MW00813910

Handbook of Essential Oils

Handbook of Essential Oils

Science, Technology, and Applications

Third Edition

Edited by
K. Hüsnü Can Başer
Gerhard Buchbauer

CRC Press is an imprint of the
Taylor & Francis Group, an **informa** business

Third edition published 2020
by CRC Press
6000 Broken Sound Parkway NW, Suite 300, Boca Raton, FL 33487-2742

and by CRC Press
2 Park Square, Milton Park, Abingdon, Oxon, OX14 4RN

CRC Press is an imprint of Taylor & Francis Group, LLC

Library of Congress Cataloging-in-Publication Data
Names: Başer, K. H. C. (Kemal Hüsnü Can), editor. | Buchbauer, Gerhard, editor.
Title: Handbook of essential oils : science, technology, and applications / [edited by]
K. Hüsnü Can Baser, Gerhard Buchbauer.
Description: Third edition. | Boca Raton : CRC Press, [2020] | Includes bibliographical references
and index.
Identifiers: LCCN 2020014781 | ISBN 9780815370963 (hardback) | ISBN 9781351246460 (ebook)
Subjects: LCSH: Essences and essential oils--Handbooks, manuals, etc.
Classification: LCC QD416.7 .H36 2020 | DDC 661/.806--dc23
LC record available at https://lccn.loc.gov/2020014781

ISBN: 9780815370963 (hbk)
ISBN: 9781351246460 (ebk)

Typeset in Times LT Std
by Nova Techset Private Limited, Bengaluru & Chennai, India

Contents

Editors

K. Hüsnü Can Başer was born on July 15, 1949, in Çankırı, Turkey. He graduated from the Eskisehir I.T.I.A. School of Pharmacy with diploma number 1 in 1972 and became a research assistant in the Pharmacognosy Department of the same school. He did his PhD in pharmacognosy between 1974 and 1978 at Chelsea College of the University of London. Upon returning home, he worked as a lecturer in pharmacognosy at the school from which he had earlier graduated and served as director of Eskisehir I.T.I.A. School of Chemical Engineering between 1978 and 1980. He was promoted to associate professorship in pharmacognosy in 1981. He served as dean of the Faculty of Pharmacy at Anadolu University (1993–2001), vice-dean of the Faculty of Pharmacy (1982–1993), head of the Department of Professional Pharmaceutical Sciences (1982–1993), head of the Pharmacognosy Department (1982–2011), member of the University Board and Senate (1982–2001; 2007), and director of the Medicinal and Aromatic Plant and Drug Research Centre (TBAM) (1980–2002) in Anadolu University.

During 1984–1994, he was appointed as the national project coordinator of Phase I and Phase II of the UNDP/UNIDO projects of the government of Turkey titled "Production of Pharmaceutical Materials from Medicinal and Aromatic Plants," through which TBAM had been strengthened. He was promoted to full professorship in pharmacognosy in 1987. After his early retirement from Anadolu University in 2011, he served as visiting professor in King Saud University in Riyadh, Saudi Arabia (2011–2015). He is currently working as head of the Pharmacognosy Department in the Faculty of Pharmacy of Near East University in Nicosia, N. Cyprus, and director of the Graduate Institute of Health Sciences of the same university. His major areas of research include essential oils, alkaloids, and biological, chemical, pharmacological, technological, and biological activity research into natural products. He is the 1995 Recipient of the Distinguished Service Medal of IFEAT (International Federation of Essential Oils and Aroma Trades) based in London, United Kingdom, and the 2005 recipient of "Science Award" (Health Sciences) of the Scientific and Technological Research Council of Turkey (TUBITAK), which are among the 18 awards he has been bestowed so far. He is listed among the 100 Turks Leading Science. He has published 827 research papers in international refereed journals, 184 papers in Turkish journals, and 141 papers in conference proceedings. He has published altogether 1212 scientific contributions as papers, books, or book chapters. According to SCI, his 594 papers were cited 10,453 times. His H-index is 47. According to Google Scholar, his publications were cited 26,540 times. His H-index is 70; i10 index is 552. His 5 books were published in Turkey, Japan, the UK, and the US (2).

More information can be found at http://www.khcbaser.com.

Gerhard Buchbauer was born in 1943 in Vienna, Austria. He studied pharmacy at the University of Vienna, from where he received his master's degree (Mag. pharm.) in May 1966. In September 1966, he assumed the duties of university assistant at the Institute of Pharmaceutical Chemistry and received his doctorate (PhD) in pharmacy and philosophy in October 1971, with a thesis on synthetic fragrance compounds. Further scientific education was practiced postdoc in the team of Professor C. H. Eugster at the Institute of Organic Chemistry, University of Zurich (1977–1978), followed by the habilitation (postdoctoral lecture qualification) in pharmaceutical chemistry with the inaugural dissertation entitled "Synthesis of Analogies of Drugs and Fragrance Compounds with Contributions to Structure-Activity-Relationships" (1979) and appointment as permanent staff of the University of Vienna and head of the first department of the Institute of Pharmaceutical Chemistry.

In November 1991, he was appointed as a full professor of pharmaceutical chemistry, University of Vienna; in 2002, he was elected as head of this institute. He retired in October 2008. He has been married since 1973 and had a son in 1974.

Among others, he is still a member of the permanent scientific committee of International Symposium on Essential Oils (ISEO); a member of the scientific committee of Forum Cosmeticum (1990, 1996, 2002, and 2008); a member of numerous editorial boards (e.g., *Journal of Essential Oil Research*, the *International Journal of Essential Oil Therapeutics*, *Scientia Pharmaceutica*); assistant editor of *Flavour and Fragrance Journal*; regional editor of *Eurocosmetics*; a member of many scientific societies (e.g., Society of Austrian Chemists, head of its working group "Food Chemistry, Cosmetics, and Tensides" [2000–2004]; Austrian Pharmaceutical Society; Austrian Phytochemical Society; vice head of the Austrian Society of Scientific Aromatherapy; and so on); technical advisor of IFEAT (1992–2008); and organizer of the 27th ISEO (September 2006, in Vienna) together with Professor Dr. Ch. Franz and senior advisor at the 50th ISEO (September 2019), organized in Vienna again.

Based on the sound interdisciplinary education of pharmacists, it was possible to establish an almost completely neglected area of fragrance and avor chemistry as a new research discipline within the pharmaceutical sciences. Our research team is the only one that conducts fragrance research in its entirety and covers synthesis, computer-aided fragrance design, analysis, and pharmaceutical/ medicinal aspects. Because of our efforts, it is possible to show and to prove that these small molecules possess more properties than merely emitting a good odor. Now, this research team has gained a worldwide scientific reputation documented by more than 450 scientific publications, about 100 invited lectures, and about 200 contributions to symposia, meetings, and congresses, as short lectures and poster presentations.

Contributors

Romana Aichinger
Department of Pharmaceutical Chemistry
Division of Clinical Pharmacy and Diagnostics
Center of Pharmacy
University of Vienna
Vienna, Austria

Adriana Arigò
Department of Chemical, Biological, Pharmaceutical and Environmental Sciences
University of Messina
Messina, Italy

Yoshinori Asakawa
Institute of Pharmacognosy
Tokushima Bunri University
Tokushima, Japan

Hugo Bovill
Ajowan Consulting
Bury St Edmunds, Suffolk, United Kingdom

W. S. Brud
Polskie Towarzystwo Aromaterapeutyczne
Warszawa, Poland

Daniel Carrillo
Tropical Research and Education Center
University of Florida
Homestead, Florida

Marie-Christine Cudlik
Department of Pharmaceutical Chemistry
Division of Clinical Pharmacy and Diagnostics
Center of Pharmacy
University of Vienna
Vienna, Austria

Jan C. R. Demyttenaere
Director, Scientific & Regulatory Affairs
EFFA/IOFI
Brussels, Belgium

Paola Dugo
Chromaleont SRL, c/o Department of Chemical, Biological, Pharmaceutical and Environmental Sciences
University of Messina
Messina, Italy

and

Unit of Food Science and Nutrition
Department of Medicine
University Campus Bio-Medico of Rome
Rome, Italy

Elaine Elisabetsky
Department of Biochemistry
Universidade Federal Do Rio Grande do Sul
Porto Alegre, Rio Grande do Sul, Brazil

Chlodwig Franz
Institute of Animal Nutrition and Functional Plant Compounds
University of Veterinary Medicine
Vienna, Austria

Darija Gajić
Department of Pharmaceutical Chemistry
Division of Clinical Pharmacy and Diagnostics
Center of Pharmacy
University of Vienna
Vienna, Austria

Susanne Hemetsberger
Department of Pharmaceutical Chemistry
Division of Clinical Pharmacy and Diagnostics
Center of Pharmacy
University of Vienna
Vienna, Austria

Eva Heuberger
Pfarrer-Lauer-Strasse
St Ingbert, Germany

Martina Höferl
Department of Pharmaceutical Chemistry
Division of Clinical Pharmacy and Diagnostics
Center of Pharmacy
University of Vienna
Vienna, Austria

Walter Jäger
Department of Pharmaceutical Chemistry
Division of Clinical Pharmacy and Diagnostics
Center of Pharmacy
University of Vienna
Vienna, Austria

Jens Jankowski (deceased)
Joh. Vögele KG
Lauffen a.N., Germany

Jan Karlsen
Department of Pharmaceutics
University of Oslo
Oslo, Norway

Paul E. Kendra
United States Department of Agriculture, Agricultural Research Service (USDA-ARS)
Subtropical Horticulture Research Station (SHRS)
Miami, Florida

Sabine Krist
Lehrstuhl für Medizinische Chemie
Medizinische Fakultät
Sigmund Freud Privat-Universität
and
Department of Pharmaceutical Chemistry
Division of Clinical Pharmacy and Diagnostics
Center of Pharmacy
University of Vienna
Vienna, Austria

Karl-Heinz Kubeczka
Untere Steigstraße
Margetshöchheim, Germany

Rosa Lemmens-Gruber
Department of Pharmacology
Center of Pharmacy
University of Vienna
Vienna, Austria

Rhiannon Lewis
Chemin Les Achaps
La Martre, France

Agnieszka Ludwiczuk
Independent Laboratory of Natural Products Chemistry, Chair and Department of Pharmacognosy
Medical University of Lublin
Lublin, Poland

Giuseppe Micalizzi
Department of Chemical, Biological, Pharmaceutical and Environmental Sciences
University of Messina
Messina, Italyand

and

Center of Sports Nutrition Science
University of Physical Education
Budapest, Hungary

Luigi Mondello
Chromaleont SRL, c/o Department of Chemical, Biological, Pharmaceutical and Environmental Sciences
University of Messina
Messina, Italy

and

Unit of Food Science and Nutrition
Department of Medicine
University Campus Bio-Medico of Rome
Rome, Italy

and

BeSep SRL, c/o Department of Chemistry, Biological, Pharmaceutical and Environmental Sciences
University of Messina
Messina, Italy

Wayne S. Montgomery
United States Department of Agriculture, Agricultural Research Service (USDA-ARS)
Subtropical Horticulture Research Station
Miami, Florida

Éva Németh-Zámbori
Department of Medicinal and Aromatic Plants
Szent István University
Budapest, Hungary

Jerome Niogret
Niogret Ecology Consulting LLC
Miami, Florida

Yoshiaki Noma
Shinkirai, Kitajima-cho
Tokushima, Japan

Johannes Novak
Institute of Animal Nutrition and Functional Plant Compounds
University of Veterinary Medicine
Vienna, Austria

Domingos S. Nunes
Department of Chemistry
Universidade Estadual de Ponta Grossa,
Ponta Grossa, Paraná, Brazil

David Owens
Carvel Research & Education Center
University of Delaware
Georgetown, Delaware

Jens-Achim Protzen
Paul Kaders GmbH
Hamburg, Germany

Klaus-Dieter Protzen
Paul Kaders GmbH
Hamburg, Germany

Robert A. Raguso
Department of Neurobiology and Behavior
Cornell University
Ithaca, New York

Catherine Regnault-Roger
Professor emeritus
Institute of Interdisciplinary Research on Environment and Materials
UPPA University of Pau et des pays de l'Adour
Pau, France

Erich Schmidt
Consultant
Essential oils
Nördlingen, Germany

Danilo Sciarrone
Department of Chemical, Biological, Pharmaceutical and Environmental Sciences
University of Messina
Messina, Italy

Charles Sell
Parsonage Farm
Church Lane
Kent, England

Isabel Charlotte Soede
Department of Pharmaceutical Chemistry
Division of Clinical Pharmacy and Diagnostics
Center of Pharmacy
University of Vienna
Vienna, Austria

Nurhayat Tabanca
United States Department of Agriculture, Agricultural Research Service (USDA-ARS)
Subtropical Horticulture Research Station (SHRS)
Miami, Florida

Sean V. Taylor
Flavor & Extract Manufacturers Association
and
The International Organization of the Flavor Industry
Washington, DC

Peter Q. Tranchida
Department of Chemical, Biological, Pharmaceutical and Environmental Sciences
University of Messina
Messina, Italy

Emanuela Trovato
Chromaleont SRL, c/o Department of Chemical, Biological, Pharmaceutical and Environmental Sciences
University of Messina
Messina, Italy

Carmen Trummer
Department of Pharmaceutical Chemistry
Division of Clinical Pharmacy and Diagnostics
Center of Pharmacy
University of Vienna
Vienna, Austria

Margita Utczás
Department of Chemical, Biological, Pharmaceutical and Environmental Sciences
University of Messina
Messina, Italy

and

Center of Sports Nutrition Science
Universityof Physical Education
Budapest, Hungary

Jürgen Wanner
Kurt Kitzing GmbH
Wallerstein, Germany

Mariosimone Zoccali
Unit of Food Science and Nutrition, Department of Medicine
University Campus Bio-Medico of Rome
Rome, Italy

1 Introduction

K. Hüsnü Can Başer and Gerhard Buchbauer

The overwhelming success of the first edition of the *Handbook of Essential Oils: Science, Technology, and Applications* had urged the publication of the second edition which was bestowed, in 2016, the *ABC James A. Duke Excellence in Botanical Literature Award* for the excellent contribution to the vast field of essential oils. This prestigious award by the *American Botanical Council* for the best book in botanical literature has prompted us to prepare a third edition of this Handbook.

As in the previous edition, updated chapters as well as completely new chapters have been included in the third edition. Some important chapters remained as such. Thus, we kept the contributions of the current Chapters 2, 5, 6, 9, 22, 23, 25, 32, and 33 as in the second edition. We skipped Chapters 15, 16, and 26 in the second edition of the *Handbook*, whereby the former Part Chapter 16, "Aromatherapy with Essential Oils", has been substituted by Rhiannon Lewis (Chapter 13). In this edition, Chapters 4, 7, 10, 26, 30, and 31 have been updated, and many new contributions have been added, covering the commonly entitled "Biological activities of…" chapters in the form of six chapters. These are "Essential Oils in Cancer Therapy" (Chapter 14), then "Antimicrobial Activity of Selected Essential Oils and Aromas" (Chapter 15), followed by "Quorum Sensing and Essential Oils" (Chapter 16), then "Essential Oils as Carrier Oils" (Chapter 27), and then two new (more chemically written) overviews, namely "Influence of Light on Essential Oil Constituents" and "Influence of Air on Essential Oil Constituents" (now Chapters 28 and 29). The new Chapter 19, entitled "Adverse Effects and Intoxication with Essential Oils" is an overview written by a pharmacologist of the University of Vienna. The former Chapter 12 now has been substituted by the updated chapter "Central Nervous System Effects of Essential Oil Compounds" (now Chapter 11) and another, newly entitled, treatise, namely "Effects of Essential Oils on Human Cognition" (now Chapter 12). "Essential Oils and Volatiles in Bryophytes" is a new chapter (Chapter 21) by Agnieszka Ludwiczuk and Yoshinori Asakawa. "Functions of Essential Oils and Natural Volatiles in Plant–Insect Interactions" (Chapter 17) was contributed by R. Raguso. "Essential Oils as Lures for Invasive Ambrosia Beetles" (Chapter 18) is yet another new contribution. A useful new chapter for GC/MS analysts is entitled "Use of Linear Retention Indices in GC/MS Libraries for Essential Oil Analysis" (Chapter 8).

Also with this third edition, we hope that many scientists, especially in the fields of essential oils in botany, chemistry, pharmacognosy, medicine, clinical aromatherapy, and other relevant aspects of these natural products, will find these contributions not only alluring for their own research but also interesting to read and to find out what manifold properties essential oils have. Especially, also in this third edition, we want to provide a strong scientific basis for essential oils and to prevent any trace of esoteric ignorance.

2 History and Sources of Essential Oil Research

Karl-Heinz Kubeczka

CONTENTS

2.1 ANCIENT HISTORICAL BACKGROUND

Plants containing essential oils have been used since furthest antiquities as spices and remedies for the treatment of diseases and in religious ceremonies because of their healing properties and their pleasant odors. In spite of the obscured beginning of the use of aromatic plants in prehistoric times to prevent, palliate, or heal sicknesses, pollen analyses of Stone Age settlements indicate the use of aromatic plants that may be dated to 10,000 BC.

One of the most important medical documents of ancient Egypt is the so-called Papyrus Ebers of about 1550 BC, a 20 m long papyrus, which was purchased in 1872 by the German Egyptologist

G. Ebers, for whom it is named, containing some 700 formulas and remedies, including aromatic plants and plant products like anise, fennel, coriander, thyme, frankincense, and myrrh. Much later, the ancient Greek physician Hippocrates (460–377 BC), who is referred to as the father of medicine, mentioned in his treatise *Corpus Hippocratium* approximately 200 medicinal plants inclusive of aromatic plants and described their efficacies.

One of the most important herbal books in history is the five-volume book *De Materia Medica*, written by the Greek physician and botanist Pedanius Dioscorides (ca. 40–90), who practiced in ancient Rome. In the course of his numerous travels all over the Roman and Greek world seeking for medicinal plants, he described more than 500 medicinal plants and respective remedies. His treatise, which may be considered a precursor of modern pharmacopoeias, was later translated into a variety of languages. Dioscorides, as well as his contemporary Pliny the Elder (23–79), a Roman natural historian, mention besides other facts turpentine oil and give some limited information on the methods in its preparation.

Many new medicines and ointments were brought from the east during the Crusades from the eleventh to the thirteenth centuries, and many herbals, whose contents included recipes for the use and manufacture of essential oil, were written during the fourteenth to the sixteenth centuries.

Theophrastus von Hohenheim, known under the name Paracelsus (1493–1541), a physician and alchemist of the fifteenth century, defined the role of alchemy by developing medicines and extracts from healing plants. He believed distillation released the most desirable part of the plant, the *Quinta essentia* or *quintessence* by a means of separating the "essential" part from the "nonessential" containing its subtle and essential constituents. The currently used term "essential oil" still refers to the theory of *Quinta essentia* of Paracelsus.

The roots of distillation methods are attributed to Arabian Alchemists centuries with Avicenna (980–1037) describing the process of steam distillation, who is credited with inventing a coiled cooling pipe to prepare essential oils and aromatic waters. The first description of distilling essential oils is generally attributed to the Spanish physician Arnaldus de Villa Nova (1235–1311) in the thirteenth century. However, in 1975, a perfectly preserved terracotta apparatus was found in the Indus Valley, which is dated to about 3000 BC and which is now displayed in a museum in Taxila, Pakistan. It looks like a primitive still and was presumable used to prepare aromatic waters. Further findings indicate that distillation has also been practiced in ancient Turkey, Persia, and India as far back as 3000 BC.

At the beginning of the sixteenth century appeared a comprehensive treatise on distillation by Hieronymus Brunschwig (ca. 1450–1512), a physician of Strasbourg. He described the process of distillation and the different types of stills in his book *Liber de arte Distillandi de compositis* (Strasbourg 1500 and 1507) with numerous block prints. Although obviously endeavoring to cover the entire field of distillation techniques, he mentions in his book only the four essential oils from rosemary, spike lavender, juniper wood, and the turpentine oil. Just before, until the Middle Ages, the art of distillation was used mainly for the preparation of aromatic waters, and the essential oil appearing on the surface of the distilled water was regarded as an undesirable by-product.

In 1551 appeared at Frankfurt on the Main the *Kräuterbuch*, written by Adam Lonicer (1528–1586), which can be regarded as a significant turning point in the understanding of the nature and the importance of essential oils. He stresses that the art of distillation is a quite recent invention and not an ancient invention and has not been used earlier.

In the *Dispensatorium Pharmacopolarum* of Valerius Cordus, published in Nuremberg in 1546, only three essential were listed; however, the second official edition of the *Dispensatorium Valerii Cordi* issued in 1592, 61 distilled oils were listed illustrating the rapid development and acceptance of essential oils. In that time, the so-called Florentine flask has already been used for separating the essential oil from the water phase.

The German J.R. Glauber (1604–1670), who can be regarded as one of the first great industrial chemists, was born in the little town Karlstadt close to Wuerzburg. His improvements in chemistry, for example, the production of sodium sulfate, as a safe laxative brought him the honor of being named Glauber's salt. In addition, he improved numerous different other chemical processes and especially new distillation devices also for the preparation of essential oils from aromatic plants. However, it lasted until the nineteenth century to get any real understanding of the composition of true essential oils.

2.2 FIRST SYSTEMATIC INVESTIGATIONS

The first systematic investigations of constituents from essential oils may be attributed to the French chemist M.J. Dumas (1800–1884) who analyzed some hydrocarbons and oxygen as well as sulfur- and nitrogen-containing constituents. He published his results in 1833. The French researcher M. Berthelot (1859) characterized several natural substances and their rearrangement products by optical rotation. However, the most important investigations have been performed by O. Wallach, an assistant of Kekule. He realized that several terpenes described under different names according to their botanical sources were often, in fact, chemically identical. He, therefore, tried to isolate the individual oil constituents and to study their basic properties. He employed together with his highly qualified coworkers Hesse, Gildemeister, Betram, Walbaum, Wienhaus, and others fractional distillation to separate essential oils and performed reactions with inorganic reagents to characterize the obtained individual fractions. The reagents he used were hydrochloric acid, oxides of nitrogen, bromine, and nitrosyl chloride—which was used for the first time by W.A. Tilden (1875)—by which frequently crystalline products had been obtained.

At that time, hydrocarbons occurring in essential oils with the molecular formula $C_{10}H_{16}$ were known, which had been named by Kekule *terpenes* because of their occurrence in turpentine oil. Constituents with the molecular formulas $C_{10}H_{16}O$ and $C_{10}H_{18}O$ were also known at that time under the generic name camphor and were obviously related to terpenes. The prototype of this group was camphor itself, which was known since antiquity. In 1891, Wallach characterized the terpenes pinene, camphene, limonene, dipentene, phellandrene, terpinolene, fenchene, and sylvestrene, which has later been recognized to be an artifact.

During 1884–1914, Wallach wrote about 180 articles that are summarized in his book *Terpene und Campher* (Wallach, 1914) compiling all the knowledge on terpenes at that time, and already in 1887, he suggested that the terpenes must be constructed from isoprene units. In 1910, he was honored with the Nobel Prize for Chemistry "in recognition of his outstanding research in organic chemistry and especially in the field of alicyclic compounds" (Laylin, 1993).

In addition to Wallach, the German chemist A. von Baeyer, who also had been trained in Kekule's laboratory, was one of the first chemists to become convinced of the achievements of structural chemistry and who developed and applied it to all of his work covering a broad scope of organic chemistry. Since 1893, he devoted considerable work to the structures and properties of cyclic terpenes (von Baeyer and Seuffert, 1901). Besides his contributions to several dyes, the investigations of polyacetylenes, and so on, his contributions to theoretical chemistry including the strain theory of triple bonds and small carbon cycles have to be mentioned. In 1905, he was awarded the Nobel Prize for Chemistry "in recognition of his contributions to the development of Organic Chemistry and Industrial Chemistry, by his work on organic dyes and hydroaromatic compounds" (Laylin, 1993). The frequently occurring acyclic monoterpenes geraniol, linalool, citral, and so on have been investigated by F.W. Semmler and the Russian chemist G. Wagner (1899), who recognized the importance of rearrangements for the elucidation of chemical constitution, especially the carbon-to-carbon migration of alkyl, aryl, or hydride ions, a type of reaction that was later generalized by H. Meerwein (1914) as Wagner–Meerwein rearrangement.

More recent investigations of J. Read, W. Hückel, H. Schmidt, W. Treibs, and V. Prelog were mainly devoted to disentangle the stereochemical structures of menthols, carvomenthols, borneols, fenchols, and pinocampheols, as well as the related ketones (see Gildemeister and Hoffmann, 1956).

A significant improvement in structure elucidation was the application of dehydrogenation of sesqui- and diterpenes with sulfur and later with selenium to give aromatic compounds as a major method, and the application of the isoprene rule to terpene chemistry, which have been very efficiently used by L. Ruzicka (1953) in Zurich, Switzerland. In 1939, he was honored in recognition of his outstanding investigations with the Nobel Prize in chemistry for his work on "polymethylenes and higher terpenes."

The structure of the frequently occurring bicyclic sesquiterpene ß-caryophyllene was for many years a matter of doubt. After numerous investigations, W. Treibs (1952) has been able to isolate the crystalline caryophyllene epoxide from the autoxidation products of clove oil, and F. Šorm et al.

(1950) suggested caryophyllene to have a four- and nine-membered ring on bases of infrared (IR) investigations. This suggestion was later confirmed by the English chemist D.H.R. Barton (Barton and Lindsay, 1951), who was awarded the Nobel Prize in Chemistry in 1969.

The application of ultraviolet (UV) spectroscopy in the elucidation of the structure of terpenes and other natural products was extensively used by R.B. Woodward in the early forties of the last century. On the basis of his large collection of empirical data, he developed a series of rules (later called the Woodward rules), which could be applied to finding out the structures of new natural substances by correlations between the position of UV maximum absorption and the substitution pattern of a diene or an α,β-unsaturated ketone (Woodward, 1941). He was awarded the Nobel Prize in Chemistry in 1965. However, it was not until the introduction of chromatographic separation methods and nuclear magnetic resonance (NMR) spectroscopy into organic chemistry that a lot of further structures of terpenes were elucidated. The almost exponential growth in our knowledge in that field and other essential oil constituents is essentially due to the considerable advances in analytical methods in the course of the last half century.

2.3 RESEARCH DURING THE LAST HALF CENTURY

2.3.1 Essential Oil Preparation Techniques

2.3.1.1 Industrial Processes

The vast majority of essential oils are produced from plant material in which they occur by different kinds of distillation or by cold pressing in the case of the peel oils from citrus fruits.

In water or hydrodistillation, the chopped plant material is submerged and in direct contact with boiling water. In steam distillation, the steam is produced in a boiler separate of the still and blown through a pipe into the bottom of the still, where the plant material rests on a perforated tray or in a basket for quick removal after exhaustive extraction. In addition to the aforementioned distillation at atmospheric pressure, high-pressure steam distillation is most often applied in European and American field stills, and the applied increased temperature significantly reduces the time of distillation. The high-pressure steam-type distillation is often applied for peppermint, spearmint, lavandin, and the like. The condensed distillate, consisting of a mixture of water and oil, is usually separated in a so-called Florentine flask, a glass jar, or more recently in a receptacle made of stainless steel with one outlet near the base and another near the top. There, the distillate separates into two layers from which the oil and the water can be separately withdrawn. Generally, the process of steam distillation is the most widely accepted method for the production of essential oils on a large scale.

Expression or cold pressing is a process in which the oil glands within the peels of citrus fruits are mechanically crushed to release their content. There are several different processes used for the isolation of citrus oils; however, there are four major currently used processes. Those are pellatrice and sfumatrice—most often used in Italy—and the Brown peel shaver as well as the FMC extractor, which are used predominantly in North and South America. For more details, see, for example, Lawrence 1995. All these processes lead to products that are not entirely volatile because they may contain coumarins, plant pigments, and so on; however, they are nevertheless acknowledged as essential oils by the International Organization for Standardization, the different pharmacopoeias, and so on.

In contrast, extracts obtained by solvent extraction with different organic solvents, with liquid carbon dioxide or by supercritical fluid extraction (SFE) may not be considered as true essential oils; however, they possess most often aroma profiles that are almost identical to the raw material from which they have been extracted. They are therefore often used in the flavor and fragrance industry and in addition in food industry, if the chosen solvents are acceptable for food and do not leave any harmful residue in food products.

2.3.1.2 Laboratory-Scale Techniques

The following techniques are used mainly for trapping small amounts of volatiles from aromatic plants in research laboratories and partly for determination of the essential oil content in plant material.

The most often used device is the circulatory distillation apparatus, basing on the publication of Clevenger in 1928 and which has later found various modifications. One of those modified apparatus described by Cocking and Middleton (1935) has been introduced in the European pharmacopoeia and several other pharmacopoeias. This device consists of a heated round-bottom flask into which the chopped plant material and water are placed and which is connected to a vertical condenser and a graduated tube, for the volumetric determination of the oil. At the bottom of the tube, a three-way valve permits to direct the water back to the flask, since it is a continuous closed-circuit distillation device, and at the end of the distillation process to separate the essential oil from the water phase for further investigations. The length of distillation depends on the plant material to be investigated; however, it is usually fixed to 3–4 h. For the volumetric determination of the essential oil content in plants according to most of the pharmacopoeias, a certain amount of xylene—usually 0.5 mL—has to be placed over the water before running distillation to separate even small droplets of essential oil during distillation from the water. The volume of essential oil can be determined in the graduated tube after subtracting the volume of the applied xylene.

Improved constructions with regard to the cooling system of the aforementioned distillation apparatus have been published by Stahl (1953) and Sprecher (1963) and, in publications of Kaiser and Lang (1951) and Mechler and Kovar (1977), various apparatus used for the determination of essential oils in plant material are discussed and depicted.

A further improvement was the development of a simultaneous distillation–solvent extraction device by Likens and Nickerson in 1964 (see Nickerson and Likens, 1966). The device permits continuous concentration of volatiles during hydrodistillation in one step using a closed-circuit distillation system. The water distillate is continuously extracted with a small amount of an organic- and water-immiscible solvent. Although there are two versions described, one for high-density and one for low-density solvents, the high-density solvent version using dichloromethane is mostly applied in essential oil research. It has found numerous applications, and several modified versions including different microdistillation devices have been described (e.g., Bicchi et al., 1987; Chaintreau, 2001).

A sample preparation technique basing on Soxhlet extraction in a pressurized container using liquid carbon dioxide as extractant has been published by Jennings (1979). This device produces solvent-free extracts especially suitable for high-resolution gas chromatography (GC). As a less time-consuming alternative, the application of microwave-assisted extraction has been proposed by several researchers, for example, by Craveiro et al. (1989), using a round-bottom flask containing the fresh plant material. This flask was placed into a microwave oven and passed by a flow of air. The oven was heated for 5 min and the obtained mixture of water and oil collected in a small and cooled flask. After extraction with dichloromethane, the solution was submitted to GC–mass spectrometry (GC-MS) analysis. The obtained analytical results have been compared with the results obtained by conventional distillation and exhibited no qualitative differences; however, the percentages of the individual components varied significantly. A different approach yielding solvent-free extracts from aromatic herbs by means of microwave heating has been presented by Lucchesi et al. (2004). The potential of the applied technique has been compared with conventional hydrodistillation showing substantially higher amounts of oxygenated compounds at the expense of monoterpene hydrocarbons.

2.3.1.3 Microsampling Techniques

2.3.1.3.1 Microdistillation

Preparation of very small amounts of essential oils may be necessary if only very small amounts of plant material are available and can be fundamental in chemotaxonomic investigations and control analysis but also for medicinal and spice plant breeding. In the past, numerous attempts have been made to minimize conventional distillation devices. As an example, the modified Marcusson device may be quoted (Bicchi et al., 1983) by which 0.2–3 g plant material suspended in 50 mL water can be distilled and collected in 100 μL analytical grade pentane or hexane. The analytical results proved to be identical with those obtained by conventional distillation.

Microversions of the distillation–extraction apparatus, described by Likens and Nickerson, have also been developed as well for high-density (Godefroot et al., 1981) and low-density solvents (Godefroot et al., 1982). The main advantage of these techniques is that no further enrichment by evaporation is required for subsequent gas chromatographic investigation.

A different approach has been presented by Gießelmann and Kubeczka (1993) and Kubeczka and Gießelmann (1995). By means of a new developed micro-hydrodistillation device, the volatile constituents of very small amounts of plant material have been separated. The microscale hydrodistillation of the sample is performed using a 20 mL crimp-cap glass vial with a Teflon®-lined rubber septum containing 10 mL water and 200–250 mg of the material to be investigated. This vial, which is placed in a heating block, is connected with a cooled receiver vial by a 0.32 mm ID fused silica capillary. By temperature-programmed heating of the sample vial, the water and the volatile constituents are vaporized and passed through the capillary into the cooled receiver vial. There, the volatiles as well as water are condensed and the essential oil collected in pentane for further analysis. The received analytical results have been compared to results from identical samples obtained by conventional hydrodistillation showing a good correlation of the qualitative and quantitative composition. Further applications with the commercially available Eppendorf MicroDistiller® have been published in several papers, for example, by Briechle et al. (1997) and Baser et al. (2001).

A simple device for rapid extraction of volatiles from natural plant drugs and the direct transfer of these substances to the starting point of a thin-layer chromatographic plate has been described by Stahl (1969a) and in his subsequent publications. A small amount of the sample (ca. 100 mg) is introduced into a glass cartridge with a conical tip together with 100 mg silica gel, containing 20% of water, and heated rapidly in a heating block for a short time at a preset temperature. The tip of the glass tube projects ca. 1 mm from the furnace and points to the starting point of the thin-layer plate, which is positioned 1 mm in front of the tip. Before introducing the glass tube, it is sealed with a silicone rubber membrane. This simple technique has proven useful for many years in numerous investigations, especially in quality control, identification of plant drugs, and rapid screening of chemical races. In addition to the aforementioned micro-hydrodistillation with the so-called TAS procedure (T, thermomicro and transfer; A, application; S, substance), several further applications, for example, in structure elucidation of isolated natural compounds such as zinc dust distillation, sulfur and selenium dehydrogenation, and catalytic dehydrogenation with palladium, have been described in the microgram range (Stahl, 1976).

2.3.1.3.2 Direct Sampling from Secretory Structures

The investigation of the essential oils by direct sampling from secretory glands is of fundamental importance in studying the true essential oil composition of aromatic plants, since the usual applied techniques such as hydrodistillation and extraction are known to produce in some cases several artifacts. Therefore, only direct sampling from secretory cavities and glandular trichomes and properly performed successive analysis may furnish reliable results. One of the first investigations with a kind of direct sampling has been performed by Hefendehl (1966), who isolated the glandular hairs from the surfaces of *Mentha piperita* and *Mentha aquatica* leaves by means of a thin film of polyvinyl alcohol, which was removed after drying and extracted with diethyl ether. The composition of this product was in good agreement with the essential oils obtained by hydrodistillation. In contrast to these results, Malingré et al. (1969) observed some qualitative differences in the course of their study on *M. aquatica* leaves after isolation of the essential oil from individual glandular hairs by means of a micromanipulator and a stereomicroscope. In the same year, Amelunxen et al. (1969) published results on *M. piperita*, who separately isolated glandular hairs and glandular trichomes with glass capillaries. They found identical qualitative composition of the oil in both types of hairs, but differing concentrations of the individual components. Further studies have been performed by Henderson et al. (1970) on *Pogostemon cablin* leaves and by Fischer et al. (1987) on *Majorana hortensis* leaves. In the latter study, significant differences regarding the oil composition of the hydrodistilled oil and the oil extracted by means of glass capillaries from the trichomes were observed. Their final conclusion was that the analysis of the respective essential oil is mainly an analysis of artifacts, formed during distillation, and the gas chromatographic analysis. Even if the investigations are performed very carefully and the successive GC has been performed by cold-on-column injection to avoid thermal stress in the injection port, significant differences of the GC pattern of directly sampled oils versus the microdistilled samples have been observed in several cases (Bicchi et al., 1985).

2.3.1.3.3 HS Techniques

Headspace (HS) analysis has become one of the very frequently used sampling techniques in the investigation of aromatic plants, fragrances, and spices. It is a means of separating the volatiles from a liquid or solid prior to gas chromatographic analysis and is preferably used for samples that cannot be directly injected into a gas chromatograph. The applied techniques are usually classified according to the different sampling principles in static HS analysis and dynamic HS analysis.

2.3.1.3.3.1 Static HS Methods In static HS analysis, the liquid or solid sample is placed into a vial, which is heated to a predetermined temperature after sealing. After the sample has reached equilibrium with its vapor (in equilibrium, the distribution of the analytes between the two phases depends on their partition coefficients at the preselected temperature, the time, and the pressure), an aliquot of the vapor phase can be withdrawn with a gas-tight syringe and subjected to gas chromatographic analysis. A simple method for the HS investigation of herbs and spices was described by Chialva et al. (1982), using a blender equipped with a special gas-tight valve. After grinding the herb and until thermodynamic equilibrium is reached, the HS sample can be withdrawn through the valve and injected into a gas chromatograph. Eight of the obtained capillary gas chromatograms are depicted in the paper of Chialva and compared with those of the respective essential oils exhibiting significant higher amounts of the more volatile oil constituents. However, one of the major problems with static HS analyses is the need for sample enrichment with regard to trace components. Therefore, a concentration step such as cryogenic trapping, liquid absorption, or adsorption on a suitable solid has to be inserted for volatiles occurring only in small amounts. A versatile and often-used technique in the last decade is solid-phase microextraction (SPME) for sampling volatiles, which will be discussed in more detail in a separate paragraph. Since different other trapping procedures are a fundamental prerequisite for dynamic HS methods, they will be considered in the succeeding text. A comprehensive treatment of the theoretical basis of static HS analysis including numerous applications has been published by Kolb and Ettre (1997, 2006).

2.3.1.3.3.2 Dynamic HS Methods The sensitivity of HS analysis can be improved considerably by stripping the volatiles from the material to be investigated with a stream of purified air or inert gas and trapping the released compounds. However, care has to be taken if grinded plant material has to be investigated, since disruption of tissues may initiate enzymatic reactions that may lead to formation of volatile artifacts. After stripping the plant material with gas in a closed vessel, the released volatile compounds are passed through a trap to collect and enrich the sample. This must be done because sample injection of fairly large sample volumes results in band broadening causing peak distortion and poor resolution. The following three techniques are advisable for collecting the highly diluted volatile sample according to Schaefer (1981) and Schreier (1984) with numerous references.

Cryogenic trapping can be achieved by passing the gas containing the stripped volatiles through a cooled vessel or a capillary in which the volatile compounds are condensed (Kolb and Liebhardt, 1986). The most convenient way for trapping the volatiles is to utilize part of the capillary column as a cryogenic trap. A simple device for cryofocusing of HS volatiles by using the first part of capillary column as a cryogenic trap has been shown in the aforementioned reference inclusive of a discussion of the theoretical background of cryogenic trapping. A similar on-column cold trapping device, suitable for extended period vapor sampling, has been published by Jennings (1981).

A different approach can be used if large volumes of stripped volatiles have to be trapped using collection in organic liquid phases. In this case, the volatiles distribute between the gas and the liquid, and efficient collection will be achieved, if the distribution factor K is favorable for solving the stripped compounds in the liquid. A serious drawback, however, is the necessity to concentrate the obtained solution prior to GC with the risk to lose highly volatile compounds. This can be overcome if a short-packed GC column is used containing a solid support coated with a suitable liquid. Novak et al. (1965) have used Celite coated with 30% silicone elastomer E-301 and the absorbed compounds were introduced into a gas chromatograph after thermal desorption. Coating with 15% silicone rubber SE 30 has been successfully used by Kubeczka (1967) with a similar device and the application of a wall-coated tubing with methyl silicone oil SF 96 has been described by Teranishi et al. (1972). A different technique has been used by Bergström (1973) and Bergström et al. (1980). They trapped the scent of flowers on Chromosorb® W coated with 10% silicon high-vacuum grease and filled a small portion of the sorbent containing the volatiles into a precolumn, which was placed in the splitless injection port of a gas chromatograph. There, the volatiles were desorbed under heating and flushed onto the GC column. In 1987, Bichi et al. applied up to 50 cm pieces of thick-film fused silica capillaries coated with a 15 μm dimethyl silicone film for trapping the volatiles in the atmosphere surrounding living plants. The plants under investigation were placed in a glass bell into which the trapping capillary was introduced through a rubber septum, while the other end of the capillary has been connected to pocket sampler. In order to trap even volatile monoterpene hydrocarbons, a capillary length of at least 50 cm and sample volume of maximum 100 mL have to be applied to avoid loss of components through breakthrough. The trapped compounds have been subsequently online thermally desorbed, cold trapped, and analyzed. Finally, a type of *enfleurage* especially designed for field experiments has been described by Joulain (1987) to trap the scents of freshly picked flowers. Around 100 g flowers were spread on the grid of a specially designed stainless steel device and passed by a stream of ambient air, supplied by an unheated portable air drier. The stripped volatiles are trapped on a layer of purified fat placed above the grid. After 2 h, the fat was collected and the volatiles recovered in the laboratory by means of vacuum distillation at low temperature.

With a third often applied procedure, the stripped volatiles from the HS of plant material and especially from flowers are passed through a tube filled with a solid adsorbent on which the volatile compounds are adsorbed. Common adsorbents most often used in investigations of plant volatiles are above all charcoal and different types of synthetic porous polymers. Activated charcoal is an adsorbent with a high adsorption capacity, thermal and chemical stability, and which is not deactivated by water, an important feature, if freshly collected plant material has to be investigated. The adsorbed volatiles can easily be recovered by elution with small amounts (10–50 μL) of carbon disulfide avoiding further concentration of the sample prior to GC analysis. The occasionally observed incomplete recovery of sample components after solvent extraction and artifact formation after thermal desorption has been largely solved by application of small amounts of special type of activated charcoal as described by Grob and Zürcher (1976). Numerous applications have been described using this special type of activated charcoal, for example, by Kaiser (1993) in a great number of field experiments on the scent of orchids. In addition to charcoal, the following synthetic porous polymers have been applied to collect volatile compounds from the HS from flowers and different other plant materials according to Schaefer (1981): Tenax® GC, different Porapak® types (e.g., Porapak P, Q, R, and T), and several Chromosorb types belonging to the 100 series. More recent developed adsorbents are the carbonaceous adsorbents such as Ambersorb®, Carboxen®, and Carbopak®, and their adsorbent properties lie between activated charcoal and the porous polymers. Especially the porous polymers have to be washed repeatedly, for example, with diethyl ether, and conditioned before use in a stream of oxygen-free nitrogen at 200°C–280°C, depending on the sort of adsorbent. The trapped components can be recovered either by thermal desorption or by solvent elution, and the recoveries can be different depending on the applied adsorbent (Cole, 1980). Another very important criterion for the selection of a suitable adsorbent for collecting HS samples is the breakthrough volume limiting the amount of gas passing through the trap.

A comprehensive review concerning HS gas chromatographic analysis of medicinal and aromatic plants and flowers with 137 references, covering the period from 1982 to 1988 has been published by Bicchi and Joulain in 1990, thoroughly describing and explaining the different methodological approaches and applications. Among other things, most of the important contributions of the Finnish research group of Hiltunen and coworkers on the HS of medicinal plants and the optimization of the HS parameters have been cited in the mentioned review.

2.3.1.3.4 Solid-Phase Microextraction

SPME is an easy-to-handle sampling technique, initially developed for the determination of volatile organic compounds in environmental samples (Arthur and Pawliszyn, 1990), and has gained, in the last years, acceptance in numerous fields and has been applied to the analysis of a wide range of analytes in various matrices. Sample preparation is based on sorption of analytes from a sample onto a coated fused silica fiber, which is mounted in a modified GC syringe. After introducing the coated fiber into a liquid or gaseous sample, the compounds to be analyzed are enriched according to their distribution coefficients and can be subsequently thermally desorbed from the coating after introducing the fiber into the hot injector of a gas chromatograph. The commercially available SPME device (Supelco Inc.) consists of a 1 cm length fused silica fiber of ca. 100 μm diameter coated on the outer surface with a stationary phase fixed to a stainless steel plunger and a holder that looks like a modified microliter syringe (Supelco, 2007). The fiber can be drawn into the syringe needle to prevent damage. To use the device, the needle is pierced through the septum that seals the sample vial. Then, the plunger is depressed lowering the coated fiber into the liquid sample or the HS above the sample. After sorption of the sample, which takes some minutes, the fiber has to be drawn back into the needle and withdrawn from the sample vial. By the same procedure, the fiber can be introduced into the gas chromatograph injector where the adsorbed substances are thermally desorbed and flushed by the carrier gas into the capillary GC column.

SPME fibers can be coated with polymer liquid (e.g., polydimethylsiloxane [PDMS]) or a mixed solid and liquid coating (e.g., Carboxen®/PDMS). The selectivity and capacity of the fiber coating can be adjusted by changing the phase type or thickness of the coating on the fiber according to the properties of the compounds to be analyzed. Commercially available are coatings of 7, 30, and 100 μm of PDMS, an 85 μm polyacrylate, and several mixed coatings for different polar components. The influence of fiber coatings on the recovery of plant volatiles was thoroughly investigated by Bicchi et al. (2000a,b). Details concerning the theory of SPME, technology, its application, and specific topics have been described by Pawliszyn (1997) and references cited therein. A number of different applications of SPME in the field of essential oil analysis have been presented by Kubeczka (1997a). An overview on publications of the period 2000–2005 with regard to HS-SPME has been recently published by Belliardo et al. (2006) covering the analysis of volatiles from aromatic and medicinal plants, selection of the most effective fibers and sampling conditions, and discussing its advantages and limitations. The most comprehensive collection of references with regard to the different application of SPME can be obtained from Supelco on CD.

2.3.1.3.5 Stir Bar Sorptive Extraction and HS Sorptive Extraction

Despite the indisputable simplicity and rapidity of SPME, its applicability is limited by the small amount of sorbent on the needle (<0.5 μL), and consequently SPME has no real opportunity to realize quantitative extraction. Parameters governing recovery of analytes from a sample are partitioning constants and the phase ratio between the sorbent and liquid or gaseous sample. Therefore, basing on theoretical considerations, a procedure for sorptive enrichment with the sensitivity of packed PDMS beds (Baltussen et al., 1997) has been developed for the extraction of aqueous samples using modified PDMS-coated stir bars (Baltussen et al., 1999).

The stir bars were incorporated into a narrow glass tube coated with a PDMS layer of 1 mm (corresponding to 55 μL for a 10 mm length) applicable to small sample volumes. Such stir bars are commercially available under the name "Twister" (Gerstel, Germany). After certain stirring time, the stir bar has to be removed, introduced into a glass tube, and transferred to thermal desorption instrument. After desorption and cryofocusing within a cooled programmed temperature vaporization (PTV) injector, the volatiles were transferred onto the analytical GC column. Comparison of SPME and the aforementioned stir bar sorptive extraction (SBSE) technique using identical phases for both techniques exhibited striking differences in the recoveries, which has been attributed to ca. 100 times higher phase ratio in SBSE than in SPME. A comprehensive treatment of SBSE, discussion of the principle, the extraction procedure, and numerous applications was recently been published by David and Sandra (2007).

A further approach for sorptive enrichment of volatiles from the HS of aqueous or solid samples has been described by Tienpont et al. (2000), referred to as HS sorptive extraction (HSSE). This technique implies the sorption of volatiles into PDMS that is chemically bound on the surface of a glass rod support. The device consists of a ca. 5 cm length glass rod of 2 mm diameter and at the last centimeter of 1 mm diameter. This last part is covered with PDMS chemically bound to the glass surface. HS bars with 30, 50, and 100 mg PDMS are commercially available from Gerstel GmbH, Mülheim, Germany. After thermal conditioning at 300°C for 2 h, the glass bar was introduced into the HS of a closed 20 mL HS vial containing the sample to be investigated. After sampling for 45 min, the bar was put into a glass tube for thermal desorption, which was performed with a TDS-2 thermodesorption unit (Gerstel). After desorption and cryofocusing within a PTV injector, the volatiles were transferred onto the analytical GC column. As a result, HSSE exceeded largely the sensitivity attainable with SPME. Several examples referring to the application of HSSE in HS analysis of aromatic and medicinal plants inclusive of details of the sampling procedure were described by Bicchi et al. (2000a).

2.3.2 Chromatographic Separation Techniques

In the course of the last half century, a great number of techniques have been developed and applied to the analysis of essential oils. A part of them has been replaced nowadays by either more effective or easier-to-handle techniques, while other methods maintained their significance and have been permanently improved. Before going into detail, the analytical facilities in the sixties of the last century should be considered briefly. The methods available for the analysis of essential oils have been at that time (Table 2.1) thin-layer chromatography (TLC), various types of liquid column chromatography (LC), and already gas–liquid chromatography (GC). In addition, several spectroscopic techniques such as UV and IR spectroscopy, MS, and ^{1}H-NMR spectroscopy have been available. In the following years, several additional techniques were developed and applied to essential oils analysis, including high-performance liquid chromatography (HPLC); different

TABLE 2.1
Techniques Applied to the Analysis of Essential Oils

Chromatographic Techniques Including Two- and Multidimensional Techniques	Spectroscopic and Spectrometric Techniques	Hyphenated Techniques
TLC	UV	GC-MS
GC	IR	GC-UV
LC	MS	HPLC-GC
HPLC	^{1}H-NMR	SFE-GC
CCC	^{13}C-NMR	GC-FTIR
SFC	NIR	GC-AES
	Raman	HPLC-MS
		SFC-GC
		GC-FTIR-MS
		GC-IRMS
		HPLC-NMR

kinds of countercurrent chromatography (CCC); supercritical fluid chromatography (SFC), including multidimensional coupling techniques, C-13 NMR, near IR (NIR), and Raman spectroscopy; and a multitude of so-called hyphenated techniques, which means online couplings of chromatographic separation devices to spectrometers, yielding valuable structural information of the individual separated components that made their identification feasible.

2.3.2.1 Thin-Layer Chromatography

TLC was one of the first chromatographic techniques and has been used for many years for the analysis of essential oils. This method provided valuable information compared to simple measurements of chemical and physical values and has therefore been adopted as a standard laboratory method for characterization of essential oils in numerous pharmacopoeias. Fundamentals of TLC have been described by Geiss (1987) and in a comprehensive handbook by Stahl (1969b), in which numerous applications and examples on investigations of secondary plant metabolites inclusive of essential oils are given. More recently, the third edition of the handbook of TLC from Shema and Fried (2003) appeared. Further approaches in TLC have been the development of high-performance TLC (Kaiser, 1976) and the application of forced flow techniques such as overpressured layer chromatography and rotation planar chromatography described by Tyihák et al. (1979) and Nyiredy (2003).

In spite of its indisputable simplicity and rapidity, this technique is now largely obsolete for analyzing such complex mixtures like essential oils, due to its low resolution. However, for the rapid investigation of the essential oil pattern of chemical races or the differentiation of individual plant species, this method can still be successfully applied (Gaedcke and Steinhoff, 2000). In addition, silver nitrate and silver perchlorate impregnated layers have been used for the separation of olefinic compounds, especially sesquiterpene hydrocarbons (Prasad et al., 1947), and more recently for the isolation of individual sesquiterpenes (Saritas, 2000).

2.3.2.2 GC

However, the separation capability of GC exceeded all the other separation techniques, even if only packed columns have been used. The exiting evolution of this technique in the past can be impressively demonstrated with four examples of the gas chromatographic separation of the essential oil from rue (Kubeczka, 1981a), a medicinal and aromatic plant. This oil was separated by S. Bruno in 1961 into eight constituents and represented one of the first gas chromatographic analyses of that essential oil. Only a few years later in 1964, separation of the same oil has been improved using a Perkin Elmer gas chromatograph equipped with a 2 m packed column and a thermal conductivity detector (TCD) operated under isothermal conditions yielding 20 separated constituents. A further improvement of the separation of the rue oil was obtained after the introduction of temperature programming of the column oven, yielding approximately 80 constituents. The last significant improvements were a result of the development of high-resolution capillary columns and the sensitive flame ionization detector (FID) (Bicchi and Sandra, 1987). By means of a 50 m glass capillary with 0.25 mm ID, the rue oil could be separated into approximately 150 constituents, in 1981. However, the problems associated with the fragility of the glass capillaries and their cumbersome installation lessened the acknowledgment of this column types, despite their outstanding quality. This has changed since flexible fused silica capillaries became commercially available, which are nearly unbreakable in normal usage. In addition, by different cross-linking technologies, the problems associated with wall coating, especially with polar phases, have been overcome, so that all important types of stationary phases used in conventional GC have been commercially available. The most often used stationary phases for the analysis of essential oils have been, and are still today, the polar phases Carbowax® 20M (DB-Wax, Supelcowax-10, HP-20M, Innowax, etc.) and 14% cyanopropylphenyl–86% methyl polysiloxane (DB- 1701, SPB-1701, HP-1701, OV-1701, etc.) and the nonpolar phases PDMS (DB-1, SPB-1, HP-1 and HP-1 ms, CPSil-5 CB, OV-1, etc.) and 5% phenyl methyl polysiloxane (DB-5, SPB-5, HP-5, CPSil-8 CB, OV-5, SE-54, etc.). Besides different column diameters of 0.53, 0.32, 0.25, 0.10, and 0.05 mm ID, a variety of film thicknesses can be purchased. Increasing column diameter and film thickness of stationary phase increases the sample capacity at the expense of separation efficiency. However, sample capacity has become important, particularly in trace analysis and with some hyphenated techniques such as GC–Fourier transform IR (GC-FTIR), in which a higher sample capacity is necessary when compared to GC-MS. On the other hand, the application of a narrow bore column with 100 μm ID and a film coating of 0.2 μm have been shown to be highly efficient and theoretical plate numbers of approximately 250,000 were received with a 25 m capillary (Lancas et al., 1988). The most common detector in GC is the FID because of its high sensitivity toward organic compounds. The universal applicable TCD is nowadays used only for fixed-gas detection because of its very low sensitivity as compared to FID, and cannot be used in capillary GC. Nitrogen-containing compounds can be selectively detected with the aid of the selective nitrogen–phosphorus detector and chlorinated compounds by the selective and very sensitive electron-capture detector, which is often used in the analysis of pesticides. Oxygen-containing compounds have been selectively detected with special O-FID analyzer even in very complex samples, which was primarily employed to the analysis of oxygenated compounds in gasoline, utilized as fuel-blending agents (Schneider et al., 1982). The oxygen selectivity of the FID is obtained by two online postcolumn reactions: first, a cracking reaction forming carbon monoxide, which is reduced in a second reactor yielding equimolar quantities of methane, which can be sensitively detected by the FID. Since in total each oxygen atom is converted to one molecule methane, the FID response is proportional to the amount of oxygen in the respective molecule. Application of the O-FID to the analysis of essential oils has been presented by Kubeczka (1991). However, conventional GC using fused silica capillaries with different stationary phases, including chiral phases, and the sensitive FID, is up to now the prime technique for the analysis of essential oils.

2.3.2.2.1 *Fast and Ultrafast GC*

Due to the demand for faster GC separations in routine work in the field of GC of essential oils, the development of fast and ultrafast GC seems worthy to be mentioned. The various approaches for fast GC have been reviewed in 1999 (Cramers et al., 1999). The most effective way to speed up GC separation without losing separation efficiency is to use shorter columns with narrow inner diameter and thinner coatings, higher carrier gas flow rates, and accelerated temperature ramps. In Figure 2.1, the conventional and fast GC separation of lime oil is shown, indicating virtually the same separation efficiency in the fast GC and a reduction in time from approximately 60 to 13 min (Mondello et al., 2000).

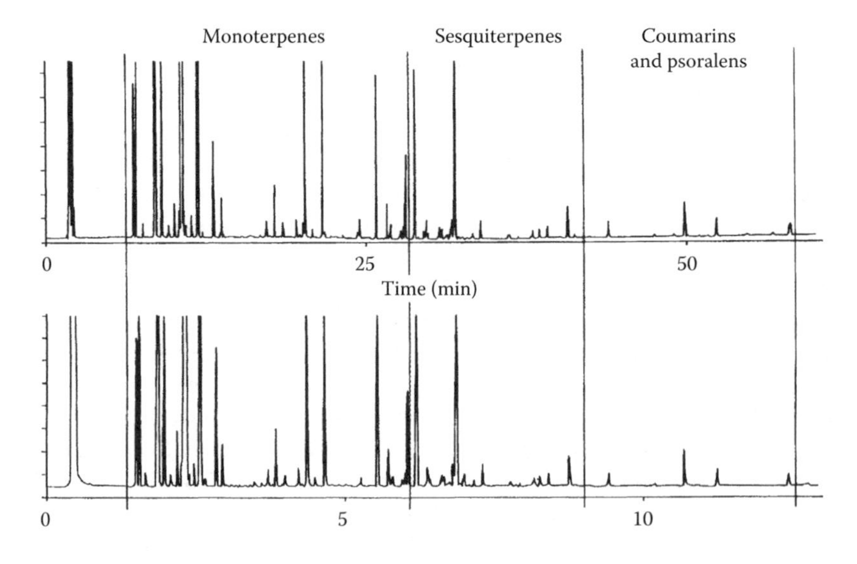

FIGURE 2.1 Comparison of conventional and fast GC separation of lime oil. (From Mondello, L. et al., *LC-GC Eur.*, 13, 495, 2000. With permission.)

TABLE 2.2
Conditions of Conventional, Fast, and Ultrafast GC

	Conventional GC	Fast GC	Ultrafast GC
Column	30 m	10 m	10–15 m
	0.25 mm ID	0.1 mm ID	0.1 mm ID
	0.25 μm film	0.1 μm film	0.1 μm film
Temperature program	50°C–350°C	50°C–350°C	45°C–325°C
	3°C/min	14°C/min	45–200°C/min
Carrier gas	H_2	H_2	H_2
	$u = 36$ cm/s	$u = 57$ cm/s	$u = 120$ cm/s
Sampling frequency	10 Hz	20–50 Hz	50–250 Hz

An ultrafast GC separation of the essential oil from lime with an outstanding reduction of time was recently achieved (Mondello et al., 2004) using a 5 m capillary with 50 μm ID and a film thickness of 0.05 μm operated with a high carrier gas velocity of 120 cm/min and an accelerated three-stage temperature program. The analysis of the essential oil was obtained in approximately 90 s, which equates to a speed gain of approximately 33 times in comparison with the conventional GC separation. However, such a separation cannot be performed with conventional GC instruments. In addition, the mass spectrometric identification of the separated components could only be achieved by coupling GC to a time-of-flight mass spectrometer. In Table 2.2, the separation parameters of conventional, fast, and ultrafast GC separation are given, indicating clearly the relatively low requirements for fast GC, while ultrafast separations can only be realized with modern GC instruments and need a significant higher employment.

2.3.2.2.2 Chiral GC

Besides fast and ultrafast GC separations, one of the most important developments in GC has been the introduction of enantioselective capillary columns in the past with high separation efficiency, so that a great number of chiral substances including many essential oil constituents could be separated and identified. The different approaches of gas chromatographic separation of chiral compounds are briefly summarized in Table 2.3. In the mid-1960s, Gil-Av published results with chiral diamide stationary phases for gas chromatographic separation of chiral compounds, which interacted with the analytes by hydrogen bonding forces (Gil-Av et al., 1965). The ability to separate enantiomers using these phases was therefore limited to substrates with hydrogen bonding donor or acceptor functions.

Diastereomeric association between chiral molecules and chiral transition metal complexes was first described by Schurig (1977). Since hydrogen bonding interaction is not essential for chiral recognition in such a system, a number of compounds could be separated, but this method was limited by the nonsufficient thermal stability of the applied metal complexes.

In 1988 König, as well as Schurig, described the use of cyclodextrin derivatives that act enantioselectively by host–guest interaction by partial intrusion of enantiomers into the cyclodextrin

TABLE 2.3
Different Approaches of Enantioselective GC

1. Chiral diamide stationary phases (Gil-Av et al., 1965)
   Hydrogen bonding interaction
2. Chiral transition metal complexation (Schurig, 1977)
   Complexation gas chromatography
3. Cyclodextrin derivatives (König et al, 1988a,b,c and Schurig and Nowotny, 1988)
   Host–guest interaction, inclusion gas chromatography

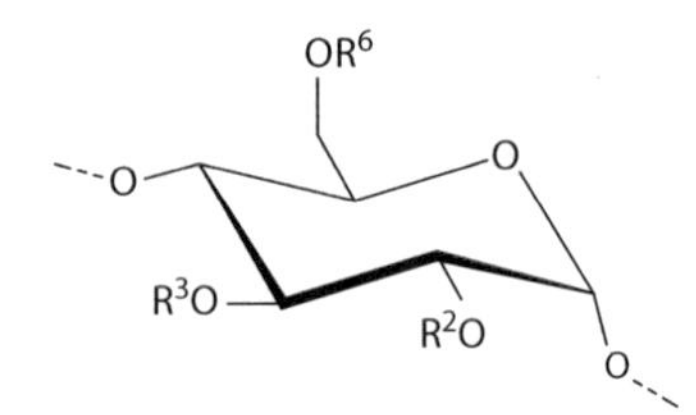

FIGURE 2.2 α-Glucose unit of a cyclodextrin.

cavity. They are cyclic α-(1–4)-bounded glucose oligomers with six-, seven-, or eight-glucose units, which can be prepared by enzymatic degradation of starch with specific cyclodextrin glucosyl transferases from different bacterial strains, yielding α-, β-, and γ-cyclodextrins, and are commercially available. Due to the significant lower reactivity of the 3-hydroxygroups of cyclodextrins, this position can be selectively acylated after alkylation of the two and six positions (Figure 2.2), yielding several nonpolar cyclodextrin derivatives, which are liquid or waxy at room temperature and which proved very useful for gas chromatographic applications.

König and coworkers reported their first results in 1988 with per-*O*-pentylated and selectively 3-*O*-acylated-2,6-di-*O*-pentylated α-, β-, and γ-cyclodextrins, which are highly stable, soluble in nonpolar solvents, and which possess a high enantioselectivity toward many chiral compounds. In the following years, a number of further cyclodextrin derivatives have been synthesized and tested by several groups, allowing the separation of a wide range of chiral compounds, especially due to the improved thermal stability (Table 2.4) (König et al., 1988a,b,c). With the application of 2,3-pentyl-6-methyl-β- and -γ-cyclodextrin as stationary phases, all monoterpene hydrocarbons commonly occurring in essential oils could be separated (König et al., 1992a). The reason for application of two different columns with complementary properties was that on one column not all enantiomers were satisfactorily resolved. Thus, the simultaneous use of these two columns provided a maximum of information and reliability in peak assignment (König et al., 1992b).

After successful application of enantioselective GC to the analysis of enantiomeric composition of monoterpenoids in many essential oils (e.g., Werkhoff et al., 1993; Bicchi et al., 1995; and references cited therein), the studies have been extended to the sesquiterpene fraction. Standard mixtures of known enantiomeric composition were prepared by isolation of individual enantiomers from numerous essential oils by preparative GC and by preparative enantioselective GC. A gas chromatographic separation of a series of isolated or prepared sesquiterpene hydrocarbon enantiomers, showing

TABLE 2.4
Important Cyclodextrin Derivatives

Research Group	Year	Cyclodextrin Derivative
Schurig and Novotny	1988	Per-*O*-methyl-β-CD
König et al.	1988c	Per-*O*-pentyl-(α,β,γ)-CD
König et al.	1988b	3-*O*-acetyl-2,6,-di-*O*-pentyl-(α,β,γ)-CD
König et al.	1989	3-*O*-butyryl-2,6-di-*O*-pentyl-(α,β)-CD
König et al.	1990	6-*O*-methyl-2,3-di-*O*-pentyl-γ-CD
Köng et al.	1990	2,6-Di-*O*-methyl-3-*O*-pentyl-(α,γ)-CD
Dietrich et al.	1992b	2,3-Di-*O*-acetyl-6-*O*-*tert*-butyl-dimethysilyl-β-CD
Dietrich et al.	1992a	2,3-Di-*O*-methyl-6-*O*-*tert*-butyl-dimethylsilyl-(β,γ)-CD
Bicchi et al.	1996	2,3-Di-*O*-ethyl-6-*O*-*tert*-butyl-dimethylsilyl-(β,γ)-CD
Takahisa and Engel	2005a	2,3-Di-*O*-methoxymethyl-6-*O*-*tert*-butyl-dimethylsilyl-β-CD
Takahisa and Engel	2005b	2,3-Di-*O*-methoxymethyl-6-*O*-*tert*-butyl-dimethylsilyl-γ-CD

the separation of 12 commonly occurring sesquiterpene hydrocarbons on a 2,6-methyl-3-pentyl-β-cyclodextrin capillary column has been presented by König et al. (1995). Further investigations on sesquiterpenes have been published by König et al. (1994). However, due to the complexity of the sesquiterpene pattern in many essential oils, it is often impossible to perform directly an enantioselective analysis by coinjection with standard samples on a capillary column with a chiral stationary phase alone. Therefore, in many cases 2D GC had to be performed.

2.3.2.2.3 Two-Dimensional GC

After preseparation of the oil on a nonchiral stationary phase, the peaks of interest have to be transferred to a second capillary column coated with a chiral phase, a technique usually referred to as "heart cutting." In the simplest case, two GC capillaries with different selectivities are serially connected, and the portion of unresolved components from the effluent of the first column is directed into a second column, for example, a capillary with a chiral coating. The basic arrangement used in 2D GC (GC-GC) is shown in Figure 2.3. By means of a valve, the individual fractions of interest eluting from the first column are directed to the second, chiral column, while the rest of the sample may be discarded. With this heart-cutting technique, many separations of chiral oil constituents have been performed in the past. As an example, the investigation of the chiral sesquiterpene hydrocarbon germacrene D shall be mentioned (Kubeczka, 1996), which was found to be a main constituent of the essential oil from the flowering herb from *Solidago canadensis*. The enantioselective investigation of the germacrene-D fraction from a GC run using a nonchiral DB-Wax capillary transferred to a 2,6-methyl-3-pentyl-β-cyclodextrin capillary exhibited the presence of both enantiomers. This is worthy to be mentioned, since in most of other germacrene D containing higher plants nearly exclusively the (−)-enantiomer can be found.

The previously mentioned 2D GC design, however, in which a valve is used to direct the portion of desired effluent from the first into the second column, has obviously several shortcomings. The sample comes into contact with the metal surface of the valve body, the pressure drop of both connected columns may be significant, and the use of only one column oven does not permit to adjust the temperature for both columns properly. Therefore, one of the best approaches to overcome these limitations has been realized by a commercially available two-column oven instrument using a Deans-type pressure balancing interface between the two columns called a "live-T connection" (Figure 2.4) providing considerable flexibility (Hener, 1990). By means of that instrument, the enantiomeric composition of several essential oils has been investigated very successfully. As an example, the investigation of the essential oil from *Lavandula angustifolia* shall be mentioned (Kreis

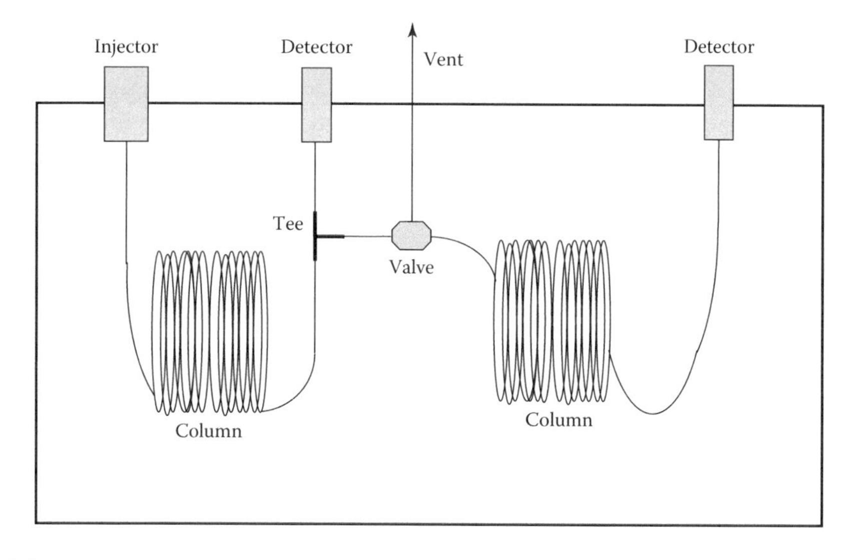

FIGURE 2.3 Basic arrangement used in 2D GC.

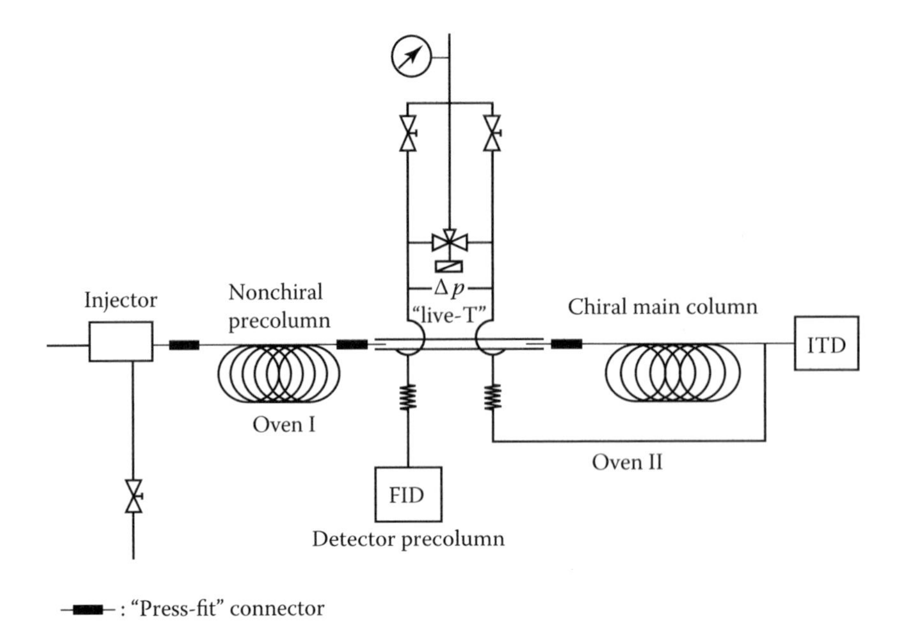

FIGURE 2.4 Scheme of enantioselective multidimensional GC with "live-T" column switching. (From Hener, U., Chirale Aromastoffe—Beiträge zur Struktur, Wirkung und Analytik, Dissertation, Goethe-University of Frankfurt/Main, Frankfurt, Germany, 1990. With permission.)

and Mosandl, 1992) showing the simultaneous stereoanalysis of a mixture of chiral compounds, which can be found in lavender oils, using the column combination Carbowax 20M as the precolumn and 2,3-di-*O*-acetyl-6-*O*-*tert*-butyldimethylsilyl-β-cyclodextrin as the main column. All the unresolved enantiomeric pairs from the precolumn could be well separated after transferring them to the chiral main column in a single run. As a result, it was found that most of the characteristic and genuine chiral constituents of lavender oil exhibit a high enantiomeric purity.

A different and inexpensive approach for transferring individual GC peaks onto a second column has been presented by Kubeczka (1997a , using an SPME device. The highly diluted organic vapor of a fraction eluting from a GC capillary in the carrier gas flow has been absorbed on a coated SPME fiber and introduced onto a second capillary. As could be demonstrated, no modification of the gas chromatograph had to be performed to realize that approach. The eluting fractions were sampled after shutting the valves of the air, of hydrogen and the makeup gas if applied. In order to minimize the volume of the detector to avoid dilution of the eluting fraction and to direct the gas flow to the fiber surface, a capillary glass tubing of 1.5 mm ID was inserted into the FID and fixed and tightened by an O-ring (Figure 2.5). At the beginning of peak elution, controlled only by time, a 100 μm PDMS fiber was introduced into the mounted glass capillary tubing and withdrawn at the end of peak elution. Afterward, the fiber within the needle was introduced into the injector of a second capillary column with a chiral stationary phase. Two examples concerning the investigation of bergamot oil have been shown. At first, the analysis of an authentic sample of bergamot oil, containing chiral linalool, and the respective chiral actetate is carried out. Both components were cut separately and transferred to an enantioselective cyclodextrin Lipodex® E capillary. The chromatograms clearly have shown that the authentic bergamot oil contains nearly exclusively the (−)-enantiomers of linalool and linalyl acetate, while the respective (+)-enantiomers could only be detected as traces. In contrast to the authentic sample, a commercial sample of bergamot oil, which was analyzed under the same conditions, exhibited the presence of significant amounts of both enantiomers of linalool and linalyl acetate indicating a falsification by admixing the respective racemic alcohol and ester.

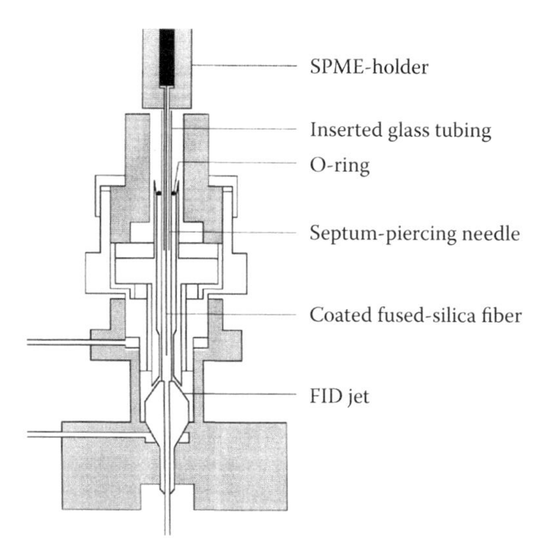

FIGURE 2.5 Cross section of an FID of an HP 5890 gas chromatograph with an inserted SPME fiber. (From Kubeczka, K.-H. *Essential Oil Symposium Proceedings*, 1997b, p. 145. With permission.)

2.3.2.2.4 Comprehensive Multidimensional GC

One of the most powerful separation techniques that has been recently applied to the investigation of essential oils is the so-called comprehensive multidimensional GC (GC × GC). This technique is a true multidimensional GC (MDGC) since it combines two directly coupled columns and importantly is able to subject the entire sample to simultaneous two-column separation. Using that technique, the need to select heart cuts, as used in conventional MDGC, is no longer required. Since components now are retained in two different columns, the net capacity is the product of the capacities of the two applied columns increasing considerably the resolution of the total system. Details regarding that technique will be given in Chapter 7.

2.3.2.3 Liquid Column Chromatography

The different types of LC have been mostly used in preparative or semipreparative scale for preseparation of essential oils or for isolation of individual oil constituents for structure elucidation with spectroscopic methods and were rarely used at that time as an analytical separation tool alone, because GC plays a central role in the study of essential oils.

2.3.2.3.1 Preseparation of Essential Oils

A different approach besides 2D GC, which has often been used in the past to overcome peak overlapping in a single GC run of an essential oil, has been preseparation of the oil with LC. The most common method of fractionation is the separation of hydrocarbons from the oxygenated terpenoids according to Miller and Kirchner (1952), using silica gel as an adsorbent. After elution of the nonpolar components from the column with pentane or hexane, the more polar oxygen-containing constituents are eluted in order of increasing polarity after applying more and more polar eluents.

A very simple and standardized fractionation in terms of speed and simplicity has been published by Kubeczka (1973) using dry-column chromatography. The procedure, which has been proved useful in numerous experiments for prefractionation of an essential oil, allows a preseparation into five fractions of increasing polarity. The preseparation of an essential oil into oxygenated constituents, monoterpene hydrocarbons, and sesquiterpene hydrocarbons, which is—depending on the oil composition—sometimes of higher practical use, can be performed successfully using reversed-phase RP-18 HPLC (Schwanbeck et al., 1982). The HPLC was operated on a semipreparative scale by stepwise elution with methanol–water 82.5:17.5 (solvent A) and pure methanol (solvent B).

The elution order of the investigated oil was according to decreasing polarity of the components and within the group of hydrocarbons to increasing molecular weight. Fraction 1 contained all oxygenated mono- and sesquiterpenoids, fraction 2 the monoterpene hydrocarbons, and fraction 3—eluted with pure methanol—the sesquiterpene hydrocarbons. A further alternative to the mentioned separation techniques is flash chromatography, initially developed by Still et al. (1978), which has often been used as a rapid form of preparative LC based on a gas- or air pressure–driven short-column chromatography. This technique, optimized for rapid separation of quantities typically in the range of 0.5–2.0 g, uses dry-packed silica gel in an appropriate column. The separation of the sample generally takes only 5–10 min and can be performed with inexpensive laboratory equipment. However, impurities and active sites on dried silica gel were found to be responsible for isomerization of a number of oil constituents. After deactivation of the dried silica gel by adding 5% water, isomerization processes could be avoided (Scheffer et al., 1976). A different approach using HPLC on silica gel and isocratic elution with a ternary solvent system for the separation of essential oils has been published by Chamblee et al. (1985). In contrast to the aforementioned commonly used offline pretreatment of a sample, the coupling of two or more chromatographic systems in an online mode offers advantages of ease of automation and usually of a shorter analysis time.

2.3.2.3.2 High-Performance Liquid Column Chromatography

The good separations obtained by GC have delayed the application of HPLC to the analysis of essential oils; however, HPLC analysis offers some advantages, if GC analysis of thermolabile compounds is difficult to achieve. Restricting factors for application of HPLC for analyses of terpenoids are the limitations inherent in the commonly available detectors and the relatively small range of k' values of liquid chromatographic systems. Since temperature is an important factor that controls k' values, separation of terpene hydrocarbons was performed at −15°C using a silica gel column and *n*-pentane as a mobile phase. Monitoring has been achieved with UV detection at 220 nm. Under these conditions, mixtures of commonly occurring mono- and sesquiterpene hydrocarbons could be well separated (Schwanbeck and Kubeczka, 1979; Kubeczka, 1981b). However, the silica gel had to be deactivated by adding 4.8% water prior to separation to avoid irreversible adsorption or alteration of the sample. The investigation of different essential oils by HPLC already has been described in the seventies of the last century (e.g., Komae and Hayashi, 1975; Ross, 1976; Wulf et al., 1978; McKone, 1979; Scott and Kucera, 1979). In the last publication, the authors have used a rather long microbore packed column, which had several hundred thousand theoretical plates. Besides relatively expensive equipment, the HPLC chromatogram of an essential oil, separated on such a column, could only be obtained at the expense of long analysis time. The mentioned separation needed about 20 h and may be only of little value in practical applications.

More recent papers with regard to HPLC separation of essential oils were published, for example, by Debrunner et al. (1995), Bos et al. (1996), and Frérot and Decorzant (2004), and applications using silver ion–impregnated sorbents have been presented by Pettei et al. (1977), Morita et al. (1983), Friedel and Matusch (1987), and van Beek et al. (1994). The literature on the use and theory of silver complexation chromatography has been reviewed by van Beek and Subrtova (1995). HPLC has also been used to separate thermally labile terpenoids at low temperature by Beyer et al. (1986), showing the temperature dependence of the separation efficiency. The investigation of an essential oil fraction from *Cistus ladanifer* using RP-18 reversed-phase HPLC at ambient temperature and an acetonitrile–water gradient was published by Strack et al. (1980). Comparison of the obtained HPLC chromatogram with the respective GC run exhibits a relatively good HPLC separation in the range of sesqui- and diterpenes, while the monoterpenes exhibited, as expected, a significant better resolution by GC. The enantiomeric separation of sesquiterpenes by HPLC with a chiral stationary phase has recently been shown by Nishii et al. (1997), using a Chiralcel® OD column.

2.3.2.4 Supercritical Fluid Chromatography

Supercritical fluids are highly compressed gases above their critical temperature and critical pressure point, representing a hybrid state between a liquid and a gas, which have physical properties intermediate between liquid and gas phases. The diffusion coefficient of a fluid is about two orders of magnitude larger and the viscosity is two orders of magnitude lower than the corresponding properties of a liquid. On the other hand, a supercritical fluid has a significant higher density than a gas. The commonly used carbon dioxide as a mobile phase, however, exhibits a low polarity (comparable to pentane or hexane), limiting the solubility of polar compounds, a problem that has been solved by adding small amounts of polar solvents, for example, methanol or ethanol, to increase mobile-phase polarity, thus permitting separations of more polar compounds (Chester and Innis, 1986). A further strength of SFC lies in the variety of detection systems that can be applied. The intermediate features of SFC between GC and LC can be profitable when used in a variety of detection systems, which can be classified in *LC-* and *GC-like* detectors. In the first case, measurement takes place directly in the supercritical medium or in the liquid phase, whereas GC-like detection proceeds after a decompression stage.

Capillary SFC using carbon dioxide as mobile phase and a FID as detector has been applied to the analysis of several essential oils and seemed to give more reliable quantification than GC, especially for oxygenated compounds. However, the separation efficiency of GC for monoterpene hydrocarbons was, as expected, better than that of SFC. Manninen et al. (1990) published a comparison of a capillary GC versus a chromatogram obtained by capillary SFC from a linalool–methyl chavicol basil oil chemotype exhibiting a fairly good separation by SFC.

2.3.2.5 Countercurrent Chromatography

CCC is according to Conway (1989) a form of liquid–liquid partition chromatography, in which centrifugal or gravitational forces are employed to maintain one liquid phase in a coil or train of chambers stationary, while a stream of a second, immiscible phase is passed through the system in contact with the stationary liquid phase. Retention of the individual components of the sample to be analyzed depends only on their partition coefficients and the volume ratio of the two applied liquid phases. Since there is no porous support, adsorption and catalytic effects encountered with solid supports are avoided.

2.3.2.5.1 Droplet Countercurrent Chromatography

One form of CCC, which has been sporadically applied to separate essential oils into fractions or in the ideal case into individual pure components, is droplet countercurrent chromatography (DCCC). The device, which has been developed by Tanimura et al. (1970), consists of 300–600 glass tubes, which are connected to each other in series with Teflon tubing and filled with a stationary liquid. Separation is achieved by passing droplets of the mobile phase through the columns, thus distributing mixture components at different ratios leading to their separation. With the development of a water-free solvent system, separation of essential oils could be achieved (Becker et al., 1981, 1982). Along with the separation of essential oils, the method allows the concentration of minor components, since relatively large samples can be separated in one analytical run (Kubeczka, 1985).

2.3.2.5.2 Rotation Locular Countercurrent Chromatography

The rotation locular countercurrent chromatography (RLCC) apparatus (Rikakikai Co., Tokyo, Japan) consists of 16 concentrically arranged and serially connected glass tubes. These tubes are divided by Teflon disks with a small hole in the center, thus creating small compartments or locules. After filling the tubes with the stationary liquid, the tubes are inclined to a 30° angle from horizontal. In the ascending mode, the lighter mobile phase is applied to the bottom of the first tube by a constant flow pump, displacing the stationary phase as its volume attains the level of the hole in the disk. The mobile phase passes through this hole and enters into the next compartment, where the process continues until the mobile phase emerges from the uppermost locule. Finally, the two phases fill approximately half of each compartment. The dissolved essential oil subsequently introduced is subjected to a multistage partitioning process that leads to separation of the individual components. Whereas gravity contributes to the phase separation, rotation of the column assembly

(60–80 rpm) produces circular stirring of the two liquids to promote partition. If the descending mode is selected for separation, the heavier mobile phase is applied at the top of each column by switching a valve. An overview on applications of RLCC in natural products isolation inclusive of a detailed description of the device and the selection of appropriate solvent systems has been presented by Snyder et al. (1984).

Comparing RLCC to the aforementioned DCCC, one can particularly stress the superior flexibility of RLCC. While DCCC requires under all circumstances a two-phase system able to form droplets in the stationary phase, the choice of solvent systems with RLCC is nearly free. So the limitations of DCCC, when analyzing lipophilic samples, do not apply to RLCC. The separation of a mixture of terpenes has been presented by Kubeczka (1985). A different method, the high-speed centrifugal CCC developed by Ito and coworkers in the mid-1960s (Ito et al., 1966), has been applied to separate a variety of nonvolatile natural compounds; however, separation of volatiles has, strange to say, until now not seriously been evaluated.

2.3.3 Hyphenated Techniques

2.3.3.1 Gas Chromatography-Mass Spectrometry

The advantage of online coupling of a chromatographic device to a spectrometer is that complex mixtures can be analyzed in detail by spectral interpretation of the separated individual components. The coupling of a gas chromatograph with a mass spectrometer is the most often used and a well-established technique for the analysis of essential oils, due to the development of easy-to-handle powerful systems concerning sensitivity, data acquisition and processing, and above all their relatively low cost. The very first application of a GC-MS coupling for the identification of essential oil constituents using a capillary column was already published by Buttery et al. (1963). In those times, mass spectra have been traced on UV recording paper with a five-element galvanometer, and their evaluation was a considerable cumbersome task.

This has changed after the introduction of computerized mass digitizers yielding the mass numbers and the relative mass intensities. The different kinds of GC-MS couplings available at the end of the seventies of the last century have been described in detail by ten Noever de Brauw (1979). In addition, different types of mass spectrometers have been applied in GC-MS investigations such as magnetic sector instruments, quadrupole mass spectrometers, ion-trap analyzers (e.g., ion-trap detector), and time-of-flight mass spectrometers, which are the fastest MS analyzers and therefore used for very fast GC-MS systems (e.g., in comprehensive multidimensional GC-MS). Surprisingly, a time-of-flight mass spectrometer was used in the very first description of a GC-MS investigation of an essential oil mentioned before. From the listed spectrometers, the magnetic sector and quadrupole instruments can also be used for selective ion monitoring, to improve sensitivity for the analysis of target compounds and for discrimination of overlapping GC peaks.

The great majority of today's GC-MS applications utilize 1D capillary GC with quadrupole MS detection and electron ionization. Nevertheless, there are substantial numbers of applications using different types of mass spectrometers and ionization techniques. The proliferation of GC-MS applications is also a result of commercially available easy-to-handle dedicated mass spectral libraries (e.g., NIST/EPA/NIH 2005; WILEY Registry 2006; MassFinder 2007; and diverse printed versions such as Jennings and Shibamoto, 1980; Joulain and König, 1998; Adams, 1989, 1995, 2007 inclusive of retention indices) providing identification of the separated compounds. However, this type of identification has the potential of producing some unreliable results, if no additional information is used, since some compounds, for example, the sesquiterpene hydrocarbons α-cuprenene and β-himachalene, exhibit identical fragmentation pattern and only very small differences of their retention index values. This example demonstrates impressively that even a good library match and the additional use of retention data may lead in some cases to questionable results, and therefore require additional analytical data, for example, from NMR measurements.

2.3.3.1.1 GC-Chemical Ionization-MS and GC-Tandem MS

Although GC-electron impact (EI)-MS is a very useful tool for the analysis of essential oils, this technique can sometimes be not selective enough and requires more sophisticated techniques such as GC-chemical ionization-MS (GC-CI-MS) and GC-tandem MS (GC-MS-MS). The application of CI-MS using different reactant gases is particularly useful, since many terpene alcohols and esters fail to show a molecular ion. The use of OH^- as a reactant ion in negative CI-MS appeared to be an ideal solution to this problem. This technique yielded highly stable quasi-molecular ions M–H, which are often the only ions in the obtained spectra of the aforementioned compounds. As an example, the EI and CI spectra of isobornyl isovalerate—a constituent of valerian oil—shall be quoted (Bos et al., 1982). The respective EI mass spectrum shows only a very small molecular ion at 238. Therefore, the chemical ionization spectra of isobornyl acetate were performed with isobutene as a reactant gas a $[C_{10}H_{17}]^+$ cation and in the negative CI mode with OH^- as a reactant gas two signals with the masses 101, the isovalerate anion, and 237 the quasi-molecular ion $[M–H]^-$. Considering all these obtained data, the correct structure of the oil constituent could be deduced. The application of isobutane and ammonia as reactant gases has been presented by Schultze et al. (1992), who investigated sesquiterpene hydrocarbons by GC-CI-MS. Fundamental aspects of chemical ionization MS have been reviewed by Bruins (1987), discussing the different reactant gases applied in positive and negative ion chemical ionization and their applications in essential oil analysis.

The utilization of GC-MS-MS to the analysis of a complex mixture will be shown in Figure 2.6. In the investigated vetiver oil (Cazaussus et al., 1988), one constituent, the norsesquiterpene ketone khusimone, has been identified by using GC-MS-MS in the collision-activated dissociation mode. The molecular ion at *m/z* 204 exhibited a lot of daughter ions, but only one of them gave a daughter ion at *m/z* 108, a fragment rarely occurring in sesquiterpene derivatives so that the presence of khusimone could be undoubtedly identified.

2.3.3.2 High-Resolution GC-FTIR Spectroscopy

A further hyphenated technique, providing valuable analytical information, is the online coupling of a gas chromatograph with a FTIR spectrometer. The capability of IR spectroscopy to provide discrimination between isomers makes the coupling of a gas chromatograph to an FTIR spectrometer

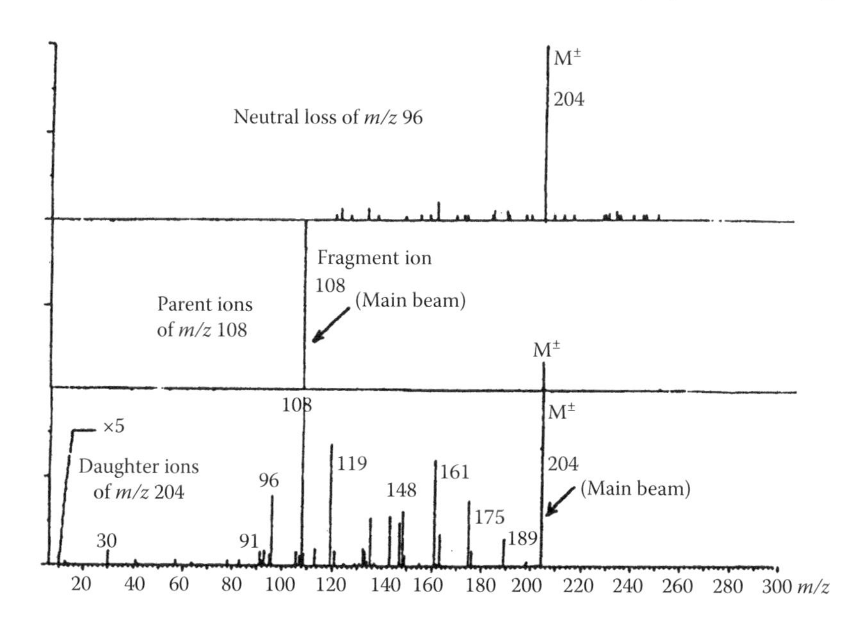

FIGURE 2.6 GC-EIMS-MS of khusimone of vetiver oil. (From Cazaussus, A. et al., *Chromatographia*, 25, 865, 1988. With permission.)

suited as a complementary method to GC/MS for the analysis of complex mixtures like essential oils. The GC/FTIR device consists basically of a capillary gas chromatograph and an FTIR spectrometer including a dedicated computer and ancillary equipment. As each GC peak elutes from the GC column, it enters a heated IR measuring cell, the so-called light pipe, usually a gold-plated glass tube with IR transparent windows. There, the spectrum is measured as an interferogram from which the familiar absorbance spectrum can be calculated by computerized Fourier transformation. After passing the light pipe, the effluent is directed back into the FID of the gas chromatograph. More detailed information on the experimental setup was given by Herres et al. (1986) and Herres (1987).

In the latter publication, for example, the vapor-phase IR spectra of all the four isomers of pulegol and dihydrocarveol are shown, which have been extracted from a GC/FTIR run. These examples convincingly demonstrate the capability of distinguishing geometrical isomers with the aid of vapor-phase IR spectra, which cannot be achieved by their mass spectra. A broad application of GC-FTIR in the analysis of essential oils, however, is limited by the lack of sufficient vapor-phase spectra of uncommon compounds, which are needed for reference use, since the spectra of isolated molecules in the vapor phase can be significantly different from the corresponding condensed-phase spectra.

A different approach has been published by Reedy et al. in 1985, using a cryogenically freezing of the GC effluent admixed with an inert gas (usually argon) onto a rotating disk maintained at liquid He temperature to form a solid matrix trace. After the separation, reflection absorption spectra can be obtained from the deposited solid trace. A further technique published by Bourne et al. (1990) is the subambient trapping, whereby the GC effluent is cryogenically frozen onto a moving IR transparent window of zinc selenide (ZnSe). An advantage of the latter technique is that the unlike larger libraries of conventional IR spectra can be searched in contrast to the limited number of vapor-phase spectra and those obtained by matrix isolation. A further advantage of both cryogenic techniques is the significant higher sensitivity, which exceeds the detection limits of a light pipe instrument by approximately two orders of magnitude.

Comparing GC/FTIR and GC/MS, advantages and limitations of each technique become visible. The strength of IR lies—as discussed before—in distinguishing isomers, whereas identification of homologues can only be performed successfully by MS. The logical and most sophisticated way to overcome these limitations has been the development of a combined GC/FTIR/MS instrument, whereby simultaneously IR and mass spectra can be obtained.

2.3.3.3 GC-UV Spectroscopy

The instrumental coupling of gas chromatograph with a rapid scanning UV spectrometer has been presented by Kubeczka et al. (1989). In this study, a UV-VIS diode-array spectrometer (Zeiss, Oberkochen, FRG) with an array of 512 diodes was used, which provided continuous monitoring in the range of 200–620 nm. By interfacing the spectrometer via fiber optics to a heated flow cell, which was connected by short heated capillaries to the GC column effluent, interferences of chromatographic resolution could be minimized. With the aid of this device, several terpene hydrocarbons have been investigated. In addition to displaying individual UV spectra, the available software rendered the analyst to define and to display individual window traces, 3D plots, and contour plots, which are valuable tools for discovering and deconvoluting gas chromatographic unresolved peaks.

2.3.3.4 Gas Chromatography-Atomic Emission Spectroscopy

A device for the coupling of capillary GC with atomic emission spectroscopy (GC-AES) has been presented by Wylie and Quimby (1989). By means of this coupling, 23 elements of a compound including all elements of organic substances separated by GC could be selectively detected providing the analyst not only with valuable information on the elemental composition of the individual components of a mixture but also with the percentages of the elemental composition. The device incorporates a microwave-induced helium plasma at the outlet of the column coupled to an optical emission spectrometer. From the 15 most commonly occurring elements in organic compounds, up to 8 could be detected and measured simultaneously, for example, C, O, N, and S, which are of importance with respect to the analysis of essential oils. The examples given in the literature (e.g., Wylie and Quimby, 1989; Bicchi et al., 1992; David and Sandra, 1992; Jirovetz et al., 1992; Schultze, 1993) indicate that the GC-AES coupling can provide the analyst with additional valuable information, which are to some extent complementary to the date obtained by GC-MS and GC-FTIR, making the respective library searches more reliable and more certain.

However, the combined techniques GC-UV and GC-AES have not gained much importance in the field of essential oil research, since UV spectra offer only low information and the coupling of a GC-AES, yielding the exact elemental composition of a component, can to some extent be obtained by precise mass measurement. Nevertheless, the online coupling GC-AES is still today efficiently used in environmental investigations.

2.3.3.5 Gas Chromatography-Isotope Ratio Mass Spectrometry

In addition to enantioselective capillary GC, the online coupling of GC with isotope-ratio MS (GC-IRMS) is an important technique in authentication of food flavors and essential oil constituents. The online combustion of effluents from capillary gas chromatographic separations to determine the isotopic compositions of individual components from complex mixtures was demonstrated by Matthews and Hayes (1978). On the basis of this work, the online interfacing of capillary GC with IRMS was later improved. With the commercially available GC-combustion IRMS device, measurements of the ratios of the stable isotopes $^{13}C/^{12}C$ have been accessible and respective investigations have been reported in several papers (e.g., Bernreuther et al., 1990; Carle et al., 1990; Braunsdorf et al., 1992, 1993; Frank et al., 1995; Mosandl and Juchelka, 1997). A further improvement was the development of the GC-pyrolysis-IRMS (GC-P-IRMS) making measurements of $^{18}O/^{16}O$ ratios and later $^{2}H/^{1}H$ ratios feasible (Juchelka et al., 1998; Ruff et al., 2000; Hör et al., 2001; Mosandl, 2004). Thus, the GC-P-IRMS device (Figure 2.7) appears today as one of the most sophisticated instruments for the appraisal of the genuineness of natural mixtures.

2.3.3.6 High-Performance Liquid Chromatography-Gas Chromatography

The online coupling of an HPLC device to a capillary gas chromatograph offers a number of advantages, above all higher column chromatographic efficiency, simple and rapid method development, simple cleanup of samples from complex matrices, and effective enrichment of the

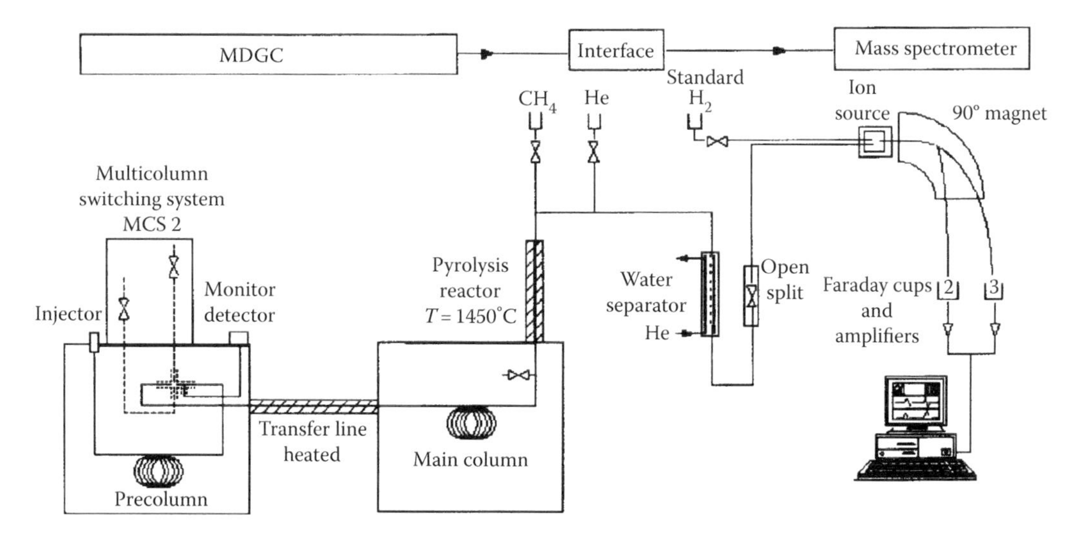

FIGURE 2.7 Scheme of an MDGC-C/P-IRMS device. (From Sewenig, S. et al., *J. Agric. Food Chem.*, 53, 838, 2005. With permission.)

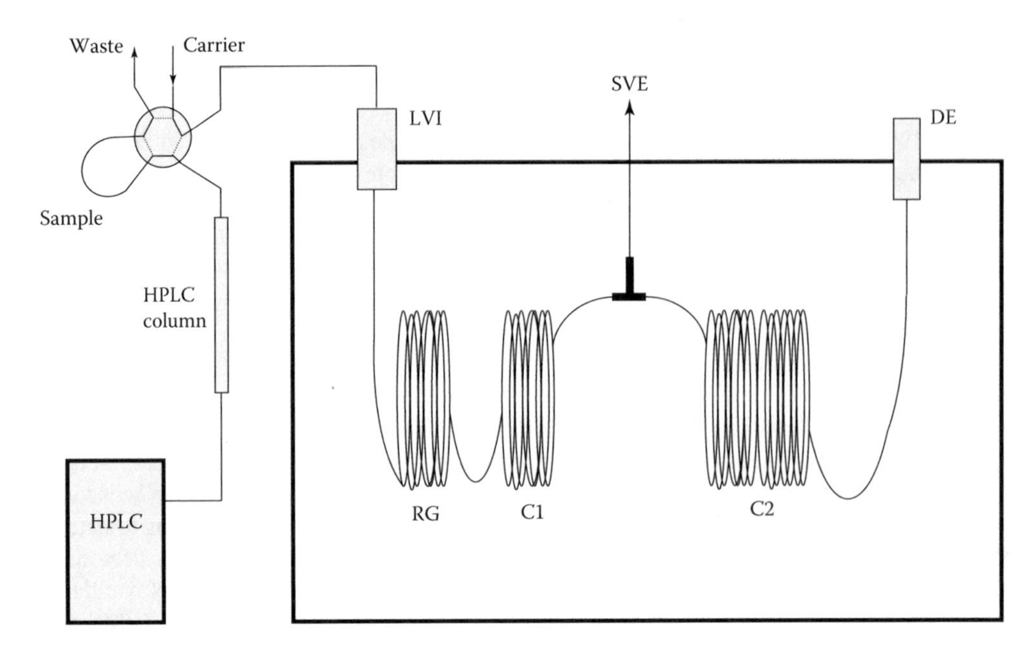

FIGURE 2.8 Basic arrangement of an HPLC-GC device with a sample loop interface. RG, retention gap; C1, retaining column; C2, analytical column; LVI, large volume injector; and SVE, solvent vapor exit.

components of interest; additionally, the entire analytical procedure can easily be automated, thus increasing accuracy and reproducibility. The commercially available HPLC-GC coupling consists of an HPLC device that is connected with a capillary gas chromatograph via an interface allowing the transfer of HPLC fractions. Two different types of interfaces have been often used. The on-column interface is a modification of the on-column injector for GC; it is particularly suited for the transfer of fairly small fraction containing volatile constituents (Dugo et al., 1994; Mondello et al., 1994a,b, 1995). The second interface uses a sample loop and allows to transfer large sample volumes (up to 1 mL) containing components with limited volatilities. Figure 2.8 gives a schematic view of such an LC-GC instrument. In the shown position of the six-port valve, the desired fraction of the HPLC effluent is stored in the sample loop, while the carrier gas is passed through the GC columns. After switching the valve, the content of the sample loop is driven by the carrier gas into the large volume injector and vaporized and enters the precolumns, where the sample components are retained and most of the solvent vapor can be removed through the solvent vapor exit. After closing this valve and increasing the GC-oven temperature, the sample components are volatilized and separated in the main column reaching the detector. The main drawback of this technique, however, may be the loss of highly volatile compounds that are vented together with the solvent. As an example of an HPLC-GC investigation, the preseparation of lemon oil with gradient elution into four fractions is quoted (Munari et al., 1990). The respective gas chromatograms of the individual fractions exhibit good separation into hydrocarbons, esters, carbonyls, and alcohols, facilitating gas chromatographic separation and identification. Due to automation of all analytical steps involved, the manual operations are significantly reduced, and very good reproducibility was obtained. In three excellent review articles, the different kinds of HPLC-GC couplings are discussed in detail, describing their advantages and limitations with numerous references cited therein (Mondello et al., 1996, 1999; Dugo et al., 2003).

2.3.3.7 HPLC-MS, HPLC-NMR Spectroscopy

The online couplings of HPLC with MS and NMR spectroscopy are further important techniques combining high-performance separation with structurally informative spectroscopic techniques, but they are mainly applied to nonvolatile mixtures and shall not be discussed in more detail here, although they are very useful for investigating plant extracts.

Some details concerning the different ionization techniques used in HPLC-MS have been presented among other things by Dugo et al. (2005).

2.3.3.8 Supercritical Fluid Extraction–Gas Chromatography

Although SFE is not a chromatographic technique, separation of mixtures can be obtained during the extraction process by varying the physical properties such as temperature and pressure to obtain fractions of different composition. Detailed reviews on the physical background of SFE and its application to natural products analysis inclusive of numerous applications have been published by Modey et al. (1995) and more recently by Pourmortazavi and Hajimirsadeghi (2007). The different types of couplings (offline and online) have been presented by several authors. Houben et al. (1990) described an online coupling of SFE with capillary GC using a programmed temperature vaporizer as an interface. Similar approaches have been used by Blanch et al. (1994) in their investigations of rosemary leaves and by Ibanez et al. (1997) studying Spanish raspberries. In both the last two papers, an offline procedure was applied. A different device has been used by Hartonen et al. (1992) in a study of the essential oil of *Thymus vulgaris* using a cooled stainless steel capillary for trapping the volatiles connected via a six-port valve to the extraction vessel and the GC column. After sampling of the volatiles within the trap, they have been quickly vaporized and flushed into the GC column by switching the valve. The recoveries of thyme components by SFE-GC were compared with those obtained from hydrodistilled thyme oil by GC exhibiting a good agreement. The SFE-GC analyses of several flavor and fragrance compounds of natural products by transferring the extracted compounds from a small SFE cell directly into a GC capillary has already been presented by Hawthorne et al. (1988). By inserting the extraction cell outlet restrictor (a 20 μm ID capillary) into the GC column through a standard on-column injection port, the volatiles were transferred and focused within the column at 40°C, followed by rapid heating to 70°C (30°C/min) and successive usual temperature programming. The suitability of that approach has been demonstrated with a variety of samples including rosemary, thyme, cinnamon, spruce needles, orange peel, and cedar wood. In a review article from Greibrokk, published in 1995, numerous applications of SFE connected online with GC and other techniques, the different instruments, and interfaces have been discussed, including the main parameters responsible for the quality of the obtained analytical results. In addition, the instrumental setups for SFE-LC and SFE-SFC couplings are given.

2.3.3.9 Supercritical Fluid Chromatography-Gas Chromatography

Online coupling of SFC with GC has sporadically been used for the investigation of volatiles from aromatic herbs and spices. The requirements for instrumentation regarding the pumps, the restrictors, and the detectors are similar to those of SFE-GC. Additional parts of the device are the separation column and the injector, to introduce the sample into the mobile phase and successively into the column. The most common injector type in SFC is the high-pressure valve injector, similar to those used in HPLC. With this valve, the sample is loaded at ambient pressure into a sample loop of defined size and can be swept into the column after switching the valve to the injection position. The separation columns used in SFC may be either packed or open tubular columns with their respective advantages and disadvantages. The latter mentioned open tubular columns for SFC can be compared with the respective GC columns; however, they must have smaller internal diameter. With regard to the detectors used in SFC, the FID is the most common applied detector, presuming that no organic modifiers have been admixed to the mobile phase. In that case, for example, a UV detector with a high-pressure flow cell has to be taken into consideration.

In a paper presented by Yamauchi and Saito (1990), cold-pressed lemon-peel oil has been separated by semipreparative SFC into three fractions (hydrocarbons, aldehydes and alcohols), and esters together with other oil constituents. The obtained fractions were afterward analyzed by capillary GC. SFC has also often been combined with SFE prior to chromatographic separation in plant volatile oil analysis, since in both techniques the same solvents are used, facilitating an online coupling. SFE and online-coupled SFC have been applied to the analysis of turmeric, the rhizomes of *Curcuma longa* L., using modified carbon dioxide as the extractant, yielding fractionation of turmerones curcuminoids in a single run (Sanagi et al., 1993). A multidimensional SFC-GC system was developed by Yarita et al. (1994) to separate online the constituents of citrus essential oils by stepwise pressure programming. The eluting fractions were introduced into a split/splitless injector of a gas chromatograph and analyzed after cryofocusing prior to GC separation. An SFC-GC investigation of cloudberry seed oil extracted with supercritical carbon dioxide was described by Manninen and Kallio (1997), in which SFC was mainly used for the separation of the volatile constituents from the low-boiling compounds, such as triacylglycerols. The volatiles were collected in a trap column and refocused before being separated by GC. Finally, an online technique shall be mentioned by which the compounds eluting from the SFC column can be completely transferred to GC, but also for selective or multistep heart-cutting of various sample peaks as they elute from the SFC column (Levy et al., 2005).

2.3.3.10 Couplings of SFC-MS and SFC-FTIR Spectroscopy

Both coupling techniques such as SFC-MS and SFC-FTIR have nearly exclusively been used for the investigation of low-volatile more polar compounds. Arpino published in 1990 a comprehensive article on the different coupling techniques in SFC-MS, which have been presented up to 1990 including 247 references. A short overview of applications using SFC combined with benchtop mass spectrometers was published by Ramsey and Raynor (1996). However, the only paper concerning the application of SFC-MS in essential oil research was published by Blum et al. (1997). With the aid of a newly developed interface and an injection technique using a retention gap, investigations of thyme extracts have been successfully performed.

The application of SFC-FTIR spectroscopy for the analysis of volatile compounds has also rarely been reported. One publication found in the literature refers to the characterization of varietal differences in essential oil components of hops (Auerbach et al., 2000). In that paper, the IR spectra of the main constituents were taken as films deposited on AgCl disks and compared with spectra obtained after chromatographic separation in a flow cell with IR transparent windows, exhibiting a good correlation.

2.3.4 Identification of Multicomponent Samples without Previous Separation

In addition to chromatographic separation techniques including hyphenated techniques, several spectroscopic techniques have been applied to investigate the composition of essential oils without previous separation.

2.3.4.1 UV Spectroscopy

UV spectroscopy has only little significance for the direct analysis of essential oils due to the inability to provide uniform information on individual oil components. However, for testing the presence of furanocoumarins in various citrus oils, which can cause photodermatosis when applied externally, UV spectroscopy is the method of choice. The presence of those components can be easily determined due to their characteristic UV absorption. In the European pharmacopoeia, for example, quality assessment of lemon oil, which has to be produced by cold pressing, is therefore performed by UV spectroscopy in order to exclude cheaper distilled oils.

2.3.4.2 IR Spectroscopy

Several attempts have also been made to obtain information about the composition of essential oils using IR spectroscopy. One of the first comprehensive investigations of essential oils was published by Bellanato and Hidalgo (1971) in the book *Infrared Analysis of Essential Oils* in which the IR spectra of approximately 200 essential oils and additionally of more than 50 pure reference components have been presented. However, the main disadvantage of this method is the low sensitivity and selectivity of the method in the case of mixtures with a large number of components and, second, the unsolvable problem when attempting to quantitatively measure individual component concentrations.

New approaches to analyze essential oils by vibrational spectroscopy using attenuated reflection (ATR) IR spectroscopy and NIR-FT-Raman spectroscopy have recently been published by Baranska et al. (2005) and numerous papers cited therein. The main components of an essential oil can be identified by both spectroscopic techniques using the spectra of pure oil constituents as references. The spectroscopic analysis is based on characteristic key bands of the individual constituents and made it, for example, possible to discriminate the oil profiles of several eucalyptus species. As can be taken from this paper, valuable information can be obtained as a result of the combined application of ATR-IR and NIR-FT-Raman spectroscopy. Based on reference GC measurements, valuable calibration equations have been developed for numerous essential oil plants and related essential oils in order to quantify the amount of individual oil constituents applying different suitable chemometric algorithms. Main advantages of those techniques are their ability to control the quality of essential oils very fast and easily and, above all, their ability to quantify and analyze the main constituents of essential oils *in situ*, that means in living plant tissues without any isolation process, since both techniques are not destructive.

2.3.4.3 Mass Spectrometry

MS and proton NMR spectroscopy have mainly been used for structure elucidation of isolated compounds. However, there are some reports on mass spectrometric analyses of essential oils. One example has been presented by Grützmacher (1982). The depicted mass spectrum (Figure 2.9) of an essential oil exhibits some characteristic molecular ions of terpenoids with masses at *m/z* 136, 148, 152, and 154. By the application of a double focusing mass spectrometer and special techniques analyzing the decay products of metastable ions, the components anethole, fenchone, borneol, and cineole could be identified, while the assignment of the mass 136 proved to be problematic.

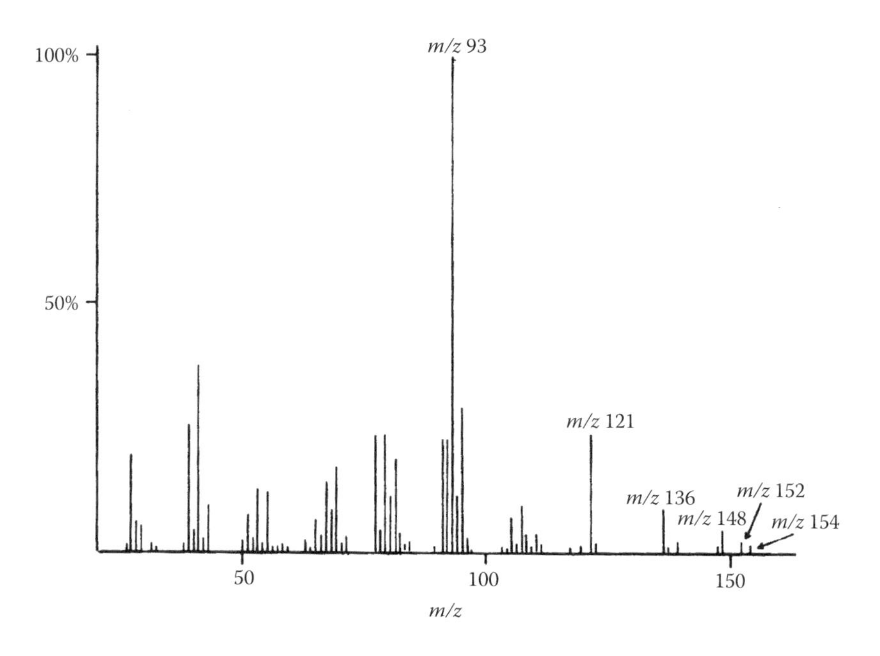

FIGURE 2.9 EI-mass spectrum of an essential oil. (From Grützmacher, H.F., 1982. In Ätherische Öle: Analytik, Physiologie, Zusammensetzung K.H. Kubeczka (ed.), pp. 1–24. Stuttgart, Germany: Georg Thieme Verlag. With permission.)

A different approach has been used by Schultze et al. (1986), investigating secondary metabolites in dried plant material by direct mass spectrometric measurement. The small samples (0.1–2 mg, depending on the kind of plant drug) were directly introduced into a mass spectrometer by means of a heatable direct probe. By heating the solid sample, stored in a small glass crucible, various substances are released depending on the applied temperature, and subsequently their mass spectra can be taken. With the aid of this technique, numerous medicinal plant drugs have been investigated and their main vaporizable components could be identified.

2.3.4.4 ^{13}C-NMR Spectroscopy

^{13}C-NMR spectroscopy is generally used for the elucidation of molecular structures of isolated chemical species. The application of ^{13}C-NMR spectroscopy to the investigation of complex mixtures is relatively rare. However, the application of ^{13}C-NMR spectroscopy to the analysis of essential oils and similar complex mixtures offers particular advantages, as have been shown in the past (Formaček and Kubeczka, 1979, 1982a; Kubeczka, 2002), to confirm analytical results obtained by GC-MS and for solving certain problems encountered with nonvolatile mixture components or thermally unstable compounds, since analysis is performed at ambient temperature.

The qualitative analysis of an essential oil is based on comparison of the oil spectrum, using broadband decoupling, with spectra of pure oil constituents, which should be recorded under identical conditions regarding solvent, temperature, and so on to ensure that differences in the chemical shifts for individual ^{13}C-NMR lines of the mixture and of the reference substance are negligible. As an example, the identification of the main constituent of celery oil is shown (Figure 2.10). This constituent can be easily identified as limonene by the corresponding reference spectrum. Minor constituents give rise to less intensive signals that can be recognized after a vertical expansion of the spectrum. For recognition of those signals, also a horizontal expansion of the spectrum is advantageous.

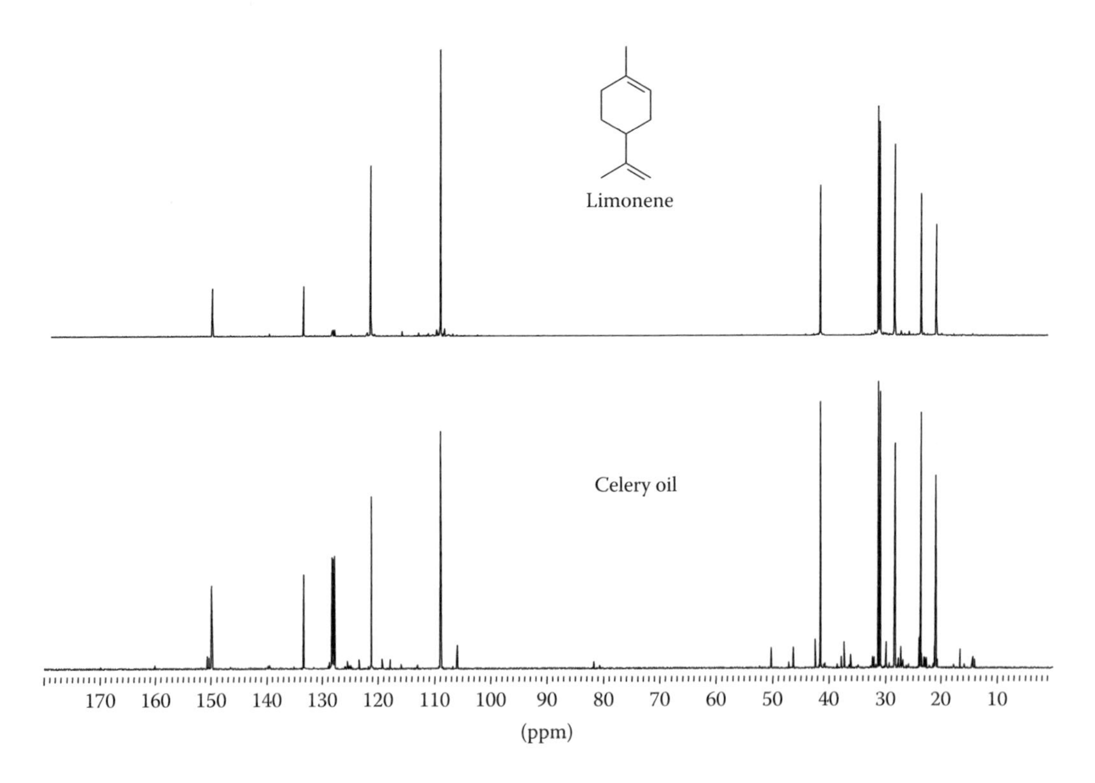

FIGURE 2.10 Identification of limonene in celery oil by ^{13}C-NMR spectroscopy.

The sensitivity of the ^{13}C-NMR technique is limited by diverse factors such as rotational sidebands, ^{13}C–^{13}C couplings, and so on, and at least by the accumulation time. For practical use, the concentration of 0.1% of a component in the entire mixture has to be seen as an interpretable limit. A very pretentious investigation has been presented by Kubeczka (1989). In the investigated essential oil, consisting of more than 80 constituents, approximately 1200 signals were counted after a horizontal and vertical expansion in the obtained broadband decoupled ^{13}C-NMR spectrum, which reflects impressively the complex composition of that oil. However, the analysis of such a complex mixture is made difficult by the immense density of individual lines, especially in the aliphatic region of the spectrum, making the assignments of lines to individual components ambiguous. Besides, qualitative analysis quantification of the individual sample components is accessible as described by Formaček and Kubeczka (1982b). After elimination of the ^{13}C-NMR signals of nonprotonated nuclei and calculation of average signal intensity per carbon atom as a measurement characteristic, it has been possible to obtain satisfactory results as shown by comparison with gas chromatographic analyses.

During the last years, a number of articles have been published by Casanova and coworkers (e.g., Bradesi et al. (1996) and references cited therein). In addition, papers dealing with computer-aided identification of individual components of essential oils after ^{13}C-NMR measurements (e.g., Tomi et al., 1995), and investigations of chiral oil constituents by means of a chiral lanthanide shift reagent by ^{13}C-NMR spectroscopy have been published (Ristorcelli et al., 1997).

REFERENCES

Adams, R. P., 1989. *Identification of Essential Oils by Ion Trap Mass Spectroscopy*. San Diego, CA: Academic Press.

Adams, R. P., 1995. *Identification of Essential Oil Components by Gas Chromatography/Mass Spectroscopy*. Carol Stream, IL: Allured Publishing Corp.

Adams, R. P., 2007. *Identification of Essential Oil Components by Gas Chromatography/Mass Spectrometry*, 4th edn. Carol Stream, IL: Allured Publishing Corp

Amelunxen, F., T. Wahlig, and H. Arbeiter, 1969. Über den Nachweis des ätherischen Öls in isolierten Drüsenhaaren und Drüsenschuppen von *Mentha piperita* L. *Z. Pflanzenphysiol.*, 61: 68–72.

Arpino, P., 1990. Coupling techniques in LC/MS and SFC/MS. *Fresenius J. Anal. Chem.*, 337: 667–685.

Arthur, C. L. and J. Pawliszyn, 1990. Solid phase microextraction with thermal desorption using fused silica optical fibres. *Anal. Chem.*, 62: 2145–2148.

Auerbach, R. H., D. Kenan, and G. Davidson, 2000. Characterization of varietal differences in essential oil components of hops (*Humulus lupulus*) by SFC-FTIR spectroscopy. *J. AOAC Int.*, 83: 621–626.

Baltussen, E., H. G. Janssen, P. Sandra, and C. A. Cramers, 1997. A novel type of liquid/liquid extraction for the preconcentration of organic micropollutants from aqueous samples: Application to the analysis of PAH's and OCP's in Water. *J. High Resolut. Chromatogr.*, 20: 395–399.

Baltussen, E., P. Sandra, F. David, and C. A. Cramers, 1999. Stir bar sorptive extraction (SBSE), a novel extraction technique for aqueous samples: Theory and principles. *J. Microcol. Sep.*, 11: 737–747.

Baranska, M., H. Schulz, S. Reitzenstein, U. Uhlemann, M. A. Strehle, H. Krüger, R. Quilitzsch, W. Foley, and J. Popp, 2005. Vibrational spectroscopic studies to acquire a quality control method of eucalyptus essential oils. *Biopolymers*, 78: 237–248.

Barton, D. H. R. and A. S. Lindsay, 1951. Sesquiterpenoids. Part I. Evidence for a nine-membered ring in caryophyllene. *J. Chem. Soc.*, 1951: 2988–2991.

Baser, K. H. C., B. Demirci, F. Demirci, N. Kirimer, and I. C. Hedge, 2001. Microdistillation as a useful tool for the analysis of minute amounts of aromatic plant materials. *Chem. Nat. Comp.*, 37: 336–338.

Becker, H., J. Reichling, and W. C. Hsieh, 1982. Water-free solvent system for droplet counter-current chromatography and its suitability for the separation of non-polar substances. *J. Chromatogr.*, 237: 307–310.

Becker, H., W. C. Hsieh, and C. O. Verelis, 1981. Droplet counter-current chromatography (DCCC). Erste Erfahrungen mit einem wasserfreien Trennsystem. *GIT Fachz. Labor. Suppl. Chromatogr.*, 81: 38–40.

Bellanato, J. and A. Hidalgo, 1971. *Infrared Analysis of Essential Oils*. London, U.K.: Heyden & Son Ltd.

Belliardo, F., C. Bicchi, C. Corsero, E. Liberto, P. Rubiolo, and B. Sgorbini, 2006. Headspace-solid-phase microextraction in the analysis of the volatile fraction of aromatic and medicinal plants. *J. Chromatogr. Sci.*, 44: 416–429.

Bergström, G., 1973. Studies on natural odoriferous compounds. *Chem. Scr.*, 4: 135–138.

Bergström, G., M. Appelgren, A. K. Borg-Karlson, I. Groth, S. Strömberg, and St. Strömberg, 1980. Studies on natural odoriferous compounds. *Chem. Scr.*, 16: 173–180.

Bernreuther, A., J. Koziet, P. Brunerie, G. Krammer, N. Christoph, and P. Schreier, 1990. Chirospecific capillary gas chromatography (HRGC) and on-line HRGC-isotope ratio mass spectrometry of γ-decalactone from various sources. *Z. Lebensm. Unters. Forsch.*, 191: 299–301.

Berthelot, M., 1859. Ueber Camphenverbindungen. *Liebigs Ann. Chem.*, 110: 367–368.

Beyer, J., H. Becker, and R. Martin, 1986. Separation of labile terpenoids by low temperature HPLC. *J. Chromatogr.*, 9: 2433–2441.

Bicchi, C., A. D. Amato, C. Frattini, G. M. Nano, E. Cappelletti, and R. Caniato, 1985. Analysis of essential oils by direct sampling from plant secretory structures and capillary gas chromatography. *J. High Resolut. Chromagtogr.*, 8: 431–435.

Bicchi, C., A. D. Amato, F. David, and P. Sandra, 1987. Direct capture of volatiles emitted by living plants. *Flavour Frag. J.*, 2: 49–54.

Bicchi, C., A. D. Amato, G. M. Nano, and C. Frattini, 1983. Improved method for the analysis of essential oils by microdistillation followed by capillary gas chromatography. *J. Chromatogr.*, 279: 409–416.

Bicchi, C., A. D. Amato, V. Manzin, A. Galli, and M. Galli, 1996. Cyclodextrin derivatives in gas chromatographic separation of racemic mixtures of volatile compounds. X. 2,3-di-*O*-ethyl-6-*O-tert*-butyl-dimethylsilyl)-β- and -γ-cyclodextrins. *J. Chromatogr. A*, 742: 161–173.

Bicchi, C., C. Cordero, C. Iori, P. Rubiolo, and P. Sandra, 2000a. Headspace sorptive extraction (HSSE) in the headspace analysis of aromatic and medicinal plants. *J. High Resolut. Chromatogr.*, 23: 539–546.

Bicchi, C., C. Cordero, and P. Rubiolo, 2000b. Influence of fibre coating in headspace solid-phase microextraction-gas chromatographic analysis of aromatic and medicinal plants. *J. Chromatogr. A*, 892: 469–485.

Bicchi, C., C. Frattini, G. Pellegrino, P. Rubiolo, V. Raverdino, and G. Tsoupras, 1992. Determination of sulphurated compounds in *Tagetes patula* cv. nana essential oil by gas chromatography with mass spectrometric, Fourier transform infrared and atomic emission spectrometric detection. *J. Chromatogr.*, 609: 305–313.

Bicchi, C. and D. Joulain, 1990. Review: Headspace-gas chromatographic analysis of medicinal and aromatic plants and flowers. *Flavour Frag. J.*, 5: 131–145.

Bicchi, C. and P. Sandra, 1987. Microtechniques in essential oil analysis. In *Capillary Gas Chromatography in Essential Oil Analysis*, P. Sandra and C. Bicchi (eds.), pp. 85–122. Heidelberg, Germany: Alfred Huethig Verlag.

Bicchi, C., V. Manzin, A. D. Amato, and P. Rubiolo, 1995. Cyclodextrin derivatives in GC separation of enantiomers of essential oil, aroma and flavour compounds. *Flavour Frag. J.*, 10: 127–137.

Blanch, G. P., E. Ibanez, M. Herraiz, and G. Reglero, 1994. Use of a programmed temperature vaporizer for off-line SFE/GC analysis in food composition studies. *Anal. Chem.*, 66: 888–892.

Blum, C., K. H. Kubeczka, and K. Becker, 1997. Supercritical fluid chromatography-mass spectrometry of thyme extracts (*Thymus vulgaris* L.). *J. Chromatogr. A*, 773: 377–380.

Bos, R., A. P. Bruins, and H. Hendriks, 1982. Negative ion chemical ionization, a new important tool in the analysis of essential oils. In *Ätherische Öle, Analytik, Physiologie, Zusammensetzung*, K. H. Kubeczka (ed.), pp. 25–32. Stuttgart, Germany: Georg Thieme Verlag.

Bos, R., H. J. Woerdenbag, H. Hendriks, J. H. Zwaving, P.A.G.M. De Smet, G. Tittel, H. V. Wikström, and J. J. C. Scheffer, 1996. Analytical aspects of phytotherapeutic valerian preparations. *Phytochem. Anal.*, 7: 143–151.

Bourne, S., A. M. Haefner, K. L. Norton, and P. R. Griffiths, 1990. Performance characteristics of a real-time direct deposition gas chromatography/Fourier transform infrared system. *Anal. Chem.*, 62: 2448–2452.

Bradesi, P., A. Bighelli, F. Tomi, and J. Casanova, 1996. L'analyse des mélanges complexes par RMN du Carbone-13—Partie I et II. *Cand. J. Appl. Spectrosc.*, 11: 15–24, 41–50.

Braunsdorf, R., U. Hener, and A. Mosandl, 1992. Analytische Differenzierung zwischen natürlich gewachsenen, fermentativ erzeugten und synthetischen (naturidentischen) Aromastoffen II. Mitt.: GC-C-IRMS-Analyse aromarelevanter Aldehyde—Grundlagen und Anwendungsbeispiele. *Z. Lebensm. Unters. Forsch.*, 194: 426–430.

Braunsdorf, R., U. Hener, S. Stein, and A. Mosandl, 1993. Comprehensive cGC-IRMS analysis in the authenticity control of flavours and essential oils. Part I: Lemon oil. *Z. Lebensm. Unters. Forsch.*, 197: 137–141.

Briechle, R., W. Dammertz, R. Guth, and W. Volmer, 1997. Bestimmung ätherischer Öle in Drogen. *GIT Lab. Fachz.*, 41: 749–753.

Bruins, A. P., 1987. Gas chromatography-mass spectrometry of essential oils, Part II: Positive ion and negative ion chemical ionization techniques. In *Capillary Gas Chromatography in Essential Oil Analysis*, P. Sandra and C. Bicchi (eds.), pp. 329–357. Heidelberg, Germany: Dr. A. Huethig Verlag.

Bruno, S., 1961. La chromatografia in phase vapore nell'identificazione di alcuni olii essenziali in materiali biologici. *Farmaco*, 16: 481–486.

Buttery, R. G., W. H. McFadden, R. Teranishi, M. P. Kealy, and T. R. Mon, 1963. Constituents of hop oil. *Nature*, 200: 435–436.

Carle, R., I. Fleischhauer, J. Beyer, and E. Reinhard, 1990. Studies on the origin of (−)-α-bisabolol and chamazulene in chamomile preparations: Part I. Investigations by isotope ratio mass spectrometry (IRMS). *Planta Med.*, 56: 456–460.

Cazaussus, A., R. Pes, N. Sellier, and J. C. Tabet, 1988. GC-MS and GC-MS-MS analysis of a complex essential oil. *Chromatographia*, 25: 865–869.

Chaintreau, A., 2001. Simultaneous distillation–extraction: From birth to maturity—Review. *Flavour Frag. J.*, 16: 136–148.

Chamblee, T. S., B. C. Clark, T. Radford, and G. A. Iacobucci, 1985. General method for the high-performance liquid chromatographic prefractionation of essential oils and flavor mixtures for gas chromatographic-mass spectrometric analysis: Identification of new constituents in cold pressed lime oil. *J. Chromatogr.*, 330: 141–151.

Chester T. L. and D. P. Innis, 1986. Separation of oligo- and polysaccharides by capillary supercritical fluid chromatography. *J. High Resolut. Chromatogr.*, 9: 209–212.

Chialva, F., G. Gabri, P. A. P. Liddle, and F. Ulian, 1982. Qualitative evaluation of aromatic herbs by direct headspace GC analysis. Application of the method and comparison with the traditional analysis of essential oils. *J. High Resolut. Chromatogr.*, 5: 182–188.

Clevenger, J. F., 1928. Apparatus for the determination of volatile oil. *J. Am. Pharm. Assoc.*, 17: 345–349.

Cocking, T. T. and G. Middleton, 1935. Improved method for the estimation of the essential oil content of drugs. *Quart. J. Pharm. Pharmacol.*, 8: 435–442.

Cole, R. A., 1980. The use of porous polymers for the collection of plant volatiles. *J. Sci. Food Agric.*, 31: 1242–1249.

Conway, W. D., 1989. *Countercurrent Chromatography—Apparatus, Theory, and Applications*. New York: VCH Inc.

Cramers, C. A., H. G. Janssen, M. M. van Deursen, and P. A. Leclercq, 1999. High speed gas chromatography: An overview of various concepts. *J. Chromatogr. A*, 856: 315–329.

Craveiro, A. A., F. J. A. Matos, J. Alencar, and M. M. Plumel, 1989. Microwave oven extraction of an essential oil. *Flavour Frag. J.*, 4: 43–44.

David, F. and P. Sandra, 1992. Capillary gas chromatography-spectroscopic techniques in natural product analysis. *Phytochem. Anal.*, 3: 145–152.

David, F. and P. Sandra, 2007. Review. Stir bar sorptive extraction for trace analysis. *J. Chromatogr. A*, 1152: 54–69.

Debrunner, B., M. Neuenschwander, and R. Benneisen, 1995. Sesquiterpenes of *Petasites hybridus* (L.) G.M. et Sch.: Distribution of sesquiterpenes over plant organs. *Pharmaceut. Acta Helv.*, 70: 167–173.

Dietrich, A., B. Maas, B. Messer, G. Bruche, V. Karl, A. Kaunzinger, and A. Mosandl, 1992a. Stereoisomeric flavour compounds, part LVIII: The use of heptakis (2,3-di-*O*-methyl-6-*O*-*tert*-butyl-dimethylsilyl)-β-cyclodextrin as a chiral stationary phase in flavor analysis. *J. High Resolut. Chromatogr.*, 15: 590–593.

Dietrich, A., B. Maas, V. Karl, P. Kreis, D. Lehmann, B. Weber, and A. Mosandl, 1992b. Stereoisomeric flavour compounds, part LV: Stereodifferentiation of some chiral volatiles on heptakis (2,3-di-*O*-acetyl-6-*O*-*tert*-butyl-dimethylsilyl)-β-cyclodextrin. *J. High Resolut. Chromatogr.*, 15: 176–179.

Dugo, G., A. Verzera, A. Cotroneo, I. S. d'Alcontres, L. Mondllo, and K. D. Bartle, 1994. Automated HPLC-HRGC: A powerful method for essential oil analysis. Part II. Determination of the enantiomeric distribution of linalol in sweet orange, bitter orange and mandarin essential oils. *Flavour Frag. J.*, 9: 99–104.

Dugo, G., P. Q. Tranchida, A. Cotroneo, P. Dugo, I. Bonaccorsi, P. Marriott, R. Shellie, and L. Mondello, 2005. Advanced and innovative chromatographic techniques for the study of citrus essential oils. *Flavour Frag. J.*, 20: 249–264.

Dugo, P., G. Dugo, and L. Mondello, 2003. On-line coupled LC–GC: Theory and applications. *LC-GC Eur.*, 16(12a): 35–43.

Dumas, M. J., 1833. Ueber die vegetabilischen Substanzen welche sich dem Kampfer nähern, und über einig ätherischen Öle. *Ann. Pharmacie*, 6: 245–258.

Fischer, N., S. Nitz, and F. Drawert, 1987. Original flavour compounds and the essential oil composition of Marjoram (*Majorana hortensis* Moench). *Flavour. Frag. J.*, 2: 55–61.

Formaček, V. and K. H. Kubeczka, 1979. Application of 13C-NMR-spectroscopy in analysis of essential oils. In *Vorkommen und Analytik ätherischer Öle*, K. H. Kubeczka (ed.), pp. 130–138. Stuttgart, Germany: Georg Thieme Verlag.

Formaček, V. and K. H. Kubeczka, 1982a. ^{13}C-NMR analysis of essential oils. In *Aromatic Plants: Basic and Applied Aspects*, N. Margaris, A. Koedam, and D. Vokou (eds.), pp. 177–181. The Hague, the Netherlands: Martinus Nijhoff Publishers.

Formaček, V. and K. H. Kubeczka, 1982b. Quantitative analysis of essential oils by ^{13}C-NMR-spectroscopy. In *Ätherische Öle: Analytik, Physiologie, Zusammensetzung*, K. H. Kubeczka (ed.), pp. 42–53. Stuttgart, Germany: Georg Thieme Verlag.
Frank, C., A. Dietrich, U. Kremer, and A. Mosandl, 1995. GC-IRMS in the authenticy control of the essential oil of *Coriandrum sativum* L. *J. Agric. Food Chem.*, 43: 1634–1637.
Frérot, E. and E. Decorzant 2004. Quantification of total furocoumarins in citrus oils by HPLC couple with UV fluorescence, and mass detection. *J. Agric. Food Chem.*, 52: 6879–6886.
Friedel, H. D. and R. Matusch, 1987. Separation of non-polar sesquiterpene olefins from Tolu balsam by high-performance liquid chromatography: Silver perchlorate impregnation of prepacked preparative silica gel column. *J. Chromatogr.*, 407: 343–348.
Gaedcke, F. and B. Steinhoff, 2000. *Phytopharmaka*. Stuttgart, Germany: Wissenschaftliche Verlagsgesellschaft (Figure 1.7).
Geiss, F., 1987. *Fundamentals of Thin-Layer Chromatography*. Heidelberg, Germany: Hüthig Verlag.
Gießelmann, G. and K. H. Kubeczka, 1993. A new procedure for the enrichment of headspace constituents versus conventional hydrodistillation. Poster presented at the *24th International Symposium on Essential Oils*, Berlin, Germany.
Gil-Av, E., B. Feibush, and R. Charles-Sigler, 1965. In *Gas Chromatography 1966*, A. B. Littlewood (ed.), 227pp. London, U.K.: Institute of Petroleum.
Gildemeister, E. and F. Hoffmann, 1956. In *Die ätherischen Öle*, W. Treibs (ed.), Vol. 1, p. 14. Berlin, Germany: Akademie-Verlag.
Godefroot, M., M. Stechele, P. Sandra, and M. Verzele, 1982. A new method fort the quantitative analysis of organochlorine pesticides and polychlorinated biphenyls. *J. High Resolut. Chromatogr.*, 5: 75–79.
Godefroot, M., P. Sandra, and M. Verzele, 1981. New method for quantitative essential oil analysis. *J. Chromatogr.*, 203: 325–335.
Greibrokk, T., 1995. Review: Applications of supercritical fluid extraction in multidimensional systems. *J. Chromatogr. A*, 703: 523–536.
Grob, K. and F. Zürcher, 1976. Stripping of trace organic substances from water: Equipment and procedure. *J. Chromatogr.*, 117: 285–294.
Grützmacher, H. F., 1982. Mixture analysis by new mass spectrometric techniques—A survey. In *Ätherische Öle: Analytik, Physiologie, Zusammensetzung*, K. H. Kubeczka (ed.), pp. 1–24. Stuttgart, Germany: Georg Thieme Verlag.
Hartonen, K., M. Jussila, P. Manninen, and M. L. Riekkola, 1992. Volatile oil analysis of *Thymus vulgaris* L. by directly coupled SFE/GC. *J. Microcol. Sep.*, 4: 3–7.
Hawthorne, S. B., M. S. Krieger, and D. J. Miller, 1988. Analysis of flavor and fragrance compounds using supercritical fluid extraction coupled with gas chromatography. *Anal. Chem.*, 60: 472–477.
Hefendehl, F. W., 1966. Isolierung ätherischer Öle aus äußeren Pflanzendrüsen. *Naturw.*, 53: 142.
Henderson, W., J. W. Hart, P. How, and J. Judge, 1970. Chemical and morphological studies on sites of sesquiterpene accumulation in *Pogostemon cablin* (Patchouli). *Phytochemistry*, 9: 1219–1228.
Hener, U., 1990. Chirale Aromastoffe—Beiträge zur Struktur, Wirkung und Analytik. *Dissertation*, Goethe-University of Frankfurt/Main, Frankfurt, Germany.
Herres, W., 1987. *HRGC-FTIR: Capillary Gas Chromatography-Fourier Transform Infrared Spectroscopy*. Heidelberg, Germany: Alfred Huethig Verlag.
Herres, W., K. H. Kubezka, and W. Schultze, 1986. HRGC-FTIR investigations on volatile terpenes. In *Progress in Essential Oil Research*, E. J. Brunke (ed.), pp. 507–528. Berlin, Germany: W. de Gruyter.
Hör, K., C. Ruff, B. Weckerle, T. König, and P. Schreier, 2001. $^{2}H/^{1}H$ ratio analysis of flavor compounds by on-line gas chromatography-pyrolysis-isotope ratio mass spectrometry (HRGC-P-IRMS): Citral. *Flavour Frag. J.*, 16: 344–348.
Houben, R. J., H. G. M. Janssen, P. A. Leclercq, J. A. Rijks, and C. A. Cramers, 1990. Supercritical fluid extraction-capillary gas chromatography: On-line coupling with a programmed temperature vaporizer. *J. High Resolut. Chromatogr.*, 13: 669–673.
Ibanez, E., S. Lopez-Sebastian, E. Ramos, J. Tabera, and G. Reglero, 1997. Analysis of highly volatile components of foods by off-line SFE/GC. *J. Agric. Food Chem.*, 45: 3940–3943.
Ito, Y., M. A. Weinstein, I. Aoki, R. Harada, E. Kimura, and K. Nunogaki, 1966. The coil planet centrifuge. *Nature*, 212: 985–987.
Jennings, W. G., 1979. Vapor-phase sampling. *J. High Resolut. Chromatogr.*, 2: 221–224.
Jennings, W. G., 1981. Recent developments in high resolution gas chromatography. In *Flavour '81*, P. Schreier (ed.), pp. 233–251. Berlin, Germany: Walter de Gruyter & Co.
Jennings, W. and T. Shibamoto, 1980. *Qualitative Analysis of Flavor and Fragrance Volatiles by Glass Capillary Gas Chromatography*. New York: Academic Press.

Jirovetz, L., G. Buchbauer, W. Jäger, A. Woidich, and A. Nikiforov, 1992. Analysis of fragrance compounds in blood samples of mice by gas chromatography, mass spectrometry, GC/FTIR and GC/AES after inhalation of sandalwood oil. *Biomed. Chromatogr.*, 6: 133–134.

Joulain, D., 1987. The composition of the headspace from fragrant flowers: Further results. *Flavour Frag. J.*, 2: 149–155.

Joulain, D. and W. A. König, 1998. *The Atlas of Spectral Data of Sesquiterpene Hydrocarbons*. Hamburg, Germany: E. B. Verlag.

Juchelka, D., T. Beck, U. Hener, F. Dettmar, and A. Mosandl, 1998. Multidimensional gas chromatography coupled on-line with isotope ratio mass spectrometry (MDGC-IRMS): Progress in the analytical authentication of genuine flavor components. *J. High Resolut. Chromatogr.*, 21: 145–151.

Kaiser, H. and W. Lang, 1951. Ueber die Bestimmung des ätherischen Oels in Drogen. *Dtsch. Apoth. Ztg.*, 91: 163–166.

Kaiser, R., 1976. *Einführung in die Hochleistungs-Dünnschicht-Chromatographie*. Bad Dürkheim, Germany: Institut für Chromatographie.

Kaiser, R., 1993. *The Scent of Orchids—Olfactory and Chemical Investigations*. Amsterdam, the Netherlands: Elsevier Science Ltd.

Kolb, B. and L. S. Ettre, 1997. *Static Headspace-Gas Chromatography: Theory and Practice*. New York: Wiley.

Kolb, B. and L. S. Ettre, 2006. *Static Headspace-Gas Chromatography: Theory and Practice*, 2nd edn. New York: Wiley.

Kolb, B. and B. Liebhardt, 1986. Cryofocusing in the combination of gas chromatography with equilibrium headspace sampling. *Chromatographia*, 21: 305–311.

Komae, H. and N. Hayashi, 1975. Separation of essential oils by liquid chromatography. *J. Chromatogr.*, 114: 258–260.

König, W. A., A. Rieck, C. Fricke, S. Melching, Y. Saritas, and I. H. Hardt, 1995. Enantiomeric composition of sesquiterpenes in essential oils. In *Proceedings of the 13th International Congress of Flavours, Fragrances and Essential Oils*, K. H. C. Baser (ed.), Vol. 2, pp. 169–180. Istanbul, Turkey: AREP Publ.

König, W. A., A. Rieck, I. Hardt, B. Gehrcke, K. H. Kubeczka, and H. Muhle, 1994. Enantiomeric composition of the chiral constituents of essential oils Part 2: Sesquiterpene hydrocarbons. *J. High Resolut. Chromatogr.*, 17: 315–320.

König, W. A., B. Gehrcke, D. Icheln, P. Evers, J. Dönnecke, and W. Wang, 1992a. New, selectively substituted cyclodextrins as stationary phases for the analysis of chiral constituents of essential oils. *J. High Resolut. Chromatogr.*, 15: 367–372.

König, W. A., D. Icheln, T. Runge, I. Pforr, and A. Krebs, 1990. Cyclodextrins as chiral stationary phases in capillary gas chromatography. Part VII: Cyclodextrins with an inverse substitution pattern—Synthesis and enantioselectivity. *J. High Resolut. Chromatogr.*, 13: 702–707.

König, W. A., D. Icheln, T. Runge, P. Evers, B. Gehrcke, and A. Krüger, 1992b. Enantioselective gas chromatography—A new dimension in the analysis of essential oils. In *Proceedings of the 12th International Congress of Flavours, Fragrances and Essential Oils*, H. Woidich and G. Buchbauer (eds.), pp. 177–186. Vienna, Austria: Austrian Association of Flavour and Fragrance Industry.

König, W. A., P. Evers, R. Krebber, S. Schulz, C. Fehr, and G. Ohloff, 1989. Determination of the absolute configuration of α-damascenone and α-ionone from black tea by enantioselective capillary gas chromatography. *Tetrahedron*, 45: 7003–7006.

König, W. A., S. Lutz, and G. Wenz, 1988a. Modified cyclodextrins—Novel, highly enantioselective stationary phases for gas chromatography. *Angew. Chem. Int. Ed. Engl.* 27: 979–980.

König, W. A., S. Lutz, G. Wenz, and E. van der Bey, 1988b. Cyclodextrins as chiral stationary phases in capillary gas chromatography II. Heptakis (3–O-acetyl-2,6-di-O-pentyl)-β-cyclodextrin. *J. High Resolut. Chromatogr. Chromatogr. Commun.*, 11: 506–509.

König, W. A., S. Lutz, P. Mischnick-Lübbecke, B. Brassat, and G. Wenz, 1988c. Cyclodextrins as chiral stationary phases in capillary gas chromatography I. Pentylated α-cyclodextrin. *J. Chromatogr.*, 447: 193–197.

Kreis, P. and A. Mosandl, 1992. Chiral compounds of essential oils XI. Simultaneous stereoanalysis of lavandula oil constituents. *Flavour Frag. J.*, 7: 187–193.

Kubeczka, K. H., 1967. Vorrichtung zur Isolierung, Anreicherung und chemischen Charakterisierung gaschromatographisch getrennter Komponenten im μg-Bereich. *J. Chromatogr.*, 31: 319–325.

Kubeczka, K. H., 1973. Separation of essential oils and similar complex mixtures by means of modifieddry-column chromatography. *Chromatographia*, 6: 106–108.

Kubeczka, K. H., 1981a. Standardization and analysis of essential oils. In *A Perspective of the Perfumes and Flavours Industry in India*, S. Jain (ed.), pp. 105–120. New Delhi, India: Perfumes and Flavours Association of India.

Kubeczka, K. H., 1981b. Application of HPLC for the separation of flavour compounds. In *Flavour 81*, P. Schreier (ed.), pp. 345–359. Berlin, Germany: Walter de Gruyter & Co.

Kubeczka, K. H., 1985. Progress in isolation techniques for essential oil constituents. In *Advances in Medicinal Plant Research*, A. J. Vlietinck and R. A. Dommisse (eds.), pp. 197–224. Stuttgart, Germany: Wissenschaftliche Verlagsgesellschaft mbH.

Kubeczka, K. H., 1989. Studies on complex mixtures: Combined separation techniques versus unprocessed sample analysis. In *Moderne Tecniche in Fitochimica*, C. Bicchi and C. Frattini (eds.), pp. 53–68. Firenze, Tuscany: Società Italiana di Fitochimica.

Kubeczka, K. H., 1991. New methods in essential oil analysis. In *Conferencias Plenarias de la XXIII Reunión Bienal de Quimica*, A. San Feliciano, M. Grande, and J. Casado (eds.), pp. 169–184. Salamanca, Spain: Universidad de Salamanca, Sección local e la R.S.E.Q.

Kubeczka, K. H., 1996. Unpublished results.

Kubeczka, K. H., 1997a. New approaches in essential oil analysis using polymer-coated silica fibers. In *Essential Oils: Basic and Applied Research*, Ch. Franz, A. Máthé, and G. Buchbauer (eds.), pp. 139–146. Carol Stream, IL: Allured Publishing Corp.

Kubeczka, K.-H., 1997b. Essential Oil Symposium Proceedings, p. 145.

Kubeczka, K.-H., 2002. *Essential Oils Analysis by Capillary Gas Chromatography and Carbon-13 NMR Spectroscopy*, 2nd completely rev. edn. Baffins Lane, England: Wiley.

Kubeczka, K. H. and G. Gießelmann, 1995. Application of a new micro hydrodistillation device for the investigation of aromatic plant drugs. Poster presented at the *43th Annual Congress on Medicinal Plant Research*, Halle, Germany.

Kubeczka, K. H., W. Schultze, S. Ebel, and M. Weyandt-Spangenberg, 1989. Möglichkeiten und Grenzen der GC-Molekülspektroskopie-Kopplungen. In *Instrumentalized Analytical Chemistry and Computer Technology*, W. Günther and J. P. Matthes (eds.), pp. 131–141. Darmstadt, Germany: GIT Verlag.

Lancas, F., F. David, and P. Sandra, 1988. CGC analysis of the essential oil of citrus fruits on 100 μm i.d. columns. *J. High Resolut. Chromatogr.*, 11: 73–75.

Lawrence, B. M., 1995. The isolation of aromatic materials from natural plant products. In *Manual of the Essential Oil Industry*, K. Tuley De Silva (ed.), pp. 57–154. Vienna, Austria: UNIDO.

Laylin, J. K., 1993. *Nobel Laureates in Chemistry, 1901–1992*. Philadelphia, PA: Chemical Heritage Foundation.

Levy, J. M., J. P. Guzowski, and W. E. Huhak, 2005. On-line multidimensional supercritical fluid chromatography/capillary gas chromatography. *J. High Resolut. Chromatogr.*, 10: 337–341.

Lucchesi, M. E., F. Chemat, and J. Smadja, 2004. Solvent-free microwave extraction of essential oil from-aromatic herbs: Comparison with conventional hydro-distillation. *J. Chromatogr. A*, 1043: 323–327.

Malingré, T. M., D. Smith, and S. Batterman, 1969. De Isolering en Gaschromatografische Analyse van de Vluchtige Olie uit Afzonderlijke Klierharen van het Labiatentype. *Pharm. Weekblad*, 104: 429.

Manninen, P. and H. Kallio, 1997. Supercritical fluid chromatography-gas chromatography of volatiles in cloudberry (*Rubus chamaemorus*) oil extracted with supercritical carbon dioxide. *J. Chromatogr. A*, 787: 276–282.

Manninen, P., M. L. Riekkola, Y. Holm, and R. Hiltunen, 1990. SFC in analysis of aromatic plants. *J. High Resolut. Chromatogr.*, 13: 167–169.

MassFinder, 2007. *MassFinder Software*, Version 3.7. Hamburg, Germany: Dr. Hochmuth Scientific Consulting.

Matthews, D. E. and J. M. Hayes, 1978. Isotope-ratio-monitoring gas chromatography-mass spectrometry. *Anal. Chem.*, 50: 1465–1473.

McKone, H. T., 1979. High performance liquid chromatography. *J. Chem. Educ.*, 56: 807–809.

Mechler, E. and K. A. Kovar, 1977. Vergleichende Bestimmungen des ätherischen Öls in Drogen nach dem Europäischen und dem Deutschen Arzneibuch. *Dtsch. Apoth. Ztg.*, 117: 1019–1023.

Meerwein, H., 1914. Über den Reaktionsmechanismus der Umwandlung von Borneol in Camphen. *Liebigs Ann.Chem.*, 405: 129–175.

Miller, J. M. and J. G. Kirchner, 1952. Some improvements in chromatographic techniques for terpenes. *Anal. Chem.*, 24: 1480–1482.

Modey, W. K., D. A. Mulholland, and M. W. Raynor, 1995. Analytical supercritical extraction of natural products. *Phytochem. Anal.*, 7: 1–15.

Mondello, L., K. D. Bartle, G. Dugo, and P. Dugo, 1994a. Automated HPLC-HRGC: A powerful method for essential oil analysis Part III. Aliphatic and terpene aldehydes of orange oil. *J. High Resolut. Chromatogr.*, 17: 312–314.

Mondello, L., K. D. Bartle, P. Dugo, P. Gans, and G. Dugo, 1994b. Automated HPLC-HRGC: A powerful method for essential oils analysis. Part IV. Coupled LC-GC-MS (ITD) for bergamot oil analysis. *J. Microcol. Sep.*, 6: 237–244.

Mondello, L., G. Dugo, and K. D. Bartle, 1996. On-line microbore high performance liquid chromatography-capillary gas chromatography for food and water analyses. A review. *J. Microcol. Sep.*, 8: 275–310.

Mondello, L., G. Zappia, G. Errante, P. Dugo, and G. Dugo, 2000. Fast-GC and Fast-GC/MS for the analysis of natural complex matrices. *LC-GC Eur.*, 13: 495–502.

Mondello, L., P. Dugo, G. Dugo, A. C. Lewis, and K. D. Bartle, 1999. Review: High-performance liquid chromatography coupled on-line with high resolution gas chromatography, State of the art. *J. Chromatogr. A*, 842: 373–390.

Mondello, L., P. Dugo, K. D. Bartle, G. Dugo, and A. Cotroneo, 1995. Automated LC-GC: A powerful method for essential oils analysis Part V. Identification of terpene hydrocarbons of bergamot, lemon, mandarin, sweet orange, bitter orange, grapefruit, clementine and Mexican lime oils by coupled HPLC-HRGC-MS (ITD). *Flavour Frag. J.*, 10: 33–42.

Mondello, L., R. Shellie, A. Casilli, P. Marriott, and G. Dugo, 2004. Ultra-fast essential oil characterization by capillary GC on a 50 μm ID column. *J. Sep. Sci.*, 27: 699–702.

Morita, M., S. Mihashi, H. Itokawa, and S. Hara, 1983. Silver nitrate impregnation of preparative silica gel columns for liquid chromatography. *Anal. Chem.*, 55: 412–414.

Mosandl, A., 2004. Authenticity assessment: A permanent challenge in food flavor and essential oil analysis. *J. Chromatogr. Sci.*, 42: 440–449.

Mosandl, A. and D. Juchelka, 1997. Advances in authenticity assessment of citrus oils. *J. Essent. Oil Res.*, 9: 5–12.

Munari, F., G. Dugo, and A. Cotroneo, 1990. Automated on-line HPLC-HRGC with gradient elution and multiple GC transfer applied to the characterization of citrus essential oils. *J. High Resolut. Chromatogr.*, 13: 56–61.

Nickerson, G. and S. Likens, 1966. Gas chromatographic evidence for the occurrence of hop oil components in beer. *J. Chromatogr.*, 21: 1–5.

Nishii, Y., T. Yoshida, and Y. Tanabe, 1997. Enantiomeric resolution of a germacrene-D derivative by chiral high-performance liquid chromatography. *Biosci. Biotechnol. Biochem.*, 61: 547–548.

NIST/EPA/NIH Mass Spectral Library 2005. Version: NIST 05. Gaithersburg, MD: Mass Spectrometry Data Center, National Institute of Standard and Technology.

Novak, J., V. Vašak, and J. Janak, 1965. Chromatographic method for the concentration of trace impurities in the atmosphere and other gases. *Anal. Chem.*, 37: 660–666.

Nyiredy, Sz., 2003. Progress in forced-flow planar chromatography. *J. Chromatogr. A*, 1000: 985–999.

Pawliszyn, J., 1997. *Solid Phase Microextraction Theory and Practice*. New York: Wiley-VCH Inc.

Pettei, M. J., F. G. Pilkiewicz, and K. Nakanishi, 1977. Preparative liquid chromatography applied to difficult separations. *Tetrahedron Lett.*, 24: 2083–2086.

Pourmortazavi, S. M. and S. S. Hajimirsadeghi, 2007. Review: Supercritical fluid extraction in plant essential and volatile oil analysis. *J. Chromatogr. A*, 1163: 2–24.

Prasad, R. S., A. S. Gupta, and S. Dev, 1947. Chromatography of organic compounds III. Improved procedure for the thin-layer chromatography of olefins on silver ion-silica gel layers. *J. Chromatogr.*, 92: 450–453.

Ramsey, E. D. and M. W. Raynor, 1996. Electron ionization and chemical ionization sensitivity studies involving capillary supercritical fluid chromatography combined with benchtop mass spectrometry. *Anal. Commun.*, 33: 95–97.

Reedy, G. T., D. G. Ettinger, J. F. Schneider, and S. Bourne, 1985. High-resolution gas chromatography/matrix isolation infrared spectrometry. *Anal. Chem.*, 57: 1602–1609.

Ristorcelli, D., F. Tomi, and J. Casanova, 1997. Enantiomeric differentiation of oxygenated monoterpenes by carbon-13 NMR in the presence of a chiral lanthanide shift reagent. *J. Magnet. Resonance Anal.*, 1997: 40–46.

Ross, M. S. F., 1976. Analysis of cinnamon oils by high-pressure liquid chromatography. *J. Chromatogr.*, 118: 273–275.

Ruff, C., K. Hör, B. Weckerle, and P. Schreier, 2000. $^{2}H/^{1}H$ ratio analysis of flavor compounds by on-line gas chromatography pyrolysis isotope ratio mass spectrometry (HRGC-P-IRMS): Benzaldehyde. *J. High Resolut. Chromatogr.*, 23: 357–359.

Ruzicka, L., 1953. The isoprene rule and the biogenesis of terpenic compounds. *Experientia*, 9: 357–396.

Sanagi, M. M., U. K. Ahmad, and R. M. Smith, 1993. Application of supercritical fluid extraction and chromatography to the analysis of turmeric. *J. Chromatogr. Sci.*, 31: 20–25.

Saritas, Y., 2000. Isolierung, Strukturaufklärung und stereochemische Untersuchungen von sesquiterpenoiden Inhaltsstoffen aus ätherischen Ölen von Bryophyta und höheren Pflanzen. *PhD dissertation*, University of Hamburg, Hamburg, Germany.

Schaefer, J., 1981. Comparison of adsorbents in head space sampling. In *Flavour '81*, P. Schreier (ed.), pp. 301–313. Berlin, Germany: Walter de Gruyter & Co.

Scheffer, J. J. C., A. Koedam, and A. Baerheim Svendsen, 1976. Occurrence and prevention of isomerization of some monoterpene hydrocarbons from essential oils during liquid–solid chromatography on silica gel. *Chromatographia*, 9: 425–432.
Schneider, W., J. C. Frohne, and H. Bruderreck, 1982. Selektive gaschromatographische Messung sauerstoffhaltiger Verbindungen mittels Flammenionisationsdetektor. *J. Chromatogr.*, 245: 71–83.
Schreier, P., 1984. *Chromatographic Studies of Biogenesis of Plant Volatiles*. Heidelberg, Germany: Alfred Hüthig Verlag.
Schultze, W., 1993. Moderne instrumentalanalytische Methoden zur Untersuchung komplexer Gemische. In *Ätherische Öle—Anspruch und Wirklichkeit*, R. Carle (ed.), pp. 135–184. Stuttgart, Germany: Wissenschaftliche Verlagsgesellschaft mbH.
Schultze, W., G. Lange, and G. Heinrich, 1986. Analysis of dried plant material directly introduced into a mass spectrometer. (Part I of investigations on medicinal plants by mass spectrometry). In *Progress in Essential Oil Research*, E. J. Brunke (ed.), pp. 577–596. Berlin, Germany: Walter de Gruyter & Co.
Schultze, W., G. Lange, and G. Schmaus, 1992. Isobutane and ammonia chemical ionization mass spectrometry of sesquiterpene hydrocarbons. *Flavour Frag. J.*, 7: 55–64.
Schurig, V., 1977. Enantiomerentrennung eines chiralen Olefins durch Komplexierungschromatographie an einem optisch aktiven Rhodium(1)-Komplex. *Angew. Chem.*, 89: 113–114.
Schurig, V. and H. P. Nowotny, 1988. Separation of enantiomers on diluted permethylated β-cyclodextrin by high resolution gas chromatography. *J. Chromatogr.*, 441: 155–163.
Schwanbeck, J. and K. H. Kubeczka, 1979. Application of HPLC for separation of volatile terpene hydrocarbons. In *Vorkommen und Analytik ätherischer Öle*, K. H. Kubeczka (ed.), pp. 72–76. Stuttgart, Germany: Georg Thieme Verlag.
Schwanbeck, J., V. Koch, and K. H. Kubeczka, 1982. HPLC-separation of essential oils with chemically bonded stationary phases. In *Essential Oils—Analysis, Physiology, Composition*, K. H. Kubeczka (ed.), pp. 70–81. Stuttgart, Germany: Georg Thieme Verlag.
Scott, R. P. W. and P. Kucera, 1979. Mode of operation and performance characteristics of microbore columns for use in liquid chromatography. *J. Chromatogr.*, 169: 51–72.
Sewenig, S., D. Bullinger, U. Hener, and A. Mosandl, 2005. Comprehensive authentication of (*E*)-α(β)-ionone from raspberries, using constant flow MDGC-C/P-IRMS and enantio-MDGC-MS. *J. Agric. Food Chem.*, 53: 838–844.
Shema, J. and B. Fried (eds.), 2003. *Handbook of Thin-Layer Chromatography*, 3rd edn. New York: Marcel Dekker.
Snyder, J. K., K. Nakanishi, K. Hostettmann, and M. Hostettmann, 1984. Application of rotation locular countercurrent chromatography in natural products isolation. *J. Liquid Chromatogr.*, 7: 243–256.
Šorm, F., L. Dolejš, and J. Pliva, 1950. *Collect. Czechoslov. Chem. Commun.*, 3: 187.
Sprecher, E., 1963. Rücklaufapparatur zur erschöpfenden Wasserdampfdestillation ätherischen Öls aus voluminösem Destillationsgut. *Dtsch. Apoth. Ztg.*, 103: 213–214.
Stahl, E., 1953. Eine neue Apparatur zur gravimetrischen Erfassung kleinster Mengen ätherischer Öle. *Microchim. Acta*, 40: 367–372.
Stahl, E., 1969a. A thermo micro procedure for rapid extraction and direct application in thin-layer chromatography. *Analyst*, 94(122):723–727.
Stahl, E. (ed.), 1969b. *Thin-Layer Chromatography. A Laboratory Handbook*, 2nd edn. Berlin, Germany: Springer.
Stahl, E., 1976. Advances in the field of thermal procedures in direct combination with thin-layer chromatography. *Acc. Chem. Res.*, 9: 75–80.
Still, W. C., M. Kahn, and A. Mitra, 1978. Rapid chromatographic technique for preparative separations with moderate resolution. *J. Org. Chem.*, 43: 2923–2925.
Strack, D., P. Proksch, and P. G. Gülz, 1980. Reversed phase high performance liquid chromatography of essential oils. *Z. Naturforsch.*, 35c: 675–681.
Supelco, 2007. *Solid Phase Microextraction CD*, 6th edn. Bellefonte, PA: Supelco.
Takahisa, E. and K. H. Engel, 2005a. 2,3-Di-*O*-methoxyethyl-6-*O*-*tert*-butyl-dimethylsilyl-β-cyclodextrin, a useful stationary phase for gas chromatographic separation of enantiomers. *J. Chromatogr. A*, 1076: 148–154.
Takahisa, E. and K. H. Engel, 2005b. 2,3-Di-*O*-methoxymethyl-6-*O*-*tert*-butyl-dimethylsilyl-γ-cyclodextrin: A new class of cyclodextrin derivatives for gas chromatographic separation of enantiomers. *J. Chromatogr. A*, 1063: 181–192.
Tanimura, T., J. J. Pisano, Y. Ito, and R. L. Bowman, 1970. Droplet countercurrent chromatography. *Science*, 169: 54–56.

ten Noever de Brauw, M.C., 1979. Combined gas chromatography-mass spectrometry: A powerful tool in analytical chemistry. *J. Chromatogr.*, 165: 207–233.

Teranishi, R., T. R. Mon, A. B. Robinson, P. Cary, and L. Pauling, 1972. Gas chromatography of volatiles from breath and urine. *Anal. Chem.*, 44: 18–21.

Tienpont, B., F. David, C. Bicchi, and P. Sandra, 2000. High capacity headspace sorptive extraction. *J. Microcol. Sep.*, 12: 577–584.

Tilden, W. A., 1875. On the action of nitrosyl chloride on organic bodies. Part II. On turpentine oil. *J. Chem. Soc.*, 28: 514–518.

Tomi, F., P. Bradesi, A. Bighelli, and J. Casanova, 1995. Computer-aided identification of individual components of essential oils using carbon-13 NMR spectroscopy. *J. Magnet. Resonance Anal.*, 1995: 25–34.

Treibs, W., 1952. Über bi- und polycyclische Azulene. XIII. Das bicyclische Caryophyllen als Azulenbildner. *Liebigs Ann. Chem.*, 576: 125–131.

Tyihák, E., E. Mincsovics, and H. Kalász, 1979. New planar liquid chromatographic technique: Overpressured thin-layer chromatography. *J. Chromatogr.*, 174: 75–81.

van Beek, T. A. and D. Subrtova, 1995. Factors involved in the high pressure liquid chromatographic separation of alkenes by means of argentation chromatography on ion exchangers: Overview of theory and new practical developments. *Phytochem. Anal.*, 6: 1–19.

van Beek, T. A., N. van Dam, A. de Groot, T. A. M. Geelen, and L. H. W. van der Plas, 1994. Determination of the sesquiterpene dialdehyde polygodial by high-pressure liquid chromatography. *Phytochem. Anal.*, 5: 19–23.

von Baeyer, A. and O. Seuffert, 1901. Erschöpfende Bromierung des Menthons. *Ber. Dtsch. Chem. Ges.*, 34: 40–53.

Wagner, G., 1899. J. Russ. Phys. Chem. Soc., 31: 690 (cited in H. Meerwein, 1914. *Liebigs Ann. Chem.*, 405: 129–175.

Wallach, O., 1914. *Terpene und Campher*, 2nd edn., Leipzig, Germany: Veit & Co.

Werkhoff, P., S. Brennecke, W. Bretschneider, M. Güntert, R. Hopp, and H. Surburg, 1993. Chirospecific analysis in essential oil, fragrance and flavor research. *Z. Lebensm. Unters. Forsch.*, 196: 307–328.

WILEY Registry, 2006. *Wiley Registry of Mass Spectral Data*, 8th edn. New York: Wiley.

Woodward, R. B. 1941. Structure and the absorption spectra of α, β-unsaturated ketones. *J. Am. Chem. Soc*,, 63: 1123–1126.

Wulf, L. W., C. W. Nagel, and A. L. Branen, 1978. High-pressure liquid chromatographic separation of the naturally occurring toxicants myristicin, related aromatic ethers and falcarinol. *J. Chromatogr.*, 161: 271–278.

Wylie, P. L. and B. D. Quimby, 1989. Applications of gas chromatography with atomic emission detector. *J. High Resolut. Chromatogr.*, 12: 813–818.

Yamauchi, Y. and M. Saito, 1990. Fractionation of lemon-peel oil by semi-preparative supercritical fluid-chromatography. *J. Chromatogr.*, 505: 237–246.

Yarita, T., A. Nomura, and Y. Horimoto, 1994. Type analysis of citrus essential oils by multidimensional supercritical fluid chromatography/gas chromatography. *Anal. Sci.*, 10: 25–29.

3 Sources of Essential Oils

Chlodwig Franz and Johannes Novak

CONTENTS

3.1 "ESSENTIAL OIL–BEARING PLANTS": ATTEMPT OF A DEFINITION

Essential oils are complex mixtures of volatile compounds produced by living organisms and isolated by physical means only (pressing and distillation) from a whole plant or plant part of known taxonomic origin. The respective main compounds are mainly derived from three biosynthetic pathways only, the mevalonate pathway leading to sesquiterpenes, the methyl-erythritol pathway leading to mono- and diterpenes, and the shikimic acid pathway *en route* to phenylpropenes. Nevertheless, there are an almost uncountable number of single substances and a tremendous variation in the composition of essential oils. Many of these volatile substances have diverse ecological functions. They can act as internal messengers, as defensive substances against herbivores, or as volatiles not only directing natural enemies to these herbivores but also attracting pollinating insects to their host (Harrewijn et al., 2001).

All plants possess principally the ability to produce volatile compounds, quite often, however, only in traces. "Essential oil plants" in particular are those plant species delivering an essential oil of commercial interest. Two principal circumstances determine a plant to be used as an essential oil plant:

1. A unique blend of volatiles like the flower scents in rose (*Rosa* spp.), jasmine (*Jasminum sambac*), or tuberose (*Polianthes tuberosa*). Such flowers produce and immediately emit the volatiles by the epidermal layers of their petals (Bergougnoux et al., 2007). Therefore, the yield is even in intensive smelling flowers very low, and besides distillation special techniques, as an example, enfleurage has to be applied to recover the volatile fragrance compounds.
2. Secretion and accumulation of volatiles in specialized anatomical structures. These lead to higher concentrations of the essential oil in the plant. Such anatomical storage structures for essential oils can be secretory idioblasts (secretory cells), cavities/ducts, or glandular trichomes (Fahn, 1979, 1988; colorfully documented by Svoboda et al, 2000).

Secretory idioblasts are individual cells producing an essential oil in large quantities and retaining the oil within the cell like the essential oil idioblasts in the roots of *Vetiveria zizanioides* that occurs within the cortical layer and close to the endodermis (Bertea and Camusso, 2002). Similar structures containing essential oils are also formed in many flowers, for example, *Rosa* sp., *Viola* sp., or *Jasminum* sp.

Cavities or ducts consist of extracellular storage space that originate either by schizogeny (created by the dissolution of the middle lamella between the duct initials and formation of an intercellular space) or by lysogeny (programmed death and dissolution of cells). In both cases, the peripheral cells are becoming epithelial cells highly active in synthesis and secretion of their products into the extracellular cavities (Pickard, 2008). Schizogenic oil ducts are characteristic for the Apiaceae family, for example, *Carum carvi*, *Foeniculum vulgare*, or *Cuminum cyminum*, but also for the Hypericaceae or Pinaceae family. Lysogenic cavities are found in Rutaceae (*Citrus* sp., *Ruta graveolens*), Myrtaceae (e.g., *Syzygium aromaticum*), and others.

Secreting trichomes (glandular trichomes) can be divided into two main categories: peltate and capitate trichomes. Peltate glands consist of a basal epidermal cell, a neck–stalk cell, and a secreting head of 4–16 cells with a large subcuticular space on the apex in which the secretion product is accumulated. The capitate trichomes possess only 1–4 secreting cells with only a small subcuticular space (Werker, 1993; Maleci Bini and Giuliani, 2006). Such structures are typical for Lamiaceae (the mint family), but also for *Pelargonium* sp.

The monoterpene biosynthesis in different species of Lamiaceae, for example, sage (*Salvia officinalis*) and peppermint (*Mentha piperita*), is restricted to a brief period early in leaf development (Croteau et al., 1981; Gershenzon et al., 2000). The monoterpene biosynthesis in peppermint reaches a maximum in 15-day-old leaves; only very low rates were observed in leaves younger than 12 days or older than 20 days. The monoterpene content of the peppermint leaves increased rapidly up to day 21, then leveled off, and kept stable for the remainder of the leaf life (Gershenzon et al., 2000).

The composition of the essential oil often changes between different plant parts. Phytochemical polymorphism is often the case between different plant organs. In *Origanum vulgare* ssp. *hirtum*, a polymorphism within a plant could even be detected on a much lower level, between different oil glands of a leaf (Johnson et al., 2004). This form of polymorphism seems to be not frequently occurring; differences in the composition between oil glands are more often related to the age of the oil glands (Grassi et al., 2004; Johnson et al., 2004; Novak et al., 2006a; Schmiderer et al., 2008).

Such polymorphisms can also be found quite frequently when comparing the essential oil composition of individual plants of a distinct species (intraspecific variation, "chemotypes") and is based on the plants" genetic background.

The differences in the complex composition of two essential oils of one kind may sometimes be difficult to assign to specific chemotypes or to differences arising in the consequence of the reactions of the plants to specific environmental conditions, for example, to different growing locations. In general, the differences due to genetic differences are much bigger than by different environmental conditions. However, many intraspecific polymorphisms are probably not yet detected or have been described only recently even for widely used essential oil crops like sage (Novak et al., 2006b).

3.2 PHYTOCHEMICAL VARIATION

3.2.1 CHEMOTAXONOMY

The ability to accumulate essential oils is not omnipresent in plants but scattered throughout the plant kingdom, in many cases, however, very frequent within—or a typical character of—certain plant families. From the taxonomical and systematic point of view, not the production of essential oils is the distinctive feature since this is a quite heterogeneous group of substances, but either the type of secretory containers (trichomes, oil glands, lysogenic cavities, or schizogenic oil ducts) or the biosynthetically specific group of substances, for example, mono- or sesquiterpenes and phenylpropenes; the more a substance is deduced in the biosynthetic pathway, the more specific it is for certain taxa: monoterpenes are typical for the genus *Mentha*, but menthol is characteristic for *M. piperita* and *Mentha arvensis* ssp. *piperascens* only; sesquiterpenes are common in the *Achillea–millefolium* complex, but only *Achillea roseoalba* (2×) and *Achillea collina* (4×) are able to produce matricine as precursor of (the artifact) chamazulene (Vetter et al., 1997). On the other hand, the phenylpropanoid eugenol, typical for cloves (*S. aromaticum*, Myrtaceae), can also be found in large amounts in distant species, for example, cinnamon (*Cinnamomum zeylanicum*, Lauraceae) or basil (*Ocimum basilicum*, Lamiaceae); as sources for anethole are known not only aniseed (*Pimpinella anisum*) and fennel (*F. vulgare*), which are both Apiaceae, but also star anise (*Illicium verum*, Illiciaceae), *Clausena anisata* (Rutaceae), *Croton zehntneri* (Euphorbiaceae), or *Tagetes lucida* (Asteraceae). Finally, eucalyptol (1,8-cineole)—named after its occurrence in *Eucalyptus* sp. (Myrtaceae)—may also be a main compound of the essential oil of galangal (*Alpinia officinarum*, Zingiberaceae), bay laurel (*Laurus nobilis*, Lauraceae), Japan pepper (*Zanthoxylum piperitum*, Rutaceae), and a number of plants of the mint family, for example, sage (*S. officinalis*, *Salvia fruticosa*, *Salvia lavandulifolia*), rosemary (*Rosmarinus officinalis*), and mints (*Mentha* sp.). Taking the aforementioned facts into consideration, chemotaxonomically relevant are (therefore) common or distinct pathways, typical fingerprints, and either main compounds or very specific even minor or trace substances (e.g., δ-3-carene to separate *Citrus grandis* from other *Citrus* sp. [Gonzalez et al., 2002]).

The plant families comprising species that yield a majority of the most economically important essential oils are not restricted to one specialized taxonomic group but are distributed among all plant classes: gymnosperms, for example, the families Cupressaceae (cedarwood, cedar leaf, juniper oil, etc.) and Pinaceae (pine and fir oils), and angiosperms, and among them within Magnoliopsida, Rosopsida, and Liliopsida. The most important families of dicots are Apiaceae (e.g., fennel, coriander, and other aromatic seed/root oils), Asteraceae or Compositae (chamomile, wormwood, tarragon oil, a.s.o), Geraniaceae (geranium oil), Illiciaceae (star anise oil), Lamiaceae (mint, patchouli, lavender, oregano, and many other herb oils), Lauraceae (litsea, camphor, cinnamon, sassafras oil, etc.), Myristicaceae (nutmeg and mace), Myrtaceae (myrtle, cloves, and allspice), Oleaceae (jasmine oil), Rosaceae (rose oil), and Santalaceae (sandalwood oil). In monocots (Liliopsida), it is substantially restricted to Acoraceae (calamus), Poaceae (vetiver and aromatic grass oils), and Zingiberaceae (e.g., ginger and cardamom).

Apart from the phytochemical group of substances typical for a taxon, the chemical outfit depends, furthermore, on the specific genotype; the stage of plant development, also influenced by environmental factors; and the plant part (see Section 3.3.2.1). Considering all these influences, chemotaxonomic statements and conclusions have to be based on comparable material, grown and harvested under comparable circumstances.

3.2.2 Inter- and Intraspecific Variation

Knowledge on biochemical systematics and the inheritance of phytochemical characters depends on extensive investigations of taxa (particularly species) and populations on single-plant basis, respectively, and several examples of genera show that the taxa do indeed display different patterns.

3.2.2.1 Lamiaceae (Labiatae) and Verbenaceae

The presumably largest genus among the Lamiaceae is *sage* (*Salvia* L.) consisting of about 900 species widely distributed in the temperate, subtropical, and tropical regions all over the world with major centers of diversity in the Mediterranean, in Central Asia, the Altiplano from Mexico throughout Central and South America, and in southern Africa. Almost 400 species are used in traditional and modern medicine, as aromatic herbs or ornamentals worldwide; among them are *S. officinalis*, *S. fruticosa*, *Salvia sclarea*, *Salvia divinorum*, *Salvia miltiorrhiza*, and *Salvia pomifera*, to name a few. Many applications are based on nonvolatile compounds, for example, diterpenes and polyphenolic acids. Regarding the essential oil, there are a vast number of mono- and sesquiterpenes found in sage but, in contrast to, for example, *Ocimum* sp. and *Perilla* sp. (also Lamiaceae), no phenylpropenes were detected.

To understand species-specific differences within this genus, the Mediterranean *S. officinalis* complex (*S. officinalis*, *S. fruticosa*, and *S. lavandulifolia*) will be confronted with the *Salvia stenophylla* species complex (*S. stenophylla*, *Salvia repens*, and *Salvia runcinata*) indigenous to South Africa: in the *S. officinalis* group, usually α- and β-thujones, 1,8-cineole, camphor, and, in some cases, linalool, β-pinene, limonene, or *cis*-sabinyl acetate are the prevailing substances, whereas in the *S. stenophylla* complex, quite often sesquiterpenes, for example, caryophyllene or α-bisabolol, are main compounds.

Based on taxonomical studies of *Salvia* spp. (Hedge, 1992; Skoula et al., 2000; Reales et al., 2004) and a recent survey concerning the chemotaxonomy of *S. stenophylla* and its allies (Viljoen et al., 2006), Figure 3.1 shows the up-to-now-identified chemotypes within these taxa. Comparing the data

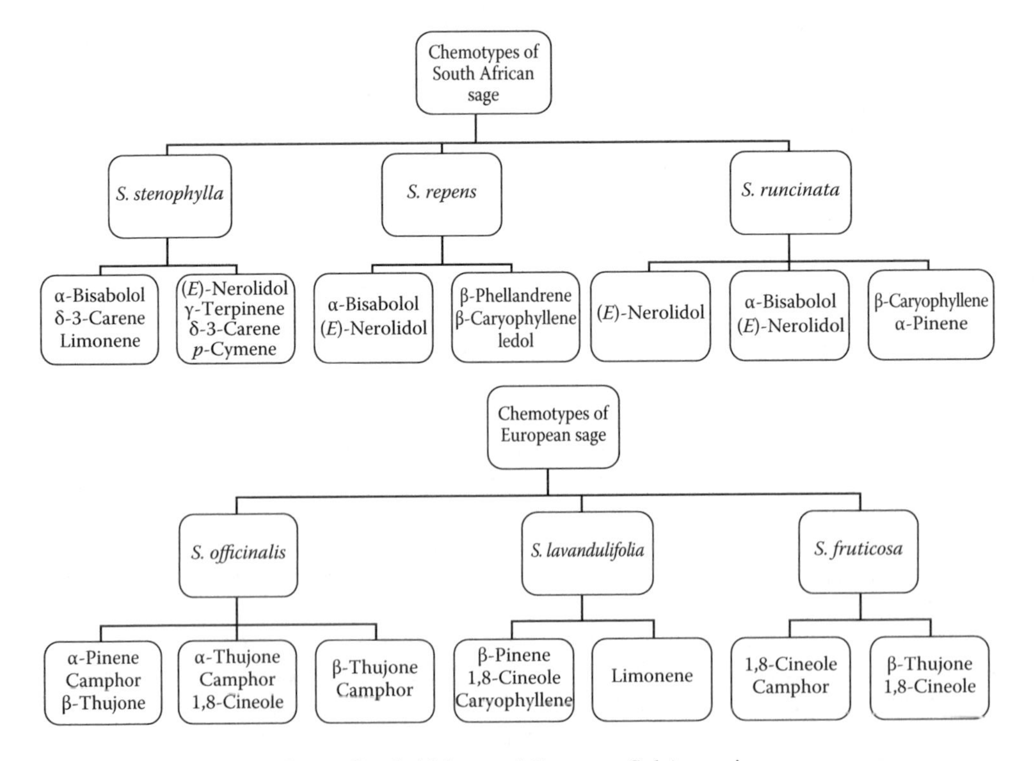

FIGURE 3.1 Chemotypes of some South African and European *Salvia* species.

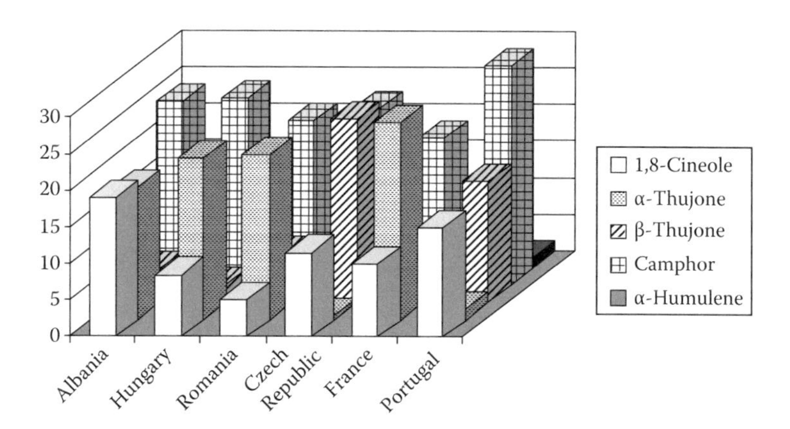

FIGURE 3.2 Composition of the essential oil of six *Salvia officinalis* origins.

of different publications, the picture is, however, not as clear as demonstrated by six *S. officinalis* origins in Figure 3.2 (Chalchat et al., 1998; Asllani, 2000). This might be due to the prevailing chemotype in a population, the variation between single plants, the time of sample collection, and the sample size. This is exemplarily shown by one *S. officinalis* population where the individuals varied in α-thujone, from 9% to 72%; β-thujone, from 2% to 24%; 1,8-cineole, from 4% to 18%; and camphor from 1% to 25%. The variation over 3 years and five harvests of one clone only ranged as follows: α-thujone 35%–72%, β-thujone 1%–7%, 1,8-cineole 8%–15%, and camphor 1%–18% (Bezzi, 1994; Bazina et al., 2002). But also all other (minor) compounds of the essential oil showed respective intraspecific variability (see, e.g., Giannouli and Kintzios, 2000).

S. fruticosa was principally understood to contain 1,8-cineole as main compound but at best traces of thujones, as confirmed by Putievsky et al. (1986) and Kanias et al. (1998). In a comparative study of several origins, Máthe et al. (1996) identified, however, a population with atypically high β-thujone similar to *S. officinalis*. Doubts on if this origin could be true *S. fruticosa* or a spontaneous hybrid of both species were resolved by extensive investigations on the phytochemical and genetic diversities of *S. fruticosa* in Crete (Karousou et al., 1998; Skoula et al., 1999). There, it was shown that all wild populations in western Crete consist of 1,8-cineole chemotypes only, whereas in the eastern part of the island, essential oils with up to 30% thujones, mainly β-thujone, could be observed. In central Crete, finally, mixed populations were found. A cluster analysis based on random amplification of polymorphic DNA (RAPD) patterns confirmed the genetic differences between the west and east Crete populations of *S. fruticosa* (Skoula et al., 1999).

A rather interesting example of diversity is *oregano*, which counts to the commercially most valued spices worldwide. More than 60 plant species are used under this common name showing similar flavor profiles characterized mainly by cymyl compounds, for example, carvacrol and thymol. With few exemptions, the majority of oregano species belong to the Lamiaceae and Verbenaceae families with the main genera *Origanum* and *Lippia* (Table 3.1). In 1989, almost all of the estimated 15,000 ton/year dried oregano originated from wild collection; today, some 7000 ha of *Origanum onites* are cultivated in Turkey alone (Baser, 2002); *O. onites* and other *Origanum* species are cultivated in Greece, Israel, Italy, Morocco, and other countries.

In comparison with sage, the genus *Origanum* is much smaller and consists of 43 species and 18 hybrids according to the actual classification (Skoula and Harborne, 2002) with main distribution areas around the Mediterranean. Some subspecies of *O. vulgare* only are also found in the temperate and arid zones of Eurasia up to China. Nevertheless, the genus is characterized by large morphological and phytochemical diversities (Kokkini et al., 1996; Baser, 2002; Skoula and Harborne, 2002).

The occurrence of several chemotypes is reported, for example, for commercially used *Origanum* species, from Turkey (Baser, 2002). In *O. onites*, two chemotypes are described, a carvacrol type and a linalool type. Additionally, a *mixed type* with both basic types mixed may occur. In

TABLE 3.1
Species Used Commercially in the World as Oregano

Family/Species	Commercial Name(s) Found in Literature
Labiatae	
Calamintha potosina Schaf.	Oregano de la sierra, oregano, origanum
Coleus amboinicus Lour. (syn. *C. aromaticus* Benth)	Oregano, oregano brujo, oregano de Cartagena, oregano de Espana, oregano Frances
Coleus aromaticus Benth.	Oregano de Espana, oregano, *Origanum*
Hedeoma floribunda Standl.	Oregano, *Origanum*
Hedeoma incona Torr.	Oregano
Hedeoma patens Jones	Oregano, *Origanum*
Hyptis albida HBK.	Oregano, *Origanum*
Hyptis americana (Aubl.) Urb. (*H. gonocephala* Gris.)	Oregano
Hyptis capitata Jacq.	Oregano, *Origanum*
Hyptis pectinata Poit.	Oregano, *Origanum*
Hyptis suaveolens (L.) Poit.	Oregano, oregano cimarron, *Origanum*
Monarda austromontana Epling	Oregano, *Origanum*
Ocimum basilicum L.	Oregano, *Origanum*
Origanum compactum Benth. (syn. *O. glandulosum* Salzm, ex Benth.)	Oregano, *Origanum*
Origanum dictamnus L. (*Majorana dictamnus* L.)	Oregano, *Origanum*
Origanum elongatum (Bonent) Emberger et Maire	Oregano, *Origanum*
Origanum floribundum Munby (*O. cinereum* Noe)	Oregano, *Origanum*
Origanum grosii Pau et Font Quer ex letswaart	Oregano, *Origanum*
Origanum majorana L.	Oregano
Origanum microphyllum (Benth) Vogel	Oregano, *Origanum*
Origanum onites L. (syn. *O. smyrneum* L.)	Turkish oregano, oregano, *Origanum*[a]
Origanum scabrum Boiss et Heldr. (syn. *O. pulchrum* Boiss et Heldr.)	Oregano, *Origanum*
Origanum syriacum L. var. *syriacum* (syn. *O. maru* L.)	Oregano, *Origanum*
Origanum vulgare L. ssp. *gracile* (Koch) letswaart (syn. *O. gracile* Koch, *O. tyttanthum* Gontscharov)	Oregano, *Origanum*
Origanum vulgare ssp. *hirtum* (Link) letswaart (syn *O. hirtum* Link)	Oregano, *Origanum*
Origanum vulgare ssp. *virens* (Hoffmanns et Link) letswaart (syn. *O. virens* Hoffmanns et Link)	Oregano, *Origanum*, oregano verde
Origanum vulgare ssp. *viride* (Boiss.) Hayek (syn. *O. viride*) Halacsy (syn. *O. heracleoticum* L.)	Greek oregano, oregano, *Origanum*[a]
Origanum vulgare L. ssp. *vulgare* (syn. *Thymus origanum* (L.) Kuntze)	Oregano, *Origanum*
Origanum vulgare L.	Oregano, orenga, Oregano de Espana
Poliomintha longiflora Gray	Oregano
Salvia sp.	Oregano
Satureja thymbra L.	Oregano cabruno, oregano, *Origanum*
Thymus capitatus (L.) Hoffmanns et Link (syn. *Coridothymus capitatus* (L.) Rchb.f.)	Spanish oregano, oregano, *Origanum*[a]
Verbenaceae	
Lantana citrosa (Small) Modenke	Oregano xiu, oregano, *Origanum*
Lantana glandulosissima Hayek	Oregano xiu, oregano silvestre, oregano, *Origanum*
Lantana hirsuta Mart et Gall.	Oreganillo del monte, oregano, *Origanum*
Lantana involucrata L.	Oregano, *Origanum*

(*Continued*)

TABLE 3.1 (*Continued*)
Species Used Commercially in the World as Oregano

Family/Species	Commercial Name(s) Found in Literature
Lantana purpurea (Jacq.) Benth. & Hook. (syn. *Lippia purpurea* Jacq.)	Oregano, *Origanum*
Lantana trifolia L.	Oregano, *Origanum*
Lantana velutina Mart. & Gal.	Oregano xiu, oregano, *Origanum*
Lippia myriocephala Schlecht. & Cham.	Oreganillo
Lippia affinis Schau.	Oregano
Lippia alba (Mill) N.E. Br. (syn. *L. involucrata* L.)	Oregano, *Origanum*
Lippia berlandieri Schau.	Oregano
Lippia cordiostegia Benth.	Oreganillo, oregano montes, oregano, *Origanum*
Lippia formosa T.S. Brandeg.	Oregano, *Origanum*
Lippia geisseana (R.A.Phil.) Soler.	Oregano, *Origanum*
Lippia graveolens HBK	Mexican oregano, oregano cimarron, oregano[a]
Lippia helleri Britton	Oregano del pais, oregano, *Origanum*
Lippia micromera Schau.	Oregano del pais, oregano, *Origanum*
Lippia micromera var. *helleri* (Britton) Moldenke	Oregano
Lippia origanoides HBK	Oregano, origano del pais
Lippia palmeri var. *spicata* Rose	Oregano
Lippia palmeri Wats.	Oregano, *Origanum*
Lippia umbellata Cav.	Oreganillo, oregano montes, oregano, *Origanum*
Lippia velutina Mart. et Galeotti	Oregano, *Origanum*
Rubiaceae	
Borreria sp.	Oreganos, oregano, *Origanum*
Scrophulariaceae	
Limnophila stolonifera (Blanco) Merr.	Oregano, *Origanum*
Apiaceae	
Eryngium foetidum L.	Oregano de Cartagena, oregano, *Origanum*
Asteraceae	
Coleosanthus veronicaefolius HBK	Oregano del cerro, oregano del monte, oregano del campo
Eupatorium macrophyllum L. (syn. *Hebeclinium macrophyllum* DC.)	Oregano, *Origanum*

[a] Oregano species with economic importance according to Lawrence (1984).

Turkey, two chemotypes of *Origanum majorana* are known, one contains *cis*-sabinene hydrate as chemotypical lead compound and is used as marjoram in cooking (*marjoramy*), while the other one contains carvacrol in high amounts and is used to distil *oregano oil* in a commercial scale. Variability of chemotypes continues also within the *marjoramy O. majorana*. Novak et al. (2002) detected in cultivated marjoram accessions additionally to *cis*-sabinene hydrate the occurrence of polymorphism of *cis*-sabinene hydrate acetate. Since this chemotype did not influence the sensorial impression much, this chemotype was not eliminated in breeding, while an *off-flavor* chemotype would have been certainly eliminated in its cultivation history. In natural populations of *O. majorana* from Cyprus besides the *classical cis*-sabinene hydrate type, a chemotype with α-terpineol as main compound was also detected (Novak et al., 2008). The two extreme *off-flavor* chemotypes in *O. majorana*, carvacrol and α-terpineol chemotypes, are not to be found anywhere in cultivated marjoram, demonstrating one of the advantages of cultivation in delivering homogeneous qualities.

The second *oregano* of commercial value—mainly used in the Americas—is *Mexican oregano* (*Lippia graveolens* HBK., Verbenaceae) endemic to California, Mexico, and throughout Central America (Fischer, 1998). Due to wild harvesting, only a few published data show essential oil contents largely ranging from 0.3% to 3.6%. The total number of up-to-now-identified essential oil compounds comprises almost 70 with the main constituents thymol (3.1%–80.6%), carvacrol (0.5%–71.2%), 1,8-cineole (0.1%–14%), and *p*-cymene (2.7%–28.0%), followed by, for example, myrcene, γ-terpinene, and the sesquiterpene caryophyllene (Lawrence, 1984; Dominguez et al., 1989; Uribe-Hernández et al., 1992; Fischer et al., 1996; Vernin et al., 2001).

In a comprehensive investigation of wild populations of *L. graveolens* collected from the hilly regions of Guatemala, three different essential oil chemotypes could be identified, a thymol, a carvacrol, and an absolutely irregular type (Fischer et al., 1996). Within the thymol type, contents of up to 85% thymol in the essential oil could be obtained and only traces of carvacrol. The irregular type has shown a very uncommon composition where no compound exceeds 10% of the oil, and also phenylpropenes, for example, eugenol and methyl eugenol, were present (Fischer et al., 1996; Fischer, 1998). In Table 3.2, a comparison of recent data is given including *Lippia alba*, commonly called *oregano* or *oregano del monte*, although carvacrol and thymol are absent from the essential oil of this species. In Guatemala, two different chemotypes were found within *L. alba*: a myrcenone and a citral type (Fischer et al., 2004). Besides it, a linalool, a carvone, a camphor (1,8-cineole), and a limonene–piperitone chemotype have been described (Dellacassa et al., 1990; Pino et al., 1997; Frighetto et al., 1998; Senatore and Rigano, 2001).

Chemical diversity is of special interest if on genus or species level both terpenes and phenylpropenes can be found in the essential oil. Most Lamiaceae preferentially accumulate mono- and sesquiterpenes in their volatile oils, but some genera produce oils also rich in phenylpropenes, among these *Ocimum* sp. and *Perilla* sp.

The genus *Ocimum* comprises over 60 species, of which *Ocimum gratissimum* and *O. basilicum* are of high economic value. Biogenetic studies on the inheritance of *Ocimum* oil constituents were reported by Khosla et al. (1989) and an *O. gratissimum* strain named *clocimum* containing 65% of eugenol in its oil was described by Bradu et al. (1989). A number of different chemotypes of basil (*O. basilicum)* have been identified and classified (Vernin et al., 1984; Marotti et al., 1996) containing up to 80% linalool, up to 21.5% 1,8-cineole, 0.3%–33.0% eugenol, and also the presumably toxic compounds methyl chavicol (estragole) and methyl eugenol in concentrations close to 50% (Elementi et al., 2006; Macchia et al., 2006).

Perilla frutescens can be classified in several chemotypes as well according to the main monoterpene components perillaldehyde, elsholtzia ketone, or perilla ketones and on the other side phenylpropanoid types containing myristicin, dillapiole, or elemicin (Koezuka et al., 1986). A comprehensive presentation on the chemotypes and the inheritance of the mentioned compounds was given by this author in Hay and Waterman (1993). In the referred last two examples, not only the sensorial but also the toxicological properties of the essential oil compounds are decisive for the (further) commercial use of the respective species" biodiversity.

Although the Labiatae family plays an outstanding role as regards the chemical polymorphism of essential oils, also in other essential oils containing plant families and genera, a comparable phytochemical diversity can be observed.

3.2.2.2 Asteraceae (Compositae)

Only a limited number of genera of the Asteraceae are known as essential oil plants, among them *Tagetes*, *Achillea*, and *Matricaria*. The genus *Tagetes* comprises actually 55 species, all of them endemic to the American continents with the center of biodiversity between 30° northern and 30° southern latitude. One of the species largely used by the indigenous population is *pericon* (*T. lucida* Cav.), widely distributed over the highlands of Mexico and Central America (Stanley and Steyermark, 1976). In contrast to almost all other *Tagetes* species characterized by the content of tagetones, this species contains phenylpropenes and terpenes. A detailed study on its diversity

TABLE 3.2
Main Essential Oil Compounds of *Lippia graveolens* and *L. alba* According to Recent Data

	L. graveolens					*L. alba*			
	Fischer et al. (1996)			Senatore and Rigano (2001)	Vernin et al. (2001)	Fischer (1998)		Senatore and Rigano (2001)	Lorenzo et al. (2001)
	Guatemala					Guatemala			
Compound	**Thymol-Type**	**Carvacrol-Type**	**Irregular Type**	**Guatemala**	**El-Salvador**	**Myrcenone-Type**	**Cineole-Type**	**Guatemala**	**Uruguay**
Myrcene	1.3	1.9	2.7	1.1	t	6.5	1.7	0.2	0.8
p-Cymene	2.7	6.9	2.8	5.5	2.1	t	t	0.7	n.d.
1,8-Cineole	0.1	0.6	5.0	2.1	t	t	**22.8**	**14.2**	1.3
Limonene	0.2	0.3	1.5	0.8	t	1.0	3.2	**43.6**	2.9
Linalool	0.7	1.4	3.8	0.3	t	4.0	2.4	1.2	**55.3**
Myrcenon	n.d.	n.d.	n.d.	n.d.	n.d.	**54.6**	3.2	n.d.	n.d.
Piperitone	n.d.	n.d.	n.d.	n.d.	n.d.	t	t	**30.6**	n.d.
Thymol	**80.6**	**19.9**	6.8	**31.6**	7.3	n.d.	n.d.	n.d.	n.d.
Carvacrol	1.3	**45.2**	1.1	0.8	**71.2**	n.d.	n.d.	n.d.	n.d.
β-Caryophyllene	2.8	3.5	8.7	4.6	9.2	2.6	1.2	1.0	9.0
α-Humulene	1.9	2.3	5.7	3.0	5.0	0.7	t	0.6	0.9
Caryophyll.-ox.	0.3	0.8	3.3	4.8	t	1.8	3.0	1.1	0.6
Z-Dihydrocarvon/*Z*-Ocimenone	n.d.	n.d.	n.d.	n.d.	n.d.	13.1	0.6	0.1	0.8
E-Dihydrocarvon	n.d.	n.d.	n.d.	n.d.	n.d.	4.9	n.d.	t	1.2

Note: n.d., Not detectable; t, traces. Main compounds in bold.

TABLE 3.3
Main Compounds of the Essential Oil of Selected *Tagetes lucida* Types (in% of dm)

Substance	Anethole Type (2)	Estragole Type (8)	Methyleugenol Type (7)	Nerolidol Type (5)	Mixed Type
Linalool	0.26	0.69	1.01	Tr.	3.68
Estragole	11.57	**78.02**	8.68	3.23	**24.28**
Anethole	**73.56**	0.75	0.52	Tr.	**30.17**
Methyleugenol	1.75	5.50	**79.80**	17.76	**17.09**
β-Caryophyllene	0.45	1.66	0.45	2.39	0.88
Germacrene D	2.43	2.89	1.90	Tr.	5.41
Methylisoeugenol	1.42	2.78	2.00	Tr.	3.88
Nerolidol	0.35	0.32	0.31	**40.52**	1.24
Spathulenol	0.10	0.16	0.12	Tr.	0.23
Carophyllene oxide	0.05	0.27	0.45	10.34	0.53

Note: Location of origin in Guatemala: (2) Cabrican/Quetzaltenango, (5) La Fuente/Jalapa, (7) Joyabaj/El Quiche, (8) Sipacapa/S. Marcos, Mixed Type: Taltimiche/San Marcos. Main compounds in bold.

in Guatemala resulted in the identification of several eco- and chemotypes (Table 3.3): anethole, methyl chavicol (estragole), methyl eugenol, and one sesquiterpene type producing higher amounts of nerolidol (Bicchi et al., 1997; Goehler, 2006). The distribution of the three main phenylpropenes in six populations is illustrated in Figure 3.3. In comparison with the plant materials investigated by Ciccio (2004) and Marotti et al. (2004) containing oils with 90%–95% estragole, only the germplasm collection of Guatemaltecan provenances (Goehler, 2006) allows to select individuals with high anethole but low to very low estragole or methyleugenol content—or with interestingly high nerolidol content, as mentioned earlier.

The genus *Achillea* is widely distributed over the northern hemisphere and consists of approximately 120 species, of which the *Achillea millefolium* aggregate (yarrow) represents a polyploid complex of allogamous perennials (Saukel and Länger, 1992; Vetter and Franz, 1996). The different taxa of the recent classification (*minor species* and *subspecies*) are morphologically and chemically to a certain extent distinct and only the diploid taxa *Achillea asplenifolia* and *A. roseoalba* as well as the tetraploids *A. collina* and *Achillea ceretanica* are characterized by proazulens, for example, achillicin, whereas the other taxa, especially 6× and 8×, contain eudesmanolides, longipinenes, germacranolides, and/or guajanolid peroxides (Table 3.4). The intraspecific variation in the proazulene content ranged from traces up to 80%; other essential oil components of the azulenogenic species are, for example,

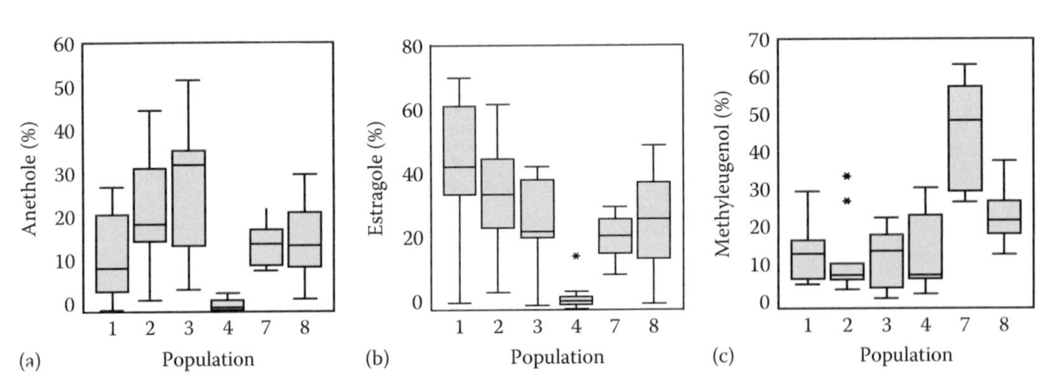

FIGURE 3.3 Variability of (a) anethole, (b) methyl chavicol (estragole), and (c) methyl eugenol in the essential oil of six *Tagetes lucida*—populations from Guatemala. * indicates erratic individuals.

TABLE 3.4
Taxa within the *Achillea-Millefolium*-Group (Yarrow)

Taxon	Ploidy Level	Main Compounds
A. setacea W. et K.	2×	Rupicoline
A. aspleniifolia Vent.	2× (4×)	**7,8-Guajanolide**
		Artabsin-derivatives
		3-Oxa-Guajanolide
A. roseo-alba Ehrend.	2×	Artabsin-derivatives
		3-Oxaguajanolide
		Matricinderivatives
A. collina Becker	4×	Artabsin-derivatives
		3-Oxaguajanolide
		Matricinderivatives
		Matricarinderivatives
A. pratensis Saukel u. Länger	4×	Eudesmanolides
A. distans ssp. Distans W. et K.	6×	Longipinenones
A. distans ssp. *styriaca*	4×	
A. tanacetifolia (stricta) W. et K.	6×	
A. mill. ssp. *sudetica*	6×	Guajanolidperoxide
A. mill. ssp. Mill. L.	6×	
A. pannonica Scheele	8× (6×)	Germacrene
		Guajanolidperoxide

Source: Franz, Ch., 2013. In *Handbuch des Arznei- u. Gewuerzpflanzenbaus*, Vol. 5, pp. 453–463. Saluplanta, Bernburg.

Note. Substances in bold are proazulenes.

α- and β-pinene, borneol, camphor, sabinene, or caryophyllene (Kastner et al., 1992). The frequency distribution of proazulene individuals among two populations is shown in Figure 3.4.

Crossing experiments resulted in proazulene being a recessive character of di- and tetraploid *Achillea* spp. (Vetter et al., 1997) similar to chamomile (Franz, 1993a,b). Finally, according to Steinlesberger (2002) also a plant-to-plant variation in the enantiomers of, for example, α- and β-pinene as well as sabinene exists in yarrow oils, which makes it even more complicated to use phytochemical characters for taxonomical purposes.

Differences in the essential oil content and composition of chamomile flowers (*Matricaria recutita*) have long been recognized due to the fact that the distilled oil is either dark blue, green, or yellow,

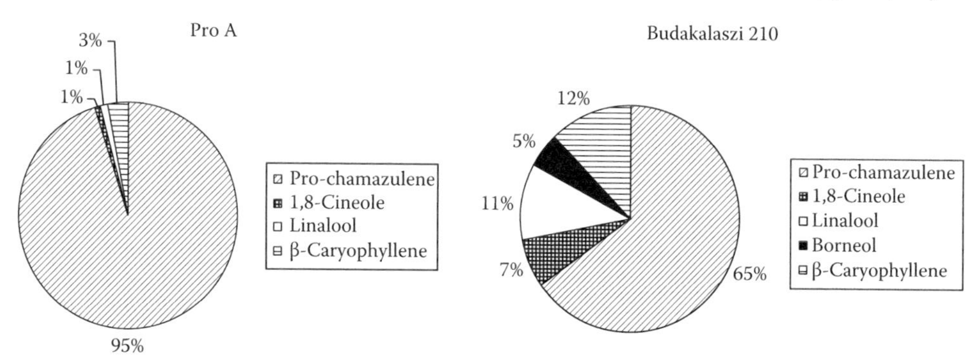

FIGURE 3.4 Frequency distribution of proazulene individuals among two *Achillea* sp. populations.

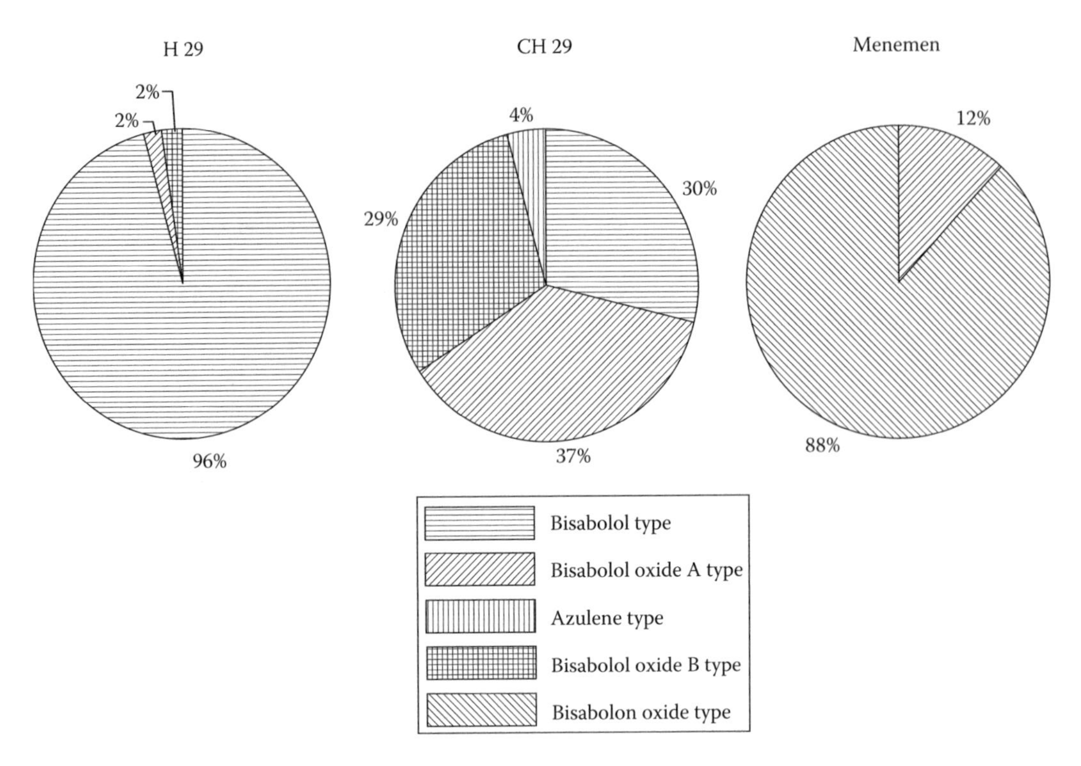

FIGURE 3.5 Frequency distribution of chemotypes in three varieties/populatios of chamomile (*Matricaria recutita* (L.) Rauschert).

depending on the prochamazulene content (matricin as prochamazulene in chamomile is transformed to the blue-colored artifact chamazulene during the distillation process). Recognizing also the great pharmacological potential of the bisabolols, a classification into the chemotypes (−)-α-bisabolol, (−)-α-bisabololoxide A, (−)-α-bisabololoxide B, (−)-α-bisabolonoxide (A), and (pro)chamazulene was made by Franz (1982, 1989a). Examining the geographical distribution revealed a regional differentiation, where an α-bisabolol—(pro) chamazulene population was identified on the Iberian peninsula; mixed populations containing chamazulene, bisabolol, and bisabololoxides A/B are most frequent in Central Europe, and prochamazulene—free bisabolonoxide populations are indigenous to southeast Europe and minor Asia. In the meantime, Wogiatzi et al. (1999) have shown for Greece and Taviani et al. (2002) for Italy a higher diversity of chamomile including α-bisabolol types. This classification of populations and chemotypes was extended by analyzing populations at the level of individual plants (Schröder, 1990) resulting in the respective frequency distributions (Figure 3.5).

In addition, the range of essential oil components in the chemotypes of one Central European population is shown in Table 3.5 (Franz, 2000).

Data on inter- and intraspecific variation of essential oils are countless, and recent reviews are known for a number of genera published, for example, in the series "Medicinal and Aromatic Plants—Industrial Profiles" (Harwood Publications, Taylor & Francis, CRC Press, respectively).

The generally observed quantitative and qualitative variations in essential oils draw the attention *i.a.* to appropriate random sampling for getting valid information on the chemical profile of a species or population. As concerns quantitative variations of a certain pattern or substance, Figure 3.6 shows exemplarily the bisabolol content of two chamomile populations depending on the number of individual plants used for sampling. At small numbers, the mean value oscillates strongly, and only after at least 15–20 individuals the range of variation becomes acceptable. Quite different appears the situation at qualitative differences, that is, *either–or variations* within populations or taxa, for example, carvacrol/thymol, α-/β-thujone/1,8-cineole/camphor, or monoterpenes/phenylpropenes.

TABLE 3.5
Grouping within a European Spontaneous Chamomile, Figures in % of Terpenoids in the Essential Oil of the Flower Heads

	Chamazulen	α-Bisabolol	α-B.-Oxide A	α-B.-Oxide B
α-Bisabolol-type				
Range	2.5–35.2	58.8–92.1	n.d.–1.0	n.d.–3.2
Mean	23.2	**68.8**	n.d.	n.d.
α-Bisabololoxide A-type				
Range	6.6–31.2	0.5–12.3	31.7–66.7	1.9–22.4
Mean	21.3	2.1	**53.9**	11.8
α-Bisabololoxide B-type				
Range	7.6–24.2	0.8–6.5	1.6–4.8	61.6–80.5
Mean	16.8	2.0	2.6	**72.2**
Chamazulene-type				
Range	76.3–79.2	5.8–8.3	n.d.–0.8	n.d.–2.6
Mean	**77.8**	7.1	n.d.	n.d.

Source: Franz, Ch., 2000. *Biodiversity and Random Sampling in Essential Oil Plants*. Lecture 31st ISEO, Hamburg, Germany.
Note: Main compounds in bold. n.d., not detected (determined).

Any random sample may give nonspecific information only on the principal chemical profile of the respective population provided that the sample is representative. This depends on the number of chemotypes, their inheritance, and frequency distribution within the population, and generally speaking, no less than 50 individuals are needed for that purpose, as it can be derived from the comparison of chemotypes in a *Thymus vulgaris* population (Figure 3.7).

The overall high variation in essential oil compositions can be explained by the fact that quite different products might be generated by small changes in the synthase sequences only. On the other hand, different synthases may be able to produce the same substance in systematically distant taxa. The different origin of such substances can be identified by, for example, the $^{12}C/^{13}C$ ratio (Mosandl, 1993). Bazina et al. (2002) stated, "Hence, a simple quantitative analysis of the essential oil composition is not necessarily appropriate for estimating genetic proximity even in closely related taxa."

3.3 IDENTIFICATION OF SOURCE MATERIALS

As illustrated by the previous paragraph, one of the crucial points of using plants as sources for essential oils is their heterogeneity. A first prerequisite for reproducible compositions is therefore an unambiguous botanical identification and characterization of the starting material. The first approach is the classical taxonomical identification of plant materials based on macro- and micromorphological features of the plant. The identification is followed by phytochemical analysis that may contribute to species identification as well as to the determination of the quality of the essential oil. This approach is now complemented by DNA-based identification.

DNA is a long polymer of nucleotides, the building units. One of four possible nitrogenous bases is part of each nucleotide, and the sequence of the bases on the polymer strand is characteristic for each living individual. Some regions of the DNA, however, are conserved on the species or family level and can be used to study the relationship of taxa (Taberlet et al., 1991; Wolfe and Liston, 1998). DNA sequences conserved within a taxon but different between taxa can therefore be used to identify a taxon (*DNA barcoding*) (Hebert et al., 2003; Kress et al., 2005). A DNA-barcoding consortium was founded in 2004 with the ambitious goal to build a barcode library for all eukaryotic life in the next 20 years (Ratnasingham and Hebert, 2007). New sequencing technologies (454, Solexa, SOLiD) enable a fast and

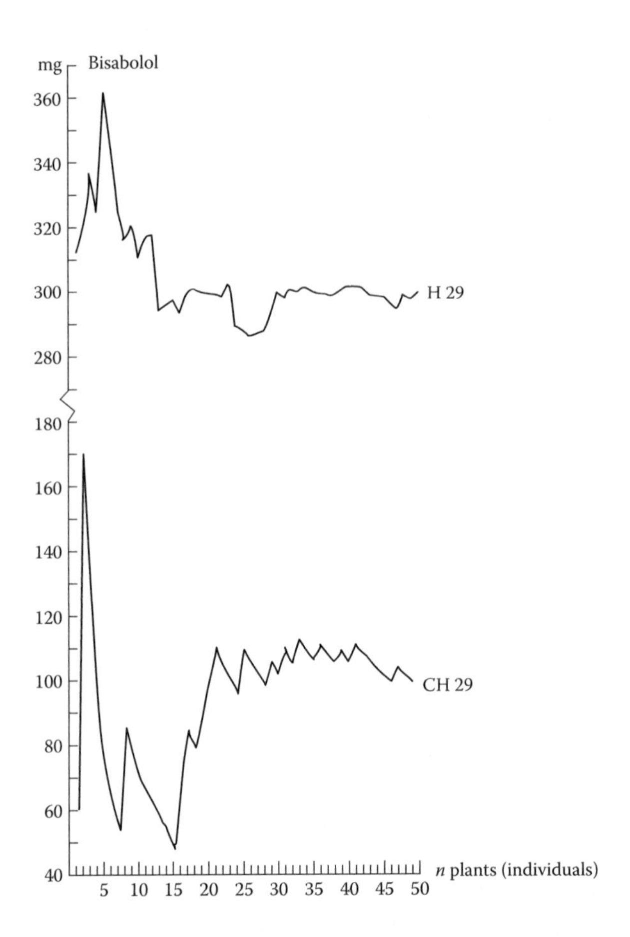

FIGURE 3.6 (−)-α-Bisabolol-content (mg/100 g crude drug) in two chamomile (*Matricaria recutita*) populations: mean value in dependence of the number of individuals used for sampling.

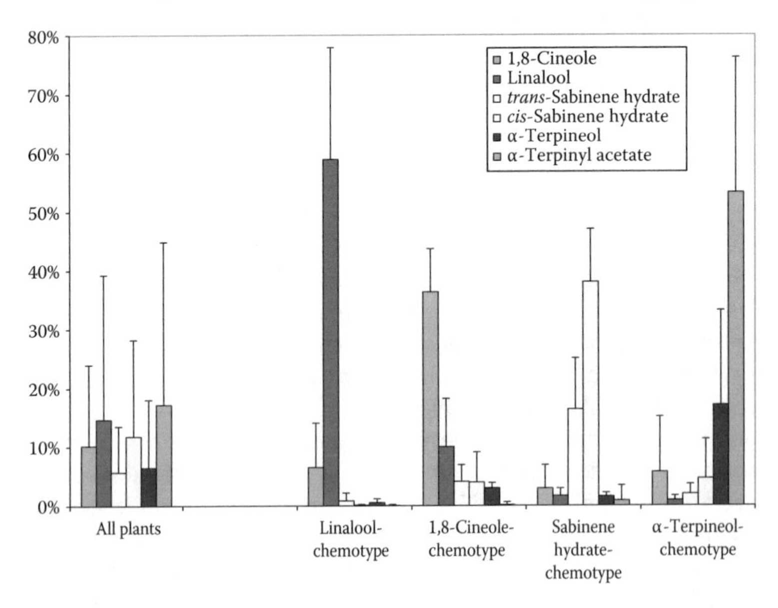

FIGURE 3.7 Mean values of the principal essential compounds of a *Thymus vulgaris* population (left) in comparison to the mean values of the chemotypes within the same population.

representative analysis, but will be applied due to their high costs in the moment only in the next phase of DNA barcoding (Frezal and Leblois, 2008). DNA barcoding of animals has already become a routine task. DNA barcoding of plants, however, is still not trivial and a scientific challenge (Pennisi, 2007).

Besides sequence information–based approaches, multilocus DNA methods (RAPD, amplified fragment length polymorphism, etc.) are complementing in resolving complicated taxa and can become a barcode for the identification of populations and cultivars (Weising et al., 2005). With multilocus DNA methods, it is furthermore possible to tag a specific feature of a plant of which the genetic basis is still unknown. This approach is called molecular markers (in *sensu strictu*) because they mark the occurrence of a specific trait like a chemotype or flower color. The gene regions visualized, for example, on an agarose gel are not the specific gene responsible for a trait but are located on the genome in the vicinity of this gene and therefore co-occur with the trait and are absent when the trait is absent. An example for such an inexpensive and fast polymerase chain reaction system was developed by Bradbury et al. (2005) to distinguish fragrant from nonfragrant rice cultivars. If markers would be developed for chemotypes in essential oil plants, species identification by DNA and the determination of a chemotype could be performed in one step.

Molecular biological methods to identify species are nowadays routinely used in feed- and foodstuffs to identify microbes, animals, and plants. Especially the discussion about traceability of genetically modified organisms (GMOs) throughout the complete chain ("from the living organism to the supermarket") has sped up research in this area (Auer, 2003; Miraglia et al., 2004). One advantage of molecular biological methods is the possibility to be used in a number of processed materials like fatty oil (Pafundo et al., 2005) or even solvent extracts (Novak et al., 2007). The presence of minor amounts of DNA in an essential oil cannot be excluded *a priori*, although distillation as separation technique would suggest the absence of DNA. However, small plant or DNA fragments could distill over or the essential oil could come in contact with plant material after distillation.

3.4 GENETIC AND PROTEIN ENGINEERING

Genetic engineering is defined as the direct manipulation of the genes of organisms by laboratory techniques, not to be confused with the indirect manipulation of genes in traditional (plant) breeding. *Transgenic or GMOs* are organisms (bacteria, plants, etc.) that have been engineered with single or multiple genes (either from the same species or from a different species), using contemporary molecular biology techniques. These are organisms with improved characteristics, in plants, for example, with resistance or tolerance to biotic or abiotic stresses such as insects, disease, drought, salinity, and temperature. Another important goal in improving agricultural production conditions is to facilitate weed control by transformed plant resistant to broadband herbicides like glufosinate. Peppermint has been successfully transformed with the introduction of the bar gene, which encodes phosphinothricin acetyltransferase, an enzyme inactivating glufosinate ammonium or the ammonium salt of glufosinate, phosphinothricin, making the plant insensitive to the systemic, broadspectrum herbicide Roundup (*Roundup Ready mint*) (Li et al., 2001).

A first step in genetic engineering is the development and optimization of transformation (gene transfer) protocols for the target species. Such optimized protocols exist for essential oil plants such as lavandin (*Lavandula* × *intermedia*; Dronne et al., 1999), spike lavender (*Lavandula latifolia*; Nebauer et al., 2000), and peppermint (*M. piperita*; Diemer et al., 1998; Niu et al., 2000).

In spike lavender, an additional copy of the 1-deoxy-D-xylulose-5-phosphate synthase gene, the first enzymatic step in the methylerythritol phosphate (MEP) pathway leading to the precursors of monoterpenes, from *Arabidopsis thaliana*, was introduced and led to an increase of the essential oil of the leaves of up to 360% and of the essential oil of flowers of up to 74% (Munoz-Bertomeu et al., 2006).

In peppermint, many different steps to alter essential oil yield and composition were already targeted (reviewed by Wildung and Croteau, 2005; Table 3.6). The overexpression of deoxyxylulose phosphate reductoisomerase (DXR), the second step in the MEP pathway, increased the essential oil yield by approximately 40% tested under field conditions (Mahmoud and Croteau, 2001). The overexpression of

TABLE 3.6
Essential Oil Composition and Yield of Transgenic Peppermint Transformed with Genes Involved in Monoterpene Biosynthesis

Gene	Method	Limonene	Mentho-Furan	Pulegone	Menthone	Menthol	Oil Yield (lb/acre)
WT	—	1.7	4.3	2.1	20.5	44.5	97.8
DXR	Overexpress	1.6	3.6	1.8	19.6	45.6	137.9
MFS	Antisense	1.7	1.2	0.4	22.7	45.2	109.7
l-3-h	Cosuppress	74.7	0.4	0.1	4.1	3.0	99.6

Source: Wildung, M.R. and R.B. Croteau, 2005. *Transgenic Res.*, 14: 365.
Note: DXR, Deoxyxylulose phosphate reductoisomerase; l-3-h, limonene-3-hydroxylase; MFS, menthofuran synthase; WT, wild type.

geranyl diphosphate synthase leads to a similar increase of the essential oil. Menthofuran, an undesired compound, was downregulated by an antisense method (a method to influence or block the activity of a specific gene). Overexpression of the menthofuran antisense RNA was responsible for an improved oil quality by reducing both menthofuran and pulegone in one transformation step (Mahmoud and Croteau, 2003). The ability to produce a peppermint oil with a new composition was demonstrated by Mahmoud et al. (2004) by upregulating limonene by cosuppression of limonene-3-hydroxylase, the enzyme responsible for the transformation of (−)-limonene to (−)-*trans*-isopiperitenol *en route* to menthol.

Protein engineering is the application of scientific methods (mathematical and laboratory methods) to develop useful or valuable proteins. There are two general strategies for protein engineering, random mutagenesis and rational design. In rational design, detailed knowledge of the structure and function of the protein is necessary to make desired changes by site-directed mutagenesis, a technique already well developed. An impressive example of the rational design of monoterpene synthases was given by Kampranis et al. (2007) who converted a 1,8-cineole synthase from *S. fruticosa* into a synthase producing sabinene, the precursor of α- and β-thujones with a minimum number of substitutions. They went also a step further and converted this monoterpene synthase into a sesquiterpene synthase by substituting a single amino acid that enlarged the cavity of the active site enough to accommodate the larger precursor of the sesquiterpenes, farnesyl pyrophosphate.

3.5 RESOURCES OF ESSENTIAL OILS: WILD COLLECTION OR CULTIVATION OF PLANTS

The raw materials for producing essential oil are resourced either from collecting them in nature (*wild collection*) or from cultivating the plants (Table 3.7).

3.5.1 Wild Collection and Sustainability

Since prehistoric times, humans have gathered wild plants for different purposes; among them are aromatic, essential oil–bearing species used as culinary herbs, spices, flavoring agents, and fragrances. With increasing demand of standardized, homogeneous raw material in the industrial societies, more and more wild species have been domesticated and systematically cultivated. Nevertheless, a high number of species are still collected from the wild due to the fact that

- Many plants and plant products are used for the subsistence of the rural population.
- Small quantities of the respective species are requested at the market only which make a systematic cultivation not profitable.
- Some species are difficult to cultivate (slow growth rate and requirement of a special microclimate).

TABLE 3.7
Important Essential Oil-Bearing Plants—Common and Botanical Names Including Family, Plant Parts Used, Raw Material Origin, and Trade Quantities of the Essential Oil

Trade Name	Species	Plant Family	Used Plant Part(s)	Wild Collection/ Cultivation	Trade Quantities[a]
Ambrette seed	*Hibiscus abelmoschus* L.	Malvaceae	Seed	Cult	LQ
Amyris	*Amyris balsamifera* L.	Rutaceae	Wood	Wild	LQ
Angelica root	*Angelica archangelica* L.	Apiaceae	Root	Cult	LQ
Anise seed	*Pimpinella anisum* L.	Apiaceae	Fruit	Cult	LQ
Armoise	*Artemisia herba-alba* Asso.	Asteraceae	Herb	Cult/wild	LQ
Asafoetida	*Ferula assa-foetida* L.	Apiaceae	Resin	Wild	LQ
Basil	*Ocimum basilicum* L.	Lamiaceae	Herb	Cult	LQ
Bay	*Pimenta racemosa* Moore	Myrtaceae	Leaf	Cult	LQ
Bergamot	*Citrus aurantium* L. ssp. *bergamia* (Risso et Poit.) Engl.	Rutaceae	Fruit peel	Cult	MQ
Birch tar	*Betula pendula* Roth. (syn. *Betula verrucosa* Erhart. *Betula alba* sensu H.J.Coste. non L.)	Betulaceae	Bark/ wood	Wild	LQ
Buchu leaf	*Agathosma betulina* (Bergius) Pillans. *A. crenulata* (L.) Pillans	Rutaceae	Leaf	Wild	LQ
Cade	*Juniperus oxycedrus* L.	Cupressaceae	Wood	Wild	LQ
Cajuput	*Melaleuca leucandendron* L.	Myrtaceae	Leaf	Wild	LQ
Calamus	*Acorus calamus* L.	Araceae	Rhizome	Cult/wild	LQ
Camphor	*Cinnamomum camphora* L. (Sieb.)	Lauraceae	Wood	Cult	LQ
Cananga	*Cananga odorata* Hook. f. et Thoms.	Annonaceae	Flower	Wild	LQ
Caraway	*Carum carvi* L.	Apiaceae	Seed	Cult	LQ
Cardamom	*Elettaria cardamomum* (L.) Maton	Zingiberaceae	Seed	Cult	LQ
Carrot seed	*Daucus carota* L.	Apiaceae	Seed	Cult	LQ
Cascarilla	*Croton eluteria* (L.) W.Wright	Euphorbiaceae	Bark	Wild	LQ
Cedarwood, Chinese	*Cupressus funebris* Endl.	Cupressaceae	Wood	Wild	MQ
Cedarwood, Texas	*Juniperus mexicana* Schiede	Cupressaceae	Wood	Wild	MQ
Cedarwood, Virginia	*Juniperus virginiana* L.	Cupressaceae	Wood	Wild	MQ
Celery seed	*Apium graveolens* L.	Apiaceae	Seed	Cult	LQ
Chamomile	*Matricaria recutita* L.	Asteraceae	Flower	Cult	LQ
Chamomile, Roman	*Anthemis nobilis* L.	Asteraceae	Flower	Cult	LQ
Chenopodium	*Chenopodium ambrosioides* (L.) Gray	Chenopodiaceae	Seed	Cult	LQ
Cinnamon bark, Ceylon	*Cinnamomum zeylanicum* Nees	Lauraceae	Bark	Cult	LQ
Cinnamon bark, Chinese	*Cinnamomum cassia* Blume	Lauraceae	Bark	Cult	LQ

(Continued)

TABLE 3.7 (*Continued*)
Important Essential Oil-Bearing Plants—Common and Botanical Names Including Family, Plant Parts Used, Raw Material Origin, and Trade Quantities of the Essential Oil

Trade Name	Species	Plant Family	Used Plant Part(s)	Wild Collection/ Cultivation	Trade Quantities[a]
Cinnamon leaf	*Cinnamomum zeylanicum* Nees	Lauraceae	Leaf	Cult	LQ
Citronella, Ceylon	*Cymbopogon nardus* (L.) W. Wats.	Poaceae	Leaf	Cult	HQ
Citronella, Java	*Cymbopogon winterianus* Jowitt.	Poaceae	Leaf	Cult	HQ
Clary sage	*Salvia sclarea* L.	Lamiaceae	Flowering herb	Cult	MQ
Clove buds	*Syzygium aromaticum* (L.) Merill et L.M. Perry	Myrtaceae	Leaf/bud	Cult	LQ
Clove leaf	*Syzygium aromaticum* (L.) Merill et L.M. Perry	Myrtaceae	Leaf	Cult	HQ
Coriander	*Coriandrum sativum* L.	Apiaceae	Fruit	Cult	LQ
Cornmint	*Mentha canadensis* L. (syn. *M. arvensis* L. f. *piperascens* Malinv. ex Holmes; *M. arvensis* L. var. *glabrata. M. haplocalyx* Briq.; *M. sachalinensis* [Briq.] Kudo)	Lamiaceae	Leaf	Cult	HQ
Cumin	*Cuminum cyminum* L.	Apiaceae	Fruit	Cult	LQ
Cypress	*Cupressus sempervirens* L.	Cupressaeae	Leaf/twig	Wild	LQ
Davana	*Artemisia pallens* Wall.	Asteraceae	Flowering herb	Cult	LQ
Dill	*Anethum graveolens* L.	Apiaceae	Herb/fruit	Cult	LQ
Dill, India	*Anethum sowa* Roxb.	Apiaceae	Fruit	Cult	LQ
Elemi	*Canarium luzonicum* Miq.	Burseraceae	Resin	Wild	LQ
Eucalyptus	*Eucalyptus globulus* Labill.	Myrtaceae	Leaf	Cult/wild	HQ
Eucalyptus, lemon-scented	*Eucalyptus citriodora* Hook.	Myrtaceae	Leaf	Cult/wild	HQ
Fennel bitter	*Foeniculum vulgare* Mill. ssp. *vulgare* var. *vulgare*	Apiaceae	Fruit	Cult	LQ
Fennel sweet	*Foeniculum vulgare* Mill. ssp. *vulgare* var. *dulce*	Apiaceae	Fruit	Cult	LQ
Fir needle, Canadian	*Abies balsamea* Mill.	Pinaceae	Leaf/twig	Wild	LQ
Fir needle, Siberian	*Abies sibirica* Ledeb.	Pinaceae	Leaf/twig	Wild	LQ
Gaiac	*Guaiacum officinale* L.	Zygophyllaceae	Resin	Wild	LQ
Galbanum	*Ferula galbaniflua* Boiss. *F. rubricaulis* Boiss.	Apiaceae	Resin	Wild	LQ
Garlic	*Allium sativum* L.	Alliaceae	Bulb	Cult	LQ
Geranium	*Pelargonium* spp.	Geraniaceae	Leaf	Cult	MQ
Ginger	*Zingiber officinale* Roscoe	Zingiberaceae	Rhizome	Cult	LQ

(Continued)

TABLE 3.7 (*Continued*)
Important Essential Oil-Bearing Plants—Common and Botanical Names Including Family, Plant Parts Used, Raw Material Origin, and Trade Quantities of the Essential Oil

Trade Name	Species	Plant Family	Used Plant Part(s)	Wild Collection/ Cultivation	Trade Quantities[a]
Gingergrass	*Cymbopogon martinii* (Roxb.) H. Wats var. *sofia* Burk	Poaceae	Leaf	Cult/wild	
Grapefruit	*Citrus* × *paradisi* Macfad.	Rutaceae	Fruit peel	Cult	LQ
Guaiacwood	*Bulnesia sarmienti* L.	Zygophyllaceae	Wood	Wild	MQ
Gurjum	*Dipterocarpus* spp.	Dipterocarpaceae	Resin	Wild	LQ
Hop	*Humulus lupulus* L.	Cannabaceae	Flower	Cult	LQ
Hyssop	*Hyssopus officinalis* L.	Lamiaceae	Leaf	Cult	LQ
Juniper berry	*Juniperus communis* L.	Cupressaceae	Fruit	Wild	LQ
Laurel leaf	*Laurus nobilis* L.	Lauraceae	Leaf	Cult/wild	LQ
Lavandin	*Lavandula angustifolia* Mill. × *L. latifolia* Medik.	Lamiaceae	Leaf	Cult	HQ
Lavender	*Lavandula angustifolia* Miller	Lamiaceae	Leaf	Cult	MQ
Lavender, Spike	*Lavandula latifolia* Medik.	Lamiaceae	Flower	Cult	LQ
Lemon	*Citrus limon* (L.) Burman fil.	Rutaceae	Fruit peel	Cult	HQ
Lemongrass, Indian	*Cymbopogon flexuosus* (Nees ex Steud.) H. Wats.	Poaceae	Leaf	Cult	LQ
Lemongrass, West Indian	*Cymbopogon citratus* (DC.) Stapf	Poaceae	Leaf	Cult	LQ
Lime distilled	*Citrus aurantiifolia* (Christm. et Panz.) Swingle	Rutaceae	Fruit	Cult	HQ
Litsea cubeba	*Litsea cubeba* C.H. Persoon	Lauraceae	Fruit/leaf	Cult	MQ
Lovage root	*Levisticum officinale* Koch	Apiaceae	Root	Cult	LQ
Mandarin	*Citrus reticulata* Blanco	Rutaceae	Fruit peel	Cult	MQ
Marjoram	*Origanum majorana* L.	Lamiaceae	Herb	Cult	LQ
Mugwort common	*Artemisia vulgaris* L.	Asteraceae	Herb	Cult/wild	LQ
Mugwort, Roman	*Artemisia pontica* L.	Asteraceae	Herb	Cult/wild	LQ
Myrtle	*Myrtus communis* L.	Myrtaceae	Leaf	Cult/wild	LQ
Neroli	*Citrus aurantium* L. ssp. *aurantium*	Rutaceae	Flower	Cult	LQ
Niaouli	*Melaleuca viridiflora*	Myrtaceae	Leaf	Cult/wild	LQ
Nutmeg	*Myristica fragrans* Houtt.	Myristicaceae	Seed	Cult	LQ
Onion	*Allium cepa* L.	Alliaceae	Bulb	Cult	LQ
Orange	*Citrus sinensis* (L.) Osbeck	Rutaceae	Fruit peel	Cult	HQ
Orange bitter	*Citrus aurantium* L.	Rutaceae	Fruit peel	Cult	LQ
Oregano	*Origanum* spp. *Thymbra spicata* L. *Coridothymus capitatus* Rechb. fil. *Satureja* spp. *Lippia graveolens*	Lamiaceae	Herb	Cult/wild	LQ
Palmarosa	*Cymbopogon martinii* (Roxb.) H. Wats var. *motia* Burk	Poaceae	Leaf	Cult	LQ
Parsley seed	*Petroselinum crispum* (Mill.) Nym. ex A.W. Hill	Apiaceae	Fruit	Cult	LQ

(*Continued*)

TABLE 3.7 (*Continued*)
Important Essential Oil-Bearing Plants—Common and Botanical Names Including Family, Plant Parts Used, Raw Material Origin, and Trade Quantities of the Essential Oil

Trade Name	Species	Plant Family	Used Plant Part(s)	Wild Collection/ Cultivation	Trade Quantities[a]
Patchouli	*Pogostemon cablin* (Blanco) Benth.	Lamiaceae	Leaf	Cult	HQ
Pennyroyal	*Mentha pulegium* L.	Lamiaceae	Herb	Cult	LQ
Pepper	*Piper nigrum* L.	Piperaceae	Fruit	Cult	LQ
Peppermint	*Mentha* x *piperita* L.	Lamiaceae	Leaf	Cult	HQ
Petitgrain	*Citrus aurantium* L. ssp. *aurantium*	Rutaceae	Leaf	Cult	LQ
Pimento leaf	*Pimenta dioica* (L.) Merr.	Myrtaceae	Fruit	Cult	LQ
Pine needle	*Pinus silvestris* L. *P. nigra* Arnold	Pinaceae	Leaf/twig	Wild	LQ
Pine needle, Dwarf	*Pinus mugo* Turra	Pinaceae	Leaf/twig	Wild	LQ
Pine silvestris	*Pinus silvestris* L.	Pinaceae	Leaf/twig	Wild	LQ
Pine white	*Pinus palustris* Mill.	Pinaceae	Leaf/twig	Wild	LQ
Rose	*Rosa* x *damascena* Miller	Rosaceae	Flower	Cult	LQ
Rosemary	*Rosmarinus officinalis* L.	Lamiaceae	Feaf	Cult/wild	LQ
Rosewood	*Aniba rosaeodora* Ducke	Lauraceae	Wood	Wild	LQ
Rue	*Ruta graveolens* L.	Rutaceae	Herb	Cult	LQ
Sage, Dalmatian	*Salvia officinalis* L.	Lamiaceae	Herb	Cult/wild	LQ
Sage, Spanish	*Salvia lavandulifolia* L.	Lamiaceae	Leaf	Cult	LQ
Sage, three lobed (Greek Turkish)	*Salvia fruticosa* Mill. (syn. *S. triloba* L.)	Lamiaceae	Herb	Cult/wild	LQ
Sandalwood, East Indian	*Santalum album* L.	Santalaceae	Wood	Wild	MQ
Sassafras, Brazilian (Ocotea cymbarum oil)	*Ocotea odorifera* (Vell.) Rohwer (*Ocotea pretiosa* [Nees] Mez.)	Lauraceae	Wood	Wild	HQ
Sassafras, Chinese	*Sassafras albidum* (Nutt.) Nees.	Lauraceae	Root bark	Wild	HQ
Savory	*Satureja hortensis* L. *Satureja montana* L.	Lamiaceae	Leaf	Cult/wild	LQ
Spearmint, Native	*Mentha spicata* L.	Lamiaceae	Leaf	Cult	MQ
Spearmint, Scotch	*Mentha gracilis* Sole	Lamiaceae	Leaf	Cult	HQ
Star anise	*Illicium verum* Hook fil.	Illiciaceae	Fruit	Cult	MQ
Styrax	*Styrax officinalis* L.	Styracaceae	Resin	Wild	LQ
Tansy	*Tanacetum vulgare* L.	Asteraceae	Flowering herb	Cult/wild	LQ
Tarragon	*Artemisia dracunculus* L.	Asteraceae	Herb	Cult	LQ
Tea tree	*Melaleuca* spp.	Myrtaceae	Leaf	Cult	LQ

(Continued)

TABLE 3.7 (*Continued*)
Important Essential Oil-Bearing Plants—Common and Botanical Names Including Family, Plant Parts Used, Raw Material Origin, and Trade Quantities of the Essential Oil

Trade Name	Species	Plant Family	Used Plant Part(s)	Wild Collection/ Cultivation	Trade Quantities[a]
Thyme	*Thymus vulgaris* L. *T. zygis* Loefl. ex L.	Lamiaceae	Herb	Cult	LQ
Valerian	*Valeriana officinalis* L.	Valerianaceae	Root	Cult	LQ
Vetiver	*Vetiveria zizanoides* (L.) Nash	Poaceae	Root	Cult	MQ
Wintergreen	*Gaultheria procumbens* L.	Ericaceae	Leaf	Wild	LQ
Wormwood	*Artemisia absinthium* L.	Asteraceae	Herb	Cult/wild	LQ
Ylang Ylang	*Cananga odorata* Hook. f. et Thoms.	Annonaceae	Flower	Cult	MQ

[a] HQ, High quantities ($>$1000 t/a); MQ, medium quantities (100–1000 t/a); LQ, low quantities ($<$100 t/a).

- Market uncertainties or political circumstances do not allow investing in long-term cultivation.
- The market is in favor of *ecologically* or *naturally* labeled wild collected material.

Especially—but not only—in developing countries, parts of the rural population depend economically on gathering high-value plant material. Less than two decades ago, almost all oregano (crude drug as well as essential oil) worldwide came from wild collection (Padulosi, 1996) and even this well-known group of species (*Origanum* sp. and *Lippia* sp.) was counted under "neglected and underutilized crops."

Yarrow (*A. millefolium s.l.*), arnica, and even chamomile originate still partly from wild collection in Central and Eastern Europe, and despite several attempts to cultivate spikenard (*Valeriana celtica*), a tiny European mountain plant with a high content of patchouli alcohol, this species is still wildly gathered in Austria and Italy (Novak et al., 1998, 2000).

To regulate the sustainable use of biodiversity by avoiding overharvesting, genetic erosion, and habitat loss, international organizations such as International Union for Conservation of Nature (IUCN), WWF/TRAFFIC, and World Health Organization (WHO) have launched together the Convention on Biological Diversity (CBD, 2001), the Global Strategy for Plant Conservation (CBD, 2002), and the Guidelines for the Sustainable Use of Biodiversity (CBD, 2004). TRAFFIC is a joint programme of World Wide Fund for Nature (WWF) and the World Conservation Union (IUCN). TRAFFIC also works in close co-operation with the Secretariat of the Convention on International Trade in Endangered Species of Wild Fauna and Flora (CITES). These principles and recommendations address primarily the national and international policy level, but provide also the herbal industry and the collectors with specific guidance on sustainable sourcing practices (Leaman, 2006). A standard for sustainable collection and use of medicinal and aromatic plants (the international standard on sustainable wild collection of medicinal and aromatic plants [ISSC-MAP]) was issued first in 2004, and its principles will be shown at the end of this chapter. This standard certifies wild-crafted plant material insofar as conservation and sustainability are concerned. Phytochemical quality cannot, however, be derived from it, which is the reason for domestication and systematic cultivation of economically important essential oil plants.

3.5.2 Domestication and Systematic Cultivation

This offers a number of advantages over wild harvest for the production of essential oils:

- Avoidance of admixtures and adulterations by reliable botanical identification.
- Better control of the harvested volumes.
- Selection of genotypes with desirable traits, especially quality.
- Controlled influence on the history of the plant material and on postharvest handling.

TABLE 3.8
Domestication Strategy for Plants of the Spontaneous Flora

1. *Studies at the natural habitat*: botany, soil, climate, growing type, natural distribution and propagation, natural enemies, pests and diseases	→	GPS to exactly localize the place
2. *Collection of the wild grown plants and seeds*: establishment of a germplasm collection, ex situ conservation, phytochemical investigation (screening)		
3. *Plant propagation*: vegetatively or by seeds, plantlet cultivation; (biotechnol.: *in vitro* propagation)	→	Biotechnol./*in vitro*
4. *Genetic improvement*: variability, selection, breeding; phytochemical investigation, biotechnology (*in vitro* techniques)	→	Biotechnol./*in vitro*
5. *Cultivation treatments*: growing site, fertilization, crop maintenance, cultivation techniques		
6. *Phytosanitary problems*: pests, diseases	→	Biotechnol./*in vitro*
7. *Duration of the cultivation*: harvest, postharvest handling, phytochemical control of the crop produced	→	Technical processes, solar energy (new techniques)
8. *Economic evaluation and calculation*	→	New techniques

Source: Modified from Franz, Ch., 1993c. Plant Res. Dev., 37: 101; Franz, Ch., 1993d. Genetic versus climatic factors influencing essential oil formation. *Proceedings of the 12th International Congress of Essential Oils, Fragrances and Flavours*, pp. 27–44, Vienna, Austria.

On the other side, it needs arable land and investments in starting material, maintenance, and harvest techniques. On the basis of a number of successful introductions of new crops a scheme and strategy of domestication was developed by this author (Table 3.8).

Recent examples of successful domestication of essential oil-bearing plants are *oregano* (Ceylan et al., 1994; Kitiki, 1997; Putievsky et al., 1997), *Lippia* sp. (Fischer, 1998), *Hyptis suaveolens* (Grassi, 2003), and *T. lucida* (Goehler, 2006). Domesticating a new species starts with studies at the natural habitat. The most important steps are the exact botanical identification and the detailed description of the growing site. National Herbaria are in general helpful in this stage. In the course of collecting seeds and plant material, a first phytochemical screening will be necessary to recognize chemotypes (Fischer et al., 1996; Goehler et al., 1997). The phytosanitary of wild populations should also be observed so as to be informed in advance on specific pests and diseases. The flower heads of wild *Arnica montana*, for instance, are often damaged by the larvae of *Tephritis arnicae* (Fritzsche et al., 2007).

The first phase of domestication results in a germplasm collection. In the next step, the appropriate propagation method has to be developed, which might be derived partly from observations at the natural habitat: while studying wild populations of *T. lucida* in Guatemala we found, besides appropriate seed set, also runners, which could be used for vegetative propagation of selected plants (Goehler et al., 1997). Wherever possible, propagation by seeds and direct sowing is however preferred due to economic reasons.

The appropriate cultivation method depends on the plant type—annual or perennial, herb, vine, or tree—and on the agroecosystem into which the respective species should be introduced. In contrast to large-scale field production of herbal plants in temperate and Mediterranean zones, small-scale sustainable agroforesty and mixed cropping systems adapted to the environment have the preference in tropical regions (Schippmann et al., 2006). Parallel to the cultivation trials dealing with all topics from plant nutrition and maintenance to harvesting and postharvest handling, the evaluation of the genetic resources and the genetic improvement of the plant material must be started to avoid developing of a detailed cultivation scheme with an undesired chemotype.

3.5.3 Factors Influencing the Production and Quality of Essential Oil-Bearing Plants

Since plant material is the product of a predominantly biological process, prerequisite of its productivity is the knowledge on the factors influencing it, of which the most important ones are

1. The already discussed intraspecific chemical polymorphism, derived from it the biosynthesis and inheritance of the chemical features, and as consequence selection and breeding of new cultivars.
2. The intraindividual variation between the plant parts and depending on the developmental stages ("morpho- and ontogenetic variation").
3. The modification due to environmental conditions including infection pressure and immissions.
4. Human influences by cultivation measures, for example, fertilizing, water supply, or pest management.

3.5.3.1 Genetic Variation and Plant Breeding

Phenotypic variation in essential oils was detected very early because of their striking sensorial properties. Due to the high chemical diversity, a continuous selection of the desired chemotypes leads to rather homogenous and reproducible populations, as this is the case with the landraces and common varieties. But Murray and Reitsema (1954) stated already that "a plant breeding program requires a basic knowledge of the inheritance of at least the major essential oil compounds." Such genetic studies have been performed over the last 50 years with a number of species especially of the mint family (e.g., *T. vulgaris*: Vernet, 1976; *Ocimum* sp.: Sobti et al., 1978; Gouyon and Vernet, 1982; *P. frutescens*: Koezuka et al., 1986, *Mentha* sp.: Croteau, 1991), of the Asteraceae/Compositae (*M. recutita*: Horn et al., 1988; Massoud and Franz, 1990), the genus (*Eucalyptus*: Brophy and Southwell, 2002; Doran, 2002), or the *V. zizanioides* (Akhila and Rani, 2002).

The results achieved by inheritance studies have been partly applied in targeted breeding as shown exemplarily in Table 3.9. Apart from the essential oil content and composition there are also other targets to be observed when breeding essential oil plants, as particular morphological characters ensuring high and stable yields of the respective plant part, resistances to pest and diseases as well as abiotic stress, low nutritional requirements to save production costs, appropriate homogeneity, and suitability for technological processes at harvest and postharvest, especially readiness for distillation (Pank, 2007; Bernáth, 2002). In general, the following breeding methods are commonly used (Franz, 1999).

3.5.3.1.1 Selection by Exploiting the Natural Variability

Since many essential oil-bearing species are in the transitional phase from wild plants to systematic cultivation, appropriate breeding progress can be achieved by simple selection. Wild collections or accessions of germplasm collections are the basis, and good results were obtained, for example, with *Origanum* sp. (Putievsky et al., 1997) in limited time and at low expenses.

Individual plants showing the desired phenotype will be selected and either generatively or vegetatively propagated (individual selection), or positive or negative mass selection techniques can be applied. Selection is traditionally the most common method of genetic improvement and the majority of varieties and cultivars of essential oil crops have this background. Due to the fact, however, that almost all of the respective plant species are allogamous, a recurrent selection is necessary to maintain the varietal traits, and this has especially to be considered if other varieties or wild populations of the same species are nearby and uncontrolled cross pollination may occur.

TABLE 3.9
Some Registered Cultivars of Essential Oil Plant

Species	Cultivar/ Variety	Country	Year of Registration	Breeding Method	Specific Characters
Achillea collina	SPAK	CH	1994	Crossing	High in proazulene
Angelica archangelica	VS 2	FR	1996	Recurrent pedigree	Essential oil index of roots: 180
Foeniculum vulgare	Fönicia	HU	1998	Selection	High anethole
Lavandula officinalis	Rapido	FR	1999	Polycross	High essential oil, high linalyl acetate
Levisticum officinale	Amor	PL	2000	Selection	High essential oil
Matricaria recutita	Mabamille	DE	1995	Tetraploid	High α-bisabolol
	Ciclo-1	IT	2000	Line breeding	High chamazulene
	Lutea	SK	1995	Tetraploid	High α-bisabolol
Melissa officinalis	Ildikó	HU	1998	Selection	High essential oil, Citral A + B, linalool
	Landor	CH	1994	Selection	High essential oil
	Lemona	DE	2001	Selection	High essential oil, citral
Mentha piperita	Todd's Mitcham	USA	1972	Mutation	Wilt resistant
	Kubanskaja	RUS	1980s	Crossing and polyploid	High essential oil, high menthol
Mentha spicata	MSH-20	DK	2000	Recurrent pedigree	High menthol, good flavor
Ocimum basilicum	Greco	IT	2000	Synthetic	Flavor
	Perri	ISR	1999	Cross-breeding	Fusarium Resistant
	Cardinal	ISR	2000	Cross-breeding	
Origanum syriacum	Senköy	TR	1992	Selection	5% essential oil, 60% carvacrol
	Carmeli	ISR	1999	Selection	Carvacrol
	Tavor	ISR	1999	Selection	Thymol
Origanum onites		GR	2000	Selfing	Carvacrol
Origanum hirtum		GR	2000	Selfing	Carvacrol
	Vulkan	DE	2002	Crossing	Carvacrol
	Carva	CH	2002	Crossing	Carvacrol
	Darpman	TR	1992	Selection	2.5% essential oil, 55% carvacrol
Origanum majorana	Erfo	DE	1997	Crossing	High essential oil,
(Majorana hortensis)	Tetrata	DE	1999	Ployploid	*cis*-Sabinene-hydrate
	G 1	FR	1998	Polycross	
Salvia officinalis	Moran	ISR	1998	Crossing	Herb yield
	Syn 1	IT	2004	Synthetic	α-Thujone
Thymus vulgaris	Varico	CH	1994	Selection	Thymol/carvacrol
	T-16	DK	2000	Recurrent pedigree	Thymol
	Virginia	ISR	2000	Selection	Herb yield

The efficacy of selection has been shown by examples of many species, for instance, of the Lamiaceae family, starting from "Mitcham" peppermint and derived varieties (Lawrence, 2007), basil (Elementi et al., 2006), sage (Bezzi, 1994; Bernáth, 2000) to thyme (Rey, 1993). It is a well-known method also in the breeding of caraway (Pank et al., 1996) and fennel (Desmarest, 1992) as well as of tropical and subtropical species such as palmarosa grass (Kulkarni, 1990), tea tree (Taylor, 1996), and eucalyptus

(Doran, 2002). At perennial herbs, shrubs, and trees clone breeding, that is, the vegetative propagation of selected high-performance individual plants, is the method of choice, especially in sterile or not type-true hybrids, for example, peppermint (*M. piperita*) or lavandin (*Lavandula* × *hybrida*). But this method is often applied also at sage (Bazina et al., 2002), rosemary (Mulas et al., 2002), lemongrass (Kulkarni and Ramesh, 1992), pepper, cinnamon, and nutmeg (Nair, 1982), and many other species.

3.5.3.1.2 Breeding with Extended Variability (Combination Breeding)

If different desired characters are located in different individuals/genotypes of the same or a closely related crossable species, crossings are made followed by selection of the respective combination products. Artificial crossings are performed by transferring the paternal pollen to the stigma of the female (emasculated) or male sterile maternal flower. In the segregating progenies individuals with the desired combination will be selected and bred to constancy, as exemplarily described for fennel and marjoram by Pank (2002b).

Hybrid breeding—common in large-scale agricultural crops, for example, maize—was introduced into essential oil plants over the last decade only. The advantage of hybrids on the one side is that the F_1 generation exceeds the parent lines in performance due to hybrid vigor and uniformity ("heterosis effect") and on the other side it protects the plant breeder by segregating of the F_2 and following generations in heterogeneous low-value populations. But it needs as precondition separate (inbred) parent lines of which one has to be male sterile and one male fertile with good combining ability.

In addition, a male fertile "maintainer" line is needed to maintain the mother line. Few examples of F_1 hybrid breeding are known especially at Lamiaceae since male sterile individuals are found frequently in these species (Rey, 1994; Langbehn et al., 2002; Novak et al., 2002; Pank, 2002a).

Synthetic varieties are based on several (more than two) well-combining parental lines or clones which are grown together in a polycross scheme with open pollination for seed production. The uniformity and performance is not as high as at F_1 hybrids but the method is simpler and cheaper and the seed quality acceptable for crop production until the second or third generation. Synthetic cultivars are known for chamomile (Franz et al., 1985), arnica (Daniel and Bomme, 1991), marjoram (Franz and Novak, 1997), sage (Aiello et al., 2001), or caraway (Pank et al., 2007).

3.5.3.1.3 Breeding with Artificially Generated New Variability

Induced mutations by application of mutagenic chemicals or ionizing radiation open the possibility to find new trait expressions. Although quite often applied, such experiments are confronted with the disadvantages of undirected and incalculable results, and achieving a desired mutation is often like searching for a needle in a haystack. Nevertheless, remarkable achievements are several colchicine-induced polyploid varieties of peppermint (Murray, 1969; Lawrence, 2007), chamomile (Czabajska et al., 1978; Franz et al., 1983a ; Repčak et al., 1992), and lavender (Slavova et al., 2004).

Further possibilities to obtain mutants are studies of the somaclonal variation of *in vitro* cultures since abiotic stress in cell and tissue cultures induces also mutagenesis. Finally, genetic engineering opens new fields and potentialities to generate new variability and to introduce new traits by gene transfer. Except research on biosynthetic pathways of interesting essential oil compounds genetic engineering, GMO's and transgenic cultivars are until now without practical significance in essential oil crops and also not (yet) accepted by the consumer.

As regards the different traits, besides morphological, technological, and yield characteristics as well as quantity and composition of the essential oil, also stress resistance and resistance to pests and diseases are highly relevant targets in breeding of essential oil plants. Well known in this respect are breeding efforts against mint rust (*Puccinia menthae*) and wilt (*Verticillium dahliae*) resulting in the peppermint varieties "Multimentha," "Prilukskaja," or "Todd's Mitcham" (Murray and Todd, 1972; Lawrence, 2007; Pank, 2007), the development of *Fusarium*-wilt and *Peronospora* resistant cultivars of basil (Dudai, 2006; Minuto et al., 2006), or resistance breeding against *Septoria petroselini* in parsley and related species (Marthe and Scholze, 1996). An overview on this topic is given by Gabler (2002).

3.5.3.2 Plant Breeding and Intellectual Property Rights

Essential oil plants are biological, cultural, and technological resources. They can be found in nature gathered from the wild or developed through domestication and plant breeding. As long as the plant material is wild collected and traditionally used, it is part of the cultural heritage without any individual intellectual property and therefore not possible to protect, for example, by patents. Even finding a new plant or substance is a discovery in the "natural nature" and not an invention since a technical teaching is missing. Intellectual property, however, can be granted to new applications that involve an inventive step. Which consequences can be drawn from these facts for the development of novel essential oil plants and new selections or cultivars?

Selection and genetic improvement of aromatic plants and essential oil crops is not only time consuming but also rather expensive due to the necessity of comprehensive phytochemical and possibly molecular biological investigations. In addition, with few exceptions (e.g., mints, lavender and lavandin, parsley but also *Cymbopogon* sp., black pepper, or cloves) the acreage per species is rather limited in comparison with conventional agricultural and horticultural crops. And finally, there are several "fashion crops" with market uncertainties concerning their longevity or half-life period, respectively. The generally unfavorable cost: benefit ratio to be taken into consideration makes essential oil plant breeding economically risky and there is no incentive for plant breeders unless a sufficiently strong plant intellectual property right (IPR) exists. Questioning "which protection, which property right for which variety?" offers two options (Franz, 2001).

3.5.3.2.1 Plant Variety Protection

By conventional methods bred plant groupings that collectively are distinct from other known varieties and are uniform and stable following repeated reproduction can be protected by way of plant breeder's rights. Basis is the International Convention for the Protection of New Varieties initially issued by UPOV (Union for the Protection of New Varieties of Plants) in 1961 and changed in 1991. A plant breeder's right is a legal title granting its holder the exclusive right to produce reproductive material of his plant variety for commercial purposes and to sell this material within a particular territory for up to 30 years (trees and shrubs) or 25 years (all other plants). A further precondition is the "commercial novelty," that is, it must not have been sold commercially prior to the filing date. Distinctness, uniformity, and stability (DUS) refer to morphological (leaf shape, flower color, etc.) or physiological (winter hardiness, disease resistance, etc.), but not phytochemical characteristics, for example, essential oil content or composition. Such "value for cultivation and use (VCU) characteristics" will not be examined and are therefore not protected by plant breeder's rights (Franz, 2001; Llewelyn, 2002; Van Overwalle, 2006).

3.5.3.2.2 Patent Protection (Plant Patents)

Generally speaking, patentable are inventions (not discoveries!) that are novel, involve an innovative step, and are susceptible to industrial application, including agriculture. Plant varieties or essentially biological processes for the production of plants are explicitly excluded from patenting. But other groupings of plants that fall neither under the term "variety" nor under "natural nature" are possible to be protected by patents. This is especially important for plant groupings with novel phytochemical composition or novel application combined with an inventive step, for example, genetic modification, a technologically new production method or a novel type of isolation (product by process protection).

Especially for wild plants and essentially allogamous plants not fulfilling DUS for cultivated varieties (cultivars) and plants where the phytochemical characteristics are more important than the morphological ones, plant patents offer an interesting alternative to plant variety protection (PVP) (Table 3.10).

In conformity with the UPOV Convention of 1991 (UPOV, 1991)

- A strong plant IPR is requested.
- Chemical markers (e.g., secondary plant products) must be accepted as protectable characteristics.

TABLE 3.10
Advantages and Disadvantages of PVP versus Patent Protection of Specialist Minor Crops (Medicinal and Aromatic Plants)

PVP	Patent
Beginning of protection: registration date	Beginning of protection: application date
Restricted to "varieties"	"Varieties" not patentable, but any other grouping of plants
Requirements: DUS = distinctness, uniformity, stability	Requirements: novelty, inventive step, industrial applicability (=NIA)
Free choice of characters to be used for DUS by PVO (Plant Variety Office)	Repeatability obligatory, product by process option
Phenotypical. Mainly morphological characters (phytochemicals of minor importance)	"Essentially biological process" not patentable
Value for cultivation and use characteristics (VCU) not protected	"Natural nature" not patentable
	Claims (e.g., phytochemical characters) depend on applicant
	Phytochemical characters and use/application (VCU) patentable

- Strong depending rights for essentially derived varieties are needed since it is easy to plagiarize such crops.
- "Double protection" would be very useful (i.e., free decision by the breeder if PVR or patent protection is applied).
- But also researchers exemption and breeders privilege with fair access to genotypes for further development is necessary.

Strong protection does not hinder usage and development; it depends on a fair arrangement only (Le Buanec, 2001).

3.5.3.3 Intraindividual Variation between Plant Parts and Depending on the Developmental Stage (*Morpho-* and *Ontogenetic Variation*)

The formation of essential oils depends on the tissue differentiation (secretory cells and excretion cavities, as discussed in Section 3.3.1) and on the ontogenetic phase of the respective plant. The knowledge on these facts is necessary to harvest the correct plant parts at the right time.

Regarding the *differences between plant parts*, it is known from cinnamon (*Cinnamomum zeylanicum*) that the root-, stem-, and leaf oils differ significantly (Wijesekera et al., 1974): only the stem bark contains an essential oil with up to 70% cinnamaldehyde, whereas the oil of the root bark consists mainly of camphor and linalool, and the leaves produce oils with eugenol as main compound. In contrast to it, eugenol forms with 70%–90% the main compound in stem, leaf, and bud oils of cloves (*S. aromaticum*) (Lawrence, 1978). This was recently confirmed by Srivastava et al. (2005) for clove oils from India and Madagascar, stating in addition that eugenyl acetate was found in buds up to 8% but in leaves between traces and 1.6% only. The second main substance in leaves as well as buds is β-caryophyllene with up to 20% of the essential oil. In *Aframomum giganteum* (Zingiberaceae), the rhizome essential oil consists of β-caryophyllene, its oxide, and derivatives mainly, whereas in the leaf oil terpentine-4-ol and pinocarvone form the principal components (Agnaniet et al., 2004).

Essential oils of the Rutaceae family, especially citrus oils, are widely used as flavors and fragrances depending on the plant part and species: in lime leaves neral/geranial and nerol/geraniol are prevailing, whereas grapefruit leaf oil consists of sabinene and β-ocimene mainly. The peel of grapefruit contains almost limonene only and some myrcene, but lime peel oil shows a composition of β-pinene, γ-terpinene, and limonene (Gancel et al., 2002). In *Phellodendron* sp., Lis et al. (2004),

Lis and Milczarek (2006) found that in flower and fruit oils limonene and myrcene are dominating; in leaf oils, in contrast, α-farnesene, β-elemol, or β-ocimene, are prevailing.

Differences in the essential oil composition between the plant parts of many Umbelliferae (Apiaceae) have exhaustively been studied by the group of Kubeczka, summarized by Kubeczka et al. (1982) and Kubeczka (1997). For instance, the comparison of the essential fruit oil of aniseed (*P. anisum*) with the oils of the herb and the root revealed significant differences (Kubeczka et al., 1986). Contrary to the fruit oil consisting of almost *trans*-anethole only (95%), the essential oil of the herb contains besides anethole, considerable amounts of sesquiterpene hydrocarbons, for example, germacrene D, β-bisabolene, and α-zingiberene. Also pseudoisoeugenyl-2-methylbutyrate and epoxi-pseudoisoeugenyl-2-methylbutyrate together form almost 20% main compounds of the herb oil, but only 8.5% in the root and 1% in the fruit oil. The root essential oil is characterized by a high content of β-bisabolene, geijerene, and pregeijerene and contains only small amounts of *trans*-anethole (3.5%). Recently, Velasco-Neguerela et al. (2002) investigated the essential oil composition in the different plant parts of *Pimpinella cumbrae* from Canary Islands and found in all above-ground parts α-bisabolol as main compound besides of δ-3-carene, limonene, and others, whereas the root oil contains mainly isokessane, geijerene, isogeijerene, dihydroagarofuran, and proazulenes—the latter is also found in *Pimpinella nigra* (Kubeczka et al., 1986). Pseudoisoeugenyl esters, known as chemosystematic characters of the genus *Pimpinella*, have been detected in small concentrations in all organs except leaves.

Finally, Kurowska and Galazka (2006) compared the seed oils of root and leaf parsley cultivars marketed in Poland. Root parsley seeds contained an essential oil with high concentrations of apiole and some lower percentages of myristicin. In leaf parsley seeds, in contrast, the content of myristicin was in general higher than apiole, and a clear differentiation between flat leaved cultivars showing still higher concentrations of apiole and curled cultivars with only traces of apiole could be observed. Allyltetramethoxybenzene as the third marker was found in leaf parsley seeds up to 12.8%, in root parsley seeds, however, in traces only. Much earlier, Franz and Glasl (1976) had published already similar results on parsley seed oils comparing them with the essential oil composition of the other plant parts (Figure 3.8). Leaf oils gave almost the same fingerprint than the seeds with high myristicin in curled leaves, some apiole in flat leaves, and higher apiole concentrations than myristicin in the leaves of root varieties. In all root samples, however, apiole dominated largely over myristicin. It is therefore possible to identify the parsley type by analyzing a small seed sample.

As shown already by Figueiredo et al. (1997), in the major number of essential oil-bearing species the oil composition differs significantly between the plant parts, but there are also plant species—as mentioned before, for example, cloves—which form a rather similar oil composition in each plant organ. Detailed knowledge in this matter is needed to decide, for instance, how exact the separation of plant parts has to be performed before further processing (e.g., distillation) or use.

Another topic to be taken into consideration is the *developmental stage* of the plant and the plant organs, since the formation of essential oils is phase dependent. In most cases, there is a significant increase of the essential oil production throughout the whole vegetative development.

And especially in the generative phase between flower bud formation and full flowering, or until fruit or seed setting, remarkable changes in the oil yield and compositions can be observed. Obviously, a strong correlation is given between formation of secretory structures (oil glands, ducts, etc.) and essential oil biosynthesis, and different maturation stages, are associated with, for example, higher rates of cyclization or increase of oxygenated compounds (Figueiredo et al., 1997).

Investigations on the ontogenesis of fennel (*F. vulgare* Mill.) revealed that the best time for picking fennel seeds is the phase of full ripeness due to the fact that the anethole content increases from <50% in unripe seeds to over 80% in full maturity (Marotti et al., 1994). In dill weed (*Anethum graveolens* L.) the content on essential oil rises from 0.1% only in young sprouts to more than 1% in herb with milk ripe umbels (Gora et al., 2002). In the herb, oil α-phellandrene prevails until the beginning of flowering with up to 50%, followed by dill ether, *p*-cymene, and limonene. The oil from green as well as ripe umbels contains, on the other hand, mainly (*S*)-carvone and (*R*)-limonene. The

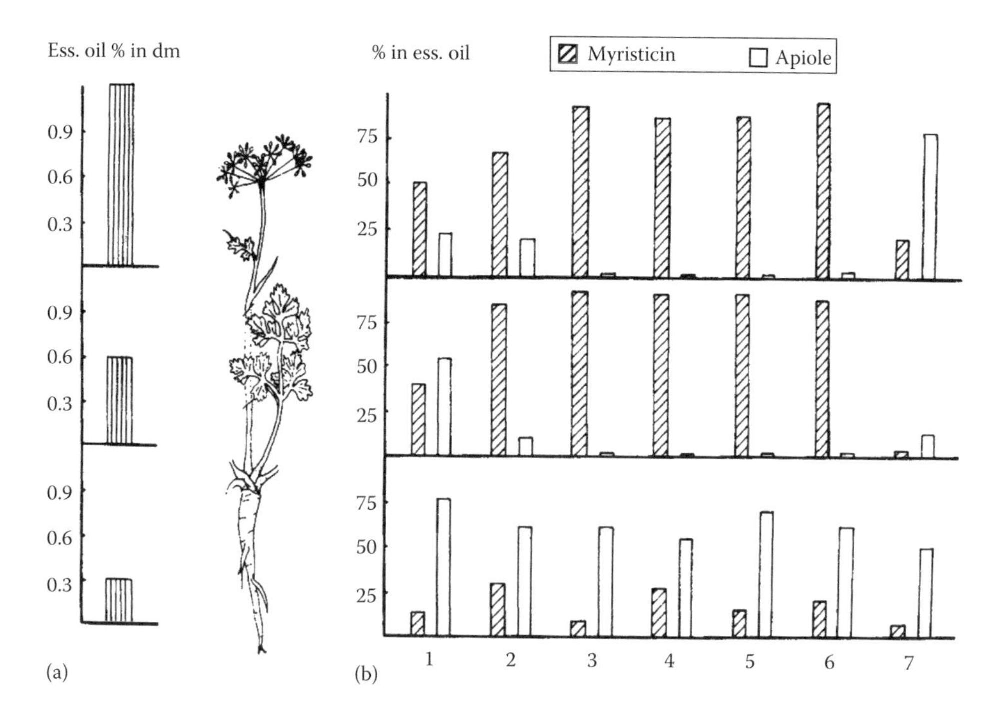

FIGURE 3.8 Differences in the essential oil of fruits, leaves, and roots of parsley cultivars (*Petroselinum crispum* (Mill.) Nyman). (a) Essential oil content, (b) content of myristicin and apiole in the essential oil. 1, 2—flat leaved cv's; 3–7—curled leaves cv's; 7—root parsley).

flavor of dill oil changes therefore dramatically, which has to be considered when determining the harvest time for distillation.

Among Compositae (Asteraceae) there are not as many results concerning ontogeny due to the fact that in general the flowers or flowering parts of the plants are harvested, for example, chamomile (*M. recutita*), yarrow (*A. millefolium s.l.*), immortelle (*Helichrysum italicum*), or wormwood (*Artemisia* sp.) and therefore the short period between the beginning of flowering and the decay of the flowers is of interest only. In chamomile (*M. recutita*), the flower buds show a relatively high content on essential oil between 0.8% and 1.0%, but the oil yield in this stage is rather low. From the beginning of flowering, the oil content increases until full flowering (all disc florets open) and decreases again with decay of the flower heads. At full bloom there is also the peak of (pro) chamazulene, whereas farnesene and α-bisabolol decrease from the beginning of flowering and the bisabololoxides rise (Franz et al., 1978). This was confirmed by Repčak et al. (1980). The essential oil of *Tagetes minuta* L. at different development stages was investigated by Worku and Bertoldi (1996). Before flower bud formation the oil content was 0.45% only, but it culminated with 1.34% at the immature seed stage. During this period *cis*-ocimene increased from 7.2% to 37.5% and *cis*-ocimenone declined from almost 40%–13.1%. Little variations could be observed at *cis*- and *trans*-tagetone only. Similar results have been reported also by Chalchat et al. (1995).

Also for *Lippia* sp. (Verbenaceae) some results are known concerning development stages (Fischer, 1998; Coronel et al., 2006). The oil content in the aerial parts increases from young buds (<1.0%) to fully blooming (almost 2.0%). But although quantitative variations could be observed for most components of the essential oils, the qualitative composition appeared to be constant throughout the growing season.

A particular situation is given with eucalypts as they develop up to five distinct types of leaves during their lifetime, each corresponding to a certain ontogenetic stage with changing oil concentrations and compositions (Doran, 2002). Usually the oil content increases from young to matured, nonlignified leaves, and is thereafter declining until leaf lignification. Almost the same curve is valid also for the 1,8-cineole concentration in the oil. But comparing the relatively extensive literature on this topic, one may conclude that the concentration at various stages of leaf maturity is determined by a complex pattern of quantitative change in individual or groups of substances, some remaining constant, some increasing, and some decreasing. Tsiri et al. (2003) investigated the volatiles of the leaves of *Eucalyptus camaldulensis* over the course of a year in Greece and found a seasonal variation of the oil concentration with a peak during summer and lowest yields during winter. The constituent with highest concentration was 1,8-cineole (25.3%–44.2%) regardless the time of harvest. The great variation of all oil compounds showed however no clear tendency, neither seasonal nor regarding leaf age or leaf position. Doran (2002) concluded therefore that genotypic differences outweigh any seasonal or environmental effects in eucalypts.

There is an extensive literature on ontogenesis and seasonal variation of Labiatae essential oils. Especially for this plant family, great differences are reported on the essential oil content and composition of young and mature leaves and the flowers may in addition influence the oil quality significantly. Usually, young leaves show higher essential oil contents per area unit compared to old leaves. But the highest oil yield is reached at the flowering period, which is the reason that most of the oils are produced from flowering plants. According to Werker et al. (1993) young basil (*O. basilicum*) leaves contained 0.55% essential oil while the content of mature leaves was only 0.13%. The same is also valid to a smaller extent for *O. sanctum*, where the essential oil decreases from young (0.54%) to senescing leaves (0.38%) (Dey and Choudhuri, 1983). Testing a number of basil cultivars mainly of the linalool chemotype, Macchia et al. (2006) found that only some of the cultivars produce methyl eugenol up to 8% in the vegetative stage. Linalool as main compound is increasing from the vegetative (10%–50%) to the flowering (20%–60%), and postflowering phase (25%–80%), whereas the second important substance eugenol reaches its peak at the beginning of flowering (5%–35%). According to the cultivars, different harvest dates are therefore recommended. In *O. sanctum*, the content of eugenol (60.3%–52.2%) as well as of methyl eugenol (6.6%–2.0%) is decreasing from young to senescent leaves and at the same time β-caryophyllene increases from 20.8% to 30.2% (Dey and Choudhuri, 1983).

As regards oregano (*O. vulgare* ssp. *hirtum*), the early season preponderance of *p*-cymene over carvacrol was reversed as the season progressed and this pattern could also be observed at any time within the plant, from the latest leaves produced (low in cymene) to the earliest (high in cymene) (Johnson et al., 2004; Figure 3.9). Already Kokkini et al. (1996) had shown that oregano contains a higher proportion of *p*-cymene to carvacrol (or thymol) in spring and autumn, whereas carvacrol/thymol prevails in the summer. This is explained by Dudai et al. (1992) as photoperiodic reaction: short days with high *p*-cymene, long days with low *p*-cymene production. But only young plants are

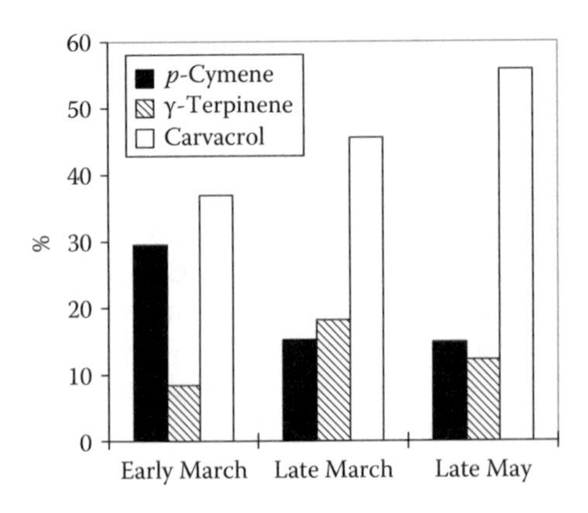

FIGURE 3.9 Average percentages and concentrations of *p*-cymene. γ-Terpinene and carvacrol at the different sampling dates of *Origanum vulgare ssp. hirtum*.

capable of making this switch, whereas in older leaves the already produced and stored oil remains almost unchanged (Johnson et al., 2004).

Presumably the most studied essential oil plant is peppermint (*M. piperita* L.). Already in the 1950s Lemli (1955) stated that the proportion of menthol to menthone in peppermint leaves changes in the course of the development toward higher menthol contents. Lawrence (2007) has just recently shown that from immature plants via mature to senescent plants the content of menthol increases (34.8%–39.9%–48.2%) and correspondingly the menthone content decreases dramatically (26.8%–17.4%–4.7%). At the same time, also an increase of menthyl acetate from 8.5% to 23.3% of the oil could be observed. At full flowering, the peppermint herb oil contains only 36.8% menthol but 21.8% menthone, 7.7% menthofuran, and almost 3% pulegone due to the fact that the flower oils are richer in menthone and pulegone and contain a high amount of menthofuran (Hefendehl, 1962). Corresponding differences have been found between young leaves rich in menthone and old leaves with high menthol and menthyl acetate content (Hoeltzel, 1964; Franz, 1972). The developmental stage depends, however, to a large extent from the environmental conditions, especially the day length.

3.5.3.4 Environmental Influences

Essential oil formation in the plants is highly dependent on climatic conditions, especially day length, irradiance, temperature, and water supply. Tropical species follow in their vegetation cycle the dry and rainy season; species of the temperate zones react more on day length, the more distant from the equator their natural distribution area is located.

Peppermint as typical long day plant needs a minimum day length (hours of day light) to switch from the vegetative to the generative phase. This is followed by a change in the essential oil composition from menthone to menthol and menthyl acetate (Hoeltzel, 1964). Franz (1981) tested six peppermint clones at Munich/Germany and at the same time also at Izmir/Turkey. At the development stage "beginning of flowering," all clones contained at the more northern site much more menthol than on the Mediterranean location, which was explained by a maximum day length in Munich of 16 h 45 min, but in Izmir of 14 h 50 min only. Comparable day length reactions have been mentioned already for oregano (Dudai et al., 1992; Kokkini et al., 1996). Also marjoram (*O. majorana* L.) was influenced not only in flower formation by day length, but also in oil composition (Circella et al., 1995). At long day treatment the essential oil contained more *cis*-sabinene hydrate. Terpinene-4-ol prevailed under short day conditions.

Franz et al. (1986) performed ecological experiments with chamomile, growing vegetatively propagated plants at three different sites, in South Finland, Middle Europe, and West Turkey. As regards the oil content, a correlation between flower formation, flowering period, and essential oil synthesis could be observed: the shorter the flowering phase, the less was the time available for oil formation, and thus the lower was the oil content. The composition of the essential oil, on the other hand, showed no qualitative change due to ecological or climatic factors confirming that chemotypes keep their typical pattern. In addition, Massoud and Franz (1990) investigated the genotype–environment interaction of a chamazulene–bisabolol chemotype. The frequency distributions of the essential oil content as well as the content on chamazulene and α-bisabolol have shown that the highest oil- and bisabolol content was reached in Egypt while under German climatic conditions chamazulene was higher. Similar results have been obtained by Letchamo and Marquard (1993). The relatively high heritability coefficients calculated for some essential oil components—informing whether a character is more influenced by genetic or other factors—confirm that the potential to produce a certain chemical pattern is genetically coded, but the gene expression will be induced or repressed by environmental factors also (Franz, 1993b,d).

Other environmental factors, for instance, soil properties, water stress, or temperature, are mainly influencing the productivity of the respective plant species and by this means the oil yield also, but have little effect on the essential oil formation and composition only (Figueiredo et al., 1997; Salamon, 2007).

3.5.3.5 Cultivation Measures, Contaminations, and Harvesting

Essential oil-bearing plants comprise annual, biennial, or perennial herbs, shrubs, and trees, cultivated either in tropical or subtropical areas, in Mediterranean regions, in temperate, or even in arid zones. Surveys in this respect are given, for instance, by Chatterjee (2002) for India, by Carruba et al. (2002) for Mediterranean environments, and by Galambosi and Dragland (2002) for Nordic countries. Nevertheless, some examples should refer to some specific items.

The *cultivation method*—if direct sowing or transplanting—and the timing influence the crop development and by that way also the quality of the product, as mentioned above. Vegetative propagation, necessary for peppermint due to its genetic background as interpecific hybrid, common in *Cymbopogon* sp. and useful to control the ratio between male and female trees in nutmeg (*Myristica fragrans*), results in homogeneous plant populations and fields. A disadvantage could be the easier dispersion of pests and diseases, as known for "yellow rot" of lavandin (*Lavandula* × *hybrida*) (Fritzsche et al., 2007). Clonal propagation can be performed by leaf or stem cuttings (Goehler et al., 1997; El-Keltawi and Abdel-Rahman, 2006; Nicola et al., 2006) or *in vitro* (e.g., Figueiredo et al., 1997; Mendes and Romano, 1997), the latter method especially for mother plant propagation due to the high costs. *In vitro* essential oil production received increased attention in physiological experiments, but has up to now no practical significance.

As regards *plant nutrition and fertilizing*, a numerous publications have shown its importance for plant growth, development, and biomass yield. The essential oil yield, obviously, depends on the plant biomass; the oil percentage is partly influenced by the plant vigor and metabolic activity. Optimal fertilizing and water supply results in better growth and oil content, for example, in marjoram, oregano, basil, or coriander (Menary, 1994), but also in delay of maturity, which causes quite often "immature" flavors.

Franz (1972) investigated the influence of nitrogen and potassium on the essential oil formation of peppermint. He could show that higher nitrogen supply increased the biomass but retarded the plant development until flowering, whereas higher potassium supply forced the maturity. With increasing nitrogen, a higher oil percentage was observed with lower menthol and higher menthone content; potassium supply resulted in less oil with more menthol and menthyl acetate. Comparable results with *R. officinalis* have been obtained by Martinetti et al. (2006), and Omidbaigi and Arjmandi (2002) have shown for *T. vulgaris* that nitrogen and phosphorus fertilization had significant effect on the herb yield and essential oil content, but did not change the thymol percentage. Also Java citronella (*Cymbopogon winterianus* Jowitt.) responded to nitrogen supply with higher herb and oil yields, but no influence on the geraniol content could be found (Munsi and Mukherjee, 1986).

Extensive pot experiments with chamomile (*M. recutita*) have also shown that high nitrogen and phosphorus nutrition levels resulted in a slightly increased essential oil content of the anthodia, but raising the potassium doses had a respective negative effect (Franz et al., 1983a,b). With nitrogen the flower formation was in delay and lasted longer; with more potassium the flowering phase was reduced, which obviously influenced the period available for essential oil production. This was confirmed by respective ^{14}C-acetate labeling experiments (Franz, 1981).

Almost no effect has been observed on the composition of the essential oil. Also a number of similar pot or field trials came to the same result, as summarized by Salamon (2007).

Salinity and salt stress get an increasing importance in agriculture especially in subtropical and Mediterranean areas. Some essential oil plants, for example, *Artemisia* sp. and *M. recutita* (chamomile) are relatively salt tolerant. Also thyme (*T. vulgaris*) showed a good tolerance to irrigation water salinity up to 2000 ppm, but exceeding concentrations caused severe damages (Massoud et al., 2002). Higher salinity reduced also the oil content, and an increase of *p*-cymene was observed. Recently, Aziz et al. (2008) investigated the influence of salt stress on growth and essential oil in several mint species. In all three mints, salinity reduced the growth severely from 1.5 g/L onward; in peppermint, the menthone content raised and menthol went down to <1.0%, in apple mint, linalool and neryl acetate decreased while myrcene, linalyl acetate, and linalyl propionate increased.

Further problems to be taken into consideration in plant production are *contaminations* with heavy metals, damages caused by pests and diseases, and *residues* of plant protection products. The most important toxic heavy metals Cd, Hg, Pb, and Zn, but also Cu, Ni, and Mn may influence the plant growth severely and by that way also the essential oil, as they may act as cofactors in the plant enzyme system. But as contaminants, they remain in the plant residue after distillation (Zheljazkov and Nielsen, 1996; Zheljazkov et al., 1997). Some plant species, for example, yarrow and chamomile accumulate heavy metals to a greater extent. This is, however, problematic for using the crude drug or for deposition of distillation wastes mainly. The same is valid for the microbial contamination of the plant material. More important in the production of essential oils are pests and diseases that cause damages to the plant material and sometimes alterations in the biosynthesis; but little is known in this respect.

In contrast to organic production, where no use of pesticides is permitted, a small number of insecticides, fungicides, and herbicides are approved for conventional herb production. The number, however, is very restricted (end of 2008 several active substances lost registration at least in Europe), and limits for residues can be found in national law and international regulations, for example, the European Pharmacopoeia. For essential oils, mainly the lipophilic substances are of relevance since they can be enriched over the limits in the oil.

Harvesting and the first steps of *postharvest handling* are the last part of the production chain of starting materials for essential oils. The harvest date is determined by the development stage or maturity of the plant or plant part, Harvesting techniques should keep the quality by avoiding adulterations, admixtures with undesired plant parts, or contaminations, which could cause "off-flavor" in the final product. There are many technical aids at disposal, from simple devices to large-scale harvesters, which will be considered carefully in Chapter 4. From the quality point of view, raising the temperature by fermentation should in general be avoided (except, in vanilla), and during the drying process further contamination with soil, dust, insects, or molds has to be avoided.

Quality and safety of essential oil-bearing plants as raw materials for pharmaceutical products, flavors, and fragrances are of highest priority from the consumer point of view. To meet the respective demands, standards and safety as well as quality assurance measures are needed to ensure that the plants are produced with care, so that negative impacts during wild collection, cultivation, processing, and storage can be limited. To overcome these problems and to guarantee a steady, affordable and

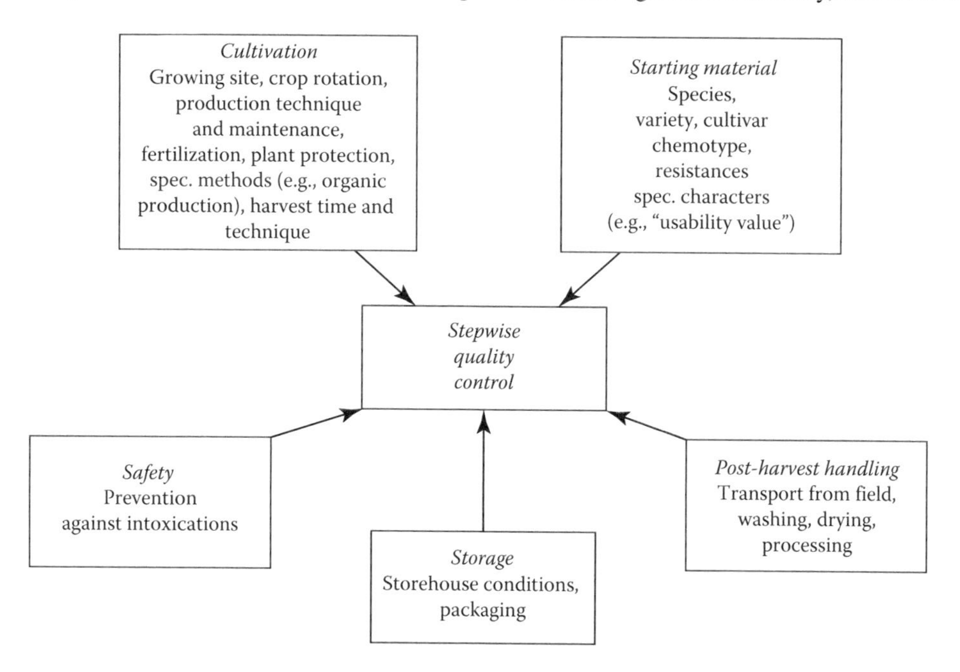

FIGURE 3.10 Main items of "good agricultural practices" (GAP) for medicinal and aromatic plants.

sustainable supply of essential oil plants of good quality (Figure 3.10), in recent years guidelines for good agricultural practices (GAP) and standards for Sustainable Wild Collection (ISSC) have been established at the national and international level.

3.6 INTERNATIONAL STANDARDS FOR WILD COLLECTION AND CULTIVATION

3.6.1 GA(C)P: Guidelines for Good Agricultural (and Collection) Practice of Medicinal and Aromatic Plants

First initiatives for the elaboration of such guidelines trace back to a roundtable discussion in Angers, France in 1983, and intensified at an International Symposium in Novi Sad 1988 (Franz, 1989b). A first comprehensive paper was published by Pank et al. (1991) and in 1998 the European Herb Growers Association (EHGA/EUROPAM) released the first version (Máthé and Franz, 1999). The actual version can be downloaded from http://www.europam.net.

In the following it was adopted and slightly modified by the European Agency for the Evaluation of Medicinal Products (EMEA), and finally as Guidelines on good agricultural and collection practices (GACP) by the WHO in 2003.

All these guidelines follow almost the same concept dealing with the following topics:

- Identification and authentication of the plant material, especially botanical identity and deposition of specimens.
- Seeds and other propagation material, respecting the specific standards and certifications.
- Cultivation, including site selection, climate, soil, fertilization, irrigation, crop maintenance, and plant protection with special regard to contaminations and residues.
- Harvest, with specific attention to harvest time and conditions, equipment, damage, contaminations with (toxic) weeds and soil, transport, possible contact with any animals, and cleaning of all equipment and containers.
- Primary processing, that is, washing, drying, distilling; cleanness of the buildings; according to the actual legal situation these processing steps including distillation—if performed by the farmer—is still part of GA(C)P; in all other cases, it is subjected to GMP (good manufacturing practice).
- Packaging and labeling, including suitability of the material.
- Storage and transportation, especially storage conditions, protection against pests and animals, fumigation, and transport facilities.
- Equipment: material, design, construction, easy to clean.
- Personnel and facilities, with special regard to education, hygiene, protection against allergens and other toxic compounds, welfare.

In the case of wild collection the standard for sustainable collection should be applied (see Section 3.3.6.2).

A very important topic is finally the *documentation* of all steps and measurements to be able to trace back the starting material, the exact location of the field, any treatment with agrochemicals, and the special circumstances during the cultivation period. Quality assurance is only possible if the traceability is given and the personnel is educated appropriately. Certification and auditing of the production of essential oil-bearing plants is not yet obligatory, but recommended and often requested by the customer.

3.6.2 ISSC-MAP: The International Standard on Sustainable Wild Collection of Medicinal and Aromatic Plants

ISSC-MAP is a joint initiative of the German Bundesamt für Naturschutz (BfN), WWF/TRAFFIC Germany, IUCN Canada, and IUCN Medicinal Plant Specialist Group (MPSG). ISSC-MAP intends to ensure the long-term survival of MAP populations in their habitats by setting principles and criteria for the management of MAP wild collection (Leaman, 2006; Medicinal Plant Specialist Group, 2007). The standard is not intended to address product storage, transport, and processing, or any issues of products, topics covered by the WHO Guidelines on GACP for Medicinal Plants (WHO, 2003). ISSC-MAP includes legal and ethical requirements (legitimacy, customary rights, and transparency), resource assessment, management planning and monitoring, responsible collection, and collection, area practices and responsible business practices. One of the strengths of this standard is that resource management not only includes target MAP resources and their habitats but also social, cultural, and economic issues.

3.6.3 FairWild

The FairWild standard (http://www.fairwild.org) was initiated by the Swiss Import Promotion Organization (SIPPO) and combines principles of FairTrade (Fairtrade Labelling Organizations International, FLO), international labor standards (International Labour Organization, ILO), and sustainability (ISSC-MAP).

3.7 CONCLUSION

This chapter has shown that a number of items concerning the plant raw material have to be taken into consideration when producing essential oils. A quality management has to be established tracing back to the authenticity of the starting material and ensuring that all known influences on the quality are taken into account and documented in an appropriate way. This is necessary to meet the increasing requirements of international standards and regulations. The review also shows that a high number of data and information exist, but sometimes without expected relevance due to the fact that the repeatability of the results is not given by a weak experimental design, an incorrect description of the plant material used, or an inappropriate sampling. On the other side, this opens the chance for many more research work in the field of essential oil-bearing plants.

REFERENCES

Agnaniet, H., C. Menut, and J.M. Bessière, 2004. Aromatic plants of tropical Africa XLIX: Chemical composition of essential oils of the leaf and rhizome of *Aframomum giganteum* K. Schum from Gabun. *Flavour Fragr. J.*, 19: 205–209.

Aiello, N., F. Scartezzini, C. Vender, L. D'Andrea, and A. Albasini, 2001. Caratterisiche morfologiche, produttive e qualitative di una nuova varietà sintetica di salvia confrontata con altre cultivar. *ISAFA Comunicaz. Ric.*, 2001(1): 5–16.

Akhila, A. and M. Rani, 2002. Chemical constituents and essential oil biogenesis in *Vetiveria zizanioides*. In *Vetiveria—The Genus Vetiveria*, M. Maffei (ed.), pp. 73–109. New York: Taylor & Francis.

Asllani, U., 2000. Chemical composition of albanian sage oil (*Salvia officinalis* L.). *J. Essent. Oil Res.*, 12: 79–84.

Auer, C.A., 2003. Tracking genes from seed to supermarket: Techniques and trends. *Trends Plant Sci.*, 8: 591–597.

Aziz, E.E., H. Al-Amier, and L.E. Craker, 2008. Influence of salt stress on growth and essential oil production in peppermint, pennyroyal and apple mint. *J. Herbs Spices Med. Plants*, 14: 77–87.

Baser, K.H.C., 2002. The Turkish *Origanum* species. In *Oregano: The Genera Origanum and Lippia*, S.E. Kintzios (ed.), pp. 109–126. New York: Taylor & Francis.

Bazina, E., A. Makris, C. Vender, and M. Skoula, 2002. Genetic and chemical relations among selected clones of *Salvia officinalis*. In *Breeding Research on Aromatic and Medicinal Plants*, C.B. Johnson and Ch. Franz (eds.), pp. 269–273. Binghampton, NY: Haworth Press.

Bergougnoux, V., J.C. Caissard, F. Jullien, J.L. Magnard, G. Scalliet, J.M. Cock, P. Hugueney, and S. Baudino, 2007. Both the adaxial and abaxial epidermal layers of the rose petal emit volatile scent compounds. *Planta*, 226: 853–866.

Bernáth, J., 2000. Genetic improvement of cultivated species of the genus Salvia. In *Sage—The Genus Salvia*, S.E. Kintzios (ed.), pp. 109–124. Amsterdam, the Netherlands: Harwood Academic Publishers.

Bernáth, J., 2002. Evaluation of strategies and results concerning genetical improvement of medicinal and aromatic plants. *Acta Hort.*, 576: 116–128.

Bertea, C.M. and W. Camusso, 2002. Anatomy, biochemistry and physiology. In *Vetiveria: Medicinal and Aromatic Plants—Industrial Profiles*, M. Maffei (ed.), Vol. 20, pp. 19–43. London, U.K.: Taylor & Francis.

Bezzi, A., 1994. Selezione clonale e costituzione di varietà di salvia (*Salvia officinalis* L.). *Atti convegno internazionale 'Coltivazione e miglioramento di piante officinali'*, pp. 97–117. Villazzano di Trento, Italy: ISAFA.

Bicchi, C., M. Fresia, P. Rubiolo, D. Monti, Ch. Franz, and I. Goehler, 1997. Constituents of *Tagetes lucida* Cav. ssp. *lucida* essential oil. *Flavour Fragr. J.*, 12: 47–52.

Bradbury, L.M.T., R.J. Henry, Q. Jin, R.F. Reinke, and D.L.E. Waters, 2005. A perfect marker for fragrance genotyping in rice. *Mol. Breed.*, 16: 279–283.

Bradu, B.L., S.N. Sobti, P. Pushpangadan, K.M. Khosla, B.L. Rao, and S.C. Gupta, 1989. Development of superior alternate source of clove oil from 'Clocimum' (*Ocimum gratissimum* Linn.). In *Proceedings of the 11th International Congress of Essential Oils, Fragrances and Flavours*, Vol. 3, pp. 97–103.

Brophy, J.J. and I.A. Southwell, 2002. Eucalyptus chemistry. In *Eucalyptus—The Genus Eucalyptus*, J.J.W. Coppen (ed.), pp. 102–160. London, U.K.: Taylor & Francis.

CBD, 2001. Convention on biological diversity. In *United Nations Environment Programme, CBD Meeting Nairobi*, Nairobi, Kenya.

CBD, 2002. Global strategy for plant conservation. In *CBD Meeting The Hague*, The Hague, the Netherlands.

CBD, 2004. Sustainable use of biodiversity. In *CBD Meeting Montreal*, Montreal, Quebec, Canada.

Carruba, A., R. la Torre, and A. Matranga, 2002. Cultivation trials of some aromatic and medicinal plants in a semiarid Mediterranean environment. *Acta Hort.*, 576: 207–214.

Ceylan, A., H. Otan, A.O. Sari, N. Carkaci, E. Bayram, N. Ozay, M. Polat, A. Kitiki, and B. Oguz, 1994. *Origanum onites L. (Izmir Kekigi) Uzerinde Agroteknik Arastirmalar, Final Report.* Izmir, Turkey: AARI.

Chalchat, J., J.C. Gary, and R.P. Muhayimana, 1995. Essential oil of *Tagetes minuta* from Rwanda and France: Chemical composition according to harvesting, location, growing stage and plant part. *J. Essent. Oil Res.*, 7: 375–386.

Chalchat, J., A. Michet, and B. Pasquier, 1998. Study of clones of *Salvia officinalis* L., yields and chemical composition of essential oil. *Flavour Fragr. J.*, 13: 68–70.

Chatterjee, S.K., 2002. Cultivation of medicinal and aromatic plants in India. *Acta Hort.*, 576: 191–202.

Ciccio, J.F., 2004. A source of almost pure methylchavicol: Volatile oil from the aerial parts of *Tagetes lucida* (Asteraceae) cultivated in Costa Rica. *Rev. de Biol. Trop.*, 52: 853–857.

Circella, G., Ch. Franz, J. Novak, and H. Resch, 1995. Influence of day length and leaf insertion on the composition of marjoram essential oil. *Flavour Fragr. J.*, 10: 371–374.

Coronel, A.C., C.M. Cerda-Garcia-Rojas, P. Joseph-Nathan, and C.A.N. Catalán, 2006. Chemical composition, seasonal variation and a new sesquiterpene alcohol from the essential oil of *Lippia integrifolia*. *Flavour Fragr. J.*, 21: 839–847.

Croteau, R., 1991. Metabolism of monoterpenes in mint (*Mentha*) species. *Planta Med.*, 57(Suppl. 1): 10–14.

Croteau, R., M. Felton, F. Karp, and R. Kjonaas, 1981. Relationship of camphor biosynthesis to leaf development in sage (*Salvia officinalis*). *Plant Physiol.*, 67: 820–824.

Czabajska, W., J. Dabrowska, K. Kazmierczak, and E. Ludowicz, 1978. Maintenance breeding of chamomile cultivar 'Zloty Lan'. *Herba Polon.*, 24: 57–64.

Daniel, G. and U. Bomme, 1991. Use of in-vitro culture for arnica (*Arnica montana* L.) breeding. *Landw. Jahrb.*, 68: 249–253.

Dellacassa, E., E. Soler, P. Menéndez, and P. Moyna, 1990. Essentail oils from *Lippia alba* Mill. N.E. Brown and *Aloysia chamaedrifolia* Cham. (Verbenaceae) from Urugay. *Flavour Fragr. J.*, 5: 107–108.

Desmarest, P., 1992. Amelioration du fenonil amier par selection recurrente, clonage et embryogenèse somatique. In *Proceedings of the Second Mediplant Conference*, pp. 19–26. Conthey, Switzerland/CH, P.

Dey, B.B. and M.A. Choudhuri, 1983. Effect of leaf development stage on changes in essential oil of *Ocimum sanctum* L. *Biochem. Physiol. Pflanzen*, 178: 331–335.

Diemer, F., F. Jullien, O. Faure, S. Moja, M. Colson, E. Matthys-Rochon, and J.C. Caissard, 1998. High efficiency transformation of peppermint (*Mentha* x *piperita* L.) with *Agrobacterium tumefaciens*. *Plant Sci.*, 136: 101–108.

Dominguez, X.A., S.H. Sánchez, M. Suárez, X Baldas, J.H., and G. Ma del Rosario, 1989. Chemical constituents of *Lippia graveolens*. *Planta Med.*, 55: 208–209.

Doran, J.C., 2002. Genetic improvement of eucalyptus. In *Eucalyptus—The Genus Eucalyptus, Medicinal and Aromatic Plants—Industrial Profiles*, J.J.W. Coppen (ed.), Vol. 22, pp. 75–101. London, U.K.: Taylor & Francis.

Dronne, S., S. Moja, F. Jullien, F. Berger, and J.C. Caissard, 1999. *Agrobacterium*-mediated transformation of lavandin (*Lavandula* × *intermedia* Emeric ex Loiseleur). *Transgenic Res.*, 8: 335–347.
Dudai, N., 2006. Breeding of high quality basil for the fresh herb market—An overview. In *International Symposium on the Labiatae*, p. 15, San Remo, Italy.
Dudai, N., E. Putievsky, U. Ravid, D. Palevitch, and A.H. Halevy, 1992. Monoterpene content of *Origanum syriacum* L. as affected by environmental conditions and flowering. *Physiol. Plant.*, 84: 453–459.
Elementi, S., R. Nevi, and L.F. D'Antuono, 2006. Biodiversity and selection of 'European' basil (*Ocimum basilicum* L.) types. *Acta Hort.*, 723: 99–104.
El-Keltawi, N.E.M. and S.S.A. Abdel-Rahman, 2006. In vivo propagation of certain sweet basil cultivars. *Acta Hort.*, 723: 297–302.
Fahn, A., 1979. *Secretory Tissues in Plants*. London, U.K.: Academic Press.
Fahn, A., 1988. Secretory tissues in vascular plants. *New Phytol.*, 108: 229–257.
Figueiredo, A.C., J.G. Barroso, L.G. Pedro, and J.J.C. Scheffer, 1997. Physiological aspects of essential oil production. In *Essential Oils: Basic and Applied Research*, Ch. Franz, A. Máthé, and G. Buchbauer (eds.), pp. 95–107. Carol Stream, IL: Allured Publishing.
Fischer, U., 1998. Variabilität Guatemaltekischer Arzneipflanzen der Gattung *Lippia* (Verbenaceae): *Lippia alba*, L. *dulcis*, L. graveolens. *Dissertation*, Veterinärmedizinischen Universität, Wien, Austria.
Fischer, U., Ch. Franz, R. Lopez, and E. Pöll, 1996. Variability of the essential oils of *Lippia graveolens* HBK from Guatemala. In *Essential Oils: Basic and Applied Research*, Ch. Franz, A. Máthé, and A.G. Buchbauer (eds.), pp. 266–269. Carol Stream, IL: Allured Publishing.
Fischer, U., R. Lopez, E. Pöll, S. Vetter, J. Novak, and Ch. Franz, 2004. Two chemotypes within *Lippia alba* populations in Guatemala. *Flavour Fragr. J.*, 19: 333–335.
Franz, Ch., 1972. Einfluss der Naehrstoffe Stickstoff und Kalium auf die Bildung des aetherischen Oels der Pfefferminze, *Mentha piperita* L. *Planta Med.*, 22: 160–183.
Franz, Ch., 1981. *Zur Qualitaet von Arznei- u. Gewuerzpflanzen*. Habil.-Schrift. Muenchen, Germany: TUM.
Franz, Ch., 1982. Genetische, ontogenetische und umweltbedingte Variabilität der Bestandteile des ätherischen Öls von Kamille (*Matricaria recutita*(L.) Rauschert). In *Aetherische Oele—Analytik, Physiologie, Zusammensetzung*, K.H. Kubeczka (ed.), pp. 214–224. Stuttgart, Germany: Thieme.
Franz, Ch., 1989a. Biochemical genetics of essential oil compounds. In *Proceedings of the 11th International Congress of Essential Oils, Fragrances and Flavours*, Vol. 3, pp. 17–25. New Delhi, India: Oxford & IBH Publishing.
Franz, Ch., 1989b. Good agricultural practice (GAP) for medicinal and aromatic plant production. *Acta Hort.*, 249: 125–128.
Franz, Ch., 1993a. Probleme bei der Beschaffung pflanzlicher Ausgangsmaterialien. In *Ätherische Öle, Anspruch und Wirklichkeit*, R. Carle (ed.), pp. 33–58. Stuttgart, Germany: Wissenschaftliche Verlagsgesellschaft.
Franz, Ch., 1993b. Genetics. In *Volatile Oil Crops*, R.K.M. Hay and P.G. Waterman (eds.), pp. 63–96. Harlow, U.K.: Longman.
Franz, Ch., 1993c. Domestication of wild growing medicinal plants. *Plant Res. Dev.*, 37: 101–111.
Franz, Ch., 1993d. Genetic versus climatic factors influencing essential oil formation. In *Proceedings of the 12th International Congress of Essential Oils, Fragrances and Flavours*, pp. 27–44. Vienna, Austria.
Franz, Ch., 1999. Gewinnung von biogenen Arzneistoffen und Drogen. In *Biogene Arzneistoffe*, 2nd edn., H. Rimpler (ed.), pp. 1–24. Stuttgart, Germany: Deutscher Apotheker Verlag.
Franz, Ch., 2000. *Biodiversity and Random Sampling in Essential Oil Plants*. Lecture 31st ISEO, Hamburg, Germany.
Franz, Ch., 2001. Plant variety rights and specialised plants. In *Proceedings of the PIPWEG 2001, Conference on Plant Intellectual Property within Europe and the Wider Global Community*, pp. 131–137. Sheffield, U.K.: Sheffield Academic Press.
Franz, Ch., 2013. Schafgarbe (*Achillea millefolium* L.). In *Handbuch des Arznei- u. Gewuerzpflanzenbaus*, Vol. 5, pp. 453–463. Saluplanta, Bernburg.
Franz, C. and H. Glasl, 1976. Comparative investigations of fruit-, leaf- and root-oil of some parsley varieties. *Qual. Plant. Plant Foods Hum. Nutr.*, 25(3/4): 253–262.
Franz, Ch., K. Hardh, S. Haelvae, E. Mueller, H. Pelzmann, and A. Ceylan, 1986. Influence of ecological factors on yield and essential oil of chamomile (*Matricaria recutita* L.). *Acta Hort.*, 188: 157–162.
Franz, Ch., J. Hoelzl, and C. Kirsch, 1983a. Influence of nitrogen, phosphorus and potassium fertilization on chamomile (*Chamomilla recutita* (L.) Rauschert). II. Effect on the essential oil. *Gartenbauwiss. Hort. Sci.*, 48: 17–22.
Franz, Ch., J. Hoelzl, and A. Voemel, 1978. Variation in the essential oil of *Matricaria chamomilla* L. depending on plant age and stage of development. *Acta Hort.*, 73: 230–238.

Franz, Ch., C. Kirsch, and O. Isaac, 1983b. Process for producing a new tetraploid chamomile variety. *German Patent DE*3423207.
Franz, Ch., C. Kirsch, and O. Isaac, 1985. Neuere Ergebnisse der Kamillenzüchtung. *Dtsch. Apoth. Ztg.*, 125: 20–23.
Franz, Ch. and Novak, J., 1997. Breeding of *Origanum* sp. In *Proceedings of the IPGRI Workshop*, Padulosi, S. (ed.), pp. 50–57. Oregano.
Frezal, L. and R. Leblois, 2008. Four years of DNA barcoding: Current advances and prospects. *Infect. Genet. Evol.*, 8: 727–736.
Frighetto, N., J.G. de Oliveira, A.C. Siani, and K. Calago das Chagas, 1998. *Lippia alba* Mill (Verbenaceae) as a source of linalool. *J. Essent. Oil Res.*, 10: 578–580.
Fritzsche, R., J. Gabler, H. Kleinhempel, K. Naumann, A. Plescher, G. Proeseler, F. Rabenstein, E. Schliephake, and W. Wradzidlo, 2007. *Handbuch des Arznei- und Gewürzpflanzenbaus: Krankheiten und Schädigungen an Arznei- und Gewürzpflanzen*, Vol. 3. Bernburg, Germany: Saluplanta e.V.
Gabler, J., 2002. Breeding for resistance to biotic and abiotic factors in medicinal and aromatic plants. In *Breeding Research on Aromatic and Medicinal Plants*, C.B. Johnson and Ch. Franz (eds.), pp. 1–12. Binghampton, NY: Haworth Press.
Galambosi, B. and S. Dragland, 2002. Possibilities and limitations for herb production in Nordic countries. *Acta Hort.*, 576: 215–225.
Gancel, A.L., D. Ollé, P. Ollitraut, F. Luro, and J.M. Brillouet, 2002. Leaf and peel volatile compounds of an interspecific citrus somatic hybrid (*Citrus aurantifolia* Swing. × *Citrus paradisi* Macfayden). *Flavour Fragr. J.*, 17: 416–424.
Gershenzon, J., M.E. McConkey, and R.B. Croteau, 2000. Regulation of monoterpene accumulation in leaves of peppermint. *Plant Physiol.*, 122: 205–213.
Giannouli, A.L. and S.E. Kintzios, 2000. Essential oils of *Salvia* spp.: Examples of intraspecific and seasonal variation. In *Sage—The Genus Salvia*, S.E. Kintzios (ed.), pp. 69–80. Amsterdam, the Netherlands: Harwood Academic Publishing.
Goehler, I., 2006. Domestikation von Medizinalpflanzen und Untersuchungen zur Inkulturnahme von *Tagetes lucida* Cav. *Dissertation*, an der Universität für Bodenkultur Wien, Wein, Austria.
Goehler, I., Ch. Franz, A. Orellana, and C. Rosales, 1997. *Propagation of Tagetes lucida Cav. Poster WOCMAP II Mendoza*. Argentina.
Gonzalez de, C.N., A. Quintero, and A. Usubillaga, 2002. Chemotaxonomic value of essential oil compounds in *Citrus* species. *Acta Hort.*, 576: 49–55.
Gora, J., A. Lis, J. Kula, M. Staniszewska, and A. Woloszyn, 2002. Chemical composition variability of essential oils in the ontogenesis of some plants. *Flavour Fragr. J.*, 17: 445–451.
Gouyon, P.H. and P. Vernet, 1982. The consequences of gynodioecy in natural populations of *Thymus vulgaris* L. *Theoret. Appl. Genet.*, 61: 315–320.
Grassi, P., 2003. Botanical and chemical investigations in *Hyptis* spp. (Lamiaceae) in El Salvador. *Dissertation*, Universität Wien, Wein, Austria.
Grassi, P., J. Novak, H. Steinlesberger, and Ch. Franz, 2004. A direct liquid, non-equilibrium solid-phase micro-extraction application for analysing chemical variation of single peltate trichomes on leaves of *Salvia officinalis*. *Phytochem. Anal.*, 15: 198–203.
Harrewijn, P., A.M. van Oosten, and P.G.M. Piron, 2001. *Natural Terpenoids as Messengers*. Dordrecht, the Netherlands: Kluwer Academic Publishers.
Hay, R.K.M. and P.G. Waterman, 1993. *Volatile Oil Crops*. Burnt Mill, U.K.: Longman Science & Technology Publications.
Hebert, P.D.N., A. Cywinska, S.L. Ball, and J.R. deWaard, 2003. Biological identifications through DNA barcodes. *Proc. R. Soc. Lond. B*, 270: 313–322.
Hedge, I.C., 1992. A global survey of the biography of the Labiatae. In *Advances in Labiatae Science*, R.M. Harley and T. Reynolds (eds.), pp. 7–17. Kew, U.K.: Royal Botanical Gardens.
Hefendehl, F.W., 1962. Zusammensetzung des ätherischen Öls von *Mentha x piperita* im Verlauf der Ontogenese und Versuche zur Beeinflussung der Ölkomposition. *Planta Med.*, 10: 241–266.
Hoeltzel, C., 1964. Über Zusammenhänge zwischen der Biosynthese der ätherischen Öle und dem photoperiodischen Verhalten der Pfefferminze (*Mentha piperita* L.). *Dissertation*, University of Tübingen, Tübingen, Germany.
Horn, W., Ch. Franz, and I. Wickel, 1988. Zur Genetik der Bisaboloide bei der Kamille. *Plant Breed.*, 101: 307–312.
Johnson, C.B., A. Kazantzis, M. Skoula, U. Mitteregger, and J. Novak, 2004. Seasonal, populational and ontogenic variation in the volatile oil content and composition of individuals of *Origanum vulgare* subsp. *hirtum*, assessed by GC headspace analysis and by SPME sampling of individual oil glands. *Phytochem. Anal.*, 15: 286–292.

Kampranis, S.C., D. Ioannidis, A. Purvis, W. Mahrez, E. Ninga, N.A. Katerelos, S. Anssour et al., 2007. Rational conversion of substrate and product specificity in a *Salvia* monoterpene synthase: Structural insights into the evolution of terpene synthase function. *Plant Cell*, 19: 1994–2005.

Kanias, G.D., C. Souleles, A. Loukis, and E. Philotheou-Panou, 1998. Statistical studies of essential oil composition in three cultivated Sage species. *J. Essent. Oil Res.*, 10: 395–403.

Karousou, R., D. Vokou, and Kokkini, 1998. Variation of *Salvia fruticosa* essential oils on the island of Crete (Greece). *Bot. Acta*, 111: 250–254.

Kastner, U., J. Saukel, K. Zitterl-Eglseer, R. Länger, G. Reznicek, J. Jurenitsch, and W. Kubelka, 1992. Ätherisches Öl—ein zusätzliches Merkmal für die Charakterisierung der mitteleuropäischen Taxa der *Achillea-millefolium*-Gruppe. *Sci. Pharm.*, 60: 87–99.

Khosla, M.K., B.L. Bradu, and R.K. Thapa, 1989. Biogenetic studies on the inheritance of different essential oil constituents of *Ocimum* species, their F1 hybrids and synthesized allopolyploids. *Herba Hung.*, 28: 13–19.

Kitiki, A., 1997. Status of cultivation and use of oregano in Turkey. In *Proceedings of the IPGRI Workshop Oregano*, S. Padulosi (ed.), pp. 122–132.

Koezuka, Y., G. Honda, and M. Tabata, 1986. Genetic control of phenylpropanoids in Perilla frutescens. *Phytochemistry*, 25: 2085–2087.

Kokkini, S., R. Karousou, A. Dardioti, N. Kirgas, and T. Lanaras, 1996. Autumn essential oils of Greek oregano (*Origanum vulgare* ssp. *hirtum*). *Phytochemistry*, 44: 883–886.

Kress, W.J., K.J. Wurdack, E.A. Zimmer, L.A. Weigt, and D.H. Janzen, 2005. Use of DNA barcodes to identify flowering plants. *PNAS*, 102: 8369–8374.

Kubeczka, K.H., 1997. The essential oil composition of *Pimpinella* species. In *Progress in Essential Oil Research*, K.H.C. Baser and N. Kirimer (eds.), pp. 35–56. Eskisehir, Turkey: ISEO.

Kubeczka, K.H., A. Bartsch, and I. Ullmann, 1982. Neuere Untersuchungen an ätherischen Apiaceen-Ölen. In *Ätherische Öle—Analytik, Physiologie, Zusammensetzung*, K.H. Kubeczka (ed.), pp. 158–187. Stuttgart, Germany: Thieme.

Kubeczka, K.H., I. Bohn, and V. Formacek, 1986. New constituents from the essential oils of *Pimpinella sp.* In *Progress in Essential Oil Research*, E.J. Brunke (ed.), pp. 279–298. Berlin, Germany: W de Gruyter.

Kulkarni, R.N., 1990. Honeycomb and simple mass selection for herb yield and inflorescence-leaf-steam-ratio in palmarose grass. *Euphytica*, 47: 147–151.

Kulkarni, R.N. and S. Ramesh, 1992. Development of lemongrass clones with high oil content through population improvement. *J. Essent. Oil Res.*, 4: 181–186.

Kurowska, A. and I. Galazka, 2006. Essential oil composition of the parsley seed of cultivars marketed in Poland. *Flavour Fragr. J.*, 21: 143–147.

Langbehn, J., F. Pank, J. Novak, and C. Franz, 2002. Influence of Selection and Inbreeding on *Origanum majorana* L. *J. Herbs Spices Med. Plants*, 9: 21–29.

Lawrence, B.M., 1978. *Essential Oils 1976–77*, pp. 84–109. Wheaton, IL: Allured Publishing.

Lawrence, B.M., 1984. The botanical and chemical aspects of Oregano. *Perform. Flavor*, 9(5): 41–51.

Lawrence, B.M., 2007. *Mint: The Genus Mentha*. Boca Raton, FL: CRC Press.

Leaman, D.J., 2006. Sustainable wild collection of medicinal and aromatic plants. In *Medicinal and Aromatic Plants*, R.J. Bogers, L.E. Craker, and D. Lange (eds.), pp. 97–107. Dordrecht, the Netherlands: Springer.

Le Buanec, B., 2001. Development of new plant varieties and protection of intellectual property: An international perspective. In *Proceedings of the PIPWEG Conference on 2001 Angers*, pp. 103–108. Sheffield, U.K.: Sheffield Academic Press.

Lemli, J.A.J.M., 1955. De vluchtige olie van *Mentha piperita* L. gedurende de ontwikkeling van het plant. *Dissertation*, University of Groningen, Groningen, the Netherlands.

Letchamo, W. and R. Marquard, 1993. The pattern of active substances accumulation in camomile genotypes under different growing conditions and harvesting frequencies. *Acta Hort.*, 331: 357–364.

Li, X., Z. Gong, H. Koiwa, X. Niu, J. Espartero, X. Zhu, P. Veronese et al., 2001. Bar-expressing peppermint (*Mentha* × *piperita* L. var. Black Mitcham) plants are highly resistant to the glufosinate herbicide Liberty. *Mol. Breed.*, 8: 109–118.

Lis, A., E. Boczek, and J. Gora, 2004. Chemical composition of the essential oils from fruits. Leaves and flowers of the Amur cork tree (*Phellodendron amurense* Rupr.). *Flavour Fragr. J.*, 19: 549–553.

Lis, A. and Milczarek, J., 2006. Chemical composition of the essential oils from fruits, leaves and flowers of *Phellodendron sachalinene* (Fr. Schmidt) Sarg. *Flavour Fragr. J.*, 21: 683–686.

Llewelyn, M., 2002. European plant intellectual property. In *Breeding Research on Aromatic and Medicinal Plants*, C.B. Johnson and Ch. Franz (eds.), pp. 389–398. Binghampton, NY: Haworth Press.

Lorenzo, D., D. Paz, P. Davies, R. Vila, S. Canigueral, and E. Dellacassa, 2001. Composition of a new essential oil type of *Lippia alba* (Mill.) N.E. Brown from Uruguay. *Flavour Fragr. J.*, 16: 356–359.

Macchia, M., A. Pagano, L. Ceccarini, S. Benvenuti, P.L. Cioni, and G. Flamini, 2006. Agronomic and phytochimic characteristics in some genotypes of *Ocimum basilicum* L. *Acta Hort.*, 723: 143–149.

Mahmoud, S.S. and R.B. Croteau, 2001. Metabolic engineering of essential oil yield and composition in mint by altering expression of deoxyxylulose phosphate reductoisomerase and menthofuran synthase. *PNAS*, 98: 8915–8920.

Mahmoud, S.S. and R.B. Croteau, 2003. Menthofuran regulates essential oil biosynthesis in peppermint by controlling a downstream monoterpene reductase. *PNAS*, 100: 14481–14486.

Mahmoud, S.S., M. Williams, and R.B. Croteau, 2004. Cosuppression of limonene-3-hydroxylase in peppermint promotes accumulation of limonene in the essential oil. *Phytochemistry*, 65: 547–554.

Maleci Bini, L. and C. Giuliani, 2006. The glandular trichomes of the Labiatae. A review. *Acta Hort.*, 723: 85–90.

Marotti, M., R. Piccaglia, B. Biavati, and I. Marotti, 2004. Characterization and yield evaluation of essential oils from different *Tagetes* species. *J. Essent. Oil Res.*, 16: 440–444.

Marotti, M., R. Piccaglia, and E. Giovanelli, 1994. Effects of variety and ontogenetic stage on the essential oil composition and biological activity of fennel (*Foeniculum vulgare* Mill.). *J. Essent. Oil Res.*, 6: 57–62.

Marotti, M., P. Piccaglia, and E. Giovanelli, 1996. Differences in essential oil composition of basil (*Ocimum basilicum* L.) of Italian cultivars related to morphological characteristics. *J. Agric. Food Chem.*, 44: 3926–3929.

Marthe, F. and P. Scholze, 1996. A screening technique for resistance evaluation to septoria blight (*Septoria petroselini*) in parsley (*Petroselinum crispum*). *Beitr. Züchtungsforsch*, 2: 250–253.

Martinetti, L., E. Quattrini, M. Bononi, and F. Tateo, 2006. Effect of the mineral fertilization and the yield and the oil content of two cultivars of rosemary. *Acta Hort.*, 723: 399–404.

Massoud, H. and C. Franz, 1990. Quantitative genetical aspects of *Chamomilla recutita* (L.) Rauschert. *J. Essent. Oil Res.*, 2: 15–20.

Massoud, H., M. Sharaf El-Din, R. Hassan, and A. Ramadan, 2002. Effect of salinity and some trace elements on growth and leaves essential oil content of thyme (*Thymus vulgaris* L.). *J. Agric. Res. Tanta Univ.*, 28: 856–873.

Máthé, A. and Ch. Franz, 1999. Good agricultural practice and the quality of phytomedicines. *J. Herbs Spices Med. Plants*, 6: 101–113.

Máthé, I., G. Nagy, A. Dobos, V.V. Miklossy, and G. Janicsak, 1996. Comparative studies of the essential oils of some species of *Sect. Salvia*. In *Proceedings of the 27th International Symposium on Essential Oils (ISEO)*, Ch. Franz, A. Máthé, and G. Buchbauer (eds.), pp. 244–247.

Medicinal Plant Specialist Group, 2007. International Standard for Sustainable Wild Collection of Medicinal and Aromatic Plants (ISSC-MAP). Version 1.0. Bundesamt für Naturschutz (BfN), MPSG/SSC/IUCN, WWF Germany, and TRAFFIC, Bonn, Gland, Frankfurt, and Cambridge. *BfN-Skripten*, 195.

Menary, R.C., 1994. Factors influencing the yield and composition of essential oils, II: Nutrition, irrigation, plant growth regulators, harvesting and distillation. In *Proceedings of the 4emes Rencontres Internationales*, pp. 116–138. Nyons, France.

Mendes, M.L. and A. Romano, 1997. In vitro cloning of *Thymus mastichina* L. field grown plants. *Acta Hort.*, 502: 303–306.

Minuto, G., A. Minuto, A. Garibaldi, and M.L. Gullino, 2006. Disease control of aromatic crops: Problems and solutions. In *International Symposium on Labiatae*, p. 33. San Remo, Italy.

Miraglia, M., K.G. Berdal, C. Brera, P. Corbisier, A. Holst-Jensen, E.J. Kok, H.J. Marvin et al., 2004. Detection and traceability of genetically modified organisms in the food production chain. *Food Chem. Toxicol.*, 42: 1157–1180.

Mosandl, A., 1993. Neue Methoden zur herkunftsspezifischen Analyse aetherischer Oele. In *Ätherische Öle—Anspruch und Wirklichkeit*, R. Carle (ed.), pp. 103–134. Stuttgart, Germany: Wissenschaftliche Verlagsgesellschaft.

Mulas, M., A.H. Dias Francesconi, B. Perinu, and E. Del Vais, 2002. Selection of Rosemary (*Rosmarinus officinalis* L.) cultivars to optimize biomass yield. In *Breeding Research on Aromatic and Medicinal Plants*, C.B. Johnson and Ch. Franz (eds.), pp. 133–138. Binghampton, NY: Haworth Press.

Munoz-Bertomeu, J., I. Arrillaga, R. Ros, and J. Segura, 2006. Up-regulation of 1-deoxy-d-xylulose-5-phosphate synthase enhances production of essential oils in transgenic spike lavender. *Plant Physiol.*, 142: 890–900.

Munsi, P.S. and Mukherjee, S.K., 1986. Response of Java citronella (*Cymbopogon winterianus* Jowitt.) to harvesting intervals with different nitrogen levels. *Acta Hort.*, 188: 225–229.

Murray, M.J., 1969. *Induced Mutations in Plants*, pp. 345–371. Vienna, Austria: IAEA.

Murray, M.J. and R.H. Reitsema, 1954. The genetic basis of the ketones carvone and menthone in *Mentha crispa* L. *J. Am. Pharm. Assoc. (Sci. Ed.)*, 43: 612–613.

Murray, M.J. and A.W. Todd, 1972. Registration of Todd's Mitcham Peppermint. *Crop Sci.*, 12: 128.

Nair, M.K., 1982. Cultivation of spices. In *Cultivation and Utilization of Aromatic Plants*, C.K. Atal and B.M. Kapur (eds.), pp. 190–214. Jammu-Tawi, India: RRL-CSIR.

Nebauer, S.G., I. Arrillaga, L. del Castillo-Agudo, and J. Segura, 2000. *Agrobacterium tumefaciens*-mediated transformation of the aromatic shrub *Lavandula latifolia*. *Mol. Breed.*, 6: 23–48.

Nicola, S., J. Hoeberechts, and E. Fontana, 2006. Rooting products and cutting timing for peppermint (*Mentha piperita* L.) radication. *Acta Hort.*, 723: 297–302.

Niu, X., X. Li, P. Veronese, R.A. Bressan, S.C. Weller, and P.M. Hasegawa, 2000. Factors affecting *Agrobacterium tumefaciens*-mediated transformation of peppermint. *Plant Cell Rep.*, 19: 304–310.

Novak, J., L. Bahoo, U. Mitteregger, and C. Franz, 2006a. Composition of individual essential oil glands of savory (*Satureja hortensis* L., Lamiaceae) from Syria. *Flavour Fragr. J.*, 21: 731–734.

Novak, J., C. Bitsch, F. Pank, J. Langbehn, and C. Franz, 2002. Distribution of the *cis*-sabinene hydrate acetate chemotype in accessions of marjoram (*Origanum majorana* L.). *Euphytica*, 127: 69–74.

Novak, J., S. Grausgruber-Gröger, and B. Lukas, 2007. DNA-Barcoding of plant extracts. *Food Res. Int.*, 40: 388–392.

Novak, J., B. Lukas, and C. Franz, 2008. The essential oil composition of wild growing sweet marjoram (*Origanum majorana* L., Lamiaceae) from Cyprus—Three chemotypes. *J. Essent. Oil Res.*, 20: 339–341.

Novak, J., M. Marn, and C. Franz, 2006b. An a-pinene chemotype in *Salvia officinalis* L. (Lamiaceae). *J. Essent. Oil Res.*, 18: 239–241.

Novak, J., S. Novak, C. Bitsch, and C. Franz, 2000. Essential oil composition of different populations of *Valeriana celtica* ssp. from Austria and Italy. *Flavour Fragr. J.*, 15: 40–42.

Novak, J., S. Novak, and C. Franz, 1998. Essential oils of rhizomes and rootlets of *Valeriana celtica* L. ssp. *norica* Vierh. from Austria. *J. Essent. Oil Res.*, 10: 637–640.

Omidbaigi, R. and A. Arjmandi, 2002. Effects of NP supply on growth, development, yield and active substances of garden thyme (*Thymus vulgaris* L.). *Acta Hort.*, 576: 263–265.

Padulosi, S. (ed.), 1996. Oregano. Promoting the conservation and use of underutilized and neglected crops. 14. *Proceedings of the IPGRI Internet Workshop on Oregano*, May 8–12, 1996, CIHEAM Valenzano (Bari). IPGRI: Rome.

Pafundo, S., C. Agrimonti, and N. Marmiroli, 2005. Traceability of plant contribution in olive oil by amplified fragment length polymorphisms. *J. Agric. Food Chem.*, 53: 6995–7002.

Pank, F., 2002a. Three approaches to the development of high performance cultivars considering the different biological background of the starting material. *Acta Hort.*, 576: 129–137.

Pank, F. 2002b. Aims and results of current medicinal and aromatic plant breeding projects. *Z. Arznei- u. Gewuerzpfl.*, 7(S): 226–236.

Pank, F., 2007. Use of breeding to customise characteristics of medicinal and aromatic plants to postharvest processing requirements. *Stewart Postharvest Rev.*, 4: 1.

Pank, F., E. Herbst, and C. Franz, 1991. Richtlinien für den integrierten Anbau von Arznei- und Gewürzpflanzen. *Drogen Rep.*, 4(S): 45–64.

Pank, F., H. Krüger, and R. Quilitzsch, 1996. Selection of annual caraway (*Carum carvi* L. var. annuum hort.) on essential oil content and carvone in the maturity stage of milky-wax fruits. *Beitr. Züchtungsforsch*, 2: 195–198.

Pank, J., H. Krüger, and R. Quilitzsch, 2007. Results of a polycross-test with annual caraway (*Carum carvi L. var. annum* hort.). *Z. Arznei- u. Gewürzpfl*, 12.

Pennisi, E., 2007. Wanted: A DNA-barcode for plants. *Science*, 318: 190–191.

Pickard, W.F., 2008. Laticifers and secretory ducts: Two other tube systems in plants. *New Phytologist*, 177: 877–888.

Pino, J.A., M. Estarrón, and V. Fuentes, 1997. Essential oil of sage (*Salvia officinalis* L.) grown in Cuba. *J. Essent. Oil Res.*, 9: 221–222.

Putievsky, E., N. Dudai and U. Ravid, 1997. Cultivation, selection and conservation of oregano species in Israel. In: Padulosi, S. (ed.) Oregano. *Proc. of the IPGRI Internat. Workshop on Oregano*, 8–12 May 1996, CIHEAM Valenzano (Bari). IPGRI: Rome.

Putievsky, E., U. Ravid, and N. Dudai, 1986. The essential oil and yield components from various plant parts of *Salvia fruticosa*. *J. Nat. Prod.*, 49: 1015–1017.

Ratnasingham, S. and P.D.N. Hebert, 2007. The barcode of life data system (http://www.barcodinglife.org). *Mol. Ecol. Notes*, 7: 355–364.

Reales, A., D. Rivera, J.A. Palazón, and C. Obón, 2004. Numerical taxonomy study of *Salvia* sect. *Salvia* (Labiatae). *Bot. J. Linnean Soc.*, 145: 353–371.

Repčak, M., P. Cernaj, and V. Oravec, 1992. The stability of a high content of a-bisabolol in chamomile. *Acta Hort.*, 306: 324–326.

Repčak, M., J. Halasova, R. Hončariv, and D. Podhradsky, 1980. The content and composition of the essential oil in the course of anthodium development in wild chamomile (*Matricaria chamomilla* L.). *Biol. Plantarum*, 22: 183–191.

Rey, C., 1993. Selection of thyme (*Thymus vulgaris* L.). *Acta Hort.*, 344: 404–407.

Rey, C., 1994. Une variete du thym vulgaire "Varico". *Rev. Suisse Vitic. Arboric. Hortic.*, 26: 249–250.

Salamon, I., 2007. Effect of the internal and external factors on yield and qualitative–quantitative characteristics of chamomile essential oil. *Acta Hort.*, 749: 45–64.

Saukel, J. and R. Länger, 1992. Die *Achillea-millefolium*-Gruppe in Mitteleuropa. *Phyton*, 32: 47–78.

Schippmann, U., D. Leaman, and A.B. Cunningham, 2006. A comparison of cultivation and wild collection of medicinal and aromatic plants under sustainability aspects. In *Medicinal and Aromatic Plants*, R.J. Bogers, L.E. Craker, and D. Lange (eds.), pp. 75–95. Dordrecht, the Netherlands: Springer.

Schmiderer, C., P. Grassi, J. Novak, M. Weber, and C. Franz, 2008. Diversity of essential oil glands of clary sage (*Salvia sclarea* L., Lamiaceae). *Plant Biol.*, 10: 433–440.

Schröder, F.J., 1990. Untersuchungen über die Variabilität des ätherischen Öles in Einzelpflanzen verschiedener Populationen der echten Kamille, *Matricaria chamomilla* L. (syn. *Chamomilla recutita* L.). *Dissertation*, TU-München-Weihenstephan, Weihenstephan, Germany.

Senatore, F. and D. Rigano, 2001. Essential oil of two *Lippia spp.* (Verbenaceae) growing wild in Guatemala. *Flavour Fragr. J.*, 16: 169–171.

Skoula, M., J.E. Abbes, and C.B. Johnson, 2000. Genetic variation of volatiles and rosmarinic acid in populations of *Salvia fruticosa* Mill. Growing in crete. *Biochem. Syst. Ecol.*, 28: 551–561.

Skoula, M., I. El-Hilalo, and A. Makris, 1999. Evaluation of the genetic diversity of *Salvia fruticosa* Mill. clones using RAPD markers and comparison with the essential oil profiles. *Biochem. Syst. Ecol.*, 27: 559–568.

Skoula, M. and J.B. Harborne, 2002. The taxonomy and chemistry of *Origanum*. In *Oregano-The Genera Origanum and Lippia*, S.E. Kintzios (ed.), pp. 67–108. London, U.K.: Taylor & Francis.

Slavova, Y., F. Zayova, and S. Krastev, 2004. Polyploidization of lavender (*Lavandula vera*) in-vitro. *Bulgarian J. Agric. Sci.*, 10: 329–332.

Sobti, S.N., P. Pushpangadan, R.K. Thapa, S.G. Aggarwal, V.N. Vashist, and C.K. Atal, 1978. Chemical and genetic investigations in essential oils of some *Ocimum* species, their F1 hybrids and synthesized allopolyploids. *Lloydia*, 41: 50–55.

Srivastava, A.K., S.K. Srivastava, and K.V. Syamasundar, 2005. Bud and leaf essential oil composition of *Syzygium aromaticum* from India and Madagascar. *Flavour Fragr. J.*, 20: 51–53.

Stanley, P.C. and J.A. Steyermark, 1976. *Flora of Guatemala: Botany*. Chicago, IL: Field Museum of Natural History.

Steinlesberger, H., 2002. Investigations on progenies of crossing exeriments of Bulgarian and Austrian yarrows (*Achillea millefolium* agg., Compositae) with focus on the enantiomeric ratios of selected Monoterpenes. *Dissertation*, University of Veterinary Medicine, Wien, Austria.

Svoboda, K.P., T.G. Svoboda, and A.D. Syred, 2000. *Secretory Structures of Aromatic and Medicinal Plants*. Middle Travelly, U.K.: Microscopix Publications.

Taberlet, P., L. Gielly, G. Pautou, and J. Bouvet, 1991. Universal primers for amplification of three non-coding regions of chloroplast DNA. *Plant Mol. Biol.*, 17: 1105–1109.

Taviani, P., D. Rosellini, and F. Veronesi, 2002. Variation for Agronomic and Essential Oil traits among wild populations of *Chamomilla recutita* (L.) Rauschert from Central Italy. In *Breeding Research on Aromatic and Medicinal Plants*, C.B. Johnson and Ch. Franz (eds.), pp. 353–358. Binghampton, NY: Haworth Press.

Taylor, R., 1996. Tea tree—Boosting oil production. *Rural Res.*, 172: 17–18.

Tsiri, D., O. Kretsi, I.B. Chinou, and C.G. Spyropoulos, 2003. Composition of fruit volatiles and annual changes in the volatiles of leaves of *Eucalyptus camaldulensis* Dehn. growing in Greece. *Flavour Fragr. J.*, 18: 244–247.

UPOV, 1991. International Convention for the Protection of New Varieties of Plants, www.upov.int/upovlex/en/conventions/1991/content.html (accessed August 21, 2015).

Uribe-Hernández, C.J., J.B. Hurtado-Ramos, E.R. Olmedo-Arcega, and M.A. Martinez-Sosa, 1992. The essential oil of *Lippia graveolens* HBK from Jalsico, Mexico. *J. Essent. Oil Res.*, 4: 647–649.

Van Overwalle, G., 2006. Intellectual property protection for medicinal and aromatic plants. In *Medicinal and Aromatic Plants*, J. Bogers, L.E. Craker, and D. Lange (eds.), pp. 121–128. Dordrecht, the Netherlands: Springer.

Velasco-Neguerela, A., J. Pérez-Alonso, P.L. Pérez de Paz, C. García Vallejo, J. Palá-Paúl, and A. Inigo, 2002. Chemical composition of the essential oils from the roots, fruits, leaves and stems of *Pimpinella cumbrae* link growing in the Canary Islands (Spain). *Flavour Fragr. J.*, 17: 468–471.

Vernet, P., 1976. Analyse génétique et écologique de la variabilité de l'essence de *Thymus vulgaris* L. *(Labiée). PhD thesis*, University of Montpellier, Montpellier, France.

Vernin, G., C. Lageot, E.M. Gaydou, and C. Parkanyi, 2001. Analysis of the essential oil of *Lippia graveolens* HBK from El Salvador. *Flavour Fragr. J.*, 16: 219–226.

Vernin, G., J. Metzger, D. Fraisse, and D. Scharff, 1984. Analysis of basil oils by GC-MS data bank. *Perform. Flavour*, 9: 71–86.

Vetter, S. and C. Franz, 1996. Seed production in selfings of tetraploid *Achillea* species (Asteraceae). *Beitr. Züchtungsforsch*, 2: 124–126.

Vetter, S., C. Franz, S. Glasl, U. Kastner, J. Saukel, and J. Jurenitsch, 1997. Inheritance of sesquiterpene lactone types within the *Achillea millefolium complex* (Compositae). *Plant Breed.*, 116: 79–82.

Viljoen, A.M., A. Gono-Bwalya, G.P.P. Kamatao, K.H.C. Baser, and B. Demirci, 2006. The essential oil composition and chemotaxonomy of *Salvia stenophylla* and its Allies *S. repens* and *S. runcinata. J. Essent. Oil Res.*, 18: 37–45.

Weising, K., H. Nybom, K. Wolff, and G. Kahl, 2005. *DNA Fingerpinting in Plants.* Boca Raton, FL: Taylor & Francis.

Werker, E., 1993. Function of essential oil-secreting glandular hairs in aromatic plants of the Lamiaceae—A review. *Flavour Fragr. J.*, 8: 249–255.

Werker E., E. Putievski, U. Ravid, N. Dudai, and I. Katzir 1993 Glandular hairs and essential oil in developing leaves of Ocimum basilicum L. (Lamiaceae). *Ann Bot* 71:43–50.

Wijesekera R., A.L. Jajewardene, and L.S. Rajapakse, 1974. Composition of the essential oils from leaves, stem bark and root bark of two chemotypes of cinnamom. *J. Sci. Food Agric.*, 25: 1211–1218.

Wildung, M.R. and R.B. Croteau, 2005. Genetic engineering of peppermint for improved essential oil composition and yield. *Transgenic Res.*, 14: 365–372.

WHO, 2003. *Guidelines on Good Agricultural and Collection Practices (GACP) for Medicinal Plants.* Geneva, Switzerland: World Health Organization.

Wogiatzi, E., D. Tassiopoulos, and R. Marquard, 1999. Untersuchungen an Kamillen-Wildsammlungen aus Griechenland. In *Fachtagg. Arznei. u. Gewürzpfl. Gießen*, pp. 186–192. Gießen, Germany: Köhler.

Wolfe, A.D. and A. Liston, 1998. Contributions of PCR-based methods to plant systematics and evolutionary biology. In *Molecular Systematics of Plants II: DNA Sequencing*, D.E. Soltis, P.S. Soltis, and J. Doyle (eds.), pp. 43–86. Dordrecht, the Netherlands: Kluwer Academic Publishers.

Worku, T. and M. Bertoldi, 1996. Essential oils at different development stages of Ethiopian *Tagetes minuta* L. In *Essential Oils: Basic and Applied Research*, Ch. Franz, A. Máthé, and G. Buchbauer (eds.), pp. 339–341. Carol Stream, IL: Allured Publishing.

Zheljazkov, V.D., N. Kovatcheva, S. Stanev, and E. Zheljazkova, 1997. Effect of heavy metal polluted soils on some qualitative and quantitative characters of mint and cornmint. In *Essential Oils: Basic and Applied Research*, Ch. Franz, A. Máthé, and G. Buchbauer (eds.), pp. 128–131. Carol Stream, IL: Allured Publishing.

Zheljazkov, V.D. and N. Nielsen, 1996. Studies on the effect of heavy metals (Cd, Pb, Cu, Mn, Zn and Fe) upon the growth, productivity and quality of lavender (*Lavandula angustifolia* Mill.) production. *J. Essent. Oil Res.*, 8: 259–274.

4 Natural Variability of Essential Oil Components

Éva Németh-Zámbori

CONTENTS

4.1 MANIFESTATION OF VARIABILITY

It is a long known fact that qualitative and quantitative composition of genuine essential oils is not a standard one. In consequence of this, they possess different quality, value, and price on the market.

As a reflection of this practical experience, in several cases, different qualities are defined for essential oils of the same species. In the International Organization for Standardization (ISO) standard series (ISO TC/54), numerous essential oils are listed in at least two (e.g., lemongrass, thyme)—but in some cases even in four (e.g., petitgrain, spearmint)—different qualities depending on the geographical source, plant organ, or main component of the oil (ISO, 2013). However, numerous other factors might contribute to the different qualities, such as variety, environment, agricultural methods, or extraction technology. In practice, the same species might be utilized for different applications based on the variable composition of its oil, like the thyme-odor type, lavender-odor type, and rose-odor type individuals of *Thymus longicaulis* subsp. *longicaulis* (Baser et al., 1993).

In several cases, the real sources of variability are hard to determine. However, for standardization of any product, it is of primary importance that the background of variability and the factors, which influence the composition of the essential oils, are detected and can be managed and controlled.

In the scientific literature, reports on variability of essential oil components are very frequently published. According to a survey on articles in the last volumes (2010–2018) of *Journal of Essential Oil Research* (Taylor & Francis Group), it can be established that more than one-third of them is evaluating biological variability at specific or intraspecific taxonomic levels or chemosyndromes due to developmental or morphological differences (Figure 4.1).

The chemical variability of the essential oils gained from different plant species varies on a large scale. Tétényi (1975) mentioned already fifty years ago that 36 families, 121 genera, 360 species are polymorphs for essential oils. This number must have increased enormously since that time because of intensive research and highly developed analytical techniques.

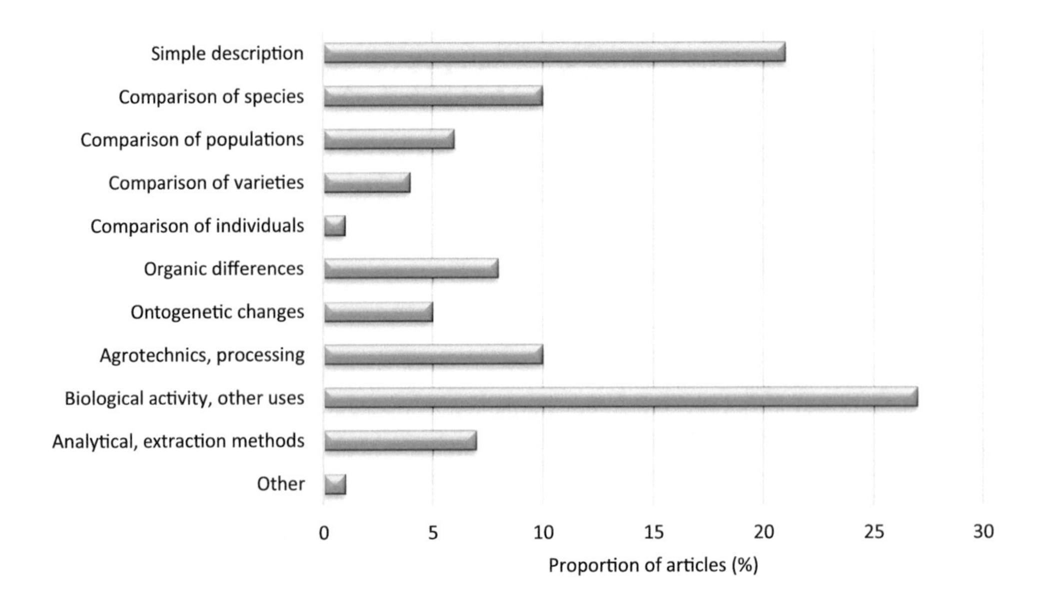

FIGURE 4.1 Distribution of topics of publications in *Journal of Essential Oil Research* between 2010 and 2018.

The backgrounds of the chemical variability of essential oil composition are usually grouped as abiotic and biotic factors. Abiotic influencing factors include the effects of the environment (exposure, soil, light intensity and length of illumination, wind, absolute and marginal temperatures, water supply as total, and frequency of precipitation) but also those in consequence of human activities (agrotechnical methods, extraction, processing, and storage). Several chapters of this book deal with these factors in detail.

The biotic/biological factors are the main topic of this chapter. Natural variability may be defined as the phenomenon when a diverse quality of essential oils is detectable as result of genetic-biological differences of the source plants. "Natural variability" is, however, a rather complex issue having many aspects, as we can see below. In this context, we deal with the essential oil spectrum, quantitative and qualitative composition of the oils, and do not discuss other chemical or physical properties of the oils. We also have to declare that although based on the accepted definition, essential oils are products produced from the plant by physical means like distillation or pressing; in this chapter, in some cases, we refer to "essential oil" also as the mixture of volatile compounds in the plant *in vivo*.

4.2 VARIABILITY AT DIFFERENT TAXONOMIC LEVELS

4.2.1 Species

Variability in the composition of essential oils has been most frequently discussed at the level of plant species and has the highest relevance from practical points of view.

A significant variation in qualitative and quantitative composition might have considerable influence on the recognition and the market value both of the drug and the essential oil itself. Besides, fluctuations in the composition of the essential oil might have significant effects on the therapeutic efficacy or sensory value of the product. Limonene seems to have a strong influence on the allelopathic property of *Tagetes minuta* (Scrivanti et al., 2003). The characteristic antioxidative property of thyme (*Thymus vulgaris*) oil is by 2.0–2.6 times higher in chemotypes containing phenolic compounds as main components (Chizzola et al., 2008). In somc phytotherapeutic preparations of chamomile, the antiphlogistic and spasmolytic effect seems to be in closest connection with the content of (–)-α-bisabolol (Schilcher, 2004). Recently, it has also been demonstrated that individual compounds may be identified as putative biomarkers to the active oils (Maree et al., 2014; Ayouniet al., 2016). Anti-quorum-sensing activity of certain essential oils may be attributed to the presence of special compounds like eugenol, geraniol geranial, menthol, and pulegone where, nevertheless, synergistic effects are of high importance (Mokhetho et al., 2018). On the other hand, adverse effects may be caused by the presence of single compounds like the carcinogen effect of *cis*-isoasarone in the essential oil of calamus (*Acorus calamus*), (Blaschek et al., 1998) or the high concentrations of thujone in wormwood (*Artemisia absinthium*) or sage (*Salvia officinalis*) oils (Lachenmeier et al., 2006).

Not each species exhibits a similar amount of variability. A huge amount of research data accumulated in the last decades proving that the incidence of diversity is one of the characteristic features of the plant species.

The well-known caraway (*Carum carvi*) seems to be an essential oil–bearing species of relatively low variability concerning the oil constituents. Nowadays, besides being a popular spice, it is a source of essential oil of excellent antimicrobial properties, but the spasmolytic and cholagogue effects justify its use in phytotherapy, too. In the oil of caraway, the ratio of the main components *S*(+) carvone and *R*(+) limonene in the oil is above 90%, most frequently above 95% (Table 4.1). Variability is manifested in most cases only in their proportions compared to each other. Minor constituents have been rarely identified and mentioned. The majority of constituents are all monoterpenes, besides the sesquiterpene β-caryophyllene and some phenolic and aliphatic compounds.

Biological variability of the oil composition seems to be more pronounced if comparing the two varieties (*Carum carvi* var. *annuum* and var. *biennis*) of caraway. In general, biennial varieties are believed to accumulate higher concentrations of total volatiles and carvone (Table 4.1). Bouwmeester and Kuijpers (1993) concluded that the restricted potential of carvone accumulation in annual varieties is the consequence of limited availability of assimilates. Indeed, the abundance of nutrients available from the more robust root system of biennial varieties might play a very important role in accumulation of secondary compounds, as the examples of other species also show (Bodor et al., 2006, 2009).

TABLE 4.1
Variability of the Main Components Carvone and Limonene in Biennial and Annual Accessions of Caraway

	Biennial		Annual	
Source of Data	**Carvone**	**Limonene**	**Carvone**	**Limonene**
Aćimović et al. (2014)	–	–	27–44	54–70
Argañosa et al. (1998)	54–57	43–45	46–50	49–53
Embong et al. (1977)	39–46	43–49	–	–
Fleischer and Fleischer (1988)	54–68	30–44	–	–
Forwick-Kreutzer et al. (2003)	52–72	–	–	–
Galambosi and Peura (1996)	47–49	39–52	–	–
Laribi et al. (2010)	–	–	76–80	13–20
Pank et al. (2008)	–	–	50–53	45–48
Puschmann et al. (1992)	47–54	–	45–52	–
Putievsky et al. (1994)	53–59	38–44	47–62	3–46
Raal et al. (2012)	44–95[a]	2–50	–	–
Sedláková et al. (2003)	72–81	18–27	–	–
Solberg et al. (2016)	14–15	69–71	–	–
Zámboriné (2005)	51–60	38–44	50–56	43–49

[a] Annual accessions might also be included.

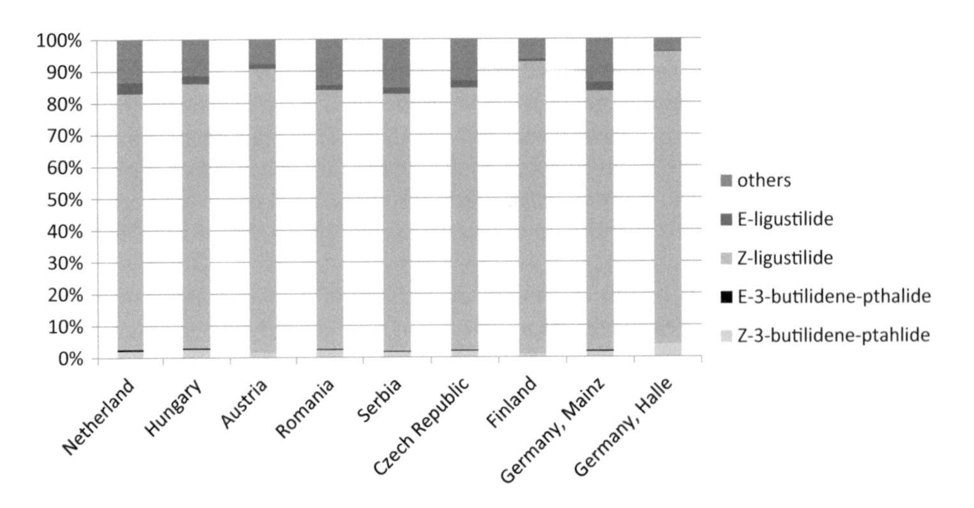

FIGURE 4.2 Distribution of characteristic phthalide components (in %) in the root oil of different European accessions of lovage (*Levisticum officinale*). (Adapted from Németh, É. et al., Unpublished.)

Based on the majority of available references, the carvone content of biennial accessions is regularly higher than the content of the annual plants. This fact may explain why biennial caraway is still in many countries in cultivation although production of the annual variety has an economic advantage based on higher seed yields and more advantageous crop rotation.

In Apiaceae species, a relatively low biological variability in essential oil composition is usual. Another example may be lovage (*Levisticum officinale*), whose aromatic volatile (essential oil) and non-volatile (mainly coumarin-type) compounds justify the application of each part of the plant as a popular spice. However, the most valuable organ is the root. The main components of the root essential oil are alkyl-phthalide type compounds, among which the most abundant ones are usually Z-ligustilide and butylidene-phthalide (Szebeni et al., 1992; Venskutonis, 1995; Novák, 2006). Only in exceptional cases have other compounds been detected as major ones, like 49% phellandrene (Scottish accession) or 26% terpinyl acetate (Dutch accession) in the investigations of Raal et al. (2008). Our own investigations on ten accessions of lovage originating from different European countries ascertained that the compositional variability is low (unpublished). The phthalides are the main components of the distilled oil practically in each accession (Figure 4.2). The presence of two isomers, E and Z, makes the pictures somewhat more diverse; however, their ratios are not significantly different in either of the accessions. In each case, the Z isomer is in multiple concentrations present more than the other one.

The seeds of the investigated accessions have been obtained from different countries and regions but—as in many cases—the real genetic origin is uncertain. Therefore, a common basic source cannot be excluded, either. However, even in this case, it might mean that lovage has a very narrow gene pool and maybe therefore possess a low chemical variability. The connection between the restricted natural distribution and small spectral variance of the oil might support the hypothesis on the development of polychemism as a tool in geographical distribution and ecological adaptation (see Section 4.5).

The Mediterranean species hyssop (*Hyssopus officinalis*) belongs to the Lamiaceae family. It is used for its spicy essential oil in the food industry and also as a strong antimicrobial agent.

Monoterpenes, which are present as the main compounds in the oil of this species (pinocamphone, isopinocamphone), are relatively seldom detected in higher quantities in essential oils of other species. Although, as the highest number of 44 components were detected in hyssop oil (Chalchat et al., 2001), the major ones are relatively uniform and found almost in each examined accession (Table 4.2). Besides the mentioned compounds, the majority of further ones are also monoterpenes,

TABLE 4.2
Main Components in the Essential Oil of Hyssop (*Hyssopus Officinalis*) According to Different References

Main Compounds (In the Row of Their Abundance)	Reference
Pinocamphone, isopinocamphone, β-pinene	Aiello et al. (2001)
Pinocarvone, isopinocamphone, β-pinene	Bernotienė and Butkienė (2010)
Pinocamphone, isopinocamphone	Chalchat et al. (2001)
Isopinocamphone, β-pinene	Danila et al. (2012)
Pinocamphone, isopinocamphone, germacrene D, pinocarvone	Galambosi et al. (1993)
Pinocamphone, isopinocamphone, β-pinene	Fraternale et al. (2004)
Terpineol, bornyl acetate, linalool	Hodzsimatov and Ramazanova (1974)
Isopinocamphone, β-pinene, pinocamphone	Joulain and Ragault (1976)
Isopinocamphone, β-pinene, pinocarvone	Kizil et al. (2010)
Isopinocamphone, pinocamphone, β-pinene	Koller and Range (1997)
Pinocamphone, β-pinene	Lawrence (1979)
Pinocamphone, isopinocamphone, β-pinene, pinocarvone	Lawrence (1992)
Isopinocamphone, pinocamphone	Mitič and Đorđević (2000)
Isopinocamphone, pinocamphone, β-pinene	Németh-Zámbori et al. (2017)
Isopinocamphone, myrtenol, β-pinene, 1,8-cineole, methyl-eugenol, limonene	Piccaglia et al. (1999)
Pinocamphone, camphor, β-pinene	Schulz and Stahl-Biskup (1991)
Isopinocamphone, 1,8-cineole, β-pinene	Tsankova et al. (1993)
1,8-Cineole, β-pinene	Vallejo et al. (1995)

products of related biosynthetic pathways (β-pinene, pinocarvone, and myrtenol). In general, it can be observed that besides some mentioned main compounds, all the others are present only in minimal concentrations (Németh-Zámbori et al., 2017). Thus, the biological variability of the herb oil of hyssop is relatively low. Only samples of the subspecies *aristatus* (Godr.) Briq., collected from three populations of Appennines, showed a different character with higher amounts of myrtenol (up to 32%), methyl-eugenol (up to 44%) and limonene (up to 15%); however, the characteristic pinane-type compounds have also been found here at different quantities (Piccaglia et al., 1999).

Summarizing the above-mentioned species, it seems to be clear that the variation in the oil composition of the above-mentioned species is principally a quantitative one. The spectrum seems to be relatively constant; changes are detectable basically in the accumulation proportions of the individual components.

On the other side, a great number of plant species can be characterized by high intraspecific variability concerning essential oil composition. In these oils, both qualitative and quantitative variations are present.

One of the most comprehensively studied genera from this respect is the genus *Achillea*. For the majority of yarrow species, a wide variability in oil composition has been detected. Based on a comprehensive literature search, in most of the species, one to three compounds have been identified as main components (Kindlovits and Németh, 2012). The evaluation is, however, not a simple one because, in most cases, the different chemical races had been detected and mentioned by different authors independently from each other. Therefore, the comparison of data is always a hard task taking into account the possible role of other influencing factors besides the genetic background.

Chamazulene is currently the most important component of the distilled oil of yarrow. In general, the proazulene accumulation potential of *A. collina* (4*n*) and its relatives *A. asplenifolia* and *A. roseo-alba* (2*n*) seems to be widely accepted (Rauchensteiner et al., 2002; Ma et al., 2010). However, even here some contradictory results can be found in the literature. In plant samples from Yugoslavia, Chalchat et al. (2000) could not identify chamazulene, but 1,8-cineole, chrysanthenon, and camphor are mentioned as main components of *A. collina*. Todorova et al. (2007) presented three chemotypes of this species (azulene-rich, azulene-poor, sesquiterpene-free types) based on the analysis of samples from six different populations in Bulgaria.

According to the literature references, the largest intraspecific variability could be devoted to *Achillea millefolium*. For this species, 19 different chemical compounds have already been mentioned in the essential oil as main components (Németh, 2005; Pecetti et al., 2012). Comparing the chamazulene content of the distilled oil, values between 0% and 85% have been detected by different authors (e.g., Figueiredo et al., 1992; Michler et al., 1992; Bélanger és Dextraze, 1993; Orav et al., 2001; Németh et al., 2007; Chou et al., 2013; etc.).

Taking into consideration only these data, we might assume that *A. millefolium* is an extremely variable species concerning its essential oil spectrum with numerous intraspecific chemical varieties. However, in this case, I would be more cautious because, in numerous references, the proper identification of the taxon is not obvious or botanical characterization is missing. The genus *Achillea* is a very complex one with species in a polyploid row, containing intrageneric sections and groups, many spontaneous hybrids, phenocopies, and aneuploid forms. Contradictory results may originate from a false definition of taxa belonging to the *A. millefolium* section only by morphological features or—on the other side—only by chromosome numbers. Similarly, investigation of non-representative samples like commercial samples, individuals of non-stable spontaneous hybrid or aneuploid character may lead to invalid information. Detailed morphological and cytological identification of any taxon belonging to the section *A. millefolium* seems to be a prerequisite for reliable chemical characterization; otherwise, comparison of the data is not really possible. In the last decade, molecular markers have also been developed for identification of certain taxa (e.g., Ma et al., 2010).

According to the above-mentioned information, unequivocal definition of the accessions showing diverse essential oil composition in the section *A. millefolium* as chemotypes could be more than questionable, and only references based on comprehensive determination of the investigated plant material can be accepted.

Nevertheless, the high variability concerning the composition of the essential oil of yarrow species is without doubt. Three chemotypes of *A. biebersteinii* were described from indigenous populations in Turkey based on the major compounds 1.8-cineole, p-cymene, camphor, piperitone and ascaridol (Toncer et al., 2010). According to Muselli et al. (2009) geographically distinct populations of *A. ligustica* also show chemically distinct characteristics. Corsican samples contain camphor (21%) and santolina alcohol (15%) as main compounds, Sardinian samples have *trans*-sabinyl acetate (18%) and *trans*-sabinol (15%), and those from Sicily can be characterized by high terpinen-4-ol (19%) and carvone (9%) accumulation. Similar results on other species are numerous.

A related species, wormwood (*Artemisia absinthium*), gained an adverse "reputation" due its thujone content and mutual side effects associated with absinthism (Lachenmeier et al., 2006). It is widely distributed in Europe and introduced also in other continents. The composition of the essential oil has been studied by several authors and highlighted that large amounts of thujones are representative only for one of the many chemotypes of *A. absinthium* while other mono- or sometimes sesquiterpenes are more frequently present as major components in the herb oil (Table 4.3). According to the investigation in the last decade, it is obvious that thujone may be not rarely even absent from the oil. Additionally, although wormwood is known among the few proazulen-containing species (Wichtl, 1997), chamazulene is only rarely and in low proportions present in the essential oil of the investigated accessions.

The real source of the polychemism in this species seems to be till now unknown. In our recent study based on the data of 12 different accessions, a connection between chemotype and habitat could not be justified in most cases. The majority of the accessions were heterogenous concerning appearance of chemotypes. The occurrence of thujone-type individuals was rather frequent in

TABLE 4.3
Main Components of Essential Oils from *Artemisia absinthium* Samples of Different Origin

Reference	Sample Origin	Determined Main Components/Chemotypes (Main Compounds in Area %)
Altunkaya et al. (2014)	Turkey	Myrcene (44%)
Arino et al. (1999)	Spain	*cis*-Chrysanthenyl acetate (31%–44%) + Cis-epoxyocimene (34%–42%)
Bagci et al. (2010)	Turkey	Chamazulene (29%)
Basta et al. (2007)	Greece	Caryophyllene-oxide (25%)
Chialva et al. (1983)	Italy	*cis*-Epoxyocimene (30%–54%) or β-thujone (41%)
	Romania	β-thujone (15%)
	France	Sabinyl acetate (32%) or chrysanthenyl acetate (42%)
	Siberia	Sabinyl acetate (85%)
Derwich et al. (2009)	Morocco	α-Thujone (40%)
Huong et al. (2018)	Spain	*cis*-Epoxyocimene (47%–76%)
	Belgium	α-Thujone (1%–52%) or β-thujone (25%–89%)
	Germany	β-Thujone (2%–85%) or *trans*-sabinyl acetate (1%–36%) or myrcene (4%–68%)
	Norway	*trans*-Sabinyl acetate (20%–78%)
	Hungary	Sabinene (2%–34%) + ß-myrcene (2%–42%)
	England	*cis*-Epoxyocimene (35%–65%) or isocitral (10%–49%) or sabinene (1%–38%)
Judzentiene and Budiene (2010)	Lithuania	*trans*-Sabinyl acetate (22%–51%) or α- and β-thujones (18%–72%)
Juteau et al. (2003)	Croatia	α-Thujone (49%) or *cis*-epoxyocimene (31%)
	France	*cis*-Chrysanthenyl acetate (34%) or *cis*-epoxyocimene (50%)
Llorens-Molina and Vacas (2015)	Spain	α-Fenchene (24%) or bornyl acetate (21%) or Myrcene (29%)
Lopes-Lutz et al. (2008)	Canada	α-Thujone (10%) + myrcene (10%) + Sabinyl acetate (26%)
Morteza-Semnani and Akbarzadeh (2005)	Iran	α-Thujone (70%)
Msaada et al. (2015)	Tunisia	Chamazulene (40%)
Mucciarelli et al. (1995)	Italy	*cis*-Epoxyocimene (25%) + *trans*-chrysanthenyl acetate (22%) + camphor (17%)
Nezhadali and Parsa (2010)	Iran	p-Cymene (10%) + camphor (15%)
Nin et al. (1995)	Italy	Terpinene-4-ol (29%)
	US	α-Thujone (70%)
Orav et al. (2006)	Greece	β-Thujone (38%)
	Estonia	β-Thujone (65%) or myrcene (30%) or Sabinyl acetate (71%)
	Russia	*cis*-Epoxyocimene (21%)
	Moldova	Myrcene (60%)
	Siberia	Sabinyl acetate (31%)
	France	Sabinyl acetate (85%)
	Armenia	Sabinyl acetate (34%)
Pino et al. (1997)	Cuba	Bornyl acetate (24%)
Rezaeinodehi and Khangholi (2008)	Iran	β-pinene (24%)
Sharopov et al. (2012)	Tajikistan	Myrcene (23%) + *cis*-chrysanthenyl acetate (18%)
Simonnet et al. (2012)	Switzerland	*cis*-Epoxyocimene (30%–40%)
Tucker et al. (1993)	US	α-Thujone (33%)

Note: +, indicates mixed chemotypes with more main components; or, indicates different chemotypes in the same accession.

TABLE 4.4
Chemotypes of *Tanacetum vulgare* According Selected References

Reference	Country	Chemotypes (Main Components)
Collin et al. (1993)	Canada	Camphor-cineole-borneol, β-thujone, chrysanthenone, dihydrocarvone
de Pooter et al. (1989)	Belgium	β-Thujone, chrysanthenyl acetate, camphor + thujone
Dragland et al. (2005)	Norway	Thujone, camphor, borneol, bornyl acetate, chrysanthenol, chrysanthenyl acetate, 1,8-cineole, α-terpineol
Forsen and Schantz (1971)	Finland	Chrysanthenyl acetate, isopinocamphone, not identified sesquiterpene
Hendrics et al. (1990)	Nether-lands	Artemisia ketone, chrysanthenol + chrysanthenyl acetate, lyratol + lyratyl acetate, β-thujone
Héthelyi et al. (1991)	Hungary	Yomogi alcohol, artemisia alcohol, davanone, lyratol + lyratyl acetate, chrysanthenol, carveol, carvone, dihydrocarvone, terpinene-4-ol, γ-campholenol, myrtenol, β-terpineol, 4-thujene-2-α-yl acetate, carvyl acetate, β-cubebene, juniper camphor, thymol, β-terpinyl acetate, linalool
Holopainen et al. (1987)	Finland	Sabinene, germacrene D
Mockute and Judzetiene (2004)	Lithuania	1,8-Cineole, artemisia ketone, camphor, α-thujone
Nano et al. (1979)	Italy	Chrysanthenyl acetate
Rohloff et al. (2004)	Norway	β-Thujone, camphor, artemisia ketone, umbellulone, chrysanthenyl acetate, chrysanthenone, chrysanthenol, 1,8-cineole
Sorsa et al. (1968)	Finland	α-pinene + tricyclene, β-pinene + sabinene, 1,8-cineole, γ-terpinene, artemisia ketone, thujone, camphor, umbellulone, borneol, humulenol

European samples, except a single one the accessions were not homogenous from this respect (Huong et al., 2018). No other works are suggesting any data on chemotype distribution except Chialva et al. (1983). Unfortunately, the plant material investigated by them included different plant parts, harvest years, samples distilled both fresh and dried, and originated either from natural habitats or from market. Under such conditions, the results cannot enable reliable conclusions, especially not in chemotaxonomic respect, although the title of the paper is suggesting this.

One of the earliest and most deeply studied plant species with respect to essential oil polymorphism has been tansy (*Tanacetum vulgare*). Formerly—due to the lack of reliable chemical-analytical investigations and systematic evaluation—it has been presented as a characteristic thujone containing species (Gildemeister and Hoffmann, 1961). Although it is true that this is the main component most frequently present in the essential oil, until today, the number of the detected main compounds in different chemotypes is near to 50. Some of these are summarized in Table 4.4. The dominant compounds are in most cases monoterpenes, but in some samples also sesquiterpene ones such as humulenole, germacrene D, or davanone were detected.

The spectrum of these monoterpenes is very wide. There are representatives of each types of the basic monoterpene skeletons except the carane group. Even if the main component itself is usually not enough for evaluation of the characteristics of the oil, tansy is a good example to illustrate the fact that the main compounds of different chemotypes may not necessarily belong to the same skeleton. It also means that they are not always products of closely related biosynthetic pathways, which might reflect a really heterogeneous genetic structure.

Large intraspecific chemical variability is by no means restricted to Asteraceae species. The genus *Thymus* comprises many species highly polymorphic for essential oil composition. Different chemotypes have been reported in at least 85 cases, mainly from the species *T. aestivus, T. herba-barona, T. hyemalis, T. mastichina, T. nitens, T. vulgaris*, and *T. zygis* (Stahl-Biskup and Sáez, 2003). For most of them, three to six intraspecific chemotypes have already been described.

Different chemotypes are often grouped as ones containing phenolic compounds and chemotypes with non-phenolic ones (Baser et al., 1993).

Common thyme, *Thymus glabrescens* Willd., is a procumbent dwarf shrub, indigenous on sunny hillsides of southeastern and central Europe. Recently, in Hungary, eight populations at different localities have been investigated and new chemotypes identified (Pluhár et al., 2008). Four chemotypes contained thymol as the main compound in the oil (15%–34%), but the second and third main compound has been different in each of them. One chemotype contained only monoterpenes as major constituents (*p*-cymene 45%, geraniol 14%, and linalyl acetate 10%) while two other ones only sesquiterpenes (germacrene D 55%, β-caryophyllene 15%, α-cubebene 51%). 1,8-cineole and thymyl acetate/carvacrol/*p*-cymene chemovarieties were described in Croatia; a terpinyl acetate chemotype was reported in Bosnia; and linalool/thymol/α-terpinyl acetate, geraniol, citronellol, and carvacrol chemovarieties were mentioned in Bulgaria (Pluhár et al., 2008). It can be established that in this species—in contrary to the formerly mentioned ones—the main compounds could be relatively well grouped based on their chemical constitution: acyclic monoterpenes, menthane skeleton group, and sesquiterpene ones. This led us to conclude that intraspecific differences in this species are primarily the results of diversity in biosynthesis at the level of terpene synthases and not in the following transformations.

Within a genus, different species may exhibit different levels of intraspecific chemical variability. The genus *Mentha* is a good example for this. Besides the best known species, *M. piperita*, there is only a small variability also in *M. pulegium*. While the first one is characterized always by the presence of menthol, the last one almost always contains pulegone as the main compound or one of the main compounds (Baser et al., 2012; Teixeira et al., 2012). The presence of piperitenon oxide in high percentages has been reported in each of the published studies for the oil composition of *M. suaveolens* (Baser et al., 1999, 2012; Božović et al., 2015). Similarly, *M. aquatica* seems to be a species of low essential oil variability. According to the available data, menthofuran has been detected in the huge majority of the investigated samples (Baser et al., 2012; Andro et al., 2013). On the other side, numerous species of the genus are really polymorphic concerning their volatile compounds. *M. longifolia, M. spicata*, *M. arvensis*, and also natural hybrids like *M. x dumetorum* exhibit a wide spectrum of essential oil compounds, and numerous chemotypes have been reported (Lawrence, 2007; Baser et al., 2012; Llorens-Molina et al., 2017).

4.2.2 Populations

During evaluation of the intraspecific essential oil variability of any species, one has to be aware of the fact that in many cases, the investigated plant material is far from a homogenous one. Although representative sampling is a prerequisite for these studies, unfortunately, this is only rarely the fact. It is still quite frequently not taken into account that different populations might reveal significant variability due to the individual differences of single plants. Description of differences among populations without referring to the individual variability within populations may lead to significant misinterpretation of data.

This is especially relevant for wild growing plants because natural populations are often heterogenous in many respects. A special difficulty is that the size of this diversity is not known either. Therefore, inadequate number of sampled individuals or bulked samples may obscure the real variability that can be demonstrated by several examples.

In natural stands of *A. crithmifolia*, considerable variability has been detected, and the level of several essential oil constituents varied on a large scale. Camphor (camphor above 50% in the oil), 1,8-cineole (this compound above 30% of the oil), and mixed-type individuals have been detected (Németh et al., 2000). It was found that the abundance of plant individuals belonging to the different chemotypes varied according to habitat. In this case, bulked samples could not tell us details about the real diversity of the stands, but individual sampling could reveal the three chemotypes present in these populations.

In a similar trial in Bulgaria, analyzing samples from seven habitats, besides camphorous- and 1,8-cineole-type individuals, an artemisia alcohol chemotype (with 24%–46% artemisia alcohol in the oil) has been described (Konakchiev and Vitkova, 2004). However, in this examination, the populations could not be characterized. and the abundance of the three chemotypes has not been described either, as only a single individual has been sampled from each habitat! Therefore, the results are only useful to provide data about the existing chemical diversity of the species but not about their frequency and distribution.

Unfortunately, some other references are even more questionable if they could give appropriate information on natural variability of this species. Bulked plant material from a Serbian population "near Niš" was characterized by high (19%) proportions of *trans*-chrysanthenyl acetate (Palić et al., 2003), while another in Greece "from Pilio mountain at the altitude of 700 m" by larger levels of α-terpineol (Tzakou et al., 1993). These data do not tell us anything about the quality of the oil of single individuals where these ratios might be much lower or higher, respectively. A single sample from a population might lead to false interpretation not only from theoretical point of view but also about practical/pharmaceutical value of these stands because the representativeness is at least questionable.

In the same genus, significant amounts of chamazulene are generally present in the essential oil of *A. collina*. Rarely can we find, however, any reference about the individual distribution of this compound inside a plant population, although collection of bulk samples may again lead to false consequences and cannot represent a basis for standard drug quality.

Table 4.5 shows that among 23 Hungarian *A. collina* populations, differences of mean values varied from 33.2% till 67.1% while the standard deviations show twelvefold differences! A population with 1.8% standard deviation ("Diósd") means, in the practice, a strongly homogenous stand where the high level of chamazulene manifests itself in almost each individual. On the other side, a population like "Alsótold," of similar mean value but with a much higher standard deviation, can be evaluated as an unstable one, less suiting even for commercial purposes. A more detailed investigation afterward revealed that the mentioned results could be traced back to individual differences. The plants in the examined wild populations of *A. collina* could be sorted in four groups based on the characteristic spectrum of the essential oil. Individuals, accumulating chamazulene in high proportions as the absolutely main component of the oil are clearly different from the ones having both β-caryophyllene and chamazulene in higher levels. Individuals of only low levels of chamazulene and having other compounds as major ones form a distinct group while the plants with essential oil lacking chamazulene are sorted in the fourth group. The evaluated mean values of the populations obviously reflect the proportion of these chemotypes (Németh et al., 2007).

As discussed above, more than twenty chemotypes of wormwood (*Artemisia absinthium*) have been described until recently in the literature. Checking the methods of the published papers, it can, however, be established that in the huge majority of the cited references, the method of sampling has only be described as follows: "aerial parts/leaves/plants were collected …" without providing any information about the number of individuals, replications, or the amount of the sample. On the other side, a paper mentioned "four different plants" which have been harvested, but in this case the low number of individual plants is surely not able to represent either the population or its variability. Intrapopulation variability has been studied only in exceptional cases. Llorens-Molina et al. (2016) presented the common occurrence of two well-distinguishable chemotypes (*cis*-beta-epoxyocimene above 70% of essential oil and *cis*-beta-epoxyocimene at 60%–70%, and with *cis*-chrysanthenyl acetate at 10%–20%) in a wild habitat in Spain. The two chemically—and presumably also genetically—distinct individuals are distinguishable only by EO analysis and do not show any external marker traits. The authors called the attention to the importance of individual monitoring during examination oil composition because of the obvious differences among plants of the same population. Similarly, the detailed study Huong et al. (2018) on 120 individual samples

TABLE 4.5
Average Values and Standard Deviations of the Essential Oil Content and its Chamazulene Level in 23 Spontaneous Hungarian *Achillea collina* Populations

	Essential Oil Content Of Flowers (% d.w.)		Chamazulene Content in Flower Oil (ess. Oil %)	
Population (origin)	**Mean**	**Std. Dv.**	**Mean**	**Std. Dv.**
Alsótold	0.55	0.49	53.6	25.3
Apc	0.27	0.06	45.7	15.4
Aszód	0.48	0.22	63.1	12.5
Balatonakali	0.36	0.11	67.1	4.2
Balatonudvari	0.29	0.07	61.0	7.2
Bokor	0.35	0.13	64.5	5.8
Csepreg	0.20	0.08	40.0	18.3
Csillebérc	0.33	0.22	60.3	7.2
Diósd	0.42	0.11	61.0	1.8
Jobbágyi	0.33	0.18	33.7	24.6
Kevélynyereg	0.30	0.13	60.7	7.6
Lupasziget	0.31	0.09	52.5	4.8
Makkoshetye	0.18	0.08	40.3	28.1
Mezőnyárád	0.27	0.14	33.2	18.9
Mikóújfalu	0.44	0.07	57.7	8.9
Nagymaros	0.71	0.36	47.8	22.9
Nagymaros	0.53	0.15	64.2	5.0
Oroszlány	0.33	0.11	60.7	2.8
Solymár	0.37	0.08	58.3	5.1
Sopron	0.37	0.29	31.3	26.9
Szigliget	0.35	0.19	47.5	19.2
Tiszavasvári	0.65	0.58	30.5	30.4
Zenta	0.47	0.12	44.7	22.4
Mean	0.39	0.24	51.3	19.3
SD value	0.334	–	25.4	–
P level	0.005	–	0.000	–

Source: Modified from Németh, É. et al. 2007. *J. Herbs, Spices Med. Plants*, 13: 57–69.

of 12 accessions provided a well-established base for chemotype definition and characterization of populations (Table 4.6). In this work, it was also demonstrated that differences among individuals manifest themselves not only in the main components but also in the total spectrum. The varying ratios of mono- and sesquiterpene compounds to each other demonstrate it very well: individuals with 89% monoterpenes and 11% sesquiterpenes represent one marginal value while, on the other side, an individual with 10% monoterpenes and 90% sesquiterpenes in the essential oil express in GC peak area percentages is the contrast.

Sampling of a population of *Thymus longicaulis* subsp. *longicaulis* in Turkey resulted in distinguishing three different chemotypes: thymol type, geraniol type, and α-terpinyl acetate types. It was shown that individuals belonging to the different chemotypes can be found near to each other even on a one-square-meter area (Baser et al., 1993).

The above examples represent quite well that in a chemically diverse species in consequence of the large plant-to-plant variability, the populations may be heterogeneous too.

TABLE 4.6
Distribution of the Identified Chemotypes in Twelve Wormwood (*Artemisia absinthium*) Accessions

Chemotype	Proportion in the Accessions (%)											
	Bel	**Eng**	**Ger0**	**Ger1**	**Ger2**	**Hum**	**HuW1**	**HuW2**	**HuW3**	**HuW4**	**Nor**	**Spa**
Pure Chemotypes (Main Component >30% of Total Gc Area)												
Thujone	**100**			30						40	20	
cis-Epoxyocimene		20					10	30				**100**
trans-Sabinyl acetate			10	10	**100**				10		40	
Sabinene		10										
ß-Myrcene				20		20	10	10	30			
Linalool										10		
cis-Chrysanthenol						10						
(*Z*)-*Iso*-citral		10		10								
Selin-11-en-4-α-ol								10				
(*E*)-Nuciferol isobutyrate			10									
Mixed Chemotypes (Two or Three Components >30% of Total Oil)												
Thujone + *cis*-epoxyocimene											10	
Thujone + *cis*-epoxyocimene + *trans*-sabinyl acetate											20	
Thujone + *trans*-sabinyl acetate											10	
Sabinene + ß-Myrcene		20	30				40		10			
ß-Myrcene + β-caryophyllene				10		10	10	20				
ß-Myrcene + (*Z*)-nuciferol isobutyrate			10									
Linalool + β-caryophyllene				10				10				
Linalool + (*Z*)-nuciferol isobutyrate										30		
β-Caryophyllene + selin-11-en-4-α-ol						20		20				
Selin-11-en-4-α-ol + (*Z*)-*iso*-citral												
Selin-11-en-4-α-ol + (*Z*)-nuciferol isobutyrate						10			10			
Other composition	0	40	40	10	0	30	30	0	40	20	0	0

Source: Modified from Huong et al. 2018. *J. Essent. Oil Res.*, 30: 421–430.

4.3 CONNECTIONS OF CHEMICAL DIVERSITY WITH OTHER PLANT CHARACTERISTICS

4.3.1 Propagation and Genetics

The homogeneity or variability of a population often stays in connection with the usual propagation method of the species. Phenotypic manifestation of diverse genetic background and appearance of different chemotypes in a plant stand can be supported by sexual propagation and cross pollination. To the contrary, vegetative propagation or autogamy enhances uniformity of the population.

Vetter and Franz (1998) proved the large degree of self-incompatibility in five *Achillea* species (*A. ceretanica, A. collina, A. pratensis, A. distans*, and *A. monticola*). While the number of seeds in cross-pollinated flowers reached 47–110 pcs, it was solely 0–11 pcs in self-pollinated ones. Our long-term practical experiences with yarrow ascertain this finding and it is in obvious coincidence with the large intraspecific chemical diversity of these species.

Xenogamy is the preferred way of fertilization in several important medicinal species. As an example, *Lamiaceae* species are cross-pollinating ones based on the morphological constitution of the flowers and the mechanism of proterandry. Beside xenogamy, geitonogamy may occur between flowers of the same plant; however, seed-set rates are much lower in this case (Putievsky et al., 1999; Németh and Székely, 2000). In some species of the same genus, both hermaphrodite and male-sterile flowers can be found. In thyme (*Thymus vulgaris*), it has been described that the latter ones occur primarily in suboptimal environments, assuring that outcrossing enhances fitness of the progenies while the hermaphrodite flower structure enables autogamy. Depending on the type of fertilization, the essential oil pattern varies characteristically (Gouyon et al., 1986).

Species that are generally propagated by vegetative methods like peppermint, tarragon, etc. do not show any or only a minimum variability among individuals. This fact sometimes is considered as an adverse phenomenon and an obstacle in effective selection and genotype improvement. Therefore, breeders usually try to increase the variability of these plants with specific methods. Mutation breeding proved to be a prosperous tool in producing wilt-resistant strains of peppermint in the US (Murray et al., 1986). Induction of polyploids by colchicine and the crossing of fertile accessions afterward has been the basics in developing the highly productive variety "Multimentha" in East-Germany (Dubiel et al., 1988). Development of new chemical varieties is endeavored today more and more by molecular genetic methods (Croteau et al., 2005; Wagner et al., 2005).

On the other side, clonal propagation is an optimal way to produce chemically homogenous populations for commercial production and processing purposes. According to my own experiences, seed sowing of tansy results in an enormous segregation of the population which is not acceptable as raw material for industrial utilization. Therefore, vegetative propagation by young shoots has been elaborated for the production of selected chemotypes (Zámboriné et al., 1987).

The chemical heterogeneity of several wild-growing populations seems to be today one of the basic motivations for introduction of economically important wild species into the agriculture and selection of their stable varieties. Breeding is going on usually parallel with development of technological methods.

Fennel has been cultivated already for many decades and selected cultivars are registered in numerous countries. The main goals of the breeding have been definitely the increase of essential oil content and stabilization of its composition. During maintenance of our cultivar "Foenipharm," we checked the most important characteristics of individual mother plants. The results show that deviations among the plants are minimal owing to the long-term breeding and variety maintenance process (Table 4.7).

Breeding of the polymorph species *Artemisia absinthium* in the Conthey Research Centre (Switzerland) resulted in a uniform variety accumulating *cis*-epoxyocymene as the main compound. After screening of more than 800 plants from 24 accessions originating from six countries, the researcher selected and stabilized the desired chemovariety (Simonnet et al., 2012).

TABLE 4.7
Fruit Characteristics of Selected Individuals in the Stock Plantation of *Foeniculum vulgare* "Foenipharm"

Plant nr.	Essential Oil (% d.w.)	Anethole (% ess. oil)	Estragole (% ess. oil)	Fenchone (% ess. oil)
1	7.91	65.13	2.34	2.14
2	6.10	61.58	2.18	2.38
3	6.34	57.88	2.09	2.07
4	4.73	57.91	2.02	2.51
5	5.59	60.14	2.17	2.41
6	4.94	61.87	2.17	2.26
7	6.33	66.30	2.38	1.87
8	4.49	67.57	2.45	1.76
9	4.91	54.84	1.98	2.53
10	4.90	69.02	2.56	1.79
CV%	18.8	7.5	8.0	12.0

Source: Németh, É., Unpublished.

Effective breeding necessitates knowledge on the genetic background; however, inheritance of volatile compounds is till now only partially detected. Earlier studies explained the presence or absence of individual volatile compounds by Mendelian genes and gene interactions.

Classical genetic studies revealed that in yarrow, azulenogenic sesquiterpene lactones are inherited through the recessive allele of a special gene (Vetter et al., 1997). Similar mechanism seems to be working in the related chamomile (*Matricaria chamomilla*), and quantitative changes may be the result of modifying polygenes (Wagner et al., 2005). Multiallelic genetic determination was stipulated for the inheritance of borneol and 1,8-cineole in *Hedeoma drummondii* (Irving and Adams, 1973) or for the inheritance of camphor in *Tanacetum vulgare* (Holopainen et al., 1987). A single gene locus may be responsible for the production of anethole and estragole, with partial dominance for high estragole content (Gross et al., 2009). Similarly, existence of chemotypes of different δ-carene levels in Scots pine (*Pinus silvestris*) are explained by the alleles of a single gene and inherited in a dominant-recessive system (Hiltunen, 1975).

Today, molecular genetic tools are more frequently involved in study of volatile compounds. In several species, function, substrate specificity, and isoforms of different terpene synthases, as well as their regulating genes, have been identified, e.g., in lavender (Tsuro and Asada, 2014), in *Origanum* species (Lukas et al., 2010), in sage (Grausgruber-Grüger et al., 2012), and in *Thuja plicata* (Foster et al., 2013).

Based on these recent findings, the genetic determination of essential oil compounds seems to be complex and could be revealed only at the metabolom level. Besides the direct regulation of the biosynthetic processes, other types of regulation interact with the formation of volatile compounds like intra- and intercellular transportation mechanisms, primarily metabolic processes, or regulation through transcription factors, which are still less known in terpenoid metabolism.

Inherited traits manifest themselves in each plant individual and this is the background of intraspecific diversity, but the appearance of variability at the population level depends also on the occurrence frequencies of corresponding genes.

4.3.2 Morphological Characteristics

Revealing the connection between chemical traits (essential oil composition) and any morphological characteristics would be of interest both from theoretical and practical points of view. External features as marker traits for oil composition would be of high importance during cultivation, breeding, or audits. Unfortunately, with some exemptions, there are no reliable data about this topic.

The leaf form, size of dissections, and color of the leaves show a great variability in *Achillea crithmifolia*. Our investigations in a controlled environment, however, revealed no connection between chemotype and leaf dissection (Németh et al., 1999). Similarly, Hofmann (1993) established, in several taxa belonging to the section *millefolium*, that morphological traits may not refer directly to specificities of essential oil composition. In *Achillea millefolium*, Gudaityté and Venskutonis (2007) tried to find a connection between the color of petals and the azulene accumulation in the flowers; however, it could not be demonstrated. A higher proazulene level was detected in connection with a higher number of internodia, narrower leaves, and ligulate flowers, but it has not been ascertained by other authors.

The shape of the leaf is very variable in case of tansy, too. According to my own observations (unpublished), some chemotypes can be distinguished from other ones based on this feature. Individuals containing the sesquiterpene davanone have shiny green, oval leaves with dense incisions, while the leaves of the chemotype accumulating thujone as the main compound are elongated, leaflets are sparsely incised but lobes are deeper, and their color grayish-green. However, similar characteristics cannot be generalized as special markers applicable for each chemotype. It is in coincidence with the opinion of Schantz and Forsén (1971), who emphasized that no characteristic connection between essential oil composition and morphology of the examined west European tansy populations could be determined.

In some cases, however, literature references seem to be contradictory in this respect. In a former publication, Hodzsimatov and Ramazanova (1974) declared that the presence of bornyl acetate, terpineol, and linalool content of the essential oil of hyssop (*Hyssopus officinalis* L.) is connected to the pink flower color. Chalchat et al. (2001) mentioned that pinocarvone is mostly present in individuals of white petal color, but Galambosi et al. (1993) found the highest pinocarvone proportions in a population of pink flowers. My own (not published) measurements and experiences showed that chemism of white, pink, or blue flowering individuals is independent from flower (petal) color, and there may be larger differences between plants with the same flower color if they originate from different accessions compared to the ones which have different petal color but have the same origin.

The investigations in 48 annual and 18 biennial caraway populations provided data about several significant correlations among oil composition and different morphological and production characteristics such as rootneck width, number of shoots, number of umbels, and seed biomass. However, no significant correlation could be found between any of these morphological features and the carvone content of the oil (Zámboriné, 2005).

In chamomile (*Chamomila recutita*), Gosztola (2012) carried out a very detailed and comprehensive analysis on Hungarian wild-growing populations from different habitats. They studied a wide range of plant characteristics and their connections. As for the oil composition, none of the most important sesquiterpenes showed any significant correlation with morphological features such as plant height and diameter of the flowers and that of the discus (Table 4.8).

By screening13 different accessions of fennel (*Foeniculum vulgare* ssp. *capillaceum* var. *vulgare*), it has been established that they represent different chemovarieties of the species (Bernáth et al., 1996). Chemovariety 1 represented by a single accession accumulating the largest concentrations of fenchone (above 30% of the oil), chemovariety 2 contained three strains characterized as methyl-chavicol-rich ones (above 20% of the oil), while the nine accessions belonging to chemovariety 3 showed high anethol contents (above 60% of the oil). Studying the connections between the main components of the oil and the morphological features, only loose or medium-strength correlations could be determined (Table 4.9). The connection between the seed size and essential oil content seems to be of the largest practical importance ($r = 0.6102$). Among the volatile components, β-pinene and limonene showed significant negative connection with leaf mass. Similarly, higher plants produced less anethole and more methyl chavicol. Although these results may be interesting, most likely, these statistical correlations have hardly any real physiological or genetic background; therefore, their universal use as markers is questionable.

TABLE 4.8
Correlation Coefficients between Main Essential Oil Components and Some Morphological Traits in Chamomile

Chemical Compound	Plant Height	Diameter of Flowers	Diameter of Discus
β-Farnesene	0.00	0.28	0.38
Bisabolol-oxide B	0.10	0.09	−0.10
α-Bisabolol	−0.34	−0.32	−0.33
Chamazulene	−0.12	−0.04	0.10
Bisabolol-oxide A	0.30	0.26	0.28
cis-Spiroether	0.20	0.13	0.09
trans-Spiroether	−0.02	0.05	0.23

Source: Modified from Gosztola, B. 2012. PhD Dissertation, Corvinus University, Budapest.

TABLE 4.9
Correlation Coefficients of Morphological and Chemical Characters of Fennel (Foeniculum vulgare) based on Investigation of Different Accessions

Morphological Feature	Essential Oil Content (mL/100 g)	Component (in % of the oil) α-Pinene	β-Pinene	Fenchone	Methyl Chavicol	Anethole	Limonene
Plant height	−0.1002	0.0866	−0.1590	0.2263	0.5927	−0.5805	0.2152
Mass of leaves	0.3531	−0.2713	−0.6458	−0.2800	−0.0768	0.2515	−0.4393
Length of seeds	0.6102	0.2334	0.0219	−0.1940	−0.3321	0.3853	−0.3229
1000 seed mass	0.4705	0.2186	0.3560	0.1711	−0.4384	0.2624	−0.0684

Source: Bernáth, J. et al. 1996. *J. Essent. Oil Res.*, 8: 247–253.

4.4 MORPHOGENETIC AND ONTOGENETIC MANIFESTATION OF THE CHEMICAL VARIABILITY

Although, until now, we discussed chemical polymorphism of the plants in general, in numerous plant species, there are well-defined deviations also between the oil composition of different plant organs: roots and aboveground parts, vegetative and generative organs, leaves and flowers. Rate and pattern of the divergences are characteristic for the species. In some cases, the transition is continuous and only the quantitative proportion of the compounds changes from the basal regions toward the apical parts beyond a relatively standard qualitative spectrum. In other species, the spectrum is suddenly changing with the differentiation of new plant organs. Besides, there are examples also for uniform composition both of the vegetative and generative organs.

Lovage (*Levisticum officinale*) always accumulates in the leaves α-terpinyl acetate as the main compound (40%–80%) beside a lower level of β-phellandrene (15%–28%). The main components of the roots 3-butylidenephthalide and Z-ligustilide are only present in lower concentrations—not rarely only in traces (Novák, 2006). The composition of the fruits is similar to that of the leaves with β-phellandrene as the main component in up to 60% (Bylaite et al., 1998).

In species where the root does not provide an official drug, data on essential oil accumulation of the underground parts are obviously much rarer. Existing data, however, show that the composition of the underground parts might be similar to, or even totally different from, that of the shoot system. In another Apiaceae species, in fennel, the difference in composition of the roots and that of the aboveground parts shows the largest deviations, while the difference between the green parts (leaves and shoots) and generative organs (flowers and fruits) is less characteristic. In the roots, the absolute main compound is dillapiol, while in the whole shoot, anethole is accumulating in the highest concentration. Composition of the essential oils from stems and leaves of fennel are qualitatively similar to that of the fruits. However, the ratio of anethole varies among organs, the vegetative green parts containing it in higher concentrations (around 90% of the oil) than flowers and fruits do (60%–68% of the oil) (Chung and Németh, 1999). The majority of the mentioned components are biosynthetically related phenylpropanoids.

Composition of the roots and that of the leaves proved to be surprisingly different both qualitatively and quantitatively in some Asteraceae species, too.

In the genus *Achillea*, the volatile composition of the root has been studied till now only in a few species. The root oil of *A. distans* contained primarily τ-cadinol, alismol, and α-cadinol (Lazarević et al., 2010), while in *A. millefolium*, epi-cubenol and the monoterpenic ester neryl isovalerate was detected in highest proportions (Lourenço et al., 1999). The sesquiterpenes (τ-cadinol) and monoterpene esters (neryl isovalerate) were also present in significant amounts in *A. lingulata* roots (Jovanović et al., 2010). Our investigation ascertained the differences also at the intraspecific level. In the root EOs of ten *A. collina* accessions, the universal main compound was 7-heptadecanone-en (28.9%–43.0%), beside other sesquiterpenes like alismol (3.1%–18.9%), β-sesquiphellandrene (0.5%–9.8%), γ-humulene (1.1%–6.8%), *cis*-cadin-4-en-7-ol (0.3%–7.4%), and β-eudesmol (0.2%–7.3%). However, monoterpenic fraction provided the smaller part of terpene constituents with only two identified compounds (neryl 2-methylbutanoate and neryl isovalerate), and by this, the root EO spectrum is considerably different from the composition of the flowering shoots (Kindlovits et al., 2018).

Intraspecific, moreover, individual differences have been detected in the volatile composition of the root oils also in the related species *Artemisia absinthium* (Llorens-Molina et al., 2016). Major components of the roots and of the aboveground organs proved to be totally different from each other. Based on leaves, three chemotypes—sabinene + myrcene, β-thujone, and new sesquiterpene type—were found in the Hungarian population and two chemotypes—(Z)-β-epoxyocimene and (Z)-β-epoxyocimene + (Z)-chrysanthemyl acetate types—were present in the Spanish one. The composition of the root volatiles was predominated by monoterpenic esters, but characteristic quantitative and qualitative differences were present among plants. No relationship could be determined between the composition of essential oil of roots and leaves of the same plant.

The results of this study on forty plants ascertain that individual variability of essential oil composition might be present in each plant organ which may lead to a confusing conclusion: based on the volatile composition of the leaves, a group of individuals might eventually be evaluated as belonging to the same chemotype, but concerning the root volatiles, they are not similar to each other. About such data there are still very few records which underlines the necessity of further investigations.

The manifestation of significant differences between the composition of the essential oil from the root and that from the shoot is, however, not a universal phenomenon. Schulz and Stahl-Biskup (1991) studied the organic diversity of the essential oil spectrum in the case of hyssop. They found that pinocamphone can be described as a universal main component in each of the roots, stems, leaves, and flowers at an accumulation proportion of 22%–60%. In case of the roots, the only characteristic difference was the presence of an unidentified—presumably—sesquiterpene type compound in 13.1%–15.6% while the spectrum of the leaves and flowers proved to be both qualitatively and quantitatively related.

The picture is somewhat similar for peppermint. According to investigations of Murray et al. (1986), components of the essential oil distilled from the stolons are highly comparable with the shoot oil. Major compounds of the stolon oil were menthofurane (46.1%), menthyl acetate (24.5%), and menthol (11.4%), which reflect only quantitative differences compared to the oil distilled from the herb or the leaves. These data refer to a relatively uniform biosynthetic process of these terpenoids in the whole shoot system developing underground or aboveground.

As a conclusion, we could declare that biological variability is manifested also in the relationship between the volatile compounds of different organs. While some species accumulate qualitatively similar compounds in each organ, some others produce different compounds; however, they result from the same biosynthetic route. At the same time, there are also species in which the volatile composition of different plant organs seems to be almost "random", and according to our present knowledge, no closer connection can be established between their biosynthetic origin.

The special composition of any plant organ is, however, not a stable phenomenon. Qualitative and quantitative compositional changes frequently occur *during ontogenesis* of the plant. These changes are either direct when the same plant organ (leaf, flower) shows an altered character during its development or they may be indirect as in variability detected due to morphogenetic changes of the shoot, such as appearance of buds, fall of the leaves. These changes—onto- and morphogenetic ones—are part of the biological variability, as they are regulated by the metabolomic processes of the plant. Consecutive expression of corresponding genes or changes at the translational or enzymatic level might result in different chemosyndromes during the plant's life.

In peppermint, it has been detected by *in vitro* enzyme activity and $^{14}CO_2$-labeling experiments that the background of these ontogenetic changes is a complex process (McConkey et al., 2000). In the first phase ("*de novo* oil biosynthetic program"), which coincides with leaf expansion and gland filling, the group of enzymes leading from geranyl diphosphate till menthone is extremely active, while about a week later, in the second period ("oil maturation program"), these enzyme activities are strongly diminished and the activity of menthone reductase increases steadily, leading to an elevated level of menthol. Early upregulation of menthone reductase in a breeding program may result in transgenic peppermint of elevated menthol accumulation potential.

In Apiaceae species, considerable compositional changes occur during the development of the seeds, which may have an influence on the quality of the drug. In caraway, a timely shift has been proved in the accumulation of the two main components carvone and limonene during seed development; therefore, their ratio is primarily depending on the ontogenetic phase. Limonene is formed right after fertilization of the flowers, which is 5–10 days later, followed by the accumulation of carvone synthesized from limonene. The timely shift is the consequence of changing enzyme activities: limonene synthase is active at the beginning while, after the mentioned time period, activity of limonene hydroxylase catalyzing the formation of carvone is increasing (Bouwmeester et al., 1998). It has been supposed that this is the rate-limiting step in enhancing carvone accumulation.

A similar phenomenon has been described for the related coriander (*Coriandrum sativum*). In the fruits of this species, linalool is the characteristic main compound, reaching more than 90% of the total essential oil in ripe seeds. At the beginning of seed development, besides linalool, the ratio of (E)-2-decenal is characteristically high (above 15%–20%) but later, during seed development, the later one decreases (Varga et al., 2012).

In sage (*Salvia officinalis*), it was detected that different types of accumulation structures (peltate glandular trichomes, capitate glandular trichomes, and ambrate resinous droplets) are present in special distribution on the leaves. Each of them has characteristic terpenoid composition (Tirillini et al., 1999). Considerable compositional differences can be found also between older and younger leaves (Grassi et al., 2004). While the younger leaves were rich in β-pinene, bornyl acetate, and sesquiterpenes, the ratio of these compounds was significantly reduced by leaf expansion (Table 4.10). On the other side, mature leaves contained three times more camphor and camphene than the newly formed ones. In this case, the changes are reflected not only among leaves of different age but also among different segments of the same leaf, too. The relatively immature regions at the

TABLE 4.10
Characteristic Main Components (%) of SPME Extracts of Sage Leaves of Different Age

			Intermediary Leaf			
Compound	**Young**	**Old**	**Base**	**Margin**	**Middle**	**Inside**
Camphene	4.5	8.1	4.7	6.5	7.1	5.7
β-Pinene	19.4	4.6	8.1	8.6	9.7	13.4
1,8-Cineole	8.0	15.5	–	–	–	–
α-Thujone	12.1	12.5	16.6	14.6	17.0	21.0
Camphor	9.7	29.1	11.9	20.0	18.2	14.9
Bornyl acetate	2.5	0.0	2.0	0.7	0.4	0.2
α-Humulene	6.0	4.7	7.4	3.5	4.2	3.6
Viridiflorene	4.7	0.5	23.3	14.0	13.6	1.5
Manool	10.4	9.4	6.2	9.2	6.4	5.2

Source: Grassi, P. J. et al. 2004. *Phytochem. Anal.*, 15: 198–203.

basal region of the leaf showed similarity to the composition of young leaves. However, there were characteristic differences observed also between the inside and the marginal regions; therefore, it seems to be likely that besides age, other factors also may influence the composition of the single oil glands on the leaf surface.

Similarly, detailed investigation of Johnson et al. (2004) revealed well measurable differences in the oil composition of leaves of different age in *Origanum vulgare* ssp. *hirtum*. Lowest levels of p-cymene were always detected in the younger leaves but, to the contrary of sage, the composition of the oil proved to be similar in older and younger parts of the same leaf. It was anticipated that in this case, the background of the differences would be the time of the leaf development and not its relative age. Lengthening days enhance the loss of p-cymene compared to carvacrol, but only the young leaves are able to make this switch. In this case, it seems that compositional diversity of leaves of different age may not be considered as simple biological variability but, more, a specific environmental response.

The detectable variability of the volatile composition may depend also on the spectrum of individual oil glands, even of the same leaf. In oregano, significant differences could be detected in the composition of individual oil glands (Johnson et al., 2004). While the majority of the oil glands yielded carvacrol and virtually no detectable amount of thymol, a small proportion of the glands produced up to 70% thymol. The oil glands of this latter type were randomly distributed on the leaf surface, and the explanation for this interesting phenomenon is still lacking.

The composition of the oil seems vary with the age, too. Although there are hardly any reports on this issue, it seems that this variability may be the result of histological transformations. Stahl-Biskup and Wichtmann (1991) detected a decrease of germacrene B, the characteristic compound (in 51%) of the seedlings, in parallel with secondary thickening of the roots and formation of secondary oil cavities. The roots of adult plants contained mainly aliphatic aldehydes up to 68% of the oil. On the contrary, the herb oil of seedling and adult plants did not differ significantly from each other.

In many cases, the detectable variability of oil composition is connected to morphological differentiation during shoot development. The emergence of the flowering stem, appearance of flowers, and development of fruits may result in qualitative and quantitative alterations. These shifts are connected not only with aging of the plant but also with changes in its organic structure.

A "classical" example for this phenomenon is the change of the compositional profile of peppermint oil during ontogenesis (Murray et al., 1986). At the beginning of shoot development, the herb contains

TABLE 4.11
Main Components (% in the Oil) of the Peppermint Oil from Top (T) and bottom (B) Parts of the Shoot during the Vegetation Period

Compound	Stem Part	July 21	Aug 3	Aug 16	Sept 7
1,8-Cineole	T	5.3	6.0	5.4	5.4
	B	5.8	4.8	4.5	4.0
Limonene	T	3.2	5.0	3.7	2.8
	B	3.0	2.3	2.7	2.1
Pulegone	T	1.1	3.4	4.6	1.0
	B	0.8	2.0	2.3	1.7
Menthofuran	T	1.4	7.6	9.9	8.2
	B	2.0	2.9	4.6	5.7
Menthone	T	34.1	30.5	28.3	19.2
	B	16.1	17.1	17.1	13.6
Menthol	T	29.9	23.3	27.1	41.5
	B	43.4	39.4	38.8	44.2
Menthyl acetate	T	3.2	3.7	2.7	3.7
	B	8.0	10.3	9.5	9.0

Source: Modified from Murray et al. 1986. In: *Flavors and Fragrances: A World Perspective*, Lawrence (eds.) et al. pp. 189–210, Amsterdam: Elsevier.

menthone (above 30%) as the main compound, while the ratio of menthol is usually the same or even lower. During shoot growth, the proportion of menthol starts to increase, and at harvest time in good-quality plant material, it reaches more than 40%. By the senescence of the leaves, elevated levels of the corresponding ester, menthyl acetate can be measured, too. Data of Table 4.11 show the mentioned changes during shoot development. On the same day, the older, bottom part (B) of the shoot is more similar to a later phenological phase (more menthol, less menthone) than the upper part (T). Besides, morphogenetic changes are reflected in the elevated values of menthofurane and pulegone on later dates as these are characteristic compounds of the flowering parts of peppermint.

The species-specific behavior in these biogenetic transformations is demonstrated by the fact that the related species *Mentha citrata* (syn. *Mentha x piperita* var. *citrata*) shows a different tendency as the main compound linalool increases by approximately 30% during flowering, while the corresponding ester, linalyl acetate, is decreasing at the same time (Malizia et al., 1996). In species possessing different chemical varieties, ontogenetic changes might be characteristic features not only to the species but also to the intraspecific chemotype.

In tansy (*Tanacetum vulgare*), Schantz et al. (1966) already described quantitative shift in the essential oil composition from budding till the end of flowering period. The main tendency of the change was an increase of the main components camphor and thujone in these two chemotypes. In my own investigation, six different chemotypes were checked in an extended period from early shooting till seed ripening (Németh et al., 1994). In five of the chemotypes containing monoterpene compounds as major components of the oil, an increase of the proportion of these components was detected. The increase is a slight and continuous one in each of the borneol-, camphor-, and 1,8-cineole-type plants, while it is a sudden and larger one in the thujone- and thujene-acetate-type individuals (Figure 4.3). Another dynamics was found in the chemotype accumulating the sesquiterpene lactone davanone in the oil as the main component. In these plants, the ratio of davanone is highest right after shooting out in spring and shows a continuous decrease after that.

Long ago, Schratz and Hörster (1970) established that during the vegetation period, significant changes can be detected in the essential oil composition of two thyme species (*Thymus vulgaris* and

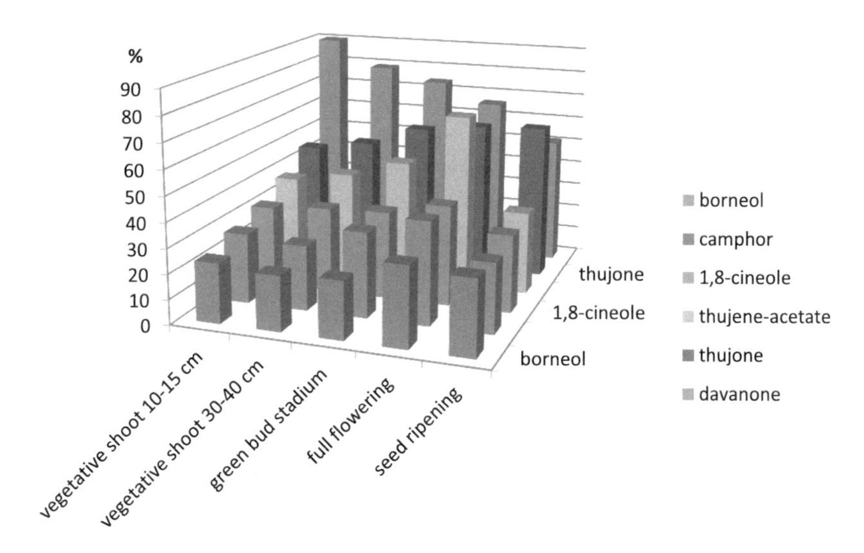

FIGURE 4.3 Proportion of the main components (in GC area %) during the vegetation period in different chemotypes of tansy (*Tanacetum vulgare*). (Adapted from Németh, É. et al. 1994. *J. Herbs, Spices Med. Plants*, 2: 85–92.)

Th. marschallianus). The dynamics of changes of the measured compounds (γ-terpinene, thymol, carvacrol, p-cymene, terpinolene, α-pinene, etc.) thorough plant development was different in each examined individual plant. The large individual variability in the components and in the dynamics of the compositional changes let the authors conclude that the overall chemical features of a genus or a species may play only a limited role for systematic characterization.

Less known and investigated are even till today the diurnal changes of volatile components. As a model plant, in *Origanum onites*, the most important compounds carvacrol and thymol fluctuate significantly during the day. Carvacrol shows a drop in early morning and in afternoon, while thymol has the lowest concentrations at 10 o'clock in the morning, and the level increases from early evening till midnight. This dynamics means 1.6–3.0 times differences in the accumulation ratios of the mentioned compounds and does not seem to be influenced severely by the phenological stage between pre-flowering and post-flowering periods (Toncer et al., 2009). More recently, Padalia et al. (2016) proved that the diurnal cycle of compositional changes of volatiles are to a large extent specific for the species. Significant quantitative and qualitative differences were registered (e.g., in *Ocimum gratissimum* and *O. kilimandscharicum*) while only small quantitative changes occurred in *O. basilicum*.

Biological variability of essential oil composition is obviously a complex system manifested in a characteristic row of chemosyndromes in the plant species or variety. The backgrounds of the evaluated changes are until now only partially detected. The spectrum of volatile compounds in the plant depends to a large extent on modifications in metabolomics processes not rarely connected to anatomical/histological changes.

Schratz and Hörster (1970) suggested already many decades ago that compositional changes associated with plant ontogenesis may be present due to the fact that only young oil glands synthesize volatile compounds, and in the fully developed leaves, secondary transformations take place. Later, it was proved that monoterpene production in peppermint (*Mentha* × *piperita* L.) is determined by the rate of biosynthesis in connection with the age of leaf glandular trichomes and is restricted to leaves 12–20 days of age (McConkey et al., 2000). According to Yamaura et al. (1992), in capitate glandular trichomes, the synthesis stops at a much earlier stage than in peltate grandular trichomes.

Today, investigations on cellular and molecular mechanisms of volatile formation are under focus. Monoterpenes are formed predominantly in the plastids via the methylerythritol phosphate

(MEP) pathway; therefore, cells rich in plastids may be rich in monoterpenes, while cells of well-developed endoplasmic reticulum may be primary sites of sesquiterpene biosynthesis based on the mevalonic acid (MVA) pathway. However, a metabolic cross talk between the plastidic MEP and cytosolic MVA pathways has been found several times (Dinesh and Nagegowda, 2010). Webb et al. (2011) concluded that in *Melaleuca alternifolia*, both pathways may contribute to the sesquiterpene formation, and the intensity of this contribution depends on other factors like the plant species, tissue, and physiological state of the plant. Although data on localization of enzymes and terpenoid biosynthetic processes are continuously accumulating, a general statement about the role of intracellular structures in detected changes of volatile profile of essential oil–bearing plants could not be established yet.

Looking for the background of biological variations of the essential oil spectrum, we have to add that synthesis of the volatile compounds is only one side. It cannot be excluded either that the biosynthesized components are specifically translocated and/or further metabolized in the cells. The best documented example for this complex compartmentation system seems to be the formation of different monoterpenic compounds of peppermint oil. Labelling studies revealed that geranyl diphosphate synthase is localized in the leucoplasts, (−)-limonene-6-hydroxylase is associated with the endoplasmic reticulum, (−)-*trans*-isopiperitenol dehydrogenase is found in the mitochondria and (+)-pulegone reductase in the cytoplasm. Thus, a well-formed subcellular compartmentation and translocation mechanism is needed in fulfilling the total biosynthetic chain. Besides, an active transport is supposed in carrying the produced components from the secretory cells into the subcuticular oil storage cavity of the peltate grandular trichomes (Kutchan, 2005). Recently, it was found that beside the linalool synthase (*LinS*) gene (responsible for the production of monoterpene linalool), a sesquiterpene synthase gene (cadinene synthase, *CadS*) was the most abundant transcript in the glandular trichomes of lavender flowers (Lane et al., 2010). This surprising result indicated that precursor supply may represent a bottleneck in the biosynthesis of sesquiterpenes in lavender flowers. Similar situations may be widespread in volatile producing plants and desire a lot of attention.

Unfortunately, molecular biological studies on volatile accumulation in connection with plant development and morphogenetic changes have been limited until today. Grausgruber-Grüger et al. (2012) investigated the connection between the accumulation of main monoterpenes and transcript levels of their synthases during the vegetation cycle of garden sage (*Salvia officinalis*). It has been established that terpene synthase mRNA expression and the level of the respective end products were in significant correlation in the cases of 1,8-cineole (correlation coefficients $r = 0.51$ and 0.67 for the two investigated cultivars) and camphor ($r = 0.75$ and 0.82), which indicated a transcriptional control of the process. The same correlation, however, could not be proved for α- and β-thujones, which shows the possible role of other regulation mechanisms in their accumulation. In lavender, the gene responsible for the formation of linalool is strongly expressed in essential oil glands of the flowers and shows the peak expression at 70% flowering stage. After that, *LinS* gene transcription decreases, but the level of the product linalool still remains high (Lane et al., 2010). The authors concluded that the production of linalool in lavender is transcriptionally regulated but other regulatory mechanisms (e.g., transcriptional upregulation of other genes, post-transcriptional regulatory mechanisms) may also be involved. Sarker and Mahmoud (2015) cloned and functionally characterized two monoterpene acetyltransferases from *Lavandula* x *intermedia* glandular trichomes which are capable of synthesizing geranyl acetate, lavandulyl acetate, and neryl acetate from their respective monoterpene substrates. They demonstrated that transcripts of their genes were significantly more abundant in flowers compared to leaf tissues, which corresponds to the higher accumulation rate of these components in the flowering parts.

Studies like the above mentioned ones represent a new focus for understanding of changing chemosyndromes in essential oil–producing species and for a potential regulation based on complex metabolomic approaches.

4.5 ORIGIN OF ESSENTIAL OIL VARIABILITY

Ontogenic changes in chemical profile of numerous taxa provide the basis for the "ontogenetic hypothesis" on the genesis of intraspecific chemical taxa. Individuals of different chemism may show different adaptation capacity to changing environmental circumstances. Thus, plants of the most appropriate compositional profile have a fitness comparing with other ones, and therefore, can be stabilized and spread through the area (Hegnauer, 1978).

Tétényi (1970) assumed that differentiated intraspecific taxa have emerged by adapting themselves to different life conditions. According to this, some alterations of ontogenetic metabolistic changes of individual development as acclimatization behavior may have occurred in certain phases characteristic of the given plant species. The special chemical differences of intraspecific taxa seem to have evolved by the stabilization of these chemical features present in consecutive phases of the ontogenesis if the deviation was of durable nature and occurred repeatedly in several progenies. Differences both in intensity and quality of metabolism in single phenological phases might be inherited, becoming taxonomic characteristics and, thus, establishing the biosynthetic basis of the existence of new chemical taxa.

It is without doubt, that chemical changes manifested in successive progenies serve frequently as a direct adaptive tool in survival of the population or individual. In other cases, however, adaptation through other—morphological, propagation-biological, phenological—features may also lead to an altered chemical profile as an indirect result.

This general statement for secondary plant compounds may be valid for essential oil compounds as well. Compounds that have some kind of adaptive value are presumably distributing to a larger extent inside the population, increasing the fitness of the plants. Many examples demonstrate already that in accordance with phylogenesis, appearance of intraspecific chemical variability may be considered as result of adaptive processes (Dudareva and Pichersky, 2008). During natural selection, the presence and accumulation level of volatile compounds have changed and show now a really colorful spectrum.

In *Thymus vulgaris,* the broad chemical variability and adaptation potential are manifested during ontogenesis. At the 1–3 months seedling stage, the plants belonging to linalool chemotype accumulate mainly phenolic compounds similarly to the carvacrol and thymol chemotypes and, only in later phenological stages, start to predominate the characteristic linalool in the oil. It has been suggested that this behavior of the plant is a chemical defense against herbivores in this young, sensitive age (Linhart and Thompson, 1995).

Seedlings of *Eucalyptus globulus* accumulate essential oil rich in pinenes while adult plants produce 1,8-cineole (eucalyptol) as the main component. Similarly, seedlings of *Cinnamomum camphora* synthesize safrole as the major volatile compound of the leaves, but later, different other compounds start to accumulate in abundance corresponding to the different intraspecific chemotype (Tétényi, 1975). Juvenile parts of *Thuja occidentalis* show multiple differences in the quantitative composition of the volatiles compared with that of the mature shoots. The accumulation rate of sabinene (25.4%) and α-pinene (26.3%) are especially high compared to the leaves of the adult plants (0.76% and 3.54%, respectively) (Gnilka et al., 2010). Interestingly, the adaptive role or other function of these changes is still not adequately declared and is worthy of further research.

Several examples demonstrate that adaptation process should be considered here in a wider sense. Production of special volatiles may enhance not only plant survival among adverse ecological conditions and assure a protection against predators, but also stimulate competitiveness and distribution.

The important role of the changing volatile profile during ontogenesis in the eco-physiological behavior of the plants has been demonstrated by the example of lavender (Guitton et al., 2010). It was found that in each individual flower, three different groups of terpenes appeared sequentially during the flowering period. The authors concluded that the terpenic volatiles in bud stadium and at the end of flowering, when flowers faded and small seeds started to develop, should serve as protective, insect-repellent compounds (e.g., ocimene, limonene, linalool). The ones, however, produced during full flowering are mainly attractive-type molecules, presumably enhancing fertilization by engaging pollinators (e.g., linalyl acetate, some sesquiterpenes).

Similarly, Muhlemann et al. (2012) proved that in snapdragon, emission of attractive volatiles—primarily myrcene, β-ocimene, linalool, and (E)-nerolidol—follows a diurnal rhythm. Molecular genetic investigations revealed that the synthesis is regulated at the level of transcription. The expression of genes encoding enzymes involved into the formation of these volatiles is increasing after anthesis and reaches its maximum when the flower is ready for pollination. This is specifically detectable in the petals of the flowers while very few changes were found in sepal expression profiles, which shows a strict organic localization corresponding to the physiological role of the changes.

Nevertheless, as mentioned above, it can be assumed that sometimes an altered chemical profile is not a tool in acclimatization but appearing as consequence of adaptation processes. This fact seems to be an explanation for the existence of variable chemotypes of *Thymus vulgaris* in different French populations. It has been proposed that the appearance of numerous intraspecific chemotypes and the wide range of main components in the essential oil may be the result of natural competition (Gouyon et al., 1986). The species is indigenous thorough the Mediterranean coasts; however, it has a relative weak competition ability, therefore autogamy can ensure survival only in an optimal environment. In these habitats, plants are hermaphrodites, and in consequence of the high rate of authogamy, they are of homozygote genetic structure. The constitution of a homozygote genotype could lead to the manifestation of recessively inherited features like the high thymol accumulation and distribution of this chemotype in the area. Nevertheless, under less favorable conditions, in order of increased competitiveness, propagation is going on through xenogamy which results in heterozygotic genetic structure. In such a genotype, the typical oil components are the dominantly inherited ones like geraniol and terpineole; therefore these chemotypes are abundantly found in the population. Based on this hypothesis, the phenotypic manifestation of genetic information, i.e., the chemosyndrome, is the indirect result of the environmental pressure.

Hybridization represents a further possible way for development of new chemotypes. Interspecific hybridization and polyploidization frequently occur, especially in neighboring and overlapping distribution areas of different taxa. Hybrid accessions of elevated fitness—either through a new chemical profile or by other advantageous features—may give rise to the distribution of these genotypes. One of the most important and widely used medicinal plants, *Achillea collina*, might have developed by this way. Based on hybridization experiments, it was suggested that this species arose by natural outcrossing and allopolyploidization of *A. setacea* and *A. asplenifolia* as their hybrid in neighboring distribution areas. Through this process, a new tetraploid species was born with a proazulene-producing potential, similarly to the diploid parent *A. asplenifolia*. Hybridization is one way for the formation of new biotyps of better adaptability and chemism (Ehrendorfer, 1963).

Besides the mentioned adaptation processes, interspecific hybridization in overlapping areas with simultaneous blooming and possibilities of crossing was mentioned as another possible evolution mechanism in the development of numerous chemical varieties of *Thymus* species, too (Stahl-Biskup and Sáez, 2003).

4.6 CHEMOTAXONOMIC ASPECTS

Essential oil variability is a target of chemotaxonomy. In connection with the complex regulation of the metabolism of terpenoids and other volatiles, chemotaxonomic aspects should be evaluated already in a more comprehensive way then it has been done for decades.

A big majority of scientific publications on intraspecific chemical variability of volatile components refer to "chemotypes". Chemotype is, in practice, a common term which may be used in each case when the exact chemical taxon cannot be defined more precisely. Formerly, the term "chemical race" was used similarly, which term unfortunately has been considered later as not unequivocal and a rather indefinitive one (Tétényi, 1975). Therefore, the use of "intraspecific chemical taxon" as a general name and the use of the accepted botanical taxon definitions (such as *forma, varietas, subspecies*, in case of cultivated species the term *cultivar* etc. with the prefix "chemo-" as special names for closely defined taxa) were suggested.

Recently, Polatoglu (2013) discussed this issue anew and proposed another definition for *chemotype*: "organisms categorized under same species, subspecies or varieties having differences in quantity and quality of their components in their whole chemical fingerprint that is related to genetic or genetic expression differences." Together with this, a new nomenclature was also suggested: a trinomial including the authors' name, the location, and the frequency distribution of the chemotype. Although there are valuable ideas behind this approach, in practice, such a description is hardly possible because of the huge amount of different investigations.

Unfortunately, at present, from the majority of literature, the range and size of chemical difference is frequently not obvious. Although there is a need for some kind of minimum criteria in defining a chemical taxon as a separate one, the range and size of divergence are not universally accepted and applied. In case of volatile components, this question is maybe even more difficult to answer than in several other secondary compounds, as essential oils consists of a huge number of individual constituents. How many compounds and which of them should be taken into consideration?

Hegnauer (1962) suggested that essential oil constituents that attain at least 1% should be considered for chemotaxonomic evaluation. However, by the quick development of analytical methods, the number of identified components has been increased enormously; thus, 1% is no longer of the same significance as it used to be.

The increase of the number of identified compounds in essential oils can be well demonstrated by comparison of the scientific articles published in the 1990–91 and 2017–18 volumes of *Journal of Essential Oil Research*. In the starting volumes, the mean number was 41 components/sample (marginal values: 10–121), while in the last volumes, it reached 68 compounds as a mean (marginal values: 21–191), which is a more than 1.5-fold increase during 20 years.

Now, there are no universally accepted criteria for delimitation of a chemical taxon. Evaluation is mostly dependent on both the spectrum and relative abundance of the components.

This approach indicates a further question if GC area percentages are appropriate and precise enough for evaluation of the significance of individual components. Area percentage is a relative value and it is changing by the total number of the identified compounds in the essential oil as a mixture, not directly dependent from the absolute accumulation quantity of the target compound. The majority of the literature still publishes area percentages, which, unfortunately, are much less adequate for chemotaxonomic or genetic/biosynthetic conclusions.

The evaluation is sometimes severely aggravated with the—seemingly—obvious question: what is, practically, an individual component? Even the determination of a chemical compound is to a certain extent dependent on the goal of the evaluation. For a relatively large group of terpenoid compounds, different enantiomers are known *in vivo*, such as (+)/(−) sabinene, (+)/(−) limonene, and (+)/(−) pinene, which usually also possess different biological activities. Usually, the plants are producing predominantly only one of the isomers, and in each case, the enantiomeric purity is characteristic for the species. In the study of Özek et al. (2010), each of the investigated eleven *Thymus* species accumulated mainly (−)-linalool (70–100%), while the studied five *Nepeta* species could be characterized by 100% presence of (+)-linalool in the hydro-distilled oil. Similarly, (1*R*)-(+)-camphor overdominated in the oils of both basil species *O. canum* and *O. kilimandscharicum* (Pragadheesh et al., 2013). In other species however, racemic mixtures of certain components were found in the genuine oils like in *Salvia microstegia* both isomers of linalool (Özek et al., 2010) or in *Lonicera iliensis* both isomers of α- and β-pinenes, sabinene and limonene (Kushnarenko et al., 2016).

The enantiomeric purity or characteristic ratio of the isomers might be a very valuable marker for authenticity control of essential oils. However, detection of the genuine constitution is often very difficult because racemization may happen during drying, distillation, storage and also by non-enzymatic reactions like autoxidation (Kreck et al., 2002). The genetic regulation of producing the corresponding isomer by a species has been less studied. In peppermint, it was proved that *in vivo* transformation of the not genuine (1S)-pulegone into (S)-isomenthone and (S)-menthofuran was carried out to a high degree (Fuchs et al., 2000), which indicates that the enzymatic processes are less stereoselective, but the end product depends to a large extent on the precursor. Different plant parts also may have specific enantiomeric distributions, which has hardly been investigated yet. (1*R*)-(+)-α-pinene is the characteristic isomer in Lithuanian accessions of *Juniperus communis* and its ratio proved to be 74%±13% in leaves while 69%±17% in unripe cones compared to the other isomer (1*S*)-(−)-α-pinene (Labokas and Ložienė, 2013).

The usage of enantiomeric distribution in chemotaxonomic evaluation is still not widespread. Further research is needed to define its role in systematics in case of different taxonomic units.

The methods of production of essential oils is a basic determining factor of its composition, the detailed discussion of which is out of the scope of this chapter and is discussed elsewhere. Nevertheless, it is important to remember the fact that even slight changes of distillation process (equipment type, distillation time, recovery method, etc.) and details of the following analysis may severely contribute to an altered spectrum and quality of the essential oil product (Zámboriné et al., 2018).

Another important aspect in selecting relevant compounds for evaluation is the well-known fact that distilled oils not rarely contain artefacts. The synthesis of these compounds not occurring in the natural oils during water distillation or storage may severely disturb the evaluation both from theoretic and practical points of view. Increased levels of terpinen-4-ol in the distilled oil of marjoram (*Majorana hortensis*) compared to the genuine volatiles (Fischer et al., 1987) or development of spathulenol from bicyclogermacrene during extraction or storage (Toyota et al., 1996) are very good examples for it. The proportions of these compounds are obviously not appropriate ones to reflect the biological variability.

Qualitative and quantitative differences may have of different significance when distinguishing taxonomical units. Denomination of qualitative and quantitative chemotypes, however, raises a further dilemma. The proportion or the absolute quantity of the accumulated volatile compounds is a continuous variable; thus, fixing a borderline between the values would be extremely difficult. A possible approach for this question is illustrated on the example of six Hungarian chemotypes of tansy (*Tanacetum vulgare*) in Table 4.12. Among the six chemotypes, the thujone, thujene acetate and davanone types may be accounted for as qualitative variants as the main compound is taking the huge majority of the total oil spectrum and these components are detected in the other types only in traces. The further three chemotypes can be considered as quantitative chemotypes as they have several components in common with each other, but definite and stable differences can be detected in the proportions of these compounds.

The example of different chemodemes (chemotype with separate distribution area) of *Achillea crithmifolia* can only be characterized properly if evaluation is carried out on the basis of three main compounds (Table 4.13). In this case, the complexity of this phenomenon can be clearly seen. Although the first two major compounds, camphor and 1,8-cineole, are deviating only in quantitative term, the third main compound seem to be definitive for the characterization of the chemotype. These three compounds are biosynthetically not closely related ones, and it can be supposed that the detectable composition is the result of diverse genetic constitution.

If discussing chemotypes, the ones of *A. crithmifolia* mentioned above may be considered as "mixed" ones because not a single compound is determining the unique chemical feature of the taxon. On the other side, we could speak about "pure" chemotypes where a single component is clearly defining the chemical profile of the taxon. Polatoglu (2013) defined the ratio of such a single component as over 50% while Huong et al. (2018) declared pure chemotypes with the main component occupying above 30% of the oil; however, unfortunately, there is not a universally accepted definition of this term.

We can conclude that *biological variability* in the case of volatile compounds is, in practice, a concentrated determination of numerous aspects: appearance of a specific spectrum of chemical compounds in plant populations, in single individuals, in different plant parts or in different periods of

TABLE 4.12
Proportion of Characteristic Compounds in the Essential Oil of Different Tansy (*Tanacetum vulgare*) Chemotypes

	Chemotype					
Compound	**Thujone**	**Thujene Acetate**	**Camphor**	**Borneol**	**1,8-Cineole**	**Davanone**
β-Pinene				3–5	7–16	
1,8-Cineole			8–10		23–31	
α-Thujone	55–64					
β-Thujone	26–32					
Thujene acetate		35–45				
Carvyl acetate		22–32				
Artemisia ketone			2–9			
Camphor			38–42	3–4	3–8	
Borneol			2–5	25–30	10–27	
Lyratol			3–5	7–10		
Bornyl acetate				30–43		
Lyratyl acetate				8–11		
Davanone						60–74
Davanol						6–13

Source: Németh, É., Unpublished.
Note: Only components are accumulating at least in 5% of total oil area are indicated.

plant life. Therefore, to check and determine the biological variability, the most proper method according to the question should be used! Characteristic requirements, marginal values, limits or intervals of quantitative features might be different for different reasons. Evaluation of a cultivar, checking the quality of a product or a drug, taxonomic studies or practical breeding each may require different assessments.

Classical chemotaxonomy emphasized that not the presence of a compound itself, but the biosynthetic processes, the potential of a plant for the formation of the given compound, should be taken into account. According to the present knowledge, this general definition seems to be oversimplified and not fully adequate anymore. Today, we have to face two approaches of chemotaxonomic considerations.

A pragmatic evaluation—although often published and mentioned otherwise—has nothing to do with real taxonomic aspects but collects and summarizes the detected chemosyndromes, i.e., compositional changes in the essential oil due to different internal or external factors. This approach is quite frequent in practice if certain quality requirements should be fulfilled. In these cases, a

TABLE 4.13
Chemotypes of *Achillea crithmifolia* in Different European Areas

Area (Country)	**Component I**	**Component II**	**Component III**	**Reference**
Bulgaria	Camphor 5%–30%	1,8-Cineole 3%–45%	Artemisia alcohol 20%–40%	Konakchiev and Vitkova (2004)
Hungary	Camphor 20%–50%	1,8-Cineole 5%–30%	Borneol 5%–10%	Németh et al. (2000)
Serbia	Camphor 30%	1,8-Cineole 30%	Chrysanthenyl acetate 20%	Palić et al. (2003)

stable composition both qualitatively and quantitatively should be assured, and from this point of view, the proper knowledge on the detectable influencing factors and possibilities for their regulation is of primary importance, indeed. That means that the range of variability should be evaluated by investigating the row of chemosyndromes that may appear under different circumstances or due to different treatments. These circumstances or treatments may consist of temperature regimes, illumination or any ecological factor or even characteristic habitat as a complex background. For this pragmatic approach, it is not necessary that biotic and abiotic factors are strictly distinguished because all of these factors may have an effect on the chemosyndromes. Modifications and control are possible in majority of cases without deeper knowledge about the biosynthetic processes or their genetic regulation. The results may be directly utilized in cultivation and production of essential oil–bearing species; however, this approach tells nothing about taxonomic relationships of the examined objects.

However, for a well-established regulation of the desired composition by breeding or evaluating the theoretical connections, it is not enough to define taxonomic relationships by simply detecting chemosyndromes. At a more sophisticated level we have to search in more detail for the physiological and genetic backgrounds or even the metabolomic connections.

At the same time, this is the aspect which seemed to be more simple some decades ago than today. As cited above, Tétényi (1970) declared that determination and separation of an intraspecific chemical taxon should be based on the divergent biosynthetic routes, on the potential of the plant for the synthesis of certain compounds instead of on the presence of any compound itself. It is justified by the fact that biosynthesis of a certain compound might take place by different routes in divergent species; therefore, simply the presence of any compound may not be proof for a taxonomic relationship. There are today already several examples for producing the same molecule as active compound by different enzymes and, thus, likely on the base of different genetic determination. Linalool synthase from *Clarkia breweri* has a different constitution than linalool synthase from *Arabidopsis* plants (Dudareva et al., 2006) and only 41% identity to the same enzyme from *Mentha citrata* (Crowell et al., 2002).

The statement that not the compound itself but its biosynthetic pathway has the taxonomic significance may be still valid. However the "pathway" should be defined as a complex metabolomics term including the genomic constitution, gene expression, transcriptional and translational regulation, enzyme activities at different levels of biosynthesis, interactions with transporters, translocation, and spatial isolation (Dudareva et al., 2006; Dinesh and Nagegowda, 2010). The result of the enzymatic reaction is determined not only by the availability and constitution of the enzyme but also by the availability of precursors and different interactions; therefore, the simple presence or absence of a certain enzyme in general cannot determine the final product.

It has been estimated that about half of all mono- and sesquiterpene synthases act as multiproduct enzymes! Numerous terpene synthases/cyclases are able to produce a wide range of terpenic skeletons. It is usually not possible to predict the product profile of terpene synthases solely on the basis of their primary structure (Tholl, 2006). Limonene cyclase frequently catalyzes the formation of myrcene and pinenes beside limonene from acyclic precursors (McCaskill and Croteau, 1998). The same synthase converts GPP into myrcene and (E)-β-ocimene while the synthase TPS1 has three acyclic sesquiterpene products: (E)-β-farnesene, (3R)-(E)-nerolidol and (E,E)-farnesol (Dudareva et al., 2006). Taxonomically unrelated species seem to have closely related pathways for the formation of main terpenoid compounds. 3-hydroxylation of the monoterpene precursor limonene by a P450 enzyme produces trans-isopiperitenol, a volatile compound characteristic for mint species (Lupien et al., 1999), while 6-hydroxylation by another P450 enzyme yields trans-carveol, which is further oxidized by nonspecific dehydrogenase to carvone, the main volatile compound of caraway, which is not a closely related species taxonomically (Bouwmeester et al., 1998).

According to McConkey et al. (2000), several enzymes of the menthol biosynthetic pathway appear to originate from widely divergent genetic resources in primarily metabolism which would make their products less significant as chemotaxonomic characteristics.

Although "the biosynthetic route" in the sense of former publications seems to be a quasi linear process resulting in a special compound, the *in vivo* plant metabolism means a very complex regulatory and operating system nowadays only partially detected. Moreover, it has been supposed that the detectable mechanisms are only a part of the reserves of the plant and plants have a resource of "hidden" biosynthetic capacities, a practically unlimited potential to produce a large array of different compounds if they are activated by novel available precursors (Lewinsohn and Gijzen, 2009).

That has been the reason why transgenic operations with single genes frequently could not result in the desired change of the volatile spectrum. Cloning of a single gene is even more unlikely to result in a substantial production of the desired volatile compound if this compound is the final product of a long metabolic pathway (Dudareva and Pichersky, 2008).

All of these factors contribute to the fact that some volatile components are really characteristic for the species or intraspecific taxon, while others are hardly to join to the taxonomic units defined by botanical features. Essential oil components should be individually evaluated if they may serve as chemotaxonomic markers for the target organism.

A very good example for this is the study of Radulović et al. (2007) about the chemotaxonomic relationships of Balkan *Achillea* species. They evaluated altogether 47 records of 23 yarrow taxa according to their volatile constituents by PCA (principal component analysis). If each component detected in the oils at least in 1% were taken into account—except for three taxa—no clear distinction could be established, presumably because of the universal presence of 1.8-cineole, camphor, and borneol in higher amounts. The rate of oxygenation in the molecule was not an appropriate sign of real taxonomic connections either, because it may be influenced by ecological/geographical factors. Finally, dealing with groups of biosynthetically related compounds instead of individual components proved to be useful: choosing the monoterpene structural types (p-menthane, bornane, pinane, etc.) as discriminative variables, the best grouping appropriate to accepted taxonomic classification could be set up.

4.7 CONSIDERATIONS FOR PROPER ASSESSMENT OF NATURAL VARIABILITY

Detection of the biological variability is usually not an easy job. Unfortunately, in practice, numerous irrelevant papers have been published about the chemical variability of essential oil–bearing species. Many articles deal with determination and simple description of the essential oil composition of a given plant material under given conditions (Figure 4.1). These types of articles might be useful to enhance the literature with new information. Mechanical assessment of such articles as a reference about the chemical variability of the given taxon is, however, a bad practice. Comparison of several independent publications about the target species is most likely not an appropriate tool for the evaluation of biological diversity either, because the plant genetic material, the habitat, environmental conditions, sampling, processing, and the analytical method itself may strongly influence the measured data, while, unfortunately, even these circumstances are rarely adequately provided. Therefore, a summary of publications—even if there is a huge number of them—is not able to reflect the biological variability of a species or any taxon.

Practically, a similar procedure is carried out if analytical results of several samples are compared in the frame of a single publication; however, samples originate from different habitats or other kind of sources. For example, a large pool of samples from different localities of Greece showed that the quantitative composition of essential oil from *Mentha pulegium* varies greatly (Kokkini et al., 2004). The most variable compound is pulegone, its proportion ranges from traces to 91% of the total oil. Fluctuations in the contents of piperitone (from traces to 97%), menthone (from traces to 53%), isomenthone (from traces to 45%), piperitenone (from traces to 40.0%) and isopiperitenone (from not detected to 23%) has been found, too. The authors emphasize, that in localities where the real Mediterranean climate dominates, the total oil content and the amounts of the more reduced products of the p-menthane biosynthetic pathway, like menthone and/or isomenthone and their derivatives, were increased. Concerning the market quality of the pennyroyal oil, these conclusion may be enough. However, the changing essential oil profile might be the consequence not only of the diverse environmental conditions from south to north, but might be the manifestation of different genotypes or even both. To conclude about the genetic-biological variability, the plant material of different habitats must be collected and investigated directly on the same plot (Llorens-Molina et al., 2016, 2017).

In any case, special care should be taken by choosing and defining a sample. Searching for the biological variability in essential oil composition, only plant material of well-defined origin should be investigated. Rather frequently, we find studies on commercial samples obtained from any market. It cannot be emphasized enough that these items are not of reproducible quality. Any statement about such samples may provide only general and/or smattering information about the characteristics of the species or taxon.

As an example, Raal et al. (2012) evaluated twenty commercial samples of caraway seeds. Although the work includes detailed analytical results on the composition of the samples, a valuable conclusion is impossible because of the inadequate definition of these samples. They originated from pharmacies and other shops of different countries but there was not enough information about the original genotype. The different age of the samples (2000–2008) also reduces the reliability of the data.

In sampling, another misunderstanding is caused by the fact that not rarely during comparison of populations, bulk samples are taken, processed and analyzed. As discussed above, individual plant variability is a very frequent and important form of biological variability, which may be hidden in such trials. As mentioned also by Franz and Novak in Chapter 3 of this book, representativeness of sample taking in a heterogeneous population is prerequisite for reliable results. This should contain a larger number of individuals (up to 50) even if the analytical methodology would make do with a small quantity of plant material. Random sampling or bulk samples may give bias and does not reflect the characteristic compositional profile of the population. Even if this type of sampling and evaluation may have its role in checking the drug quality of a commercial plantation when harvesting results in bulk material, but does not give an answer to the question about biological diversity.

In many cases, a single population is practically a mixture of individuals belonging to different chemotypes and their ratio is most often unknown. As it is reviewed in *Achillea crithmifolia* in Section 4.2.2, in each habitat, different chemotypes may be found in different proportions (Németh et al., 2000). If this fact is not respected, the published results provide only some kind of "analytical mean" of the different chemotypes instead of characterizing the diversity of the taxon.

In summary, relevant data about natural variability should be based on individual plant samples, which have been taken and analyzed in appropriate replications. Figure 4.1 shows the low proportion (approximately 1%) of recent publications dealing with individual differences compared to other topics.

In the case of investigating perennial species, the weather conditions of the growth period (vegetation year) and the age of the plantation can hardly be separated. The problem is similar at different harvest times within the same year. Reliable information about the characteristics of the taxon can be obtained only by consequent sampling, keeping other circumstances constant and continuing the trials for a longer period, if necessary.

By taking two samples in October 2007 and in June 2011, Guimarães et al. (2012) wanted to establish the influence of the collection period on the concentrations of the essential oil components in *Mikania glauca*. Even if the sampling would have been representative, there are several problems, which make the evaluation more than questionable. The two random sampling times do not only mean that four years passed but the plants became older, too. Phenological phase of the plants might have been different in October and in June, two different periods of the year. Besides, weather conditions in the growth period or during sampling may also have an effect on the analytical results. The information is useful in showing that quality of the leaves is not stable, but no exact conclusion about the real influencing factors and therefore not about their possible regulation is available from the data.

There are also some further problems to mention. Unfortunately, in the literature quite frequently, the effect of different factors or treatments is interpreted as a direct influencing factor on the essential oil composition; however, the basis of the change is the biological variability of the plants.

Caraway may be a good example for this. Formerly, a negative correlation between wind velocity and loss of essential oil and carvone was anticipated as the consequence of increased volatilization, thus, practically, an environmental factor influencing the oil composition. However, Bouwmeester (1998) suggested that the loss of carvone is in connection with seed shedding enhanced by the wind. Shedding results in loss especially of the older/more ripen seeds while the premature, younger ones remain on the stalks. Due to the delayed activity of limonene synthase, in these premature fruits. carvone usually did not reach its maximum level. In this case, the biological variability connected to the special biological process has been mistaken and evaluated as only an environmental effect.

Our investigations in caraway revealed another interesting finding about the effect of row distance on the essential oil components of the seeds (Valkovszki and Zámbori-Németh, 2010). It was established, that the wider row distance (48 cm) resulted in 10%–25% loss of carvone in the oil compared to narrower spacing (Figure 4.4). The difference is more pronounced in fertilized plots where optimal nutrient supply is assured for each plant individual. For the farmers, it is worth considering if quality may be improved by this way, but looking for the backgrounds of this phenomenon, the picture seems to be more complex and the influence of the studied treatments only indirect. In a wider spacing, the plant develops more umbels but seeds in higher order umbels ripen several days—sometimes even weeks—later, than older ones. It has been already explained that highest carvone content is detected in the fully ripen seeds. In consequence of large proportion of higher order umbels in plots of wider spacing, most likely a big ratio of seeds still did not reach at harvest the phenological stage of maximal carvone content.

Agrotechnical methods in general act indirectly through influencing growth dynamics, developmental characteristics, and organic proportions, which further on determine the accumulation of special compounds.

Propagation method seems to have an effect on essential oil quality in many cases. Zheljazkov et al. (1996) detected differences in the main components of peppermint oil in plantations propagated by different methods. Menthol content of the oil was highest and menthyl-acetate level was the lowest in plots established by rhizomes in autumn (Table 4.14). The investigated clones reacted not uniformly, the variety "Zephir," for example, showed the highest pulegone content in plots propagated by rhizomes in summer, while all the other cultivars had the lowest proportions in this treatment. The results, however, should be evaluated as an indirect consequence of the propagation method, as it is most unlikely that planting itself has any effect on the biosynthesis of terpenoids in the newly developed shoots several months after transplanting. However, propagation method may

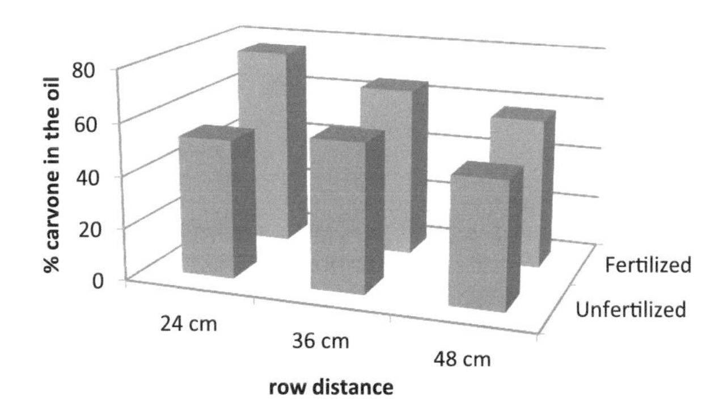

FIGURE 4.4 Carvone content (in GC area %) of caraway (*Carum carvi*) seed oil in plots of different row distances and fertilization treatments. (Adapted from Valkovszki, N. and Zámbori-Németh, É. 2010. *Acta Aliment. Hung.*, 40: 235–246.)

TABLE 4.14
Main Essential Oil Components (In Area %) of Some Peppermint Cultivars after Different Propagation Methods

Variety	Propagation Method	Menthol	Neo-Menthol	Menthyl Acetate	Pulegone
'No. 1'	Rooted cuttings	42.3	7.16	10.70	0.53
	Summer rhizomes	44.3	6.41	9.64	0.18
	Autumn rhizomes	47.8	5.84	8.38	0.86
'No. 101'	Rooted cuttings	53.4	4.24	9.08	0.29
	Summer rhizomes	53.8	4.48	10.40	0.16
	Autumn rhizomes	60.5	4.46	8.63	0.22
'Zephir'	Rooted cuttings	59.3	1.80	9.81	0.54
	Summer rhizomes	58.9	2.33	9.76	1.06
	Autumn rhizomes	59.1	1.82	8.93	0.61
'Mentolna 18'	Rooted cuttings	62.4	2.34	7.36	0.17
	Summer rhizomes	62.1	1.95	7.18	0.09
	Autumn rhizomes	63.8	2.10	6.71	0.14

Source: Modified from Zheljazkov, V. et al. 1996. *J. Essent. Oil Res.*, 8: 35–45.

have an influence on the phenological phases, appearance of flowers, and size and number of leaves. Besides, accelerated or prolonged development due to the different propagation method and time means different weather conditions for the growing period. All of these factors are able to influence the accumulation of volatiles in peppermint, which may reflect itself in the obtained data. The results are important in optimization of agrotechniques under the given circumstances, but obviously, do not represent the primary background of the detected variability.

Not rarely, essential oil plants are propagated experimentally by *in vitro* methods. According to Ibrahim et al. (2011), tissue culture provides a fast breeding technology for tarragon (*Artemisia dracunculus*). They report that in the plantation installed with plantlets obtained from callus cultures and organogenesis, estragole concentration was significantly reduced. According to our knowledge, tissue culture from different plant parts as explants is hardly able to change the genotype of the plant, except by somatic mutations which is, however, not reproducible during the technology. On the other side, volatile concentrations may be influenced by an eventual structural (tissue constitution and essential oil–accumulating organelles) change of the plant which, however, is already an indirect effect and should be evaluated accordingly.

Thus, it is important to know that practically all of the agrotechnical interventions have effects on the growth and development of the plantation. Planting, irrigation, pruning, and fertilization influence the physiological processes of the plants but do not necessarily represent the direct influencing factors of volatile formation.

REFERENCES

Aćimović, M. V., S. I. Oljača, V. V. Tešević, M. M. Todosijević, and J. N. Djisalov. 2014. Evaluation of caraway essential oil from different production areas of Serbia. *Hort. Sci. (Prague)*, 41: 122–130.

Aiello, N., F. Scartezzini, L. D'Andrea, A. Albasini, and P. Rubiolo. 2001. *Reserach on St. John's Wort, Hyssop, Rosemary and Common Golden Rod at Different Planting Density in Trentino.* Publications of ISAFA Research Centre, pp. 23–31.

Altunkaya, A., B. Yıldırım, K. Ekici, and Ö. Terzioğlu 2014. Determining essential oil composition, antibacterial and antioxidant activity of water wormwood extracts. *Gida*, 39: 17–24.

Andro, A. R., I. Boz, M. Zamfirache, and I. Burzo 2013. Chemical composition of essential oils from *Mentha aquatica* L. at different moments of the ontogenetic cycle. *J. Med. Plants Res.*, 7: 47.

Argañosa, G. C., F. W. Sosulski, and A. E. Slinkard. 1998. Seed yields and essential oils of annual and biennial caraway (*Carum carvi* L.) grown in Western Canada. *J. Herbs Spices Med. Plants*, 6: 9–17.
Arino, A., I. Arberas, G. Renobales, S. Arriaga, and J. B. Dominguez. 1999. Seasonal variation in wormwood (*Artemisia absinthium* L.) essential oil composition. *J. Essent. Oil Res.*, 11: 619–622.
Ayouni, K., M. Berboucha-Rahmani, H. K. Kim, D. Atmani, R. Verpoorte, and Y. H. Choi. 2016. Metabolomic tool to identify antioxidant compounds of *Fraxinus angustifolia* leaf and stem bark extracts. *Ind. Crop. Prod.*, 88: 65–77.
Bagci, E., M. Kursat, and Civelek, S. 2010. Essential oil composition of the aerial parts of two *Artemisia* species (*A. vulgaris* and *A. absinthium*) from East Anatolian region. *J. Essent. Oil Bearing Plants*, 13: 66–72.
Baser, K. H. C., M. Kürkçüoğlu, B. Demirci, T. Özek, and G. Tarımcılar. 2012. Essential oils of *Mentha* species from Marmara region of Turkey. *J. Essent. Oil Res.*, 24: 265–272.
Baser, K. H. C., M. Kürkçüoğlu, G. Tarımcılar, and G. Kaynak. 1999. Essential oils of *Mentha* species from Northern Turkey. *J. Essent. Oil Res.*, 11: 579–588.
Baser, K. H. C., T. Özek, N. Kirimer, and G. Tümen. 1993. The occurrence of three chemotypes of *Thymus longicaulis* C. Presl. subsp. *longicaulis* in thc same population. *J. Essent. Oil Res.*, 5: 291–295.
Basta, A., O. Tzakou, and M. Couladis. 2007. Chemical composition of *Artemisa absinthium* L. from Greece. *J. Essent. Oil Res.*, 19: 316–318.
Bélanger, A. and Dextraze, L. 1993. Variability of chamazulene within *Achillea millefolium* L. *Acta Horticult. Nr.*, 330, 141–145.
Bernáth, J., É. Németh, A. Katta, and É. Héthelyi. 1996. Morphological and chemical evaluation of fennel (*Foeniculum vulgare* Mill.) populations of different origin. *J. Essent. Oil Res.*, 8: 247–253.
Bernotienė, G. and R. Butkienė. 2010. Essential oils of *Hyssopus officinalis* L. cultivated in East Lithuania. *Chemija.*, 21: 135–138.
Blaschek, W., R. Hänsel, K. Keller, J. Reichling, H. Rimpler, and G. Schneider. 1998. *Hagers Handbuch der pharmazeutischen Praxis, Folgeband 2.* Drogen: A-K, Berlin-Heidelberg: Springer Verlag.
Bodor, Zs., É. Németh, and K. Csalló. 2006. Produktionspotenzial ein- und zweijahrigen Formen des Muskatellersalbeis (*Salvia sclarea* L.) und Einfluss unterschiedlicher Aussaatzeiten. *Z. Arznei- und Gewürzpflanzen.*, 11: 40–47.
Bodor, Zs., É. Németh, K. Csalló, and Sz. Sárosi. 2009. Einfluss unterschiedlicher Aussaatzeiten und Standorte auf die Produktion der Königskerzensorte 'Napfény' (*Verbascum phlomoides*). *Z. Arznei- und Gewürzpflanzen.*, 14: 32–36.
Bouwmeester, H. J. 1998. Regulation of essential oil formation in caraway. In: *Caraway – The Genus Carum*, É. Németh (ed.), pp. 83–101, Amsterdam: Harwood Academic Publishers.
Bouwmeester, H. J. and A. M. Kuijpers. 1993. Relationship between assimilate supply and essential oil accumulation in annual and biennial caraway. *J. Essent. Oil Res.*, 5: 143–152.
Bouwmeester, H. J., J. Gershenzon, M. C. J. M. Koning, and R. Croteau. 1998. Biosynthesis of the monoterpenes limonene and carvone in the fruit of caraway. *Plant Physiol.*, 117: 901–912.
Božović, M., A. Pirolli, and R. Ragno. 2015. *Mentha suaveolens* Ehrh. (Lamiaceae) essential oil and its main constituent piperitenone oxide: Biological activities and chemistry. *Molecules*, 20: 8605–8633.
Bylaite, E., R. P. Venskutonis, and J. P. Roozen. 1998. Influence of harvesting time of volatile components in different anatomical parts of lovage (*Levisticum officinale* KOCH.). *J. Agric. Food Chem.*, 46: 3735–3740.
Chalchat, J., D. Adamovic, and M. S. Gorunovic. 2001. Composition of oils of three cultivated forms of *Hyssopus officinalis* endemic in Yugoslavia: F. *albus*, F. *cyaneus* and F. *ruber*. *J. Essent. Oil Res.*, 13: 419–421.
Chalchat, J., M. S. Gorunovic, D. Petrovic, and V. V. Zlatkovic. 2000. Aromatic plants of Yugoslavia. II. Chemical composition of oils of *Achillea clavenae, A. collina, A. lingulata. J. Essent. Oil Res.*, 12: 7–10.
Chialva, F., P. A. P. Liddle, and G. Doglia. 1983. Chemotaxonomy of wormwood (*Artemisia absinthium* L.), *Z. Lebensm. Unters. Forsch.*, 176: 363–366.
Chizzola, R., H. Michitsch, and C. Franz. 2008. Antioxidative properties of *Thymus vulgaris* leaves: Comparison of different extracts and essential oil chemotypes. *J. Agric. Food Chem.*, 56: 6897–6904.
Chou S. T., H. Y. Peng, J. C. Hsu, C. C. Lin, and Y. Shih 2013. *Achillea millefolium* L. essential oil inhibits LPS-induced oxidative stress and nitric oxide production in RAW 264.7 macrophages. *Int. J. Mol. Sci.*, 14: 12978–12993.
Chung, H. G., É. Németh, 1999: Studies on the essential oil of different fennel (*Foeniculum vulgare* Mill.) populations during ontogeny. *Internat. J. Hort. Sci.*, 5: 27–30.
Collin, G. J., H. Deslauriers, N. Pageau, and M. Gagnon. 1993. Essential oil of tansy (*Tanacetum vulgare* L.) of Canadian origin. *J. Essent. Oil Res*, 5: 629–638.

Croteau, R. B., R. M. Davis, K. L. Ringer, and M. R. Wildung. 2005. (–)-Menthol biosynthesis and molecular genetics. *Naturwissenschaften.*, 92: 562–577.

Crowell, A. L., D. C. Williams, E. M. Davis, M. R. Wildung, and R. Croteau. 2002. Molecular cloning and characterization of a new linalool synthase. *Arch. Biochem. Biophys.*, 405: 112–21.

Danila, D., C. Stefanache, A. Spac, E. Gille, and U. Stanescu. 2012. Studies on the variation of the volatile oil composition in experimental variants of *Hyssopus officinalis*. In: *Program and Book of Abstracts, 43rd ISEO Lisbon*, p. 105.

de Pooter, H. L., J. Vermeesch, and N. S. Schamp. 1989. The essential oils of *Tanacetum vulgare* L. and *Tanacetum parthenium* (L.) Schultz-Bip. *J. Essent. Oil Res.*, 1: 9–13.

Derwich, E., Z. Benziane, and A. Boukir. 2009. Chemical compositions and insecticidal activity of essential oils of three plants *Artemisia* sp. (*A. herba-alba, A. absinthium* and *A. Pontica* (Morocco). *Electronic J. of Environ. Agricult. Food Chem.*, 8: 1202–1211.

Dinesh, A. and D. A. Nagegowda. 2010. Plant volatile terpenoid metabolism: Biosynthetic genes, transcriptional regulation and subcellular compartmentation. *FEBS Lett.*, 584: 2965–2973.

Dragland, S., J. Rohloff, R. Mordal, T. H. Iversen. 2005. Harvest regimen optimalization and essential oil production in five tansy (*Tanacetum vulgare* L.) genotypes under Northern climate. *J. Agric. Food Chem.*, 53: 4946–53.

Dubiel, E., M. Herold, F. Pank, W. Schmidt, and M. Stein. 1988. 30 Jahre "Multimentha" (*Mentha piperita* L.). *Drogenreport.*, 1: 31–64.

Dudareva, N. F. Negre, D. A. Nagegowda, and I. Orlova. 2006. Plant volatiles: Recent advances and future perspectives. *Critical Rev. Plant Sci.*, 25: 417–440.

Dudareva, N. and E. Pichersky. 2008. Metabolic engineering of plant volatiles. *Curr. Opin. Biotechnol.*, 19: 181–189.

Ehrendorfer, F. 1963. Probleme, Methoden und Ergebnisse der experimetellen Systematik. *Planta Med.*, 3: 234–251.

Embong, M. B., D. Hadziyeu, and S. Molnar. 1977. Essential oils from species grown in Alberta, caraway oil (*Carum carvi*). *Can. J. Plant Sci.*, 57: 543–549.

Figueiredo, C., J. M. S. Barroso, M. S. Pais, and J. C. Scheffer. 1992. Composition of the essential oils from leaves and flowers of *Achillea millefolium*. *Flavour Fragr. J.*, 7: 219–222.

Fischer, N., S. Nitz, and F. Drawert. 1987. Original flavour compounds and the essential oil composition of marjoram (*Majorana hortensis*). *Flavour Fragr. J.*, 2: 55–61.

Fleischer, A. and Z. Fleischer. 1988. The essential oil of annual *Carum carvi* L. grown in Israel. In: *Flavours and Fragrances – A World Perspective*, B. M. Lawrence (ed.), pp. 33–40, Amsterdam: Elsevier Sci. Publ. B.V.

Forsen, K. and M. Schantz. 1971. Neue Hauptbestandteile im ätherischen Öl des Rainfarns in Finnland. *Arch. Pharmaz.*, 304: 944–952.

Forwick-Kreutzer, J., B. M. Moseler, R. Wingender, and J. Wunder. 2003. Intraspecific diversity of wild plants and their importance for nature conservation and agriculture. *Mitt. Biol. Bundesanst. Land-Forstwirtsch.*, 393: 210–215.

Foster, A. J., D. E. Hall, L. Mortimer, S. Abercromby, R. Gries, J. Bohlmann, J. Russell, and J. Mattson. 2013. Identification of genes in *Thuja plicata* foliar terpenoid defenses. *Plant Physiol.*, 161: 1993–2004.

Fraternale, D., D. Ricci, F. Epifano, and M. Curini. 2004. Composition and antifungal activity of two essential oils of hyssop. *J. Essent. Oil Res.*, 16: 617–622.

Fuchs, S., T. Beck, and A. Mosandl. 2000. Biogeneseforschung ätherischer Öle mittels SPME-enentio-MDGC/MS. *GIT Labor-Fachzeitschrift.*, 4: 358–362.

Galambosi, B. and P. Peura. 1996. Agrobotanical features and oil content of wild and cultivated forms of caraway (*Carum carvi* L.). *J. Essent. Oil Res.*, 8: 389–397.

Galambosi, B., K. P. Svoboda, A. G. Deans, and É. Héthelyi. 1993. Agronomical and phytochemical investigation of *Hyssopus officinalis*. *Agric. Sci. Finl.*, 2: 293–302.

Gildemeister, E. and F. Hoffmann. 1961. *Die ätherischen Öle*. Vol. 7., 661. Berlin: Akademie Verlag.

Gnilka, R., A. Szumny, and Cz. Wawrzeńczyk. 2010. Efficient method of isolation of pure (–)-α- and (+)-β-thujone from *Thuja occidentalis* essential oil. In: *Program and Book of Abstracts, 41st ISEO Wroclaw*, S. Lochynski and Cz. Wawreńczyk, (ed.), p. 69.

Gosztola, B. 2012. Morphological and chemical diversity of different chamomile (*Matricaria recutita* L.) populations of the Great Hungarian Plain. (In Hungarian). PhD Dissertation, Corvinus University of Budapest.

Gouyon, P. H., Ph. Vernet, J. L. Guillerm, and G. Valdeyron. 1986. Polymorphism and environment: The adaptive value of the oil polymorphism in *Thymus vulgaris* L. *Heredity*, 57: 59–66.

Grassi, P., J. Novak, H. Steinlesberger, and Ch. Franz. 2004. A direct liquid, non-equilibrium solid-phase micro-extraction application for analysing chemical variation of single peltate trichomes on leaves of *Salvia officinalis*. *Phytochem. Anal.*, 15: 198–203.

Grausgruber-Grüger, S., C. Schmiderer, R. Steinborn, and J. Novak. 2012. Seasonal influence on gene expression of monoterpene synthases in *Salvia officinalis* (Lamiaceae), *J. Plant Physiol.*, 169: 353–359.

Gross, M., E. Lewinsohn, Y. Tadmor, E. Bar, N. Dudai, Y. Cohen, and J. Friedman. 2009. The inheritance of volatile phenylpropenes in bitter fennel (*Foeniculum vulgare* Mill. var. vulgare, Apiaceae) chemotypes and their distribution within the plant. *Biochem. Syst. Ecol.*, 37: 308–316.

Gudaitytė, O. and P. R. Venskutonis. 2007. Chemotypes of *Achillea millefolium* transferred from 14 different locations in Lithuania to the controlled environment., *Biochem. Syst. Ecol.*, 35: 582–592.

Guimarães, L. G., M. G. Cardoso, L. F. Silva et al. 2012. Chemical analyses of the essential oils from leaves of *Mikania glauca* Mart. ex Baker. *J. Essent. Oil Res.*, 24: 599–604.

Guitton, Y., F. Nicole, S. Moja, T. Bednabdelkader, N. Valot, S. Legrand, F. Julien, and L. Legendre. 2010. Lavender inflorescence – a model to study regulation of terpene synthesis. *Plant Signal. Behav.*, 5–6: 749–775.

Hegnauer, R. 1962. *Chemotaxonomie der Pflanzen*, Vol. 1, p. 114, Basel: Birkhäuser Verlag.

Hegnauer, R. 1978. Die systemische Bedeutung der ätherischen Öle. *Dragoco Rep.*, 24: 203–230.

Hendrics, H., D. J. D. van der Elst, F. M. S. van Putten, and R. Bos. 1990. The essential oil of Dutch tansy. *J. Essent. Oil Res.*, 2: 155–162.

Héthelyi, É., P. Tétényi, B. Dános, and I. Koczka. 1991. Phytochemical and antimicrobial studies on the essential oils of the *Tanacetum vulgare* clones by GC/MS. *Herba Hung.*, 30: 82–88.

Hiltunen, R. 1975. Variation and inheritance of some monoterpenes in *Pinus silvestris*. *Planta Med.*, 28: 315–323.

Hodzsimatov, K. H. and N. Ramazanova. 1974. Some biological characteristics of essential oil accumulation and composition in hyssop cultivated in Taskent region. (In Russian). *Rastit. Resursi*, 11: 238–242.

Hofmann, L. 1993. Einfluss von Genotyp, Ontogenese und äusseren Faktoren auf pflanzenbauliche Merkmale sowie ätherische Öle und Flavonoide von Klonen der Schafgarbe (*Achillea millefolium* Aggregat). PhD Dissertation, Technische Universität, München.

Holopainen, M., R. Hiltunen, J. Lokki, K. Forsen, and M. Schantz. 1987. Model for the genetic control of thujone, sabinene and umbellulone in tansy (*Tanacetum vulgare*). *Hereditas*, 106: 205–208.

Huong, N. T., Sz. Tavaszi-Sárosi, J. A. Llorens-Molina, and É. Zámborine-Németh. 2018. Compositional variability in essential oils of twelve wormwood (*Artemisia absinthium* L.) accessions. *J. Essent. Oil Res.*, 30: 421–430.

Ibrahim, A. K., A. A. Safwat, S. E. Khattab, and F. M. El Sherif. 2011. Efficient callus induction, plant regeneration and estragole estimation in tarragon (*Artemisia dracunculus* L.). *J. Essent. Oil Res.*, 23: 16–20.

International Organization for Standardization. 2013. Standards catalogue, ISO/TC 54 – Essential oils. https://www.iso.org/committee/48956/x/catalogue/ (accessed December. 2018).

Irving, S. and R. P. Adams. 1973. Genetic and biosynthetic relationships of monoterpene. In: *Terpenoids, structure, biogenesis and distribution – Recent Advances in Phytochemistry*, Runeckless, E. C. and T. J. Marby (eds.), Vol. 6. pp. 187–214, New York – London: Academic Press.

Johnson, C. B., A. Kazantzis, M. Skoula, U. Mitteregger, and J. Novak. 2004. Seasonal, populational ad otogenetic variation in the volatile oil content and composition of individuals of *Origanum vulgare* subsp. *hirtum* assessed by GC headspace analysis and by SPME sampling of individual oil glands. *Phytochem. Anal.*, 15: 286–292.

Joulain, D. and M. Ragault. 1976. Sur quelques nouvaux constituants de l'huile essentielle *d'Hyssopus officinalis* L. *Rivista Ital.*, 58: 129–131.

Jovanović, O., N. Radulovits, R. Palić, and B. Zlatković. 2010. Root essential oil of *Achillea lingulata* Waldst. & Kit. (Asteraceae). *J. Essent. Oil Res.*, 22: 336–339.

Judzetiene, A. and J. Budiene. 2010. Compositional variation in essential oils of wild *Artemisia absintium* from Lithuania. *J. Essent. Oil Bearing Plants*, 13: 275–285.

Juteau, F., I. Jerkovic, V. Masotti, M. Milos, J. Mastelic, J. M. Bessière, and J. Viano. 2003. Composition and antimicrobial activity of the essential oil of *Artemisia absinthium* from Croatia and France. *Planta Med.*, 69: 158–161.

Kindlovits, S. and É. Németh. 2012. Sources of variability of yarrow (*Achillea* spp.) essential oil. *Acta Aliment.*, 41: 92–103.

Kindlovits, S., Sz. Sárosi, K. Inotai, G. Petrović, G. Stojanović, and É. Németh. 2018. Phytochemical characteristics of root volatiles and extracts of *Achillea collina* Becker genotypes. *J. Essent. Oil Res.*, 30: 330–340.

Kizil, S., N. Hasami, V. Tolan, E. Kilinc, and H. Karatas. 2010. Chemical composition, antimicrobial and antioxidant activities of hyssop (*Hyssopus officinalis* L.) essential oil. *Plant Soil Environ.*, 49: 277–282.

Kokkini, S., E. Hanlidou, R. Karousou, and T. Lanaras. 2004. Clinal variation of *Mentha pulegium* essential oils along the climatic gradient of Greece. *J. Essent. Oil Res.*, 16: 588–593.

Koller, W. D. and P. Range. 1997. Geruchsprägende Inhaltsstoffe von Fenchel und Ysop, *Z. Arzn. Gew. Pfl.*, 2: 73–80.

Konakchiev, A. and A. Vitkova. 2004. Essential oil composition of *Achillea crithmifolia* Waldst. et Kit. *J. Essent. Oil Bearing Plants*, 7: 32–36.

Kreck, M., A. Scharrer, S. Bilke, and A. Mosandl. 2002. Enantioselective analysis of monoterpene compounds in essential oils stir bar sorptive extraction (SBSE)-enantio-MDGC-MS. *Flavour Fragr. J.*, 17: 32–40.

Kushnarenko, S. V., L. N. Karasholakova, G. Ozek, K. T. Abidkulova, and N. M. Mukhitdinov. 2016. Investigation of essential oils from three natural populations of *Lonicera iliensis*. *Chem. Nat. Compd.*, 52: 751–753.

Kutchan, T. M. 2005. A role for intra- and intercellular translocation in natural product biosynthesis. *Curr. Opin. Biotechnol.*, 8: 292–300.

Labokas, J. and K. Ložienė. 2013. Variation of essential oil yield and relative amounts of enantiomers of α-pinene in leaves and unripe cones of *Juniperus communis* L. growing wild in Lithuania. *J. Essent. Oil Res.*, 25: 244–250.

Lachenmeier, D. W., J. Emmert, T. Kuballa, and G. Sartor. 2006. Thujone: Cause of absinthism? *Forensic Sci Internat.*, 158: 1–8.

Lane A., A. Boecklemann, G. N. Woronuk, L. Sarker, and S. S. Mahmoud. 2010. A genomics resource for investigating regulation of essential oil production in *Lavandula angustifolia*. *Planta*, 231: 835–845.

Laribi, B., K. Kouki, A. Mougou, and B. Marzouk. 2010. Fatty acid and essential oil composition of three Tunisian caraway (*Carum carvi* L.) seed ecotypes. *J. Sci. Food Agric.*, 90: 391–396. DOI: 10.1002/jsfa.3827

Lawrence, B. M. 1979. *Progress in Essential Oils 1979–80*. Wheaton: Allured Publishing Corp.

Lawrence, B. M. 1992. Progress in essential oils. *Perfumer and Flav.*, 17: 54–55.

Lawrence, B. M. 2007. Oil composition of other *Mentha* species. In: *Mint. The Genus Mentha – Medicinal and Aromatic Plants – Industrial Profiles*, B. M. Lawrence (ed.), pp. 217–232, Boca Raton: CRC Press.

Lazarević J., N. Radulović, B. Zlatković, and R. Palić. 2010. Composition of *Achillea distans* Willd. subsp. *distans* root essential oil. *Nat. Prod. Res.*, 24: 718–731.

Lewinsohn, E. and M. Gijzen. 2009. Phytochemical diversity: The sounds of silent metabolism. *Plant Science*, 176: 161–169.

Linhart, Y. B. and J. D. Thompson. 1995. Terpene-based selective herbivory by *Helix aspersa* (Mollusca) on *Thymus vulgaris* (Labiatae). *Oecologia*, 102: 126–132.

Llorens-Molina, J. A., V. Castell, S. Vacas, and M. Verdeguer. 2017. TLC-GC/MS Method for identifying and selecting valuable essential oil chemotypes from wild populations of *Mentha longifolia* L. *Nat. Volatiles and Essent. Oils*, 4: 49–61.

Llorens-Molina, J. A. and S. Vacas. 2015. Seasonal variations in essential oil of aerial parts and roots of an *Artemisia absinthium* L. population from a Spanish area with supramediterranean climate (Teruel, Spain). *J. Essent. Oil Res.*, 27: 395–405.

Llorens-Molina, J. A., S. Vacas, V. Castell, and É. Németh-Zámboriné. 2016. Variability of essential oil composition of wormwood (*Artemisia absinthium* L.) affected by plant organ. *J. Essent. Oil Res.*, 29: 11–21.

Lopes-Lutz, D., D. S. Alviano, C. S. Alviano, and P. P. Kolodziejczyk. 2008. Screening of chemical composition, antimicrobial and antioxidant activities of *Artemisia* essential oils. *Phytochemistry.*, 69: 1732–1738.

Lourenço, P. M. L., A. C. Figueiredo, J. G. Barroso, L. G. Pedro, M. M. Oliveira, S. G. Deans, and J. J. C. Scheffer. 1999. Essential oil from hairy root cultures and from plant roots of *Achillea millefolium*. *Phytochemistry*, 51: 637–642.

Lukas, B., R. Samuel, and J. Novak. 2010. Oregano or marjoram? The enzyme γ-terpinene synthase affects chemotype formation in the genus *Origanum*. *Israel J. Plant Sci.*, 58: 211–220.

Lupien, S., F. Karp, M. Wildung, and R. Croteau. 1999. Regiospecific cytochrome P450 limonene hydroxylases from mint (*Mentha*) species: CDNA isolation, characterization, and functional expression of (–)-4S-limonene-3-hydroxylase and (–)-4S-limonene-6-hydroxylase. *Arch. Biochem. Biophys.*, 368: 181–192.

Ma, J. X., Y. N. Li, C. Vogl, F. Ehrendorfer, and Y. P. Guo. 2010. Allopolyploid speciation and ongoing backcrossing between diploid progenitor and tetraploid progeny lineages int he *Achillea millefolium* species complex: Analyses of single copy nuclear genes and genomic AFLP. *BMC Evol. Biol.*, 10: 100.

Malizia, R. A., S. Molli, D. A. Cardell, and J. A. Retamar. 1996. Essential oil of *Mentha citrata* grown in Argentina. Variation in the composition and yield at full- and post- flowering. *J. Essent. Oil Res.*, 8: 347–349.

Maree, J., G. P. P. Kamatou, S. Gibbons, A. M. Viljoen, and S. Van Vuuren. 2014. The application of GC-MS combined with chemometrics for the identification of antimicrobial compounds from selected commercial oils. *Chemometr. Intell. Lab.*, 130: 172–181.

McCaskill, D. and R. Croteau. 1998. Bioengineering terpenoid production: Current problems. *Trends Biotechnol.*, 16: 189–203.

McConkey, M., J. Gershenzon, and R. Croteau. 2000. Developmental regulation of monoterpene biosynthesis in the grandular trichomes of peppermint (*Mentha* x *piperita* L.). *Plant Physiol.*, 122: 215–223.

Michler, B., A. Preitschopf, P. Erhard, and C. Arnold. 1992. *Achillea millefolium*: Zusammenhänge zwischen Standortfaktoren, Ploidiegrad, Vorkommen von Proazulenen und Gehalt an Chamazulen im ätherischen Öl. *PZ-Wissenschaft*, 137: 23–29.

Mitić, V. and S. Đorđević. 2000. Essential oil composition of *Hyssopus officinalis* L. cultivated in Serbia. *Facta Universitatis*, 2: 105–108. http://casopisi.junis.ni.ac.rs/index.php/FUPhysChemTech

Mockute D. and Judzetiene, A. 2004. Composition of the essential oils of *Tanacetum vulgare* L. growing wild in Vilnius district (Lithuania). *J. Essent. Oil Res.*, 16: 550–553.

Mokhetho, K. C., M. SAndasi, A. Ahmad, G. P. Kamatou, and A. Viljoen. 2018. Identification of potential anti-quorum sensing compounds in essential oils: A gas-chromatography-based metabolomics approach. *J Essent. Oil Res.*, 30: 399–408.

Morteza-Semnani, K. and M. Akbarzadeh. 2005. Essential oils composition of Iranian *Artemisia absinthium* L. and *Artemisia scoparia* Waldst. et Kit. *J. Essent. Oil Res.*, 17: 321–322.

Msaada, K., N. Salem, O. Bachrouch, S. Bousselmi, S. Tammar, A. Alfaify. and Marzouk, B. 2015. Chemical composition and antioxidant and antimicrobial activities of wormwood (*Artemisia absinthium* L.) essential oils and phenolics. *J. Chemistry.*, 2015: 1–12. http://dx.doi.org/10.1155/2015/804658

Mucciarelli, M., R. Caramiello, M. Maffei. and F. Chialva. 1995. Essential oils from some *Artemisia* species growing spontaneously in North-West Italy. *Flavour Frag. J.*, 10: 25–32.

Muhlemann, J. K., H. Maeda, C. Y. Chang et al. 2012. Developmental changes in the metabolic network of snapdragon flowers. *PLOS One*, 7: e40381. https://doi.org/10.1371/journal.pone.0040381

Murray, M. J. P., Marble, D., Lincoln, and F. W., Hefendehl. 1986. Peppermint oil quality differences and the reasons for them. In: *Flavors and Fragrances: A World Perspective*, B. M. Lawrence, B. D. Mookherjee, and B. D. Willis (eds.), pp. 189–210, Amsterdam: Elsevier Publishers.

Muselli A., M. Pau, J. M. Desjobert, M. Foddai, M. Usai, and J. Costa. 2009. Volatile constituents of *Achillea ligustica* all. by HS-SPME/GC/GC-MS. Comparison with essential oils obtained by hydrodistillation from Corsica and Sardinia. *Cromatographia*, 69: 575–e40585.

Nano, G. M., C. Bicchi, C. Frattini, and M. Gallino. 1979. Wild piemontese plants. *Planta Med.*, 35: 270–274.

Németh, É. 2005. Essential oil composition of species in the genus *Achillea*. *J. Essent. Oil Res.*, 17: 501–512.

Németh, É., J. Bernáth, and É. Héthelyi. 2000. Chemotypes and their stability in *Achillea crithmifolia* W. et K. populations. *J. Essent. Oil Res.*, 12: 53–58.

Németh, É., J. Bernáth, and Zs. Pluhár. 1999. Variability of selected characters of production biology and essential oil accumulation in populations of *Achillea crithmifolia*. *Plant Breed.*, 118: 263–267.

Németh, É., J. Bernáth, and G. Tarján. 2007. Quantitative and qualitative studies of essential oils of Hungarian *Achillea* populations. *J. Herbs, Spices Med. Plants*, 13: 57–69.

Németh, É., É. Héthelyi, and J. Bernáth. 1994. Comparison studies on *Tanacetum vulgare* L.chemotypes. *J. Herbs, Spices Med. Plants*, 2: 85–92.

Németh, É. and G. Székely. 2000. Floral biology of medicinal plants II. Lamiaceae species. *Internat. J. Hort. Sci.*, 6: 137–140.

Németh-Zámbori, É., P. Rajhárt, and K. Inotai. 2017. Effect of genotype and age on essential oil and total phenolics in hyssop (*Hyssopus officinalis* L.). *J. Appl. Bot. Food Qual.*, 90, 25–30.

Nezhadali, A. and M. Parsa. 2010. Study of the volatile compounds in *Artemisia absinthium* from Iran using HS/SPME/GC/MS. *Adv. Appl. Sci. Res.*, 1: 174–179.

Nin, S. P. and M. Bosetto. 1995. Quantitative determination of some essential oil components of selected *Artemisia absinthium* plants. *J. Essent. Oil Res.*, 7: 271–277. *Not. Bot. Hort. Agrobot. Cluj*, 38: 99–103.

Novák, I. 2006. Possibilities for improving production and quality of lovage (*Levisticum officinale* Koch.) drugs. PhD Dissertation, Corvinus University of Budapest.

Orav, A., T. Kailas, and K. Ivask. 2001. Composition of the essential oil from *Achillea millefolium* from Estonia. *J. Essent. Oil Res.*, 13: 290–294.

Orav, A., A. Raal, E. Arak, M. Muurisepp, and T. Kailas, T. 2006. Composition of the essential oil of *Artemisia absinthium* L. of different geographical origin. *Proc. of Estonian Acad. of Sci. Chem.*, 55: 155.

Özek, T., N. Tabanca, F. Demirci, D. E. Wedge, and K. H. C. Baser. 2010. Enantiomeric distribution of some linalool containing essential oils and their biological activities. *Rec. Nat. Prod.* 4:180–192.

Padalia, R. C., R. S. Verma, and A. Chauhan. 2016. Diurnal variations in aroma profile of *Ocimum basilicum* L., *O. gratissiumum* L., *O. americanum* L., *O. kilimandscharicum* Guerke. *J. Essent. Oil Res.*, 29: 248–261.

Palić, R., G. Stojanović, T. Nasković, and N. Ranelović. 2003. Composition and antibacterial activity of *Achillea crithmifolia* and *Achillea nobilis* essential oils. *J. Essent. Oil Res.*, 15: 434–437.

Pank, F., H. Krüger, and R. Quilitzsch. 2008. Ergebnisse zwanzigjähriger rekurrenter Selektion zur Steigerung des Aetherischöl-gehaltes von einjährigem Kümmel (*Carum carvi* L. var. *annuum* hort). *Z. Arzn. Gew. Pfl.*, 13: 24–28.

Pecetti, L., A. Tava, M. Romani, R. Cecotti, and M. Mella. 2012. Variation in terpene and linear chain hydrocarbon content in yarrow (*Achillea millefolium* L.) germplasm from the Rhaetian Alps, Italy. *Chem. Biodivers.*, 9: 2282–2295.

Piccaglia, R., L. Pace, and E. Tammaro. 1999. Characterisation of essential oils from three Italian ecotypes of hyssop (*Hyssopus officinalis* ssp. *artistatus* Briq.). *J. Essent. Oil Res.*, 11: 693–699.

Pino, J. A., A. Rosado, and V. Fuentes. 1997. Chemical composiition of the essential oil of *Artemisia absinthium* L. from Cuba. *J. Essent. Oil Res.*, 9: 87–89.

Pluhár, Zs., Sz. Sárosi, I. Novák, and G. Kutta. 2008. Essential oil polymorphism of Hungarian common thyme (*Thymus glabrescens* Willd.) populations. *Nat. Prod. Commun.*, 3: 1151–1154.

Polatoglu, K. 2013. "Chemotypes"- A fact that should not be ignored in natural product studies. *The Natural Products J.*, 3: 10–14. http://www.eurekaselect.com/107926

Pragadheesh, V. S., A. Saroj, A. Yadav, A. Samad, and C. S. Chanotiya. 2013. Compositions, enantiomer characterization and antifungal activity of two *Ocimum* essential oils. *Ind. Crops Prod.*, 50: 333–337.

Puschmann, G., V, Stephani, and D. Fritz. 1992. Investigations on the variability of caraway (*Carum carvi* L.). *Gartenbauwissenschaft*, 57: 275–277.

Putievsky E., A. Paton, E. Lewinsohn, U. Ravid, D. Haimovich, I. Katzir, D. Saadi, and N. Dudai. 1999. Crossability and relationship between morphological and chemical varieties of *Ocimum basilicum* L. *J. Herbs Spices Med. Plants*, 6: 11–24.

Putievsky, E., U. Ravid, N. Dudai, and I. Katzir. 1994. A new cultivar of caraway (*Carum carvi* L.) and its essential oil. *J. Herbs, Spices Med. Plants*, 2: 85–90.

Raal, A., E. Arak, and A. Orav. 2012. The content and composition of the essential oil found in *Carum carvi* L. commercial fruits obtained from different countries. *J. Essent. Oil Res.*, 24: 53–59.

Raal, A., E. Arak, A. Orav, T. Kailas, and M. Muurisepp. 2008. Composition of the essential oil of *Levisticum Officinale* W. D. J. Koch from some European countries. *J. Essent. Oil Res.*, 20: 318–322.

Radulović, N., B. Zlatković, R. Palić, and G. Stojanović. 2007. Chemotaxonomic significance of the Balkan *Achillea* volatiles. *Nat. Prod. Commun.*, 2: 453–474.

Rauchensteiner F., S. Nejati, I. Werner, S. Glasl, J. Saukel, J. Jurenits, and W. Kubelka. 2002. Determination of taxa of the *Achillea millefolium* group and *Achillea crithmifolia* by morphological and phytochemical methods I. Characterisation of Central European taxa. *Sci. Pharm.*, 70: 199–230.

Rezaeinodehi, A. and S. Khangholi. 2008. Chemical composition of the essential oil of *Artemisia absinthium* growing wild in Iran. *Pakistan J. Biol. Sci.*, 11: 946–949.

Rohloff, J., S. Dragland, and R. Mordal. 2004. Chemotypical variation of tansy (*Tanacetum vulgare* L.) from 40 different location in Norway. In: *Program and Book of Abstracts, 34th ISEO*, Würzburg, p. 113.

Sarker L. S. and S. S. Mahmoud. 2015. Cloning and functional characterization of two monoterpene acetyltransferases from glandular trichomes of *L. x intermedia*. *Planta*, 242: 709–719.

Schratz E. and H. Hörster. 1970. Zusammensetzung des ätherischen Öles von *Thymus vulgaris* and *Thymus marschallianus* in Abhängigkeitt von Blattalter und Jahreszeit. *Planta Med.*, 19: 161–175.

Schantz, M. and K. Forsén. 1971. Begleitstoffe in verschiedenen Chemotypen von *Chrysanthemum vulgare* L. Bernh. I. Reine Thujon- und Campher-Typen, Farmaseut. *Aikakauslehti*, 80: 122–131.

Schantz, M., M. Jarvi, and R. Kaartinen. 1966. Die Veränderungen des ätherischen Öles während der Entwicklung der Blütenkörbchen von *Chrysanthemum vulgare*. *Planta Med.*, 4: 421–435.

Schilcher, H. 2004. *Wirkungsweise und Anwendungsformen der Kamillenblüten*. Berlin: Berliner Medizinische Verlagsanstalt GmbH.

Schulz, G. and E. Stahl-Biskup. 1991. Essential oils and glycosidic bound volatile from leaves, stems, flowers and roots of *Hyssopus officinalis*. *Flavour Fragr. J.*, 6: 69–73.

Scrivanti, L. R., M. P. Zunino, and J. A. Zygadlo. 2003. *Tagetes minuta* and *Schinus areira* essential oils as allelopathic agents. *Biochem. Syst. Ecol.*, 31: 563–572.

Sedláková, J., B. Kocourková, L. Lojková, and V. Kubáň. 2003. Determination of essential oil content in caraway (*Carum carvi* L.) species by means of supercritical fluid extraction. *Plant Soil Environ.*, 49(6):277–282.

Sharopov, S. F., V. Sulaimonova, and W. N. Setzer. 2012. Composition of the essential oil of *Artemisia absinthium* from Tajikistan. *Rec. Nat. Prod.*, 6: 127–134.

Simonnet, X., M. Qennoz, E. Capella, O. Panero, and I. Tonutti. 2012. Agricultural and phytochemical evaluation of *Artemisia absinthium* hybrids. *Acta Hort.*, 955: 169–172.

Solberg, S. O., M. Göransson, M. A. Petersen, F. Yndgaard, and S. Jeppson. 2016. Caraway essential oil composition and morphology: The role of location and genotype. *Biochem. Syst. Ecol.*, 66: 351–357.

Sorsa, M., M. Schantz, J. Lokki, and K. Forsén. 1968. Variability of essential oil components in *Chrysanthemum vulgare* L. in Finland. *Ann. Acad. Scient. Fennicae, Series A. IV. Biologica*, 135: 1–12.

Stahl-Biskup, E. and F. Sáez. 2003. *Thyme – The genus Thymus. Medicinal and Aromatic Plants- Industrial Profiles*. London: Taylor and Francis.

Stahl-Biskup, E. and E. M. Wichtmann. 1991. Composition of the essential oils from roots of some *Apiaceae* in relation to the development of their oil duct systems. *Flavour Fragr. J.*, 6: 249–255.

Szebeni, Zs., B. Galambosi, and Y. Holm. 1992. Growth, yield and essential oil content of lovage grown in Finland. *J. Essent. Oil Res.*, 4: 375–380.

Teixeira, B., A. Marques, C. Ramos, I. Batista, C. Serrano, O. Matos, N. R. Neng, J. M. F. Nogueira, J. A. Saraiva, and M. L. Nunes. 2012. European pennyroyal (*Mentha pulegium*) from Portugal: Chemical composition of essential oil and antioxidant and antimicrobial properties of extracts and essential oil. *Ind. Crops Prod.*, 36: 81–87.

Tétényi, P. 1970. *Infraspecific Chemical Taxa of Medicinal Plants*. Budapest: Akadémiai Kiadó.

Tétényi, P. 1975. Homology of biosynthetic routes: Base of chemotaxonomy. *Herba Hung.*, 14: 37–42.

Tholl, D. 2006. Terpene synthases and the regulation, diversity and biological roles of terpene metabolism. *Curr. Opin. Biotechnol.*, 9: 297–304.

Tirillini, B., A. Ricci, and R. Pellegrino. 1999. Secretion constituents of leaf glandular trichomes of *Salvia officinalis* L. *J. Essent. Oil Res.*, 11: 565–569.

Todorova M., A. Trendafilova, B. Mikhova, A. Vitkova, and H. Duddeck. 2007. Chemotypes in *Achillea collina* based on sesquiterpene lactone profile. *Phytochemistry*, 68: 1722–1730.

Toncer, O., S. Basbag, S. Karaman, E. Diraz, and M. Basbag. 2010. Chemical composition of the essential oils of some *Achillea* species growing wild in Turkey. *Int. J. Agric. Biol.*, 12: 527–530.

Toncer, O., S. Karaman, S. Kilil, and E. Diraz. 2009. Changes in essential oil composition of Oregano (*Origanum onies* L.) due to diurnal variations at different developmental stages. *Not. Bot. Hort. Agrobot. Cluj.*, 37: 177–181.

Toyota, M., H. Koyama, M. Mizutani, and Y. Asakawa. 1996. (−)-*ent*-Spathulenol isolated from liverworts is an artefact. *Phytochemistry*, 41: 1347–1350.

Tsankova, E. T., A. N. Konakchiev, and A. M. Genova. 1993. Chemical composition of the essential oils of two *Hyssopus officinalis* taxa. *J. Essent. Oil Res.*, 5: 609–611.

Tsuro M. and S. Asada. 2014. Differential expression of limonene synthase gene affects production and composition of essential oils in leaf and floret of transgenic lavandin (*Lavandula x intermedia* Emerice x Loisel.). *Plant Biotech. Rep.*, 8: 193–201.

Tucker, A. O., M. J. Maciarello, and G. Sturtz. 1993. The essential oils of *Artemisia* "Powis Castle" and its putative parents, *A. absinthium* and *A. arborescens*. *J. Essent. Oil Res.*, 5: 239–242.

Tzakou, O., A. Loukis, and N. Argyriadou. 1993. Volatile constituents of *Achillea crithmifolia* flowers from Greece. *J. Essent. Oil Res.*, 5: 345–346.

Valkovszki, N. J. and É. Zámbori-Németh. 2010. Effects of growing conditions on content and composition of the essential oil of annual caraway (*Carum carvi* L. var. *annua*). *Acta Aliment. Hung.*, 40: 235–246.

Vallejo, M. C. G., J. G. Herraiz, M. J. Pérez-Alonso, and A. Velasco-Negueruela. 1995. Volatile oil of *Hyssopus officinalis* L. from Spain. *J. Essent. Oil Res.*, 7: 567–568.

Varga, L., B. Berhardt, B. Gosztola, Sz. Sárosi, and É. Németh-Zámbori. 2012. Essential oil accumulation during ripening process of selected *Apiaceae* species. In: *Proc. 7th CMAPSEEC Subotica*, Z. Dajic-Stevanovic, and D. Radanovic (eds.), p. 389–396, Belgrade: Inst. Med. Plant Res. "Dr Josif Pančić".

Venskutonis, P. R. 1995. Essential oil composition of some herbs cultivated in Lithuania. In: *Proc. 13th Internat. Congress of Flavours, Fragrances and Essential Oils*. Vol. 2. Istanbul, Turkey, pp. 108–123.

Vetter, S. and Ch. Franz. 1998. Samenbildung bei Kreuzungen und Selbstungen mit polyploiden *Achillea* Arten. *Z. Arzn. Gew. Pfl.*, 3: 11–14.

Vetter, S., Ch. Franz, S. Glasl, U. Kastner, J. Saukel, and J. Jurenits. 1997. Inheritance of sesquiterpene lactone types within the *Achillea millefolium* complex (Compositae). *Plant Breed.*, 116: 79–82.

Wagner, C., W. Friedt, R. A. Marquard, and F. Ordon. 2005. Molecular analyses on the genetic diversity and inheritance of (–)-α-bisabolol and chamzilene content in tetraploid chamomile (*Chamomilla recutita* (L.) Rausch.). *Plant Sci.*, 169: 917–927.

Webb, H., K. Carsten, L. Rob, J. Hamill, and W. Foley. 2011. The regulation of quantitative variation of foliar terpenes in medicinal tea tree *Melaleuca alternifolia*. *BMC Proceedings*, 5(Suppl. 7): O20.

Wichtl, M. 1997. *Teedrogen und Phytopharmaka*. Stuttgart: Wissenschaftliche Verlagsgesellschaft.

Yamaura, T., S. Tanaka, M. Tabata. 1992. Localization of the biosynthesis and accumulation of monoterpenoids in glandular trichomes of thyme. *Planta Med.*, 58: 153–158.

Zámboriné, N. É. 2005. Methods of taxonomic investigations on caraway (Carum carvi L.), practical importance of single plant caharacteristics. In: *Hungarian. "Widening of the variety spectrum in horticulture" Special issue of Kertgazdaság*, pp. 209–220.

Zámboriné, N. É., D. É. Kertészné, and P. Tétényi. 1987. Propagation of *Tanacetum vulgare* L. *Herba Hung.*, 26: 137–143.

Zámboriné N. É., K. Ruttner, and P. Radácsi. 2018. Small changes in the distillation method result in variable quality of yarrow (*Achillea collina*) essential oil. *Facta Universitatis*, 16(Special issue): 1. http://casopisi.junis.ni.ac.rs/index.php/FUPhysChemTech

Zheljazkov, V., B. Yankov, V. Topalov. 1996. Comparison of three methods of mint propagation and their effects on the yield of fresh material and essential oil. *J. Essent. Oil Res.*, 8: 35–45.

5 Production of Essential Oils

Erich Schmidt

CONTENTS

5.1 INTRODUCTION

5.1.1 General Remarks

Essential oils have become an integral part of everyday life. They are used in a great variety of ways: as food flavorings, as feed additives, as flavoring agents by the cigarette industry, and in the compounding of cosmetics and perfumes. Furthermore, they are used in air fresheners and deodorizers as well as in all branches of medicine such as in pharmacy, balneology, massage, and homeopathy. A more specialized area will be in the fields of aromatherapy and aromachology. In recent years, the importance of essential oils as biocides and insect repellents has led to a more detailed study of their antimicrobial potential. Essential oils are also good natural sources of substances with commercial potential as starting materials for chemical synthesis.

Essential oils have been known to mankind for hundreds of years, even millenniums. Long before the fragrances themselves were used, the important action of the oils as remedies was recognized. Without the medical care as we enjoy in our time, self-healing was the only option to combat parasites or the suffering of the human body. Later on essential oils were used in the preparation of early cosmetics, powders, and soaps. As the industrial production of synthetic chemicals started and increased during the nineteenth century, the production of essential oils also increased owing to their importance to our way of life.

The quantities of essential oils produced around the world vary widely. The annual output of some essential oils exceeds 35,000 tons, while that of others may reach only a few kilograms. Some production figures, in metric tons, based on the year 2008 are shown in Table 5.1.

TABLE 5.1
Production Figures of Important Essential Oils (2008)

Essential Oil	Production in Metric Tons (2008)	Main Production Countries
Orange oils	51,000	United States, Brazil, Argentina
Cornmint oil	32,000	India, China, Argentina
Lemon oils	9200	Argentina, Italy, Spain
Eucalyptus oils	4000	China, India, Australia, South Africa
Peppermint oil	3300	India, United States, China
Clove leaf oil	1800	Indonesia, Madagascar
Citronella oil	1800	China, Sri Lanka
Spearmint oils	1800	United States, China
Cedarwood oils	1650	United States, China
Litsea cubeba oil	1200	China
Patchouli oil	1200	Indonesia, India
Lavandin oil Grosso	1100	France
Corymbia citriodora	1000	China, Brazil, India, Vietnam

Source: Perfumer & Flavorist. 2009. A preliminary report on the world production of some selected essential oils and countries, Vol. 34, January 2009 pp. 38–44, Perfumer & Flavorist Carol Stream, IL 60188-2403.

Equally wide variations also occur in the monetary value of different essential oils. Prices range from $1.80/kg for orange oil to $120,000.00/kg for orris oil. The total annual value of the world market is of the order of several billions of USD. A large, but variable, labor force is involved in the production of essential oils. While, in some cases, harvesting and oil production will require just a few workers, other cases will require manual harvesting and may require multiple working steps. Essential oil production either from wild-growing or from cultivated plants is possible almost anywhere, excluding the world's coldest, permanently snow-covered regions. It is estimated that the global number of plant species is of the order of 300,000. About 10% of these contain essential oils and could be used as a source for their production. All continents possess their own characteristic flora with many odor-producing species. Occasionally, these plants may be confined to a particular geographical zone such as *Santalum album* to India and Timor in Indonesia, *Pinus mugo* to the European Alps, or *Abies sibirica* to the Commonwealth of Independent States (CIS, former Russia). For many countries, mainly in Africa and Asia, essential oil production is their main source of exports. Essential oil export figures for Indonesia, Sri Lanka, Vietnam, and even India are very high.

Main producer countries are found in every continent. In Europe, the center of production is situated in the countries bordering the Mediterranean Sea: Italy, Spain, Portugal, France, Croatia, Albania, and Greece, as well as middle-eastern Israel, all of which produce essential oils in industrial quantities. Among Central European countries, Bulgaria, Romania, Hungary, and Ukraine should be mentioned. The huge Russian Federation spread over much of Eastern Europe and Northern Asia has not only nearly endless resources of wild-growing plants but also large areas of cultivated land. The Asian continent with its diversity of climates appears to be the most important producer of essential oils. China and India play a major role followed by Indonesia, Sri Lanka, and Vietnam. Many unique and unusual essential oils originate from the huge Australian continent and from neighboring New Zealand and New Caledonia. Major essential oil–producing countries in Africa include Morocco, Tunisia, Egypt, and Algeria with Ivory Coast, South Africa, Ghana, Kenya, Tanzania, Uganda, and Ethiopia playing a minor role. The important spice-producing islands of Madagascar, the Comoros, Mayotte, and Réunion are situated along the eastern coast of the African continent. The American continent is also one of the biggest essential oil producers. The United States, Canada, and Mexico possess a wealth of natural aromatic plant material. In South America, essential oils are

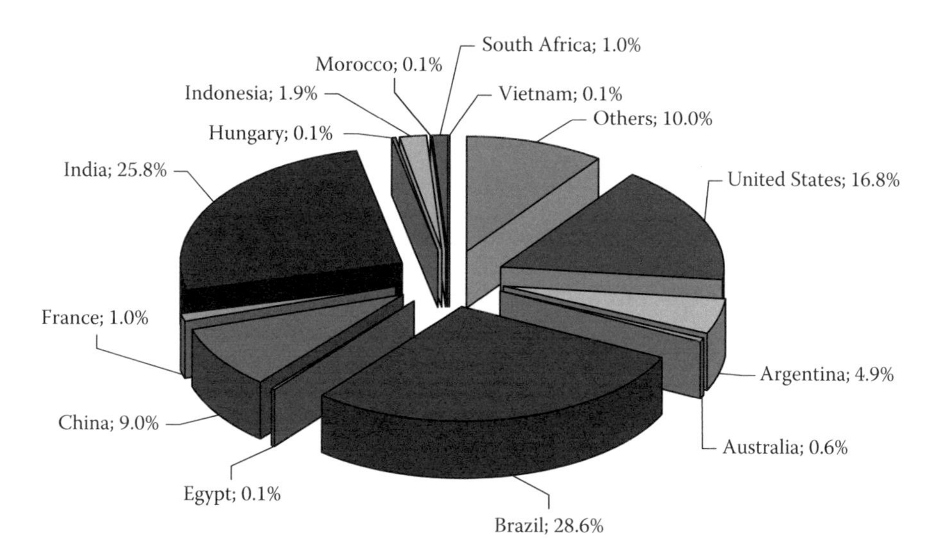

FIGURE 5.1 Production countries and essential oil production worldwide (2008). (Adapted from Perfumer & Flavorist. 2009. A preliminary report on the world production of some selected essential oils and countries, Vol. 34, January 2009 pp. 38–44, Perfumer & Flavorist Carol Stream, IL 60188-2403.)

produced in Brazil, Argentina, Paraguay, Uruguay, Guatemala, and the Island of Haiti. Apart from the aforementioned major essential oil–producing countries, there are many more, somewhat less important ones, such as Germany, Taiwan, Japan, Jamaica, and the Philippines. Figure 5.1 shows production countries and essential oil production worldwide (2008).

Cultivation of aromatic plants shifted during the last two centuries. From 1850 to 1950, the centers of commercial cultivation of essential oil plants have been the Provence in France, Italy, Spain, and Portugal. With the increase of labor costs, this shifted to the Mediterranean regions of North Africa. As manual harvesting proved too expensive for European conditions, and following improvements in the design of harvesting machinery, only those crops that lend themselves to mechanical harvesting continued to be grown in Europe. In the early 1990s, even North Africa proved too expensive, and the centers of cultivation moved to China and India. At the present time, manual-handling methods are tending to become too costly even in China, and thus India remains as today's center for the cultivation of fragrant plant crops (Dey et al., 2001).

5.1.2 Definition and History

Not all odorous extracts of essential oil–bearing plants comply with the International Organization for Standardization (ISO) definition of an "essential oil." An essential oil as defined by the ISO in document ISO 9235.2: 2013—aromatic natural raw materials—vocabulary is as follows.

Product obtained from a natural raw material (2.19) of plant origin, by steam distillation, by mechanical processes from the epicarp of citrus fruits, or by dry distillation, after separation of the aqueous phase—if any—by physical processes. Note 1 to entry: The essential oil can undergo physical treatments which do not result in any significant change inits composition (e.g. filtration, decantation, centrifugation).

An alternative definition of essential oils, established by Professor Dr. Gerhard Buchbauer of the Institute of Pharmaceutical Chemistry, University of Vienna, includes the following suggestion: "Essential oils are more or less volatile substances with more or less odorous impact, produced either by steam distillation or dry distillation or by means of a mechanical treatment from one single species" (Buchbauer et al., 1994). This appears to suggest that mixing several different plant species within the production process is not allowed. As an example, the addition of lavandin plants to lavender plants will yield a natural essential oil but not a natural lavender essential oil. Likewise, wild-growing varieties of *Thymus* will not result in a thyme oil as different chemotypes will totally change the composition of the oil. It follows that blending of different chemotypes of the same botanical species is inadmissible as it will change the chemical composition and properties of the final product. However, in view of the global acceptance of some specific essential oils, there will be exceptions. For example, oil of geranium, ISO/DIS 4730, is obtained from *Pelargonium* × ssp., for example, from hybrids of uncertain parentage rather than from a single botanical species (ISO/DIS, 4731, 2005). It is a well established and important article of commerce and may, thus, be considered to be an acceptable exception. In reality, it is impossible to define "one single species" as many essential oils being found on the market come from different plant species. Even in ISO drafts, it is confirmed that various plants are allowed. There are several examples like rosewood oils, distilled from *Aniba rosaeodora* and *Aniba parviflora*, two different plant species. The same happens with the oil of gum turpentine from China, where mainly *Pinus massoniana* will be used, beside other *Pinus* species. Eucalyptus provides another example: oils produced in Portugal have been produced from hybrids such as *Eucalyptus globulus* ssp. *globulus* × *Eucalyptus globulus* ssp. *bicostata* and *Eucalyptus globulus* ssp. *globulus* × *Eucalyptus globulus* ssp. *Eucalyptus globulus* ssp. *pseudoglobulus*. These subspecies were observed from various botanists as separate species. The Chinese eucalyptus oils coming from the Sichuan province are derived from *Cinnamomum longipaniculatum*. Oil of *Melaleuca* (*terpinen-4-ol type*) is produced from *Melaleuca alternifolia* and in smaller amounts also from *Melaleuca linariifolia* and *Melaleuca dissitiflora*. For the future, this definition must be discussed on the level of ISO rules.

Products obtained by other extraction methods, such as solvent extracts, including supercritical carbon dioxide extracts, concretes or pomades, and absolutes as well as resinoids, and oleoresins are *not essential oils* as they do not comply with the earlier mentioned definition. Likewise, products obtained by enzymic treatment of plant material do not meet the requirements of the definition of an essential oil. There exists, though, at least one exception that ought to be mentioned. The well-known "essential oil" of wine yeast, an important flavor and fragrance ingredient, is derived from a microorganism and not from a plant.

In many instances, the commercial terms used to describe perfumery products as essential oils are either wrong or misleading. So-called artificial essential oils, nature-identical essential oils, reconstructed essential oils, and in some cases even essential oils complying with the constants of pharmacopoeias are merely synthetic mixtures of perfumery ingredients and have nothing to do with pure and natural essential oils.

Opinions differ as to the historical origins of essential oil production. According to some, China has been the cradle of hydrodistillation, while others point to the Indus culture (Levey, 1959; Zahn, 1979). On the other hand, some reports also credit the Arabs as being the inventors of distillation. Some literature reports suggest that the earliest practical apparatus for water distillation has been dated from the Indus Culture of some 5000 years ago. However, no written documents have been found to substantiate these claims (Levey, 1955; Zahn, 1979). The earliest documented records of a method and apparatus of what appears to be a kind of distillation procedure were published by Levy from the high culture of Mesopotamia (Levey, 1959). He described a kind of cooking pot from Tepe Gaure in northeastern Mesopotamia, which differed from the design of cooking pots of that period. It was made of brown clay, 53 cm in diameter and 48 cm high. Its special feature was a channel between the raised edges. The total volume of the pot was 37 L and that of the channel was 2.1 L. As the pot was only half-filled when in use, the process appears to represent a true distillation. While the Arabs appear to be, apart from the existence of the pot discovered in Mesopotamia, the inventors of hydrodistillation, we ought to go back 3000 years BC.

The archaeological museum of Texila in Pakistan has on exhibit a kind of distillation apparatus made of burnt clay. At first sight, it really has the appearance of a typical distillation apparatus, but it is more likely that at that time, it was used for the purification of water (Rovesti, 1977). Apart from that, the assembly resembles an eighteenth-century distillation plant (Figure 5.2). It was

FIGURE 5.2 Reconstruction of the distillation plant from Harappa.

again Levy who demonstrated the importance of the distillation culture. Fire was known to be of greatest importance. Initial heating, the intensity of the heat, and its maintenance at a constant level right down to the cooling process were known to be important parameters. The creative ability to produce natural odors points to the fact that the art of distillation was a serious science in ancient Mesopotamia. While the art of distillation had been undergoing improvements right up to the eighth century, it was never mentioned in connection with essential oils, merely with its usefulness for alchemical or medicinal purposes ("Liber servitorius" of Albukasis). In brief, concentration and purification of alcohol appeared to be its main reason for being in existence, its "raison d'être" (Koll and Kowalczyk, 1957).

The Mesopotamian art of distillation had been revived in ancient Egypt as well as being expanded by the expression of citrus oils. The ancient Egyptians improved these processes largely because of their uses in embalming. They also extracted, in addition to myrrh and storax, the exudates of certain East African coastal species of *Boswellia*, none of which are of course essential oils. The thirteenth century Arabian writer Ad-Dimaschki also provided a description of the distillation process, adding descriptions of the production of distilled rose water as well as of the earliest improved cooling systems. It should be understood that the products of these practices were not essential oils in the present accepted sense but merely fragrant distilled water extracts exhibiting the odor of the plant used.

The next important step in the transfer of the practice of distillation to the Occident, from ancient Egypt to the northern hemisphere, was triggered by the crusades of the Middle Ages from the twelfth century onward. Hieronymus Brunschwyk listed in his treatise *The True Art to Distil* about 25 essential oils produced at that time. Once again, one should treat the expression "essential oils" with caution; it would be more accurate to refer to them as "fragrant alcohols" or "aromatic waters." Improvements in the design of equipment led to an enrichment in the diversity of essential oils derived from starting materials such as cinnamon, sandalwood, and also sage and rosemary (Gildemeister and Hoffmann, 1931).

The first evidence capable to discriminate between volatile oils and odorous fatty oils was provided in the sixteenth century. The availability of printed books facilitated "scientists" seeking guidance on the distillation of essential oils. While knowledge of the science of essential oils did not increase during the seventeenth century, the eighteenth century brought about only small progress in the design of equipment and in refinements of the techniques used. The beginning of the nineteenth century brought about progresses in chemistry, including wet analysis, and restarted again, chiefly in France, in an increased development of hydrodistillation methods. Notwithstanding the "industrial" production of lavender already in progress since the mid-eighteenth century, the real breakthrough occurred at the beginning of the following century. While until then the distillation plant was walled in, now the first moveable apparatus appeared. The "Alambique Vial Gattefossé" was easy to transport and placed near the fields. It resulted in improved product quality and reduced the length of transport. These stills were fired with wood or dried plant material. The first swiveling still pots had also been developed that facilitated the emptying of the still residues. These early stills had a capacity of about 50–100 kg of plant material. Later on, their capacity increased to 1000–1200 kg. At the same time, cooling methods were also improved. These improvements spread all over the northern hemisphere to Bulgaria, Turkey, Italy, Spain, Portugal, and even to northern Africa. The final chapter in the history of distillation of plant material came about with the invention of the "alembic à bain-marie," technically speaking a double-walled distillation plant. Steam was not only passed through the biomass, but was also used to heat the wall of the still. This new method improved the speed of the distillation as well as the quality of the top notes of the essential oils thus produced.

The history of the expression of essential oils from the epicarp of citrus fruits is not nearly as interesting as that of hydrodistillation. This can be attributed to the fact that these expressed fragrance concentrates were more readily available in antiquity as expression could be effected by implements made of wood or stone. The chief requirement for this method was manpower, and that was available in unlimited amount. The growth of the industry led to the invention of new mechanical machinery, followed by automation and reduction of manpower. But this topic will be dealt with later on.

5.1.3 Production

Before dealing with the basic principles of essential oil production, it is important to be aware of the fact that the essential oil we have in our bottles or drums is not necessarily identical with what is present in the plant. It is wishful thinking, apart for some rare exceptions, to consider an "essential oil" to be the "soul" of the plant and thus an exact replica of what is present in the plant. Only expressed oils that have not come into contact with the fruit juice and that have been protected from aerial oxidation may meet the conditions of a true plant essential oil. The chemical composition of distilled essential oils is not the same as that of the contents of the oil cells present in the plant or with the odor of the plants growing in their natural environment. Headspace technology, a unique method allowing the capture of the volatile constituents of oil cells and thus providing additional information about the plant, has made it possible to detect the volatile components of the plant's "aura." One of the best examples is rose oil. A nonprofessional individual examining pure and natural rose oil on a plotter, even in dilution, will not recognize its plant source. The alteration caused by hydrodistillation is remarkable as plant material in contact with steam undergoes many chemical changes. Hot steam contains more energy than, for example, the surface of the still. Human skin that has come into contact with hot steam suffers tremendous injuries, while short contact with a metal surface at 100°C results merely in a short burning sensation. Hot steam will decompose many aldehydes, and esters may be formed from acids generated during the vaporization of certain essential oil components. Some water-soluble molecules may be lost by solution in the still water, thus altering the fragrance profile of the oil.

Why do so many plants produce essential oils? Certainly neither to regale our nose with pleasant fragrances of rose or lavender nor to heighten the taste (as taste is mostly related to odor) of ginger, basil, pepper, thyme, or oregano in our food! Nor to cure diseases of the human body or influence human behavior! Most essential oils contain compounds possessing antimicrobial properties, active against viruses, bacteria, and fungi. Often, different parts of the same plant, such as leaves, roots, and flowers, may contain volatile oils of different chemical composition. Even the height of a plant may play a role. For example, the volatile oil obtained from the gum of the trunk of *Pinus pinaster* at a height of 2 m will contain mainly pinenes and significant car-3-ene, while oil obtained from the gum collected at a height of 4 m will contain very little or no car-3-ene. The reason for this may be protection from deer that browse the bark during the winter months. Some essential oils may act not only as insect repellents but even prevent their reproduction. In many cases, it has been shown that plants attract insects that in turn assist in pollinating the plant. It has also been shown that some plants communicate through the agency of their essential oils. Sometimes, essential oils are considered to be simply metabolic waste products! This may be so in the case of eucalypts as the oil cells present in the mature leaves of *Eucalyptus* species are completely isolated and embedded deeply within the leaf structure. In some cases, essential oils act as germination inhibitors thus reducing competition by other plants (Porter, 2001).

Essential oil yields vary widely and are difficult to predict. The highest oil yields are usually associated with balsams and similar resinous plant exudations, such as gurjun, copaiba, elemi, and Peru balsam, where they can reach 30%–70%. Clove buds and nutmeg can yield between 15% and 17% of essential oil, while other examples worthy of mention are cardamom (about 8%), patchouli (3.5%) and fennel, star anise, caraway seed, and cumin seed (1%–9%). Much lower oil yields are obtained with juniper berries, where 75 kg of berries are required to produce 1 kg of oil, sage (about 0.15%), and other leaf oils such as geranium (also about 0.15%). Rose petals in 4000 kg will yield 1 kg of oil, and 1000 kg of bitter orange flowers is required for the production of also just 1 kg of oil. The yields of expressed fruit peel oils, such as bergamot, orange, and lemon, vary from 0.2% to about 0.5%.

A number of important agronomic factors have to be considered before embarking on the production of essential oils, such as climate, soil type, influence of drought and water stress, and stresses caused by insects and microorganisms, propagation (seed or clones), and cultivation practices. Other important factors include precise knowledge on which part of the biomass is to be used, location of the oil cells within the plant, timing of harvest, method of harvesting, storage, and preparation of the biomass prior to essential oil extraction (Yanive and Palevitch, 1982).

5.1.4 Climate

The most important variables include temperature, number of hours of sunshine, and frequency and magnitude of precipitations. Temperature has a profound effect on the yield and quality of the essential oils, as the following example of lavender will show. The last years in the Provence, too cold at the beginning of growth, were followed by very hot weather and a lack of water. As a result, yields decreased by one-third. The relationship between temperature and humidity is an additional important parameter. Humidity coupled with elevated temperatures produces conditions favorable to the proliferation of insect parasites and, most importantly, microorganisms. This sometimes causes plants to increase the production of essential oil for their own protection. Letchamo have studied the relationship between temperature and concentration of daylight on the yield of essential oil and found that the quality of the oil was not influenced (Letchamo et al., 1994). Herbs and spices usually require greater amounts of sunlight. The duration of sunshine in the main areas of herb and spice cultivation, such as the regions bordering the Mediterranean Sea, usually exceeds 8 h/day. In India, Indonesia, and many parts of China, this is well in excess of this figure, and two or even three crops per year can be achieved. Protection against cooling and heavy winds may be required. Windbreaks provided by rows of trees or bushes and even stone walls are particularly common in southern Europe. In China, the *Litsea cubeba* tree is used for the same purpose. In colder countries, the winter snow cover will protect perennials from frost damage. Short periods of frost with temperatures below −10°C will not be too detrimental to plant survival. However, long exposure to heavy frost at very low subzero temperatures will result in permanent damage to the plant ensuing from a lack of water supply.

5.1.5 Soil Quality and Soil Preparation

Every friend of a good wine is aware of the influence of the soil on the grapes and finally on the quality of the wine. The same applies to essential oil–bearing plants. Some crops, such as lavender, thyme, oregano, and clary sage require meager but lime-rich soils. The Jura Chalk of the Haute

Provence is destined to produce a good growth of lavender and is the very reason for the good quality and interesting top note of its oils compared with lavender oils of Bulgarian origin growing on different soil types (Meunier, 1985). Soil pH affects significantly oil yield and oil quality. Figueiredo et al. found that the pH value "strongly influences the solubility of certain elements in the soil. Iron, zinc, copper and manganese are less soluble in alkaline than in acidic soils because they precipitate as hydroxides at high pH values" (Figueiredo et al., 2005). It is essential that farmers determine the limits of the elemental profile of the soil. Furthermore, the spacing of plantings should ensure adequate supply with essential trace elements and nutrients. Selection of the optimum site coupled with a suitable climate plays an important role as they will provide a guarantee for optimum crop and essential oil quality.

5.1.6 Water Stress and Drought

It is well known to every gardener that lack of water, as well as too much water, can influence the growth of plants and even kill them. The tolerance of the biomass to soil moisture should be determined in order to identify the most appropriate site for the growing of the desired plant. Since fungal growth is caused by excess water, most plants require well-drained soils to prevent their roots from rotting and the plant from being damaged, thus adversely affecting essential oil production. Lack of water, for example, dryness, exerts a similar deleterious influence. Flowers are smaller than normal and yields drop. Extreme drought can kill the whole plant as its foliage dries closing down its entire metabolism.

5.1.7 Insect Stress and Microorganisms

Plants are living organisms capable of interacting with neighbor plants and warning them of any incipient danger from insect attack. These warning signals are the result of rapid changes occurring in their essential oil composition, which are then transferred to their neighbors who in turn transmit this information on to their neighbors forcing them to change their oil composition as well. In this way, the insect will come into contact with a chemically modified plant material, which may not suit its feeding habits thus obliging it to leave and look elsewhere. Microorganisms can also significantly change the essential oil composition as shown in the case of elderflower fragrance. Headspace gas chromatography coupled with mass spectroscopy (GC/MS) has shown that linalool, the main constituent of elderflowers, was transformed by a fungus present in the leaves, into linalool oxide. The larvae of Cécidomye (*Thomasissiana lavandula*) damage the lavender plant with a concomitant reduction of oil quality. Mycoplasmose and the fungus *Armillaria mellea* can affect the whole plantation and totally spoil the quality of the oil.

5.1.8 Location of Oil Cells

As already mentioned, the cells containing essential oils can be situated in various parts of the plant. Two different types of essential oil cells are known, superficial cells, for example, glandular hairs located on the surface of the plant, common in many herbs such as oregano, mint, and lavender, and cells embedded in plant tissue, occurring as isolated cells containing the secretions (as in citrus fruit and eucalyptus leaves) or as layers of cells surrounding intercellular space (canals or secretory cavities), for example, resin canals of pine. Professor Dr. Johannes Novak (Institute of Applied Botany, Veterinary University, Vienna) has shown impressive pictures and pointed out that the chemical composition of essential oils contained in neighboring cells (oil glands) could be variable but that the typical composition of a particular essential oil was largely due to the averaging of the enormous number of individual cells present in the plant (Novak, 2005). It has been noted in a publication entitled *Physiological Aspects of Essential Oil Production* that individual oil glands do not always secrete the same type of compound and that the process of secretion can be different

(Kamatou et al., 2006). Different approaches to distillation are dictated by the location of the oil glands. Preparation of the biomass to be distilled, temperature, and steam pressure affect the quality of the oil produced.

5.1.9 Types of Biomass Used

Essential oils can occur in many different parts of the plant. They can be present in flowers (rose, lavender, magnolia, bitter orange, and blue chamomile) and leaves (cinnamon, patchouli, petitgrain, clove, perilla, and laurel); sometimes the whole aerial part of the plant is distilled (*Melissa officinalis*, basil, thyme, rosemary, marjoram, verbena, and peppermint). The so-called fruit oils are often extracted from seed, which forms part of the fruit, such as caraway, coriander, cardamom, pepper, dill, and pimento. Citrus oils are extracted from the epicarp of species of *Citrus*, such as lemon, lime, bergamot, grapefruit, bitter orange as well as sweet orange, mandarin, clementine, and tangerine. Fruit or perhaps more correctly berry oils are obtained from juniper and *Schinus* species. The well-known bark oils are obtained from birch, cascarilla, cassia, cinnamon, and massoia. Oil of mace is obtained from the aril, a fleshy cover of the seed of nutmeg (*Myristica fragrans*). Flower buds are used for the production of clove oil. Wood and bark exudations yield an important group of essential oils such as galbanum, incense, myrrh, mastix, and storax, to name but a few. The needles of conifers (leaves) are a source of an important group of essential oils derived from species of *Abies*, *Pinus*, and so on. Wood oils are derived mostly from species of *Santalum* (sandalwood), cedar, amyris, cade, rosewood, agarwood, and guaiac. Finally, roots and rhizomes are the source of oils of orris, valerian, calamus, and angelica.

What happens when the plant is cut? Does it immediately start to die as happens in animals and humans? The water content of a plant ranges from 50% to over 80%. The cutting of a plant interrupts its supply of water and minerals. Its life-sustaining processes slow down and finally stop altogether. The production of enzymes stops, and autooxidative processes start, including an increase in bacterial activity leading to rotting and molding. Color and organoleptic properties, such as fragrance, will also change usually to their detriment. As a consequence of this, unless controlled drying or preparation is acceptable options, treatment of the biomass has to be prompt.

5.1.10 Timing of the Harvest

The timing of the harvest of the herbal crop is one of the most important factors affecting the quality of the essential oil. It is a well-documented fact that the chemical composition changes throughout the life of the plant. Occasionally, it can be a matter of days during which the quality of the essential oil reaches its optimum. Knowledge of the precise time of the onset of flowering often has a great influence on the composition of the oil. The chemical changes occurring during the entire life cycle of Vietnamese *Artemisia vulgaris* have shown that 1,8-cineole and β-pinene contents before flowering were below 10% and 1.2%, respectively, whereas at the end of flowering, they reached values above 24% and 10.4% (Nguyen et al., 2004). These are very large variations indeed occurring during the plant's short life span. In the case of the lavender life cycle, the ester value of the oil is the quality-determining factor. It varies within a wide range and influences the value of the oil. As a rule of thumb, it is held that its maximum value is reached at a time when about two-thirds of the lavender flowers have opened and, thus, that harvesting should commence. In the past, growers knew exactly when to harvest the biomass. These days, the use of a combination of microdistillation and GC techniques enables rapid testing of the quality of the oil and thus the determination of the optimum time for harvesting to start. Oil yields may in some cases be influenced by the time of harvesting. One of the best examples is rose oil. The petals should be collected in the morning between 6 a.m. and 9 a.m. With rising day temperatures, the oil yield will diminish. In the case of oil glands embedded within the leaf structure, such as in the case of eucalypts and pines, oil yield and oil quality are largely unaffected by the time of harvesting.

5.1.11 Agricultural Crop Establishment

The first step is, in most cases, selection of plant seed that suits best the requirements of the product looked for. Preparation of seedbeds, growing from seed, growing and transplanting of seedlings, and so on should follow well-established agricultural practices. The spacing of rows has to be considered (Kassahun et al., 2011; based on example of peppermint leaves). For example, dill prefers wider row spacing than anise, coriander, or caraway (Novak, 2005). The time required before a crop can be obtained depends on the species used and can be very variable. Citronella and lemongrass may take 7–9 months from the time of planting before the first crop can be harvested, while lavender and lavandin require up to 3 years. The most economical way to extract an essential oil is to transport the harvested biomass directly to the distillery. For some plants, this is the only practical option. *M. officinalis* ("lemon balm") is very prone to drying out and thus to loss of oil yield. Some harvested plant material may require special treatment of the biomass before oil extraction, for example, grinding or chipping, breaking or cutting up into smaller fragments, and sometimes just drying. In some cases, fermentation of the biomass should precede oil extraction. Water contained within the plant material can be named as chemically, physicochemically, and mechanically bound water (Grishin et al., 2003). According to these authors, only the mechanically bound water, which is located on the surface and the capillaries of plants, can be reduced. Drying can be achieved simply by spreading the biomass on the ground where wind movement affects the drying process. Drying can also be carried out by the use of appropriate drying equipment. Drying, too, can affect the quality of the essential oil. Until the middle of the 1980s, cut lavender and lavandin have been dried in the field (Figure 5.3), a process requiring about 3 days. The resulting oils exhibited the typical fine, floral odor; however, oil yields were inferior to yields obtained with fresh material. Compared with the present-day procedure with container harvesting and immediate processing (the so-called vert-broyé), this quality of the oil is greener and harsher and requires some time to harmonize. However, yields are better, and one step in the production process has been eliminated. Clary sage is a good example demonstrating the difference between oils distilled from fresh plant material on the one hand and dried plant material on the other. The chemical differences are clearly shown in Table 5.2. Apart from herbal biomass, fruits and seed may also have to be dried before distillation.

FIGURE 5.3 Lavender drying on the field.

TABLE 5.2
Differences in the Composition of the Essential Oil of Clary Sage Manufactured Fresh and Dried

Component	"Vert Broyee" (%)	Traditional (%)
Myrcene	0.9–1.0	0.9–1.1
Limonene	0.2–0.4	0.3–0.5
Ocimene *cis*	0.3–0.5	0.4–0.6
Ocimene *trans*	0.5–0.7	0.8–1.0
Copaene alpha	0.5–0.7	1.4–1.6
Linalool	13.0–24.0	6.5–13.5
Linalyl acetate	56.0–70.5	62.0–78.0
Caryophyllene beta	1.5–1.8	2.5–3.0
Terpineol alpha	1.0–5.0	Max. 2.1
Neryl acetate	0.6–0.8	0.7–1.0
Germacrene D	1.1–7.5	1.5–12
Geranyl acetate	1.4–1.7	2.2–2.5
Geraniol	1.4–1.7	1.2–1.5
Sclareol	0.4–1.8	0.6–2.8
Minor changes		
Middle changes		
Big changes		

These include pepper, coriander, cloves, and pimento berries, as well as certain roots such as vetiver, calamus, lovage, and orris. Clary sage is harvested at the beginning of summer but distilled only at the end of the harvesting season.

Seeds and fruits of the families Apiaceae, Piperaceae, and Myristicaceae usually require grinding up prior to steam distillation. In many cases, the seed has to be dried before comminution takes place. Celery, coriander, dill, ambrette, fennel, and anise belong to the Apiaceae. All varieties of pepper belong to the Piperaceae while nutmeg belongs to the Myristicaceae. The finer the material is ground, the better will be the oil yield and, owing to shorter distillation times, also the quality of the oil. In order to reduce losses of volatiles by evaporation during the comminution of the seed or fruit, the grinding can also be carried out under water, preferably in a closed apparatus. Heartwood samples, such as those of *S. album*, *Santalum spicatum*, and *Santalum austrocaledonicum*, have to be reduced to a very fine powder prior to steam distillation in order to achieve complete recovery of the essential oil. In some cases, coarse chipping of the wood is adequate for efficient essential oil extraction. This includes cedarwood, amyris, rosewood, birch, guaiac, linaloe, cade, and cabreuva.

Plant material containing small branches as well as foliage, which includes pine needles, has to be coarsely chopped up prior to steam distillation. Examples of such material are juniper branches, *M. alternifolia*, *Corymbia citriodora*, *P. mugo*, *P. pinaster*, *Pinus sylvestris*, *Pinus nigra*, *Abies alba*, *A. sibirica*, and *Abies grandis*, as well as mint and peppermint. Present-day mechanized harvesting methods automatically affect the chopping up of the biomass. This also reduces the volume of the biomass, thus increasing the quantity of material that can be packed into the still and making the process more economical.

It appears that the time when the seed is sown influences both oil yield and essential oil composition, for example, whether it is sown in spring or autumn. Important factors affecting production of plant material are application of fertilizers, herbicides, and pesticides and the availability and kind of pollination agents. In many essential oil–producing countries, no artificial nitrogen, potassium, or phosphorus fertilizers are used. Instead, both in Europe and overseas, the biomass left over after steam distillation is spread in the fields as an organic fertilizer. Court et al. reported, from the field tests conducted with peppermint, that an increase in fertilizer affects plant oil yield. However, higher doses did not result in further increases in oil yield or in changes in the oil composition (Court et al., 1993). Herbicides and pesticides do not appear to influence either oil yield or oil composition. The accumulation of pesticide residues in essential oils has a negative influence on their quality and on their uses. The yield and quality of essential oils are also influenced by the timing and type of pollinating agent. If the flower is ready for pollination, the intensity of its fragrance and the amount of volatiles present are at their maximum. If on the other hand the weather is too cold at the time of flowering, pollination will be adversely affected, and transformation to fruit is unlikely to take place. Such an occurrence has a very significant effect on the plant's metabolism and finally on its essential oil. Grapevine cultivators use the following trick to attract pollinators to their vines. A rose flower placed at the end of each grapevine row attracts pollinators who then also pollinate the unattractive flowers of the grapevine.

5.1.12 Propagation from Seed and Clones

Plants can be grown from seed or propagated asexually by cloning. Lavender plants raised from seed are kept for 1 year in pots before transplanting into the field. It then takes another 3 years before the plantation yields enough flowers for commercial harvesting and steam distillation. Plants of any species raised from seed will exhibit wide genetic variations among the progeny, as exist between the members of any species propagated by sexual means, for example, humans. In the case of lavender (*Lavandula angustifolia*), the composition of the essential oil from individual plants varies from plant to plant or, more precisely, from one genotype to the other. Improvement of the crop by selective breeding of those genotypes that yield the most desirable oil is a very slow process requiring years to accomplish. Charles Denny, who initiated the Tasmanian lavender industry in 1921, selected within 11 years 487 genotypes from a source of 2500 genotypes of *L. angustifolia* for closer examination, narrowing them down to just 13 strains exhibiting large yields of superior oil. Finally, four of these genotypes were grown on a large scale and mixed together in what is called "comunelles." The quality of the oils produced was fairly constant from year to year, both in their physicochemical properties and in their olfactory characteristics (Denny, 1995, private information to the author).

Cloning is the preferred method for the replication of plants having particular, usually commercially desirable, characteristics. Clones are obtained from buds or cuttings of the same individual, and the essential oils, for example, obtained from them are the same, or very similar to those of the parent. Cloning procedures are well established but may vary in their detail among different species. One important advantage of clones is that commercial harvesting may be possible after a shorter time as compared with plantations grown from seed. One risk does exist though. If the mother plant is diseased, all clones will also be affected, and the plantation would have to be destroyed.

No field of agriculture requires such a detailed and comprehensive knowledge of botany and soil science as well as of breeding and propagation methods, harvesting methods, and so on as that of the cultivation of essential oil–bearing plants. The importance of this is evident from the very large amount of scientific research carried out in this field by universities as well as by industry.

5.1.13 Commercial Essential Oil Extraction Methods

There are three methods in use. Expression is probably the oldest of these and is used almost exclusively for the production of *Citrus* oils. The second method, hydrodistillation or steam distillation, is the most commonly used one of the three methods, while dry distillation is used only rarely in some very special cases.

5.1.14 Expression

Cold expression, for example, expression at ambient temperature without the involvement of extraneous heat, was practiced long before humans discovered the process of distillation, probably

TABLE 5.3
Important Essential Oil Production from Plants of the Rutaceae Family

Botanical Term	Expressed	Distilled	Used Plant Parts
Citrus aurantifolia (Christm.) Swingle	Lime oil	Lime oil distilled	Pericarp, fruit juice, or crushed fruits
Citrus aurantium L., syn. *Citrus amara* Link, syn. *Citrus bigaradia* Loisel, syn. *Citrus vulgaris* Risso	Bitter orange oil	Neroli oil, bitter orange petitgrain oil	Flower, pericarp, leaf, and twigs with sometimes little green fruits
Citrus bergamia (Risso et Poit.), *Citrus aurantium* L. ssp. *bergamia* (Wight et Arnott) Engler	Bergamot oil	Bergamot petitgrain oil	Pericarp, leaf, and twigs with sometimes little green fruits
Citrus hystrix DC., syn. *Citrus torosa* Blanco	Kaffir lime oil, combava	Kaffir leaves oil	Pericarp, leaves
Citrus latifolia Tanaka	Lime oil Persian type		Pericarp
Citrus limon (L.) Burm. *f.*	Lemon oil	Lemon petitgrain oil	Flower, pericarp, leaf, and twigs with sometimes little green fruits
Citrus reticulate Blanco syn. *Citrus nobilis* Andrews	Mandarin oil	Mandarin petitgrain oil	Flower, pericarp, leaf, and twigs with sometimes little green fruits
Citrus sinensis (L.) Osbeck, *Citrus djalonis* A. Chevalier	Sweet orange oil		Pericarp
Citrus × *paradisi* Macfad.	Grapefruit oil		Pericarp

because the necessary tools for it were readily available. Stones or wooden tools were well suited to breaking the oil cells and freeing their fragrant contents. This method was used almost exclusively for the production of *Citrus* peel oils. *Citrus* and the allied genus *Fortunella* belong to the large family Rutaceae. *Citrus* fruits used for the production of the oils are shown in Table 5.3. *Citrus* fruit cultivation is widely spread all over the world with a suitable climate. Oils with the largest production include orange, lemon, grapefruit, and mandarin. Taking world lemon production as an example, the most important lemon-growing areas in Europe are situated in Italy and Spain with Cyprus and Greece being of much lesser importance. Nearly 90% of all lemon fruit produced originates from Sicily where the exceptionally favorable climate enables an almost around-the-year production. There is a winter crop from September to April, a spring crop from February to May, and a summer crop from May to September. The Spanish harvest calendar is very similar. Other production areas in the northern hemisphere are in the United States, particularly in Florida, Arizona, and California, and in Mexico. In the southern hemisphere, large-scale lemon producers are Argentina, Uruguay, and Brazil. Lemon production is also being developed in South Africa, Ivory Coast, and Australia. China promises to become a huge producer of lemon in the future.

The reason for extracting citrus oils from fruit peel using mechanical methods is the relative thermal instability of the aldehydes contained in them. Fatty, for example, aliphatic, aldehydes such as heptanal, octanal, nonanal, decanal, and dodecanal are readily oxidized by atmospheric oxygen, which gives rise to the formation of malodorous carboxylic acids. Likewise, terpenoid aldehydes such as neral, geranial, citronellal, and perillaldehyde as well as the α- and β-sinensals are sensitive to oxidation. Hydrodistillation of citrus fruit yields poor quality oils owing to chemical reactions that can be attributed to heat and acid-initiated degradation of some of the unstable fruit volatiles. Furthermore, some of the terpenic hydrocarbons and esters contained in the peel oils are also sensitive to heat and oxygen. One exception to this does exist. Lime oil of commerce can be either cold pressed or steam distilled. The chemical composition of these two types of oil as well as their odors differs significantly from each other. The expressed citrus peel is normally treated

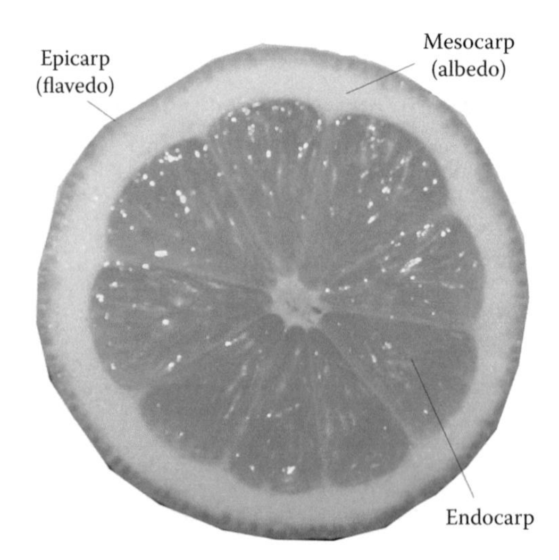

FIGURE 5.4 Parts of a citrus fruit.

with hot steam in order to recover any essential oil still left over in it. The products of this process, consisting mainly of limonene, are used in the solvent industry. The remaining peel and fruit flesh pulp are used as cattle feed.

The oil cells of citrus fruit are situated just under the surface in the epicarp, also called flavedo, in the colored area of the fruit. Figure 5.4 is a cross section of the different parts of the fruit also showing the juice cells present in the fruit. An essential oil is also present in the juice cells. However, the amount of oil present in the juice cells is very much smaller than the amount present in the flavedo; also their composition differs from each other.

Until the beginning of the twentieth century, industrial production of cold-pressed citrus oils was carried out manually. One has to visualize huge halls with hundreds of workers, men and women, seated on small chairs handling the fruit. First of all, the fruit had to be washed and cut into two halves. The pulp was then removed from the fruit using a sharp-edged spoon, called the "rastrello," and after, the peel was soaked in warm water. The fruit peel was now manually turned inside out so that the epicarp was on the inside and squeezed by hand to break the oil glands, and the oil soaked up with a sponge. The peel was now turned inside out once again and wiped with the sponge and the sponge squeezed into a terracotta bowl, the "concolina." After decantation, the oil was collected in metal containers. This was an extremely laborious process characterized by substantial oil losses. A later improvement of the fruit peel expression process was the "scodella" method. The apparatus was a metallic hemisphere lined inside with small spikes, with a tube attached at its center. The fruit placed inside the hemisphere was rotated while being squeezed against the spikes thus breaking the oil cells. The oil emulsion, containing some of the wax coating the fruit, flowed into the central tube was collected, and the oil was subsequently separated by centrifugation.

Neither of these methods, even when used simultaneously, was able to satisfy the increased demand for fruit peel oils at the start of the industrial era. The quantity of fruit processed could be increased, but the extraction methods were time wasting and the oil yields too low. With the advent of the twentieth century, the first industrial machinery was developed. Today the only systems of significance in use for the industrial production of peel oils can be classified into four categories: "sfumatrici" machines and "speciale sfumatrici," "pellatrici" machines, "food machinery corporation (FMC) whole fruit process," and "brown oil extractors (BOEs)" (Arnodou, 1991).

It is important to be aware of the fact that the individual oil glands within the epicarp are not connected to neighboring glands. The cell walls of these oil glands are very tough, and it is believed that the oil they contain is either a metabolic waste product or a substance protecting the plant from being browsed by animals.

FIGURE 5.5 "Pellatrici" method. The spiked Archimedes screw with lemons, washed with water.

The machines used in the "sfumatrici" methods consist in principle of two parts, a fixed part and a moveable part. The fruit is cut into two and the flesh is removed. In order to extract the oil, the citrus peel is gently squeezed, by moving it around between the two parts of the device and rinsing off the squeezed-out oil with a jet of water. The oil readily separates from the liquid on standing and is collected by decantation. Since the epicarp may contain organic acids (mainly citric acid followed by malic and oxalic acid), it is occasionally soaked in lime solution in order to neutralize the acids present. Greater concentrations of acid could alter the quality of the oil. Degradation of aldehydes is also an important consideration. In the "special sfumatrici" method, the peel is soaked in the lime solution for 24 h before pressing. By means of a metallic chain drawn by horizontal rollers with ribbed forms, the technical process is finished. The oils obtained by these methods may have to be "wintered," for example, refrigerated in order to freeze out the peel waxes that are then filtered off.

In the "pellatrici" method, the peel oil is removed during the first step and the fruit juice in the second step (Figure 5.5). In the first step, the fruit is fed through a slowly turning Archimedean screw-type valve. The screw is covered with numerous spikes that will bruise the oil cells in the epicarp and initiate the flow of oil. The oil is, once again, removed by means of a jet of water. The fruit is finally carried to a fast-rotating, spiked, roller carpet where the remaining oil cells, located deeper within the epicarp, are bruised and their oil content recovered, thus resulting in maximum oil yield. The process involves centrifugation, filtration, and "wintering" as previously mentioned.

The "brown process" (Reeve, 2005) is used mainly in the United States and in South America, but less in Europe. The BOE (Figure 5.6) is somewhat similar to the machinery of the "pellatrici" method. A device at the front end controls the quantity of fruit entering the machine. The machine itself consists of numerous pairs of spiked rollers turning in the same direction, as well as moving horizontally, thus reaching all oil cells. The spiked rollers as well as the fruit are submerged in water for easy transport. Any residual water and oil adhering to the fruit are removed by a special system of rollers and added to the oil emulsion generated on the first set of rollers. Any solid particles are then removed by passing it through a fine sieve. The emulsion is then centrifuged and the aqueous phase recycled. The BOE is manufactured in V4A steel to avoid contact with iron.

The most frequently used type of extractor is the FMC in-line. It is assumed that in the United States, more than 50% of extractors are of the FMC type (Figure 5.7). Other large producer countries, such as Brazil and Argentina, use exclusively FMC extractors. The reason for this is the design of the machinery, as fruit juice and oil are produced in one step without the two coming into contact with each other. The process requires prior grading of the fruit as the cups used in this process are designed for different sizes of fruit. An optimum fruit size is important as bigger fruit would be over

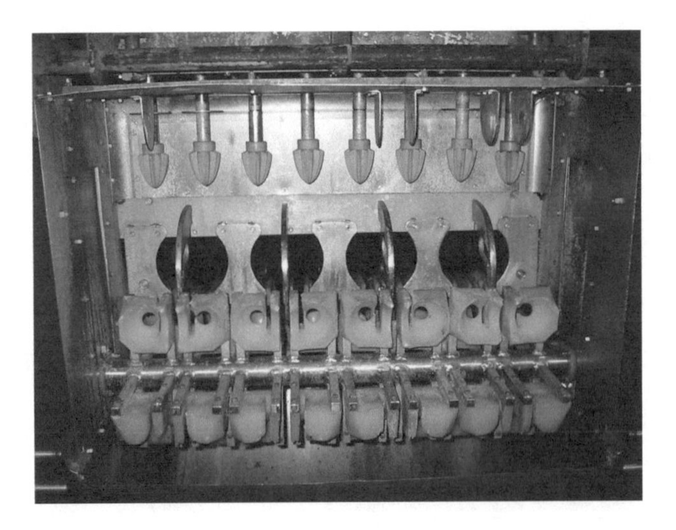

FIGURE 5.6 "Brown" process. A battery of eight juice squeezers waiting for fruits.

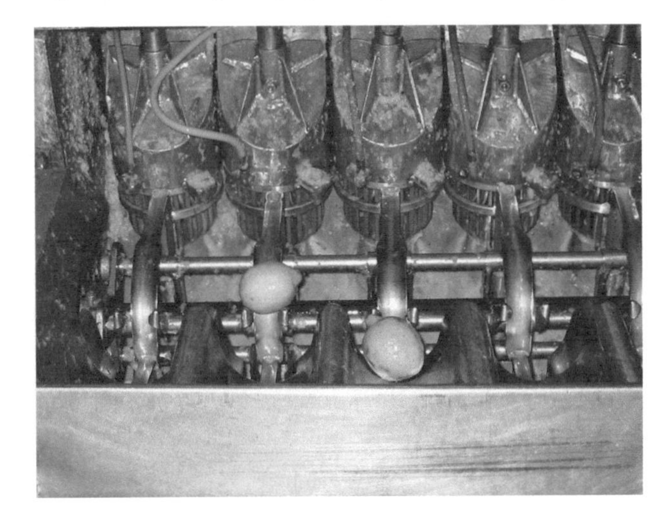

FIGURE 5.7 Food machinery corporation extractor.

squeezed and some essential oil carried over into the juice making it bitter. On the other hand, if the fruit were too small, the yield of juice would be reduced. Different frame sizes allow treating 3, 5, or 8 fruits at the same time. This technique was revolutionary in its concept and works as follows: the fruit is carried to, and placed into, a fixed cup. Another cup, bearing a mirror image relationship to the fixed cup, is positioned exactly above it. Both cups are built of intermeshing jaws. The moveable cup is lowered toward the fixed cup thus enclosing the fruit. At the same time, a circular knife cuts a hole into the bottom of the fruit. When pressure is applied to the fruit, the expressed juice will exit through the cut hole on to a mesh screen and be transported to the juice manifold, while at the same time, the oil is squeezed out of the surface of the peel. As before, the oil is collected using a jet of water. The oil–water emulsion is then separated by centrifugation.

An examination of the developments in the design of citrus fruit processing machinery shows quite clearly that the quality of the juice was more important than the quality of the oil, the only exception being oil of bergamot. Nevertheless, oil quality improved during the last decades and complies with the requirements of ISO standards. The expressed pulp of the more valuable kind fruit is very often treated with high-pressure steam to recover additional amounts of colorless oils of variable composition. The kinds of fruit treated in this manner are bergamot, lemon, and mandarin.

5.1.15 Steam Distillation

Steam or water distillation is unquestionably the most frequently used method for the extraction of essential oil from plants. The already mentioned history of steam distillation and the long-standing interest of mankind in extracting the fragrant and useful volatile constituents of plants testify to this. Distillation plants of varying design abound all over the world. While in some developing countries traditional and sometimes rather primitive methods are still being used (Figure 5.8), the essential oils produced are often of high quality. Industrialized countries employ technologically more evolved and complex equipment, computer aided with in-process analysis of the final product. Both of these very different ways of commercial essential oil production provide excellent quality oils. One depends on skill and experience, the other on superior technology and expensive equipment. It should be borne in mind that advice by an expert on distillation is a prerequisite for the production of superior quality oils. The term "distillation" is derived from the Latin "distillare," which means "trickling down." In its simplest form, distillation is defined as "evaporation and subsequent condensation of a liquid." All liquids evaporate to a greater or lesser degree, even at room temperature. This is due to thermally induced molecular movements within the liquid resulting in some of the molecules being ejected into the airspace above them (diffusing into the air). As the temperature is increased, these movements increase as well, resulting in more molecules being ejected, for example, in increased evaporation. The definition of an essential oil, ISO 9235, item 3.1.1 is "... product obtained from vegetable raw material—either by distillation with water or steam" and in item 3.1.2 "... obtained with or without added water in the still" (ISO/DIS, 9235.2, 1997, p. 2). This means that even "cooking" in the presence of water represents a method suitable for the production of essential oils. The release of the essential oil present in the oil glands (cells) of a plant is due to the bursting of the oil cell walls caused by the increased pressure of the heat-induced expansion of the oil cell contents. The steam flow acts as the carrier of the essential oil molecules. The basic principle of either water or steam distillation is a limit value of a liquid–liquid–vapor system. The theory of hydrodistillation is the following. Two nonmiscible liquids (in our case water and essential oil) A and B form two separate

FIGURE 5.8 Bush distillation device, opened.

phases. The total vapor pressure of that system is equal to the sum of the partial vapor pressures of the two pure liquids:

$$\rho = \rho_A + \rho_B \quad (\rho \text{ is total vapor pressur of the system})$$

With complete nonmiscibility of both liquids, r is independent of the composition of the liquid phase. The boiling temperature of the mixture (T_M) lies below the boiling temperatures (T_A and T_B) of liquids A and B. The proportionality between the quantity of each component and the pressure in the vapor phase is given in the formula

$$\frac{N_{oil}}{N_{water}} = \frac{P_{oil}}{P_{water}}$$

where

N_{oil} is the number of moles of the oil in the vapor phase
N_{water} is the number of moles of water in the vapor phase

It is nearly impossible to calculate the proportions as an essential oil is a multicomponent mixture of variable composition.

The simplest method of essential oil extraction is by means of hydrodistillation, for example, by immersion of the biomass in boiling water. The plant material soaks up water during the boiling process, and the oil contained in the oil cells diffuses through the cell walls by means of osmosis. Once the oil has diffused out of the oil cells, it is vaporized and carried away by the stream of steam. The volatility of the oil constituents is not influenced by the rate of vaporization but does depend on the degree of their solubility in water. As a result, the more water-soluble essential components will distil over before the more volatile but less water-soluble ones. The usefulness of hydrodiffusion can be demonstrated by reference to rose oil. It is well known that occasionally some of the essential oil constituents are not present as such in the plant but are artifacts of the extraction process. They can be products of either enzymic splitting or chemical degradation, occurring during the steam distillation, of high-molecular-weight and thus nonvolatile compounds present in the plants. These compounds are often glycosides. The main constituents of rose oil, citronellol, geraniol, and nerol are products of a fermentation that takes place during the water-distillation process.

Hydrolysis of esters to alcohols and acids can occur during steam distillation. This can have serious implications in the case of ester-rich oils, and special precautions have to be taken to prevent or at least to limit the extent of ester degradation. The most important examples of this are lavender or lavandin oils rich in linalyl acetate and cardamom oil rich in α-terpinyl acetate. Chamazulene, a blue bicyclic sesquiterpene, present in the steam-distilled oil of German chamomile, *Chamomilla recutita* (L.) Rauschert, flower heads, is an artifact resulting from matricin by a complex series of chemical reactions: dehydrogenation, dehydration, and ester hydrolysis. As chamazulene is not a particularly stable compound, the deep-blue color of the oil can change to green and even yellow on aging.

The design of a water/steam distillation plant at its simplest, sometimes called "false-bottom apparatus," is as follows: a still pot (a mild steel drum or similar vessel) is fitted with a perforated metal plate or grate, fixed above the intended level of the water, and a lid with a gooseneck outlet. The lid has to be equipped with a gasket or a water seal to prevent steam leaks. The steam outlet is attached to a condenser, for example, a serpentine placed in a drum containing cold water. An oil collector (Florentine flask) placed at the bottom end of the serpentine separates the oil from the distilled water (Figure 5.9). The whole assembly is fixed on a brick fireplace. A separate water inlet is often provided to compensate for water used up during the process. The biomass is placed inside the still pot above the perforated metal plate, and sufficient biomass should be used to completely fill the still pot. The fuel used is firewood. This kind of distillation plant was extensively used at the end of the nineteenth

FIGURE 5.9 Old distillation apparatus modernized by electric heating.

century, mainly for field distillations. A disadvantage of this system was that in some cases excessive heat imparted a burnt smell to the oils. Furthermore, when the water level in the still dropped too much, the plant material could get scorched. Till today, there is a necessity to clean the distillation vessel after two cycles with water to avoid burning notes in the essential oil. In any case, the quality of oils obtained in this type of apparatus was very variable and varied with each distillation. A huge improvement to this process was the introduction of steam generated externally. The early steam generators were very large and unwieldy, and the distillation plant could no longer be transported in the field. The biomass had now to be transported to the distillation plant, unlike with the original type of distillation plant. Originally, the generator was fuelled with dry, extracted biomass. Today, gas or fuel oil is used. The delivery of steam can be carried out in various ways. Most commonly, the steam is led directly into the still through its bottom. Overheating is thus avoided and the biomass is heated rapidly. It also allows regulation of steam quantity and pressure and reduces distillation time and improves oil quality. In another method, the steam is injected in a spiraling motion. This method is more effective as the steam comes into contact with a greater surface of the biomass. The velocity of steam throughput and the duration of the distillation depend on the nature of the biomass. It can vary from 100 kg/h in the case of seed and fruits to 400 kg/h for clary sage. The duration of the distillation can vary from about 20 min for lavender flowers (Denny, 1995, personal communication) to 700 min for dried angelica root. The values quoted are for a 4 m^3 still pot (Omidbaigi, 2005). Specialists on distillation found a formula that distillation can be stopped when the ratio of oil to water coming from condenser will achieve 1:40. In all cases of hydrodistillation, the distillation water is recovered and reused for steam generation. In a cohobation, the aqueous phase of the distillate is continuously reintroduced into the still pot. In this method, any essential oil constituents emulsified or dissolved in the water are captured, thus increasing total oil yield. There is one important exception: in the case of rose oil, the distillation water is collected and redistilled *separately* in a second step. The "floral water" contains increased amounts of β-phenylethyl alcohol, up to 15%, whereas its maximum permissible content in rose oil is 3%. The reason for this is its significant solubility in water, ca. 2%.

The distillation of rose oil is an art in itself as not only quality but also quantity plays an important role. It takes two distillation cycles to produce between 200 and 280 g of rose oil. Jean-François

Arnodou describes its manufacture as follows (Arnodou, 1991): the still pot is loaded with 400 kg of rose petals and 1600 L of water. The contents are heated until they boil and steam distilled. Approximately, the quantity of flowers used is then distilled. That action will last about 2–3 h. Specially designed condensers are required in order to obtain a good quality. The condensing system comprises a tubular condenser followed by a second cooler to allow the oil to separate. The oil is collected in Florentine-type oil separators. About 300 L of the oil-saturated still waters are then redistilled in a separate still in order to recover most of the oil contained in them. Both oils are mixed together and constitute the rose oil of commerce. BIOLANDES described in 1991 the whole process, which uses a microprocessor to manage parameters such as pressure and temperature, regulated by servo-controlled pneumatic valves.

A modern distillation plant consists of the biomass container (still pot), a cooling system (condenser), an oil separator, and a high-capacity steam generator. The kettle (still pot) looks like a cylindrical vertical storage tank with steam pipes located at the bottom of the still. Perforated sievelike plates are often used to separate the plant charge and prevent compaction, thus allowing the steam unimpeded access to the biomass. The outlet for the oil-laden steam is usually incorporated into the design of the usually hemispherical, hinged still pot lid. The steam is then passed through the cooling system, either a plate heat exchanger or a surface heat exchanger, such as a cold-water condenser. The usually liquid condensate is separated into essential oil and distillation water in an appropriate oil separator such as a Florentine flask. The distillation water may, in some cases, be redistilled, and the remaining essential oil is recovered, dried, and stored. Figure 5.10 shows a cross section of such a still.

The following illustrations show different parts of an essential oil production plant. Figure 5.11 shows a battery of four production units in the factory. Each still has a capacity of 3000–5000 L. Owing to their large size, the upper half of the stills is on the level as shown, while the lower half is situated on the lower level. Figure 5.12 shows open stills and displays the steam/oil vapor outlets on the underside of the lids leading to the cooling units. On the right side of the illustration, one can see the perforated plate used to prevent clumping of the biomass. Several such perforated plates, up to 12, depending on the type of biomass, are used to prevent clumping. Spacers on the central upright control the optimum distance between these plates for improved steam penetration. Figure 5.13 shows the unloading of the still. Unloading is much faster than the loading process where the biomass is compacted either manually or by means of tractor wheel (Figure 5.14). This type of loading is

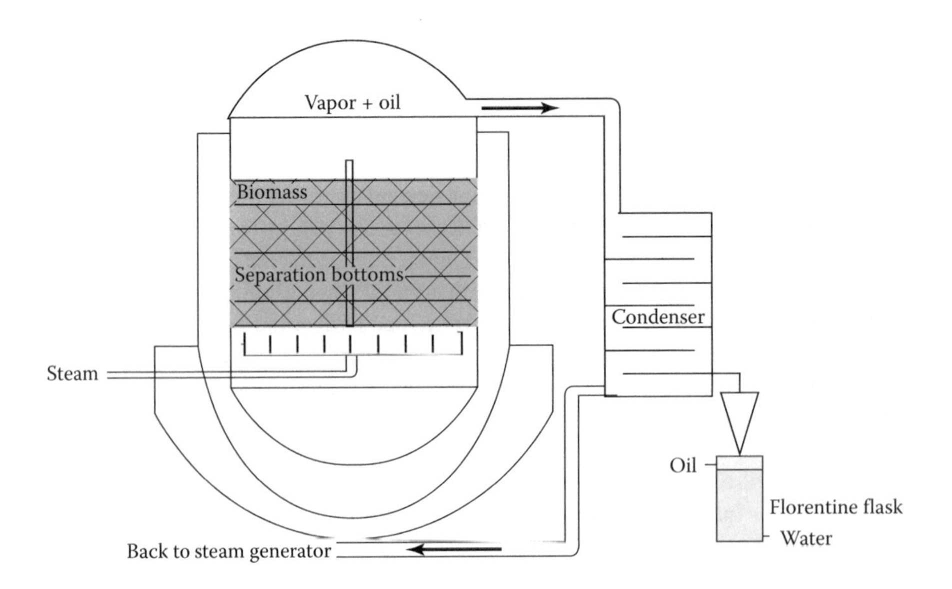

FIGURE 5.10 Cross section of a hydrodistillation plant.

FIGURE 5.11 Battery of four distillation units.

FIGURE 5.12 Open kettle with opening for vapor and oil.

called "open mouth" loading. Figure 5.15 shows the cooling unit. The cold water enters the tank equipped with a coil condenser. The cooling water is recycled so that no water is wasted. The two main types of industrially used condensers are the following. The earliest was the coil condenser that consisted of a coiled tube fixed in an open vessel of cold water with cold water entering the tank from the bottom and leaving at the top. The oil-rich steam is passed through the coils of the condenser from the top end. The second type of condenser is the pipe bundle condenser where the steam is passed through several vertical tubes immersed in a cold-water tank. The tubes have on the inside walls horizontal protuberances that slow down the rate of the steam flow and thus result in more effective cooling. Figure 5.16 shows the inside of a Florentine flask where the oil is separated from the water. Most essential oils are lighter than water and thus float on top of the water. Some essential oils have a specific gravity >1, for example, they are heavier than water thus collecting at

FIGURE 5.13 Unloading a kettle.

FIGURE 5.14 Loading a kettle and pressing by concreted tractor wheel.

the bottom of the collection vessel. A modified design of the Florentine flask for such oils is shown in Figure 5.17. Figure 5.18 shows oil in the presence of turbid distillation water. The liquid phase is contaminated with biomass matter, and the oil has to be filtered. The capacity of the still pot depends on the biomass. Weights vary from 150 to 650 kg/m^3. Wilted and dried plants are much lighter than seeds and fruits or dried roots that can be very heavy.

A very special case is the production of the essential oils of ylang-ylang from the fresh flowers of *Cananga odorata* (Lam.) Hook. f. et Thomson forma *genuina*. The hydro-distillation process is

FIGURE 5.15 Cooling unit.

FIGURE 5.16 Inner part of a Florentine flask.

started and after a certain time the obtained oil is saved. With ongoing distillation, this procedure is repeated three times to achieve at least four separate fractions. The chemical composition of the first fraction is characterized by a high concentration of *p*-cresol methylether, methyl benzoate, benzyl acetate, linalool, and *E*-cinnamyl acetate. The second fraction contains less of those volatiles but an increased amount of geraniol, geranyl acetate, and β-caryophyllene. The third fraction contains higher boiling substances such as germacrene-D, (*E,E*)-α-farnesene, (*E,E*)-farnesol, benzyl benzoate, (*E,E*)-farnesyl acetate, and benzyl salicylate. Of course, smaller quantities of the lower boiling components are also present. This kind of fractionation has been practiced for a long time. At the same time, the whole oil, obtained by a single distillation is available as "ylang-ylang complete." This serves as an example of the importance the duration of the distillation can have on the quality of the oil.

Raw materials occurring in the form of hard grains have to be comminuted, for example, ground up before water distillation. This is carried out in the presence of water, such as in a wet-grinding turbine, and the water is used later during the distillation. The stills themselves are equipped with blade stirrers ensuring thorough mixing and particularly dislodging oil particles or biomass articles

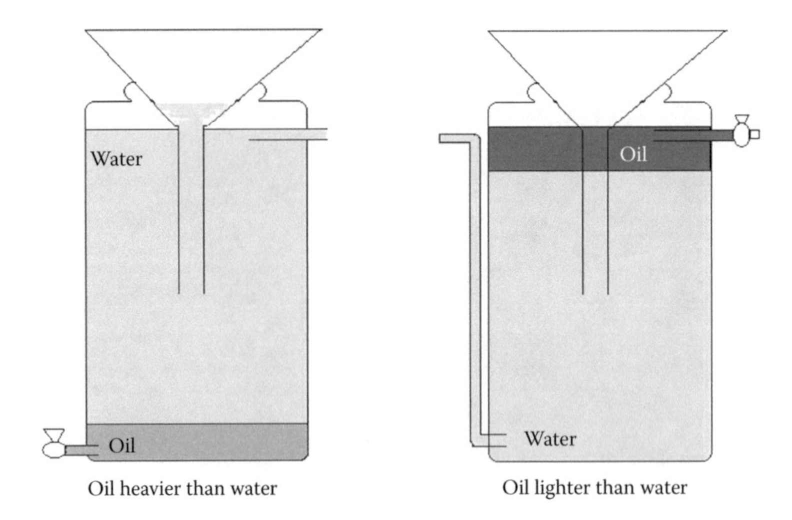

FIGURE 5.17 Two varieties of Florentine flasks.

FIGURE 5.18 Oil and muddy water in the Florentine flask.

sticking to the walls of the still, the consequence of which can be burning and burnt notes. Dry grinding is likely to result in a significant loss of volatiles. Pepper, coriander, cardamom, celery seed, and angelica seed as well as roots, cumin, caraway, and many other seeds and fruits are treated in this manner. The process used in all these cases is called "turbo distillation." The ratio oil/condensate is very low when this method is used, and it is for that reason that turbo distillation uses a fractionating column to enrich the volatiles. This also assists in preventing small particles of biomass passing into the condenser and contaminating the oil. As in many other distillation and rectification units, cold traps are installed to capture any very volatile oil constituents that may be present. This water-distillation procedure is also used for gums such as myrrh, olibanum, opopanax, and benzoin.

Orris roots are also extracted by water distillation. However, in this case, the distillation has to be carried out under conditions of slightly elevated pressure. This is achieved by means of a reflux column filled with Raschig rings. This is important as the desired constituents, the irones, exhibit very high boiling points. It is noteworthy that in this case, there is no cooling of the vapors, as not only the irones but also the long-chain hydrocarbons will immediately be transported to the top of the column. Figure 5.19 shows a Florentine flask with the condensed oil/water emerging at a

FIGURE 5.19 Orris distillation, Florentine flask at nearly 98°C.

temperature of nearly 98°C. Orris oil or orris butter (note that the term orris "concrete" is incorrect, as the process is not a solvent extraction) is one of the few essential oils that are, at least partly, solid at room temperature. Depending on its *trans*-anethole content, rectified star anise oil is another example of this nature.

A relatively new technique that saves time in loading and unloading of the biomass is the "on-site" or "container" distillation. The technique is very simple as the container that is used to pick up the biomass and transport it to the distillery serves itself as the still pot. The first plant crops treated in this way were peppermint and mint, clary sage, lavandin grosso, *L. angustifolia*, *Eucalyptus polybractea*, and tea tree. In its simplest form, the mobile still assembly is composed of the following components: a tractor is coupled to an agricultural harvester that cuts the plant material and delivers it directly into the still pot (or vat) via a chute. The still pot (vat) is permanently fixed onto a trailer that is coupled to the harvester. Once the still pot is completely filled, it is towed by the tractor into the factory where it is uncoupled and attached to the steam supply and condenser and distillation commences. Presupposition for a proper working of the container as vat is a perfect insulation. Every loss of steam and heat will guide to worse quality and diminished quantity. Lids will have to be placed properly and fixed by clamps. The tractor and harvester are attached to an empty still pot and the process is repeated. The design, shape, and size of the still pot as well as the type of agricultural harvester depend on the type of plant crop, the size of the plantation, the terrain, and so on. The extracted biomass can be used as mulch or, after drying, as fuel for the steam boiler. The unloading is automated using metal chains running over the tubes with steam valves. This method requires less manpower and thus reduces labor costs. Loading and unloading costs are minimal. It lends itself best to fresh biomass, lavender and lavandin, mallee eucalyptus, tea tree, and so on. It may not be as useful for the harvesting and distillation of mint and peppermint as these crops have to be wilted before oil extraction. Figures 5.20 through 5.22 show the harvesting of mallee eucalyptus, containers in processing, and the whole site of container distillation. Figure 5.23 gives a view into the interior of a container.

Another interesting distillation method has been developed by the LBP Freising, Bavaria, Germany. The plant consists of two tubes, each 2 m long and 25 cm in diameter, open at the top. The

FIGURE 5.20 Harvesting blue mallee with distillation container.

FIGURE 5.21 A battery of containers to be distilled, one opened to show the biomass.

tubes are attached vertically to a central axis that can be rotated. One tube is connected, hydraulically or mechanically, to the steam generator and on top to a condenser. During the distillation of the contents of the tube, which may take 25–40 min depending on the biomass, the other tube can be loaded. When the distillation of the first tube has been completed, the tubes are rotated around the axis and distillation of the second tube commenced. The first distilled tube can now be emptied and reloaded. The only disadvantage of this type of apparatus is the small size. Only 8.5–21 kg of biomass can be treated. This system has been developed for farmers intending to produce small quantities of essential oil. The apparatus is transportable on a truck and will work satisfactorily provided a supply of power is available.

Most commercially utilized essential oil distillation methods, except the mobile still on-site methods, suffer from high labor costs. Apart from harvesting the biomass, three to four laborers will be required to load and unload the distillation pots, regulating steam pressure and temperature, and so on. The loading and distribution of the biomass in the distillation vessel may not be homogeneous.

FIGURE 5.22 Distillation plant with container technique.

FIGURE 5.23 View inside the distillation container, showing the steam tubes and the metallic grid.

This will adversely affect the steam flow through the biomass by channeling, for example, the steam passing through less compacted areas and thus not reaching other more compacted areas. This will result in lower oil yields and perhaps even alter the composition of the oil. In times of high energy costs, the need for consequent recovery has to be considered. Given the demand for greater quantities of essential oils, the question is how to achieve it and at the same time improve the quality of the oils. For this, several considerations have to be taken into account. The first is how to process large quantities of biomass in a given time. Manpower has to be decreased as it still is the most important factor affecting costs. The biomass as a whole has to be treated uniformly to ensure higher oil yields and more constant and thus better oil composition. How can energy costs and water requirements be reduced in an ecologically acceptable way? The answer to this was the development of continuous distillation during the last years of the twentieth century. Until then, all distillation processes were discontinuous. Stills had to be loaded, the distillation stopped, and stills unloaded. The idea was to develop a process where the steam production was continuous with permanent unchanged parameters. This was achieved by the introduction of an endless screw that fed the plant material slowly into the still pot from the top and removed the exhausted plant material from the bottom at the same speed. The plant material moves against the flow of dry steam entering the still from the bottom. In this fashion, all of the biomass comes into contact with the steam ensuring optimum essential oil extraction.

The earliest of these methods is known as the "Padova system." It consists of a still pot 6 m high and about 1.6 m in diameter (Arnodou, 1991). Its total volume is about 8 m^3. The feeding of the still with the plant biomass as well as its subsequent removal is a continuous process. The plant material is delivered via a feed hopper situated at the top of the still. Before entering the still, it is compressed and cut by a rotating knife to ensure a more uniform size. Finally, a horizontal moving cone regulates the quantity of biomass entering the still. The biomass that enters the still moves in the opposite direction to that of the steam. The steam saturated with essential oil vapors is then passed into the cooling system. The exhausted plant material is simultaneously removed by means of an Archimedes screw. This type of plant was originally designed for the distillation of wine residues. A different system is provided by the "chimie fine aroma process continuous distillation" method. Once again, the plant material is delivered via a hopper to several interconnected tubes. These tubes are slightly inclined and connected to each other. The biomass is carried slowly through the tubes, by means of a worm screw, in a downward direction. Steam is injected at the end of the last tube and is directed upward in the opposite direction to that of the movement of the plant material. The essential oil–laden steam is deflected near the point of entrance of the biomass, into the condenser. The exhausted plant material is unloaded by another worm screw located near the point of the steam entrance to the system.

Texarome, a big producer of cedarwood oil and related products holds a patent on another continuous distillation system. In contrast to other systems, the biomass is conveyed pneumatically within the system. It is a novel system spiked with new technology of that time. Texan cedarwood oil is produced from the whole tree, branches, roots, and stumps. Cedarwood used in Virginia uses exclusively branches, stumps, sawdust, and other waste for oil production; wood is used mainly for furniture making. The wood is passed through a chipper and then through a hammer mill. The dust is collected by means of a cyclone. Any coarse dust is reground to the desired size. The dust is now carried via a plug feeder to the first contactor where superheated steam in reverse flow exhausts it in a first step and following that in a similar second step at the next contactor. The steam and oil vapors are carried into a condenser. The liquid distillate is then separated in Florentine flasks. This process does all transport entirely by pneumatic means. The recycling of cooling water and the use of the dried plant matter as a fuel contribute to environmental requirements (Arnodou, 1991). In the 1990s, the BIOLANDES company designed its own system of continuous distillation. The reason for this was BIOLANDES" engagement in the forests of South West France. The most important area of pine trees (*P. pinaster* Sol.) supplying the paper industry exists between Bordeaux and Biarritz. Twigs and needles have been burnt or left to rot to assist with reforestation with new trees. These needles contain a fine essential oil very similar to that of the dwarf pine oil (*P. mugo* Turra.). Compared to other needle oils, dwarf pine oil is very expensive and greatly appreciated. The oil was produced by a discontinuous distillation but as demand rose, new and improved methods were required. First of all, the collection of the branches had to be improved. A tractor equipped with a crude grinder and a ground wood storage box follows the wood and branch cutters and transports it to the nearby distillation unit where the biomass is exhausted via a continuous distillation process. In contrast to the earlier described methods, the BIOLANDES continuous distillation process operates somewhat differently. The plant material is carried by mechanical means from the storage to the fine cutter and via an Archimedes screw to the top of the distillation pot. The plant material is now compressed by another vertical screw and transported into a chamber that is then hermetically closed on its back but opening at the front. Biomass is falling down allowing the countercurrent passage of hot steam through it. The steam is supplied through numerous nozzles. Endless screws at the bottom of the still continuously dispose of the exhausted biomass. Oil-laden steam is channeled from the top of the still into condenser and then the oil separator.

It is well known that clary sage yields an essential oil on hydrodistillation. However, a very important component of this oil, sclareol, is usually recovered in only very small quantity when this method is used; the reason for this being its very high boiling point. Sclareol can be recovered in very high yield and quality by extraction with volatile solvents. Consequently, BIOLANDES has incorporated an extraction step in its system (Figure 5.24). Any waste biomass, whether of extracted or nonextracted material, is used for energy production or, mostly, for composting. The energy recovery management distinguishes this system from all other earlier described processes. In all of the latter, large amounts of cold water are required to condense the essential oil–laden steam. This results in significant wastage of water as well as in latent energy losses. The BIOLANDES system recovers this latent heat. Hot water from the condenser is carried into an aerodynamic radiator. Air used as the transfer gas takes up the energy of the hot water, cooling it down, so that it can be recycled to the condenser. The hot air is then used to dry about one-third of the biomass waste that is used as an energy source for steam and even electricity production. In other words, this system is energetically self-sufficient. Furthermore, since it is fully automated, it results in constant quality products. A unit comprising two stills of 7.5 m^3 capacity can treat per hour 3 ton of pine needles, 1.5 ton of juniper branches, and 0.25 ton of cistus branches (Arnodou, 1991). The advantages are once again short processing time of large amounts of biomass, reduced labor costs, and near-complete energy sufficiency. All operations are automated and water consumption is reduced to a minimum. The system can also operate under slight pressure thus improving the recovery of higher boiling oil constituents.

The following is a controversial method for essential oil extraction by comparison with classical hydrodistillation methods. In this method, the steam enters the distillation chamber from the top passes through the biomass in the still pot (e.g., the distillation chamber) and percolates into the condenser located below it. Separation of the oil from the aqueous phase occurs in a battery of Florentine flasks. It is claimed that this method is very gentle and thus suitable for the treatment of sensitive plants. The biomass is held in the still chamber (e.g., still pot) on a grid that allows easy disposal of the spent plant matter at the completion of the distillation. The whole apparatus is

FIGURE 5.24 Scheme of the BIOLANDES continuous production unit. (A) biomass; (B), distillation vat; (C) condenser; (D) Florentine flask; (E) extraction unit; (F) solvent recovery; and (G) exhausted biomass.

relatively small, distillation times are reduced, and there is less chance of the oil being overheated. It appears that this method is fairly costly and thus likely to be used only for very-high-priced biomass.

Recently, microwave-assisted hydrodistillation methods have been developed, so far mainly in the laboratory or only for small-scale projects. Glass vessels filled with biomass, mainly herbs and fruits or seeds, are heated by microwave power. By controlling the temperature at the center of the vessel, dry heat conditions are established at about 100°C. As the plant material contains enough water, the volatiles are evaporated together with the steam solely generated by the microwave heat and can be collected in a suitably designed condenser/cooling system. In this case, changes in the composition of the oil will be less pronounced than in oil obtained by conventional hydrodistillation. This method has attracted interest owing to the mild heat to which the plant matter is exposed. Kosar reported improvements in the quality of microwave-extracted fennel oil due to increases in the yields of its oxygenated components (Kosar et al., 2007).

Very different products can result from the dry distillation of plant matter. ISO Standard 9235 specifies in Section 3.1.4 that products of dry distillation, for example, "... obtained by distillation without added water or steam" are in fact essential oils (ISO/DIS, 9235.2, 1997, p. 2). Dry distillation involves heating in the absence of aerial oxygen, normally in a closed vessel, preventing combustion. The plant material is thus decomposed to new chemical substances. Birch tar from the wood exudate of *Betula pendula* Roth. and cade oil from the wood of *Juniperus oxycedrus* L. are manufactured in this way. Both oils contain phenols, some of which are recognized carcinogens. For this reason, the production of these two oils is no longer of any commercial importance, though very highly rectified and almost phenol-free cade oils do exist.

Some essential oils require rectification. This involves redistillation of the crude oil in order to remove certain undesirable impurities, such as very small amounts of constituents of very low volatility, carried over during the steam or water distillation (such as high-molecular-weight phenols, leaf wax components) as well as small amounts of very volatile compounds exhibiting an undesirable odor, and thus affecting the top note of the oils, such as sulfur compounds (dimethyl sulfide present in crude peppermint oil), isovaleric aldehyde (present in *E. globulus* oil), and certain nitrogenous compounds (low-boiling amines, sulfides, mercaptans, and polysulfides). In some cases, rectification can also be used to enrich the essential oil in a particular component such as 1,8-cineole in low-grade eucalyptus oil. Rectification is usually carried out by redistillation under vacuum to avoid overheating and thus partial decomposition of the oil's constituents. It can also be carried out by steaming. Commonly rectified oils include eucalyptus, clove, mint, turpentine, peppermint, and patchouli. In the case of patchouli and clove oils, rectification improves their, often unacceptably, dark color.

Fractionation of essential oils on a commercial scale is carried out in order to isolate fractions containing a particular compound in very major proportions and occasionally even individual essential oil constituents in a pure state. In order to achieve the required separation, fractionations are conducted under reduced pressure (e.g., under vacuum) to prevent thermal decomposition of the oil constituents, using efficient fractionating columns. A number of different types of fractionating columns are known, but the one most commonly used in laboratory stills or small commercial stills is a glass or stainless steel column filled with Raschig rings. Raschig rings are short, narrow diameter, rings made of glass or any other chemically inert material. Examples of compounds produced on a commercial scale are citral (a mixture of geranial and neral) from *L. cubeba*, 1,8-cineole from eucalyptus oil (mainly *E. polybractea* and other cineole-rich species) as well as from *Cinnamomum camphora* oil, eugenol from clove leaf oil, α-pinene from turpentine, citronellal from citronella oil, linalool from Ho-oil, geraniol from palmarosa oil, and so on. A small-scale high-vacuum plant used for citral production is shown in Figure 5.25. The reflux ratio, for example, the amount of distillate separated and the amount of distillate returned to the still, determines the equilibrium conditions of the vapors near the top of the fractionation column, which are essential for good separation of the oil constituents.

Apart from employing fractional distillation, with or without the application of a vacuum, some essential oil constituents are also obtained on a commercial scale by freezing out from the essential oil, followed by centrifugation at below freezing point of the desired product. Examples are menthol

FIGURE 5.25 High-vacuum rectification plant in small scale. The distillation assembly is composed of a distillation vessel (1) of glass, placed in an electric heating collar (2). The vessel is surmounted by a jacketed fractionation column (3), packed with glass spirals or Raschig rings, of such a height as to achieve maximum efficiency (e.g., have the maximum number of theoretical plates capable of being achieved for this type of apparatus). The reflux ratio is automatically regulated by a device (4), which also includes the head condenser (5) and a glass tube that leads the product to another condenser (6), from there to both receivers (7). The vacuum pump unit is placed on the right (8).

from *Mentha* species (this is usually further purified by recrystallization with a suitable solvent); *trans*-anethole from anise oil, star anise oil, and particularly fennel seed oil; and 1,8-cineole from cineole-rich eucalyptus oils.

Most essential oils are complex mixtures of terpenic and sesquiterpenic hydrocarbons and their oxygenated terpenoid and sesquiterpenoid derivatives (alcohols, aldehydes, ketones, esters, and occasionally carboxylic acids), as well as aromatic (benzenoid) compounds such as phenols, phenolic ethers, and aromatic esters. So-called "terpeneless" and "sesquiterpeneless" essential oils are commonly used in the flavor industry. Many terpenes are bitter in taste, and many, particularly the terpenic hydrocarbons, are poorly soluble or even completely insoluble in water–ethanol mixtures. Since the hydrocarbons rarely contribute anything of importance to their flavoring properties, their removal is a commercial necessity. They are removed by the so-called washing process, a method used mostly for the treatment of citrus oils. This process takes advantage of the different polarities of individual essential oil constituents. The essential oil is added to a carefully selected solvent (usually a water–ethanol solution) and the mixture partitioned by prolonged stirring. This removes some of the more polar oil constituents into the water–ethanol phase (e.g., the solvent phase). Since a single partitioning step is not sufficient to effect complete separation, the whole process has to be repeated several times. The water–ethanol fractions are combined and the solvent removed. The residue contains now very much reduced amounts of hydrocarbons but has been greatly enriched in the desired polar oxygenated flavor constituents, aldehydes such as octanal, nonanal, decanal, hexenal, geranial, and neral; alcohols such as nerol, geraniol, and terpinen-4-ol; oxides such as 1,8-cineole and 1,4-cineole; and esters and sometimes carboxylic acids. Apart from water–ethanol mixtures, hexane or light petroleum fractions (sometimes called "petroleum ether") have sometimes also been added as they will enhance the separation process. However, these are highly flammable liquids, and care has to be taken in their use.

"Folded" or "concentrated" oils are citrus oils from which some of the undesirable components (usually limonene) have been removed by high-vacuum distillation. In order to avoid thermal degradation of the oil, temperatures have to be kept as low as possible. Occasionally, a solvent is used as a "towing" agent to keep the temperature low.

Another, more complex, method for the concentration of citrus oils is a chromatographic separation using packed columns. This method allows a complete elimination of the unwanted hydrocarbons. This method, invented by Erich Ziegler, uses columns packed with either silica or aluminum oxide. The oil is introduced onto the column and the hydrocarbons eluted by means of a suitable nonpolar solvent of very low boiling point. The desirable polar citrus oil components are then washed out using a polar solvent (Ziegler, 1982).

Yet another valuable flavor product of citrus fruits is the "essence oil." The favored method for the transport of citrus juice is in the form of a frozen juice concentrate. The fruit juice is partly dehydrated by distilling off under vacuum the greater part of the water and frozen. Distilling off the water results in significant losses of the desirable volatiles responsible for the aroma of the fruit. These volatiles are captured in several cold traps and constitute the "aqueous essence" or "essence oil" that has the typical fruity and fresh fragrance, but slightly less aldehydic than that of the oil. This oil is used to enhance the flavor of the reconstituted juice obtained by thawing and dilution with water of the frozen concentrate.

Producing essential oils today is, from a marketing point of view, a complex matter. As in the field of other finished products, the requirements of the buyer or producer of the consumer product must be fulfilled. The evaluation of commercial aspects of essential oil production is not an easy task and requires careful consideration. There is no sense in producing oils in oversupply. Areas of short supply, depending on climatic or political circumstances should be identified and acted upon. As in other industries, global trends are an important tool and should continually be monitored. For example, which are the essential oils that cannot be replaced by synthetic substitutes such as patchouli oil or blue chamomile oils? A solution to this problem can lie in the breeding of suitable plants. For example, a producer of a new kind "pastis," the traditional aperitif of France, wants to introduce a new flavor with a rosy note in the fennel component of the flavor. This will require the study and identification of oil constituents with "rosy" notes and help biologists to create new botanical varieties by genetic crossing, for example, by genetic manipulation, of suitable target plant species. Any new lines will first be tested in the laboratory and then in field trials. Test distillations will be carried out and the chemical composition of the oils determined. Agronomists and farmers will be involved in all agricultural aspects of the projects: soil research and harvesting techniques. Variability of all physicochemical aspects of the new strains will be evaluated. At this point, the new types of essential oils will be presented to the client. If the client is satisfied with the quality of the oils, the first larger plantations shall be established, and consumer market research will be initiated. If everything has gone to plan, that is, all technical problems have been successfully resolved and the finished product has met with the approval of the consumers, large-scale production can begin. This example describes the current way of satisfying customer demands.

Global demand for essential oils is on the increase. This also generates some serious problems for which immediate solutions may not easily be found. The first problem is the higher demand for certain essential oils by some of the world's very major producers of cosmetics. They sometimes contract oil quantities that can be of the order of 70% of world production. This will not only raise the price but also restrict consumer access to certain products. From this arises another problem. Our market is to some extent a market of copycats. How can one formulate the fragrance of a competitor's product without having access to the particular essential oil used by him or her, particularly as this oil may have other functions than just being a fragrance, such as certain physiological effects on both the body as well as the mind? Lavender oil from *L. angustifolia* is a calming agent as well as possessing anti-inflammatory activity. No similar or equivalent natural essential oil capable of replacing it is known. Another problem affecting the large global players is ensuring the continuing availability of raw material of the required quality needed to satisfy market demand. This is clearly an almost impossible demand as nobody can assure that climatic conditions required for optimum growth of a particular essential oil crop will remain unchanged. Another problem may be the farmer himself or herself. Sometimes, it may be financially more worthwhile for the farmer to cultivate other than essential oil plant crops. All these factors may have some detrimental effects on the availability of essential oils. Man's responsibility for the continued health of the environment may also be one of the reasons for the disappearance of an essential oil from the market. Sandalwood (species of *Santalum*, but mainly Indian *S. album*) requires in some cases up to 100 years to regenerate to a point where they are large enough to be harvested. This and their uses in religious ceremonies have resulted in significant shortages of Indian oil. Owing to the large monetary value of Indian sandalwood oil, indiscriminate cutting of the wood has just about entirely eliminated it from native forests in Timor (Indonesia). Sandalwood oils of other origins are available, *S. spicatum* from western Australia and *S. austrocaledonicum* from New Caledonia and Vanuatu. However, their wood oils differ somewhat in odor as well as in chemical composition from genuine Indian oils.

Some essential oils are disappearing from the market owing to the hazardous components they contain and are, therefore, banned from most applications in cosmetics and detergents. These oil components, all of which are labeled as being carcinogenic, include safrole, asarone, methyl eugenol, and elemicin. Plant diseases are another reason for essential oil shortages as they, too, can be affected by a multitude of diseases, some cancerous, which can completely destroy the total crop. For example, French lavender is known to suffer from a condition whereby a particular protein causes a decrease in the growth of the lavender plants. This process could only be slowed down by cultivation at higher altitudes. In the middle of the twentieth century, lavender has been cultivated in the Rhône valley at an altitude of 120 m. Today, lavender is growing only at altitudes around 800 m. The growing shoots of lavender plants are attacked by various pests, in particular the larvae of Cécidomye (*T. lavandula*) that, if unchecked, will defoliate the plants and kill them. Some microorganisms such as *Mycoplasma* and a fungus *A. mellea* can cause serious damage to plantations. At the present time, the use of herbicides and pesticides is an unavoidable necessity. Wild-growing plants are equally prone to attack by insect pests and plant diseases.

The progression from wild-growing plants to essential oil production is an environmental problem. In some developing countries, damage to the natural balance can be traced back to overexploitation of wild-growing plants. Some of these plants are protected worldwide, and their collection, processing, and illegal trading are punishable by law. In some Asian countries, such as in Vietnam, collection from the wild is state controlled and limited to quantities of biomass accruing from natural regeneration.

The state of technical development of the production in the developing countries is very variable and depends largely on the geographical zone they are located in. Areas of particular relevance are Asia, Africa, South America, and Eastern Europe. As a rule, the poorer the country, the more traditional and less technologically sophisticated equipment is used. Generally, standards of the distillation apparatus are those of the 1980s. At that time, the distillation equipment was provided and installed by foreign aid programs with European and American know-how. Most of these units are still in existence and, owing to repairs and improvements by local people, in good working order. Occasionally, primitive equipment has been locally developed, particularly when the state did not provide any financial assistance. Initially, all mastery and expertise of distillation techniques came from Europe, mainly from France. Later on, that knowledge was acquired and transferred to their countries by local people who had studied in Europe. They are no longer dependent on foreign know-how and able to produce oils of constant quality. Conventional hydrodistillation is still the main essential oil extraction method used, one exception being hydrodiffusion often used in Central America, mainly Guatemala and El Salvador, and Brazil in South America. The construction of the equipment is carried out in the country itself and makes the producer independent from higher-priced imports. Steam is generated by oil-burning generators only in the vicinity of cities. In country areas, wood or dried spent biomass is used. As in all other essential oil–producing countries, the distillation plants are close to the cultivation areas. Wild-growing plants are collected, provided the infrastructure exists for their transport to the distillation plant. For certain specific products, permanent fixed distillation plants are used. A forward leap in the technology will be only possible if sufficient investment funds became available in the future. Essential oil quantities produced in those countries are not small, and important specialities such as citral-rich ginger oil from Ecuador play a role on the world market. It should be a compulsory requirement that developing countries treat their wild-growing plant resources with the utmost care. Harvesting has to be controlled to avoid their disappearance from the natural environment and quantities taken adjusted to the ability of the environment to spontaneously regenerate. On the other hand, cultivation will have to be handled with equal care. The avoidance of monoculture will prevent leaching the soil of its nutrients and guard the environment from possible insect propagation. Balanced agricultural practices will lead to a healthy environment and superior quality plants for the production of essential oils.

The following are some pertinent remarks on the now-prevailing views of "green culture" and "organically" grown plants for essential oil production. It is unjustified to suggest that such products are of better quality or greater activity. Comparisons of chemical analyses of "bio-oils," for example, oils from "organically" grown plants, and commercially produced oils show absolutely no differences, qualitative or quantitative, between them. While the concept of pesticide- and fertilizer-free agriculture is desirable and should be supported, the huge worldwide consumption of essential oils could never be satisfied by bio-oils.

Finally, some remarks as to the concept of honesty are attached to the production of natural essential oils. During the last 30 years or so, adulteration of essential oils could be found every day. During the early days, cheap fatty oils (e.g., peanut oil) were used to cut essential oils. Such adulterations were easily revealed by means of placing a drop of the oil on filter paper and allowing it to evaporate (Karg, 1981). While an unadulterated essential oil will evaporate completely or at worst leave only a trace of nonvolatile residue, a greasy patch indicates the presence of a fatty adulterant. As synthetic components of essential oils became available around the turn of the twentieth century, some lavender and lavandin oils have been adulterated by the addition of synthetic linalool and linalyl acetate to the stills before commencing the distillation of the plant material. With the advent of improved analytical methods, such as GC and GC/MS, techniques of adulterating essential oil were also refined. Lavender oil can again serve as an example. Oils distilled from mixtures of lavender and lavandin flowers mimicked the properties of genuine good-quality lavender oils. However, with the introduction of chiral GC techniques, such adulterations were easily identified and the genuineness of the oils guaranteed. This also allowed the verification of the enantiomeric distribution of monoterpenes, monoterpenoid alcohols, and esters present in essential oils. Nuclear magnetic resonance is probably one of the best, but also one of the most expensive, methods available for the authentication of naturalness and will be cost-effective only with large-batch quantities or in the case of very expensive oils. In the future, 2D GC (GC/GC) will provide the next step for the control of naturalness of essential oils.

Another important aspect is the correct botanical source of the essential oil. This can perhaps best be discussed with reference to eucalyptus oil of the 1,8-cineole type. Originally, before commercial eucalyptus oil production commenced in Australia, eucalyptus oil was distilled mainly from *E. globulus* Labill. trees introduced into Europe (mainly Portugal and Spain [ISO Standard 770]). It should be noted that this species exists in several subspecies: *E. globulus* ssp. *bicostata* (Maiden, Blakely, & J. Simm.) Kirkpatr., *E. globulus* Labill. ssp. *globulus*., *E. globulus* ssp. *pseudoglobulus* (Naudin ex Maiden) Kirkpatr., and *E. globulus* ssp. *maidenii* (F. Muell.) Kirkpatr. It has been shown that the European oils were in fact mixed oils of some of these subspecies and of their hybrids (report by H.H.G. McKern of ISO/TC 54 meeting held in Portugal in 1966). The European Pharmacopoeia Monograph 0390 defines eucalyptus oil as the oil obtained from *E. globulus* Labill., *Eucalyptus fruticetorum* F. von Mueller Syn. *E. polybractea* R.T. Baker (this is the correct botanical name), *Eucalyptus smithii* R.T. Baker, and other species of *Eucalyptus* rich in 1,8-cineole. The Council of Europe's book *Plants in Cosmetics*,

Vol. 1, page 127, confuses the matter even further. It entitles the monograph as *E. globulus* Labill. et al. species, for example, and includes any number of unnamed *Eucalyptus* species. The *Pharmacopoeia of the Peoples Republic of China* (English Version, Vol. 1) 1997 goes even further defining eucalyptus oil as the oil obtained from *E. globulus* Labill. and *C. camphora* as well as from other plants of those two families. ISO Standard 3065—Oil of Australian eucalyptus—80%–85% cineole content, simply mentions that the oil is distilled from the appropriate species. The foregoing passage simply shows that Eucalyptus oil does not necessarily have to be distilled from a single species of *Eucalyptus*, for example, *E. globulus*, although suggesting that it is admissible to include 1,8-cineole-rich *Cinnamomum* oils is incorrect and unrealistic. This kind of problem is not unusual or unique. For example, the so-called English lavender oil, considered by many to derive from *L. angustifolia*, is really, in the majority of cases, the hybrid lavandin (Denny, 1995, personal communication).

Another pertinent point is how much twig and leaf material can be used in juniper berry oil. In Indonesia, it is common practice to space individual layers of patchouli leaves in the distillation vessel with twigs of the gurjun tree. Gurjun balsam present in the twigs contains an essential oil that contaminates the patchouli oil. Can this be considered to constitute an adulteration or simply a tool required for the production of the oil?

5.1.16 Concluding Remarks

As mentioned at the beginning, essential oils do have a future. In spite of regulatory limitations, dangerous substance regulations, and dermatological concerns, and problems with pricing the world production of essentials oil will increase. Essential oils are used in a very large variety of fields. They are an integral constituent of fragrances used in perfumes and cosmetics of all kinds, skin softeners to shower gels and body lotions, and even to "aromatherapy horse care massage oils." They are widely used in the ever-expanding areas of aromatherapy or, better, aromachology. Very large quantities of natural essential oils are used by the food and flavor industries for the flavoring of small goods, fast foods, ice creams, beverages, both alcoholic and nonalcoholic soft drinks, and so on. Their medicinal properties have been known for many years and even centuries. Some possess antibacterial or antifungal activity, while others may assist with the digestion of food. However, as they are multicomponent mixtures of somewhat variable composition, the medicinal use of whole oils has contracted somewhat, the reason being that single essential oil constituents were easier to test for effectiveness and eventual side effects. Despite all that, the use of essential oils is still "number one" on the natural healing scene. With rising health care and medicine costs, self-medication is on the increase and with it a corresponding increase in the consumption of essential oils. Parallel to this, the increase in various esoteric movements is giving rise to further demands for pure natural essential oils.

In the field of agriculture, attempts are being made at the identification of ecologically more friendly natural biocides, including essential oils, to replace synthetic pesticides and herbicides. Essential oils are also used to improve the appetite of farm animals, leading to more rapid increases in body weight as well as to improved digestion.

Finally, some very cheap essential oils or oil components such as limonene, 1,8-cineole, and the pinenes are useful as industrial solvents, while phellandrene-rich eucalyptus oil fractions are marketed as industrial perfumes for detergents and the like.

In conclusion, a "golden future" can be predicted for that useful natural product: the "essential oil"!

ACKNOWLEDGMENTS

The author thanks first of all Dr. Erich Lassak for his tremendous support, for so many detailed information, and also for some pictures; Klaus Dürbeck for some information about production of essential oils in development countries; Dr. Tilmann Miritz, Miritz Citrus Ingredients, for Figures 5.4 through 5.6; Bernhard Mirwald for Figure 5.9; and Tim Denny from Bridestowe Estate, Lilydale, Tasmania, for Figures 5.20 through 5.23.

REFERENCES

Arnodou, J. F. 1991. The taste of nature; industrial methods of natural products extraction. *Presented at a Conference organized by the Royal Society of Chemistry in Canterbury*, Canterbury, U.K., July 16–19, 1991.

Buchbauer, G., W. Jäger, L. Jirovetz, B. Nasel, C. Nasel, J. Ilmberger, and H. Diertrich. 1994. Aromatherapy Research: Studies on the Biological Effects of Fragrance Compounds and Essential Oils upon Inhalation. *25th International Symposium on Essential Oils*. Grasse, France.

Court, W.A., R. C. Roy, R. Pocs, A. F. More, and P. H. White. 1993. Effects of harvest date on the yield and quality of the essential oil of peppermint. *Can. J. Plant. Sci.*, 73: 815–824.

Denny, T. 1995. Bridestowe estates. Tasmania, Private information to the author.

Dey, D., R. Gaudile, K. Goad. 2001. Essential oils Industry, Alberta Agriculture, Food and Rural Development, Goverment of Alberta, 2001–2007.

Figueiredo, A. C., J. G. Barroso, L. G. Pedro, and J. J. C. Scheffer. 2005. Physiological aspects of essential oil production. *Plant Sci.*, 169(6): 1112–1117.

Gildemeister, E. and F. Hoffmann. 1931. *Die Ätherische Öle*. Miltitz. Germany: Verlag Schimmel & Co.

Grishin, A. M., A. N. Golovanov, and S. V. Rusakov. 2003. Evaporation of free water and water bound with forest combustibles under isothermal conditions. *J. Eng. Phys. Thermophys.*, 76(5): 1.

ISO/DIS 4731. 2005. *Oil of Geranium*. Geneva, Switzerland: International Standard Organisation.

ISO/DIS 9235.2. 1997. *Aromatic Natural Raw Materials—Vocabulary*. Geneva, Switzerland: International Standard Organisation.

Kamatou, G. P. P., R. L. van Zyl, S. F. van Vuuren, A. M. Viljoen, A. C. Figueiredo, J. G. Barroso, L. G. Pedro, and P. M. Tilney. 2006. Chemical composition, leaf trichome types and biological activities of the essential oils of four related salvia species indigenous to Southern Africa. *J. Essent. Oil Res.*, 18(Special edition): 72–79.

Karg, J. E. 1981. Das Geschäft mit ätherischen Ölen. *SÖFW, Seifen-Öle-Fette-Wachse*, 107(5/1981): 121–124.

Kassahun, M. T., J. da Silva, S. A. Mekonnen. 2011. Agronomic Characters, Leaf and essential Oil Yield of Peppermint (*Mentha piperita L.*) as Influenced by Harvesting Age and Row Spacing, Medicinal and Aromatic Plant Science and Biotechnology 5(1), 49–53.

Koll, N., and W. Kowalczyk. 1957. *Fachkunde der Parfümerie und Kosmetik*. Leipzig, Germany: Fachbuchverlag.

Kosar, M., T. Özek, M. Kürkcüoglu, and K. H. C. Baser. 2007. Comparison of microwave-assisted hydrodistillation and hydrodistillation methods for the fruit essential oils of *Foeniculum vulgare. J. Essent. Oil Res.*, 19: 426–429.

Letchamo, W., R. Marquard, J. Hölzl, and A. Gosselin. 1994. The selection of Thymus vulgaris cultivars to grow in Canada. *Angewandte Botanik*, 68: 83–88.

Levey, M. 1955. Evidences of ancient distillation, sublimation and extraction in Mesopotamia. *Centaurus*, 4(1): 23–33.

Levey, M. 1959. *Chemistry and Chemical Technology in Ancient Mesopotamia*. Amsterdam, Netherlands: Elsevier.

Meunier, C. 1985. *Lavandes & Lavandins*. Aix-en-Provence, France: ÉDISUD.

Nguyen, T. P. T., T. T. Nguyen, M. H. Tran, H. T. Tran, A. Muselli, A. Bighelli, V. Castola, and J. Casanova. 2004. *Artemisia vulgaris* L. from Vietnam, chemical variability and composition of the oil along the vegetative life of the plant. *J. Essent. Oil Res.*, 16: 358–361.

Novak, J. 2005. Genetics of monoterpenes in the genera Origanum and Salvia, lecture held on the 35th International Symposium on *Essential Oils*, Budapest, Hungary.

Omidbaigi, R. 2005. Processing of essential oil plants. In: *Processing, Analysis and Application of Essential Oils*. Dr. Leopold Jirovetz, Prof. Dr. Gerhard Buchbauer, Har Krishan Bhalla & Sons (eds.), Dehradun, India.

Perfumer & Flavorist. 2009. A preliminary report on the world production of some selected essential oils and countries, Vol. 34, January 2009 pp. 38–44, Perfumer & Flavorist Carol Stream, IL 60188-2403.

Porter, N. 2001. *Crop and Food Research*. Christchurch: Crop & Foodwatch Research, No. 39, October.

Reeve, D. 2005. A cultivated zest. *Perf. Flav.*, 30(3): 32–35.

Rovesti, P. 1977. Die Destillation ist 5000 Jahre alt. *Dragoco Rep.*, 3: 49–62.

Yanive, Z., and D. Palevitch. 1982. Effect of drought on the secondary metabolites of medicinal and aromatic plants. In: *Cultivation and Utilization of Medicinal Plants*, Atal, C.V. and B.M. Kapur (eds.). Jammu Tawi, India: CSIR.

Zahn, J. 1979. *Nichts neues mehr seit Babylon*. Hamburg, Germany: Hoffmann und Campe.

Ziegler, E. 1982. *Die natürlichen und künstlichen Aromen*. Heidelberg, Germany: Alfred Hüthig Verlag, pp. 187–188.

6 Chemistry of Essential Oils

Charles Sell

CONTENTS

6.1 INTRODUCTION

The term "essential oil" is a contraction of the original "quintessential oil." This stems from the Aristotelian idea that matter is composed of four elements: fire, air, earth, and water. The fifth element, or quintessence, was then considered to be spirit or life force. Distillation and evaporation were thought to be processes of removing the spirit from the plant, and this is also reflected in our language since the term "spirits" is used to describe distilled alcoholic beverages such as brandy, whiskey, and eau de vie. The last of these again shows reference to the concept of removing the life force from the plant. Nowadays, of course, we know that, far from being spirit, essential oils are physical in nature and composed of complex mixtures of chemicals. One thing that we do see from the ancient concepts is that the chemical components of essential oils must be volatile since they are removed by distillation. In order to have boiling points low enough to enable distillation, and atmospheric pressure steam distillation in particular, the essential oil components need to have molecular weights below 300 Da (molecular mass relative to hydrogen = 1) and are usually fairly hydrophobic. Within these constraints, nature has provided an amazingly rich and diverse range of chemicals (Lawrence, 1985; Hay and Waterman, 1993) but there are patterns of molecular structure that give clues to how the molecules were constructed. These synthetic pathways have now been confirmed by experiment and will serve to provide a structure for the contents of this chapter.

6.2 BASIC BIOSYNTHETIC PATHWAYS

The chemicals produced by nature can be classified into two main groups. The primary metabolites are those that are universal across the plant and animal family and constitute the basic building blocks of life. The four subgroups of primary metabolites are proteins, carbohydrates, nucleic acids, and lipids. These families of chemicals contribute little to essential oils although some essential oil components are degradation products of one of these groups, lipids being the most significant. The secondary metabolites are those that occur in some species and not others, and they are usually classified into terpenoids, shikimates, polyketides, and alkaloids. The most important as far as essential oils are concerned are the terpenoids and the shikimates are the second. There are a number of polyketides of importance in essential oils but very few alkaloids. Terpenoids, shikimates, and polyketides will therefore be the main focus of this chapter.

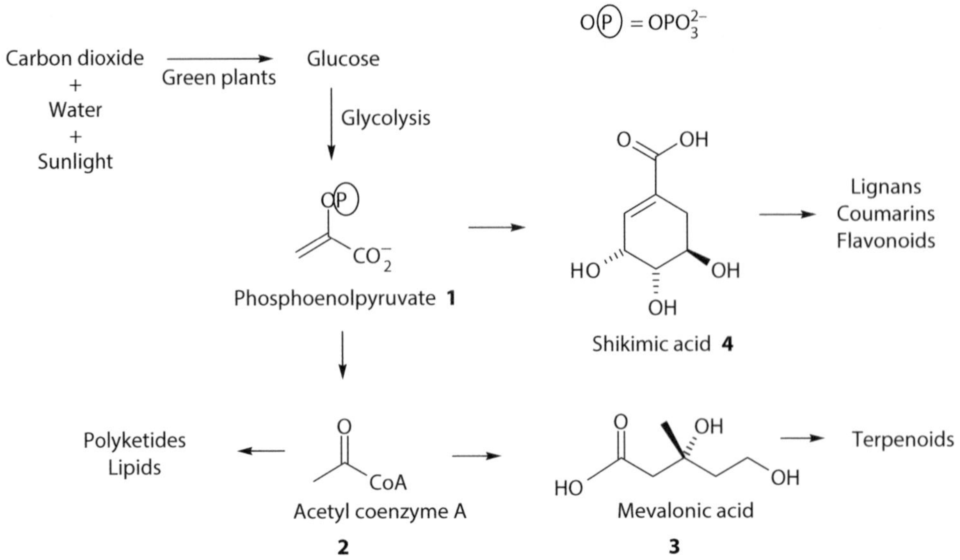

FIGURE 6.1 General pattern of biosynthesis of secondary metabolites.

The general scheme of biosynthetic reactions (Bu'Lock, 1965; Mann et al., 1994) is shown in Figure 6.1. Through photosynthesis, green plants convert carbon dioxide and water into glucose. Cleavage of glucose produces phosphoenolpyruvate (**1**), which is a key building block for the shikimate family of natural products. Decarboxylation of phosphoenolpyruvate gives the two-carbon unit of acetate and this is esterified with coenzyme-A to give acetyl CoA (**2**). Self-condensation of this species leads to the polyketides and lipids. Acetyl CoA is also a starting point for synthesis of mevalonic acid (**3**), which is the key starting material for the terpenoids. In all of these reactions and indeed all the natural chemistry described in this chapter, nature uses the same reactions that chemists do (Sell, 2003). However, nature's reactions tend to be faster and more selective because of the catalysts it uses. These catalysts are called enzymes, and they are globular proteins in which an active site holds the reacting species together. This molecular organization in the active site lowers the activation energy of the reaction and directs its stereochemical course (Matthews and van Holde, 1990; Lehninger, 1993).

Many enzymes need cofactors as reagents or energy providers. Coenzyme-A has already been mentioned earlier. It is a thiol and is used to form thioesters with carboxylic acids. This has two effects on the acid in question. First, the thiolate anion is a better leaving group than alkoxide and so the carbonyl carbon of the thioester is reactive toward nucleophiles. Second, the thioester group increases the acidity of the protons adjacent to the carbonyl group and therefore promotes the formation of the corresponding carbanions. In biosynthesis, a key role of adenosine triphosphate (ATP) is to make phosphate esters of alcohols (phosphorylation). One of the phosphate groups of ATP is added to the alcohol to give the corresponding phosphate ester and adenosine diphosphate. Another group of cofactors of importance to biosynthesis includes pairs such as NADP/NADPH, TPN/TPNH, and DPN/DPNH. These cofactors contain an *N*-alkylated pyridine ring. In each pair, one form comprises an *N*-alkylated pyridinium salt and the other the corresponding *N*-alkyl-1,4-dihydropyridine. The two forms in each pair are interconverted by gain or loss of a hydride anion and therefore constitute redox reagents. In all of the cofactors mentioned here, the reactive part of the molecule is only a small part of the whole. However, the bulk of the molecule has an important role in molecular recognition. The cofactor docks into the active site of the enzyme through recognition, and this holds the cofactor in the optimum spatial configuration relative to the substrate.

6.3 POLYKETIDES AND LIPIDS

The simplest biosynthetic pathway to appreciate is that of the polyketides and lipids (Bu'Lock, 1965; Mann et al., 1994). The key reaction sequence is shown in Figure 6.2. Acetyl CoA (**2**) is carboxylated to give malonyl CoA (**5**) and the anion of this attacks the CoA ester of a fatty acid. Obviously, the fatty acid could be acetic acid, making this a second molecule of acetyl CoA. After decarboxylation, the product is a b-ketoester with a backbone that is two carbon atoms longer than the first fatty acid. Since this is the route by which fatty acids are produced, it explains why fatty acids are mostly even numbered. If the process is repeated with this new acid as the feedstock, it can be seen that various poly-oxoacids can be built up, each of which will have a carbonyl group on every alternate carbon atom, hence the name polyketides. Alternatively, the ketone function can be reduced to the corresponding alcohol, and then eliminated, and the double bond hydrogenated. This sequence of reactions gives a higher homologue of the starting fatty acid, containing two more carbon atoms in the chain. Long chain fatty acids, whether saturated or unsaturated, are the basis of the lipids.

There are three main paths by which components of essential oils and other natural extracts are formed in this family of metabolites: condensation reactions of polyketides, degradation of lipids, and cyclization of arachidonic acid.

Figure 6.3 shows how condensation of polyketides can lead to phenolic rings. Intramolecular aldol condensation of the tri-keto-octanoic acid and subsequent enolization leads to orsellinic acid (**6**). Polyketide phenols can be distinguished from the phenolic systems of the shikimates by the fact that the former usually retain evidence of oxygenation on alternate carbon atoms, either as acids, ketones, phenols, or as one end of a double bond. The most important natural products containing polyketide phenols are the extracts of oakmoss and tree moss (*Evernia prunastri*). The most significant in odor terms is methyl 3-methylorsellinate (**7**) and ethyl everninate (**8**), which is usually also present in reasonable quantity. Atranol (**9**) and chloratranol (**10**) are minor components but they are skin sensitizers and so limit the usefulness of oakmoss and tree moss extracts, unless they are removed from them. Dimeric esters of orsellinic and everninic acids and analogues also exist in mosses. They are known as depsides and hydrolysis yields the monomers, thus increasing the odor of the sample. However, some depsides, such as atranorin (**11**), are allergens and thus contribute to safety issues with the extracts.

The major metabolic route for fatty acids involves b-oxidation and cleavage giving acetate and a fatty acid with two carbon atoms less than the starting acid, that is, the reverse of the biosynthesis reaction. However, other oxidation routes also exist and these give rise to new metabolites that were

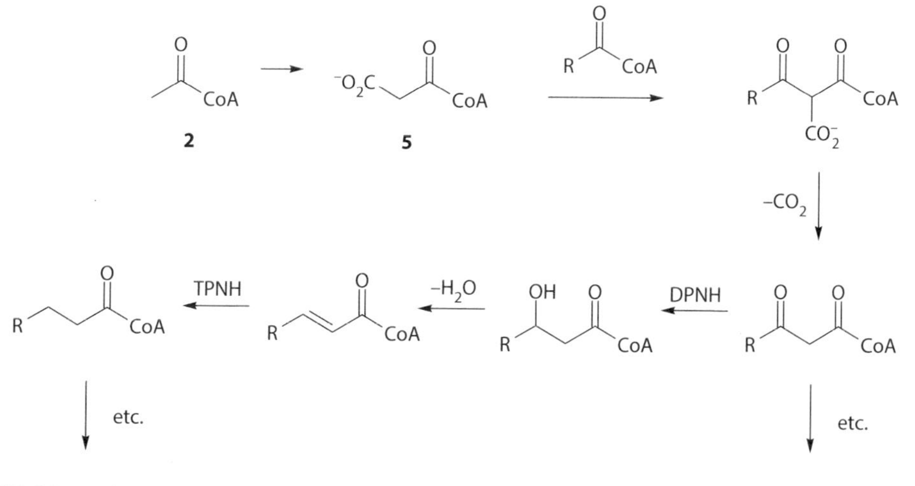

FIGURE 6.2 Polyketide and lipid biosynthesis.

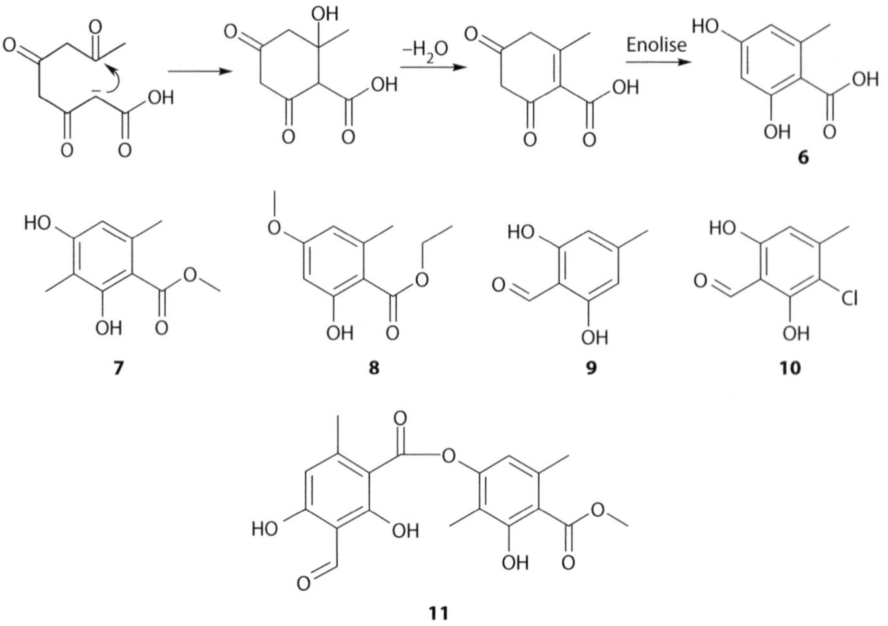

FIGURE 6.3 Polyketide biosynthesis and oakmoss components.

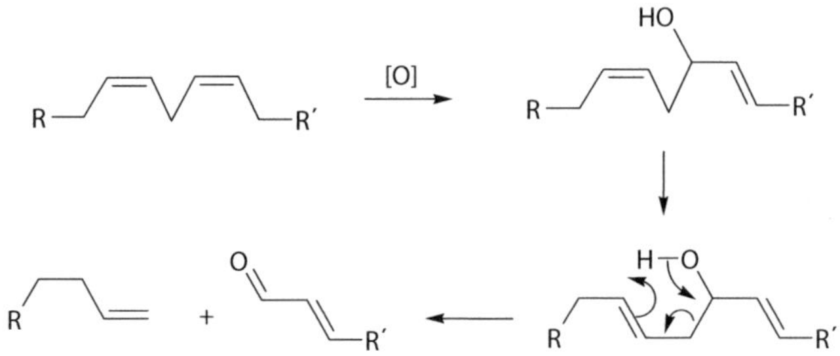

FIGURE 6.4 Fragmentation of polyunsaturated fats to give aldehydes.

not on the biosynthetic pathway. For example, Figure 6.4 shows how allylic oxidation of a dienoic acid and subsequent cleavage can lead to the formation of an aldehyde.

Allylic oxidation followed by lactonization rather than cleavage can, obviously, lead to lactones. Reduction of the acid function to the corresponding alcohols or aldehydes is also possible as are hydrogenation and elimination reactions. Thus, a wide variety of aliphatic entities are made available. Some examples are shown in Figure 6.5 to illustrate the diversity that exists. The hydrocarbon (*E,Z*)-1,3,5-undecatriene (**12**) is an important contributor to the odor of galbanum. Simple aliphatic alcohols and ethers are found, the occurrence of 1-octanol (**13**) in olibanum and methyl hexyl ether (**14**) in lavender being examples. Aldehydes are often found as significant odor components of oils, for example, decanal (**15**) in orange oil and (*E*)-4-decenal (**16**) in caraway and cardamom. The ketone 2-nonanone (**17**) that occurs in rue and hexyl propionate (**18**), a component of lavender, is just one of a plethora of esters that are found. The isomeric lactones γ-decalactone (**19**) and δ-decalactone (**20**) are found in osmanthus (Essential Oils Database, 2006). Acetylenes also occur as essential oil components, often as polyacetylenes such as methyl deca-2-en-4,6,8-triynoate (**21**), which is a component of *Artemisia vulgaris*.

Arachidonic acid (**22**) is a polyunsaturated fatty acid that plays a special role as a synthetic intermediate in plants and animals (Mann et al., 1994). As shown in Figure 6.6, allylic oxidation at

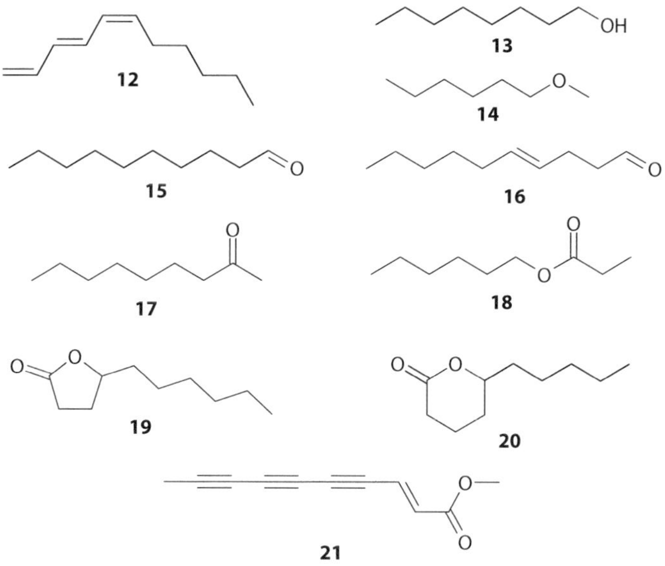

FIGURE 6.5 Some lipid-derived components of essential oils.

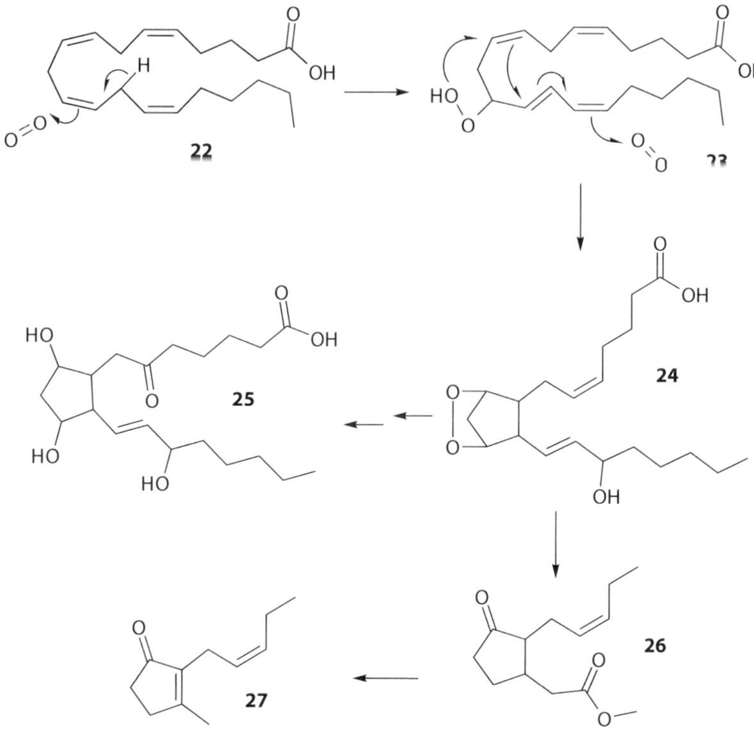

FIGURE 6.6 Biosynthesis of prostaglandins and jasmines.

the 11th carbon of the chain leads to the hydroperoxide (**23**). Further oxidation (at the 15th carbon) with two concomitant cyclization reactions gives the cyclic peroxide (**24**). This is a key intermediate for the biosynthesis of prostaglandins such as 6-ketoprostaglandin F_{1a} (**25**) and also for methyl jasmonate (**26**). The latter is the methyl ester of jasmonic acid, a plant hormone, and is a significant odor component of jasmine, as is jasmone (**27**), a product of degradation of jasmonic acid.

6.4 SHIKIMIC ACID DERIVATIVES

Shikimic acid (**4**) is a key synthetic intermediate for plants since it is the key precursor for both the flavonoids and lignin (Bu'Lock, 1965; Mann et al., 1994). The flavonoids are important to plants as antioxidants, colors, protective agents against ultraviolet light, and the like, and lignin is a key component of the structural materials of plants, especially woody tissues. Shikimic acid is synthesized from phosphoenolpyruvate (**1**) and erythrose 4-phosphate (**28**), as shown in Figure 6.7, and thus its biosynthesis starts from the carbohydrate pathway. Its derivatives can usually be recognized by the characteristic shikimate pattern of a six-membered ring with either a one- or three-carbon substituent on position one and oxygenation in the third, and/or fourth, and/or fifth positions. However, the oxygen atoms of the final products are not those of the starting shikimate since these are lost initially and then replaced.

Figure 6.8 shows some of the biosynthetic intermediates stemming from shikimic acid (**4**) and which are of importance in terms of generating materials volatile enough to be essential oil

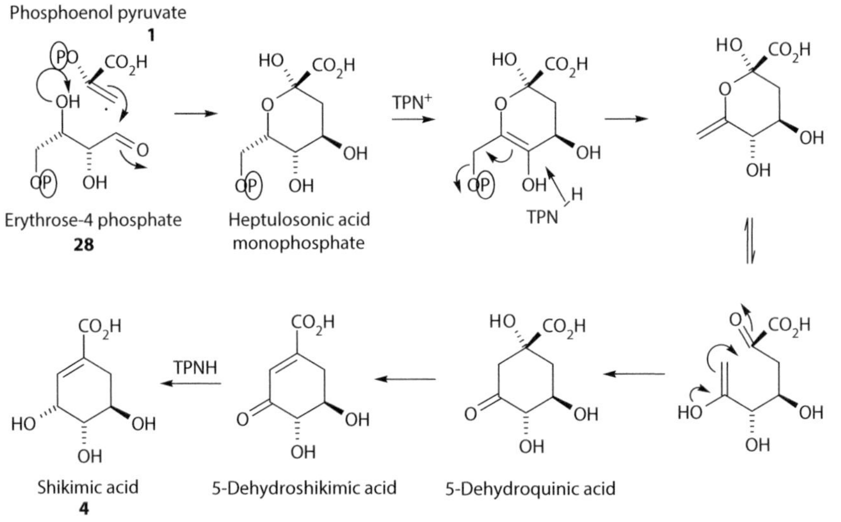

Phosphoenol pyruvate 1, Erythrose-4 phosphate 28, Heptulosonic acid monophosphate, TPN^+, TPN, TPNH, Shikimic acid 4, 5-Dehydroshikimic acid, 5-Dehydroquinic acid

FIGURE 6.7 Biosynthesis of shikimic acid.

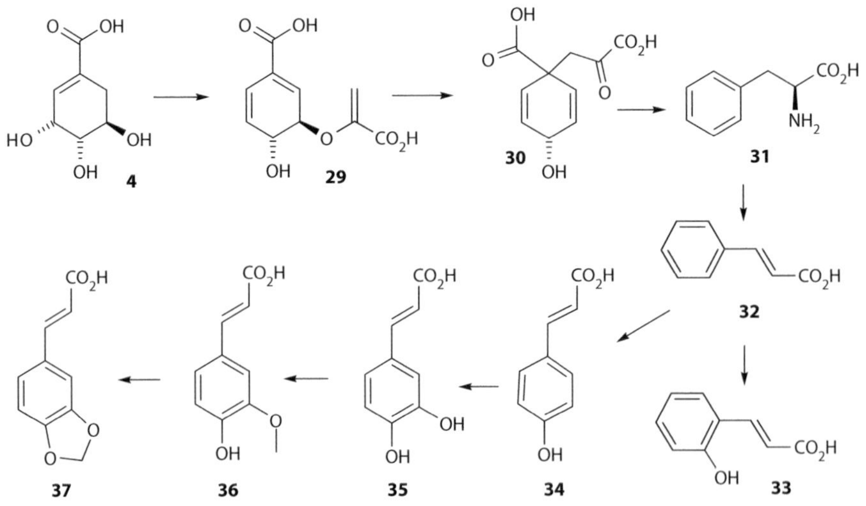

FIGURE 6.8 Key intermediates for shikimic acid.

components. Elimination of one of the ring alcohols and reaction with phosphoenolpyruvate (**1**) gives chorismic acid (**29**) that can undergo an oxy-Cope reaction to give prephenic acid (**30**). Decarboxylation and elimination of the ring alcohol now gives the phenylpropionic acid skeleton. Amination and reduction of the ketone function gives the essential amino acid phenylalanine (**31**), whereas reduction and elimination leads to cinnamic acid (**32**). Ring hydroxylation of the latter gives the isomeric *o*- and *p*-coumaric acids, (**33**) and (**34**), respectively. Further hydroxylation gives caffeic acid (**35**) and methylation of this gives ferulic acid (**36**). Oxidation of the methyl ether of the latter and subsequent cyclization gives methylenecaffeic acid (**37**). In shikimate biosynthesis, it is often possible to arrive at a given product by different sequences of the same reactions, and the exact route used will depend on the genetic makeup of the plant.

Aromatization of shikimic acid, without addition of the three additional carbon atoms from phosphoenolpyruvate, gives benzoic acid derivatives. Benzoic acid itself occurs in some oils and its esters are widespread. For example, methyl benzoate is found in tuberose, ylang ylang, and various lilies. Even more common are benzyl alcohol, benzaldehyde, and their derivatives (Günther, 1948; Gildemeister and Hoffmann, 1956; Arctander, 1960; Essential Oils Database, 2006). Benzyl alcohol occurs in muguet, jasmine, and narcissus, for example, and its acetate is the major component of jasmine oils. The richest sources of benzaldehyde are almond and apricot kernels, but it is also found in a wide range of flowers, including lilac, and other oils such as cassia and cinnamon. Hydroxylation or amination of benzoic acid leads to further series of natural products and some of the most significant, in terms of odors of essential oils, are shown in Figure 6.9. *o*-Hydroxybenzoic acid is known as salicylic acid (**38**) and both it and its esters are widely distributed in nature. For instance, methyl salicylate (**39**) is the major component (about 90% of the volatiles) of wintergreen and makes a significant contribution to the scents of tuberose and ylang ylang although only present at about 10% in the former and less than 1% in the latter. *o*-Aminobenzoic acid is known as anthranilic acid (**40**). Its methyl ester (**41**) has a very powerful odor and is found in such oils as genet, bitter orange flower, tuberose, and jasmine. Dimethyl anthranilate (**42**), in which both the nitrogen and acid functions have been methylated, occurs at low levels in citrus oils. *p*-Hydroxybenzoic acid has been found in vanilla and orris but much more common is the methyl ester of the corresponding aldehyde, commonly known as anisaldehyde (**44**). As the name suggests, the latter one is an important component of anise and it is also found in oils such as lilac and the smoke of agar wood. The corresponding alcohol, anisyl alcohol (**45**), and its esters are also widespread components of essential oils.

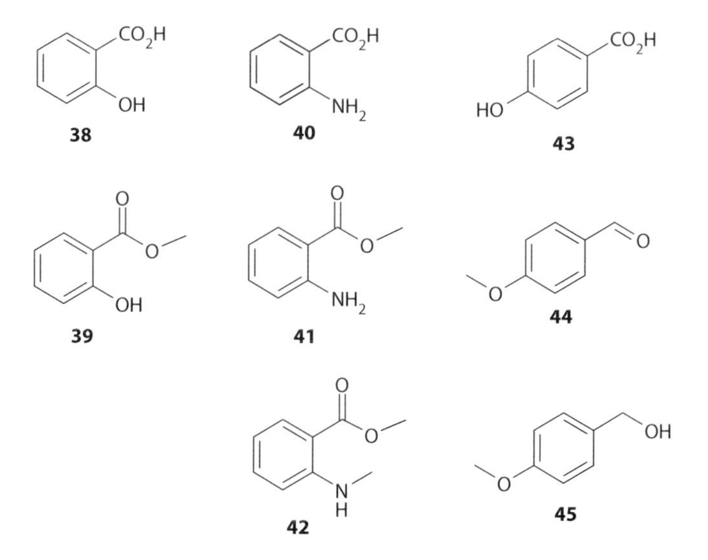

FIGURE 6.9 Hydroxy- and aminobenzoic acid derivatives.

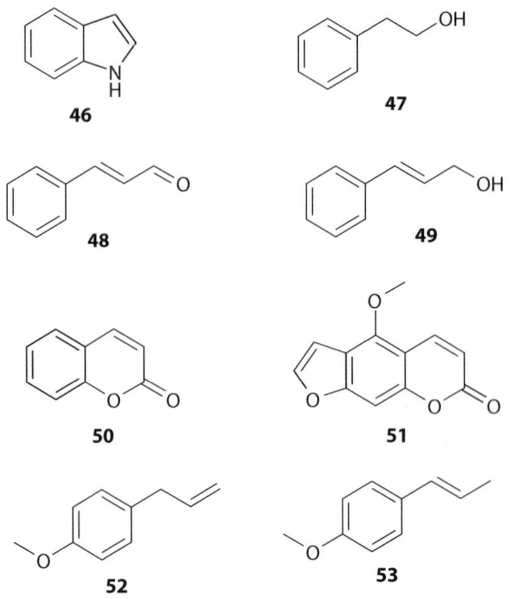

FIGURE 6.10 Some shikimate essential oil components.

Indole (**46**) and 2-phenylethanol (**47**) are both shikimate derivatives. Indole is particularly associated with jasmine. It usually occurs in jasmine absolute at a level of about 3%–5% and makes a very significant odor contribution to it. However, it does occur in many other essential oils as well. 2-Phenylethanol occurs widely in plants and is especially important for rose where it usually accounts for one-third to three-quarters of the oil. The structures of both are shown in Figure 6.10.

Figure 6.10 also shows some of the commonest cinnamic acid–derived essential oil components. Cinnamic acid (**32**) itself has been found in, for example, cassia and styrax, but its esters, particularly the methyl ester, are more frequently encountered. The corresponding aldehyde, cinnamaldehyde (**48**), is a key component of cinnamon and cassia and also occurs in some other oils. Cinnamyl alcohol (**49**) and its esters are more widely distributed, occurring in narcissus, lilac, and a variety of other oils. Lactonization of *o*-coumaric acid (**33**) gives coumarin (**50**). This is found in new mown hay to which it gives the characteristic odor. It is also important in the odor profile of lavender and related species and occurs in a number of other oils. Bergapten (**51**) is a more highly oxygenated and substituted coumarin. The commonest source is bergamot oil, but it also occurs in other sources, such as lime and parsley. It is phototoxic and consequently constitutes a safety issue for oils containing it.

Oxygenation in the *p*-position of cinnamic acid followed by methylation of the phenol and reduction of the acid to alcohol with subsequent elimination of the alcohol gives estragole (also known as methyl chavicol (**52**) and anethole (**53**)). Estragole is found in a variety of oils, mostly herb oils such as basil, tarragon, chervil, fennel, clary sage, anise, and rosemary. Anethole occurs in both the (*E*)- and (*Z*)-forms, the more thermodynamically stable (*E*)-isomer (shown in Figure 6.10) is the commoner, the (*Z*)-isomer is the more toxic of the two. Anethole is found in spices and herbs such as anise, fennel, lemon balm, coriander, and basil and also in flower oils such as ylang ylang and lavender.

Reduction of the side chain of ferulic acid (**36**) leads to an important family of essential oil components, shown in Figure 6.11. The key material is eugenol (**53**), which is widespread in its occurrence. It is found in spices such as clove, cinnamon, and allspice, herbs such as bay and basil, and in flower oils including rose, jasmine, and carnation. Isoeugenol (**54**) is found in basil, cassia, clove, nutmeg, and ylang ylang. Oxidative cleavage of the side chain of shikimates to give benzaldehyde derivatives is common and often significant, as it is in this case, where the product

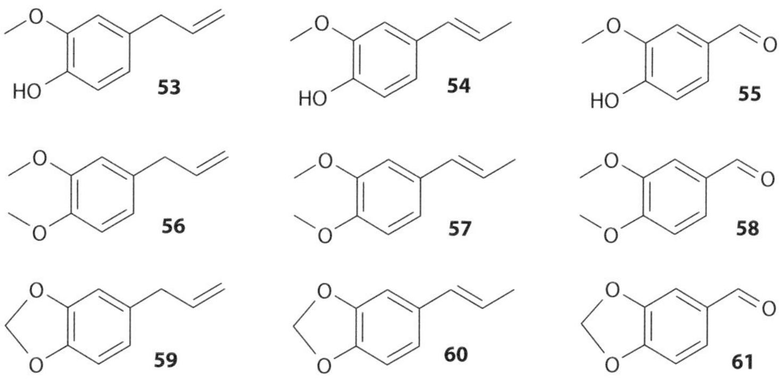

FIGURE 6.11 Ferulic acid derivatives.

is vanillin (**55**). Vanillin is the key odor component of vanilla and is therefore of considerable commercial importance. It also occurs in other sources such as jasmine, cabreuva, and the smoke of agar wood. The methyl ether of eugenol, methyl eugenol (**56**), is very widespread in nature, which, since it is the subject of some toxicological safety issues, creates difficulties for the essential oils business. The oils of some *Melaleuca* species contain up to 98% methyl eugenol, and it is found in a wide range of species including pimento, bay, tarragon, basil, and rose. The isomer, methyl isoeugenol (**57**), occurs as both (*E*)- and (*Z*)-isomers, the former being slightly commoner. Typical sources include calamus, citronella, and some narcissus species. Oxidative cleavage of the side chain in this set of substances produces veratraldehyde (**58**), a relatively rare natural product. Formation of the methylenedioxy ring, via methylenecaffeic acid (**37**), gives safrole (**59**), the major component of sassafras oil. The toxicity of safrole has led to a ban on the use of sassafras oil by the perfumery industry. Isosafrole (**60**) is found relatively infrequently in nature. The corresponding benzaldehyde derivative, heliotropin (**61**), also known as piperonal, is the major component of heliotrope.

6.5 TERPENOIDS

The terpenoids are, by far, the most important group of natural products as far as essential oils are concerned. Some authors, particularly in older literature, refer to them as terpenes, but this term is nowadays restricted to the monoterpenoid hydrocarbons. They are defined as substances composed of isoprene (2-methylbutadiene) units. Isoprene (**62**) is not often found in essential oils and is not actually an intermediate in biosynthesis, but the 2-methylbutane skeleton is easily discernable in terpenoids. Figure 6.12 shows the structures of some terpenoids. In the case of geraniol (**63**), one end of one isoprene unit is joined to the end of another making a linear structure (2,6-dimethyloctane). In guaiol (**64**), there are three isoprene units joined together to make a molecule with two rings. It is easy to envisage how the three units were first joined together into a chain and then formation of bonds from one point in the chain to another produced the two rings. Similarly, two isoprene units were used to form the bicyclic structure of a-pinene (**65**).

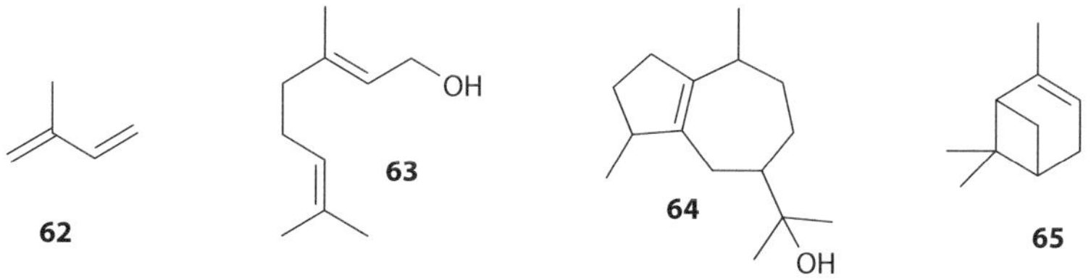

FIGURE 6.12 Isoprene units in some common terpenoids.

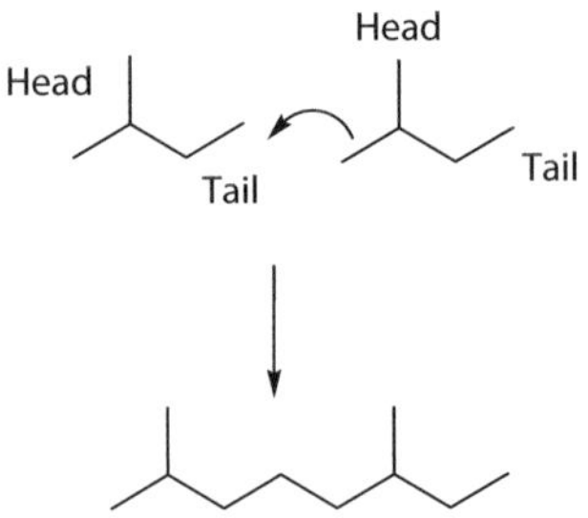

FIGURE 6.13 Head-to-tail coupling of two isoprene units.

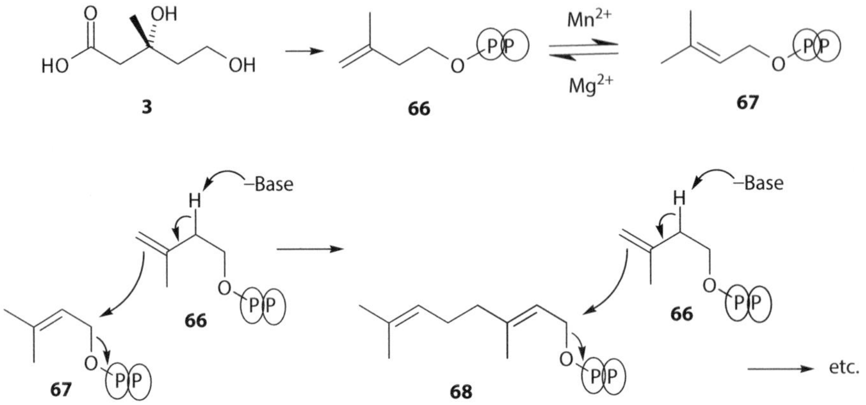

FIGURE 6.14 Coupling of C5 units in terpenoid biosynthesis.

The direction of coupling of isoprene units is almost always in one direction, the so-called head-to-tail coupling. This is shown in Figure 6.13. The branched end of the chain is referred to as the head of the molecule and the other as the tail.

This pattern of coupling is explained by the biosynthesis of terpenoids (Bu'Lock, 1965; Croteau, 1987; Mann et al., 1994). The key intermediate is mevalonic acid (**3**), which is made from three molecules of acetyl CoA (**2**). Phosphorylation of mevalonic acid followed by elimination of the tertiary alcohol and concomitant decarboxylation of the adjacent acid group gives isopentenyl pyrophosphate (**66**). This can be isomerized to give prenyl pyrophosphate (**67**). Coupling of these two 5-carbon units gives a 10-carbon unit, geranyl pyrophosphate (**68**), as shown in Figure 6.14, and further additions of isopentenyl pyrophosphate (**66**) lead to 15-, 20-, 25-, and so on carbon units.

It is clear from the mechanism shown in Figure 6.14 that terpenoid structures will always contain a multiple of five carbon atoms when they are first formed. The first terpenoids to be studied contained 10 carbon atoms per molecule and were called monoterpenoids. This nomenclature has remained and so those with 5 carbon atoms are known as hemiterpenoids; those with 15, sesquiterpenoids; those with 20, diterpenoids; and so on. In general, only the hemiterpenoids, monoterpenoids, and sesquiterpenoids are sufficiently volatile to be components of essential oils. Degradation products of higher terpenoids do occur in essential oils, so they will be included in this chapter.

6.5.1 Hemiterpenoids

Many alcohols, aldehydes, and esters, with a 2-methylbutane skeleton, occur as minor components in essential oils. Not surprisingly, in view of the biosynthesis, the commonest oxidation pattern is that of prenol, that is, 3-methylbut-2-ene-1-ol. For example, the acetate of this alcohol occurs in ylang ylang and a number of other oils. However, oxidation has been observed at all positions. Esters such as prenyl acetate give fruity top notes to oils containing them, and the corresponding thioesters contribute to the characteristic odor of galbanum.

6.5.2 Monoterpenoids

Geranyl pyrophosphate (**68**) is the precursor for the monoterpenoids. Heterolysis of its carbon–oxygen bond gives the geranyl carbocation (**69**). In natural systems, this and other carbocations discussed in this chapter do not exist as free ions but rather as incipient carbocations held in enzyme active sites and essentially prompted into cation reactions by the approach of a suitable reagent. For the sake of simplicity, they will be referred to here as carbocations. The reactions are described in chemical terms but all are under enzymic control, and the enzymes present in any given plant will determine the terpenoids it will produce. Thus, essential oil composition can give information about the genetic makeup of the plant. A selection of some of the key biosynthetic routes to monoterpenoids (Devon and Scott, 1972) is shown in Figure 6.15.

Reaction of the geranyl carbocation with water gives geraniol (**63**) that can subsequently be oxidized to citral (**71**). Loss of a proton from (**69**) gives myrcene (**70**) and this can be isomerized to other acyclic hydrocarbons. An intramolecular electrophilic addition reaction of (**69**) gives the monocyclic carbocation (**72**) that can eliminate a proton to give limonene (**73**) or add water to give a-terpineol (**74**). A second intramolecular addition gives the pinyl carbocation (**75**) that can lose a proton to give either α-pinene (**65**) or β-pinene (**76**). The pinyl carbocation (**75**) is also reachable directly from the menthyl carbocation (**72**). Carene (**77**), another bicyclic material, can be produced

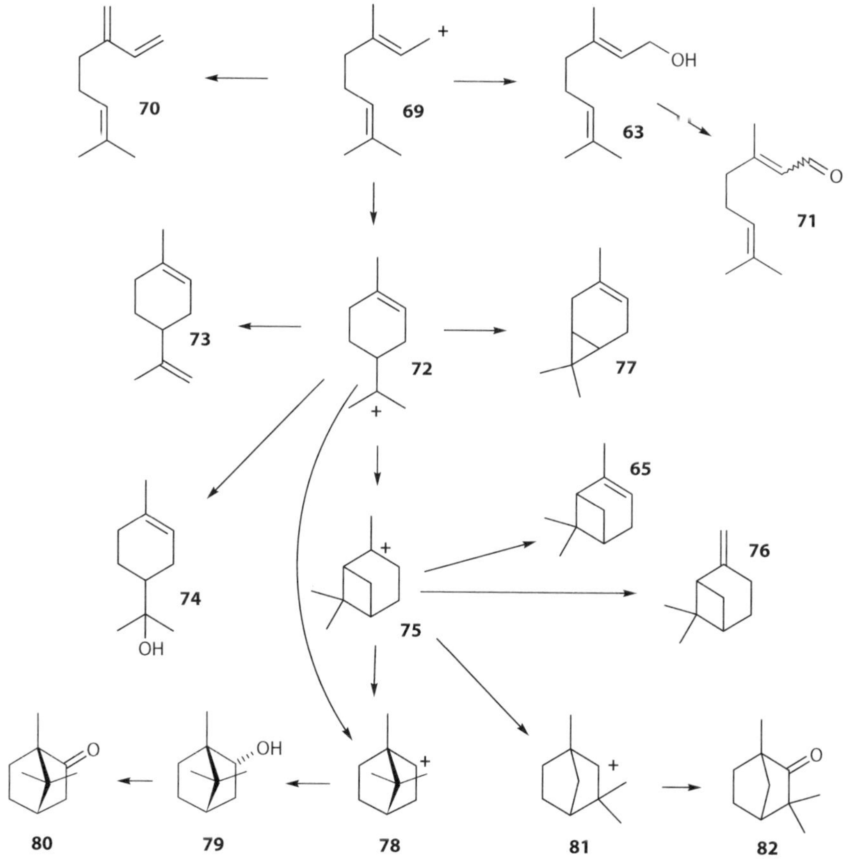

FIGURE 6.15 Formation of monoterpenoid skeletons.

through similar reactions. Wagner–Meerwein rearrangement of the pinyl carbocation (**75**) gives the bornyl carbocation (**78**). Addition of water to this gives borneol (**79**) and this can be oxidized to camphor (**80**). An alternative Wagner–Meerwein rearrangement of (**75**) gives the fenchyl skeleton (**81**) from which fenchone (**82**) is derived.

Some of the more commonly encountered monoterpenoid hydrocarbons (Günther, 1948; Gildemeister and Hoffmann, 1956; Arctander, 1960; Essential Oils Database, 2006; Sell, 2007) are shown in Figure 6.16. Many of these can be formed by dehydration of alcohols, and so their presence in essential oils could be as artifacts arising from the extraction process. Similarly, *p*-cymene (**83**) is one of the most stable materials of this class and can be formed from many of the others by appropriate cyclization and/or isomerization and/or oxidation reactions and so its presence in any essential oil could be as an artifact.

Myrcene (**70**) is very widespread in nature. Some sources, such as hops, contain high levels and it is found in most of the common herbs and spices. All isomers of α-ocimene (**84**), β-ocimene (**85**), and allo-ocimene (**86**) are found in essential oils, the isomers of β-ocimene (**85**) being the most frequently encountered. Limonene (**73**) is present in many essential oils, but the major occurrence is in the citrus oils that contain levels up to 90%. These oils contain the dextrorotatory (*R*)-enantiomer, and its antipode is much less common. Both α-phellandrene (**87**) and β-phellandrene (**88**) occur widely in essential oils. For example, (−)-α-phellandrene is found in Eucalyptus dives and (*S*)-(−)-b-phellandrene in the lodgepole pine, Pinus contorta. *p*-Cymene (**83**) has been identified in many essential oils and plant extracts and thyme and oregano oils are particularly rich in it. α-Pinene (**65**), β-pinene (**76**), and 3-carene (**77**) are all major constituents of turpentine from a wide range of pines, spruces, and firs. The pinenes are often found in other oils, 3-carene less so. Like the pinenes, camphene (**89**) is widespread in nature.

Simple hydrolysis of geranyl pyrophosphate gives geraniol, (*E*)-3,7-dimethylocta-2,6-dienol (**63**). This is often accompanied in nature by its geometric isomer, nerol (**90**). Synthetic material is usually a mixture of the two isomers and when interconversion is possible, the equilibrium mixture comprises about 60% geraniol (**63**) and 40% nerol (**90**). The name geraniol is often used to describe a mixture of geraniol and nerol. When specifying the geometry of these alcohols, it is better to use the modern (*E*)/(*Z*) nomenclature as the terms *cis* and *trans* are somewhat ambiguous in this

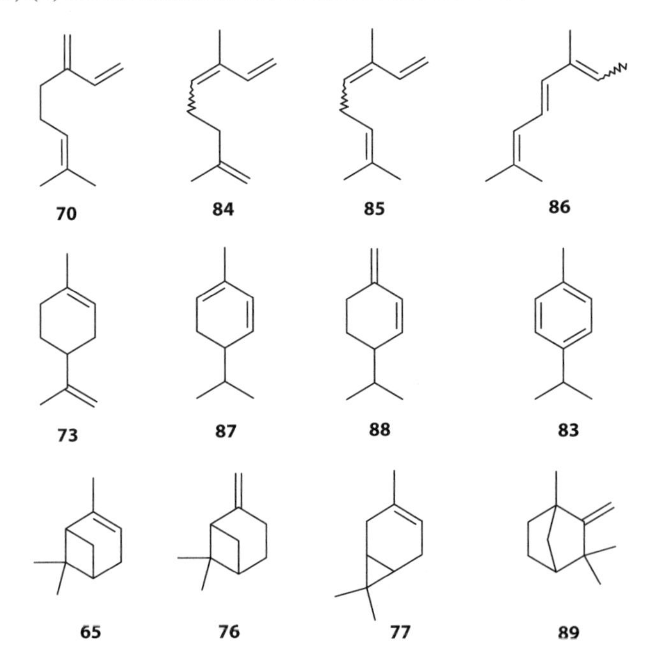

FIGURE 6.16 Some of the more common terpenoid hydrocarbons.

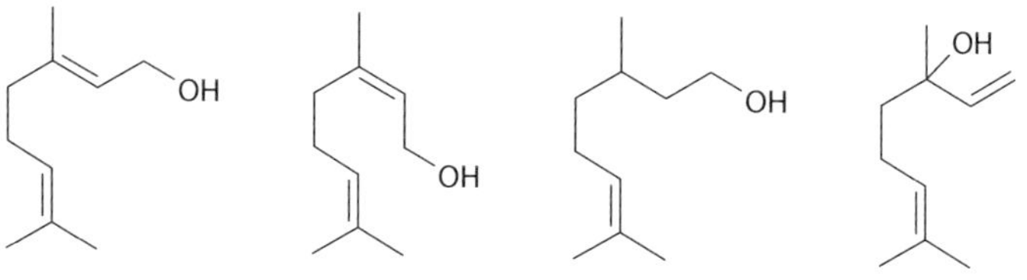

FIGURE 6.17 Key acyclic monoterpenoid alcohols.

case and earlier literature is not consistent in their use. Both isomers occur in a wide range of essential oils, geraniol (**63**) being particularly widespread. The oil of *Monarda fistulosa* contains over 90% geraniol (**63**) and the level in palmarosa is over 80%. Geranium contains about 50% and citronella and lemongrass each contain about 30%. The richest natural sources of nerol include rose, palmarosa, citronella, and davana, although its level in these is usually only in the 10%–15% range. Citronella and related species are used commercially as sources of geraniol, but the price is much higher than that of synthetic material. Citronellol (**91**) is a dihydrogeraniol and occurs widely in nature in both enantiomeric forms. Rose, geranium, and citronella are the oils with the highest levels of citronellol. Geraniol, nerol, and citronellol, together with 2-phenylethanol, are known as the rose alcohols because of their occurrence in rose oils and also because they are the key materials responsible for the rose odor character. Esters (the acetates in particular) of all these alcohols are also commonly encountered in essential oils (Figure 6.17).

Allylic hydrolysis of geranyl pyrophosphate produces linalool (**92**). Like geraniol, linalool occurs widely in nature. The richest source is ho leaf, the oil of which can contain well over 90% linalool. Other rich sources include linaloe, rosewood, coriander, freesia, and honeysuckle. Its acetate is also frequently encountered and is a significant contributor to the odors of lavender and citrus leaf oils.

Figure 6.18 shows a selection of cyclic monoterpenoid alcohols. α-Terpineol (**74**) is found in many essential oils as is its acetate. The isomeric terpinen-4-ol (**93**) is an important component of Ti tree oil, but its acetate, surprisingly, is more widely occurring, being found in herbs such as marjoram and rosemary. l-Menthol (**94**) is found in various mints and is responsible for the cooling effect of oils containing it. There are eight stereoisomers of the menthol structure, l-menthol is the commonest in nature and also has the strongest cooling effect. The cooling effect makes menthol and mint oils valuable commodities, the two most important sources being corn mint (*Mentha arvensis*) and peppermint (*Mentha piperita*). Isopulegol (**95**) occurs in some species including *Eucalyptus citriodora* and citronella. Borneol (*endo*-1,7,7-trimethylbicyclo[2.2.1]heptan-2-ol) (**79**)

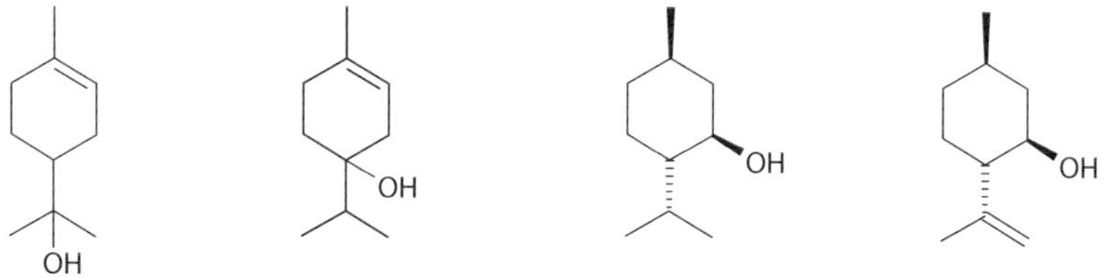

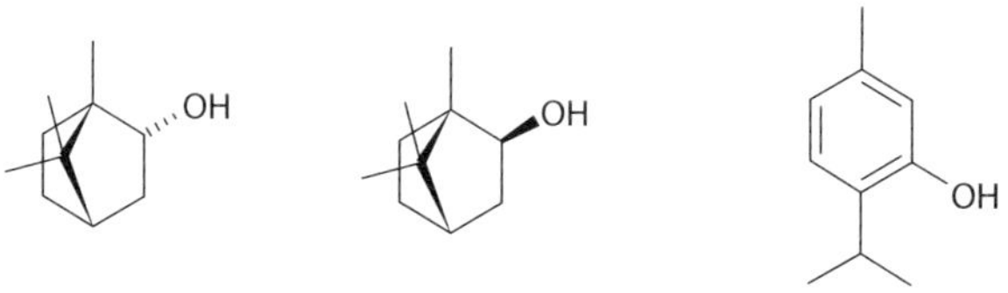

FIGURE 6.18 Some cyclic monoterpenoid alcohol.

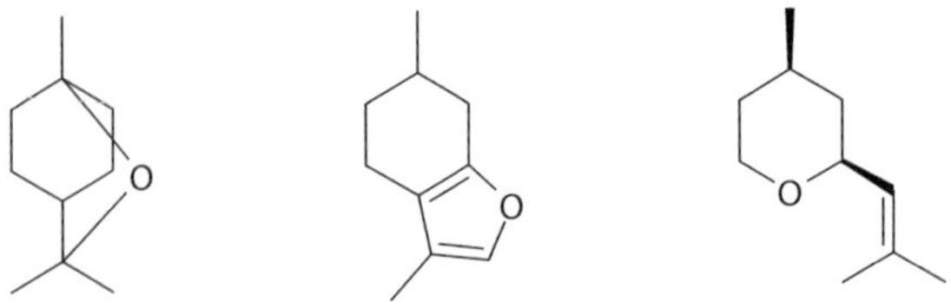

FIGURE 6.19 Some monoterpenoid ethers.

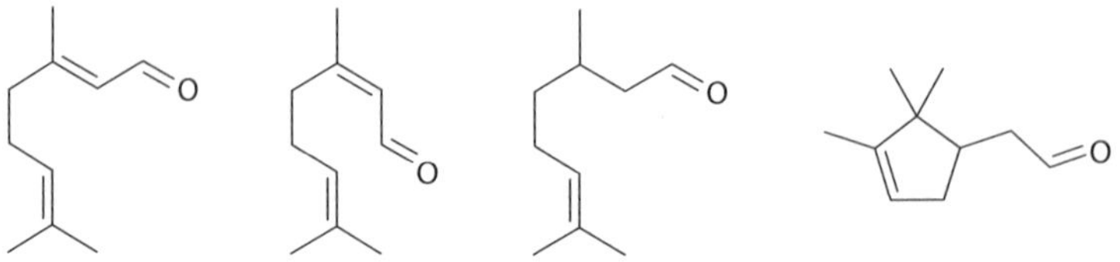

FIGURE 6.20 Some monoterpenoid aldehydes.

and esters thereof, particularly the acetate, occur in many essential oils. Isoborneol (*exo*-1,7,7-trimethylbicyclo[2.2.1]heptan-2-ol) (**96**) is less common; however, isoborneol and its esters are found in quite a number of oils. Thymol (**97**), being a phenol, possesses antimicrobial properties, and oils, such as thyme and basil, which find appropriate use in herbal remedies. It is also found in various *Ocimum* and *Monarda* species.

Three monoterpenoid ethers are shown in Figure 6.19. 1,8-Cineole (**98**), more commonly referred to simply as cineole, comprises up to 95% of the oil of *Eucalyptus globulus* and about 40%–50% of cajeput oil. It also can be found in an extensive range of other oils and often as a major component. It has antibacterial and decongestant properties and consequently, eucalyptus oil is used in various paramedical applications. Menthofuran (**99**) occurs in mint oils and contributes to the odor of peppermint. It is also found in several other oils. Rose oxide is found predominantly in rose and geranium oils. There are four isomers, the commonest being the levorotatory enantiomer of *cis*-rose oxide (**100**). This is also the isomer with the lowest odor threshold of the four.

The two most significant monoterpene aldehydes are citral (**71**) and its dihydro analogue citronellal (**103**), both of which are shown in Figure 6.20. The word citral is used to describe a mixture of the two geometric isomers geranial (**101**) and neral (**102**) without specifying their relative proportions. Citral occurs widely in nature, both isomers usually being present, the ratio between them usually being in the 40:60 to 60:40 range. Lemongrass contains 70%–90% citral and the fruit of *Litsea cubeba* contains about 60%–75%. Citral also occurs in *Eucalyptus staigeriana*, lemon balm, ginger, basil, rose, and citrus species. It is responsible for the characteristic smell of lemons although lemon oil usually contains only a few percent of it. Citronellal (**103**) also occurs widely in essential oils. *E. citriodora* contains up to 85% citronellal, and significant amounts are also found in some chemotypes of *L. cubeba*, citronella Swangi leaf oil, and *Backhousia citriodora*. Campholenic aldehyde (**104**) occurs in a limited range of species such as olibanum, styrax, and some eucalypts. Material produced from a-pinene (**65**) is important as an intermediate for synthesis.

Figure 6.21 shows some of the commoner monoterpenoid ketones found in essential oils. Both enantiomers of carvone are found in nature, the (*R*)-(−)- (usually referred to as L-carvone) (**105**) being the commoner. This enantiomer provides the characteristic odor of spearmint (*Mentha cardiaca*, *Mentha gracilis*, *Mentha spicata*, and *Mentha viridis*), the oil of which usually contains 55%–75% of L-carvone. The (*S*)-(+)-enantiomer (**106**) is found in caraway at levels of 30%–65% and in dill at 50%–75%. Menthone is fairly common in essential oils particularly in the mints, pennyroyal, and sages, but lower levels are also found in oils such as rose and geranium. The L-isomer is commoner

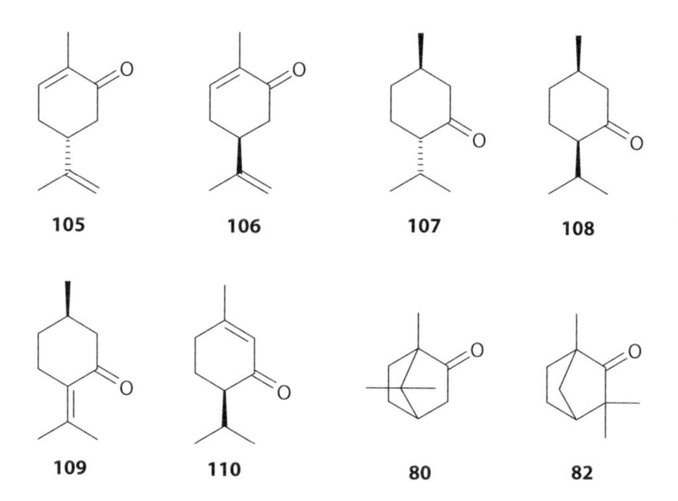

FIGURE 6.21 Some monoterpenoid ketones.

than the *d*-isomer. Isomenthone is the *cis*-isomer and the two interconvert readily by epimerization. The equilibrium mixture comprises about 70% menthone and 30% isomenthone. The direction of rotation of plane-polarized light reverses on epimerization, and therefore *l*-menthone (**107**) gives D-isomenthone (**108**). (+)-Pulegone (**109**) accounts for about 75% of the oil of pennyroyal and is also found in a variety of other oils. (−)-Piperitone (**110**) also occurs in a variety of oils, the richest source being *E. dives*. Both pulegone and piperitone have strong minty odors. Camphor (**80**) occurs in many essential oils and in both enantiomeric forms. The richest source is the oil of camphor wood, but it is also an important contributor to the odor of lavender and of herbs such as sage and rosemary. Fenchone (**82**) occurs widely, for example, in cedar leaf and lavender. Its laevorotatory enantiomer is an important contributor to the odor of fennel.

6.5.3 Sesquiterpenoids

By definition, sesquiterpenoids contain 15 carbon atoms. This results in their having lower volatilities and hence higher boiling points than monoterpenoids. Therefore, fewer of them (in percentage terms) contribute to the odor of essential oils, but those that do often have low-odor thresholds and contribute significantly as endnotes. They are also important as fixatives for more volatile components.

Just as geraniol (**63**) is the precursor for all the monoterpenoids, farnesol (**111**) is the precursor for all the sesquiterpenoids. Its pyrophosphate is synthesized in nature by the addition of isopentenyl pyrophosphate (**66**) to geranyl pyrophosphate (**68**) as shown in Figure 6.14, and hydrolysis of that gives farnesol. Incipient heterolysis of the carbon–oxygen bond of the phosphate gives the nascent farnesyl carbocation (**112**), and this leads to the other sesquiterpenoids, just as the geranyl carbocation does to monoterpenoids. Starting from farnesyl pyrophosphate, the variety of possible cyclic structures is much greater than that from geranyl pyrophosphate because there are now three double bonds in the molecule. Similarly, there is also a greater scope for further structural variation resulting from rearrangements, oxidations, degradation, and so on (Devon and Scott, 1972). The geometry of the double bond in position 2 of farnesol is important in terms of determining the pathway used for subsequent cyclization reactions, and so these are best discussed in two blocks.

Figure 6.22 shows a tiny fraction of the biosynthetic pathways derived from (*Z*,*E*)-farnesyl pyrophosphate. Direct hydrolysis leads to acyclic sesquiterpenoids such as farnesol (**111**) and nerolidol (**113**). However, capture of the carbocation (**112**) by the double bond at position 6 gives a cyclic structure that of the bisabolane skeleton (**114**), and quenching of this with water gives bisabolol (**115**). A hydrogen shift in (**114**) leads to the isomeric carbocation (**116**) that still retains the bisabolane skeleton. Further cyclizations and rearrangements take the molecule through various

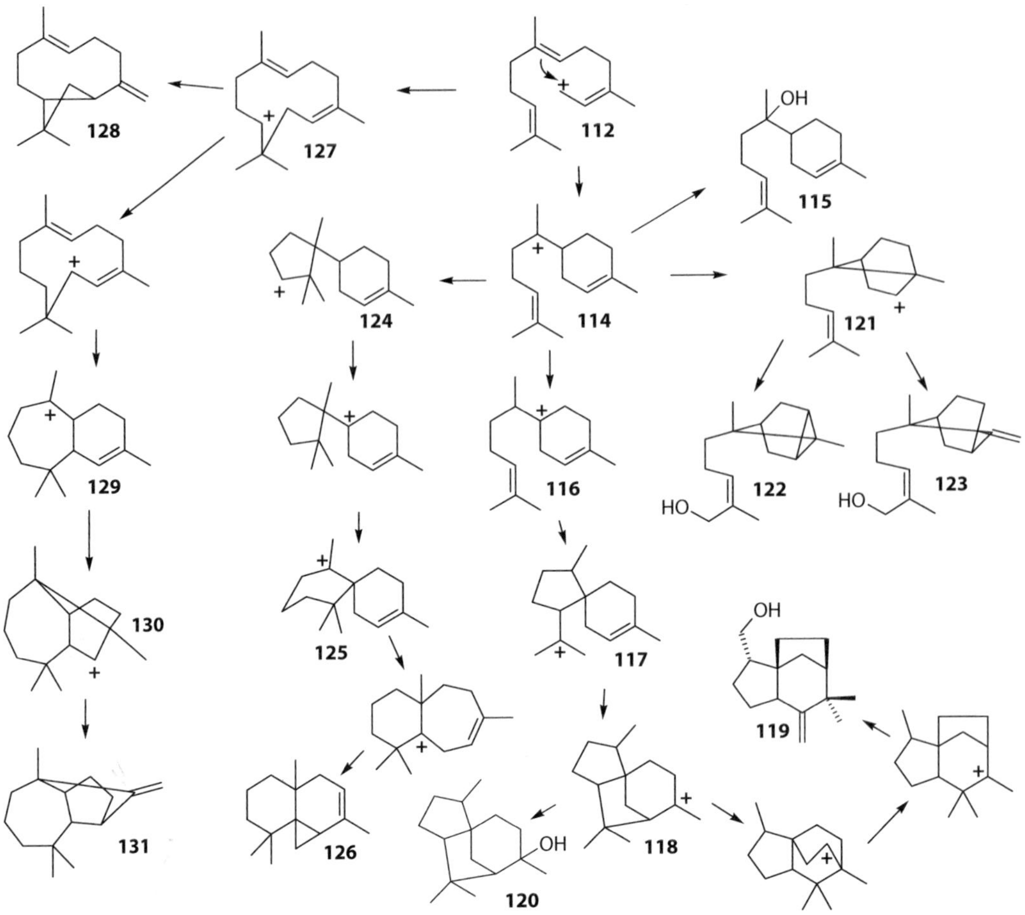

FIGURE 6.22 Some biosynthetic pathways from (*Z*,*E*)-farnesol.

skeletons, including those of the acorane (**117**) and cedrane (**118**) families, to the khusane family, illustrated by khusimol (**119**) in Figure 6.22. Obviously, a wide variety of materials can be generated along this route, an example being cedrol (**120**) formed by reaction of cation (**118**) with water. The bisabolyl carbocation (**114**) can also cyclize to the other double bonds in the molecule leading to, *inter alia*, the campherenane skeleton (**121**) and hence α-santalol (**122**) and β-santalol (**123**), or, via the cuparane (**124**) and chamigrane (**125**) skeletons, to compounds such as thujopsene (**126**). The carbocation function in (**112**) can also add to the double bond at the far end of the chain to give the *cis*-humulane skeleton (**127**). This species can cyclize back to the double bond at carbon 2 before losing a proton, thus giving caryophyllene (**128**). Another alternative is for a series of hydrogen shifts, cyclizations, and rearrangements to lead it through the himachalane (**129**) and longibornane (**130**) skeletons to longifolene (**131**).

Figure 6.23 shows a few of the many possibilities for biosynthesis of sesquiterpenoids from (*E*,*E*)-farnesyl pyrophosphate. Cyclization of the cation (**132**) to C-11, followed by loss of a proton gives all *trans*- or α-humulene (**133**), whereas cyclization to the other end of the same double bond gives a carbocation (**134**) with the germacrane skeleton. This is an intermediate in the biosynthesis of odorous sesquiterpenes such as nootkatone (**135**) and α-vetivone (**137**). β-Vetivone (**137**) is synthesized through a route that also produces various alcohols, for example, (**138**) and (**139**), and an ether (**140**) that has the eudesmane skeleton. Rearrangement of the germacrane carbocation (**134**) leads to a carbocation (**141**) with the guaiane skeleton, and this is an intermediate in the synthesis

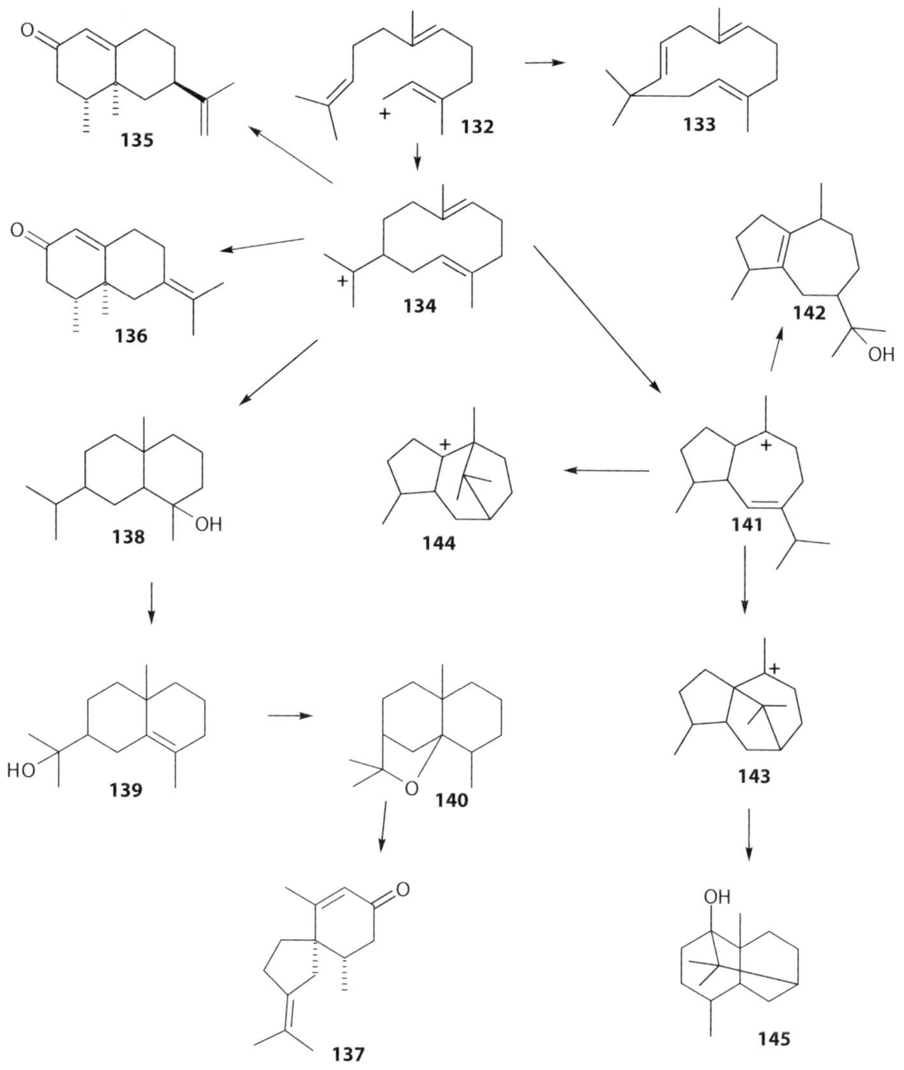

FIGURE 6.23 Some biosynthetic pathways from (*E,E*)-farnesol.

of guaiol (**142**). Carbocation (**141**) is also an intermediate in the biosynthesis of the a-patchoulane (**143**) and b-patchoulane (**144**) skeletons and of patchouli alcohol (**145**).

All four isomers of farnesol (**111**) are found in nature and all have odors in the muguet and linden direction. The commonest is the (*E,E*)-isomer that occurs in, among others, cabreuva and ambrette seed, while the (*Z,E*)-isomer has been found in jasmine and ylang ylang, the (*E,Z*)-isomer in cabreuva, rose, and neroli, and the (*Z,Z*)-isomer in rose. Nerolidol (**113**) is the allylic isomer of farnesol and exists in four isomeric forms: two enantiomers each of two geometric isomers. The (*E*)-isomer has been found in cabreuva, niaouli, and neroli oils among others and the (*Z*)-isomer in neroli, jasmine, ho leaf, and so on. Figure 6.24 shows the structures of farnesol and nerolidol with all of the double bonds in the *trans*-configuration.

α-Bisabolol (**115**) is the simplest of the cyclic sesquiterpenoid alcohols. If farnesol is the sesquiterpenoid equivalent of geraniol and nerolidol of linalool, then α-bisabolol is the equivalent of α-terpineol. It has two chiral centers and therefore exists in four stereoisomeric forms, all of which occur in nature. The richest natural source is *Myoporum crassifolium* Forst., a shrub from New Caledonia, but α-bisabolol can be found in many other species including chamomile, lavender,

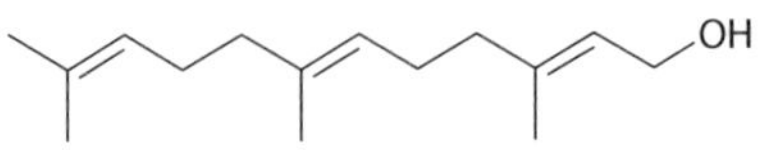

111

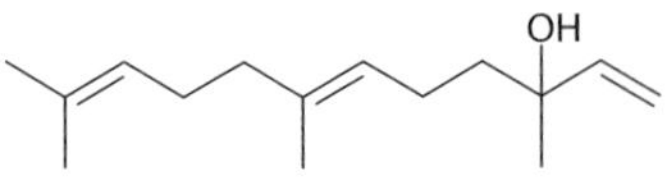

113

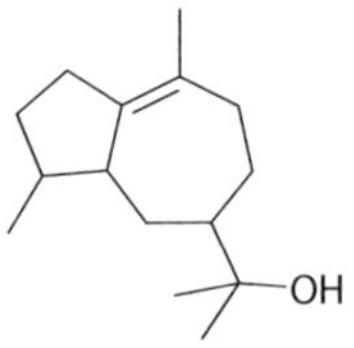

147

FIGURE 6.24 Some sesquiterpenoid alcohols.

and rosemary. It has a faint floral odor and anti-inflammatory properties and is responsible, at least in part, for the related medicinal properties of chamomile oil.

The santalols (**122**) and (**123**) have more complex structures and are the principal components of sandalwood oil. Cedrol (**120**) is another complex alcohol, but it is more widely occurring in nature than the santalols. It is found in a wide range of species, the most significant being trees of the *Juniperus, Cupressus*, and *Thuja* families. Cedrene (**146**) occurs alongside cedrol in cedarwood oils. Cedrol is dehydrated to cedrene in the presence of acid, and so the latter can be an artifact of the former and the ratio of the two will often depend on the method of isolation. Thujopsene (**126**) also occurs in cedarwood oils, usually at a similar level to that of cedrol/cedrene, and it is found in various other oils also. Caryophyllene (**128**) and α-humulene (the all *trans*-isomer) (**133**) are widespread in nature, cloves being the best-known source of the former and hops of the latter. The ring systems of these two materials are very strained making them quite reactive chemically, and caryophyllene, extracted from clove oil as a by-product of eugenol production, is used as the starting material in the synthesis of several fragrance ingredients. Longifolene (**131**) also possesses a strained ring system. It is a component of Indian turpentine and is therefore readily available as a feedstock for fragrance ingredient manufacture.

Guaiacwood oil is the richest source of guaiol (**142**) and the isomeric bulnesol (**147**), but both are found in other oils, particularly guaiol that occurs in a wide variety of plants. Dehydration and dehydrogenation of these give guaiazulene (**148**), which is used as an anti-inflammatory agent. Guaiazulene is also accessible from α-gurjunene (**149**), the major component of gurjun balsam. Guaiazulene is blue in color as is the related olefin chamazulene (**150**). The latter occurs in a variety of oils, but it is particularly important in chamomile to which it imparts the distinctive blue tint (Figure 6.25).

Vetiver and patchouli are two oils of great importance in perfumery (Williams, 1996, 2004). Both contain complex mixtures of sesquiterpenoids, mostly with complex polycyclic structures (Sell, 2003). The major components of vetiver oil are a-vetivone (**136**), b-vetivone (**137**), and khusimol (**119**), but the most important components as far as odor is concerned are minor constituents such as khusimone (**151**), zizanal (**152**), and methyl zizanoate (**153**). Nootkatone (**154**) is an isomer of a-vetivone and is an important odor component of grapefruit. Patchouli alcohol (**145**) is the major constituent of patchouli oil but, as is the case also with vetiver, minor components are more important for the odor profile. These include norpatchoulenol (**155**) and nortetrapatchoulol (**156**) (Figure 6.26).

The molecules of chamazulene (**150**), khusimone (**151**), norpatchoulenol (**155**), and nortetrapatchoulol (**156**) each contain only 14 carbon atoms in place of the normal 15 of sesquiterpenoids. They are all degradation products of sesquiterpenoids. Degradation, either by enzymic action or from environmental chemical processes, can be an important factor in generating essential oil components. Carotenoids are a family of tetraterpenoids characterized by having a tail-to-tail fusion between two diterpenoid fragments. In the case of β-carotene (**157**), both ends of the chain have been cyclized to form cyclohexane rings. Degradation of the central part of the chain

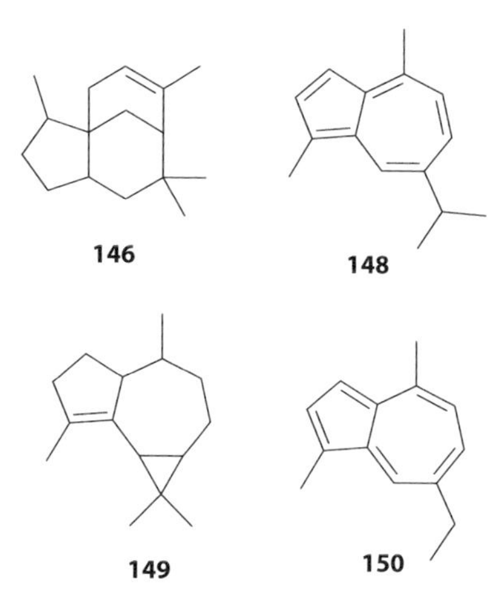

FIGURE 6.25 Some sesquiterpenoid hydrocarbons.

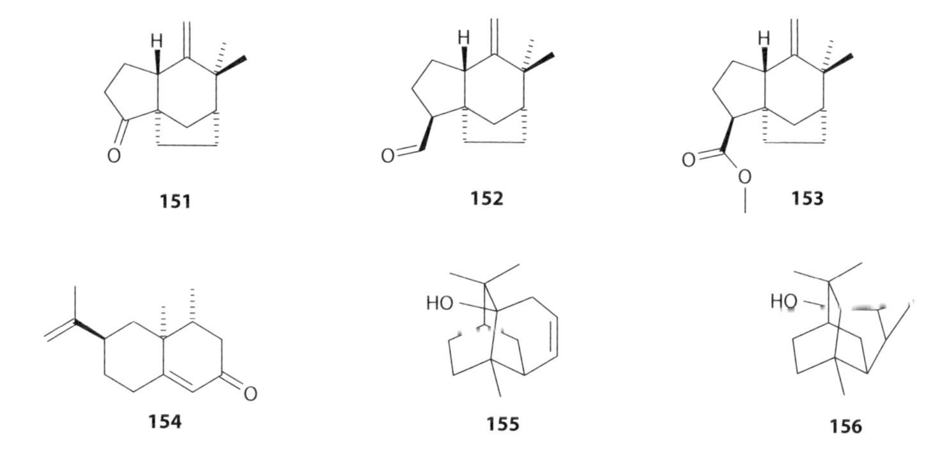

FIGURE 6.26 Components of vetiver, patchouli, and grapefruit.

leads to a number of fragments that are found in essential oils and the two major families of such are the ionones and damascones. Both have the same carbon skeleton, but in the ionones (Sell, 2003; Essential Oils Database, 2006), the site of oxygenation is three carbon atoms away from the ring, and in damascones oxygenation is found at the chain carbon next to the ring (Figure 6.27).

The ionones occur naturally in a wide variety of flowers, fruits, and leaves, and are materials of major importance in perfumery (Günther, 1948; Gildemeister and Hoffmann, 1956; Arctander, 1960; Essential Oils Database, 2006; Sell, 2007). About 57% of the volatile components of violet flowers are a- (**158**) and b-ionones (**159**), and both isomers occur widely in nature. The damascones are also found in a wide range of plants. They usually occur at a very low level, but their very intense odors mean that they still make a significant contribution to the odors of oils containing them. The first to be isolated and characterized was β-damascenone (**160**), which was found at a level of 0.05% in the oil of the Damask rose. Both β-damascenone (**160**) and the α- (**161**) and β-isomers (**162**) have since been found in many different essential oils and extracts. In the cases of safranal (**163**) and cyclocitral (**165**), the side chain is degraded even further leaving only one of its carbon atoms attached to the cyclohexane ring. About 70% of the volatile component of saffron is safranal and it makes a significant contribution to its odor. Other volatile carotenoid degradation products that occur in essential oils and contribute to their odors include the theaspiranes (**165**), vitispiranes (**166**), edulans (**167**), and dihydroactinidiolide (**168**).

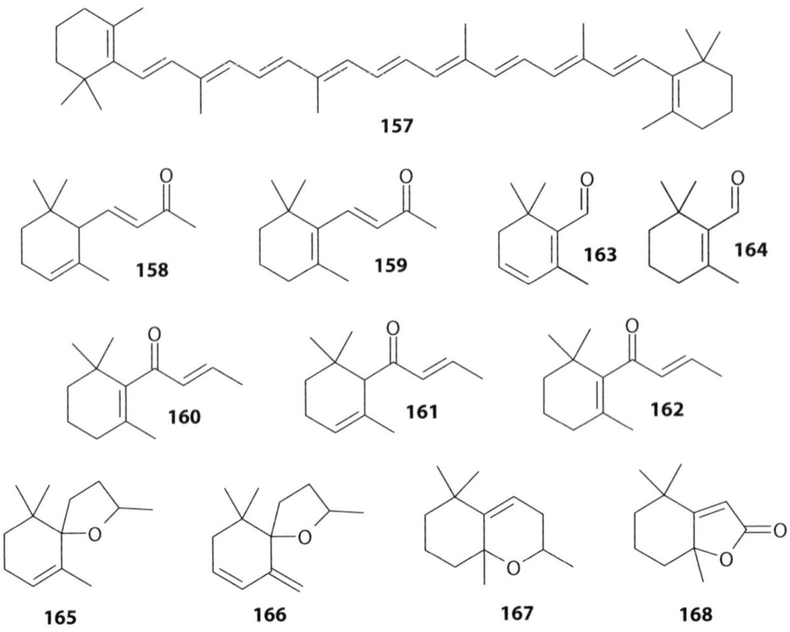

157

158 159 163 164

160 161 162

165 166 167 168

FIGURE 6.27 Carotenoid degradation products.

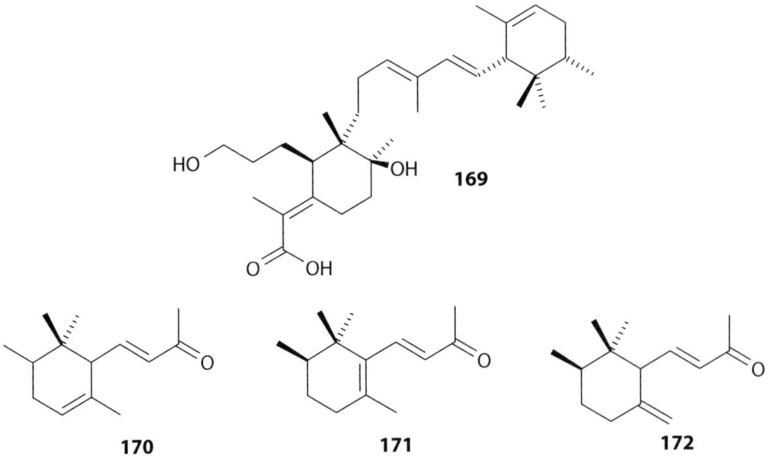

HO OH 169 O OH

170 171 172

FIGURE 6.28 Iripallidal and the irones.

The similarity in structure between the ionones and the irones might lead to the belief that the latter are also carotenoid derived. However, this is not the case as the irones are formed by degradation of the triterpenoid iripallidal (**169**), which occurs in the rhizomes of the iris. The three isomers, α- (**170**), β- (**171**), and γ-irone (**172**), are all found in iris and the first two in a limited number of other species (Figure 6.28).

6.6 SYNTHESIS OF ESSENTIAL OIL COMPONENTS

It would be impossible, in a volume of this size, to review all of the reported syntheses of essential oil components and so the following discussion will concentrate on some of the more commercially important synthetic routes to selected key substances. In the vast majority of cases, there is a balance between routes using plant extracts as feedstocks and those using petrochemicals. For some materials, plant-derived and petrochemical-derived equivalents might exist in economic competition, while for others, one source is more competitive. The balance will vary over time and the market will respond accordingly. Sustainability of production routes is a complex issue and easy assumptions might be totally incorrect. Production and extraction of plant-derived feedstocks often requires considerable expenditure of energy in fertilizer production, harvesting, and processing, and so it is quite possible that production of a material derived from a plant source would use more mineral oil than the equivalent derived from petrochemical feedstocks.

Figure 6.29 shows some of the plant-derived feedstocks used in the synthesis of lipids and polyketides (Sell, 2006). Rapeseed oil provides erucic acid (**173**) that can be ozonolyzed to give brassylic acid (**174**) and heptanal (**175**), both useful building blocks. The latter can also be obtained, together with undecylenic acid (**176**), by pyrolysis of ricinoleic acid (**177**) that is available from castor oil. Treatment of undecylenic acid (**176**) with acid leads to movement of the double bond along the chain and eventual cyclization to give γ-undecalactone (**178**), which has been found in narcissus oils. Aldol condensation of heptanal (**175**) with cyclopentanone, followed by Baeyer–Villiger oxidation, gives δ-dodecalactone (**179**), identified in the headspace of tuberose. Such aldol reactions, followed by appropriate further conversions, are important in the commercial production of analogues of methyl jasmonate (**26**) and jasmone (**27**).

Ethylene provides a good example of a petrochemical feedstock for the synthesis of lipids and polyketides. It can be oligomerized to provide a variety of alkenes into which functionalization can be introduced by hydration, oxidation, hydroformylation, and so on. Of course, telomerization can be used to provide functionalized materials directly.

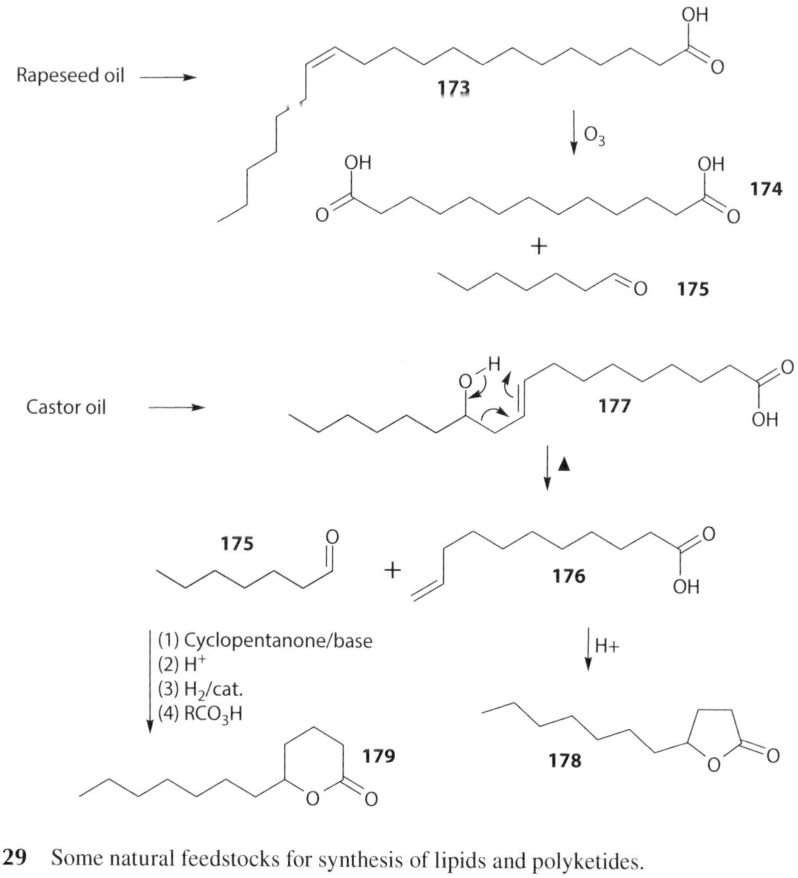

FIGURE 6.29 Some natural feedstocks for synthesis of lipids and polyketides.

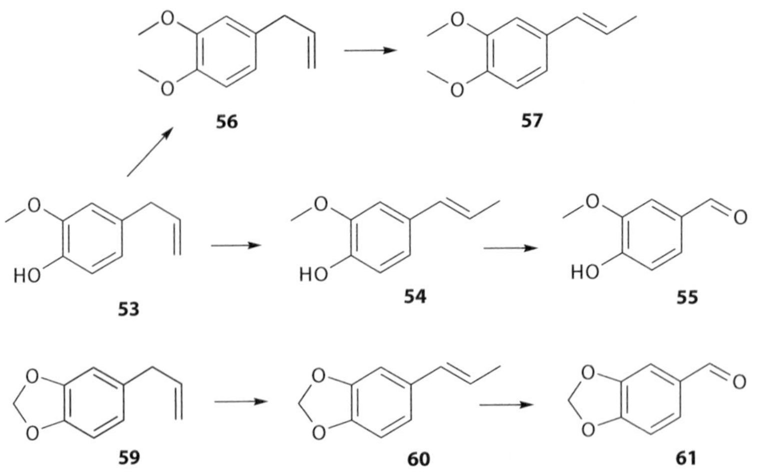

FIGURE 6.30 Shikimates from eugenol and safrole.

Eugenol (**53**) (e.g., clove oil) and safrole (**59**) (e.g., sassafras) are good examples of plant-derived feedstocks that are used in the synthesis of other shikimates. Methylation of eugenol produces methyl eugenol (**56**) and this can be isomerized using acid or metal catalysts to give methyl isoeugenol (**57**). Similarly, isomerization of eugenol gives isoeugenol (**54**), and oxidative cleavage of this, for example, by ozonolysis, gives vanillin (**55**). This last sequence of reactions, when applied to safrole, gives isosafrole (**60**) and heliotropin (**61**). All of these conversions are shown in Figure 6.30.

Production of shikimates from petrochemicals for commercial use mostly involves straightforward chemistry (Arctander, 1969; Däniker, 1987; Bauer and Panten, 2006; Sell, 2006). Nowadays, the major starting materials are benzene (**180**) and toluene (**181**), which are both available in bulk from petroleum fractions. Alkylation of benzene with propylene gives cumene (**182**), the hydroperoxide of which fragments to give phenol (**183**) and acetone. Phenol itself is an important molecular building block and further oxidation gives catechol (**184**). Syntheses using these last two materials will be discussed in the succeeding text. Alkylation of benzene with ethylene gives ethylbenzene, which is converted to styrene (**185**) via autoxidation, reduction, and elimination in a process known as styrene monomer/propylene oxide (SMPO) process. The epoxide (**186**) of styrene serves as an intermediate for 2-phenylethanol (**47**) and phenylacetaldehyde (**187**), both of which occur widely in essential oils. 2-Phenylethanol is also available directly from benzene by Lewis acid–catalyzed addition of ethylene oxide and as a by-product of the SMPO process. Currently, the volume available from the SMPO process provides most of the requirement. All of these processes are illustrated in Figure 6.31.

Phenol (**183**) and related materials, such as guaiacol (**188**), were once isolated from coal tar, but the bulk of their supply is currently produced from benzene via cumene as shown in Figure 6.31. The use of these intermediates to produce shikimates is shown in Figure 6.32. In principle, anethole (**53**) and estragole (methyl chavicol) (**52**) are available from phenol, but in practice, the demand is met by extraction from turpentine. Carboxylation of phenol gives salicylic acid (**38**) and hence serves as a source for the various salicylate esters. Formylation of phenol by formaldehyde, in the presence of a suitable catalyst, has now replaced the Reimer–Tiemann reaction as a route to hydroxybenzaldehydes. The initial products are saligenin (**189**) and *p*-hydroxybenzyl alcohol (**190**), which can be oxidized to salicylaldehyde (**191**) and *p*-hydroxybenzaldehyde (**192**), respectively. Condensation of salicylaldehyde with acetic acid/acetic anhydride gives coumarin (**50**) and *O*-alkylation of *p*-hydroxybenzaldehyde gives anisaldehyde (**44**). As mentioned earlier, oxidation of phenol provides a route to catechol (**184**) and guaiacol (**188**). The latter is a precursor for vanillin, and catechol also provides a route to heliotropin (**61**) via methylenedioxybenzene (**193**).

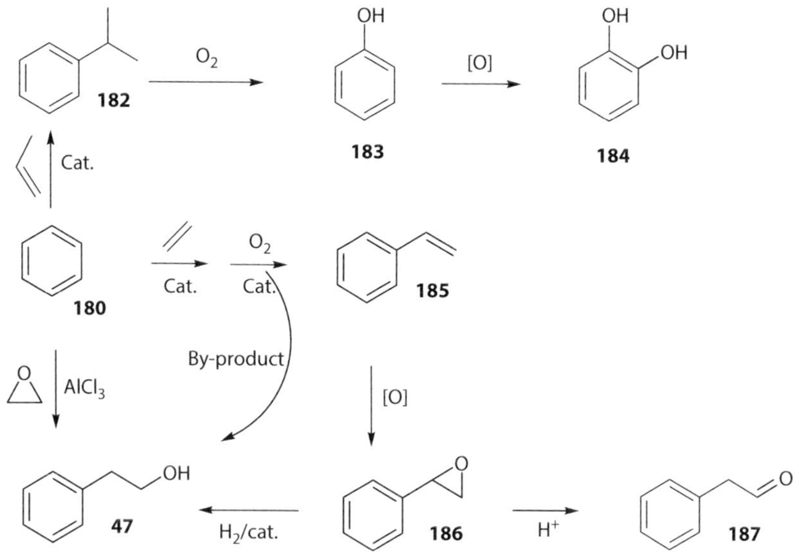

FIGURE 6.31 Benzene as a feedstock for shikimates.

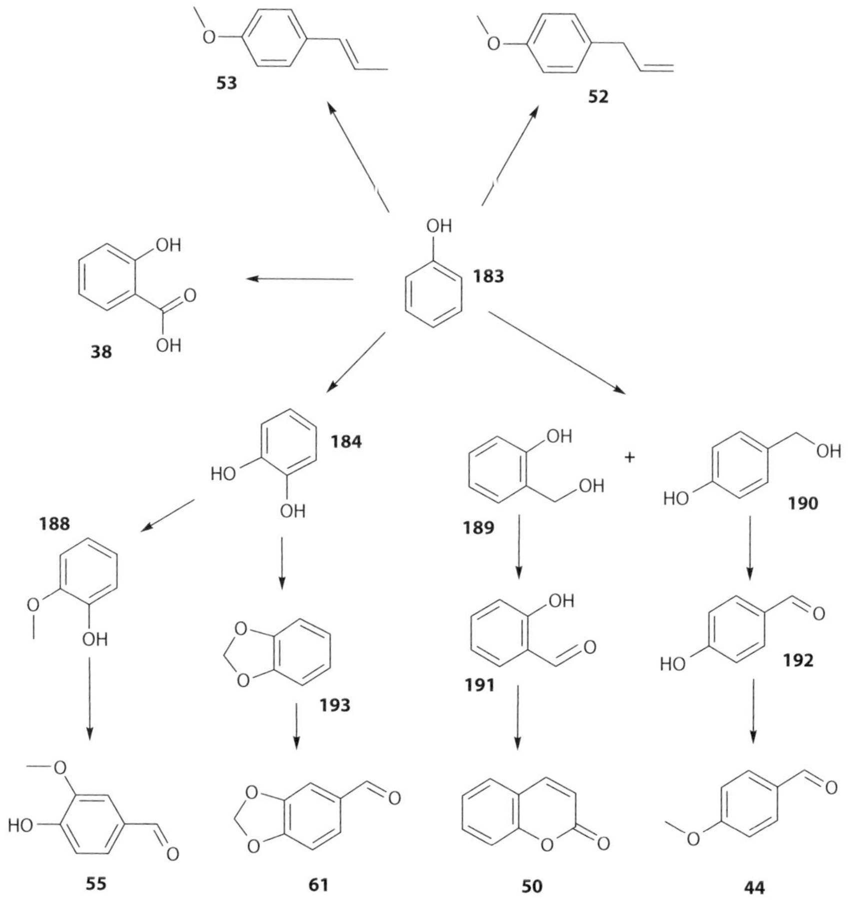

FIGURE 6.32 Synthesis of shikimates from phenol.

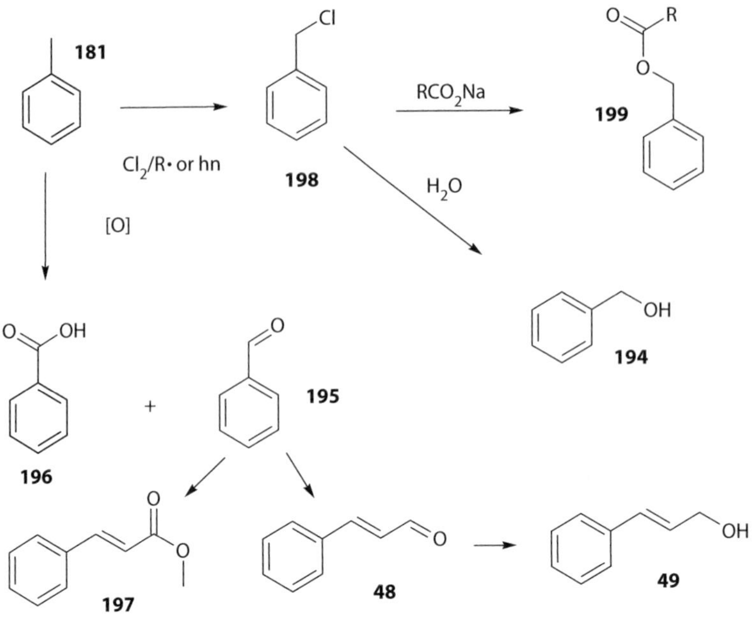

FIGURE 6.33 Shikimates from toluene.

Oxidation of toluene (**181**) with air or oxygen in the presence of a catalyst gives benzyl alcohol (**194**), benzaldehyde (**195**), or benzoic acid (**196**) depending on the chemistry employed. The demand for benzoic acid far exceeds that for the other two oxidation products and so such processes are usually designed to produce mostly benzoic acid with benzaldehyde as a minor product. For the fragrance industry, benzoic acid is the precursor for the various benzoates of interest, while benzaldehyde, through aldol-type chemistry, serves as the key intermediate for cinnamate esters (such as methyl cinnamate (**197**)) and cinnamaldehyde (**48**). Reduction of the latter gives cinnamyl alcohol (**49**) and hence, through esterification, provides routes to all of the cinnamyl esters. Chlorination of toluene under radical conditions gives benzyl chloride (**198**). Hydrolysis of the chloride gives benzyl alcohol (**194**), which can, in principle, be esterified to give the various benzyl esters (**199**) of interest. However, these are more easily accessible directly from the chloride by reaction with the sodium salt of the corresponding carboxylic acid. All of these conversions are shown in Figure 6.33.

Methyl anthranilate (**41**) is synthesized from either naphthalene (**200**) or *o*-xylene (**201**) as shown in Figure 6.34. Oxidation of either starting material produces phthalic acid (**202**). Conversion of this

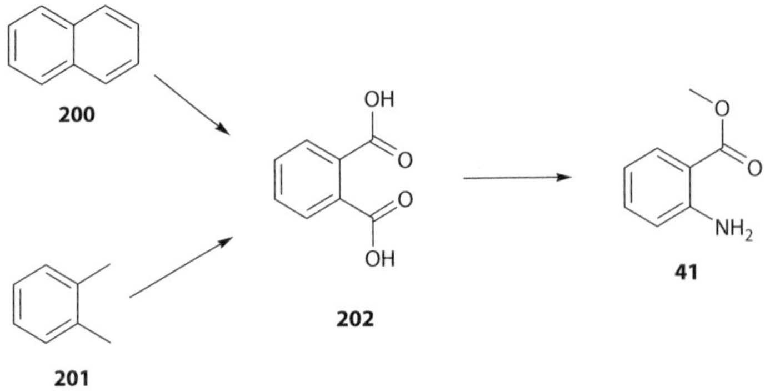

FIGURE 6.34 Synthesis of methyl anthranilate.

diacid to its imide, followed by the Hoffmann reaction, gives anthranilic acid, and the methyl ester can then be obtained by reaction with methanol.

In volume terms, the terpenoids represent the largest group of natural and nature identical fragrance ingredients (Däniker, 1987; Sell, 2007). The key materials are the rose alcohols [geraniol (**63**)/nerol (**90**), linalool (**23**), and citronellol (**91**)], citronellal (**103**), and citral (**71**). Interconversion of these key intermediates is readily achieved by standard functional group manipulation. Materials in this family serve as starting points for the synthesis of a wide range of perfumery materials including esters of the rose alcohols. The ionones are prepared from citral by aldol condensation followed by cyclization of the intermediate y-ionones.

The sources of the aforementioned key substances fall into three main categories: natural extracts, turpentine, and petrochemicals. The balance depends on economics and also on the product in question. For example, while about 10% of geraniol is sourced from natural extracts, it is only about 1% in the case of linalool. Natural grades of geraniol are obtained from the oils of citronella, geranium, and palmarosa (including the variants jamrosa and dhanrosa). Citronella is also used as a source of citronellal. Ho, rosewood, and linaloe were used as sources of linalool, but conservation and economic factors have reduced these sources of supply very considerably. Similarly, citral was once extracted from *L. cubeba* but overharvesting has resulted in loss of that source.

Various other natural extracts are used as feedstocks for the production of terpenoids as shown in Figure 6.35. Two of the most significant ones are clary sage and the citrus oils (obtained as by-products of the fruit juice industry). After distillation of the oil from clary sage, sclareol (**203**) is extracted from the residue, and this serves as a starting material for naphthofuran (**204**), known under trade names such as Ambrofix, Ambrox, and Ambroxan. The conversion is shown in Figure 6.35. Initially, sclareol is oxidized to sclareolide (**205**). This was once effected using oxidants such as permanganate and dichromate, but nowadays, the largest commercial process uses a biotechnological oxidation. Sclareolide is then reduced using lithium aluminum hydride, borane, or similar reagents and the resulting diol is cyclized to the naphthofuran. D-Limonene (**73**) and valencene (**206**) are both extracted from citrus oils. Reaction of D-limonene with nitrosyl chloride gives an adduct that is rearranged to the oxime of L-carvone, and subsequent hydrolysis produces the free ketone (**105**). Selective oxidation of valencene gives nootkatone (**135**).

Turpentine is obtained by tapping of pine trees and this product is known as gum turpentine. However, a much larger commercial source is the so-called crude sulfate turpentine, which is

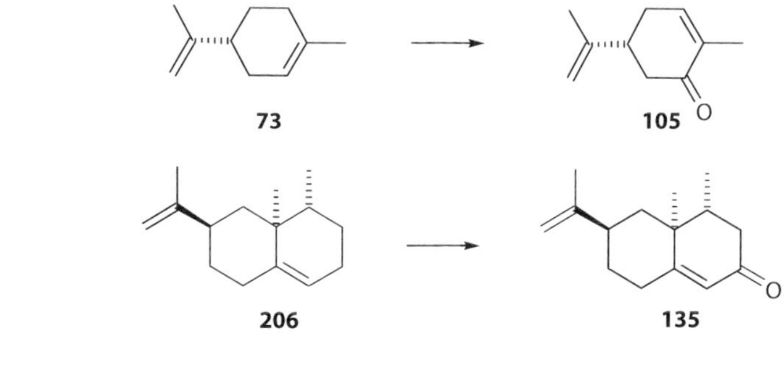

FIGURE 6.35 Partial synthesis of terpenoids from natural extracts.

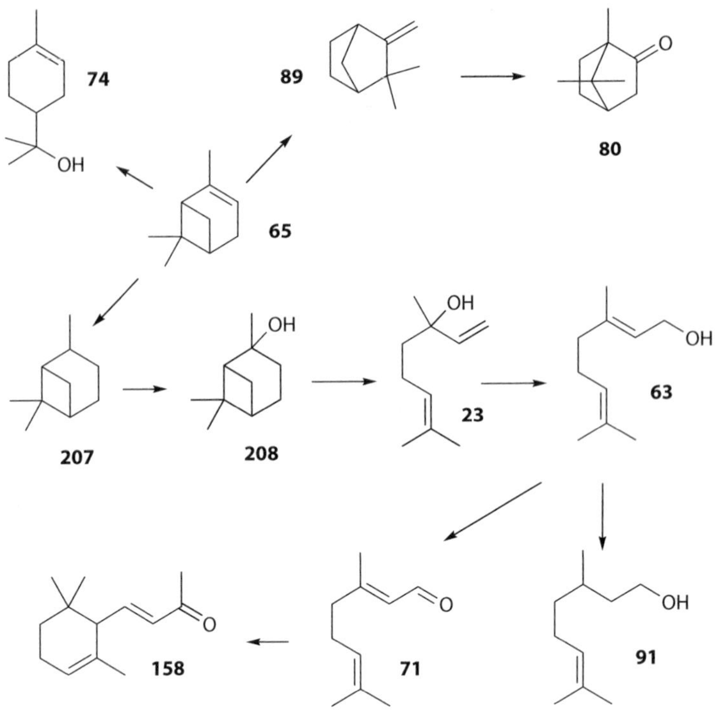

FIGURE 6.36 Products from α-pinene.

obtained as a by-product of the Kraft paper process. The major components of turpentine are the two pinenes with a-pinene (**65**) predominating. Turpentine also serves as a source of *p*-cymene (**83**) and, as mentioned earlier, the shikimate anethole (**53**) (Zinkel and Russell, 1989).

Figure 6.36 shows some of the major products manufactured from α-pinene (**65**) (Sell, 2003, 2007). Acid-catalyzed hydration of a-pinene gives α-terpineol (**74**), which is the highest tonnage material of all those described here. Acid-catalyzed rearrangement of α-pinene gives camphene (**89**) and this, in turn, serves as a starting material for production of camphor (**80**). Hydrogenation of α-pinene gives pinane (**207**), which is oxidized to pinanol (**208**) using air as the oxidant. Pyrolysis of pinanol produces linalool (**23**) and this can be rearranged to geraniol (**63**). Hydrogenation of geraniol gives citronellol (**91**), whereas oxidation leads to citral (**71**). The major use of citral is not as a material in its own right, but as a starting material for production of ionones, such as a-ionone (**158**) and vitamins A, E, and K.

Some of the major products manufactured from β-pinene (**76**) are shown in Figure 6.37. Pyrolysis of β-pinene gives myrcene (**70**) and this can be *hydrated* (not in one step but in a multistage process) to give geraniol (**63**). The downstream products from geraniol are then the same as those described in the preceding paragraph and shown in Figure 6.36. Myrcene is also a starting point for D-citronellol (**209**), which is one of the major feedstocks for the production of L-menthol (**94**) as will be described in the succeeding text.

Currently, there are two major routes to terpenoids that use petrochemical starting materials (Sell, 2003, 2007). The first to be developed is an improved version of a synthetic scheme demonstrated by Arens and van Dorp in 1948. The basic concept is to use two molecules of acetylene (**210**) and two of acetone (**211**) to build the structure of citral (**71**). The route, as it is currently practiced, is shown in Figure 6.38. Addition of acetylene (**210**) to acetone (**211**) in the presence of base gives methylbutynol (**212**), which is hydrogenated, under Lindlar conditions, to methylbutenol (**213**). The second equivalent of acetone is introduced as the methyl ether of its enol form, that is, methoxypropene (**214**). This adds to methylbutenol, and the resultant adduct undergoes a Claisen

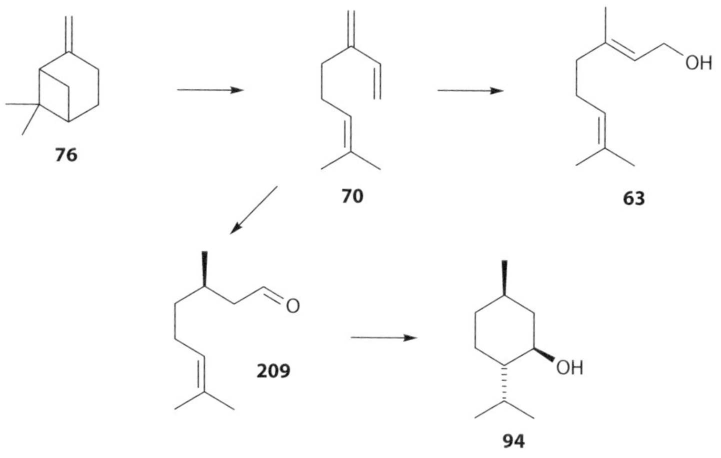

FIGURE 6.37 Products from β-pinene.

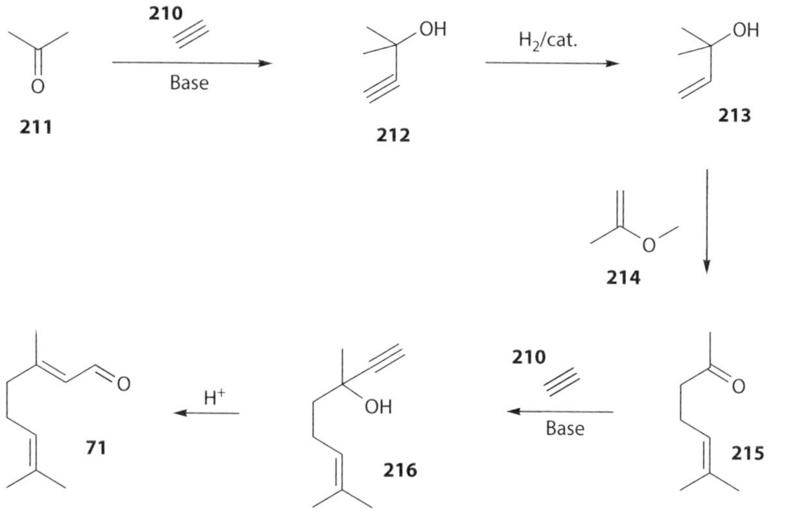

FIGURE 6.38 Citral from acetylene and acetone.

rearrangement to give methylheptenone (**215**). Base catalyzed addition of the second acetylene to this gives dehydrolinalool (**216**), which can be rearranged under acidic conditions to give citral (**71**). Hydrogenation of dehydrolinalool under Lindlar conditions gives linalool (**23**) and thus opens up all the routes to other terpenoids as described earlier and illustrated in Figure 6.36.

The other major route to citral is shown in Figure 6.39. This starts from isobutene (**217**) and formaldehyde (**218**). The ene reaction between these produces isoprenol (**219**). Isomerization of isoprenol over a palladium catalyst gives prenol (**220**) and aerial oxidation over a silver catalyst gives prenal (senecioaldehyde) (**221**). When heated together, these two add together to form the enol ether (**222**), which then undergoes a Claisen rearrangement to give the aldehyde (**223**). This latter molecule is perfectly set up (after rotation around the central bond) for a Cope rearrangement to give citral (**71**). Development chemists have always striven to produce economic processes with the highest overall yield possible thus minimizing the volume of waste and hence environmental impact. This synthesis is a very good example of the fruits of such work. The reaction scheme uses no reagents, other than oxygen, employs efficient catalysts, and produces only one by-product, water, which is environmentally benign.

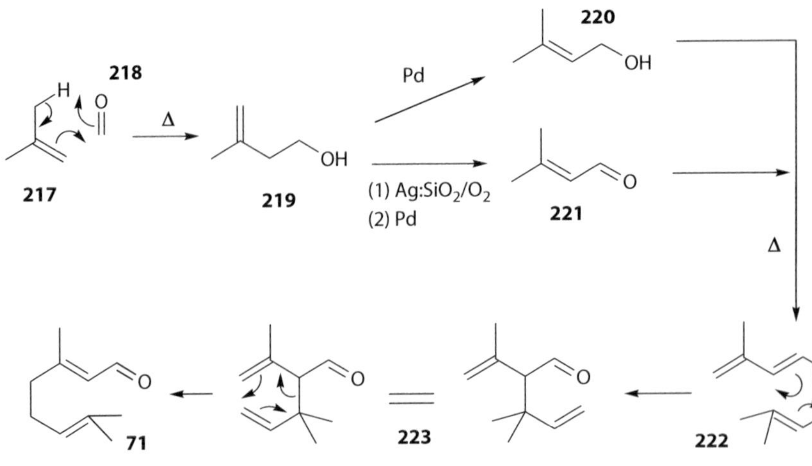

FIGURE 6.39 Citral from isobutylene and acetone.

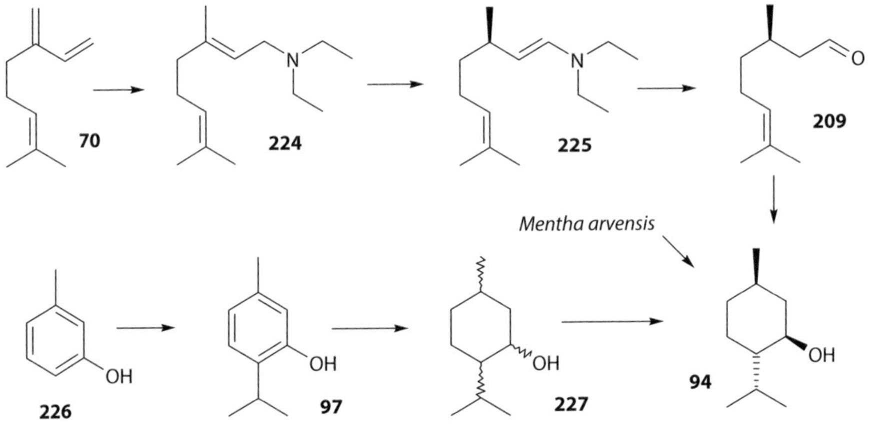

FIGURE 6.40 Competing routes to L-menthol.

The synthesis of L-menthol (**94**) provides an interesting example of different routes operating in economic balance. The three production routes in current use are shown in Figure 6.40. The oldest and simplest route is extraction from plants of the *Mentha* genus and *M. arvensis* (corn mint) in particular. This is achieved by freezing the oil to force the L-menthol to crystallize out. Diethylamine can be added to myrcene (**70**) in the presence of base and rearrangement of the resultant allyl amine (**224**) using the optically active catalyst ruthenium (*S*)-BINAP perchlorate gives the homochiral enamine (**225**). This can then be hydrolyzed to *d*-citronellol (**209**). The chiral center in this molecule ensures that, on acid-catalyzed cyclization, the two new stereocenters formed possess the correct stereochemistry for conversion, by hydrogenation, to give L-menthol as the final product. Starting from the petrochemically sourced *m*-cresol (**226**), propenylation gives thymol (**97**), which can be hydrogenated to give a mixture of all eight stereoisomers of menthol (**227**). Fractional distillation of this mixture gives racemic menthol. Resolution was originally carried out by fractional crystallization, but recent advances include methods for the enzymic resolution of the racemate to give L-menthol.

Estimation of the long-term sustainability of each of these routes is complex and the final outcome is far from certain. In terms of renewability of feedstocks, *m*-cresol might appear to be at a disadvantage against mint or turpentine. However, as the world's population increases, use of agricultural land will come under pressure for food production, hence increasing pressure on mint cultivation and turpentine, hence, myrcene is a by-product of paper manufacture and is therefore vulnerable to trends in paper recycling and "the paperless office." In terms of energy consumption, and hence current dependence on petrochemicals, the picture is also not as clear as might be imagined. Harvesting and processing of mint requires energy and, if the crop is grown in the same field over time, fertilizer is required and this is produced by the very energy-intensive Haber process. The energy required to turn trees in a forest into pulp at a sawmill is also significant and so turpentine supply will also be affected by energy prices. No doubt, the skills of process chemists will be of increasing importance as we strive to make the best use of natural resources and minimize energy consumption (Baser and Demirci, 2007).

REFERENCES

Arctander, S., 1960. *Perfume and Flavour Materials of Natural Origin*. Elizabeth, NJ: Steffen Arctander. (Currently available from Allured Publishing Corporation.)

Arctander, S., 1969. *Perfume and Flavor Chemicals (Aroma Chemicals)*. Montclair, NJ: Steffen Arctander. (Currently available from Allured Publishing Corporation.)

Baser, K.H.C. and F. Demirci, 2007. Chemistry of essential oils. In *Flavours and Fragrances: Chemistry, Bioprocessing and Sustainability*, R.G. Berger (ed.), pp. 43–86, Berlin, Germany: Springer.

Bauer, K. and J. Panten, 2006. *Common Fragrance and Flavor Materials: Preparation, Properties and Uses*. New York: Wiley-VCH.

Bu'Lock, J.D., 1965. *The Biosynthesis of Natural Products*. New York: McGraw-Hill.

Croteau, R., 1987. Biosynthesis and catabolism of monoterpenoids. *Chem. Rev.*, 87: 929.

Däniker, H.U., 1987. *Flavors and Fragrances (Worldwide)*. Stamford, CA: SRI International.

Devon, T.K. and A.I. Scott, 1972. *Handbook of Naturally Occurring Compounds, Vol. 2, The Terpenes*. New York: Academic Press.

Essential Oils Database, 2006. *Boelens Aromachemical Information Systems*. Canton, GA: Leffingwell & Associates. http://www.leffingwell.com/baciseso.htm (accessed August 20, 2015).

Gildemeister, E. and F. Hoffmann, 1956. *Die Ätherischen Öle*. Berlin, Germany: Akademie-Verlag.

Günther, E., 1948. *The Essential Oils*. New York: D van Nostrand.

Hay, R.K.M. and P.G. Waterman (eds.), 1993. *Volatile Oil Crops: Their Biology, Biochemistry and Production*. London, U.K.: Longman.

Lawrence, B.M., 1985. A review of the world production of essential oils. *Perfumer Flavorist*, 10(5): 1.

Lehninger, A.L., 1993. *Principles of Biochemistry*. New York: Worth.

Mann, J., R.S. Davidson, J.B. Hobbs, D.V. Banthorpe, and J.B. Harbourne, 1994. *Natural Products: Their Chemistry and Biological Significance*. London, U.K.: Longman.

Matthews, C.K. and K.E. van Holde. 1990. *Biochemistry*. Redwood City, CA: Benjamin/Cummings.

Sell, C.S., 2003. *A Fragrant Introduction to Terpenoid Chemistry*. Cambridge, U.K.: Royal Society of Chemistry.

Sell, C.S. (ed.), 2006. *The Chemistry of Fragrances from Perfumer to Consumer*, 2nd ed. Cambridge, U.K.: Royal Society of Chemistry.

Sell, C.S., 2007. Terpenoids. In *Kirk-Othmer Encyclopedia of Chemical Technology*, Kirk-Othmen (Ed.), 5th ed. New York: Wiley.

Williams, D.G., 1996. *The Chemistry of Essential Oils*. Weymouth, U.K.: Micelle Press.

Williams, D.G., 2004. *Perfumes of Yesterday*. Weymouth, U.K.: Micelle Press.

Zinkel, D.F. and J. Russell (eds.), 1989. *Naval Stores*. New York: Pulp Chemicals Association, Inc.

7 Analysis of Essential Oils

Adriana Arigò, Mariosimone Zoccali, Danilo Sciarrone, Peter Q. Tranchida, Paola Dugo, and Luigi Mondello

CONTENTS

7.1 INTRODUCTION

The production of essential oils was industrialized in the first half of the nineteenth century, owing to an increased demand for these matrices as perfume and flavor ingredients (Rowe, 2005). As a consequence, the need to perform their systematic investigation also became unprecedented. It is interesting to point out that in the second edition of Parry's monograph, published in 1908, about 90 essential oils were listed, and very little was known about their composition (Parry, 1908). Further important contributions to the essential oil research field were made by Semmler (Semmler, 1907), Gildemeister and Hoffmann (Gildemeister and Hoffman, 1950), Finnemore (Finnemore, 1926), and Guenther (Guenther, 1972). Obviously, it is unfeasible to cite all the researchers involved in the progress of essential oil analysis.

As is widely acknowledged, the composition of essential oils is mainly represented by monoterpene and sesquiterpene hydrocarbons and their oxygenated (hydroxyl and carbonyl) derivatives, along with aliphatic aldehydes, alcohols, and esters. Terpenes can be considered as the most structurally varied class of plant natural products, derived from the repetitive fusion of branched five-carbon units (isoprene units) (Croteau et al., 2000). In this respect, analytical methods applied in the characterization of essential oils have to account for a great number of molecular species. Moreover, it is also of great importance to highlight that an essential oil chemical profile is closely related to the extraction procedure employed and, hence, the choice of an appropriate extraction method becomes crucial. On the basis of the properties of the plant material, the following extraction techniques can be applied: steam distillation (SD), possibly followed by rectification and fractionation, solvent extraction (SE), fractionation of solvent extracts, maceration, expression (cold pressing of citrus peels), *enfleurage*, supercritical fluid extraction (SFE), pressurized-fluid extraction, simultaneous distillation-extraction (SDE), Soxhlet extraction, microwave-assisted hydrodistillation (MAHD), dynamic (DHS) and static (SHS) headspace (HS) techniques, solvent-assisted flavor evaporation (SAFE), solid-phase microextraction (SPME), and direct thermal desorption (DTD), among others.

Apart from the great interest in performing systematic studies on essential oils, there is also the necessity to trace adulterations, mainly in economically important essential oils. As can be observed with almost all commercially available products, market changes occur rapidly, affecting individual plants or industrial processes. In general, market competition, along with the limited interest of consumers with regards to essential oil quality, may induce producers to adulterate their commodities by the addition of products of lower value. Different types of adulterations can be encountered: (a) the simple addition of natural and/or synthetic compounds, with the aim of generating an oil characterized by specific quality values, such as density, optical rotation, residue percentage, ester value, etc. or (b) refined sophistications in the reconstitution and counterfeiting of commercially valuable oils. In the latter case, natural and/or synthetic compounds are added to enhance the market value of an oil, attempting to maintain the qualitative, or even quantitative, composition of natural essential oils, and making adulteration detection a troublesome task. Consequently, the exploitation of modern analytical methodologies, such as gas chromatography (GC) and related hyphenated techniques, is practically unavoidable.

As a consequence of diffused illegal practice in the production of essential oils, there has been an enhanced request for legal standards of commercial purity, while essential oils were included as herbal drugs in pharmacopoeias (*The International Pharmacopoeia*; *Japanese Pharmacopeia*; *British Pharmacopoeia*; *The United States Pharmacopoeia*; *European Pharmacopoeia*) and also in a compendium denominated as *Martindale: The Complete Drug Reference* (formerly named as *Martindale's: The Extra Pharmacopoeia*) (Martindale, 2007). In view of the need for standardized methodologies, these pharmacopoeias commonly include the descriptions of several tests, processes, and apparatus. In addition, various international standard regulations have been introduced in which the characteristics of specific essential oils are described, and the botanical source and physicochemical requirements are reported. Such standardized information was created to facilitate the assessment of quality; for example, ISO 3761:1997 specifies that for Brazilian rosewood essential oil (*Aniba rosaeodora* Ducke), an alcohol content in the 84%–93% range, determined as linalool, is required (ISO 3761:1997). Moreover, guidelines for the analysis of essential oils are also available; for example, for the measurement of the refractive index (ISO 280: 1198) and optical rotation (ISO 592: 1998), as also for gas chromatography, analysis using capillary columns (ISO chromatography, ISO 8432:1987) (TC 57 Essential Oils). The French Standards Association (Association Française de Normalisation - AFNOR) also develops norms and standard methods dedicated to the essential oils research field, with the aim of assessing quality in relation to specific physical, organoleptic, chemical and chromatographic characteristics (AFNOR).

The present contribution provides an overview on the classical and modern analytical techniques commonly applied to characterize essential oils. Modern techniques will be focused on chromatographic analyses, including theoretical aspects and applications. For modern analytical techniques, most recent advances with respect to the information reported in the previous edition of this book are here discussed.

7.2 CLASSICAL ANALYTICAL TECHNIQUES

The thorough study of essential oils is based on the relationship between their physical and chemical properties, and it is completed by the assessment of organoleptic qualities. The earliest analytical methods applied in the investigation of an essential oil were commonly focused on quality aspects, concerning mainly two properties, namely identity and purity (Simões and Spitzer, 1999).

The following techniques are commonly applied to assess essential oil physical properties: specific gravity (SG), which is the most frequently reported physicochemical property and is a special case of relative density, $[\rho]^{T(°C)}$, defined as the ratio of the densities of a given oil and of water when both are at identical temperatures. The attained value is characteristic for each essential oil and commonly ranges between 0.696 and 1.118 at 15°C (Gildemeister and Hoffman, 1950). In cases in which the determinations were made at different temperatures, conversion factors can be used to normalize data.

The measurement of optical rotation, $[\alpha]_D^{20}$, either dextrorotatory or laevorotatory, is also widely recognized. Optical activity is determined by using a polarimeter, with the angle of rotation depending on a series of parameters, such as oil nature, the length of the column through which the light passes, the applied wavelength, and the temperature. The degree and direction of rotation are of great importance for purity assessments, since they are related to the structures and the concentration of chiral molecules in the sample. Each optically active substance has its own specific rotation, as defined in Biot's law:

$$[\alpha]_\lambda^T = \frac{\alpha_\lambda^T}{c \cdot l}$$

where α is the optical rotation at a temperature T expressed in degrees Celsius, l is the optical path length in dm, λ is the wavelength, and c is the concentration in g/100 mL. It is worthy of note that a standard 100 mm tube is commonly used; in cases in which darker or lighter colored oils are analyzed, longer or shorter tubes are used, respectively, and the rotation should be extrapolated for a 100 mm-long tube. Moreover, prior to the measurement, the essential oil should be dried out with anhydrous sodium sulfate and filtered.

The determination of the refractive index, $[\eta]_D^{20}$, also represents a characteristic physical constant of an oil, usually ranging from 1.450 to 1.590. This index is represented by the ratio of the sine of the angle of incidence (i) to the sine of the angle of refraction (e) of a beam of light passing from a less dense to a denser medium, such as from air to the essential oil:

$$\sin i / \sin e = N/n$$

where N and n are, respectively, the indices of the more and the less dense medium. The Abbé type refractometer, equipped with a monochromatic sodium light source, is recommended for routine essential oil analysis; the instrument is calibrated through the analysis of distilled water at 20°C, producing a refractive index of 1.3330. In cases in which the measurement is performed at a temperature above or below 20°C, a correction factor per degree must be added or subtracted, respectively (Bosart, 1937).

A further procedure which can be applied for the purity assessment of essential oils is based on water solubility; the test, which reveals the presence of polar substances, such as alcohols, glycols and their esters, and glycerin acetates, is carried out as follows: the oil is added to a saturated solution of sodium chloride, which after homogenization is divided in two phases; the volume of the oil, which is the organic phase, should remain unaltered; volume reduction indicates the presence of water-soluble substances. On the other hand, the solubility, or immiscibility, of an essential oil in ethanol reveals much on its quality. Considering that essential oils are slightly soluble in water and are miscible with ethanol, it is simple to determine the number of volumes of water-diluted ethanol required for the complete solubility of one volume of oil; the analysis is carried out at 20°C, if the oil is liquid at this temperature. It must be emphasized that oils rich in oxygenated compounds are more readily soluble in dilute ethanol than those richer in hydrocarbons. Moreover, aged or improperly stored oils frequently present decreased solubility (Guenther, 1972).

The investigation on the solubility of essential oils in other media is also widely accepted, such as the evaluation of the presence of water by means of a simple procedure: the addition of a volume of essential oil to an equal volume of carbon disulfide or chloroform; in the case that the oil is rich in oxygenated constituents, it may contain dissolved water, generating turbidity. A further solubility test, in which the oil is dissolved in an aqueous solution of potassium hydroxide, is applied to oils containing molecules with phenolic groups; finally, the incomplete dissolution of oils rich in aldehydes in a dilute bisulfite solution may denote the presence of impurities.

The estimation of melting and congealing points, as well as the boiling range of essential oils, is also of great importance for identity and purity assessments. Melting point evaluations are a valuable modality to control essential oil purity, since a large number of molecules generally comprised in essential oils melt within a range of 0.5°C or, in the case of decomposition, over a narrow temperature range. On the other hand, the determination of the congealing point is usually applied in cases where the essential oil consists mainly of one molecule, such as the oil of cloves which contains about 90% of eugenol. In the latter case, such a test enables the evaluation of the percentage-amount of the abundant compound. At the congealing point, crystallization occurs accompanied by heat liberation, leading to a rapid increase in temperature which is then stabilized at the so-called congealing point. A further purity evaluation method is represented by the boiling range determination, through which the percentage of oil which distills below a certain temperature or within a temperature range is investigated.

An additional test usually performed in essential oil analysis is the evaporation residue, in which the percentage of the oil that is not released at 100°C is determined. In the specific case of cold-pressed citrus oils, this test enables purity assessment, since a lower amount of residue in an expressed oil may indicate an addition of less valuable distilled volatile components to the oil; an increased residue amount reveals the possible presence ofterpenes with higher molecular weights, through the addition of single compounds (or other essential oils), or of heavier oils, such as rosin oil, cheaper citrus oils or by directly using the citrus oil residue. An example consists of the addition of lime oil to sophisticate lemon oils. In oxidized or polymerized oils, the presence of less-volatile compounds is common; in this case a simple test may be carried out by applying a drop of oil on a piece of filter paper; if a transparent spot persists for a period over 24 h, the oil is most probably degraded. Furthermore, the residue can be subjected to acid and saponification number analyses; for instance, the addition of rosin oil would increase the acid number since this oil, differently from other volatile oils, is characterized by the presence of complex acids. By definition, the acid number is the number of milligrams of potassium hydroxide required to neutralize the free acids contained in 1 g of an oil. This number is preserved in cases in which the essential oil has been carefully dried and stored in dark and airtight recipients. As commonly observed, the acid number increases along the aging process of an oil; oxidation of aldehydes and hydrolysis of esters trigger the increase of the acid number.

To compensate for the reduction of the non-volatile residue resulting from a dilution with volatile diluents such as citrus terpenes, an appropriate amount of a non-volatile adulterant (e.g., castor oil or distillation residue from citrus oil) was added. In this case, evaporation residue resulted of limited value for the detection of the adulteration. Cold-pressed citrus oils present a typical UV spectrum with a maximum of absorption at 315 nm, while distilled oils are transparent in the same region. This difference can offer a method (known as CD method) to detect distilled oils in cold-pressed ones (Sale, 1953).

Classical methodologies have been also widely applied to assess essential oil chemical properties (Guenther, 1972; Simões and Spitzer, 1999), such as the determination of the presence of halogenated hydrocarbons and heavy metals. The former investigation is exploited to reveal the presence of halogenated compounds, commonly added to the oils for adulteration purposes. Several tests have been developed for halogen detection, with the Beilstein method (Beilstein, 1872) being the one most reported. In practice, a copper wire is cleaned and heated in a Bunsen burner flame to form a coating of copper (II) oxide. It is then dipped in the sample to be tested and once again heated in the flame. A positive test is indicated by a green flame caused by the formation of a copper halide. Attention is be paid to positive or inconclusive results, since they may be induced by trace amounts of organic acids, nitrogen-containing compounds (Guenther, 1972), or salts (Panda, 2003). An alternative to the Beilstein method is the sodium fusion test, in which the oil is first mineralized, and in the case that halogenated hydrocarbons are present, a residue of sodium halide is formed, which is soluble in nitric acid, and precipitates as the respective silver halide by the addition of a small amount of silver nitrate solution (Simões and Spitzer, 1999). With regard to the detection of heavy metals, several tests are described to investigate and ensure the absence, especially, of copper and lead. One method is based on the extraction of the essential oil with a diluted hydrochloric acid solution, followed by the formation of an aqueous phase to which a buffered thioacetamide solution is added. The latter reagent leads to the formation of sulfite ions that are used in the detection of heavy metals.

The determination of esters derived from phthalic acid is also of great interest for the toxicity evaluation of an essential oil. Considering that esters commonly contained in essential oils are derived from monobasic acids, at first, saponification is carried out through the addition of an ethanolic potassium hydroxide solution. The formed potassium phthalate, which is not soluble in ethanol, generates a crystalline precipitate (Simões and Spitzer, 1999).

The use of qualitative information alone is not sufficient to correctly characterize an essential oil, and quantitative data are of extreme importance. Classical methods are generally focused on chemical groups, and the assessment of quantitative information through titration is widely applied (e.g., for the acidimetric determination of saponified terpene esters). Saponification can be performed with heat, and in the case readily saponified esters are to be investigated, in the cold, and afterwards the alkali excess is titrated with aqueous hydrochloric acid; thereafter, the ester number can be calculated. A further test is the determination of terpene alcohols by acetylating with acetic anhydride; part of the acetic anhydride is consumed in the reaction and can be quantified through titration of acetic acid with sodium hydroxide. The percentage of alcohol can then be calculated. The latter method is applied when the alcoholic constituents of an essential oil are not well known, in the case these are established, the oil is saponified and the ester number of the acetylated oil is calculated and used to estimate the free alcohol content.

Other chemical classes worthy of mention are aldehydes and ketones that may be investigated through different tests. The bisulfite method is recommended for essential oils rich in aldehydic compounds, such as lemongrass, bitter almond, and cassia, while the neutral sulfite test is more suitable for ketone-rich oils, such as spearmint, caraway, and dill oils. For essential oils presenting small amounts of aldehydes and ketones, the hydroxylamine method, or its modification the Stillman–Reed method, are the most indicated methods (Panda, 2003). In the latter case the aldehyde and ketone contents are determined through the addition of a neutralized hydroxylamine hydrochloride solution and subsequent titration with standardized acid (Stillman–Reed method) (Stillman and Reed, 1932); in the former analytical procedure, the aldehyde and ketone content is established through the addition of a hydroxylamine hydrochloride solution, followed by neutralization with the reaction products, that is, alkali of the hydrochloric acid. These methods may be applied in the determination of citral in citrus oils and carvone in caraway oil. With regard to the determination of phenols, such as eugenol in clove oil or thymol and carvacrol in thyme oil, the test is commonly made through the addition of potassium hydroxide solutions, forming water-soluble salts. It has to be pointed out that besides phenols, other constituents are soluble in alkali solutions and in water (Guenther, 1972; Panda, 2003).

Essential oils are also often analyzed by means of chromatographic methods. In general, the principle of chromatography is based on the distribution of the constituents to be separated between two immiscible phases; one of these is a stationary bed (stationary phase) with a large surface area, while the other is a mobile phase which percolates through that stationary bed in a definite direction (Poole, 2003). Planar chromatography may be referred to as a classical method for essential oil analysis, being well-represented by thin layer chromatography (TLC) and paper chromatography (PC). In both techniques, the stationary phase is distributed as a thin layer on a flat support, in PC being self-supporting, while in TLC coated on a glass, plastic, or metal surface; the mobile phase is allowed to ascend through the layer by capillary forces. TLC is a fast and inexpensive method for identifying substances and testing the purity of compounds, being widely used as a preliminary technique providing valuable information for subsequent analyses (Falkenberg et al., 1999). Separations in TLC involve the distribution of one or a mixture of substances between a stationary phase and a mobile phase. The stationary phase is a thin layer of adsorbent (usually silica gel or alumina) coated on a plate. The mobile phase is a developing solvent which travels up the stationary phase, carrying the samples with it. Components of the samples will separate on the stationary phase according to their stationary phase-mobile phase affinities (Wagner et al., 2003). In practice, a small quantity of the sample is applied near one edge of the plate, and its position is marked with a pencil. The plate is then positioned in a developing chamber with one end immersed in the developing solvent, the mobile phase, avoiding the direct contact of the sample with the solvent. When the mobile phase reaches about two-thirds of the plate length, the plate is removed and dried, the solvent front is traced, and the separated components are located. In some cases, the spots are directly visible, but in others, they must be visualized by using methods applicable to almost all organic samples, such as the use of a solution of iodine or sulfuric acid, both of which react with organic compounds yielding dark products. The use of an ultraviolet (UV) lamp is also advisable, especially if a substance which aids in the visualization of compounds is incorporated into the plate, as is the case of many commercially available TLC plates. Data interpretation is made through the calculation of the ratio of fronts (R_f) value for each spot, which is defined as

$$R_f = \frac{Z_S}{Z_{St}}$$

where Z_S is the distance from the starting point to the center of a specific spot, and Z_{St} is the distance from the starting point to the solvent front (Wagner et al., 2003; Hahn-Deinstrop, 2000). A concise review on TLC has been made by J. Sherma (Sherma, 2000).

The R_f value is characteristic for any given compound on the same stationary phase using the identical mobile phase. Hence, known R_f values can be compared to those of unknown substances to aid in their identification (Wagner et al., 2003). On the other hand, separations in paper chromatography involve the same principles as those in TLC, differing in the use of a high-quality filter paper as stationary phase instead of a thin adsorbent layer, by the increased time requirements, and by poorer resolution. It is worth highlighting that TLC has largely replaced PC in contemporary laboratory practice (Poole, 2003).

Essential oils can be characterized by their organoleptic properties, an assessment which involves human subjects as measuring tools. These procedures present an immediate problem, linked to the innate variability between individuals, not only as a result of their previous experiences and expectations, but also to their sensitivity (Richardson, 1999). In this respect, individuals are selected and screened for specific anosmia, as proposed by Friedrich et al. (Friedrich et al., 2001). In the case no insensitivities are found, the panelists are introduced to two sensorial properties, quality and intensity. Odor quality is described according to the odor families, while intensity is measured through the rating of a sensation based on an intensity interval scale. The assessment of an essential oil odor can be performed through its addition to filter paper strips and subsequent evaluation by the panelists. Considering that each volatile compound is characterized by a different volatility, the evaluation of the paper strip in different periods of time enables the classification of the odors in top, middle, and bottom notes (Curtis and Williams, 2001). In addition, the olfactive assessment during the determination of the evaporation residue is also of significance, since by-notes of low-boiling adulterants or contaminants may be detected as the oil vaporizes, and the odor of the final hot residue can reveal the addition of high-boiling compounds. Olfactive analysis is also valuable after the determination of phenols in essential oils, by studying the non-phenolic portion (Guenther, 1972).

Noteworthy is that the use of the earliest analytical techniques for the systematic study of essential oils, such as specific gravity, relative density, optical activity, and refractive index, or melting, congealing, and boiling points determinations, are generally applied for the assessment of pure compounds and may be extended to evaluate essential oils composed of a major compound. Classical methods cannot be used as stand-alone methods and need to be combined with modern analytical techniques, especially gas chromatography, for the assessment of essential oil authenticity and quality.

7.3 MODERN ANALYTICAL TECHNIQUES

Most of the methods applied in the analysis of essential oils rely on chromatographic procedures, which enable component separation and identification. However, additional confirmatory evidence is required for reliable identification, avoiding equivocated characterizations.

In the early stages of research in the essential oil field, attention was devoted to the development of methods in order to acquire deeper knowledge on the profiles of volatiles; however, this analytical task was made troublesome due to the complexity of these real-world samples. Over the last decades, the aforementioned research area has benefited from the improvements in instrumental analytical chemistry, especially in the chromatographic area, and nowadays, the number of known constituents has drastically increased.

The primary objective in any chromatographic separation is always the complete resolution of the compounds of interest in the minimum time. To achieve this task, the most suitable analytical column (dimension and stationary phase type) has to be used, and adequate chromatographic parameters must be applied to limit peak enlargement phenomena. A good knowledge of chromatographic theory is, indeed, of great support for the method optimization process, as well as for the development of innovative techniques.

In gas chromatographic (GC) analysis, the compounds to be analyzed are vaporized and eluted by the mobile gas phase, the carrier gas, through the column. The analytes are separated on the basis of their relative vapor pressures and affinities for the stationary bed. On the other hand, in liquid chromatographic (LC) analysis, the compounds are eluted by a liquid mobile phase consisting of a solvent or a mixture of solvents, the composition of which may vary during the analysis (gradient elution), and are separated according to their affinities for the stationary bed. In general, the volatile fraction of an essential oil is analyzed by GC, while the non-volatile by LC.

At the outlet of the chromatography column, the analytes emerge separated in time. The analytes are then detected and a signal is recorded generating a chromatogram, which is a signal vs. time graphic ideally with peaks presenting a Gaussian distribution-curve shape. The peak area and height are a function of the amount of solute present, and its width is a function of band spreading in the column (Ettre and Hinshaw, 1993), while its retention time can be related to the solute's identity. Hence, the information contained in the chromatogram can be used for qualitative and quantitative analysis.

7.3.1 The Use of Gas Chromatography and Gas Chromatography-Mass Spectrometry (GC-MS), and Linear Retention Indices in Essential Oil Analysis

The analysis of essential oils by means of gas chromatography began in the 1950s, when Prof. Liberti (Liberti and Conti, 1956) started analyzing citrus essential oils only a few years after James and Martin first described gas-liquid chromatography (GLC), commonly referred to as gas chromatography (GC) (James and Martin, 1952), a milestone in the evolution of instrumental chromatographic methods.

After its introduction, gas chromatography developed at a phenomenal rate, growing from a simple research novelty to a highly sophisticated instrument. Moreover, the current-day requirements for high resolution and trace analysis are satisfied by modern column technology. In particular, inert, thermostable, and efficient open-tubular columns are available, along with associated selective detectors and injection methods, which allow on-column injection of liquid and thermally labile samples. The development of robust fused-silica columns, characterized by superior performances to that of glass columns, brings open-tubular GC columns within the scope of almost every analytical laboratory.

At present, essential oil GC analyses are more frequently performed on capillary columns, which, after their introduction, rapidly replaced packed GC columns. In general, packed columns support larger sample-size ranges, from tenths of a microliter up to 20 μL, and thus, the dynamic range of the analysis can be enhanced. Trace-level components can be easily separated and quantified without preliminary fractionation or concentration. On the other hand, the use of packed columns leads to lower resolution due to the higher pressure drop per unit length. Packed columns need to be operated at higher column flow rates, since their low permeability requires high pressures to significantly improve resolution (Scott, 2001). It is worthy of note that since the introduction of fused-silica capillary columns, considerable progress has been made in column technology, and a great number of papers regarding GC applications on essential oils have been published.

The choice of the capillary column in an essential oil GC analysis is of great importance for the overall characterization of the matrix; the stationary phase chemical nature and film thickness, as well as the column length and internal diameter, are to be considered. In general, essential oil GC analysis js carried out on 25–50 m columns, with 0.20–0.32 mm internal diameters, and 0.25 μm stationary phase film thickness. It must be noted that the degree of separation of two components on two distinct stationary phases can be drastically different. As it is well known, non-polar columns produce boiling-point separations, while on polar stationary phases, compounds are resolved according to their polarity. Considering that essential oil components such as terpenes and their oxygenated derivatives frequently present similar boiling points, these elute in a narrow retention-time range on a non-polar column. In order to overcome this limit, the analytical method can be modified by applying a slower oven temperature rate to widen the elution range of the oil or by using a polar stationary phase, as oxygenated compounds are more retained than hydrocarbons. However, choosing different stationary phases may provide little improvement as resolution can be improved for a series of compounds but new coelutions can also be generated.

Considering gas chromatographic analyses using flame ionization (FID), thermal conductivity (TCD), or other detectors, which do not provide structural information of the analyzed molecules, retention data (more precisely, retention indices) are used as the primary criterion for peak assignment. The most thoroughly studied, diffused, and accepted retention index calculation methods are based on the logarithmic-based equation developed by Kováts in 1958 (Kováts, 1958) for isothermal conditions and on the equation propounded by van den Dool and Kratz in 1963 (van den Dool and Kratz, 1963), which does not use the logarithmic form and is used in the case of temperature-programming conditions.

In addition, peak assignment can be greatly facilitated by the combined use of gas chromatography and mass spectrometry (GC-MS).

Mass spectrometry (MS) can be defined as the study of systems through the formation of gaseous ions, with or without fragmentation, which are then characterized by their mass to charge ratios (*m/z*) and relative abundances (Todd, 1995). The analyte may be ionized thermally, by an electric field or by impacting energetic electrons, ions, or photons.

During the past decade, there has been a tremendous growth in popularity of mass spectrometers as a tool for routine analytical experiments as well as fundamental research. This is due to a number of features including relatively low cost, simplicity of design and extremely fast data acquisition rates. Although the sample is destroyed by the mass spectrometer, the technique is very sensitive and only low amounts of material are used in the analysis.

The most frequent and simple identification method in GC-MS consists of the comparison of the acquired unknown mass spectra with those contained in a reference MS library. Often, the combined use of GC-MS data search and retention indices filters can represent a very powerful tool for reliable identification of volatile components in essential oils.

For detailed discussion on this specific subject, the reader can refer to Chapter 8 of this book.

7.3.2 Fast Gas Chromatography for Essential Oil Analysis

Nowadays in daily routine work, apart from increased analytical sensitivity, demands are also made on the efficiency in terms of speed of the laboratory equipment. Regarding the rapidity of analysis, two aspects need to be considered: (a) the costs in terms of time required, for example, as is the case in quality control analysis, and (b) the efficiency of the utilized analytical equipment.

When compared to conventional GC, the primary objective of fast GC is to maintain sufficient resolving power in a shorter time, by using adequate columns and instrumentation in combination with optimized run conditions to provide 3–10-times faster analysis times (Korytár et al., 2002; Cramers et al., 1999; Cramers and Leclercq, 1999). The technique can be accomplished by manipulating a number of analysis parameters, such as column length, column internal diameter (ID), stationary phase, film thickness, carrier gas, linear velocity, oven temperature, and ramp rate. Fast GC is typically performed using short, 0.10 mm or 0.18 mm ID capillary columns with hydrogen carrier gas and rapid oven temperature ramp rates. In general, capillary gas chromatographic analysis may be divided in three groups, based solely on column internal diameter types; namely, as conventional GC when 0.25 mm ID columns are applied, fast GC using 0.10–0.18 mm ID columns, and ultrafast GC for columns with an ID of 0.05 mm or less. In addition, GC analyses times between 3 and 12 min can be defined as "fast," between 1 and 3 min as "very fast," and below 1 min as "ultrafast." Fast GC requires instrumentation provided with high split ratio injection systems because of low sample column capacities, increased inlet pressures, rapid oven heating rates, and fast electronics for detection and data collection (Mondello et al., 2003).

The application of two methods, conventional (30 m $\times$ 0.25 mm ID, 0.25 μm d_f column) and fast (10 m $\times$ 0.10 mm ID, 0.10 μm d_f column), on five different citrus essential oils (bergamot, mandarin, lemon, and bitter and sweet oranges) has been reported (Mondello et al., 2003). The fast method allowed the separation of almost the same compounds as the conventional analysis, while quantitative data showed good reproducibility. The effectiveness of the fast GC method, through the use of narrow-bore columns, was demonstrated. An ultrafast GC lime essential oil analysis was also performed on a 5 m $\times$ 50 μm capillary column with 0.05 μm stationary phase film thickness (Mondello et al., 2004). The total analysis time of this volatile essential oil was less than 90 s; a chromatogram is presented in Figure 7.1.

Another technique, ultrafast module-GC (UFM-GC) with direct resistively heated narrow-bore columns, has been applied to the routine analysis of four essential oils of differing complexities: chamomile, peppermint, rosemary, and sage (Bicchi et al., 2004). All essential oils were analyzed by conventional GC with columns of different lengths; namely, 5 m and 25 m, with a 0.25 mm ID, and by fast GC and UFM-GC with narrow bore columns (5 m $\times$ 0.1 mm ID). Column performances were evaluated and compared through the Grob test, separation numbers, and peak capacities. UFM-GC was successful in the qualitative and quantitative analysis of essential oils of different compositions with analysis times between 40 s and 2 min vs. 20–60 min required by conventional GC. UFM-GC allows a drastically reduced analysis time, although the very high column heating rates may lead to changes in selectivity compared to conventional GC and changes that are more marked than those of classical fast GC. In a further work, the same researchers (Bicchi et al., 2005) stated that in UFM-GC experiments, the appropriate flow choice can compensate, in part, for the loss of separation capability due to the heating rate increase.

Besides the numerous fast GC applications on citrus essential oils, other oils have also been subjected to analysis, such as rose oil by means of ultrafast GC (Mondello, 2004) and very fast GC (Tranchida et al., 2005), both using narrow-bore columns. Rosemary and chamomile oils have been investigated by means of fast GC on two short conventional columns of distinct polarity

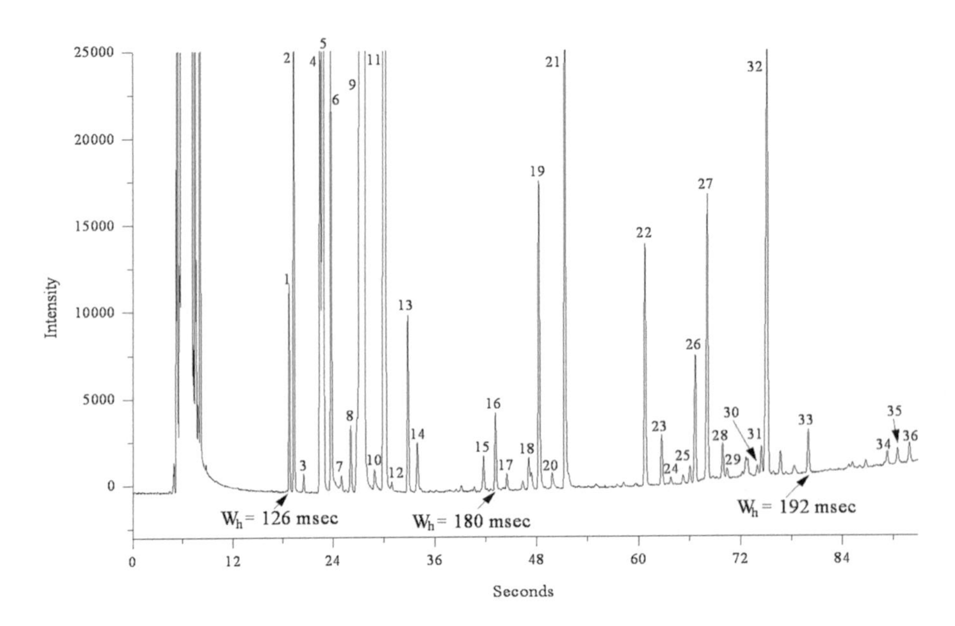

FIGURE 7.1 Fast GC analysis of a lime essential oil on a 5 m × 5 mm (0.05 μm film thickness) capillary column, applying fast temperature programming. The peak widths of three components are marked to provide an illustration of the high efficiency of the column, even under extreme operating conditions. (Mondello, L. et al.: Ultra-fast essential oil characterization by capillary GC on a 50 μm ID column, *J. Sep. Sci.*, 27: 699–702. 2004. Copyright Wiley-VCH Verlag GmbH & Co. KGaA. Reproduced with permission.)

(5 m × 0.25 mm ID) (Bicchi et al., 2001). The latter oil has also been analyzed through fast HS-SPME-GC on a narrow-bore column (Rubiolo et al., 2006). Fast and very fast GC analyses on narrow-bore columns have also been carried out on patchouli and peppermint oils (Proot et al., 1986). Recently, a rapid plant volatile screening using HS-SPME and person-portable GC-MS (PGC) was carried out. The PGC was equipped with a low thermal mass (LTM) narrow bore (0.1 mm ID) column and a toroidal ion trap-mass spectrometer (IT-MS). Fast GC analysis allowed near-real-time measurement of leaf volatiles released from the different plants under study (Wong et al., 2019).

7.3.3 Gas Chromatography-Olfactometry for the Assessment of Odor-Active Components of Essential Oils

The discriminatory capacity of the mammalian olfactory system is such that thousands of volatile chemicals are perceived as having distinct odors. It is accepted that the sensation of odor is triggered by highly complex mixtures of volatile molecules, mostly hydrophobic, and usually occurring in trace level concentrations (ppm or ppb). These volatiles interact with odorant receptors of the olfactive epithelium located in the nasal cavity. Once the receptor is activated, a cascade of events is triggered to transform the chemical-structural information contained in the odorous stimulus into a membrane potential (Firestein, 1992; Firestein, 2001), which is projected to the olfactory bulb and then transported to higher regions of the brain (Malnic et al., 1999) where the translation occurs.

It is known that only a small portion of the large number of volatiles occurring in a fragrant matrix contributes to its overall perceived odor (Grosch, 1994; van Ruth, 2001). Further, these molecules do not contribute equally to the overall flavor profile of a sample; hence, a large GC peak area, generated by a chemical detector, does not necessarily correspond to high odor intensities because of differences in intensity/concentration relationships.

The description of a gas chromatograph modified for the sniffing of its effluent to determine volatile odor activity, was first published in 1964 by Fuller, Steltenkamp and Tisserand (Fuller et al., 1964).

In general, gas chromatography-olfactometry (GC-O) is carried out on a standard GC that has been equipped with a sniffing port, also denominated olfactometry port or transfer line, in substitution of, or in addition to, the conventional detector. When a flame FID or a mass spectrometer is also used, the analytical column effluent is split and transferred to the conventional detector and to the human nose. GC-O was a breakthrough in analytical aroma research, enabling the differentiation of a multitude of volatiles, previously separated by GC into odor-active and non-odor-active, related to their existing concentrations in the matrix under investigation. Moreover, it is a unique analytical technique which associates the resolution power of capillary GC with the selectivity and sensitivity of the human nose.

GC-O systems are often used in addition to either an FID or a mass spectrometer. With regard to detectors, splitting column flow between the olfactory port and a mass spectral detector provides simultaneous identification of odor-active compounds. Another variation is to use an in-line, non-destructive detector such as a TCD (Nishimura, 1995) or a photo-ionization detector (PID) (Wright, 1997). Especially when working with GC-O systems equipped with detectors that do not provide structural information, retention indexes are commonly associated to odor description supporting peak assignment.

Over the last decades, GC-O has been extensively used in essential oil analysis in combination with sophisticated olfactometric methods; the latter were developed to collect and process GC-O data and, hence, to estimate the sensory contribution of a single odor-active compound. The odor-active compounds of essential oils extracted from citrus fruits (*Citrus* sp.), such as orange, lime, and lemon, were among the first character impact compounds identified by flavor chemists (McGorrin, 2002).

GC-O methods are commonly classified in four categories: dilution, time-intensity, detection frequency, and posterior intensity methods. Dilution analysis, the most applied method, is based on successive dilutions of an aroma extract until no odor is perceived by the panelists. This procedure, usually performed by a reduced number of assessors, is mainly represented by CHARM (combined hedonic aroma response method) (Acree et al., 1984), developed by Acree and co-workers, and AEDA (aroma extraction dilution analysis), first presented by Ullrich and Grosch (Ullrich and Grosch, 1987). The former method has been applied to the investigation of two sweet orange oils from different varieties, one Florida Valencia and the other Brazilian Pera (Gaffney et al., 1996). The intensities and qualities of their odor-active components were assessed. CHARM results indicated for both the oils that the most odor-active compounds are associated with the polar fraction compounds: straight chain aldehydes (C_8-C_{14}), β-sinensal, and linalool presented the major CHARM responses. On the other hand, AEDA has been used to investigate the odor-active compounds responsible for the characteristic odors of juzu oil (*Citrus junos* Sieb. ex Tanaka) (Song et al., 2000a) and dadai (*Citrus aurantium* L. var. *cyathifera* Y. Tanaka) (Song et al., 2000b) cold-pressed essential oils.

Time-intensity methods, such as OSME (Greek word for odor), are based on the immediate recording of the intensity as a function of time by moving the cursor of a variable resistor (McDaniel et al., 1992). An interesting application of the time intensity approach was demonstrated for cold-pressed grapefruit oil (Lin and Rouseff, 2001), in which 38 odor-active compounds were detected and, among these, 22 were considered as aroma impact compounds. A comparison between the grapefruit oil GC chromatogram and the corresponding time-intensity aromagram for that sample is shown in Figure 7.2.

A further approach, the detection frequency method (Linssen et al., 1993; Pollien et al., 1997), uses the number of evaluators detecting an odor-active compound in the GC effluent as a measure of its intensity. This GC-O method is performed with a panel composed of numerous and untrained evaluators; 8–10 assessors are a good agreement between low variation of the results and analysis time. It must be added that the results attained are not based on real intensities and are limited by the scale of measurement. An application of the detection frequency method was reported for the evaluation of leaf- and wood-derived essential oils of Brazilian rosewood (*Aniba rosaeodora* Ducke) essential oils by means of enantioselective (Es)-GC-O analyses (d'Acampora Zellner et al., 2006).

Another GC-O technique, the posterior intensity method (Casimir and Whitfield, 1978), proposes the measurement of a compound odor intensity and its posterior scoring on a previously determined scale. This posterior registration of the perceived intensity may cause a considerable variance between

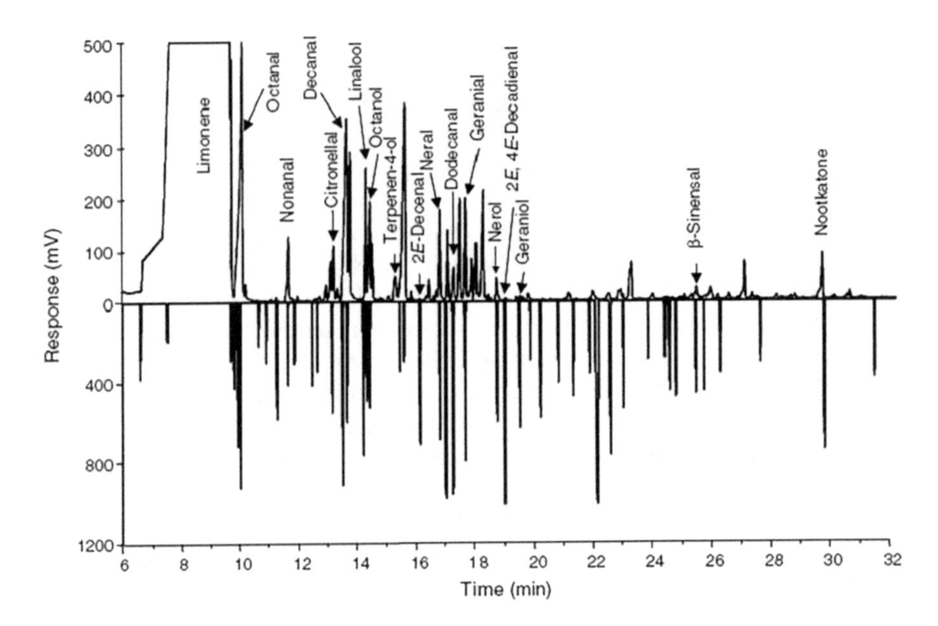

FIGURE 7.2 GC-FID chromatogram with some components identified by means of MS (top) and a time-intensity aromagram of grapefruit oil (bottom). The separation was performed on a polyethylene glycol column (30 m × 0.32 mm ID, 0.25 μm film thickness). (Lin, J., and R. L. Rouseff: Characterization of aroma-impact compounds in cold-pressed grapefruit oil using time–intensity GC–olfactometry and GC–MS. *Flav. Fragr. J.*, 16: 457–463. 2001. Copyright Wiley-VCH Verlag GmbH & Co. KGaA. Reproduced with permission.)

assessors. The attained results may generally be well correlated with detection frequency method results and, to a lesser extent, with dilution methods. In the abovementioned research performed on the essential oils of Brazilian rosewood, this method was also used to give complementary information on the intensity of the linalool enantiomers (d'Acampora Zellner et al., 2006).

Other GC-O applications are also reported in the literature using the so-called peak-to-odor impression correlation, method in which the olfactive quality of an odor-active compound perceived by a panelist is described. The odor-active compounds of the essential oils of black pepper (*Piper nigrum*) and Ashanti pepper (*Piper guineense*) were assessed applying the afore cited correlation method (Jirovetz et al., 2002). The odor profile of the essential oils of leaves and flowers of *Hyptis pectinata* (L.) Poit. were also investigated by using the peak-to-odor impression correlation (Jirovetz and Ngassoum, 1999).

The choice of the GC-O method is of extreme importance for the correct characterization of a matrix, since the application of different methods to an identical real sample can distinctly select and rank the odor-active compounds according to their odor potency and/or intensity. Commonly, detection frequency and posterior intensity methods result in similar odor intensity/concentration relationships, while dilution analyses investigate and attribute odor potencies. The number of studies on the characterization of odor-active components in plant and food samples and essential oils is enormous, and it is impossible to report every single reference. Recent review articles on this specific topic are available (Chin and Marriott, 2015; Giungato et al., 2018).

7.3.4 Gas Chromatographic Enantiomeric Characterization of Essential Oils

Capillary gas chromatography is currently the method of choice for enantiomer analysis of essential oils, and enantioselective GC (Es-GC) has become an essential tool for stereochemical analysis, mainly after the introduction of cyclodextrin (CD) derivatives as chiral stationary phases (CSPs) in 1983 by Sybilska and Koscielski, at the University of Warsaw, for packed columns (Sybilska and Koscielski, 1983), and applied to capillary columns in the same decade (Schurig and Nowotny, 1988;

König, 1991). Moreover, Nowotny et al. first proposed diluting CD derivatives in moderately polar polysiloxane (OV-1701) phases to provide them with good chromatographic properties and a wider range of operative temperatures (Nowotny et al., 1989).

The advantage on the application of Es-GC lies mainly in its high separation efficiency and sensitivity, simple detection, unusually high precision and reproducibility, and also the need for a small amount of sample. Moreover, its main use is related with the characterization of the enantiomeric composition and the determination of the enantiomeric excess (*ee*) and/or ratio (ER) of chiral research chemicals, intermediates, metabolites, flavors and fragrances, drugs, pesticides, fungicides, herbicides, pheromones, etc. Information on *ee* or ER is of great importance to characterize natural flavor and fragrance materials, such as essential oils, since the obtained values are useful tools, or even "fingerprints," for the determination of their quality, applied extraction technique, geographic origin, biogenesis, and also authenticity (Dugo et al., 1992).

A great number of essential oils have already been investigated by means of Es-GC using distinct CSPs; unfortunately, a universal chiral selector with widespread potential for enantiomer separation is not available, and thus effective optical separation of all chiral compounds present in a matrix may be unachievable on a single chiral column. In 1997, Bicchi et al. (1997) reported the use of columns that addressed particular chiral separations, noting that certain CSPs preferentially resolved certain enantiomers. Thus, a 2,3-di-*O*-ethyl-6-*O*-*tert*-butyldimethylsilyl-β-CD on polymethylphenylsiloxane (PS086) phase allowed the characterization of lavender and citrus oils containing linalyl oxides, linalool, linalyl acetate, borneol, bornyl acetate, α-terpineol, and *cis*- and *trans*-nerolidol. On the other hand, peppermint oil was better analyzed by using a 2,3-di-*O*-methyl-6-*O*-*tert*-butyldimethylsilyl-β-CD on PS086 phase, and especially for α- and β-pinene, limonene, menthone, isomenthone, menthol, isomenthol, pulegone, and methyl acetate. König (König, 1998) performed an exhaustive investigation of the stereochemical correlations of terpenoids, concluding that when using a heptakis (6-*O*-methyl-2,3-di-*O*-penthyl)-β-CD and octakis (6-*O*-methyl-2,3-di-*O*-penthyl)-γ-CD in polysiloxane, the presence of both enantiomers of a single compound is common for monoterpenes, less common for sesquiterpenes, and never observed for diterpenes.

Substantial improvements in chiral separations have been extensively published in the field of chromatography. At present, over 100 stationary phases with immobilized chiral selectors are available (Curtis and Williams, 2001), presenting increased stability and extended lifetime. It can be affirmed that enantioselective chromatography has now reached a high degree of sophistication. To better characterize an essential oil, it is advisable to perform Es-GC analysis on at least two, or better three, columns coated with different CD derivatives. This procedure enables the separation of more than 85% of the racemates that commonly occur in these matrices (Bicchi et al., 1999), while the reversal of enantiomer elution order can take place in several cases. The analyst must be aware of some practical aspects prior to an Es-GC analysis: as is well accepted, variations in linear velocity can affect the separation of enantiomeric pairs; resolution (R_S) can be improved by optimizing the gas linear velocity, a factor of high importance in cases of difficult enantiomer separation. Satisfactory resolution requires $R_S \geq 1$, and baseline resolution is obtained when $R_S \geq 1.5$ (Lee et al., 1984). Resolution can be further improved by applying slow temperature ramp rates (1°C to 2°C/min is frequently suggested). Moreover, according to the chiral stationary phase used, the initial GC oven temperature can affect peak width; initial temperatures of 35°C–40°C are recommended for the most column types. Furthermore, attention should be devoted to the column sample capacity, which varies with different compounds; overloading results in broad tailing peaks and reduced enantiomeric resolution. The troublesome separation and identification of enantiomers due to the fact that each chiral molecule splits into two chromatographic signals for each existing stereochemical center is also worthy of note. As a consequence, the increase in complexity of certain regions of the chromatogram may lead to imprecise *ee* and/or ER values. In terms of retention time repeatability, and also reproducibility, it can be affirmed that good results are being achieved with commercially available chiral columns. A recent study on the role of substituents in cyclodextrin derivatives for enantioselective gas chromatography analysis of chiral terpenoids in the essential oil of *Mentha spicata* has been published (Pragadheesh et al., 2015). The study gives details on the most appropriate CD derivative for the separation of carvone enantiomers and thus the developed method could be ideal for the analysis of all limonene-carvone reach spearmint essential oils.

The retention-index calculation of optically active compounds can be considered as a troublesome issue due to complex inclusion complexation retention mechanisms on CD stationary phases; if a homologous series, such as the *n*-alkanes, is used, the hydrocarbons randomly occupy positions in the chiral cavities. As a consequence, *n*-alkanes can be considered as unsuitable for retention index determinations. Nevertheless, other reference series can be employed on CD stationary phases, such as linear chain fatty acid methyl esters (FAMEs) and fatty acid ethyl esters (FAEEs). However, retention indices are seldom reported for optically active compounds, and publications refer to retention times rather than indices.

The innovations in Es-GC analysis have not only concerned the development and applications of distinct CSPs, but also the development of distinct enantioselective analytical techniques, such as Es-GC-mass spectrometry (Es-GC-MS), Es-GC-olfactometry (Es-GC-O), enantioselective multidimensional gas chromatography (Es-MDGC), Es-MDGC-MS, Es-GC hyphenated to isotopic ratio mass spectrometry (Es-GC-IRMS), Es-MDGC-IRMS.

It is obvious that an enantioselective separation in combination with MS detection presents the additional advantage of qualitative information. Notwithstanding, a difficulty often encountered is that related to peak assignment, due to the similar fragmentation pattern of isomers. The reliability of Es-GC-MS results can be increased by using an effective tool, namely retention indices. It can be assumed that in the enantioselective recognition of optically active isomers in essential oils, mass spectra can be exploited to locate the two enantiomers in the chromatogram, and the linear retention index (LRI) when possible, enables their identification (Rubiolo et al., 2007).

In addition, the well-known property of odor-activity recognized for several isomers, can be assessed by means of Es-GC-MS-O and can represent an outstanding tool for precise enantiomer characterization (see Figure 7.3). As demonstrated by Mosandl and his group (Lehmann et al., 1995), Es-GC-O is a valid tool for the simultaneous stereo-differentiation and olfactive evaluation of the volatile optically active components present in essential oils. It is worthwhile to point out

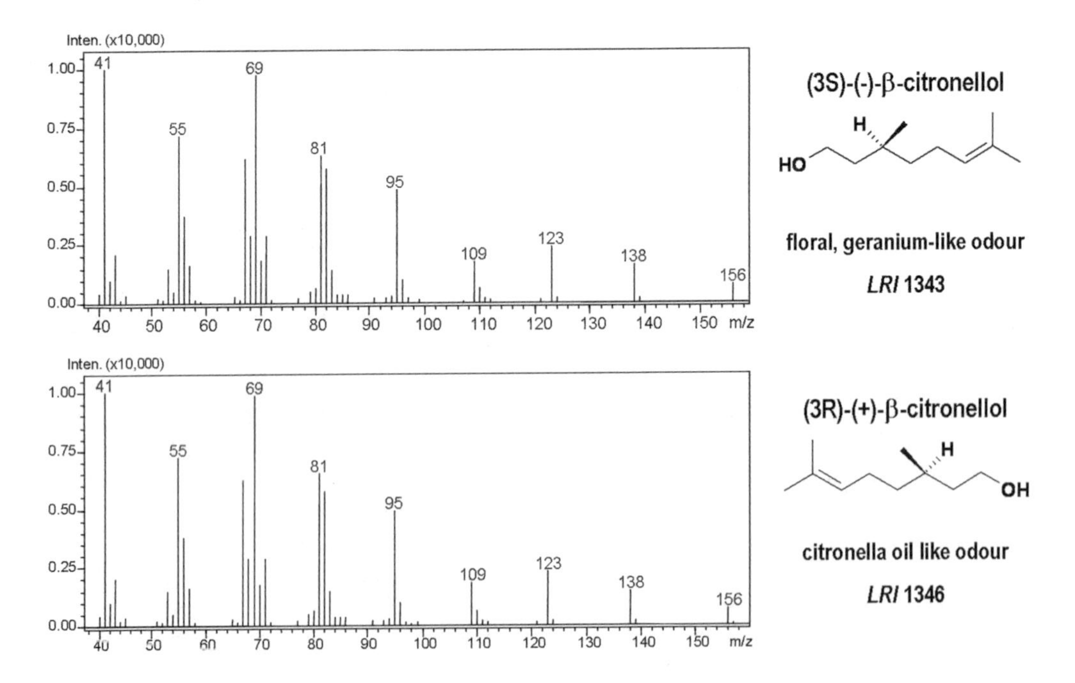

FIGURE 7.3 Representation of the mass spectra similarity of β-citronellol enantiomers.

that the preponderance of one of the enantiomers, defined by the enantiomeric excess, results in a characteristic aroma (Boelens et al., 1993) and is of great importance for the olfactive characterization of the sample.

A recent review article on the authenticity of essential oils summarizes analytical methodologies adopted for the detection of adulteration, including, among others, chiral gas chromatography and isotope-ratio mass spectrometry (Do et al., 2015). Es-GC is a cheap, sensitive technique, but at the same time, no universal stationary phase is available.

7.3.5 Multidimensional Gas Chromatography (MDGC)

7.3.5.1 Multidimensional Gas Chromatographic Techniques (MDGC)

Conventional GC-FID and/or GC-MS are commonly employed for the assessment of essential oil quality, through the determination of the qualitative and quantitative profile and of the enantiomeric excess (ee) or ratio (ER) of volatile chiral compounds. In spite of the considerable advances made in instrumentation, the detection of all the constituents of an essential oil still represents an extremely difficult task. In fact, gas chromatograms relative to complex mixtures are frequently characterized by several overlapping compounds: well-known examples are octanal and α-phellandrene, as well as limonene and 1,8-cineole, coeluted on 5% diphenyl–95% dimethylpolysiloxane stationary phases. On the other hand, insufficient resolution is observed between citronellol and nerol or geraniol and linalyl acetate. Whereas, the overlapping of monoterpene alcohols and esters with sesquiterpene hydrocarbons is frequently reported on polyethylene glycol stationary phase (Mondello et al., 2010). Hence, the straightforward identification of a component in such mixtures may be a cumbersome challenge. The use of MDGC, especially in combination with a third MS dimension, greatly increases the analytical potential of the technique (Tranchida et al., 2012). In MDGC, key fractions of a sample are selected from the first column and re-injected onto a second one differing in selectivity, where ideally, they should be fully resolved thanks to an increased total peak capacity, calculated as the peak capacity of column 1 plus the peak capacity of column 2 (or more) multiplied for the number of heart-cuts selected. Deans-switch, twin-oven MDGC system are nowadays used in multiple-cut analysis, involving two chromatographic columns of different polarities, but generally of identical dimensions, with both GC systems commonly equipped with detectors. Furthermore, when heart-cut operations are not carried out, the primary column elutes normally in the first dimension (^{1}D) GC system, while heart-cut fractions are chromatographically resolved on the secondary column (Deans, 1968). MDGC is a useful approach for the fractionation of compounds of particular interest in a specific sample; one of its major application areas is chiral analysis, exploiting using a conventional column as ^{1}D and a chiral stationary phase in ^{2}D (Sciarrone et al., 2010a; Casilli et al., 2014; Wong et al., 2015; Hong et al., 2017; Bonaccorsi et al., 2012; Dugo et al., 2012; Schipilliti et al., 2012). Other arrangements have consisted of a ^{2}D achiral column (Do et al., 2015; Sciarrone et al., 2011; Sciarrone et al., 2010b) coupled to an isotopic ratio MS (IRMS) for evaluation of genuineness and geographical origin (Juchelka et al., 1998), as well as for preparative purposes (Sciarrone et al., 2015a,b, 2012, 2013, 2014, 2016, 2017, 2019; Pantò et al., 2015; Ochiai and Sasamoto, 2011; Ball et al., 2012). Enantio-multidimensional GC (eMDGC), has been reported for the first time by Schomburg et al., achieving the chiral separation of four racemic menthol isomers and the two α-ionol isomers in the form of their isopropylurethane derivatives by using XE-60-(S)-valines–(S)-α-phenylethylamide as the ^{2}D stationary phase (Schomburg et al., 1984). The employment of eMDGC in essential oil analysis has become a very useful tool to overcome the drawbacks of eGC (Elbashir et al., 2018). However, heart-cut systems have a common drawback, related to the design of the transfer system employed, such that each cut causes a slight variation in the backpressure of the first dimension, leading to shifted retention times of the next eluting components. A fully automated, multidimensional, double-oven MDGC system able to overcome this issue has been presented by Mondello et al. (Mondello et al., 2008): the main feature of such a transfer device consisted in the capability to generate the same ^{1}D backpressure in both standby and cut conditions, allowing the transfer of substantially unlimited heart-cut fractions (Figure 7.4).

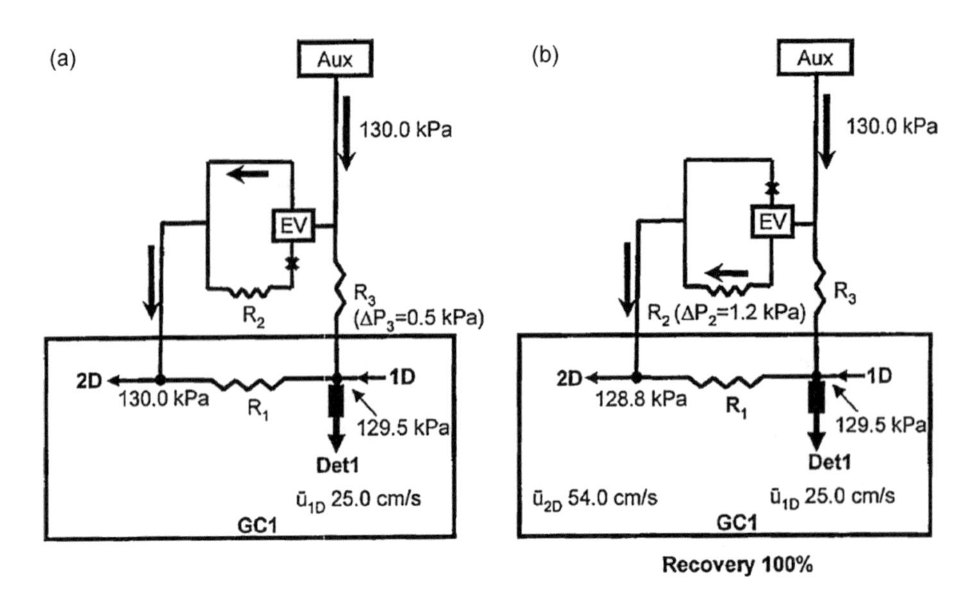

FIGURE 7.4 Schemes of the MDGC interface in (A) standby and (B) cut mode oil. (Reprinted from *J. Chromatogr. A*, 1217. Sciarrone, D. et al., Evaluation of tea tree oil quality and ascaridole: A deep study by means of chiral and multi heart-cuts multidimensional gas chromatography system coupled to MS Detection. 6422–6427. Copyright 2010a, with permission from Elsevier.)

In both operational modes, a constant pressure is supplied from an external pressure control unit to a fused-silica restrictor (R_3) and to a two-way solenoid valve (V), both located outside the GC oven. The two metal branches connected to the valve are characterized by another fused-silica restrictor (R_2) at one side, producing a pressure drop slightly higher than that generated by R_3 ($\Delta P_2 > \Delta P_3$), while no restriction is present on the other side. In the standby mode, a lower pressure is generated on the side of the first dimension respect to the second dimension branch, where it passes through the solenoid valve without any restriction. In such a configuration, analytes eluting from the first column are directed to FID1 simulating a monodimensional analysis. Once the solenoid valve is activated, the transfer device passes to the cutting mode: the pressure on the first-dimension side of the interface remains unchanged, while the pressure on the second-dimension side becomes lower than that of the first dimension. Under such conditions, the primary-column eluate is free to reach the second capillary. The same system has been used for the determination of the enantiomeric distribution of monoterpene hydrocarbons and monoterpene alcohols in the essential oils of mandarin (Sciarrone et al., 2010a), pistachio (Lo Presti et al., 2008), and sandalwood (Sciarrone et al., 2011). The quali/quantitative assessment with respect to normative limits was also carried out in tea tree (Sciarrone et al., 2010b). In Figure 7.5, the monodimensional chiral analysis of mandarin oil is compared to the second dimension chiral separation achieved with the MDGC approach (Sciarrone et al., 2010a). Well-resolved peaks of components present in large amounts, and also of the minor compounds, were attained through the partial transfer of the major concentrated components. For most of the enantiomers, the ee values achieved in eGC and eMDGC were in good agreement, with slight variations probably due to the use of different instruments and columns. However, considerable variations were observed for camphene and linalool, due to the presence of coelutions, while the ee values of camphor and citronellal were calculated only in the multidimensional applications. The latter measurements were achieved thanks to the possibility to overload the primary apolar column by injection of a higher sample amount. Next, only selected slices of the more abundant components were transferred to the secondary chiral column, while in the case of trace components, the entire peaks were selected for the heart-cut. An MDGC approach with a polar and chiral column set, coupled to mass spectrometry and olfactometry port, was applied for the analysis of citrus hybrid peel extract (Casilli et al., 2014). An MDGC-MS method for the

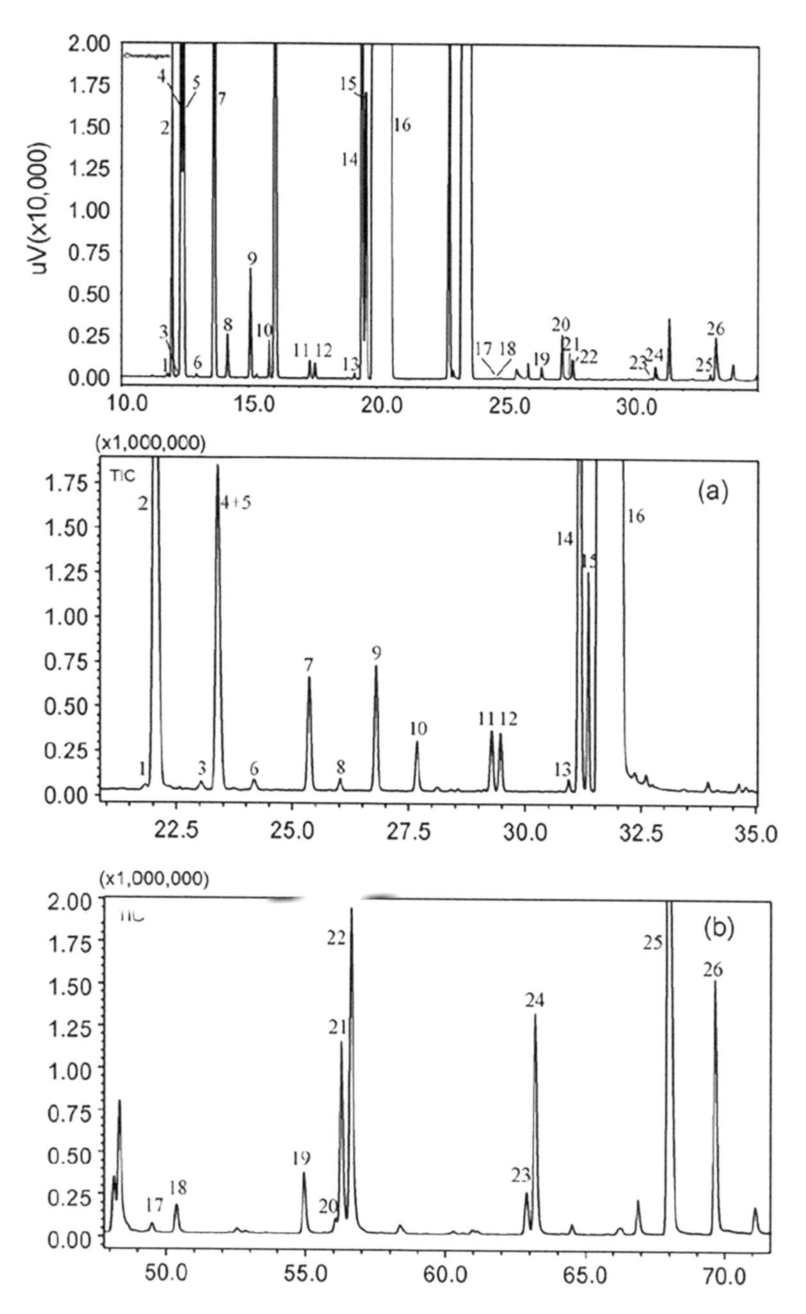

FIGURE 7.5 Comparison of eGC and eMDGC chiral analysis of mandarin essential oil. (Reprinted from *J. Chromatogr. A*, 1217. Sciarrone, D. et al., Thorough evaluation of the validity of conventional enantio-gas chromatography in the analysis of volatile chiral compounds in mandarin essential oil: A comparative investigation with multidimensional gas chromatography. 1101–1105. Copyright 2010b, with permission from Elsevier.)

determination of the enantiomeric distribution of selected compounds in lime oils was also reported, namely α-thujene, camphene, β-pinene, sabinene, α-phellandrene, β-phellandrene, limonene, linalool, terpinen-4-ol, and α-terpineol (Bonaccorsi et al., 2012). Gas chromatography–combustion–isotope ratio mass spectrometry and enantioselective multidimensional gas chromatography were exploited complementarily for petitgrain oils landmark and lime essential oil assessment (Schipilliti et al., 2012; Bonaccorsi et al., 2012).

7.3.5.2 Multidimensional Preparative Gas Chromatography

When the isolation of specific volatile components is required because of the lack of standard components or to allow the structure elucidation of unknown molecules, preparative gas chromatography could be considered as an option. Nevertheless, monodimensional prep-GC is commonly considered a time-consuming and "tricky" technique since the isolation of volatile components from the gas stream is not an easy task. Furthermore, a series of shortcomings related to the unsatisfactory degree of purity of the fraction collected have prevented the widespread use of this technique. When the sample is characterized by a high level of complexity (as is often the case of natural samples), the presence of coeluting interferences is very common and requires particular precautions, which affect the time requested for the collection of a certain amount. The first issue is related to the reduced amount of neat, or diluted sample, which can be injected in each run, in order not to impair the efficiency of the system by overloading the GC column. Moreover, the collection of sufficient amounts of chemicals for further evaluations would require a long time. The use of conventional 0.25 mm ID columns requires hundreds of injections to afford the collection of a sufficient amount (μg-mg) due to their reduced sample capacity. For such a reason, mega-bore columns (0.32–0.53 mm ID) are generally employed in prep-GC systems since they are able to manage higher sample amounts, even at the price of a reduced efficiency compared to micro-bore capillaries (0.1–0.25 mm ID). Thanks to the improvement in separation power, heart-cutting multidimensional preparative chromatography (MDGC-prep) has demonstrated effectiveness to allow the collection of target compounds from complex samples (Sciarrone et al., 2015a). With such an approach, by the coupling of mega-bore columns with different selectivity in the heart-cut mode, it is possible to achieve a separation system able to afford higher purity of the compounds purification, with higher sample amounts, and finally with higher throughput. Furthermore, the multidimensional approach can be effectively used to obtain increased productivity, thanks to the possibility to overload the first chromatographic dimension by the injection of a high sample amount. After a ^{1}D pre-purification step, the ^{2}D column (or even ^{3}D) would allow to purify the fraction transferred before the collection, reducing the number of runs required. Recently, MDGC systems have been successfully coupled to prep-GC (Ochiai and Sasamoto, 2011; Ball et al., 2012; Sciarrone et al., 2012, 2013, 2014, 2016, 2017, 2019). Sciarrone et al. reported a three-dimensional MDGC-prep system, exploiting an apolar ^{1}D column, a medium-polarity ^{2}D wax column, and an ionic liquid stationary phase as ^{3}D, with similar polarity but different in selectivity, each one equipped with a three-restrictors Deans switch device (Figure 7.6). The collection of components present in a range between 10% and 30% in the sample was successfully reported, allowing for further investigations of unknown molecules by NMR, MS, and fourier transform infrared (FTIR). The same group described a further improvement of the system, enabling the collection of low-level components in the sample (1%–10%), obtained including an LC pre-separation step before the three-dimensional GC system. Due to the higher sample capacity of the packed LC column, it was possible to inject 10–20-fold sample amounts than a direct-GC injection (actually limited by the injector liner internal volume). The use of an LC system operated in normal phase allowed the isolation of fractions of eluate containing the concentrated components of interest. The latter were on-line transferred to the GC large-volume injector in order to evaporate the mobile phase (organic) before the start of the GC run. The procedure demonstrated the capability to reduce to about 1/10–1/20 the time to collect a milligram of components.

7.3.5.3 Multidimensional Gas Chromatography Coupled to Isotope Ratio Mass Spectrometry (MDGC-IRMS)

Enantioselectivity and isotope discrimination during biosynthesis are well recognized principles of authenticity evaluation. As above discussed, enantioselective multidimensional gas chromatography (MDGC) with the combination of a non-chiral pre-separation column and a chiral main column has proved to be the method of choice for the enantioselective analysis of chiral compounds from complex matrices. In the same way, isotope ratio mass spectrometry (IRMS) is well established and

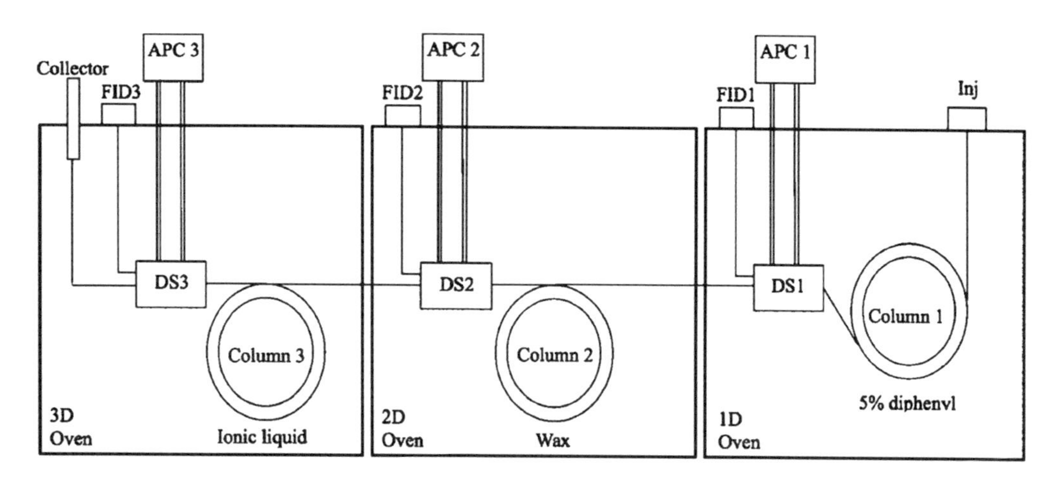

FIGURE 7.6 Triple Deans switch multidimensional prep-GC system scheme oil. (Reprinted from *Anal. Chim. Acta*, 785. Sciarrone, D. et al., Rapid collection and identification of a novel component from *Clausena lansium* Skeels leaves by means of three-dimensional preparative gas chromatography and nuclear magnetic resonance/infrared/mass spectrometric analysis. 119–125. Copyright 2013, with permission from Elsevier.)

widespread in different research fields for the determination of the isotopic ratio of GC-separated compounds. In the case of carbon, the molecules are oxidized to CO_2 into a combustion furnace (C) and transferred to Faraday cups for *m/z* 44, 45, and 46, by using a magnetic sector. The results can be often related to specific geographical origin and easily discriminate natural from synthetic compounds, allowing highlighting of possible adulterations or frauds, especially in the field of essential oils (Do et al., 2015). Obviously, calibration is of great importance in isotope ratio measurements. Two calibration procedures are commonly used in GC-C-IRMS measurements, the first dealing with the use of standard compounds co-injected and combusted along with the analytes. The second procedure uses standard CO_2 gas directly introduced into the ion source. The complexity of natural samples often precludes to find even a few seconds retention time window free of peaks for the selected standards, which in turn must be selected for each matrix. Thus the use of standard CO_2 gas, directly introduced into the ion source before and after each sample, has become the most practiced procedure in such cases. Monodimensional gas chromatographic separation is often unable to afford the complete separation of minor components from complex matrices, while MDGC, thanks to its higher peak capacity, gives the highest performance in direct sample cleanup. When using an IRMS system coupled to GC, in mono- or multidimensional systems, the chromatographic conditions have to be carefully optimized in order to avoid a multitude of erroneous judgements (Nitz et al., 1992). Unlike what happens when using a GC hyphenated to a quadrupole mass spectrometer (GC-QMS), where coeluting compounds can still be identified and quantified in the selected ion monitoring (SIM) or extracted ion mode, excellent chromatographic resolution is mandatory for GC-C-IRMS analysis, since the isotopic ratio of the CO_2 generated from coeluted peaks cannot be treated in the same way. Baseline-to-baseline integration over an entire peak is moreover required for accurate measurement of its isotopic composition, since column fractionation would retain in a slightly different way the heavier isotopes, eluted in the first part of the peak, and the lighter ones, eluted in the tail (Ricci et al., 1994). As a consequence, the use of MDGC approaches coupled to IRMS detection appears to be the best choice. The hyphenation of a multidimensional GC system in heart-cut mode to IRMS is not a novel concept even if, probably due to technical difficulties, only few papers have appeared in the literature (Juchelka et al., 1998). An MDGC-C-IRMS system was presented for the origin-specific analysis of flavor and fragrance compounds, exploiting the $^{13}C/^{12}C$ ratio of the detected enantiomers. The authors used a double-oven system equipped with a "live switching" coupling piece (live T-piece) as the transfer system between an apolar-chiral column set. The results showed no significant variations in the $^{13}C/^{12}C$ ratio of the standard injected, measured in mono- and multidimensional conditions. Nowadays, the use of Deans switch devices based on a pressure balancing effect, generally affected by retention time shifts of the next eluting components (see Section 7.3.5.1), exposes to the risk of isotopic fractionation resulting from the incomplete transfer of peaks during consecutive heart-cuts. The recent Deans switch systems, able to maintain a constant ^{1}D backpressure, represent a new frontier to be explored for MDGC-C-IRMS users, and the development of new applications in the field of essential oils analysis is foreseeable.

7.3.6 Comprehensive Two-Dimensional Gas Chromatography and Multidimensional Liquid-Gas Chromatography

7.3.6.1 Analysis of Essential Oils through Comprehensive Two-Dimensional Gas Chromatography (GC × GC)

Comprehensive two-dimensional gas chromatography (GC × GC) was first described in 1991 (Liu and Phillips, 1991), with it taking nearly 10 years to be applied to the analysis of an essential oil (Dimandja et al., 2000). Since that first research, various investigations have been published, also regarding complex samples related to essential oils, such as perfumes. In many cases, the use of such a powerful GC methodology appeared to be justified, while for others its use seemed to be an excess in analytical terms.

Briefly, a GC × GC analysis is performed by using two capillary columns, each coated with a different type of stationary phase. Most commonly, a conventional column (e.g., 30 m × 0.25 mm ID) with a non-polar stationary phase is used in the first dimension (^{1}D), while a short mid-polarity micro-bore one (e.g., 1–2 m × 0.10 mm ID) is used in the second dimension (^{2}D). A special transfer device, defined modulator, is situated between the two analytical dimensions and works in a continuous manner throughout the analysis. Cryogenic modulators (a form of thermal modulation) are by far the most popular GC × GC transfer devices. The modulator first accumulates fractions from the ^{1}D and then releases them onto the ^{2}D, where rapid separations occur. The accumulation-release process occurs in a sequential manner, with its time duration defined as the modulation period, usually lasting in the range of 4–8 s. The modulation period is also equivalent to the timeframe of each rapid separation on the second column. A raw GC × GC chromatogram is nothing more than a continuous stream of rapid ^{2}D separations, positioned side-by-side along the retention-time axis. For the fundamental objective of two-dimensional (2D) visualization, peak integration, and identification, the use of dedicated software is necessary. Compared to GC, thermal modulation GC × GC offers enhanced resolving power, selectivity, and sensitivity. The reader is directed to the literature for more details on this powerful GC technology (Marriot et al., 2012).

In general, an essential oil can be subjected to a GC analysis for basically three analytical objectives: (a) untargeted analysis, (b) pre-targeted analysis and (c) fingerprinting. A series of examples will be shown, illustrating the need (or not) of comprehensive 2D GC in such types of investigations.

7.3.6.1.1 Untargeted Analyses

The untargeted analysis of an essential oil, by using a conventional low-polarity or mid-polarity capillary column, is performed for objectives which may relate to pure research curiosity, or to define the state of preservation. The use of mass spectrometry (MS) is mandatory for peak identification, advisably with the support of linear retention index (LRI) data. The use of relative retention information is important because many terpenes are characterized by a similar structure, a factor that hinders reliable MS differentiation. The generation of two sets of LRI data, by using two parallel columns with a chemically different stationary phase, has been reported for a long time (Bicchi et al., 1988). Such an option, in itself, highlights the requirement of an increased GC resolving power in untargeted analyses.

Gas chromatography-olfactometry (GC-O) can also be considered a method used for untargeted analyses, even though the information provided is different with respect to GC-MS. For such a reason, it is a common occurrence to split the column effluent between a mass spectrometer and an olfactometry port to attain complementary types of information (Delahunty et al., 2006). Again, the non-sufficient resolving power of a single GC column can hinder the attainment of reliable GC-O-MS essential oil information, calling for the use of a multidimensional (MD) GC method. However, the enhanced separation power of heart-cutting MDGC, and not GC × GC, is the preferred choice in such instances (Delahunty et al., 2006).

In general, the separation power of GC × GC is most impressive in untargeted investigations, with essential oils making no exception. The first GC × GC separations involving essential oils appeared in 2000: Dimandja et al. used a thermal sweeper (a now obsolete modulator), and flame ionization detection (FID), for the analysis of peppermint and spearmint oils (Dimandja et al., 2000). A rather curious set of columns was used (the first column was shorter than the second): non-polar, 1 m × 0.1 mm ID × 3.5 μm d_f + medium-polarity 2 m × 0.1 mm ID × 0.5 μm d_f. A side-by-side comparison was made with GC-MS results, attained by using a non-polar 30 m column: three times more peaks were detected in the GC × GC-FID analysis of peppermint oil compared to GC-MS (89 vs. 30). Even though not fully expressed, the analytical power of GC × GC was evident in this initial untargeted experiment.

In a recent study, Filippi et al. carried out an in-depth qualitative and quantitative investigation of four samples of vetiver oil (Haiti, Indonesia, Brazil, La Réunion) by using GC × GC combined with single-quadrupole (Q) MS and GC × GC-FID, as well as GC-QMS (Filippi et al., 2013). A total ion current (TIC) GC-QMS chromatogram, relative to Haitian vetiver oil, is shown in Figure 7.7A, with a total number of 101 peaks detected with a signal-to-noise ratio (*s/n*) > 10. The same sample was analyzed by using GC × GC-QMS, with the number of detected peaks (*s/n* > 10) reaching 535, even though the injector split ratio was doubled compared to the GC-QMS analysis (Figure 7.7B). It is noteworthy, however, that only 117 compounds were identified and then subjected to quantification by using GC × GC-FID. As a consequence, there was still a great deal of concealed information, despite the high number of detected essential oil constituents. In general, such a misbalance is often observed when using GC × GC-MS.

The research of Filippi et al. highlighted the enhanced separation power and sensitivity of GC × GC. An additional emphasized advantage was the formation of elution patterns of specific compound classes (e.g., ketones, hydrocarbons, alcohols, aldehydes, etc.), increasing the reliability of identification. Finally, as will be discussed later, GC × GC generates a true analytical fingerprint possessing a wealth of exploitable information.

The on-line multidimensional combination of high-performance liquid chromatography and gas chromatography (LC-GC) has a proven usefulness within the context of essential oil analysis (Marriott et al., 2001). The LC dimension is used to perform the separation of chemical classes of compounds on the basis of polarity. Fractions with a reduced complexity can then be subjected to GC analysis, usually by injecting large sample volumes.

The LC-GC concept was extended to the off-line combination of LC and GC × GC-QMS (LC//GC × GC-QMS) by Tranchida et al. (Tranchida et al., 2013b), who exploited the first analytical dimension for the separation of hydrocarbons and O-containing compounds in sweet orange and bergamot essential oils. In the bergamot oil analysis, for instance, approximately 350 peaks were detected with 195 analytes tentatively identified (53 hydrocarbons and 142 oxygenated constituents), against 64 compounds (31 hydrocarbons and 33 oxygenated constituents) using direct GC-QMS.

Flow modulation (FM) is an interesting alternative to cryogenic modulation due to the limited hardware and operational costs, and to the capability to modulate with efficiency compounds with a high volatility (e.g., $\leq C_6$), as well as a low one (e.g., $\geq C_{30}$). On the other hand, FM cannot match the higher peak capacity and sensitivity provided by cryogenic modulation (CM). A further recognized disadvantage of FM is the presumed requirement of very high ^{2}D gas flows (e.g., ≥20 mL min^{-1}), a problematic issue when using mass spectrometry (Edwards et al., 2011). However, it has also been demonstrated that through fine method optimization, it is possible to carry out FM GC × GC applications at gas flows within the range 6–8 mL min^{-1} (Tranchida et al., 2014). Several commercially available MS systems can handle such gas flows.

Costa et al. used FM GC × GC-QMS for the untargeted analysis of *Artemisia arborescens* L. leaf essential oil (Costa et al., 2016). The column set consisted of a low-polarity 20 m × 0.18 mm

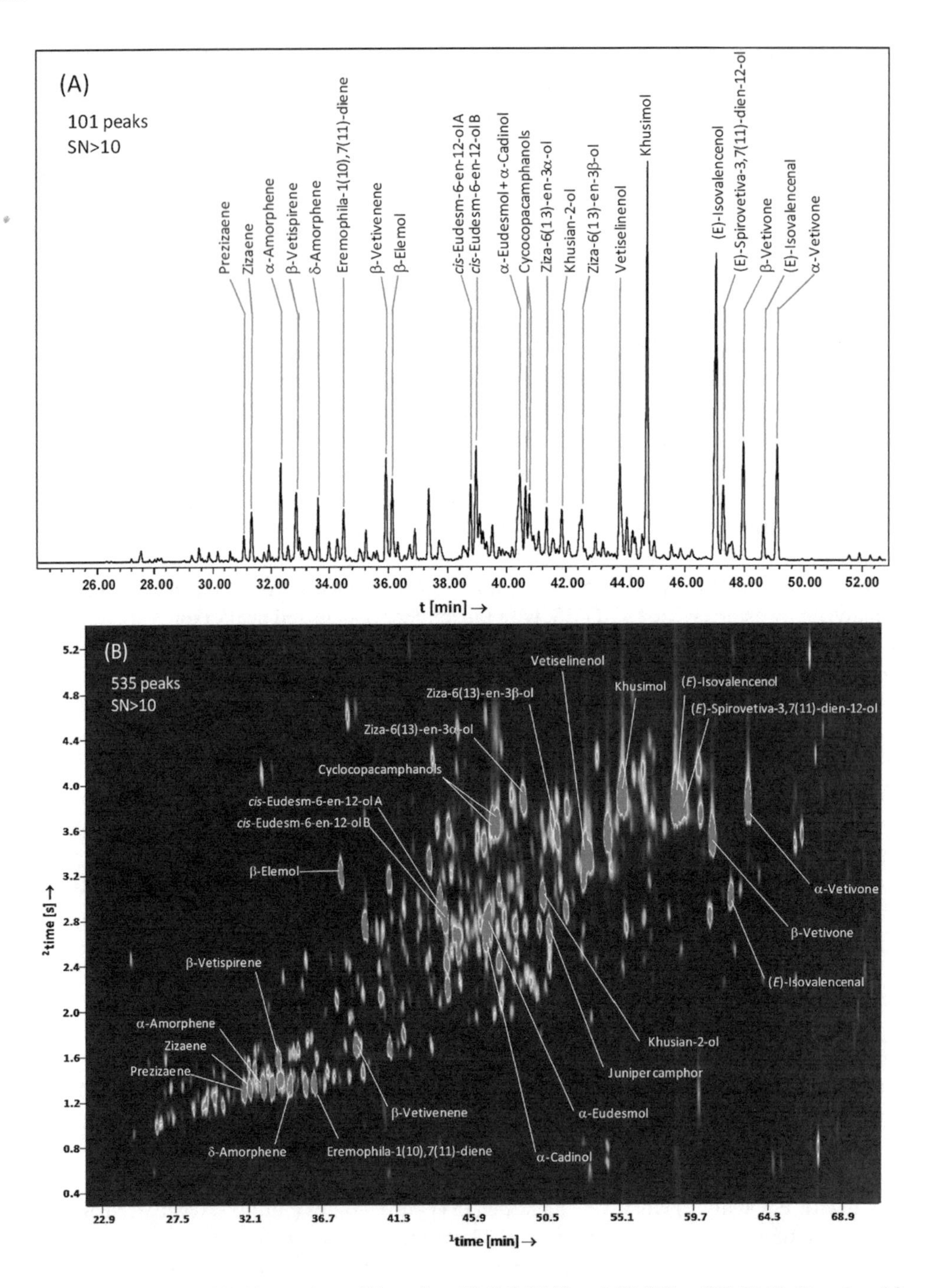

FIGURE 7.7 Analysis of Haitian vetiver oil by using (A) GC-QMS and (B) GC × GC-QMS. (Reprinted from *J. Chromatogr. A.*, 1288. Filippi, J. J. et al., Qualitative and quantitative analysis of vetiver essential oils by comprehensive two-dimensional gas chromatography and comprehensive two-dimensional gas chromatography/mass spectrometry. 127–148. Copyright 2013, with permission from Elsevier.)

ID × 0.18 μm d_f ¹D and a mid-polarity 10 m × 0.32 mm ID × 0.20 μm d_f ²D. The use of rather long wide bore ²D columns in FM analyses is an obliged choice, related not only to the high gas flows (in this case 8 mL min^{-1}), but also to the wide chromatography bands released from the modulator. In fact, differently from CM no solute re-concentration effects occur. Figure 7.8 illustrates three

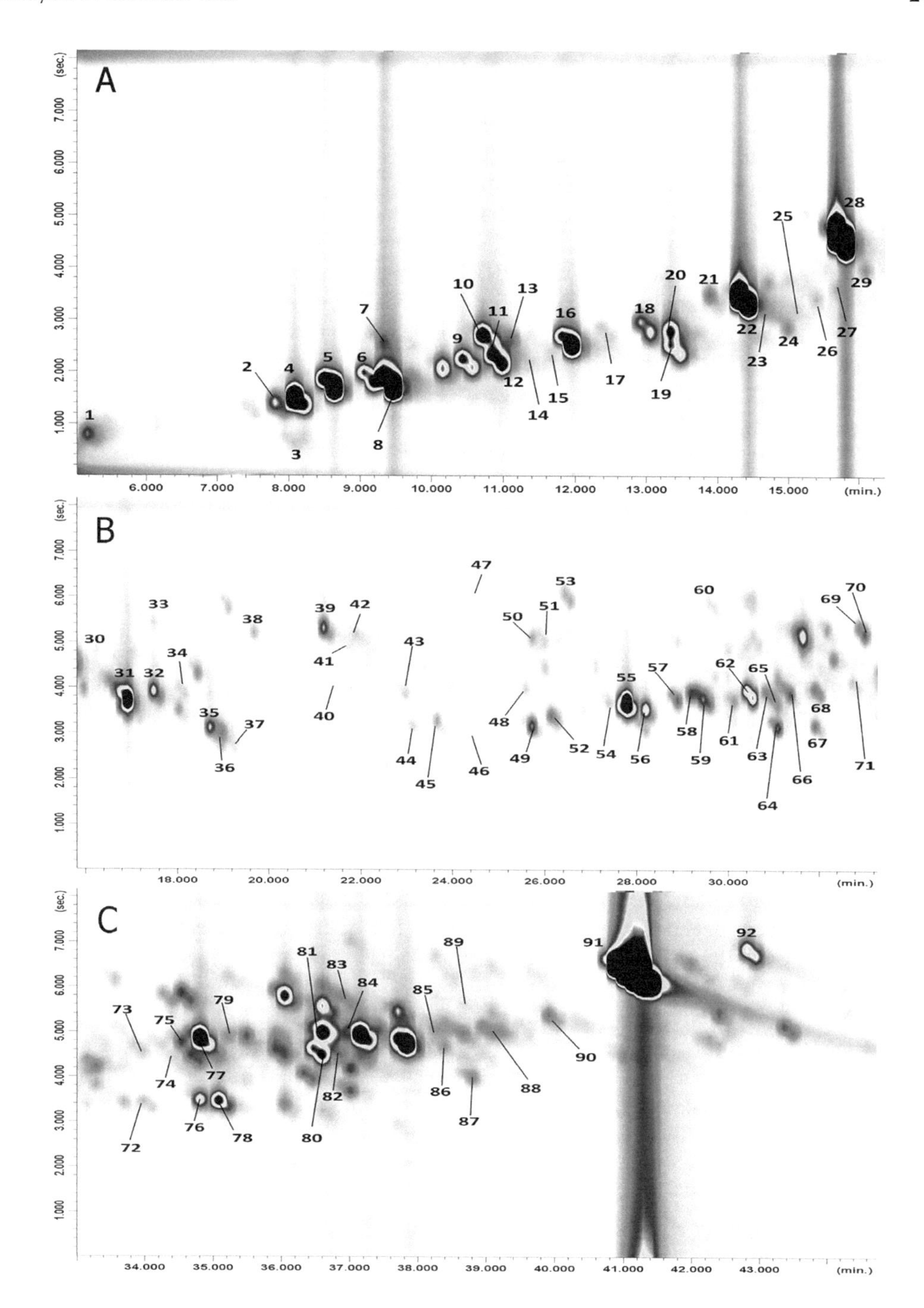

FIGURE 7.8 Three chromatogram expansions (A-C) relative to the FM GC × GC-QMS analysis of *Artemisia arborescens* L. leaf essential oil. (Reprinted from *J. Pharm. Biomed. Anal.*, 117. Costa, R. et al., 2016. Phytochemical screening of *Artemisia arborescens* L. by means of advanced chromatographic techniques for identification of health-promoting compounds. 499–509. Copyright 2016, with permission from Elsevier.)

2D expansions, extrapolated from the entire chromatogram, relative to the FM GC × GC-QMS analysis of *Artemisia arborescens* L. leaf essential oil. Mass spectral database searching, with the support of LRI information, enabled the tentative identification of 92 compounds (about two times more were detected).

7.3.6.1.2 Pre-targeted Analyses

With regard to the pre-targeted analysis of essential oils, there are a series of different possibilities, with the use of GC × GC often questionable. For example, pre-targeted essential oil analysis can relate to the determination of phytosanitary compounds, plasticizers, antioxidants, etc. (Di Bella et al., 2006; Garland et al., 1999). It is the specificity and sensitivity of MS which play a fundamental role in such an applicational field. Even so, the use of GC × GC has been described: Tranchida et al. exploited FM GC × GC, hyphenated with triple quadrupole (QqQ) MS, for the determination of the antioxidants butylated hydroxyanisole (BHA) and butylated hydroxytoluene (BHT), as well as the fungicide *o*-phenylphenol (OPP), in mandarin essential oil (Tranchida et al., 2013a). The QqQ mass spectrometer was a novel device and was capable of rapid switching between the untargeted (scan) and targeted (MSMS) modes during the same analysis. The main aim of the study was to evaluate the QqQMS instrument under challenging analytical conditions. However, the capability of the mass spectrometer to provide two distinct types of information, practically simultaneously, does potentially justify the use of GC × GC in this type of application.

A further example of pre-targeted analysis, very useful to confirm (or not) genuineness, is that involving the determination of the enantiomeric composition of specific chiral constituents. Target analyte elution on shape-selective GC stationary phases (enantio-GC), in particular those containing cyclodextrins, represents in this case the most important step of the analytical chain. Enantio-GC essential oil separations are usually carried out by using a single column or, more preferably, on the second dimension of a heart-cutting MDGC instrument (Marriot et al., 2001). Even though both enantio-GC × GC and GC × enantio-GC studies have been reported (Shellie et al., 2001; Shellie and Marriott, 2002), such approaches have not found popularity. On one hand, the modulation process can degrade ^{1}D enantiomer resolution, while on the other, column lengths in the ^{2}D are usually too short to provide satisfactory degrees of separation.

The combination of either conventional GC or heart-cutting MDGC, with isotope ratio mass spectrometry (IRMS), has a proven usefulness for the measurement of target analyte isotopic ratios (Schipilliti et al., 2011; Juchelka et al., 1998). Again, MDGC-IRMS represents the preferred choice for the pre-targeted analysis of essential oils, also because the IRMS combustion chamber generally leads to extensive band broadening, causing a resolution loss. Isotopic ratio determinations are performed to guarantee authenticity or to define a geographical origin. To the best of the present authors' knowledge, the use of GC × GC-IRMS for the analysis of essential oils has not yet been reported.

7.3.6.1.3 Fingerprinting

Fingerprinting is a form of untargeted analysis, inasmuch that interest is devoted to the chromatography profile, as a whole or in part, rather than to peak-to-peak assignment. The concept of fingerprinting is two-dimensional in its nature and so it matches well with GC × GC. In fact, an intimate bond can be created between a highly detailed 2D chromatogram and a specific type of sample, preferably by using adequate data treatment software tools. Fine and reliable GC × GC between-sample differentiation can be achieved on the basis of a specific industrial process (e.g., type of roasting), of geographical origin, or of sensorial properties (e.g., oxidized vegetable oils) (Cordero et al., 2010; Purcaro et al., 2014).

The GC × GC fingerprinting of essential oils has been rarely reported; in one of such rare instances, Cordero et al. used FM GC × GC with dual detection (FID/QMS) for the fingerprinting of four vetiver oils, namely Java, Brazil, Bourbon, and Haiti (Cordero et al., 2015). Prior to visualization, thresholds were set for *s/n* values (>25) and peak volumes (>30,000). Figure 7.9A illustrates the result for the Haitian sample, highlighting also the formation of chemical-class elution patterns. Following alignment, a comparative visualization was made between the Bourbon and Haitian vetiver oils. Regions with a red and green color highlight positive and negative detector response differences, respectively, for the Haitian oil (Figure 7.9B). A further software function allowed a more clear between-oil differentiation: 315 peaks were aligned, with those characterized by the largest peak volume difference (CV% >50%) pinpointed by a yellow circle (Figure 7.9C).

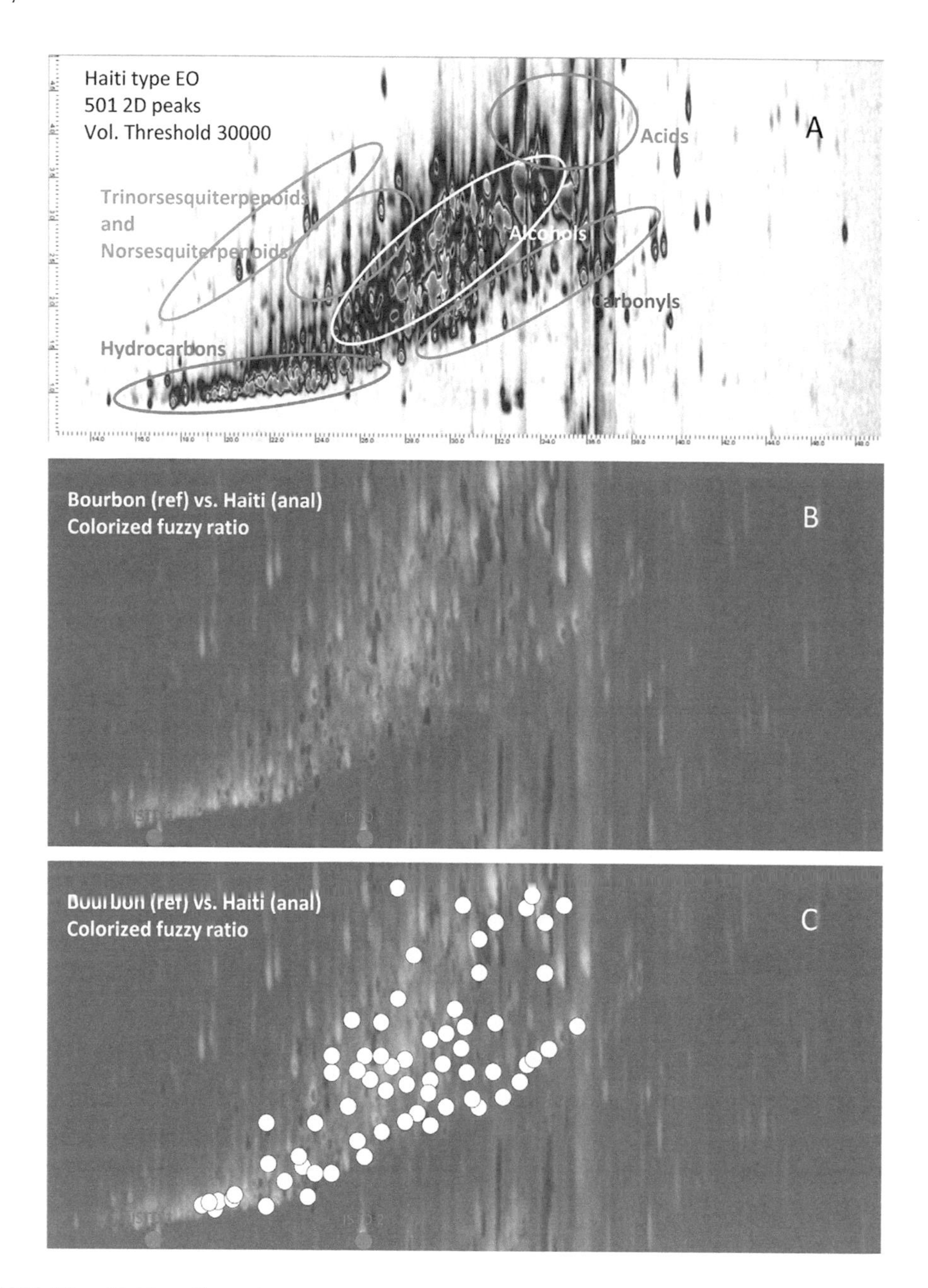

FIGURE 7.9 (A) Two-dimensional chromatogram of Haitian vetiver oil; (B) visual comparison between Bourbon and Haitian vetiver oils; yellow circles indicate compounds with the largest quantitative differences between the four vetiver oils. (Reprinted from *J. Chromatogr. A*, 1417. Cordero, C. et al., Potential of the reversed-inject differential flow modulator for comprehensive two-dimensional gas chromatography in the quantitative profiling and fingerprinting of essential oils of different complexity. 79–95. Copyright 2015, with permission from Elsevier.)

For instance, β-vetivenene, a sesquiterpene hydrocarbon, was contained in the highest amounts in the Java and Brazilian oils, while the Haitian oil was characterized, by far, by the highest concentration of (E)-isovalencenol, an oxygenated sesquiterpene. The availability of such detailed information, accompanied by the use of suitable statistical processes, results in a powerful sample classification approach.

7.3.6.2 On-Line and Off-Line Coupled Liquid Chromatography-Gas Chromatography

The analysis of very complex mixtures is often troublesome due to the variety of chemical classes to which the sample components belong to and to their wide range of concentrations. As such, several compounds cannot be resolved by monodimensional GC. In this respect, less complex and more homogeneous mixtures can be attained by the fractionation of the matrix by means of LC prior to GC separation. The multidimensional LC-GC approach, combines the selectivity of the LC separation with the high efficiency and sensitivity of GC separation, enabling the separation of compounds with similar physicochemical properties in samples characterized by a great number of chemical classes.

For the highly volatile components, commonly present in essential oils, the most adequate transfer technique is partially concurrent eluent evaporation (Grob, 1987). In this technique, a retention gap is installed, followed by a few meters of pre-column and the analytical capillary GC column, both with an identical stationary phase, for the separation of the LC fractionated components. A vapor exit is placed between the pre-column and the analytical column, allowing partial evaporation of the solvent. Hence, column and detector overloading are avoided. This transfer technique can be applied to the analysis of GC components with a boiling point of at least 50°C higher than the solvent. The interface proposed by Grob was improved in 2009 by Biedermann and Grob, introducing the so-called Y-interface in order to reduce the memory effect (Biedermann and Grob, 2009). In the last few years, different transfer devices have been developed to transfer the fractions to the GC system (Purcaro et al., 2013; Biedermann and Grob, 2012).

The composition of citrus essential oils has been elucidated by means of LC-GC, and the development of new methods for the study of single classes of components has been well reported. The aldehyde composition in sweet orange oil has been investigated (Mondello et al., 1994a), as also industrial citrus oil mono- and sesquiterpene hydrocarbons (Mondello et al., 1995) and the enantiomeric distribution of monoterpene alcohols in lemon, mandarin, sweet orange, and bitter orange oils (Dugo et al., 1994a,b).

The hyphenation of LC-GC systems to mass spectrometric detectors has also been reported for the analyses of neroli (Mondello et al., 1994b), bitter and sweet oranges, lemon, and petitgrain mandarin oils (Mondello et al., 1996). It has to be highlighted, that the preliminary LC separation, which reduces mutual component interference, greatly simplifies MS identification.

7.3.7 Liquid Chromatography, Liquid Chromatography Hyphenated to Mass Spectrometry, and Multidimensional Liquid Chromatographic Technique in the Analysis of Essential Oils

Essential oils generally contain only volatile components, since their preparation is performed by steam distillation. Citrus oils, extracted by cold-pressing machines, are an exception, containing more than 200 volatile and non-volatile components. The non-volatile fraction, constituting 1%–10% of the oil, is represented mainly by hydrocarbons, fatty acids, sterols, carotenoids, waxes, and oxygen heterocyclic compounds (coumarins, psoralens, and polymethoxylated flavones) (Di Giacomo and Mincione, 1994). The latter can have an important role in the identification of a cold-pressed oil and in the control of both quality and authenticity (Di Giacomo and Mincione, 1994; Dugo et al., 1996; Mc Hale and Sheridan, 1989; Mc Hale and Sheridan, 1988), when the information attained by means of GC is not sufficient. The analysis of these compounds is usually performed by means of liquid chromatography (LC), also referred to as high performance liquid chromatography (HPLC), in normal-phase (NP-HPLC) or reversed-phase (RP-HPLC) applications (Dugo and Di Giacomo, 2002).

Nowadays, the analysis of oxygen heterocyclic compounds plays a fundamental role in the characterization and quality control of citrus essential oils; therefore, rapid analytical strategies have been developed. Fast methods were validated in order to perform a rapid quality control and determine furocoumarins in essences, according to the European Regulation EC No. 1223/2009, which sets the maximum amount of furocoumarins admitted in finished cosmetic products containing citrus essential oils (EC No. 1223/2009). New HPLC methods are based on the use of short columns, which ensure fast separations and guarantee a high separation efficiency thanks to the small dimension of the packaging particles. Fused-core columns are the most employed; the superficially porous structure of their particles reduce the analytes path, and consequently the molecule dispersion and retention times. These columns were employed to study the chemical composition variability of citrus essences depending on seasoning and production procedure (Dugo et al., 2010; Dugo et al., 2011; Dugo et al., 2012; Russo et al., 2012). A satisfactory separation of 38 targets was achieved by using an Ascentis Express C18 column (50 × 4.6 mm ID with particle size of 2.7 μm) in ten minutes, or also in shorter time (three minutes), maintaining a good compromise between chromatographic resolution and analysis time in RP-HPLC-photo dyode array detector PDA Figure 7.10 (Russo et al., 2015). This method was applied to analyze seven citrus essential oils (lime, lemon, bergamot, bitter orange, grapefruit, mandarin, and sweet orange). The method developed was further improved by combining the use of LRIs (using alkyl aryl ketones) and UV spectral libraries for reliable identification of oxygen heterocyclic components in citrus essential oils (Arigò et al., 2019). In some cases, the oxygen heterocyclic components proved to be more effective as markers in adulteration detection than the volatile components (Fan et al., 2015). More recent methods are based on HPLC systems coupled to MS detectors (Mehl et al., 2014; Mehl et al., 2015; Donato et al., 2014; Masson et al., 2016). This approach allows the rapid identification and characterization of oxygen heterocyclic compounds in citrus oils, and detection of some minor components for the first time in some oils was reported (Dugo et al., 1999). In addition, markers to monitor authenticity and possible adulteration were identified.

Furthermore, the high selectivity and sensitivity of MS detection, mainly atmospheric pressure chemical ionization (APCI)-positive ionization in *Multiple Reaction Monitoring* acquisition mode, allows the determination of oxygen heterocyclic compounds in finished cosmetic products, as imposed by the European regulation cited above. A new ionization approach was employed to analyze polymethoxyflavones isolated from the non-volatile residue of mandarin essential oil. This technique consists of a nano-LC-EI-MS configuration made of a nano prominence HPLC system coupled to a GC-QMS, working in electron impact ionization mode (Russo et al., 2016).

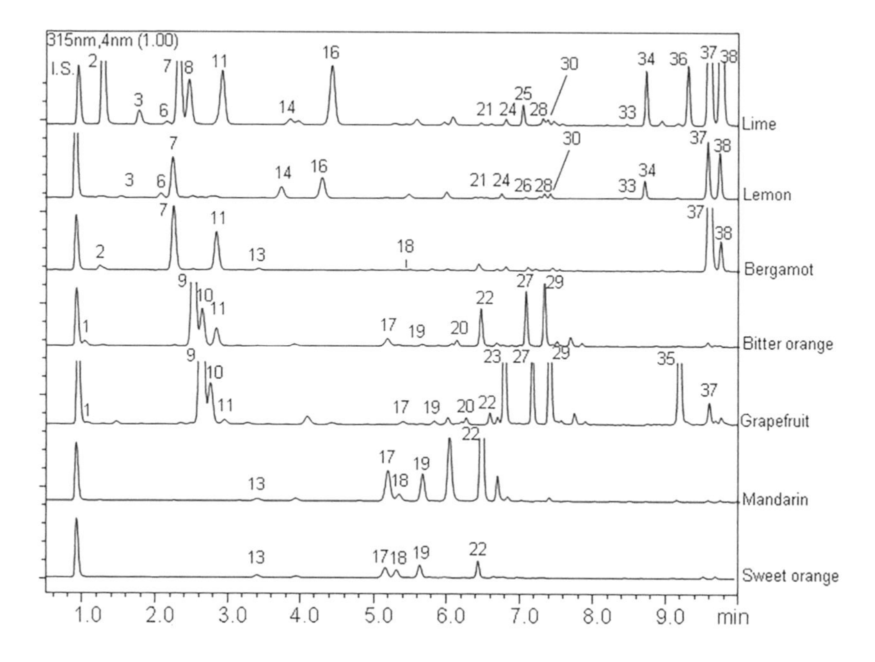

FIGURE 7.10 Reverse-phase high-performance liquid chromatography (RP-HPLC) chromatograms of oxygen heterocyclic compounds present in Citrus essential oils. (Reprinted with permission of Taylor & Francis Group, from Russo, M. et al., 2015. *J. Essent. Oil Res.*, 27: 307–315.)

Apart from citrus oils, other essential oils have also been analyzed by means of liquid chromatography, such as the blackcurrant bud (Píry and Pribela, 1994), *Nigella sativa* essential oil (Abdel-Reheem et al., 2014); cinnamon, caraway, and cardamom fruit oils (Porel et al., 2014); *Zanthoxylum zanthoxyloides* Lam. fruit oil (Tine et al., 2017), and mint (Hawryt et al., 2017) and rice leaf essential oils (Minh et al., 2019).

The most recent applications of liquid chromatography in the analysis of essential oils are mainly related to the isolation of interesting compounds (Russo et al., 2016; Gu et al., 2016; Padhan et al., 2017; Piochon-Gauthier et al., 2014), to the analysis of thermolabile components (Quassinti et al., 2014; Maggi et al., 2015), or to characterize the oils obtained through innovative and green techniques (Khajenoori et al., 2015).

Recent applications regard the determination of several pesticides in essential oils by HPLC coupled to tandem mass spectrometry (Fillatre et al., 2014; Fillatre et al., 2016).

High performance liquid chromatography (HPLC) has acquired a role of great importance in food analysis, as demonstrated by the wide variety of applications reported. Single LC column chromatographic processes have been widely applied for sample profile elucidation, providing satisfactory degrees of resolving power; however, whenever highly complex samples require analysis, a monodimensional HPLC system can prove to be inadequate. Moreover, peak overlapping may occur even in the case of relatively simple samples, containing components with similar properties.

The basic principles of MDGC are also valid for multidimensional LC (MDLC). The most common use of MDLC separation is the pre-treatment of a complex matrix in an off-line mode. The off-line approach is very easy, but presents several disadvantages: it is time-consuming, operationally intensive, and difficult to automate and to reproduce. Moreover, sample contamination or formation of artefacts can occur. On the other hand, on-line MDLC, though requiring specific interfaces, offers the advantages of ease of automation and greater reproducibility in a shorter analysis time. In the on-line heart-cutting system, the two columns are connected by means of an interface, usually a switching valve, which allows the transfer of fractions of the first column effluent onto the second column.

In contrast to comprehensive gas chromatography (GC × GC), the number of comprehensive liquid chromatography (LC × LC) applications reported in the literature are much less. It can be affirmed that LC × LC presents a greater flexibility when compared to GC × GC since the mobile phase composition can be adjusted in order to obtain enhanced resolution (Dugo et al., 2006a). Comprehensive HPLC systems, developed and applied to the analysis of food matrices, have employed the combination of either NP × RP or RP × RP separation modes. However, it is worthy of note that the two separation mechanisms exploited should be as orthogonal as possible, so that no or little correlation exists between the retention of compounds in both dimensions.

A typical comprehensive two-dimensional HPLC separation is attained through the connection of two columns by means of an interface (usually a high pressure switching valve), which entraps specific quantities of first-dimension eluate and directs it onto a secondary column. This means that the first column effluent is divided into “cuts” which are transferred continuously to the second dimension by the interface. The type of interface depends on the methods used, although multiport valve arrangements have been the most frequently employed.

Various comprehensive HPLC systems have been developed and proved to be effective for the separation of complex sample components, and in the resolution of a number of practical problems. In fact, the very different selectivities of the various LC modes enable the analysis of complex mixtures with minimal sample preparation. However, comprehensive HPLC techniques are complicated by the operational aspects of switching effectively from one operation step to another, by data acquisition and interpretation issues. Therefore, careful method optimization and several related practical aspects should be considered.

In the most common approach, a micro-bore LC column in the first and a conventional column in the second dimension are used. In this case, an 8-, 10-, or 12-port valve equipped with two sample loops (or trapping columns) is used as an interface. A further approach foresees the use of a conventional LC column in the first, and two conventional columns in the second, dimension. One or two valves that allow transfers from the first column to two parallel secondary columns (without the use of storage loops) are used as interface.

One of the best examples of the application of comprehensive NPLC × RPLC in essential oil analysis is represented by the analysis of oxygen heterocyclic components in cold-pressed lemon oil, by using normal-phase with a micro-bore silica column in the first dimension and a monolithic C18 column in the second dimension with a 10-port switching valve as interface (Dugo et al., 2004). In Figure 7.11, an NPLC × RPLC separation of the oxygen heterocyclic fraction of a lemon oil sample is presented. Oxygen heterocyclic components (coumarins, psoralens, and polymethoxylated flavones) represent the main part of the non-volatile fraction of cold-pressed citrus oils. Their structures and substituents have an important role in the characterization of these oils. Positive peak identification of these compounds was obtained by both the relative location of the peaks in the 2D plane, which varied in relation to their chemical structure and by characteristic UV spectra. In a later experiment, a similar setup was used for a citrus oil extract composed of lemon and orange oil (François et al., 2006). The main difference with respect to the earlier published work (Dugo et al., 2004) was the employment of a bonded phase (diol) column in the first dimension. Under optimized LC conditions, the high degree of orthogonality between the NP and RP systems tested resulted in increased ^{2}D peak capacity.

The NPLC × RPLC approach has been also developed for the analysis of carotenoids, pigments mainly distributed in plant-derived foods, especially in orange and mandarin essential oils (Dugo et al., 2006b; Dugo et al., 2008). In terms of structures, food carotenoids are polyene hydrocarbons, characterized by a C_{40} skeleton that derives from eight isoprene units. They present an extended conjugated double bond (DB) system that is responsible for the yellow, orange, or red colors in plants and are notable for their wide distribution, structural diversity, and various functions. Carotenoids are usually classified in two main groups: hydrocarbon carotenoids, known as carotenes (e.g., β-carotene and lycopene), and oxygenated carotenoids, known as xanthophylls (e.g., β-cryptoxanthin and lutein). The elucidation of carotenoid patterns is particularly challenging because of the complex composition of carotenoids in natural matrices, their great structural diversity, and their extreme instability. An innovative comprehensive dual gradient elution HPLC system was employed using

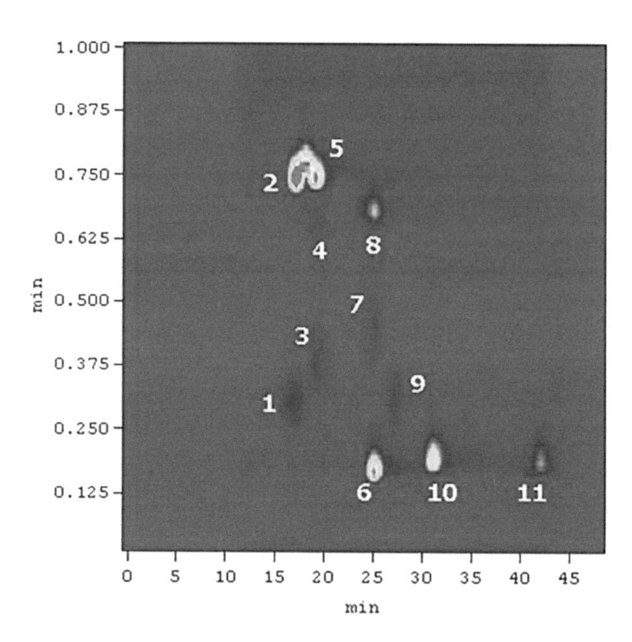

FIGURE 7.11 Comprehensive NP (adsorption)-LC × RPLC separation of the oxygen heterocyclic fraction of a lemon oil sample (for peak identification see Dugo et al., 2004). (Reprinted with permission from Dugo, P. et al., 2004. Comprehensive two-dimensional normal-phase (adsorption)-reversed-phase liquid chromatography. *Anal. Chem.*, 76: 2525–2530. Copyright 2004, American Chemical Society.)

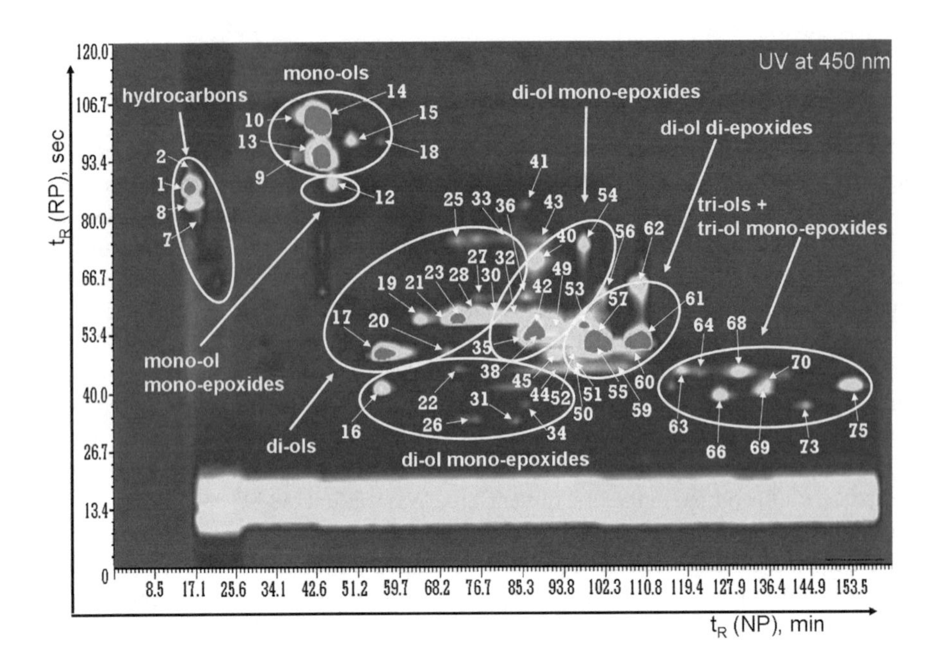

FIGURE 7.12 Contour plot of the comprehensive HPLC analyses of carotenoids present in sweet orange essential oil with peaks and compound classes indicated (for peak identification see Dugo et al., 2006b). (Reprinted with permission from Dugo, P. et al., 2006b. Elucidation of carotenoid patterns in citrus products by means of comprehensive normal-phase × reversed-phase liquid chromatography. *Anal. Chem.*, 78: 7743–7750. Copyright 2006b, American Chemical Society.)

an NPLC × RPLC setup, composed of silica and C18 columns in the first and second dimensions, respectively. Free carotenoids in orange essential oil and juice (after saponification), were identified by combining the two-dimensional retention data with UV-visible spectra (Dugo et al., 2006b) obtained by using a PDA detector (Figure 7.12). The carotenoid fraction of a saponified mandarin oil has been studied by means of comprehensive LC, in which a ^{1}D micro-bore silica column was applied for the determination of free carotenoids and a cyanopropyl column for the separation of esters; a monolithic column was used in the ^{2}D (Dugo et al., 2008). Detection was performed by connecting a photodiode array detection (DAD) system in parallel with an MS detection system operated in the APCI positive-ion mode. Thus, the identification of free carotenoid and carotenoid esters was carried out by combining the information provided by the DAD and MS systems and the peak positions in the 2D chromatograms.

7.4 GENERAL CONSIDERATIONS ON ESSENTIAL OIL ANALYSIS

As evidenced by the numerous techniques described in the present contribution, chromatography, especially gas chromatography, has evolved into the dominant method for essential oil analysis. This is to be expected because the complexity of the samples must be unraveled by some type of separation, before the sample constituents can be measured and characterized; in this respect, gas chromatography provides the greatest resolving power for most of these volatile mixtures.

In the past, a vast number of investigations have been carried out on essential oils, and many of these natural ingredients have been investigated following the introduction of GC-MS, which marked a real turning point in the study of volatile molecules. Es-GC also represented a landmark in the detection of adulterations, and in the cases where the latter technique could fail, gas chromatography hyphenated to isotope ratio mass spectrometry by means of a combustion interface (GC-IRMS) has proved to be a valuable method to evaluate the genuineness of natural product components. In addition, the introduction of GC-O was a breakthrough in analytical aroma research, enabling the differentiation of a multitude of volatiles in odor-active and non-odor-active compounds, according to their existing concentrations in a matrix. The investigation of the non-volatile fraction of essential oils, by means of LC and its related hyphenated techniques, contributed greatly toward the progress of the knowledge on essential oils. Many extraction techniques have also been developed, boosting the attained results. Moreover, the continuous demand for new synthetic compounds reproducing the sensations elicited by natural flavors triggered analytical investigations toward the attainment of information on scarcely known properties of well-known matrices.

REFERENCES

Abdel-Reheem, M. A. T., M. M. Oraby, and S. K. Alanazi, 2014. *Nigella sativa* essential oil constituents and its antimicrobial, cytotoxic and necrotic replies. *J. Pure Appl. Microbiol.*, 8: 389–398.

Acree, T. E., J. Barnard, and D. Cunningham, 1984. A procedure for the sensory analysis of gas chromatographic effluents. *Food Chem.*, 14: 273–286.

AFNOR- Association Française de Normalisation, Saint-Denis, http://www.afnor.org/portail.asp (accessed January 11, 2008).

Arigò, A., F. Rigano, G. Micalizzi, P. Dugo, and L. Mondello, 2019. Oxygen heterocyclic compound screening in *Citrus* essential oils by linear retention index approach applied to liquid chromatography coupled to photodiode array detector. *Flav. Fragr. J.*, 34: 349–364. in press.

Ball, G. I., L. Xu, A. P. Mc Nichol, and L. I. Aluwihare, 2012. A two-dimensional, heart-cutting preparative gas chromatograph facilitates highly resolved single-compound isolations with utility towards compound-specific natural abundance radiocarbon (14C) analyses. *J. Chromatogr. A.*, 1220: 122–131.

Beilstein, F., 1872. Ueber den Nachweis von Chlor, Brom und Jod in organischen Substanzen. *Ber. Dtscg. Chem. Ges.*, 5: 620–621.

Bicchi, C., C. Brunelli, C. Cordero, P. Rubiolo, M. Galli, and A. Sironi, 2004. Direct resistively heated column gas chromatography (ultrafast module-GC) for high-speed analysis of essential oils of differing complexities. *J. Chromatogr. A*, 1024: 195–207.

Bicchi, C., C. Brunelli, C. Cordero, P. Rubiolo, M. Galli, and A. Sironi, 2005. High-speed gas chromatography with direct resistively-heated column (ultra fast module-GC)-separation measure (*S*) and other chromatographic parameters under different analysis conditions for samples of different complexities and volatilities. *J. Chromatogr. A*, 1071: 3–12.

Bicchi, C., C. Brunelli, M. Galli, and A. Sironi, 2001. Conventional inner diameter short capillary columns: An approach to speeding up gas chromatographic analysis of medium complexity samples. *J. Chromatogr. A*, 931: 129–140.

Bicchi, C., A. D'Amato, V. Manzin, and P. Rubiolo, 1997. Cyclodextrin derivatives in GC separation of racemic mixtures of volatiles. Part XI. Some applications of cyclodextrin derivatives in GC Enantioseparations of essential oil components. *Flav. Fragr. J.*, 12: 55–61.

Bicchi, C., A. D'Amato, and P. Rubiolo, 1999. Cyclodextrin derivatives as chiral selectors for direct gas chromatographic separation of enantiomers in the essential oil, aroma and flavour fields. *J. Chromatogr. A*, 843: 99–121.

Bicchi, C., C. Frattini, G. M. Nano, and A. D'Amato, 1988. On column injection-dual channel analysis of essential oil. *J. High Resol. Chromatogr.*, 11: 56–60.

Biedermann, M., and K. Grob, 2009. Memory effects with the on-column interface for on-line coupled high performance liquid chromatography-gas chromatography: The Y-interface. *J. Chromatogr. A*, 1216: 8652–8658.

Biedermann, M., and K. Grob, 2012. On-line coupled high performance liquid chromatography-gas chromatography for the analysis of contamination by mineral oil. Part 1: Method of analysis. *J. Chromatogr. A*, 1255: 56–75.

Boelens, M. H., H. Boelens, and L. J. van Gemert, 1993. Sensory properties of optical isomers. *Perf. Flav.*, 18: 2–16.

Bonaccorsi, I., D. Sciarrone, L. Schipilliti, P. Dugo, L. Mondello, and G. Dugo, 2012. Multidimensional enantio gas chromatography/mass spectrometry and gas chromatography-combustion-isotopic ratio mass spectrometry for the authenticity assessment of lime essential oils (*C. aurantifolia* Swingle and *C. latifolia* Tanaka). *J. Chromatogr. A*, 1226: 87–95.

Bosart, L. W. 1937. Perfumery Essential Oil Record, 28, 95.

British Pharmacopoeia BP 2008, 2007. The Stationery Office (TSO), Norwich.
Casilli, A., E. Decorzant, A. Jaquier, and E. Delort, 2014. Multidimensional gas chromatography hyphenated to mass spectrometry and olfactometry for the volatile analysis of citrus hybrid peel extract. *J. Chromatogr. A*, 1373: 169–178.
Casimir, D. J., and F. B. Whitfield, 1978. Flavour impact values, a new concept for assigning numerical values for the potency of individual flavour components and their contribution to the overall flavour profile, *Ber. Int. Fruchtsaftunion.*, 15: 325.
Chin, S. T., and P. J. Marriott, 2015. Review of the role and methodology of high resolution approaches in aroma analysis. *Anal. Chim. Acta*, 854: 1–12.
Cordero, C., E. Liberto, C. Bicchi et al., 2010. Profiling food volatiles by comprehensive two-dimensional gas chromatography coupled with mass spectrometry: Advanced fingerprinting approaches for comparative analysis of the volatile fraction of roasted hazelnuts (*Corylus avellana* L.) from different origins. *J. Chromatogr. A*, 1217: 5848–5858.
Cordero, C., P. Rubiolo, L. Cobelli et al., 2015. Potential of the reversed-inject differential flow modulator for comprehensive two-dimensional gas chromatography in the quantitative profiling and fingerprinting of essential oils of different complexity. *J. Chromatogr. A*, 1417: 79–95.
Costa, R., S. Ragusa, M. Russo et al., 2016. Phytochemical screening of *Artemisia arborescens* L. by means of advanced chromatographic techniques for identification of health-promoting compounds. *J. Pharm. Biomed. Anal.*, 117: 499–509.
Cramers, C. A., H. G. Janssen, M. M. van Deursen, and P. A. Leclercq, 1999. High-speed gas chromatography: An overview of various concepts. *J. Chromatogr. A*, 856: 315–329.
Cramers, C. A., and P. A. Leclercq, 1999. Strategies for speed optimisation in gas chromatography: An overview, *J. Chromatogr. A*, 842: 3–13.
Croteau R., Kutchan, T. M., and N. G. Lewis, 2000. Natural Products (Secondary Metabolites). In *Biochemistry & Molecular Biology of Plants*, 1st ed., B. Buchanan, W. Gruissen, and R. Jones (eds.). New Jersey: ASPB and Wiley, chap. 24.
Curtis, T., and Williams, D. G., 2001. *Introduction to Perfumery*, 2nd ed. New York: Micelle Press, chap. 3.
d'Acampora Zellner, B., M. Lo Presti, L. E. Barata, P. Dugo, G. Dugo, and L. Mondello, 2006. Evaluation of leaf-derived extracts as an environmentally sustainable source of essential oils by using gas chromatography-mass spectrometry and enantioselective gas chromatography-olfactometry. *Anal. Chem.*, 78: 883–890.
Deans, D. R. 1968. A new technique for heart cutting in gas chromatography. *Chromatographia*, 1: 18–22.
Delahunty, C. M., G. Eyres, and J.-P. Dufour, 2006. Gas chromatography-olfactometry. *J. Sep. Sci.*, 29: 2107–2125.
Di Bella, G., L. Serrao, F. Salvo, V. Lo Turco, M. Croce, and G. Dugo, 2006. Pesticide and plasticizer residues in biological citrus essential oils from 2003–2004. *Flav. Fragr. J.* 21: 497–501.
Di Giacomo, A., and B. Mincione, 1994. *Gli Olii Essenziali Agrumari in Italia*. Baruffa editore.
Dimandja, J-M. D., S. B. Stanfill, J. Grainger, and D. G. Patterson, 2000. Application of comprehensive two-dimensional gas chromatography (GC × GC) to the qualitative analysis of essential oils. *J. High Resolut. Chromatogr.*, 23: 208–214.
Do, T. K. T., F. Hadjji-Minaglou, S. Antoniotti, and X. Fernandez, 2015. Authenticity of essential oils. *Trends Anal. Chem.*, 66: 146–157.
Donato, P., I. Bonaccorsi, M. Russo, and P. Dugo, 2014. Determination of new bioflavonoids in bergamot (*Citrus bergamia*) peel oil by liquid chromatography coupled to tandem ion trap-time of flight mass spectrometry. *Flav Fragr. J.*, 29, 131–136.
Dugo, G., Lamonica, G., Cotroneo, A., Stagno d'Alcontres, I., Verzera, A., Donato, M.G., Dugo, P., and G. Licandro, 1992. High resolution gas chromatography for detection of adulterations of citrus cold-pressed essential oils. *Perf. & Flav.*, 17: 57–74.
Dugo, G., I. Bonaccorsi, D. Sciarrone et al., 2012. Characterization of cold-pressed and processed bergamot oils by using GC-FID, GC-MS, GC-C-IRMS, enantio-GC, MDGC, HPLC and HPLC-MS-IT-TOF. *J. Essent. Oil Res.*, 24: 93–117.
Dugo G., and A. Di Giacomo, 2002. *Citrus: The Genus Citrus*, 1st ed. Book Chapter, Taylor & Francis Group.
Dugo, P., I. Bonaccorsi, C. Ragonese et al., 2011. Analytical characterization of mandarin (*Citrus* deliciosa Ten.) essential oil. *Flav. Frag. J.*, 26: 34–46.
Dugo, P., O. Favoino, R. Luppino, G. Dugo, and L. Mondello, 2004. Comprehensive two-dimensional normal-phase (adsorption)-reversed-phase liquid chromatography. *Anal. Chem.*, 76: 2525–2530.
Dugo, P., M. M. Fernandez, A. Cotroneo, and L. Mondello, 2006a. Optimization of a comprehensive two-dimensional normal-phase and reversed phase-liquid chromatography system. *J. Chromatogr. Sci.*, 44: 561–565.

Dugo, P., M. Herrero, T. Kumm, D. Giuffrida, G. Dugo, and L. Mondello, 2008. Comprehensive normal-phase × reversed-phase liquid chromatography coupled to photodiode array and mass spectrometry detectors for the analysis of free carotenoids and carotenoid esters from Mandarin. *J. Chromatogr. A*, 1189: 196–206.

Dugo, P., L. Mondello, E. Cogliandro, A. Verzera, and G. Dugo, 1996. On the genuineness of citrus essential oils. 51. Oxygen heterocyclic compounds of bitter orange oil (*Citrus aurantium* L.). *J. Agric. Food Chem.*, 44: 544–549.

Dugo, P., L. Mondello, E. Sebastiani, R. Ottanà, G. Errante, and G. Dugo, 1999. Identification of minor oxygen heterocyclic compounds of citrus essential oils by liquid chromatography-atmospheric pressure chemical ionisation mass spectrometry. *J. Liquid Chrom. & Rel. Tech.*, 22: 2991–3005.

Dugo, P., C. Ragonese, M. Russo et al., 2010. Sicilian lemon oil: Composition of volatile and oxygen heterocyclic fractions and enantiomeric distribution of volatile components. *J. Sep. Sci.*, 33: 3374–3385.

Dugo, P., V. Škeříková, T. Kumm et al., 2006b. Elucidation of carotenoid patterns in citrus products by means of comprehensive normal-phase × reversed-phase liquid chromatography. *Anal. Chem.*, 78: 7743–7750.

Dugo, G., A. Verzera, A. Cotroneo, I. Stagno d'Alcontres, L. Mondello, and K. D. Bartle, 1994a. Automated HPLC-HRGC: A powerful method for essential oil analysis. Part II. Determination of the enantiomeric distribution of Linalol in sweet orange, bitter orange and mandarin essential oils. *Flav. Fragr. J.*, 9: 99–104.

Dugo, G., A. Verzera, A. Trozzi, A. Cotroneo, L. Mondello, and K. D. Bartle, 1994b. Automated HPLC-HRGC: A powerful method for essential oils analysis. Part I. Investigation on enantiomeric distribution of monoterpene alcohols of lemon and mandarin essential oils. *Essenz. Deriv. Agrum.*, 44: 35–44.

Edwards, M., A. Mostafa, and T. Górecki, 2011. Modulation in comprehensive two-dimensional gas chromatography: 20 years of innovation. *Anal. Bioanal. Chem.*, 401: 2335–2349.

Elbashir, A. A., and H. Y. Aboul-Enein, 2018. Multidimensional gas chromatography for chiral analysis. *Critical Rev. Anal. Chem.*, 48: 416–427.

Ettre, L. S., and J. V. Hinshaw, 1993. *Basic relationships of Gas Chromatography*, 1st ed. Cleveland: Advanstar Data, chap. 4.

Regulation (ec) no 1223/2009 of the European Parliament and of the Council of 30 November 2009 on Cosmetic products (recast) (text with eea relevance). European Parliament, *Off. J. Eur. Commun.* L 342 (2009) 59.

European Pharmacopoeia, 6th ed. Strasburg: European Directorate for the Quality of Medicines & Healthcare (Edqm), 2007.

Falkenberg, M. B., R. I. Santos, and C. M. O. Simões, 1999. Introdução à Análise Fitoquímica. In *Farmacognosia: Da planta ao medicamento*, C. M. O. Simões et al. (ed.). 1999. Editora da UFSC and Editora da Universidade/UFRGS: Florianópolis, chap. 10.

Fan, H., Q. Wu, J. E. Simon, S. Lou, and C. Ho, 2015. Authenticity analysis of citrus essential oils by HPLC-UV-MS on oxygenated heterocyclic components. *J. Food and Drug Anal.*, 23: 30–39.

Filippi, J. J., E. Belhassen, N. Baldovini, H. Brevard, and U. J. Meierhenrich, 2013. Qualitative and quantitative analysis of vetiver essential oils by comprehensive two-dimensional gas chromatography and comprehensive two-dimensional gas chromatography/mass spectrometry. *J. Chromatogr. A.*, 1288: 127–148.

Fillâtre, Y., D. Rondeau, A. Daguin, and P. Y. Communal, 2016. A workflow for multiclass determination of 256 pesticides in essential oils by liquid chromatography tandem mass spectrometry using evaporation and dilution approaches: Application to lavandin, lemon and cypress essential oils. *Talanta,* 149: 178–186.

Fillâtre, Y., D. Rondeau, A. Daguin, A. Jadas-Hecart, and P. Y. Communal, 2014. Multiresidue determination of 256 pesticides in lavandin essential oil by LC/ESI/sSRM: Advantages and drawbacks of a sampling method involving evaporation under nitrogen. *Anal. & Bioanal. Chem.*, 406: 1541–1550.

Finnemore, H., 1926. *The Essential Oils*, 1st ed. London: Ernest Benn.

Firestein, S., 1992. Electrical signals in olfactory transduction. *Curr. Opin. Neurobiol.*, 2: 444–448.

Firestein, S., 2001. How the olfactory system makes sense of scents. *Nature*, 413: 211–218.

François, I., A. de Villiers, and P. Sandra, 2006. Considerations on the possibilities and limitations of comprehensive normal phase-reversed phase liquid chromatography (NPLC × RPLC). *J. Sep. Sci.*, 29: 492–498.

Friedrich J. E., T. E. Acree, and E. H. Lavin, 2001. Selecting standards for gas chromatography-olfactometry. In *Gas Chromatography-Olfactometry: The State of the Art*, J. V. Leland et al. (eds.). Washington, D.C.: American Chemical Society, 2001, chap. 13.

Fuller, G. H., R. Seltenkamp, and G. A. Tisserand, 1964. The gas chromatograph with human sensor: Perfumer model. *Ann. NY Acad. Sci.* 116: 711–724.

Gaffney, B. M., M. Havekotte, B. Jacobs, L. Costa, 1996. Charm analysis of two *Citrus sinensis* peel oil volatiles. *Perf. Flav.*, 21: 1.

Garland, S. M., R. C. Menary, and N. W. Davies, 1999. Dissipation of propiconazole and tebuconazole in peppermint crops (*Mentha piperita* (Labiatae)) and their residues in distilled oils. *J. Agric. Food Chem.*, 47: 294–298.

Gildemeister, E., and F. R. Hoffman, 1950. *Die Ätherischen Öle*, Volume One, 3rd ed. Leipzig: Verlag Von Schimmel & Co.

Giungato, P., A. Di Gilio, J. Palmisani et al. 2018. Synergistic approaches for odor active compounds monitoring and identification: State of the art, integration, limits and potentialities of analytical and sensorial techniques. *Trends Anal. Chem.*, 107: 116–129.

Grob, K., 1987. On-line coupled HPLC-HRGC. In *Proc. 8th International Symposium on Capillary Chromatography*, Riva del Garda, Italy.

Grosch, W., 1994. Determination of potent odourants in foods by aroma extract dilution analysis (AEDA) and calculation of odour activity values (OAVs). *Flav. Fragr. J.*, 9: 147–158.

Guenther, E., 1972. *The Essential Oils - Volume One: History-Origin in Plants Production-Analysis, Reprint of 1st ed.* (1948). Florida: Krieger Publishing Company.

Gu, X., Y. Zhao, K. Li et al., 2016. Differentiation of volatile aromatic isomers and structural elucidation of volatile compounds in essential oils by combination of HPLC separation and crystalline sponge method. *J. Chromatogr. A*, 1474: 130–137.

Hahn-Deinstrop, E., 2000. *Applied Thin Layer Chromatography: Best Practice and Avoidance of Mistakes*, 2nd ed. Weinheim: Wiley-VCH, chap. 1.

Hawrył, H. M., M. A. Hawrył, R. Świeboda, K. Stępak, M. Niemiec, and M. Waksmundzka-Hajnos, 2017. Characterization of mint essential oils by high-performance liquid chromatography. *Anal. Lett.*, 50: 2078–2089.

Hong, J. H., N. Khan, N. Jamila et al., 2017. Determination of Volatile Flavour Profiles of Citrus spp. Fruits by SDE-GC–MS and Enantiomeric composition of chiral compounds by MDGC–MS. *Phytochem. Anal.*, 28: 392–403.

ISO 3761–1997, 1997. International Organization for Standardization, Geneva, http://www.iso.org/iso/iso_catalogue/catalogue_tc/catalogue_tc_browse.htm?commid=48956 (accessed December 15, 2007).

James, A. T., and A. J. P. Martin, 1952. Gas-liquid partition chromatography: The separation and micro-estimation of volatile fatty acids from formic acid to dodecanoic acid. *Biochem. J.*, 50: 679–690.

Japanese Pharmacopoeia JP XV. 2007, 15th ed., Yakuji Nippo, Ltd., Tokyo, 2007.

Jirovetz, L., G. Buchbauer, M. B. Ngassoum, and M. Geissler, 2002. Aroma compound analysis of *Piper nigrum* and *Piper guineense* essential oils from Cameroon using solid-phase microextraction-gas chromatography, solid-phase microextraction-gas chromatography-mass spectrometry and olfactometry. *J. Chromatogr. A*, 976: 265–275.

Jirovetz, L., and M. B. Ngassoum, 1999. Olfactory evaluation and CG/MS analysis of the essential oil of leaves and flowers of *Hyptis pectinata* (L.) Poit. From Cameroon. *SoFW J.*, 125: 35.

Juchelka, D., T. Beck, U. Hener, F. Dettmar, and A. Mosandl, 1998. Multidimensional gas chromatography coupled on-line with isotope ratio mass spectrometry (MDGC-IRMS): Progress in the analytical authentication of genuine flavor components. *J. High Resolut. Chromatogr.*, 21: 145–151.

Khajejenori, M., A. H. Asl, and M. H. Eikani, 2015. Optimization of subcritical water extraction of pimpinella anisum Seeds. *Journal of Essential Oils Bearing Plants*, 18: 1310–1320.

König, W. A., 1991. *Gas Chromatographic Enantiomer Separation with Modified Cyclodextrins*, 1st ed. Heidelberg: Hüthig.

König, W. A., 1998. Enantioselective capillary gas chromatography in the investigation of stereochemical correlations of terpenoids. *Chirality*, 10: 499–504.

Korytár, P., H. G. Janssen, E. Matisová, and U. A. T. Brinkman, 2002. Practical fast gas chromatography: Methods, instrumentation and applications, *TrAC*, 21, 558–572.

Kováts, E., 1958. Gas-chromatographische charakterisierung organischer verbindungen. teil 1: Retentionsindices aliphatischer halogenide, alkohole, aldehyde und ketone, *Helv. Chim. Acta*, 41, 1915–1932.

Lee, M. L., F. J. Yang, and K. D. Bartle, 1984. *Open Tubular Column Gas Chromatography*, 1st ed. New York: Wiley & Sons, chap. 2.

Lehmann, D., A. Dietrich, U. Hener, and A. Mosandl, 1995. Stereoisomeric flavour compounds LXX, 1-p-menthene-8-thiol: Separation and sensory evaluation of the enantiomers by enantioselective gas chromatography/olfactometry. *Phytochem. Anal.*, 6: 255–257.

Liberti, A., and G. Conti, 1956. Possibilità di applicazione della cromatografia in fase gassosa allo studio della essenza. In *Proc. 1° Convegno Internazionale di Studi e Ricerche sulle Essenze*, Reggio Calabria, Italy.

Lin, J., and R. L. Rouseff, 2001. Characterization of aroma-impact compounds in cold-pressed grapefruit oil using time–intensity GC–olfactometry and GC–MS. *Flav. Fragr. J.*, 16: 457–463.

Linssen, J. P. H., J. L. G. M. Janssens, J. P. Roozen, M. A. Posthumus, 1993. Combined gas chromatography and sniffing port analysis of volatile compounds of mineral water packed in polyethylene laminated packages. *Food Chem.*, 46: 367–371.

Liu, Z., and J. B. Phillips, 1991. Comprehensive two-dimensional gas chromatography using an on-column thermal modulator interface. *J. Chromatogr. Sci.*, 29: 227–231.

Lo Presti, M., D. Sciarrone, M. L. Crupi et al., 2008. Evaluation of the volatile and chiral composition in *Pistacia lentiscus* L. essential oil. *Flav. Fragr. J.*, 23: 249–257.

Maggi, F., F. Papa, C. Giuliani et al. 2015. Essential oil chemotypification secretory structures of the neglected vegetable *Smyrnium olusatrum* L. (Apiaceae) growing in central Italy. *Flavour Fragr. J.*, 30: 139–159.

Malnic, B., J. Hirono, T. Sato, and L. B. Buck, 1999. Combinatorial receptor codes for odors. *Cell*, 96: 713–723.

Marriott, P. J., S.-T. Chin, B. Maikhunthod, H. G. Schmarr, and S. Bieri, 2012. Multidimensional gas chromatography. *TrAC*, 34: 1–21.

Marriott, P. J., R. Shellie, and C. Cornwall, 2001. Gas chromatographic technologies for the analysis of essential oils. *J. Chromatogr. A.*, 936: 1–22.

Martindale: *The Complete Drug Reference*, 2007. 35th ed., Sweetman, S., Ed., Pharmaceutical Press, London.

Masson, J., E. Liberto, J. C. Beolor, H. Brevard, C. Bicchi, and P. Rubiolo, 2016. Oxygenated heterocyclic compounds to differentiate *Citrus* spp. essential oils through metabolomics strategies. *Food Chem.*, 206: 223–233.

McDaniel, M. R., R. Miranda-Lopez, B. T. Watson, N. J. Micheals, and L. M. Libbey, 1992. Pinot noir aroma: A sensory/gas chromatographic approach. In *Flavors and Off-Flavors (Developments in Food Science* Vol. 24), G. Charalambous (ed.). Amsterdam: Elsevier Science Publishers, pp. 23–26.

McGorrin, R. J., 2002. Character impact compounds: Flavors and off-flavors in foods. In *Flavor, Fragrance, and Odor Analysis*, R. Marsili (ed.), 1st ed. New York: Marcel Dekker, chap. 14.

Mc Hale, D., and J. B. Sheridan, 1988. Detection of adulteration of cold-pressed lemon oil. *Flav. Fragr. J.*, 3: 127–133.

Mc Hale, D., and J. B. Sheridan, 1989. The oxygen heterocyclic compounds of citrus peel oils. *J. Essent. Oil Res.*, 1: 139–149.

Mehl F., G. Marti, J. Boccard et al., 2014. Differentiation of lemon essential oil based on volatile and non-volatile fractions with various analytical techniques: A metabolomics approach. *Food Chem.*, 143: 325–335.

Mehl F., G. Marti, P. Merle et al., 2015. Integrating metabolomics data from multiple analytical platforms for a comprehensive characterization of lemon essential oils. *Flav. Fragr. J.*, 30: 131–138.

Minh, T. N., T. D. Xuan, T. M. Van, Y. Andriana, T. D. Khanh, and H. D. Tran, 2019. Phytochemical analysis of potential biological activities of essential oil from rice leaf. *Molecules*, 24: 546–558.

Mondello, L., 2004. Determinazione della Composizione e Individuazione delle Adulterazioni degli Olii Essenziali mediante Ultrafast-GC. In *Qualità e sicurezza degli Alimenti*. Milan: Morgan Edizioni Scientifiche, 113–116.

Mondello, L., K. D. Bartle, G. Dugo, and P. Dugo, 1994a. Automated HPLC-HRGC: A powerful method for essential oil analysis. Part III. Aliphatic and terpene aldehydes of orange oil. *J. High Resol. Chromatogr.*, 17: 312–314.

Mondello, L., A. Casilli, P. Q. Tanchida, L. Cicero, P. Dugo, and G. Dugo. 2003. Comparison of fast and conventional GC analysis for citrus essential oils, *J. Agric. Food Chem.*, 51: 5602–5606.

Mondello, L., A. Casilli, P. Q. Tranchida, D. Sciarrone, P. Dugo, and G. Dugo, 2008. Analysis of allergens in fragrances by using multiple heart-cut multidimensional gas chromatography-mass spectrometry. *LC-GC Europe*, 21: 130–137.

Mondello, L., R. Costa, D. Sciarrone, and G. Dugo, 2010. The chiral compound of citrus oils. In *Citrus Oils: Composition, Advanced Analytical Techniques, Contaminants, and Biological Activity*, Giovanni Dugo, and Luigi Mondello (eds.). Taylor & Francis Group, LLC, CRC Press.

Mondello, L., P. Dugo, K. D. Bartle, G. Dugo, and A. Cotroneo, 1995. Automated HPLC-HRGC: A powerful method for essential oils analysis. Part V. Identification of terpene hydrocarbons of bergamot, lemon, mandarin, sweet orange, bitter orange, grapefruit, clementine and Mexican lime oils by coupled HPLC-HRGC-MS(ITD). *Flav. Fragr. J.*, 10: 33–42.

Mondello, L., P. Dugo, K. D. Bartle, B. Frere, and G. Dugo, 1994b. On-line high performance liquid chromatography coupled with high resolution gas chromatography and mass spectrometry (HPLC-HRGC-MS) for the analysis of complex mixtures containing highly volatile compounds. *Chromatog.*, 39: 529–538.

Mondello, L., P. Dugo, G. Dugo and K. D. Bartle, 1996. On-line HPLC-HRGC-MS for the analysis of natural complex mixtures. *J. Chromatogr. Sci.*, 34: 174–181.

Mondello, L., R. Shellie, A. Casilli, P. Q. Tranchida, P. Marriot, and G. Dugo, 2004. Ultra-fast essential oil characterization by capillary GC on a 50 μm ID column, *J. Sep. Sci.*, 27: 699–702.

Nishimura, O., 1995. Identification of the characteristic odorants in fresh rhizomes of ginger *(Zingiber officinale* Roscoe) using aroma extract dilution analysis and modified multidimensional gas chromatography-mass spectroscopy, *J. Agric. Food Chem.*, 43: 2941–2945.

Nitz, S., B. Weinreich, and F. Drawert, 1992. Multidimensional gas chromatography-isotope ratio mass-spectrometry (MDGC-IRMS) A. System description and technical requirements. *J. High Resol. Chromatogr.*, 15: 387–391.

Nowotny, H. P., D. Schmalzing, D. Wistuba, and V. Schurig, 1989. Extending the scope of enantiomer separation on diluted methylated β-cyclodextrin derivatives by high-resolution gas chromatography. *J. High Res. Chromatogr.*, 12: 383–393.

Ochiai, N., and K. J. Sasamoto, 2011. Selectable one-dimensional or two-dimensional gas chromatography-olfactometry/mass spectrometry with preparative fraction collection for analysis of ultra-trace amounts of odor compounds. *J. Chromatogr. A.*, 1218: 3180–3185.

Padhan, D., S. Pattnaik, A. K. Behera, 2017. Growth-arresting activity of acmella essential oil and its isolated component D-limonene (1,8 P-Mentha diene) against *Trichophyton rubrum* (microbial type culture collection 296). *Pharmacognosy Magazine*, 13: 555–560.

Panda, H., 2003. *Essential Oils Handbook*, 1st ed. New Delhi: National Institute of Industrial Research, chap. 110.

Pantò, S., D. Sciarrone, M. Maimone et al., 2015. Performance evaluation of a versatile multidimensional chromatographic preparative system based on three-dimensional gas chromatography and liquid chromatography–two-dimensional gas chromatography for the collection of volatile constituents. *J. Chromatogr. A,* 1417: 96–103.

Parry, E. J., 1908. *The Chemistry of Essential Oils and Artificial Perfumes*, 2nd ed. London: Scott, Greenwood & Son.

Piochon-Gauthier, M., J. Legault, M. Sylvestre, and A. Pichette, 2014. The essential oil of *Populus balsamifera* buds: Its chemical composition and cytotoxic activity. *Natural Prod. Comm.,* 9: 257–260.

Píry, J., and A. Príbela, 1994. Application of high-performance liquid chromatography to the analysis of the complex volatile mixture of blackcurrant buds (*Ribes nigrum* L.). *J. Chromatogr. A,* 665: 105–109.

Pollien P., A. Ott, F. Montigon, M. Baumgartner, R. Muñoz-Box, and A. Chaintreau, 1997. Hyphenated headspace-gas chromatography-sniffing technique: Screening of impact odorants and quantitative aromagram comparisons. *J. Agric. Food Chem.*, 45: 2630–2637.

Poole, C. F., 2003. *The Essence of Chromatography*, 1st ed. Amsterdam: Elsevier.

Porel, A., Y. Sanyal, and A. Kundu, 2014. Simultaneous HPLC determination of 22 components of essential oils; method robustness with experimental design. *Indian J. Pharm. Sci*, 76: 19–30.

Pragadheesh, V. S., A. Yadav, and C. S. Chanotiya, 2015. Role of substituents in cyclodextrin derivatives for enantioselective gas chromatographic separation of chiral terpenoids in the essential oils of Mentha spicata. *J. Chromatogr. B*, 1002: 30–41.

Proot, M., P. Sandra, and E. Geeraert, 1986. Resolution of triglycerides in capillary SFC as a function of column temperature, *J. High Res. Chromatogr. Commun.*, 9: 189–192.

Purcaro, G., C. Cordero, E. Liberto, C. Bicchi, and L. S. Conte, 2014. Toward a definition of blueprint of virgin olive oil by comprehensive two-dimensional gas chromatography. *J. Chromatogr. A,* 1334: 101–111.

Purcaro, G., M. Zoccali, and P. Q. Tranchida et al., 2013. Comparison of two different multidimensional liquid-gas chromatography interfaces for determination of mineral oil saturated hydrocarbons in foodstuffs. *Anal. Bioanal. Chem.,* 405: 1077–1084.

Quassinti, L., F. Maggi, L. Barboni et al., 2014. Wild celery (*Smyrnium olusatrum* L.) oil and isofuranodiene induce apoptosis in human colon carcinoma cells. *Fitoterapia,* 97: 133–141.

Ricci, M. P., D. A. Merritt, K. H. Freeman, J. M. Hayes, 1994. Acquisition and processing of data for isotope-ratio-monitoring mass spectrometry. *Org. Geochem.,* 21: 561–571.

Richardson, A., 1999. Measurement of fragrance perception. In *The Chemistry of Fragrances*, D. H. Pybus and C. S. Sell (eds.). Cambridge: Royal Society of Chemistry, chap. 8.

Rowe, D. J., 2005. Introduction. In *Chemistry and Technology of Flavors and Fragrances*, D. J. Rowe (ed.). Oxford: The Blackwell Publishing, chap. 1.

Rubiolo, P., F. Belliardo, C. Cordero, E. Liberto, B. Sgorbini, and C. Bicchi, 2006. Headspace-solid-phase microextraction fast GC in combination with principal component analysis as a tool to classify different chemotypes of chamomile flower-heads (*Matricaria recutita* L.). *Phytochem. Anal.*, 17: 217–225.

Rubiolo, P., E. Liberto, C. Cagliero et al., 2007. Linear retention indices in enantioselective GC-Mass Spectrometry (Es-GC-MS) as a tool to identify enantiomers in flavour and fragrance fields. In *38th International Symposium on Essential Oils*, Graz.

Russo, M., I. Bonaccorsi, R. Costa, R. Trozzi, P., Dugo, and L. Mondello, 2015. Reduced time HPLC analyses for fast quality control of citrus essential oil. *J. Essent. Oil Res.,* 27: 307–315.

Russo, M., F. Rigano, A. Arigò et al., 2016. Rapid isolation, reliable characterization, and water solubility improvement of polymethoxyflavones from cold-pressed mandarin essential oil. *J. Sep. Sci.* 2016, 39, 2018–2027.
Russo, M., G. Torre, C. Carnovale, I. Bonaccorsi, L. Mondello, and P. Dugo, 2012. A new HPLC method developed for the analysis of oxygen heterocyclic compounds in Citrus essential oils. *J. Essent. Oil Res.*, 24, 119–129.
Sale, J. W., 1953. Analysis of lemon oils. *J. Assoc. Offic. Agr. Chemists*, 36: 112–119.
Schipilliti, L., I. Bonaccorsi, D. Sciarrone, L. Dugo, L. Mondello, and G. Dugo, 2012. Determination of petitgrain oils landmark parameters by using gas chromatography–combustion–isotope ratio mass spectrometry and enantioselective multidimensional gas chromatography. *Anal. Bioanal. Chem.*, 405: 679–690.
Schipilliti, L., G. Dugo, L. Santi, P. Dugo, and L. Mondello, 2011. Authentication of bergamot essential oil by gas chromatography-combustion-isotope ratio mass spectrometer (GC-C-IRMS). *J. Essent. Oil Res.*, 23(2): 60–71.
Schurig, V., and H. P. Nowotny, 1988. Separation of enantiomers on diluted permetylated β-cyclodextrin by high-resolution gas chromatography. *J. Chromatogr.*, 441: 155–163.
Sciarrone, D., R. Costa, C. Ragonese et al., 2011. Application of a multidimensional GC system with simultaneous mass spectrometric and FID detection to the analysis of sandalwood oil. *J. Chromatogr. A*, 1218: 137–142.
Sciarrone, D., D. Giuffrida, A. Rotondo et al., 2017. Quali-quantitative characterization of the volatile constituents in *Cordia verbenacea* D.C. essential oil exploiting advanced chromatographic approaches and nuclear magnetic resonance analysis. *J. Chromatogr. A*. 1524: 246–253.
Sciarrone, D., S. Pantò, F. Cacciola, R. Costa, P. Dugo, and L. Mondello, 2015a. Advanced preparative techniques for the collection of pure components from essential oils. *Nat. Volatiles & Essent. Oils*, 2: 1–15.
Sciarrone, D., S. Pantò, P. Donato, and L. Mondello, 2016. Improving the productivity of a multidimensional chromatographic preparative system by collecting pure chemicals after each chromatographic dimension. *J. Chromatogr. A*, 1475: 80–85.
Sciarrone, D., S. Pantò, C. Ragonese, P. Dugo, and L. Mondello, 2015b. Evolution and status of preparative gas chromatography as a green sample-preparation technique. *TrAC*, 71: 65–73.
Sciarrone, D., S. Pantò, C. Ragonese, P. Q. Tranchida, P. Dugo, and L. Mondello, 2012. Increasing the isolated quantities and purities of volatile compounds by using a triple deans-switch multi-dimensional preparative gas chromatographic system with an apolar-wax-ionic liquid stationary-phase combination. *Anal. Chem.*, 84: 7092–7098.
Sciarrone, D., S. Pantò, A. Rotondo et al., 2013. Rapid collection and identification of a novel component from *Clausena lansium* Skeels leaves by means of three-dimensional preparative gas chromatography and nuclear magnetic resonance/infrared/mass spectrometric analysis. *Anal. Chim. Acta*, 785: 119–125.
Sciarrone, D., S. Pantò, P. Q. Tranchida, P. Dugo, and L. Mondello, 2014. Rapid isolation of high solute amounts using an online four-dimensional preparative system: Normal phase-liquid chromatography coupled to methyl siloxane–ionic liquid–wax phase gas chromatography. *Anal. Chem.*, 86: 4295–4301.
Sciarrone, D., C. Ragonese, C. Carnovale et al., 2010a. Evaluation of tea tree oil quality and ascaridole: A deep study by means of chiral and multi heart-cuts multidimensional gas chromatography system coupled to MS Detection. *J. Chromatogr. A*, 1217: 6422–6427.
Sciarrone, D., A. Schepis, G. De Grazia et al., 2019. Collection and identification of an unknown component from *Eugenia uniflora* essential oil exploiting a multidimensional preparative three-GC system employing apolar, mid-polar and ionic liquid stationary phases. *Faraday Discuss.* DOI:10.1039/C8FD00234G.
Sciarrone, D., L. Schipilliti, C. Ragonese et al., 2010b. Thorough evaluation of the validity of conventional enantio-gas chromatography in the analysis of volatile chiral compounds in mandarin essential oil: A comparative investigation with multidimensional gas chromatography. *J. Chromatogr. A*, 1217: 1101–1105.
Schomburg, G., H. Husmann, E. Hübinger, and W. A. König, 1984. Multidimensional capillary gas chromatography-enantiomeric separation of selected cuts using a chiral second column. *J. High Resolut. Chromatogr.*, 7: 404–410.
Scott, R. P. W., 2001. *Gas Chromatography*, Chrom Ed. Series, http://www.chromatography-online.org/ (accessed December 15, 2007).
Semmler, F. W., 1906–1907. *Die Ätherischen Öle*, Vol. I to IV, 3rd ed. Leipzig: Verlag von Veit.
Shellie, R., and P. J. Marriott, 2002. Comprehensive two-dimensional gas chromatography with fast enantioseparation. *Anal. Chem.*, 74: 5426–5430.
Shellie, R., P. Marriott, and C. Cornwell, 2001. Application of comprehensive two-dimensional gas chromatography (GC × GC) to the enantioselective analysis of essential oils. *J. Sep. Sci.*, 24: 823–830.

Sherma, J., 2000. Thin-layer chromatography in food and agricultural analysis, *J. Chromatogr. A*, 880: 129–147.
Simões, C. M. O., and V. Spitzer, 1999. Óleos voláteis. In *Farmacognosia: Da planta ao medicamento*, C.M.O. Simões et al. (ed.), 1999. Florianópolis: Editora da UFSC and Editora da Universidade/UFRGS, chap. 18.
Song, H. S., M. Sawamura, T. Ito, A. Ido, and H. Ukeda, 2000a. Quantitative determination and characteristic flavour of daidai (*Citrus aurantium* L. var. *cyathifera* Y. Tanaka) peel oil. *Flav. Fragr. J.*, 15: 323–328.
Song, H. S., M. Sawamura, T. Ito, K. Kawashimo, and H. Ukeda, 2000b. Quantitative determination and characteristic flavour of *Citrus junos* (yuzu) peel oil. *Flav. Fragr. J.,* 15: 245–250.
Stillman, R. C., and R. M. Reed, 1932. Perfumery Essential Oil Record, 23: 278.
Sybilska, D., and T. Koscielski, 1983. β-cyclodextrin as a selective agent for the separation of o-, m- and p-xylene and ethylbenzene mixtures in gas-liquid chromatography. *J. Chromatogr. A.*, 261: 357–362.
TC 57 Essential Oils, International Organization for Standardization, Geneva, http://www.iso.org/iso/iso_catalogue/catalogue_tc/catalogue_tc_browse.htm?commid=48956, (accessed December 15, 2007).
The International Pharmacopoeia (IntPh), 2007. Pharmacopoeia, 4th ed. Geneva: World Health Organization Press.
The United States Pharmacopoeia USP/NF 2008, 2007. Maryland: United States Pharmacopoeia Convention Inc.
Tine, Y., A. Diop, W. Diatta et al., 2017. Chemical diversity and antimicrobial activity of volatile compounds from Zanthoxylum zanthoxyloides Lam. According to compounds classes, plant organs and Senegalese sample locations. *Chem. Biodivers.*, 14.
Todd, J. F. J., 1995. Recommendations for nomenclature and symbolism for mass spectroscopy. *Int. J. Mass Spectrom. Ion Process,* 142: 211–240.
Tranchida, P. Q., A. Casilli, G. Dugo, L. Mondello, and P. Dugo. 2005. Fast gas chromatographic analysis with a 0.05 mm ID micro-bore capillary column, *G.I.T. Lab. J.*, 9: 22.
Tranchida, P. Q., Franchina, F. A., Dugo, P., and Mondello, L. 2014. Use of greatly-reduced gas flows in flow-modulated comprehensive two-dimensional gas chromatography-mass spectrometry. *J. Chromatogr. A*, 1359: 271–276.
Tranchida, P. Q., F. A. Franchina, M. Zoccali et al., 2013a. Untargeted and targeted comprehensive two-dimensional GC analysis using a novel unified high-speed triple quadrupole mass spectrometer. *J. Chromatogr. A,* 1278: 153–159.
Tranchida, P. Q., D. Sciarrone, P. Dugo, and L. Mondello, 2012. Heart-cutting multidimensional gas chromatography: Recent evolution, applications, and future prospects. *Anal. Chim. Acta,* 716: 66–75.
Tranchida, P. Q., M. Zoccali, I. Bonaccorsi, P. Dugo, L. Mondello, and G. Dugo, 2013b. The off-line combination of high performance liquid chromatography and comprehensive two-dimensional gas chromatography–mass spectrometry: A powerful approach for highly detailed essential oil analysis. *J. Chromatogr. A,* 1305: 276–284.
Ullrich, F., and W. Grosch, 1987. Identification of the most intense volatile flavour compounds formed during autoxidation of linoleic acid, *Z. Lebensm. Unters. Forsch.*, 184: 277–282.
van den Dool, H., and P. D. Kratz, 1963. A generalization of the retention index system including linear temperature programmed gas-liquid chromatography. *J. Chromatogr. A*, 11, 463–471.
van Ruth, S. M., 2001. Methods for gas chromatography-olfactometry: A review. *Biomolec. Eng.*, 17: 121–128.
Wagner, H., S. Bladt, and V. Rickl, 2003. *Plant Drug Analysis: A Thin Layer Chromatography Atlas*, 2nd ed. Heidelberg: Springer Verlag, p. 1.
Wong, Y. F., R. N. West, S. T. Chin, and P. J. Marriot, 2015. Evaluation of fast enantioselective multidimensional gas chromatography methods for monoterpenic compounds: Authenticity control of Australian tea tree oil. *J. Chromatogr. A*, 1406: 307–315.
Wong Y. F., D. Yan, R. A. Shellie, D. Sciarrone, and P. J. Marriot, 2019. Rapid plant volatiles screening using headspace SPME and person-portable gas chromatography–mass spectrometry, *Chromatographia,* 82: 297–305.
Wright, D. W., 1997. Application of multidimensional gas chromatography techniques to aroma analysis. In *Techniques for Analyzing Food Aroma (Food Science and Technology)*, R. Marsili (ed.), 1st ed. New York: Marcel Dekker, chap. 5.

8 Use of Linear Retention Indices in GC-MS Libraries for Essential Oil Analysis

Emanuela Trovato, Giuseppe Micalizzi, Paola Dugo, Margita Utczás, and Luigi Mondello

CONTENTS

8.1 INTRODUCTION

The continuous fundamental and difficult challenge of the modern science of separation consists of providing adequate chromatographic techniques that allow resolving mixtures of compounds into less complex mixtures or ultimately into pure components with rapid and effective separations, improving at the same time the resolving power, and allowing highly selective or sensitive detections.

Essential oils are very complex mixtures consisting of monoterpene and sesquiterpene hydrocarbons, their oxygenated derivatives, and aliphatic oxygenated compounds. The technique most used for essential oil analysis and characterization is GC-MS. Many terpenes have identical mass spectra due to similarities in the initial molecule, or in the fragmentation patterns and rearrangements after ionization, and this raises difficulties in the GC-MS peak identification of these samples. The combining of retention time information with mass spectra results may support the identification by MS library search.

In a GC separation, each compound is characterized by a different retention behavior on a specific column, expressed by absolute retention time, retention factor, and the relative retention time. The absolute retention time (t_R) varies depending on mobile phase flow, temperature program, column phase ratio and length, and also between consecutive runs on the same column and under the same analytical conditions, and for these reasons, it is not so useful for identification purposes.

The need to express gas chromatographic retention data in a standardized system has long been recognized and an enormous effort has been devoted to this purpose. James and Martin (James and Martin, 1952), in their first publication on gas–liquid chromatography in 1952 mentioned that the retention volume of a pure substance on a certain gas chromatographic column is a characteristic value, which could be used for the identification of sample components.

In 1955, Littlewood et al. (1955) attempted to make retention data independent of temperature, stationary phase film thickness, carrier gas flow rate, and pressure drop across the column, introducing the specific retention volume; but this parameter was not practically applied. Instead, from the many proposals, the use of relative retention data expressed with respect to the retention of a standard substance has found consensus. Since gas chromatographic separations are performed over a wide range of temperatures and on stationary phases of different polarities, no single substance would fulfill the role of a universal standard. In general, the most thoroughly studied, diffused, and accepted retention index calculation methods are calculated applying the equation developed by Kováts in 1958 (Kováts, 1958) for isothermal analysis conditions and the equation propounded by van den Dool and Kratz in 1963, (van den Dool and Kratz, 1963) in the case of temperature programming analysis conditions, while values derived from the former equation are usually called retention index (I) or Kováts index (KI). Values calculated using the latter approach are commonly denominated in literature as retention index (I), linear retention index (LRI) or programmed-temperature retention index (PTRI).

This chapter provides an overview of the theoretical aspects and applications of retention indices to gas chromatographic analysis of essential oils. Gas chromatographic retention parameters and retention index calculations are described; furthermore, a critical viewpoint is discussed with respect to the indiscriminate use of retention indices from different sources.

8.2 RETENTION INDEX THEORIES

As is well known and reported by Poole, the retention of analytes in gas chromatographic capillary columns results from the differential distribution (partition) of the solutes between the stationary liquid and the mobile gas phases (Poole, 2003). A compound's retention behavior on a specific column is characterized by three parameters: retention time (t_R), retention factor (k), and relative retention (r).

Retention time (t_R) is expressed as the time that the solute spends in the mobile phase, that is the unretained peak time (t_M), summed to the adjusted retention time (t'_R), which corresponds to the time the solute spends distributed into the stationary phase (see Equation 8.1):

$$t_R = t_M + t'_R \tag{8.1}$$

The magnitude of retention depends on the partition coefficient, or distribution constant (K), which is defined as the ratio of the equilibrium concentrations of a solute in the stationary (C_S) and mobile phases (C_m) during partitioning in the column:

$$K_D = C_S / C_m \tag{8.2}$$

Consequently, the greater is the K value for a sample component, the higher is its solubility and the longer its retention in the stationary phase.

Generally, the affinity of a solute for the stationary phase depends on its vapor pressure and the activity coefficient of the solute in that phase. The column's ability to differentiate two solutes, eluting them with different retention times depends from differences between those two properties.

Retention factor (k) quantifies the ratio of the time spent in the stationary phase and that spent in the mobile phase, as expressed by the following equation:

$$k = t'_R / t_M \tag{8.3}$$

where t'_R is the adjusted retention time and t_M the unretained peak time.

Relative retention (r) expresses the degree of separation between two peaks, a standard (st) and the solute of interest, not necessarily in adjacent positions. Relative retention can be expressed using standard solute retention factors or adjusted retention times. The latter are used in Equation 8.4, where i refers to an individual solute. The relative retention of solutes eluting after the standard will be >1, while that of compounds eluting prior to the standard <1:

$$r_i = t'_{Ri} / t'_{R(St)} \tag{8.4}$$

Retention time is not useful for peak identification, since the retention time is influenced by the applied linear velocity, temperature, phase ratio, and column length. Compared to retention time, retention factor is more advantageous due to the inclusion of the unretained peak time, contouring changes caused by linear velocity discrepancies and column length differences. On the other hand, relative retention depends also on the phase ratio, so if an identical reference peak is chosen, and the same column temperature is applied, the results are comparable. However, for peaks eluting far from the reference peak, the accuracy of this information can be degraded.

Kováts (1958), to overcome this limit, introduced a retention index (I) system in which a homologous series of normal paraffins were applied as reference peaks. In isothermal GC conditions, as is well known, this homologous series elutes with retention times increasing exponentially. Under such conditions, a semilogarithmic relationship exists between the adjusted retention times (t'_{Ri}) of the n-paraffins and their carbon numbers (c_n), as expressed in Equation 8.5, where a and b are proportionality constants:

$$\log t'_{Ri} = a \cdot c_n + b \tag{8.5}$$

Each analyte is referenced in terms of its position between the two n-paraffins that bracket its retention time. The index calculation is based on a linear interpolation of the carbon chain length of the two bracketing paraffins. In the original definition, the retention index of a particular substance was calculated using only n-paraffins with even carbon atoms as references (as represented in Equation 8.6):

$$I_s^{st.ph.}(T) = 200(\log X_S - \log X_Z / \log X_{(Z+2)} - \log X_Z) + 100z \tag{8.6}$$

where I is the isothermal retention index at temperature T, s is the compound of interest, $st.ph.$ is the stationary phase, and X is, for instance, the retention time used for the calculation, while z and $z + 2$ are n-alkanes with z and $z + 2$ carbon numbers, respectively. By definition, the retention index of the n-paraffins is equal to 100 times their carbon number for any given stationary phase and at any given column temperature, for example, n-C_8 has an index of 800.

At first, the use of even-number n-paraffins was preferred, assuming the possible occurrence of an oscillation in the chromatographic properties of successive numbers of the complete n-paraffin series. Later, it was experimentally confirmed that this statement was erroneous and Kováts redefined the first and proposed another fundamental equation:

$$I_s^{st.ph.}(T) = 100[z + (\log X_S - \log X_Z / \log X_{(Z+1)} - \log X_Z)] \tag{8.7}$$

where z and $z + 1$ are n-alkanes with z and $z + 1$ carbon numbers, respectively. In this equation, the closer setting of the reference paraffins enables a more accurate evaluation of a compound's behavior in the uniform scale.

In general, the retention index system is a combination of two chromatographic parameters, relative retention (Equation 8.4) and retention volume (V_R), which can be measured experimentally from the mobile phase volumetric flow rate (F_C) and the retention time, as shown in Equation 8.8:

$$V_R = t_R \cdot F_C \tag{8.8}$$

However, several drawbacks of this system could be observed. As reported by Castello and Testini, a comparison between retention indices calculated on columns of distinct stationary phases does not allow distinguishing whether the difference in the values is related to an increased retention of the substance of interest or to a decreased retention of the members of the reference series on the stationary phase in question, or vice versa (Castello and Testini, 1996).

In the equation proposed by Kováts, the retention indices refer to data obtained under isothermal elution conditions, while in temperature-programmed conditions, the series of *n*-paraffins elutes in a linear mode.

In the latter case, a constant increment is added for each successive peak to the retention time of its predecessor, instead of a nonlinear increment, as could be observed under isothermal conditions. Equation 8.9 expresses a similar relationship between the programmed-temperature retention times ($\log t_R^T$) of the *n*-paraffins and their carbon numbers (c_n):

$$\log t_R^T = a' \cdot c_n + b' \tag{8.9}$$

where a' and b' are constants depending on the stationary phase and on the nature of the chemical group bound to the alkyl chain.

The calculation of retention index in programmed-temperature conditions is based on Equation 8.10 proposed by van den Dool and Kratz (1963), which does not use the logarithmic form:

$$I(T) = 100\left[z + \left(t_{R_i}^T - t_{R_z}^T / t_{R_{z+1}}^T - t_{R_z}^T\right)\right] \tag{8.10}$$

The indices calculated by Equation 8.10 are commonly denominated as linear retention indices (LRIs). When series of *n*-paraffins containing only odd or even carbon atoms as references are used, Equation 8.11 must be applied:

$$I(T) = 100n + \left(t_{R_i}^T - t_{R_z}^T / t_{R_{z+1}}^T - t_{R_z}^T\right) + 100z \tag{8.11}$$

where *n* represents the difference in carbon atom number of the two *n*-paraffins that bracket a solute's retention time.

The retention indices calculated for isothermal and temperature-programmed conditions commonly present proximate values, but are not identical. It is important for the reproducibility of temperature-programmed retention indices that working variables, such as stationary phase film thickness, carrier gas flow rate, and linear temperature-programming rate being standardized. In fact, there is the possibility that I^T values vary according to the applied temperature rate and the initial temperature.

In 1974, van den Dool described a method called the g-pack value (van den Dool, 1974) for the characterization of GC columns using relative retention indices, as proposed by Kováts. The indices were determined for limonene, linalool, linalyl acetate, acetophenone, naphthalene, and cinnamyl alcohol, considered as representative for the substances commonly detected in essential oils, using *n*-paraffins with even carbon atoms from C_8 to C_{24} as references. The g-pack value may decrease due to modifications in the selectivity of a stationary phase, such as those deriving from column bleeding, while the modifications of the stationary phase due to oxidation may increase a g-pack value, but is not affected by alterations of the carrier gas flow rate, temperature-programming rate and initial temperature. To evaluate the dependence of the values from stationary phase, the g-pack values were calculated for two stationary phases, a polyethylene glycol representative of polar columns and a 100% dimethyl polysiloxane, representative of non-polar columns. Since the elution of homologous series under temperature-programmed conditions was not linear, van den Dool suggested the use of both even and odd carbon-numbered *n*-alkanes to reduce possible errors.

The continuous interest and study on retention mechanisms, and also on retention index systems, led to the introduction of several other concepts compared to the well-known proposed by Kováts and, later by van den Dool and Kratz, some of which are shortly reported here.

Evans and Smith (Evans and Smith, 1961b, 1962) suggested a molecular retention index, derived from the equation proposed by Kováts, which was defined as the molecular weight of a hypothetical n-alkane presenting the identical retention of the compound being investigated. The formula of that hypothetical n-alkane equivalent to the target compound is $C_I H_{2I'+2}$ and will possess an effective molecular weight according to the following relation:

$$M_e = 14.026I' + 2.016 = 0.1406I + 2.016 \quad (8.12)$$

where $I = 100I'$, whereas I' is the number of carbon atoms and M_e is the molecular weight of the hypothetical n-alkane. The difference between that hypothetical and the real molecular weight may be a useful parameter to correlate retention and chemical structure. In the same year, Evans (Evans and Smith, 1961a,b) proposed a further system, defined as the theoretical nonane value, determining first the retention value as proposed by Kováts and then transferring to a system in which n-nonane is the standard.

However, the accuracy of the determination decreased when compounds with retention values far from that of the reference standard were analyzed.

Robinson (Robinson and Odell, 1971) developed another system, the standard retention index, based on the equation proposed by Kováts. In that system, the reference parameters are the boiling points of the target analyte and the reference standards rather than the retention time or the adjusted retention time. Novák (Novák and Ružiková, 1974) developed the generalized retention index, which also consisted of a modification of the equation proposed by Kováts, which may be applied also when homologous series other than n-alkanes are used as references. Zenkevich created a further system (Zenkevich and Ioffem, 1988) with the identical name, generalized retention index, that combines the logarithmic system of Kováts with the retention indices based on the linear relationship between I and $(t'_R + q\log t'_R)$, where t'_R is the adjusted retention time and q a coefficient related to the analytical conditions. Berezkin reported (Berezkin, 1974), the invariant retention index (I_0), based on the fact that the retention indices should be independent of the effects of the stationary phase, proposing the following relationships between the retention indices and the invariant retention index:

$$I = I_0 + a/P_L \quad (8.13)$$

$$I = I_0 + b/k_S \quad (8.14)$$

$$I = I_0 + c/V_L \quad (8.15)$$

where a, b, and c are constants; P_L is the percentage of stationary phase present in the column; V_L is the volume of stationary phase in the column; and k_S is the capacity factor of the reference substance.

Rasanen et al. (1995) measured the homologous retention indices of alkaline drugs through software able to automatically assign the so-called secondary retention index standard series (I^*) by a pattern recognition algorithm; the I^* standards were then identified as members of a pattern rather than individual peaks. The same software has been used to calculate the I^* values of other peaks through linear interpolation using absolute retention times, to compare the obtained values to those contained in libraries reporting the identified compounds.

The development of the unified retention index (UI_T) concept was developed by Dimov (1985) after the investigation on the variation in the retention index of hydrocarbons on squalene, which is largely attributed to random errors. The unified retention index (UI_T) may be defined as a statistical treatment of data by means of linear regression (see Equation 8.16):

$$UI_T = UI_0 + (dUi/dT)T \quad (8.16)$$

where UI_0 is the value of UI_T at 0°C and dUi/dT is the index increment with the analysis temperature (the slope of the curve).

Since the index itself and the temperature increment of the unified retention index are statistical values, they are therefore more reliable than the individual experimental retention values.

It is also worth noting that it is characterized by a standard deviation (SD), and the calculation of the confidence interval at any desired level is possible.

Škrbic considered the unified retention index not only a convenient mode of determining the retention index value at any temperature within the investigated range, but also as an evaluation method of the proper column temperature for the analysis of complex mixtures, demonstrating that the elution sequence of the sample components can change (Škrbic, 1997). Harangi proposed the virtual carbon number for the identification of compounds without applying any index calculation, but using solely the retention behavior (Harangi, 2003). According to this system, for the calculation of retention indices, a homologous series is not necessary, since the analyzed matrix certainly contains a series of compounds with well-known retention indices, not necessarily belong to a homologous series that could be used as index references. The index of the other compounds shall then be calculated using virtual carbon numbers of the reference compounds instead of the carbon numbers of the *n*-alkanes.

Blumberg proposed a further method for compound identification using only retention behavior without index calculation, the retention time locking (RTL) for programmed-temperature analysis (Blumberg and Klee, 1998). In this method, the inlet pressure is adjusted to provide an identical retention time for the same compound in any system equipped with the same nominal column. The RTL method applies constant pressure, and only small deviations in column size and stationary phase loading parameters are tolerated; moreover, it cannot be used under constant flow-rate conditions.

Stránský et al. (2006) proposed reduced Kováts index (RKI) for retention data determination of wax esters. RKIs are calculated based on Equation 8.17, where NCA is the number of carbon atoms:

$$RKI = I - 100 \cdot NCA \tag{8.17}$$

These values should express the influence of the presence and position of the ester functionality or position of double bond on the stationary phase (Zellner et al., 2008). However, it has to be pointed out that although used for temperature-programmed GC analyses, retention index calculations were logarithmic-based.

A further index system, the molecular topological index, proposed by Balaban (1982), has been shown to be a very important structural parameter for describing the chromatographic behavior of a compound. Heinzen and Yunes investigated the determination of molecular topological indices and their correlation with retention indices of linear alkylbenzene isomers with C_{10} to C_{14} linear alkyl chains (Heinzen and Yunes, 1996).

The prediction of retention index has attracted a great deal of interest, as also the extrapolation of values of an analyte's specific physical properties that could be correlated through its retention index. The development of predictive relationships would represent a key for the reliability of retention data. More general approaches to retention index prediction are based on the generation of topological, geometric, and electronic molecular descriptors, which are subsequently fitted to the retention index using multiple linear regression or artificial neural networks. Bruchmann et al. (1993) investigated neural networks that were explored to predict GC retention index data based on electrotopographic indexes of monofunctional compounds, such as acyclic and cyclic monoterpenes and a mixed set of monosubstituted compounds and terpenes. According to the authors, predictions by neural networks are generally in good agreement with predictions done by multiple linear regression techniques. In addition, the prediction of retention indices of homologous series was proved to be effective, as demonstrated by Junkes et al. (2002).

More recently, Veselinović et al. (2017) have presented another method for the prediction of retention indices based on the Monte Carlo method. In this study, the authors used, for the first time, descriptors based on simplified molecular-input line-entry system (SMILES) and local graph invariants including the number of paths and valence shells for successfully developing a predictive quantitative structure-property relationships (QSPR) model for gas chromatography retention indices prediction.

8.3 LINEAR RETENTION INDICES PRESENT IN THE DATABASES

The combined use of the mass spectrum and the RI information revolutionized the spectral database construction. In fact, an increasing number of mass spectral collections found on the market nowadays also contain RI values. However, it is important to point out that the combined use of RI values can facilitate the correct peak assignment of the problematic compounds with identical mass spectrum, only if a reliable database has been selected.

In the commercially available databases and libraries, different types of RI can be found, with several names (Kovàts indices, KI; arithmetic indices, AI; normal alkane RI; linear retention indices, LRI), which may lead to confusion in the comparison of the experimental value with the one present in the database. For a reliable identification, the most adequate one should be selected, according to the analytical conditions (stationary phase, homologue series, temperature program). The most used LRI values are calculated on non-polar or semi-non-polar stationary phases against *n*-alkane homologue series. Thanks to the stability of this type of stationary phase and its intensive interaction with the hydrocarbons, this combination provides the most stable LRI values. In the essential oil field, polar columns are often applied for a better separation of critical pairs, or to have a further parameter to achieve a highly reliable identification, confirmed using three independent analytical information (MS spectrum, polar and non-polar LRI values). Even though in some commercial database LRI values calculated on polar column against *n*-alkanes can be found, it is well known that due to the poor solubility of alkanes in this type of stationary phase, a fluctuating behavior can be observed in the LRI values. Considering this fact, instead of the *n*-alkanes, further homologue series were proposed, such as 2-alkanones, alkyl ethers, alkyl halides, alkyl acetates, fatty acid methyl (FAMEs) and ethyl esters (FAEEs) (Shibamoto, 1987; Castello, 1999; Zellner et al., 2008), and para-substituted *n*-alkylphenols (Mjøs et al., 2006). In the flavor and fragrance analysis, the most used polar homologue series is the FAEEs series, resulting in more stable LRI values on polar columns than using *n*-alkanes. It is worthy to note that the stability of the polar stationary phases is much lower than the non- or semi-non-polar phases. Thus, according to the column aging, notable shift (10–50 LRI units) can be observed in the calculated LRI values in the case of some components (e.g., esters, alcohols, ketones). This can be controlled using a suitable mixture of column age sensible compounds to evaluate the actual state of the stationary phase used.

8.4 ACCURACY OF RETENTION INDICES

Even using the correct analytical combination (column, homologue series, conditions) and the most suitable library LRI for our analysis, the reliability of the reported values is undoubtedly the base of a correct identification. There are three main types of databases: only MS spectral collections, MS databases including LRI values from different sources and obtained in different statistical ways, and MS libraries containing experimental LRI values obtained under well-defined conditions.

For a preliminary, tentative identification, MS databases can be used resulting a questionable candidate for the target compound. According to the fragmentation and the molecular ion, the main structure of the molecule can be estimated, but without any further information, it could not be confirmed.

Widely used databases, containing huge number of MS spectra and LRI values, have their limitations. In particular, the reported LRI values are not experimentally obtained. The first calculated LRI values were predicted according to the suggestion of Stein et al. (2007). The predicted retention indices are calculated in isothermal conditions considering the contribution of different

TABLE 8.1
Comparison between Average Experimental and Predicted RI Values of Essential Oil Compounds

Common Name	CAS	Molecular Weight	Molecular Formula	$LRI_{exp,average}$	$LRI_{predicted}$	ΔLRI
Diacetone alcohol	123-42-2	116	C6 H12 O2	833	845	−13
n-Hexanol	111-27-3	102	C6 H14 O	865	860	5
Isoamyl acetate	123-92-2	130	C7 H14 O2	870	820	50
2,5-Dimethyl-pyrazine	123-32-0	108	C6 H8 N2	910	894	16
2,3-Dimethyl-pyrazine	5910-89-4	108	C6 H8 N2	915.5	894	22
α-Thujene	2867-05-2	136	C10 H16	925.5	902	24
Sabinene	3387-41-5	136	C10 H16	970.5	897	74
β-Pinene	127-91-3	136	C10 H16	976	943	33
δ-3-Carene	13466-78-9	136	C10 H16	1008.5	948	61
Isopentyl butyrate	106-27-4	158	C9 H18 O2	1055	1019	36
Terpinolene	586-62-9	136	C10 H16	1086	1052	34
para-Cymenene	1195-32-0	132	C10 H12	1091	1073	18
Linalyl formate	115-99-1	182	C11 H18 O2	1213	1270	−57
Neral	106-26-3	152	C10 H16 O	1236.5	1174	63
Bornyl acetate	92618-89-8	196	C12 H20 O2	1286	1277	9
α-Longipinene	39703-24-7	204	C15 H24	1351	1403	−52
α-Copaene	138874-68-7	204	C15 H24	1376	1430	−54
trans-Geranyl acetate	105-87-3	196	C12 H20 O2	1380	1352	28
β-Dihydroionone	17283-81-7	194	C13 H22 O	1436	1418	18
trans-α-Amylcinnamaldehyde	78605-96-6	202	C14 H18 O	1667	1663	4

Note: Average experimental LRI values were calculated from the data of two databases, namely FFNSC 3 (Shimadzu, Kyoto, Japan and John Wiley & Sons, Hoboken, NJ, USA) and Adams library 4th ed. (FarHawk, Fineview, NY, USA), containing experimental LRI values obtained in programmed-temperature conditions on an SLB-5 ms and a HP-5 stationary phases, respectively. Predicted Kovàts RI values were reported for non-polar stationary phases in the NIST 17 spectral database.

chemical groups, and they are given as the estimated value with a confidence interval. These values can differ significantly from the real LRI of a compound (Table 8.1); for non-polar column data, the declared median absolute prediction error was 46, while for polar one, it was 65 units (Stein et al., 2007). Due to the poor accuracy, using these estimated LRI values results in only a slight improvement with respect to the match obtained by MS only.

Also cited RI values are reported from different references, mainly on non-polar columns (different brands of 100% dimethyl polysiloxane: DB-1, RTX-1, CP sil 5 CB, etc.). Zellner et al. demonstrated in their review that even well-known compounds, like limonene and linalool, can have significantly different LRIs reported in scientific journals (Zellner et al., 2008). It can be caused on the one hand by the analytical condition differences and on the other hand by the misidentified compounds present in the literature. Considering these facts, the described data collections can provide more information about the compound, but an accurate and reliable peak assignment cannot be obtained with them. Furthermore, for certain compounds, average RI can be found in the databases on non-polar, semi-non-polar and polar stationary phases against *n*-alkanes. The values are given as the median of different RI values obtained on chromatographic columns with similar polarity, the standard deviation and the number of the data used for the calculation, as it is shown in Table 8.2. Not surprisingly, the best SD values (average: 4 LRI units) can be observed in the case of the non-polar and semi-non-polar

TABLE 8.2
Summary of Median LRI Values on Non-polar (NP), Semi-Non-polar (SNP), and Polar (P) Stationary Phases with Standard Deviation (SD) and Number of Used Data (N°) Reported in NIST 17 Spectral Database, and Comparison between Average Experimental LRI Calculated from the Data of FFNSC 3 and Adams Library 4th ed. and Median LRI Values of Essential Oil Compounds

Common Name	Molecular Formula	$LRI_{SNP, exp\ average}$	ΔLRI_{SNP}	LRI_{NP}	SD	N°	LRI_{SNP}	SD	N°	LRI_{P}	SD	N°
Diacetone alcohol	C6 H12 O2	833	−6	816	4	22	838	8	27	1358	14	34
n-Hexanol	C6 H14 O	865	−3	854	5	175	868	4	223	1355	7	347
Isoamyl acetate	C7 H14 O2	870	−6	859	4	82	876	2	100	1122	7	168
2,5-Dimethyl-pyrazine	C6 H8 N2	910	−7	889	4	73	917	7	69	1320	11	130
2,3-Dimethyl-pyrazine	C6 H8 N2	916	−11	897	4	60	926	7	44	1344	9	93
α-Thujene	C10 H16	926	−4	925	3	352	929	2	488	1028	7	224
Sabinene	C10 H16	971	−4	967	4	451	974	2	619	1124	8	387
β-Pinene	C10 H16	976	−3	973	5	587	979	2	849	1112	7	518
δ-3-Carene	C10 H16	1009	−3	1006	5	212	1011	2	336	1147	7	192
Isopentyl butyrate	C9 H18 O2	1055	−2	1040	2	23	1056	4	39	1259	5	24
Terpinolene	C10 H16	1086	−2	1079	3	413	1088	2	607	1283	7	412
para-Cymenene	C10 H12	1091	1	1074	4	84	1090	2	106	1444	11	95
Linalyl formate	C11 H18 O2	1213	−2	1206	1	7	1215	16	9	1570	9	5
Neral	C10 H16 O	1237	−4	1218	5	125	1240	3	168	1680	13	161
Bornyl acetate	C12 H20 O2	1286	1	1270	5	217	1285	3	328	1581	11	200
α-Longipinene	C15 H24	1351	−2	1355	5	23	1353	3	63	1525	17	15
α-Copaene	C15 H24	1376	0	1376	4	377	1376	2	698	1492	7	371
trans-Geranyl acetate	C12 H20 O2	1380	−3	1361	2	201	1382	3	206	1752	11	199
β-Dihydroionone	C13 H22 O	1436	—	—	—	—	—	—	—	—	—	—
trans-α-Amylcinnamaldehyde	C14 H18 O	1667	5	1615	16	3	1662	10	3	2247	0	1

column type, firstly due to their high stability in long-term use, secondly thanks to the high number of data present in the scientific literature, and thirdly because of the perfect compatibility between the stationary phase and the homologue series. However, SD values for polar columns, with poly(ethylene glycol) were more than twice higher (average: 9 LRI units), mainly due to the lack of the latter requirement. As previously mentioned, on the polar wax columns, the *n*-alkanes have a fluctuating behavior resulting in very variable LRI even in the same conditions; thus, their use in this combination is not preferred. Furthermore, the deviation can be influenced by the wide range of average molecular weights of poly(ethylene glycols) and the variable level of cross-linking.

Comparing the median results with average experimental LRI values, mainly a good correspondence was obtained (±4 LRI units), but in some critical cases with similar MS fragmentation—for example, in isomer 2,5-dimethyl-pyrazine and 2,3-dimethyl-pyrazine or terpene (sabinene and β-pinene)—identification problems can occur thanks to the slight retention differences. Summarizing, the LRI obtained on non-polar and semi-non-polar columns, for a lot of type of unknown essential oil compound, can be a good choice to have a second "dimension" in the identification, even demonstrating lower reliability than experimental LRI databases.

The experimental LRI value approach is substantially different. The LRI values are determined from pure analytical standards or isolated and characterized essential oil constituents under well-specified chromatographic conditions in parallel analysis. On the same column type, they are perfectly comparable regardless of the brand and exact quality of the stationary phase. An important evidence of this fact is shown in Table 8.3, where LRI values are compared from two MS databases

TABLE 8.3
Comparison between Experimental LRI of Essential Oil Constituents from the FFNSC 3 (LRI, SLB-5 ms, 2015) and Adams Library 4th ed. (AI, HP-5, 2007)

Common Name	Molecular Formula	LRI_{FFNSC}	AI_{ADAMS}	$LRI_{average}$	Reproducibility Interlaboratory
Diacetone alcohol	C6 H12 O2	834	831	833	2.1
n-Hexanol	C6 H14 O	867	863	865	2.8
Isoamyl acetate	C7 H14 O2	871	869	870	1.4
2,5-Dimethyl-pyrazine	C6 H8 N2	912	908	910	2.8
2,3-Dimethyl-pyrazine	C6 H8 N2	916	915	916	0.7
α-Thujene	C10 H16	927	924	926	2.1
Sabinene	C10 H16	972	969	971	2.1
β-Pinene	C10 H16	978	974	976	2.8
δ-3-Carene	C10 H16	1009	1008	1009	0.7
Isopentyl butyrate	C9 H18 O2	1057	1052	1055	3.5
Terpinolene	C10 H16	1086	1086	1086	0.0
para-Cymenene	C10 H12	1093	1089	1091	2.8
Linalyl formate	C11 H18 O2	1212	1214	1213	1.4
Neral	C10 H16 O	1238	1235	1237	2.1
Bornyl acetate	C12 H20 O2	1285	1287	1286	1.4
α-Longipinene	C15 H24	1352	1350	1351	1.4
α-Copaene	C15 H24	1378	1374	1376	2.8
trans-Geranyl acetate	C12 H20 O2	1380	1379	1380	0.7
β-Dihydroionone	C13 H22 O	1438	1434	1436	2.8
trans-α-Amylcinnamaldehyde	C14 H18 O	1667	1667	1667	0.0

Note: AI: Arithmetic retention index based on the calculation of van den Dool and Kratz (van den Dool and Kratz, 1963); actually. it is equal to LRI.

containing experimental RI values on semi-non-polar columns from different brands acquired with almost one decade of difference. As can be noticed, the obtained LRI values remained the same according to the calculated interlaboratory reproducibility (less than 4 LRI units in every case), even using two different temperature programs. These observations prove that this approach provides an excellent accuracy of the confirmatory identification parameter. Using the databases containing experimental LRI values gives the possibility of a correct double confirmed peak assignment even using different columns with the same polarity and different temperature gradient.

To reach the best available LRI matching the use of a selected database with experimental LRIs combined with the same analytical conditions (column, temperature program) described in it is recommended. According to the results obtained (Table 8.4) from an experiment on three inter-day repetitions under the analytical conditions established by the FFNSC 3, the repeatability intralaboratory was less than 1, and the difference between the obtained average LRI and those present in the library was not higher than 3 in every case. Considering that in a normal essential oil analysis, the average peak width at half height in terms of LRI is about 2 units, completing with the excellent repeatability using a cutoff window of ±5 LRI units can be suitable for a reliable identification, excluding the wrong candidates for the target molecule.

Some of the experimental databases are boosted with multi-LRI features, thus they contain different LRIs obtained under various conditions; for example, isothermal and programmed temperature in the Adams library, or on different stationary phases with different homologue series; for example, 5 LRI values on Equity-1 (Merck KGaA, Darmstadt, Germany, part#: 28046-U), SLB-5 ms (Merck KGaA,

TABLE 8.4
Repeatability of Experimental LRI Values under Standardized Conditions (SLB-5 ms, C_7-C_{30} *n*-alkanes, Programmed Temperature from 40°C to 350°C at 3°C/min)

Common Name	Molecular Formula	LRI_{lib}	LRI_{exp1}	LRI_{exp2}	LRI_{exp3}	$LRI_{average}$	Repeatability Intralaboratory	ΔLRI
Diacetone alcohol	C6 H12 O2	834	837	836	836	836	0.6	2
n-Hexanol	C6 H14 O	867	868	868	869	868	0.6	1
Isoamyl acetate	C7 H14 O2	871	874	875	874	874	0.6	3
2,5-Dimethyl-pyrazine	C6 H8 N2	912	909	910	910	910	0.6	−2
2,3-Dimethyl-pyrazine	C6 H8 N2	916	915	915	916	915	0.6	−1
α-Thujene	C10 H16	927	926	926	927	926	0.6	−1
Sabinene	C10 H16	972	972	972	973	972	0.6	0
β-Pinene	C10 H16	978	977	978	978	978	0.6	0
δ-3-Carene	C10 H16	1009	1009	1009	1010	1009	0.6	0
Isopentyl butyrate	C9 H18 O2	1057	1054	1055	1055	1055	0.6	−2
Terpinolene	C10 H16	1086	1087	1087	1087	1087	0.0	1
para-Cymenene	C10 H12	1093	1090	1091	1091	1091	0.6	−2
Linalyl formate	C11 H18 O2	1212	1211	1211	1211	1211	0.0	−1
Neral	C10 H16 O	1238	1239	1238	1239	1239	0.6	1
Bornyl acetate	C12 H20 O2	1285	1285	1285	1285	1285	0.0	0
α-Longipinene	C15 H24	1352	1351	1352	1352	1352	0.6	0
α-Copaene	C15 H24	1378	1379	1378	1378	1378	0.6	0
trans-Geranyl acetate	C12 H20 O2	1380	1377	1378	1378	1378	0.6	−2
β-Dihydroionone	C13 H22 O	1438	1435	1434	1435	1435	0.6	−3
trans-α-Amylcinnamaldehyde	C14 H18 O	1667	1666	1667	1667	1667	0.6	0

Note: LRI_{lib}: LRI present in FFNSC 3, $LRI_{exp,i}$: LRI obtained from the intra-day repeated analysis, ΔLRI: $LRI_{average}-LRI_{lib}$.

Darmstadt, Germany, part#: 28471-U) and Supelcowax-10 (Merck KGaA, Darmstadt, Germany, part#: 24079) with *n*-alkanes (Merck KGaA, Darmstadt, Germany, part#: 49451-U), FAMEs (Merck KGaA, Darmstadt, Germany, part#: 49453-U), and FAEEs (Merck KGaA, Darmstadt, Germany, part#: 49454-U) in the FFNSC 3. Especially the latter provides a great identification tool for the essential oil compounds. Using the MS spectral similarity and two LRIs obtained on two columns with different polarity (one polar and one non-polar) and suitable homologue series, three orthogonal analytical data with high accuracy can be achieved, resulting in an univocal peak assignment also for problematic isomers, terpenes, and pheromones.

8.5 RETENTION INDEX COMPILATIONS

Some of the most important retention index compilations are available in printed format, such as published as books and widely used as references, as those authored by Jennings and Shibamoto (1980) (Huethig Verlag: Heidelberg, Germany) containing MS spectra and KI on methyl silicone and on PEG stationary phase of more than 1150 flavor and fragrance compounds and by Adams (2007) (FarHawk, Fineview, NY, US), including spectral information and KI and AI of 2205 essential oil constituents. A further available library book is the *Sadtler Standard Gas Chromatography Retention Index Library*, reporting multi-RI on three different columns of over 2000 compounds (Sadtler Research Laboratories, 1984). An important reference dataset is the Joulain and König hydrocarbon sesquiterpene MS library of 307 components along with their RI calculated on a 25 m fused-silica capillary column coated with 100% dimethyl polysiloxane (Joulain and König, 1998). Worth noting is the compilation and evaluation of Babushok of the average or median RIs with standard deviations and confidence intervals of over 500 frequently reported essential oil constituents from the National Institute of Standards and Technology (NIST) data collection (Babushok et al., 2011). RI databases also can be available from the internet. The most known online RI datasets are the LRI & Odor Database (Mottram), the Flavornet (Acree and Arn, 2004), and the Pherobase (El-Sayed, 2003). The latter is the largest database of behavior-modifying chemicals with over 30,000 entries, among them about 3500 semiochemicals. Furthermore, related KIs are also reported according to cited values on different stationary phases in scientific papers. The website includes an interactive feature for the calculation of LRI of the target molecule and then allows an automatic search in the database according to the LRI filter window setting. Flavornet (Acree and Arn, 2004) lists 738 odorants along with their chromatographic and sensory properties. Chromatographic data include KI calculated with *n*-alkanes and ethyl esters on four different columns. The first-mentioned database reports over 9000 LRI values of more than 5000 odor compounds identified in food samples. The listed LRIs calculated on different stationary phases with *n*-alkanes were obtained from the scientific literature. On the website, a search feature helps the user, providing filters for different column types, molecular weight, and LRI within the filter window.

On the market, several MS databases can be found in user-friendly electronic format; however, only a few of them also include LRI information, particularly experimental LRI data. The most known of them are the NIST MS 17 (National Institute of Standards and Technology, Gaithersburg, MD, US), the Mass Finder 4 (Detlev H., Hochmuth Scientific Consulting, Hamburg, Germany), Adams Library 4th ed. (FarHawk, Fineview, NY, US), and FFNSC 3 (Shimadzu, Kyoto, Japan and John Wiley & Sons, Hoboken, NJ, US). The first-mentioned MS library is a comprehensive data collection, which contains 404,045 retention indices for 99,400 compounds, including 72,361 of which are in the EI library. As previously mentioned, these RI values comprise different type of RIs, and they are extracted from scientific literature or predicted, average, or median values. The latter three databases are pronouncedly dedicated to the flavor and fragrance field, containing mainly hydrocarbon and oxygenated mono- and sesquiterpenes, other essential oil constituents, and synthetic odor and flavor compounds. The Mass Finder 4 library reports the MS spectra of approximately 2000 terpenoids and related constituents commonly found in essential oils, along with their substance name, molecular formula, graphical chemical structure, and accurate experimental retention indices on non-polar stationary phase (DB-1), providing a highly reliable compound search. The Adams library is the digital version of the book mentioned previously, listing exclusively essential oil components with their experimentally obtained RI values. Thanks to the electronic format, it can be widely used for automatic search of real natural sample constituents. The innovative FFNSC 3 library with multi-LRI feature is commercialized in two electronic formats, in Shimadzu format for Shimadzu instrumentation and in NIST format compatible with further producers, by Shimadzu and Wiley, respectively. As the name implies, it contains the MS spectra of about 3500 synthetic and natural flavor and fragrance compounds and their related CAS registry number, systematic and common name, molecular weight and formula, and five different experimental LRIs on three stationary phases calculated with *n*-alkanes, FAMEs, or FAEEs. Due to the multi-LRI option, this library provides an excellent tool for the confirmed identification by three analytical data of isomers or terpenes.

A promising future prospective, an LRI database in the combination of FTIR spectral information of 1500 flavor and fragrance compounds, was introduced recently (Utczás et al., 2018). The library contains basic information of the listed compounds and eight experimental LRI values on non-polar, semi-non-polar and polar stationary phases calculated with *n*-alkanes, FAMEs, and FAEEs in all scientifically correct combinations. Furthermore, it contains the high-quality FTIR spectra of each molecule acquired by a novel analytical technique, GC coupled with condensed phase FTIR. Thanks to the solid phase deposition in the FTIR after the separation step, the centrifugal distortion of the molecules present in gas phase can be avoided, providing excellent spectral resolution. Thus, this technique can be a complementary system to the GC-MS-LRI using the FTIR spectrum with two orthogonal LRI or it can be also coupled with a normal GC-MS, giving the possibility to use two different spectral data and one LRI value in the identification procedure. Summarizing, this innovative library can be a solution for the identification of compounds with similar MS spectral fragmentation and retention behaviors, such as in the case of pheromone isomers.

The determination of the quality or, in the case of natural products, of biological or geographical origin and authenticity is highly based on the enantiomeric ratio of certain compounds. The use of GC-MS in this field is linked with the introduction of chiral stationary phases with cyclodextrin chiral selectors in the late 1980s (König et al., 1988; Schurig and Novotny, 1988). According to the MS spectrum, the two enantiomers cannot be distinguished due to the identical fragmentation; thus, further information is necessary for their determination. However, LRI can be a useful tool as a secondary identification parameter; careful circumspection is required. Noteworthy, a GC-MS library was proposed by Liberto et al. containing enantiomers of 134 racemates including their LRI on four different cyclodextrin derivatives calculated with *n*-alkanes. An effective operation in the correct enantiomer determination was demonstrated using an adequate retention index allowance in the automatized search. The multi-LRI feature provided a solution to choose the most suitable stationary phase for the separation of usually poorly or inseparable enantiomers (Liberto et al., 2008).

8.6 RESEARCH SOFTWARE

The introduction of suitable spectral searching software significantly boosted the identification of the compounds rather than comparing the spectra of the sample with those in the literature. The most widely used MS spectral searching software is the NIST MS Search 2.0 (National Institute of Standards and Technology). The major part of the chromatographic post run software contains a built-in connection with the NIST MS Search. The software is able to compare with its specific searching algorithm the target and the library spectra, presenting also a differential comparison between the two spectra and also a statistical analysis about the possible candidates considering 1000 as similarity for the perfect match (Figure 8.1).

Furthermore, the combination of NIST MS Search with AMDIS (Automated Mass Spectral Deconvolution and Identification System) provides the possibility to the automatic use of the LRI information also (Figure 8.2). After conversion in .msl (AMDIS format), the MS spectral libraries including LRI values can be applied easily. Using a "calibration" run of a homologue series, the

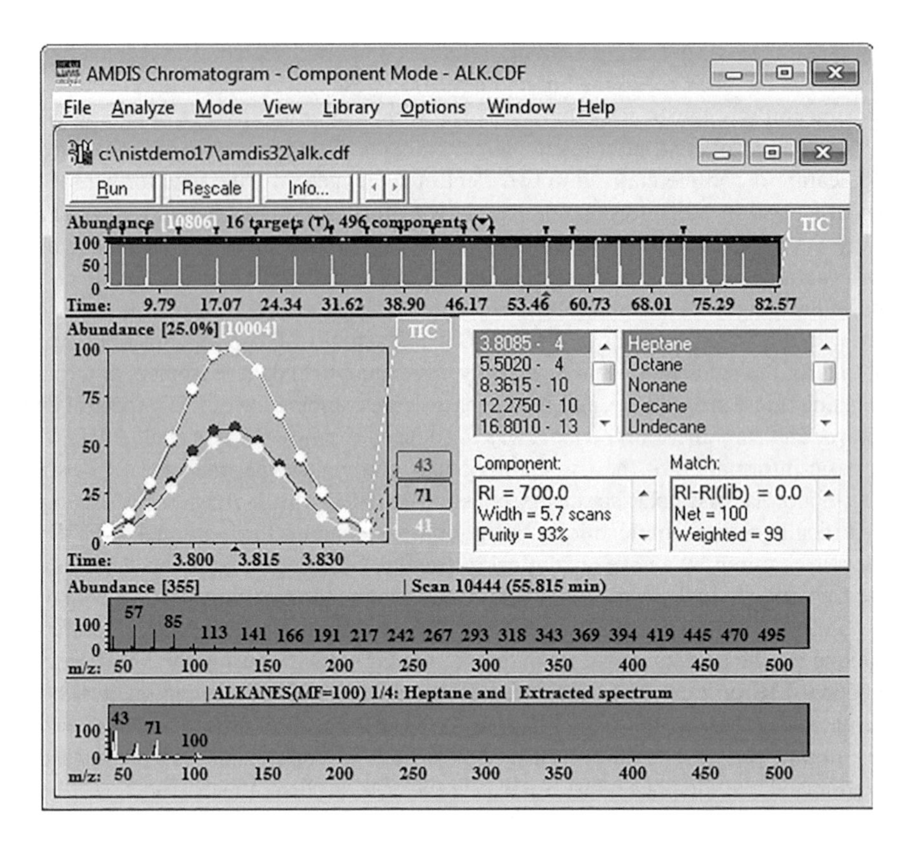

FIGURE 8.1 NIST MS Search 2.0 MS spectral search features in the identification of a cold pressed *Citrus aurantifolia* oil.

software can automatically identify the members and assign the conventional LRI values, then calculate the LRI of the analyzed compounds and search according to MS spectral information and LRI filter.

Zhang et al. developed a method to employ NIST 2008 retention index database information for molecular retention matching via constructing a set of empirical distribution functions (DFs) of the absolute retention index deviation to its mean value. The DF information was further implemented into a software program called iMatch. The performance of iMatch was evaluated using experimental data of a mixture of standards and metabolite extract of rat plasma with spiked-in standards. About 19% of the molecules identified by ChromaTOF were filtered out by iMatch from the identification list of mass spectral matching, while all of the spiked-in standards were preserved. The analysis results demonstrated that using the RI values, via constructing a set of DFs, can improve the spectral matching-based identifications by reducing a significant portion of false positives (Zhang et al., 2011).

NIST MS Search can be used in combination with further post-run software. One of these with an excellent LRI feature is ChromatoPlus Spectra 2018 (Chromaleont SRL, Messina, Italy). This program is suitable for complete quantitative and qualitative analysis. The qualitative analysis has two connected parts. The MS spectral search is carried out by the NIST MS Search using an MS library in NIST MS Search format, while an integrated algorithm in ChromatoPlus Spectra calculates the LRI of the target compounds and then filters the results obtained in the MS spectral search (Figure 8.3) according to the LRI library (.lribin format) and the LRI allowance set.

For this process, after the analysis of the reference homologue series, the data file should be elaborated and the conventional LRI values should be assigned to the identified peaks in the software. Prior to the search, the previously integrated homologue series data file should be loaded, library and suitable LRI should be selected, and searching requirements (MS similarity, LRI filter window)

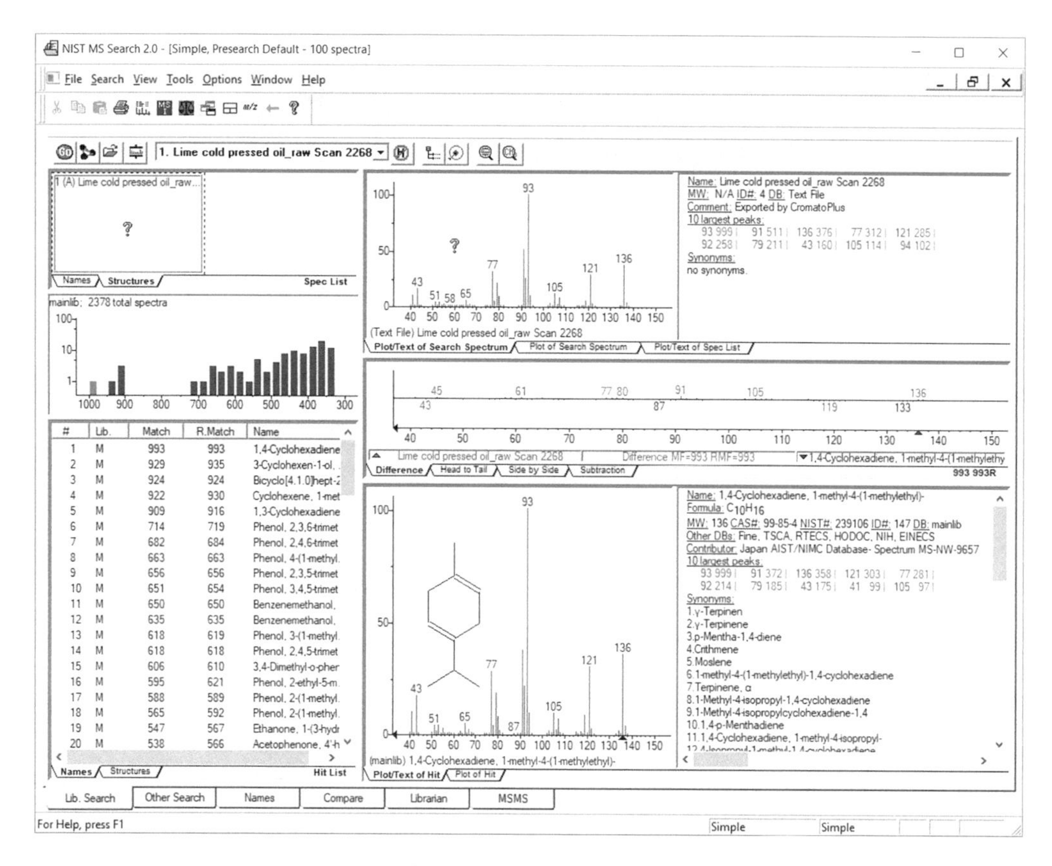

FIGURE 8.2 Automatic identification of the members of the *n*-alkane homologue series and assignment of conventional LRIs by AMDIS.

should be established (Figure 8.4). Using the appropriate settings, the combined MS and LRI search, a single match can be achieved, removing the wrong candidates from the primary list.

A further very useful post-run software, dedicated to the Shimadzu instrumentation, is the GC-MS Solution (latest version 4.41). It is worth noting that the files coming from other instrumentation can be elaborated transforming them in the universal.cdf format. Basically, this program works similarly to the previous one; the main difference is the integrated MS and LRI feature. This unique software provides a complete post-run operation, including integration, automatic LRI calculation, univocal identification based on MS, and LRI information (Figure 8.5). For the identification process, the steps are the same; the advantage is that it is even more user-friendly due to the unified functions.

8.7 APPLICATION OF RETENTION INDICES IN GC-FID AND GC-MS ESSENTIAL OIL ANALYSES

The composition of essential oils is characterized by a wide diversity of components belonging to several chemical classes and is mainly represented by mono- and sesquiterpene hydrocarbons and their oxygenated (hydroxyl and carbonyl) derivatives, along with aliphatic aldehydes, alcohols, and esters. Since the early stages of research in this field, attention was devoted to the development of methods to acquire deeper knowledge on the profiles of the essential oils that was made troublesome due to the complexity of those samples. The utilization of retention indices, when combined with other characterization methods, such as gas chromatography (GC) coupled to mass spectrometry and related hyphenated techniques, was shown to be an effective tool for the identification of flavor and fragrance compounds.

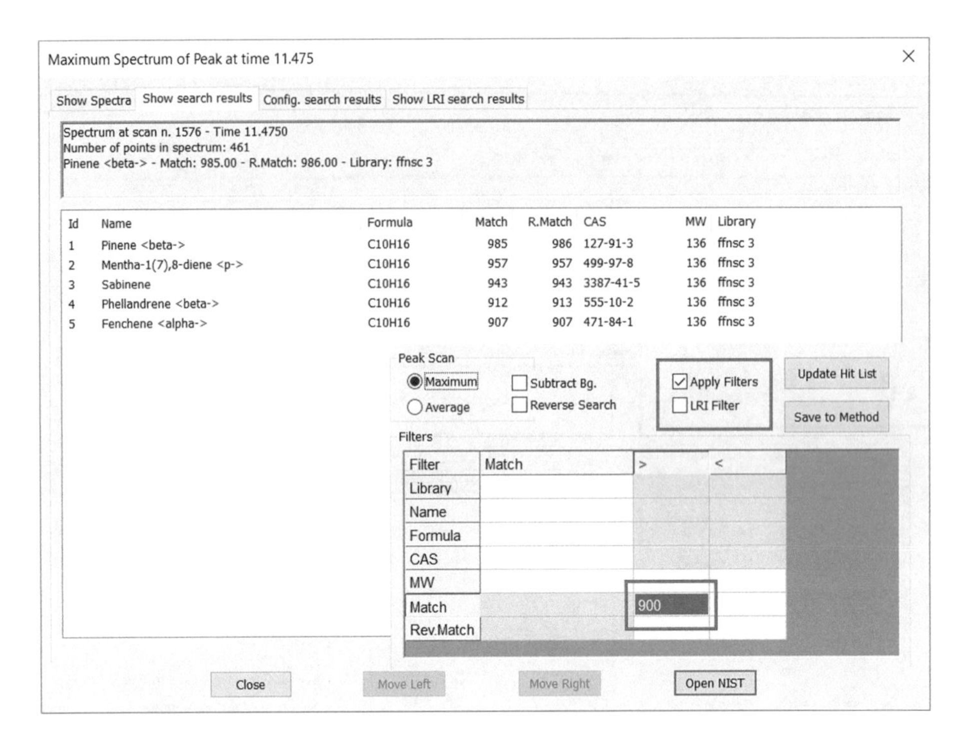

FIGURE 8.3 Automatic compound search carried out by ChromatoPlus Spectra 2018 using 900 of MS spectral similarity without LRI filter according to the FFNSC 3 library in the analysis of *Pistacia vera*. All the listed candidates meet the requirements.

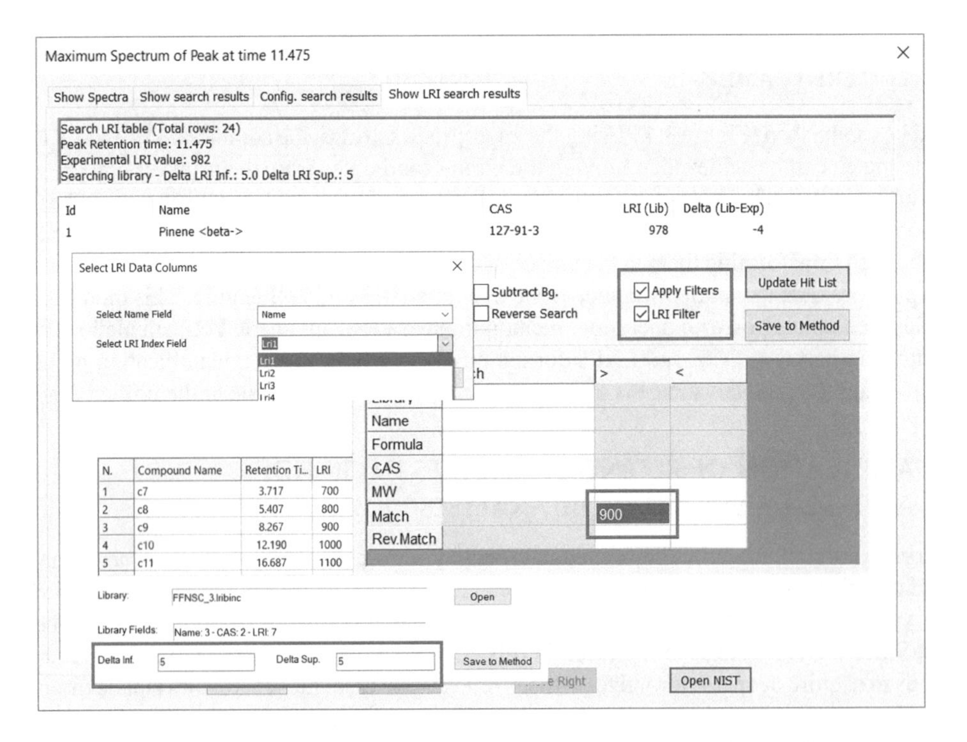

FIGURE 8.4 Combined use of 900 of MS spectral similarity and ±5 LRI units of LRI filter window in ChromatoPlus Spectra 2018 providing a single match for the target compound in the analysis of *Pistacia vera* according to the FFNSC 3 library.

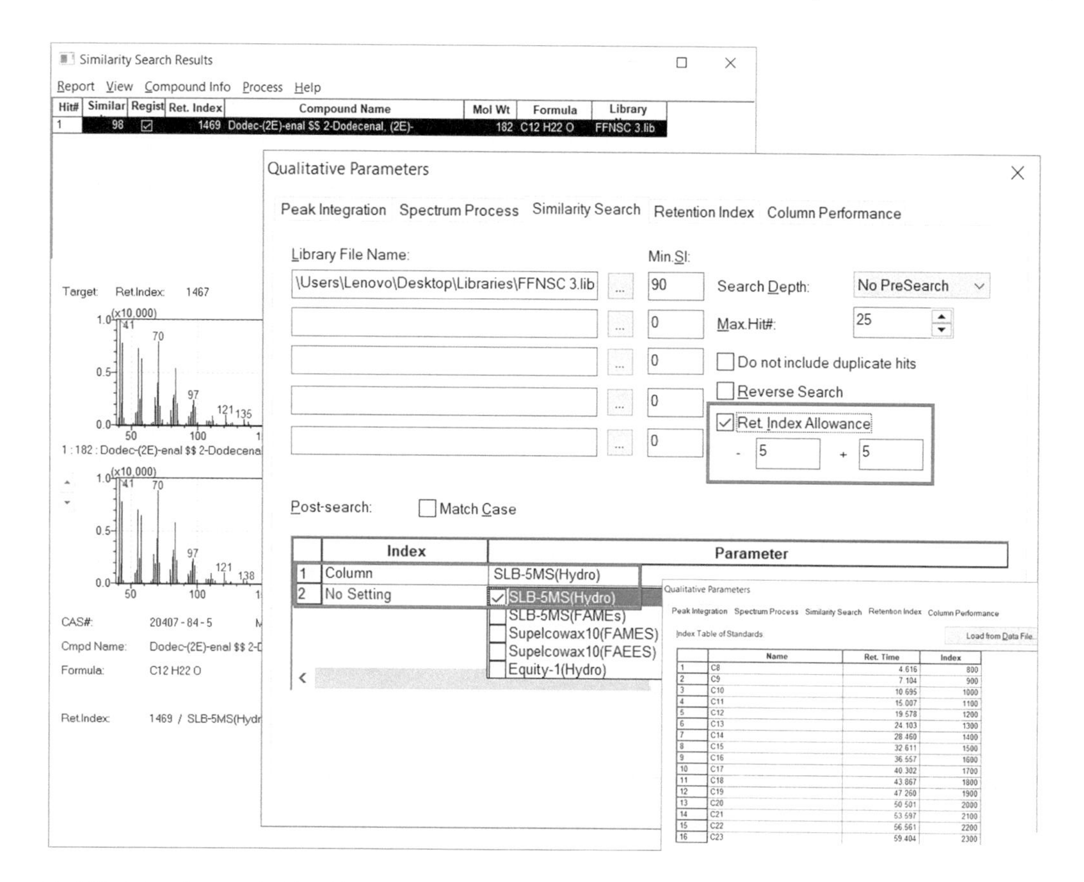

FIGURE 8.5 Automatic identification of a target compound in *Coriandrum sativum* leaf oil using 90% of MS similarity and ±5 LRI units LRI allowance by GC/MS Solution according to the FFNSC 3 library. The LRI values were calculated on the SLB-5 ms stationary phase according to *n*-alkanes (Hydro).

Considering gas chromatographic analyses using flame ionization (FID), thermal conductivity (TCD) or other detectors, which do not provide structural information of the analyzed molecules, retention indices are used as the primary criterion for peak assignment. Moreover, with the growing popularity of GC coupled to mass spectrometry (GC-MS), a large number of studies has used mass spectral information for peak identification; the most frequent and simple identification method consisting of the comparison of the acquired unknown mass spectra with those contained in a reference MS library. However, as is well-known, isomers or similar substances, when analyzed by means of GC-MS, can be identified in an inaccurate or unreliable manner, a drawback which is often observed in essential oil analyses.

The composition of essential oils is mainly represented by terpenes with mass spectra usually very similar, and peak assignment becomes difficult and sometimes impracticable. In order to increase the reliability of the analytical results and to address the qualitative determination of compositions of complex samples by GC-MS, retention indices can be a very useful tool. The use of retention indices in conjunction with the structural information provided by GC-MS is widely accepted and routinely used to confirm the identity of compounds. The analysis and identification of *Cordia verbenacea* essential oil (Sciarrone et al., 2017) is shown in Figure 8.6 and Table 8.5, respectively.

In some cases, even when using GC-MS, a further method may still be required to satisfactorily characterize a compound, such as nuclear magnetic resonance (NMR) or infrared (IR) spectroscopy.

Generally, when associated with its non-polar GC retention index, the mass spectrum of a given sesquiterpene is usually sufficient for its identification, provided the reference data have been recorded using authentic samples. Indeed, for this class of compounds, there would be no need to adopt a polar stationary phase, which could even lead to misinterpretations caused by possible changes in

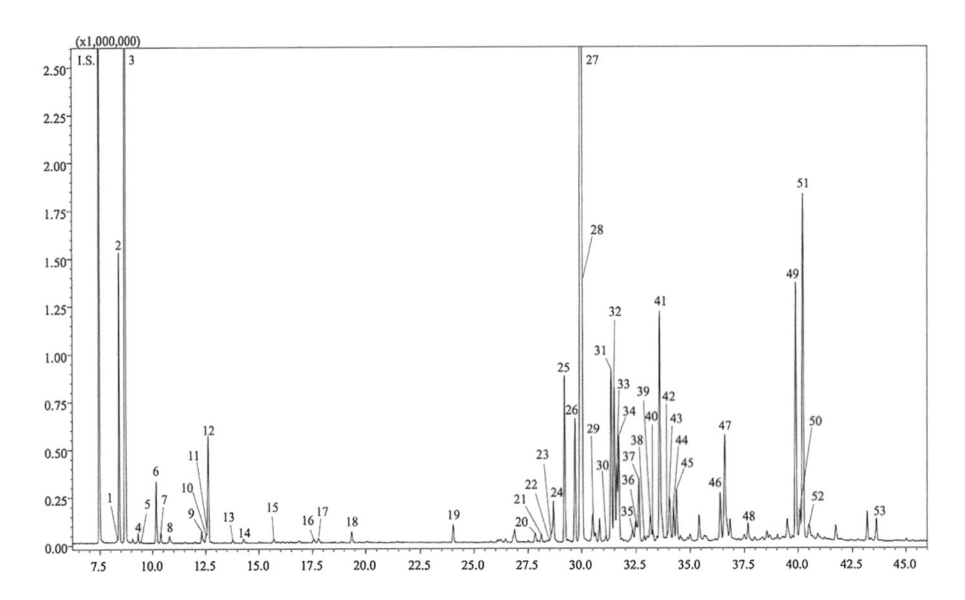

FIGURE 8.6 GC/MS chromatogram of *Cordia verbenacea* essential oil. For peak identification refer to Table 8.5. Reproduced with kind permission from Elsevier. (From Sciarrone D. et al. 2017. *J. Chromatogr. A*, 1524: 246–253. With permission.)

TABLE 8.5
Volatile Compounds of *Cordia verbenacea* Essential Oil Identified by GC/MS Solution Using FFNSC 3, Adams Library 4th edition and Wiley and NIST 14, MS database (LRI_{lib}) and Experimental (LRI_{exp}) LRI Values, Database Spectral Similarity (Sim%), Response Factor (R.F.), Absolute Content as g/100 g, Relative Abundances (as Area%), Repeatability (CV%), and Chemical Class (Class)

No	Compound	LRI_{lit}	LRI_{exp}	Sim%	R.F.	g/100g	Area%	CV%	Class
1	3-Methylapopinene	927	923	94	1	0.03	0.03	3.07	M
2	α-Thujene	927	925	97	1	2.59	2.69	0.42	M
3	α-Pinene	933	934	97	1	24.48	25.32	0.24	M
4	Camphene	954	950	96	1	0.09	0.09	2.67	M
5	Thuja-2,4(10)-diene	953	954	95	1	0.01	0.01	8.19	M
6	Sabinene	972	972	97	1	0.64	0.66	1.02	M
7	β-Pinene	979	978	95	1	0.12	0.12	1.49	M
8	Myrcene	991	989	92	1	0.12	0.13	6.21	M
9	p-Cymene	1025	1025	96	1	0.13	0.13	1.76	M
10	Limonene	1028	1030	92	1	0.1	0.1	2.34	M
11	β-Phellandrene	1031	1029	91	1	0.13	0.13	7.28	M
12	Eucalyptol	1032	1032	89	1.3	1.41	1.12	1.17	MO
13	γ-Terpinene	1058	1059	93	1	0.04	0.04	5.11	M
14	*cis*-Sabinene hydrate	1069	1071	91	1.3	0.08	0.08	0.92	MA
15	*trans*-Sabinene hydrate	1099	1102	90	1.3	0.09	0.08	1.16	MA
16	*trans*-Pinocarveol	1141	1142	95	1.3	0.13	0.1	7.15	MA
17	*trans*-Verbenol	1145	1147	91	1.3	0.12	0.11	3.21	MA
18	Terpinen-4-ol	1184	1182	91	1.3	0.23	0.18	0.99	MA
19	Bornyl acetate	1285	1285	95	1.6	0.34	0.22	0.3	ME

(Continued)

TABLE 8.5 (*Continued*)
Volatile Compounds of *Cordia verbenacea* Essential Oil Identified by GC/MS Solution Using FFNSC 3, Adams Library 4th edition and Wiley and NIST 14, MS database (LRI_{lib}) and Experimental (LRI_{exp}) LRI Values, Database Spectral Similarity (Sim%), Response Factor (R.F.), Absolute Content as g/100 g, Relative Abundances (as Area%), Repeatability (CV%), and Chemical Class (Class)

No	Compound	LR_{llit}	LRI_{exp}	Sim%	R.F.	g/100g	Area%	CV%	Class
20	Cyclosativene	1367	1371	93	1	0.14	0.15	1.34	S
21	α-Copaene	1375	1378	92	1	0.11	0.11	0.92	S
22	β-Bourbonene	1382	1384	92	1	0.02	0.02	4.76	S
23	7-*epi*-Sesquithujene	1387	1388	93	1	0.16	0.17	1.31	S
24	β-Elemene	1390	1391	92	1	0.63	0.71	2.87	S
25	Funenbrene	1403	1403	98	1	2.17	2.24	0.05	S
26	*cis*-α-Bergamotene	1416	1415	97	1	1.65	1.71	0.33	S
27	α-Santalene	1418	1422	95	1	17.3	17.9	0.13	S
28	(*E*)-Caryophyllene	1424	1422	94	1	7.4	7.66	0.25	S
29	*trans*-α-Bergamotene	1432	1435	92	1	0.45	0.46	0.7	S
30	*epi*-β-Santalene	1446	1447	92	1	0.09	0.1	3.34	S
31	Sesquisabinene	1455	1455	98	1	2.65	2.74	0.15	S
32	α-Humulene	1454	1458	97	1	2.13	2.21	0.07	S
33	β-Santalene	1459	1461	96	1	0.93	0.96	2.19	S
34	9-*epi*-(*E*)-Caryophyllene	1464	1463	97	1	1.43	1.48	1.15	S
35	γ-Curcumene	1482	1478	92	1	0.14	0.15	5.49	S
36	Germacrene D	1480	1483	92	1	0.23	0.23	1.11	S
37	*trans*-β-Bergamotene	1483	1485	95	1	0.92	0.95	0.42	S
38	β-Selinene	1492	1491	95	1	0.06	0.06	1.58	S
39	Bicyclogermacrene	1497	1498	95	1	0.46	0.47	1.09	S
40	γ-Muurolene	1497	1501	95	1	0.14	0.15	1.13	S
41	β-Bisabolene	1508	1509	95	1	3.36	3.48	0.34	S
42	Cubebol	1518	1519	93	1.3	0.28	0.22	1.46	SA
43	δ-Cadinene	1518	1521	93	1	0.83	0.86	1.72	S
44	β-Sesquiphellandrene	1523	1525	93	1	0.46	0.48	1.25	S
45	(*E*)-γ-Bisabolene	1528	1528	93	1	0.72	0.74	1.34	S
46	Spatulenol	1576	1580	90	1.3	1.12	0.89	1.9	SA
47	Caryophyllene oxide	1587	1585	90	1.5	3.76	2.59	0.5	SO
48	Humulene epoxide II	1613	1614	94	1.5	0.5	0.35	1.36	SO
49	(*E*)-α-Bergamotenal		1671		1.3	4.76	4.46	0.42	SAld
50	(*Z*)-α-Santalol	1676	1678	95	1.3	0.84	0.79	0.2	SA
51	(*E*)-α-Santalal		1680		1.3	6.29	5.89	0.5	SAld
52	(*Z*)-α-trans-Bergamotol	1688	1688	92	1.3	0.83	0.66	1.52	SA
53	(*Z*)-α-Santalol acetate	1778	1776	94	1.5	0.53	0.36	4.2	SE
Monoterpene hydrocarbons						28.48	29.45		
Sesquiterpene hydrocarbons						44.58	46.19		
Oxygenated monoterpenes						2.4	1.89		
Oxygenated sesquiterpenes						18.91	16.21		

Note: Compounds isolated by using prep-MDGC system and characterized by using NMR spectroscopy.*Abbreviations:* M, monoterpene; S, sesquiterpene; Ald, aldehyde; E, ester; O, oxide; A, alcohol.

the retention behavior of sesquiterpene hydrocarbons as a result of column aging or deterioration, and depending on the manufacturer. Moreover, in agreement with the authors' experience, for each individual sesquiterpene, it is necessary to pay great attention to the retention index and the mass spectrum registration, since many compounds with rather similar mass spectra may be eluted in a narrow range; on a non-polar column, in fact, more than 160 compounds can be eluted between two successive reference *n*-alkanes.

The application of the retention index system has been used in other gas chromatographic techniques, such as olfactometry, that allow an analytical aroma screening. In fact, it enables the differentiation of a multitude of volatiles, previously separated by GC, into odor-active and non-odor-active compounds, related to their concentrations existing in the matrix under investigation. The retention index when GC-O systems are equipped with detectors that do not provide structural information is commonly associated to its odor description, supporting reliable peak assignment.

8.8 GAS CHROMATOGRAPHIC ENANTIOMER CHARACTERIZATION SUPPORTED BY RETENTION INDICES

As reported by Dugo et al. (1992), the main application of enantioselective GC (Es-GC) to flavor and fragrance materials concerns the precise determination of enantiomeric compositions and ratios, which characterize the matrix, determining its quality and, in the case of natural materials, its geographic origin, biogenesis, authenticity, and extraction technique applied (Dugo et al., 1992).

The calculation of linear retention indices of optically active compounds can be problematic due to complex retention mechanisms on chiral stationary phases (CSPs) (Liberto et al., 2008).

For many years, GC-MS has been the benchmark technique for the qualitative analysis of flavor and fragrance volatiles. However, when considering Es-GC-MS, the identification of a target compound can be hindered due to the similar fragmentation pattern of enantiomers. The automation of library search algorithms incorporating linear retention indices as part of the match criteria can also be extended to enantioselective GC-MS analysis, assisting in the discrimination of enantiomers. An enantioselective MS library is characterized by two features, the mass spectrum, which could be useful to determine the position of enantiomers in a chromatogram, and the retention index, which can be considered as a tool to discriminate these isomers. Moreover, considering that a universal chiral selector with widespread capability for enantiomer separation is not available, and thus effective optical separation of all chiral compounds present in a matrix may be unachievable on a single chiral column, the use of a set of columns coated with distinct CD derivatives as chiral selectors is necessary.

8.9 RETENTION INDICES APPLIED TO MULTIDIMENSIONAL GAS CHROMATOGRAPHIC ANALYSIS

As is well known, the determination of the structure of a component directly from complex multicomponent mixtures such as essential oils is difficult, if not impossible, due to many overlapping peaks in the gas chromatograms, but the use of multidimensional gas chromatography (MDGC) makes the analysis more reliable, providing improved resolution and analyte peak capacity and reducing interfering chemical background.

Moreover, the application of first and second dimension's retention indices on the respective phases can provide additional supporting information to reject incorrectly matched peaks. As reported by Nolvachai et al. in 2017 (Nolvachai et al., 2017), linear retention indices on the first phases ${}^{1}I$ and on the second phases ${}^{2}I$ can be calculated directly from retention time (t_R) of a target analyte normalized to the reference compound series scale according to the van den Dool and Kratz relationship in temperature (T)-programmed separation (van den Dool and Kratz, 1963).

For isothermal ${}^{2}D$ separation (which applies to the fast ${}^{2}D$ separation in GC $\times$ GC), ${}^{2}I$ can be calculated according to Kovats index as reported by M. Jiang et al. (2015). Alternatively, I values of a peak can also be approximated according to the plot of reference compound t_R values vs. carbon number, as suggested by Chin et al. (2012).

The use of a retention index system in comprehensive analysis has also been widely discussed in the literature and is still an emerging topic of research. Actually, there are only available comprehensive sets of retention indices for a few GC phase types, and to have the additional information provided by index values, only a few phases can be used.

To calculate I values for peaks of interest is necessary to inject under the same experimental conditions reference compounds whose elution times span the retention times of the peaks of interest. In heart-cut MDGC, as reported by Nolvachai et al. (2017), s alkane positions in ^1D can be obtained directly by injection of alkanes into MDGC with detection at the end of the ^1D column. Another approach to obtain reference compound position in ^2D consists of experimental injection of reference compounds onto the system followed by heart-cutting and then trapping at the inlet of the ^2D column. The heart-cut alkanes and sample regions are analyzed under the exact same conditions, and ^2I values may be calculated.

In GC × GC, ^1t_R can be approximated from the time of the modulated peak with maximum response and subtraction from the ^2t_R value. The generation of reference compound positions in ^2D can be difficult, but they can be predicted based on 1DGC isothermal data using the same experimental column with corrected flow calculation, such as for prediction of saturated FAME positions in GC × GC, as suggested by Kulsing et al. (2014) and Nolvachai et al. (2015). In GC × GC with T-programmed separation, calculation of ^2I cannot be directly estimated by comparation of target analyte ^2t_R with reference literature, since ^2t_R of compounds vary according to the prevailing T at the elution point. In 2006 and in 2008, Bieri et al. (Bieri and Marriott, 2006, 2008) investigated the variation of alkane retention time with T exploiting their "isovolatility" relationships in order to calculate ^2I values, by multiple injections of the reference compound mixture into the ^2D column.

8.10 CONCLUDING REMARKS

The use of the combination of spectral information along with single or multi-LRI values is a consolidated method not only in normal chromatographic analysis but also in comprehensive applications or in chiral separations. A reliable identification procedure is undoubtably unimaginable without the support of almost one LRI; nevertheless the right choice requires caution. Firstly, the appropriate selection of the LRI type and homologue series according to the analytical parameters (polarity of the stationary phase, temperature program) is a crucial point in the accuracy of the calculated LRIs. Secondly, for the LRI comparison, the use of trustworthy data is fundamental. Among the multitude of literature information and commercially available datasets, the most suitable and reliable one has to be chosen. In particular, the use of the experimental LRI values as reference along with the application of the specified analytical conditions can boost the reliability of the compound identification. Thanks to the widely applied experimental LRI databases, purchasing all the standards or running in different conditions the same analysis for the correct reproduction of the cited LRI values can be avoided. Thirdly, exploiting the automatized possibilities provided by the searching software, the time-consuming LRI calculation and identification can be replaced with a fast and reliable peak assignment. Even though, using the appropriate settings in the software, spectral and LRI search can result a very repeatable and good quality match reducing the competence demand of the laboratory staff dealing with the routine identification, an in-depth expertise is still a requirement for a correct characterization of complex essential oil samples.

REFERENCES

Acree, T. and H. Arn, 2004. *Flavornet*. New York: Cornell University, NYSAES. http://www.flavornet.org/index.html (accessed April 19, 2019).

Adams, R. P., 2007. *Identification of Essential Oil Components by gas Chromatography/Mass Spectrometry*, 4th ed. NY, USA: FarHawk, Fineview.

Babushok, V. I., P. J. Linstrom, and I. G. Zenkevich, 2011. Retention indices for frequently reported compounds of plant essential oils. *J. Phys. Chem. Ref. Data.*, 40(4): 043101–043147. https://aip.scitation.org/doi/pdf/10.1063/1.3653552?classpdf.

Balaban, A. T., 1982. Highly discriminating distance-based topological index. *Chem. Phys. Lett.*, 89: 399–404.

Berezkin, V. G., 1974. Application of gas chromatography in petrochemistry and petroleum refining. *J. Chromatogr. A*, 91: 559–582.

Bieri, S., and P. J. Marriott, 2006. Generating multiple independent retention index data in dual-secondary column comprehensive two-dimensional gas chromatography. *Anal. Chem.*, 78: 8089–8097.

Bieri, S., and P. J. Marriott, 2008. Dual-injection system with multiple injections for determining bidimensional retention indexes in comprehensive two-dimensional gas chromatography. *Anal. Chem.*, 80: 760–768.

Blumberg, L. M., and M. Klee, 1998. Method translation and retention time locking in partition GC. *Anal. Chem.*, 70: 3828–3839.

Bruchmann, A., P. Zinn, and C. M. Haffer, 1993. Prediction of gas chromatographic retention index data by neural networks. *Anal. Chim. Acta*, 283: 869–880.

Castello, G., 1999. Retention index systems: Alternatives to the n-alkanes as calibration standards. *J. Chromatogr. A*, 842: 51–64.

Castello, G., and G. Testini, 1996. Determination of retention indices of polychlorobiphenyls by using other compounds detectable by electron-capture detection or selected polychlorobiphenyls as reference series. *J. Chromatogr. A*, 741: 241–249.

Chin, S. T., G. T. Eyres, and P. J. Marriott, 2012. System design for integrated comprehensive and multidimensional gas chromatography with mass spectrometry and olfactometry. *Anal. Chem.*, 84: 9154–9162.

Dimov, N. J., 1985. Unified retention index of hydrocarbons separated on squalane. *J. Chromatogr. A*, 347: 366–374.

Dugo, G., G. Lamonica, A. Cotroneo et al. 1992. High resolution gas chromatography for detection of adulterations of citrus cold-pressed essential oils. *Perfumer Flavorist*, 17: 57–74.

El-Sayed, A. M. 2003. *The Pherobase: Database of Insect Pheromones and Semiochemicals*. Lincoln, New Zealand: Hort Research. http://www.pherobase.com (accessed April 19, 2019).

Evans, M. B., and J. F. Smith, 1961a. Gas-liquid chromatography in qualitative analysis: Part I. An interpolation method for the prediction of retention data. *J. Chromatogr.*, 5: 300–307.

Evans, M. B., and J. F. Smith, 1961b. Prediction of retention data in gas-liquid chromatography from molecular formulæ. *Nature*, 190: 905–906.

Evans, M. B., and J. F. Smith, 1962. Gas-liquid chromatography in qualitative analysis: Part. III. The constancy of ΔMe values within homologous series and relations within the periodic table. *J. Chromatogr.*, 8: 303–307.

Harangi, J., 2003. Retention index calculation without n-alkanes-the virtual carbon number. *J. Chromatogr. A*, 993: 187–195.

Heinzen, V. E. F., and R. A. Yunes, 1996. Using topological indices in the prediction of gas chromatographic retention indices of linear alkylbenzene isomers. *J. Chromatogr. A*, 719: 462–467.

James, A. T., and A. J. P. Martin, 1952. Gas-liquid partition chromatography: The separation and micro-estimation of volatile fatty acids from formic acid to dodecanoic acid. *Biochem. J.*, 50: 679–690.

Jennings, W., and Shibamoto, T. 1980. *Qualitative Analysis of Flavour and Fragrance Volatiles by Glass Capillary Gas Chromatography*, 1st ed. New York, NY: Academic Press.

Jiang, M., C. Kulsing, Y. Nolvachai, and P. J. Marriott, 2015. Two-dimensional retention indices improve component identification in comprehensive two-dimensional gas chromatography of saffron. *Anal. Chem.*, 87: 5753–5761.

Joulain, D., and W. A. König, 1998. *The Atlas of Spectral Data of Sesquiterpene Hydrocarbons*. Hamburg: E.B.-Verlag.

Junkes, B. S., R. D. M. C. Amboni, V. E. F. Heinzen, R. A. Yunes, 2002. Use of a semi-empirical topological method to predict the chromatographic retention of branched alkenes. *Chromatographia*, 55: 75–80.

König, W. A., S. Lutz, and G. Wenz, 1988. Modified cyclodextrins? Novel, highly enantioselective stationary phases for gas chromatography. *Angew. Chem. Int. Ed.*, 27: 979–980.

Kováts, E., 1958. Gas-chromatographische charakterisierung organischer verbindungen. teil 1: Retentions indices aliphatischer halogenide, alkohole, aldehyde und ketone. *Helv. Chim. Acta*, 41: 1915–1932.

Kulsing, C., Y. Nolvachai, A. X. Zeng, S. T. Chin, B. Mitrevski, and P. J. Marriott, 2014. From molecular structures of ionic liquids to predicted retention of fatty acid methyl esters in comprehensive two-dimensional gas chromatography. *ChemPlusChem*, 79: 790–797.

Liberto, E., C. Cagliero, B. Sgorbini et al. 2008. Enantiomer identification in the flavour and fragrance fields by "interactive" combination of linear retention indices from enantioselective gas chromatography and mass spectrometry. *J. Chromatogr. A*, 1195: 117–126.

Littlewood, A. B., C. S. G. Phillips, and D. T. Price, 1955. The chromatography of gases and vapours. Part V. Partition analyses with columns of silicone 702 and of tritolyl phosphate. *J. Chem. Soc.*, 1480–1489.

Mjøs, S. A., S. Meier, and S. Boitsov, 2006. Alkylphenol retention indices. *J. Chromatogr. A*, 1123: 98–105.

Mottram, R. The LRI and Odour Database. Flavour Research Group, School of Food Biosciences, University of Reading, UK. http://www.odour.org.uk/index.html (accessed April 19, 2019).

Nolvachai, Y., C. Kulsing, and P. J. Marriott, 2015. Thermally sensitive behavior explanation for unusual orthogonality observed in comprehensive two-dimensional gas chromatography comprising a single ionic liquid stationary phase. *Anal. Chem.*, 87: 538–544.

Nolvachai, Y., C. Kulsing, and P. J. Marriott, 2017. Multidimentional gas chromatography in food analysis. *Trends Analyt. Chem.*, 96: 124–137.

Novák, J., and J. Ružičková, 1974. Generalization of the gas chromatographic retention index system. *J. Chromatogr.*, 91: 79–88.

Poole, C. F., 2003. *The Essence of Chromatography*, 1st ed. Amsterdam: Elsevier.

Rasanen, I., I. Ojanperä, and E. Vuori, 1995. Comparison of four homologous retention index standard series for gas chromatography of basic drugs. *J. Chromatogr. A*, 693: 69–78.

Robinson, P. G., and A. L. Odell, 1971. A system of standard retention indices and its uses the characterisation of stationary phases and the prediction of retention indices. *J. Chromatogr.*, 57: 1–10.

Sadtler Research Laboratories. 1984. *The Sadtler Standard Gas Chromatography Retention Index Library.* Philadelphia, PA: Sadtler Research Laboratories.

Schurig, V., and H. P. Novotny, 1988. Separation of enantiomers on diluted permethylated β-cyclodextrin by high-resolution gas chromatography. *J. Chromatogr. A*, 441: 155–163.

Sciarrone D., D. Giuffrida, A. Rotondo et al. 2017. Quali-quantitative characterization of the volatile constituents in *Cordia verbenacea* D.C. essential oil exploiting advanced chromatographic approaches and nuclear magnetic resonance analysis. *J. Chromatogr. A*, 1524: 246–253.

Shibamoto, T. 1987. In *In Capillary Gas Chromatography in Essential Oil Analysis*, C. Bicchi and P. Sandra (eds.), Heidelberg: Huethig Verlag.

Škrbic, B. D., 1997. Unified retention concept-statistical treatment of Kováts retention index. *J. Chromatogr. A*, 764: 257–264.

Stein, S. E., V. I. Babushok, R. L. Brown, and P. J. Linstrom, 2007. Estimation of Kováts retention indices using group contributions. *J. Chem. Inf. Model.*, 47: 975–980.

Stránský, K., M. Zarovúcka, I. Valterová, and Z. Wimmer, 2006. Gas chromatographic retention data of wax esters. *J. Chromatogr. A*, 1128: 208–219.

Utczás, M., E. Trovato, and L. Mondello, 2018. Gas chromatography coupled with condensed phase FTIR: A novel and reliable technique for flavor and fragrance analysis. *Presented at the 42nd International Symposium on Capillary Chromatography Symposium*, Riva del Garda, Italy.

Van den Dool, H., 1974. *Standardization of GC Analysis of Essential Oils.* Rotterdam: Proefschrift, Rijksuniversiteit te Groningen.

Van den Dool, H., and P. D. Kratz, 1963. A generalization of the retention index system including linear temperature programmed gas-liquid partition chromatography. *J. Chromatogr.*, 11: 463–471.

Veselinović, A. M., D. Velimorović, B. Kaličanin, A. Toropova, A. Toropov, and J. Veselinović, 2017. Prediction of gas chromatographic retention indices based on Monte Carlo method. *Talanta*, 168: 257–262.

Zellner, B. A., C. Bicchi, P. Dugo, P. Rubiolo, G. Dugo, and L. Mondello, 2008. Linear retention indices in gas chromatographic analysis: A review. *Flavour. Fragr. J.*, 23: 297–314.

Zenkevich, I. G., and B. V. Ioffem, 1988. System of retention indices for a linear temperature programming regime. *J. Chromatogr.*, 439: 185–194.

Zhang, J., A. Fang, B. Wang et al. 2011. iMatch: A retention index tool for analysis of gas chromatography-mass spectrometry data. *J. Chromatogr. A*, 1218: 6522–6530.

9 Safety Evaluation of Essential Oils

Constituent-Based Approach Utilized for Flavor Ingredients—An Update

Sean V. Taylor

CONTENTS

9.1 INTRODUCTION

Based on their action on the human senses, plant-derived essential oils have functioned as sources of food, preservatives, medicines, symbolic articles in religious and social ceremonies, and remedies to modify behavior. In many cases, essential oils and extracts gained widespread acceptance as multifunctional agents due to their strong stimulation of the human gustatory (taste) and olfactory (smell) senses. Cinnamon oil exhibits a pleasing warm spicy aftertaste, characteristic spicy aroma, and preservative properties that made it attractive as a food flavoring and fragrance. Four millennia ago, cinnamon oil was the principal ingredient of a holy ointment mentioned in Exodus 32:22–26. Because of its perceived preservative properties, cinnamon and cinnamon oil were sought by Egyptians for embalming. According to Dioscorides (Dioscorides, 50 AD), cinnamon was a breath freshener, would aid in digestion, would counteract the bites of venomous beasts, reduced inflammation of the intestines and the kidneys, and acted as a diuretic. Applied to the face, it was purported to remove undesirable spots. It is not surprising, then, that in 1000 BC, cinnamon was more expensive than gold.

It is not unexpected that cultures throughout history ascribed essential oils and extracts with healing and curative powers, and their strong gustatory and olfactory impacts continue to spur desirable emotions in humans, resulting in their often considerable economic value and cultural importance. However, this cultural importance and widespread demand and (when available) use often occurred with only a limited understanding or acknowledgment of the toxic effects associated with high doses of these plant products. The *natural* origin of these products and their long history of use by humans have, in part, mitigated concerns as to whether these products are efficacious or whether they are safe under conditions of intended use (Arctander, 1969). The adverse effects resulting from the human use of pennyroyal oil as an abortifacient or wild germander as a weight control agent are reminders that no substance is inherently safe independent of considerations of dose. In the absence of information concerning efficacy and safety, recommendations for the quantity and quality of natural products, including essential oils, to be consumed as a medicine remain ambiguous. However, when the intended use is, for example, as a flavor that is subject to governmental regulation, effective and safe levels of use are defined by fundamental biological limits and careful risk assessment.

Flavors derived from essential oils (heretofore known as flavors) are complex mixtures that act directly on the gustatory and olfactory receptors in the mouth and nose leading to taste and aroma responses, respectively. Saturation of these receptors by the individual chemicals within the flavors occurs at very low levels in animals. Hence, with few exceptions, the effects of flavors are self-limiting. The evolution of the human diet is tightly tied to the function of these receptors. Taste and aroma not only determine what we eat but often allow us to evaluate the quality of food and, in some cases, identify unwanted contaminants. The principle of self-limitation taken together with the long history of use of essential oils as flavors in food creates initial conditions upon which has been concluded that these complex mixtures are safe under intended conditions of use. In the United States, the conclusion by the U.S. Food and Drug Administration (21 CFR Sec. 182.10, 182.20, 482.40, and 182.50) that certain oils are "generally recognized as safe" (GRAS) for their intended use was based, in large part, on these two considerations. In Europe and Asia, the presumption of "safe under conditions of use" has been bestowed on essential oils based on similar considerations.

For other intended uses such as dietary supplements or direct food additives, a traditional toxicology approach has been used to demonstrate the safety of essential oils. This relies on performing toxicity tests on laboratory animals, assessing intake and intended use, and determining adequate margins of safety between estimated daily intake by humans and toxic levels resulting from animal studies. Given the constantly changing marketplace and the consumer demand for new and interesting products, however, many new intended uses for these complex mixtures are regularly created, and the exact composition of the essential oil or extract may slightly vary based on processing and desired characteristics. The resources necessary to test all complex mixtures for each intended use are simply not economical. Ultimately, for essential oils that are complex mixtures of chemicals being sold into a competitive marketplace, an approach where each mixture is tested is effective only when specifications for the composition and purity are clearly defined and adequate quality controls are in place for the continued commercial use of the oil. In the absence of such specifications, the results of toxicity testing apply specifically and only to the complex mixture tested. Recent safety evaluation approaches (Schilter et al., 2003) suggest that a multifaceted decision tree approach can be applied to prioritize natural products and the extent of data required to demonstrate safety under conditions of use. The latter approach offers many advantages, both economic as well as scientific, over more traditional approaches. Nevertheless, various levels of resource-intensive toxicity testing of an essential oil are required in this approach.

9.2 CONSTITUENT-BASED EVALUATION OF ESSENTIAL OILS

The chemical constitution of a natural product is fundamental to understanding the product's intended use as well as factors that would affect its safety. Recent advances in analytical methodology have made intensive investigation of the chemical composition of a natural product economically feasible and even routine. High-throughput instrumentation necessary to perform extensive qualitative and quantitative analysis of complex chemical mixtures and to evaluate the variation in the composition of the mixture is now a reality. In fact, analytical tools needed to chemically characterize these complex mixtures are becoming more cost effective, while the cost of traditional toxicology is becoming more cost intensive. Based on the wealth of existing chemical and biological data on the constituents of essential oils and similar data on essential oils themselves, it is possible to validate a constituent-based safety evaluation of an essential oil.

As noted earlier, it is scientifically valid to evaluate the safety of a natural mixture based on its chemical composition. Fundamentally, it is the interaction between one or more molecules in the natural product and macromolecules (proteins, enzymes, etc.) that yield the biological response, regardless of whether it is a desired functional effect such as a pleasing taste or a potential toxic effect such as liver necrosis. Many of the advertised beneficial properties of ephedra are based on the presence of the central nervous system stimulant ephedrine. So too, the gustatory and olfactory properties of coriander oil are, in part, based on the binding of linalool, benzyl benzoate, and other molecules to the appropriate receptors. It is these molecular interactions of chemical constituents that ultimately determine conditions of use.

9.3 SCOPE OF ESSENTIAL OILS: USED AS FLAVOR INGREDIENTS

9.3.1 Plant Sources

Essential oils, as products of distillation, are mixtures of mainly low-molecular-weight chemical substances. Sources of essential oils include components (e.g., pulp, bark, peel, leaf, berry, blossom) of fruits, vegetables, spices, and other plants. Essential oils are prepared from food and nonfood sources. Many of the approximately 100 essential oils used as flavoring ingredients in food are derived directly from food (i.e., lemon oil, basil oil, and cardamom oil); far fewer are extracts from plants that are not normally consumed as food (e.g., cedar leaf oil or balsam fir oil).

Whereas an essential oil is typically obtained by steam distillation of the plant or plant part, an oleoresin is produced by extraction of the same with an appropriate organic solvent. The same volatile constituents of the plant isolated in the essential oil are primarily responsible for aroma and taste of the plant as well as the subsequent extract or oleoresin. Hence, borneol, bornyl acetate, camphor, and other volatile constituents in rosemary oil can provide a flavor intensity as potent as the mass of dried rosemary used to produce the oil. A few exceptions include cayenne pepper, black pepper, ginger, paprika, and sesame seeds, which contain key nonvolatile flavor constituents (e.g., gingerol and zingerone in ginger). These nonvolatile constituents are often higher molecular weight, hydrophilic substances that would be lost during distillation in the preparation of an essential oil, but they remain present in the oleoresin or extract. For economic reasons, crude essential oils are often produced via distillation at the source of the plant raw material and subsequently further processed at modern flavor facilities.

9.3.2 Processing of Essential Oils for Flavor Functions

Because essential oils are a product of nature, environmental and genetic factors will impact the chemical composition of the plant. Factors such as species and subspecies, geographical location, harvest time, plant part used, and method of isolation all affect the chemical composition of the crude material separated from the plant. Variability in the composition of the crude essential oil as isolated from nature has been the subject of much research and development since plant yields of essential oils are major economic factors in crop production.

However, the crude essential oil that arrives at the flavor processing plant is not normally used as such. The crude oil is often subjected to a number of processes that are intended to increase purity and to produce a product with the intended flavor characteristics. Some essential oils may be distilled and cooled to remove natural waxes and improve clarity, while others are distilled more than once (i.e., rectified) to remove undesirable fractions or to increase the relative content of certain chemical constituents. Some oils are dry or vacuum distilled. Normally, at some point during processing, the essential oil is evaluated for its technical function as a flavor. This evaluation typically involves analysis (normally by GLC or liquid chromatography) of the composition of the essential oil for chemical constituents that are markers for the desired technical flavor effect. For an essential oil such as cardamom oil, levels of target constituents such as terpinyl acetate, 1,8-cineole, and limonene are markers for technical viability as a flavoring substance. Based on this initial assessment, the crude essential oil may be blended with other sources of the same oil or chemical constituents isolated from the oil to reach target ranges for key constituent markers that reflect flavor function. The mixture may then be further rectified by distillation. Each step of the process is driven by flavor function. Therefore, the chemical composition of product to be marketed may be significantly different from that of the crude oil. Also, the chemical composition of the processed essential oil is more consistent than that of the crude batches of oil isolated from various plant harvests. The range of concentrations for individual constituents and for groups of structurally related constituents in an essential oil are dictated, in large part, by the requirement that target levels of critical flavor imparting constituents – essentially, principal flavor components – must be maintained.

9.3.3 Chemical Composition and Congeneric Groups

In addition to the key chemical constituents that are the principal flavor components within an essential oil or extract and that allow the natural complex mixture to achieve the technical flavor effect, an essential oil found on the market will normally contain many other chemical constituents, some having little or no flavor function. However, the chemical constituents of essential oils are not infinite in structural variation. Because they are derived from higher plants, these constituents are formed via one of four or five major biosynthetic pathways: lipoxygenase oxidation of lipids, shikimic acid, isoprenoid (terpenoid), and photosynthetic pathways. In ripening vegetables, lipoxygenases oxidize polyunsaturated fatty acids, eventually yielding low-molecular-weight aldehydes (2-hexenal), alcohols (2,6-nonadienol), and esters, many exhibiting flavoring properties. Plant amino acids phenylalanine and tyrosine are formed via the shikimic acid pathway and can subsequently be deaminated, oxidized, and reduced to yield important aromatic substances such as cinnamaldehyde and eugenol. The vast majority of constituents detected in commercially viable essential oils are terpenes (e.g., hydrocarbons [limonene], alcohols [menthol], aldehydes [citral], ketones [carvone], acids, and esters [geranyl acetate]) that are formed via the isoprene pathway (Roe and Field, 1965). Since all of these pathways operate in plants, albeit to different extents depending upon the species, season, and growth environment, many of the same chemical constituents are present in a wide variety of essential oils.

A consequence of having a limited number of plant biosynthetic pathways is that structural variation of chemical constituents in an essential oil is limited. Essential oils typically contain 5–10 distinct chemical classes or congeneric groups. Some congeneric groups, such as aliphatic terpene hydrocarbons, contain upward of 100 chemically identified constituents. In some essential oils, a single constituent (e.g., citral in lemongrass oil) or congeneric group of constituents (e.g., hydroxyallylbenzene derivatives, eugenol and eugenyl acetate, in clove bud oil) comprises the majority of the mass of the essential oil. In others, no single congeneric group predominates. For instance, although eight congeneric groups comprise >98% of the composition of oil of *Mentha piperita* (peppermint oil), greater than 95% of the oil is accounted for by three chemical groups: (1) terpene aliphatic and aromatic hydrocarbons; (2) terpene alicyclic secondary alcohols, ketones, and related esters; and (3) terpene 2-isopropylidene–substituted cyclohexanone derivatives and related substances.

The formation and members of a congeneric group are chosen based on a combination of structural features and known biochemical fate. Substances with a common carbon skeletal structure and functional

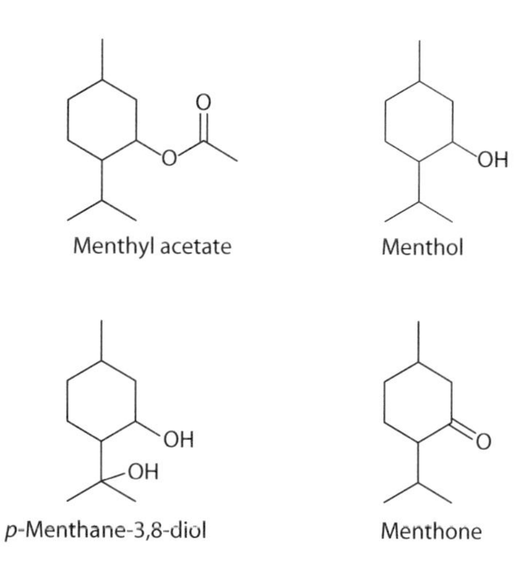

FIGURE 9.1 Congeneric groups are formed by members sharing common structural and metabolic features, such as the group of 2-isopropylidene-substituted cyclohexanone derivatives and related substances.

groups that participate in common pathways of metabolism are assigned to the same congeneric group. For instance, menthyl acetate hydrolyzes prior to absorption to yield menthol, which is absorbed and is interconvertible with menthone in fluid compartments (e.g., the blood). Menthol is either conjugated with glucuronic acid and excreted in the urine or undergoes further hydroxylation mainly at C8 to yield a diol that is also excreted, either free or conjugated. Despite the fact that menthyl acetate is an ester, menthol is an alcohol, menthone a ketone, and 3,8-menthanediol a diol, they are structurally and metabolically related (Figure 9.1). Therefore, all are members of the same congeneric group.

In the case of *M. piperita*, the three principal congeneric groups listed earlier have different metabolic options and possess different organ-specific toxic potential. The congeneric group of terpene aliphatic and aromatic hydrocarbons is represented mainly by limonene and myrcene. The second and most predominant congeneric group is the alicyclic secondary alcohols, ketones, and related esters that include d-menthol, menthone, isomenthone, and menthyl acetate. Although the third congeneric group contains alicyclic ketones similar in structure to menthone, it is metabolically quite different in that it contains an exocyclic isopropylidene substituent that undergoes hydroxylation principally at the C9 position, followed by ring closure and dehydration to yield a heteroaromatic furan ring of increased toxic potential. In the absence of a C4–C8 double bond, neither menthone nor isomenthone can participate in this intoxication pathway. Hence, they are assigned to a different congeneric group.

The presence of a limited number of congeneric groups in an essential oil is critical to the organization of constituents and subsequent safety evaluation of the oil itself. Members of each congeneric group exhibit common structural features and participate in common pathways of pharmacokinetics and metabolism and exhibit similar toxicologic potential. Recent guidance on chemical grouping has been published (OECD, 2014). If the mass of the essential oil (>95%) can be adequately characterized chemically and constituents assigned to well-defined congeneric groups, the safety evaluation of the essential oil can be reduced to (1) a safety evaluation of each of the congeneric groups comprising the essential oil and (2) a *sum of the parts* evaluation of the all congeneric groups to account for any chemical or biological interactions between congeneric groups in the essential oil under conditions of intended use. Validation of such an approach lies in the stepwise comparison of the dose and toxic effects for each key congeneric group with similar equivalent doses and toxic effects exhibited by the entire essential oil. Using such an approach, the scientifically independent but industry-sponsored Expert Panel of the U.S. Flavor and Extract Manufacturers Association (FEMA) has conducted safety evaluations on a number of natural flavoring complexes, many of which are essential oils and extracts, under the auspices of the GRAS concept (Smith et al., 2003, 2004). Without question, other approaches have been developed (Meek et al., 2011).

Potential interactions between congeneric groups can, to some extent, be analyzed by an in-depth comparison of the biochemical and toxicologic properties of different congeneric groups in the essential oil. For some representative essential oils that have been the subject of toxicology studies, a comparison of data for the congeneric groups in the essential oil with data on the essential oil itself (congeneric groups together) is a basis for analyzing for the presence or absence of interactions. Therefore, the impact of interaction between congeneric groups is minimal if the levels of and endpoints for toxicity of congeneric groups (e.g., tertiary terpene alcohols) are similar to those of the essential oil (e.g., coriander oil).

Since composition plays such a critical role in the evaluation, analytical identification requirements are also critical to the evaluation. Complete chemical characterization of the essential oil may be difficult or economically unfeasible based on the small volume of essential oil used as a flavor ingredient. In these few cases, mainly for low-volume essential oils, the unknown fraction may be appreciable and a large number of chemical constituents will not be identified. However, if the intake of the essential oil is low or significantly less than its intake from consumption of food (e.g., thyme) from which the essential oil is derived (e.g., thyme oil), there should be no significant concern for safety under conditions of intended use. For those cases in which chemical characterization of the essential oil is limited but the volume of intake is more significant, it may be necessary to perform additional analytical work to decrease the number of unidentified constituents or, in other cases, to perform selected toxicity studies on the essential oil itself. A principal goal of the safety evaluation of essential oils is that no congeneric groups that have significant human intakes should go unevaluated.

9.3.4 Chemical Assay Requirements and Chemical Description of Essential Oil

The safety evaluation of an essential oil initially involves specifying the biological origin, physical and chemical properties, and any other relevant identifying characteristics. An essential oil produced under good manufacturing practices should be of an appropriate purity (quality), and chemical characterization should be complete enough to guarantee a sufficient basis for a thorough safety evaluation of the essential oil under conditions of intended use. Because the evaluation is based primarily on the actual chemical composition of the essential oil, full specifications used in a safety evaluation will necessarily include not only information on the origin of the essential oil (commercial botanical sources, geographical sources, plant parts used, degree of maturity, and methods of isolation) and physical properties (specific gravity, refractive index, optical rotation, solubility, etc.) but also chemical assays for a range of essential oils currently in commerce.

9.3.4.1 Intake of the Essential Oil

Based on current analytical methodology, it is possible to identify literally hundreds of constituents in an essential oil and quantify the constituents to part per million levels. But is this necessary or desirable? From a practical point of view, the level of analysis for constituents should be directly related to the level of exposure to the essential oil. The requirements to identify and quantify constituents for use of 2,000,000 kg of peppermint oil annually should be far greater than that for use of 2000kg of coriander oil or 50 kg of myrrh oil annually. Also, there is a level at which exposure to each constituent is so low that there is no significant risk associated with intake of that substance. A conservative no significant risk level of 1.5 μg/day (0.0015 mg/day or 0.000025 mg/kg/day) has been adopted by regulatory authorities as a level at which the human cancer risk is below one in one million – this is commonly referred to as the "threshold of regulation" (FDA, 2005). Therefore, if consumption of an essential oil results in an intake of a constituent that is less than 1.5 μg/day, there should be no requirement to identify and quantify that constituent.

Determining the level to which the constituents of an essential oil should be identified depends upon estimates of intake of food or flavor additives. These estimates are traditionally calculated using a *volume-based* or a *menu–census* approach. A volume-based approach assumes that the total annual volume of use of a substance reported by an industry is distributed over a portion of the population consuming that substance. A menu–census approach is based on the concentration of the substance (essential oil) added to each flavor, the amount of flavor added to each food category, the portion of food consumed daily, and the total of all exposures across all food types. Although the latter is quite accurate for food additives consumed at higher levels in a wide variety of food such as food emulsifiers, the former method provides an efficient and conservative approximation of intake, if a fraction of the total population is assumed to consume all of the substance.

For the World Health Organization (WHO) and the U.S. FDA, intake is calculated using a method known as the *per capita* intake (PCI × 10) method, or alternatively known as the maximized survey-derived intake (MSDI) (Rulis et al., 1984; Woods and Doull, 1991). The MSDI method assumes that only 10% of the population consumes the total annual reported volume of use of a flavor ingredient. This approximation provides a practical and cost-effective approach to the estimation of intake for flavoring substances. The annual volumes of flavoring agents are relatively easy to obtain by industry-wide surveys, which can be performed on a regular basis to account for changes in food trends and flavor consumption. A recent poundage survey of U.S. flavor producers was collected in 2005 and published by FEMA in 2007 (Adams et al., 2007). Similar surveys were conducted in recent years in Europe (EFFA, 2005) and Japan (JFFMA, 2002).

Calculation of intake using the MSDI method has been shown to result in conservative estimates of intake and thus is appropriate for safety evaluation. Over the last three decades, two comprehensive studies of flavor intake have been undertaken. One involved a detailed dietary analysis (DDA) of a panel of 12,000 consumers who recorded all foods that they consumed over a 14-day period, and the flavoring ingredients in each food were estimated by experience flavorists to estimate intake of each flavoring substance (Hall, 1976; Hall and Ford, 1999). The other study utilized a robust full stochastic model (FSM) to estimate intake of flavoring ingredients by typical consumers in the United Kingdom (Lambe et al., 2002). The results of the data-intensive DDA method and the model-based FSM support the use of PCI data as a conservative estimate of intake.

With regard to essential oils, the MSDI method provides overestimates of intake for oils that are widely distributed in food. The large annual volume of use reported for essential oils such as orange oil, lemon oil, and peppermint oil indicate widespread use in a large variety of foods resulting in consumption of these oils by significantly more than 10% of the population. Citrus flavor is pervasive in a multitude of foods and beverages. Therefore, for selected high-volume essential oils, a simple *PCI* rather than a *per capita* × 10 intake may be more appropriate. However, the intake of the congeneric groups and the group of unidentified constituents for these high-volume oils is still estimated by the MSDI method.

An alternative approach to MSDI now employed in the evaluation of flavoring substances by the WHO/UN Food and Agricultural Organization Joint Expert Committee on Food Additives (JECFA) relies on a modification of the traditional menu–census approach. The JECFA single portion exposure technique (SPET) identifies the highest intake from a single food category and assigns it as the overall intake for a flavoring substance. This is calculated from the concentration of the flavoring substance within a food category multiplied by the daily portion size for that food category. At JECFA, the highest intake from either MSDI or SPET is now used to assign an intake to a flavoring substance. JECFA has concluded that both methods, MSDI and SPET, can provide complementary and useful information regarding the uses of flavoring substances. To date, SPET has not been used for the safety evaluation of natural complex mixtures at JECFA, as the flavoring agent evaluations at JECFA have focused thus far on chemically defined substances.

9.3.4.2 Analytical Limits on Constituent Identification

As described earlier, the analytical requirements for detection and identification of the constituents of an essential oil are set by the intake of the oil and by the conservative assumption that constituents with intakes less than 1.5 μg/day will not need to be identified. For instance, if the annual volume of use of coriander oil in the United States is 10,000 kg, then the estimated daily PCI of the oil is

$$\frac{10{,}000\ \text{kg/year} \times 10^9\ \mu\text{g/kg}}{365\ \text{days/year} \times 31{,}000{,}000\ \text{persons}} = 883\ \mu\text{g coriander oil/person/day}$$

Based on the intake of coriander oil (883 μg/day), any constituent present at greater than 0.17% would need to be chemically characterized and quantified:

$$\frac{1.5\ \mu g/day}{978\ \mu g/day} \times 100 = 0.17\%$$

For the vast majority of essential oils, meeting these characterization requirements does not require exotic analytical techniques and the identification of the constituents is of a routine nature. However, what would the requirements be for very high-volume essential oils, such as orange oil, cold pressed (567,000 kg), or peppermint oil (1,229,000 kg)? In these cases, a practical limit must be applied and can be justified based on the concept that the intake of these oils is widespread and far exceeds the 10% assumption of MSDI. Based on current analytical capabilities, 0.10% or 0.05% could be used as a reasonable limit of detection, with the lower level used for an essential oil that is known or suspected to contain constituents of higher toxic potential (e.g., methyl eugenol in basil).

9.3.4.3 Intake of Congeneric Groups

Once the analytical limits for identification of constituents have been met, it is key to evaluate the intake of each congeneric group from consumption of the essential oil. A range of concentration of each congeneric group is determined from multiple analyses of different lots of the essential oil used in flavorings. The intake of each congeneric group is determined from mean concentrations (%) of constituents recorded for each congeneric group. For instance, for peppermint oil the alicyclic secondary alcohol/ketone/related ester group may contain (−)-menthol, (−)-menthone, (−)-menthyl acetate, and isomenthone in mean concentrations of 43.0%, 20.3%, 4.4%, and 0.40%, respectively, with the result that that congeneric group accounts for 68.1% of the oil. It should be emphasized that although members in a congeneric group may vary among the different lots of oil, the variation in concentration of congeneric groups in the oil is relatively small.

Routinely, the daily *PCI* of the essential oil is derived from the annual volumes reported in industry surveys (NAS, 1965, 1970, 1975, 1982, 1987; Lucas et al., 1999; JFFMA, 2002; EFFA, 2005). If a conservative estimate of intake of the essential oil is made using a volume-based approach such that a defined group of constituents are set for each essential oil, target constituents can be monitored in an ongoing quality control program, and the composition of the essential oil can become one of the key specifications linking the product that is distributed in the marketplace to the chemically based safety evaluation.

Limited specifications for the chemical composition of some essential oils to be used as food flavorings are currently listed in the Food Chemicals Codex (FCC, 2013). For instance, the chemical assay for cinnamon oil is given as "not less than 80%, by volume, as total aldehydes." Any specification developed related to this safety evaluation procedure should be consistent with already published specifications including FCC and ISO standards. However, based on chemical analyses for the commercially available oil, the chemical specification or assay can and should be expanded to

1. Specify the mean of concentrations for congeneric groups with confidence limits that constitute a sufficient number of commercial lots constituting the vast majority of the oil.
2. Identify key constituents of intake >1.5 μg/day in these groups that can be used to efficiently monitor the quality of the oil placed into commerce over time.
3. Provide information on trace constituents that may be of a safety concern.

For example, given its most recent reported annual volume (1060 kg) (Harman et al., 2013), it is anticipated that a chemical specification for lemongrass oil would include (1) greater than 98.7% of the composition chemically identified; (2) not more than 92% aliphatic terpene primary alcohols, aldehydes, acids, and related esters, typically measured as citral; and (3) not more than 15%

aliphatic terpene hydrocarbons, typically measured as myrcene. The principal goal of a chemical specification is to provide sufficient chemical characterization to ensure safety of the essential oil from use as a flavoring. From an industry standpoint, the specification should be sufficiently descriptive as to allow timely quality control monitoring for constituents that are responsible for the technical flavor function. These constituents should also be representative of the major congeneric group or groups in the essential oil. Also, monitored constituents should include those that may be of a safety concern at sufficiently high levels of intake of the essential oil (e.g., pulegone). The scope of a specification should be sufficient to ensure safety in use but not impose an unnecessary burden on industry to perform ongoing analyses for constituents unrelated to the safety or flavor of the essential oil.

9.4 SAFETY CONSIDERATIONS FOR ESSENTIAL OILS, CONSTITUENTS, AND CONGENERIC GROUPS

9.4.1 Essential Oils

9.4.1.1 Safety of Essential Oils: Relationship to Food

The close relationship of natural flavor complexes to food itself has made it difficult to evaluate the safety and regulate the use of essential oils. In the United States, the Federal Food Drug and Cosmetic Act recognizes that a different, lower standard of safety must apply to naturally occurring substances in food than applies to the same ingredient intentionally added to food. For a substance occurring naturally in food, the act applies a realistic standard that the substance must "… not ordinarily render it [the food] injurious to health" (21 CFR 172.30). For added substances, a much higher standard applies. The food is considered to be adulterated if the added substance "… may render it [the food] injurious to health" (21 CFR 172.20). Essential oils used as flavoring substances occupy an intermediate position in that they are composed of naturally occurring substances, many of which are intentionally added to food as individual chemical substances. Because they are considered neither a direct food additive nor a food itself, no current standard can be easily applied to the safety evaluation of essential oils.

The evaluation of the safety of essential oils that have a documented history of use in foods starts with the presumption that they are safe based on their long history of use over a wide range of human exposures without known adverse effects. With a high degree of confidence, one may presume that essential oils derived from food are likely to be safe. Annual surveys of the use of flavoring substances in the United States (NAS, 1965, 1970, 1975, 1981, 1987; Lucas et al., 1999; Harman et al., 2013; 21 CFR 172.510) in part, document the history of use of many essential oils. Conversely, confidence in the presumption of safety decreases for natural complexes that exhibit a significant change in the pattern of use or when novel natural complexes with unique flavor properties enter the food supply. Recent consumer trends that have changed the typical consumer diet have also changed the exposure levels to essential oils in a variety of ways. As one example, changes in the use of cinnamon oil in low-fat cinnamon pastries would alter intake for a specialized population of eaters. Also, increased international trade has coupled with a reduction in cultural cuisine barriers, leading to the introduction of novel plants and plant extracts from previously remote geographical locations. Osmanthus absolute (FEMA No. 3750) and Jambu oleoresin (FEMA No. 3783) are examples of natural complexes recently used as flavoring substances that are derived from plants not indigenous to the United States and not commonly consumed as part of a Western diet. Furthermore, the consumption of some essential oils may not occur solely from intake as flavoring substances; rather, they may be regularly consumed as dietary supplements with advertised functional benefits. These impacts have brought renewed interest in the safety evaluation of essential oils. Although the safety evaluation of essential oil must still rely heavily on knowledge of the history of use, a flexible science-based approach would allow for rigorous safety evaluation of different uses for the same essential oil.

9.4.2 Safety of Constituents and Congeneric Groups in Essential Oils

It is well established that when consumed in high quantities, some plants do indeed exhibit toxicity. Historically, humans have used plants as poisons (e.g., hemlock), and many of the intended medicinal uses of plants (pennyroyal oil as an abortifacient) have produced undesirable toxic side effects. High levels of exposure to selected constituents in the plant or essential oil (i.e., pulegone in pennyroyal oil) have been associated with the observed toxicity. However, with regard to flavor use, experience through long-term use and the predominant self-limiting impact of flavorings on our senses have restricted the amount of a plant or plant part that we use in or on food.

Extensive scientific data on the most commonly occurring major constituents in essential oils have not revealed any results that would give rise to safety concerns at low levels of exposure. Chronic studies have been performed on more 30 major chemical constituents (menthol, carvone, limonene, citral, cinnamaldehyde, benzaldehyde, benzyl acetate, 2-ethyl-1-hexanol, methyl anthranilate, geranyl acetate, furfural, eugenol, isoeugenol, etc.) found in many essential oils. The majority of these studies were hazard determinations that were sponsored by the National Toxicology Program, and they were normally performed at dose levels many orders of magnitude greater than the daily intakes of these constituents from consumption of the essential oil. Even at these high intake levels, the majority of the constituents show no carcinogenic potential (Smith et a., 2005b). In addition to dose/exposure, for some flavor ingredients, the carcinogenic potential that was assessed in the study is related to several additional factors including the mode of administration, species and sex of the animal model, and target organ specificity. In the vast majority of studies, the carcinogenic effect occurs through a nongenotoxic mechanism in which tumors form secondary to preexisting high-dose, chronic organ toxicity, typically to the liver or kidneys. Selected subgroups of structurally related substances (e.g., aldehydes, terpene hydrocarbons) are associated with a single-target organ and tumor type in a specific species and sex of rodent (i.e., male rat kidney tumors secondary to alpha-2u-globulin neoplasms with limonene in male rats) or using a single mode of administration (i.e., forestomach tumors that arise due to high doses of benzaldehyde and hexadienal given by gavage).

Given their long history of use, it appears unlikely that there are essential oils consumed by humans that contain constituents not yet studied that are weak nongenotoxic carcinogens at chronic high-dose levels. Even if there are such cases, because of the relatively low intake (Lucas et al., 1999) as constituents of essential oils, these yet-to-be-discovered constituents would be many orders of magnitude less potent than similar levels of aflatoxins (found in peanut butter), the polycyclic heterocyclic amines (found in cooked foods), or the polynuclear aromatic hydrocarbons (also found in cooked foods). There is nothing to suggest that the major biosynthetic pathways available to higher plants are capable of producing substances such that low levels of exposure to the substance would result in a high level of toxicity or carcinogenicity.

The toxic and carcinogenic potentials exhibited by constituent chemicals in essential oils can largely be equated with the toxic potential of the congeneric group to which that chemical belongs. A comparison of the oral toxicity data (JECFA, 2004) for limonene, myrcene, pinene, and other members of the congeneric group of terpene hydrocarbons shows similar low levels of toxicity with the same high-dose target organ endpoint (kidney) in animal studies, of which the relevance to humans is unlikely. Likewise, dietary toxicity and carcinogenicity data (JECFA, 2001) for cinnamyl alcohol, cinnamaldehyde, cinnamyl acetate, and other members of the congeneric group of 3-phenyl-1-propanol derivatives show similar toxic and carcinogenic endpoints. The safety data for the congeneric chemical groups that are found in vast majority of essential oils have been reviewed (Adams et al., 1996, 1997, 1998, 2004; JECFA 1997, 1998, 1999, 2000a, b, 2001, 2003, 2004; Newberne et al., 1999; Smith et al., 2002a, b). Available data for different representative members in each of these congeneric groups support the conclusion that the toxic and carcinogenic potential of individual constituents adequately represent similar potentials for the corresponding congeneric group.

The second key factor in the determination of safety is the level of intake of the congeneric group from consumption of the essential oil. Intake of the congeneric group will, in turn, depend upon the variability of the chemical composition of the essential oil in the marketplace and on the conditions of use. As discussed earlier, chemical analysis of the different batches of oil obtained from the same and different manufacturers will produce a range of concentrations for individual constituents in each congeneric group of the essential oil. The mean concentration values (%) for constituents are then summed for all members of the congeneric group. The total % determined for the congeneric group is multiplied by the estimated daily intake (PCI $\times$ 10) of the essential oil to provide a conservative estimate of exposure to each congeneric group from consumption of the essential oil.

In some essential oils, the intake of one constituent, and therefore, one congeneric group, may account for essentially all of the oil (e.g., linalool in coriander oil, citral in lemongrass oil, benzaldehyde in bitter almond oil). In other oils, exposure to a variety of congeneric groups over a broad concentration range may occur. As noted earlier, cardamom oil is an example of such an essential oil. Ultimately, it is the relative intake and the toxic potential of each congeneric group that is the basis of the congeneric group-based safety evaluation. The combination of relative intake and toxic potential will prioritize congeneric groups for the safety evaluation. Hypothetically, a congeneric group of increased toxic potential that accounts for only 5% of the essential oil may be prioritized higher than a congeneric group of lower toxic potential accounting for 95%.

The following guide and examples therein are intended to more fully illustrate the principles described earlier that are involved in the safety evaluation of essential oils. Fermentation products, process flavors, substances derived from fungi, microorganisms, or animals, and direct food additives are explicitly excluded. The guide is designed primarily for application to essential oils and extracts for use as flavoring substances. The guide is a tool to organize and prioritize the chemical constituents and congeneric groups in an essential oil in such a way as to allow a detailed analysis of their chemical and biological properties. This analysis as well as consideration of other relevant scientific data provides the basis for a safety evaluation of the essential oil under conditions of intended use. Validation of the approach is provided, in large part, by a detailed comparison of the doses and toxic effects exhibited by constituents of the congeneric group with the equivalent doses and effects provided by the essential oil. This methodology, with some variations based on expert judgment as appropriate, has been in use for more than 10 years through the FEMA GRAS program, in safety evaluations conducted by the FEMA Expert Panel (Smith et al., 2004).

9.5 GUIDE AND EXAMPLE FOR THE SAFETY EVALUATION OF ESSENTIAL OILS

9.5.1 INTRODUCTION

The guide is a procedure involving a comprehensive evaluation of the chemical and biological properties of the constituents and congeneric groups of an essential oil. Constituents in, for instance, an essential oil that are of known structure are organized into congeneric groups that exhibit similar metabolic and toxicologic properties. The congeneric groups are further classified according to levels (Structural Classes I, II, and III) of toxicologic concern using a decision tree approach (Cramer et al., 1978; Munro et al., 1996b). Based on intake data for the essential oil and constituent concentrations, the congeneric groups are prioritized according to intake and toxicity potential. The procedure ultimately focuses on those congeneric groups that, due to their structural features and intake, may pose some significant risk from the consumption of the essential oil. Key elements used to evaluate congeneric groups include exposure, structural analogy, metabolism, and toxicology, which includes toxicity, carcinogenicity, and genotoxic potential (Oser and Hall, 1977; Oser and Ford, 1991; Woods and Doull, 1991; Adams et al., 1996, 1997, 1998, 2004; Newberne et al., 1999; Smith et al., 2002a,b, 2004). Throughout the analysis of these data, it is essential that professional judgment and expertise be applied to complete the safety evaluation of the essential oil. As an example of how a typical evaluation process for an essential oil is carried out according to this guide, the safety evaluation for flavor use of corn mint oil (*Mentha arvensis*) is outlined in the following text.

9.5.2 Elements of the Guide for the Safety Evaluation of the Essential Oil

9.5.2.1 Introduction

In Step 1 of the guide, the evaluation procedure estimates intake based on industry survey data for each essential oil. It then organizes the chemically identified constituents that have an intake >1.5 μg/day into congeneric groups that participate in common pathways of metabolism and exhibit a similar toxic potential. In Steps 2 and 3, each identified chemical constituent is broadly classified according to toxic potential (Cramer et al., 1978) and then assigned to a congeneric group of structurally related substances that exhibit similar pathways of metabolism and toxicologic potential.

Before the formal evaluation begins, it is necessary to specify the data (e.g., botanical, physical, chemical) required to completely describe the product being evaluated. In order to effectively evaluate an essential oil, attempted complete analyses must be available for the product intended for the marketplace from a number of flavor manufacturers. Additional quality control data are useful, as they demonstrate consistency in the chemical composition of the product being marketed. A Technical Information Paper drafted for the particular essential oil under consideration organizes and prioritizes these data for efficient sequential evaluation of the essential oil.

In Steps 8 and 9, the safety of the essential oil is evaluated in the context of all congeneric groups and any other related data (e.g., data on the essential oil itself or for an essential oil of similar composition). The procedure organizes the extensive database of information on the essential oil constituents in order to efficiently evaluate the safety of the essential oil under conditions of use. It is important to stress, however, that the guide is not intended to be nor in practice operates as a rigid checklist. Each essential oil that undergoes evaluation is different, and different data will be available for each. The overriding objective of the guide and subsequent evaluation is to ensure that no significant portion of the essential oil should go unevaluated.

9.5.2.2 Prioritization of Essential Oil According to Presence in Food

In Step 1, essential oils are prioritized according to their presence or absence as components of commonly consumed foods (Step 1). This question evaluates the relative intake of the essential oil as an intentionally added flavoring substance versus its intake as a component part of food. Many essential oils are isolated from plants that are commonly consumed as a food. Little or no safety concerns should exist for the intentional addition of the essential oil to the diet, if intake of the oil from consumption of traditional foods (garlic) substantially exceeds intake as an intentionally added flavoring substance (garlic oil). In many ways, the first step applies the concept of "long history of safe use" to essential oils. That is, if exposure to the essential oil occurs predominantly from consumption of a normal diet, a conclusion of safety is straightforward. Step 1 of the guide clearly places essential oils that are consumed as part of a traditional diet on a lower level of concern than those oils derived from plants that are either not part of the traditional diet or whose intake is not predominantly from the diet. The first step also mitigates the need to perform comprehensive chemical analysis for essential oils in those cases where intake is low and occurs predominantly from consumption of food. An estimate of the intake of the essential oil is based on the most recent poundage available from flavor industry surveys and the assumption that the essential oil is consumed by only 10% of the population for an oil having a survey volume <50,000 kg/year and 100% of the population for an oil having a survey volume >50,000 kg/year. In addition, the detection limit for constituents is determined based on the daily PCI of the essential oil.

9.5.2.2.1 Corn Mint Oil

To illustrate the type of data considered in Step 1, consider corn mint oil. Corn mint oil is produced by the steam distillation of the flowering herb of *M. arvensis*. The crude oil contains upward of 70% (−)-menthol, some of which is isolated by crystallization at low temperature. The resulting dementholized oil is corn mint oil. Although produced mainly in Brazil during the 1970s and 1980s, corn mint oil is now produced predominantly in China and India. Corn mint has a more stringent taste compared to that of peppermint oil, *M. piperita*, but can be efficiently produced and is used as a more cost-effective substitute. Corn mint oil isolated from various crops undergoes subsequent *clean up*, further distillation, and blending to produce the finished commercial oil. Although there may be significant variability in the concentrations of individual constituents in different samples of crude essential oil, there is far less variability in the concentration of constituents and congeneric groups in the finished commercial oil. The volume of corn mint oil reported in the most recent U.S. poundage survey is 446,000 kg/year (Harman et al., 2013), which is approximately 25% of the potential market of peppermint oil. Because corn mint oil is a high-volume essential oil, it is highly likely that the entire population consumes the annual reported volume, and therefore, the daily PCI is calculated based on 100% of the population (310,000,000). This results in a daily PCI of approximately 3.9 mg/person/day (0.066 mg/kg bw/day) of corn mint oil:

$$\frac{446{,}000 \text{ kg/year} \times 10^9 \ \mu\text{g/kg}}{365 \text{ days/year} \times 310 \times 10^6 \text{ persons}} = 3942 \ \mu\text{g/person/day}$$

Based on the intake of corn mint oil (3942 μg/day), any constituent present at greater than 0.038% would need to be chemically characterized and quantified:

$$\frac{1.5 \ \mu\text{g/day}}{3942 \ \mu\text{g/day}} \times 100 = 0.038\%$$

9.5.2.3 Organization of Chemical Data: Congeneric Groups and Classes of Toxicity

In Step 2, constituents are assigned to one of three structural classes (I, II, or III) based on toxic potential (Cramer et al., 1978). Class I substances contain structural features that suggest a low order of oral toxicity. Class II substances are clearly less innocuous than Class I substances but do not contain structural features that provide a positive indication of toxicity. Class III substances contain structural features (e.g., an epoxide functional group, unsubstituted heteroaromatic derivatives) that permit no strong presumption of safety and in some cases may even suggest significant toxicity. For instance, the simple aliphatic hydrocarbon, limonene, is assigned to structural Class I, while elemicin, which is an allyl-substituted benzene derivative with a reactive benzylic/allylic position, is assigned to Class III. Likewise, chemically unidentified constituents of the essential oil are automatically placed in Structural Class III, since no presumption of safety can be made.

The toxic potential of each of the three structural classes has been quantified (Munro et al., 1996a). An extensive toxicity database has been compiled for substances in each structural class. The database covers a wide range of chemical structures, including food additives, naturally occurring substances, pesticides, drugs, antioxidants, industrial chemicals, flavors, and fragrances. Conservative no observable effect levels (fifth percentile NOELs) have been determined for each class. These fifth percentile NOELs for each structural class are converted to human exposure threshold levels by applying a 100-fold safety factor and correcting for mean bodyweight (60/100). The human exposure threshold levels are referred to as thresholds of toxicological concern (TTC). With regard to flavoring substances, the TTCs are even more conservative, given that the vast majority of NOELs for flavoring substances are above the 90th percentile. These conservative TTCs have since been adopted by the WHO and Commission of the European Communities for use in the evaluation of chemically identified flavoring agents by JECFA and the European Food Safety Authority (EFSA) (JECFA, 1997; EC, 1999).

Step 3 is a key step in the guide. It organizes the chemical constituents into congeneric groups that exhibit common chemical and biological properties. Based on the well-recognized biochemical pathways operating in plants, essentially all of the volatile constituents found in essential oils, extracts, and oleoresins belong to well-recognized congeneric groups. Recent reports (Maarse et al., 1992, 1994, 2000; Njissen et al., 2003) of the identification of new naturally occurring constituents indicate that newly identified substances fall into existing congeneric groups. The Expert Panel, JECFA, and the EC have acknowledged that individual chemical substances can be evaluated in the context of their respective congeneric group (JECFA, 1997; EC, 1999; Smith et al., 2005a, b). The congeneric group approach provides the basis for understanding the relationship between the biochemical fate of members of a chemical group and their toxicologic potential. Within this framework, the objective is to continuously build a more complete understanding of the absorption, distribution, metabolism, and excretion of members of the congeneric group and their potential to cause systemic toxicity. Within the guidelines, the structural class of each congeneric group is assigned based on the highest structural class of any member of the group. Therefore, if an essential oil contained a group of furanone derivatives that were variously assigned to Structural Classes II and III, in the evaluation of the oil, the congeneric group would, in a conservative manner, be assigned to Class III.

The types and numbers of congeneric groups in a safety evaluation program are, by no means, static. As new scientific data and information become available, some congeneric groups are combined while others are subdivided. This has been the case for the group of alicyclic secondary alcohols and ketones that were the subject of a comprehensive scientific literature review in 1975 (FEMA, 1975). Over the last two decades, experimental data have become available indicating that a few members of this group exhibit biochemical fate and toxicologic potential inconsistent with that for other members of the same group. These inconsistencies, almost without exception, arise at high-dose levels that are irrelevant to the safety evaluation of low levels of exposure to flavor use of the substance. However, given the importance of the congeneric group approach in the safety assessment program, it is critical to resolve these inconsistencies. Additional metabolic and toxicologic studies may be required to distinguish the factors that determine these differences. Often the effect of dose and a unique structural feature results in utilization of a metabolic activation pathway not utilized by other members of a congeneric group. Currently, evaluating bodies including JECFA, EFSA, and the FEMA Expert Panel have classified flavoring substances into the same congeneric groups for the purpose of safety evaluation.

In Steps 5, 6, and 7, each congeneric group in the essential oil is evaluated for safety in use. In Step 5, an evaluation of the metabolism and disposition is performed to determine, under current conditions of intake, whether the group of congeneric constituents is metabolized by well-established detoxication pathways to yield innocuous products. That is, such pathways exist for the congeneric group of constituents in an essential oil, and safety concerns will arise only if intake of the congeneric group is sufficient to saturate these pathways potentially leading to toxicity. If a significant intoxication pathway exists (e.g., pulegone), this should be reflected in a higher decision tree class and lower TTC threshold. At Step 6 of the procedure, the intake of the congeneric group relative to the respective TTC for one of the three structural classes (1800 μg/day for Class I; 540 μg/day for Class II; 90 μg/day for Class III; see Table 9.1) is evaluated. If the intake of the congeneric group is less than the threshold for the respective structural class, the intake of the congeneric group presents no significant safety concerns. The group passes the first phase of the evaluation and is then referred to Step 8, the step in which the safety of the congeneric group is evaluated in the context of all congeneric groups in the essential oil.

If, at Step 5, no sufficient metabolic data exist to establish safe excretion of the product or if activation pathways have been identified for a particular congeneric group, then the group moves to Step 7, and toxicity data are required to establish safe use under current conditions of intake. There are examples where low levels of xenobiotic substances can be metabolized to reactive substances. In the event that reactive metabolites are formed at low levels of intake of naturally occurring substances, a detailed analysis of dose-dependent toxicity data must be performed. Also, if the intake of the congeneric group is greater than the human exposure threshold (suggesting metabolic saturation may occur), then toxicity data are also required. If, at Step 7, a database of relevant toxicological data for a representative member or members of the congeneric group indicates that a sufficient margin of safety exists for the intake of the congeneric group, the members of that congeneric group are concluded to be safe under conditions of use of the essential oil. The congeneric group then moves to Step 8.

TABLE 9.1
Structural Class Definitions and Their Human Intake Thresholds

Class	Description	Fifth Percentile NOEL (mg/kg/day)	Human Exposure Threshold (TTC)[a] (μg/day)
I	Structure and related data suggest a low order of toxicity. If combined with low human exposure, they should enjoy an extremely low priority for investigation. The criteria for adequate evidence of safety would also be minimal. Greater exposures would require proportionately higher priority for more exhaustive study.	3.0	1800
II	Intermediate substances. They are less clearly innocuous than those of Class I, but do not offer the basis either of the positive indication of toxicity or of the lack of knowledge characteristic of those in Class III.	0.91	540
III	Permit no strong initial presumptions of safety or that may even suggest significant toxicity. They thus deserve the highest priority for investigation. Particularly when per capita intake is high of a significant subsection of the population that has a high intake, the implied hazard would then require the most extensive evidence for safety in use.	0.15	90

[a] The human exposure threshold was calculated by multiplying the fifth percentile NOEL by 60 (assuming an individual weighs 60 kg) and dividing by a safety factor of 100.

In the event that insufficient data are available to evaluate a congeneric group at Step 7, or the currently available data result in margins of safety that are not sufficient, the essential oil cannot be further evaluated by this guide and must be set aside for further considerations.

Use of the guide requires scientific judgment at each step of the sequence. For instance, if a congeneric group that accounted for 20% of a high-volume essential oil was previously evaluated and found to be safe under intended conditions of use, the same congeneric group found at less than 2% of a low-volume essential oil does not need to be further evaluated.

Step 8 considers additivity or synergistic interactions between individual substances and between the different congeneric groups in the essential oil. As for all other toxicological concerns, the level of exposure to congeneric groups is relevant to whether additive or synergistic effects present a significant health hazard. The vast majority of essential oils are used in food in extremely low concentrations, which therefore results in very low intake levels of the different congeneric groups within that oil. Moreover, major representative constituents of each congeneric group have been tested individually and pose no toxicological threat even at dose levels that are orders of magnitude greater than normal levels of intake of essential oils from use in traditional foods. Based on the results of toxicity studies both on major constituents of different congeneric groups in the essential oil and on the essential oil itself, it can be concluded that the toxic potential of these major constituents is representative of that of the oil itself, indicating the likely absence of additivity and synergistic interaction. In general, the margin of safety is so wide and the possibility of additivity or synergistic interaction so remote that combined exposure to the different congeneric groups and the unknowns are considered of no health concern, even if expert judgment cannot fully rule out additivity or synergism. However, case-by-case considerations are appropriate. Where possible combined effects might be considered to have toxicological relevance, additional data may be needed for an adequate safety evaluation of the essential oil.

Additivity of toxicologic effect or synergistic interaction is a conservative default assumption that may be applied whenever the available metabolic data do not clearly suggest otherwise. The extensive database of metabolic information on congeneric groups (JECFA, 1997, 1998, 1999, 2000a, b, 2001, 2003, 2004) that are found in essential oils suggests that the potential for additive effects and synergistic interactions among congeneric groups in essential oils is extremely low. Although additivity of effect is the approach recommended by NAS/NRC committees (NRC, 1988, 1994) and regulatory agencies (EPA, 1988), the Presidential Commission of Risk Assessment and Risk Management recommended (Presidential Commission, 1996) that "For risk assessments involving multiple chemical exposures at low concentrations, without information on mechanisms, risks should be added. If the chemicals act through separate mechanisms, their attendant risks should not be added but should be considered separately." Thus, the risks of chemicals that act through different mechanisms, that act on different target systems, or that are toxicologically dissimilar in some other way should be considered to be independent of each other. The congeneric groups in essential oils are therefore considered separately.

Further, the majority of individual constituents that comprise essential oils are themselves used as flavoring substances that pose no toxicological threat at doses that are magnitudes greater than their level of intake from the essential oil. Rulis (1987) reported that "The overwhelming majority of additives present a high likelihood of having safety assurance margins in excess of 10^5." He points out that this is particularly true for additives used in the United States at less than 100,000 lb/year. Because more than 90% of all flavoring ingredients are used at less than 10,000 lb/year (Hall and Oser, 1968), this alone implies intakes commonly many orders of magnitude below the no-effect level. Nonadditivity thus can often be assumed. As is customary in the evaluation of any substance, high-end data for exposure (consumption) are used, and multiple other conservatisms are employed to guard against underestimation of possible risk. All of these apply to complex mixtures as well as to individual substances.

9.5.2.3.1 Corn Mint Oil Congeneric Groups

In corn mint oil, the principal congeneric group is composed of terpene alicyclic secondary alcohols, ketones, and related esters, as represented by the presence of (−)-menthol, (−)-menthone, (+)-isomenthone, (−)-menthyl acetate, and other related substances. Samples of triple-distilled commercial corn mint oil may contain up to 95% of this congeneric group. The biochemical and biological fate of this group of substances has been previously reviewed (Adams et al., 1996; JECFA, 1999). Key data on metabolism, toxicity, and carcinogenicity are cited in the following text (Table 9.2) in order to complete the evaluation. Although constituents in this group are effectively detoxicated via conjugation of the corresponding alcohol or ω-oxidation followed by conjugation and excretion (Williams, 1940; Madyastha and Srivatsan, 1988; Yamaguchi et al., 1994), the intake of the congeneric group (3745 μg/person/day or 3.75 mg/person/day, see Table 9.2) is higher than the exposure threshold of (540 μg/person/day or 0.540 mg/person/day for Structural Class II. Therefore, toxicity data are required for this congeneric group. In both short- and long-term studies (NCI, 1979; Madsen et al., 1986), menthol, menthone, and other members of the group exhibit NOAELs at least 1000 times the daily PCI (*eaters only*) (3.75 mg/person/day or 0.062 mg/kg bw/day) of this congeneric group resulting from intake of the essential oil. For members of this group, numerous *in vitro* and *in vivo* genotoxicity assays are consistently negative (Florin et al., 1980; Heck et al., 1989; Sasaki et al., 1989; Muller, 1993; Zamith et al., 1993; Rivedal et al., 2000; NTP Draft, 2003a). Therefore, the intake of this congeneric group from consumption of *M. arvensis* is not a safety concern.

Although it is a constituent of corn mint oil and is also a terpene alicyclic ketone structurally related to the aforementioned congeneric group, pulegone exhibits a unique structure (i.e., 2-isopropylidenecyclohexanone) that participates in a well-recognized intoxication pathway (see Figure 9.2) (McClanahan et al., 1989; Thomassen et al., 1992; Adams et al., 1996; Chen et al., 2001) that leads to the formation of menthofuran. This metabolite subsequently oxidizes and ring opens to yield a highly reactive 2-ene-1,4-dicarbonyl intermediate that reacts readily with proteins resulting in hepatotoxicity at intake levels at least two orders of magnitude less than no observable effect levels for structurally related alicyclic ketones and secondary alcohols (menthone, carvone, and menthol). Therefore, pulegone

TABLE 9.2
Safety Evaluation of Corn Mint Oil, *Mentha arvensis*[a]

Congeneric Group	Step 2. Decision Tree Class (TTC, μg/Person/Day)	Step 3. High% from Multiple Commercial Samples	Step 4. Intake, μg/Person/Day	Step 5. Metabolism Pathways	Step 6. Intake of Congeneric Group or Total of Unidentified Constituents Group < TTC for Class?	Step 7. Relevant Toxicity Data if Intake of Group > TTC
Secondary alicyclic saturated and unsaturated alcohol/ketone/ketal/ester (e.g., menthol, menthone, isomenthone, menthyl acetate)	II (540)	95	3745	1. Glucuronic acid conjugation of the alcohol followed by excretion in the urine.2 ω-Oxidation of the side-chain substituents to yield various polyols and hydroxyacids and excreted as glucuronic acid conjugates.	No, 3745 > 540 μg/person/day	NOEL of 600,000 μg/kg/kg bw/day for menthol (103-week dietary study in mice) (NCI, 1979) NOEL of 400,000 μg/kg/kg bw/day for menthone (28-day gavage study in rats) (Madsen et al., 1986)
Aliphatic terpene hydrocarbon (e.g., limonene, pinene)	I (1800)	8	315	1. ω-Oxidation to yield polar hydroxy and carboxy metabolites excreted as glucuronic acid conjugates.	Yes, 315 <1800 μg/person/day	Not required
2-Isopropylidene cyclohexanone and metabolites (e.g., pulegone)	III (90)	2	79	1. Reduction to yield menthone or isomenthone, followed by hydroxylation of ring or side-chain positions and then conjugation with glucuronic acid. 2. Conjugation with glutathione in a Michael-type addition leading to mercapturic acid conjugates that are excreted or further hydroxylated and excreted. 3. Hydroxylation catalyzed by cytochrome P-450 to yield a series of ring- and side-chain-hydroxylated pulegone metabolites, one of which is a reactive 2-ene-1,4-dicarbonyl derivative. This intermediate is known to form protein adducts leading to enhanced toxicity. (Austin et al., 1988)	Yes, 79 < 90 μg/person/day	Not required. But NOAEL of 9375 μg/kg bw/day for pulegone (90 day gavage study in rats) (NTP, 2002)

[a] Based on daily per capita intake of 3942 μg/person/day for corn mint oil.

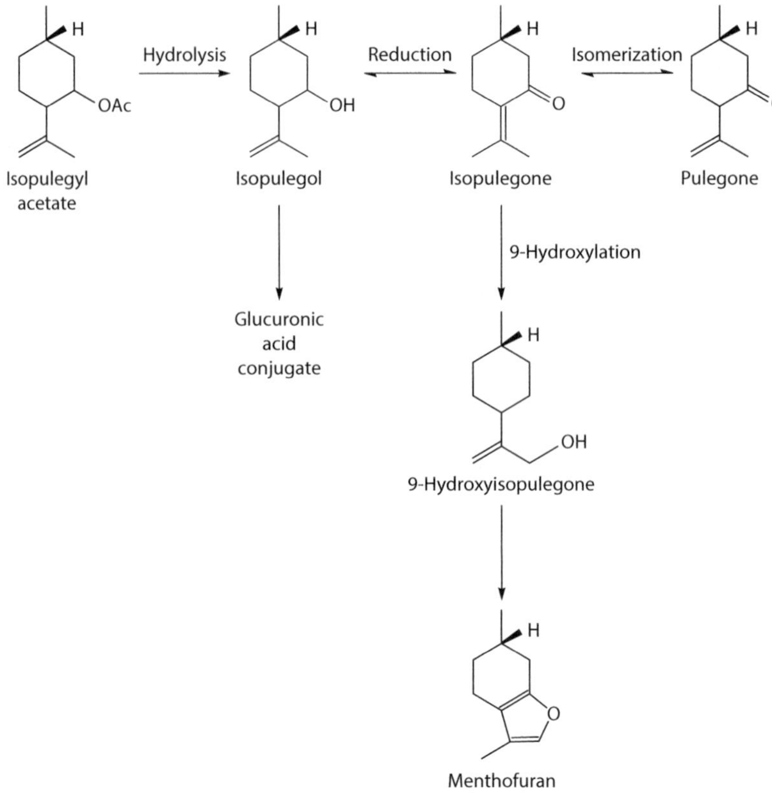

FIGURE 9.2 Metabolism of isopulegone, pulegone, and isopulegyl acetate.

and its metabolite (menthofuran), which account for <2% of commercial corn mint oil, are considered separately in the guide. In this case, the daily PCI of 79 μg/person/day (1.3 μg/kg bw/day) does not exceed the 90 μg/day threshold for Class III. However, a 90-day study on pulegone (NTP, 2002) showed a NOAEL (9.375 mg/kg bw/day) that is approximately 7200 times the intake of pulegone and its metabolites as constituents of corn mint oil. Also, in a 28-day study with peppermint oil (*M. piperita*) containing approximately 4% pulegone and menthofuran, a NOAEL of 200 mg/kg bw/day for male rats and a NOAEL of 400 mg/kg bw/day for female rats were established, which corresponds to a NOAEL of 8 mg/kg bw/day for pulegone and menthofuran (Serota, 1990). In a 90-day study with a mixture of *M. piperita* and *M. arvensis* oils (Splindler and Madsen, 1992; Smith et al., 1996), a NOAEL of 100 mg/kg bw/day was established, which corresponds to a NOAEL of 4 mg/kg bw/day for pulegone and menthofuran.

The only other congeneric group that accounts for >2% of the composition of corn mint oil is a congeneric group of terpene hydrocarbons ((+) and (−)-pinene, (+) limonene, etc.). Although these may contribute up to 8% of the oil, upon multiple redistillations during processing, the hydrocarbon content can be significantly reduced (<3%) in the finished commercial oil. Using the 8% figure to determine a conservative estimate of intake, the intake of terpene hydrocarbons is 315 μg/person/day (5.26 μg/kg bw/day). This group is predominantly metabolized by cytochrome P450-catalyzed hydroxylation, conjugation, and excretion (Ishida et al., 1981; Madyastha and Srivatsan, 1987; Crowell et al., 1994; Poon et al., 1996; Vigushin et al., 1998; Miyazawa et al., 2002). The daily PCI of 315 μg/person/day is less than the exposure threshold (1800 μg/person/day) for Structural Class I. Although no additional data would be required to complete the evaluation of this group, NOAELs (300 mg/kg bw/day) from long-term studies (NTP, 1990) on principal members of this group are orders of magnitude greater than the daily *PCI* (*eaters only*) of terpene hydrocarbons (0.088 mg/kg bw/day). Therefore, all congeneric groups in corn mint oil are considered safe for use when consumed in corn mint oil.

Finally, the essential oil itself is evaluated in the context of the combined intake of all congeneric groups and any other related data in Step 8. Interestingly, members of the terpene alicylic secondary alcohols, ketones, and related esters, multiple members of the monoterpene hydrocarbons, and peppermint oil itself show a common nephrotoxic effect recognized as alpha-2u-globulin nephropathy. The microscopic evidence of histopathology of the kidneys for male rats in the mint oil study is consistent with the presence of alpha-2u-globulin nephropathy. In addition, a standard immunoassay for detecting the presence of alpha-2u-globulin was performed on kidney sections from male and female rats in the mint oil study (Serota, 1990). Results of the assay confirmed the presence of alpha-2u-globulin nephropathy in male rats (Swenberg and Schoonhoven, 2002). This effect is found only in males rats and is not relevant to the human health assessment of corn mint oil. Other toxic interactions between congeneric groups are expected to be minimal given that the NOELs for the congeneric groups and those for finished mint oils are on the same order of magnitude.

Based on the aforementioned assessment and the application of the scientific judgment, corn mint oil is concluded to be *GRAS* under conditions of intended use as a flavoring substance. Given the criteria used in the evaluation, recommended specifications should include the following chemical assay:

1. Less than 95% alicyclic secondary alcohols, ketones, and related esters, typically measured as (−)-menthol
2. Less than 2% 2-isopropylidenecyclohexanones and their metabolites, measured as (−)-pulegone
3. Less than 10% monoterpene hydrocarbons, typically measured as limonene

9.5.3 Summary

The safety evaluation of an essential oil is performed in the context of all available data for congeneric groups of identified constituents and the group of unidentified constituents, data on the essential oil or a related essential oil, and any potential interactions that may occur in the essential oil when consumed as a flavoring substance.

The guide provides a chemically based approach to the safety evaluation of an essential oil. The approach depends on a thorough quantitative analysis of the chemical constituents in the essential oil intended for commerce. The chemical constituents are then assigned to well-defined congeneric groups that are established based on extensive biochemical and toxicologic information, and this is evaluated in the context of intake of the congeneric group resulting from consumption of the essential oil. The intake of unidentified constituents considers the consumption of the essential oil as a food, a highly conservative toxicologic threshold, and toxicity data on the essential oil or an essential oil of similar chemical composition. The flexibility of the guide is reflected in the fact that high intake of major congeneric groups of low toxicologic concern will be evaluated along with low intake of minor congeneric groups of significant toxicological concern (i.e., higher structural class). The guide also provides a comprehensive evaluation of all congeneric groups and constituents that account for the majority of the composition of the essential oil. The overall objective of the guide is to organize and prioritize the chemical constituents of an essential oil in order that no reasonably possible significant risk associated with the intake of essential oil goes unevaluated.

REFERENCES

Adams, T.B., Cohen, S., Doull, J. et al. 2004. The FEMA GRAS assessment of cinnamyl derivatives used as flavor ingredients. *Food Chem. Toxicol.* 42: 157–185.

Adams, T.B., Doull, J., Goodman, J.I. et al. 1997. The FEMA GRAS assessment of furfural used as a flavor ingredient. *Food Chem. Toxicol.* 35: 739–751.

Adams, T.B., Greer, D.B., Doull, J. et al. 1998. The FEMA GRAS assessment of lactones used as flavor ingredients. *Food Chem. Toxicol.* 36: 249–278.

Adams, T.B., Hallagan, J.B., Putman, J.M. et al. 1996. The FEMA GRAS assessment of alicyclic substances used as flavor ingredients. *Food Chem. Toxicol.* 34: 763–828.

Adams, T.B., McGowen, M.M., Williams, M.C. et al. 2007. The FEMA GRAS assessment of aromatic substituted secondary alcohols, ketones and related esters used as flavor ingredients. *Food Chem. Toxicol.* 45: 171–201.

Arctander, S. 1969. *Perfume and Flavor Chemicals*, Vol. 1. Rutgers University, Montclair, NJ (1981).

Austin, C.A., Shephard, E.A., Pike, S.F., Rabin, B.R., and Phillips, I.R. 1988. The effect of terpenoid compounds on cytochrome P-450 levels in rat liver. *Biochem. Pharmacol.* 37(11): 2223–2229.

Chen, L., Lebetkin, E.H., and Burka, L.T. 2001. Metabolism of (R)-(+)-pulegone in F344 rats. *Drug Metabol. Dispos.* 29(12): 1567–1577.

Cramer, G., Ford, R., and Hall, R. 1978. Estimation of toxic hazard – A decision tree approach. *Food Cosmet. Toxicol.* 16: 255–276.

Crowell, P., Elson, C.E., Bailey, H., Elegbede, A., Haag, J., and Gould, M. 1994. Human metabolism of the experimental cancer therapeutic agent d-limonene. *Cancer Chemother. Pharmacol.* 35: 31–37.

Dioscorides (50 AD) *Inquiry into Plants and Growth of Plants – Theophrastus.* De Materia Medica.

EPA (U.S. Environmental Protection Agency). 1988. *Technical Support Document on Risk Assessment of Chemical Mixtures.* EPA-600/8-90/064. U.S. Environmental Protection Agency, Office of Research and Development, Washington, DC.

European Communities (EC). 1999. Commission of European Communities Regulation No. 2232/96.

European Flavour and Fragrance Association (EFFA). 2005. *European Inquiry on Volume Use.* Private communication to the Flavor and Extract Manufacturers Association (FEMA), Washington, DC.

Flavor and Extract Manufacturers Association (FEMA). 1975. *Scientific Literature Review of Alicyclic Substances Used as Flavor Ingredients.* U.S. National Technical Information Services, PB86-1558351/LL, FEMA, Wahington, DC.

Florin, I., Rutberg, L., Curvall, M., and Enzell, C.R. 1980. Screening of tobacco smoke constituents for mutagenicity using the Ames test. *Toxicology* 18: 219–232.

Food Chemical Codex (FCC). 2013. *Food Chemicals Codex* (9th edn.). United States Pharmacopeia (USP), Rockville, MD.

Food and Drug Administration. 2005. Threshold of regulation for substances used in food-contact articles. 21 CFR 170.39.

Hall, R.L. 1976. Estimating the distribution of daily intakes of certain GRAS substances. Committee on GRAS list survey – Phase III. National Academy of Sciences/National Research Council, Washington, DC.

Hall, R.L. and Ford, R.A. 1999. Comparison of two methods to assess the intake of flavoring substances. *Food Addit. Contam.* 16: 481–495.

Hall, R.L. and Oser, B.L. 1968. Recent progress in the consideration of flavoring substances under the Food Additives Amendment. *Food Technol.* 19(2): 151.

Harman, C.L., Lipman, M.D., and Hallagan, J.B. 2013. *Flavor and Extract Manufacturers Association of the United States 2010 Poundage and Technical Effects Survey.* Flavor and Extract Manufacturers Association, Washington, DC.

Heck, J.D., Vollmuth, T.A., Cifone, M.A., Jagannath, D.R., Myhr, B., and Curren, R.D. 1989. An evaluation of food flavoring ingredients in a genetic toxicity screening battery. *Toxicologist* 9(1): 257.

Ishida, T., Asakawa, Y., Takemoto, T., and Aratani, T. 1981. Terpenoids Biotransformation in Mammals III: Biotransformation of alpha-pinene, beta-pinene, 3–carene, carane, myrcene, and p-cymene in rabbits. *J. Pharm. Sci.* 70: 406–415.

Japanese Flavor and Fragrance Manufacturers Association (JFFMA). 2002. Japanese inquiry on volume use. Private communication to the Flavor and Extract Manufacturers Association (FEMA), Washington, DC.

JECFA. 1997. Evaluation of certain food additives and contaminants. Forty-sixth Report of the Joint FAO/WHO Expert Committee on Food Additives. World Health Organization, WHO Technical Report Series 868.

JECFA. 1998. Evaluation of certain food additives and contaminants. Forty-seventh Report of the Joint FAO/WHO Expert Committee on Food Additives. WHO Technical Report Series 876. World Health Organization, Geneva, Switzerland.

JECFA. 1999. Procedure for the Safety Evaluation of Flavouring Agents. Evaluation of certain food additives and contaminants. Forty-ninth Report of the Joint FAO/WHO Expert Committee on Food Additives. World Health Organization, WHO Technical Report Series 884.

JECFA. 2000a. Evaluation of certain food additives and contaminants. Fifty-first report of the Joint FAO/WHO Expert Committee on Food Additives. WHO Technical Report Series No. 891. World Health Organization, Geneva, Switzerland.

JECFA. 2000b. Evaluation of certain food additives and contaminants. Fifty-third report of the Joint FAO/WHO Expert Committee on Food Additives. WHO Technical Report Series No. 896. World Health Organization, Geneva, Switzerland.

JECFA. 2001. Evaluation of certain food additives and contaminants. Fifty-fifth report of the Joint FAO/WHO Expert Committee on Food Additives. WHO Technical Report Series No. 901. World Health Organization, Geneva, Switzerland.

JECFA. 2003. Evaluation of certain food additives and contaminants. Fifty-ninth report of the Joint FAO/WHO Expert Committee on Food Additives. WHO Technical Report Series No. 913. World Health Organization, Geneva, Switzerland.

JECFA. 2004. Evaluation of certain food additives and contaminants. Sixty-first report of the Joint FAO/WHO Expert Committee on Food Additives. World Health Organization, Geneva, Switzerland.

Lambe, J., Cadby, P., and Gibney, M. 2002. Comparison of stochastic modelling of the intakes of intentionally added flavouring substances with theoretical added maximum daily intakes (TAMDI) and maximized survey-derived daily intakes (MSDI). *Food Addit. Contam.* 19(1): 2–14.

Lucas, C.D., Putnam, J.M., and Hallagan, J.B. 1999. *Flavor and Extract Manufacturers Association (FEMA) of the United States.* 1995 Poundage and Technical Effects Update Survey. Washington, DC.

Maarse, H., Visscher, C.A., Willemsens, L.C., and Boelens, M.H. 1992, 1994, 2000. *Volatile Components in Food-Qualitative and Quantitative Data.* Centraal Instituut Voor Voedingsonderzioek TNO, Zeist, the Netherlands.

Madsen, C., Wurtzen, G., and Carstensen, J. 1986. Short-term toxicity in rats dosed with menthone. *Toxicol. Lett.* 32: 147–152.

Madyastha, K.M. and Srivatsan, V. 1987. Metabolism of beta-myrcene *in vivo* and *in vitro*: Its effects on rat-liver microsomal enzymes. *Xenobiotica* 17(5): 539–549.

Madyastha, K.M. and Srivatsan, V. 1988. Studies on the metabolism of l-menthol in rats. *Drug Metab. Dispos.* 16: 765.

McClanahan, R.H., Thomassen, D., Slattery, J.T., and Nelson, S.D. 1989. Metabolic activation of (R)-(+)-pulegone to a reactive enonal that covalently binds to mouse liver proteins. *Chem. Res. Toxicol.* 2: 349–355.

Meek, M.E., Boobis, A.R., Crofton, K.M., Heinemeyer, G., Van Raaij, M., and Vickers, C. 2011. Risk assessment of combined exposure to multiple chemicals: A WHO/IPCS framework. *Reg. Toxicol. Pharmacol.* 60: S1–S14.

Miyazawa, M., Shindo, M., and Shimada, T. 2002. Sex differences in the metabolism of (+)- and (–)-limonene enantiomers to carveol and perillyl alcohol derivatives by cytochrome P450 enzymes in rat liver microsomes. *Chem. Res. Toxicol.* 15(1): 15–20.

Muller, W. 1993. Evaluation of mutagenicity testing with *Salmonella typhimurium* TA102 in three different laboratories. *Environ. Health Perspect. Suppl.* 101: 33–36.

Munro, I., Ford, R., Kennepohl, E., and Sprenger, J. 1996a. Correlation of structural class with no-observed-effect-levels: A proposal for establishing a threshold of concern. *Food Chem. Toxicol.* 34: 829–867.

Munro, I.C., Ford, R.A., Kennepohl, E., and Sprenger, J.G. 1996b. Thresholds of toxicological concern based on structure-activity relationships. *Drug Metab. Rev.* 28(1/2): 209–217.

National Academy of Sciences (NAS). 1965, 1970, 1975, 1981, 1982, 1987. *Evaluating the Safety of Food Chemicals.* National Academy of Sciences, Washington, DC.

National Cancer Institute (NCI). 1979. Bioassay of dl-menthol for possible carcinogenicity. National Technical Report Series No. 98. U.S. Department of Health, Education and Welfare, Bethesda, MD.

National Research Council (NRC). 1988. *Complex Mixtures: Methods for In Vivo Toxicity Testing.* National Academy Press, Washington, DC.

National Research Council (NRC). 1994. *Science and Judgment in Risk Assessment.* National Academy Press, Washington, DC.

National Toxicology Program (NTP). 1990. *Carcinogenicity and toxicology studies of d-limonene in F344/N rats and B6C3F1 mice.* NTP-TR-347. U.S. Department of Health and Human Services. NIH Publication No. 90-2802. National Toxicology Program, Research Triangle Park, NC.

National Toxicology Program (NTP). 2002. *Toxicity studies of pulegone in B6C3F1 mice and rats (Gavage studies).* Battelle Research Laboratories, Study No. G004164-X. Unpublished Report. National Toxicology Program, Research Triangle Park, NC.

National Toxicology Program (NTP). 2003a. *Draft report on the Initial study results from a 90-day toxicity study on beta-myrcene in mice and rats.* Study number C99023 and A06528. National Toxicology Program, Research Triangle Park, NC.

Newberne, P., Smith, R.L., Doull, J. et al. 1999. The FEMA GRAS assessment of *trans*-anethole used as a flavoring substance. *Food Chem. Toxicol.* 37: 789–811.

Nijssen, B., van Ingen-Visscher, K., and Donders, J. 2003. *Volatile Compounds in Food 8.1. Centraal Instituut Voor Voedingsonderzioek TNO,* Zeist, the Netherlands. http://www.voeding.tno.nl/vcf/VcfNavigate.cfm.

Organization for Economic Cooperation and Development (OECD). 2014. *Guidance on Grouping of Chemicals,* 2nd edn. OECD Environment, Health and Safety Publications, Series on Testing and Assessment, No. 194. OECD, Paris, France.

Oser, B. and Ford, R. 1991. FEMA Expert Panel: 30 years of safety evaluation for the flavor industry. *Food Technol.* 45(11): 84–197.
Oser, B. and Hall, R. 1977. Criteria employed by the Expert Panel of FEMA for the GRAS evaluation of flavoring substances. *Food Cosmet. Toxicol.* 15: 457–466.
Poon, G., Vigushin, D., Griggs, L.J., Rowlands, M.G., Coombes, R.C., and Jarman, M. 1996. Identification and characterization of limonene metabolites in patients with advanced cancer by liquid chromatography/mass spectrometry. *Drug Metab. Dispos.* 24: 565–571.
Presidential Commission on Risk Management and Risk Assessment. 1996. *Risk assessment and risk management in regulatory decision making.* Final Report, Vols. 1 and 2. Presidential Commission on Risk Management, Washington, DC.
Rivedal, E., Mikalsen, S.O., and Sanner, T. 2000. Morphological transformation and effect on gap junction intercellular communication in Syrian Hamster Embryo Cells Screening Tests for Carcinogens Devoid of Mutagenic Activity. *Toxicol. In Vitro* 14(2): 185–192.
Roe, F. and Field, W. 1965. Chronic toxicity of essential oils and certain other products of natural origin. *Food Cosmet. Toxicol.* 3: 311–324.
Rulis, A.M. 1987. *De Minimis* and the threshold of regulation. In: Felix, C.W. (Ed.), *Food Protection Technology.* Lewis Publishers Inc., Chelsea, MI, pp. 29–37.
Rulis, A.M., Hattan, D.G., and Morgenroth, V.H. 1984. FDA's priority-based assessment of food additives. I. Preliminary results. *Reg. Toxicol. Pharmacol.* 26: 44–51.
Sasaki, Y.F., Imanishi, H., Ohta, T., and Shirasu, Y. 1989. Modifying effects of components of plant essence on the induction of sister-chromatid exchanges in cultured Chinese hamster ovary cells. *Mutat. Res.* 226: 103–110.
Schilter, B., Andersson, C., Anton, R., Constable, A., Kleiner, J., O'Brien, J., Renwick, A.G., Korver, O., Smit, F., and Walker, R. 2003. Guidance for the safety assessment of botanicals and botanical preparations for use in food and food supplements. *Food Chem. Toxicol.* 41: 1625–1649.
Serota, D. 1990. *28-Day toxicity study in rats.* Hazelton Laboratories America, HLA Study No. 642-477. Private Communication to FEMA. Unpublished Report.
Smith, R.L., Adams, T.B., Doull, J. et al. 2002a. Safety assessment of allylalkoxybenzene derivatives used as flavoring substances – Methyleugenol and estragole. *Food Chem. Toxicol.* 40: 851–870.
Smith, R.L., Cohen, S.M., Doull, J. et al. 2005a. Criteria for the safety evaluation of flavoring substances the expert panel of the flavor and extract manufacturers association. *Food Chem. Toxicol.* 43: 1141–1177.
Smith, R.L., Cohen, S., Doull, J., Feron, V.J., Goodman, J.I., Marnett, L.J., Portoghese, P.S., Waddell, W.J., Wagner, B.M., and Adams, T.B. 2003. Recent progress in the consideration of flavor ingredients under the Food Additives Amendment. 21 GRAS Substances. *Food Technol.* 57: 46.
Smith, R.L., Cohen, S., Doull, J., Feron, V.J., Goodman, J.I., Marnett, L.J., Portoghese, P.S., Waddell, W.J., Wagner, B.M., and Adams, T.B. 2004. Safety evaluation of natural flavour complexes. *Toxicol. Lett.* 149: 197–207.
Smith, R.L., Cohen, S.M., Doull, J. et al. 2005b. A procedure for the safety evaluation of natural flavor complexes used as ingredients in food: Essential oils. *Food Chem. Toxicol.* 43: 345–363.
Smith, R.L., Doull, J., Feron, V.J. et al. 2002b. The FEMA GRAS assessment of pyrazine derivatives used as flavor ingredients. *Food Chem. Toxicol.* 40: 429–451.
Smith, R.L., Newberne, P., Adams, T.B., Ford, R.A., Hallagan, J.B., and the FEMA Expert Panel. 1996. GRAS flavoring substances 17. *Food Technol.* 50(10): 72–78, 80–81.
Splindler, P. and Madsen, C. 1992. Subchronic toxicity study of peppermint oil in rats. *Toxicol. Lett.* 62: 215–220.
Swenberg, J. and Schoonhoven, R. 2002. Private communication to FEMA.
Thomassen, D., Knebel, N., Slattery, J.T., McClanahan, R.H., and Nelson, S.D. 1992. Reactive intermediates in the oxidation of menthofuran by cytochrome P-450. *Chem. Res. Toxicol.* 5: 123–130.
Vigushin, D., Poon, G.K., Boddy, A., English, J., Halbert, G.W., Pagonis, C., Jarman, M., and Coombes, R.C. 1998. Phase I and pharmacokinetic study of D-limonene in patients with advanced cancer. *Cancer Chemother. Pharmacol.* 42(2): 111–117.
Williams, R.T. 1940. Studies in detoxication. 7. The biological reduction of l-Menthone to d-neomenthol and of d-isomenthone to d-isomenthol in the rabbit. The conjugation of d-neomenthol with glucuronic acid. *Biochem. J.* 34: 690–697.
Woods, L. and Doull, J. 1991. GRAS evaluation of flavoring substances by the Expert Panel of FEMA. *Regul. Toxicol. Pharmacol.* 14(1): 48–58.
Yamaguchi, T., Caldwell, J., and Farmer, P.B. 1994. Metabolic fate of [^{3}H]-l-menthol in the rat. *Drug Metab. Dispos.* 22: 616–624.
Zamith, H.P., Vidal, M.N.P., Speit, G., and Paumgartten, F.J.R. 1993. Absence of genotoxic activity of beta-myrcene in the *in vivo* cytogenetic bone marrow assay. *Braz. J. Med. Biol. Res.* 26: 93–98.

10 Metabolism of Terpenoids in Animal Models and Humans

Walter Jäger and Martina Höferl

CONTENTS

10.1 INTRODUCTION

Terpenoids are main constituents of plant-derived essential oils. Because of their pleasant odor, they are widely used in the food, fragrance, and pharmaceutical industries. Furthermore, in traditional medicine, terpenoids are also well known for their anti-inflammatory, antibacterial, antifungal, antitumor, and sedative activities. Although large amounts are used in the industry, the knowledge about their biotransformation in humans is still scarce. Yet, metabolism of terpenoids can lead to the formation of new biotransformation products with unique structures and often different flavor and biological activities compared to the parent compounds. All terpenoids easily enter the human body by oral absorption, penetration through the skin, or inhalation, very often leading to measurable blood concentrations. A number of different enzymes, however, readily metabolize these compounds to more water-soluble molecules. Although nearly every tissue has the ability to metabolize drugs, the liver is the most important organ of drug biotransformation. In general, metabolic biotransformation occurs at two major categories called Phase I and Phase II reactions (Spatzenegger and Jäger, 1995). Phase I concerns mostly cytochrome P450 (CYP)-mediated oxidation, as well as reduction and hydrolysis. Phase II is a further step where a Phase I product is completely transformed to high water solubility. This is done by attaching already highly water-soluble endogenous entities such as sugars (glucuronic acids) or salts (sulfates) to the Phase I intermediate and forming a Phase II final product. It is not always necessary for a compound to undergo both Phases I and II; indeed, for many terpenoids, one or the other is enough to eliminate these volatile plant constituents. In the following concise review, special emphasis will be put on metabolism of selected mono- and sesquiterpenoids, not only in animal and *in vitro* models but also in humans.

10.2 METABOLISM OF MONOTERPENES

10.2.1 Borneol

Borneol is a component in many essential oils; for example, oils of Pinaceae, *Salvia officinalis*, *Rosmarinus officinalis*, and *Artemisia* species (Bornscheuer et al., 2014). *Cinnamomum camphora* chemotype borneol and *Blumea balsamifera*, which are rich in (+)- and (–)-borneol, respectively, are used as sources for preparation of bingpian, a drug of traditional Chinese medicine (Zhao et al., 2012). Moreover, borneol is used to give soaps, perfumes, and other products a scent of spruce needles. *In vitro* studies with rat liver microsomes could provide evidence for four metabolites (Figure 10.1) (Zhang et al., 2008). The main metabolite, camphor, could also be detected in rat plasma (Sun et al., 2014).

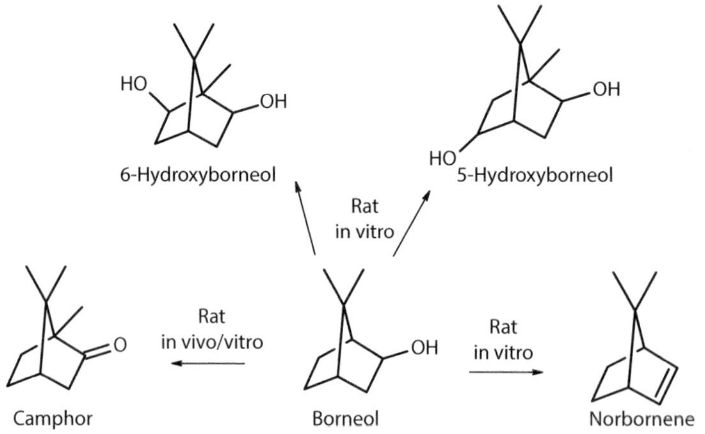

FIGURE 10.1 Proposed metabolism of borneol in rat liver microsomes. (Adapted from Zhang et al. 2008 *J. Chromatogr. Sci.* 46:419–423.)

10.2.2 Camphene

Camphene is found in higher concentrations in the essential oils of common coniferous trees (e.g., *Abies alba* or *Tetraclinis articulata*), in the rhizome of *Zingiber officinalis*, and in *Salvia officinalis* and *Rosmarinus officinalis* (Bornscheuer et al., 2014). Up to now, there is only one publication on various biotransformation products in the urine of rabbits after its oral administration. As shown in Figure 10.2, camphene is metabolized into two diastereomeric glycols (camphene-2,10-glycols). Their formation obviously involves two isomeric epoxide intermediates, which are hydrated by epoxide hydrolase. The monohydroxylated camphene and tricyclene derivatives were apparently formed through the non-classical cation intermediate (Ishida et al., 1979). So far, there are no studies available about the biotransformation of camphene in humans.

10.2.3 Camphor

(+)-Camphor is extracted from the wood of *Cinnamomum camphora*, a tree endemic to Southeast Asia. Furthermore, it is also one of the major constituents of the essential oils of *Salvia officinalis* and *Rosmarinus officinalis*. Camphor is commercially used as a moth repellent and antiseptic in cosmetics (aftershaves, face tonics, mouthwash etc.) and pharmaceutically in ointments for treatment of rheumatic pains and coughs (O'Neil, 2006; Bornscheuer et al., 2014). In dogs, rabbits, and rats, camphor is extensively metabolized whereat the major hydroxylation products are 5-*endo*-and 5-*exo*-hydroxycamphor (Leibman and Ortiz, 1973). A small amount was also identified as 3-*endo*-hydroxycamphor (Figure 10.3). Both 3- and 5-bornane groups can be further reduced to 2,3- and 2,5-bornanedione. Minor biotransformation steps also involve the reduction of camphor to borneol and isoborneol. Interestingly, all hydroxylated camphor metabolites are further conjugated in a Phase II reaction with glucuronic acid. Camphor is extensively metabolized by human liver microsomes to 5-*exo*-hydroxycamphor (Leibman and Ortiz, 1973; Gyoubu and Miyazawa, 2007). In an *in vitro*

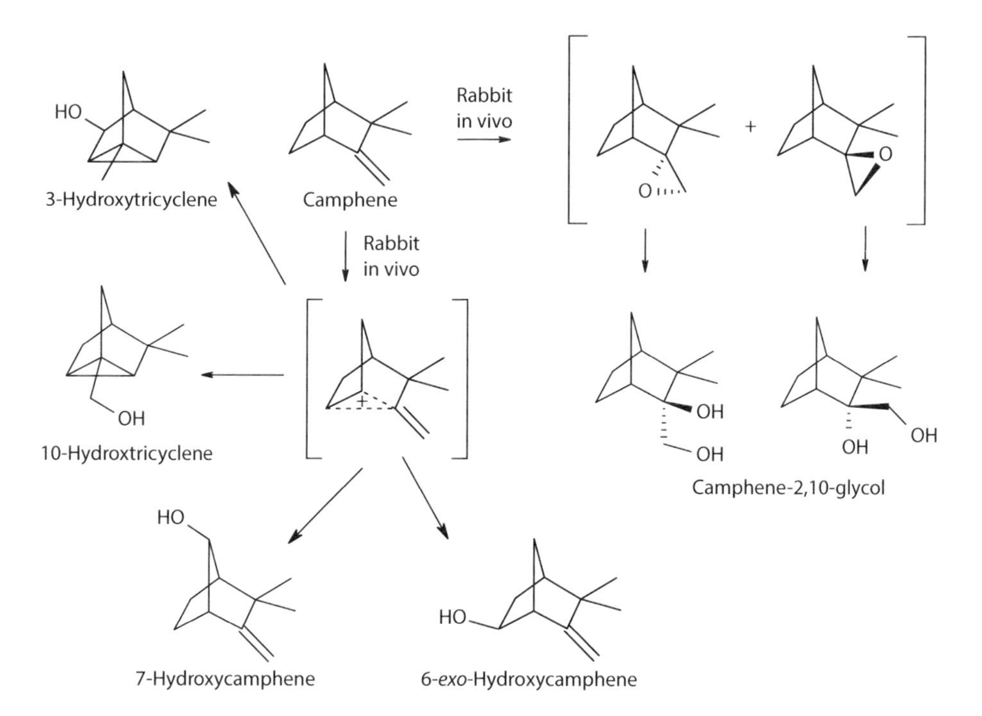

FIGURE 10.2 Urinary excretion of camphene metabolites in rabbits. (Adapted from Ishida et al. 1979. *J. Pharm. Sci.* 68:928–930.)

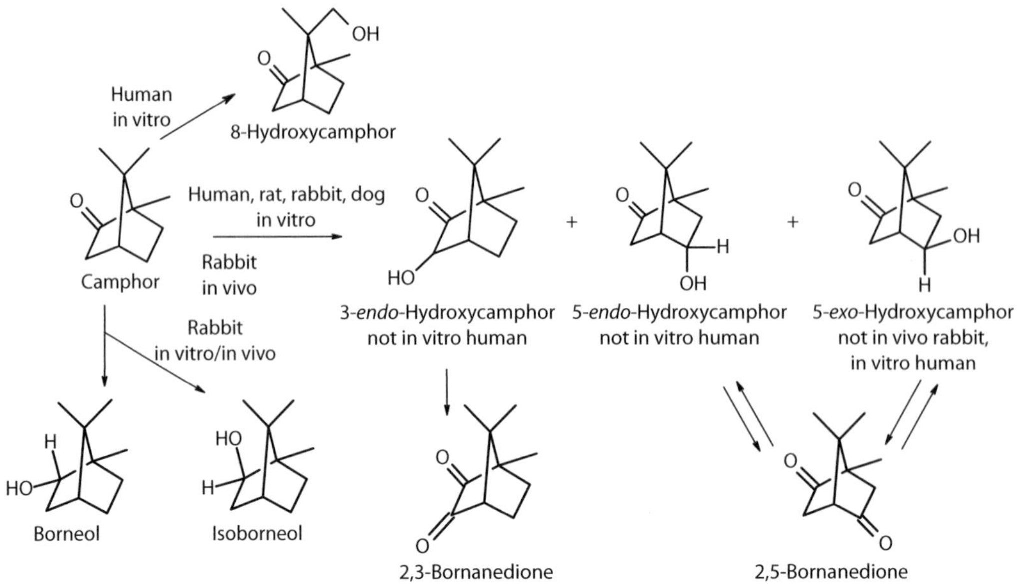

FIGURE 10.3 Metabolisms of camphor in dogs, rabbits, and rats. (Adapted from Leibmann and Ortiz. 1972. *Drug Metab. Dispos.* 1:543–551.)

experiment using *Salmonella typhimurium* expressing human CYP2A6 and NADPH-P450 reductase, 5-*exo*-hydroxycamphor was found as a metabolite of camphor, together with 8-hydroxycamphor (Nakahashi and Miyazawa, 2011).

10.2.4 3-Carene

3-Carene is found in various Pinaceae essential oils: (+)-3-Carene is a major compound of *Pinus palustris* essential, and (−)-3-carene of *Pinus sylvestris*. It is used as raw material in perfumery (Bornscheuer et al., 2014). In rabbits, 3-carene is metabolized into 3-caren-9-ol and presumably via 3-caren-10-ol, and further oxidized into carboxylic and dicarboxylic acid products (Ishida et al., 1981, 2005). Another metabolic pathway takes place via opening of the methylene bridge to *m*-mentha-4,6-dien-8-ol with subsequent dehydrogenation of the cyclohexene ring to *m*-cymen-8-ol (Ishida, 2005). *In vitro* experiments with human liver microsomes revealed 3-caren-10-ol and 3-carene epoxide as metabolites (Figure 10.4). Hydroxylation was catalyzed by CYP2B6, CYP2C19, and CYP2D6, whereas epoxidation could be attributed to CYP1A2 (Duisken et al., 2005). Interestingly, neither of these two metabolites, nor the supposed hydrolysis product of the epoxide, 3-carene-3,4-diol, could be detected in rabbit or human urine, yet (Ishida, 2005; Schmidt et al., 2013, 2015). A recent study (Schmidt et al., 2015) identified chaminic acid as a new metabolite in human urine after oral intake of 3-carene (Figure 10.4). Moreover, Schmidt et al. suggested that dihydrochaminic acid and carene-3,4,9-triol are additional metabolites of 3-carene in humans (Schmidt et al., 2015).

10.2.5 Carvacrol

Carvacrol is used as disinfectant and found in high concentrations in the essential oils of, for example, *Thymus vulgaris* chemotype carvacrol, *Origanum vulgare*, *Majorana hortensis*, or *Satureja hortensis* (Bornscheuer et al., 2014). In rat, only small amounts of unchanged carvacrol were excreted 24 h after oral application. As β-glucuronidase and sulfatase were used for sample preparation before GC analysis, carvacrol might also be excreted as its glucuronide and sulfate, respectively. Both of the aliphatic groups present undergo extensive metabolism, whereas aromatic

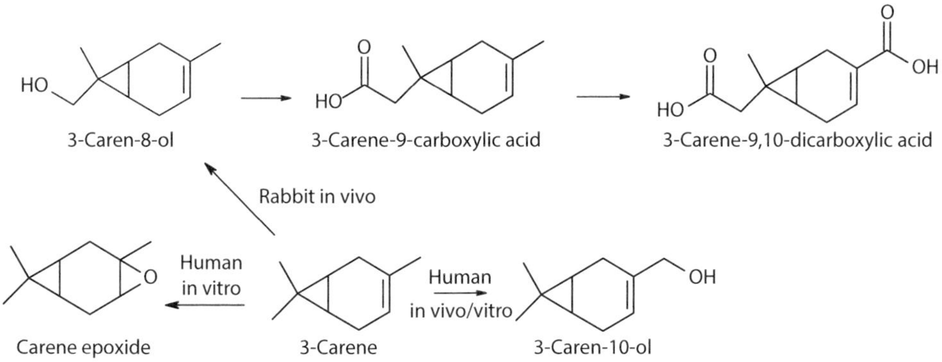

FIGURE 10.4 Proposed metabolism of 3-carene in rabbits, humans, and human liver microsomes. (Adapted from Ishida et al. 1981. *J. Pharm. Sci.* 70:406–415; Duisken et al. 2005. *Curr. Drug Metab.* 6:593–601; Schmidt et al. 2015. *Arch. Toxicol.* 89:381–392.)

hydroxylation to 2-hydroxycarvacrol is only a minor important pathway for carvacrol. Further oxidation of 7-hydroxycarvacrol results in isopropylsalicylic acid (Austgulen et al., 1987) (Figure 10.5). Carvacrol is metabolized by recombinant human CYP1A2, CYP2A6 and CYP2B6 (Dong et al., 2012b). An *in vitro* study with human microsomes demonstrated that recombinant UGT1A9 was mainly responsible for glucuronidation in liver, and rUGT1A7 in intestinal microsomes, forming monoglucuronated metabolites (Dong et al., 2012a).

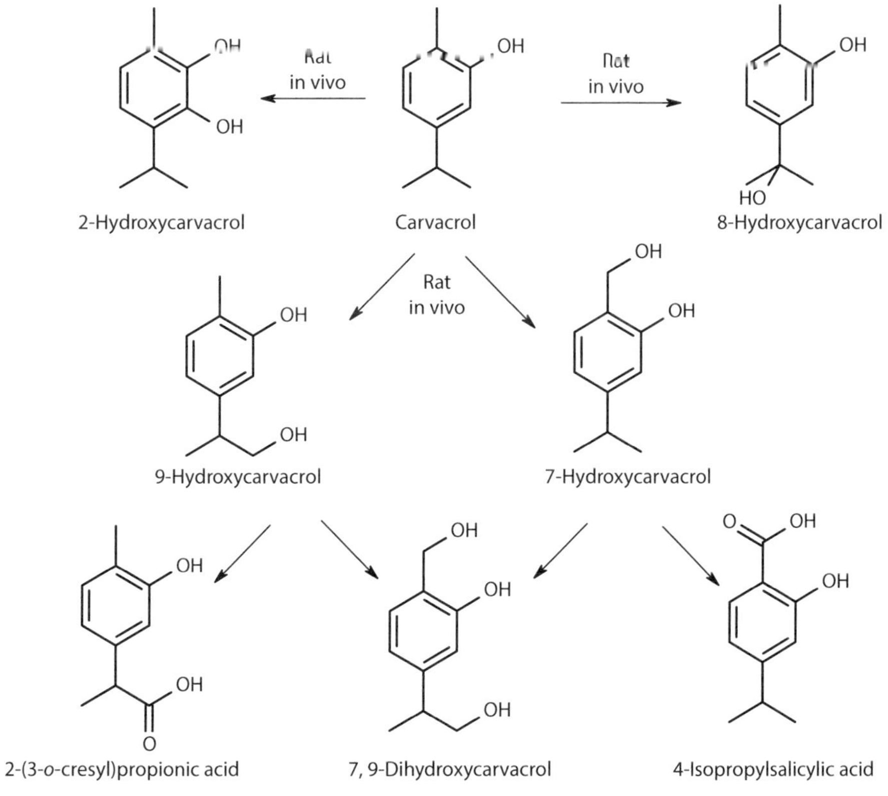

FIGURE 10.5 Metabolism and urinary excretion of carvacrol in rats. (Adapted from Austgulen et al. 1987. *Pharmacol. Toxicol.* 61:98–102.)

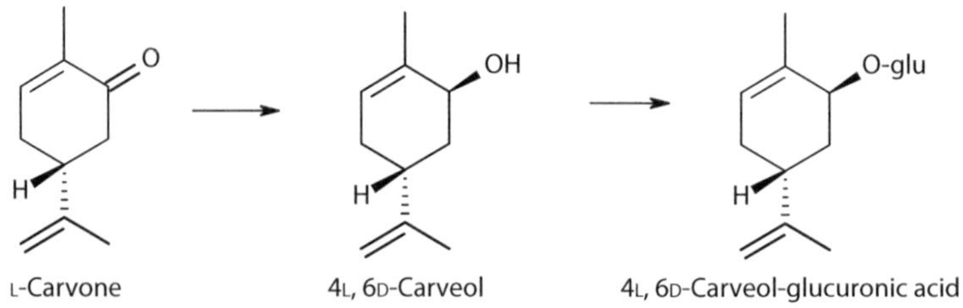

L-Carvone 4L, 6D-Carveol 4L, 6D-Carveol-glucuronic acid

FIGURE 10.6 Metabolic pathway of (R)-(–)-carvone in healthy subjects. (Adapted from Jäger et al., 2000. *J. Pharm. Pharmacol.* 52:191–197.)

10.2.6 Carvone

The (R)-(–)- and (S)-(+)-enantiomers of the monoterpene ketone carvone are found in various plants. While (S)-(+)-carvone is the main constituent of the essential oil *Carum carvi* and *Anethum graveolens*, (R)-(–)-carvone is found in the oil of *Mentha spicata* var. *crispa* (O'Neil, 2006; Bornscheuer et al., 2014). Because of minty odor and taste, large amounts of (R)-(–)-carvone are frequently added to toothpastes, mouthwashes, and chewing gums. (S)-(+)-carvone possesses the typical caraway aroma and is mainly used as a flavor compound in food industry. Due to its spasmolytic effect, (S)-(+)-carvone is also used as stomachic and carminative (Jäger et al., 2001). After separate topical applications of (R)-(–)- and (S)-(+)-carvone, both enantiomers are rapidly absorbed, resulting in significantly higher maximal plasma concentrations (C_{max}) and areas under the blood concentrations time curves (AUC) for (S)-(+)- compared to (R)-(–)-carvone. As demonstrated in Figure 10.6, analysis of control and ß-glucuronidase-pretreated urine samples only revealed stereoselective metabolism of (R)-(–)-carvone but not of (S)-(+)-carvone to (4R,6S)-(–)-carveol and (4R,6S)-(–)-carveol glucuronide, indicating that stereoselectivity in Phases I and II metabolism has significant effects on (R)-(–)- and (S)-(+)-carvone pharmacokinetics (Jäger et al., 2000) (Figure 10.6). Contrary to the study of Jäger et al., (2000), Engel could not demonstrate any differences in the formation of metabolites after peroral application of (R)-(–)- and (S)-(+)-carvone to human volunteers which may be due to the separation of biotransformation products on a nonchiral gas chromatography column. As shown in Figure 10.7, besides carveol, several metabolites could be identified in the urine samples (Engel, 2001).

10.2.7 1,4-Cineole

1,4-Cineole is a flavor constituent of *Citrus aurantiifolia* and *Piper cubeba* (Bornscheuer et al., 2014). *In vitro* and *in vivo* animal studies demonstrated extensive biotransformation of this monoterpene, strongly suggesting biotransformation in the human body, too. After oral application to rabbits, four neutral and one acidic metabolite could be isolated from urine (Asakawa et al., 1988). Using rat and human liver microsomes, however, only 2-hydroxylation could be observed, indicating species-related differences in 1,4-cineole metabolism (Miyazawa et al., 2001b) (Figure 10.8).

10.2.8 1,8-Cineole

1,8-Cineole is widely distributed in plants and found in high concentrations in the essential oils of *Eucalyptus globulus* and *Laurus nobilis*. It is extensively used in cosmetics and for ointments against cough, muscular pain, and rheumatism (O'Neil, 2006; Bornscheuer et al., 2014). In rat liver microsomes, 1,8-cineole is predominantly metabolized to 3-hydroxy-1,8-cineole and, to a lesser extent, to 2- and 9-hydroxy-1,8-cineole, respectively. (Miyazawa et al., 2001a). In the urine of rabbits and koalas, various 7- and 9-oxydated metabolites were found (Boyle et al., 2001). In human liver

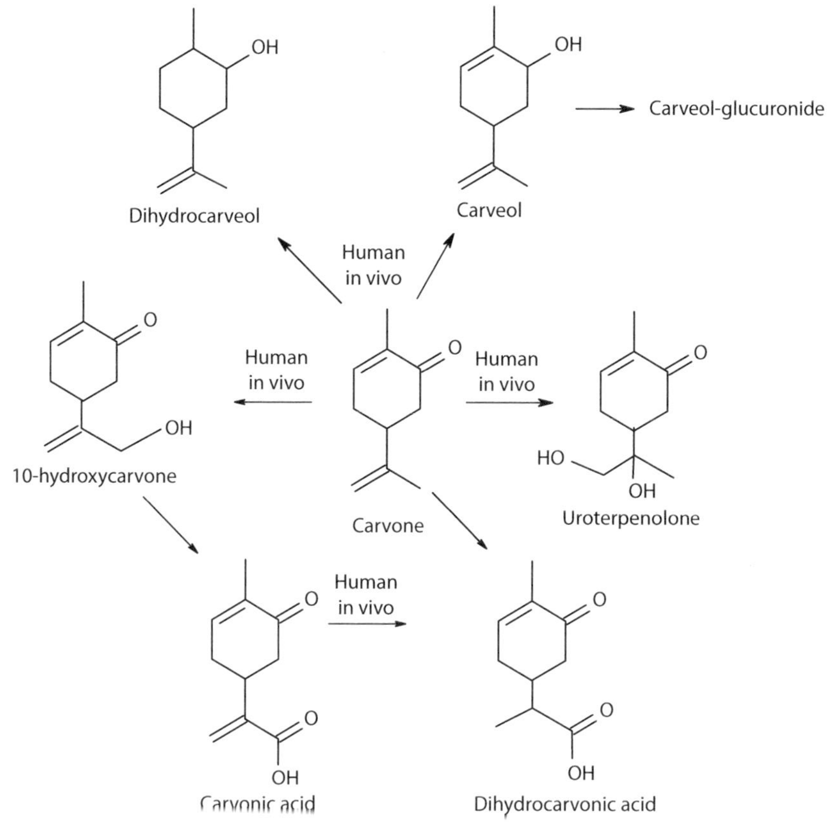

FIGURE 10.7 Proposed metabolic pathway of (R)-(−)- and (S)-(+)-carvone in healthy volunteers. (Adapted from Engel. 2001. *J. Agric. Food Chem.* 49:4069–4075.)

microsomes, however, only the 2- and 3-hydroxylated products catalyzed by the isoenzyme CYP3A4 could be identified (seen Figure 10.9. (Miyazawa and Shindo, 2001; Miyazawa et al., 2001a). Both metabolites could also be also found in the urine of three human volunteers after oral administration of a cold medication containing 1,8-cineole (Miyazawa et al., 2001b); these metabolites are therefore excellent biomarkers for the 1,8-cineole intake in humans. Besides these main biotransformation products, Horst and Rychlik could also identify small amounts of 7- and 9-hydroxy-1,8-cineole in human urine (Horst and Rychlik, 2010).

10.2.9 Citral

Both natural and synthetic citral are isomeric mixtures of geranial and neral, in which geranial is usually the predominant isomer. Major amounts are found in the essential oils of *Cymbopogon* sp., *Backhousia citriodora*, *Litsea cubeba*, *Verbena officinalis*, or *Melissa officinalis*. Moreover, it is found in many citrus oils. Because of its intense lemon aroma, citral has been extensively used for flavoring food, cosmetics, and detergents (O'Neil, 2006; Bornscheuer et al., 2014). Studies in rats have shown that citral is rapidly metabolized into several acids and a biliary glucuronide and excreted, with urine as the major route of elimination of citral, followed by expired air and feces. As demonstrated in Figure 10.10, seven urinary metabolites were isolated and identified (Diliberto et al., 1990). Based on the rat study mentioned above, extensive biotransformation of citral in human subjects is highly suggested.

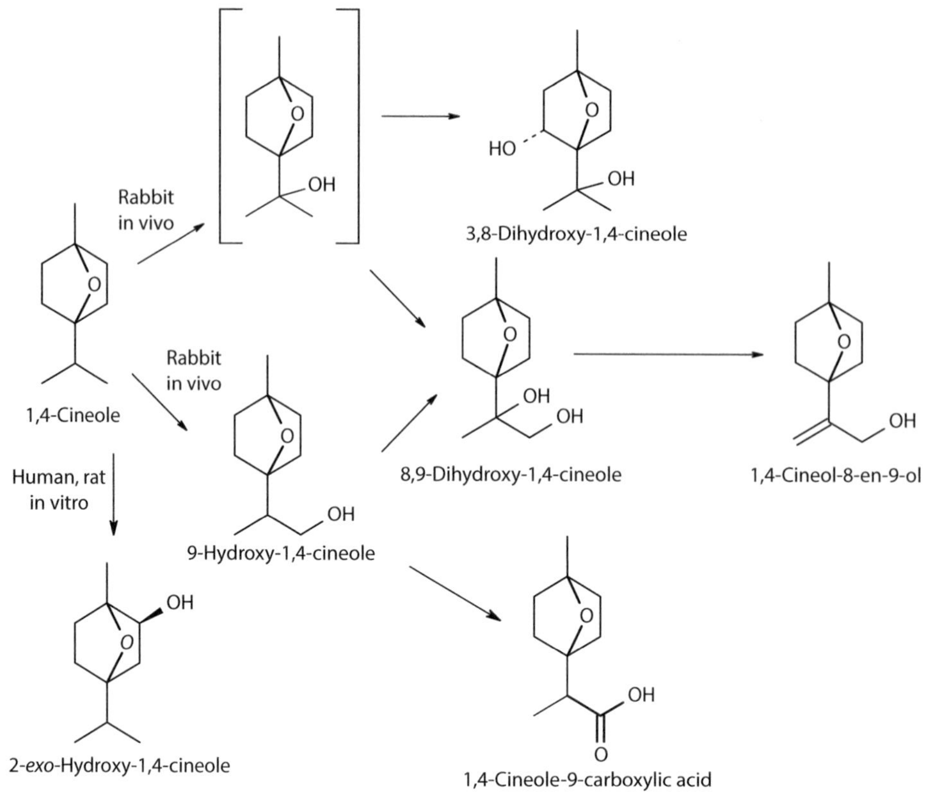

FIGURE 10.8 Proposed metabolism of 1,4-cineole in rabbits and in rat and human liver microsomes. (Adapted from Asakawa et al. 1988. *Xenobiotica* 18:1129–1134; Miyazawa et al. 2001a. *Xenobiotica* 31: 713–723.)

10.2.10 Citronellal

Citronellal is a monocyclic monoterpene aldehyde with high concentrations found in the essential oils of *Corymbia citriodora*, *Melissa officinalis*, and various *Cymbopogon* species. It is used for perfuming soaps and other products (O'Neil, 2006; Bornscheuer et al., 2014). Only one study described biotransformation of citronellal in rabbits. Ishida et al. could isolate three neutral metabolites of (+)-citronellal in the urine of rabbits (Figure 10.11). An additional acidic metabolite was formed as the result of regioselective oxidation of the aldehyde and dimethyl allyl groups (Ishida et al., 1989). Based on animal data, metabolism of citronellal is also expected in humans.

10.2.11 p-Cymene

p-Cymene is found in many essential oils; for example, *Carum carvi* and *Thymus vulgaris*. It is used as a fragrance compound in perfumery (Bornscheuer et al., 2014). The main *p*-cymene metabolite found *in vitro* using liver microsomes of brushtail possum, koala, and rat was cuminyl alcohol, accompanied by its oxidation product, cumic acid (Pass et al., 2002). Various hydroxylated and carboxylated *p*-cymene metabolites were found in rabbits (Matsumoto et al., 1992). *In vitro* assays with human recombinant CYP resulted in several metabolites, thymol, cuminyl alcohol, and cuminaldehyde (Figure 10.12). In human blood and urine samples only thymol and its conjugates, thymol glucuronide and sulfate, could be identified (Meesters et al., 2009).

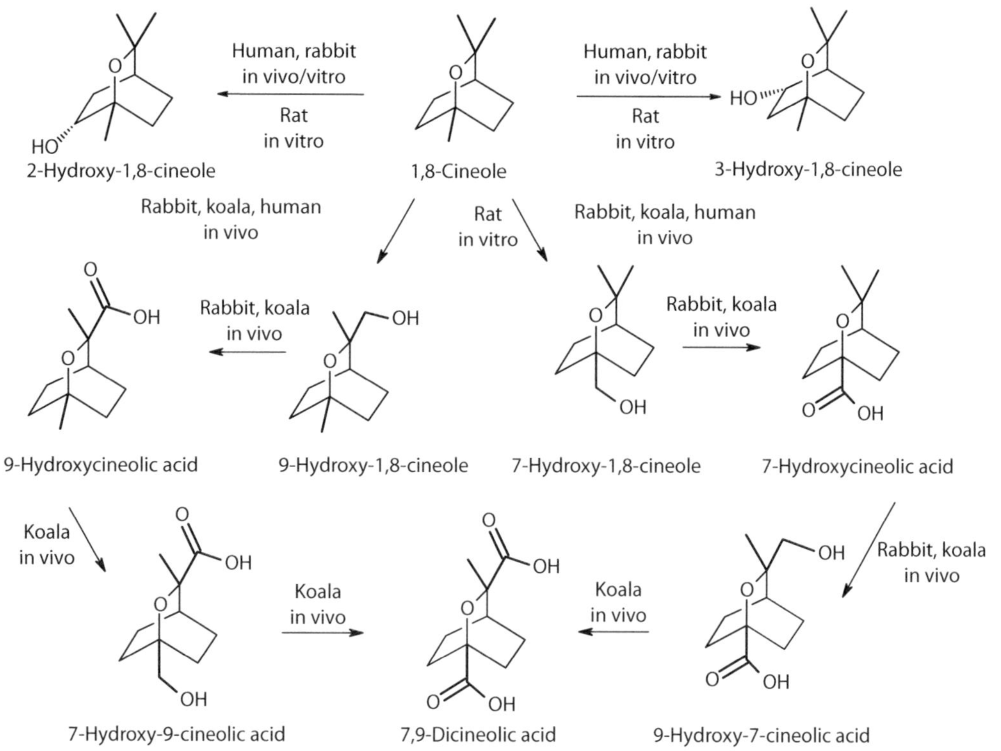

FIGURE 10.9 Proposed metabolism of 1,8-cineole *in vitro* (rat and human liver microsomes) and *in vivo* (rabbits, koalas, and humans). (Adapted from Miyazawa et al., 2001b. *Drug Metab. Dispos.*, 29: 200–205; Boyle, R. et al., 2001. *Comp. Biochem. Physiol. C*, 129: 385–395; Shipley, L. A et al., 2012. *J. Chem. Ecol.*, 38: 1178–1189.)

10.2.12 Fenchone

(+)-Fenchone is found in notable concentrations in the essential oil of *Foeniculum vulgare*, whereas (–)-fenchone is a component of *Thuja occidentalis* essential oil. (+)-Fenchone is used as a food flavor and as a carminative (Bornscheuer et al., 2014). A study (Miyazawa and Gyoubu, 2006) investigated the biotransformation of (+)-fenchone in human liver microsomes into 6-*exo*-hydroxyfenchone, 6-*endo*-hydroxyfenchone, and 10-hydroxyfenchone (Figure 10.13). Metabolism of (–)-fenchone resulted in similar hydroxylated products (Miyazawa and Gyoubu, 2007). There are currently no data about metabolism of this compound in humans. However, in *Salmonella typhimurium* expressing human CYP2D6 and NADPH-P450 reductase, (+)-fenchone was metabolized into 6-*exo*- and 6-*endo*-hydroxyfenchone (Nakahashi et al., 2013).

10.2.13 Geraniol

Geraniol is a major component in the essential oils of *Geranium graveolens*, *Cymbopogon martinii*, and other *Cymbopogon* species. It has a rose-like odor and is commonly used in perfumes and cosmetics and as a flavor (O'Neil, 2006; Bornscheuer et al., 2014). Several metabolites could be identified in rat urine after oral administration (Chadha and Madyastha, 1984). Geraniol can be either metabolized via 8-hydroxygeraniol and 8-carboxygeraniol to Hildebrandt acid or directly oxidized to geranic acid and 3-hydroxycitronellic acid (Figure 10.14). The observed selective oxidation of the C-8 in geraniol also occurs in higher plants as the first step in the biosynthesis of indole alkaloids.

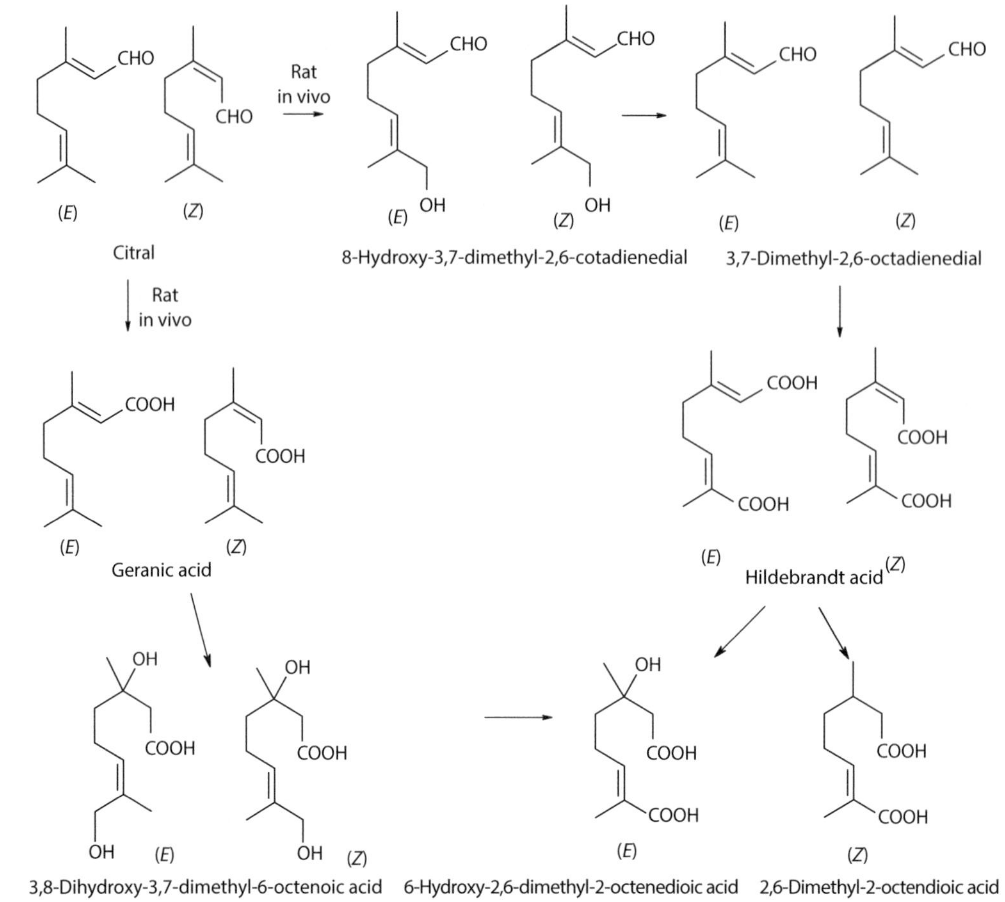

FIGURE 10.10 Proposed metabolism of citral in rats. (Adapted from Diliberto et al. 1990. *Drug Metab. Dispos.* 18:886–875.)

When incubated with human CYPs found in skin, geraniol was metabolized not only into the aldehydes neral and geranial, but also into epoxides like 2,3-epoxygeraniol, 6,7-epoxygeraniol, and 6,7-epoxygeranial (Hagvall et al., 2008).

10.2.14 Limonene

Limonene is one of the most common terpenes found in aromatic plants. The (+)-isomeric form is more abundantly present in plants than the racemic mixture and the (–)-isomeric form. (+)-Limonene has an orange odor and is a major constituent of citrus peel oils such as *Citrus aurantium* sp.

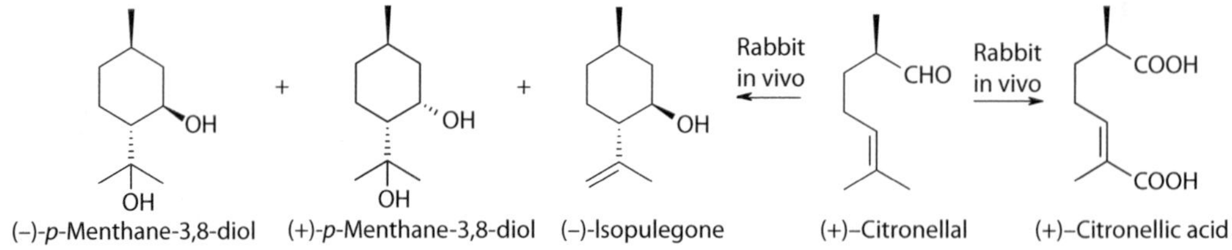

FIGURE 10.11 Proposed metabolism of citronellal in rabbits. (Adapted from Ishida et al. 1989. *Xenobiotica* 19:843–855.)

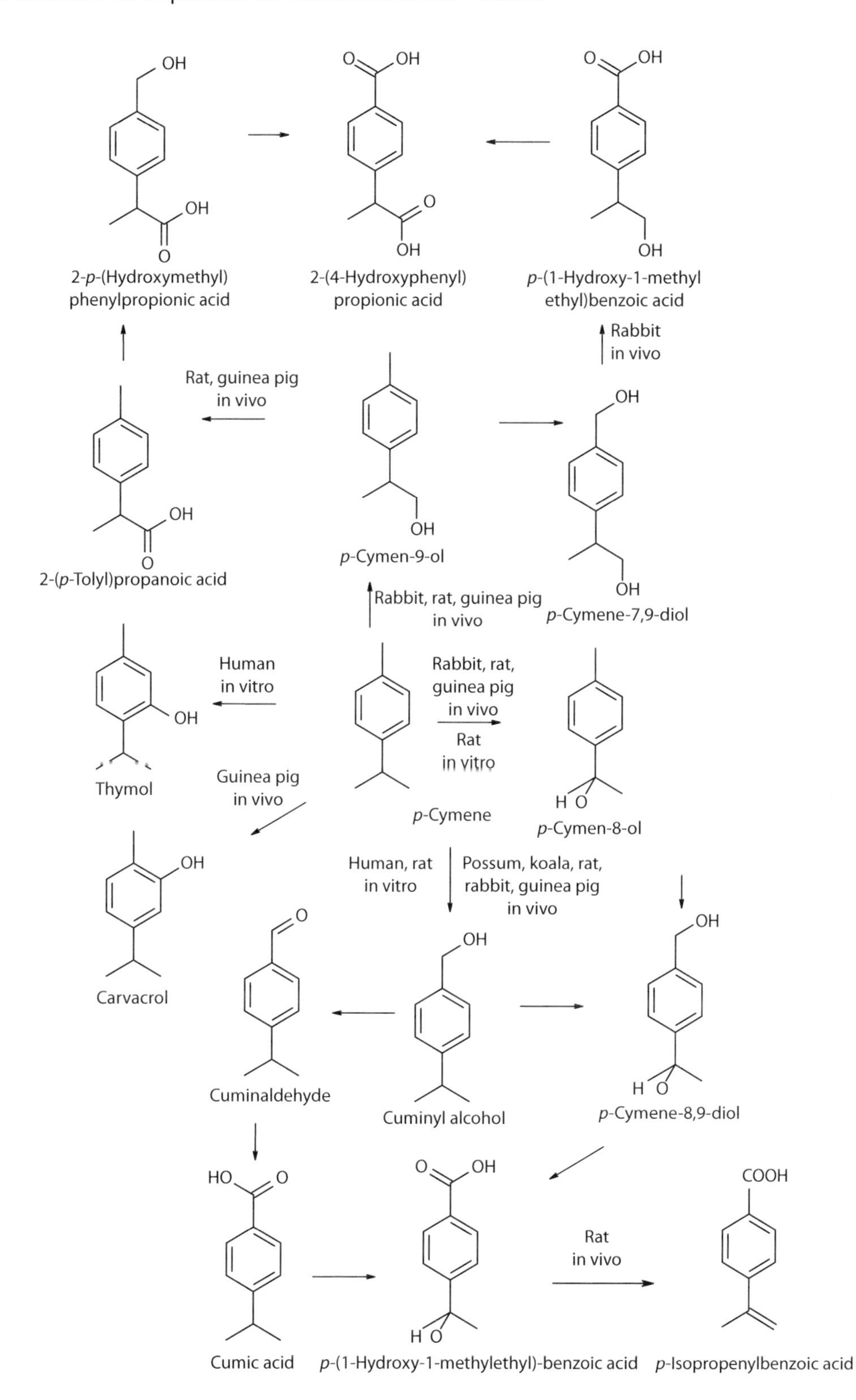

FIGURE 10.12 Proposed metabolism of *p*-cymene in brushtail possum, koala, rabbit, rat, and human volunteers. (Adapted from Matsumoto et al. 1992. *Chem. Pharm. Bull.* 40:1721–1726; Meesters et al. 2009. *Xenobiotica* 39:663–671; Pass et al. 2002. *Xenobiotica* 32:383–397.)

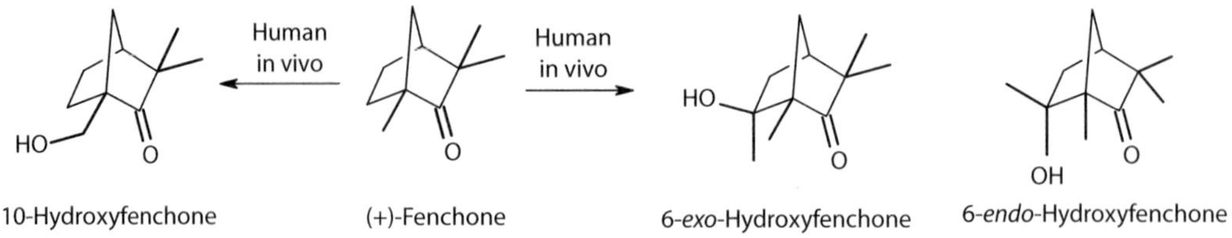

FIGURE 10.13 Proposed metabolism of fenchone in human liver microsomes. (Adapted from Miyazawa and Gyoubu 2006. *Biol. Pharm. Bull* 29:2354–2358; Nakahashi et al. 2013. *J. Oleo Sci* 62:293–296.)

aurantium and *Citrus limon*, whereas (+)-limonene is found in the essential oil of *Abies procera* and dipentene in turpentine oil. (+)-Limonene is extensively used as fragrance in perfumery and household products (O'Neil, 2006; Bornscheuer et al., 2014). Several research groups have successfully described the biotransformation of (+)-limonene *in vitro* (rat and human liver microsomes) and *in vivo* (rat, mice, guinea pigs, dogs, rabbits, and human volunteers and patients). As shown in Figure 10.15, (+)-limonene is extensively biotransformed to several metabolites, whereat in humans, the main biotransformation products are perillyl alcohol; perillic acid and its isomer; *cis-* and *trans*-dihydroperillic acid; *cis-* and *trans*-carveol; limonene-1,2-diol; limonene-10-ol; uroterpenol; several glucuronides of perillic acid; and limonene-10-ol (Crowell et al., 1992; Miyazawa et al., 2002; Shimada et al., 2002; Schmidt and Göen, 2017b).

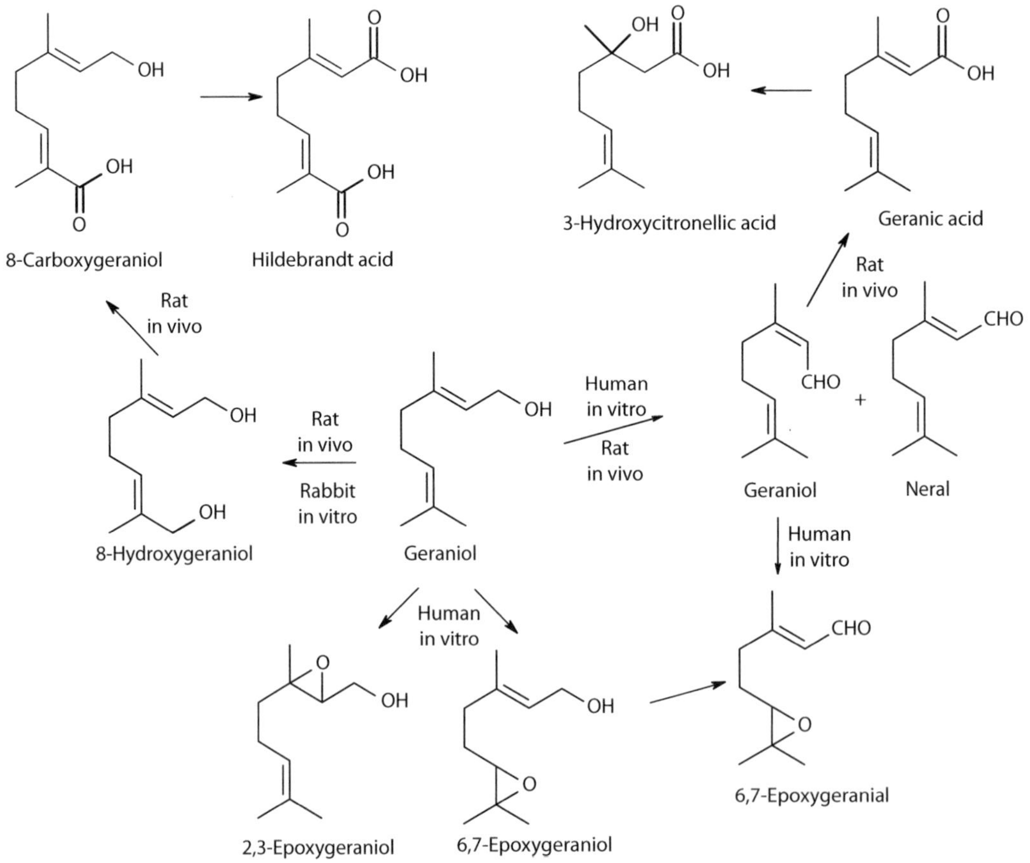

FIGURE 10.14 Proposed metabolism of geraniol in rats. (Adapted from Chadha and Madyastha 1984. *Xenobiotica* 14:365–374; Hagvall et al. 2008. *Toxicol. Appl. Pharmacol.* 233:308–313.)

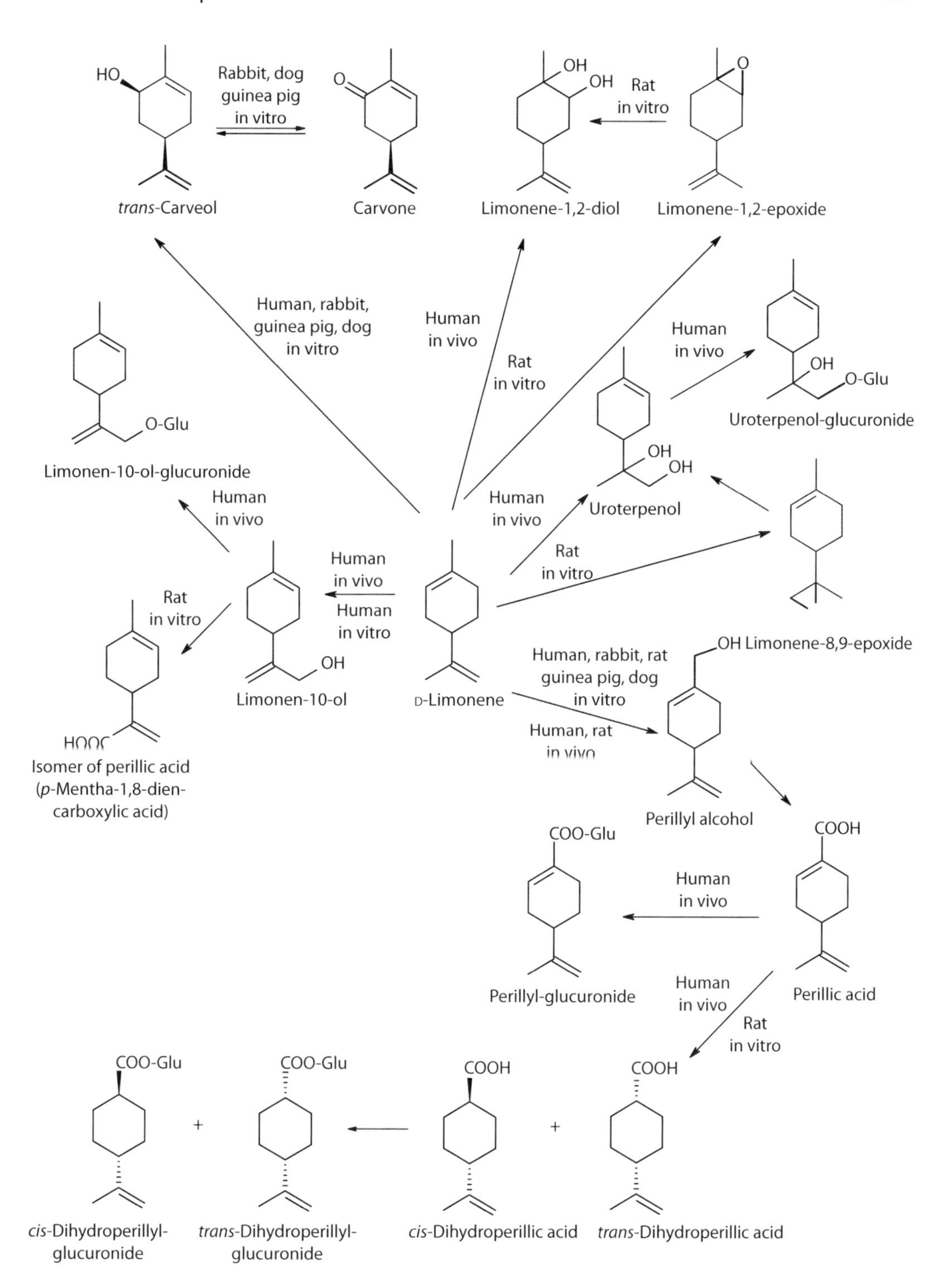

FIGURE 10.15 Proposed metabolism of (+)-limonene in rats, rabbits, guinea pigs, dogs, and humans. (Adapted from Crowell et al. 1992. *Cancer Chemother. Pharmacol* 31:205–212; Miyazawa et al. 2002. *Drug Metab. Dispos.* 30:602–607; Shimada et al. 2002. *Drug Metab. Pharmacokin.* 17:507–515; Schmidt and Göen 2017. *Arch. Toxicol.* 91:1175–1185.)

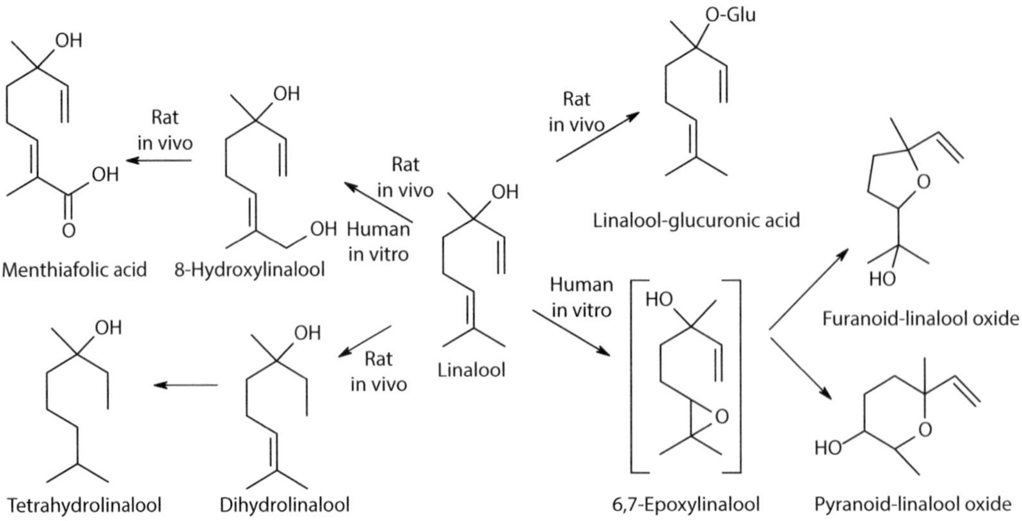

FIGURE 10.16 Proposed metabolism of linalool in rats. (Adapted from Chadha and Madyastha 1984. *Xenobiotica* 14:365–374; Meesters et al. 2007. *Xenobiotica* 37:604–617.)

10.2.15 Linalool

(–)-Linalool is the major compound of the essential oils of *Aniba rosaeodora*, *Cinnamomum camphora* leaves, *Bursera delpechiana*, and *Lavandula angustifolia*. It has a fresh, light floral odor and is used in large quantities in perfumery as well as soap and detergent products. (+)-Linalool is found in *Coriandrum sativum* and neroli absolue (*Citrus aurantium* ssp. *aurantium*) (O'Neil 2006; Bornscheuer et al., 2014). In rat, linalool is metabolized by CYP isoenzymes to dihydro- and tetrahydrolinalool and to 8-hydroxylinalool, which is further oxidized to 8-carboxylinalool (Figure 10.16). CYP-derived metabolites are then converted to glucuronide conjugates (Chadha and Madyastha 1984). Meesters et al. additionally demonstrated that 6,7-epoxyidation by recombinant human CYP2D6 resulted in cyclic ethers, and pyranoid- and furanoid-linalool oxides. The proposed intermediate product, 6,7-epoxylinalool, may cause allergic reactions. The enzymatic oxidation of linalool to 8-hydroxylinalool was catalyzed by CYP2C19 and CYP2D6 (Meesters et al. 2007).

10.2.16 Linalyl Acetate

Due to its pleasant odor, (–)-linalyl acetate is used as an ingredient in perfumes and cosmetic products. It is a major compound in the essential oils of petitgrain (*Citrus aurantium* spp. *aurantium*), *Citrus bergamia*, and *Lavandula angustifolia* (O'Neil, 2006; Bornscheuer et al., 2014). As an ester, linalyl acetate is hydrolyzed *in vivo* by carboxylesterases or esterases to linalool (Figure 10.17), which

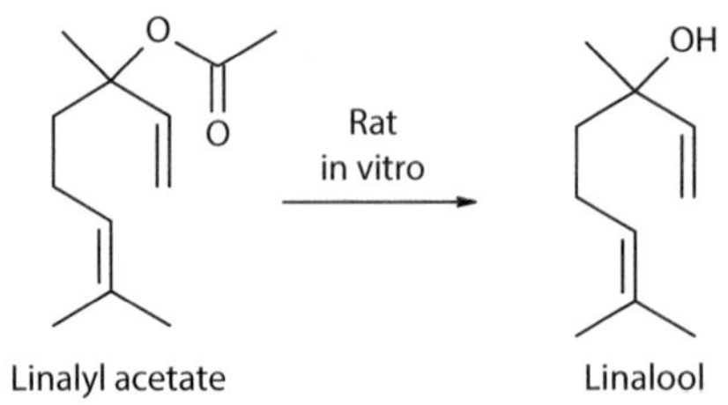

FIGURE 10.17 Proposed metabolism of linalyl acetate in rats. (Adapted from Bickers, D. et al., 2003. *Food Chem. Toxicol.* 41:919–942.)

is then further metabolized to numerous oxidized products (see metabolism of linalool) (Bickers et al., 2003).

10.2.17 Menthofuran

(+)-Menthofuran is a constituent of peppermint oils (2%–10%) (Bornscheuer et al., 2014). It has been demonstrated that menthofuran contributes to the nephrotoxicity of pulegone (Khojasteh et al., 2010). Treatment of rat liver slices with toxic concentrations of menthofuran produced several monohydroxylated metabolites, mintlactone and 7-hydroxymintlactone. Glutathione conjugates of 7-hydroxymintlactone and menthofuran were also identified (Figure 10.18). The metabolites could also be detected *in vivo* in rat urine. These reactions are catalyzed by human CYP1A2, 2B6, 2C19, 2E1, and 3A4 (Khojasteh-Bakht et al., 1999; Lassila et al., 2016). *In vitro* experiments with human liver S9 fraction also demonstrate the formation of several menthofuran-glutathione conjugates (Lassila et al., 2016).

10.2.18 Menthol

(–)-Menthol is a major component of the essential oils of *Mentha piperita* and *M. var. piperascens*; besides, it is also found in other mint oils. It has a pleasant typical minty odor and taste, and is widely used to flavor liqueurs and confectionaries and in perfumery, the tobacco industry, toiletries, oral hygiene products, lotions, and hair tonics. Due to its locally anesthetic, antipruritic, stomachic, and carminative properties, it is used in many indications—for example, topically against coughs, rhinitis, mild burns, insect bites and muscle aches and internally against dyspeptic symptoms (O'Neil, 2006; Bornscheuer et al., 2014). After oral administration to human volunteers, menthol is rapidly metabolized and only menthol glucuronide can be measured in plasma or urine (Kaffenberger and Doyle, 1990). Interestingly, unconjugated menthol is only detected after a transdermal application. In rats, however, hydroxylation at C-7 and at C-8 and C-9 of the isopropyl moiety form a series a mono- and dihydroxymenthols and carboxylic acids, some of which are excreted in part as glucuronic acid conjugates. Additional metabolites are mono- and or dihydroxylated menthol derivatives (Figure 10.19). Similar to humans, the main metabolite in rats is, again, menthol glucuronide (Madyastha and Srivatsan, 1988b; Gelal et al., 1999; Spichiger et al., 2004). *In vitro* experiments using human liver microsomes have revealed that the metabolite *p*-menthane-3,8-diol is mainly produced by CYP2A6 (Miyazawa et al., 2011).

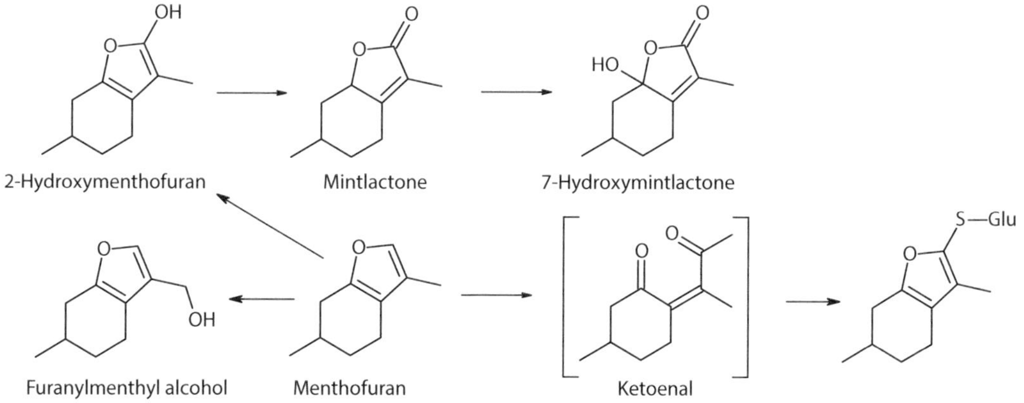

FIGURE 10.18 Proposed metabolism menthofuran in rats. (Adapted from Khojasteh et al. 2010. *Chem. Res. Toxicol.* 23:1824–1832.)

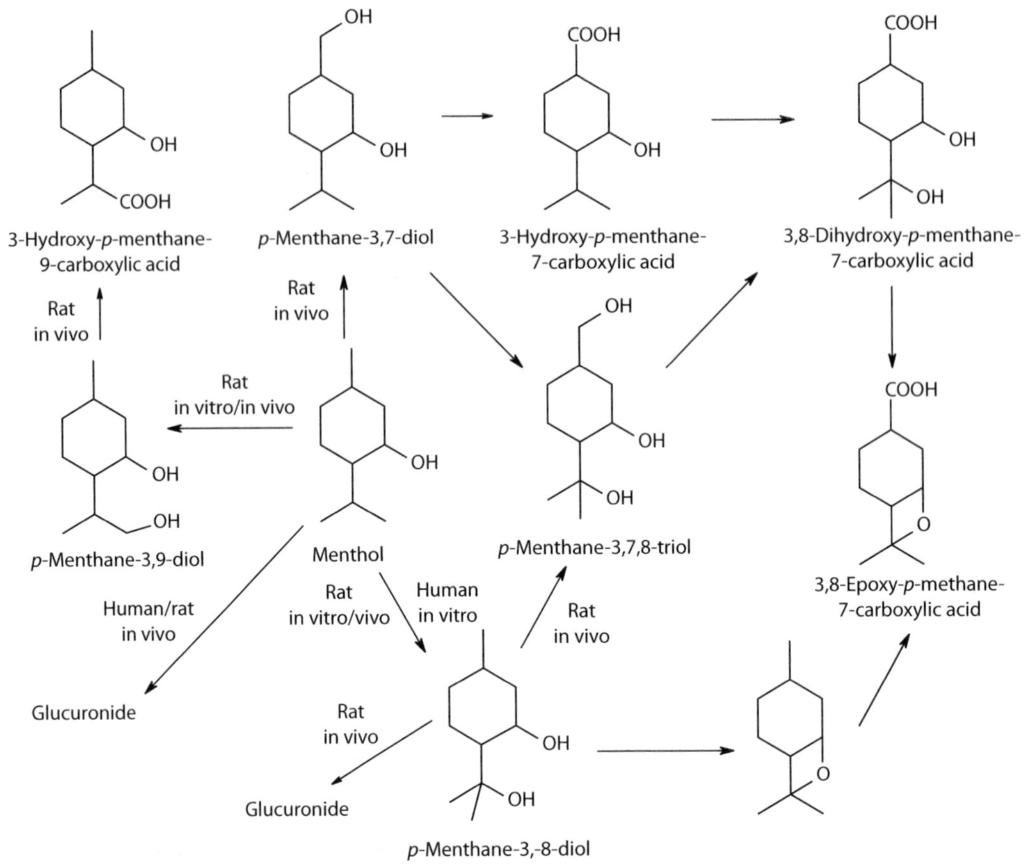

FIGURE 10.19 Proposed metabolism of menthol in rats and humans. (Adapted from Madyastha and Srivatsan 1988. *Environ. Contam. Toxicol.* 41:17–25; Gelal et al. 1999. *Clin. Pharmacol. Ther.* 66:128–235; Spichinger et al. 2004. *J. Chromatogr. B* 799:111–117.)

10.2.19 Menthone

(–)-Menthone is the main compound of *Mentha arvensis* essential oil and a minor compound of pennyroyal and peppermint oils and is used in perfumery (Bornscheuer et al., 2014). In human liver microsomes, (–)-menthone is metabolized into (+)-neomenthol and 7-hydroxymenthone (Figure 10.20) (Miyazawa and Nakanishi, 2006).

10.2.20 Myrcene

Myrcene is the major constituent of the essential oil of *Humulus lupulus and Levisticum officinale* (Bornscheuer et al., 2014). After oral application, several metabolites in the urine of rabbits have been identified, whereby the formation of the two glycols may be due to the hydration of the corresponding epoxides formed as intermediates (Ishida et al., 1981). The formation of uroterpenol may proceed via limonene, derived from myrcene in the acidic conditions of rabbit stomachs (Figure 10.21).

10.2.21 α- and β-Pinene

The constitutional isomers α- and β-pinene are both major constituents of pine resins and found in many essential oils; for example, *Pinus species*, *Piper nigrum*, and *Juniperis communis*. Interestingly,

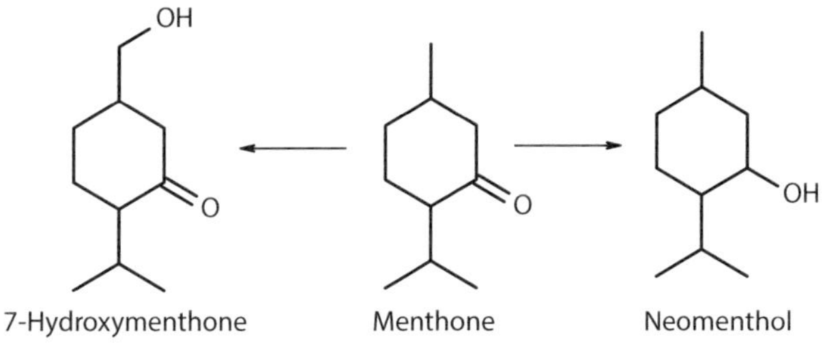

FIGURE 10.20 Proposed metabolism of menthone in rats liver microsomes. (Adapted from Miyazawa and Nakanishi 2006. *Biosci. Biotechnol. Biochem.* 70:1259–1261.)

α-pinene is more common in European pines, whereas β-pinene is more common in North America (O'Neil, 2006; Bornscheuer et al., 2014). As shown in Figure 10.22, metabolism of α- and β-pinene in humans leads to the formation of *trans*- and *cis*-verbenol and myrtenol, respectively (Schmidt et al., 2013; Schmidt and Göen, 2017a). The main urinary metabolite of α-pinene in rabbits is *trans*-verbenol; the minor biotransformation products are myrtenol and myrtenic acid. The main urinary metabolite of β-pinene in rabbits, however, is *cis*-verbenol, indicating stereoselective hydroxylation (Ishida et al., 1981; Eriksson and Levin, 1996).

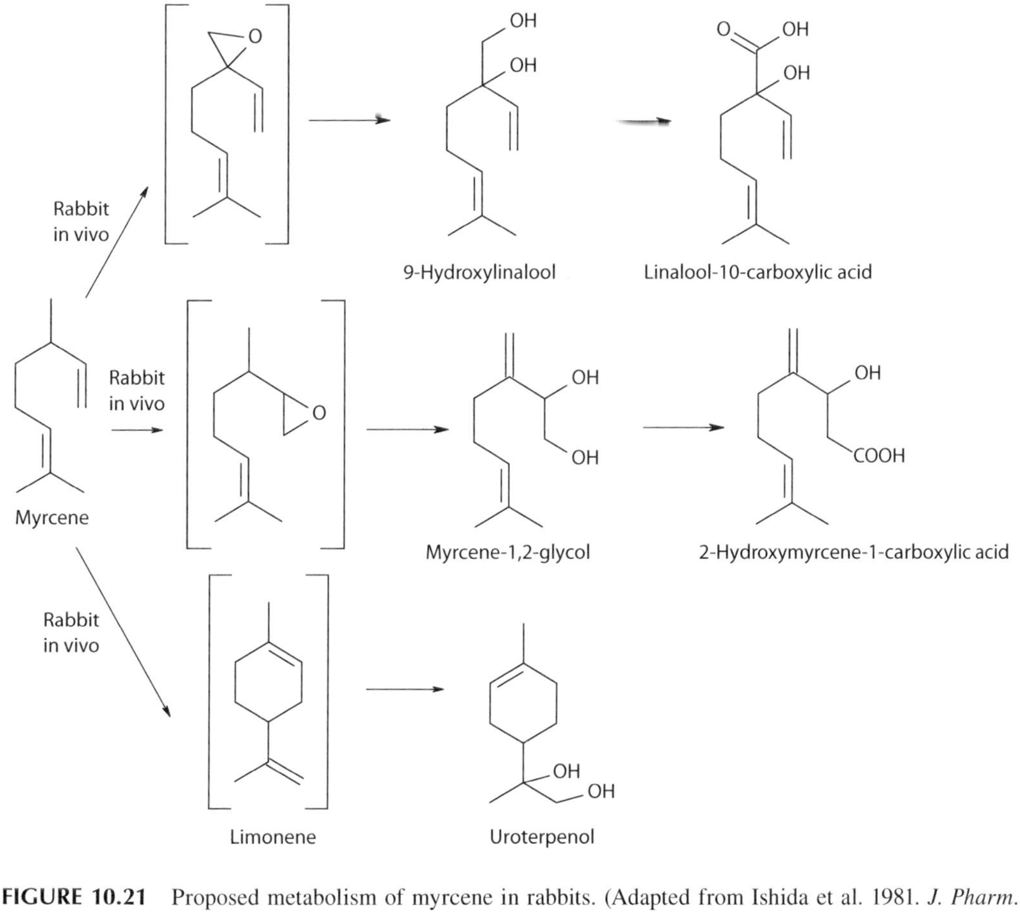

FIGURE 10.21 Proposed metabolism of myrcene in rabbits. (Adapted from Ishida et al. 1981. *J. Pharm. Sci.* 70:406–415.)

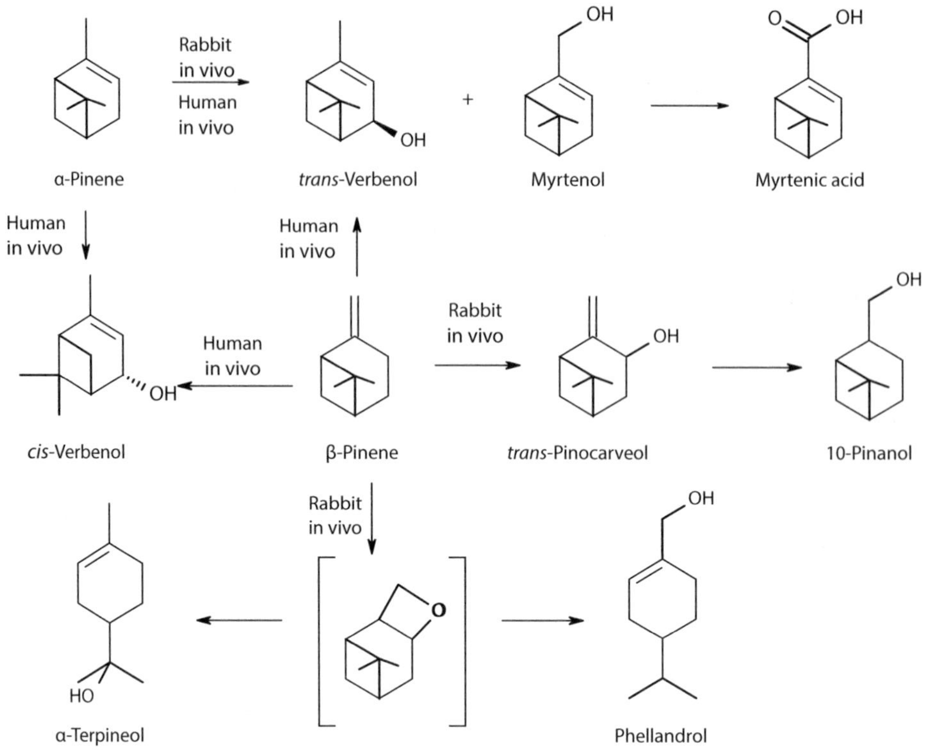

FIGURE 10.22 Proposed metabolism of α- and β-pinene in rabbits and humans. (Adapted from Ishida et al.,1981. *J. Pharm. Sci.* 70:406–415; Eriksson and Levin 1996. *J. Chromatogr. B* 677:85–98; Schmidt and Göen 2017. *Arch. Toxicol.* 91:677–687.)

10.2.22 Pulegone

(R)-(+)-Pulegone is present in essential oils of Lamiaceae. *Hedeoma pulegioides* and *Mentha pulegium*, both commonly called pennyroyal, contain essential oils that are chiefly pulegone (O'Neil, 2006; Bornscheuer et al., 2014). Pennyroyal herb had been used for inducing menstruation and abortion. In higher doses, however, it may result in central nervous system toxicity, gastritis, hepatic and renal failure, pulmonary toxicity, and death. Commercially available, pennyroyal oils have a pulegone content >80% and are both hepatotoxic and pneumotoxic in mice (Engel, 2003). Though pulegone has been used for flavoring food and oral hygiene products, the pulegone content in foods and beverages is restricted by EU law. At nontoxic concentrations, pulegone is oxidized selectively at the 10-position and further metabolized into menthofuran. *In vitro* data show that this metabolic pathway is mainly catalyzed by human CYP2E1, and to a lesser extent by CYP1A2 and CYP 2C19 (Khojasteh-Bakht et al., 1999). Alternatively, it may be reduced to menthone, which has been detected in trace levels in urine samples. It might be possible that pulegone is also reduced at the carbonyl group first. Consequently, pulegol is either reduced very efficiently to menthol or rearranged to 3-p-menthen-8-ol (Engel, 2003) (Figure 10.23). In rats, three major pathways have been identified: (a) hydroxylation followed by glucuronidation, (b) reduction to menthone and hydroxylation, and (c) conjugation with glutathione and further metabolism (Chen et al., 2001; Ferguson et al., 2007).

10.2.23 α-Terpineol

α-Terpineol, a monocyclic monoterpene alcohol, is found in the essential oil sources such as neroli, petitgrain, *Melaleuca alternifolia*, and *Origanum vulgare* (Bornscheuer et al., 2014). Based on its

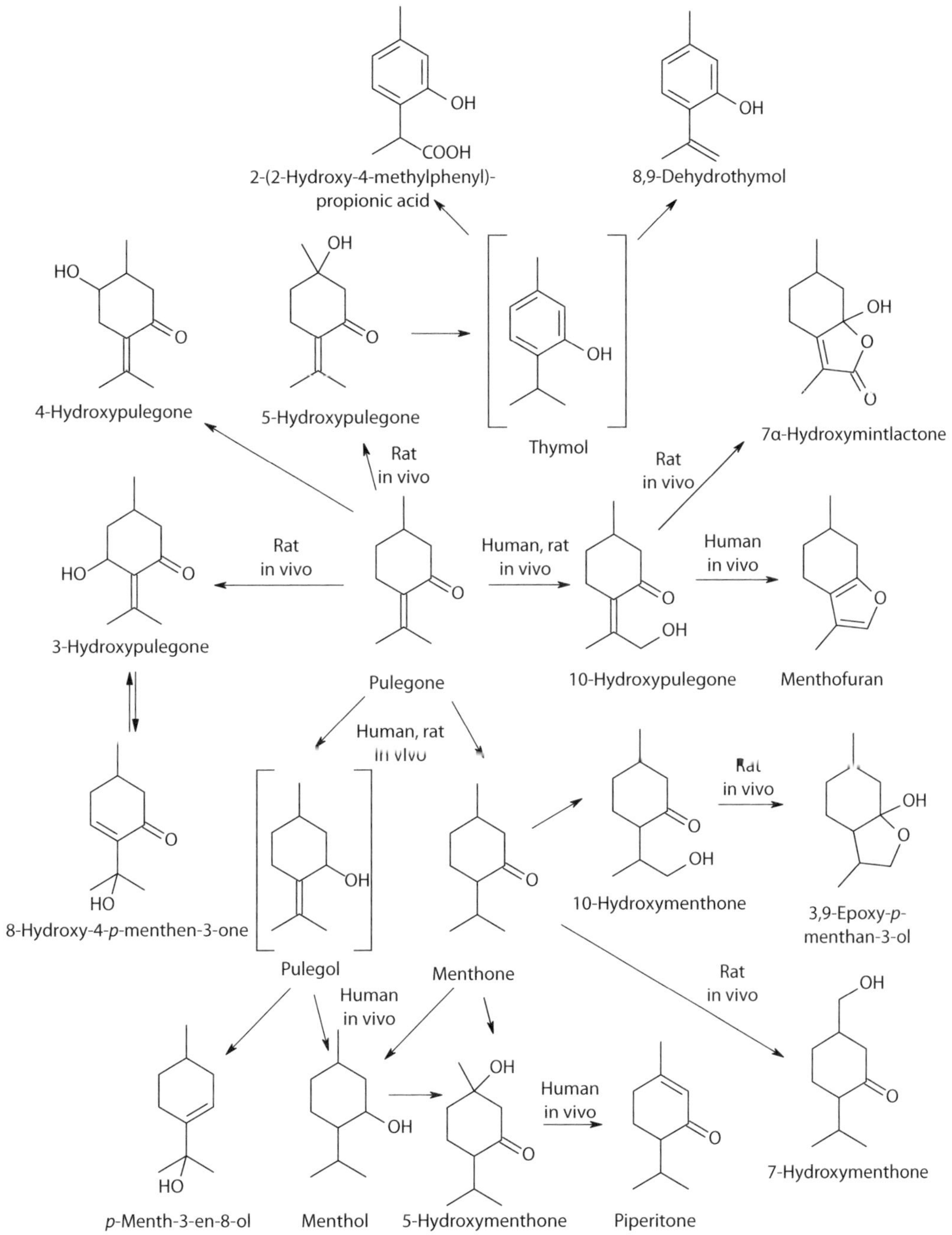

FIGURE 10.23 Proposed metabolism of pulegone in humans and rats. (Adapted from Engel 2003. *J. Agric. Food Chem.* 51:6589–6597; Chen et al. 2001. *Drug Metab. Dispos.* 29:1567–1577.)

pleasant lilac-like odor, α-terpineol is widely used in the manufacture of perfumes, cosmetics, soaps, and antiseptic agents. After oral administration to rats (600 mg/kg body weight), α-terpineol is metabolized to p-menthane-1,2,8-triol, probably formed from the epoxide intermediate. Notably, allylic methyl oxidation and the reduction of the 1,2-double bond are the major routes for the biotransformation of α-terpineol in rat (Figure 10.24). Although allylic oxidation of C-1 methyl

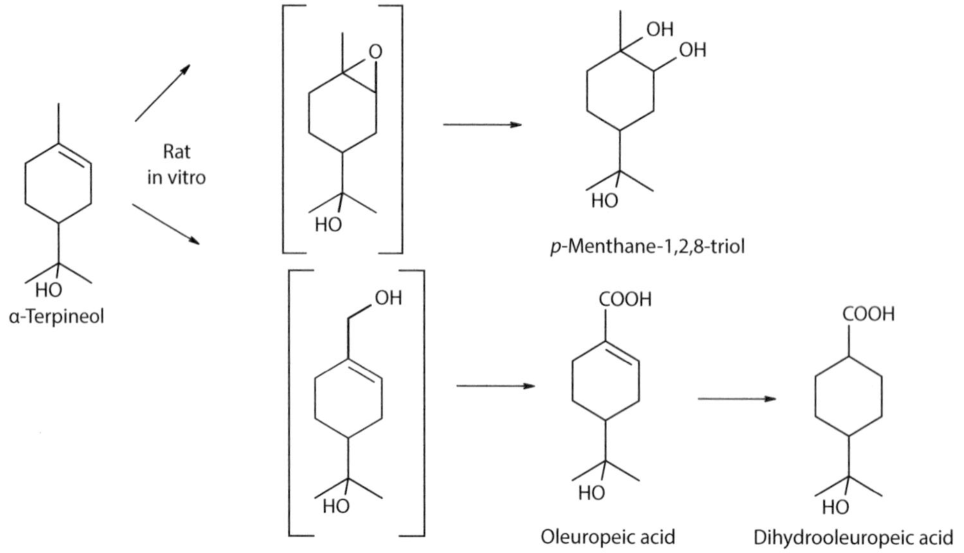

FIGURE 10.24 Proposed metabolism of α-terpineol in rats. (Adapted from Madyastha and Srivatsan 1988. *Environ. Contam. Toxicol.* 41:17–25.)

seems to be the major pathway, the alcohol *p*-ment-1-ene-7,8-diol could not be isolated from the urine samples. Probably, this compound is accumulated and is readily further oxidized to oleuropeic acid (Madyastha and Srivatsan, 1988a).

10.2.24 Terpinen-4-ol

(–)-Terpinen-4-ol is the main compound of *Melaleuca alternifolia* essential oil, which is widely used in cosmetic and pharmaceutical products due to its antiseptic properties (Bornscheuer et al., 2014). As shown in Figure 10.25, terpinen-4-ol is oxidized to *p*-menth-1-en-4,8-diol and 1,2-epoxy-*p*-menthan-4-ol. Further experiments with human recombinant CYPs identified CYP3A4, CYP 2A6, and to a lesser extent CYP1A2 as the main enzymes involved in the formation of terpinene-4-ol (Haigou and Miyazawa, 2012).

10.2.25 α- and β-Thujone

α-Thujone and β-thujone are bicyclic monoterpenes that differ in the stereochemistry of the C-4 methyl group. The isomer ratio depends on the plant source, with high content of α-thujone in *Thuja occidentalis* and β-thujone in the essential oils of *Tanacetum vulgare*, *Salvia officinalis*, and *Artemisia absinthium*. Due to their neurotoxicity, maximum permissible amounts for thujones in

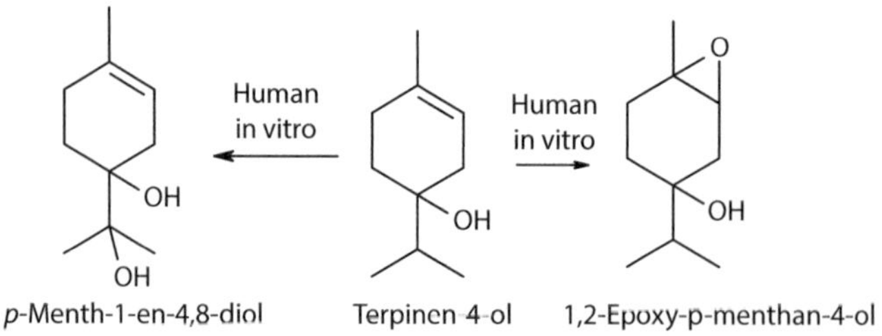

FIGURE 10.25 Proposed metabolism of terpinen-4-ol in human liver microsomes. (Adapted from Haigou and Miyazawa 2012. *J. Oleo Sci.* 61:35–43.)

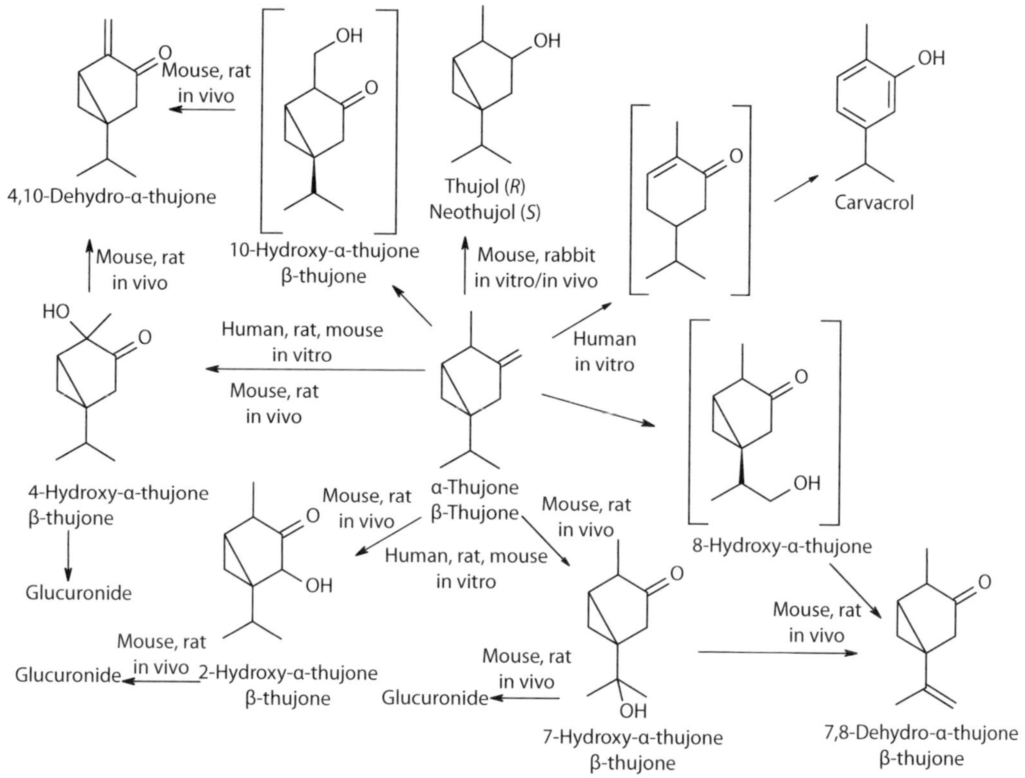

FIGURE 10.26 Proposed *in vitro* and *in vivo* metabolism of α- and β-thujone in rats. (Adapted from Ishida et al. 1989. *Xenobiotica* 19:843–855; Höld et al. 2000. *Environ. Chem. Toxicol.* 97:3826–3831; Höld et al. 2001. *Chem. Res. Toxicol.* 14:589–595.)

foods and beverages have been defined in the EU (O'Neil, 2006; Bornscheuer et al., 2014). α-Thujone is best known as the active ingredient of the alcoholic beverage absinthe, which was a very popular European drink in the 1800s. *In vivo* rabbit, rat, and mouse models conformed to the *in vitro* data using rat, mice, and human liver microsomes where α- and β-thujone are extensively metabolized to six hydroxy-thujones and three dehydrothujones (Figure 10.26). In Phase II reactions, several metabolites are further conjugated with glucuronic acid (Ishida et al., 1989; Höld et al., 2000, 2001). Recent studies have shown that the main human metabolites 4- and 7-hydroxy-a-thujone are formed by CYP2A, CYP2B6, CYP2D6, and CYP3A4 (Abass et al., 2011). CYP3A4 seems to be involved in the formation of most of the thujone metabolites in human models (Jiang and Ortiz de Montellano, 2009). Based on the *in vitro* and *in vivo* data, biotransformation should also be pronounced in humans after the intake of α- and β-thujone.

10.2.26 Thymol

Thymol is the major constituent of the essential oils of *Thymus vulgaris* and *Origanum vulgare*. It is medically used in preparations against bronchitis and oral infections. Due to its antiseptic properties, it is also an ingredient of mouthwashes and toothpastes (O'Neil, 2006; Bornscheuer et al., 2014). Although after oral application to rats, large quantities were excreted unchanged or as their glucuronide and sulfate conjugates, extensive oxidation of the methyl and isopropyl groups also occurred (Austgulen et al., 1987), resulting in the formation of derivatives of benzyl alcohol and 2-phenylpropanol and their corresponding carboxylic acids. Ring hydroxylation was only a minor reaction in rats (Figure 10.27), whereas in humans, several ring-hydroxylated products were found

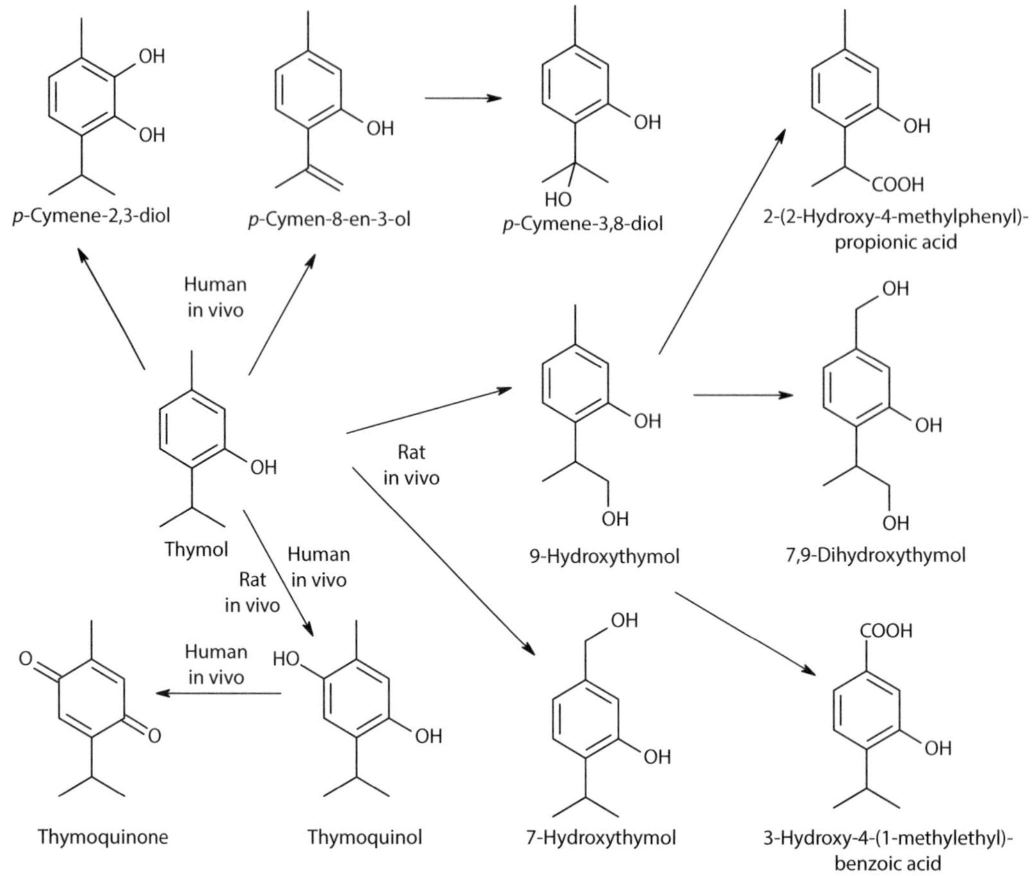

FIGURE 10.27 Proposed metabolism of thymol in humans and rats. (Adapted from Austgulen et al. 1987. *Pharmacol. Toxicol.* 61:98–102; Thalhamer et al. 2011. *J. Pharm. Biomed. Anal.* 56:64–69; Pisarčíková et al. 2017. *Biomed. Chromatogr.* 31:e3881.)

after oral application. All metabolites could be detected in human urine samples that were treated with glucuronidase prior to analysis (Thalhamer et al., 2011). CYP1A2, CYP2A6, and CYP2D6 were shown to be the main enzymes responsible for metabolizing thymol in human liver microsomes (Dong et al., 2012b). Thymol is excreted as glucuronide and sulfate into human urine which can be also assumed for its metabolites (Kohlert et al., 2002). After feeding broiler chickens various amounts of thymol, high concentrations of thymol sulfate and thymol glucuronide could be detected in the duodenal wall; decreasing concentrations of thymol sulfate were also found in plasma and liver samples (Pisarčíková et al., 2017).

10.3 METABOLISM OF SESQUITERPENES

10.3.1 Caryophyllene

(–)-β-Caryophyllene is a common sesquiterpene with a clove- or turpentine-like odor. It can be found in the essential oils of *Syzygium aromaticum*, *Piper nigrum*, and *Humulus lupulus*. It is used as a flavoring substance; for example, for chewing gums (O'Neil, 2006; Bornscheuer et al., 2014). The biotransformation of (–)-β-caryophyllene in rabbits yielded the main metabolite (10S)-(–)-14-hydroxycaryophyllen-5,6-oxide and the minor biotransformation product caryophyllene-5,6-oxide-2,12-diol. The formation of the minor metabolites is easily explained via the regioselective

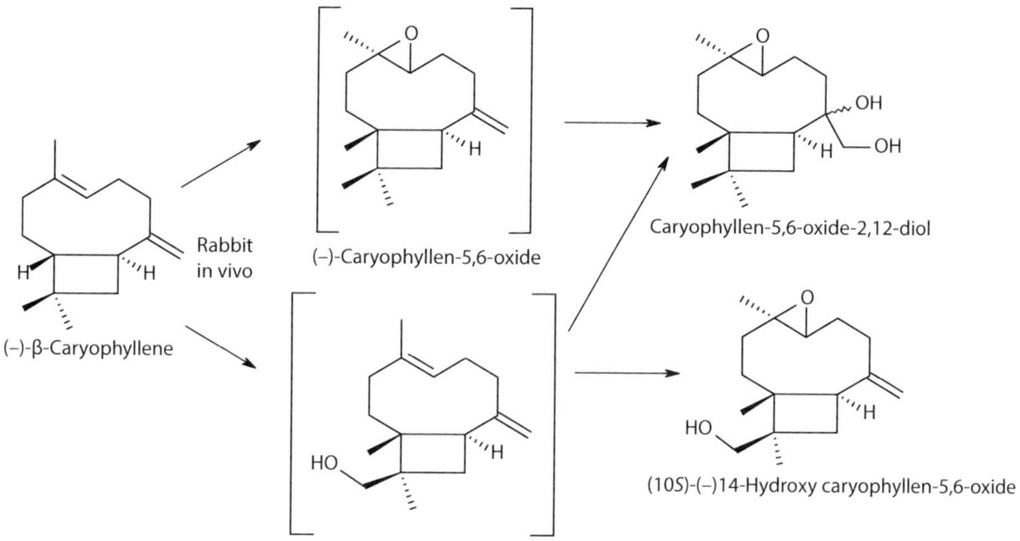

FIGURE 10.28 Proposed metabolism of β-caryophyllene in rabbits. (Adapted from Asakawa et al. 1981. *J. Pharm. Sci.* 70:710–711; Asakawa et al. 1986. *Xenobiotica* 16:753–767.)

hydroxylation of the diepoxide intermediate (Asakawa et al., 1981, 1986). Based on *in vivo* data from rabbits, extensive biotransformation in humans is highly suggested after oral administration (Figure 10.28).

10.3.2 Farnesol

Farnesol is present in many essential oils such as *Cymbopogon* species and neroli. It is used in perfumery and soaps to emphasize the odors of sweet floral perfumes and due to its fixative and antibacterial properties (O'Neil, 2006; Bornscheuer et al., 2014). Interestingly, it is also produced in humans where it acts on numerous nuclear receptors (Joo and Jetten, 2010). *In vitro* studies using recombinant drug metabolizing enzymes and human liver microsomes have shown that CYP isoenzymes participate in the metabolism of farnesol to 12-hydroxyfarnesol (Figure 10.29). Subsequently, farnesol and its metabolite are glucuronidated (DeBarber et al., 2004; Staines et al., 2004).

10.3.3 Longifolene

Longifolene is primarily found in Indian turpentine oil, which is commercially extracted from *Pinus roxburghii* (chir pine) (Bornscheuer et al., 2014). In rabbits, longifolene is metabolized as follows:

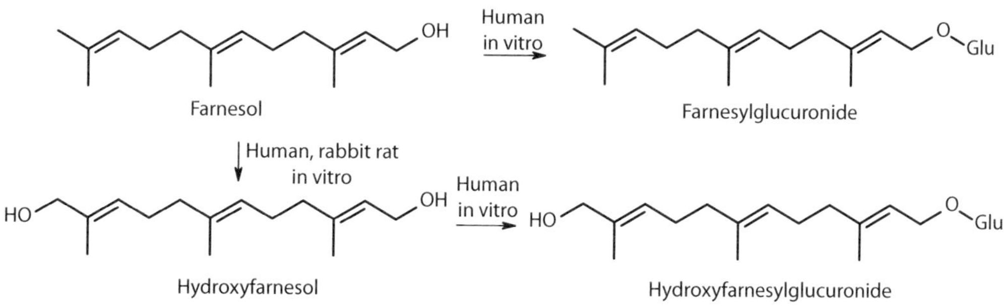

FIGURE 10.29 Proposed metabolism of farnesol in human liver microsomes. (Adapted from DeBarber et al. 2004. *Biochim. Biophys. Acta* 1682:18–27; Staines et al. 2004. *Biochem. J.* 384:637–645.)

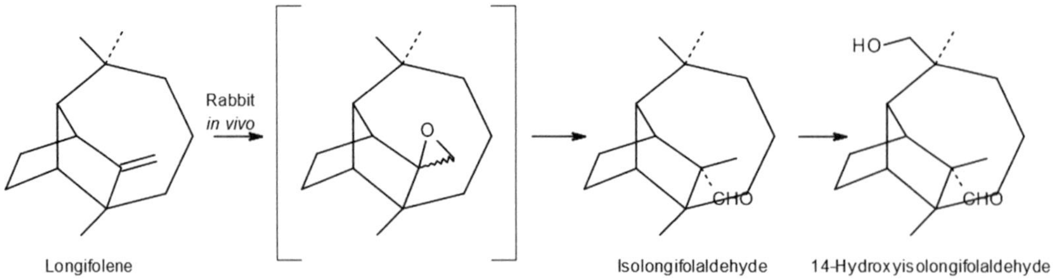

Rabbit *in vivo*

Longifolene Isolongifolaldehyde 14-Hydroxyisolongifolaldehyde

FIGURE 10.30 Proposed metabolism of (+)-longifolene in rabbits. (Adapted from Asakawa et al. 1986. *Xenobiotica* 16:753–767.)

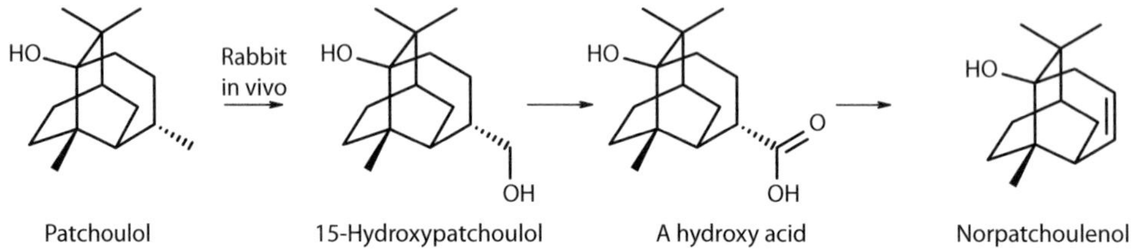

Rabbit in vivo

Patchoulol 15-Hydroxypatchoulol A hydroxy acid Norpatchoulenol

FIGURE 10.31 Proposed metabolism of patchoulol in rabbits and dogs. (Adapted from Bang et al. 1975. *Tetrahedron Lett.* 26:2211–2214; Ishida 2005. *Chem. Biodivers.* 2:569–590.)

attack on the *exo*-methylene group from the *endo*-face to form its epoxide followed by isomerization of the epoxide to a stable endo-aldehyde. Then, rapid CYP-catalyzed hydroxylation of this *endo*-aldehyde occurs (Asakawa et al., 1986) (Figure 10.30).

10.3.4 Patchoulol

(–)-Patchoulol is the major active ingredient and the most odor-intensive component of patchouli oil, the volatile oil of *Pogostemon cablin*, one of the most important raw materials of perfumery, which is also used for its insect repellent activity (Bornscheuer et al., 2014). In rabbits and dogs, patchoulol is hydroxylated at the C-15 position, yielding a diol that is subsequently oxidized to a carboxylic acid. After decarboxylation and oxidation, the 3,4-unsaturated norpatchoulen-1-ol is formed, which also has a characteristic odor (Figure 10.31). All these urinary metabolites are also found as glucuronides, explaining their excellent water solubility (Bang et al., 1975; Ishida, 2005).

REFERENCES

Abass, K., P. Reponen, S. Mattila et al. 2011. Metabolism of α-thujone in human hepatic preparations *in vitro*. *Xenobiotica*, 41: 101–111.

Asakawa, Y., T. Ishida, M. Toyota et al. 1986. Terpenoid biotransformation in mammals. IV Biotransformation of (+)-longifolene, (-)-caryophyllene, (-)-caryophyllene oxide, (-)-cyclocolorenone, (+)-nootkatone, (-)-elemol, (-)-abietic acid and (+)-dehydroabietic acid in rabbits. *Xenobiotica*, 16: 753–767.

Asakawa, Y., and Z. Taira, T. Takemoto et al. 1981. X-ray crystal structure analysis of 14-hydroxycaryophyllene oxide, a new metabolite of (-)-caryophyllene, in rabbits. *Journal of Pharmaceutical Sciences*, 70: 710–711.

Asakawa, Y., M. Toyota, and T. Ishida. 1988. Biotransformation of 1,4-cineole, a monoterpene ether. *Xenobiotica*, 18: 1129–1134.

Austgulen, L. T., E. Solheim, and R. R. Scheline. 1987. Metabolism in rats of *p*-cymene derivatives: Carvacrol and thymol. *Pharmacology & Toxicology*, 61: 98–102.

Bang, L., G. Ourisson, and P. Teisseire. 1975. Hydroxylation of patchoulol by rabbits. Hemi-synthesis of nor-patchoulenol, the odour carrier of patchouli Oil. *Tetrahedron Letters*, 16: 2211–2214.

Bickers, D., P. Calow, H. Greim et al. 2003. A toxicologic and dermatologic assessment of linalool and related esters when used as fragrance ingredients. *Food and Chemical Toxicology: An International Journal Published for the British Industrial Biological Research Association*, 41: 919–942.

Bornscheuer, U., W. Streit, B. Dill et al., (eds.). 2014. *Römpp online 4.0*. Stuttgart, Germany: Georg Thieme Verlag KG.

Boyle, R., S. McLean, W. Foley et al. 2001. Metabolites of dietary 1,8-cineole in the male koala (*Phascolarctos cinereus*). *Comparative Biochemistry and Physiology Part C: Toxicology & Pharmacology*, 129: 385–395.

Chadha, A., and K. M. Madyastha. 1984. Metabolism of geraniol and linalool in the rat and effects on liver and lung microsomal enzymes. *Xenobiotica*, 14: 365–374.

Chen, L.-J., E. H. Lebetkin, and L. T. Burka. 2001. Metabolism of (R)-(+)-pulegone in F344 rats. *Drug Metabolism and Disposition*, 29: 1567–1577.

Crowell, P. L., S. Lin, E. Vedejs et al. 1992. Identification of metabolites of the antitumor agent d-limonene capable of inhibiting protein isoprenylation and cell growth. *Cancer Chemotherapy and Pharmacology*, 31: 205–212.

DeBarber, A. E., L. A. Bleyle, J.-B. O. Roullet et al. 2004. Omega-hydroxylation of farnesol by mammalian cytochromes p450. *Biochimica et Biophysica Acta*, 1682: 18–27.

Diliberto, J. J., P. Srinivas, D. Overstreet et al. 1990. Metabolism of citral, an alpha,beta-unsaturated aldehyde, in male F344 rats. *Drug Metabolism and Disposition: The Biological Fate of Chemicals*, 18: 866–875.

Dong, R.-H., Z.-Z. Fang, L.-L. Zhu et al. 2012a. Identification of UDP-glucuronosyltransferase isoforms involved in hepatic and intestinal glucuronidation of phytochemical carvacrol. *Xenobiotica*, 42: 1009–1016.

Dong, R.-H., Z.-Z. Fang, L.-L. Zhu et al. 2012b. Identification of CYP isoforms involved in the metabolism of thymol and carvacrol in human liver microsomes (HLMs). *Die Pharmazie*, 67: 1002–1006.

Duisken, M., D. Benz, T. H. Peiffer et al. 2005. Metabolism of delta(3)-carene by human cytochrome p450 enzymes: Identification and characterization of two new metabolites. *Current Drug Metabolism*, 6: 593–601.

Engel, W. 2001. *In vivo* studies on the metabolism of the monoterpenes S-(+)- and R-(-)-carvone in humans using the metabolism of ingestion-correlated amounts (MICA) approach. *Journal of Agricultural and Food Chemistry*, 49: 4069–4075.

Engel, W. 2003. *In vivo* studies on the metabolism of the monoterpene pulegone in humans using the metabolism of ingestion-correlated amounts (MICA) approach: explanation for the toxicity differences between (S)-(−)- and (R)-(+)-pulegone. *Journal of Agricultural and Food Chemistry*, 51: 6589–6597.

Eriksson, K., and J. O. Levin. 1996. Gas chromatographic-mass spectrometric identification of metabolites from alpha pinene in human urine after occupational exposure to sawing fumes. *Journal of Chromatography. B, Biomedical Applications*, 677: 85–98.

Ferguson, L.-J. C., E. H. Lebetkin, F. B. Lih et al. 2007. 14C-labeled pulegone and metabolites binding to α2u-globulin in kidneys of male F-344 rats. *Journal of Toxicology and Environmental Health, Part A*, 70: 1416–1423.

Gelal, A., P. Jacob 3rd, L. Yu et al. 1999. Disposition kinetics and effects of menthol. *Clinical Pharmacology and Therapeutics*, 66: 128–135.

Gyoubu, K., and M. Miyazawa. 2007. In vitro metabolism of (-)-camphor using human liver microsomes and CYP2A6. *Biological and Pharmaceutical Bulletin*, 30: 230–233.

Hagvall, L., J. M. Baron, A. Börje et al. 2008. Cytochrome P450-mediated activation of the fragrance compound geraniol forms potent contact allergens. *Toxicology and Applied Pharmacology*, 233: 308–313.

Haigou, R., and M. Miyazawa. 2012. Metabolism of (+)-terpinen-4-ol by cytochrome P450 enzymes in human liver microsomes. *Journal of Oleo Science*, 61: 35–43.

Höld, K. M., N. S. Sirisoma, and J. E. Casida. 2001. Detoxification of alpha- and beta-Thujones (the active ingredients of absinthe): Site specificity and species differences in cytochrome P450 oxidation *in vitro* and *in vivo*. *Chemical Research in Toxicology*, 14: 589–595.

Höld, K. M., N. S. Sirisoma, T. Ikeda et al. 2000. Alpha-thujone (the active component of absinthe): Gamma-aminobutyric acid type A receptor modulation and metabolic detoxification. *Proceedings of the National Academy of Sciences of the United States of America*, 97: 3826–3831.

Horst, K., and M. Rychlik. 2010. Quantification of 1,8-cineole and of its metabolites in humans using stable isotope dilution assays. *Molecular Nutrition & Food Research*, 54: 1515–1529.

Ishida, T. 2005. Biotransformation of terpenoids by mammals, microorganisms, and plant-cultured cells. *Chemistry & Biodiversity*, 2: 569–590.

Ishida, T., Y. Asakawa, T. Takemoto et al. 1979. Terpenoid biotransformation in mammals. II: Biotransformation of dl-camphene in rabbits. *Journal of Pharmaceutical Sciences*, 68: 928–930.

Ishida, T., Y. Asakawa, T. Takemoto et al. 1981. Terpenoids biotransformation in mammals III: Biotransformation of alpha-pinene, beta-pinene, pinane, 3-carene, carane, myrcene, and p-cymene in rabbits. *Journal of Pharmaceutical Sciences*, 70: 406–415.

Ishida, T., M. Toyota, and Y. Asakawa. 1989. Terpenoid biotransformation in mammals. V. Metabolism of (+)-citronellal, (+-)-7–hydroxycitronellal, citral, (-)-perillaldehyde, (-)-myrtenal, cuminaldehyde, thujone, and (+-)-carvone in rabbits. *Xenobiotica*, 19: 843–855.

Jäger, W., M. Mayer, P. Platzer et al. 2000. Stereoselective metabolism of the monoterpene carvone by rat and human liver microsomes. *The Journal of Pharmacy and Pharmacology*, 52: 191–197.

Jäger, W., M. Mayer, G. Reznicek et al. 2001. Percutaneous absorption of the montoterperne carvone: Implication of stereoselective metabolism on blood levels. *The Journal of Pharmacy and Pharmacology*, 53: 637–642.

Jiang, Y., and P. R. Ortiz de Montellano. 2009. Cooperative effects on radical recombination in CYP3A4-catalyzed oxidation of the radical clock β-thujone. *ChemBioChem*, 10: 650–653.

Joo, J. H., and A. M. Jetten. 2010. Molecular mechanisms involved in farnesol-induced apoptosis. *Cancer Letters*, 287: 123–135.

Kaffenberger, R. M., and M. J. Doyle. 1990. Determination of menthol and menthol glucuronide in human urine by gas chromatography using an enzyme-sensitive internal standard and flame ionization detection. *Journal of Chromatography*, 527: 59–66.

Khojasteh, S. C., S. Oishi, and S. D. Nelson. 2010. Metabolism and toxicity of menthofuran in rat liver slices and in rats. *Chemical Research in Toxicology*, 23: 1824–1832.

Khojasteh-Bakht, S. C., W. Chen, L. L. Koenigs et al. 1999. Metabolism of (R)-(+)-pulegone and (R)-(+)-menthofuran by human liver cytochrome P-450s: Evidence for formation of a furan epoxide. *Drug Metabolism and Disposition*, 27: 574–580.

Kohlert, C., G. Schindler, R. W. März et al. 2002. Systemic availability and pharmacokinetics of thymol in humans. *The Journal of Clinical Pharmacology*, 42: 731–737.

Lassila, T., S. Mattila, M. Turpeinen et al. 2016. Tandem mass spectrometric analysis of S- and N-linked glutathione conjugates of pulegone and menthofuran and identification of P450 enzymes mediating their formation. *Rapid Communications in Mass Spectrometry: RCM*, 30: 917–926.

Leibman, K. C., and E. Ortiz. 1973. Mammalian metabolism of terpenoids. I. Reduction and hydroxylation of camphor and related compounds. *Drug Metabolism and Disposition: The Biological Fate of Chemicals*, 1: 543–551.

Madyastha, K. M., and V. Srivatsan. 1988a. Biotransformations of alpha-terpineol in the rat: Its effects on the liver microsomal cytochrome P-450 system. *Bulletin of Environmental Contamination and Toxicology*, 41: 17–25.

Madyastha, K. M., and V. Srivatsan. 1988b. Studies on the metabolism of l-menthol in rats. *Drug Metabolism and Disposition: The Biological Fate of Chemicals*, 16: 765–772.

Matsumoto, T., T. Ishida, T. Yoshida et al. 1992. The enantioselective metabolism of *p*-cymene in rabbits. *Chemical & Pharmaceutical Bulletin*, 40: 1721–1726.

Meesters, R. J. W., M. Duisken, and J. Hollender. 2007. Study on the cytochrome P450-mediated oxidative metabolism of the terpene alcohol linalool: Indication of biological epoxidation. *Xenobiotica*, 37: 604–617.

Meesters, R. J. W., M. Duisken, and J. Hollender. 2009. Cytochrome P450-catalysed arene-epoxidation of the bioactive tea tree oil ingredient *p*-cymene: Indication for the formation of a reactive allergenic intermediate? *Xenobiotica*, 39: 663–671.

Miyazawa, M., and K. Gyoubu. 2006. Metabolism of (+)-fenchone by CYP2A6 and CYP2B6 in human liver microsomes. *Biological & Pharmaceutical Bulletin*, 29: 2354–2358.

Miyazawa, M., and K. Gyoubu. 2007. Metabolism of (-)-fenchone by CYP2A6 and CYP2B6 in human liver microsomes. *Xenobiotica*, 37: 194–204.

Miyazawa, M., S. Marumoto, T. Takahashi et al. 2011. Metabolism of (+)- and (-)-menthols by CYP2A6 in human liver microsomes. *Journal of Oleo Science*, 60: 127–132.

Miyazawa, M., and K. Nakanishi. 2006. Biotransformation of (-)-menthone by human liver microsomes. *Bioscience, Biotechnology, and Biochemistry*, 70: 1259–1261.

Miyazawa, M., and M. Shindo. 2001. Biotransformation of 1,8-cineole by human liver microsomes. *Natural Product Letters*, 15: 49–53.

Miyazawa, M., M. Shindo, and T. Shimada. 2001a. Oxidation of 1,8-cineole, the monoterpene cyclic ether originated from *Eucalyptus polybractea*, by cytochrome P450 3A enzymes in rat and human liver microsomes. *Drug Metabolism and Disposition*, 29: 200–205.

Miyazawa, M., M. Shindo, and T. Shimada. 2001b. Roles of cytochrome P450 3A enzymes in the 2-hydroxylation of 1,4-cineole, a monoterpene cyclic ether, by rat and human liver microsomes. *Xenobiotica*, 31: 713–723.

Miyazawa, M., M. Shindo, and T. Shimada. 2002. Metabolism of (+)- and (−)-limonenes to respective carveols and perillyl alcohols by CYP2C9 and CYP2C19 in human liver microsomes. *Drug Metabolism and Disposition*, 30: 602–607.

Nakahashi, H., and M. Miyazawa. 2011. Biotransformation of (-)-camphor by *Salmonella typhimurium* OY1002/2A6 expressing human CYP2A6 and NADPH-P450 reductase. *Journal of Oleo Science*, 60: 545–548.

Nakahashi, H., N. Yagi, and M. Miyazawa. 2013. Biotransformation of (+)-fenchone by *Salmonella typhimurium* OY1002/2A6 expressing human CYP2A6 and NADPH-P450 reductase. *Journal of Oleo Science*, 62: 293–296.

O'Neil, M. J., (eds.). 2006. *The Merck Index: An Encyclopedia of Chemicals, Drugs, and Biologicals*. Whitehouse Station, NJ: Wiley.

Pass et al. 2002. Microsomal metabolism and enyzme kinetics of the terpene *p*-cymene in the common brushtail possum (*Trichosurus vulpecula*), koala (*Phascolarctos cinereus*) and rat. *Xenobiotica* 32: 383–397.

Pisarčíková, J., V. Oceľová, Š. Faix et al. 2017. Identification and quantification of thymol metabolites in plasma, liver and duodenal wall of broiler chickens using UHPLC-ESI-QTOF-MS. *Biomedical Chromatography*, 31: e3881.

Schmidt, L., V. N. Belov, and T. Göen. 2013. Sensitive monitoring of monoterpene metabolites in human urine using two-step derivatisation and positive chemical ionisation-tandem mass spectrometry. *Analytica Chimica Acta*, 793: 26–36.

Schmidt, L., V. N. Belov, and T. Göen. 2015. Human metabolism of Δ^3-carene and renal elimination of Δ^3-caren-10-carboxylic acid (chaminic acid) after oral administration. *Archives of Toxicology*, 89: 381–392.

Schmidt, L., and T. Göen. 2017a. Human metabolism of α-pinene and metabolite kinetics after oral administration. *Archives of Toxicology*, 91: 677–687.

Schmidt, L., and T. Göen. 2017b. R-Limonene metabolism in humans and metabolite kinetics after oral administration. *Archives of Toxicology*, 91: 1175–1185.

Shimada, T., M. Shindo, and M. Miyazawa. 2002. Species differences in the metabolism of (+)- and (-)-limonenes and their metabolites, carveols and carvones, by cytochrome P450 enzymes in liver microsomes of mice, rats, guinea pigs, rabbits, dogs, monkeys, and humans. *Drug Metabolism and Pharmacokinetics*, 17: 507–515.

Spatzenegger, M., and W. Jäger. 1995. Clinical importance of hepatic cytochrome P450 in drug metabolism. *Drug Metabolism Reviews*, 27: 397–417.

Spichiger, M., R. C. Mühlbauer, and R. Brenneisen. 2004. Determination of menthol in plasma and urine of rats and humans by headspace solid phase microextraction and gas chromatography–mass spectrometry. *Journal of Chromatography B*, 799: 111–117.

Staines, A. G., P. Sindelar, M. W. H. Coughtrie et al. 2004. Farnesol is glucuronidated in human liver, kidney and intestine *in vitro*, and is a novel substrate for UGT2B7 and UGT1A1. *The Biochemical Journal*, 384: 637–645.

Sun, X.-M., Q.-F. Liao, Y.-T. Zhou et al. 2014. Simultaneous determination of borneol and its metabolite in rat plasma by GC-MS and its application to pharmacokinetics study. *Journal of Pharmaceutical Analysis*, 4:345–350.

Thalhamer, B., W. Buchberger, and M. Waser. 2011. Identification of thymol phase I metabolites in human urine by headspace sorptive extraction combined with thermal desorption and gas chromatography mass spectrometry. *Journal of Pharmaceutical and Biomedical Analysis*, 56: 64–69.

Zhang, R., C. Liu, T. Huang et al. 2008. *In vitro* characterization of borneol metabolites by GC-MS upon incubation with rat liver microsomes. *Journal of Chromatographic Science*, 46: 419–423.

Zhao, J., Y. Lu, S. Du et al. 2012. Comparative pharmacokinetic studies of borneol in mouse plasma and brain by different administrations. *Journal of Zhejiang University. Science B*, 13: 990–996.

11 Central Nervous System Effects of Essential Oil Compounds

Elaine Elisabetsky and Domingos S. Nunes

CONTENTS

11.1 OVERVIEW

The purpose of this chapter is to provide a critical review of the scientific product in the area of psychopharmacology of essential oils (EOs). The chapter is based on the literature review of EOs reported to act as anxiolytics, hypnotics, anticonvulsants, antidepressants, and neuroprotectors, especially in experimental models. Other psychopharmacology and/or neurological effects (including analgesia, antipsychotic, hallucinogenic, and other effects), as well as pharmacokinetic and toxicology aspects are beyond the scope of this chapter and have been discussed in Chapters 10, 17, and 26.

Clinical observations focusing on the effects of EOs* are informative and should be encouraged (Höferl et al., 2016; Soto-Vásquez and Alvarado-Garcia, 2018; Wang and Heinbockel, 2018), but well-substantiated conclusions require accurately characterized samples to allow comparisons among laboratories and sound clinical trials. In this context, it is disappointing that hundreds of clinical observations by aromatherapists and naturopaths are not well documented to facilitate the design of observational clinical studies. Although not conclusive, observational studies are of great value to inform on specific therapies and/or the design of more rigorous clinical trials (Graz et al., 2007; Graz et al., 2010; Nunes et al., 2010; Peers et al., 2014).

Because EOs are complex mixtures of dozens of compounds (the very feature that makes them unique), the sample variability makes biological experiments with EOs difficult to reproduce. Without reproducibility, the robustness of results is weakened or at least questionable (Janero, 2016). Moreover, the understanding of reliable effects and its underlying mechanism of action are key features to translational research. Hoping that this review and constructive critics will optimize pharmacological innovation in EOs research, we chose to focus on the effects of compounds isolated from EOs.

11.2 INTRODUCTION

11.2.1 Translatability and Reproducibility

Translatability refers to the translation of experimental results in the field of biomedicine to safe and effective therapies. Despite the academic efforts and the substantial investment of capital in drug development by the pharmaceutical industry, innovations in this area are rare (Wehling, 2009). The question of translational research is complex and has many dimensions (Mullane et al., 2014), but unavoidable issues include hypothesis-driven research, the suitability of the methods employed to answer the question in place (Goodman et al., 2016), and the proper execution of validated methods (Kimmelman et al., 2014). As in all fields of biomedical inquiry, the question of translatability is applicable EO research, and an overall evaluation of research designs and methods is needed to attain meaningful improvements in the area (Ioannidis, 2018). This review shows key features to translatability (e.g., kinetics, validated target, mechanism of action, toxicity) (Wehling, 2009) are often not consider, ultimately weakening drug development in the area.

Replicability† in science has been abundantly discussed (Ioannidis, 2018). Alarming data show that, despite the rigorous methodologies common to science and scientific methods, 70% of the researchers interviewed responded that they could not replicate colleagues' experiments (Baker, 2016). It is not uncommon that even the laboratory that originated the data cannot reproduce its own results. The crisis sets in when it was estimated that as much as 75% of biomedical research may be lost (unreproducible), which severely decreases its value and leads to huge investments by the pharmaceutical industry on projects that fail (Goodman et al., 2016). Underlying the lack of experimental reproducibility, among others, are sample size (N), false positives, biases of

* Pharmacological activities of EOs and associated plant species have been reviewed elsewhere (Nunes et al., 2010; Löscher and Schmidt, 2011; Sousa, 2012; Peers et al., 2014; Baker, 2016).

† Replicability has been defined as "the ability of a researcher to duplicate the results of a prior study if the same procedures are followed but new data are collected"; reproducibility "refers to the ability of a researcher to duplicate results of a prior study using the same materials as were used by the original investigator." (Bollen et al., 2015)

various natures, misuse of statistics, absence of adequate randomization methods, lack of blinded observation, and observations made by a single observer (Kimmelman et al., 2014; Goodman et al., 2016). Considering the time, financial resources, and number of animals involved in biomedical research, it is fair to claim ethical and moral obligations to improve the quality and reliability of preclinical studies (Peers et al., 2014). As stated by Ioannidis (Ioannidis, 2005), "... the high rate of non-replication (lack of confirmation) of research discoveries is a consequence of the convenient, yet ill-founded strategy of claiming conclusive research findings solely on the basis of a single study assessed by formal statistical significance, typically for a p-value less than 0.05."

The literature review shows that many compounds from EOs have been subjected to just one study; though they have been included in the review, its conclusions must be viewed with caution until replication is available. It is also noteworthy that some studies evaluate a given property of a compound by using one experimental model, whereas others use a battery of models. Another common limitation is the postulation of a mechanism of action based on the reversal of a given effect by an antagonist in a single dose experiment (Kenakin et al., 2014). Single dose experiments, while perfectly accepted to suggest the involvement of a target in the mechanism of action, must be further corroborated with proper evidence for a conclusive definition. As in other areas of biomedical research, more often than not, conclusion on promising results or therapeutic claims are fairly unsubstantiated.

11.3 REVIEW METHODOLOGY

11.3.1 Literature Review

For this chapter, bibliographical references were collected in the online database Web of Science, since 1945 up to now (December 2018). An initial survey was done by searching in the field "issue" the following association of the keywords: "essential oils" AND "antidepressant OR antidepressive OR anxyolitic OR anxiolytic OR anticonvulsivant OR anticonvulsive OR anticonvulsant OR neuroprotector OR neuroprotection OR hypnotic OR epilepsy." The resulting 785 reference abstracts were acquired and read to establish an initial list of 61 substances isolated from EOs reported to have pharmacological studies using experimental models for anxiolytic, hypnotic, anticonvulsant, antidepressant, and/or neuroprotector effects.

A second survey followed on each of these 61 substances, this time searching in the field "issue" the association of the keywords "compound name" AND "antidepressant OR antidepressive OR anxyolitic OR anxiolytic OR anticonvulsivant OR anticonvulsive OR anticonvulsant OR neuroprotector OR neuroprotection OR hypnotic OR epilepsy." From the resulting 875 abstracts, about 300 were selected for in-depth analysis.

This analysis led us to reduce the potential initial 61 compounds to the 34 compounds discussed in this chapter. Reports of activity published as abstracts at meetings, online bulletins, or devoid of DOI and/or another main stream identification were excluded. Compounds with reported activity in the areas of interest of this chapter but reporting frail data (e.g., unusual methods and/or known methods improperly conducted, inadequate samples sizes, lacking positive and/or negative controls, unvalidated endpoint observations, unacceptable statistics) were also left out.*

A brief description of the pathology, the mechanism of action of major marketed drugs, and a brief description of the mostly used methods used for evaluation are given for each pharmacology topic in order to facilitate the reader appraisal on the aspects reported for each compound.

* 1-(4-(3,5-Dibutyl-4-hydroxybenzyl) piperazin-1-yl)-2-methoxyethan-1-one, 1-nitro-2-phenylethane, 3-butylidene-4,5-dihydrophtalide, α-atlantone, benzylalcohol, carvone, cedrol, citral, citronellal, cyano-carvone, dimethoxytoluene, eucalyptol, geranial, hydroxycarvone, N-isopropylanthranilate, N-methylanthranilate, methylisoeugenol, myrcene, neral, α-phellandrene, sarisan, sulcatylacetate, α-terpineol, terpinen-4-ol, α-turmerone, β-turmerone, (-)-verbenone.

11.3.2 Identification of EOs and/or their Compounds with Psychopharmacologic Action

The need for hypothesis-driven research, instead of fishing expeditions in search for a hypothesis (often based on the available results) has been thoroughly discussed (Ioannidis, 2018), as well as its implications for pharmacology and drug development (Kenakin et al., 2014). One approach in the search for psychoactive compounds from EOs is to formulate hypotheses on pharmacological properties based on the medicinal use of aromatic species in traditional systems of medicine. Buchbauer and colleagues (Buchbauer et al., 1993) showed the effects of inhalation of several EOs and their compounds in mice mobility. Since sedative-like effects (decreased mobility) were observed in low concentrations, the data justify the efficacy of pure fragrance compounds and/or their EOs used as mild sedatives in folk medicine. Examples include coumarin for the soporific property of hay and linalool and linalyl acetate for sandalwood oil, neroli oil, and especially lavender oil. By correlating the effects on mobility with compounds concentrations retrieved in plasma, their lipophilic characters, and detection thresholds, the authors concluded that chemical structures and functional groups of the compounds are likely to determine their biological effects.

Another example is the ethnopharmacology approach to search for anticonvulsants in Amazonian traditional medicine (Elisabetsky and Brum, 2003). Based on the descriptions of symptoms and diseases by healers and users among *caboclo* or *riberinhos* in the State of Pará (Brazil), a list of conditions that could refer to epilepsy-like conditions and associated home medicines was generated. The most frequently mentioned formula was the combination of the juices (obtained by mechanical pressure) from *arruda* (*Ruta graveolens* L., Rutaceae), *cipó pucá* (*Cissus sicyoides* L., Vitaceae), *catinga de mulata* (*Aeollanthus suaveolens* Mart. Ex Spreng., Lamiaceae) leaves, and a teaspoon of *gergelim preto* (*Sesamun indicum* L., Pedaliaceae) seeds. The formula, prepared as reported, slowed (but did not inhibit) pentylenetetrazole (PTZ)-induced seizures in mice. *A. suaveolens*, an African species never studied before, as the popular name suggests, is strongly aromatic, and its EO proved to have anticonvulsant properties (Coelho de Souza et al., 1997). The main components in the EO were (*E*)-β-farnesene (37.75%), δ-decene-2-lactone (20.6%–44.3%), linalyl acetate (11.32%), linalool 10.49%), and δ-decanolactone (0.37%–3.02%) (Elisabetsky and Brum, 2003). Although the study of (*E*)-β-farnesene is limited, we found that linalyl acetate was devoid of activity (Coelho de Souza et al., 1997) and that linalool and γ-decanolactone were active in several hypno-sedative and seizure animal models (Pereira et al., 1997). Several linalool-producing plants are or have been used in traditional medical systems for purposes evocative of central sedative effects, most notably lavender baths or inhalation.

Because neurotransmitter systems are present in neuronal circuitries implicated in more than a neuropsychiatric condition, plants used traditionally for one purpose may act through a pharmacological mechanism applicable to other pharmacological areas. A plant species used to "improve sleep" generates the hypothesis of hypnotic, anxiolytic and anticonvulsant properties, as the same mechanism of action may apply to all. Antidepressants are used for analgesics and/or anxiolytic purposes, and insults to central nervous system (CNS) apparently share the commonality of generating oxidative stress and neuroinflammation (Hurley and Tizabi, 2013). Unfortunately, one finds in the literature reference to "a plant traditionally used" without detailed use or references to original field data, or even statements unsubstantiated by the original report (often the wrong information is thereafter used repetitively after the first misquotation). An accurate ethnopharmacology is obviously necessary for this approach.

11.4 COMPOUNDS FROM EOs WITH PSYCHOPHARMACOLOGY POTENTIAL

The components isolated from EOs belong to several chemical classes, such as monoterpenes, sesquiterpenes, and phenylpropanoids. There are not enough studies on structure-activity relationships that would facilitate the search for new psychoactive EO compounds. In the absence of ethnopharmacological information, the major advantage of using EOs in the search for new drugs with central activity would be the availability of a wide variety of natural substances and their molecular characteristics.

There are enormous difficulties in the way to find pharmacologically useful drugs (Lipinski, 2000), but absorption, distribution, metabolism, and excretion were proposed as simple filters in the search of active substances in large databases (Clark, 1999; Lipinski et al., 2012). Of particular relevance to this chapter is the estimative that over 98% of all small-molecule drugs do not cross the blood–brain barrier (BBB) (Pardridge, 2005).

Lipinsky's "rule of five" describes a set of requirements for a drug to be orally absorbed (Lipinski et al., 2012): (1) molecular weight, MW <500; (2) polar surface area (Clark, 1999), PSA <140 $Å^2$; (3) octanol/water partition coefficient (Leo, 1993), LogP <5; (4) no more than five hydrogen bond donors (HBD); and (5) no more than 10 hydrogen bond acceptors (HBA). The HBD and HBA values are additives, and in the rule of five, the sum of both types of hydrogen bonding must be equal or less than 10. In this chapter we will use the molecular descriptors TPSA (Clark, 1999) and XLogP3 (Cheng et al., 2007) instead of PSA and LogP, respectively. As can be seen in Table 11.1, the psychoactive volatiles (PAVs) that originated from EOs easily conform to the conditions placed by the rule of five.

MW represents the molecule size. Above the 400–500 g/mol limit, the size factor (molecular volume) becomes impeditive for substance absorption through membranes (Wessel et al., 1998; Lipinski and Hopkins, 2004). None of the PAVs exceeds this limit. Theoretical studies applied to the compounds in Table 11.1 can use more precise descriptors, such as molecular volume (Molinspiration, 2019) or molecular complexity (PubChem, 2019).

TPSA is the topological sum of the surfaces of the atoms that form the polar chemical clusters in a molecular structure, usually formed by atoms of nitrogen, oxygen, and the attached hydrogens. TPSA values are used to search for substances of interest in large databases or to carry out theoretical studies to refine the knowledge about structure/activity relationships within small groups of similar structures. The values of TPSA, found in several online databases (Molinspiration, 2019; PubChem, 2019), can also be correlated to oral absorption in a general sense, TPSA <140 $Å^2$ (Ertl et al., 2000), or specifically to permeability in CACO-2 monolayers (Artursson et al., 1996), and BBB penetration (values below 60 $Å^2$) (Palm et al., 1997; Shityakov et al., 2013).

In Table 11.1, there are only two violations to the maximum of five allowed by the rule of five for XLogP3 (phytol **4** and oleamide **34**), though only one violation to the rule does not preclude adequate oral absorption and/or effectiveness (Lipinski, 2000).

HBD and the HBA are parameters related to BBB penetration and the interactions of the substance with the active site. Synthetic small-molecule drugs can present values up to 8–10 for the sum of HBD and HBA, but this is a strong limiting factor to the number of clinically useful compounds since the number and strength of the hydrogen bonds can significantly alter lipophilicity (Waring, 2010).

The molecular descriptors are more useful to predict absorption than activity. Numerous CNS inactive natural or synthetic volatile compounds show molecular descriptors similar to those of PAV compounds **1** to **34** (Table 11.1). Even using MW <400 and a total value of HBD plus HBA below 8–10 (to ensure adequate size and high lipophilicity), the database searches will still result in large numbers of substances that cross the BBB but have no affinity for the active sites that ensure psychopharmacological activities. Pardridge (2005) points out that affective disorders, chronic pain, and epilepsy (Pardridge, 2005) are CNS disorders that respond to drugs that exhibit MW <400 and high lipid solubility, two major characteristics of PAVs. The selected list of PAVs in Table 11.1 includes substances from different chemical classes and subclasses that were selected after careful review of published results on experiments suggesting anxiolytic, hypnotic, antiepileptic, antidepressant, and/or neuroprotective activities. It is interesting to draw a quick comparison between the molecular descriptors of these PAVs and those of some synthetic psychoactive drugs. Table 11.2 shows a selection of clinically useful synthetic drugs used as anxiolytics, hypnotics, antiepileptics, antidepressants, and/or neuroprotectors displaying variable chemical structures and mechanisms of action. The synthetic drugs are listed in increasing order of MW, between 144.21 (valproic acid) and 385.51 (buspirone). Six out of 19 drugs present MW >300 (fluoxetine, zolpidem, alprazolam, topiramate, buspirone, and trazodone), representing a violation of the rule for penetrating the

TABLE 11.1
Lipinski's Molecular Descriptors for the Psychoactive Volatile Compounds

Compound	MW	TPSA	XLogP3	HBD	HBA
		Alcohols			
Myrtenol **1**	152.23	20.2	1.6	1	1
Geraniol **2**	154.25	20.2	2.9	1	1
Farnesol **3**	222.37	20.2	4.8	1	1
Phytol **4**	296.53	20.2	**8.2**[a]	1	1
Fragranol **5**	154.25	20.2	2.9	1	1
Citronellol **6**	156.27	20.2	3.2	1	1
Borneol **7**	154.25	20.2	2.7	1	1
Isopulegol **8**	154.25	20.2	3.0	1	1
Linalool **9**	154.25	20.2	2.7	1	1
Nerolidol **10**	222.37	20.2	4.6	1	1
Bisabolol **11**	222.37	20.2	3.8	1	1
		Phenols and Aromatic Methylethers			
Thymol **12**	150.22	20.2	3.3	1	1
Carvacrol **13**	150.22	20.2	3.1	1	1
Eudesmin **14**	386.44	55.4	2.9	0	6
Methyleugenol **15**	178.23	18.5	2.5	0	2
α-Asarone **16**	208.25	27.7	3.0	0	3
		Hydrocarbons			
α-Pinene **17**	136.23	0.0	2.8	0	0
β-Pinene **18**	136.23	0.0	3.1	0	0
Limonene **19**	136.23	0.0	3.4	0	0
Caryophyllene **20**	204.35	0.0	4.4	0	0
		Carbonyl Compounds			
Decursinol angelate **21**	328.36	61.8	3.7	0	5
Fragranyl acetate **22**	196.25	26.3	3.2	0	2
Fragranyl benzoate **23**	275.37	26.3	4.9	0	2
Benzyl benzoate **24**	212.25	26.3	4.0	0	2
γ-Decanolactone **25**	170.25	26.3	2.7	0	2
Safranal **26**	150.22	17.1	2.1	0	1
Thymoquinone **27**	164.20	34.1	2.0	0	2
Ar-Turmerone **28**	216.32	17.1	4.0	0	1
Dehydrofukinone **29**	218.34	17.1	4.1	0	1
		Epoxides			
Curzerene **30**	216.32	13.1	4.6	0	1
Epoxycarvone **31**	166.22	29.6	1.6	0	2
Linalool oxide **32**	170.25	29.5	1.4	1	2
Limonene epoxide **33**	152.23	12.5	2.5	0	1
		Nitrogenated Compound			
Oleamide **34**	281.48	43.1	**6.6**[a]	1	1

Source: Data from https://pubchem.ncbi.nlm.nih.gov.

Abbreviations: Molecular weight, MW (g/mol); topological polar surface area, TPSA (Å^2); octanol/water partition coefficient, XLogP3; number of hydrogen bond donors, HDB; and hydrogen bond acceptors, HBA.

[a] Rule of five violations by compounds **4** and **34**.

TABLE 11.2
Lipinski's Molecular Descriptors for Some Clinically Useful Drugs

Compound	MW g/mol	TPSA	XLogP3	HBD	HBA
Valproic acid	144.21	37.3	2.8	1	2
Levetiracetam	170.21	63.4	−0.3	1	2
Selegiline	187.28	3.2	2.8	0	1
Velanfaxine	177.40	32.7	2.9	1	3
Zonisamide	212.22	94.6	0.2	1	5
Pentobarbital	226.27	75.3	2.1	2	3
Riluzole	234.19	76.4	3.6	1	7
Felbamate	238.24	105	0.6	2	4
Phenytoin	252.27	58.2	2.5	2	2
Diazepam	284.07	32.7	3.0	0	2
Amitryptiline	277.41	3.2	5.0	0	1
Maproptiline	277.41	12.0	4.6	1	1
Fluoxetine	309.33	21.3	4.0	1	5
Zolpidem	307.39	37.6	2.5	0	2
Topiramate	339.35	124.0	−0.8	1	9
Alprazolam	308.76	43.1	2.1	0	3
Buspirone	385.51	69.6	2.6	0	6
Tradozone	371.86	42.4	2.8	0	4

Source: Data from https://pubchem.ncbi.nlm.nih.gov.

BBB, especially because the molecular weight is usually associated with an elevated number of heteroatoms, decreasing lipophilicity.

Among the PAVs in Table 11.1, eudesmin **14** and decursinol angelate **21** have MWs above 300, exactly because of their higher number of oxygens, raising these compounds TPSA values to 55.4 and 61.8, respectively, the higher among PAVs. An elevated number of oxygens increases the MW and may significantly alter the molecular volume, increasing the importance of intramolecular hydrogen bonds that can decrease the volume. Nevertheless, PAVs, natural or synthetic, in general possess few oxygens in their structures and thus present elevated lipophilicity, with average XLogP3 values circa 3.0 and always above zero for compounds **1–34**. Of note, the increase in carbon chain from C10 to C20 in a molecule with only one oxygen can elevate the value of XLogP3, causing absorption difficulties.

Among the six drugs with MWs >300 in Table 11.2, topiramate presents 10 heteroatoms. As a result, its lipophilicity is low, with XLogP3=−0.8 the lowest in the list. Moreover, its high TPSA (124 Å^2) value and a total HBD/HBA number of 10 are two violations to the rule that could prevent its absorption trough the BBB or could have withdrawn it from clinical trials (Shin et al., 2018).

A relatively higher polarity (lower values for XLogP3 and higher for the HBD/HBA sum) seems to be a general feature for the molecules of topiramate, levetiracetam, zonisamide and felbamate, all with high values for TPSA >60 Å^2. On the other hand, among PAVs at Table 11.1, eudesmin **14** and decursinol angelate **21** present relatively higher values for MW, TPSA and HBD/HBA summation than the other PAVs and closer to the values found for synthetic drugs. However, contrary to the drugs in clinic, compounds **14** and **21** also have higher values for XLogP3, suggesting a greater lipophilicity in relation to the synthetic compounds and the other natural PAVs.

The comparison of the main molecular characteristics of synthetic drugs (Table 11.2) and PAVs (Table 11.1) clearly shows some differences or trends of the two types of small-molecules drugs. The majority of PAVs have XLogP3 values around 3.0, and the values for all other molecular descriptors are relatively lower than those found for synthetic drugs. The amide group has been frequently used as a polar group in the synthetic psychoactive molecules, but only one amide (oleamide **34**) is found in the list of PAVs. The amide group presents two canonical forms, which leads to a resonance hybrid with separation of charges, δ^+ on the nitrogen atom and δ^- on the oxygen, resulting in two types of strong hydrogen bonds, one with donor and the another with acceptor character. None of the other chemical functions present in PAV molecules have this marked amide characteristic.

Although the molecular descriptors are only general data on the molecular characteristics of the active substances, a careful structure-activity analysis may point to a common path for the development of new synthetic and natural psychoactive drugs or provide general rules to narrow the searches in compound libraries. It may be possible to find a mid-term of molecular characteristics between PAVs and synthetic compounds looking for or creating molecular structures with the correct balance between the forces of the polar groups and the size of the non-polar areas. Even with the few data available for analysis, the limits that underpin the concept of PAVs of natural origin are given: (a) the compound possesses only one chemical function in its structure, making clear the difference between a small polar area and a vast non-polar area in the molecule, guaranteeing a good level of lipophilicity, and (b) the compound contains 0–3 heteroatoms, generally oxygens only, therefore presenting TPSA <60 Å^2; compounds rarely present MW above 300 Da and only positive values for XLogP3, predominantly values between 1 and 5. The number of molecules that fit the concept of PAV is limited.

Even considering that various structures may be created by synthetic chemistry using this model, real advantages would arise from exploring the general characteristics in the structural designs of PAVs.

An important characteristic is that chemicals structures of PAVs are formed by few and well-known basic blocks present in plant secondary metabolites, as in the terpenoids C10, C15, and C20 and the arylpropanoids C6 and C3. Even in relation to their effectiveness as drugs, the PAVs may demonstrate that there is no need for molecules with strong polar bonds to obtain marked biological effects, since the special forms and molecular properties of natural compounds have good chances of being recognized by the organisms' protein active sites (Buchbauer et al., 1993). Among the synthetic compounds listed, one finds a variety of structural scaffolds but containing only one heteroatom and therefore only one simple chemical function: two secondary amides (maprotiline and fluoxetine) and two tertiary amines (amitriptyline and trazodone). These four compounds are clinically used as antidepressants, acting through mechanisms of action that alter synaptic concentration of 5HT/NOR, just 5HT, or just NOR. All four compounds present structures that are closer to the concept discussed for PAVs: the secondary or tertiary amine functions are used as a weakly polar group attached to the molecule at a distant tip from the large non-polar area. This is an example of convergence among the concepts underpinning the synthetic and PAV psychoactive compounds.

The small niche of PAVs is here defined, but it is obviously necessary to scrutinize the individual molecular characteristics that underlie their mechanisms of action. To that end, the chemical structures of PAVs **1–34** were divided in six groups containing characteristic organic functions: alcohols, phenols and aromatic methyl ethers, hydrocarbons, carbonylic compounds, monoterpene epoxides and furanes, and nitrogenated compounds. Certainly, as research in EOs progresses, new PAV structures containing other chemical functions—such as carboxylic acids, nitro-compounds and amines—may join the PAV list.

11.4.1 Alcohols

Table 11.1 shows 11 primary, secondary, or tertiary alcohols, with cyclic or open-chain structures, with or without an α,β-unsaturation to the alcohol function (**1** to **11**, Figure 11.1). With the exception of compounds **2** and **3**, all other alcohols on the list have at least one chiral center. In addition, **2**, **3**, **4**, and **10** have *cis*/*trans* isomers, which raise significantly the number of isomers and molecular conformation to be considered in theoretical studies.

Among the four primary α,β-unsaturated alcohols **1–4**, myrtenol **1** (C10, monoterpene) presents the lower XLogP3 due to its rigid structure and small volume in comparison with the open chain present in the others.

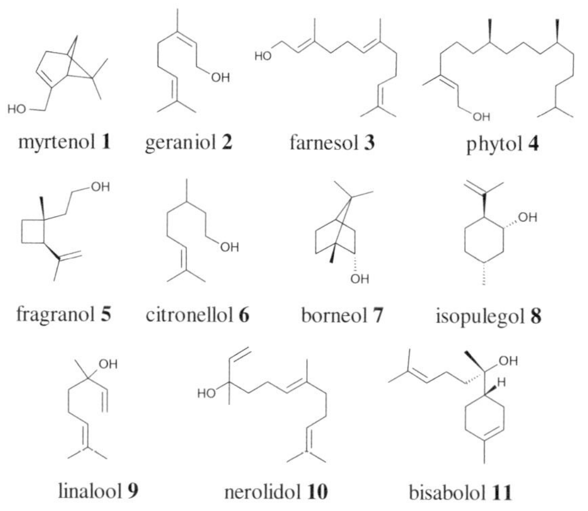

myrtenol **1** geraniol **2** farnesol **3** phytol **4**

fragranol **5** citronellol **6** borneol **7** isopulegol **8**

linalool **9** nerolidol **10** bisabolol **11**

FIGURE 11.1 Psychopharmacologic active alcohols.

Geraniol **2** (C10, monoterpene), farnesol **3** (C15, sesquiterpene), and phytol **4** (C20, diterpene) form a good series for structure-activity-mechanism of action comparative studies. These are three primary alcohols with open chains that present an α,β-unsaturation to the carbon that bears the hydroxyl group. All three present an identical unit isopentenol that forms the polar extremity of the chain and identical isoprene units (in **2** and **3**) or saturation (**4**) in the non-polar extremities. The MWs raise in the same sequence (**2** < **3** < **4**) of their XLogP3, with accompany increases in lipophilicity, whereas the other molecular descriptors remain in the same values.

The primary α,β-saturated alcohols fragranol **5** and citronellol **6** (C10, monoterpenes) present very similar Lipinski's molecular descriptors, which make them exemplary for studying the effects of variations in the open carbonic chains. The study is even more interesting due to the occurrence of chiral centers in the **5** and **6** structures.

Borneol **7** has very average values for its molecular descriptors in comparison to the other alcohols in the series (Table 11.1), with an uncommon property of facilitating other drugs to cross the BBB, information arising from traditional Chinese medicine and now fairly well studied (Zhang et al., 2015, 2017; Yu et al., 2013a,b; He et al., 2018).

An interesting experiment related the binding effectiveness of the two (+) and (−) borneol isomers and isoborneol to the $GABA_A$ receptors (but not specifically the benzodiazepine [BDZ] site) to the configuration of the hydroxyl in these structures. Only, at the isomer (+) of borneol **7**, the hydroxyl is positioned *exo* in the opposite direction to the geminal dimethyl bridge, while (−)-borneol and isoborneol possess the hydroxyl group in the *endo* position. The isomers with *endo* hydroxyl bind less effectively to the $GABA_A$ receptor (Yu et al., 2013a,b; Zhang et al., 2017).

Borneol **7** and isopulegol **8** (C10, monoterpenes) are α,β-saturated secondary alcohols presenting small, rigid, cyclic structures. Borneol **7** has a totally rigid structure with a small molecular volume (MV = 165.72), followed closely by myrtenol **1** (MV = 160.07) and isopulegol **8** (MV = 171.55) (Molispiration, 2019). Molecules with higher or total rigidity are useful to reveal details of a given active site architecture, since a rigid structure does not adapt to the site by simple bonds twists. Totally rigid structures present only one conformation and are already in their minimum conformational energy, of special interest to structure-activity relationship studies because they may reveal precise information on the geometry of the active site to which they bind, with better chances of representing entirely new models for specific pharmacological activities.

A couple of compounds that also wait for comparative structure-activity-mechanism of action studies is formed by linalool **9** (C10, monoterpene) and nerolidol **10** (C15, sesquiterpene). These are tertiary alcohols with α,β-unsaturation to the carbon that bears the hydroxyl group. The extremities of the open chains of these two alcohols are identical: a tertiary hydroxyl on one side and an isoprenic C5 unity on the other. Both present an asymmetric center in the tertiary carbon and may be present in EOs with only one optic isomer (*R* or *S*) or as a mixture of the two. Another factor to consider is that nerolidol **10** has *cis-trans* isomerism in the double bond C6–C7, which increases the number of possible isomers and conformers.

The third tertiary alcohol, bisabolol **11**, has its α,β-saturated hydroxyl group positioned in the middle of the structure, between an isoprenic unit and a six-member ring. Though **11** presents little structural similarity with **9** or **10** or the remaining alcohol in the series, it should not be difficult to find similar natural molecules forming adequate groups for structure-activity relationships, since **11** contains all the carbon scaffold of α-terpineol linked to an isoprenic unit.

11.4.2 Phenols and Aromatic Methyl Ethers

Thymol **12** and carvacrol **13** are two small C10 phenolics presenting identical MW, TPSA, HBD, HBA, the same molecular planar shape, and XLogP3 values of 3.3 and 3.1, respectively (Table 11.1). The small difference between partition coefficients of these two phenols is caused by the two possible positions for the phenolic hydroxyl in the same carbon structure. Both compounds have only one rotatable bond that links the isopropyl group and the aromatic ring, giving these molecules the possibility to present a low energy conformation with all heavy atoms in the same plan. Several natural small-molecule phenols, including **12** and **13** and several flavones, presented anxiolytic properties likely to be mediated by $GABA_A$ signaling (Wang et al., 2017). The structures of **12**, **13**, and various small phenolics show molecular descriptors similar to the PAV propofol, including TPSA, XLogP3, MW, and MV (Molispiration, 2019; PubChem, 2019), as well as a 100% planar form. Flavonoids are a vast class of non-volatile natural phenolic compounds with a classic C15 carbon scaffold, many 100% planar, like the flavones, which also present anxiolytic activity associated with their affinity for the BDZ site in $GABA_A$ receptors (Paladini et al., 1999) (Figure 11.2).

Methyleugenol **15** and α-asarone **16** are methyl ethers of the corresponding phenols and also show the planar characteristic of **12** and **13**, able to accommodate in one plan all of its heavy atoms. In the molecules **14**, **15**, and **16**, there are no hydroxyls, and the oxygens may act in hydrogen bonds only as acceptors, but all of the compounds **12** and **16** show XLogP3 circa 3.0.

The chemical structure of eudesmin **14** corresponds to a dimer of methyleugenol **15**, being also 100% planar in the major part of its structure. Among the synthetic psychoactive drugs, there are many examples of molecules that are planar in the major part of theirs structures, such as the carbamazepine and the barbiturates. As mentioned, eudesmin **14** presents relatively high values of TPSA and the HBD/HBA sum as compared to other PAVs (Table 11.1), without diminishing its lipophilicity (XLogP3 = 2.9) even when compared to the similar phenols and aromatic methyl ethers in its group. In contrast, synthetic psychoactive drugs in general present higher values for their molecular descriptors, with lower values for XLogP3 and consequent lower lipophilicity in comparison to PAVs (Table 11.2).

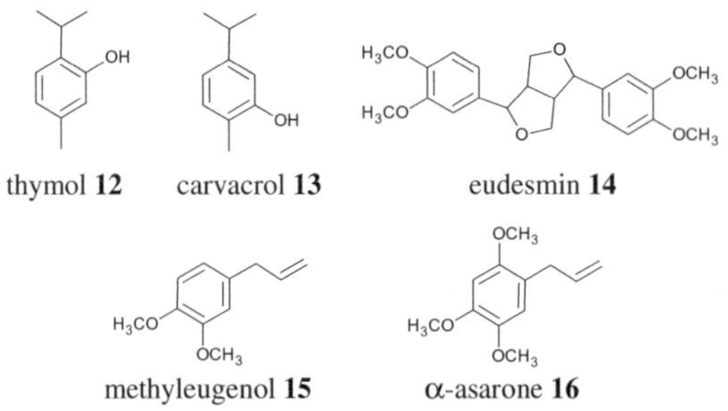

thymol **12** carvacrol **13** eudesmin **14**

methyleugenol **15** α-asarone **16**

FIGURE 11.2 Psychopharmacologic active phenols and aromatic methyl ethers.

11.4.3 Hydrocarbons

The hydrocarbons series, the monoterpenes α-pinene **17**, β-pinene **18**, and limonene **19**, and the sesquiterpene caryophyllene **20** (Figure 11.3) are devoid of heteroatoms and present zero value for TPSA, HBD and HBA. All are cyclic molecules of reduced volume, with XLogP3 between 2.8 and 4.4, not particularly lipophilic. In structural terms, the four compounds of this group represent the very opposite of the model followed by the other PAVs and synthetic psychoactive drugs, being very non-polar since lacking the heteroatom. They remain as markers of a frontier of centrally active PAVs with TPSA equal to zero.

Caryophyllene **20** is among the few substances with sufficiently adequate shape and measures to present binding to the CB2 receptor (Galdino et al., 2012; Bahi et al., 2014). A reasonable analysis on the similarities of the structures of **20**, **34**, and cannabidiol requires theoretic calculations.

11.4.4 Carbonyl Compounds

Among the nine carbonyl compounds **21** through **29** of this survey, only fragranol acetate **22** and γ-decanolactone **25** did not present α,β-unsaturation to the oxygenated function carbons (Figure 11.4), a factor that affects the main molecular characteristics of this group in relation to the electronic behavior of different carbonyls. In the α,β-saturated carbonyls of esters **22** and **25**, the resulting resonance hybrid presents a charge separation with δ^- over the carbonyl oxygen and δ^+ over the other oxygen.

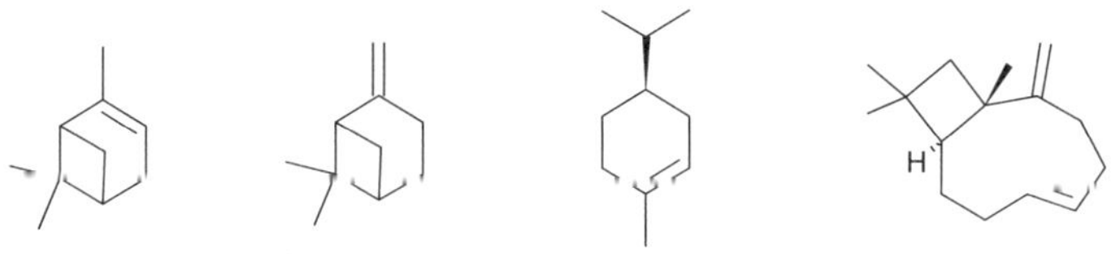

α-pinene **17** β-pinene **18** limonene **19** caryophyllene **20**

FIGURE 11.3 Psychopharmacologic active terpenoid hydrocarbons.

decursinol angelate **21** fragranol acetate **22** fragranol benzoate **23** benzyl benzoate **24**

γ-decanolactone **25** safranal **26** thymoquinone **27**

ar-turmerone **28** dehydrofukinone **29**

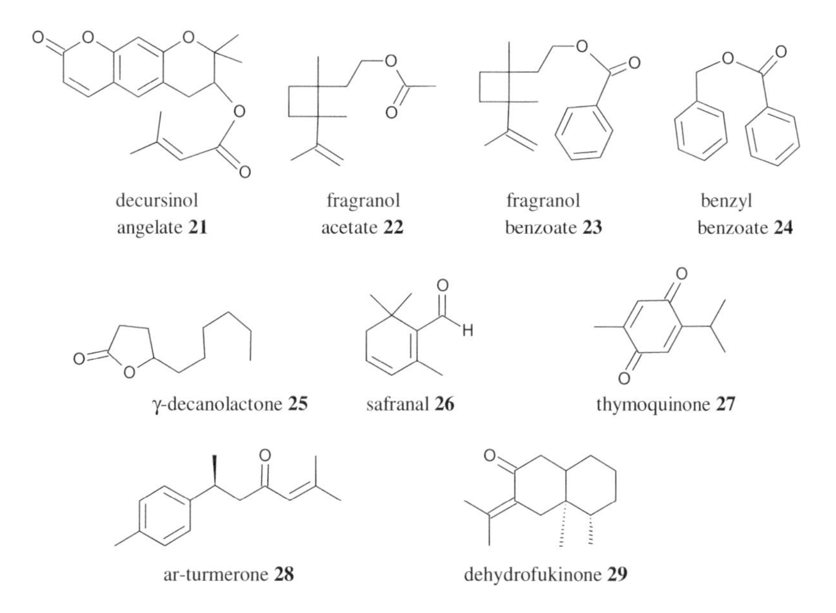

FIGURE 11.4 Psychopharmacologic active carbonyl compounds (aldehyde, ketones, esters, and lactones).

The two most important characteristics of the α,β-unsaturated carbonyl of compounds **21**, **23**, **24**, and **26–29** are the electronic resonance that renders planar different extensions of these molecular structures and the negative charge over the carbonyl oxygen, stronger in **22** and **25**, which possess α,β-saturated carbonyls. The carbonyl compounds group presents as a structural characteristic a bigger planar proportion of the molecule, highlighting the structure of compounds **21**, **24**, **26**, **27**, and **28**, with different capacities as HB acceptors.

11.4.5 Monoterpene Epoxides and Furanes

Curzerene **30** (C15, furan ring), epoxycarvone **31** (C10, epoxide), linalool oxide **32** (C15, cyclic C4-ether), and limonene epoxide **33** (C10, epoxide) are cyclic ethers of different kinds, sizes, and formats. Compounds **30–33** have two or three chiral centers and present four possible optical isomers each, but few conformations to consider.

The sesquiterpene **30** is a small, bicyclic molecule with high lipophilicity (XLogP3 = 4.6), with its oxygen involved in the aromaticity of the furan ring, resulting in a structure with no more than a weak interaction as an HB acceptor. The trisubstituted furan ring in **30** extends that planar part of this molecule, forcing half of the 16 heavy atoms to remain at a single plane. Judging by the potency observed *in vivo* in various psychopharmacology tests, the simple structure of **30** seems to be efficient in binding to the BDZ site at the $GABA_A$ receptor, causing strong effects in very low doses (Abbasi et al., 2017).

The epoxides **31** and **33** present higher polarity than their original counterparts limonene **19** and carvone. The epoxycarvone **31** and linalool oxide **32** are exceptions to the one chemical function rule for PAVs since these compounds present two oxygenated chemical functions resulting in the lowest XLogP3 (1.4 and 1.6, respectively) of the list (Figure 11.5).

11.4.6 Nitrogenated Compounds

The oleamide **34**, the primary amide of oleic acid, is the only PAV of this list that contains a nitrogenated chemical function. The amide group per se is fairly polar and able to form strong hydrogen bonds; nevertheless, **34** contains 20 carbons in an open chain, guaranteeing its high lipophilicity. It is a natural PAV with various well-studied activities many times compared to the anandamide (De Petrocellis et al., 2000; Leggett et al., 2004), which continues to represent an innovative model for the study of bioactivity in a large series of structures (Figure 11.6).

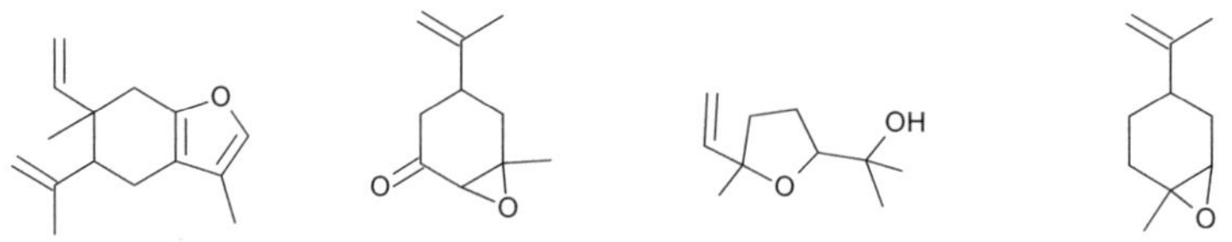

curzenere **30** epoxycarvone **31** linalool oxide **32** limonene epoxide **33**

FIGURE 11.5 Psychopharmacologic active terpenoid epoxides of several types.

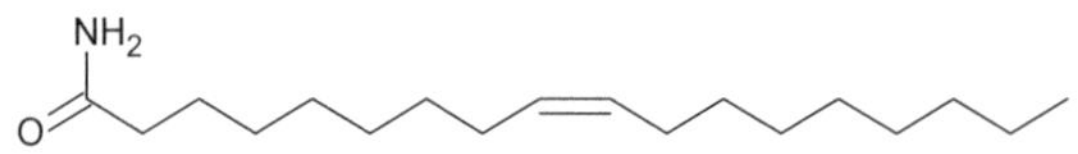

oleamide **34**

FIGURE 11.6 Structure of oleamide **34**, the only psychopharmacologic active volatile amide in the list.

TABLE 11.3
Overview of Psychopharmacological Properties of 34 Compounds from Essential Oils

Compound	Anxiolytic	Hypnotic	Anticonvulsant	Antidepressive	Neuroprotector
Myrtenol **1**	X	–	–	–	–
Geraniol **2**	–	–	–	X	X
Farnesol **3**	–	–	–	–	X
Phytol **4**	X	X	X	–	–
Fragranol **5**	X	–	–	–	–
Citronellol **6**	–	–	X	–	–
Borneol **7**	X	–	X	–	X
Isopulegol **8**	X	–	–	–	–
Linalool **9**	X	–	X	X	–
Nerolidol **10**	X	–	X	–	X
Bisabolol **11**	X	–	–	–	–
Thymol **12**	X	–	–	X	X
Carvacrol **13**	X	–	X	X	X
Eudesmin **14**	–	–	X	–	–
Methyleugenol **15**	X	–	X	–	–
α-Asarone **16**	X	X	X	X	X
α-Pinene **17**	X	X	–	–	–
β-Pinene **18**	–	–	–	X	–
Limonene **19**	X	X	–	X	–
Caryophyllene **20**	X	–	X	X	X
Decursinol angelate **21**	–	X	–	–	–
Fragranyl acetate **22**	X	–	–	–	–
Fragranyl benzoate **23**	X	–	–	–	–
Benzyl benzoate **24**	X	–	–	–	–
γ-Decanolactone **25**	X	X	X	–	–
Safranal **26**	X	X	X	–	X
Thymoquinone **27**	X	–	X	–	X
Ar-Turmerone **28**	–	–	X	–	–
Dehydrofukinone **29**	X	X	X	–	–
Curzerene **30**	–	–	X	–	–
Epoxycarvone **31**	–	–	X	–	–
Linalool oxide **32**	X	–	–	–	–
Limonene epoxide **33**	X	–	–	–	–
Oleamide **34**	X	X	X	X	–

Notes: –, no references; X, at least one reference.

Table 11.3 presents an overview of the psychopharmacogy properties of the essential oil compounds selected under the rational described above. The following sections provide a detailed analysis of the data obtained through our survey.

11.5 COMPOUNDS FROM EOS WITH ANXIOLYTIC PROPERTIES

11.5.1 Assessing Anxiolytic Properties

Anxiety-related disorders are among the most common psychiatric illnesses. Anxiety is not a unitary disease, but a complex phenomenon; it can be divided into state or trait, normal and pathological. There are different types of pathological anxiety, various etiological origins, and different neural circuits, neurochemical systems, and brain areas seem to be involved (Brooks and Stein, 2015). Anxiety can accompany various psychiatric conditions not necessarily accompanying the progress of the primary disease, and thus anxiety often needs to be addressed in its own right.

11.5.1.1 Methods

Like humans, different animal species express anxiety and fear in response to environment circumstances. The criteria for validating experimental models of anxiety have been reviewed (Belzung and Lemoine, 2011), and particular models are more suitable to answer particular questions. The most commonly used models to screen for anxiolytic properties are (Santos et al., 2017a) the elevated pus maze (EPM), the light and dark box (LD or LDB), and the whole board (WB). In these tests, subjects under the effect of anxiolytics spend more time in the open arms (EPM) or light compartment (LD) and execute more head dips (WB), all of which present some level of insecurity or conflict between the instinct defensive and exploratory behaviors. The open field (OF) test is highly sensitive but very unspecific, with several observation endpoints prone to interference of drugs other than hypno-sedatives; it becomes somewhat more specific as an anxiety test when central versus periphery ambulation is assessed.

These tests can be skewed by drugs that affect locomotion, which is either measured in the same test (total number of entries in open and closed arms in the EPM, total number of crossing between compartments in the LD, and ambulation between holes in the WB) or in a separate OF test or activity cage.

Adverse effects of hypno-sedative drugs include sedation, memory impairments, decreased alertness, and slowed reaction time, which can be assessed in specific models (the rotarod being the most common). While diazepam is the most common positive control, muscimol (agonist at GABA receptor) or flumazenil (antagonist at the benzodiazepine site in the GABA receptor) are used to check for the participation of $GABA_A$ in the mechanism of action of the test drug.

11.5.1.2 Mechanisms of Action

Most anxiolytic drugs act through 5HT, GABA or noradrenergic systems. While the effects of GABA modulating agents (benzodiazepines) are acute, those of selective serotonin reuptake inhibitors (SSRI) or noradrenaline and serotonin reuptake (SNRI) requires chronic administration (O'Donnell and Shelton, 2011; Shelton and Od, 2011).

The balance between neuronal excitation and inhibition is vital for ordinary brain function and critical for CNS disorders. The balance is achieved by the proper activation of the γ-aminobutyric acid (GABA) receptor system (major inhibitory CNS system) and blockade of neuronal voltage-gated sodium channels (Na^+ channels) and/or glutamate-mediated excitability. Type A GABA ($GABA_A$) are pentameric proteins that form Cl^--permeable ion channels. Widely distributed in the CNS, these receptors constitute the primary inhibitory control of neural activity, under normal or pathophysiological conditions. $GABA_A$ receptors possess up to 19 subunits, composing heteromeric receptors associated with physiological and pharmacological properties. Of the $GABA_A$ receptors in the adult brain, 43% show the subunit combination $\alpha1\beta2\gamma22$, an important molecular target for hypno-sedative drugs used for anxiety, insomnia, and epilepsies (McKernan and Whiting, 1996). Given that the specific roles of subunits in different diseases and/or adverse effects are still not fully understood, novel $GABA_A$ receptor modulators are continuously investigated (Ding et al., 2014).

11.5.2 Anxiolytic Properties of EO Compounds

An overview of the anxiolytic properties of the compounds discussed in this chapter is presented at Table 11.4. Phytol **4** showed anxiolytic-like effect on 3 mice models, in which its effects were reversed by flumazenil, without affecting motor coordination (Costa et al., 2014). Isopulegol **8** seems to be a psycholeptic, acting on mice models of anxiety, potentiating barbital sleep and increasing immobility in the tail suspension test (TST) (Silva et al., 2007), though data do not show consistent dose-response. Nerolidol **10** showed anxiolytic-like effect in two mice models without affecting motor coordination (Goel et al., 2016). Acute bisabolol **11** induces anxiolytic-like effects in mice, possibly related to $GABA_A$ receptors (Tabari and Tehrani, 2017). Thymol **12** showed anxiolytic-like effects, congruent to the author's hypothesis that natural small-molecule phenols are, in general, anxiolytics (Wang et al., 2017). Despite characterization as a $GABA_A$ activator (Ding et al., 2014), methyleugenol **15** did not

TABLE 11.4
Anxyolitic Properties of Compounds from Essential Oils

Compound	Behavioral models										Mechanisms of action						Active dose (mg/kg)	Via	Use
	EPM	LD	HB	OF	OF center	RR	MB	Fear recall	SI	NSF	Antioxidant	GABAa	GLU	5HT1a	CB	COR			
Myrtenol **1**	Y	Y	–	N	–	–	–	–	–	–	–	Y	–	–	–	–	25–75	ip	Y
Phytol **4**	Y	Y	–	N	–	N	Y	–	–	–	–	Y	–	–	–	–	25–75	ip	N
Fragranol **5**	–	Y	–	Y	–	–	–	–	–	–	–	–	–	–	–	–	50–150	po	N
Borneol **7**	Y	Y	–	–	Y	–	–	Y	–	–	–	Y	–	–	–	–	1 nM	ip	Y
Isopulegol **8**	Y	–	Y	N	–	N	–	–	–	–	–	Y	–	–	–	–	25–50	ip	N
Linalool **9**	Y	Y	Y	Y	–	Y	–	–	–	–	–	Y	–	N	–	–	30, 1%–3% 200 μL 8.6–86 μg	ipinhaled icv	Y
Nerolidol **10**	Y	–	–	–	Y	N	–	–	–	–	–	–	–	–	–	–	12–50	ip	N
Bisabolol **11**	Y	–	–	Y	–	N	–	–	–	–	–	Y	–	N	–	–	0.5–1	ip	S
Thymol **12**	Y	–	–	–	–	–	–	–	–	–	–	–	–	–	–	–	20	po	N
Carvacrol **13**	Y	–	–	N	–	N	–	–	–	–	–	Y	–	–	–	–	12.5–50	po	N
Methyleugenol **15**	N	–	N	N	–	–	–	–	N	–	–	–	–	–	–	–	0.025– 0.25 μL/kg	po	Y
α-Asarone **16**	Y	Y	Y	Y	Y	–	Y	–	–	Y	–	Y	Y	–	–	Y	200×21d 3.5–28×6d 2–20×7d	ip po po	Y
α-pinene **17**	Y	–	–	–	–	–	–	–	–	–	–	–	–	–	–	–	10μL/L	inhaled	N
Limonene **19**	Y N	Y N	–	N Y	–	Y	–	–	–	–	–	N	–	–	–	–	1%–2% 200 3.4–6.7 mg/L	inhaled ipinhaled	Y
Caryophyllene **20**	Y	Y	Y	–	Y	N	Y	–	–	–	–	N	–	N	Y	–	50–200	ip po	N
Fragranyl acetate **22**	–	Y	–	N	–	–	–	–	–	–	–	–	–	–	–	–	50–150	po	N
Fragranyl benzoate **23**	–	Y	–	N	–	–	–	–	–	–	–	–	–	–	–	–	50–150	po	N
Benzyl benzoate **24**	Y	Y	–	N	–	–	–	–	–	–	–	–	–	Y	–	–	2%	inhaled	1%–4%
γ-Decanolactone **25**	N	–	–	Y	–	–	–	–	–	–	–	–	–	–	–	–	0.1–0.3	ip	N
Safranal **26**	Y	–	–	Y	Y	–	–	–	–	–	–	–	–	–	–	–	0.05–0.35 mL/ kg	ip	Y
Thymoquinone **27**	Y	Y	–	N	–	–	–	–	Y	–	Y	Y	–	–	–	–	10–20	ip	N
Dehydrofukinone **29**	–	–	–	Y	–	–	–	–	–	–	–	–	–	–	–	–	10–50 mg/L	water	N
Linalool oxide **32**	Y	Y	–	–	–	Y	–	–	–	–	–	–	–	–	–	–	0.65%–5%	inhaled	N
Limonene epoxide **33**	Y	–	–	Y	–	Y	Y	–	–	–	Y	–	–	–	–	–	25–75	ip	N
Oleamide **34**	Y	Y	Y	Y	–	–	–	–	Y	–	–	–	–	–	–	–	10–20 5	ip	N

Notes: Y, positive effects; N, negative effects; –, no data.

Abbreviations: EPM, elevated plus maze; LD, light/dark test; HB, Hole board; OF, open field; RR, rotaod; MB, murbe burryin; SI, social interaction; NSF, novel suppress feeding; CB, cannabinoid; COR, corticosterone; Use, compatible traditional use.d,days; icv, intracerebroventricular; ip, intraperitoneally; po, per oz; *, unless stated otherwise; #, in a model of arsenic exposure.

show anxiolytic-like effects (Norte et al., 2005). Inhaled α-pinene **17** (8.6 mg/L, 90 min/day for 1, 3, and 5 days) showed anxiolytic-like effects in the EPM (Satou et al., 2014). β-pinene **18** diminished ambulation in the OF but had no effects on EPM or muscle tone (Guzmán-Gutiérrez et al., 2012). γ-Decanolactone **25** did not show anxiolytic effects (Viana et al., 2007). No replication or further investigation were found for the alleged anxiolytic-like properties of these compounds.

Electrophysiologic analysis showed that myrtenol **1** is a positive allosteric modulator of δ-$GABA_A$ receptors, modulating tonic and phasic inhibition in neurons that express these receptors (Kessler et al., 2014). This property is compatible with the sedative effects of volatile oils that contain significant amounts of these compounds, as well as the reported anxiolytic effects in mice models (Moreira et al., 2014).

Fraganol **5**, fragranyl acetate **22**, and fragranyl benzoate **23** showed anxiolytic properties in the LDT and OF in mice in doses roughly equivalent to 17% of the ED50 for the EO (Radulović et al., 2012). While this preliminary study suggests anxiolytic properties, it is difficult to predict how these compounds are metabolized and which metabolites reach the CNS, since it is not unlikely that fragranyl acetate and fragranyl benzoate can be metabolized into fragranol.

Borneol **7** (intrahippocampal infusion) attenuated contextual and cued fear expression (assessed at 24 h and 7 d intervals) and behaved as anxiolytic in the OF, EPM, and LD. It facilitates GABAergic currents measured in hippocampal slices, an effect blocked by bicuculine, suggesting the involvement of GABA receptors in its mechanism of action (Cao et al., 2018).

Linalool **9**, one of lavender EO major components, showed sedative (Linck et al., 2010) and anxiolytic-like effects in mice (Norte et al., 2005; Silva et al., 2007; Kessler et al., 2014; Tabari and Tehrani, 2017), rats (Yamada et al., 2005), and neonate chicks (Gastón et al., 2016) in several experimental models. Effects were documented following inhalation, intraperitoneal (ip), and/or intracerebroventricular (icv) administrations, and consistently did not affect motor coordination. Harada et al. (2018) proved that anxiolytic-like effects are the result of olfactory input since they absent in anosmic mice. They concluded that olfactory input is likely to activate central $GABA_A$ (but not $5HT_{1A}$)-dependent anxiolytic circuits which is compatible with binding data showing that linalool does not act directly on $GABA_A$ receptors (Brum et al., 2001). The data corroborates the anxiolytic effects of lavender EO odor repetitively documented in different animal species (see de Sousa et al., 2015).

Carvacrol **13** showed anxiolytic-like effects in the EPM, without apparent sedation (barbital sleep and ambulation) or muscle relaxation (rotarod). Effects were reversed by flumazenil (Melo et al., 2009).

α-Asarone **16** showed anxiolytic-like effects when given orally to mice in several well-established models (Liu et al., 2015). Corroborating its anxiolytic-like profile, it also showed anxiolytic-like effects in rat models when given before corticosterone (21 days); **16** prevented the reductions in BDNF, TrkB, and tyrosine hydroxylase expression in the hippocampus and locus coeruleus, respectively. The authors concluded that the anxiolytic-like effect is achieved by modifying the central noradrenergic system and BDNF function (Lee et al., 2014).

In a model of chronic inflammatory pain (complete Freund's adjuvant-induced) known to be anxiogenic, **16** had anxiolytic effects without being analgesic. The study showed that the model induced an imbalance between excitatory (upregulated GluR1 and NR2A glutamate receptors) and inhibitory (downregulated GABAA-α2 and GABAA-γ2 receptors) neurotransmission in the basolateral amygdala, a key area for anxiety. One-week treatment with **16** reversed the altered expressions of these proteins. Electrophysiological recordings showed that **16** treatment restored the excitation/inhibition balance by enhancing GABAergic neurotransmission and decreasing glutamatergic neurotransmission in the basolateral amygdala (Tian et al., 2017).

Limonene **19** is the main component of *Citrus junos*, widely used in food and cosmetic formulas due to its citric smell. It is also present in some medicinal species, as in *Lippia alba* (Viana et al., 2000). Systemic (ip) **19** showed sedative and muscle relaxant activity in higher doses, but not anxiolytic effects in the EPM (do Vale et al., 2002; Yu et al., 2013a,b). Inhaled (90 min) limonene showed an anxiolytic-like effect in LD and EPM (Satou et al., 2012). Another study reported anxiolytic-like effects of inhaled (7 min) limonene in the EPM, not reversible by flumazenil (Lima et al., 2013).

Caryophyllene **20** orally administered showed anxiolytic-like effects in several mice models without affecting motor coordination. The effect seems to be independent of 5-HT1A or benzodiazepine receptors, compatible with its CB2 agonist profile (Galdino et al., 2012). A second study, with caryophyllene administered ip, replicated its anxiolytic-like effects in some of the models and extended to others, corroborating the involvement of the CB2 receptor in its mechanism of action, showing that the effects were abolished by previous administration of the CB2 receptor antagonist AM630 (Bahi et al., 2014).

Benzyl benzoate **24** (the major component of *Cananga odorata* EO) showed anxiolytic activities after inhalation in doses that do not affect locomotion. The inhalation decreased the ratio of 5-HIAA/5-HT in the hippocampus (but not striatum or prefrontal cortex) but did not change that of DOPAC/DA. The authors concluded that the anxiolytic effect of benzyl benzoate (as well as *C. odorata* EO) might be based on the 5-HTnergic and, indirectly, on DAnergic pathways (Zhang et al., 2016).

Safranal **26** showed anxiolytic effects in doses devoid of effects on motor activity or coordination (Hosseinzadeh and Noraei, 2009), compatible with the hypno-sedative profile better discussed in Section 11.6.2.1.

Thymoquinone **27** showed anxiolytic-like activity in unstressed mice, but only higher doses in stressed mice (Gilhotra and Dhingra, 2011). Because thymoquinone attenuated the immobilization-induced increase in plasma nitrite levels and a decrease in GABA content, the authors suggested these effects are related to antianxiety-like effect in stressed mice.

Dehydrofukinone **29** induced sedation in catfish (less validated than the zebrafish) with a concentration 20-fold lower than that required for anesthesia in the same model; **29** reversed the acute, but not long-term, rise in cortisol levels in stressed animals. Because the effect was synergic with diazepam and partially reversed by flumazenil, the authors suggested the involvement of $GABA_A$ receptors (Garlet et al., 2016).

Linalool oxide **32** is a minor component of EOs or can be produced by oxidizing linalool. Inhaled linalool oxide showed anxiolytic-like effect in mice models (Souto-Maior et al., 2011), though the inhalation method does not allow accurate estimations of inhaled concentrations.

Limonene epoxide **33** induced sedative/anxiolytic effects sensitive to flumazenil in much smaller doses than the reported to have an (ip) LD50 of 4.0 g/kg; higher doses affected motor coordination (de Almeida et al., 2012). The same group latter corroborated the anxiolytic effect of limonene epoxide in a different model, with single and repeated (14 days) doses, again sensitive to flumazenil. Antioxidant properties were shown *in vitro* (formation of nitrite ion, hydroxyl radical, and reactive substances to thiobarbituric acid) and *in vivo* (decreased lipid peroxidation and nitrites levels and increased catalase and superoxide dismutase activity in the hippocampus), suggesting a combined mechanism of action (de Almeida et al., 2014).

Oleamide **34** is structurally related to the cannabinoid endogenous agonist anandamide. Though a CB1 agonist, its interaction with the cannabinoid system is still not fully understood; reports also show interactions with other neurotransmitter systems, including $GABA_A$, 5HT and DA (Fedorova et al., 2001). In addition, **34** showed an anxiolytic-like effect in several models in stressed (socially isolated) or non-stressed (group housed) mice (Fedorova et al., 2001; Wei et al., 2007).

11.6 HYPNOTIC PROPERTIES OF COMPOUNDS FROM EOs

11.6.1 Assessing Hypnotic Properties

Insomnia refers to disturbance of sleep onset, maintenance, or quality, a condition that affects 6%–20% of the general population in different countries (Ohayon, 2002). Insomnia is considered the second most prevalent mental disorder (Blanken et al., 2019), existing as an independent disorder, a side effect of drugs, or a symptom related to a variety of somatic or mental disorders (Riemann et al., 2017). Though acute insomnia may arise from acute stress or cognitive or emotional increased arousal, a chronic continuation of sleep troubles contrasting with the sleep history of the afflicted patient is of considerable morbidity and prone to treatment.

Hypnotics are drugs that induce drowsiness and facilitate the onset and maintenance of a sleep state from which the recipient can be easily aroused (Mihic and Harris, 2011). An ideal hypnotic would induce sleep quickly, maintain the sleep architecture similar to the physiologic pattern, and be devoid of after effects upon waking.

11.6.1.1 Methods

The most common test for identifying a hypnotic effect is the potentiation of barbital-induced sleep in mice: different doses of the test compound are given in combination with a fixed dose of pentobarbital; the latency to sleep and total sleep time are measured. Sleeping time is defined by the time elapsed between the loss and recover of the righting reflex. EEG analysis allows for a detailed analysis of sleep architecture, including times of rapid eye movement (REM) and non-REM (NREM) and rounds of wakefulness.

11.6.1.2 Mechanisms of Action

GABA receptors play a crucial role in sleep. $GABA_A$ α1 subunit is associated with sedation, the α2/3 subunits with anxiety, and the α5 with temporal and spatial memory (Draguhn et al., 1990). Benzodiazepines promote the binding of GABA to the $GABA_A$ subtype of GABA receptors. Novel receptor agonists, referred to as Z compounds, are agonists on the benzodiazepine site of the $GABA_A$ receptor, selective to $GABA_A$ receptors that contain the α1 subunit.

The melatonergic and cannabinoid systems are also part of the endogenous sleep-promoting systems and represent new drug targets.

11.6.2 Hypnotic Properties of EO Compounds

An overview of the hypnotic properties of the compounds discussed in this chapter is presented at Table 11.5. Phytol **4** did not show an effect on barbital sleep (Costa et al., 2014). α-Asarone **16** improved the quality of sleep in normal and sleep-deprived rats (Radhakrishnan et al., 2017). Limonene **19** potentiated barbital sleep only in the highest dose tested, whereas muscle relaxation was already present in the lowest (do Vale et al., 2002). γ-Decanolactone **25** potentiated barbiturate sleep when given ip or po, in doses much smaller than the LD50 (Coelho de Souza et al., 1997). No additional data on the hypnotic properties was found for any of these compounds.

α-Pinene **17** administered orally was compared to zolpidem using electrophysiology (vigilance states, rapid eye movement sleep [REMS] and non-REM sleep [NREMS]), molecular modeling (homology and docking), and behavior (potentiation of pentobarbital sleep) (Yang et al., 2016). Compatible with the hypno-sedative data reported with inhaled **17**, its oral administration prolonged pentobarbital sleep and decreased its latency. Both effects seem to be dose-dependent and were reversed by flumazenil. Neither **17** nor zolpidem affected REMs. The analysis of time course of NREMS, REMS, and Wake for 24 h showed that unlike zolpidem, α-pinene induced NREMS without causing adverse effects after sleep induction. **17** and zolpidem increased the number of state transitions (Wake to NREMS and NREMS to Wake); no effects were seen on stage transitions (NREMS to REMS or REMS to NREMS). This sleep architect analysis shows that α-pinene enhances the quantity of sleep, decreasing sleep latency and increasing NREMS time. EEG power density analysis suggests that, unlike zolpidem, α-pinene **17** increased the quantity of sleep without compromising sleep intensity.

The combined evidence from EEG (where flumazenil completely abolished **17** hypnotic effect), patch clamp experiments (**17** increased the decay time constant of inhibitory post synaptic currents in hippocampal CA1 pyramidal neurons), and docking analysis demonstrated that **17** acts as partial modulator at the BZD binding site. The sleep quantity, but not intensity, was positively correlated with efficacy, potency, and binding energy. Authors conclude that though α-pinene

TABLE 11.5
Hypnotic Properties of Compounds from Essential Oils

Compound	Behavioral models					Mechanism of action				Active dose (mg/kg)	Via	Use
	Barbital latency	Barbital time	Motor coor- dination	Catfish	EEG	GABAa	GLU	5HT	Eletrophysiology			
Phytol **4**	Y	N	N	–	–	Y	–	–	–	75	ip	N
α-Asarone **16**	–	–	–	–	Y	–	–	–	–	10–80	ip	Y
α-Pinene **17**	Y	Y	–	–	Y	Y	–	–	–	25–100	po	N
Limonene **19**	Y	Y	–	–	–	–	–	–	–	200	ip	Y
Decursinol angelate **21**	Y	Y	–	–	Y	Y	–	–	–	10–50	po	Y
γ-Decanolactone **25**	–	Y	Y	–	–	–	–	–	–	300–400 500	ip po	N
Safranal **26**	Y	Y	–	–	Y	–	–	–	–	180–360 0.05–0.35 mL/kg	po ip	Y
Dehydrofukinone **29**	–	–	–	Y	–	–	–	–	–	5–50 mg/L	Water	N
Oleamide **34**	–	–	–	–	Y	Y	–	–	Y	4.1–20 μM 10 2.8–5.6 μg 25 μg 50–200	ip icv icv ip	N

Notes: Y, positive effects; N, negative effects; –, no data.
Abbreviations: EEG, eletroencephalogram; GABAa, g-amynobutiric acid; receptor subtype A; GLU, glutamate; 5HT, serotonin; Use, compatible traditional use; Ip, intraperitoneal; icv, intracerebroventricular; po, per oz; d, days.

showed lower binding energy and efficacy than zolpidem, when considering sleep quantity and intensity, the effectiveness of monoterpene **17** for sleep seems to be better than zolpidem.

A rather detailed study showed that decursinol angelate **21** orally given to mice dose dependently suppressed spontaneous locomotion and potentiated latency and duration in hypnotic and sub-hypnotic doses of pentobarbital-induced sleep (Woo et al., 2017). Noteworthy, **21** increased the number of animals that slept with sub-hypnotic pentobarbital treatment. EEG analysis in rats showed that **21** modified sleep architecture, reducing sleep/wake cycles, increasing non-rapid eye movement (NREM) and rapid eye movement (REM). In addition, **21** increased intracellular Cl^- influx in primary cultured of rat hypothalamic cells, resulting in hyperpolarization and non-selectively activated the subunits of $GABA_A$ receptors and markedly increased protein expression of GAD65/67, a key enzyme in GABA synthesis. The body of data is consistent with hypnosis induced trough the GABAergic system.

Safranal **26** potentiated the effects of hypnotic and sub-hypnotic doses of pentobarbital in mice, shortening the latency for and increasing the duration of NREM sleep (Hosseinzadeh and Noraei, 2009). Safranal per se did not affect baseline sleep wave profiles (Liu et al., 2012). The analysis of c-fos expression suggested that safranal activated the sleep-promoting neurons in the ventrolateral preoptic area (critical for NREM sleep promotion), with consequent inhibition of the wakefulness-promoting neurons in the tuberomammillary nucleus.

Oleamide **34** is part of a brain lipid family that induces sleep (Cravatt et al., 1995). Systemic and icv administration of **34** to rats and mice markedly decreased sleep latency without changing other sleep parameters (Basile et al., 1999; Mendelson, 2001). When the major metabolizing enzyme for **34** (fatty acyl amide hydrolase, type I) is administered intraventricularly, sleep latency is reduced and total sleep is increased (Mendelson, 2001). In one study, acute and the subchronic (15 days) centrally administered oleamide increased REMS in rats, without affecting waking and NREM sleep. No abstinence effects on the sleep–waking cycle were observe after drug cessation (Herrera-Solís et al., 2010). Another study found that **50** administered systemically increased slow-wave sleep and decreased wakefulness and sleep latency, with no effects on REM sleep. The oleamide-induced increase in slow-wave sleep was prevented by 5-HT reuptake inhibitors (Yang et al., 2003). Research has found that **34** acts as a full cannabinoid receptor agonist (Leggett et al., 2004), and at least part of its hypnotic action involves the CB1 receptor system (Mendelson and Basile, 1999). A very well thought of and elegant series of experiments generated data to show that the *cis* isomer of oleamide is a stereoselective modulator of voltage-gated Na^+ channel and $GABA_A$ receptor complex (Verdon et al., 2000). The mechanism is consistent with the hypno-sedative effects observed *in vivo* for *cis*-oleamide and its anticonvulsant effects (Verdon et al., 2000). Therefore, **34** effects on CB1 are compounded with other properties, such as the allosteric modulation of other receptors and fatty acid amide hydrolase inhibition.

11.7 COMPOUNDS REPORTED TO POSSESS ANTICONVULSANT EFFECTS

Epilepsy is characterized by the propensity of the brain to have abnormally and spontaneously synchronized neuronal discharges (Schulze-Bonhage, 2017). These discharges result in transient brain dysfunction (seizures) that may occur in a variety of ways (with or without motor manifestations, with or without loss of consciousness, etc.), depending on the brain areas affected or involved (Fisher et al., 2014). Epilepsy has miscellaneous etiologies and affects an estimated 50 million people worldwide (Berg et al., 2010). Though anticonvulsive drugs, alone or in combination, allow most patients to remain seizure free, an estimated 30% are refractory to existing treatments ("difficult to treat epilepsy" or "refractory epilepsy"). Even with the introduction of about 15 new (second-generation) anticonvulsant drugs in the clinic since 1989 (Reimers et al., 2018), there are more advantages in terms of tolerability (relevant in any chronic disease) than in efficacy (Schulze-Bonhage, 2017). Among the reasons for lack of advancement in this area are the candidate selection process and the role of the most commonly used screening models.

The classic experimental models used in the search for new drugs are not models of chronic epilepsy but of acute convulsions induced by electric current (such as the maximum electroshock, MES) or convulsive agents (especially pentylenetetrazol, PTZ).

11.7.1 Assessing Antiepileptic Properties

11.7.1.1 Methods

The two most commonly used seizure models in the search for anticonvulsant properties of drug candidates are the maximal electroshock (MES) and the pentylenetetrazol (PTZ)-induced seizures (Löscher, 2011). The validated MES model is the induction of clonic-tonic convulsions by a short-duration (0.2 second) electrical stimulus (50 mA in mice and 150 mA in rats) transcorneal (or transauricular). The parameter to be observed is the tonic extension of the hind limbs, typically during 30 min after electroshock induction. Drugs that prevent the onset of convulsions thus induced may have clinical utility for generalized tonic-clonic seizures. However, some of the newer anticonvulsants with significant clinical value (vigabatrin, tiagabine, and leviracetam) have not been identified by these models (false negatives), raising justifiable criticism (Löscher and Schmidt, 2011).

The validated PTZ model is the induction of clonic seizures of at least 5 s duration within 30 min of PTZ administration, used at a dose that induces such seizures in at least 97% of the non-drug treated animals (control). This test has predictive value for non-convulsive epileptic seizures (absence or myoclonic), although several drugs useful for this purpose had no effect in this test (false negatives). Drugs that protect animals from MES and PTZ, i.e., inhibit the onset of seizures (tonic extension of the hind limbs for MES and 5 s clones for PTZ) are considered active in these models. Latency for the first seizure is not a validated parameter, being highly variable and poorly reproducible (Swinyard and Kupferberg, 1985; Löscher, 2011).

New candidates for anticonvulsant/antiepileptic drugs are also tested in different seizure (bicuculine, picrotoxin, strychnine, quinolinic acid, kainate, NMDA, etc.) or epilepsy (kindling, pilocarpine status epilepticus, etc.) models and animal species. It is necessary at a later stage to consider long-term studies with chronic administration of the candidate, especially since tolerance is relatively common among anticonvulsant drugs.

11.7.1.2 Mechanisms of Action

Drugs used in the management of epilepsy most often increase the ability to inhibit abnormal neuronal activity, by acting on Na, K, Ca, or channels and/or enhancing GABA-mediated neuronal inhibition. There still gaps in the understanding on the mechanism of action of the newer anticonvulsant agents that include GABA uptake inhibitors, GABA metabolism inhibitors, antagonists of glutamate receptors (NMDA and AMPA/kainate), binding to the synaptic SV2A and $\alpha 2\delta$ proteins (implicated in neurotransmitter release), modulation of hyperpolarization-activated cyclic nucleotide–gated channels, and inhibitors of brain carbonic anhydrase (that modulates neuronal excitability) (McNamara, 2011). Finally, cannabinoids are known to have pro- and anticonvulsive effects. Overall, it seems that cannabidiol is mostly devoid of the psychoactive effects associated with the endocannabinoid system and is a promising well-tolerated therapeutic agent for the treatment of seizures, particularly in patients with treatment resistant forms of epilepsy. Significant gaps still exist on the mechanism, safety, and efficacy of cannabinoids, especially for long-term use (Rosenberg et al., 2015).

It is argued that instead of focusing on the synaptic aspects involved in seizures, it would be useful to develop models that target the epileptogenic tissue and the process that transforms normal tissue into epileptic (such as structural changes, synaptic plasticity, cellular homeostasis, glial changes, changes in the blood–brain barrier, and/or inflammatory processes) (Schulze-Bonhage, 2017).

11.7.2 Anticonvulsant Properties of EO Compounds

An overview of the anticonvulsant properties of the compounds discussed in this chapter is presented at Table 11.6. Phytol **4** showed anticonvulsant effects, reversed by flumazenil, in one mouse model (Costa et al., 2012). Borneol **7** (ip) protected mice from PTZ and MEs convulsions; the effect at PTZ were blocked by flumazenil (Quintans-Jr et al., 2010). Safranal **26** protected mice against PTZ, an

TABLE 11.6
Anticonvulsant Properties of Compounds from Essential Oils

Compound	Models											Mechanisms of action					Active dose (mg/kg)	Via	Use
	PTZ				MES	PILO	KA	QA	NMDA	6hz	EEP	GABAa	GLU	INF	OS	APO			
	ip/sc	iv	KD	ZF															
Phytol **4**	–	–	–	–	–	Y	–	–	–	–	–	N	–	–	–	–	25–75	ip	N
Citronellol **6**	Y	–	–	–	Y	–	–	–	–	–	Y	–	–	–	–	–	400	ip	N
Borneol **7**	Y	–	–	–	–	–	–	–	–	–	–	Y	–	–	–	–	50–200	ip	N
Linalool **9**	Y	–	Y	–	Y	–	–	Y	Y	–	–	N	Y	–	–	–	0.3–5 μM, 0.3–1 mM, 50–400	icv ip	Y
Nerolidol **10**	–	–	Y	–	–	–	–	–	–	–	–	–	–	–	Y	–	12.5–50d/21d	ip	N
Carvacrol **13**	Y	–	–	–	Y	–	–	–	–	–	–	N	–	–	–	–	200	ip	N
Eudesmin **14**	Y	–	Y	–	Y	Y	–	–	–	Y	–	Y	Y	–	Y	Y	5–20 5–20/d × 21d	ip	N
Methyleugenol **15**	–	–	–	–	–	–	–	–	–	–	Y	–	–	–	–	–	EC50 367 μM	#	N
α-Asarone **16**	N	–	–	–	N	Y	Y	–	N	–	–	Y	Y	Y	Y	–	30–60 50–200 × 2d/28d 100	ip po ip	Y
Caryophyllene **20**	N	–	–	–	Y	–	Y	–	–	–	–	–	–	–	Y	–	30 50–100/d × 7d	ip ip	N
γ-Decanolactone **25**	Y	–	–	–	Y	–	–	–	–	–	–	–	Y	Y	–	–	250–600 2000–3000	ip po	Y
Safranal **26**	Y	–	–	–	–	–	–	–	–	–	–	Y	–	–	–	–	145–291	ip	Y
Thymoquinone **27**	N Y	–	Y	–	N Y	Y	Y	–	–	Y	–	–	–	Y	Y	–	10–20/100 10 200–400μM	ip po icv	N
Ar-Turmerone **28**	–	–	Y	Y	–	–	–	–	–	Y	–	N	–	Y	Y	–	46 μM 05–50	## ip	Y
Dehydrofukinone **29**	N	–	–	–	–	–	–	–	–	–	Y	Y	–	–	–	–	10–100 μM	#	N
Curzerene **30**	Y	–	–	–	–	–	–	–	–	–	–	Y	–	–	–	–	0.1–0.4	ip	N
Epoxycarvone **31**	Y	–	–	–	Y	–	–	–	–	–	Y	Y	–	–	Y	–	300–400	ip	N
Oleamide **34**	–	–	–	–	–	–	Y	–	–	–	Y	Y	–	–	–	Y	0.5–10 × 4d 64 μM	po	N

Notes: Y, positive effects; N, negative effects; **–**, no data.

Abbreviations: PTZ, pentiletetrazol; KD, kindling; ZF, zebra ish larvae; MES, maximum eletroshock; PILO, Li-pilocarpine; KA, kainate; QA, quinolinic acid; NMDA, n-methyl-aspartate; EEP, electrophysiology; GABAa, GABA receptor subtype A; GLU, glutamate; INF, inflammation; OS, oxidative stress; APO, apoptosis; Use, compatible traditional use; d, days; ip, intraperitoneally; po, per oz; #, *in vitro*; ##, per well.

effect reversed by flumazenil (Hosseinzadeh and Sadeghnia, 2007). No further corroboration was found in the literature for these preliminary data.

Citronellol **6** protected 80% of mice from MES and PTZ seizures; it dose-dependently decreased the amplitude of action potential (no effect on repolarization) in rat sciatic nerve preparation, suggesting that this was at least one component of the mechanism of anticonvulsive action (de Sousa et al., 2006).

Linalool **9** showed anticonvulsant properties (ip and icv) in various experimental models of convulsion and/or epilepsy in mice, including PTZ-kindling and those related to the glutamatergic system (NMDA- and quinolinic acid-induced convulsions) (Elisabetsky et al., 1999; Elisabetsky and Brum, 2003; McNamara, 2011; Rosenberg et al., 2015). Neurochemical analyzes identified a complex anticonvulsive mechanism of action, where linalool acts primarily as a modulator of glutamate transmission: antagonist to NMDA receptors (glutamate binding) and noncompetitive antagonist of dizocilpine (NMDA antagonist) in cortical membranes (Elisabetsky et al., 1999), reducing potassium-stimulated (but not basal) glutamate release and glutamate uptake (Brum et al., 2001). Linalool does not seem to interfere with $GABA_A$ (muscimol binding or GABA release) (Brum et al., 2001). Leal-Cardoso and colleagues (Leal-Cardoso et al., 2010), using sciatic nerve preparations and intact neurons dissociated from the dorsal root ganglion, found that linalool reversibly blocks sciatic nerve excitability in a dose-dependent manner, decreasing the amplitude of action potential, and blocking the generation of action potentials without affecting resting membrane potential. Further investigation is needed to reveal if these effects observed in peripheral nerves are relevant to the anticonvulsant activity characterized *in vivo* for systemic linalool.

Nerolidol **10** suppressed the progression of PTZ-kindling (reduced seizure severity), comparable to sodium valproate in the highest studied dose. The kindling-induced shifts in the neurochemical parameters include the depletion of NE, DA, and 5-HT; enhanced activity of acetylcholinesterase; and oxidative stress. Nerolidol restored levels of amines and reduced the oxidative stress (decreased TBARS, nitrite levels, GSH levels, and CAT activity) in the cortex and hippocampus (Kaur et al., 2016). As depression and memory deficits are major challenges for epileptic patients, the effects of **10** on TST and inhibitory avoidance (step down) were investigated: nerolidol reduced depressive-like behavior and the memory deficit induced by PTZ-kindling. These protective effects are likely to be associated with the antioxidant property of nerolidol, and its inhibition of kindling-induced increased acetylcholinesterase activity.

Eudesmin **14** protected mice against PTZ and MES, in doses that diminished ambulation and potentiated pentobarbital sleep. In PTZ-kindling epileptic rats, **14** increased GABA and decreased GLU (whole brain), gradually decreasing to control levels. Eudesmin upregulated the expressions of GAD65, GABAA, and Bcl-2 and downregulated caspase-3, all altered in PTZ-kindled rats (Liu et al, 2015). The modulation of GLU and GABA and the antiapoptotic effects are of interest for long-term antiepileptic effects (Jiang et al., 2015).

Methyleugenol **15** was characterized as a $GABA_A$ receptor activator means of electrophysiology; **15** induced concentration-dependent GABA-mediated, Cl^--permeable current in cultured hippocampal neurons and significantly sensitized GABA- (but not glutamate- or glycine-) induced currents; these currents were inhibited by picrotoxin ($GABA_A$ channel blocker) and bicuculline ($GABA_A$ competitive antagonist) but not strychnine (glycine receptor inhibitor). Though no *in vivo* tests were found, the data show that methyleugenol potentially regulates neuronal excitability in the CNS, which is of relevance to counteract epileptic seizures (Ding et al., 2014).

Chellian and colleagues (2017) review the pharmacological effects of asarones, based on indexed references between 1960 and 2017. Reported pharmacological effects include antidepressant, anxiolytic, anti-Alzheimer's, antiparkinsonian, antiepileptic, anticancer, antihyperlipidemic, antithrombotic, anticholestatic, and radioprotective activities across multiple molecular targets, which still require confirmation and characterization. The pharmacokinetic assessment shows poor bioavailability and short plasma half-life in rodents, with metabolism mainly through cytochrome P450. Of note, toxicological studies suggest that α- (and β-) asarone can cause hepatomas, mutagenicity, genotoxicity, and teratogenicity.

α-Asarone **16** did not fully protect mice against PTZ-induced and picrotoxin ($GABA_A$ antagonists) or MES seizures (Chen et al., 2013; Huang et al., 2013); on the contrary, it potentiated picrotoxin-induced seizures in rats (Sharma et al., 1961). Non-toxic doses of **16** (60 mg/kg and less) were ineffective against GABAergic and glutamatergic-seizures or MES (Pages et al., 2010), but protected mice against magnesium deficiency-dependent audio seizures, a model that responds to antioxidant/anti-inflammatory compounds (Pages et al., 2010). Moreover, **16** was effective in the lithium-pilocarpine model in rats (Huang et al., 2013; Miao et al., 2013). In this study, α-asarone, in a brain concentration of 34.9 ± 2.1 μg/mL (in line with the *in vitro* dose range), attenuated learning and memory deficits and neuroinflammation induced by pilocarpine. The protection was attributed to NF-κB pathway inhibition in microglia, suggesting that **20** may be useful to temporal lobe epilepsy and other conditions in which microglia-mediated neuroinflammation is a compounding trigger (Liu et al., 2017). Though positive neuromodulation of **16** on $GABA_A$ receptors was verified by electrophysiology (Chellian et al., 2017), indicating the GABAergic system as a target, there was no inhibition of GABA uptake or GABA transaminase (Wang et al., 2014) and no effects on muscimol or flunitrazepam binding. α-Asarone protected kainate- but not NMDA-induced seizures in mice (Pages et al., 2010; Huang et al., 2013). Binding studies confirmed the antagonistic effect of α-asarone at NMDA receptors (Cho et al., 2002).

In summary, data indicate that α-asarone might be useful as adjuvant to synthetic anticonvulsant activity, its effects resulting from a complex mechanism of action that includes activation of the GABAergic system, increased glutamate uptake, antagonism of the NMDA receptor, tonic inhibition of the Nav1.2 chains, and antioxidant and anti-inflammatory properties (Pages et al., 2010; Huang et al., 2013; Wang et al., 2014; Liu et al., 2015; Chellian et al., 2017).

β-caryophyllene **20**, a CB2 agonist, was active against MES and kainite-induced status epilepticus, but not against PTZ. The attenuation of seizure intensity in the kainite model was accompanied by suppression of the status epilepticus-induced lipid peroxidation in the hippocampus (Tchekalarova et al., 2018).

γ-Decanolactone **25** ip or po was effective against PTZ in mice, and also against MES when given ip. The effects were dose-dependent, and the ED50 was almost fourfold lower than the LD50 (Galdino et al., 2012); **25** dose-dependently inhibited the binding of glutamate to rat cortical membranes, with 96.8% inhibition obtained with 5 mM (Pereira et al., 1997). The anticonvulsant activity was confirmed in the mouse PTZ-kindling model, where **25** attenuated severity and seizure progression, accompanied by decreased neuroinflammation and damage to brain DNA possibly related to its antagonism to NMDA-like glutamatergic receptors (Viana et al., 2007).

Among the PAVs that act as anticonvulsants, thymoquinone seems to be the best studied. Its effects were replicated in several laboratories and various models, including those with higher translational value. Results were associated with neurochemical substrates, and a couple of preliminary clinical observations are available.

Thymoquinone **27** administered icv to rats provided partial protection against PTZ, reversible by flumazenil (Hosseinzadeh et al., 2005); **27** showed partial protective effects against PTZ and MES and lower the ED50 for valproate, which could be useful to minimize valproate adverse effects (Raza, 2006). Thymoquinone protected mice against PTZ-kindling model, which induced increased NO production, oxidative stress, and cerebral glutamate levels. Mice submitted to PTZ-kindling exhibit cognitive deficits in the Water Maze and inhibitory avoidance. In the same study, the gene expression and NO protein synthetase isoforms were examined (RT-PCR and histochemistry). The authors concluded that thymoquinone attenuates cognitive deficits, oxidative stress, and overproduction of NO (Abdel-Zaher et al., 2017); **27** attenuated the incidence and severity of seizures and cognitive deficits induced by lithium-pilocarpine (Shao et al., 2016, 2017). The effects are likely to be related to its antioxidant (Nrf2, HO-1, and SOD) and anti-inflammatory effects (decreased expression of TNF-α and COX-2) assessed in cortex and hippocampus. It is noteworthy that beneficial effects of thymoquinone were seen in the lithium-pilocarpine and PTZ-kindling models, of higher translational value.

Temporal lobe epilepsy is one of the most common among the "difficult to treat" epilepsies, with almost a third of patients without adequate response to available treatments. In a rat intrahippocampal kainate experimental model of temporal lobe epilepsy, orally given thymoquinone decreased epileptic activity, reducing the increase in malondialdehyde (but not nitrite and nitrate) levels and SOD activity. Of note, **27** attenuated the hippocampal CA1 and CA3 neuronal loss and the signs of damage to the dentate nucleus, possibly suggestive of neuroprotective activity against the epileptic insult. The neuroprotection may be associated with the antioxidant and anti-inflammatory activities observed for this compound in other studies (Dariani et al., 2013).

Though appropriate clinical trials are required for a reliable definition of the therapeutic profile of this compound, a pilot trial yielded promising results (Akhondian et al., 2011). In this pilot double-blinded crossover clinical trial with children with refractory epilepsy, thymoquinone was administered as an adjunctive therapy and its effects on frequency of seizures compared with placebo: adjunctive thymoquinone (0.5 mg/kg for 4 weeks) was well tolerated and effective to a statistically and clinically significant extent. Relevant to the translational potential of **27** as an antiepileptic drug, though a complete and critical long-term toxicological study is not yet available, anti-inflammatory effects were observed without major toxic or adverse effects (Hosseinzadeh et al., 2005; Raza, 2006; Alyoussef and Al-Gayyar, 2016; Gökce et al., 2016; Abdel-Zaher et al., 2017). Mice given up to 90 mg/kg (in drinking water) for 90 days did not show meaningful toxicological changes (Badary et al., 1998).

Ar-turmerone **28** was active in a PTZ model in zebrafish larvae (Orellana-Paucar et al., 2012), the iv PTZ perfusion test in mice, and the 6 Hz psychomotor model (NMRI mice) that models partial epilepsy (Orellana-Paucar et al., 2013). In addition, **28** modulated two seizure-related genes (*c-fos* and *BDNF*) expression patterns in zebrafish. There were no effects on motor function and/or balance with doses 500 times higher than the effective dose in the 6 Hz model. The quantification of **28** in brain revealed good absorption after ip injection and the ability to cross the BBB and to stay in the brain for a suitable time. Authors speculate that the protective effect against PTZ-induced lipid peroxidation and oxidative stress-induced DNA damage (Ribeiro et al., 2005) and the anti-neuroinflammatory effects of ar-turmerone would contribute to reduce the neuronal loss that accompanies convulsions, as in the case of zonisamide (Willmore, 2005).

Electrophysiology studies in cortical synaptosomes showed that dehydrofukinone **29** modulates synaptosomal membrane potential and depolarization-evoked calcium influx; the effects were sensitive to flumazenil. *In vivo* evaluation showed that **29** was ineffective against PTZ (Garlet et al., 2017).

Smyrnium cordifolium is used in Iranian traditional medicine for the treatment of anxiety and insomnia; curzerene **30** is the main component (65.26%) of its EO. It has been shown that **30** afforded 100% protection against PTZ at 0.4 mg/kg with an ED50 of 0.25 ± 0.09 mg/kg. Flumazenil and naloxone diminished curzerene protection against PTZ in circa 50%, suggesting participation of the GABAergic and (kappa) opioid receptors in **30** anticonvulsant activity (Abbasi et al., 2017).

Epoxycarvone **31** anticonvulsant activity seems to involve the GABAergic system and its ability to reduce neuronal activity through voltage-dependent Na^+ channels (de Almeida et al., 2008). Its antioxidant activity may also be of relevance. Four stereoisomers identical to the natural compound were tested: as expected, no significant differences were found in comparison to epoxycarvone (Salgado et al., 2015).

Oleamide **34** prevented the kainate-induced seizures excitotoxic damage, a model of temporal lobe epilepsy. Neurochemical analysis suggests that oleamide neuroprotection against excitotoxicity-induced neuronal death and behavioral seizures is, at least in part, via inhibition of the calpain protease (Nam et al., 2017); **34** had no effects on epileptiform discharges induced by 4-aminopyridine (4AP) in slices of rat hippocampus and visual cortex (Dougalis et al., 2004).

11.8 ANTIDEPRESSANT PROPERTIES

Depression is one of the top diseases in the global disease burden rank, estimated to affect as many as 350 million people worldwide (Ledford, 2014). It is generally subdivided into major and bipolar depression, in which antidepressant drugs are clinically employed.

11.8.1 Assessing Antidepressant Properties

11.8.1.1 Methods

The facts that depression is multifactorial; presents affective, cognitive and homeostatic abnormalities; is presented in enormously varied ways; and lacks a clear biomarker present major difficulties for validating animal models (see Berton et al., 2012; Nestler and Hyman, 2010 for excellent discussions).

The forced swimming (FST) and tail suspension (TST) tests are considered tests of predictive value and are based on "behavior despair," in which animals under the acute effect of antidepressants increase the time struggling to escape (Nestler and Hyman, 2010). The unpredictable chronic mild stress (UCMS) is considered a model with construct and face validity, though it is time consuming and of difficult replication (internal and external); more attuned with the clinic, it responds only to chronic antidepressant treatment. A major problem with the UCMS is the enormous variability in protocols, complicating the comparison of data obtained in different laboratories with a given compound and/or efficacy among different compounds (Strekalova and Steinbusch, 2010). The most common observation endpoints include body weight change, sucrose preference (or the surrogate splash test), and changes in fur, all of which can be associated with anhedonia.

11.8.1.2 Mechanisms of Action

The second-generation antidepressants include the selective serotonin reuptake inhibitors (SSRIs) and the selective serotonin-norepinephrine uptake inhibitors (SNRIs). Relatively selective norepinephrine uptake inhibitors are also in the market. First-generation antidepressants include the monoamine oxidase inhibitors (MAOIs) and the tricyclic antidepressants (TCAs), which enhance monoaminergic transmission either by inhibiting its intraneuronal metabolism or its reuptake from the synaptic cleft, respectively. Second-generation agents have fewer side effects than the older agents, though limitations in efficacy and extended latency for symptoms amelioration are common to all (Shelton and Od, 2011).

11.8.2 Antidepressant Properties of OE Compounds

An overview of the antidepressant properties of the compounds discussed in this chapter is presented at Table 11.7. Methyleugenol **15** did not show meaningful effects in the FST (Norte et al., 2005). In a model of neuropathic pain (L5 and L6 spare nerve injury, SNI), limonene **19** decreased the SNI-induced increase in immobility in the FST (Piccinelli et al., 2015). No additional data are available.

Geraniol **2** attenuated the decreased body weight gain and sucrose preference and increased FST and TST immobility time and corticosterone levels induced by the UCMS. Geraniol and fluoxetine reversed and/or minimized the UCMS-induced neuroinflammation in the prefrontal cortex of mice Deng et al., 2015b).

Linalool **9** showed antidepressant effects in mice and rats (Nestler and Hyman, 2010; Strekalova and Steinbusch, 2010; Berton et al., 2012; dos Santos et al., 2018). The effect is consistent with some linalool-bearing species used traditionally, such as the Mexican Mazahua people who use *L. glaucescens* along with other species to prepare a decoction used to treat sadness, nervousness, anger, and *susto* ("fright") (Guzmán-Gutiérrez et al., 2012). Serotonin postsynaptic 5-HT1A receptors and $\alpha 2$ adrenergic receptors were implicated in the antidepressant-like effects (Guzmán-Gutiérrez et al., 2015).

Thymol **12** presented consistent antidepressant-like effects in mice UCMS. All behavioral parameters during UCMS and the immediately followed TST and FST were attenuated by thymol. Thymol restored the UCMS-induced decrease in 5HT and NOR, excessive cortisol, and inhibited the activation of proinflammatory cytokines (Deng et al., 2015a).

Oral carvacrol **13** showed antidepressant-like effects in mice FST and TST. This antidepressant-like effect was prevented by SCH23390 or sulpiride, but not p-chlorophenylalanine and prazosin, suggesting involvement of the dopaminergic (but not serotonergic and noradrenergic) system (Melo et al., 2011). The effects on FST were not observed in rats with oral higher doses (Zotti et al.,

TABLE 11.7
Antidepressant Properties of Compounds from Essential Oils

Compound	Behavioral models							Mechanism of action					Active dose (mg/kg)	Via	Use
	FST	TST	UCMS					5HT	NOR	DA	Neuro inflam- mation	CB			
			Body Weight	Sucrose Preference	Splash Test	COR	NSF								
Geraniol **2**	–	–	Y	Y	–	Y	–	–	–	–	Y	–	20–40/d × 3weeks	po	N
Linalool **9**	Y	Y	–	–	Y	–	–	Y	Y	N	–	–	30–200	ip	Y
Thymol **12**	Y	Y	Y	Y	–	Y	–	Y	Y	–	Y	–	15–30	po	N
Carvacrol **13**	Y	Y	–	–	–	–	–	N Y	N	Y	–	–	12.5–50	po	N
													12.5/d × 7d	po	
α-Asarone **16**	Y	Y	–	–	–	–	–	Y	Y	–	–	–	15–20	Ip	N
β-Pinene **18**	Y	–	–	–	–	–	–	Y	Y	Y	–	–	54–173 100 × 3	ip ip	N
Limonene **19**	Y[a]	–	–	–	–	–	–	–	–	–	–	–	10/d × 15d	po	N
Caryophyllene **20**	Y	Y	–	–	–	–	Y	–	–	–	–	Y	50	ip	N
Oleamide **34**	Y	–	Y	Y	–	–	–	–	–	–	–	Y	10–20	ip	N
													5/d × 2weeks		

Notes: Y, positive effects; N, negative effects; –, no data.

Abbreviations: FST, forced swimming test; TST, tail suspension test; UCMS, unpredictable chronic mild stress; COR, corticosterone; NSF, novel suppress feeding; 5HT, serotonin; NOR, norepineprhine; DA, dompamine; CB, Cannabinoid; Use, compatible traditional use; ip, intraperitoneal; po, per oz; d, days; [a], after neuropathic pain.

2013). The study showed that a 7-day treatment with doses compatible with the first study increased dopamine and serotonin levels in the prefrontal cortex and hippocampus; the higher acute doses ineffective in the FST increased dopamine (but decreased serotonin) in both brain areas.

α-Asarone **16** produced some effect in FST and TST (Han et al., 2013), involving the noradrenergic (α1 and α2 adrenoceptors) and serotonergic (particularly 5-HT1A receptors) systems.

β-pinene **17** diminished immobility in the FST, though it also decreased ambulation in OF (Guzmán-Gutiérrez et al., 2012). The antidepressive-like effect was replicated and extended by the same group, in which the use of antagonists suggests **17** effects are mediated by $5HT_{1A}$ serotonin receptors (WAY100635), β adrenoceptors (propranolol but not yohimbine), and D_1 dopamine receptors (SCH23390) (Guzmán-Gutiérrez et al., 2015).

β-caryophyllene **20** systemically administered produced antidepressant-like activity in three well-established models with predictive validity, and the use of the CB2 receptor antagonist AM630 provided evidence that the CB2 receptor is involved in mechanism of action of **20** (Bahi et al., 2014).

Oleamide **34**, a CB1 receptor agonist, was effective in the FST; the effects were reversed by CB1 and CB2 antagonists and potentialized by nicotine and scopolamine (Kruk-Slomka et al., 2015). Also, **34** was also effective in the UCMS, partially reversing the lingering effects of UCMS after two weeks cessation (during which treatments were installed) in body weight and sucrose preference (Ge et al, 2015).

11.9 NEUROPROTECTOR PROPERTIES OF COMPOUNDS FROM EOs

Neurodegeneration is considered a multifactorial process. The process of neuronal injury seems to result from a complex interaction of genetic and environmental influences that affect specific neuron populations (Standaert and Roberson, 2011) that are altered in specific disorders. Despite specificities, it has become clear than all brain insults lead to oxidative stress and neuroinflammation, that in an inextricable way feed each other and ultimately contribute to cell death (Khoshnam et al., 2017).

11.9.1 Assessing Neuroprotective Properties

11.9.1.1 Methods

Alzheimer's disease (AD) is often modeled by icv administration of amyloid peptides in rodents, causing measurable memory deficits and changes in neuronal tissue (Saito and Saido, 2018). Parkinson's disease (PD) can be modeled by toxins specific to nigrostriatal neurons (e.g., MPTP), locally administered dopamine toxins (e.g., 6-hydroxydopamine), or other neurotoxins (e.g., rotenone), causing motor disturbances and changes in neuronal tissue (Gubellini and Kachidian, 2015). Various methods of depriving the brain or neuronal cells of oxygen and glucose (oxygen-glucose deprivation, OGD) *in vitro* (cultures), *ex vivo* (brain slices), or *in vivo* (artery occlusion models) mimic ischemia or stroke and provide the possibility to follow cellular events and changes in biochemical pathways that follow (Hansel et al., 2015). Modifications include disturbances of cell calcium homeostasis; depletion of adenine nucleotides; activation of phospholipases, proteases, and endonucleases; and the generation of free radicals (ROS). All of these steps and related pathways are considered potential targets for drug intervention and constitute endpoints in the assessment neuroprotective effects of drugs (Bellaver et al., 2015). Multiple sclerosis (MS) is characterized by inflammation, demyelination, axonal loss, and gliosis. The experimental autoimmune encephalomyelitis (EAE) is the most commonly used MS model (Constantinescu et al., 2011).

Because mitigating oxidative stress and neuroinflammatory responses to brain insults may affect the progression of neurodegeneration from various origins (acute insults or chronic disorders), antioxidant and anti-inflammatory drugs that are effective as such at pertinent brain areas have become relevant in the search for neuroprotectors (Gilgun-Sherki et al., 2002).

11.9.1.2 Mechanisms of Action

Though the symptoms of neurodegenerative disorders are treated with various drugs, few are genuine neuroprotectors. Riluzole, used in amyotrophic lateral sclerosis, is often used as a positive control for neuroprotection; it acts by inhibiting the release of glutamate and blocking postsynaptic glutamate receptors (NMDA and kainate).

11.9.2 Neuroprotective Properties of EO Compounds

An overview of the neuroprotective properties of the compounds discussed in this chapter is presented at Table 11.8. The antioxidant and anti-inflammatory properties of farnesol **3** were the basis for investigating neuroprotective effects of this compound. In a model of inflammation-driven oxidative stress, mice orally treated with **3** convincingly showed attenuated degenerative changes in the cortex and hippocampus of lipopolysaccharide[LPS]-treated mice. The analysis of several complementary parameters suggested that regulation of cellular antioxidant defense systems and intrinsic apoptotic cascade are part of **3** neuroprotective property (Santhanasabapathy and Sudhandiran, 2015).

Geraniol **2** attenuated markers of oxidative stress, neurotransmission, and mitochondrial function in a rat model of acrylamide-induced neuroinflammation (Prasad, 2014a). Oxidative markers were improved in unstressed *Drosophilla* exposed to acrylamide (Prasad, 2014 b).

Borneol **7** protected cortical neurons in primary culture submitted to OGD followed by reperfusion through various signaling pathways, including mitochondria viability, oxidative stress (RO and iNO/NO), neuroinflammation (NF-kB), and apoptosis (caspases 3 and 9) (Liu et al., 2011).

Nerolidol **10** showed protective effects in the rotenone-induced neurodegeneration model of PD (Javed et al., 2016). Rotenone induced a marked state of oxidative stress (reduced SOD, CAT, and GHS and increased MDA), neuroinflammation (increased proinflammatory cytokines IL-1β, IL-6, and TNF-α and mediators COX-2 and iNOS) in rat brain tissues. Consistently, increased activated astrocytes (GFAP), microglia (Iba 1) and loss of dopamine (DA) neurons in the substantia *nigra pars compacta* and dopaminergic nerve fibers in the striatum were observed. As expected from its well-documented antioxidant capacity, nerolidol increased the level of SOD, CAT, and GSH, and decreased that of MDA. Nerolidol also inhibited the release of proinflammatory cytokines and inflammatory mediators and prevented glial cell activation and the loss of dopaminergic neurons and nerve fibers. Overall, data suggest that nerolidol attenuates rotenone-induced dopaminergic neurodegeneration through its antioxidant and anti-inflammatory properties.

General anesthetics are considered candidates to reduce the brain damage that follows ischemic events due to their ability to reduce the brain metabolic demand, antagonize glutamate excitotoxicity, and enhance inhibitory transmission (Kawaguchi et al., 2005). It has been suggested that the neuroprotective properties of the anesthetic propofol against ischemia are also associated with its antioxidant properties (Lee et al., 2005). Thymol **12**, a structural analogue of propofol, showed partial protection against an injury model in cortical neurons culture, though not correlated with its antioxidant capacity (Delgado-Marín et al., 2017). Treatment with **12** and carvacrol **13** reversed the cognitive deficits induced by β-amyloid and scopolamine in doses much smaller than the lethal dose (Azizi et al., 2012). Corroborating its neuroprotective properties, carvacrol reduced the neurological deficits and the infarct volume in a mouse cerebral ischemia/reperfusion model (middle cerebral artery occlusion, 75 min of ischemia and 24 h of reperfusion). The protection was verified up to 2 h after reperfusion when **13** was administered ip, whereas when given icv, it lasted for 6 hours. Results suggested the involvement of the PI3 K/Akt pathway, a critical survival mediator after cerebral ischemia (Zhao et al., 2006), but the roles other pathways cannot be ruled out. Though not investigated in the same study, carvacrol antioxidant properties are likely to be part of the mechanism of action (Guimarães et al., 2010).

α-Asarone **16** effectively ameliorated learning and memory deficits in a systemic LPS neuroinflammation model in mice, which also caused microglial activation, neuronal damage in the hippocampus. The protective effects were mediated through the inhibition of proinflammatory

TABLE 11.8
Neuroprotector Properties of Compounds from Essential Oils

Compound	Insult models													Mechanism of action					Active dose (mg/kg)	Via	Use
	HR culture	HR in vivo		MSI	ACD	LPS	QA	H_2O_2	RTN	ECM	Chemical	Memory deficits		SO	INF	MTC	APO	Glia			
		Beh	IV									SCL	β-AM								
Geraniol **2**	–	–	–	–	Y	–	–	–	–	–	–	–	–	Y	Y	Y	–	–	100/d × 4 weeks	po	N
Farnesol **3**	–	–	–	–	–	Y	–	–	–	–	–	–	–	Y	Y	–	Y	–	100 × 28d	po	N
Borneol **7**	Y	–	–	–	–	–	–	–	–	–	–	–	–	Y	–	Y	–	–	0.001–0.3 μM		N
Nerolidol **10**	–	–	–	–	–	–	–	–	Y	–	–	–	–	Y	Y	–	–	Y	50/d/4 weeks	ip	N
Thymol **12**	–	–	–	–	–	–	–	Y	–	–	–	Y	Y	N	–	–	–	–	100–500 μM	ip	N
																			0.5–2		
Carvacrol **13**	–	Y	Y	–	–	–	–	–	–	–	–	Y	Y	–	–	–	Y	–	25–50	ip	N
																			0.5–2		
α-Asarone **16**	–	–	–	–	–	Y	–	–	–	–	–	–	Y	Y	Y	Y	Y	Y	30 × 3d	po	Y
																			10	ip	
																			10 × 28d	po	
																			10 × 16d	po	
Caryophyllene **20**	–	–	–	–	–	–	–	–	–	Y	–	–	–	Y	Y	–	–	Y	50(2 × /day)	po	N
Safranal **26**	–	Y	Y	–	–	–	Y	–	–	–	–	–	–	Y	–	–	–	–	145–291	ip	N
																			72–145	ip	
																			0.5/d × 30d	ip	
																			363–727	ip	
Thymoquinone **27**	–	–	Y	Y	–	–	–	–	–	–	Y	–	–	Y	Y	–	Y	–	2.5–5 × 3	po	N
																			30 × 3 weeks	po	
																			5	po	
																			50 × 12 weeks	po	
																			20 × month	po	

Abbreviations: HR = hypoxia/reperfusion; Beh = behavior; IV = infarct volume; MSI = mild spinal injury; ACD = acrylamide-induced neurodegeneration; LPS = lypopoly sacharide; QA = quinolinic acid; ECM = encephalo myelitis H2O2 = hydrogene peroxidase, RTN = rotenone; SCL = scopolamine; b-AM = Beta- amyloid peptide; OS = oxidative stress; INF = inflammation; MTC = mitochondria; APO = apoptosis Y = positive effects; N = negative effects; - = no data; d = days; ip = intraperitoneal; po = per oz.

cytokines and microglial activation (Shin et al., 2014). Possibly related to its effects on neurogenesis, it was show that **16** modulates microglial morphological dynamics (Cai et al., 2016). α-Asarone promoted neuroprogenitor cells proliferation *in vivo* and *in vitro* (Mao et al., 2015), elegantly validating the traditional use of *A. tatarinowii*. Authors suggests that **16** is a drug candidate for promoting neurogenesis against cognitive decline associated with aging and neurodegenerative disorders, extending the findings of neuroprotective effects of *Rhizoma acori graminei* extracts against PC12 cells treated with $A\beta_{(1-40)}$ (Irie and Keung, 2003). α-Asarone prevented cognitive deficits, activation of astrocytes (GFAP immunoreactivity), nitrite levels in the hippocampus and temporal cortex, and neural damage to hippocampus CA1 in a rat icv model of induced AD (Limón et al., 2009). Additionally, stimulation of glutamate uptake and inhibition of the glutamate transporter EAAC1-mediated current by α-asarone could contribute to reduced excitotoxicity (Gu et al., 2010). Authors suggest that the combined inhibitory effects on increased intracellular calcium flux, neuroinflammation, and oxidative stress are related to the cognitive status preservation and that **16** should be better scrutinized as a drug candidate to AD.

β-caryophyllene **20** showed convincing evidence of neuroprotection in a murine model (EAE) of MS (Sadeghnia et al., 2013). Administered orally twice daily since the immune attack, β-caryophyllene attenuated the phenotypic and pathological parameters of the model, including *in vivo* and *ex vivo* analysis in brain areas. The study suggests that the mechanisms underlying **20** immunomodulatory effects are linked to its ability to inhibit microglial cells, CD4+ and CD8+ T lymphocytes, as well as protein expression of proinflammatory cytokines. In addition, it diminished axonal demyelination and modulated Th1/Treg immune balance through the activation of CB2 receptor. The multifaceted mechanism of action, adequate to the complex disease the model aims to replicate, together with the body of positive results reported in this study, call for replication and a curve-effect study.

Evidence for a neuroprotective effect of safranal **26** comes from a series of studies performed by one research group. A rat model of transient global cerebral ischemia (four-vessel-occlusion for 20 min) (Sadeghnia et al., 2013) found that **26** significantly reduced hippocampus malonaldehyde (MDA) levels, thiobarbituric acid reactive substances (TBARS), total sulfhydryl (SH) groups, and the antioxidant capacity of hippocampus (FRAP). No direct evidence of reduced infarct size was provided. Additional evidence of neuroprotection came from quinolinic acid-induced oxidative damage to rat hippocampus. Safranal inhibited the quinolinic acid-induced lipid peroxidation and oxidative DNA damage and improved antioxidant and thiol redox status; no direct measure of tissue damaged was investigated (Sadeghnia et al., 2013). Administration of **26** for a month attenuated the increased lipid peroxidation and decreased GSH brain content and restored SOD and GST activities of aging rats (Samarghandian et al., 2015). In a more comprehensive study in a rat model of transient focal ischemia (middle cerebral artery occlusion for 30 min, followed by 24 h of reperfusion), safranal attenuated the neurobehavioral deficit (paw asymmetry), neuronal loss (hippocampus CA1 and CA3 areas), and oxidative stress (increased lipid peroxidation, depletion of total sulfhydryl content, and antioxidant power in the brain tissue). The protective effects of **26** were more apparent in low doses applied up to 6 h after the stroke induction and attributed to its antioxidant effects (Sadeghnia et al., 2017). Because anti-inflammatory effects were reported for **26** (Limón et al., 2009; Gu et al., 2010; Tamaddonfard et al, 2013; Zhu and Yang, 2014), this property could be a contributing mechanism to its neuroprotective effect.

The benefits of thymoquinone **27** in neurodegenerative conditions and brain tumors have been reviewed (Elmaci and Altinoz, 2016). Thymoquinone attenuated the anxiogenic changes in a model of arsenic exposure (11 days). The effects were attributed to the ability of **27** to attenuate hippocampal oxidative stress and proinflammatory cytokine levels induced by the exposure (Firdaus et al., 2018), suggesting that thymoquinone may be useful to counteract episodic arsenic intoxication. Thymoquinone co-administration with toluene prevented most of the histopathological changes induced by chronic toluene exposure (Kanter, 2011). Thymoquinone also attenuated markers of lead-induced neurotoxicity in rats, including apoptosis, excitotoxicity, neurotransmitter alterations, damage to mitochondria, cerebrovascular endothelial cells, astroglia, and oligodendroglia (Radad et al., 2014). In a rat model of transient global ischemia (common carotid arteries occlusion followed by reperfusion), thymoquinone decreased neuronal death in hippocampus CA1 and returned malondialdehyde, glutathione, catalase, and superoxide dismutase to control levels (Al-Majed et al., 2006). In another rat model of cerebral ischemia reperfusion (four-vessel-occlusion method for 20 min), **27** attenuated the lipid peroxidation process during ischemia/reperfusion in rat hippocampus (Hosseinzadeh et al., 2007).

Finally, in a model of mild contusion injury to rat spinal cord, thymoquinone reduced the injury, ameliorated the behavioral score, and decrease the water content in spinal cord tissue. The protective effects of thymoquinone were attributed to its activation of PPAR-γ and the PI3 K/Akt pathways (Chen et al., 2018).

11.10 CONCLUDING REMARKS AND PERSPECTIVES

Though comprehensive, this review does not expect to be all-inclusive. Of the compounds here reviewed, 25 were reported as anxiolytics, 18 as anticonvulsants, 9 as hypnotics, 9 as antidepressants, and 10 as neuroprotectors. Because anxiolytics, hypnotics, and anticonvulsants often share the same mechanism of action, it was somewhat expected that the same compounds were present in these three areas. Only α-asarone **16** is present in the five areas. Perhaps anxiolytics also reflect an area easier to be recognized and described by users of traditional systems of medicine where aromatic medicinal plants are used. The anti-inflammatory activity of medicinal species used for pain and/or inflammation may prompt neuroprotection screening. Certainly, this analysis is biased by whatever was reported instead of a full data set in which all compounds were studied for all areas.

Another limitation in an overall view of the field is that the smaller the studies conducted in a scientific field, the less likely the research findings are to be true (Ioannidis, 2018). Some compounds seem to have been studied once and/or by one research group only, and data should be viewed as preliminary. On the contrary, linalool **9**, asarone **16**, and thymoquinone **27** have fairly substantiated data.

Though pharmacokinetics is beyond the scope of this chapter, differentiation of *in vitro* vs. *in vivo*, ip vs. oral, ip vs. icv, and inhaled vs. systemic effects is necessary. Advantages of inhalation cannot be fully explored until accurate calculations of actual inhaled concentrations in experimental models is provided, or reproducibility and predictability is weakened. Consistent proofs that whatever the effects of interest (receptor modulation, anti-inflammatory, antioxidant, etc.) can be traced to brain areas relevant to the pathophysiology are ultimately needed to define the appropriate drug exposure to produce the desired effect. Experiments including the analysis of effects in biomarkers in brain areas were found almost exclusively in the neuroprotector area.

Though the ratios of ED50s/LD50s are eventually reported, in general, toxicity studies are lacking. Long-term toxicity and tolerance studies will certainly be necessary, especially for epilepsy and neurodegenerative disorders.

If, on one hand, the lipophilic nature of compounds from EOs is advantageous to penetrate the BBB and reach the relevant targets, on the other, poor water solubility presents challenges for pharmaceutical formulation. Improvements in drug delivery, such as attempts to develop complexes (Santos et al., 2017) and/or nanotechnology (Conte et al., 2017) to improve bioavailability, should become increasingly important in the field.

In terms of drug development, accumulating data to compounds studied by different research groups, in different models, and subjacent mechanisms of action, are more likely to succeed if geared to research and development (Wehling, 2009; Wendler and Wehling, 2017).

REFERENCES

Abbasi, N., Mohammadpour, S., Karimi, E., Aidy, A., Karimi, P., Azizi, M., and Asadollahi, K. Protective effects of *Smyrnium cordifolium boiss* essential oil on pentylenetetrazol-induced seizures in mice: Involvement of benzodiazepine and opioid antagonists. *Journal of Biological Regulators and Homeostatic Agents* 2017, 31, 683–689.

Abdel-Zaher, A.O., Farghaly, H.S.M., Farrag, M.M.Y., Abdel-Rahman, M.S., and Abdel-Wahab, B.A. A potential mechanism for the ameliorative effect of thymoquinone on pentylenetetrazole-induced kindling and cognitive impairments in mice. *Biomedicine & Pharmacotherapy* 2017, 88, 553–561.

Akhondian, J., Kianifar, H., Raoofziaee, M., Moayedpour, A., Toosi, M.B., and Khajedaluee, M. The effect of thymoquinone on intractable pediatric seizures (pilot study). *Epilepsy Research* 2011, 93, 39–43.

Alberti, T., Barbosa, W., Vieira, J., Raposo, N., and Dutra, R. (−)-β-Caryophyllene, a CB2 receptor-selective phytocannabinoid, suppresses motor paralysis and neuroinflammation in a murine model of multiple sclerosis. *International Journal of Molecular Sciences* 2017, 18, 691.

Al-Majed, A.A., Al-Omar, F.A., and Nagi, M.N. Neuroprotective effects of thymoquinone against transient forebrain ischemia in the rat hippocampus. *European Journal of Pharmacology* 2006, 543, 40–47.

Alyoussef, A., and Al-Gayyar, M.M.H. Thymoquinone ameliorated elevated inflammatory cytokines in testicular tissue and sex hormones imbalance induced by oral chronic toxicity with sodium nitrite. *Cytokine* 2016, 83, 64–74.

Artursson, P., Palm, K., and Luthman, K. Caco-2 monolayers in experimental and theoretical predictions of drug transport. *Advanced Drug Delivery Reviews* 1996, 22, 67–84.

Azizi, Z., Ebrahimi, S., Saadatfar, E., Kamalinejad, M., and Majlessi, N. Cognitive-enhancing activity of thymol and carvacrol in two rat models of dementia. *Behavioural Pharmacology* 2012, 23, 241–249.

Badary, O.A., Al-Shabanah, O.A., Nagi, M.N., Al-Bekairi, A.M., and Elmazar, M.M.A. Acute and subchronic toxicity of thymoquinone in mice. *Drug Development Research* 1998, 44, 56–61.

Bahi, A., Al Mansouri, S., Al Memari, E., Al Ameri, M., Nurulain, S.M., and Ojha, S. β-Caryophyllene, a CB2 receptor agonist produces multiple behavioral changes relevant to anxiety and depression in mice. *Physiology & Behavior* 2014, 135, 119–124.

Baker, M. 1,500 scientists lift the lid on reproducibility. *Nature* 2016, 533, 452–454.

Basile, A.S., Hanuš, L., and Mendelson, W.B. Characterization of the hypnotic properties of oleamide. *Neuro Report* 1999, 10, 947–951.

Bellaver, B., Souza, D.G., Bobermin, L.D., Gonçalves, C.-A., Souza, D.O., and Quincozes-Santos, A. Guanosine inhibits LPS-induced pro-inflammatory response and oxidative stress in hippocampal astrocytes through the heme oxygenase-1 pathway. *Purinergic Signalling* 2015, 11, 571–580.

Belzung, C., and Lemoine, M. Criteria of validity for animal models of psychiatric disorders: Focus on anxiety disorders and depression. *Biology of Mood & Anxiety Disorders* 2011, 1, 9.

Berg, A.T., Berkovic, S.F., Brodie, M.J., Buchhalter, J., Cross, J.H., van Emde Boas, W., Engel, J. et al. Revised terminology and concepts for organization of seizures and epilepsies: Report of the ILAE Commission on Classification and Terminology, 2005–2009. *Epilepsia* 2010, 51, 676–685.

Berton, O., Hahn, C.-G., and Thase, M.E. Are we getting closer to valid translational models for major depression? *Science* 2012, 338, 75–79.

Blanken, T.F., Benjamins, J.S., Borsboom, D., Vermunt, J.K., Paquola, C., Ramautar, J., Dekker, K. et al. Insomnia disorder subtypes derived from life history and traits of affect and personality. *The Lancet Psychiatry* 2019, 6(2), 151–163.

Bollen, K., Cacioppo, J.T., Kaplan, R.M., Krosnick, J.A., Olds, J.L., and Dean, H. *Report of the Subcommittee on the Replicability in Science Advisory Committee to the National Science Foundation Directorate for Social, Behavioral, & Economic Sciences, 2015.* Retrieved from https://www.nsf.gov/sbe/AC_Materials/SBE_Robust_and_Reliable_Research_Report.pdf

Brooks, S.J., and Stein, D.J. A systematic review of the neural bases of psychotherapy for anxiety and related disorders. *Dialogues in Clinical Neuroscience* 2015, 17, 261–279.

Brum, L.F., Elisabetsky, E., and Souza, D. Effects of linalool on [(3)H]MK801 and [(3)H] muscimol binding in mouse cortical membranes. *Phytotherapy Research* 2001, 15, 422–425.

Buchbauer, G., Jirovetz, L., Jäger, W., Plank, C., and Dietrich, H. Fragrance compounds and essential oils with sedative effects upon inhalation. *Journal of Pharmaceutical Sciences* 1993, 82, 660–664.

Cai, Q., Li, Y., Mao, J., and Pei, G. Neurogenesis-Promoting natural Product α-Asarone modulates morphological dynamics of activated microglia. *Frontiers in Cellular Neuroscience* 2016, 10.

Cao, B., Ni, H.-Y., Li, J., Zhou, Y., Bian, X.-L., Tao, Y., Cai, C.-Y. et al. (+)-Borneol suppresses conditioned fear recall and anxiety-like behaviors in mice. *Biochemical and Biophysical Research Communications* 2018, 495, 1588–1593.

Chellian, R., Pandy, V., and Mohamed, Z. Pharmacology and toxicology of α- and β-Asarone: A review of preclinical evidence. *Phytomedicine* 2017, 32, 41–58.

Chen, Q.-X., Miao, J.-K., Li, C., Li, X.-W., Wu, X.-M., and Zhang, X. Anticonvulsant activity of acute and chronic treatment with a-Asarone from *Acorus gramineus* in seizure models. *Biological and Pharmaceutical Bulletin* 2013, 36, 23–30.

Chen, Y., Wang, B., and Zhao, H. Thymoquinone reduces spinal cord injury by inhibiting inflammatory response, oxidative stress and apoptosis via PPAR-γ and PI3K/Akt pathways. *Experimental and Therapeutic Medicine* 2018, 15(6), 4987–4994.

Cheng, T., Zhao, Y., Li, X., Lin, F., Xu, Y., Zhang, X., Li, Y., Wang, R., and Lai, L. Computation of octanol–water partition coefficients by guiding an additive model with knowledge. *Journal of Chemical Information and Modeling* 2007, 47, 2140–2148.

Cho, J., Kim, Y.H., Kong, J.-Y., Yang, C.H., and Park, C.G. Protection of cultured rat cortical neurons from excitotoxicity by asarone, a major essential oil component in the rhizomes of *Acorus gramineus*. *Life Sciences*. 2002, 71, 591–599.

Clark, D.E. Rapid calculation of polar molecular surface area and its application to the prediction of transport phenomena. 1. Prediction of intestinal absorption. *Journal of Pharmaceutical Sciences* 1999, 88, 807–814.

Coelho de Souza, G.P., Elisabetsky, E., Nunes, D.S., Rabelo, S.K., and Nascimento da Silva, M. Anticonvulsant properties of gamma-decanolactone in mice. *Journal of Ethnopharmacology* 1997, 58, 175–181.

Constantinescu, C.S., Farooqi, N., O'Brien, K., and Gran, B. Experimental autoimmune encephalomyelitis (EAE) as a model for multiple sclerosis (MS): EAE as model for MS. *British Journal of Pharmacology* 2011, 164, 1079–1106.

Conte, R., Marturano, V., Peluso, G., Calarco, A., and Cerruti, P. Recent advances in nanoparticle-mediated delivery of anti-inflammatory phytocompounds. *International Journal of Molecular Sciences* 2017, 18, 709.

Costa, D.A., de Oliveira, G.A.L., Lima, T.C., dos Santos, P.S., de Sousa, D.P., and de Freitas, R.M. Anticonvulsant and antioxidant effects of cyano-carvone and its action on acetylcholinesterase activity in mice hippocampus. *Cellular and Molecular Neurobiology* 2012, 32, 633–640.

Costa, J.P., de Oliveira, G.A.L., de Almeida, A.A.C., Islam, M.T., de Sousa, D.P., and de Freitas, R.M. Anxiolytic-like effects of phytol: Possible involvement of GABAergic transmission. *Brain Research* 2014, 1547, 34–42.

Cravatt, B.F., Prospero-Garcia, O., Siuzdak, G., Gilula, N.B., Henriksen, S.J., Boger, D.L., and Lerner, R.A. Chemical characterization of a family of brain lipids that induce sleep. *Science* 1995, 268, 1506–1509.

Dariani, S., Baluchnejadmojarad, T., and Roghani, M. Thymoquinone attenuates astrogliosis, neurodegeneration, mossy fiber sprouting, and oxidative stress in a model of temporal lobe epilepsy. *Journal of Molecular Neuroscience* 2013, 51, 679–686.

de Almeida, A.A.C., Costa, J.P., de Carvalho, R.B.F., de Sousa, D.P., and de Freitas, R.M. Evaluation of acute toxicity of a natural compound (+)-limonene epoxide and its anxiolytic-like action. *Brain Research* 2012, 1448, 56–62.

de Almeida, A.A.C., de Carvalho, R.B.F., Silva, O.A., de Sousa, D.P., and de Freitas, R.M. Potential antioxidant and anxiolytic effects of (+)-limonene epoxide in mice after marble-burying test. *Pharmacology Biochemistry and Behavior* 2014, 118, 69–78.

de Almeida, R.N., de Sousa, D.P., Nóbrega, F.F. de F., Claudino, F. de S., Araújo, D.A.M., Leite, J.R., and Mattei, R. Anticonvulsant effect of a natural compound α,β-epoxy-carvone and its action on the nerve excitability. *Neuroscience Letters* 2008, 443, 51–55.

De Petrocellis, L., Melck, D., Bisogno, T., and Di Marzo, V. Endocannabinoids and fatty acid amides in cancer, inflammation and related disorders. *Chemistry and Physics of Lipids* 2000, 108, 191–209.

de Sousa, D., Hocayen, P., Andrade, L., and Andreatini, R. A systematic review of the anxiolytic-like effects of essential oils in animal models. *Molecules* 2015, 20, 18620–18660.

de Sousa, D.P., Gonçalves, J.C.R., Quintans-Júnior, L., Cruz, J.S., Araújo, D.A.M., and de Almeida, R.N. Study of anticonvulsant effect of citronellol, a monoterpene alcohol, in rodents. *Neuroscience Letters* 2006, 401, 231–235.

Delgado-Marín, L., Sánchez-Borzone, M., and García, D.A. Neuroprotective effects of GABAergic phenols correlated with their pharmacological and antioxidant properties. *Life Sciences* 2017, 175, 11–15.

Deng, X.-Y., Li, H.-Y., Chen, J.-J., Li, R.-P., Qu, R., Fu, Q., and Ma, S.-P. Thymol produces an antidepressant-like effect in a chronic unpredictable mild stress model of depression in mice. *Behavioural Brain Research* 2015a, 291, 12–19.

Deng, X.-Y., Xue, J.-S., Li, H.-Y., Ma, Z.-Q., Fu, Q., Qu, R., and Ma, S.-P. Geraniol produces antidepressant-like effects in a chronic unpredictable mild stress mice model. *Physiology & Behavior* 2015b, 152, 264–271.

Ding, J., Huang, C., Peng, Z., Xie, Y., Deng, S., Nie, Y.-Z., Xu, T.-L., Ge, W.-H., Li, W.-G., and Li, F. Electrophysiological characterization of methyleugenol: A novel agonist of GABA(A) receptors. *ACS Chemical Neuroscience* 2014, 5, 803–811.

do Vale, T.G., Furtado, E.C., Santos, J.G., and Viana, G.S.B. Central effects of citral, myrcene and limonene, constituents of essential oil chemotypes from *Lippia alba* (Mill.) N.E. Brown. *Phytomedicine* 2002, 9, 709–714.

dos Santos, É.R.Q., Maia, C.S.F., Fontes Junior, E.A., Melo, A.S., Pinheiro, B.G., and Maia, J.G.S. Linalool-rich essential oils from the Amazon display antidepressant-type effect in rodents. *Journal of Ethnopharmacology* 2018, 212, 43–49.

Dougalis, A., Lees, G., and Ganellin, C.R. The sleep lipid oleamide may represent an endogenous anticonvulsant: An *in vitro* comparative study in the 4-aminopyridine rat brain-slice model. *Neuropharmacology* 2004, 46, 541–554.

Draguhn, A., Verdorn, T.A., Ewert, M., Seeburg, P.H., and Sakmann, B. Functional and molecular distinction between recombinant rat GABAA receptor subtypes by Zn2+. *Neuron* 1990, 5, 781–788.

Elisabetsky, E., and Brum, L.F. Linalool as active component of traditional remedies: Anticonvulsant properties and mechanism of action. *Curare* 2003, 26, 237–244.

Elisabetsky, E., Brum, L.F., and Souza, D.O. Anticonvulsant properties of linalool in glutamate-related seizure models. *Phytomedicine* 1999, 6, 107–113.

Elmaci, I., and Altinoz, M.A. Thymoquinone: An edible redox-active quinone for the pharmacotherapy of neurodegenerative conditions and glial brain tumors. A short review. *Biomedicine & Pharmacotherapy* 2016, 83, 635–640.

Ertl, P., Rohde, B., and Selzer, P. Fast calculation of molecular polar surface area as a sum of fragment-based contributions and its application to the prediction of drug transport properties. *Journal of Medicinal Chemistry* 2000, 43, 3714–3717.

Fedorova, I., Hashimoto, A., Fecik, R.A., Hedrick, M.P., Hanus, L.O., Boger, D.L., Rice, K.C., and Basile, A.S. Behavioral evidence for the interaction of oleamide with multiple neurotransmitter systems. *Journal of Pharmacology and Experimental Therapeutics* 2001, 299, 332–342.

Firdaus, F., Zafeer, M.F., Ahmad, M., and Afzal, M. Anxiolytic and anti-inflammatory role of thymoquinone in arsenic-induced hippocampal toxicity in Wistar rats. *Heliyon* 2018, 4, e00650.

Fisher, R.S., Acevedo, C., Arzimanoglou, A., Bogacz, A., Cross, J.H., Elger, C.E., Engel, J. et al. ILAE Official Report: A practical clinical definition of epilepsy. *Epilepsia* 2014, 55, 475–482.

Galdino, P.M., Nascimento, M.V.M., Florentino, I.F., Lino, R.C., Fajemiroye, J.O., Chaibub, B.A., de Paula, J.R., de Lima, T.C.M., and Costa, E.A. The anxiolytic-like effect of an essential oil derived from *Spiranthera odoratissima* A. St. Hil. leaves and its major component, β-caryophyllene, in male mice. *Progress in Neuro-Psychopharmacology and Biological Psychiatry* 2012, 38, 276–284.

Garlet, Q.I., Pires, L. da C., Milanesi, L.H., Marafiga, J.R., Baldisserotto, B., Mello, C.F., and Heinzmann, B.M. (+)-Dehydrofukinone modulates membrane potential and delays seizure onset by GABAa receptor-mediated mechanism in mice. *Toxicology and Applied Pharmacology* 2017, 332, 52–63.

Garlet, Q.I., Pires, L.C., Silva, D.T., Spall, S., Gressler, L.T., Bürger, M.E., Baldisserotto, B., and Heinzmann, B.M. Effect of (+)-dehydrofukinone on GABAA receptors and stress response in fish model. *Brazilian Journal of Medical and Biological Research* 2016, 49.

Gastón, M.S., Cid, M.P., Vázquez, A.M., Decarlini, M.F., Demmel, G.I., Rossi, L.I., Aimar, M.L., and Salvatierra, N.A. Sedative effect of central administration of *Coriandrum sativum* essential oil and its major component linalool in neonatal chicks. *Pharmaceutical Biology* 2016, 54, 1954–1961.

Ge, L., Zhu, M., Yang, J.-Y., Wang, F., Zhang, R., Zhang, J.-H., Shen, J., Tian, H.-F., and Wu, C.-F. Differential proteomic analysis of the anti-depressive effects of oleamide in a rat chronic mild stress model of depression. *Pharmacology Biochemistry and Behavior* 2015, 131, 77–86.

Gilgun-Sherki, Y., Rosenbaum, Z., Melamed, E., and Offen, D. Antioxidant therapy in acute central nervous system injury: Current state. *Pharmacological Review* 2002, 54, 271–284.

Gilhotra, N., and Dhingra, D. Thymoquinone produced antianxiety-like effects in mice through modulation of GABA and NO levels. *Pharmacological Reports* 2011, 63, 660–669.

Goel, R., Kaur, D., and Pahwa, P. Assessment of anxiolytic effect of nerolidol in mice. *Indian Journal of Pharmacology* 2016, 48, 450.

Gökce, E.C., Kahveci, R., Gökce, A., Cemil, B., Aksoy, N., Sargon, M.F., Kısa, U. et al. Neuroprotective effects of thymoquinone against spinal cord ischemia-reperfusion injury by attenuation of inflammation, oxidative stress, and apoptosis. *Journal of Neurosurgery: Spine* 2016, 24(6), 949–59.

Goodman, S.N., Fanelli, D., and Ioannidis, J.P.A. What does research reproducibility mean? *Science Translational Medicine* 2016, 8, 341ps12–341ps12.

Graz, B., Elisabetsky, E., and Falquet, J. Beyond the myth of expensive clinical study: Assessment of traditional medicines. *Journal of Ethnopharmacology* 2007, 113, 382–386.

Graz, B., Falquet, J., and Elisabetsky, E. Ethnopharmacology, sustainable development and cooperation: The importance of gathering clinical data during field surveys. *Journal of Ethnopharmacology* 2010, 130, 635–638.

Gu, Q., Du, H., Ma, C., Fotis, H., Wu, B., Huang, C., and Sochwarz, W. Effects of α-Asarone on the glutamate transporter EAAC1 in *Xenopus* oocytes. *Planta Medica* 2010, 76, 595–598.

Gubellini, P., and Kachidian, P. Animal models of Parkinson's disease: An updated overview. *Revue Neurologique* 2015, 171, 750–761.

Guimarães, A.G., Oliveira, G.F., Melo, M.S., Cavalcanti, S.C.H., Antoniolli, A.R., Bonjardim, L.R., Silva, F.A. et al. Bioassay-guided Evaluation of Antioxidant and Antinociceptive Activities of Carvacrol: Antioxidant and antinociceptive effects of carvacrol. *Basic & Clinical Pharmacology & Toxicology* 2010, 107, 949–957.

Guzmán-Gutiérrez, S.L., Bonilla-Jaime, H., Gómez-Cansino, R., and Reyes-Chilpa, R. Linalool and β-pinene exert their antidepressant-like activity through the monoaminergic pathway. *Life Sciences* 2015, 128, 24–29.

Guzmán-Gutiérrez, S.L., Gómez-Cansino, R., García-Zebadúa, J.C., Jiménez-Pérez, N.C., and Reyes-Chilpa, R. Antidepressant activity of Litsea glaucescens essential oil: Identification of β-pinene and linalool as active principles. *Journal of Ethnopharmacology* 2012, 143, 673–679.

Han, P., Han, T., Peng, W., and Wang, X.-R. Antidepressant-like effects of essential oil and asarone, a major essential oil component from the rhizome of *Acorus tatarinowii*. *Pharmaceutical Biology* 2013, 51, 589–594.

Hansel, G., Tonon, A.C., Guella, F.L., Pettenuzzo, L.F., Duarte, T., Duarte, M.M.M.F., Oses, J.P., Achaval, M., and Souza, D.O. Guanosine protects against cortical focal Ischemia. Involvement of inflammatory response. *Molecular Neurobiology* 2015, 52, 1791–1803.

Harada, H., Kashiwadani, H., Kanmura, Y., and Kuwaki, T. Linalool odor-induced anxiolytic effects in mice. *Frontiers in Behavioral Neuroscience* 2018, 12.

He, Q., Liu, J., Liang, J., Liu, X., Li, W., Liu, Z., Ding, Z., and Tuo, D. Towards improvements for penetrating the blood–brain barrier—recent progress from a material and pharmaceutical perspective. *Cells* 2018, 7, 24.

Herrera-Solís, A., Vásquez, K.G., and Prospéro-García, O. Acute and subchronic administration of anandamide or oleamide increases REM sleep in rats. *Pharmacology Biochemistry and Behavior* 2010, 95, 106–112.

Höferl, M., Hütter, C., and Buchbauer, G. A pilot study on the physiological effects of three essential oils in humans. *Natural product communications* 2016, 11, 1561–1564.

Hosseinzadeh, H., and Noraei, N.B. Anxiolytic and hypnotic effect of *Crocus sativus* aqueous extract and its constituents, crocin and safranal, in mice. *Phytotherapy Research* 2009, 23, 768–774.

Hosseinzadeh, H., Parvardeh, S., Asl, M.N., Sadeghnia, H.R., and Ziaee, T. Effect of thymoquinone and *Nigella sativa* seeds oil on lipid peroxidation level during global cerebral ischemia-reperfusion injury in rat hippocampus. *Phytomedicine* 2007, 14, 621–627.

Hosseinzadeh, H., Parvardeh, S., Nassiri-Asl, M., and Mansouri, M.-T. Intracerebroventricular administration of thymoquinone, the major constituent of *Nigella sativa* seeds, suppresses epileptic seizures in rats. *Medical Science Monitor* 2005, 11, BR106–BR110.

Hosseinzadeh, H., and Sadeghnia, H.R. Protective effect of safranal on pentylenetetrazol-induced seizures in the rat: Involvement of GABAergic and opioids systems. *Phytomedicine* 2007, 14, 256–262.

Huang, C., Li, W.-G., Zhang, X.-B., Wang, L., Xu, T.-L., Wu, D., and Li, Y. Alpha-asarone from *Acorus gramineus* alleviates epilepsy by modulating A-Type GABA receptors. *Neuropharmacology* 2013, 65, 1–11.

Hurley, L.L., and Tizabi, Y. Neuroinflammation, neurodegeneration, and depression. *Neurotoxicity Research* 2013, 23, 131–144.

Ioannidis, J.P.A. Meta-research: Why research on research matters. *PLOS Biology* 2018, 16, e2005468.

Ioannidis, J.P.A. Why most published research findings are false. *PLOS Medicine* 2005, 2, e124.

Irie, Y., and Keung, W.M. Rhizoma acori graminei and its active principles protect PC-12 cells from the toxic effect of amyloid-β peptide. *Brain Research* 2003, 963, 282–289.

Janero, D.R. The reproducibility issue and preclinical academic drug discovery: Educational and institutional initiatives fostering translation success. *Expert Opinion on Drug Discovery* 2016, 11, 835–842.

Javed, H., Azimullah, S., Abul Khair, S.B., Ojha, S., and Haque, M.E. Neuroprotective effect of nerolidol against neuroinflammation and oxidative stress induced by rotenone. *BMC Neuroscience* 2016, 17.

Jiang, Y., Yu, B., Fang, F., Cao, H., Ma, T., and Yang, H. Modulation of chloride channel functions by the plant lignan compounds kobusin and eudesmin. *Frontiers in Plant Science* 2015, 6.

Kanter, M. Protective effects of thymoquinone on the neuronal injury in frontal cortex after chronic toluene exposure. *Journal of Molecular Histology* 2011, 42, 39–46.

Kaur, D., Pahwa, P., and Goel, R.K. Protective effect of nerolidol against pentylenetetrazol-induced kindling, oxidative stress and associated behavioral comorbidities in mice. *Neurochemical Research* 2016, 41, 2859–2867.

Kawaguchi, M., Furuya, H., and Patel, P.M. Neuroprotective effects of anesthetic agents. *Journal of Anesthesia* 2005, 19, 150–156.

Kenakin, T., Bylund, D.B., Toews, M.L., Mullane, K., Winquist, R.J., and Williams, M. Replicated, replicable and relevant–target engagement and pharmacological experimentation in the 21st century. *Biochemical Pharmacology* 2014, 87, 64–77.

Kessler, A., Sahin-Nadeem, H., Lummis, S.C.R., Weigel, I., Pischetsrieder, M., Buettner, A., and Villmann, C. GABA $_A$ receptor modulation by terpenoids from *Sideritis* extracts. *Molecular Nutrition & Food Research* 2014, 58, 851–862.

Khoshnam, S.E., Winlow, W., Farzaneh, M., Farbood, Y., and Moghaddam, H.F. Pathogenic mechanisms following ischemic stroke. *Neurological Sciences* 2017, 38, 1167–1186.

Kimmelman, J., Mogil, J.S., and Dirnagl, U. Distinguishing between exploratory and confirmatory preclinical research will improve translation. *PLOS Biology* 2014, 12, e1001863.

Kruk-Slomka, M., Michalak, A., and Biala, G. Antidepressant-like effects of the cannabinoid receptor ligands in the forced swimming test in mice: Mechanism of action and possible interactions with cholinergic system. *Behavioural Brain Research* 2015, 284, 24–36.

Leal-Cardoso, J.H., da Silva-Alves, K.S., Ferreira-da-Silva, F.W., dos Santos-Nascimento, T., Joca, H.C., de Macedo, F.H.P., de Albuquerque-Neto, P.M. et al. Linalool blocks excitability in peripheral nerves and voltage-dependent Na+ current in dissociated dorsal root ganglia neurons. *European Journal of Pharmacology* 2010, 645, 86–93.

Ledford, H. Medical research: If depression were cancer. *Nature* 2014, 515, 182–184.

Lee, B., Sur, B., Yeom, M., Shim, I., Lee, H., and Hahm, D.-H. Alpha-Asarone, a major component of *Acorus gramineus*, attenuates corticosterone-induced anxiety-like behaviours via modulating TrkB signaling process. *The Korean Journal of Physiology & Pharmacology* 2014, 18, 191.

Lee, H., Jang, Y.-H., and Lee, S.-R. Protective effect of propofol against Kainic acid-induced lipid peroxidation in mouse brain homogenates: Comparison with trolox and melatonin. *Journal of Neurosurgical Anesthesiology* 2005, 17, 144–148.

Leggett, J.D., Aspley, S., Beckett, S.R.G., D'Antona, A.M., Kendall, D.A., and Kendall, D.A. Oleamide is a selective endogenous agonist of rat and human CB $_1$ cannabinoid receptors. *British Journal of Pharmacology* 2004, 141, 253–262.

Leo, A.J. Calculating log Poct from structures. *Chemical Reviews* 1993, 93, 1281–1306.

Lima, N.G.P.B., De Sousa, D.P., Pimenta, F.C.F., Alves, M.F., De Souza, F.S., Macedo, R.O., Cardoso, R.B., de Morais, L.C.S.L., Melo Diniz, M. de F.F., and de Almeida, R.N. Anxiolytic-like activity and GC–MS analysis of (R)-(+)-limonene fragrance, a natural compound found in foods and plants. *Pharmacology Biochemistry and Behavior* 2013, 103, 450–454.

Limón, I.D., Mendieta, L., Díaz, A., Chamorro, G., Espinosa, B., Zenteno, E., and Guevara, J. Neuroprotective effect of alpha-asarone on spatial memory and nitric oxide levels in rats injected with amyloid-β(25–35). *Neuroscience Letters* 2009, 453, 98–103.

Linck, V.M., da Silva, A.L., Figueiró, M., Caramão, E.B., Moreno, P.R.H., and Elisabetsky, E. Effects of inhaled linalool in anxiety, social interaction and aggressive behavior in mice. *Phytomedicine* 2010, 17, 679–683.

Lipinski, C., and Hopkins, A. Navigating chemical space for biology and medicine. *Nature* 2004, 432, 855–861.

Lipinski, C.A. Drug-like properties and the causes of poor solubility and poor permeability. *Journal of Pharmacological and Toxicological Methods* 2000, 44, 235–249.

Lipinski, C.A., Lombardo, F., Dominy, B.W., and Feeney, P.J. Experimental and computational approaches to estimate solubility and permeability in drug discovery and development settings. *Advanced Drug Delivery Reviews* 2012, 64, 4–17.

Liu, H., Lai, X., Xu, Y., Miao, J., Li, C., Liu, J., Hua, Y., Ma, Q., and Chen, Q. α-Asarone attenuates cognitive deficit in a pilocarpine-induced status epilepticus rat model via a decrease in the nuclear factor-κB activation and reduction in microglia neuroinflammation. *Frontiers in Neurology* 2017, 8.

Liu, H., Song, Z., Liao, D.-G., Zhang, T.-Y., Liu, F., Zhuang, K., Luo, K., Yang, L., He, J., and Lei, J.-P. Anticonvulsant and Sedative Effects of Eudesmin isolated from *Acorus tatarinowii* on mice and rats: Antiepileptic effects of eudesmin. *Phytotherapy Research* 2015, 29, 996–1003.

Liu, R., Zhang, L., Lan, X., Li, L., Zhang, T.-T., Sun, J.-H., and Du, G.-H. Protection by borneol on cortical neurons against oxygen-glucose deprivation/reperfusion: Involvement of anti-oxidation and anti-inflammation through nuclear transcription factor κappaB signaling pathway. *Neuroscience* 2011, 176, 408–419.

Liu, Z., Xu, X.-H., Liu, T.-Y., Hong, Z.-Y., Urade, Y., Huang, Z.-L., and Qu, W.-M. Safranal enhances non-rapid eye movement sleep in pentobarbital-treated mice: Safranal enhances NREM sleep in mice. *CNS Neuroscience & Therapeutics* 2012, 18, 623–630.

Löscher, W. Critical review of current animal models of seizures and epilepsy used in the discovery and development of new antiepileptic drugs. *Seizure* 2011, 20, 359–368.

Löscher, W., and Schmidt, D. Modern antiepileptic drug development has failed to deliver: Ways out of the current dilemma: Ways out of the current dilemma with new AEDs. *Epilepsia* 2011, 52, 657–678.

Mao, J., Huang, S., Liu, S., Feng, X.-L., Yu, M., Liu, J., Sun, Y.E. et al. A herbal medicine for Alzheimer's disease and its active constituents promote neural progenitor proliferation. *Aging Cell* 2015, 14, 784–796.

McKernan, R.M., and Whiting, P.J. Which GABAA-receptor subtypes really occur in the brain? *Trends in Neurosciences*. 1996, 19, 139–143.

McNamara, J.O. Pharmacotherapy of the Epilepsies. In *Goodman & Gilman's: The Pharmacological Basis of Therapeutics*, L L., Brunton, B.A., Chabner, B.C., Knollmann eds., New York: MacGrow Hill, 2011, Chapter 22, pp. 609–628 ISBN 978-0-07-162442-8.

Melo, F.H.C., Moura, B.A., de Sousa, D.P., de Vasconcelos, S.M.M., Macedo, D.S., Fonteles, M.M. de F., Viana, G.S. de B., and de Sousa, F.C.F. Antidepressant-like effect of carvacrol (5-isopropyl-2-methylphenol) in mice: Involvement of dopaminergic system: Involvement of dopaminergic system. *Fundamental & Clinical Pharmacology* 2011, 25, 362–367.

Melo, F.H.C., Venâncio, E.T., De Sousa, D.P., De França Fonteles, M.M., De Vasconcelos, S.M.M., Viana, G.S.B., and De Sousa, F.C.F. Anxiolytic-like effect of Carvacrol (5-isopropyl-2-methylphenol) in mice: Involvement with GABAergic transmission: Anxiolytic-like effect of carvacrol. *Fundamental & Clinical Pharmacology* 2009, 24, 437–443.

Mendelson, W. The hypnotic actions of the fatty acid amide, oleamide. *Neuropsychopharmacology* 2001, 25, S36–S39.

Mendelson, W.B., and Basile, A.S. The hypnotic actions of oleamide are blocked by a cannabinoid receptor antagonist. *Neuroreport* 1999, 10, 3237–3239.

Miao, J.-K., Chen, Q.-X., Li, C., Li, X.-W., Wu, X.-M., and Zhang, X. Modulation effects of a-asarone on the GABA homeostasis in the lithium-pilocarpine model of temporal Lobe Epilepsy. *International Journal of Pharmacology* 2013, 9, 24–32.

Mihic, S.J., Harris, R.A. Hypnotics and sedatives. In *Goodman & Gilman's: The Pharmacological Basis of Therapeutics*, L.L., Brunton, B.A., Chabner, B.C., Knollmann eds., New York: MacGrow Hill, 2011, Chapter 17, pp. 457–479 ISBN 978-0-07-162442-8.

Molisnpiration. 2019. https://molinspiration.com, assessed in November 2019.

Moreira, M.R.C., Salvadori, M.G. da S.S., de Almeida, A.A.C., de Sousa, D.P., Jordán, J., Satyal, P., de Freitas, R.M., and de Almeida, R.N. Anxiolytic-like effects and mechanism of (−)-myrtenol: A monoterpene alcohol. *Neuroscience Letters* 2014, 579, 119–124.

Mullane, K., Winquist, R.J., and Williams, M. Translational paradigms in pharmacology and drug discovery. *Biochemical Pharmacology* 2014, 87, 189–210.

Nam, H.Y., Na, E.J., Lee, E., Kwon, Y., and Kim, H.-J. Antiepileptic and neuroprotective effects of oleamide in rat striatum on kainate-induced behavioral seizure and excitotoxic damage via calpain inhibition. *Frontiers in Pharmacology* 2017, 8.

Nestler, E.J., and Hyman, S.E. Animal models of neuropsychiatric disorders. *Nature Neuroscience* 2010, 13, 1161–1169.

Norte, M.C.B., Cosentino, R.M., and Lazarini, C.A. Effects of methyl-eugenol administration on behavioral models related to depression and anxiety, in rats. *Phytomedicine* 2005, 12, 294–298.

Nunes, D.S., Linck, V., Silva, A.L., Figueiró, M., Elisabetsky, E. Psychopharmacology of Essential oils. In *Handbook of Essential Oils: Science, Technology, and Applications*, K. Hüsnü Can Baser and G. Buchbauer eds., London: CRC Press/Taylor and Francis, 2010, Chapter 10.2, pp. 297–314 ISBN 978-1-4200-6315-8.

O'Donnell, J.M., and Shelton, R.C. Drug therapy of depression and anxiety disorders. In: *Goodman & Gilman's: The Pharmacological Basis of Therapeutics*, L.L. Brunton, B.A. Chabner, B.C. Knollmann eds. 12th edition, pp. 412–415. ISBN-13: 978-0-07-162442-8.

Ohayon, M.M. Epidemiology of insomnia: What we know and what we still need to learn. *Sleep Medicine Reviews* 2002, 6, 97–111.

Orellana-Paucar, A.M., Afrikanova, T., Thomas, J., Aibuldinov, Y.K., Dehaen, W., de Witte, P.A.M., and Esguerra, C.V. Insights from Zebrafish and mouse models on the activity and safety of Ar-turmerone as a potential drug candidate for the treatment of Epilepsy. *PLOS ONE* 2013, 8, e81634.

Orellana-Paucar, A.M., Serruys, A.-S.K., Afrikanova, T., Maes, J., De Borggraeve, W., Alen, J., León-Tamariz, F. et al. Anticonvulsant activity of bisabolene sesquiterpenoids of *Curcuma longa* in zebrafish and mouse seizure models. *Epilepsy & Behavior* 2012, 24, 14–22.

Pages, N., Maurois, P., Delplanque, B., Bac, P., Stables, J.P., Tamariz, J., Chamorro, G., and Vamecq, J. Activities of α-asarone in various animal seizure models and in biochemical assays might be essentially accounted for by antioxidant properties. *Neuroscience Research* 2010, 68, 337–344.

Paladini, A.C., Marder, M., Viola, H., Wolfman, C., Wasowski, C., and Medina, J.H. Flavonoids and the central nervous system: From forgotten factors to potent anxiolytic compounds. *Journal of Pharmacy and Pharmacology* 1999, 51, 519–526.

Palm, K., Stenberg, P., Luthman, K., and Artursson, P. Polar molecular surface properties predict the intestinal absorption of drugs in humans. *Pharmaceutical Research* 1997, 14, 568–571.

Pardridge, W.M. The blood-brain barrier: Bottleneck in brain drug development. *NeuroRX* 2005, 2, 3–14.

Peers, I.S., South, M.C., Ceuppens, P.R., Bright, J.D., and Pilling, E. Can you trust your animal study data? *Nature Reviews Drug Discovery* 2014, 13, 560–560.

Pereira, P., Elisabetsky, E., and Souza, D.O. Effect of gamma-decanolactone on glutamate binding in the rat cerebral cortex. *Neurochemical Research* 1997, 22, 1507–1510.

Piccinelli, A.C., Santos, J.A., Konkiewitz, E.C., Oesterreich, S.A., Formagio, A.S.N., Croda, J., Ziff, E.B., and Kassuya, C.A.L. Antihyperalgesic and antidepressive actions of (R)-(+)-limonene, α-phellandrene, and essential oil from *Schinus terebinthifolius* fruits in a neuropathic pain model. *Nutritional Neuroscience* 2015, 18, 217–224.

Prasad, S.N., and Muralidhara, M. Mitigation of acrylamide-induced behavioral deficits, oxidative impairments and neurotoxicity by oral supplements of geraniol (a monoterpene) in a rat model. *Chemico-Biological Interactions* 2014a, 223, 27–37.

Prasad, S.N., and Muralidhara, M. Neuroprotective effect of geraniol and curcumin in an acrylamide model of neurotoxicity in *Drosophila melanogaster*: Relevance to neuropathy. *Journal of Insect Physiology* 2014b, 60, 7–16.

PubChem. 2019. https://pubchem.ncbi.nlm.nih.gov, assessed in November 2019.

Quintans-Jr, L.J., Guimarães, A.G., and Araújo, B.E.S. Carvacrol, (-)-borneol and citral reduce convulsant activity in rodents. *African Journal of Biotechnology* 2010, 9, 6566–6572.

Radad, K., Hassanein, K., Al-Shraim, M., Moldzio, R., and Rausch, W.-D. Thymoquinone ameliorates lead induced brain damage in Sprague Dawley rats. *Experimental and Toxicologic Pathology* 2014, 66, 13–17.

Radhakrishnan, A., Jayakumari, N., Kumar, V.M., and Gulia, K.K. Sleep promoting potential of low dose α-asarone in rat model. *Neuropharmacology* 2017, 125, 13–29.

Radulović, N.S., Dekić, M.S., Ranđelović, P.J., Stojanović, N.M., Zarubica, A.R., and Stojanović-Radić, Z.Z. Toxic essential oils: Anxiolytic, antinociceptive and antimicrobial properties of the yarrow *Achillea umbellata Sibth.* et Sm. (Asteraceae) volatiles. *Food and Chemical Toxicology* 2012, 50, 2016–2026.

Raza, M. Beneficial interaction of thymoquinone and sodium valproate in experimental models of epilepsy: Reduction in hepatotoxicity of valproate. *Scientia Pharmaceutica* 2006, 74, 159–173.

Reimers, A., Berg, J.A., Larsen Burns, M., Brodtkorb, E., Johannessen, S.I., and Johannessen Landmark, C. Reference ranges for antiepileptic drugs revisited: A practical approach to establish national guidelines. *Drug Design, Development and Therapy* 2018, 12, 271–280.

Ribeiro, M.C.P., de Ávila, D.S., Schneider, C.Y.M., Hermes, F.S., Furian, A.F., Oliveira, M.S., Rubin, M.A., Lehmann, M., Krieglstein, J., and Mello, C.F. α-Tocopherol protects against pentylenetetrazol- and methylmalonate-induced convulsions. *Epilepsy Research* 2005, 66, 185–194.

Riemann, D., Baglioni, C., Bassetti, C., Bjorvatn, B., Dolenc Groselj, L., Ellis, J.G., Espie, C.A. et al. European guideline for the diagnosis and treatment of insomnia. *Journal of Sleep Research* 2017, 26, 675–700.

Rosenberg, E.C., Tsien, R.W., Whalley, B.J., and Devinsky, O. Cannabinoids and Epilepsy. *Neurotherapeutics* 2015, 12, 747–768.

Sadeghnia, H.R., Kamkar, M., Assadpour, E., Boroushaki, M.T., and Ghorbani, A. Protective effect of safranal, a constituent of *Crocus sativus*, on quinolinic acid-induced oxidative damage in rat hippocampus. *Iranian Journal of Basic Medical Sciences* 2013, 16, 73–82.

Sadeghnia, H.R., Shaterzadeh, H., Forouzanfar, F., and Hosseinzadeh, H. Neuroprotective effect of safranal, an active ingredient of *Crocus sativus*, in a rat model of transient cerebral ischemia. *Folia Neuropathologica* 2017, 3, 206–213.

Saito, T., and Saido, T.C. Neuroinflammation in mouse models of Alzheimer's disease. *Clinical and Experimental Neuroimmunology* 2018, 9, 211–218.

Salgado, P., da Fonsêca, D., Braga, R., de Melo, C., Andrade, L., de Almeida, R., and de Sousa, D. Comparative anticonvulsant study of epoxycarvone stereoisomers. *Molecules* 2015, 20, 19660–19673.

Samarghandian, S., Azimi-Nezhad, M., and Samini, F. Preventive effect of safranal against oxidative damage in aged male rat brain. *Experimental Animals* 2015, 64, 65–71.

Santhanasabapathy, R., and Sudhandiran, G. Farnesol attenuates lipopolysaccharide-induced neurodegeneration in Swiss albino mice by regulating intrinsic apoptotic cascade. *Brain Research* 2015, 1620, 42–56.

Santos, P., Herrmann, A.P., Benvenutti, R., Noetzold, G., Giongo, F., Gama C.S., Piato, A.L., Elisabetsky, E. Anxiolytic properties of N-acetylcysteine in mice. *Behavioural Brain Research*. 2017a,317,461–469.

Santos, P.S., Souza, L.K.M., Araújo, T.S.L., Medeiros, J.V.R., Nunes, S.C.C., Carvalho, R.A., Pais, A.C.C., Veiga, F.J.B., Nunes, L.C.C., and Figueiras, A. Methyl-β-cyclodextrin inclusion complex with β-Caryophyllene: Preparation, characterization, and improvement of pharmacological activities. *ACS Omega* 2017b, 2, 9080–9094.

Satou, T., Kasuya, H., Maeda, K., and Koike, K. Daily inhalation of α-Pinene in mice: effects on behavior and organ accumulation: Daily inhalation of α-Pinene in mice. *Phytotherapy Research* 2014, 28, 1284–1287.

Satou, T., Miyahara, N., Murakami, S., Hayashi, S., and Koike, K. Differences in the effects of essential oil from *Citrus junos* and (+)-limonene on emotional behavior in mice. *Journal of Essential Oil Research* 2012, 24, 493–500.

Schulze-Bonhage, A.A. 2017 review of pharmacotherapy for treating focal epilepsy: Where are we now and how will treatment develop? *Expert Opinion on Pharmacotherapy* 2017, 18, 1845–1853.

Shao, Y., Feng, Y., Xie, Y., Luo, Q., Chen, L., Li, B., and Chen, Y. Protective effects of thymoquinone against convulsant activity induced by lithium-pilocarpine in a model of status epilepticus. *Neurochemical Research* 2016, 41, 3399–3406.

Shao, Y.Y., Li, B., Huang, Y.M., Luo, Q., Xie, Y.M., Chen, Y.H. Thymoquinone attenuates brain injury via an antioxidative pathway in a status epilepticus rat model. *Transl Neurosci.* 2017, Mar 25, 8, 9–14.

Sharma, J.D., Dandiya, P.C., Baxter, R.M., and Kandel, S.I. Pharmacodynamical effects of asarone and β-asarone. *Nature* 1961, 192, 1299–1300.

Shelton, R.C., and Od., J.M. Drug therapy of depression and anxiety disorders. In *Goodman & Gilman's: The Pharmacological Basis of Therapeutics*, L.L., Brunton, B.A., Chabner, B.C., Knollmann eds., New York: Mac Grow Hill, 2011, Chapter 15, pp. 412–415 ISBN 13: 978-0-07-162442-8.

Shin, J.-W., Cheong, Y.-J., Koo, Y.-M., Kim, S., Noh, C.-K., Son, Y.-H., Kang, C., and Sohn, N.-W. α-Asarone ameliorates memory deficit in lipopolysaccharide-treated mice via suppression of pro-inflammatory cytokines and microglial activation. *Biomolecules & Therapeutics* 2014, 22, 17–26.

Shin, M., Jang, D., Nam, H., Lee, K.H., and Lee, D. Predicting the absorption potential of chemical compounds through a deep learning approach. *IEEE/ACM Transactions on Computational Biology and Bioinformatics* 2018, 15, 432–440.

Shityakov, S., Neuhaus, W., Dandekar, T., and Förster, C. Analysing molecular polar surface descriptors to predict blood-brain barrier permeation. *International Journal of Computational Biology and Drug Design* 2013, 6, 146.

Silva, M.I.G., de Aquino Neto, M.R., Teixeira Neto, P.F., Moura, B.A., do Amaral, J.F., de Sousa, D.P., Vasconcelos, S.M.M., and de Sousa, F.C.F. Central nervous system activity of acute administration of isopulegol in mice. *Pharmacology Biochemistry and Behavior* 2007, 88, 141–147.

Sousa, F.C.F., Oliveira, I.C.M., Fernandes, M.L., De Sousa, D.P. Advances in the Research of Essential Oils: Anxiolytic and Sedative Activity. In: De Sousa, D.P. (Org.). *Medicinal Essential Oils: Chemical, Pharmacological and Therapeutic Aspects.* 1 ed., New York: Nova Science Publishers, 2012, vol. 1, pp. 123–140.

Souto-Maior, F.N., Carvalho, F.L. de, Morais, L.C.S.L. de, Netto, S.M., de Sousa, D.P., and Almeida, R.N. de Anxiolytic-like effects of inhaled linalool oxide in experimental mouse anxiety models. *Pharmacology Biochemistry and Behavior* 2011, 100, 259–263.

Standaert, D.G., and Roberson, E.D. Treatment of central nervous system degenerative disorders. In *Goodman & Gilman's: The Pharmacological Basis of Therapeutics*, L.L., Brunton, B.A., Chabner, B.C., Knollmann eds., New York: Mac Grow Hill, 2011, Chapter 22, pp. 609–628 ISBN 978-0-07-162442-8.

Strekalova, T., and Steinbusch, H.W.M. Measuring behavior in mice with chronic stress depression paradigm. *Progress in Neuro-Psychopharmacology and Biological Psychiatry* 2010, 34, 348–361.

Swinyard, E.A., and Kupferberg, H.J. Antiepileptic drugs: Detection, quantification, and evaluation. *Federation Proceedings* 1985, 44, 2629–2633.

Tabari, M.A., and Tehrani, M.A.B. Evidence for the involvement of the GABAergic, but not serotonergic transmission in the anxiolytic-like effect of bisabolol in the mouse elevated plus maze. *Naunyn-Schmiedeberg's Archives of Pharmacology* 2017, 390, 1041–1046.

Tamaddonfard, E., Farshid, A.-A., Eghdami, K., Samadi, F., and Erfanparast, A. Comparison of the effects of crocin, safranal and diclofenac on local inflammation and inflammatory pain responses induced by carrageenan in rats. *Pharmacological Reports* 2013, 65, 1272–1280.

Tchekalarova, J., da Conceição Machado, K., Gomes Júnior, A.L., de Carvalho Melo Cavalcante, A.A., Momchilova, A., and Tzoneva, R. Pharmacological characterization of the cannabinoid receptor 2 agonist, β-caryophyllene on seizure models in mice. *Seizure* 2018, 57, 22–26.

Tian, J., Tian, Z., Qin, S., Zhao, P., Jiang, X., and Tian, Z. Anxiolytic-like effects of α-asarone in a mouse model of chronic pain. *Metabolic Brain Disease* 2017, 32, 2119–2129.

Verdon, B., Zheng, J., Nicholson, R.A., Ganelli, C.R., and Lees, G. Stereoselective modulatory actions of oleamide on GABA $_A$ receptors and voltage-gated Na^+ channels *in vitro*: A putative endogenous ligand for depressant drug sites in CNS. *British Journal of Pharmacology* 2000, 129, 283–290.

Viana, C.C.S., de Oliveira, P.A., Brum, L.F. da S., Picada, J.N., and Pereira, P. Gamma-decanolactone effect on behavioral and genotoxic parameters. *Life Sciences* 2007, 80, 1014–1019.

Viana, G.S. de B., Vale, T.G. do, Silva, C.M.M., and Matos, F.J. de A. Anticonvulsant activity of essential oils and active principles from chemotypes of *Lippia alba* (Mill.) N.E. Brown. *Biological & Pharmaceutical Bulletin* 2000, 23, 1314–1317.

Wang, X., Chen, Y., Wang, Q., Sun, L., Li, G., Zhang, C., Huang, J., Chen, L., and Zhai, H. Support for natural small-molecule phenols as anxiolytics. *Molecules* 2017, 22, 2138.

Wang, Z.-J., and Heinbockel, T. Essential oils and their constituents targeting the GABAergic system and sodium channels as treatment of neurological diseases. *Molecules* 2018, 23, 1061.

Wang, Z.-J., Levinson, S.R., Sun, L., and Heinbockel, T. Identification of both GABAA receptors and voltage-activated Na+ channels as molecular targets of anticonvulsant α-asarone. *Frontiers in Pharmacology* 2014, 5.

Waring, M.J. Lipophilicity in drug discovery. *Expert Opinion on Drug Discovery* 2010, 5, 235–2248.

Wehling, M. Assessing the translatability of drug projects: What needs to be scored to predict success? *Nature Reviews Drug Discovery* 2009, 8, 541–546.

Wei, X.Y., Yang, J.Y., Dong, Y.X., and Wu, C.F. Anxiolytic-like effects of oleamide in group-housed and socially isolated mice. *Progress in Neuro-Psychopharmacology and Biological Psychiatry* 2007, 31, 1189–1195.

Wendler, A., and Wehling, M. Translatability score revisited: Differentiation for distinct disease areas. *Journal of Translational Medicine* 2017, 15.

Wessel, M.D., Jurs, P.C., Tolan, J.W., and Muskal, S.M. Prediction of human intestinal absorption of drug compounds from molecular structure. *Journal of Chemical Information and Computer Sciences* 1998, 38, 726–735.

Willmore, L.J. Antiepileptic drugs and neuroprotection: Current status and future roles. *Epilepsy & Behavior* 2005, 7, 25–28.

Woo, J.H., Ha, T.-W., Kang, J.-S., Hong, J.T., and Oh, K.-W. Potentiation of decursinol angelate on pentobarbital-induced sleeping behaviors via the activation of $GABA_A$-ergic systems in rodents. *The Korean Journal of Physiology & Pharmacology* 2017, 21, 27.

Yamada, K., Mimaki, Y., and Sashida, Y. Effects of inhaling the vapor of *Lavandula burnatii* super-derived essential oil and linalool on plasma adrenocorticotropic hormone (ACTH), catecholamine and gonadotropin levels in experimental menopausal female rats. *Biological and Pharmaceutical Bulletin* 2005, 28, 378–379.

Yang, H., Woo, J., Pae, A.N., Um, M.Y., Cho, N.-C., Park, K.D., Yoon, M., Kim, J., Lee, C.J., and Cho, S. Pinene, a major constituent of pine tree oils, enhances non-rapid eye movement sleep in mice through GABAA-benzodiazepine receptors. *Molecular Pharmacology* 2016, 90, 530–539.

Yang, J.-Y., Wu, C.-F., Wang, F., Song, H.-R., Pan, W.-J., and Wang, Y.-L. The serotonergic system may be involved in the sleep-inducing action of oleamide in rats. *Naunyn-Schmiedeberg's Archives of Pharmacology* 2003, 368, 457–462.

Yu, B., Ruan, M., Cui, X., Guo, J.-M., Xu, L., and Dong, X.-P. Effects of borneol on the pharmacokinetics of geniposide in cortex, hippocampus, hypothalamus and striatum of conscious rat by simultaneous brain microdialysis coupled with UPLC–MS. *Journal of Pharmaceutical and Biomedical Analysis* 2013a, 77, 128–132.

Yu, B., Ruan, M., Dong, X., Yu, Y., and Cheng, H. The mechanism of the opening of the blood–brain barrier by borneol: A pharmacodynamics and pharmacokinetics combination study. *Journal of Ethnopharmacology* 2013b, 150, 1096–1108.

Zhang, N., Zhang, L., Feng, L., and Yao, L. The anxiolytic effect of essential oil of *Cananga odorata* exposure on mice and determination of its major active constituents. *Phytomedicine* 2016, 23, 1727–1734.

Zhang, Q.-L., Fu, B.M., and Zhang, Z.-J. Borneol, a novel agent that improves central nervous system drug delivery by enhancing blood–brain barrier permeability. *Drug Delivery* 2017, 24, 1037–1044.

Zhang, Q., Wu, D., Wu, J., Ou, Y., Mu, C., Han, B., and Zhang, Q. Improved blood–brain barrier distribution: Effect of borneol on the brain pharmacokinetics of kaempferol in rats by *in vivo* microdialysis sampling. *Journal of Ethnopharmacology* 2015, 162, 270–277.

Zhao, H., Sapolsky, R.M., and Steinberg, G.K. Phosphoinositide-3-kinase/Akt survival signal pathways are implicated in neuronal survival after stroke. *Molecular Neurobiology* 2006, 34, 249–270.

Zhu, K.-J., and Yang, J.-S. Anti-allodynia effect of safranal on neuropathic pain induced by spinal nerve transection in rat. *International Journal of Clinical and Experimental Medicine* 2014, 7, 4990–4996.

Zotti, M., Colaianna, M., Morgese, M., Tucci, P., Schiavone, S., Avato, P., and Trabace, L. Carvacrol: From ancient flavoring to neuromodulatory agent. *Molecules* 2013, 18, 6161–6172.

12 Effects of Essential Oils on Human Cognition

Eva Heuberger

CONTENTS

12.1 INTRODUCTION

A number of attempts have been made to unravel the effects of natural essential oils (EOs) and fragrances on the human central nervous system. Among these attempts, two major lines of research have been followed to identify psychoactive, particularly stimulating and sedative, effects of fragrances. On the one hand, researchers have investigated the influence of EOs and fragrances on brain potentials, which are indicative of the arousal state of the human organism by means of neurophysiological methods. On the other hand, behavioral studies have elucidated the effects of EOs and fragrances on basic and higher cognitive functions, such as alertness and attention, learning and memory, or problem solving. The scope of the following section is to give a broad overview about the current knowledge in these fields. Much of the research reviewed has been carried out in healthy populations, and only recently, investigators have started to focus on clinical aspects of the administration of fragrances and EOs. However, since the latter topic is covered in Chapter 13 of this volume, it is omitted here in the interest of space.

Olfaction differs from other senses in several ways. First, in humans and many other mammals, the information received by peripheral olfactory receptor cells is mainly processed in brain areas located ipsilaterally to the stimulated side of the body, whereas in the other sensory systems, it is transferred to the contralateral hemisphere. Second, in contrast to the other sensory systems, olfactory information reaches a number of cortical areas without being relayed in the thalamus (Kandel et al. 1991; Zilles and Rehkämpfer 1998; Wiesmann et al. 2001) (Figure 12.1).

Owing to this missing thalamic control, as well as to the fact that the olfactory system presents anatomical connections and overlaps with brain areas involved in emotional processing—such as the amygdala, hippocampus and prefrontal cortex of the limbic system (Reiman et al. 1997; Davidson and Irwin 1999; Phan et al. 2002; Bermpohl et al. 2006)—the effects of odorants on the organisms are supposedly exerted not only via pharmacological but also via psychological mechanisms. In humans and probably also in other mammals, psychological factors may be based on certain stimulus features, such as odor valence (Baron and Thomley 1994), on semantic cues; for example, memories

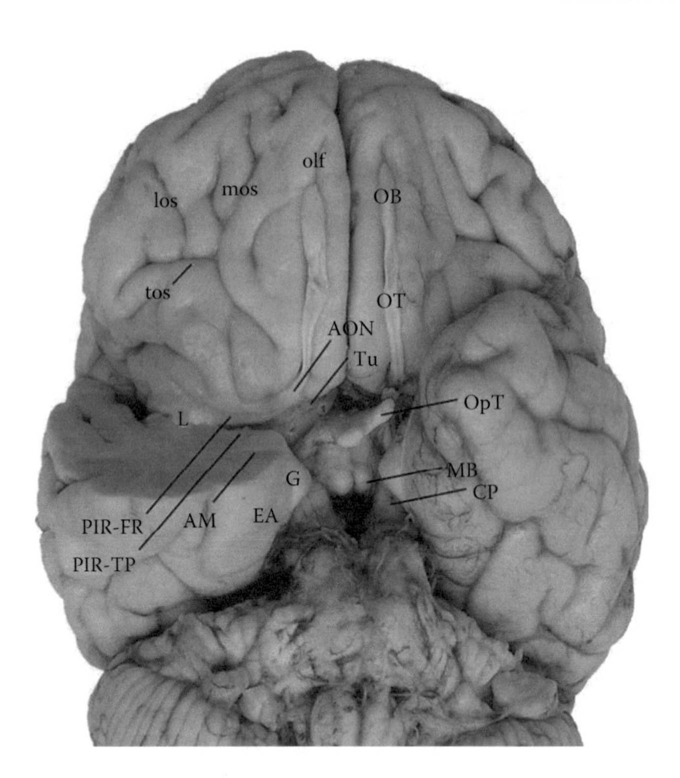

FIGURE 12.1 (See color insert.) Macroscopic view of the human ventral forebrain and medial temporal lobes, depicting the olfactory tract, its primary projections, and surrounding non-olfactory structures. The right medial temporal lobe has been resected horizontally through the mid-portion of the amygdala (AM) to expose olfactory cortex. AON, anterior olfactory nucleus; CP, cerebral peduncle; EA, entorhinal area; G, gyrus ambiens; L, limen insula; los, lateral olfactory sulcus; MB, mammillary body; mos, medial olfactory sulcus; olf, olfactory sulcus; PIR-FR, frontal piriform cortex; OB, olfactory bulb; OpT, optic tract; OT, olfactory tract; tos, transverse olfactory sulcus; Tu, olfactory tubercle; PIR-TP, temporal piriform cortex. Figure prepared with the help of Dr. Eileen H. Bigio, Dept. of Pathology, Northwestern University Feinberg School of Medicine, Chicago, IL. (Taken with permission from Gottfried, J. A. and Zald, D. A. [2005] *Brain Research Reviews* 50:287–304.)

and experiences associated with a particular odor, or on placebo effects related to the expectation of certain effects (Jellinek 1997). None of the latter mechanisms is substance- (i.e., odorant-) specific, but their effectiveness depends on cognitive mediation and control.

Many odorants stimulate not only the olfactory system via the first cranial nerve (*N. olfactorius*) but also the trigeminal system via the fifth cranial nerve (*N. trigeminus*), which enervates the nasal mucosa. The trigeminal system is part of the body's somatosensory system and mediates mechanical- and temperature-related sensations, such as itching and burning or warmth and cooling sensations. Trigeminal information reaches the brain via the trigeminal ganglion and the ventral posterior nucleus of the thalamus. The primary cortical projection area of the somatosensory system is the contralateral postcentral gyrus of the parietal lobe (Zilles and Rehkämpfer 1998). The reticular formation in the brain stem is part of the reticular activating system (RAS) (Figure 12.2) and receives collaterals from the trigeminal system. Thus, trigeminal stimuli have direct effects on arousal. Utilizing this direct connection, highly potent trigeminal stimulants, such as ammonia and menthol, have been used in the past in smelling salts to awaken people who fainted.

It has been shown in experimental animals that due to their lipophilic properties, fragrances do not only penetrate the skin (Hotchkiss 1998), but also the blood–brain barrier (Buchbauer et al. 1993). Also, odorants have been found to bind to several types of brain receptors (Aoshima and Hamamoto 1999; Elisabetsky et al. 1999; Okugawa et al. 2000), and it has been suggested that these

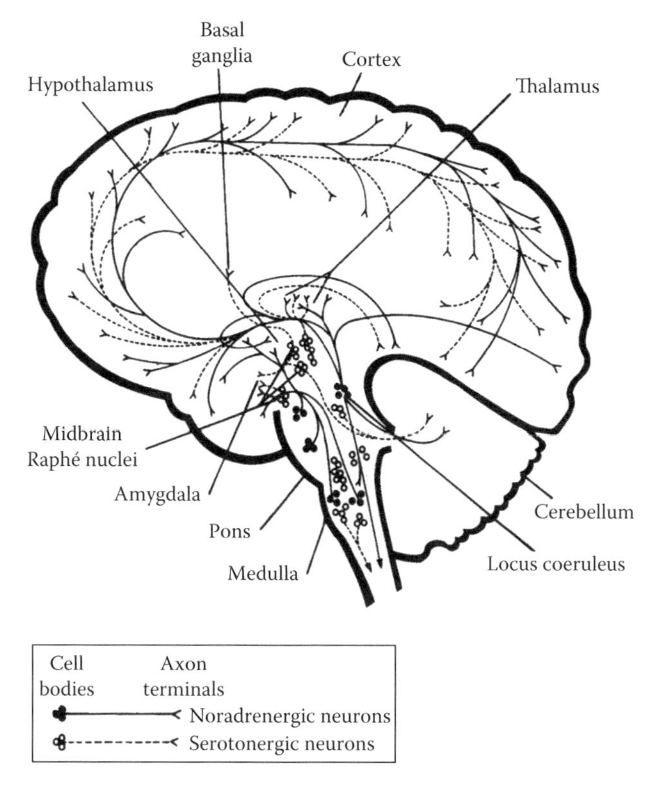

FIGURE 12.2 Schematic of the reticular activating system (RAS) with noradrenergic and serotonergic connections (Taken with permission from Grilly D.M., *Drugs & Human Behavior*, Allyn & Bacon, Boston, 2002.)

odorant–receptor interactions are responsible for psychoactive effects of fragrances in experimental animals. With regard to these findings, it is important to note that Heuberger and co-workers have observed differential effects of fragrances as a function of chirality (Heuberger et al. 2001). It seems likely that such differences in effectiveness are related to enantiomeric selectivity of receptor proteins. However, the question remains whether effects of fragrances on human arousal and cognition rely on a similar psychopharmacological mechanism.

12.2 ACTIVATION AND AROUSAL: DEFINITION AND NEUROANATOMICAL CONSIDERATIONS

Activation, or arousal, refers to the ability of an organism to adapt to internal and external challenges (Schandry 1989). Activation is an elementary process, which serves in the preparation for overt activity. Nevertheless, it does not necessarily result in overt behavior (Duffy 1972). Activation varies in degree and may be described along a continuum from deep sleep to overexcitement. Early theoretical accounts of activation have emphasized physiological responses as the sole measurable correlate of arousal. Current models, however, consider physiological, cognitive, and emotional activity as observable consequences of activation processes. It has been shown that arousal processes within each of these three systems (i.e., physiological, cognitive, and emotional) can occur to varying degrees so that the response of one system need not be correlated linearly to that of the other systems (Baltissen and Heimann 1995).

It has long been established that the RAS, which comprises the reticular formation with its sensory afferents and widespread hypothalamic, thalamic, and cortical projections, plays a crucial role in the control of both phasic and tonic activation processes (Becker-Carus 1981; Schandry 1989).

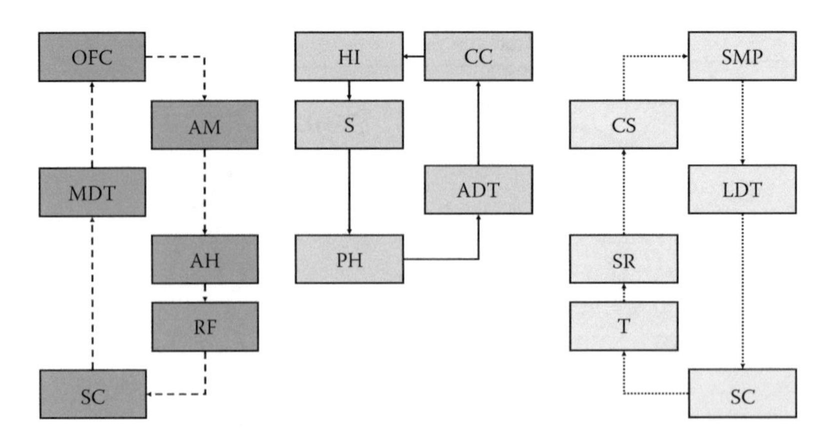

FIGURE 12.3 (See color insert.) Control of activation processes. OFC, orbitofrontal cortex; AM, amygdala; MDT, medial dorsal thalamus; AH, anterior hypothalamus; RF, reticular formation; SC, spinal cord; HI, hippocampus; CC, cingulate cortex; S, septum; ADT, anterior dorsal thalamus; PH, posterior hypothalamus; SMP, sensory-motor projections; CS, corpus striatum; LDT, lateral dorsal thalamus; SR, subthalamic regions; T, tectum. Orange, structures of the arousal network; green, structures of the effort network; blue, structures of the activation network. (Adapted from Pribram, K. H. and McGuinness, D. 1975. *Psychological Review* 82 (2):116–149.)

Pribram and McGuinness (1975) distinguish three separate but interacting neural networks in the control of activation (Figure 12.3). The arousal network involves the amygdala and related frontal cortical structures and regulates phasic physiological responses to novel incoming information. The activation network centers on the basal ganglia of the forebrain and controls the tonic physiological readiness to respond. Finally, the effort network, which comprises hippocampal circuits, coordinates the arousal and activation networks. Noradrenergic projections from the locus coeruleus, which is located within the dorsal wall of the rostral pons, are particularly important in the regulation of circadian alertness, the sleep–wake rhythm, and the sustainment of alertness (alerting) (Pedersen et al. 1998; Aston-Jones et al. 2001). On the other hand, tonic alertness seems to be dependent on cholinergic (Baxter and Chiba 1999; Gill et al. 2000) frontal and inferior parietal thalamic structures of the right hemisphere (Sturm et al. 1999). Other networks that are involved in the control of arousal and attentional functions are found in posterior parts of the brain (e.g., the parietal cortex, superior colliculi, and posterior-lateral thalamus), as well as in anterior regions, (e.g., the cingulate and prefrontal cortices) (Posner and Petersen 1990; Paus 2001).

12.3 INFLUENCE OF ESSENTIAL OILS AND FRAGRANCES ON BRAIN POTENTIALS INDICATIVE OF AROUSAL

12.3.1 Spontaneous EEG Activity

Recordings of spontaneous electroencephalographic (EEG) activity during the administration of EOs and fragrances have been used widely to assess stimulant and sedative effects of these substances. Given that the alpha and beta bands, sometimes also the theta band of the EEG, are thought to be most indicative of central arousal processes, particular attention has been paid to changes within these bands in response to olfactory stimulation. Alpha waves are slow brain waves within a frequency range of 8–13 Hz and amplitudes between 5 and 100 μV. They typically occur over posterior areas of the brain in an awake but relaxed state, especially with closed eyes. The alpha rhythm disappears immediately when subjects open their eyes and when cognitive activity is required, for example, when external stimuli are processed or tasks are solved. This phenomenon is often referred to as alpha block or desynchronization. Simultaneously with the alpha block, faster brain waves occur, such as beta waves with smaller amplitudes (2–20 μV) and frequencies between 14 and 30 Hz. The beta rhythm, which is most evident frontally, is characteristic of alertness, attention, and arousal. In contrast, theta waves are very slow brain waves occurring in fronto-temporal areas with amplitudes between 5 and 100 μV in the frequency range between 4 and 7 Hz. Although the theta rhythm is most commonly associated with drowsiness and light sleep, some researchers found theta activity to correlate with memory processes (Grunwald et al. 1999; Hoedlmoser et al. 2007) and creativity (Razumnikova 2007). Other authors found correlations between theta activity and ratings of anxiety and tension (Lorig and Schwartz 1988). With regard to animal olfaction, it has been proposed that the theta rhythm generated by the hippocampus is concomitant to sniffing and allows for encoding and integration of olfactory information with other cognitive and motor processes (Kepecs et al. 2006).

Recordings of the spontaneous EEG allow deriving a large number of measures. Time- (index) or voltage- (power) based rates of typical frequency bands, as well as ratios between certain frequency bands (e.g., between the alpha and the beta band or the theta and the beta band), within a selected time interval are most commonly used to quantify EEG patterns. Period analysis quantifies the number of waves which occur in the various frequency bands within a distinct time interval of the EEG record and is supposed to be more sensitive to task-related changes than spectral analysis (Lorig 1989). Coherence and neural synchrony are parameters that describe the covariation of a given signal at different electrodes. These measures inform on the functional link between brain areas (Oken et al. 2006).

The pattern of the spontaneous EEG varies with the arousal level of the CNS. Thus, different states of consciousness, such as sleep, wakefulness, or meditation, can be distinguished by their characteristic EEG patterns. For instance, an increase of central activation is typically characterized by a decrease in alpha and an increase in beta activity (Schandry 1989). More precisely, a decline of alpha and beta power together with a decrease of alpha index and an increase of beta and theta activity have been observed under arousing conditions (i.e., in a mental calculation and a psychosocial stress paradigm) (Walschburger 1976). In addition, when subjects are maximally attentive, frequencies in the alpha band are attenuated and activity in the beta and even higher frequency bands can be observed. On the other hand, fatigue and performance decrements in situations requiring high levels of attention are often associated with increases in theta and decreases in beta activity (Oken et al.2006). Drowsiness and the onset of sleep are characterized by an increase in slow, and a decrease in fast, EEG waves. However, high activity in the alpha band, particularly in the range between 7 and 10 Hz, is not indicative of low arousal states of the brain, such as relaxation, drowsiness, and the onset of sleep. It rather seems to be a component of selective neural inhibition processes, which are necessary for a number of cognitive processes, such as perception, attention, and memory (Miller and O'Callaghan 2006; Palva and Palva 2007).

Changes of spontaneous EEG activity accompany a wide range of cognitive as well as emotional brain processes. Moreover, "EEG measurements [...] do not tell investigators what the brain is doing" (Lorig 1989, p. 93). Thus, it is somewhat naïve to interpret changes that are induced by the application of an odorant as a result of a single and specific process, particularly when no other correlates of the process of interest are assessed. Nevertheless, this is exactly the approach that has been taken by many researchers to identify stimulant or relaxing effects of odors. The simplest, but probably most problematic, setup for such experiments in terms of interpretation of the results is the comparison of spontaneous EEG activity in response to odorants with a no-odor baseline. Using this design, Sugano (Sugano 1992) observed increased EEG alpha activity after inhalation of α-pinene, 1,8-cineole, lavender, sandalwood, musk, and eucalyptus odors. Considering that traditional aromatherapy discriminates these fragrances by their psychoactive effects—for example, lavender is assigned relaxing properties while eucalyptus is supposedly stimulant (Valnet 1990)—these findings are at least rather curious. Also, Ishikawa and co-workers recorded the spontaneous EEG in 13 Japanese subjects while drinking either lemon juice with a supplement of lemon odor or lemon juice without it (Ishikawa et al. 2002). It was shown that alpha power, indicative of increased relaxation, was enhanced by supplementation with lemon odor. Again, with regard to aromatherapeutic accounts of lemon EO, this result is somewhat counterintuitive, even more so as, in the same study, the juice supplemented with lemon odor increased spontaneous locomotion in experimental animals. Haneyama and Kagatani tested a fragrant spray made from extracts of Chinese spikenard roots (*Nardostachys chinensis* Batalin [Valerianaceae]) in butylene glycol and found increased alpha activity in subjects under stress (Haneyama and Kagatani 2007). This finding was interpreted by the authors as demonstrating a sedative effect of the extract. However, it is unknown how stress was induced in the subjects and how it was measured. Similarly, Ishiyama (Ishiyama 2000) concluded from measurements of the frequency fluctuation patterns of the alpha band that smelling a blend of terpene compounds typically found in forests induced feelings of refreshment and relaxation in human subjects, although without proper description of how these feelings were assessed. In a very recent paper, Kim and co-workers (Kim et al. 2018) studied the effect after inhalation of the EO of Indian chrysanthemum (*Chrysanthemum indicum* L. [Asteraceae]), which can be found in several mountain regions of South Korea, on blood pressure and EEG in ten healthy adult male and female participants. EEG recordings were taken for three minutes during the inhalation of the EO and compared to a three-minute baseline recording obtained previously while subjects were at rest with their eyes closed. Unfortunately, it remains unclear how the EO was applied, which concentration was used, and whether or not subjects were instructed to close their eyes during the inhalation of the EO. This study found widespread but insignificant increases in relative theta and alpha waves as well as decreases in relative beta and gamma waves. Also, decreases in blood pressure and heart rate after inhalation of the EO compared to the previous control condition were described that were significant at an intra-individual level. These findings were interpreted as demonstrating a mentally and physically relaxing effect of the EO of Indian chrysanthemum, although the study design does not allow to exclude order effects, and the observed changes in EEG activity were overall insignificant. According to the authors, two major compounds of the EO (camphor and 1,8-cineole) were associated with the odor profile of Indian chrysanthemum in a sniffing test. This is a noteworthy finding, since in other studies on attention and memory (see Section 12.4), these components have been associated with activation rather than with relaxation.

Inconclusive findings as those described above are not unexpected with such simple experimental designs as there are several problems associated with these kinds of experiments. First, a no-odor baseline is often inappropriate, as it does not control for cognitive activity of the subject. For instance, subjects might be puzzled by the fact that they do not smell anything, eventually focusing attention to the search for an odor. This may lead to quite high arousal levels rather than the intended resting brain state. This was the case with Lorig and Schwarzt (Lorig and Schwartz 1988), who tested changes of spontaneous EEG in response to spiced apple, eucalyptus, and lavender fragrances diluted in an odorless base: contrary to the authors' expectations, alpha activity in the no-odor condition was less than during odor presentation.

Considering the various mechanisms outlined by Jellinek (Jellinek 1997) by which fragrances influence human arousal and behavior, another inherent problem of such simple designs is that little is known about how subjects process stimulus-related information—for example, the pleasantness or intensity of an odorant—and whether or not higher cognitive processes related to the odorant are initiated by the stimulation. For instance, subjects might be able to identify and label some odors but might fail to do so with others; similarly, some odors might trigger the recall of associated memories while others might not. In order to assess psychoactive effects of EOs and fragrances, it seems necessary to control for these factors; for instance, by assessing additional variables which inform on the subject's perception of primary and secondary stimulus features and which correlate to the subject's cognitive or emotional arousal state. In the above mentioned study, Lorig and Schwarzt (Lorig and Schwartz 1988) collected ratings of intensity and pleasantness of the tested fragrances as well as subjective ratings of a number of affective states in addition to the EEG recordings. Analysis of the amount of EEG theta activity revealed that the spiced apple odor produced more relaxation than lavender and eucalyptus; the analysis of the secondary variables suggested that this relaxing effect was correlated with subjective estimates of anxiety and tension. As to the EEG patterns, similar results were observed when subjects imagined food odors and practiced relaxation techniques. Thus, the authors conclude that the relaxing effect of spiced apple was probably related to its association with food. These cognitive influences also seem to be a plausible explanation for the increase in alpha power by lemon odor in the Ishikawa et al. study (Ishikawa, Miyake, and Yokogoshi 2002). Other studies related to food odors were conducted by Kaneda and colleagues (Kaneda et al. 2005; Kaneda et al. 2006, 2011). These authors investigated the influence of smelling beer flavors on the frequency fluctuation of alpha waves in frontal areas of the brain. The results showed relaxing effects of the aroma of hop extracts as well as of linalool, geraniol, ethyl acetate, and isoamyl acetate, but not of humulene and myrcene. In addition, in the right hemisphere, these fluctuations were correlated with subjective estimates of arousal and with the intensity of the hop aroma. Lee et al. (Lee et al. 1994) also found evidence for differential EEG patterns as a function of odor intensity for citrus, lavender, and a floral odors. A ten-minute exposure to the weaker intensity of the citrus fragrance in comparison to lavender odor increased the rate of occipital alpha. Moreover, there was a general trend for citrus to be rated as more comfortable than the other fragrances. In contrast, the higher intensity of the floral fragrance increased the rate of occipital beta more than lavender odor.

Several authors have shown influences of odor pleasantness and familiarity on changes of the spontaneous EEG. For instance, Kaetsu and colleagues (Kaetsu et al. 1994) reported that pleasant odors increased alpha activity, while unpleasant ones decreased it. In a study on the effects of lavender and jasmine odor on electrical brain activity (Yagyu 1994), it was shown that changes in the alpha, beta, and theta bands in response to these fragrances were similar when subjects rated them as pleasant, while lavender and jasmine odor led to distinct patterns when they were rated as unpleasant. Increases of alpha activity in response to pleasant odors might be explained by altered breathing patterns since it has been demonstrated that pleasant odors induce deeper in- and exhalations than unpleasant odors and that this form of breathing by itself increases activity in the alpha band (Lorig 2000). Masago and co-workers (Masago et al. 2000) tested the effects of lavender, chamomile, sandalwood, and eugenol fragrances on ongoing EEG activity and self-ratings of comfort and found a significant positive correlation between the degree of comfort and the odorants' potency to decrease alpha activity in parietal and posterior temporal regions. In relation to the previously described investigations, this finding is rather difficult to explain, although it differs from the other studies in that it differentiated between electrode sites rather than reporting merely global changes in electrical brain activity. Therefore, this result suggests that topographical differences in electrical brain activity induced by fragrances may be important and need further investigation. In fact, differences in hemispheric localization of spontaneous EEG activity in response to pleasant and unpleasant fragrances, respectively, seem to be quite consistent. While pleasant odors induced higher activation in left frontal brain regions, unpleasant ones led to bilateral and widespread activation (Kim and Watanuki 2003) or no differences were observed when an unpleasant odor (valerian) was compared to a no-odor control condition (Kline et al. 2000). Another interesting finding in the study of Kim et al. (Kim and Watanuki 2003) was that EEG activity in response to the tested fragrances was observed when subjects were at rest but vanished after performing a mental task. The importance of distinguishing EEG activity arising from different areas of the brain is highlighted by an investigation by Van Toller and co-workers (Van Toller et al. 1993). These authors recorded alpha wave activity at 28 sites of the scalp immediately after the exposure to a number of fragrances covering a range of different odor types and hedonic tones at iso-intense concentrations; the odorants had to be rated in terms of pleasantness, familiarity, and intensity. It was shown that in posterior regions of the brain, changes in alpha activity in response to these odors compared to an odorless blank were organized in distinct topographical maps. Moreover, alpha activity in a set of electrodes at frontal and temporal sites correlated with the psychometric ratings of the fragrances.

More recently, a research group from Thailand screened a variety of EOs used in aromatherapy for their effects on electrical brain activity and mood after inhalation. All experiments were conducted in healthy human subjects and employed the same A-B-C design, that is, a no-odor control condition (A) was followed by a condition (B) in which a carrier oil without EO was presented and finally by the experimental condition (C) in which the EO of interest diluted in the carrier was administered. The EOs under study were jasmine (Sayowan et al. 2013), rosemary (Sayorwan et al. 2013), citronella (Sayowan et al. 2012), and lavender (Sayorwan et al. 2012). In light of the above-mentioned influence of hedonic odor evaluation on EEG activity, it is important to mention that in all of these investigations, subjects were only allowed to participate if they rated the tested odors as highly to moderately pleasant. The authors were able to confirm the effect described by traditional aromatherapy for every single EO. In regard to the inconsistent findings reported by others, the perfect conformity between observed effects and aromatherapeutic claims is quite remarkable. The selection of subjects based on their hedonic preferences does not seem to offer a fully satisfactory explanation since there shouldn't have emerged differences between the EOs if the effects relied solely on mechanisms involving hedonic odor valence. Perhaps other factors, such as familiarity with the odors and, even more importantly, expectations of their effects (Lorig and Roberts 1990; Jellinek 1997), have influenced the physiological response of the study participants. Thus, it would have been worthwhile for the authors to assess these potentially confounding influences.

A very interesting approach to the study of EO effects on EEG activity and how they might be related to olfactory perception has been reported by Seo et al. (Seo et al. 2016). This group compared human EEG activity in response to the EO of Korean fir (*Abies koreana* Wilson [Pinaceae]) after unilateral and bilateral nasal inhalation in twenty healthy male and female volunteers. A defined amount (10 μL) of the EO was presented on paper strips in a randomized sequence either to both nostrils or solely to the left and right nostril, respectively, at a distance of 3 cm. In the unilateral conditions, the other nostril was blocked with a cotton plug. In each condition, EEG recordings were taken during 45 s before and during EO inhalation. Between conditions, subjects were allowed to rest for 3 min. During the whole procedure, subjects were asked to sit quietly with eyes closed and to breathe normally. The results indicated significant differences in EEG activity pattern during EO inhalation compared to baseline as a function of condition and electrode position. In short, bilateral inhalation increased alpha wave activity as well as beta wave activity in left frontal and right parietal regions, supporting a relaxing effect. Unilateral exposure, however, affected EEG activity quite differently: while inhalation through the right nostril increased alertness and attention—as shown by a reduction in theta wave activity in parietal areas—left-nostril stimulation was associated with decreased beta and theta wave activity in frontal and parietal brain regions. The authors suggested that differences in air-flow between the two nostrils may have driven these differences in physiological effect. Although Seo and colleagues collected information on some psychophysical features of the Korean fir oil (e.g., pleasantness), the observed changes of electrical brain activity were not correlated with such subjective ratings. However, the relation between subjective evaluation and physiological effects could be highly informative since differences in air-flow between the nostrils, the so-called nasal cycle, have been associated with differences in perception (Sobel et al. 1999).

As to influences of the familiarity of or experience with fragrances, Kawano (Kawano 2001) reported that the odors of lemon, lavender, patchouli, marjoram, rosemary, and sandalwood increased alpha activity over occipital electrode sites in subjects to whom these fragrances were well known. On the other hand, lag times between frontal and occipital alpha phase were shorter in subjects less experienced with the fragrances, indicating that these subjects were concentrating more on smelling—and probably identifying—the odorants. These findings were confirmed in an investigation comparing professional perfume researchers, perfume salespersons, and general workers (Min et al. 2003). This study showed that measures of cortico-cortical connectivity (i.e., the averaged cross mutual information content) in odor processing were more pronounced in frontal areas with perfume researchers, whereas with perfume salespersons and general workers, a larger network of posterior temporal, parietal, and frontal regions was activated. These results could be due to a greater involvement of orbitofrontal cortex neurons in perfume researchers who exhibit high sophistication in discriminating and identifying odors. Moreover, it was shown that the value of the averaged cross mutual information content was inversely related to preference in perfume researchers and perfume salespersons, but not in general workers.

As pointed out above, the administration of EOs and fragrances to naïve subjects can lead to cognitive processes, which are unknown to the investigator and sometimes even the subject. Nevertheless, these processes will affect spontaneous EEG activity. Some researchers have sought to solve this problem by engaging subjects in a secondary task while the influence of the odorant of interest was assessed. This procedure does not only draw the subject's attention away from the odor stimulus but also provides the desired information about his/her arousal state. Another benefit of such experimental designs is that the task may control for the subject's arousal state if a certain amount of attention is required to perform it. Measurement of changes of alpha and theta activity in the presence or absence of 1,8-cineole, methyl jasmonate and *trans*-jasmin lactone in subjects who performed a simple visual task showed that the increase in slow wave activity was attenuated by 1,8-cineole and methyl jasmonate, while augmented by *trans*-jasmin lactone (Nakagawa et al. 1992). At least in the case of 1,8-cineole, these findings are supported by results from experimental animals and humans, indicating activating effects of this odorant (Kovar et al. 1987; Nasel et al. 1994; Bensafi et al. 2002). An investigation of the effects of lemon odor on EEG alpha, beta, and theta activity showed that the odor reduced power in the lower alpha range while it increased power in the higher alpha, lower beta, and lower theta bands (Krizhanovs'kii et al. 2004). These findings of increased arousal were in agreement with better performance in a cognitive task. In addition, it was shown that inhalation of the lemon fragrance was most effective during rest and in the first minutes of the cognitive task, but wore off after less than 10 min. In several experiments, the group of Sugawara demonstrated complex interactions between electrical brain activity induced by the exposure to fragrances, sensory profiling, and various types of tasks (Sugawara et al. 2000; Satoh and Sugawara 2003). In one study, they showed that the odor of peppermint, in contrast to basil, was rated less favorable on a number of descriptors and reduced the magnitude of beta waves after, as compared to before, performance of a cognitive task. In a similar investigation, these authors showed that the sensory evaluation as well as changes in spontaneous EEG activity in response to the odors of the linalool enantiomers differed as a function of the molecular structure and the kind of task. For instance, R-(–)-linalool was rated more favorable and led to larger decreases of beta activity after listening to natural sounds than before. In contrast, after, as compared to before, cognitive effort R-(–)-linalool was rated as less favorable and tended to increase beta power. A similar pattern was found for RS-(±)-linalool, whereas for S-(+)-linalool, the pattern was different, particularly with regard to EEG activity.

In the study of Yagyu (Yagyu 1994), the effects of lavender and jasmine fragrances on the performance in a critical flicker fusion and an auditory reaction time task were assessed in addition to changes of the ongoing EEG. In contrast to the EEG findings, lavender decreased performance in both tasks, independent of its hedonic evaluation. Jasmine, however, had no effect on task performance. The EEG changes in response to these odorants might well explain their effects on performance: lavender induced decreases of activity in the beta band, which is associated with states of low attention, regardless of being rated as pleasant or unpleasant; jasmine, on the other hand, increased EEG beta activity when it was judged unpleasant but lowered beta activity when judged pleasant, so that overall its effect on performance leveled out. The effects of lavender and rosemary fragrances on electrical brain activity, mood states, and math computations were investigated by Diego and Field and co-workers (Diego et al. 1998; Field et al. 2005). These investigations showed that the exposure to lavender increased beta power, elevated feelings of relaxation, reduced feelings of depression, and improved both speed and accuracy in the cognitive task. In contrast, rosemary odor decreased frontal alpha and beta power, decreased feelings of anxiety, increased feelings of relaxation and alertness, and increased speed in the math computations. The EEG results were interpreted as indicating increased drowsiness in the lavender group and increased alertness in the rosemary group; however, the behavioral data showed performance improvement and similar mood ratings in both groups. These findings suggest that different electrophysiological arousal patterns may still be associated with similar behavioral arousal patterns, emphasizing the importance of collecting additional endpoints to evaluate psychoactive effects of EOs and fragrances.

12.3.2 Contingent Negative Variation

The contingent negative variation (CNV) is a slow, negative-event-related brain potential, which is generated when an imperative stimulus is preceded by a warning stimulus. It reflects expectancy and preparation (Walter et al. 1964). The amplitude of the CNV is correlated to attention and arousal (Tecce 1972). Since changes of the magnitude and latency of CNV components have long been associated with the effects of psychoactive drugs (Kopell et al. 1974; Ashton et al. 1977), measurement of the CNV has also been used to evaluate psychostimulant and sedating effects of EOs and fragrances. In a pioneering investigation, Torii and colleagues (Torii et al. 1988) measured CNV magnitude changes evoked by a variety of EOs, such as jasmine, lavender, and rose oil, in male subjects. CNV was recorded at frontal, central, and parietal sites after the presentation of an odorous or blank stimulus in the context of a cued reaction time paradigm. In addition, physiological markers of arousal (i.e., skin potential level and heart rate) were measured. Results showed that at frontal sites, the amplitude of the early negative shift of the CNV was significantly altered after the presentation of odor stimuli, and that these changes were mostly congruent with stimulating and sedative properties reported for the tested oils in the traditional aromatherapy literature. In contrast to other psychoactive substances, such as caffeine or benzodiazepines, presentation of the EOs affected neither physiological parameters nor reaction times. The authors concluded that the EOs tested influenced brain waves "almost exclusively" while having no effects on other indicators of arousal.

Subsequently, CNV recordings have been used by other researchers on a variety of EOs and fragrances to establish effects of odors on the human brain along the activation–relaxation continuum. For instance, Sugano (Sugano 1992), in the aforementioned study, demonstrated that α-pinene, sandalwood, and lavender odor increased the magnitude of the CNV in healthy young adults, whereas eucalyptus reduced it. It is interesting to note, however, that all of these odors—despite their differential influence on the CNV—increased spontaneous alpha activity in the same experiment. An increase of CNV magnitude was also observed with the EO from pine needles (Manley 1993) which was interpreted as a stimulating effect. Aoki (Aoki 1996) investigated the influence of odors from several coniferous woods—hinoki (*Chamaecyparis obtusa* [Siebold & Zucc.] Endl. [Cupressaceae]), sugi (*Cryptomeria japonica* D. Don [Cupressaceae]), akamatsu (*Pinus densiflora* Siebold & Zucc. [Pinaceae]), hiba (*Thujopsis dolabrata* var. *hondai* Siebold & Zucc. [Cupressaceae]), Alaska cedar (*Chamaecyparis nootkatensis* [D. Don] Spach [Cupressaceae]), Douglas fir (*Pseudotsuga manziesii* [Mirbel] Franco [Pinaceae]), and Western red cedar (*Thuja plicata* Donn [Cupressaceae])—on the CNV and found conflicting effects: the amplitude of the early CNV component at central sites was decreased by these wood odors, and the alpha/beta-wave ratio of the EEG increased. Moreover, the decrease of CNV magnitude was correlated with the amount of α-pinene in the tested wood odors. Also, Sawada and co-workers (Sawada et al. 2000) measured changes of the early component of the CNV in response to stimulation with terpenes found in the EOs of woods and leaves. These authors noticed a reduction of the CNV magnitude after the administration of α-pinene, Δ-3-carene, and bornyl acetate. However, a more recent investigation (Hiruma et al. 2002) showed that hiba (*Thujopsis dolabrata* Siebold & Zucc. [Cupressaceae]) odor increased the CNV magnitude at frontal and central sites and shortened reaction times to the imperative stimulus in female subjects. These authors thus concluded that the odor of hiba heightened the arousal level of the CNS.

Although the CNV is believed to be largely independent of individual differences, such as age, sex, or race (Manley 1997), there seem to be cognitive influences which need not be considered when interpreting the effects of odor stimuli on the CNV. Lorig and Roberts (Lorig and Roberts 1990) repeated the study by Torii and colleagues (Torii et al. 1988), but they introduced a new variable into the paradigm: subjects were exposed to the original two odors—that is, lavender and jasmine, referred to as odor A and odor B, respectively—as well as to a mixture of the two fragrances. However, in half of the trials in which the mixture was administered, subjects were led to believe that they received a low concentration of odor A while in the other half of the trials, they thought that they would be exposed to a low concentration of odor B. In none of the four conditions were subjects given the correct odor names. As in the Torii et al. study, lavender reduced the amplitude of the CNV, whereas jasmine increased it. When the mixture was administered, the CNV magnitude, however, decreased when subjects believed that they would receive a low concentration of lavender, but increased when they thought they were inhaling a low concentration of jasmine. This means that the alteration of the CNV amplitude was not solely related to the substance that had been administered but also to the expectation of the subjects. Another point made by Lorig and Roberts (Lorig and Roberts 1990) is that in their study, self-report data indicated that lavender was actually rated as more arousing than jasmine. Since low CNV amplitudes are not only associated with low arousal but also with high arousal in the context of distraction (Travis and Tecce 1998), the lavender odor might in fact have led to higher arousal levels than jasmine, even though the CNV magnitude was smaller with lavender. Other authors have noted that CNV changes might not only reflect effects of odor stimuli but also anticipation, expectancy, and the emotional state of the subjects who are exposed to these odorants (Hiruma et al. 2005). The involvement of these and other cognitive factors might well explain why the findings of CNV changes in response to odorants are rather inconsistent.

12.4 EFFECTS OF ESSENTIAL OILS AND FRAGRANCES ON SELECTED BASIC AND HIGHER COGNITIVE FUNCTIONS

Psychoactive effects of odorants at the cognitive level have been explored in humans using a large number of methods. A variety of testing procedures ranging from simple alertness or mathematical tasks to tests that assess higher cognitive functions, such as memory or creativity, have been employed to study stimulant or relaxing/sedating effects of EOs and fragrances. Nevertheless, the efficiency of odorants is commonly defined by changes in performance in such tasks as a function of the exposure to fragrances.

12.4.1 ALERTNESS AND ATTENTION

A number of studies are available on the influence of fragrances on attentional functions. The integrity and the level of the processing efficiency of the attentional systems is a fundamental prerequisite of all higher cognitive functions. Attentional functions can be divided into four categories: alertness, selective attention, divided attention, and vigilance (Posner and Rafal 1987; Keller and Groemminger 1993; Sturm 1997). Alertness is the most basic form of attention and is intrinsically dependent on the general level of arousal. Selective attention describes the ability to focus on relevant stimulus information while non-relevant features are neglected; divided attention describes the ability to concomitantly process several stimuli from different sensory modalities. Vigilance refers to the sustainment of attention over longer time periods. Since the critical stimuli typically occur only rarely in time, vigilance can be seen as a counterforce against increasing fatigue in boring situations. Vigilance is crucial in everyday life, in situations like long-distance driving (particularly at night), working in assembly lines, or monitoring a radar screen (e.g., in air traffic control).

In a pioneering study, Warm and colleagues (Warm et al. 1991) investigated the influence of peppermint and muguet odors on human visual vigilance. Peppermint, which was rated stimulant, was expected to increase task performance, while muguet, rated as relaxing, was expected to impair it. After intermittent inhalation, none of these fragrances increased processing speed in the task, but subjects in both odorant conditions detected more targets than a control group receiving unscented air. On the other hand, neither fragrance influenced subjective mood or judgements of workload. Gould and Martin (Gould and Martin 2001) studied the effects of bergamot and peppermint EOs on human sustained attention. Again, peppermint was expected to improve performance, while bergamot was characterized as relaxing by an independent sample of subjects and was thus expected to have a deteriorating effect on vigilance performance. However, only bergamot had a significant influence in the anticipated direction; that is, subjects in this condition detected fewer targets than subjects in the peppermint or a no-odor control condition, which was probably related to the subjects' expectation of a relaxing effect.

The influence of the inhalation of a number of EOs and fragrances on performance in basic attentional tasks was assessed by Heuberger and Ilmberger (2010) and Ilmberger and co-workers (Ilmberger et al. 2001). In the first experiment, several EOs with presumed activating effects were inhaled by human subjects during an alertness task. Contrary to the authors' hypotheses, the results suggested that these Eos, when compared to an odorless control, did not increase the speed of information processing. Even more unexpectedly, motor learning was impaired in the groups that received EOs. According to the authors, this effect was due to distraction induced by the strong odor stimuli. Their interpretation was supported by the reaction times that tended to be higher in the EO treated groups than in the corresponding control groups. Alternatively, the authors argued that a ceiling effect might be responsible for the observed effects. Given that healthy subjects with intact attentional systems already perform at optimal levels of information processing in such basic tasks, it seems likely that activating EOs cannot enhance performance any further. Similarly, performance of healthy subjects may be too robust to be influenced by deactivating fragrances. In the second study, subjects inhaled several activating and sedating fragrances while engaging in a vigilance task. Again, the expected effects were not observed. For one compound (linalyl acetate), even the opposite effect on vigilance performance was demonstrated. Further analyses showed that this effect was strongly correlated to subjective estimates of odor pleasantness.

Schneider (Schneider 2016) explored the effects of two EO mixtures on selective attention. Healthy human subjects inhaled the odors via a special nasal inhalator. The author used the d2 test, a paper-and-pencil test in which several targets have to be marked while neglecting a number of distractors. A fixed number of items has to be processed in blocks in a given time interval. After time has expired, participants must stop processing the current block whether or not they have reached its end and start with the next block. In this study, eight blocks of items had to be processed. In-between blocks, seven 1 min breaks were introduced during which subjects inhaled either one of the two mixtures (treatment groups) or practiced a technique of personal choice to re-focus on the task (control group). The EO mixtures called "focus" and "alert" consisted of peppermint, rosemary, and cinnamon EO and lemon, peppermint, rosemary, grapefruit, black pepper, and basil oil, respectively. During each break, two inhalations were taken, alternatingly through one nostril while the other one was closed with a finger. Subjects were instructed to inhale through the open nostril, hold their breath for a little while, and exhale through their mouth. The results of this study indicated that subjects in both odor conditions performed better than the control group with regard to both speed and accuracy. The author interpreted these findings as indicating a positive effect of the odor mixtures. However, since pre-treatment baseline performance measures were not assessed in this study, it is not clear whether the results really show an increase in selective attention performance induced by the EO mixtures or just *a priori* group differences. Even more ambiguously, when the task was divided into two halves and performance in the first four blocks was compared to that in the second four blocks, it appeared that the strongest improvement took place in the no-odor control group. According to the author, this finding proved that the observed overall improvement in performance in the odor groups was not due to a stronger learning effect. However, one could also argue that inhalation of the EO mixtures attenuated learning over time. Although this study yielded promising results, their interpretation is strongly constrained by the abovementioned methodological weaknesses of the investigation.

The observation that the effectiveness of fragrances is dependent on task complexity is supported by investigations on the EO of peppermint (Ho and Spence 2005), lavender, and rosemary (Moss et al. 2003). These fragrances rather affected performance in difficult tasks or in tasks testing higher cognitive functions than in simple ones testing basic functions. Another interesting finding of the study of Ilmberger et al. (2001) was that changes in performance were correlated with subjective ratings of characteristic odor properties, particularly with pleasantness and efficiency. Similar results were obtained in another study for the EO of peppermint (Sullivan et al. 1998) which showed that in a vigilance task, subjects benefited most from the effects of this fragrance when they experienced the task as quite difficult and thought that the EO had a stimulant effect. In addition, the studies of

Ilmberger and co-workers clearly demonstrated effects of expectation, that is, a placebo effect, as correlations between individual task performance and odor ratings were not only revealed in the EO groups but also in the no-odor control groups. This finding was in part supported by a recent study by Babulka and co-workers (Babulka et al. 2017). Their investigation showed no effects of inhaled lavender or rosemary EO or a placebo pill with stimulant suggestion in comparison to a no-treatment control condition on sustained attention, self-rated alertness, or heart rate. However, both EOs and the placebo pill induced positive expectations about cognitive performance and self-rated alertness. Correlation analyses demonstrated that the expected changes predicted perceived (but not observed) changes with respect to commission errors and self-rated alertness. Thus, the authors concluded that expectations played a critical role in the perceived effects of the tested EOs. Nevertheless, a study by Moss and Oliver (Moss and Oliver 2012) strongly supports the existence of pharmacological factors involved in the effects of inhaled rosemary EO on human attentional performance. These authors studied several tasks with varying cognitive load and found significant correlations between a number of performance measures and plasma levels of 1,8-cineole, which is one of the main compounds of rosemary EO.

The effects of a pleasant and an unpleasant blend of fragrances on selective attention was studied by Gilbert and colleagues (Gilbert et al. 1997). No influence of either fragrance blend was found on attentional performance, but the authors observed a sex-specific effect of suggesting the presence of ambient odors. In the presence of a pleasant or no odor in the testing room, male subjects performed better when they were led to believe that no odor was present. Female subjects, however, performed better when they thought that they were exposed to an odorant under the same conditions. No such interaction was found in the unpleasant fragrance condition. These data again emphasize that, in addition to hedonic preferences, expectation of an effect may crucially influence the effects of odorants on human performance and that these factors may affect women and men differently.

Millot and co-workers evaluated the influence of pleasant (lavender oil) and unpleasant (pyridine) ambient odors on performance in a visual or auditory alertness task and in a divided attention task (Millot et al. 2002). The results showed that in the alertness task, both fragrances compared to an unscented control condition improved performance by shortening reaction times. These findings were irrespective of the modality of the alertness task and independent of the hedonic valence of the odors. However, none of the odorants exerted any influence on performance in the selective attention task in which subjects had to attend to auditory stimuli while neglecting visual ones. The authors concluded that pleasant odors enhance task performance by decreasing subjective feelings of stress (i.e., by reducing over-arousal), while unpleasant fragrances increase activation from suboptimal to optimal levels, thus having the same beneficial effects on cognitive performance. With this explanation, the authors, however, presume that subjects in their experimental groups started from dissimilar arousal levels, which seems rather unlikely given that subjects were assigned to these groups at random. Moreover, cognitive performance should be affected by alterations of the arousal level more readily with increasing task difficulty. Thus, the interpretation given by the authors does not thoroughly explain why reaction times were influenced by the odorants in the simple alertness task but not in the more sophisticated selective attention task.

Degel and Köster (Degel and Köster 1999) exposed healthy subjects to either lavender, jasmine, or no fragrance. Subjects were unaware of the presence of odorants. They had to perform a mathematical test, a letter-counting (i.e., selective attention) task, and a creativity test. The authors expected a negative effect on performance of lavender and a positive effect of jasmine. The results, however, showed that lavender decreased the error rate in the selective attention task, whereas jasmine increased the number of errors in the mathematical test. Ratings of odor valence collected after testing demonstrated that lavender was judged more pleasant than jasmine, independent of which odor had been presented during the testing. As subjects did not know that a fragrance had been administered, implicit evaluation of odor pleasantness probably influenced their performance. This relation is supported by the fact that subjects who were not able to correctly identify the odors preferentially associated pictures of the room they had been tested in with the odor that had been present during testing. Improvement of performance as a result of the inhalation of lavender EO has also been reported in another investigation (Sakamoto et al. 2005). In this study, subjects completed five sessions of a visual vigilance task involving tracking of a moving target. During phases of rest in-between sessions, they were exposed to lavender, jasmine, or no aroma. As estimated from the performance decrement in the control group, fatigue was highest and arousal was lowest in the last session. By contrast, tracking speed increased and tracking error decreased in the lavender group in the last session when compared to the no-aroma group. Jasmine had no effect on task performance. The authors argued that lavender aroma decreased arousal during the resting period and hence helped to achieve optimal levels for the following task period. Since no secondary variables indicative of arousal or of subjective evaluation of aroma quality were assessed in this investigation, no inferences can be made on the mechanisms underlying the observed effects. Diego and co-workers in the aforementioned investigation studied the influence of lavender and rosemary EOs after a three-minute inhalation period on a mathematical task (Diego et al. 1998). In contrast to the authors' expectations, both odorants positively affected performance by increasing calculation speed, although only lavender improved calculation accuracy. In addition, subjects in both fragrance groups reported to be more relaxed. Those in the lavender group had less depressed mood, while those in the rosemary group felt more alert and had lower state anxiety scores. These findings were interpreted as indicating over-arousal caused by rosemary EO, which led to an increase of calculation speed at the cost of accuracy. In contrast, lavender EO seemed to have reduced the subjects' arousal level and thus led to better performance than rosemary EO. However, since subjects in both fragrance groups felt more relaxed—but obviously only the lavender group benefited from this increase in relaxation—this is a somewhat unsatisfying explanation for the observed results.

Evidence for the influence of physicochemical odorant properties on visual information processing was supplied by Michael and colleagues (Michael et al. 2005). These authors found that the exposure to both allyl isothiocyanate (AIC), a mixed olfactory/trigeminal stimulus, and 2-phenyl ethyl alcohol (2-PEA), a pure olfactory stimulant, impaired performance in a highly demanding visual attention task. The task involved reaction to a target as well as neglecting a distractor that appeared at different time intervals. In trials without a distractor, only 2-PEA significantly increased the reaction times of healthy subjects; in trials with a distractor, subjects reacted more slowly in both odor conditions as compared to the no-odor control condition. However, AIC impaired performance independent of the interval between distractor and target, whereas 2-PEA only had a negative effect when the interval between target and distractor was short. While 2-PEA seemed to have led to performance decrements by decreasing subjects' arousal levels, AIC as a strong trigeminal irritant seemed to have shifted attention toward the distractor stimuli. A similar observation has also been made for the annoying odor propionic acid (Hey et al. 2009). In this study, the error rate in a response-inhibition task increased as a function of odorant concentration, suggesting a relationship between cognitive distraction and sensory annoyance.

Differences in effectiveness of fragrances as a function of the route of administration were explored by Heuberger and co-workers (Heuberger et al. 2008). These authors investigated the influence of two monoterpenes, 1,8-cineole and (±)-linalool, on performance in a visual sustained-attention task. The fragrances were applied for 20 min by inhalation and dermal application, respectively. 1,8-Cineole was expected to induce activation and improve task performance while (±)-linalool was considered sedating/relaxing, thus impairing performance. Since one of the aims of the study was to assess fragrance effects that were not mediated by stimulation of the olfactory system, inhalation of the odorants was prevented in the dermal application conditions. In each condition, subjects rated their mood and well-being. In addition, ratings of odor pleasantness, intensity, and effectiveness were assessed in the inhalation conditions. Regarding performance on the vigilance task, no difference was observed in the inhalation condition between the fragrance groups and a control group, which had received odorless air. However, 1,8-cineole increased feelings of relaxation and calmness, whereas (±)-linalool led to increased vigor and mood. In addition, individual performance was correlated to the pleasantness of the odor and to expectations of its effect. In contrast, in the dermal application conditions, subjects having received 1,8-cineole performed faster than those having received (±)-linalool. These findings were interpreted as indicating the involvement of different mechanisms after inhalation and non-olfactory administration of fragrances. This study suggests that psychological effects are predominant when fragrances are applied by means of inhalation, that is, when the sense of smell is stimulated. On the other hand, pharmacological effects of odorants become evident when processing of odor information is prevented.

Along the same lines, two studies investigated the effects of Spanish sage (*Salvia lavandulifolia* Vahl [Lamiaceae]) on performance in a number of cognitive tasks after oral application or inhalation. In the first study by Kennedy and colleagues (Kennedy et al. 2011), participants who received a single oral dose of 50 μL of the EO of *S. lavandulifolia* reacted faster in a simple reaction-time task than subjects who received a placebo. By contrast, this facilitating effect on attention speed was not detected in the second study when subjects inhaled the EO of *S. lavandulifolia* (Moss et al. 2010). In a very recent double-blind, placebo-controlled investigation in healthy adult human subjects (Kennedy et al. 2018), the effect of different dosages of orally administered peppermint (*Mentha x piperita* L. [Lamiaceae]) EO on cognitive performance and mood measures was explored. Each participant received a capsule containing either 50 μL or 100 μL EO or a placebo capsule containing vegetable oil. The capsules were administered in random order on three separate days with a wash-out period of one week after each administration. In each condition, subjects completed several individual cognitive tests as well as a cognitive demand battery testing attention and executive and memory functions. Moreover, several mood measures were assessed. The cognitive and mood measurements were taken before as well as 1 h, 3 h, and 6 h after ingestion of each capsule. The results of this study demonstrated that administration of the higher dose of peppermint EO, as compared to both the lower dose and the placebo, improved cognitive performance (e.g., sustained attention) assessed by means of the battery but not by the individual tests. Similarly, self-rated mental fatigue as a component of the cognitive battery was attenuated by the higher dose of oral peppermint EO, but individual mood measures remained unchanged. According to the authors, this finding is owed to a higher sensitivity of the test battery or, alternatively, to greater statistical power to detect treatment induced changes. In addition, the authors showed *in vitro* that the tested EO had anticholinergic, calcium regulatory, $GABA_A$ receptor, and nicotinic receptor binding properties and speculated that these properties may be related to the observed effects of peppermint EO on cognitive improvement and fatigue.

12.4.2 Learning and Memory

During the past few years, effects of EOs and fragrances on memory functions and learning have been explored in a growing number of studies. While learning can briefly be defined as "a process through which experience produces a lasting change in behavior or mental processes" (Zimbardo et al. 2003, p. 206), memory is a cognitive system composed of three separate subsystems or stages that cooperate closely to encode, store, and retrieve information. Sensory memory constitutes the first of the three memory stages and is responsible for briefly retaining sensory information. The second stage, working memory, transitorily preserves recent events and experiences. Long-term memory, the third subsystem, has the highest capacity of all stages and stores information based on meaning associated with the information (Zimbardo et al. 2003). A basic form of learning which has been identified as a potent mediator of fragrance effects in humans (Jellinek 1997) is conditioning—that is, the (conscious or unconscious) association of a stimulus with a specific response or behavior. For instance, Epple and Herz (Epple and Herz 1999) demonstrated that children who were exposed to an odorant during the performance of an insolvable task performed worse on other, solvable tasks when the same odorant was presented again. In contrast, no such impairment was observed when no odor or a different odor was presented. These results were interpreted as demonstrating negative olfactory conditioning. Along the same lines, Chu (Chu 2008) was able to show positive olfactory conditioning. In his study, children successfully performed a cognitive task deemed insolvable in the presence of an ambient odor. When they were re-exposed to the same odorant, performance on other tasks improved significantly in comparison to another group of children who received a different odor. Since the children in Chu's investigation were described in school reports as underachieving and lacking self-confidence, one might speculate that only children with these specific attributes benefit from the influence of fragrances. This seemed to be confirmed by a study of Kerl (Kerl 1997), who found that ambient odors of lavender and jasmine did not improve memory functions in school children in general. However, lavender tended to increase performance in the memory task in children with high anxiety levels, which may have been related to the stress-relieving properties of lavender. On the other hand, jasmine impaired memory performance in lethargic children, and this impairment was correlated with the children's rating of the odor's hedonic valence. This result might indicate that lethargic children were distracted by the presence of an odorant they liked.

Several studies examined the influence of inhaled EOs on a number of memory-related variables in adults (Moss et al. 2003, 2008; Tildesley et al. 2005; Moss et al. 2010, 2018). These authors reported that lavender reduced the quality of memory and rosemary increased it, while both EOs reduced the speed of memory when compared to a no-odor control condition. At the same time, rosemary increased alertness in comparison to both the control and the lavender group, but exposure to the odorants led to higher contentedness than no scent. Similarly, peppermint enhanced memory quality while ylang-ylang impaired it and also reduced processing speed. Ratings of mood showed that peppermint increased alertness while ylang-ylang decreased alertness but increased subjective calmness. In a third experiment, oral administration of Spanish sage (*Salvia lavandulifolia* Vahl [Lamiaceae]) improved both quality and speed of memory and increased subjective ratings of alertness, calmness, and contentedness. In a fourth study, the effects of inhalation of two species of sage, *S. officinalis* L. (Lamiaceae) and *S. lavandulifolia* Vahl (Lamiaceae), were compared. This investigation demonstrated differences between the two species in regard to their impact on the quality of memory. In comparison to a no-odor control, only *S. officinalis* aroma improved endpoints related to quality of memory. In contrast, the impact of the two species on subjective measures of mood was identical. Compared to the no-odor control, both species increased alertness but had no effect on calmness and contentedness. Moreover, there were no differences between the two species with respect to odor hedonics or intensity.

The comparison between oral administration and inhalation of *S. lavandulifolia* EO highlights differences in effectiveness in relation to the route of administration that are in accordance with the authors own observations (Hongratanaworakit et al. 2004; Heuberger et al., 2006; Heuberger et al. 2008; Ambrosch et al.2018). Most interestingly, new research on memory suggests that these differences in the effectiveness of fragrances on overt behavior can be traced back to functional changes at the neuronal level. In the study by Ambrosch and co-workers, the effects of 1,8-cineole and (−)-linalool on the neuronal substrate of working memory and cognitive performance were investigated. The authors measured brain activation by means of functional magnetic resonance imaging (fMRI) in young, healthy subjects who performed a working memory task, that is, the 1-back task. In this task, subjects monitored numbers on a screen and had to decide as fast as possible whether two consecutive numbers on the screen were the same or different. Subjects were assigned randomly to one of four groups and received either 1,8-cineole or (−)-linalool by inhalation or dermal application, respectively. During the latter, sensory evaluation of the odorants was precluded in that clean air was delivered to the subjects via a face mask. Administration of the fragrances was initiated about 15 min before subjects performed the cognitive task to allow for the buildup of detectable plasma levels (Friedl et al. 2010, 2015). Brain activation and performance in the working memory task during the administration of the fragrances was compared to appropriate placebo conditions. The authors found significant brain activation in response to dermal administration of 1,8-cineole and (−)-linalool but not after inhalation. Moreover, consistent with the authors' hypotheses, brain areas related to working memory and attention were activated in the 1,8-cineole dermal condition, specifically in the frontal cortex, the anterior cingulate, and the precuneus. In contrast, dermal (−)-linalool led to significant activation of subcortical limbic brain areas, such as the basal ganglia and the posterior cingulate gyrus, particularly in male participants. The former area has been associated with working memory while the latter is part of the default mode network. Thus, the authors suggested that activation of this brain area was indicative of a relaxing effect of dermal (−)-linalool. In terms of cognitive performance, no effects were observed except that subjects in the (−)-linalool dermal condition, particularly males, committed less incorrect responses in the odorant trial than in the placebo trial. Probably, dermal (−)-linalool facilitated working memory performance by inducing a relaxed state. Although the authors identified several limitations of their study, they concluded that dermal application of 1,8-cineole and (−)-linalool, as opposed to inhalation, resulted in the activation of distinct neuronal networks which indicated differential pharmacological effects of these two fragrances.

Odors have been claimed to be powerful cues for contextual memory. The beneficial influence of an odorant being present in the learning phase on successive retrieval of information has been shown, for instance, by Morgan (Morgan 1996). In this study, subjects were exposed or not exposed to a fragrance during the encoding of words unrelated to odor. Recall of the learned material was tested in three unannounced sessions 15 min apart, as well as five days after the learning phase. The results showed that performance in those groups which had not been exposed to an odorant in the learning phase declined continuously over time, whereas it remained stable in those groups that had learned with ambient odor present. In addition, subjects who had learned under odor exposure performed significantly better when the odor was present during recall than those who had not received an odorant during the learning phase. These findings show that odorants in the encoding phase may serve as cues for later recall of the stored information. Recently, similar results have been reported in three-month-old infants (Suss, Gaylord, and Fagen 2012). Schwabe and Wolf (Schwabe and Wolf 2009) showed that the detrimental effects of stress experienced before memory retrieval can be attenuated by odor context cues presented during the encoding phase. Enhancement of declarative memory by olfactory context cues has even been found in adults who were presented with odorants while asleep (Rasch et al. 2007). Moreover, a recent study showed that memory consolidation during sleep is accelerated by odors (Diekelmann et al. 2012). On the other hand, fear memory can be extinguished selectively when odors are presented during sleep, which have served as contextual cues during a learning phase in the awake state (Hauner et al. 2013).

According to a study by Walla and co-workers, it seems to be crucial whether or not an odorant in the encoding phase of a mnemonic task is consciously perceived and processed (Walla et al. 2002). These authors found differences in brain activation in a word-recognition task as a function of conscious versus unconscious olfactory processing in the encoding phase. In other words, when odorants were presented and consciously perceived during the learning phase, word recognition was more likely negatively affected than when the odor was not consciously processed. In addition, the same researchers demonstrated that word recognition performance was significantly poorer when the odorants were presented simultaneously with the words as opposed to continuously during the encoding phase and when semantic (deep), as opposed to non-semantic (shallow), encoding was required. These effects can be explained by a competition of processing resources in brain areas engaged in both language and odor processing (Walla et al. 2003a). Similar results were observed in an experiment involving the encoding of faces with and without odorants present in the learning phase (Walla et al. 2003b). Again, recognition accuracy was impaired when an odor was presented simultaneously with the face stimulus during encoding.

Reichert et al. took a somewhat different approach (Reichert et al. 2017). These authors examined the effects of congruent vs. incongruent contextual odors on the encoding and recognition of abstract line drawings and the underlying functional brain networks. In this study, two groups of subjects were presented simultaneously with an olfactory stimulus (lavender) via an olfactometer while they saw abstract line drawings that had to be encoded on a computer screen. In the subsequent retrieval conditions, the olfactory context was either the same (lavender, congruent group) or different (vanilla, incongruent group). A third group received only odorless air during encoding and retrieval. These authors did not find any differences in behavioral retrieval performance between the three groups, which might be attributed to the noisy and demanding environment of the fMRI scanner. Also, as pointed out above by Walla and co-workers, processing and timing issues may be responsible for the lack of memory enhancement in this study. In contrast, differences in the recruitment of functional brain networks were observed that point to the involvement of the olfactory cortex when recognition is successful and the odor context is congruent during encoding and retrieval of memory.

As discussed above, subjective experience of valence seems to modulate the influence of fragrances on cognition but it seems that not all cognitive functions are affected in the same way and to the same degree. Hackländer and Bermeitinger (Hackländer and Bermeitinger 2017) tackled the question whether olfactory-enhanced memory is contingent on affective congruency between the olfactory cue and the to-be-remembered material. In their study, three independent groups of subjects received a pleasant, an unpleasant, or no odor stimulus immediately before the presentation of 90 words that had to be encoded and that were either positive, negative, or neutral in regard to valence. The results of this study showed that in comparison to the no-odor group, recognition of the verbal material was enhanced in both odor groups irrespective of the affective congruency of the olfactory stimuli and the words; that is, both the pleasant and the unpleasant odorant increased word retrieval. In contrast, when investigating short-term memory, Danuser and co-workers (Danuser et al. 2003) found no effects of pleasant olfactory stimuli, whereas unpleasant odorants reduced the performance of healthy subjects, probably by distracting them. Martin and Chaudry (Martin and Chaudry 2014) described a detrimental effect of unpleasant ambient odor compared to pleasant odor on working memory and explained this finding by a shift in attention (distraction) caused by the unpleasant stimulus. Interestingly, in their study, one specific aspect of working memory—spatial span—was particularly affected by the exposure to odorants. However, it is not quite clear why spatial span was more sensitive to the effects of ambient odors than other functions of working memory. Also, Habel and co-workers studied the effect of neutral and unpleasant olfactory stimulation on the performance of a working memory task (Habel et al. 2007). These authors found that malodors significantly deteriorated working memory, but only in about half of the subjects. It was also shown that subjects in the affected group differed significantly in brain activation patterns from those in the unaffected group, the latter showing stronger activation in fronto-parieto-cerebellar networks associated with working memory. In contrast, subjects whose performance was impaired by the unpleasant odor showed greater activation in areas associated with emotional processing, such as the temporal and medial frontal cortex. The authors concluded that individual differences exist for the influence of fragrances on working memory and that unaffected subjects were better able to counteract the detrimental effect of unpleasant odor stimuli. This interpretation receives strong support by a recent investigation by Nordin and co-workers (Nordin et al. 2017). This group reported that exposure to unpleasant, as opposed to pleasant, olfactory stimuli impaired the ability to focus on an imagined cognitive task and could thus explain the negative impact of unpleasant ambient odors on cognitive performance that has been found in many studies.

Leppanen and Hietanen (Leppanen and Hietanen 2003) tested the recognition speed of happy and disgusted facial expressions when pleasant or unpleasant odorants were presented during the recognition task. This study showed that pleasant olfactory stimuli had no particular influence on the speed of recognition of emotional facial expressions, that is, happy faces were recognized faster than disgusted faces. This result was also observed when no odorant was administered. In the unpleasant condition, however, the advantage for recognizing happy faces disappeared. In the view of the authors, these findings demonstrate that unpleasant odorants may modulate emotion-related brain structures, which form the perceptual representation of facial expressions. Walla and colleagues supplied evidence that performance in a face recognition task was only affected when conscious odor processing took place (Walla et al. 2005). In this study, two olfactory stimuli (2-PEA and dihydrogen sulfide [H_2S]), a trigeminal stimulus (carbon dioxide [CO_2]), or no odor were presented briefly and simultaneously to the presentation of faces. The results showed that irrespective of their valence, the pure odorants improved recognition performance whereas CO_2 decreased it. In addition, only CO_2, which is associated with painful sensations, was processed consciously by the participants of this investigation.

The effect of expectancy on implicit learning was demonstrated in a recent study by Colagiuri et al. (Colagiuri et al. 2011). Expectancy was manipulated in this experiment by providing to the subjects positive, negative, or no information about a possible effect of an odor present during a visual search task. Unknown to the subjects, the search task comprised a contingency: on half of the trials, the location of the target was cued by a distinct spatial configuration of the distractors. While neither the mere presence of the odor nor the manipulation had any influence on the participants' awareness of the contingency, reaction time on cued trials was affected by the information subjects had received on the effect of the odor. Participants who had received the negative information reacted more slowly than those who had received no information. Those who had received the positive information reacted faster than the other groups. These findings strongly resemble the effects of expectation observed on EEG measures such as the CNV and basic cognitive tasks.

With regard to the content of memory, some researchers have claimed a special relationship between autobiographical—that is, personally meaningful, episodic—memories and fragrances. As a result of this special link, it has been observed that memories evoked by olfactory cues are often older, more vivid, more detailed, and more affectively toned than those cued by other sensory stimuli (Chu and Downes 2002; Goddard et al. 2005; Willander and Larsson 2007). This phenomenon has been explained by the peculiar neuroanatomical connection of the memory systems with the emotional systems. Evidence for this hypothesis has for instance been supplied by the group of Herz (Herz and Cupchik 1995; Herz 2004; Herz et al. 2004) who demonstrated that presentation of odorants resulted in more emotional memories than presentation of the same cue in auditory or visual form. The authors also showed that if the odor cue was hedonically congruent with the item that had to be remembered, memory for associated emotional experience was improved. Moreover, personally salient fragrance cues were associated with higher functional activity in emotion-related brain regions, such as the amygdala and the hippocampus.

12.4.3 Other Cognitive Tasks

The study of Degel and Köster described earlier (Degel and Köster 1999) showed that under certain conditions, odorants may influence attentional performance even when subjects are unaware of their presence. According to an investigation by Holland and co-workers, the unnoticed presence of odorants may also affect everyday behavior and higher cognitive functions (Holland et al. 2005). The authors reported that subliminal concentrations of a citrus-scented cleaning product increased identification of cleaning-related words in a lexical decision task. Moreover, subjects exposed to the subliminal odor listed cleaning-related activities more frequently when asked to describe planned activities during the day and kept their environment tidier during an eating task.

The effects of pleasant suprathreshold fragrances on other higher cognitive functions were, for instance, investigated by Baron (Baron 1990). In this study, subjects were exposed to pleasant or neutral ambient odors while solving a clerical coding task and negotiating about monetary issues with a fellow participant. Before performing these tasks, subjects indicated self-set goals and self-efficacy. Following the tasks, subjects rated the experimental rooms in terms of pleasantness and comfort, as well as their mood. In addition, they were asked which conflict management strategies they would adopt in the future. Although neither fragrance had a direct effect on performance in the clerical task, subjects in the pleasant odor condition set higher goals and adopted a more efficient strategy in the task than those in the neutral odor condition. In addition, male subjects in the pleasant-odor condition rated themselves as more efficient than those in the neutral-odor condition. In the negotiation task, subjects in the pleasant-odor condition set higher monetary goals and made more concessions than those in the neutral-odor condition. Moreover, subjects in the pleasant-odor condition were in a better mood and reported planning to handle future conflicts less often through confrontation and avoidance. Thus, this study showed that pleasant ambient fragrances offer a potential to create a more comfortable work environment and diminish aggressive behavior in situations involving competition. Similar findings were reported in a recent study by Sellaro and colleagues (Sellaro et al. 2014), who examined the effects of an arousing ambient odorant (peppermint) and a calming odorant (lavender) on interpersonal trust. The authors used the Trust Game as an experimental paradigm in which one participant (the trustor) transfers money to another participant (the trustee). The authors reported that in comparison to a no-odor control condition and the arousing-odor condition, the trustors transferred significantly more money to the trustees in the calming odor condition, which was probably due to a more inclusive cognitive-control state induced by the calming odor of lavender.

Gilbert and co-workers examined the effect of a pleasant and an unpleasant blend of fragrances on the same clerical coding task as in the study by Baron (Baron 1990) but were unable to show any influence of either fragrance mix on task performance (Gilbert et al. 1997). However, subjects exposed to unpleasant odorants believed that these odorants had negative effects on their performance in simple and difficult mathematical and verbal tasks (Knasko 1993).

Also, Ludvigson and Rottman (Ludvigson and Rottman 1989), studying the influence of lavender and clove EOs in ambient air in comparison to a no-odor control condition, found that lavender impaired performance in a mathematical reasoning task, while clove was devoid of effects. However, the lavender effect was only observed in the first of two sessions held one week apart. Also, subjects in the lavender condition rated the experimental conditions more favorably, while clove odor decreased subjects' willingness to return to the second session. Moreover, subjects who were exposed to an odor in one of the two sessions were generally less willing to return and had worse moods than subjects who never received an odor. To further complicate things, these odorant by session interactions were related to personality factors in a highly complex manner.

Another study that demonstrated rather complex effects of the administration order of olfactory stimuli was reported by Gaygen and Hedge (Gaygen and Hedge 2009). These authors found detrimental effects of a commercial air freshener containing fragrances rated as pleasant on certain aspects of word recognition in a lexical decision task only when the odorants were applied in the second of two independent sessions that was preceded by a no-odor condition. In contrast no effects were found when the odor was applied in the first session and the no-odor condition in the second. Probably, the observed effect is attributable to the influence of a novel cue (odor in session two) in an otherwise familiar context (experimental context without odor) that allocated attentional resources.

More recently, Finkelmeyer and co-workers (Finkelmeyer et al. 2010) investigated the effect of mood states induced by olfactory stimuli on inhibitory control performance. In their study, subjects performed the Stroop color-word interference task, either in the presence of an unpleasant odor (H_2S) generating a negative affective state or an emotionally neutral odor (eugenol). In this task, color names are written in colored ink and the task of the subject is to attend to the color of the ink and ignore the color name. Typically, naming of the ink color is more difficult and reaction times are higher when the color of the ink and the color name are incongruent. Contrary to the authors' expectations, presentation of the negative odor facilitated cognitive processing and reduced Stroop interference, while the neutral odor had no effects on performance. In the opinion of the authors, this finding can be explained by increasing cognitive control as a result of mood-congruent processing. Saito and co-workers presented still different results (Saito et al. 2018). In their study, subjects who were exposed to the honey-like floral odor of MCMP, a blend of four different odorants, during a fatigue-inducing task (a 2-back task) performed more accurately in color-word congruent trials than in a corresponding no-odor control condition. Interestingly, MCMP activated a similar set of olfactory receptors as the so-called Hex-Hex Mix, a blend of *cis*-3-hexenol and *trans*-2-hexenal, which has also been shown to reduce fatigue and improve cognitive performance. Thus, the authors suggested that the anti-fatigue effects of MCMP and Hex-Hex Mix were probably exerted directly through the peripheral olfactory system.

12.5 CONCLUSIONS

The research reviewed in this chapter demonstrates that in the study on the effects of EOs and fragrances on human arousal and cognition, incoherent findings are the rule rather than the exception. This means that still more work must be done until scientists will understand in detail how the effects of EOs are exerted. However, only by thoroughly understanding the mechanisms will we be able to predict precisely which effects will be elicited by administering a particular EO. One reason for the lack of consistent results may be that in many human studies, clear associations between constituents of EOs and observed effects are still missing. While the relationship between the observed effects of an EO and its composition or dosage may not always be linear, exact specifications about the origin, composition, and concentration of the tested oils seems compulsory to establish clear pharmacological profiles and dose–response curves for specific oils. If researchers attended to these issues more carefully, it would be easier to compare the results from different studies and to generalize findings.

Another aspect that clearly contributes to inconclusive findings is the involvement of a variety of mechanisms of action in the effects of EOs on human arousal and cognition. In a very valuable review on the assessment of olfactory processes with electrophysiological techniques, Lorig (Lorig 2000) pointed out that EEG changes induced by odorants have to be interpreted with great care, since factors other than direct odor effects may be responsible for changes, particularly when relaxing effects reflected by the induction of slow-wave activity are concerned. Physiological processes, such as altered breathing patterns in response to pleasant versus unpleasant odors, cognitive factors (e.g., as a consequence of expectancy or the processing of secondary stimulus features), or an inappropriate baseline condition can lead to changes of the EEG pattern, which are quite unrelated to any psychoactive effect of the tested odorant. Well-designed paradigms are thus necessary to control for cognitive influences that might mask substance-specific effects of fragrances. Also, while EEG and other electrophysiological techniques are highly efficient to elucidate fragrance effects on the central nervous system in the time domain, we know only a little about spatial aspects of such effects. Brain imaging techniques, such as fMRI, have proved valuable to address these questions.

When evaluating psychoactive effects of EOs and fragrances on human cognitive functions, the results should be interpreted just as cautiously as those of electrophysiological studies, as similar confounding factors, ranging from influences of stimulus-related features (e.g., pleasantness) to expectation of fragrance effects and even personality traits, may be influencing the observed outcome. In regard to higher cognitive functioning, such as language or emotional processing, conscious as opposed to sub- or unconscious processing of odor information seems to differentially affect performance due to differences in the utilization of shared neuronal resources. Even seemingly small variations in experimental setup—for example, the timing of stimulus presentation or stimulus duration—appear to have significant impact on the observed results. Thus, it seems worthwhile to measure additional parameters which are indicative of (subjective) stimulus information processing and emotional arousal if hypotheses are being built on direct (pharmacological) and cognitively mediated (psychological) odor effects on human behavior. Moreover, comparison of different forms of application that involve or exclude stimulation of the olfactory system, such as inhalative vs. non-inhalative (e.g., dermal or oral), administration have proved useful in the distinction of pharmacological from psychological mechanisms. Eventually, sophisticated techniques and elaborated designs will serve to enlarge our understanding of psychoactive effects of EOs and fragrances in humans.

REFERENCES

Ambrosch, S., C. Duliban, H. Heger, E. Moser, E. Laistler, C. Windischberger, and E. Heuberger. 2018. Effects of 1,8-Cineole and (–)-Linalool on Functional Brain Activation in a Working Memory Task. *Flavour and Fragrance Journal*. 33(3): 235–244.

Aoki, Hiroyuki. 1996. Effect of odors from coniferous woods on contingent negative variation (CNV). *Zairyo*. 45(4):397–402.

Aoshima, H. and K. Hamamoto. 1999. Potentiation of GABAA receptors expressed in Xenopus oocytes by perfume and phytoncid. *Bioscience, Biotechnology, and Biochemistry*. 63(4):743–748.

Ashton, H., J. E. Millman, R. Telford, and J. W. Thompson. 1977. The use of event-related slow potentials of the human brain as an objective method to study the effects of centrally acting drugs [proceedings]. *Neuropharmacology*. 16(7–8):531–532.

Aston-Jones, G., S. Chen, Y. Zhu, and M. L. Oshinsky. 2001. A neural circuit for circadian regulation of arousal. *Nature Neuroscience*. 4(7):732–738.

Babulka, P., T. Berkes, R. Szemerszky, and F. Köteles. 2017. No effects of rosemary and lavender essential oil and a placebo pill on sustained attention, alertness, and heart rate. *Flavour and Fragrance Journal*. 32(4):305–311.

Baltissen, R. and H. Heimann. 1995. Aktivierung, Orientierung und Habituation bei Gesunden und psychisch Kranken. In *Biopsychologie Von Streß und Emotionalen Reaktionen - Ansätze Interdisziplinärer Forschung*, edited by F. Debus, G. Erdmann, and K.W. Dallus, 233–245. Göttingen: Hogrefe-Verlag für Psychologie.

Baron, R. A. 1990. Environmentally induced positive affect: Its impact on self-efficacy, task performance, negotiation, and conflict. *Journal of Applied Social Psychology*. 20(5, Pt 2):368–384.

Baron, R. A. and J. Thomley. 1994. A whiff of reality: Positive affect as a potential mediator of the effects of pleasant fragrances on task performance and helping. *Environment and Behavior*. 26(6):766–784.

Baxter, M. G. and A. A. Chiba. 1999. Cognitive functions of the basal forebrain. *Current Opinion in Neurobiology*. 9(2):178–183.

Becker-Carus, C. 1981. *Grundriß der Physiologischen Psychologie*. Heidelberg: Quelle & Meyer.

Bensafi, M., C. Rouby, V. Farget, B. Bertrand, M. Vigouroux, and A. Holley. 2002. Autonomic nervous system responses to odours: The role of pleasantness and arousal. *Chemical Senses*. 27(8):703–709.

Bermpohl, F., A. Pascual-Leone, A. Amedi, L. B. Merabet, F. Fregni, N. Gaab, D. Alsop, G. Schlaug, and G. Northoff. 2006. Dissociable networks for the expectancy and perception of emotional stimuli in the human brain. *NeuroImage*. 30(2):588–600.

Buchbauer, G., L. Jirovetz, M. Czejka, Ch. Nasel, and H. Dietrich. 1993. New results in aromatherapy research. *Paper read at 24th International Symposium on Essential Oils*, July 21–24, At TU Berlin, Germany.

Chu, S. 2008. Olfactory Conditioning of Positive Performance in Humans. *Chemical Senses*. 33(1):65–71.

Chu, S. and J. J. Downes. 2002. Proust nose best: Odors are better cues of autobiographical memory. *Memory and Cognition*. 30(4):511–518.

Colagiuri, B., E. J. Livesey, and J. A. Harris. 2011. Can expectancies produce placebo effects for implicit learning? *Psychonomic Bulletin & Review*. 18(2):399–405.

Danuser, B., D. Moser, T. Vitale-Sethre, R. Hirsig, and H. Krueger. 2003. Performance in a complex task and breathing under odor exposure. *Human Factors*. 45(4):549–562.

Davidson, R. J. and W. Irwin. 1999. The functional neuroanatomy of emotion and affective style. *Trends in Cognitive Sciences*. 3(1):11–21.

Degel, J. and E. P. Köster. 1999. Odors: Implicit Memory and Performance Effects. *Chemical Senses*. 24(3):317–325.

Diego, M. A., N. A. Jones, T. Field, M. Hernandez-Reif, S. Schanberg, C. Kuhn, V. McAdam, R. Galamaga, and M. Galamaga. 1998. Aromatherapy positively affects mood, EEG patterns of alertness and math computations. *International Journal of Neuroscience*. 96(3–4):217–224.

Diekelmann, S., S. Biggel, B. Rasch, and J. Born. 2012. Offline consolidation of memory varies with time in slow wave sleep and can be accelerated by cuing memory reactivations. *Neurobiology of Learning and Memory*. 98(2):103–111.

Duffy, E. 1972. Activation. In *Handbook of Psychophysiology*, edited by N. S. Greenfield, and R. A. Sternbach, 577–622. New York: Holt, Rinehart and Winston Inc.

Elisabetsky, E., L. F. Brum, and D. O. Souza. 1999. Anticonvulsant properties of linalool in glutamate-related seizure models. *Phytomedicine*. 6(2):107–113.

Epple, G. and R. S. Herz. 1999. Ambient odors associated to failure influence cognitive performance in children. *Developmental Psychobiology*. 35(2):103–107.

Field, T., M. Diego, M. Hernandez-Reif, W. Cisneros, L. Feijo, Y. Vera, K. Gil, D. Grina, and Q. C. He. 2005. Lavender fragrance cleansing gel effects on relaxation. *International Journal of Neuroscience*. 115(2):207–222.

Finkelmeyer, A., T. Kellermann, D. Bude, T. Niessen, M. Schwenzer, K. Mathiak, and M. Reske. 2010. Effects of aversive odour presentation on inhibitory control in the Stroop colour-word interference task. *BMC Neuroscience*. 11:131.

Friedl, S. M., E. Heuberger, K. Oedendorfer, S. Kitzer, L. Jaganjac, I. Stappen, and G. Reznicek. 2015. Quantification of 1,8-cineole in human blood and plasma and the impact of liner choice in head-space chromatography. *Current Bioactive Compounds*. 11(1):49–155.

Friedl, S. M., K. Oedendorfer, S. Kitzer, G. Reznicek, G. Sladek, and E. Heuberger. 2010. Comparison of liquid-liquid partition, HS-SPME and static HS GC/MS analysis for the quantification of (-)-linalool in human whole blood samples. *Natural Product Communications*. 5(9):1447–1452.

Gaygen, D. E. and A. Hedge. 2009. Effect of acute exposure to a complex fragrance on lexical decision Performance. *Chemical Senses*. 34(1):85–91.
Gilbert, A. N., S. C. Knasko, and J. Sabini. 1997. Sex differences in task performance associated with attention to ambient odor. *Archives of Environmental Health*. 52(3):195–199.
Gill, T. M., M. Sarter, and B. Givens. 2000. Sustained visual attention performance-associated prefrontal neuronal activity: Evidence for cholinergic modulation. *Journal of Neuroscience*. 20(12):4745–4757.
Goddard, L., L. Pring, and N. Felmingham. 2005. The effects of cue modality on the quality of personal memories retrieved. *Memory*. 13(1):79–86.
Gottfried, J. A. and D. A. Zald. 2005. On the scent of human olfactory orbitofrontal cortex: meta-analysis and comparison to non-human primates. *Brain Research Reviews*. 50:287–304.
Gould, A. and G. N. Martin. 2001. A good odour to breathe? The effect of pleasant ambient odour on human visual vigilance. *Applied Cognitive Psychology*. 15:225–232.
Grilly, D. M. 2002. *Drugs & Human Behavior*, Allyn & Bacon, Boston.
Grunwald, M., T. Weiss, W. Krause, L. Beyer, R. Rost, I. Gutberlet, and H. J. Gertz. 1999. Power of theta waves in the EEG of human subjects increases during recall of haptic information. *Neuroscience Letters*. 260(3):189–192.
Habel, U., K. Koch, K. Pauly et al. 2007. The influence of olfactory-induced negative emotion on verbal working memory: Individual differences in neurobehavioral findings. *Brain Research*. 1152:158–170.
Hackländer, R. P. M. and C. Bermeitinger. 2017. Olfactory context-dependent memory and the effects of affective congruency. *Chemical Senses*. 42(9):777–788.
Haneyama, H. and M. Kagatani. 2007. Sedatives containing Nardostachys chinensis extracts. *CAN*. 147:197215.
Hauner, K. K., J. D. Howard, C. Zelano, and J. A. Gottfried. 2013. Stimulus-specific enhancement of fear extinction during slow-wave sleep. *Nature Neuroscience*. 16(11):1553–1555.
Herz, R. S. 2004. A naturalistic analysis of autobiographical memories triggered by olfactory visual and auditory stimuli. *Chemical Senses*. 29(3):217–224.
Herz, R. S. and G. C. Cupchik. 1995. The emotional distinctiveness of odor-evoked memories. *Chemical Senses*. 20(5):517–528.
Herz, R. S., J. Eliassen, S. Beland, and T. Souza. 2004. Neuroimaging evidence for the emotional potency of odor-evoked memory. *Neuropsychologia*. 42(3):371–378.
Heuberger, E, T Hongratanaworakit, C Bohm, R Weber, and G Buchbauer. 2001. Effects of chiral fragrances on human autonomic nervous system parameters and self-evaluation. *Chemical Senses*. 26(3):281–292.
Heuberger, E., T. Hongratanaworakit, and G. Buchbauer. 2006. East Indian sandalwood and α-santalol odor increase physiological and self-rated arousal in humans. *Planta Medica*. 72(9):792–800.
Heuberger, E. and J. Ilmberger. 2010. The influence of essential oils on human vigilance. *Natural Product Communications*. 5(9):1441–191446.
Heuberger, E., J. Ilmberger, E. Hartter, and G. Buchbauer. 2008. Physiological and behavioral wffects of 1,8-cineol and (±)-linalool: A comparison of inhalation and massage aromatherapy. *Natural Product Communications*. 3(7):1103–1110.
Hey, K., S. Juran, M. Schaeper, S. Kleinbeck, E. Kiesswetter, M. Blaszkewicz, K. Golka, T. Bruening, and C. van Thriel. 2009. Neurobehavioral effects during exposures to propionic acid-An indicator of chemosensory distraction? *NeuroToxicology*. 30(6):1223–1232.
Hiruma, T., T. Matuoka, R. Asai, Y. Sato, N. Shinozaki, T. Sutoh, T. Nashida, T. Ishiyama, H. Yabe, and S. Kaneko. 2005. Psychophysiological basis of smells. *Seishin Shinkeigaku Zasshi*. 107(8):790–801.
Hiruma, T., H. Yabe, Y. Sato, T. Sutoh, and S. Kaneko. 2002. Differential effects of the hiba odor on CNV and MMN. *Biological Psychology*. 61(3):321–331.
Ho, C. and C. Spence. 2005. Olfactory facilitation of dual-task performance. *Neuroscience Letters*. 389(1):35–40.
Hoedlmoser, K, M Schabus, W Stadler, P Anderer, G Kloesch, C Sauter, W Klimesch, and J Zeitlhofer. 2007. EEG Theta-Aktivität während deklarativem Lernen und anschließendem REM-Schlaf im Zusammenhang mit allgemeiner Gedächtnisleistung. *Klinische Neurophysiologie*. 38:P304.
Holland, R. W., M. Hendriks, and H. Aarts. 2005. Smells like clean spirit. Nonconscious effects of scent on cognition and behavior. *Psychological Science*. 16(9):689–693.
Hongratanaworakit, T, E Heuberger, and G Buchbauer. 2004. Evaluation of the effects of East Indian sandalwood oil and a-santalol on humans after transdermal absorption. *Planta Medica*. 70(1):3–7.
Hotchkiss, S. A. M. 1998. Absorption of Fragrance Ingredients Using In Vitro Models with Human Skin. In *Fragrances: Beneficial and Adverse Effects*, edited by P. J. Frosch, J. D. Johansen, and I. R. White, 124–135. Berlin Heidelberg: Springer-Verlag.
Ilmberger, J, E Heuberger, C Mahrhofer, H Dessovic, D Kowarik, and G Buchbauer. 2001. The influence of essential oils on human attention. I: Alertness. *Chemical Senses*. 26(3):239–245.

Ishikawa, S., Y. Miyake, and H. Yokogoshi. 2002. Effect of lemon odor on brain neurotransmitters in rat and electroencephalogram in human subject. *Aroma Research*. 3(2):126–P130.

Ishiyama, S. 2000. Aromachological effects of volatile compounds in forest. *Aroma Research*. 1(4):15–21.

Jellinek, J. S. 1997. Psychodynamic odor effects and their mechanisms. *Cosmetics & Toiletries*. 112(9):61–71.

Kaetsu, I., T. Tonoike, K. Uchida, K. Sutani, M. Hanada, T. Ioku, K. Saruwatari, T. Kai, and N. Kanesaka. 1994. Effect of controlled release of odorants on electroencephalogram during mental activity. *Proceedings of the International Symposium on Controlled Release of Bioactive Materials*. 21ST:589–590.

Kandel, E. R., J. H. Schwartz, and T. M. Jessel. 1991. *Principles of Neural Science*, 3rd ed. Englewood Cliffs: Prentice-Hall International Inc.

Kaneda, H., H. Kojima, M. Takashio, and T. Yoshida. 2005. Relaxing effect of hop aromas on human. *Aroma Research*. 6(2):164–170.

Kaneda, H., H. Kojima, and J. Watari. 2006. Effect of beer flavors on changes in human feelings. *Aroma Research*. 7(4):342–347.

Kaneda, H., H. Kojima, and J. Watari. 2011. Novel psychological and neurophysiological significance of beer aromas. Part I: Measurement of changes in human emotions during the smelling of hop and ester aromas using a measurement system for brainwaves. *Journal of the American Society Brewing Chemists*. 69(2):67–74.

Kawano, K. 2001. Meditation-like effects of aroma observed in EEGs with consideration of each experience. *Aroma Research*. 2(1):30–46.

Keller, I. and O. Groemminger. 1993. Aufmerksamkeit. In *Neuropsychologische Diagnostik*, edited by D. Y. von Cramon, N. Mai, and W. Ziegler, 65–90. Weinheim: VCH.

Kennedy, D. O., F. L. Dodd, B. C. Robertson, E. J. Okello, J. L. Reay, A. B. Scholey, and C. F. Haskell. 2011. Monoterpenoid extract of sage (*Salvia lavandulaefolia*) with cholinesterase inhibiting properties improves cognitive performance and mood in healthy adults. *Journal of Psychopharmacology (London, U. K.)*. 25(8):1088–1100.

Kennedy, D., E. Okello, P. Chazot, M. J. Howes, S. Ohiomokhare, P. Jackson, C. Haskell-Ramsay, J. Khan, J. Forster, and E. Wightman. 2018. Volatile terpenes and brain function: Investigation of the cognitive and mood effects of *Mentha x piperita* L. essential oil with *in vitro* properties relevant to central nervous system function. *Nutrients*. 10(8): 1029.

Kepecs, A., N. Uchida, and Z. F. Mainen. 2006. The sniff as a unit of olfactory processing. *Chemical Senses*. 31(2):167–179.

Kerl, S. 1997. Zur olfaktorischen Beeinflussbarkeit von Lernprozessen. *Dragoco Report*. 44:45–59.

Kim, D. S., Y. M. Goo, J. Cho et al. 2018. Effect of volatile organic chemicals in *Chrysanthemum indicum* Linne on blood pressure and electroencephalogram. *Molecules*. 23(8): pii: E2063.

Kim, Y.-K. and S. Watanuki. 2003. Characteristics of electroencephalographic responses induced by a pleasant and an unpleasant odor. *Journal of Physiological Anthropology and Applied Human Science*. 22(6):285–291.

Kline, J. P., G. C. Blackhart, K. M. Woodward, S. R. Williams, and G. E. Schwartz. 2000. Anterior electroencephalographic asymmetry changes in elderly women in response to a pleasant and an unpleasant odor. *Biological Psychology*. 52(3):241–250.

Knasko, S. C. 1993. Performance, mood, and health during exposure to intermittent odors. *Archives of Environmental Health*. 48(5):305–308.

Kopell, B. S., W. K. Wittner, D. T. Lunde, L. J. Wolcott, and J. R. Tinklenberg. 1974. The effects of methamphetamine and secobarbital on the contingent negative variation amplitude. *Psychopharmacology*. 34(1):55–62.

Kovar, K. A., B. Gropper, D. Friess, and H. P. Ammon. 1987. Blood levels of 1,8-cineole and locomotor activity of mice after inhalation and oral administration of rosemary oil. *Planta Medica*. 53(4):315–318.

Krizhanovs'kii, S. A., I. H. Zima, M. Yu Makarchuk, N. H. Piskors'ka, and A. O. Chernins'kii. 2004. Effect of citrus essential oil on the attention level and electrophysical parameters of human brain. *Physics of the Alive*. 12(1):111–120.

Lee, C. F., T. Katsuura, S. Shibata, Y. Ueno, T. Ohta, S. Hagimoto, K. Sumita, A. Okada, H. Harada, and Y. Kikuchi. 1994. Responses of electroencephalogram to different odors. *Annals of Physiological Anthropology*. 13(5):281–291.

Leppanen, J. M. and J. K. Hietanen. 2003. Affect and face perception: Odors modulate the recognition advantage of happy faces. *Emotion*. 3(4):315–326.

Lorig, T. S. 1989. Human EEG and odor response. *Progress in Neurobiology*. 33(5–6):387–398.

Lorig, T. S. 2000. The application of electroencephalographic techniques to the study of human olfaction: A review and tutorial. *International Journal of Psychophysiology*. 36(2):91–104.

Lorig, T. S. and M. Roberts. 1990. Odor and cognitive alteration of the contingent negative variation. *Chemical Senses*. 15(5):537–545.

Lorig, T. S. and G. E. Schwartz. 1988. Brain and odor: I. Alteration of human EEG by odor administration. *Psychobiology*. 16(3):281–284.

Ludvigson, H. W. and T. R. Rottman. 1989. Effects of ambient odors of lavender and cloves on cognition, memory, affect and mood. *Chemical Senses*. 14(4):525–536.

Manley, C. H. 1993. Psychophysiological effect of odor. *Critical Reviews In Food Science and Nutrition*. 33(1):57–62.

Manley, C. H. 1997. Psychophysiology of odor. *Rivista Italiana EPPOS*. (Spec. Num., 15th Journees Internationales Huiles Essentielles, 1996):375–386.

Martin, G. N. and A. Chaudry. 2014. Working memory performance and exposure to pleasant and unpleasant ambient odor: Is spatial span special? *The International Journal of Neuroscience*. 124(11):806–811.

Masago, R., T. Matsuda, Y. Kikuchi, Y. Miyazaki, K. Iwanaga, H. Harada, and T. Katsuura. 2000. Effects of inhalation of essential oils on EEG activity and sensory evaluation. *Journal of Physiological Anthropology and Applied Human Science*. 19(1):35–42.

Michael, G. A., L. Jacquot, J. L. Millot, and G. Brand. 2005. Ambient odors influence the amplitude and time course of visual distraction. *Behavioral Neuroscience*. 119(3):708–715.

Miller, D. B. and J. P. O'Callaghan. 2006. The pharmacology of wakefulness. *Metabolism*. 55(Supplement 2):S13–S19.

Millot, J. L., G. Brand, and N. Morand. 2002. Effects of ambient odors on reaction time in humans. *Neuroscience Letters*. 322(2):79–82.

Min, B.-C., S.-H. Jin, I.-H. Kang, H. L. Dong, K. K. Jin, T. L. Sang, and K. Sakamoto. 2003. Analysis of mutual information content for EEG responses to odor stimulation for subjects classified by occupation. *Chemical Senses*. 28(9):741–749.

Morgan, C. L. 1996. Odors as cues for the recall of words unrelated to odor. *Perceptual and Motor Skills*. 83(3 Pt 2):1227–1234.

Moss, M., J. Cook, K. Wesnes, and P. Duckett. 2003. Aromas of rosemary and lavender essential oils differentially affect cognition and mood in healthy adults. *International Journal of Neuroscience*. 113(1):15–38.

Moss, M., S. Hewitt, L. Moss, and K. Wesnes. 2008. Modulation of cognitive performance and mood by aromas of peppermint and ylang-ylang. *International Journal of Neuroscience*. 118(1):59–77.

Moss, M. and L. Oliver. 2012. Plasma 1,8-cineole correlates with cognitive performance following exposure to rosemary essential oil aroma. *Therapeutic Advances in Psychopharmacology*. 2(3):103–113.

Moss, L., M. Rouse, K. A. Wesnes, and M. Moss. 2010. Differential effects of the aromas of *Salvia* species on memory and mood. *Human Psychopharmacology*. 25(5):388–396.

Moss, M., E. Smith, M. Milner, and J. McCready. 2018. Acute ingestion of rosemary water: Evidence of cognitive and cerebrovascular effects in healthy adults. *Journal of Psychopharmacology*. 32(12):1319–1329.

Nakagawa, M., H. Nagai, and T. Inui. 1992. Evaluation of drowsiness by EEGs. Odors controlling drowsiness. *Fragrance Journal*. 20(10):68–72.

Nasel, C., B. Nasel, P. Samec, E. Schindler, and G. Buchbauer. 1994. Functional imaging of effects of fragrances on the human brain after prolonged inhalation. *Chemical Senses*. 19(4):359–364.

Nordin, S., L. Aldrin, A. S. Claeson, and L. Andersson. 2017. Effects of negative affectivity and odor valence on hemosensory and symptom perception and perceived ability to focus on a cognitive task. *Perception*. 46(3–4):431–446.

Oken, B. S., M. C. Salinsky, and S. M. Elsas. 2006. Vigilance, alertness, or sustained attention: Physiological basis and measurement. *Clinical Neurophysiology*. 117(9):1885–1901.

Okugawa, H., R. Ueda, K. Matsumoto, K. Kawanishi, and K. Kato. 2000. Effects of sesquiterpenoids from Oriental incenses on acetic acid-induced writhing and D_2 and 5-HT_{2A} receptors in rat brain. *Phytomedicine*. 7(5):417–422.

Palva, S. and J. M. Palva. 2007. New vistas for [alpha]-frequency band oscillations. *Trends in Neurosciences*. 30(4):150–158.

Paus, T. 2001. Primate anterior cingulate cortex: Where motor control, drive and cognition interface. *Nature Reviews Neuroscience*. 2(6):417–424.

Pedersen, C. A., B. N. Gaynes, R. N. Golden, D. L. Evans, and J. J. Jr. Haggerty. 1998. Neurobiological aspects of behavior. In *Human Behavior: An Introduction for Medical Students*, edited by A. Stoudemire, 403–472. Philadelphia, New York: Lippincott-Raven Publishers.

Phan, K. L., T. W., Stephan F. Taylor, and I. Liberzon. 2002. Functional Neuroanatomy of Emotion: A Meta-Analysis of Emotion Activation Studies in PET and fMRI. *NeuroImage*. 16(2):331–348.

Posner, M. I. and S. E. Petersen. 1990. The attention system of the human brain. *Annual Review of Neuroscience*. 13:25–42.

Posner, M. I. and R. D. Rafal. 1987. Cognitive Theories of Attention and the Rehabilitation of Attentional Deficits. In *Neuropsychological Rehabilitation*, edited by M. J. Meier, A. L. Benton, and L. Diller, 182–201. London: Churchill-Livingston.

Pribram, K. H. and McGuinness, D. 1975. Arousal, activation, and effort in the control of attention. *Psychological Review*, 82(2), 116–149.

Rasch, B., C. Buchel, S. Gais, and J. Born. 2007. Odor cues during slow-wave sleep prompt declarative memory consolidation. *Science*. 315(5817):1426–1429.

Razumnikova, O. M. 2007. Creativity related cortex activity in the remote associates task. *Brain Research Bulletin*. 73(1–3):96–102.

Reichert, J. L., M. Ninaus, W. Schuehly, C. Hirschmann, D. Bagga, and V. Schopf. 2017. Functional brain networks during picture encoding and recognition in different odor contexts. *Behavioural Brain Research*. 333:98–108.

Reiman, E. M., R. D. Lane, G. L. Ahern, G. E. Schwartz, R. J. Davidson, K. J. Friston, L. S. Yun, and K. Chen. 1997. Neuroanatomical correlates of externally and internally generated human emotion. *American Journal of Psychiatry*. 154(7):918–925.

Saito, N., E. Yamano, A. Ishii, M. Tanaka, J. Nakamura, and Y. Watanabe. 2018. Involvement of the olfactory system in the induction of anti-fatigue effects by odorants. *PLOS ONE*. 13(3):e0195263.

Sakamoto, R., K. Minoura, A. Usui, Y. Ishizuka, and S. Kanba. 2005. Effectiveness of aroma on work efficiency: Lavender aroma during recesses prevents deterioration of work performance. *Chemical Senses*. 30(8):683–691.

Satoh, T. and Y. Sugawara. 2003. Effects on humans elicited by inhaling the fragrance of essential oils: Sensory test, multi-channel thermometric study and forehead surface potential wave measurement on basil and peppermint. *Analytical Sciences*. 19(1):139–146.

Sawada, K., R. Komaki, Y. Yamashita, and Y. Suzuki. 2000. Odor in forest and its physiological effects. *Aroma Research*. 1(3):67–71.

Sayorwan, W., N. Ruangrungsi, T. Piriyapunyporn, T. Hongratanaworakit, N. Kotchabhakdi, and V. Siripornpanich. 2013. Effects of inhaled rosemary oil on subjective feelings and activities of the nervous system. *Sci Pharm*. 81(2):531–542.

Sayorwan, W., V. Siripornpanich, T. Piriyapunyaporn, T. Hongratanaworakit, N. Kotchabhakdi, and N. Ruangrungsi. 2012. The effects of lavender oil inhalation on emotional states, autonomic nervous system, and brain electrical activity. *Journal of the Medical Association of Thailand*. 95(4):598–606.

Sayowan, W., V. Siripornpanich, T. Hongratanaworakit, N. Kotchabhakdi, and N. Ruangrungsi. 2013. The effects of jasmine oil inhalation on brainwave activities and emotions. *Journal of Health Research*. 27(2):73–77.

Sayowan, W., V. Siripornpanich, T. Piriyapunyaporn, T. Hongratanaworakit, N. Kotchabhakdi, and N. Ruangrungsi. 2012. The harmonizing effects of citronella oil on mood states and brain activities. *Journal of Health Research*. 26(2):69–75.

Schandry, R. 1989. *Lehrbuch Der Psychophysiologie*, 2. Auflage ed. Weinheim: Psychologie Verlags Union.

Schneider, R. 2016. Direct application of specially formulated scent compositions (AromaStick®) prolongs attention and enhances visual scanning speed. *Applied Cognitive Psychology*. 30(4):650–654.

Schwabe, L. and O. T. Wolf. 2009. The context counts: Congruent learning and testing environments prevent memory retrieval impairment following stress. *Cognitive Affective & Behavioral Neuroscience*. 9(3):229–236.

Sellaro, R., W. W. van Dijk, C. R. Paccani, B. Hommel, and L. S. Colzato. 2014. A question of scent: Lavender aroma promotes interpersonal trust. *Frontiers in Psychology*. 5:1486.

Seo, M., K. Sowndhararajan, and S. Kim. 2016. Influence of binasal and uninasal inhalations of essential oil of *Abies koreana* twigs on electroencephalographic activity of human. *Behavioural Neurology*. 2016:9250935.

Sobel, N., R. M. Khan, A. Saltman, E. V. Sullivan, and J. D. E. Gabrieli. 1999. Olfaction: The world smells different to each nostril. *Nature*. 402(6757):35.

Sturm, W. 1997. Aufmerksamkeitsstoerungen. In *Klinische Neuropsychologie*, edited by W. Hartje, and K. Poeck, 283–289. Stuttgart: Thieme.

Sturm, W., A. de Simone, B. J. Krause, K. Specht, V. Hesselmann, I. Radermacher, H. Herzog, L. Tellmann, H. W. Muller-Gartner, and K. Willmes. 1999. Functional anatomy of intrinsic alertness: Evidence for a fronto-parietal-thalamic-brainstem network in the right hemisphere. *Neuropsychologia*. 37(7):797–805.

Sugano, H. 1992. Psychophysiological studies of fragrance. In *Fragrance: The Psychology and Biology of Perfume*, edited by S. Van Toller, and G. H. Dodd, 221–228. Barking: Elsevier Science Publishers Ltd.

Sugawara, Y., C. Hara, T. Aoki, N. Sugimoto, and T. Masujima. 2000. Odor distinctiveness between enantiomers of linalool: Difference in perception and responses elicited by sensory test and forehead surface potential wave measurement. *Chemical Senses.* 25(1):77–1484.

Sullivan, T. E, J. S Warm, B. K Schefft, W. N. Dember, M. W O'Dell, and S. J. Peterson. 1998. Effects of olfactory stimulation on the vigilance performance of individuals with brain injury. *Journal of Clinical and Experimental Neuropsychology.* 20(2):227–236.

Suss, C., S. Gaylord, and J. Fagen. 2012. Odor as a contextual cue in memory reactivation in young infants. *Infant Behavior & Development.* 35(3):580–583.

Tecce, J. J. 1972. Contingent negative variation (CNV) and psychological processes in man. *Psychological Bulletin.* 77(2):73–108.

Tildesley, N. T. J., D. O. Kennedy, E. K. Perry, C. G. Ballard, K. A. Wesnes, and A. B. Scholey. 2005. Positive modulation of mood and cognitive performance following administration of acute doses of *Salvia lavandulaefolia* essential oil to healthy young volunteers. *Physiology & Behavior.* 83(5):699–709.

Torii, S., H. Fukada, H. Kanemoto, R. Miyanchi, Y. Hamauzu, and M. Kawasaki. 1988. Contingent negative variation (CNV) and the psychological effects of odour. In *Perfumery - The Psychology and Biology of Fragrance*, edited by S. Van Toller, and G. H. Dodd, 107–120. London New York: Chapman and Hall.

Travis, F. and J. J. Tecce. 1998. Effects of distracting stimuli on CNV amplitude and reaction time. *International Journal of Psychophysiology.* 31(1):45–50.

Valnet, J. 1990. *The Practice of Aromatherapy.* Rochester: Inner Traditions.

Van Toller, S., J. Behan, P. Howells, M. Kendal-Reed, and A. Richardson. 1993. An analysis of spontaneous human cortical EEG activity to odors. *Chemical Senses.* 18(1):1–16.

Walla, P., B. Hufnagl, J. Lehrner, D. Mayer, G. Lindinger, L. Deecke, and W. Lang. 2002. Evidence of conscious and subconscious olfactory information processing during word encoding: A magnetoencephalographic (MEG) study. *Brain Research. Cognitive Brain Research.* 14(3):309–316.

Walla, P., B. Hufnagl, J. Lehrner, D. Mayer, G. Lindinger, H. Imhof, L. Deecke, and W. Lang. 2003a. Olfaction and depth of word processing: A magnetoencephalographic study. *NeuroImage.* 18(1):104–116.

Walla, P., B. Hufnagl, J. Lehrner, D. Mayer, G. Lindinger, H. Imhof, L. Deecke, and W. Lang. 2003b. Olfaction and face encoding in humans: A magnetoencephalographic study. *Brain Research Cognitive Brain Research.* 15(2):105–115.

Walla, P., D. Mayer, L. Deecke, and W. Lang. 2005. How chemical information processing interferes with face processing: A magnetoencephalographic study. *NeuroImage.* 24(1):111–117.

Walschburger, P. 1976. Zur Beschreibung von Aktivierungsprozessen: Eine Methodenstude zur psychophysiologischen Diagnostik. *Dissertation*, Philosophische Fakultät, Albert-Ludwigs-Universität, Freiburg (Breisgau).

Walter, W. G., R. Cooper, V. J. Aldridge, W. C. McCallum, and A. L. Winter. 1964. Contingent negative variation: An electric sign of sensorimotor association and expectancy in the human brain. *Nature.* 203:380–384.

Warm, J. S., W. N. Dember, and R. Parasuraman. 1991. Effects of olfactory stimulation on performance and stress in a visual sustained attention task. *Journal of the Society of Cosmetic Chemists.* 42:199–210.

Wiesmann, M, I Yousry, E Heuberger, A Nolte, J Ilmberger, G Kobal, T A Yousry, B Kettenmann, and T P Naidich. 2001. Functional magnetic resonance imaging of human olfaction. *Neuroimaging Clinics of North America.* 11(2):237–250.

Willander, J. and M. Larsson. 2007. Olfaction and emotion: The case of autobiographical memory. *Memory and Cognition.* 35(7):1659–1663.

Yagyu, T. 1994. Neurophysiological findings on the effects of fragrance: Lavender and jasmine. *Integrative Psychiatry.* 10:62–67.

Zilles, K. and G. Rehkämpfer. 1998. *Funktionelle Neuroanatomie*, 3. Aufl. ed. Berlin, Heidelberg, New York: Springer-Verlag.

Zimbardo, P. G., A. L. Weber, and R. L. Hohnson. 2003. *Psychology - Core Concepts*, 4th ed. Boston: Allyn and Bacon.

13 Aromatherapy
An Overview and Global Perspectives

Rhiannon Lewis

CONTENTS

13.1 INTRODUCTION

The goal of this chapter is to present a concise overview of the practice of aromatherapy around the world based on the author's 30-year experience as aromatherapist, educator, and editor of an international aromatherapy journal. To this day, a degree of confusion exists as to what the titles "aromatherapy" and "aromatherapist" actually mean; it is the author's intention to help bring clarity by offering some historical context, defining different aromatherapy styles, and painting a broad perspective of this profession as it is currently practiced. This chapter has certain limitations; for example, it confines information concerning aromatherapeutic applications to humans only, reports only on recent work and studies that mention aromatherapy, and omits a detailed overview of the historical use of aromatics for health and well-being. Furthermore, as the pharmacological and psychological effects of essential oils are the subject of other chapters in this text, in-depth detail on the mechanism of action of aromatherapy interventions is also limited to avoid repetition.

13.2 THE IMAGE OF AROMATHERAPY: TREAT OR TREATMENT?

For many, the word aromatherapy conjures up images of scented candles, bubble baths, and luxury pampering rather than an established and increasingly structured profession that offers direct and measurable health benefits as well as valuable psychosocial support. The former fluffy image further compounds existing skepticism among doctors or other healthcare providers and the medical establishment in general, as well as researchers. The increasing banality of the word over the past 30 years has also led to members of the scientific community distancing themselves from using the term even if their work involves the therapeutic benefits of essential oils in ways that aromatherapists typically employ. Dunning, in her book *Essential Oils in Therapeutic Care* (2007; page 6), explains:

> Aromatherapy is a confusing term that may not adequately convey the complexity or therapeutic benefits of essential oils… In fact, overuse of the term "aromatherapy" for commercial reasons may have obscured its therapeutic applications, which makes it more difficult for skeptical conventional practitioners to take aromatherapy seriously.

This misperception has led some aromatherapy practitioners to try and differentiate themselves by conveying a more a health-orientated or professional stance using titles such as "Essential Oil Therapist," "Medical Aromatherapist," "Advanced Aromatherapist," "Clinical Aromatherapist," "Aromatologist," and so on. For the most part, these titles are not clearly defined, are loosely used, and can vary widely in their meaning even between practitioners. This chapter thus aims to offer clarity through a better understanding of what aromatherapy means, how it originated, and what it currently has to offer.

13.3 DEFINING AROMATHERAPY: IS THERE CONSENSUS?

The debate as to what constitutes aromatherapy stems from a general misunderstanding, and sometimes disagreement, of the word itself and is compounded by a lack of an internationally recognized definition.

The word aromatherapy (*aromathérapie*) was first used before the second world war by René-Maurice Gattefossé (1881–1950), a French chemist, engineer, and perfumer who wrote a number of texts detailing the use and benefits of essential oils, perfumes, and aromatic extracts in cosmetology and pharmacy. For Gattefossé, *l'aromathérapie* was a science that merited closer attention as he both observed and confirmed his own findings along with those of his colleagues concerning the physiological benefits and medicinal potentials of essential oils (in particular, for antisepsis, cosmetology, and wound repair). In 1937, he published *Aromathérapie-les huiles essentielles hormones végétales*, where the term *aromathérapie* was used for the first time (Gattefossé, 1937).

Despite being the author of the name and describing various physiological potentials of essential oils and essences via inhalation, diffusion, and external application, over the ensuing years, the meaning of aromathérapie/aromatherapy has at times been interpreted differently from that of Gattefossé himself. For example, some researchers have proposed it literally to mean a "therapy by aroma," thereby limiting aromatherapy to the use of essential oils and other volatile substances by inhalation only (Buchbauer et al., 1991). For example, Buchbauer and colleagues (1994) proposed a universal definition of aromatherapy:

> Therapeutic use of fragrances or of volatile substances to cure and to mitigate or to prevent diseases, infection and indispositions only by means of inhalation…

This proposition was not widely adopted at the time but did contribute to the rise in more rigorous international studies concerning the measurable psychological, behavioral, and physiological effects of inhaled essential oils which continues today and is detailed in other chapters of this book.

It could also be argued that Buchbauer et al.'s definition is partly allied to but goes further than that of "aromachology." This term was established by the Sense of Smell Institute in 1989 (SOSI: the research and education division of the Fragrance Foundation) and refers to the measurable and reproducible effect of inhaled aromas/fragrances (not necessarily of natural origin) on mind, mood, and behavior. Aromatherapy and aromachology are, however, distinct from one another in that the former is concerned with both physiological and psychological benefits through the appropriate selection and administration of essential oils.

The main understanding of aromatherapy today includes more than the inhalation of the fragrance/aroma of essential oils and aromatic extracts. As will be discussed in this chapter, the majority of practicing aromatherapists around the world are those that are trained in the UK Holistic Aromatherapy style, using a range of administration methods with an emphasis on topical application (often combined with bodywork techniques such as massage) as well as inhalation.

A more current definition of aromatherapy therefore might be a therapy that uses essential oils and other aromatic plant extracts for their body-mind benefits, without confining the therapy exclusively to effects obtained via their aroma, although this is, of course, a very important aspect of how the therapy works. In modern day aromatherapy, essential oils and related substances are currently used via a number of administration routes depending on the clinical expertise of the practitioner (see Figure 13.1).

13.4 AROMA THERAPY OR AROMA CARE?

To add complexity to the situation, between countries and different actors, the interpretation of the word "therapy" also gives rise to differing opinions and emphasis.

For example, the word therapy might be interpreted as:

- "The treatment of physical, mental, or social disorders or disease" (http://www.collinsdictionary.com/dictionary/english/therapy),
- "A treatment that helps someone feel better, grow stronger, etc., especially after an illness" (http://dictionary.cambridge.org/dictionary/british/therapy)
- "Healing power or quality"
- (http://www.thefreedictionary.com/therapy).

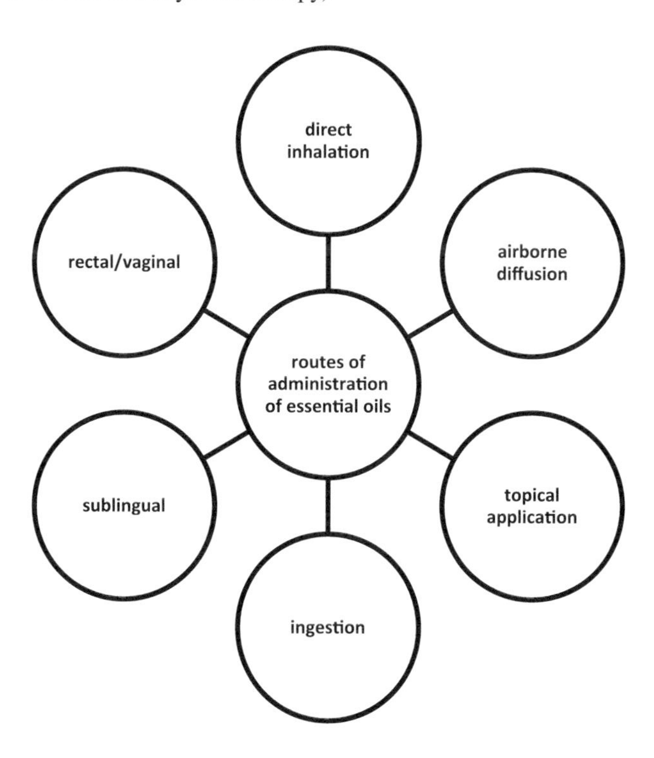

FIGURE 13.1 Possible routes of essential oil administration in aromatherapy.

For those in European countries such as France, Germany, Austria, and Belgium, the former interpretation is emphasized (therapy=treatment of disorder or disease), with the conclusion that any person providing a "therapy" such as aromatherapy is required to have a medical qualification as it constitutes a medical act. Thus, non-medical practitioners who use essential oils may use a different term to describe their work; for example, in Austria, the term "aroma care" and "aroma practitioner" are used to describe what would otherwise be called aromatherapy and aromatherapist in the United Kingdom.

In countries such as the United Kingdom, Canada, and Australia, the latter interpretations of "therapy" may be more appropriate and do not refer to direct treatment of disease but more to the aspect of promoting health, preventing decline, and improving well-being. This concept of helping to restore psychological and physiological balance using essential oils is a core tenet of the British holistic style of aromatherapy. Promoting health and well-being and providing complementary care alongside orthodox medical practice does not require a formal medical qualification. This is reflected in the profile of aromatherapists practicing in these countries, most of whom do not possess a medical qualification but who have followed professional training in aromatherapy that includes basic anatomy, physiology, and pathology as part of the syllabus.

With the above in mind, a clearer understanding may be thus gleaned from reading the definitions of aromatherapy provided by a number of professional associations/organizations around the world that support aromatherapists as healthcare professionals and who provide information to the general public (see Table 13.1).

13.5 AROMATHERAPY: SCIENTIFIC OR ENERGETIC?

Adding another layer of complexity, a certain number of holistic aromatherapy practitioners and educators underpin their teaching and practice with theory based on traditional medicine systems including traditional Chinese medicine and Ayurveda that are based on energetic/elemental/vibrational concepts. According to many, the ethereal, vibrational nature of essential oils enables them to influence and harmonize the energy field of the individual, thereby promoting a more balanced health and emotional state. Using the main concepts of the traditional medicine system in question, the therapist is able to conduct a holistic health assessment to ascertain a pattern of imbalance and then match the client's needs with the appropriate selection and administration of essential oils. Many holistic aromatherapists practice multiple therapies (massage, reflexology, acupressure, reiki, etc.) and adapt their care and treatment style to the person's unique needs.

Other aromatherapy practitioners (predominantly health professionals) underpin their teaching and practice with evidence-based data and with emphasis on the chemical composition of essential oils, the bioactivity of their individual components, route of administration, galenics, and dose. This more allopathic approach is adopted in many healthcare settings.

In reality, in the practice of aromatherapy, there is overlap between these differing approaches, depending on the clinical expertise of the practitioner and the goals of treatment (Figure 13.2). These will be explored in more detail below.

13.6 AN AROMATIC EVOLUTION: DIFFERENT AROMATHERAPY STYLES

Since Gattefossé's day, the use of essential oils for therapeutic purposes has been expanded, evolved, adapted, and exported to a number of different countries and, over time, has also been shaped by cultural influences as well as by the scope of practice and competencies of the practitioners themselves.

The predominant aromatherapy styles are identified in Figure 13.3, and these will be discussed in turn.

1. Subtle aromatherapy
2. Self-care aromatherapy
3. Holistic aromatherapy
4. Cosmetic aromatherapy
5. Clinical aromatherapy
6. Medical aromatherapy

TABLE 13.1
International Definitions of Aromatherapy

Organization	Definition
The International Federation of Aromatherapy UK www.ifaroma.org	"Aromatherapy is an ancient therapeutic treatment that enhances well-being, relieves stress and helps in the rejuvenation and regeneration of the human body."
The Alliance of International Aromatherapists USA www.alliance-aromatherapists.org	"Aromatherapy refers to the inhalation and topical application of true, authentic essential oils from aromatic plants to restore or enhance health, beauty and well-being."
The International Aromatherapy and Aromatic Medicine Association. AUSTRALIA www.iaama.org.au	"Aromatherapy is the evidence-based, therapeutic use of essential oils to treat, influence or modify the mind, body and spirit by aromatherapists (professionally qualified therapists) to promote health and well-being."
The National Association for Holistic Aromatherapy USA www.naha.org	"Aromatherapy, also referred to as Essential Oil therapy, can be defined as the art and science of utilizing naturally extracted aromatic essences from plants to balance, harmonize and promote the health of body, mind and spirit."
The Allied Health Professions Council of South Africa SOUTH AFRICA www.ahpcsa.co.za	"Therapeutic Aromatherapy is a multifaceted, non-invasive system of treatment that uses aromatic plants extracts such as volatile essential oils and hydrosols, via a range of application modes, for therapeutic purposes, in order to facilitate the restoration of health…"
The Austrian Society for Scientific Aromatherapy and Aroma Care (ASsAAC) AUSTRIA www.oegwa.at	"Aroma care is the purposeful, qualified use of natural high quality essential oils, fatty vegetable oils, hydrolats and other aroma care products in professional health care and nursing of the sick. Aroma care ranks among complementary care methods. It serves the promotion and preservation of health and well-being, as well as health maintenance and preventive measures.""
The International Federation of Professional Aromatherapists UK www.ifparoma.org	"Aromatherapy uses the volatile aromatic plant essences, known as essential oils, to treat ill-health and help maintain good health."
Forum Essenzia GERMANY www.forum-essenzia.org (excerpt from the 2019 position paper for aroma care in clinical inpatient facilities)	"Aroma care and therapy strive for a holistic view of the patient. Essential oils or ready-to-use aroma care products help mental and alleviate physical complaints integrated with conventional medicine."
VAGA Professional association for commercial aroma practitioners and aroma care specialists AUSTRIA www.aromapraktiker.eu	"Aroma care is a recognised complementary method of health and healthcare. It supports well-being and relaxation, personal hygiene, improving breathing, digestion and elimination and creating a pleasant indoor ambience."
Canadian Federation of Aromatherapists CANADA www.cfacanada.com	"Aromatherapy, also referred to as essential oil therapy, can be defined as the art and science of utilizing naturally extracted aromatic essences from plants to balance, harmonize, and promote the health of body, mind, and spirit. It seeks to unify physiological, psychological, and spiritual processes to enhance an individual's innate healing process."

13.6.1 Subtle Aromatherapy

As mentioned above, many practitioners integrate aspects of traditional medicine systems (TCM, Ayurveda, Unani Tibb) with their use of essential oils with the goal of helping restore energetic balance to the person, thereby enhancing health and well-being (Miller and Miller, 1995; Mojay,

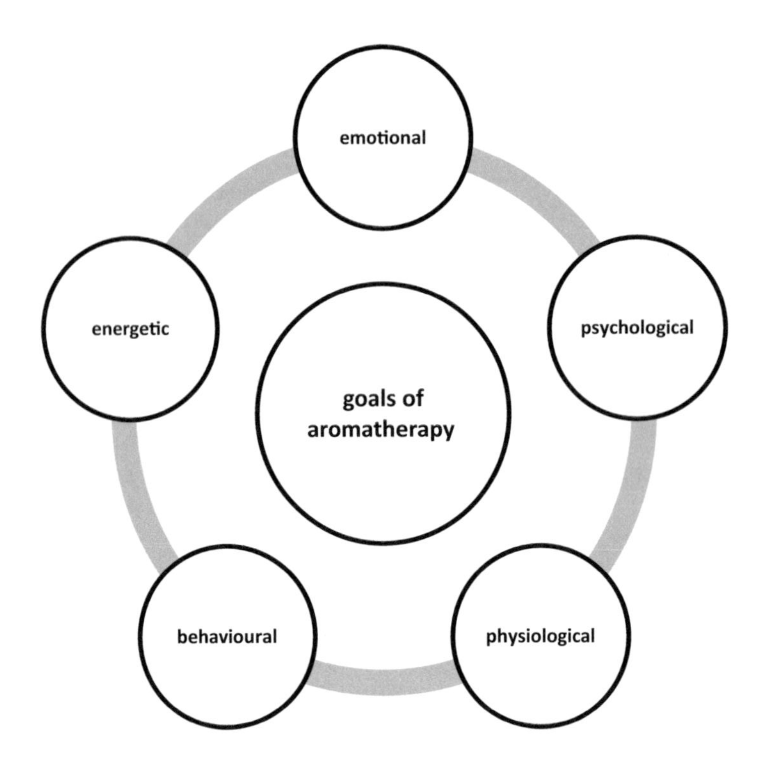

FIGURE 13.2 Different goals of aromatherapy.

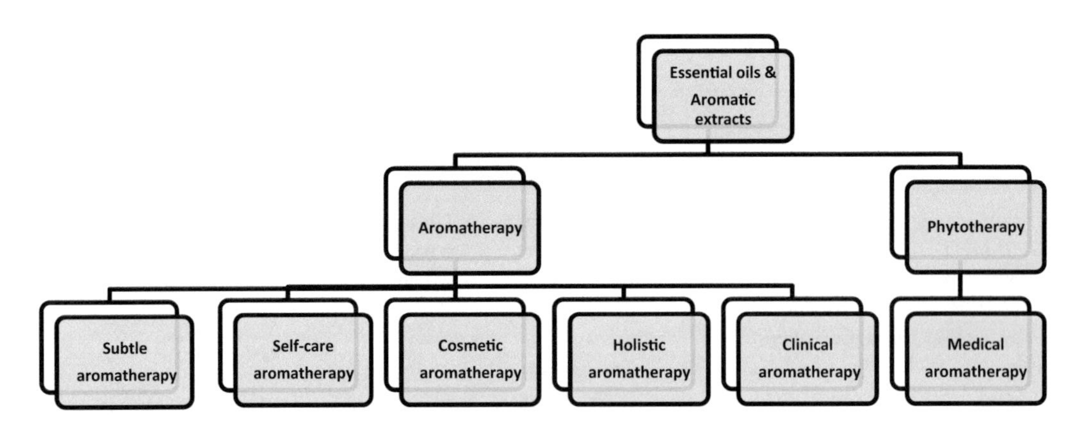

FIGURE 13.3 Different styles of aromatherapy.

2000; Holmes, 2001, 2016; Irani, 2001; Zafar et al., 2007). Others may integrate their use of essential oils with other subtle non-physical approaches such as reiki or other healing modalities without being linked to religious belief systems.

One of the first books to be published on the subtle aromatherapy style (*Subtle Aromatherapy* by Davis [1991]) elaborated on chakras, dowsing, crystals, flower remedies, and meditation in combination with essential oils, and variations of these techniques continue today, with terms such as vital force, prana, chi, vibration, spirit, soul, and energy being used interchangeably. In this gentle style of aromatherapy, there is a tendency to use low doses of essential oils in topical or inhaled form with an emphasis on the fragrance impact of essential oils to balance body, mind, and spirit. A recent example of the subtle use of essential oils is the practice of soul midwifery, where low doses of essential oils are used along with other healing modalities to assist the person in the dying phase and where essential oils are regarded as sacred tools of transformation (Warner, 2013, and 2018). Essential oils are also increasingly used in mindfulness and meditation practices (Godfrey, 2018) to reduce stress and promote balance. It may be argued that much of the perceived benefit of essential oils in subtle aromatherapy is obtained through the impact of fragrance on the psyche.

Critics of aromatherapy often refer to this style of aromatherapy as being "new age," unscientific, and nonsensical; however, it does represent one facet of how some therapists use essential oils for health and healing and has been part of the aromatherapy world for over 30 years. Furthermore, using vitalistic principles to underpin a therapeutic intervention is not an approach that is exclusive to aromatherapy.

13.6.2 Self-Care Aromatherapy

At its most basic level, essential oils can be incorporated into the home environment for general improvement of family health and well-being as well as contributing to a fragrant ambience. A plethora of affordable self-help aromatherapy texts are available in all the main languages and are of variable accuracy and quality. Some make unsubstantiated health claims and promises, others offer simple and safe advice for improved health and well-being. Most offer do-it-yourself "recipes" for a range of non-life-threatening common ailments and self-care applications. Public awareness of essential oils and their potential benefits for the family is increasing worldwide, but confusion as to what truly constitutes aromatherapy as a therapeutic intervention remains. In the home situation, essential oils are mainly used in bathing, airborne diffusion, household cleaning, and topical applications for skin and body care—prepared according to suggested recipes provided by essential oil suppliers and authors.

Since the 1990s, a worrying trend in self-care aromatherapy has emerged with the promotion of ingestion of essential oils by auto-medication as well as the use of undiluted or high doses of topically applied essential oils to actively treat disease states with little guidance or support. This trend is driven by network marketing, largely targeting the family unit including babies and children and carries significant risk of adverse reactions. This trend will be addressed later in the Section 13.10.2 as it poses a threat not only to the consumer but also to the future of aromatherapy as a profession.

13.6.3 Holistic Aromatherapy

In the United Kingdom, along with herbal medicine, homeopathy, and therapies such as massage and reflexology, aromatherapy is among the most popular forms of complementary medicine used by the public (Posadzki et al., 2013). The exportation of aromatherapy from France to the UK in the 1950s and 1960s gave rise to a therapy style that is allied with the cosmetic and holistic benefits of aromatherapy and includes massage or similar touch techniques as a main means of essential oil delivery (Harris, 2003; Bensouilah, 2005; Jenkins, 2006). Pioneers of this "Anglo Saxon" style include individuals such as Marguerite Maury (Maury, 1989) and Micheline Arcier, neither of whom were doctors or pharmacists but who were directly influenced and supported by key French figures such as Dr. Jean Valnet, considered the father of modern day aromatherapy. Maury developed a "softer" form of aromatherapy compared to the French medical style from which it originated, with a focus on whole-person health and well-being rather than being disease-orientated.

This holistic Anglo Saxon style of aromatherapy has since been exported internationally and uses, for the most part, the benefits of inhaled and topically applied blends of essential oils in dosage ranges of 1%–3% concentration, often delivered within a traditional treatment sequence that includes a health and well-being consultation, selection of an individualized blend of essential oils, and its administration, coupled with body massage or other touch techniques (Harris, 2003). Emphasis is placed on the "individual prescription" and the need for the overall fragrance of the blend to be harmonious and pleasing to the client. The selection of essential oils for treatment places more weight on their fragrance notes and client preferences than their chemical composition and bioactivity. The full treatment duration varies from one to one and one-half hours and usually includes aftercare advice, self-care, and support. A full series of treatments averages six sessions.

Public perception of holistic aromatherapy as a "feel good" therapy, being useful predominantly for stress, relaxation, as a "pick me up," or for general aches and pains may have contributed to the continued perception of traditional holistic aromatherapy as a nice but not integral complement to medical care (Furnham, 2000; Emslie et al., 2002; Bishop et al., 2008).

In summary, the holistic aromatherapy style emphasizes a personalized approach to improving health and well-being through body-mind balance and engages the client actively in their own care. The combined effect of the aroma of the essential oil(s), their topical application, and the psychophysiological benefits of touch all contribute to treatment efficacy. This personalized and holistic approach has enjoyed a longstanding reputation for being beneficial for anxiety and depression, pain, and well-being, as well as for enhancing quality of life.

13.6.4 Cosmetic Aromatherapy

Closely allied with holistic aromatherapy and currently in vogue with the rising popularity of the spa concept and similar well-being settings, aromatherapy has grown significantly within these sectors (Petersen, 2012). In many spas and beauty clinics, aromatherapy use is often product-based more than offering individualized aromatherapy treatments; ready-made blends are delivered with a range of massage styles as well as specific cosmetic treatments such as peels, scrubs, wraps, and face masks by aestheticians and massage therapists.

With a focus on physical appearance, aromatherapy is frequently offered as a complementary treatment in haircare, nail, and beauty salons, offering aromatherapy facial and body treatments for the purpose of resolving troubled skin and for beautifying the complexion. Once again, for the most part, these treatments use ready-formulated aromatherapy blends, and the cosmetician is rarely a qualified professional aromatherapist. Limited aromatherapy education is provided by many cosmetology institutes, with focus on aesthetician training and skin benefits of essential oils.

In the case where the cosmetician is trained in aromatherapy, the following definition of aesthetic/cosmetic aromatherapy may apply (Falsetto, 2010):

> A cosmetic aromatherapist is trained in the art and science of adding essential oils to cosmetic bases, such as lotions and oils, for use in skin care and hair care. In addition, many cosmetic aromatherapists work in beauty salons and spas, using custom aromatherapy blends in spa and beauty treatments, such as massages and facials…

Essential oils also play an important part in therapeutic phytocosmetology as defined by Goetz (2007): using plant extracts including essential oils not only to repair skin changes via topical applications but also using plant extracts as medicine (via ingestion) to bring about skin improvement. This global skin health concept is shared by others (Baudoux and Zhiri, 2003) and for many years, aromatherapy has figured in texts concerning phytocosmetic applications such as that by D'Amelio (1999). For qualified aromatherapists seeking to specialize in skin care via topical applications, several texts offer evidence-based information. Bensouilah and Buck (2006) offer detail on the both psychological and physiological effects of essential oils and their positive impact on skin health along with detail on essential oil safety that is particularly relevant to skin applications. The more recent text by Sade (2017) offers guidance on creating cosmetic formulations using essential oils as well as fixed oils for skin care.

13.6.5 Clinical Aromatherapy

Over the past 30 years, aromatherapists and healthcare professionals, including nurses and midwives, have sought to integrate essential oils into clinical environments, with a goal to improve quality of life and health status of patients. This style has grown out of holistic aromatherapy and has evolved in parallel with the rise in complementary therapy care provision in clinical settings such as hospitals, residential homes, and hospices, as well as in response to increasing public demand. The goal of clinical aromatherapy is to accompany the patient in their journey with their disease alongside allopathic care, helping them with symptoms related to the disease or its treatment, as well as enhancing well-being and quality of life.

Delivering aromatherapy to fragile patients with complex health needs who are poly-medicated generally demands a further depth of aromatherapy knowledge and skills, as well as a sound evidence base to underpin aromatic interventions. As a result, specialized aromatherapy training opportunities exist for health practitioners which enable them to successfully and safely incorporate essential oils in these clinical settings and to be able to measure and evaluate specific outcomes.

In many clinical settings, nurses deliver aromatherapy alongside their regular care. However, depending on the country in question, a nursing or allied health professional qualification may not always be a prerequisite for aromatherapy delivery. For example, in the UK, the majority of aromatherapists working in clinical settings are neither nurses nor allied health professionals (Lewis, 2015). However, these aromatherapists often undertake further training (either privately or provided in-house by the clinical establishment) to enable them to work safely and confidently with sick patients and are usually provided with peer support and mentoring. In cancer care and palliative care environments, these issues are especially pertinent (Mackereth et al., 2009, 2010; Carter et al., 2010).

In order to implement aromatherapy successfully in clinical settings, the need for an evidence-based approach is paramount. To this end, in 2004, a specialist publication was launched to meet the needs of practitioners working in these environments. To date, the *International Journal of Clinical Aromatherapy* is the only journal in the English language dedicated to reporting of clinical trials, pilot studies, case reports, case series, literature reviews, and specialist papers on themes relevant to the clinical aromatherapist (www.ijca.net).

What actually distinguishes clinical aromatherapy from holistic aromatherapy, however, is a subject of ongoing debate as this newly emerging style has yet to be recognized, and the term "clinical aromatherapy" is widely used without clear understanding. In 2016, an attempt was made to clarify and present opinions from around the world in a special edition of the *International Journal of Clinical Aromatherapy* entitled "Defining Clinical Aromatherapy". This edition was prompted by participant surveys conducted at an international conference for clinical aromatherapy in 2012 and 2014 (Lewis, 2015), as well as a UK-based survey conducted by Carter et al. (2009). Here, Carter et al. defined the clinical aromatherapist as "a skilled and knowledgeable practitioner, who assesses, prescribes, applies and reviews the use of essential oils with patients".

As clinical aromatherapy has increased around the world across a wide range of healthcare settings, much work has been accomplished and documented in these environments by experienced educators and aromatherapists such as Dr. Jane Buckle, author of *Clinical Aromatherapy: Essential Oils in Healthcare* (2015), who over many years has trained approximately 2000 nurses in the use of aromatherapy in hospitals and other clinical settings. Other publications on integrating aromatherapy in healthcare settings have become available that overlap with some medical aromatherapy applications (Dunning, 2007; Wabner and Beier, 2009; Tavares, 2011; Price and Price, 2012; Kerkhof Knapp Hayes, 2015; Steflitsch et al., 2013; Zimmermann, 2018).

It is in the clinical sector of aromatherapy care where most research has been conducted and therefore where most evidence of efficacy exists. Three areas where aromatherapy is particularly advanced include

1. Women's health and midwifery care (Burns, 2005; Burns et al., 2000, 2007; Simkin and Bolding, 2004; Fanner, 2005; Bastard and Tiran, 2006; Imura et al., 2006; Antoniak, 2008; Chitty, 2009; Tillett and Ames, 2010; Conrad and Adams, 2012; Igarishi, 2013; Kazemzadeh et al., 2016)
2. Elderly care, especially dementia (Brooker et al., 1997; Smallwood et al., 2001; Ballard et al., 2002; Holmes et al., 2002; Thorgrimsen et al., 2003; Bowles et al., 2005; Lee, 2005; Lin et al., 2007; Lee, 2008; Nguyen and Paton, 2008; Johannessen, 2013; Grace, 2015; Yang et al., 2015; Press-Sandler et al., 2016)
3. Cancer/palliative care (Wilkinson et al., 1999; Cawthorne and Carter, 2000; Stringer, 2000; DeValois and Clarke, 2001; Louis and Kowalski, 2002; Schwan, 2004; Wilkinson et al., 2007; Curry et al., 2008; Imanishi et al., 2009; Dyer et al., 2014; Clemo-Crosby et al., 2018).

TABLE 13.2
Symptoms that Respond to Clinical Aromatherapy Interventions

Symptom	Evidence
Nausea	Geiger, (2005), De Pradier (2006), Dyer et al. (2010), Ferruggiari et al. (2012), Hines et al. (2012), Lua et al. (2012), Hunt et al. (2013), Hodge et al. (2014), McIlvoy et al. (2015), Sites et al. (2014), Cronin et al. (2015), Lua et al. (2015)
Pain	Dyer (2004), Yip and Tam (2008), Bagheri-Nesami et al. (2014), Karaman et al. (2016), Seyyed- Rasooli et al., (2016), Bikmoradi et al. (2017), Yayla and Ozdemir (2019), Ziyaeifard et al. (2017)
Stress, anxiety, and depression	Morris, (2008), Imanishi et al. (2009), Yim et al. (2009), Lee et al. (2011), Bagheri-Nesami et al. (2014), Hur et al. (2014), Karaman et al. (2016), Seyyed-Rasooli et al. (2016), Sanchez-Vidana et al. (2017), Yayla and Ozdemir (2017), Ziyaeifard et al. (2017), Lewis (2018)
Wound malodour	Warnke et al. (2004), Mercier and Knevitt, (2005), Warnke et al. (2006), Stringer et al. (2014)
Symptoms of insomnia	Lewith et al. (2005), Field et al. (2008), Chen et al. (2012), Chien et al. (2012), Fismer and Pilkington, (2012), Johannessen, (2013), Lillehei and Halcon, (2014), Hwang and Shin (2015), Lillehei et al. (2015), Dyer et al. (2016), Karadag et al. (2017), Muz and Tasci, (2017)

Clinical aromatherapy interventions are often focused on symptom management. Main symptoms where an accumulating evidence base exists for clinical aromatherapy interventions are listed in Table 13.2. Other symptoms that respond well to clinical aromatherapy but for whom there is less robust evidence (predominantly in the form of case reports) include skin breakdown/wound management, oral mucositis, breathlessness, and constipation.

In the clinical aromatherapy sector, patient-reported outcome measures such as MYMOP or MYCaW (developed for cancer supportive care) (Paterson, 1996; Paterson et al., 2007) are becoming the most commonly used research tools for collecting quantitative and qualitative data, documenting outcomes from the patient's perspective. These are easy to use and not time-consuming tools for evaluating service provision and are particularly adapted for assessing complementary therapies such as aromatherapy.

Since 2010, there has been a significant rise in the integration of essential oils in French hospitals, supported by foundations such as Gattefossé and Pierre Fabre who offer prizes and grants for hospital-based initiatives and those in elderly care settings. Whilst this phenomenon is largely driven by doctors, pharmacists, and nurses, the goals and methods of administration of essential oils more closely mirrors that of the clinical aromatherapy style (symptom management, well-being, quality of life, etc.) than the French medical aromatherapy approach (curing disease).

13.6.6 Medical Aromatherapy

Within France and other countries such as Belgium, aromatherapy remains largely in the hands of those qualified in pharmacy and medicine and, here, "legitimate aromatherapy" is seen to constitute a medical act, based on orthodox diagnosis and prescription and dispensed by a pharmacist. Medical aromatherapy/aromatic medicine is more allied to herbal medicine/ phytotherapy than the other forms of aromatherapy described in this chapter. The "official" training in aromatherapy is university-based, and to complete all levels of studies, one has to be a medical professional (doctor, pharmacist, nurse, midwife, veterinarian) and the training is offered as part of optional post-graduate continuing education. There remains no provision for a recognized university diploma in aromatherapy for non-medical personnel in France. A selection of French universities currently offering diplomas in phytoaromatherapy or clinical aromatherapy are listed in Table 13.3.

TABLE 13.3
Examples of Universities in France Offering Training in Aromatherapy or Phytoaromatherapy

University	Diploma and Study Duration	Participants
University of Strasbourg Faculty of Pharmacy	Diploma of Clinical Aromatherapy 105 hours (of which 35 are home-based learning)	Doctors, pharmacists, midwives, nurses, physiotherapists—priority given to those working in a hospital setting
University of Rennes 1 Faculty of Biological and Pharmaceutical Sciences	Diploma of Aromatherapy 105 hours (of which 25 are home-based learning)	Pharmacists, doctors, veterinarians or 4th-year students of these specialties
University of Bourgogne Department of Continuing Professional Development	Diploma of Aromatherapy 55 hours (of which 13 are home study and 3 are e-learning)	Doctors, pharmacists and pharmacy technicians, nurses, midwives
Paris Descartes University The Continuing Education department at the faculty of Pharmacy	Diploma of Phytotherapy and Aromatherapy 80 hours	Pharmacists, pharmacy technicians, doctors, veterinarians, dentists, midwives
University of Franche-Comte	Diploma of Phytotherapy and Aromatherapy 80 hours (of which 8 are home study)	Doctors and pharmacists
University of Montpellier 1 Department of Science, Technology and Health	Diploma of Phytotherapy and Aromatherapy 80 hours	Pharmacists, pharmacy technicians, doctors, veterinarians, dentists, midwives

Medical aromatherapy has more focus on the pharmacological actions of essential oils and their constituents than the psychological impact of fragrant essential oils that is so favored by the holistic aromatherapy style. Following a medical consultation, diagnosis is then coupled with information on essential oil chemistry, bioactivity, pharmacokinetics, toxicology, and galenics in order to formulate the most effective prescription for the patient. Combinations of plant extracts (phytoaromatherapy) are frequently prescribed together. The routes of administration are not limited to topical and inhalation; internal routes such as oral, sublingual, rectal, and vaginal are also commonly used (Figure 13.1.). Main pathologies targeted by the medical approach are those that are infectious or inflammatory in nature.

Outside of mainland Europe, doctors and pharmacists generally have little exposure to or opportunity to extend their knowledge and skills in aromatherapy. A recent paper (Esposito et al., 2014) discusses the success of the implementation of an elective course in Aromatherapy Science for Doctor of Pharmacy (PharmD) students at Sullivan University College of Pharmacy (Louisville, Kentucky, US) to enable pharmacists to advise customers and be more informed in potential drug–essential oil interactions. This type of program, if extended to other pharmacy colleges and medical schools would go some way to preparing a new generation of healthcare professionals who are informed in complementary/alternative medicines such as aromatherapy and herbal medicine. As essential oils are sold in pharmacies and aromatherapy is increasingly being used in hospitals and primary care settings, it is extremely important for doctors and pharmacists to become better informed about the potentials and risks of aromatherapy delivery in these environments.

A few medical publishers engage in publication of reference texts concerning aromatherapy that meet the needs of doctors and pharmacists. *Aromatherapy Science* (Lis-Balchin, 2006) published by Pharmaceutical Press was one of the first to provide pharmacists, general practitioners, nurses, and other healthcare professionals with reliable, scientifically based information on this growing discipline. Since then, others have followed, such as the text (in French) by Kaloustian and Hadji-Minaglou (2013).

13.7 DIFFERENT CULTURES, DIFFERENT AROMATHERAPY STYLES

With the internationalization of aromatherapy, it quickly becomes apparent that each country and culture has adapted techniques and approaches to best meet the needs of the public and the profession. A few examples are given below.

In the United Kingdom and Ireland, the holistic and cosmetic styles of aromatherapy as described earlier in the chapter prevail; most practitioners are working to deliver aromatherapy care for health, beauty, and well-being. Aromatherapy as a business is not lucrative for practitioners; many supplement their income by practicing several therapies or work part-time in a completely different occupation (Lewis, 2015). This self-regulating profession is represented by two aromatherapy-specific bodies (IFA and IFPA, see Table 13.1). Suitably qualified therapists may also register with the UK voluntary regulator for complementary health professionals. regulatory body, the Complementary and Natural Healthcare Council (www.chnc.org.uk).

Aromatherapy in the United States has evolved quite differently from Continental Europe and the United Kingdom. Aromatherapy as a business model predominates, with numerous aromatherapists undertaking training with a specific goal to develop their own product range upon completion or to develop and sell essential oils. Unlike the Anglo Saxon style, aromatherapy massage is less well established due to different massage licensing laws between states. In US medical settings, care is delivered primarily by nurse-aromatherapy practitioners (Buckle, 2003; Johnson et al., 2016). Two well-established aromatherapy organizations support practitioners. Public awareness of aromatherapy in the US has risen since the 1990s due to aggressive promotion through network marketing by two leading US essential oil companies; this has both advantages and disadvantages; on the one hand, self-care aromatherapy is widely promoted as a natural health solution; on the other, it is often the case that essential oil use is uncontrolled, inappropriate, and potentially hazardous to the uninformed user.

In Canada, the predominant aromatherapy style is holistic, closely following the UK style, being therapy orientated rather than product driven. Aromatherapy is well represented across the country with two well-established aromatherapy organizations. In Quebec, the French medical style of aromatherapy is predominant.

Aromatherapy on Continental Europe is highly variable according to country, but generally to be able to call themselves an aromatherapist, practitioners are required to be medical professionals or to have completed training to the level of naturopath in order to practice (Germany, Holland, for example) or be in the medical or paramedical profession (France, Belgium, for example). Most interventions, even when prescribed by medical personnel, remain via diffusion/ inhalation and/ or external applications; in limited cases, essential oils are prescribed orally or rectally by doctors.

Aromatherapy in Asian countries largely follows the holistic Anglo Saxon style with modifications based on culture and commercial interests. Interestingly, despite a long history in alternative approaches such as with traditional Japanese "Kampo" medicine or traditional Chinese medicine, both of which use aromatic and medicinal plants for medicinal benefit, when it comes to the practice of aromatherapy or the use of essential oils in these countries, it is more often the Western style that is followed, with emphasis on essential oils that originate in different parts of the world rather than from indigenous aromatic plants. Currently, Asian countries are highly influenced by the main US-based network marketing companies.

Aromatherapy in Australia has been influenced by both the Anglo Saxon and French medical aromatherapy styles. In recent years, an Advanced Diploma in Aromatic Medicine has been recognized as a profession with educational standards and recognition being established at a national level since 2009. Practitioners with this diploma are permitted to use essential oils internally (oral, rectal) even without a formal medical qualification. This is the first country worldwide to have obtained a government accredited training package in this therapy, although the future of this accreditation system is currently under threat. Australian aromatherapists typically employ treatment approaches that may use external doses of essential oils that may be higher than the Anglo Saxon style, as well as some internal dosing methods such as suppositories and oral ingestion.

13.8 EVALUATING EFFICACY: DOES IT WORK?

As a person-centered and complex therapy that offers a number of possible mechanisms of action, evaluating the effectiveness a typical aromatherapy treatment demands that the researcher understands the multifaceted nature of essential oils: that they are both, at the same time and inseparably, fragrant and pharmacologically active and thus capable of exerting varying activity based on individual responses.

When it comes to direct and specific physiological effects such as accelerating wound healing or treating an infection, then it becomes clear that the pharmacological action of the selected essential oil(s) dominates their effects. However, when aromatherapy is used for more complex issues such as management of chronic pain or assisting the patient to regain a healthier sleep pattern, then the effect of the aromatherapy intervention will be due to both psychosocial and physiological factors.

The moment a pharmacologically active agent also engages the senses, it becomes impossible to completely dissociate psychological from physiological effects. As all essential oils by their very nature are fragrant, the hedonics and semantics of the aroma are likely to directly influence the therapeutic outcome (Moss et al., 2006; Robbins and Broughan, 2007). For these reasons, it is extremely difficult to objectively measure the effects of aromatherapy when it is delivered externally or inhaled, as the sense of smell is completely subjective and can color and modulate any physiological responses that ensue. What is more, the general context of an aromatherapy treatment and the client–therapist relationship will also exert an influence.

Therefore, the complexity of chemical components of an essential oil do not entirely explain the resultant global effects of aromatherapy. Rather than trying to get around this issue in terms of research, it may be necessary to embrace the fact that aromatherapy by its very nature is complex and multifaceted and that it is this "totum" that leads to the effects experienced by the patient (see Figure 13.4).

Finding ways to evaluate the psycho-pharmaco-physiological effects of the whole treatment is challenging but not impossible (Kohara et al., 2004; Takeda, 2008; Xu et al., 2008). Qualitative and mixed methods research designs are more likely to yield a more accurate evaluation of efficacy in the real aromatherapy context but are less popular forms of research design than the gold standard of the randomized controlled trial (RCT) (Verhoef et al., 2002; Fox et al., 2013).

The difficulty of conducting an RCT in aromatherapy, taking into account the complexity of the therapy, may account in part for the disappointing findings in the increasing number of meta-analyses and systematic reviews. These often conclude that there is insufficient evidence to confirm the benefits of aromatherapy in areas such as geriatric care, specific symptom management, and midwifery (Smith et al., 2011; Hines et al., 2012; Forrester et al., 2014) despite the widespread and longstanding use of aromatherapy in these settings, with positive reporting from therapists, staff, and patients. The consistent observation from authors of reviews and meta-analyses is that very few studies meet the pre-specified eligibility criteria or have sufficient data or numbers and many suffer from poor design, and thus, the conclusion is almost invariably that there is insufficient evidence to support aromatherapy use for the topic studied. By means of example, in 2000, Cooke and Ernst conducted a systematic review of aromatherapy, concluding that "There is no published literature that provides a sound rationale for the use of aromatherapy massage as a medical intervention. In the absence of hard efficacy data for lasting and relevant health effects, it is probably best considered as a pleasant diversion for those who can afford it and are prepared to pay for it" (Cooke and Ernst, 2000)

Furthermore, in 2012, Lee et al. conducted a widespread overview of a number of systematic reviews that have been published on aromatherapy; only 10 out of 201 potentially relevant studies met their eligibility/inclusion criteria (Lee et al., 2012). They concluded that while there was some encouraging evidence as to its benefits for psychological health and pain, (largely due to aromatherapy's relaxing effect), "due to a number of caveats, the evidence is not sufficiently convincing that aromatherapy is an effective therapy for any condition."

Research tools for complex interventions that consider the whole person and that involve high patient participation have been helped in recent years, with the development of validated questionnaires such as MYCaW (Measure Yourself Concerns and Wellbeing) and MYMOP (Measure Your Medical

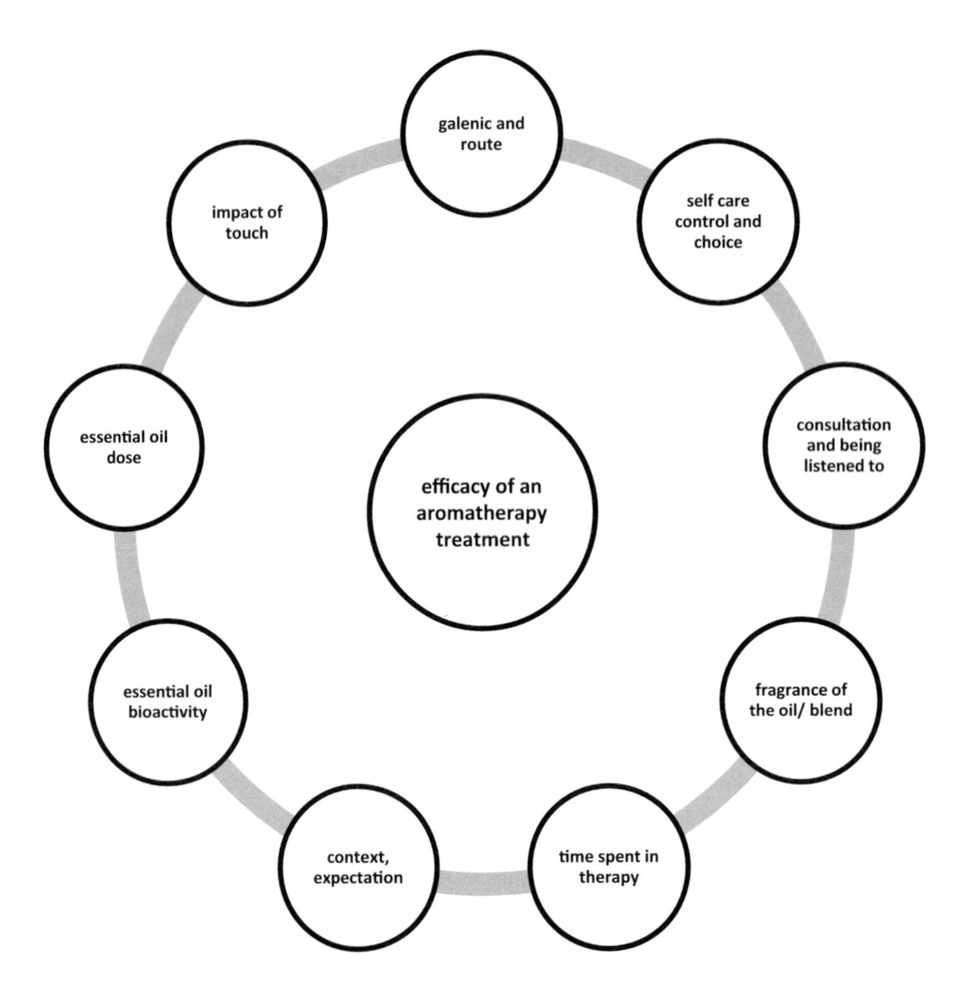

FIGURE 13.4 Aromatherapy as a complex intervention that is hard to evaluate.

Outcome Profile) for evaluating outcomes of complementary therapies in cancer support and similar settings (Paterson, 1996; Paterson et al., 2007). These collate both quantitative and qualitative data from patient-reported outcome measures. As a result, over the past 10 years, it has become easier to document outcomes from the patient's perspective rather than that of the clinician or researcher. This appears to be a better fit generally for complementary therapies that are, by their nature, focused more on whole-person outcomes than those that are disease or symptom specific. An increasing number of clinical settings that offer complementary therapies now routinely use these patient-reported outcome measures (Dyer et al., 2014; Harrington et al., 2012).

13.9 AROMATHERAPY: AN EVOLVING THERAPY

Since its origins, aromatherapy has progressed in its scope, depth, and rigor, and this continues today. With improved reporting of benefits has also come a desire to extend knowledge and skills to continue to improve therapeutic efficacy. Examples of three areas where aromatherapy has made recent progress include:

1. Aroma-alone strategies using personalized aroma inhaler devices
2. Inclusion of hydrosols/hydrolats (distilled plant waters) in aromatherapy applications
3. Use of aromatic extracts such as those obtained by supercritical carbon dioxide extraction.

13.9.1 Personalized Aroma Inhaler Devices

In cancer and palliative care, an increase in the use of aroma-only applications of aromatherapy using personalized aroma inhalers has led to a number of positive service evaluations (Dyer et al., 2008; Dyer et al., 2010; Stringer and Donald, 2010; Hackman et al., 2012; Dyer et al., 2014, 2016). This simple and cost-effective strategy (introduced to cancer and palliative care by the author in 2004) enables patients to maintain a locus of control, providing a welcome and pleasant distraction, while reinforcing any hands-on aromatherapy treatments that may have been given by the therapist. Aroma inhaler use is particularly used in these settings for anxiety, sleep disorders, and nausea management.

13.10 AROMATHERAPY: A SAFE HEALING MODALITY?

In professional aromatherapy training, aromatherapists receive education on key safety issues as part of their basic studies. A comprehensive text (Tisserand and Young, 2013) exists as a key resource.

In an aromatherapeutic treatment context, there are two main issues to consider:

1. Safety to the therapist and those who handle essential oils
2. Safety to the patient/consumer

13.10.1 Safety to the Therapist

Due to their extended exposure to and handling of essential oils, in terms of risk, the aromatherapist is more vulnerable to potential adverse events over the course of their career than the patient receiving aromatherapeutic care. Several cases of occupational hazard among aromatherapists have been reported (Selvaag et al., 1995; Bleasel et al., 2002; Crawford et al., 2004; Boonchai et al., 2007), with the greatest risks reported being contact allergic dermatitis following exposure to these fragranced products. Additionally, the aromatherapist's workplace (often small treatment rooms with inadequate ventilation) may contain high levels of volatile organic compounds that can over time lead to adverse health consequences (Huang et al., 2012), and other staff members may report adverse effects in response to the aromas of essential oils (Tysoe, 2000).

13.10.2 Safety to the Patient/Consumer

Given the approximately 40-year history of holistic aromatherapy delivery, reporting of adverse events by patients connected to professional aromatherapy practice are relatively scarce and idiosyncratic. This may be due to under-reporting, as no globally recognized system of reporting is in place to date but may also suggest that providing aromatherapists continue to use essential oils along the same traditional lines as outlined in this chapter, the risk/benefit ratio is highly favorable toward aromatherapy being both a safe and effective modality.

Two valuable international databases of adverse reactions to essential oils exist, but they are not cross-linked or associated with an official public health body. Public and professional awareness of these databases is low, and they are run and published by private institutions:

- The Injury Reporting Database (collated by the Atlantic Institute of Aromatherapy and published through Aromatherapy United— www.aromatherapyunited.org)
- The Adverse Reaction Database (collated and published by the Tisserand Institute— www.tisserandinstitute.org/safety/adverse-reaction-database/)

According to P&S Market Research in their report on the aromatherapy market 2014–2025, the global aromatherapy market generated 1.2 billion US dollars in 2017, and the trend continues upward. As more people use essential oils, there will undoubtedly be more reporting of essential oil harm. This global aromatherapy "craze" is mostly promoted and encouraged by large essential oil companies using network marketing whose goal is to sell quantity of product (individual essential oils and essential oil blends); thus the recommended doses are often elevated (topical and oral), putting the user at further risk of adverse effects of essential oils. This evolving trend is not widely supported or endorsed by the professional aromatherapy community; in fact, to the contrary, it has generated much anxiety and heated debate around issues of public safety, and some professional aromatherapy organizations have released position papers to denounce these practices. Thus, the accidental exposure/misuse of essential oils under the guise of "aromatherapy" by an uninformed/misinformed public is currently posing a greater and more disturbing risk. For vulnerable populations such as pregnant women, children, babies, the elderly, or those taking multiple medications, home care and self-treatment with essential oils could have potentially disastrous consequences (Sibbritt et al., 2014). The majority of adverse reactions detailed in the above-mentioned databases attest to the risk of this current surge in aromatherapy popularity for using high concentrations of essential oils.

On the other hand, given the longstanding safe history of aromatherapy use, it has not been helpful to have potential risks sensationalized and exaggerated in one recent systematic review of aromatherapy (Posadzki et al., 2012) in which the authors suggested that the risk/benefit ratio of aromatherapy was unhealthy, with risk being potentially serious and benefit being basically nil. In this paper, which continues to be widely consulted by health professionals, out of 40 cases/case series of so-called aromatherapy hazard, at least 10 cases did not relate to aromatherapy practice at all and should not have passed the authors' own pre-specified eligibility criteria. This highlights another pressing need: that researchers conducting large-scale reviews of the literature concerning complementary and alternative therapies should possess at least a basic knowledge or understanding of the therapy in question.

13.11 CONCLUSION: MOVING TOWARD GREATER INTEGRATION

Aromatherapy has come a long way since the early 1960s, with a steady evolution in the quality of education, the professionalism of publications, a more rigorous evaluation, and documentation of risks and benefits, as well as greater awareness and acceptance by the medical establishment. The distinction between differing aromatherapy styles is also becoming clearer, despite the continued debate as to what defines the therapy and who should practice it. This all bodes well for the continued integration of aromatherapy into mainstream medical care. This is in line with the global move toward integrative healthcare in accordance with the WHO's Traditional Medicine Strategy 2014–2023 that encourages a broader vision of person-centered healthcare. The otherwise controversial Smallwood report (2005), compiled by economist Christopher Smallwood, highlighted important areas in health service provision where complementary therapies such as aromatherapy have a role to play as well as demonstrate efficacy. These areas are where "effectiveness gaps" exist in orthodox care, providing the opportunity for complementary therapies to improve patient health and well-being outcomes. These include chronic and complex diseases where cure is not the main goal, as well as palliative care. The place of aromatherapy in these situations is well established and demonstrated in surveys of both therapist and patient experiences and expectations (Long et al., 2001; Osborn et al., 2001).

In December 2018, the All-Party Parliamentary Group for Integrated Healthcare (PGIH) in the UK published an important 31-page document entitled "Integrated Healthcare: Putting the Pieces Together". This report collates the submissions made by 113 organizations and stakeholders that responded to their enquiry that was launched in 2017. Within this document, aromatherapy was mentioned with respect to cancer care. The opening statement of this document reads

> The future of healthcare lies in our health system recognising that physical, emotional and mental health are intrinsically linked. And that only by treating a patient as a whole person can we tackle the root cause of illness and deal with the problem of patients presenting with multiple and complex conditions. (Williams, 2018)

As a multifaceted therapy that is capable of exerting an influence on mind, body and emotions, aromatherapy has an increasingly secure future in healthcare provision. For its continued evolution (in all its styles and aspects), what is needed is further clarity on aromatherapy scope and practice, greater collaboration and consensus at an international level, establishment of international training standards, a greater public profile, a coordinated system of adverse reporting, and more rigorous evaluation of treatment outcomes.

REFERENCES

Antoniak P. 2008. Essential oil therapy with a client experiencing post-partum psychosis: A case study. *Int J Clin Aromather*, 5(1): 8–11.

Bagheri-Nesami M, Espahbodi F, Nikkah A, Shorofi SA, and Charati JY. 2014. The effects of lavender aromatherapy on pain following needle insertion into a fistula in hemodialysis patients. *Complement Ther Clin Pract*, 20(1): 1–4.

Ballard CG, O'Brien JT, Reichelt K, and Perry EK. 2002. Aromatherapy as a safe and effective treatment for the management of agitation in severe dementia: The results of a double-blind, placebo-controlled trial with *Melissa*. *J Clin Psychiatry*, 63: 553–558.

Bastard J, and Tiran D. 2006. Aromatherapy and massage for antenatal anxiety: Its effect on the fetus. *Complement Ther Clin Pract*, 12(1): 48–54.

Baudoux D, and Zhiri A. 2003. *Volume 2: Dermatologie. Collection Aromatherapie Professionnellement*. Luxemburg: Inspir S.A.

Bensouilah J. 2005. The history and development of modern British aromatherapy. *Int J Aromather*, 15(2): 134–140.

Bensouilah J, and Buck P. 2006. *Aromadermatology. Aromatherapy in the Treatment and Care of Common Skin Conditions*. Oxford: Radcliffe Publishing.

Bikmoradi A, Khaleghverdi M, Seddighi I, Moradkhani S, Soltanian A, and Cheraghi F. 2017. Effect of inhalation aromatherapy with lavender essence on pain associated with intravenous catheter insertion in preschool children: A quasi-experimental study. *Complement Ther Clin Pract*, 28: 85–91.

Bishop FL, Yardley L, and Lewith GT. 2008. Treat or treatment: A qualitative study analyzing patients' use of complementary and alternative medicine. *Am J Public Health*, 98(9): 1700–1705.

Bleasel N, Tate B, and Rademaker M. 2002. Allergic contact dermatitis following exposure to essential oils. *Australas J Dermatol*, 43: 211–213.

Boonchai W, Iamtharachai P, and Sunthonpalin P. 2007. Occupational allergic contact dermatitis from essential oils in aromatherapists. *Contact Dermatitis*, 56(3): 181–182.

Bowles EJ, Cheras P, Stevens J, and Myers S. 2005. A survey of aromatherapy practices in aged care facilities in Northern NSW, Australia. *Int J Aromather*, 15: 42–50.

Brooker DJR, Snape M, Johnson E, Ward D, and Payne M. 1997. Single case evaluation of the effects of aromatherapy and massage on disturbed behaviour in severe dementia. *Br J Clin Psychol*, 36: 287–296.

Buchbauer G, and Jirovetz L. 1994. Aromatherapy—use of fragrances and essential oils as medicaments. *Flavour Fragr J*, 9(5): 217–222.

Buchbauer G Jirovetz, L. Jager, W Dietrich, H, and Plank, C 1991. Aromatherapy: Evidence for sedative effects of the essential oil of lavender after inhalation. *Z Naturforsch Teil C*, 46(11–12): 1067–1072.

Buckle J. 2003. Aromatherapy in the USA. *Int J Aromather*, 13(1): 42–46.

Buckle J. 2015. *Clinical Aromatherapy: Essential oils in Healthcare*. St Louis: Elsevier.

Burns E. 2005. Aromatherapy in childbirth: Helpful for mother – what about baby? *Int J Clin Aromather*, 2(2): 36–38.

Burns E, Blamey C, Ersser SJ, Lloyd AJ, and Barnetson L. 2000. The use of aromatherapy in intrapartum midwifery practice. An observational study. *Comp Ther Nurs Midwifery*, 6(1): 33–34.

Burns E, Zobbi V, Panzeri D, Oskrochi R, and Regalia A. 2007. Aromatherapy in childbirth: A pilot randomised controlled trial. *Br J Obstet Gynecol*, 114: 838–844.

Carter A, Mackereth P, and Stringer J. 2010. Aromatherapy in Cancer Care; do aromatherapists in cancer care need specific training to do this work? *In Essence*, 9: 20–22.

Carter A, Mackereth P, Tavares M, and Donald G. 2009. Take me to a clinical aromatherapist: An exploratory survey of delegates to the first Clinical Aromatherapy Conference, Manchester UK. *IJCA*, 6(1): 3–8.

Cawthorne A, and Carter A. 2000. Aromatherapy and its application in cancer and palliative care. *Comp Ther Nurs Midwifery*, 6: 83–86.

Chen JH, Chao YH, Lu SF, Shiung TF, and Chao YF. 2012. The effectiveness of Valerian acupressure on the sleep of ICU patients: A randomized clinical trial. *Int J Nursing Studies*, 48(8): 913–920.

Chien LW, Cheng SL, and Liu CF. 2012. The effect of lavender aromatherapy on autonomic nervous system in midlife women with insomnia. *Evid Based Complement Alternat Med*, 2012: 740813.

Chitty A. 2009. Review of evidence: Complementary therapies in pregnancy. *New Digest*, 46: 20–26.
Clemo-Crosby AC, Day J, Stidston C, McGinley S, Powell RJ. 2018. Aromatherapy Massage for Breast Cancer Patients: A Randomised Controlled Trial. *J Nurs Womens Health*, 3:144. doi: 10.29011/2577-1450.100044.
Conrad P, and Adams C. 2012. The effects of clinical aromatherapy for anxiety and depression in the high risk postpartum woman—A pilot study. *Complement Ther Clin Pract*, 18(3): 164–168.
Cooke B, and Ernst E. 2000. Aromatherapy: A systematic review. *Brit J Gen Pract*, 50(45): 493–496.
Crawford GH, Katz KA, Ellis E, and James WD. 2004. Use of aromatherapy products and increased risk of hand dermatitis in massage therapists. *Arch Dermatol*, 140(8): 991–996.
Cronin SN, Odom-Forren J, Roberts H, Thomas M, Williams S, and Wright MI. 2015. Effects of controlled breathing with or without aromatherapy in the treatment of postoperative nausea. *J Perianesth Nurs*, 30(5): 389–397.
Curry SV, Donaghy K, and Hughes, CM. 2008. Aromatherapy massage for improving well-being of carers of patients with cancer: A pilot double-blind randomised controlled clinical trial. *Int J Clin Aromather*, 5(2): 9–16.
D'Amelio FS. 1999. *Botanicals. A phytocosmetic Desk Reference*. Boca Raton: CRC Press LLC.
Davis P. 1991. *Subtle Aromatherapy*. Safron Walden: CW Daniels.
De Pradier E. 2006. A trial mixture of three essential oils in the treatment of postoperative nausea and vomiting. *Int J Aromatherapy*, 16(1): 15–20.
De Valois B, and Clarke E. 2001. A retrospective assessment of 3 years of patient audit for an aromatherapy massage service for cancer patients. *Int J Aromather*, 11(3): 134–143.
Dunning T. 2007. *Essential Oils in Therapeutic Care*. Melbourne: Australia Scholarly publishing.
Dyer J. 2004. The use of aromatherapy massage for the relief of pain in cancer patients. *Int J Clin Aromather*, 1(1): 8–10.
Dyer J, Cleary L, McNeill S, Ragsdale-Lowe M, and Osland C. 2016. The use of aromasticks to help sleep problems: A patient experience survey. *Complement Ther Clin Pract*, 22: 51–58.
Dyer J, Cleary L, Ragsdale-Loew M, McNeill, and Osland C. 2014. The use of aromasticks at a cancer centre: A retrospective audit. *Complement Ther Clin Pract*, 20(4): 203–206.
Dyer J, Ragsdale-Lowe M, Cardoso M, McNeill S, and Cleary L. 2010. The use of aromasticks for nausea in a cancer hospital. *Int J Clin Aromather*, 7(2): 3–6.
Dyer J, Ragsdale-Lowe M, McNeill S, and Tratt L. 2008. A snapshot survey of current practice: The use of aromasticks for symptom management. *Int J Clin Aromather*, 5(2): 17–21.
Emslie MJ, Campbell MK, and Walker KA. 2002. Changes in public awareness of, attitudes to, and use of complementary therapy in North East Scotland: Surveys in 1993 and 1999. *Complement Ther Med*, 10(3): 148–153.
Esposito ER, Bystrek MV, and Klein JS. 2014. An elective course in aromatherapy science. *Am J Pharm Educ*, 78(4): 70.
Falsetto S. 2010. http://sedonaaromatherapie.com/blog/2010/11/01/the-work-of-a-cosmetic-aromatherapist/ Accessed online December 3, 2018.
Fanner F. 2005. The use of aromatherapy for pain management through labour. *Int J Clin Aromather*, 2(1): 10–14.
Ferruggiari L, Ragione B, Rich ER, and Lock K. 2012. The effect of aromatherapy on postoperative nausea in women undergoing surgical procedures. *J Perianesth Nurs*, 27(4): 246–251.
Field T, Field T, Cullen C, Largie S, Diego M, Schanberg S, and Kuhn C. 2008. Lavender bath oil reduces stress and crying and enhances sleep in very young infants. *Early Hum Dev*, 84(6): 399–401.
Fismer M, and Pilkington K. 2012. Lavender and sleep: A systematic review of the evidence. *Eur J Integr Med*, 4: e436–e447.
Forrester LT, Maayan N, Orrell M, Spector AE, Buchan LD, and Soares-Weiser K. 2014. Aromatherapy for dementia. *Cochrane Database Syst Rev*, (2). Art. No.: CD003150. DOI: 10.1002/14651858.CD003150.pub2.
Fox P, Butler M, Coughlan B, Murray M, Boland N, Hanan T, Murphy H, Forrester PO' Brien M, and O' Sullivan N. 2013. Using a mixed methods research design to investigate complementary alternative medicine (CAM) use among women with breast cancer in Ireland. *Eur J Oncol Nurs*, 17(4): 490–497.
Furnham A. 2000. How the public classify complementary medicine: A factor analytic study. *Complement Ther Med*, 8(2): 82–27.
Gattefossé R-M. 1937. *Aromathérapie- Les Huiles Essentielles Hormones Végétales*. Librairie des Sciences Girardot.
Geiger JL. 2005. The essential oil of ginger, *Zingiber officinale*, and anaesthesia. *Int J Aromather*, 15(1): 7–14.
Godfrey HD. 2018. *Essential oils for Mindfulness and Meditation*. Rochester Vermont: Healing Arts Press.
Goetz P. 2007. *La Phytocosmetologie Therapeutique*. France: Springer-Verlag.
Grace U-M. 2015. Reducing anxiety and restlessness in institutionalised elderly care patients in Finland: A qualitative update on four years of treatment. *IJCA*, 10(1): 22–28.

Hackman E, Mackereth P, Maycock P, Orrett L, and Stringer J. 2012. Expanding the use of aromasticks for surgical and day care patients. *Int J Clin Aromather*, 8(1–2): 10–15.
Harrington JE, Baker BS, and Hoffman CJ. 2012. Effect of an integrated support programme on the concerns and wellbeing of women with breast cancer: A national service evaluation. *Complement Ther Clin Pract*, 18(1): 10–15.
Harris R. 2003. Anglo-Saxon aromatherapy: Its evolution and current situation. *Int J Aromather*, 13(1): 9–17.
Hines S, Steels E, Chang A, and Gibbons K. 2012. Aromatherapy for treatment of post-operative nausea and vomiting. *Cochrane Database Syst Rev*, 4: CD007598.
Hodge NS, McCarthy MS, and Pierce RM. 2014. A prospective randomized study of the effectiveness of aromatherapy for the relief of postoperative nausea and vomiting. *J Perianesth Nurs*, 29(1): 5–11.
Holmes C, Hopkins V, Hensford C, MacLaughlin V, Wilkinson D, and Rosenvinge H. 2002. Lavender oil as a treatment for agitated behaviour in severe dementia: A placebo controlled study. *Int J Geriatr Psychiat*, 17(4): 305–308.
Holmes P. 2001. *Clinical Aromatherapy: Using Essential Oils for Healing Body and Soul*. USA: Tiger Lily Press.
Holmes P. 2016. *Aromatica: A Clinical Guide to Essential Oil Therapeutics*. London & Philadelphia: Singing Dragon.
Huang H-L, Tsai T-J, Hsu N-Y, Lee C-C, Wu P-C, and Su H-J. 2012. Effects of essential oils on the formation of formaldehyde and secondary organic aerosols in an aromatherapy environment. *Build Environ*, 57: 120–125.
Hunt R, Dienemann J, Norton HJ, Hartley W, Hudgens A, Stern T, and Divine G. 2013. Aromatherapy as a treatment for postoperative nausea: A randomized trial. *Anaesth Analg*, 117(3): 597–604.
Hur MH, Song JA, Lee J, and Lee, MS. 2014. Aromatherapy for stress reduction in healthy adults: A systematic review and meta-analysis of randomized clinical trials. *Maturitas*, 79(4): 362–369.
Hwang E, and Shin S. 2015. The effects of aromatherapy on sleep improvement: A systematic literature review and meta-analysis. *J Altern Complement Med*, 21(2): 61–68.
Igarashi T. 2013. Physical and psychological effects of aromatherapy inhalation on pregnant women: A randomized controlled trial. *J Altern Complement Med*, 19: 805–810.
Imanishi J, Kuriyama H, Shigemori I, Watanabe S, Aihara Y, Kita M, and Sawai K et al. 2009. Anxiolytic effect of aromatherapy massage in patients with breast cancer. *eCAM*. 6(1): 123–128.
Imura M, Misao H, and Ushijima H. 2006. The psychological effects of aromatherapy massage in healthy postpartum mothers. *J Midwifery Women's Health*, 51: e21–e27.
Irani F. 2001. *The Magic of Ayurveda Aromatherapy*. West Pennant Hills, Australia: Subtle Energies.
Jenkins S. 2006. Modern British aromatherapy – The way forward: Education and regulation. *IJA*, 16(2): 85–88.
Johannessen B. 2013. Nurses experience of aromatherapy use with dementia patients experiencing disturbed sleep patterns. An action research project. *Complement Ther Clin Pract*, 19(4):209–213. http://dx.doi.org/10.1016/j.ctcp.2013.01.003
Johnson JR, Rivard RL, Griffin KH, Kolste AK, Joswiak D, Kinney ME, and Dusek JA. 2016. The effectiveness of nurse-delivered aromatherapy in an acute care setting. *Complement Ther Med*, 25: 164–169.
Kaloustian J, and Hadji-Minaglou F (editors). 2013. *La connaissance des huiles essentielles : Qualitologie et Aromathérapie*. Entre science et tradition pour une application médicale raisonnée Collection Phytotherapie pratique.
Karadag E, Samancioglu B, Ozden D, and Bakir E. 2017. Effects of aromatherapy on sleep quality and anxiety of patients. *Br Assoc Criti Care Nur*, 22(2): 105–112.
Karaman T, Karaman S, Dogru S, Tapar H, Sahin A, Suren M, Arici S, and Kaya Z. 2016. Evaluating the efficacy of lavender aromatherapy on peripheral venous cannulation pain and anxiety: A prospective, randomized study. *Complement Ther Clin Pract*, 23: 64–68.
Kazemzadeh R, Nikjou R, Rostamnegad M, and Norouzi H. 2016. Effect of lavender aromatherapy on menopause hot flushing: A crossover randomized clinical trial. *J Chin Med Assoc*, 79(9): 489–492.
Kerkhof Knapp-Hayes M. 2015. *Complementary Nursing in End of Life Care*. Netherlands: Kicozo.
Kohara H, Miyauchi T, Suehiro Y, Ueoka H, Takeyama H, and Morita T. 2004. Combined modality treatment of aromatherapy, footsoak, and reflexology relieves fatigue in patients with cancer. *J Palliative Med*, 7(6): 791–796.
Lee MS, Choi J, Posadzki P, Ernst E. 2012. Aromatherapy for health care: An overview of systematic reviews. *Maturitas*, 1(3): 257–260.
Lee SY. 2005. The effect of lavender aromatherapy on cognitive function, emotion, and aggressive behavior of elderly with dementia. *Taehan Kanho Hakhoe*, 35: 303–312 (in Korean with an English abstract).

Lee SY. 2008. The effect of aromatherapy hand massage on cognitive function, sleep disturbance and problematic behaviors of elderly with dementia. *J Korean Clin Nurs Res*, 14: 115–126. (Article in Korean.)
Lee Y, Wu Y, Tsan H, Leung A, and Cheung W. 2011. A systematic review on the anxiolytic effects of aromatherapy in people with anxiety symptoms. *J Altern Complem Med*, 17(2): 101–108.
Lewis R. 2015. Towards defining clinical aromatherapy: The essence of Botanica. *Int J Clin Aroma*, 10(1): 35–47.
Lewis R. 2018. Clinical aromatherapy spotlight: Acute procedural anxiety in cancer care. *Int J Clin Aromather*, 13(1): 19–31.
Lewith GT, Godfrey AD, and Prescott P. 2005. A single-blinded, randomised pilot study evaluating the aroma of *Lavandula angustifolia* as a treatment for mild insomnia. *J Alt Complement Med*, 11(4): 631–637.
Lillehei A, and Halcón L. 2014. A systematic review of the effect of inhaled essential oils on sleep. *J Altern Complement Med*, 20(6): 441–451.
Lillehei AS, Halcón LL, Savik K, and Reis R. 2015. Effect of inhaled lavender and sleep hygiene on self-reported sleep issues: A randomized controlled trial. *J Altern Complement Med*, 21(7): 430–438.
Lin PW, Chan W, Ng BF, and Lam LC. 2007. Efficacy of aromatherapy (*Lavandula angustifolia*) as an intervention for agitated behaviour in Chinese older persons with dementia: A cross-over randomized trial. *Int J Geriatr Psychiatr*, 22: 405–410.
Lis-Balchin M. 2006. *Aromatherapy Science. A guide for Health Professionals*. London: Pharmaceutical Press.
Long, L, Huntley A, and Ernst E. 2001. Which Complementary Therapies benefit which conditions. A survey of the opinions of 223 professional organisations. *Complement Ther Med*, 9: 178–185.
Louis M, and Kowalski SD. 2002. Use of aromatherapy with hospice patients to decrease pain, anxiety and depression and to promote an increased sense of well-being. *Am J Hospice Palliat Care*, 19(6): 381–386.
Lua PL, Salihah N, and Mazlan N. 2015. Effects of inhaled ginger aromatherapy on chemotherapy-induced nausea and vomiting and health-related quality of life in women with breast cancer. *Comp Ther Med*, 23(3): 396–404.
Lua PL, and Zakaria NS. 2012. A brief review of current scientific evidence involving aromatherapy use for nausea and vomiting. *J Altern Complement Med*, 18(6): 534–540.
Mackereth P, Carter A, Parkin S, Stringer J, Caress A, Todd C, Long A, and Roberts D. 2009. Complementary Therapist's training and cancer care: A multi-site survey. *Eur J Oncol Nurs*, 13: 330–335.
Mackereth P, Parkin S, Donald G, and Antcliffe N. 2010. Clinical supervision and complementary therapists: An exploration of the rewards and challenges of cancer care. *Complement Ther Clin Pract*, 16(3): 143–148.
Maury M. 1989. *The Secret of Life and Youth: Regeneration Through Essential Oils: A Modern Alchemy*. England: MacDonald and Co.
McIlvoy L, Richmer L, Kramer D, Jackson R, Shaffer L, Lawrence J, and Inman K. 2015. The efficacy of aromatherapy in the treatment of post discharge nausea in patients undergoing outpatient abdominal surgery. *J Perianesth Nurs*, 30(5): 383–388.
Mercier D, and Knevitt A. 2005. Using topical aromatherapy for the management of fungating wounds in a palliative care unit. *J Wound Care*, 14(10): 497–501.
Miller L, and Miller B. 1995. *Ayurveda and Aromatherapy. The Earth Essential Guide to Ancient Wisdom and Modern Healing*. USA: Lotus Press.
Mojay G. 2000. *Aromatherapy for Healing the Spirit: Restoring Emotional and Mental Balance with Essential Oils*. Randolf, USA: Healing Arts Press – Inner Traditions.
Morris, N. 2008. The effects of lavender (*Lavandula angustifolia*) essential oil baths on stress and anxiety. *Int J Clin Aromather*, 5(1): 3–7.
Moss M, Howarth R, Wilkinson L, and Wesnes K. 2006. Expectancy and the aroma of Roman chamomile influence mood and cognition in healthy volunteers. *IJA*, 16(2): 63–73.
Muz G, and Taşcı S. 2017. Effect of aromatherapy via inhalation on the sleep quality and fatigue level in people undergoing hemodialysis. *Appl Nurs Res*, 37: 28–35.
Nguyen Q, and Paton C. 2008. The use of aromatherapy to treat behavioural problems in dementia. *International Journal of Geriatric*, 23(4): 337–346.
Osborn CE, Barlas P, Baxter GD, and Barlow JH. 2001. Aromatherapy: Survey of common practice in the management of rheumatic disease symptoms. *Comp Ther Med*, 9: 62–67.
P&S Market Research. https://www.psmarketresearch.com/market-analysis/aromatherapy-market.
Paterson C. 1996. Measuring outcomes in primary care: A patient generated measure, MYMOP, compared with the SF-36 health survey. *BMJ*, 312(7037): 1016–1020.
Paterson C, Thomas K, Manasse A, Cooke H, and Peace G. 2007. Measure Yourself Concerns and Wellbeing (MYCaW): An individualised questionnaire for evaluating outcome in cancer support care that includes complementary therapies. *Complement Ther Med*, 15(1): 38–45.
Petersen D. 2012. What's hot and what's not: US trends in aromatherapy essential oil choices. Paper presented at: *Asian Aroma Ingredients Congress & Expo*, Bali, Indonesia.

Posadzki P, Alotaibi A, and Ernst E. 2012. Adverse effects of aromatherapy: A systematic review of case reports and case series. *Int J Risk Saf Med*, 24: 147–161.

Posadzki P, Watson LK, Alotaibi A, and Ernst E. 2013. Prevalence of use of complementary and alternative medicine (CAM) by patients/consumers in the UK: Systematic review of surveys. *Clin Med*, 13(2): 126–131.

Press-Sandler O, Freud T, Volkov I, Peleg R, and Press Y. 2016. Aromatherapy for the treatment of patients with behavioral and psychological symptoms of dementia: A descriptive analysis of RCTs. *J Altern Complem Med*, 22(6): 422–428.

Price S, and Price L. 2012. *Aromatherapy for Health Practitioners*, 4 ed. Churchill Livingstone; Elsevier.

Robbins G, and Broughan C. 2007. The effects of manipulating participant expectations of an essential oil on memory through verbal suggestion. *Int J Essential Oil Ther*, 1: 56–60.

Sade D. 2017. *The Aromatherapy Beauty Guide: Using the Science of Carrier and Essential Oils to Create Natural Personal Care Products*. Toronto: Robert Rose.

Sánchez-Vidaña DI, Ngai SP, He W, Chow JK, Lau BW, and Tsang HW. 2017. The effectiveness of aromatherapy for depressive symptoms: A systematic review. *Evid-Based Complement Alternat Med*., 2017:5869315 https://www.hindawi.com/journals/ecam/2017/5869315/ Accessed March 7, 2017.

Schwan R. 2004. Integrative palliative aromatherapy care program at San Diego Hospice and Palliative Care. *Int J Clin Aromather*, 1(2): 5–9.

Selvaag E, Holm J-O, and Thune P. 1995. Allergic contact dermatitis in an aromatherapist with multiple sensitisations to essential oils. *Contact Dermatitis*, 33(5): 354–355.

Seyyed-Rasooli A, Salehi F, Mohammadpoorasl A, Goljaryan S, Seyyedi Z, and Thomson B. 2016. Comparing the effects of aromatherapy massage and inhalation aromatherapy on anxiety and pain in burn patients: A single-blind randomized clinical trial. *Burns*, 42(8): 1774–1780.

Sibbritt DW, Catling CJ, Adams J, Shaw, AJ, and Homer CSE. 2014. The self-prescribed use of aromatherapy oils by pregnant women. *Women Birth*, 27(10): 41–45.

Simkin P, Bolding A. 2004. Update on nonpharmacologic approaches to relieve labor pain and prevent suffering. *J Midwifery Women's Health* 49(6): 480–504.

Sites DSS, Johnson NT, Miller JA, Torbush PH, Hardin JS, Knowles SS, Nance J, Fox TH, and Tart RC. 2014. Controlled breathing with or without peppermint aromatherapy for postoperative nausea and/or vomiting symptom relief: A randomized controlled trial. *J Perianesth Nurs*, 29(1): 12–19.

Smallwood C. 2005. The role of complementary and alternative medicine in the NHS. An investigation into the potential contribution of mainstream complementary therapies to healthcare in the UK. http://www.getwelluk.com/uploadedFiles/Publications/SmallwoodReport.pdf

Smallwood J, Brown R, Coulter F, Irvine E, and Copland C. 2001. Aromatherapy and behaviour disturbances in dementia: A randomized controlled trial. *Int J Geriatr Psychiatry*, 16: 1010–1013.

Smith CA, Collins CT, and Crowther CA. 2011. Aromatherapy for pain management in labour. *Cochrane Database Syst Rev*, (7). Art. No.: CD009215. DOI: 10.1002/14651858.CD009215.

Steflitsch W, Wolz D, and Buchbauer G. 2013. *Aromatherapie in Wissenschaft und Praxis*. Wiggensbach: Stadelmann Verlag.

Stringer J. 2000. Massage and aromatherapy on a leukaemia unit. *Comp Ther Nurs Midwifery*, 6: 72–76.

Stringer J, and Donald G. 2010. Aromasticks in cancer care: An innovation not to be sniffed at. *Complement Ther Clin Pract*, 11(11): 1222–1229.

Stringer J, Donald G, Knowles R, and Warn P. 2014. The symptom management of fungating malignant wounds using a novel essential oil cream. *Wounds UK*, 10: 54–59.

Takeda H, Tsujita J, Kaya M, Takemura M, and Oku Y. 2008. Differences between the physiologic and psychologic effects of aromatherapy body treatment. *J Altern Complement Med*, 14(6): 655–661.

Tavares M. 2011. *Integrating clinical aromatherapy in specialist palliative care. The use of Essential Oils for Symptom Management*. Self-published. www.aromapac.ca

Thorgrimsen L, Spector A, Wiles A, and Orrell M. 2003. Aroma therapy for dementia. *Cochrane Database Syst Rev*, (3). Art. No.: CD003150. DOI: 10.1002/14651858.CD003150.

Tillett J, and Ames D. 2010. The uses of aromatherapy in women's health. *J Perinat Neonatal Nurs*, 24(3): 238–245.

Tisserand R, and Young R. 2013. *Essential Oil Safety*. 2nd ed. Churchill Livingstone. Elsevier.

Tysoe P. 2000. The effect on staff of essential oil burners in extended care settings. *Int J Nurs Pract*, 6: 110–112.

Verhoef MJ, Casebeer AL, and Hilsden RJ. 2002. Assessing Efficacy of Complementary Medicine: Adding Qualitative Research Methods to the "Gold Standard". *J Altern Complem Med*, 8(3): 275–281.

Wabner D, and Beier C (editors). 2009. *Aromatherapie. Grundlagen, Wirkprinzipien, Praxis*. München: Elsevier Urban & Fischer.

Warner F. 2013. *The Soul Midwives' Handbook*. London: Hay House.

Warner F. 2018. *Sacred Oils. Working with 20 Precious Oils to Heal Spirit and Soul*. London: Hay House.

Warnke PH, Sherry E, Russo PAJ, Acil Y, Wiltfang J, Sivananthan S, Sprengel M, Roldan JC, Bredee JP, and Springe ING. 2006. Antibacterial essential oils in malodorous cancer patients: Clinical observations in 30 patients. *Phytomed*, 13(7): 463–467.

Warnke PH, Terheyden H, Acil Y, Springer IN, Sherry E, Reynolds M, Russo PA, Bredee JP, and Podschun R. 2004. Tumour smell reduction with antibacterial essential oils. *Cancer*, 100(4): 879–880.

Wilkinson S, Aldridge J, Salmon I, Cain E, and Wilson B. 1999. An evaluation of aromatherapy massage in palliative care. *Palliat Med*, 13: 409–417.

Wilkinson S, Love S, Westcombe A, Gambles M, Burgess C, Cargill A, Young T, Mayer E, and Ramirez A. 2007. Effectiveness of aromatherapy massage in the management of anxiety and depression in patients with cancer: A multicentre randomised controlled trial. *Am J Clin Oncol*, 25(5): 532–539.

Williams M. 2018. Integrated Healthcare: Putting the Pieces Together. All-Party Parliamentary Group for Integrated Healthcare. http://icamhub.com/wp-content/uploads/2019/01/PGIH-Report-Download.pdf. Accessed February 13, 2019.

Xu F, Uebaba K, Ogawa H, Tatsuse T, Wang BH, Hisajima T, and Venkatraman S. 2008. Pharmaco-physio-psychologic effect of Ayurvedic oil-dripping treatment using an essential oil from Lavandula angustifolia. *J Altern Complement Med*, 14(8): 947–956.

Yang MH, Lin LC, Wu SC, Chiu JH, Wang PN, and Lin JG. 2015. Comparison of the efficacy of aroma-acupressure and aromatherapy for the treatment of dementia-associated agitation. *BMC Complement Altern Med*, 15: 93.

Yayla EM, and Ozdemir L. 2019. Effect of inhalation aromatherapy on procedural pain and anxiety after needle insertion into an implantable central venous port catheter: A quasi-randomized controlled pilot study. *Cancer Nurs*, 42(1): 35–41.

Yim V, Ng A, Tsang H, and Leung A. 2009. A review on the effects of aromatherapy for patients with depressive symptoms. *J Altern Complem Med*, 15(2): 187–195.

Yip YB, and Tam ACY. 2008. An experimental study on the effectiveness of massage with aromatic ginger and orange essential oil for moderate-to-severe knee pain among the elderly in Hong Kong. *Compl Ther Med*, 16: 131–138.

Zafar S, Hifzul K, Ahsan MT, Siddiqui AI, and Munawwar H. 2007. Conference Paper: Evidence Based Unani Aromatherapy: Rationale and Relevance. *International Conference on Research Advances and Traditional Medicine: Challenges and Opportunities*. New Delhi.

Zimmermann E. 2018. *Aromatherapie fur Pflege- und Heilberufe*. Stuttgart: Haug Verlag.

Ziyaeifard M, Zahedmehr A, Ferasatkish R, Faritous Z, Alavi M, Alebouyeh MR, Dehdashtian, E, Ziyaeifard, P, and Yousefi Z. 2017. Effects of lavender oil inhalation for anxiety and pain in patients undergoing coronary angioplasty. *Iran Heart J*, 18(1): 47–50.

14 Essential Oils in Cancer Therapy

Carmen Trummer and Gerhard Buchbauer

CONTENTS

14.1 INTRODUCTION

Cancer is a growing health problem and is one of the main reasons for the cause of death worldwide (Manjamalai et al., 2012). In 2012, 14.1 million new cases appeared and the most prevalent diagnosed cancers were lung (1.82 million), breast (1.67 million), and colorectal (1.36 million). The most common cancer-related deaths were caused by lung cancer (1.6 million), liver cancer (745,000), and stomach cancer (723,000) (Ferlay et al., 2015).

Nature is a rich source of biological and chemical diversity, and most of the anticancer drugs available in the market are obtained from medicinal plants (Shan et al., 1999). During recent years, essential oils (EOs) have come into the focus for the treatment of cancer. EOs are hydrophobic liquids containing volatile aroma compounds from plants and have been used since ancient time for their health properties. Currently, a lot of *in vitro* studies with EOs against different cancer cell lines are being carried out with promising results. EOs could be evolving to be a potential new medicinal resource. The research for EOs as new anticancer agents ought to be continued.

14.2 CANCER CELL LINES

A cell line is a population of cells descended from a single cell containing the same genetic material (Alberts et al., 2002). Human cancer-derived cell lines are substantial models that are commonly used in laboratories to examine the biology of cancer and to evaluate the therapeutic efficacy of anticancer agents (Sharma et al., 2010). The first cultured cancer cell line was HeLa (taken from

Henrietta Lacks) in 1951 (Scherer et al., 1953). From that time on, hundreds of cancer cell lines have been developed and propagated. There are two different types of production processes, either *in vitro* as monolayer cultures or *in vivo* as xenografts in mice (Mattern et al., 1988).

14.3 TESTS TO ASSESS CYTOTOXIC ACTIVITY

Several methods are used to investigate the cytotoxic, antiproliferative and anticancer activities of EOs. Three of the most important assays are the MTT assay, the TUNEL assay, and flow cytometry and are described below.

14.3.1 MTT Assay

The 3-(4,5-dimethylthiazol-2-yl)-2,5-diphenyl-tetrazolium bromide (MTT) assay is a colorimetric assay used for measuring cell survival and proliferation (Mosmann , 1983). The method is based on the metabolic reduction of soluble MTT by mitochondrial enzyme activity of viable tumor cells. The product, an insoluble, colored formazan; can be quantified spectrophotometrically (Momtazi et al., 2017); and depends on cell type, cellular metabolism, and incubation time with MTT (Mosmann, 1983). This provides an estimate of cell viability.

14.3.2 TUNEL Assay

Terminal deoxynucleotidyl transferase dUTP nick end labeling (TUNEL) is a method to investigate cell death by detecting apoptotic DNA fragmentation. The assay relies on the use of terminal deoxynucleotidyl transferase (TdT), an enzyme that catalyzes attachment of deoxynucleotides, tagged with a fluorochrome or another marker, to 3′-hydroxyl termini of DNA double strand breaks (http://en.wikipedia.org/wiki/TUNEL.assay).

14.3.3 Flow Cytometry

Flow cytometry is a routinely used method to detect and measure physical and chemical characteristics of a population of cells. This method allows study of cellular populations with high precision. Some important measurable parameters are apoptosis (quantification, measurement of DNA degradation, caspase activity) and cell viability (Picot et al., 2012).

14.4 EOs AGAINST DIVERSE HUMAN CANCER CELL LINES

EOs possess remarkable cytotoxic, antiproliferative, and antitumor properties on several cancer cell lines. In laboratory practice, it is usual to test one EO against different cancer cell lines. In this chapter, studies are presented that deal with the effect of one EO on a range of different cancer cell lines.

Sharopov et al. (2017) evaluated the cytotoxic activity of *Foeniculum vulgare* (Lin.) Mill. (Apiaceae) EO from Tajikistan against five different human cancer cell lines (HeLa [human cervical cancer[, Caco-2 [human colorectal adenocarcinoma], MCF-7 [human breast adenocarcinoma], CCRF-CEM [human T-lymphoblast leukemia], and CEM/ADR5000 [Adriamycin-resistant leukemia]). The plant and its EO are often used in traditional medicine as carminative, digestive, and diuretic (Mimica-Dukic et al., 2003), and the major components are *trans*-anethole (36.8%), α-ethyl-*p*-methoxy-benzyl alcohol (9.1%), *p*-anisaldehyde (7.7%), carvone (4.9%), 1-phenyl-penta-2,4-diyne (4.8%), and fenchyl butanoate (4.2%). Because the oil is rich in lipophilic secondary metabolites, it can cross cell membranes by free diffusion. A lot of constituents carry a reactive carbonyl group, which can react with amino groups of amino acid in proteins or in nucleotides. It was found that the cytotoxic effect is due to the lipophilic properties of the EO. But also, the alkylating properties of the main components *trans*-anethole and *p*-anisaldehyde are important factors for their efficacy. The calculated IC_{50} values were 207 mg/L for HeLa, 75 mg/L for Caco-2, 59 mg/L for MCF-7, 32 mg/L for CCRF-CEM, and 165 mg/L for CEM/ADR5000 cell lines (Sharopov et al., 2017). Supplementary, the main compound *trans*-anethole was investigated for its cytotoxicity in RC-37 cells. The IC_{50} value was 100 mg/L (Astani et al., 2011), and it showed cell death and loss of cellular ATP and adenine nucleotide pools (Nakagawa, 2003). These findings indicate that anethole could be the cause of the overall cytotoxicity of the EO (Sharopov et al., 2017).

In traditional medicine, the EOs of *Teucrium alopecurus* L. (Lamiaceae) have been used for their anti-inflammatory properties (Guesmi et al., 2018), and the phenolic and terpenic components from *Teucrium* species are known for their ability to treat obesity, hypercholesterolemia, and diabetes (Rajabalian, 2008).

Guesmi et al. (2018) evaluated the EO of *T. alopecurus*, from aerial parts, for its anticancer activities. The EO exhibited, dose dependently, proliferation-inhibition effects on HCT-116, U266, SCC4, Panc28, KBM5, and MCF-7 cells. Furthermore, the EO showed the efficacy to suppress the growth of the carcinoma cells and it induced apoptosis. It was found that the anticancer effects are possibly due to their phenolic and/or sesquiterpene content (+)-limonene, α-bisabolol, humulene, thymol, and (+)-epi-bicyclosesquiphellandrene.

Dahham et al. (2015) investigated the cytotoxic effect of β-caryophyllene from the EO of *Aquilaria crassna* Pierre (Thymelaeaceae), which is a plant with diverse traditional medicinal properties. The sesquiterpene was tested for its inhibitory effect on proliferation against seven human cancer cell lines and two normal cell lines, using the MTT test. β-Caryophyllene exhibited among the tested cancer cells an antiproliferative effect against HCT 116 (colon cancer, $IC_{50} = 19\ \mu M$), PANC-1 (pancreatic cancer, $IC_{50} = 27\ \mu M$), and HT29 (colon cancer, $IC_{50} = 63\ \mu M$) cells. The compound showed moderate activities against ME-180, PC3, K562, and MCF-7, and low toxicity against the normal cell lines 3T3-L1 and RGC-5. The authors found that the EO induces chromatin condensation and DNA fragmentation in HCT 116 cells. For untreated HCT 116 cells, the apoptotic indices were 1.7% ± 0.04%; after the 24 h treatment with β-caryophyllene, it increased to 58.4% ± 7%.

Loizzo et al. (2007) studied the cytotoxic properties of EOs from Labiatae and Lauraceae families and their ability to inhibit human tumor cell growth. The authors cultured the human amelanotic melanoma cell line C32, renal cell adenocarcinoma ACHN, hormone-dependent prostate carcinoma LNCaP, and the human breast cancer cell line MCF-7 in different media and tested the cytotoxicity using the sulforhodamine B (SRB) assay. *Laurus nobilis* L. (Lauraceae) fruit oil showed significant anticancer effects on the amelanotic melanoma cells (IC50 75.45 μg/mL) and the renal adenocarcinoma cells (IC_{50} 78.24 μg/mL). In contrast, the cytotoxic activity of *L. nobilis* leaf oil on both cell lines was less (IC_{50} 202.62 μg/mL for ACHN and C32 with IC_{50} of 209.69 μg/mL). Interestingly, the EO of *Sideritis perfoliata* L. (Labiatae) was also found to have high activity on both cell lines. The IC_{50} value was 100.90 mg/mL for C32 and 98.58 μg/mL for ACHN. The EO of *Salvia officinalis* L. (Labiatae) also showed interesting growth-inhibitory effects on renal cell adenocarcinoma cells with an IC_{50} of 100.70 μg/mL. Likewise, when *Pistacia palestina* L. (Labiatae) EO was applied, growth inhibition appeared on ACHN and C32 cells (IC_{50} of 204.70 and 356.98 μg/mL, respectively). All the oils showed no reaction with human breast cancer cells (MCF-7) and hormone-dependent prostate carcinoma cells (LNCaP).

Also, several identified compounds of the EOs were tested for their anticancer activity in *in vitro* experiments (1,8-cineole, limonene, linalool, β-caryophyllene, and α-humulene). Linalool was active against amelanotic melanoma (IC_{50} 23.16 μg/mL) and renal cell adenocarcinoma (IC_{50} of 23.77 μg/mL). β-caryophyllene exhibited cytotoxic activity against ACHN and C32 cell lines (IC_{50} 21.81 and 20.10 μg/mL, respectively). Among the tested constituents, the sesquiterpene α-humulene showed the highest cytotoxicity tested on the LNCaP cells. The calculated IC_{50} was 11.24 μg/mL. But it showed no results against the MCF-7, C32, and ACHN cell lines (IC50 >50 μg/mL).

Lippia citriodora Pal. (Verbenaceae), commonly known as lemon verbena, is native to South and Central America and was brought to Europe in the seventeenth century (Quirantes-Piné et al., 2013). The plant was traditionally used for its antispasmodic, diuretic, and sedative properties and to relief gastrointestinal symptoms (Fitsiou et al., 2018). The authors analyzed the antiproliferative activity of *L. citriodora* grown in Greece on a panel of cancer cell lines. To determine the cell viability, SRB or MTT assay were employed. Citral, which is the sum of the two isomers neral (*cis*-citral) (17.2%) and geranial (*trans*-citral) (26.4%), was identified as the main component, followed by nerol (8.0%), geraniol (5.7%), and spathulenol (3.3%). The cells got treated with increasing concentrations of the EO fraction (0.64–920 μg/mL) or citral (0.63–900 μg/mL) for 72 h.

The oil fraction showed the most potent cytotoxic effects against the A375 (melanoma) cells (EC_{50} = 9.1 ± 0.6 μg/mL). Similar viability levels were discovered against HepG2 (hepatocellular carcinoma), MCF-7 (breast adenocarcinoma), and Caco2 (colon adenocarcinoma) cells (EC_{50} = 74±2.8, 89 ± 1.4 and 71 ± 2.6 mg/mL, respectively). Only less cytotoxic activity was observed against THP-1 cells (leukemic monocytes) (EC_{50} = 11 ± 3.6 μg/mL).

Citral demonstrated a significantly stronger cytotoxic effect against all tested cell lines than the oil. Among the cancer cell lines, MCF-7 cells were the most sensitive (EC_{50} =1.3 ± 0.19 μg/mL) when subjected to citral. Followed by Caco2 and HepG2 cells (EC_{50} = 3.7 ± 0.21 μg/mL and 7 ± 0.35 μg/mL, respectively). This effect could be due to an antagonistic effect between the components of the EO.

An *in vitro* antitumor cytotoxicity assay of *Artemisia herba-alba* Asso. (Asteraceae) EOs demonstrated antiproliferative properties against P815 (murine mastocytoma cell line) and BSR (kidney carcinoma cell line of hamsters) cancer cell lines (Tilaoui et al., 2015). *A. alba* is a dwarf shrub and the EOs from the leaves and aerial parts mainly contain oxygenated sesquiterpenes, while the capitulum oil is mainly composed by monoterpenes. The results showed that all EOs possess a significant cytotoxic effect against the tested cell lines in a dose-dependent manner but leaf and capitulum EOs were more active than EOs obtained from aerial parts.

Nagappan et al. (2011) evaluated the antiproliferative effects of three carbazole alkaloids (mahanine, mahanimbicine, and mahanimbine) and the EO from the leaves of *Murraya koenigii* L. (Rutaceae) against human breast (MCF-7), human cervical (HeLa), and murine leukemia cell lines (P388). From the EO of *M. koenigii*, 34 aromatic volatile constituents were identified. The two sesquiterpene hydrocarbons, β-caryophyllene (19.5%) and α-humulene (15.2%), represented the main volatile metabolites. To evaluate the anticancer activity, cells were treated with all compounds and the EOs dissolved in DMSO to a concentration of 30 μg/mL. The carbazole alkaloids and the EOs exhibit antiproliferative effects against all three tested cell lines in a dose-dependent manner: the higher the concentration of the tested compound, the lower the cell viability. Mahanimbine showed the most significant cytotoxic effects with IC_{50} values of 2.12, 5.00, and 1.98 μg/mL in the MCF-7, P388, and HeLa cell lines, respectively. These results could be important for the development of a new antitumor lead metabolite.

Croton matourensis Aubl. (Euphorbiaceae) is a medicinal plant and is popularly known as "orelha de burro" or "maravuvuia" in Brazil (Lima et al., 2018). The tree has been used in folk medicine for the treatment of infections, fractures, and colds (de Sousa Trinidade and Alves Lameira , 2014). An experimental study was carried out to evaluate the *in vitro* cytotoxic and *in vivo* antitumor effects from the leaf EO of *C. matourensis* collected from the Amazon rainforest. The scientists examined the following cell lines: MCF-7 (breast adenocarcinoma), HCT116 (colon carcinoma), HepG2 (hepatocellular carcinoma), HL-60 (promyelocytic leukemia), and human non-cancer cell line MRC-5 (lung fibroblasts). The EO was obtained by hydrodistillation and major components identified were β-caryophyllene (12.4%), thunbergol (11.7%), cembrene (7.1), *p*-cymene (5.0%), and β-elemene (4.9%). To measure cell proliferation and cytotoxicity, the Alamar blue assay, which is a non-toxic alternative to the MTT-assay, was used. The most promising anticancer potential effect was on HL-60, with an IC_{50} value of 17.8 μg/mL. Other inhibiting concentration 50% values were 23.3 μg/mL for MCF-7, 28.9 μg/mL for HCT116, 28.5 μg/mL for HepG2, and 25.8 μg/mL for MRC-5. Moreover, the *in vivo* antitumor activity of the EO from the leaves of *C. matourensis* was tested in C.B-17 severe combined immunodeficient (SCID) mice with HepG2 cell xenografts. The animals got doses of 40 and 80 mg/kg/day of the EO injected for 21 consecutive days. At the end of the treatment, the tumor mass inhibition rates of the EO were 34.6%–55.9%. Interestingly, the positive control 5-fluorouracil (10 mg/kg/day) reached a tumor weight reduction of 44.2%. In conclusion, these results indicated the *in vitro* and *in vivo* anticancer properties of this plant (Lima et al., 2018).

In the present study, varying to the others before, a range of ten EOs were investigated for their *in vitro* toxicology toward A-549 (human lung carcinoma), PC-3 (human prostate carcinoma), and MCF-7 (human breast cancer) cell lines (Zu et al., 2010). The Chinese scientists used the EOs of mint (*Mentha spicata* L., Lamiaceae), ginger (*Zingiber officinale* Rosc., Zingiberaceae), lemon (*Citrus limon* Burm. f., Rutaceae), grapefruit (*Citrus paradisi* Macf., Rutaceae), jasmine (*Jasminum grandiflora* L., Oleaceae), *Lavandula angustifolia* Mill., (Lamiaceae), chamomile (*Matricaria chamomilla* L., Asteraceae), thyme (*Thymus vulgaris* L., Lamiaceae), rose (*Rosa damascena* Mill., Rosaceae), and cinnamon (*Cinnamomum zeylanicum* Nees, Lauraceae). The cells were exposed to increasing concentrations of the EOs, and cell viability was detected by MTT assay. Principally, a dose-dependent decrease in the survival of the tumor cell lines has been revealed. Most of the EOs showed significant cytotoxicities against A549 cells. But the EO of mint showed no effect on A549 cells with the tested concentration of 0.002%–0.2% (v/v). All EOs exhibited strong cytotoxic properties against PC-3 cells at a concentration of 0.20% (v/v). In comparison to the other EOs, cinnamon, thyme, chamomile, and jasmine EOs exhibited the strongest cytotoxicities toward MCF-7 cells. Interestingly the EO of *T. vulgaris* displayed the strongest cytotoxicity against all three human cancer cells. The IC_{50} values for *T. vulgaris* EO were 0.01% for PC-3, 0.01% for A549, and 0.03% (v/v) against MCF-7 cells (Zu et al., 2010).

14.5 EOs AGAINST BREAST CANCER

Breast cancer is the most frequently diagnosed cancer in woman, with 485,000 deaths per year worldwide (Zhang et al., 2018). Several factors such as gender, family history, diet, use of alcohol, lifestyle, and endocrine aspects are associated with breast tumors, but it is unclear which factor is most important in breast cancer pathogenesis (Abdulkareem, 2013).

This chapter concerns different types of plants and their EOs that retain antitumor properties. Most of them are reported in *in vitro* cell line models.

Ortiz et al. (2016) examined the cytotoxic and genotoxic effects of sandalwood (*Santalum album* L. Santalaceae) EO. The EO which is obtained from trees from the *Santalum* genus is also used in the cosmetic, perfume, and food industries. For their studies, they used breast adenocarcinoma (MCF-7) and nontumorigenic breast epithelial (MCF-10 A) cells. *(Z)-α*-Santalol (25.3%) is one of the main components of the 300 identified chemical constituents of sandalwood EO. Other main active constituents were (*Z*)-nuciferol (18.3%), (*E*)-*β*-santalol (11.0%), and (*E*)-nuciferol (10.5%). The authors tested eight different concentrations of sandalwood EO and all showed a decrease in cell viability. The calculated IC_{50} value for the MCF-7 cell line was 8.03 μg/mL, and for the MCF-10A cell line, it was 12.3 mg/mL.

It was found that the EO of *Decatropis bicolor* (Zucc.) Radlk (Rutaceae) can induce apoptosis of the MDA-MB-231 breast cancer cell line. Estanislao Gómez et al. (2016) analyzed the effect from a pharmacological perspective. The plant is also well known in traditional medicine in some communities in Mexico. The authors used the MTT assay to investigate the cytotoxic activities. There were no cytotoxic results found for the aqueous extract, but the ethanolic, acetonic, and hexanic extracts showed a cytotoxic effect. The authors used different concentrations of the oily extract and different times of incubation to evaluate the cell viability. The oil was able to decline the cell viability 15%, 36%, 60%, and 77% using doses of 40, 60, 80, and 100 μg/mL, respectively, after 24 h of incubation. Additionally, after 48 h of treatment, the cell viability decreased 35%, 69%, 88%, and 91% with 40, 60, 80, and 100 μg/mL, respectively. Finally, the greatest cytotoxic effect was observed after 72 h of incubation; the cell viability reduction was more than 85% with doses of 60, 80, and 100 μg/mL. The IC_{50} value was 53.81 ± 1.7 μg/mL. The positive control paclitaxel (0.25 μg/mL) showed 30%, 40%, and 53% of viability reduction at 24, 48, and 72 h, respectively, while the DA-MB-231 cells without treatment retained a viability of 100%.

Zhong et al. (2018) found that the EO of Rhizoma Curcumae (*Curcuma longa* L., *Curcuma xanthorrhiza* L.) (Zingiberaceae), and the main bioactive compounds (furanodienone and furanodiene) exhibited significant inhibitory effects on viability of doxorubicin-resistant MCF-7 breast cancer cells. The authors tested the cell viability after a 48-h treatment with doxorubicin using the MTT assay. The EO, furanodienone, and furanodiene showed IC_{50} values of 77.0 μg/mL, 52.1 μM, and 69.6 μM, respectively. Combined effects of furanodiene and doxorubicin were also investigated after 24 h of treatment. The best synergistic inhibitory effects on the viability were observed with high concentrations of furanodiene combined with low concentrations of doxorubicin (Zhong et al., 2018). Moreover, it was found that furanodiene induced extrinsic and intrinsic apoptosis associated to the AMPK-dependent and NF-κB-independent pathways in doxorubicin-resistant MCF-7 cells (Zhong et al., 2016).

Frankincense EO has long been known for its anti-rheumatism, anti-inflammatory (Poeckel and Werz, 2006), antibacterial, and antifungal activities (Syrovets et al., 2005). Frankincense (*Boswellia carteri* Birdw., Burseraceae), pine needle (Pinaceae), and geranium (Geraniaceae) EOs were found to be capable of suppressing cell viability, proliferation, migration, and invasion in human breast cancer MCF-7 cells (Ren et al., 2017). Different dilutions of these three EOs were used to observe the cell viability (frankincense, 1:1000–1:4000; pine needle, 1:1000–1:4000; geranium, 1:1000–1:8000). After 48 h of treatment, cell viability decreased in a dose-dependent manner measured by the CCK-8 assay. Interestingly, at an oil dilution of 1:1000, no MCF-7 cells stayed viable. The concentration that inhibited cell vitality by 50% (IC_{50}) of frankincense EO was 42.8 μg/mL. For pine needle and geranium EO, the calculated IC_{50} values were 90.2 and 73.9 μg/mL, respectively. In addition, the EO stimulation causes morphological changes of the MCF-7 cells. The author also found that the EOs were able to decrease the colony size and colony forming capacity of the breast cancer cells, in comparison to the control groups. These results demonstrated that the EO treatment had an impact on breast cancer cell proliferation. Furthermore, flow cytometry showed that the three EOs induced apoptosis but there were no effects on cell cycle progression. Another important finding was that Frankincense, pine needle and geranium EOs affected the AMPK/mTOR pathway. Thus, the EOs showed an influence on the progression of breast cancer cells.

Pallenis spinosa (L.) de Cassini (Asteraceae) is an annual herbaceous plant (Al-Eisawi, 1982) and has long been used for the treatment of eczema, rheumatism, diabetes, headaches, and infections (Cavero and calvo, 2015). The *P. spinosa* EO of flowers and leaves was studied for its cytotoxicity against MCF-7 and MDA-MB-231 breast adenocarcinomas. Saleh et al. (2017) found out that the predominant constituents in the flower oil are sesquiterpenes (96.4%). While the leaf sample mainly contained oxygenated sesquiterpenes (51.0%) and sesquiterpene hydrocarbons (34.0%). The highest cytotoxic activity against MCF-7 and MDA-MB-231 was found in the flower EO. Increasing concentrations of the EO were applied to the cells for 48 h. The calculated IC_{50} values for the flower oil were 0.25 ± 0.03 μg/mL against MCF-7 and 0.21 ± 0.03 μg/mL against MDA-MB-231. And the IC_{50} values from the leaf oil against MCF-7 were 2.4 ± 0.5 and 1.5 ± 0.1 μg/mL against MDA-MB-231, respectively. These findings indicate that *P. spinose* EO shows potent natural anticancer compounds.

Rosmarinus officinalis L. (Lamiaceae) is commonly known as rosemary and is widespread in the Mediterranean region. 1,8-Cineole is the main component of the EO of *R. officinalis*, followed by α-pinene, camphene, β-myrcene, borneol, and camphor. *R. officinalis* EO from Tunisia was found to have some anticancer activities. The authors used the human cancer cell line MCF-7 and the HeLa cell line for their studies. The cells were exposed to increasing concentrations of the EO and tested by MTT assay. The EO indicated proliferation inhibition on both cell lines, but the inhibition was significantly stronger in the HeLa cells (IC_{50} value of 0.011 μL/mL) than in the MCF-7 cells (IC_{50} value of 0.253 μL/mL) (Jardak et al., 2017). These results indicate that the EO of *R. officinalis* is a potent anticancer agent and these activities are due to the main components, 1,8-cineole, camphor, and α-pinene (Moteki et al., 2002).

Blepharocalix salicifolius (Kunth.) O. Berg. (Myrtaceae) is an aromatic spice of the Myrtaceae family which is widespread in South America. The EO of this plant was found to possess potent cytotoxic properties. As major constituents, bicyclogermacrene (17.5%), globulol (14.1%), viridiflorol (8.8%), γ-eudesmol (7.9%), and α-eudesmol (6.9%) were established. The authors found out that this EO showed cytotoxic effects against the MDA-MB-231 (46.60 μg/mL) breast cancer cell line by impairing the cellular metabolism of the cancer cells (Furtado et al., 2018).

The tiny evergreen tree *Tamarix aphylla* (L.) (Tamaricaceae) is native in the Middle East and across Africa and Asia. *T. aphylla* H. Karst EO of the aerial parts was found to have cytotoxic effects on human breast adenocarcinoma (MCF-7), colorectal adenocarcinoma (Caco-2), pancreatic carcinoma (Panc-1) cancer cell lines, and normal human fibroblasts in a dose-dependent manner. The EO showed the strongest inhibitory effects on MCF-7 cells, with IC_{50} values for *T. aphylla* aqueous extract and ethanol extract of 2.17 ± 0.10 and 26.65 ± 3.09 μg/mL. In comparison with the control drug cisplatin, *T. aphylla* aqueous extract showed also a comparable cytotoxic effect with an IC_{50} value of 1.17 ± 0.13 μg/mL. Unfortunately, both extracts showed, in the tested concentrations, no cytotoxic effects against Panc-1 or Caco-2 cell lines (Alhourani et al., 2018).

The aldehyde component citral (3,7 dimethyl-2,6-octadien-1-al) is a major compound in several EOs, especially of citrus fruits, lemongrass and ginger. Citral was analyzed for its potential cytotoxic activity *in vitro* against MDA-MB-231 breast cancer cells using the MTT assay. Therefore, spheroids of MDA-MB-231 breast cancer cells were treated with different concentrations of citral (Nigjeh et al., 2018). Spheroids are a type of *in vitro* cultured cell in a three-dimensional form. (Ho et al., 2012) Compared to the monolayer of MDA-MB-231 cells, citral inhibited the growth of the MDA-MB-231 spheroids at a lower IC_{50} value. Additionally, an apoptosis study was performed and citral showed early and late apoptotic changes in a dose-dependent manner. Moreover, a gene expression study demonstrated that citral can suppress the self-renewal capacity of spheroids and down regulates the Wnt/β-catenin pathway. Further studies should be performed to investigate the potential of citral as an alternative treatment (Nigjeh et al., 2018).

The aromatic shrub *Hedyosmum sprucei* Solms (Chloranthaceae) is native in the tropical Andes of Ecuador and Peru (Todzia, 1988). *H. sprucei* EO was evaluated for its cytotoxic effects on the breast adenocarcinoma cell line MCF 7 and the lung adenocarcinoma cell line A549. The EO was characterized by a high amount of germacrene-D (23.2%) followed by β-caryophyllene (15.5%), δ-cadinene (5.5%), α-copaene (5.1%), and α-phellandrene (3.5%). The data were less interesting on the A549 cell line. But against the MCF-7 (breast cancer) cell line, the EO showed promising cytotoxic results. After 48 h of the treatment, the calculated IC_{50} value was 32.76 ± 4.92 μg/mL, and after 72 h, the IC_{50} was 33.64 ± 0.43 μg/mL (Guerrini et al., 2016).

It has been demonstrated that terpenes such as germacrene-D and β-caryophyllene performed cytotoxic activity against breast cancer cells. Hence, the high content of these main compounds in *H. sprucei* EO may explain the cytotoxic properties toward MCF-7 (Legault and Pichette, 2007).

14.6 EOs AGAINST PROSTATE CANCER

Prostate cancer is the most commonly diagnosed cancer in men overall and is one of the leading causes of death from cancer in older men (American Cancer Society, 2017). Various treatments like androgen-deprivation therapy or chemotherapy are available (Stavridi et al., 2010), but the research to find more effective and less toxic drugs has become necessary.

It was found that three EOs of Lebanese *Salvia* (*Salvia aurea* L., *S. judaica* L., and *S. viscosa* L.) (Lamiaceae) showed significant anticancer activities against human prostate cancer cells (DU-145) (Russo et al., 2018). The EOs from the aerial parts were used, and the authors treated the cells for 72 h with different concentrations. The EOs with caryophyllene oxide as the main component were able to reduce the growth of human prostate cancer cells. They also activated an apoptotic process and increased the reactive oxygen species generation.

Bayala et al. (2014) investigated the effects of EOs from the leaves of *Ocimum basilicum* L. (Lamiaceae), *O. americanum* L. (Lamiaceae), *Hyptis spicigera* Jacq. (Lamiaceae), *Lippia multiflora* Mold. (Verbenaceae), *Ageratum conyzoides* L. (Asteraceae), *Eucalyptus camaldulensis* Dehn.

(Myrtaceae), and *Zingiber officinale* Rosc. (Zingiberaceae) from Burkina Faso against prostate cancer cell lines, namely LNCaP and PC3. The results showed that *O. basilicum*, *Z. officinale*, *L. multiflora*, and *A. conyzoides* possess significant antiproliferative effects on both cell lines, whereas *O. americanum*, *H. spicigera*, and *E. camaldulensis* exhibited no activity. Tests showed that *A. conyzoides* (IC_{50} 0.35 mg/mL) and *Z. officinale* (IC_{50} 0.38 mg/mL) possessed the best antiproliferative effects on the LNCaP cell line. The EOs of *L. multiflora* and *O. basilicum* showed the most significant anticancer effects on the PC-3 cell line (IC_{50} 0.30 mg/mL).

Lavandula angustifolia L. (Lamiaceae) is the most cultivated lavender species and has been used for the treatment of gastrointestinal disorders, irritability, and rheumatism. Zhao et al. (2016) studied the therapeutic anticancer effects of lavender EO, linalool, and linalyl acetate in a mouse xenograft model. A cell suspension of human prostate cancer cell line PC-3 was injected into male nude mice. As soon as the xenograft tumors reached a size of ~30 mm^3, the mice got medicated with solvent control, lavender EO, linalool, or linalyl acetate (200 mg/kg). After 4 weeks of treatment, the tumor growth was reduced in comparison with the control group. For lavender EO ($P < 0.001$), linalool ($P < 0.001$), and linalyl acetate ($P = 0.016$). The author group suggested that lavender EO and linalool are promising sources for therapeutic agents for the prostate cancer treatment.

Carvacol is a natural-bioactive monoterpenoid that is present in many EOs. It is extracted from thyme or other herbs, spices, vegetables, or fruits (Suntres et al., 2015). Luo et al. (2016) found that carvacol blocked transient receptor potential melastatin-like 7 currents (TRPM7) in prostate cancer cells. TRPM7 belongs to the melastatin-like transient receptor potential (TRPM) subfamily and is overexpressed in a lot of cancer tissues and cell lines. The EO compound carvacol showed a reduction of cell proliferation, migration, and invasion of PC-3 and DU145 cells. Carvacol was also able to decrease the protein expression of MMP-2, p-Akt, and p-ERK.

14.7 EOs AGAINST LIVER CANCER

Liver cancer is the sixth most common cancer worldwide (Ferlay et al., 2015) and has a high recurrence rate and poor prognosis (Liu et al., 2017).

The leading cause of liver cancer is cirrhosis due to hepatitis C, hepatitis B, or alcohol. Most of the new cases occur in less-developed countries, with a high incidence in Asia and Africa (Ferlay et al., 2015). It is essential to elucidate new therapeutic strategies, and EOs are a potent source for new active components. *Eupatorium adenophorum* Spreng. (Asteraceae) is a well-known medicinal weed and is traditionally used in treating fever, desensitization, traumatism, and phyma in China (Yunnan Institute of Materia Medica, 1975). The EO of *E. adenophorum* Spreng., which is mainly composed of sesquiterpenes, was evaluated for its anticancer activities (Chen et al., 2018). It exhibited *in vitro* hepatocellular carcinoma cell apoptosis and also inhibited the growth of HepG2 xenografts. The tests showed that the EO decreased the ratio of Bcl-2/Bax, while the activation of caspase-9 and -3 increased. This may be due to an activation of the mitochondrial apoptotic pathway. The treatment with *E. adenophorum* Spreng. EO inhibits the proliferation of HepG2, Hep3B, and SMMC-7721 cells. All cells were incubated with 0, 5, 10, 30, 50, 100, 150, and 200 μg/mL of *E. adenophorum* Spreng EO for 48 h. The inhibiting concentration 50% (IC_{50}) values were 17.74 ± 1.92 μg/mL for HepG2, 49.56 ± 5.01 μg/mL for Hep3B, and 39.20 ± 3.37 μg/mL for SMMC-7721 cells (Chen et al., 2018).

In the present study, *in vitro* tests of *Siegesbeckia orientalis* L., *S. glabrescens* L. and *S. pubescens* (Xi-Xian) (Asteraceae) EOs showed antitumor properties against Hep3B (liver) cancer cells and against HeLa (cervical) cells (Gao et al., 2018). Hydrodistillation was used to obtain the EOs, and by gas chromatography–mass spectrometry the authors identified 148 compounds (56 in *S. orientalis*, 62 in *S. glabrescens*, and 59 in *S. pubescens*). The main compounds of each EO were different. Caryophyllene oxide showed the highest contribution on cancer cell cytotoxicity. The results showed that the EO of *S. pubescens* was more effective than the EOs of *S. orientalis* and *S. glabrescens*. Therefore, *S. pubescens* exerted the strongest cytotoxicity, with IC_{50} values of 38.10 μg/mL for Hep3B and 37.72 μg/mL for HeLa. Interestingly, the IC_{50} value of the positive control 5-fluorouracil was lower compared with the three EOs. It was also found that the *S. orientalis* ethanol extract exhibited anticancer activity against A549 (lung cancer), HepG2 (hepatoma), MDA-MB-231 (breast cancer), and especially on LNCaP (prostate cancer) cell lines. These findings indicate that the EOs are a promising source for anticancer agents.

Murraya paniculata L (Rutaceae) is known as orange jessamine (Zhang et al., 2011) and possess anti-inflammatory, antibiotic, and analgesic effects (Selestino-Neta et al., 2016). In a study about the effects of β-caryophyllene and *M. paniculata* EO, Selestino-Neta et al. (2016) studied the cytotoxic activity against hepatoma cancer cells using the colorimetric MTT assay. Concentrations between 7.8 and 500 μg/mL were used. The cytotoxicity of the EO was expressed as IC_{50} values and was for tumorous cells of hepatocytes 63.73 μg/mL. Interestingly, at the highest tested concentration of 500 μg/mL, β-caryophyllene showed the lowest activity. The mechanism of action of the EOs against cancer cells works possibly due to a synergistic effect among all the components of the EO.

Glandora rosmarinifolia (Ten.) D.C. Thomas (Boraginaceae) EO was evaluated for its *in vitro* antitumor activity. Different hepatocellular carcinoma cell lines (HA22T/VGH, HepG2, Hep3B) and triple negative breast cancer cell lines (SUM 149, MDA-MB-231) were examined. The results showed that the EO exerts significant antitumor effects on all tested cell lines and induces cell growth inhibition in a concentration-dependent manner (Poma et al., 2018).

Another study was conducted to appraise the anticancer activity of three *Plectranthus* species (*P. cylindraceus* Hocst. ex Benth., *P. asirensis* JRI Wood and *P. barbatus* Andrews) (Lamiaceae) grown and collected in Saudi Arabia (Mothana et al., 2018). MTT assay was used to determine the cytotoxic activity against HeLa (human cervical cancer), HepG2 (human hepatocellular liver carcinoma), and HT-29 (human colon cancer) cell lines. The EO of these species exhibited promising anticancer activities against all cell lines. The IC_{50} values ranged from 3.88 to 7.51 μg/mL. *P. cylindraceus* EO showed the strongest inhibitory effect.

The EOs of carvacrol and rosemary were analyzed for their ability to induce apoptosis in human hepatoma HepG2 cells and their effect on the cell cycle (Melušová et al., 2015). A previous study demonstrated that both EOs induced growth inhibition on HepG2 and BHNF-1 human cell lines. HepG2 cells were treated with different concentrations of the EOs for 24 h and the cell cycle phases were investigated by flow cytometry. An increasing accumulation of cells in the G1 phase was depicted in a dose-dependent manner. In addition, it was found that carvacrol and rosemary decreased the number of viable cells by increasing concentrations. Both compounds induced formation of apoptotic bodies and conducted shrinking of the cytoplasmic membrane.

14.8 EOS AGAINST LUNG CANCER

Lung cancer, also known as lung carcinoma, is considered as a major global health problem. It is the most common cancer-related cause of death in males and the second most frequent cause of death in females after breast cancer (Ramalingam and Belani, 2008). Most lung cancer cases are due to tobacco smoking and air pollution. Worldwide in 2012, a total of 1.8 million cases of lung cancer were reported and 1.6 million resulted in death (World Cancer Report, 2014a,b) These data suggest that it's important to find new therapeutic agents.

The monoterpene carvacrol is widespread in many EOs of the Lamiaceae plant family. For example, it is present in the EO fraction of *Origanum, Satureja*, *Thymbra, Thymus*, and *Corydothymus*. Koparal and Zeytinoglu (2003) investigated the effects of carvacrol on a human non-small cell lung cancer (NSCLC) cell line, A549. Usually the A549 cells exhibit a polygonal shape and a sheet-like pattern in the culture, which is like the epithelial origin. The authors treated the cells with four different concentrations of the oil (100, 250, 500, and 1000 μM) for 24 h; 100 μM of carvacrol did not show any effects, but the treatment with 250 μM of carvacrol lead to morphological shape changes of the cells. At concentrations of 500 and 1000 μM, carvacrol exerted apoptotic characteristics, and morphological changes could be described. The phenolic compound carvacrol is also well known for its antibacterial, antifungal, analgesic, and antioxidant activities.

The annual herb *Nigella sativa* L. (Ranunculaceae), commonly known as black seed, was reported to have antioxidant, anti-inflammatory, and antibacterial properties. Al-Sheddi et al. (2014), investigated the cytotoxic activity of *N. sativa* seed EO and extract against the human lung cancer cell line A-549, using the MTT assay and the neutral red uptake (NRU) assay. The data revealed that the EO induces a decrease in cell viability in a concentration-dependent manner.

Populus alba L. (Salicaceae) and *Rosmarinus officinalis* L. (Lamiaceae) EO of leaves and flowers were evaluated for their cytotoxic effects (Gezici et al., 2017). The EOs of these medicinal and aromatic plants were obtained by hydrodistillation, and about 300 compounds were identified. The used cell lines were A549 (human lung adenocarcinoma), H1299 (human non-small cell lung cancer), MCF-7 (human breast adenocarcinoma), and non-tumor HUVEC cells. *R. officinalis* exerted stronger inhibitory effects than *P. alba*. Calculated IC_{50} values for *R. officinalis* were ranging from 3.06 to 7.38 μg/mL and for *P. alba* from 12.05 to 28.16 μg/mL. The highest activity was achieved on A459 and H1299 cells, while only moderate activity was exhibited against MCF-7 cells. The significant growth inhibition effect of *R. officinalis* may be attributed to the presence of carnosol, methyl carnosate, carnosic, and rosmarinic acids.

Tridax procumbens L. (Asteraceae) is a long-used traditional medical plant (Manjamalai et al., 2012). The following study showed that EO of *T. procumbens* L. is capable of suppressing lung metastasis by the B16F-10 cell line in C57BL/6 mice (20–25 g). In comparison with the untreated mice, the group which was treated with the EO showed an inhibition of tumor nodule formation of 71.7%. It was also found, that the EO was able to inhibit the formation of tumor directed new blood vessels (39.5%). The authors also studied apoptosis using the TUNEL assay. In comparison with cancer alone, the treated cells showed an increase of apoptotic cells (Manjamalai et al., 2012).

Navel orange is a type of citrus which is cultivated in many countries around the world and Gannan in Jiangxi Province is the top navel orange producing area in China (Eldahshan and Halim, 2016). Yang et al. (2017) studied the EO from Gannan navel orange peel, with 74.6% of limonene, for its anticancer activities. The EO was able to inhibit the proliferation of a human lung cancer cell line A549 and prostate cancer cell line 22RV-1. They used the MTT assay to evaluate the effect of different concentrations on cell viability. Concentrations between 6.25 and 200 μg/mL showed the best inhibition of proliferation in both cell lines. The higher the concentration of the navel orange EO, the better was the inhibition.

14.9 EOs AGAINST MELANOMA

Melanoma is a malignant proliferation of melanocytes (pigment containing cells) and it typically occurs in the skin. About 133,000 new cases appear each year worldwide, and 80% of melanoma is caused by ultraviolet damage (World Cancer Report, 2014a,b). This chapter deals with the antitumor activity of EOs and their compounds against melanoma.

Salvia officinalis L. and *Thymus vulgaris* L. are well-known medicinal aromatic plants of the Lamiaceae family. Since ancient times, they have been recognized for their ability to treat inflammations, physical and mental fatigue, nervousness, cough, and skin ulceration (Bouaziz et al., 2009). Alexa et al., 2018 examined the *in vitro* antiproliferative activity of the EO of *S. officinalis* and *T. vulgaris* on two melanoma cell lines, namely A375 human melanoma and B164A5 mouse melanoma. As main components of *S. officinalis*, EOs identified were β-caryophyllene (25.4%), camphene (14.1%), eucalyptol (14.0%), and β-pinene (11.2%). In *T. vulgaris*, EO γ-terpinene (68.4%) represents the main component followed by thymol (24.7%). The results of this study showed that *S. officinalis* EO was more effective in decreasing the cell viability of the two tested cell lines. The tests concentration of 100 μg/mL resulted an inhibition ratio of 30% for B164A5 cell line and 27% for A375 cell line. However, the testes concentration of 100 μg/mL *T. vulgaris* EO showed an inhibition ratio of 30% for the B164A5 cell line and 27% for the A375 cell line. Also, the mixture of the two EOs showed promising results, and further studies should be conducted (Alexa et al., 2018).

The following study demonstrated the anticancer properties, of the EO from *Pituranthos tortuosus* Benth. and Hook f. ex Asch. and Schweinf. (Apiaceae) against melanoma. The oil was isolated from the aerial parts of *P. tortuosus*, and major constituents identified were sabinene (24.2%), α-pinene (18.0%), limonene (16.1%), and terpinen-4-ol (7.2%) (Krifa et al., 2015). The cytotoxicity against B16F10 melanoma cells was tested. The cells were seeded, and after 24 h, different concentrations of the EO were added. After 48 h of incubation, the maximum of proliferation inhibition was detected at the concentration of 400 μg/mL (91.3% ± 3.6%). The calculated IC_{50} was 80 μg EO/mL after 48 h. The authors concluded that *P. tortuosus* could be an important medicinal resource in cancer treatment.

The Verbenaceae family, including *Lippia*, is native in tropical and subtropical distributions (Trease and Evans, 1983) and is often used by local citizens to relieve gastrointestinal and respiratory diseases (Morton, 1981). The EO of *Lippia alba* Mill. (Verbenaceae) turned out to exert significant anticancer activities against B16F10Nex2 (murine melanoma) and A549 (human lung adenocarcinoma) cell lines (Santos et al., 2016). The EO was characterized by a high amount of the isomeric monoterpenes nerol/geraniol (27.1%) and citral (21.9%), 6-methyl-5-heptene-2-one (12.0%), and *β*-caryophyllene (9.3%). Interestingly, compared to the standard drugs cisplatin and paclitaxel, the oil possessed lower IC_{50} values. *L. alba* EO inhibited the proliferation of B16F10Nex2 cells with an IC_{50} value of 45.8 μg/mL, while at a concentration of 63.9 μg/mL, the EO inhibited only 50% of the A549 cell proliferation. Cisplatin showed an IC_{50} value of 52.8 μg/mL against B16F10Nex2, and paclitaxel illustrated an IC_{50} value of 84.3 μg/mL against A549. In conclusion, the EO of *L. alba* displayed promising anticancer potential and deserves further studies because it showed lower IC_{50} values in comparison to standard drugs in the tested cells.

14.10 EOs AGAINST COLORECTAL CANCER

Most of cancers occurring in the colon and rectum are adenocarcinomas, with 945,000 new cases diagnosed each year. (World Cancer Report, 2014a,b) The following studies show the great potential of EOs for the development of new chemotherapeutics for the treatment of colorectal cancer.

Melaleuca alternifolia (Maid. and Betch.) Cheel, also known as the tea tree, belongs to the botanical family Myrtaceae. The main component of the EO is terpinen-4-ol, and it was studied for its antitumoral activity against colorectal cancer cell lines HCT116 and RKO (Nakayama et al., 2017). To evaluate the effect of terpinen-4-ol *in vivo*, a xenograft model was used. The authors implanted HCT116 cells (2 × 106 cells per mouse) into 14 mice and injected 200 mg/kg of terpinen-4-ol. The EO exhibited apoptotic cell death in HCT116 and RKO cells via reactive oxygen species. This happened in a dose-dependent manner. Compared with the control group, the EO also showed an inhibition of the proliferation of HCT116 xenografts. All these effects were achieved without damaging normal cells.

The EO of *Piper aequale* L. (Piperaceae) was studied for its sensitivity against human colorectal carcinoma (HCT-116) and human gastric tumor (ACP 03) cell lines (da Silva et al., 2017). δ-Elemene (19.0%), β-pinene (15.6%) and α-pinene (12.6%) were identified as the main compounds. The cells were treated for 72 h with the EO. In all tested concentrations (0.75–3.0 μg/mL), the EO induced apoptosis in the ACP 03 cells. The cytotoxicity was expressed as IC_{50} values. The IC_{50} value for human gastric tumor (ACP 03) attained 1.54 and 8.69 μg/mL for human colorectal carcinoma (HCT-116).

Euphorbia macrorrhiza C.A. Mey. ex Ledeb. (Euphorbiaceae), a perennial herb which is widespread in West Siberia, Kazakhstan, and North China, was evaluated for its anticancer properties against Caco-2 cells (human colorectal carcinoma) (Lin et al., 2012). Many species of the genus *Euphorbia* have been considered throughout the ages to have a wealth of healing properties. For example, *Euphorbia* medicinal plants have been used for the treatment of skin diseases, gonorrhea, migraine, intestinal parasites, and warts (Singla and Kamla, 1990). Chemical profiles for *E. macrorrhiza* EO of the aerial parts and roots demonstrated that acorenone B (16.7% and 25.8%) is the major compound, followed by (+)-cycloisosativene (15.0% and 12.4%), 3α-hydroxy-5β-androstane (10.6% and 5.5%), and copaene (7.4% and 6.3%). After 24 h of treatment with different concentrations of the EO, MTT assay was conducted to quantify cell proliferation. The strongest inhibitory effect on Caco-2 cells was shown by EO of the roots with an IC_{50} value of 11.86 μg/mL. The data of this study suggested that *E. macrorrhiza* EO is a good natural source to produce new antitumor drugs (Lin et al., 2012).

Pinus koraiensis Sieb. and Zucc, (Pinaceae) is an evergreen tree widespread in Korea, China, and eastern Russia. The EO of *P. koraiensis* contains a lot of different components, including camphene (21.1%), D-limonene (21.0%), α-pinene (16.7%) ,and borneol (11.5%) (Kim et al., 2012). Cho et al. (2014) analyzed the effects of *P. koraiensis* EO, obtained from the leaves, on HCT116 colorectal cancer cells. *P. koraiensis* EO has been found to reduce cell proliferation on HCT116 cells through G1 arrest. This happens without affecting normal cells. Additionally, *P. koraiensis* EO was able to suppress PAK1 expression in a dose-dependent manner, thereby a decrease in ERK and AKT phosphorylation and beta-catenin expression occurred. This study demonstrated the potent anticancer activity of *P. koraiensis* EO, which may be a novel chemotherapeutic agent for the treatment of colorectal cancer.

14.11 EOs AGAINST EHRLICH CARCINOMA

Mesosphaerum sidifolium (L'Hérit.) Harley and J.F.B.Pastore (Lamiaceae) has been used as a folk remedy in different countries to treat stomach disorders and headaches, besides its use as expectorant, carminative, and tonic (Matos, 1999). Rolim et al. (2017) evaluated the antitumor activity of *M. sidifolium* EO. The authors used aerial parts of *M. sidifolium* (L'Hérit.) Harley and J.F.B.Pastore (syn. *Hyptis umbrosa*), which were collected in Brazil. Fenchone (24.8%) is the main compound followed by cubebol (6.9%), limonene (5.4%), spathulenol (4.5%), β-caryophyllene (4.6%), and α-cadinol (4.7%). For the evaluation of the *in vivo* antitumor activity, Ehrlich tumor cells (5–7 days old) were implanted in female mice. As control group, 5-fluoruracil (25 mg/kg) was analyzed. After 9 days of treatment with the EO of *M. sidifolium* (50, 100, or 150 mg/kg) and its major component (30 or 60 mg/kg), the results abounded. In comparison with the control group, the EO of *M. sidifolium* (100 or 150 mg/kg) induced a significant reduction in tumor volume and weight and tumor cell total count. For fenchone (60 mg/kg) also, a decrease in all analyzed parameters was found. The survival of the mice increased for both concentrations.

14.12 EOs AGAINST GLIOBLASTOMA

The EO of *Mentha crispa* L. (Lamiaceae) and its major constituent rotundifolone and a series of six related monoterpenes were tested for their anticancer effects on the human U87MG glioblastoma cell line (Turkez et al., 2018). Glioblastoma (GBM) is one of the most widespread and aggressive brain tumors. *M. crispa* EO, 1,2-perillaldehyde epoxide (EPER1), and perillaldehyde (PALD) were more effective than other tested compounds with IC_{50} values of 16.26, 15.09, and 14.88 μg/mL, respectively. It was also found that the EO increased the expression of BRAF, EGFR (epidermal growth factor receptor), KRAS (kirsten rat sarcoma), NFκB1 (nuclear factor NF-kappa-B), NFκB1A, NFκB2, PIK3CA (phosphatidylinositol-4,5-bisphosphate 3-kinase), PIK3R (phosphatidylinositol 3-kinase regulatory), PTEN (phosphatase and tensin homolog), and TP53 (tumor protein p53) genes. Furthermore, the expression of some genes, for example AKT1 (RAC-alpha serine/threonine-protein kinase), AKT2 (RAC-beta serine/threonine-protein kinase), FOS, and RAF1 (proto-oncogene c-RAF) decreased. See Table 14.1.

TABLE 14.1
Essential Oils and Their Anticancer Properties: A Collection in Tabular Form

EO/EO Constituent	Anticancer	Year	Reference
Foeniculum vulgare (Lin.) Mill. (Apiaceae)	HeLa Caco-2 MCF-7 CCRF-CEM CEM/ADR5000	2017	Sharopov et al. (2017)
Teucrium Alopecurus Lin. (Lamiaceae)	HCT-116 U266 SCC4 Panc28 KBM5 MCF-7	2008	Guesmi et al. (2018)
β-caryophyllene of *Aquilaria crassna* (Thymelaeaceae)	HCT-116 PANC-1 HT29 ME-180 PC-3 K562 MCF-7	2015	Dahham et al. (2015)
Laurus nobilis Lin. (Lauraceae)	C32 ACHN	2007	Loizzo et al. (2007)
Sideritis perfoliata (Labiatae)	C32 ACHN	2007	Loizzo et al. (2007)
Salvia officinalis Lin. (Labiatae)	ACHN	2007	Loizzo et al. (2007)
Pistacia palestina (Labiatae)	ACHN C32	2007	Loizzo et al. (2007)
Lippia citriodora Pal. (Verbenaceae)	A375 HepG2 MCF-7 Caco2	2018	Fitsiou et al. (2018)
Artemisia alba Asso (Asteraceae)	P815 BSR	2015	Tilaoui et al. (2015)
Murraya koenigii Lin. (Rutaceae)	MCF-7 HeLa P388	2011	Nagappan et al. (2011)
Croton matourensis Aubl. (Euphorbiaceae)	HL-60 MCF-7 HCT-116 HepG2 MRC-5	2018	Lima et al. (2018)
Zingiber officinale Rosc. (Zingiberaceae)	A549 PC-3 MCF-7	2010	Zu et al. (2010)
Mentha spicata Lin. (Lamiaceae)	PC-3 MCF-7	2010	Zu et al. (2010)
Citrus limon Burm.f. (Rutaceae)	A549 PC-3 MCF-7	2010	Zu et al. (2010)

(*Continued*)

TABLE 14.1 (*Continued*)
Essential Oils and Their Anticancer Properties: A Collection in Tabular Form

EO/EO Constituent	Anticancer	Year	Reference
Citrus paradisi Macf. (Rutaceae)	A549 PC-3 MCF-7	2010	Zu et al. (2010)
Jasminum grandiflora Lin. (Oleaceae)	A549 PC-3 MCF-7	2010	Zu et al. (2010)
Lavandula angustifolia Mill. (Lamiaceae)	A549 PC-3 MCF-7	2010	Zu et al. (2010)
Matricaria chamomilla Lin. (Asteraceae)	A549 PC-3 MCF-7	2010	Zu et al. (2010)
Thymus vulgaris Lin. (Lamiaceae)	A549 PC-3 MCF-7	2010	Zu et al. (2010)
Rosa damascena Mill. (Rosaceae)	A549 PC-3 MCF-7	2010	Zu et al. (2010)
Cinnamomum zeylanicum Pre. (Lauraceae)	A549 PC-3 MCF-7	2010	Zu et al. (2010)
Santalum album Lin. (Santalaceae)	MCF-7 MCF-10 A	2016	Ortiz et al. (2016)
Decatropis bicolor (Zucc.) Radlk (Rutaceae)	MDA-MB-231	2016	Estanislao Gómez et al. (2016)
Curcuma longa Lin., *Curcuma xanthorrhiza* Lin (Zingiberaceae)	MCF-7	2018	Zhong et al. (2018)
Frankincense (Burseraceae), *pine needle* (Pinaceae) and *geranium* (Geraniaceae)	MCF-7	2018	Ren et al. (2017)
Pallenis spinosa (L.) de Cassini (Asteraceae)	MCF-7 MDA-MB-231	2017	Saleh et al. (2017)
Rosmarinus officinalis Lin. (Lamiaceae)	MCF-7 HeLa	2017	Jardak et al. (2017)
Blepharocalix salicifolius (Myrtaceae)	MDA-MB-231	2018	Furtado et al. (2018)
Tamarix aphylla Lin. (Tamaricaceae)	MCF-7 Caco-2 Panc-1	2018	Alhourani et al. (2018)
Citral	MDA-MB-231	2018	Nigjeh et al. (2018)
Hedyosmum sprucei Solms (Chloranthaceae)	MCF-7	2018	Guerrini et al. (2016)
Salvia aurea Lin., *S. Judaica* Lin. and *S. viscosa* Lin. (Lamiaceae)	DU-145	2018	Russo et al. (2018)
Ocimum basilicum Lin. (Lamiaceae),	LNCaP PC-3	2014	Bayala et al. (2014)
Lippia multiflora Mold. (Verbenaceae)	LNCaP PC-3	2014	Bayala et al. (2014)
Ageratum conyzoides Lin. (Asteraceae)	LNCaP PC-3	2014	Bayala et al. (2014)
Zingiber officinale Rosc. (Zingiberaceae)	LNCaP PC-3	2014	Bayala et al. (2014)

(*Continued*)

TABLE 14.1 (*Continued*)
Essential Oils and Their Anticancer Properties: A Collection in Tabular Form

EO/EO Constituent	Anticancer	Year	Reference
Lavandula angustifolia Mill. (Lamiaceae)	PC-3	2016	Zhao et al. (2016)
Carvacol	PC-3 DU145	2016	Luo et al. (2016)
Eupatorium adenophorum Spreng. (Asteraceae)	HepG2 Hep3B SMMC-7721	2018	Chen et al., (2018)
Siegesbeckia orientalis, *S. glabrescens* and *S. pubescens* (Asteraceae)	Hep3B HeLa	2018	Gao et al. (2018)
β-caryophyllene of *Murraya paniculata* Lin. (Rutaceae)	Hepa 1c1c7	2016	Selestino-Neta et al. (2016)
Glandora rosmarinifolia (Boraginaceae)	HA22T/VGH HepG2 Hep3B SUM 149 MDA-MB-231	2018	Poma et al. (2018)
Plectranthus cylindraceus Hocst. ex Benth., *P. asirensis* JRI Wood and *P. barbatus* Andrews (Lamiaceae)	HeLa HepG2 HT-29	2018	Mothana et al. (2018)
Carvacrol	HepG2	2015	Melušová et al. (2015)
Rosmarinus officinalis Lin. (Lamiaceae)	HepG2	2015	Melušová et al. (2015)
Carvacrol	A549	2003	Koparal and Zeytinoglu (2003)
Nigella Sativa Lin. (Ranunculaceae)	A549	2014	Al-Sheddi et al. (2014)
Populus alba Lin. (Salicaceae)	A549 H1299	2017	Gezici et al. (2017)
Rosmarinus officinalis Lin. (Lamiaceae)	A549 H1299	2017	Gezici et al. (2017)
Tridax procumbens Lin. (Asteraceae)	B16F-10	2012	Manjamalai et al. (2012)
Citrus sinensis Osbeck cv. Newhall (Rutaceae)	A549 22RV-1	2017	Yang et al. (2017)
Salvia officinalis Lin. (Lamiaceae)	A375 B164A5	2018	Alexa et al. (2018)
Thymus vulgaris Lin. (Lamiaceae)	A375 B164A5	2018	Alexa et al. (2018)
Pituranthos tortuosus (Apiaceae)	B16F10	2015	Krifa et al. (2015)
Lippia alba Mill. (Verbenaceae)	B16F10Nex2 A549	2016	Santos et al., 2016
Melaleuca alternifolia (Maid. and Betch.) Cheel (Myrtaceae)	HCT-116 RKO	2017	Nakayama et al. (2017)
Piper aequale Lin. (Piperaceae)	HCT-116 ACP 03	2017	da Silva et al. (2017)
Euphorbia macrorrhiza (Euphorbiaceae)	Caco-2	2012	Lin et al. (2012)
Pinus koraiensis Sieb. and Zucc. (Pinaceae)	HCT-116	2014	Cho et al. (2014)
Mesosphaerum sidifolium (Lamiaceae)	Ehrlich carcinoma	2017	Rolim et al. (2017)
Mentha crispa Lin. (Lamiaceae)	U87MG	2018	Turkez et al. (2018)

14.13 CONCLUSION

Historically, medicinal plants and their EOs were primarily used in folk medicine. However, today EOs are getting more and more in the focus for their therapeutic properties against cancer. The results of the present investigations revealed the potential of EOs against a variety of human cancer types. They showed promising results against breast, prostate, liver, lung, and colorectal cancer. In addition, important studies against melanoma, Ehrlich carcinoma, and glioblastoma have been published. Most of the studies are executed *in vitro* by cell line culture. Only a few of them are carried out *in vivo* or in xenograft models. To sum up, the EOs and their ingredients were capable to inhibit the proliferation of tumor cell lines and exhibited significant cytotoxic properties. Some EOs achieved comparable results with standard drugs. It has been shown that the mechanism of action could not yet been proved in all EOs. The selected EOs possess therapeutic properties and the research on new anticancer agents and their mechanism of action ought to be continued. Furthermore, more *in vivo* studies with EOs should be conducted to analyze whether the efficacy is also possible *in vivo.*

REFERENCES

Abdulkareem, I. H. 2013. A review on aetio-pathogenesis of breast cancer. *J. Genet. Syndr. Gene Ther.*, 4, 1–5.

Alberts, B., A. Johnson, J. Lewis et al. 2002. *Molecular Biology of the Cell.* 4th ed. Garland Science, New York; Table 8-3, Some Landmarks in the Development of Tissue and Cell Culture.

Al-Eisawi, D. M. 1982. List of Jordan vascular plants. *Amman*, 152, 79–182.

Alexa, E., R. M. Sumalan, C. Danciu et al. 2018. Synergistic antifungal, allelopatic and anti-proliferative potential of *Salvia officinalis* L., and *Thymus vulgaris* L. EOs. *Molecules*, 23(1), 185.

Alhourani, N., V. Kasabri, Y. Bustanji, R. Abbassi, and M. Hudaib. 2018. Potential Antiproliferative Activity and Evaluation of EO Composition of the Aerial Parts of Tamarix aphylla (L.) H. Karst.: A Wild Grown Medicinal Plant in Jordan. *Evid. Based Complement. and Alternat. Med.*, 2018, 9363868.

Al-Sheddi, E. S., N. N. Farshori, M. M. Al-Oqail, Musarrat, J., A. A. Al-Khedhairy, and M. A. Siddiqui 2014. Cytotoxicity of *Nigella sativa* seed oil and extract against human lung cancer cell line. *Asian Pac. J. Cancer Prev.*, 15(2), 983–987.

American Cancer Society. 2017. Available online: http://www.cancer.org/, accessed December 21, 2018.

Astani, A., J. Reichling, and P. Schnitzler. 2011. Screening for antiviral activities of isolated compounds from essential oils. *Evid. Based Complement. Alternat. Med.*, 2011, 253643.

Bayala, B., I. H. Bassole, C. Gnoula et al. 2014. Chemical composition, antioxidant, anti-inflammatory and anti-proliferative activities of EOs of plants from Burkina Faso. *PLOS ONE*, 9(3), e92122.

Bouaziz, M., T. Yangui, S. Sayadi, and A. Dhouib. 2009. Disinfectant properties of EOs from *Salvia officinalis* L. cultivated in Tunisia. *Food Chem. Toxicol.*, 47(11), 2755–2760.

Cavero, R. Y., and M. I. Calvo 2015. Medicinal plants used for musculoskeletal disorders in Navarra and their pharmacological validation. *J. Ethnopharmacol.*, 168, 255–259.

Chen, H., B. Zhou, J. Yang et al. 2018. EO derived from *Eupatorium adenophorum* Spreng. Mediates anticancer effect by Inhibiting STAT3 and AKT activation to induce apoptosis in hepatocellular carcinoma. *Front Pharmacol.*, 9, 483.

Cho, S. M., E. O. Lee, S. H. Kim, and H. J. Lee 2014. EO of *Pinus koraiensis* inhibits cell proliferation and migration via inhibition of p21-activated kinase 1 pathway in HCT116 colorectal cancer cells. BMC Complement. *Med. Rev.*, 14(1), 275.

Dahham, S. S., Y. M. Tabana, M. A. Iqbal, M. B. Ahamed, M. O. Ezzat, A. S. Majid, and A. M. Majid 2015. The anticancer, antioxidant and antimicrobial properties of the sesquiterpene β-caryophyllene from the EO of Aquilaria crassna. *Molecules*, 20(7), 11808–11829.

da Silva, J. K., R. da Trindade, N. S. Alves, P. L. Figueiredo, J. Maia, and W. N. Setzer 2017. EOs from neotropical piper species and their biological activities. *Int. Mol. Sci.*, 18(12), 2571.

de Sousa Trindade, M. J., and O. Alves Lameira. 2014. Species from the Euphorbiaceae family used for medicinal purposes in Brazil. *Revista Cubana de Plantas Medicinales*, 19(4), 292–309.

Eldahshan, O. A., and A. F. Halim. 2016. Comparison of the composition and antimicrobial activities of the EOs of green branches and leaves of Egyptian navel orange (*Citrus sinensis* (L.) Osbeck var. Malesy). *Chem. Biodivers.*, 13(6), 681–685.

Estanislao Gómez, C. C., A. Aquino Carreño, D. G. Pérez Ishiwara et al. 2016. *Decatropis bicolor* (Zucc.) Radlk EO induces apoptosis of the MDA-MB-231 breast cancer cell line. *BMC Complement. Altern. Med.*, 16, 266.

Ferlay, J., I. Soerjomataram, R. Dikshit, S. Eser, C. Mathers, M. Rebelo, and F. Bray. 2015. Cancer incidence and mortality worldwide: Sources, methods and major patterns in GLOBOCAN 2012. *Int. J. Cancer*, 136(5), E359–E386.

Fitsiou, E., G. Mitropoulou, K. Spyridopoulou, M. Vamvakias, H. Bardouki, A. Galanis, and A. Pappa. 2018. Chemical composition and evaluation of the biological properties of the EO of the dietary phytochemical *Lippia citriodora*. *Molecules*, 23(1), 123, 1–11.

Furtado, F. B., B. C. Borges, T. L. Teixeira et al. 2018. Chemical composition and bioactivity of EO from *Blepharocalyx salicifolius*. *Int. J. Mol. Sci.*, 19(1), 33.

Gao, X., J. Wei, L. Hong, S. Fan, G. Hu, and J. Jia. 2018. Comparative analysis of chemical composition, anti-inflammatory activity and antitumor activity in EOs from *Siegesbeckia orientalis*, *S. glabrescens* and *S. pubescens* with an ITS sequence analysis. *Molecules*, 23(9), 2185, 1–11.

Gezici, S., N. Sekeroglu, and A. Kijjoa. 2017. *In vitro* anticancer activity and antioxidant properties of EOs from *Populus alba* L. and *Rosmarinus officinalis* L. from South Eastern Anatolia of Turkey. *Indian J. Pharm. Educ. Res.*, 51(3), S498–S503.

Guerrini, A., G. Sacchetti, A. Grandini, A. Spagnoletti, M. Asanza, and L. Scalvenzi. 2016. Cytotoxic effect and TLC bioautography-guided approach to detect health properties of Amazonian *Hedyosmum sprucei* EO. *Evid. Based Complement Alternat. Med.*, 2016, 1638342.

Guesmi, F., A. K. Tyagi, S. Prasad, and A. Landoulsi. 2018. Terpenes from EOs and hydrolate of *Teucrium alopecurus* triggered apoptotic events dependent on caspases activation and PARP cleavage in human colon cancer cells through decreased protein expressions. *Oncotarget.*, 9(64), 32305–32320.

Ho, W. Y., S. K. Yeap, C. L. Ho, R. A. Rahim, and N. B. Alitheen 2012. Development of multicellular tumor spheroid (MCTS) culture from breast cancer cell and a high throughput screening method using the MTT assay. *PLOS ONE*, 7(9), e44640.

Jardak, M., J. Elloumi-Mseddi, S. Aifa, and S. Mnif. 2017. Chemical composition, anti-biofilm activity and potential cytotoxic effect on cancer cells of *Rosmarinus officinalis* L. EO from Tunisia. *Lipids Health Dis.*, 16(1), 190.

Kim, J. H., H. J. Lee, S. J. Jeong, M. H. Lee, and S. H. Kim 2012. EO of *Pinus koraiensis* leaves exerts antihyperlipidemic effects via up-regulation of low-density lipoprotein receptor and inhibition of acyl-coenzyme A: Cholesterol acyltransferase. *Phytother. Res.*, 26(9), 1314–1319.

Koparal, A. T., and M. Zeytinoğlu. 2003. Effects of carvacrol on a human non-small cell lung cancer (NSCLC) cell line, A549. *Cytotechnology*, 43(1–3), 149–154.

Krifa, M., El Mekdad, H., N. Bentouati, A. Pizzi, K. Ghedira, M. Hammami, and L. Chekir-Ghedira. 2015. Immunomodulatory and anticancer effects of Pituranthos tortuosus EO. *Tumor Biol.*, 36(7), 5165–5170.

Legault, J., and A. Pichette. 2007. Potentiating effect of β-caryophyllene on anticancer activity of α-humulene, isocaryophyllene and paclitaxel. *J. Pharm. Pharmacol.*, 59(12), 1643–1647.

Lima, E., R. Alves, T. Anunciação, V. Silva, L. Santos, M. Soares, and D. Bezerra. 2018. Antitumor effect of the EO from the leaves of *Croton matourensis* Aubl. (Euphorbiaceae). *Molecules*, 23(11), 2974, 1–10.

Lin, J., J. Dou, J. Xu, and H. A. Aisa 2012. Chemical composition, antimicrobial and antitumor activities of the EOs and crude extracts of *Euphorbia macrorrhiza*. *Molecules*, 17(5), 5030–5039.

Liu, L., J. Z. Liao, X. X. He, and P. Y. Li 2017. The role of autophagy in hepatocellular carcinoma: Friend or foe. *Oncotarget*, 8(34), 57707–e57722.

Loizzo, M. R., R. Tundis, F. Menichini, A. M. Saab, G. A. Statti, and F. Menichini. 2007. Cytotoxic activity of EOs from Labiatae and Lauraceae families against *in vitro* human tumor models. *Anticancer Res.*, 27(5A), 3293–3299.

Luo, Y, J. Y. Wu, M. H. Lu, Z. Shi, N. Na, J. M. Di. 2016. Carvacrol alleviates prostate cancer cell proliferation, migration, and invasion through regulation of PI3K/Akt and MAPK signaling pathways. *Oxid. Med. Cell Longev.*, 1469693, 1–9.

Manjamalai, A., M. J. Kumar, and V. M. Grace. 2012. EO of *Tridax procumbens* L induces apoptosis and suppresses angiogenesis and lung metastasis of the B16F-10 cell line in C57BL/6 mice. *Asian Pac. J. Cancer Prev.*, 13(11), 5887–5895.

Matos, F. J. A. 1999. *Plantas da Medicina Popular do Nordeste: Propriedades atribuídas e confirmadas.* Editora da UFC, Fortaleza, p. 80.

Mattern, J., M. Bak, E. W. Hahn, and M. Volm. 1988. Human tumor xenografts as model for drug testing. *Cancer Metastasis Rev.*, 7(3), 263–284.

Melušová, M., S. Jantová, and E. Horváthová. 2015. Carvacrol and rosemary oil at higher concentrations induce apoptosis in human hepatoma HepG2 cells. *Interdiscip. Toxicol.*, 7(4), 189–94.

Mimica-Dukic, N., S. Kujundyic, M. Sokovic, and M. Couladis. 2003. EO composition and antifungal activity of *Foeniculum vulgare* Mill. Obtained by different distillation conditions. *Phytother. Res.*, 17, 368–371.

Momtazi, A. A., O. Askari-Khorasgani, E. Abdollahi, H. Sadeghi-Aliabadi, F. Mortazaeinezhad, and A. Sahebkar. 2017. Phytochemical analysis and cytotoxicity evaluation of *Kelussia odoratissima* Mozaff. *J. Acupunct. Meridian Stud.*, 10(3), 180–186.

Morton, J. F. 1981. Atlas of medicinal plants of middle America: Bahamas to Yucatan. Thomas, C. C. (Ed.). Springfield, IL, US.

Mosmann, T. 1983. Rapid colorimetric assay for cellular growth and survival: Application to proliferation and cytotoxicity assays. *J. Immunol. Methods*, 65(1–2), 55–63.

Moteki, H., H. Hibasami, Y. Yamada et al. 2002. Specific induction of apoptosis by 1,8–cineole in two human leukemia cell lines, but not a in human stomach cancer cell line. *Oncol. Rports*, 9(4), 757–760.

Mothana, R. A., J. M. Khaled, O. M. Noman et al. 2018. Phytochemical analysis and evaluation of the cytotoxic, antimicrobial and antioxidant activities of EOs from three *Plectranthus* species grown in Saudi Arabia. *BMC Complement Altern. Med.* 18(1), 237, 1–9.

Nagappan, T., P. Ramasamy, M. E. A. Wahid, T. C. Segaran, and C. S. Vairappan. 2011. Biological activity of carbazole alkaloids and EO of Murraya koenigii against antibiotic resistant microbes and cancer cell lines. *Molecules*, 16(11), 9651–9664.

Nakagawa, Y. T. S. 2003. Cytotoxic and xenoestrogenic effects via biotransformation of trans-anethole on isolated rat hepatocytes and cultured mcf-7 human breast cancer cells. *Biochem. Pharmacol.*, 66, 63–73

Nakayama, K., S. Murata, H. Ito et al. 2017. Terpinen-4-ol inhibits colorectal cancer growth via reactive oxygen species. *Oncol Lett.*, 14(2), 2015–2024.

Nigjeh, S. E., S. K. Yeap, N. Nordin, B. Kamalideghan, H. Ky, and R. Rosli. 2018. Citral induced apoptosis in MDA-MB-231 spheroid cells. *BMC Complement. Altern. Med.*, 18(1), 56, 1–9.

Ortiz, C., L. Morales, M. Sastre, W. E. Haskins, and J. Matta. 2016. Cytotoxicity and genotoxicity assessment of sandalwood EO in human breast cell lines MCF-7 and MCF-10A. *Evid. Based Complement. Altern. Med.*, 3696232, 1–11.

Picot, J., C. L. Guerin, C. Le Van Kim, and C. M. Boulanger. 2012. Flow cytometry: Retrospective, fundamentals and recent instrumentation. *Cytotechnology*, 64(2), 109–130.

Poeckel, D., and O. Werz. 2006. Boswellic acids: Biological actions and molecular targets. *Curr. Med. Chem.*, 13, 3359–3369.

Poma, P., M. Labbozzetta, M. Notarbartolo et al. 2018. Chemical composition, *in vitro* antitumor and pro-oxidant activities of *Glandora rosmarinifolia* (Boraginaceae) EO. *PLOS ONE*, 13(5): e0196947.

Quirantes-Piné, R., M. Herranz-Lopez, L. Funes, I. Borrás-Linares, and V. Micol, A. Segura-Carretero, and A. Fernández-Gutiérrez. 2013. Phenylpropanoids and their metabolites are the major compounds responsible for blood-cell protection against oxidative stress after administration of *Lippia citriodora* in rats. *Phytomedicine*, 20(12), 1112–1118.

Rajabalian, S. 2008. Methanolic extract of *Teucrium polium* L. potentiates the cytotoxic and apoptotic effects of anticancer drugs of vincristine, vinblastine and doxorubicin against a panel of cancerous cell lines. *Exp. Oncol.*, 30, 133–138.

Ramalingam, S., and C. Belani. 2008. Systemic chemotherapy for advanced non-small cell lung cancer: Recent advances and future directions. *The Oncol.*, 13(Supplement 1), 5–13.

Ren, P., X. Ren, L. Cheng, and L. Xu. 2017. Frankincense, pine needle and geranium EOs suppress tumor progression through the regulation of the AMPK/mTOR pathway in breast cancer. *Oncol. Rep.*, 39(1), 129–137.

Rolim, T. L., D. R. P. Meireles, T. M. Batista, T. K. G. de Sousa, V. M. Mangueira, R. A. de Abrantes, and M. S. da Silva. 2017. Toxicity and antitumor potential of *Mesosphaerum sidifolium* (Lamiaceae) oil and fenchone, its major component. *BMC Complement. Altern. Med.*, 17(1), 347, 1–10.

Russo, A., V. Cardile, A. Graziano, R. Avola, M. Bruno, and D. Rigano. 2018. Involvement of Bax and Bcl-2 in induction of apoptosis by EOs of three Lebanese *Salvia* species in human prostate cancer cells. *Int. J. Molecular Sci.*, 19(1), 292, 1–9.

Saleh, A. M., M. A. Al-Qudah, A. Nasr, S. A. Rizvi, A. Borai, and M. Daghistani. 2017. Comprehensive analysis of the chemical composition and *in vitro* cytotoxic mechanisms of *Pallines spinosa* flower and leaf EOs against breast cancer cells. *Cellular Physiol. Biochem.*, 42(5), 2043–2065.

Santos, N., R. C. Pascon, M. A. Vallim, C. R. Figueiredo, M. G. Soares, J. Lago, and P. Sartorelli. 2016. Cytotoxic and antimicrobial constituents from the EO of *Lippia alba* (Verbenaceae). *Medicines* (Basel, Switzerland), 3(3), 22, 1–40.

Scherer, W. F., J. T. Syverton, and G. O. Gey. 1953. Studies on the propagation *in vitro* of poliomyelitis viruses: IV. Viral multiplication in a stable strain of human malignant epithelial cells (strain HeLa) derived from an epidermoid carcinoma of the cervix. *J. Experimental Med.*, 97(5), 695–710.

Selestino Neta, M. C., C. Vittorazzi, A. C. Guimarães, J. D. Martins, M. Fronza, D. C. Endringer, and R. Scherer. 2016. Effects of β-caryophyllene and *Murraya paniculata* EO in the murine hepatoma cells and in the bacteria and fungi 24-h time-kill curve studies. *Pharm. Biol.*, 55(1), 190–197.

Shan, B., J. C. Medina, E. Santha, W. P. Frankmoelle, T. C. Chou, R. M. Learned, ... and T. Rosen. 1999. Selective, covalent modification of β-tubulin residue Cys-239 by T138067, an antitumor agent with *in vivo* efficacy against multidrug-resistant tumors. *Proc. Natl. Acad. Sci.*, 96(10), 5686–5691.

Sharma, S. V., D. A. Haber, and J. Settleman. 2010. Cell line-based platforms to evaluate the therapeutic efficacy of candidate anticancer agents. *Nature Rev. Cancer*, 10(4), 241, 2–3.

Sharopov, F., A. Valiev, P. Satyal et al. 2017. Cytotoxicity of the EO of Fennel (*Foeniculum vulgare*) from Tajikistan. *Foods*, 6(9), 73, 2–8.

Singla, A. K., and P. Kamla. 1990. Phytoconstituents of Euphorbia species. *Fitoterapia*, 41(6), 483–516.

Stavridi, F., E. M. Karapanagiotou, and K. N. Syrigos 2010. Targeted therapeutic approaches for hormone-refractory prostate cancer. *Cancer Treatment Rev.*, 36(2), 122–130.

Suntres, Z. E., J. Coccimiglio, and M. Alipour. 2015. The bioactivity and toxicological actions of carvacrol. *Critical Rev. Food Sci. Nutr.*, 55(3), 304–318.

Syrovets, T., B. Büchele, C. Krauss, Y. Laumonnier, and T. Simmet. 2005. Acetyl-boswellic acids inhibit lipopolysaccharide-mediated TNF-alpha induction in monocytes by direct interaction with IkappaB kinases. *J. Immunol.*, 174, 498–506.

Tilaoui, M., H. A. Mouse, A. Jaafari, and A. Zyad. 2015. Comparative phytochemical analysis of EOs from different biological parts of *Artemisia herba alba* and their cytotoxic effect on cancer cells. *PLOS ONE*, 10(7), e0131799.

Todzia, C. A. 1988. Monograph 48. Chloranthaceae: Hedyosmum. *Flora Neotrop.*, 48, 1–138.

Trease, G. E., and W. C. Evans, 1983. *W.C. Pharmacognosy*, 12th edn. Bailliere & Tindall, London, UK.

Turkez, H., O. O. Tozlu, T. C. Lima, de Brito, A., and D. P. de Sousa. 2018. A comparative evaluation of the cytotoxic and antioxidant activity of *Mentha crispa* EO, its major constituent rotundifolone, and analogues on human glioblastoma. *Oxid. Med. Cell. Longevity*, 2018, 2083923, doi:10.1155/2018/2083923

Wikipedia: https://en.wikipedia.org/wiki/TUNEL.assay, accessed January 5, 2019.

World Cancer Report. 2014a. World Health Organization, Chapter 5. 1. ISBN 9283204298.

World Cancer Report. 2014b. World Health Organization. 2014. pp. Chapter 5.14. ISBN 978-9283204299.

Yang, C., H. Chen, H. Chen, B. Zhong, X. Luo, and J. Chun. 2017. Antioxidant and anticancer activities of EO from Gannan navel orange peel. *Molecules*, 22(8), 1391.

Yunnan Institute of Materia Medica. 1975. *List of Medicinal Plants in Yunnan*. Yunnan People's Publishing House, Yunnan.

Zhang, J. Y., N. Li, Y. Y. Che, Y. Zhang, S. X. Liang, M. B. Zhao, and P. F. Tu 2011. Characterization of seventy polymethoxylated flavonoids (PMFs) in the leaves of *Murraya paniculata* by on-line high-performance liquid chromatography coupled to photodiode array detection and electrospray tandem mass spectrometry. *J. Pharm. Biomed. Anal.*, 56(5), 950–961.

Zhang, W., W. Yu, G. Cai et al. 2018. A new synthetic derivative of cryptotanshinone KYZ3 as STAT3 inhibitor for triple-negative breast cancer therapy. *Cell Death Dis.*, 9(11), 1098, 1–10. Published Oct 27, 2018.

Zhao, Y., R. Chen, Y. Wang, C. Qing, W. Wang, and Y. Yang. 2016. *In vitro* and *in vivo* efficacy studies of lavender angustifolia eo and its active constituents on the proliferation of human prostate cancer. *Integr. Cancer Ther.*, 16(2), 215–226.

Zhong, Z., H. Yu, S. Wang, Y. Wang, and L. Cui. 2018. Anti-cancer effects of Rhizoma Curcumae against doxorubicin-resistant breast cancer cells. *Chin Med.*, 13(44), 1–9, doi:10.1186/s13020-018-0203-z.

Zhong, Z.-F., W. Tan, W.W. Qiang et al. 2016. Furanodiene alters mitochondrial function in doxorubicin-resistant MCF-7 human breast cancer cells in an AMPK-dependent manner. *Mol. Biosyst.*, 12(5), 1626–1637.

Zu, Y., H. Yu, L. Liang, Y. Fu, T. Efferth, X. Liu, and N. Wu. 2010. Activities of ten EOs towards Propionibacterium acnes and PC-3, A-549 and MCF-7 cancer cells. *Molecules*, 15(5), 3200–3210.

15 Antimicrobial Activity of Selected Essential Oils and Aroma Compounds against Airborne Microbes

Sabine Krist

CONTENTS

15.1 INTRODUCTION

Airborne microbes surround humans 24 h a day and are a hitherto underestimated cause of risk for human health. Air disinfectants used today are not optimal to decrease airborne microbes because of their toxicity to humans and their unpleasant smell (Moore and Kaczmarek, 1991). The development of an easy and safe way to reduce the amount of airborne microbes in locations where people gather or where foods are processed is thus required. Essential oils are complex mixtures of plant secondary metabolites, mostly composed of terpenoids. In plant systems, they act as defense compounds against microbes, herbivores, and other ecological factors (Baby and Verughese, 2009). Antimicrobial effects of essential oils and aroma compounds have been described in a number of studies (Horne et al., 2001; Inouye et al., 2001; Cimanga et al., 2002; Gadalla and Hassan, 2004; Moliszewska, 2003; Rhayour et al., 2003; Bagamboula et al., 2004; Bennis et al., 2004; Zambonelli et al., 2004; Ahmad et al., 2014;

Sadeghi et al., 2015; Chouhan et al., 2017), but very little is yet known about the antimicrobial effects of scents on airborne microbes. The majority of studies bring the aroma compounds or essential oils and the microbes into direct contact via methods such as disc diffusion, well diffusion, agar dilution, or broth dilution and thus evaluate their antimicrobial activity. However, studies on vapors of aroma compounds on microbes are rare, and even the few that do exist only use small-scale methods (Bouaziz et al., 2009). This is reason enough to focus on measurements of antimicrobial activity of vapors of essential oils and their main aroma compounds, which were performed by my research group.

15.2 METHODS

15.2.1 General

The air samples were taken with an RCS Air Sampler, purchased from Biotest AG, Dreieich, Germany. The RCS Air Sampler uses inertial impaction to collect the airborne microbes. The microbes were impacted on commercially available agar strips, which were incubated after sampling in an incubator. According to the manufacturer's specifications, the sampling volume of the RCS Air Sampler is 280 L/min and the separation volume is 40 L/min for particles with a diameter of 4 μm. Agar strips TC (Art. No. 941105050) for determination of total microbial counts, obtained from Biotest AG, Dreieich, Germany, were used as culture media. Colony forming units (CFU/m^3) were calculated after incubation at 30°C for 48 hours using the formula in Figure 15.1.

Afterward, the average reduction of germ count (see Figure 15.2), the standard deviation (SD; see Figure 15.3) and the mean reduction (AD; see Figure 15.4) were calculated, where n is the number of measurements and x is the result of measurements.

The statistical comparisons between control and each volatile compound, as well as the value comparison prior and after fragrance diffusion, were performed using Student's t-test; $p < 0.05$ was considered to be significant.

For the identification of the airborne microbes, the colonies were inoculated from the agar strips, transferred into standard I nutrient broth (Merck, Darmstadt, Germany) and cultivated for 24 h at 30°C. A Gram stain using crystal violet, iodine/alcohol, and safranin (all Merck, Darmstadt,

$$\text{CFU/m}^3 = \frac{\text{Counted colonies} \cdot 25}{\text{Sampling time (min)}}$$

FIGURE 15.1 Calculation of the total microbial count in air.

$$\bar{x} = \frac{1}{n}\sum_{i=1}^{n} x_i = \frac{x_1 + x_2 + \cdots + x_n}{n}$$

FIGURE 15.2 Calculation of the average reduction of germ count.

$$\sigma = \sqrt{\frac{n\sum x^2 - \left(\sum x\right)^2}{n(n-1)}}$$

FIGURE 15.3 Calculation of the standard deviation σ.

$$\text{Mean-deviation} = \frac{1}{n}\sum (x - \bar{x})$$

FIGURE 15.4 Calculation of the mean deviation.

Germany) was performed to group bacteria. The bacteria were isolated on standard I nutrient agar (Merck, Darmstadt, Germany), examined with a light microscope (Alphaphot 2 YS"-H, Nikon, Japan), and identified by performing a respective API-test (bioMérieux, Marcy ÍEtoile, France).

15.2.2 Studies in Small Examination Rooms

For the antimicrobial studies in small examination rooms, the experiments were carried out in restrooms at the Center of Pharmacy, University of Vienna, Austria. The air volume of every washroom was 8.0 m^3. The temperature was 24°C and the air humidity was 46%.

First, stock emulsions (o/w) of the aroma compounds and the essential oils were prepared in concentrations 1:100 (emulsion/suspension in distilled water); 4.0 mg were taken from each stock emulsion and filled into atomizer bottles (10 mL), which had been purchased from VWR International GmbH, Vienna, Austria (Art. No. 215-6270). The spray angle was 35°, and 0.05 mL was atomized by one spraying performance. By atomizing 4.0 mg of diluted aroma-chemicals/essential oils in the testing rooms, a concentration of 5.0 mg/m^3 was achieved.

After determining the total microbial count with the RCS Air Sampler in the testing room by sampling air for 2 min (blank values), the aroma compounds or essential oils were vaporized. Fifteen minutes later, the total microbial count was measured again with the same experimental setup. Ten measurements were taken for each aroma substance or essential oil and for the blank values, respectively. After incubation of the agar strips and enumeration of colonies, the total germ count in the air in CFU/m^3 and the average bacterial count decrease in the air were calculated.

15.2.2.1 Examined Essential Oils

The following essential oils, all of which have a known antimicrobial activity, were tested against airborne microbes by applying the method described above:

- Australian sandalwood essential oil (Jirovetz et al., 2006a)
  Source plant: *Santalum spicatum* A. DC. (Santalaceae)
- Cabreuva wood oil (Wanner et al., 2010)
  Source plant: *Myrocarpus frondosus* and *Myrocarpus fastigiatus* (Fabaceae)
- Carnation oil *Oleum Caryophyllorum foliorum* (Lopez et al., 2005)
  Source plant: *Eugenia caryophyllata* Thurnberg (Myrtaceae)
- Dill herb essential oil, *Oleum Anethi e herba* (Jirovetz et al., 2004)
  Source plant: *Anethum graveolens* L. (Apiaceae)
- Dill seed essential oil, *Oleum Seminis Anethi aethereum* (Jirovetz et al., 2004)
  Source plant: *Anethum graveolens* L. (Apiaceae)
- East Indian sandalwood oil, *Oleum Ligni Santali* (Jirovetz et al., 2006a)
  Source plant: *Santalum album* L. (Santalaceae)
- Jasmin essential oil, *Oleum Jasmini* (Jirovetz et al., 2007)
  Source plant: *Jasminum grandiflorum* L. (Oleaceae)
- Noble fir cone oil, *Abietis fructuum aetheroleum, syn. Oleum Templini* (Yang et al., 2009)
  Source plant: *Albis alba* Mill. (Pinaceae)
- Palmarosa essential oil, *Oleum Palmarosae* (Jirovetz et al., 2006b)
  Source plant: *Cymbopogon martinii* (Roxb.) J.F. Wats var. motia (Poaceae)
- Pine essential oil (Jirovetz et al., 2005) source plant: *Pinus pinaster* (Pinaceae)
- Rosemary essential oil, *Rosmarini aetheroleum*, *Oleum Rosmarini* (Nieto, 2017)
  Source plant: *Rosmarinus officinalis* L. (Lamiaceae)
- Spruce needle essential oil, *Oleum Abietis sibiricae* (Donaldson et al., 2005)
  Source plant: *Abietis sibirica* L. (Pinaceae)
- Sweet orange essential oil, *Aurantii dulcis aetheroleum* (Geraci et al., 2017)
  Source plant: *Citrus sinensis* L. (Rutaceae)

- Tea tree essential oil, *Melaleucae aetheroleum*, *Melaleucae alternifolia aetheroleum* (Brun et al., 2019)
  Source plant: *Melaleucae alternifolia* (Myrtaceae)
- West Indian sandalwood essential oil (Jirovetz et al., 2006a)
  Source plant: *Amyris balsamifera* L. (Rutaceae)

15.2.2.2 Examined Aroma Compounds

The following aroma compounds, all of which have a known antimicrobial activity, were tested against airborne microbes by applying the method described above:

- α-Bisabolol CAS[23089-26-1] (De Lucca et al., 2011)
- Benzyl benzoate CAS[120-51-4] (Essien et al., 2018)
- Bornyl acetate CAS[76-49-3] (Ebani et al., 2018)
- (+)-Camphene CAS[79-92-5] (Cutillas et al., 2017)
- δ-3-Carene CAS[13466-78-9] (Yousefi et al., 2017)
- Carvacrol CAS[499-75-2] (Knowles et al., 2005)
- (+)-Carvone CAS[2244-16-8] (Aggarwal et al., 2002)
- (−)-Carvone CAS[6485-40-1] (Aggarwal et al., 2002)
- β-Caryophyllene CAS[87-44-5] (Schmidt et al., 2005)
- 1,8-Cineole CAS[470-82-6] (Sonboli et al., 2006)
- *trans*-Cinnamaldehyde CAS[104-55-2] (Ferhout et al., 1999)
- Citral CAS[5392-40-5] (Pattnaik et al., 1997)
- (−)-Citronellal CAS[5949-05-3] (Saibabu et al., 2017)
- *p*-Cymene CAS[99-87-6] (Kisko and Roller, 2005)
- α-2,5-Dimethylstyrene CAS [1195-32-0] (Fayyaz et al., 2015)
- Eugenol CAS[97-53-0] (Marchese et al., 2017)
- *trans,trans*-Farnesol CAS[106-28-5] (Wanner et al., 2010)
- Geraniol CAS[106-24-1] (Lei et al., 2019)
- Geranyl acetate CAS[105-87-3] (Jirovetz et al., 2006b)
- Isophytol CAS[505-32-8] (Tao et al., 2013)
- (+)-Limonene CAS[5989-27-5] (Aggarwal et al., 2002)
- (−)-Limonene CAS[5989-54-8] (Aggarwal et al., 2002)
- Linalool CAS[78-70-6] (Park et al., 2012)
- Nerolidol CAS[7212-44-4] (Brehm-Stecher and Johnson, 2003)
- *cis*-Nerolidol CAS[3790-78-1] (Lee et al., 2014)
- (−)-Perillaldehyde CAS[18031-40-8] (Friedman et al., 2002)
- α-Pinene CAS[7785-26-4] (Santoyo et al., 2005)
- β-Pinene CAS[18172-67-3] (Rivas da Silva et al., 2012)
- Phytol (mixture of isomers) CAS[150-86-7] (Ghaneian et al., 2015)
- Santalol (α-, β-isomers) and CAS[1103-45-1] (Schmidt et al., 2005)
- α-Terpinene CAS[99-86-5] (Ahmad et al., 2014)
- γ-Terpinene CAS[99-85-4] (Ahmad et al., 2014)
- Terpinen-4-ol (racemate) CAS[562-74-3] (Jirovetz et al., 2005)
- Terpineol (mixture of α-, β-, and γ-Terpineol) CAS[8006-39-1] (Jirovetz et al., 2005)
- Terpinolene CAS[586-62-9] (Hinou et al., 1989)
- Thymol CAS[89-83-8] (Bagamboula et al., 2004)

15.2.3 Studies in Large Examination Rooms

For the antimicrobial studies in large examination rooms, the experiments were carried out

A. In a lecture room at the Center of Pharmacy, University of Vienna, Austria, using an air washer in order to vaporize the aroma compounds.

B. In a lecture room at the Center of Pharmacy, University of Vienna, Austria, using a room diffuser in order to vaporize the aroma compounds.

C. In pharmacy showrooms in Vienna, using a room diffuser in order to vaporize the aroma compounds.

For test method **A**, an air washer (Venta, Type LW24, with a capacity of 7 L water) was fixed on the desk at the center of the lecture room (Figure 15.5, point 1). The air volume of the lecture room was 168 m^3. During testing, the doors and windows were kept closed. The temperature was 24°C and the air humidity was 44%. Experiments were carried out only once a day, right after the end of a lecture. The wide constancy of germ count in the air of the lecture room was ascertained by a large number of preliminary tests. After the measurements, the lecture room was aerated by opening the windows for 30 min. Due to this daily procedure, uniform starting conditions were ensured.

First of all, the air samples in this room were taken with the RCS Air Sampler for 8 min at each of the five well-defined measuring points (Figure 15.5, measuring points 1–5) in order to get the blank values. Then, 0.84 g of volatile compounds were added to 7 L distilled water in the air washer and vaporized. The achieved concentration of aroma compounds was 5.0 mg/m^3. One hour later, the air samples were collected, again following the same procedure. The measurements were performed ten times with each aroma compound and the blanks.

For test method **B**, the same lecture room as for method **A** was used. The volatile compounds were vaporized with a room diffuser (Venta-AirScenter, Type RB 10). The room diffuser was fixed on the desk at the center of the lecture room (Figure 15.5, point 1). The air samples in the test room were taken with the RCS Air Sampler for 8 min at each of the five well-defined measuring points to get the blank values (Figure 15.5, measuring points 1–5). Then, volatile aroma compounds were sprayed with the room diffuser for 5 h. The air samples were collected again by following the same procedure. The measurements were performed ten times for each aroma compound and the blanks.

For test method **C**, the aroma compounds and essential oils were also vaporized with a room diffuser (Airwick Symphonia fragrant plug). This time, three pharmacy show rooms, all located in Vienna, were used as test rooms. These test rooms had an average air volume of 84 m^3 and an average air humidity of 48%. The room diffuser was fixed at the center of the showroom. The air samples were taken at two well-defined measuring points (one next to the entrance, one right in the middle of the

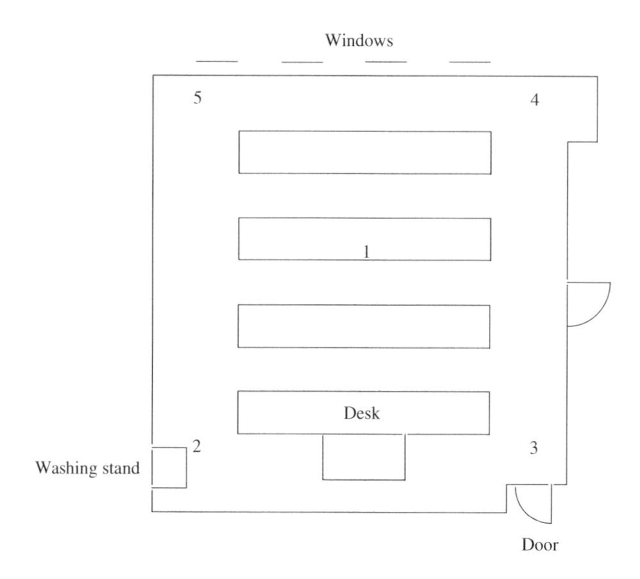

FIGURE 15.5 Large examination room with air washer or room diffuser and measuring points. The air washer/room diffuser was fixed on the desk at measuring point 1. Point 1: 0 m away from the air washer/room diffuser and 1 m height. Point 2–5: 5 m away from the air washer/room diffuser and 1 m height.

showroom). In the morning, the initial values were taken by sampling air for 8 min at both measuring points. Afterward, the essential oils and aroma compounds were sprayed with the room diffuser for 9 h. Air samples were taken again after 3, 6, and 9 h. To obtain the blank values, the same measuring procedure as described above was applied, but without spraying an aroma compound or an essential oil.

15.2.3.1 Examined Essential Oils

The following essential oils, both of which have a known antimicrobial activity, were tested against airborne microbes by applying the method described above:

- Sweet orange essential oil, *Aurantii dulcis Aetheroleum* (Geraci et al., 2017) source plant: *Citrus sinensis* L. (Rutaceae) **C**
- Noble fir cone oil, *Abietis fructuum aetheroleum, syn. Oleum Templini* (Yang et al., 2009) source plant: *Abies alba* Mill. (Pinaceae) **C**

15.2.3.2 Examined Aroma Compounds

The following aroma compounds, all of which have a known antimicrobial activity, were tested against airborne microbes by applying the method described above:

- Carvacrol CAS[499-75-2] (Knowles et al., 2005) **A, B**
- 1,8-Cineole CAS[470-82-6] (Sonboli et al., 2006) **A, B**
- *trans*-Cinnamaldehyde CAS[104-55-2] (Ferhout et al., 1999) **A**
- Citral CAS[5392-40-5] (Pattnaik et al., 1997) **A**
- (−)-Citronellal CAS[5949-05-3] (Saibabu et al., 2017) **A**
- Eugenol CAS[97-53-0] (Marchese et al., 2017) **A**
- Geraniol CAS[106-24-1] (Lei et al., 2019) **A**
- (+)-Limonene CAS[5989-27-5] (Aggarwal et al., 2002) **A, C**
- (−)-Limonene CAS[5989-54-8] (Aggarwal et al., 2002) **A, C**
- Linalool CAS[78-70-6] (Park et al., 2012) **A, B**
- (−)-Perillaldehyde CAS[18031-40-8] (Friedman et al., 2002) **A, B**
- Terpineol (mix. of α-, β-, and γ-Terpineol) CAS[8006-39-1] (Jirovetz et al., 2005) **A, B**
- γ-Terpinene CAS[99-85-4] (Ahmad et al., 2014) **A**

15.3 RESULTS

15.3.1 Studies in Small Examination Rooms

15.3.1.1 Results for Essential Oils

TABLE 15.1
Average Reduction of Germ Count (CFU) by Essential Oils in Small Examination Rooms

Essential oil	Australian sandalwood oil	Cabreuva oil	Carnation oil
Average germ count reduction in %	48.28 (Wosny, 2007)	25.60 (Banovac, 2012)	49.36 (Pasierb, 2007)
Essential oil	Dill herb essential oil	Dill seed essential oil	East Indian sandalwood oil
Average germ count reduction in %	40.24 (Zaussinger, 2008)	35.17 (Zaussinger, 2008)	37.97 (Wosny, 2007)
Essential oil	Jasmin essential oil	Noble fir cone oil	Palmarosa essential oil
Average germ count reduction in %	38.89 (Pasierb, 2007)	27.87 (Banovac, 2012)	43.52 (Pasierb, 2007)
Essential oil	Pine essential oil	Rosemary essential oil	Spruce needle essential oil
Average germ count reduction in %	25.59 (Kretz, 2007)	48.28 (Zaussinger, 2008)	51.50 (Pasierb, 2007)
Essential oil	Sweet orange essential oil	Tea tree essential oil	West Indian sandalwood essential oil
Average germ count reduction in %	27.22 (Banovac, 2012)	19.93 (Kretz, 2007)	43.60 (Wosny, 2007)

15.3.1.2 Results for Aroma Compounds

TABLE 15.2
Average Reduction of Germ Count (CFU) by Aroma Compounds in Small Examination Rooms

Aroma compound	α-Bisabolol	Benzyl benzoate	Bornyl acetate
Average germ count reduction in %	46.32 (Krist et al., 2015)	49.52 (Pasierb, 2007)	45.43 (Pasierb, 2007)
Aroma compound	(+)-Camphene	δ-3-Carene	Carvacrol
Average germ count reduction in %	55.61 (Pasierb, 2007)	44.17 (Pasierb, 2007)	30.90 (Feichtinger, 2005)
Aroma compound	(+)-Carvon	(−)-Carvon	β-Caryophyllene
Reduction of CFU (%)	36.16 (Zaussinger, 2008)	37.67 (Zaussinger, 2008)	33.25 (Zaussinger, 2008)
Aroma compound	1,8-Cineole	*trans*-Cinnamaldehyde	Citral
Average germ count reduction in %	44.18 (Kretz, 2007)	43.77 (Krist et al., 2007)	51.40 (Strobl, 2005)
Aroma compound	(−)-Citronellal	*p*-Cymene	α-2,5-Dimethylstyrene
Average germ count reduction in %	46.30 (Strobl, 2005)	28.98 (Kretz, 2007)	43.80 (Kretz, 2007)
Aroma compound	Eugenol	*trans,trans*-Farnesol	Geraniol
Average germ count reduction in %	61.94 (Krist et al., 2007)	37.26 (Krist et al., 2015)	52.09 (Pasierb, 2007)
Aroma compound	Geranyl acetate	Isophytol	(+)-Limonene
Average germ count reduction in %	45.68 (Pasierb, 2007)	60.63 (Pasierb, 2007)	44.79 (Zaussinger, 2008)
Aroma compound	(−)-Limonene	Linalool	Nerolidol
Average germ count reduction in %	55.42 (Zaussinger, 2008)	56.28 (Krist et al., 2007)	18.35 (Krist et al., 2015)
Aroma compound	*cis*-Nerolidol	(−)-Perillaldehyde	α-Pinene
Average germ count reduction in %	31.89 (Krist et al., 2015)	35.60 (Strobl, 2005)	49.75 (Pasierb, 2007)
Aroma compound	β-Pinene	Phytol (mixture of isomers)	Santalol (α-, β- isomers)
Average germ count reduction in %	24.07 (Zaussinger, 2008)	48.87 (Pasierb, 2007)	43.59 (Wosny, 2007)
Aroma compound	α-Terpinene	γ-Terpinene	Terpinen-4-ol (racemate)
Average germ count reduction in %	26.41 (Kretz, 2007)	16.28 (Kretz, 2007)	24.09 (Kretz, 2007)
Aroma compound	Terpineol (mixture of α-, β-, and γ-Terpineol)	Terpinolene	Thymol
Average germ count reduction in %	41.06 (Feichtinger, 2005)	17.59 (Kretz, 2007)	54.27 (Krist et al., 2007)

15.3.2 Studies in Large Examination Rooms

15.3.2.1 Results for Essential Oils

TABLE 15.3
Average Reduction of Germ Count (CFU) by Essential Oils in Large Examination Rooms

Essential oils	Sweet orange essential oil	Noble fir cone oil
Average germ count reduction in %	45.67 (Karbasiyan, 2012)	45.11 (Karbasiyan, 2012)
Applied Method	C	C

15.3.2.2 Results for Aroma Compounds

TABLE 15.4
Average Reduction of Germ Count (CFU) by Aroma Compounds in Large Examination Rooms

Aroma compound	Carvacrol	1,8-Cineol	*trans*-Cinnnam-aldehyde
Average germ count reduction in %	37.5 (Krist et al., 2008)	64 (Sato et al., 2007)	45 (Sato et al., 2006)
Applied Method	B	A	A
Aroma compound	Citral	(−)-Citronellal	Eugenol
Average germ count reduction in %	17 (Sato et al., 2006)	30 (Sato et al., 2006)	13 (Sato et al., 2006)
Applied Method	A	A	A
Aroma compound	Geraniol	(+)-Limonene	(−)-Limonene
Average germ count reduction in %	46 (Sato et al., 2007)	2 (Karbasiyan, 2012)	39 (Karbasiyan, 2012)
Applied Method	A	C	C
Aroma compound	Linalool	(−)-Perillaldehyde	Terpineol
Average germ count reduction in %	53 (Sato et al., 2007)	53 (Sato et al., 2006)	68 (Sato et al., 2007)
Applied Method	A	A	A
Aroma compound	γ-Terpinene		
Average germ count reduction in %	40 (Sato et al., 2007)		
Applied Method	A		

15.3.3 Identified Airborne Microbes

The identification procedure of the airborne microbes led to the following results: Approximately 60% of the examined colonies turned out to be cocci (*Micrococcus luteus, Micrococcus* sp., *Staphylococcus epidermidis*), approximately 25% were bacilli (*Bacillus cereus* ssp. *mycoides*, *Bacillus* sp.), and the rest consisted of miscellaneous bacterial colonies such as *Corynebacterium* sp. Only very few fungi were detectable (*Penicillium* sp.).

15.4 DISCUSSION

Due to the fact that all currently used air disinfectants have disadvantages and based on the well-known antimicrobial activity of essential oils and aroma compounds (Pattnaik et al., 1997; Swamy et al., 2016), the idea of investigating the effect of these volatiles on airborne microbes took form. Since there were no current studies concerning the antimicrobial properties of aroma compounds and essential oils on airborne microbes available, new methods of examination had to be developed. Not only the selected essential oils themselves, but also their main components, have been tested by applying the same methods.

A reduction of the total microbial count in the air was achieved with each of the tested essential oils and aroma compounds. Out of the tested essential oils, spruce needle essential oil, carnation oil, Australian sandalwood oil, rosemary essential oil, sweet orange oil, and noble fir cone oil were the most successful in diminishing airborne microbes, with an average germ count reduction of approximately 50%. The tested main components of the essential oils were even more effective in reducing airborne microbes. The most successful aroma compounds were terpineol, 1,8-cineol, eugenol and isophytol, with an average germ count reduction of more than 60%, followed by linalool, (−)-limonene, thymol, (−)-perillaldehyde, and geraniol, which performed an average germ count reduction on airborne microbes of more than 50%. The majority of these volatiles showed comparable results in the small and large examination rooms; see, for example, the results for carvacrol, *t*-cinnamaldehyde, geraniol, or linalool.

Contrary to the developed method, "traditional" air disinfection with, for example, formaldehyde or triethylene glycol requires high concentrations of these chemicals and partly longer residence time to achieve disinfection of indoor air. Formaldehyde, for example, is used at concentrations of 5 g/m^3 for air disinfection and has to be left on for 6 h (Wallhaeusser, 1995). In contrast, the tested essential oils and aroma compounds allow persons to stay in the room during air disinfection, which is not possible when using formaldehyde or triethylene glycol because of their toxicity. As essential oils have "GRAS"-status (generally recognized as safe by the FDA), leaving no harmful effect, and as antimicrobial resistance against synthetic antimicrobial agents is increasing to a degree, which has to be called frightening (Lorenzi et al., 2009; Mittal et al., 2019), the method presented in this chapter is a new, promising alternative for lowering microbial air contamination in rooms.

15.5 CONCLUSION

The convenient safe method to decrease airborne microbes by essential oils or aroma compounds presents an additional or even alternative method to currently used air disinfectants, especially in places where people gather, such as lecture halls, theatres, stations, airports, rest homes, or hospitals.

ACKNOWLEDGEMENT AND DEDICATION

I would like to thank Prof. Dr. Gerhard Buchbauer for his perpetual support throughout my previous scientific career.

This work has been created partly by affiliation with the Sigmund-Freud University of Vienna.

This work is dedicated to my son Florens Emmanuel Krist (born November 7, 2016).

REFERENCES

Aggarwal, K. K., S. P. S. Khanuja, A. Ahmad, T. R. S. Kumar, V. K. Gupta, and S. Kumar. 2002. Antimicrobial activity profiles of the two enantiomeres of limonene and carvone isolated from the oils Mentha spicata and Anethum sowa. *Flavour Fragr. J.*, 17(1): 59–63.

Ahmad, A., S. van Vuuren, and A. Viljoen. 2014. Unravelling the Complex Antimicrobial Interaction of Essential Oils. The Case of *Thymus vulgaris* (Thyme). *Molecules*, 19(3): 2896–2910.

Baby, S., and G. Varughese. 2009. Essential Oils and New Antimicrobial Strategies. In: *New Strategies Combating Bacterial Infection*, I. Ahmad, and F. Aqil (eds.), pp. 165–203.

Bagamboula, C. F., M. Uyttendaele, and J. Debrevere. 2004. Inhibitory effect of thyme and basil essential oils, carvacrol, thymol, estragol, linalool and p-cymene towards *Shigella sonnei* and *S. flexneri*. *Food Microbiol.*, 21: 33–42.

Banovac, D. 2012. Antimikrobielle Wirkung ausgewählter flüchtiger Verbindungen und ätherischer Öle auf luftgetragene Keime. *Diplomarbeit*, University of Vienna.

Bennis, S., F. Chamin, T. Bouchikhi, and A. Remmal. 2004. Surface alteration of *Saccharomyces cerevisiae* induced by thymol and eugenol. *Lett. Appl. Microbiol.*, 38: 454–458.

Bouaziz, M., T. Yangui, S. Sayadi, and A. Dhouib. 2009. Disinfectant properties of essential oils from *Saliva officinalis* L. cultivated in Tunisia. *Food Chem. Toxicol.*, 47(11): 2755–60.

Brehm-Stecher, B. F., and E. A. Johnson. 2003. Sensitization of *Staphylococcus aureus* and *Escherichia coli* to antibiotics by the sesquiterpenoids nerolidol, farnesol, bisabolol, and apritone. *Antimicrob. Agents Chemother.*, 47(10): 3357–3360.

Brun, P., G. Bernabè, R. Filippini, and A. Piovan. 2019. *In Vitro* antimicrobial activities of commercially available tea tree (*Melaleuca alternifola*) essential oils. *Curr. Microbiol.*, 76(1):108–116.

Chouhan, S., K. Sharma, and S. Guleria. 2017. Antimicrobial activity of some essential oils – present status and future perspectives. *Medicines (Basel)*, 4(3): 58.

Cimanga, K., K. Kambu, L. Tona, S. Apers, T. De Bruyne, N. Hermans, J. Totté, L. Pieters, and A. J. Vlietinck. 2002. Correlation between chemical composition and antimicrobial activity of essential oils of some aromatic medicinal plants growing in the Democratic Republic of Congo. *J. Ethnopharmacology*, 79: 213–220.

Cutillas, A. B., A. Carrasco, R. Martinez-Gutierrez, V. Tomas, and J. Tudela. 2017. Composition and antioxidant, antienzymatic and antimicrobial activities of volatile molecules from Spanish *Salvia lavandulifolia* (Vahl) essential oils. *Molecules*, 22(8): 1382.

De Lucca, A. J., A. Pauli, H. Schilcher, T. Sien, D. Bhatnagar, and T. J. Walsh. 2011. Fungicidal and bactericidal properties of bisabolol and dragostanol. *J. Essent. Oil Res.*, 23(3): 47–54.

Donaldson, J. R., S. L. Warner, R. G. Cates, and D. G. Young. 2005. Assessment of antimicrobial activity of fourteen essential oils when using dilution and diffusion methods. *Pharm. Biol.*, 43(8): 687–695.

Ebani, V. V., S. Nardoni, F. Bertelloni, S. Giovanelli, B. Ruffoni, C. D'Ascenzi, L. Pistelli, and F. Mancianti. 2018. Activity of *Salvia dolomitica* and *Salvia somalensis* essential oils against bacteria, molds and yeasts. *Molecules*, 23(2): 396.

Essien, E. E., P. S. Thomas, R. Ascrizzi, W. N. Setzer, and G. Flamini. 2018. *Senna occidentalis* (L.) link and *Senna hirsuta* (L.) H. S. Irwin & Barneby: Constituents of fruit essential oils and antimicrobial activity. *Nat. Prod. Res.*, 18: 1–4.

Ferhout, H., J. Bohatier, J. Guillot, and J. C. Chalchat. 1999. Antifungial Activity of Selected Essential Oils, Cinnamaldehyde and Carvacrol aganinst *Malassezia furfur* and *Candida albicans*. *J. Essent. Oil Res.*, 11(1): 119–129.

Fayyaz, N., A. M. Sani, and M. N. Najafi. 2015. Antimicrobial activity and composition of essential oil from *Echinophora platyloba*. *J. Ess. Oil Bear. Plant.*, 18(5): 1157–1164.

Feichtinger, Y. 2005. Wirkung ausgewählter alkoholischer Duftstoffe auf die Luftkeimzahl. *Diplomarbeit*, University of Vienna.

Friedman, M., P. R. Henika, and R. E. Mandrell. 2002. Bactericidal activities of plant essential oils and some of their isolated constituents against *Camphylobacter jejuni*, *Escherichia coli*, *Listeria monocytogenes*, and *Salmonella enterica*. *J. Food Prot.*, 65(10): 1545–1560.

Gadalla, M. A., and H. M. Hassan. 2004. The use of natural preservatives in cosmetics preparation. I – Minimum inhibitory concentrations of essential oils against isolated fungi from some cosmetic products. *Egypt. J. Biotech.*, 16: 308–316.

Geraci, A., V. Di Stefano, E. Di Martino, D. Schillaci, and R. Schicchi. 2017. Essential oil components of orange peels and antimicrobial activity. *Nat. Prod. Res.*, 31(6): 653–659.

Ghaneian, M. T., M. H. Ehrampoush, A. Jebali, S. Hekmatimoghaddam, and M. Mahmoudi. 2015. Antimicrobial activity, toxicity and stability of phytol as a novel surface disinfectant. *Environ. Health Eng. Manag. J.*, 2(1): 13–16.

Hinou, J. B., C. E. Harvala, and E. B. Hinou. 1989. Antimicrobial activity screening of 32 common constituents of essential oils. *Pharmazie*, 44(4): 302–303.

Horne, D., M. Holm, C. Oberg, S. Chao, and D. G. Young. 2001. Antimicrobial effects of essential oils on *Streptococcus pneumoniae*. *J. Essent. Oil Res.*, 13: 387–392.

Inouye S., T. Takizawa, and H. Yamaguchi. 2001. Antibacterial activity of essential oils and their major constituents against respiratory tract pathogens by gaseous contact. *J. Antimicrob. Chemother.*, 47(5): 565–573.

Jirovetz, L., G. Buchbauer, Z. Denkova, A. S. Stoyanova, I. Murgov, V. Gearon, S. Birkbeck, E. Schmidt, and M. Geissler. 2006a. Comparative study on the antimicrobial activities of different sandalwood essential oils of various origin. *FFJ*, 21: 465–468.

Jirovetz, L., G. Buchbauer, Z. Denkova, A. S. Stoyanova, I. Murgov, E. Schmidt, and M. Geissler. 2005. Antimicrobial testings and gas chromatographic analysis of pure oxygenated monoterpenes 1,8-cineole, α-terpineol, terpinen-4–ol and camphor as well as target compounds in essential oils of pine (*Pinus pinaster*), rosemary (*Rosmarinus officinalis*), tea tree (*Melaleuca alternifolia*). *Scientia Pharmaceutica*, 73(1): 27–39.

Jirovetz, L., G. Buchbauer, G. Schweiger, Z. Denkova, A. Slavchev, A. Stoyanova, E. Schmidt, and M. Geissler. 2007. Chemical composition, olfactory evaluation and antimicrobial activities of *Jasminum grandiflorum* L. absolute from India. *Nat. Prod. Com.*, 2(4): 407–412.

Jirovetz, L., G. Buchbauer, A. S. Stoyanova, Z. Denkova, and I. Murgov. 2004. Antimicrobial testings and chiral phase gas chromatographic analysis of dill oils and related key compounds. *Ernährung*, 28(6): 257–260.

Jirovetz, L., G. Eller, G. Buchbauer, E. Schmidt, Z. Denkova, A. S. Stoyanova, R. Nikolova, and M. Geissler. 2006b. Chemical composition, antimicrobial activities and odor descriptions of some essential oils with characteristic floral-rosy scent and their principal aroma compounds. *Recent Res. Dev. Agro. Hort.*, 2: 1–12.

Karbasiyan, Z. 2012. Antimikrobielle Wirkung von Terpenen auf luftgetragene Keime in Apothekenverkaufsräumen/Arztpraxis. *Diplomarbeit*, University of Vienna.

Kisko, G., and S. Roller. 2005. Carvacrol and *p*-cymene inactivate Escherichia coli O157:H7 in apple juice. *BMC Microbiol.*, 5: 36.

Knowles, J. R., S. Roller, D. B. Murray, and A. S. Naidu. 2005. Antimicrobial action of carvacrol at different stages of dual-species biofilm development by *Staphylococcus aureus* and *Salmonella enterica* serovar Typhimurium. *Appl. Environ. Microbiol.*, 71(2): 797–803.

Kretz, H. 2007. Zur Wirkung der ätherischen Öle von Pinus pinaster und Melaleuca alternifolia auf die Luftkeimzahl. *Diplomarbeit*, University of Vienna.

Krist, S., D. Banovac, N. Tabanca, D. E. Wedge, V. K. Gochev, J. Wanner, E. Schmidt, and L. Jirovetz. 2015. Antimicrobial activity of nerolidol and its derivatives against airborne microbes and further biological activities. *Nat. Prod. Comm.*, 10(1): 143–148.

Krist, S., L. Halwachs, G. Sallaberger, and G. Buchbauer. 2007. Effects of scents on airborne microbes, part I: Thymol, eugenol, *trans*-cinnamaldehyde and linalool. *Flavour Fragr. J.*, 22: 44–48.

Krist, S., K. Sato, S. Glasl, M. Hoeferl, and J. Saukel. 2008. Antimicrobial effect of vapours of terpineol, (R)-(−)-linalool, carvacrol, (S)-(−)-perillaldehyde and 1,8–cineole on airborne microbes using a room diffuser. *Flavour Fragr. J.*, 23: 353–356.

Lee, K., J. H. Lee, S. I. Kim, M. H. Cho, and J. Lee. 2014. Anti-biofilm, anti-hemolysis, and anti-virulence activities of black pepper, cananga, myrrh oils, and nerolidol against *Staphylococcus aureus*. *Appl. Microbiol. Biotechnol.*, 98(22): 9447–9457.

Lei, Y., P. Fu, X. Jun, and P. Cheng. 2019. Pharmacological properties of geraniol – A review. *Planta Med.*, 85(1): 48–55.

Lopez, P., C. Sanchez, R. Battle, and C. Nerin. 2005. Solid- and vapour-phase antimicrobial activities of six essential oils: Susceptibility of selected foodborne bacterial and fungal strains. *J. Agric. Food Chem.*, 53(17): 6939–6946.

Lorenzi V., A. Muselli, A. F. Bernardini, L. Berti, J. M. Pagès, L. Amaral, and J. M. Bolla. 2009. Geraniol restores antibiotic activities against multidrug-resistant isolates from gram-negative species. *Antimicrob. Agents Chemother*, 53: 2209.

Marchese, A., R. Barbieri, E. Coppo, I. E. Orhan, M. Daglia, S. F. Nabavi, M. Izadi, M. Abdollahi, S. M. Nabavi, and M. Ajami. 2017. Antimicrobial activity of eugenol and essential oils containing eugenol: A mechanistic viewpoint. *Crit. Rev. Microbiol.*, 43(6): 668–689.

Mittal, R. P., A. Rana, and V. Jaitak. 2019. Essential oils: An impending substitute of synthetic antimicrobial agents to overcome antimicrobial resistance. *Curr. Drug Targets*, 20(6):605–624.

Moliszewska, E. B., 2003. Antifungal properties of some ropenylbenzenes. *B. Pol. Acad. Sci. Biol. Sci.*, 51(3): 229–235.

Moore R. M., and R. G. Kaczmarek. 1991. Occupational hazards to health care workers: Diverse, ill-defined, and not fully appreciated. *Am. J. Infect. Control.*, 18: 316–327.

Nieto, G. 2017. Biological activities of three essential oils of the Lamiaceae family. *Medicines (Basel)*, 4(3): 63.

Park, S.-N., Y. K. Lim, M. O. Freire, E. Cho, D. Jin, and J.-K. Kook. 2012. Antimicrobial effect of linalool and α-terpineol against periodontopathic and cariogenic bacteria. *Anaerobe*, 18(3): 369–372.

Pasierb, P. 2007. Über die antimikrobielle Wirkung von ätherischem Fichtennadelöl, Jasminöl, Nelkenblattöl und Palmarosaöl auf luftgetragene Keime. *Diplomarbeit*, University of Vienna.

Pattnaik, S., V. R. Subramanyam, M. Bapaji, and C. R. Kole. 1997. Antibacterial and antifungal activity of aromatic constituents of essential oils. *Microbios*, 89(358): 39–46.

Rhayour, K., T. Bouchikhi, A. Tantaoui-Elaraki, K. Sendide, and A. Remmal. 2003. The mechanism of bactericidal action of oregano and clove essential oils and of their phenolic major components on *Escherichia coli* and *Bacillus subtilis*. *J. Essent. Oil Res.*, 15: 286–292.

Rivas da Silva, A. C., P. M. Lopes, M. M. Barros de Azevedo, D. C. Costa, C. S. Alviano, and D. S. Alviano. 2012. Biological activities of α-pinene and β-pinene enantiomers. *Molecules*, 17(6): 6305–6316.

Sadeghi, E., A. Dargahi, A. Mohammadi, F. Asadi, and S. Sahraee. 2015. Antimicrobial effect of essential oils: A systematic review. *Food Hygiene*, 5(2): 1–26.

Saibabu, V., S. Singh, M. A. Ansari, Z. Fatima, and S. Hameed. 2017. Insights into the intracellular mechanisms of citronellal in *Candida albicans*: Implications for reactive oxygen species-mediated necrosis, mitochondrial dysfunction, and DNA damage. *Rev. Soc. Bras. Med. Trop.*, 50(4): 524–529.

Santoyo S., S. Cacero, L. Jaime, E. Ibanez, F. J. Senorans, and G. Reglero. 2005. Chemical composition and antimicrobial activity of *Rosmarinus officinalis* L. essential oil obtained via supercritical fluid extraction. *J. Food Prot.*, 68(4): 790–795.

Sato, K., S. Krist, and G. Buchbauer. 2006. Antimicrobial Effect of *trans*-cinnamaldehyde, (−)-perillaldehyde, (−)-citronellal, citral, eugenol and carvacrol on airborne microbes using an airwasher. *Biol. Pharm. Bull.*, 29(11): 2292–2294.

Sato, K., S. Krist, and G. Buchbauer. 2007. Antimicrobial effect of vapours of geraniol, (R)-(−)-linalool, terpineol, γ-terpinene and 1,8–cineole on airborne microbes using an airwasher. *Flavour Fragr. J.*, 22: 435–437.

Schmidt, E., L. Jirovetz, G. Buchbauer, Z. Denkova, A. Stoyanova, I. Murgov, and M. Geissler. 2005. Antimicrobial testings and gas chromatographic analyses of aroma chemicals. *J. Essent. Oil Bear. Pl.*, 8(1): 99–106.

Sonboli, A., B. Babakhani, and A. R. Mehrabian. 2006. Antimicrobial activity of six constituents of essential oil from *Salvia*. *J. Biosci.*, 61(3/4): 160–164.

Strobl, M. N. 2005. Wirkung ausgewählter Terpenaldehyde auf die Luftkeimzahl. *Diplomarbeit*, University of Vienna.

Swamy, M. K., M. S. Akhtar, and U. R. Sinniah. 2016. Antimicrobial properties of plant essential oils against human pathogens and their mode of action: An updated review. *Evid. Based Complement. Alternat. Med.*, 2016(3): 1–21.

Tao, R., C. Z. Wang, and Z. W. Kong. 2013. Antibacterial/antifungal activity and synergistic interactions between polyprenols and other lipids isolated from *Ginkgo biloba* L. leaves. *Molecules*, 18(2): 2166–2182.

Wallhaeusser, K. H. 1995. *Praxis der Sterilisation, Desinfektion-Konservierung*. Stuttgart, New York: Georg Thieme Verlag.

Wanner, J., E. Schmidt, S. Bail, L. Jirovetz, G. Buchbauer, V. Gochev, T. Girova, T. Atanasova, and A. Stoyanova. 2010. Chemical composition and antimicrobial activity of selected essential oils and some of their main components. *Nat. Prod. Comm.*, 5(9): 1359–1364.

Wosny, M. 2007. Antimikrobielle Wirkung verschiedener Sandelholzöle und alpha- und beta-Santalol auf luftgetragene Keime. *Diplomarbeit*, University of Vienna.

Yang, S. A., S. K. Jeon, E. J. Lee, N. K. Im, K. H. Jhee, S. P. Lee, and I. S Lee. 2009. Radical scavenging activity of essential oil of silver fir (*Abis alba*). *J. Clin. Biochem. Nutr.*, 44(3): 253–259.

Yousefi, K., S. Hamedeyazdan, D. Hodaei, F. Lotfipour, B. Baradaran, M. Orangi, and F. Fathiazad. 2017. An *in vitro* ethnopharmacological study on *Prangos ferulacea*: A wound healing agent. *Bioimpacts*, 7(2): 75–82.

Zambonelli, A., A. Zechini D'Aulerio, A. Severi, S. Benvenuti, L. Maggi, and A. Bianchi. 2004. Chemical composition and fungicidal activity of commercial essential oils of *Thymus vulgaris* L. *J. Essent. Oil Res.*, 16: 69–74.

Zaussinger, M. 2008. Antimikrobielle Wirkung einiger ätherischer Öle von Rosmarinus officinalis und Anethum graveolens gegenüber luftgetragenen Keimen. *Diplomarbeit*, University of Vienna.

16 Quorum Sensing and Essential Oils

Isabel Charlotte Soede and Gerhard Buchbauer

CONTENTS

16.1 INTRODUCTION

16.1.1 Cell-to-Cell Communication

Bacteria have the ability to communicate with each other by producing signal molecules, which they release into their environment. The fundamentals of cell-to-cell communication can be subdivided in three stages: detection, signal transduction, and response. Signal molecules are being detected when binding to extracellular receptors or intracellular receptor proteins. Ligand-binding leads to a change of confirmation on the receptor, which then causes an intracellular signaling cascade. This cascade then leads to a specific reaction—for example, the activation of genes (Campbell and Reece, 2009).

16.1.2 Quorum Sensing: Definition

Quorum sensing (QS) describes the ability of bacteria to sense the number of bacteria in their environment by using signal molecules, also called autoinducers. This way of communication is important to coordinate group behavior depending on the number of bacteria in the environment of a bacterial population. As soon as a critical external concentration of signal molecules ("quorum") is reached, the population reacts by modifying its gene expression. This means the production of these signals is only effective when being secreted from a large number of cells (Waters and Bassler, 2005).

Various species of bacteria have developed different signaling molecules for QS, which means that a direct communication between different classes is not necessarily possible. If there is an interspecies communication, it is called "cross talk." It is also possible for one species of bacteria to have several different QS signals that lead to different reactions—whereby the main reason for this kind of cell-to-cell communication appears to be the adaption to changing environmental conditions or the defense of toxic substances or other organisms, as scientists of the Centre for Biomolecular Sciences of the University of Nottingham found (Jensen et al., 2008).

Examples of cell reactions caused by QS are bioluminescence, the expression of virulence factors, proliferation, sporulation, mating, and biofilm formation (Bassler, 2002). Most research on QS-signaling circuits refers to systems based on either acyl homoserine lactones (AHLs) or peptides as signal molecules.

16.2 QS IN BACTERIA

16.2.1 Gram-Negative Bacteria

16.2.1.1 The LUXRI -System

The research on QS began in the early 1970s, when Nealson, Platt, and Hastings found out that *Vibrio fischeri* and *Vibrio harveyi* produce signal molecules when growing. As soon as a critical concentration of these molecules is reached, luminescence of the bacteria is activated. This team was also the first to assign the term "autoinducer" for QS molecules (QSM) (Nealson et al., 1970). The term "quorum sensing" was first mentioned in 1995 by Fuqua, Winans, and Greenberg (1994) in the *Journal of Bacteriology*: their research on QS was based on the question of why the Gram-negative marine bacterium *V. fischeri* contains high levels of luciferase only when a large number of individuals are present (Greenberg, 1997).

In nature, *V. fischeri* populations live—inter alia—in the light organs of the squid *Euprymna scolopes*. Visik et al. found out that this symbiosis is essential for the bacteria to reach the necessary cell density for the synthesis of luciferase (Visick et al., 2000). Eberhard et al. isolated and identified the QS signal of *V. fischeri* in 1981—the signal 3-O-C6-homoserine-lactone—N-(3-oxohexanoyl)-L-homoserinlactone—(HSL or AHL) is an N-acyl-lactone (see Figure 16.1) (Eberhard et al., 1981).

Engelbrecht and Silverman found out that the luciferase operon is controlled by the expression of the proteins LUXI and LUXR, whereas LUXI is necessary for the synthesis of AHL and LUXR is a specific cytoplasmic receptor for the autoinducer and therefore regulates the response to the autoinducer (Engebrecht and Silverman, 1984). Enzymes such as luciferase, being part of the luminescence system in bacteria, are encoded by the *lux* (Latin for light) genes. Hence, proteins with names starting with LUX indicate their correspondence to luminescence genes (*lux*) (Meighen, 1993). Stevens et al. proved that as soon as a critical concentration is reached, LUXR specifically binds the QS molecule and activates an operon, which then leads to the transcription of the luciferase (Stevens et al., 1994).

AHL molecules differ in their ability to permeate cell membranes. Whitehead et al. suggested that this is caused by their structural variability (acyl side chain) (Whitehead et al., 2001). For example, Kaplan and Greenberg compared the cellular and external concentrations of autoinducer and found out that *V. fischeri* cells are freely permeable for the molecules—the autoinducer can diffuse into

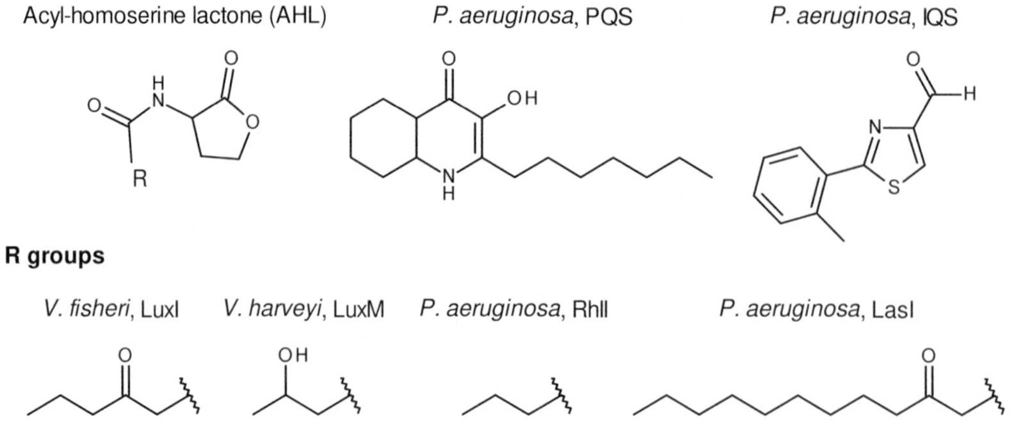

Acyl-homoserine lactone (AHL) *P. aeruginosa*, PQS *P. aeruginosa*, IQS

R groups

V. fisheri, LuxI *V. harveyi*, LuxM *P. aeruginosa*, RhlI *P. aeruginosa*, LasI

FIGURE 16.1 *V. fishery, V. harvey* and *P. aeruginosa* QS signals. (Adapted and newly drawn from Waters CM and Bassler BL, 2005.)

and out of the cell (Kaplan and Greenberg, 1985)—whereas Pearson et al. showed that there is an active efflux pump for HSL in *P. aeruginosa*, which produces—among others—a 3OC12-HSL as an AHL QS molecule (see Figure 16.1/Lasl) (Pearson et al., 1999).

In summary, this signaling circuit consists of a synthase-regulator complex, which is dependent on an AHL and when activated, leads to the expression of specific genes, typically for Gram-negative bacteria. There are over 50 species using AHL molecules as autoinducers. These systems are referred to as LUXRI-homologous systems, whereas the nomenclature consists of the genes/corresponding proteins controlling the QS circuit (e.g., las/LAS) and other letters indicating the respective function (e.g., I = AHL synthesizing protein, R = AHL receptor), resulting in a QS system designated as, for example, lasRI in *P. aeruginosa* (Bassler, 2002). Other bacteria using the LUXRI –homologous system for QS are, for instance, *Aeromonas hydrophila*, *A. salmonicida*, and *Chromobacterium violaceum* (Greenberg, 1997). To give an example of QS systems being partly homologous to LUXRI, the QS systems of *V. fischeri* and *P. aeruginosa* are being presented. As *C. violaceum* is a typical strain used for QS detection, its QS-system is also being introduced.

16.2.1.2 *Vibrio harveyi*

Besides the LUXRI system, Gram-negative bacteria also use other signaling pathways; one example is *V. harveyi*, a marine Gram-negative bacterium that also uses QS for bioluminescence. *V. harveyi* produces three different autoinducers:

- Cao and Meighen identified an N-acyl-homoserine-lactone, as an autoinducer (3-hydroxy-C4-HSL [Figure 16.1]) named harveyi autoinducer 1 (HAI-1) (Cao and Meighen, 1989). However, it was observed that the genes involved in the QS circuit are not homologous to the ones in the LUXRI system described earlier (Bassler et al., 1993).
- The second autoinducer discovered was named AI-2 (autoinducer 2), a furanosyl borate di-ester (Bassler et al., 1994), (Chen et al., 2002). Surette et al. identified the genes for luxS which is responsible for the production of AI-2 (Surette et al., 1999). LUXS binds periplasmatic to LUXP; this complex then interacts with a membrane-bound histidine kinase—LuxQ. Lux Q, just like LUXN has the same response regulator domains like those in two-component signal transduction systems in Gram-positive bacteria (Bassler et al., 1994).
- Henke and Bassler discovered a third QS system in *V. harveyi*—being named after the bacterial strain in which the system was primary discovered: *Vibrio cholerae*, a bacterium causing the disease cholera. The autoinducer (AI) is called *cholerae* autoinducer 1 (CAI-1), which is produced by CqsA (Cqs = *cholerae* quorum-sensing autoinducer) and binds to CqsS (=sensor) (Henke and Bassler, 2004).

16.2.1.3 *Vibrio parahaemolyticus*

QS in *V. parahaemolyticus* is poorly characterized, as many isolates seem to have silenced QS. What is known is that OpaR is a LuxR homologue existing in this organism. OpaR regulates surface sensing and type III secretion system. The signaling cascade was shown to be highly similar to *V. harveyi*. Swarming motility in *V. parahaemolyticus* is regulated by OpaR and can therefore be used to detect a QS-inhibition (QSI) (Huan Liu et al., 2013; Banu et al., 2018).

16.2.1.4 *Pseudomonas aeruginosa*

P. aeruginosa can cause severe and persistent nosocomial and wound infections as they are often resistant to several antibiotics. Affected patient groups of those nosocomial infections are, for example, cystic fibrosis patients, immune-compromised elderly or long-term intubated patients. It has been shown that QS plays a role in virulence gene expression; for instance. by murine models with pneumonia, showing reduced virulence of the bacteria when deleting QS genes. As *P. eruginosa* is such an important human pathogen, it is often used as a model organism for QS research.

P. aeruginosa has several virulence mechanisms, namely iron scavenging (proteases, siderophores), cytotoxicity (pyocyanin, exotoxin A), antibiotic resistance (efflux pumps, modifying enzymes), immune evasion (elastase, alkaline protease), biofilm development (alginate, rhamnolipids), and motility (flagella, type IV pili).

Some of the genes regulating those virulence mechanisms are known to be regulated by QS examples are elastase, proteases, hydrogen cyanide, exotoxin A, secretion proteins, catalase, rhamnolipid, pyocyanin, lectins, acylated HSLs, and pyoverdin. It has also been shown that QSMs of *P. aeruginosa* influence the production of several eukaryotic cytokines (IL-8, IL-12, TNF-β), which may also influence its pathogenicity (Smith and Iglewski, 2003a,b; Lee and Zhang, 2014).

There are four main QS systems identified in *P. aeruginosa*, whereas (1) and (2) are LUXRI homologous systems. The QS-systems are closely connected to each other, showing a complex communication network including various components mutually influencing each other.

- The LasRI system, named after its elastase-regulating properties, possesses the AI N-(3-oxododecanoyl)-homoserine lactone (OdDHL) (Figure 16.1).
- The RhlIR system_designated after the RhlR. RhlR works as the AI receptor of this system and is a protein encoded by a rhamnolipid synthase gene cluster. The belonging AIs of this QS system are N-butyryl-L-homoserine-lactones (C4-HSL = BHL) (Figure 16.1) (Lee and Zhang, 2014).

Both of the pathways were shown to be regulated by various mechanisms; examples are QscR (quorum-sensing-control repressor) (Chugani et al., 2001) and VqsR (virulence and quorum-sensing regulator) (Juhas et al., 2004), being LuxR homologues. QscR inhibits both pathways by forming LasR-QSM/RhIR-QSM complexes; on the other hand, it binds to OdDHL and is activated its own regulon. VqsR positively activates the LasRI system and is regulated by the LasR-QSM complex. Just like QscR and VqsR, there are many other super-regulators which reflects the complexity of QS regulation in *P. aeruginosa* (Lee and Zhang, 2014).

Virulence factors controlled by the Las system include, for example, elastase-causing degradation of different matrix proteins (collagen, elastin) and protease-causing disruption of epithelial barriers. A gene (*rhlAB*) controlled by the RhlIR system codes for rhamnolipids and causes necrosis of host macrophages and polymorphonuclear lymphocytes (Lee and Zhang, 2014). Reimmann et al. also showed the correlation among OdDHL presence, virulence factors (pyocyanin, elastase, rhamnolipids, hydrogen cyanide), and swarming motility, as both the virulence factors and motility decreased when OdDHL was reduced (Reimmann et al., 2002).

Quinolone-based QS: The main QSM was found to be 2-heptyl-3-hydroxy-4-quinolone (Figure 16.1), also known as PQS—the *Pseudomonas* quinolone signal; the corresponding receptor is

PqsR (*Pseudomonas* quinolone signal receptor). Research also determined a connection between the quinolone system and LasR, as part of the PQS synthesis cluster is controlled by LasR and turn-controlled of PQS by the LasR-OdDHL. Quinolone-based QS was shown to influence biofilm formation and virulence-factor production (pyocyanin, elastase, PA-IL lectin, rhamnolipids) as null mutation of the PQS system leads to reduced biofilm formation and virulence-factor production, respectively (Lee and Zhang, 2014).

Integrated Quorum Sensing Signal (IQS) based QS: The AI is 2-(2-hydroxyphenyl)-thiazole-4-carbaldehyde (Figure 16.1), belonging to a new class of QSM. Disruption of this system causes reduced production of the QSM PQS and BHL, and also, pyocyanin, rhamnolipids, and elastase are decreased (Lee and Zhang, 2014).

Summarized, there is a clear connection between virulence and QS, and therefore, inhibition of QS pathways could lead to reduced virulence in *P. aeruginosa* being a promising approach for new treatments of infections caused by those bacteria. It is worth a mention that as explained above, QS plays a role in biofilm formation but is only one of several regulating factors such as swarming motility, rhamnolipids, and siderophores (Kalia et al., 2015).

16.2.1.5 *Chromobacterium violaceum*

Given the fact that in most of the research done on EOs as QS inhibitors (QSI), different strains of *C. violaceum* were being used as the sensor, its QS system is presented here. *C. violaceum* is an aquatic bacterium growing in soil and water which can cause abscesses and bacteremia in humans. The QS system consists of LuxRI homologues, namely CviIR (Cv = *Chromobacterium violaceum*).

Research showed that antagonists of natural AHLs can reduce the pathogenicity by means of confirmation-modification of CviR of *C. violaceum* on the nematode *Caenorhabditis elegans*, allowing the presumption that QSI may also reduce virulence in of *C. violaceum* in human infections. One well-studied gene product controlled by QS in *C. violaceum* is the purple pigment violacein. The violacein gene cluster consists out of four genes: vioABCD, whereas the vioA promotor is directly controlled by CviR. To detect QS inhibition, different strains of *C. violaceum* are being used; they are listed in the "Materials and Methods" section (Mcclean et al., 1997; August et al., 2000).

16.2.2 Gram-Positive Bacteria

Gram-positive bacteria use autoinducing peptides (AIPs) as signals. AIPs cannot freely diffuse through the membrane; they bind on extracellular histidine kinase receptors. This signaling way is classified as a "two-component" type, as it consists of a sensor and a response-regulator protein. As soon as the molecule binds, the kinase activity of the receptor is activated, which leads to a phosphorylation cascade resulting in the binding of a responsive regulator, a DNA-binding regulator controlling the transcription of QS genes. Bacteria using this kind of signaling path are *Bacillus subtilis, Streptococcus pneumonaiae*, and *Staphylococcus aureus*. For the excretion of AIPs, bacteria often express ATP-binding cassette (ABC)-transporter (Kleerebezem et al., 1997).

16.2.2.1 *Staphylococcus aureus*

S. aureus is a commensal and an opportunistic pathogen causing mild to life-threatening infections in both humans and animals. Infections caused by this pathogen are linked to several virulence factors and biofilm formation. The QS system in *S. aureus* is encoded by the accessory gene regulator (agr). Agr contains two operons, whereas one includes agrBDCA, which triggers the production of RNAII transcript. The other operon is responsible for transcription of RNAIII, which is in turn considered as the effector molecule of the agr locus (Yarwood and Schlievert, 2003). Consequently, the transcriptional level of RNAIII can be used to determine QSI activity of different EOs against *S. aureus*.

16.3 QS IN FUNGI

Although there has been done a lot of research on QS in bacteria in the last decades, QS in fungi was not investigated until 2001, when Hornby et al. discovered the extracellular QS-molecule farnesol in the dimorphic fungus *Candida albicans* (Hornby et al., 2001). *C. albicans* is one of the most frequent human fungal pathogens causing, for example, catheter-related infections (Ramage et al., 2002).

In yeast-mycelium fungi, there is a direct correlation between cell morphology—either formation of budding yeasts (high cell density) or germ tubes (low cell density)—on initial cell density. This dependence is called the "inoculum size effect," firstly mentioned by Kulkarni and Nickerson in 1980 (Kulkarni and Nickerson, 1980). The isoprenoid farnesol in *C. albicans* causes a shift in morphology, resulting in inhibition of mycelial growth and formation of budding yeasts. Hornby et al. set up the hypothesis that farnesol may be a QSM as its physical properties are suitable, in contrast to other extracellular molecules produced by *C. albicans* (geranylgeraniol, geraniol). These characteristics include the solubility in water, the lipophilicity, and the fact that farnesol is produced continuously during cell growth in an amount proportional to cell mass (Hornby et al., 2001).

Ramage et al. (Ramage et al., 2002) showed a biofilm-inhibiting effect of farnesol, the inhibition was shown to be dependent on the time cells had to adhere to a surface before farnesol was added. They hypothesized that in the presence of high farnesol concentrations, mycelial development is inhibited and therefore biofilm formation is reduced. The cause of this may be the detachment of cells in order to colonize new substrate areas and to prevent overpopulation (Ramage et al., 2002; Albuquerque and Casadevall, 2012).

In 2001, Oh et al. identified farnesoic acid as another autoregulatory substance also inhibiting filamentous growth in a strain of *C. albicans* (Oh et al., 2001). Cho et al. showed that tyrosol is an autoregulatory molecule in *C. albicans*. Its effects are the shortening of the lag phase of diluted cultures, as well as the opposite effect of farnesol regarding germ tube formation—it accelerates the conversion from yeasts to mycelium (Cho et al., 2008).

16.3.1 EOs as Biofilm Inhibitors in *C. albicans*?

As mentioned above, farnesol, a sesquiterpene-alcohol, inhibits biofilm formation due to morphological changes. As terpenoid structures are also constituents of EOs, maybe EOs interfere with the fungal QS system similarly and are therefore able to inhibit biofilm formation. Agarwal et al. positively tested several plant oils on their biofilm-inhibiting activity, although it has not been clarified if this has any connection to QS (Agarwal et al., 2008).

16.4 QS IN *SACCHAROMYCES CEREVISIAE*

Chen et al. showed that nitrogen starvation in *S. cerevisiae* results in the production of aromatic alcohols functioning as QSM. These molecules again lead to a morphogenetic switch from budding yeasts to the filamentous form (Chen and Fink, 2006):

Besides *C. albicans and S. cerevisiae*, QS activities in other species (*Ceratocystis ulmi, Cryptococcus neoformans, Uromyces phaseoli, Neurospora crassa*, etc.) have been described, but research on which molecules are responsible and how the signaling pathways work is still at the beginning and processes are not yet understood (Hogan, 2006; Albuquerque and Casadevall, 2012).

16.5 QS IN VIRUSES

In 2017, Erez et al. discovered a QS-like system—also called "the arbitrium system"—in viruses (phages) of the SPbeta group to coordinate lysis–lysogeny decisions. When infecting the host *Bacillus*, a peptide consisting of six amino acids is released. In subsequent infections, the concentration of the peptide is measured by progeny phages which lysogenize when a certain level is reached (Erez et al., 2017).

16.6 QS AND BIOFILM FORMATION

A biofilm is "a thin layer of microorganisms adhering to the surface of a structure, which may be organic or inorganic, together with the polymers that they secrete" (Miller-Keane Encyclopedia and Dictionary of Medicine, 2003).

In several species, QS is involved in biofilm development; however, the research on this topic is still at the beginning and there is only little knowledge about mutual influence between the biofilm (formation, pathogenic potential, structure) and the QS system (onset of gene expression, interspecies signaling) (Parsek and Greenberg, 2005). In several studies presented in the next sections, research groups found QS inhibitors to be potent biofilm inhibitors too. Although the relation cannot always be found; for example, Kim et al. showed a significant biofilm inhibition of EHEC cells by bay, clove, and pimento berry oils and eugenol. When looking at the transcriptional changes by clove oil and eugenol in more detail, the group showed that the expression of QS -elated genes (*luxS*, *luxR*, and *tnaA*) was not appreciably affected (Kim et al., 2016).

Biofilms cannot only be found attached on medical devices and implants, including, for instance, catheters, orthopedic implants, or cardiac pacemakers, but also adhere to endothelial or epithelial lining in lungs, intestinal or vaginal mucus layers, or teeth. Biofilm formation can result in significantly increased resistance against antibiotics and host defenses and can therefore lead to chronic inflammations or impaired wound healing (Bryers, 2009).

Other fields where biofilms cause worries are agriculture, fishery (fish—pathogenic bacteria like *Aeromonas hydrophila or A. salmonicida)* and water treatment plants (especially, biofilms on reverse osmosis membranes built by the species *Legionella* in interaction with mainly γ-proteobacteria cause economic losses). As biofilms occur ubiquitous, they are generally associated with deterioration of materials they populate (Kalia, 2013). Biofilm formation can be divided in three stages: adhesion to a surface, maturation of the biofilm meaning cell-division, and production an extracellular matrix in order to protect the bacteria from environmental stresses. The last stage is dispersal to allow bacteria to escape the biofilm community (Toole et al., 2000).

Whereas Solano et al. believe that QS is most reasonable in the last step of biofilm disassembly, as the signal density reaches sufficient concentrations only then (Solano and Echeverz, 2014), research has shown that QS influences all stages of biofilm formation (Parsek and Greenberg, 2005). In human pathogens *Staphylococcus aureus* (Yarwood and Schlievert, 2003), *Helicobacter pylori* (Cole et al., 2004), and *Salmonella enterica* (Prouty et al., 2002) QS plays a role in the attachment phase of biofilm formation (Costerton et al., 1995). An example for an AHL-based QS system responsible for biofilm maturation is *Aeromonas hydrophila* (Lynch et al., 2002). Lynch et al. found out that the plant pathogen *Xanthomonas campestris* uses a QS system which is involved in biofilm dispersal (Lynch et al., 2002).

16.7 QS INHIBITORS (QSIs)

As described earlier, QS is involved in biofilm formation and production of virulence factors, and the inhibition of this way of communicating is therefore a promising target for new antibiotics. Kalia summarized the possible targeting points of QSI (with regard to AHL QS-systems): activity reduction of the AHL; receptor protein or synthase inhibition of the AI production; degradation of the QSM; and usage of QSM analogues, decoy receptors, or antibodies targeting the AHL receptors.

To effectively select QS molecules, it has been proposed that QSI should be highly specific small molecules—preferably longer than the native AHL—to efficiently reduce QS-regulated gene expression. There should preferably be no adverse effects, neither on bacteria nor host. QSM should be chemically stable and, as far as possible, resistant to the host's metabolic degradation. Meeting all these criteria, the chance of resistance formation is reduced, the risk of harming beneficial bacteria in the present of the host is low, and due to the small molecules, the risk of antigenicity is low too (Kalia, 2013). A term often used when discussing QSI is "quorum quenching," meaning the quench or interference with QS systems, as further discussed in the next paragraph (Dong et al., 2007). There can be made a rough division between natural (e.g., prokaryotic, animal-based, fungus based, plant based) and synthetic (e.g., modified AHL molecules, antagonists of receptor-ligand interactions) QS inhibitors (Kalia, 2013).

16.7.1 Practical Applicability of QS Inhibitors

QS inhibition is already used successfully to control membrane fouling by biofilms in water-purification plants. In this area, the most commonly applied method is quorum quenching, meaning the enzymatic degradation of AHLs by using, for instance, purified QQ enzyme obtained from porcine species or bacteria that produce QQ enzymes (Bouayed et al., 2016).

Clinical applicability of QS inhibitors seems to be a more challenging task, as typical drug discovery obstacles such as the lack of potency in animal models, delivery, and stability of the agents delay or impede the practical use of QSI. Until now, it was difficult to say at what point of infection QS inhibition would be of value or if a use as prophylactic agent could be helpful. Whiteley, Diggle, and Greenberg ultimately assume that those agents will be likely to occur in combinational therapies with conventional antimicrobials (Whiteley et al., 2017).

An example was the effective treatment of *P. aeruginosa*-infected mice with halogenated furanones causing attenuated virulence through QSI in the murine lungs (Hentzer et al., 2003). Although, furanones only have a limited practical benefit due to their lack of stability and their cytotoxicity (Riedel et al., 2005). This is a potential area for less toxic substances like EOs or their components. Especially, chronic lung infections caused by *P. aeruginosa* represent an inviting target for QSI (Whiteley et al., 2017). Due to the diverse components of EOs, screening for QSI is an applicable model for further tests, though they are often done with single components. In 2018, Haque et al. summarized developments in the field of QS inhibition. Many of the mentioned active components are of natural origin and are often major constituents of EOs—for example, leugenol, farnesol, carvacrol, and menthol (Haque et al., 2018). A step toward clinical trials is using infected *Caenorhabditis elegans* as a preclinical model to detect QS inhibition and consequences for the organism when being treated with EOs (Husain et al., 2013).

16.8 EOs AS QS INHIBITORs

An EO was defined by the International Organization for Standardization (ISO –rule, 9235 Geneva, 1997: 2013) as "a product obtained from a natural raw material of plant origin, by steam distillation, by mechanical processes from the epicarp of citrus fruits, or by dry distillation, after separation of the aqueous phase—if any—by physical processes." EOs represent a great reservoir of possible new therapeutics as a single oil may contain more than 50 components. The research on EOs as QSIs in human medicine, as well as in the food industry, plant pathology, and so on, has been growing in the last years (Kerekes et al., 2013; Nazzaro et al., 2013). EOs already are widely used because of their antibacterial properties—in, for example, antiseptics and preservatives (e.g., food preservation) (Burt, 2004). The main advantages of EOs used as antimicrobials are that their mammalian toxicity is low, methods of obtaining of EOs are relatively easy, and they are environmentally friendly as they are degraded quickly in water and soil (Kerekes et al., 2013).

As mentioned before, Kalia et al. suggested, that QSI should be preferably small molecules with longer side chains than the natural AHLs. EOs are small molecules and there can be found constituents with longer acyl chains than those of some natural AHLs (Kalia, 2013). When looking at examples of EO components like the acyclic monoterpene citral compared to a typical AHL-AI, namely N-hexanoyl-homoserine lactone (C-6-HSL) produced by *C. violaceum*, there is a clear resemblance, from which one might assume that EO components could bind to AHL receptors and therefore function as antagonists. Especially those acyclic monoterpenes show resemblance due to their partly branched side chain as well as sesquiterpenes like farnesol; whereas on the first sight, cyclic components like menthol or the phenylpropanoid eugenol show less resemblance.

Several EOs have been tested on their anti-QS activity, most of them using strains of *C. violaceum* or *P. aeruginosa* as sensor organisms. This means research on EO as QSI is currently concentrated on AHL-based quorum sensing. It is not without relevance to mention that especially biofilms are problematic in human infections or in industry. Current QSI tests on *C. violaceum* concentrate only on the mechanism of signal inhibition causing less pigment production, not including information on biofilm formation. Those assays can therefore be considered as rough screening methods for inhibitors of luxRI-based QS, but it is not possible to give a presumption on whether the EO would inhibit biofilm formation or inhibit QS in other organisms with similar QS systems too. Research done on the QS inhibitory effect of EOs on *P. aeruginosa* are more advanced, as they quantify virulence-factor expression/swarming motility and biofilm formation. As it is known that QS plays a major role in these areas and even some of the associated genes are known, this seems to be a more applicable model, as the tested factors are jointly responsible for the pathogenicity of *P. aeruginosa* infections.

16.8.1 Principle of QSI Activity Testing of EOs

EOs are chosen, and usually, different concentrations are being tested. As many of the EOs are known to inhibit bacterial growth, the research team often used subMIC-concentrations to detect QS independently from growth inhibition. Other methods to observe QS independently are concurrent cell viability tests.

In order to identify the components of the used EOs, most of the research teams did GC-MS analyses. As EOs differ in their composition depending on factors such as collecting point, growth conditions, and production method, this is an important step before the following testing of the oils. An alternative method is to use EOs complying to a pharmacopoeia, as quality and percentages of components have to meet certain criteria. In order to prove an inhibitory activity of the oils, different bacterial strains with known QS systems can be used, showing a certain reaction when QS is inhibited. This can be, for example, less pigment production (*C. violaceum*), reduced virulence-factor production (*P. aeruginosa*), or reduced fluorescence.

16.8.2 Materials and Methods

This section gives an overview on possible methods to detect QS inhibition of EOs and is based on those used in the discussed research results in the following section; however, no warranty on completeness or accuracy of the given data is taken over. Primary, sensor strains and their belonging QS-inhibition detection methods are listed. A more detailed characterization of some strains is given subsequent to the table. Afterward, common QS inhibition assays are presented (see Table 16.1).

16.8.2.1 Sensor Strains

16.8.2.1.1 Chromobacterium violaceum Strains

ATCC12472: This wild-type strain produces and responds to the AHLs by producing the pigment violacein. In the presence of a QS inhibitor, violacein production is inhibited or decreases, which can be evaluated as QSI activity. The major AHL is 3-hydroxy-C10-HSL (Morohoshi et al., 2008).

VIR07: VIR07 is a mutant strain of ATCC12472 in which the gene for AHL production (cviI, LUXi-homologue) was deleted. Violacein production is inhibited by C6–C8 HSLs and C4HSLs. Violacei production is induced by long chain HSLs C10–C16 (mainly C10) (Morohoshi et al., 2008).

ATCC31532: This wild-type strain has N-hexanoyl-L-homoserine lactone (C6-HSL) as QSM and produces violacein in the presence of this QSM (production of antibiotics, protease, chitinase are controlled by QS too) (McLean et al., 2004).

CV026: CV026 is a mutant of the wild-type strain ATCC31532 which is unable to produce AHL-molecules on its own, and therefore, exogenous C6-HSL molecules have to be added. Violacein production is induced by C4–C8 HSLs, whereas C10–C14 HSLs inhibit it (McClean et al., 1997; Morohoshi et al., 2008).

CV ATCC31592: Alvarez et al. used this strain as a biosensor to test the EO of *Hyptis dilatata* via violacein production (Alvarez et al., 2015).

TABLE 16.1
Sensor Strains

Sensor Strain	Detection Method
C. violaceum (CV) ATCC12472	Violacein
CV VIR07	Violacein
CV ATCC31532	Violacein
CV ATCC 31592	Violacein
CV026	Violacein
CV CIP 103350T	Violacein
QSIS 1	Green fluorescence
Pseudomonas putida F117 + (pRK-C12)	Green fluorescence
E. coli MT102 + (pJBA132)	Green fluorescence
E. coli [pSB401]	Luminescence
E. coli [pSB1075]	Luminescence
E. coli DH5α (pJN105L + pSC11)	β-galactosidase
E. coli + pKDT17	β-galactosidase
P. aeruginosa PAO1	Virulence factors, biofilm, swarming motility
P. aeruginosa qsc 119	β-galactosidase
P. aeruginosa HT5	Virulence factors, biofilm
P. aeruginosa PAO1-JP2	Virulence factors
Aeromonas hydrophila	Virulence factors, biofilm
Staphylococcus aureus ATCC 25923	Transcriptional level of RNAIII
Vibrio parahaemolyticus ATCC 17802	Biofilm, virulence factors, swarming/swimming motility

CV CIP 103350T: This is a wild-type strain producing violacein in the presence of this QSM.

QSIS 1: This strain is one of the constructed strains used for the QSI selector method A. It was developed by Rasmussen et al. (2005). Based on a *luxI-gfp*-reporter, the *luxI* promotor region was fused to *ohlA* from *S. liquefaciens*. If a QSI is present (and 3-oxo-C6-HSL), cells can survive (Rasmussen et al., 2005).

16.8.2.1.2 Pseudomonas putida F117 (pRK-C12) + E. coli MT102 (pJBA132)

These strains are not able to produce AHL on their own. They sense AHLs with a sensor plasmid. This plasmid expresses a promotor controlled by AHL and is fused to a *gfp* gene. The LuxR-receptor protein is specific for long-chain AHLs (C12) in *P. putida* and for short-chain AHLs (C6) in *E. coli* (Jaramillo-Colorado et al., 2012).

16.8.2.1.3 E. coli (pSB401) and (pSB1075)

These strains contain an N-acyl-HSL-sensor plasmid, constructed by Winson et al. (1998).

PsB401: The plasmid contains a fusion of the genes *luxRI′* (*Photobacterium fischeri*) and *luxCDABE* (*Photorhabdus luminescence*). The strains show bioluminescence in response to short-chain AHLs.

PsB1075: The plasmid contains a fusion of the genes *lasRI′* (*P. aeruginosa PAO1*) and *luxCDABE* (*Photorhabdus luminescence*). The strain shows bioluminescence in response to long-chain AHLs (Winson et al., 1998; Yap et al., 2014).

16.8.2.1.4 E. coli DH5α harboring pJN105L and pSC11

This strain contains *lasR* and *plasI:lacZ* and therefore can be used as bio-reporter strain, as their activity depends on the presence of AHL molecules. Kalia et al. (2015) used this strain to detect 3-oxo-C12-HSL produced by PAO1.

16.8.2.1.5 E.coli Harboring pKDT17

This is another sensor strain containing a plasmid with a *lasB′-lacZ* reporter showing lasB promoter activity and *lasR* being controlled by the lac promoter (Pearson et al., 1994). Again, β-galactosidase production is dependent on the presence of AHL molecules.

16.8.2.1.6 P. aeruginosa: Strains

PAO1: PAO1 is a wild-type strain of *P. aeruginosa*; its QS systems were explained previously.

PAO1-JP2 is an isogenetic mutant strain of wild-type PAO1. In this strain, two QS systems are silenced: *lasI* and *rhlI* (Ganesh and Vittal Rai, 2016).

qsc 119: This strain is a mutant that is not able to produce AHLs on its own. When external AHL is added, the strain reacts by producing β-galactosidase (Luciardi et al., 2016).

16.8.2.1.7 HT5 ("Hospital Tucumán 5")

This strain was used by Luciardi et al. (2016) to test the effect of mandarin EO on QS and virulence-factor production. It is a strain isolated from a patient with food poisoning which is resistant to several antibiotics. As reference strain, they used *P. aeruginosa* ATC 27853.

16.8.2.1.8 Aeromonas hydrophila

This pathogen has an LUXRI-homologous QS-system and produces C-4 and C-6-HSL (mainly C4-HSL) as QS-molecules. *A. hydrophila* builds biofilms on different surfaces. Lynch et al. (2002) showed the correlation between QS and biofilm structure, showing less-differentiated biofilm architecture in AHL-deficient *ahyI* mutant strains, although other differences (height, surface coverage) were not observed.

16.8.2.1.9 Pseudomonas flourescens KM121—an Indirect Sensor Strain

P. fluorescens can be used in combination with CV026. CV026 produces violacein if AHL is produced by *P. fluorescens*. When examining the AHL profile, Myszka et al. found out that this strain produces C6-HSL, 3-oxo-C6 HSL, and 3-OH-C8-HSL (Myszka et al., 2016).

16.8.2.2 QS-Inhibition Detection Assays

16.8.2.2.1 Disc Diffusion Assay

McLean et al. (2004) developed a simple screening test for bacteria or plant material (leaf, stem, flower) for its QS-inhibitory potential (AHL-dependent signaling). For this method, they used either *Pseudomonas aurofaciens* 30–84, CV ATCC 12472, or CV026 as indicator strains. They basically incubated the test bacteria or placed the plant material on a Petri plate (Lysogeny broth LB) and overlaid it with previously incubated indicator strains (LB broth/LB broth + antibiotics). After incubating, the color was examined again. A lack of pigment production indicated QSI activity of the used bacteria/plant material. Hence, QSI was detected by a colorless, opaque halo (McLean et al., 2004).

Adonizio et al. (2006) adjusted this method for testing different plant extracts. They used CV12472 and CV026 as sensor strains and referring to a standard disc diffusion test by Bauer et al. (1966). The extracts (20 μL) were loaded on sterile paper discs (6 mm diameter) and placed on the overnight cultures of indicator strains. They were incubated overnight. QSI was measured via measuring the colorless but viable cells around the disc (in mm). In any case, a possible antimicrobial effect should be included in the investigations to make sure QS inhibition is independent from growth inhibition. This agar-diffusion method, slightly adjusted (incubation time, diameter of the discs, wells instead of discs, volume of EO) became a common and fast way to test QSI potential of EOs (Adonizio et al., 2006; Choo et al., 2006).

16.8.2.2.2 Flask Incubation Assay

To determine the QSI potential of vanilla extract, Choo et al. (2006) used this assay in 2006. CV026 was incubated for 16–18 h and then inoculated in Erlenmeyer flasks containing either LB/LB with HHL (N-hexanoyl-HSL) or LB with HHL and the potential QSI. Afterwards, flasks were incubated in a shaken incubator for 24 h (27°C, 150 rev/min). QSI was measured then via the Blosser and Gray method to quantify violacein production by absorbance-measurement at 585 nm. They obtained violacein after the flask incubation assay by taking 1 mL of each flask, vortexing it and dissolving the resulting insoluble pellet containing violacein in DMSO after discarding the supernatant (Blosser and Gray, 2000; Choo et al., 2006).

Snoussi et al. (2018) modified the classic flask incubation assay by using 96-well plates instead of Erlenmeyer flasks. Wells were filled with LB broth inoculated with ATCC 12472. After incubation for 24 hours, growth (optical density) in each well was measured by a multimode plate reader (OD_{600}). In the next step, contents of the wells were transferred to Eppendorf tubes where they were centrifuged in order to collect the cells. Violacein was then extracted with water-saturated n-butanol by centrifugation, violacein was separated from cells, and was then quantified spectrophotometrically at OD_{585} (Snoussi et al., 2018).

16.8.2.2.3 The QSI Selector (QSIS)

This assay, developed by Rasmussen et al. (2005) is another method making it possible to screen for QSI activity of EOs. Method A, established in *E.coli* is based on a gene encoding a lethal protein being fused to a promotor controlling QS. This leads to cell death in the presence of AHL signal molecules; therefore cell growth is only possible when QSIs are present. Method B bases on a repressor-controlled antibiotic-resistance gene. This repressor is controlled by a QS-regulated promotor. If AHL is present, the expression of the gene is inhibited; leading to an antibiotic effect and cell growth is inhibited. In presence of QSI, bacteria can grow. To detect QSI, they used, for example, a *luxI-gfp* reporter in QSIS1 (see "Sensor Strains" section) (Rasmussen et al., 2005).

16.8.2.2.4 Fluorescence-Based QSI Detection

Jaramillo-Colorado et al. (2012) used this method for the detection of QSI, which is based on the research of Wagner-Döbler et al. (Wagner-Döbler et al., 2005; Jaramillo-Colorado et al., 2012). The group screened several α-proteobacteria for AHL-production by using a sensor strain which carries plasmids consisting of *gfp*-reporter constructs (*luxR-gfp*). LuxR works as a reporter protein, initiating fluorescence by activating the transcription of gfp (green fluorescence protein). They used two sensor strains: *P. putida* F117 (pRK-C12) and *E. coli* MT102 (pJBA132) (see "Sensor Strains" section). Fluorescence was measured at an excitation wavelength of 485 nm.

16.8.2.2.5 Bioluminescence-Based QSI Detection

Winson et al. (1998) developed this method to investigate AHL-based QS. The authors constructed reporter strains with plasmid vectors containing *luxR*, *lasR*, or *rh1R* in fusion with *lux-I* homologues and *luxCDABE* from *Photorabdus luminescens* ATCC29999. In the presence of specific AHLs, the strains show bioluminescence via expression of *luxCDABE.*

16.8.2.2.6 β-Galactosidase-Based QSI Detection

An example of using β-galactosidase with the encoding gene LacZ was an assay carried out by Luciardi et al., they used the reporter strain *P. aeruginosa qscl19* to detect inhibition of QS by mandarin EO. External AHLs were obtained from *P. aeruginosa* ATCC27853 or HT5, being wild type strains. After incubating with different concentrations of EO, β-galactosidase was measured spectrophotometrically (Luciardi et al., 2016).

16.8.2.2.7 Quantitative Real-Time PCR (RT-PCR)

Sharifi et al. (2018) used RT-PCR to detect QS inhibition in *Staphylococcus aureus* strains by detecting expressional changes of QS genes (RNAIII). To do so, biofilms were grown with and without EOs. RNA was then isolated by using commercial RNA extraction and purification kits. RNA was then reverse transcribed to cDNA. This cDNA was used for quantitative real-time PCR to detect expression levels of *RNAIII.* Primer pairs used for RT-PCR assay are, for example, *hld* gene (delta-hemolysin gene), which is part of the *RNAIII* gene and 16S (rRNA) (Sharifi et al., 2018).

16.8.2.2.8 Swarming Motility of P. aeruginosa.

A method to test swarming motility was described by Vattem et al. (2007); they used 40 μL of the test substances (dietary phytochemical extracts) on 0.3% LB-agar plates. Then they inoculated the plates with PAO1. After incubating, they measured the swarming motility by determining the diameter of the motility swarms. Another assay, being the basis for Kalia et al. (2015), who tested the effect of cinnamon oil on swarming motility of PAO1 was done by Krishnan et al. (2012). The authors cleaned the test extract on pre-warmed swarm plates containing Bacto-agar, glucose, and yeast extract. They point-inoculated the plates with PAO1. After incubation, swarming was measured by lengths of dendrites on the swarm plates (Vattem et al., 2007).

16.8.2.2.9 Swimming Motility of P. fluorescens/PAO1

Myszka et al. (2016) used this assay to determine flagella-based motility of *P. fluorescens* by measuring the distance bacteria move from an inoculation point in a certain incubation period. Swimming motility is a bacterial movement from individual cells, rotating their flagella to move in liquid environments. Swarming however describes a multicellular movement of bacteria in biofilms over solid surfaces through flagella. Bala et al. (2011) described a similar method to detect swimming motility of PAO1, assessing the zone diameter of swim (in mm). Similar assays as described above were used to detect swimming and swarming motility in *Vibrio parahaemolyticus* (Packiavathy et al., 2013).

To measure the impact of EOs on QS-dependent virulence-factor production, different assays can be used, mainly based on an appropriate chemical processing followed by absorbance measurements.

16.8.2.2.10 LasA Staphylolytic Assay of P. aeruginosa

P. aeruginosa produces several extracellular protease enzymes which are related to the virulence of the pathogen. One of them is the QS-controlled protease LasA, which is involved in degradation of the host protein elastin and shows staphylolytic activity. Consequently, LasA- inhibition can be assessed by measuring the efficacy of culture supernatants of *P. aeruginosa* to lyse boiled *S. aureus* cells (Ganesh and Vittal Rai, 2016).

16.8.2.3 Biofilm Inhibition

The following list includes examples of biofilm assays as they were carried out in the results discussed in the following sections and does not claim to be complete.

16.8.2.3.1 Biofilm Visualization by Crystal Violet Assay

Crystal violet assay is an indirect quantification method of death cells. Crystal violet binds to protein and DNA on detached cells. Reduced cell attachment because of cell death leads to reduced violet staining in a culture. This simple assay can be used for the effect of EO on biofilms (Feoktistova et al., 2016).

16.8.2.3.2 Biofilm Assay by Le Thi

This assay was used by Myszka et al. (Myszka et al., 2016) to detect adhesion of *P. fluorescens* on stainless-steel plates. After removing unattached cells with PBS (phosphate-buffered saline), bacteria are stained with acridine-orange and then observed in a fluorescence microscope to characterize the biofilm (Le Thi et al., 2001).

16.8.2.3.3 Microtiter Plate Assay (MtP) to Detect Inhibition of Biofilm Formation

A certain inoculum of bacteria is added to 96-well plates containing test agents in different concentrations and TSB (tryptic soy broth) and incubated for 24 hours. After that, the biofilm is rinsed twice with sterile phosphate buffer saline (PBS) to remove non-adherent cells. The biofilm equaling attached bacteria are then stained with safranin for 10 min and rinsed afterward to remove the dye. After adding a sodium chloride solution and sonicating the plate to re-suspend the bacterial cells, absorbance is measured at OD490. Biofilm reduction is then calculated by comparing the results with a negative control. The same assay can be used to detect the effect of preformed biofilms. In that case, bacteria is grown in the wells for 24 h and EOs are added afterward. After another incubation time of 24 h, the biofilm is measured using a microtiter plate reader as described above (Sieniawska et al., 2013).

16.8.2.3.4 Scanning Electron Microscopy (SEM)

SEM is a common method to visualize biofilms. There are slightly varying methods to prepare samples. For example, bacteria can be grown in microtiter plates with glass coverslips. They are then incubated with potential biofilm inhibitors for certain amounts of time. Then, samples are fixed with buffered glutaraldehyde and dehydrated afterward. After drying, samples are glued on stubs and coated with gold in order to examine them by SEM (Sharifi et al., 2018).

16.8.2.4 Other Assays

16.8.2.4.1 C. Elegans Paralytic and Fast-Killing Assay

Depending on the medium *C. elegans* is grown on, *P. aeruginosa* either kills the nematode within a few hours (fast killing, in rich, high-salt media) through mainly QS-controlled virulence factors or via paralysis (brain–heart infusion, BHI) regulated by LasR and Rh1R QS regulators. By using PAO1-JP2 as a control (*lasI* and *rhII* mutant), it is possible to relate survival/paralysis of infected *C. elegans* with QS inhibition of its pathogen (Ganesh and Vittal Rai, 2016).

16.8.3 Research on EOs as QSI

In this chapter, research done on EOs will be presented, sorted by publishing year. To achieve a better overview and comparability, contents and outcomes will be presented for each study. After that, information on the used EOs (components, extraction method, and concentrations of EOs, if specified), sensor strains, and methods are listed. Detailed data on QS inhibition of the oils are then shown in charts. If additional research was done not being directly traced to QS, these results are explained after the charts. Tables summarizing all the discussed papers will be shown in the next section. The presented research results are concentrated on QS-related results; therefore, other results concerning, for example, the effect on bacterial growth are not discussed. Many investigations include the effect of EOs on biofilm formation. These results are mentioned but it is important to consider that biofilm formation is a complex mechanism where QS is only one of many signaling systems playing a role here. Hence, relating to the studies discussed in this work, one must not conclude a direct cause and effect relationship between QS inhibition of the oils and decreased biofilm formation; it is often more likely an inhibition caused by EOs, whereas the mechanism of inhibition is not known or not investigated in respective papers.

16.8.3.1 Clove, Cinnamon, Peppermint, and Lavender Oil

Khan et al. evaluated the anti-QS activity in 20 different EOs and showed QSI-activity in four of them. The inhibition zone was, with clove EO, 19 mm (CV12472) and 17 (CV026 + C-6-AHL); with cinnamon EO, 12 and 11 mm, resp.; with peppermint EO, 11 and 10 mm, resp.; and with lavender EO, 11 and 10 mm, resp. Negatively tested EOs were *Apium graveolens, Citrus limon, C. paradisi* and *C. sinensis, Cymbopogon citratus* and *C. martini, Foeniculum vulgare, Eucalyptus* sp., *Myristica fragrans, Olea europea, Petroselinum crispum, Rosmarinus officinalis, Santalum album, Thymus vulgaris, Trachyspermum ammi, Zea mays*, and *Zingiber officinale* (Khan et al., 2009).

EOs: subMICs
Sensor strains: CV12472, CVO26
AHL producing strain: CV31532 (C6-AHL)
Performed assay: Disc diffusion assay: (CV 12472)/(CV026 + CV12472)

16.8.3.1.1 *Further investigations on clove oil*

In addition to these analyses, Khan et al. performed further research on clove oil (0.04–0.20 (v/v%). In order to prove that the violacein production inhibition is not caused by the antibiotic effect of clove oil, quantitative violacein measurements in clove-oil treated and non-treated cultures were done concurrently with cell viability tests in both of the cultures. The test showed that clove oil works as QSI up to a concentration of 0.12%; higher concentrations lead to growth inhibition. Using GC-MS, the team revealed eugenol as the main component of clove oil but they discovered, that pure eugenol has no effect on the inhibition of pigment production in CV026. Hence, other components like α/β-caryophyllene could be the effective substances (Khan et al., 2009).

16.8.3.1.2 *Effect of clove oil on swarming motility in PAO1*

Kahn et al. analyzed the effect of clove oil on swarming motility in a *P. aeruginosa strain.* The research revealed that subMICs of clove oil show a concentration-dependent effect on the reduction of swarming motility in PAO1, whereas the main component of clove oil, eugenol, exerted no effect (Khan et al., 2009).

16.8.3.2 Rose, Lavender, Geranium and Rosemary Oil

Szabo et al. (2010) investigated the effect of different EOs on bacterial growth and QS in CV026 using either *E. coli* ATTC 31298 or Ezf 10 as AHL-producing strains. The following EO were used, most of them being already known as inhibitors of bacterial growth: rose, lavender, chamomile, orange, eucalyptus, geranium, citrus, and rosemary oil. QS inhibition was proved in rose, lavender, geranium, and rosemary when using pure oil in both combinations, whereas eucalyptus oil only inhibited violacein production in the CV026 + *E. coli* ATTC 31298 combination and citrus oil only inhibited the CV026 + Ezf 10–17 combination. The most potent QS inhibitor was rose oil. Interestingly, rose oil showed a comparable QSI activity in the 10% dilution as the concentrated oil on CV026 + *E. coli* ATTC 31298 combination, whereas the other tested oils had a higher QSI potential when being higher concentrated (Szabo et al., 2010). Negatively tested oils were *Geranium robertianum, Citrus sinensis*, and *Rosmarinus officinalis* (see Table 16.2).

EOs: Quality of the Hungarian pharmacopoeia (Ph. Hg. VII.), concentrated or diluted (10%) in DMSO
Sensor strain: CV026
AHL-Producing strains: *E. coli* ATTC 31298 and Ezf 10–17 as, a strain isolated from a grapevine crown gall tumor
Performed assay*:* Disc diffusion assay

TABLE 16.2
Inhibition of QS Regulated Violacein Production in *C. violaceum* Strains (Inhibition Zone Size, in mm)

EO	CV026 + *E. coli* ATTC 31298 10%/100% EO	CV026 + Ezf 10–17 10%/100% EO
Rosa damascena	15–20/18 mm	10/20 mm
Eucalyptus globulus	0/6	3/0
Citrus sinensis	0/0	2.5/16
Rosmarinus officinalis	0/18	3/18
Lavandula angustifolia	0/10	5/15
Geranium robertianum	0/13	3/15

Source: Szabó, M. Á. et al. 2010. *Phytother. Res.* 24: 782–786.

16.8.3.3 Oils of Piper Species

Olivero et al. (2011) investigated the anti-QS activity of different piper species EOs (*P. bogotense*, *P. brachypodom*, and *P. bredemeyeri*). In this study, an IC_{50} for QS inhibition was defined, representing the concentration of EOs leading to a 50% reduction of violacein production compared to the amount produced by a fully induced CV026 strain. The investigation revealed the highest anti-QS-activity in *P. bredemeyeri*, following *P. brachypodom* and *P. bogotense*. The EO concentrations for IC_{50} values also had a minor effect on cell-growth inhibition, but these results did not differ significantly from the control cells. The research group concluded QSI was basically independent from cell growth inhibition (Olivero et al., 2011).

16.8.3.3.1 Connection between Molecular Structure and QS Inhibition

The main components of the oils (<10%) are listed below:

- *P. bredemeyeri*: sabinene/β-pinene and α-pinene
- *P. brachypodom*: *trans*-β-caryophyllene and caryophyllene oxide
- *P. bogotense: trans*-sabinene hydrate and α-phellandrene

The major components of *P. bredemeyeri* oil (with the exception of α-pinene which also can be found in *P. bogotense*) do not appear in other *Piper* species and may therefore be partly responsible for the high activity of this oil. Olivero et al. also made the hypothesis that those components may be competitive inhibitors because of their similarity in size and polarity with HHLs (Olivero et al., 2011).

EOs: Plant material was collected in different Columbian states; oils were produced by microwave-assisted hydro-distillation. For the EO of *P. bogotense* and *P. bredemeyeri*, herbs were used; for *P. brachypodom* oil, the whole plant was processed. Different dilutions with DMSO were tested. To evaluate the connection between EOs and their constituents, the team refers to literature in which components of the oils have been analyzed (Olivero et al., 2011).

The QS regulated violacein production in *C. violaceum* revealed an IC_{50} (μg/mL) (concentration of the EO leading to a 50% reduction of the production) with the EO of *P. bogotense* of 93.1 (77.0–112.6), with the EO of *P. brachypodom* of 513.8 (429.6–614.4) and with the EO of *P. bredemeyeri* of 45.6 (41.2–50.5) (Olivero et al., 2011).

Sensor Strain: CV026 + C6-HSL
Performed assay: Flask incubation assay

16.8.3.4 EOs from Columbian Plants

Jaramillo-Colorado et al. (2012) investigated the anti-QS activity of different Columbian plants on a *Pseudomonas putida*, as well as on an *E. coli*, strain. As the chemical composition of *Lippia alba* oils depends on the collection location of the plants, tests were carried out with oils from different collection sites. Additionally, the team differed between oils extracted by means of microwave-assisted hydro-distillation (MWHD) or hydro-distillation (HD). In concentrations of 1.2 mg/mL, anti-QS activity was measured for some of the *L. alba* oils, *Ocotea* sp., and *Elettaria cardamomum* in *P. putida* (the highest activity inhibition was reached by a the geranial/neral chemotype of *L. alba*), whereas only two of the limonene/carvone chemotypes showed a low inhibitory activity of 9% and 1%, respectively). All of the oils were active QS inhibitors in *E. coli*, the most active was an oil of *L. alba* chemotype 2 and *Ocotea* sp. (Jaramillo-Colorado et al., 2012).

16.8.3.4.1 Connection between Molecular Structure and QS Inhibition

When taking a look at the main components of the investigated oils, it is apparent that CT1 is a more active QS inhibitor of C12-HSL-regulated QS. The research group hypothesized that this is maybe because of the linear molecular structure of the components. Limonene and carvone are cyclic molecules and therefore show more similarity to C6-AHLs. QSI of different constituents of EOs were also investigated separately. Carvone, geranyl-acetate, α-pinene ,and citral were investigated.

TABLE 16.3
Inhibition of QS Regulated Fluorescence in *E. coli* and *P. putida* Strains (% of QS-Decrease)

EOs	*P. putida (C12-AHL)*	*E. coli (C6-AHL)*
Lippia alba (1) CT 1	65%	18%
L. alba (2) CT 2	–	15%–84%
L. alba (3) CT 2	1%	72%–78%
L. alba (4) CT2	–	61%–62%
L. alba (5) CT2	9%	44%–72%
Swinglea glutinosa	–	28%–73%
Myntotachys mollis	–	56%–65%
Ocotea sp.	25%	71%–76%
Elettaria cardamomum	31%	21%–22%
Zingiber officinale	–	34%–69%

Source: Jaramillo-Colorado, B. et al. 2012. *Nat. Prod. Res.* 26 (12): 1075–86.

Only citral was active in *P. putida* (long-chain AHL); all of the other molecules only showed activity in *E. coli* (short-chain AHL.). This result coincides with the tests on the EOs as, again, the linear citral inhibits long-chain-AHL-dependent QS, and branched structures only inhibit QS in the short-chain-AHL-controlled *E. coli* strain (see Table 16.3) (Jaramillo-Colorado et al., 2012).

EOs: Several *L. alba* oils 1–5 (MWHD/HD) belonging to chemotype 1 with the main components geranial and neral or chemotype 2 with limonene and carvone as main components; *Swinglea glutinosa* (MWHD) *Myntotachys mollis* (HD) Ocotea sp. (MWHD), *E. cardamomum* (HD), and *Zingiber officinale* (HD) components of the oils were characterized via GC-MS
Sensor strains: *E. coli* + pJBA132 (C6). *P. putida* + pRK12 (C12)
Performed assay: Fluorescence-based QSI assay on *P. putida* + *E. coli*

16.8.3.5 Tea Tree and Rosemary Oils

Alvarez et al. (2012) tested tea tree oil and rosemary oil on their QSI potential using disc diffusion assay and flask incubation assay. Tested concentrations were 0.125 μg/mL, 0.5 μg/mL, and 1 μg/mL. Interestingly, QS inhibition was inversely dose-dependent, meaning higher concentrations were less effective. The lowest concentration inhibited violacein production by about 80%. MQSIC (minimum QS inhibitory concentration) was estimated and revealed a high QSI activity of both oils, whereas MQSIC concentrations did not affect bacterial growth (Alvarez et al., 2012). The EO of *M. alternifolia* exerted a QS-inhibition zone of 15–19 mm, a QS decrease of 80%, and with MQSI (μg/mL), a value of 0.21 (0.13–0.29). The inhibition zone with the EO of *R. officinalis* was 15–19 mm, the QS decrease was 80% and the MQSI value was 0.21 (0.14–0.27).

EOs: *Melaleuca alternifolia* (Maiden and Betche) Cheel, *Rosmarinus officinalis*
Sensor Strain: *C. violaceum* ATCC 12472
Performed Assays: Disc diffusion assay and flask incubation assay

16.8.3.6 Clary Sage, Juniper, Lemon, and Marjoram Oils

Kerekes et al. (2013) investigated EOs of clary sage, juniper, lemon, and marjoram on their QSI potential. The decision to use those oils was made because of their lack of phenolic compounds. The oils themselves, as well as the main constituent of each oil, were tested. All of the oils inhibited QS in the tested bacterial strains, except lemon and limonene showing only very little or no effect. Interestingly, the parent oils showed better inhibitory activity than their major components, opposite to the biofilm inhibitory potential that was also tested in this study (Kerekes et al., 2013). The inhibition zone with 1/2/3 μL EO of *S. sclarea* amounts to 10/20/25 mm and with 1/2/3 μL of the main constituent, namely linalool, 10/15/20 mm. The inhibition zone values with 1/2/3 μL EO of *J. communis* were 0/15/20 mm, and with the main constituent, namely α-pinene, were 0.1/0.1/0.1 mm. Finally, the inhibition zone with the same amounts of the EO of *O. majorana* comes to 10/15/20 mm and with the main component, namely terpinen-4-ol, to 0/15/20 mm.

EOs: 1/2/3 μL, components were determined by the producer of the oils
Sensor strain: CV6269
Performed assay: Disc diffusion assay: CV6269
Negatively tested EO: Citrus lemon (limonene) (Kerekes et al., 2013)

16.8.3.6.1 Effect on Biofilm Formation in Different Organisms

Biofilm-inhibition in *Bacillus cereus* var. mycoides 0042, *E.* coli 0582, and *P. putida* 291 and in the yeast *Pichia anomaila* 8061Mo was detected via crystal violet assay; additionally, a structural analysis of the biofilm modifications with scanning electron microscopy (SEM) has been done. The team revealed that only marjoram EO has an inhibitory effect on biofilms of *B. cereus*, whereas all of the tested oils inhibit biofilm formation in *E. coli*. Biofilms formed by *P. anomaila* are inhibited by all tested oils and also by α-pinene and terpinen-4-ol, whereas only marjoram and its main component terpinen-4-ol inhibited biofilms in *P. putida*; the other oils had too-high MICs to allow further investigations. When examining mixed biofilm (*E. coli* and *P. putida* 1:1), marjoram and cinnamon oils showed inhibitory effects (Kerekes et al., 2013).

16.8.3.6.2 Connection between Molecular Structure and QS Inhibition

The oils investigated in this study mainly contain terpenoids. As mentioned above, their inhibitory potential is comparable; only lemon oil did not show any activity. When correlating these results to the main components of the oils, it seems that terpene alcohols (terpineol, linalool) possess a higher activity than monoterpenes (limonene and α-pinene). Then again, it is apparent that the multi-component character of EOs should be included in such observations as α-pinene, being that the main component of juniper oil has no significant QS-inhibiting effect by itself, whereas the oil was proved to be a potent QS-inhibitor, leading to the conclusion that other components have to be responsible for this effect (Kerekes et al., 2013).

16.8.3.7 Clove, Rose, Chamomile, and Pine Turpentine Oils

Eris and Ulusoy (2013) analyzed the QSI potential of different EOs and determined the highest inhibitory potential in clove, rose, chamomile, and pine turpentine EOs. They started their investigations by screening the oils by the QSIS method, and only the oils showing the highest activity in this assay were submitted to further tests. Forty-two essential and edible oils were screened by the QSIS method, 12 of them showed positive results. The activity was classified as weak (*), medium (**), or strong (***) positive, or negative, whereas the most active substances underwent further tests. The team performed an agar-plate assay, but results are not discussed here, as only pictures were published, indicating the QSI activity of rose, clove, pine turpentine, and chamomile EO on CV026; as quantification of the results is missing, those values are not mentioned in the Table 16.4. The flask incubation method revealed the highest QS inhibition for clove and rose oil. Concerning biofilm formation, the study showed a significant reduction in PCI (*P. aeruginosa* clinical isolate) for rose, clove, pine turpentine, and chamomile oil, but interestingly, the oils did not show a significant change in PAO1. Negatively tested EOs were *Cedrus* sp., *Citrus aurantium*, *C. bergamia*, *C. limon*, *C. reticulata*, *Eucalyptus globulus*, *Laurus nobilis*, *Lilium candium*, *Melissa officinalis*, *Mentha piperita*, *Myroxylon pereirae*, *Nigella sativa*, *Pinus* sp., *Punica granatum*, *Rosmarinus officinalis*, *Santalum album*, *Viola odorata*, and *Vitis vinifera*. Negatively tested EOs on biofilm-formation in PAO1 were *Eugenia caryophyllata*, *Matricaria recutita*, *Pinus sylvestris*, and *Rosa damascena* (Eris and Ulusoy, 2013).

TABLE 16.4
Inhibition of QS-Regulated Fluorescence in QSIS1; Violacein Production in CV026, VIR07, and CV12472 (% of QS-Decrease); and Biofilm-Inhibition in PCI (% of Biofilm-Decrease) by EOs

EO	QSIS1	CV026	VIR07	CV12472	PCI
Rosa damascena	***	80%	70%	43%	53%
Eugenia caryophyllata	***	80%	72%	39%	52%
Matricaria recutita	**	75%	61%	32%	58%
Cinnamomum zeylanicum	**				
Cymbopogon nardus	*				
Juniperus communis	*				
Myrtus communis	*				
Pinus sylvestris	**	67%	65%	25%	59%
Salvia officinalis	*				
Thymus vulgaris	*				
Vanilla fragrans	**				
Lavandula angustifolia	*				

Source: Eris, R., S. Ulusoy. 2013. *J. EO Bear.* Plants 16 (2): 126–135.

EO: Ethanol dilutions of filter-sterilized EOs; MICs were not evaluated but biomonitor strain cell counts treated with EOs were compared to those without treatment, and no significant deviations could be determined.
Sensor strains: QSIS1, CV026, VIR07, ATCC12472, PAO1, PCI
Performed assays:

- QSIS: QSIS1
- Disc diffusion assay: CV026 + C6AHL, VIR07 + C12AHL, ATCC1247
- Flask incubation assay: CV026, VIR07, ATCC12472
- Biofilm-inhibition via crystal violet assay: PAO1, PCI

Inhibition of QS-regulated fluorescence in QSIS1, violacein production in CV026, VIR07, and CV12472 (% of QS-decrease) and biofilm-inhibition in PCI (% of biofilm decrease) by EOs (Eris and Ulusoy, 2013)

16.8.3.8 Clove Oil

Husain et al. (2013) investigated clove oil, which also was already proved to be a potent QSI in *C. violaceum* strains and an inhibitor of swarming motility in PAO1 in 2009 (Khan et al., 2009). To understand the inhibition of clove oil more precisely, the team used the *E. coli* strain MG4 pKDT17 to find out if the QSI effect of clove oil is due to inhibition of the LasR/RhlR system. They proved this assumption as AHL-concentrations were decreased significantly (65%), being shown by the reduced β-galactosidase activity of the sensor strain. The team continued the research on the effects on PAO1, proving an inhibitory effect on QS-related virulence factors as well as on swimming motility (Husain et al., 2013). SubMICs of clove oil additionally led to enhanced survival of a PAO1 infected *Caenorhabditis elegans*. Clove-oil treatment led to decreased biofilm formation and virulence-factor production (protease, exopolysaccharide [EPS]) in *A. hydrophila*. The inhibition of QS-regulated ß-galactosidase production in PAO1 by clove oil was 65% of QS decrease (Husain et al., 2013).

EO: subMICs of *Syzigium aromaticum* EO
Sensor strains: PAO1, *Aeromonas hydrophila WAF-79* (isolated from hospital wastewater), *E. coli* MG4 pKDT17

Performed assays:

- β-galactosidase—assay (*E. coli* MG4 pKDT17*)* to quantify AHLs produced by PAO1
- Virulence-factor production: protease, chitinase, pyocyanin LasB, EPS
- Swimming motility of PAO1
- Biofilm assay via crystal violet staining: PAO1
- Survival of *C. elegans* infected with PAO1

16.8.3.9 Oregano and Carvacrol

Alvarez et al. (2014) investigated the QSI activity of Mexican oregano EO. QSI via flask incubation assay was quantified as percentage of violacein reduction. As exact inhibition values were not published, the listed results are rough estimates. Concentrations used were 0.0156 mg/mL to 0.125 mg/mL, showing a reduction of violacein production of ~50% even in the lowest concentration and about 80% for 0.0625%. In the highest concentration, cell viability was reduced too, whereas the other concentrations did not affect viability significantly (Alvarez et al., 2014). Using a concentration of the EO of *L. graveolens* Kunt of (a) 6.25 mg/mL, (b) 12.5 mg/mL, and (c) 25 mg/mL, an inhibition zone of 4.6 ± 0.29 mm, then 5.6 ± 0.29 mm, and finally 5.8 ± 1.04 mm was achieved. Using 0.0156% EO of *L. graveolens* Kunt, a QSI decrease of <50% was obtained; then with 0.0312%, a decrease of >60 was noticed; the concentration of 0.0625% revealed a >75% decrease; and finally the concentration of 0.125% EO yielded a QSI-decrease of >95%.

EOs: *Lippia graveolens* Kunt
Sensor strain: ATCC12472
Performed assays: Disc diffusion assay, flask incubation method with concurrent cell-viability testing (Alvarez et al., 2014)

16.8.3.10 Lemongrass and Cinnamon Oils

Mukherji and Prabhune (2014) performed tests on various EOs and proved lemongrass oil and cinnamon oil to be active QS inhibitors. The group was the first one to adapt the EOs chemically in order to change their polarity by converting them into EO sophorolipids (= glycolipids) produced by a *Candida bombicola* strain to obtain oils that are more active in aquatic environments. They tested the EOs themselves, testing both the EOs with added oleic acid sophorolipid (OASL) and the EO sophorolipids. The test results revealed a higher QS inhibition when adding oleic acid sophorolipid into the growth medium, caused by its function as an emulsifier, even to of the inactive oils showed QSI activity then (basil, ylang ylang). After converting the oils to their respective EO sophorolipids, all of the tested oils showed inhibitory activity. The outcome of the study indicates that QSI is not only on account of constituents of the EOs, but physical properties play an important role too (Mukherji and Prabhune., 2014).

EOs: Main components are indicated but no reference on this data is mentioned
Sensor strain: CV026 + C6HSL
Performed assay: Disc diffusion assay

Both EOs of *Cymbopogon citratus* (20%) and *Cinnamomum verum* (10%) revealed an inhibition zone of 15 mm. The negatively tested EOs were *Boswellia carteri*, *Cananga odorata*, *Citrus bergamia*, *Citrus sinensis*, *Cymbopogon nardus*, *Eucalyptus* ssp., *Melaleuca alternifolia*, *Mentha piperita*, *Ocimum basilicum*, and *Rosmarinus officinalis*.

16.8.3.11 EOs from South American Species

The research done by Pellegrini et al. (2014), who tested several oils from South American plant species, came to the result that *Salvia officinalis*, *Artemisia annua*, *Lepechinia floribunda*, *Schinus molle*, *Satureja odora*, and *Minthostachys mollis* are QS inhibitors. To quantify the results of the flask incubation assay, MQSIC was defined as the concentration of EO that results in 50% inhibition of QS (measured by means of the violacein production). *S. molle* (inhibition zone size [IZS] 50 ± 1.2 mm, MQSIC 0.005%), *M. mollis* (IZS >90 ± 0.1 mm, 0.0649%) and *S. odora* showed the lowest MQSIC-value (IZS 90 ± 1 mm, QSI = highest QS inhibition). The other EOs revealed the following data: *S. officinalis* (IZS >90 ± 0.1 mm, 0.0259% MQSIC), *A. annua* (IZS 70 ± 0.2 mm, 0.0138%), and *L. floribunda* (QSI = Growth inhibition and 0.0137% MQSI). With the exception of *S. odora* when being tested by flask incubation assay and *L. floribunda* tested via disc diffusion assay, all oils showed QS inhibition independently from cell-growth inhibition. Hence, *S. odora* and *L. floribunda* QS inhibition was just due to an antimicrobial effect, and the results of the respective methods can therefore not be counted as positive (Pellegrini et al., 2014).

EOs: Extraction by steam distillation, components were analyzed in previous reports via solid-phase micro-extraction (SPME) coupled to GC-MS, undiluted oils were used for Disc diffusion test, Flask incubation assay was performed with different concentrations.
Sensor Strain: ATCC1247
Performed assays: Disc diffusion assay, flask incubation assay with concurrent cell viability testing

16.8.3.12 Lavender Oil

Lavender oil inhibits QS in *E. coli* psB1075 and *E. coli* pSB401, these sensor strains react to the presence of either long-chain or short-chain AHLs by light production. Yap et al. (2014) proved the oil to be an inhibitor of the long-chain-sensitive strain *E. coli* psB107. The team did not determine MICs of the oil but they tested the bactericidal effect of the used concentrations, and no changes of cell survival were detected. Lavender oil was negatively tested on *E Coli* [pSB401]+ 3-oxo-C6-HSL (Yap et al., 2014).

EO: 0.01%–0.025%, components were analyzed via GC-MS, main components are linalyl anthranilate (38.4%) and linalool (34.6%)
Sensor strains: *E. coli* [pSB401] + 3-oxo-C6-HSL or and [pSB1075] +3-oxo-C12-HSL: *L. angustifolia* showed against both strains no effect
Performed assay: Luminescence-based QSI assay (quantified as reduction of total light production)

16.8.3.13 *Lippia alba* Oils

Olivero-Verbel et al. (2014) continued the research on *Lippia alba* EOs, which have already been investigated in 2012 (Jaramillo-Colorado et al., 2012). The authors also divided them in different groups depending on their extraction-method (MWDH/HD) and their chemotype (geranial/neral = CT1, limonene/carvone = CT2). In a difference from the research discussed before, this group also investigated the effect on cell growth of the tested oil concentrations and found that QSI is independent from the effect on cell growth. They found that the most active QS-inhibiting *L. alba* oils were those with high concentrations of geranial/neral (CT1), followed by the oils with limonene/carvone (CT2) as main constituents. In other words, CT1 is a more active QSI than CT2 in CV026 and ATCC31532, responding to C6HSL. As mentioned before, Jaramillo-Colorado et al. came to the result that CT2 is more active in the C6-AHL-sensible *E. coli* strain (Olivero-Verbel et al., 2014). The obtained data show the corresponding IC_{50} (μg/mL) values: *L. alba* (A) CT2: 66.91 (55.89–80.11), $R^2 = 0.95$; *L. alba* (B) CT2: 2.24 (1.98–2.54), $R^2 = 0.88$; *L. alba* (C) CT1: 0.62 (0.53–0.72), $R^2 = 0.85$; *L. alba* (D) CT1 (leaves) 15.79 (14.3–17.42), $R^2 = 0.82$; and *L. alba* (E) CT2: 30.32 (27.27–33.71), $R^2 = 0.84$.

EOs: MWHD/HD oils from different Columbian locations, the components were characterized via GC-MS; 0.01–300 mg/mL EO dissolved in DMSO were used for the assay
Sensor strains: CV026 + C6-HSL
Performed assay: Flask incubation assay, determination of IC_{50} values

16.8.3.14 *Ferula* and *Dorema* Oils

Sepahi et al. (2015) investigated the EOs of *Ferula* and *Dorema*, both belonging to the Apiaceae family on their QSI potential. Both of the oils were active, *Ferula* oil even fully inhibited violacein production in CV026, although the exact diameters of the inhibitory halos weren't published. Regarding QS in *P. aeruginosa*, the team showed reduced production of different QS-controlled virulence factors (elastase, pyoverdine, pyocyanin), as well as inhibition of biofilm formation in PAO1 (Sepahi et al., 2015).

EO: Extraction by hydro-distillation. The EO of *Ferula foetida* L. inhibited elastase, pyoverdine, pyocyanin, and biofilm formation, whereas *Dorema aucheri* Boiss EO did not inhibit pyocyanin and biofilm formation. However both EOs inhibited the QS-regulated violacein production in CV026.
Sensor Strains: CV026, PAO1.

Performed assays:

- Disc diffusion assay: CV026
- Virulence Factor Production: PAO1

16.8.3.15 *Hyptis dilatata* Oil

Hyptis dilatata, being traditionally used for the treatment of wound infections of cattle in Columbia was proved to be a QS inhibitor by Alvarez et al. (2015). The main components of the oil are monoterpenes (2-ß-pinene, camphor, 1-ß-pinene), sesquiterpenes (mainly aromadendrene), and oxygenated derivates of these substance classes. As inhibition diameters and used oil concentrations are not published, a comparison to other EOs is not possible (Álvarez et al., 2015).

EO: Extraction by steam distillation, GC-MS analysis of the components, methanol solutions of the oils were used
Sensor Strain: CVATCC 31592
Performed assay: Disc diffusion assay

16.8.3.16 Peppermint Oil and Menthol

Husain et al. (2015) tested peppermint oil and its main component menthol on QSI activity. Additionally, they investigated the effect of these substances on biofilm formation in *P. aeruginosa* and *A. hydrophila,* as well as their inhibitory potential on QS-regulated virulence factors. This research group was the same team that revealed the QSI activity of clove, cinnamon, peppermint, and lavender in 2009. As peppermint oil was proved to be an active QS inhibitor, they decided to continue research on this oil. Besides carrying out tests mentioned above, the team also performed molecular docking analysis of different constituents of the oil (menthone, limonene, menthol, menthyl-acetate) to an AHL–binding site of LASR having the natural autoinducer 3-oxo-C12-AHL. Using different subMICs of the oil, a maximal violacein reduction of 83.3% of was recorded in CV026, showing no significant difference in cell viability in comparison to non-treated cultures. Peppermint oil led to a decrease of LasB elastase, protease, chitinase, and pyocyanin activity in PAO1. The oil also inhibited swarming motility of PAO1 and decreased production of EPS. Biofilm reduction was shown too. The oil also showed reduction of protease, EPS production, and biofilm formation in *A. hydrophila*. Molecular docking of peppermint oil constituents revealed menthol as the best docked structure (based on binding energy scores), followed by isopulegol and 1-hydroxyaceton. Non-cyclic aliphatic constituents showed low affinities. When investigating the effects of menthol on QSI, a maximum reduction of violacein production of 85% was reached (400 μg/mL). Just like the oil itself, biofilm reduction and reduced production of different virulence factors in the tested strains was shown too (Husain et al., 2015).

EO: subMICs, GC-MS was carried out to analyze the components
Sensor Strains: CV026, PAO1, *Aeromonas hydrophila* (WAF-38)
Performed assays: (Husain et al., 2015)

- Flask incubation method + cell viability testing: CV026
- Biofilm inhibition via crystal violet assay: PAO1, *A. hydrophila*
- Virulence-factor inhibition: LasB elastase, protease, chitinase, and pyocyanin in PAO1, protease and exo-polymeric substance (EPS) in *A. hydrophila*
- Swarming motility: PAO1

Inhibited QS-associated processes in PAO1 by peppermint oil: Biofilm formation, elastase/protease/pyoverdine/chitinase/EPS-production, swarming motility (Husain et al., 2015).
Inhibited QS-associated processes in *A. hydrophila* by peppermint oil: Biofilm formation, protease/pyoverdine/chitinase/EPS production, swarming motility (Husain et al., 2015).

16.8.3.17 Cinnamon Oil

Cinnamon oil was tested by Kalia et al. (2015) regarding QS inhibition (CV026, PAO1) and its effect on biofilm formation, virulence-factor production, and swarming activity (PAO1). Whereas QS inhibition is often quantified via flask incubation assay, the authors additionally applied a β-galactosidase assay in order to quantify AHL produced by PAO1. Cinnamon oil already tested positively on QSI in 2009 by Khan et al. (2009) using the sensor strains CV026 and CV12472 (disc diffusion assay). The oil revealed a concentration-dependent inhibition of violacein in CV026 with a maximum inhibition of 78% (0.6 μL/mL). 3-oxo-C12HSL-production in PAO1 was reduced by 38% (0.2 μL/mL), higher concentrations showed higher inhibition, but that was most likely reducible to growth inhibition. All the other tested virulence factors as well as swarming motility were inhibited. Biofilm formation was reduced by 31%; further microscopy analysis showed changes in surface topology, integrity, structure, EPS, cell viability, and DNA content (Kalia et al., 2015).

EO: subMICs, *C. verum*
Sensor strains: CV026, PAO1 (GFP-tagged), *E. coli* DH5α pJN105L (containing l*asR* gene) and pSC11 (PlasI:lacZ)
Performed assays:

- Flask incubation assay: CV026
- β-galactosidase assay: To quantify 3-oxo-C12HSL produced by PAO1
- Virulence-factor production: pyocyanin, alginate, protease in PAO1
- Swarming motility of PAO1
- Biofilm assay (crystal violet staining); further biofilm analysis by scanning electron microscopy and confocal laser scanning microscopy

Inhibited QS-associated processes in PAO1 by cinnamon oil: biofilm formation, pyocyanin/alginate/protease-production, swarming motility (Kalia et al., 2015)

16.8.3.18 Coriander and (S)-(+)-Linalool

A study carried out by Duarte et al. (2016) proved that coriander EO, as well as its main component (*S*)-(+)-linalool, possesses a concentration-dependent QSI potential. Disc diffusion assay revealed a higher inhibitory potential of (*S*)-(+)-linalool (69.5 ± 0.71 mm QSI radius, mean ± standard deviation $p < 0.05$) than the coriander EO itself (63.5 ± 2.12, mean ± standard deviation $p < 0.05$), proving that this component is mainly responsible for the inhibitory effect of the oil. Additionally, both of the substances inhibit biofilm formation (5 μL/mL >90%) in *Campylobacter jejuni* and *Campylobacter coli* strains, whereas the oil is the more active inhibitor (Duarte et al., 2016).

EO: Pure or dilutions in DMSO
Sensor Strain: ATCC12472
Performed assays: Disc diffusion assay, flask incubation assay

16.8.3.19 Cinnamon Oil

Cinnamomum verum (cort.) EO, containing mainly *trans*-cinnamaldehyde (72.8%), benzyl alcohol (12.5%), and eugenol (6.6%) (GC-MS results) was tested on the ability to inhibit QS measured by bioluminescence of two *E. coli* strains as listed below. Concentrations of 0.01%, 0.0075%, and 0.005% significantly inhibited bioluminescence dose-dependently, whereas no antimicrobial effect was observed at these concentrations (Yap et al., 2015).

EO: C. verum (cort.) EO
Sensor strains: *E. coli* [pSB401] and *E. coli* [pSB1075] supplemented with C6-HSL/3-oxo-C12-HSL
Performed assay: Bioluminescence assay

16.8.3.20 Spice Oil Nano-Emulsions

Venkadesaperumal et al. (2016) investigated the QSI potential of *Cuminum cyminum*, *Piper nigrum*, and *Foeniculum vulgare*. In contrast to the other QSI tests done by other groups who performed assays with pure or diluted oils, this team used nano-emulsions of the EOs, formulated by ultrasonic emulsification of Tween 80 and water. One reason the team mentioned for using these emulsions instead of pure EOs is that EOs lose their bactericidal activity after certain dilutions because of changes in droplet structures. Other advantages are the high stability, high solubility, low turbidity, and enlarged surface, allowing a rapid penetration of nano-emulsions into the cell. All of the tested oils showed inhibitory activity when using disc diffusion assay and flask incubation assay; results are shown below (rough estimates for flask incubation assay are given for results without exact data). Additionally to QS-inhibition assays, the team investigated the effect of the nano-emulsions on biofilm formation and EPS production in *E. coli*, *Klebsiella pneumoniae*, *Salmonella typhimurium*, and *C. violaceum* strains, showing a concentration-dependent inhibitory potential of the oils (Venkadesaperumal et al., 2016).

EOs: subMICs (30 and 50 μL/mL) of EOs in different emulsion ratios
Bacterial strains: CV026 + C6HSL
Performed assays: Flask incubation assay and disc diffusion assay (30 and 50 μL/mL), and using 50 μL/mL the percentage reduction of violacein amounted to 70% with *C. cyminum*, >55% with *P. nigerum*, and 75.7% with *F. vulgare*.

16.8.3.21 Eucalyptus Oils

Eucalyptus globulus revealed an inhibition zone size at 10 mm QSI radius and >90% inhibition of violacein at 5 μL/mL, and also *E. radiata* inhibited QS, concentration dependent, in CV ATCC12472 (20 mm QSI radius and >90% inhibition of violacein at 5 μL/mL) (Luís et al., 2016).

EOs: Extraction by hydro-distillation (leaves, small branches), GC-MS analysis of the components
Sensor strain: ATCC12472
Performed assays: Disc diffusion assay, flask incubation assay

16.8.3.22 Mandarin Oil

Luciardi et al. (2016) investigated mandarin oil and the main component limonene on its effects on QS, virulence factors, and biofilm formation in *P. aeruginosa* strains. This research group differed between two oils depending on their production method, which was either cold-pressing (= EOP) or cold pressing followed by steam distillation (= EOPD). EOP, EOPD, and limonene reduced AHL, concentration dependent, in both of the tested strains. AHL-reduction was shown to be higher than cell growth inhibition (CGI). An interesting fact the group mentioned in the paper was the increased AHL production in the presence of the antibiotic azithromycin, showing a possible connection between bacterial stress and AHL production. Concerning biofilm formation, the oils and limonene were concentration-dependent inhibitors. The team also showed a reduced elastase-B production (>70%)

TABLE 16.5
Inhibition of QS Regulated β-Galactosidase-Production in ATCC 27853 and HT5 Quantified by *P. aeruginosa* qsc119 (% of QS-Decrease) by Mandarin Oil and Below Inhibition of Biofilm-Formation in ATCC27853 and HT5 by Mandarin Oil (% of Decrease[a])

	ATCC27853	HT5
EOP 4 mg/mL	49% (CGI: 17%)	38% (CGI: 33%)
EOPD 4 mg/mL	50% (no cell CGI)	37% (CGI: 25%)
Limonene	34% (no CGI)	30% (CGI: 24%)
EOP 4 mg/mL	82%	73%
EOPD 4 mg/mL	88%	92%
Limonene 4 mg/mL	49%	58%

Source: Luciardi, M.C. et al. 2016. *LWT - Food Sci. Technol.* 68: 373–380.
Note: CGI = Cell Growth Inhibition.
[a] Only highest inhibition values listed.

in all of the tested strains and with all tested substances. The analysis of the components revealed monoterpenes as main components (97.66%, EOP/96.58 EOPD mainly represented by limonene, followed by γ-terpinene, myrcene, and α-pinene). When comparing EOP and EOPD, EOPD contained more oxygenated monoterpenes and sesquiterpene fractions. The research team believes that EOPD showed a more pronounced biofilm inhibition because of the higher amount of linalool, α-terpineol, γ-terpinene, and terpinolene, respectively (See Table 16.5) (Luciardi et al., 2016).

EO: subMICs of *Citrus reticulata* oils (EOP/EOPD), GC-MS analysis of constituents was carried out
Sensor strains: *P. aeruginosa* ATCC27853 and HT5 (Hospital Tucuman, strain resistant to several antibiotics) as AHL producers, *P. aeruginosa* qsc 119 as sensor strain
Performed assays:

- β-galactosidase assay (*P. aeruginosa* qsc 119) to quantify AHLs produced by ATCC27853 and HT5
- Biofilm assay (crystal violet staining): ATCC27853, HT5
- Further tests on the biofilm were done on metabolic activity, bacterial viability, and the relation to growth inhibition
    - Elastase-B-activity assay: ATCC27853, HT5

16.8.3.23 Thyme Oil, Thymol and Carvacrol

The influence of *Thymus vulgare* EO (0.1 to 0.5 μL/mL) and its main components carvacrol and thymol (0.001 to 0.01 μL/mL) on QSI, flagella gene (*fglA*, responsible for motility)-expression (AHL-related), and biofilm formation in a *P. fluorescens* strain was investigated by Myszka et al. (2016). The reason for the team to investigate the oil was the fact that *P. fluorescens* biofilms are a concern of the food industry as mature biofilms are difficult to remove, and available antimicrobial agents often show proper disinfection only when staying in contact to the surface for a long time period. As the flagellum may be the first part of the bacterium to have contact to surfaces and its production is controlled by autoinducers, the team hypothesized that this is a possible point of blocking to inhibit biofilm formation. Violacein production was reduced significantly for all tested substances; 90% (EO), 80% (carvacrol), and 78% (thymol) reduction was shown in the 72 h cultures. When analyzing the AHLs, treated cultures showed no AHL molecules after treatment with the EO/thymol/

carvacrol (72 h), whereas reference samples contained C6-HSL, 3-oxo-C6 HSL, 3-OH-C8-HSL, and 3-OH-C8-HSL. The investigation on the effect of subMICs of the oil and its main components on the gene expression of *flgA* in *P. fluorescens* revealed inhibitory effects of all the tested substances; the EO had the highest inhibitory effect on both gene expression and motility. Regarding the biofilm, formation was inhibited by all the tested substances, and the EO again showed the highest inhibitory effect (Myszka et al., 2016).

EO: subMICs of *T. vulgare* EO (leaves) produced by hydro-distillation (0.1–0.5 μL/mL), GC-MS analysis was done to figure out the components of the oil; main EO components (<5%) were thymol (55.42), carvacrol (6.84%), and *p*-cymene (5.33%)
Sensor strain: *P. fluorescens* KM121, isolated from stainless-steel surface in a food process plant as AHL-producing strain, CV026 as biosensor to measure AHL production from KM121
Performed assays:

- Flask incubation assay for AHL quantification, AHL profile was analyzed
- Swimming motility assay + *flgA*-gene expression in KM121
- Biofilm assay (acridine-orange staining, examination of the biofilm by fluorescence microscopy)

Inhibited QS-associated processes in KM121 by thyme oil: Biofilm formation, swimming motility, *flgA*-gene expression (Myszka et al., 2016).

16.8.3.24 *Murraya koenigii* CO_2-Extract—A Preclinical Infectious Model

Ganesh and Rai (2016) investigated the effect of *Murraya koenigii EO* on *P. aeruginosa*-infected *Caenorhabditis elegans*. The team used several assays in order to evaluate QS inhibition, as well as survival of the nematodes.

In this study, the EO was gained using supercritical fluid CO_2 extraction and therefore is rather to be classified as a CO_2 extract. Therefore, these results will not be listed in tables. As this is one of only a few studies testing the anti-QS effect *in vivo* on *C. elegans*, it is worth a mention, although results will not be listed in the overview tables in Section 16.9 (Ganesh and Vittal Rai, 2016).

The study showed a significant inhibition of two QS-controlled processes: pyocyanin production (64.2%,) and staphylolytic activity (63.1 ± 1.2%) after 60 minutes. Non-treated worms showed 100% mortality after 4 h when being infected with *P. aeruginosa*. A paralytic assay showed that 70% of treated nematodes were still alive after 3 h (55%–60% after 4 hours). The concentrations of *Murraya koenigii EO* were neither toxic nor did they affect lifespan or survival of non-infected *C. elegans*. In a fast-killing assay, 60% of treated nematodes were still alive when being treated with the EO. *C. elegans* infected with a *lasI* and *rhlI* mutant strain of *P. aeruginosa*, PAO1-JP2, showed 90% survival after 4 h, showing a QS-dependent virulence (Ganesh and Rai, 2016).

EO: subMICs of *Murraya koenigii* EO
Sensor strains: PAO1, PAO1-JP2 (*Staphylococcus aureus* ATCC 25923 for staphylolytic assay)
Test organism: *C. elegans* wild-type Bristol strain N2
Performed assays:

- Pyocyanin inhibition assay: Quantitative assessment of pyocyanin (phenazin, a QS-controlled virulence factor) inhibition in *P. aeruginosa* PAO1 treated with *Murraya koenigii* EO
- LasA staphylolytic assay: To detect QS-dependent protease LasA production
- Paralytic assay and fast-killing assay in *P. aeruginosa*-infected *C. elegans*

16.8.3.25 Lavender, Rosemary, and Eucalyptus Oils

Luìs et al. (2017) assessed the antioxidant, antibacterial, and anti-QS activities of eight commercial EOs. Rosemary, eucalyptus, and lavender of different origins were shown to be active QSIs. The chemical composition of EOs was analyzed by GC-MS, therefore the authors were able to recognize interrelations between the activity of the oils and their components. For example, *L. angustifolia* originating from France has linalool and linalyl acetate as main constituents and inhibited QS. The plant originating from Croatia contains mainly linalool and linalyl formate and did not show any effect (Luís et al., 2017). The data for the inhibition zone size using the sensor strain CV ATCC12472 were as follows: *E. citriodora* EO from China: 8.36 ± 0.93 mm, *L. officinalis* EO from Croatia: 4.84 ± 0.08 mm, *L. angustifolia* EO from France: 8.60 ± 0.56 mm, *L. angustifolia* EO from France at a high altitude: 5.01 ± 0.43 mm, and *Rosmarinus officinalis* EO from Tunisia: 8.69 ± 0.81 mm.

EOs: GC-MS-analyzed EOs, acquired commercially in a local pharmacy in Portugal and in Corsic. EOs were obtained by hydro-distillation from fresh or dried aerial plant parts. 10 μL were used for QS-inhibition assay. Negatively tested EOs were *Eucalyptus smithii* from Australia, *E. staigeriana* (also from Australia), *L. angustifolia* (from Croatia), *Cymbopogon citratus* (from South Africa), and *Juniperus communis* (from Bosnia-Herzegovina).
Sensor strains: *C. violaceum* ATCC 12472
Performed assay: Disc diffusion assay

16.8.3.26 Green Cardamom EO

Abdullah et al. (2017) investigated the QSI effect of green cardamom (*Elletaria cardamomum*) on *C. violaceum*. Main compounds include terpinyl acetate (38.4%), 1,8-cineole (28.71%), and linalyl acetate (8.42%). Concentrations of 0.625 and 0.313 mg/mL inhibited QS significantly (about 75%–80% of QS decrease) with very little effect on growth.

EO: Green cardamom fruits were purchased at a local market in Pakistan. The acquired EO was analyzed by GC-MS. SubMIC concentrations were used.
Sensor Strain: *Chromobacterium violaceum* ATCC 12472
Performed assay: Flask incubation assay

16.8.3.27 Corsican *Mentha suaveolens* ssp. *insularis* Oil and Others

Poli et al. (2018) tested the QS inhibitory potential of 12 EOs on *C. violaceum*. Minimal inhibitory concentration (MIC), as well as minimal QS inhibiting concentration (MQSIC), was determined. *cis-cis-p*-Menthenolide and pulegone (main components of *Mentha suaveolens* ssp. *insularis*) as well as minthlactone (isomer of *cis-cis-p*-menthenolide), were further tested on QS inhibition and biofilm degradation. *Cedrus atlantica* EO showed a very high MQSIC (6.00 mg/mL), meaning a high concentration is necessary to inhibit QS. Hence, this oil is less effective than other oils like *M. suaveolens* ssp. and *Myrtus communis*, which were the most potent inhibitors of the study (Poli et al., 2018).

EOs in this study were obtained using steam distillation. The group noticed varying QSI activity depending on the isolation method of EOs. Their assumption is a higher activity due to more oxygenated compounds when using microwave-assisted distillation compared to steam distillation. MQSIC/MIC ratios were calculated, whereby a higher ratio represents a higher anti-QS activity. By using this parameter, it is visible that *Xanthoxylum armatum* was the least efficient EO, as the effect is likely due to antimicrobial action. The same applies to *Calamintha nepeta* ssp. *nepeta*, *Citrus clementina*, *C. atlantica*, and *Lavandula stoechas*. Higher values (16.32) represent relevant QS inhibition. All in all, *M. suaveolens* ssp. *insularis* showed the highest anti-QS activity.

Poli et al. also found out that *cis-cis-p*-menthenolide has a considerably higher MQSIC/MIC ratio than the other main component of *M. suaveolens* ssp. *insularis* EO and can therefore be made primarily responsible for the anti-QS effect. An interesting finding was the significant difference in activity from *cis-cis-p*-menthenolide compared to its isomer minthlactone, as their structural difference is only a double-bond position (exo-cyclic or intra-cyclic). SEM images of *C. violaceum* grown under normal conditions were compared with *M. suaveolens* ssp. *insularis* EO, *cis-cis-p*-menthenolide at MIQSC, and minthlactone at MIQSC and confirmed the previous results. The pictures confirmed the previous

TABLE 16.6
Inhibition of QS Regulated Violacein Production in *C. Violaceum* by EOs

EO	MQSIC (mg/mL)	MQSIC/MIC Ratio	Major Compounds
Mentha suaveolens ssp. Insularis	0.1(512)	32	Pulegone (44.4%), *cis-cis-p*-menthenolide (27.3%)
Pulgeone	0.1	4	
cis-cis-p-Menthenolide	0.05	64	
Mintlactone	0.4	128	
Citrus limon	0.4	16	Limonene (66.4%), γ-terpinene (10.1%)
Eucalyptus polybractea	0.05	16	p-cymene (25.5%), cryptone (11.42%)
Helichrysum italicum	0.8	16	Neryl acetate (39.6%), α-curcumene (7.6%)
Myrtus communis	0.1	16	α-pinene (52.9%), 1,8-cineole (20.6%) α-pinene
Cedrus atlantica	6.0	8	α-pinene (55%), himachalol (8.3%)
Lavandula stoechas	0.2	8	Fenchone (34.9%), camphor (28.9%)
Calamintha nepeta ssp. nepeta	0.2	4	Pulegone (49.0%), menthone (21.5%)
Citrus clementina	0.1	4	Sabinene (31.4%), linalool (20.4%) geranial
Cymbopogon citratus	0.2	4	Geranial (44.0%), neral (31.1%) (*E*)-anethole
Foeniculum vulgare	0.2	4	(*E*)-anethole (75.7%), limonene (8.5%)
Xanthoxylum armatum	0.2	2	Linalool (52.9%), limonene (15.0%)

Source: Adapted from Poli et al. 2018. *Molecules*. 23(9). doi:10.3390/molecules23092125.

results. The EO, as well as *cis-cis-p*-menthenolide, did not affect growth, but the matrix was degraded and fragmented. Minthlactone visibly reduced population density (see Table 16.6).

EOs: 12 EOs obtained by steam distillation, components were analyzed by GC. Dilutions of EOs ranging from 50.00 mg/mL (dilution: 1) to 1.25×10^{-2} mg/mL (dilution: 4096), obtained by a two-fold serial dilution were used. EOs were mainly purchased in Corsica (France).
Sensor strains*:* *C. violaceum* wild-type strain CIP 103350T.
Performed assays: Flask incubation assay, anti-biofilm assay (SEM).

16.8.3.28 South African Oils Plus a Gas Chromatography-Based Metabolomics Approach

Mokhetho et al. (2018) tested 40 commercial EOs of therapeutic relevance in South Africa using the disc diffusion method and the flask incubation assay, along with the MIC assay (see Table 16.6). GC-MS was used to profile EO components. The major components of one active EO were tested on anti-QS (AQS) activity. Additionally, the team combined those components in the same proportion as they would occur in the natural oil and tested the activity of the combination. By changing percentages of the constituents, possible synergistic, additive, and antagonistic interactions were assessed. To link the chemical profiles and biological activity of the EOs, an untargeted metabolomics approach was used (SIMCA-P + 14.0).

Several of the tested oils exhibited good antimicrobial activity (zone of inhibition, ZOI) including lemon, juniper, eucalyptus, lavender, labrador tea, verbena, pennyroyal, spikenard, sweet marjoram, and rosemary oil, although all of them were less effective than eugenol (see Table 16.7). Seventeen EOs showed moderate AQS activity (zone of turbidity = 3.00–5.50 mm) compared to eugenol (ZOT = 12.0 mm). ZOT-values were generally lower therefore tested oils have most probably a greater ability to inhibit growth than to reduce QS. Results of Flask incubation assay along with

TABLE 16.7
Antimicrobial and QSI Activity of 40 Essential Oils against *C. violaceum*

EO/Positive Control	ZOI (mm)	ZOT (mm)	MIC (mg/mL)	MQSIC (mg/mL)
Eugenol (Positive control)	9.5 ± 0.4	12.0 ± 0.0	0.1	0.1
Artemisia dracunculus L.	4.8 ± 0.4	0.5 ± 0.7	0.1	0.6
Chamaemelum nobile (L.) All	0.5 ± 0.7	4.3 ± 0.4	>0.5	0.5
Citrus limon (L.) Osbeck (1)	1.3 ± 1.1	5.5 ± 1.1	0.3	0.1
Citrus limon (L.) Osbeck (2)	5.3 ± 0.4	1.0 ± 0.0	0.3	0.1
Citrus reticulata Blanco (1)	1.5 ± 0.7	4.0 ± 0.0	0.3	0.1
Citrus reticulata Blanco (2)	0.8 ± 0.4	3.3 ± 0.4	>0.5	0.5
Citrus sinensis (L.) Osbeck	1.3 ± 1.1	3.3 ± 0.4	0.5	0.3
Cymbopogon citratus (DC.) Stapf	0.8 ± 1.1	3.0 ± 0.7	0.1	0.1
Cymbopogon flexuosus (Nees ex Steud.)	0.3 ± 0.4	3.3 ± 0.4	0.1	0.1
Cymbopogon giganteus Chiov.	3.5 ± 1.4	2.0 ± 0.7	0.5	0.1
Cymbopogon martini (Roxb.) W.Watson	0.8 ± 0.4	3.5 ± 1.4	0.3	0.1
Cymbopogon winterianus Jowitt	1.0 ± 0.7	2.3 ± 1.1	0.3	0.1
Eucalyptus dives Schauer	2.5 ± 0.0	3.0 ± 0.7	0.3	0.1
Eucalyptus globulus Labill.	7.3 ± 0.4	1.0 ± 0.7	0.5	0.1
Eugenia caryophyllus (Spreng.)	0.5 ± 0.7	3.3 ± 0.4	>0.5	N/A
Gaultheria procumbens L.	0.5 ± 0.7	3.0 ± 1.4	0.1	0.1
Inula graveolens (L.) Desf.	0.3 ± 0.4	3.3 ± 1.1	>0.5	0.5
Juniperus communis L. (1)	3.3 ± 0.4	1.3 ± 1.1	0.5	0.3
Juniperus communis L. (2)	2.3 ± 0.4	3.0 ± 0.7	0.5	0.3
Juniperus oxycedrus L.	6.3 ± 1.1	0.0	>0.5	nd
Lavandula buchii var. tolpidifolia (Svent.)(1)	7.8 ± 0.4	0.0	0.5	nd
Lavandula buchii var. tolpidifolia (Svent)(2)	5.7 ± 0.4	1.0 ± 0.0	0.5	0.1
Lavandula buchii var. tolpidifolia (Svent.)(3)	0.3 ± 0.4	0.5 ± 0.0	0.3	0.1
Ledum groenlandicum Oeder	6.8 ± 0.4	0.0	0.5	nd
Lippia citriodora (Palau) Kunth	6.5 ± 0.7	0.5 ± 0.0	>0.5	0.5
Mentha arvensis L.	0.5 ± 0.7	1.0 ± 0	0.1	0.1
Mentha pulegium L.	5.3 ± 0.4	2.3 ± 0.4	0.3	0.1
Mentha × *citrata* Ehrh	2.8 ± 0.4	4.0 ± 0.7	0.5	0.5
Mentha × *piperita* L.	3.8 ± 0.4	1.0 ± 0.0	0.5	0.3
Myrtus communis L. (1)	0.5 ± 0.7	4.0 ± 1.4	0.3	0.1
Nardostachys jatamansi (D.Don) DC.	7.0 ± 0.7	1.3 ± 0.4	>0.5	nd
Origanum compactum Benth.	0.3 ± 0.4	2.5 ± 0.0	0.1	0.1
Origanum majorana L. (1)	6.5 ± 0.7	0.5 ± 0.0	>0.5	0.5
Origanum majorana L. (2)	4.5 ± 0.7	0.8 ± 0.4	0.1	0.1
Pinus ponderosa Douglas ex C.Lawson	1.3 ± 0.4	3.0 ± 0.0	0.3	0.1
Rosmarinus officinalis L.	6.0 ± 0.7	0.5 ± 0.0	>0.5	0.5
Satureja montana L.	0.4 ± 0.4	2.3 ± 0.4	0.1	0.1
Styrax benzoin Dryand	0.8 ± 0.4	3.5 ± 0.7	>0.5	nd
Thymus vulgaris L. (1)	3.3 ± 0.4	2.3 ± 1.1	0.1	0.1
Thymus vulgaris L. (2)	0.0	2.3 ± 0.0	0.0	0.1

Source: Adapted from Mokhetho et al., 2018. *J. Essent. Oil Res.* 30 (6): 399–408.
Abbreviation: nd = not detected.

MIC-assay revealed good antimicrobial activity with MICs <0.5 mg/mL. The majority of those oils also inhibited QS at low concentrations. QSI was higher with increasing concentrations. The most potent QS inhibitors of this assay (0.1 mg/mL) were EOs of *C. limon, C. flexuosus, C. martini, E. caryophyllus, G. procumbens, O. compactum* and *T. vulgaris*. Interestingly, *Juniperus oxycedrus* and *Picea mariana* led to an enhancement of QS.

Moketho et al. (2018) used GC-MS-based metabolomics to identify active components in a complex data matrix. A similar approach was already used to detect biomarkers responsible for antimicrobial activity by Maree et al. (2014). As the team wanted to find out which EO constituents are associated with good AQS activity, they separated the EOs into poorly active and highly active classes by OPLS-D, whereas GC-MS data were merged with percentage inhibition at 0.25 mg/mL concentrations. Five *Cymbopogon* spp. oils were distinctly separated into the active class. In order to obtain possible pure compounds for further studies, chemical constituents occurring in the active oils were identified using an S-plot showing X-variables highly correlated to the two oil classes at the extreme ends. The authors postulated geranial, geraniol, menthol, eugenol ,and pulegone as biomarker molecules being responsible for AQS.

Additional research was pursued on *Cymbopogon* spp., as it showed the best AQS activity (see Table 16.8). Therefore, a GC-MS from *C. martin* was carried out in order to identify its main constituents. Those compounds were then tested individually and in combination on their AQS activity. Geraniol (81.1%), geranyl acetate (8.3%), and linalool (3.0%) accounted for 92.4% of the total EO. Interestingly, geranyl acetate alone enhanced QS in all four concentrations tested (0.50, 0.25, 0.13, 0.06 mg/mL). The combination of linalool, geraniol, and geranyl acetate inhibited QS more effectively than the individual components. In lower concentrations, the activity even exceeded eugenol, which was the positive control. These findings highlighted the importance of interaction between components of EOs (Mokhetho et al., 2018).

EOs. 40 EOs, purchased in Pranarôm (Belgium); 20 µL of a 1 mg/mL EO solution was used for disc diffusion assay

Sensor strain: *C. violaceum* ATCC 12472

Performed assays:

- Disc diffusion assay: positive control, eugenol. ZOI was defined as the diameter of clear area surrounding the well, ZOT as zone of turbidity (= QS-inhibition)
- Flask incubation method: 0.06, 0.13, 0.25 and 0.50 mg/mL EO; MIC and MQSIC were determined
- Multivariate data analysis to correlate EO profiles to AQS activity

TABLE 16.8

Percentage QSI in *C. violaceum* of *Cymbopogon Martini* EO and Combinations of its Major Constituents

EO Constituents	0.50 mg/mL	0.25 mg/mL	0.13 mg/mL	0.06 mg/mL
Cymbopogon martini	95.00	94.67	89.34	70.00
Eugenol (+ control)	95.00	95.00	84.00	60.00
Linalool + geraniol + geranyl acetate	93.00	93.00	92.00	90.00
Geraniol + linalool	92.00	91.00	83.00	61.00
Geraniol	90.00	88.00	83.00	60.00
Geranyl acetate	−55.00	−54.00	−49.00	−48.00
Linalool	95.00	70.00	55.00	53.00

Source: Adapted from Mokhetho et al., 2018. *J. Essent. Oil Res.* 30 (6): 399–408.

16.8.3.29 *Thymus daenensis* and *Satureja hortensis* Oils

Sharifi et al. (2018) investigated the effect of *Satureja hortensis* and *Thymus daenensis* EO on biofilm formation, biofilm disruption, planktonic growth, and QS on some *Staphylococcus aureus* isolates. GC-MS revealed the main compounds of *T. daenensis* EO are carvacrol (40.69%), γ-terpinene (30.28%), and α-terpinene (5.52%). The main components of *S. hortensis* EO of the study are thymol (41.28%), γ terpinene (37.63%), *p*-cymene (12.2%), and α-terpinene (3.52%). All the tested subMIC concentrations of both oils (except MIC/16 of *S. hortensis*) significantly inhibited biofilm formation. Biofilm inhibition was dose-dependent, whereas *T. daenensis* was more potent. Preformed biofilms were significantly disrupted by both of the oils at MIC/2 and MIC/4. *T. daenensis* even significantly disrupted the biofilm at MIC/8. SEM confirmed those results, as treated samples with MIC/2 decreased the number of adherent bacterial cells.

Real-time PCR revealed that MIC/2 (0.0625 μL/mL) concentrations of S. hortensis significantly down-regulated *hld*-gene expression. On the other hand, MIC/2 of *T. daenensis* EO had the opposite effect, as the *hld*-gene was upregulated by 1.2-fold, though this result was not statistically significant ($p > 0.05$) This result indicates an anti-biofilm activity of *T. daenensis* EO independent from *hld*-gene expression (Sharifi et al., 2018).

Bacterial strains: *S. aureus* isolates from respiratory patients, dairy cattle with subclinical mastitis, and milk and food samples. Using culture characteristics, biochemical tests, and PCR, isolates were identified. Results were calculated as mean of 10 isolates. Additionally, *S. aureus* ATCC 25923 was used as reference strain. Additionally, *S. epidermidis* ATCC 12228, a non-biofilm-producer strain, was used in all the experiments.

EOs: EO components were determined by GC-MS. Plant material was obtained from a university in Iran; EOs were then obtained by hydro-distillation. MIC and MBC (minimal bactericidal concentration) were quantified. Experiments were carried out using subMICs.

Performed assays:

- Microtiter plate test (mtP) to assess activity against biofilm formation and biofilm eradication as earlier described. (Sieniawska et al., 2013) MIC/2 to MIC/16 concentration was used.
- SEM: To assess antibiofilm-activity.
- Quantitative real-time RT-polymerase chain reaction (PCR) of *hdl* (RNAIII transcript) to detect QSI: MIC/2 concentrations of EOs were used, assay was performed as described before. EO treated biofilm formed by *S. aureus* ATCC 25923 was used to measure *hld*-gene expression.
- Tetrazolium-based colorimetric assay (MTT) to detect cytotoxicity.

Inhibition of QS (*hdl* expression inhibition in *S. aureus* ATCC 25923) *S. hortensis* MIC/2.
Negatively tested oil: *T. daenensis*

16.8.3.30 *Carum copticum* EO

Snoussi et al. (2018) tested the effect of *Carum copticum EO. C. copticum* fruits are widely used in India as a spice in curry powder and are known to have several areas of application due to their antibacterial, antifungal, anti-inflammatory, anti-hypertensive, and antitussive efficacy. The authors investigated the constituents of the EO by GC-EIMS. Additionally, polyphenols of the methanolic extract were analyzed. *C. copticum* EO, as well as the methanolic extract, was positively tested on antibacterial, antifungal, and antioxidative activity. The main components of the EO included oxygenated monoterpenes (53.0%), with thymol accounting for the greatest share. The monoterpene hydrocarbons *p*-cymene and γ-terpinene together accounted for about 44% of the oil. Disc diffusion test with CV026 revealed an anti-QS activity of the EO, although the diameter of inhibition is not mentioned. Microplate assay with ATCC12472 revealed an IC_{50} value for violacein inhibition of 0.23 mg/mL, whereas the MIC was 0.6 mg/mL (Snoussi et al., 2018).

EOs: Dried seeds were purchased in Jeddah, a city in Saudi Arabia. EO was obtained by hydro-distillation. GC-EIMS (gas chromatography electron ionization mass spectrometry) was performed to identify components of the oil, MIC, MBC, and MFC (minimal fungicidal concentration).
Sensor strains: ATCC 12472 (microplate assay) and CV026 (disc diffusion assay)
Performed assays:

- DPPH assay: To measure antioxidant activity
- Disc diffusion assay: As described before, slightly adjusted; 2 μL of the EO was used.
- Micro-plate assay: Adjusted flask incubation assay as described before, spectrophotometric quantification of violacein, IC_{50} was recorded

16.8.3.31 *Cinnamomum tamala* oil Combined with DNase

Banu et al. (2018) tested the antibiofilm and QSI potential of *Cinnamomum tamala* EO from India, as well as its main compounds cinnamaldehyde and linalool, on the pathogen *Vibrio parahaemolyticus*, which can cause food-borne infections when consuming raw sea foods. Additionally, synergistic effects of EO together with DNase and marine bacterial DNase on preformed biofilms were investigated. eDNA (environmental DNA) is seen to be as an important component of biofilms to ensure structural integrity in *V. parahaemolyticus*. Enzymes like DNase can therefore weaken biofilms and, as a result, can improve the efficacy of antibacterial agents by improving biofilm penetration. The authors also tested linalool and cinnamaldehyde on their preservative efficacy in prawns infected with *V. parahaemolyticus* and proved them to reduce bacterial load and inhibit lipid peroxidation equal to the standard food preservative sodium benzoate. Also, SEM analysis revealed that the compounds do not damage muscle tissue of the tested prawns, which makes them a possible new food preservative.

Looking at biofilm disruption, results revealed a higher efficacy of the EO combined with DNases than the compounds alone to disrupt preformed biofilms or biofilm formation. *C. tamala* EO significantly inhibited EPS and alginate production dose-dependently (subMICs were used). Due to these findings, Banu et al. expected that the application of antibiotics in combination with EOs reducing EPS and alginate to possibly enhance penetration through biofilms and improve their efficacy that way. The similar assumption was made for eDNA inhibition, as again the combination of EO with DNases lead to reduced eDNA accumulation. Swarming and swimming was measured using subMICs (0.5% v/v) of the oil. Both types of motility were significantly inhibited in a dose-dependent manner by the oil as well as by the two major compounds of the oil. As swarming motility is QS dependent, this result can be evaluated as QS inhibition.

All in all, the work of this author group demonstrated multiple fields of application of C. tamala EO and an efficient way to combine EO/EO compounds in order inhibit biofilms, virulence factor production, and bacterial motility (Banu et al., 2018).

Sensor strains: *V. parahaemolyticus* (ATCC 17802)
EO: Oil was obtained *C. tamala* leaves collected in Almora, Uttarkhand, India. Components were identified by GC-MS. MIC was determined in the study.
DNase: Commercially acquired
Marine Bacterial DNase: isolated from *Vibrio alginolyticus*
Performed assays: Biofilm assays (test agents: EO, DNases, and combinations)

- Crystal violet assay: To detect young biofilm inhibition of EO as well as the effect on preformed biofilms
- Visualization of biofilm inhibition by light microscopy, confocal laser scanning microcopy
- eDNA staining by fluorescence microscopy: To detect DNA present in a biofilm

Other assays:

- Inhibition of virulence factors (test agent: EO): EPS, alginate (embedded in the EPS of biofilms)
- Swarming and swimming motility assay (test agents: EO, cinnamaldehyde, and linalool)

- Bacteriological analysis of tiger prawns (*P. vannamei*), including measurement of the initial bacterial load, bactericidal effect, preservative effect of linalool, cinnamaldehyde, ciprofloxacin, and sodium benzoate
- SEM of linalool or cinnamaldehyde combined with DNase to prove that muscle tissue of the prawns is not damaged.

16.9 OVERVIEW OF THE RESULTS

This section gives an overview of the results discussed in the Section 16.8 (see Tables 16.9 through 16.13). If oils were tested in different concentrations below their minimal inhibitory concentrations (MICs) of the sensor strains, the label "subMIC" is used. If no MIC was evaluated in the respective study or higher concentrations were used, concentrations or volumes are shown. In the case of a study not clearly stating the concentration, cells are marked by N/A. Methods used are discussed in Section 16.8.2.

Results given for disc diffusion (DD) test quantify QSI by measuring colorless (but viable) cells around EO-impregnated paper discs (in mm). It is important to mention that several different concentrations of EOs were used in the studies, therefore values cannot always be compared. Additionally, some publications do not clearly state whether the diameter contains or excludes the antimicrobial diameter (in case of exceeding subMIC concentrations). Results from other assays equal the reduction of violacein/fluorescence/luminescence/β-galactosidase, depending on the used method. Most of the oils were tested in different concentrations; these tables only lists the maximal reductions at the respective concentrations of EOs. If both MQSIC and other inhibition values are published, MQSIC is cited. For cases not indicating exact values, N/A is cited.

16.9.1 Positively Tested Oils

16.9.1.1 Disc Diffusion Assay

TABLE 16.9
Essential Oils Inhibiting Quorum Sensing Detected by Disc Diffusion Assay

EO	Concentration	Sensor Strain	Diameter of QS Inhibition (mm)	Reference
Artemisia annua L.	100%	ATCC12472	70 ± 0.2	Mukherji and Prabhune (2014)
Artemisia dracunculus L.	1 mg/mL (20 μL)	ATCC 12472	0.5 ± 0.7	Mokhetho et al. (2018)
Carum copticum L.	100% (2 μL)	CV026	N/A	Snoussi et al. (2018)
Chamaemelum nobile (L.) *All*	1 mg/mL (20 μL)	ATCC 12472	4.3 ± 0.4	Mokhetho et al. (2018)
Cinnamomum verum J.Presl	subMICs	CV026	11	Khan et al. (2009)
Cinnamomum verum J.Presl	subMICs	ATCC12472	12	Khan et al. (2009)
Cinnamomum verum J.Presl	100 mg/mL	CV026	15	Mukherji and Prabhune. (2014)
Citrus limon (L.) Osbeck *(1)*	1 mg/mL (20 μL)	ATCC 12472	5.5 ± 1.1	Mokhetho et al. (2018)
Citrus limon (L.) Osbeck *(2)*	1 mg/mL (20 μL)	ATCC 12472	1.0 ± 0.0	Mokhetho et al. (2018)
Citrus reticulata Blanco *(1)*	1 mg/mL (20 μL)	ATCC 12472	4.0 ± 0.0	Mokhetho et al. (2018)
Citrus reticulata Blanco *(2)*	1 mg/mL (20 μL)	ATCC 12472	3.3 ± 0.4	Mokhetho et al. (2018)
Citrus sinensis (L.) Osbeck	1 mg/mL (20 μL)	ATCC 12472	3.3 ± 0.4	Mokhetho et al. (2018)
Coriandrum sativum L.	100%	ATCC12472	63.50 ±2.12	Duarte et al. (2016)
Cuminum cyminum L.	subMICs	CV026	N/A	Venkadesaperumal et al. (2016)
Cymbopogon citratus (DC.) Stapf	200 mg/mL	CV026	15	Mukherji and Prabhune (2014)

(Continued)

TABLE 16.9 (*Continued*)
Essential Oils Inhibiting Quorum Sensing Detected by Disc Diffusion Assay

EO	Concentration	Sensor Strain	Diameter of QS Inhibition (mm)	Reference
Cymbopogon citratus (DC.) Stapf	1 mg/mL (20 μL)	ATCC 12472	3.0 ± 0.7	Mokhetho et al. (2018)
Cymbopogon flexuosus (Nees ex Steud.) W.Watson	1 mg/mL (20 μL)	ATCC 12472	3.3 ± 0.4	Mokhetho et al. (2018)
Cymbopogon giganteus Chiov.	1 mg/mL (20 μL)	ATCC 12472	2.0 ± 0.7	Mokhetho et al. (2018)
Cymbopogon martini (Roxb.) W.Watson	1 mg/mL (20 μL)	ATCC 12472	3.5 ± 1.4	Mokhetho et al. (2018)
Cymbopogon winterianus Jowitt	1 mg/mL (20 μL)	ATCC 12472	2.3 ± 1.1	Mokhetho et al. (2018)
Dorema aucheri Boiss.	N/A	CV026	N/A	Sepahi et al. (2015)
Eucalyptus citriodora Hook *(China)*	10 μL	ATCC 12472	8.36 ± 0.93	Luís et al. (2017)
Eucalyptus dives Schauer	1 mg/mL (20 μL)	ATCC 12472	3.0 ± 0.7	Mokhetho et al. (2018)
Eucalyptus globulus Labill.	N/A	ATCC12472	10	Luís et al. (2016)
Eucalyptus globulus Labill.	1 mg/mL (20 μL)	ATCC 12472	1.0 ± 0.7	Mokhetho et al. (2018)
Eucalyptus radiata Sieber ex DC.	N/A	ATCC12472	20	Luís et al. (2016)
Eugenia caryophyllata Thunb.	1 mg/mL (20 μL)	ATCC 12472	3.3 ± 0.4	Mokhetho et al. (2018)
Ferula asafoetida L.	N/A	CV026	N/A	Sepahi et al. (2015)
Foeniculum vulgare Mill.	subMICs	CV026	N/A	Venkadesaperumal et al., (2016)
Gaultheria procumbens L.	1 mg/mL (20 μL)	ATCC 12472	3.0 ± 1.4	Mokhetho et al. (2018)
Geranium robertianum L.	100%	CV026+ATTC 31301	13	Szabó et al. (2010)
Geranium robertianum L.	100%	CV026+Ezf 10–20	15	Szabó et al. (2010)
Hyptis dilatata Benth.	30/60 μL (diluted)	CV31592	N/A	Álvarez et al. (2015)
Inula graveolens (L.) Desf.	1 mg/mL (20 μL)	ATCC 12472	3.3 ± 1.1	Mokhetho et al. (2018)
Juniperus communis L.	3 μL	CV6269	20	Kerekes et al. (2013)
Juniperus communis L. *(1)*	1 mg/mL (20 μL)	ATCC 12472	1.3 ± 1.1	Mokhetho et al. (2018)
Juniperus communis L. *(2)*	1 mg/mL (20 μL)	ATCC 12472	3.0 ± 0.7	Mokhetho et al. (2018)
Lavandula angustifolia Mill.	subMICs	CV026	10	Khan et al. (2009)
Lavandula angustifolia Mill.	subMICs	ATCC12472	11	Khan et al. (2009)
Lavandula angustifolia Mill. *(France)*	10 μL	ATCC 12472	8.60 ± 0.56	Luís et al. (2017)
Lavandula angustifolia Mill. *(High altitude, France)*	10 μL	ATCC 12472	5.01 ± 0.43	Luís et al. (2017)
Lavandula buchii Webb *var. tolpidifolia* Svent. *(3)*	1 mg/mL (20 μL)	ATCC 12472	0.5 ± 0.0	Mokhetho et al. (2018)
Lavandula buchii Webb *var. tolpidifolia (Svent)* M. C. León Labill. *(2)*	1 mg/mL (20 μL)	ATCC 12472	1.0 ± 0.0	Mokhetho et al. (2018)
Lavandula officinalis Chaix ex Vill. *(Croatia)*	10 μL	ATCC 12472	4.84 ± 0.08	Luís et al. (2017)

(*Continued*)

TABLE 16.9 (*Continued*)
Essential Oils Inhibiting Quorum Sensing Detected by Disc Diffusion Assay

EO	Concentration	Sensor Strain	Diameter of QS Inhibition (mm)	Reference
Lippia citriodora (Palau) *Kunth*	1 mg/mL (20 μL)	ATCC 12472	0.5 ± 0.0	Mokhetho et al. (2018)
Lippia graveolens Kunt	25 mg/mL	ATCC12472	5.8±1.04	Alvarez et al. (2014)
Melaleuca alternifolia (Maiden & Betche) Cheel	N/A	ATCC 12472	15–19	María V. Alvarez et al. (2012)
Mentha × citrata Ehrh	1 mg/mL (20 μL)	ATCC 12472	4.0 ± 0.7	Mokhetho et al. (2018)
Mentha × piperita L.	1 mg/mL (20 μL)	ATCC 12472	1.0 ± 0.0	Mokhetho et al. (2018)
Mentha arvensis L.	1 mg/mL (20 μL)	ATCC 12472	1.0 ± 0	Mokhetho et al. (2018)
Mentha piperita L.	subMICs	CV026	10	Khan et al. (2009)
Mentha piperita L.	subMICs	ATCC12472	11	Khan et al. (2009)
Mentha pulegium L.	1 mg/mL (20 μL)	ATCC 12472	2.3 ± 0.4	Mokhetho et al. (2018)
Minthostachys mollis (Kunth) Griseb.	pure EO/N/A	ATCC12472	>90 ± 0.1	Pellegrini et al. (2014)
Myrtus communis L. *(1)*	1 mg/mL (20 μL)	ATCC 12472	4.0 ± 1.4	Mokhetho et al. (2018)
Nardostachys jatamansi (D.Don) DC.	1 mg/mL (20 μL)	ATCC 12472	1.3 ± 0.4	Mokhetho et al. (2018)
Origanum compactum Benth.	1 mg/mL (20 μL)	ATCC 12472	2.5 ± 0.0	Mokhetho et al. (2018)
Origanum majorana L.	100% 3 μL	CV6269	20	Szabó et al. (2010)
Origanum majorana L. *(1)*	1 mg/mL (20 μL)	ATCC 12472	0.5 ± 0.0	Mokhetho et al. (2018)
Origanum majorana L. *(2)*	1 mg/mL (20 μL)	ATCC 12472	0.8 ± 0.4	Mokhetho et al. (2018)
Pinus ponderosa Douglas ex C. Lawson	1 mg/mL (20 μL)	ATCC 12472	3.0 ± 0.0	Mokhetho et al. (2018)
Piper nigrum L.	50 μL/mL	CV026	N/A	Venkadesaperumal et al. (2016)
Rosa damascena Mill.	100%	CV026+ATTC 31298	18	Szabó et al. (2010)
Rosa damascena Mill.	100%	CV026+Ezf 10–17	20	Szabó et al. (2010)
Rosmarinus officinalis L. *(Tunisia)*	10 μL	ATCC 12472	8.69 ± 0.81	Luís et al. (2017)
Rosmarinus officinalis L	100%	CV026+ATTC 31303	18	Szabó et al. (2010)
Rosmarinus officinalis L	100%	CV026+Ezf 10–22	18	Szabó et al. (2010)
Rosmarinus officinalis L.	1 mg/mL (20 μL)	ATCC 12472	0.5 ± 0.0	Mokhetho et al. (2018)
Rosmarinus officinalis L.	N/A	ATCC 12472	15–19	Alvarez et al. (2012)
Salvia officinalis L.	100%	ATCC12472	>90 ± 0.1	Pellegrini et al. (2014)
Salvia sclarea L.	3 μL	CV6269	25	Szabó et al. (2010)
Satureja montana L.	1 mg/mL (20 μL)	ATCC 12472	2.3 ± 0.4	Mokhetho et al. (2018)
Satureja odora (Gris.) Epl.	100%	ATCC12472	90 ±1	Pellegrini et al. (2014)
Schinus molle L.	100%	ATCC12472	50±1.2	Pellegrini et al. (2014)
Styrax benzoin Dryand	1 mg/mL (20 μL)	ATCC 12472	3.5 ± 0.7	Mokhetho et al. (2018)
Syzigium aromaticum (L.) Merr. & L.M. Perry	subMICs	CV026	17	Khan et al. (2009)
Syzigium aromaticum (L.) Merr. & L.M. Perry	subMICs	ATCC12472	19	Khan et al. (2009)
Thymus vulgaris L. *(1)*	1 mg/mL (20 μL)	ATCC 12472	2.3 ± 1.1	Mokhetho et al. (2018)
Thymus vulgaris L. *(2)*	1 mg/mL (20 μL)	ATCC 12472	2.3 ± 0.0	Mokhetho et al. (2018)

16.9.1.2 Flask Incubation Assay

TABLE 16.10
Essential Oils Inhibiting Quorum Sensing Detected by Flask Incubation Assay

EO	Concentration	Sensor strain	% Inhibition	Reference
Artemisia annua L.	MQSIC = 0.0138%	ATCC12472	500%	Pellegrini et al. (2014)
Artemisia dracunculus L.	MQSIC = 0.6 mg/mL	ATCC 12472	60.0%	Mokhetho et al. (2018)
Calamintha nepeta (L.) Savi *ssp. nepeta*	MQSIC = 0.2 mg/mL	CV CIP 103350T	50.0%	Poli et al. (2018)
Carum copticum L.	MQSIC = 0.23 mg/ML	ATCC 12472	50.0%	Snoussi et al. (2018)
Cedrus atlantica (Endl.) Manetti ex Carrière	MQSIC = 6.0 mg/mL	CV CIP 103350T	50.0%	Poli et al. (2018)
Chamaemelum nobile (L.) *All*	MQSIC = 0.5 mg/mL	ATCC 12472	50.0%	Mokhetho et al. (2018)
Cinnamomum verum J.Presl	subMICs	CV026	78.0%	Kalia et al. (2015)
Citrus clementina Hort ex Tan	MQSIC = 0.1 mg/mL	CV CIP 103350T	50.0%	Poli et al. (2018)
Citrus limon (L.) Osbeck	MQSIC = 0.4 mg/mL	CV CIP 103350T	50.0%	Poli et al. (2018)
Citrus limon (L.) Osbeck *(1)*	MQSIC = 0.1 mg/mL	ATCC 12472	10.0%	Mokhetho et al. (2018)
Citrus limon (L.) Osbeck *(2)*	MQSIC = 0.1 mg/mL	ATCC 12472	10.0%	Mokhetho et al. (2018)
Citrus reticulata Blanco *(1)*	MQSIC = 0.1 mg/mL	ATCC 12472	10.0%	Mokhetho et al. (2018)
Citrus reticulata Blanco *(2)*	MQSIC = 0.5 mg/mL	ATCC 12472	50.0%	Mokhetho et al. (2018)
Citrus sinensis (L.) Osbeck	MQSIC = 0.3 mg/mL	ATCC 12472	30.0%	Mokhetho et al. (2018)
Citrus sinensis L.	100%	CV026+Ezf 10–21	1500.0%	Szabó et al. (2010)
Coriandrum sativum L.	5 μL/mL	ATCC12472	>90%	Duarte et al. (2016)
Cuminum cyminum L.	50 μL/mL	CV026	>70%	Venkadesaperumal et al. (2016)
Cymbopogon citratus (DC.) Stapf.	MQSIC = 0.2 mg/mL	CV CIP 103350T	50.0%	Poli et al.
Cymbopogon citratus (DC.) Stapf	MQSIC = 0.1 mg/mL	ATCC 12472	10.0%	Mokhetho et al. (2018)
Cymbopogon flexuosus (Nees ex Steud.) W. Watson	MQSIC = 0.1 mg/mL	ATCC 12472	10.0%	Mokhetho et al. (2018)
Cymbopogon giganteus Chiov.	MQSIC = 0.1 mg/mL	ATCC 12472	10.0%	Mokhetho et al. (2018)
Cymbopogon martini (Roxb.) W. Watson	MQSIC (mg/mL) = 0.1 0.5 mg/mL	ATCC 12472	50% 95%	Mokhetho et al. (2018)
Cymbopogon winterianus Jowitt	MQSIC = 0.1 mg/mL	ATCC 12472	50.0%	Mokhetho et al. (2018)
Elettaria cardamomum (L.) Maton	subMICs	ATCC 12472	≈75%–80%	Abdullah et al. (2017)
Eucalyptus dives Schauer	MQSIC = 0.1 mg/mL	ATCC 12472	50.0%	Mokhetho et al. (2018)
Eucalyptus globulus Labill.	5 μL/mL	ATCC12472	>90%	Luís et al. (2016)
Eucalyptus globulus Labill.	100%	CV026+ATTC 31300	6 mm	Szabó et al. (2010)
Eucalyptus globulus Labill.	MQSIC = 0.1 mg/mL	ATCC 12472	50.0%	Mokhetho et al. (2018)
Eucalyptus polybractea R.T. Baker	MQSIC = 0.05 mg/mL	CV CIP 103350T	50.0%	Poli et al. (2018)

(*Continued*)

TABLE 16.10 (*Continued*)
Essential Oils Inhibiting Quorum Sensing Detected by Flask Incubation Assay

EO	Concentration	Sensor strain	% Inhibition	Reference
Eucalyptus radiata Sieber ex DC.	5 μL/mL	ATCC12472	>90%	Luís et al. (2016)
Eugenia caryophyllata Thunb.	50 mL (dil.)	CV026	80.0%	Eris and Ulusoy (2013)
Eugenia caryophyllata Thunb.	50 mL (dil.)	VIR07	72.0%	Eris and Ulusoy (2013)
Eugenia caryophyllata Thunb.	50 mL (dil.)	ATCC12472	39.0%	Eris and Ulusoy (2013)
Foeniculum vulgare Mill.	50 μL/mL	CV026 + C6HSL	75.7%	Sepahi et al. (2015)
Foeniculum vulgare Mill.	MQSIC = 0.2 mg/mL	CV CIP 103350T	50.0%	Poli et al.
Gaultheria procumbens L.	MQSIC = 0.1 mg/mL	ATCC 12472	50.0%	Mokhetho et al. (2018)
Helichrysum italicum (Roth) G. Don fil.	MQSIC = 0.8 mg/mL	CV CIP 103350T	50.0%	Poli et al.
Inula graveolens (L.) *Desf.*	MQSIC = 0.5 mg/mL	ATCC 12472	50.0%	Mokhetho et al. (2018)
Juniperus communis L. *(1)*	MQSIC = 0.3 mg/mL	ATCC 12472	50.0%	Mokhetho et al. (2018)
Juniperus communis L. *(2)*	MQSIC = 0.3 mg/mL	ATCC 12472	50.0%	Mokhetho et al. (2018)
Lippia alba (Mill.) N.E.Br. ex Britton & P.Wilson *(C) CT1*	MQSIC = 0.62 μg/mL	CV026 + C6HSL	50.0%	Olivero-Verbel et al. (2014)
Lippia alba (Mill.) N.E.Br. ex Britton & P.Wilson *(A) CT2*	MQSIC = 66.91 μg/mL	CV026 + C6HSL	50.0%	Olivero-Verbel et al. (2014)
Lippia alba (Mill.) N.E.Br. ex Britton & P.Wilson *(D) CT1*	MQSIC = 15.79 μg/mL	CV026 + C6HSL	50.0%	Olivero-Verbel et al. (2014)
Lippia alba (Mill.) N.E.Br. ex Britton & P.Wilson *(E) CT2*	MQSIC = 30.32 μg/mL	CV026 + C6HSL	50.0%	Olivero-Verbel et al. (2014)
Lippia alba (Mill.) N.E.Br. ex Britton & P.Wilson *(B) CT2*	lMQSIC = 2.24 μg/mL	CV026 + C6HSL	50.0%	Olivero-Verbel et al. (2014)
Lavandula angustifolia L.	100%	CV026 + ATTC 31299	10 mm	Szabó et al. (2010)
Lavandula angustifolia L.	100%	CV026 + Ezf 10–18	15 mm	Szabó et al. (2010)
Lavandula buchii Webb *var. tolpidifolia (Svent.)* M. C. León *(3)*	MQSIC = 0.1 mg/mL	ATCC 12472	50.0%	Mokhetho et al. (2018)
Lavandula buchii Webb *var. tolpidifolia (Svent)* M. C. León *(2)*	MQSIC = 0.1 mg/mL	ATCC 12472	50.0%	Mokhetho et al. (2018)
Lavandula stoechas L.	MQSIC = 0.2 mg/mL	CV CIP 103350T	50.0%	Poli et al. (2018)
Lepechinia floribunda (Benth.) Epling	MQSIC = 0.0137%	ATCC12472	50.0%	Pellegrini et al. (2014)
Linalool	5 μL/mL	ATCC12472	>90%	Duarte et al. (2016)
Lippia citriodora (Palau) Kunth	MQSIC = 0.5 mg/mL	ATCC 12472	50.0%	Mokhetho et al. (2018)
Lippia graveolens Kunt	0.13%	ATCC12472	>95%	Alvarez et al. (2014)

(*Continued*)

TABLE 16.10 (*Continued*)
Essential Oils Inhibiting Quorum Sensing Detected by Flask Incubation Assay

EO	Concentration	Sensor strain	% Inhibition	Reference
Matricaria recutita L.	50 mL (dil.)	CV026	67.0%	Eris and Ulusoy (2013)
Matricaria recutita L.	50 mL (dil.)	VIR07	65.0%	Eris and Ulusoy (2013)
Matricaria recutita L.	50 mL (dil.)	ATCC12472	25.0%	Eris and Ulusoy (2013)
Melaleuca alternifolia (Maiden & Betche) Cheel	0.125 μg/mL MQSIC = 0.21 μg/mL	ATCC 12472	80% 50%	Alvarez et al. (2012)
Mentha × citrata Ehrh	MQSIC = 0.5 mg/mL	ATCC 12472	50.0%	Mokhetho et al. (2018)
Mentha × piperita L.	MQSIC = 0.3 mg/mL	ATCC 12472	50.0%	Mokhetho et al. (2018)
Mentha arvensis L.	MQSIC = 0.1 mg/mL	ATCC 12472	50.0%	Mokhetho et al. (2018)
Mentha piperita L.	subMIC	CV026	83.3%	Husain et al. (2015)
Mentha pulegium L.	MQSIC = 0.1 mg/mL	ATCC 12472	50.0%	Mokhetho et al. (2018)
Mentha suaveolens Ehrh. *ssp. insularis* (Req.) Greuter	MQSIC = 0.1 mg/mL	CV CIP 103350T	50.0%	Poli et al. (2018)
Minthostachys mollis (Kunth) Griseb.	MQSIC = 0.0649%	ATCC12472	50.0%	Pellegrini et al. (2014)
Myrtus communis L.	MQSIC = 0.1 mg/mL	CV CIP 103350T	50.0%	Poli et al. (2018)
Myrtus communis L. *(1)*	MQSIC = 0.1 mg/mL	ATCC 12472	50.0%	Mokhetho et al. (2018)
Origanum compactum Benth.	MQSIC = 0.1 mg/mL	ATCC 12472	50.0%	Mokhetho et al. (2018)
Origanum majorana L. *(1)*	MQSIC = 0.5 mg/mL	ATCC 12472	50.0%	Mokhetho et al. (2018)
Origanum majorana L. *(2)*	MQSIC = 0.1 mg/mL	ATCC 12472	50.0%	Mokhetho et al. (2018)
Pinus ponderosa Douglas ex C.Lawson	MQSIC = 0.1 mg/mL	ATCC 12472	50.0%	Mokhetho et al. (2018)
Pinus sylvestris L.	50 mL (dil.)	CV026	75.0%	Eris and Ulusoy (2013)
Pinus sylvestris L.	50 mL (dil.)	VIR07	61.0%	Eris and Ulusoy (2013)
Pinus sylvestris L.	50 mL (dil.)	CV12472	32.0%	Eris and Ulusoy (2013)
Piper bogotense C. DC.	MQSIC = 93.1 μL/mL	CV026 + C6AHL	50.0%	Olivero et al. (2011)
Piper brachypodon C. DC.	MQSIC = 513.8 μL/mL	CV026 + C6AHL	50.0%	Olivero et al. (2011)
Piper bredemeyeri J. Jacq.	MQSIC = 45.6 μL/mL	CV026 + C6AHL	50.0%	Olivero et al. (2011)
Piper nigrum L.	subMICs	CV026 +C6HSL	>55%	Venkadesaperumal et al. (2016)
Rosa damascena Mill.	50 mL (dil.)	CV026	80.0%	Eris and Ulusoy (2013)
Rosa damascena Mill.	50 mL (dil.)	VIR07	70.0%	Eris and Ulusoy (2013)
Rosa damascena Mill.	50 mL (dil.)	CV12472	43.0%	Eris and Ulusoy (2013)
Rosmarinus officinalis L.	MQSIC = 0.5 mg/mL	ATCC 12472	50.0%	Mokhetho et al.
Rosmarinus officinalis L.	0.125 μg/mL MQSIC = 0.21 μg/mL	ATCC 12472	80% 50%	Alvarez et al. (2012)
Salvia officinalis L.	MQSIC = 0.0259%	ATCC1247	50.0%	Pellegrini et al. (2014)
Satureja montana L.	MQSIC = 0.1 mg/mL	ATCC 12472	50.0%	Mokhetho et al. (2018)
Schinus molle L.	MQSIC = 0.005%	ATCC12472	50.0%	Pellegrini et al. (2014)
Syzigium aromaticum (L.) Merr. & L.M. Perry	subMICs	ATCC12472	78.4%	Khan et al. (2009)
Thymus vulgaris L.	subMICs	CV026	90.0%	Myszka et al. (2016)
Thymus vulgaris L. *(1)*	MQSIC = 0.1 mg/mL	ATCC 12472	50.0%	Mokhetho et al. (2018)
Thymus vulgaris L. *(2)*	MQSIC = 0.1 mg/mL	ATCC 12472	50.0%	Mokhetho et al. (2018)
Xanthoxylum armatum DC.	MQSIC = 0.2 mg/mL	CV CIP 103350T	50.0%	Poli et al. (2018)

16.9.1.3 QS Inhibiting EOs Evaluated by Other Assays

TABLE 16.11
QS Inhibiting EOs Evaluated by Other Assays

EO	Concentration	Assay	Sensor Strain	QS Inhibition	Reference
Cinnamomum verum	subMICs	ß-Galactosidase	PAO1	38.0%	M. Kalia et al. (2015)
Cinnamomum verum	subMIC	Fluorescence	*E. Coli* [pSB1075] + 3-oxo-C12-HSL	N/A	Yap et al. (2014)
Cinnamomum verum	subMIC	Fluorescence	*E. coli* [pSB401] + C6-HSL	N/A	Yap et al. (2014)
Cinnamomum zeylanicum	50 mL	QSIS	QSIS4	N/A	Eris and Ulusoy (2013)
Citrus reticulata	4 mg/mL	ß-Galactosidase	ATCC27853	49.0%	Luciardi et al. (2016)
Citrus reticulata	4 mg/mL	ß-Galactosidase	ATCC27853	50.0%	Luciardi et al. (2016)
Citrus reticulata	4 mg/mL	ß-Galactosidase	HT5	38.0%	Luciardi et al. (2016)
Citrus reticulata	4 mg/mL	ß-Galactosidase	HT5	37.0%	Luciardi et al. (2016)
Cymbopogon nardus	50 mL	QSIS	QSIS5	N/A	Eris and Ulusoy (2013)
Elettaria cardamomum	1.2 mg/mL	Fluorescence	*P. putida* + pRK12 (C12)	31.0%	Jaramillo-Colorado et al. (2012)
Elettaria cardamomum	1.2 mg/mL	Fluorescence	*E. coli* + pJBA132 (C6)	21%–22%	Jaramillo-Colorado et al. (2012)
Eugenia caryophyllata	50 mL	QSIS	QSIS2	N/A	Eris and Ulusoy (2013)
Juniperus communis	50 mL	QSIS	QSIS6	N/A	Eris and Ulusoy (2013)
L.alba (3) CT 2	1.2 mg/mL	Fluorescence	*P. putida* + pRK12 (C12)	1.0%	Jaramillo-Colorado et al. (2012)
L.alba (3) CT 2	1.2 mg/mL	Fluorescence	*E. coli* + pJBA132 (C6)	72%–78%	Jaramillo-Colorado et al. (2012)
L.alba (2) CT 2	1.2 mg/mL	Fluorescence	*E.coli* + pJBA132 (C6)	15%–84%	Jaramillo-Colorado et al. (2012)
L.alba (4) CT2	1.2 mg/mL	Fluorescence	*E. coli* + pJBA132 (C6)	61%–62%	Jaramillo-Colorado et al. (2012)
L.alba (5) CT2	1.2 mg/mL	Fluorescence	*P. putida* + pRK12 (C12)	9.0%	Jaramillo-Colorado et al. (2012)
L.alba (5) CT2	1.2 mg/mL	Fluorescence	*E. coli* + pJBA132 (C6)	44%–72%	Jaramillo-Colorado et al. (2012)
Lavandula angustfolia	50 mL	QSIS	QSIS12	N/A	Eris and Ulusoy (2013)
Lavandula angustifolia	N/A	Luminescence	*E. Coli* [pSB1075] + 3-oxo-C12-HSL	N/A	Yap et al. (2014)
Lippia alba (1) CT 1	1.2 mg/mL	Fluorescence	*P. putida* + pRK12 (C12)	65.0%	Jaramillo-Colorado et al. (2012)

(Continued)

TABLE 16.11 (*Continued*)
QS Inhibiting EOs Evaluated by Other Assays

EO	Concentration	Assay	Sensor Strain	QS Inhibition	Reference
Lippia alba (1) CT 1	1.2 mg/mL	Fluorescence	*E. coli* + pJBA132 (C6)	18.0%	Jaramillo-Colorado et al. (2012)
Matricaria recutita	50 mL	QSIS	QSIS3	N/A	Eris and Ulusoy (2013)
Myntotachys mollis	1.2 mg/mL	Fluorescence	*E. coli* + pJBA132 (C6)	56%–65%	Jaramillo-Colorado et al. (2012)
Myrtus communis	50 mL	QSIS	QSIS7	N/A	Eris and Ulusoy (2013)
Ocotea sp.	1.2 mg/mL	Fluorescence	*P. putida* + pRK12 (C12)	25.0%	Jaramillo-Colorado et al. (2012)
Ocotea sp.	1.2 mg/mL	Fluorescence	*E. coli* + pJBA132 (C6)	71%–76%	Jaramillo-Colorado et al. (2012)
Pinus sylvestris	50 mL	QSIS	QSIS8	N/A	Eris and Ulusoy (2013)
Rosa damascena	50 mL	QSIS	QSIS1	N/A	Eris and Ulusoy (2013)
Salvia officinalis	50 mL	QSIS	QSIS9	N/A	Eris and Ulusoy (2013)
Satureja hortensis	subMIC	Hdl expression	*S. aureus* ATCC 25923	N/A	Sharifi et al. (2018)
Swinglea glutinosa	1.2 mg/mL	Fluorescence	*E. coli* + *pJBA132* (C6)	28%–73%	Jaramillo-Colorado et al. (2012)
Syzigium aromaticum	subMICs	ß-Galactosidase	PAO1	56.0%	Husain et al. (2013)
Thymus vulgaris	50 mL	QSIS	QSIS10	N/A	Eris and Ulusoy (2013)
Vanilla fragrans	50 mL	QSIS	QSIS11	Fluorescence	Eris and Ulusoy (2013)
Zingiber officinale	1.2 mg/mL	Fluorescence	*E. coli* + pJBA132 (C6)	34%–69%	Jaramillo-Colorado et al. (2012)

16.9.2 Negatively Tested Oils

TABLE 16.12
Essential Oils Showing no QSI Effect in the Respective Studies

EO	Assay	Sensor Strain	Reference
Cymbopogon martini	DD	CV026 + C6AHL	Khan et al. (2009)
Cymbopogon martini	DD	CV12472	Khan et al. (2009)
Apium graveolens	DD	CV026 + C6AHL	Khan et al. (2009)
Apium graveolens	DD	CV12472	Khan et al. (2009)
Boswellia carteri	DD	CV026 + C6AHL	Mukherji and Prabhune (2014)
Cananga odorata	DD	CV026 + C6AHL	Mukherji and Prabhune (2014)
Cedrus sp.	QSIS	QSIS1	Eris and Ulusoy (2013)
Citrus aurantium	QSIS	QSIS1	Eris and Ulusoy (2013)
Citrus bergamia	QSIS	QSIS1	Eris and Ulusoy (2013)
Citrus bergamia	DD	CV026 + C6AHL	Mukherji and Prabhune. (2014)
Citrus limon	DD	CV026 + C6AHL	Khan et al. (2009)
Citrus limon	DD	CV12472	Khan et al. (2009)
Citrus limon	DD	CV6269	Szabó et al. (2010)
Citrus limonum	QSIS	QSIS1	Eris and Ulusoy (2013)
Citrus paradisi	DD	CV026 + C6AHL	Khan et al. (2009)
Citrus paradisi	DD	CV12472	Khan et al. (2009)
Citrus reticulata	QSIS	QSIS1	Eris and Ulusoy. (2013)
Citrus sinensis	DD	CV026 + C6AHL	Khan et al. (2009)
Citrus sinensis	DD	CV12472	Khan et al. (2009)
Citrus sinensis	DD	CV026 + C6AHL	Mukherji and Prabhune (2014)
Citrus sinensis	V	CV026 +ATTC 31302	Szabó et al. (2010)
Cymbopogon citratus	DD	CV026 + C6AHL	Khan et al. (2009)
Cymbopogon citratus	DD	CV12472	Khan et al. (2009)
Cymbopogon citratus (South Africa)	DD	ATCC 12475	Luís et al. (2017)
Cymbopogon nardus	DD	CV026 + C6AHL	Mukherji and Prabhune (2014)
Eucalyptus globulus	QSIS	QSIS1	Eris and Ulusoy (2013)
Eucalyptus globulus	V	CV026 + Ezf 10–19	Szabó et al. (2010)
Eucalyptus smithii (Australia)	DD	ATCC 12472	Luís et al. (2017)
Eucalyptus sp	DD	CV026 + C6AHL	Mukherji and Prabhune (2014)
Eucalyptus sp.	DD	CV026 + C6AHL	Khan et al. (2009)
Eucalyptus sp.	DD	CV12472	Khan et al. (2009)
Eucalyptus staigeriana (Australia)	DD	ATCC 12473	Luís et al. (2017)
Eugenia caryophyllus	V	ATCC 12472	Mokhetho et al. (2018)
Foeniculum vulgare	DD	CV026 + C6AHL	Khan et al. (2009)
Foeniculum vulgare	DD	CV026 + C6AHL	Khan et al. (2009)
Foeniculum vulgare	DD	CV12472	Khan et al. (2009)
Foeniculum vulgare	DD	CV12472	Khan et al. (2009)
Foeniculum vulgare Miller	QSIS	QSIS1	Eris and Ulusoy (2013)
Jasminum officinale	QSIS	QSIS1	Eris and Ulusoy (2013)
Juniperus communis (Bosnia)	DD	ATCC 12476	Luís et al. (2017)
Juniperus communis	V	CV026 + ATTC 31302/Ezf 10–20	Szabó et al. (2010)
Juniperus oxycedrus	DD	ATCC 12472	Mokhetho et al. (2018)

(Continued)

TABLE 16.12 (*Continued*)
Essential Oils Showing no QSI Effect in the Respective Studies

EO	Assay	Sensor Strain	Reference
Juniperus oxycedrus	V	ATCC 12473	Mokhetho et al. (2018)
L.alba (2) CT 2	F	*P. putida* + pRK12 (C12)	Jaramillo-Colorado et al. (2012)
L.alba (4) CT2	F	*P. putida* + pRK12 (C12)	Jaramillo-Colorado et al. (2012)
Laurus nobilis	QSIS	QSIS1	Eris and Ulusoy (2013)
Lavandula angustifolia	F	*E. coli* [pSB401] + 3-oxo-C6-	Yap et al. (2014)
Lavandula angustifolia (Croatia)	DD	ATCC 12474	Luís et al. (2017)
Lavandula buchii var. tolpidifolia (1)	DD	ATCC 12472	Mokhetho et al. (2018)
Lavandula buchii var. tolpidifolia (1)	V	ATCC 12474	Mokhetho et al. (2018)
Ledum groenlandicum	DD	ATCC 12472	Mokhetho et al. (2018)
Ledum groenlandicum	V	ATCC 12475	Mokhetho et al. (2018)
Lilium candium	QSIS	QSIS1	Eris and Ulusoy (2013)
Matricaria recutica L.,	V	CV026 + ATTC 31302/Ezf 10–21	Szabó et al. (2010)
Melaleuca alternifolia	DD	CV026 + C6AHL	Mukherji and Prabhune (2014)
Melissa officinalis	QSIS	QSIS1	Eris and Ulusoy (2013)
Mentha piperita	DD	CV026+ C6AHL	Mukherji and Prabhune (2014)
Mentha piperita	QSIS	QSIS1	Eris and Ulusoy (2013)
Myntotachys mollis	F	*P. putida* + pRK12 (C12)	Jaramillo-Colorado et al. (2012)
Myristica fragrans	DD	CV026 + C6AHL	Khan et al. (2009)
Myristica fragrans	DD	CV12472	Khan et al. (2009)
Myroxylon pereirae	QSIS	QSIS1	Eris and Ulusoy (2013)
Nardostachys jatamansi	V	ATCC 12473	Mokhetho et al. (2018)
Nigella sativa	QSIS	QSIS1	Eris and Ulusoy (2013)
Ocimum basilicum	QSIS	QSIS1	Eris and Ulusoy. (2013)
Ocimum basilicum	DD	CV026 + C6AHL	Mukherji and Prabhune (2014)
Olea europa	QSIS	QSIS1	Eris and Ulusoy (2013)
Olea europaea	DD	CV026 + C6AHL	Khan et al. (2009)
Olea europaea	DD	CV12472	Khan et al. (2009)
Orange	V	CV026 + ATTC 31302/Ezf 10–19	Szabó et al. (2010)
Petroselinum crispum	DD	CV026 + C6AHL	Khan et al. (2009)
Petroselinum crispum	DD	CV12472	Khan et al. (2009)
Pinus sp.	QSIS	QSIS1	Eris and Ulusoy (2013)
Punica granatum	QSIS	QSIS1	Eris and Ulusoy (2013)
Rosmarinus off.	DD	CV026 + C6AHL	Mukherji and Prabhune (2014)
Rosmarinus officinalis	DD	CV026 + C6AHL	Khan et al. (2009)
Rosmarinus officinalis	DD	CV12472	Khan et al. (2009)
Rosmarinus officinalis	QSIS	QSIS1	Eris and Ulusoy (2013)
Santalum album	DD	CV026 + C6AHL	Khan et al. (2009)
Santalum album	DD	CV12472	Khan et al. (2009)
Santalum album	QSIS	QSIS1	Eris and Ulusoy (2013)
Styrax benzoin	V	ATCC 12474	Mokhetho et al. (2018)
Swinglea glutinosa	F	*P. putida* + pRK12 (C12)	Jaramillo-Colorado et al. (2012)
T. daenensis	*hdl* expression	*S. aureus* ATCC 25923	Sharifi et al. (2018)

(Continued)

TABLE 16.12 (*Continued*)
Essential Oils Showing no QSI Effect in the Respective Studies

EO	Assay	Sensor Strain	Reference
Thymus vulgaris	DD	CV026 + C6AHL	Khan et al. (2009)
Thymus vulgaris	DD	CV12472	Khan et al. (2009)
Trachyspermum ammi	DD	CV026+ C6AHL	Khan et al. (2009)
Trachyspermum ammi	DD	CV12472	Khan et al. (2009)
Viola odorata	QSIS	QSIS1	Eris and Ulusoy (2013)
Vitis vinifera	QSIS	QSIS1	Eris and Ulusoy (2013)
Zea mays	DD	CV026 + C6AHL	Khan et al. (2009)
Zea mays	DD	CV12472	Khan et al. (2009)
Zingiber officinale	F	*P. putida* + pRK12 (C12)	Jaramillo-Colorado et al. (2012)
Zingiber officinale	DD	CV026 + C6AHL	Khan et al. (2009)
Zingiber officinale	DD	CV12472	Khan et al. (2009)
Zingiber officinale	QSIS	QSIS1	Eris and Ulusoy (2013)

Abbreviations: DD = Disc diffusion assay, V = Flask incubation assay including spectrophotometric violacein quantification, QSIS: QS-inhibitor-selector assay, β: β-galactosidase assay, F: Fluorescence-based assay, L: Luminescence-based assay, hdl-expression-assay

16.9.3 EOs Tested on the Inhibition of QS-Related Processes

TABLE 16.13
Essential Oils Tested on the Inhibition of QS-Related Processes

EO	Sensor Strain	Positive	Negative	Reference
Cinnamomum tamala	*V. parahaemolyticus* (ATCC 17802)	Biofilm eDNA EPS Alginate, Swarming Swimming		Farisa Banu et al. (2018)
Cinnamomum verum	PAO1	Biofilm, protease Pyocyanin Alginate Swarming motility		M. Kalia et al. (2015)
Citrus reticulata EOP	ATCC27853	Biofilm Elastase B		Luciardi et al. (2016)
Citrus reticulata EOP	HT5	Biofilm Elastase B		Luciardi et al. (2016)
Citrus reticulata EOPD	ATCC27853	Biofilm Elastase B		Luciardi et al. (2016)
Citrus reticulata EOPD	HT5	Biofilm Elastase B		Luciardi et al. (2016)
Dorema aucheri	PAO1	Biofilm Elastase, Pyoverdine	Pyocyanin	Sepahi et al. (2015)
Eugenia caryophyllata	PAO1		Biofilm	Eris and Ulusoy (2013)
Eugenia caryophyllata	PCI	Biofilm		Eris and Ulusoy (2013)
Ferula asafoetida	PAO1	Biofilm Elastase Pyoverdine Pyocyanin		Sepahi et al. (2015)

(*Continued*)

TABLE 16.13 (*Continued*)
Essential Oils Tested on the Inhibition of QS-Related Processes

EO	Sensor Strain	Positive	Negative	Reference
Matricaria recutita	PAO1		Biofilm	Eris and Ulusoy (2013)
Matricaria recutita	PCI	Biofilm		Eris and Ulusoy (2013)
Mentha piperita	PAO1	Biofilm Elastase Protease Pyoverdine Pyocyanin Chitinase EPS Swarming		Husain et al. (2015)
Mentha piperita	*Aeromonas hydrophila*	Biofilm Protease EPS		Husain et al. (2015)
Pinus sylvestris	PAO1		Biofilm	Eris and Ulusoy (2013)
Pinus sylvestris	PCI	Biofilm		Eris and Ulusoy (2013)
Rosa damascena	PAO1		Biofilm	Eris and Ulusoy (2013)
Rosa damascena	PCI	Biofilm		Eris and Ulusoy (2013)
Satureja hortensis	*Staphyloccus aureus*	Biofilm Planctonic growth		Sharifi et al. (2018)
Syzigium aromaticum	PAO1	Biofilm Swarming		Khan et al. (2009)
Syzigium aromaticum	*Aeromonas hydrophila*	Biofilm Rotease EPS		Husain et al. (2013)
Syzigium aromaticum	PAO1	Biofilm Elastase Protease Pyocyanin Chitinase EPS Swimming		Husain et al. (2013)
Thymus daenensis	*Staphylococcus aureus*	Biofilm Planktonic growth		Sharifi et al. (2018)
Thymus vulgare	*Pseudomonas fluorescens*	Biofilm Swimming *fg1A* gene-expression		Myszka et al. (2016)

16.10 DISCUSSION

As presented before, several EOs where tested positively on QS inhibition or inhibition of QS-related processes like biofilm formation and virulence-factor production in several different sensor strains. The large variety of assays used to detect QS inhibition of EOs and deviations in their execution makes it difficult to compare the activity of the oils. To illustrate this point, the aspects of the studies complicating comparability are present here.

16.10.1 The QS Activity of EOs Can Only Be Valued Individually for the Respective Assay

QS inhibitors detected by the disc diffusion assay using *C. violaceum* CV026 or ATCC12472 as sensor strains with an inhibition of >20 mm can be accounted as extremely sensitive. EOs belonging to this group are *Artemisia annua, Coriandrum sativum, Eucalyptus radiata, Juniperus communis, Minthostachys mollis, Origanum majorana, Salvia sclarea, Satureja odora*, and *Schinus molle.*

15–19 mm inhibition can still be valued as sensitive and was visible in the following oils: *Cinnamomum verum, Coriandrum sativum, Geranium robertianum, Melaleuca alterinifolia, Rosa damascena* L., *Rosmarinus officinalis, Syzygium aromaticum, Cinnamomum verum, Eucalyptus globulus, Geranium robertianum, Lavandula angustifolia*, and *Mentha piperata* EOs showed a moderate effect of 9–14 mm QSI. Repetitions emerging in different groups are due to diverging results in the studies, which is not surprising as oils often were obtained in different countries resulting in altered constituents, and varying amounts of EOs were used depending on the study.

When looking at the results from the flask incubation assay, the introduction of MQSIC (concentration of EO leading to a 50% reduction of violacein production) allows a better comparability of EOs, whereas a low MQSIC reveals high QSI potential. EOs revealing a very low MQSIC of >0.1 mg/mL are different *Lippia alba* hemotypes, *Piper bredemeyeri, Piper bogotense, Schinus molle*, and *Eucalyptus polybractea.* MQSIC of 0.1 mg/mL was measured, for example, by thyme, oregano, myrtle, lavender, eucalyptus, and different mint oils. When comparing results of those different assays, it is visible that in some cases, EOs show a QSI activity in several assays like *Cinnamomum verum* which was proved to be active in the disc diffusion assay, flask incubation assay (both *C. violaceum* strains), ß-galactosidase-assay, and fluorescence assay (both *E. coli* strains). Although some oils show rather conflicting results depending on the assay. An example is oregano oil (*Lippia graveolens* Kunt) which was highly active in the flask incubation assay but has a rather low activity when being tested by disc diffusion assay (Alvarez et al., 2014).

16.10.2 An EO Being an Active Inhibitor of AHL-Based QS in One Strain is Not Necessarily Active in Another Strain Also Producing AHLs

Due to the varying side chains of AHLs in different organisms, EOs show different activities depending on the QS system of the tested organism, which was demonstrated by Yap et al. (2014) when investigating the effect of lavender oil on either short chain or long chain AHL sensitive sensor strains. For one thing, this aspect is important to consider when conducting studies on EOs, as the respective AHLs have to be defined. This ambiguity exists, for example, in the study done by Szabo et al. (2010), as the research team did not specify the AHLs produced by *E. coli* ATTC 31298 and Ezf 10–17; they just referred to an unpublished work by Szegedi. This point should be considered before assuming that specific EOs already tested positively might inhibit QS in other bacteria with AHL-based QS.

16.10.3 EOs Often Not Only Inhibit QS but Also Inhibit Growth

As mentioned before, the disc diffusion assay makes a clear differentiation between QS inhibition and growth inhibition, possible because a strain which is only QS inhibited is colorless and turbid whereas a clear halo indicates cell death. The assays often include concurrent cell viability tests in order to individually determine QS inhibition or EOs are used in subMIC concentrations.

16.10.4 Extraction Method and Collecting Site of the EOs Influences Test Results

Constituents of EOs vary significantly depending on their production method, resulting in different QS activity, which was shown, for example, by Jaramillo-Colorado et al. (2012) when investigating clove oil either extracted by microwave-assisted hydro-distillation or merely hydro-distillation. The relevance of the collecting site regarding constituents of the EOs was illustrated when comparing different EOs from *Lippia alba* (Jaramillo-Colorado et al., 2012).

As mentioned in Section 16.8, EOs are defined according to their extraction method, therefore it is quite relevant to include this information in publications, although this specification is often not mentioned in the discussed studies. This aspect should be considered when interpreting study results. An example is a study published by Ganesh and Vittal Rai (2015) in which they discussed the anti-QS activity of EOs. However, they used supercritical CO_2 extraction to obtain the oils, which is an extraction method not in accordance to the ISO definition of EOs (Ganesh and Vittal Rai, 2015). GC-MS-analysis can also give a clue on the extraction method, as oil ingredients may indicate an extraction method not being rule consistent. This was demonstrated by a study conducted by Eris and Ulusoy (2013) as they analyzed the components of the tested oils by GC-MS. The main clove oil components mentioned in the paper where the solvents propylene glycol (73.2%) and diethylene glycol (16.0%). They also found propylene glycol in chamomile oil. The team also mentioned "vanilla EO," although vanilla does not contain an EO (Eris and Ulusoy, 2013).

16.11 CONCLUSION

QS systems in different organisms vary widely; research on those pathways is especially concentrated on AHL-based QS, which may be due to the fact that many pathogenic bacteria possess LUXRI-homologous QS systems. As QS regulates a wide range of bacterial functions such as biofilm formation, virulence-factor production, or swarming motility, QS inhibitors represent a group of active substances, possibly having multifaceted effects on bacterial populations differing markedly from classical antibiotic targets. Research on EOs on QS inhibitors only started a few years ago with *in vitro* studies on *C. violaceum*. Given the fact that EOs already are well studied and are safe substances which are easy to gain, they meet important criteria for possible new drugs. *In vitro* studies on different sensor strains, all based on AHLs as autoinducers revealed a significant QS-inhibitory potential of many different EOs. Not only did several oils inhibit QS itself but also they inhibit QS-related processes. However, methods of detecting QS inhibition are widely, differing from each other, and therefore prevent a transparent comparison of the EOs. In summary, EOs are potent QS inhibitors in the tested sensor strains, but research on this topic is still at its beginning. As step toward clinical studies is using the preclinical infection model *Caenorhabditis elegans* (Husain et al., 2013; Ganesh and Vittal Rai,2016). Closer investigations are necessary to, for example, understand the exact inhibition mechanisms or to define the effective components of the oils

16.12 ABBREVIATIONS

Abbreviation	Definition
"I" in, e.g., LuxI/LasI	AI synthesis protein
"R" in, e.g., LuxR/LasR	AI receptor protein
ABC	ATP-binding cassette
agr	Accessory gene regulator
AHL or HSL	Acyl homoserine lactone
AI	Autoinducer
AI-2	*V. harveyi* autoinducer 2, a furanosyl borate diester
AIP	Autoinducer peptide
ATCC12472	*C. violaceum* wild type strain
ATCC31532	*C. violaceum* wild type strain
BHL	N-butyryl-L-homoserine-lactones
CAI-1	*V. cholerae* autoinducer 1
CGI	Cell growth inhibition
CqsA	*V. cholerae* quorum-sensing autoinducer
CqsS	*V. cholerae* quorum-sensing sensor
CT	Chemotype
CV ATCC31592	*C. violaceum* strain
CV026	*C. violaceum* mutant strain of ATCC31532
CviIR	QS system in *C. violaceum*
DD	Disc diffusion assay
EO(s)	Essential oil(s)

EPS	Exopolysacharide
F	Flourescence based assay
GC-EIMS	Gas chromatography electron ionization mass spectrometry
Gfp	Green fluorescence protein
HAI-1	*V. harveyi* autoinducer 1, an AHL
HD	Hydro-distillation
HHL	N-hexanoyl-HSL
HT5	"Hospital Tucuman 5," a *P. aeruginosa* strain isolated from a patient
IC_{50}	Concentration of EO leading to a 50% reduction of violacein production
IQS	Integrated quorum sensing signal
IZS	Inhibition zone size
L	Luminescence-based assay
LasRI	*P. aeruginosa* QS system, designated after ist elastase-regulating properties
LB agar	Luria Bertani agar/lysogeny broth, a growth medium for bacteria
LUX	Proteins coded by luminescence genes (*lux*)
lux	Luminescence genes (*lux* = Latin for light)
LUXRI homologous systems	QS systems homologous to LUXRI-QS in e.g. *V.fisheri* using AHL as Ais
MBC	Minimal bactericidal concentration
MFC	Minimal fungicidal concentration
MIC	Minimal inhibitory concentration
MQSIC	Minimum QS inhibitory concentration; the minimal EO concentration, inhibiting violacein production of *C. violaceum* by at least 50%
MWHD	Microwave assisted hydrodestillation
OD	Dptical density
OdDHL	N-(3-oxododecanoyl)-homoserine lactone
PAO1	Wild-type strain of *P. aeruginosa*
PBS	Phosphate buffered saline, a buffer solution
PCI	*P. aeruginosa* clinical isolate
PQS	pseudomonas quinolone signal
PqsR	PQS receptor
QS	Quorum sensing
QscR	QS-control repressor
QSI	QS inhibition/QS inhibitor(s)
QSIS	QSI selector, a QS-detection method
QSIS 1	QS strain constructed via QSI-selector method A
QSM	Quorum sensing molecule(s)
RhlIR	*P. aeruginosa* QS system, designated after RhlR
RhlR	Protein encoded by a rhamnolipid synthase gene cluster, AI receptor of BHL
RT-PCR	Real-time polymerase chain reaction
SEM	Scanning electron microscopy
SPME	Solid-phase microextraction
ß	ß-galactosidase assay
subMIC	Concentrations below MIC
V	Flask incubation assay
VIR07	*C. violaceum* mutant strain of ATCC12472
VqsR	virulence and QS regulator
ZOI	Zone of inhibition = antimicrobial activity DD
ZOT	Zone of turbidity = QSI DD

REFERENCES

Abdullah, A. A., M. S. Butt, M. Shalid, Q. Huang. 2017. Evaluating the antimicrobial potential of green cardamom essential oil focusing on quorum sensing inhibition of *Chromobacterium violaceum*. *J. Food Sci. Technol.* 54 (8): 2306–2315.

Adonizio, A. L., K. Downum, B. C. Bennett, K. Mathee. 2006. Anti-quorum sensing activity of medicinal plants in southern Florida. *J. Ethnopharmacol.* 105 (3): 427–435.

Agarwal, V., L. Priyanka, V. Pruthi et al. 2008. Prevention of *Candida albicans* biofilm by plant oils. *Mycopathologia* 165 (1): 13–19.

Albuquerque, P., A. Casadevall. 2012. Quorum sensing in fungi--a Review. *Med. Mycol.* 50 (4): 337–345.

Álvarez, F., E. Tello, K. Bauer et al. 2015. Cytotoxic and antimicrobial diterpenes from *Hyptis dilatata*. *Curr. Bioact. Compd.* 11 (3): 189–197.

Alvarez, M. V., M. R. Moreira, A. Ponce. 2012. Antiquorum sensing and antimicrobial activity of natural agents with potential use in food. *J. Food Saf.* 32 (3): 379–387.

Alvarez, M. V., L. A. Ortega-Ramirez, M. M. Gutierrez-Pacheco et al. 2014. Oregano essential oil-pectin edible films as anti-quorum sensing and food antimicrobial agents. *Front. Microbiol* 5 (12): 1–7.

August, P. R., T. H. Grossman, C. Minor et al. 2000. Sequence analysis and functional characterization of the violacein biosynthetic pathway from *Chromobacterium violaceum*. *J. Mol. Microbiol Biotechnol.* 2 (4): 513–519.

Bala, A., R. Kumar, K. Harjai. 2011. Inhibition of quorum sensing in *Pseudomonas aeruginosa* by azithromycin and its effectiveness in urinary tract infections." *J Med. Microbiol.* 60: 300–306.

Banu, S. F., D. Rubini, R. Murugan et al. 2018. Exploring the antivirulent and sea food preservation efficacy of essential oils combined with DNase on *Vibrio parahaemolyticus*. *Lwt-Food Science and Technology* 95: 107–115.

Bassler, L. Bonnie. 2002. Small talk: Cell-to-cell communication in bacteria. *Cell* 109 (4): 421–424.

Bassler B. L., M. Wright, R. E. Showalter, M. R. Silverman. 1993. Intercellular signalling in *Vibrio harveyi*: Sequence and function of genes regulating expression of luminescence. *Mol. Microbiol.* 9 (4): 773–786.

Bassler B. L., M. Wright, M. R. Silverman. 1994. Multiple signalling systems controlling expression of luminescence in *Vibrio harveyi*: Sequence and function of genes encoding a second sensory pathway. *Mol. Microbiol.* 13 (2): 273–286.

Bauer A. W., W. M. Kirby, J. C. Sherris, M. Turck. 1966. Antibiotic susceptibility testing by a standardized single disk method. *Tech. Bull. Regist. Med. Technol.* 6 (3): 49–52.

Blosser, R. S., K. M. Gray. 2000. Extraction of Violacein from *Chromobacterium violaceum* provides a new quantitative bioassay for N-acyl homoserine lactone autoinducers. *J. Microbiol. Methods.* 40 (1): 47–55.

Bouayed, N., N. Dietrich, C. Lafforgue et al. 2016. Process-oriented review of bacterial quorum quenching for membrane biofouling mitigation in membrane bioreactors (MBRs). *Membranes* 6 (4): 52–58.

Bryers, J. D. 2009. Medial biofilms. *Biotechnol. Bioeng.* 2008. 100 (1): 1–18.

Burt, S. A. 2004. Essential oils: Their antibacterial properties and potential applications in foods - a review. *Int. J. Food Microbiol.* 94 (3): 223–253; cited from: Gracia-Valenzuela, M. H., C. Orozco-Medina, C. Molina-Maldonado et al. 2012. Antibacterial effect of essential oregano oil (*Lippia berlandieri*) on pathogenic bacteria of shrimp *Litopenaeus vannamei*. *Hidrobiologica* 22(3): 201–206.

Campbell, N. A., J. B. Reece J. B. 2009. *Biologie*, 8. Aktualisierte Auflage. Verlag Pearson Studium, München, pp 280–287. ISBN 978-3-8273-7287-1, 280–87.

Cao, J. G., E. A. Meighen. 1989. Purification and structural identification of an autoinducer for the luminescence system of *Vibrio harveyi*. *J. Biol. Chem.* 264 (36): 21670–21676.

Chen, H., G. R. Fink. 2006. Feedback control of morphogenesis in fungi by aromatic alcohols. *Gene Development.* 20: 1150–1161.

Chen X., S. Schauder, N. Potier et al., 2002. Structural identification of a bacterial quorum-sensing signal containing boron. *Nature* 415 (6871): 545–549.

Cho, T., T. Aoyama, M. Toyoda et al. 2008. Farnesol as a quorum-sensing molecule in *Candida albicans*. *Japanese Journal of Medical Mycology* 49 (4): 281–286.

Choo, J. H., Y. Rukayadi, J. K. Hwang et al. 2006. Inhibition of bacterial quorum sensing by vanilla extract. *Lett. Appl. Microbiol.* 42 (6): 637–641.

Chugani, S. A., M. Whiteley, K. M. Lee et al. 2001. QscR, a modulator of quorum-sensing signal synthesis and virulence in *Pseudomonas aeruginosa*. *Proc. Natl. Acad. Sci. U. S. A.* 98 (5): 2752–2757.

Cole, S. P., J. Harwood, R. Lee et al. 2004. Characterization of monospecies biofilm formation by *Helicobacter pylori*. *J. Bacteriol.* 186 (10): 3124–3132.

Costerton, W. J., Z. Lewandowski, D. E. Caldwell et al. 1995. Microbial biofilms. *Annu. Rev. Microbiol.* 49 (1): 711–745.

Dong, Y. H., L. Y. Wang, L. H. Zhang et al. 2007. Quorum-quenching microbial infections: Mechanisms and implications. *Philos. Trans. R. Soc. Lond. Series B, Biol. Sci.* 362 (1483): 1201–1211.
Duarte, A., A. Luis, M. Oleastro, F. C. Domingues. 2016. Antioxidant properties of coriander essential oils and linalool and their potential to control *Campylobacter* spp. *Food Control* 61: 115–122.
Eberhard, A. A., L. Burlingame, C. E. Eberhard et al. 1981. Structural identification of autoinducer of *Photobacterium fischeri* luciferase. *Biochemistry* 20 (9): 2444–2449.
Engebrecht, J., M. Silverman. 1984. Identification of genes and gene products necessary for bacterial bioluminescence. *Proc. Natl. Acad. Sci. U. S. A.* 81 (13): 4154–4158.
Erez, Z., I. Steinberger-Levy, M. Shamir et al. 2017. Communication between viruses guides lysis-lysogeny decisions. *Nature* 541 (7638): 488–493.
Eris, R., S. Ulusoy. 2013. Rose, clove, chamomile essential oils and pine turpentine inhibit quorum sensing in *Chromobacterium violaceum* and *Pseudomonas aeruginosa*. *J. EO Bear. Plants* 16 (2): 126–135.
Feoktistova, M., P. Geserick, M. Leverkus et al. 2016. Crystal violet assay for determining viability of cultured cells. *Cold Spring Harb. Protoc.* 2016 (4): pdb.prot087379. doi:10.1101/pdb.prot087379.
Fuqua, W. C., S. C. Winans, E. P. Greenberg. 1994. Quorum sensing in bacteria: The LuxR-Luxl family of cell density-responsive transcriptional regulators. *J. Bacteriol.* 176 (2), 269–275.
Ganesh, P. S., R. Vittal Rai. 2015. Evaluation of anti-bacterial and anti-quorum sensing potential of essential oils extracted by supercritical CO_2 method against *Pseudomonas aeruginosa*. *J. EO Bear. Plants* 18 (2): 264–275.
Ganesh, P. S., R. Vittal Rai. 2016. Inhibition of quorum-sensing-controlled virulence factors of *Pseudomonas aeruginosa* by *Murraya koenigii* essential oil: A study in a *Caenorhabditis elegans* infectious model. *J. Med. Microbiol.* 65 (12): 1528–1535.
Greenberg, E. P. 1997. Quorum sensing in gram-negative bacteria. *ASM News*. 63 (7): 371–377.
Haque, S., F. Ahmad, S. A. Dar et al. 2018. Developments in strategies for quorum sensing virulence factor inhibition to combat bacterial drug resistance. *Microb. Pathog.* 121 (March): 293–302.
Henke, J. M., B. L. Bassler. 2004. Three parallel quorum-sensing systems regulate gene expression in *Vibrio harveyi*. *J. Bacteriol.* 186 (20): 6902–6914.
Hentzer, M., H. Wu, J. B. Andersen et al. 2003. Attenuation of *Pseudomonas aeruginosa* virulence by quorum-sensing inhibitors. *EMBO J.* 22 (15): 3803–3815.
Hogan, D. 2006. Talking to themselves: Autoregulation and quorum sensing in fungi. *Eukaryot. Cell* 5 (4): 613–619.
Hornby, J. M., E. C. Jensen, A. D. Lisec et al. 2001. Quorum sensing in the dimorphic fungus *Candida albicans* is mediated by farnesol. *Appl. Environ. Microbiol.* 67 (7): 2982–2992.
Huan Liu, S. Srinivas, X. He. 2013. Quorum sensing in *Vibrio* and its relevance to bacterial virulence. *J. Bacteriol. Parasitol.* 04 (03). doi:10.4172/2155-9597.1000172.
Husain, F. M., I. Ahmad, A. Asif et al. 2013. Influence of clove oil on certain quorum-sensing-regulated functions and biofilm of *Pseudomonas aeruginosa* and *Aeromonas hydrophila*. *J. Biosci.* 38 (5): 835–44.
Husain, F. M., I. Ahmad, M. S. Khan et al. 2015. Sub-MICs of *Mentha piperita* EO and menthol Inhibits AHL mediated quorum sensing and biofilm of gram-negative bacteria. *Front. Microbiol.* 6: 420–432.
ISO/9235.2. ISO-Rule 1997, Aromatic Natural Raw Materials-Vocabulary, Geneva, Switzerland: International Standard Organisation, 2013, Standards catalogue, ISO/TC54-Essential Oils, https://www.iso.org/obp/ui/#iso:std:iso:9235:ed-2:v1:en:term:2.11.
Jaramillo-Colorado, B., J. Olivero-Verbel, E. E. Stashenko et al. 2012. Anti-quorum sensing activity of essential oils from Colombian plants. *Nat. Prod. Res.* 26 (12): 1075–86.
Jensen, R. O., K. Winzer, S. R. Clarke et al. 2008. Differential recognition of *Staphylococcus aureus* quorum-sensing signals depends on both extracellular loops1 and 2 of the transmembrane sensor AgrC. *J. Mol. Biol.*, 381: 300–309.
Juhas, M., L. Wiehlmann, B. Huber et al. 2004. Global regulation of quorum sensing and virulence by VqsR in *Pseudomonas aeruginosa*. *Microbiology* 150 (4): 831–841.
Kalia, V. C. 2013. Quorum sensing inhibitors: An overview. *Biotechnol. Adv.* 31 (2): 224–245.
Kalia, M., V. Kumar Yadav, P. Kumar Singh et al. 2015. Effect of cinnamon oil on quorum sensing-controlled virulence factors and biofilm formation in *Pseudomonas aeruginosa*. *PLoS ONE* 10 (8): 1–18.
Kaplan, H. B., E. P. Greenberg. 1985. Diffusion of autoinducer is involved in regulation of the *Vibrio fischeri* luminescence system. *J. Bacteriol.* 163 (3): 1210–1214.
Kerekes, E. B., E. Deak, M. Tako et al. 2013. Anti-biofilm forming and anti-quorum sensing activity of selected essential oils and their main components on food-related micro-organisms. *J. Appl. Microbiol.* 115 (4): 933–942.
Khan, M. S. A., M. Zahin, S. Hasan et al. 2009. Inhibition of quorum sensing regulated bacterial functions by plant essential oils with special reference to clove oil. *Lett. Appl. Microbiol.* 49 (3): 354–360.

Kim, Y.-G., J. H. Lee, G. Gwon et al. 2016. Essential oils and eugenols inhibit biofilm formation and the virulence of *Escherichia coli* O157:H7. *Sci. Rep.* 6: 1–11.

Kleerebezem, M., L. E. Quadri, O. P. Kuipers et al. 1997. Quorum sensing by peptide pheromones and two-component signal-transduction systems in gram-positive bacteria. *Mol. Microbiol.* 24 (5): 895–904.

Krishnan, T., W.-F. Yin, K.-G. Chan et al. 2012. Inhibition of quorum sensing-controlled virulence factor production in pseudomonas aeruginosa PAO1 by Ayurveda spice clove (Syzygium aromaticum) bud extract. *Sensors* 12 (4): 4016–4030.

Kulkarni, R. K., K. W. Nickerson. 1980. Nutritional control of dimorphism in *Ceratocystis ulmi*. *Exp. Mycol.* 5 (2): 149–154.

Le Thi, T. T., C. Prigent-Combaret, C. Dorel, P. Le Lejeune. 2001. First stages of biofilm formation: Characterization and quantification of bacterial functions involved in colonization process. *Methods Enzymology* 336: 152–159.

Lee, J., L. Zhang. 2014. The hierarchy quorum sensing network in *Pseudomonas aeruginosa*. *Protein Cell* 6 (1): 26–41.

Luciardi, M. C., M. Amparo-Blázquez, E. Cartagena et al. 2016. Mandarin essential oils inhibit quorum sensing and virulence factors of *Pseudomonas aeruginosa*. *LWT - Food Sci. Technol.* 68: 373–380.

Luís, A., A. Duarte, J. Gominho et al. 2016. Chemical composition, antioxidant, antibacterial and anti-quorum sensing activities of *Eucalyptus globulus* and *Eucalyptus radiata* EOs. *Ind. Crops Prod.* 79: 274–282.

Luís, A., A. Duarte, L. Pereira et al. 2017. Chemical profiling and evaluation of antioxidant and anti-microbial properties of selected commercial essential oils: A comparative study. *Medicines* 4 (2): 36.

Lynch, M. J., S. Swift, D. F. Kirke et al. 2002. The regulation of biofilm development by quorum sensing in *Aeromonas hydrophila*. *Environ. Microbiol.* 4 (1): 18–28.

Maree, J., G. P. Kamatou, S. Gibbons et al. 2014. The application of GC-MS combined with chemometrics for the identification of antimicrobial compounds from selected commercial EOs. *Chemometr. Intell. Lab. Syst.* 130: 172–81.

McClean, K. H., M. K. Winson, L. Fish et al. 1997. Quorum sensing and *Chrornobacteriurn violaceurn*: Exploitation of violacein production and inhibition for the detection of N-acyl homoserine lactones. *Microbiology* 143 (1997): 3703–11.

McLean, R. J. C., L. S. Pierson 3rd, C. Fuqua et al. 2004. A simple screening protocol for the identification of quorum signal antagonists. *J. Microbiol. Methods* 58 (3): 351–360.

Meighen, E. A. 1993. Bacterial bioluminescence: Organization regulation and application of the *lux* genes. *FASEB J.* 7 (11): 1016–1022.

Miller-Keane Encyclopedia and Dictionary of Medicine, *Nursing, and Allied Health*, Seventh Edition. (2003). Retrieved from http://Medical-Dictionary.Thefreedictionary.Com/Biofilm, Accessed July 22, 2016.

Mokhetho, K. C., M. Sandasi, A. Ahmad et al. 2018. Identification of potential anti-quorum sensing compounds in EOs: A gas chromatography-based metabolomics approach. *J. Essent. Oil Res.* 30 (6): 399–408.

Morohoshi, T., M. Kato, K. Fukamachi et al. 2008. N -acylhomoserine lactone regulates violacein production in *Chromobacterium violaceum* type strain ATCC12472. *FEMS Microbiol. Lett.* 279: 124–130.

Mukherji, R., A. Prabhune. 2014. Novel glycolipids synthesized using plant essential oils and their application in quorum sensing inhibition and as antibiofilm agents. *Scientific World J.* 2014 (3): 890709. doi:10.1155/2014/890709.

Myszka, K., M. T. Schmidt, M. Majcher et al. 2016. Inhibition of quorum sensing-related biofilm of *Pseudomonas fluorescens* KM121 by *Thymus vulgare* essential oils and its major bioactive compounds. *Int. Biodeterior. Biodegrad.* 114: 252–59.

Nazzaro, F., F. Fratianni, L. DeMartino et al. 2013. Effect of essential oils on pathogenic bacteria. *Pharmaceuticals* 6 (12): 1451–74. J. W. Hastings.

Nealson, K. H., T. Platt et al. 1970. Cellular control of the synthesis and activity of the bacterial luminescent system. *J. Bacteriol.* 104 (1): 313–322.

Oh, K. B., H. Miyazawa, T. Naito et al. 2001. Purification and characterization of an autoregulatory substance capable of regulating the morphological transition in *Candida albicans*. *Proc. Natl. Acad. Sci. U. S. A.* 98 (8): 4664–4668.

Olivero, J. T., N. P. C. Pajaro, E. Stashenko. 2011. Antiquorum sensing activity of essential oils isolated from different species of the genus *Piper*. *Vitae,* 18 (1): 77–82.

Olivero-Verbel, J., A. Barreto-Maya, A. Bertel-Sevilla et al. 2014. Composition, anti-quorum sensing and antimicrobial activity of essential oils from *Lippia alba*. *Braz. J. Microbiol.* 45 (3): 759–767.

Packiavathy, I., Pitchaikani Sasikumar, S. K. Pandian, A. Veera Ravi. 2013. Prevention of quorum-sensing-mediated biofilm development and virulence factors production in *Vibrio* spp. by curcumin. *Appl. Microbiol. Biotechnol.* 97 (23): 10177–10187.

Parsek, M. R., E. P. Greenberg. 2005. Sociomicrobiology: The connections between quorum sensing and biofilms. *Trends Microbiol.* 13 (1): 27–33.

Pearson, J. P., K. M. Gray, L. Passador et al. 1994. Structure of the autoinducer required for expression of *Pseudomonas aeruginosa* virulence genes. *Proc. Natl. Acad. Sci. U. S. A.* 91 (1): 197–201.

Pearson, J. P., C. van Delden, B. H. Iglewski et al. 1999. Active efflux and diffusion are involved in transport of pseudomonas aeruginosa cell-to-cell signals. *J. Bacteriol.* 181 (4): 1203–1210.

Pellegrini, M. C., M. V. Alvarez, A. G. Ponce et al. 2014. Anti-quorum sensing and antimicrobial activity of aromatic species from South America. *J. Essent. Oil Res.* 26 (6): 458–465.

Poli, J.-P., E. Guinoiseau, D. de Rocca Serra et al. 2018. Anti-quorum sensing activity of 12 EOs on *Chromobacterium violaceum* and specific action of *cis-cis-p*-menthenolide from Corsican *Mentha Suaveolens* ssp. *Insularis*. *Molecules*. 23 (9): E2125.

Prouty, A. M., W. H., Schwesinger, J. S. Gum. 2002. Biofilm formation and interaction with the surfaces of gallstones by *Salmonella* spp. *Society* 70 (5): 2640–2649.

Ramage, G., S. P. Saville, B. L. Wickers et al. 2002. Inhibition of *Candida albicans* biofilm formation by farnesol, a quorum-sensing molecule. *Appl. Environ. Microbiol.* 68 (11): 5459–5463.

Rasmussen, T. B., T. Bjarnsholt, M. E. Skindersoe et al. 2005. Screening for quorum-sensing inhibitors (QSI) by use of a novel genetic system, the QSI selector. *J. Bacteriol.* 187 (5): 1799–1814.

Reimmann, C., N. Ginet, L. Michel et al. 2002. Genetically programmed autoinducer destruction reduces virulence gene expression and swarming motility in *Pseudomonas aeruginosa* PAO1. *Microbiology* 148 (4): 923–932.

Riedel, K., S. Schoenmann, L. Eberl. 2005. Quorum sensing in plant-associated bacteria. *BioSpektrum* 11: 1385–1388.

Sepahi, E., S. Tarighi, F. S. Ahmadi, A. Bagheri. 2015. Inhibition of quorum sensing in *Pseudomonas aeruginosa* by two herbal EOs from Apiaceae family. *J. Microbiol.* 53 (2): 176–180.

Sharifi, A., A. Mohammadzadeh, T. Salehi Zahraei, P. Mahmoodi, 2018. Antibacterial, antibiofilm and antiquorum sensing effects of *Thymus daenensis* and *Satureja hortensis* essential oils against *Staphylococcus aureus* isolates. *J. Appl. Microbiol.* 124 (2): 379–388.

Sieniawska, E., R. Los, T. Baj et al. 2013. Antimicrobial efficacy of *Mutellina purpurea* essential oil and α-pinene against *Staphylococcus epidermidis* grown in planktonic and biofilm cultures. *Ind. Crops Prod.* 51. 152–157.

Smith, R. S., B. H. Iglewski. 2003a. *P. aeruginosa* quorum-sensing systems and virulence. *Curr. Opin. Microbiol.* 6 (1): 56–60.

Smith, R. S., Iglewski, B. H. 2003b. *Pseudomonas aeruginosa* quorum sensing as a potential antimicrobial target. *J. Clin. Investig.* 112 (10): 1460–1465.

Snoussi, M., E. Noumi, R. Punchappady-devasya et al. 2018. Antioxidant properties and anti-quorum sensing potential of *Carum copticum* essential oil and phenolics against *Chromobacterium violaceum*. *J. Food Sci. Technol.* 55 (8): 2824–2832.

Solano, C., M. Echeverz. 2014. Biofilm dispersion and quorum sensing. *Curr. Opin. Microbiol.* 18 (1): 96–104.

Stevens, A. M., K. M. Dolan, E. P. Greenberg. 1994. Synergistic binding of the *Vibrio fischeri* LuxR transcriptional activator domain and RNA polymerase to the Lux promoter region." *Proc. Natl. Acad. Sci. U. S. A.* 91 (26): 12619–12623.

Surette, M. G., M. B. Miller, B. L. Bassler. 1999. Quorum sensing in *Escherichia coli*, *Salmonella typhimurium*, and *Vibrio harveyi*: A new family of genes responsible for autoinducer production. *Proc. Natl. Acad. Sci. U. S. A.* 96 (4): 1639–1644.

Szabó, M. Á., G. Z. Varga, J. Hohmann et al. 2010. Inhibition of quorum-sensing signals by EOs. *Phytother. Res.* 24:782–786.

Toole, G. O., H. B. Kaplan, R. Kolter. 2000. Biofilm formation as microbial development. *Annu. Rev. Microbiol.* 54:49–79.

Vattem, D. A., K. Mihalik, S. H. Crixell, R. J. McLean. 2007. Dietary phytochemicals as quorum sensing inhibitors. *Fitoterapia* 78 (4): 302–310.

Venkadesaperumal, G., S. Rucha, K. Sundar, P. K. Shetty. 2016. Anti-quorum sensing activity of spice oil nanoemulsions against food borne pathogens. *LWT - Food Sci. Technol.* 66. Elsevier Ltd: 225–231.

Visick, K. L., J. Foster, J. Doino et al. 2000. *Vibrio fischeri lux* genes play an important role in colonization and development of the host light organ. *J. Bacteriol.* 182 (16): 4578–4586.

Wagner-Döbler, I., V. Thie, L. Eberl et al. 2005. Discovery of complex mixtures of novel long-chain quorum sensing signals in free-living and host-associated marine alphaproteobacteria. *Chembiochem* 6 (12): 2195–2206.

Waters, C. M., B. L. Bassler. 2005. Quorum sensing: Cell-to-cell communication in bacteria. *Annu. Rev. Cell Dev. Biol.* 21 (1): 319–346.

Whitehead, N. A., A. M. L. Barnard, H. Slater et al. 2001. Quorum-sensing in gram-negative bacteria. *FEMS Microbiol. Rev.* 25 (4): 365–404.

Whiteley, M., S. P. Diggle, E. P. Greenberg et al. 2017. Progress in and promise of bacterial quorum sensing research. *Nature* 551 (7680): 313–320.

Winson, M. K., S. Swift, L. Fish et al. 1998. Construction and analysis of LuxCDABE-based plasmid sensors for investigating N-acyl homoserine lactone-mediated quorum sensing. *FEMS Microbiol. Lett.* 163 (2): 185–192.

Yap, P. S. X., T. Krishnan, B. C. Yiap et al. 2014. Membrane disruption and anti-quorum sensing effects of synergistic interaction between *Lavandula angustifolia* (lavender oil) in combination with antibiotic against plasmid-conferred multi-drug-resistant *Escherichia coli*. *J. Appl. Microbiol.* 116 (5): 1119–1128.

Yap, P. S. X., T. Krishnan, K. G. Khan, S. H. E. Lim. 2015. Antibacterial mode of action of *Cinnamomum verum* Bark EO, alone and in combination with piperacillin, against a multi-drug-resistant *Escherichia coli* strain. *J. Microbiol. Biotechnol.* 25 (8): 1299–1306.

Yarwood, J. M., P. M. Schlievert. 2003. Quorum sensing in *Staphylococcus* infections. *J. Clin. Investig.* 112 (11): 1620–1625.

17 Functions of Essential Oils and Natural Volatiles in Plant-Insect Interactions

Robert A. Raguso

CONTENTS

17.1 INTRODUCTION

17.1.1 Brief Historical Overview of Essential Oil (EO) Functional Ecology

Essential oils (EOs) provide a glimpse into the hidden world of plant metabolic diversity, revealed through analytical chemistry. EO diversity is manifested at all levels of organization, ranging from the contents of glandular trichomes or floral tissues within a plant, to the physiological plasticity in EO emissions induced by biotic or abiotic stresses, to heritable EO variation (chemotypes) across the geographic distribution of a given species, and (finally), to species-specific variation in EO bouquets between related species. EOs have played important historical roles in establishing the conceptual foundations of chemosystematics (Rodman et al., 1981; Adams, 1998) as well as plant defense and coevolutionary theory (Raguso et al., 2015a), and improvements in analytical and statistical methods have drawn EOs closer to the mainstream of ecological research (Kallenbach et al., 2014; Raguso et al., 2015b). More recent research has led to the rapidly growing field of herbivore-induced plant responses, in which herbivore damage may dramatically alter a plant's constitutive EO composition, along with non-volatile chemical defenses, physical defenses and extra-floral nectar. Herbivore induction often triggers systemic release of a fundamentally different EO blend, with the potential to attract carnivores whose activities may reduce herbivore damage and thus (indirectly) constitute a form of defense (Karban and Baldwin, 1997; Dicke and Baldwin, 2010). There is a growing realization that herbivore-induced responses impact flowers and below-ground plant tissues, as well as vegetative shoots and leaves, revealing unexpected richness in chemically mediated ecological interactions across the lives of plants (Johnson et al., 2015; Papadopoulou and van Dam, 2017). The primary goal of this chapter is to outline progress in our understanding of the biological functions of EOs, with an emphasis on complex insect–plant interactions, especially pollination. This outline will focus on natural sources of EO variation, in contrast to the historical emphasis on domesticated and horticultural plants in the voluminous literature on EO chemistry, especially in flowers.

17.1.2 Challenges in Identifying EO Functional Roles and Selective Pressures

Chemical ecologists face at least three major challenges in their quest to identify the natural functions of EOs in plants. The first challenge is to determine which volatile compounds are released into the headspace of living plants, where they are most likely to mediate biological interactions. Natural volatile composition and dosage are important because attraction, repellence, toxicity, growth inhibition, and other functions often are determined by quantitative relationships among EO blend components (Dötterl and Vereecken, 2010; Galen et al., 2011). This largely methodological challenge has been addressed using non-invasive headspace sorption techniques and sensitive analytical methods. These approaches now extend beyond the traditional coupled gas chromatography-mass spectrometry (GC-MS) to include hyphenated GC systems (Mondello et al., 2008; Mitrevski and Marriott, 2012), direct volatile sampling via proton-transfer-reaction mass spectrometry (PTR-MS; Riffell et al., 2014; Farré-Armengol et al., 2016) and tissue surface chemical mapping using matrix-assisted laser desorption/ionization (MALDI) mass spectrometry (Kaspar et al., 2011; Shroff et al., 2015).

The second major challenge for chemical ecologists is to identify the ecological roles played by EOs, which are often complex and unexpected. This challenge has deep roots in the work of early 20th-century scientists (Clements and Long, 1923; Knoll, 1926), whose clever behavioral assays decoupled EOs and other chemical stimuli from floral color, shape, and texture, challenging living pollinators to respond to experimentally modified floral traits (Borg-Karlson, 1990). A major conceptual change has come with the recognition that a full spectrum of organisms (from microbes to browsing ungulates) interact with flowers and, with it, the realization that bioassays should be extended to these organisms as well as to pollinators (Junker et al., 2013). Recent insights on the functional ecology of flowers have been gained through novel genetic tools (e.g., inbred lines, hybrids, gene silencing; Galliot et al., 2006; Kessler et al., 2008) and more sophisticated experimental manipulation of floral stimuli (e.g., augmenting 3D-printed flowers with EO extracts; Policha et al., 2016; Nordström et al., 2017).

The third major challenge, critical to ecological and evolutionary theory, has been to measure the selective forces acting upon and shaping EOs as phenotypic traits. Progress in this area has been impaired by the difficulty of measuring and experimentally manipulating complex chemical blends as traits, extending their natural variation while gaining a clearer understanding of their inheritance. The genetic tools described above have helped substantially in this regard, when combined with untargeted approaches that measure the impacts of floral antagonists (herbivores, florivores, nectar- and pollen-thieves) and microbes as well as pollinators, typically using fruit- or seed-production as the ultimate measure of fitness (Kessler et al., 2008, 2013). It remains daunting to dissect highly complex chemical traits, such as pine oleoresins or glandular trichome exudates, which may contain tens to hundreds of related compounds resulting from interactions among several genetic loci, through the action of their encoded biosynthetic enzymes (Pichersky and Raguso, 2018). In some plant families (e.g., Asteraceae, Lamiaceae, Verbenaceae), floral and vegetative EOs may be quite similar due to shared classes of glandular trichomes covering the entire plant (Manan et al., 2016; Tissier et al., 2017). This review begins with a discussion of plants in the mint family, well known as sources of culinary spices and herbal medicines (Bozin et al., 2006).

17.2 AROMATIC MEDICINAL HERBS WITH MULTIPLE EO FUNCTIONS

17.2.1 Glandular Plants of Mediterranean Biomes

All of the world's major biomes contain plants with medicinal properties. Of these, the Mediterranean biome, defined by cool, wet winters and hot, dry summers and bounded between 30°–45° N or S latitudes (Cowling et al., 1996), contains multitudes of aromatic plants, many of which have been cultivated for their medicinal and/or culinary properties from prehistoric times (Kantsa et al., 2015). The *maquis* shrublands of the Mediterranean basin show a remarkable resemblance to the South African *fynbos*, Californian *chaparral*, Chilean *matorral*, and Australian *mallee* in their physiognomy and ecology, being open, fire-adapted habitats with unusually high degrees of local endemism. Indeed, these five Mediterranean climate regions combined account for less than 3% of the world's land surface yet contain nearly almost 20% of the world's floristic diversity (Rundel et al., 2018).

In southern Europe, the *maquis* transitions into a coastal scrub habitat typified by low-nutrient limestone soils dominated by aromatic, resinous shrubby plants in the Lamiaceae family, including some of the best-known sources of culinary herbs and commercially valuable EOs (e.g., *Lavandula*, *Salvia*, *Rosmarinus*, and *Origanum*). In southern France, this calcareous coastal scrub, the *garrigue*, is home to *Thymus vulgaris*, noteworthy for its qualitative variation in EO composition, manifested in a fine-scale mosaic of population-level differences perceptible to the human nose (Gouyon et al., 1986). Careful genetic crosses and GC-MS analyses led to the assignment of individual plants to six different chemical phenotypes or "chemotypes," each dominated by a single volatile oxygenated monoterpenoid compound, with epistatic relationships dictated by their biosynthetic pathway dynamics (Vernet et al., 1986). These chemotypes include acyclic monoterpene alcohols (geraniol and linalool), followed by cyclic terpene alcohols (alpha terpineol and 4-thujanol), and phenolic terpene alcohols (thymol and carvacrol; Linhart and Thompson, 1999).

17.2.2 Focal Studies on *Thymus vulgaris* EO Chemotypes in the *Garrigue*

17.2.2.1 Biotic Interactions in a Geographic Mosaic

Careful experimental studies have revealed a fascinating landscape of ecological functions and selective forces shaping the mosaic of thyme EO chemotypes across southern France. Depending on location, thyme foliage is assailed by a variety of herbivores, ranging from *Arima marginata*, a chrysomelid beetle specialized on aromatic plants, to more generalized invertebrates (*Helix* snails and *Leptophyes* grasshoppers) and vertebrate browsers (domesticated or feral sheep and goats) (Linhart and Thompson, 1999). Interestingly, no single EO chemotype is an ideal herbivore deterrent; linalool-dominated plants are least attractive to sheep and grasshoppers but are most vulnerable to snails and beetles, whereas the phenolic chemotypes (thymol and carvacrol) are repellent to most animals and microbes but are vulnerable to grasshopper attack (Linhart and Thompson, 1995, 1999). Local differences in biotic interactions would favor balanced polymorphism in EO chemotypes, especially given that these traits are under relatively simple Mendelian genetic control. An additional level of biotic interactions involves competition between *T. vulgaris* and neighboring plants, particularly grasses, which could crowd or overtop these short, slowly growing plants in the open landscape of the *garrigue* (Tarayre et al., 1995; Linhart et al., 2005). The phenolic terpenes emerge as the most effective allelopathic agents against the germination or survival of competitor grasses such as *Brachypodium phoenicoides* and *Bromus* spp., whether tested as aqueous leachates from *T. vulgaris* foliage or as dilute EO distillates (Linhart et al., 2014). However, the impacts of growing beneath or alongside a canopy of thyme leaves are complicated by the positive effects that the plants may have on surrounding soil nutrients, which may counter-balance direct allelopathy due to EOs.

17.2.2.2 Geographic Variation and Abiotic Stress

The biotic interactions described above would appear to favor thymol- or carvacrol-dominated EO chemotypes of *Thymus vulgaris* across the *garrigue*, with minor exceptions. Instead, one finds a patchwork of thyme populations in which non-phenolic chemotypes may be dominant, and the thymol and carvacrol chemotypes appear to partition microhabitat by soil type and slope aspect (Gouyon et al., 1986). Additional studies indicate that abiotic (thermal) factors associated with elevational gradients provide the most powerful check against a selective sweep by the phenolic chemotypes of thyme EOs and may explain why sharp clinal variation is maintained despite considerable gene flow (Linhart and Thompson, 1999). Thymol and carvacrol chemotypes are rare at higher elevations where winter temperatures are coldest and show demonstrably lower ability (as seedlings) to survive and regrow after early winter freezing temperatures (−10°C), in strong contrast to their superior ability to endure hot, dry summers (Amiot et al., 2005). Reciprocal transplant experiments support the association of specific thyme chemotypes with differences in thermal tolerance (Thompson et al., 2007). However, if the climate of southern France continues to warm (without explosive growth of *Leptophyes* grasshopper populations), the phenolic EO chemotypes of *T. vulgaris* would be predicted to spread beyond the lowest elevations of the *garrigue*.

17.2.2.3 Thymol in Nectar, Impacts on Bee Health

Thymol and carvacrol are not limited to the essential oil of *Thymus vulgaris*. Surveys of the EOs of other *Thymus* species, of related *Origanum* and other mint-family plants reveal a complex mosaic of terpenoid-dominated EOs from Portugal (Miguel et al., 2004; Figueiredo et al., 2008) to Turkey (Tümen et al., 1995; Bagci and Başer, 2005) and north to the Baltic Sea (Mockute and Bernotiene, 2001). Despite the low solubility of the phenolic terpenes in aqueous medium, thymol has been reported as a constituent of many floral nectars, including those of *Thymus capitatus* and *T. serpyllum*, which are important nectar sources for diverse bee groups across the Mediterranean region (Petanidou and Vokou, 1993; Petanidou and Smets, 1996), as well as the honey bee (*Apis mellifera*) and its associated honey industry (Alissandrakis et al., 2007; Karabagias et al., 2014). Recent research has focused on the non-caloric value that thymol, eugenol, and other EO substances might provide to bees foraging in diverse floral habitats, especially those infected with internal parasites. The antimicrobial potential of thymol was noted in earlier studies of the utility of *T. vulgaris* EO against a spectrum of microbial agents at low dosages (Hammer et al., 1999; Rota et al., 2008) and in reduction of infection by *Nosema ceranae*, a microsporidian fungal pathogen of honey bees, along with increased survivorship of bees fed a thymol syrup (Costa et al., 2010).

A serious problem for social bees with generalized foraging patterns (*Bombus terrestris* in Europe, *B. impatiens* in North America) is the spread of the trypanosome *Crithidia bombi*, an obligate gut parasite of bumble bees that is transmitted among bees through infected feces (Schmid-Hempel and Durrer, 1991). In addition to direct impacts on bee survival and reproduction, infection by *C. bombi* also reduces the ability of bees to learn different floral traits in association with sugar rewards (Gegear et al., 2006). One promising development is that thymol was found to inhibit the growth of *C. bombi* when applied in dosages comparable to their natural occurrence in *T. vulgaris* floral nectar (5–8 ppm), presumably by disrupting the parasite's cell and mitochondrial membranes (Palmer-Young et al., 2016). Subsequent research indicates that the prophylactic effects of nectar thymol on *C. bombi* are synergized when eugenol, another widespread component of floral EOs and nectars, is present, suggesting that diversified nectar meals may help bees to mitigate trypanosome infection (Palmer-Young et al., 2017b). However, strains of thymol-resistant *C. bombi* may evolve quickly, even when multiple EO components are used in biocontrol (Palmer-Young et al., 2017a). We are just beginning to understand how non-sugar nectar components, including but not limited to volatile EO constituents, impact the health of bees and other flower-visiting animals (Richardson et al., 2015).

17.3 FUNCTIONAL ECOLOGICAL LINKS BETWEEN EOs AND RESINS

17.3.1 Oleoresin in Conifers and Multi-product TPS Enzymes

The preceding discussion of EO antimicrobial properties and the evolution of pathogen resistance calls to mind the fundamental relationship between terpenoid EOs and the non-volatile terpenoid compounds present in the oleoresins of coniferous trees. In the oleoresin of pines and other conifers, volatile "turpentine" (monoterpene and sesquiterpene EO blends) helps to solubilize and mobilize non-volatile "rosin" (terpenoid resin acids and gums), which acts as both a physical and chemical defense by sealing wounds and overwhelming bark beetles and associated fungal pathogens (Paine et al., 1997). Oleoresin is an ancient plant defense, whose prevalence in prehistoric forests is evidenced by fossilized amber deposits, whose chemical composition may still be analyzed (Pereira et al., 2009). Early biochemical studies of oleoresin components in grand fir (*Abies grandis*) trees revealed that individual terpene synthase enzymes and their structural genes (the *tps* superfamily) often are responsible for the biosynthesis of a single major product and numerous minor products (Steele et al., 1998a). It is thought that the generation of complex and variable EO blends from relatively few enzymes represents a coevolutionary response to the selective pressures exerted by bark beetles (Scolytidae), making it more difficult for them to perceive aggregation pheromones or evolve resistance to a shifting target of toxic oleoresin blends (Phillips and Croteau, 1999; Raffa, 2001). Controlled physiological assays using grand fir saplings demonstrated that wounding induces the expression of C10 (monoterpene), C15 (sesquiterpene), and C20 (diterpene acid)-related *tps* genes in specialized resin ducts, beginning with almost immediate monoterpene synthase activity, followed by coordinated expression of sesquiterpene and diterpene acid-related *tps* genes and their associated biosynthetic enzymes (Steele et al., 1998b). Thus, the EOs and non-volatile resin components of oleoresin in coniferous trees are biosynthetically and functionally linked.

17.3.2 EOs, Resins, and TPS Enzymes in Angiosperms

Terpene synthases are responsible for terpenoid biosynthesis in angiosperms as well as gymnosperms, and the "one gene, multiple products" pleiotropic relationship first described for *tps* genes in fir trees is common to many flowering plant lineages (Chen et al., 2011). For example, the cineole synthase enzyme, responsible for the biosynthesis of 1,8-cineole (eucalyptol), produces from 7 to 10 additional monoterpenoids in *Salvia officinalis* (Lamiaceae), the common sage (Wise et al., 1998), in *Arabidopsis thaliana* (Cruciferae), a self-pollinating weed and model for genetic research (Chen et al., 2004) and across *Nicotiana* sect. *Alatae* (Solanaceae), a group of wild tobacco species from southern Brazil (Fähnrich et al., 2012). In addition to coniferous trees, many angiosperms also demonstrate links between EOs, gums, and resins in ways that have been valued since human antiquity. The family Burseraceae includes aromatic trees whose resins—frankincense, myrrh, and copal—were used as forms of currency, to honor royalty and to invoke the sacred throughout the ancient world (Pichersky and Raguso, 2018). Neotropical members of this family (*Bursera* and *Protium*) have become model systems for the study of coevolutionary "arms races," in which specialized beetles and other herbivores drive the evolution of more complex EO blends and resins by evolving counter-adaptations to simpler blends. The diversification of terpenoid EO blends in *Bursera* and *Protium* is especially pronounced among related species that share a habitat, presumably under selective pressure to avoid attracting the herbivores using the related species as host plants (Becerra, 2007; Fine et al., 2013). Molecular analyses reveal high levels of *tps* gene duplication in *Protium*, providing a likely mechanism for the diversification of EOs and resin components (Zapata and Fine, 2013).

17.3.3 Resins as Resources for Nest-Making Bees

17.3.3.1 Resin Collection from Aromatic Tropical Trees

Like humans, bees also value plant resins, but for more practical reasons—to use as antimicrobial and waterproof building materials for their nests. Honey bees mix naturally collected plant waxes and resins with their own glandular secretions to produce *propolis*, which has been valued for its medicinal and sacred properties from ancient to modern times (Burdock, 1998; Salatino et al., 2011). Beyond honey bees, literally thousands of social and solitary bee species utilize plant resins as nest-building materials. This is especially true in tropical rainforests, where an effective bees' nest should shed the constant rain, repel or exclude predators, and prevent pathogenic microbial growth (Drescher et al., 2014). Leonhardt et al. (2011) studied such a community, calculating food-web networks to explore the resin-foraging relationships between 16 species of meliponine (stingless) bees and 15 tree species in a lowland dipterocarp rainforest in Borneo. Although the resin samples were chemically diverse (1117 total compounds), only 113 of these compounds were incorporated into bee nests, featuring sesquiterpenoids and triterpenes but lacking the monoterpene and diterpene components of the original resins. Behavioral experiments in the same habitat revealed that stingless bees are attracted to the resin of a rainforest conifer, *Agathis borneensis*, through olfactory responses to the volatile monoterpene and sesquiterpene EO extracts of the resin (Leonhardt et al., 2010). Stingless bees appear to learn specific ratios of these volatile components, just as honey bees are conditioned on quantitative ratios in the scents of rewarding flowers (Wright and Schiestl, 2009).

17.3.3.2 Resin Collection from *Dalechampia* and *Clusia* Flowers

In Central and South America, the one-sided relationship between bees and their resin sources is leveraged into a plant-pollinator mutualism by two plant genera whose flowers produce resins as floral rewards, instead of nectar (Armbruster, 1984). Many of the approximately 120 species of *Dalechampia* (Euphorbiaceae) secrete a floral resin of triterpene ketones and alcohols (e.g., 11a,12a-Epoxi-3-oxo-D-friedooleanan-14-ene (**1**); de Araújo et al., 2007), which is collected for nest building by female orchid bees (euglossines), along with stingless bees and megachilid bees that may collect both resin and pollen (Armbruster et al., 2009). Like poinsettia and other spurges, *Dalechampia* plants combine small monoecious (single sex on the same plant) flowers with showy bracts in a pseudanthial "blossom," which serves as the unit of pollinator attraction (Figure 17.1A). Phenotypic selection studies revealed that bees primarily utilize visual cues when foraging for *Dalechampia* resins. In Gabon, West Africa, *D. ipomoifolia* produces colorless floral resins, and its megachilid bee pollinators show foraging preferences for colorful bracts whose size is correlated with resin gland area and reward quantity, selecting for morphologically integrated flowers that optimize pollen placement onto stigmas by bees while collecting resins (Armbruster et al., 2005; Figure 17.1B). In contrast, the tropical Mexican species *D. schottii* has small, inconspicuous bracts but produces blue-colored resins. Female orchid bee pollinators (*Euglossa* sp.) ignored floral bracts of *D. schottii* as a proxy for reward quality and their foraging preferences resulted in direct selection on resin gland area (Bolstad et al., 2010). Phylogenetic analyses reveal that resin-based pollination is a shared-ancestral condition in the genus *Dalechampia*, with several independent, derived shifts to pollination by EO-collecting male orchid bees (see Section 17.4.1) in species that emit monoterpenes from the stigmas of female flowers (Armbruster, 1997). The coevolutionary patterns of resin deployment in defense of leaves, developing fruits, and male flowers in conjunction with the targeting of these tissues by the larvae of specialized herbivorous butterflies suggests that resin-based plant defenses served as a pre-adaptation for the evolution of resin-based pollination in *Dalechampia* (Armbruster et al., 2009).

The genus *Clusia* (Guttiferae) contains about 150 dioecious (single-sex flowers on different plants) epiphyte and tree species, in which male flowers offer pollen and resins whereas female flowers only provide resins from sterile staminodes (Armbruster, 1984; Figure 17.1C). As for *Dalechampia*, the primary pollinators of *Clusia* flowers are female orchid bees and stingless (meliponine) bees that collect floral resins for nest building across tropical America. However, the floral resins of *Clusia* are not terpene-based at all, consisting of poly-isoprenylated benzophenones such as grandone (**2**), which are also found in the fruits (Figure 17.1D), roots and leaves of *Clusia* trees (de Oliveira et al., 1996). Unsurprisingly, *Clusia* resins offer antimicrobial as well as structural benefits, with evidence for targeted toxicity against gram-staining *Paenobacillus* bacteria with pathogenic impacts on honey bees (Lokvam and Braddock, 1999). The phenolic resins of *Clusia* flowers show no obvious volatile fraction, unlike the tree-produced terpenoid resins and oleoresins described above. Instead, the visually showy flowers of *Clusia* produce floral EOs with diverse chemical composition (aliphatic

FIGURE 17.1 Plants with relationships between essential oils, resins, and resin-collecting bees. Panels a–d: Neotropical plants with resin-secreting flowers. (A) Blossom of *Dalechampia tiliifolia* from Gamboa, Panama, with showy bracts and a glistening resin gland (center). Insert (**1**) shows an oxygenated triterpene component of *Dalechampia* resin (11a,12a-Epoxi-3-oxo-D-friedooleanan-14-ene). (B) Female *Euglossa* orchid bee (arrow) collecting resin—note body placement resulting in pollen transfer. (C) *Clusia uvitana* from La Selva biological station, Costa Rica, with resin glands at the flower's center. (D) Maturing fruits of a *Clusia* tree, Incachaca, Bolivia, with resin darkening at the site of a wound (arrow). Insert (**2**) shows grandone, a poly-isoprenylated benzophenone component of *Clusia* resin. Panels e–g: Resinous, purple-flowered Mediterranean plants with sesquiterpene-dominated essential oils and prominent roles as bee-attractants in plant-pollinator networks in Greece. (E,F) *Cistus x purpurea*, resinous foliage and flower. Insert (**3**) shows Labd-13-ene-8a,15-diol, an oxygenated diterpene from *Cistus creticus* resin. (G) *Lavandula stoechas*, cultivated in central California, with a similar Mediterranean climate. (Photos by R.A. Raguso.)

and aromatic compounds as well as terpenoids), which attract stingless bees in field trials (Nogueira et al., 2001). As with herbivore-avoidance in *Bursera* and *Protium* trees, species-specific floral EOs may enhance the fitness of co-occurring *Clusia* trees, in this case by promoting reproductive isolation through scent-mediated constancy by bees.

17.3.4 *Cistus* and *Lavandula*: Pollinator Hubs in the *Phrygana*

As discussed in Section 17.2, the Mediterranean region is rich in aromatic plants and bee diversity, including numerous flowering as well as non-flowering (e.g., cypress and pine tree) sources of propolis (Potts et al., 2006). Thus, plant-pollinator relationships in this region might be expected to include chemical links to resinous plants. Many of the signature resin components of propolis in Crete and mainland Greece are labdane-type diterpenes (Popova et al., 2010), which typify the commercially valuable *labdanum* resin produced by *Cistus creticus* shrubs (Cistaceae). As in grand fir, the expression of *tps* genes and associated enzymes (e.g., copal-8-ol diphosphate synthase) that mediate the synthesis of labdane-related diterpenes (e.g., Labd-13-ene-8a,15-diol (**3**)) in *Cistus* is localized to glandular trichomes (vs. resin ducts) and is induced by mechanical wounding (Falara et al., 2010; Figure 17.1E).

Cistus cretica is a characteristic shrub of the *phrygana*, the Greek equivalent of the French *garrigue*, where it blooms for several months and is visited by a broad spectrum of bees. Recent studies on the Greek island of Lesvos explored the complex relationships between *C. cretica* and 40 additional species of co-flowering plants, their "community volatilome" of 351 total

EO compounds and 168 identified insect-pollinator species. These studies were remarkable for their community-wide reach, their unbiased statistical approach, and the resulting non-random relationships identified between insect visitation, floral EOs, and color, as coded through bee- and butterfly-specific visual-perception models (Kantsa et al., 2017). *Cistus cretica*, along with an aromatic mint-family plant, *Lavandula stoechas*, emerged as dominant "hubs" in the plant-pollinator network (Figure 17.1F,G), combining purple floral color with distinctive sesquiterpene EOs, dense floral displays, and extended blooming periods; all properties that underscored strong statistical relationships with native bees, especially megachilids (Kantsa et al., 2018). *Lavandula stoechas* provides comparable services as a cornucopian floral resource for solitary bees along the coast of southern Spain (Herrera, 1997), as does the related *Lavandula latifolia* in upland Iberian habitats (Herrera, 2005). Future studies should explore the connection between solitary bee pollination, resin collection, and floral attraction to sesquiterpene-dominated EOs in *Cistus* and *Lavandula* plants (Figure 17.1).

17.4 EO-MEDIATED POLLINATOR SPECIALIZATION: LESSONS FROM AROIDS

17.4.1 Perfume-Collecting Male Orchid Bees as a Pollination Niche

The study of floral volatiles as pollinator attractants began over 50 years ago with examinations of highly specialized pollination systems in orchids, the most diverse family of angiosperm plants. The success of these early studies hinged on field experiments that revealed how EOs mediate the behavioral responses of pollinators, including a diverse guild of over 600 species of New World orchids pollinated by male euglossine (orchid) bees (Williams and Whitten, 1983). Neotropical orchids in several genera (e.g., *Stanhopea, Gongora, Catasetum*) produce floral EOs in such high quantities that they are harvested in the liquid phase by male bees (Gerlach and Schill, 1991; Eltz et al., 1999), which use EO blends collected from flowers and other sources in pheromone-like reproductive displays (Pokorny et al., 2017). Studies of the perfume blends collected by males of 15 sympatric *Euglossa* bee species in Panama revealed strong divergence of chemical composition, again in species-specific patterns that would enhance reproductive isolation if used by female bees for mate choice and recognition (Zimmermann et al., 2009). Thus, the reproductive biology and behavior of male orchid bees has driven the evolution of an EO-based, specialized pollination system in orchids, just as the reproductive biology of female orchid bees has driven the evolution of resin-based, specialized pollination systems in *Dalechampia* and *Clusia* plants.

Not all plants pollinated by EO-seeking male orchid bees are orchids. For example, males of several *Eulaema* species collect trans-carvone oxide, an unusual EO component in flowers, resulting in convergent evolution among several species of *Catasetum* orchids along with *Dalechampia spathulata* (Euphorbiaceae; Whitten et al., 1986), *Unonopsis stipitata* (Annonaceae) (Teichert et al., 2009), *Gloxinia perennis* (Gesneriaceae; Gerlach and Schill, 1991), and *Anthurium* and *Spathiphyllum* species (Araceae; Schwerdtfeger et al., 2002). These last two families—the Gesneriaceae and Araceae—provide a useful comparison when considering how EOs contribute to plant-pollinator diversification. Both families are pan-tropical and diverse, with growth forms varying from understory plants to epiphytes. Despite its high diversity in floral form and color, pollinator affinities, and a preponderance of trichomes, the Gesneriaceae or "African violet" family (160 genera, 3300 spp.) is noteworthy for its lack of floral scent, outside of *Gloxinia* and related genera (Möller and Clark, 2013). In contrast, the Araceae (arum family or "aroids") show comparable diversification (126 genera, 3300 spp.) with a much greater emphasis on EO chemistry, associated with multiple evolutionary shifts between pollinator classes and from rewarding mutualisms to deception (Bröderbauer et al., 2012; Chartier et al., 2014). The following section outlines several of these transitions in the Araceae, as a model for understanding how volatile chemistry can mediate specialized pollination even when floral morphology is relatively conservative.

17.4.2 Scent-Driven Pollinator Modes in the Araceae

Pollinator diversity in aroids is surprising, given the apparent constraints in their floral *bauplan*. All species in this family place small, single-sex flowers ("florets") onto a dense spike inflorescence (a "spadix") subtended by a bract (a "spathe"), which may be colorful and, in some species, may form a chamber (Bröderbauer et al., 2012; Figure 17.2). Only a few aroids provide nectar as a floral reward, and many species present male and female floral functions that are separated in space and/or time. When the spadix includes sterile tissue (lacking florets), this "appendix" often is involved in EO production and thermogenesis (Skubatz et al., 1996). Recent studies have used phylogenetic trait reconstruction to map the evolution of plant-pollinator interactions (Chartier et al., 2014) and insect-trapping mechanisms (i.e., within floral chambers; Bröderbauer et al., 2012) in aroids. These analyses, combined with a growing body of literature on floral volatiles and pollinator attraction in aroids, paint an intriguing picture of EO diversification associated with distinctive pollinator niches (Gibernau and Vignola, 2016). The primitive condition in aroids is inferred from basal lineages with

FIGURE 17.2 Floral diversification in the Araceae. (A) *Lysichiton americanum*, with yellow banner spathe, indole (**4**) and rove beetle pollinators on the spadix, near Seattle, Washington, US. (B) Bee-pollinated *Anthurium* sp. with ipsdienol (**5**) and (C) scarab beetle pollinated *Philodendron radiatum,* with p-methoxystyrene (**6**), both at La Selva biological station, Costa Rica. (D) The spathe of *Xanthosoma mexicana* (Barro Colorado Island, Panama) forms a chamber below the spadix, emitting cis-jasmone (**7**) and providing a mating site for cyclocephaline scarabs. (E) Giant spathe of *Amorphophallus titanum* opening after dusk, and (F) margin between male florets and sterile appendix (arrow), the source of fetid dimethyltrisulfide (**8**) and heat in *Amorphophallus konjak*, both cultivated at Cornell University, Ithaca, NY, US. (Photos by R.A. Raguso.)

hermaphroditic, pollen-rewarding flowers, including extant "skunk cabbage" genera (*Lysichiton* and *Symplocarpus*) from temperate North America and East Asia. *Lysichiton americanum* produces a bright-yellow spathe scented with C9 and C11 aliphatic alkenes and indole (**4**), which attracts staphylinid (rove) beetles that mate and eat pollen while aggregating on the spadix (Brodie et al., 2018; Figure 17.2A). Indole is a widespread floral volatile with context-dependent functions, as it is present in herbivore dung and deceptive dung-mimicking flowers (Urru et al., 2011; Schiestl and Dötterl, 2012), in jasmine and other night-blooming, hawkmoth-pollinated flowers (Bischoff et al., 2015), and in *Stanhopea* orchids pollinated by male orchid bees in the genus *Eulema* (Williams and Whitten, 1983). However, in the community context of springtime temperate forests in northwestern North America, indole paired with bright yellow attracts one species of rove beetle, *Pelecomalium testaceum*, the pollinator of *Lysichiton americanum* (Brodie et al., 2018).

Nearly one-third of all aroid species belong to the genus *Anthurium*, with the full diversity of pollinators that might be expected of a large genus (Croat, 1980). Spathes range from waxy and brilliantly colored (scarlet in the bird-pollinated *A. sanguineum*; Kraemer and Schmitt, 1999) to flimsy, green, and reflexed in the orchid bee-pollinated *A. rubrinervium*, from which bees collect EOs directly from the spadix (Hentrich et al., 2007; Figure 17.2B). *Anthurium rubrinervium* produces ipsdienol (**5**), a monoterpene alcohol that attracts male *Euglossa* bees to orchids in tropical rainforests but is better known as an aggregation pheromone for scolytid bark beetles (*Ips pini*), a major pest in temperate pine forests (Whitten et al., 1988). Floral EOs remain poorly known for most of the approximately 1000 species of *Anthurium*, but the examples above illustrate a theme common to aroids and orchids: the olfactory preferences and perfume-gathering behavior of male orchid bees represent a niche dimension that can be exploited by flowers, resulting in convergent evolution of EOs, including carvone oxide for *Eulema*-pollinated plants and ipsdienol for *Euglossa*-pollinated plants. The greater antiquity of orchid bee lineages suggests that orchids and aroids evolved more recently, recruiting euglossine bees as specialized pollinators when they gained the biosynthetic capability to produce different EO blends (Ramírez et al., 2011).

The evolution of monoecy in aroids is associated with protogyny (maturation of female flowers before male flowers) and the modification of the spathe into a chamber, with the capacity to trap and detain pollinators among female-phase flowers until (hours later) male flowers shed pollen onto departing insects (Chartier et al., 2014). Trap inflorescences have evolved independently in several aroid genera and play a major role in the "mating mutualism" pollination mode of the large (about 400 sp.) genus *Philodendron* (Bröderbauer et al., 2012). In the best studied species, *P. selloum*, the large white, waxy spathe opens at dusk with a burst of fragrance and heat associated with the highest respiratory rate measured for a flowering plant (Seymour, 2001). Large dynastine scarab beetles are attracted by the scent and spathe, enter the warm floral chamber, and spend the evening mating in a warm, protected space (Gottsberger and Silberbauer-Gottsberger, 1991; Figure 17.2C,D). Although the EO blend of *P. selloum* is complex (over 80 compounds), three abundant and antennally sensitive volatiles (4-methoxystyrene (**6**), 3,4-dimethoxystyrene, and cis-jasmone (**7**)) are sufficient to attract the specific beetle pollinator, *Erioscelis emarginata* (Dötterl et al., 2012). Like orchid bees, scarab beetles are abundant, ancient, species-rich, and have well-developed chemical-communication systems using methoxylated aromatic compounds, aliphatic acyloins, and fatty acid-derived esters (Schiestl and Dötterl, 2012). Beside aroids, the mating mutualism trap flower or "love hotel" pollination syndrome includes some of the most spectacular flowers of the New World tropics, such as the Victoria water lily (Nymphaeaceae), giant *Magnolia* flowers (Magnoliaceae), Panama hat plants (Cyclanthaceae), and rind-like *Annona* flowers (Annonaceae), all of which detain scarab beetle pollinators in chambers, produce heat, and emit strong fragrances at nightfall (Gottsberger, 1990; Ervik and Knudsen, 2003). The EO components that attract these beetles are diverse (jasmones, ionones, terpenoids), interesting (methoxylated aromatics and chiral aliphatic acyloins), and sometimes novel in a floral context (4-methyl-5-vinyl thiazole), common to *Caladium bicolor* and four *Annona* species pollinated by *Cyclocephala* scarab beetles (Maia et al., 2012).

Finally, aroids are well known practitioners of brood-site deception, in which the chamber inflorescence traps saprophytic flies or beetles with fetid odors that mimic feces, carrion, fungi, or rotting fruit, decaying substrates that are used by the larvae of these insects for sustenance (Urru et al., 2011; Jürgens et al., 2013). As with sexual deception, the effectiveness of brood-site deception is thought to hinge on the scarcity of quality resources, resulting in powerful selective pressure to respond to cues that reliably indicate such resources (Schiestl and Dötterl, 2012). Deceptive pollination has evolved at least five times in the Araceae, such that fecal- and urine-mimicry dominate in the genus *Arum* (indicated by indole, skatole, cresol, and 2-heptanone) and carrion mimicry is well represented in *Amorphophallus* (associated with S-volatiles and thermogenesis; Kite et al., 1998; Figure 17.2E,F). Carrion mimicry in *Helicodiceros muscivoros* requires calliphorid (blow) flies to enter the floral trap during the female phase, which is accomplished through emitting dimethyl di- and tri-sulfides (**8**) as fly attractants and producing heat from the hairy, tail-like appendix, which flies land upon before walking into the chamber (Angioy et al., 2004). Twenty years after their initial studies, Kite and Hetterscheid (2017) expanded their initial GC-MS analyses to a total of 92 species of the Old World tropical genus *Amorphophallus* (about 200 spp.) cultivated in a common greenhouse, extending our knowledge far beyond the spectacular *A. titanum*, which has become a global sensation in botanical gardens but remains poorly understood in nature (Barthlott et al., 2009). The patterns emerging from Kite and Hetterscheid's study serve as a fitting summary of the major themes outlined in this chapter: (1) not all species of *Amorphophallus* produce fetid odors, as EO composition (and human-perceived odor quality) varies markedly across the genus with geography, rather than presenting variations on a theme of gaseous sulfur malodor; (2) EOs often co-vary with inflorescence color, with carrion mimicry associated with sulfides and darker (blood red) colors; (3) species sharing distinctive odorants (e.g., trimethylamine, the odor of rotting fish) recur in several clades and are not closest relatives; and (4) conversely, closely related species with shared geographic distributions differ markedly in floral scent and other characteristics, suggesting selection to reduce pollinator competition and/or enhance reproductive isolation through pollinator specialization. The lessons learned from the EOs of aroids reinforce the larger themes running through this chapter and highlight that competing selective forces—mating success, reproductive isolation, avoidance of predators and pathogens, abiotic stress, and community context—shape the ecological functions of plant essential oils.

REFERENCES

Adams, R. P. 1998. The leaf essential oils and chemotaxonomy of *Juniperus* sect. *Juniperus*. *Biochem. Syst. Ecol*., 26: 637–645.

Alissandrakis, E., P. A. Tarantilis, P. C. Harizanis, and M. Polissiou. 2007. Comparison of the volatile composition in thyme honeys from several origins in Greece. *J. Agric. Food Chem*., 55: 8152–8157.

Amiot, J., Y. Salmon, C. Collin, and J. D. Thompson. 2005. Differential resistance to freezing and spatial distribution in a chemically polymorphic plant *Thymus vulgaris*. *Ecol. Lett*., 8: 370–377.

Angioy, A. M., M. C. Stensmyr, I. Urru, M. Puliafito, I. Collu, and B. S. Hansson. 2004. Function of the heater: The dead horse arum revisited. *Proc. R. Soc. B*., 271: S13–S15.

de Araújo, M. R. S., M. A. S. Lima, and E. R. Silveira. 2007. Triterpenes and steroids of *Dalechampia pernambucensis* Baill. *Biochem. Syst. Ecol*., 35: 311–313.

Armbruster, W. S. 1984. The role of resin in angiosperm pollination: Ecological and chemical considerations. *Am. J. Bot*., 71: 1149–1160.

Armbruster, W. S. 1997. Exaptations link evolution of plant–herbivore and plant–pollinator interactions: A phylogenetic inquiry. *Ecology*, 78: 1661–1672.

Armbruster, W. S., L. Antonsen, and C. Pélabon. 2005. Phenotypic selection on *Dalechampia* blossoms: Honest signaling affects pollination success. *Ecology*, 86: 3323–3333.

Armbruster, W. S., J. Lee, and B. G. Baldwin. 2009. Macroevolutionary patterns of defense and pollination in *Dalechampia* vines: Adaptation, exaptation, and evolutionary novelty. *Proc. Natl Acad. Sci. USA*, 106: 18085–18090.

Bagci, E., and K. H. C. Başer. 2005. Study of the essential oils of *Thymus haussknechtii* Velen and *Thymus kotschyanus* Boiss. et Hohen var. *kotschyanus* (Lamiaceae) taxa from the eastern Anatolian region in Turkey. *Flavour Fragr. J.*, 20: 199–202.

Barthlott, W., J. Szarzynski, P. Vlek, W. Lobin, and N. Korotkova. 2009. A torch in the rain forest: Thermogenesis of the titan arum (*Amorphophallus titanum*). *Plant Biol.*, 11: 499–505.

Becerra, J. X. 2007. The impact of herbivore–plant coevolution on plant community structure. *Proc. Natl Acad. Sci. USA*, 104: 7483–7488.

Bischoff, M., R. A. Raguso, A. Jürgens, and D. R. Campbell. 2015. Context-dependent reproductive isolation mediated by floral scent and color. *Evolution*, 69: 1–13.

Bolstad, G. H., W. S. Armbruster, C. Pélabon, R. Pérez-Barrales, and T. F. Hansen. 2010. Direct selection at the blossom level on floral reward by pollinators in a natural population of *Dalechampia schottii*: Full-disclosure honesty? *New Phytol.*, 188: 370–384.

Borg-Karlson, A. K. 1990. Chemical and ethological studies of pollination in the genus *Ophrys* (Orchidaceae). *Phytochemistry*, 29: 1359–1387.

Bozin, B., N. Mimica-Dukic, N. Simin, and G. Anackov. 2006. Characterization of the volatile composition of essential oils of some Lamiaceae spices and the antimicrobial and antioxidant activities of the entire oils. *J. Agric. Food Chem.*, 54: 1822–1828.

Brodie, B. S., A. Renyard, R. Gries, H. Zhai, S. Ogilvie, J. Avery, and G. Gries. 2018. Identification and field testing of floral odorants that attract the rove beetle *Pelecomalium testaceum* (Mannerheim) to skunk cabbage, *Lysichiton americanus* (L.). *Arthropod-Plant Inte.*, 12: 591–599.

Bröderbauer, D., A. Diaz, and A. Weber. 2012. Reconstructing the origin and elaboration of insect-trapping inflorescences in the Araceae. *Am. J. Bot.*, 99: 1666–1679.

Burdock, G. A. 1998. Review of the biological properties and toxicity of bee propolis (propolis). *Food Chem. Toxicol.*, 36: 347–363.

Chartier, M., M. Gibernau, and S. S. Renner. 2014. The evolution of pollinator–plant interaction types in the Araceae. *Evolution*, 68: 1533–1543.

Chen, F., D. K. Ro, J. Petri, J. Gershenzon, J. Bohlmann, E. Pichersky, and D. Tholl. 2004. Characterization of a root-specific *Arabidopsis* terpene synthase responsible for the formation of the volatile monoterpene 1, 8–cineole. *Plant Physiol.*, 135: 1956–1966.

Chen, F., D. Tholl, J. Bohlmann, and E. Pichersky. 2011. The family of terpene synthases in plants: A mid-size family of genes for specialized metabolism that is highly diversified throughout the kingdom. *Plant J.*, 66: 212–229.

Clements, F. E., and F. L. Long. 1923. *Experimental Pollination: An Outline of the Ecology of Flowers and Insects (No. 336)*. Carnegie Institution of Washington.

Costa, C., M. Lodesani, and L. Maistrello. 2010. Effect of thymol and resveratrol administered with candy or syrup on the development of *Nosema ceranae* and on the longevity of honeybees (*Apis mellifera* L.) in laboratory conditions. *Apidologie*, 41: 141–150.

Cowling, R. M., P. W. Rundel, B. B. Lamont, M. K. Arroyo, and M. Arianoutsou. 1996. Plant diversity in Mediterranean-climate regions. *Trends Ecol. Evol.*, 11: 362–366.

Croat, T. B. 1980. Flowering behavior of the neotropical genus *Anthurium* (Araceae). *Am. J. Bot.*, 67: 888–904.

Dicke, M., and I. T. Baldwin. 2010. The evolutionary context for herbivore-induced plant volatiles: Beyond the "cry for help". *Trends Plant Sci.*, 15: 167–175.

Dötterl, S., A. David, W. Boland, I. Silberbauer-Gottsberger, and G. Gottsberger. 2012. Evidence for behavioral attractiveness of methoxylated aromatics in a dynastid scarab beetle-pollinated Araceae. *J. Chem. Ecol.*, 38: 1539–1543.

Dötterl, S., and N. J. Vereecken. 2010. The chemical ecology and evolution of bee–flower interactions: A review and perspectives. *Can. J. Zool.*, 88: 668–697.

Drescher, N., H. M. Wallace, M. Katouli, C. F. Massaro, and S. D. Leonhardt. 2014. Diversity matters: How bees benefit from different resin sources. *Oecologia*, 176: 943–953.

Eltz, T., W. M. Whitten, D. W. Roubik, and K. E. Linsenmair. 1999. Fragrance collection, storage, and accumulation by individual male orchid bees. *J. Chem. Ecol.*, 25: 157–176.

Ervik, F., and J. T. Knudsen. 2003. Water lilies and scarabs: Faithful partners for 100 million years? *Biol. J. Linn. Soc.*, 80: 539–543.

Fähnrich, A., A. Brosemann, L. Teske, M. Neumann, and B. Piechulla. 2012. Synthesis of "cineole cassette" monoterpenes in *Nicotiana* section *Alatae*: Gene isolation, expression, functional characterization and phylogenetic analysis. *Plant Mol. Biol.*, 79: 537–553.

Falara, V., E. Pichersky, and A. K. Kanellis. 2010. A copal-8-ol diphosphate synthase from the angiosperm *Cistus creticus* subsp. *creticus* is a putative key enzyme for the formation of pharmacologically active, oxygen-containing labdane-type diterpenes. *Plant Physiol.*, 154: 301–310.

Farré-Armengol, G., J. Peñuelas, T. Li, P. Yli-Pirilä, I. Filella, J. Llusia, and J. D. Blande. 2016. Ozone degrades floral scent and reduces pollinator attraction to flowers. *New Phytol.*, 209: 152–160.

Figueiredo, A. C., J. G. Barroso, L. G., Pedro, L. Salgueiro, M. G. Miguel, and M. L. Faleiro. 2008. Portuguese *Thymbra* and *Thymus* species volatiles: Chemical composition and biological activities. *Curr. Pharm. Des.*, 14: 3120–3140.

Fine, P. V. A., M. R. Metz, J. Lokvam et al. 2013. Insect herbivores, chemical innovation, and the evolution of habitat specialization in Amazonian trees. *Ecology*, 94: 1764–1775.

Galen, C., R. Kaczorowski, S. L. Todd, J. Geib, and R. A. Raguso. 2011. Dosage-dependent impacts of a floral volatile compound on pollinators, larcenists, and the potential for floral evolution in the alpine skypilot *Polemonium viscosum*. *Am. Nat.*, 177: 258–272.

Galliot, C., J. Stuurman, and C. Kuhlemeier. 2006. The genetic dissection of floral pollination syndromes. *Curr. Opin. Plant Biol.*, 9: 78–82.

Gegear, R. J., M. C. Otterstatter, and J. D. Thompson. 2006. Bumble-bee foragers infected by a gut parasite have an impaired ability to utilize floral information. *Proc. R. Soc. B*, 273: 1073–1078.

Gerlach, G., and R. Schill. 1991. Composition of orchid scents attracting euglossine bees. *Bot. Acta*, 104: 385–391.

Gibernau, M., and R. D. S. Vignola. 2016. Pollinators and visitors of aroid inflorescences III–phylogenetic & chemical insights. *Aroideana*, 39: 4–22.

Gottsberger, G. 1990. Flowers and beetles in the South American tropics. *Bot. Acta*, 103: 360–365.

Gottsberger, G., and I. Silberbauer-Gottsberger. 1991. Olfactory and visual attraction of *Erioscelis emarginata* (Cyclocephalini, Dynastinae) to the inflorescences of *Philodendron selloum* (Araceae). *Biotropica*, 23: 23–28.

Gouyon, P. H., P. Vernet, J. L. Guillerm, and G. Valdeyron. 1986. Polymorphisms and environment: The adaptive value of the oil polymorphisms in *Thymus vulgaris* L. *Heredity*, 57: 59–66.

Hammer, K. A., C. F. Carson, and T. V. Riley. 1999. Antimicrobial activity of essential oils and other plant extracts. *J. Appl. Microbiol.*, 86: 985–990.

Hentrich, H., R. Kaiser, and G. Gottsberger. 2007. Floral scent collection at the perfume flowers of *Anthurium rubrinervium* (Araceae) by the kleptoparasitic orchid bee *Aglae caerulea* (Euglossini). *Ecotropica*, 13: 149–155.

Herrera, C. M. 2005. Plant generalization on pollinators: Species property or local phenomenon? *Am. J. Bot.*, 92: 13–20.

Herrera, J. 1997. The role of colored accessory bracts in the reproductive biology of *Lavandula stoechas*. *Ecologyv* 78: 494–504.

Johnson, M. T., S. A. Campbell, S. A., and S. C. Barrett. 2015. Evolutionary interactions between plant reproduction and defense against herbivores. *Annu. Rev. Ecol. Evol. Syst.*, 46: 191–213.

Junker, R. R., N. Blüthgen, T. Brehm, J. Binkenstein, J. Paulus, H. M. Schaefer, and M. Stang. 2013. Specialization on traits as basis for the niche-breadth of flower visitors and as structuring mechanism of ecological networks. *Funct. Ecol.*, 27: 329–341.

Jürgens, A., S. L. Wee, A. Shuttleworth, and S. D. Johnson. 2013. Chemical mimicry of insect oviposition sites: A global analysis of convergence in angiosperms. *Ecol. Lett.*, 16: 1157–1167.

Kallenbach, M., Y. Oh, E. J. Eilers, D. Veit, I. T. Baldwin, and M. C. Schuman. 2014. A robust, simple, high-throughput technique for time-resolved plant volatile analysis in field experiments. *Plant J.*, 78: 1060–1072.

Kantsa, A., R. A. Raguso, A. G. Dyer, S. P. Sgardelis, J. M. Olesen, and T. Petanidou. 2017. Community-wide integration of floral colour and scent in a Mediterranean scrubland. *Nat. Ecol. Evol.*, 1: 1502–1510.

Kantsa, A., R. A. Raguso, A. G. Dyer, J. M. Olesen, T. Tscheulin, and T. Petanidou. 2018. Disentangling the role of floral sensory stimuli in pollination networks. *Nat. Commun.*, 9: 1041. https://www.nature.com/articles/s41467-018-03448-w

Kantsa, A., S. Sotiropoulou, M. Vaitis, and T. Petanidou. 2015. Plant volatilome in Greece: A review on the properties, prospects, and chemogeography. *Chem. Biodivers.*, 12: 1466–1480.

Karabagias, I. K., M. V. Vavoura, A. Badeka, S. Kontakos, and M. G. Kontominas. 2014. Differentiation of Greek thyme honeys according to geographical origin based on the combination of phenolic compounds and conventional quality parameters using chemometrics. *Food Anal. Methods*, 7: 2113–2121.

Karban, R., and I. T. Baldwin. 1997. *Induced Responses to Herbivory*. Chicago: Univ Chicago Press. 319p.

Kaspar, S., M. Peukert, A. Svatos, A. Matros, and H. P. Mock. 2011. MALDI-imaging mass spectrometry–an emerging technique in plant biology. *Proteomics*, 11: 1840–1850.

Kessler, D., C. Diezel, D. G. Clark, T. A. Colquhoun, and I. T. Baldwin. 2013. Petunia flowers solve the defence/apparency dilemma of pollinator attraction by deploying complex floral blends. *Ecol. Lett.*, 16: 299–306.

Kessler, D., K. Gase, and I. T. Baldwin. 2008. Field experiments with transformed plants reveal the sense of floral scents. *Science*, 321: 1200–1202.

Kite, G. C., W. L. A. Hetterscheid, M. J. Lewis et al. 1998. Inflorescence odours and pollinators of *Arum* and *Amorphophallus* (Araceae). In: S.J. Owens and P.J. Rudall (Editors). *Reproductive Biology*, pp. 295–315. Royal Botanic Gardens. Kew.

Kite, G. C., and W. L. Hetterscheid. 2017. Phylogenetic trends in the evolution of inflorescence odours in *Amorphophallus*. *Phytochemistry*, 142: 126–142.

Knoll, F. R. 1926. Die Arum-blütenstände und ihre Besucher. *Abhandlungen der Zoologisch–Botanischen Gesellschaft Wien*, 12: 382–481.

Kraemer, M., and U. Schmitt. 1999. Possible pollination by hummingbirds in *Anthurium sanguineum* Engl. (Araceae). *Plant Syst. Evol.*, 217: 333–335.

Leonhardt, S. D., S. Zeilhofer, N. Blüthgen, and T. Schmitt. 2010. Stingless bees use terpenes as olfactory cues to find resin sources. *Chem. Senses*, 35: 603–611.

Leonhardt, S. D., T. Schmitt, and N. Blüthgen, 2011. Tree resin composition, collection behavior and selective filters shape chemical profiles of tropical bees (Apidae: Meliponini). *PLOS ONE*, 6: e23445. https://doi.org/10.1371/journal.pone.0023445

Linhart, Y. B., and J. D. Thompson. 1995. Terpene-based selective herbivory by *Helix aspersa* (Mollusca) on *Thymus vulgaris* (Labiatae). *Oecologia*, 102: 126–132.

Linhart, Y. B., and J. D. Thompson. 1999. Thyme is of the essence: Biochemical polymorphism and multi-species deterrence. *Evol. Ecol. Res.*, 1: 151–171.

Linhart, Y. B., K. Keefover-Ring, K. A. Mooney, B. Breland, and J. D. Thompson. 2005. A chemical polymorphism in a multitrophic setting: Thyme monoterpene composition and food web structure. *Am. Nat.*, 166: 517–529.

Linhart, Y. B., P. Gauthier, K. Keefover-Ring, and J. D. Thompson. 2014. Variable phytotoxic effects of *Thymus vulgaris* (Lamiaceae) terpenes on associated species. *Int. J. Plant Sci.*, 176: 20–30.

Lokvam, J., and J. F. Braddock. 1999. Anti-bacterial function in the sexually dimorphic pollinator rewards of *Clusia grandiflora* (Clusiaceae). *Oecologia*, 119: 534–540.

Maia, A. C. D., S. Dötterl, R. Kaiser et al. 2012. The key role of 4-methyl-5-vinylthiazole in the attraction of scarab beetle pollinators: A unique olfactory floral signal shared by Annonaceae and Araceae. *J. Chem. Ecol.*, 38: 1072–1080.

Manan, A. A., R. M. Taha, E. E. Mubarak, and H. Elias. 2016. In vitro flowering, glandular trichomes ultrastructure, and essential oil accumulation in micropropagated *Ocimum basilicum* L. *In Vitro Cell. Dev. Biol. Plant*, 52: 303–314.

Miguel, G., M. Simoes, A. C. Figueiredo, J. G. Barroso, L. G., Pedro, and L. Carvalho. 2004. Composition and antioxidant activities of the essential oils of *Thymus caespititius, Thymus camphoratus* and *Thymus mastichina*. *Food Chem.*, 86: 183–188.

Mitrevski, B., and P. J. Marriott. 2012. Novel hybrid comprehensive 2D–multidimensional gas chromatography for precise, high-resolution characterization of multicomponent samples. *Anal. Chem.*, 84: 4837–4843.

Mockute, D., and G. Bernotiene. 2001. The α-terpenyl acetate chemotype of essential oil of *Thymus pulegioides* L. *Biochem. Syst. Ecol.*, 29: 69–76.

Möller, M., and J. L. Clark. 2013. The state of molecular studies in the family Gesneriaceae: A review. *Selbyana*, 31: 95–125.

Mondello, L., P. Q. Tranchida, P. Dugo, and G. Dugo. 2008. Comprehensive two-dimensional gas chromatography-mass spectrometry: A review. *Mass Spectrom. Rev.*, 27: 101–124.

Nogueira, P. C. D. L., V. Bittrich, G. J. Shepherd, A. V. Lopes, and A. J. Marsaioli. 2001. The ecological and taxonomic importance of flower volatiles of *Clusia* species (Guttiferae). *Phytochemistry*, 56: 443–452.

Nordström, K., J. Dahlbom, V. S. Pragadheesh et al. 2017. In situ modeling of multimodal floral cues attracting wild pollinators across environments. *Proc. Natl Acad. Sci. USA*, 114: 13218–13223.

de Oliveira, C. M., A. Porto, V. Bittrich, I. Vencato, and A. J. Marsaioli. 1996. Floral resins of *Clusia* spp.: Chemical composition and biological function. *Tetrahedron Lett.*, 37: 6427–6430.

Paine, T. D., K. F. Raffa, and T. C. Harrington. 1997. Interactions among scolytid bark beetles, their associated fungi, and live host conifers. *Annu. Rev. Entomol.*, 42: 179–206.

Palmer-Young, E. C., B. M. Sadd, and L. S. Adler. 2017a. Evolution of resistance to single and combined floral phytochemicals by a bumble bee parasite. *J. Evol. Biol.*, 30: 300–312.

Palmer-Young, E. C., B. M. Sadd, P. C. Stevenson, R. E. Irwin, and L. S. Adler. 2016. Bumble bee parasite strains vary in resistance to phytochemicals. *Sci. Rep.*, 6: 37087. https://www.nature.com/articles/srep37087

Palmer-Young, E. C., B. M. Sadd, R. E. Irwin, and L. S. Adler. 2017b. Synergistic effects of floral phytochemicals against a bumble bee parasite. *Ecol. Evol.*, 7: 1836–1849.

Papadopoulou, G. V., and N. M. van Dam. 2017. Mechanisms and ecological implications of plant-mediated interactions between belowground and aboveground insect herbivores. herbivores. *Ecol. Res.*, 32: 13–26.

Pereira, R., I. de Souza Carvalho, B.R. Simoneit, and D. de Almeida Azevedo. 2009. Molecular composition and chemosystematic aspects of Cretaceous amber from the Amazonas, Araripe and Recôncavo basins, Brazil. *Org. Geochem.*, 40: 863–875.

Petanidou, T., and D. Vokou. 1993. Pollination ecology of Labiatae in a phryganic (East Mediterranean) ecosystem. *Am. J. Bot.*, 80: 892–899.

Petanidou, T., and E. Smets. 1996. Does temperature stress induce nectar secretion in Mediterranean plants? *New Phytol.*, 133: 513–518.

Phillips, M. A., and R. B. Croteau. 1999. Resin-based defenses in conifers. *Trends Plant Sci.*, 4: 184–190.

Pichersky, E., and R. A. Raguso. 2018. Why do plants produce so many terpenoid compounds? *New Phytol.*, 220: 692–702.

Pokorny, T., I. Vogler, R. Losch et al. 2017. Blown by the wind: The ecology of male courtship display behavior in orchid bees. *Ecology*, 98: 1140–1152.

Policha, T., A. Davis, M. Barnadas, B. T. Dentinger, R. A. Raguso, and B. A. Roy. 2016. Disentangling visual and olfactory signals in mushroom-mimicking *Dracula* orchids using realistic three-dimensional printed flowers. *New Phytol.*, 210: 1058–1071.

Popova, M. P., K. Graikou, I. Chinou, and V. S. Bankova. 2010. GC-MS profiling of diterpene compounds in Mediterranean propolis from Greece. *J. Agric. Food Chem.*, 58: 3167–3176.

Potts, S. G., T. Petanidou, S. Roberts, C. O'Toole, A. Hulbert, and P. Willmer. 2006. Plant-pollinator biodiversity and pollination services in a complex Mediterranean landscape. *Biol. Conserv.*, 129: 519–529.

Raffa, K. F. 2001. Mixed messages across multiple trophic levels: The ecology of bark beetle chemical communication systems. *Chemoecology*, 11: 49–65.

Raguso, R. A., A. A. Agrawal, A. E. Douglas, G. Jander, A. Kessler, K. Poveda, and J. S. Thaler. 2015a. The raison d'être of chemical ecology. *Ecology*, 96: 617–630.

Raguso, R. A., J. N. Thompson, and D. R. Campbell. 2015b. Improving our chemistry: Challenges and opportunities in the interdisciplinary study of floral volatiles. *Nat. Prod. Rep.*, 32: 893–903.

Ramírez, S. R., T. Eltz, M. K. Fujiwara et al. 2011. Asynchronous diversification in a specialized plant-pollinator mutualism. *Science*, 333: 1742–1746.

Richardson, L. L., L. S. Adler, A. S. Leonard et al. 2015. Secondary metabolites in floral nectar reduce parasite infections in bumblebees. *Proc. R. Soc. B.*, 282: 20142471.

Riffell, J. A., E. Shlizerman, E. Sanders et al. 2014. Flower discrimination by pollinators in a dynamic chemical environment. *Science*, 344: 1515–1518.

Rodman, J. E., A. R. Kruckeberg, and I. A. Al-Shehbaz. 1981. Chemotaxonomic diversity and complexity in seed glucosinolates of *Caulanthus* and *Streptanthus* (Cruciferae). *Syst. Bot.*, 6: 197–222.

Rota, M. C., A. Herrera, R. M. Martínez, J. A. Sotomayor, and M. J. Jordán. 2008. Antimicrobial activity and chemical composition of *Thymus vulgaris, Thymus zygis* and *Thymus hyemalis* essential oils. *Food Control*, 19: 681–687.

Rundel, P. W., M. T. Arroyo, R. M. Cowling et al. 2018. Fire and plant diversification in Mediterranean-climate regions. *Front. Plant Sci.*, 9: 851.

Salatino, A., C. C. Fernandes-Silva, A. A. Righi, and M. L. F. Salatino. 2011. Propolis research and the chemistry of plant products. *Nat. Prod. Rep.*, 28: 925–936.

Schiestl, F. P., and S. Dötterl. 2012. The evolution of floral scent and olfactory preferences in pollinators: Coevolution or pre-existing bias? *Evolution*, 66: 2042–2055.

Schmid-Hempel, P., and S. Durrer. 1991. Parasites, floral resources and reproduction in natural populations of bumblebees. *Oikos*, 62: 342–350.

Schwerdtfeger, M., G. Gerlach, and R. Kaiser. 2002. Anthecology in the neotropical genus *Anthurium* (Araceae): A preliminary report. *Selbyana*, 23: 258–267.

Seymour, R. S. 2001. Diffusion pathway for oxygen into highly thermogenic florets of the arum lily *Philodendron selloum*. *J. Exp. Bot.*, 52: 1465–1472.

Shroff, R., K. Schramm, V. Jeschke et al. 2015. Quantification of plant surface metabolites by matrix-assisted laser desorption–ionization mass spectrometry imaging: Glucosinolates on *Arabidopsis thaliana* leaves. *Plant J.*, 81: 961–972.

Skubatz, H., D. D. Kunkel, W. N. Howald, R. Trenkle, and B. Mookherjee. 1996. The *Sauromatum guttatum* appendix as an osmophore: Excretory pathways, composition of volatiles and attractiveness to insects. *New Phytol.*, 134: 631–640.

Steele, C. L., J. Crock, J. Bohlmann, and R. Croteau. 1998a. Sesquiterpene synthases from grand fir (*Abies grandis*) comparison of constitutive and wound-induced activities, and cDNA isolation, characterization, and bacterial expression of δ-selinene synthase and γ-humulene synthase. *J. Biol. Chem.*, 273: 2078–2089.

Steele, C. L., S. Katoh, J. Bohlmann, and R. Croteau. 1998b. Regulation of oleoresinosis in grand fir (*Abies grandis*): Differential transcriptional control of monoterpene, sesquiterpene, and diterpene synthase genes in response to wounding. *Plant Physiol.*, 116: 1497–1504.

Tarayre, M., J. D. Thompson, J. Escarré, and Y. B. Linhart. 1995. Intra-specific variation in the inhibitory effects of *Thymus vulgaris* (Labiatae) monoterpenes on seed germination. *Oecologia*, 101: 110–118.

Teichert, H., S. Dötterl, B. Zimma, M. Ayasse, and G. Gottsberger. 2009. Perfume-collecting male euglossine bees as pollinators of a basal angiosperm: The case of *Unonopsis stipitata* (Annonaceae). *Plant Biol.*, 11: 29–37.

Thompson, J. D., P. Gauthier, J. Amiot et al. 2007. Local adaptation in a chemically polymorphic plant: The relative performance of *Thymus vulgaris* chemotypes in a spatially heterogeneous environment. *Ecol. Monogr.*, 77: 421–439.

Tissier, A., J. A. Morgan, and N. Dudareva. 2017. Plant volatiles: Going "in" but not "out" of trichome cavities. *Trends Plant Sci.*, 22: 930–938.

Tümen, G., N. Kirimer, and K. H. C. Başer. 1995. Composition of the essential oil of *Thymus* species growing in Turkey. *Chemistry of Natural Compounds*, 31: 42–47.

Urru, I., M. C. Stensmyr, and B. S. Hansson. 2011. Pollination by brood-site deception. *Phytochemistry*, 72: 1655–1666.

Vernet, P., R. H. Gouyon, and G. Valdeyron. 1986. Genetic control of the oil content in *Thymus vulgaris* L: A case of polymorphism in a biosynthetic chain. *Genetica*, 69: 227–231.

Whitten, W. M., H. G. Hills, and N. H. Williams. 1988. Occurrence of ipsdienol in floral fragrances. *Phytochemistry*, 27: 2759–2760.

Whitten, W. M., N. H. Williams, W. S. Armbruster, M. A. Battiste, L. Strekowski, L., and N. Lindquist. 1986. Carvone oxide: An example of convergent evolution in euglossine pollinated plants. *Syst. Bot.*, 11: 222–228.

Williams, N. H., and W. M. Whitten. 1983. Orchid floral fragrances and male euglossine bees: Methods and advances in the last sesquidecade. *Biol. Bull.*, 164: 355–395.

Wise, M. L., T. J. Savage, E. Katahira, and R. Croteau. 1998. Monoterpene synthases from common sage (*Salvia officinalis*) cDNA isolation, characterization, and functional expression of (+)-sabinene synthase, 1, 8–cineole synthase, and (+)-bornyl diphosphate synthase. *J. Biol. Chem.*, 273: 14891–14899.

Wright, G. A., and F. P. Schiestl. 2009. The evolution of floral scent: The influence of olfactory learning by insect pollinators on the honest signaling of floral rewards. *Funct. Ecol.*, 23: 841–851.

Zapata, F., and P. V. Fine. 2013. Diversification of the monoterpene synthase gene family (TPSb) in *Protium*, a highly diverse genus of tropical trees. *Mol. Phylogenet. Evol.*, 68: 432–442.

Zimmermann, Y., S. R. Ramírez, and T. Eltz. 2009. Chemical niche differentiation among sympatric species of orchid bees. *Ecology*, 90: 2994–3008.

18 Essential Oils as Lures for Invasive Ambrosia Beetles

Paul E. Kendra, Nurhayat Tabanca, Wayne S. Montgomery, Jerome Niogret, David Owens, and Daniel Carrillo

CONTENTS

18.1 INTRODUCTION

Ambrosia beetles in the weevil subfamily Scolytinae (Coleoptera: Curculionidae) are wood borers that have developed close symbiotic relationships with fungi (Hulcr and Stelinski, 2017). Upon dispersal, adult females carry fungal conidia (spores) in specialized storage organs known as mycangia. Colonizing females excavate galleries in the xylem of host trees, inoculate the gallery walls with conidia, and subsequently cultivate fungal gardens to provide food for themselves and their progeny. Most species of ambrosia beetle will colonize only physiologically stressed or dying trees, and host-seeking females locate vulnerable trees by their elevated emissions of ethanol (Miller and Rabaglia, 2009; Ranger et al., 2010). As such, these beetles can be considered beneficial insects that initiate decomposition of woody material and contribute to the maintenance of healthy forest ecosystems.

Unfortunately, with significant increases in global trade in recent decades, exotic ambrosia beetles have become one of the most frequently intercepted insects at ports of entry (Rabaglia et al., 2008). Although benign or minor pests in their native range, a few species and their fungal symbionts have emerged as serious threats to forests and agriculture following establishment in new environments and exploitation of novel hosts (Hulcr and Dunn, 2011). This is the case with the redbay ambrosia beetle, *Xyleborus glabratus* Eichhoff (Fraedrich et al., 2008; Kendra et al., 2013a) and a group of cryptic species morphologically similar to the tea shot-hole borer, *Euwallacea fornicatus* Eichhoff (Kasson et al., 2013; O'Donnell et al., 2015). These ambrosia beetles are endemic to Southeast Asia, but are now established in portions of the US, where they vector fungal pathogens responsible for vascular wilt diseases in commercially grown avocado (*Persea americana* Mill.) and numerous native American trees.

A critical component for successful pest management is the identification of effective attractants, followed by development of lures for early pest detection. This process is facilitated by an understanding of the unique aspects of the insect's chemical ecology. First, as with other ambrosia beetles within the tribe Xyleborini, species-specific sex pheromones are not used by *X. glabratus* or *Euwallacea* spp.; this is because new females usually mate with their flightless male siblings prior to dispersal from the natal gallery. Second, atypical of ambrosia beetles, neither *X. glabratus* or *Euwallacea* spp. are strongly attracted to ethanol. These beetles function ecologically as primary colonizers, capable of attacking healthy unstressed trees. As a result, they are attracted to the volatile terpenoids naturally emitted from the wood of host trees (i.e., kairomones used for host location). Since essential oils consist of concentrated plant terpenoids, they have provided an ideal substrate for development of lures for ambrosia beetle pests (Hanula and Sullivan, 2008; Kendra et al., 2012c, 2014b, 2015b, 2016a, 2018; Owens et al., 2017). This chapter will summarize the succession of essential oil lures used for *X. glabratus* over the last decade, outline development of the current lure which is highly enriched in (–)-α-copaene, present chemical analysis of the α-copaene lure, and describe its recent applications for detection of *Euwallacea* in Florida avocado groves.

18.2 REDBAY AMBROSIA BEETLE

18.2.1 Background

Females of *X. glabratus* are the primary vectors of the fungus that causes laurel wilt, a lethal disease of trees and woody shrubs within the family Lauraceae, and a disease not observed in nature until the establishment of *X. glabratus* in the US (Fraedrich et al., 2008; Hanula et al., 2008) (Figure 18.1). Although multiple fungal species have been isolated from the mycangia of *X. glabratus*, the beetle's predominant symbiont, *Raffaelea lauricola* T. C. Harr., Fraedrich and Aghayeva, is the confirmed pathogen (Harrington et al., 2008, 2010). Introduction of *R. lauricola* into susceptible hosts triggers secretion of gels and the formation of parenchymal tyloses within xylem vessels (Ploetz et al., 2012).

FIGURE 18.1 (See color insert.) Female (A) and male (B) of the redbay ambrosia beetle, *Xyleborus glabratus*, vector of the fungal pathogen that causes laurel wilt. (C) Cross-section through the trunk of a swampbay tree, *Persea palustris*, showing extensive galleries excavated by redbay ambrosia beetles. (D) Silkbay tree, *Persea humilis*, killed by laurel wilt in Highlands County, FL, US.

This defensive response can be extreme, resulting in decreased water transport, systemic wilt, and ultimately death of the infected tree.

Since initially detected near a maritime port in Georgia in 2002, *X. glabratus* has continued to spread throughout the southeastern US. It is now established in nine states (USDA-FS, 2018) where laurel wilt has reached epidemic levels (reviewed by Kendra et al., 2013a; Hughes et al., 2015; Ploetz et al., 2017b). Throughout this range, primary hosts are native *Persea* species, including redbay (*Persea borbonia* [L.] Spreng.), swampbay (*P. palustris* [Raf.] Sarg.), and silkbay (*P. humilis* Nash), and as of 2017, an estimated 300 million native bay trees have been killed (Hughes et al., 2017). In southern Florida, avocado trees are also readily attacked. Although *X. glabratus* clearly transported the *R. lauricola* pathogen southward into commercial groves, the progression of laurel wilt in this agroecosystem has unfolded quite differently than that in native forests. Apparently avocado is a poor host for *X. glabratus*, limiting its reproduction (Brar et al., 2013); in contrast, established species of ambrosia beetle successfully colonize avocado trees stressed by laurel wilt. This situation has led to an unprecedented lateral transfer of *R. lauricola* to at least nine additional beetle species, which now comprise a community of secondary vectors of the laurel-wilt pathogen in avocado groves (Carrillo et al., 2014; Ploetz et al., 2017c). Because of the severe impact of laurel wilt, it has been identified as a high-priority disease by the USDA National Plant Disease Recovery System (USDA-ARS, 2018), and separate recovery plans have been drafted for the disease in forest hosts (Hughes et al., 2015) and in avocado (Ploetz et al., 2017a).

18.2.2 Host Attractants

All known hosts of *X. glabratus* in the US are members of the Lauraceae family; therefore, research on kairomone attractants has focused on this taxonomic group. Kendra et al., (2014a) conducted a comparative study of nine lauraceous species (including avocado cultivars representative of each of the three botanical races) to determine boring preferences and in-flight attraction of *X. glabratus* as related to phytochemical emissions from host wood. Emissions of α-copaene, α-cubebene, α-humulene, and calamenene (all sesquiterpene hydrocarbons) were positively correlated with attraction to Lauraceae. Of these compounds, α-copaene and α-humulene had been correlated previously with attraction to three additional avocado cultivars, as well as to wood from lychee (*Litchi chinensis* Sonn.; Sapindaceae) (Kendra et al., 2011a). Lychee is not a reproductive host of *X. glabratus*, but particular cultivars will attract females and initiate boring behavior owing to sesquiterpene emission profiles similar to those of the Lauraceae (Kendra et al., 2011a, 2011b, 2013b). In addition, independent research identified eucalyptol (1,8 cineole; a monoterpene ether) as another host-based attractant (Kuhns et al., 2014).

For all these terpenoids, olfactory chemoreception has been confirmed in female *X. glabratus* using freshly dissected antennae and newly developed electroantennography (EAG) techniques designed to accommodate the minute antennae of this species (Figure 18.2) (Kendra et al., 2012a, 2014a). Comparative EAG recordings have been useful for assessing potential attractants, with good correlation between amplitude of antennal responses and relative attraction observed in the field. The current hypothesis is that the initial step in the host-location process consists of in-flight detection (by antennal olfactory receptors) of a mixture of volatile terpenoids characteristic of the Lauraceae (i.e., a long-range "signature bouquet" indicative of an appropriate host) (Kendra et al., 2011a, 2014a; Niogret et al., 2011a). This is followed by a series of cues presented in sequential order, including mid-range visual cues and integration of various short-range cues (olfactory, gustatory, contact chemosensory, tactile) which reinforce the message that a suitable host has been located, triggering a switch from "host-location" behavior to "host-acceptance and boring" behavior (Kendra et al., 2014a and references therein).

18.2.3 Field Lures

Although identification of specific attractant chemicals has been useful for understanding the chemical ecology of *X. glabratus*, production of field lures using synthetic sesquiterpenes poses

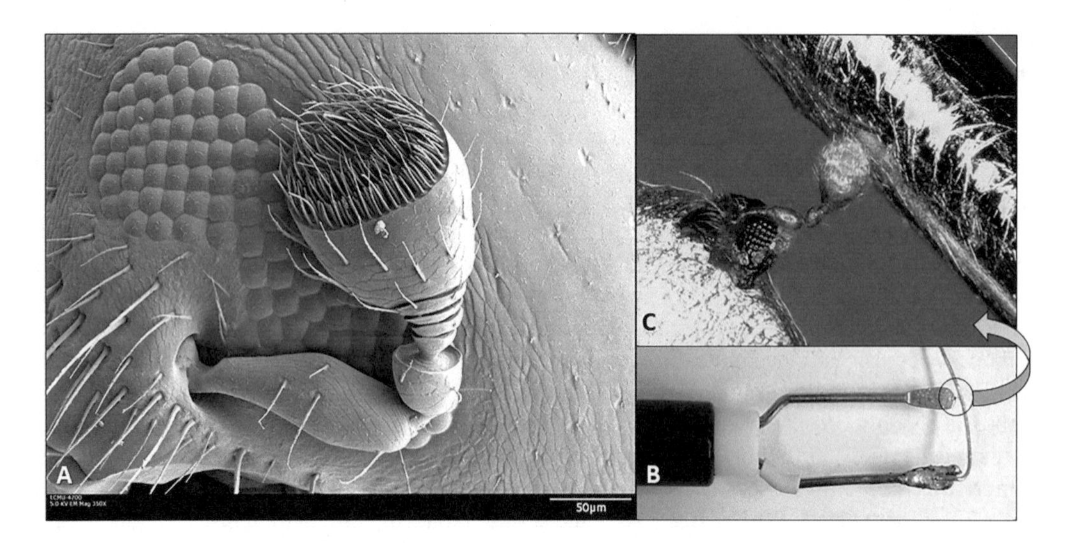

FIGURE 18.2 **(See color insert.)** (A) Scanning electron micrograph of female redbay ambrosia beetle antenna (length = 0.4 mm), showing concentration of olfactory sensilla on the distal end of antennal club. (B) Two-pronged gold electroantennography (EAG) electrode modified to accommodate minute antennae by soldering a flexible gold wire to one prong. (C) Magnified view of the EAG antennal preparation, with head capsule mounted to the prong (indifferent electrode) and back of antennal club in contact with wire (different electrode).

a challenge. Many sesquiterpenes, particularly α-copaene, are difficult to synthesize, not readily available, and prohibitively costly in quantities sufficient for trap deployment (Flath et al., 1994). Thus, development of economical lures has relied on the use of plant-derived essential oils (whole or distilled oil products) naturally high in attractive sesquiterpenes.

18.2.3.1 Manuka and Phoebe Oils

Crook et al., (2008) identified two essential oils attractive to the emerald ash borer, *Agrilus planipennis* Fairmaire (Coleoptera: Buprestidae). These consisted of manuka oil (extracted from *Leptospermum scoparium* Forst. and Forst.; Myrtaceae), and phoebe oil (obtained from *Phoebe porosa* Mez; Lauraceae). When first tested in Georgia and South Carolina, these two oils were also found to be good baits for trapping *X. glabratus* (Hanula and Sullivan, 2008). However, when commercial manuka and phoebe oil lures were evaluated in Florida in 2009–2010, both were found to be fairly non-specific, capturing a variety of non-target Scolytinae (Kendra et al., 2011b, 2012c). Moreover, the manuka oil lures were not competitive with host *Persea* wood (Figure 18.3A) and had a field life of only 2–3 weeks due to rapid depletion of sesquiterpenes under Florida field conditions (Figure 18.3C). In contrast, phoebe oil lures were competitive with host wood (Figure 18.3), captured significantly more *X. glabratus* than manuka oil lures (Figure 18.3B), and despite rapid loss of terpenoids initially, continued to release low levels of attractive sesquiterpenes for at least 10 weeks (Figure 18.3C). In addition, EAG response to phoebe oil was comparable to that of *Persea* wood, and both were significantly higher than EAG response to manuka oil (Kendra et al., 2012a). Unfortunately, shortly after completion of the field studies by Kendra et al., (2012c), phoebe oil lures were no longer available commercially due to a limited supply of source trees in Brazil. Therefore, the suboptimal manuka oil lure became the standard for detection of *X. glabratus* in the US.

18.2.3.2 Cubeb Oil

The poor performance (and increasing cost) of manuka oil lures prompted research to evaluate additional essential oils as potential alternatives. Field tests in 2012 (Kendra et al., 2013a, 2014b) compared captures of female *X. glabratus* among seven essential oils. Treatments included whole oil preparations of manuka and phoebe oils, plus five new oils: cubeb, ginger root, angelica seed, tea

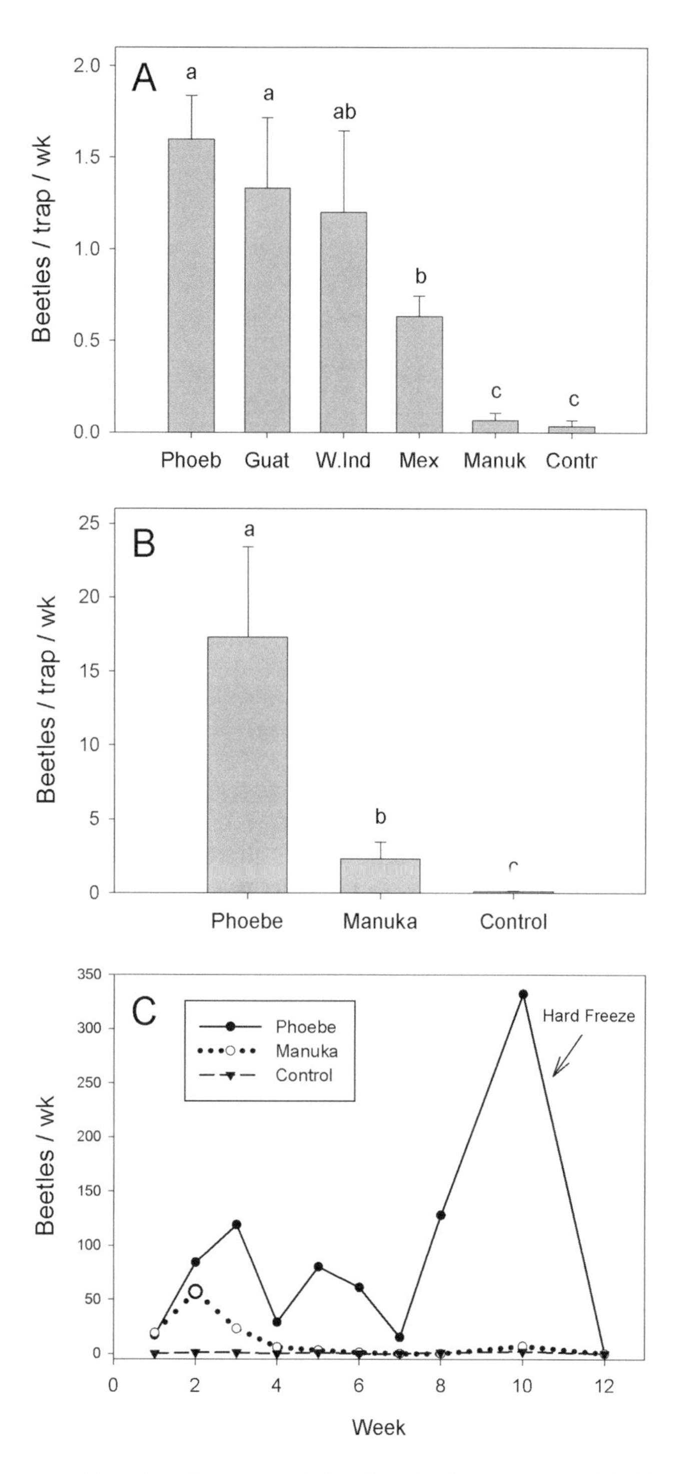

FIGURE 18.3 Captures of female redbay ambrosia beetle in field tests conducted in Florida, US. (A) Mean (±SE) captures in an 8-wk test (Alachua County) with commercial phoebe oil lure (Phoeb), commercial manuka oil lure (Manuk), wood bolts of three avocado cultivars: "Brooks Late," Guatemalan race (Guat); "Simmonds," West Indian race (W. Ind); "Seedless Mexican," Mexican race (Mex), and an unbaited control (Contr). (Adapted from Kendra et al., 2011a. *J. Chem. Ecol.*, 37: 932–942.) (B) Mean (±SE) captures and (C) summed weekly captures in a 12-wk test (Highlands County) with commercial oil lures and an unbaited control. Bars topped with the same letter are not significantly different. (Adapted from Kendra, P. E. et al., 2012c. *J. Econ. Entomol.*, 105: 659–669.)

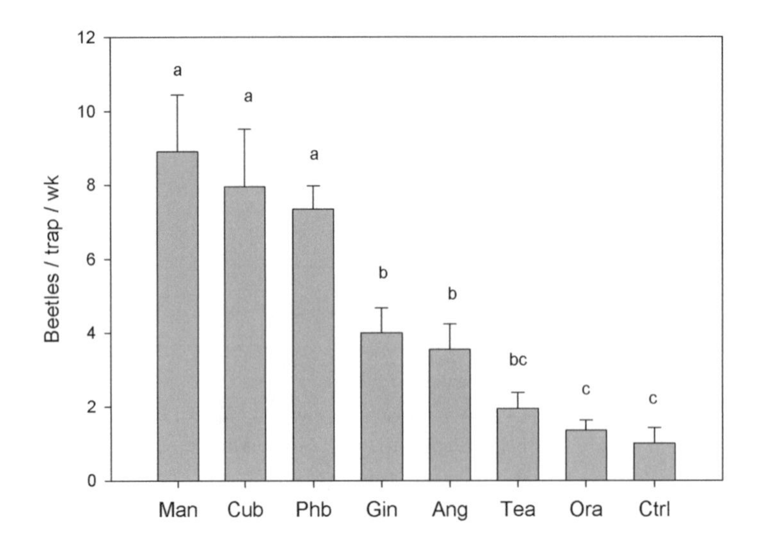

FIGURE 18.4 Mean (±SE) captures of female redbay ambrosia beetle with seven essential oils deployed in a 4-wk field test conducted in Highlands County, FL, USA. Treatments consisted of manuka (Man), cubeb (Cub), phoebe (Phb), ginger root (Gin), angelica seed (Ang), tea tree (Tea), Valencia orange (Ora) oils, and an unbaited control trap (Ctrl). A membrane-based dispenser was used to prepare the lures, and each lure was loaded with 5 mL of neat oil. Bars topped with the same letter are not significantly different. (Adapted from Kendra, P. E. et al., 2013a. *Am. J. Plant. Sci.*, 4: 727–738.)

tree, and Valencia orange oils. These latter oils were chosen based on their sesquiterpene content, which included constituents correlated with captures of *X. glabratus* previously (Kendra et al., 2011a, 2012c). In addition, they all had been reported to attract males of the Mediterranean fruit fly, *Ceratitis capitata* Wied. (Diptera: Tephritidae), another pest that responds positively to substrates containing α-copaene (Flath et al., 1994; Niogret et al., 2011b). Of the new oils evaluated, ginger root and angelica seed oils were moderately attractive, but whole cubeb oil (extracted from berries of tailed pepper, *Piper cubeba* L.; Piperaceae) was identified as a strong new attractant for *X. glabratus* (Figure 18.4) (Kendra et al., 2013a, 2014b).

In 2013, a commercial formulation of cubeb oil became available, which consisted of a plastic bubble lure containing a proprietary oil product distilled to remove monoterpenoids, thereby enriching the sesquiterpene content (Synergy Semiochemicals Corp., Burnaby, BC, Canada). The cubeb bubble lure captured significantly more *X. glabratus* than the commercial manuka oil lure in several independent field trials (Hanula et al., 2013; Kendra et al., 2014b). In EAG analyses, olfactory responses to the cubeb lure and phoebe lure were equivalent, and both were significantly greater than response obtained with the manuka oil lure (Kendra et al., 2014b).

Further evaluations (Kendra et al., 2015b) indicated that field attraction of cubeb oil lures was significantly better than that of phoebe oil lures (Figure 18.5A), and the bubble lures had a field life of at least 12 weeks (Figure 18.5B). Temporal analysis of cubeb lure emissions revealed that its superior longevity was due to extended low release of sesquiterpenes, primarily α-copaene (Figure 18.6A) and α-cubebene (Figure 18.6B) (Kendra et al., 2015b). The bubble dispenser used for the new cubeb lure has a much lower surface area-to-volume ratio than the flat rectangular design of the manuka oil lure (Figure 18.7). This difference, coupled with a thicker release membrane, has resulted in a much improved delivery system for sustained release of sesquiterpene attractants.

18.2.3.3 Copaiba Oil and Enriched α-Copaene Oil

After the cubeb bubble lure replaced the manuka lure for detection of *X. glabratus*, research efforts (in collaboration with Synergy Semiochemicals Corp.) focused on elucidation of the attractive components found in cubeb oil. Since this essential oil is composed of a complex mixture of

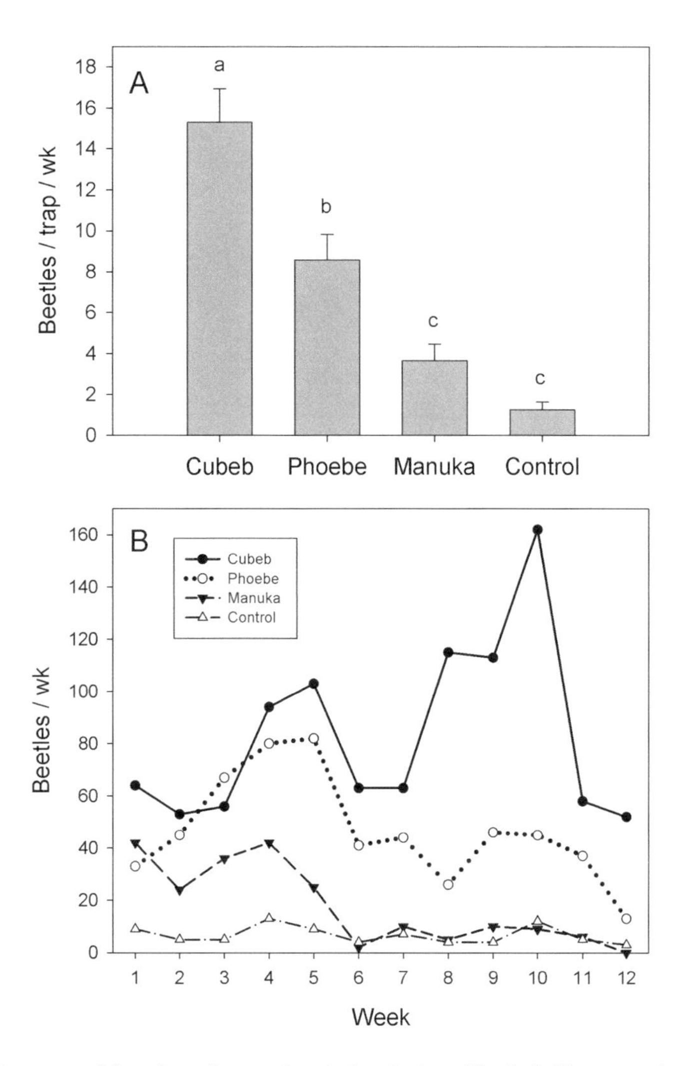

FIGURE 18.5 Captures of female redbay ambrosia beetle in a 12-wk field test conducted in Highlands County, FL, US. (A) Mean (±SE) captures and (B) summed weekly captures obtained with commercial oil lures and an unbaited control. Bars topped with the same letter are not significantly different. (Adapted from Kendra, P. E. et al., 2015b. *J. Econ. Entomol.*, 108: 350–361.)

terpenoids, hydro-distillation was used to separate whole cubeb oil into multiple fractions, based on boiling point of the terpenoid constituents. These fractions were then evaluated through EAG analyses and binary-choice bioassays to quantify olfactory and behavioral responses, respectively. Results indicated that fractions with high α-copaene and α-cubebene content were the most bioactive for female *X. glabratus* (Kendra et al., 2016a), supporting previous hypotheses regarding the importance of these kairomones (Kendra et al., 2011a, 2012c, 2014a, 2014b, 2015b; Niogret et al., 2013). Based on this observation, two additional essential oil products were evaluated, both formulated in slow-release bubble dispensers. The first lure contained whole copaiba oil (extracted from species of *Copaifera* L.; Leguminosae), an essential oil low in α-cubebene, but with twice as much α-copaene as whole cubeb oil. In field trials (Figure 18.8A), the prototype copaiba oil lure captured equal numbers of *X. glabratus* as the commercial cubeb oil lure (Kendra et al., 2016a).

The second prototype lure contained a proprietary essential oil product that was highly distilled to achieve 50% α-copaene content. In multiple field tests, this enriched α-copaene lure captured significantly more *X. glabratus* than either the cubeb oil lure or the copaiba oil lure (Figure 18.8A)

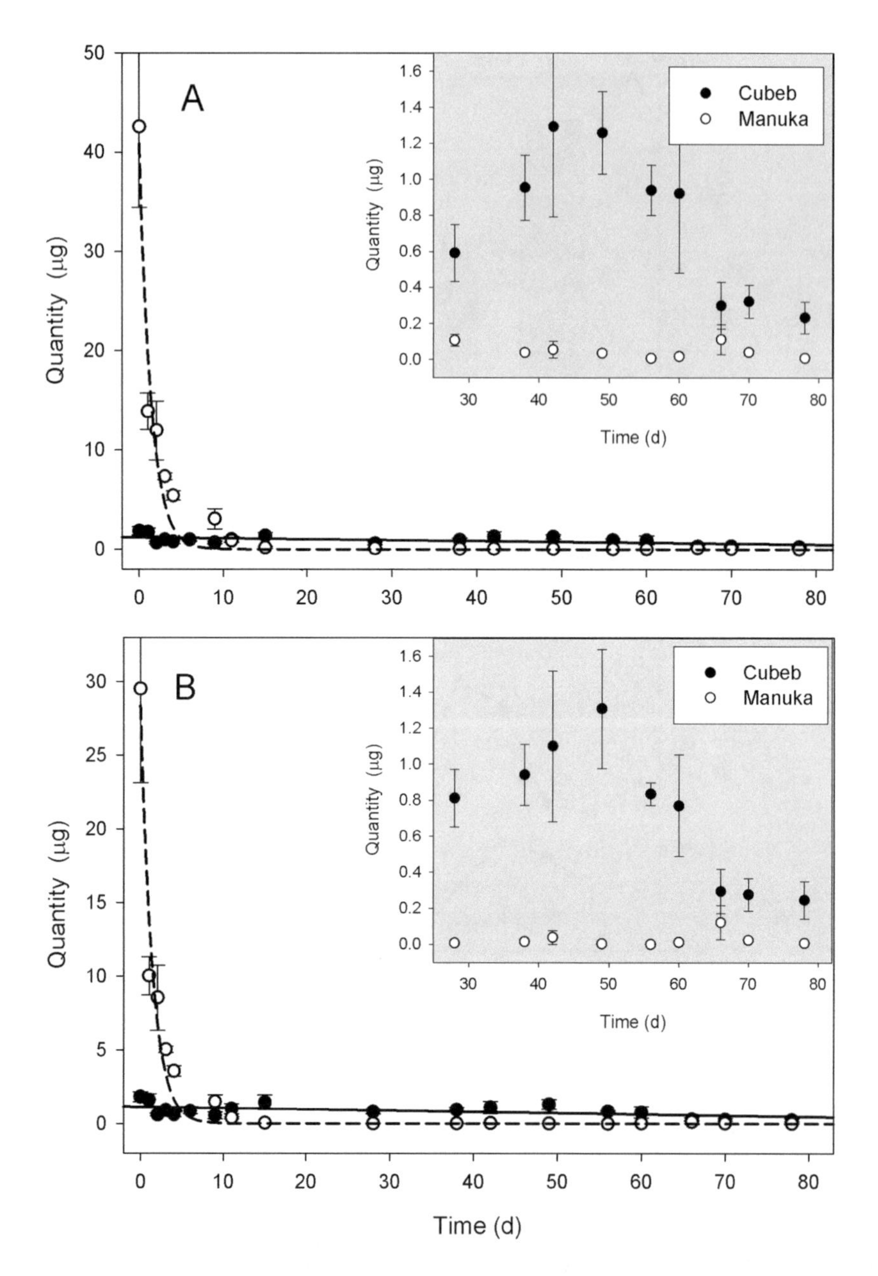

FIGURE 18.6 Emissions of (A) α-copaene and (B) α-cubebene quantified over time from commercial oil lures field-deployed for 12 wk in Miami-Dade County, FL, US. Inset enhances the scale for emissions beginning at 4 wk, the point at which manuka lures lost efficacy for attraction of redbay ambrosia beetle in field tests. Volatiles were isolated by super-Q collection, analyzed by GC-MS, and identified by comparison of Kovats retention index with synthetic chemicals. (Adapted from Kendra, P. E. et al., 2015b. *J. Econ. Entomol.*, 108: 350–361.)

and displayed field longevity of at least three months (Figure 18.8B) (Kendra et al., 2016a, 2016b). EAG response to the α-copaene lure was also significantly higher than response recorded with either the cubeb oil lure or the copaiba oil lure (Kendra et al., 2016a). This research confirmed the role of α-copaene as a primary host-location cue, and identified the 50% α-copaene lure as an improved tool for more sensitive detection of *X. glabratus* than that provided by the cubeb oil lure. The α-copaene bubble lure is now produced commercially and has been adopted by state and federal action agencies for use in *X. glabratus* monitoring programs in the US.

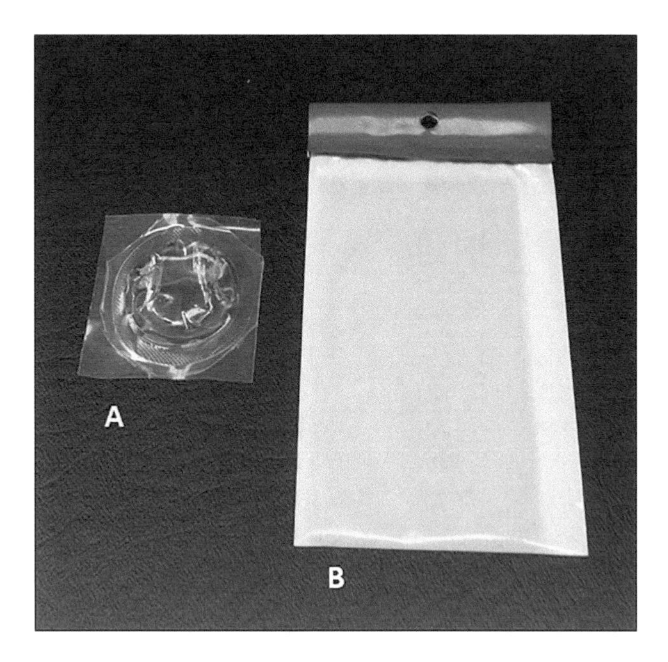

FIGURE 18.7 **(See color insert.)** Commercially available lures for redbay ambrosia beetle. (A) Cubeb bubble lure (Synergy Semiochemicals Corp.; Burnaby, BC, Canada). (B) Manuka oil lure (ChemTica Internacional; Heredia, Costa Rica).

18.3 CHEMICAL ANALYSIS OF THE α-COPAENE LURE

18.3.1 Method

Three replicates of the α-copaene lure (Synergy Semiochemicals Corp., Product # 3302, Lot # 160511) were analyzed by gas chromatography-mass spectrometry (GC-MS) using a 5975B mass-selective detector (Agilent Technologies, Santa Clara, SA, US) and equipped with a DB-5 column (30 m, 0.25 mm ID, film thickness 0.25 μm; Agilent Technologies). The oven temperature was programmed from 60°C (1.3 min) to 246°C at a rate of 3°C/min. Helium was used as the carrier gas at a flow rate of 1.3 mL/min. The PTV injector and ion source temperature were 200°C and 230°C, respectively. Mass spectra were recorded at 70 eV. The mass range was *m/z* 35–450 and scan rate was set at 2.8 scans/second. Data acquisition and processing were done using software Mass Hunter B.07.02 (Agilent Technologies).

Chemical identification of lure components was based on (a) linear retention indices (LRI), which were calculated using *n*-alkanes on the DB-5 column according to the calculation method developed by van den Dool and Kratz (1963); (b) comparison of the mass spectra with published literature values (MassFinder, 2004; Adams, 2007; FFNSC-3, 2015; NIST, 2017; Wiley, 2017); (c) comparison with our own library "SHRS Essential Oil Constituents-DB-5," which was built with authentic standards and components of known essential oils analyzed on the DB-5 column; and (d) comparison with reference compounds purchased from Fluka Chemical Co., Buchs, SG, Switzerland (δ-cadinene, Cas # 483-76-1; α-copaene. Cas # 3856-25-5; β-elemene, Cas # 515-13-9; valencene, Cas # 4630-07-3), Sigma-Aldrich Ltd, St. Louis, MO, US (alloaromadendrene, Cas # 25246-27-9; β-caryophyllene, Cas # 87-44-5; caryophyllene oxide, Cas # 1139-30-6, cyclosativene, Cas # 22469-52-9; and α-humulene, Cas # 6753-98-6), and α-cubebene (Cas # 17699-14-8) from Bedoukian Research, Inc., Danbury, CT, US.

18.3.2 Results

Fifteen compounds were identified from the α-copaene enriched oil lure, and relative percentages of the components are shown in Table 18.1. The analysis indicated that sesquiterpene hydrocarbons were

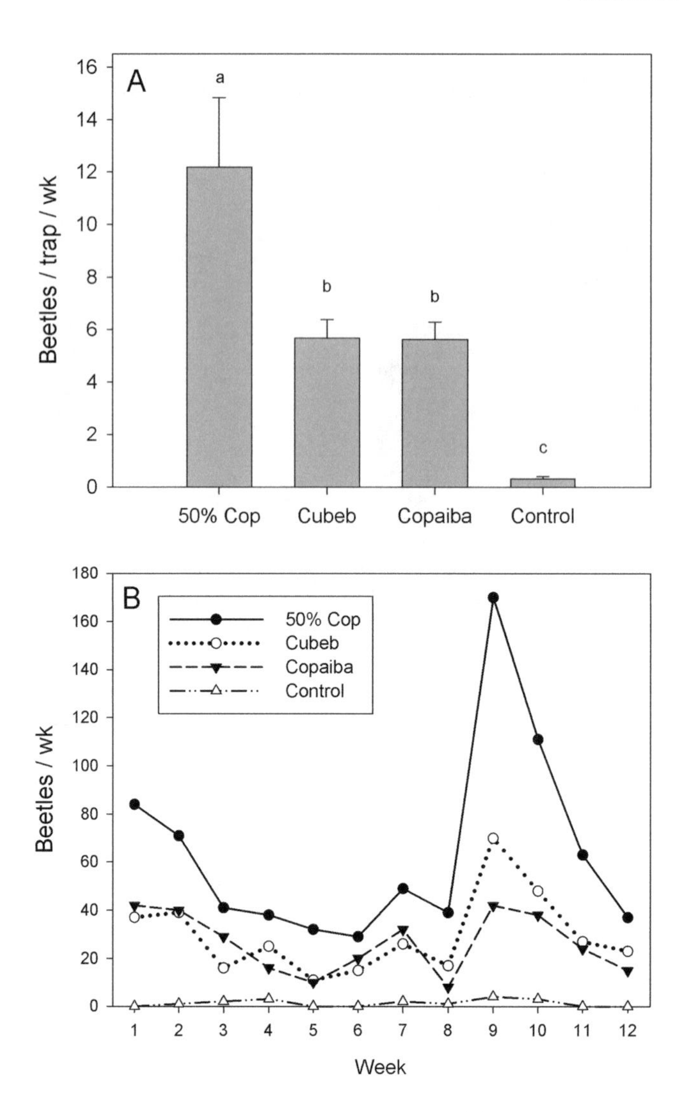

FIGURE 18.8 Captures of female redbay ambrosia beetle in a 12-wk field test conducted in Highlands County, FL, US. (A) Mean (±SE) captures and (B) summed weekly captures obtained with commercial cubeb oil lure, prototype copaiba oil lure, prototype 50% α-copaene lure (50% cop), and an unbaited control trap. Bars topped with the same letter are not significantly different. (Adapted from Kendra, P. E. et al, 2016a. *J. Pest Sci.*, 89: 427–438.)

the dominant constituents, comprising 99.3% of the lure contents; only one oxygenated sesquiterpene was detected, caryophyllene oxide (0.7%). Major sesquiterpene hydrocarbons consisted of α-copaene (60.2%), β-caryophyllene (18.7%), and δ-cadinene (10.3%), with lesser amounts of α-cubebene (2.4%), α-humulene (3.2%), and γ-muurolene (1.1%). These results are consistent with those reported by Owens et al. (2018b) for a different lot of α-copaene lures analyzed on a DB-5MS column. Although α-copaene is the primary attractant, other sesquiterpene components (part of the laurel bouquet, Kendra et al., 2014a) may be contributing to the overall attraction of this lure.

Owens et al. (2018b) also evaluated the enantiomeric distribution of α-copaene on an Rt-βDEXse column (2,3-di-*O*-ethyl-6-*O*-*tert*-butyl dimethylsilyl-β-cyclodextrin; Restek Corp., Bellefonte, PA, US), which indicated that the (–)-enantiomer was predominant (99.91%) in the commercial lure, with a very high enantiomeric excess of 99.82% (Figure 18.9). Knowledge of enantiomer distributions is an important consideration when developing management programs for insect pests. Although

TABLE 18.1
Chemical Composition of α-copaene Enriched Lure (n = 3)

#	LRI_{exp}[a1]	LRI_{Lit}[b2]	Compound	%	Identification method[c1]
1	1335	1335	δ-Elemene	0.27 ± 0.02	a, b, c
2	1348	1345	α-Cubebene	2.42 ± 0.11	a, b, c, d
3	1363	1369	Cyclosativene	0.01 ± 0.00	a, b, c, d
4	1374	1374	α-Copaene	60.21 ± 0.09	a, b, c, d
5	1388	1387	β-Cubebene	0.31 ± 0.01	a, b, c
6	1390	1389	β-Elemene	0.25 ± 0.02	a, b, c, d
7	1407	1409	α-Gurjunene	0.84 ± 0.03	a, b, c
8	1416	1417	β-Caryophyllene	18.72 ± 0.10	a, b, c, d
9	1450	1452	α-Humulene	3.19 ± 0.02	a, b, c, d
10	1456	1458	Alloaromadendrene	0.81 ± 0.05	a, b, c, d
11	1474	1478	γ-Muurolene	1.07 ± 0.07	a, b, c
12	1488	1496	Valencene	0.43 ± 0.02	a, b, c, d
13	1498	1500	α-Muurolene	0.54 ± 0.02	a, b, c
14	1521	1522	δ-Cadinene	10.26 ± 0.12	a, b, c, d
15	1580	1582	Caryophyllene oxide	0.66 ± 0.01	a, b, c, d

[a1] LRI was calculated against *n*-alkanes based on DB-5 column.
[b1] LRI values from Adams library (2007).
[c1] Identification method given in Section 18.3.1.

X. glabratus and *E. nr. fornicatus* (see Section 18.4.3.2 below) are highly attracted to (–)-α-copaene in this essential oil lure, the (+)–α-copaene isolated from angelica seed oil (*Angelica archangelica* L.) was found to be more attractive to males of *C. capitata* then its (–)- enantiomer (Flath et al., 1994).

18.4 *EUWALLACEA* SHOT-HOLE BORERS

18.4.1 Background

The type species, *E. fornicatus*, is a polyphagous beetle, but it is best documented as a pest of cultivated tea, *Camellia sinensis* (L.) Kuntze (Theaceae), in Sri Lanka, India, and elsewhere in Asia (Hazarika et al., 2009) (Figure 18.10). Females typically colonize small-diameter branches which reduces production of tender foliage (flushes), the commodity harvested commercially. In recent years, beetles morphologically indistinguishable from *E. fornicatus* have emerged as pests of avocado, woody ornamentals, and native trees outside of Asia. Pest populations have become established in the US (Rabaglia et al., 2006; Eskalen et al., 2013), Israel (Mendel et al., 2012), Australia (Campbell and Geering, 2011), Mexico (García-Avila et al., 2016), and, most recently, South Africa (Paap et al., 2018). Molecular investigations have concluded that beetles from different geographic regions possess sufficient genetic diversity to constitute a cryptic species complex; therefore, all members are conservatively referred to as *E.* near *fornicatus* currently, pending taxonomic separation (Kasson et al., 2013, O'Donnell et al., 2015). The primary fungal symbionts are species within the genus *Fusarium* (Hypocreales: Nectriaceae) (Freeman et al., 2012). Unlike the *R. lauricola* pathogen that systemically colonizes host xylem, *Fusarium* spp. fungi remain localized and destroy the functional xylem in close proximity to beetle galleries. With avocado trees, beetle attacks tend to be concentrated at the base of branches, so the localized lesions essentially girdle the branch, impede water conduction, and eventually kill the branch. This newly described vascular disease has been termed *Fusarium* dieback (Freeman et al., 2012).

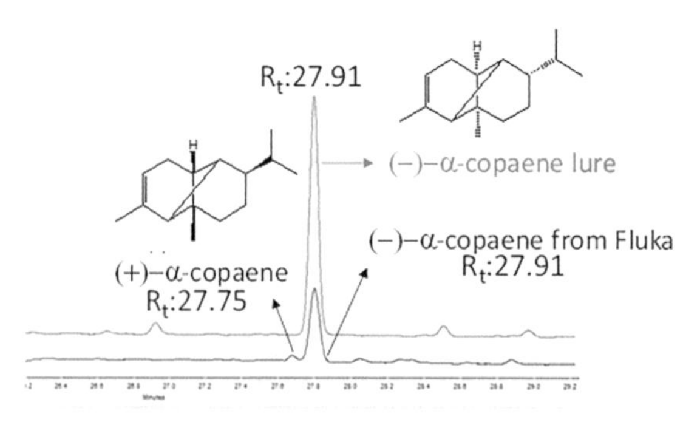

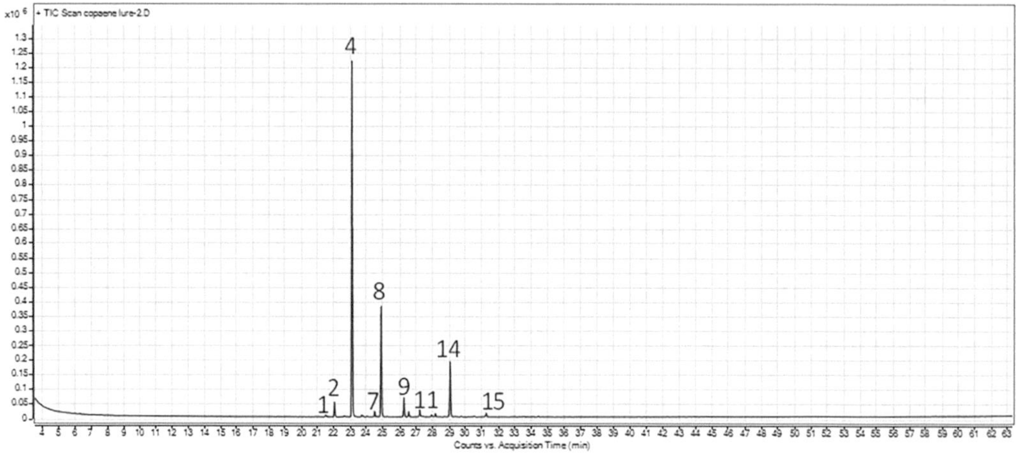

FIGURE 18.9 GC-MS chromatogram of the α-copaene enriched lure and enantiomeric distribution of α-copaene using an Rt-βDEXse column.

FIGURE 18.10 **(See color insert.)** Female (A) and male (B) of *Euwallacea* nr. *fornicatus*, vector of the fungal pathogen that causes *Fusarium* dieback. (C) Avocado tree (*Persea americana*) in Miami-Dade County, FL, US, with multiple attacks by *E.* nr. *fornicatus* concentrated at the base of the branch.

In the US, two distinct species of *E.* nr. *fornicatus* now exist in California, commonly referred to as the polyphagous shot-hole borer, first detected in the vicinity of Los Angeles, and the Kuroshio shot-hole borer, found farther south near San Diego (Eskalen et al., 2013; Boland, 2016). These beetles are attributed with high mortality of a variety of phylogenetically diverse species, with the most severe impact on avocado, box elder (*Acer negundo* L., Sapindaceae), castor bean (*Ricinus communis* L., Euphorbiaceae), California live oak (*Quercus agrifolia* Née, Fagaceae), and red willow (*Salix* laevigata Bebb, Salicaceae). A third distinct population, more closely aligned with tea shot-hole borer (Stouthamer et al., 2017), is now established in southern Florida where avocado is the preferred host (Carrillo et al., 2016; Kendra et al., 2017; Owens et al., 2018b). That state's avocado industry, already heavily impacted by laurel wilt, now faces a new threat with *Fusarium* dieback. In addition to avocado, reported hosts of Florida *E.* nr. *fornicatus* include native species like swampbay and wild tamarind (*Lysiloma latisiliquum* [L.] Benth., Fabaceae), ornamental royal poinciana (*Delonix regia* [Boj. ex Hook] Raf., Fabaceae), naturalized exotics like woman's tongue tree (*Albizia lebbeck* [L.] Bentham, Fabaceae), and fruit crops such as mango (*Mangifera indica* L., Anacardiaceae) and soursop (*Annona muricata* L., Annonaceae) (Rabaglia et al., 2006; Carrillo et al., 2012, 2016; Owens et al., 2018a).

18.4.2 Host Attractants

Unlike the situation with *X. glabratus*, little attention has been directed toward identification of specific host-based attractants for the *Euwallacea* pests. This is not unexpected, given the extensive host range exhibited by members of the *E.* nr. *fornicatus* complex. For example, hosts of the tea shot-hole borer are reported to include species from more than 35 plant families in Asia (Danthanarayana, 1968), and the polyphagous shot-hole borer, as implied by its common name, has even more documented hosts in California, with species represented from more than 50 plant families (Eskalen et al., 2013).

To the authors' knowledge, only one published report has addressed host attractants of *E. fornicatus*; this was in Sir Lanka and was restricted to tea (Karunaratne et al., 2008). In olfactometer tests, females were attracted to the dominant volatiles known from tea, including eugenol, hexanol, α- and β pinene, geraniol, and methyl salicylate. In addition, attraction was increased when these volatiles were combined with ethanol. With *E.* nr. *fornicatus* in Florida field trials, ethanol appeared to be a weak attractant when combined with quercivorol lures (see Section 18.4.3.1 below) (Carrillo et al., 2015; Kendra et al., 2015a); however, with polyphagous shot-hole borer, ethanol was found to be a repellent in tests conducted in California (Dodge et al., 2017) and in Israel (Byers et al., 2018).

18.4.3 Field Lures

18.4.3.1 Quercivorol

Current monitoring for *E.* nr. *fornicatus* utilizes a commercial lure containing quercivorol, (1*S*, 4*R*)-*p*-menth-2-en-1-ol (Kashiwagi et al., 2006), a proposed fungal volatile emitted by *Fusarium* spp. (i.e., a food-based attractant) (Cooperband et al., 2017). These lures were first found attractive to host-seeking females in Florida (Carrillo et al., 2015), with comparable results later obtained with *Euwallacea* populations in California (Dodge et al., 2017) and Israel (Byers et al., 2017). A recent analysis (Owens et al., 2018b) of the commercial lure indicated it contains a mixture of four isomers of *p*-menth-2-en-1-ol, and the (1*S*,4*R*)-enantiomer is not the major component. However, this lure is commonly referred to as the "quercivorol lure" in entomological literature, but it is unclear exactly which isomer(s) is/are responsible for attraction of *E.* nr. *fornicatus*.

18.4.3.2 Enriched α-Copaene Lure

In 2015, while conducting field tests for *X. glabratus* in a Florida avocado grove, a serendipitous discovery was made. Females of *E.* nr. *fornicatus* also appeared to be attracted to the enriched α-copaene lure (Kendra et al., 2015a). More focused evaluations in 2016 demonstrated that α-copaene is equal in attraction to quercivorol and that the combination of semiochemicals results in additive to synergistic increases in trap capture (Figure 18.11) (Kendra et al., 2017). EAG analyses confirmed olfactory chemoreception of both compounds, with a higher response elicited with the combination of

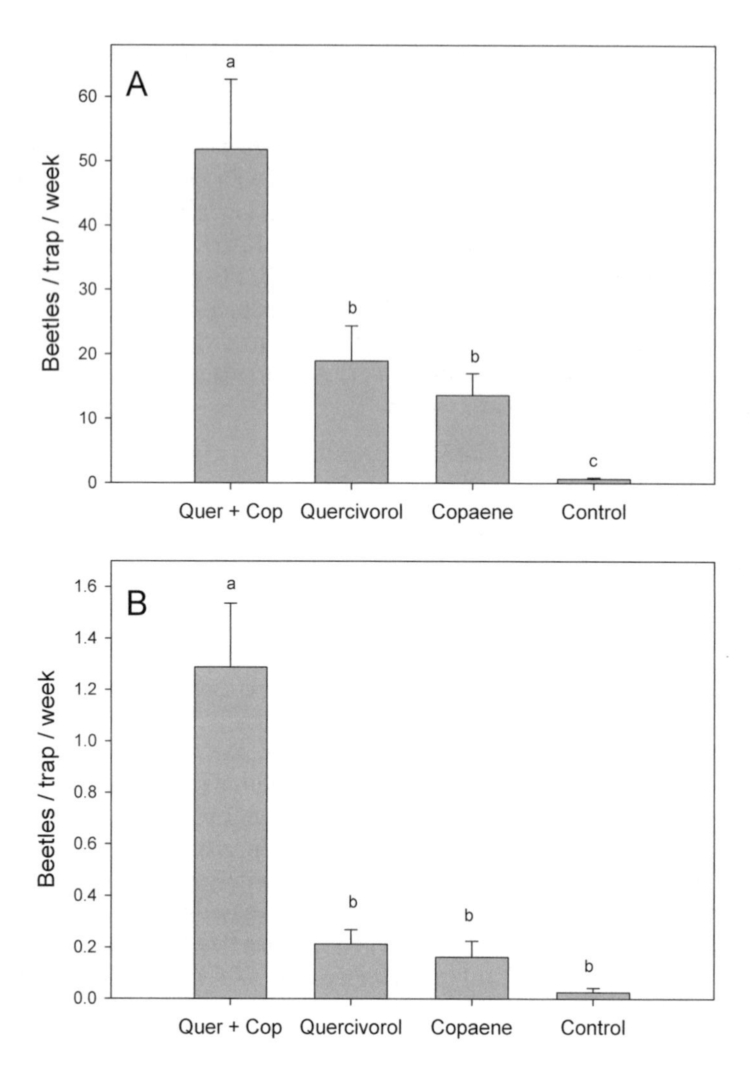

FIGURE 18.11 Mean (±SE) captures of female *Euwallacea* nr. *fornicatus* in field tests conducted in commercial avocado groves in Miami-Dade County, FL, US. (A) Results obtained at a site free of laurel wilt with high beetle populations. (B) Results from a site with advanced laurel wilt and very low numbers of *E.* nr. *fornicatus*. Lure treatments consisted of a quercivorol bubble lure and an α-copaene bubble lure, deployed separately and in combination. Bars topped with the same letter are not significantly different. (Adapted from Kendra, P. E. et al, 2017. *PLOS ONE*, 12: e0179416.)

volatiles (Figure 18.12). This two-component lure, combining a food attractant with a host attractant, is the most effective lure identified to date for *E.* nr. *fornicatus* in Florida, with field longevity of 12 weeks, minimal attraction of non-target ambrosia beetles, and an effective sampling range of ~30 m (Owens et al., 2018b, 2019). In early 2018, this combination lure was adopted by SAGARPA (Secretaría de Agricultura, Ganadería, Desarrollo Rural, Pesca y Alimentación) in Mexico in survey programs for both *E.* nr. *fornicatus* and *X. glabratus* in high-risk areas, including ports, international borders, and avocado-production regions (DGSV-CNRF, 2018).

18.5 OTHER APPLICATIONS

Although recognized as valuable tools for early pest detection, essential oils have played another important role in our investigations of ambrosia beetles. In 2011, a method was developed for field

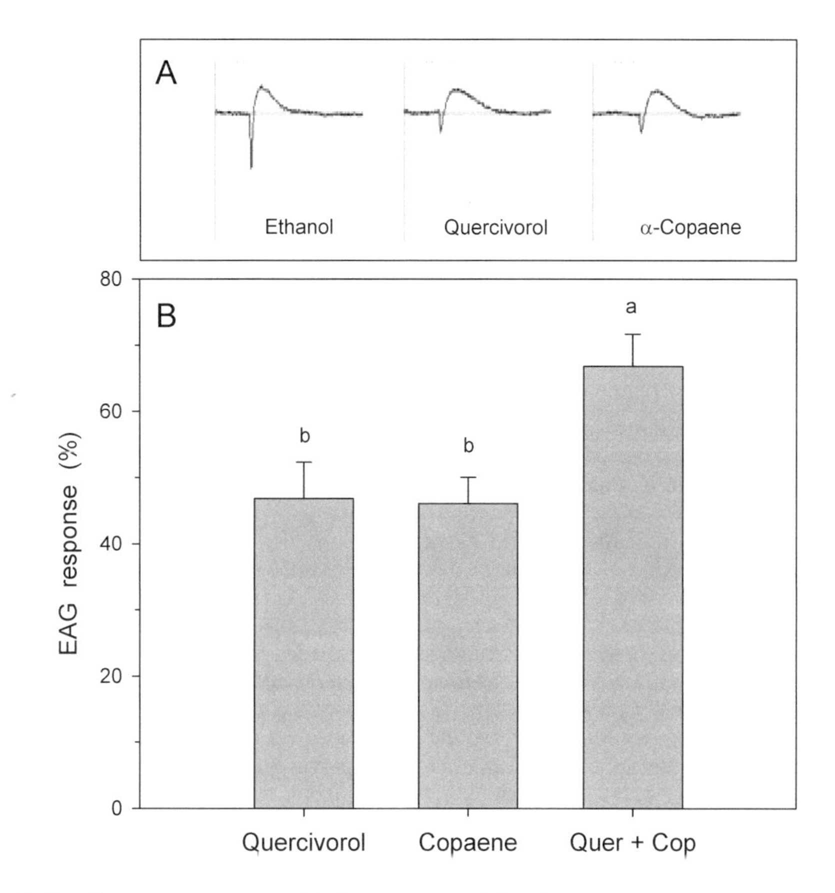

FIGURE 18.12 Electroantennogram (EAG) responses of female *Euwallacea* nr. *fornicatus* to volatiles emitted from commercial lures containing quercivorol and α-copaene. (A) Representative EAG recordings obtained with single excised antennae. (B) Comparative EAG responses (mean ± SE) to fixed 2-mL doses of volatiles. Responses to lure volatiles are expressed as normalized percentages relative to the standard reference compound (ethanol, 2-mL saturated vapor). Bars topped with the same letter are not significantly different. (Adapted from Kendra, P. E. et al, 2017. *PLOS ONE*, 12: e0179416.)

collection of live, host-seeking female *X. glabratus* using host wood and essential oils as bait (Kendra et al., 2012b). As an alternative to establishing a laboratory colony of *X. glabratus*, this method provided an immediate supply of adult beetles for research use, on an "as needed" basis. Furthermore, females obtained in this manner were behaviorally and physiologically in "host-seeking mode," the perfect cohort for evaluation of host-based attractants in controlled tests.

The method consisted of spreading a white cotton sheet in a clearing of a forest exhibiting laurel wilt, placing freshly cut *Persea* or lychee wood in the center of the sheet, and suspending essential oil lures (initially manuka and/or phoebe) from a tripod positioned above the wood. At intervals of 15–20 minutes, new wood was added to the sheet and lures were fanned to generate a pulsed release of attractive host volatiles. This effectively lured in female ambrosia beetles from the surrounding woods. As they landed, beetles were collected by hand with a soft brush and stored in plastic boxes with moist tissue until needed for experimentation.

This simple collection method provided the foundation for numerous projects since 2011, not only advancing our knowledge of *X. glabratus*, but of many other scolytine beetles, including pest and non-pest species. An immediate benefit of real-time collections was the documentation of species-specific flight windows. This was observed initially for female *X. glabratus,* which initiated flight much earlier than other species, allowing for selective capture of this target species (Kendra et al., 2012b). Subsequently, flight windows were recorded for *Xyleborus affinis* Eichhoff and *X.*

ferrugineus (Fabricius) (Kendra et al., 2012a), eight species of *Hypothenemus* (Johnson et al., 2016), *E.* nr. *fornicatus* (Kendra et al., 2017), and the community of ambrosia beetles found in Florida avocado groves affected by laurel wilt (Menocal et al., 2018).

Field-caught females greatly accelerated comparative evaluations of host species and essential oil attractants through EAG analyses (Kendra et al., 2012a, 2014b, 2016a), choice and no-choice behavioral bioassays (Kendra et al., 2013b, 2014a, 2016a), and field cage release-recapture experiments (Kendra et al., 2015b, 2016b). In addition, the collection of in-flight ambrosia beetles facilitated collaborations with plant pathologists to study the fungal symbionts transported in female mycangia (Bateman et al., 2015; Campbell et al., 2016; Ploetz et al., 2017c).

REFERENCES

Adams, R. P., 2007. *Identification of Essential Oil Components by Gas Chromatography/Mass Spectrometry*, 4th ed. Carol Stream: Allured Publishing Corp.

Bateman, C., P. E. Kendra, R. Rabaglia, and J. Hulcr, 2015. Fungal symbionts in three exotic ambrosia beetles, *Xylosandrus amputatus*, *Xyleborinus andrewesi*, and *Dryoxylon onoharaense* (Coleoptera: Curculionidae: Scolytinae: Xyleborini) in Florida. *Symbiosis*, 66: 141–148.

Boland, J. M., 2016. The impact of an invasive ambrosia beetle on the riparian habitats of the Tijuana River Valley, California. *Peer J*, 4: e2141.

Brar, G. S., J. L. Capinera, P. E. Kendra, S. Mclean, and J. E. Peña, 2013. Life cycle, development, and culture of *Xyleborus glabratus* (Coleoptera: Curculionidae: Scolytinae). *Florida Entomol*, 96: 1158–e1167.

Byers, J. A., Y. Maoz, and A. Levi-Zada, 2017. Attraction of the *Euwallacea* sp. near *fornicatus* (Coleoptera: Curculionidae) to quercivorol and to infestation in avocado. *J. Econ. Entomol.*, 110: 1512–1517.

Byers, J. A., Y. Maoz, D. Wakarchuk, D. Fefer, and A. Levi-Zada, 2018. Inhibitory effects of semiochemicals on the attraction of an ambrosia beetle *Euwallacea* nr. *fornicatus* to quercivorol. *J. Chem. Ecol.*, 44: 565–575.

Campbell, P., and A. Geering, 2011. Biosecurity capacity building for the Australian avocado industry – Laurel wilt. *Proceedings VII World Avocado Congress 2011 (Actas VII Congreso Mundial del Aguacate 2011)*. Cairns, Australia. 5–9 September 2011. http://www.avocadosource.com/WAC7/Section_03/CampbellPaul2011.pdf

Campbell, A. S., R. C. Ploetz, T. J. Draeden, P. E. Kendra, and W. S. Montgomery, 2016. Geographic variation in mycangial communities of *Xyleborus glabratus*. *Mycologia*, 108: 657–667.

Carrillo, D., R. E. Duncan, and J. E. Peña, 2012. Ambrosia beetles (Coleoptera: Curculionidae: Scolytinae) that breed in avocado wood in Florida. *Florida Entomol.*, 95: 573–579.

Carrillo, D., R. E. Duncan, J. N. Ploetz, A. F. Campbell, R. C. Ploetz, and J. E. Peña, 2014. Lateral transfer of a phytopathogenic symbiont among native and exotic ambrosia beetles. *Plant Pathol.*, 63: 54–62.

Carrillo, D., T. Narvaez, A. A. Cossé, R. Stouthamer, and M. Cooperband, 2015. Attraction of *Euwallacea* nr. *fornicatus* (Coleoptera: Curculionidae: Scolytinae) to lures containing quercivorol. *Florida Entomol.*, 98: 780–782.

Carrillo, D., L. F. Cruz, P. E. Kendra, T. I. Narvaez, W. S. Montgomery, A. Monterroso, C. De Grave, and M. F. Cooperband, 2016. Distribution, pest status, and fungal associates of *Euwallacea* nr. *fornicatus* in Florida avocado groves. Special Issue: Invasive Insect Species. *Insects*, 7: 55.

Cooperband, M. F., A. A. Cossé, T. H. Jones, D. Carrillo, K. Cleary, I. Canlas, and R. Stouthamer, 2017. Pheromones of three ambrosia beetles in the *Euwallacea fornicatus* species complex: Ratios and preferences. *Peer J.*, 5: e3957.

Crook, D. J., A. Khirimian, J. A. Francese, I. Fraser, T. M. Poland, A. J. Sawyer, and V. C. Mastro, 2008. Development of a host-based semiochemical lure for trapping emerald ash borer, *Agrilus planipennis* (Coleoptera: Buprestidae). *Environ. Entomol.*, 37: 356–365.

Danthanarayana, W., 1968. The distribution and host-range of the shot-hole borer (*Xyleborus fornicatus* Eichh.) of tea. *Tea Quarterly.*, 39: 61–69.

DGSV-CNRF, 2018. Dirección General de Sanidad Vegetal — Centro Nacional de Referencia Fitosanitaria. Manuales operativos de plagas cuarentenarias. Plan de acción para la vigilancia y aplicación de medidas de control contra complejos ambrosiales reglamentados en México: *Xyleborus glabratus- Raffaelea lauricola* y *Euwallacea* sp.—*Fusarium euwallaceae*. http://www.sinavef.senasica.gob.mx/SIRVEF/ManualesOperativos.aspx

Dodge, C., J. Coolidge, M. Cooperband, A. Cossé, D. Carrillo, and R. Stouthamer, 2017. Quercivorol as a lure for the polyphagous and Kuroshio shot hole borers, *Euwallacea* spp. nr. *fornicatus* (Coleoptera: Scolytinae), vectors of *Fusarium* dieback. *Peer J.*, 5: e3656.

Eskalen, A., R. Stouthamer, S. C. Lynch, P. F. Rugman-Jones, M. Twizeyimana, A. Gonzalez, and T. Thibault, 2013. Host range of *Fusarium* dieback and its ambrosia beetle (Coleoptera: Scolytinae) vector in southern California. *Plant Dis.*, 97: 938–951.

FFNSC-3, 2015. *Flavors and Fragrances of Natural and Synthetic Compounds 3*. Hoboken, NJ, USA: Mass Spectral Database, Scientific Instrument Services Inc.

Flath, R. A., R. T. Cunningham, T. R. Mon, and J. O. John, 1994. Male lures for Mediterranean fruit fly (*Ceratitis capitata* Wied.): Structural analogs of α-copaene. *J. Chem. Ecol.*, 20: 2595–2609.

Fraedrich, S. W., T. C. Harrington, R. J. Rabaglia, M. D. Ulyshen, A. E. Mayfield III, J. L. Hanula, J. M. Eickwort, and D. R. Miller, 2008. A fungal symbiont of the redbay ambrosia beetle causes a lethal wilt in redbay and other Lauraceae in the southeastern USA. *Plant Dis.*, 92: 215–224.

Freeman, S., A. Protasov, M. Sharon, K. Mohotti, M. Eliyahu, N. Okon-Levy, M. Maymon, and Z. Mendel, 2012. Obligate feed requirement of *Fusarium* sp. Nov., an avocado wilting agent, by the ambrosia beetle *Euwallacea* aff. *fornicata*. *Symbiosis*, 58: 245–251.

García-Avila, C. D. J., F. J. Trujillo-Arriaga, J. A. López-Buenfil, R. González-Gómez, D. Carrillo, L. F. Cruz, I. Ruiz-Galván, A. Quezada-Salinas, and N. Acevedo-Reyes, 2016. First report of *Euwallacea* nr. *fornicatus* (Coleoptera: Curculionidae) in Mexico. *Florida Entomol.*, 99: 555–556.

Hanula, J. L., and B. Sullivan, 2008. Manuka oil and phoebe oil are attractive baits for *Xyleborus glabratus* (Coleoptera: Curculionidae: Scolytinae), the vector of laurel wilt. *Environ. Entomol.*, 37: 1403–1409.

Hanula, J. L., A. E. Mayfield III, S. W. Fraedrich, and R. J. Rabaglia, 2008. Biology and host associations of redbay ambrosia beetle, *Xyleborus glabratus* (Coleoptera: Curculionidae: Scolytinae), exotic vector of laurel wilt killing redbay (*Persea borbonia*) trees in the southeastern United States. *J. Econ. Entomol.*, 101: 1276–1286.

Hanula, J. L., B. T. Sullivan, and D. Wakarchuk, 2013. Variation in manuka oil lure efficacy for capturing *Xyleborus glabratus* (Coleoptera: Curculionidae: Scolytinae), and cubeb oil as an alternative attractant. *Environ. Entomol.*, 42: 333–340.

Harrington, T. C., S. W. Fraedrich, and D. N. Aghayeva, 2008. Raffaelea lauricola, a new ambrosia beetle symbiont and pathogen on the Lauraceae. *Mycotaxon*, 104: 399–404.

Harrington, T. C., D. N. Aghayeva, and S. W. Fraedrich, 2010. New combinations of *Raffaelea*, *Ambrosiella*, and *Hyalorhinocladiella*, and four new species from the redbay ambrosia beetle, *Xyleborus glabratus*. *Mycotaxon*, 111: 337–361.

Hazarika, L. K., M. Bhuyan, and B. N. Hazarika, 2009. Insect pests of tea and their management. *Ann. Rev. Entomol.*, 54: 267–284.

Hughes, M. A., J. A. Smith, R. C. Ploetz et al., 2015. Recovery plan for laurel wilt on redbay and other forest species caused by *Raffaelea lauricola* and disseminated by *Xyleborus glabratus*. *Plant Health Prog.*, 16: 173–210.

Hughes, M. A., J. J. Riggins, F. H. Koch, A. I. Cognato, C. Anderson, J. P. Formby, T. J. Dreaden, R. C. Ploetz, and J. A. Smith, 2017. No rest for the laurels: Symbiotic invaders cause unprecedented damage to southern USA forests. *Biol. Invasions.*, 19: 2143–2157.

Hulcr, J. and R. R. Dunn, 2011. The sudden emergence of pathogenicity in insect-fungus symbioses threatens naïve forest ecosystems. *Proc. R. Soc. B. Biol. Sci.*, 278: 2866–2873.

Hulcr, J. and L. L. Stelinski, 2017. The ambrosia symbiosis: From evolutionary ecology to practical management. *Annu. Rev. Entomol.*, 52: 285–303.

Johnson, A. J., P. E. Kendra, J. Skelton, and J. Hulcr, 2016. Species diversity, phenology, and temporal flight patterns of *Hypothenemus* pygmy borers (Coleoptera: Curculionidae: Scolytinae) in South Florida. *Environ. Entomol.*, 45: 627–632.

Karunaratne, W. S., V. Kumar, J. Pettersson, and N. S. Kumar, 2008. Response of the shot-hole borer of tea, *Xyleborus fornicatus* (Coleoptera: Scolytidae) to conspecifics and plant semiochemicals. *Acta Agr. Scand. B—Soil Plant Sci.*, 58: 345–351.

Kashiwagi, T., T. Nakashima, S-I Tebayashi, and C-S Kim, 2006. Determination of the absolute configuration of quercivorol, (1*S*,4*R*)-*p*-menth-2-en-1-ol, an aggregation pheromone of the ambrosia beetle *Platypus quercivorus* (Coleoptera: Platypodidae). *Biosci. Biotech. Biochem.*, 70: 2544–2546.

Kasson, M. T., K. O'Donnell, A. P. Rooney et al., 2013. An inordinate fondness for *Fusarium*: Phylogenetic diversity of fusaria cultivated by ambrosia beetles in the genus *Euwallacea* on avocado and other plant hosts. *Fungal Genet. Biol.*, 56: 147–157.

Kendra, P. E., W. S. Montgomery, J. Niogret, J. E. Peña, J. L. Capinera, G. Brar, N. D. Epsky, and R. R. Heath, 2011a. Attraction of the redbay ambrosia beetle, *Xyleborus glabratus*, to avocado, lychee, and essential oil lures. *J. Chem. Ecol.*, 37: 932–942.

Kendra, P. E., J. S. Sanchez, W. S. Montgomery, K. E. Okins, J. Niogret, J. E. Peña, N. D. Epsky, and R. R. Heath, 2011b. Diversity of Scolytinae (Coleoptera: Curculionidae) attracted to avocado, lychee, and essential oil lures. *Florida Entomol.*, 94: 123–130.

Kendra, P. E., W. S. Montgomery, J. Niogret, M. A. Deyrup, L. Guillén, and N. E. Epsky, 2012a. *Xyleborus glabratus*, *X. affinis*, and *X. ferrugineus* (Coleoptera: Curculionidae: Scolytinae): Electroantennogram responses to host-based attractants and temporal patterns in host-seeking flight. *Environ. Entomol.*, 41: 1597–1605.

Kendra, P. E., W. S. Montgomery, J. S. Sanchez, M. A. Deyrup, J. Niogret, and N. D. Epsky, 2012b. Method for collection of live redbay ambrosia beetles, *Xyleborus glabratus* (Coleoptera: Curculionidae: Scolytinae). *Florida Entomol.*, 95: 513–516.

Kendra, P. E., J. Niogret, W. S. Montgomery, J. S. Sanchez, M. A. Deyrup, G. E. Pruett, R. C. Ploetz, N. D. Epsky, and R. R. Heath, 2012c. Temporal analysis of sesquiterpene emissions from manuka and phoebe oil lures and efficacy for attraction of *Xyleborus glabratus* (Coleoptera: Curculionidae: Scolytinae). *J. Econ. Entomol.*, 105: 659–669.

Kendra, P. E., W. S. Montgomery, J. Niogret, and N. D. Epsky, 2013a. An uncertain future for American Lauraceae: A lethal threat from redbay ambrosia beetle and laurel wilt disease (A review). Special Issue: The Future of Forests. *Am. J. Plant. Sci.*, 4: 727–738.

Kendra, P. E., R. C. Ploetz, W. S. Montgomery, J. Niogret, J. E. Peña, G. S. Brar, and N. D. Epsky, 2013b. Evaluation of *Litchi chinensis* for host status to *Xyleborus glabratus* (Coleoptera: Curculionidae: Scolytinae) and susceptibility to laurel wilt disease. *Florida Entomol.*, 96: 1442–1453.

Kendra, P. E., W. S. Montgomery, J. Niogret, G. E. Pruett, A. E. Mayfield III, M. MacKenzie, M. A. Deyrup, G. R. Bauchan, R. C. Ploetz, and N. D. Epsky, 2014a. North American Lauraceae: Terpenoid emissions, relative attraction and boring preferences of redbay ambrosia beetle, *Xyleborus glabratus* (Coleoptera: Curculionidae: Scolytinae). *PLOS ONE*, 9: e102086.

Kendra, P. E., W. S. Montgomery, J. Niogret, E. Q. Schnell, M. A. Deyrup, and N. D. Epsky, 2014b. Evaluation of seven essential oils identifies cubeb oil as most effective attractant for detection of *Xyleborus glabratus*. *J. Pest Sci.*, 87: 681–689.

Kendra, P. E., T. I. Narvaez, W. S. Montgomery, and D. Carrillo, 2015a. Ambrosia beetle communities in forest and agricultural ecosystems with laurel wilt disease (D3524). *53rd Annual Meeting of the Entomological Society of America*, Minneapolis, MN, USA 15–18 November 2015. https://www.esa.confex.com/esa/2015/webprogram/Session27045.html

Kendra, P. E., J. Niogret, W. S. Montgomery, M. A. Deyrup, and N. D. Epsky, 2015b. Cubeb oil lures: Terpenoid emissions, trapping efficacy, and longevity for attraction of redbay ambrosia beetle (Coleoptera: Curculionidae: Scolytinae). *J. Econ. Entomol.*, 108: 350–361.

Kendra, P. E., W. S. Montgomery, M. A. Deyrup, and D. Wakarchuk, 2016a. Improved lure for redbay ambrosia beetle developed by enrichment of α-copaene content. *J. Pest Sci.*, 89: 427–438.

Kendra, P. E., W. S. Montgomery, E. Q. Schnell, M. A. Deyrup, and N. D. Epsky, 2016b. Efficacy of α-copaene, cubeb, and eucalyptol lures for detection of redbay ambrosia beetle (Coleoptera: Curculionidae: Scolytinae). *J. Econ. Entomol.*, 109: 2428–2435.

Kendra, P. E., D. R. Owens, W. S. Montgomery, T. I. Narvaez, G. R. Bauchan, E. Q. Schnell, N. Tabanca, and D. Carrillo, 2017. α-Copaene is an attractant, synergistic with quercivorol, for improved detection of *Euwallacea* nr. *fornicatus* (Coleoptera: Curculionidae: Scolytinae). *PLOS ONE*, 12: e0179416.

Kendra, P. E., W. S. Montgomery, J. Niogret, N. Tabanca, D. Owens, and N. D. Epsky, 2018. Utility of essential oils for development of host-based lures for *Xyleborus glabratus* (Coleoptera: Curculionidae: Scolytinae), vector of laurel wilt. Special Issue: Research for Natural Bioactive Products. *Open Chem.*, 16: 393–400.

Kuhns, E. H., X. Martini, Y. Tribuiani, M. Coy, C. Gibbard, J. Peña, J. Hulcr, and L. L. Stelinski, 2014. Eucalyptol is an attractant of the redbay ambrosia beetle, *Xyleborus glabratus*. *J. Chem. Ecol.*, 40: 355–362.

MassFinder, 2004. *MassFinder Software, Version 3.* Hamburg: Dr. Hochmuth Scientific Consulting.

Mendel, Z., A. Protasov, M. Sharon, A. Zveibil, S. Ben Yehuda, K. O'Donnell, R. Rabaglia, M. Wysoki, and S. Freeman, 2012. An Asian ambrosia beetle *Euwallacea fornicatus* and its novel symbiotic fungus *Fusarium* sp. pose a serious threat to the Israeli avocado industry. *Phytoparasitica*, 40: 235–238.

Menocal, O., P. E. Kendra, W. S. Montgomery, J. H. Crane, and D. Carrillo, 2018. Vertical distribution and daily flight periodicity of ambrosia beetles (Coleoptera: Curculionidae) in Florida avocado orchards affected by laurel wilt. *J. Econ. Entomol.*, 111: 1190–1196.

Miller, D. R., and R. J. Rabaglia, 2009. Ethanol and (-)-α-pinene: Attractant kairomones for bark and ambrosia beetles in the southeastern U.S. *J. Chem. Ecol.*, 35: 435–448.

Niogret, J., P. E. Kendra, N. D. Epsky, and R. R. Heath, 2011a. Comparative analysis of terpenoid emissions from Florida host trees of the redbay ambrosia beetle, *Xyleborus glabratus* (Coleoptera: Curculionidae: Scolytinae). *Florida Entomol.*, 94: 1010–1017.

Niogret, J., W. S. Montgomery, P. E. Kendra, R. R. Heath, and N. D. Epsky, 2011b. Attraction and electroantennogram responses of male Mediterranean fruit fly (Diptera: Tephritidae) to volatile chemicals from *Persea, Litchi*, and *Ficus* wood. *J. Chem. Ecol.*, 37: 483–491.

Niogret, J., N. C. Epsky, R. J. Schnell, E. J. Boza, P. E. Kendra, and R. R. Heath, 2013. Terpenoid variations within and among half-sibling avocado trees, *Persea americana* Mill. (Lauraceae). *PLOS ONE*, 8: e73601.

NIST, 2017. *NIST/EPA/NIH Mass Spectral Library, Version: NIST 17.* Mass Spectrometry Data Center. Gaithersburg, MD, USA: National Institute of Standard and Technology.

O'Donnell, K., S. Sink, R. Libeskind-Hadas et al., 2015. Discordant phylogenies suggest repeated host shifts in the *Fusarium-Euwallacea* ambrosia beetle mutualism. *Fungal Genet. Biol.*, 82: 277–290.

Owens, D., W. S. Montgomery, T. I. Narvaez, M. A. Deyrup, and P. E. Kendra, 2017. Evaluation of lure combinations containing essential oils and volatile spiroketals for detection of host-seeking *Xyleborus glabratus* (Coleoptera: Curculionidae: Scolytinae). *J. Econ. Entomol.*, 110: 1596–1602.

Owens, D., L. F. Cruz, W. S. Montgomery, T. I. Narvaez, E. Q. Schnell, N. Tabanca, R. E. Duncan, D. Carrillo, and P. E. Kendra, 2018a. Host range expansion and increasing damage potential of *Euwallacea* nr. *fornicatus* (Coleoptera: Curculionidae) in Florida. *Florida Entomol.*, 101: 229–e73236.

Owens, D., P. E. Kendra, N. Tabanca, T. I. Narvaez, W. S. Montgomery, E. Q. Schnell, and D. Carrillo, 2018b. Quantitative analysis of contents and volatile emissions from α-copaene and quercivorol lures, and longevity for attraction of *Euwallacea* nr. *fornicatus* in Florida. Special Issue: Invasive insect pests of forests and urban trees: Pathways, early detection, and management. *J. Pest Sci.*, 91: doi 10.1007/s10340-018-0960-6.

Owens, D., M. Seo, W. S. Montgomery, M. J. Rivera, L. L. Stelinski, and P. E. Kendra, 2019. Dispersal behavior of *Euwallacea* nr. *fornicatus* (Coleoptera: Curculionidae: Scolytinae) in avocado groves and estimation of lure sampling range. *Agr. Forest Entomol.*, 21: 199–208.

Paap, T., Z. W. de Beer, C. Migliorini, W. J. Nel, and M. J. Wingfield, 2018. The polyphagous shot hole borer (PSHB) and its fungal symbiont *Fusarium euwallaceae*: A new invasion in South Africa. *Australasian Plant Pathol.*, 47: 231–237.

Ploetz, R. C., J. M. Pérez-Martínez, J. A. Smith, M. Hughes, T. J. Dreaden, S. A. Inch, and Y. Fu, 2012. Responses of avocado to laurel wilt, caused by *Raffaelea lauricola*. *Plant Pathol.*, 61: 801–808.

Ploetz, R. C., M. A. Hughes, P. E. Kendra et al., 2017a. Recovery plan for laurel wilt of avocado, caused by *Raffaelea lauricola*. *Plant Health Prog.*, 18: 51–77.

Ploetz, R. C., P. E. Kendra, R. A. Choudhury, J. Rollins, A. Campbell, K. Garrett, M. Hughes, and T. Dreaden, 2017b. Laurel wilt in natural and agricultural ecosystems: Understanding the drivers and scales of complex pathosystems. Special Issue: Forest Pathology and Plant Health. *Forests*, 8: 48.

Ploetz, R. C., J. L. Konkol, T. Narvaez, R. E. Duncan, R. J. Saucedo, A. Campbell, J. Mantilla, D. Carrillo, and P. E. Kendra, 2017c. Presence and prevalence of *Raffaelea lauricola*, cause of laurel wilt, in different species of ambrosia beetle in Florida USA. *J. Econ. Entomol.*, 110: 347–354.

Rabaglia, R. J., S. A. Dole, and A. I. Cognato, 2006. Review of American Xyleborina (Coleoptera: Curculionidae: Scolytinae) occurring north of Mexico, with an illustrated key. *Ann. Entomol. Soc. Amer.*, 99: 1034–1056.

Rabaglia, R., D. Duerr, R. Acciavatti, and I. Ragenovich, 2008. *Early Detection and Rapid Response for Non-Native Bark and Ambrosia Beetles*. Washington, D. C.: US Department of Agriculture, Forest Service, Forest Health Protection, http://www.fs.fed.us/foresthealth/publications/EDRRProjectReport.pdf

Ranger, C. M., M. E. Reding, A. B. Persad, and D. A. Herms, 2010. Ability of stress-related volatiles to attract and induce attacks by *Xylosandrus germanus* and other ambrosia beetles (Coleoptera: Curculionidae, Scolytinae). *Agr. Forest Entomol.*, 12: 177–185.

Stouthamer, R., P. Rugman-Jones, P. Q. Thu et al., 2017. Tracing the origin of a cryptic invader: Phylogeography of the *Euwallacea fornicatus* (Coleoptera: Curculionidae: Scolytinae) species complex. *Agr. Forest Entomol.*, 19: 366–375.

USDA-ARS, 2018. United States Department of Agriculture, Agricultural Research Service, National Plant Disease Recovery System. https://www.ars.usda.gov/crop-production-and-protection/plant-diseases/docs/npdrs

USDA-FS, 2018. United States Department of Agriculture, Forest Service, Southern Regional Extension Forestry. Distribution of Counties with Laurel Wilt Disease as of September 20, 2018. https://www.fs.usda.gov/Internet/FSE_DOCUMENTS/fseprd571973.pdf

van den Dool, H, and P. D. Kratz, 1963. A generalization of the retention index system including linear temperature programmed gas-liquid partition chromatography. *J. Chromatogr., A.*, 11: 463–471.

Wiley, 2017. *Wiley Registry of Mass Spectral Data*, 11th edition. Ringoes, NJ, USA: Scientific Instrument Services, Inc..

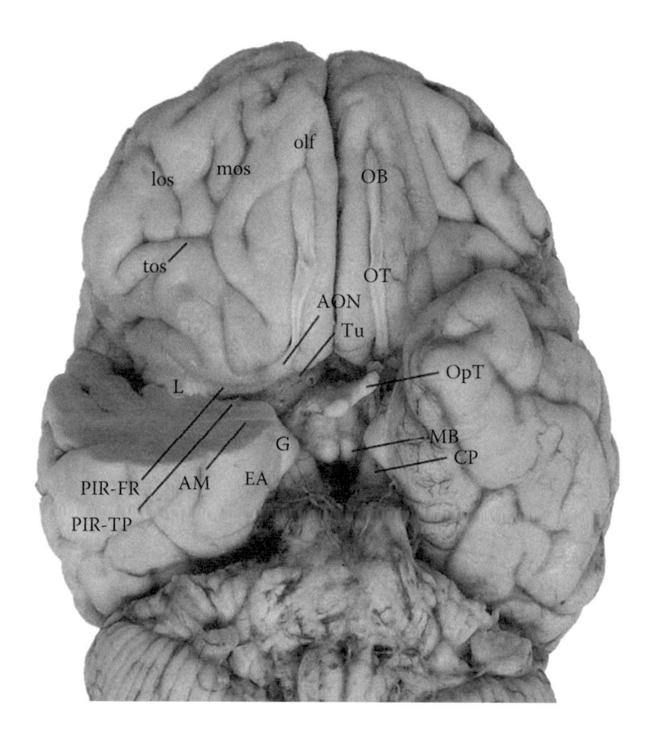

FIGURE 12.1 Macroscopic view of the human ventral forebrain and medial temporal lobes, depicting the olfactory tract, its primary projections, and surrounding non-olfactory structures. The right medial temporal lobe has been resected horizontally through the mid-portion of the amygdala (AM) to expose olfactory cortex. AON, anterior olfactory nucleus; CP, cerebral peduncle; EA, entorhinal area; G, gyrus ambiens; L, limen insula; los, lateral olfactory sulcus; MB, mammillary body; mos, medial olfactory sulcus; olf, olfactory sulcus; PIR-FR, frontal piriform cortex; OB, olfactory bulb; OpT, optic tract; OT, olfactory tract; tos, transverse olfactory sulcus; Tu, olfactory tubercle; PIR-TP, temporal piriform cortex. Figure prepared with the help of Dr. Eileen H. Bigio, Dept. of Pathology, Northwestern University Feinberg School of Medicine, Chicago, IL. (Taken with permission from Gottfried, J. A. and Zald, D. A. [2005] *Brain Research Reviews* 50:287–304.)

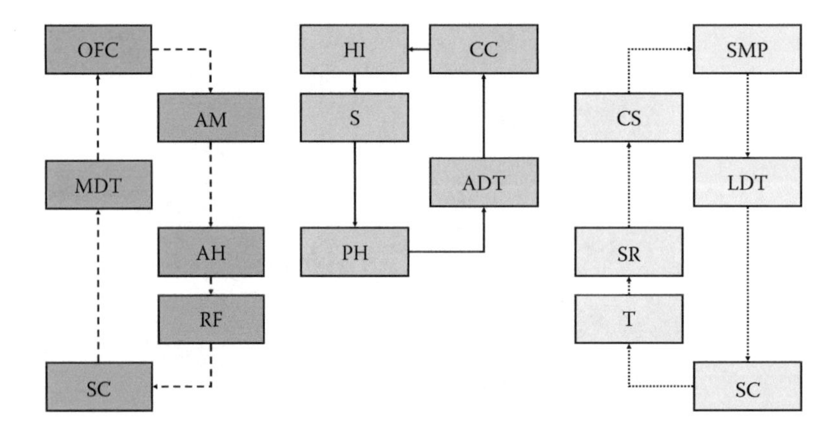

FIGURE 12.3 Control of activation processes. OFC, orbitofrontal cortex; AM, amygdala; MDT, medial dorsal thalamus; AH, anterior hypothalamus; RF, reticular formation; SC, spinal cord; HI, hippocampus; CC, cingulate cortex; S, septum; ADT, anterior dorsal thalamus; PH, posterior hypothalamus; SMP, sensory-motor projections; CS, corpus striatum; LDT, lateral dorsal thalamus; SR, subthalamic regions; T, tectum. Orange, structures of the arousal network; green, structures of the effort network; blue, structures of the activation network. (Adapted from Pribram, K. H. and McGuinness, D. 1975. *Psychological Review* 82 (2):116–149.)

FIGURE 18.1 Female (A) and male (B) of the redbay ambrosia beetle, *Xyleborus glabratus,* vector of the fungal pathogen that causes laurel wilt. (C) Cross-section through the trunk of a swampbay tree, *Persea palustris*, showing extensive galleries excavated by redbay ambrosia beetles. (D) Silkbay tree, *Persea humilis*, killed by laurel wilt in Highlands County, FL, US.

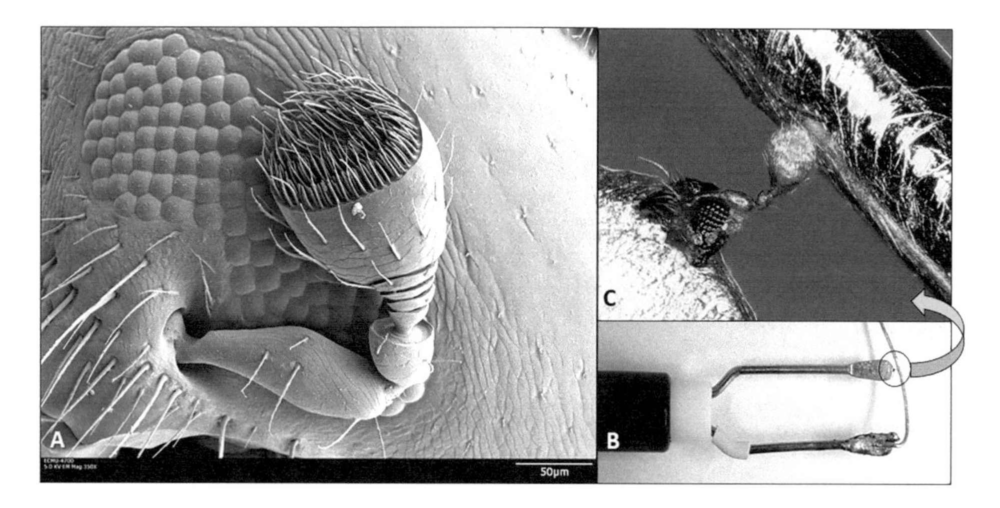

FIGURE 18.2 (A) Scanning electron micrograph of female redbay ambrosia beetle antenna (length = 0.4 mm), showing concentration of olfactory sensilla on the distal end of antennal club. (B) Two-pronged gold electroantennography (EAG) electrode modified to accommodate minute antennae by soldering a flexible gold wire to one prong. (C) Magnified view of the EAG antennal preparation, with head capsule mounted to the prong (indifferent electrode) and back of antennal club in contact with wire (different electrode).

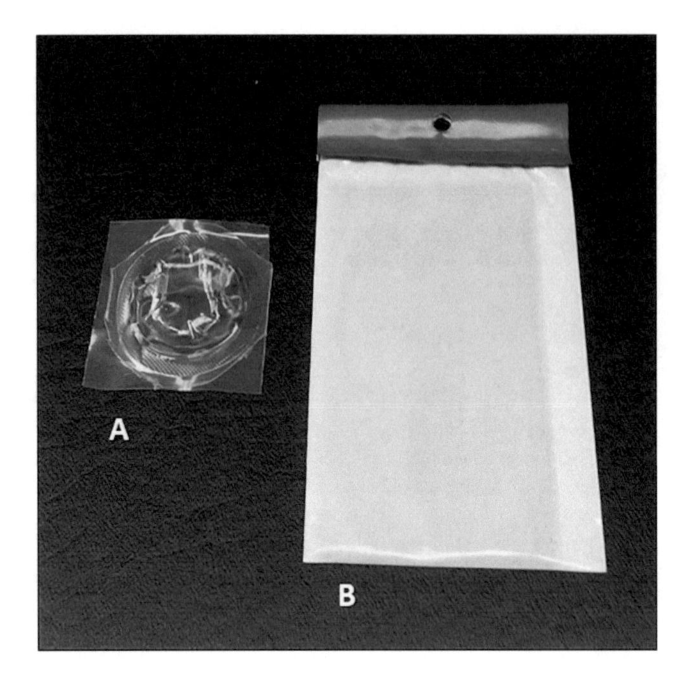

FIGURE 18.7 Commercially available lures for redbay ambrosia beetle. (A) Cubeb bubble lure (Synergy Semiochemicals Corp.; Burnaby, BC, Canada). (B) Manuka oil lure (ChemTica Internacional; Heredia, Costa Rica).

FIGURE 18.10 Female (A) and male (B) of *Euwallacea* nr. *fornicatus*, vector of the fungal pathogen that causes *Fusarium* dieback. (C) Avocado tree (*Persea americana*) in Miami-Dade County, FL, US, with multiple attacks by *E.* nr. *fornicatus* concentrated at the base of the branch.

FIGURE 21.1 Representatives of bryophytes: (a) moss, (b) liverwort, (c) hornwort.

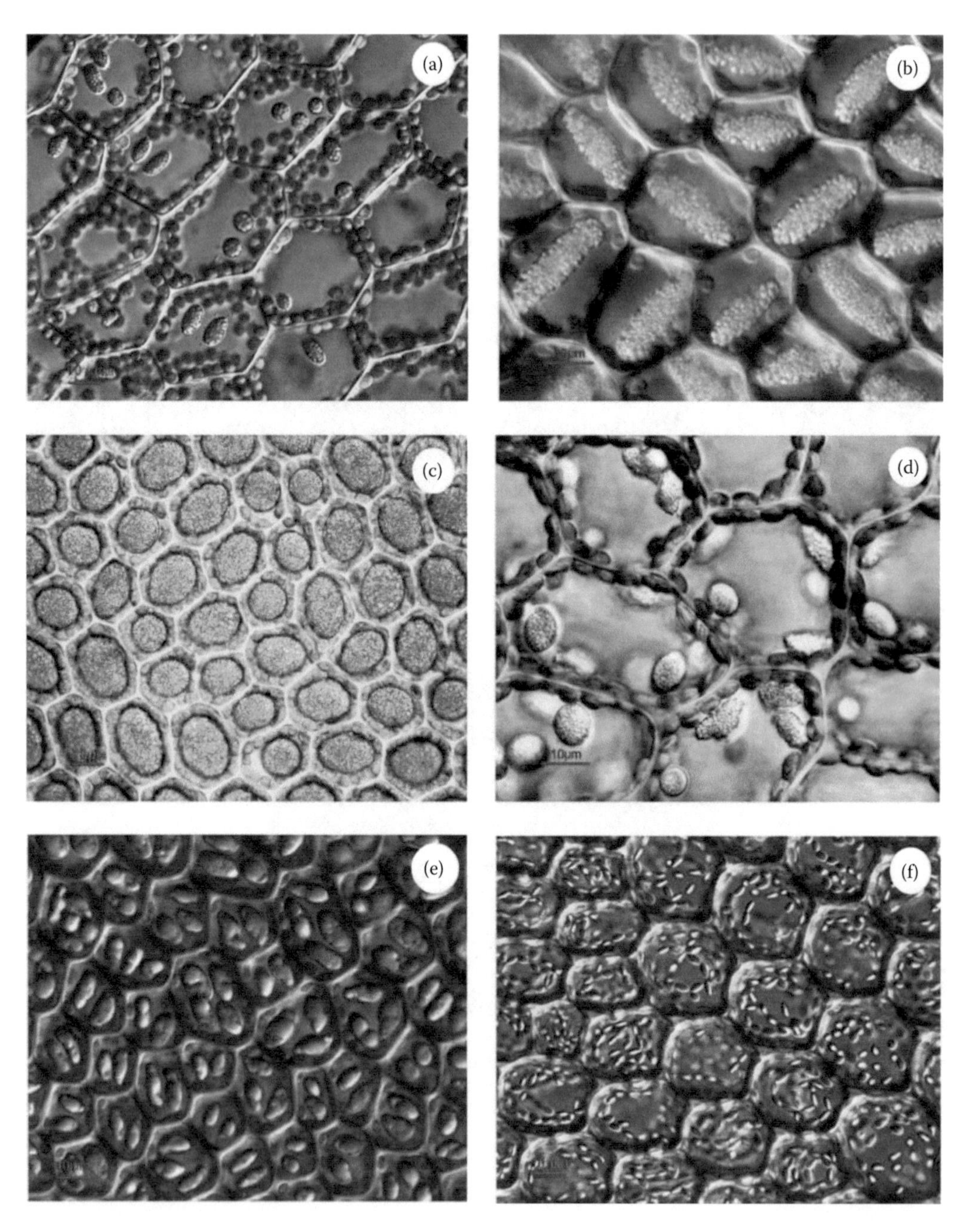

FIGURE 21.2 Different types of oil bodies in the cells of liverworts: (a) *Calypogeia azurea*, (b) *Cheilolejeunea anthocarpa*, (c) *Radula constricta*, (d) *Solenostoma truncatum*, (e) *Bazzania tridens*, and (f) *Trocholejeunea sandvicensis*. (From He et al. 2013. *Crit. Rev. Plant Sci.*, 32: 293–302. With permission.)

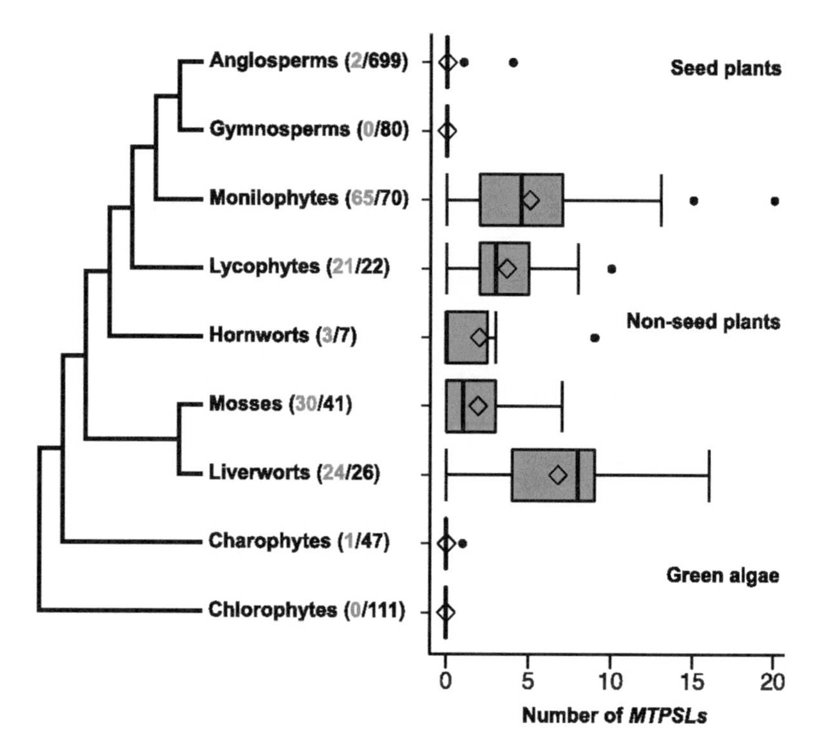

FIGURE 21.10 Distribution of MTPSL genes identified from the transcriptomes of 1103 plant species. The numbers in parentheses represent the number of transcriptomes containing putative MTPSLs (in red) and total transcriptomes analyzed in each lineage (in black). (From Jia et al. 2016. *Proc. Natl. Acad. Sci. USA*, 113: 12328–12333. With permission.)

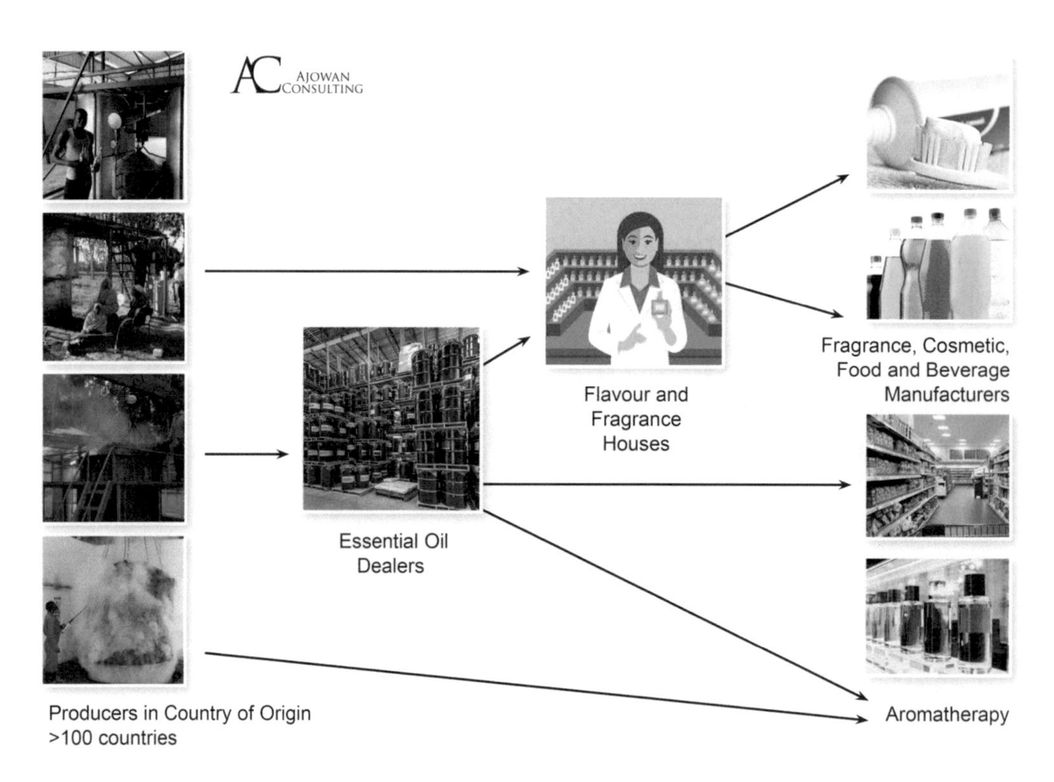

FIGURE 30.3 The essential oil trade flows.

19 Adverse Effects and Intoxication with Essential Oils

Rosa Lemmens-Gruber

CONTENTS

19.1 INTRODUCTION

Studies on the toxicity of essential oils are comprehensive due to the fact that essential oils are composed of numerous compounds. In most cases, knowledge about intoxication with essential oils is based on case studies. In the absence of data for whole oil, insight may be derived from data for individual components. However, it is noteworthy to keep in mind that data for essential oils may not necessarily apply to individual components, and vice versa data for components may not apply to the whole oil.

In most cases, essential oils are safe in use except in overdosage, wrong application route, and in hypersensitive persons. As essential oils are highly lipophilic, they are absorbed readily, and thus, also after topical application, systemic reactions may occur. Noteworthy, lipophilic constituents of essential oils may also pass the blood–brain and placental barriers, and therefore the application during pregnancy and lactation has to be done carefully. It is well documented that the toxicity of essential oils is concentration-dependent, and thus most unwanted side effects can be avoided by application of low doses. This advice, however, is not relevant for hypersensitive people, because allergic reactions can occur independently of the concentration.

Adverse effects attributed to compounds of essential oils can be initiated by oxidative metabolism and conversion of components to reactive intermediates. Oxidation products formed by exposure to light and/or air may play an important role. Thus, formation of reactive compounds should be prevented by adequate storage. Further causes for intolerance and intoxication comprise impurity and contamination, adulteration, and wrong declaration of the essential oil.

19.1.1 General Side Effects

Common side effects of essential oils include irritation of skin and mucosa. Besides these effects, for some furocoumarin-containing essential oils, photosensitivity has been reported. Noteworthy, in sensitive persons, signs of allergy on skin and in the respiratory tract have been reported in several publications, including following examples.

In a five-year study with 3065 patients suffering from contact dermatitis, 16.6% patients were found to be allergic to a fragrance mix. In these patients, the most frequent allergens were identified as isoeugenol (57.9%), eugenol (55.4%), cinnamyl alcohol (34.4%), and oak moss (24.2%) (Turić et al., 2011).

A 53-year-old patient who used volatile essential oils in aroma lamps over years suffered from relapsing eczema contact allergy to various essential oils. Sensitization was due to previous exposure to lavender, jasmine, and rosewood, and also, laurel, eucalyptus, and pomerance showed positive tests, although without known previous exposure. A diagnosis of allergic airborne contact dermatitis could be established (Schaller and Korting, 1995).

Vicks VapoRub is a commonly used inhalant ointment that helps relieve symptoms of upper respiratory tract infections. It contains several plant substances, including turpentine oil, eucalyptus oil, and cedar leaf oil, which can potentially irritate or sensitize the skin, as well as camphor, menthol, nutmeg oil, and thymol. Many reports describe allergic contact dermatitis to the various constituents in Vicks VapoRub ointment (Noiles and Pratt, 2010).

19.2 CAMPHOR AND CAMPHOR-CONTAINING ESSENTIAL OILS

The dietary exposure (herbs such as basil, coriander, marjoram, rosemary, and sage) to camphor was estimated to be 1.5 mg/person/day and thus is safe (Council of Europe, 2001). However, the essential oil from the so-called camphor tree (*Cinnamomum camphora*) contains up to 84% d-camphor and may be the source for intoxication. For the use of camphor as cough suppressant and decongestant, a limit of 11% allowable camphor in consumer products is set. The complex sensory properties of camphor can be explained by the modulation of several temperature-activated ion channels from the TRP family, such as TRPV3 (Moqrich et al., 2005), TRPV1 (Xu et al., 2005), TRPM8 (Selescu et al., 2013), and TRPA1 (Alpizar et al., 2013).

Numerous reports have described the consequences of overdosage/poisoning with essential oils or their constituents like camphor (see inchem.org), eucalyptol, and thujone. Camphor is a well-known toxin responsible for thousands of poisonings per year. A review of the American Association of Poison Control Centers reports about almost 10,000 cases of intoxication with camphor-containing products in the period between 1990 and 2003 in the US (Manoguerra et al., 2006). Nonetheless, camphor still can be found in many over-the-counter remedies. Severe intoxication is mostly caused by accidental oral use or mistake with castor oil. After ingestion, symptoms of burning sensation in the mouth, nausea, and vomiting occur within 5–15 min (Committee on Drugs, 1978). But significant resorption of camphor may also occur after substantial cutaneous and inhalative application as the risk for intoxication depends on concentration, rate of absorption, exposition, and mixture of components. Locally, it can produce irritation of skin, eyes, and mucous membranes of the respiratory tract. Even usual topical application of camphor may cause contact eczema (Committee on Drugs, 1994).

As camphor is easily absorbed in the gastrointestinal tract, the toxidrome manifests within minutes and includes gastrointestinal, hepato-, nephro-, and neurotoxic symptoms (Jimenez et al., 1983; Von Bruchhausen et al., 1993), as well as pulmonary and cardiac effects (Santos and Cabot, 2015). Severe ingestions may progress to seizures, apnea, and coma. Most individuals are no longer symptomatic after 24–48 h, but physiologic derangement may persist for far longer in some instances (Santos and Cabot, 2015).

19.2.1 Case Reports

Especially children react sensitively when essential oils are ingested accidentally. Flaman et al. (2001) reported intoxications in 251 children after ingestion of essential oil or products. Out of these 251 children, 50 ingested eucalyptus oil, 18 camphorated oil, 93 a vaporizing liquid, and 90 Vicks VapoRub. The most common symptoms were cough, lethargy, and vomiting, whereas two children had seizures but recovered, and even one child died.

19.2.1.1 Dose Range for Oral Intoxication

In humans admitted to the hospital in a state of acute intoxication after ingestion of 6–10 g camphor, camphor hydroxylated in the positions 3, 5, and 8 (or 9) were identified as major metabolites in the urine. Hydroxylated intermediates were subsequently oxidized to the corresponding ketones and carboxylic acids, the latter being conjugated with glucuronic acid (Köppel et al., 1982). Incubation of human liver microsomes with l-camphor resulted in the formation of 5-exo-hydroxycamphor as the only oxidation product (Gyoubu and Miyazawa, 2007). Among the 11n enzymes tested, only CYP2A6 was found to hydroxylate l-camphor. The important role of CYP2A6 was supported by the good correlation between contents of CYP2A6 and rates of formation of 5-exo-hydroxycamphor in liver microsomes from nine human samples (The EFSA Journal, 2008).

In infants, a dosage of 0.5–1.0 g (Siegel and Wason, 1986), respectively a lowest lethal dose of 70 mg/kg body weight (inchem.org), may cause death or severe intoxication, probably due to immature hepatic metabolism (Uc et al., 2000). This dose range confirms previous reports. According to Smith and Margolis (1954), as little as 1 g camphor ingested in one teaspoonful of camphorated oil (20% camphor in cottonseed oil) was fatal in a 19-month-old child.

Ragucci et al. (2007) presented the case of a 10-year-old boy who was admitted to the emergency room with symptoms of lethargy, nausea, vomiting, and rigors. He had chewed three over-the-counter cold remedy transdermal patches containing 4.7% (95.4 mg/patch) camphor and 2.6% menthol as active ingredients approximately 24 h before admission to the hospital. Assuming a body weight of 30 kg, this would correspond to an intake of camphor of approximately 10 mg/kg body weight.

On the basis of these and similar data, a probable lethal dose was estimated to be in the range of 50–500 mg/kg body weight (Gleason et al., 1969; Phelan, 1976). In sensitive people, clinically insignificant signs of toxicity may be observed already at 5 mg/kg body weight, whereas clinically manifest toxicity may require doses higher than 30 mg/kg body weight, and indeed, the study of Geller et al. (1984) demonstrates a large variation in sensitivity of humans to the acute toxicity of camphor.

19.2.1.2 Neurotoxic Effects

Neurotoxic effects include symptoms such as excitement, hallucinations, delirium, tremors, and convulsions (Opdyke, 1978). Camphor and camphor-containing essential oils have been included in the list of potential epileptogenics (Burkhard et al., 1999; Spinella, 2001).

In a retrospective case series from 2007 to 2014, the patient-reported ingestion rate of camphor was 1.5–15 g. In 30 patients, nausea and vomiting occurred in 73.3% cases, and tonic-clonic seizure was seen in 40% of patients. Mean exposure time was significantly longer in patients who experienced seizure (Rahimi et al., 2017).

Confirming the occurrence of seizures in animal toxicity tests, convulsions have been observed, especially in children following ingestion and also dermal or inhalation exposure of camphor- and thujone-containing products or eucalyptus oil (Craig, 1953; Millet et al., 1981; Melis et al., 1989; Committee on Drugs, 1994; Gouin and Patel, 1996; Burkhard et al., 1999; Woolf, 1999; Flaman et al., 2001; Ruha et al., 2003; Stafstrom, 2007; Khine et al., 2009; Halicioglu et al., 2011; see Bazzano et al., 2017). In lethal cases of camphor intoxication, postmortem necrosis of neurons was detected, similar to lesions observed in animal experiments (Smith and Margolis, 1954).

The risk of serious neurological side effects in young children such as convulsions, accompanied by apnea and asystole, has led the Committee for Medicinal Products for Human Use (CHMP) to implement contraindication for the use of suppositories containing terpenic constituents of essential oils in children under 30 months old (EMA, 2012).

19.2.1.3 Effects Following Inhalative Application

Camphor-containing products are almost exclusively used topically or for inhalation purpose. Nonetheless, topical and inhalative application is not free of risk due to the narrow therapeutic index of camphor. In infants, vapors of camphor and eucalyptus oil my cause laryngospasm with apnea and asystole (Kratschmer–Holmgren reflex). But, also, adults suffering from bronchial asthma or airway hypersensitivity may react with bronchoconstriction (inchem.org). No critical threshold could be determined for induction of laryngospasm.

Vicks VapoRub (VVR) that contains 5.46% camphor and 1.35% eucalyptus oil is often used to relieve symptoms of chest congestion. A report about a toddler in whom severe respiratory distress developed after VVR was applied directly under her nose led Abanses et al. (2009) to study the effect of VVR on mucociliary function in an animal model of airway inflammation. They could document a VVR-induced stimulation of mucin secretion (63% above control) and tracheal mucociliary transport velocity in the lipopolysaccharide-inflamed ferret airway. Findings were similar to the acute inflammatory stimulation observed with exposure to irritants and may lead to mucus obstruction of small airways and increased nasal resistance (Abanses et al., 2009).

19.3 EUCALYPTUS OIL

Human data are derived from accidental intoxications, which have been reported following ingestion of eucalyptus oil. However, it is difficult to draw general conclusions from reports on accidental intoxications with eucalyptus oil, because in many cases the amounts ingested could only be estimated roughly and the precise composition of the ingested product was not reported. Distilled eucalyptus oil contains at least 70% 1,8-cineole (eucalyptol) and a complex mixture of other compounds. Whitman and Ghazizadeh (1994) argue that the component hydrocyanic acid (prussic acid) may contribute to the toxicity of eucalyptus oil.

19.3.1 Case Reports

19.3.1.1 Dose Range for Oral Intoxication

Reports about lethal doses of eucalyptus oil vary markedly (De Vincenzi et al., 2002). Although it is reported that in adults, less than 2.5 mL is usually asymptomatic, transient coma has followed ingestion of 1 mL, and adult fatalities have been recorded with as little as 3.5 mL (Hindle, 1994). In general, minor depression of consciousness is presumed if more than 3 mL of 100% oil is ingested and significant central nervous system depression with more than 5 mL (Darben et al., 1998). Death is commonly seen after ingestion of 30 mL, but also has occurred after 4–5 mL (Patel and Wiggins, 1980; Flaman et al., 2001; Karunakara and Jyotirmanju, 2012). In particular, infants are prone to severe intoxication. In 27 infant patients who ingested known doses of eucalyptus oil, 10 had no effects after a mean of 1.7 mL, 11 had minor poisoning after a mean of 2.0 mL, five had moderate poisoning after a mean of 2.5 mL and one had major poisoning after 7.5 mL (Tibballs, 1995).

19.3.1.2 Intoxication after Ingestion

The toxic symptoms are rapid in onset, which include a burning sensation in the mouth and throat, abdominal pain, and spontaneous vomiting (Patel and Wiggins, 1980; Flaman et al., 2001). The initial central nervous system effects are giddiness, ataxia, and disorientation followed by loss of consciousness occurring in 10–15 min (Flaman et al., 2001; Karunakara and Jyotirmanju, 2012). Oil of eucalyptus was capable of producing severe cardiovascular, respiratory, and central nervous system manifestations after ingestion of as little as 4 mL by an adult (Gurr, 1965).

In particular, ingestion of eucalyptus oil causes significant morbidity in infants and young children. There are several reports which document severe intoxication following oral application of eucalyptus oil. For example, accidental ingestion of eucalyptus oil by a 3-year-old boy caused profound central nervous system depression within 30 minutes and rapid recovery after gastric lavage (Patel and Wiggins, 1980).

Out of 109 children (mean age 23.5 months) with eucalyptus oil ingestion, 59% of them were symptomatic; 28% had depression of conscious state, ranging from drowsiness to unconsciousness after ingesting 5–10 mL. Vomiting occurred in 37%, ataxia in 15%, and pulmonary disease in 11%. No treatment was given for 12% (Tibballs, 1995).

Convulsions are rare in adults, but Kumar et al. (2015) reported two cases of 3- and 6-year-old boys, who presented with status epilepticus within 10 min of accidental ingestion of 10 mL of eucalyptus oil. They had four and eight episodes of tonic-clonic convulsions, respectively, without known previous history of seizures. The children recovered completely within 20 h (Kumar et al., 2015). Similarly, Karunakara and Jyotirmanju (2012) reported two cases, one of a 2-year-old female child with convulsions who accidentally ingested around 5–10 mL eucalyptus oil. Immediately after ingestion of the oil, the child had two episodes of vomiting, and within 20 minutes generalized tonic-clonic convulsions started. In the second case, a 2-year, 7-month old male child, who was treated for two days with 5 mL of eucalyptus oil as a home remedy for cold, started having generalized tonic-clonic convulsions within 10 minutes after oral application (Karunakara and Jyotirmanju, 2012).

In contrast, Webb and Pitt (1993) identified 42 cases of oral eucalyptus oil poisoning in children under 14 years of age between 1984 and 1991 with less severe symptoms. Thirty-three children (80%) were asymptomatic, although four of these children ingested more than 30 mL of eucalyptus oil. Only two children had clear symptoms or clinical signs. Thus, the authors speculate that eucalyptus oil may be a less toxic compound than previously thought (Webb and Pitt, 1993).

Due to lack of a specific antidote, the management of eucalyptus oil poisoning is mainly supportive and symptomatic. The main risk is aspiration following vomiting and depression of the central nervous system. Therefore, emesis is contraindicated (Patel and Wiggins, 1980; Flaman et al., 2001; Karunakara and Jyotirmanju, 2012). The role of activated charcoal is controversial (Flaman et al., 2001), because once ingested, eucalyptus oil is readily distributed throughout the body and efforts at elimination by using activated charcoal is unlikely to help significantly. In a case of severe intoxication with prolonged coma, successful management of intoxication was enabled by the use of mannitol, hemodialysis, and peritoneal dialysis (Gurr, 1965).

Although reports about dosages causing toxic effects are divergent, already relatively small amounts of eucalyptus oil are also reported to be fatal, notably in children in whom severe toxicity may predispose to the development of status epilepticus. Thus, eucalyptus oil should never be given orally.

19.3.1.3 Intoxication after Topical Application and Inhalation

Eucalyptus oil is well documented as being extremely toxic if ingested. But Darben et al. (1998) also present a case of systemic eucalyptus oil intoxication from topical application of a home remedy for urticaria, containing eucalyptus oil. The 6-year-old girl presented with slurred speech, ataxia, and muscle weakness progressing to unconsciousness following the widespread application. Six hours following removal of the topical preparation her symptoms had resolved (Darben et al., 1998).

19.3.2 Eucalyptol (1,8-Cineole)

The case reports on acute toxicity in humans refer to the ingestion of eucalyptus oil and not to eucalyptol as such. Distilled eucalyptus oil typically contains at least 70% eucalyptol (1,8-cineole) and a complex mixture of other compounds. Thus, the contribution of eucalyptol and other constituents to the reported intoxications with eucalyptus oil do not provide information for adequate estimates of toxic dose levels for eucalyptol. Moreover, toxicological data available on eucalyptol are rather limited. However, similar to camphor essential oil, eucalyptol also produced electrocortical seizure activity in rat brain (Culic et al., 2009; Kolassa, 2013), which supports the context between eucalyptol and convulsions in children intoxicated with eucalyptus oil.

Eucalyptol is rapidly absorbed from the gastrointestinal tract. In a subacute toxicity study, no significant differences in body weight and relative organ weight between the control group and eucalyptol treatment groups were found. The histopathological examinations showed dose-related granular degeneration and vacuolar degeneration in liver and kidney tissue after administration of high doses. The electron microscopy assays indicated that the influence of eucalyptol on the target organ at the subcellular level were mainly on the mitochondria, endoplasmic reticulum, and other membrane-type structures of liver and kidney (Hu et al., 2014; Xu et al., 2014).

A limited long-term study with mice did not show treatment-related effects, including effects on tumor incidence. The study was performed with males only and the histopathological examination was limited to only a few organs. No evidence for genotoxicity has been found in bacterial tests. In Chinese hamster ovary cells, chromosomal aberrations were not induced, and sister chromatid exchanges were only observed at cytotoxic doses (Opinion of the Scientific Committee on Food on eucalyptol, EU, 2002).

Acute and subacute oral toxicity studies of eucalyptol in Kunming mice revealed an LD_{50} value (95% CI) of 3849 mg/kg (Xu et al., 2014). Even if eucalyptol were responsible for the acute toxicity of eucalyptus oil, the estimated daily intake of eucalyptol from food would be much lower than the amount tentatively assumed to be present in the lowest lethal doses of eucalyptus oil reported (Opinion of the Scientific Committee on Food on eucalyptol, EU, 2002).

19.4 THUJONE-CONTAINING ESSENTIAL OILS

Thujones are renowned as the possible psychoactive and toxic principles of absinthe (Olsen, 2000). Thujone is a common name for two naturally occurring monoterpene diastereomeric ketones, (−)-α-thujone and (+)-β-thujone, that can be found in essential oils of sage (*Salvia officinalis* L.), absinthe wormwood (*Artemisia absinthium* L.), eastern arborvitae (*Thuja occidentalis* L.), and tansy (*Tanacetum vulgare* L.) in different quantities (Blagojevic et al., 2006). These plants are wild-growing and ornamental plant species with ethnopharmacological usages. Their essential oils are important components of numerous flavoring, perfumery, cosmetic, and pharmaceutical products. However, their potential toxic effects, in particular the known neurotoxicity of thujones, are the main reasons why some countries impose restriction upon their utilization.

19.4.1 Case Reports

Precautions should be taken into account because there are several case reports in which sage oil induced the development of vertical nystagmus, hyperreflexia, clonic spasms, and tonic-clonic convulsions, with muscle dystonia before seizures and postictal muscle weakness occurred (Burkhard et al., 1999; Halicioglu et al., 2011). The convulsive properties of a diluted essential oil (1:99 dilution, and repeating this dilution 30 times) containing thujones were observed in a 7-month-old baby (Stafstrom, 2007). In adults, the oil of *A. absinthium* caused convulsions and paralysis, probably due to its content of thujones, camphor, and 1,8-cineole (see Radulović et al., 2017). α-Thujone in absinthe and herbal medicines is a rapid-acting and readily detoxified modulator of the GABA-gated chloride channel (Höld et al., 2000). Although all of the described neurotoxic effects could be attributed to the presence of thujones in these oils, numerous other oil constituents could also cause the observed neurotoxicity (Radulović et al., 2017).

19.5 PEPPERMINT OIL

The main component of peppermint oil (*Oleum Menthae piperitae*) is menthol (Table 19.1).

19.5.1 Adulterations

Dementholized essential oil of *Mentha arvensis* L. var. *piperascens* Malinv., which is mainly produced in Brazil and China, has to be labelled correctly as mint oil or Japanese mint oil. The composite of the oil is comparable to peppermint oil of *Mentha x piperita* L., but it is much cheaper. Synthetic

TABLE 19.1
Content of Components of Peppermint Oil

Content (%)	Component
~45	Menthol
~24	Menthon
1–9	Menthofurane
~5	Eucalyptol (1,8-cineol)
~4	Menthyl acetate
~3	Isomenthon
~1.5	Limonene
~1	Pulegon
<1	Carvon

menthol and menthyl acetate might be added to essential oils of insufficient quality. Occasionally, peppermint oil is mingled with the essential oil of spearmint *Mentha spicata* var. *crispa*, which contains up to 10% carvone, which can be used as an indicator for adulteration, because peppermint oil should contain less than 1% carvone. Peppermint oil may also be adulterated with the essential oil of pennyroyal (*Mentha pulegium*), which is characterized by an extremely high content of pulegone. This adulteration is important to be identified due to the pronounced hepatotoxicity of pulegone. Thus, the content of pulegone in peppermint oil should not exceed 1%, and accordingly, no confirmed cases of liver damage caused by peppermint oil or mint oil have been reported so far.

19.5.2 Case Reports

Documented case reports indicate that especially children are prone to unwanted side effects induced by menthol or menthol-containing essential oils and products. Inhalation of menthol can cause apnea and laryngoconstriction in susceptible individuals (Gardiner, 2000). The excessive inhalation of mentholated preparations may also cause nausea, anorexia, cardiac problems, ataxia, confusion, euphoria, nystagmus, and diplopia as reported in the case of a 13-year-old boy who inhaled 5 mL of olbas oil (containing 200 mg menthol) instead of the recommended few drops (O'Mullane et al., 1982).

Direct application of peppermint oil to the nasal area or chest to infants should be avoided because of the risk of apnea, laryngeal and bronchial spasms, acute respiratory distress with cyanosis, and even respiratory arrest (Blake et al., 1993; Wyllie and Alexander, 1994). When studying the incidence of respiratory reactions to stimulation of the nasal and propharyngeal mucosa in 44 newborn premature infants, the inhalation of menthol fumes or application of drops to the nasal mucosa caused transient respiratory arrest or reduced respiration rate in 43% of infants (Javorka et al., 1980). Thus, in children, menthol-containing products should not be applied directly to the nostrils (Dost and Leiber, 1967). There are also reports that menthol can cause jaundice in newborn babies, and in some cases, this has been linked to a glucose-6-phosphatase dehydrogenase deficiency (Olowe and Ransome-Kuti, 1980).

Another report in children documents facial edema, shortness of breath, and cyanosis after application of menthol-containing cologne to the face in an otherwise healthy 2-month-old infant (Arikan-Ayyildiz et al., 2012). Perfumes or colognes may aggravate respiratory symptoms in patients with asthma, and they are also targeted as one of the most common causes of cosmetic allergic contact dermatitis (Kumar et al., 1995; Lessenger, 2001). Immediate hypersensitivity reactions to menthol range from urticaria and rhinitis to asthma (Kawane, 1996; Marlowe, 2003; Andersson and Hindsen, 2007), but also one case of anaphylaxis induced by menthol-containing toothpaste has been reported in a metamizole-allergic woman (Paiva et al., 2010).

A near fatal case due to ingestion of toxic doses of oral peppermint oil in suicidal intention was reported by Nath et al. (2012). The authors describe the case of a patient who arrived in a comatose state and in shock. She was managed with mechanical ventilation and inotropes. Her vital parameters reached normal within 8 hours and she became conscious by 24 hours. The side effects of peppermint oil are considered to be mild, but this case report warns that ingestion of oral toxic doses of peppermint oil could be dangerous.

Adverse reactions to orally applied peppermint oil capsules are rare but can include hypersensitivity reaction, contact dermatitis, abdominal pain, heartburn, perianal burning, bradycardia, and muscle tremor (Parys, 1983; Nash et al., 1986; Weston, 1987; Wilkinson and Beck, 1994; Sainio and Kanerva, 1995). In 12 patients, contact sensitivity to menthol and peppermint occurred, with oral symptoms including burning-mouth syndrome, recurrent oral ulceration, or a lichenoid reaction (Morton et al., 1995).

Some precautions have to be taken especially in patients with specific organ dysfunction. For example, peppermint oil is contraindicated in obstruction of the bile ducts, gallbladder inflammation, severe liver damage, hiatal hernia, and kidney stones (Gardiner, 2000). Patients with achlorhydria should use peppermint oil only in enteric-coated capsules (Rees et al., 1979), and patients with gastrointestinal reflux should be aware that peppermint may worsen reflux symptoms (Gardiner, 2000).

19.5.3 Menthol

In contrast to eucalyptol, camphor, and thujone, menthol is quickly metabolized after oral administration. Because of the high first-pass glucuronidation of menthol, no significant systemic, but only local, effects can be expected after enteral absorption of commonly used doses of menthol (Kolassa, 2013).

Menthol is classified by the US Food and Drug Administration (FDA) as safe and effective as a topical over-the-counter (OTC) product. The FDA has approved concentrations of menthol of up to 16% for OTC external use; their safety profile has been demonstrated by *in vitro* and *in vivo* studies and most investigations reveal a low potential for toxicity in humans (Api et al., 2016a). A chronic toxicity study with Fischer rats and B6C3F1 mice (50/sex/dose), which received DL-menthol with their diet, revealed no clinical effects, signs of toxicity, histopathological changes in major organs and tissues, or increased incidences of tumors compared to controls (National Cancer Institute, 1979). Moreover, most published studies report no genotoxic or mutagenic effects for menthol (see Kamatou et al., 2013). It is suggested that some contradictory outcomes of studies may have resulted from the use of different stereoisomers of menthol (Liu et al., 2013).

19.5.3.1 Cooling vs. Irritating Effect

Menthol is well known for its cooling effect or sensation when it is inhaled, chewed, consumed or applied to the skin (Yosipovitch et al., 1996) due to its ability to activate the cold-sensitive transient receptor potential cation channel melastatin family member 8 (TRPM8) (Peier et al., 2002) and transient receptor potential ankyrin 1 (TRPA1) (Karashima et al., 2007), thereby increasing calcium flux through the channels (Farco and Grundmann, 2013). However, in contrast to low concentrations, which cause cooling sensation, after topical application of menthol, high concentrations may cause irritation and local anesthesia (Eccles, 1994).

19.5.3.2 Menthol-Induced Analgesia

Menthol acts as a nonselective analgesic agent through multiple peripheral and central pain targets. Besides the menthol-induced cooling sensation by activation of TRPM8, Liu et al. (2013) could demonstrate that TRPM8 is also a principal mediator of menthol-induced analgesia of acute and inflammatory pain. In addition, direct activation of γ-aminobutyric acid type A ($GABA_A$) receptors (Zhang et al., 2008) and blockade of Na^+ and Ca^{2+} channels in dorsal horn neurons (Pan et al., 2012) and peripheral neurons (Swandulla et al., 1987; Sidell et al., 1990; Gaudioso et al., 2012) play prominent roles in analgesia. Data indicate that both menthol and peppermint oil exert Ca^{2+}-channel-blocking properties also in other tissue which may corroborate, for example, their usage in the treatment of irritable bowel syndrome (Hawthorn et al., 1988). Menthol was also found to inhibit nicotinic acetyl choline receptors and serotonin gated ion channels, which are known to contribute to pain signaling (Heimes et al., 2011; Hans et al., 2012).

However, L-menthol with TRPA1 and other targets may also have pro-algesic and inflammatory effects. Indeed, topical menthol application at high concentrations can be accompanied by skin irritation, and menthol inhalation can exacerbate asthma in some patients, both conditions in which TRPA1 has a documented role (Bautista et al., 2006; Caceres et al., 2009). As L-menthol was found to interact with a series of ion channels, these interactions may contribute to analgesia but may also underlie the irritating and pro-inflammatory side effects observed in many patients treated with these agents (Liu et al., 2013).

19.5.3.3 Menthol and Tobacco-Related Chemicals

Aside from its benefits, menthol has been shown to inhibit mucosal recognition of nicotine and toxic cigarette components, thus potentially leading to toxic effects when consuming mentholated cigarette brands (Farco and Grundmann, 2013). Menthol enhances penetration of the tobacco carcinogen nitrosonornicotine and nicotine through buccal and floor of mouth mucosa *in vitro*, even after short exposure (Squier et al., 2010). Studies conducted in order to evaluate the effects of menthol in cigarettes showed that (−)-menthol is an inhibitor of CYP2A6, and that smoking mentholated cigarettes might lead to inhibition of the metabolism of nicotine and other tobacco-related chemicals, which could be the reason for the increased risk of lung toxicity when smoking mentholated cigarettes (MacDougall et al., 2003).

However, in a rat model, cigarette-smoke-induced inflammatory responses were not observed or were much lower after exposure to mentholated tobacco products (Kogel et al., 2016; Oviedo et al., 2016). Based on these findings, it is suggested that L-menthol, through TRPM8, is a strong suppressor of respiratory irritation responses, even during highly noxious exposures to cigarette smoke or smoke irritants. This feature could bear the risk that L-menthol, as a cigarette additive, may promote smoking initiation and nicotine addiction (Ha et al., 2015). Moreover, menthol, when administered with nicotine, showed evidence of psychoactive properties by affecting brain activity and behavior compared to nicotine administration alone (Thompson et al., 2017).

19.5.3.4 Menthol and Dermal Penetration

Menthol is one of the most effective terpenes used to enhance the dermal penetration of pharmaceuticals and other agents. However, the mechanism by which terpenes enhance drug permeation through the skin is not completely understood (Nair, 2001; Kamatou et al., 2013). Nonetheless, this has to be kept in mind when applying drugs to the skin in combination with menthol or menthol-rich essential oils, which might cause unwanted topical and systemic drug effects due to the increased penetration rate.

19.6 PENNYROYAL OIL

Pennyroyal oil, derived from *Mentha pulegium* or *Hedeoma pulegoides*, is a highly toxic agent containing high concentrations of R(+)-pulegon (62%–97%) and its metabolite menthofuran. It may cause both hepatic and neurologic injury if ingested. A potential source of pennyroyal oil is certain mint teas mistakenly used as home remedies to treat minor ailments and colic in infants. While no apparent evidence exists for beneficial activities, there are strong data supporting the hepatotoxic effects of pennyroyal oil (Bakerink et al., 1996; Chen et al. 2003). Even small quantities can cause acute liver and lung damage, while high doses (15 mL) of the oil even may result in death (Dietz and Bolton, 2007). If at all, it only should be applied topically.

19.6.1 Case Reports

Hepatic and neurologic injury developed in two infants after ingestion of mint tea. Examination of the mint plants, from which the teas were brewed, indicated that they contained the toxic agent pennyroyal oil. In one of the children, lethal fulminant liver failure with cerebral edema and necrosis developed. This infant had a positive proof only for menthofuran. The other infant, who was positive for both pulegone and menthofuran, developed hepatic dysfunction and a severe epileptic encephalopathy (Bakerink et al., 1996).

Menthofuran and its metabolites were found in relatively small amounts in the urine of six human volunteers. However, menthofuran was present in the serum of two individuals, hours after ingestion of a large amount of pennyroyal oil (Anderson et al., 1996). In a fatally poisoned patient, 18 ng/mL of pulegone and 1 ng/mL of menthofuran were found in serum analyzed at 26 h postmortem, 72 h following acute ingestion. In another case, 40 ng/mL menthofuran were found in serum with no detectable pulegone levels, 10 h after ingestion.

A literature review of cases of human intoxication with pennyroyal oil indicates that ingestion of 10 mL (corresponding to approximately 5.4–9 g pulegone, approximately 90–150 mg/kg body weight for a 60 kg person; calculated with a relative density of 0.9 as for peppermint oil) resulted in moderate to severe toxicity, and ingestion of more than 15 mL (corresponding to approximately 8–13 g pulegone, approximately 130–215 mg/kg body weight for a 60 kg person) resulted in death. The clinical pathology was characterized by massive centrilobular necrosis of the liver, pulmonary edema and internal hemorrhage (European Commission Opinion of the Scientific Committee on Food on pulegone and menthofuran, 2002).

19.6.2 Pulegone and Menthofuran

There are no formal toxicokinetic studies performed in humans. A few studies analyzed serum levels and/or urinary excretion of pulegon and metabolites, and tentative metabolic pathways have been uncovered. Pulegone and menthofuran are absorbed from the gastrointestinal tract, but there are no studies available to estimate oral bioavailability. There are no studies on dermal penetration, but the use of pulegone as a dermal absorption enhancer seems to suggest that it may be absorbed. Furthermore, there are no inhalation studies available (EMA, 2014).

In vivo metabolism of pulegone is extremely complex, with numerous metabolites in the urine and bile of treated animals (Chen et al., 2003). At lower doses than 80 mg/kg body weight, urinary phase 2 conjugates of hydroxylated metabolites predominate, which involves reduction of pulegone to menthone or isomenthone followed by hydroxylation in ring or side chain and subsequent conjugation with glucuronic acid. For a second pathway, conjugation with glutathione is reported (The EFSA Journal, 2005). At higher doses, the major bioactivation step involves the oxidation of pulegone by cytochrome P450 enzymes CYP2E1, CYP1A2, and CYP2C19 to its metabolite menthofuran (Gordon et al., 1987; Thomassen et al., 1990; Khojasteh-Bakht et al., 1999; Sztajnkrycer et al., 2003). Menthofuran is metabolized by the same human liver CYP enzymes involved in the metabolism of pulegone and additionally by CYP2A6. Menthofuran inhibits human CYP2A6 irreversibly, possibly by covalent adduction (Khojasteh-Bakht et al., 1998). Analysis of bile of treated rats showed the presence of a glutathione conjugate of menthofuran, which confirms that menthofuran is an *in vivo* metabolite of pulegone and that further metabolites of menthofuran are formed, including the hepatotoxic compound γ-ketenal (8-pulegone aldehyde) and menthofuran epoxide. These latter intermediates are either further metabolized and form glutathione conjugates or, at high concentrations and/or glutathione depletion, covalently bind to cellular macromolecules, causing hepatic injury (Madyastha and Moorthy, 1989; McClanahan et al., 1989; Anderson et al., 1996; Khojasteh-Bakht et al., 1999; Chen et al., 2001). In addition, binding to hepatic proteins might be facilitated by another metabolite of pulegone, *p*-cresol, which has been shown to be a glutathione depletory (Madyastha and Raj, 2002). Evidence from animal experiments suggests that *N*-acetylcysteine provides at least partial protection from the hepatotoxic effects of pennyroyal oil (Anderson et al., 1996) because it counteracts glutathione depletion and thus promotes inactivation via this conjugation pathway.

(*R*)-(+)-menthofuran, which may be formed as a metabolite of (R)-(+)-pulegone at high doses, exhibits qualitatively similar toxicity in laboratory rodents. However, in a recent human metabolism study on six human volunteers, menthofuran and its metabolites were only found in relatively small amounts in the urine (Engel, 2003). Nevertheless, also in humans, the formation of menthofuran as an intermediate leading to other reactive compounds, such as γ-ketoenal, cannot be excluded (The EFSA Journal, 2005).

Nephropathy seems to be sex- and species-specific as nephrotoxicity was only observed in male rats. The reason for that was speculated to be binding to α2μ-globulin in these animals (Spindler and Madsen, 1992).

19.6.3 Precautions

No quantitative data concerning absorption of pulegone and menthofuran through the skin exist, although it is known that pulegone has been used as a penetration enhancer. It is to ensure that the sum of pulegone and menthofuran within the daily dose is less than 3.5 mg for adults. The short-term use of a maximum 14 days is restricted to intact skin.

The value of 20 mg/kg body weight per day, based on the National Toxicology Program chronic study, is taken as a lowest-observed-adverse-effect level (LOAEL) value. It is possible to use a safety factor of 300 (not 100, because LOAEL was the lowest significant effect level). Consequently the acceptable exposure is 0.07 mg/kg body weight per day, which is close to the current acceptable daily intake (ADI) value of 0.1 mg/kg body weight per day. The daily dose for an adult of 50 kg body weight is thus 3.5 mg/person/day. The intake (pulegone + menthofuran) of 3.5 mg/person/day (even if the limit presents the overall intake from all sources) can be accepted for herbal medicinal products as short-term intake (maximum 14 days) (EMA, 2014). If at all, pennyroyal oil only should be applied topically.

19.7 WINTERGREEN OIL

Wintergreen essential oil is usually obtained from the leaves of *Gaultheria* species, and in trace amounts, wintergreen is used as a food flavoring. The main constituent of the essential oil of wintergreen is methyl salicylate.

19.7.1 Methyl Salicylate

Methyl salicylate is widely available in high concentrations as a component in many over-the-counter brands of ointments, lotions, liniments, rubefacient, and medicated oils intended for topical application to relieve joint and muscular pain (Davis, 2007). Among the most potent forms is wintergreen oil with a content of approximately 98% methyl salicylate. Methyl salicylate in oil of wintergreen is an infrequent cause of salicylate poisoning but is the most dangerous salicylate formulation by strength. One teaspoon of wintergreen oil contains approximately 6 g of methyl salicylate, which is equivalent to approximately 20 aspirin tablets of 300 mg. Childhood fatalities may occur after ingestion of as little as 4 mL of oil of wintergreen (Ellenhorn and Barceloux, 1988); 30 mL of wintergreen oil is equivalent to 55.7 g of aspirin (Johnson, 1985). This conversion illustrates the potency and potential toxicity of oil of wintergreen, even in small quantities. When applied to large areas of skin, topical salicylic acid may cause sufficient dermal absorption to produce toxic serum salicylate levels. Seneviratne et al. (2015) and Lucas (2000) reported four cases of accidental intoxication with small amounts of methyl salicylate, whereof three ended fatally.

Intoxication with salicylates causes nausea and vomiting, hyperthermia, tinnitus, hyperventilation, and primary respiratory alkalosis, followed by metabolic acidosis, coagulopathy, and, finally, vasomotor collapse, respiratory depression, and coma. In addition, essential oil of wintergreen can also produce allergy-like symptoms and asthma, and it can also affect blood clotting. Thus, taking wintergreen oil along with warfarin, aspirin, or any other oral anticoagulant can increase the chances of bruising and bleeding.

19.8 TEA TREE OIL (MELALEUCA OIL)

Tea tree oil, also known as maleleuca oil, is a complex mixture of terpene hydrocarbons and tertiary alcohols with at least 30% terpinen-4-ol and maximal 15% 1,8-cineole. The essential oil is distilled mainly from plantation stands of the Australian native plant *Melaleuca alternifolia* (Maiden and Betche) Cheel of the Myrtaceae family. Occasionally, essential oils are named tea tree oil, but originate from other *Melaleuca* species, such as *Melaleuca leucadendra* (cajuput oil) or *Melaleuca viridiflora* (niauli oil).

Tea tree oil is widely used as an alternative antimicrobial and anti-inflammatory agent (de Souza et al., 2017; Najafi-Taher et al., 2018; Casarin et al., 2018). However, relatively limited data are available on the safety and toxicity of the oil. In many reports on the safety and toxicity of tea tree oil, the composition of the oil is not stated (Hammer et al., 2006). Particularly in the presence of oxygen, light, and high temperature, the oil degrades to decreased contents of α-terpinen, γ-terpinen, and terpinols, while the content of *p*-cymene increases tenfold. Oxidation processes produce peroxides, endoperoxides, and epoxide, which contribute to toxicity.

19.8.1 Toxicity Following Oral Exposure

Oral exposure to tea tree oil can be toxic as reported for experimental studies in rats (LD_{50} 1.9–2.6 mL/kg, Russell, 1999) and from limited cases of human poisoning, especially in children. Two case reports about children less than two years tell about reversible clinical symptoms of drunkenness due to central nervous system depression after ingestion of about 10 mL pure tea tree oil (Jacobs and Hornfeldt, 1994; Del Beccaro, 1995). However, another report of a 4-year-old boy who ingested two teaspoons of 100% pure tea tree oil describes worse symptoms of ataxia followed by unconsciousness, unresponsiveness, and required intubation (Morris et al., 2003). An adult patient who drank approximately half a tea cup of tea tree oil (corresponds to 0.5–1.0 mL/kg body weight) was comatose for 12 h and hallucinatory for another 36 h. Abdominal pain and diarrhea continued for approximately six weeks (Seawright, 1993). A 60-year-old man who ingested one and a half teaspoons of tea tree oil in order to prevent a cold presented with swollen feet and face and red rash all over his body, which disappeared after one week (Elliott, 1993). Thus, oral application of tea tree oil should not be recommended. 100% pure tea tree oil has to be labelled that it must be kept out of the reach of children, is packaged with a child-resistant cap, and is labelled "not to be taken internally."

19.8.2 Toxicity Following Dermal Exposure

The irritant capacity of tea tree oil has been investigated in small (20 patients) and larger (>300 volunteers) trials at concentrations ranging from 1% to 100% in different formulations and time span. Patients did not demonstrate any or only weak irritant reactions (see Hammer et al., 2006).

However, numerous case reports of contact allergy have been published (see Hammer et al., 2006). The reactions occurred in response to 100% pure tea tree oil as well as lower concentrations in various formulations. The frequency of allergic reaction reported in numerous trials ranged from 0.1% to 10.7% (Hammer et al., 2006). Patch testing of allergic patients with tea tree oil components showed that they reacted mostly to the sesquiterpenoid fractions but not the pure monoterpenes (Southwell et al., 1997). Probably the oxidation products formed during prolonged storage are the main allergens (Hausen, 2004), while newly distilled tea tree oil seems to have a relatively low sensitizing capacity. The most important allergens formed could be terpinolene, α-terpinene, ascardiole, and 1,2,4-trihydroxymethane (Hausen et al., 1999; Hausen, 2004)

19.8.3 Systemic Reactions

Toxicity following dermal application of inappropriately high doses of melaleuca oil to cats or dogs treated for fleas has been described. Animals had typical signs of depression, weakness, un-coordination and muscle tremors (Villar et al., 1994), and especially cats experienced severe symptoms of hypothermia, incoordination, dehydration, and trembling. One cat which had pre-existing renal damage even died after three days (Bischoff and Guale, 1998). Only one report of a systemic reaction to tea tree oil in humans is published (Mozelsio et al., 2003). Following dermal application, a 38-year-old man suffered immediate flushing, pruritus, a constricted throat, and lightheadedness. His reaction, however, was not due to an IgG or IgE response.

Henley et al. (2007) investigated possible causes of gynecomastia in three prepubertal boys who repeatedly applied lavender and tea tree oil topically and who were otherwise healthy and had normal serum concentrations of endogenous steroids. In all three boys, gynecomastia coincided with the topical application of products that contained lavender and tea tree oils. Studies in human cell lines confirmed the findings of estrogenic and antiandrogenic activities of both essential oils. Gynecomastia resolved in each patient shortly after the use of products containing these oils was discontinued (Henley et al., 2007). However, this matter is discussed controversially (Carson et al., 2014).

19.8.4 Ototoxicity

Tea tree oil has been suggested as an effective treatment for a number of microorganisms commonly associated with otitis externa and otitis media, but it is possible ototoxicity has only been evaluated in a single study by Zhang and Robertson (2000) in guinea pigs. The authors showed that concentrations up to 2% tea tree oil may be safe for use within the ear, whereas higher concentrations applied for a relatively short time were, to some extent, ototoxic to the high-frequency region of the cochlea (Zhang and Robertson, 2000).

19.8.5 Developmental Toxicity

No studies of potential developmental toxicity following exposure to tea tree oil have been published yet. However, an embryo- and fetotoxicity study with α-terpinene, a constituent of tea tree oil with a content of approximately 9%, has demonstrated significant toxicity in a rat model (Araujo et al., 1996), with delayed ossification and skeletal malformations. These limited data suggest that tea tree oil might be embryo- and fetotoxic, although only if ingested at relatively high concentrations.

19.8.6 *In Vitro* Toxicity

Several cytotoxicity studies have been performed in numerous cell lines and showed that tea tree oil and/or its components are cytotoxic at concentrations between 20 and 2700 μg/mL (Hammer et al., 2006). However, for topically applied products, data on percutaneous penetration and *in vivo* disposition including metabolism are needed for any relevant assessment of a potential *in vivo* risk. Furthermore, in bacterial reverse mutation assays, tea tree oil and its components were not mutagenic (Hammer et al., 2006) and not genotoxic (Pereira et al., 2014).

To conclude, if oxidation products can be avoided, the available literature suggests that tea tree oil can be used topically in diluted form by the majority of individuals without adverse effects.

19.9 SASSAFRAS OIL

The main constituent of sassafras oil is safrole, which is also present in small amounts in a number of spices. Sassafras oil is extracted from the bark and roots of the tree *Sassafras albidum*. It has had a traditional and widespread use as a natural diuretic, as well as a remedy against urinary tract disorders or kidney problems until safrole was discovered to be hepatotoxic and weakly carcinogenic (Fennell et al., 1984; Rietjens et al., 2005). Thus, the FDA banned the use of sassafras oil as a food and flavoring additive because of the high content of safrole and its proven carcinogenic effects. However, pure sassafras oil is still available online and also in some health-food stores.

19.9.1 Safrole

Safrole has been demonstrated to be genotoxic and carcinogenic. It produces liver tumors in mice and rats by oral administration, and it produces also liver and lung tumors in male infant mice by subcutaneous application. It has been shown that safrole is able to induce chromosome aberrations, sister chromatid exchanges, and DNA adducts in hepatocytes of F344 rats exposed *in vivo* (Daimon et al., 1998).

However, the carcinogenic potency appears to be relatively low and dependent on the metabolism. Safrole is metabolically activated through the formation of intermediates, and these are able to directly react with DNA. Two bioactivation pathways of safrole to potentially hepatotoxic intermediates have been reported. One involves P450-catalyzed hydroxylation, thereby producing 1′-hydroxysafrole, and on further conjugation with sulfate, generating a reactive sulfate ester. This ester creates a highly reactive carbocation, which alkylates DNA. The second pathway involves P450-catalyzed hydroxylation of the methylenedioxy ring and finally formation of reactive *p*-quinone methide. Both pathways could explain the genotoxic effects of safrole. DNA adducts have been identified *in vitro* and *in vivo* (see Dietz and Bolton, 2005).

In vitro experiments confirm the genotoxic effects of safrole. Therefore, the existence of a threshold cannot be assumed and the EU Scientific Committee on Food could not establish a safe exposure limit. Consequently, reductions in the exposure and restrictions in the use levels by health authorities are justified (Opinion of the Scientific Committee on Food on the safety of the presence of safrole, 2002).

19.10 CLOVE OIL (*Oleum Caryophylli, Caryophylli floris aetheroleum*)

Clove oil is the common name for an extract from the flower buds of *Syzygium aromaticum* (L.) Merrill et L. M. Perry, also known as *Caryophyllus aromaticus* L., *Eugenia aromatica* (L.) Baill., *Eugenia caryophyllus* (Spreng.) Bullock & S. G. Harrison, and *Eugenia caryophyllata* Thunberg. It is a complex mixture of chemical substances, the main component being eugenol. The essential oil is widely used and well known for its medicinal properties. Traditional uses of clove oil include use in dental care, as an antiseptic and analgesic, where the undiluted oil may be rubbed on the gums to treat toothache. It is active against oral bacteria associated with dental caries and periodontal disease (Cai and Wu, 1996) and effective against a large number of other bacteria (Chaieb et al., 2007). The major component of clove oil is usually considered to be eugenol with a content of 88.6% (Chaieb et al., 2007).

19.10.1 Case Reports

Recent growth in aromatherapy sales has been accompanied by an unfortunate increase in accidental poisoning from these products. Clove oil warrants special attention, because it has hepatotoxic effects at high concentrations.

Eugenol is a material commonly used in dentistry with few reported side effects. It is not, however, a bio-friendly material when in contact with oral soft tissues. It can produce both local irritative and cytotoxic effects, as well as hypersensitivity reactions, as reported by Sarrami et al. (2002) in two cases and by Barkin et al. (1984), who reported one case with a serious allergic reaction. Also, an allergic reaction after use of a eucalyptol-containing anti-inflammatory cream was reported (Vilaplana and Romaguera, 2000).

A case report about a 24-year-old woman who spilled a small amount of clove oil on her face in an attempt to relieve a toothache is described by Nelson et al. (2011). She experienced permanent infraorbital anesthesia and anhidrosis.

Ingestion of 1–2 teaspoon of clove oil in a 2-year-old boy resulted in metabolic acidosis, coma, seizures, hypoglycemia, and liver failure (Ford et al., 2001).

Also, Janes et al. (2005) described the development of fulminant hepatic failure in a 15-month-old boy after ingestion of 10 mL clove oil. The hepatic impairment resolved after intravenous administration of *N*-acetylcysteine, and recovery occurred over the next four days.

19.10.2 Eugenol, Isoeugenol

For some data, isoeugenol was used as a read-across for eugenol, because both belong to the generic class of phenols with common structures and the only differ in the position of the double bond in the alkene chain. It is assumed that the structural differences do not essentially change the physicochemical properties nor raise any additional structural alerts. In addition, they are predicted to have similar metabolites. Therefore, the toxicity profiles are expected to be similar.

19.10.2.1 Repeated Dose Toxicity

The margin of exposure for eugenol is adequate for the repeated dose toxicity endpoint at the current level of use.

From a dietary 13-week subchronic toxicity study conducted in rats, the no-observed-adverse-effect level (NOAEL) for repeated dose toxicity was determined to be 300 mg/kg/day for eugenol and 37.5 mg/kg/day for isoeugenol (Api et al., 2016b), based on reduced body weights.

The Research Institute for Fragrance Materials (RIFM) and the US National Toxicology Program (US NTP) concluded that hepatocellular tumors observed following eugenol administration were considered to be associated with the dietary administration of eugenol, but because of the lack of a dose-response effect in male mice and the marginal combined increases in female mice, there was equivocal evidence of carcinogenicity (Api et al., 2016c). In rats, it was not carcinogenic. Similarly,

TABLE 19.2
Repeated Dose Toxicity

	Eugenol	Isoeugenol
Total systemic exposure (mg/kg/day)	0.019	0.0065 Dermal uptake 38.4%
Lowest dose level in mouse NTP study	>23,600 times higher	>11,500 times higher
NOAEL (mg/kg/day)	300	37.5
Margin of Exposure (mg/kg/day)	300/0.019 = 15789	37.5/0.0004 = 93,000

the US NTP reported that isoeugenol is hepatocarcinogenic in male mice at 75 mg/kg/day and equivocally carcinogenic in female mice and male rats (thymoma, mammary gland carcinoma) at 300 mg/kg/day (see Api et al., 2016b).

However, hepatotoxicity might have played a role in the development of the hepatic tumors in B6C3F1 mice, which are known to be sensitive to the development of liver tumors by non-genotoxic mechanisms. Thus, it is assumed that these data are not relevant to human risk, as the increase in the incidence of tumors in male B6C3F1 mice reflects the impact of high-dose liver damage to an organ already prone to spontaneous development of liver neoplasms (Haseman et al., 1986) (Table 19.2).

A margin of exposure (MOE) of >100 is deemed acceptable.

19.10.2.2 Developmental and Reproductive Toxicity

A gavage developmental toxicity study with isoeugenol, conducted in rats, revealed intrauterine growth retardation and delayed skeletal ossification at maternally toxic dosages (George et al., 2001) (Table 19.3). Reproductive toxicity of isoeugenol was confirmed in rats in a gavage multigenerational continuous breeding study based on a decreased number of male pups per litter during the F0 cohabitation and decreased male and female pup weights during the F1 cohabitation (Layton et al., 2001) (Table 19.3). Although estimated values for margin of exposure (MOE) suggest no developmental and reproductive toxicity of isoeugenol.

The developmental toxicity data on eugenol are insufficient. There are no reproductive toxicity data on eugenol. Taking the NOAEL of 230 mg/kg/day from the read across analogue isoeugenol and considering 22.6% absorption from skin contact and 100% from inhalation, a MOE of 12,105 was estimated. As a MOE of >100 is deemed acceptable, the margin of exposure for eugenol is adequate for the developmental and reproductive toxicity endpoints at the current level of use.

19.10.2.3 Skin Sensitization

The available data demonstrate that eugenol is a weak sensitizer with a weight-of-evidence no-expected-sensitization-induction level (WoE NESIL) of 5900 mg/cm^2 (Gerberick et al., 2009).

TABLE 19.3
Developmental Toxicity

	Eugenol	Isoeugenol
Gavage developmental toxicity study	No data	NOAEL: 500 mg/kg/day MOE: 1,250,000
Gavage multigenerational continuous breeding study	No data	NOAEL: 230 mg/kg/day MOE: 575,000

19.10.3 Methyleugenol

Methyleugenol, the methyl ether of eugenol, is a natural constituent of numerous essential oils in very low amounts. However, it is not allowed to be added to cosmetic products and beverages as it has been listed by the National Toxicology Program's Report on Carcinogens as reasonably anticipated to be a human carcinogen. This finding is based on the observation of increased incidence of malignant tumors at multiple tissue sites in experimental animals of different species. High doses of methyleugenol (at least 30 mg/kg body weight for 25 days) cause autoinduction of 1′-hydroxylation by P450 cytochromes, with formation of the proximate carcinogen 1′-hydroxymethyleugenol (Jeurissien et al., 2006). Methyleugenol and its two metabolites 1-hydroxymethyleugenol and 2′,3′-epoxymethyleugenol induced unscheduled DNA synthesis (UDS) *in vitro*. In addition, methyleugenol formed DNA adducts *in vitro* and *in vivo* (Cartus et al., 2012; Herrmann et al., 2013). Sipe et al. (2014) showed that both methyleugenol and eugenol readily undergo peroxidative metabolism *in vitro* to form free radicals. Due to auto-oxidation of methyleugenol, commercial products contain 10–30 mg/L hydroperoxide. Spectroscopic studies show that the hydroperoxide is not a good substrate for catalase, which suggests that glutathione peroxidase may be important in the inhibition of this pathway *in vivo*. Thus, the authors suggest that peroxidase metabolism may contribute to the observed carcinogenicity of methyleugenol in extrahepatic tissues (Sipe et al., 2014).

Since methyleugenol has been demonstrated to be genotoxic and carcinogenic, the existence of a threshold cannot be assumed and the EU Scientific Committee on Food could not establish a safe exposure limit. Consequently, reductions in exposure and restrictions in use levels are indicated (European Commission Opinion of the Scientific Committee on Food, 2001).

19.11 BERGAMOT OIL

The main components of volatile fraction of bergamot oil, a cold-pressed essential oil from the peel of bergamot (*Citrus bergamia* Risso et Poiteau) fruit, are limonene (25%–53%), linalyl acetate (15%–40%), linalool (2%–20%), γ-terpinene, and β-pinene (Sawamura et al., 2006; see Navarra et al., 2015). In the non-volatile fraction, the main compounds are coumarins and furanocoumarins such as bergapten (5-methoxypsoralen) and bergamottin (5-geranyloxypsoralen) (Dugo et al., 2000).

The bergamot essential oil is produced in relatively small quantities and thus it is rather expensive. This makes it particularly subject to adulteration (Schipilliti et al., 2011) either by adding distilled essences of poor quality and low cost, for example of bitter orange and bergamot mint, or by adding mixtures of natural or synthetic terpenes. An increase in limonene accompanied by a decrease in linalool hints at adulteration with cheaper citrus oils. In addition, bergamot oil should not contain more than 1.5% *p*-cymene, which is an indicator for degradation and may cause dry skin and redness of eyes and skin, as well as drowsiness and nausea.

In several studies, application of bergamot oil directly to the skin was shown to have a concentration-dependent phototoxic effect after exposure to UV light, presumably due to bergapten and also bergamottin (Kaddu et al., 2001; Kejlova et al., 2007). Cutaneous lesions developed gradually within 48–72 h when aromatherapy with bergamot oil was applied 2–3 days prior to ultraviolet (UV) exposure (Kaddu et al., 2001). Thus, for applications on areas of skin exposed to sunshine, excluding bath preparations, soaps, and other products which are washed off the skin, bergamot oil should not be used such that the level in the consumer products exceeds 0.4% (International Fragrance Association, IFRA). Cases have become much more rare since the introduction of psoralen-free bergamot oil.

Earl Grey tea is black tea flavored with bergamot essential oil. In one case study, a patient who consumed four liters of Earl Grey tea per day, suffered from paresthesias, fasciculations, and muscle cramps (Finsterer, 2002). The author explained the adverse effects in this patient by the effect of bergapten, which is reported to be an axolemmal potassium channel blocker (Bohuslavizki et al., 1994; Wulff et al., 1998; During et al., 2000). This may lead to hyperexcitability of the axonal membrane, causing fasciculations and muscle cramps.

19.12 ESSENTIAL OILS OF NUTMEG AND OTHER SPICES

Myristicin (methoxysafrole) and elemecin, along with safrole, are present in small amounts in the essential oil of nutmeg and, to an even lesser extent, in other spices such as parsley and dill. The essential oil *Myristicae aetheroleum* is used in the form of balsam, pastilles, drops, and massage oil for the treatment of common cold, digestive disorders, and rheumatism.

The main components myristicin and elemecin are psychoactive compounds with anticholinergic activity causing dry mouth and skin, obstipation, urinary retention, and tachycardia. Intoxication with high concentrations may cause severe hypotension and liver injury, as well as central nervous symptoms such as memory disturbances, visual distortions, excitation, confusion, anxiety, and hallucinations. Although misuse is voluntary, intoxication is invariably accidental. Although the risks of nutmeg intoxication after voluntary use are known, some people still use it for a low-cost recreational-drug alternative (Demetriades et al., 2005).

In case reports (e.g., Abernethy and Becker, 1992; Sangalli and Chiang, 2000; Stein et al., 2001; Kelly and Clarke, 2003; McKenna et al., 2004), intoxications with nutmeg had effects that varied from person to person. With a slow onset of symptoms, the effects persisted for several hours. After-effects even lasted up to several days.

19.13 CONCLUSION

To conclude, in most cases, essential oils are safe in use except in overdosage, wrong application route, and in hypersensitive persons:

- Toxicity of essential oils is concentration-dependent, and thus most unwanted side effects can be avoided by application of low doses and restriction to small skin areas.
- Hypersensitive people should avoid essential oils because allergic reactions occur independently of the applied concentration and route of application.
- Adequate storage of essential oils is mandatory, because oxidation products formed by exposure to light and/or air may play an important role in toxicity.
- Most severe intoxications occur in children following accidental ingestion.

Subjects of special concern are the

- Epileptogenic effect of camphor-, eucalyptol-, and thujone-containing essential oils
- Severe hepatotoxicity of pennyroyal oil due to pulegone and menthofuran, and at high concentrations of clove oil due to eucalyptol
- Salicylate intoxication with wintergreen oil
- Phototoxicity of coumarin- and furocoumarin-containing bergamot oil
- Application in pregnant women, because components of essential oils easily cross the placental barrier and may harm the fetus and neonates
- The use particularly in infants and young children as they are prone to severe intoxication

REFERENCES

Abanses, J.C., S. Arima, and B.K. Rubin. 2009. Vicks VapoRub induces mucin secretion, decreases ciliary beat frequency, and increases tracheal mucus transport in the ferret trachea. *Chest*, 135: 143–148.

Abernethy, M.K., and L.B. Becker. 1992. Acute nutmeg intoxication. *Am. J. Emerg. Med.*, 10: 429–430.

Alpizar, Y.A., M. Gees, A. Sanchez, A. Apetrei, T. Voets, B. Nilius, and K. Talavera. 2013. Bimodal effects of cinnamaldehyde and camphor on mouse TRPA1. *Pflugers Arch.*, 465: 853–864.

Anderson, I.B., W.H. Mullen, J.E. Meeker, S.C. Khojasteh Bakht, S. Oishi, S.D. Nelson, and P.D. Blanc. 1996. Pennyroyal toxicity: Measurement of toxic metabolite levels in two cases and review of the literature. *Ann. Intern. Med.*, 124: 726–734.

Andersson, M., and M. Hindsen. 2007. Rhinitis because of toothpaste and other menthol-containing products. *Allergy*, 62: 336–337.

Api, A.M., D. Belsito, S. Bhatia et al. 2016a. RIFM fragrance ingredient safety assessment, 2-Hydroxy-α,α,4-trimethylcyclohexanemethanol, CAS Registry Number 42822–86–6. *Food Chem. Toxicol.*, 97S: S209–S215.

Api, A.M., D. Belsito, S. Bhatia et al. 2016b. RIFM fragrance ingredient safety assessment, isoeugenol, CAS Registry Number 97–54–1. *Food Chem. Toxicol.*, 97S: S49–S56.

Api, A.M., D. Belsito, S. Bhatia et al. 2016c. RIFM fragrance ingredient safety assessment, eugenol, CAS Registry Number 97–53–0. *Food Chem. Toxicol.*, 97S: S25–S37.

Araujo, I.B., C.A.M. Souza, R.R. De-Carvalho, S.N. Kuriyama, R.P. Rodrigues, R.S. Vollmer, E.N. Alves, and F.J.R. Paumgartten. 1996. Study of the embryo foetotoxicity of a-terpinene in the rat. *Food Chem. Toxicol.*, 34: 477–482.

Arikan-Ayyildiz, Z., F. Akgül, Ş. Yılmaz, D. Özdemir, and N. Uzuner. 2012. Anaphylaxis in an infant caused by menthol-containing cologne. *Allergol. Immunopathol. (Madr)*, 40: 198.

Bakerink, J.A., S.M. Gospe Jr., R.J. Dimand, and M.W. Eldridge. 1996. Multiple organ failure after ingestion of pennyroyal oil from herbal tea in two infants. *Pediatrics*, 98: 944–947.

Barkin, M.E., J.P. Boyd, and S. Cohen. 1984. Acute allergic reaction to eugenol. *Oral Surg. Oral Med. Oral Pathol.*, 57: 441–442.

Bautista, D.M., S.E. Jordt, T. Nikai, P.R. Tsuruda, A.J. Read, J. Poblete, E.N. Yamoah, A.I. Basbaum, and D. Julius. 2006. TRPA1 mediates the inflammatory actions of environmental irritants and proalgesic agents. *Cell*, 124: 1269–1282.

Bazzano, A.N., C. Var, F. Grossman, and R.A. Oberhelman. 2017. Use of camphor and essential oil balms for infants in Cambodia. *J. Trop. Pediatr.*, 63: 65–69.

Bischoff, K., and F. Guale. 1998. Australian tea tree (*Melaleuca alternifolia*) oil poisoning in three purebred cats. *J. Vet. Diagn. Invest.*, 10: 208–210.

Blagojevic, P., N. Radulovic, R. Palic, and G. Stojanovic. 2006. Chemical composition of the essential oils of Serbian wild-growing *Artemisia absinthium* and *Artemisia vulgaris*. *J. Agr. Food Chem.*, 54: 4780–4789.

Blake, K.D., C.R. Fertleman, and M.A. Meates. 1993. Dangers of common cold treatments in children [letter] [published erratum appears in Lancet 1993; 341(8848):842]. *Lancet*, 341: 640.

Bohuslavizki, K.H., W. Hänsel, A. Kneip, E. Koppenhöfer, E. Niemöller, and K. Sanmann. 1994. Mode of action of psoralens, benzofurans, acridinons, and coumarins on the ionic currents in intact myelinated nerve fibres and its significance in demyelinating diseases. *Gen. Physiol. Biophys.*, 13: 309–328.

Burkhard, P.R., K. Burkhardt, C.A. Haenggeli, and T. Landis. 1999. Plant induced seizures: Reappearance of an old problem. *J. Neurol.*, 246: 667–670.

Caceres, A.I., M. Brackmann, M.D. Elia et al. 2009. A sensory neuronal ion channel essential for airway inflammation and hyperreactivity in asthma. *Proc. Natl. Acad. Sci. USA*, 106: 9099–9104.

Cai L., and C.D. Wu. 1996. Compounds from syzygium aromaticum possessing growth inhibitory activity against oral pathogens. *J. Nat. Prod.* 59(10), 987–990.

Carson, C.F., R. Tisser, and T. Larkman. 2014. Lack of evidence that essential oils affect puberty. *Reprod. Toxicol.*, 44: 50–51.

Cartus, A.T., K. Herrmann, L.W. Weishaupt, K.-H. Merz, W. Engst, H. Glatt, and D. Schrenk. 2012. Metabolism of methyleugenol in liver microsomes and primary hepatocytes: Pattern of metabolites, cytotoxicity, and DNA-adduct formation. *Toxicol. Sci.*, 129: 21–34.

Casarin, M., J. Pazinatto, R.C.V. Santos, and F.B. Zanatta. 2018. *Melaleuca alternifolia* and its application against dental plaque and periodontal diseases: A systematic review. *Phytother. Res.*, 32: 230–242.

Chaieb, K., H. Hajlaoui, T. Zmantar, A.B. Kahla-Nakbi, M. Rouabhia, K. Mahdouani, and A. Bakhrouf. 2007. The chemical composition and biological activity of clove essential oil, *Eugenia caryophyllata* (*Syzigium aromaticum* L. Myrtaceae): A short review. *Phytother. Res.*, 21: 501–506.

Chen, L.J., E.H. Lebetkin, and L.T. Burka. 2003. Metabolism of (R)-(+)-menthofuran in Fischer-344 rats: Identification of sulfonic acid metabolites. *Drug Metab. Dispos.*, 31: 1208–1213.

Committee on Drugs. 1978. Camphor: Who needs it? American Academy of Pediatrics. *Pediatrics*, 62: 404–406.

Committee on Drugs. 1994. Camphor revisited. Focus on toxicity. American Academy of Pediatrics. *Pediatrics*, 94: 127–128.

Council of Europe. 2001. Committee of Experts on Flavouring Substances of the Council of Europe. *48th Meeting*, Strasbourg, 2–6 April 2001.

Craig, J.O. 1953. Poisoning by the volatile oils in childhood. *Arch. Dis. Child.*, 28: 475–483.

Culic, M., G. Kekovic, G. Grbic, L. Martac, M. Sokovic, J. Podgorac, and S. Sekulic. 2009. Wavelet and fractal analysis of rat brain activity in seizures evoked by camphor essential oil and 1,8-cineole. *Gen. Physiol. Biophys.*, 28: 33–40.

Daimon, H., S. Sawada, S. Asakura, and F. Sagani. 1998. In vivo genotoxicity and DNA adduct levels in the liver of rats treated with safrole. *Carcinogenesis*, 19: 141–146.

Darben, T., B. Cominos, and C.T. Lee. 1998. Topical eucalyptus oil poisoning. *Australas. J. Dermatol.*, 39: 265–267.

Davis, J.E. 2007. Are one or two dangerous? Methyl salicylate exposure in toddlers. *J. Emerg. Med.*, 32: 63–69.

Del Beccaro, M.A. 1995. Melaleuca oil poisoning in a 17-month-old. *Vet. Hum Toxicol.*, 37: 557–558.

Demetriades, A.K., P.D. Wallman, A. McGuiness, and M.C. Gavalas. 2005. Low cost, high risk: Accidental nutmeg intoxication. *Emerg. Med. J.*, 22: 223–225.

De Souza, M.E., D.J. Clerici, C.M. Verdi et al. 2017. Antimicrobial activity of *Melaleuca alternifolia* nanoparticles in polymicrobial biofilm *in situ*. *Microb. Pathog.*, 113: 432–437.

De Vincenzi, M., M. Silano, A. De Vincenzi, F. Maialetti, and B. Scazzocchio. 2002. Constituents of aromatic plants: Eucalyptol. *Fitoterapia*, 73: 269–275.

Dietz, B., and J.L. Bolton. 2007. Botanical dDietary sSupplements gGone bBad. *Chem. Res. Toxicol.*, 20: 586–590.

Dost, S., and B. Leiber. 1967. *Menthol and Menthol containing External Remedies. Use, Mode of Effect and Tolerance in Children*. Stuttgart: George Thieme Verlag.

Dugo, P., L. Mondello, L. Dugo, R. Stancanelli, and G. Dugo. 2000. LC-MS for the identification of oxygen heterocyclic compounds in citrus essential oils. *J. Pharm. Biomed. Anal.*, 24: 147–150.

Düring, T., F. Gerst, W. Hansel, H. Wulff, and E. Koppenhofer. 2000. Effects of three alkoxypsoralens on voltage gated ion channels in Ranvier nodes. *Gen. Physiol. Biophys.*, 19: 345–364.

Eccles, R. 1994. Menthol and related cooling compounds. *J. Pharm. Pharmacol.*, 46: 618–630.

Ellenhorn, M.J., and D.G. Barceloux. 1988. *Medical Toxicology - Diagnosis and Treatment of Human Poisoning*. New York, NY: Elsevier Science Publishing Co., Inc., p. 562.

Elliott, C. 1993. Tea tree oil poisoning. *Med. J. Aust.*, 159: 830–831.

EMA. 2012. EMA/67070/2012. Assessment report for suppositories containing terpenic derivatives. 20 January 2012. Procedure number: EMEA/H/A-1284 (http://www.ema.europa.eu/docs/en_GB/document_library/Referrals_document/Terpenic_31/WC500122526.pdf)

EMA. 2014. EMA/HMPC/138386/2005. Rev. 1 Committee on Herbal Medicinal Products (HMPC) Public statement on the use of herbal medicinal products containing pulegone and menthofuran. November 24, 2014. (http://www.ema.europa.eu/docs/en_GB/document_library/Public_statement/2014/12/WC500179556.pdf)

Engel, W. 2003. In vivo studies on the metabolism of the monoterpene pulegone in humans using the metabolism of ingestion-correlated amounts (MICA) approach: Explanation for the toxicity differences between (S)-(−) and (R)-(+)-pulegone. *J. Agric. Food Chem.*, 51: 6589–6597.

European Commission Opinion of the Scientific Committee on Food on Methyleugenol (4-Allyl-1,2-dimethoxybenzene). 2001. SCF/CS/FLAV/FLAVOUR/4 ADD1 FINAL (http://europa.eu.int/comm/food/fs/sc/scf/index_en.html)

European Commission Opinion of the Scientific Committee on Food on pulegone and menthofuran. 2002. SCF/CS/FLAV/FLAVOUR/3 ADD2 Final (https://ec.europa.eu/food/sites/food/files/safety/docs/sci-com_scf_out133_en.pdf)

Farco, J.A., and O. Grundmann. 2013. Menthol - Pharmacology of an important naturally medicinal "cool". *Mini Rev. Med. Chem.*, 13: 124–131.

Fennell, T.R., J.A. Miller, and E.C. Miller. 1984. Characterization of the biliary and urinary glutathione and N-acetylcysteine metabolites of the hepatic carcinogen 1′-hydroxysafrole and its 1′-oxo metabolite in rats and mice. *Cancer Res.*, 44: 3231–3240.

Finsterer, J. 2002. Earl Grey tea intoxication. *Lancet*, 359(9316): 1484.

Flaman, Z., S. Pellechia-Clark, B. Bailey, and M. McGuigan. 2001. Unintentional exposure of young children to camphor and eucalyptus oils. *Paediatr. Child. Health*, 6: 80–83.

Ford, M.D., K.A. Delaney, L.J. Ling, and T. Erickson. 2001. *Clinical Toxicology*. Philadelphia, PA: W.B. Saunders Company, p. 346.

Gardiner, P. 2000. Peppermint (Mentha piperita). The Longwood Herbal Task Force and The Center for Holistic Pediatric Education and Research (http://www.mcp.edu/herbal/default.htm)

Gaudioso, C., J. Hao, M.F. Martin-Eauclaire, M. Gabriac, and P. Delmas. 2012. Menthol pain relief through cumulative inactivation of voltage-gated sodium channels. *Pain*, 153: 473–484.

Geller, R.J., D.A. Spyker, L.K. Garretson, and A.D. Rogol. 1984. Camphor toxicity: Development of a triage strategy. *Vet. Hum. Toxicol.*, 26(Suppl 2): 8–10.

George, J.D., C.J. Price, M.C. Marr, C.B. Myers, and G.D. Jahnke. 2001. Evaluation of the developmental toxicity of isoeugenol in Sprague-Dawley (CD) rats. *Toxicol. Sci. Former Fundam. Appl. Toxicol.*, 60: 112–120.

Gerberick G.F., J.A. Troutman, L.M. Foertsch, J.D. Vassallo, M. Quijano, R.L.M. Dobson, C. Goebel, and J-P. Lepoittevin. 2009. Investigation of peptide reactivity of pro-hapten skin snsitizers using a peroxidase-peroxide oxidation system. *Toxico.l Sci.* 112(1), 164–174.
Gleason, M.N., R.E. Gosselin, H.C. Hodge, and R.P. Smith. 1969. *Clinical Toxicology of Commercial Products*, 3rd ed. Baltimore: Williams & Williams Co, pp. 56–57.
Gordon, W.P., A.C. Huitric, C.L. Seth, R.H. McClanahan, and S.D. Nelson. 1987. The metabolism of the abortifacient terpene, (R)-(+)-pulegone, to a proximate toxin, menthofuran. *Drug Metab. Dispos.*, 15: 589–594.
Gouin, S., and H. Patel. 1996. Unusual cause of seizure. *Pediatr. Emerg. Care*, 12: 298–300.
Gurr, F.W. 1965. Scroggie JG: Eucalyptus oil poisoning treated by dialysis and mannitol infusion, with an appendix on the analysis of biological fluids for alcohol and eucalyptol. *Australas. Ann. Med.*, 14: 238–249.
Ha, M.A., G.J. Smith, J.A. Cichocki, L. Fan, Y.S. Liu, A.I. Caceres, S.E. Jordt, and J.B. Morris. 2015. Menthol attenuates respiratory irritation and elevates blood cotinine in cigarette smoke exposed mice. *PLOS ONE*, 10: e0117128.
Halicioglu, O., G. Astarcioglu, I. Yaprak, and H. Aydinlioglu. 2011. Toxicity of *Salvia officinalis* in a newborn and a child: An alarming report. *Pediatr. Neurol.*, 45: 259–260.
Hammer, K.A., C.F. Carson, T.V. Riley, and J.B. Nielsen. 2006. A review of the toxicity of *Melaleuca alternifolia* (tea tree) oil. *Food Chem. Toxicol.*, 44: 616–625.
Hans, M., M. Wilhelm, and D. Swandulla. 2012. Menthol suppresses nicotinic acetylcholine receptor functioning in sensory neurons via allosteric modulation. *Chem. Senses*, 37: 463–469.
Haseman, J.K., J.S. Winbush, and M.W. O'Donnell Jr. 1986. Use of dual control groups to estimate false positive rates in laboratory animal carcinogenicity studies. *Fundam. Appl. Toxicol.*, 7: 573–584.
Hausen, B.M. 2004. Evaluation of the main contact allergens in oxidized tea tree oil. *Dermatitis*, 15: 213–214.
Hausen, B.M., J. Reichling, and M. Harkenthal. 1999. Degradation products of monoterpenes are the sensitizing agents in tea tree oil. *Am. J. Contact Dermat.*, 10: 68–77.
Hawthorn, M., J. Ferrante, E. Luchowski, A. Rutledge, X.Y. Wei, and D.J. Triggle. 1988. The actions of peppermint oil and menthol on calcium channel dependent processes in intestinal, neuronal and cardiac preparations. *Aliment. Pharmacol. Ther.*, 2: 101–118.
Heimes, K., F. Hauk, and E.J. Verspohl. 2011. Mode of action of peppermint oil and (−)-menthol with respect to 5-HT3 receptor subtypes: Binding studies, cation uptake by receptor channels and contraction of isolated rat ileum. *Phytother. Res.*, 25: 702–708.
Henley, D.V., N. Lipson, K.S. Korach, and C.A. Bloch. 2007. Prepubertal gynecomastia linked to lavender and tea tree oils. *N. Engl. J. Med.*, 356: 479–485.
Herrmann, K., F. Schumacher, W. Engst, K.E. Appel, K. Klein, U.M. Zanger, and H. Glatt. 2013. Abundance of DNA adducts of methyleugenol, a rodent hepatocarcinogen, in human liver samples. *Carcinogenesis*, 34: 1025–1030.
Hindle, R.C. 1994. Eucalyptus oil ingestion. *N. Z. Med. J.*, 107: 185–186.
Hold, K.M., N.S. Sirisoma, T. Ikeda, T. Narahashi, and J.E. Casida. 2000. α-Thujone (the active component of absinthe): γ-Aminobutyric acid type A receptor modulation and metabolic detoxification. *Proc. Natl. Acad. Sci. USA*, 97: 3826–3831.
Hu, Z., R. Feng, F. Xiang et al. 2014. Acute and subchronic toxicity as well as evaluation of safety pharmacology of eucalyptus oil-water emulsions. *Int. J. Clin. Exp. Med.*, 7: 4835–4845.
inchem.org (http://www.inchem.org/documents/pims/pharm/camphor.htm#SectionTitle)
Jacobs, M.R., and C.S. Hornfeldt. 1994. Melaleuca oil poisoning. *J. Toxicol. Clin. Toxicol.*, 32: 461–464.
Janes, S.E., C.S. Price, and D. Thomas. 2005. Essential oil poisoning: N-acetylcysteine for eugenol-induced hepatic failure and analysis of a national database. *Eur. J. Pediatr.*, 164: 520–522.
Javorka, K., Z. Tomori, and L. Zavarská. 1980. Protective and defensive airway reflexes in premature infants. *Physiol. Bohemoslov.*, 29: 29–35.
Jeurissien, S.M.F., J.J.P. Bogaards, M.G. Boersma et al. 2006. Human cytochrome P450 enzymes of importance for the bioactivation of methyleugenol to the proximate carcinogen 1′-hydroxymethyleugenol. *Chem. Res. Toxicol.*, 19: 111–116.
Jimenez, J.F., A.L. Brown, W.C. Arnold, and W.J. Byrne. 1983. Chronic Camphor Ingestion Mimicking Reye's Syndrome. *Gastroenterology*, 84: 394–398.
Johnson, P.N. 1985. Methyl salicylate/aspirin equivalence. *Vet. Hum. Toxicol.*, 26: 317–318.
Kaddu, S., H. Kerl, and P. Wolf. 2001. Accidental bullous phototoxic reactions to bergamot aromatherapy oil. *J. Am. Acad. Dermatol.*, 45: 458–461.
Kamatou, G.P., I. Vermaak, A.M. Viljoen, and B.M. Lawrence. 2013. Menthol: A simple monoterpene with remarkable biological properties. *Phytochemistry*, 96: 15–25.

Karashima, Y., N. Damann, J. Prenen, K. Talavera, A. Segal, T. Voets, and B. Nilius. 2007. Bimodal action of menthol on the transient receptor potential channel TRPA1. *J. Neurosci.*, 27: 9874–9884.

Karunakara, B.P., and C.S. Jyotirmanju. 2012. Eucalyptus oil poisoning in children. *J. Pediatr. Sci.*, 4: e132.

Kawane, H. 1996. Menthol and aspirin induced asthma. *Respir. Med.*, 90: 247.

Kejlova, K., D. Jirova, H. Bendova, H. Kandarova, Z. Weidenhoffer, H. Kolarova, and M. Liebsch. 2007. Phototoxicity of bergamot oil assessed by *in vitro* techniques in combination with human patch tests. *Toxicol. In Vitro*, 21: 1298–1303.

Kelly, B.D., and M.M. Clarke. 2003. Nutmeg and psychosis. *Schizophr. Res.*, 60: 95–96.

Khine, H., D. Weiss, N. Graber, R.S. Hoffman, N. Esteban-Cruciani, and J.R. Avner. 2009. A cluster of children with seizures caused by camphor poisoning. *Pediatrics*, 123: 1269–1272.

Khojasteh-Bakht, S.C., L.L. Koenigs, R.M. Peter, W.F. Trager, and S.D. Nelson. 1998. (R)-(+)-Menthofuran is a potent, mechanism-based inactivator of human liver cytochrome P450 2A6. *Drug Metab. Dispos.*, 26: 701–704.

Khojasteh-Bakht, S.C., S.D. Nelson, and W.M. Atkins. 1999. Glutathione S-transferase catalyzes the isomerization of (R)-2-hydroxymenthofuran to mintlactones. *Arch. Biochem. Biophys.*, 370: 59–65.

Kogel, U., B. Titz, W.K. Schlage et al. 2016. Evaluation of the Tobacco Heating System 2.2. Part 7: Systems toxicological assessment of a mentholated version revealed reduced cellular and molecular exposure effects compared with mentholated and non-mentholated cigarette smoke. *Regul. Toxicol. Pharmacol.*, 81(Suppl 2): S123–S138.

Kolassa, N. 2013. Menthol differs from other terpenic essential oil constituents. *Regul. Toxicol. Pharmacol.*, 65: 115–118.

Köppel, C., J. Tenczer, Th. Schirop, and K. Ibe. 1982. Camphor poisoning. Abuse of camphor as a stimulant. *Arch. Toxicol.*, 51: 101–106.

Kumar, K.J., S. Sonnathi, C. Anitha, and M. Santhoshkumar. 2015. Eucalyptus oil poisoning. *Toxicol. Int.*, 22(1): 170–171.

Kumar, P., V.M. Caradonna-Graham, S. Gupta, X. Ci, P.N. Rao, and J. Thompson. 1995. Inhalation challenge effects of perfume scent strips in patients with asthma. *Ann. Allergy Asthma Immunol.*, 75: 429–433.

Layton, K.A., G.W. Wolfe, Y. Wang, J. Bishop, and R.E. Chaping. 2001. Reproductive effects of isoeugenol in Sprague-Dawley rats when assessed by the continuous breeding protocol. *Toxicologist*, 60(1): 384.

Lessenger, J.E. 2001. Occupational acute anaphylactic reaction to assault by perfume spray in the face. *J. Am. Board Fam. Pract.*, 14: 137–140.

Liu, B., L. Fan, S. Balakrishna, A. Sui, J.B. Morris, and S.E. Jordt. 2013. TRPM8 is the principal mediator of menthol-induced analgesia of acute and inflammatory pain. *Pain*, 154(10): 2169–2177.

Lucas, G.N. 2000. Acute drug poisoning in children. *Sri Lanka J. Child Health*, 29: 45–48.

MacDougall, J.M., K. Fandrick, X. Zhang, S.V. Serafin, and J.R. Cashman. 2003. Inhibition of human liver microsomal (S)-nicotine oxidation by (–)-menthol and analogues. *Chem. Res. Toxicol.*, 16: 988–993.

Madyastha, K.M., and B. Moorthy. 1989. Pulegone mediated hepatotoxicity: Evidence for covalent binding of R(+)-[14C]pulegone to microsomal proteins *in vitro*. *Chem. Biol. Interact.*, 72: 325–333.

Madyastha, K.M., and C.P. Raj. 2002. Stereoselective hydroxylation of 4-methyl-2-cyclohexenone in rats: Its relevance to R-(+)-pulegone-mediated hepatotoxicity. *Biochem. Biophys. Res. Commun.*, 297(2): 202–205.

Manoguerra, A.S., A.R. Erdman, P.M. Wax et al. 2006. Camphor Poisoning: An evidence-based practice guideline for out-of-hospital management. *Clin. Toxicol (Phila)*, 44(4): 357–370.

Marlowe, K.F. 2003. Urticaria and asthma exacerbations after ingestion of menthol-containing lozanges. *Am. J. Health Syst. Pharm.*, 60: 1657–1659.

McClanahan, R.H., D. Thomassen, J.T. Slattery, and S.D. Nelson. 1989. Metabolic activation of (R)-(+)-pulegone to a reactive enonal that covalently binds to mouse liver proteins. *Chem. Res. Toxicol.*, 2: 349–355.

McKenna, A., S.P. Nordt, and J. Ryan. 2004. Acute nutmeg poisoning. *Eur. J. Emerg. Med.*, 11(4): 240–241.

Melis, K., A. Bochner, and G. Janssens. 1989. Accidental nasal eucalyptol and menthol instillation. *Eur. J. Pediatr.*, 148: 786–787.

Millet, Y., J. Jouglard, M.D. Steinmetz, P. Tognetti, P. Joanny, and J. Arditti. 1981. Toxicity of some essential plant oils. Clinical and experimental study. *Clin. Toxicol.*, 18: 1485–1498.

Moqrich, A., S.W. Hwang, T.J. Earley, M.J. Petrus, A.N. Murray, K.S. Spencer, M. Andahazy, G.M. Story, and A. Patapoutian. 2005. Impaired thermosensation in mice lacking TRPV3, a heat and camphor sensor in the skin. *Science*, 307(5714): 1468–1472.

Morris, M.C., A. Donoghue, J.A. Markowitz, and K.C. Osterhoudt. 2003. Ingestion of tea tree oil (Melaleuca oil) by a 4-year-old boy. *Pediatr. Emerg. Care*, 19: 169–171.

Morton, C.A., J. Garioch, P. Todd, P.J. Lamey, and A. Forsyth. 1995. Contact sensitivity to menthol and peppermint in patients with intra-oral symptoms. *Contact Dermatitis*, 32: 281–284.

Mozelsio, N.B., K.E. Harris, K.G. McGrath, and L.C. Grammer. 2003. Immediate systemic hypersensitivity reaction associated with topical application of Australian tea tree oil. *Allergy Asthma Proc.*, 24: 73–75.

Nair, B. 2001. Final report on the safety assessment of *Mentha piperita* (peppermint) oil, *Mentha piperita* (peppermint) leaf extract, *Mentha piperita* (peppermint) leaf, and *Mentha piperita* (peppermint) leaf water. *Int. J. Toxicol.*, 20(Suppl 3): 61–73.

Najafi-Taher, R., B. Ghaemi, S. Kharazi, S. Rasoulikoohi, and A. Amani. 2018. Promising antibacterial effects of silver nanoparticle-loaded tea tree oil nanoemulsion: A synergistic combination against resistance threat. *AAPS PharmSciTech*, 19(3): 1133–1140.

Nash, P., S.R. Gould, and D.E. Bernardo. 1986. Peppermint oil does not relieve the pain of irritable bowel syndrome. *Br. J. Clin. Pract.*, 40: 292–293.

Nath, S.S., C. Pandey, and D. Roy. 2012. A near fatal case of high dose peppermint oil ingestion - Lessons learnt. *Indian J. Anaesth.*, 56(6): 582–584.

National Cancer Institute. 1979. Carcinogenesis, Technical Report Series No. 98. Bioassay of dl-menthol for possible carcinogenicity (https://ntp.niehs.nih.gov/ntp/htdocs/lt_rpts/tr098.pdf)

Navarra, M., C. Mannucci, M. Delbò, and G. Calapai. 2015. *Citrus bergamia* essential oil: From basic research to clinical application. *Front. Pharmacol.*, 6: 36.

Nelson, L., N. Lewin, M.A. Howland, R. Hoffman, L. Goldfrank, and N. Flomenbaum. 2011. *Goldfrank's Toxicologic Emergencies*, 9th ed. New York, N.Y.: McGraw-Hill, p. 625.

Noiles, K., and M. Pratt. 2010. Contact dermatitis to Vicks VapoRub. *Dermatitis*, 21(3): 167–169.

Olowe, S.A., and O. Ransome-Kuti. 1980. The risk of jaundice in glucose-6-phosphate dehydrogenase deficient babies exposed to menthol. *Acta Paediatr Scand*, 69: 341–345.

Olsen, R.W. 2000. Absinthe and g-aminobutyric acid receptors. *P. Natl. Acad. Sci. USA*, 97: 4417–4418.

O'Mullane, N.M., P. Joyce, S.V. Kamath, M.K. Tham, and D. Knass. 1982. Adverse CNS effects of menthol-containing olbas oil. *Lancet*, 1(8281): 1121.

Opdyke, D.L.J. 1978. Camphor USP in monographs on fragrance raw materials. *Food Cosmet .Toxicol.*, 16(Suppl 1): 665–1669.

Opinion of the Scientific Committee on Food on eucalyptol. 2002. (https://ec.europa.eu/food/sites/food/files/safety/docs/sci-com_scf_out126_en.pdf)

Opinion of the Scientific Committee on Food on the safety of the presence of safrole. 2002. (https://ec.europa.eu/food/sites/food/files/safety/docs/sci-com_scf_out116_en.pdf)

Oviedo, A., S. Lebrun, U. Kogel et al. 2016. Evaluation of the Tobacco Heating System 2.2. Part 6: 90–day OECD 413 rat inhalation study with systems toxicology endpoints demonstrates reduced exposure effects of a mentholated version compared with mentholated and non-mentholated cigarette smoke. *Regul. Toxicol. Pharmacol.*, 81(Suppl 2): S93–S122.

Paiva, M., S. Piedade, and A. Gaspar. 2010. Toothpaste-induced anaphylaxis caused by mint (Mentha) allergy. *Allergy*, 65: 1201–1202.

Pan, R., Y. Tian, R. Gao, H. Li, X. Zhao, J.E. Barrett, and H. Hu. 2012. Central mechanisms of menthol-induced analgesia. *J. Pharmacol. Exp. Ther.*, 343(3): 661–672.

Parys, B.T. 1983. Chemical burns resulting from contact with peppermint oil mar: A case report. *Burns Incl. Therm Inj.*, 9: 374–375.

Patel, S., and J. Wiggins. 1980. Eucalyptus oil poisoning. *Arch. Dis. Child.*, 55(5): 405–406.

Peier, A.M., A. Moqrich, A.C. Hergarden et al. 2002. A TRP channel that senses cold stimuli and menthol. *Cell*, 108(5): 705–715.

Pereira, T.S., J.R. de Sant'anna, E.L. Silva, A.L. Pinheiro, and M.A. de Castro-Prado. 2014. In vitro genotoxicity of Melaleuca alternifolia essential oil in human lymphocytes. *J. Ethnopharmacol.*, 151(2): 852–857.

Phelan, W.J. 1976. Camphor poisoning: Over-the-counter dangers. *Pediatrics*, 57, 428–431.

Radulović, N.S., M.S. Genčić, N.M. Stojanović, P.J. Randjelović, Z.Z. Stojanović-Radić, and N.I. Stojiljković. 2017. Toxic essential oils. Part V: Behaviour modulating and toxic properties of thujones and thujone-containing essential oils of *Salvia officinalis* L., *Artemisia absinthium* L., *Thuja occidentalis* L. and *Tanacetum vulgare* L. *Food Chem. Toxicol.*, 105: 366–369.

Ragucci, K.R., P.R. Trangmar, J.G. Bigby, and T.D. Detar. 2007. Camphor ingestion in a 10-year-old male. *South. Med. J.*, 100: 204–207.

Rahimi, M., F. Shokri, H. Hassanian-Moghaddam, N. Zamani, A. Pajoumand, and S. Shadnia. 2017. Severe camphor poisoning, a seven-year observational study. *Environ. Toxicol. Pharmacol.*, 52: 8–13.

Rees, W.D., B.K. Evans, and J. Rhodes. 1979. Treating irritable bowel syndrome with peppermint oil. *Br. Med. J.*, 2: 835–836.

Rietjens, I.M., M.J. Martena, M.G. Boersma, W. Spiegelenberg, and G.M. Alink. 2005. Molecular mechanisms of toxicity of important food-borne phytotoxins. *Mol. Nutr. Food Res.*, 49: 131–158.

Ruha, A.M., K.A. Graeme, and A. Field. 2003. Late seizure following ingestion of Vicks VapoRub. *Acad. Emerg. Med.*, 10: 691.

Russell, M. 1999. Toxicology of tea tree oil. In: Southwell, I., and Lowe, R. (Eds.), *Tea Tree: The Genus Melaleuca*. Amsterdam: Harwood Academic Publishers, pp. 191–201.

Sainio, E.L., and L. Kanerva. 1995. Contact allergens in toothpastes and a review of their hypersensitivity. *Contact Dermatitis*, 33: 100–105.

Sangalli, B.C., and W. Chiang. 2000. Toxicology of nutmeg abuse. *J. Toxicol. Clin. Toxicol.*, 38(6): 671–678.

Santos, C.D., and J.C. Cabot. 2015. Persistent effects after camphor ingestion: A case report and literature review. *J. Emerg. Med.*, 48(3): 298–304.

Sarrami, N., M.N. Pemberton, M.H. Thornhill, and E.D. Theaker. 2002. Adverse reactions associated with the use of eugenol in dentistry. *Br. Dent. J.*, 193(5): 257–259.

Sawamura, M., Y. Onishi, J. Ikemoto, N.T.M. Tu, and N.T.L. Phi. 2006. Characteristic odour components of bergamot (*Citrus bergamia* Risso) essential oil. *Flavour Fragr. J.*, 21(4): 609–615.

Schaller, M., and H.C. Korting. 1995. Allergic airborne contact dermatitis from essential oils used in aromatherapy. *Clin. Exp. Dermatol.*, 20(2): 143–145.

Schipilliti, L., P. Dugo, L. Santi, G. Dugo, and L. Mondello. 2011. Authentication of bergamot essential oil by gas-chromatography-combustion-isotope ratio mass spectrometer (GC-C-IRMS). *J. Essent. Oil Res.*, 23: 60–671.

Seawright, A. 1993. Tea tree oil poisoning - comment. *Med. J. Aust.*, 159: 831.

Selescu, T., A.C. Ciobanu, C. Dobre, G. Reid, and A. Babes. 2013. Camphor activates and sensitizes transient receptor potential melastatin 8 (TRPM8) to cooling and icilin. *Chem. Senses*, 38: 563–575.

Seneviratne, M.P., S. Karunarathne, A.H. de Alwis, A.H.N. Fernando, and R. Fernando. 2015. Accidental methyl salicylate poisoning in two adults. *Ceylon Med. J.*, 60: 65.

Sidell, N., M.A. Verity, and E.P. Nord. 1990. Menthol blocks dihydropyridine-insensitive Ca^{2+} channels and induces neurite outgrowth in human neuroblastoma cells. *J. Cell. Physiol.*, 142: 410–419.

Siegel, E., and S. Wason. 1986. Camphor toxicity. *Pediatr. Clin. North Am.*, 33(2): 375–379.

Sipe, H.J. Jr., O.M. Lardinois, and R.P. Mason. 2014. Free radical metabolism of methyleugenol and related compounds. *Chem. Res. Toxicol.*, 27(4): 483–489.

Smith A.G., and G. Margolis. 1954. Camphor poisoning; Anatomical and pharmacological study; report of a fatal case; experimental investigation of protective action of barbiturate. *Am J Pathol*, 30(5), 857–869.

Southwell, I.A., S. Freeman, and D. Rubel. 1997. Skin irritancy of tea tree oil. *J. Essent. Oil Res.*, 9: 47–852.

Spindler, P., and C. Madsen. 1992. Subchronic toxicity study of peppermint oil in rats. *Toxicol. Lett.*, 62(2–3): 215–220.

Spinella, M. 2001. Herbal medicines and epilepsy: The potential for benefit and adverse effects. *Epilepsy Behav.*, 2: 524–532.

Squier, C.A., M.J. Mantz, and P.W. Wertz. 2010. Effect of menthol on the penetration of tobacco carcinogens and nicotine across porcine oral mucosa *ex vivo*. *Nicotine Tob. Res.*, 12(7): 763–767.

Stafstrom, C.E. 2007. Seizures in a 7-month-old child after exposure to the essential plant oil thuja. *Pediatr. Neurol.*, 37: 446–448.

Stein, U., H. Greyer, and H. Hentschel. 2001. Nutmeg (myristicin) poisoning - Report on a fatal case and a series of cases recorded by a poison information centre. *Forensic. Sci. Int.*, 118(1): 87–90.

Swandulla, D., E. Carbone, K. Schäfer, and H.D. Lux. 1987. Effect of menthol on two types of Ca currents in cultured sensory neurons of vertebrates. *Pflugers Arch.*, 409: 52–59.

Sztajnkrycer, M.D., E.J. Otten, G.R. Bond, C.J. Lindsell, and R.J. Goetz. 2003. Mitigation of pennyroyal oil hepatotoxicity in the mouse. *Acad. Emerg. Med.*, 10: 1024–1028.

The EFSA Journal. 2005. 298:1–32. Opinion of the Scientific Panel on Food Additives, Flavourings, Processing Aids and Materials in contact with Foods on a request from the Commission on Pulegone and Menthofuran in flavourings and other food ingredients with flavouring properties. Question number EFSA-Q-2003–119 (http://www.efsa.europa.eu/sites/default/files/scientific_output/files/main_documents/298.pdf)

The EFSA Journal. 2008. 729:1–15. Camphor in flavourings and other food ingredients with flavouring properties Opinion of the Scientific Panel on Food Additives, Flavourings, Processing Aids and Materials in Contact with Food on a request from the Commission (http://www.efsa.europa.eu/sites/default/files/scientific_output/files/main_documents/729.pdf)

Thomassen, D., J.T. Slattery, and S.D. Nelson. 1990. Menthofuran dependent and independent aspects of pulegone hepatotoxicity: Roles of glutathione. *J. Pharmacol. Exp. Ther.*, 253: 567–572.

Thompson, M.F., G.L. Poirier, M.I. Dávila-García et al. 2017. Menthol enhances nicotine-induced locomotor sensitization and *in vivo* functional connectivity in adolescence. *J. Psyhopharmacol.*, 32(3): 332–343.

Tibballs, J. 1995. Clinical effects and management of eucalyptus oil ingestion in infants and young children. *Med. J. Aust.*, 163(4): 177–180.

Turić, P., J. Lipozencić, V. Milavec-Puretić, and S.M. Kulisić. 2011. Contact allergy caused by fragrance mix and *Myroxylon pereirae* (balsam of Peru)--A retrospective study. *Coll. Antropol.*, 35(1): 83–87.

Uc, A., W.P. Bishop, and K.D. Sanders. 2000. Camphor hepatotoxicity. *South. Med. J.*, 93(6): 596–8.

Vilaplana, J., and C. Romaguera. 2000. Allergic contact dermatitis due to eucalyptol in an anti-inflammatory cream. *Contact Dermatitis*, 43(2): 118.

Villar, D., M.J. Knight, S.R. Hansen, and W.B. Buck. 1994. Toxicity of melaleuca oil and related essential oils applied topically on dogs and cats. *Vet. Hum. Toxicol.*, 36: 139–142.

Von Bruchhausen, F., G. Dannhardt, S. Ebel, and A. Frahm. 1993. *Hagers Handbuch der Pharmazeutischen Praxis*. Bd. 7. Stoffe A-D, 5. Auflage, Springer Verlag Berlin Heidelberg, 649–650.

Webb, N.J.A., and W.R. Pitt. 1993. Eucalyptus oil poisoning in childhood: 41 cases in south-east Queensland. *J. Paediatr. Child Health*, 29(5): 368–371.

Weston, C.F. 1987. Anal burning and peppermint oil [letter]. *Postgrad. Med. J.*, 63: 717.

Whitman, B.W., and H. Ghazizadeh. 1994. Eucalyptus oil: Therapeutic and toxic aspects of pharmacology in humans and animals. *J. Paediatr. Child Health*, 30(2): 190–191.

Wilkinson, S.M., and M.H. Beck. 1994. Allergic contact dermatitis from menthol in peppermint. *Contact Dermatitis*, 30: 42–43.

Woolf, A. 1999. Essential oil poisoning. *Clin. Toxicol.*, 37: 721–727.

Wulff, H., H. Rauer, T. Düring, C. Hanselmann, K. Ruff, A. Wrisch, S. Grissmer, and W. Hänsel. 1998. Alkoxypsoralens, novel nonpeptide blockers of Shaker-type K+ channels: Synthesis and photoreactivity. *J. Med. Chem.*, 41(23): 4542–4549.

Wyllie, J.P., and F.W. Alexander. 1994. Nasal instillation of "olbas oil" in an infant. *Arch. Dis. Child.*, 70: 357–358.

Xu, H., N.T. Blair, and D.E. Clapham. 2005. Camphor activates and strongly desensitizes the transient receptor potential vanilloid subtype 1 channel in a vanilloid-independent mechanism. *J. Neurosci.*, 25(39): 8924–8937.

Xu, J., Z.Q. Hu, C. Wang et al. 2014. Acute and subacute toxicity study of 1,8-cineole in mice. *Int. J. Clin. Exp. Pathol.*, 7(4): 1495–1501.

Yosipovitch, G., C. Szolar, X.Y. Hui, and H. Maibach. 1996. Effect of topically applied menthol on thermal, pain and itch sensations and biophysical properties of the skin. *Arch. Dermatol. Res.*, 288(5–6): 245–248.

Zhang, S.Y., and D. Robertson. 2000. A study of tea tree oil ototoxicity. *Audiol. Neurootol.*, 5: 64–68.

Zhang, X.B., P. Jiang, N. Gong, X.L. Hu, D. Fei, Z.Q. Xiong, L. Xu, and T.L. Xu. 2008. A-type GABA receptor as a central target of TRPM8 agonist menthol. *PLOS ONE*, 3: e3386.

20 Adulteration of Essential Oils

Erich Schmidt and Jürgen Wanner

CONTENTS

20.1 INTRODUCTION

20.1.1 General Remarks

Requirements of governmental bodies in the use of natural products increased tremendously in the last years. Not only the directions by the flavor regulation but also the demand of the consumer organizations presses the industry and producers to supply solely natural products, often desired in "bio" quality. The green responsibility for natural products brought synthetic aroma chemicals in a poor light. Regarding the mass market for cosmetic products, genuineness in fragrances is very limited because of the sources of raw material. On the other side, prices for raw materials increase since years for natural flavor and fragrance ingredients. The acceptance of fragrance compounds containing essential oils and natural aroma chemicals is a way to fulfill the consumer demand for safe, effective, and too affordable finished products. But even today with that claims and in spite of fabulous analytical equipment and highly sophisticated methods, many adulterated essential oils are found on the market. The question arises, what are the motivations for such a serious condition? This and above all, the manner of adulteration and the matching analysis methods shall be treated within this chapter.

Essential oils are constituents of around 30,000 species of plants around the world. Only a few of them are used in today's flavor, cosmetic, animal feeding, and pharmaceutical industry as well as in aromatherapy. Observing the product range of producers and dealers, about 250–300 essential oils are offered in varying quantities. Within those oils, 150 can be characterized as important oils in quantity and price.

Consumption of these essential oils worldwide is tremendous. The data published by Lawrence (2009) show a total quantity of more than 120,000 metric tons. A monetary value can barely be calculated, as prices rise and fall over in a year; exchange rates in foreign currency and disposals within the dealer's community falsify the result. Gambling with natural raw materials and the finished essential oil takes place with all the important ones. Comparison of production figures and those of export and import statistic will vary sometimes obviously; discrepancy appears in higher amounts. Reunion Island's National Institute of Statistics and Economic Studies (INSEE, 2008–2009) reports for the year 2002 the low quantity of 0,4 to kg of "vetiver oil Bourbon" exported; the sold quantity in the market was more than ten times higher. From 2003 until today, no vetiver oil is produced on Reunion Island. The essential oils traded were proven pure and natural but were coming from other production areas.* However, "Bourbon" quality generated a much higher price. It must be assumed that the value of essential oils worldwide will be far above $10 billion and this dimension invites to make some extra money out of it.

Screening essential oil qualities from the market with high-end analytical equipment leads to detection of many adulterated essential oils. Alarming is the fact that these adulterations appear not only within the consumer market but also in industry and trade. The questions arise again, what is the explanation for such a behavior and what are the reasons? Adulterations are subject of many publications with high impact; most of them came out in the beginning of the nineties, when new analytical methods were developed and applied. However, adulteration of essential oils began a long time ago. This chapter deals with adulteration starting in history up to the present. High criminal energy must be responsible for such sacrileges. Financial advantages, market shares, monopole status, and sometimes simply a sportive action are only some reasons.

20.2 DEFINITION AND HISTORY

Observing the significant number of publications about essential oils using the Internet, the results show lack of knowledge, finding "extracted" essential oils, distilled ones by "supercritical fluids," distilled Jasmine oil, therapeutic essential oil, and so on. It is remarkable that in scientific publications of serious publishers, there is the same confusion and ignorance about the term "essential oil." Even in regulatory and institutional papers, this term is used falsely and unprofessionally. Essential oils are clearly defined by the International Standard Organization (ISO) in Draft International Standard (DIS) ISO, 9235, 2013 (former 9235.2)—Aromatic natural raw materials—Vocabulary. This DIS war, revised 2013, tightened the strict definition for an essential oil:

> Essential oil: Product obtained from a natural vegetable raw material of plant origin, by steam distillation, by mechanical process from the epicarp of *Citrus* fruits, or by dry distillation, after separation of the aqueous phase—if any—by physical process.

Further it is mentioned that steam distillation can be carried out with or without added water in a still. No change has been made with the note that "the essential oil can undergo physical treatments (e.g., filtrations, decantation, centrifugation) which do not result in any significant change in its composition." This is a tightening of the old definition. Natural raw material also was updated in the definition: "Natural raw material of vegetal, animal or microbiologic origin, as such, obtained by physical, enzymatic or microbiological processes, or obtained by traditional preparation processes (e.g., extraction, distillation, heating, torrefaction, fermentation)." These definitions show clearly the importance of standards to avoid any adulteration during all stages of collection, production, and trade. Terms like "pure or natural essential oils" now are misleading as an essential oil has to fulfill this demand. Other regulatory elements are used for standardization like the pharmacopoeias around the world. Furthermore medicinal authorities and animal feeding product industry settled own regulatory standards. In the field of aromatherapy, no standards are settled although just in that field, the existence of ISO standards or pharmacopoeias as base should be implemented. The

* Based on the authors' private information and experiences.

use of natural cosmetics is reported to result in a higher demand for oils but the quantity available on the market was and is insufficient to satisfy this developing market. Today many aromatic chemicals from "natural" source, produced by enzymatic reaction from sometimes not natural starters by microorganisms, help the producers to fill up that shortage. But are these chemicals really "natural"?

Adulteration is defined as "making impure or inferior by adding foreign substances to something" or "is a legal term meaning that a product fails to meet international, federal or state standards" (The Free Dictionary n.d. [a]). In this context all kinds of adulterations are enclosed like poisonous, economical, and unethical ones. In the United States, FDA regulations for food and cosmetics are the basis of law; in Europe consumer protection and food regulations as well as the cosmetic regulation have to be applied. Adulterations must not always be deliberated with criminal or unethical background. Environmental pollution like herbicides or pesticides cannot be prevented; these are unavoidable "adulterations." Negligently adulteration, basing on the lack of knowledge, still happens today. By selection of the botanical raw material, very often mistakes are done. Another instance is the missing of suitable production facilities and again, insufficient professionalism of the staff working there. The vocabulary for adulteration of essential oils is fanciful: extending, stretching, cutting, bouquetting, and rounding up to even "sophistication."

The history of adulteration reaches back to the turn of nineteenth to twentieth century. It is ongoing hand in hand with the establishment of chemistry and the development of synthetic aromatic chemicals. Already in 1834 it was possible to isolate cinnamaldehyde from cinnamon bark oil and as a result in 1856, it was synthesized. It is worth to know that Otto Wallach, a basic scientist for terpene chemistry at the University of Göttingen, already started in 1884 with the clarification of terpene structures. In 1885 he defined the contribution for the structure of pinenes, limonene, 1,8-cineole, dipentene, terpineol, pulegol, caryophyllene, and cadinene. The development of synthetic aroma chemicals found in essential oils and extracts started with the synthesis of vanillin in 1874 by W. Haarmann and F. Tiemann. Literature gives only a few information about adulteration and the authors" research is going back to publications from 1919. The company Schimmel & Co. in Leipzig, Germany, was founded in 1829 under the name Spahn & Büttner. In 1879 Schimmel was the first company to establish an industrial laboratory for the production of essential oils. Famous well-known scientists like Wallach (Isolation of Muscon) and Eduard Gildemeister (the famous books of Gildemeister and Hofmann) were there at work. In 1909 Schimmel started publishing the famous "Schimmel-Berichte" (Schimmel reports). These reports about progress in research on essential oils, containing composition of essential oils, isolation of chemicals, cultivation of fragrance plants, and adulteration of essential oils, were discussed. In the report from Schimmel-Berichte (April/October, 1919), the following adulterations were reported: "Bergamot oil with phthalates and terpenes, bitter almond oil with raw benzaldehyde, cassia oil with phthalates, cinnamon oil Ceylon with camphor oil, lemon oil with water, lavender oil with phthalates and terpineol, peppermint oil with glycerol esters, menthol, phthalates and spirit, sandalwood oil with benzyl alcohol, star anise oil with fatty oils and rose oil with palmarosa oil and spermaceti."

That points out clearly that already at that time adulteration was omnipresent. Schimmel was the first company to create a synthetic neroli oil on the market. The cause of that was the establishment of a syndicate comprising of the South French growers and, what is more, frost and storm damages during orange flower blooming. In 1920, again in one of the "Schimmel-chronicles" (Schimmel-Bericht, April/May, 1920), the adulteration of lemon oil is reported: "Skilled mingled, with terpenes, sesquiterpenes and citral, the determination with specific gravity and optical rotation are not sufficient to detect adulteration in lemon oil." As a solution for that problem, preparation of as much as possible fractions, tested for the solution in ethanol of 89% Vol., was proposed. Rochussen (1920) reported that several oils were adulterated, as other than the permitted parts of the plant were added. The utilization of turpentine oil was the most used component for adulteration of those essential oils, bearing large amounts of terpenes. Mineral oils (paraffin, petrolatum, benzene, and Vaseline [petrolatum]) were reported to adjust the change of physical data forced by other adulteration chemicals. For the same purpose cedarwood, gurjun balsam, and copaiba balsam oils were applied because of their weak odor.

At that time, physical constants like optical rotation, density, refraction, boiling point, freezing point, residue on evaporation, ester value, acid value, carbonyl value, and water content could give satisfying data about the examined oils. In addition the so-called wet chemistry could give some more answers. Detection reaction, titration, photometry, and gravimetric analysis were used to achieve a more precise result about the genuineness of an essential oil (detailed information will follow). Bovill (2000) published in Perfumer & Flavorist an original text of E.W. Bovill, RC Treatt, London, about the essential oil market in 1934. One key sentence was as follows: "Essential oils, largely because they are liquids, are easy to adulterate in ways which are difficult to detect and – it must be admitted they very frequently adulterated." At that time it was obvious that adulteration was a fight likely as between a thief and a policeman. Previously, adulteration was done by producers and brokers and customers had to believe the words about purity and nativeness. The only key of trustfulness could be the relationship between producers/traders and clients by permanent control in the harvest and production time locally. Ohloff (1990) confirmed that through hydrodistillation in industrial scale, progress was made in synthesizing compounds with effective methods. Camphor was one of the important chemical components (not only for the flavor and fragrance use but as starting substance for important "plastic" products), together with borneol. Starting with the synthesis of important aromatic chemicals at accessible prices like linalool, geraniol, phenylethanol, and the esters, the adulteration of essential oils started progressively. In a flash, essential oil components were available in sufficient quantities and at prices much cheaper than the essential oil itself. Another event was the problem of the two world wars. Especially in World War II, the acquisition of raw materials of natural or synthetic kind was no more possible at all. The military machine used nearly all raw material resources, chemists were forced to establish substitutes for essential oils, and that resulted in several new synthetic aroma chemicals, which now could be used to stretch or to substitute an essential oil. Thus, the ingenuity of corruption by essential oil producers increased rapidly, and even with general application of chromatographic analysis in industrial laboratories beginning in the 1980s, falsifications could hardly be detected. The authors noticed in the course of a visit to Provence area and producers of lavender and lavandin oil that starting with the analysis by gas chromatography using packed columns in 1979 led to the result that compositions of samples of essential oils were not adequate to literature of that time. By informing the producers about the establishment of chromatographic equipment, the quality was immediately rising. But not all was satisfying. So a "touristic" tour to the facilities of the producers was decided during harvest and production time, end of August 1981. Starting in Puimoisson and with the help of a little blue empty perfume bottle ("flacon montre," used at that time for lavender perfumes), the staff was asked for a little bit of lavender oil. By giving them some Francs, he had the opportunity to watch around the distillation facility. Together with the oil sample, information about special observations was noticed. The same happened in facilities of Riez, Montguers, Simiane, Richerenches, Rosans, Remuzat, Banon, and Apt, and from all production units oil samples could be collected and these were analyzed in the laboratory by gas chromatography. As a result, only one from 10 samples was in accordance with the requirements. The others were adulterated mainly by synthetic linalool and linalyl acetate (recognized by dihydro- and dehydrolinalool as well as dihydro- and dehydrolinalyl acetate) but also with too high contents of limonene, borneol, and camphor (by adding lavandin oil), far too little amounts of lavandulol and lavandulyl acetate. Adulteration by adding those chemicals after distillation could not have happened, as the samples he received were filled directly from the Florentine flask. Finally he was in luck to observe that before closing the cover of the plant, linalyl acetate was spread by means of a portable pump with tank directly on the lavender. The cover was closed and the essential oil now running from the Florentine flask could be seriously called as natural. Interesting was the observation of the distillation facilities. In four cases drums with linalool and linalyl acetate from BASF (Badische Anilin- & Soda-Fabrik) could be detected. Concerning linalool and linalyl acetate, it is worth to mention that 1 year later other drums were occurring from a producer in Bulgaria produced by another synthesis pathway without de- and dihydro components.* At that time it was a customary practice, as reported by Touche et al. (1981), to use relationships of *cis*-β-ocimene to *trans*-β-ocimene (R1) and *trans*-β-ocimene to octanone-3 (R2) as well as linalool + linalyl acetate to lavandulol + lavandulyl acetate (R3). Statistically collected values over many years guaranteed the correct relations.

Another observation was the use of lavender bushels together with some lavandin bushels. This could be transparently seen; they were larger in size and the leaves were characteristic of Lavandin Grosso. At least, there were other chemicals in a rack, such as camphor and geraniol, borneol, terpinen-4-ol, and β-caryophyllene. What should these chemicals be for? He was told that these are for special clients that are so-called comunelles, meaning a mixture of several charges from different farmers but from the same plant was established. The last production unit had constrained not only a laboratory but also a production unit with many chemicals. From that time on the author did no more trust any producer or trade companies.*

Today such a way of working is no more possible. Ordering an essential oil means, with all possible parameters, the properties are scanned and tested and standards must be fulfilled. In spite of all loyalty of a producer/customer relationship, only analytical, physical, and sensory examination is inevitable.

20.3 ADULTERATION

20.3.1 Unintended Adulteration

The term "adulteration" must be understood as a negative concept. It is in nearly all cases a criminal and unethical behavior. But there will be the possibility that it is an act of unwanted adulteration without any intent and has nothing to do with dishonesty. Reasons can be the lack of knowledge, bad equipment, wrong treatment of the plant material, not allowed methods for distillation and no good manufacturing practice. Of course, all these things should be avoided, but they happen.

Selection of correct plant material is the basic for the production of essential oils. It is clearly confirmed in every ISO standard, from which botanical source the oil has to be produced. When a cultivation of the biomass is possible, it must be guaranteed that the correct species is used. By using wild collected plants, the possible risk of collecting similar species or chemotypes cannot be excluded. As an example, essential oil of *Thymus serpyllum* L. from Turkey origin was sent for analysis. As a result, the oil was absolutely different from experience and literature. Thymol and carvacrol were too low, and linalool was higher as well as geraniol and geranyl acetate: What could have happened? People collecting the wild thyme took everything that look like a thyme plant without looking for the chemotypes (which is hardly possible by visual selection). Higher quantities of linalool and geraniol chemotypes were present. The oil was natural at all, but it is not to be used in flavor application or aromatherapy. A further example is the petitgrain oil: The petitgrain from Paraguay is not a bitter orange plant as such. *Citrus aurantium* L. is the true bitter orange plant used for production of the essential oil from peel, but also from leaves to obtain petitgrain oil "bigarade." The oil from Paraguayan plants is derived from *Citrus sinensis* L. Pers. × *Citrus aurantium* L. ssp. *amara* var. *pumila*. This is a hybrid from a sweet and bitter orange. The composition of that oil is different from the first because of its higher content of limonene, *trans*-β-ocimene, and mainly linalyl acetate (ISO/DIS, 8901 and ISO/DIS, 3064 2015). The Paraguayan oil was sold as bitter orange oil; however, curiously, the oil of petitgrain from Paraguay was much cheaper. A further mistake can be the inappropriate treatment of the biomass. Fresh distilled plant material results in a different final product than the oil from dried or fermented material. The question arises, is it a desired procedure or a mistake or an omission? Good manufacturing practice (GMP) is a standard application procedure in technologically developed countries, but what about emerging nations or those without any technical progress? Can such essential oils be traded or used?

* Based on the authors' private information and experiences.

Hydrodistillation is nowadays a highly developed, automated and computerized method to produce essential oils. Applying computer programs for detection- and production parameters observing all conditions. Starting with the plant material (moisture content, degree of maturity, oil content) followed by temperature, steam quantity, steam pressure, cooling temperature, and subsequent treatment, all is handled electronically. In spite of such a technology, mistakes can happen as in the case of distilling *Melissa officinalis* L. from a biological culture. When analyzing that oil by gas chromatography–mass spectrometry (GC-MS), a content of more than 5% of thymol was detected and confirmed three times. After discussing the problem with the producer, he guaranteed that only balm, fresh cut, was used. Contamination in the analyzed sample could be excluded as the second sample showed identical results. After checking all distillation equipment, the reason was found. Before producing Melissa oil, thyme was distilled and he discovered a small chamber within his own constructed cooling unit. There remained about 50–60 mL of oil and it was mixed up with balm oil in the course of hydrodistillation. This is a great financial loss for the producer taking account of the price.* Adulteration can also be done by, for example, adding basic solution to the vessel while distilling. Sometimes it is used to avoid artifacts formation from the plant acids. By raw materials with high-grade acids, for example, barks like massoia, this is sometimes applied. The neutralization of the pH value inhibits the ester formation or degradation reactions in the course of the processing. The use of cohobation is another fact for discussion. This method uses the recovery of volatiles from distillation of water, remaining solved and without separation by Florentine flask. As long as this water is returned to the distillation vessel, there is no problem at all. For some oils like thyme oil, it is of importance to use cohobation, because of the high content of phenols, which dissolve not insignificantly in distilled water. Venskutonis (2002) reports that "cohobation minimizes the loss of oxygenated compounds but increases the risk of hydrolysis and degradation." Similar is the situation with rose oil. Here the cohobation is an important step to receive a genuine reflection of rose odor. Treating the distillation water with salt (NaCl) or pH adjusting to remove plant acids seems to be discussed, as it will still remain an essential oil. By extracting the distillation water with any solvents, which then will be removed by distillation in vacuum, will lead to an extract. This extract, added to the distilled essential oil will result in the loss of the term "essential oil" by definition.! Codistillation is reported for several "essential oils." Mainly sandalwood oils are mentioned to be extracted and redistilled. Valder (2003) reported that the production method of sandalwood oil is sometimes the hexane extraction followed by codistillation with propylene glycol and finally a rectification is done. Baldovini (2010) stated that this method is not used anymore as it does not longer comply with ISO rules. Rajeswara (2007) tested the effect of codistillation using crop weed with several plant materials like citronella, lemongrass, palmarosa, *Corymbia citriodora*, and basil. He observed the increase of yields and also changes in composition. The author was able to observe a similar effect, analyzing bio lavender oil. The oil differed in composition from standards with higher ester content, higher oxidation compounds, and less 3-octanone. The cultivation was bio, without removing any weed. The percentage of weed was esteemed to be more than 30%. The oil did not match ISO standard.* Contaminated containers for production and transport are another possibility of adulteration. Reused drums, not cleaned, are often the cause for worst damage of the essential oil. Plastic containers contain phthalates as plasticizers and the hydrocarbons solve these. Phthalates contaminate even in smallest traces the oil for the use in natural finished products.

20.3.2 Intentional Adulteration

If anybody adulterates willfully and knowingly, it is a criminal act and never a wangle or minor offense. The target to cheat a customer by supplying an essential oil being not conform to any standard is a felony. There is no excuse for such an unethical behavior. Again, until now, adulteration can be found on essential oil market. Terms like cutting, stretching, blending, bouquetting, or sophistication try to moderate this act. The causes for adulteration are manifold and often related with economical and environmental reasons.

* Based on the authors' private information and experiences.

20.3.3 Prices

The cutting of essential oil happens in most cases because of the price. They are subject to falling and increasing prices, depending on crops and the demand but also speculation. The last is often the reason why essential oils come to market with adulteration in every dimension. All economical ambition is the increase of profit, the more the better to fulfill shareholders demand. Essential oil market makes no distinction as it is the same as the law of supply and demand. Speculation was always a reason to ameliorate the process. By hoarding raw materials, the profit can go up in high levels. Lavender oil is an example for that. Most of the farmers in France keep parts of the production in the cellars to wait as far as the new crop starts, until the price seems to be the best. Another example from the authors' experience was the competitive battle with cornmint oil in the 1980s. China and Brazil were the two countries with the highest production at that time. Both increased the cultivation area with the result that in the next season prices fall down tremendously because of oversupply. So Brazil decided to decrease the cultivations and in the next year, prices were going up like a rocket, as China too had decreased production, and demand of the market could not be fulfilled. At that time, most menthol was produced from cornmint and consumption was rising. Stabilization was achieved by supplement of sufficient synthetic menthol and India coming to the market as third production continent. Nowadays speculation has another dimension. Likely nearly all foodstuffs and plant crops are merchandized at the stocks and of course this has influence on the oil market. Increased prices in essential oil market cannot be transferred simply onto the industry of finished products. Mass market and global players will not accept any price advance, so the only solution is cutting or stretching. This situation leads to a competitive market, where adulterated essential oils with certificate of naturalness are sold.

20.3.4 Availability

The market changes from monopole situation for suppliers to buyers market within a few months. Shortage of biomass leads to a rapid increase of prices. The causes will be dryness in the production area, flood during harvest time, and ice and frost like in the case of citrus fruits in California or Florida. Tempests are crucial for destruction of stretches of land as well as deletion caused by insects or microbes. Annual rainfalls in tropic regions and lack of sunshine in Mediterranean region are reasons for partial failure. Unavailability or decreasing availability is a high risk for global players in the fine perfumery industry. The costs of launching a perfume worldwide exceed the amount of $100 million easily and one can imagine that the lack of an important natural raw material and therefore, unavailability of that perfume can lead to ruinous losses. In consequence of that, natural essential oils and also extracts are no more or only in small quantities added to the perfume compounds. It is easy to replace a natural product by a compound. Lemon oil, bergamot oil, rose oil, lavender oil, and many others can be substituted by basic synthetic compositions. Only some special oils like vetiver, sandalwood, patchouli, and orris are never perfect to compensate.

20.3.5 Demand of Clients

The various fields of applications sometimes cause the client industries to require special essential oils. One of the reasons is of course the price. The end-consumer industry settles the prices for the market. Competitive prices have to be adjusted and the problem for the use of essential oils is that prices can change from crop to crop or even from week to week.

20.3.6 Regulations

A huge number of regulations have been established for the protection of consumers worldwide. Cosmetic regulation, flavor regulation, and REACH (*R*egistration, *E*valuation, *A*uthorisation and

Restriction of Chemicals; the European chemical law) extend to the Globally Harmonized System of Classification and Labeling of Chemicals (GHS). All these provisions cover essential oils, but not per se. They are dealt with their contents. Most oils contain one or more chemicals, classified as toxic, allergic, sensitizing, or carcinogenic. Consequently these oils are no more or only in very limited quantities applicable. As an example, the methyl eugenol must be mentioned. Methyl eugenol is a native ingredient of distilled rose oil. Between 1.5% and 3.5% is found in this essential oil. Considering the European Cosmetic Regulation, the maximum dose in finished leave-on products might not exceed 0.001%, which results in the fact that rose oil is no more applicable in cosmetic products. The same pertains to pimento oil from leaves and berries, oil of myrtle, magnolia flower oil, laurel leaf oil, and bay oil. The demand of the industry is rose oil with nearly no or only traces of methyl eugenol. Applying various methods like fractioning and remixing and chemical reactions it was tempted to fulfill the regulation. The result was not as desired, as the odor was not a rose anymore and the finished product could not be named as "essential oil" or "natural."

To avoid an increase of cases of allergic contact dermatitis in the public, many essential oils nowadays have to labeled as "sensitizing—can cause allergic reaction"! Cosmetic producers as well as discounters with private labels avoid naming and labeling and thus, force the replacement of natural products. Essential oil producers offer natural oils with reduced values of sensitizers and this is not possible. Maybe the essential oils are treated in some way, but then are no more "natural," according to ISO standard.

20.3.7 Aging

It is impossible to characterize aging as adulteration without intent. Changes over storing, either correct or false, always result in quantitative modification of the composition. The formation of oxidative compounds like epoxides, peroxides, and hyperoxides is a dangerous effect. Terpenoids show the properties to be volatile but also to be thermolabile. The mentioned oxidative substances show the negative reaction activities coming in contact with the skin. Sensitizing and allergic reactions are consecutive. Countless components of essential oils are subject to oxidation reaction with atmospheric oxygen. Citrus oils implicate such chemicals. Dugo and Mondello (2011) confirm that the reaction of degradation of sabinene, limonene, γ-terpinene, neral, and geranial in lemon oil results in the enlargement of p-cymene, *cis*-8,9-limonene oxide, (*E*)-dihydro carvone, (*Z*)-carveol, 2,3-epoxy geraniol, and carvone. Particularly the increase of p-cymene in citrus essential oils is a distinct detection for heterodyning and aging. Sawamura et al. (2004) demonstrated in a poster compositional changes in commercial mandarin essential oil and detected in a 12-month test the decrease of limonene (60%) and myrcene (nearly 100%) and the increase of (*Z*)-carveol (from 0.0% to 2.8%) and carvones (from 0.2% to 3.0%). Brophy et al. (1989) could observe degradation in terpinen-4-ol, γ-terpinene, and β-pinene but rise of p-cymene and α-terpineol. Examination of peroxide value seems not to be the suitable tool to recognize aging of essential oils, as some of the reactions are reversible. Eucalyptus oil, rosemary oil, and of course citrus oils have demonstrated that warming of these oils in a closed system at 40°C reduces obviously the peroxide value. The most suitable method of recognizing aging is analysis by GC-MS. It has to be the responsibility of producers and traders to take care for the safe quality of their essential oils.

20.3.8 Cupidity

This of course is the main reason for adulteration. Cupidity is defined as "strong desire, especially for possessions or money; greed" (The Free Dictionary n.d. [b]). Although this is the main reason, here is not the place to discuss anything about cupidity as it is human fault.

20.3.9 Simple Sports?

The knowledge about absence of qualified analytical equipment on the side of the client seduces producers as well as the trade to cut essential oils. With the development of new chromatographic methods and an increase in sensitivity as well as selectivity, the inventiveness of the producers shows high wealth of ideas. Again, the race between a thief and a police could have been stopped by the costs of investments in analytical systems, but there seems no end by observing bad qualities on the market.

20.4 POSSIBLE ADULTERATIONS FOR ESSENTIAL OILS

20.4.1 Water

Adding of water is very simple and cheap. This is not possible for all essential oils but is likely for those possessing compounds with high affinity of binding to water. Conifer varieties and citrus oils are examples. Siberian pine (*Abies sibirica* Ledeb.) was supplied in August at 25°C–30°C. Visually no water could be detected. In January the oil came from stock (−5°C) and the quantity of water could be esteemed to 8%. Is that adulteration or natural behavior? Rajeswara Rao et al. (2002) mentioned water contents up to 20%, but that level seems too high by following GMP. Responsible for this effect are monoterpenes, and like conifer oils, citrus oils contain higher quantities of these compounds. Citrus expression techniques use a lot of water to spray away the oil/wax emulsion. Centrifugation will separate water and waxes from essential oil. Unfortunately also aldehydes and alcohols contained in the water/wax phase are removed. Cotroneo et al. (1987) observed this effect by comparing oils from manual sponge method without water and the industrial technical methods. An easy method to detect higher value of water in citrus oil is the visible cloudiness and the deposit of waxes when cooled down to 10°C. A validated method to detect the water content in essential oils is the Karl Fischer titration (ISO, 11021, 1999).

20.4.2 Ethanol

Mostafa (1990) reports that ethanol is the main alcohol used in moderate quantities to dilute essential oils. Ethanol is a component of rose oil. The process of water/steam distillation forms alcohol in certain quantities. In the chromatographic profile in ISO, 9842, 2003, essential oil of rose, the value of ethanol is specified from 0% to 7%, depending on the origin (maxima: Bulgaria 2%, Turkey 7%, and Morocco 3%). The question arises whether the alcohol is coming directly from the process or if the distilled water is washed and extracted by alcohol and this extract is added to the essential oil. If it is the case it is no more an essential oil. Rose oil is a very expensive product and 1% or 2% of added ethanol increases the profit.

20.4.3 Fatty Oils or Mineral Oils

As mentioned in the introduction, adulteration with fatty oils in history is well known. Rochussen (1920) confirms that and mentions paraffin, petrolatum, and castor oil. The easiest way to detect was the blotting paper test. Essential oils containing fatty oils leave behind a lasting grease spot. By testing the solubility in ethanol of 90%, a blur occurs, showing the presence of fatty oils. Only castor oil is soluble in ethanol but can be detected by evaluation of nonvolatile residue content method. Since more than three decades, such an adulteration became no more apparent.

20.4.4 High Boiling Glycols

For a long time such adulterations remained undiscovered. High boiling materials like polyethylene glycols could not be detected with normal GC-MS method, even at 280°C column temperature. It takes many hours before these chemicals leave the column and give a long, small hill in chromatogram. Not only expensive essential oils like sandalwood, vetiver, and orris were cut with those chemicals but also extracts like rose absolute. By happenstance the author found the stretching of this rose absolute. Typically this product is of deep red color and its use in white cosmetic cream was impossible. A high vacuum distillation was performed to remove the color, which really seemed to be not a problem. The end temperature was 200°C in high vacuum but still about 20% remained in the retort. Never a component of rose absolute with that boiling point was detected or mentioned in the literature before.* At least in, 1974, Peyron et al. reported about the use of hexylene glycol up to 40% in vetiver and cananga oils (Peyron et al., 1974). By treatment with water, separation by rectification, and liquid gas chromatography, it could be detected. In 1991, John et al. reported the thin-layer chromatography method for detection. This method is simple and effective and can be used for routine analysis.

20.4.5 Oils from Other Parts of the Same Species or Other Species with Similar Essential Oil Composition

This is a very simple method to stretch. Clove leaves are easier to harvest and bring similar oil as the stem or bud oil. Clove bud oil is nearly three times higher in price than clove leaf oil. Therewith the concocting of that oil makes sense. Pimento berry oil and leaf oil are closely related in composition. The price of the berry oil is double the leaf oil; consequently, adulteration is interesting. By GC-MS method and regarding minor components, this immixture will be detected. Eucalyptus oil from China is derived from *Cinnamomum longepaniculatum* (Gamble) N. Chao ex H.W. Li 1,8-cineole type. Either the whole oil sold as eucalyptus oil or *Eucalyptus* sp. added is clearly not allowed and must be seen as adulteration.

20.4.6 Related Botanical Species

Something of the kind can only happen in the origin, where biomass is growing and distillation is proceeded. Some plants have relatives, which then might have comparable oil compositions. Ylang oil is produced from the flowers of *Cananga odorata* (Lam.) Hook. F. Thomson forma *genuina*. The other form of *Cananga odorata* is cananga oil with *Cananga odorata* (Lam.) Hook. F. Thomson forma *macrophylla* as plant source. On the market, cananga oil is cheaper than ylang oil. Instead of mixing the distilled oils of both, the flowers are distilled together. Furthermore flowers of climbing ylang-ylang *Artabotrys uncinatus* (Lam.) Merill are sometimes added before distillation. When producing patchouli oil, the vessel is filled with leaves, and to avoid that they stick together during processing, branches of the gurjun tree (*Dipterocarpus alatus* Roxb. Ex G. Don and *Dipterocarpus turbinatus* C.F. Gaertn.) are added to avoid that. As a result, the oil of patchouli is contaminated with α-Gurjunene (up to 3%),which is not part of a pure patchouli oil (private information to the author).

20.4.7 Fractions of Essential Oils

This is one of the most applied adulteration methods. Fractions arise in all cases, when essential oils or extracts are concentrated, washed out, and rectified or are residues of removal from centrifugation and distillation as well as from recovery of water streams of expressed citrus peels. Heads and tails from rectification are added to similar essential oils, like peppermint terpenes to mint oil. Essential oils that are high in terpene content, cheap, and available on the market are citrus, eucalyptus, *Litsea cubeba*, cornmint and mint, petitgrain, spearmint, vetiver, lavender and lavandin, cedarwood, citronella, clove, and *Corymbia citriodora* terpenes. Especially citrus terpenes from various species are appropriate for stretching. Dugo (1993) recognized the contamination of bitter orange oil by the use of sweet orange and lemon oils and its terpenes. By applying liquid chromatography (LC), high-resolution gas chromatography (HRGC), and GC-MS methods, they confirmed the addition of less than 3% of these components. Limonene is the main component of many citrus oils. On the other side, limonene is a component of nearly all essential oils in variable quantities. Limonene is a chiral

* Based on the authors' private information.

terpene; chirality describes a basic property of nature. Some molecules show the property to have three-dimensional structures that possess a chiral center. These molecules show a basic property of nature, to look like image and mirror image, but both cannot be aligned. It is a matter of common knowledge that chemical compounds in essential oils tend to be chiral and the preference of one form can be observed. In contrary synthetic molecules are racemic, meaning both forms of the molecule are balanced. This is a real fingerprint for detection of adulterations with terpenes owing different chiral properties. On normal GC columns a separation of chiral molecules is impossible. With the use of columns coated with modified cyclodextrins, those mirror-imaged molecules will be separated and appear as two different signals. In 1995 bergamot essential oils, sold in 10 mL bottles to end users, were subject for investigation of the German journal ÖKO Test (ÖKO Test, 1995). This magazine, dealing with naturality and safety of consumer products, charged the University of Frankfurt, Prof. Mosandl, to check the authenticity of the oils by chiral separation of linalool and linalyl acetate. Thirty-seven of fifty-three samples could be confirmed not being natural and "pure." In a private letter of Prof. König (University of Hamburg) to the author, he confirmed: "Linalool and linalyl acetate are present in bergamot oil only in pure R-enantiomers. That for bergamot oil—unnatural—(*S*)-linalyl acetate cannot be found even in traces." The invention of this technique by König, Mosandl, Casabianca, and Dugo, improved by Mondello, Bicchi, and Rubiolo, was a huge step to diminish the cases of adulterations in essential oils. Limonene occurs in D-, L-, and DL-form in nature. Limonene in sweet orange oil is always in D-form; most of the pine oils show pure L-forms. For the detection of stretching, other terpenes and terpene alcohols are used like α- and β-thujene, α- and β-pinene, α- and β-camphene, α- and β-sabinene, δ3-carene, linalool, β-borneol, myrtenol, linalyl acetate, menthol, *cis*- and *trans*-terpineol, 1-phenylethanol, nerolidol, terpinen-4-ol, and many more. Whenever the chiral distribution differs from normal proportion, a manual intervention was made. However, the last sentence has to be relativized carefully. Chiral composition of camphor and borneol in natural essential oil of rosemary is very variable, depending on the origin! Another method of stretching is the use of distilled residues of expressed citrus peels. After citrus production process, some quantities of essential oil components still can be found in the "waste products." In addition, Reeve et al. (2002) report of the use of distilled mandarin oil, produced from recovered water streams from spiking and pressing. Because of the contact with acidic juice, the composition of such an oil shows significant difference in composition from peel oil. This oil can be sold as "distilled" mandarin oil, but never be used to be blended with pressed oils. Such a handling must be seen as adulteration and is not covered by ISO definitions.

20.4.8 Natural Isolates

Single chemicals can be derived from essential oils by methods of fractioning and rectification. As this procedure is from a natural source, of course, this chemical must be characterized as natural. Numerous components are offered on the market like citral from *Litsea cubeba* oil, geraniol from palmarosa oil, linalool from ho oil, coriander oil or lavandin, pinenes from different *Pinus* species, citronellal from *Corymbia citriodora*, cedrol from cedarwood, or even the santalols from different sandalwood species. All these are added to "finish" essential oils. As already mentioned before, synthetic chemicals can no more be applied as enantiomeric separation is a state of the art today and will convict the matter of fact of adulteration. By using synthesized, correct chiral compounds, the detection is hardly to be recognized but with NMR method, but this is an expensive analysis.

20.4.9 Chemically Derived Synthetic Compounds, Which Are Proved to Appear in Nature

Compounds formerly named "nature identical" are to be defined by law since some years as synthetic products. These molecules have been found in nature by analytical proof and are published in an authorized scientific journal. The term "nature identical" is no more valid and allowed in Europe in relation to flavor and fragrance substances. Such molecules are identical with those appearing in nature but are produced by a synthetic process. These processes contain undesired by-products. The use of such synthetic compounds is easy to detect, as by-products from manufacturing can easily be detected by GC-MS systems. On the other hand, chiral separation will help to confirm adulteration.

20.4.10 Steam Distilled Residues from Expression

After an expression process of citrus oils either in exhausted peels or in centrifugation residues, carryovers of the volatiles will remain. These can be removed by distillation with high-strung steam, and oils acquired are colorless and still have the smell of the starting material. Components are similar those of the expressed oils but contain for reason of the production method some oxidized chemicals. Nevertheless these oils are used to adulterate the expressed oils, and with that process the naturality per definition by ISO standard is lost. The main ingredient is limonene, followed by myrcene and γ-terpinene. Traces of aldehydes still can be found. Such adding can be detected by observing higher values of oxidation compounds.

20.4.11 Enzymatically Produced Chemicals (Natural by Law)

Enzyme is defined as "Any of numerous proteins or conjugated proteins produced by living organisms and functioning as biochemical catalysts" (The Free Dictionary n.d. [c]). Enzymes of microorganisms in a medium with added nutrients produced such molecules. Isolation has to be done by any physical process to isolate the desired molecule designed by the biosystem. Although this manufacturing process is authorized within the applicable legal requirements for the use of such enzymes in the EU, it is somewhat uncertain. These processes will as a rule not happen in nature. Within a plant, production of hydrocarbons takes place without microorganisms by conversion of a starting material and by enzymatic reaction within a cell system. The question arises if that molecule from microorganisms can really be named as natural, particularly if it is generated from "any" starters. In the authors" mind, adding such a compound must also be named as adulteration within the definition of essential oil, which does not allow such a process and any additives.

At least it must be mentioned that the mixture of essential oils from the same species but from different geographical sources cannot be called an adulteration. They fulfill the requirements of standards as long as the specific provenance is not laid down in the specification. Lavender oil from France can be mixed up with Chinese, Bulgarian, and Russian origin and is still lavender oil (*Lavandula angustifolia* Mill.), but may not be sold as "French lavender." Furthermore if an essential oil from the same species but from another geographical area is recognizably different in composition, it must be mentioned in standards and certificates. One example is geranium oil from North Africa compared to other origins. 10-epi-γ-Eudesmol is only present in North African oil. The cause for this is adaptation to different conditions in a longer period.

20.5 METHODS TO DETECT ADULTERATIONS

20.5.1 Organoleptic Methods

20.5.1.1 Appearance and Color

Appearance is the visual aspect of a thing or person. In this case it is the appearance of the "essential oil," starting with color, going on with mobility, and finally with the odor itself. The color of essential oils is dependent on the starting material. Citrus oils are colored weak yellowish with lemon oil, light green to darker green with bergamot oil, and orange to brownish red with orange and mandarin oils. The color is dependent on the degree of ripeness of the fruits but they alter with storage and influence of light and warmth. Hereby the age of such an essential oil can be detected. Colors too will appear in hydrodistilled essential oils. The normal case for these oils is colorlessness accompanied by mobile fluidness. Weak yellow color can appear in oils like cardamom, rosewood, tarragon, or turpentine. Oils of lavender and lavandin are pale yellow and mobile; sandalwood oil is almost colorless to golden yellow but viscous. Lemongrass and citronella Java-type oils are pale yellow to yellowish brown; the latter is slightly viscous. Geranium oil has various shades, starting from amber yellow to greenish yellow. Yellow to light brown is the rectified oil of clove leaf, but the crude oil is black. Vetiver and patchouli oils are yellow to reddish brown; both are viscous or even highly viscous. The oils of thyme (*Thymus vulgaris* L. and *Thymus zygis* (Loefl.) L.) are red. The color results from a reaction between thymol and iron of the still. Divergent colors have to be observed critical as these can be a result of aging. Citrus oils can be clear at ambient temperature but becomes cloudy at lower temperatures. Sometimes waxes can undergo precipitation by storage under 4°C.

20.5.1.2 Odor

The odor is the most important factor for the application of essential oils. The highest quantity of usage is in the flavor and fragrance industry. Hatt et al. (2011) confirmed that the human sense of scent is the most important, even if the visual is dominating today. Smelling influences the human being by affecting the limbic system and is responsible for feelings and instincts, to trigger hormonal effects and activate pheromone production. The fact, that the human species has 350 gene receptors for smelling but only 4 for seeing will demonstrate the vital necessity of this sense.

In times of high sophisticated analytical technique with highest resolutions, the human sense of smell is not replaceable. Especially small traces of aromatic chemicals with very low odor threshold are quickly identifiable. Examples are vanillin from vanilla beans and maltol, found in malt and caramel. For perfumer beginners, a single chemical is easy to recognize again. Essential oils are multicomponent mixtures, varying sometimes in wider limits. In spite of electronic noses a well-trained perfumer can recognize more nuances. Common human nose might be duped by adulterations, but not those of professional perfumers. At least the sense of taste is used to recognize adulteration. A good example is rose oil. Part of a drop applied to the tongue and mixture with synthetic geraniol or citronellol reflects in soapy and bitter taste.*

20.5.1.3 Physical–Chemical Methods

These are inevitable in spite of many analytical certificates according to regulations. In history these values have been the only reference points to confirm naturalness. Although the bandwidth is often too wide, the values can be used as simple and easy to establish proof-samples. Gross mistakes and adulteration can be detected as well as aging of the oil. A series of ISO standards have been established especially for the quality control of essential oils:

Relative density (*ISO 279*): This is defined as the relation between the mass of a defined volume of substance at 20°C and the mass of a comparable volume of water at the same temperature.

Refractive index (*ISO 280*): This is the ratio of the sine of the angle of refraction, when a ray of light is passing from one medium into another. Three decimals are obligatory for the result.

Optical rotation (*ISO 592*): This is the property of defined substances, to turn the plane of polarized light.

Residue on evaporation (*ISO, 4715*), 1978: This is the residue of the essential oil after vaporization on the water quench, using defined conditions and is expressed in percent (*m/m*).

Determination of acid value (*1242*): This value shows the number of milligrams of potassium hydroxide required to neutralize free acid in one gram of essential oil.

Miscibility with ethanol (*ISO 875*), 1999: For that key figure, a defined quantity of an essential oil is added to an also defined mixture of distilled water and ethanol at 20°C. As result a blur occures visual. The value before obscuration is the value for a clear solution.

* Based on the authors' private experience.

Determination of ethanol content (*ISO 17494*): This is the method of detecting the quantity of ethanol contained in an essential oil by GC analysis on a suitable column.

Determination of the carbonyl value—free hydroxylamine method (*ISO 1271*): The determination is done by converting the carbonyl compounds to oximes by reaction with free hydroxylamine liberated in a mixture of hydroxylammonium chloride and potassium hydroxide.

Determination of water content—Karl Fischer method (*ISO 11021*): A reagent (Karl Fischer reagent without pyridine) is used for the reaction of the absorbed water from an essential oil. The reagent is produced by titration, using a Karl Fischer apparatus. The final result of the reaction is obtained by an electronic method.

Determination of phenol content (*ISO, 1272*), 2000: The transformation of phenolic compounds into their alkaline phenol esters, then soluble in aqueous phase, in a defined volume of essential oil is proceeded. The volume of unabsorbed portion is measured.

Determination of peroxide value (*ISO 3960*): The peroxide value is the quantity, expressed as oxygen, to oxidize potassium iodide under specific conditions, divided through the mass of testing substance.

Determination of freezing point (*ISO 1041*): This is the highest temperature during freezing of an undercooled liquid.

Iodine number: Kumar and Madaan (1979) report in their chapter the use of iodine number. This method is a measure of the unsaturation of a substance (as an oil or fat) expressed as the number of grams of iodine or equivalent halogen absorbed by 100 g of the substance (Merriam-Webster, n.d.). By using the iodine monobromide-mercuric acetate reagent for iodination, a much better result could be achieved, as reported by Kumar. With this method adulterations on essential oils can be successfully detected.

Thin-layer chromatography: This method belongs to wet chemistry (see Section 20.18.5.2).

All earlier-mentioned methods are still involved in the concerning standards and are useful tools to confirm naturalness and purity of essential oils.

20.5.1.4 Calculation of Relationship Coefficient

This method might be derided by some people, but once a relationship is recognized, it will help to detect mixtures with natural fractions or isolates. Touche et al. (1981) presented reference factors for identification of adulteration in French lavender. The ratio between (*Z*)-β-ocimene and (*E*)-β-ocimene (R_1), (*E*)-β-ocimene and 3-Octanone (R_2), and linalool + linalyl acetate to lavandulol + lavandulyl acetate had been determined using GC data of production values over more than 5 years. Schmidt (2003) showed in accordance to that the differentiation of the minor components of lavender oils as the possibility to detect even the provenance of the oil. Of course such a calculation of relationship can be done for many other essential oils. As long as typical minor substances are present, it is a valuable tool for genuineness control.

20.5.2 Analytical Methods

20.5.2.1 General Tests

Before the availability of sophisticated analytical instrumentation, only rather simple tests could be carried out to detect adulterations in essential oils (EOs), and it can be assumed that at that time many falsifications went undetected and adulterations were widespread. To determine the identity and purity of EOs, several elementary methods are described in the European Pharmacopoeia (Ph. Eur.) and ISO Standards (ISO/TC 54).

Physicochemical properties like relative density (Ph.Eur. 10th ed. 2.2.5, ISO 279), refractive index (Ph.Eur. 10th ed 2.2.6, ISO 280), optical rotation (Ph.Eur. 10th ed. 2.2.7, ISO 592), and flash point (ISO/DIS 3679:2015) should be within certain limits and numerical values can be found in the literature, on the Internet, and in ISO standards. Solubility tests can indicate the presence of contaminants (solubility in ethanol ISO 875), for instance, a turbidity of a solution of an EO in carbon disulfide is an indication of water (ISO 11021:1999). To detect fatty oil adulterations, a drop of the EO is placed on a filter paper and after 24 h no visible grease spot must be left behind (Ph. Eur. 7.0/2.08.07.00). Determination of the evaporation residue serves the same purpose and detects nonvolatile or low volatile compounds (ISO, 4715). An instrumentally more demanding method would be thermogravimetry.

Basic chemical tests can also provide useful information on the properties of EOs. Ester and saponification value (ISO, 7660), content of free and total alcohol (ISO, 1241), carbonyl value (ISO, 1279, 1996), and phenol content (ISO, 1272, 2000) are helpful indicators of identity and purity of EOs.

Last but not the least sensory evaluation is important especially in perfume houses and should be carried out by experts or a panel of fragrance professionals most effectively done with the help of reference samples of known status.

But some parameters alone determined by the earlier-mentioned methods tell us nothing about the composition of the tested EO. Therefore, separation of the oil into their individual components is necessary.

20.5.2.2 Thin-Layer Chromatography

One of the oldest and easiest separation methods is thin-layer chromatography (TLC), which nevertheless has the disadvantage of having a quite low separation efficiency. A drop of the EO is placed at one end of a plate (glass, aluminum foil) coated with a thin layer of an adsorbent (silica gel, diatomaceous earth, aluminum oxide, cellulose) as the stationary phase and placed in a glass chamber filled with a small amount of an appropriate solvent as a mobile phase. The solvent is drawn up the plate through capillary forces and separates the oil into several fractions. These fractions or spots can be made visible by derivatization with a chromophoric reagent or by inspection under UV light. A reference oil of known purity and composition must be analyzed simultaneously or under the same conditions for comparison (Ph. Eur. 7.0/2.02.26.00); (Jaspersen-Schib and Flueck, 1962; Phokas, 1965; Atal and Shah, 1966; Sen et al., 1974; Kubeczka and Bohn, 1985; Nova et al., 1986; John et al., 1991).

20.5.2.3 Gas Chromatography (GC, GLC, HRGC, GC-FID, GC-MS)

The analysis of natural products can be demanding especially in the field of EO research since an EO can contain 300 or more components at a concentration ranging from more than 90% to a few ppm or less of the total oil amount. A separation technique of high efficiency is needed and gas liquid chromatography (GLC) or simply GC is one of the most importantly used analytical methods for volatile compounds and is especially suited in EO and fragrance analysis. With the introduction of capillary columns, the separation performance increased dramatically and theoretical plate numbers of more than 1×10^5 are commonplace and provide gas chromatograms of high resolution (HRGC).

A small amount of the EO (nanograms to micrograms) is injected with a microliter syringe into a closed evaporation chamber (injector) held at elevated temperature (>200°C) where the sample evaporates and the vapor is transferred by an inert carrier gas (N_2, H_2, He, Ar) as the mobile phase to a quartz capillary column of an internal diameter ranging from 0.05 to 0.53 mm and a length between a few meters up to a hundred meters or more. The inner wall of the capillary is coated with a thin film of a liquid phase (polysiloxane, polyglycol) in which separation takes place through permanent absorption and desorption of the sample compounds between liquid and mobile phase. The column is mounted in an oven either held at constant temperature or more often programmed at a constant rising temperature rate and permanently flushed with the carrier gas. The substances pass through the column and leave at a reproducible time (retention time) and then need to be registered by an appropriate detector. The most common and versatile detector for organic compounds is the flame ionization detector (FID) in which the eluted substances are burnt in a tiny hot hydrogen/air flame producing charged carbon ions that are collected on an anode. The resulting current is registered after amplification by an electrometer. Amplitude and signal (peak) area are proportional to the amount of the corresponding compound. Retention times depend on many device parameters, and therefore, instrument-independent retention indices calculated according to Kovats (1958) or Van Den Dool and Kratz (1963) are used instead and can be compared with literature data for substance identification. To verify identification, a second analysis on a column of different polarity must be performed, and retention times or indices on both columns must match the data from the assumed compound or of those from a reference substance. This two-column approach is also useful to detect peak overlapping, which occurs quite often on single-column GC analysis of natural complex substances. Substance amounts are calculated from the individual peak areas as a percentage of the total peak areas of all substances multiplied by a substance-dependent correction factor, which accounts for the different sensitivities of each compound to FID detection.

These correction factors should be determined at least for the major compound found in an EO relative to a standard substance (preferably a *n*-paraffin) whose factor is typically set to 1. Unfortunately many EO compounds are not available commercially in pure form or easily synthesized, so for convenience all correction factors are set to 1 to a first approximation.

Chromatographic profiles and percentage composition of all important commercial essential oils and many oil-bearing plants can be found in the ISO standards and the literature of essential oil research (Formácek, 2002; Kubezka and Formácek, 2002; Adams, 2007), and genuine EOs should fit into this image within certain limits. Chemical profiling in this manner can detect adulterations if additional or missing peaks are found or percentage composition deviates substantially. Addition of synthetic compounds can in some cases be disclosed by the presence of by-products or impurities left behind from the synthesis pathway. Dihydro linalool, for example, is an intermediate in the industrial production of linalool and is always present in small amounts ($\ll$1%), and if a trace of this substance is found in an EO, it is a clear indicator of an adulteration with synthetic linalool.

However, fingerprinting of EOs by GC with FID detection alone cannot reveal the chemical identity of detected peaks if any deviations from the expected profiles occur. In that case further information is needed besides the chromatographic data, and mass spectrometry provides detailed information on the structure of the separated compounds. High-resolution gas chromatography coupled with a mass spectrometric detector (HRGC-MS or simply GC-MS), most commonly quadrupole ion trap detectors in EO analysis, together with sophisticated chromatographic software and special mass spectral libraries of essential oil components (Königet al., 2004; Adams, 2007; Mondello, 2011) separates and identifies most components of an EO. Since some classes of compounds show very similar mass spectra, like some groups of mono- and sesquiterpenes, retention indices must be taken into account as a second criterion for an unequivocal identification. Usually a GC-MS run is performed for identifying the EO components and a second GC-FID run for peak area, respectively, for percentage composition determination. Normally this is done on two different instruments. Identical capillary columns must be used in both GCs, and device parameters must be adjusted properly to obtain closely similar chromatographic profiles for both detectors to facilitate peak allocation between the two chromatograms. Peaks identified in the mass spectrometry (MS) chromatogram must be correctly assigned in the FID chromatogram for peak integration. Sometimes this is proving difficult since separation on two instruments is never exactly equal especially if one column ends up in a high vacuum (MS) and the other at atmospheric pressure (FID). Therefore, a series of closely eluting peaks may not be resolved in the same manner on both columns. To overcome this problem, an FID-MS splitter can be used since here the separation takes place on one GC column and the effluent is split to both detectors, MS and FID. To detect peak overlapping, the same procedure should be undertaken on another capillary column of different polarity. In EO analysis GC separations are preferably performed on 95% dimethylpolysiloxane/5% diphenylpolysiloxane and on Carbowax 20 M columns. In the end you come up with two FID and two MS chromatograms, each of the two carried on different capillary columns. These results in two analyses that should for the most part coincide and accept the overlapping peaks on the other GC column. EO analysis performed in this way confirms the chemical composition of an EO and detects adulterations with exogenic substances if they are amenable by GC, that is, volatile. Adulterations with gas chromatographic undetectable substances like very high boiling vegetable or mineral oils can be disclosed by a change in the percentage composition or a too low total peak area. Specific marker compounds or a diastereomeric isomer distribution can also be used for authentication of EO (Teisseire, 1987).

In some unusual EOs not all detected peaks can be identified by MS library search because of missing entries but, with the necessary experience, it can often be estimated in which group of natural substances these peaks belong by looking at the mass spectrometric pattern. All in all an extensive experience in EO analysis and a thorough knowledge of the scientific literature is needed to evaluate the authenticity of an EO and detect adulterations. If this is the case the analyst should be in a position to disclose falsifications with cheaper or inappropriate EOs, synthetic natural substances, solvents, or other unnatural chemicals.

20.5.2.4 Chiral Analysis

Many EOs contain substances with asymmetric carbon atoms, that is, there are two molecular forms of these molecules that behave like image und mirror image and cannot be brought into alignment (Busch and Busch, 2006). This is the reason why EOs rotate the oscillating plane of light and the resulting overall optical rotation is measured by a polarimeter. However, the enantiomeric distribution of single substances cannot be determined in this way because separation is needed in the first place. This will be enabled by GC, and additional separation of optically active compounds into their enantiomers can be achieved if a chiral selector is added to the stationary phase of the GC column. Cyclodextrin derivatives have proven to be very useful and chiral capillary columns with different optically active cyclodextrin selectors are commercially available. The enantiomeric separation should be done at rather low temperature since the interaction between the chiral molecules and the chiral selector is rather weak; therefore, the temperature gradient of the GC program should be low and the carrier gas flow is higher than the optimal flow. If the appropriate chiral GC column is used, the enantiomeric distribution of certain EO components can be determined, and since nature often prefers one enantiomer over the other, the determination of the enantiomeric excess is a valuable tool to detect adulterations by synthetic components that are racemic and change the enantiomeric ratio. The optical purity of authentic EOs can be found in the literature (Mosandl, 1998; Busch and Busch, 2006; Dugo and Mondello, 2011).

20.5.2.5 GC-GC and GC×GC (Two-Dimensional Gas Chromatography, ^{2}D GC)

Gas chromatography is one of the most widely deployed methods in analytical chemistry to investigate organic sample material due to its simple ease of use, the ready availability of sophisticated inexpensive instrumentation, and the large amount of qualitative and quantitative information that can be retrieved if the appropriate configuration is employed. Especially the high separation efficiency for volatiles makes GC very suitable to investigate complex mixtures and sample matrices. But for some applications, the separation performance is not sufficient when it comes to very complex mixtures like odors, flavors, crude oil products, and foodstuff. Co-elution with other analytes or sample matrix elements causes problems in detection and quantitation especially when the analytes differ greatly in their concentration. This problem can be solved by cutting out the co-eluting part of the chromatogram and a subsequent second chromatographic separation of the excised effluent preferably on a stationary phase of different polarity. This technique, called heart-cutting or two-dimensional GC (^{2}D GC or GC-GC), is done with the help of a diverting valve or a Deans switch. The sought-after substances are then resolved in the second GC column.

An even more elaborate technique recently developed not only cuts out one or several parts of the first GC column but virtually cuts the whole 1D chromatogram into small equal pieces (each several seconds long), refocuses each effluent, and separates it on a short second GC column within a very short time (several seconds). Refocusing, which here means stopping the effluent of the first column for several seconds onto a very small area and then releasing the focused substances into the second column, is done with a cryogenic modulator, and this procedure is repeated until all substances are eluted from the first GC column. The overall separation efficiency is calculated by multiplying the theoretical plate numbers of the two columns, which results in rather large figures and a separation efficiency, which cannot be achieved by simple GC alone. Instruments and the appropriate data acquisition software for this comprehensive ^{2}D GC (or GC × GC) are now commercially available. Instead of a 1D chromatogram, a 2D contour plot is generated, and since the "half-widths" of substances are very small, detection must proceed very fast, so in case a mass spectrometer is used as a detector, a fast scanning quadrupole or a time-of-flight MS (TOF-MS) must be chosen. Such an instrument arrangement like GC × GC-TOF-MS leaves almost nothing to be desired for EO analysis (Marsili, 2010).

20.5.2.6 ^{13}C NMR (Nuclear Magnetic Resonance)

Another useful method to validate the identity of an EO offers the spectroscopy of the magnetic properties of ^{13}C nuclei. ^{13}C NMR spectroscopy is a very useful technique regularly used to elucidate the structure of individual substances and has been applied by Kubeczka and Formácek (2002) to a large number of essential oils and individual reference compounds. Even though it is not very common to apply this technique to substance mixtures, as it is with an EO, the spectrum of a genuine oil is very distinct and characterized by the chemical shifts, signal multiplicities, and intensities of all components. Functional groups can be identified by chemical shifts and provide information on chemical structures of individual substances within the oil. No separation of the oil into its components is necessary, and therefore, the approach is simply measuring a solution of the oil in an NMR tube and comparing the spectrum with literature data or reference oils and compounds. Unfortunately NMR instruments are rarely found in traditional analytical laboratories, and thus, this method has not gained the significance in the field of EO analysis it deserves.

20.6 IMPORTANT ESSENTIAL OILS AND THEIR POSSIBLE ADULTERATION

20.6.1 Ambrette Seed Oil

This is a very expensive oil. The main ingredients are (*E*,*E*)-farnesyl acetate (60%) and ambrettolide (8.5%). Synthetic ambrettolide is used as a fixative in perfumery, is nearly odorless but has exalting properties. Compared to natural ambrette seed oil, the price of ambrettolide is only 10%. Detection of naturality can be done by isotope ratio mass spectrometry (IRMS).

20.6.2 Amyris Oil

ISO standard 3525 shows character and data for this oil. Blending (professional term within dealers and perfumers for adulterating) is done by Virginia cedarwood oil, α-terpineol, and copaiba balm. Also, elemol distilled from elemi resin is used. Detection is done by GC-MS.

20.6.3 Angelica Oils

The fruit oil contains up to 76% of β-phellandrene and α-pinene (13%) as main ingredients. As no chiral values for the β-phellandrene are described in literature, it must be assumed that adulteration is done by this compound. β-Phellandrene is naturally available by geranyl diphosphate cycling (BRENDA, BRaunschweig ENzyme DAtabase), 2007. For adulteration copaiva balm, gurjun balsam, lovage root oil, and amyris oil were used in the past. Cheap α-pinene can also be used to "improve" the composition.

20.6.4 Anise Fruit Oil

ISO standard 3475 shows character and data for that oil. This oil often is produced not only from the fruits but also from the whole aerial part. Values of *cis*- and *trans*-anethole are then reduced in smaller amount. Adulteration with star anise oil can be easily detected. Anise fruit oil does not contain any foeniculin, but star anise oil contains up to 3% (ISO 11016). On the other hand pseudo-isoeugenol-2-methylbutyrate is a component of anise fruit oil (0.3%–2%, ISO 3475) and does not appear in star anise oil. The midratio of *cis*-anethole to *trans*-anethole in anise seed oil in relation to star anise oil is 0.3%:0.6%. Synthetic anethole, fennel terpene limonene (80°), and terpineol were used for blending.

20.6.5 Armoise Oil

This oil is often blended with α- and β-thujones from cheaper sources like *Thuja orientalis*. Furthermore camphor or white camphor oil and camphene are used. GC-MS method is used to detect these adulterations.

20.6.6 Basil Oils

ISO standard 11043 shows character and data for the methyl chavicol–type oil. Synthetic methyl chavicol is used to adulterate that oil and can be detected by NMR. Basil oil from linalool type will be adulterated by synthetic linalool but can easily be detected by chiral separation. Casabianca (1996a) found the minimum value *R*-(−)-linalool with 99.8% and is congruent with the authors' results.

20.6.7 Bergamot Oil

ISO standard 3520 shows character and data for that oil. As mentioned in the text, this oil was adulterated with synthetic linalool and linalyl acetate but also with bergamot terpenes and distilled oil from bergamot peel residues. Blending was done with limonene (80°), nerol, geranyl acetate, petitgrain oil Paraguay, rosewood oil, and citral from *Litsea cubeba* oil. This oil was subject to many studies with chiral separation. König et al. (1992); Mosandl et al. (1991); Dugo et al. (1992) and Casabianca (1996a), reported similar values for linalool and linalyl acetate. (*R*)-(−)-linalyl acetate is present always in purity higher than 99.0% and (*R*)-(−)-linalool with a minimum of 99.0% too. (*S*-pinene value varies between α)-(−)- 68.0% and 71.1%, (*S*-pinene between 91.1% and 92.6%β)-(−)-, and (*R*)-(+)-limonene between 97.4% and 98.0% (Juchelka, 1996a). Blending is done by terpinyl acetate, citral synthetic, *n*-decanal, *n*-nonanal, nerol, limonene (80°), and sweet orange terpenes. Dugo and Mondello (2011) published the following data for chiral ratios: (*S*)-thujene (0.5%–1.3%):(α)-(+)-(*R*)-thujene (98.7%–99.5%); (α)-(−)-(*R*)-(+)-α-pinene (31.0%–36.1%):(*S*)-(−)-α-pinene (63.9–69.0); (1*S*,4*R*)-(−)-camphene (85.7%–92.7%):(1*R*,4*S*)-(+)-camphene (7.3%–14.3%); (*R*)-(+)-β-pinene (7.6%–10.3%):(*S*)-(−)-β-pinene (89.7%–92.4%); (*R*)-(+)-sabinene (13.7%–19.8%):(*S*)-(−) (80.2%–86.3%); (*R*)-(−)-α-phellandrene (43.1%–54.7%):(*S*)-(+)-α-phellandrene (45.3%–56.9%); (*R*)-(−)-β-phellandrene (24.1%–36.9%):(*S*)-(+)-β-phellandrene (63.1%–75.9%); (*S*)-(−)-limonene (1.2%–2.1%):(*R*)-(+)-limonene (97.9%–98.8%); (*R*)-(−)-linalool (97.8%–99.5%):(*S*)-(+)-linalool (0.5%–2.2%); (*S*)-(−)-citronellal (>98%):(*R*)-(+)-citronellal (<2%); (*R*)-(−)-linalyl acetate (99.1%–99.9%):(*S*)-(+)-linalyl acetate (0.1%–1.0%); (*S*)-(+)-terpinen-4-ol (22.4%–44.7%):(*R*)-(−)-terpinen-4-ol (55.3%–77.6%); (*S*)-(−)-α-terpineol (14.0%–68.5%):(*R*)-(+)-α-terpineol (31.5%–86.0%); (*S*)-(−)-citronellol (12.0%–20.0%):(*R*)-(+)-citronellol (80.0%–88.0%); (*S*)-(−)-terpinyl acetate (36.0%–44.0%):(*R*)-(+)-terpinyl acetate (56.0%–64.0%).

20.6.8 Bitter Orange Oil

ISO standard 3517 shows character and data for this oil. In the past, sweet orange oil was used as well as orange terpenes and distilled bitter orange residues from production for adulteration. Limonene of high purity from other citrus fruits was also applied. Dugo (2011) shows the similarity of the components between bitter and sweet orange oil. McHale et al. (1983) report the use of grapefruit oil, as it contains higher concentrations of coumarins and psoralens. Today the adding of purified components from other citrus sources is used. Mingling up with sweet orange oil can be detected by measuring the δ-3-carene and camphene content. As bitter orange oil contains only traces of δ-3-carene, the sweet oil goes up to 0.1%. Also, the ratio δ-3-carene/camphene can be used for detection (Dugo et al., 1992). Chiral values for components are reported by Dugo and Mondello (2011): (*R*)-(+)-α-pinene (89.7%–97.4%):(*S*)-(−)-α-pinene (2.6–10.3); (1*S*,4*R*)-(−)-camphene (35.8%–47.6%):(1*R*,4*S*)-(+)-camphene (52.4%–64.2%); (*R*)-(+)-β-pinene (6.1%–7.9%):(*S*)-(−)-β-pinene (92.1%–93.9%); (*R*)-(+)-sabinene (49.4%–80.6%):(*S*)-(−)-sabinene (19.4%–50.6%); (*R*)-(−)-α-phellandrene (60.1%–74.9%):(*S*)-(+)-α-phellandrene (25.1%–39.9%); (*R*)-(−)-β-phellandrene (0.6%–5.7%):(*S*)-(+)-β-phellandrene (94.3%–99.4%); (*S*)-(−)-limonene (0.5%–0.8%):(*R*)-(+)-limonene (99.2%–99.5%); (*R*)-(−)-linalool (61.2%–89.8%):(*S*)-(+)-linalool (10.2%–38.8%); (*S*)-(−)-citronellal 42.5%:(*R*)-(+)-citronellal (57.5%); (*R*)-(−)-linalyl acetate (99.2%–99.4%):(*S*)-(+)-linalyl acetate (0.6%–0.8%); (*S*)-(+)-terpinen-4-ol (67.5%–71.5%):(*R*)-(−)-terpinen-4-ol (28.5%–32.5%); (*S*)-(−)-α-terpineol (6.6%–29.8%):(*R*)-(+)-α-terpineol (93.4%–70.2%).

20.6.9 Bitter Orange Petitgrain Oil

ISO standard, 8901, 2010 shows character and data for this oil. This is also known as "petitgrain oil bigarade." Adulteration is done by using petitgrain oil from Paraguay, heads and tails from fractioning of that oil (to receive the petitgrain oil "terpene free"), and rectified orange terpenes and limonene. Blending is done with synthetic linalool, linalyl acetate, geraniol, geranyl acetate, and α-terpineol. Detection is done by GC-MS as well as by multidimensional enantiomeric separation systems. Juchelka (1996a,b) published the following chiral ratio values: (*R*)-(−)-linalyl acetate (98.0%–99.1%):(*S*)-(+)-linalyl acetate (0.2%–2.0%); (*R*)-(−)-linalool (66.4%–90.2%):(*S*)-(+)-linalool (9.8%–33.6%); (*S*)-(−)-α-terpineol (26.4%–28.4%):(*R*)-(+)-α-terpineol (71.6%–73.6%). Dugo and Mondello (2011) reported following chiral ratios: (*R*)-(+)-α-pinene (6.7%–12.0%): (*S*)-(−)-α-pinene (93.3%–88.0%); (*R*)-(+)-β-pinene (0.1%–1.1%): (*S*)-(−)-β-pinene (98.8%–99.9%); (*S*)-(−)-limonene (29.2%–39.2%): (*R*)-(+)-limonene (60.8%–70.8%); (*R*)-(−)-linalool (66.4%–90.2%):(*S*)-(+)-linalool (9.8%–33.6%); (*R*)-(−)-linalyl acetate (93.4%–99.1%):(*S*)-(+)-linalyl acetate (0.9%–6.6%); (*S*)-(+)-terpinen-4-ol (47.9%–67.4%):(*R*)-(−)-terpinen-4-ol (32.6%–52.1%); (*S*)-(−)-α-terpineol (27.5%–28.4%):(*R*)-(+)-α-terpineol (71.6%–72.5%).

20.6.10 Cajeput Oil

The oil obtained from *Melaleuca cajuputi* Powell is the natural variety. To adulterate this oil, other species like *Eucalyptus* ssp. or *Cinnamomum camphora* (1,8-cineole type) are used. Eucalyptus terpenes, α-phellandrene, and α-terpineol were used for blending. Detection is done by GC/MS with the smaller components like α-selinene, α-humulene, and α-terpineol.

20.6.11 Camphor Oil

This oil is produced from *C. camphora* Sieb. wood in China and Taiwan. Many chemotypes and also varieties from *Cinnamomum* species exist, but the true camphor oil is meant. This oil is rarely adulterated because of the cheap price but, fractioned and enriched in camphor up to 90%, is used to mix up camphor containing oils like rosemary, Spanish sage, and spike lavender. Smallest traces of safrole (a carcinogenic compound) shows the use of *C. camphora* higher boiling fractions.

20.6.12 Cananga Oil

ISO standard 3523 shows character and data for the oil, produced from *C. odorata* (Lam.) Hook. F. et Thomson forma *macrophylla*. It can be adulterated by linalool from lavandin oil showing nearly equal chiral values and β-caryophyllene from clove oil fractions. Blending is done with benzyl acetate, linalool, α-terpineol, geraniol, α-terpinyl formate, methyl benzoate, and allo-ocimene, all synthetically produced further from Virginia cedarwood oil, copaiva balm oil, clove leaf oil, and gurjun balsam oil. Cananga oil is used to adulterate ylang-ylang oil, as yield is higher and the flowers have a higher distribution in nature. Detection is done by GC/MS and by multidimensional chiral separation. Bernreuther et al. (1991) mentions the ratio for linalool with (*S*)-(+) 2% and (*R*)-(–) 98%.

20.6.13 Caraway Oil

ISO standard 8896 shows technical data for that oil. Synthetic carvone is used for adulteration. European pharmacopoeia requires contents of (+)-limonene 35.5%–45.0%, (+)-carvone 50.0%–65.0%, and (–)-carvone with a maximum of 0.7% for a pure quality. ISO 8896 standard evaluated a content of *cis*-dihydrocarvone from 0.3% to 1.2%.

20.6.14 Cardamom Oil

ISO standard 4733 shows character and data for that oil. This oil is produced mainly in India, Sri Lanka, and Guatemala. Adulteration is done by 1,8-cineole from eucalyptus or camphor oil, α-terpinyl acetate, and linalool. The adding of 1,8-cineole is hard to be discovered; linalool ratio is reported by Casabianca (2011) to be between 7% of (*R*)-(–)-linalool to 93% of (*S*)-(+)-linalool and 100.0% (*R*)-(–)-linalyl acetate to 0.0% (*S*)-(+)-linalyl acetate. The presence of δ-terpinyl acetate seems to show adulteration as to nowadays it is only detected in clary sage oil. δ-Terpinyl acetate is a side product in synthetic terpinyl acetate.

20.6.15 Cassia Oil

ISO standard 3216 shows character and data for the Chinese type oil. Main component is *trans*-cinnamaldehyde. Synthetic cinnamaldehyde as well as coumarin is used for adulteration. This oil is often used for the adulteration of cinnamon bark oil. That is easy to detect, as coumarin is not a component of that oil. If *o*-methoxy cinnamaldehyde is found in cinnamon bark oil, it is a sign for adulteration with cassia oil. The naturality of the cinnamaldehyde can be detected by the combination of GC–combustion– IRMS (GC-C-IRMS) and GC-P-IRMS.

20.6.16 Cedar Leaf Oil

Adulteration is done by adding thujones from other species. According to Ravid (1992), the chiral value of (–)-fenchone is 100%. The thujones were found to be (–)-α-thujone 90% and (+)-α-thujone 10% (Gnitka, 2010).

20.6.17 Cedarwood Oils

ISO standard 9843 shows character and data for Virginia cedarwood oil. This is obtained from *Juniperus virginiana* L., whereas the Texas oil is distilled from *Juniperus ashei* J. Buchholz. If these two oils are mixed together, the status of essential oil per definition is lost. Another adulteration is adding cedrol of cheaper cedarwood from Chinese oil (*Chamaecyparis funebris* (Endl.) Franco). Production is done by cooling down the oil and separating the crystals by filtration. As by-products, the terpenes will also serve as adulterations for other cedarwood oils. This is only valid for cedarwood oils coming from the species *Juniperus*. Recognition of adulterations is done by GC-MS.

20.6.18 Celery Seed Oil

ISO standard 3760 shows character and data for this oil. Blending is done with α-terpineol, limonene from orange oil, rectified copaiba oil, lovage root oil, and amyris oil. Detection is best done by GC-MS.

20.6.19 Chamomile Oil Blue

ISO standard 19332 shows character and data for that oil. Adulteration is done mainly by synthetic α-bisabolol, chamazulene, and (*E*)-βfarnesene. The method developed by Carle et al. (1990) and Carle (1996) using $\delta^{13}C$ and δD isotopes with IRMS is a real tool for naturality assay.

20.6.20 Chamomile Oil Roman

This essential oil, distilled from *Chamaemelum nobile* (L.) All., possesses a series of angelates in various concentrations. Bail (2009) published the following values for angelates and specific esters: isobutyl angelate 32.1%, 2-methylbutyl angelate 16.2%, isobutyl isobutyrate 5.3%, methyl 2-methylbutyrate 1.9%, prenyl acetate 1.4%, 2-methylbutyl 2-methylbutyrate 1.2%, and 2-methylbutyl acetate 1.2%. As most of these compounds are available as synthetic chemicals, adulteration can be done easily. Detection is done by the combination of GC-C-IRMS.

20.6.21 Cinnamon Bark Oil

Only in some pharmacopoeia monographs this oil is dealt with a chromatic profile. Adulteration is done once by adding cassia oil. This can easily be detected by any quantity of coumarin, never to be found in pure cinnamon bark oil. On the other hand, adulteration by synthetic cinnamaldehyde will be detected by the combination of GC-C-IRMS and GC-P-IRMS as the assessment of synthetic and natural cinnamaldehyde is possible. Sewenig et al. (2003) reported that the chiral ratio of linalool is (*R*)-(−) 95% and (*S*)-(+) 5%.

20.6.22 Cinnamon Leaf Oil

ISO standard 3524 shows character and data for this oil. As eugenol is the main component, rectified clove leaf oil or isolated eugenol from this oil is used. Also β-caryophyllene, synthetic benzyl benzoate, and *trans*-cinnamaldehyde are added. Blending is done by α-terpineol. Casabianca (1998) reported the ratio of (*R*)-(−)-linalool and (*S*)-(+)-linalool to be 64.0% and 36.0%, respectively.

20.6.23 Citronella Oil

ISO standard 3849 shows character and data for the Sri Lanka–type oil and ISO standard 3848 for the Java-type oil. Synthetic citronellal, citronellol, and geraniol were used. In addition citronella oil terpenes were used to cover such adulterations. Lawrence (1996a) mentions the ratio of citronellal in Java type with *R*-(+) enantiomer is 90%. Casabianca (1996a,b) found chiral ratio of (*R*)-(+)-citronellol and (*S*)-(−)-citronellol to be 75.0%–79.0% and 21.0%–25.0%, respectively.

20.6.24 Clary Sage Oil

The chiral ratio of (*R*)-(−)-linalool and (*S*)-(+)-linalool is 80.6%–94.0% and 6.0%–19.4%, respectively, and of (*R*)-(−)-linalyl acetate and (*S*)-(+)-linalyl acetate is 93.0%–98.1% and 1.9%–7.0%, respectively, dependent on fresh or ensilaged biomass, found by Casabianca (1996a,b).

20.6.25 Clove Oils

ISO standards 3141, 3142, and 3143 dealing with the oil from buds, leaves, and stems show character and data for these oils. The best and most expensive oil is from buds. Adulterations with oils from leaves and stem are used for adulteration. For those other oils, residues from isolation of natural eugenol are applied. Sometimes β-caryophyllene from other sources was used too. Detection can be made by GC-MS analysis.

20.6.26 Coriander Fruit Oil

ISO standard 3516 shows character and data for this oil. It was subject to adulteration with linalool, as it is the main component. By using chiral separation, Braun and Franz (2001) found the chiral ratio for linalool as follows: (*S*)-(+) from 64.8% to 87.3% and (*R*)-(−) from 12.7% to 35.2%, while Casabianca (1996a,b) found (*S*)-(+) 87% to (*R*)-(−) 13%. Chiral ratio in limonene according to Casabianca (1996a,b) is (*R*)-(+) 62.0%–93.0% and (*S*)-(−) 7.0%–38.0%.

20.6.27 Corymbia Citriodora Oil

ISO standard 3044 shows character and data for this oil. Adulteration is done by synthetic citronellal as well as by citronellol. Detection is done by GC-MS but better by multidimensional chiral separation. The chiral ratio of linalool is (*R*)-(−) 100.0% and (*S*)-(+) 0.0%, citronellal (*R*)-(+) 55.6%–57.2% and (*S*)-(−) 42.8%–44.4%, and citronellol (*R*)-(+) 49.0%–54.0% and (*S*)-(−) 46.0%–51.0%, found by Casabianca (1996a,b).

20.6.28 Corn Mint Oil

ISO standard 9776 shows character and data for this oil. Adulterations were carried out using synthetic menthol, menthyl acetate and menthone, terpenes from fractioning of corn mint oil, and limonene. Detection is done by enantiomeric separation by GC-MS. Mosandl (2000) reported the following chiral ratios: (1*S*)-(α)-pinene (56.5%–73.5%):(1*R*)-(α)-pinene (26.5%–43.5%); (+)-(1*S*)-(−)-β-pinene (49.1%–55.6%):(1*R*)-(+)-β-pinene (44.6%–50.9%); (1*S*)-(−)-limonene (98.1%–99.9%):(1*R*)-(+)-Limonene (0.1%–1.9%); (1*R*,3*R*,4*S*)-(−)-menthol (min 99.9%):(1*S*,3*S*,4*R*)-(+)-menthol (max 0.1%); and (4*R*)-(−)-piperitone 21.0%:(4*S*)-(+)-piperitone <0.1%.

20.6.29 Cumin Fruit Oil

ISO standard 9776 shows character and data for this oil. The main components are cumin aldehyde, p-mentha-1.3-dien-7al, p-mentha-1,4-dien-7al, γ-terpinene, and β-pinene. Blending is done by orange terpenes, p-cymene, and piperitone. Detection is done by GC-MS and by the combination of GC-C-IRMS and GC-P-IRMS as assessment of synthetic and natural cuminaldehyde is possible.

20.6.30 Cypress Oil

Adulteration is done by either turpentine oil or α- and β-pinene. Further, blending is done using δ-3-carene and cedrol from cedarwood Chinese type. AFNOR (1992) presents data in the standard NF T 75-254. Using chiral GC as his method of analysis, Casabianca (1996a,b) determined that the enantiomeric ratio of α-thujene in cypress oil was as follows: (1*R*)-(+)-α-thujene (45%):(1*S*)-(−)-α-thujene (55%).

20.6.31 Dill Oils

Dill weed oil is dominated by α-phellandrene, limonene, and carvone. Dill ether and the absence of dill apiol are further criteria for that oil. Dill seed oil contains mainly carvone and dihydrocarvone. Adulteration is done using phellandrenes, distilled limonene coming from orange terpenes, synthetic carvone, and dihydrocarvone. Detection is done by 2D enantiomeric separation. Lawrence (1996a,b) reports, here and elsewhere.> the following ratios for dill seed oil: (+)-limonene 98.4%:(−)-limonene 1.6%; (+)-carvone 98.7%:(−)-carvone 1.3%; (+)-*trans*-carveol 33.3%:(−)-*trans*-carveol 66.7%; and (+)-*cis*-carveol 100%:(−)-*cis*-carveol 0%. The authors own findings from biocultivated oil was (+)-carvone 98.4%:(−)-carvone 1.6%; (*S*)-(−)-α-pinene 4.0%:(*R*)-(+)-α-pinene 96.0%; (+)-limonene

95.4%:(−)-limonene 4.6%; (*S*)-(−)-β-phellandrene 0%:(*S*)-(+)-β-phellandrene 100%; and (*R*)-(−)-α-phellandrene 100%:(*R*)-(+)-α-phellandrene 0%.

20.6.32 Dwarf Pine Oil

ISO standard 21093 shows character and data for this oil. In the past, addition of turpentine oil was used for adulteration; later the essential oil of the needles of *Pinus maritima* was used, as the composition showed very close data of compounds. Chiral analysis gives helpful results. Kreis et al. (1991) reported chiral ratio in monoterpenes as follows: (1*R*,5*R*)-(+)-α-pinene (44%):(1*S*,5*S*)-(−)-α-pinene (56%); (1*R*,5*R*)-(+)-β-pinene (2%):(1*S*,5*S*)-(−)-β-pinene (98%); and (4*R*)-(+)-limonene (40%):(4*S*)-(−)-limonene (60%).

20.6.33 Elemi Oil

ISO standard 10624 shows character and data for this oil. Elemol and elemicin are the lead compounds. As minor component 10-epi-γ-eudesmol must be detected between 0.2% and 0.3% to ensure naturness. Adulteration is done by limonene, α-phellandrene, and sabinene. Detection is done by GC-MS.

20.6.34 Eucalyptus Oil

ISO standard 770 shows character and data for this oil. Schmidt (2010) reported the Chinese eucalyptus oil coming from Sichuan province and is derived from *C. longepaniculatum* (Gamble) N. Chao ex H.W. Li. This must be labeled and it is not correct to mix this up with true eucalyptus varieties. Adulteration is done by 1,8-cineole from various *Cinnamomum* varieties. The detection is not easy as all these adulterations are natural.

20.6.35 Fennel Oil Sweet

Although this oil should be produced solely from fruits, more often the whole aerial parts are used. Adulteration is done by synthetic anethole or from other sources like star anise oil. Blending is done too with star anise oil and limonene +60°. Analysis is done by GC-MS or by multidimensional enantiomeric separation. According to Ravid (1992), the chiral ratio of (+)-fenchone is 100%–0%.

20.6.36 Fennel Oil Bitter

ISO standard 17412 shows character and data for this oil. It is often mixed up with the herb oil derived from bitter fennel and sweet fennel produced from the whole aerial parts of the plant. Blending is done by sweet fennel oil, star anise oil, and fennel terpenes. Analysis is done by GC-MS or by multidimensional enantiomeric separation. Ravid reported that the chiral ratio of (+)-fenchone is 100%:0%. Casabianca (1996b) confirms a chiral ratio of (4*R*)-(+)-α-phellandrene 100%:(4*S*)-(−)-α-phellandrene 0%.

20.6.37 Geranium Oils

ISO standard 4731 shows character and data for that oil. This important and high-price product is and was often the target for adulteration. Blending was done with synthetic geraniol, citronellol, limonene—60°, terpinyl formate, synthetic rhodinol, α-terpineol cristallin, and distilled bergamot oil. Lawrence (1996a,b) showed results of chiral separation of citronellol, *cis*- and *trans*-rose oxide, menthone, isomenthone, and linalool for geranium oils from different origins. The chiral ratio of (*R*)-(−) linalool and (*S*)-(+)-linalool is 42.0%–55.0% and 45.0%–58.0% found by Casabianca (1996b). In 1996 he published the following chiral data: (from various origins)

(*R*)-(−)-citronellol (18.0%–43.0%) and (*S*)-(+)-citronellol (50.0%–82.0%); (2*R*,4*S*)-(+)-*cis*-rose oxide (24.0%–38.0%):(2*S*,4*R*)-(−)-*cis*-rose oxide (62.0%–76.0%); (2*R*,4*R*)-(−) *trans*-rose oxide (70.0%–76.0%):(2*S*,4*S*)-(+)-*trans*-rose oxide (72.0%–76.0%); (1*S*,4*R*)-(+)-menthone (>99.0%):(1*R*,4*S*)-(−)-menthone (<1%); and (1*S*,4*S*)-(−)-isomenthone (>99.0%):(1*R*,4*R*)-(+)-isomenthone (<1.0%); Lawrence (1999a) confirmed further the values: (2*R*,4*S*)-(+)-*cis*-rose oxide (35.5%–49.3%):(2*S*,4*R*)-(−)-*cis*-rose oxide (50.7%–64.5%); (2*R*,4*R*)-(−) *trans*-rose oxide (49.2%–62.4%):(2*S*,4*S*)-(+)-*trans*-rose oxide (37.6%–50.8%); and (2*S*,4*R*)-(−)-*cis*-rose oxide ketone (50.5%–62.6%):(2*R*,4*S*)-(+)-*cis*-rose oxide ketone (37.4%–49.5%).

20.6.38 Grapefruit Oil

ISO standard 3053 shows character and data for this oil. Pure oils possess as marker the compound nootkatone from traces up to 0.8%, depending on the fruit status. This compound is used for blending, together with *n*-octanal, *n*-nonanal, *n*-decanal, and synthetic citral. Adulteration is performed by orange terpenes and distilled grapefruit residues from expression and limonene—80°. Detection must be done exclusively by multidimensional enantiomeric separation. Dugo and Mondello (2011) published the following chiral data: (*R*)-(−)-α-pinene (0.3%–0.8%):(*S*)-(+)-α-pinene (99.2%–99.7%); (*R*)-(+)-β-pinene (62.0%–76.8%):(*S*)-(−)-β-pinene (23.2%–38.0%); (*R*)-(+)-sabinene (98.4%–98.5%):(*S*)-(−)-sabinene (1.5%–1.6%); (*S*)-(−)-limonene (0.5%–0.6%):(*R*)-(+)-limonene (98.4%–98.5%); (*R*)-(−)-linalool (32.0%–43.0%):(*S*)-(+)-linalool (57.0%–68.0%); (*S*)-(−)-citronellal (16.6%–21.4%):(*R*)-(+)-citronellal (57.0%–68.0%); (*S*)-(−)-α-terpineol (1.2%–3.3%):(*R*)-(+)-α-terpineol (96.7%–99.8%); and (*S*)-(+)-carvone:(*R*)-(−)-carvone 34.8%.

20.6.39 Juniper Berry Oil

ISO standard 8897 shows character and data for this oil. *J. communis* oil is often mixed up with *J. oxycedrus*. As marker for that the myrcene content is rising up. Real markers are germacrene D- and δ-cadinene. The sesquiterpene fraction gives more information. Further on, addition of fractions of juniper berry oil from rectification as well as adding juniper branches oil is made. Kartnig et al. (1999) published some chiral data comparing self-distilled and commercial qualities of juniper berry α- and β-pinene, limonene, and terpinen-4-ol oils. Chirality was recognized as useful components for quality control of that oil. Mosandl et al. (1991) report a ratio for (*S*)-(−)-α-pinene 77%:(*R*-)-(+)-α-pinene 23%.

20.6.40 Lavandin Oils

ISO standard 3054 and 8902 shows character and data for Abrial and Grosso lavandin oil. This oil is adulterated by adding acetylated lavandin, lavandin distilled heads and tails, camphor oil white, Spanish sage oil, and spike lavender oil. Blending is done by terpinyl acetate, turpentine oil, methyl α-terpineol and ethyl amyl ketone, hexyl ketone, and geranyl acetate, all from synthetic source. Chiral ratio of linalool is from (*R*)-(−) 64.8% to 87.3% and (*S*)-(+) from 12.7% to 35.2%. Chiral ratio of linalyl acetate according to the findings of Casabianca (1996a) is (*R*)-(−) from 98.1% to 100.0% and (*S*)-(+) from 0.0% to 1.0%). Renaud (2001) reported chiral ratios for linalool and linalyl acetate as follows: (3*R*)-(−)-linalool (94.5%–98.2%):(3*S*)-(+)-linalool (1.8%–5.5%) and (3*R*)-(−)-linalyl acetate (99%–100%):(3*S*)-(+)-linalyl acetate (0%–1%). Lawrence (1996a) published the following chiral ratios: (3*R*)-(−)-linalyl acetate (98.3%–100%):(3*S*)-(+)-linalyl acetate (0.0%–1.7%); (3*R*)-(−)-linalool (95.0%–96.6%):(3*S*)-(+)-linalool (3.4%–5.0%). (2*S*,5*S*)-*trans*-linalool oxide (4.2%–23.3%):(2*R*,5*R*)-*trans*-linalool oxide (76.7%–95.8%); (2*R*,5*S*)-*cis*-linalool oxide (82.9%–95.8%):(2*S*,5*R*)-*trans*-linalool oxide (4.2%–17.1%); (*R*)-(−)-lavandulol (96.2%–99.0%):(*S*)-(+)-lavandulol (1.0%–3.8%); and (*R*)-(−)-terpinen-4-ol (1.6%–10.9%):(*S*)-(+)-terpinen-4-ol (89.1%–98.4%).

20.6.41 Lavender Oil

ISO standard 3515 shows character and data for oils from various origins. Most applied adulterations were already discussed in the text. Chiral separations showed the most effective results in analysis. The ratio of linalool is (*R*)-(−) 98.0%–100.0% and (*S*)-(+) 0.0%–2.0% found by Casabianca (1996a). He too found that the chiral ratio of linalyl acetate is (*R*)-(−) from 98.0% to 100.0% and (*S*)-(+) from 0.0% to 2.0%. Stoyanova reports the following data for Bulgarian lavender oil: (3*R*)-(−)-linalyl acetate (100%):(3*S*)-(+)-linalyl acetate (0%); (3*R*)-(−)-linalool (95.0%–96.6%):(3*S*)-(+)-linalool (3.4%–5.0%); and (3*R*)-(+)-camphor (27.4%–52.2%):(3*S*)-(−)-camphor (47.8%–78.6%). Kreis and Mosandl (1992a,) published the following data: (2*S*,5*S*)-*trans*-linalool oxide (3.9%–23.3%):(2*R*,5*R*)-*trans*-linalool oxide (76.7%–96.1%); (2*R*,5*S*)-*cis*-linalool oxide (82.9%–95.8%):(2*S*,5*R*)-*trans*-linalool oxide (4.2%–17.1%); (*R*)-(−)-lavandulol (89.8%–100%):(*S*)-(+)-lavandulol (0%–10.2%); (*R*)-(−)-terpinen-4-ol (0%–10.9%):(*S*)-(+)-terpinen-4-ol (89.1%–100%); and (3*R*)-(−)-linalool (95.1%–98.2%):(3*S*)-(+)-linalool (18%–4.9%).

20.6.42 Lemon Oil

ISO standard 855 shows character and data for that oil. Adulteration is done by distilled lemon oil from residues, orange terpenes or limonene from orange terpenes, and synthetic citral from *Litsea cubeba* oil. Lemon oil washed as residues from production of terpene-free oil is preferably used, as these contain still all components of the pure lemon oil. Also lemon terpenes and heads of distilled grapefruit oils could be found. Blending is done by using synthetic decanal, nonanal, octanal, and citronellal from *Corymbia citriodora* oil. Detection is made by GC-MS and mainly by multidimensional enantiomeric separation with various methods (see part of methods). Mondello (1998) reports some constituents with chiral ratios as follows: (*R*)-(+)-β-pinene 6.3%:(*S*)-(−)-β-pinene 93.7%; (*R*)-(+)-sabinene 14.9%:(*S*)-(−)-sabinene 85.1%; (*S*)-(−)-limonene 1.6%:(*R*)-(+)-limonene 98.4%; (*S*)-(+)-terpinen-4-ol 24.7%:(*R*)-(−)-terpinen-4-ol 75.3%; and (*S*)-(−)-α-terpineol 75.2%:(*R*)-(+)-α-terpineol 75.2%. Further on, Dugo and Mondello (2011) gave the following data: (*R*)-(+)-α-pinene (25.5%–31.5%):(*S*)-(−)-α-pinene (68.5%–74.5%); (1*S*,4*R*)-(−)-camphene (86.2%–92.4%):(1*R*,4*S*)-(+)-camphene (7.6%–13.8%); (*S*)-(−)-β-pinene (93.2%–95.7%):(*R*)-(+)-β-pinene (4.3%–6.8%); (*R*)-(+)-sabinene (12.4%–15.0%):(*S*)-(−)-sabinene (85.0%–87.6%); (*R*)-(−)-α-phellandrene (46.9%–52.6%):(*S*)-(+)-α-phellandrene (47.4%–53.1%); (*R*)-(−)-β-phellandrene (31.1%–53.9%):(*S*)-(+)-β-phellandrene (46.1%–68.9%); (*S*)-(−)-limonene (1.4%–1.6%):(*R*)-(+)-limonene 98.4%–98.6%); (*R*)-(−)-linalool (52.0%–74.5%):(*S*)-(+)-linalool (25.5%–48.0%); (*S*)-(−)-citronellal (89.5%–94.8%):(*R*)-(+)-citronellal (5.2%–10.5%); (*S*)-(+)-terpinen-4-ol (12.0%–26.2%):(*R*)-(−)-terpinen-4-ol (73.8%–88.0%); and (*S*)-(−)-α-terpineol (66.4%–82.0%):(*R*)-(+)-α-terpineol (18.0%–33.6%).

20.6.43 Lemongrass Oil

ISO standard 4718 shows character and data for this oil. Adulteration is done by adding synthetic citral or citral from *Litsea cubeba* oil. Blending is done with addition of geranyl acetate and 6-methyl-5-heptene-2-one. Detecting is done by GC-MS and multidimensional chiral separation. Wang et al. (1995) reported chiral ratio for linalool to be (3*S*)-(+)-linalool 30.9%:(3R)-(−)-linalool 69.1% and (*R*)-(−)-linalool 58.0%:(*S*)-(+)-linalool 42.0%.

20.6.44 Lime Oil Distilled

ISO standard 3809 shows character and data for that oil. Adulteration is done by adding limonene from different sources, synthetic terpineol and γ-terpinene from lime terpenes as well as from heads of the production of terpene-free lemon oil. Detection must be done by multidimensional chiral separation. Dugo and Mondello (2011) report the following data for chiral ratios: (*S*)-(−)-β-pinene (96.0%–96.8%) (*R*)-(+)-β-pinene (3.2%–4.0%); (*S*)-(−)-limonene (5.5%–8.7%):(*R*)-(+)-limonene (91.3%–94.5%); (*R*)-(−)-linalool (49.8%–80.0%):(*S*)-(+)-linalool (50.0%–50.2%); (*S*)-(+)-terpinen-4-ol (42.3%–45.0%):(*R*)-(−)-terpinen-4-ol (55.0%–57.7%); and (*S*)-(−)-α-terpineol (53.3%–56.8%):(*R*)-(+)-α-terpineol (46.7%–43.2%).

20.6.45 Lime Oil Expressed

ISO standard 23954 shows character and data for this oil. Blending is done by adding limonene from different sources, citral from *Litsea cubeba* oil and γ-terpinene from lime terpenes as well as from heads of the production of terpene-free lemon oil. Detection is done by GC-MS, looking for absence of δ-3-carene (not even in traces) and multidimensional enantiomeric separation with various methods (see part of methods). Dugo and Mondello (2011) reported the following chiral ratios: (*S*-)-(−)-(β)-pinene (98.7%–90.9%):(*R*)-(+)-β-pinene (9.1%–10.3%); (*R*)-(+)-sabinene (18.2%–23.4%):(*S*)-(−)-sabinene (76.6%–81.8%); (*S*)-(−)-limonene (0.4%–2.7%):(*R*)-(+)-limonene (97.3%–99.6%); (*R*)-(−)-linalool (54.4%–69.3%):(*S*)-(+)-linalool (30.7%–45.6%); (*S*)-(+)-terpinen-4-ol (18.6%–24.9%):(*R*)-(−)-terpinen-4-ol (75.1%–81.4%); and (*S*)-(−)-α-terpineol (74.5%–80.8%):(*R*)-(+)-α-terpineol (19.2%–25.5%).

20.6.46 *Litsea cubeba* Oil

ISO standard 3214 shows character and data for this oil. Adulteration is done by adding synthetic citral and can be detected by GC-MS by checking the values for geranial and neral but also for isogeranial and isoneral.

20.6.47 Mandarin Oil

ISO standard 3528 shows character and data for this oil. α-Sinensal is a marker for mandarin essential oil. Adulteration is made by synthetic methyl-*n*-methyl anthranilate, as well as methyl anthranilate. Orange terpenes and limonene (80°), dipentene, citronellal, and citral are used for blending. Detection is done by GC-MS but improved results show multidimensional chiral separation. Dugo and Mondello (2011) reported the following chiral data: (*S*)-(+)-α-thujene (0.3%–1.9%):(*R*)-(−)-α-thujene (98.1%–99.7%); (*R*)-(+)-α-pinene (41.7%–54.5%):(*S*)-(−)-α-pinene (45.5–58.3); (1*S*,4*R*)-(−)-camphene (31.8%–72.6%):(1*R*,4*S*)-(+)-camphene (27.4%–68.2%); (*R*)-(+)-β-pinene (87.8%–99.1%):(*S*)-(−)-β-pinene (0.9%–12.2%); (*R*)-(+)-sabinene (71.3%–83.5%):(*S*)-(−) (16.5%–28.7%); (*R*)-(−)-α-phellandrene (44.3%–55.0%):(*S*)-(+)-α-phellandrene (45.0%–55.7%); (*R*)-(−)-β-phellandrene (0.4%–3.0%):(*S*)-(+)-β-phellandrene (97.0%–99.6%); (*S*)-(−)-limonene (1.5%–2.91%):(*R*)-(+)-limonene (97.1%–98.5%); (1*R*,4*R*)-(+)-camphor (17.0%–36.5%):(1*S*4*R*)-(−)-camphor (63.5%–83.0%); (*R*)-(−)-linalool (13.0%–22.7%):(*S*)-(+)-linalool (77.3%–87.0%); (*S*)-(−)-citronellal 3.9%–9.2%):(*R*)-(+)-citronellal 90.8%–96.1%); (*S*)-(+)-terpinen-4-ol (9.5%–23.8%):(*R*)-(−)-terpinen-4-ol (76.2%–90.5%); and (*S*)-(−)-α-terpineol (66.1%–75.9%):(*R*)-(+)-α-terpineol (24.1%–33.9%).

20.6.48 Melissa Oil (Lemon Balm)

This expensive essential oil was and is often the target of adulteration: citronella oil (*Cymbopogon winterianus* or *Cymbopogon nardus*), lemongrass oil, lemon oil citral, and geraniol rose oxides, natural and synthetic. Lawrence (1996a,b) reports the chiral values of (*R*)-(+)-citronellal in that oil being 97.2%–98.2%. Also the value for (*R*)-(+)-methyl citronellate is at a minimum of 99.0% By using GC analysis with different chiral GLC phases, Schultze (1993) confirmed citronellal from other sources as follows: lemongrass oil (*S*)-(−) 30%–55%, (*R*)-(+) 70%–45%; citronella oil (*S*)-(−) 10%–15%, (*R*)-(+) 90%–85%; and catnip oil (*S*)-(−) 98%–99.9%, (*R*)-(+) 2%–0.1%. Schultze also published δ-values to confirm naturality.

20.6.49 *Mentha citrata* Oil

The chiral ratio of linalool is (*R*)-(–) 42.0%–55.0% and (*S*)-(+) 45.0%–58.0% found by Casabianca (1996a). Chiral ratio of linalyl acetate according to the findings of Casabianca (1996a,b) is (*R*)-(–) from 98.6% to 99.0% and (*S*)-(+) from 1.0% to 1.4%.

20.6.50 Mountain Pine Oil

ISO standard 21093 shows character and data for that oil. *Pinus maritima* shows nearly identical values and adulteration is hardly to recognize. As long as the price of the oil of *P. maritima* was one-fifth of the *Pinus mugo*, it was sold in large quantities as mountain pine oil. Now prices are nearly identical and mixing up makes no sense anymore. Adulteration is done by α-pinene, β-pinene, δ-3-carene, (–)-limonene, myrcene, β-phellandrene, and *l*-bornyl acetate from various sources.

20.6.51 Neroli Oil

ISO standard 3517 shows character and data for that oil. Adulteration is made by geraniol from palmarosa oil, linalool from rose wood oil, orange oil sweet, citral from *Litsea cubeba* oil, and petitgrain oil Paraguay. Blending is done by methyl-*n*-methyl anthranilate, methyl anthranilate, synthetic phenyl ethyl alcohol, and synthetic indol. Detection is done by GC-MS and multidimensional chiral separation. Casabianca (1996a) found the chiral ratio for linalool with (*S*)-(+) 12.0–13.8 and (*R*)-(–) 86.2%–88.0% and for linalyl acetate with (*S*)-(+) 1.8–5.0 and (*R*)-(–) 95.0%–98.2%. Dugo and Mondello (2011) reported the following chiral ratios: (*R*)-(+)-α-pinene (2.2%–13.6%):(*S*)-(–)-α-pinene (86.4%–97.8%); (*R*)-(+)-β-pinene (0.1%–0.8%):(*S*)-(–)-β-pinene (99.2%–99.9%); (*S*)-(–)-limonene (1.9%–6.9%):(*R*)-(+)-limonene (93.1%–98.1%); (*R*)-(–)-linalool (70.8%–81.5%):(*S*)-(+)-linalool (18.4%–29.2%); (*R*)-(–)-linalyl acetate (95.4%–98.2%):(*S*)-(+)-linalyl acetate (1.8%–4.6%); (*S*)-(+)-terpinen-4-ol (36.0%–37.6%):(*R*)-(–)-terpinen-4-ol (62.4%–64.0%); (*S*)-(–)-α-terpineol (28.1%–39.8%):(*R*)-(+)-α-terpineol (60.2%–71.9%); and (*R*)-(–)-(*E*)-nerolidol (0.4%–1.8%):(*S*)-(+)-(*E*)-nerolidol (98.2–99.6).

20.6.52 Nutmeg Oil

ISO standard 3215 shows character and data for this oil. Adulteration is done by monoterpenes α- and β-pinene, sabinene from different sources, and α- and β-phellandrene as well as synthetic linalool, terpinen-4-ol and α-terpineol. In the past, safrole from sassafras oil was used for blending. Detection is done by GC-MS, looking for the quantities of safrole and myristicin; in addition 2D chiral separation is recommended. König et al. (1992) reported the following chiral ratios: (+)-α-thujene 10.3%:(–) α-thujene 89.7%; (–)-α-pinene 79.3:(+)-α-pinene 20.7%; (–)-camphene 100%:(+)-camphene 0%; (+)-β-pinene 41.9%:(–)-β-pinene 58.1%; (–)-α-phellandrene 0%:(+)-α-phellandrene 100%; (+)-sabinene 89.5%:(–)-sabinene 10.5%; (+)-δ-3-carene 0%:(–)-δ(–)-3-carene 100%; (–)-β-phellandrene 7.7%:(+)-β-phellandrene 92.3%; and (–)-limonene 60.9%:(+)-limonene 39.1%.

20.6.53 Orange Oil Sweet

ISO 3140 shows character and data for this oil. Adulteration is done by adding orange terpenes or purified limonene. Casabianca (1996a,b) found chiral ratio for linalool with (*S*)-(+) 86 to (*R*)-(–) 14. Hara et al. (1999) reported chiral ratios of several hydrocarbon components as follows: (1*R*,5*R*)-(+)-α-pinene (99.6%–99.7%):(1*S*,5*S*)-(–)-α-pinene (0.3%–0.4%); (1*R*,5*R*)-(+)-β-pinene (66.1%–77.8%):(1*S*,5*S*)-(–)-β-pinene (22.2%–33.9%); (4*R*)-(+)-limonene (99.1%–99.4%):(4*S*)-(–)-limonene (0.6%–0.9%); (3*S*)-(+)-linalool (82.5%–96.3%):(3*R*)-(–)-linalool (3.7%–17.5%); (4*R*)-(+)-α-terpineol (97.0%–98.4%):(4*S*)-(–)-α-terpineol (1.9%–3.0%); and (3*R*)-(+)-citronellal (31.3%–87.8%):(3*S*)-(–)-citronellal (12.2%–68.7%).

20.6.54 Origanum Oil

This oil is from *Origanum vulgare* L. ssp. *hirtum* (Link) Ietsw. Adulteration is done with synthetic thymol and carvacrol or with limonene from different sources. The chiral ratio of linalool is (*R*)-(−) 82.0% and (*S*)-(+) 18.0% found by Casabianca (1996a,b).

20.6.55 Palmarosa Oil

ISO 4727 shows character and data for this oil. As the main component is geraniol, adulteration is done by synthetic geraniol. Detection must be done by the combination of GC-C-IRMS and GC-P-IRMS as the assessment of synthetic and natural geraniol is possible.

20.6.56 Parsley Oil

ISO standard 3527 shows character and data for this oil. Adulteration is done by turpentine oil or pure α-pinene, β-pinene, and elemicin from elemi resinoid. 1,2,3,4-Tetramethoxy-5-allyl benzene is a key compound up to 12% as well as apiol. Blending is done by celery grain oil, nutmeg oil, and carrot seed oil. Detection is done by GC-MS system.

20.6.57 Pine Oil Siberian

ISO 10869 shows character and data for this oil. Adulteration is done by turpentine oil, *l*-bornyl acetate synthetic, terpinyl acetate, and Virginia cedarwood oil. Detection is done either by GC-MS or chiral separation. Ochocka (2002) determined the enantiomeric ratios of four monoterpene hydrocarbons in *A. sibirica* oils (Korean source): (1*R*,5*R*)-(+)-α-pinene 25.4%:(1*S*,5*S*)-(−)-α-pinene 74.6%; (1*R*,5*R*)-(+)-β-pinene 13.2%:(1*S*,5*S*)-(−)-β-pinene 86.8%; (4*R*)-(+)-limonene 6.8%:(4*S*)-(−)-limonene 93.2%; and (3*R*)-(+)-camphene 4.8%:(3*S*)-(−)-camphene 95.2%.

20.6.58 Patchouli Oil

ISO standard 3757 shows character and data for this oil. This very complex oil is adulterated by gurjun balm oil (see text). Blending is done by patchouli terpenes, cedarwood oil, pepper oil, white camphor oil, and guaiac wood oil. Detection is done by GC-MS.

20.6.59 Pepper Oil

ISO standard 3061 shows character and data for this oil. Blending is done with turpentine oil, α-phellandrene from other sources, limonene from orange terpenes, and clove leaf oil terpenes. The chiral ratio of linalool is (*R*)-(−) 81.0%–89.0% and (*S*)-(+) 11.0%–19.0% found by Casabianca (1996a); König et al. (1992) reported the following chiral ratios: (+)-α-thujene 100%:(−)-α-thujene 0%; (−)-α-pinene 74.6%:(+)-α-pinene 25.4%; (−)-camphene 66.6%:(+)-camphene 33.4%; (+)-β-pinene 2.6%:(−)-β-pinene 97.4%; (+)-δ3-carene 2.6%:(−)-δ3-carene 97.4%; (−)-α-phellandrene 0%:(+)-α-phellandrene 100%; (−)-β-phellandrene 100%; and (−)-limonene 61.7%:(+)-limonene 38.3%.

20.6.60 Peppermint Oil

ISO standard 856 shows character and data for that oil. Adulteration is done by synthetic menthol, menthol from cornmint oil, and fractions of peppermint terpenes. Detection is done by chiral separation using 2D enantiomeric columns on GC-MS system. The chiral ratio of 3-octanol is (*R*)-(−) 94.1%–100.0% and (*S*)-(+) 0.0%–5.9% according to values found by Casabianca (1996b); Mosandl

(2000) reports the following ratios: (1*S*)-(−)-α-pinene (45.1%–68.1%):(1*R*)-(+)-α-pinene (31.9%–54.9%); (1*S*)-(−)-β-pinene (41.7%–53.6%):(1*R*)-(+)-β-pinene (46.4%–58.3%); (4*S*)-(−)-limonene (74.4%–98.3%):(4*R*)-(+)-limonene (1.7%–25.6%); (1*R*,3*R*,4*S*)-(−)-menthol (min 99.9%):(1*S*,3*S*,4*R*)-(+)-menthol (max 0.1%); and (4*R*)-(−)-piperitone 2.0%–13.0%:(4*S*)-(+)-piperitone 87.0%–98.0%.

20.6.61 Petitgrain Oil Paraguay Type

ISO standard 25157 shows character and data for this oil. A marker for natural quality is (Z,Z)-farnesol with 2.0%–3.5%. Adulteration is done by limonene (80°) and rectified orange terpenes. The chiral ratio of linalool is (*R*)-(−) 71.4%–73.3% and (*S*)-(+) 26.7%–28.6% found by Casabianca (1996a).

20.6.62 Pimento Oils

ISO standard 4729 shows character and data for the oil from *Pimento dioica* (L.) Merr. (Pimento oil Jamaica type). ISO standard 3045 shows character and data for the oil of *Pimenta racemosa* (Mill.) J.W. Moore (Bay oil). The difference between the oils is the content of eugenol and myrcenol. Pimento berry oil shows a content of more than 80% of eugenol and traces of myrcene, while bay oil contains maximum 56% of eugenol but up to 30% of myrcene. Eugenol from cinnamon leaf oil is added as adulteration for both oils and myrcene from other sources. Detection is done by GC-MS.

20.6.63 Rose Oil

ISO standard 9842 shows character and data for oils from various sources. Blending is done with synthetic phenyl ethyl alcohol, synthetic rhodinol, and geraniol from palmarosa oil. Geranium oil, ylang oil, rose absolute, and palmarosa oil are used for "finishing." Detection has to be done by GC-MS system and by chiral separation with multidimensional GC-MS. Kreis and Mosandl (1992) report the enantiomeric ratios for (*S*)-(−)-citronellol with >99%; (2*S*,4*R*)-*cis*-rose oxide as well as (2*R*,4*R*)-*cis*-rose oxide show a purity higher than 99.5%.

20.6.64 Rosemary Oil

ISO standard 1342 shows character and data for oils from various sources. Turpentine oil, synthetic camphor, and limonene from orange terpenes are used to blend this oil. Adulteration is done by 1,8-cineole from eucalyptus or white camphor oil. Detection is usually done by GC-MS but also by multidimensional chiral separation. Enantiomeric ratio of linalool is reported by Casabianca (1996a) to be (*R*)-(−) 23 to (*S*)-(+) 77. Kreis (1991) published the following values of chiral separation: (−)-borneol (84.6%–97.8%):(+)-borneol (2.2%–15.4%) and (−)-isoborneol (29.1%–53.8%):(+)-isoborneol (46.2%–70.9%). König et al. (1992) reported the following chiral ratios: (+)-α-thujene 33.7%:(−)-α-thujene 66.3%; (−)-α-pinene 41.7%:(+)-α-pinene 58.3%; (−)-β-pinene 84.4%:(+)-β-pinene 15.6%; (−)-camphene 83.3%:(+)-camphene 16.7%; (−)-α-phellandrene 0%:(+)-α-phellandrene 100%; (−)-β-phellandrene 2.5%:(+)-β-phellandrene 97.5%; and (−)-limonene 64.8%:(+)-limonene 35.2%. The authors own results analyzing pure oils showed the following ratios: (+)-β-pinene 26.0%:(−)-β-pinene 74.0%; (−)-α-pinene 42.0%:(+)-α-pinene 58.0%; (*S*)-(−)-sabinene 66.5%:(*R*)-(+)-sabinene 33.5%; and (*S*)-(−)-camphene 31.4%:(*R*)-(+)-camphene 68.6%.

20.6.65 Rosewood Oil

ISO standard 3761 shows character and data for this oil. Blending is done by synthetic linalool, α-terpineol, geraniol, and heads of rosewood oil. Eremophilane is a marker according to Lawrence (1999a,b) with values of 0.3%–0.9%. Detection is done by GC-MS on a chiral column. Casabianca (1996a) found chiral ratio for linalool with (*S*)-(+) 50.0%–51% and (*R*)-(−) 49.0%–50%. Lawrence

(1999a,b) reported the following chiral ratios: (3*R*)-(−)-linalool 10.0%:(3*S*)-(+)-linalool 90.0%. (For information, rosewood leaf oil ratio is (3*R*)-(−)-linalool 22.2%:(3*S*)-(+)-linalool 77.8%).

20.6.66 Sage Oil (*Salvia officinalis*)

ISO standard 9909 shows character and data for this oil. As the main components α-thujone β-thujone are known, adulteration is done with thuja oil or cedar leaf oil. β-Caryophyllene, 1,8-cineole, and borneol from other sources are used. α-Humulene is a marker with up to 12% total content. Detection is done by GC-MS analysis.

20.6.67 Sage Oil Spanish Type

ISO standard 3526 shows character and data for this oil derived from *Salvia lavandulifolia* Vahl. Adulteration is done by eucalyptus oil, camphor oil chemotypes like 1,8-cineole and camphor, α-pinene, and limonene from different sources. Blending is done with synthetic linalyl acetate und terpinyl acetate. Detection can be achieved by using GC and GC-MS analytical equipment.

20.6.68 Sandalwood Oil

ISO standard 3518 shows character and data for the oil derived from *Santalum album* L. This very expensive oil with source from India is no longer sold on the market these days. Today other varieties from various sources like New Caledonia (*Santalum austrocaledonicum* Vieill. and *Santalum spicatum* (R.Br.) A. DC) from Australia are used in flavor and fragrance industry. α- and β-Santalols are the main components and responsible for the fragrance. Adulteration will take place with cheaper varieties, coming from Australia. This is very simple to detect as *Santalum spicatum* contains *cis*-nuciferol in high amounts. The same can be seen with *S. austrocaledonicum*, where the *cis*-lanceol can be found in high amounts compared to the other oils. Braun et al. (2005) found in analyzing *S. album* oil the following chiral substances and values: (1*R*,4*R*,5*S*)-α-acorenol (0.22%), (1*R*,4*R*,5*R*)-β-acorenol (0.11%), (1*R*,4*S*,5*S*)-epi-α-acorenol (0.13%), and (1*R*,4*S*,5*R*)-epi-β-acorenol (<0.01%).

20.6.69 Spearmint Oils

ISO 3033-1-3033.4 shows character and data for these oils from various varieties and hybrids. Adulteration is done by synthetic levo-carvone. Detection is done by enantiomeric separation by GC-MS. Coleman et al. (2002), Lawrence and Cole (2011) report the chiral ratio of monoterpenes in native spearmint to be (1*R*,5*R*)-(+)-α-pinene-(40.3%):(1*S*,5*S*)-(−)-α-pinene (59.7%); (1*R*)-(+)-camphene (99.9%):(1*S*)-(−)-camphene (0.1%); (1*R*,5*R*)-(+)-β-pinene (48.7%):(1*S*,5*S*)-(−)-β-pinene (51.3%); and (4*R*)-(+)-limonene (1.9%):(4*S*)-(−)-limonene (98.1%). Further on, Nakamoto et al. (1996) gave results about *cis*- and *trans*-carveol as follows: (2*R*,4*S*)-(+)-*trans*-carveol (15%):(2*S*,4*R*)-(−)-trans-carveol (85%) and (2*S*,4*S*)-(+)-cis-carveol (4%):(2*R*,4*R*)-(−)-*cis*-carveol (96%). Mosandl (2000) reports the following chiral ratios: (1*S*-)-(−)-α-pinene (59.7%–62.4%):(1*R*)-(+)-α-pinene (37.6%–40.3%); (1*S*)-(−)-β-pinene (51.3%–52.1%):(1*R*)-(+)-α-pinene (98.1%–98.8%); (−)-camphene (<0.1%):(+)-camphene (>99.9%), and (4*S*)-(+)-carvone (99.1%–99.9%):(4*R*)-(−)-carvone (0.1%–0.9%).

20.6.70 Spike Lavender Oil

ISO standard 4719 shows character and data for that oil. Adulteration is done by white camphor oil, 1,8-cineole distilled from eucalyptus oil, synthetic camphor, and linalool. Blending is done with terpenes from eucalyptus oil, turpentine oil, *n*-bornyl acetate, lavandin, rosemary oil, HO leaf oil, and α-terpineol. Detection can be made by GC-MS and by multidimensional chiral separation. Ravid (1992) mentions the chiral ratio of terpinen-4-ol as (4S)-(+)-terpinen-4-ol 93%:(4R)-(−)-terpinen-4-ol 7%.

20.6.71 Star Anise Oil

ISO standard 11016 shows character and data for the oil. This oil is not the target for adulteration but is itself an oil for the adulteration of oils containing *trans*-anethole. For the detection of possible adulteration by limonene and α-pinene trace components, *cis*- and *trans*-bergamotene and foeniculin are markers with quantities according to the standard.

20.6.72 Tarragon Oil

ISO standard 10115 shows character and data for this oil. Adulteration and blending are done by synthetic anethole, eugenol from cinnamon leaf oil, synthetic estragole, and *cis*- and *trans* ocimene from other sources. Detection is done either by GC-MS or by 2D chiral separation. The chiral ratio of linalool is (*R*)-(−) 80.0%–90.0% and (*S*)-(+) 10.0%–20.0% found by Casabianca (1996a).

20.6.73 Tea Tree Oil

ISO standard 4730 shows character and data for this oil. Adulteration is done by adding α-pinene and β-pinene from different sources and γ-terpinene from terpenes resulting from citrus concentration production. Synthetic terpinen-4-ol is also used. Detection is done by GC-MS and by using multidimensional chiral analytical equipment. Cornwell et al. (1995) reported the following chiral ratios for steam distillation of whole branches: (+)-sabinene 58%:(−)-sabinene 42%; (+)-α-pinene 91%:(−)-α-pinene 9%; (+)-α-phellandrene 49%:(−)-α-phellandrene 51%; (+)-β-phellandrene 71%:(−)-β-phellandrene 29%; (+)-limonene 62%:(−)-limonene 38%; (+)-terpinen-4-ol 65%:terpinen-4-ol 35%; and (+)-α-terpineol 80%:(−)-α (−)-terpineol 20%.

20.6.74 Thyme Oil

ISO standard 14715 shows character and data for the oil of *T. zygis* (Loefl.) L. For real thyme oil (*T. vulgaris* L.), no ISO standard is available. Origanum oil and marjoram oil from Spain were used for adulteration in earlier times. Also synthetic thymol and carvacrol were applied components. Pure essential oil of thyme is red, as most of the steel vessels contain iron and this reacts with the thymol. White thyme oil, sold in high quantities, is mostly blended and synthetic. By using multidimensional chiral analytical equipment starting in 1990, Herner et al., 1990 reported the following enantiomeric ratios: (1*S*)-(−)-α-pinene 89%:(1*R*)-(+)-α-pinene 11%; (1*S*)-(−)-β-pinene 96%:(1*R*)-(+)-β-pinene 4%; and (4*S*)-(−)-limonene 70%:(4*R*)-(+)-limonene 30%. Kreis et al. (1991) reported the chiral ratio of borneol as follows: (−)-borneol (98.1%–99.6%):(+)-borneol (0.4%–1.9%). The chiral ratio of linalool is (*R*)-(−) 94.5%–99.0% and (*S*)-(+) 1.0%–5.5%. Casabianca (1996a) showed the following value for chiral ratios of linalyl acetate, although this is a minor component: (3*R*)-(−)-linalyl acetate (93.8%–99.2%):(3*S*)-(−)-linalyl acetate (0.8%–6.2%).

20.6.75 Turpentine Oil

ISO standard 21389 shows character and data for this oil. It is distilled from *Pinus massoniana* Lamb. and produced in China. Adulteration is rarely observed, but this oil is a source for the production of α-pinene and β-pinene. Both compounds are used for the adulteration of various essential oils containing these monoterpenes in higher value. Another type of turpentine oil is distilled from *Pinus pinaster* Aiton but is less of interest for adulterations as the quantities produced are minor.

20.6.76 Vetiver Oil

ISO standard 4716 shows character and data for this oil. Limonene chiral ratio is (*R*)-(−) with 100%. The identical value was reported by Möllenbeck et al. (1997) for α-terpineol. He too found a ratio of linalool to be (*R*)-(−) 80% and (*R*)-(+) with 20% as well as terpinen-4-ol to be (*R*)-(−) 66% and (*S*)-(+) with 34%. Because of use GC×GC-TOF for detection it will be impossible to adulterate this essential oil.

20.6.77 Ylang-Ylang Oils

ISO standard 3063 shows character and data for this oil, distilled from the flowers of *C. odorata* (Lam.) Hook. F. Thomson forma genuina. Adulteration is done either by one of the fractions of cananga oil (extra, first, second, or third) or with various fractions like heads and tails. The mingling up of various fractions like extra, I, II, or III will not touch the purity or naturality of that oils as long as all are from the same botanical source. *C. odorata* (Lam.) Hook. F. Thomson forma *macrophylla* is used for adulteration. Blending is done by linalyl acetate, benzyl acetate, synthetic geraniol from various sources, methyl benzoate, benzyl alcohol, methyl salicylate, all synthetic and bay leaf oil, cedarwood terpenes, lavandin residues, and traces of ethyl vanillin (synthetic). Detection is done by GC/MS and by multidimensional chiral separation. Casabianca (1996a) reported the chiral ratio for linalool to be (*S*)-(+) 1.6–3.0 and (*R*)-(−) 97.0%–98.4%.

REFERENCES

Adams R.P. 2007. *Identification of Essential Oil Components by Gas Chromatography/Mass Spectrometry*, 4th edn., Allured Publishing Corp., Carol Stream, IL.

AFNOR (Association française de normalisation). 1992. NF T 75-254, Huile essentielle de cyprès, Tour Europe 92049 Paris La Défense Cedex.

Atal C.K., and K.C. Shah. 1966. TLC patterns of some volatile oils and crude drugs and their adulterants. *Indian J. Pharm.*, 28(6), 162–163.

Bail S. May/June 2009. Antimicrobial activities of Roman chamomile oil from France and its main compounds. *J. Essent. Oil Res.*, 21, 283–285.

Baldovini N. 2010. Phytochemistry of the heartwood from fragrant *Santalum* species: A review. *Flavour Fragr. J.*, 26, 7–26.

Bernreuther A., and Schreierc P., 1991. Multidimensional gas chromatography/mass spectrometry: A powerful tool for the direct chiral evaluation of aroma compounds in plant tissue II. Linalool in essential oils and fruits. *Phytochem. Anal.*, 2, 167–170.

Bovill E.W. 2000. The essential oil market in 1934. published by Lawrence B. *Perfum. Flavor.*, 25, 22–32.

Braun M., and Franz C. 2001. Chirale säulen decken Verfälschungen auf. *Pharm. Ztg.*, 146, 2493–2499.

Braun N. et al. 2005. Santalum spicatum (R.Br) DC. (Santalaceae)—Nor-helifolenal and acorenol isomers: Isolation and biogenetic considerations. *J. Essent. Oil Res.*, 15, 381–386.

BRENDA. 2007. *The Comprehensive Enzyme Information System*. Technische Universität Braunschweig, Braunschweig, Germany. http://www.brenda-enzymes.org/php/ (accessed February 2014).

Brophy J.J. et al. 1989. Gas chromatographic quality control for oil of *Melaleuca* terpinen-4-ol type. *J. Agric. Food Chem.*, 37, 1330–1335.

Busch K.W., and Busch M.A. 2006. *Chiral Analysis*. Elsevier B.V., Amsterdam, The Netherlands.

Carle R. 1996. Kamillenöl—Gewinnung und Qualitätsbeurteilung, Deutsche Apotherker Zeitung, 136. Jahrgang, Nr. 26, pp. 17–28.

Carle R. et al. 1990. Studies on the origin of (−)-α-bisabolol and chamazulene in chamomile preparations. Part 1. Investigations by isotope ratio mass spectrometry (IRMS). *Plant Med.*, 56, 456–460.

Casabianca H. 1996a. Le Point Sur L'Analyse Chirale Du Linalool Et De L'Acetate De Linalyle Dans Diverses Plantes. *Rivista Italiana Eppos*, 7, spi, pp. 227–243.

Casabianca H., Graff J. B., Faugier V., and Fleig Grenier C. 1998. Enantiomeric Distribution Studies of Linalool and Linalyl Acetate. A Powerful Tool for Authenticity Control of Essential Oils. *J. High Resolut. Chromatogr.* 21(2), 107–112.

Casabianca H. 1996b. Analyses Chirales Et Isotopiques Des Principaux Constituants De Roses Et De Geraniums. *Rivista Italiana Eppos*, 7, spi, pp. 244–261.

Coleman W.M., Lawrence B.M., and Cole S.K.. 2002. Semiquantitative determination of off-notes in mint oils by solid-phase microextraction. *J. Chromatogr. Sci.*, 40, 133–139.

Cornwell C.P. et al. November/December 1995. Incorporation of oxygen-18 into Terpinen-4-ol from the $H_2^{18}O$ steam distillates of *Melaleuca alternifolia* (Tea Tree). *J. Essent. Oil Res.*, 7, 613–620.

Cotroneo, A. et al. 1987. Sulla genuinità delle essenze agrumarie. Nota XVI. Differenze quantitative nella composizione di essenze di limone estratte a macchina e di essenze estratte manualmente senza uso di acqua. *Essenz. Deriv. Agrum.*, 57, 220–235.

Dugo G. et al. September/October 1992. High resolution gas chromatography for detection of adulterations of citrus cold-pressed essential oils. *Perf. Flavor.*, 17, 57–74.

Dugo G. 1993. On the genuineness of citrus essential oils. *Flavour Fragr. J.*, 8, 25–33.

Dugo D.L. 2011. *Citrus Oils, Advanced Analytical Techniques, Contaminant, and Biological Activity.* CRC Press, Taylor & Francis Group, Boca Raton, FL, p. 245.

Dugo G., and Mondello L. 2011. *Composition of the Volatile Fraction of Citrus Peel Oils in Citrus Oils—Composition, Advanced Analytical Techniques, Contaminants, and Biological Activity.* Taylor & Francis Group, Boca Raton, FL, p. 147.

Formácek V. 2002. *Essential Oil Analysis by Capillary Gas Chromatography and Carbon-13 NMR Spectroscopy.* John Wiley & Sons, Ltd., Chichester, U.K.

Gnitka R. 2010. Efficient method of isolation of pure (−)-α- and (+)-β-thujone from Thuja occidentalis essential oil. Poster presented at the *Symposium on Essential Oils*, Wroclaw, Poland.

Hara F. et al. 1999. The analysis of some chiral components in citrus volatile compounds. *Proceedings 43rd TEAC Meeting*, Oita, Japan, pp. 360–362.

Hatt H. et al. 2011. Wo Düfte ihren Anfang nehmen, Spektrum der Wissenschaft. *Gehirn und Geist*, Spektrum der Wissenschaft. Gehirn und Geist, (3), 38–241.

Herner U. et al. 1990. Enantiomeric distribution of α-pinene and β-pinene and limonene in essential oils and extracts. Part 2. Oils perfumes and cosmetics. *Flavour. Fragr. J.*, 5, 201–205.

ISO 3679:2015-06. Determination of flash no-flash and flash point — Rapid equilibrium closed cup method, Geneva, Switzerland, http://www.iso.org/"www.iso.org

ISO. International Organization for Standardization TC 54, Standard catalogue, ISO/TC 54, 2015. http://www.iso.org/iso/home/store/catalogue_tc/catalogue_tc_browse.htm?commid=48956 (accessed April 2014).

ISO/DIS 279, Essential oils—Determination of relative density at 20 degrees C - Reference method, 1998, Geneva, Switzerland, http://www.iso.org

ISO/DIS 280, Essential oils—Determination of refractive index, 1998, Geneva, Switzerland, http://www.iso.org

ISO/DIS 592, Essential oils—Determination of optical rotation, 1998, Geneva, Switzerland, http://www.iso.org

ISO/DIS 875©. Essential oils—Evaluation of miscibility in ethanol, ISO. 1999, Geneva, Switzerland, http://www.iso.org (accessed January 2014).

ISO/DIS 1241©. Essential oils—Determination of ester values, before and after acetylation, and evaluation of the contents of free and total alcohols, ISO, 1996, Geneva, Switzerland, http://www.iso.org (accessed January 2014).

ISO/DIS 11021, Essential oils—Determination of water content—Karl Fischer method, 1999, Geneva, Switzerland, http://www.iso.org

ISO/DIS 1272©. 2000. Essential oils—Determination of content of phenols, ISO, 2000, Geneva, Switzerland, http://www.iso.org (accessed January 2014).

ISO/DIS 1279©. 1996. Essential oils—Determination of carbonyl value—Potentiometric methods using hydroxylammonium chloride, ISO, 1996, Geneva, Switzerland, http://www.iso.org (accessed January 2014).

ISO/DIS 3064©. Essential oil of petitgrain, Paraguayan type (Citrus aurantium L. var. Paraguay (syn. Citrus aurantium var. bigaradia Hook f.)), ISO, 2010, Geneva, Switzerland, http://www.iso.org (accessed January 2014).

ISO/DIS 3679, Determination of flash no-flash and flash point—Rapid equilibrium closed cup method, ISO, (accessed march 2015).

ISO/DIS 4715©. Essential oils—Quantitative evaluation of residue on evaporation, ISO, 1978, Geneva, Switzerland, http://www.iso.org (accessed January 2014).

ISO/DIS 7660©. 1983. Essential oils—Determination of ester value of oils containing difficult-to-saponify esters, ISO, 1983, Geneva, Switzerland, http://www.iso.org (accessed January 2014).

ISO/DIS 8901©. Oil of bitter orange petitgrain, cultivated (Citrus aurantium L.), ISO, 2010, Geneva, Switzerland, http://www.iso.org (accessed January 2014).

ISO/DIS 9235. Aromatic natural raw materials—Vocabulary, ISO, 2013, Geneva, Switzerland, http://www.iso.org.

ISO/DIS 9842. Oil of rose (Rosa × damascena Miller), 2003, ISO International.

ISO/DIS 11021©. Essential oils—Determination of water content—Karl Fischer method, ISO, 1999, Geneva, Switzerland, http://www.iso.org (accessed January 2014).

Jaspersen-Schib R., and H. Flueck. 1962. Identification and purity determination of essential oils by thin layer chromatography. *Congr. Sci. Farm. Conf. Commun.*, 21, Pisa, Italy, 1961, pp. 608–614.

John, M.D. et al. 1991. Detection of adulteration of polyethylene glycols in oil of sandalwood. *Indian Perf.*, 35(4), 186–187.
Juchelka D. 1996a. Authenticity profiles of bergamot oil. *Pharmazie*, 51(6), 418.
Juchelka D. 1996b. Chiral compounds of essential oils. XX. Chirality evaluation and authenticity profiles of neroli and petitgrain oils. *J. Essent. Oil Res.*, 8, 487–497.
Kartnig T. et al. 1999. Gaschromatographische Untersuchungen an Ätheroleum Juniperi unter besonderer Berücksichtigung der Trennung enantiomerer Komponenten. *Sci. Pharm.*, 67, 77–82.
König W.A. et al. 1992. Enantiomeric composition of the chiral constituents in essential oils. Part 1: Monoterpene hydrocarbons. *J. High Res. Chromatogr.*, 15(3), 184–189.
Koenig W.A.D., Joulain D.H., Hochmuth S.A., Robertet, and G. Hochmuth. 2004. *Terpenoids and Related Constituents of Essential Oils.* MassFinder 3: Convenient and Rapid Analysis of GCMS, Hamburg, Germany.
Kovats E. 1958. Characterization of organic compounds by gas chromatography. Part 1: Retention indices of aliphatic halides, alcohols, aldehydes and ketones *Helv. Chim. Acta*, 41, 1915.
Kreis P. et al. 1991. Chirale Inhaltsstoffe ätherischer Öle. Deutsche Apotheker Zeitung, Nr. 39 pp. 1984–1987.
Kreis P., Juchelka D., Motz C., and Mosandl A. 1991, Chirale Inhaltsstoffe aÅNtherischer OÅNle. IX: Stereodifferenzierung von Borneol, Isoborneol und Bornylacetat. *Dtsch. Apoth. Ztg.*, 131, 1984–1987.
Kreis P., and Mosandl A. 1992a. Chiral compounds of essential oils. Part XI. Simultaneous stereoanalysis of *Lavandula* oil constituents. *Flavour Fragr. J.*, 7, 187–193.
Kreis P., and Mosandl A. 1992b. Chiral compounds of essential oils. Part XII. Authenticity control of rose oils, using enantioselective multidimensional gas chromatography. *Flavour Fragr. J.*, 7, 199–201.
Kubeczka K.H., and Bohn I. 1985. Pimpinella root and its adulteration. Detection of adulteration by thin-layer and gas chromatography. Structure revision of the principal components of the essential oils. *Deutsche Apotheker Zeitung*, 125(8), 399–402.
Kubezka K.H., and Formácek V. 2002. *Essential Oil Analysis by Capillary Gas Chromatography and Carbon-13 NMR Spectroscopy.* John Wiley & Sons, Ltd., Chichester, U.K.
Kumar and Madaan 1979, *Handbook of Herbs and Spices*, Chapter 3. 335, p 48, Woodhead Publishing Ltd.
Lawrence B. May/June 1996a. Progress in essential oils, Vol. 21, November/December 1996. *Perfum. Flavor.*, 21, 64.
Lawrence B. November/December 1996b. Dill oil, Progress in essential oils, Vol. 21, May/June. *Perfum. Flavor.*, 21, 59.
Lawrence B. May/June 1999a. Progress in essential oils. *Perfum. Flavor.*, 24, 2–7.
Lawrence B. January/February 1999b. Progress in essential oils. *Perfum. Flavor.*, 11.
Lawrence B. 2009. A preliminary report on the world production of some selected essential oils and countries. *Perfum. Flavor.*, 34, 38–44.
Marsili R. 2010. *Flavor, Fragrance and Odor Analysis*, 2nd edn. Ray Marsili, CRC Press, Taylor & Francis Group, Boca Raton, FL.
McHale D. et al. February/March 1983. Detection of adulteration of cold-pressed bitter orange oil. *Perfum. Flavor.*, 8, 40–41.
Merriam-Webster. n.d. http://www.merriam-webster.com/dictionary/iodine%20number.
Möllenbeck S., König T., Schreier P., Schwab W., Rajaonarivony J., and Ranarivelo L. 1997. Chemical composition and analyses of enantiomeres of essential oils from Madagascar. *Flavour Frag. J.*, 12, 63–69.
Mondello L. April 1998. Multidimensional tandem capillary gas chromatography system for the analysis of real complex samples. Part I: Development of a fully automated tandem gas chromatography system. *J. Chromatogr. Sci.*, 36, 206.
Mondello L. 2011. *Flavour and Fragrance Natural and Synthetic Compounds 2*, 2nd edn. John Wiley & Sons, Ltd., Chichester, U.K.
Mosandl A. et al. 1991. Stereoisomeric flavor compounds 48. Chirospecific analysis of natural flavors and essential oils using multidimensional gas chromatography. *J. Agric. Food Chem.*, 39, 1131–1134.
Mosandl A. 1998. Enantioselective analysis, in: *Flavourings*, Edited by Ziegler E., Ziegler H., Wiley-VCH Verlag GmbH & Co. KGaA, Weinheim, Germany.
Mosandl A. 2000. Authenticity assessment of essential oils—The current state and the future. *Lecture hold on the 31st International Symposium on Essential Oils*, Hamburg, Germany.
Mostafa M. M., Gomaa M. A., and El-Masry M. H. (1990a), 'Physico-chemical properties as atool to detect adulteration of some Egyptian volatile oils', *Egyptian J Food SCi*, 16(1/2), 63–77.
Nakamoto et al. 1996. Enantiomeric distributions of carveols in grapefruit, orange and spearmint oils. *IFEAT Proceedings International Conference of Aromas and Essential Oils*, IFEAT Secretariat, Tel Aviv, Israel, pp. 36–50.

Nova D., Karmazin M., and Buben I. 1986. Anatomical and chemical discrimination between the roots of various varieties of parsley (*Petroselinum crispum* Mill./A. W. Hill.) and parsnip (*Pastinaca sativa* L. ssp. sativa). *Cesko-Slovenska Farmacie*, 35(8), 363–366.

Ochocka J.R. 2002. Determination of enantiomers of terpene hydrocarbons in essential oils obtained from species of Pinus and Abies. *Pharm. Biol.*, 40, 395–399.

Ohloff G. 1990. *Riechstoffe und Geruchssinn: Die molekulare Welt der Düfte*. Springer-Verlag, Berlin, Germany.

ÖKO Test. 1995. Gepanschte Seelen, Oktober/95, ÖKO-TEST Verlag GmbH, Frankfurt, Germany, http://www.germany.ru/wwwthreads/files/1646–4014361-__196_t___246_le___214_kotest.pdf.

Peyron L. et al. 1974. Abnormal presence of hexylene glycol in commercial essential oils of vetiver and cananga. *Plantes Medicinales et Phytotherapie* (1975), 9(3), 192–203.

Ph.Eur, European Pharmacopoeia 7.4. 2012. European Directorate for the Quality of Medicines & HealthCare http://www.edqm.eu/en/edqm-homepage-628.html.

Phokas G. 1965. Thin-layer chromatography of the essential oil of *Melissa officinalis* and some of its adulterants. *Pharm. Deltion Epistemonike Ekdosis*, 5(1–2), 9–16.

Rajeswara Rao B.R. et al. 2002. Water soluble fractions of rose-scented geranium (*Pelargonium* species) essential oil. *Bioresource Technol.*, 84, 243–246.

Rajeswara B.R. 2007. Effect of crop-weed mixed distillation on essential oil yield and composition of five aromatic crops. *JEOBP*, 10(2), 127–132.

Ravid U. 1992. Chiral GC analysis of enantiomerically pure Fenchone in essential oils. *Flav. Fragr. J.*, 7, 169–172.

Reeve D. et al. July/August 2002. Riding the citrus trail: When is a mandarin a tangerine. *Perf. Flavor.*, 27, 20–22.

Renaud E.N.C. 2001. Essential oil quantity and composition from 10 cultivars of organically grown lavender and lavandin. *J. Essent. Oil Res.*, 13, 269–273.

Reunion Islands National Institute of Statistics (INSEE). 2008–2009. 10.1: Revenus et production agricoles, 10.1.3—Production végétale, INSEE-RÉUNION—TER 2008–2009, p. 185, http://www.insee.fr/fr/insee_regions/reunion/themes/dossiers/ter/ter2008_production_vegetale.pdf.

Rochussen F. 1920. *Ätherische Öle und Riechstoffe*. Walter de Gruyter & Co., Berlin, Germany, pp. 19, 23, 41, 69–72.

Sawamura M. et al. 2004. Compositional changes in commercial mandarin essential oil for aromatherapy, Poster presented at the *35th International Symposium on Essential Oils*, Giardini Naxos, Italy.

Schimmel-Bericht & Co. Edition. April/May 1919. Schimmel & Co., Leipzig, Germany, pp. 4, 13, 15, 22, 23, 36, 41–43, 48, 51, 52, 66.

Schimmel-Bericht & Co. Edition. April/October 1920. Schimmel & Co., Leipzig, Germany, pp. 30–31.

Schmidt E. 2010. Production of essential oils, in: *Handbook of Essential Oils*, Edited by Baser K.H.C. and G. Buchbauer, CRC Press, Taylor & Francis Group, Boca Raton, FL.

Schmidt E. July/August 2003. The characteristics of lavender oils from Eastern Europe, *JEOR*, 28, 48–160.

Schultze W. 1993. Differentiation of original lemon balm oil (*Melissa officinalis*) from several lemonlike smelling oils by chirospecific GC analysis of citronellal and isotope ratio mass spectrometry. *Planta Med..*, 59(Suppl.), A635.

Sen A.R., Sen Gupta P., and Ghose Dastidar N. 1974. Detection of *Curcuma zedoaria* and *C. aromatica* in *C. longa* (turmeric) by thin-layer chromatography. *Analyst* (Cambridge, United Kingdom), *Analyst*, 99(1176), 153–155.

Sewenig S. et al. 2003. Online determination of $^{2}H/^{1}H$ and $^{13}C/^{12}C$ isotope ratios of cinnamaldehyde from different sources using gas chromatography isotope mass spectrometry. *Eur. Food Res. Technol.*, 217(5), 444–448.

Teisseire P. 1987. Industrial quality control of essential oils by capillary GC, in: *Capillary Gas Chromatography in Essential Oil Analysis*, Edited by P. Sandra, and Bicchi C., Huethig, Basel, NY.

The Free Dictionary. n.d. (a). http://www.thefree dictionary.com/adulteration (accessed February 2014).

The Free Dictionary. n.d. (b). http://www.thefreedictionary.com/cupidity (accessed February 2014).

The Free Dictionary. n.d. (c). http://www.thefreedictionary.com/enzymatic (accessed February 2014).

Touche J. et al. 1981. Maillettes et lavandes fines francaises, Rivista Italiana E.P.P.O.S., LXIII - n. 6 – settembre-ottobre, pp. 320–323.

Valder C. 2003. Western Australian sandalwood oil—New constituents of *Santalum spicatum* (R.Br) A DC. (Santalaceae). *J. Essent. Oil Res.*, 15, 178–186.

Van Den Dool H., and Kratz P.D. 1963. A generalization of the retention index system including linear temperature programmed gas—liquid partition chromatography. *J. Chromatogr.*, 11, 463–471.

Venskutonis P.R. 2002. *Thyme—The Genus Thyme*. Taylor & Francis, New York, p. 226.

Wang X.-H. et al. 1995. The direct chiral separation of some optically active compounds in essential oils by multidimensional gas chromatography. *J. Chromatogr. Sci.*, 33, 22–25.

21 Essential Oils and Volatiles in Bryophytes

Agnieszka Ludwiczuk and Yoshinori Asakawa

CONTENTS

21.1 INTRODUCTION

There are about 20,000 species of bryophytes in the world. The bryophytes, including liverworts (Marchantiophyta), hornworts (Anthocerotophyta), and mosses (Bryophyta) (Figure 21.1), are small non-vascular, spore-forming plants that are placed taxonomically between algae and pteridophytes (Asakawa et al., 2013b). Bryophytes are considered to be the oldest terrestrial plants. Most plant biologists believe that they evolved from a single algal ancestor in the Charophyta. Among bryophytes, liverworts are the earliest embryophyte lineage, and are the sister of all other land plants. Hornworts have been described as the sister to vascular plants, while mosses were placed between the liverworts and the hornworts (Qiu et al., 2006). The key features that probably enabled early plants to thrive on land are the rigid cell wall; several plastids per cell; metabolites enabling protection, for example, to UVB radiation; and the mutualistic interaction with fungi to gain access to inorganic nutrients (Rensing, 2018).

Bryophytes can grow on various substrates (soil, humus, peat, rocks, stones, leaves, and bark of trees) and can be found almost everywhere, from the Arctic regions by the temperate zone, to the hot and humid tropics, where they have reached a huge diversity. Some of the bryophytes are also known as aquatic plants, since they can leave under the water surface. In some landscapes, such as boreal peatlands, montane tropical rainforests, and unglaciated parts of Antarctica, they are an important component of ecosystem biomass and play significant roles in ecosystem functioning. Many of them are pioneers on bare and distributed habitats, helping other plants to gain a foothold (Bednarek-Ochyra et al., 2000; Crandall-Stotler et al., 2008, 2009; Vanderpoorten and Goffinet, 2009). As the first inhabitants of terrestrial habitats, they were frequently exposed to adverse environmental conditions, such as pathogen attack, insect predation, and UV injury (Xie and Lou, 2009; Whitehead

FIGURE 21.1 (**See color insert.**) Representatives of bryophytes: (a) moss, (b) liverwort, (c) hornwort.

et al., 2018). In general, bryophytes display a low morphological complexity, but a high degree of chemical diversification (Asakawa, 1982, 1995; Asakawa et al., 2013b; 2013c; Ludwiczuk and Asakawa, 2014), suggesting that secondary metabolites, and especially terpenoids, may play an important role in bryophyte–environment interactions (Chen et al., 2018; Whitehead et al., 2018).

Compared with angiosperms, bryophytes are seldom utilized by insects, animals, and humans as food. These plants are assumed to be unpalatable, and three classes of mechanisms have been suggested as possible barriers of herbivory on bryophytes: chemical defenses, low digestibility, and low nutrient content (Haines and Renwick, 2009). However, a number of bryophytes have been widely use as medicinal plants in China and North America as decoctions or crushed, and the resulting powder mixed with oil, to cure burns, bruises, external wounds, fractures, snake bites, convulsions, uropathy, pneumonia, neurasthenia, etc. (Glime, 2007; Asakawa, 2008; Harris, 2008; Asakawa et al., 2013a). This chapter reviews current knowledge of terpenoid secondary metabolites and other volatile components in bryophytes from three perspectives: chemical diversity, chemotaxonomy, and biological functions.

21.2 TERPENOIDS AND OTHER VOLATILES FROM BRYOPHYTES

Among the bryophytes, the liverworts are an extremely rich source of terpenoids and other volatiles; however, a few mosses and hornworts are also known to produce such kinds of components. A number of bryophytes are known to emit volatile terpenoids and simple aromatic compounds when crushed (Asakawa, 1982, 1995; Asakawa et al., 2013b;c). In case of liverworts, the characteristic odor is associated with constituents of oil bodies. Oil bodies occur only in liverworts and are unique intracellular organelles bound by unit membranes that are structurally asymmetric like the tonoplast and are filled with osmiophilic globules suspended in a matrix of carbohydrates and proteins (Figure 21.2) (Suire, 2000).

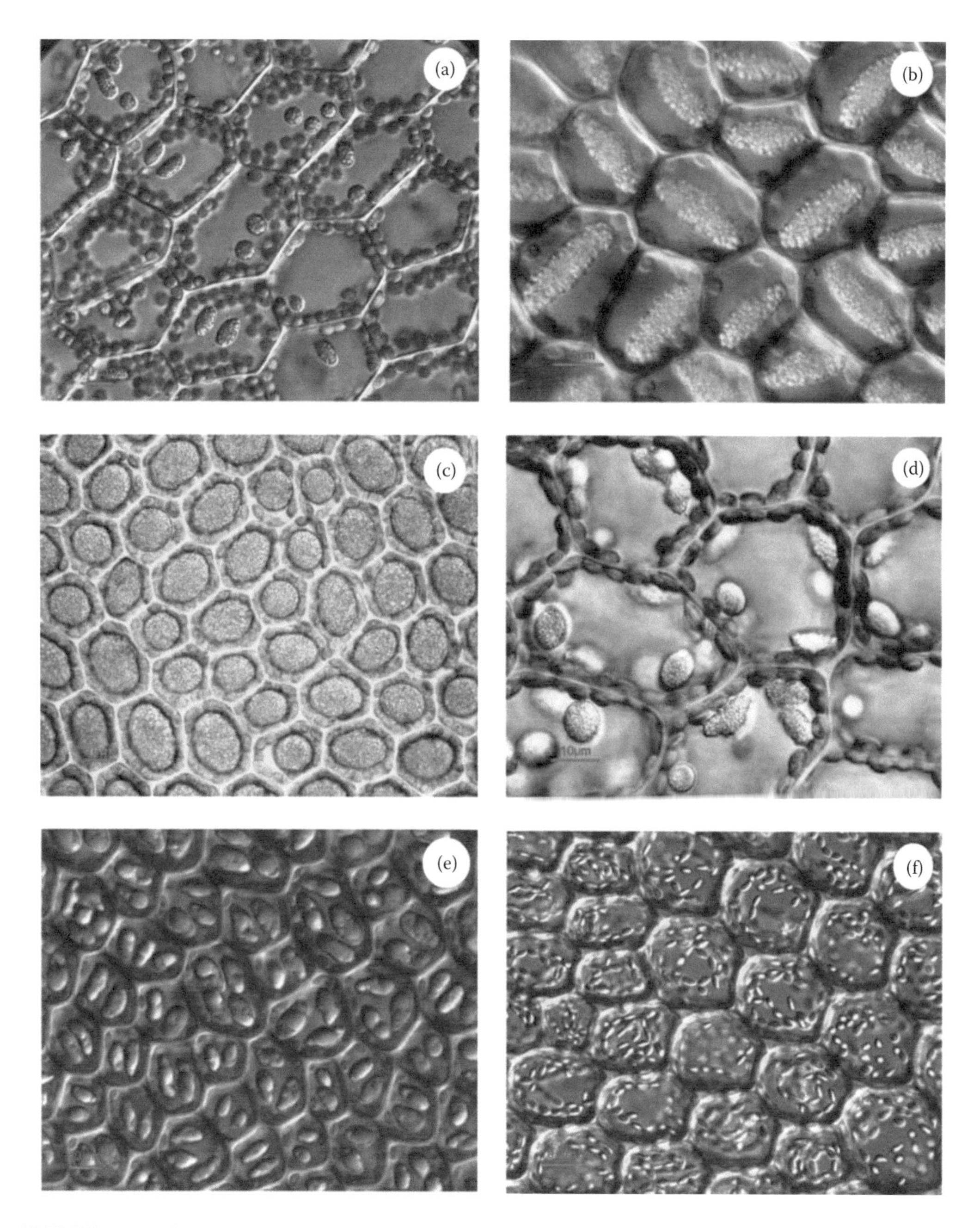

FIGURE 21.2 (**See color insert.**) Different types of oil bodies in the cells of liverworts: (a) *Calypogeia azurea*, (b) *Cheilolejeunea anthocarpa*, (c) *Radula constricta*, (d) *Solenostoma truncatum*, (e) *Bazzania tridens*, and (f) *Trocholejeunea sandvicensis*. (From He et al. 2013. *Crit. Rev. Plant Sci.*, 32: 293–302. With permission.)

They occur in both gametophyte and sporophyte generations (Duckett, 1986), and are absent from fewer than 10% of liverwort species. Variations occur in oil body size, shape, color, number, and distribution among taxa (Crandall-Stotler et al., 2008, 2009). The contents of the oil bodies are easily extracted with solvent; for example, by use of ultrasonic bath or just distillate the essential oil (Asakawa et al., 2013b; Asakawa and Ludwiczuk, 2013). The literature review shows that most of the scientific work on bryophyte chemistry concerns research on volatile compounds found in extracts than on essential oil (EO) composition. This is due to the fact that bryophytes are morphologically small plants and it is difficult to collect a sufficiently large amount of plant material for research (Ludwiczuk and Asakawa, 2014).

21.2.1 Chemical Diversity of Natural Volatiles in Bryophytes

The diversity of volatile compounds in bryophytes, especially terpenoids, is enormously rich. While many terpenoids are observed in both bryophytes and seed plants, some are unique to bryophytes.

21.2.1.1 Liverwort Components

The first studies concerning liverworts components were published in 1905 by Karl Müller (1905). He believed that the characteristic odor of many liverworts is associated with oil-body constituents, which are mono- and sesquiterpenes and corresponding alcohols, and noticed the odor of various liverwort species, for example, *Jungermannia obovata* (carrot), *Lophozia bicrenata* (cedar oil), *Geocalyx* sp. (turpentine), and *Leptolejeunea* sp. (licorice) (Müller, 1905; Harrison, 1983). Since that time, over the next 50 years, phytochemical studies of the liverworts has been neglected completely, mostly because of the difficulty of obtaining a large quantity of plant material for chemical analysis. Isolating pure compounds and determining chemical structures with small amounts of plant material became possible and easier due to the development of analytical methods. Thus, reports dealing with isolation, structure elucidation, and chemistry of terpenoids and volatile aromatic compounds of the liverworts in the early 1970s and latter have increased dramatically. Over the last 40 years, more than 1600 compounds belonging to the terpenoids have been reported from this plant group (Asakawa, 1982, 1995; Asakawa et al., 2013b; He et al., 2013).

When liverworts are crushed, an intense mushroom-like, sweet woody, or seaweed scent is emitted. The presence of 1-octen-3-ol (**1**) and its acetate is responsible for the mushroom-like scent of a number of liverworts. Generally, 1-octen-3-yl acetate is more abundant than the free alcohol. A Tahitian liverwort, *Cyathodium foetidisimum* emits an incredibly unpleasant odor, for which the presence of skatole (**2**) is responsible (Ludwiczuk et al., 2009; Sakurai et al., 2018). The stink bug smell of the New Zealand *Chiloscyphus pallidus* is attributable to (*Z*)-, and (*E*)-pent-2-enal (**3**), (*Z*)-dec-2-enal (**4**), and (*E*)-dec-2-enal (Toyota and Asakawa, 1994). *Cheilolejeunea imbricata* produces strong milky smelling (*R*)-dodec-2-en-1,5-olide (**5**) and (*R*)-tetradec-2-en-1,5-olide. A very tiny liverwort, *Leptolejeunea elliptica* emits sweet mold-like odor, which is mainly due to the presence of 1-ethyl-4-methoxybenzene (**6**) (Toyota et al., 1997).

Liverworts are also known to produce dictyopterenes, which are known as sex pheromones of marine brown algae, such as *Ectocarpus siliculosus*, *Dictyopteris membranacea*, and *Cutleria multifida* (Kajiwara et al., 1989; Boland, 1995). These pheromones are lipophilic, volatile acetogenins consisting of C_8 or C_{11} linear or monocyclic hydrocarbons or their epoxides (Kobayashi and Ishibashi, 1999). The liverwort species *Fossombronia angulosa*, collected in Greece, produced dictyopterene (**7**) as the major volatile component, as well as dictyotene (**8**) and multifidene (**9**) (Ludwiczuk et al., 2008). Dictyotene (**8**) and (*E*)-ectocarpene (**10**) were also found in the Tahitian *Chandonanthus hirtellus* as the major volatiles (Ludwiczuk et al., 2009).

Monoterpenoids are compounds found in the essential oils obtained from many aromatic plants. These compounds contribute to the flavor and aroma of the plant from which they are extracted. However, this group of terpenoids is not dominant in terms of liverworts components. The available literature data show that monoterpenoids are usually present in complex thalloid liverworts. These terpenes are the major volatile components found in the liverworts belonging to genera of *Asterella*, *Conocephalum*, and *Wiesnerella*, and the most characteristic compounds are sabinene (**11**), α-pinene (**12**), limonene (**13**), and acetate derivatives of borneol (**14**), myrtenol, nerol (**15**), and geraniol (Ludwiczuk et al., 2008; Ludwiczuk and Asakawa, 2015). Selected fragrant components found in liverworts are presented in Figure 21.3.

The most characteristic and diverse group of volatile components present in the extracts as well as in the essential oils from liverworts are sesquiterpenoids. The first well-documented study on liverwort

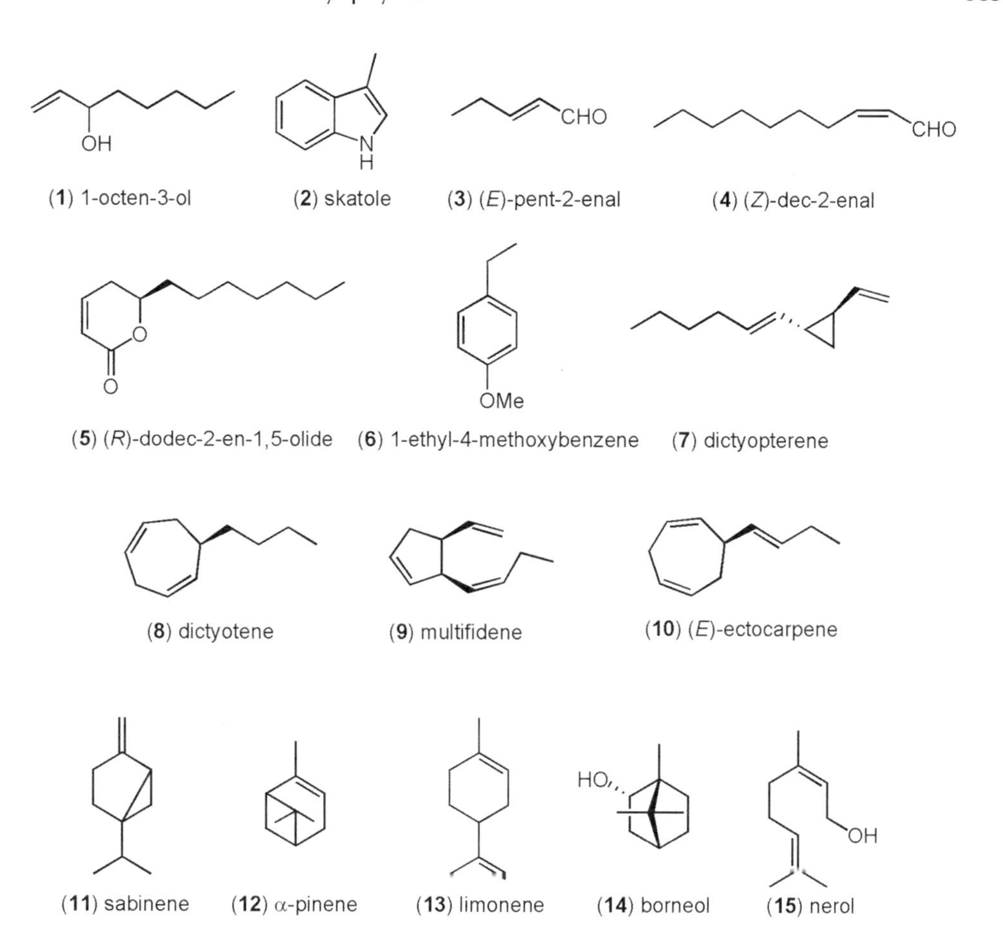

FIGURE 21.3 Fragrant components from liverworts.

components was conducted by Huneck and Klein (1967). These authors indicated the presence of (−)-longiborneol (**16**) and (−)-longifolene (**17**) in the EO from the leafy liverwort *Scapania undulata*. Andersen and coworkers (1977a) confirmed the presence of both components in the EO of *S. undulata* and additionally identified longipinanol, α- and β-longipinene, longicyclene from the longifolane group and also anastreptene (**18**), β-barbatene (**19**), caryophyllene, α- and γ-himachalene, and β-chamigrene. Up to now, about 900 compounds have been isolated and/or detected in the Marchantiophyta (Asakawa et al., 2013). Liverworts often produce unique sesquiterpenoids with carbon skeletons which are not isolated from higher plants, and the absolute configuration of most of the liverwort sesquiterpenoids correspond to the enantiomer of those of higher plants. For example, the (+)-enantiomer of cadina-3,5-diene was isolated from *Conocephalum conicum*, while the (−)-enantiomer was present in the angiosperms, *Piper cubeba* and *Leptospermum scoparium* (Manuka, or New Zealand tea tree) (Melching et al., 1997). The liverwort, *Reboulia hemisphaerica* produces (+)-thujopsene, while the (−)-enantiomer is a major component of cedar wood oil (Asakawa et al., 2013b). Exceptions to this include frullanolide, which occurs in both enantiomeric forms in the genus *Frullania* (Knoche et al., 1969; Asakawa, 1982), and β-caryophyllene, which is found in *Pellia endiviifolia*, *P. epiphylla*, and *Metzgeria conjugata* as a mixture of (+) and (−) forms (Fricke et al., 1995).

Sesquiterpene components occurring in the liverworts belong to more than 60 different skeletal groups, among which the eudesmane and aromadendrane skeletons are most prevalent. Others, like cuparane, pinguisane, and barbatane (= gymnomitrane) are also quite common. Liverworts are plants in which one can find relatively rare groups of naturally occurring compounds. These are *seco*-africanes (e.g., **20**), noraristolanes, 1,10-*seco*- (e.g., **21**) and 2,3-*seco*-aromadendranes

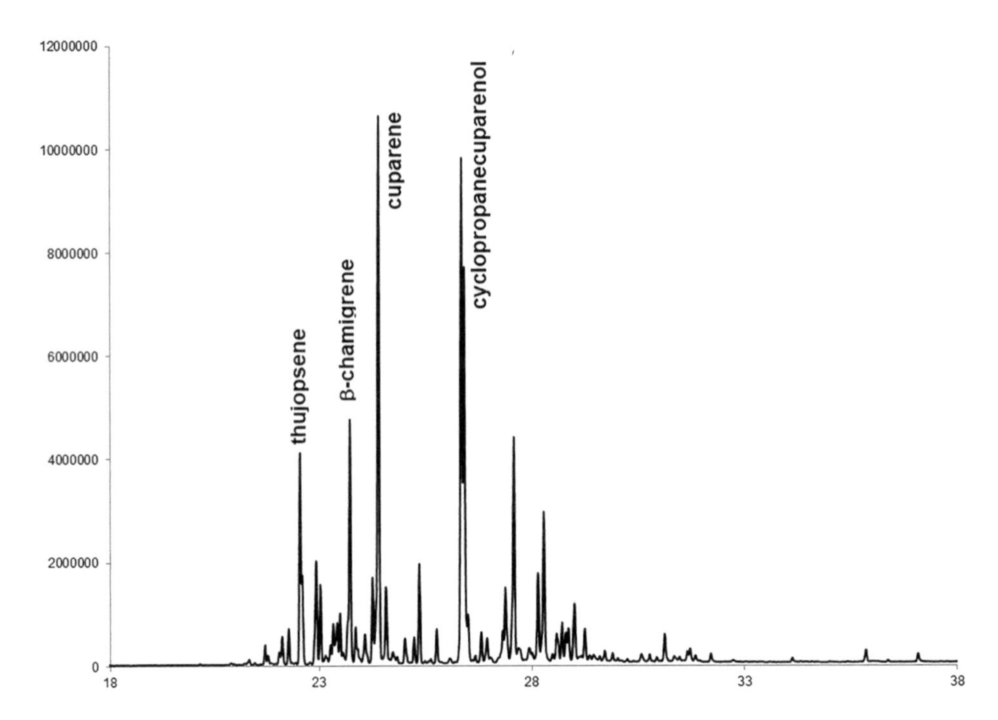

FIGURE 21.4 Gas chromatogram of the essential oil of *Marchantia polymorpha* subsp. *ruderalis*.

(e.g., **22**), *seco*-cuparanes, neotrifaranes, chenopodanes, and ricciocarpanes, among others. To date, dumortane- (e.g., **23**) and pinguisane-type (e.g., **24, 25**) sesquiterpenoids have only been found in liverworts (Asakawa, 1982, 1995; Asakawa et al., 2013b). The example of gas chromatogram of essential oil characteristic by the presence of sesquiterpenoids obtained from *Marchantia polymorpha* subsp. *ruderalis* is presented on Figure 21.4. Structures of selected sesquiterpenoids characteristic for liverwort are shown in Figure 21.5.

These spore-forming plants are also rich sources of a number of different skeletal diterpenoids. In the essential oils obtained from liverwort species, diterpene hydrocarbons and simple alcohols can be found. Volatile extracts are more rich in the diterpenic compounds. The most prevalent are clerodane, kaurane, and labdane skeletons. Compounds of spiroclerodane, 5,10-*seco*-clerodane (e.g., **26**), 9,10-*seco*-clerodane, *epi*-homoverrucosane, *seco*-infuscane, infuscane, *abeo*-labdane (e.g., **27**), and sacculatane (e.g., **28, 29**) types are representatives of diterpenoids found only in the liverworts.

Among rare, naturally occurring diterpenoids, the liverworts biosynthesize dolabellanes (e.g., **30**), fusicoccanes (e.g., **31**), cembranes (e.g., **32**), vibsanes, neodenudatanes, verticillanes (e.g., **33**), viscidanes, and prenylguaianes (Asakawa, 1982, 1995; Asakawa et al., 2013b). Some of the characteristic diterpenoids present in liverworts are presented in Figure 21.6.

Liverworts belonging to the genus *Radula* are very exceptional spore-forming plants. These produce a number of volatile bibenzyl compounds. The most characteristic are prenylated bisbibenzyls (e.g., **34**). Compounds with a 2,2-dimethylchromane ring skeleton (e.g., **35**) and dihydrooxepin skeleton (e.g., **36**) are also prevalent (Asakawa, 1982, 1995; Asakawa et al., 2013b). 3-Methoxy bibenzyl (**37**) was the major EO component from *R. complanata* and *R. lindenbergiana*. Interestingly, some *Radula* species produce bibenzyl cannabinoids (e.g., **38**). All cannabinoids isolated from *Radula* thus far belong to one of three types: (a) the *o*-cannabichromene , (b) *o*-cannabicyclol , and (c) tetrahydrocannabinol, with the *o*-cannabichromene type as the most prevalent (Asakawa et al., 2013b). The presence of radulanin H (**36**), a compound with a didihydrooxepin skeleton, was also confirmed in the EO of *R. lindenbergiana* (Figueiredo et al., 2009) (Figure 21.7). The content of terpenoids in *Radula* species is generally very low, but there are some exceptions; for example,

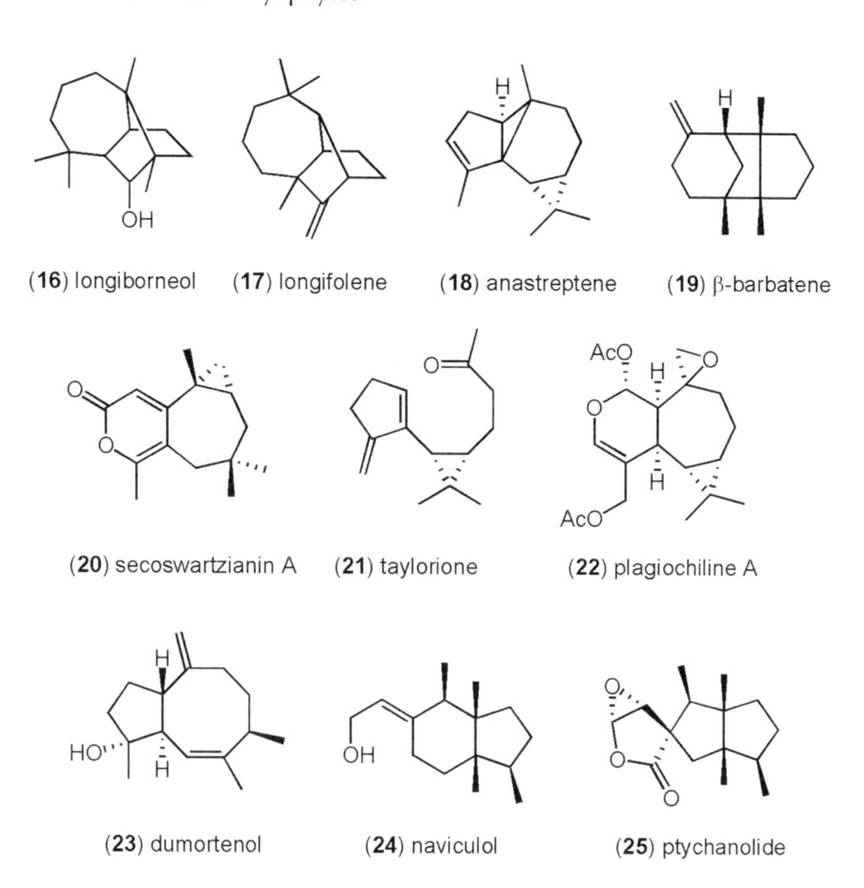

FIGURE 21.5 Structures of selected sesquiterpenoids present in liverworts.

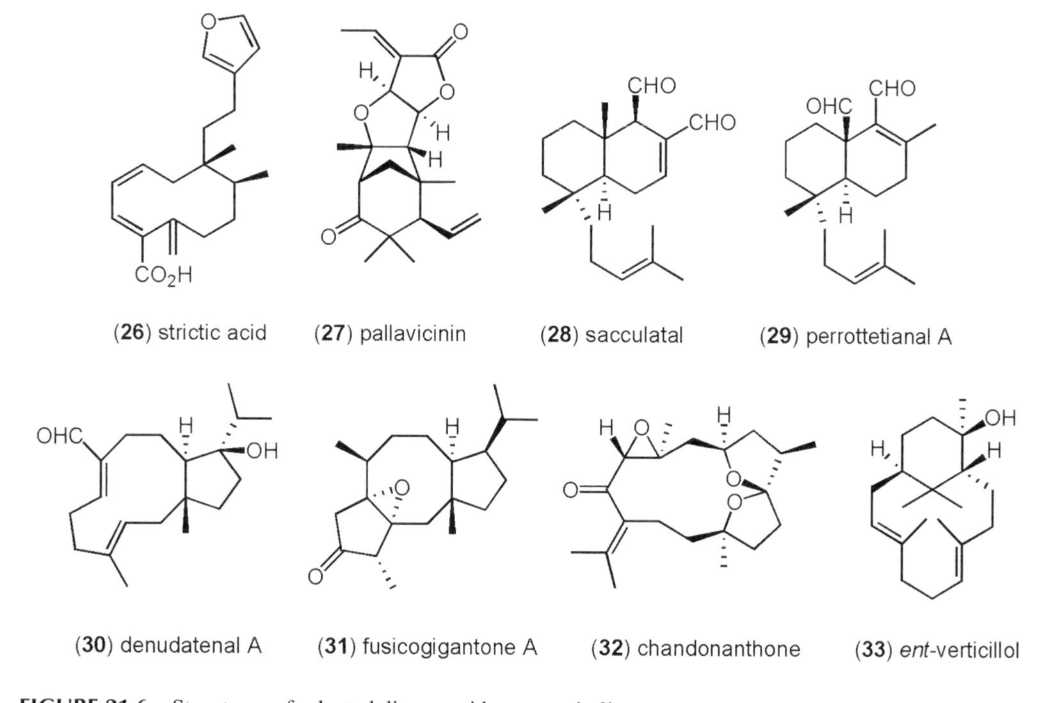

FIGURE 21.6 Structures of selected diterpenoids present in liverworts.

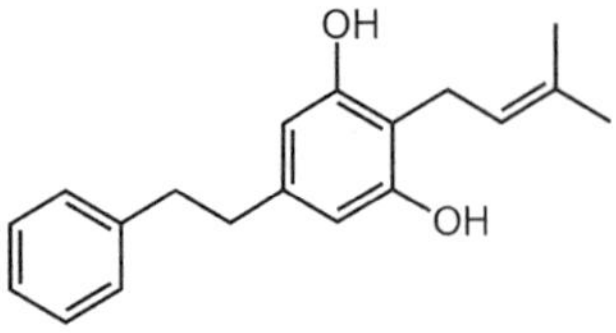

(**34**) 3,5-dihydroxy-4-(3-methyl-2-butenyl)bibenzyl

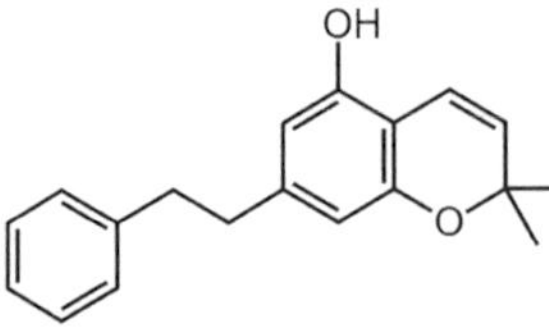

(**35**) 2,2-dimethyl-5-hydroxy-7-(2-phenylethyl)chromene

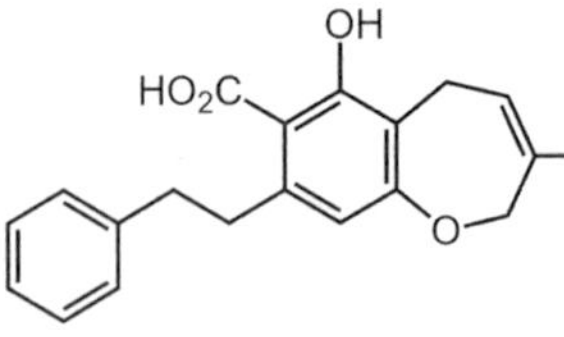

(**36**) radulanin H

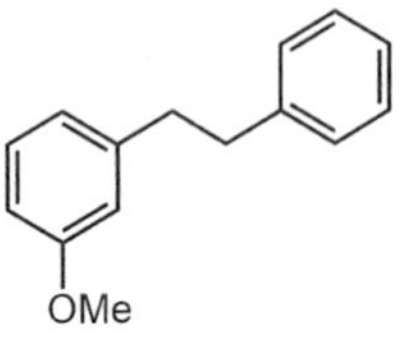

(**37**) 3-methoxybibenzyl

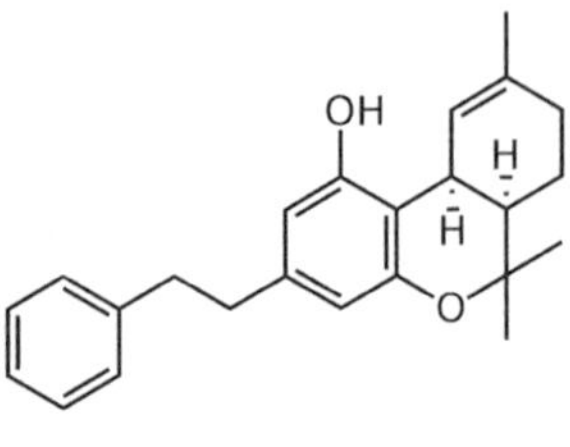

(**38**) perrottetinene

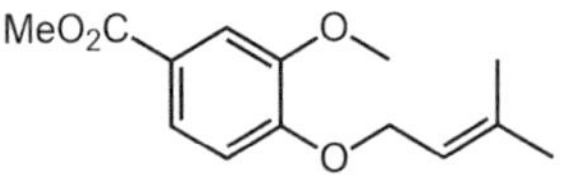

(**39**) trichocolein

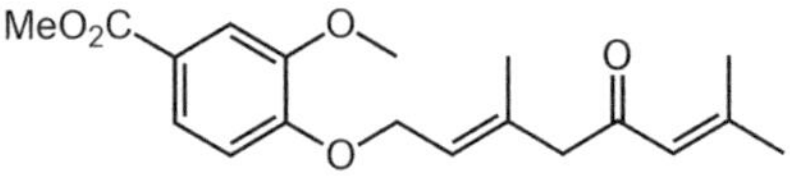

(**40**) tomentellin

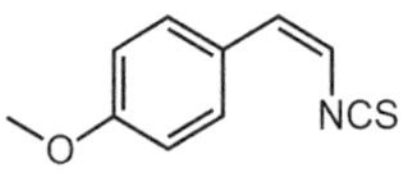

(**41**) (*Z*)-coriandrin

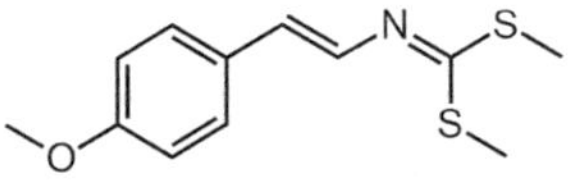

(**42**) (*E*)-*O*-methyltridentatol A

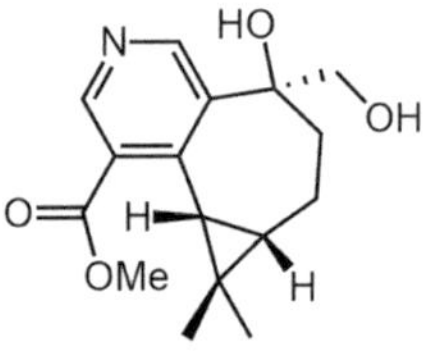

(**43**) plagiochianin B

FIGURE 21.7 Bibenzyls characteristic for Radula species (**34–38**), methyl benzoates with prenyl ether group present in genus Trichocolea (**39**, **40**), and nitrogen-containing compounds (**41–43**).

R. perrottetii. From the essential oil of the Japanese collection, the presence of a huge amount of bisabola-2,6,11-triene was confirmed. From this essential oil, two viscidane diterpenoids were also isolated (Tesso et al., 2005).

Another, very interesting liverwort genus is *Trichocolea*, best represented in tropics. All of the chemically studied specimens produce volatile methyl benzoates with a prenyl ether group, like trichocolein (**39**) or tomentellin (**40**) (Figure 21.7), together with a simple vanillic acid methyl ester. Some of the *Wettsteina* species are known to biosynthesize volatile isocoumarin and naphthalene derivatives (Asakawa et al., 2013b).

The occurrence of nitrogen- and sulfur-containing compounds among liverworts is very rare (Asakawa et al., 2013b; Ludwiczuk and Asakawa, 2014). There are just a few examples of such components in these spore-forming plants. Examples include skatole (**2**), detected in the Tahitian *Cyathodium foetidissimum* (Ludwiczuk et al., 2009; Sakurai et al., 2018), and the coriandrins (e.g., **41**) and methyl tridentatols (e.g., **42**) isolated from the essential oil obtained from the Spanish collection of *Corsinia coriandrina* (von Reuβ and König, 2005). The same components were detected in the volatile extracts from this liverwort species collected in Turkey, Spain, and France (Asakawa et al., 2018). An exceptional example is the pyridine-type aromadendrane alkaloid, plagiochianin B (**43**) isolated from the Chinese liverwort, *Plagiochila duthiana* (Han et al., 2018) (Figure 21.7).

21.2.1.2 Volatile Components in Mosses and Hornworts

In comparison to the liverworts, the other two groups of the bryophytes, mosses and hornworts, are not rich in volatile compounds. It is mainly because these spore-forming plants do not have oil bodies. The distribution of terpenoids, simple benzoic, cinnamic, and phthalic acid derivatives, as well as coumarins, have been investigated for several species of mosses and hornworts. The mentioned group of compounds were detected in mosses, including *Mnium, Plagiomnium, Homalia, Hypnum, Plagiothecium, Brachythecium, Bryum, Syntrichia, Breutelia, Leptodontium, Macromitrium, Campylopus, Rhacocarpus, Thuidium,* and *Taxiphyllum* species (Asakawa, 1995; Ozdemir et al., 2010; Asakawa et al., 2013b; Valarezo et al., 2018), and in the hornworts from *Anthoceros* species (Sonwa and König, 2003; Xiong et al., 2018). Figure 21.8 presents terpenoids detected and/or isolated from mosses and hornworts.

The most common monoterpenoid detected in mosses was β-cyclocitral (**44**), while in hornworts, α-pinene (**12**) and limonene (**13**) were present. Other compounds—for example, β-pinene (**46**), myrcene, α-phellandrene (**45**), terpinolene, and camphor (**47**)—were also detected as frequent constituents (Saritas et al., 2001; Sonwa and König, 2003; Valarezo et al., 2018; Xiong et al., 2018). The Japanese *Plagiomnium acutum* produces *ent*-sesquiterpene hydrocarbons, β-cedrene (**48**), α-cedrene, and α-acoradiene (**49**) (Toyota et al., 1998a). The three new sesquiterpene hydrocarbons,

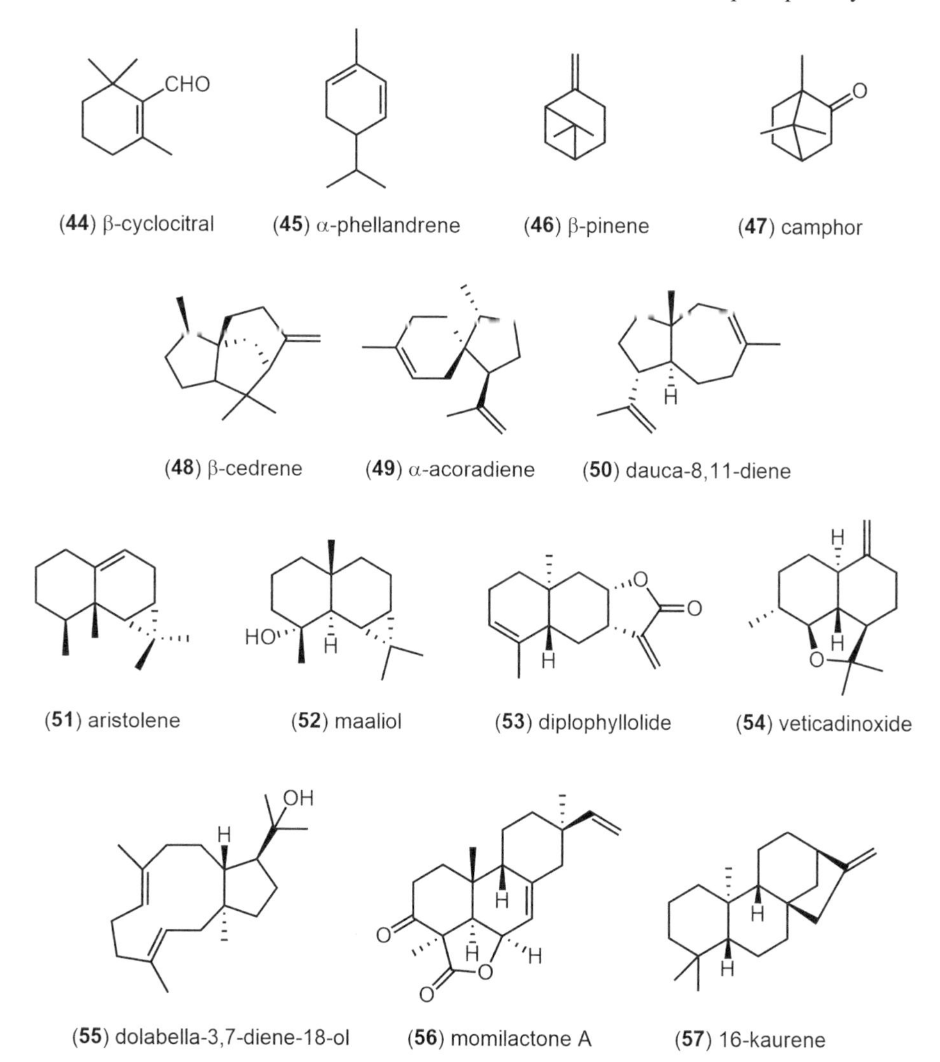

FIGURE 21.8 Terpenoids detected and/or isolated from mosses and hornworts.

(+)-10-*epi*-muurola-4,11-diene, (−)-1,2-dihydro-α-cuparenone, and (+)-dauca-8,11-diene (**50**), were isolated as major components from the essential oil *of Mnium hornum* and *Plagiomnium undulatum* by preparative GC (Saritas et al., 2001). The most characteristic components present in the essential oils hydro-distilled from mosses originated from Ecuador were sesquiterpenoids, and among them, α- and β-selinene, slina-3,11-dien-6α-ol, *epi*-α-muurulol, and α-cadinol were the major terpenoids (Valarezo et al., 2018). Sonwa and König (2003) identified several sesquiterpenoids—for example, aristolene (**51**), maaliol (**52**), diplophyllolide (**53**) and veticadinoxide (**54**)—in the EO of the hornwort *Anthoceros caucasicus.* Xiong and coworkers (2018) also showed the presence of sesquiterpenoids in another two *Anthoceros* species, *A. punctatus* and *A. agrestis.* These were, for example, α-amorphene, bicyclogermacrene, β-bisabolene, or β-cubebene (Xiong et al., 2018). The presence of diterpenoids was also confirmed in mosses and hornworts. A dolabellane-type compound, (+)-dolabella-3,7-dien-18-ol (**52**) ,was isolated from the Japanese moss *Plagiomnium acutum* (Toyota et al., 1998a. Two pimaranes, momilactones A (**56**) and B, were found in *Hypnum plumaeforme.* Both components were previously identified as phytoalexins in rice (Nozaki et al., 2007). In the essential oil obtained from the hornwort *A. caucasicus,* the presence of 16-kaurene (**57**) and isoabienol was confirmed (Sonwa and König, 2003). Gas chromatogram of the essential oil obtained from *Anthoceros caucasicus* is presented in Figure 21.9.

21.2.2 Microbial Terpene Synthase-Like Genes in Bryophytes

Embryophytes, informally called land plants and including bryophytes, lypophytes, ferns, gymnosperms, and angiosperms, are believed to have evolved from charophycean green algae. Among embryophytes, bryophytes are considered to be the oldest terrestrial plants, and among them, it has been suggested that the liverworts are the earliest embryophyte lineage and are the sisters of all other land plants (Qiu et al., 2006). After the transition from the aquatic environment, land plants have continued to originate evolutionary innovations, and one of them was the ability to produce an enormous diversity of secondary metabolites. These metabolites are generally believed

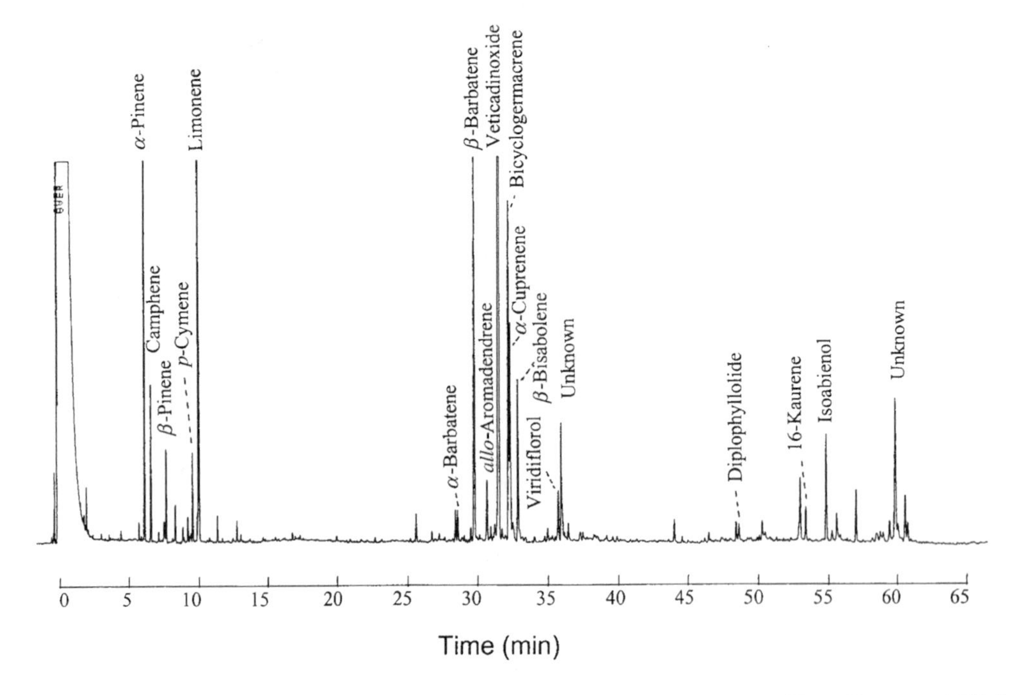

FIGURE 21.9 Gas chromatogram of the essential oil of *Anthoceros caucasicus*. (From Sonwa and König, 2003. *Flavour Fragr. J.*, 18: 286–289. With permission.)

to be important for defense against herbivores and microbial pathogens or resistance to the abiotic stresses (Gershenzon and Dudareva, 2007; Vickers et al., 2009; Chen et al., 2018; Jia et al., 2018).

Among the secondary metabolites made by plants, terpenoids are the largest and structurally most diverse (Ludwiczuk et al., 2017). The diversity of terpenoids in bryophytes, particularly in liverworts, is enormously rich (Asakawa et al., 2013b). While many terpenoids are observed in both bryophytes and seed plants, some are unique to bryophytes. Based on the existing literature data, it is just the beginning to understand the molecular and biochemical basis underlying terpenoid biosynthesis in bryophytes. The first bryophyte terpene synthase (TPS) gene was isolated from *Physcomitrella patens*, a moss used as a model organism for studies on plant evolution. This gene, *PpCPS/KS*, encodes a bifunctional enzyme that has both copalyl diphosphate synthase (CSP) and kaurene synthase (KS) activities (Hayashi et al., 2006; Chen et al., 2018). Unlike *Physcomitrella patens*, the genome of the liverwort *Marchantia polymorpha* contains several typical plant *TPS* genes, including both bifunctional and monofunctional enzymes based on domain analysis (Bowman et al., 2017).

Compared to seed plants, which have only one type of terpene synthase genes, so-called typical plant terpene synthase genes, bryophytes employ not only typical plant terpene synthase genes but also another class of terpene synthase genes called microbial terpene synthase-like (MTPSL) genes for terpenoid biosynthesis (Jia et al., 2018). MTPSLs are structurally and phylogenetically more related to bacterial and fungal terpene synthases than to typical plant terpene synthases. They are widely distributed in nonseed plants, including bryophytes, lycophytes, and ferns, but absent in seed plants and green algae. The conclusion that *MTPSL* genes are widely distributed in nonseed plants was drawn from the analysis of transcriptomes. From 166 analyzed plants, *MTPSL* genes were identified from 24 species of liverworts, 30 species of mosses, three species of hornworts, 21 species of lycophytes, and 65 species of ferns (Jia et al., 2016). Distribution of *MTPSL* genes identified from the transcriptomes of 1103 plant species is presented on Figure 21.10.

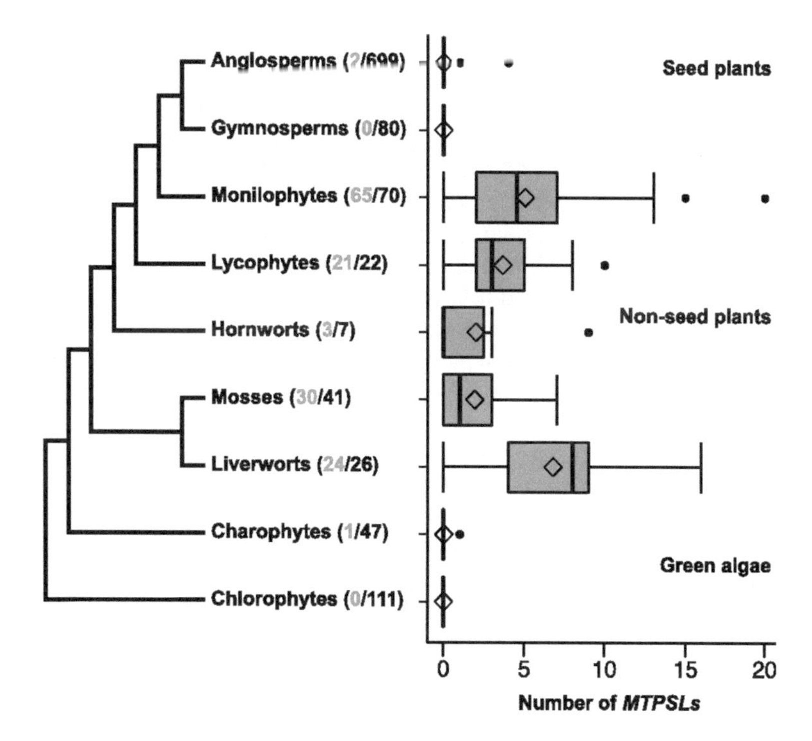

FIGURE 21.10 (**See color insert.**) Distribution of MTPSL genes identified from the transcriptomes of 1103 plant species. The numbers in parentheses represent the number of transcriptomes containing putative MTPSLs (in red) and total transcriptomes analyzed in each lineage (in black). (From Jia et al. 2016. *Proc. Natl. Acad. Sci. USA*, 113: 12328–12333. With permission.)

Biochemical studies suggest the MTPSLs are largely responsible for the terpenoid diversity in bryophytes, particularly mono- and sesquiterpenoids. A typical plant TPS function in the nonseed plants is catalysis of the biochemical reactions for the production of diterpenoids (Chen et al., 2018; Jia et al., 2018).

The wide distribution of *MTPSL* genes in nonseed plants and their absence in green algae raises questions about their biological function. It seems that these were necessary in transition from aquatic to terrestrial environments. MTPSLs may have helped early land plants to resist the biotic and abiotic stresses caused by the new environment. The good example are liverworts that produce and store terpenoids in the oil bodies that have been suggested to function in drought resistance and herbivore defense (Chen et al., 2018; Jia et al., 2018). However, much more remains to be done to investigate the biological functions of their products and how they have influenced the evolution of the *MTPSL* gene family in nonseed land plants.

21.3 SIGNIFICANCE FOR CHEMOTAXONOMIC STUDIES

Successful phytochemical work with any plant material must begin with careful preparation of plant samples, and proper botanical identification is of key concern. Correct identification of plant material guarantees the reproducibility of the phytochemical research, and carelessness at this stage of an investigation may greatly reduce the scientific value of the overall study (Jones and Kinghorn, 2006). In comparison to the higher plants, the identification of bryophytes, and especially liverworts, is challenging. It is because of their small size, their often microscopic or even chemical distinguishing features, and their enormous diversity (Crandall-Stotler et al., 2008). Taxonomy of the liverworts traditionally based on morphological studies of specimens collected in herbaria. Liverworts are very sensitive to environmental changes caused by humans, which induce alteration of their genetic and morphological features. It is believed that about 20% of the described liverwort species are synonyms. Furthermore, as plants with simple morphological organization, they have an evolutionary potential of hidden genetic variation. Characteristic for that group of plants is the existence of cryptic species (= sibling species), which exhibit significant genetic differences, but they are morphologically indistinguishable (Saw, 2001). The recent development of molecular techniques has helped to solve many problems of liverworts taxonomy; however, still many questions remain unanswered. Although genetic-based studies are very useful to assess phylogenetic relationships among plant taxa at higher levels of the taxonomic hierarchy, at the genus level, there is not such an ideal molecular character to assess relationships. At this level, the number of chemical features available from plant secondary metabolites can be much higher, and these can provide sufficient data to be used for cladistic analyses (Zidorn, 2008).

The major purpose of chemosystematic studies is searching for single components or group of compounds, which allow proper botanical identification of the liverwort species, and their classification. Based on the available literature data concerning chemotaxonomy of the liverworts, there are two directions of such kinds of studies (Asakawa et al., 2013b; c; Ludwiczuk and Asakawa, 2014, 2015, 2017):

- Chromatographic fingerprinting of the liverworts growing under different climatic and geographical conditions for the presence of chemical markers, and
- Searching for the chemical relationships between liverwort species classified in one genus, within one liverwort species, and between liverworts species and genera classified in one family.

As was shown in Section 21.2.1.1, liverworts have yielded a rich array of secondary metabolites. Many of these compounds are characterized by unique structures, and some of them have not been found in any other plants, fungi, or marine organisms. Gas chromatographic profiling of the volatile extracts received from the liverworts has been applied with success in differentiating liverworts species and also used to resolve the taxonomic problems at the genus and family levels. Due to the fact that liverworts are morphologically very small and it is difficult to collect a sufficient amount of plant material, there is little research data on the composition of the essential oils obtained from these plants. Nevertheless, the available data indicate that, as in the case of aromatic plants, the components present in essential oils obtained from the liverworts can be used in chemosystematic studies of this plant group (Ludwiczuk and Asakawa, 2014, 2015, 2017).

21.3.1 Chemotaxonomic Value of Essential Oils

The studies on liverworts conducted in the 1970s mainly concerned essential oils (Benesova et al., 1971; Matsuo, 1971; Andersen et al., 1973, 1977b; a; Ohta et al., 1977). Published results showed that EO components can be utilized for chemosystematic studies of liverworts and suggested that the liverworts belonging to the orders Metzgeriales (simple thalloid) and Marchantiales (complex thalloid) can be distinguished from the leafy liverworts (order Jungermanniales) by the presence of β-barbatene (**19**) and anastreptene (**18**). Analysis of EO constituents showed that longipinane, longifolane- (e.g., **17**) and longibornane-type (e.g., **16**) sesquiterpenoids are characteristic components found in *Scapania undulata*, cadinanes for *Conocephalum conicum*, bazzananes for genus *Bazzania*, and eudesmanolides in the genus *Diplophyllum* (Andersen et al., 1973; 1977b; a; Ohta et al., 1977).

The recent studies indicate that among the investigated essential oils obtained from liverworts, those obtained from the following species are of chemotaxonomic value: *Conocephalum conicum*, *Reboulia hemisphaerica*, and *Lepidozia fauriana*, as well as from the genera of *Asterella*, *Radula*, *Plagiochila*, *Marsupella*, *Mylia*, and *Lophocolea* (Ludwiczuk and Asakawa, 2015, 2017). Characteristic components found in the essential oils obtained from liverwort species are presented in Figure 21.11.

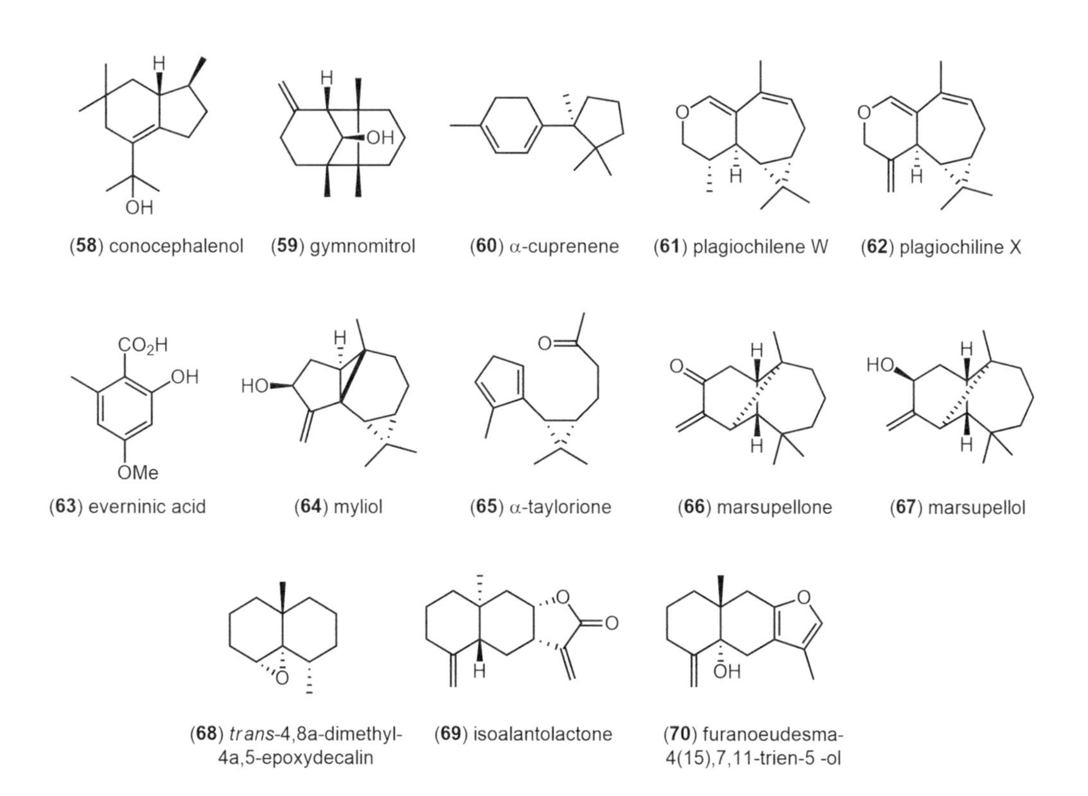

FIGURE 21.11 Characteristic compounds found in essential oils from liverwort species.

Conocephalum conicum is one of the most common liverworts in the northern hemisphere. Molecular studies have shown that this liverwort is a complex of six cryptic species (A, C, F, J, L, and S), which are indistinguishable by morphological features (Odrzykoski and Szweykowski, 1991). The GC-MS fingerprinting of ether extracts obtained from samples representing cryptic species A, F, J, L, and S (recently described as *C. salebrosum*) indicated the existence of chemical polymorphism between these species and the possibility to distinguish them based on the volatile components. For each of the analyzed cryptic species, the chemical markers were described (Figure 21.12) (Ludwiczuk et al., 2013a).

Melching and König (1999) investigated the essential oil and dichloromethane extracts obtained from the European *C. conicum* belonging to L-type. They showed that although different peak areas were observed in the hydro-distillation and solvent extraction in gas chromatograms, all the components were present in both samples. One of the major components present both in essential oil and extract was conocephalenol (**58**), which is in accordance with the data described by Ludwiczuk et al. (2013a).

The GC-MS fingerprinting of the ether extracts of the Japanese liverwort *Reboulia hemisphaerica* showed the existence of three chemotypes. One is characterized by the presence of cyclomyltaylane and chamigrane sesquiterpenoids, the second chemotype contained mainly gymnomitranes, while the third one produced aristilanes and gymnomitranes as the most characteristic compounds (Toyota et al., 1999; Ludwiczuk et al., 2008). There is only one scientific paper which concerns chemical composition of EO hydro-distilled from this liverwort species. Warmers and König (1999) showed that gymnomitranes and cuparanes are the major components present in the essential oil obtained from the European specimen, as well as in solvent extract. The most characteristic components were (–)-gymnomitr-3(15),4-diene, (+)-gymnomitrol (**59**), cuparene, and (-)-α-cuprenene (**60**). These data are very similar to those obtained from the ether extract of *R. hemisphaerica* collected in Japan and belonging to the second chemotype (Ludwiczuk et al., 2008).

The EOs obtained from three collections of *Lepidozia fauriana* growing in Taiwan showed that this species is not homogenous. Two of the analyzed specimens were characterized by the presence of chiloscyphane sesquiterpenoids, while amorphanes were the most characteristic components present in the third collection (Paul et al., 2001). These data are in accordance with the other reports concerning chemistry of this liverwort species (Shu et al., 1994; Shy and Wu, 2006). Shy and Wu (2006) divided *L. fauriana* into three chemotypes: amorphane (I), chiloscyphane (II), and eudesmane (III). In the essential oil of another Taiwanese *Lepidozia* species, *L. vitrea*, the eudesmane-type sesquiterpenoids were the major components (Paul et al., 2001). The same components were isolated from the ether extract from the Japanese specimen (Toyota et al., 1996).

Based on the chemical composition of EO from liverwort species, it is also possible to draw some chemotaxonomic conclusions at the genus level. One of the examples is genus *Plagiochila*. The species classified in this genus can be simply divided into two groups characterized by either the presence or absence of 2,3-seco-aromadendrane-type sesquiterpenoids. For both groups, several chemotypes can be distinguished (Asakawa et al., 2013b). The essential oil from the German liverwort *P. asplenioides* showed the presence of 2,3-*seco*-aromadendranes. Plagiochilines W (**61**) and X (**62**) were detected among other components of this EO (Adio and Konig, 2005). Based on such results, this liverwort was classified as chemotype I, which is the most characteristic for *Plagiochila species* (Asakawa et al., 2013b).

The analysis of essential oil composition of four *Plagiochila* species from Madeira showed that these species are characterized by the absence of 2,3-seco-aromadendrane-type sesquiterpenoids (Figueiredo et al., 2005). *P. bifaria* produces a large amount of evernic acid methyl ester (**63**), as well as the 9,10-dihydrophenanthrenes and methyl benzoates. The presence of methyl benzoates were also confirmed in another two *Plagiochila* species, *P. retrorsa* and *P. stricta* (Figueiredo et al., 2005). Based on these data, the mentioned three *Plagiochila* species belong to type XII of the *Plagiochila*, which is characterized by the occurrence of everninic acid methyl ester (**63**) together with a wide range of other aromatic compounds, especially 9,10-dihydrophenanthrenes (Asakawa et al., 2013b). In the essential oil obtained from the fourth species collected in Madeira, *P. maderensis*, the presence of methyl everninate (**63**) was also confirmed (Figueiredo et al., 2005).

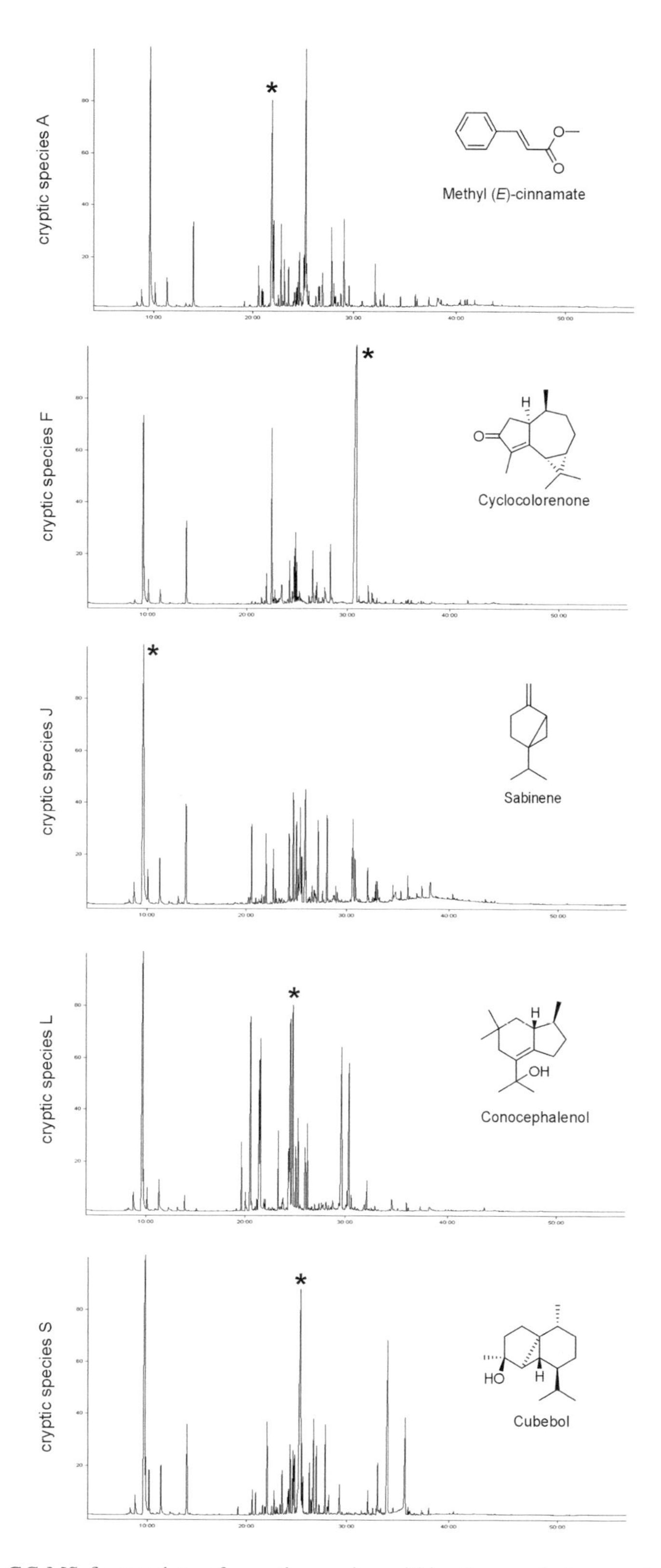

FIGURE 21.12 GC-MS fingerprints of cryptic species within *Conocephalum conicum* complex with identified marker compounds. (From Ludwiczuk and Asakawa, 2014. *J. AOAC Int.*, 97: 1234–1243.)

However, based on the volatile extract composition, this liverwort was classified as chemotype VI, which produces bibenzyls as the major and most characteristic components (Asakawa et al., 2013b). Rycroft et al. (2004) have analyzed the solvent extracts obtained from two collections of *P. maderensis* and detected 4-hydroxy-3′-methoxybibenzyl as the major volatile component, but this compound was absent in the essential oil from this liverwort species (Figueiredo et al., 2005).

Based on the terpenoid chemistry, two chemotypes of *Mylia* species have been discovered so far. Type I (*M. nuda, M. taylorii*) contains aromadendranes and seco-aromadendranes, while type II (*M. anomala, M. verrucosa*) produces mainly cyathane-type diterpenoids (Asakawa et al., 2013b). *Mylia taylorii* is known as a rich source of unique sesquiterpenoids, with rare skeletons like *ent*-5,10-cycloaromadendranes and *ent*-1,10-seco-aromadendranes (Asakawa, 1995; Asakawa et al., 2013b). The presence of these types of sesquiterpenoids were confirmed in the essential oils obtained from the German, Austrian, and Canadian specimens. These were myli-4(15)-ene, myli-4(15)-en-3-one, and myliol (**64**) among *ent*-5,10-cycloaromadendranes, and taylorione (**21**), α-*taylorione* (**65**), and 3-acetoxytaylorione among *ent*-1,10-seco-aromadendranes (von Reuβ et al., 2004). The Taiwanese *M. nuda* is characterized by having a very similar chemical composition to that of *M. taylorii*. 5,10-Cycloaromadendranes and 1,10-seco-aromadendranes have been detected in this species (von Reuβ et al., 2004).

The chemical analysis of the solvent extracts of the German and Russian *Marsupella emarginata* showed that longipinane-type sesquiterpenoids have been found as the major and most characteristic components (Nagashima et al., 1993). These data were confirmed by Adio et al. (2002). GC-MS analysis of the hydro-distillation product from *M. emerginata* revealed the presence of longipinanes—for example, marsupellone (**66**) and marsupellol (**67**)—together with barbatane-type sesquiterpenoids—for example, 5β-acetoxygymnomitr-3(15)-ene and barbatenal (Adio et al., 2002; Adio and König, 2007). In the EO from the Austrian *M. aquatica*, the most characteristic components were amorphane-type sesquiterpenoids (Adio et al., 2002, 2007). The chemical profile of EO from the Austrian *M. alpina* is very different from other *Marsupella* species since it produces the eudesmane sesquiterpenoids, with *ent*-diplophyllolide (**53**) being the main compound (Adio et al., 2002). These data showed that each of the *Marsupella* species produces its own peculiar compounds, and the analysis of EO components can be used in identification of these liverworts.

The liverworts belonging to genus *Lophocolea* are characterized by characteristic fragrance. From the hydro-distillate of *L. heterophylla*, homomonoterpene alcohols, including 2-methylisoborneol, were isolated (Tabacchi et al., 1992). The same compound was also found in the *n*-hexane extract of this liverwort species (Toyota et al., 1990). The highly fragrant compound occurring in the EO of L. bidentata was epoxy-trinoreudesmane sesquiterpene (**68**) (Rieck et al., 1995). The most characteristic components occurring in the *Lophocolea* species are furanoeudesmanes and 12,8-eudesmanolides (Asakawa et al., 2013b). Both types of compounds were found in the essential oils hydro-distilled from *Lophocolea* species. *ent*-Isoalantolactone (**69**) (Tabacchi et al., 1992) and furanoeudesma-4(15),7,11-trien-5α-ol (**70**) (Rieck and König, 1996) were detected in *L. heterophylla*, while the presence of diplophyllolide (**53**) were confirmed in *L. bidentata* (Tabacchi et al., 1992; Rieck et al., 1995).

21.3.2 Chemotaxonomic Value of Volatile Extracts

Compounds occurring in the essential oils constitute a powerful tool for studying chemical differences between or within liverwort species for which the characteristic components are mono- and sesquiterpenoids, as well as some volatile aromatic compounds. The use of the essential oil components in chemosystematic studies has, however, some limitations. There are many liverworts species for which the characteristic components are highly oxygenated sesquiterpenoids and diterpenoids. Such compounds cannot be obtained by steam or hydro-distillation of plant material. For these species, the fingerprinting of solvent extracts is the method of choice. The examples of such liverworts are *Barbilophozia floerkei*, *Chandonanthus hirtellus*, *Odontoshisma denudatum*, the genera of *Trichocolea*, *Porella*, and *Thysananthus*, and the Lejeuneaceae family (Asakawa et al., 2013b; Ludwiczuk and Asakawa, 2015, 2017).

The investigations of the ether extracts of the liverwort *Barbilophozia floerkei* showed that dolabellane- and fusicoccane-type diterpenoids are the most characteristic for this species (Asakawa, 1995; Asakawa et al., 2013b). None of these compounds were found in the essential oil from this species. The characteristic chemical constituents of this liverwort are trinorsesquiterpenes, (+)-1,2,3,6-tetrahydro-1,4-dimethylazulene, and (–)-2,3,3a,4,5,6-hexahydro-1,4-dimethylazulen-4-ol, along with the known trinoranastreptenes that have been isolated together with a number of common sesqui- and monoterpenoids (Adio and König, 2007).

Diterpenoids of cembrane, as well as dolabellane, fusicoccane, and verticillane skeleton, are characteristic components found in *Chandonanthus hirtellus*. Among them, cembranes are considered to be chemical markers of this liverwort species, and among them, chandonanthone (**71**) is the major, and most characteristic, metabolite of the volatile extracts obtained from different collections of this liverwort species (Ludwiczuk et al., 2009; Ludwiczuk and Asakawa, 2010; Asakawa et al., 2013b).

Odontoshisma denudatum is the liverwort species which produce *ent*-vibsanes and neodenudatenones. These diterpenoids have not been detected in or been isolated from any other plants (Asakawa et al., 2013b). Another liverwort species, characteristic by the presence of diterpenoids, is *Jackiella javanica*. This species produce exclusively *ent*-verticillane-type diterpenoids as the major components. Verticillanes are very rare in nature. They are considered as putative biosynthetic precursors of the taxanes (Koepp et al., 1995). While these compounds are also related biogenetically to the cembranes (Karlsson et al., 1978; Basar et al., 2001), up to now, no cembranoids have been detected in *J. javanica*. Cembranoids together with *ent*-verticillol (**33**) and its epimer have been found in Tahitian *Chandonanthus hirtellus* (Ludwiczuk et al., 2009; Komala et al., 2010b).

The genus *Trichocolea* is best represented in the tropics, and many of them—*T. hatcheri, T. lanata, T. mollissima, T. pluma, T. tomentella*, collected in Japan, Taiwan, New Zealand, East Malaysia, and French Polynesia—have been investigated chemically. All studied specimens produce methyl benzoates with a prenyl ether group, and these compounds are significant chemical markers of the genus *Trichocolea*. Trichocolein (**39**), tomentellin (**40**), and the simple vanillic acid methyl ester are the major components occurring in this genus (Asakawa et al., 2013b). Occasionally, some species biosynthesize their own characteristic metabolites. *T. pluma* originating from Taiwan, beside trichocolein, produces the labdane-type diterpene alcohol, *ent*-3α-hydroxylabda-8(17),(12*E*),14-triene (Chang and Wu, 1987), while from the New Zealand *T. mollissima* diterpenoids of the isopimarane series have been isolated (Lorimer et al., 1997b; Nagashima et al., 2003b).

Analysis of the chemical composition of *Porella* liverworts, showed that the greatest chemical diversity exhibited, was the sesqui- and diterpenoids present in these plants (Asakawa, 1982, 1995; Asakawa et al., 2013b). The complexity of the chemical composition of these two groups of compounds was used for chemotaxonomic studies. For the description of interspecific relationships the following compounds were used (Ludwiczuk et al., 2011):

- Among sesquiterpenoids: africanes, aromadendranes, drimanes, elemanes, germacranes, guaianes, monocyclofarnesanes, pinguisanes and santalanes;
- Among diterpenoids: sacculatanes, kauranes and labdanes.

On the basis of the sesqui- and diterpenoid composition, all investigated *Porella* species were divided into six chemotypes, and then the results were compared with DNA-based studies conducted by Hentschel and colleagues (2007). The data are shown in Table 21.1.

The comparison of the genetics-based classification with the recognized chemotypes shows some striking correlations, especially in the case of chemotypes I, III, and IV. Chemotype III and IV occur only in species of lineage B, whereas species belonging to chemotype I are exclusive to members of lineage A and molecular clade A1 (with the exception of *Porella fauriei*). In addition to the genetic and chemical similarities, the members of clade A1 are recognized by the glossy color of the plants; other species of *Porella* are usually dull in color. Based on the recognize similarity of morphological, genetic, and chemical features, a new section within the genus *Porella*, called *Porella*

TABLE 21.1
Correlation between Molecular Classification and Chemistry of *Porella* Species (Ludwiczuk et al., 2011)

Molecular Classification According to Hentschel et al. (2007)	Chemotype (Ludwiczuk et al., 2011)
Clade A1 (69%)	
P. arboris-vitae	I
P. vernicosa	I
P. gracillima	I
P. obtusata	I
P. canariensis	I
P. roellii	I
Clade A2 (98%)	
P. densifolia	V
P. stephaniana	II
Clade A3 (<50%)	
P. japonica	IV
Clade A4 (<50%)	
P. fauriei	I
Clade B1 (97%)	
P. platyphylla	III
P. cordaeana	V
P. navicularis	III
Clade B2 (89%)	
P. acutifolia	IV
P. camphylophylla	II
P. perrottetiana	II
Clade B3 (99%)	
P. caespitans	VI
P. subobtusa	VI
Clade B4 (100%)	
P. swartziana	VI
Clade B5 (81%)	
P. grandiloba	III

section *Vernicosae* Ludwiczuk, Nagashima, Gradstein and Asakawa, sect. nov. was created. The characteristic feature of this new section, in addition to the glossy color of the thalli, was the pungent taste of liverworts included in this section, due to the presence of polygodial (**72**), a drimane-type sesquiterpene dialdehyde (Ludwiczuk et al., 2011).

An example of problems with taxonomic classification of liverworts is the genus *Thysananthus* belonging to the Lejeuneaceae family. Sukkharak and coworkers (2011) made an attempt to explain inter- and intraspecific relationships of *Thysananthus* species, based on their chemical and morphological characteristics. Chromatographic fingerprinting of the volatile extracts obtained from 20 liverwort specimens collected in Thailand and Malaysia, belonging to four species, *T. comosus*, *T. convolutus*, *T. retusus*, and *T. spathulistipus*, indicated variability of the chemical composition depending on liverwort species. Studies have shown that the chemical constitution of *T. retusus* is rather different from that of the other *Thysananthus* species. The major components present in these liverworts were pinguisane-type sesquiterpenoids; for example, α-pinguisene (**73**), and deoxopinguisone (**75**). The chemical data were supportive of the classification of *Thysananthus* into two subgenera, namely *Thysananthus* and *Sandeanthus* (Gradstein, 1992). The data also showed the chemical heterogeneity in *T. convolutus* collected in Thailand and Malaysia. The chemical dissimilarity of these samples is reflected in their morphology, as the samples from Malaysia have entire leaves, whereas Thai samples have toothed leaves. *T. convolutus* is a polymorphic species with respect to the leaf dentation. Our data support the resurrection of *T. gottschei* as a separate taxon, previously considered a synonym of *T. convolutus* (Sukkharak et al., 2011).

The Lejeuneaceae is the largest family of liverworts, with at least 1000 species, including the already-mentioned *Thysananthus* species. Due to the high degree of morphological homoplasticity within the Lejeuneaceae, the division of this family into natural subunits has been considered to be notoriously difficult. At present, the division of the Lejeuneaceae into two subfamilies, Ptychanthoideae and Lejeuneoideae, is widely adopted (Gradstein, 2013). GC-MS fingerprinting of representatives of this liverwort family, collected in Thailand, Malaysia, Japan, and Ecuador, was performed by Ludwiczuk and colleagues (2013a; b). The ether extracts of 31 liverwort samples indicated variability of the chemical composition of the terpenoids depending on liverwort species. Each of the analyzed liverworts was characterized by the presence of particular compounds; however, there are some characteristic components that can be used as chemical markers of this liverwort family. These are the pinguisane- (**73**–**75**) and monocyclofarnesane-type (**76**, **77**) sesquiterpenoids and the fusicoccane-type diterpenoids (**78**, **79**), together with the sesquiterpene hydrocarbon, isolepidozene (**80**). All the characteristic components were present in the liverworts belonging to the subfamily Ptychanthoideae. Subfamily Lejeuneoideae was characterized by the absence of the mentioned components, with the exception of β-pinguisene (**74**), which was found only in plants classified in the subtribe Cololejeuneinae (Figure 21.13). Although the chemotaxonomic conclusions presented by

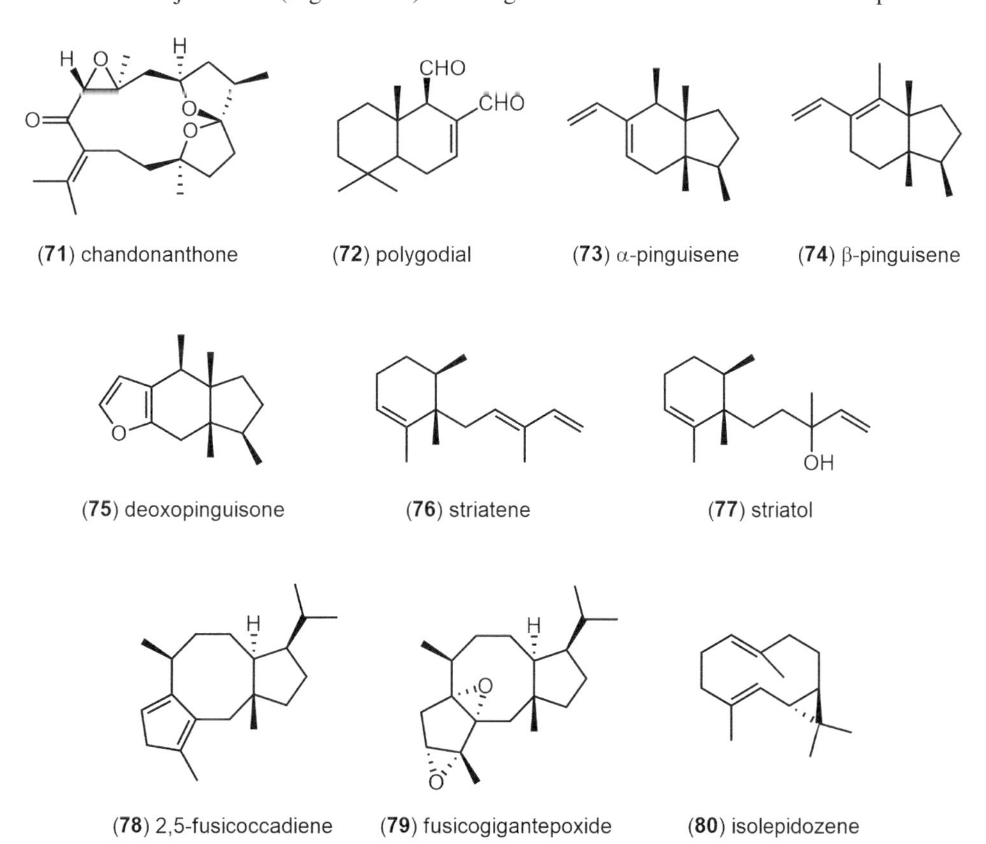

FIGURE 21.13 Structures of terpenoids characteristic for volatile extracts from liverworts.

Ludwiczuk et al. (2013a; b) were based on a quite-small sample size (16 liverwort species), it can be confirmed that chemical variation of the volatile compounds may be chemosystematically important for the whole Lejeuneaceae family. Results from the chemical investigations can help to understand the relationships among the taxa within this family, but especially can be used as additional support for molecular studies.

21.4 BIOLOGICAL FUNCTIONS

As the first inhabitants of terrestrial habitats, bryophytes were frequently exposed to adverse environmental conditions, such as pathogen attack, insect predation, and UV injury (Xie and Lou, 2009; Whitehead et al., 2018). As it was shown earlier (Section 21.2.1), these spore-forming plants display a high degree of chemical diversification (Asakawa, 1982, 1995; Asakawa et al., 2013b, 2013c), suggesting that secondary metabolites, and especially terpenoids, may play an important role in bryophyte-environment interactions (Chen et al., 2018; Whitehead et al., 2018). The bryophyte components are also known to have interesting biological properties such as antibacterial, antifungal, antiviral, neurotropic, insect repellent, muscle relaxing, cytotoxic, and apoptosis-inducing activities, among others (Asakawa, 1982, 1995; Asakawa et al., 2013b, 2013c; Asakawa and Ludwiczuk, 2018; Ludwiczuk and Asakawa, 2019).

21.4.1 Chemical Defense

Bryophyte metabolites, especially terpenoids, are a kind of chemical weapon that is necessary for these small plants to survive in the environment since they lack mechanical protection like higher vascular plants. Bryophytes are colonial organisms that are usually grown in humid locations where they form mats and cushions over soil, rocks, or on the trunks and leaves of vascular plants. Many kinds of invertebrates inhabit such bryophyte colonies. Despite this, there is little evidence of feedant activity on the bryophytes. These small, ubiquitous plants are also not infected by either bacteria or fungi (Asakawa, 2008; Chen et al., 2018). Some examples of terpenoids with antifeedant activity are shown in Figure 21.14.

It has been demonstrated that terpenoids present in liverworts and in other bryophytes show antimicrobial and antifungal effects. Lipophilic extracts of several liverworts, such as *Bazzania*, *Frullania*, *Marchantia*, *Conocephalum*, *Porella*, and *Plagiochila* spp. show antibacterial and antifungal activity (Asakawa, 1982, 1995, 2008; Asakawa et al., 2013b; Asakawa and Ludwiczuk, 2018, Ludwiczuk and Asakawa, 2019). Several antimicrobials have been isolated from the mentioned liverwort genera. In the leafy liverwort, *Bazzania trilobata*, several antifungal gymnomitrane-, calamenane- and drimane-type sesquiterpenoids were found (Scher et al., 2004). These components were tested against a range of phytopathogenic fungi, including *Botrytis cinerea*, *Cladosporium cucumerinum*, *Pyricularia oryzae*, *Phytophthora infestans*, and *Septoria tritici*. Gymnomitrol (**81**) showed inhibition against *P. infestans* at IC_{50} 0.1 μg/mL. 5-Hydroxycalamenene (**82**) showed inhibitory activity against *P. oryzae* at IC_{50} 1.7 μg/mL while 7-hydroxycalamenene (**83**) had antifungal activity against *P. oryzae*, *C. cucumerinum*, and *S. tritici* at IC_{50} 4.1, 10.0, and 11.8 μg/mL, respectively. 7-Hydroxycalamenene was also tested for *in vivo* activity against *Plasmopara viticola* on grape vine leaves and showed inhibitory activity at a concentration of 250 ppm. The infection was reduced from 100% in the control to 30% in the treated plants in a greenhouse. Drimenol (**84**) inhibited the growth of *C. cucumerinum* at concentrations of 6.6 μg/mL. Drimenal (**85**) exhibited potent activity against *P. infestans* and *S. tritici*, with an IC_{50} value of <0.3 and 17.6 μg/mL, respectively (Scher et al., 2004).

Sesquiterpene lactones, dehydrocostus lactone (**86**), acetyltriflocusolide lactone, and 11-αH-dihydrodehydrocostus lactone, characteristic for *Targionia lorbeeriana*, showed antifungal activity against *Cladosporium cucumerinum* with MIC values of 0.5, 10, and 3 μg/mL, respectively. Dehydrocostus lactone (**86**) also showed larvicidal activity against *Aedes aegypti*, with an LC_{100}

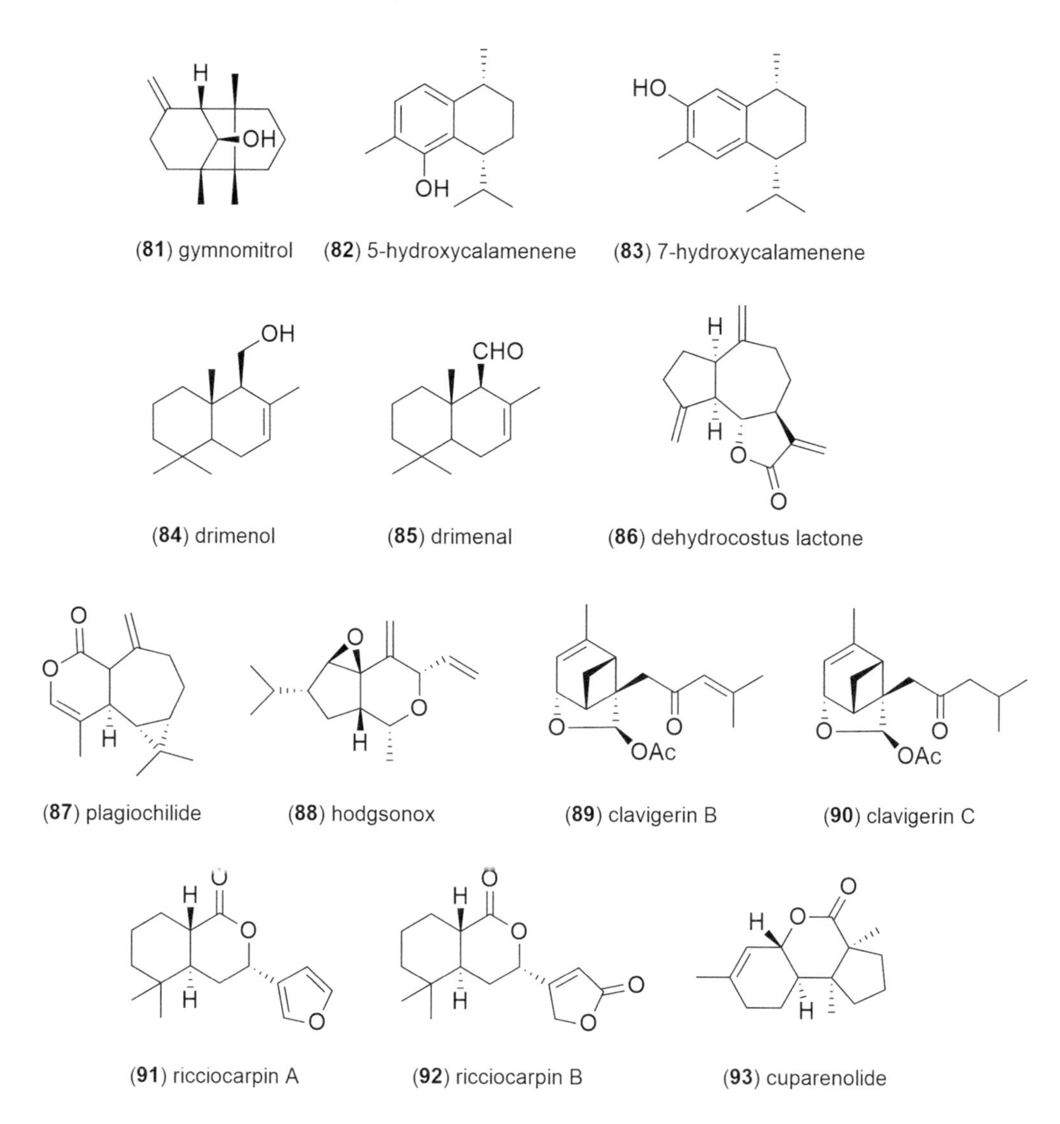

FIGURE 21.14 Compounds with antifeedant activity found in liverworts.

value of 12.5 ppm and antifungal activity against *C. albicans* (MIC 5 μg/mL) using a bioautographic TLC method (Neves et al., 1999).

The phytochemical studies on *Plagiochila* species showed that plagiochiline A (**22**), an representative of 2,3-seco-aromadendranes widely occurring in these liverworts, is a strong antifeedant against the African armyworm, *Spodoptera exempta*, at 1–10 ng/cm^2 in a choice leaf disk test for 2 hr (Asakawa et al., 1980). Another compound isolated from *Plagiochila* species, plagiochilide (**87**), killed brown planthopper, *Nilaparvata lugens* (Delphacidae) at 100 μg/mL (Asakawa, 1995). Plagiochiline A (**22**) and two fusicoccane diterpenoids, fusicogigantones A (**31**) and B isolated form Argentine *Plagiochila diversifolia* and *P. bursata* showed strong insecticidal activity against *Spodoptera frugiperda*. These compounds, incorporated to the larval diet at 100 mg/g, reduced the larval growth and produced larval mortality at early instars (Ramirez et al., 2010, 2017).

Polygodial (**72**) from the *Porella vernicosa* complex was the most active metabolite among all drimanes tested for antifeedant activity against aphids (Asakawa et al., 1988). Polygodial killed mosquito larvae at a concentration of 40 ppm and had mosquito repellent activity which was stronger than the commercially available DEET (Asakawa, 1995; Asakawa and Ludwiczuk, 2018).

Hodgsonox (**88**), isolated from the New Zealand liverwort *Lepidolaena hogdsoniae* is toxic to larvae of the blowfly *Lucilia cuprina*, with a LC_{50} value of 0.27 mg/mL (Ainge et al., 2001). Clavigerins B (**89**) and C (**90**), compounds occurring in another *Lepidolaena* species, *L. clavigera*, have significant antifeedant activities against *Anthrenocerus australis* (0.026% for **89** and 0.052% for **90**) and *Tineola bisselliella* (0.1% for both compounds). They showed similar efficiency as the well-known insect antifeedant azadirachtin (Perry et al., 2003).

Cuparane- and monocyclofarnesane-type sesquiterpenoids occurring in the liverwort *Ricciocarpos natans* have been tested against the snail *Biomphalaria glabrata*. Among tested compounds, ricciocarpin A (**91**) was the most toxic with an LC_{100} of 11 ppm, while ricciocarpin B (**92**) caused a significant reduction of the activity to an LC_{100} of 43 ppm. Cuparenolide (**93**) showed molluscicidal activity with LC_{100} at 32 ppm (Wurzel et al., 1990).

21.4.2 Allelopathic Activity

It is well known that higher plants do not want to grow around bryophytes. Van Tooren (1990) found that the numbers of emerging seedlings of angiosperms were reduced up to 30% in the presence of a bryophyte layer. Asakawa (1982) reported that most crude extracts of bryophytes, especially those containing pungent substances, show inhibitory activity against germination and root elongation. Several plant growth regulatory terpenoids have been isolated from liverworts (Figure 21.15). One of them is the pungent sesquiterpene dialdehyde, (–)-polygodial (**72**), which inhibits the germination and root elongation of rice in the husk at 100 ppm. At a concentration of less than 25 ppm, it dramatically promotes root elongation of rice (Asakawa, 1982). The extracts obtained from liverworts species (e.g.e, *Plagiochila*, *Lepidozia*, and *Marchantia*) can inhibit the germination of rice seedlings (Asakawa, 1990). Isobicyclogermacrenal (**94**) and lepidozenal (**95**) from *Lepidozia vitrea* were tested against rice seedlings. Both compounds completely inhibited the growth of leaves and roots at concentrations of 50 and 250 ppm, respectively. The 50% growth inhibition (I_{50}) of leaves and roots was observed for isobicyclogermacrenal (**94**) at 7 ppm (Matsuo et al., 1984a). Vitrenal (**96**) isolated form *L. vitrea* was tested for growth regulatory activity using rice seedlings and lettuce hypocotyls and showed strong inhibition (I_{50}=18 ppm) (Matsuo et al., 1984b; Kodama et al., 1986). Kato-Noguchi and Seki (2010) showed that the intact stems of the moss *Rhynchostegium pallidifolium* and its secondary metabolite, 3-hydroxy-β-ionone (**97**), inhibit the hypocotyl and root growth of cress (*Lepidium sativum*).

21.4.3 Medical Uses

There is much less knowledge available about medicinal properties of bryophytes. An ancient method of determining the medicinal properties of plants in the concept of Paracelsus, which deals with resemblance of plant body parts to shape and structure of organs in the human or animal body for which it is remedial. As per above philosophy, liverworts (e.g.) were used to cure hepatic disorders (Bowman, 2016). Similarly, the expressed oil from the moss *Polytrichum commune*, which bears hairy calyptra and called hair cup moss, was used by the women of ancient time on their hair (Glime, 2007). Different

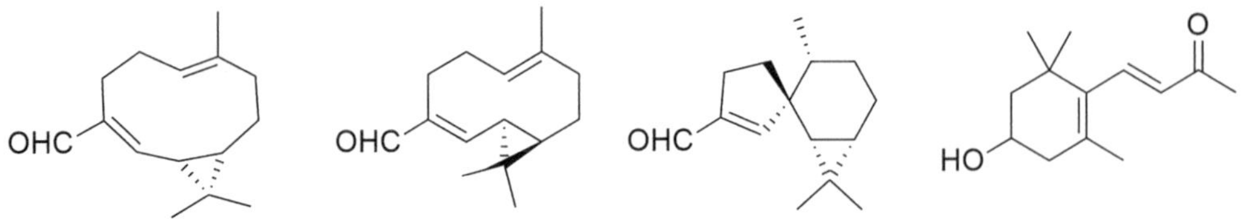

OHC OHC OHC O HO

(**94**) isobicyclogermacrenal (**95**) lepidozenal (**96**) vitrenal (**97**) 3-hydroxy-β-ionone

FIGURE 21.15 Terpenoids with allelopathic activity from bryophytes.

ethnic groups used the bryophyte species to cure various ailments in their daily lives. The most popular use was for treating skin diseases, but also as antipyretic, antimicrobial, and wound-healing remedies (Chandra et al., 2017). Chinese traditional medicine names 40 kinds of bryophytes that have been also used to treat illnesses of the cardiovascular system, tonsillitis, bronchitis, tympanitis, or cystitis (Saxena and Harinder, 2004). *Rhodobryum* species are commonly used to treat diseases, including heart palpitations, chest tightness, and neurasthenia. *Rhodobryum roseum* elicited a significant cardioprotective effect by augmentation of the endogenous antioxidants and inhibition of lipid peroxidation of the membranes. Significant myocardial necrosis, increased serum marker enzymes (lactate dehydrogenase, glutamate oxaloacetic transaminase, and creatine kinase) by isoproterenol, were reversed by pretreatment with ethanolic extract of *R. roseum* (Hu et al., 2009).

Phytochemists and biochemists have isolated a vast number of biologically active organic compounds from bryophytes which are of potential use in the pharmaceutical industry (Figure 21.16). Metabolites

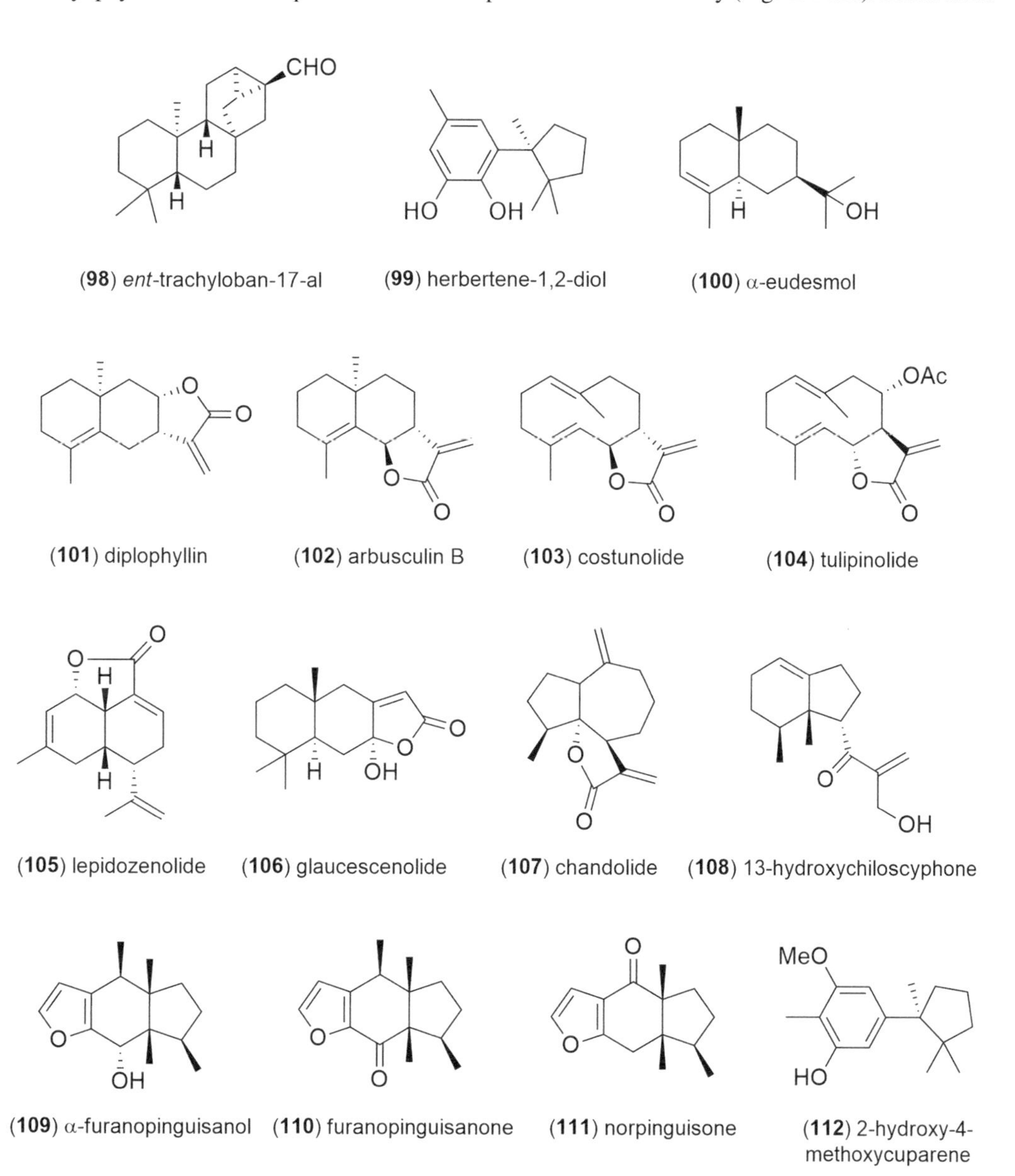

FIGURE 21.16 Terpenoids isolated from liverworts with pharmacological properties.

occurring in these tiny plants show the following biological properties: antibiotic, neurotropic, muscle-relaxing, as well as cytotoxic, and apoptosis-inducing activities (Asakawa, 1982, 1995, 2008; Asakawa et al., 2013b; Asakawa and Ludwiczuk, 2018).

It has been demonstrated that terpenoids present in liverworts and in other bryophytes have antibiotic effects. A thallose liverwort, *Pellia endiviifolia*, is known to biosynthesize the diterpene dialdehyde called sacculatal (**28**). This compound showed potent antibacterial activity against *Streptococcus mutans* (a causative organism of dental caries), exhibiting a LD_{50} value of 8 μg/mL (Asakawa, 2008). In the search for new antituberculosis lead compounds from bryophytes, Scher et al. (2010) isolated several trachylobane diterpenoids from the liverwort *Jungermannia exsertifolia* subsp. *cordifolia,* among which *ent*-trachyloban-17-al (**98**) showed the most significant activity against the virulent *Mycobacterium tuberculosis* H37Rv strain, with a MIC_{90} value of 24 μg/mL.

The herbertane sesquiterpenoids isolated from the Madagascan species *Mastigophora diclados* were tested against a *Staphylococcus aureus* strain using an agar diffusion method. These sesquiterpenoids showed weaker activity than the standard antibiotics chloramphenicol (zone of inhibition, 22 mm) and kanamycin (23 mm). Herbertene-1,2-diol (**99**) showed antibacterial activity (13 mm), but its dimmer, mastigophorene C, displayed stronger activity (17 mm) (Harinantenaina and Asakawa, 2004).

A sesquiterpene alcohol, α-eudesmol (**100**) isolated from *Porella stephaniana*, showed highly promising antitrypanosomal activity against *Trypanosoma brucei* strain GUTat 3.1, with an ED_{50} value of 0.10 μg/mL. This activity was stronger than that of the commercially used antitrypanosomal drugs eflornithine and suramin (ED_{50} 2.27 and 1.5 μg/mL, respectively) (Otoguro et al., 2011). Plagiochiline A (**22**), from the Peruvian *Plagiochila disticha*, was reported to show antileishmanial activity against *Leishmania amazonensis* axenic amastigotes with an IC_{50} of 7.1 μM and trypanocidal activity against *Trypanosoma cruzi* trypomastigotes at an MIC 14.5 μM (Aponte et al., 2010).

Many species of bryophytes have been shown to possess antitumor activity. The first antitumor active compound, diplophyllin (**101**), was reported in 1977 from the liverworts *Diplophyllum albicans* and *D. taxifolium* (Ohta et al., 1977). This compound shows significant activity (ED_{50} 4–16 μg/mL) against cervical carcinoma cell line HeLa (KB cell). The antitumor sesquiterpenoids costunolide and tulipinolide have also been isolated from the liverworts *Conocephalum japonicum*, *Frullania monocera*, *F. tamarisci*, *Lepidozia vitrea*, *Marchantia polymorpha*, *Plagiochila semidecurrens*, *Porella japonica*, and *Wiesnerella denudata* (Asakawa, 1981, 1982; Matsuo et al., 1981a; b). Other compounds that showed the cytotoxicity for KB cells were plagiochilline A (**22**) (Asakawa, 1990), diplophyllolide (**53**), α-herbertenol, herbertene-1,2-diol (**99**) (Komala et al., 2010a), and sacculatal (**28**) (Asakawa, 2008). Germacrane- and pinguisane-type sesquiterpenoids from the Indonesian and Tahitian *Frullania* sp. and Japanese *Porella perrottetiana* also were found to be active against cervical adenocarcinoma (Komala et al., 2011).

Antileukemic activity has also been demonstrated in several compounds from leafy liverworts. Biologically directed isolation of the ethanol extract from the New Zealand liverworts *Clasmatocolea vermicularis* and *Chiloscyphus subporosa* led to the isolation of (−)-diplophyllolide (**53**), as the major cytotoxic component. This compound showed activity against P388 murine leukemia cells, with an IC_{50} value of 0.4 μg/mL (Lorimer et al., 1997a). Three more sesquiterpene lactones isolated from liverworts, arbusculin B (**102**) and costunolide (**103**) from *Hepatostolonophora paucistipula* and lepidozenolide (**105**) from *Lepidozia fauriana*, showed cytotoxic activity against P388 cells, with IC_{50} values of 1.1, 0.7, and 2.1, μg/mL, respectively (Shu et al., 1994; Baek et al., 2003). Other liverwort components active against P388 were marsupellone (**66**) and its acetoxy derivative form *Marsupella emarginata*, plagiochilline A (**22**) from *Plagiochila ovalifolia*, as well as glaucescenolide (**106**) from *Schistochila glaucescens* (Nagashima et al., 1993; Toyota et al., 1998b; Scher et al., 2002).

There are also some scientific data concerning the activity against human leukemia cell lines HL-60. From the Tahitian liverwort *Chandonanthus hirtellus*, a new sesquiterpene lactone, chandolide (**107**), was isolated. This compound showed cytotoxic activity with an IC_{50} value of 5.3 μg/mL (Komala et al., 2010b). Tulipinolide (**104**) obtained from another Tahitian liverwort,

Frullania species, showed cytotoxicity against the HL-60 cell line at IC_{50} 4.6 μM. The Japanese liverwort *Porella perrottetiana* produces cytotoxic lactone, $4\alpha,5\beta$-epoxy-8-*epi*-inunolide (IC_{50} 8.5 μM) (Komala et al., 2011). Among diterpenoids, it is worth mentioning that *ent*-16-kauren-15-one derivatives together with rearranged kaurenes named jungermannenones A–D isolated from the New Zealand *Jungermannia* species. These kauranes showed cytotoxic activity against HL-60 cells at IC_{50} values from 0.4 to 7.0 μM (Nagashima et al., 2003a; Nagashima et al., 2005).

Polygodial (**72**) widely occurring in *Porella vernicosa* complex, as well as sacculatal (**28**) isolated from *Pellia endiviifolia,* showed cytotoxic activity against a human melanoma cell line, A375 (IC_{50} value range 2–4 μg/mL) (Asakawa, 2008), while 13-hydroxychiloscyphone (**108**) obtained from *Chiloscyphus rivularis* collected in Oregon showed cytotoxic activity against A-549 lung carcinoma cells (IC_{50} value 2.0 μg/mL) (Wu et al., 1997).

There are some literature data that indicate the occurrence of anti-inflammatory and immunomodulatory tepenoids among liverwort species. Two pinguisane-type sesquiterpenoids, α-furanopinguisanol (**109**) and furanopinguisanone (**110**) isolated from the Serbian *Porella cordaeana* were studied on the viability of splenocytes, as a possible mode of regulation of innate and immune inflammatory responses. It was shown that α-furanopinguisanol (**109**) induced a blast-like transformation of splenocytes in higher concentration (10^{-4} M), while in lower ones (10^{-8} to 10^{-6} M), it acted as a cytotoxic agent. Furanopinguisanone (**110**) exerted prominent cytotoxicity in all tested concentrations (Radulovic et al., 2016). Two other pinguisane sesqiterpenoids, norpinguisone (**111**) and its ester, isolated form *Porella densifolia* were screened for anti-inflammatory activities by the model of lipopolysaccharide (LPS)-induced nitric oxide (NO) production with RAW264.7 cells. Both compounds showed the inhibition at IC_{50}=1.7 μM and IC_{50}=45.5 μM, respectively (Quang and Asakawa, 2010).

Harinantenaina and coworkers (2007) showed that herbertane sesquiterpenoids isolated from *Mastigophora diclados* and *Herbertus sakuraii*, as well as cuparenoids from *Lejeunea aquatica* and *Bazzania decrescens*, showed inhibition of lipopolysaccharide (LPS)-induced production of NO. The highest inhibition was attributed to 2-hydroxy-4-methoxycuparene (**112**, IC_{50}=4.1 μM). The strong inhibitory activity of this compound was related to the inhibition of LPS-induced iNOS mRNA (Harinantenaina et al., 2007).

21.5 CONCLUSIONS

Bryophytes represent a diverse group of thc first green, lower-land plants to develop during the process of evolution. There are described more than 20,000 species, but only a few percentages of them have been studied chemically. It was shown that these tiny plants contain numerous very interesting, from the chemistry point of view, terpenoids and other volatiles. The discovery of microbial terpene synthase-like (MTPSL) genes in nonseed plants suggests that these are largely responsible for the terpenoid diversity in bryophytes, particularly mono- and sesquiterpenoids. Among the bryophytes, the liverworts, as the earliest embryophyte lineage, produce a wide array of terpenoids. Over the last 40 years, more than 1600 terpenoids were described (Asakawa, 1982, 1995; Asakawa et al., 2013b). Many of them are characterized by unprecedented structures, and some, like pinguisanes or sacculatanes, have not been found in any other organisms.

An enormous diversity of terpenoids and other volatiles present in liverworts is used for chemotaxonomic studies. Compounds occurring in the essential oils and volatile extracts constitute a powerful tool for studying chemical differences between or within liverwort species, genera, or families. Additionally, results from the chemical investigations, together with the other biological or genetic information, can help to understand real relationships among the taxa.

Existing literature data show that terpenoid secondary metabolites present in bryophytes are involved in many biochemical and ecological processes, especially as defenses against biotic stresses such as insects and microbial pathogens. It is also well known that higher plants do not want to grow around bryophytes. It was reported that most crude extracts of bryophytes, especially those containing pungent and bitter substances, show inhibitory activity against germination and root elongation. Several plant growth regulatory terpenoids have been isolated from bryophytes.

Bryophytes are also an interesting source of active metabolites having pharmacological activity. In ancient times, bryophytes have been used as herbal medicines in various parts of the world. Dioscorides ascribed medicinal properties to *Marchantia polymorpha*, especially its antimicrobial activity (Dodoens, 1578). Nowadays, a vast number of biologically active compounds were isolated from bryophytes, and these are of potential use in the pharmaceutical industry. It has been demonstrated that bryophyte extract and isolated compounds have effects on human pathogenic fungi and may cure skin diseases, but currently are not sold for that purpose. However, a patent has been obtained to cure fungal infections of horses with *Ceratodon purpureus* and *Bryum argenteum* (Frahm, 2004; Glime, 2007). Several medical uses seem promising, such as antileukemic properties and anticancer agents.

It is, however, necessary to mention that a very small fraction of bryophytes has been tested for their pharmacological efficacy. Although the exact mode of action of some of described bioactive compounds remains unknown, bryophytes could serve as an attractive candidate for therapeutic properties. Further work on isolation, characterization, structural elucidation, pharmacological evaluation, determination of mode of action, and clinical trial of these active principles could open an exciting aspect of future drug-development programs.

REFERENCES

Adio, A.M., and König, W.A. 2005. Sesquiterpene constituents from the essential oil of the liverwort *Plagiochila asplenioides*. *Phytochemistry*, 66: 599–609.

Adio, A.M., and König, W.A. 2007. Sesquiterpenoids and norsesquiterpenoids from three liverworts. *Tetrahedron: Asymmetry*, 18: 1693–1700.

Adio, A.M., Paul, C., König, W.A., and Muhle, H. 2002. Volatile components from European liverworts *Marsupella emarginata, M. aquatica and M. alpina*. *Phytochemistry*, 61: 79–91.

Adio, A.M., von Reuβ, S.H., Paul, C., Muhle, H., and König, W.A. 2007. Sesquiterpenoid constituents of the liverwort *Marsupella aquatica*. *Tetrahedron: Asymmetry*, 18: 1245–1253.

Ainge, G.D., Gerard, P.J., Hinkley, S.F.R., Lorimer, S.D., and Weavers, R.T. 2001. Hodgsonox, a new class of sesquiterpene from the liverwort *Lepidolaena hodgsoniae*. Isolation directed by insecticidal activity. *J. Org. Chem.*, 66: 2818–2821.

Andersen, N.H., Bissonette, P., Liu, C.-B., Shunk, B., Ohta, Y., Tseng, C.L.W., Moore, A., and Huneck, S. 1977a. Sesquiterpenes of nine European liverworts from the genera, *Anastrepta, Bazzania, Jungermannia, Lepidozia* and *Scapania*. *Phytochemistry*, 16: 1731–1751.

Andersen, N.H., Costin, C.R., Kramer, C.M. Jr., Ohta, Y., and Huneck, S. 1973. Sesquiterpenes of *Barbilophozia* species. *Phytochemistry*, 12: 2709–2716.

Andersen, N.H., Ohta, Y., Liu, C.-B., Kramer, C.M., Allison, K., and Huneck, S. 1977b. Sesquiterpenes of thalloid liverworts of the genera *Conocephalum, Lunularia, Metzgeria* and *Riccardia*. *Phytochemistry*, 16: 1727–1729.

Aponte, J.C., Yang, H., Vaisberg, A.J., Castillo, D. et al. 2010. Cytotoxic and anti-infective sesquiterpenes present in *Plagiochila disticha* (Plagiochilaceae) and *Ambrosia peruviana* (Asteraceae). *Planta Med.*, 76: 705–707.

Asakawa, Y. 1981. Biologically active substances obtained from bryophytes. *J. Hattori Bot. Lab.*, 50: 123–142.

Asakawa, Y. 1982. Chemical constituents of the Hepaticae. In: Herz, W., Grisebach, H., Kirby, G.W. (eds.). *Progress in the Chemistry of Organic Natural Products*. Springer, Vienna, vol. 42, pp. 1–285.

Asakawa, Y. 1990. Biologically active substances from bryophytes. In: Chopra, R.N., Bhatla, S.C. (eds.). *Bryophytes Development: Physiology and Biochemistry*. CRC Press, Boca Raton, pp. 259–287.

Asakawa, Y. 1995. Chemical constituents of the bryophytes. In: Herz, W., Kirby, G.W., Moore, R.E., Steglich, W., Tamm, Ch. (eds.). *Progress in the Chemistry of Organic Natural Products*. Springer, Vienna, vol. 65, pp. 1–618.

Asakawa, Y. 2008. Liverworts – Potential source of medicinal compounds. *Curr. Pharm. Design*, 14: 3067–3088.

Asakawa, Y., Baser, K.H.C., Erol, B., von Reuβ, S., Konig, W.A., Ozenoglu, H., and Gokler, I. 2018. Volatile components of some selected Turkish liverworts. *Nat. Prod. Commun.* 2018, 13: 899–902.

Asakawa, Y., Dawson, G.W., Griffiths, D.C., Lallemand et al. 1988. Activity of drimane antifeedants and related compounds against aphids and comparative biological effects and chemical reactivity of (–)- and (+)-polygodial. *J. Chem. Ecol.*, 14: 1845–1855.

Asakawa, Y., and Ludwiczuk, A. 2013. Bryophytes: Liverworts, Mosses, and Hornworts: Extraction and Isolation Procedures. In: Roessner, U., Dias, D.A. (eds.). *Metabolomics Tools for Natural Product Discovery: Methods and Protocols*, Humana Press, New York, USA, Vol. 1055, pp. 1–20.

Asakawa, Y., and Ludwiczuk, A. 2018. Chemical constituents of bryophytes: Structures and biological activity. *J. Nat. Prod.*, 81: 641–660.

Asakawa, Y., Ludwiczuk, A., and Hashimoto, T. 2013a. Cytotoxic and antiviral compounds from bryophytes and inedible fungi. *J. Pre-Clin. Clin. Res.*, 7: 73–85.

Asakawa, Y., Ludwiczuk, A., and Nagashima, F. 2013b. Chemical constituents of bryophytes: Bio- and chemical diversity, biological activity, and chemosystematics. In: Kinghorn, A.D., Falk, H., Kobayashi, J. (eds.). *Progress in the Chemistry of Organic Natural Products*. Springer-Verlag, Vienna, vol. 95, pp. 1–796.

Asakawa, Y., Ludwiczuk, A., and Nagashima, F. 2013c. Phytochemical and biological studies of bryophytes. *Phytochemistry*, 91: 52–80.

Asakawa, Y., Toyota, M., Takemoto, T., Kubo, I., and Nakanishi, K. 1980. Insect antifeedant secoaromadendrane-type sesquiterpenes from *Plagiochila* species. *Phytochemistry*, 19: 2147–2154.

Baek, S.-H., Perry, N.B., and Lorimer, S.D. 2003. *ent*-Costunolide from the liverwort *Hepatostolonophora paucistipula*. *J. Chem. Res. (S)*, 14–15.

Basar, S., Koch, A., and König, W.A. 2001. A verticillane-type diterpene from *Boswelia carterii* essential oil. *Flavour Fragr. J.*, 16: 315–318.

Bednarek-Ochyra, H., Vana, J., Ochyra, R., and Smith, R.I.L. 2000. *The liverwort flora of Antarctica*. Polish Academy of Sciences, Cracow.

Benesova, V., Sedmera, P., Herout, V., and Sorm, F. 1971. The structure of a tetracyclic sesquiterpenic alcohol from liverwort *Mylia taylorii* (Hook.) Gray. *Tetrahedron Lett.*, 12: 2679–2682.

Boland, W. 1995. The chemistry of gamete attraction: Chemical structures, biosynthesis, and (a)biotic degradation of algal pheromones. *Proc. Natl. Acad. Sci. U. S. A.*, 92: 37–43.

Bowman, J.L. 2016. A brief history of *Marchantia* from Greece to genomics. *Plant Cell Physiol.*, 57: 210–229.

Bowman, J.L., Kohchi, T., Yamato, K.T., Jenkins et al. 2017. Insights into land plant evolution garnered from the *Marchantia polymorpha* genome. *Cell*, 171: 287–304. e215.

Chandra, S., Chandra, D., Barh, A., Pankaj, and Pandey, R.K. 2017. Bryophytes: Hoard of remedies, an ethno-medicinal review. *J. Trad. Complement. Med.*, 7: 94–98.

Chang, S.J., and Wu, C.-L. 1987. A new labdane alcohol from the liverwort *Trichocolea pluma*. *Huaxue*, 45: 142–149.

Chen, F., Ludwiczuk, A., Wei, G., Chen, X., Crandall-Stotler, B., and Bowman, J.L. 2018. Terpenoid secondary metabolites in bryophytes: Chemical diversity, biosynthesis and biological functions. *Crit. Rev. Plant Sci.*, 37, 210–231. https://doi.org/10.1080/07352689.2018.1482397

Crandall-Stotler, B., Stotler, R.E., and Long, D.G. 2008. Morphology and classification of the Marchantiophyta. In: Goffinet, B., Shaw, A.J. (eds.) *Bryophyte Biology: Second Edition*. Cambridge University Press, Cambridge, UK, pp. 1–54.

Crandall-Stotler, B., Stotler, R.E., and Long, D.G. 2009. Phylogeny and classification of the Marchantiophyta. *Edin. J. Bot.*, 66: 155–198.

Dodoens, R. 1578. *A Nievve Herball* (English translation by H. Lyte). London, G. Deves.

Duckett, J. 1986. Ultrastructure in bryophyte systematics and evolution: An evaluation. *J. Bryol.*, 14: 25–42.

Figueiredo, A.C., Sim-Sim, M., Barroso, J.G., Pedro, L.G., Esquivel, M.G., Fontinha, S., Martins, S., Lobo, C., and Stech, M. 2009. Liverwort *Radula* species from Portugal: Chemotaxonomical evaluation of volatiles composition. *Flavour Fragr. J.*, 24: 316–325.

Figueiredo, A.C., Sim-Sim, M., Costa, M.M., Barroso, J.G., Pedro, L.G., Esquivel, M.G., Gutierres, F., Lobo, C., and Fontinha, S. 2005. Comparison of the essential oil composition of four *Plagiochila* species: P. bifaria, P. maderensis, P. retrorsa and P. stricta. *Flavour Fragr. J.*, 20: 703–709.

Frahm, J.-P. 2004. Recent developments of commercial products from bryophytes. *The Bryologist*, 107: 277–283.

Fricke, C., Rieck, A., Hardt, I.H., König, W.A., and Muhle, H. 1995. Identification of (+)-β-caryophyllene in essential oils of liverworts by enantioselective gas chromatography. *Phytochemistry*, 39: 1119–1121.

Gershenzon, J., and Dudareva, N. 2007. The function of terpene natural products in the natural world. *Nat. Chem. Biol.*, 3: 408–414.

Glime, J.M. 2007. Economic and ethnic uses of bryophytes. In: Zander, R.H. (ed.), *Flora of North America North of Mexico*. Oxford University Press, New York, Oxford, vol. 27, pp. 14–41.

Gradstein, S.R. 1992. The genera *Thysananthus*, *Dendrolejeunea*, and *Fulfordianthus* gen. nov. (Studies on Lejeuneaceae subfamily Ptychnthoideae XXI). *The Bryologist*, 95: 42–51.

Gradstein, S.R. 2013. A classification of Lejeuneaceae (Marchantiophyta) based on molecular and morphological evidence. *Phytotaxa*, 100: 6–20.

Haines, W.P., and Renwick, J.A.A. 2009. Bryophytes as food: Comparative consumption and utilization of mosses by a generalist insect herbivore. *Entomol. Exp. Appl.*, 133: 296–306.

Han, J.-J., Zhang, J.-Z., Zhu, R.-X., Li, Y., Qiao, Y.-N., Gao, Y., Jin, X.-Y., Chen, W., Zhou, J.-C., and Lou, H.-X. 2018. Plagiochianins A and B, two *ent*-2,3-seco-aromadendrane derivatives from the liverwort *Plagiochila duthiana*. *Org. Lett.*, 20: 6550–6553.

Harinantenaina, L., and Asakawa, Y. 2004. Chemical constituents of Malagasy liverworts, part II: Mastigophoric acid methyl ester of biogenetic interest from *Mastigophora diclados* (Lepicoleaceae subf. Mastigophoroideae). *Chem. Pharm. Bull.*, 52: 1382–1384.

Harinantenaina, L., Quang, D.N., Nishizawa, T., Hashimoto, T., Kohichi, C., Soma, G.-I., and Asakawa, Y. 2007. Bioactive compounds from liverworts: Inhibition of lipopolysaccharide-induced inducible NOS mRNA in RAW 264.7 cells by herbertenoids and cuparenoids. *Phytomedicine*, 14: 486–491.

Harris, E.S.J. 2008. Ethnobryology: Traditional uses and folk classification of bryophytes. *The Bryologist*, 111: 169–218.

Harrison, L.J. 1983, Secondary metabolites of the Hepaticae. PhD Thesis, University of Glasgow, United Kingdom, pp. 1–241.

Hayashi, K., Kawaide, H., Notomi, M., Sakigi, Y., Matsuo, A., and Nozaki, H. 2006. Identification and functional analysis of bifunctional *ent*-kaurane synthase from the moss *Physcomitrella patens*. *FEBS Lett.*, 580: 6175–6181.

He, X., Sun, Y., Zhu, R.-L. 2013. The oil bodies of liverworts: Unique and important organelles in land plants. *Crit. Rev. Plant Sci.*, 32: 293–302.

Hentschel, J., Zhu, R.-L., Long, D.G., Davison, P.G., Schneider, H., Gradstein, S.R., and Heinrichs, J. 2007. A phylogeny of *Porella* (Porellaceae, Jungermanniopsida) based on nuclear and chloroplast DNA sequences, *Mol. Phylogen. Evol.*, 45: 693–705.

Hu, Y., Guo, D.H., Liu, P., Rahman, K., Wang, D.X., and Wang, B. 2009. Antioxidant effects of a *Rhodobryum roseum* extract and its active components in isoproterenol-induced myocardial injury in rats and cardiac myocytes against oxidative stress-triggered damage. *Pharmazie*, 64: 53–57.

Huneck, S., and Klein, E. 1967. Inhaltsstoffe der Moose-III. Uber die vergleichende gas- und diinnschichtchromatographische untersuchung der atherischen Ole einiger Lebermoose und isolierung von (-)-Longifolen und (-)-Longibomeol aus *Scapania undulata* (L.) Dum. *Phytochemistry*, 6: 383–390.

Jia, Q., Kollner, T.G., Gershenzon, J., and Chen, F. 2018. MTPSLs: New terpene synthases in nonseed plants. *Trends Plant Sci.*, 23: 121–128.

Jia, Q., Li, G., Kollner, T.G., Fu J. et al. 2016. Microbial-type terpene synthase genes occur widely in nonseed land plants, but not in seed plants. *PNAS*, 113: 12328–12333.

Jones, W.P., and Kinghorn, A.D. 2006. Extraction of Plant Secondary Metabolites. In: Sarker, S.D., Latif, Z., Gray, A.I. (eds.), *Natural Products Isolation*. 2nd ed. (Methods of Biotechnology Series, No. 20), Humana Press, Totowa, New Jersey, pp. 323–351.

Kajiwara, T., Hatanaka, A., Tanaka, Y., Kawai, T., Ishihara, M., Tsuneya, T., and Fujimura, T. 1989. Volatile constituents from marine brown algae of Japanese *Dictyopteris*. *Phytochemistry*, 28: 636–639.

Karlsson, B., Pilotti, A.-M., Söderholm, A.-C., Norin, T., Sundin, S., and Sumimoto, M. 1978. The structure and absolute configuration of verticillol, a macrocyclic diterpene alcohol from the wood of *Sciadopitys verticillata* Sieb. et Zucc. *(Taxodiaceae)*. *Tetrahedron*, 34: 2349–2354.

Kato-Noguchi, H., and Seki, T. 2010. Allelopathy of the moss *Rhynchostegium pallidifolium* and 3-hydroxy-β-ionone. *Plant Signal. Behav.*, 5: 702–704.

Knoche, H., Ourisson, G., Perold, G.W., Foussereau, J., and Maleville, J. 1969. Allergenic component of a liverwort: A sesquiterpene lactone. *Science*, 166: 239–240.

Kobayashi, J., and Ishibashi, M. 1999. Marine Natural Products and Marine Chemical Ecology. In: Barton, D., Nakanishi, K., Meth-Cohn, O., (eds.), *Comprehensive Natural Products Chemistry*. Elsevier, Amsterdam, vol. 5, p 420.

Kodama, M., Tambunan, U.S.F., and Tsunoda, T. 1986. Total synthesis of (-)-vitrenal and its biological activity. *Tetrahedron Lett.*, 27: 1197–1200.

Koepp, A.E., Hezari, M., Zajicek, J., Stofer Vogel, B., LaFevert, R.E., Lewis, N.G., and Croteau, R. 1995. Cyclization of geranylgeranyl diphosphate to taxa-4(5),11(12)-diene is the committed step of taxol biosynthesis in pacific yew. *J. Biol. Chem.*, 270: 8686–8690.

Komala, I., Ito, T., Nagashima, F., Yagi, Y., and Asakawa, Y. 2010a. Cytotoxic, radical scavenging and antimicrobial activities of sesquiterpenoids from the Tahitian liverwort *Mastigophora diclados* (Brid.) Nees (Mastigophoraceae). *J. Nat. Med.*, 64: 417–422.

Komala, I., Ito, T., Nagashima, F., Yagi, Y., and Asakawa, Y. 2011. Cytotoxic bibenzyls, germacrane- and pinguisane-type sesquiterpenoids from the Indonesian, Tahitian and Japanese liverworts. *Nat. Prod. Commun.*, 6: 303–309.

Komala, I., Ito, T., Nagashima, F., Yagi, Y., Kawahata, M., Yamaguchi, K., and Asakawa, Y. 2010b. Zierane sesquiterpene lactone, cembrane and fusicoccane diterpenoids, from the Tahithian liverwort *Chendonanthus hirtellus*. *Phytochemistry*, 71: 1387–1394.

Lorimer, S.D., Burges, E.J., and Perry, N.B. 1997a. Diplophyllolide: A cytotoxic sesquiterpene lactone from the liverworts *Clasmatocolea vermicularis* and *Chiloscyphus subporosa*. *Phytomedicine*, 4: 261–263.

Lorimer, S.D., Perry, N.B., Burges, E.J., and Foster, L.M. 1997b. 1-Hydroxyditerpenes from two New Zealand liverworts, *Paraschistochila pinnatifolia* and *Trichocolea molissima*. *J. Nat. Prod.*, 60: 421–424.

Ludwiczuk, A., and Asakawa, Y. 2010. Chemosystematics of selected liverworts collected in Borneo. *Trop. Bryol.*, 31: 33–42.

Ludwiczuk, A., and Asakawa, Y. 2014. Fingerprinting of secondary metabolites of liverworts: Chemosystematic approach. *J. AOAC Int.*, 97: 1234–1243.

Ludwiczuk, A., and Asakawa, Y. 2015. Chemotaxonomic value of essential oil components in liverwort species. *A Review. Flavour Frag. J.*, 30: 189–196.

Ludwiczuk, A., and Asakawa, Y. 2017. GC/MS fingerprinting of solvent extracts and essential oils obtained from liverwort species. *Nat. Prod. Commun.*, 12: 1301–1305.

Ludwiczuk, A., and Asakawa, Y. 2019. Bryophytes as a source of bioactive volatile terpenoids – A review. *Food Chem. Toxicol.*, 132: 110649; https://doi.org/10.1016/j.fct.2019.110649

Ludwiczuk, A., Gradstein, S.R., Nagashima, F., and Asakawa, Y. 2011. Chemosystematics of *Porella* (Marchantiophyta, Porellaceae). *Nat. Prod. Commun.*, 6: 315–321.

Ludwiczuk, A., Komala, I., Pham, A., Bianchini, J.-P., Raharivelomanana, P., and Asakawa, Y. 2009. Volatile components from selected Tahitian liverworts. *Nat. Prod. Commun.*, 4: 1387–1392.

Ludwiczuk, A., Nagashima, F., Gradstein, S.R., and Asakawa, Y. 2008. Volatile components from selected Mexican, Ecuadorian, Greek, German, and Japanese liverworts. *Nat. Prod. Commun.*, 3: 133–140.

Ludwiczuk, A., Odrzykoski, I.J., and Asakawa. Y. 2013a. Identification of cryptic species within liverwort *Conocephalum conicum* based on the volatile components. *Phytochemistry*, 95: 234–241.

Ludwiczuk, A., Skalicka-Woźniak, K., and Georgiev, M.I. 2017. Terpenoids. In: Badal, S., Delgoda, R. (eds.) *Pharmacognosy Fundamentals, Applications and Strategy*. Elsevier, Academic Press, Amsterdam, pp. 233–266.

Ludwiczuk, A., Sukkharak, P., Gradstein, R., Asakawa, Y., and Głowniak, K. 2013b. Chemical relationships between liverworts of the family Lejeuneaceae (Porellales, Jungermanniopsida). *Nat. Prod. Commun.*, 8: 1515–1518.

Matsuo, A. 1971. Structure of bazzanene. *Tetrahedron*, 27: 2757–2764.

Matsuo, A., Atsumi, K., Nakayama, M., and Hayashi, S. 1981a. Structure of *ent*-2,3-seco-alloaromadendrane sesquiterpenoids having plant growth inhibitory activity from *Plagiochila semidecurrens* (liverwort). *J. Chem. Soc. Perkin Trans*, 1: pp. 2816–2824.

Matsuo, A., Kubota, N., Nakayama, M., and Hayashi, S. 1981b. -(-)-Lepidozenal, a sesquiterpenoid with a novel trans-fused bicyclo[8.1.0] undecane system from the liverwort *Lepidozia vitrea*. *Chem. Lett.*, pp. 1097–1100.

Matsuo, A., Nakayama, N., and Nakayama, M. 1984a. Structures and conformations of (-)-isobicyclogermacrenal and (-)-lepidozenal, two key sesquiterpenois of the *cis*- and *trans*-10,3-bicyclic ring system from the liverwort *Lepidozia vitrea*: X-ray crystal structure analysis of the hydroxyl derivative of (-)-isobicyclogermacrenal. *J. Chem. Soc. Perkin Trans.*, 1: pp. 203–214.

Matsuo, A., Uto, S., Nozaki, H., and Nakayama, M. 1984b. Structure and absolute configuration of (+)- vitrenal, a novel carbon skeletal sesquiterpenoid having plant-growth-inhibitory activity from the liverwort *Lepidozia vitrea*. *J. Chem. Soc. Perkin Trans.*, 1: pp. 215–221.

Melching, S., Bülow, N., Wihstutz, K., Jung, S., and König, W.A. 1997. Natural occurrence of both enantiomers of cadina-3, 5–diene and δ-amorphene. *Phytochemistry*, 44: 1291–1296.

Melching, S., and Konig, W.A. 1999. Sesquiterpenes from the essential oil of the liverwort *Conocephalum conicum*. *Phytochemistry*, 51: 517–523.

Müller, K. 1905. Beitrag zur Kenntnis der ätherischen Öle bei Lebermoosen. *Hoppe-Seyler's Z. Physiol. Chem.*, 45: 299–319.

Nagashima, F., Kasai, W., Kondoh, M., Fuji, M., Watanabe, Y., Braggins, J.E., and Asakawa, Y. 2003a. New *ent*-kaurene-type diterpenoids possessing cytotoxicity from the New Zealand liverwort *Jungermannia* species. *Chem. Pharm. Bull.*, 51: 1189–1192.

Nagashima, F., Kondoh, M., Fuji, M., Takaoka, S., Watanabe, Y., and Asakawa, Y. 2005. Novel cytotoxic kaurane-type diterpenoids from the New Zealand liverwort *Jungermannia* species. *Tetrahedron*, 61: 4531–4544.

Nagashima, F., Murakami, M., Takaoka, S., and Asakawa, Y. 2003b. *ent*-Isopimarane-type diterpenoids from the New Zealand liverwort *Trichocolea molissima*. *Phytochemistry*, 64: 1319–1325.

Nagashima, F., Ohi, Y., Nagai, T., Tori, M., Asakawa, Y., and Huneck, S. 1993. Terpenoids from some German and Russian liverworts. *Phytochemistry*, 33: 1445–1448.

Neves, M., Morais, R., Gafner, S., Stoeckli-Evans, H., and Hostettmann, K. 1999. New sesquiterpene lactones from the Portuguese liverwort *Targionia lorbeeriana*. *Phytochemistry*, 50: 967–972.

Nozaki, H., Hayashi, K.-I., Nishimura, N., Kawaide, H., Matsuo, A., and Takaoka, D. 2007. Momilactone A and B as allelochemicals from moss *Hypnum plumaeforme*: First occurrence in Bryophytes. *Biosci. Biotechnol. Biochem.*, 71: 3127–3130.

Odrzykoski, I.J., and Szweykowski, J. 1991. Genetic differentiation without concordant morphological divergence in the thallose liverwort *Conocephalum conicum*. *Plant Syst. Evol.*, 178: 135–151.

Ohta, Y., Andersen, N.H., and Liu, C.-B. 1977. Sesquiterpene constituents of two liverworts of genus *Diplophyllum*: Novel eudesmanolides and cytotoxicity studies for enantiomeric methylene lactones. *Tetrahedron*, 33: 617–628.

Otoguro, K., Iwatsuki, M., Ishiyama, M., Namatame, M., Nishihara-Tukashima, A., Kiyohara, H., Hashimoto, T., Asakawa, Y., Omura, S., and Yamada, H. 2011. In vitro antitrypanosomal activity of plant terpenes against Trypanosoma brucei. *Phytochemistry*, 72: 2024–2030.

Ozdemir, T., Ucuncu, O., Cansu, T.B., Kahriman, N., and Yayli, N. 2010. Volatile constituents in mosses (*Brachythecium albicans* (Hedw.) Schimp., *Bryum pallescens* Schleich. ex Schwagr and Syntrichia intermedia Brid.) grown in Turkey. *Asian J. Chem.*, 22: 7285–7290.

Paul, C., Konig, W.A., and Wu, C.-L. 2001. Sesquiterpenoid constituents of the liverworts *Lepidozia fauriana* and *Lepidozia vitrea*. *Phytochemistry*, 58: 789–798.

Perry, N.B., Burgess, E.J., Foster, L.M., and Gerard, P.J. 2003. Insect antifeedant sesquiterpene acetals from the liverwort *Lepidolaena clavigera*. *Tetrahedron Lett.*, 44: 1651–1653.

Qiu, Y.-L., Li, L., Wang, B., Chen, Z. et al. 2006. The deepest divergences in land plants inferred from phylogenomic evidence. *Proc. Natl. Acad. Sci. USA*, 103: 15511–15516.

Quang, D.N., and Asakawa, Y. 2010. Chemical constituents of the Vietnamese liverwort *Porella densifolia*. *Fitoterapia*, 81: 659–661.

Radulovic, N.S., Filipovic, S.I., Zlatkovic, D.B., Dordevic, M.R., Stojanovic, N.M., Randjelovic, P.J., Mitic, K.V., Jevtovic-Stoimenov, T.M., and Randelovic, V.N. 2016. Immunomodullatory pinguisane-type sesquiterpenes from the liverwort *Porella cordaeana* (Porellaceae): The "new old" furanopinguisanol and its oxidation product exert mutually different effects on rat splenocytes. *RSC Adv.*, 6: 41847–41860.

Ramirez, M., Kamiya, N., Popich, S., Asakawa, Y., and Bardon, A. 2010. Insecticidal constituents from the Argentine liverwort *Plagiochila bursata*. *Chem. Biodivers.*, 7: 1855–1861.

Ramirez, M., Kamiya, N., Popich, S., Asakawa, Y., and Bardon, A. 2017. Constituents of the Argentine liverwort *Plagiochila diversifolia* and their insecticidal activities. *Chem. Biodivers.*, 14: e1700229; doi: 10.1002/cbdv.201700229.

Rensing, S.A. 2018. Great moments in evolution: The conquest of land by plants. *Curr. Opin. Plant Biol.*, 42: 49–54.

Rieck, A., Bülow, N., and König, W.A. 1995. An epoxy-trinoreudesmane sesquiterpene from the liverwort *Lophocolea bidentata*. *Phytochemistry*, 40: 847–851.

Rieck, A., and König, W.A. 1996. Furano-eudesma 4(15),7,11-trien-5α-ol from the liverwort *Lophocolea heterophylla*. *Phytochemistry*, 43: 1055–1056.

Rycroft, D.S., Groth, H., and Heinrichs, J. 2004. Reinstatement of *Plagiochila maderensis* (Jungermanniopsida: Plagiochilaceae) based on chemical evidence and nrDNA ITS sequences. *J. Bryol.*, 26: 37–45.

Sakurai, K., Tomiyama, K., Kawakami, Y., Yaguchi, Y., and Asakawa, Y. 2018. Characteristic scent from the Tahitian liverwort, *Cyathodium foetidissimum*. *J. Oleo Sci.*, 67: 1265–1269.

Saritas, Y., Mekem Sonwa, M., Iznaguen, H., König, W.A., Muhle, H., and Mues, R. 2001. Volatile constituents in mosses (Musci). *Phytochemistry*, 57: 443–457.

Saw, J. 2001. Biogeographic patterns and cryptic speciation in bryophytes. *J. Biogeogr.*, 28: 253–261.

Saxena, D.K., and Harinder 2004. Uses of Bryophytes. *Resonance*, 9: 56–65. https://doi.org/10.1007/BF02839221

Scher, J.M., Burgess, E.J., Lorimer, S.D., and Perry, N.B. 2002. A cytotoxic sesquiterpene and unprecedented sesquiterpene-bisbibenzyl compounds from the liverwort *Schistochila glaucescens*. *Tetrahedron*, 58: 7875–7882.

Scher, J.M., Schinkovitz, A., Zapp, J., Wang, Y., Franzblau, S.G., and Beker, H. 2010. Structure and anti-TB activity of trachylobanes from the liverwort *Jungermannia exsertifolia ssp. cordifolia*. *J. Nat. Prod.*, 73: 656–663.

Scher, J.M., Speakman, J.B., Zapp, J., and Becker, H. 2004. Bioactivity guided isolation of antifungal compounds from the liverwort *Bazzania trilobata* (L.) S.F. Gray. *Phytochemistry*, 65: 2583–2588.

Shu, Y.-F., Wie, H.-C., and Wu, C.-L. 1994. Sesquiterpenoids from liverworts *Lepidozia vitrea* and *Lepidozia fauriana*. *Phytochemistry*, 58: 789–798.

Shy, H.-S., and Wu, C.-L. 2006. Eudesmane-type sesquiterpenoids from the liverwort *Lepidozia fauriana*. *J. Asian Nat. Prod. Res.*, 8: 723–731.

Sonwa, M., and König, W.A. 2003. Chemical constituents of the essentials oil of the hornwort *Anthoceros caucasicus*. *Flavour Fragr. J.*, 18: 286–289.

Suire, C. 2000. A comparative, transmission-electron microscopic study on the formation of oil-bodies in liverworts. *J. Hattori Bot. Lab.*, 89: 209–232.

Sukkharak, P., Ludwiczuk, A., Asakawa, Y., and Gradstein, S.R. 2011. Studies on the genus *Thysananthus* (Marchantiophyta, Lejeuneaceae). 3. Terpenoids in selected species of *Thysananthus* and in *Dendrolejeunea fruticosa*. *Cryptogam. Bryol.*, 32: 199–209.

Tabacchi, R., Joulain, D., Huneck, S., and Herout, V. 1992. In: Woidich, H., Buchbauer, G. (eds.), *Proceedings of the 12th International Congress of Flavours, Fragrances and Essential Oils*, Austrian Association of Flavour and Fragrance Industry, Vienna, pp. 125–134.

Tesso, H., Konig, W.A., and Asakawa, Y. 2005. Composition of the essential oil of the liverwort *Radula perrottetii* of Japanese origin. *Phytochemistry*, 66: 941–949.

Toyota, M., and Asakawa, Y. 1994. Volatile constituents of the liverwort *Chiloscyphus pallidus* (Mitt.) Engel & Schuster. *Flavour Fragr. J.*, 9: 237–240.

Toyota, M., Asakawa, Y., and Phram, J.-P. 1990. Homomono- and sesquiterpenoids from the liverwort *Lophocolea heterophylla*. *Phytochemistry*, 29: 2334–2337.

Toyota, M., Kimura, K., and Asakawa, Y. 1998a. Occurrence of *ent*-sesquiterpene in the Japanese moss – *Plagiomnium acutum*: First isolation and identification of *ent*-sesqui- and dolabellane-type diterpenoids from the Musci. *Chem. Pharm Bull.*, 46: 1488–1489.

Toyota, M., Konoshima, M., and Asakawa, Y. 1999. Terpenoid constituents of the liverwort *Reboulia hemisphaerica*. *Phytochemistry*, 52: 105–112.

Toyota, M., Koyama, H., and Asakawa, Y. 1997. Volatile components of the liverworts *Archilejeunea olivacea, Cheilolejeunea imbricata* and *Leptolejeunea elliptica*. *Phytochemistry*, 44: 1262–1264.

Toyota, M., Nakaishi, E., and Asakawa, Y. 1996. Eudesmane-type sesquiterpenoids from the liverwort *Lepidozia vitrea*. *Phytochemistry*, 41: 833–836.

Toyota, M., Tanimura, K., and Asakawa, Y. 1998b. Cytotoxic 2,3-secoaromadendrane-type sesquiterpenoids from the liverwort *Plagiochila ovalifolia*. *Planta Med.*, 64: 462–464.

Valarezo, E., Vidal, V., Calva, J., Jaramillo, S.P., Febres, J.D., and Benitez, A. 2018. Essential oil constituents of mosses species from Ecuador. *J. Essent. Oil-Bear. Plants*, 21: 189–197.

van Tooren, B.F. 1990. Effects of a bryophyte layer on the emergence of chalk grassland species. *Acta Oecol.*, 11: 155–163.

Vanderpoorten, A., and Goffinet, B. 2009. *Introduction to Bryophytes*. Cambridge University Press, Cambridge UK.

Vickers, C.E., Gershenzon, J., Lerdau, M.T., and Loreto, F. 2009. A unified mechanism of action for volatile isoprenoids in plant abiotic stress. *Nat. Chem. Biol.*, 5: 283–291.

von Reuβ, S.H., and König, W.A. 2005. Olefinic isothiocyanates and iminodithiocarbonates form the liverwort *Corsinia coriandrina*. *Eur. J. Org. Chem*, 2005: 1184–1188.

von Reuβ, S.H., Wu, C.-L., Muhle, H., and König, W.A. 2004. Sesquiterpene constituents from the essential oils of the liverworts *Mylia taylorii* and *Mylia nuda*. *Phytochemistry*, 65: 2277–2291.

Warmers, U., and Konig, W.A. 1999. Gymnomitrane-type sesquiterpenoids of the liverworts *Gymnomitron obtusum* and *Reboulia hemishaerica*. *Phytochemistry*, 52: 1501–1505.

Whitehead, J., Wittemann, M., and Cronberg, N. 2018. Allelopathy in bryophytes – a review. *Lindbergia*, 41: doi: 10.25227/linbg.01097.

Wu, C., Gunatilaka, A.A.L., McCabe, F.L., Johnson, R.K., Spjut, R.W., and Kingston, D.G.I. 1997. Bioactive and other sesquiterpenes from *Chiloscyphus rivularis*. *J. Nat. Prod.*, 60: 1281–1286.

Wurzel, G., Becker, H., Eicher, T., and Tiefensee, K. 1990. Molluscicidal properties of constituents from the liverwort *Ricciocarpos natans* and of synthetic lunularic acid derivatives. *Planta Med.*, 56: 444–445.

Xie, C.-F., and Lou, H.-X. 2009. Secondary metabolites in Bryophytes: An ecological aspect. *Chem. Biodivers.*, 6: 303–312.

Xiong, W., Fu, J., Kollner, T.G., Chen, X., Jia, Q., Guo, H., Qian, P., Guo, H., Wu, G., and Chen, F. 2018. Biochemical characterization of microbial type terpene synthases in two closely related species of hornworts, *Anthoceros punctatus* and *Anthoceros agrestis*. *Phytochemistry*, 149: 116–122.

Zidorn, C. 2008. Plant Chemosystematics. In: Waksmundzka-Hajnos, M., Sherma, J., Kowalska, T. (eds.), *Thin Layer Chromatography in Phytochemistry*. CRC Press, Boca Raton, pp. 77–101.

22 Biotransformation of Monoterpenoids by Microorganisms, Insects, and Mammals

Yoshiaki Noma and Yoshinori Asakawa

CONTENTS

22.1 INTRODUCTION

A large number of monoterpenoids have been detected in or isolated from essential oils and solvent extracts of fungi, algae, liverworts, and higher plants, but the presence of monoterpenoids in fern is negligible. Vegetables, fruits, and spices contain monoterpenoids; however, their fate in human and other animal bodies has not yet been fully investigated systematically. The recent development of analytical instruments makes it easy to analyze the chemical structures of very minor components, and the essential oil chemistry field has dramatically developed.

Since monoterpenoids, in general, show characteristic odor and taste, they have been used as cosmetic materials and food additives and often for insecticides, insect repellents, and attractant drugs. In order to obtain much more functionalized substances from monoterpenoids, various chemical reactions and microbial transformations of commercially available and cheap synthetic monoterpenoids have been carried out. On the other hand, insect larva and mammals have been used for direct biotransformations of monoterpenoids to study their fate and safety or toxicity in their bodies.

The biotransformation of α-pinene (**4**) by using the black fungus *Aspergillus niger* was reported by Bhattacharyya et al. (1960) half a century ago. During that period, many scientists studied the biotransformation of a number of monoterpenoids by using various kinds of bacteria, fungi, insects, mammals, and cultured cells of higher plants. In this chapter, the microbial transformation of monoterpenoids using bacteria and fungi is discussed. Furthermore, the biotransformation by using insect larva, mammals, microalgae, as well as suspended culture cells of higher plants is also summarized. In addition, several biological activities of biotransformed products are also represented. At the end of this chapter, the metabolite pathways of representative monoterpenoids for further development on biological transformation of monoterpenoids are demonstrated.

22.2 METABOLIC PATHWAYS OF ACYCLIC MONOTERPENOIDS

22.2.1 Acyclic Monoterpene Hydrocarbons

22.2.1.1 Myrcene

The microbial biotransformation of myrcene (**302**) was described with *Diplodia gossypina* ATCC 10936 (Abraham et al., 1985). The main reactions were hydroxylation, as shown in Figure 22.1. On oxidation, myrcene (**302**) gave the diol (**303**) (yield up to 60%) and also a side product (**304**) that possesses one carbon atom less than the parent compound, in yields of 1%–2%.

One of the publications dealing with the bioconversion of myrcene (Busmann and Berger, 1994) described its transformation to a variety of oxygenated metabolites, with *Ganoderma applanatum*, *Pleurotus flabellatus*, and *Pleurotus sajor-caju* possessing the highest transformation activities. One of the main metabolites was myrcenol (**305**) (2-methyl-6-methylene-7-octen-2-ol), which gives a fresh, flowery impression and dominates the sensory impact of the mixture (see Figure 22.1).

β-Myrcene (**302**) was converted by common cutworm larvae, *Spodoptera litura*, to give myrcene-3,(10)-glycol (**308**) via myrcene-3,(10)-epoxide (**307**) (Figure 22.2) (Miyazawa et al., 1998).

22.2.1.2 Citronellene

(−)-Citronellene (**309**) and (+)-citronellene (**309′**) were biotransformed by the cutworm *S. litura* to give (3*R*)-3,7-dimethyl-6-octene-1,2-diol (**310**) and (3*S*)-3,7-dimethyl-6-octene-1,2-diol (**310′**), respectively (Takeuchi and Miyazawa, 2005) (Figure 22.3).

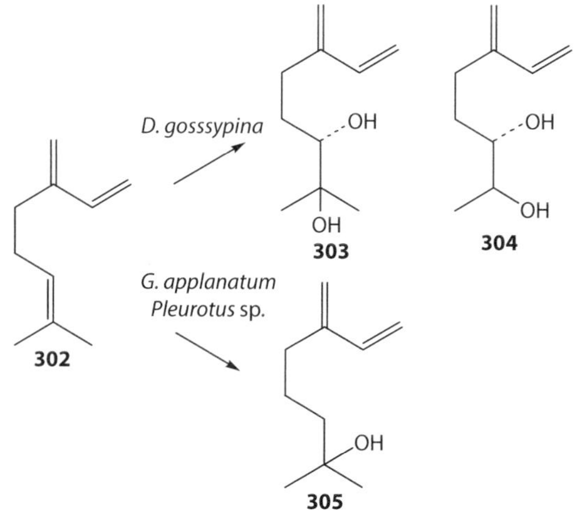

FIGURE 22.1 Biotransformation of myrcene (**302**) by *Diplodia gossypina* (Abraham et al., 1985), *Ganoderma applanatum*, and *Pleurotus* sp. (Modified from Busmann, D. and Berger, R.G., *J. Biotechol.*, 37, 39, 1994.)

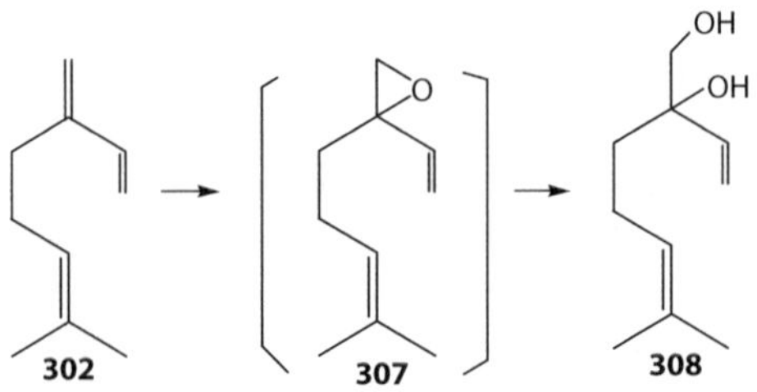

FIGURE 22.2 Biotransformation of myrcene (**302**) by *Spodoptera litura*. (Modified from Miyazawa, M. et al. Biotransformation of β-myrcene by common cutworm larvae, *Spodoptera litura* as a biocatalyst, *Proceedings of 42nd TEAC*, 1998, pp. 123–125.)

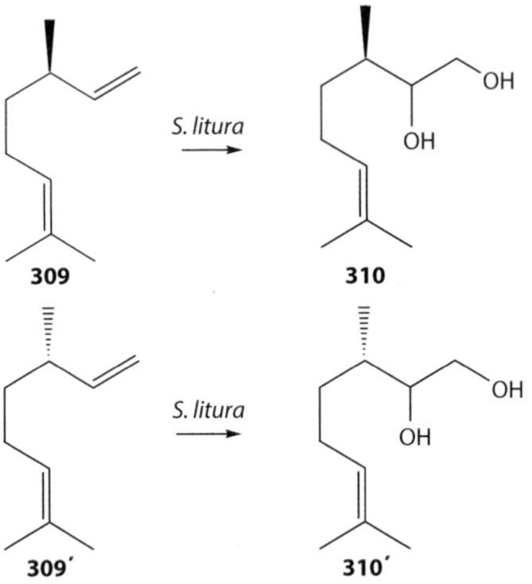

FIGURE 22.3 Biotransformation of (−)-citronellene (**309**) and (+)-citronellene (**309′**) by *Spodoptera litura*. (Modified from Takeuchi, H. and Miyazawa, M., Biotransformation of (−)- and (+)-citronellene by the larvae of common cutworm (*Spodoptera litura*) as biocatalyst, *Proceedings of 49th TEAC*, 2005, pp. 426–427.)

22.2.2 Acyclic Monoterpene Alcohols and Aldehydes

22.2.2.1 Geraniol, Nerol, (+)- and (−)-Citronellol, Citral, and (+)- and (−)-Citronellal

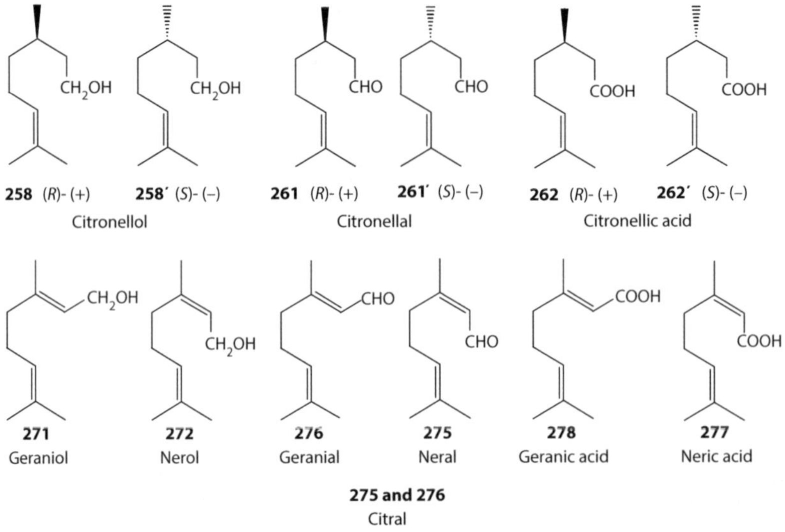

The microbial degradation of the acyclic monoterpene alcohols citronellol (**258**), nerol (**272**), geraniol (**271**), citronellal (**261**), and citral (equal mixture of **275** and **276**) was reported in the early part of 1960 (Seubert and Remberger, 1963; Seubert et al., 1963; Seubert and Fass, 1964a,b). *Pseudomonas citronellolis* metabolized citronellol (**258**), citronellal (**261**), geraniol (**271**), and geranic acid (**278**). The metabolism of these acyclic monoterpenes is initiated by the oxidation of the primary alcohol group to the carboxyl group, followed by the carboxylation of the C-10 methyl group (β-methyl) by a biotin-dependent carboxylase (Seubert and Remberger, 1963). The carboxymethyl group is eliminated at a later stage as acetic acid. Further degradation follows the β-oxidation pattern. The details of the pathway are shown in Figure 22.4 (Seubert and Fass, 1964a).

The microbial transformation of citronellal (**261**) and citral (**275** and **276**) was reported by way of *Pseudomonas aeruginosa* (Joglekar and Dhavlikar, 1969). This bacterium, capable of utilizing citronellal (**261**) or citral (**275** and **276**) as the sole carbon and energy source, has been isolated from soil by the enrichment culture technique. It metabolized citronellal (**261**) to citronellic acid (**262**) (65%), citronellol (**258**) (0.6%), dihydrocitronellol (**259**) (0.6%), 3,7-dimethyl-1,7-octanediol (**260**) (1.7%), and menthol (**137**) (0.75%) (Figure 22.5). The metabolites of citral (**275** and **276**) were geranic acid (**278**) (62%), 1-hydroxy-3,7-dimethyl-6-octen-2-one (**279**) (0.75%), 6-methyl-5- heptenoic acid (**280**) (0.5%), and 3-methyl-2-butenoic acid (**286**) (1%) (Figure 22.5). In a similar way, *Pseudomonas convexa* converted citral (**275** and **276**) to geranic acid (**278**) (Hayashi et al., 1967). The biotransformation of citronellol (**258**) and geraniol (**271**) by *P. aeruginosa*, *P. citronellolis*, and *Pseudomonas mendocina* was also reported by another group (Cantwell et al., 1978).

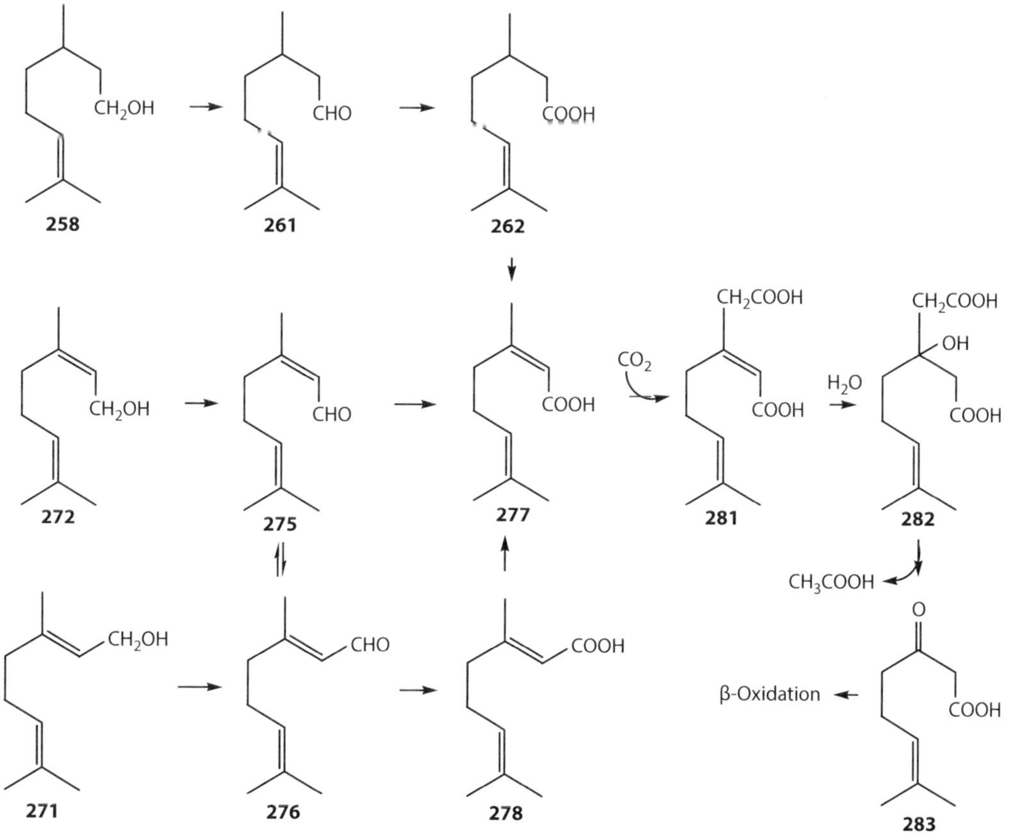

FIGURE 22.4 Biotransformation of citronellol (**258**), nerol (**272**), and geraniol (**271**) by *Pseudomonas citronellolis*. (Modified from Madyastha, K.M., *Proc. Indian Acad. Sci. (Chem. Sci.)*, 93, 677, 1984.)

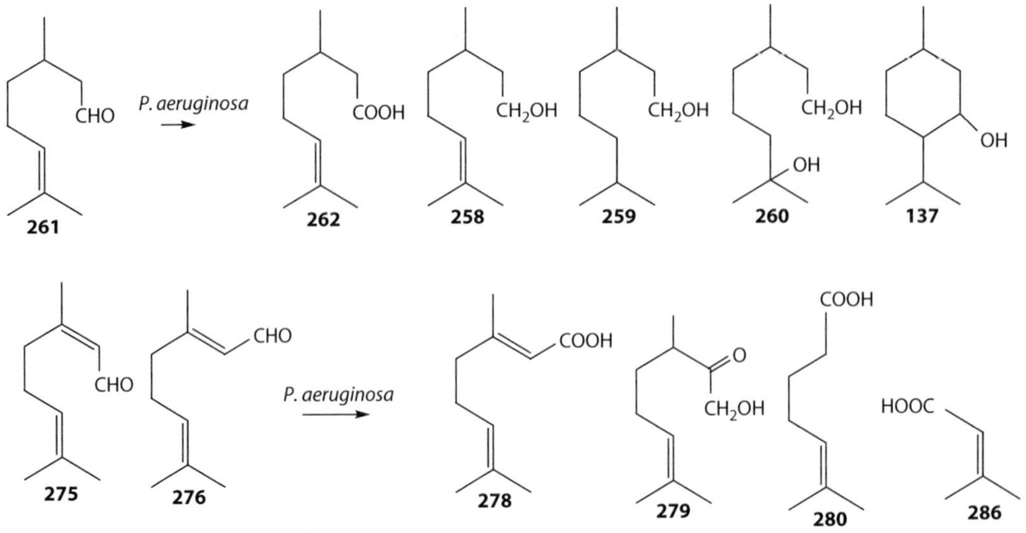

FIGURE 22.5 Biotransformation of citronellal (**261**) and citral (**275** and **276**) by *Pseudomonas aeruginosa*. (Modified from Joglekar, S.S. and Dhavlikar, R.S., *Appl. Microbiol.*, 18, 1084, 1969.)

A research group in Czechoslovakia patented the cyclization of citronellal (**261**) with subsequent hydrogenation to menthol by *Penicillium digitatum* in 1952. Unfortunately the optical purities of the intermediates pulegol and isopulegol were not determined, and presumably the resulting menthol was a mixture of enantiomers. Therefore, it cannot be excluded that this extremely interesting cyclization is the result of a reaction primarily catalyzed by the acidic fermentation conditions and only partially dependent on enzymatic reactions (Babcka et al., 1956) (Figure 22.6).

Based on previous data (Madyastha et al., 1977; Rama Devi and Bhattacharyya, 1977a), two pathways for the degradation of geraniol (**271**) were proposed by Madyastha (1984) (Figure 22.7). Pathway A involves an oxidative attack on the 2,3-double bond, resulting in the formation of an epoxide. Opening of the epoxide yields the 2,3-dihydroxygeraniol (**292**), which upon oxidation forms 2-oxo, 3-hydroxygeraniol (**293**). The ketodiol (**293**) is then decomposed to 6-methyl-5-hepten-2-one (**294**) by an oxidative process. Pathway B is initiated by the oxidation of the primary alcoholic group to geranic acid (**278**), and further metabolism follows the mechanism as proposed earlier for *P. citronellolis* (Seubert and Remberger, 1963; Seubert et al., 1963). In the case of nerol (**272**), the

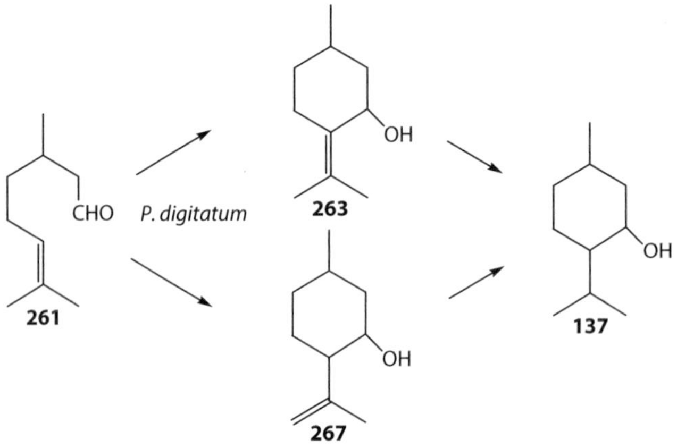

FIGURE 22.6 Biotransformation of citronellal to menthol by *Penicillium digitatum*. (Modified from Babcka, J. et al. Patent 56-9686b.)

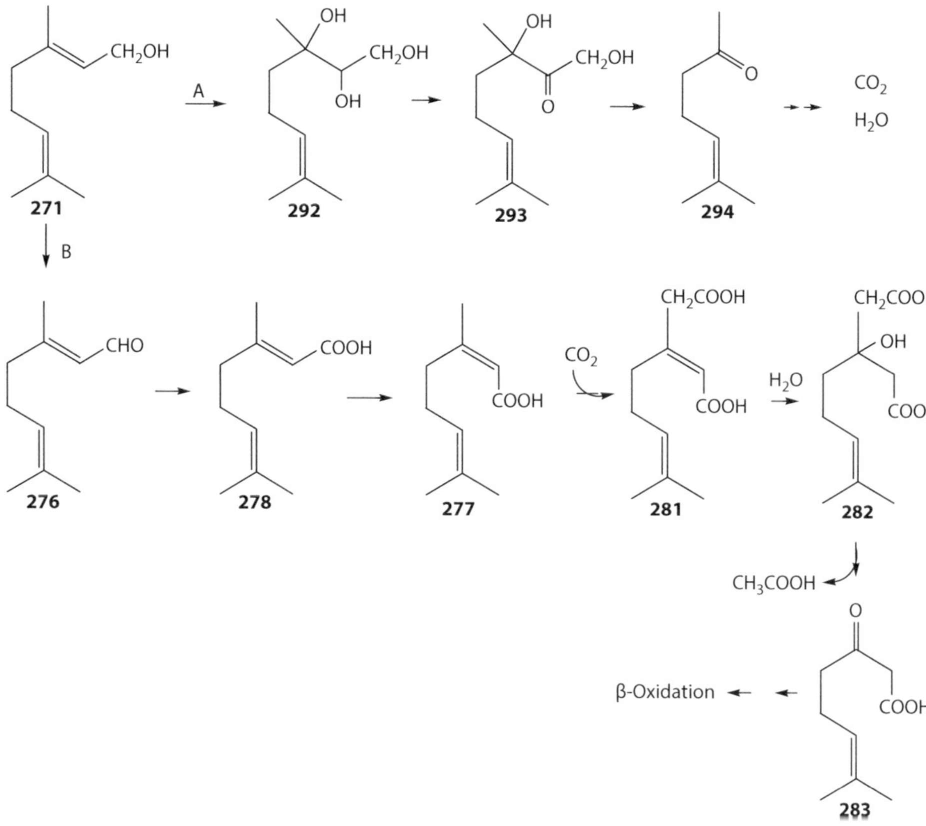

FIGURE 22.7 Metabolism of geraniol (**271**) by *Pseudomonas incognita*. (Modified from Madyastha, K.M., *Proc. Indian Acad. Sci.* (*Chem. Sci.*), 93, 677, 1984.)

Z-isomer of geraniol (**271**), degradative pathways analogous to pathways A and B as in geraniol (**271**) are observed. It was also noticed that *Pseudomonas incognita* metabolizes acetates of geraniol (**271**), nerol (**272**), and citronellol (**258**) much faster than their respective alcohols (Madyastha and Renganathan, 1983).

Euglena gracilis Z. converted citral (**275** and **276**, 56:44, peak area in gas chromatograph [GC]) to geraniol (**271**) and nerol (**272**), respectively, of which geraniol (**271**) was further transformed to (+)- and (−)-citronellol (**258** and **258′**). On the other hand, when either geraniol (**271**) or nerol (**272**) was added, both compounds were isomerized to each other, and then, geraniol (**271**) was transformed to citronellol. These results showed that *Euglena* could distinguish between the stereoisomers geraniol (**271**) and nerol (**272**) and hydrogenated geraniol (**271**) selectively. (+)-, (−)-, and (±)-Citronellal (**261**, **261′**, and **261** and **261′**) were also transformed to the corresponding (+)-, (−)-, and (±)-citronellol (**258**, **258′**, and **258** and **258′**) as the major products and (+)-, (−)-, and (±)-citronellic acids (**262**, **262′**, and **262** and **262′**) as the minor products, respectively (Noma et al., 1991b) (Figure 22.8).

Dunaliella tertiolecta also reduced citral (geranial (**276**) and neral (**275**)=56:44) and (+)-, (−)-, and (±)-citronellal (**261**, **261′**, and **261** and **261′**) to the corresponding alcohols: geraniol (**271**), nerol (**272**), and (+)-, (−)-, and (±)-citronellol (**258**, **258′**, and **258** and **258′**) (Noma et al., 1991a, 1992a).

Citral (a mixture of geranial (**276**) and neral (**275**), 56:44 peak area in GC) is easily transformed to geraniol (**271**) and nerol (**272**), respectively, of which geraniol (**32**) is further hydrogenated to (+)-citronellol (**258**) and (−)-citronellol (**258′**). Geranic acid (**278**) and neric acid (**277**) as the minor products are also formed from **276** and **275**, respectively. On the other hand, when either **271** or **272** is used as a substrate, both compounds are isomerized to each other, and then **271** is

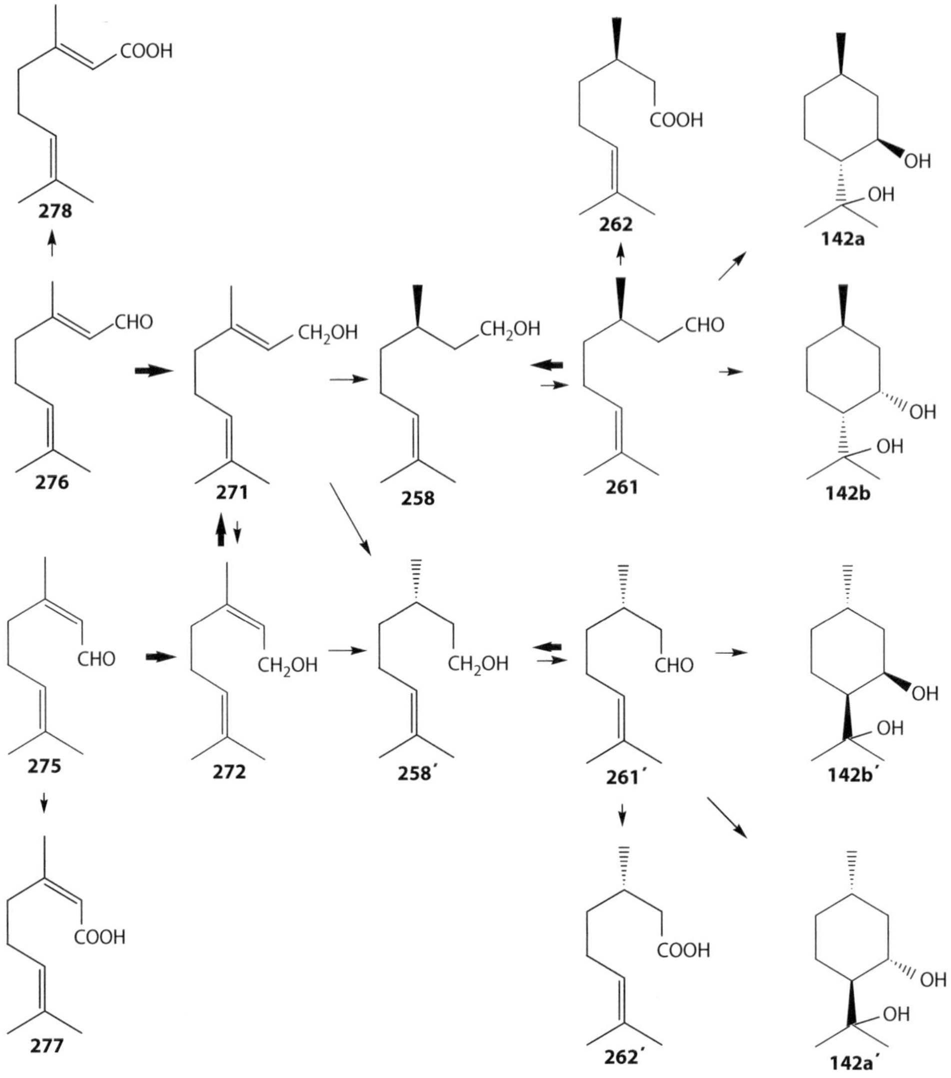

FIGURE 22.8 Metabolic pathways of citral (**275** and **276**) and its metabolites by *Euglena gracilis* Z. (Modified from Noma, Y. et al. *Phytochemistry*, 30, 1147, 1991a.)

transformed to citronellol (**258** or **258′**). These results showed the *Euglena* could distinguish between the stereoisomers **271** and **272** and hydrogenated selectively **271** to citronellol (**258** or **258′**). (+)-, (−)-, and (±)-Citronellal (**261**, **261′**, and equal mixture of **261** and **261′**) are also transformed to the corresponding citronellol and *p*-menthane-*trans*- and *cis*-3,8-diols (**142a, b, a′** and **b′**) as the major products, which are well known as mosquito repellents and plant growth regulators (Nishimura et al., 1982; Nishimura and Noma, 1996), and (+)-, (−)-, and (±)-citronellic acids (**262**, **262′**, and equal mixture of **262** and **262′**) as the minor products, respectively.

Streptomyces ikutamanensis, Ya-2-1, also reduced citral (geranial (**276**) and neral (**275**)=56:44)) and (+)-, (−)-, and (±)-citronellal (**261**, **261′**, and **261** and **261′**) to the corresponding alcohols: geraniol (**271**), nerol (**272**), and (+)-, (−)-, and (±)-citronellol (**258**, **258′**, **258** and **258′**). Compounds **271** and **272** were isomerized to each other. Furthermore, terpene alcohols (**258′**, **272**, and **271**) were epoxidized to give 6,7-epoxygeraniol (**274**), 6,7-epoxynerol (**273**), and 2,3-epoxycitronellol (**268**). On the other hand, (+)- and (±)-citronellol (**258** and **258** and **258′**) were not converted at all (Noma et al., 1986) (Figure 22.9).

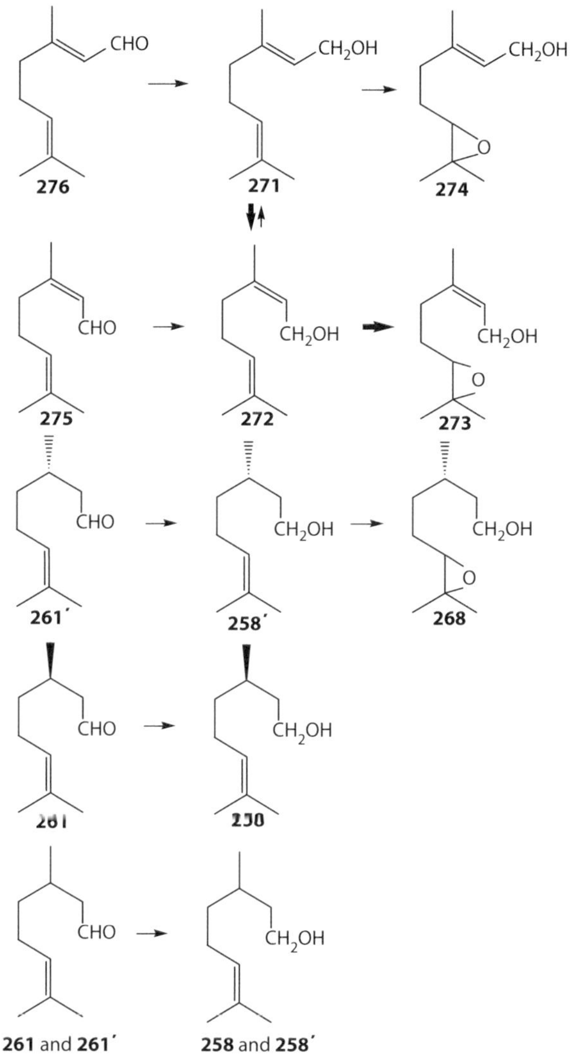

FIGURE 22.9 Reduction of terpene aldehydes and epoxidation of terpene alcohols by *Streptomyces ikutamanensis* Ya-2-1. (Modified from Noma, Y. et al. Reduction of terpene aldehydes and epoxidation of terpene alcohols by S. *ikutamanensis*, Ya-2-1, *Proceedings of 30th TEAC*, 1986, pp. 204–206.)

A strain of *A. niger*, isolated from garden soil, was able to transform geraniol (**271**), citronellol (**258** and **258′**), and linalool (**206**) to their respective 8-hydroxy derivatives. This reaction was called "ω-hydroxylation" (Madyastha and Krishna Murthy, 1988a,b).

Fermentation of citronellyl acetate with *A. niger* resulted in the formation of a major metabolite, 8-hydroxycitronellol, accounting for approximately 60% of the total transformation products, accompanied by 38% citronellol. Fermentation of geranyl acetate with *A. niger* gave geraniol and 8-hydroxygeraniol (50% and 40%, respectively, of the total transformation products).

One of the most important examples of fungal bioconversion of monoterpene alcohols is the biotransformation of citral by *Botrytis cinerea*. *B. cinerea* is a fungus of high interest in winemaking (Rapp and Mandery, 1988). In an unripe state of maturation, the infection of grapes by *B. cinerea* is very much feared, as the grapes become moldy (*gray rot*). With fully ripe grapes, however, the growth of *B. cinerea* is desirable; the fungus is then called "noble rot" and the infected grapes deliver famous sweet wines, such as Sauternes of France or Tokaji Aszu of Hungary (Brunerie et al., 1988).

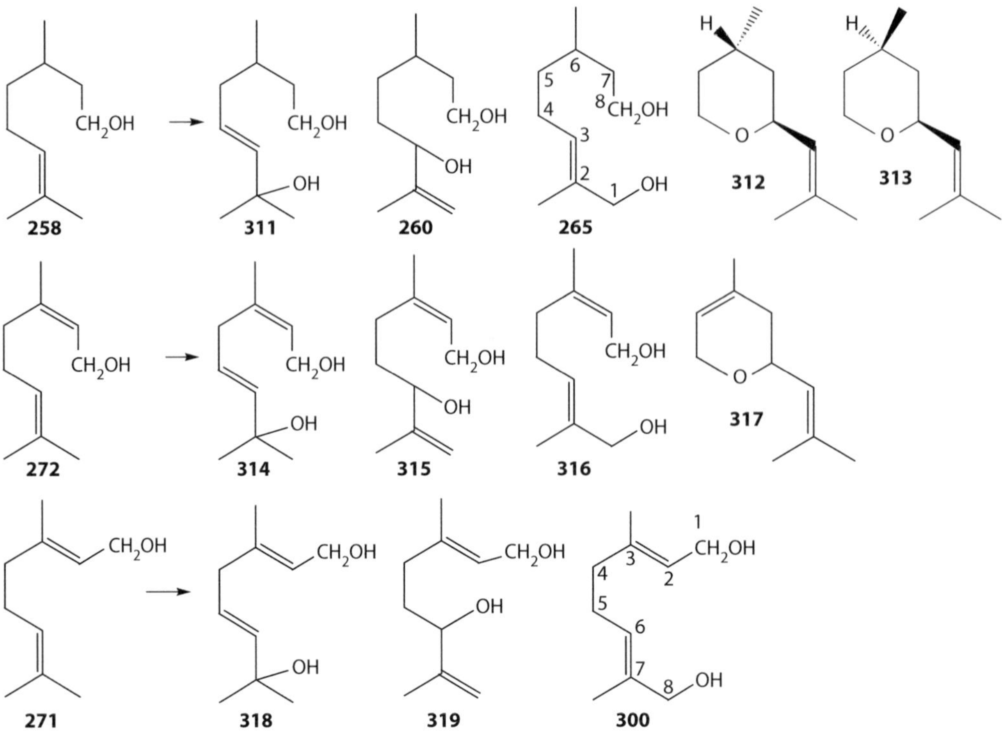

FIGURE 22.10 Biotransformation of geraniol (**271**), nerol (**272**), and citronellol (**258**) by *Botrytis cinerea*. (Modified from Miyazawa, M. et al. *Nat. Prod. Lett.*, 8, 303, 1996a.)

One of the first reports in this area dealt with the biotransformation of citronellol (**258**) by *B. cinerea* (Brunerie et al., 1987a, 1988). The substrate was mainly metabolized by ω-hydroxylation. The same group also investigated the bioconversion of citral (**275** and **276**) (Brunerie et al., 1987b). A comparison was made between grape must and a synthetic medium. When using grape must, no volatile bioconversion products were found. With a synthetic medium, biotransformation of citral (**275** and **276**) was observed yielding predominantly nerol (**272**) and geraniol (**271**) as reduction products and some ω-hydroxylation products as minor compounds. Finally, the bioconversion of geraniol (**271**) and nerol (**272**) was described by the same group (Bock et al., 1988). When using grape must, a complete bioconversion of geraniol (**271**) was observed mainly yielding ω-hydroxylation products.

The most important metabolites from geraniol (**271**), nerol (**272**), and citronellol (**258**) are summarized in Figure 22.9. In the same year, the biotransformation of these monoterpenes by *B. cinerea* in model solutions was described by another group (Rapp and Mandery, 1988). Although the major metabolites found were ω-hydroxylation compounds, it is important to note that some new compounds that were not described by the previous group were detected (Figure 22.9). Geraniol (**271**) was mainly transformed to (2*E*,5*E*)-3,7-dimethyl-2,5-octadiene-1,7-diol (**318**), (*E*)-3,7-dimethyl-2,7-octadiene-1,6-diol (**319**), and (2*E*,6*E*)-2,6-dimethyl-2,6-octadiene-1,8-diol (**300**) and nerol (**272**) to (2*Z*,5*E*)-3,7-dimethyl-2,5-octadiene-1,7-diol (**314**), (*Z*)-3,7-dimethyl-2,7-octadiene-1,6-diol (**315**), and (2*E*,6*Z*)-2,6-dimethyl-2,6-octadiene-1,8-diol (**316**). Furthermore, a cyclization product (**318**) that was not previously described was formed. Finally, citronellol (**258**) was converted to *trans*- (**312**) and *cis*-rose oxide (**313**) (a cyclization product not identified by the other group), (*E*)-3,7-dimethyl-5-octene-1,7-diol (**311**), 3,7-dimethyl-7-octene-1,6-diol (**260**), and (*E*)-2,6-dimethyl-2-octene-1,8-diol (**265**) (Miyazawa et al., 1996a) (Figure 22.10).

One of the latest reports in this area described the biotransformation of citronellol by the plant pathogenic fungus *Glomerella cingulata* to 3,7-dimethyl-1,6,7-octanetriol (Miyazawa et al., 1996a).

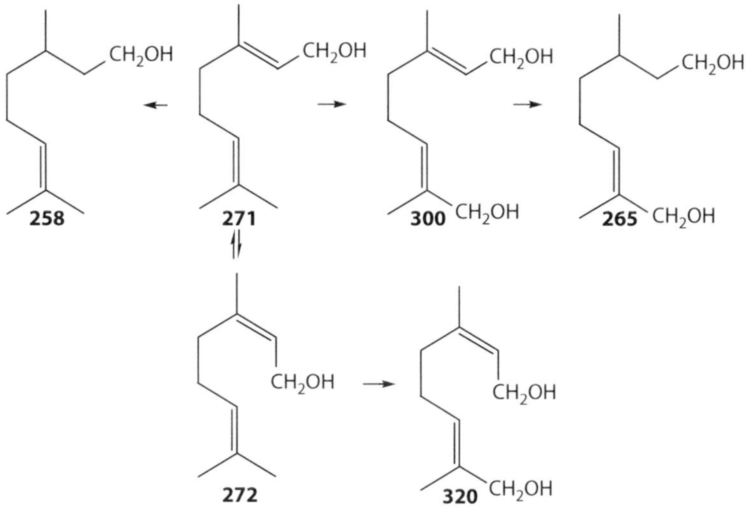

FIGURE 22.11 The biotransformation of geraniol (**271**) and nerol (**272**) by *Catharanthus roseus*. (Modified from Hamada, H. and Yasumune, H., The hydroxylation of monoterpenoids by plant cell biotransformation, *Proceedings of 39th TEAC*, 1995, pp. 375–377.)

The ability of fungal spores of *P. digitatum* to biotransform monoterpene alcohols, such as geraniol (**271**) and nerol (**272**) and a mixture of the aldehydes, that is, citral (**276** and **275**), has only been discovered very recently by Demyttenaere and coworkers (Demyttenaere et al., 1996, 2000; Demyttenaere and De Pooter, 1996, 1998). Spores of *P. digitatum* were inoculated on solid media. After a short incubation period, the spores germinated and a mycelial mat was formed. After 2 weeks, the culture had completely sporulated and bioconversion reactions were started. Geraniol (**271**), nerol (**272**), or citral (**276** and **275**) was sprayed onto the sporulated surface culture. After 1 or 2 days, the period during which transformation took place, the cultures were extracted. Geraniol and nerol were transformed into 6-methyl-5-hepten-2-one by sporulated surface cultures of *P. digitatum*. The spores retained their activity for at least 2 months. An overall yield of up to 99% could be achieved.

The bioconversion of geraniol (**271**) and nerol (**272**) was also performed with sporulated surface cultures of *A. niger*. Geraniol (**271**) was converted to linalool (**206**), α-terpineol (**34**), and limonene (**68**), and nerol (**272**) was converted mainly to linalool (**206**) and α-terpineol (**34**) (Demyttenaere et al., 2000).

The biotransformation of geraniol (**271**) and nerol (**272**) by *Catharanthus roseus* suspension cells was carried out. It was found that the allylic positions of geraniol (**271**) and nerol (**272**) were hydroxylated and reduced to double bond and ketones (Figure 22.11). Geraniol (**271**) and nerol (**272**) were isomerized to each other. Geraniol (**271**) and nerol (**272**) were hydroxylated at C10 to 8-hydroxygeraniol (**300**) and 8-hydroxynerol (**320**), respectively. 8-Hydroxygeraniol (**300**) was hydrogenated to 10-hydroxycitronellol (**265**). Geraniol (**271**) was hydrogenated to citronellol (**258**) (Hamada and Yasumune, 1995).

Cyanobacterium converted geraniol (**271**) to geranic acid (**278**) via geranial (**276**), followed by hydrogenation to give citronellic acid (**262**) via citronellal (**261**). Furthermore, the substrate **271** was isomerized to nerol (**272**), followed by oxidation, reduction, and further oxidation to afford neral (**275**), citronellal (**261**), citronellic acid (**262**), and nerolic acid (**277**) (Kaji et al., 2002; Hamada et al., 2004) (Figure 22.12).

Plant suspension cells of *C. roseus* converted geraniol (**271**) to 8-hydroxygeraniol (**300**). The same cells converted citronellol (**258**) to 8- (**265**) and 10-hydroxycitronellol (**264**) (Hamada et al., 2004) (Figure 22.13).

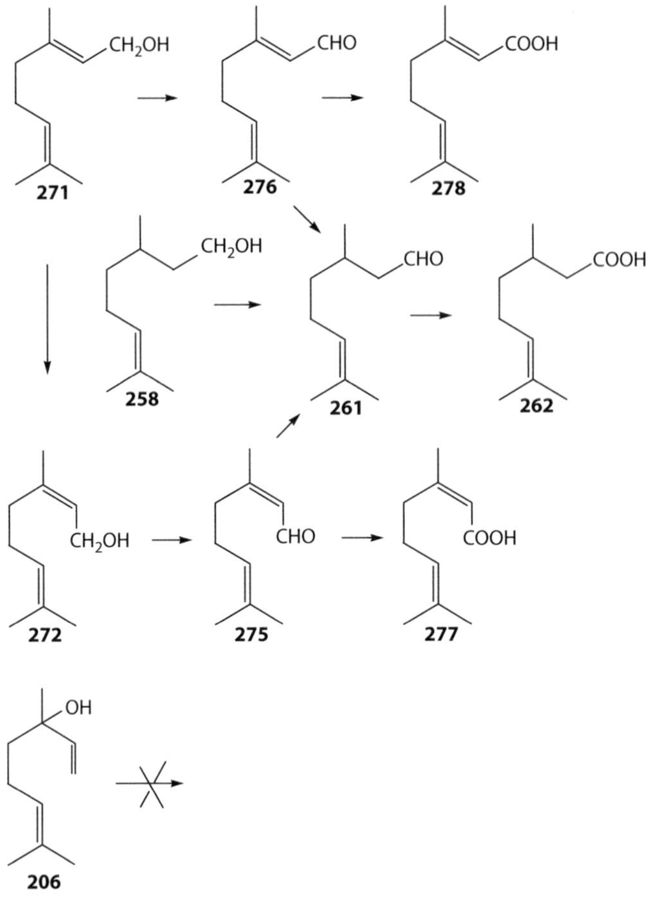

FIGURE 22.12 Biotransformation of geraniol (**271**) and citronellol (**258**) by cyanobacterium.

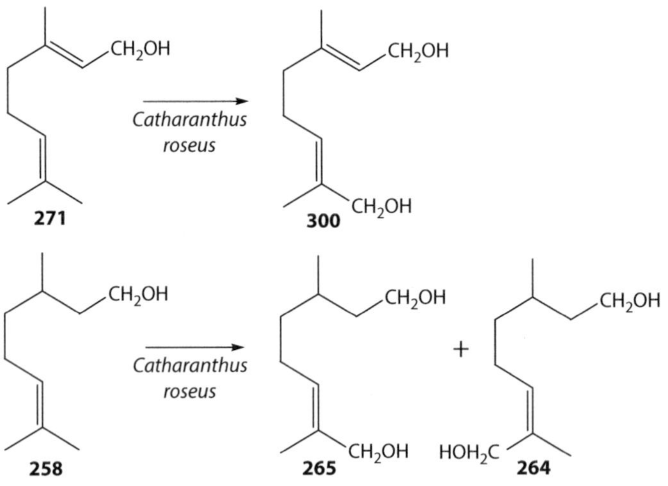

FIGURE 22.13 Biotransformation of geraniol (**271**), citronellol (**258**), and linalool (**206**) by plant suspension cells of *Catharanthus roseus*. (Modified from Hamada, H. et al. Biotransformation of acyclic monoterpenes by biocatalysts of plant cultured cells and *Cyanobacterium, Proceedings of 48th TEAC*, 2004, pp. 393–395.)

Nerol (**272**) was converted by the insect larvae *S. litura* to give 8-hydroxynerol (**320**), 10-hydroxynerol (**321**), 1-hydroxy-3,7-dimethyl-(2*E*,6*E*)-octadienal (**322**), and 1-hydroxy-3,7-dimethyl-(2*E*,6*E*)-octadienoic acid (**323**) (Takeuchi and Miyazawa, 2004) (Figure 22.14).

22.2.2.2 Linalool and Linalyl Acetate

(+)-Linalool (**206**) [(*S*)-3,7-dimethyl-1,6-octadiene-3-ol] and its enantiomer (**206′**) [(*R*)-3,7-dimethyl-1,6-octadiene-3-ol] occur in many essential oils, where they are often the main component. (*S*)-(+) Linalool (**206**) makes up 60%–70% of coriander oil. (*R*)-(−)-linalool (**206′**), for example, occurs at a concentration of 80%–85% in Ho oils from *Cinnamomum camphora*; rosewood oil contains ca. 80% (Bauer et al., 1990).

C. roseus converted (+)-linalool (**206**) to 8-hydroxylinalool (**219**) (Hamada et al., 2004) (Figure 22.15).

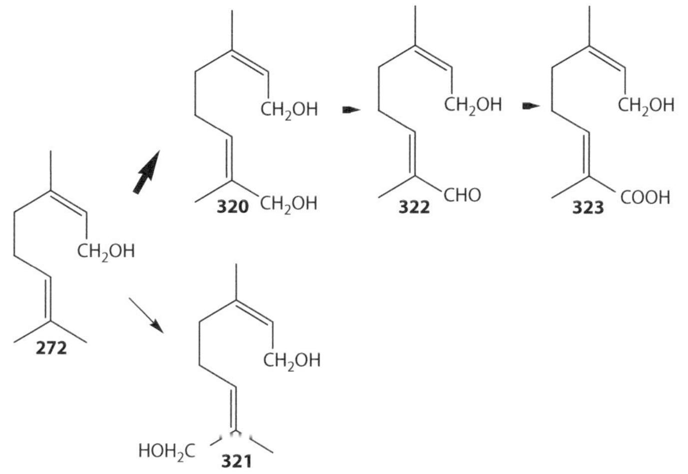

FIGURE 22.14 Biotransformation of nerol (**272**) by *Spodoptera litura*. (Modified from Takeuchi, H. and Miyazawa, M., Biotransformation of nerol by the larvae of common cutworm (*Spodoptera litura*) as a biocatalyst, *Proceedings of 48th TEAC*, 2004, pp. 399–400.)

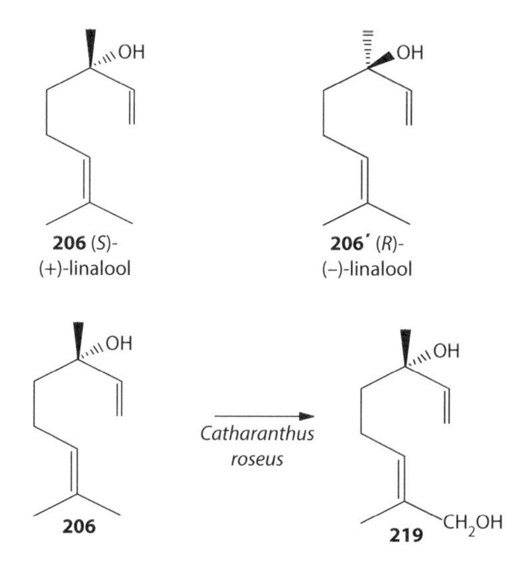

FIGURE 22.15 Biotransformation of linalool (**206**) by plant suspension cells of *Catharanthus roseus*. (Modified from Hamada, H. et al. Biotransformation of acyclic monoterpenes by biocatalysts of plant cultured cells and *Cyanobacterium*, *Proceedings of 48th TEAC*, pp. 393–395, 2004.)

The biodegradation of (+)-linalool (**206**) by *Pseudomonas pseudomallei* (strain A), which grows on linalool as the sole carbon source, was described in 1973 (Murakami et al., 1973) (Figure 22.16).

Madyastha et al. (1977) isolated a soil Pseudomonad, *P. incognita*, by the enrichment culture technique with linalool as the sole carbon source. This microorganism, the "linalool strain" as it was called, was also capable of utilizing limonene (**68**), citronellol (**258**), and geraniol (**271**) but failed to grow on citral (**275** and **276**), citronellal (**261**), and 1,8-cineole (**122**). Fermentation was carried out with shake cultures containing 1% linalool (**206**) as the sole carbon source. It was suggested by the authors that linalool (**206**) was metabolized by at least three different pathways of biodegradation (Figure 22.17). One of the pathways appeared to be initiated by the specific oxygenation of C-8 methyl group of linalool (**206**), leading to 8-hydroxylinalool (**219**), which was further oxidized to linalool-8-carboxylic acid (**220**). The presence of furanoid linalool oxide (**215**) and 2-methyl-2-vinyltetrahydrofuran-5-one (**216**) as the unsaturated lactone in the fermentation medium suggested

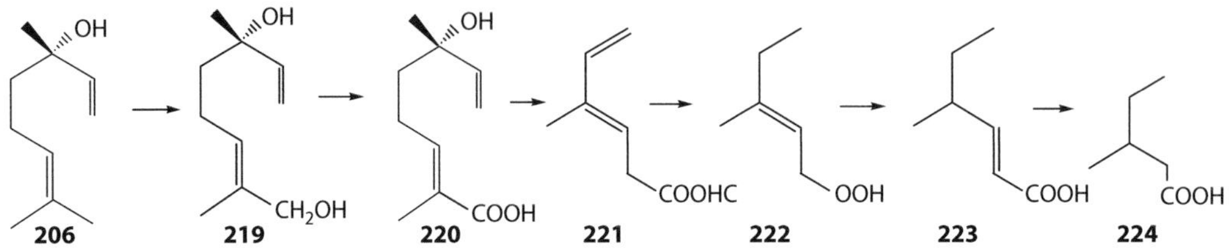

FIGURE 22.16 Degradative metabolic pathway of (+)-linalool (**206**) by *Pseudomonas pseudomallei*. (Modified from Murakami, T. et al. *Nippon Nogei Kagaku Kaishi*, 47, 699, 1973.)

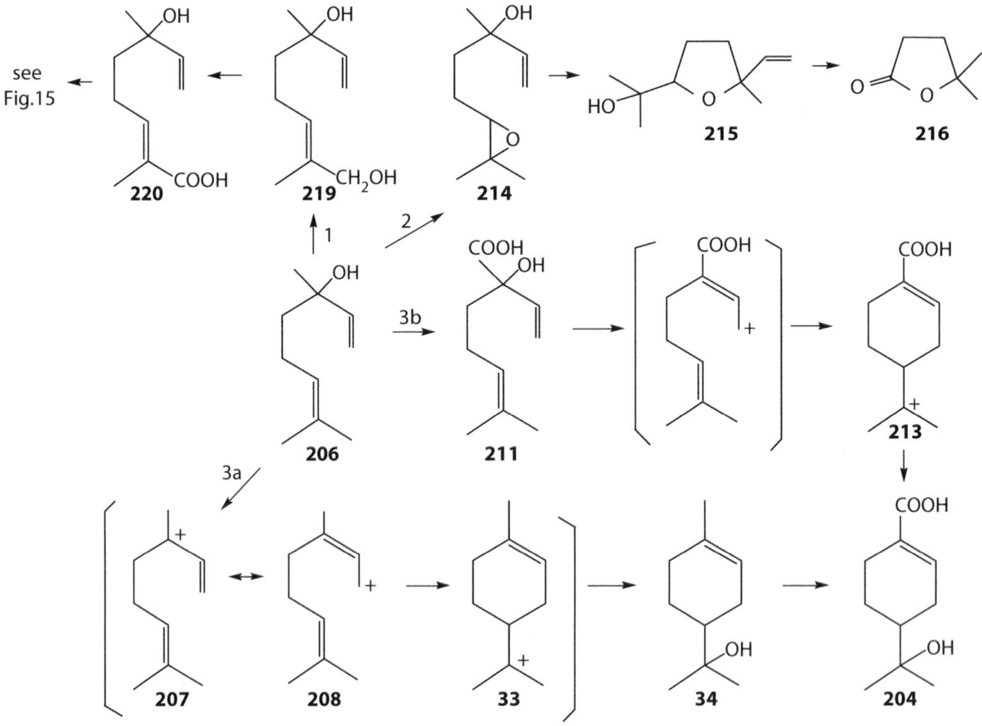

FIGURE 22.17 Biotransformation of linalool (**206**) by *Pseudomonas incognita* (Madyastha et al., 1977) and *Streptomyces albus* NRRL B1865. (Modified from David, L. and Veschambre, H., *Tetrahadron Lett.*, 25, 543, 1984.)

another mode of utilization of linalool (**206**). The formation of these compounds was believed to proceed through the epoxidation of the 6,7-double bond giving rise to 6,7-epoxylinalool (**214**), which upon further oxidation yielded furanoid linalool oxide (**215**) and 2-methyl-2-vinyltetrahydrofuran-5-one (**216**) (Figure 22.17).

The presence of oleuropeic acid (**204**) in the fermentation broth suggested a third pathway. Two possibilities were proposed: (3a) water elimination giving rise to a monocyclic cation (**33**), yielding α-terpineol (**34**), which upon oxidation gave oleuropeic acid (**204**), and (3b) oxidation of the C-10 methyl group of linalool (**206**) before cyclization, giving rise to oleuropeic acid (**204**). This last pathway was also called the "prototropic cyclization" (Madyastha, 1984).

Racemic linalool (**206** and **206′**) is cyclized into *cis*- and *trans*-linalool oxides by various microorganisms such as *Streptomyces albus* NRRL B1865, *Streptomyces hygroscopicus* NRRL B3444, *Streptomyces cinnamonensis* ATCC 15413, *Streptomyces griseus* ATCC 10137, and *Beauveria sulfurescens* ATACC 7159 (David and Veschambre, 1984) (Figure 22.17).

A. niger isolated from garden soil biotransformed linalool and its acetates to give linalool (**206**), 2,6-dimethyl-2,7-octadiene-1,6-diol (8-hydroxylinalool (**219a**), α-terpineol (**34**), geraniol (**271**), and some unidentified products in trace amounts (Madyastha and Krishna Murthy, 1988a,b).

The biotransformation of linalool (**206**) by *B. cinerea* was carried out and identified transformation products such as (*E*)-(**219a**) and (*Z*)-2,6-dimethyl-2,7-octadiene-1,6-diol (**219b**), *trans*- (**215a**) and *cis*-furanoid linalool oxide (**215b**), *trans*- (**217a**) and *cis*-pyranoid linalool oxide (**217b**) (Figure 22.18) and their acetates (**217a-Ac, 217b-Ac**), 3,9-epoxy-*p*-menth-1-ene (**324**) and 2-methyl-2-vinyltetrahydrofuran-5-one (**216**) (unsaturated lactone) (Bock et al., 1986) (Figure 22.19). Quantitative analysis, however, showed that linalool (**206**) was predominantly (90%) metabolized to (*E*)-2,6-dimethyl-2,7-octadiene-1,6-diol (**219a**) by *B. cinerea*. The other compounds were only found as by-products in minor concentrations.

The bioconversion of (*S*)-(+)-linalool (**206**) and (*R*)-(−)-linalool (**206′**) was investigated with *D. gossypina* ATCC 10936 (Abraham et al., 1990). The biotransformation of (±) linalool (**206** and **206′**) by *A. niger* ATCC 9142 with submerged shaking culture yielded a mixture of *cis*- (**215b**) and *trans*-furanoid linalool oxide (**215a**) (yield 15%–24%) and *cis*- (**217b**) and *trans*-pyranoid linalool oxide (**217a**) (yield 5%–9%) (Demyttenaere and Willemen, 1998). The biotransformation of (*R*)-(−)-linalool (**206a**) with *A. niger* ATCC 9142 yielded almost pure *trans*-furanoid linalool oxide (**215a**) and *trans*-pyranoid linalool oxide (**217a**) (ee>95) (Figure 22.20). These conversions were purely biocatalytic, since in acidified water (pH<3.5) almost 50% linalool (**206**) was recovered unchanged, and the rest was evaporated. The biotransformation was also carried out with growing surface cultures.

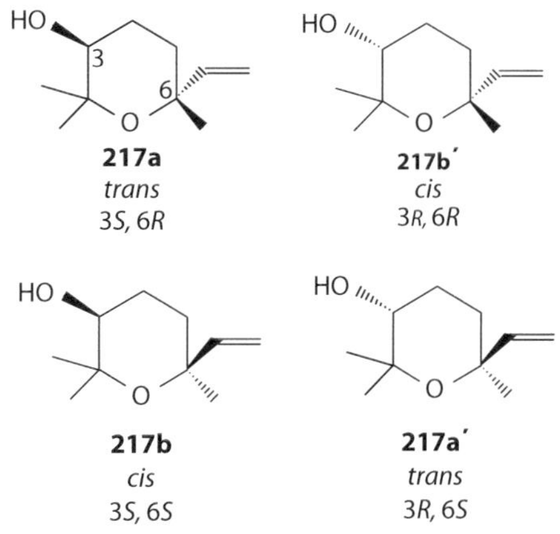

FIGURE 22.18 Four stereoisomers of pyranoid linalool oxides.

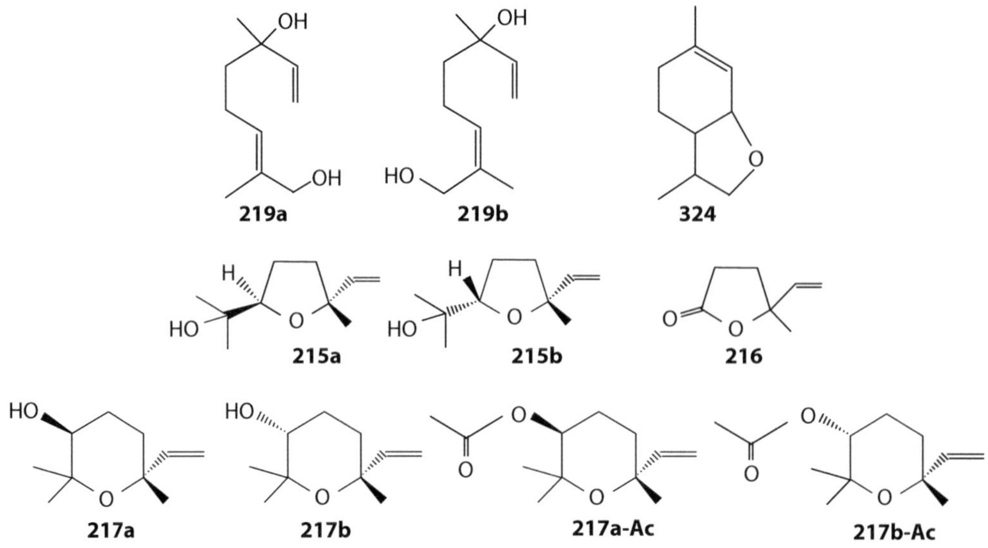

FIGURE 22.19 Biotransformation products of linalool (**206**) by *Botrytis cinerea*. (Modified from Bock, G. et al. *J. Food Sci*., 51, 659, 1986.)

S. ikutamanensis Ya-2-1 also converted (+)- (**206**), (–)- (**206′**), and racemic linalool (**206** and **206′**) via corresponding 2,3-epoxides (**214** and **214′**) to *trans*- and *cis*-furanoid linalool oxides (**215a, b, a′** and **b′**) (Noma et al., 1986) (Figure 22.21). The absolute configuration at C-3 and C-6 of *trans*- and *cis*-linalool oxides is shown in Figure 22.22.

Biotransformation of racemic *trans*-pyranoid linalool oxide (**217a** and **a′**) and racemic *cis*-linalool-pyranoid (**217b** and **b′**) has been carried out using fungus *G. cingulata* (Miyazawa et al., 1994b). *trans*- and *cis*-Pyranoid linalool oxide (**217a** and **217b**) were transformed to *trans*- (**217a′-1**) and *cis*-linalool oxide-3-malonate (**217b′-1**), respectively. In the biotransformation of racemic *cis*-linalool oxide-pyranoid, (+)-(3*R*,6*R*)-*cis*-pyranoid linalool oxide (**217a** and **a′**) was converted to (3*R*,6*R*)-pyranoid-*cis*-linalool oxide-3-malonate (**217a′-1**). (–)-(3*S*, 6*S*)-*cis*-Pyranoid linalool oxide-pyranoid (**217a′**) was not metabolized. On the other hand, in the biotransformation of racemic *trans*-pyranoid linalool oxide (**217b** and **b′**), (–)-(3*R*,6*S*)-*trans*-linalool oxide (**217b′**) was transformed to (3*R*,6*S*)-*trans*-linalool oxide-3-malonate (**217b′-1**) (Figure 22.23). (+)-(3*S*,6*S*)-*trans*-Pyranoid-linalool oxide (**217b**) was not metabolized. These facts showed that *G. cingulata* recognized absolute configuration of the secondary hydroxyl group at C-3. On the basis of this result, it has become apparent that the optical resolution of racemic pyranoid linalool oxide proceeded in the biotransformation with *G. cingulata* (Miyazawa et al., 1994b).

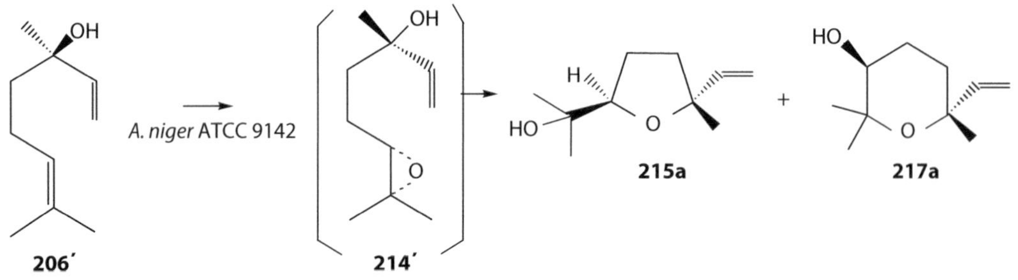

FIGURE 22.20 Biotransformation of (*R*)-(–)-linalool (**206′**) by *Aspergillus niger* ATCC 9142. (Modified from Demyttenaere, J.C.R. and Willemen, H.M., *Phytochemistry*, 47, 1029, 1998.)

206a
S. ikutamanensis, Ya-2-1
214a
215a
215b
206b
214b
215a′
215b′
206ab
214ab

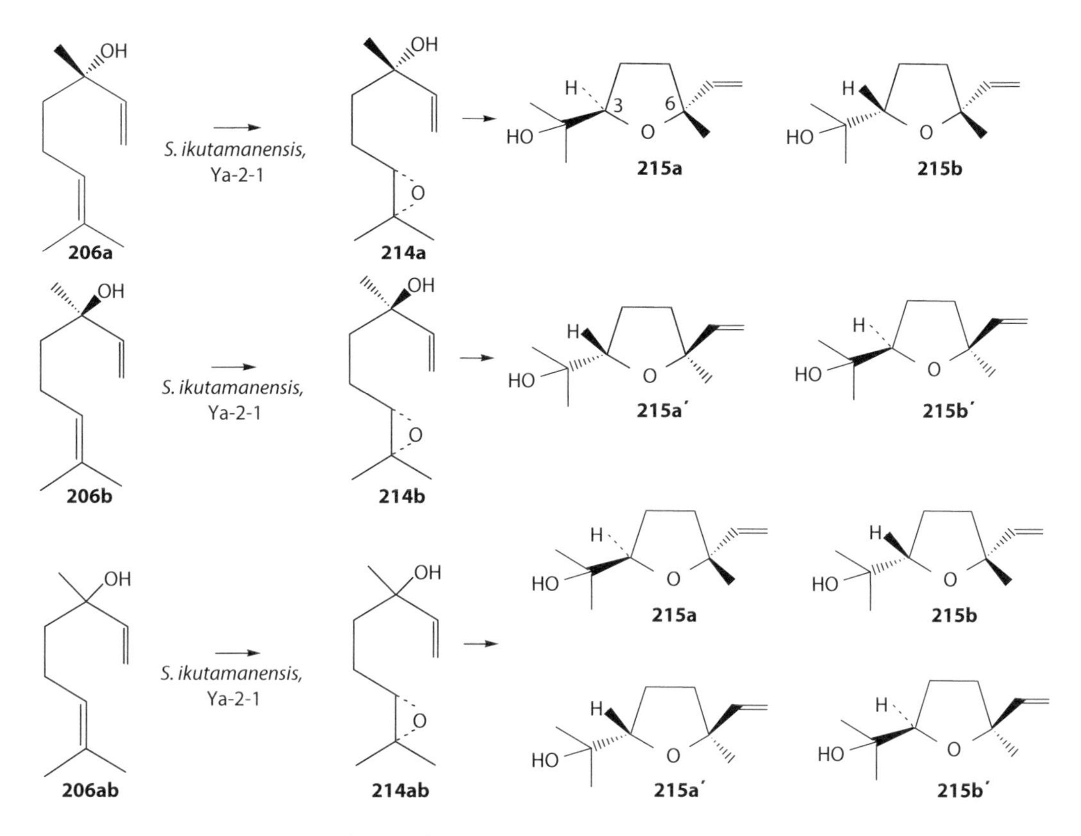

FIGURE 22.21 Metabolic pathway of (+)-(**206**), (−)-(**206′**), and racemic linalool (**206** and **206′**) by *Streptomyces ikutamanensis* Ya-2-1. (Modified from Noma, Y. et al. Reduction of terpene aldehydes and epoxidation of terpene alcohols by S. *ikutamanensis*, Ya-2-1, *Proceedings of 30th TEAC*, 1986, pp. 204–206.)

Linalool (**206**) and tetrahydrolinalool (**325**) were converted by suspension cells of *C. roseus* to give 1-hydroxylinalool (**219**) from linalool (**206**) and 3,7-dimethyloctane-3,5-diol (**326**), 3,7-dimethyloctane-3,7-diol (**327**), and 3,7-dimethyloctane-3,8-diol (**328**) from tetrahydrolinalool (**325**) (Hamada and Furuya, 2000; Hamada et al., 2004) (Figure 22.24).

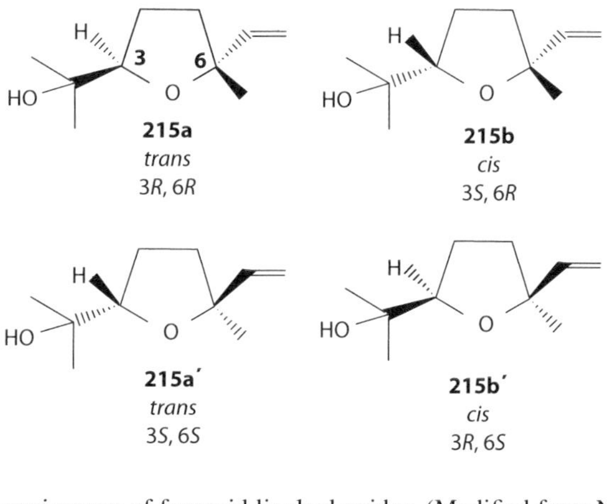

215a
trans
3R, 6R
215b
cis
3S, 6R
215a′
trans
3S, 6S
215b′
cis
3R, 6S

FIGURE 22.22 Four stereoisomers of furanoid linalool oxides. (Modified from Noma, Y. et al. Reduction of terpene aldehydes and epoxidation of terpene alcohols by S. *ikutamanensis*, Ya-2-1, *Proceedings of 30th TEAC*, 1986, pp. 204–206.)

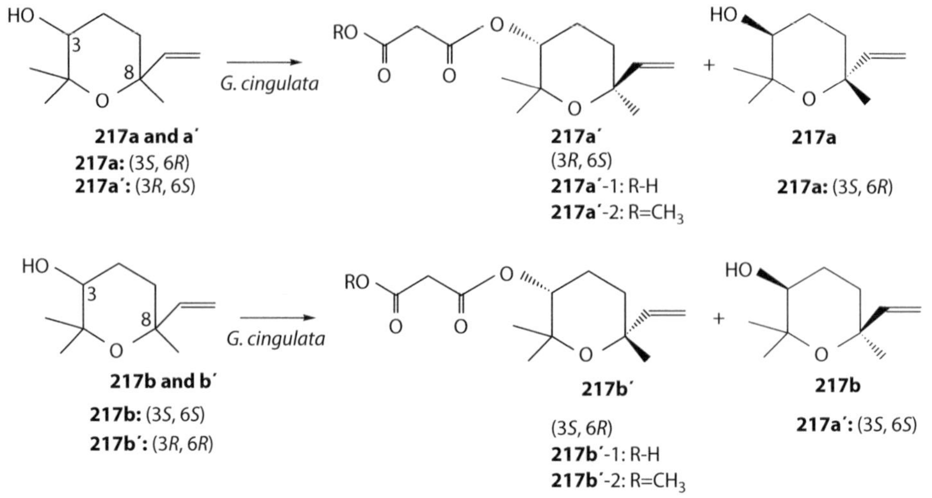

FIGURE 22.23 Biotransformation of racemic *trans*-linalool oxide-pyranoid (**217a** and **a′**) and racemic *cis*-linalool-pyranoid (**217b** and **b′**) by *Glomerella cingulata*. (Modified from Miyazawa, M. et al. Biotransformation of linalool oxide by plant pathogenic microorganisms, *Glomerella cingulata*, *Proceedings of 38th TEAC*, 1994a, pp. 101–102.)

(±)-Linalyl acetate (**206-Ac**) was hydrolyzed to (+)-(*S*)-linalool (**206**) and (±)-linalyl acetate (**206-Ac**) by *Bacillus subtilis*, *Trichoderma S*, *Absidia glauca*, and *Gibberella fujikuroi* as shown in Figure 22.25. But (±)-dihydrolinalyl acetate (**469-Ac**) was not hydrolyzed by the aforementioned microorganisms (Oritani and Yamashita, 1973a).

22.2.2.3 Dihydromyrcenol

Dihydromyrcenol (**329**) was fed by *S. litura* to give 1,2-epoxydihydromyrcenol (**330**) as a main product and 3β-hydroxydihydromyrcenol (**331**) as a minor product. Dihydromyrcenyl acetate (**332**) was converted to 1,2-dihydroxydihydromyrcenol acetate (**333**) (Murata and Miyazawa, 1999) (Figure 22.26).

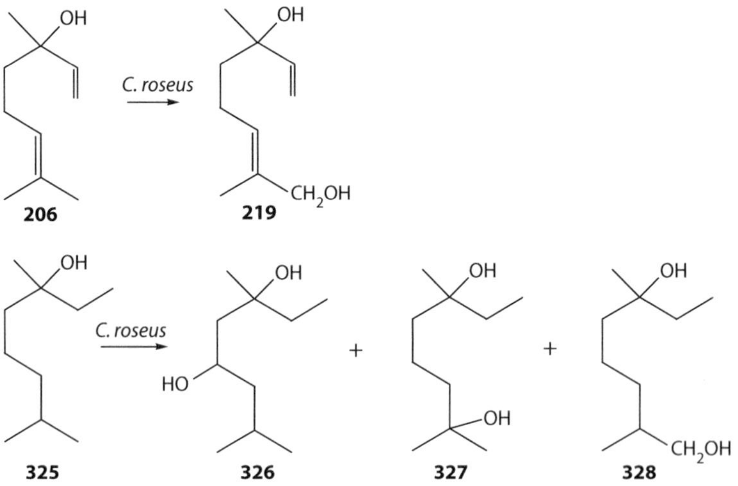

FIGURE 22.24 Biotransformation of linalool (**206**) and tetrahydrolinalool (**325**) by *Catharanthus roseus*. (Modified from Hamada, H. and Furuya, T., Hydroxylation of monoterpenes by plant suspension cells, *Proceedings of. 44th TEAC*, pp. 167–168, 2000; Hamada, H. et al. Biotransformation of acyclic monoterpenes by biocatalysts of plant cultured cells and *Cyanobacterium*, *Proceedings of 48th TEAC*, 2004, pp. 393–395.)

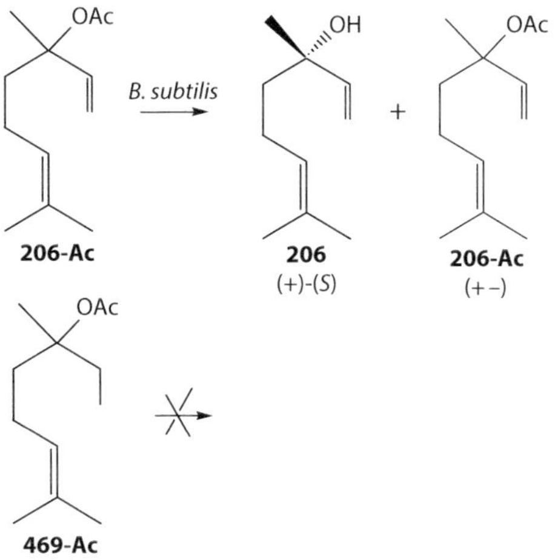

FIGURE 22.25 Hydrolysis of (±)-linalyl acetate (**206-Ac**) by microorganisms. (Modified from Oritani, T. and Yamashita, K., *Agric. Biol. Chem.*, 37, 1923, 1973a.)

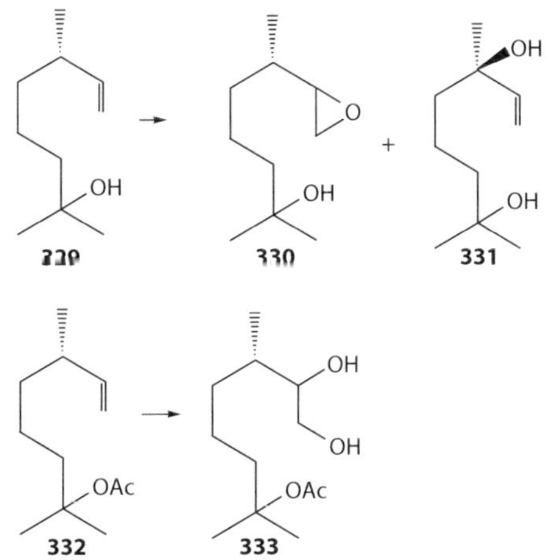

FIGURE 22.26 Biotransformation of dihydromyrcenol (**329**) and dihydromyrcenyl acetate (**332**) by *Spodoptera litura*. (Modified from Murata, T. and Miyazawa, M., Biotransformation of dihydromyrcenol by common cutworm larvae, *Spodoptera litura* as a biocatalyst, *Proceedings of 43rd TEAC*, 1999, pp. 393–394.)

22.3 METABOLIC PATHWAYS OF CYCLIC MONOTERPENOIDS

22.3.1 Monocyclic Monoterpene Hydrocarbon

22.3.1.1 Limonene

68 (*R*)-limonene **68′** (*S*)-limonene

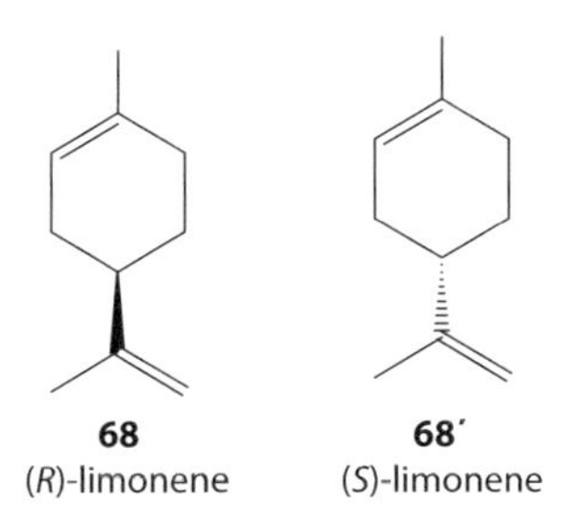

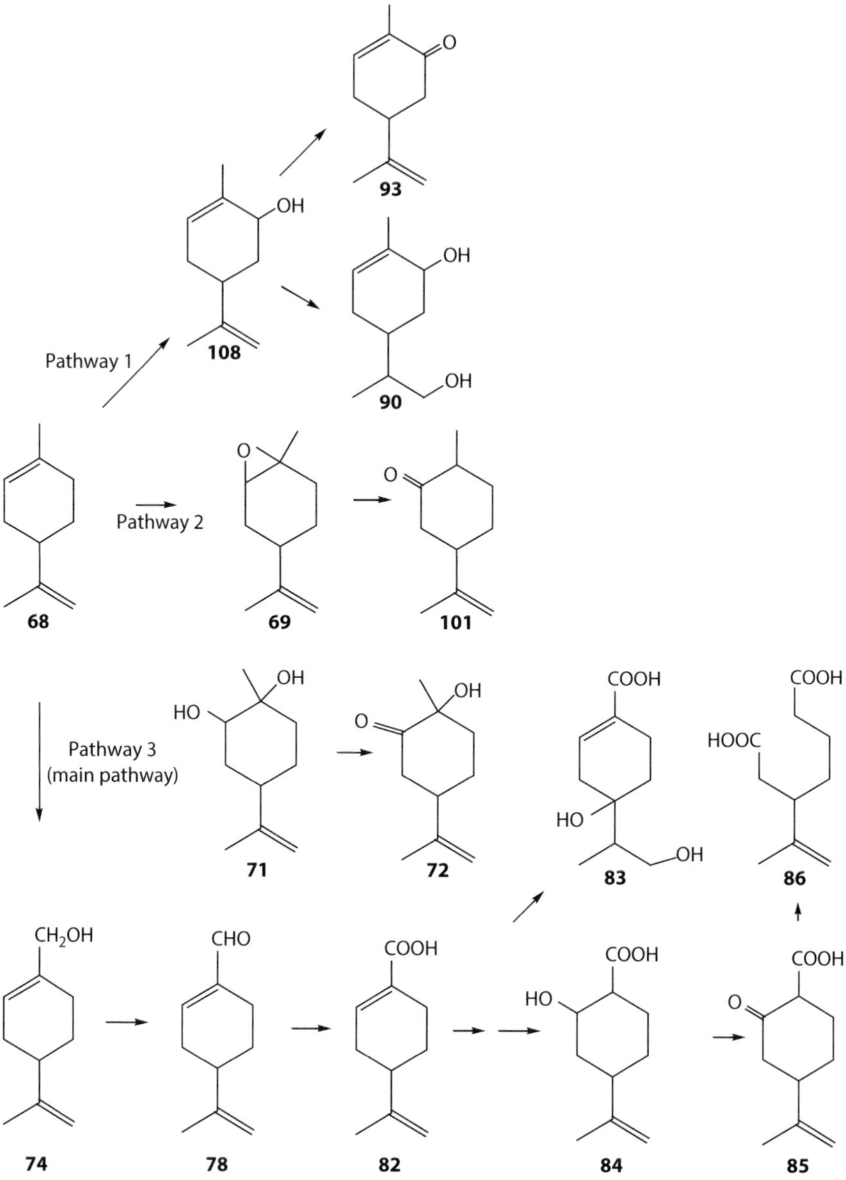

FIGURE 22.27 Pathways for the degradation of limonene (**68**) by a soil Pseudomonad sp. strain (L). (Modified from Krasnobajew, V., in: *Biotechnology*, Kieslich, K., ed., vol. 6a, Verlag Chemie, Weinheim, Germany, 1984, pp. 97–125.)

Limonene is the most widely distributed terpene in nature after α-pinene (**4**) (Krasnobajew, 1984). (4*R*)-(+)-Limonene (**68**) is present in citrus peel oils at a concentration of over 90%; a low concentration of the (4*S*)-(−)-limonene (**68′**) is found in oils from the *Mentha* species and conifers (Bauer et al., 1990). The first microbial biotransformation on limonene was carried out by using a soil Pseudomonad. The microorganism was isolated by the enrichment culture technique on limonene as the sole source of carbon (Dhavalikar and Bhattacharyya, 1966). The microorganism was also capable of growing on α-pinene (**4**), β-pinene (**1**), 1-*p*-menthene (**62**), and *p*-cymene (**178**). The optimal level of limonene for growth was 0.3%–0.6% (v/v) although no toxicity was observed at 2% levels. Fermentation of limonene (**68**) by this bacterium in a mineral salt medium resulted in the formation of a large number of neutral and acidic products such as dihydrocarvone (**64**), carvone (**61**), carveol (**60**), 8-*p*-menthene-1,2-*cis*-diol (**65b**), 8-*p*-menthen-1-ol-2-one (**66**), 8-*p*-menthene-1,2-*trans*-diol (**65a**), and 1-*p*-menthene-6,9-diol (**62**). Perillic acid (**69**), β-isopropenyl pimaric acid (**72**), 2-hydroxy-8-*p*-menthen-7-oic acid (**70**), and 4,9-dihydroxy-1-*p*-menthen-7-oic acid (**73**) were isolated and identified as acidic compounds. Based on these data, three distinct pathways for the catabolism of limonene (**68**) by the soil Pseudomonad were proposed by Dhavalikar et al. (1966), involving allylic oxygenation (pathway 1), oxygenation of the 1,2-double bond (pathway 2), and progressive oxidation of the 7-methyl group to perillic acid (**82**) (pathway 3) (Figure 22.27) (Krasnobajew, 1984). Pathway 2 yields (+)-dihydrocarvone (**101**) via intermediate limonene epoxide (**69**) and 8-*p*-menthen-1-ol-2- one (**72**) as oxidation product of limonene-1,2-diol (**71**). The third and main pathway leads to perillyl alcohol (**74**), perillaldehyde (**78**), perillic acid (**82**), constituents of various essential oils and used in the flavor and fragrance industry (Fenaroli, 1975), 2-oxo-8-*p*-menthen-7-oic acid (**85**), β-isopropenyl pimaric acid (**86**), and 4,9-dihydroxy-1-*p*-menthene-7-oic acid (**83**).

(+)-Limonene (**68**) was biotransformed via limonene-1,2-epoxide (**69**) to 8-*p*-menthene 1,2-*trans*-diol (**71b**). On the other hand, (+)-carvone (**93**) was biotransformed via (−)-isodihydrocarvone (**101b**) and 1α-hydroxydihydrocarvone (**72**) to (+)-8-*p*-menthene-1,2-*trans*-diol (**71a**) (Noma et al., 1985a,b) (Figure 22.28). A soil Pseudomonad formed 1-hydroxydihydrocarvone (**72**) and

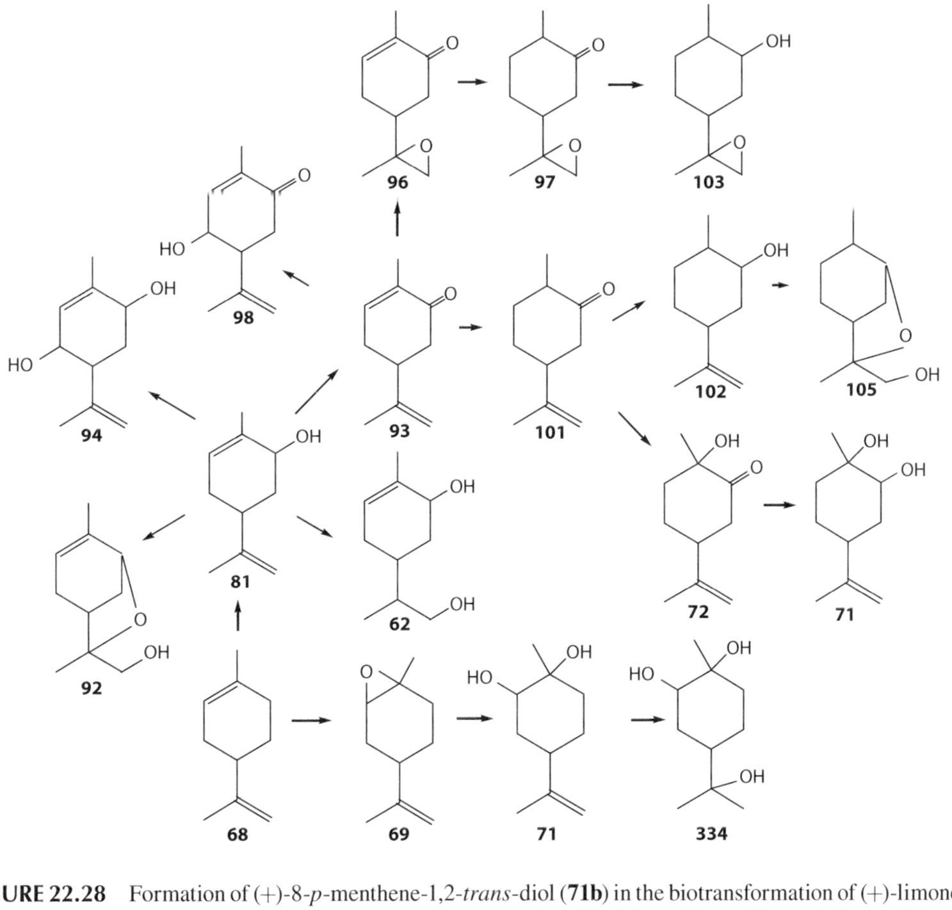

FIGURE 22.28 Formation of (+)-8-*p*-menthene-1,2-*trans*-diol (**71b**) in the biotransformation of (+)-limonene (**68**) and (+)-carvone (**93**) by *Aspergillus niger* TBUYN-2. (Modified from Noma, Y. et al. *Annual Meeting of Agricultural and Biological Chemistry*, Sapporo, p. 68; Noma, Y. et al. Biotransformation of carvone. 6. Biotransformation of (−)-carvone and (+)-carvone by a strain of *Aspergillus niger*, *Proceedings of 29th TEAC*, 1985a, pp. 235–237.)

8-*p*-menthene-1,2-*trans*-diol (**71b**) from (+)-limonene (**68**). Dhavalikar and Bhattacharyya (1966) considered that the formation of 1-hydroxydihyFdrocarvone (**66**) is from dihydrocarvone (**64**).

Pseudomonas gladioli was isolated by an enrichment culture technique from pine bark and sap using a mineral salt broth with limonene as the sole carbon source (Cadwallander et al., 1989; Cadwallander and Braddock, 1992). Fermentation was performed during 4–10 days in shake flasks at 25°C using a pH 6.5 mineral salt medium and 1.0% (+)-limonene (**68**). Major products were identified as (+)-α-terpineol (**34**) and (+)-perillic acid (**82**). This was the first report of the microbial conversion of limonene to (+)-α-terpineol (**34**).

The first data on fungal bioconversion of limonene (**68**) date back to the late 1960s (Kraidman et al., 1969; Noma, 2007). Three soil microorganisms were isolated on and grew rapidly in mineral salt media containing appropriate terpene substrates as sole carbon sources. The microorganisms belonged to the class Fungi Imperfecti, and they had been tentatively identified as *Cladosporium* species. One of these strains, designated as *Cladosporium* sp. T_7, was isolated on (+)-limonene (**68a**). The growth medium of this strain contained 1.5 g/L of *trans*-limonene-1,2-diol (**71a**). Minor quantities of the corresponding *cis*-1,2-diol (**71b**) were also isolated. The same group isolated a fourth microorganism from a terpene-soaked soil on mineral salt media containing (+)-limonene as the sole carbon source (Kraidman et al., 1969). The strain, *Cladosporium*, designated T_{12}, was capable of converting (+)-limonene (**68a**) into an optically active isomer of α-terpineol (**34**) in yields of approximately 1.0 g/L.

α-Terpineol (**34**) was obtained from (+)-limonene (**68**) by fungi such as *P. digitatum*, *Penicillium italicum*, and *Cladosporium* and several bacteria (Figure 22.29). (+)-*cis*-Carveol (**81b**), (+)-carvone (**93**) (an important constituent of caraway seed and dill-seed oils) (Fenaroli, 1975; Bouwmester et al., 1995), and 1-*p*-menthene-6,9-diol (**90**) were also obtained by *P. digitatum* and *P. italicum*. (+)-(*S*)-Carvone (**93**) is a natural potato sprout–inhibiting, fungistatic, and bacteriostatic compound (Oosterhaven et al., 1995a,b). It is important to note that (−)-carvone (**93′**, the *spearmint flavor*) was not yet described in microbial transformation (Krasnobajew, 1984). However, the biotransformation of limonene to (−)-carvone (**93′**) was patented by a Japanese group (Takagi et al., 1972). *Corynebacterium* species grown on limonene was able to produce about 10 mg/L of 99% pure (−)-carvone (**93′**) in 24–48 h.

Mattison et al. (1971) isolated *Penicillium* sp. cultures from rotting orange rind that utilized limonene (**68**) and converted it rapidly to α-terpineol (**34**). Bowen (1975) isolated two common citrus molds, *P. italicum* and *P. digitatum*, responsible for the postharvest diseases of citrus fruits. Fermentation of *P. italicum* on limonene (**68**) yielded *cis*- (**81b**) and *trans*-carveol (**81a**) (26%) as the main products, together with *cis*- and *trans*-*p*-mentha-2,8-dien-1-ol (**73**) (18%), (+)-carvone (**93′**) (6%), *p*-mentha-1,8-dien-4-ol (**80**) (4%), perillyl alcohol (**74**) (3%), and 8-*p*-menthene-1,2-diol (**71**) (3%). Conversion of **68** by *P. digitatum* yielded the same products in lower yields (Figure 22.29).

The biotransformation of limonene (**68**) by *A. niger* is a very important example of fungal bioconversion. Screening for fungi capable of metabolizing the bicyclic hydrocarbon terpene α-pinene (**4**) yielded a strain of *A. niger* NCIM 612 that was also able to transform limonene (**68**) (Rama Devi and Bhattacharyya, 1978). This fungus was able to carry out three types of

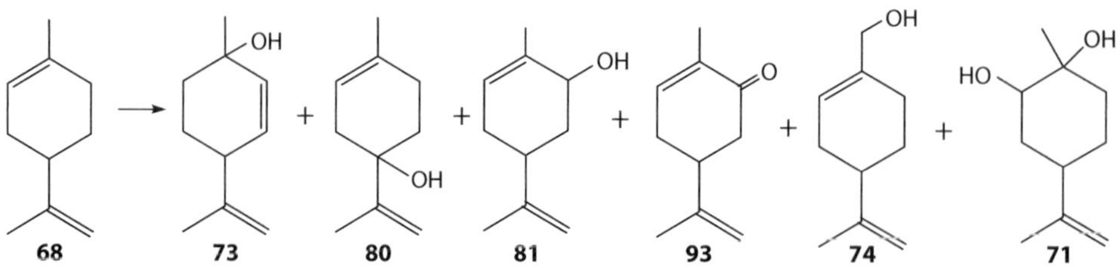

FIGURE 22.29 Biotransformation products of limonene (**68**) by *Penicillium digitatum* and *Penicillium italicum*. (Modified from Bowen, E.R., *Proc. Fla. State Hortic. Soc.*, 88, 304, 1975.)

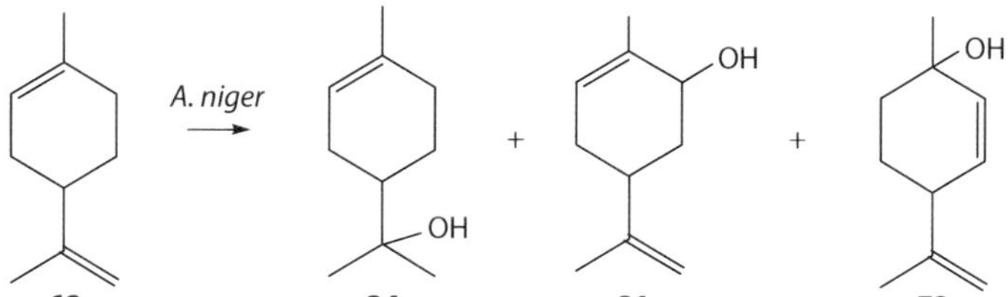

FIGURE 22.30 Biotransformation of limonene (**68**) by *Aspergillus niger* NCIM 612. (Modified from Rama Devi, J. and Bhattacharyya, P.K., *J. Indian Chem. Soc.*, 55, 1131, 1978.)

oxygenative rearrangements α-terpineol (**34**), carveol (**81**), and *p*-mentha-2,8-dien-1-ol (**73**) (Rama Devi and Bhattacharyya, 1978) (Figure 22.30). In 1985, Abraham et al. (1985) investigated the biotransformation of (*R*)-(+)-limonene (**68a**) by the fungus *P. digitatum*. A complete transformation for the substrate to α-terpineol (**34**) by *P. digitatum* DSM 62840 was obtained with 46% yield of pure product.

The production of glycols from limonene (**68**) and other terpenes with a 1-menthene skeleton was reported by *Corynespora cassiicola* DSM 62475 and *D. gossypina* ATCC 10936 (Abraham et al., 1984). Accumulation of glycols during fermentation was observed. An extensive overview on the microbial transformations of terpenoids with a 1-*p*-menthene skeleton was published by Abraham et al. (1986).

The biotransformation of (+)-limonene (**68**) was carried out by using *Aspergillus cellulosae* M-77 (Noma et al., 1992d) (Figures 22.31 and 22.32). It is important to note that (+)-limonene (**68a**) was mainly converted to (+)-isopiperitenone (**111**) (19%) as new metabolite, (1*S*,2*S*,4*R*)-(+)-limonene-1,2-*trans*-diol (**71a**) (21%), (+)-*cis*-carveol (**81b**) (5%), and (+)-perillyl alcohol (**74**) (12%) (Figure 22.32).

(+)-Limonene (**68**) was biotransformed by a kind of citrus pathogenic fungi, *P. digitatum* (Pers.; Fr.) Sacc. KCPYN., to isopiperitenone (**111%**, 7% GC ratio), 2α-hydroxy-1,8-cineole (**125b**, 7%), (+)-limonene-1,2-*trans*-diol (**71a**, 6%), and (+)-*p*-menthane-1β,2α,8-triol (**334%**, 45%) as main products and (+)-*trans*-sobrerol (**95a**, 2%), (+)-*trans*-carveol (**81a**), (+)-carvone (**93**), (−)-isodihydrocarvone (**101b**), and (+)-*trans*-isopiperitenol (**110a**) as minor products (Noma and Asakawa, 2006a, 2007a) (Figure 22.33). The metabolic pathways of (+)-limonene by *P. digitatum* are shown in Figure 22.34.

On the other hand, (−)-limonene (**68′**) was also biotransformed by a kind of citrus pathogenic fungi, *P. digitatum* (Pers.; Fr.) Sacc. KCPYN., to give isopiperitenone (**111′**), 2α-hydroxy-1,8-cineole (**125b′**), (−)-limonene-1,2-*trans*-diol (**71′**), and *p*-menthane-1,2,8-triol (**334′**) as main products together with (+)-*trans*-sobrerol (**80′**), (+)-*trans*-carveol (**81a′**), (−)-carvone (**93′**), (−)-dihydrocarvone (**101a′**), and (+)-isopiperitenol (**110a′**) as minor products (Noma and Asakawa, 2007b) (Figure 22.35.)

Newly isolated unidentified red yeast, *Rhodotorula* sp., converted (+)-limonene (**68**) mainly to (+)-limonene-1,2-*trans*-diol (**71a**), (+)-*trans*-carveol (**81a**), (+)-*cis*-carveol (**81b**), and (+)-carvone (**93′**) together with (+)-limonene-1,2-*cis*-diol (**71b**) as minor product (Noma and Asakawa, 2007b) (Figure 22.36).

Cladosporium sp. T_7 was cultivated with (+)-limonene (**68**) as the sole carbon source; it converted **68** to *trans*-*p*-menthane-1,2-diol (**71a**) (Figure 22.36) (Mukherjee et al., 1973).

On the other hand, the same red yeast converted (−)-limonene (**68′**) mainly to (−)-limonene-1,2-*trans*-diol (**71a′**), (−)-*trans*-carveol (**81a′**), (−)-*cis*-carveol (**81b′**), and (−)-carvone (**93′**) together with (−)-limonene-1,2-*cis*-diol (**71b′**) as minor product (Noma and Asakawa, 2007b) (Figure 22.37).

The biotransformation of (+)- and (−)-limonene (**68** and **68′**), (+)- and (−)-α-terpineol (**34** and **34′**), (+)- and (−)-limonene-1,2-epoxide (**69** and **69′**), and caraway oil was carried out by citrus pathogenic fungi *Penicillium* (Pers.; Fr.) Sacc. KCPYN and newly isolated red yeast, a kind of *Rhodotorula* sp. *P. digitatum* KCPYN converted limonenes (**68** and **68′**) to the corresponding isopiperitone (**111** and **111′**), 1α-hydroxy-1,8-cineole (**125b** and **125b′**), limonene-1,2-*trans*-diol

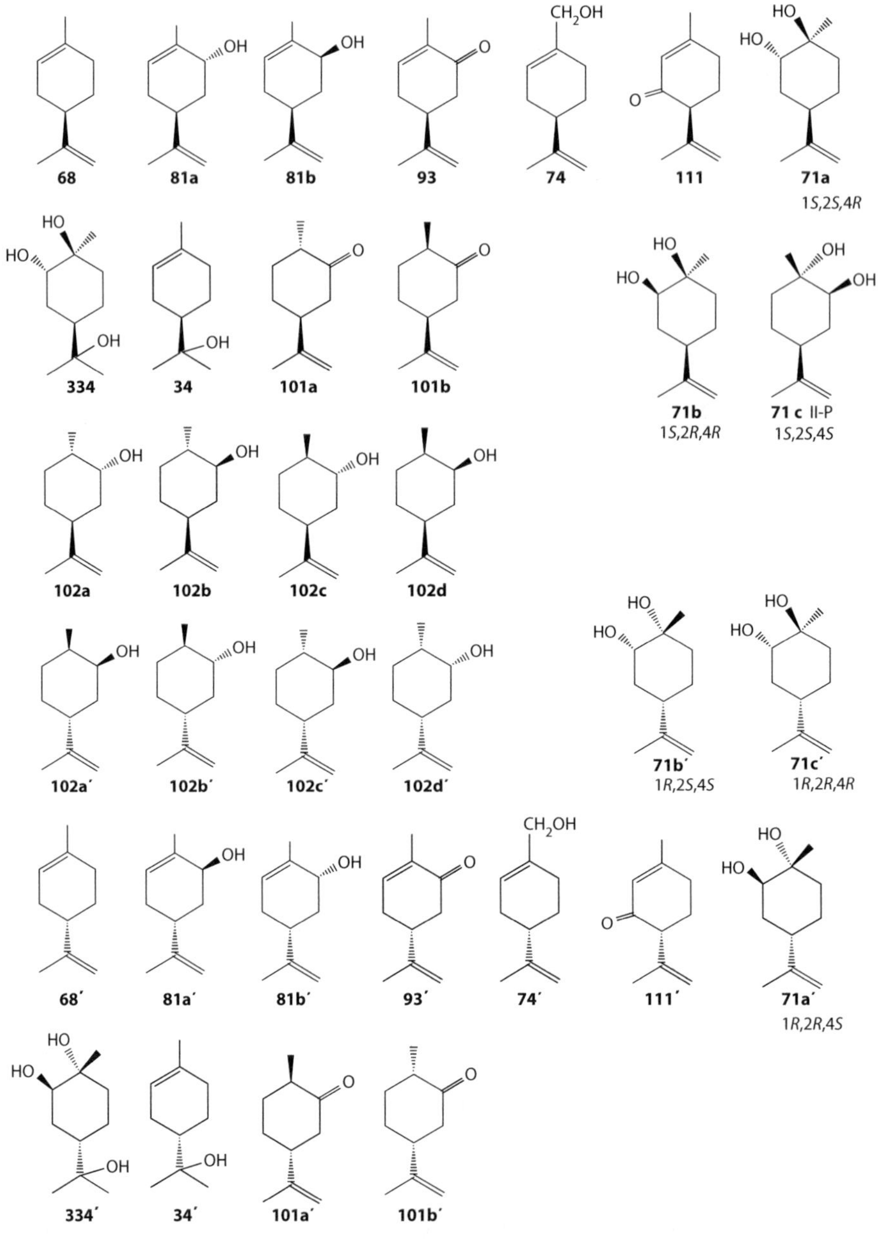

FIGURE 22.31 (+)- and (−)-limonenes (**68** and **68′**) and related compounds.

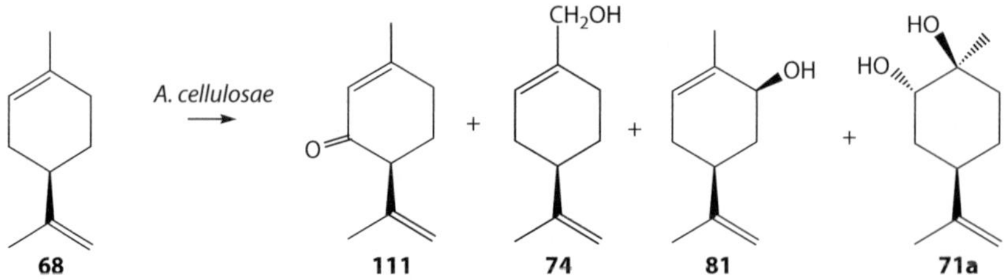

FIGURE 22.32 Biotransformation of (+)-limonene (**68**) by *Aspergillus cellulosae* IFO 4040. (Modified from Noma, Y. et al. *Phytochemistry*, 31, 2725, 1992d.)

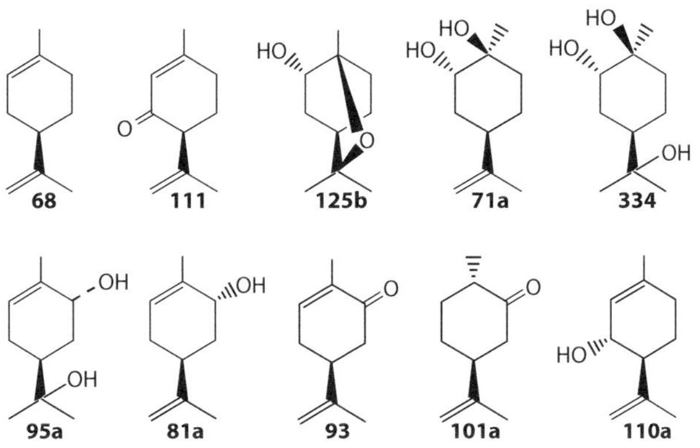

FIGURE 22.33 Metabolites of (+)-limonene (**68**) by a kind of citrus pathogenic fungi, *Penicillium digitatum* (Pers.; Fr.) Sacc. KCPYN. (Modified from Noma, Y. and Asakawa, Y., Biotransformation of (+)-limonene and related compounds by *Citrus* pathogenic fungi, *Proceedings of 50th TEAC*, pp. 431–433, 2006a; Noma, Y. and Asakawa, Y., Biotransformation of limonene and related compounds by newly isolated low temperature grown *citrus* pathogenic fungi and red yeast, *Book of Abstracts of the 38th ISEO*, 2007a, p. 7.)

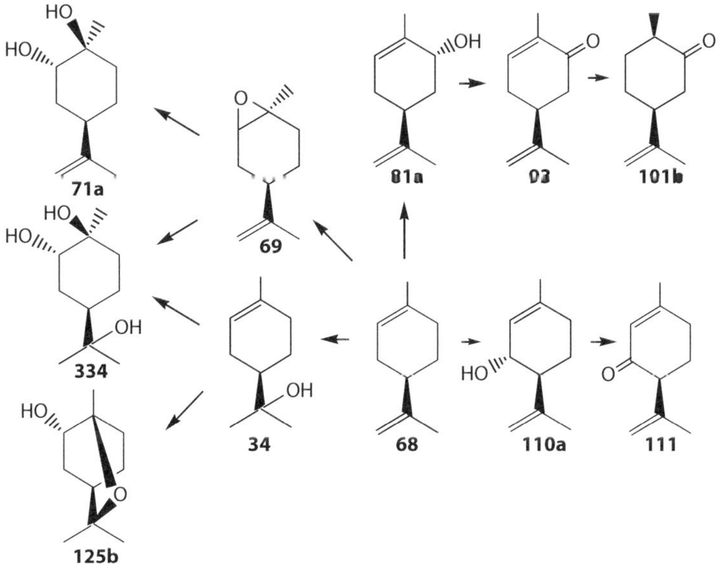

FIGURE 22.34 Biotransformation of (+)-limonene (**68**) by citrus pathogenic fungi, *Penicillium digitatum* (Pers.; Fr.) Sacc. KCPYN. (Modified from Noma, Y. and Asakawa, Y., Biotransformation of (+)-limonene and related compounds by *Citrus* pathogenic fungi, *Proceedings of 50th TEAC*, pp. 431–433, 2006a; Noma, Y. and Asakawa, Y., Biotransformation of limonene and related compounds by newly isolated low temperature grown *citrus* pathogenic fungi and red yeastm, *Book of Abstracts of the 38th ISEO*, 2007a, p. 7.)

(**71a** and **71a′**), *p*-menthane-1,2,8-triol (**334** and **334′**), and *trans*-sobrerol as main products. (+)- and (−)-α-Terpineol (**34** and **34′**) were the precursors of 2α-hydroxy-1,8-cineole (**125b** and **b′**) and *p*-menthane-1,2,8-triol (**334**). (+)- and (−)-Limonene-1,2-epoxide (**69** and **69′**) were also the precursor of limonene-1,2-*trans*-diol (**71a**). *Rhodotorula* sp. also biotransformed (+)- and (−)-limonene (**68** and **68′**) to the corresponding *trans*- and *cis*-carveols (**81a** and **b**) as main products. This microbe also converted caraway oil, equal mixture of (+)-limonene (**68**) and (+)-carvone (**93**). (+)-Limonene

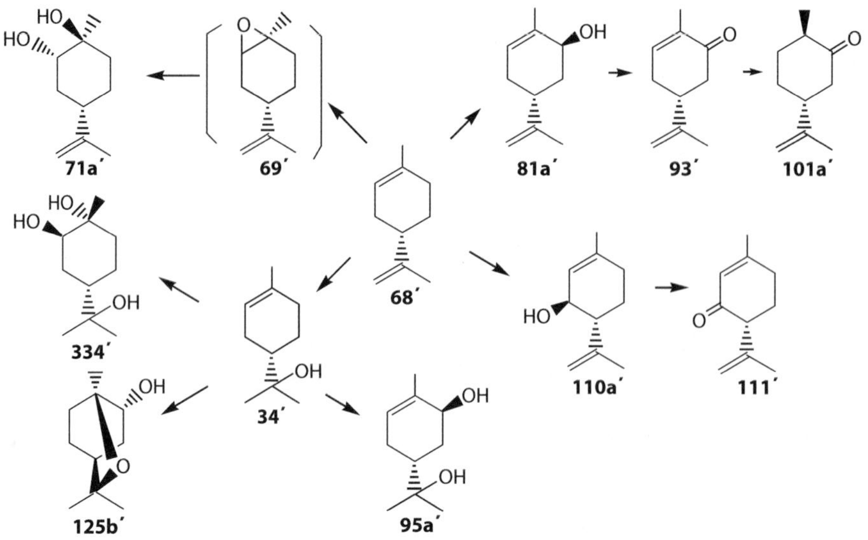

FIGURE 22.35 Biotransformation of (−)-limonene (**68′**) by citrus pathogenic fungi, *Penicillium digitatum* (Pers.; Fr.) Sacc. KCPYN. (Modified from Noma, Y. and Asakawa, Y., Microbial transformation of limonene and related compounds, *Proceedings of 51st TEAC*, 2007b, pp. 299–301.)

(**68**) disappeared and (+)-carvone (**93**) was produced and accumulated in the cultured broth (Noma and Asakawa, 2007b).

(4*S*)-(−)- (**68′**) and (4*R*)-(+)-Limonene (**68**) and their epoxides (**69** and **69′**) were incubated by cyanobacterium. It was found that the transformation was enantio- and regioselective. Cyanobacterium biotransformed only (4*S*)-limonene (**68′**) to (−)-*cis*- (**81b′**, 11.1%) and (−)-*trans*-carveol (**81a′**, 5%) in low yield. On the other hand, (4*R*)-limonene oxide (**69**) was converted to limonene-1,2-*trans*-diol (**71a′**) and 1-hydroxy-(+)-dihydrocarvone (**72a′**). However, (4*R*)-(+)-limonene (**68**) and (4*S*)-limonene oxide (**69′**) were not converted at all (Figure 22.38) (Hamada et al., 2003).

(+)-Limonene (**68**) was fed by *S. litura* to give (+)-limonene-7-oic acid (**82**), (+)-limonene-9-oic acid (**70**), and (+)-limonene-8,9-diol (**79**); (−)-limonene (**68′**) was converted to (−)-limonene-7-oic

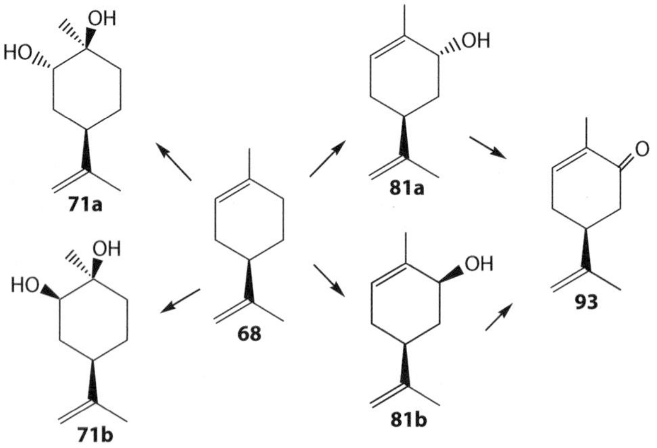

FIGURE 22.36 Biotransformation of (+)-limonene (**68**) by red yeast, *Rhodotorula* sp. and *Cladosporium* sp. T_7. (Modified from Mukherjee, B.B. et al. *Appl. Microbiol.*, 25, 447, 1973; Noma, Y. and Asakawa, Y., Microbial transformation of limonene and related compounds, *Proceedings of 51st TEAC*, 2007b, pp. 299–301.)

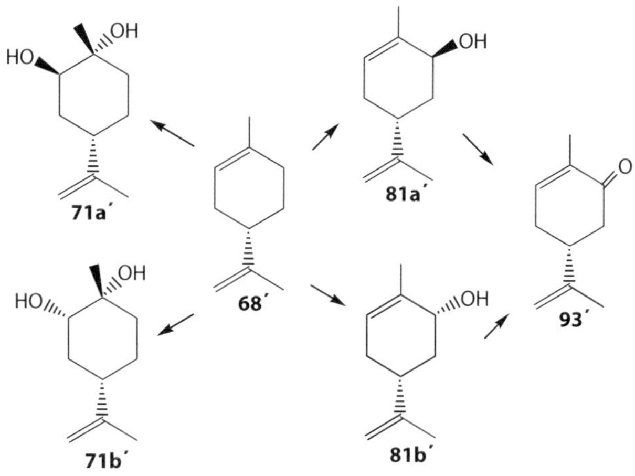

FIGURE 22.37 Biotransformation of (−)-limonene (**68′**) by a kind of *Rhodotorula* sp. (Modified from Noma, Y. and Asakawa, Y., Microbial transformation of limonene and related compounds, *Proceedings of 51st TEAC*, 2007b, pp. 299–301.)

acid (**82′**), (−)-limonene-9-oic acid (**70′**), and (−)-limonene-8,9-diol (**79′**) (Figure 22.39) (Miyazawa et al., 1995a).

Kieslich et al. (1985) found a nearly complete microbial resolution of a racemate in the biotransformation of (±)-limonene by *P. digitatum* (DSM 62840). The (*R*)-(+)-limonene (**68**) is converted to the optically active (+)-α-terpineol, $[\alpha]_D=+99°$, while the (*S*)-(−)-limonene (**68′**) is presumably adsorbed onto the mycelium or degraded via unknown pathways (Kieslich et al., 1985) (Figure 22.40).

(4*S*)- and (4*R*)-Limonene epoxides (**69a′** and **a**) were biotransformed by cyanobacterium to give 8-*p*-menthene-1α,2β-ol (**71a**, 68.4%) and 1α-hydroxy-8-*p*-menthen-2-one (**72**, 31.6%) (Hamada et al., 2003) (Figure 22.41).

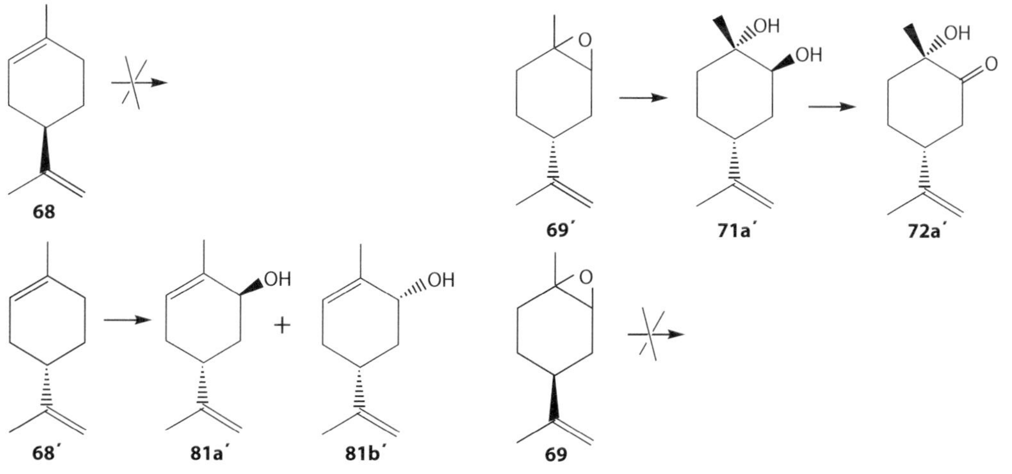

FIGURE 22.38 Biotransformation of (+)- and (−)-limonene (**68** and **68′**) and limonene epoxide (**69** and **69′**) by cyanobacterium. (Modified from Hamada, H. et al. Enantioselective biotransformation of monoterpenes by *Cyanobacterium*, *Proceedings of 47th TEAC*, 2003, pp. 162–163.)

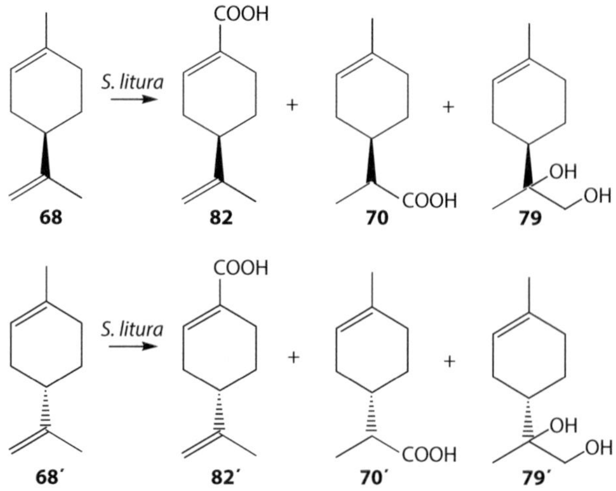

S. litura

68 82 70 79

68′ 82′ 70′ 79′

FIGURE 22.39 Biotransformation of (+)-limonene (**68**) and (−)-limonene (**68′**) by *Spodoptera litura*. (Modified from Miyazawa, M. et al. Biotransformation of terpinene, limonene and α-phellandrene in common cutworm larvae, *Spodoptera litura* Fabricius, *Proceedings of 39th TEAC*, 1995a, pp. 362–363.)

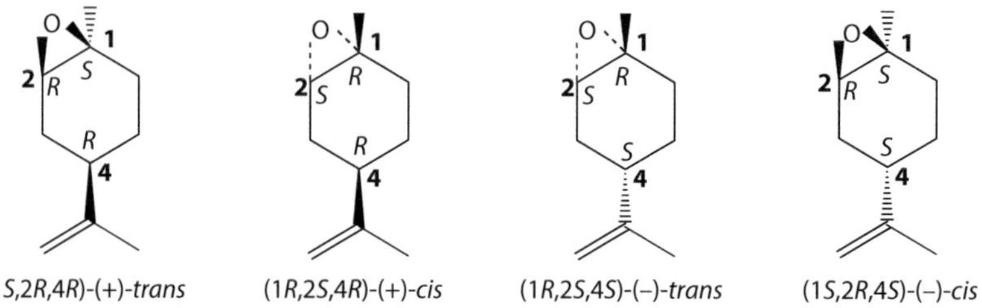

(1*S*,2*R*,4*R*)-(+)-*trans* (1*R*,2*S*,4*R*)-(+)-*cis* (1*R*,2*S*,4*S*)-(−)-*trans* (1*S*,2*R*,4*S*)-(−)-*cis*

The mixture of (+)-*trans*- (**69a**) and *cis*- (**69b**) and the mixture of (−)-*trans*-(**69a′**) and *cis*-limonene-1,2-epoxide (**69b′**) were biotransformed by citrus pathogenic fungi *P. digitatum* (Pers.; Fr.) Sacc. KCPYN to give (1*R*,2*R*,4*R*)-(−)-*trans*-(**71a**) and

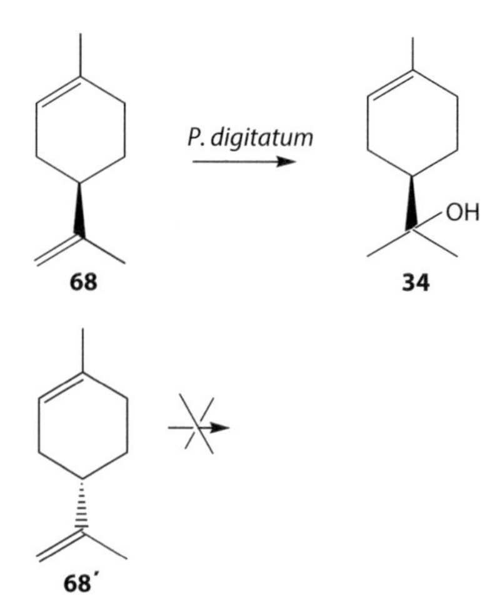

FIGURE 22.40 Microbial resolution of racemic limonene (**68** and **68′**) and the formation of optically active α-terpineol by *Penicillium digitatum*. (Modified from Kieslich, K. et al. In: *Topics in flavor research*, R.G. Berger, S. Nitz, and P. Schreier, eds., Marzling Hangenham, Eichborn, 1985, pp. 405–427.)

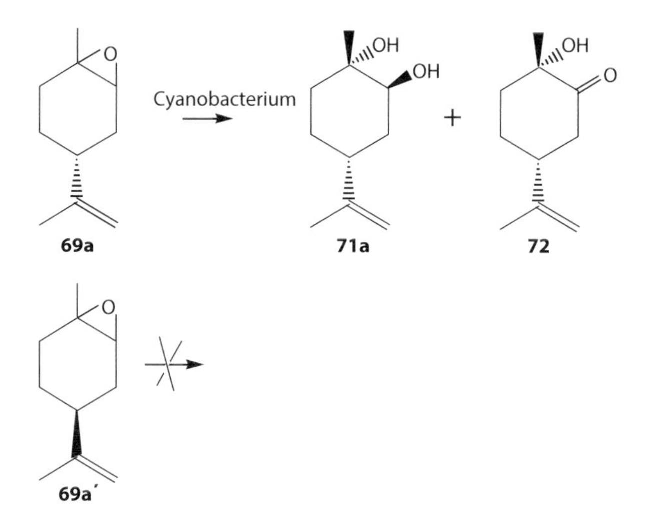

FIGURE 22.41 Enantioselective biotransformation of (4*S*)-(**69a′**) and (4*R*)-limonene epoxides (**69a**) by cyanobacterium. (Modified from Hamada, H. et al. Enantioselective biotransformation of monoterpenes by *Cyanobacterium*, *Proceedings of 47th TEAC*, 2003, pp. 162–163.)

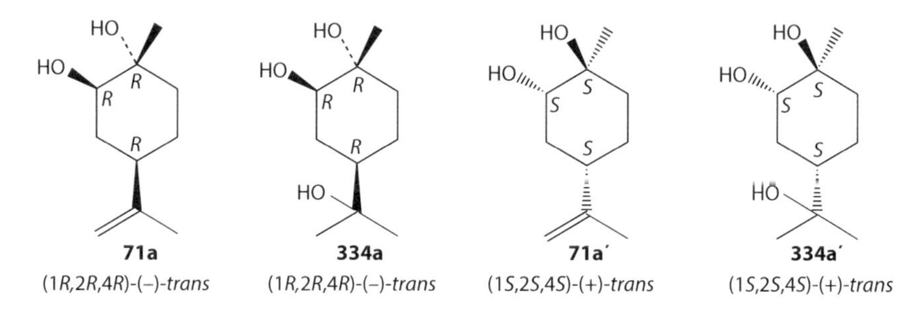

FIGURE 22.42 Biotransformation of (+)-*trans*-(**69a**) and *cis*-(**69b**) and (−)-*trans*-(**69a′**) and *cis*-limonene-1,2-epoxide (**69b′**) by citrus pathogenic fungi, *Penicillium digitatum* (Pers.; Fr.) Sacc. KCPYN, and their metabolites. (Modified from Noma, Y. and Asakawa, Y., Microbial transformation of limonene and related compounds, *Proceedings of 51st TEAC*, 2007b, pp. 299–301.)

(1*S*,2*S*,4*S*)-(+)-8-*p*-menthene-1,2-*trans*-diol (**71a′**) and (−)-*p*-menthane-1,2,8-triols (**334a** and **334a′**) (Noma and Asakawa, 2007b) (Figure 22.42).

Biotransformation of 1,8-cineole (**122**) by *A. niger* gave racemic 2α-hydroxy-1,8-cineole (**125b** and **b′**) (Nishimura et al., 1982). When racemic 2α-hydroxy-1,8-cineole (**125b** and **b′**) was biotransformed by *G. cingulata*, only (−)-2α-hydroxy-1,8-cineole (**125b′**) was selectively esterified with malonic acid to give its malonate (**125b′**-Mal). The malonate was hydrolyzed to give optical pure **125b′** (Miyazawa et al., 1995b). On the other hand, citrus pathogenic fungi *P. digitatum* biotransformed limonene (**68**) to give optical pure **125b** (Noma and Asakawa, 2007b) (Figure 22.43).

When monoterpenes, such as limonene (**68**), α-pinene (**4**), and 3-carene (**336**), were administered to the cultured cells of *Nicotiana tabacum*, they were converted to the corresponding epoxides enantio- and stereoselectively. The enzyme (p38) concerning with the epoxidation reaction was purified from the cultured cells by cation-exchange chromatography. The enzyme had not only epoxidation activity but also peroxidase activity. Amino acid sequence of p38 showed 89% homology in their 9 amino acid overlap with horseradish peroxidase (Yawata et al., 1998) (Figure 22.44). It was found that limonene and carene were converted to the corresponding epoxides in the presence of hydrogen peroxide and *p*-cresol by a radical mechanism with the peroxidase. (*R*)-limonene (**68**),

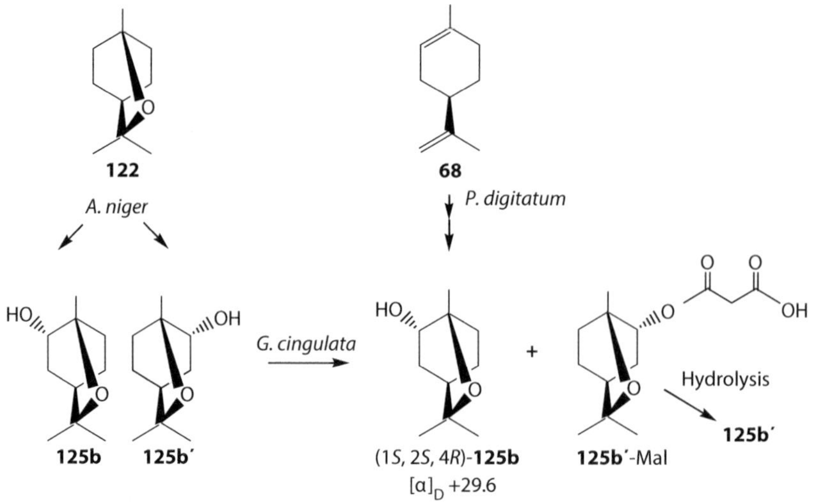

FIGURE 22.43 Formation of optical pure (+)- and (−)-2α-hydroxy-1,8-cineole (**125b** and **b′**) from the biotransformation of 1,8-cineole (**122**) and (+)-limonene (**68**) by citrus pathogenic fungi, *Penicillium digitatum* (Pers.; Fr.) Sacc. KCPYN and *Aspergillus niger* TBUYN-2. (Modified from Nishimura, H. et al. *Agric. Biol. Chem.*, 46, 2601, 1982; Miyazawa, M. et al. Biotransformation of 2-endo-hydroxy-1,4-cineole by plant pathogenic microorganism, *Glomerella cingulata*, *Proceedings of 39th TEAC*, pp. 352–353, 1995b; Noma, Y. and Asakawa, Y., Microbial transformation of limonene and related compounds, *Proceedings of 51st TEAC*, 2007b, pp. 299–301.)

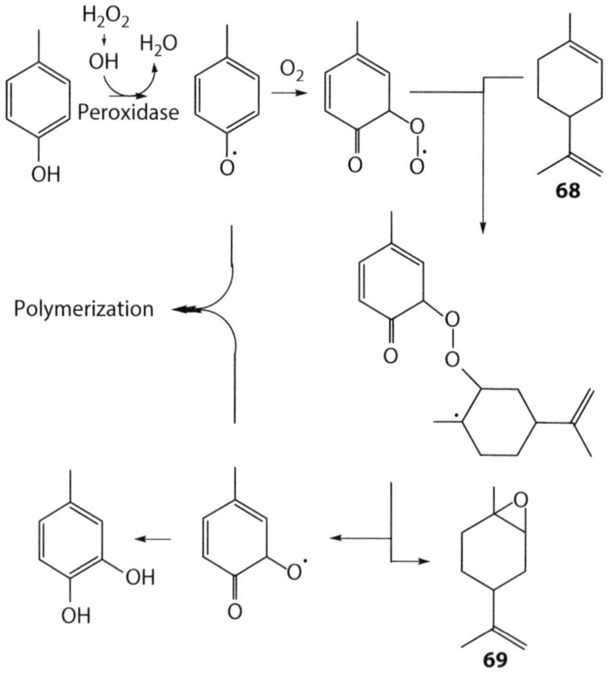

FIGURE 22.44 Proposed mechanism for the epoxidation of (+)-limonene (**68**) with p38 from the cultured cells of *Nicotiana tabacum*. (Modified from Yawata, T. et al. Epoxidation of monoterpenes by the peroxidase from the cultured cells of *Nicotiana tabacum*, *Proceedings of 42nd TEAC*, 1998, pp. 142–144.)

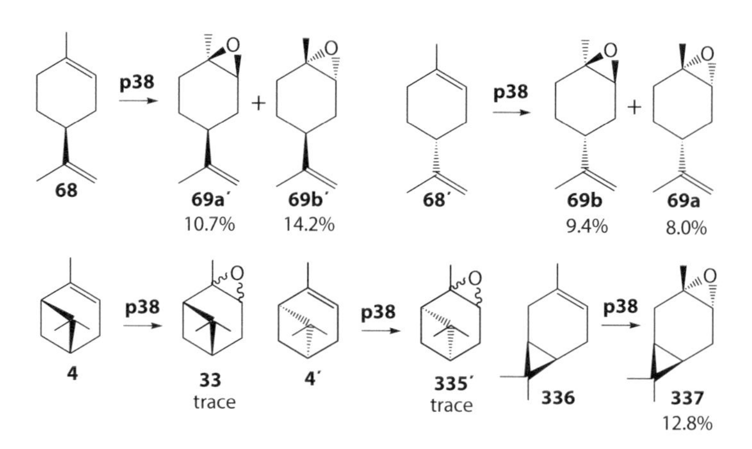

FIGURE 22.45 Epoxidation of limonene (**68**), α-pinene (**4**), and 3-carene (**336**) with p38 from the cultured cells of *Nicotiana tabacum*. (Modified from Yawata, T. et al. Epoxidation of monoterpenes by the peroxidase from the cultured cells of *Nicotiana tabacum*, *Proceedings of 42nd TEAC*, 1998, pp. 142–144.)

(*S*)-limonene (**68′**), (1*S*,5*R*)-α-pinene (**4**), (1*R*,5*R*)-α-pinene (**4**), and (1*R*,6*R*)-3-carene (**336**) were oxidized by cultured cells of *N. tabacum* to give corresponding epoxides enantio- and stereoselectively (Yawata et al., 1998) (Figure 22.45).

22.3.1.2 Isolimonene

S. litura converted (1*R*)-*trans*-isolimonene (**338**) to (1*R*,4*R*)-*p*-menth-2-ene-8,9-diol (**339**) (Miyazawa et al., 1996b) (Figure 22.46).

22.3.1.3 *p*-Menthane

Hydroxylation of *trans*- and *cis*-*p*-menthane (**252a** and **b**) by microorganisms is also very interesting from the viewpoint of the formation of the important perfumes such as (−)-menthol (**137b′**) and (−)-carvomenthol (**49b′**), plant growth regulators, and mosquito repellents such as *p*-menthane-*trans*-3,8-diol (**142a**), *p*-menthane-*cis*-3,8-diol (**142b**) (Nishimura and Noma, 1996), and *p*-menthane-2,8-diol (**93**) (Noma, 2007). *P. mendocina* strain SF biotransformed **252b** stereoselectively to *p*-*cis*-menthan-1-ol (**253**) (Tsukamoto et al., 1975) (Figure 22.47).

On the other hand, the biotransformation of the mixture of *p*-*trans*- (**252a**) and *cis*-menthane (**252b**) (45:55, peak area in GC) by *A. niger* gave *p*-*cis*-menthane-1,9-diol (**254**) via *p*-*cis*-menthan-1-ol (**253**). No metabolite was obtained from **252a** at all (Noma et al., 1990) (Figure 22.47).

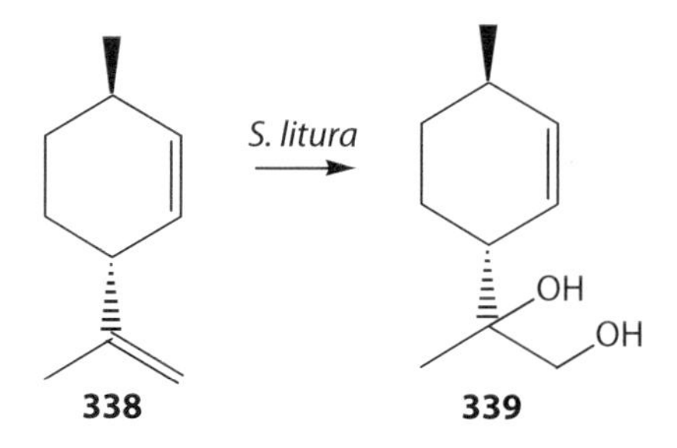

FIGURE 22.46 Biotransformation of (1*R*)-*trans*-isolimonene (**338**) by *Spodoptera litura*. (Modified from Miyazawa, M. et al. Biotransformation of *p*-menthanes using common cutworm larvae, *Spodoptera litura* as a biocatalyst, *Proceedings of 40th TEAC*, 1996b, pp. 80–81.)

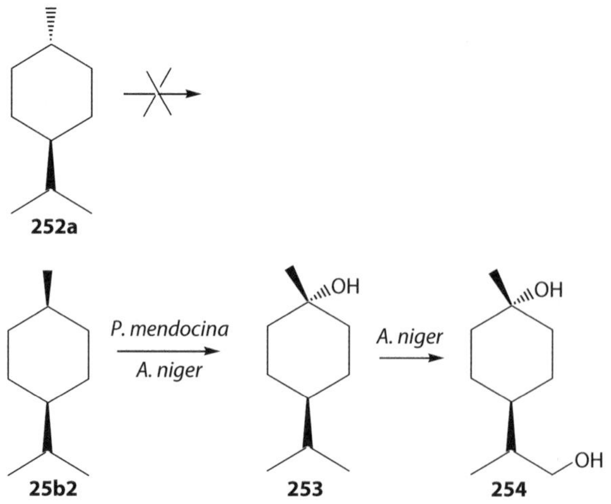

FIGURE 22.47 Biotransformation of the mixture of *trans*-(**252a**) and *cis*-*p*-menthane (**252b**) by *Pseudomonas mendocina* SF and *Aspergillus niger* TBUYN-2. (Modified from Tsukamoto, Y. et al. Microbiological oxidation of *p*-menthane 1. Formation of formation of *p*-*cis*-menthan-1-ol, *Proceedings of 18th TEAC*, pp. 24–26, 1974; Tsukamoto, Y. et al. *Agric. Biol. Chem.*, 39, 617, 1975; Noma, Y., *Aromatic Plants from Asia their Chemistry and Application in Food and Therapy*, L. Jiarovetz, N.X. Dung, and V.K. Varshney, Har Krishan Bhalla & Sons, Dehradun, India, pp. 169–186, 2007.)

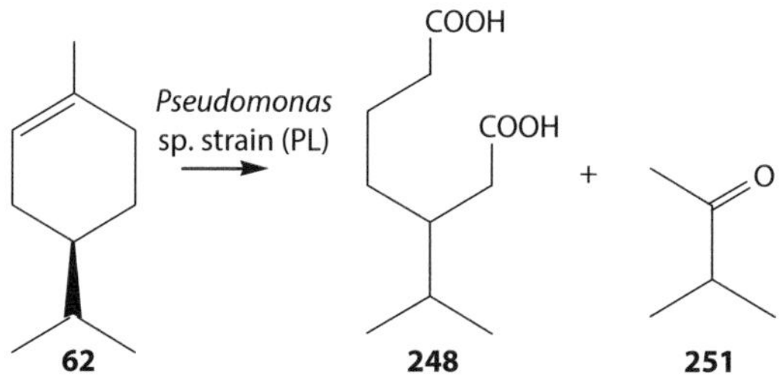

FIGURE 22.48 Biodegradation of (4*R*)-1-*p*-menthene (**62**) by *Pseudomonas* sp. *strain* (PL). (Modified from Hungund, B.L. et al. *Indian J. Biochem.*, 7, 80, 1970.)

22.3.1.4 1-*p*-Menthene

Concentrated cell suspension of *Pseudomonas* sp. strain (PL) was inoculated to the medium containing 1-*p*-menthene (**62**) as the sole carbon source. It was degraded to give β-isopropyl pimelic acid (**248**) and methylisopropyl ketone (**251**) (Hungund et al., 1970) (Figure 22.48).

As shown in Figure 22.49, *S. litura* converted (4*R*)-*p*-menth-1-ene (**62**) at C-7 position to (4*R*)-phellandric acid (**65**) (Miyazawa et al., 1996b). On the other hand, when *Cladosporium* sp. T_1 was cultivated with (+)-limonene (**68**) as the sole carbon source, it converted **62′** to *trans*-*p*-menthane-1,2-diol (**54**) (Mukherjee et al., 1973).

22.3.1.5 3-*p*-Menthene

When *Cladosporium* sp. T_8 was cultivated with 3-*p*-menthene (**147**) as the sole carbon source, it was converted to *trans*-*p*-menthane-3,4-diol (**141**) as shown in Figure 22.50 (Mukherjee et al., 1973).

22.3.1.6 α-Terpinene

α-Terpinene (**340**) was converted by *S. litura* to give α-terpinene-7-oic acid (**341**) and *p*-cymene-7-oic acid (**194**, cuminic acid) (Miyazawa et al., 1995a) (Figure 22.51).

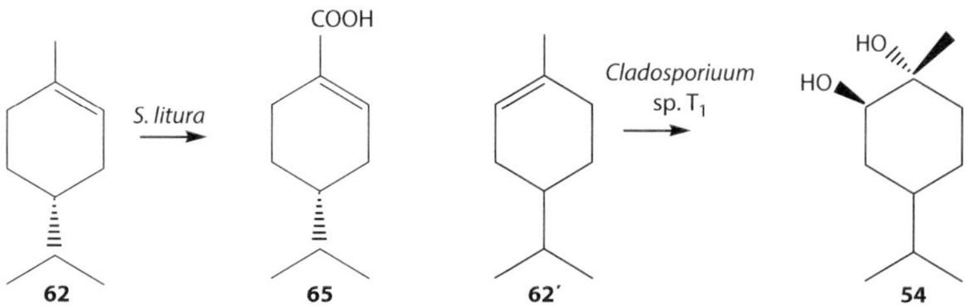

FIGURE 22.49 Biotransformation of (4*R*)-*p*-menth-1-ene (**62**) by *Spodoptera litura* and *Cladosporium* sp. T_1. (Modified from Miyazawa, M. et al. Biotransformation of *p*-menthanes using common cutworm larvae, *Spodoptera litura* as a biocatalyst, *Proceedings of 40th TEAC*, pp. 80–81, 1996b; Mukherjee, B.B. et al. *Appl. Microbiol.*, 25, 447, 1973.)

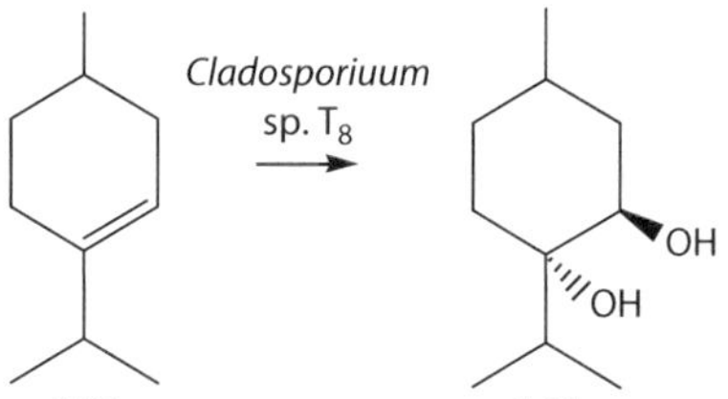

FIGURE 22.50 Biotransformation of *p*-menth-3-ene (**147**) by *Cladosporium* sp. T_8. (Modified from Mukherjee, B.B. et al. *Appl. Microbiol.*, 25, 447, 1973.)

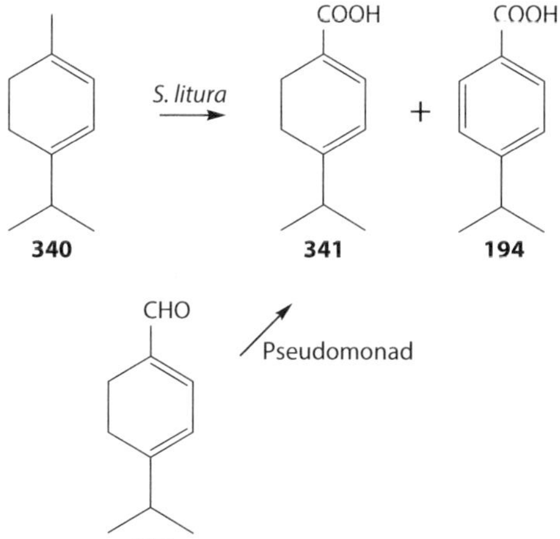

FIGURE 22.51 Biotransformation of α-terpinene (**340**) by *Spodoptera litura* and *p*-mentha-1,3-dien-7-al (**463**) by a soil Pseudomonad. (Modified from Kayahara, H. et al. *J. Ferment. Technol.*, 51, 254, 1973; Miyazawa, M. et al. Biotransformation of terpinene, limonene and α-phellandrene in common cutworm larvae, *Spodoptera litura* Fabricius, *Proceedings of 39th TEAC*, 1995a, pp. 362–363.)

A soil Pseudomonad has been found to grow with *p*-mentha-1,3-dien-7-al (**463**) as the sole carbon source and to produce α-terpinene-7-oic acid (**341**) in a mineral salt medium (Kayahara et al., 1973) (Figure 22.51).

22.3.1.7 γ-Terpinene

γ-Terpinene (**344**) was converted by *S. litura* to give γ-terpinene-7-oic acid (**345**) and *p*-cymene-7-oic acid (**194**, cuminic acid) (Miyazawa et al., 1995a) (Figure 22.52).

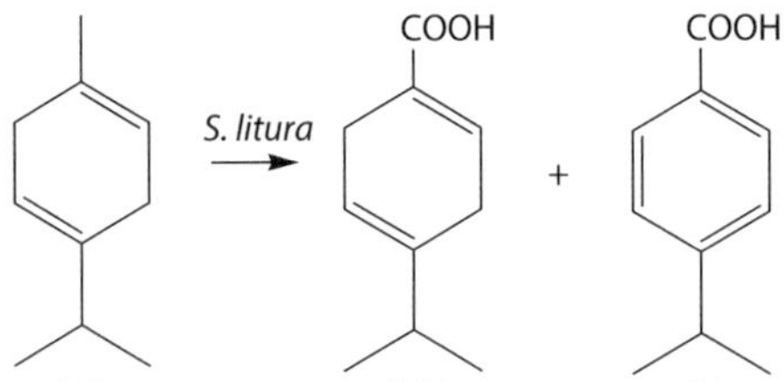

FIGURE 22.52 Biotransformation of γ-terpinene (**344**) by *Spodoptera litura*. (Modified from Miyazawa, M. et al. Biotransformation of terpinene, limonene and α-phellandrene in common cutworm larvae, *Spodoptera litura* Fabricius. *Proceedings of 39th TEAC*, 1995a, pp. 362–363.)

22.3.1.8 Terpinolene

Terpinolene (**346**) was converted by *A. niger* to give (1*R*)-8-hydroxy-3-*p*-menthen-2-one (**347**), (1*R*)-1,8-dihydroxy-3-*p*-menthen-2-one (**348**), and 5β-hydroxyfenchol (**350b′**). In the case of *C. cassiicola*, it was converted to terpinolene-1,2-*trans*-diol (**351**) and terpinolene-4,8-diol (**352**). Furthermore, in the case of rabbit, terpinolene-9-ol (**353**) and terpinolene-10-ol (**354**) were formed from **346** (Asakawa et al., 1983). *S. litura* also converted **346** to give 1-*p*-menthene-4,8-diol (**352**), cuminic acid (**194%**, 29% main product), and terpinolene-7-oic acid (**357**) (Figure 22.53).

22.3.1.9 α-Phellandrene

α-Phellandrene (**355**) was converted by *S. litura* to give α-phellandrene-7-oic acid (**356**) and *p*-cymene-7-oic acid (**194**, cuminic acid) (Miyazawa et al., 1995a) (Figure 22.54).

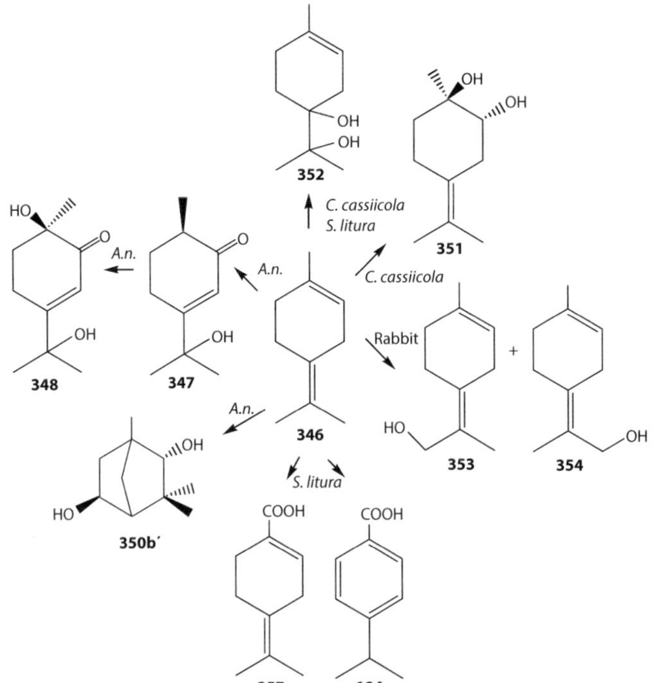

FIGURE 22.53 Biotransformation of terpinolene (**346**) by *Aspergillus niger* (Asakawa et al., 1991), *Corynespora cassiicola* (Abraham et al., 1985), rabbit (Asakawa et al., 1983), and *Spodoptera litura*. (Modified from Miyazawa, M. et al. Biotransformation of terpinene, limonene and α-phellandrene in common cutworm larvae, *Spodoptera litura* Fabricius. *Proceedings of 39th TEAC*, 1995a, pp. 362–363.)

22.3.1.10 *p*-Cymene

Pseudomonas sp. strain (PL) was cultivated with *p*-cymene (**178**) as the sole carbon source to give cumyl alcohol (**192**), cumic acid (**194**), 3-hydroxycumic acid (**196**), 2,3-dihydroxycumic acid (**197**), 2-oxo-4-methylpentanoic acid (**201**), 9-hydroxy-*p*-cymene (**189**), and *p*-cymene-9-oic acid (**190**) as shown in Figure 22.55 (Madyastha and Bhattacharyya, 1968; Yamada et al., 1965). On the other hand, *p*-cymene (**178**) was converted regiospecifically to cumic acid (**194**) by *Pseudomonas* sp., *Pseudomonas desmolytica*, and *Nocardia salmonicolor* (Madyastha and Bhattacharyya, 1968) (Figure 22.56).

p-Cymene (**178**) is converted to thymoquinone (**358**) and analogues, **179** and **180**, by various kinds of microorganisms (Demirci et al., 2007) (Figure 22.57).

22.3.2 Monocyclic Monoterpene Aldehyde

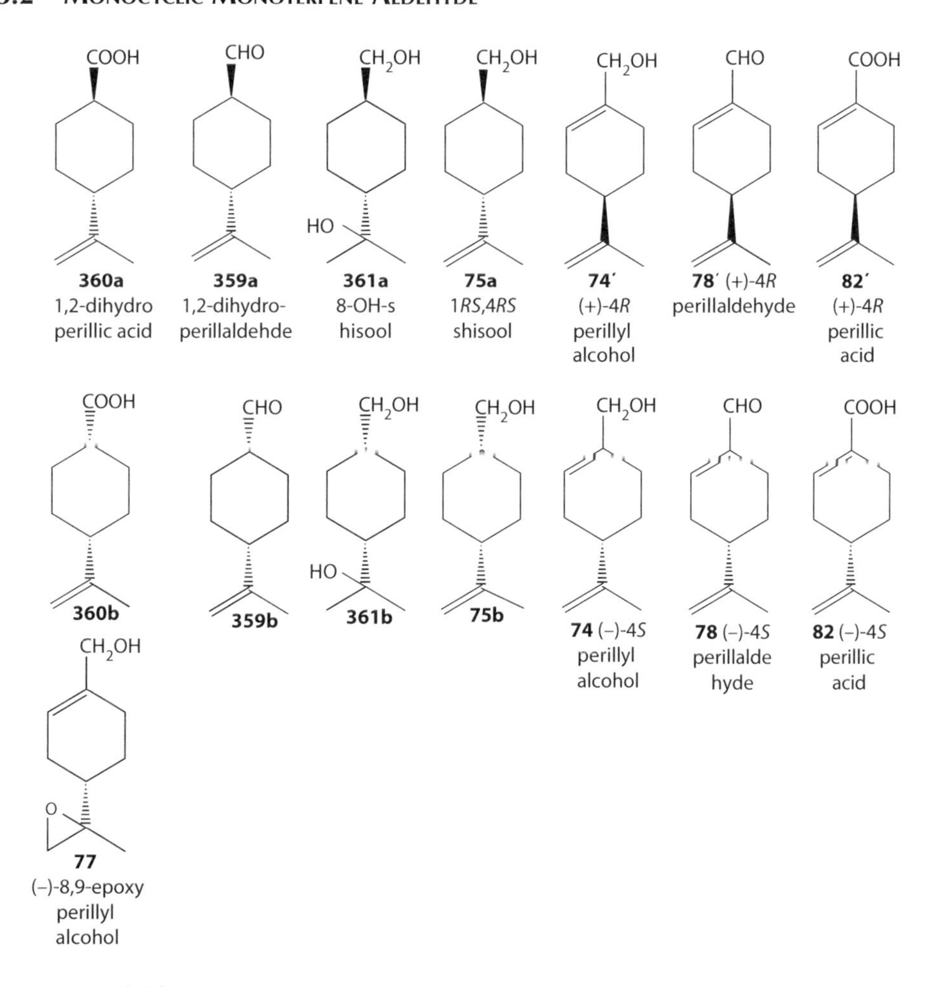

22.3.2.1 Perillaldehyde

Biotransformation of (−)-perillaldehyde (**78**), (+)-perillaldehyde (**78′**), (−)-perillyl alcohol (**74**), *trans*-1,2-dihydroperillaldehyde (**359a**) and *cis*-1,2-dihydroperillaldehyde (**359b**), and *trans*-shisoic acid (**360a**) and *cis*-shisoic acid (**360b**) was carried out by *E. gracilis* Z. (Noma et al., 1991b), *D. tertiolecta* (Noma et al., 1991a, 1992a), *Chlorella ellipsoidea* IAMC-27 (Noma et al., 1997), *S. ikutamanensis* Ya-2-1 (Noma et al., 1984, 1986), and other microorganisms (Kayahara et al., 1973) (Figure 22.58).

(−)-Perillaldehyde (**78**) is easily transformed to give (−)-perillyl alcohol (**74**) and *trans*-shisool (**75a**), which is well known as a fragrance, as the major product and (−)-perillic acid (**82**) as the minor product.

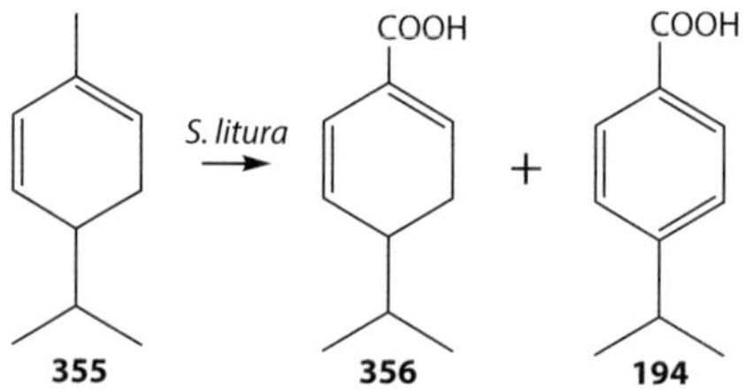

FIGURE 22.54 Biotransformation of α-phellandrene (**355**) by *Spodoptera litura*. (Modified from Miyazawa, M. et al. Biotransformation of terpinene, limonene and α-phellandrene in common cutworm larvae, *Spodoptera litura* Fabricius. *Proceedings of 39th TEAC*, 1995a, pp. 362–363.)

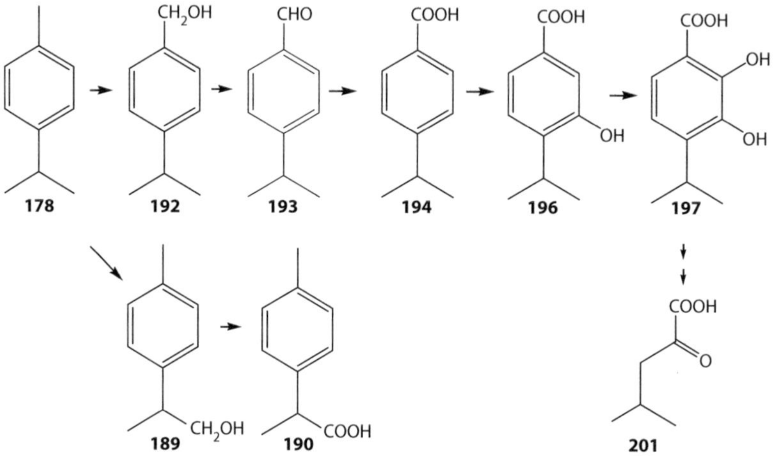

FIGURE 22.55 Biotransformation of *p*-cymene (**178**) by *Pseudomonas* sp. strain (PL). (Modified from Madyastha, K.M. and Bhattacharyya, P.K., *Indian J. Biochem*., 5, 161, 1968.)

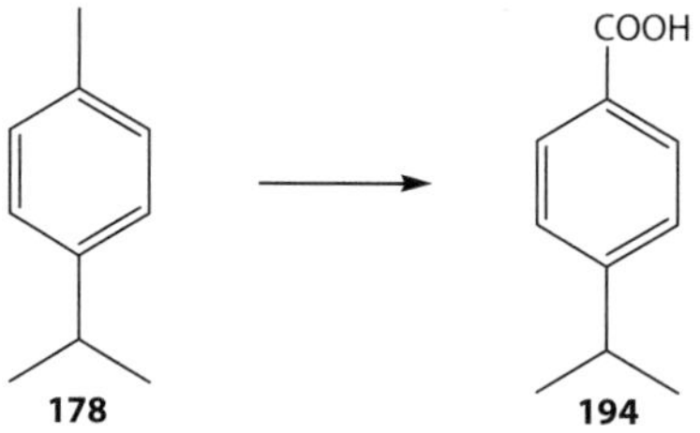

FIGURE 22.56 Biotransformation of *p*-cymene (**178**) to cumic acid (**194**) by *Pseudomonas* sp., *Pseudomonas desmolytica*, and *Nocardia salmonicolor*. (Modified from Yamada, K. et al. *Agric. Biol. Chem*., 29, 943, 1965; Madyastha, K.M. and Bhattacharyya, P.K., *Indian J. Biochem*., 5, 161, 1968; Noma, Y., 2000. unpublished data.)

(−)-Perillyl alcohol (**74**) is also transformed to *trans*-shisool (**75a**) as the major product with *cis*-shisool (**75b**) and 8-hydroxy-*cis*-shisool (**361b**). Furthermore, *trans*-shisool (**75a**) and *cis*-shisool (**75b**) are hydroxylated to 8-hydroxy-*trans*-shisool (**361a**) and 8-hydroxy-*cis*-shisool (**361b**), respectively. *trans*-1,2-Dihydroperillaldehyde (**359a**) and *cis*-1,2-dihydroperillaldehyde (**359b**) are also transformed to **75a** and **75b** as the major products and *trans*-shisoic acid (**360a**) and *cis*-shisoic acid (**360b**) as the minor products, respectively. Compound **360a** was also formed from **75a**. In the biotransformation of (±)-perillaldehyde (**74** and **74′**), the same results were obtained as described in the case of **74**. In the case of *S. ikutamanensis* Ya-2-1, (−)-perillaldehyde (**78**) was converted to (−)-perillic acid (**82**), (−)-perillyl alcohol (**74**), and (−)-perillyl alcohol-8,9-epoxide (**77**), which was the major product.

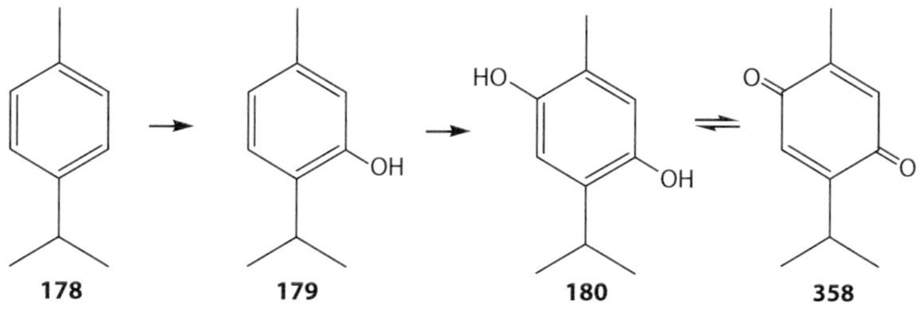

HO OH OH O O

178 179 180 358

FIGURE 22.57 Biotransformation of *p*-cymene (**178**) to thymoquinone (**358**) and analogues by microorganisms. (Modified from Demirci, F. et al. Biotransformation of *p*-cymene to thymoquinone, *Book of Abstracts of the 38th ISEO*, SL-1, 2007, p. 6.)

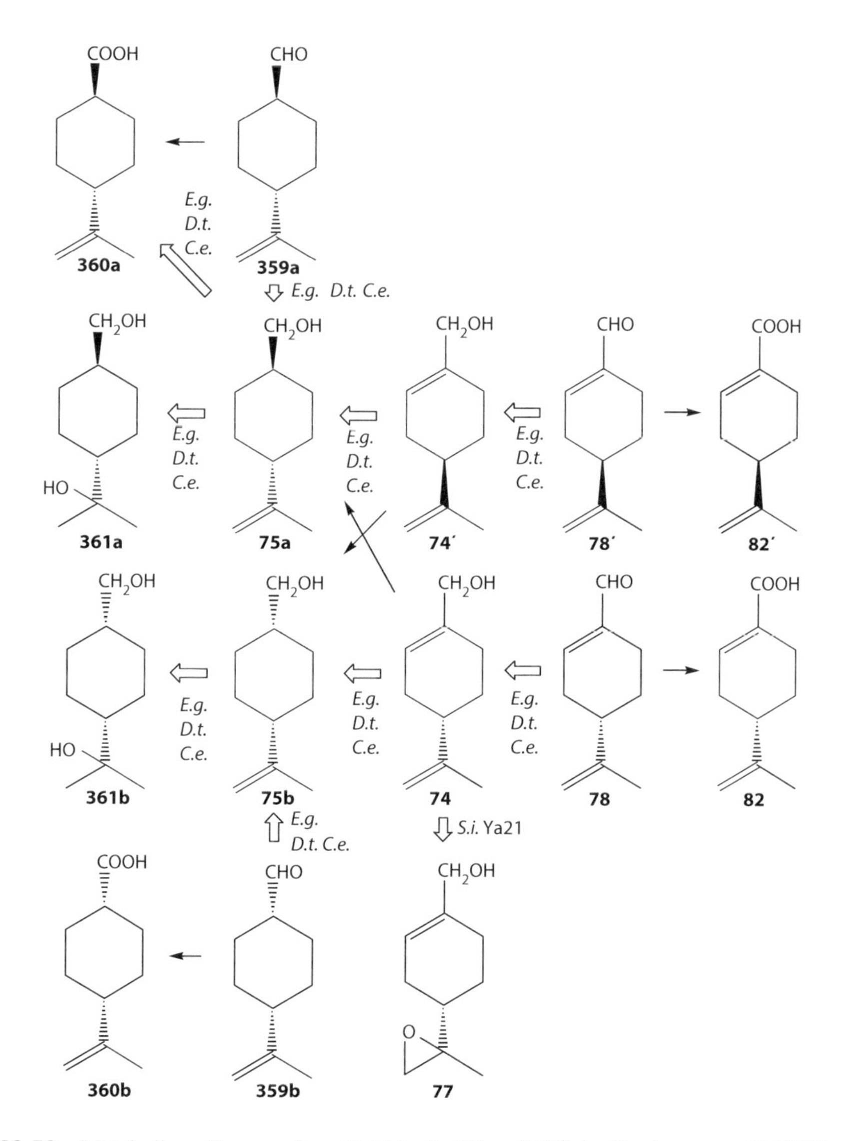

FIGURE 22.58 Metabolic pathways of perillaldehyde (**78** and **78′**) by *Euglena gracilis* Z (Noma et al., 1991b), *Dunaliella tertiolecta* (Noma et al., 1991a, 1992a), *Chlorella ellipsoidea* IAMC-27 (Noma et al., 1997), *Streptomyces ikutamanensis* Ya-2-1 (Noma et al., 1984, 1986), a soil Pseudomonad (Kayahara et al., 1973), and rabbit (Ishida et al., 1981a).

A soil Pseudomonad has been found to grow with (–)-perillaldehyde (**78**) as the sole carbon source and to produce (–)-perillic acid (**82**) in a mineral salt medium (Kayahara et al., 1973).

On the other hand, rabbit metabolized (–)-perillaldehyde (**78**) to (–)-perillic acid (**82**) along with minor shisool (**75a**) (Ishida et al., 1981a).

22.3.2.2 Phellandral and 1,2-Dihydrophellandral

Biotransformation of (–)-phellandral (**64**), *trans*-tetrahydroperillaldehyde (**362a**), and *cis*-tetrahydroperillaldehyde (**362b**) was carried out by microorganisms (Noma et al., 1986, 1991a,b, 1997). (–)-Phellandral (**64**) was metabolized mainly via (–)-phellandrol (**63**) to *trans*-tetrahydroperillyl alcohol (**66a**). *trans*-Tetrahydroperillaldehyde (**362a**) and *cis*-tetrahydroperillaldehyde (**362b**) were also transformed to *trans*-tetrahydroperillyl alcohol (**66a**) and *cis*-tetrahydroperillyl alcohol (**66b**) as the major products and *trans*-tetrahydroperillic acid (**363a**) and *cis*-tetrahydroperillic acid (**363b**) as the minor products, respectively (Figure 22.59).

22.3.2.3 Cuminaldehyde

Cumin aldehyde (**193**) is transformed by *Euglena* (Noma et al., 1991b), *Dunaliella* (Noma et al., 1991a), and *S. ikutamanensis* (Noma et al., 1986) to give cumin alcohol (**192**) as the major product and cuminic acid (**194**) as the minor product (Figure 22.60).

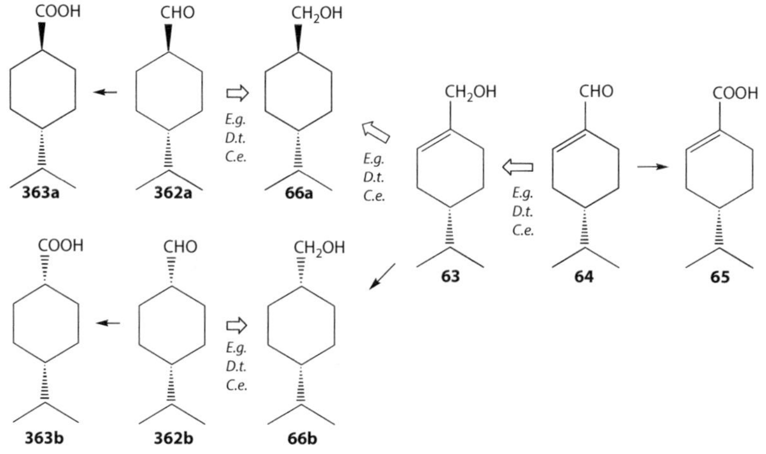

FIGURE 22.59 Metabolic pathways of (–)-phellandral (**64**) by microorganisms. (Modified from Noma, Y. et al. Reduction of terpene aldehydes and epoxidation of terpene alcohols by *S. ikutamanensis*, Ya-2-1, *Proceedings of 30th TEAC*, 1986, pp. 204–206; Noma, Y. et al. *Phytochemistry*, 30, 1147, 1991a; Noma, Y. et al. Biotransformation of monoterpenes by photosynthetic marine algae, *Dunaliella tertiolecta*, *Proceedings of 35th TEAC*, 1991b, pp. 112–114; Noma, Y. et al. Biotransformation of terpenoids and related compounds by *Chlorella* species, *Proceedings of 41st TEAC*, 1997, pp. 227–229.)

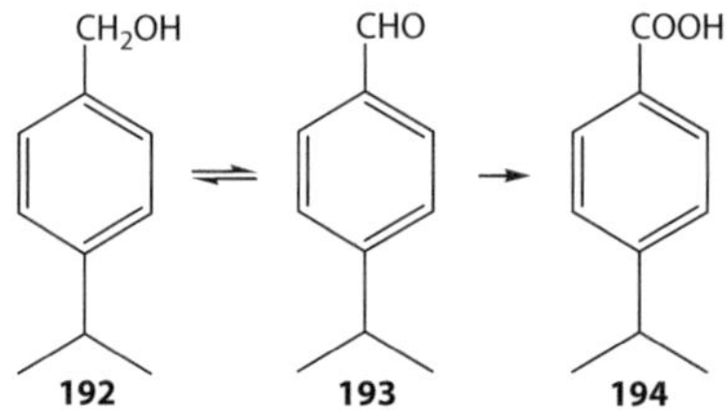

FIGURE 22.60 Metabolic pathway of cumin aldehyde (**193**) by microorganism. (Modified from Noma, Y. et al. Reduction of terpene aldehydes and epoxidation of terpene alcohols by *S. ikutamanensis*, Ya-2-1, *Proceedings of 30th TEAC*, 1986, pp. 204–206; Noma, Y. et al. *Phytochemistry*, 30, 1147, 1991a; Noma, Y. et al. Biotransformation of monoterpenes by photosynthetic marine algae, *Dunaliella tertiolecta, Proceedings of 35th TEAC*, 1991b, pp. 112–114.)

22.3.3 Monocyclic Monoterpene Alcohol

22.3.3.1 Menthol

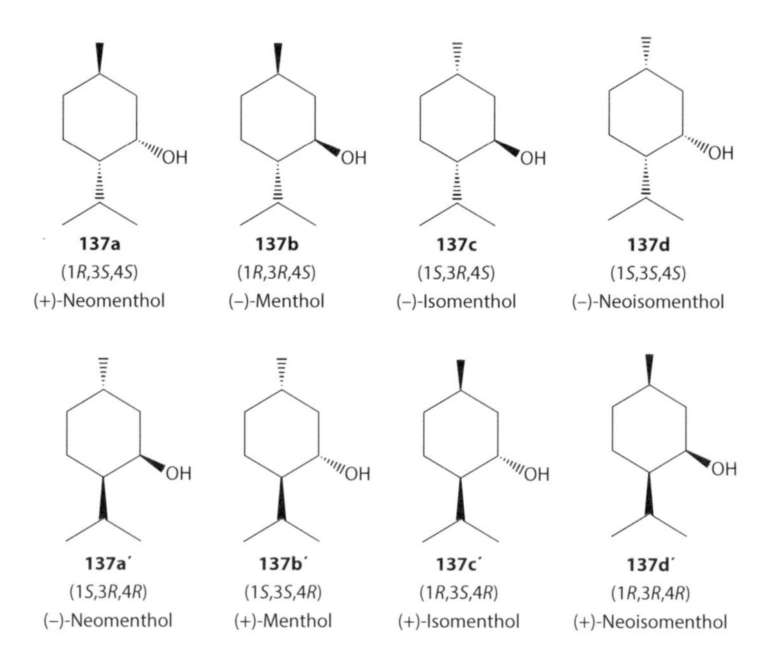

Menthol (**137**) is one of the rare naturally occurring monocyclic monoterpene alcohols that have not only various physiological properties, such as sedative, anesthetic, antiseptic, gastric, and antipruritic, but also characteristic fragrance (Bauer et al., 1990). There are in fact eight isomers with a menthol (*p*-menthan-3-ol) skeleton; (−)-menthol (**137b**) is the most important one, because of its cooling and refreshing effect. It is the main component of peppermint and cornmint oils obtained from the *Mentha piperita* and *Mentha arvensis* species. Many attempts have been made to produce (−)-menthol (**137b**) from inexpensive terpenoid sources, but these sources also unavoidably yielded the (±)-isomers (**137b** and **137b′**): isomenthol (**137c**), neomenthol (**137a**), and neoisomenthol (**137d**) (Krasnobajew, 1984). Japanese researchers have been active in this field, maybe because of the large demand for (−)-menthol (**137b**) in Japan itself, that is, 500 t/year (Janssens et al., 1992). Indeed, most literature deals with the enantiomeric hydrolysis of (±)-menthol (**137b** and **137b′**) esters to optically pure *l*-menthol (**137b**). The asymmetric hydrolysis of (±)-menthyl chloroacetate by an esterase of *Arginomonas non-fermentans* FERM-P-1924 has been patented by the Japanese Nippon Terpene Chemical Co. (Watanabe and Inagaki, 1977a,b). Investigators from the Takasago Perfumery Co. Ltd. claim that certain selected species of *Absidia*, *Penicillium*, *Rhizopus*, *Trichoderma*, *Bacillus*, *Pseudomonas*, and others asymmetrically hydrolyze esters of (±)-menthol isomers such as formates, acetates, propanoates, caproates, and esters of higher fatty acids (Moroe et al., 1971; Yamaguchi et al., 1977) (Figure 22.61).

Numerous investigations into the resolution of the enantiomers by selective hydrolysis with microorganisms or enzymes were carried out. Good results were described by Yamaguchi et al. (1977) with the asymmetric hydrolysis of (±)-methyl acetate by a mutant of *Rhodotorula mucilaginosa*, yielding 44 g of (−)-menthol (**137b**) from a 30% (±)-menthyl acetate mixture per liter of cultured medium for 24 h. The latest development is the use of immobilized cells of *Rhodotorula minuta* in aqueous saturated organic solvents (Omata et al., 1981) (Figure 22.62).

Besides the hydrolysis of menthyl esters, the biotransformation of menthol and its enantiomers has also been published (Shukla et al., 1987; Asakawa et al., 1991). The fungal biotransformation of (−)- (**137b**) and (+)-menthols (**137b′**) by *A. niger* and *A. cellulosae* was described (Asakawa et al., 1991). *A. niger* converted (−)-menthol (**137b**) to 1- (**138b**), 2- (**140b**), 6- (**139b**), 7- (**143b**), 9-hydroxymenthols (**144b**), and the mosquito repellent-active 8-hydroxymenthol (**142b**), whereas

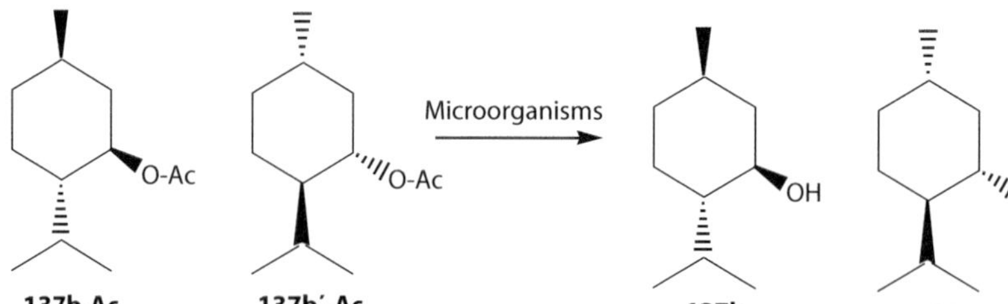

FIGURE 22.61 Asymmetric hydrolysis of racemic menthyl acetate (**137b-Ac** and **137b′-Ac**) to obtain pure (−)-menthol (**137b**). (Modified from Watanabe, Y. and Inagaki, T., Japanese Patent 77.12.989. No. 187696x, 1977a; Watanabe, Y. and Inagaki, T., Japanese Patent 77.122.690. No. 87656g, 1977b; Moroe, T. et al. Japanese Patent, 2.036. 875. no. 98195t, 1971; Oritani, T. and Yamashita, K., *Agric. Biol. Chem.*, 37, 1695, 1973b.)

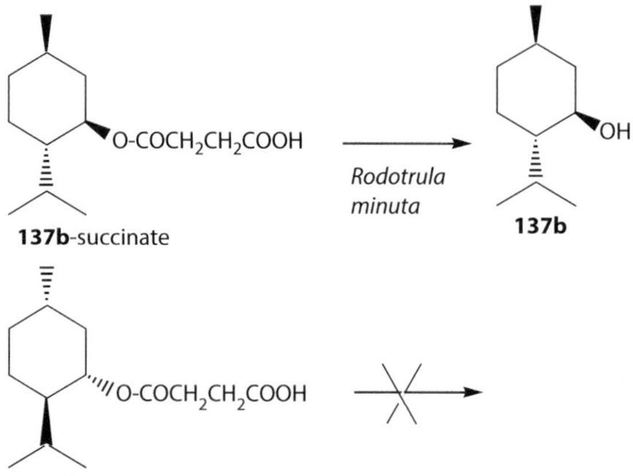

FIGURE 22.62 Asymmetric hydrolysis of racemic menthyl succinate (**137b-** and **137b′-succinates**) to obtain pure (−)-menthol (**137b**). (Modified from Yamaguchi, Y. et al. *J. Agric. Chem. Soc. Jpn.*, 51, 411, 1977.)

(+)-menthol (**137b′**) was smoothly biotransformed by the same microorganism to 7-hydroxymenthol (**143b**). The bioconversion of (+)- (**137a′**) and (−)-neomenthol (**137a**) and (+)-isomenthol (**137c′**) by *A. niger* was studied later by Takahashi et al. (1994), mainly giving hydroxylated products. Noma and Asakawa (1995) reviewed the schematic menthol hydroxylation in detail.

Incubation of (−)-menthol (**137b**) with *Cephalosporium aphidicola* for 12 days yielded 10-acetoxymenthol (**144bb-Ac**), 1α-hydroxymenthol (**138b**), 6α-hydroxy-menthol (**139bb**), 7-hydroxymenthol (**143b**), 9-hydroxymenthol (**144ba**), and 10-hydroxymenthol (**144bb**) (Atta-ur-Rahman et al., 1998) (Figure 22.63).

A. niger TBUYN-2 converted (−)-menthol (**137b**) to 1α- (**138b**), 2α- (**140b**), 4β- (**141b**), 6α- (**139bb**), 7- (**143b**), 9-hydroxymenthols (**144ba**), and the mosquito repellent-active 8-hydroxymenthol (**142b**) (Figure 22.64). *A. cellulosae* M-77 biotransformed (−)-menthol (**137b**) to 4β-hydroxymenthol (**141b**) predominantly. The formation of **141b** is also observed in *A. cellulosae* IFO 4040 and *Aspergillus terreus* IFO 6123, but its yield is much less than that obtained from **137b** by *A. cellulosae* M-77 (Asakawa et al., 1991) (Table 22.1).

On the other hand, (+)-menthol (**137b′**) was smoothly biotransformed by *A. niger* to give 1β-hydroxymenthol (**138b′**), 6β-hydroxymenthol (**139ba′**), 2β-hydroxymenthol (**140ba′**), 4α-hydroxymenthol (**141b′**), 7-hydroxymenthol (**143b′**), 8-hydroxymenthol (**142b′**), and 9-hydroxymenthol (**144ba′**) (Figure 22.65) (Table 22.2).

S. litura converted (−)- and (+)-menthols (**137b** and **137b′**) that gave the corresponding 10-hydroxy products (**143b** and **143b′**) (Miyazawa et al., 1997a) (Figure 22.66).

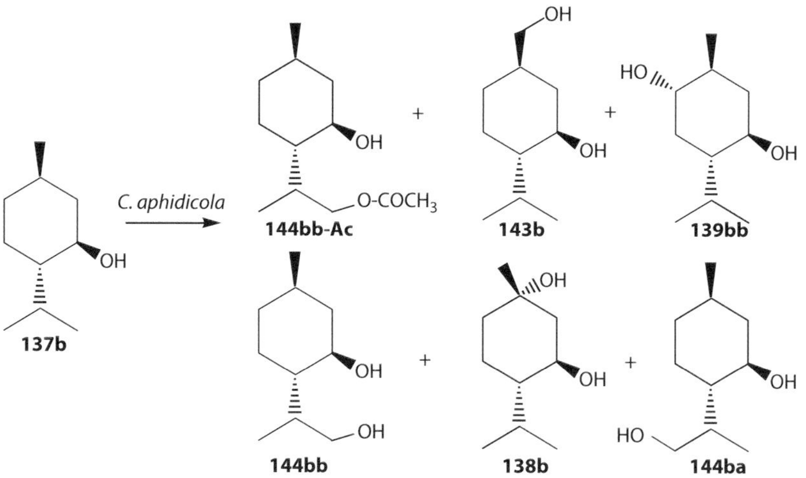

FIGURE 22.63 Biotransformation of (−)-menthol (**137b**) by *Cephalosporium aphidicola*. (Modified from Atta-ur-Rahman, M. et al. *J. Nat. Prod.*, 61, 1340, 1998.)

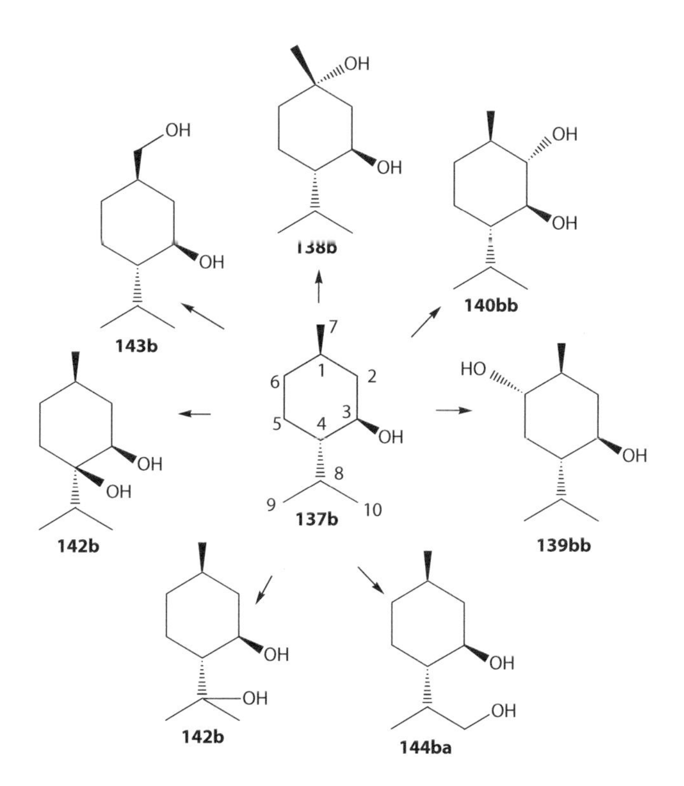

FIGURE 22.64 Metabolic pathways of (−)-menthol (**137b**) by *Aspergillus niger*. (Modified from Asakawa, Y. et al. *Phytochemistry*, 30, 3981, 1991.)

(−)-Menthol (**137b**) was glycosylated by *Eucalyptus perriniana* suspension cells to (−)-menthol diglucoside (**364**, 26.6%) and another menthol glycoside. On the other hand, (+)-menthol (**137b′**) was glycosylated by *E. perriniana* suspension cells to (+)-menthol di- (**364′**, 44.0%) and triglucosides (**365**, 6.8%) (Hamada et al., 2002) (Figure 22.67).

(−)-Menthol (**137b**) and its enantiomer (**137b′**) were converted to their corresponding 8-hydroxy derivatives (**142b** and **142b′**) by human CYP2A6 (Nakanishi and Miyazawa, 2005) (Figure 22.68).

TABLE 22.1
Metabolites of (−)-Menthol (137b) by Various *Aspergillus* spp. (Static Culture)

Microorganisms	138b	142b	139bb	143b	139bb	144ba	141b
A. awamori IFO 4033	+[a]	++	−	+	++	+++	−
A. fumigatus IFO 4400	−	+	−	+	+	+	−
A. sojae IFO 4389	++	+	+	−	−	++++	−
A. usami IFO 4338	−	−	−	+	−	+++	−
A. cellulosae M-77	+	−	−	+	−	++	++++
A. cellulosae IFO 4040	−	+	−	−	−	++	++
A. terreus IFO 6123	+	+	+	−	+	+	−
A. niger IFO 4049	−	+	−	+	−	+++	−
A. niger IFO 4040	−	+	−	+++	−	+++	−
A. niger TBUYN-2	+	++	+	+	++	++	−

[a] Symbols +, ++, +++, etc., are relative concentrations estimated by GC-MS.

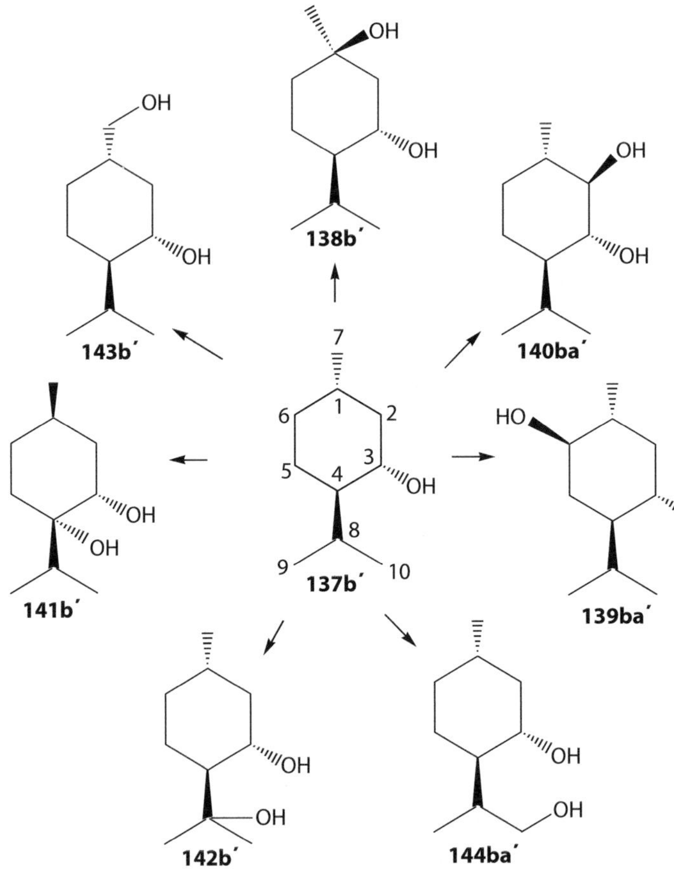

FIGURE 22.65 Metabolic pathways of (+)-menthol (**137b′**) by *Aspergillus niger.* (Modified from Noma, Y. et al. Microbiological conversion of menthol. Biotransformation of (+)-menthol by a strain of Aspergillus niger. *Proceedings of 33rd TEAC*, 1989, pp. 124–126; Asakawa, Y. et al. *Phytochemistry*, 30, 3981, 1991.)

TABLE 22.2
Metabolites of (+)-Menthol (137b′) by Various *Aspergillus* spp. (Static Culture)

Microorganisms	138b′	142b′	140ba′	143b′	139ba′	144ba′	141b′
A. awamori IFO 4033	+[a]	++	−	+++	−	+++	−
A. fumigates IFO 4400	+	++	−	+	−	++	−
A. sojae IFO 4389	+	++	−	−	−	+++	−
A. usami IFO 4338	+	−	−	+	−	+++	−
A. cellulosae M-77	−	+	−	−	−	++	++++
A. cellulosae IFO 4040	+	+	−	−	++	+	+
A. terreus IFO 6123	+	+++	+	+	+	++	−
A. niger IFO 4049	+	−	−	−	+	+++	−
A. niger IFO 4040	+	++	−	+	−	++	−
A. niger TBUYN-2	++	+	−	+++++	+	+	−

[a] Symbols +, ++, +++, etc., are relative concentrations estimated by GC-MS.

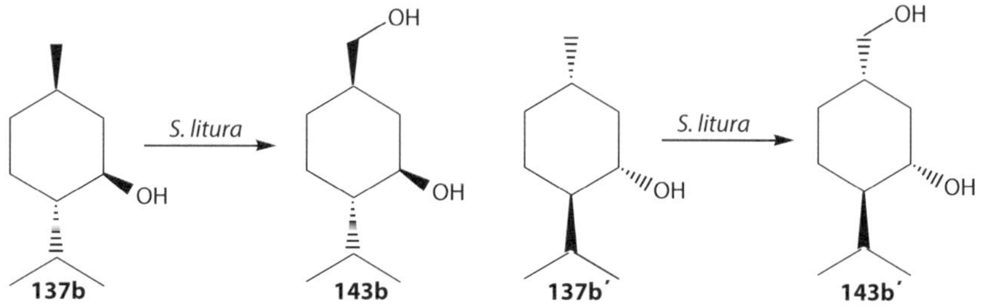

FIGURE 22.66 Biotransformation of (−)- (**137b**) and (+)-menthol (**137b′**) by *Spodoptera litura*. (Modified from Miyazawa, M. et al. Biotransformation of (−)-menthol and (+)-menthol by common cutworm Larvae, *Spodoptera litura* as a biocatalyst, *Proceedings of 41st TEAC*, 1997a, pp. 391–392.)

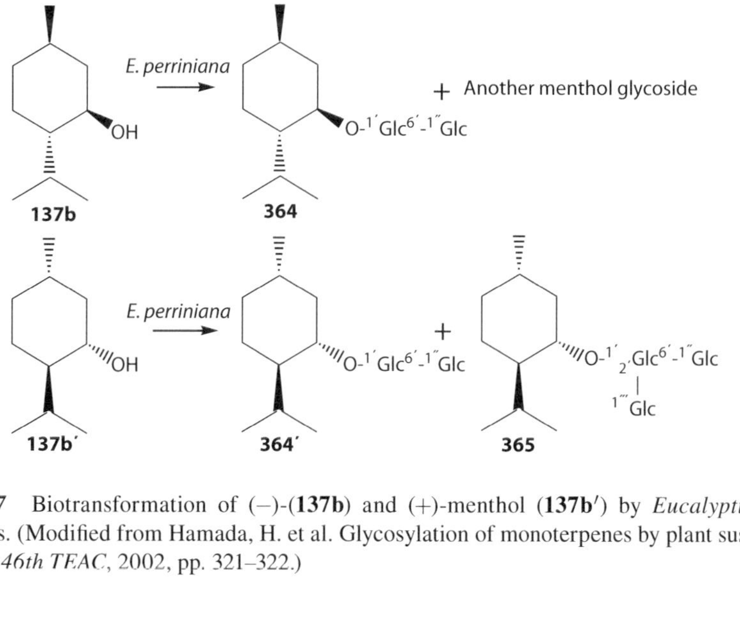

FIGURE 22.67 Biotransformation of (−)-(**137b**) and (+)-menthol (**137b′**) by *Eucalyptus perriniana* suspension cells. (Modified from Hamada, H. et al. Glycosylation of monoterpenes by plant suspension cells, *Proceedings of 46th TEAC*, 2002, pp. 321–322.)

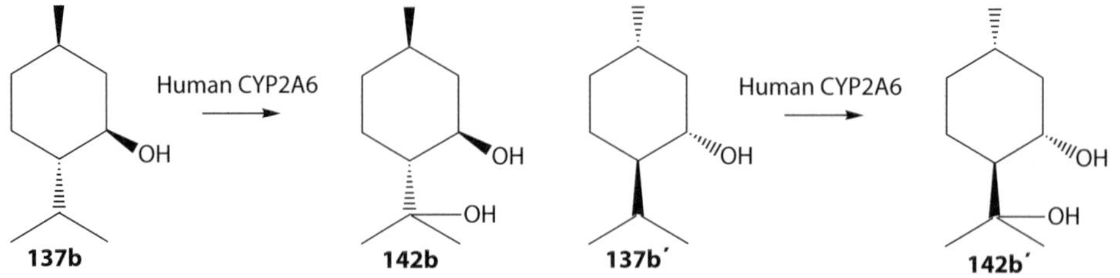

FIGURE 22.68 Biotransformation of (−)-menthol (**137b**) and its enantiomer (**137b′**) by human CYP2A6. (Modified from Nakanishi, K. and Miyazawa, M., Biotransformation of (+)- and (−)- menthol by liver microsomal humans and rats, *Proceedings of 49th TEAC*, 2005, pp. 423–425.)

By various assays, cytochrome P450 molecular species responsible for the metabolism of (−)- (**137b**) and (+)-menthol (**137b′**) was determined to be CYP2A6 and CYP2B1 in human and rat, respectively. Also, kinetic analysis showed that K and V_{max} values for the oxidation of (−)- (**137b**) and (+)-menthol (**137b′**) recombinant CYP2A6 and CYP2B1 were determined to be 28 μM and 10.33 nmol/min/nmol P450 and 27 μM and 5.29 nmol/min/nmol P450, 28 μM and 3.58 nmol/min/nmol P450, and 33 μM and 5.3 nmol/min/nmol P450, respectively (Nakanishi and Miyazawa, 2005) (Figure 22.68).

22.3.3.2 Neomenthol

(+)-Neomenthol (**137a**) is biotransformed by *A. niger* TBUYN-2 to give five kinds of diols (**138a**, **143a**, **144aa**, **144ab**, and **142a**) and two kinds of triols (**145a** and **146a**) as shown in Figure 22.69 (Takahashi et al., 1994).

(−)-Neomenthol (**137a′**) is biotransformed by *A. niger* to give six kinds of diols (**140a′**, **139a′**, **143a′**, **144aa′**, **144ab′**, and **142a′**) and a triol (**146a′**) as shown in Figure 22.70 (Takahashi et al., 1994).

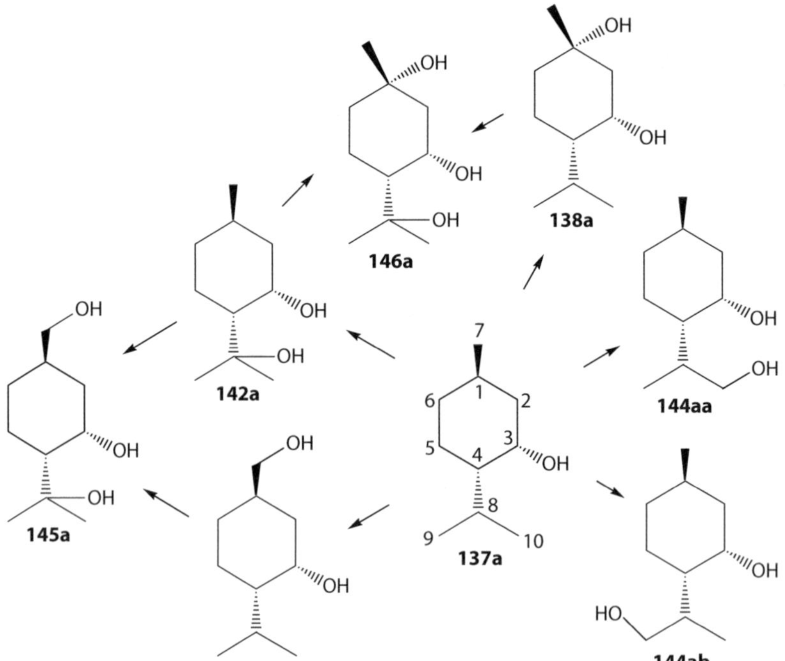

FIGURE 22.69 Metabolic pathways of (+)-neomenthol (**137a**) by *Aspergillus niger*. (Modified from Takahashi, H. et al. *Phytochemistry*, 35, 1465, 1994.)

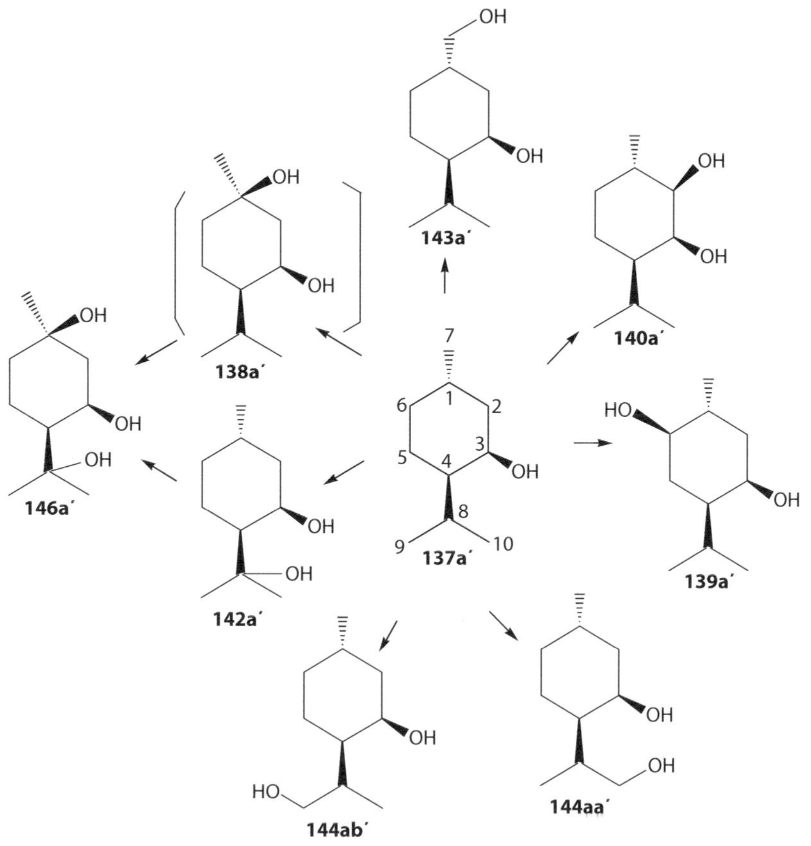

FIGURE 22.70 Metabolic pathways of (−)-neomenthol (**137a′**) by *Aspergillus niger*. (Modified from Takahashi, H. et al. *Phytochemistry*, 35, 1465, 1994.)

22.3.3.3 (+)-Isomenthol

(+)-Isomenthol (**137c**) is biotransformed to give two kinds of diols such as 1β-hydroxy- (**138c**) and 6β-hydroxyisomenthol (**139c**) by *A. niger* (Takahashi et al., 1994) (Figure 22.71).

(±)-Isomenthyl acetate (**137c-Ac** and **137c′-Ac**) was asymmetrically hydrolyzed to (−)-isomenthol (**137c**) with (+)-isomenthol acetate (**137c′-Ac**) by many microorganisms and esterases (Oritani and Yamashita, 1973b) (Figure 22.72).

22.3.3.4 Isopulegol

(−)-Isopulegol (**366**) was biotransformed by *S. litura* larvae to give 7-hydroxy-(−)-isopulegol (**367**), 9-hydroxy-(−)-menthol (**144ba**), and 10-hydroxy-(−)-isopulegol (**368**). On the other hand,

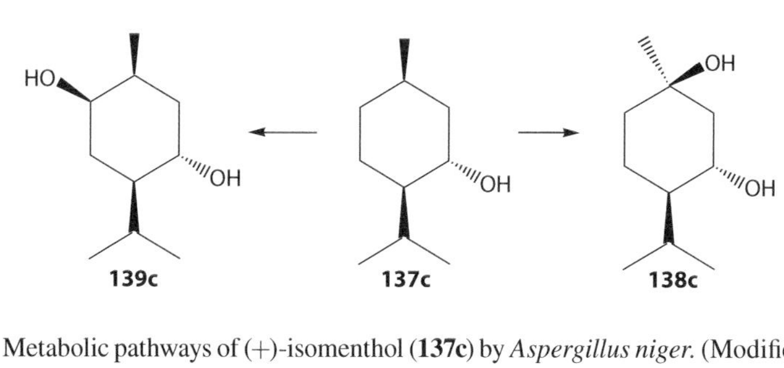

FIGURE 22.71 Metabolic pathways of (+)-isomenthol (**137c**) by *Aspergillus niger*. (Modified from Takahashi, H. et al. *Phytochemistry*, 35, 1465, 1994.)

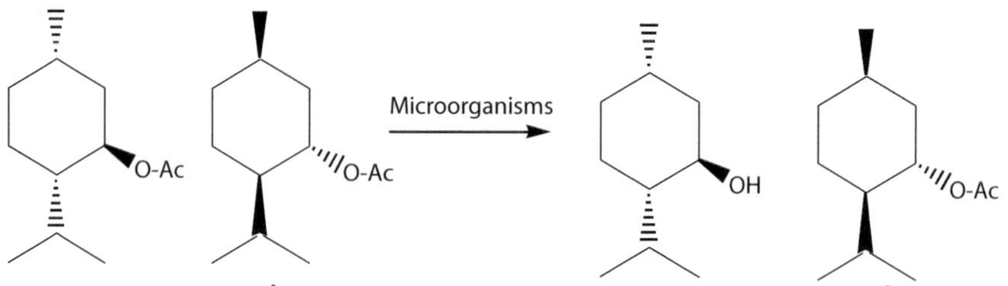

FIGURE 22.72 Microbial resolution of (±)-isomenthyl acetate (**137c-Ac** and **137c′-Ac**) by microbial esterase. (Modified from Oritani, T. and Yamashita, K., *Agric. Biol. Chem.*, 37, 1695, 1973b.)

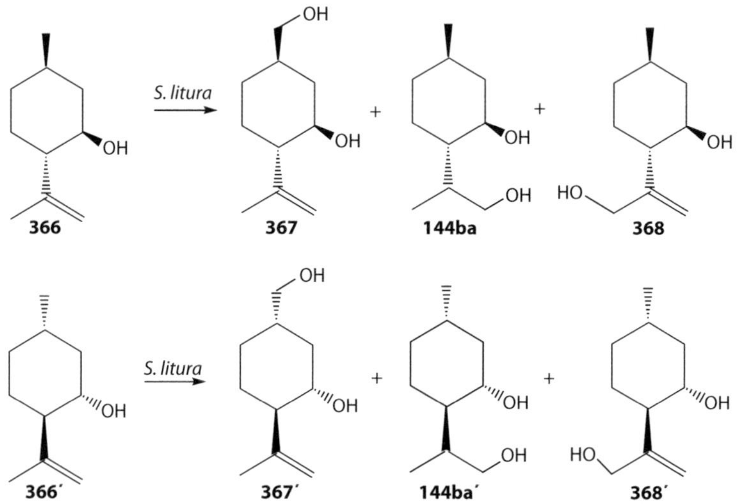

FIGURE 22.73 Biotransformation of (−)-(**366**) and (+)-isopulegol (**366′**) by *Spodoptera litura*. (Modified from Ohsawa, M. and Miyazawa, M., Biotransformation of (+)- and (−)-isopulegol by the larvae of common cutworm (*Spodoptera litura*) as a biocatalyst, *Proceedings of 45th TEAC*, 2001, pp. 375–376.)

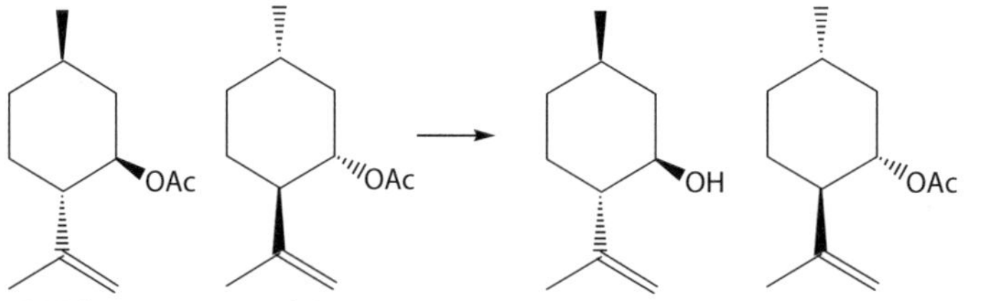

FIGURE 22.74 Microbial resolution of (±)-isopulegyl acetate (**366-Ac** and **366′-Ac**) by microorganisms. (Modified from Oritani, T. and Yamashita, K., *Agric. Biol. Chem.*, 37, 1687, 1973c.)

(+)-isopulegol (**366′**) was biotransformed by the same larvae in the same manner to give 7-hydroxy-(+)-isopulegol (**367′**), 9-hydroxy-(+)-menthol (**144ba′**), and 10-hydroxy-(+)-isopulegol (**368′**) (Ohsawa and Miyazawa, 2001) (Figure 22.73).

Microbial resolution of (±)-isopulegyl acetate (**366-Ac** and **366′-Ac**) was studied by microorganisms. (±)-Isopulegyl acetate (**366-Ac** and **366′-Ac**) was hydrolyzed asymmetrically to give a mixture of (−)-isopulegol (**366**) and (+)-isopulegyl acetate (**366′-Ac**) (Oritani and Yamashita, 1973c) (Figure 22.74).

22.3.3.5 α-Terpineol

P. pseudomallei strain T was cultivated with α-terpineol (**34**) as the sole carbon source to give 8,9-epoxy-*p*-menthan-1-ol (**58**) via epoxide (**369**) and diepoxide (**57**) as intermediates (Hayashi et al., 1972) (Figure 22.75).

(+)-α-Terpineol (**34**) was formed from (+)-limonene (**34**) by citrus pathogenic *P. digitatum* (Pers.; Fr.) Sacc. KCPYN, which was further biotransformed to *p*-menthane-1β,2α,8-triol (**334**), 2α-hydroxy-1,8-cineole (**125b**), and (+)-*trans*-sobrerol (**95a**) (Noma and Asakawa, 2006a, 2007a) (Figure 22.76). *Penicillium* sp. YuzuYN also biotransformed **34** to **334**. Furthermore, *A. niger* Tiegh CBAYN and *C. roseus* biotransformed **34** to give **95a** and (+)-oleuropeyl alcohol (**204**), respectively (Hamada et al., 2001; Noma and Asakawa, 2006a, 2007a) (Figure 22.76).

OH OH
342, 4*S*-(+) **342′,** 4*R*-(−)

Gibberella cyanea DSM 62719 biotransformed (−)-α-terpineol (**34′**) to give *p*-menthane-1-β,2α,8-triol (**334′**), 2α-hydroxy-1,8-cineole (**125b′**), 1,2-epoxy-α-terpineol (**369′**), (−)-oleuropeyl alcohol (**204′**), (−)-*trans*-sobrerol (**95a′**), and *cis*-sobrerol (**95b′**) (Abraham et al., 1986) (Figure 22.76). In cases of *P. digitatum* (Pers. Fr.) Sacc. KCPYN, *Penicillium* sp. YuzuYN, and *A. niger* Tiegh CBAYN, **34′** was biotransformed to give **369′**, **95a′**, and **334′**, respectively (Noma and Asakawa, 2006a, 2007a) (Figure 22.77). *C. roseus* biotransformed **34′** to give **95a′** and **204′** (Hamada et al., 2001) (Figure 22.77).

22.3.3.6 (−)-Terpinen-4-ol

G. cyanea DSM 62719 biotransformed (*S*)-(−)-terpinen-4-ol (**342**) (1-*p*-menthen-4-ol) to give 2α-hydroxy-1,4-cineole (**132b**), 1-*p*-menthene-4α,6-diol (**372**), and *p*-menthane-1β,2α,4α-triol (**371**) (Abraham et al., 1986). On the other hand, *A. niger* TBUYN-2 also biotransformed (−)-terpinen-4-ol (**342**) to give 2α-hydroxy-1,4-cineole(**132b**) and (+)-*p*-menthane-1β,2α,4α-triol (**371**) (Noma

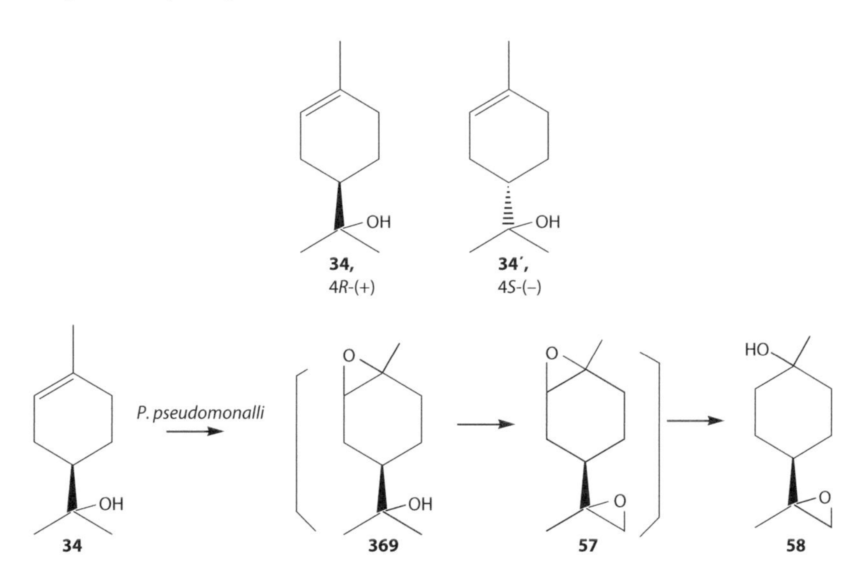

FIGURE 22.75 Biotransformation of (+)-α-terpineol (**34**) to 8,9-epoxy-*p*-menthan-1-ol (**58**) by *Pseudomonas pseudomallei* strain T. (Modified from Hayashi, T. et al. *Biol. Chem.*, 36, 690, 1972.)

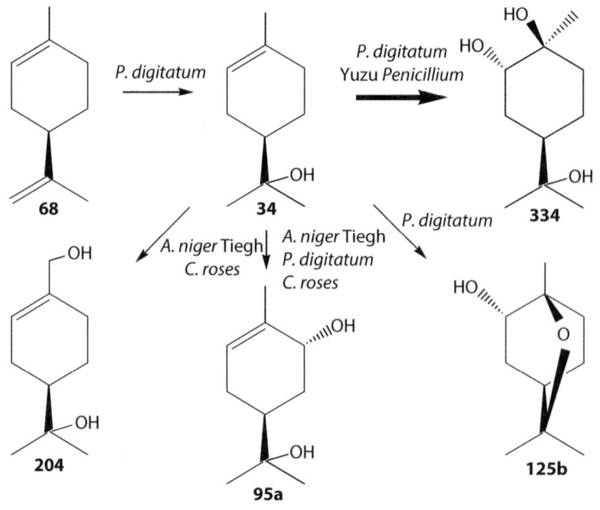

FIGURE 22.76 Biotransformation of (+)-α-terpineol (**34**) by citrus pathogenic fungi, *Penicillium digitatum* (Pers.; Fr.) Sacc. KCPYN, *Penicillium* sp. YuzuYN, and *Aspergillus niger* Tiegh CBAYN. (Modified from Noma, Y. and Asakawa, Y., Biotransformation of (+)-limonene and related compounds by *Citrus* pathogenic fungi, *Proceedings of 50th TEAC*, pp. 431–433, 2006a; Noma, Y. and Y. Asakawa, Biotransformation of limonene and related compounds by newly isolated low temperature grown *citrus* pathogenic fungi and red yeast, *Book of Abstracts of the 38th ISEO*, 2007a, p. 7.)

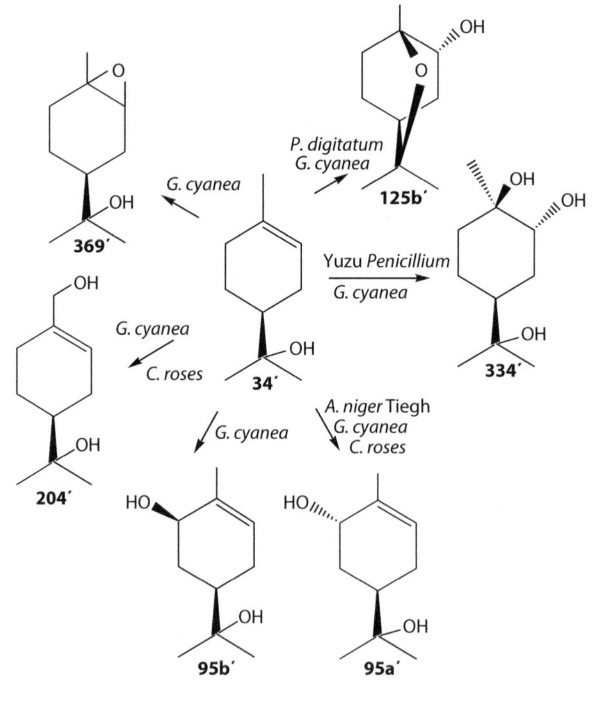

FIGURE 22.77 Biotransformation of (−)-α-terpineol (**34′**) by *Gibberella cyanea* DSM 62719, *Penicillium digitatum* (Pers. Fr.) Sacc. KCPYN, *Penicillium* sp. YuzuYN, and *Aspergillus niger* Tiegh CBAYN. (Modified from Abraham, W.-R. et al. *Appl. Microbiol. Biotechnol.*, 24, 24, 1986; Noma, Y. and Asakawa, Y., Biotransformation of (+)-limonene and related compounds by *Citrus* pathogenic fungi, *Proceedings of 50th TEAC*, 2006a, pp. 431–433; Noma, Y. and Asakawa, Y., Biotransformation of limonene and related compounds by newly isolated low temperature grown *citrus* pathogenic fungi and red yeast, *Book of Abstracts of the 38th ISEO*, 2007a, p. 7.)

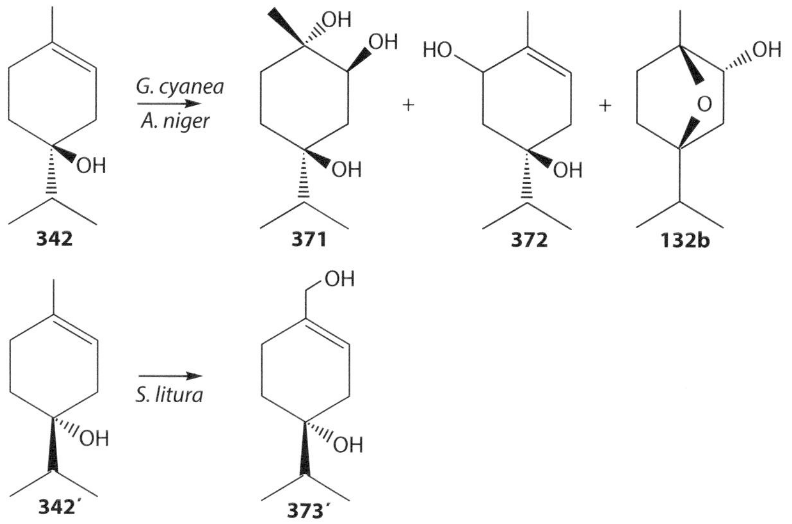

FIGURE 22.78 Biotransformation of (−)-terpinen-4-ol (**342**) by *Gibberella cyanea* DSM 62719, *Aspergillus niger* TBUYN-2, and *Spodoptera litura*. (Modified from Abraham, W.-R. et al. *Appl. Microbiol. Biotechnol.*, 24, 24, 1986; Kumagae, S. and Miyazawa, M., *Proceedings of 43rd TEAC*, 1999, pp. 389–390; Noma, Y. and Asakawa, Y., Microbial transformation of limonene and related compounds, *Proceedings of 51st TEAC*, 2007b, pp. 299–301.)

and Asakawa, 2007b) (Figure 22.78). On the other hand, *S. litura* biotransformed (*R*)-terpinen-4-ol (**342′**) to (4*R*)-*p*-menth-1-en-4,7-diol (**373′**) (Kumagae and Miyazawa, 1999) (Figure 22.78).

22.3.3.7 Thymol and Thymol Methyl Ether

Thymol (**179**) was converted at the concentration of 14% by *Streptomyces humidus*, Tu-1 to give (1*R*,2*S*)- (**181a**) and (1*R*,2*R*)-2-hydroxy-3-*p*-menthen-5-one (**181b**) as the major products (Noma et al., 1988b) (Figure 22.79). On the other hand, in a *Pseudomonas*, thymol (**179**) was biotransformed to 6-hydroxy- (**180**), 7-hydroxy- (**479**), 9-hydroxy- (**480**), 7,9-dihydroxythymol (**482**), thymol-7-oic acid (**481**), and thymol-9-oic acid (**483**) (Chamberlain and Dagley, 1968) (Figure 22.79).

Thymol methyl ether (**459**) was converted by fungi *A. niger*, *Mucor ramannianus*, *Rhizopus arrhizus*, and *Trichothecium roseum* to give 7-hydroxy- (**460**) and 9-hydroxythymol methyl ether (**461**) (Demirci et al., 2001) (Figure 22.79).

22.3.3.8 Carvacrol and Carvacrol Methyl Ether

When cultivated in a liquid medium with carvacrol (**191**), as a sole carbon source, the bacterial isolated from savory and pine consumed the carvacrol in the range of 19%–22% within 5 days of cultivation. The fungal isolates grew much slower and after 13 days of cultivation consumed 7.1%–11.4% carvacrol (**191**). Pure strains belonging to the bacterial genera of bacterium, *Bacillus* and *Pseudomonas*, as well as fungal strain from *Aspergillus*, *Botrytis*, and *Geotrichum* genera, were also tested for their ability to grow in medium containing carvacrol (**191**). Among them, only in *Bacterium* sp. and *Pseudomonas* sp. carvacrol (**191**) uptake was monitored. Both *Pseudomonas* sp. 104 and 107 consumed the substrate in the amount of 19%. These two strains also exhibited the highest cell mass yield and the highest productivity (1.1 and 1.2 g/L/day) (Schwammle et al., 2001).

Carvacrol (**191**) was biotransformed to 3-hydroxy- (**470**), 9-hydroxy (**471**), 7-hydroxy- (**475**), and 8-hydroxycarvacrol (**474**), 8,9-dehydrocarvacrol (**473**), carvacrol-9-oic acid (**472**), carvacrol-7-oic acid (**476**), and 8,9-dihydroxycarvacrol (**477**) by rats (Ausgulen et al., 1987) and microorganisms (Demirci, 2000) including *T. roseum* and *Cladosporium* sp. (Figure 22.80). Furthermore, carvacrol methyl ether (**191-Me**) was converted by the same fungi to give 7-hydroxy- (**475-Ac**) and 9-hydroxycarvacrol methyl ether (**471-Me**) and 7,9-dihydroxycarvacrol methyl ether (**478**) (Demirci, 2000) (Figure 22.80).

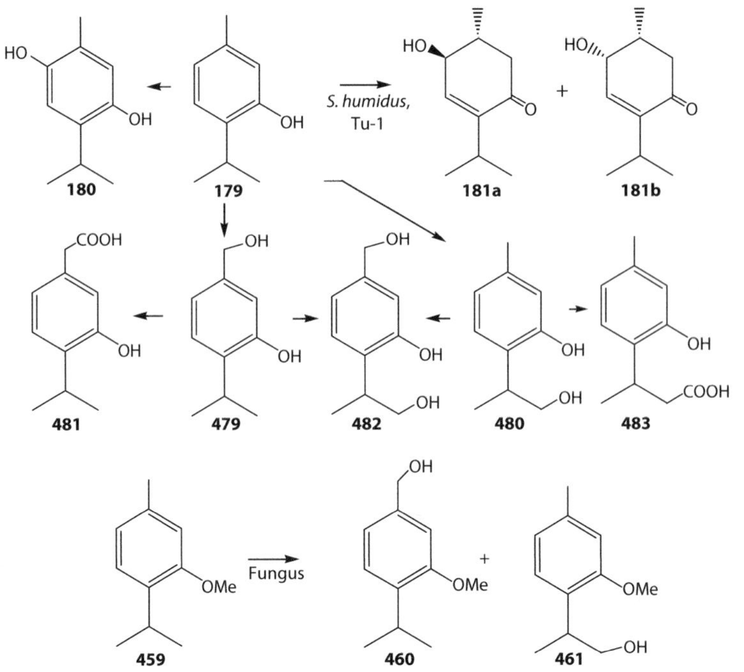

FIGURE 22.79 Biotransformation of thymol (**179**) and thymol methyl ether (**459**) by actinomycetes *Streptomyces humidus*, Tu-1, and fungi *Aspergillus niger*, *Mucor ramannianus*, *Rhizopus arrhizus*, and *Trichothecium roseum*. (Modified from Chamberlain, E.M. and Dagley, S., *Biochem. J.*, 110, 755, 1968; Noma, Y. et al. Microbial transformation of thymol formation of 2-hydroxy-3-*p*-menthen-5-one by *Streptomyces humidus*, Tu-1, *Proceedings of 28th TEAC*, 1988a, pp. 177–179; Demirci, F. et al. The biotransformation of thymol methyl ether by different fungi. *Book of Abstracts of the XII Biotechnology Congr.*, 2001, p. 47.)

22.3.3.9 Carveol

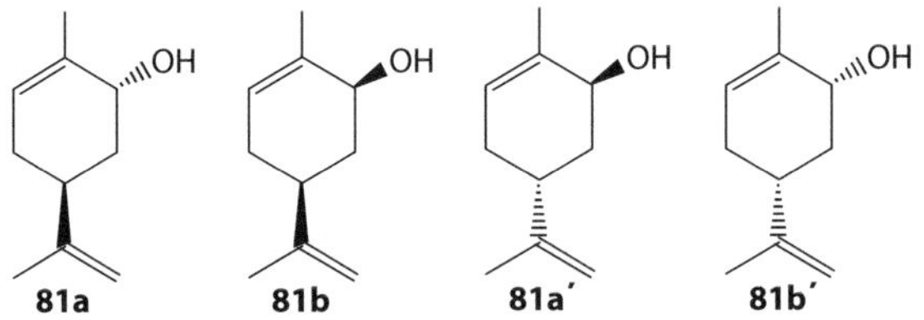

At first, soil Pseudomonad biotransformed (+)-limonene (**68**) to (+)-carvone (**93**) and (+)-1-*p*-menthene-6,9-diol (**90**) via (+)-*cis*-carveol (**81b**) as shown in Figure 22.81 (Dhavalikar and Bhattacharyya, 1966; Dhavalikar et al., 1966).

Second, *Pseudomonas ovalis* strain 6-1 (Noma, 1977) biotransformed the mixture of (−)-*cis*-carveol (**81b′**) and (−)-*trans*-carveol (**81a′**) (94:6, GC ratio) to (−)-carvone (**93'**) (Noma, 1977), which was further metabolized reductively to give (+)-dihydrocarvone (**101a′**), (+)-isodihydrocarvone (**101b′**), (+)-neodihydrocarveol (**102a**), and (−)-dihydrocarveol (**102b**) (Noma et al., 1984). Hydrogenation at C1, 2-position did not occur, but the dehydrogenation at C6-position occurred to give (−)-carvone (**93**) (Figure 22.82).

On the other hand, in *Streptomyces* A-5-1 and *Nocardia* 1-3-11, which were isolated from soil, (−)-carvone (**93′**) was reduced to give mainly (−)-*trans*-carveol (**81a′**) and (−)-*cis*-carveol (**81b′**), respectively. On the other hand, (−)-*trans*-carveol (**81a′**) and (−)-*cis*-carveol (**81b′**) were dehydrogenated to give **93′** by strain 1-3-11 and other microorganisms (Noma et al., 1986). The reaction between *trans*- and *cis*-carveols (**81a′** and **81b′**) and (−)-carvone (**93′**) is reversible (Noma, 1980) (Figure 22.82).

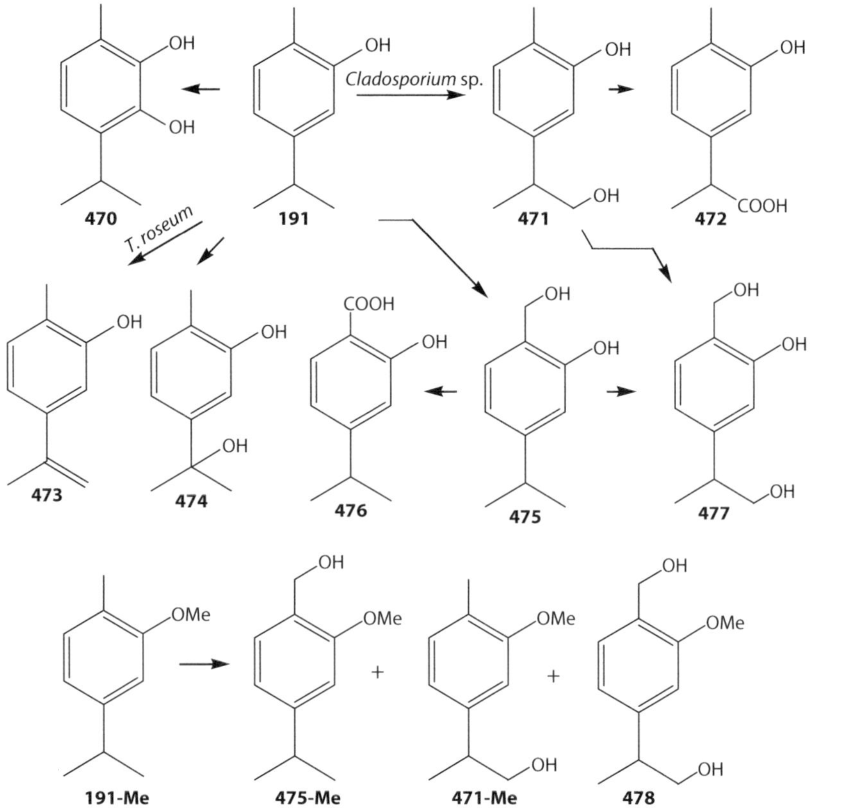

FIGURE 22.80 Biotransformation of carvacrol (**191**) and carvacrol methyl ether (**191-Me**) by rats. (Modified from Ausgulen, L.T. et al. *Pharmacol. Toxicol.*, 61, 98, 1987) and microorganisms (Modified from Demirci, F., *Microbial transformation of bioactive monoterpenes*. Ph.D. thesis, Anadolu University, Eskisehir, Turkey, 2000, pp. 1–137.)

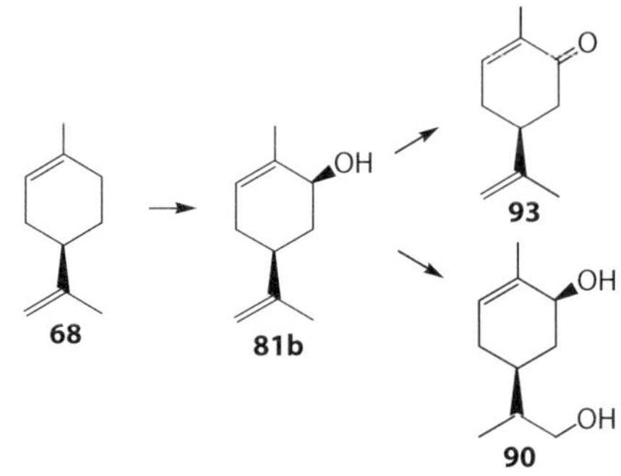

FIGURE 22.81 Proposed metabolic pathway of (+)-limonene (**68**) and (+)-*cis*-carveol (**81b**) by soil Pseudomonad. (Modified from Dhavalikar, R.S. and Bhattacharyya, P.K., *Indian J. Biochem.*, 3, 144, 1966; Dhavalikar, R.S. et al. *Indian J. Biochem.*, 3, 158, 1966.)

Third, the investigation for the biotransformation of the mixture of (−)-*trans*- (**81a′**) and (−)-*cis*-carveol (**81b′**) (60:40 in GC ratio) was carried out by using 81 strains of soil actinomycetes. All actinomycetes produced (−)-carvone (**93′**) from the mixture of (−)-*trans*- (**81a′**) and (−)-*cis*-carveol (**81b′**) (60:40 in GC ratio). However, 41 strains of actinomycetes converted (−)-*cis*-carveol (**81b′**) to give (4*R*,6*R*)-(+)-6,8-oxidomenth-1-en-9-ol (**92a′**), which is named as bottrospicatol after the name

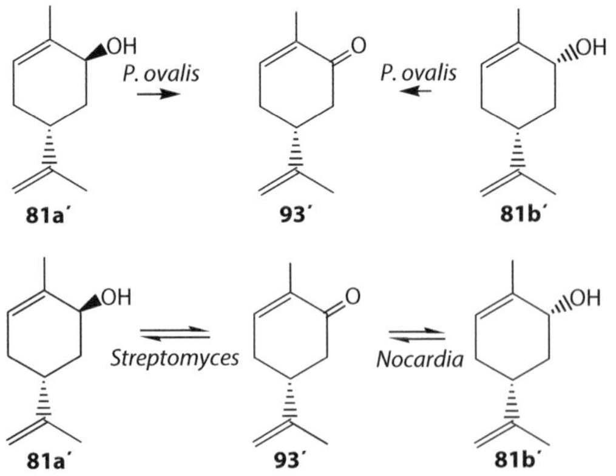

FIGURE 22.82 Biotransformation of (−)-*trans*-(**81a′**) and (−)-*cis*-carveol (**81b′**) (6:94, GC ratio) by *Pseudomonas ovalis* strain 6-1, *Streptomyces* A-5-1, and *Nocardia* 1-3-11. (Modified from Noma, Y., *Nippon Nogeikagaku Kaishi*, 51, 463, 1977; Noma, Y., *Agric. Biol. Chem.*, 44, 807, 1980.)

of the microorganism *Streptomyces bottropensis* [Bottro], and (−)-*cis*-carveol (**81b′**) containing *Mentha spicata* [spicat] and alcohol [ol] (Nishimura et al., 1983a) (Figure 22.83).

(+)-Bottrospicatol (**92a′**) was prepared by epoxidation of (−)-carvone (**93′**) with *m*CPBA to (−)-carvone-8,9-epoxide (**96′**), followed by stereoselective reduction with $NaBH_4$ to alcohol, which was immediately cyclized with 0.1 N H_2SO_4 to give diastereomixture of bottrospicatol (**92a′** and **b′**) (Nishimura et al., 1983a) (Figure 22.84).

Further investigation showed *S. bottropensis* SY-2-1 (Noma and Iwami, 1994) has different metabolic pathways for (−)-*trans*-carveol (**81a′**) and (−)-*cis*-carveol (**81b′**). That is, *S. bottropensis* SY-2-1 converted (−)-*trans*-carveol (**81a′**) to (−)-carvone (**93′**), (−)-carvone-8,9-epoxide (**96′**), (−)-5β-hydroxycarvone (**98a′**), and (+)-5β-hydroxyneodihydrocarveol (**100aa′**) (Figure 22.85). On the other hand, *S. bottropensis* SY-2-1 converted (−)-*cis*-carveol (**81b′**) to give (+)-bottrospicatol (**92a′**) and (−)-5β-hydroxy-*cis*-carveol (**94ba′**) as main products together with (+)-isobottrospicatol (**92b′**) as the minor product as shown in Figure 22.85.

In the metabolism of *cis*-carveol by microorganisms, there are four pathways (pathways 1–4) as shown in Figure 22.86. At first, *cis*-carveol (**81**) is metabolized to carvone (**93**) by C2 dehydrogenation (Noma, 1977, 1980) (pathway 1). Second, *cis*-carveol (**81b**) is metabolized via epoxide as intermediate

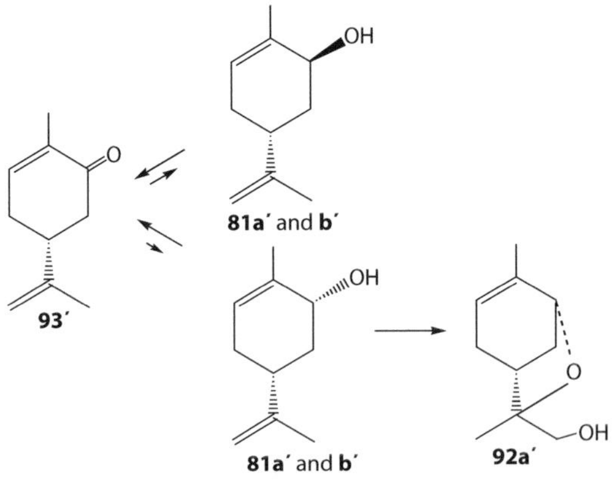

FIGURE 22.83 The metabolic pathways of *cis* carveol (**81b′**) by *Pseudomonas ovalis* strain 6-1. (Modified from Noma, Y., *Nippon Nogeikagaku Kaishi*, 51, 463, 1977) and *Streptomyces bottropensis* SY-2-1 and other microorganisms (Modified from Noma, Y. et al. *Agric. Biol. Chem.*, 46, 2871, 1982; Nishimura, H. et al. Biological activity of bottrospicatol and related compounds produced by microbial transformation of (−)-*cis*-carveol towards plants, *Proceedings of 27th TEAC*, 1983a, pp. 107–109.)

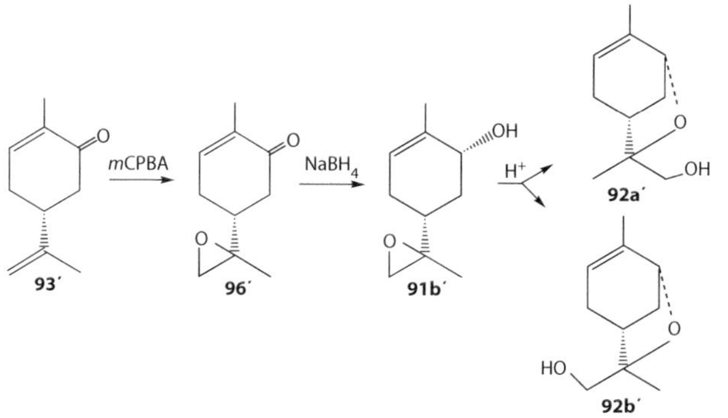

FIGURE 22.84 Preparation of (+)-bottrospicatol (**92a′**) and (+)-isobottrospicatol (**92b′**) from (−)-carvone (**93′**) with *m*CPBA. (Modified from Nishimura, H. and Noma, Y., *Biotechnology for Improved Foods and Flavors*, G.R. Takeoka et al. ACS Symp. Ser. 637, pp. 173–187, 1996. American Chemical Society, Washington, DC.)

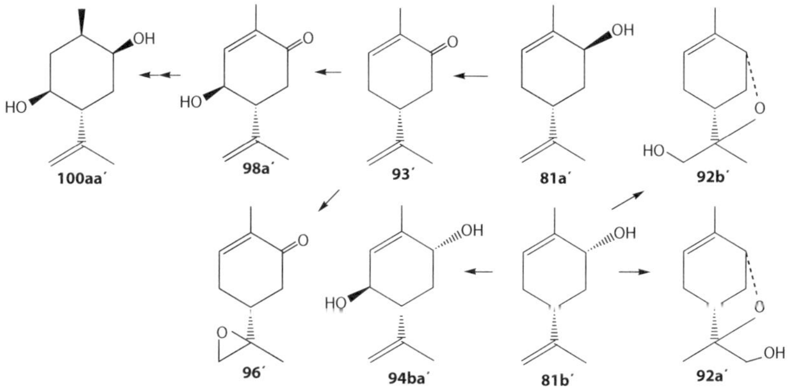

FIGURE 22.85 Biotransformation of (−)-*trans*- (**81a′**) and (−)-*cis*-carveol (**81b′**) by *Streptomyces bottropensis* SY-2-1 and *Streptomyces ikutamanensis* Ya-2-1. (Modified from Noma, Y. et al. *Agric. Biol. Chem.*, 46, 2871, 1982; Noma, Y. and Nishimura, H., Microbiological conversion of carveol. Biotransformation of (−)-*cis*-carveol and (+)-*cis*-carveol by *S. bottropensis*, Sy-2-1, *Proceedings of 28th TEAC*, 1984, pp. 171–173; Noma, Y. and H. Nishimura, *Agric. Biol. Chem.*, 51, 1845, 1987.)

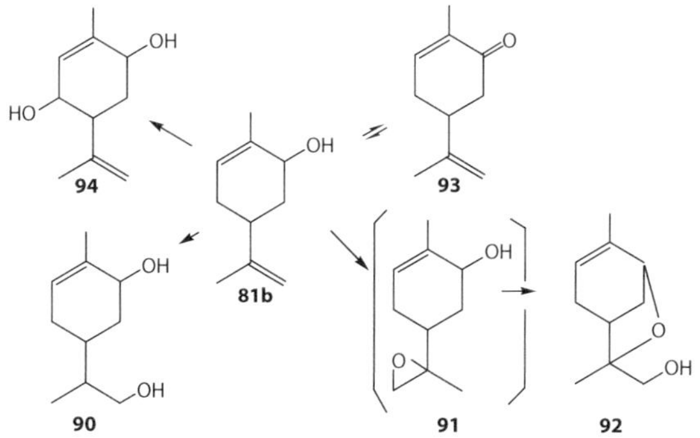

FIGURE 22.86 General metabolic pathways of carveol (**81**) by microorganisms. (Modified from Noma, Y. et al. *Agric. Biol. Chem.*, 46, 2871, 1982; Noma, Y. and Nishimura, H., Microbiological conversion of carveol. Biotransformation of (−)-*cis*-carveol and (+)-*cis*-carveol by *S. bottropensis*, Sy-2-1, *Proceedings of 28th TEAC*, pp. 171–173, 1984; Noma, Y. and Nishimura, H., *Agric. Biol. Chem.*, 51, 1845, 1987; Nishimura, H. and Noma, Y., *Biotechnology for Improved Foods and Flavors*, G.R. Takeoka, et al. ACS Symp. Ser. 637, pp.173–187. American Chemical Society, Washington, DC, 1996.)

TABLE 22.3
Effects of (−)-*cis*- (81b′) and (−)-*trans*-Carveol (81a′) Conversion Products by *Streptomyces bottropensis* SY-2-1 on the Germination of Lettuce Seeds

Compounds	Germination Rate (%) 24 h	48 h
(−)-Carvone (**93′**)	47	89
(+)-Bottrospicatol (**92′**)	3	48
(−)-Carvone-8,9-epoxide (**96′**)	2	77
5β-Hydroxyneodihydrocarveol (**102aa′**)	86	96
5β-Hydroxycarvone (**98a′**)	91	96
Control	95	96

Note: Concentration of each compound was adjusted at 200 ppm.

to bottrospicatol (**92**) by rearrangement at C2 and C8 (Noma et al., 1982; Nishimura et al., 1983a,b; Noma and Nishimura, 1987) (pathway 2). Third, *cis*-carveol (**81b**) is hydroxylated at C5 position to give 5-hydroxy-*cis*-carveol (**94**) (Noma and Nishimura, 1984) (pathway 3). Finally, *cis*-carveol (**81b**) is metabolized to 1-*p*-menthene-2,9-diol (**90**) by hydroxylation at C9 position (Dhavalikar and Bhattacharyya, 1966; Dhavalikar et al., 1966) (pathway 4).

Effects of (−)-*cis*- (**81b′**) and (−)-*trans*-carveol (**81a′**) conversion products by *S. bottropensis* SY-2-1 on the germination of lettuce seeds were examined, and the result is shown in Table 22.3. (+)-Bottrospicatol (**92′**) and (−)-carvone-8,9-epoxide (**96′**) showed strong inhibitory activity for the germination of lettuce seeds.

S. bottropensis SY-2-1 has also different metabolic pathways for (+)-*trans*-carveol (**81a**) and (+)-*cis*-carveol (**81b**) (Noma and Iwami, 1994). That is, *S. bottropensis* SY-2-1 converted (+)-*trans*-carveol (**81a**) to (+)-carvone (**93**), (+)-carvone-8,9-epoxide (**96**), and (+)-5α-hydroxycarvone (**98a**) (Noma and Nishimura, 1982, 1984) (Figure 22.87). On the other hand, *S. bottropensis* SY-2-1 converted (+)-*cis*-carveol (**81b**) to give (−)-isobottrospicatol (**92b**) and (+)-5-hydroxy-*cis*-carveol

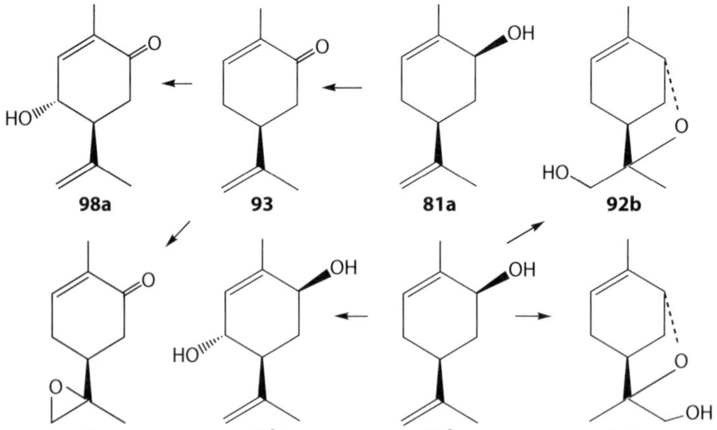

FIGURE 22.87 Metabolic pathways of (+)-*trans*- (**81a**) and (+)-*cis*-carveol (**81b**) by *Streptomyces bottropensis* SY-2-1. (Modified from Noma, Y. and Nishimura, H., *Agric. Biol. Chem.*, 51, 1845, 1987; Nishimura, H. and Noma, Y., *Biotechnology for Improved Foods and Flavors*, G.R. Takeoka, et al. ACS Symp. Ser. 637, pp.173–187. American Chemical Society, Washington, DC, 1996.)

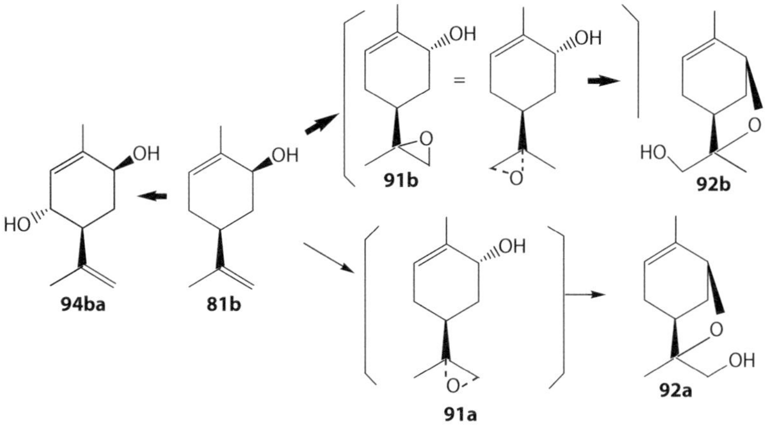

FIGURE 22.88 Metabolic pathways of (+)-*cis*-carveol (**81b**) by *Streptomyces bottropensis* SY-2-1 and *Streptomyces ikutamanensis* Ya-2-1. (Modified from Noma, Y. and Nishimura, H., *Agric. Biol. Chem.*, 51, 1845, 1987; Nishimura, H. and Noma, Y., *Biotechnology for Improved Foods and Flavors*, G.R. Takeoka et al. ACS Symp. Ser. 637, pp. 173–187. American Chemical Society, Washington, DC, 1996.)

(**94b**) as the main products and (−)-bottrospicatol (**92a**) as the minor product as shown in Figure 22.88 (Noma et al., 1980; Noma and Nishimura, 1987; Nishimura and Noma, 1996).

Biological activities of (+)-bottrospicatol (**92a′**) and related compounds for plant's seed germination and root elongation were examined toward barnyard grass, wheat, garden cress, radish, green foxtail, and lettuce (Nishimura and Noma, 1996).

Isomers and derivatives of bottrospicatol were prepared by the procedure shown in Figure 22.89. The chemical structure of each compound was confirmed by the interpretation of spectral data. The effects of all isomers and derivatives on the germination of lettuce seeds were compared. The germination inhibitory activity of (+)-bottrospicatol (**92a′**) was the highest of isomers. Interestingly,

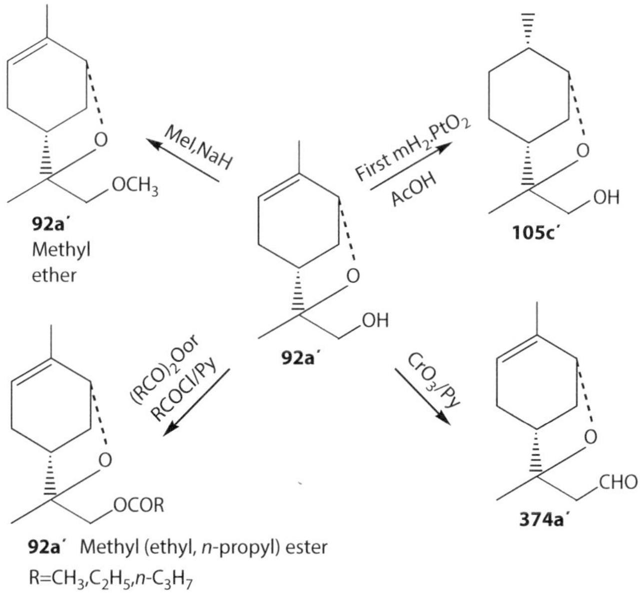

FIGURE 22.89 Preparation of (+)-bottrospicatol (**92a′**) derivatives. (Modified from Nishimura, H. and Noma, Y., *Biotechnology for Improved Foods and Flavors*, G.R. Takeoka et al. ACS Symp. Ser. 637, pp. 173–187. American Chemical Society, Washington, DC, 1996.)

(−)-isobottrospicatol (**92b**) was not effective even in a concentration of 500 ppm. (+)-Bottrospicatol methyl ether (**92a′**-methyl ether) and esters [**92a′**-methyl (ethyl and *n*-propyl) ester] exhibited weak inhibitory activities. The inhibitory activity of (−)-isodihydrobottrospicatol (**105c′**) was as high as that of (+)-bottrospicatol (**92a′**). Furthermore, an oxidized compound, (+)-bottrospicatal (**374a′**), exhibited higher activity than (+)-bottrospicatol (**92a′**). So, the germination inhibitory activity of (+)-bottrospicatal (**374a′**) against several plant seeds, lettuce, green foxtail, radish, garden cress, wheat, and barnyard grass was examined. The result indicates that (+)-bottrospicatal (**374a′**) is a selective germination inhibitor as follows: lettuce > green foxtail > radish > garden cress > wheat > barnyard grass.

Enantio- and diastereoselective biotransformation of *trans*- (**81a** and **81a′**) and *cis*-carveols by *E. gracilis* Z. (Noma and Asakawa, 1992) and *Chlorella pyrenoidosa* IAM C-28 was studied (Noma et al., 1997).

In the biotransformation of racemic *trans*-carveol (**81a** and **81a′**), *C. pyrenoidosa* IAM C-28 showed high enantioselectivity for (−)-*trans*-carveol (**81a′**) to give (−)-carvone (**93′**), while (+)-*trans*-carveol (**81a**) was not converted at all. The same *C. pyrenoidosa* IAM C-28 showed high enantioselectivity for (+)-*cis*-carveol (**81b**) to give (+)-carvone (**93**) in the biotransformation of racemic *cis*-carveol (**81b** and **81b′**). (−)-*cis*-Carveol (**81b′**) was not converted at all. The same phenomenon was observed in the biotransformation of mixture of (−)-*trans*- and (−)-*cis*-carveol (**81a′** and **81b′**) and the mixture of (+)-*trans*- and (+)-*cis*-carveol (**81a** and **81b**) as shown in Figure 22.90. The high enantioselectivity and the high diastereoselectivity for the dehydrogenation of (−)-*trans*- and (+)-*cis*-carveols (**81a** and **81b′**) were shown in *E. gracilis* Z. (Noma and Asakawa, 1992), *C. pyrenoidosa* IAM C-28 (Noma et al., 1997), *N. tabacum*, and other *Chlorella* spp.

On the other hand, the high enantioselectivity for **81a′** was observed in the biotransformation of racemic (+)-*trans*-carveol (**81a**) and (−)-*trans*-carveol (**81a′**) by *Chlorella sorokiniana* SAG to give (−)-carvone (**93′**).

It was considered that the formation of (−)-carvone (**93′**) from (−)-*trans*-carveol (**81a′**) by diastereo- and enantioselective dehydrogenation is a very interesting phenomenon in order to produce mosquito repellent (+)-*p*-menthane-2,8-diol (**50a′**) (Noma, 2007).

(4*R*)-*trans*-Carveol (**81a′**) was converted by *S. litura* to give 1-*p*-menthene-6,8,9-triol (**375**) (Miyazawa et al., 1996b) (Figure 22.91).

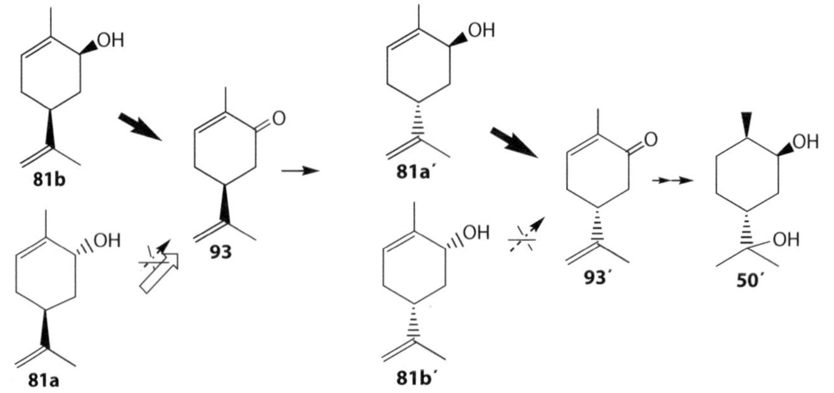

FIGURE 22.90 Enantio- and diastereoselective biotransformation of *trans*- (**81a** and **a′**) and *cis*-carveols (**81b** and **b′**) by *Euglena gracilis* Z and *Chlorella pyrenoidosa* IAM C-28. (Modified from Noma, Y. and Asakawa, Y., *Phytochemistry*, 31, 2009, 1992; Noma, Y. et al. Biotransformation of terpenoids and related compounds by *Chlorella* species, *Proceedings of 41st TEAC*, 1997, pp. 227–229.)

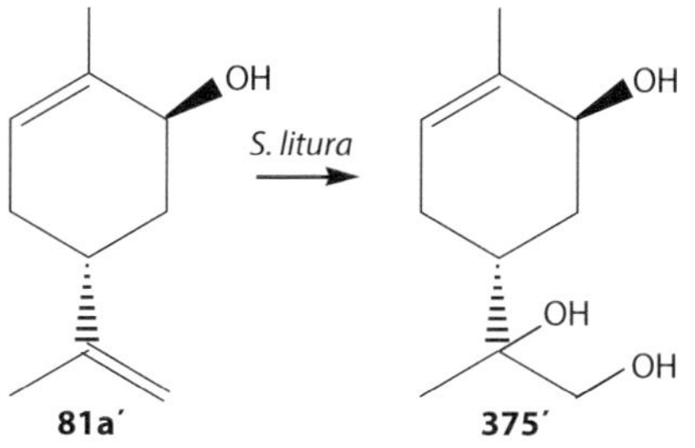

OH
S. litura
OH
OH
OH
81a′
375′

FIGURE 22.91 Biotransformation of (4*R*)-*trans*-carveol (**81a′**) by *Spodoptera litura*. (Modified from Miyazawa, M. et al. Biotransformation of *p*-menthanes using common cutworm larvae, *Spodoptera litura* as a biocatalyst, *Proceedings of 40th TEAC*, 1996b, pp. 80–81.)

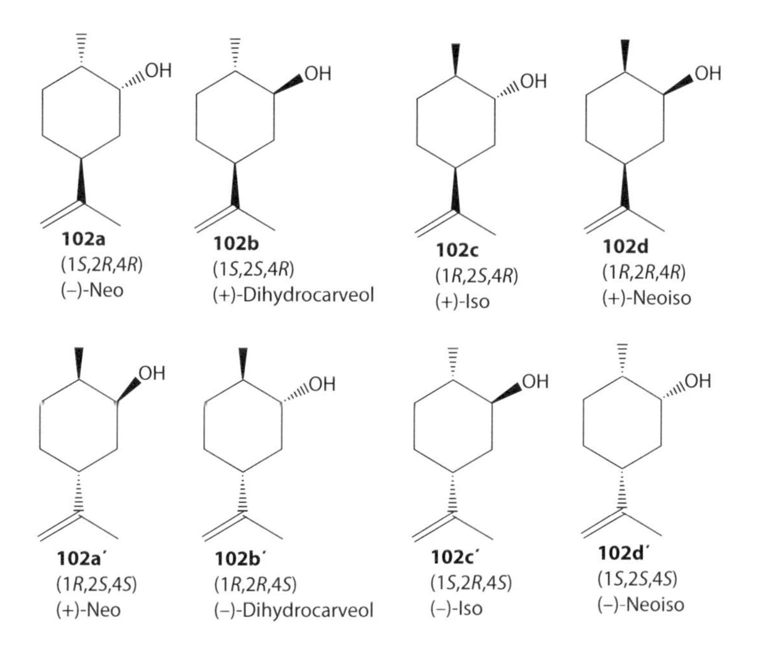

FIGURE 22.92 Chemical structure of eight kinds of dihydrocarveols.

22.3.3.10 Dihydrocarveol

(+)-Neodihydrocarveol (**102a′**) was converted to *p*-menthane-2,8-diol (**50a′**), 8-*p*-menthene-2, 8-diol (**107a′**), and *p*-menthane-2,8,9-triols (**104a′** and **b′**) by *A. niger* TBUYN-2 (Noma et al., 1985a,b; Noma and Asakawa, 1995) (Figures 22.92 and 22.93). In the case of *E. gracilis* Z., mosquito repellent (+)-*p*-menthane-2,8-diol (**50a′**) was formed stereospecifically from (–)-carvone (**93′**) via (+)-dihydrocarvone (**101a′**) and (+)-neodihydrocarveol (**102a′**) (Noma et al., 1993; Noma, 1988, 2007). (–)-Neodihydrocarveol (**102a**) was also easily and stereospecifically converted by *E. gracilis* Z. to give (–)-*p*-menthane-2,8-diol (**50a**) (Noma et al., 1993).

On the other hand, *A. glauca* converted (–)-carvone (**93′**) stereospecifically to give (+)-8-*p*-menthene-2,8-diol (**107a′**) via (+)-dihydrocarvone (**101a′**) and (+)-neodihydrocarveol (**102a′**) (Demirci et al., 2004) (Figure 22.93).

(+)- (**102b**) and (–)-Dihydrocarveol (**102b′**) were converted by 10 kinds of *Aspergillus* spp. to give mainly (+)- (**107b′**) and (–)-10-hydroxydihydrocarveol (**107b**, 8-*p*-menthene-2,10-diol) and (+)-(**50b′**) and (–)-8-hydroxydihydrocarveol (**50b**, *p*-menthane-2,8-diol), respectively (Figure 22.94). The metabolic pattern of dihydrocarveols is shown in Table 22.4.

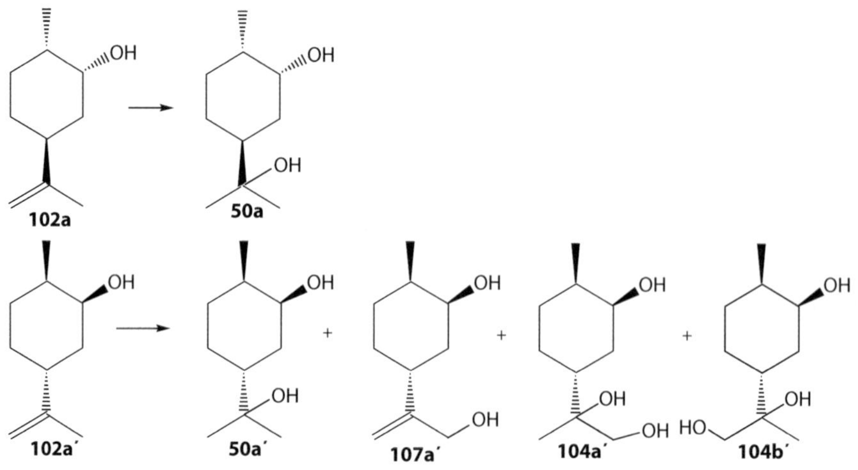

FIGURE 22.93 Biotransformation of (−)- and (+)-neodihydrocarveol (**102a** and **a′**) by *Euglena gracilis* Z, *Aspergillus niger* TBUYN-2, and *Absidia glauca*. (Modified from Noma, Y. et al. *Annual Meeting of Agricultural and Biological Chemistry*, Sapporo, 1985a, p. 68; Noma, Y. et al. Biotransformation of carvone. 6. Biotransformation of (−)-carvone and (+)-carvone by a strain of *Aspergillus niger*, *Proceedings of 29th TEAC*, 1985b, pp. 235–237; Noma, Y. et al. Formation of 8 kinds of *p*-menthane-2,8-diols from carvone and related compounds by *Euglena gracilis* Z. Biotransformation of monoterpenes by photosynthetic microorganisms. Part VIII, *Proceedings of 37th TEAC*, 1993, pp. 23–25; Noma, Y., *Aromatic Plants from Asia their Chemistry and Application in Food and Therapy*, L. Jiarovetz, N.X. Dung, and V.K. Varshney, pp. 169–186, Har Krishan Bhalla & Sons, India, 2007; Noma, Y. and Asakawa, Y., *Biotechnology in Agriculture and Forestry, Vol. 33. Medicinal and Aromatic Plants VIII*, Y.P.S. Bajaj, ed., pp. 62–96, Springer, Berlin, Germany, 1995; Demirci, F. et al. *Naturforsch*., 59c, 389, 2004.)

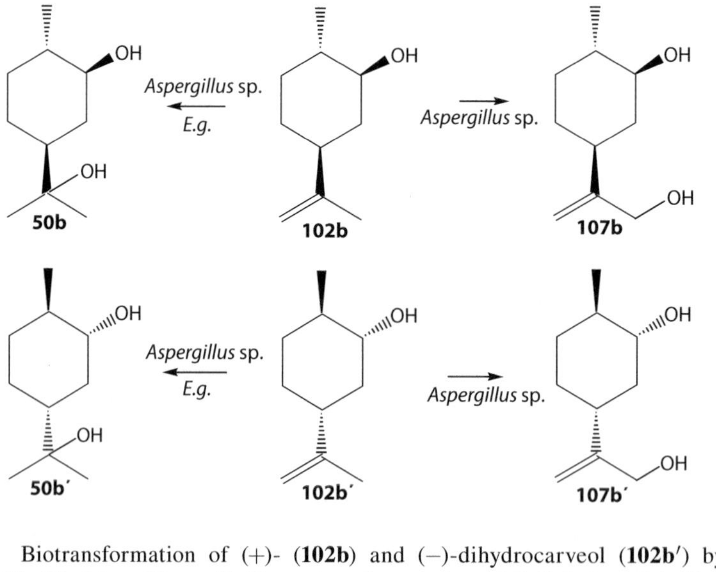

FIGURE 22.94 Biotransformation of (+)- (**102b**) and (−)-dihydrocarveol (**102b′**) by 10 kinds of *Aspergillus* spp. (Modified from Noma, Y., Formation of *p*-menthane-2,8-diols from (−)-dihydrocarveol and (+)-dihydrocarveol by *Aspergillus* spp. *The Meeting of Kansai Division of The Agricultural* and *Chemical Society of Japan*, Kagawa, 1988, p. 28) and *Euglena gracilis* Z (Modified from Noma, Y. et al. Formation of 8 kinds of *p*-menthane-2,8-diols from carvone and related compounds by *Euglena gracilis* Z. Biotransformation of monoterpenes by photosynthetic microorganisms. Part VIII, *Proceedings of 37th TEAC*, 1993, pp. 23–25.)

TABLE 22.4
Metabolic Pattern of Dihydrocarveols (102b and 102b′) by 10 Kinds of *Aspergillus* spp.

	Compounds					
Microorganisms	**107b′**	**50b′**	**C.r. (%)**	**107b**	**50b**	**C.r. (%)**
A. awamori, IFO 4033	0	98	99	3	81	94
A. fumigatus, IFO 4400	0	14	34	+	6	14
A. sojae, IFO 4389	0	47	59	1	50	85
A. usami, IFO 4338	0	32	52	+	5	7
A. cellulosae, M-77	0	27	52	+	7	14
A. cellulosae, IFO 4040	0	30	55	1	5	8
A. terreus, IFO 6123	0	79	92	+	18	46
A. niger, IFO 4034	0	29	49	+	8	12
A. niger, IFO 4049	4	50	67	9	34	59
A. niger, TBUYN-2	29	68	100	30	53	100

Note: C.r.—conversion ratio.

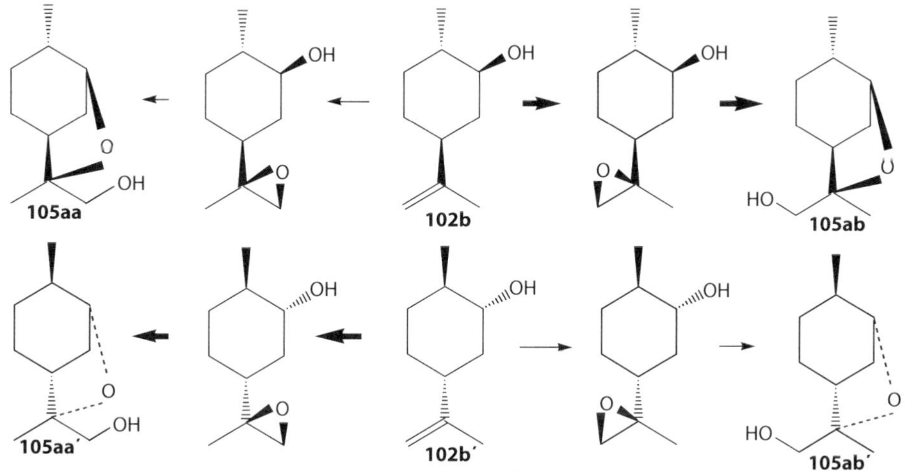

FIGURE 22.95 Biotransformation of (+)- (**102b**) and (−)-dihydrocarveol (**102b′**) by *Streptomyces bottropensis* SY-2-1. (Modified from Noma, Y., *Kagaku to Seibutsu*, 22, 742, 1984.)

In the case of the biotransformation of *S. bottropensis* SY-2-1, (+)-dihydrocarveol (**102b**) was converted to (+)-dihydrobottrospicatol (**105aa**) and (+)-dihydroisobottrospicatol (**105ab**), whereas (−)-dihydrocarveol (**102b′**) was metabolized to (−)-dihydrobottrospicatol (**105aa′**) and (−)-dihydroisobottrospicatol (**105ab′**). (+)-Dihydroisobottrospicatol (**105ab**) and (−)-dihydrobottrospicatol (**105aa′**) are the major products (Noma, 1984) (Figure 22.95).

E. gracilis Z. converted (−)-iso- (**102c**) and (+)-isodihydrocarveol (**102c′**) to give the corresponding 8-hydroxyisodihydrocarveols (**50c** and **50c′**), respectively (Noma et al., 1993) (Figure 22.96).

In the case of the biotransformation of *S. bottropensis* SY-2-1, (−)-neoisodihydrocarveol (**102d**) was converted to (+)-isodihydrobottrospicatol (**105ba**) and (+)-isodihydroisobottrospicatol (**105bb**), whereas (+)-neoisodihydrocarveol (**102d′**) was metabolized to (−)-isodihydrobottrospicatol (**105ba′**) and (−)-isodihydroisobottrospicatol (**105bb′**). (+)-Isodihydroisobottrospicatol (**105bb**) and (−)-isodihydrobottrospicatol (**105ba′**) are the major products (Noma, 1984) (Figure 22.97).

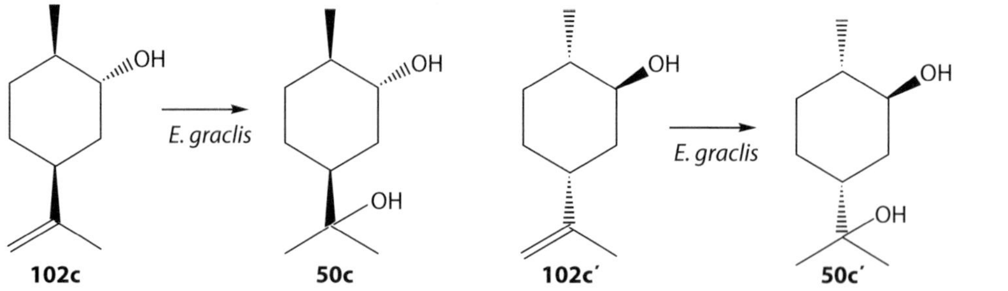

FIGURE 22.96 Biotransformation of (+)-iso- (**102c**) and (−)-dihydrocarveol (**102c′**) by *Euglena gracilis* Z. (Modified from Noma, Y. et al. Formation of 8 kinds of *p*-menthane-2,8-diols from carvone and related compounds by *Euglena gracilis* Z. Biotransformation of monoterpenes by photosynthetic microorganisms. Part VIII, *Proceedings of 37th TEAC*, 1993, pp. 23–25.)

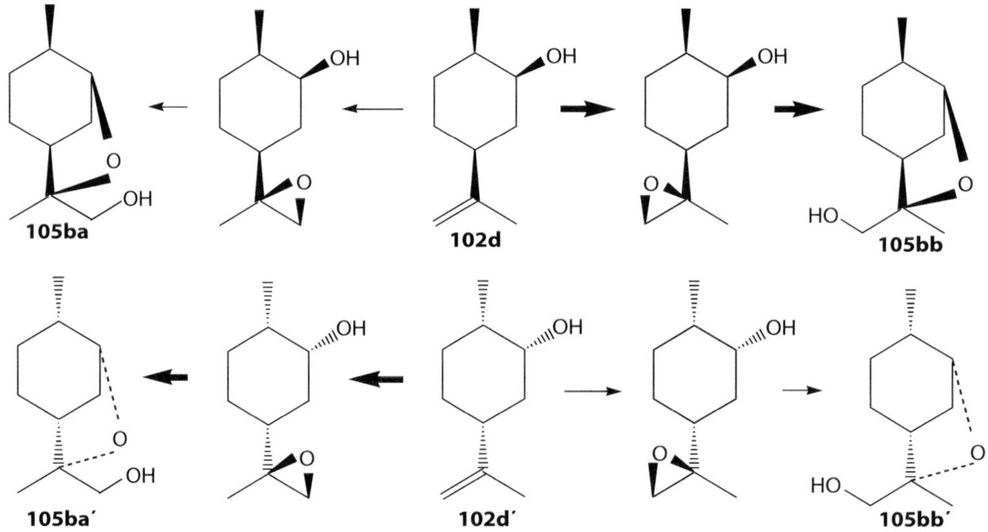

FIGURE 22.97 Formation of dihydroisobottrospicatols (**105**) from neoisodihydrocarveol (**102d** and **d′**) by *Streptomyces bottropensis* SY-2-1. (Modified from Noma, Y., *Kagaku to Seibutsu*, 22, 742, 1984.)

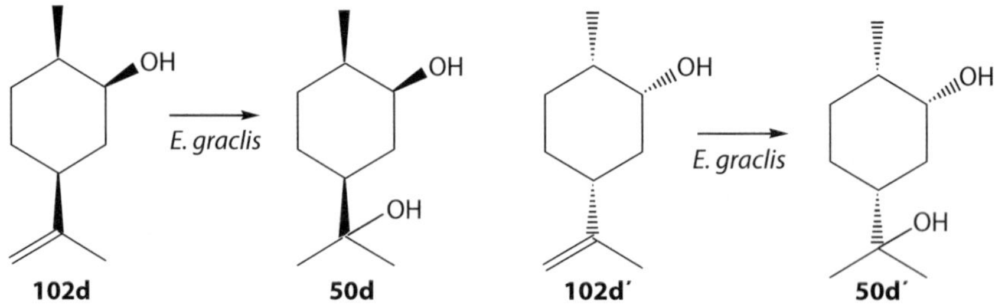

FIGURE 22.98 Biotransformation of (+)- (**102c**) and (−)-neoisodihydrocarveol (**102c′**) by *Euglena gracilis* Z. (Modified from Noma, Y. et al. Formation of 8 kinds of *p*-menthane-2,8-diols from carvone and related compounds by *Euglena gracilis* Z. Biotransformation of monoterpenes by photosynthetic microorganisms. Part VIII, *Proceedings of 37th TEAC*, 1993, pp. 23–25.)

E. gracilis Z. converted (−)- (**102d**) and (+)-neoisodihydrocarveol (**102d′**) to give the corresponding 8-hydroxyneoisodihydrocarveols (**50d** and **50d′**), respectively (Noma et al., 1993) (Figure 22.98).

Eight kinds of 8-hydroxydihydrocarveols (**50a–d** and **50a′–d′**; 8-*p*-menthane-2,8-diols) were obtained from carvone (**93** and **93′**), dihydrocarvones (**101a–b** and **101a′–b′**), and dihydrocarveols (**102a–d**, **102a′–d′**) by *E. gracilis* Z. as shown in Figure 22.99 (Noma et al., 1993).

22.3.3.11 Piperitenol

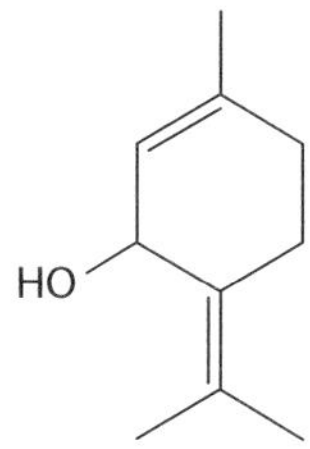

HO

458

Incubation of piperitenol (**458**) with *A. niger* gave a complex metabolites whose structures have not yet been determined (Noma, 2000).

22.3.3.12 Isopiperitenol

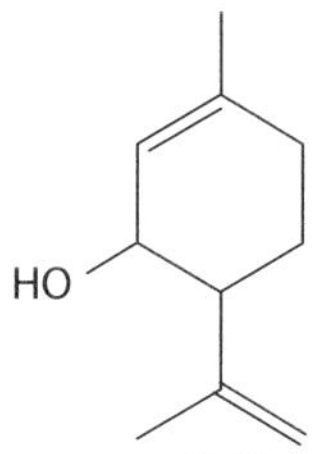

HO

110

Piperitenol (**458**) was metabolized by *A. niger* to give a complex alcohol mixtures whose structures have not yet been determined (Noma, 2000).

22.3.3.13 Perillyl Alcohol

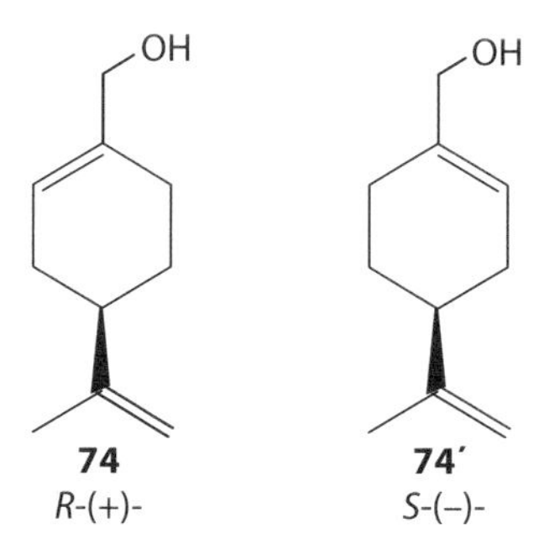

(−)-Perillyl alcohol (**74′**) was epoxidized by *S. ikutamanensis* Ya-2-1 to give 8,9-epoxy-(−)-perillyl alcohol (**77′**) (Noma et al., 1986) (Figure 22.100).

(−)-Perillyl alcohol (**74′**) was glycosylated by *E. perriniana* suspension cells to (−)-perillyl alcohol monoglucoside (**376′**) and diglucoside (**377′**) (Hamada et al., 2002; Yonemoto et al., 2005) (Figure 22.101).

Furthermore, 1-perillyl-β-glucopyranoside (**376**) was converted into the corresponding oligosaccharides (**377–381**) using a cyclodextrin glucanotransferase (Yonemoto et al., 2005) (Figure 22.102).

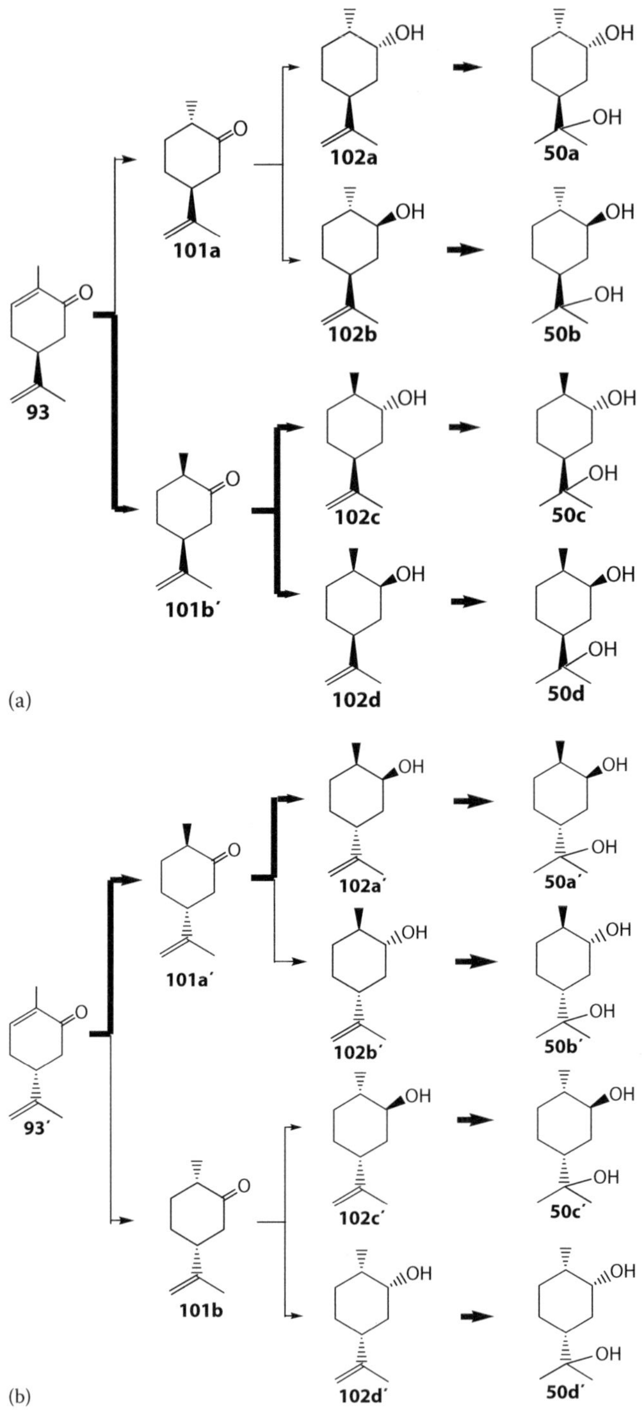

FIGURE 22.99 Formation of eight kinds of 8-hydroxydihydrocarveols (**50a–50d**, **50a′–50d′**), dihydrocarvones (**101a–101b** and **101a′–101b′**), and dihydrocarveols (**102a–102d** and **102a′–102d′**) from (+)- (**93**) and (–)-carvone (**93′**) by *Euglena gracilis* Z.: (a) denotes 8-hydroxydihydrocarveols and (b) denotes dihydrocarveols. (Modified from Noma, Y. et al. Formation of 8 kinds of *p*-menthane-2,8-diols from carvone and related compounds by *Euglena gracilis* Z. Biotransformation of monoterpenes by photosynthetic microorganisms. Part VIII, *Proceedings of 37th TEAC*, 1993, pp. 23–25.)

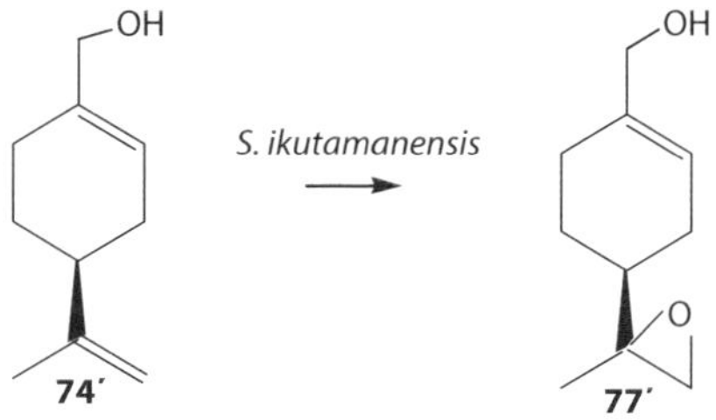

S. ikutamanensis

74′ 77′

FIGURE 22.100 Biotransformation of (−)-perillyl alcohol (**74′**) by *Streptomyces ikutamanensis* Ya-2-1. (Modified from Noma, Y. et al. Reduction of terpene aldehydes and epoxidation of terpene alcohols by *S. ikutamanensis*, Ya-2-1, *Proceedings of 30th TEAC*, 1986, pp. 204–206.)

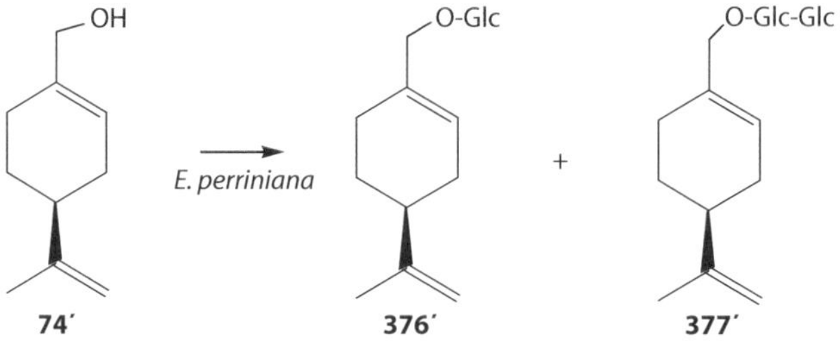

E. perriniana

74′ 376′ 377′

FIGURE 22.101 Biotransformation of (−)-perillyl alcohol (**74′**) by *Eucalyptus perriniana* suspension cell. (Modified from Hamada, H. et al. Glycosylation of monoterpenes by plant suspension cells, *Proceedings of 46th TEAC*, 2002, pp. 321–322; Yonemoto, N. et al. Preparation of (−)-perillyl alcohol oligosaccharides, *Proceedings of 49th TEAC*, 2005, pp. 108–110.)

22.3.3.14 Carvomenthol

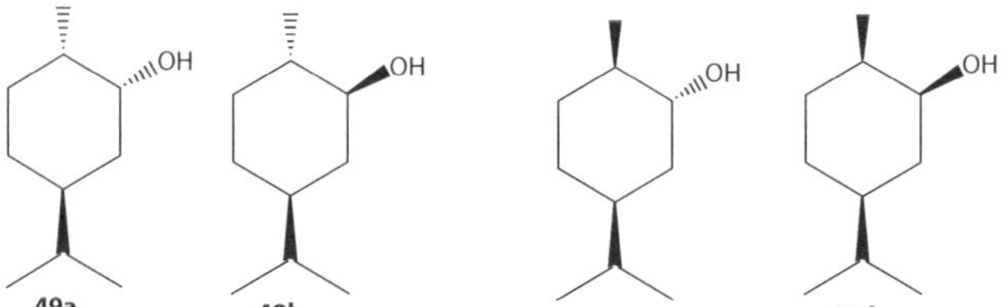

49a (1S,2R,4R) (−)-Neo
49b (1S,2S,4R) (+)-Carvomenthol
49c (1R,2S,4R) (−)-Iso
49d (1R,2R,4R) (−)-Neoiso

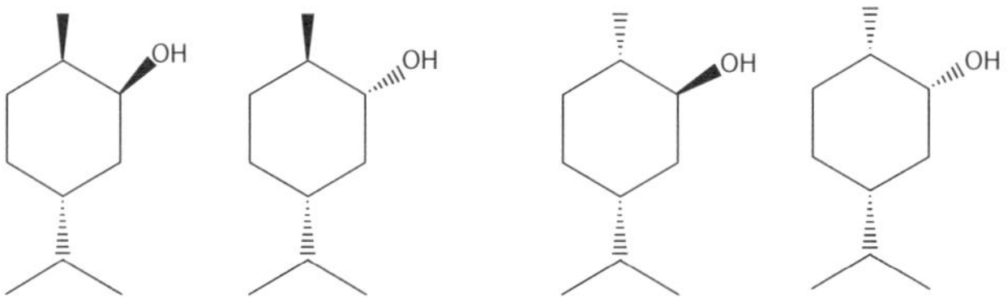

49a′ (1R,2S,4S) (+)-Neo
49b′ (1R,2R,4S) (−)-Carvomenthol
49c′ (1S,2R,4S) (+)-Iso
49d′ (1S,2S,4S) (+)-Neoiso

(+)-Iso- (**49c**) and (+)-neoisocarvomenthol (**49d**) were formed from (+)-carvotanacetone (**47**) via (−)-isocarvomenthone (**48b**) by *P. ovalis* strain 6-1, whereas (+)-neocarvomenthol (**49a′**) and (−)-carvomenthol (**49b′**) were formed from (−)-carvotanacetone (**47′**) via (+)-carvomenthone (**48a′**) by the same bacteria, of which **48b**, **48a′**, and **49d** were the major products (Noma et al., 1974a) (Figure 22.103).

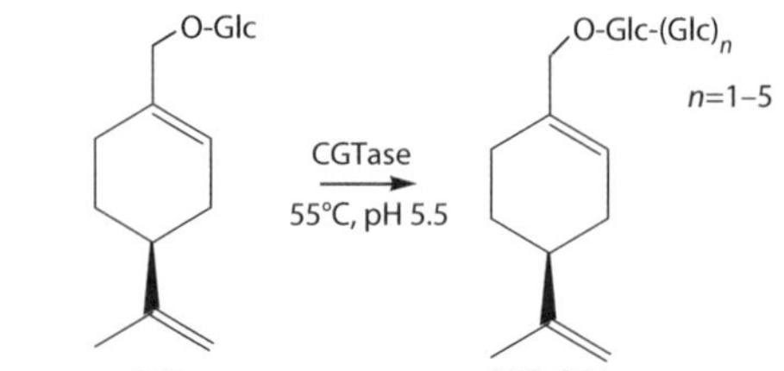

FIGURE 22.102 Biotransformation of (−)-perillyl alcohol monoglucoside (**376**) by CGTase. (Modified from Yonemoto, N. et al. Preparation of (−)-perillyl alcohol oligosaccharides, *Proceedings of 49th TEAC*, pp. 108–110, 2005.)

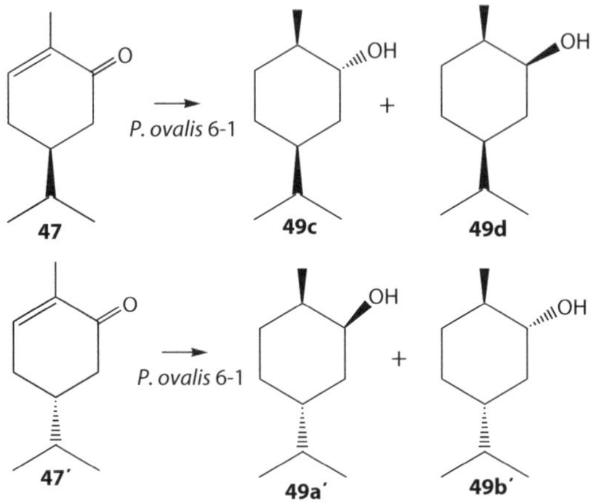

FIGURE 22.103 Formation of (−)-iso- (**49c**), (−)-neoiso- (**49d**), (+)-neo- (**49a′**), and (−)-carvomenthol (**49b′**) from (+)- (**47**) and (−)-carvotanacetone (**47′**) by *Pseudomonas ovalis* strain 6-1. (Modified from Noma, Y. et al. *Agric. Biol. Chem.*, 38, 1637, 1974a.)

Microbial resolution of carvomenthols was carried out by selected microorganisms such as *Trichoderma S* and *B. subtilis* var. *niger* (Oritani and Yamashita, 1973d). Racemic carvomenthyl acetate, racemic isocarvomenthyl acetate, and racemic neoisocarvomenthyl acetate were asymmetrically hydrolyzed to (−)-carvomenthol (**49b′**) with (+)-carvomenthyl acetate, (−)-isocarvomenthol (**49c**) with (+)-isocarvomenthyl acetate, and (+)-neoisocarvomenthol (**49d′**) with (−)-neoisocarvomenthyl acetate, respectively; racemic neocarvomenthyl acetate was not hydrolyzed (Oritani and Yamashita, 1973d) (Figure 22.104).

22.3.4 Monocyclic Monoterpene Ketone

22.3.4.1 α, β-Unsaturated Ketone

22.3.4.1.1 Carvone

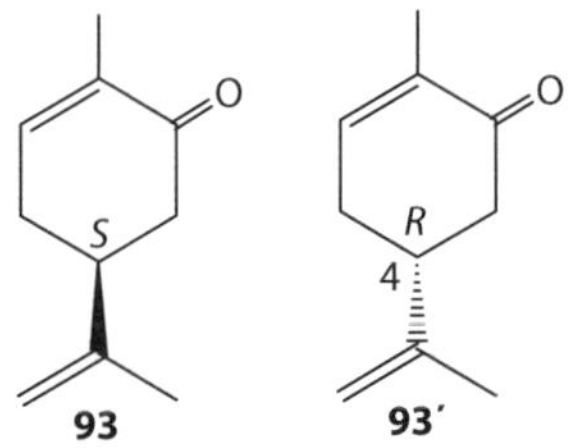

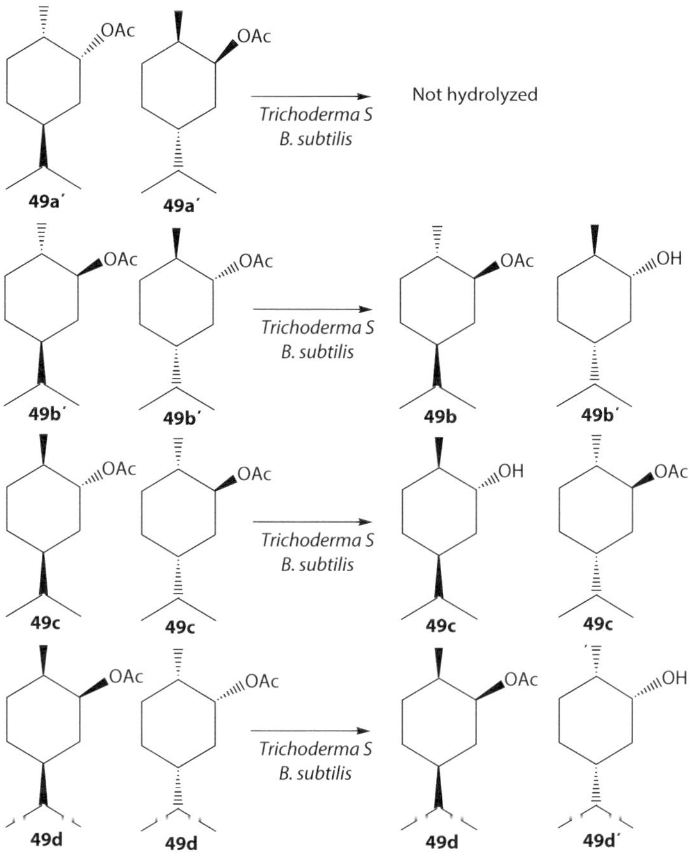

Not hydrolyzed

Trichoderma S
B. subtilis

49a′ 49a′

49b′ 49b′ 49b 49b′

49c 49c 49c 49c

49d 49d 49d 49d′

Racemic acetates

FIGURE 22.104 Microbial resolution of carvomenthols by *Trichoderma S* and *Bacillus subtilis* var. *niger*. (Modified from Oritani, T. and Yamashita, K., *Agric. Biol. Chem.*, 37, 1691, 1973d.)

Carvone occurs as (+)-carvone (**93**), (−)-carvone (**93′**), or racemic carvone. (*S*)-(+)-Carvone (**93**) is the main component of caraway oil (ca. 60%) and dill oil and has a herbaceous odor reminiscent of caraway and dill seeds. (*R*)-(−)-Carvone (**93′**) occurs in spearmint oil at a concentration of 70%–80% and has a herbaceous odor similar to spearmint (Bauer et al., 1990).

The distribution of carvone convertible microorganisms is summarized in Table 22.5. When ethanol was used as a carbon source, 40% of bacteria converted (+)- (**93**) and (−)-carvone (**93'**). On the other hand, when glucose was used, 65% of bacteria converted carvone. In the case of yeasts, 75% converted (+)- (**93**) and (−)-carvone (**93′**). Of fungi, 90% and 85% of fungi converted **93** and **93′**, respectively. In actinomycetes, 56% and 90% converted **93** and **93′**, respectively.

Many microorganisms except for some strains of actinomycetes were capable of hydrogenating the C=C double bond at C-1, 2 position of (+)- (**93**) and (−)-carvone (**93′**) to give mainly (−)-isodihydrocarvone (**101b**) and (+)-dihydrocarvone (**101a′**), respectively (Noma and Tatsumi, 1973; Noma et al., 1974b; Noma and Nonomura, 1974; Noma, 1976, 1977) (Figure 22.105) (Tables 22.6 and 22.7).

Furthermore, it was found that (−)-carvone (**93′**) was converted via (+)-isodihydrocarvone (**101b′**) to (+)-isodihydrocarveol (**102c′**) and (+)-neoisodihydrocarveol (**102d′**) by some strains of actinomycetes (Noma, 1979a,b). (−)-Isodihydrocarvone (**101b**) was epimerized to (−)-dihydrocarvone (**101a**) after the formation of (−)-isodihydrocarvone (**101b**) from (+)-carvone (**93**) by the growing cells, the resting cells, and the cell-free extracts of *Pseudomonas fragi* IFO 3458 (Noma et al., 1975).

TABLE 22.5
Distribution of (+)- (93) and (−)-Carvone (93′) Convertible Microorganisms

Microorganisms	Number of Microorganisms Used	Numbers of Carvone Convertible Microorganisms	Ratio (%)
Bacteria	40	16 (ethanol, **93**)	40
		16 (ethanol, **93′**)	40
		26 (glucose, **93**)	65
		26 (glucose, **93′**)	65
Yeasts	68	51 (**93**)	75
		51 (**93′**)	75
Fungi	40	34 (**93**)	85
		36 (**93′**)	90
Actinomycetes	48	27 (**93**)	56
		43 (**93′**)	90

Source: Noma, Y. et al. Formation of 8 kinds of *p*-menthane-2,8-diols from carvone and related compounds by *Euglena gracilis* Z. Biotransformation of monoterpenes by photosynthetic microorganisms. Part VIII, *Proceedings of 37th TEAC*, 1993, pp. 23–25.

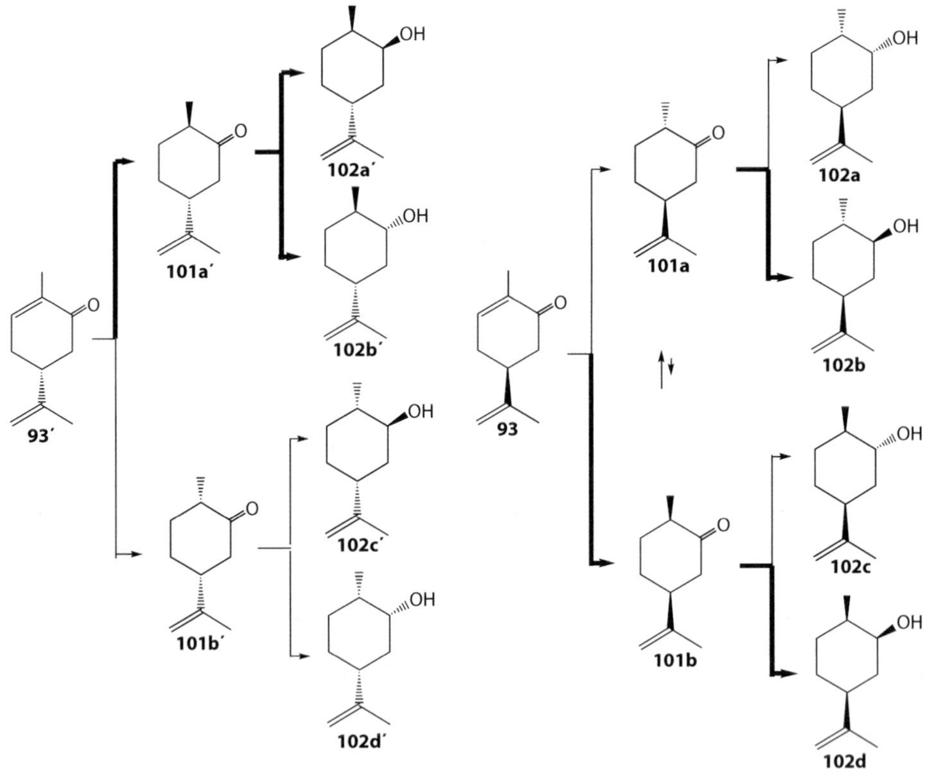

FIGURE 22.105 Biotransformation of (+)- (**93**) and (−)-carvone (**93′**) by various kinds of microorganisms. (Modified from Noma, Y. and Tatsumi, C., *Nippon Nogeikagaku Kaishi*, 47, 705, 1973; Noma, Y. et al. *Agric. Biol. Chem.*, 38, 735, 1974b; Noma, Y. et al. Microbial transformation of carvone, *Proceedings of 18th TEAC*, pp. 20–23, 1974c; Noma, Y. and Nonomura, S., *Agric. Biol. Chem.*, 38, 741, 1974; Noma, Y., *Ann. Res. Stud. Osaka Joshigakuen Junior College*, 20, 33, 1976; Noma, Y., *Nippon Nogeikagaku Kaishi*, 51, 463, 1977.)

TABLE 22.6
Ratio of Microorganisms That Carried Out the Hydrogenation of C = C Double Bond of Carvone by *Si* Plane Attack toward Microorganisms That Converted Carvone

Microorganisms	Ratio (%)
Bacteria	100[a]
	96[b]
Yeasts	74
Fungi	80
Actinomycetes	39

[a] When ethanol was used.
[b] When glucose was used.

TABLE 22.7
Summary of Microbial and Chemical Hydrogenation of (−)-Carvone (93′) for the Formation of (+)-Dihydrocarvone (101a′) and (+)-Isodihydrocarvone (101b′)

	Compounds	
Microorganisms	**101a′**	**101b′**
Amorphosporangium auranticolor	100	0
Microbispora rosea IFO 3559	86	0
Bacillus subtilis var. *niger*	85	13
Bacillus subtilis IFO 3007	67	11
Pseudomonas polycolor IFO 3918	75	15
Pseudomonas graveolens IFO 3460	74	17
Arthrobacter pascens IFO 121139	73	12
Pichia membranifaciens IFO 0128	70	16
Saccharomyces ludwigii IFO 1043	69	18
Alcaligenes faecalis IAM B-141-1	70	13
Zn-25% KOH–EtOH	73	27
Raney-10% NaOH	71	19

Source: Noma, Y., *Ann. Res. Stud. Osaka Joshigakuen Junior College*, 20, 33, 1976.

Consequently, the metabolic pathways of carvone by microorganisms were summarized as the following eight groups (Figure 22.105):

Group 1: (−)-Carvone (**93′**)- (+)-dihydrocarvone (**101a′**)-(+)-neodihydrocarveol (**102a′**)
Group 2: **93′**–**101a′**-(−)-Dihydrocarveol (**102b′**)
Group 3: **93′**–**101a′**-**102a′** and **102b′**
Group 4: **93′**-(+)-Isodihydrocarvone (**101b′**)–**102c′** and **102d′**
Group 5: (+)-Carvone (**93**)-(−)-isodihydrocarvone (**101b**)-(−)-neoisodihydrocarveol (**102d**)
Group 6: **93**–**101b**–**102c**
Group 7: **93**–**101b**–**102c** and **102d**
Group 8: **93**–**101b**–**101a**

The result of the mode action of both the hydrogenation of carvone and the reduction for dihydrocarvone by microorganism is as follows. In bacteria, only two strains were able to convert (−)-carvone (**93′**) via (+)-dihydrocarvone (**101a′**) to (−)-dihydrocarveol (**102b′**) as the major product (group 3, when ethanol was used as a carbon source, 12.5% of (−)-carvone (**93′**) convertible microorganisms belonged to this group, and when glucose was used, 8% belonged to this group) (Noma and Tatsumi, 1973; Noma et al., 1975), whereas when (+)-carvone (**93**) was converted, one strain converted it to a mixture of (−)-isodihydrocarveol (**102c**) and (−)-neoisodihydrocarveol (**102d**) (group 7%, 6% and 4% of **93** convertible bacteria belonged to this group, when ethanol and glucose were used, respectively), and four strains converted it via (−)-isodihydrocarvone (**101b**) to (−)-dihydrocarvone (**101a**) (group 8%, 6% and 15% of (+)-carvone (**93′**) convertible bacteria belonged to this group, when ethanol and glucose were used, respectively.) (Noma et al., 1975). In yeasts, 43% of carvone convertible yeasts belong to group 1%, 14% to group 2%, and 33% to group 3 (of this group, three strains are close to group 1) and 12% to group 5%, 4% to group 6%, and 27% to group 7 (of this group, three strains are close to group 5 and one strain is close to group 6). In fungi, 51% of fungi metabolized (−)-carvone (**93′**) by way of group 1% and 3% via group 3, but there was no strain capable of metabolizing (−)-carvone (**93′**) via group 2, whereas 20% of fungi metabolized (+)-carvone (**93**) via group 5% and 29% via group 7, but there was no strain capable of metabolizing (+)-carvone (**93**) via group 6. In actinomycetes, (−)-carvone (**93′**) was converted to dihydrocarveols via group 1 (49%), group 2 (0%), group 3 (9%), and group 4 (28%), whereas (+)-carvone (**93**) was converted to dihydrocarveols via group 5 (7%), group 6 (0%), group 7 (19%), and group 8 (0%).

Furthermore, (+)-neodihydrocarveol (**102a′**) stereospecifically formed from (−)-carvone (**93′**) by *A. niger* TBUYN-2 was further biotransformed to mosquito repellent (1*R*,2*S*,4*R*)-(+)-*p*-menthane-2,8-diol (**50a′**), (1*R*,2*S*,4*R*)-(+)-8-*p*-menthene-2,10-diol (**107a′**), and the mixture of (1*R*,2*S*,4*R*,8*S*/*R*)-(+)-*p*-menthane-2,8,9-triols (**104aa′** and **104ab′**), while *A. glauca* ATCC 22752 gave **107a′** stereoselectively from **102a′** (Demirci et al., 2001) (Figure 22.106).

On the other hand, (−)-carvone (**93′**) was biotransformed stereoselectively to (+)-neodihydrocarveol (**102a′**) via (+)-dihydrocarvone (**101a′**) by a strain of *A. niger* (Noma and Nonomura, 1974), *E. gracilis* Z. (Noma et al., 1993), and *Chlorella miniata* (Gondai et al., 1999). Furthermore, in *E. gracilis* Z., mosquito repellent (1*R*,2*S*,4*R*)-(+)-*p*-menthane-2,8-diol (**50a′**) was obtained stereospecifically from (−)-carvone (**93′**) via **101a′** and **102a′** (Figure 22.107).

As the microbial method for the formation of mosquito repellent **50a′** was established, the production of (+)-dihydrocarvone (**101a′**) and (+)-neodihydrocarveol (**102a′**) as the precursor of

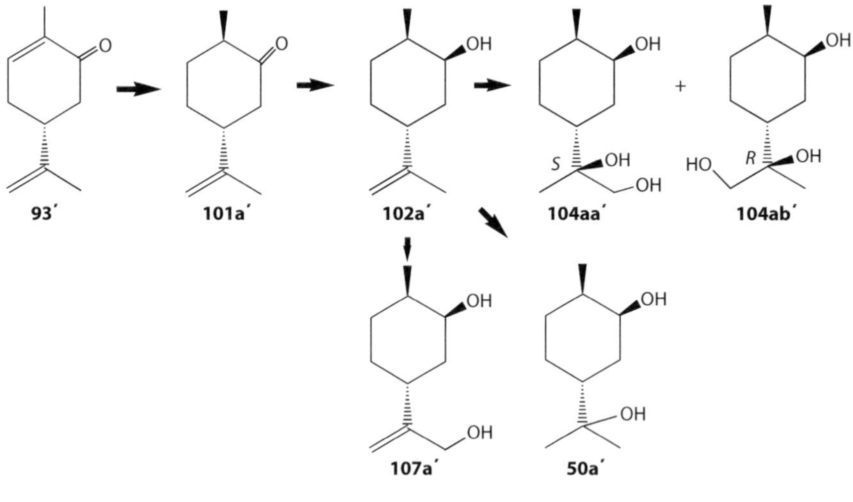

FIGURE 22.106 Metabolic pathways of (−)-carvone (**93′**) by *Aspergillus niger* TBUYN-2 and *Absidia glauca* ATCC 22752. (Modified from Demirci, F. et al. The biotransformation of thymol methyl ether by different fungi, *Book of Abstracts of the XII Biotechnology Congr.*, p. 47, 2001.)

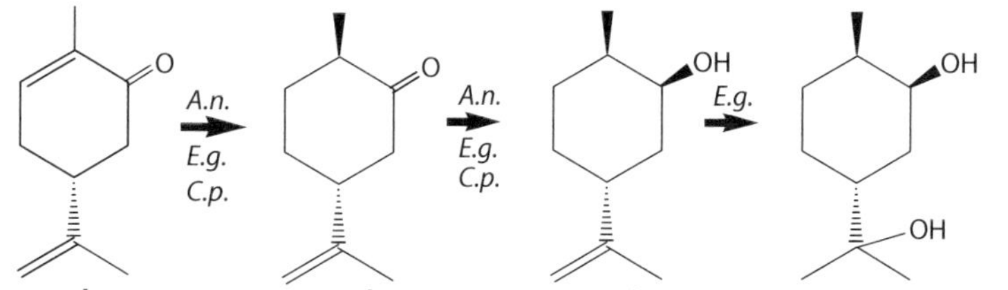

FIGURE 22.107 Metabolic pathway of (−)-carvone (**93′**) by *Aspergillus niger*, *Euglena gracilis* Z, and *Chlorella miniata*. (Modified from Noma, Y. and Nonomura, S., *Agric. Biol. Chem.*, 38, 741, 1974; Noma, Y. et al. Formation of 8 kinds of *p*-menthane-2,8-diols from carvone and related compounds by *Euglena gracilis* Z. Biotransformation of monoterpenes by photosynthetic microorganisms. Part VIII, *Proceedings of 37th TEAC*, 1993, pp. 23–25; Gondai, T. et al. Asymmetric reduction of enone compounds by *Chlorella miniata*, *Proceedings of 43rd TEAC*, 1999, pp. 217–219.)

mosquito repellent **50a′** was investigated by using 40 strains of bacteria belonging to *Escherichia*, *Aerobacter*, *Serratia*, *Proteus*, *Alcaligenes*, *Bacillus*, *Agrobacterium*, *Micrococcus*, *Staphylococcus*, *Corynebacterium*, *Sarcina*, *Arthrobacter*, *Brevibacterium*, *Pseudomonas*, and *Xanthomonas* spp.; 68 strains of yeasts belonging to *Schizosaccharomyces*, *Endomycopsis*, *Saccharomyces*, *Schwanniomyces*, *Debaryomyces*, *Pichia*, *Hansenula*, *Lipomyces*, *Torulopsis*, *Saccharomycodes*, *Cryptococcus*, *Kloeckera*, *Trigonopsis*, *Rhodotorula*, *Candida*, and *Trichosporon* spp.; 40 strains of fungi belonging to *Mucor*, *Absidia*, *Penicillium*, *Rhizopus*, *Aspergillus*, *Monascus*, *Fusarium*, *Pullularia*, *Keratinomyces*, *Oospora*, *Neurospora*, *Ustilago*, *Sporotrichum*, *Trichoderma*, *Gliocladium*, and *Phytophthora* spp.; and 48 strains of actinomycetes belonging to *Streptomyces*, *Actinoplanes*, *Nocardia*, *Micromonospora*, *Microbispora*, *Micropolyspora*, *Amorphosporangium*, *Thermopolyspora*, *Planomonospora*, and *Streptosporangium* spp.

As a result, 65% of bacteria, 75% of yeasts, 90% of fungi, and 90% of actinomycetes converted (−)-carvone (**93′**) to (+)-dihydrocarvone (**101a′**) or (+)-neodihydrocarveol (**102a′**) (Figures 22.105 and 22.108). Many microorganisms are capable of converting (−)-carvone (**93′**) to (+)-neodihydrocarveol (**102a′**) stereospecifically. Some of the useful microorganisms are listed in

TABLE 22.8
Summary of Microbial and Chemical Reduction of (−)-Carvone (93′) for the Formation of (+)-Neodihydrocarveol (102a′)

	Compounds					
Microorganisms	**101a′**	**101b′**	**102a′**	**102b′**	**102c′**	**102d′**
Torulopsis xylinus IFO 454	0	0	100	0	0	0
Monascus anka var. *rubellus* IFO 5965	0	0	100	0	0	0
Fusarium anguioides Sherbakoff IFO 4467	0	0	100	0	0	0
Phytophthora infestans IFO 4872	0	0	100	0	0	0
Kloeckera magna IFO 0868	0	0	98	2	0	0
Kloeckera antillarum IFO 0669	19	4	72	0	0	0
Streptomyces rimosus	+	0	98	0	0	0
Penicillium notatum Westling IFO 464	6	2	92	0	0	0
Candida pseudotropicalis IFO 0882	17	4	79	0	0	0
Candida parapsilosis IFO 0585	16	4	80	0	0	0
$LiAlH_4$	0	0	17	67	2	13
Meerwein–Ponndorf–Verley reduction	0	0	29	55	9	5

Source: Noma, Y., *Ann. Res. Stud. Osaka Joshigakuen Junior College*, 20, 33, 1976.

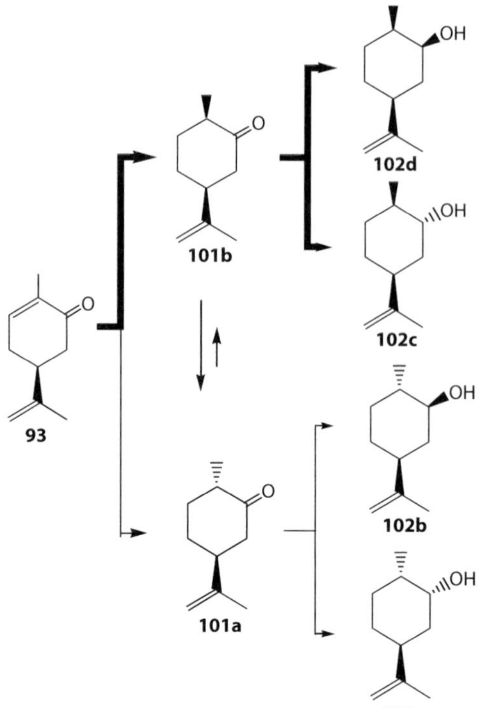

FIGURE 22.108 Metabolic pathways of (+)-carvone (**93**) by *Pseudomonas ovalis* strain 6-1 and other many microorganisms. (Modified from Noma, Y. et al. *Agric. Biol. Chem.*, 38, 735, 1974b.)

Tables 22.7 and 22.8. There is no good chemical method to obtain (+)-neodihydrocarveol (**102a′**) in large quantity. It was considered that the method utilizing microorganisms is a very useful means and better than the chemical synthesis for the production of mosquito repellent precursor (+)-neodihydrocarveol (**102a′**).

(−)-Carvone (**93**) was biotransformed by *A. niger* TBUYN-2 to give mainly (+)-8-hydroxyneodihydrocarveol (**50a′**), (+)-8,9-epoxyneodihydrocarveol (**103a′**), and (+)-10-hydroxyneodihydrocarveol (**107a′**) via (+)-dihydrocarvone (**101a′**) and (+)-neodihydrocarveol (**102a′**). *A. niger* TBUYN-2 dehydrogenated (+)-*cis*-carveol (**81b**) to give (+)-carvone (**93**), which was further converted to (−)-isodihydrocarvone (**101b**). Compound **101b** was further metabolized by four pathways to give 10-hydroxy-(−)-isodihydrocarvone (**106b**), (1*S*,2*S*,4*S*)-*p*-menthane-1,2-diol (**71d**) via 1α-hydroxy-(−)-isodihydrocarvone (**72b**) as intermediate, (−)-isodihydrocarveol (**102c**), and (−)-neoisodihydrocarveol (**102d**). Compound **102d** was further converted to isodihydroisobottrospicatol (**105bb**) via 8,9-epoxy-(−)-neoisodihydrocarveol (**103d**); compound **105′** was a major product (Noma et al., 1985a) (Figure 22.109).

In the case of the plant pathogenic fungus, *A. glauca* (−)-carvone (**93′**) was metabolized to give the diol 10-hydroxy-(+)-neodihydrocarveol (**107a′**) (Nishimura et al., 1983b).

(+)-Carvone (**93**) was converted by five bacteria and one fungus (Verstegen-Haaksma et al., 1995) to give (−)-dihydrocarvone (**101a**), (−)-isodihydrocarvone (**101b**), and (−)-neoisodihydrocarveol (**102d**). Sensitivity of the microorganism to (+)-carvone (**93**) and some of the products prevented yields exceeding 0.35 g/L in batch cultures. The fungus *Trichoderma pseudokoningii* gave the highest yield of (−)-neoisodihydrocarveol (**102d**) (Figure 22.110). (+)-Carvone (**93**) is known to inhibit fungal growth of *Fusarium sulphureum* when it was administered via the gas phase (Oosterhaven et al., 1995a,b). Under the same conditions, the related fungus *Fusarium solani* var. *coeruleum* was not inhibited. In liquid medium, both fungi were found to convert (+)-carvone

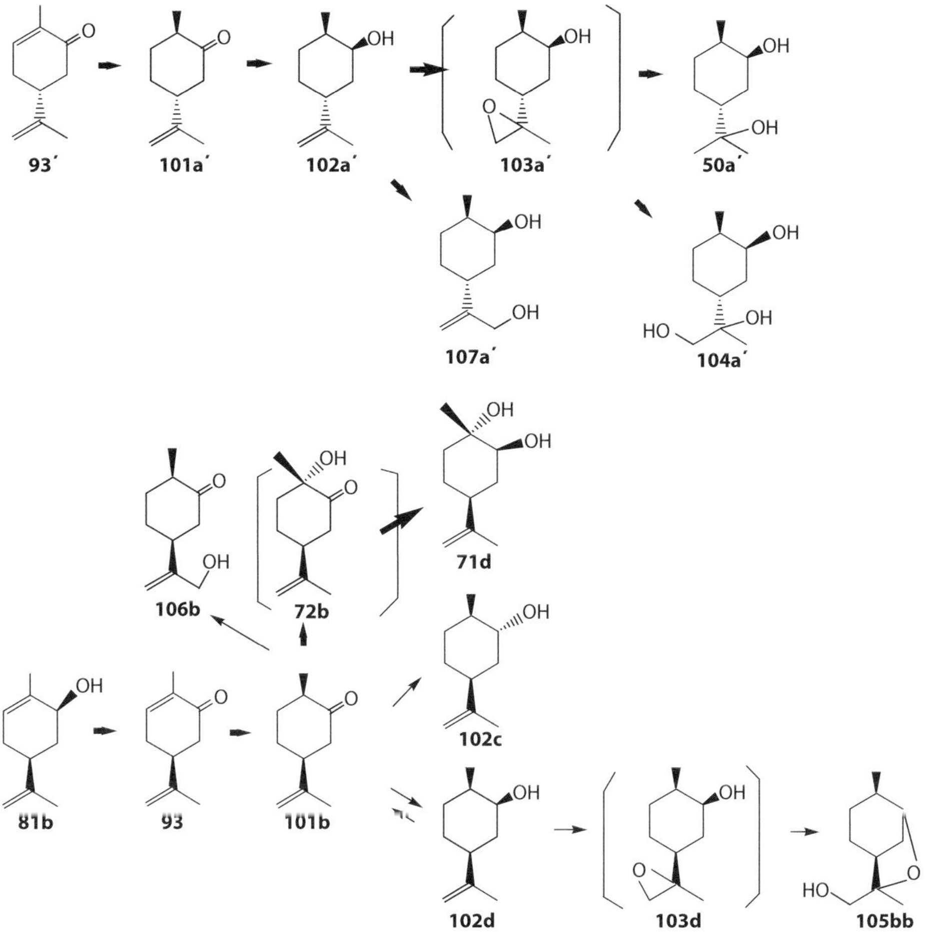

FIGURE 22.109 Possible main metabolic pathways of (−)-carvone (**93′**) and (+)-carvone (**93**) by *Aspergillus niger* TBUYN-2. (Modified from Noma, Y. et al. Biotransformation of (−)-carvone and (+)-carvone by *Aspergillus* spp., *Annual Meeting of Agricultural and Biological Chemistry*, Sapporo, Japan, 1985a, p. 68.)

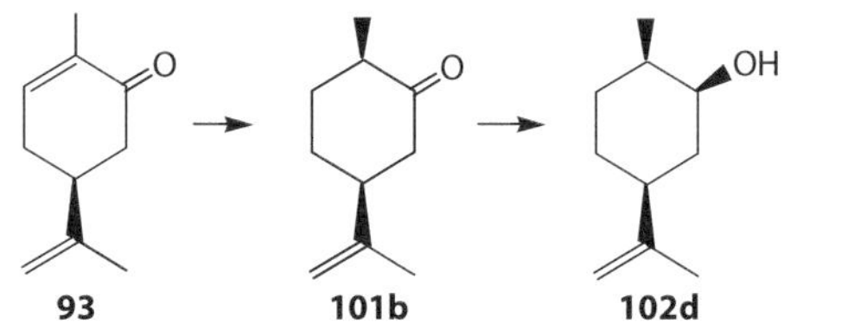

FIGURE 22.110 Biotransformation of (+)-carvone (**93**) by *Trichoderma pseudokoningii*. (Modified from Verstegen-Haaksma, A.A. et al. *Ind. Crops Prod.*, 4, 15, 1995.)

(**93**), with the same rate, mainly to (−)-isodihydrocarvone (**101b**), (−)-isodihydrocarveol (**102c**), and (−)-neoisodihydrocarveol (**102d**).

22.3.4.1.1.1 Biotransformation of Carvone to Carveols by Actinomycetes The distribution of actinomycetes capable of reducing carbonyl group of carvone containing α, β-unsaturated ketone to (−)-*trans*- (**81a′**) and (−)-*cis*-carveol (**81b′**) was investigated. Of 93 strains of actinomycetes, 63 strains were capable of converting (−)-carvone (**93′**) to carveols. The percentage of microorganisms that produced carveols from (−)-carvone (**93′**) to total microorganisms was about 71%. Microorganisms that produced carveols were classified into three groups according to the formation of (−)-*trans*-carveol (**81a′**) and (−)-*cis*-carveol (**81b′**): group 1, (−)-carvone-**81b′** only; group 2, (−)-carvone-**81a′** only; and group 3, (−)-carvone mixture of **81a′** and **81b′**. Three strains belonged to group 1 (4.5%), 34 strains belonged to group 2 (51.1%), and 29 strains belonged to group 3 (44%; of this group two strains were close to group 1 and 14 strains were close to group 2).

Streptomyces A-5-1 isolated from soil converted (−)-carvone (**93′**) to **101a′**–**102d′** and (−)-*trans*-carveol (**81a′**), whereas *Nocardia*, 1-3-11 converted (−)-carvone (**93′**) to (−)-*cis*-carveol (**81b′**) together with **101a′**–**81a′** (Noma, 1980). In the case of *Nocardia*, the reaction between **93′** and **81a′** was reversible and the direction from **81a′** to **93′** is predominantly (Noma, 1979a,b, 1980) (Figure 22.111).

(−)-Carvone (**93′**) was metabolized by actinomycetes to give (−)-*trans*- (**81a′**) and (−)-*cis*-carveol (**81b′**) and (+)-dihydrocarvone (**101a′**) as reduced metabolites. Compound **81b′** was further metabolized to (+)-bottrospicatol (**92a′**). Furthermore, **93′** was hydroxylated at C-5 position and C-8, 9 position to give 5β-hydroxy-(−)-carvone (**98a′**) and (−)-carvone-8,9-epoxide (**96′**), respectively. Compound **98a′** was further metabolized to 5β-hydroxyneodihydrocarveol (**100aa′**) via 5β-hydroxydihydrocarvone (**99a′**) (Noma, 1979a,b, 1980) (Figure 22.111).

Metabolic pattern of (+)-carvone (**93**) is similar to that of (−)-carvone (**93′**) in *S. bottropensis*. (+)-Carvone (**93**) was converted by *S. bottropensis* to give (+)-carvone-8,9-epoxide (**96**) and (+)-5α-hydroxycarvone (**98a**) (Figure 22.112). (+)-Carvone-8,9-epoxide (**96**) has light sweet aroma and has strong inhibitory activity for the germination of lettuce seeds (Noma and Nishimura, 1982).

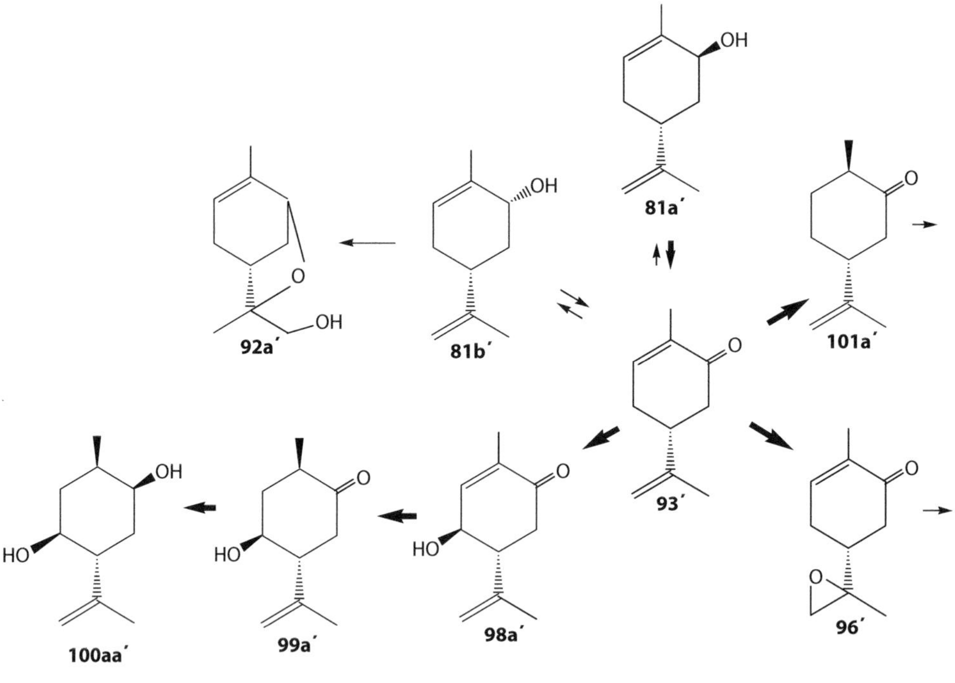

FIGURE 22.111 Metabolic pathways of (−)-carvone (**93′**) by *Streptomyces bottropensis* SY-2-, *Streptomyces ikutamanensis* Ya-2-1, *Streptomyces* A-5-1, and *Nocardia* 1-3-11. (Modified from Noma, Y., *Nippon Nogeikagaku Kaishi*, 53, 35, 1979a; Noma, Y., *Ann. Res. Stud. Osaka Joshigakuen Junior College*, 23, 27, 1979b; Noma, Y., *Agric. Biol. Chem.*, 44, 807, 1980; Noma, Y. and Nishimura, H., Biotransformation of ( )-carvone and (|)-carvone by *S. ikutamanensis* Ya-2-1, *Book of Abstracts of the Annual Meeting of Agricultural and Biological Chemical Society*, p. 390, 1983a; Noma, Y. and Nishimura, H., Biotransformation of carvone. 5. Microbiological transformation of dihydrocarvones and dihydrocarveols. *Proceedings of 27th TEAC*, 1983b, pp. 302–305.)

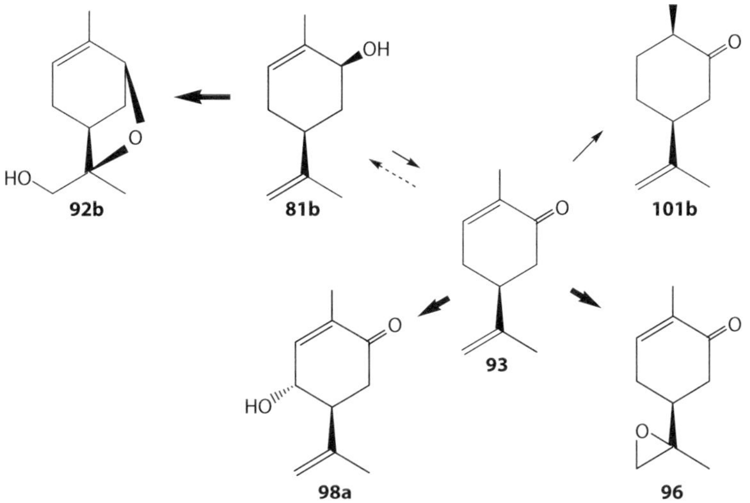

FIGURE 22.112 Metabolic pathways of (+)-carvone (**93′**) by *Streptomyces bottropensis* SY-2-1 and *Streptomyces ikutamanensis* Ya-2-1. (Modified from Noma, Y. and Nishimura, H., Biotransformation of carvone. 4. Biotransformation of (+)-carvone by *Streptomyces bottropensis*, SY-2-1, *Proceedings of 26th TEAC*, 1982, pp. 156–159l; Noma, Y. and Nishimura, H., Biotransformation of (−)-carvone and (+)-carvone by *S. ikutamanensis* Ya-2-1, *Book of Abstracts of the Annual Meeting of Agricultural and Biological Chemical Society*, 1983a, p. 390; Noma, Y. and Nishimura, H., Biotransformation of carvone. 5. Microbiological transformation of dihydrocarvones and dihydrocarveols, *Proceedings of 27th TEAC*, 1983b, pp. 302–305; Noma, Y., *Kagaku to Seibutsu*, 22, 742, 1984.)

The investigation of (−)-carvone (**93′**) and (+)-carvone (**93**) conversion pattern was carried out by using rare actinomycetes. The conversion pattern was classified as follows (Figure 22.113):

Group 1: Carvone (**93**), dihydrocarvones (**101**), dihydrocarveol (**102**), dihydrocarveol-8,9-epoxide (**103**), dihydrobottrospicatols (**105**), 5-hydroxydihydrocarveols (**100**)
Group 2: Carvone (**93**), carveols (**89**), bottrospicatols (**92**), 5-hydroxy-*cis*-carveols (**12**)
Group 3: Carvone (**93**), 5-hydroxycarvone (**98**), 5-hydroxyneodihydrocarveols (**15**)
Group 4: Carvone (**93**), carvone-8,9-epoxides (**96**)

Of 50 rare actinomycetes, 22 strains (44%) were capable of converting (−)-carvone (**93′**) to give (−)-carvone-8,9-epoxide (**96′**) via pathway 4 and (+)-5β-hydroxycarvone (**98a′**), (+)-5α-hydroxycarvone (**98b′**), and (+)-5β-hydroxyneodihydrocarveol (**100aa′**) via pathway 3 (Noma and Sakai, 1984).

On the other hand, in the case of (+)-carvone (**93**) conversion, 44% of rare actinomycetes were capable of converting (+)-carvone (**93**) to give (+)-carvone-8,9-epoxide (**96**) via pathway 4 and (−)-5α-hydroxycarvone (**98a**), (−)-5β-hydroxycarvone (**98b**), and (−)-5α-hydroxyneodihydrocarveol (**100aa**) via pathway 3 (Noma and Sakai, 1984).

22.3.4.1.1.2 Biotransformation of Carvone by Citrus Pathogenic Fungi, A. niger Tiegh TBUYN Citrus pathogenic *A. niger* Tiegh (CBAYN) and *A. niger* TBUYN-2 hydrogenated C=C double bond at C-1, 2 position of (+)-carvone (**93**) to give (−)-isodihydrocarvone (**101b**) as the major product together with a small amount of (−)-dihydrocarvone (**101a**), of which **101b** was further metabolized through two kinds of pathways as follows: One is the pathway to give (+)-1α-hydroxyn eoisodihydrocarveol (**71**) via (+)-1α-hydroxyisodihydrocarvone (**72**) and the other one is the pathway to give (+)-4α-hydroxyisodihydrocarvone (**378**) (Noma and Asakawa, 2008) (Figure 22.114).

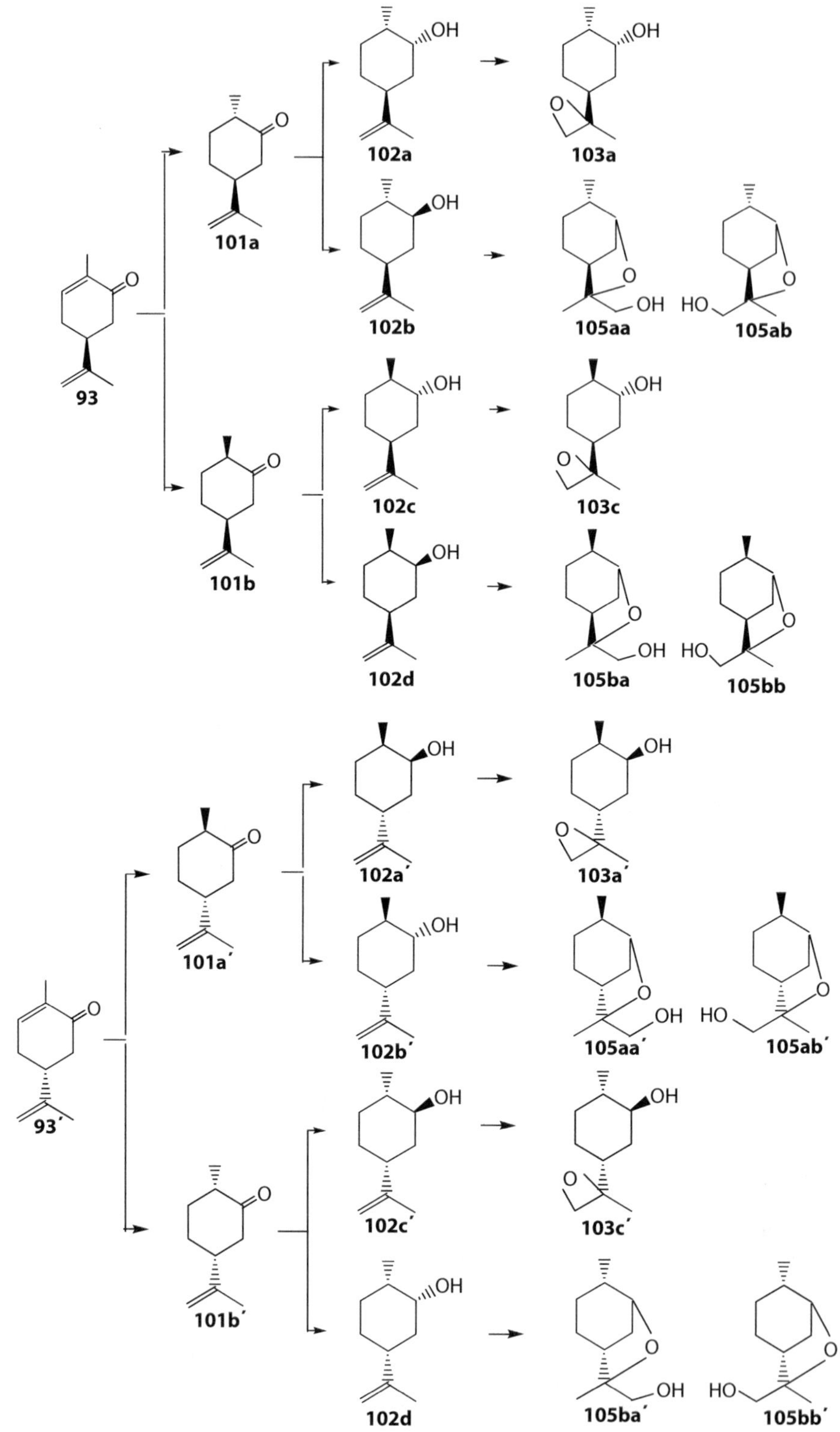

FIGURE 22.113 Metabolic pathways of (+)- (**93**) and (−)-carvone (**93′**) and dihydrocarveols (**102a-d** and **102a′-d′**) by *Streptomyces bottropensis* SY-2-1 and *Streptomyces ikutamanensis* Ya-2-1. (Modified from Noma, Y., *Kagaku to Seibutsu*, 22, 742, 1984.)

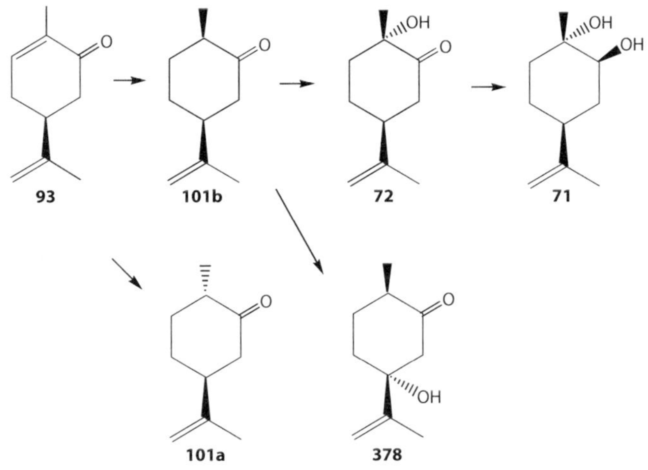

FIGURE 22.114 Metabolic pathways of (+)-carvone (**93**) by citrus pathogenic fungi, *Aspergillus niger* Tiegh CBAYN and *A. niger* TBUYN-2. (Modified from Noma, Y. and Asakawa, Y., New metabolic pathways of (+)-carvone by Citrus pathogenic *Aspergillus niger* Tiegh CBAYN and *A. niger* TBUYN-2, *Proceedings of 52nd TEAC*, 2008, pp. 206–208.)

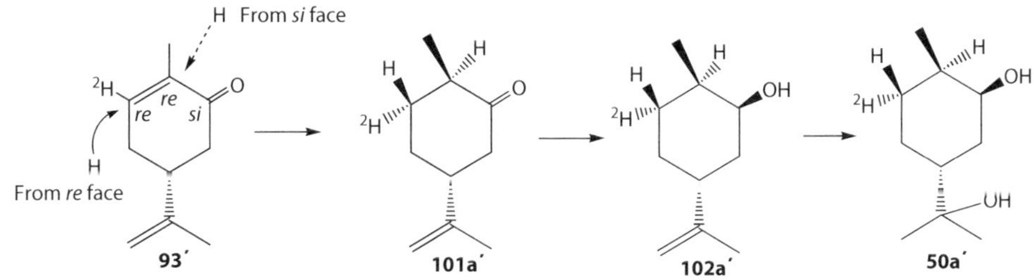

FIGURE 22.115 The stereospecific hydrogenation of the C=C double bond of α,β-unsaturated ketones, the reduction of saturated ketone, and the hydroxylation by *Euglena gracilis* Z. (Modified from Noma, Y. et al. Biotransformation of [6-2H]-(−)-carvone by *Aspergillus niger, Euglena gracilis* Z and *Dunaliella tertiolecta, Proceedings of 39th TEAC*, 1995, pp. 367–368; Noma, Y. and Asakawa, Y., *Euglena gracilis* Z: Biotransformation of terpenoids and related compounds, in Bajaj, Y.P.S., (ed.), *Biotechnology in Agriculture and Forestry*, vol. 41, Medicinal and Aromatic Plants X, Springer, Berlin, Germany, 1998, pp. 194–237.)

The biotransformation of enones such as (−)-carvone (**93′**) by the cultured cells of *C. miniata* was examined. It was found that the cells reduced stereoselectively the enones from *si* face at α-position of the carbonyl group and then the carbonyl group from *re* face (Figure 22.115).

Stereospecific hydrogenation occurs independent of the configuration and the kinds of the substituent at C-4 position, so that the methyl group at C-1 position is fixed mainly at *R*-configuration. [2-^{2}H]-(−)-Carvone ([2-^{2}H]-**93′**) was synthesized in order to clear up the hydrogenation mechanism at C-2 by microorganisms. Compound [2-^{2}H]-**93** was also easily biotransformed to [2-^{2}H]-8-hydroxy-(+)-neodihydrocarveol (**50a′**) via [2-^{2}H]-(+)-neodihydrocarveol (**102a′**). On the basis of ^{1}H-NMR spectral data of compounds **102a′** and **50a′**, the hydrogen addition of the carbon–carbon double bond at the C_1 and C_2 position by *A. niger* TBUYN-2, *E. gracilis* Z., and *D. tertiolecta* occurs from the *si* face and *re* face, respectively, that is, *anti* addition (Noma et al., 1995) (Figure 22.115) (Table 22.9).

22.3.4.1.1.3 Hydrogenation Mechanisms of C=C Double Bond and Carbonyl Group In order to understand the mechanism of the hydrogenation of α-, β unsaturated ketone of (−)-carvone (**93′**) and the reduction of carbonyl group of dihydrocarvone (**101a′**) (−)-carvone (**93′**), (+)-dihydrocarvone (**101a′**) and the analogues of (−)-carvone (**93′**) were chosen and the conversion of the analogues

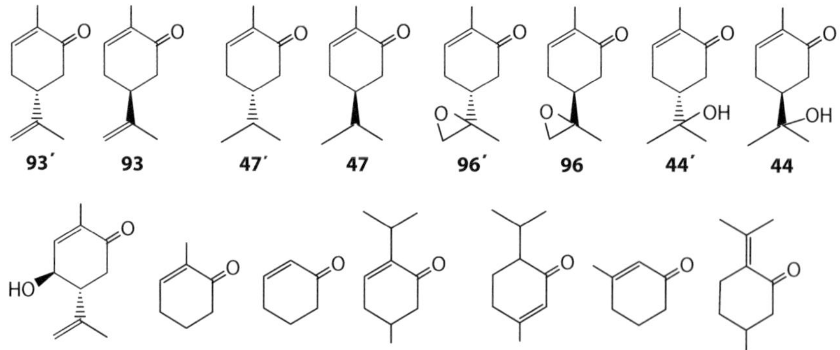

FIGURE 22.116 Substrates used for the hydrogenation of C = C double bond with *Pseudomonas ovalis* strain 6-1, *Streptomyces bottropensis* SY-2-1, *Streptomyces ikutamanensis* Ya-2-1, and *Euglena gracilis* Z.

was carried out by using *P. ovalis* strain 6-1. As the analogues of carvone (**93** and **93′**), (−)- (**47′**) and (+)-carvotanacetone (**47**), 2-methyl-2-cyclohexenone (**379**), the mixture of (−)-*cis*- (**81b′**) and (−)-*trans*-carveol (**81a′**), 2-cyclohexenone, racemic menthenone (**148**), (−)-piperitone (**156**), (+)-pulegone (**119**), and 3-methyl-2-cyclohexenone (**381**) were chosen. Of these analogues, (−)- (**47′**) and (+)-carvotanacetone (**47**) were reduced to give (+)-carvomenthone (**48a′**) and (−)-isocarvomenthone (**48b′**), respectively. 2-Methyl-2-cyclohexenone (**379**) was mainly reduced to (−)-2-methylcyclohexanone. But other compounds were not reduced.

The efficient formation of (+)-dihydrocarvone (**101a**), (−)-isodihydrocarvone (**101b′**), (+)-carvomenthone (**48a**), (−)-isocarvomenthone (**48b′**), and (−)-2-methylcyclohexanone from (−)-carvone (**93**), (+)-carvone (**93′**), (−)-carvotanacetone (**47**), (+)-carvotanacetone (**47′**), and 2-methyl-2-cyclohexenone (**379**) suggested at least that C=C double bond conjugated with carbonyl group may be hydrogenated from behind (*si* plane) (Noma et al., 1974b; Noma, 1977) (Figure 22.116).

22.3.4.1.1.4 What Is Hydrogen Donor in the Hydrogenation of Carvone to Dihydrocarvone? What Is Hydrogen Donor in Carvone Reductase? Carvone reductase prepared from *E. gracilis* Z., which catalyzes the NADH-dependent reduction of the C=C bond adjacent to the carbonyl group, was characterized with regard to the stereochemistry of the hydrogen transfer into the substrate. The reductase was isolated from *E. gracilis* Z. and was found to reduce stereospecifically the C=C double bond of carvone by *anti* addition of hydrogen from the *si* face at α-position to the carbonyl group and the *re* face at β-position (Tables 22.9 and 22.10). The hydrogen atoms participating in the

TABLE 22.9
Summary for the Stereospecificity of the Reduction of the C = C Double Bond of [2-2H]-(−)-Carvone ([2-2H]-93) by Various Kinds of Microorganisms

Microorganisms	Stereochemistry at C-2H of Compounds	
	102a	50a
Aspergillus niger TBUYN-2	β	
Euglena gracilis Z	β	β
Dunaliella tertiolecta	β	
The cultured cells of *Nicotiana tabacum* (Suga et al., 1986)	β	

TABLE 22.10
Purification of the Reductase from *Euglena gracilis* Z.

	Total Protein (mg)	Total Activity Unit × 104	Sp. Act. Units per Gram Protein	Fold
Crude extract	125	2.2	1.7	1
DEAE Toyopearl	7	1.5	21	12
AF-Blue Toyopearl	0.1	0.03	30	18

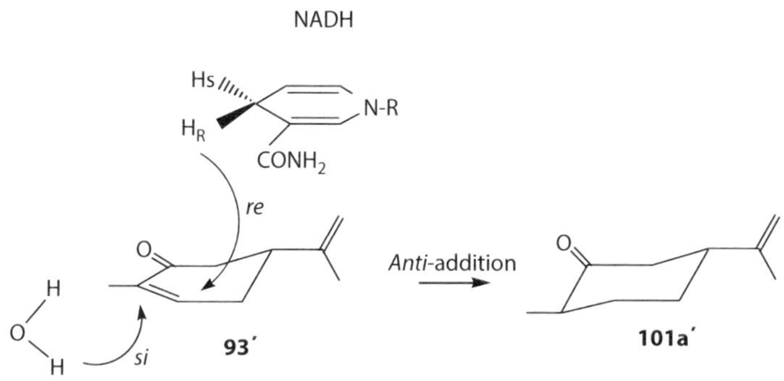

FIGURE 22.117 Stereochemistry in the reduction of (−)-carvone (**93′**) by the reductase from *Euglena gracilis* Z. (Modified from Shimoda, K. et al. *Phytochemistry*, 49, 49, 1998.)

enzymatic reduction at α- and β-position to the carbonyl group originate from the medium and the *pro*-4*R* hydrogen of NADH, respectively (Shimoda et al., 1998) (Figure 22.117).

In the case of biotransformation by using cyanobacterium, (+)- (**93**) and (−)-carvone (**93′**) were converted with a different type of pattern to give (+)-isodihydrocarvone (**101b′**, 76.6%) and (−)-dihydrocarvone (**101a**, 62.2%), respectively (Kaji et al., 2002) (Figure 22.118). On the other hand, *Catharanthus rosea*–cultured cell biotransformed (−)-carvone (**93′**) to give 5β-hydroxy-(+)-neodihydrocarveol (**100aa′**, 57.5%), 5α-hydroxy-(+)-neodihydrocarveol (**100ab′**, 18.4%), 5α-hydroxy-(−)-carvone (**98b′**), 4β-hydroxy-(−)-carvone (**384′**, 6.3%), 10-hydroxycarvone (**390′**), 5β-hydroxycarvone (**98′**), 5β-hydroxyneodihydrocarveol (**100ab′**), 5β-hydroxyneodihydrocarveol (**100aa′**), and 5α-hydroxydihydrocarvone (**99b′**) as the metabolites as shown in Figure 22.119, whereas (+)-carvone (**93**) gave 5α-hydroxy-(+)-carvone (**98a**, 65.4%) and 4α-hydroxy-(+)-carvone (**384**, 34.6%) (Hamada and Yasumune, 1995; Hamada et al., 1996; Kaji et al., 2002) (Figure 22.119) (Table 22.11).

(−)-Carvone (**93′**) was incubated with cyanobacterium, enone reductase (43 kDa) isolated from the bacterium, and microsomal enzyme to afford (+)-isodihydrocarvone (**101b′**) and (+)-dihydrocarvone (**101a′**). Cyclohexenone derivatives (**379** are **385**) were treated in the same enone reductase with

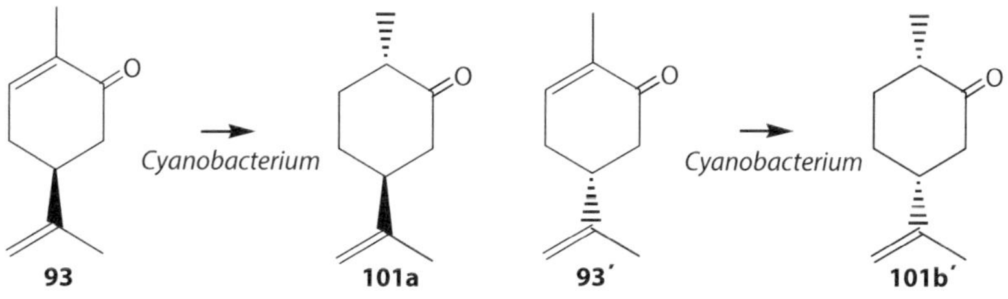

FIGURE 22.118 Biotransformation of (−)- and (+)-carvone (**93** and **93′**) by cyanobacterium. (Modified from Kaji, M. et al. Glycosylation of monoterpenes by plant suspension cells, *Proceedings of 46th TEAC*, 2002, pp. 323–325.)

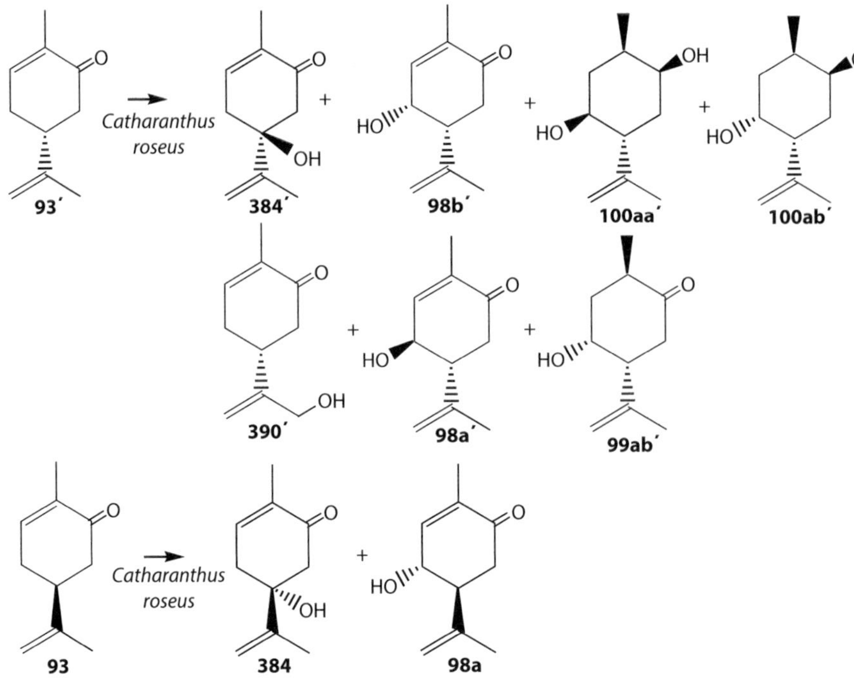

FIGURE 22.119 Biotransformation of (+)- and (−)-carvone (**93** and **93′**) by *Catharanthus roseus*. (Modified from Hamada, H. and Yasumune, H., The hydroxylation of monoterpenoids by plant cell biotransformation, *Proceedings of 39th TEAC*, pp. 375–377, 1995; Hamada, H. et al. The hydroxylation and glycosylation by plant catalysts, *Proceedings of 40th TEAC*, 1996, pp. 111–112; Kaji, M. et al. Glycosylation of monoterpenes by plant suspension cells, *Proceedings of 46th TEAC*, 2002, pp. 323–325.)

TABLE 22.11
Enantioselectivity in the Reduction of Enones (379 and 385) by Enone Reductase

Microsomal Enzyme	Substrate	Product	ee	Configuration[a]
−	**379**	**382a**	>99	R
−	**385**	**386a**	>99	R
+	**379**	**382b**	85	S
+	**385**	**386b**	80	S

[a] Preferred configuration at α-position to the carbonyl group of the products.

microsomal enzyme to give the dihydro derivative (**382a**, **386a**) with *R*-configuration in excellent *ee* (over 99%) and the metabolites (**382b**, **386b**) with *S*-configuration in relatively high *ee* (85% and 80%) (Shimoda et al., 2003) (Figure 22.120).

In contrast, almost all the yeasts tested showed reduction of carvone, although the enzyme activity varied. The reduction of (−)-carvone (**93′**) was often much faster than the reduction of (+)-carvone (**93**). Some yeasts only reduced the carbon–carbon double bond to yield the dihydrocarvone isomers (**101a′** and **b′** and **101a** and **b**) with the stereochemistry at C-1 with *R*-configuration, while others also reduced the ketone to give the dihydrocarveols with the stereochemistry at C-2 always with *S* for

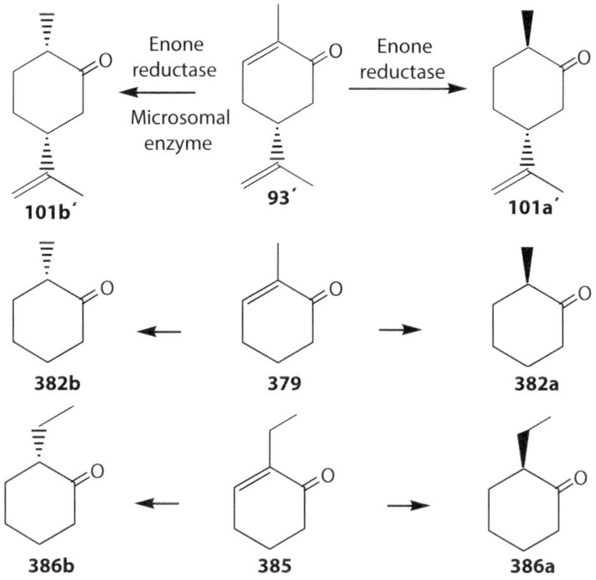

Enone reductase
Microsomal enzyme
Enone reductase
101b′ 93′ 101a′
382b 379 382a
386b 385 386a

FIGURE 22.120 Biotransformation of 2-methyl-2-cyclohexenone (**379**) and 2-ethyl-2-cyclohexenone (**385**) by enone reductase.

(−)-carvone (**93′**), but sometimes *S* and sometimes *R* for (+)-carvone (**93**). In the case of (−)-carvone (**93′**) yields increased up to 90% within 2 h (van Dyk et al., 1998).

22.3.4.1.2 Carvotanacetone

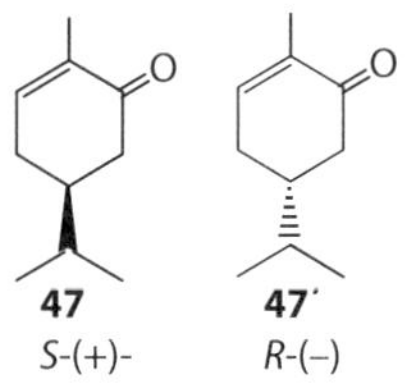

47 *S*-(+)-
47′ *R*-(−)

In the conversion of (+)- (**47**) and (−)-carvotanacetone (**47′**) by *P. ovalis* strain 6-1, (−)-carvotanacetone (**47′**) is converted stereospecifically to (+)-carvomenthone (**48a′**) an the latter compound is further converted to (+)-neocarvomenthol (**49a′**) and (−)-carvomenthol (**49b′**) in small amounts, whereas (+)-carvotanacetone (**47**) is converted mainly to (−)-isocarvomenthone (**48b**) and (−)-neoisocarvomenthol (**49d**), forming (−)-carvomenthone (**48a**) and (−)-isocarvomenthol (**49c**) in small amounts as shown in Figure 22.121 (Noma et al., 1974a).

Biotransformation of (−)-carvotanacetone (**47**) and (+)-carvotanacetone (**47′**) by *S. bottropensis* SY-2-1 was carried out (Noma et al., 1985c).

As shown in Figure 22.122, (+)-carvotanacetone (**47**) was converted by *S. bottropensis* SY-2-1 to give 5β-hydroxy-(+)-neoisocarvomenthol (**139db**), 5α-hydroxy-(+)-carvotanacetone (**51a**), 5β-hydroxy-(−)-carvomenthone (**52ab**), 8-hydroxy-(+)-carvotanacetone (**44**), and 8-hydroxy-(−)-carvomenthone (**45a**), whereas (−)-carvotanacetone (**47′**) was converted to give 5β-hydroxy(−)-carvotanacetone (**51a′**) and 8-hydroxy-(−)-carvotanacetone (**44′**).

A. niger TBUYN-2 converted (−)-carvotanacetone (**47′**) to (+)-carvomenthone (**48a′**), (+)-carvomenthone (**49a′**), diastereoisomeric *p*-menthane-2,9-diols [**55aa′** (8*R*) and **55ab′** (8*S*) in the ratio of 3:1], and 8-hydroxy-(+)-neocarvomenthol (**102a′**). On the other hand, the same fungus converted (+)-carvotanacetone (**47**) to (−)-isocarvomenthone (**48b**), 1α-hydroxy-(+)-neoisocarvomenthol (**54**)

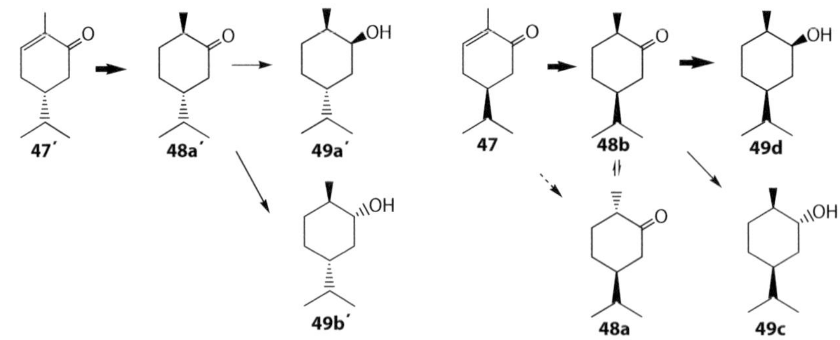

FIGURE 22.121 Metabolic pathways of (−)-carvotanacetone (**47′**) and (+)-carvotanacetone (**47**) by *Pseudomonas ovalis* strain 6-1. (Modified from Noma, Y. et al. *Agric. Biol. Chem.*, 38, 1637, 1974a.)

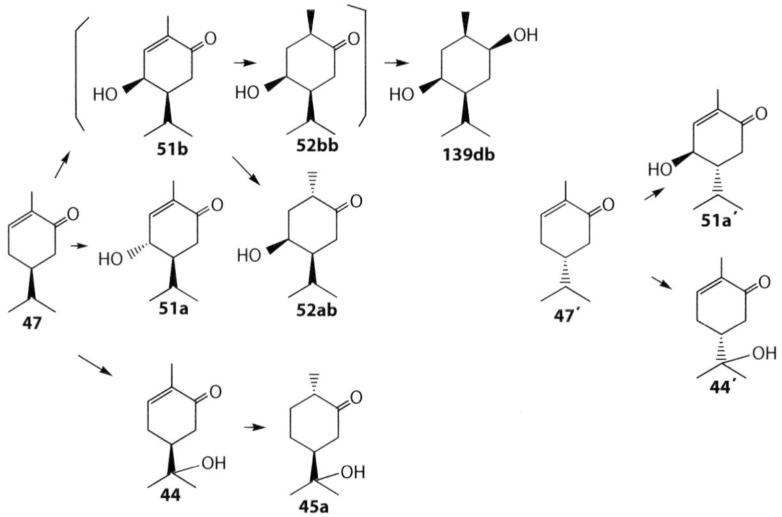

FIGURE 22.122 Proposed the metabolic pathways of (+)-carvotanacetone (**47**) and (−)-carvotanacetone (**47′**) by *Streptomyces bottropensis* SY-2-1. (Modified from Noma, Y. et al. Microbiological conversion of (−)-carvotanacetone and (+)-carvotanacetone by *S. bottropensis* SY-2-1, *Proceedings of 29th TEAC*, 1985c, pp. 238–240.)

via 1α-hydroxy-(+)-isocarvomenthone (**53**), and 8-hydroxy-(−)-isocarvomenthone (**45b**) as shown in Figure 22.123 (Noma et al., 1988a).

22.3.4.1.3 Piperitone

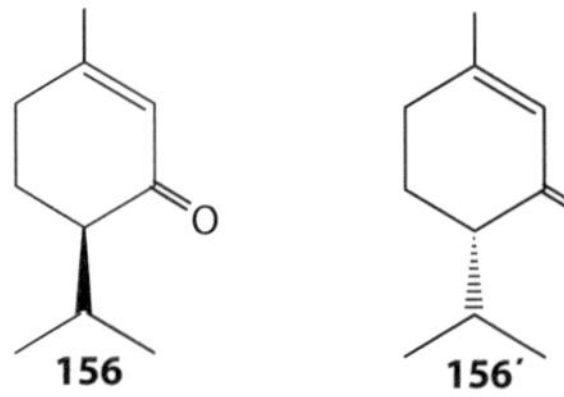

156 *R*-(−) 156′ *S*-(+)

A large number of yeasts were screened for the biotransformation of (−)-piperitone (**156**). A relatively small number of yeasts gave hydroxylation products of (−)-piperitone (**156**). Products obtained from (−)-piperitone (**156**) were 7-hydroxypiperitone (**161**), *cis*-6-hydroxypiperitone (**158b**), *trans*-6-hydroxypiperitone (**158a**), and 2-isopropyl-5-methylhydroquinone (**180**). Yields for the

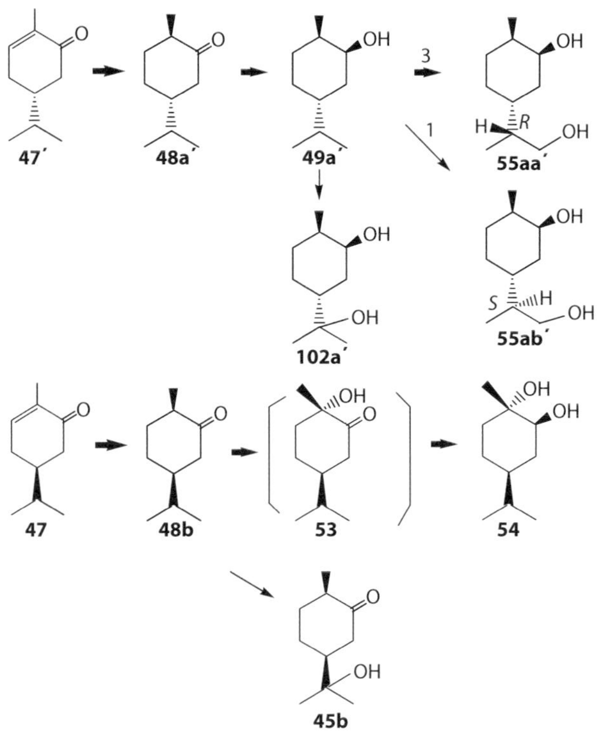

FIGURE 22.123 Proposed metabolic pathways of (−)-carvotanacetone (**47**) and (+)-carvotanacetone (**47′**) by *Aspergillus niger* TBUYN-2. (Modified from Noma, Y. et al. Microbiological conversion of (−)-carvotanacetone and (+)-carvotanacetone by a strain of *Aspergillus niger, Proceedings of 32nd TEAC*, 1988b, pp. 146–148.)

hydroxylation reactions varied between 8% and 60%, corresponding to the product concentrations of 0.04–0.3 g/L. Not one of the yeasts tested reduced (−)-piperitone (**156**) (van Dyk et al., 1998). During the initial screen with (−)-piperitone (**156**), only hydroxylation products were obtained. The hydroxylation products (**161**, **158a**, and **158b**) obtained with nonconventional yeasts from the genera *Arxula*, *Candida*, *Yarrowia*, and *Trichosporon* have recently been described (van Dyk et al., 1998) (Figure 22.124).

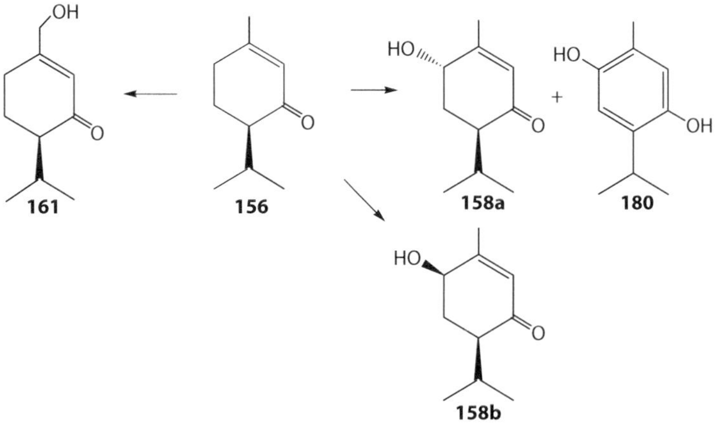

FIGURE 22.124 Hydroxylation products of (*R*)-(−)-piperitone (**156**) by yeast. (Modified from van Dyk, M.S. et al. *J. Mol. Catal. B: Enzym.*, 5, 149, 1998.)

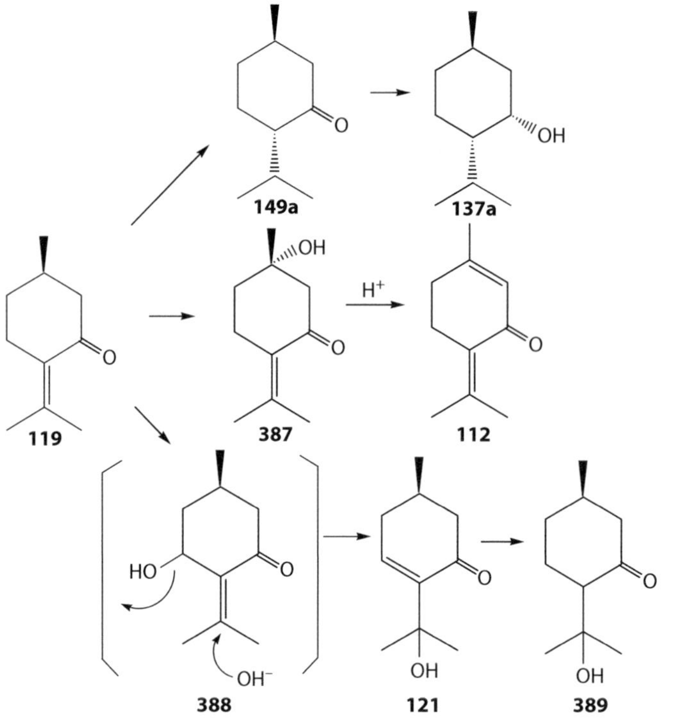

FIGURE 22.125 Biotransformation of (+)-pulegone (**119**) by *Aspergillus* sp., *Botrytis allii*, and *H. isolate* (UOFS Y-0067). (Modified from Miyazawa, M. et al. *Chem. Express*, 6, 479, 1991a; Miyazawa, M. et al. *Chem. Express*, 6, 873, 1991b; Ismaili-Alaoui, M. et al. *Tetrahedron Lett.*, 33, 2349, 1992; van Dyk, M.S. et al. *J. Mol. Catal. B: Enzym.*, 5, 149, 1998.)

22.3.4.1.4 Pulegone

(*R*)-(+)-Pulegone (**119**), with a mint-like odor monoterpene ketone, is the main component (up to 80%–90%) of *Mentha pulegium* essential oil (pennyroyal oil), which is sometimes used in beverages and food additive for human consumption and occasionally in herbal medicine as an abortifacient drug. The biotransformation of (+)-pulegone (**119**) by fungi was investigated (Ismaili-Alaoui et al., 1992). Most fungal strains grown in a usual liquid culture medium were able to metabolize (+)-pulegone (**119**) to some extent in a concentration range of 0.1–0.5 g/L; higher concentrations were generally toxic, except for a strain of *Aspergillus* sp. isolated from mint leaves infusion, which was able to survive to concentrations of up to 1.5 g/L. The predominant product was generally l-hydroxy-(+)-pulegone (**384**) (20%–30% yield). Other metabolites were present in lower amounts (5% or less) (see Figure 22.125). The formation of 1-hydroxy-(+)-pulegone (**387**) was explained by hydroxylation at a tertiary position. Its dehydration to piperitenone (**112**), even under the incubation conditions, during isolation or derivative reactions precluded any tentative determination of its optical purity and absolute configuration.

Botrytis allii converted (+)-pulegone (**119**) to (−)-(1*R*)-8-hydroxy-4-*p*-menthen-3-one (**121**) and piperitenone (**112**) (Miyazawa et al., 1991b,c). *Hormonema isolate* (UOFS Y-0067) quantitatively reduced (+)-pulegone (**119**) and (−)-menthone (**149a**) to (+)-neomenthol (**137a**) (van Dyk et al., 1998) (Figure 22.125).

Biotransformation by the recombinant reductase and the transformed *Escherichia coli* cells were examined with pulegone, carvone, and verbenone as substrates (Figure 22.126). The recombinant reductase catalyzed the hydrogenation of the exocyclic C=C double bond of pulegone (**119**) to give menthone derivatives (Watanabe et al., 2007) (Tables 22.12 and 22.13).

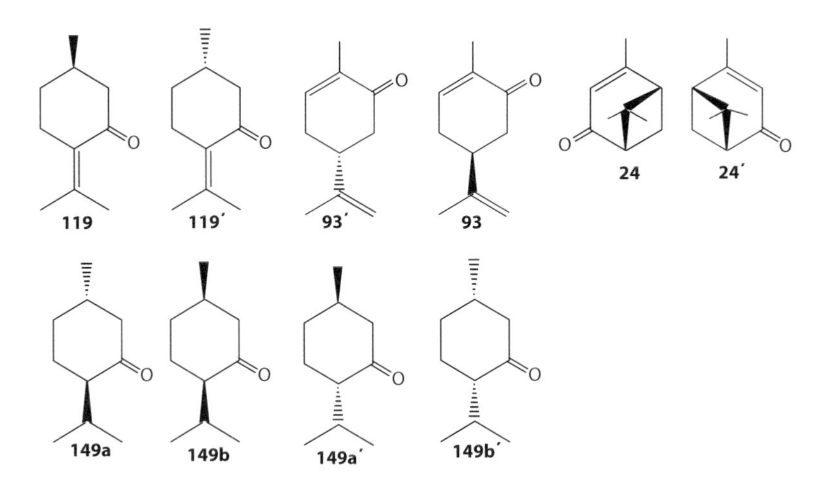

FIGURE 22.126 Chemical structures of substrate reduced by the recombinant pulegone reductase and the transformed *Escherichia coli* cells.

TABLE 22.12
Substrate Specificity in the Reduction of Enones with the Recombinant Pulegone Reductase

Entry No. (Reaction Time)	Substrates	Products	Conversions (%)
1 (3 h)	(*R*)-(+)-Pulegone **(119)**	(1*R*, 4*R*)-Isomenthone **(149b)**	4.4
2 (12 h)	(*R*)-Pulegone **(119)**	(1*S*, 4*R*)-Menthone **(149a)**	6.8
3 (3 h)	(*S*)-(−)-Pulegone **(119′)**	(1*R*, 4*R*)-Isomenthone **(149b)**	14.3
4 (12 h)	(*S*)-Pulegone **(119′)**	(1*S*, 4*R*)-Menthone **(149a)**	15.7
5 (12 h)	(*R*)-(−)-Carvone **(93′)**	(1*S*, 4*S*)-Isomenthone **(149b′)**	0.3
6 (12 h)	(*S*)-(+)-Carvone **(93)**	(1*R*, 4*S*)-Menthone **(149a′)**	0.5
7 (12 h)	(1*S*, 5*S*)-Verbenone **(24)**	(1*S*, 4*S*)-Isomenthone **(149b′)**	1.6
8 (12 h)	(1*R*, 5*R*)-Verbenone **(24′)**	(1*R*, 4S)-Menthone **(149a′)**	2.1
		—	N.d.
		—	N.d.
		—	N.d.
		—	N.d.

Note: N.d., denotes not detected.

TABLE 22.13
Biotransformation of Pulegone (119 and 119′) with the Transformed *Escherichia coli* Cells[a]

Substrates	Products	Conversion (%)
(*R*)-(+)-Pulegone **(119)**	(1*R*, 4*R*)-Isomenthone **(149b)**	26.8
(*S*)-(−)-Pulegone **(119′)**	(1*S*, 4*R*)-Menthone **(149a)**	30.0
	(1*S*, 4*S*)-Isomenthone **(149b′)**	32.3
	(1*R*, 4*S*)-Menthone **(149a′)**	7.1

[a] Reaction times of the transformation reaction are 12 h.

22.3.4.1.5 Piperitenone and Isopiperitenone

Piperitenone (**112**) is metabolized to 5-hydroxypiperitenone (**117**), 7-hydroxypiperitenone (**118**), and 7,8-dihydroxypiperitenone (**157**). Isopiperitenol (**110**) is reduced to give isopiperitenone (**111**), which is further metabolized to piperitenone (**112**), 7-hydroxy- (**113**), 10-hydroxy- (**115**), 4-hydroxy- (**114**), and 5-hydroxyisopiperitenone (**116**). Compounds **111** and **112** are isomerized to each other. Pulegone (**119**) was metabolized to **112**, 8,9-dehydromenthenone (**120**) and 8-hydroxymenthenone (**121**) as shown in the biotransformation of the same substrate using *B. allii* (Miyazawa et al., 1991c) (Figure 22.127).

H. isolate (UOFS Y-0067) reduced (4*S*)-isopiperitenone (**111**) to (3*R*,4*S*)-isopiperitenol (**110**), a precursor of (–)-menthol (**137b**) (van Dyk et al., 1998) (Figure 22.128).

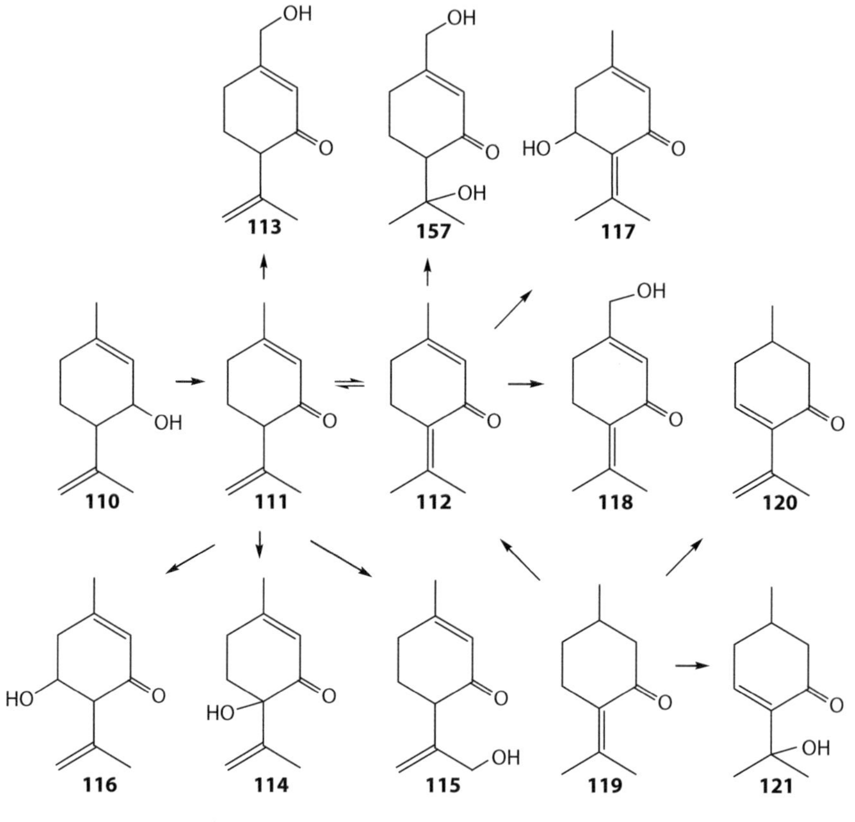

FIGURE 22.127 Biotransformation of isopiperitenone (**111**) and piperitenone (**112**) by *Aspergillus niger* TBUYN-2. (Modified from Noma, Y. et al. Biotransformation of isopiperitenone, 6-gingerol, 6-shogaol and neomenthol by a strain of *Aspergillus niger, Proceedings of 37th TEAC*, 1992c, pp. 26–28.)

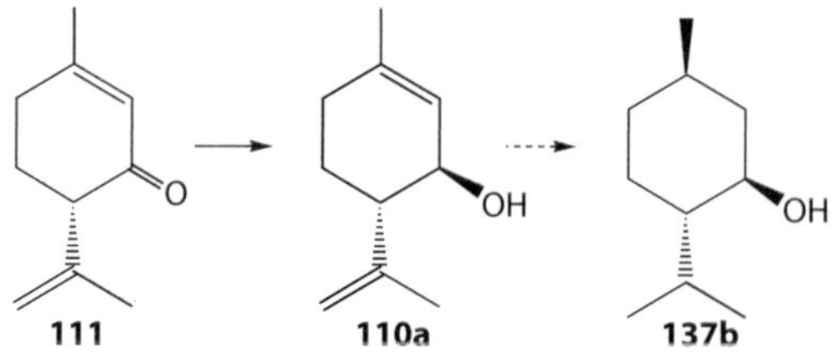

FIGURE 22.128 Biotransformation of isopiperitenone (**111**) by *H. isolate* (UOFS Y-0067). (Modified from van Dyk, M.S. et al. *J. Mol. Catal. B: Enzym.*, 5, 149, 1998.)

22.3.4.2 Saturated Ketone

22.3.4.2.1 Dihydrocarvone

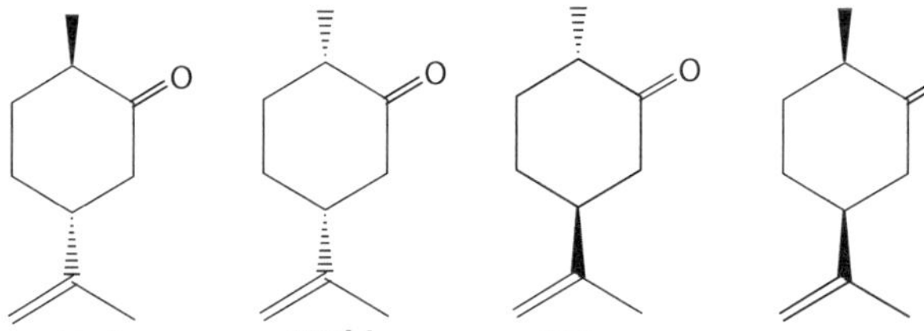

101a′ (1*R*,4*S*) (+)
101b′ (1*S*,4*S*) (+)-Iso
101a (1*S*,4*R*) (−)
101b (1*R*,4*R*) (−)-Iso

In the reduction of saturated carbonyl group of dihydrocarvone by microorganism, (+)-dihydrocarvone (**101a′**) is converted stereospecifically to either (+)-neodihydrocarveol (**102a′**) or (−)-dihydrocarveol (**102b′**) or nonstereospecifically to the mixture of **102a′** and **102b′**, whereas (−)-isodihydrocarvone (**101b**) is converted stereospecifically to either(−)-neoisodihydrocarveol (**102d**) or (−)-isodihydrocarveol (**102c**) or nonstereospecifically to the mixture of **102c** and **102d** by various microorganisms (Noma and Tatsumi, 1973; Noma et al., 1974c; Noma and Nonomura, 1974; Noma, 1976, 1977).

(+)-Dihydrocarvone (**101a′**) and (+)-isodihydrocarvone (**101b′**) are easily isomerized chemically to each other. In the microbial transformation of (−)-carvone (**93′**), the formation of (+)-dihydrocarvone (**101a′**) is predominant. (+)-Dihydrocarvone (**101a′**) was reduced to both/either (+)-neodihydrocarveol (**102a′**) and/or (−)-dihydrocarveol (**102b**), whereas in the biotransformation of (+)-carvone (**93**), (+)-isodihydrocarvone (**101b**) was formed predominantly. (+)-Isodihydrocarvone (**101b**) was reduced to both (+)-isodihydrocarveol (**102c**) and (+)-neoisodihydrocarveol (**102d**) (Figure 22.129).

However, *P. fragi* IFO 3458, *Pseudomonas fluorescens* IFO 3081, and *Aerobacter aerogenes* IFO 3319 and IFO 12059 formed (−)-dihydrocarvone (**101a**) predominantly from (+)-carvone (**93**).

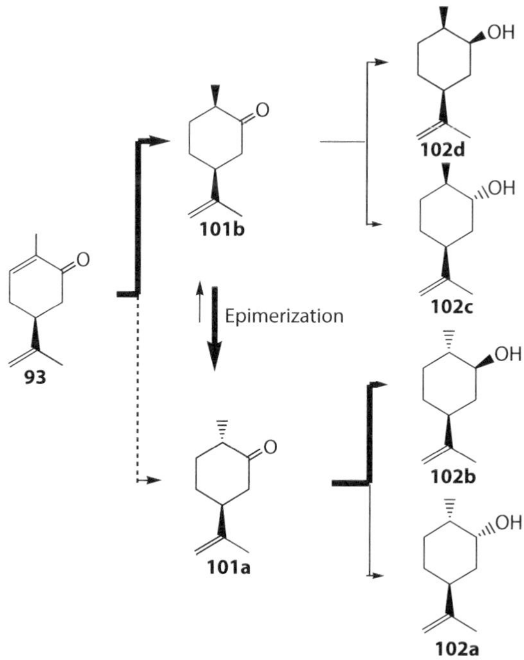

93 **101b** **101a** **102d** **102c** **102b** **102a** Epimerization

FIGURE 22.129 Proposed metabolic pathways of (+)-carvone (**93**) and (−)-isodihydrocarvone (**101b**) by *Pseudomonas fragi* IFO 3458. (Modified from Noma, Y. et al. *Agric. Biol. Chem.*, 39, 437, 1975.)

In the time course study of the biotransformation of (+)-carvone (**93**), it appeared that predominant formation of (−)-dihydrocarvone is due to the epimerization of (−)-isodihydrocarvone (**101b′**) by epimerase of *P. fragi* IFO 3458 (Noma et al., 1975).

22.3.4.2.2 Isodihydrocarvone Epimerase

22.3.4.2.2.1 Preparation of Isodihydrocarvone Epimerase The cells of *P. fragi* IFO 3458 were harvested by centrifugation and washed five times with 1/100 M KH_2PO_4–Na_2HPO_4 buffer (pH 7.2). Bacterial extracts were prepared from the washed cells (20 g from 3 L medium) by sonic lysis (Kaijo Denki Co., Ltd., 20Kc., 15 min, at 5°C–7°C) in 100 mL of the same buffer. Sonic extracts were centrifuged at 25, 500 *g* for 30 min at −2°C. The opalescent yellow supernatant fluid had the ability to convert (−)-isodihydrocarvone (**101b**) to (−)-dihydrocarvone (**101a**). On the other hand, the broken cell preparation was incapable of converting (−)-isodihydrocarvone (**101b**) to (−)-dihydrocarvone (**101a**). The enzyme was partially purified from this supernatant fluid about 56-fold with heat treatment (95°C–97°C for 10 min), ammonium sulfate precipitation (0.4–0.7 saturation), and DEAE-Sephadex A-50 column chromatography.

The reaction mixture consisted of a mixture of (−)-isodihydrocarvone (**101b**) and (−)-dihydrocarvone (**101a**) (60:40 or 90:10), 1/30 M KH_2PO_4–Na_2HPO_4 buffer (pH 7.2), and the crude or partially purified enzyme solution. The reaction was started by the addition of the enzyme solution and stopped by the addition of ether. The ether extract was applied to analytical gas–liquid chromatography (GLC) (Shimadzu GC-4A 10% PEG-20M, 3 m × 3 mm, temperature 140°C–170°C at the rate of 1°C/min, N_2 35 mL/min), and epimerization was assayed by measuring the peak areas of (−)-isodihydrocarvone (**101b**) and (−)-dihydrocarvone (**101a**) in GLC before and after the reaction.

The crude extract and the partially purified preparation were found to be very stable to heat treatment; 66% and 36% of the epimerase activity remained after treatment at 97°C for 60 and 120 min, respectively (Noma et al., 1975).

A strain of *A. niger* TBUYN-2 hydroxylated at C-1 position of (−)-isodihydrocarvone (**101b**) to give 1α-hydroxyisodihydrocarvone (**72b**), which was easily and smoothly reduced to (1*S*, 2*S*, 4*S*)-(−)-8-*p*-menthene-1,2-*trans*-diol (**71d**), which was also obtained from the biotransformation of (−)-*cis*-limonene-1,2-epoxide (**69**) by microorganisms and decomposition by 20% HCl (Figure 22.127) (Noma et al., 1985a,b). Furthermore, *A. niger* TBUYN-2 and *A. niger* Tiegh (CBAYN) biotransformed (−)-isodihydrocarvone (**101b**) to give (−)-4α-hydroxyisodihydrocarvone (**378b**) and (−)-8-*p*-menthene-1,2-*trans*-diol (**71d**) as the major products together with a small amount of 1α-hydroxyisodihydrocarvone (**72b**) (Noma and Asakawa, 2008) (Figure 22.130).

22.3.4.2.3 Menthone and Isomenthone

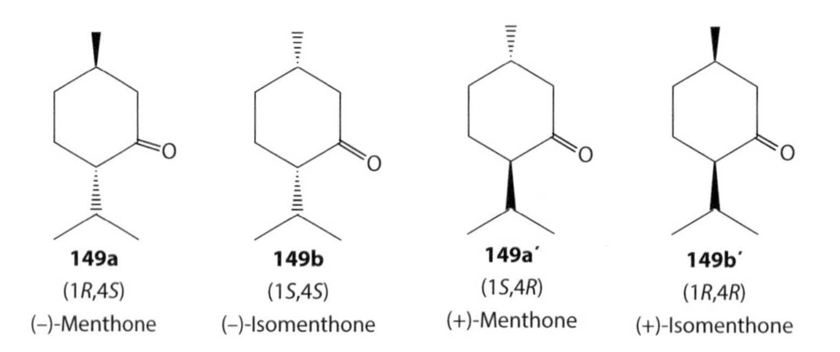

The growing cells of *P. fragi* IFO 3458 epimerized 17% of racemic isomenthone (**149b** and **b′**) to menthone (**149a** and **a′**) (Noma et al., 1975). (−)-Menthone (**149a**) was converted by *P. fluorescens* M-2 to (−)-3-oxo-4-isopropyl-1-cyclohexanecarboxylic acid (**164a**), (+)-3-oxo-4-isopropyl-1-cyclohexanecarboxylic acid (**164b**), and (+)-3-hydroxy-4-isopropyl-1-cyclohexanecarboxylic acid (**165ab**). On the other hand, (+)-menthone (**149a′**) was converted to

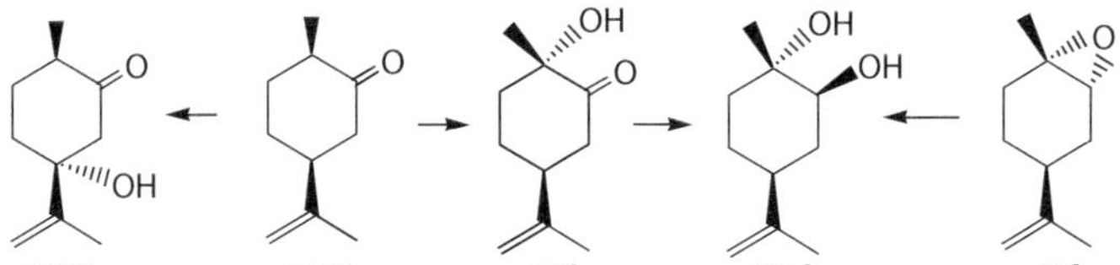

378b 101b 72b 71d 69b

FIGURE 22.130 Biotransformation of (+)-carvone (**93**), (−)-isodihydrocarvone (**101b**), and (−)-*cis*-limonene-1,2-epoxide (**69b**) by *Aspergillus niger* TBUYN-2 and *A. niger* Tiegh (CBAYN). (Modified from Noma, Y. et al. Biotransformation of (−)-carvone and (+)-carvone by Aspergillus spp., *Annual Meeting of Agricultural and Biological Chemistry*, Sapporo, Japan, 1985a, p. 68; Noma, Y. and Asakawa, Y., New metabolic pathways of (+)-carvone by Citrus pathogenic Aspergillus niger Tiegh CBAYN and A. niger TBUYN-2, *Proceedings of 52nd TEAC*, 2008, pp. 206–208.)

give (+)-3-oxo-4-isopropyl-1-cyclohexane carboxylic acid (**164a′**) and (−)-3-oxo-4-isopropyl-1-cyclohexane carboxylic acid (**164b′**). Racemic isomenthone (**149b** and **b′**) was converted to give racemic 1-hydroxy-1-methyl-4-isopropylcyclohexane-3-one (**150**), racemic piperitone (**156**), racemic 3-oxo-4-isopropyl-1-cyclohexene-1-carboxylic acid (**162**), 3-oxo-4-isopropyl-1-cyclohexane carboxylic acid (**164b**), 3-oxo-4-isopropyl-1-cyclohexane carboxylic acid (**164a**), and (+)-3-hydroxy-4-isopropyl-1-cyclohexane carboxylic acid (**165ab**) (Figure 22.131).

Soil plant pathogenic fungi *Rhizoctonia solani* 189 converted (−)-menthone (**149a**) to 4β-hydroxy(−)-menthone (**392%**, 29%) and 1 α, 4 β-dihydroxy-(−)-menthone (**393%**, 71%) (Nonoyama et al., 1999) (Figure 22.131). (−)-Menthone (**149a**) was transformed by *S. litura* to give 7-hydroxymenthone (**151a**), 7-hydroxyneomenthol (**165c**), and 7-hydroxy-9-carboxymenthone (**394a**) (Hagiwara et al., 2006) (Figure 22.132). (−)-Menthone (**149a**) gave 7-hydroxymenthone (**151a**) and (+)-neomenthol (**137c**) by human liver microsome (CYP2B6). Of 11 recombinant human P450 enzymes (expressed in *Trichoplusia ni* cells) tested, CYP2B6 catalyzed oxidation of (−)-menthone (**149a**) to 7-hydroxymenthone (**151a**) (Nakanishi and Miyazawa, 2004; Miyazawa et al., 1992b,c) (Figure 22.132).

22.3.4.2.4 Thujone

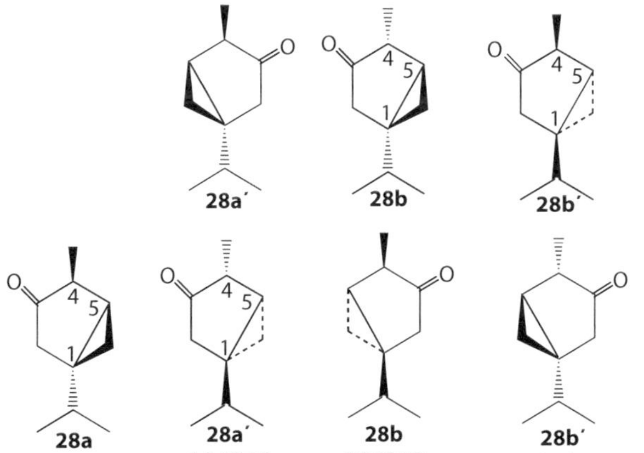

28a′ 28b 28b′

28a 1*S*,4*S*,5*R* (+)-3-
28a′ 1*R*,4*R*,5*S* (−)-3-
28b 1*S*,4*R*,5*R* (−)-3-iso
28b′ 1*R*,4*S*,5*S* (+)-3-iso

β-Pinene (**1**) is metabolized to 3-thujone (**28**) via α-pinene (**4**) (Gibbon and Pirt, 1971). α-Pinene (**4**) is metabolized to give thujone (**28**). Thujone (**28**) was biotransformed to thujoyl alcohol (**29**) by *A. niger* TBUYN-2 (Noma, 2000). Furthermore, (−)-3-isothujone (**28b**) prepared from *Armoise* oil was biotransformed by plant pathogenic fungus *B. allii* IFO 9430 to give 4-hydroxythujone (**30**) and 4,6-dihydroxythujone (**31**) (Miyazawa et al., 1992a) (Figure 22.133).

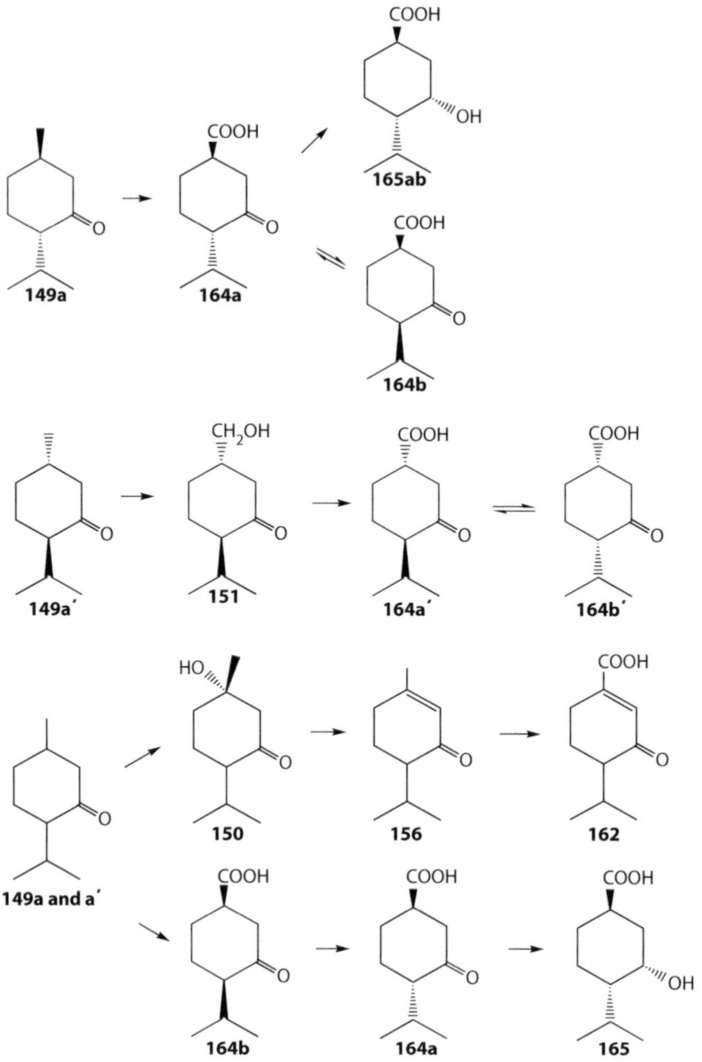

FIGURE 22.131 Biotransformation of (−)- (**149a**) and (+)-menthone (**149a′**) and racemic isomenthone (**149b** and **149b′**) by *Pseudomonas fluorescens* M-2. (Modified from Sawamura, Y. et al. Microbiological oxidation of p-menthane 1. Formation of formation of p-cis-menthan-1-ol, *Proceedings of 18th TEAC*, 1974, pp. 27–29.)

22.3.4.3 Cyclic Monoterpene Epoxide

22.3.4.3.1 1,8-Cineole

1,8-Cineole (**122**) is a main component of the essential oil of *Eucalyptus radiata* var. *australiana* leaves, comprising ca. 75% in the oil, which corresponds to 31 mg/g fr.wt. leaves (Nishimura et al., 1980).

The most effective utilization of **122** is very important in terms of renewable biomass production. It would be of interest, for example, to produce more valuable substances, such as plant growth regulators, by the microbial transformation of **122**. The first reported utilization of **122** was presented by MacRae et al. (1979), who showed that it was a carbon source for *Pseudomonas flava* growing on *Eucalyptus* leaves. Growth of the bacterium in a mineral salt medium containing **122** resulted in the oxidation at the C-2 position of **122** to give the metabolites (1*S*,4*R*,6*S*)-(+)-2α-hydroxy-1,8-cineole (**225a**), (1*S*,4*R*,6*R*)-(−)-2β-hydroxy-1,8-cineole (**125a**), (1*S*,4*R*)-(+)-2-oxo-1,8-cineole (**126**), and (−)-(*R*)-5,5-dimethyl-4-(3′-oxobutyl)-4,5-dihydrofuran-2(3H)-one (**128**) (Figure 22.134).

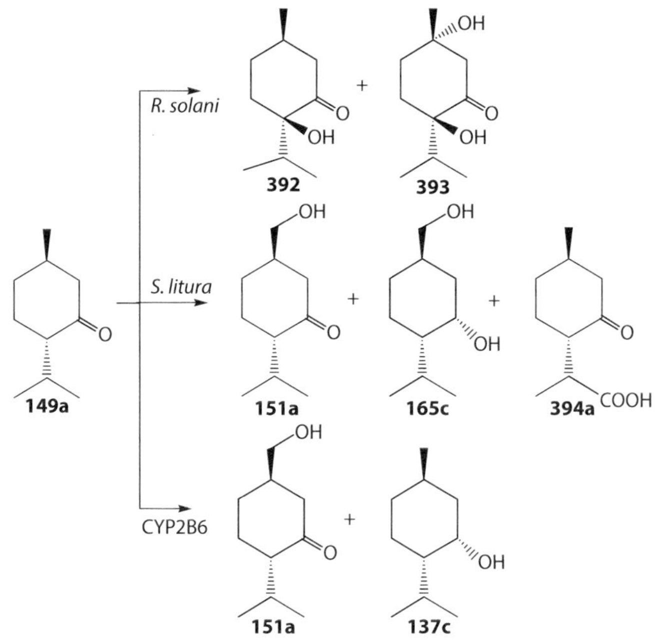

FIGURE 22.132 Metabolic pathway of (−)-menthone (**149a**) by *Rhizoctonia solani* 189, *Spodoptera litura*, and human liver microsome (CYP2B6). (Modified from Nonoyama, H. et al. Biotransformation of (−)-menthone using plant parasitic fungi, Rhizoctonia solani as a biocatalyst, Proceedings of 43rd TEAC, 1999, pp. 387–388; Nakanishi, K. and Miyazawa, M., Biotransformation of (−)-menthone by human liver microsomes, Proceedings of 48th TEAC, 2004, pp. 401–402; Hagiwara, Y. et al. Biotransformation of (+)- and (−)-menthone by the larvae of common cutworm (Spodoptera litura) as a biocatalyst, Proceedings of 50th TEAC, 2006, pp. 279–280.)

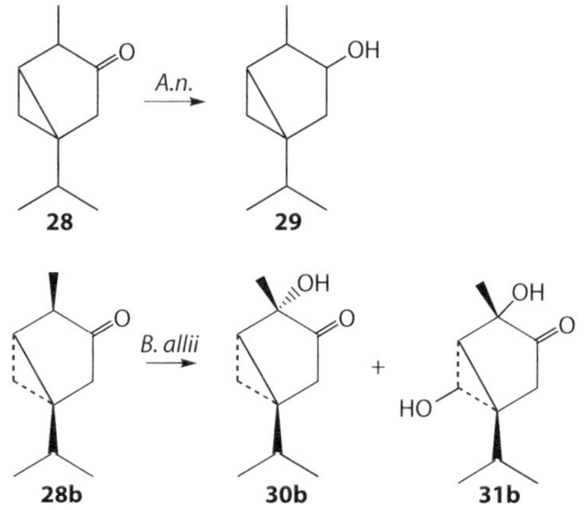

FIGURE 22.133 Biotransformation of (−)-3-isothujone (**28b**) by *Aspergillus niger* TBUYN-2 and plant pathogenic fungus *Botrytis allii* IFO 9430. (Modified from Gibbon, G.H. and Pirt, S.J., FEBS Lett., 18, 103, 1971; Miyazawa, M. et al. Biotransformation of thujone by plant pathogenic microorganism, Botrytis allii IFO 9430, Proceedings of 36th TEAC, 1992a, pp. 197–198.)

S. bottropensis SY-2-1 biotransformed 1,8-cineole (**122**) stereochemically to (+)-2α-hydroxy-1,8-cineole (**125b**) as the major product and (+)-3α-hydroxy-1,8-cineole (**123b**) as the minor product. Recovery ratio of 1,8-cineole metabolites as ether extract was ca. 30% in *S. bottropensis* SY-2-1 (Noma and Nishimura, 1980, 1981) (Figure 22.135).

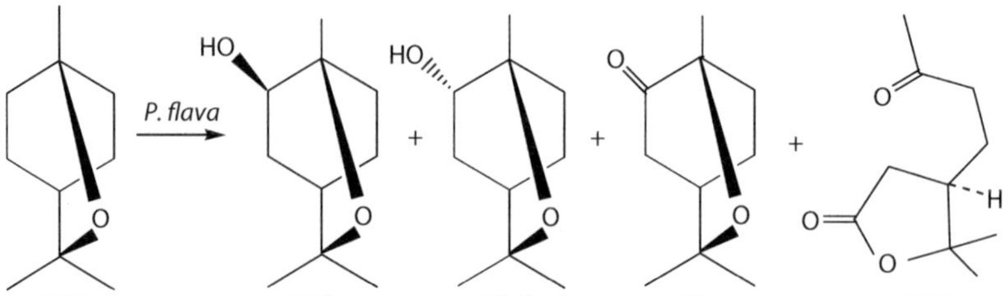

FIGURE 22.134 Biotransformation of 1,8-cineole (**84**) by *Pseudomonas flava*. (Modified from MacRae, I.C. et al. Aust. J. Chem., 32, 917, 1979.)

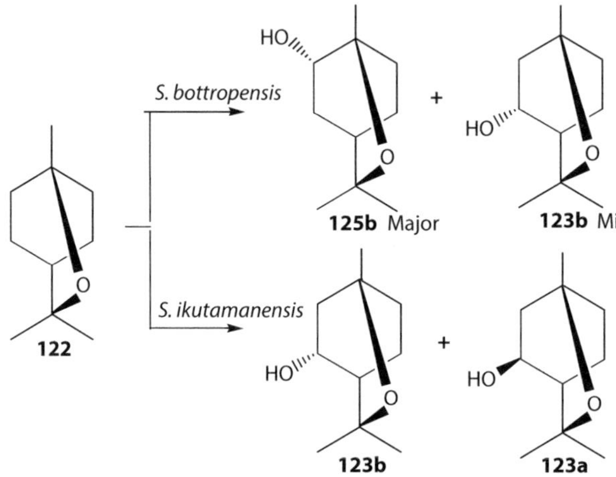

FIGURE 22.135 Biotransformation of 1,8-cineole (**122**) by *Streptomyces bottropensis* SY-2-1 and *Streptomyces ikutamanensis* Ya-2-1. (Modified from Noma, Y. and Nishimura, H., Microbiological transformation of 1,8-cineole. Oxidative products from 1,8-cineole by *S. bottropensis*, SY-2-1, *Book of abstracts of the Annual Meeting of Agricultural and Biological Chemical Society*, 1980, p. 28; Noma, Y. and Nishimura, H., Microbiological transformation of 1,8-cineole. Production of 3β-hydroxy-1,8-cineole from 1,8-cineole by *S. ikutamanensis*, Ya-2-1, *Book of Abstracts of the Annual Meeting of Agricultural and Biological Chemical Society*, 1981, p. 196.)

In the case of *S. ikutamanensis* Ya-2-1, 1,8-cineole (**122**) was biotransformed regioselectively to give (+)-3α-hydroxy-1,8-cineole (**123b**, 46%) and (+)-3β-hydroxy-1,8-cineole (**123b**, 29%) as the major product. Recovery ratio as ether extract was ca. 8.5% in *S. ikutamanensis*,Ya-2-1 (Noma and Nishimura, 1980, 1981) (Figure 22.135).

When (+)-3α-hydroxy-1,8-cineole (**123b**) was used as substrate in the cultured medium of *S. ikutamanensis* Ya-2-1, (+)-3β-hydroxy-1,8-cineole (**123a**, 32%) was formed as the major product together with a small amount of (+)-3-oxo-1,8-cineole (**126a**, 1.6%). When (+)-3β-hydroxy-1,8 cineole (**123a**) was used, (+)-3-oxo-1,8-cineole (**126a**, 9.6%) and (+)-3α-hydroxy-1,8-cineole (**123b**, 2%) were formed. When (+)-3-oxo-1,8-cineole (**126a**) was used, (+)-3α-hydroxy- (**123b**, 19%) and (+)-3β-hydroxy-1,8-cineole (**123a**, 16%) were formed.

Based on the aforementioned results, it is obvious that (+)-3β-hydroxy-1,8-cineole (**123b**) is formed mainly in the biotransformation of 1,8-cincole (**122**), (+)-3α-hydroxy-1,8-cineole (**123b**), and (+)-3-oxo-1,8-cineole (**126a**) by *S. ikutamanensis* Ya-2-1. The production of (+)-3β-hydroxy-1,8-cineole (**123b**) is interesting, because it is a precursor of mosquito repellent, *p*-menthane-3,8-diol (**142aa′**) (Noma and Nishimura, 1981) (Figure 22.136).

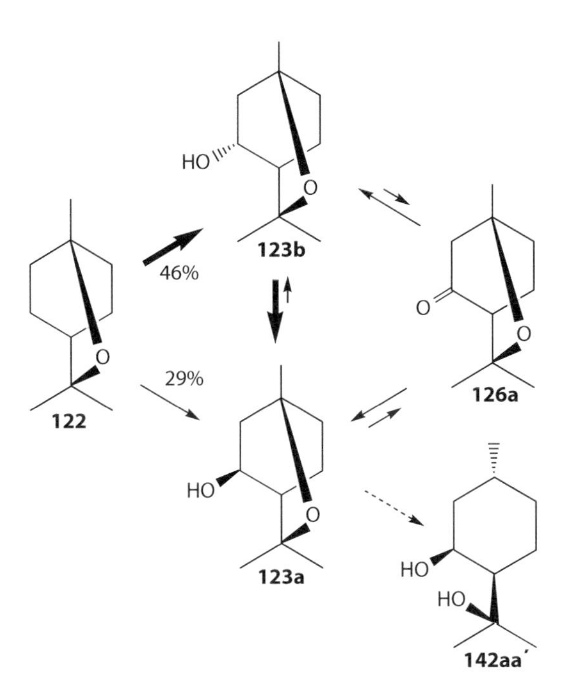

FIGURE 22.136 Biotransformation of 1,8-cineole (**122**), (+)-3α-hydroxy-1,8-cineole (**123b**), (+)-3β-hydroxy-1,8-cineole(**123a**), and (+)-3-oxo-1,8-cineole (**126a**) by *Streptomyces ikutamanensis* Ya-2-1. (Modified from Noma, Y. and Nishimura, H., Microbiological transformation of 1,8-cineole. Production of 3β-hydroxy-1,8-cineole from 1,8-cineole by *S. ikutamanensis*, Ya-2-1, *Book of Abstracts of the Annual Meeting of Agricultural and Biological Chemical Society*, 1981, p. 196.)

When *A. niger* TBUYN-2 was cultured in the presence of 1,8-cineole (**122**) for 7 days, it was transformed to three alcohols [racemic 2α-hydroxy-1,8-cineoles (**125b** and **b′**), racemic 3α-hydroxy- (**123b** and **b′**), and racemic 3β-hydroxy-1,8-cineoles (**123a** and **123a′**)] and two ketones [racemic 2-oxo- (**126** and **126′**) and racemic 3-oxo-1,8-cineoles (**124** and **124′**)] (Figure 22.135). The formation of 3α-hydroxy- (**123b** and **b′**) and 3β-hydroxy-1,8-cineoles (**123a** and **123a′**) is of great interest not only due to the possibility of the formation of *p*-menthane-3,8-diol (**142** and **142′**), the mosquito repellents, and plant growth regulators that are synthesized chemically from 3α-hydroxy- (**123b** and **b′**) and 3β-hydroxy-1,8-cineoles (**123a** and **123a′**), respectively, but also from the viewpoint of the utilization of *E. radiata* var. *australiana* leaves oil as biomass. An Et_2O extract of the culture broth (products and **122** as substrate) was recovered in 57% of substrate (w/w) (Nishimura et al., 1982; Noma et al., 1996) (Figure 22.137).

Plant pathogenic fungus *Botryosphaeria dothidea* converted 1,8-cineole (**122**) to optical pure (+)-2α-hydroxy-1,8-cineole (**125b**) and racemic 3α-hydroxy-1,8-cineole (**123b** and **b′**), which were oxidized to optically active 2-oxo- (**126**) (100% ee) and racemic 3-oxo-1,8-cineole (**124** and **124′**), respectively (Table 22.14). The cytochrome P450 inhibitor 1-aminobenzotriazole inhibited the hydroxylation of the substrate (Noma et al., 1996) (Figure 22.138). *S. litura* also converted 1,8-cineole (**122**) to give three secondary alcohols (**123b, 125a,** and **b**) and two primary alcohols (**395** and **127**) (Hagiwara and Miyazawa, 2007). *Salmonella typhimurium* OY1001/3A4 and NADPH-P450 reductase hydroxylated 1,8-cineole (**122**) to 2β-hydroxy-1,8-cineole (**125a**, $[\alpha]_D + 9.3$, 65.3% ee) and 3β-hydroxy-1 to 8-cineole (**123a**, $[\alpha]_D$–27.8, 24.7% ee) (Saito and Miyazawa, 2006).

Extraction of the urinary metabolites from brushtail possums (*Trichosurus vulpecula*) maintained on a diet of fruit impregnated with 1,8-cineole (**122**) yielded *p*-cresol (**129**) and the novel C-9 oxidated products 9-hydroxy-1,8-cineole (**127a**) and 1,8-cineole-9-oic acid (**462a**) (Flynn and Southwell, 1979; Southwell and Flynn, 1980) (Figure 22.139).

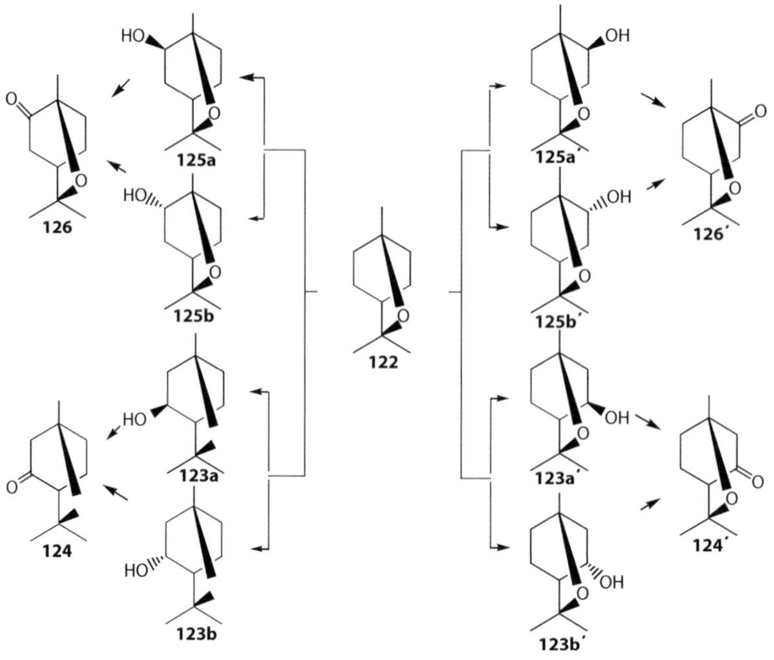

FIGURE 22.137 Biotransformation of 1,8-cineole (**122**) by *Aspergillus niger* TBUYN-2. (Modified from Nishimura, H. et al. *Agric. Biol. Chem.*, 46, 2601, 1982; Noma, Y. et al. Biotransformation of 1,8-cineole. Why do the biotransformed 2α- and 3α-hydroxy-1,8-cineole by *Aspergillus niger* have no optical activity? *Proceedings of 40th TEAC*, 1996, pp. 89–91.)

TABLE 22.14
Stereoselectivity in the Biotransformation of 1,8-Cineole (122) by *Aspergillus niger*, *Botryosphaeria dothidea*, and *Pseudomonas flava*

	Products
Microorganisms	**125a and a′, 125b and b′, 123b and b′, 123a and a′**
Aspergillus niger TBUYN-2	2:43:49:6
	50:50:41:59
Botryosphaeria dothidea	4:59:34:3
	100:0:53:47
Pseudomonas flava	
	29:71:0:0
	100:0

Source: Noma, Y. et al. Biotransformation of 1,8-cineole. Why do the biotransformed 2α- and 3α-hydroxy-1,8-cineole by *Aspergillus niger* have no optical activity? *Proceedings of 40th TEAC*, 1996, pp. 89–91.

1,8-Cineole (**122**) gave 2β-hydroxy-1,8-cineole (**125a**) by CYP450 human and rat liver microsome. Cytochrome P450 molecular species responsible for metabolism of 1,8-cineole (**122**) was determined to be CYP3A4 and CYP3A1/2 in human and rat, respectively. Kinetic analysis showed that K_m and V_{max} values for the oxidation of 1,8-cineole (**122**) by human and rat treated with pregnenolone-16α-carbonitrile recombinant CYP3A4 were determined to be 50 μM and 90.9 nmol/min/nmol P450, 20 μM and 11.5 nmol/min/nmol P450, and 90 μM and 47.6 nmol/min/nmol P450, respectively (Shindo et al., 2000).

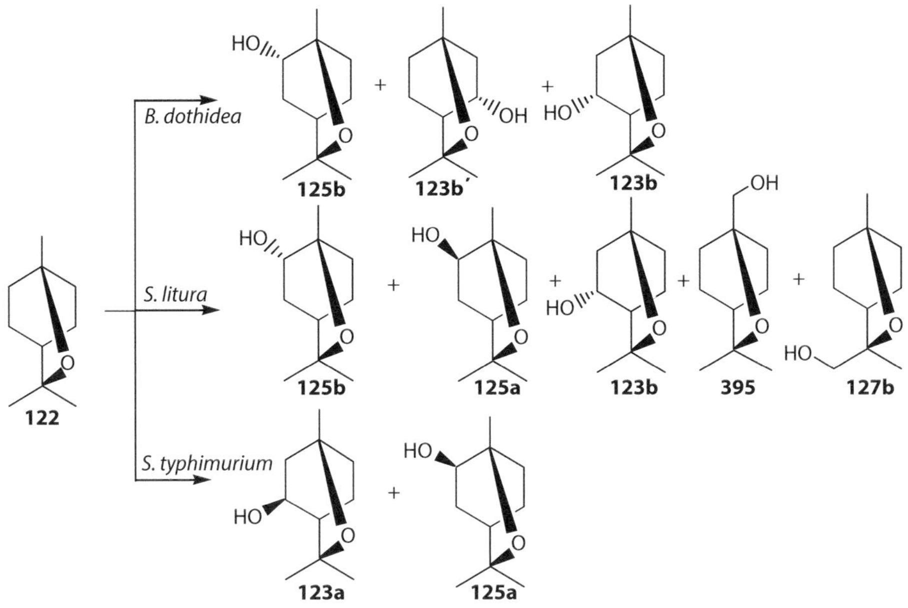

FIGURE 22.138 Biotransformation of 1,8-cineole (**122**) by *Botryosphaeria dothidea*, *Spodoptera litura*, and *Salmonella typhimurium*. (Modified from Noma, Y. et al. Biotransformation of 1,8-cineole. Why do the biotransformed 2α- and 3α-hydroxy-1,8-cineole by *Aspergillus niger* have no optical activity? *Proceedings of 40th TEAC*, 1996, pp. 89–91; Saito, H. and Miyazawa, M., Biotransformation of 1,8-cineole by *Salmonella typhimurium* OY1001/3A4, *Proceedings of 50th TEAC*, pp. 275–276, 2006; Hagiwara, Y. and Miyazawa, M., Biotransformation of cineole by the larvae of common cutworm (*Spodoptera litura*) as a biocatalyst, *Proceedings of 51st TEAC*, 2007, pp. 304–305.)

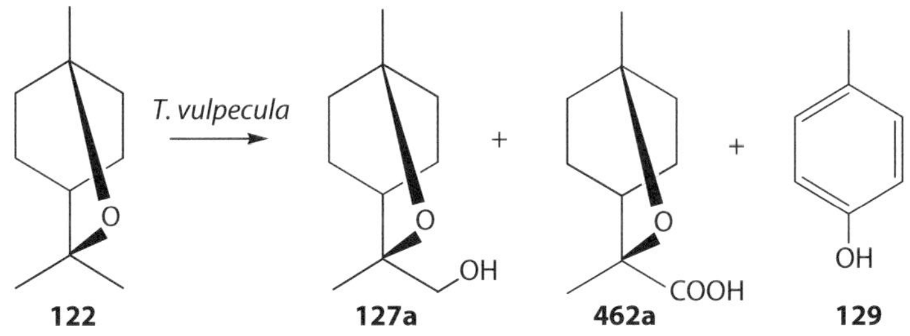

FIGURE 22.139 Metabolism of 1,8-cineole in *Trichosurus vulpecula*. (Modified from Southwell, I.A. and Flynn, T.M., *Xenobiotica*, 10, 17, 1980.)

Microbial resolution of racemic 2α-hydroxy-1,8-cineoles (**125b** and **b′**) was carried out by using *G. cingulata*. The mixture of **125b** and **b′** was added to a culture of *G. cingulata* and esterified to give after 24 h (1*R*,2*R*,4*S*)-2α-hydroxy-1,8-cineole-2-yl-malonate (**130b′**) in 45% yield (ee 100%). The recovered alcohol showed 100% ee of the (1*S*,2*S*,4*R*)-enantiomer (**125b**) (Miyazawa et al., 1995c). On the other hand, optically active (+)-2α-hydroxy-1,8-cineole (**125b**) was also formed from (+)-limonene (**68**) by a strain of citrus pathogenic fungus *P. digitatum* (Saito and Miyazawa, 2006; Noma and Asakawa, 2007a) (Figure 22.140).

Esters of racemic 2α-hydroxy-1,8-cineole (**125b** and **b′**) were prepared by a convenient method (Figure 22.141). Their odors were characteristic. Then products were tested against antimicrobial activity and their microbial resolution was studied (Hashimoto and Miyazawa, 2001) (Table 22.15).

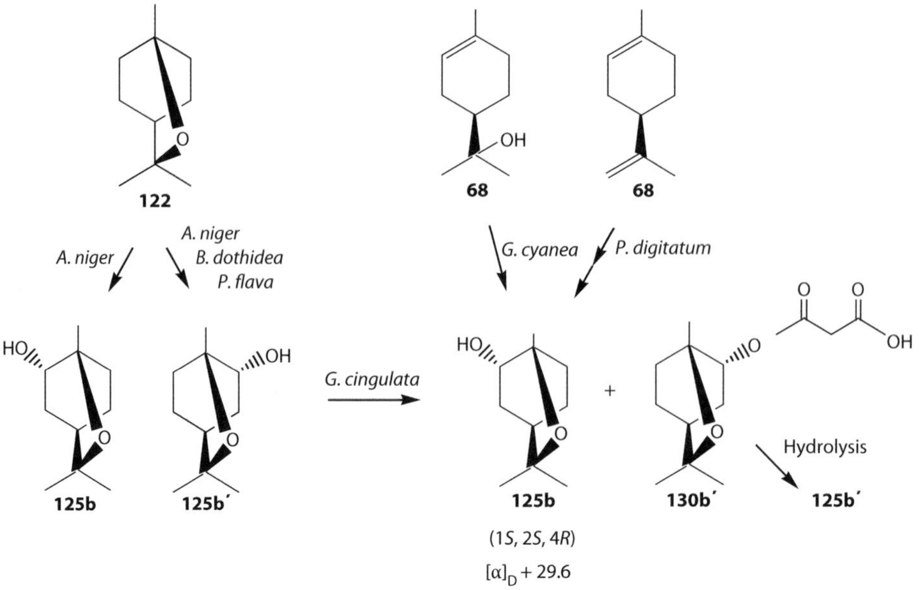

122
A. niger
A. niger
B. dothidea
P. flava
125b
125b′
G. cingulata
68
68
G. cyanea
P. digitatum
125b
(1S, 2S, 4R)
[α]D + 29.6
130b′
Hydrolysis
125b′

FIGURE 22.140 Formation of 2α-hydroxy-1,8-cineoles (**125b** and **b′**) from 1,8-cineole (**122**) and optical resolution by *Glomerella cingulata* and *Aspergillus niger* TBUYN-2 and **125b′** from (+)-limonene (**68**) by *Penicillium digitatum*. (Modified from Nishimura, H. et al. *Agric. Biol. Chem.*, 46, 2601, 1982; Abraham, W.-R. et al. *Appl. Microbiol. Biotechnol.*, 24, 24, 1986; Miyazawa, M. et al. Biotransformation of 2-*endo*-hydroxy-1,4-cineole by plant pathogenic microorganism, *Glomerella cingulata*, *Proceedings of 39th TEAC*, 1995b, pp. 352–353; Noma, Y. et al. Reduction of terpene aldehydes and epoxidation of terpene alcohols by *S. ikutamanensis*, Ya-2-1, *Proceedings of 30th TEAC*, 1986, pp. 204–206; Noma, Y. and Asakawa, Y., Biotransformation of limonene and related compounds by newly isolated low temperature grown *citrus* pathogenic fungi and red yeast, *Book of Abstracts of the 38th ISEO*, 2007a, p. 7.)

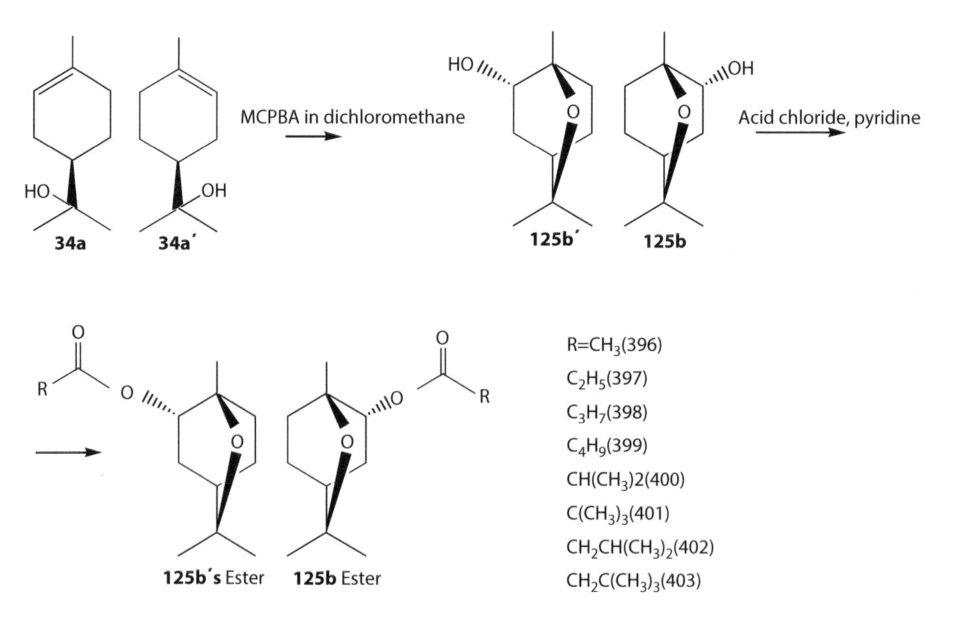

FIGURE 22.141 Chemical synthesis of esters of racemic 2α-hydroxy-1,8-cineole (**125b** and **b′**). (Modified from Hashimoto Y. and Miyazawa, M., Microbial resolution of esters of racemic 2-*endo*-hydroxy-1,8-cineole by *Glomerella cingulata*, *Proceedings of 45th TEAC*, 2001, pp. 363–365.)

TABLE 22.15
Yield and Enantiomer Excess of Esters of Racemic 2α-Hydroxy-1,8-Cineole (125b and b′) on the Microbial Resolution by *Glomerella cingulata*

	0 h	24 h		48 h	
Compounds	**% ee**	**% ee**	**Yield (%)**	**% ee**	**Yield (%)**
396	(−)36.3	(+)85.0	24.0	(+)100	14.1
397	(−)36.9	(+)73.8	18.6	(+)100	8.6
398	(−)35.6	(+)33.2	13.7	(+)75.4	3.5
399	(−)36.8	(+)45.4	14.4	(+)100	2.3
400	(−)35.4	(−)21.4	25.2	(+)20.6	8.0
401	(−)36.7	(−)37.8	31.5	(−)40.6	15.2
402	(−)36.1	(−)29.8	46.8	(−)15.0	24.0
403	(−)36.3	(−)37.6	72.2	(−)39.0	36.9

Source: Hashimoto, Y. and Miyazawa, M., Microbial resolution of esters of racemic 2-*endo*-hydroxy-1,8-cineole by *Glomerella cingulata*, *Proceedings of 45th TEAC*, 2001, pp. 363–365.

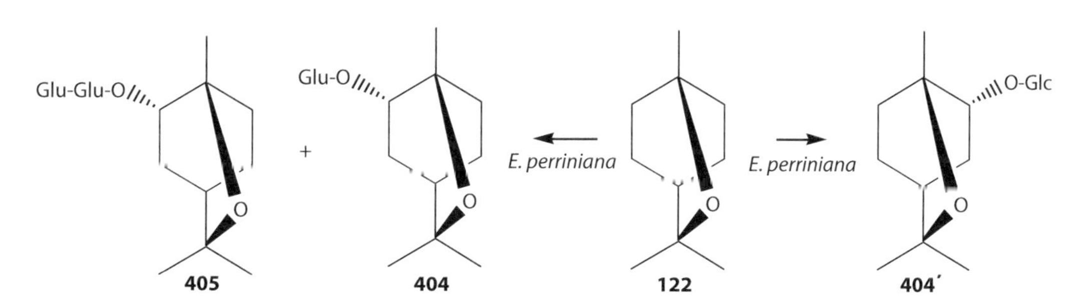

FIGURE 22.142 Biotransformation of 1,8-cineole (**122**) by *Eucalyptus perriniana* suspension cell. (Modified from Hamada, H. et al. Glycosylation of monoterpenes by plant suspension cells, *Proceedings of 46th TEAC*, 2002, pp. 321–322.)

1,8-Cineole (**122**) was glucosylated by *E. perriniana* suspension cells to 2α-hydroxy-1,8 cineole monoglucoside (**404**, 16.0% and **404′**, 16.0%) and diglucosides (**405**, 1.4%) (Hamada et al., 2002) (Figure 22.142).

22.3.4.3.2 1,4-Cineole

Regarding the biotransformation of 1,4-cineole (**131**), *S. griseus* transformed it to 8-hydroxy-1,4-cineole (**134**), whereas *Bacillus cereus* transformed 1,4-cineole (**131**) to 2α-hydroxy-1,4 cineole (**132b**, 3.8%) and 2β-hydroxy-1,4-cineoles (**132a**, 21.3%) (Liu et al., 1988) (Figures 22.143 and 22.144). On the other hand, a strain of *A. niger* biotransformed 1,4-cineole (**131**) regiospecifically to 2α-hydroxy-1,4-cineole (**132b**) (Miyazawa et al., 1991e) and (+)-3α-hydroxy-1,4-cineole (**133b**) (Miyazawa et al., 1992d) along with the formation of 8-hydroxy-1,4-cineole (**134**) and 9-hydroxy-1,4 cineole (**135**) (Miyazawa et al., 1992e) (Figure 22.144).

Microbial optical resolution of racemic 2α-hydroxy-1,4-cineoles (**132b** and **b′**) was carried out by using *G. cingulata* (Liu et al., 1988). The mixture of 2α-hydroxy-1,4-cineoles (**132b** and **b′**) was added to a culture of *G. cingulata* and esterified to give after 24 h (1*R*,2*R*,4*S*)-2α-hydroxy-1,4-cineole-2-yl-malonate (**136′**) in 45% yield (ee 100%). The recovered alcohol showed an ee of 100% of the (1*S*,2*S*,4*R*)-enantiomer (**132b**). On the other hand, optically active (+)-2α-hydroxy-1,4-cineole

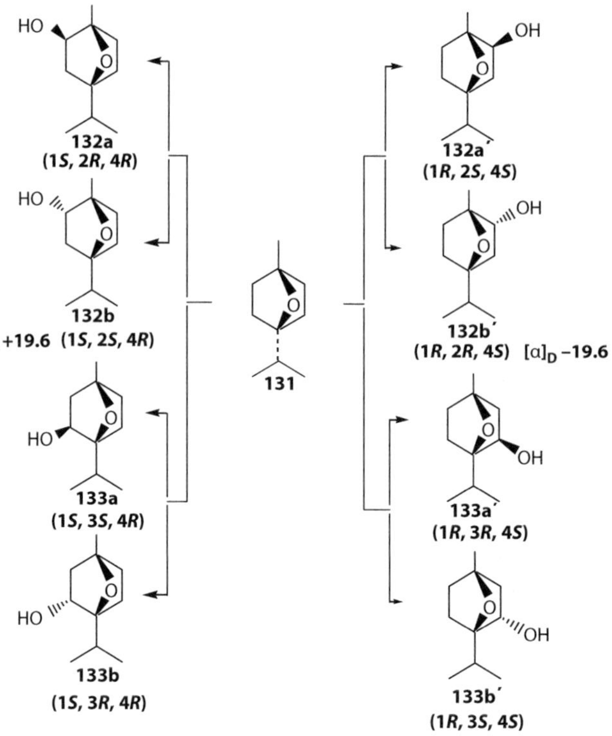

FIGURE 22.143 Metabolic pathways of 1,4-cineole (**131**) by microorganisms.

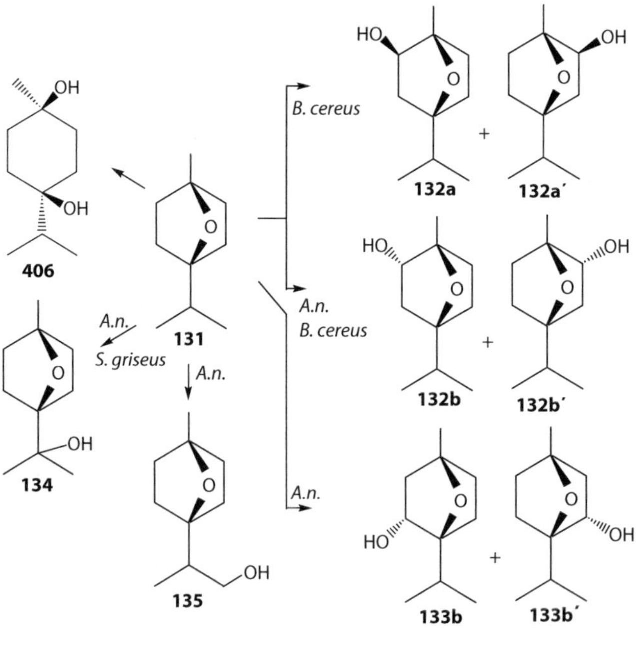

FIGURE 22.144 Metabolic pathways of 1,4-cineole (**131**) by *Aspergillus niger* TBUYN-2, *Bacillus cereus*, and *Streptomyces griseus*. (Modified from Liu, W. et al. *J. Org. Chem.*, 53, 5700, 1988; Miyazawa, M. et al. *Chem. Express*, 6, 771, 1991c; Miyazawa, M. et al. *Chem. Express*, 7, 305, 1992b; Miyazawa, M. et al. *Chem. Express*, 7, 125, 1992c; Miyazawa, M. et al. Biotransformation of 2-endo-hydroxy-1,4-cineole by plant pathogenic microorganism, Glomerella cingulata, *Proceedings of 39th TEAC*, 1995b, pp. 352–353.)

131
342
A. niger A. niger
A. niger
G. cyanea
G. cingulata
HO OH HO O OH
132b 132b′
rac-132b and b′
$[\alpha]_D$ +0.13
132b
(1S, 2S, 4R)
$[\alpha]_D$ +19.6
136′
132b′
(1R, 2R, 4S)
$[\alpha]_D$ −19.6

FIGURE 22.145 Formation of optically active 2α-hydroxycineole from 1,4-cineole (**131**) and terpinen-4-ol (**342**) by *Aspergillus niger* TBUYN-2, *Gibberella cyanea*, and *Glomerella cingulata*. (Modified from Abraham, W.-R. et al. *Appl. Microbiol. Biotechnol.*, 24, 24, 1986; Miyazawa, M. et al. *Chem. Express*, 6, 771, 1991c; Miyazawa, M. et al. Biotransformation of 2-*endo*-hydroxy-1,4-cineole by plant pathogenic microorganism, *Glomerella cingulata*, *Proceedings of 39th TEAC*, 1995b, pp. 352–353; Noma, Y. and Asakawa, Y., Microbial transformation of limonene and related compounds, *Proceedings of 51st TEAC*, 2007b, pp. 299–301.)

(**132b**) was also formed from (−)-terpinen-4-ol (**342**) by *G. cyanea* DSM (Abraham et al., 1986) and *A. niger* TBUYN-2 (Noma and Asakawa, 2007b) (Figure 22.145).

22.4 METABOLIC PATHWAYS OF BICYCLIC MONOTERPENOIDS

22.4.1 Bicyclic Monoterpene

22.4.1.1 α-Pinene

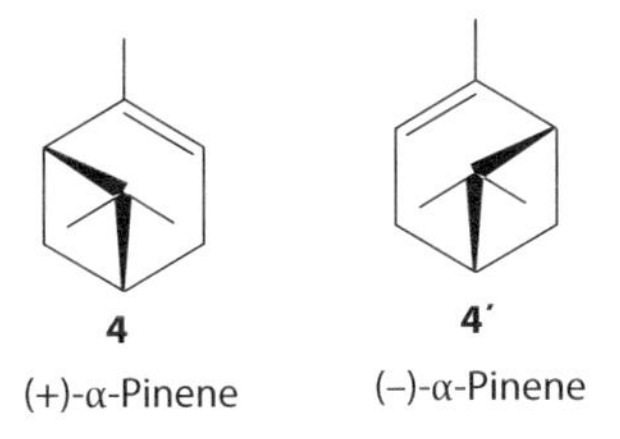

α-Pinene (**4** and **4′**) is the most abundant terpene in nature and obtained industrially by fractional distillation of turpentine (Krasnobajew, 1984). (+)-α-Pinene (**4**) occurs in oil of *Pinus palustris* Mill. at concentrations of up to 65% and in oil of *Pinus caribaea* at concentrations of 70% (Bauer et al., 1990). On the other hand, *P. caribaea* contains (−)-α-pinene (**4′**) at the concentration of 70%–80% (Bauer et al., 1990).

The biotransformation of (+)-α-pinene (**4**) was investigated by *A. niger* NCIM 612 (Bhattacharyya et al., 1960; Prema and Bhattachayya, 1962). A 24 h shake culture of this strain metabolized 0.5% (+)-α-pinene (**4**) in 4–8 h. After the fermentation of the culture broth contained (+)-verbenone (**24**) (2%–3%), (+)-*cis*-verbenol (**23b**) (20%–25%), (+)-*trans*-sobrerol (**43a**) (2%–3%), and (+)-8-hydroxycarvotanacetone (**44**) (Bhattacharyya et al., 1960; Prema and Bhattachayya, 1962) (Figure 22.146).

The degradation of (+)-α-pinene (**4**) by a soil *Pseudomonas* sp. (PL strain) was investigated by Hungund et al. (1970). A terminal oxidation pattern was proposed, leading to the formation of organic

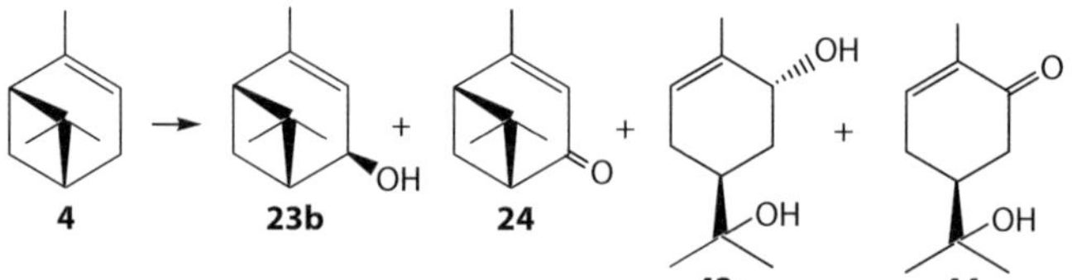

FIGURE 22.146 Biotransformation of (+)-α-pinene (**4**) by *Aspergillus niger* NCIM 612. (Modified from Bhattacharyya, P.K. et al. *Nature*, 187, 689, 1960; Prema, B.R. and Bhattachayya, P.K., *Appl. Microbiol.*, 10, 524, 1962.)

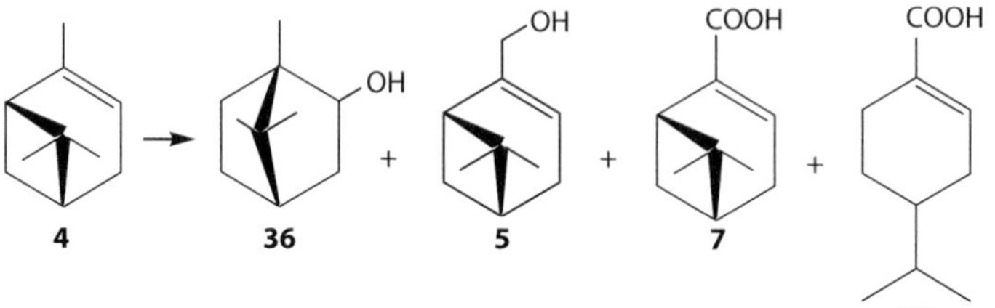

FIGURE 22.147 Biotransformation of (+)-α-pinene (**4**) by *Pseudomonas* sp. (PL strain). (Modified from Shukla, O.P., and Bhattacharyya, P.K., *Indian J. Biochem.*, 5, 92, 1968.)

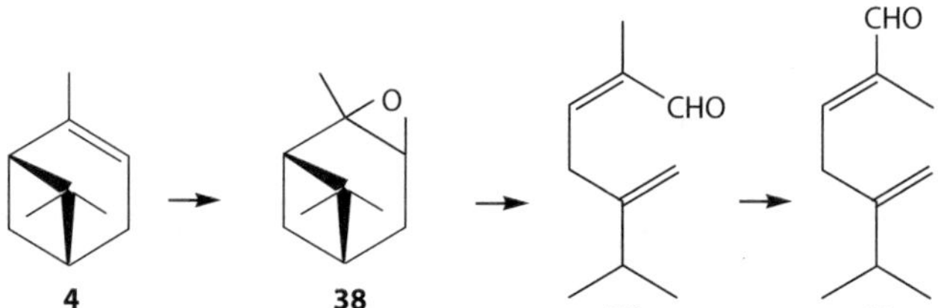

FIGURE 22.148 Biotransformation of (+)-α-pinene (**4**) by *Pseudomonas fluorescens* NCIMB 11671. (Modified from Best, D.J. et al. *Biocatal. Biotransform.*, 1, 147, 1987.)

acids through ring cleavage. (+)-α-Pinene (**4**) was fermented in shake cultures by a soil *Pseudomonas* sp. (PL strain) that is able to grow on (+)-α-pinene (**4**) as the sole carbon source, and borneol (**36**), myrtenol (**5**), myrtenic acid (**84**), and α-phellandric acid (**65**) (Shukla and Bhattacharyya, 1968) (Figure 22.147) were obtained.

The degradation of (+)-α-pinene (**4**) by *P. fluorescens* NCIMB11671 was studied, and a pathway for the microbial breakdown of (+)-α-pinene (**4**) was proposed as shown in Figure 22.148 (Best et al., 1987; Best and Davis, 1988). The attack of oxygen is initiated by enzymatic oxygenation of the 1,2-double bond to form α-pinene epoxide (**38**), which then undergoes rapid rearrangement to produce a unsaturated aldehyde, occurring as two isomeric forms. The primary product of the reaction (*Z*)-2-methyl-5-isopropylhexa-2,5-dien-1-al (**39**, isonovalal) can undergo chemical isomerization to the *E*-form (novalal, **40**). Isonovalal (**39**), the native form of the aldehyde, possesses citrus, woody, and spicy notes, whereas novalal (**40**) has woody, aldehydic, and cyclone notes. The same biotransformation was also carried out by *Nocardia* sp. strain P18.3 (Griffiths et al., 1987a,b).

Pseudomonas PL strain and PIN 18 degraded α-pinene (**4**) by the pathway proposed in Figure 22.149 to give two hydrocarbon, limonene (**68**) and terpinolene (**346**), and neutral metabolite, borneol (**36**). A probable pathway has been proposed for the terminal oxidation of β-isopropylpimelic acid (**248**) in the PL strain and PIN 18 (Shukla and Bhattacharyya, 1968).

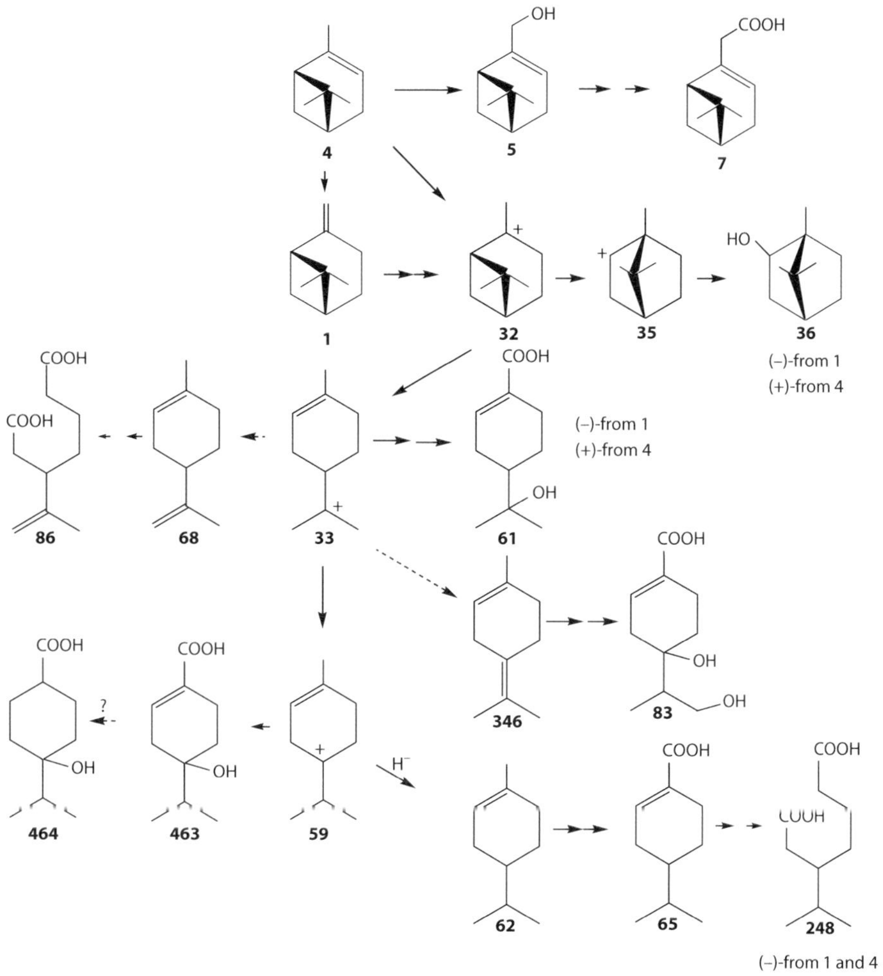

FIGURE 22.149 Metabolic pathways of degradation of α- and β-pinene by a soil Pseudomonad (PL strain) and *Pseudomonas* PIN 18. (Modified from Shukla, O.P. and Bhattacharyya, P.K., *Indian J. Biochem.*, 5, 92, 1968.)

Pseudomonas PX 1 biotransformed (+)-α-pinene (**4**) to give (+)-*cis*-thujone (**29**) and (+)-*trans*-carveol (**81a**) as major compounds. Compounds **81a**, **171**, **173**, and **178** have been identified as fermentation products (Gibbon and Pirt, 1971; Gibbon et al., 1972) (Figure 22.150).

A. niger TBUYN-2 biotransformed (−)-α-pinene (**4′**) to give (−)-α-terpineol (**34′**) and (−)-*trans*-sobrerol (**43a′**) (Noma et al., 2001). The mosquitocidal (+)-(1*R*,2*S*,4*R*)-1-*p*-menthane-2,8-diol (**50a′**) was also obtained as a crystal in the biotransformation of (−)-α-pinene (**4′**) by *A. niger* TBUYN-2 (Noma et al., 2001; Noma, 2007) (Figure 22.151).

(1*R*)-(+)-α-Pinene (**4**) and its enantiomer (**4′**) were fed to *S. litura* to give the corresponding (+)- and (−)-verbenones (**24** and **24′**) and (+)- and (−)-myrtenols (**5** and **5′**) (Miyazawa et al., 1996c) (Figure 22.152).

(−)-α-Pinene (**4′**) was treated in human liver microsomes CYP2B6 to afford (−)-*trans*-verbenol (**23′**) and (−)-myrtenol (**5′**) (Sugie and Miyazawa, 2003) (Figure 22.153).

In rabbit, (+)-α-pinene (**4**) was metabolized to (−)-*trans*-verbenols (**23**) as the main metabolites together with myrtenol (**5**) and myrtenic acid (**7**). The purities of (−)-verbenol (**23**) from (−)- (**4′**), (+)- (**4**), and (+/−)-α-pinene (**4** and **4′**) were 99%, 67%, and 68%, respectively. This means that the biotransformation of (−)-**4′** in rabbit is remarkably efficient in the preparation of (−)-*trans*-verbenol (**23a**) (Ishida et al., 1981b) (Figure 22.154).

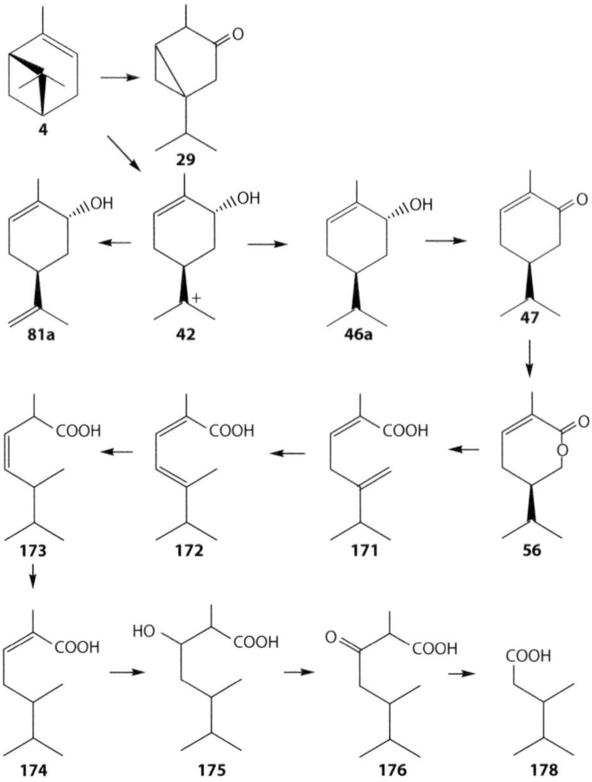

FIGURE 22.150 Proposed metabolic pathways for (+)-α-pinene (**4**) degradation by *Pseudomonas* PX 1. (Modified from Gibbon, G.H. and Pirt, S.J., FEBS Lett., 18, 103, 1971; Gibbon, G.H. et al. Degradation of α-pinene by bacteria, *Proceedings of IV IFS: Fermentation Technology Today*, 1972, pp. 609–612.)

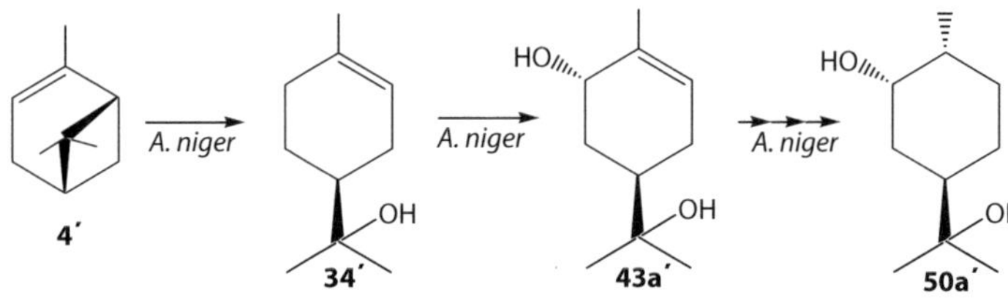

FIGURE 22.151 Biotransformation of (−)-α-pinene (**4**) by *Aspergillus niger* TBUYN-2. (Modified from Noma, Y. et al. Microbiological transformation of β-pinene, *Proceedings of 45th TEAC*, 2001, pp. 88–90.)

(−)-α-Pinene (**4′**) was biotransformed by the plant pathogenic fungus *B. cinerea* to afford 3α-hydroxy-(−)-β-pinene (**2a′**, 10%), 8-hydroxy-(−)-α-pinene (**434′**, 12%), 4β-hydroxy-(−)-pinene-6-one (**468′**, 16%), and (−)-verbenone (**24′**) (Farooq et al., 2002) (Figure 22.155).

22.4.1.2 β-Pinene

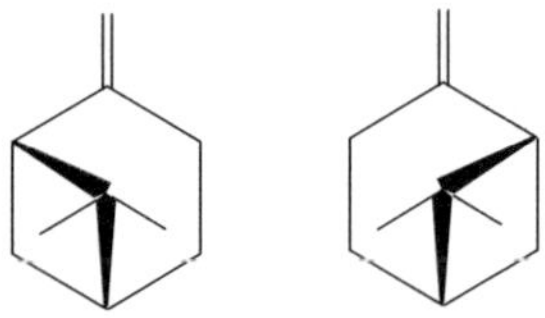

1 (+)-β-Pinene

1′ (−)-β-Pinene

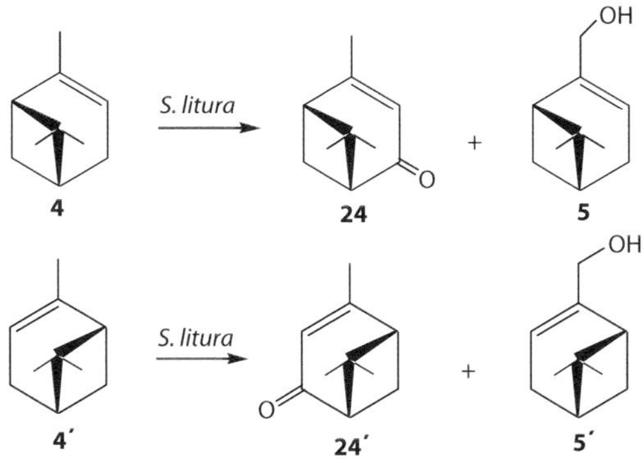

FIGURE 22.152 Biotransformation of (+)- (**4**) and (−)-α-pinene (**4′**) by *Spodoptera litura*. (Modified from Miyazawa, M. et al. Biotransformation of pinanes by common cutworm larvae, *Spodoptera litura* as a biocatalyst, *Proceedings of 40th TEAC*, 1996c, pp. 84–85.)

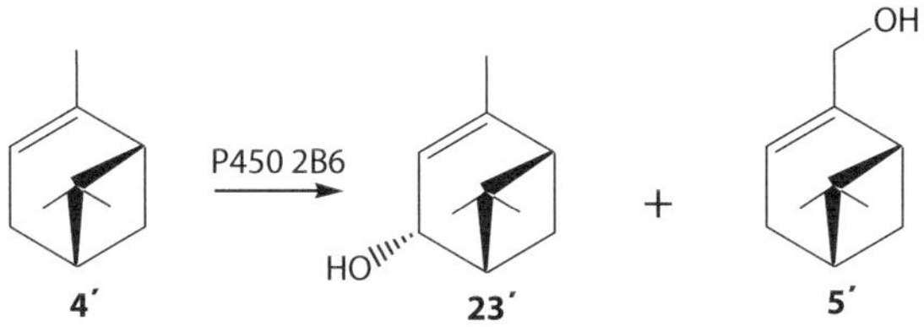

FIGURE 22.153 Biotransformation of (−)-α-pinene (**4′**) by human liver microsomes CYP2B6. (Modified from Sugie, A. and Miyazawa, M., Biotransformation of (−)-α-pinene by human liver microsomes, *Proceedings of 47th TEAC*, 2003, pp. 159–161.)

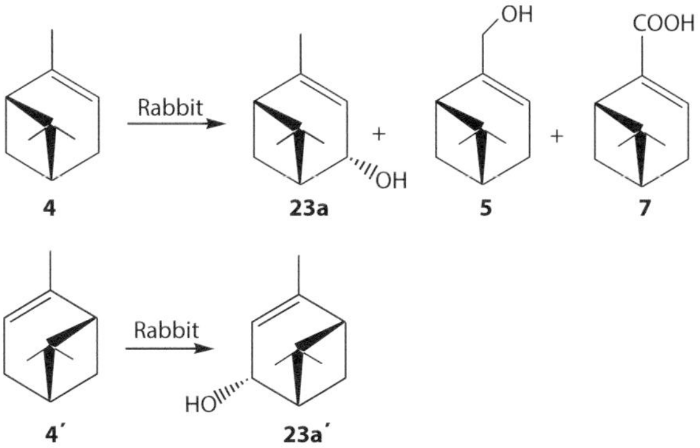

FIGURE 22.154 Biotransformation of α-pinene by rabbit. (Modified from Ishida, T. et al. *J. Pharm. Sci.*, 70, 406, 1981b.)

(+)-β-Pinene (**1**) is found in many essential oils. Optically active and racemic β-pinenes are present in turpentine oils, although in smaller quantities than (+)-α-pinene (**4**) (Bauer et al., 1990).

Shukla et al. (1968) obtained a similarly complex mixture of transformation products from (−)-β-pinene (**1′**) through degradation by a *Pseudomonas* sp/(PL strain). On the other hand, Bhattacharyya and Ganapathy (1965) indicated that *A. niger* NCIM 612 acts differently and more specifically on the pinenes by preferably oxidizing (−)-β-pinene (**1′**) in the allylic position to form the interesting products pinocarveol (**2′**) and pinocarvone (**3′**), besides myrtenol (**5′**) (see Figure 22.156). Furthermore, the conversion of (−)-β-pinene (**1′**) by *Pseudomonas putida arvilla* (PL strain) gave borneol (**36′**) (Rama Devi and Bhattacharyya, 1978) (Figure 22.156).

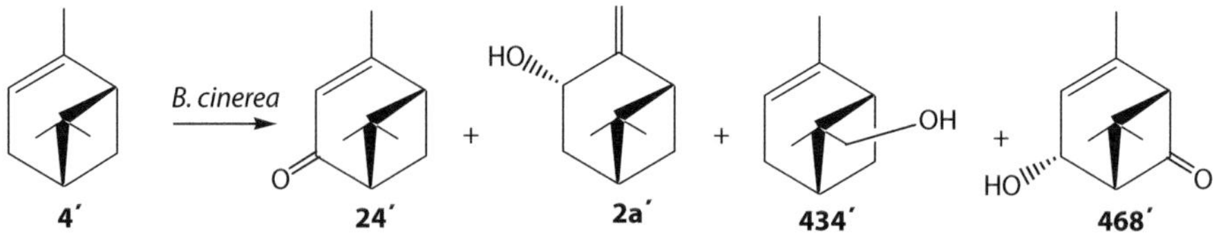

FIGURE 22.155 Microbial transformation of (−)-α-pinene (**4′**) by *Botrytis cinerea*. (Modified from Farooq, A. et al. *Z. Naturforsch.*, 57c, 686, 2002.)

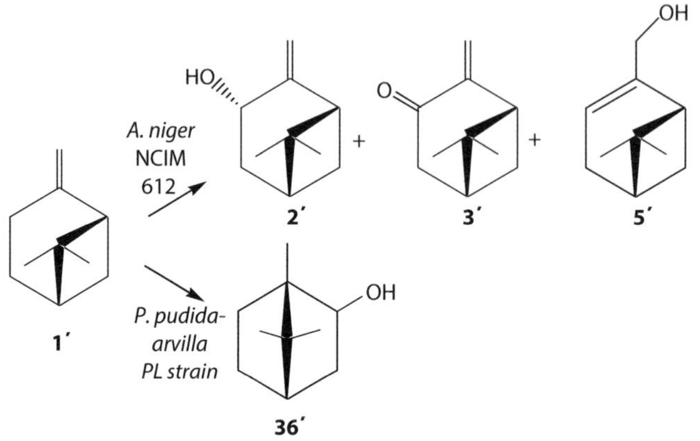

FIGURE 22.156 Biotransformation of (−)-β-pinene (**1′**) by *Aspergillus niger* NCIM 612 and *Pseudomonas putida arvilla* (PL strain). (Modified from Bhattacharyya, P.K. and Ganapathy, K., *Indian J. Biochem.*, 2, 137, 1965; Rama Devi, J. and Bhattacharyya, P.K., *J. Indian Chem. Soc.*, 55, 1131, 1978.)

P. pseudomallei isolated from local sewage sludge by the enrichment culture technique utilized caryophyllene as the sole carbon source (Dhavalikar et al., 1974). Fermentation of (−)-β-pinene (**1′**) by *P. pseudomallei* in a mineral salt medium (Seubert's medium) at 30°C with agitation and aeration for 4 days yielded camphor (**37′**), borneol (**36a′**), isoborneol (**36b′**), α-terpineol (**34′**), and β-isopropyl pimelic acid (**248′**) (see Figure 22.154). Using modified Czapek Dox medium and keeping the other conditions the same, the pattern of the metabolic products was dramatically changed. The metabolites were *trans*-pinocarveol (**2′**), myrtenol (**5′**), α-fenchol (**11′**), á-terpineol (**34′**), myrtenic acid (**7′**), and two unidentified products (see Figure 22.157).

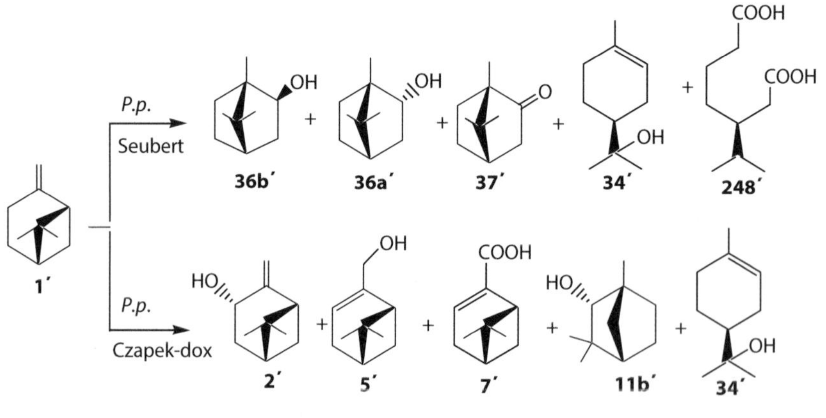

FIGURE 22.157 Biotransformation of (−)-β-pinene (**1′**) by *Pseudomonas pseudomallei*. (Modified from Dhavalikar, R.S. et al. *Dragoco Rep.*, 3, 47, 1974.)

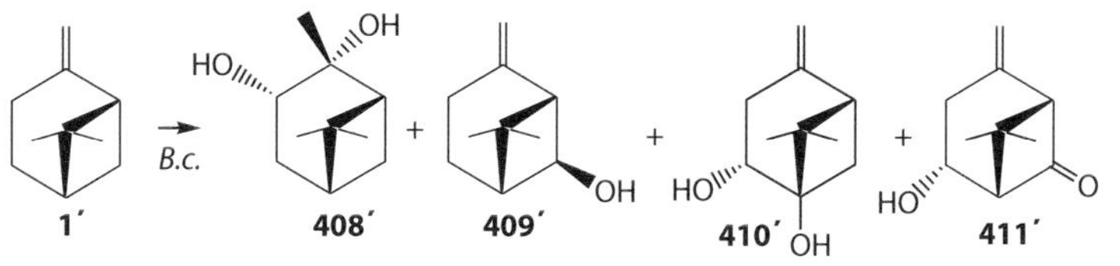

FIGURE 22.158 Biotransformation of (−)-β-pinene (**1′**) by *Botrytis cinerea*. (Modified from Farooq, A. et al. *Z. Naturforsch*., 57c, 686, 2002.)

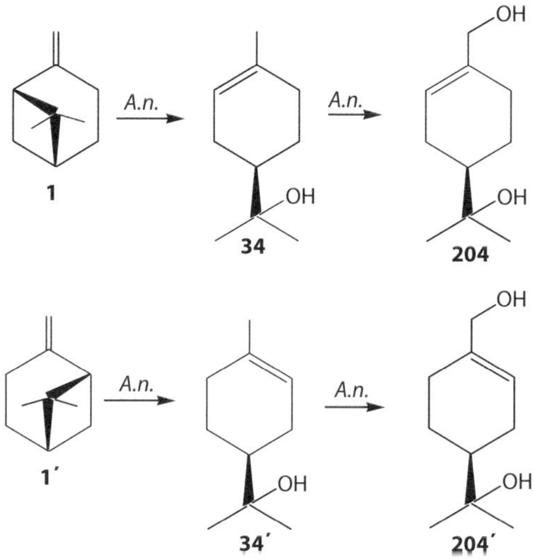

FIGURE 22.159 Biotransformation of (+)- (**1**) and (−)-β-pinene (**1′**) by *Aspergillus niger* TBUYN-2. (Modified from Noma, Y. et al. Microbiological transformation of β-pinene, *Proceedings of 45th TEAC*, 2001, pp. 88–90.)

(−)-β-Pinene (**1′**) was converted by plant pathogenic fungi, *B. cinerea*, to give four new compounds such as (−)-pinane-2α,3α-diol (**408′**), (−)-6β-hydroxypinene (**409′**), (−)-4α,5-dihydroxypinene (**410′**), and (−)-4α-hydroxypinene-6-one (**411′**) (Figure 22.158).

This study progressed further biotransformation of (−)-pinane-2α,3α-diol (**408′**) and related compounds by microorganisms as shown in Figure 22.158.

As shown in Figure 22.159, (+)- (**1**) and (−)-β-pinenes (**1′**) were biotransformed by *A. niger* TBUYN-2 to give (+)-α-terpineol (**34**) and (+)-oleuropeyl alcohol (**204**) and their antipodes (**34′** and **204′**), respectively. The hydroxylation process of α-terpineol (**34**) to oleuropeyl alcohol (**204**) was completely inhibited by 1-aminotriazole as a cytochrome P450 inhibitor.

(−)-β-Pinene (**1′**) was at first biotransformed by *A. niger* TBUYN-2 to give (+)-*trans*-pinocarveol (**2a′**) (**274**). (+)-*trans*-Pinocarveol (**2a′**) was further transformed by three pathways: First, (+)-*trans*-pinocarveol (**2a′**) was metabolized to (+)-pinocarvone (**3′**), (−)-3-isopinanone (**413′**), (+)-2α-hydroxy-3-pinanone (**414′**), and (+)-2α,5-dihydroxy-3-pinanone (**415′**). Second, (+)-*trans*-pinocarveol (**2a′**) was metabolized to (+)-6β-hydroxyfenchol (**349ba′**), and third, (+)-*trans*-pinocarveol (**2a′**) was metabolized to (−)-6β,7-dihydroxyfenchol (**412ba′**) via epoxide and diol as intermediates (Noma and Asakawa, 2005a) (Figure 22.160).

(−)-β-Pinene (**1′**) was metabolized by *A. niger* TBUYN-2 with three pathways as shown in Figure 22.154 to give (−)-α-pinene (**4′**), (−)-α-terpineol (**34′**), and (+)-*trans*-pinocarveol (**2a′**). (−)-α-Pinene (**4′**) is further metabolized by three pathways. At first, (−)-α-pinene (**4′**) was metabolized via (−)-α-pinene epoxide (**38′**), *trans*-sobrerol (**43a′**), (−)-8-hydroxycarvotanacetone (**44′**), and (+)-8-hydroxycarvomenthone (**45a**) to (+)-*p*-menthane-2,8-diol (**50a′**), which was also metabolized

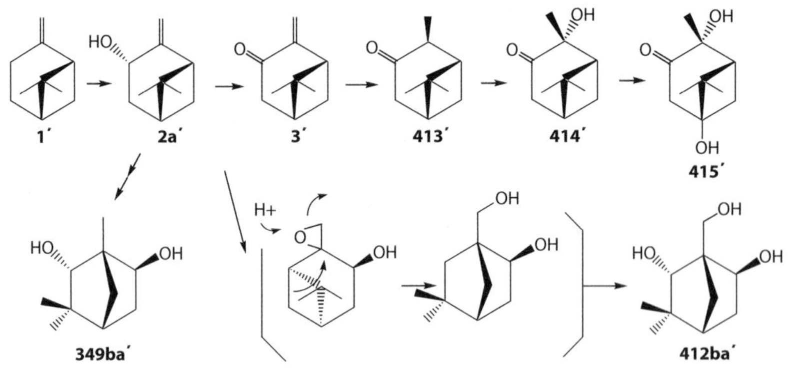

FIGURE 22.160 The metabolism of (−)-β-pinene (**1′**) and (+)-*trans*-pinocarveol (**2a′**) by *Aspergillus niger* TBUYN-2. (Modified from Noma, Y. and Asakawa, Y., New metabolic pathways of β-pinene and related compounds by *Aspergillus niger, Book of Abstracts of the 36th ISEO*, 2005a, p. 32.)

in (−)-carvone (**93′**) metabolism. Second, (−)-α-pinene (**4′**) is metabolized to myrtenol (**83′**), which is metabolized by rearrangement reaction to give (−)-oleuropeyl alcohol (**204′**). (−)-α-Terpineol (**34′**), which is formed from β-pinene (**1′**), was also metabolized to (−)-oleuropeyl alcohol (**204′**), and (+)-*trans*-pinocarveol (**2a′**), formed from (−)-β-pinene (**1′**), was metabolized to pinocarvone (**3′**), 3-pinanone (**413′**), 2α-hydroxy-3-pinanone (**414′**), 2α,5-dihydroxy-3-pinanone (**415′**), and 2α,9-dihydroxy-3-pinanone (**416′**). Furthermore, (+)-*trans*-pinocarveol (**2a′**) was metabolized by rearrangement reaction to give 6β-hydroxyfenchol (**349ba′**) and 6β,7-dihydroxyfenchol (**412ba′**) (Noma and Asakawa, 2005a) (Figure 22.161).

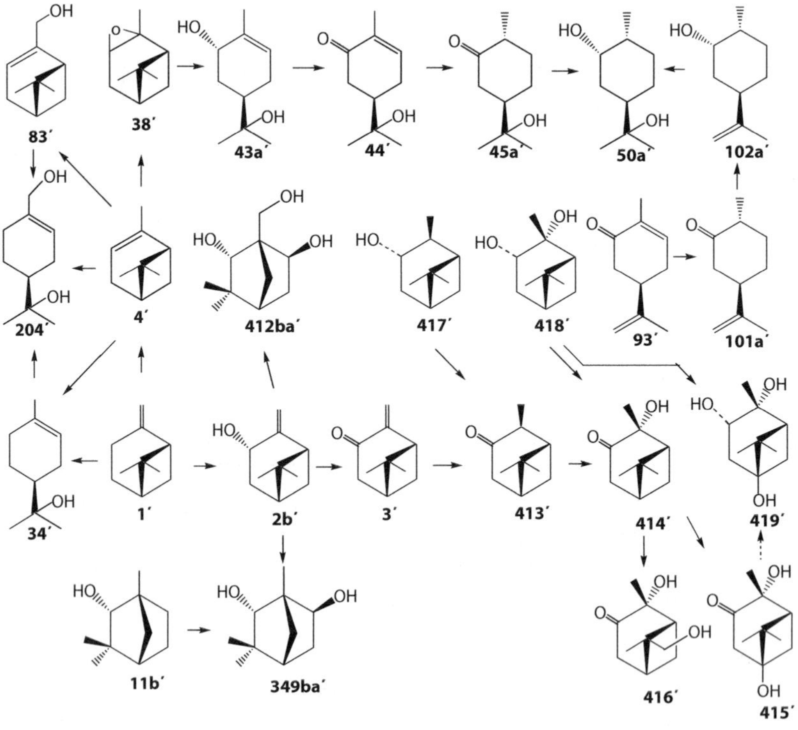

FIGURE 22.161 Biotransformation of (−)-β-pinene (**1′**), (−)-α-pinene (**4′**), and related compounds by *Aspergillus niger* TBUYN-2. (Modified from Noma, Y. and Asakawa, Y., New metabolic pathways of β-pinene and related compounds by *Aspergillus niger, Book of Abstracts of the 36th ISEO*, 2005a, p. 32.)

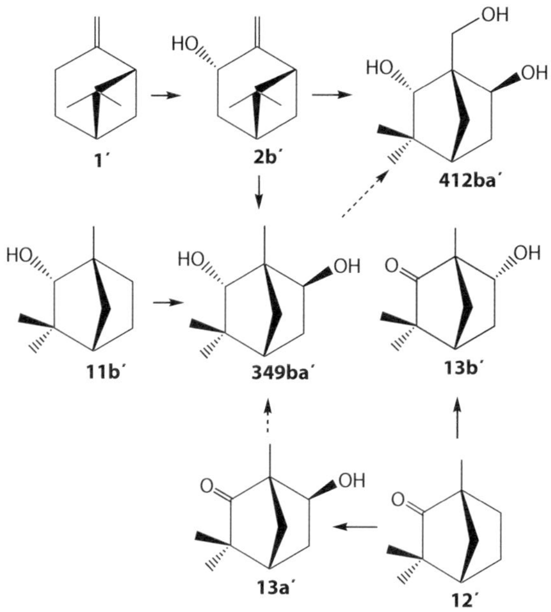

FIGURE 22.162 Relationship of the metabolism of (–)-β-pinene (**1′**), (+)-fenchol (**11′**), and (–)-fenchone (**12′**) by *Aspergillus niger* TBUYN-2. (Modified from Noma, Y. and Asakawa, Y., New metabolic pathways of β-pinene and related compounds by *Aspergillus niger, Book of Abstracts of the 36th ISEO*, 2005a, p. 32.)

(–)-β-Pinene (**1′**) was metabolized by *A. niger* TBUYN-2 to give (+)-*trans*-pinocarveol (**2a′**), which was further metabolized to 6β-hydroxyfenchol (**349ba′**) and 6β, 7-dihydroxyfenchol (**412ba′**) by rearrangement reaction (Noma and Asakawa, 2005a) (Figure 22.162). 6β-Hydroxyfenchol (**349ba′**) was also obtained from (–)-fenchol (**11b′**). (–)-Fenchone was hydroxylated by the same fungus to give 6β- (**13a′**) and 6α-hydroxy-(–)-fenchone (**13b′**). There is a close relationship between the metabolism of (–)-β-pinene (**1′**) and those of (–)-fenchol (**11′**) and (–)-fenchone (**12′**).

(–)-β-Pinene (**1′**) and (–)-α-pinene (**4′**) were isomerized to each other. Both are metabolized via (–)-α-terpineol (**34'**) to (–)-oleuropeyl alcohol (**204′**) and (–)-oleuropeic acid (**61′**). (–)-Myrtenol (**5′**) formed from (–)-α-pinene (**1′**) was further metabolized via cation to (–)-oleuropeyl alcohol (**204′**) and (–)-oleuropeic acid (**61′**). (–)-α-Pinene (**4′**) is further metabolized by *A. niger* TBUYN-2 via (–)-α-pinene epoxide (**38′**) to *trans*-sobrerol (**43a′**), (–)-8-hydroxycarvotanacetone (**44′**), (+)-8-hydroxycarvomenthone (**45a**), and mosquitocidal (+)-*p*-menthane-2,8-diol (**50a′**) (Bhattacharyya et al., 1960; Noma et al., 2001, 2002, 2003) (Figure 22.163).

The major metabolites of (–)-β-pinene (**1′**) were *trans*-10-pinanol (myrtanol) (**8ba′**) (39%) and (–)-1-*p*-menthene-7,8-diol (oleuropeyl alcohol) (**204′**) (30%). In addition, (+)-*trans*-pinocarveol (**2a′**) (11%) and (–)-α-terpineol (**34′**) (5%) and verbenol (**23a** and **23b**) and pinocarveol (**2a′**) were oxidation products of α-(**4**) and β-pinene (**1**), respectively, in bark beetle, *Dendroctonus frontalis*. (–)-*cis*- (**23b′**) and (+)-*trans*-verbenols (**23a′**) have pheromonal activity in *Ips paraconfusus* and *Dendroctonus brevicomis*, respectively (Ishida et al., 1981b) (Figure 22.164).

22.4.1.3 (±)-Camphene

Racemate camphene (**437** and **437′**) is a bicyclic monoterpene hydrocarbon found in *Liquidambar* species, *Chrysanthemum*, *Zingiber officinale*, *Rosmarinus officinalis*, and among other plants. It was administered into rabbits. Six metabolites, camphene-2,10-glycols (**438a, 438b**), which were the major metabolites, together with 10-hydroxytricyclene (**438c**), 7-hydroxycamphene (**438d**), 6-exo-hydroxycamphene (**438e**), and 3-hydroxytricyclene (**438f**), were obtained (Ishida et al., 1979). On the basis of the production of the glycols (**438a** and **438b**) in good yield, these alcohols might be formed

FIGURE 22.163 Metabolic pathways of (−)-β-pinene (**1′**) and related compounds by *Aspergillus niger* TBUYN-2. (Modified from Bhattacharyya, P.K. et al. Nature, 187, 689, 1960; Noma, Y. et al. Microbiological transformation of β-pinene, Proceedings of 45th TEAC, 2001, pp. 88–90; Noma, Y. et al. Stereoselective formation of (1R, 2S, 4R)-(+)-p-menthane-2,8-diol from α-pinene, *Book of Abstracts of the 33rd ISEO*, 2002, p. 142; Noma, Y. et al. Biotransformation of (+)- and (−)-pinane-2,3-diol and related compounds by *Aspergillus niger, Proceedings of 47th TEAC*, 2003, pp. 91–93.)

through their epoxides as shown in Figure 22.165. The homoallyl camphene oxidation products (**438c–f**) apparently were formed through the nonclassical cation as the intermediate.

22.4.1.4 3-Carene and Carane

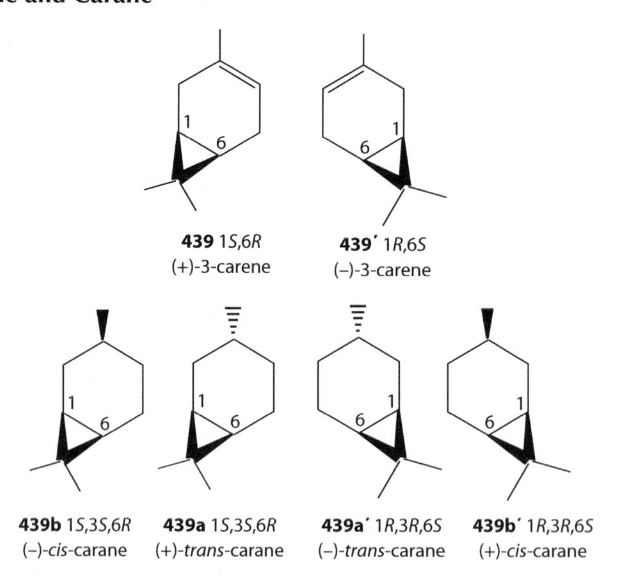

(+)-3-Carene (**439**) was biotransformed by rabbits to give *m*-mentha-4,6-dien-8-ol (**440**) (71.6%) as the main metabolite together with its aromatized *m*-cymen-8-ol (**441**). The position of C-5 in the substrate is thought to be more easily hydroxylated than C-2 by enzymatic systems in the rabbit liver. In addition to the ring opening compound, 3-carene-9-ol (**442**), 3-carene-9-carboxylic acid

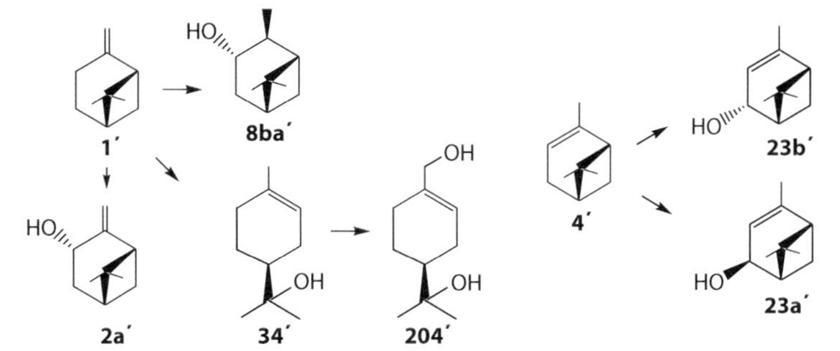

FIGURE 22.164 Metabolism of β-pinene (**1**) by bark beetle, *Dendroctonus frontalis*. (Modified from Ishida, T. et al. *J. Pharm. Sci*., 70, 406, 1981b.)

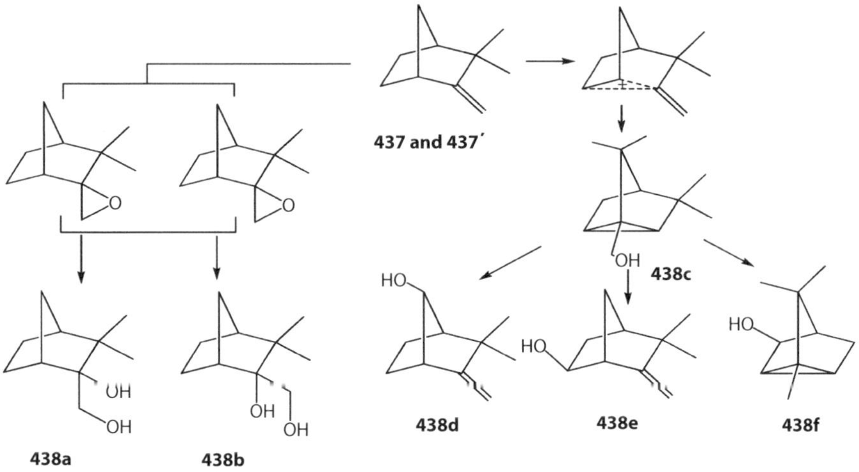

FIGURE 22.165 Biotransformation of (±)-camphene (**437** and **437′**) by rabbits. (Modified from Ishida, T. et al. *J. Pharm. Sci*., 68, 928, 1979.)

(**443**), 3-carene-9,10-dicarboxylic acid (**445**), chamic acid, and 3-caren-10-ol-9-carboxylic acid (**444**) were formed. The formation of such compounds is explained by stereoselective hydroxylation and carboxylation of *gem*-dimethyl group (Ishida et al., 1981b) (Figure 22.166). In the case of (–)-*cis*-carane (**446**), two C-9 and C-10 methyl groups were oxidized to give dicarboxylic acid (**447**) (Ishida et al., 1981b) (Figure 22.166).

3-(+)-Carene (**439**) was converted by *A. niger* NC 1M612 to give either hydroxylated compounds of 3-carene-2-one or 3-carene-5-one, which was not fully identified (Noma et al., 2002) (Figure 22.167).

22.4.2 Bicyclic Monoterpene Aldehyde

22.4.2.1 Myrtenal and Myrtanal

6 (+)- **6′** (–)- Myrtenal

435b (+)-*cis*- Binihiol

435a (–)-*trans*- **435b′** (–)-*cis*- **435a′** (+)-*trans*- Myrtanal

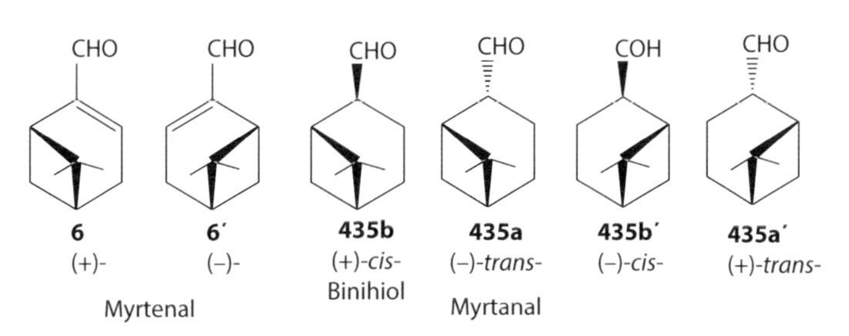

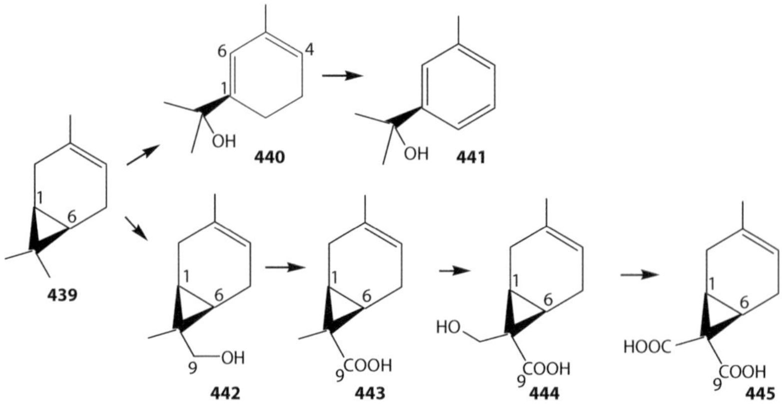

FIGURE 22.166 Metabolic pathways of (+)-3-carene (**439**) by rabbit (Modified from Ishida, T. et al. *J. Pharm. Sci.*, 70, 406, 1981b). 3-(+)-Carene (**439**) was converted by *Aspergillus niger* NC 1M612 to give either hydroxylated compounds of 3-carene-2-one or 3-carene-5-one, which was not fully identified (Figure 22.167). (Modified from Noma, Y. et al. Stereoselective formation of (1*R*, 2*S*, 4*R*)-(+)-*p*-menthane-2,8-diol from α-pinene, *Book of Abstracts of the 33rd ISEO*, 2002, p. 142.)

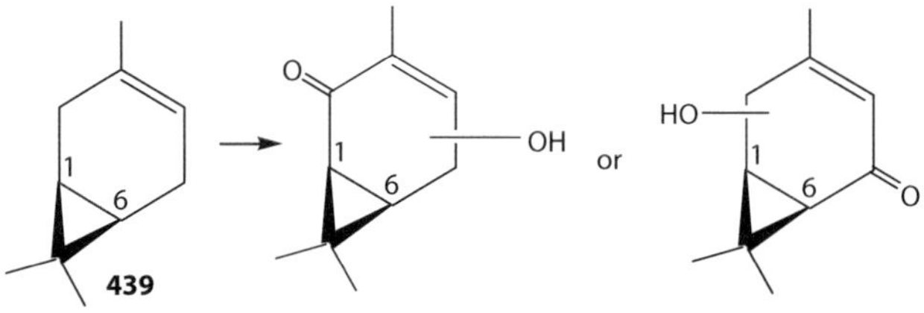

FIGURE 22.167 Metabolic pathways of (+)-3-Carene (**439**) by *Aspergillus niger* NC 1M612. (Modified from Noma, Y. et al. Stereoselective formation of (1*R*, 2*S*, 4*R*)-(+)-*p*-menthane-2,8-diol from α-pinene, *Book of Abstracts of the 33rd ISEO*, 2002, p. 142.)

E. gracilis Z. biotransformed (–)-myrtenal (**6′**) to give (–)-myrtenol (**5′**) as the major product and (–)-myrtenoic acid (**7′**) as the minor product. However, further hydrogenation of (–)-myrtenol (**5′**) to *trans*- and *cis*-myrtanol (**8a** and **8b**) did not occur even at a concentration less than ca. 50 mg/L. (*S*)-*trans*- and (*R*)-*cis*-myrtanal (**435a′** and **435b′**) were also transformed to *trans*- and *cis*-myrtanol (**8a′** and **8b′**) as the major products and (*S*)-*trans*- and (*R*)-*cis*-myrtanic acid (**436a′** and **436b′**) as the minor products, respectively (Noma et al., 1991b) (Figure 22.168).

In the case of *A. niger* TBUYN-2, *Aspergillus sojae*, and *Aspergillus usami*, (–)-myrtenol (**5′**) was further metabolized to 7-hydroxyverbenone (**25′**) as a minor product together with (–)-oleuropeyl alcohol (**204′**) as a major product (**279, 280**). (–)-Oleuropeyl alcohol (**204′**) is also formed from (–)-α-terpineol (**34**) by *A. niger* TBUYN-2 (Noma et al., 2001) (Figure 22.168).

Rabbits metabolized myrtenal (**6′**) to myrtenic acid (**7′**) as the major metabolite and myrtanol (**8a′** or **8b′**) as the minor metabolite (Ishida et al., 1981b) (Figure 22.168).

22.4.3 Bicyclic Monoterpene Alcohol

22.4.3.1 Myrtenol

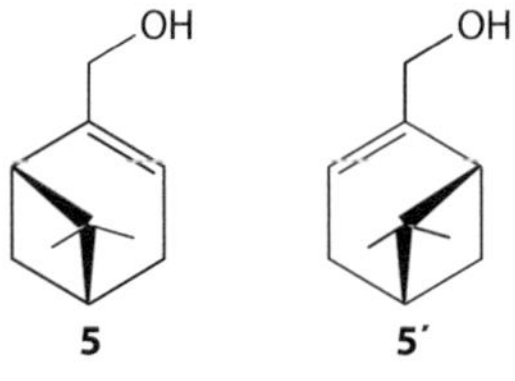

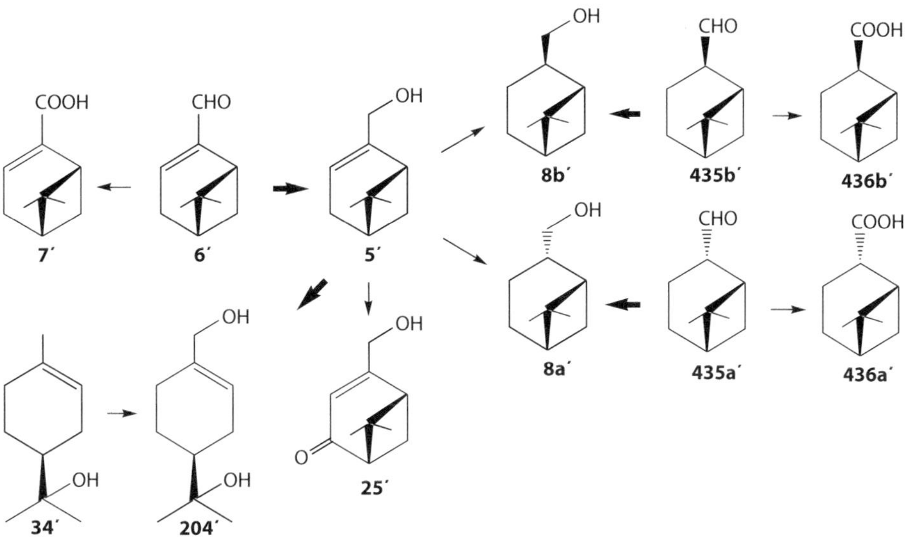

COOH CHO OH OH CHO COOH
7′ 6′ 5′ 8b′ 435b′ 436b′
OH CHO COOH
8a′ 435a′ 436a′
OH OH OH OH O
34′ 204′ 25′

FIGURE 22.168 Biotransformation of (–)-myrtenal (**6′**) and (+)-*trans*- (**435a′**) and (–)-*cis*-myrtanal (**435b′**) by microorganisms. (Modified from Noma, Y. et al. *Phytochemistry*, 30, 1147, 1991a; Noma, Y. and Asakawa, Y., Microbial transformation of (–)-myrtenol and (–)-nopol, *Proceedings of 49th TEAC*, 2005b, pp. 78–80; Noma, Y. and Asakawa, Y., Biotransformation of β-pinene, myrtenol, nopol and nopol benzyl ether by *Aspergillus niger* TBUYN-2, *Book of Abstracts of the 37th ISEO*, 2006b, p. 144.)

(–)-Myrtenol (**5′**) was biotransformed mainly to (–)-oleuropeyl alcohol (**204′**), which was formed from (–)-α-terpineol (**34′**) as a major product by *A. niger* TBUYN-2. In the case of *A. sojae* IFO 4389 and *A. usami* IFO 4338, (–)-myrtenol (**5′**) was metabolized to 7-hydroxyverbenone (**25′**) as a minor product together with (–)-oleuropeyl alcohol (**204′**) as a major product (Noma and Asakawa, 2005b) (Figure 22.169).

22.4.3.2 Myrtanol

S. litura converted (–)-*trans*-myrtanol (**8a**) and its enantiomer (**8a′**) to give the corresponding myrtanic acid (**436** and **436′**) (Miyazawa et al., 1997b) (Figure 22.170).

22.4.3.3 Pinocarveol

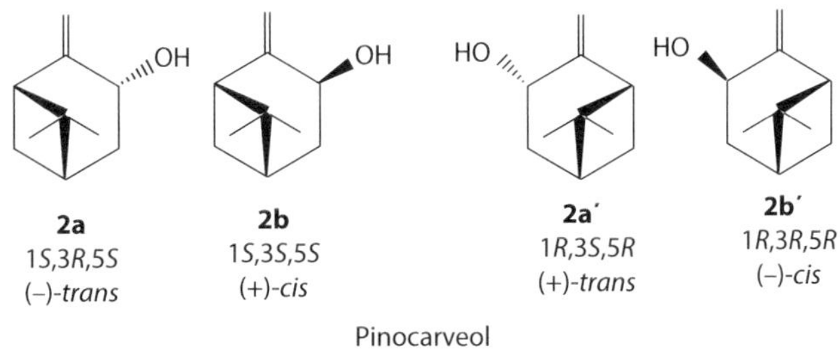

OH OH HO HO
2a 1*S*,3*R*,5*S* (–)-*trans*
2b 1*S*,3*S*,5*S* (+)-*cis*
2a′ 1*R*,3*S*,5*R* (+)-*trans*
2b′ 1*R*,3*R*,5*R* (–)-*cis*

Pinocarveol

(+)-*trans*-Pinocarveol (**2a′**) was biotransformed by *A. niger* TBUYN-2 to the following two pathways. That is, (+)-*trans*-pinocarveol (**2a′**) was metabolized via (+)-pinocarvone (**3′**), (–)-3-isopinanone (**413′**), and (+)-2α-hydroxy-3-pinanone (**414′**) to (+)-2α,5-dihydroxy-3-pinanone (**415′**) (pathway 1). Furthermore, (+)-*trans*-pinocarveol (**2a′**) was metabolized to epoxide followed by rearrangement reaction to give 6β-hydroxyfenchol (**349ba′**) and 6β,7-dihydroxyfenchol (**412ba′**) (Noma and Asakawa, 2005a) (Figure 22.171). *S. litura* converted (+)-*trans*-pinocarveol (**2a′**) to (+)-pinocarvone (**3′**) as a major product (Miyazawa et al., 1995b) (Figure 22.171).

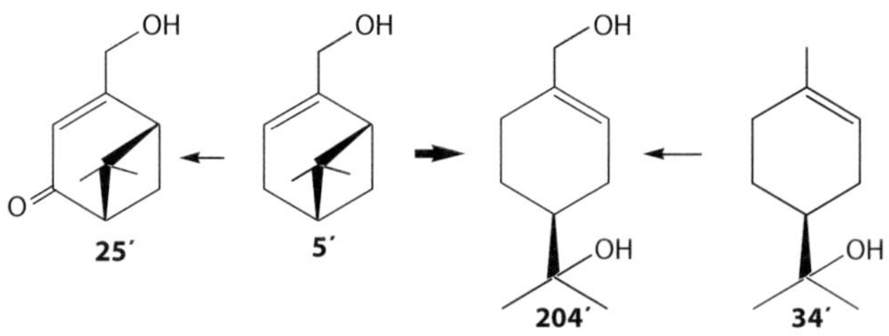

FIGURE 22.169 Biotransformation of (−)-myrtenol (**5′**) and (−)-α-terpineol (**34′**) by *Aspergillus niger* TBUYN-2. (Modified from Noma, Y. and Asakawa, Y., Microbial transformation of (−)-myrtenol and (−)-nopol, *Proceedings of 49th TEAC*, 2005b, pp. 78–80.)

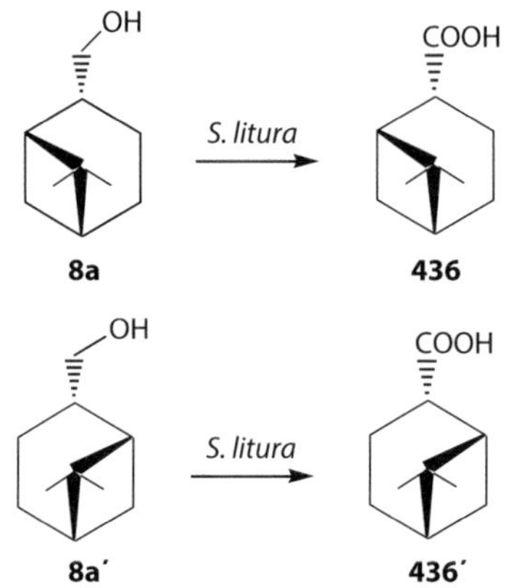

FIGURE 22.170 Biotransformation of (−)-*trans*-myrtanol (**8a**) and its enantiomer (**8a′**) by *Spodoptera litura*. (Modified from Miyazawa, M. et al. Biotransformation of (+)-trans myrtanol and (−)-trans-myrtanol by common cutworm Larvae, Spodoptera litura as a biocatalyst, *Proceedings of 41st TEAC*, 1997b, pp. 389–390.)

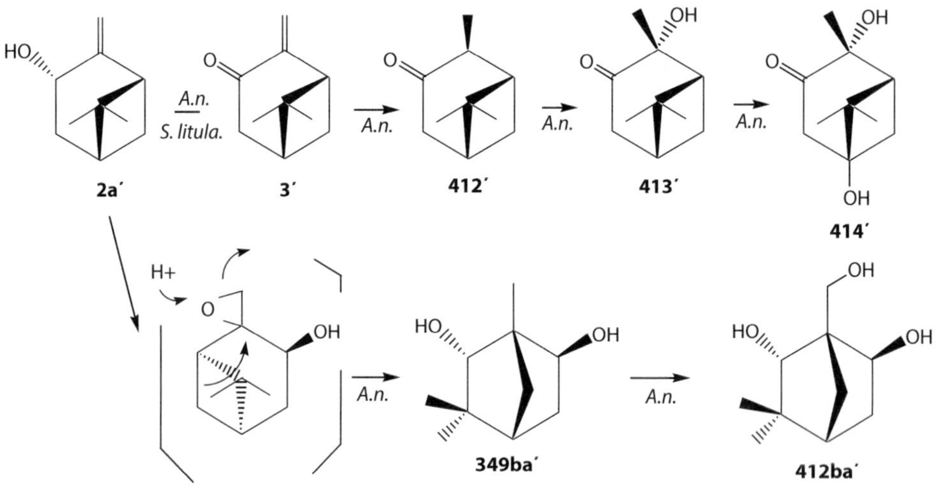

FIGURE 22.171 Biotransformation of (+)-*trans*-pinocarveol (**2a′**) by *Aspergillus niger* TBUYN-2 and *Spodoptera litura*. (Modified from Miyazawa, M. et al. Biotransformation of *trans*-pinocarveol by plant pathogenic microorganism, *Glomerella cingulata*, and by the larvae of common cutworm, *Spodoptera litura* Fabricius, *Proceedings of 39th TEAC*, 1995c, pp. 360–361; Noma, Y. and Asakawa, Y., New metabolic pathways of β-pinene and related compounds by *Aspergillus niger*, *Book of Abstracts of the 36th ISEO*, 2005a, p. 32.)

22.4.3.4 Pinane-2,3-Diol

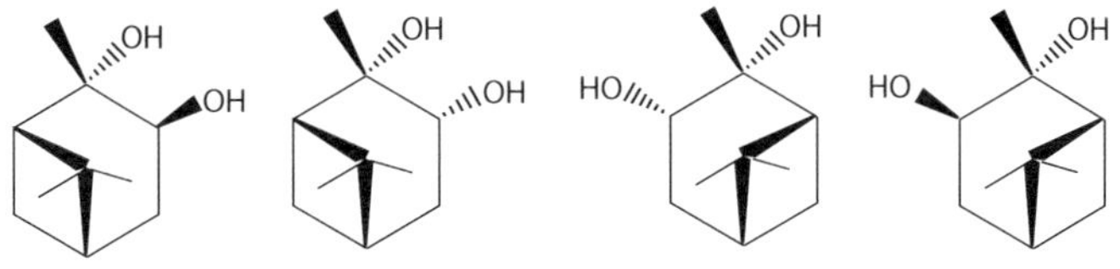

418aa ()-Pinane-2,3-diol 1*S*,2*S*,3*S*,5*S*
418ab (+)-Pinane-2,3-diol 1*S*,2*S*,3*R*,5*S*
418ab′ (–)-Pinane-2,3-diol 1*R*,2*R*,3*S*,5*R*
418aa′ ()-Pinane-2,3-diol 1*R*,2*R*,3*R*,5*R*

This results led us to study the biotransformation of (–)-pinane-2,3-diol (**418ab′**) and (+)-pinane-2,3 diol (**418ab**) by *A. niger* TBUYN-2. (–)-Pinane-2,3-diol (**418ab′**) was easily biotransformed to give (–)-pinane-2,3,5-triol (**419ab′**) and (+)-2,5-dihydroxy-3-pinanone (**415a′**) as the major products and (+)-2-hydroxy-3-pinanone (**414a′**) as the minor product.

On the other hand, (+)-pinane-2,3-diol (**418ab**) was also biotransformed easily to give (+)-pinane-2,3,5-triol (**419ab**) and (–)-2,5-dihydroxy-3-pinanone (**415a**) as the major products and (–)-2-hydroxy-3-pinanone (**414a**) as the minor product (Noma et al., 2003) (Figure 22.172). *G. cingulata* transformed (–)-pinane-2,3-diol (**418ab′**) to a small amount of (+)-2α-hydroxy-3-pinanone (**414ab′**, 5%) (Kamino and Miyazawa, 2005), whereas (+)-pinane-2,3-diol (**418ab**) was transformed to a small amount of (–)-2α-hydroxy-3-pinanone (**414ab**, 10%) and (–)-3-acetoxy-2α-pinanol (**433ab-Ac**, 30%) (Kamino et al., 2004) (Figure 22.172).

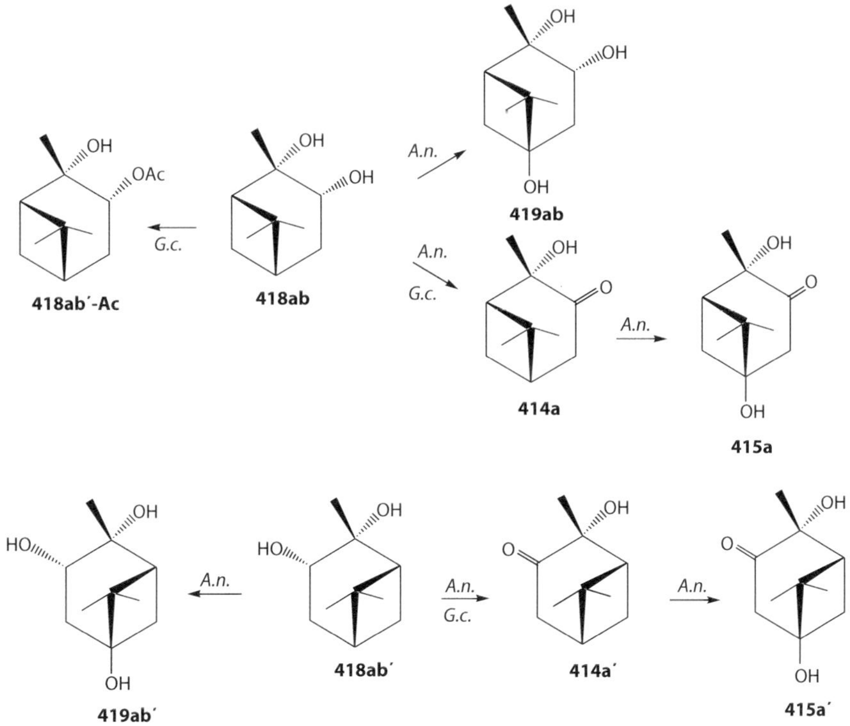

FIGURE 22.172 Biotransformation of (+)-pinane-2,3-diol (**418ab′**) and (–)-pinane-2,3-diol (**418ab′**) by *Aspergillus niger* TBUYN-2(**276**)] and *Glomerella cingulata*. (Modified from Noma, Y. et al. Biotransformation of (+)- and (–)-pinane-2,3-diol and related compounds by *Aspergillus niger, Proceedings of 47th TEAC*, 2003, pp. 91–93; Kamino, F. et al. Biotransformation of (1*S*,2*S*,3*R*,5*S*)-(+)-pinane-2,3-diol using plant pathogenic fungus, *Glomerella cingulata* as a biocatalyst, *Proceedings of 48th TEAC*, 2004, pp. 383–384; Kamino, F. and Miyazawa, M., Biotransformation of (+)-and (–)-pinane-2,3-diol using plant pathogenic fungus, *Glomerella cingulata* as a biocatalyst, *Proceedings of 49th TEAC*, 2005, pp. 395–396.)

22.4.3.5 Isopinocampheol (3-Pinanol)

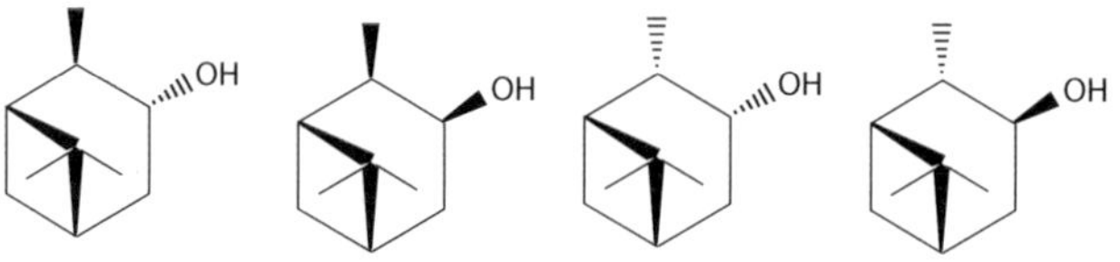

420 ba (−)-isopino 1*R*,2*R*,3*R*,5*S*
420bb (+)-neoiso 1*R*,2*R*,3*S*,5*S*
420aa (−)-neo 1*R*,2*S*,3*R*,5*S*
420ab (+)-pinocampherol 1*R*,2*S*,3*S*,5*S*

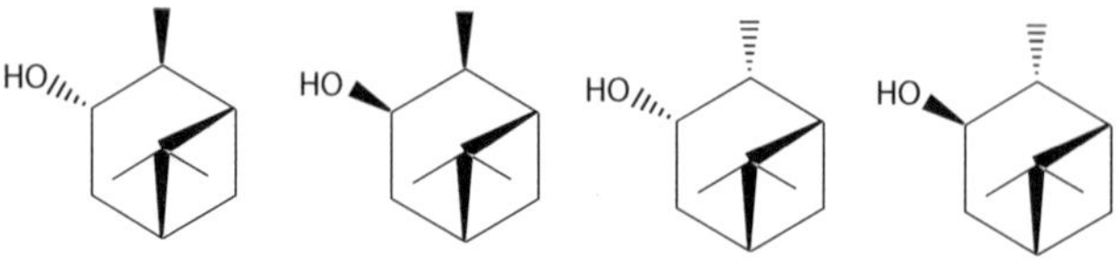

420ba′ (+)- 1*S*,2*S*,3*S*,5*R*
420bb′ (−)- 1*S*,2*S*,3*R*,5*R*
420aa′ (+)-neo 1*S*,2*R*,3*S*,5*R*
420ab′ 1*S*,2*R*,3*R*,5*R*

22.4.3.5.1 Chemical Structure of (−)-Isopinocampheol (420ba) and (+)-Isopinocampheol (420ba′)

Biotransformation of isopinocampheol (3-pinanol) with 100 bacterial and fungal strains yielded 1-, 2-, 4-, 5-, 7-, 8-, and 9-hydroxyisopinocampheol besides three rearranged monoterpenes, one of them bearing the novel isocarene skeleton. A pronounced enantioselectivity between (−)- (**420ba**) and (+)-isopinocampheol (**420ba′**) was observed. The phylogenetic position of the individual strains could be seen in their ability to form the products from (+)-isopinocampheol (**420ba′**). The formation of 1,3-dihydroxypinane (**421ba′**) is a domain of bacteria, while 3,5- (**415ba′**) or 3,6-dihydroxypinane (**428baa′**) was mainly formed by fungi, especially those of the phylum *Zygomycotina*. The activity of *Basidiomycotina* toward oxidation of isopinocampheol was rather low. Such informations can be used in a more effective selection of strains for screening (Wolf-Rainer, 1994) (Figure 22.173).

(+)-Isopinocampheol (**420ba′**) was metabolized to 4β-hydroxy-(+)-isopinocampheol (**424′**), 2β-hydroxy-(+)-isopinocampheol acetate (**425ba′-Ac**), and 2α-methyl,3-(2-methyl-2-hydroxypropyl)-cyclopenta-1β-ol (**432′**) (Wolf-Rainer, 1994) (Figure 22.174).

(−)-Isopinocampheol (**420ba**) was converted by *S. litura* to give (1*R*,2*S*,3*R*,5*S*)-pinane-2,3-diol (**418ba**) and (−)-pinane-3,9-diol (**423ba**), whereas (+)-isopinocampheol (**420ba′**) was converted to (+)-pinane-3,9-diol (**423ba′**) (Miyazawa et al., 1997c) (Figure 22.175).

(−)-Isopinocampheol (**420ba**) was biotransformed by *A. niger* TBUYN-2 to give (+)-(1*S*,2*S*,3*S*,5*R*)-pinane-3,5-diol (**422ba**, 6.6%), (−)-(1*R*,2*R*,3*R*,5*S*)-pinane-1,3-diol (**421ba**, 11.8%), and pinane-2,3-diol (**418ba**, 6.6%), whereas (+)-isopinocampheol (**420ba′**) was biotransformed by *A. niger* TBUYN-2 to give (+)-(1*S*,2*S*,3*S*,5*R*)-pinane-3,5-diol (**422ba′**, 6.3%) and (−)-(1*R*,2*R*,3*R*,5*S*)-pinane-1,3,-diol (**421ba′**, 8.6%) (Noma et al., 2009) (Figure 22.176). On the other hand, *G. cingulata* converted (−)-(**420ba**) and (+)-isopinocampheol (**420ba′**) mainly to (1*R*,2*R*,3*S*,4*S*,5*R*)-3,4-pinanediol (**484ba**) and (1*S*,2*S*,3*S*,5*R*,6*R*)-3,6-pinanediol (**485ba′**), respectively, together with (**418ba**), (**422ba**), (**422ba′**), and (**486ba′**) as minor products (Miyazawa et al., 1997c) (Figure 22.176). Some similarities exist between the main metabolites with *G. cingulata* and *R. solani* (Miyazawa et al., 1997c) (Figure 22.176).

22.4.3.6 Borneol and Isoborneol

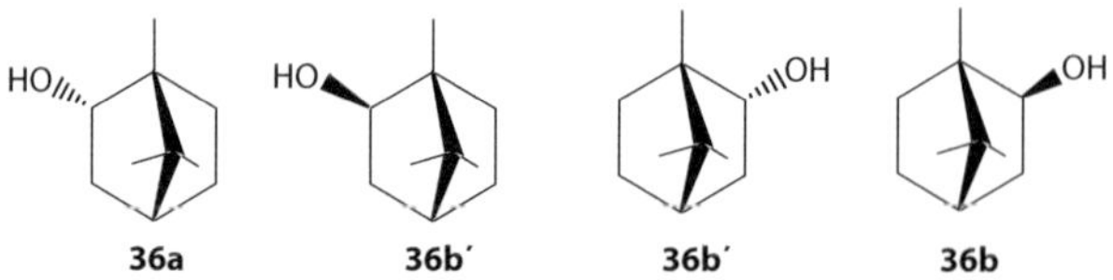

36a (1*R*,2*S*)-(+)-borneol (36a)
36b′ (1*R*,2*R*)-(−)-isoborneol (36b′)
36b′ (1*S*,2*R*)-(−)-borneol (36a′)
36b (1*S*,2*S*)-(+)-isoborneol (36b)

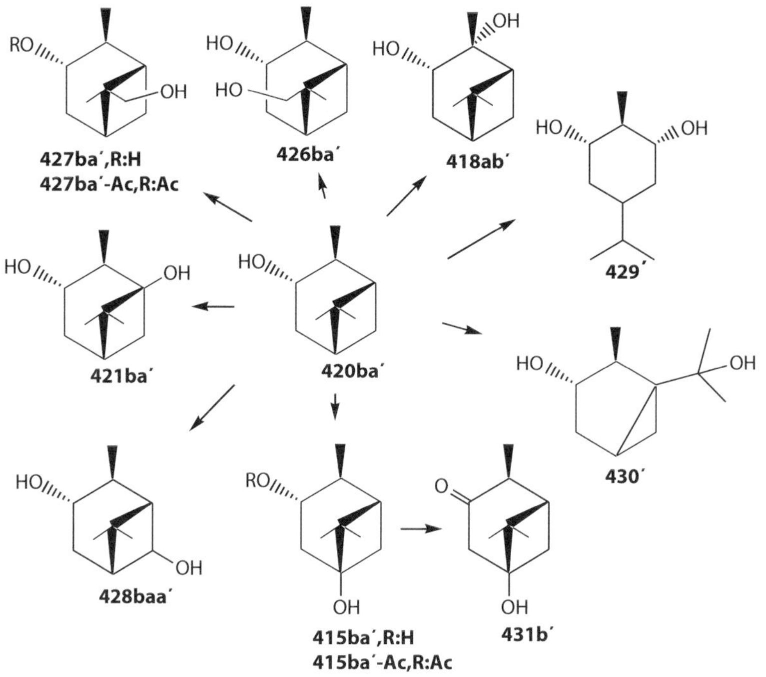

FIGURE 22.173 Metabolic pathways of (+)-isopinocampheol (**420ba′**) by microorganisms. (Modified from Wolf-Rainer, A., *Z. Naturforsch.*, 49c, 553, 1994.)

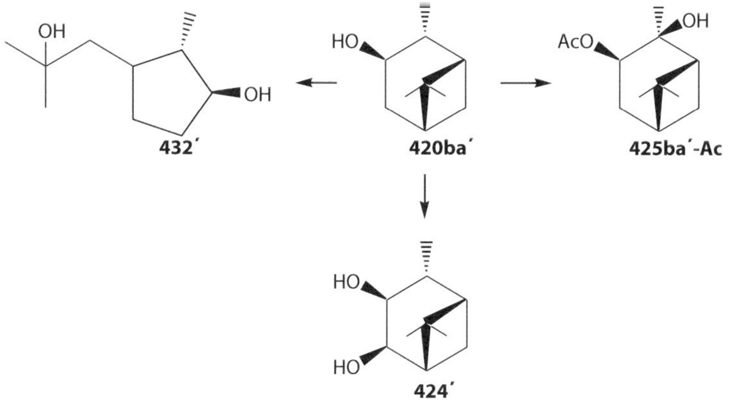

FIGURE 22.174 Metabolic pathways of (+)-isopinocampheol (**420ba′**) by microorganisms. (Modified from Wolf-Rainer, A., *Z. Naturforsch.*, 49c, 553, 1994.)

(–)-Borneol (**36a′**) was biotransformed by *P. pseudomallei* strain H to give (–)-camphor (**37′**), 6-hydroxycamphor (**228′**), and 2,6-diketocamphor (**229′**) (Hayashi et al., 1969) (Figure 22.177).

E. gracilis Z. showed enantio- and diastereoselectivity in the biotransformation of (+)- (**36a**), (–)- (**36a′**), and (±)-racemic borneols (equal mixture of **36a** and **36a′**) and (+)- (**36b**), (–)- (**36b′**), and (±)-isoborneols (equal mixture of **36b** and **36b′**). The enantio- and diastereoselective dehydrogenation for (–)-borneol (**36a′**) was carried out to give (–)-camphor (**37′**) at ca. 50% yield (Noma et al., 1992d; Noma and Asakawa, 1998). The conversion ratio was always ca. 50% even at different kinds of concentration of (–)-borneol (**36a′**). When (–)-camphor (**37′**) was used as a substrate, it was also converted to (–)-borneol (**36a′**) in 22% yield for 14 days. Furthermore, (+)-camphor (**37**) was also

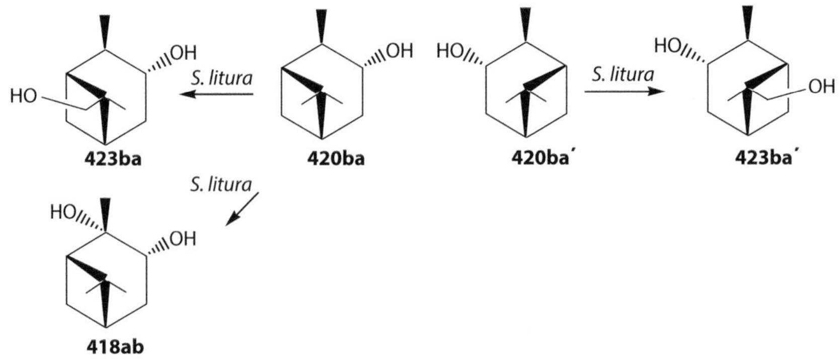

FIGURE 22.175 Biotransformation of (−)- (**420ba**) and (+)-isopinocampheol (**420ba′**) by *Spodoptera litura*. (Modified from Miyazawa, M. et al. *Phytochemistry*, 45, 945, 1997c.)

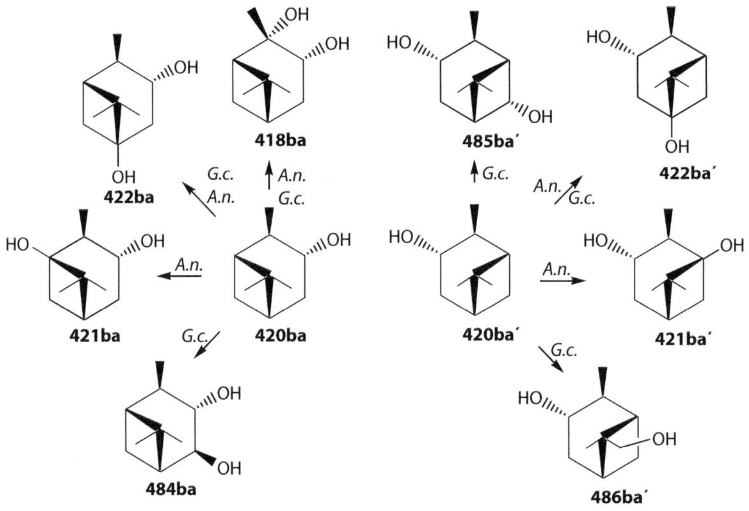

FIGURE 22.176 Biotransformation of (−)- (**420ba**) and (+)-isopinocampheol (**420ba′**) by *Aspergillus niger* TBUYN-2 and *Glomerella cingulata*. (Modified from Miyazawa, M. et al. *Phytochemistry*, 45, 945, 1997c; Noma, Y. et al. Unpublished data, 2009.)

reduced to (+)-borneol (**36a**) in 4% and 18% yield for 7 and 14 days, respectively (Noma et al., 1992d; Noma and Asakawa, 1998) (Figure 22.178).

(+)- (**36a**) and (−)-Borneols (**36a′**) were biotransformed by *S. litura* to (+)- (**370a**) and (−)-bornane-2,8-diols (**370a′**), respectively (Miyamoto and Miyazawa, 2001) (Figure 22.179).

22.4.3.7 Fenchol and Fenchyl Acetate

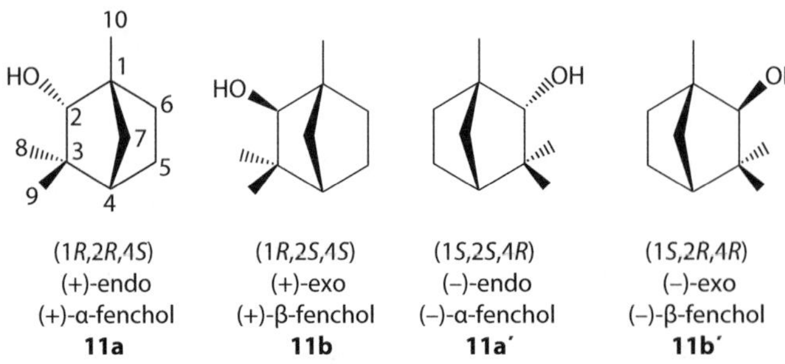

(1*R*,2*R*,4*S*) (+)-endo (+)-α-fenchol **11a**

(1*R*,2*S*,4*S*) (+)-exo (+)-β-fenchol **11b**

(1*S*,2*S*,4*R*) (−)-endo (−)-α-fenchol **11a′**

(1*S*,2*R*,4*R*) (−)-exo (−)-β-fenchol **11b′**

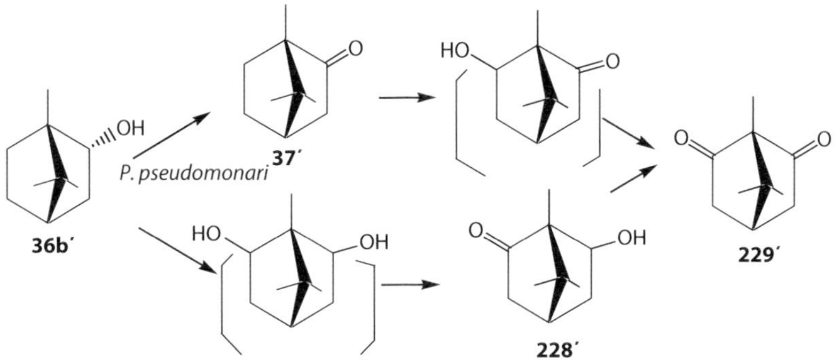

FIGURE 22.177 Biotransformation of (−)-borneol (**36a′**) by *Pseudomonas pseudomallei* strain. (Modified from Hayashi, T. et al. *J. Agric. Chem. Soc. Jpn.*, 43, 583, 1969.)

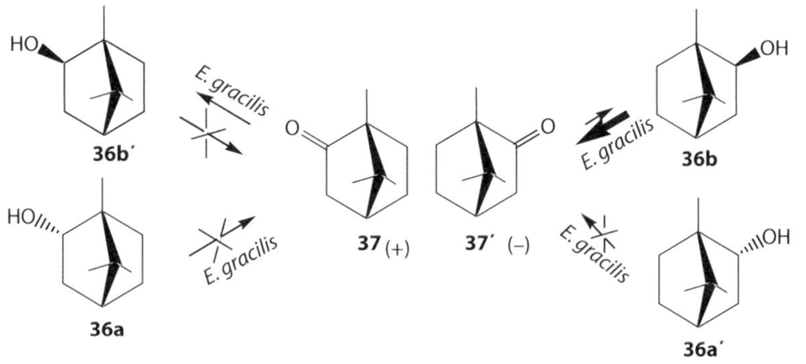

FIGURE 22.178 Enantio- and diastereoselectivity in the biotransformation of (+)- (**36a**) and (−)-borneols (**36a′**) by *Euglena gracilis* Z. (Modified from Noma, Y. et al. Biotransformation of terpenoids and related compounds, *Proceedings of 36th TEAC*, 1992b, pp. 199–201; Noma, Y. and Asakawa, Y., *Euglena gracilis* Z: Biotransformation of terpenoids and related compounds, in Bajaj, Y.P.S. (ed.), *Biotechnology in Agriculture and Forestry*, Vol. 41, Medicinal and Aromatic Plants X, Springer, Berlin, Germany, 1998, pp. 194–237.)

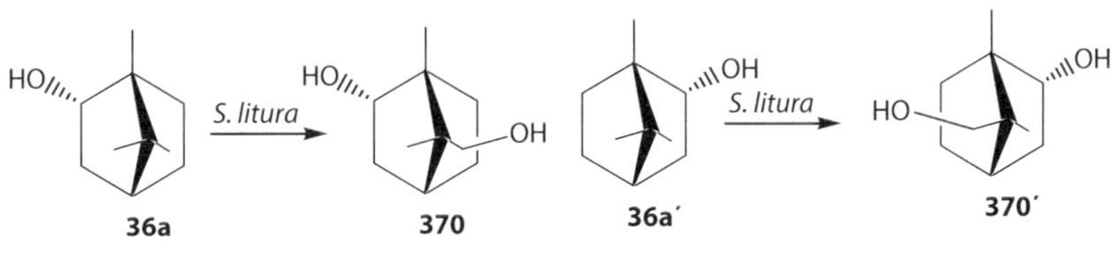

FIGURE 22.179 Biotransformation of (+)- (**36a**) and (−)-borneols (**36a′**) by *Spodoptera litura*. (Modified from Miyamoto, Y. and Miyazawa, M., Biotransformation of (+)- and (−)-borneol by the larvae of common cutworm (*Spodoptera litura*) as a biocatalyst, *Proceedings of 45th TEAC*, 2001, pp. 377–378.)

(1*R*,2*R*,4*S*)-(+)-Fenchol (**11a**) was converted by *A. niger* TBUYN-2 and *A. cellulosae* IFO 4040 to give (−)-fenchone (**12**), (+)-6β-hydroxyfenchol (**349ab**), (+)-5β-hydroxyfenchol (**350ab**), and 5α-hydroxyfenchol (**350aa**) (Noma and Asakawa, 2005a) (Figure 22.180). The larvae of common cutworm, *S. litura*, converted (+)-fenchol (**11a**) to (+)-10-hydroxyfenchol (**467a**), (+)-8-hydroxyfenchol (**465a**), (+)-6β-hydroxyfenchol (**349ab**), and (−)-9-hydroxyfenchol (**466a**) (Miyazawa and Miyamoto, 2004) (Figure 22.180).

(+)-*trans*-Pinocarveol (**2**), which was formed from (−)-β-pinene (**1**), was metabolized by *A. niger* TBUYN-2 to 6β-hydroxy-(+)-fenchol (**349ab**) and 6β,7-dihydroxy-(+)-fenchol (**412ba′**). (−)-Fenchone (**12**) was also metabolized to 6α-hydroxy- (**13b**) and 6β-hydroxy-(−)-fenchone (**13a**).

FIGURE 22.180 Biotransformation of (+)-fenchol (**11a**) by *Aspergillus niger* TBUYN-2, *Aspergillus cellulosae* IFO 4040, and the larvae of common cutworm, *Spodoptera litura*. (Modified from Miyazawa, M. and Miyamoto, Y., *Tetrahadron*, 60, 3091, 2004; Noma, Y. and Asakawa, Y., New metabolic pathways of β-pinene and related compounds by *Aspergillus niger, Book of Abstracts of the 36th ISEO*, 2005a, p. 32.)

(+)-Fenchol (**11**) was metabolized to 6β-hydroxy-(+)-fenchol (**349ab**) by *A. niger* TBUYN-2. The relationship of the metabolisms of (+)-*trans*-pinocarveol (**2**), (–)-fenchone (**12**), and (+)-fenchol (**11**) by *A. niger* TBUYN-2 is shown in Figure 22.181 (Noma and Asakawa, 2005a).

(+)-α-Fencyl acetate (**11a-Ac**) was metabolized by *G. cingulata* to give (+)-5-β-hydroxy-α-fencyl acetate (**350a-Ac**, 50%) as the major metabolite and (+)-fenchol (**11a**, 20%) as the minor metabolite (Miyazato and Miyazawa, 1999). On the other hand, (–)-α-fencyl acetate (**11a′-Ac**) was metabolized to (–)-5-β-hydroxy-α-fencyl acetate (**350a′-Ac**, 70%) and (–)-fenchol (**11a′**, 10%) as the minor metabolite by *G. cingulata* (Miyazato and Miyazawa, 1999) (Figure 22.182).

22.4.3.8 Verbenol

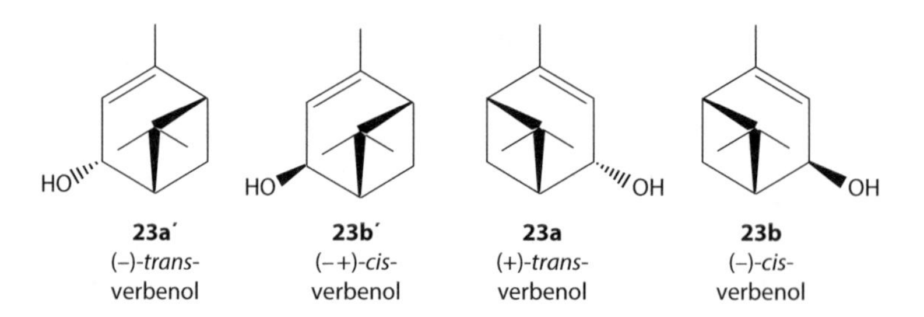

(–)-*trans*-Verbenol (**23a′**) was biotransformed by *S. litura* to give 10-hydroxyverbenol (**451a′**). Furthermore, (–)-verbenone (**24′**) was also biotransformed in the same manner to give 10-hydroxyverbenone (**25′**) (Yamanaka and Miyazawa, 1999) (Figure 22.183).

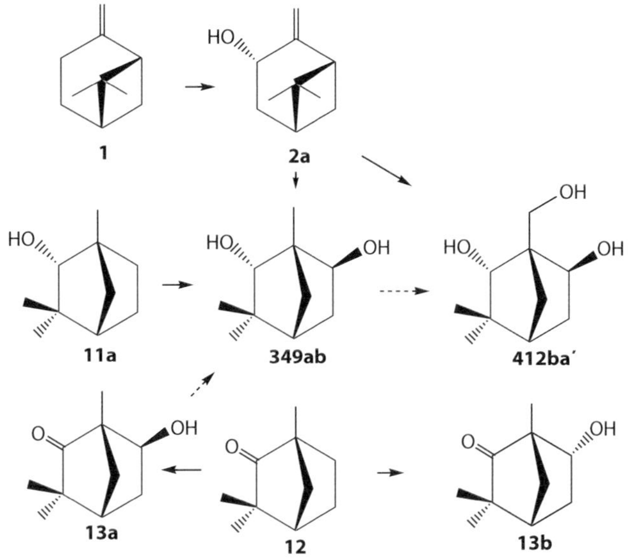

1 → 2a (HO) → 349ab (HO, OH) ⇢ 412ba′ (HO, OH, OH)
11a → 349ab
12 → 13a (O, OH); 12 → 13b (O, OH)

FIGURE 22.181 Metabolism of (+)-*trans*-pinocarveol (**2**), (−)-fenchone (**12**), and (+)-fenchol (**11**) by *Aspergillus niger* TBUYN-2. (Modified from Noma, Y. and Asakawa, Y., New metabolic pathways of β-pinene and related compounds by *Aspergillus niger, Book of Abstracts of the 36th ISEO*, 2005a, p. 32.)

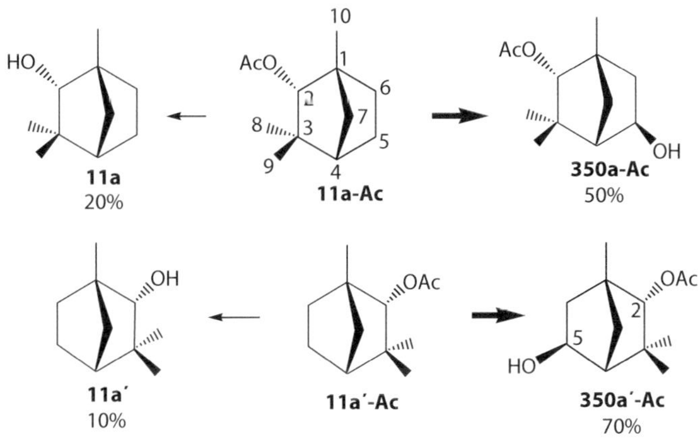

11a 20% ← 11a-Ac → 350a-Ac 50%
11a′ 10% ← 11a′-Ac → 350a′-Ac 70%

FIGURE 22.182 Biotransformation of (+)- (**11a-Ac**) and (−)-α-fencyl acetate (**11a′-Ac**) by *Glomerella cingulata*. (Modified from Miyazato, Y. and Miyazawa, M., Biotransformation of (+)- and (−)-α-fenchyl acetated using plant parasitic fungus, *Glomerella cingulata* as a biocatalyst, *Proceedings of 43rd TEAC*, 1999, pp. 213–214.)

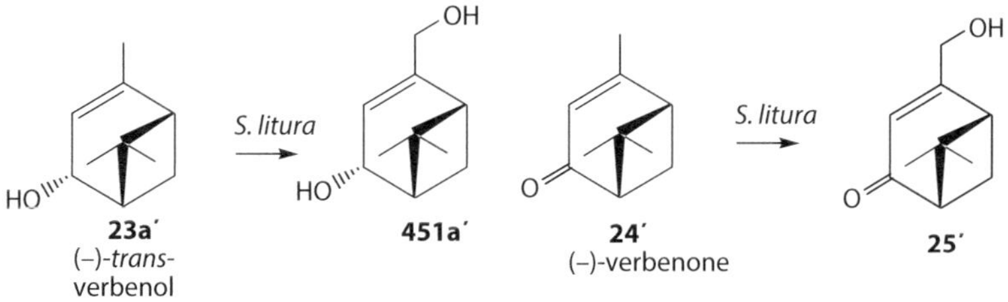

23a′ (−)-*trans*-verbenol → (*S. litura*) 451a′
24′ (−)-verbenone → (*S. litura*) 25′

FIGURE 22.183 Metabolism of (−)-*trans*-verbenol (**23a′**) and (−)-verbenone (**24′**) by *Spodoptera litura*. (Modified from Yamanaka, T. and Miyazawa, M., Biotransformation of (−)-*trans*-verbenol by common cutworm larvae, *Spodoptera litura* as a biocatalyst, *Proceedings of 43rd TEAC*, 1999, pp. 391–392.)

FIGURE 22.184 Biotransformation of (–)-nopol (**452′**) by *Aspergillus niger* TBUYN-2, *Aspergillus sojae* IFO 4389, and *Aspergillus usami* IFO 4338. (Modified from Noma, Y. and Asakawa, Y., Microbial transformation of (–)-myrtenol and (–)-nopol, *Proceedings of 49th TEAC*, 2005b, pp. 78–80; Noma, Y. and Asakawa, Y., Microbial transformation of (–)-nopol benzyl ether, *Proceedings of 50th TEAC*, 2006c, pp. 434–436.)

22.4.3.9 Nopol and Nopol Benzyl Ether

Biotransformation of (–)-nopol (**452′**) was carried out at 30°C for 7 days at the concentration of 100 mg/200 mL medium by *A. niger* TBUYN-2, *A. sojae* IFO 4389, and *A. usami* IFO 4338. (–)-Nopol (**452′**) was incubated with *A. niger* TBUYN-2 to give 7-hydroxymethyl-1-*p*-menthen-8-ol (**453′**). In cases of *A. sojae* IFO 4389 and *A. usami* IFO 4338, (–)-nopol (**452′**) was metabolized to 3-oxonopol (**454′**) as a minor product together with 7-hydroxymethyl-1-*p*-menthen-8-ol (**453′**) as a major product (Noma and Asakawa, 2005b, 2006c) (Figure 22.184).

Biotransformation of (–)-nopol benzyl ether (**455′**) was carried out at 30°C for 8–13 days at the concentration of 277 mg/200 mL medium by *A. niger* TBUYN-2, *A. sojae* IFO 4389, and *A. usami* IFO 4338. (–)-Nopol benzyl ether (**455′**) was biotransformed by *A. niger* TBUYN-2 to give 4-oxonopol-2′, 4′-dihydroxy benzyl ether (**456′**), and (–)-oxonopol (**454′**). 7-Hydroxymethyl-1-*p*-menthen-8-ol benzyl ether (**457′**) was not formed at all (Figure 22.185). 4-Oxonopol-2′,4′-dihydroxybenzyl ether (**456′**) shows strong antioxidative activity (IC_{50} 30.23 μM). The antioxidative activity of 4-oxonopol-2′,4′-dihydroxybenzyl ether (**456′**) is the same as that of butyl hydroxyl anisole (BHA) (Noma and Asakawa, 2006b,c).

Citrus pathogenic fungi *A. niger* Tiegh (CBAYN) also transformed (–)-nopol (**452′**) to (–)-oxonopol (**454′**) and 4-oxonopol-2′,4′-dihydroxybenzyl ether (**456′**) (Noma and Asakawa, 2006b,c) (Figure 22.186).

22.4.4 Bicyclic Monoterpene Ketones

22.4.4.1 α-, β-Unsaturated Ketone

22.4.4.1.1 Verbenone

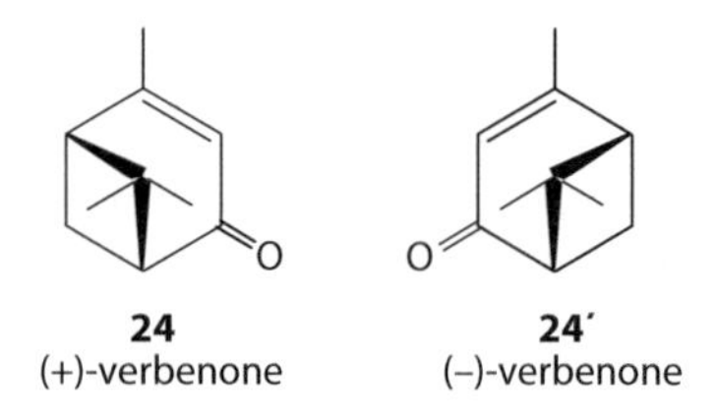

(–)-Verbenone (**24′**) was hydrogenated by reductase of *N. tabacum* to give (–)-isoverbanone (**458b′**) (Suga and Hirata, 1990; Shimoda et al., 1996, 1998, 2002; Hirata et al., 2000) (Figure 22.187).

22.4.4.1.2 Pinocarvone

3 (–)-Pinocarvone **3′** (+)-Pinocarvone

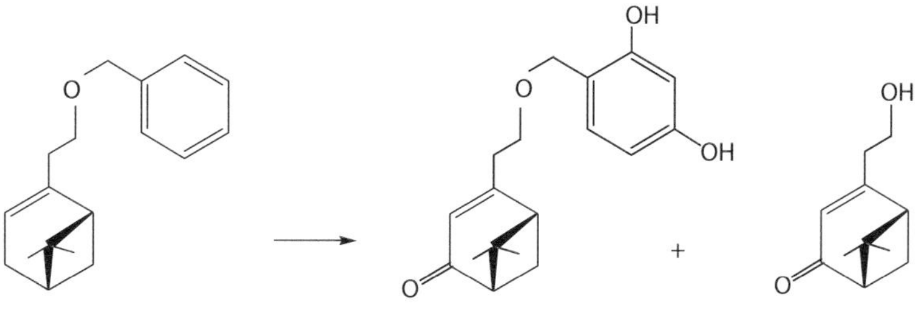

455′ 456′ 454′

FIGURE 22.185 Biotransformation of (–)-nopol benzyl ether (**455′**) by *Aspergillus niger* TBUYN-2. (Modified from Noma, Y. and Asakawa, Y., Biotransformation of β-pinene, myrtenol, nopol and nopol benzyl ether by *Aspergillus* niger TBUYN-2, *Book of Abstracts of the 37th ISEO*, 2006b, p. 144; Noma, Y. and Asakawa, Y., Microbial transformation of (−)-nopol benzyl ether, *Proceedings of 50th TEAC*, 2006c, pp. 434–436.)

455′ 457 456′ 454′

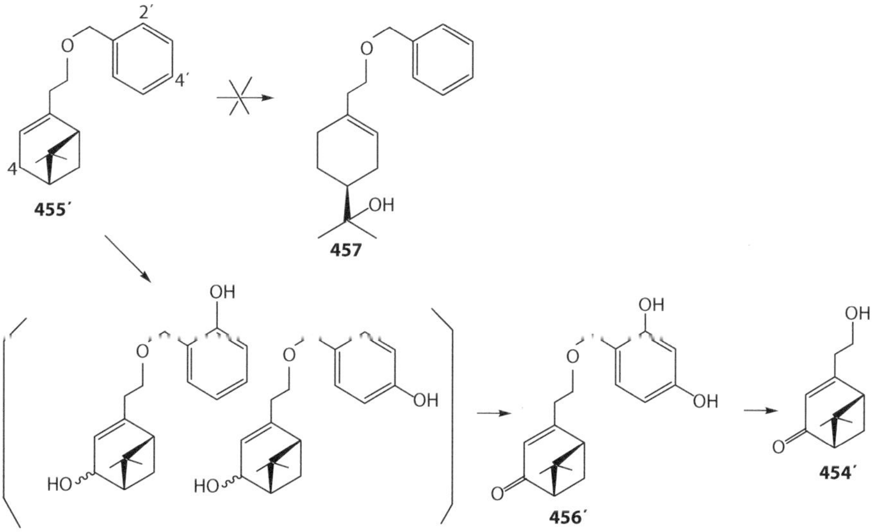

FIGURE 22.186 Proposed metabolic pathways of (–)-nopol benzyl ether (**455′**) by microorganisms. (Modified from Noma, Y. and Asakawa, Y., Biotransformation of β-pinene, myrtenol, nopol and nopol benzyl ether by *Aspergillus niger* TBUYN-2, *Book of Abstracts of the 37th ISEO*, 2006b, p. 144; Noma, Y. and Asakawa, Y., Microbial transformation of (−)-nopol benzyl ether, *Proceedings of 50th TEAC*, 2006c, pp. 434–436.)

A. niger TBUYN-2 transformed (+)-pinocarvone (**3′**) to give (–)-isopinocamphone (**413b′**), 2α-hydroxy-3-pinanone (**414b′**), and 2α, 5-dihydroxy-3-pinanone (**415b′**) together with small amounts of 2α, 10-dihydroxy-3-pinanone (**416b′**) (Noma and Asakawa, 2005a) (Figure 22.188).

22.4.4.2 Saturated Ketone

22.4.4.2.1 Camphor

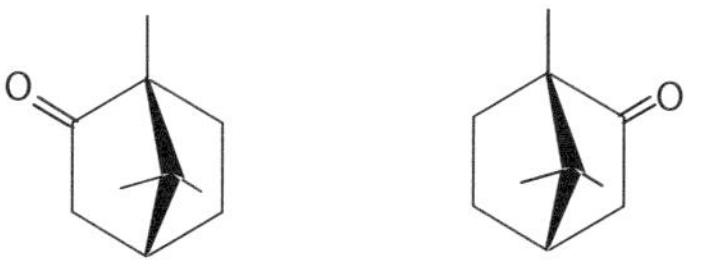

37 (1*R*)-(+)-camphor **(37)**

37′ (1*S*)-(−)-camphor **(37′)**

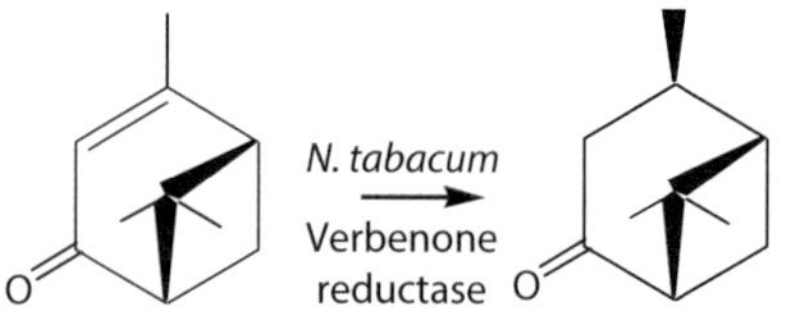

FIGURE 22.187 Hydrogenation of (–)-verbenone (**24′**) to (–)-isoverbanone (**458b′**) by verbenone reductase of *Nicotiana tabacum*. (Modified from Suga, T. and Hirata, T., Phytochemistry, 29, 2393, 1990; Shimoda, K. et al. J. Chem. Soc., Perkin Trans., 1, 355, 1996; Shimoda, K. et al. Phytochemistry, 49, 49, 1998; Shimoda, K. et al. Bull. Chem. Soc. Jpn., 75, 813, 2002; Hirata, T. et al. Chem. Lett., 29, 850, 2000.)

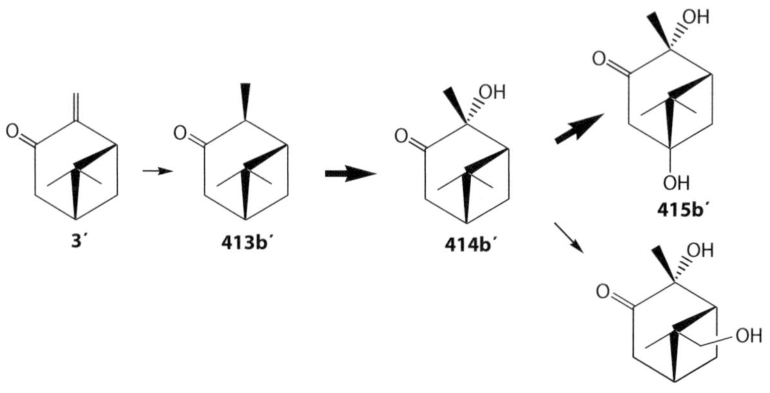

FIGURE 22.188 Biotransformation of (+)-pinocarvone (**3′**) by *Aspergillus niger* TBUYN-2. (Modified from Noma, Y. and Asakawa, Y., New metabolic pathways of β-pinene and related compounds by *Aspergillus niger, Book of Abstracts of the 36th ISEO*, 2005a, p. 32.)

(+)- (**37**) and (–)-Camphor (**37′**) are found widely in nature, of which (+)-camphor (**37**) is more abundant. It is the main component of oils obtained from the camphor tree *C. camphora* (Bauer et al., 1990). The hydroxylation of (+)-camphor (**37**) by *P. putida* C_1 was described (Abraham et al., 1988). The substrate was hydroxylated exclusively in its 5-exo- (**235b**) and 6-exo- (**228b**) positions.

Although only limited success was achieved in understanding the catabolic pathways of (+)-camphor (**37**), key roles for methylene group hydroxylation and biological Baeyer–Villiger monooxygenases in ring cleavage strategies were established (Trudgill, 1990). A degradation pathway of (+)-camphor (**37**) by *P. putida* ATCC 17453 and *Mycobacterium rhodochrous* T_1 was proposed (Trudgill, 1990).

The metabolic pathway of (+)-camphor (**37**) by microorganisms is shown in Figure 22.189. (+)-Camphor (**37**) is metabolized to 3-hydroxy- (**243**), 5-hydroxy- (**235**), 6-hydroxy- (**228**), and 9-hydroxycamphor (**225**) and 1,2-campholide (**237**). 6-Hydroxycamphor (**228**) is degradatively metabolized to 6-oxocamphor (**229**) and 4-carboxymethyl-2,3,3-trimethylcyclopentanone (**230**), 4-carboxymethyl-3,5,5-trimethyltetrahydro-2-pyrone (**231**), isohydroxycamphoric acid (**232**), isoketocamphoric acid (**233**), and 3,4,4-trimethyl-5-oxo-*trans*-2-hexenoic acid (**234**), whereas 1,2-campholide (**237**) is also degradatively metabolized to 6-hydroxy-1,2-campholide (**238**), 6-oxo-1,2-campholide (**239**), and 5-carboxymethyl-3,4,4-trimethyl-2-cyclopentenone (**240**), 6-carboxymethyl-4,5,5-trimethyl-5,6-dihydro-2-pyrone (**241**), and 5-carboxymethyl-3,4,4-trimethyl-2-heptene-1,7-dioic acid (**242**). 5-Hydroxycamphor (**235**) is metabolized to 6-hydroxy-1,2-campholide (**238**), 5-oxocamphor (**236**), and 6-oxo-1,2-campholide (**239**). 3-Hydroxycamphor (**243**) is also metabolized to camphorquinone (**244**) and 2-hydroxyepicamphor (**245**) (Bradshaw et al., 1959;

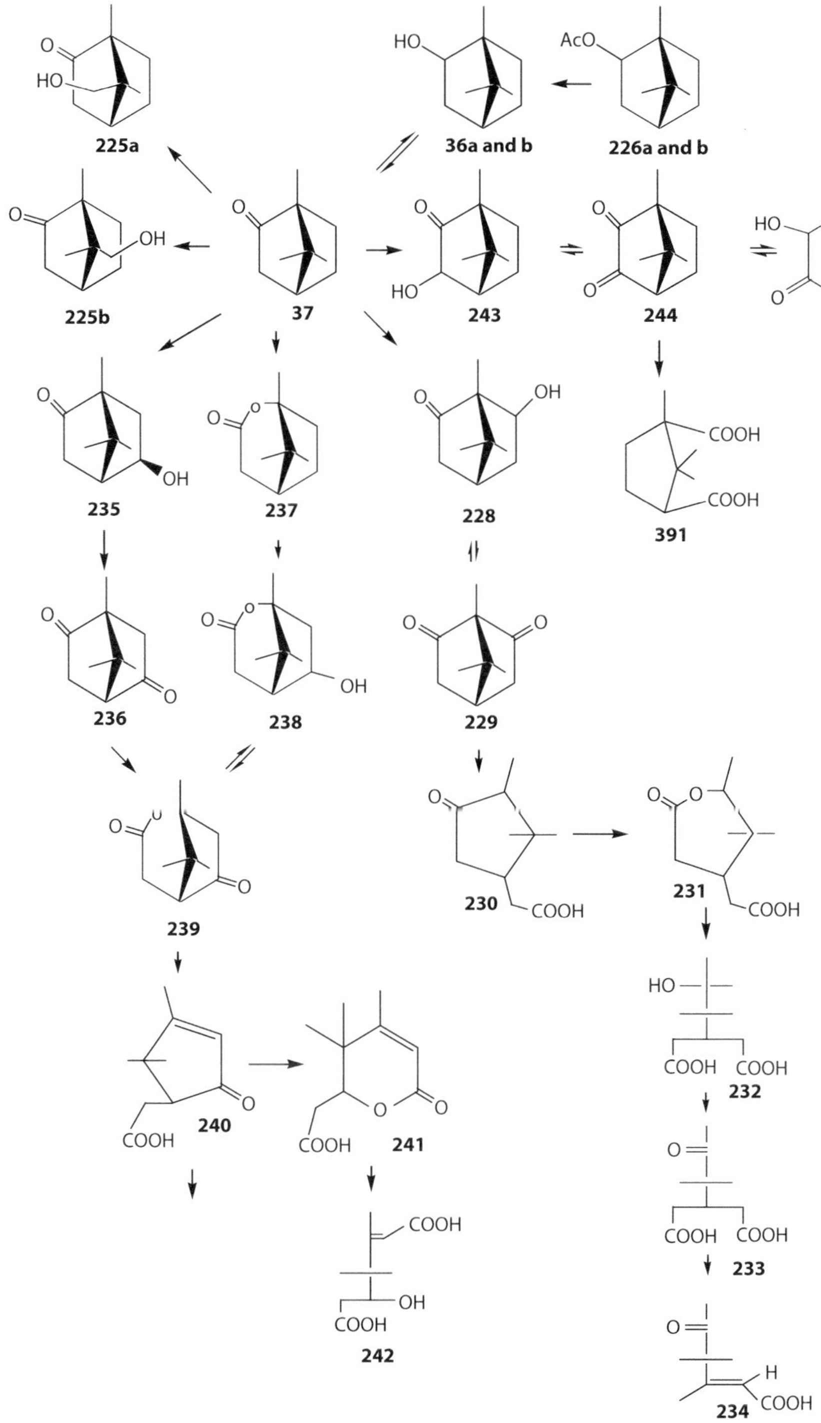

FIGURE 22.189 Metabolic pathways of (+)-camphor (**37**) by *Pseudomonas putida* and *Corynebacterium diphtheroides*. (Modified from Bradshaw, W.H. et al. *J. Am. Chem. Soc.*, 81, 5507, 1959; Conrad, H.E. et al. *Biochem. Biophys. Res. Commun.*, 6, 293, 1961; Conrad, H.E. et al. *J. Biol. Chem.*, 240, 495, 1965a; Conrad, H.E. et al. *J. Biol. Chem.*, 240, 4029, 1965b; Gunsalus, I.C. et al. *Biochem. Biophys. Res. Commun.*, 18, 924, 1965; Chapman, P.J. et al. *J. Am. Chem. Soc.*, 88, 618, 1966; Hartline, R.A. and Gunsalus, I.C., *J. Bacteriol.*, 106, 468, 1971; Oritani, T. and Yamashita, K., *Agric. Biol. Chem.*, 38, 1961, 1974.)

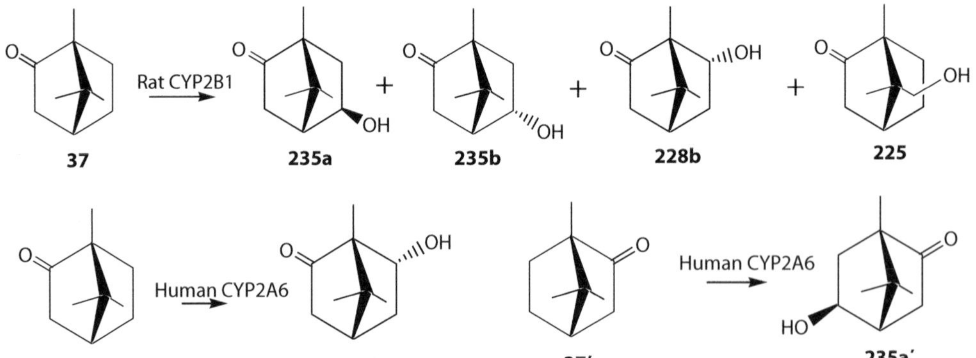

37 235a 235b 228b 225

37 228b 37′ 235a′

FIGURE 22.190 Biotransformation of (+)-camphor (**37**) by rat P450 enzyme (above) and (+)- (**37**) and (–)-camphor (**37′**) by human P450 enzymes.

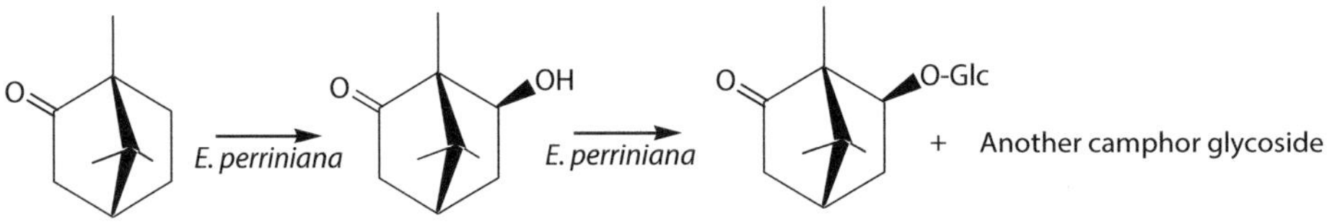

37 228 3 new

FIGURE 22.191 Biotransformation of (+)-camphor (**37**) by *Eucalyptus perriniana* suspension cell.

Conrad et al., 1961, 1965a,b; Gunsalus et al., 1965; Chapman et al., 1966; Hartline and Gunsalus, 1971; Oritani and Yamashita, 1974) (Figure 22.189).

Human CYP2A6 converted (+)-camphor (**37**) and (–)-camphor (**37′**) to 6-*endo*-hydroxycamphor (**228a**) and 5-*exo*-hydroxycamphor (**235b**), while rat CYP2B1 did 5-*endo*- (**235a**), 5-*exo*- (**235b**), and 6-*endo*-hydroxycamphor (**228a**) and 8-hydroxycamphor (**225**) (Gyoubu and Miyazawa, 2006) (Figure 22.190).

(+)-Camphor (**37**) was glycosylated by *E. perriniana* suspension cells to (+)-camphor monoglycoside (3 new, 11.7%) (Hamada et al., 2002) (Figure 22.191).

22.4.4.2.2 Fenchone

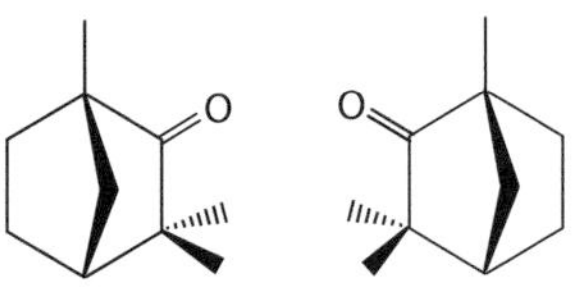

12 (+)- 12′ (–)-

(+)-Fenchone (**12**) was incubated with *Corynebacterium* sp. (Chapman et al., 1965) and *Absidia orchidis* (Pfrunder and Tamm, 1969a) give 6β-hydroxy- (**13a**) and 5β-hydroxyfenchones (**14a**) (Figure 22.191). On the other hand, *A. niger* biotransformed (+)-fenchone (**12**) to (+)-6α- (**13b**) and (+)-5α-hydroxyfenchones (**14b**) (Miyazawa et al., 1990a,b) and 5-oxofenchone (**15**), 9-formylfenchone (**17b**), and 9-carboxyfenchone (**18b**) (Miyazawa et al., 1990a,b) (Figure 22.192).

Furthermore, *A. niger* biotransformed (–)-fenchone (**12′**) to 5α-hydroxy- (**14b′**) and 6α-hydroxyfenchones (**13b′**) (Yamamoto et al., 1984) (Figure 22.193).

(+)- and (–)-Fenchone (**12** and **12′**) were converted to 6β-hydroxy- (**13a**, **13a′**), 6α-hydroxyfenchone (**13b**, **13b′**), and 10 hydroxyfenchone (**4**, **4′**) by P450. Of the 11 recombinant human P450 enzymes

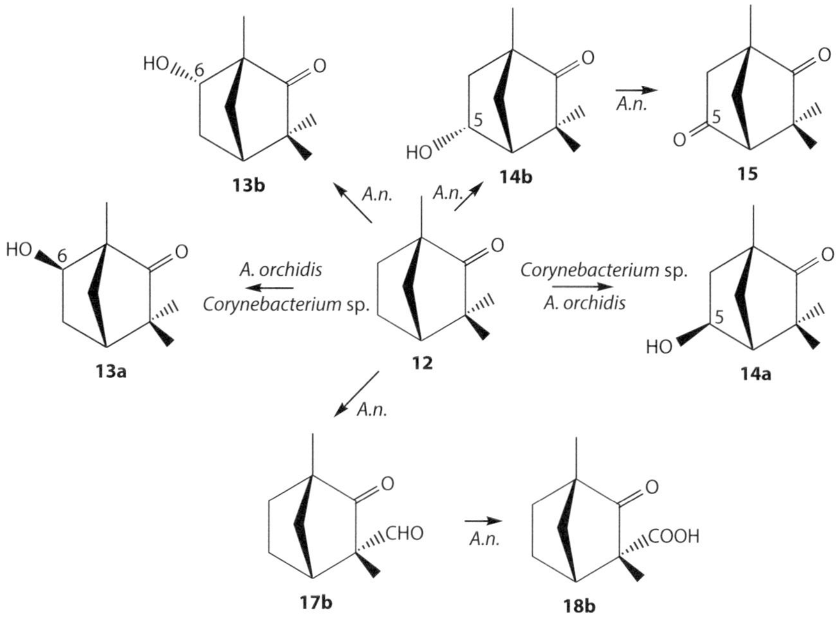

FIGURE 22.192 Metabolic pathways of (+)-fenchone (**12**) by *Corynebacterium* sp., *A. orchidis*, and *Aspergillus niger* TBUYN-2. (Modified from Chapman, P.J. et al. *Biochem. Biophys. Res. Commun.*, 20, 104, 1965; Pfrunder, B. and Tamm, Ch., *Helv. Chim. Acta.*, 52, 1643, 1969a; Miyazawa, M. et al. *Chem. Express*, 5, 237, 1990a; Miyazawa, M. et al. *Chem. Express*, 5, 407, 1990b.)

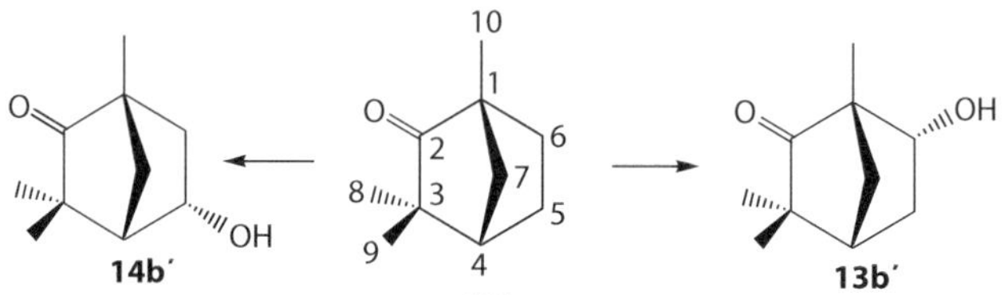

FIGURE 22.193 Metabolic pathways of (−)-fenchone (**12′**) by *Aspergillus niger* TBUYN-2. (Modified from Yamamoto, K. et al. Biotransformation of *d*- and *l*-fenchone by a strain of *Aspergillus niger*, *Proceedings of 28th TEAC*, 1984, pp. 168–170.)

tested, CYP2A6 and CYP2B6 catalyzed oxidation of (+)- (**12**) and (−)-fenchone (**12′**) (Gyoubu and Miyazawa, 2005) (Figure 22.194).

22.4.4.2.3 3-Pinanone (Pinocamphone and Isopinocamphone)

413a 1*R*2*S*5*S* (+)-pino-camphone	**413b** 1*R*,2*R*,5*S* (+)-isopino-camphone	**413a′** 1*S*2*R*5*R* (−)-pino-camphone	**413b′** 1*S*2*S*5*R* (−)-isopino-camphone

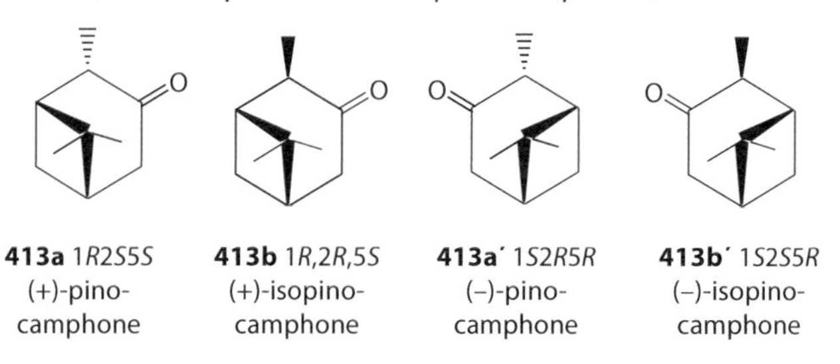

(+)- (**413**) and (−)-Isopinocamphone (**413′**) were biotransformed by *A. niger* to give (−)- (**414**) and (+)-2-hydroxy-3-pinanone (**414′**) as the main products, respectively, which inhibit strongly

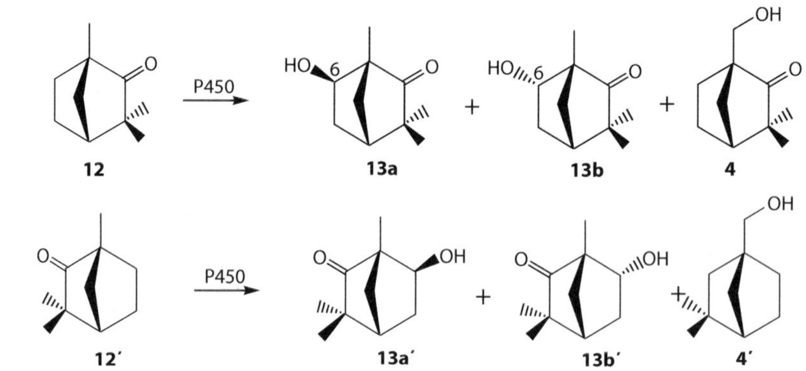

FIGURE 22.194 Biotransformation of (+)-fenchone (**12**) and (−)-fenchone (**12′**) by P450 enzymes. (Modified from Gyoubu, K. and Miyazawa, M., Biotransformation of (+)- and (−)-fenchone by liver microsomes, *Proceedings of 49th TEAC*, 2005, pp. 420–422.)

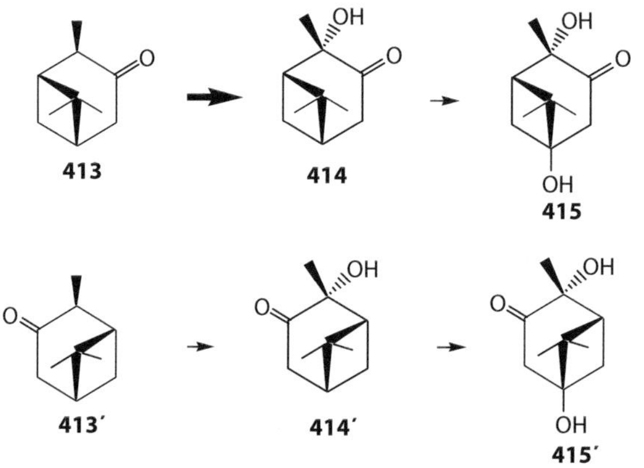

FIGURE 22.195 Biotransformation of (+)-isopinocamphone (**413b**) and (−)-isopinocamphone (**413b′**) by *Aspergillus niger* TBUYN-2. (Modified from Noma, Y. et al. Biotransformation of (+)- and (−)-pinane-2,3-diol and related compounds by *Aspergillus niger*, *Proceedings of 47th TEAC*, pp. 91–93, 2003; Noma, Y. et al. Biotransformation of (+)- and (−)-3-pinanone by *Aspergillus niger*, *Proceedings of 48th TEAC*, 2004, pp. 390–392.)

germination of lettuce seeds, and (−)- (**415**) and (+)-2,5-dihydroxy-3-pinanone (**415′**) as the minor components, respectively (Noma et al., 2003, 2004) (Figure 22.195).

22.4.4.2.4 2-Hydroxy-3-Pinanone

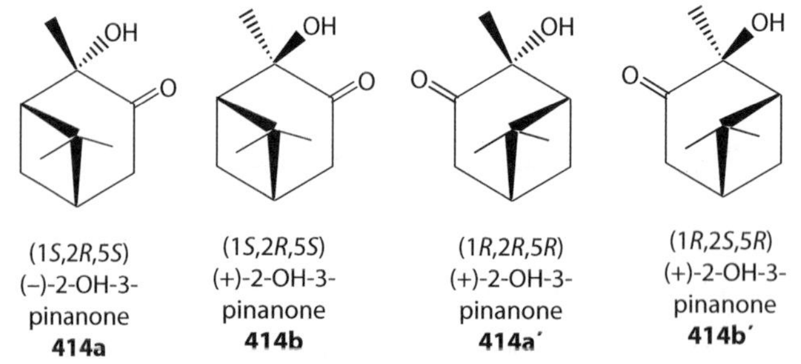

(−)-2α-Hydroxy-3-pinanone (**414**) was incubated with *A. niger* TBUYN-2 to give (−)-2α, 5-dihydroxy-3-pinanone (**415**) predominantly, whereas the fungus converted (+)-2α-hydroxy-3-pinanone (**414′**)

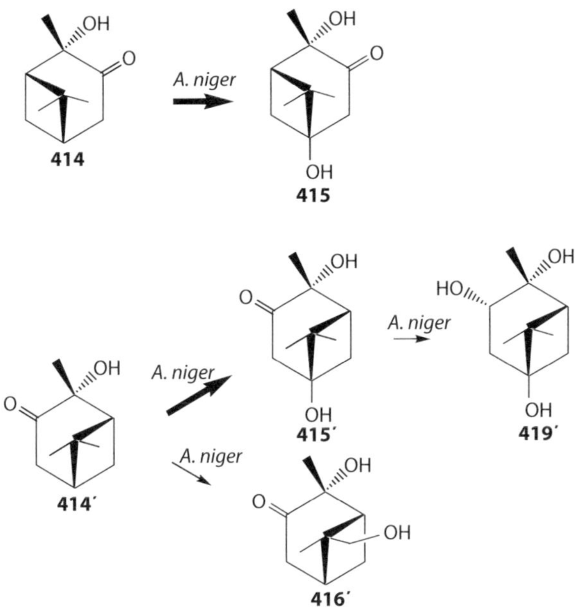

FIGURE 22.196 Biotransformation of (−)- (**414**) and (+)-2-hydroxy-3-pinanone (**414′**) by *Aspergillus niger* TBUYN-2. (Modified from Noma, Y. et al. Biotransformation of (+)- and (−)-pinane-2,3-diol and related compounds by *Aspergillus niger*, *Proceedings of 47th TEAC*, pp. 91–93, 2003; Noma, Y. et al. Biotransformation of (+)- and (−)-3-pinanone by *Aspergillus niger*, *Proceedings of 48th TEAC*, 2004, pp. 390–392.)

mainly to 2α, 5-dihydroxy-3-pinanone (**415′**), 2α,9-dihydroxy-3-pinanone (**416′**), and (−)-pinane-2α,3α,5-triol (**419ba′**) (Noma et al., 2003, 2004) (Figure 22.196).

A. niger TBUYN-2 metabolized β-pinene (**1**), isopinocamphone (**414b**), 2α-hydroxy-3-pinanone (**414a**), and pinane-2,3-diol (**419ab**) as shown in Figure 22.197. On the other hand, *A. niger* TBUYN-2 and *B. cinerea* metabolized β-pinene (**1′**), isopinocamphone (**414b′**), 2α-hydroxy-3-pinanone (**414a′**),

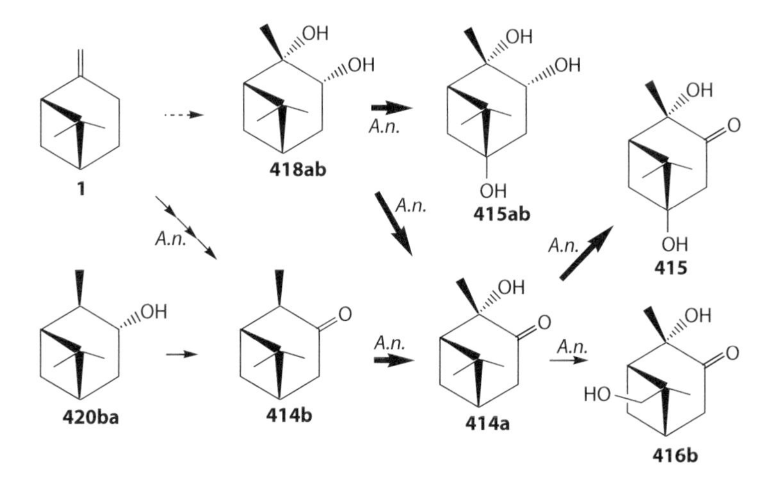

FIGURE 22.197 Relationship of the metabolism of β-pinene (**1**), isopinocamphone (**414b**), 2α-hydroxy-3-pinanone (**414a**), and pinane-2,3-diol (**419ab**) in *Aspergillus niger* TBUYN-2. (Modified from Noma, Y. et al. Biotransformation of (+)- and (−)-pinane-2,3-diol and related compounds by *Aspergillus niger*, *Proceedings of 47th TEAC*, 2003, pp. 91–93; Noma, Y. et al. Biotransformation of (+)- and (−)-3-pinanone by *Aspergillus niger*, *Proceedings of 48th TEAC*, 2004, pp. 390–392.)

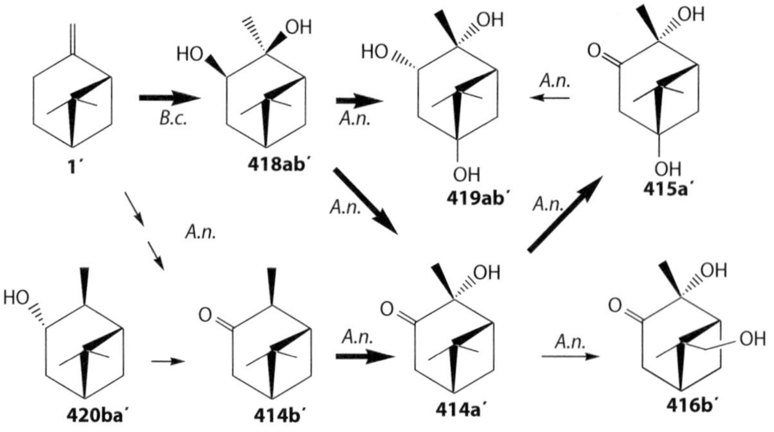

FIGURE 22.198 Relationship of the metabolism of β-pinene (**1′**), isopinocamphone (**414b′**), 2α-hydroxy-3-pinanone (**414a′**), and pinane-2,3-diol (**419ab′**) in *Aspergillus niger* TBUYN-2 and *Botrytis cinerea*. (Modified from Noma, Y. et al. Biotransformation of (+)- and (−)-pinane-2,3-diol and related compounds by *Aspergillus niger, Proceedings of 47th TEAC*, 2003, pp. 91–93; Noma, Y. et al. Biotransformation of (+)- and (−)-3-pinanone by *Aspergillus niger, Proceedings of 48th TEAC*, 2004, pp. 390–392.)

and pinane-2,3-diol (**419ab′**) as shown in Figure 22.198. The relationship of the metabolism of β-pinene (**1, 1′**), isopinocamphone (**414b, 414b′**), 2α-hydroxy-3-pinanone (**414a, 414a′**), and pinane-2,3-diol (**419ab, 419ab′**) in *A. niger* TBUYN-2 and *B. cinerea* is shown in Figures 22.197 and 22.198.

22.4.4.2.4.1 Mosquitocidal and Knockdown Activity Knockdown and mortality activity toward mosquito, *Culex quinquefasciatus*, was carried out for the metabolites of (+)- (**418ab**) and (−)-pinane-2,3-diols (**418ab′**) and (+)- and (−)-2-hydroxy-3-pinanones (**414** and **414′**) by Dr. Radhika Samarasekera, Industrial Technology Institute, Sri Lanka. (−)-2-Hydroxy-3-pinanone (**414′**) showed the mosquito knockdown activity and the mosquitocidal activity at the concentration of 1% and 2% (Table 22.16).

22.4.4.2.4.2 Antimicrobial Activity The microorganisms were refreshed in Mueller–Hinton broth (Merck) at 35°C–37°C and inoculated on Mueller–Hinton agar (Mast Diagnostics, Merseyside, United Kingdom) media for preparation of inoculum. *E. coli* (NRRL B-3008), *P. aeruginosa* (ATCC 27853),

TABLE 22.16
Knockdown and Mortality Activity toward Mosquito[a]

Compounds	Knockdown (%)	Mortality (%)
(+)-2,5-Dihydroxy-3-pinanone (**415%**, 2%)	27	20
(−)-2,5-Dihydroxy-3-pinanone (**415′**, 2%)	NT	7
(+)-2-Hydroxy-3-pinanone (**414%**, 2%)	40	33
(−)-2-Hydroxy-3-pinanone (**414′**, 2%)	100	40
(−)-2-Hydroxy-3-pinanone (**414′**, 1%)	53	7
(+)-Pinane-2,3,5-triol (**419%**, 2%)	NT	NT
(−)-Pinane-2,3,5-triol (**419%**, 2%)	13	NT
(+)-Pinane-2,3-diol (**418%**, 2%)	NT	NT
(−)-Pinane-2,3-diol (**418′**, 2%)	NT	NT

[a] The results are against *Culex quinquefasciatus*.

TABLE 22.17
Biological Activity of Pinane-2,3,5-Triol (419 and 419′), 2,5-Dihydroxy-3-Pinanone (415 and 415′), and 7-Hydroxymethyl-1-*p*-Menthene-8-ol (453′) toward MRSA

	MIC (mg/mL)							
	Compounds					Control		
Microorganisms	**419**	**415′**	**415**	**419′**	**453′**	**ST1**	**ST2**	**ST3**
Escherichia coli	0.5	0.5	0.25	0.5	0.25	0.007	0.0039	Nt
Pseudomonas aeruginosa	0.5	0.125	0.125	0.25	0.25	0.002	0.0078	Nt
Enterobacter aerogenes	0.5	0.5	0.25	0.5	1.00	0.007	0.0019	Nt
Salmonella typhimurium	0.25	0.125	0.125	0.25	0.25	0.01	0.0019	Nt
Candida albicans	0.5	0.125	0.125	0.25	1.00	NT	NT	0.0625
Staphylococcus epidermidis	0.5	0.5	0.25	0.5	1.00	0.002	0.0009	NT
MRSA	0.25	0.125	0.125	0.25	0.125	0.5	0.031	NT

Source: Iscan (2005, unpublished data).
Note: MRSA, methicillin-resistant *Staphylococcus aureus*; NT, not tested; ST1, ampicillin Na (sigma); ST2, chloramphenicol (sigma); ST3, ketoconazole (sigma).

Enterobacter aerogenes (NRRL 3567), *S. typhimurium* (NRRL B-4420), *Staphylococcus epidermidis* (ATCC 12228), methicillin-resistant *Staphylococcus aureus* (MRSA, clinical isolate, Osmangazi University, Faculty of Medicine, Eskisehir, Turkey), and *Candida albicans* (clinical isolate, Osmangazi University, Faculty of Medicine, Eskisehir, Turkey) were used as pathogen test microorganisms. Microdilution broth susceptibility assay (*R1, R2*) was used for the antimicrobial evaluation of the samples. Stock solutions were prepared in DMSO (Carlo-Erba). Dilution series were prepared from 2 mg/mL in sterile distilled water in microtest tubes from where they were transferred to 96-well microtiter plates. Overnight grown bacterial and candial suspensions in double strength Mueller–Hilton broth (Merck) was standardized to approximately 10^8 CFU/mL using McFarland No. 0.5 (10^6 CFU/mL for *C. albicans*). A volume of 100 μL of each bacterial suspension was then added to each well. The last row containing only the serial dilutions of samples without microorganism was used as negative control. Sterile distilled water, medium, and microorganisms served as a positive growth control. After incubation at 37°C for 24 h, the first well without turbidity was determined as the minimal inhibition concentration (MIC), and chloramphenicol (sigma), ampicillin (sigma), and ketoconazole (sigma) were used as standard antimicrobial agents (Amsterdam, 1997; Koneman et al., 1997) (Table 22.17).

22.5 SUMMARY

22.5.1 Metabolic Pathways of Monoterpenoids by Microorganisms

About 50 years is over since the hydroxylation of α-pinene (**4**) was reported by *A. niger* in 1960 (Bhattacharyya et al., 1960). During these years, many investigators have studied the biotransformation of a number of monoterpenoids by using various kinds of microorganisms. Now we summarize the microbiological transformation of monoterpenoids according to the literatures listed in the references including the metabolic pathways (Figures 22.199 through 22.206) for the further development of the investigation on microbiological transformation of terpenoids.

Metabolic pathways of β-pinene (**1**), α-pinene (**4**), fenchol (**11**), fenchone (**12**), thujone (**28**), carvotanacetone (**47**), and sobrerol (**43**) are summarized in Figure 22.199. In general, β-pinene (**1**) is metabolized by six pathways. At first, β-pinene (**1**) is metabolized via α-pinene (**4**) to many metabolites such as myrtenol (**5**) (Shukla et al., 1968; Shukla and Bhattacharyya, 1968), verbenol (**23**) (Bhattacharyya et al., 1960; Prema and Bhattachayya, 1962), and thujone (**28**) (Gibbon and

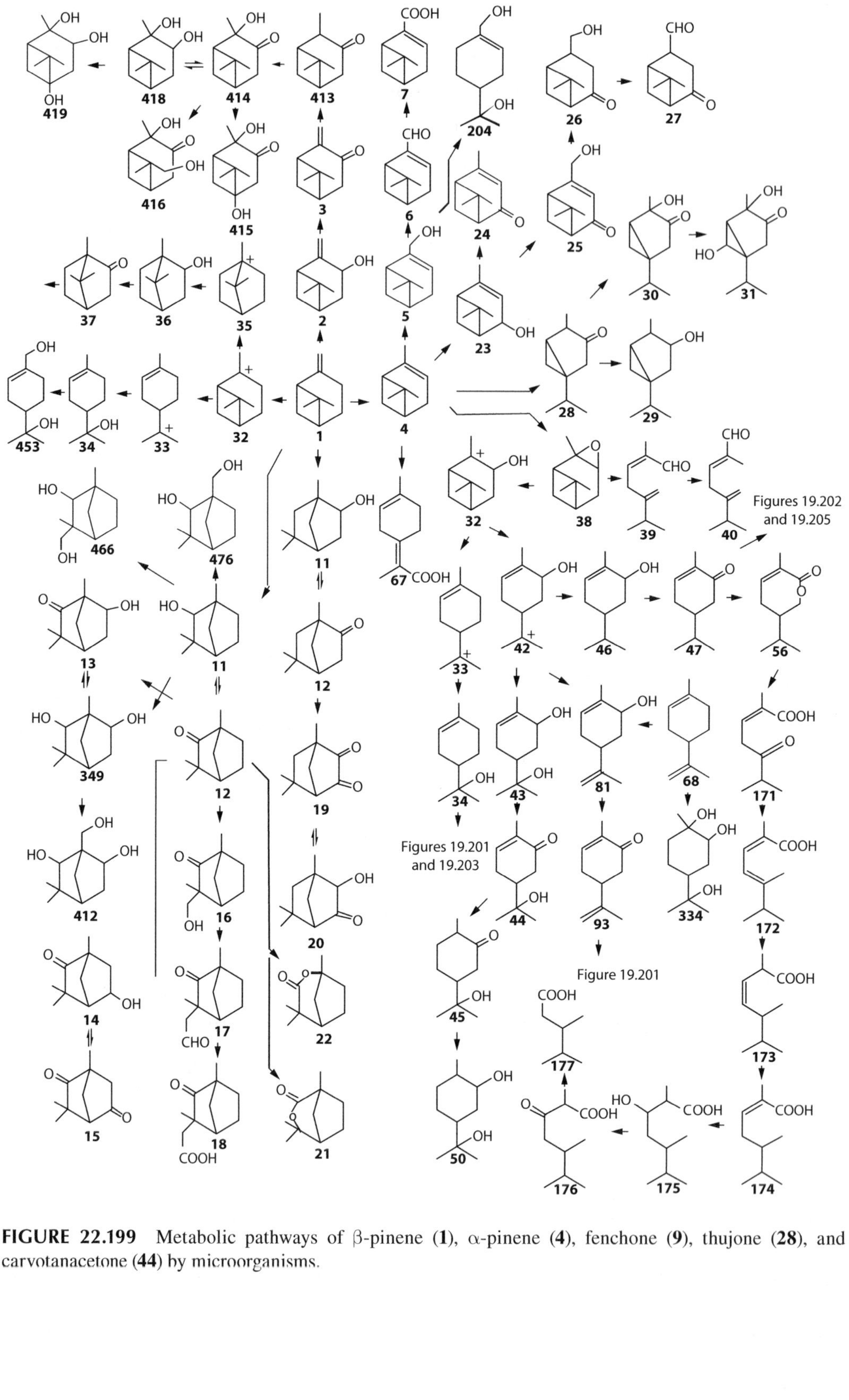

FIGURE 22.199 Metabolic pathways of β-pinene (**1**), α-pinene (**4**), fenchone (**9**), thujone (**28**), and carvotanacetone (**44**) by microorganisms.

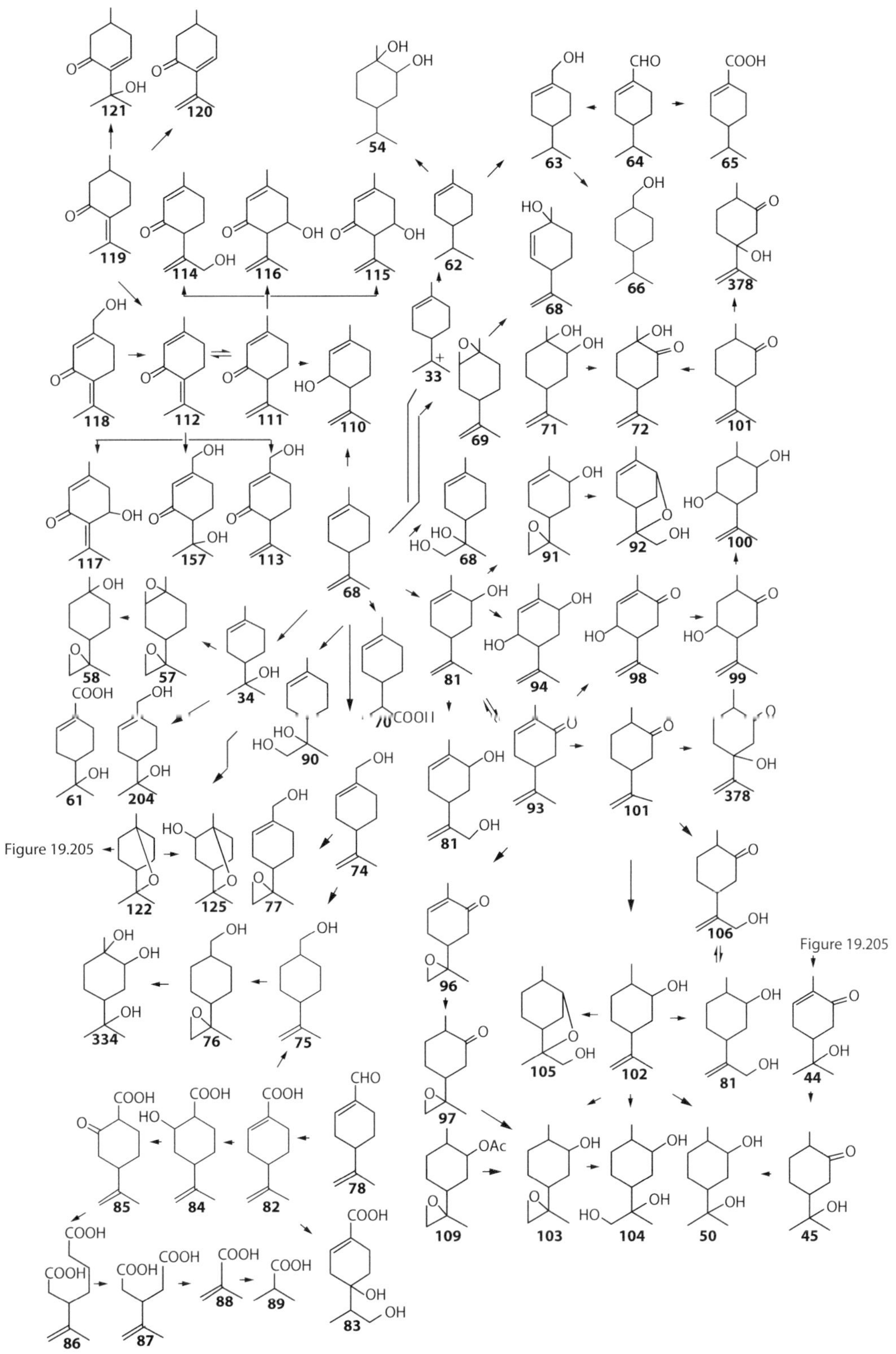

FIGURE 22.200 Metabolic pathways of limonene (**68**), perillyl alcohol (**74**), carvone (**93**), isopiperitenone (**111**), and piperitenone (**112**) by microorganisms.

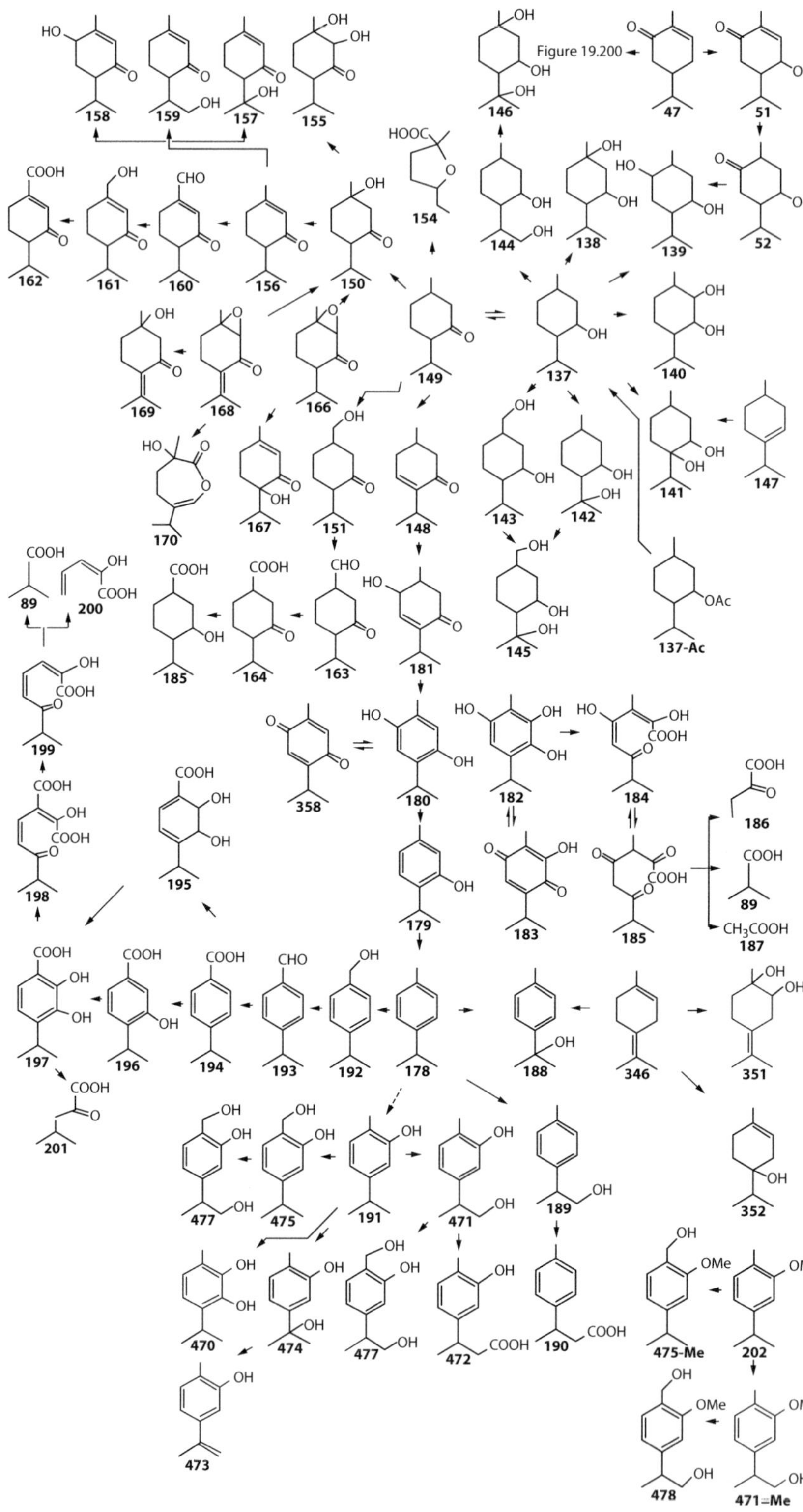

FIGURE 22.201 Metabolic pathways of menthol (**137**), menthone (**149**), *p*-cymene (**178**), thymol (**179**), carvacrol methyl ether (**201**), and carvotanacetone (**47**) by microorganisms and rabbit.

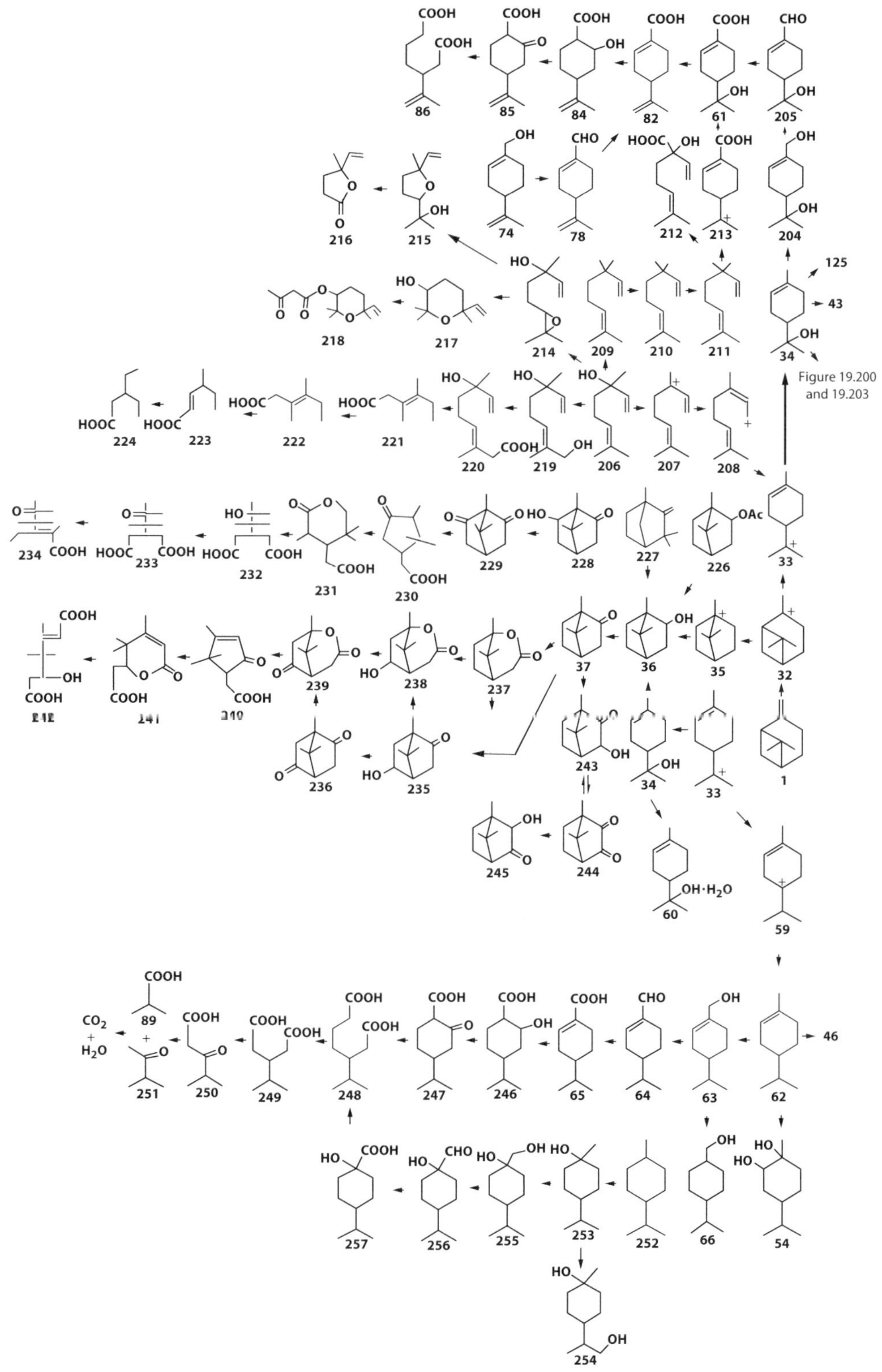

FIGURE 22.202 Metabolic pathway of borneol (**36**), camphor (**37**), phellandral (**64**), linalool (**206**), and *p*-menthane (**252**) by microorganisms.

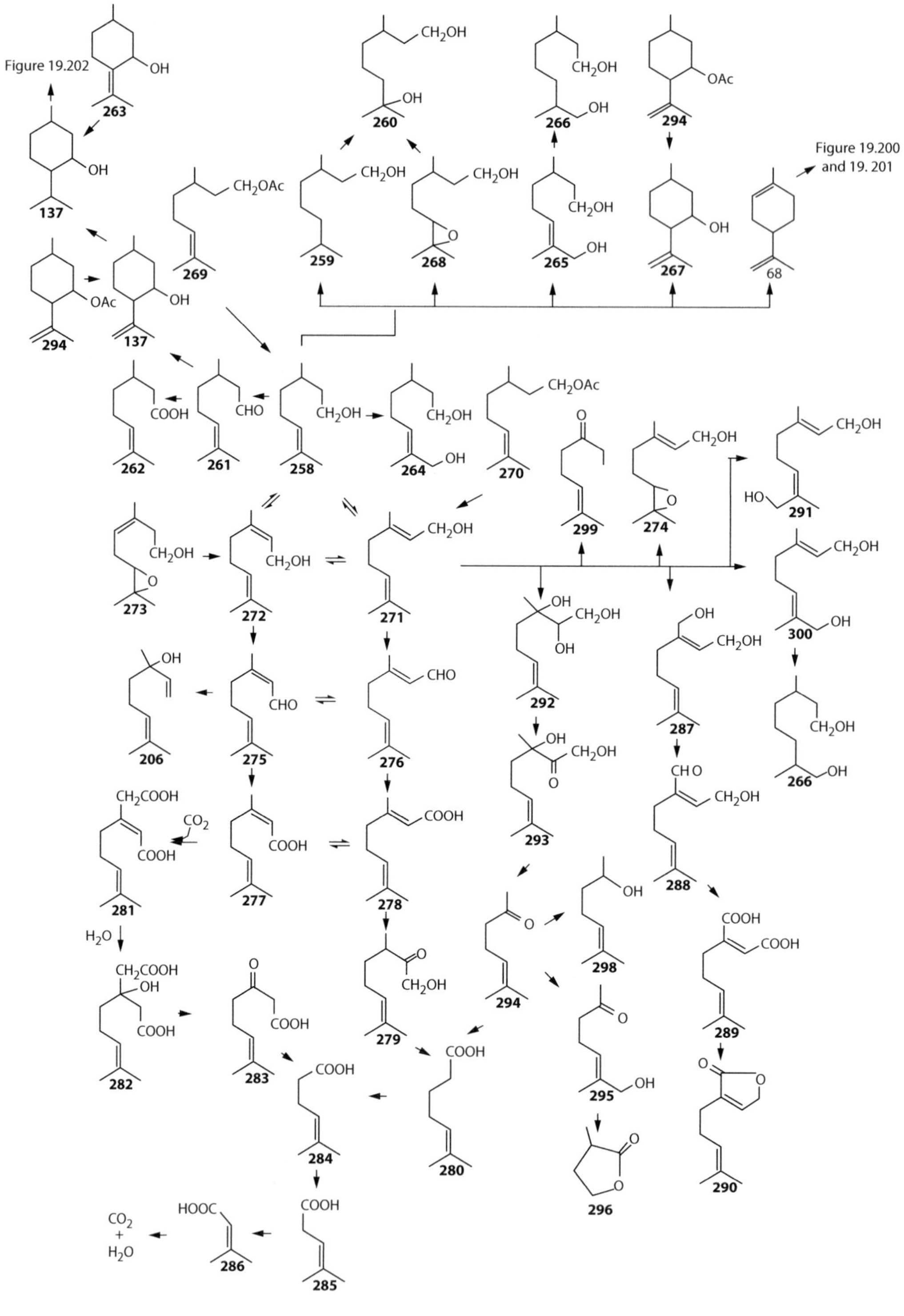

FIGURE 22.203 Metabolic pathways of citronellal (**258**), geraniol (**271**), nerol (**272**), and citral (**275** and **276**) by microorganisms.

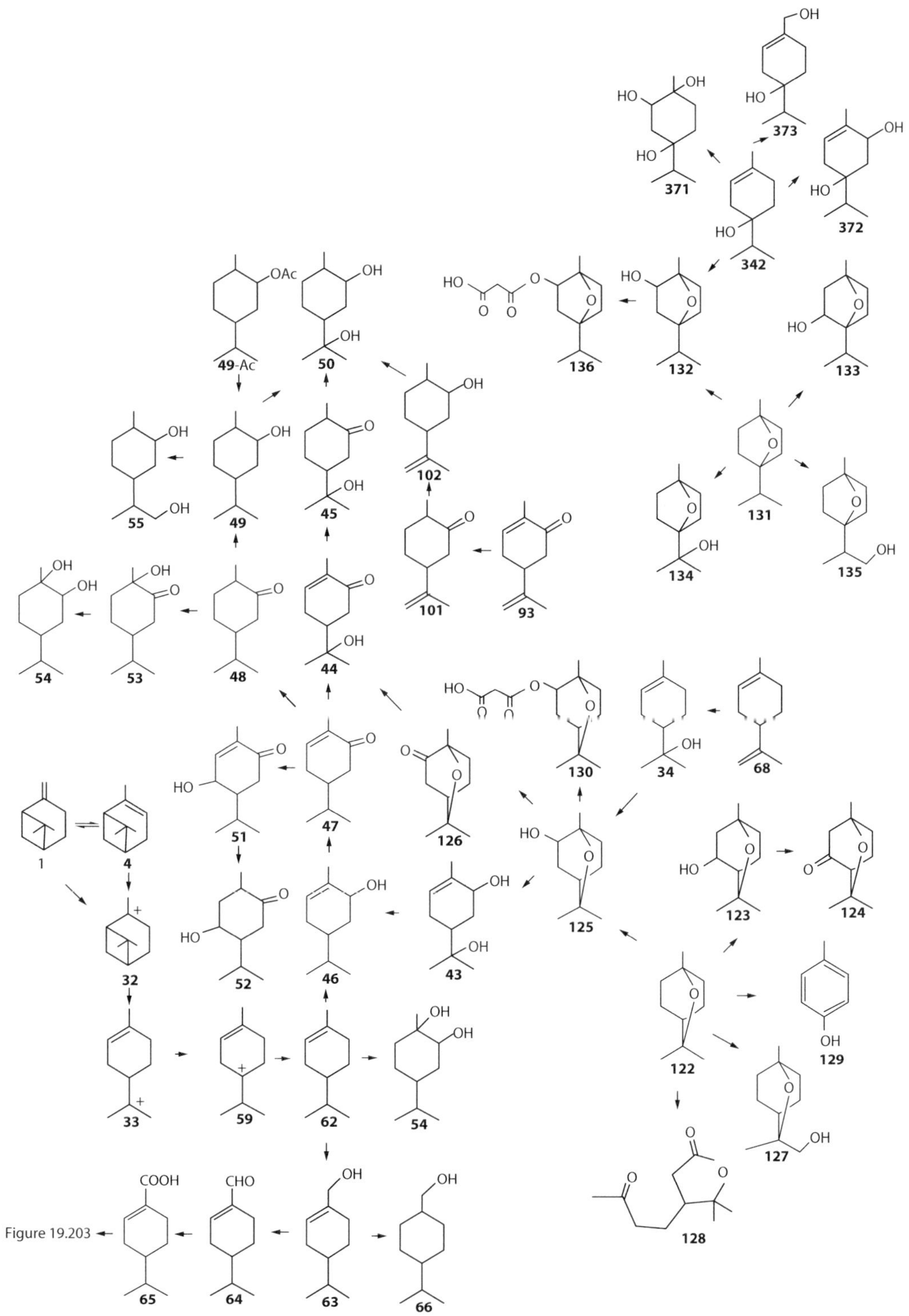

FIGURE 22.204 Metabolic pathways of 1,8-cineole (**122**), 1,4-cineole (**131**), phellandrene (**62**), and carvotanacetone (**47**) by microorganisms.

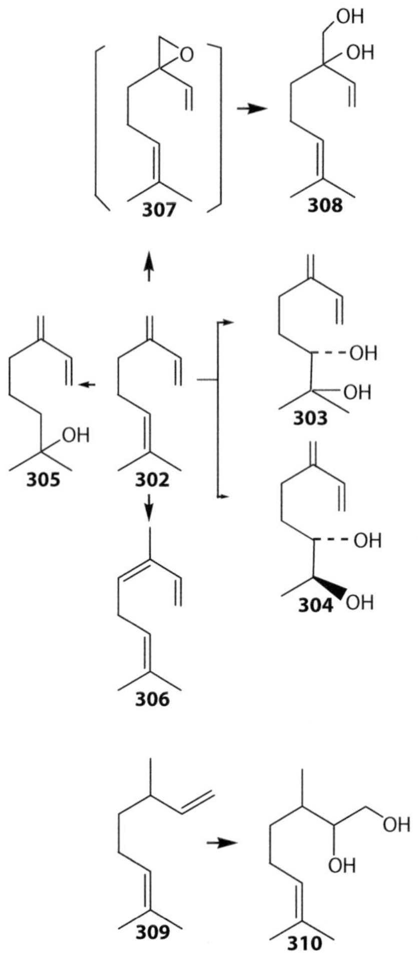

FIGURE 22.205 Metabolic pathways of myrcene (**302**) and citronellene (**309**) by rat and microorganisms.

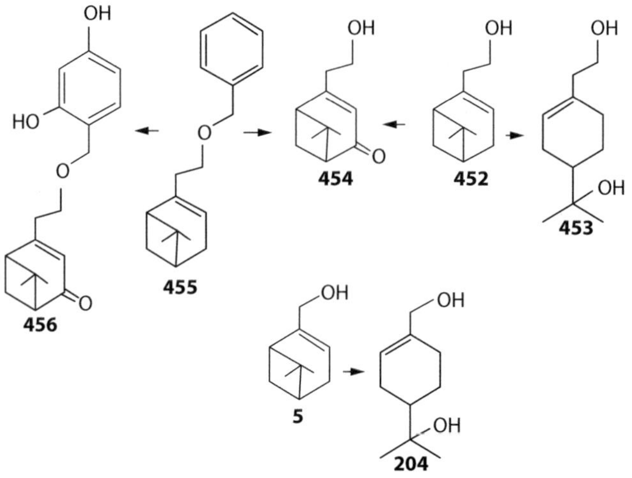

FIGURE 22.206 Metabolic pathways of nopol (**452**) and nopol benzyl ether (**455**) by microorganisms.

Pirt, 1971). Myrtenol (**5**) is further metabolized to myrtenal (**6**) and myrtenic acid (**7**). Verbenol (**23**) is further metabolized to verbenone (**24**), 7-hydroxyverbenone (**25**), 7-hydroxyverbenone (**26**), and 7-formyl verbanone (**27**). Thujone (**28**) is further metabolized to thujoyl alcohol (**29**), 1-hydroxythujone (**30**), and 1,3-dihydroxythujone (**31**). Second, β-pinene (**1**) is metabolized to pinocarveol (**2**) and pinocarvone (**3**) (Ganapathy and Bhattacharyya, unpublished data). Pinocarvone (**3**) is further metabolized to isopinocamphone (**413**), which is further hydroxylated to give 2-hydroxy-3-pinanone (**414**). Compound **414** is further metabolized to give pinane-2,3-diol (**419**), 2,5-dihydroxy- (**415**), and 2,9-dihydroxy-3-pinanone (**416**). Third, β-pinene (**1**) is metabolized to α-fenchol (**11**) and fenchone (**12**) (Dhavalikar et al., 1974), which are further metabolized to 6-hydroxy- (**13**) and 5-hydroxyfenchone (**14**), 5-oxofenchone (**15**), fenchone-9-al (**17**), fenchone-9-oic acid (**18**) via 9-hydroxyfenchone (**16**), 2,3-fencholide (**21**), and 1,2-fencholide (**22**) (Pfrunder and Tamm, 1969a,b; Yamamoto et al., 1984; Christensen and Tuthill, 1985; Miyazawa et al., 1990a,b). Fenchol (**12**) is also metabolized to 9-hydroxyfenchol (**466**) and 7-hydroxyfenchol (**467**), 6-hydroxyfenchol (**349**), and 6,7-dihydroxyfenchol (**412**). Fourth, β-pinene (**1**) is metabolized via fenchoquinone (**19**) to 2-hydroxyfenchone (**20**) (Pfrunder and Tamm, 1969b; Gibbon et al., 1972). Fifth, β-pinene (**1**) is metabolized to α-terpineol (**34**) via pinyl cation (**32**) and 1-*p*-menthene-8-cation (**33**) (Saeki and Hashimoto, 1968, 1971; Hosler, 1969; Hayashi et al., 1972). α-Terpineol (**34**) is metabolized to 8,9-epoxy-1-*p*-methanol (**58**) via diepoxide (**57**), terpin hydrate (**60**), and oleuropeic acid (**204**) (Saeki and Hashimoto, 1968, 1971; Shukla et al., 1968; Shukla and Bhattacharyya, 1968; Hosler, 1969; Hungund et al., 1970; Hayashi et al., 1972). As shown in Figure 22.202, oleuropeic acid (**204**) is formed from linalool (**206**) and α-terpineol (**34**) via **204**, **205**, and **213** as intermediates (Shukla et al., 1968; Shukla and Bhattacharyya, 1968; Hungund et al., 1970) and degradatively metabolized to perillic acid (**82**), 2-hydroxy-8-*p*-menthen-7-oic acid (**84**), 2-oxo-8-*p*-menthen-7-oic acid (**84**), 2-oxo-8-*p*-menthen-1-oic acid (**85**), and β-isopropyl pimelic acid (**86**) (Shukla et al., 1968; Shukla and Bhattacharyya, 1968; Hungund et al., 1970). Oleuropeic acid (**204**) is also formed from β-pinene (**1**) via α-terpineol (**34**) as the intermediate (Noma et al., 2001). Oleuropeic acid (**204**) is also formed from myrtenol (**5**) by rearrangement reaction by *A. niger* TBUYN-2 (Noma and Asakawa, 2005b). Finally, β-pinene (**1**) is metabolized to borneol (**36**) and camphor (**37**) via two cations (**32** and **35**) and to 1-*p*-menthene (**62**) via two cations (**33** and **59**) (Shukla and Bhattacharyya, 1968). 1-*p*-Menthene (**62**) is metabolized to phellandric acid (**65**) via phellandrol (**63**) and phellandral (**64**), which is further degradatively metabolized through **246–251** and **89** to water and carbon dioxide as shown in Figure 22.204 (Shukla et al., 1968). Phellandral (**64**) is easily reduced to give phellandrol (**63**) by *Euglena* sp. and *Dunaliella* sp. (Noma et al., 1984, 1986, 1991a,b, 1992b). Furthermore, 1-*p*-menthene (**62**) is metabolized to 1-*p*-menthen-2-ol (**46**) and *p*-menthane-1,2-diol (**54**) as shown in Figure 22.204. Perillic acid (**82**) is easily formed from perillaldehyde (**78**) and perillyl alcohol (**74**) (Figure 22.19) (Swamy et al., 1965; Dhavalikar and Bhattacharyya, 1966; Dhavalikar et al., 1966; Ballal et al., 1967; Shima et al., 1972; Kayahara et al., 1973). α-Terpineol (**34**) is also formed from linalool (**206**). α-Pinene (**4**) is metabolized by five pathways as follows: First, α-pinene (**4**) is metabolized to myrtenol (**5**), myrtenal (**6**), and myrtenoic acid (**7**) (Shukla et al., 1968; Shukla and Bhattacharyya, 1968; Hungund et al., 1970; Ganapathy and Bhattacharyya, unpublished results). Myrtenal (**6**) is easily reduced to myrtenol (**5**) by *Euglena* and *Dunaliella* spp., (Noma et al., 1991a,b, 1992b). Myrtanol (**8**) is metabolized to 3-hydroxy- (**9**) and 4-hydroxymyrtanol (**10**) (Miyazawa et al., 1994a). Second, α-pinene (**4**) is metabolized to verbenol (**23**), verbenone (**24**), 7-hydroxyverbenone (**25**), and verbanone-7-al (**27**) (Bhattacharyya et al., 1960; Prema and Bhattachayya, 1962; Miyazawa et al., 1991a). Third, α-pinene (**4**) is metabolized to thujone (**28**), thujoyl alcohol (**29**), 1-hydroxy- (**30**), and 1,3-dihydroxythujone (**31**) (Gibbon and Pirt, 1971; Miyazawa et al., 1992a; Noma, 2000). Fourth, α-pinene (**4**) is metabolized to sobrerol (**43**) and carvotanacetol (**46**, 1-*p*-menthen-2-ol) via α-pinene epoxide (**38**) and two cations (**41** and **42**). Sobrerol (**43**) is further metabolized to 8-hydroxycarvotanacetone (**44**, carvonhydrate), 8-hydrocarvomenthone (**45**), and *p*-menthane-2,8-diol (50) (Prema and Bhattachayya, 1962; Noma, 2007). In the metabolism of sobrerol (**43**), 8-hydroxycarvotanacetone (**44**), and 8-hydroxycarvomenthone (**45**) by *A. niger* TBUYN-2, the formation of *p*-menthane-2,8-diol (**50**) is very highly enantio- and diastereoselective in the reduction of 8-hydroxycarvomenthone (Noma, 2007). 8-Hydroxycarvotanacetone (**44**) is a common metabolite from sobrerol (**43**) and carvotanacetone (**47**). That is, carvotanacetone (**47**) is metabolized to carvomenthone (**48**), carvomenthol (**49**), 8-hydroxycarvomenthol (**50**), 5-hydroxycarvotanacetone (**51**), 8-hydroxycarvotanacetone (**44**), 5-hydroxycarvomenthone (**52**), and 2,3-lactone (**56**) (Gibbon and Pirt, 1971; Gibbon et al., 1972; Noma et al., 1974a, 1985c, 1988a). Carvomenthone (**48**) is metabolized to **45**, 8-hydroxycarvomenthol (**50**), 1-hydroxycarvomenthone (**53**), and *p*-menthane-1,2-diol (**54**) (Noma et al., 1985b, 1988a). Compound **52** is metabolized to 6-hydroxymenthol (**139**), which is the common metabolite of menthol (**137**) (see Figure 22.201). Carvomenthol (**49**) is metabolized to 8-hydroxycarvomenthol (**50**) and *p*-menthane-2, 9-diol (**55**). Finally, α-pinene (**4**) is metabolized to borneol (**36**) and camphor (**37**) via **32** and **35** and to phellandrene (**62**) via **32** and two cations (**33** and **59**) as mentioned in the metabolism of β-pinene (**1**). Carvotanacetone (**47**) is also metabolized degradatively to 3,4-dimethylvaleric acid (**177**) via **56** and **158–163** as shown in Figure 22.201 (Gibbon and Pirt, 1971; Gibbon et al., 1972). α-Pinene (**4**) is also metabolized to 2-(4-methyl-3-cyclohexenylidene)-propionic acid (**67**) (Figure 22.199).

Metabolic pathways of limonene (**68**), perillyl alcohol (**74**), carvone (**93**), isopiperitenone (**111**), and piperitenone (**112**) are summarized in Figure 22.199. Limonene (**68**) is metabolized by eight pathways. That is, limonene (**68**) is converted into α-terpineol (**34**) (Savithiry et al., 1997), limonene-1,2-epoxide (**69**), 1-*p*-menthene-9-oic acid (**70**), perillyl alcohol (**74**), 1-*p*-menthene-8,9-diol (**79**), isopiperitenol (**110**), *p*-mentha-1,8-diene-4-ol (**80**, 4-terpineol), and carveol (**81**) (Dhavalikar and Bhattacharyya, 1966; Dhavalikar et al., 1966; Bowen, 1975; Noma et al., 1982, 1992b; Miyazawa et al., 1983; Savithiry et al., 1997; Van der Werf et al., 1997, 1998b; Van der Werf and de Bont, 1998a). Dihydrocarvone (**101**), limonene-1,2-diol (**71**), 1-hydroxy-8-*p*-menthene-2-one (**72**), and *p*-mentha-2,8-diene-1-ol (**73**) are formed from limonene (**68**) via limonene epoxide (**69**) as intermediate. Limonene (**68**) is also metabolized via carveol (**78**), limonene-1,2-diol (**71**), carvone (**93**), 1-*p*-menthene-6,9-diol (**95**), 8,9-dihydroxy-1-*p*-menthene (**90**), α-terpineol (**34**), 2α-hydroxy-1,8-cineole (**125**), and *p*-menthane-1,2,8-triol (**334**). Bottrospicatol (**92**) and 5-hydroxycarveol (**94**) are formed from *cis*-carveol by *S. bottropensis* SY-2-1 (Noma et al., 1982; Nishimura et al., 1983a; Noma and Asakawa, 1992; Noma et al., 1982). Carvyl acetate and carvyl propionate (both are shown as **106**) are hydrolyzed enantio- and diastereoselectively to carveol (**78**) (Oritani and Yamashita, 1980; Noma, 2000). Carvone (**93**) is metabolized through four pathways as follows: First, carvone (**93**) is reduced to carveol (**81**) (Noma, 1980). Second, it is epoxidized to carvone-8,9-epoxide (**96**), which is further metabolized to dihydrocarvone-8,9-epoxide (97), dihydrocarveol-8,9-epoxide (**103**), and menthane-2,8,9-triol (**104**) (Noma et al., 1980; Noma and Nishimura, 1982; Noma, 2000). Third, **93** is hydroxylated to 5-hydroxycarvone (**98**), 5-hydroxydihydrocarvone (**99**), and 5-hydroxydihydrocarveol (**100**) (Noma and Nishimura, 1982). Dihydrocarvone (**101**) is metabolized to 8-*p*-menthene-1,2-diol (**71**) via l-hydroxydihydrocarvone (**72**), 10-hydroxydihydrocarvone (**106**), and dihydrocarveol (**102**), which is metabolized to 10-hydroxydihydrocarveol (**107**), *p*-menthane-2,8-diol (**50**), dihydrocarveol-8,9-epoxide (**100**), *p*-menthane-2,8,9-triol (**104**), and dihydrobottrospicatol (**105**) (Noma et al., 1985a,b). In the biotransformation of (+)-carvone by plant pathogenic fungi, *A. niger* Tiegh, isodihydrocarvone (**101**) was metabolized to 4-hydroxyisodihydrocarvone (**378**) and 1-hydroxyisodihydrocarvone (**72**) (Noma and Asakawa, 2008). 8,9-Epoxydihydrocarvyl acetate (**109**) is hydrolyzed to 8,9-epoxydihydrocarveol (**103**). Perillyl alcohol (**74**) is metabolized through three pathways to shisool (**75**), shisool-8,9-epoxide (**76**), perillyl alcohol-8,9-epoxide (**77**), perillaldehyde (**78**), perillic acid (**82**), and 4,9-dihydroxy-1-*p*-menthen-7-oic acid (**83**). Perillic acid (**82**) is metabolized degradatively to **84–89** as shown in Figure 22.200 (Swamy et al., 1965; Dhavalikar and Bhattacharyya, 1966; Dhavalikar et al., 1966; Ballal et al., 1967; Shukla et al., 1968; Shukla and Bhattacharyya, 1968; Shima et al., 1972; Kayahara et al., 1973; Hungund et al., 1970). Isopiperitenol (**110**) is reduced to isopiperitenone (**111**), which is metabolized to 3-hydroxy- (**115**), 4-hydroxy- (**116**), 7-hydroxy- (**113**), and 10-hydroxyisopiperitenone (**114**) and piperitenone (**112**). Compounds isopiperitenone (**111**) and piperitenone (**112**) are isomerized to each other (Noma et al.,

1992c). Furthermore, piperitenone (**112**) is metabolized to 8-hydroxypiperitone (**157**) and 5-hydroxy- (**117**) and 7-hydroxypiperitenone (**118**). Pulegone (**119**) is metabolized to **112**, 8-hydroxymenthenone (**121**), and 8,9-dehydromenthenone (**120**).

Metabolic pathways of menthol (**137**), menthone (**149**), thymol (**179**), and carvacrol methyl ether (**202**) are summarized in Figure 22.201. Menthol (**137**) is generally hydroxylated to give 1-hydroxy- (**138**), 2-hydroxy- (**140**), 4-hydroxy- (**141**), 6-hydroxy- (**139**), 7-hydroxy- (**143**), 8-hydroxy- (**142**), and 9-hydroxymenthol (**144**) and 1,8-dihydroxy- (**146**) and 7,8-dihydroxymenthol (**148**) (Asakawa et al., 1991; Takahashi et al., 1994; Van der Werf et al., 1997). Racemic menthyl acetate and menthyl chloroacetate are hydrolyzed asymmetrically by an esterase of microorganisms (Brit Patent, 1970; Moroe et al., 1971; Watanabe and Inagaki, 1977a,b). Menthone (**149**) is reductively metabolized to **137** and oxidatively metabolized to 3,7-dimethyl-6-hydroxyoctanoic acid (**152**), 3,7-dimethyl-6-oxooctanoic acid (**153**), 2-methyl-2,5-oxidoheptenoic acid (**154**), 1-hydroxymenthone (**150**), piperitone (**156**), 7-hydroxymenthone (**151**), menthone-7-al (**163**), menthone-7-oic acid (**164**), and 7-carboxylmenthol (**165**) (Sawamura et al., 1974). Compound **156** is metabolized to menthone-1,2-diol (**155**) (Miyazawa et al., 1991e, 1992d, e). Compound **148** is metabolized to 6-hydroxy- (**158**), 8-hydroxy- (**157**), and 9-hydroxypiperitone (**159**), piperitone-7-al (**160**), 7-hydroxypiperitone (**161**), and piperitone-7-oic acid (**162**) (Lassak et al., 1973). Compound **149** is also formed from menthenone (**148**) by hydrogenation (Mukherjee et al., 1973), which is metabolized to 6-hydroxymenthenone (**181**, 6-hydroxy-4-*p*-menthen-3-one). 6-Hydroxymenthenone (**181**) is also formed from thymol (**179**) via 6-hydroxythymol (**180**). 6-Hydroxythymol (**180**) is degradatively metabolized through **182–185** to **186**, **187**, and **89** (Mukherjee et al., 1974). Piperitone oxide (**166**) is metabolized to 1-hydroxymenthone (**150**) and 4-hydroxypiperitone (**167**) (Lassak et al., 1973; Miyazawa et al., 1991d,e). Piperitenone oxide (**168**) is metabolized to 1-hydroxymenthone (**150**), 1-hydroxypulegone (**169**), and 2,3-seco-*p*-menthalacetone-3-en-1-ol (**170**) (Lassak et al., 1973; Miyazawa et al., 1991e). *p*-Cymene (**178**) is metabolized to 8-hydroxy- (**188**) and 9-hydroxy-p-cymene (**189**), 2-(4-methylphenyl)-propanoic acid (**190**), thymol (**179**), and cumin alcohol (**192**), which is further converted degradatively to *p*-cumin aldehyde (**193**), cumic acid (**194**), *cis*-2,3-dihydroxy-2,3-dihydro-*p*-cumic acid (**195**), 2,3-dihydroxy-*p*-cumic acid (**197**), **198–200,** and **89** as shown in Figure 22.3 (Chamberlain and Dagley, 1968; DeFrank and Ribbons, 1977a,b; Hudlicky et al., 1999; Noma, 2000). Compound **197** is also metabolized to 4-methyl-2-oxopentanoic acid (**201**) (DeFrank and Ribbons, 1977a). Compound **193** is easily metabolized to 192 and **194** (Noma et al., 1991b. Carvacrol methyl ether (**202**) is easily metabolized to 7-hydroxycarvacrol methyl ether (**203**) (Noma, 2000).

Metabolic pathways of borneol (**36**), camphor (**37**), phellandral (**64**), linalool (**206**), and *p*-menthane (**252**) are summarized in Figure 22.202. Borneol (**36**) is formed from β-pinene (**1**), α-pinene (**4**), **34**, bornyl acetate (**226**), and camphene (**229**), and it is metabolized to **36**; 3-hydroxy- (**243**), 5-hydroxy- (**235**), 6-hydroxy- (**228**), and 9-hydroxycamphor (**225**); and 1,2-campholide (**23**). Compound **228** is degradatively metabolized to 6-oxocamphor (**229**) and **230–234**, whereas **237** is also degradatively metabolized to 6-hydroxy-1,2-campholide (**238**), 6-oxo-1,2-campholide (**239**), and **240–242**. 5-Hydroxycamphor (**235**) is metabolized to **238**, 5-oxocamphor (**236**), and 6-oxo-1,2-campholide (**239**). Compound **243** is also metabolized to camphorquinone (**244**) and 2-hydroxyepicamphor (**245**) (Bradshaw et al., 1959; Conrad et al., 1961, 1965a,b; Gunsalus et al., 1965; Chapman et al., 1966; Hartline and Gunsalus, 1971; Oritani and Yamashita, 1974). 1-*p*-Menthene (**62**) is formed **1** and **4** via three cations (**32**, **33**, and **59**) and metabolized to phellandrol (**63**) (Noma et al., 1991b) and *p*-menthane-1,2-diol (**54**). Compound **63** is metabolized to phellandral (**64**) and 7-hydroxy-*p*-menthane (**66**). Compound **64** is furthermore metabolized degradatively to CO_2 and water via phellandric acid (**65**), **246–251**, and **89** (Dhavalikar and Bhattacharyya, 1966; Dhavalikar et al., 1966; Ballal et al., 1967; Shukla et al., 1968; Shukla and Bhattacharyya, 1968; Hungund et al., 1970). Compound **64** is also easily reduced to phellandrol (**63**) (Noma et al., 1991b, 1992a). *p*-Menthane (**252**) is metabolized via 1-hydroxy-*p*-menthane (**253**) to *p*-menthane-1,9-diol (**254**) and *p*-menthane-1,7-diol (**255**) (Tsukamoto et al., 1974, 1975; Noma et al., 1990). Compound **255** is degradatively metabolized via **256–248** to CO_2 and water through the degradation pathway of phellandric acid (**65**,

246–251, and **89**) as aforementioned. Linalool (**206**) is metabolized to α-terpineol (**34**), camphor (**37**), oleuropeic acid (**61**), 2-methyl-6-hydroxy-6-carboxy-2,7-octadiene (**211**), 2-methyl-6-hydroxy-6-carboxy-7-octene (**199**), 5-methyl-5-vinyltetrahydro-2-furanol (**215**), 5-methyl-5-vinyltetrahydro-2-furanone (**216**), and malonyl ester (**218**). 1-Hydroxylinalool (**219**) is metabolized degradatively to 2,6-dimethyl-6-hydroxy-*trans*-2,7-octadienoic acid (**220**), 4-methyl-*trans*-3, 5-hexadienoic acid (**221**), 4-methyl-*trans*-3,5-hexadienoic acid (**222**), 4-methyl-*trans*-2-hexenoic acid (**223**), and isobutyric acid (**224**). Compound **206** is furthermore metabolized via **213** to **61**, **82**, and **84**–**86** as shown in Figure 22.2 (Mizutani et al., 1971; Murakami et al., 1973; Madyastha et al., 1977; Rama Devi and Bhattacharyya, 1977a,b; Rama Devi et al., 1977; David and Veschambre, 1985; Miyazawa et al., 1994a,b).

Metabolic pathways of citronellol (**258**), citronellal (**261**), geraniol (**271**), nerol (**272**), citral [neral (**275**) and geranial (**276**)], and myrcene (**302**) are summarized in Figure 22.203 (Seubert and Fass, 1964a,b; Hayashi et al., 1968; Rama Devi and Bhattacharyya, 1977a,b). Geraniol (**271**) is formed from citronellol (**258**), nerol (**272**), linalool (**206**), and geranyl acetate (**270**) and metabolized through 10 pathways. That is, compound **271** is hydrogenated to give citronellol (**258**), which is metabolized to 2,8-dihydroxy-2,6-dimethyl octane (**260**) via 6,7-epoxycitronellol (**268**), isopulegol (**267**), limonene (**68**), 3,7-dimethyloctane-1,8-diol (**266**) via 3,7-dimethyl-6-octene-1,8-diol (**265**), **267**, citronellal (**261**), dihydrocitronellol (**259**), and nerol (**272**). Citronellyl acetate (**269**) and isopulegyl acetate (**301**) are hydrolyzed to citronellol (**258**) and isopulegol (**267**), respectively. Compound **261** is metabolized via pulegol (**263**) and isopulegol (**267**) to menthol (**137**). Compound **271** and **272** are isomerized to each other. Compound **272** is metabolized to **271**, **258**, citronellic acid (**262**), nerol-6-,7-epoxide (**273**), and neral (**275**). Compound **272** is metabolized to neric acid (**277**). Compounds **275** and **276** are isomerized to each other. Compound **276** is completely decomposed to CO_2 and water via geranic acid (**278**), 2,6-dimethyl-8-hydroxy-7-oxo-2-octene (**279**), 6-methyl-5-heptenoic acid (**280**), 7-methyl-3-oxo-6-octenoic acid (**283**), 6-methyl-5-heptenoic acid (**284**), 4-methyl-3-heptenoic acid (**284**), 4-methyl-3-pentenoic acid (**285**), and 3-methyl-2-butenoic acid (**286**). Furthermore, compound **271** is metabolized via 3-hydroxymethyl-2,6-octadiene-1-ol (**287**), 3-formyl-2,6-octadiene-1-ol (**288**), and 3-carboxy-2,6-octadiene-1-ol (**289**) to 3-(4-methyl-3-pentenyl)-3-butenolide (**290**). Geraniol (**271**) is also metabolized to 3,7-dimethyl-2,3-dihydroxy-6-octen-1-ol (**292**), 3,7-dimethyl-2-oxo-3-hydroxy-6-octen-1-ol (**293**), 2-methyl-6-oxo-2-heptene (**294**), 6-methyl-5-hepten-2-ol (**298**), 2-methyl-2-heptene-6-one-1-ol (**295**), and 2-methyl-γ-butyrolactone (**296**). Furthermore, **271** is metabolized to 7-methyl-3-oxo-6-octanoic acid (**299**), 7-hydroxymethyl-3-methyl-2,6-octadien-1-ol (**291**), 6,7-epoxygeraniol (**274**), 3,7-dimethyl-2,6-octadiene-1,8-diol (**300**), and 3,7-dimethyloctane-1,8-diol (**266**).

Metabolic pathways of 1,8-cineole (**122**), 1,4-cineole (**131**), phellandrene (**62**), carvotanacetone (**47**), and carvone (**93**) by microorganisms are summarized in Figure 22.204.

1,8-Cineole (**112**) is biotransformed to 2-hydroxy- (**125**), 3-hydroxy- (**123**), and 9-hydroxy-1,8-cineole (**127**), 2-oxo- (**126**) and 3-oxo-1,8-cineole (**124**), lactone [**128**, (*R*)-5,5-dimethyl-4-(3′-oxobutyl)-4,5-dihydrofuran-2-(3H)-one], and *p*-hydroxytoluene (**129**) (MacRae et al., 1979; Nishimura et al., 1982; Noma and Sakai, 1984). 2-Hydroxy-1,8-cineole (**125**) is further converted into 2-oxo-1,8-cineole (**126**), 1,8-cineole-2-malonyl ester (**130**), sobrerol (**43**), and 8-hydroxycarvotanacetone (**44**) (Miyazawa et al., 1995b). 2-Hydroxy-1,8-cineole (**125**) and 2-oxo-1,8-cineole (**126**) are also biodegraded to sobrerol (**43**) and 8-hydroxycarvotanacetone (**44**), respectively. 2-Hydroxy-1,8-cineole (**125**) was esterified to give malonyl ester (**130**). 2-Hydroxy-1,8-cineole (**125**) was formed from limonene (**68**) by citrus pathogenic fungi *P. digitatum* (Noma and Asakawa, 2007b). 1,4-Cineole (**131**) is metabolized to 2-hydroxy- (**132**), 3-hydroxy- (**133**), 8-hydroxy- (**134**), and 9-hydroxy-1,4-cineole (**135**). Compound **132** is also esterified to malonyl ester (**136**) as well as **125** (Miyazawa et al., 1995b). Terpinen-4-ol (342) is metabolized to 2-hydroxy-1,4-cineole (**132**), 2-hydroxy- (**372**) and 7-hydroxyterpinene-4-ol (**342**), and *p*-menthane-1,2,4-triol (**371**) (Abraham et al., 1986; Kumagae and Miyazawa, 1999; Noma and Asakawa, 2007a). Phellandrene (**62**) is metabolized to carvotanacetol (**46**) and phellandrol (**63**). Carvotanacetol (**46**) is further metabolized through the metabolism of carvotanacetone (**47**). Phelandrol (**63**) is also metabolized to give phellandral (**64**), phellandric acid (**65**), and 7-hydroxy-*p*-menthane (**66**). Phellandric acid (**65**) is completely degraded to carbon dioxide and water as shown in Figure 22.202.

Metabolic pathways of myrcene (**302**) and citronellene (**309**) by microorganisms and insects are summarized in Figure 22.205. β-Myrcene (**302**) was metabolized with *D. gossypina* ATCC 10936 (Abraham et al., 1985) to the diol (**303**) and a side product (**304**). β-Myrcene (**302**) was metabolized with *G. applanatum*, *P. flabellatus*, and *P. sajor-caju* to myrcenol (**305**) (2-methyl-6-methylene-7-octen-2-ol) and **306** (Busmann and Berger, 1994).

β-Myrcene (**302**) was converted by common cutworm larvae, *S. litura*, to give myrcene-3,(10)-glycol (**308**) via myrcene-3,(10)-epoxide (**307**) (Miyazawa et al., 1998). Citronellene (**309**) was metabolized by cutworm *S. litura* to give 3,7-dimethyl-6-octene-1,2-diol (**310**) (Takeuchi and Miyazawa, 2005). Myrcene (**302**) is metabolized to two kinds of diols (**303** and **304**), myrcenol (**305**), and ocimene (**306**) (Seubert and Fass, 1964a,b; Abraham et al., 1985). Citronellene (**309**) was metabolized to (**310**) by *S. litura* (Takeuchi and Miyazawa, 2005).

Metabolic pathways of nopol (**452**) and nopol benzyl ether (**455**) by microorganisms are summarized in Figure 22.206. Nopol (**452**) is metabolized mainly to 7-hydroxyethyl-α-terpineol (453) by rearrangement reaction and 3-oxoverbenone (454) as minor metabolite by *Aspergillus* spp. including *A. niger* TBUYN-2 (Noma and Asakawa, 2006b,c). Myrtenol (**5**) is also metabolized to oleuropeic alcohol (**204**) by rearrangement reaction. However, nopol benzyl ether (**455**) was easily metabolized to 3-oxoverbenone (**454**) and 3-oxonopol-2′,4′-dihydroxybenzylether (**456**) as main metabolites without rearrangement reaction (Noma and Asakawa, 2006c).

22.5.2 Microbial Transformation of Terpenoids as Unit Reaction

Microbiological oxidation and reduction patterns of terpenoids and related compounds by fungi belonging to *Aspergillus* spp. containing *A. niger* TBUYN-2 and *E. gracilis* Z. are summarized in Tables 22.18 and 22.19, respectively. Dehydrogenation of secondary alcohols to ketones, hydroxylation of both nonallylic and allylic carbons, oxidation of olefins to form diols and triols via epoxides, reduction of both saturated and α,β-unsaturated ketones, and hydrogenation of olefin conjugated with the carbonyl group were the characteristic features in the biotransformation of terpenoids and related compounds by *Aspergillus* spp.

Compound names: **1**, β-pinene; **2**, pinocarveol; **3**, pinocarvone; **4**, α-pinene; **5**, myrtenol; **6**, myrtenal; **7**, myrtenoic acid; **8**, myrtanol; **9**, 3-hydroxymyrtenol; **10**, 4-hydroxymyrtanol; **11**, α-fenchol; **12**, fenchone; **13**, 6-hydroxyfenchone; **14**, 5-hydroxyfenchone; **15**, 5-oxofenchone; **16**, 9-hydroxyfenchone; **17**, fenchone-9-al; **18**, fenchone-9-oic acid; **19**, fenchoquinone; **20**, 2-hydroxyfenchone; **21**, 2,3-fencholide; **22**, 1,2-fencholide; **23**, verbenol; **24**, verbenone; **25**, 7-hydroxyverbenone; **26**, 7-hydroxyverbenone; **27**, verbanone-4-al; **28**, thujone; **29**, thujoyl alcohol; **30**, 1-hydroxythujone; **31**, 1,3-dihydroxythujone; **32**, pinyl cation; **33**, 1-*p*-menthene-8-cation; **34**, α-terpineol; **35**, bornyl cation; **36**, borneol; **37**, camphor; **38**, α-pinene epoxide; **39**, isonovalal; **40**, novalal; **41**, 2-hydroxypinyl cation; **42**, 6-hydroxy-1-*p*-menthene-8-cation; **43**, *trans*-sobrerol; **44**, 8-hydroxycarvotanacetone; (carvonehydrate); **45**, 8-hydrocarvomenthone; **46**, 1-*p*-menthen-2-ol; **47**, carvotanacetone; **48**, carvomenthone; **49**, carvomenthol; **50**, 8-hydroxycarvomenthol; **51**, 5-hydroxycarvotanacetone; **52**, 5-hydroxycarvomenthone; **53**, 1-hydroxycarvomenthone; **54**, *p*-menthane-1,2-diol; **55**, *p*-menthane-2,9-diol; **56**, 2,3-lactone; **57**, diepoxide; **58**, 8,9-epoxy-1-*p*-methanol; **59**, 1-*p*-menthene-4-cation; **60**, terpin hydrate; **61**, oleuropeic acid (8-hydroxyperillic acid); **62**, 1-*p*-menthene; **63**, phellandrol; **64**, phellandral; **65**, phellandric acid; **66**, 7-hydroxy-*p*-menthane; **67**, 2-(4-methyl-3-cyclohexenylidene)-propionic acid; **68**, limonene; **69**, limonene-1,2-epoxide; **70**, 1-*p*-menthene-9-oic acid; **71**, limonene-1,2-diol; **72**, 1-hydroxy-8-*p*-menthene-2-one; **73**, 1-hydroxy-*p*-menth-2,8-diene; **74**, perillyl alcohol; **75**, shisool; **76**, shisool-8,9-epoxide; **77**, perillyl alcohol-8,9-epoxide; **78**, perillaldehyde; **79**, 1-*p*-menthene-8,9-diol; **80**, 4-hydroxy-*p*-menth-1,8-diene (4-terpineol); **81**, carveol; **82**, perillic acid; **83**,

TABLE 22.18
Microbiological Oxidation and Reduction Patterns of Monoterpenoids by *Aspergillus niger* TBUYN-2

	Microbiological Oxidation	
Oxidation of Alcohols	**Oxidation of Primary Alcohols to Aldehydes and Acids**	
	Oxidation of secondary alcohols to ketones	(−)-*trans*-Carveol (**81a′**), (+)-*trans*-carveol (**81a**), (−)-*cis*-carveol (**81b′**), (+)-*cis*-carveol (**81b**), 2α-hydroxy-1,8 cineole (**125b**), 3α-hydroxy-1,8-cineole (**123b**), 3β-hydroxy-1,8-cineole (**123**a)
Oxidation of aldehydes to acids	Hydroxylation of nonallylic carbon	(−)-Isodihydrocarvone (**101c′**), (−)-carvotanacetone (**47′**), (+)-carvotanacetone (**47**), *cis-p*-menthane (**252**), 1α-hydroxy-*p*-menthane (**253**), 1,8-cineole (**122**), 1,4-cineole (**131**), (+)-fenchone (**12**), (−)-fenchone (**12′**), (−)-menthol (**137b′**), (+)-menthol (**137b**), (−)-neomenthol (**137a**), (+)-neomenthol (**137a**), (+)-isomenthol (**137c**)
	Hydroxylation of allylic carbon	(−)-Isodihydrocarvone (**101b**), (+)-neodihydrocarveol (**102a′**), (−)-dihydrocarveol (**102b′**), (+)-dihydrocarveol (**102b**), (+)-limonene (**68**), (−)-limonene (**68′**)
Oxidation of olefins	Formation of epoxides and oxides	
	Formation of diols	(+)-Neodihydrocarveol (**102a′**), (+)-dihydrocarveol (**102b**), (−)-dihydrocarveol (**102b′**), (+)-limonene (**68**), (−)-limonene (**68′**)
	Formation of triols	(+)-Neodihydrocarveol (**102a′**)
Lactonization		
	Microbiological Reduction	
Reduction of aldehydes to alcohols		
Reduction of ketones to alcohols	Reduction of saturated ketones	(+)-Dihydrocarvone (**101a′**), (−)-isodihydrocarvone (**101b**), (+)-carvomenthone (**48a′**), (−)-isocarvomenthone (**48b**)
	Reduction of α,β-unsaturated ketones	
Hydrogenation of olefins	Hydrogenation of olefin conjugated with carbonyl group	(−)-Carvone (**93′**), (+)-carvone (**93**), (−)-carvotanacetone (**47′**), (+)-carvotanacetone (**47**)
	Hydrogenation of olefin not conjugated with a carbonyl group	

TABLE 22.19
Microbiological Oxidation, Reduction, and Other Reaction Patterns of Monoterpenoids by *Euglena gracilis Z*

	Microbiological Oxidation	
Oxidation of Alcohols	**Oxidation of Primary Alcohols to Aldehydes and Acids**	
	Oxidation of secondary alcohols to ketones	(−)-*trans*-Carveol (**81a′**), (+)-*cis*-carveol (**81b**), (+)-isoborneol (**36b**).[a]
Oxidation of aldehydes to acids		Myrtenal (**6**), myrtanal, (−)-perillaldehyde (**78**), *trans*- and *cis*-1,2-dihydroperillaldehydes (**261a** and **261b**), (−)-phellandral (**64**), *trans*- and *cis*-tetrahydroperillaldehydes, cuminaldehyde (**193**), (+)- and (−)-citronellal (**261** and **261′**).[b]
Hydroxylation	Hydroxylation of nonallylic carbon	(+)-Limonene (**68**), (−)-limonene (**68′**).
	Hydroxylation of allylic carbon	
Oxidation of olefins	Formation of epoxides and oxides	
	Formation of diols	
	Formation of triols	(+)- and (−)-Neodihydrocarveol (**102a′** and **a**), (−)- and(+)-dihydrocarveol (**102b′** and **b**), (+)- and (−)-isodihydrocarveol (**102c′** and **c**), (+)- and (−)-neoisodihydrocarveol (**102d′** and **d**).
Lactonization		
	Microbiological Reduction	
Reduction of aldehydes to alcohols	Reduction of terpene aldehydes to terpene alcohols	Myrtenal (**6**), myrtanal, (−)-perillaldehyde (**78**), *trans*- and *cis*-1,2-dihydroperillaldehydes (**261a** and **261b**), phellandral (**64**), *trans*- and *cis*-1,2-dihydroperillaldehydes (**261a** and **261b**), *trans*- and *cis*-tetrahydroperillaldehydes, cuminaldehyde (**193**), citral (**275** and **276**), (+)- (**261**) and (−)-citronellal (**261′**).
	Reduction of aromatic and related aldehydes to alcohols	
	Reduction of aliphatic aldehydes to alcohols	
Reduction of ketones to alcohols	Reduction of saturated ketones	(+)-Dihydrocarvone (**101a′**), (−)-isodihydrocarvone (**101b**), (+)-carvomenthone (**48a′**), (−)-isocarvomenthone (**48b**), (+)-dihydrocarvone-8,9-epoxides (**97a′**), (+)-isodihydrocarvone-8,9-epoxides (**97b′**), (−)-dihydrocarvone-8,9-epoxides (**97a**).
	Reduction of α,β-unsaturated ketones	
	Hydrogenation of olefins Hydrogenation of olefin conjugated with carbonyl group	(−)-Carvone (**93′**), (+)-carvone (**93**), (−)-carvotanacetone (**47′**), (+)-carvotanacetone (**47**), (−)-carvone-8,9-epoxides (**96′**), (+)-carvone-8,9-epoxides (**96**).
	Hydrogenation of olefin not conjugated with a carbonyl group	

(Continued)

TABLE 22.19 (*Continued*)
Microbiological Oxidation, Reduction, and Other Reaction Patterns of Monoterpenoids by *Euglena gracilis Z*

	Hydrolysis	
Hydrolysis	Hydrolysis of ester	(+)-*trans*- and *cis*-Carvyl acetates (**108a** and **b**), (−)-*cis*-carvyl acetate (**108b′**), (−)-*cis*-carvyl propionate, geranyl acetate (**270**).
	Hydration	
Hydration	Hydration of C = C bond in isopropenyl group to tertiary alcohol	(+)-Neodihydrocarveol (**102a′**), (−)-dihydrocarveol (**102b′**), (+)-isodihydrocarveol (**102c′**), (+)-neoisodihydrocarveol (**102d′**), (−)-neodihydrocarveol (**102a**), (+)-dihydrocarveol (**102b**), (−)-isodihydrocarveol (**102c**), (−)-neoisodihydrocarveol (**102d**), *trans*- and *cis*-shisools (**75a** and **75b**).
	Isomerization	
Isomerization		Geraniol (**271**), nerol (**272**).

[a] Diastereo- and enantioselective dehydrogenation is observed in carveol, borneol, and isoborneol.
[b] Acids were obtained as minor products.

4,9-dihydroxy-1-*p*-menthene-7-oic acid; **84**, 2-hydroxy-8-*p*-menthen-7-oic acid; **85**, 2-oxo-8-*p*-menthen-7-oic acid; **86**, β-isopropyl pimelic acid; **87**, isopropenylglutaric acid; **88**, isobutenoic acid; **89**, isobutyric acid; **90**, 1-*p*-menthene-8,9-diol; **91**, carveol-8,9-epoxide; **92**, bottrospicatol; **93**, carvone; **94**, 5-hydroxycarveol; **95**, 1-*p*-menthene-6,9-diol; **96**, carvone-8,9-epoxide; **97**, dihydrocarvone-8,9-epoxide; **98**, 5-hydroxycarvone; **99**, 5-hydroxydihydrocarvone; **100**, 5-hydroxydihydrocarveol; **101**, dihydrocarvone; **102**, dihydrocarveol; **103**, dihydrocarveol-8,9-epoxide; **104**, *p*-menthane-2,8,9-triol; **105**, dihydrobottrospicatol; **106**, 10-hydroxydihydrocarvone; **107**, 10-hydroxydihydrocarveol; **108**, carvyl acetate and carvyl propionate; **109**, 8,9-epoxydihydrocarvyl acetate; **110**, isopiperitenol; **111**, isopiperitenone; **112**, piperitenone; **113**, 7-hydroxyisopiperitenone; **114**, 10-hydroxyisopiperitenone; **115**, 4-hydroxyisopiperitenone; **116**, 5-hydroxyisopiperitenone; **117**, 5-hydroxypiperitenone; **118**, 7-hydroxypiperitenone; **119**, pulegone; **120**, 8,9-dehydromenthenone; **121**, 8-hydroxymenthenone; **122**, 1,8-cineole; **123**, 3-hydroxy-1,8-cineole; **124**, 3-oxo-1,8-cineole; **125**, 2-hydroxy-1,8-cineole; **126**, 2-oxo-1,8-cineole; **127**, 9-hydroxy-1,8-cineole; **128**, lactone (*R*)-5,5-dimethyl-4-(3′-oxobutyl)-4,5-dihydrofuran-2-(3H)-one; **129**, *p*-hydroxytoluene; **130**, 1,8-cineole-2-malonyl ester; **131**, 1,4-cineole; **132**, 2-hydroxy-1,4-cineole; **133**, 3-hydroxy-1,4-cineole; **134**, 8-hydroxy-1,4-cineole; **135**, 9-hydroxy-1, 4-cineole; **136**, 1,4-cineole-2-malonyl ester; **137**, menthol; **138**, 1-hydroxymenthol; **139**, 6-hydroxymenthol; **140**, 2-hydroxymenthol; **141**, 4-hydroxymenthol; **142**, 8-hydroxymenthol; **143**, 7-hydroxymenthol; **144**, 9-hydroxymenthol; **145**, 7,8-dihydroxymenthol; **146**, 1,8-dihydroxymenthol; **147**, 3-*p*-menthene; **148**, menthenone; **149**, menthone; **150**, 1-hydroxymenthone; **151**, 7-hydroxymenthone; **152**, 3,7-dimethyl-6-hydroxyoctanoic acid; **153**, 3,7-dimethyl-6-oxooctanoic acid; **154**, 2-methyl-2,5-oxidoheptenoic acid; **155**, menthone-1,2-diol; **156**, piperitone; **157**, 8-hydroxypiperitone; **158**, 6-hydroxypiperitone; **159**, 9-hydroxypiperitone; **160**, piperitone-7-al; **161**, 7-hydroxypiperitone; **162**, piperitone-7-oic acid; **163**, menthone-7-al; **164**, menthone-7-oic acid; **165**, 7-carboxylmenthol; **166**, piperitone oxide; **167**, 4-hydroxypiperitone; **168**, piperitenone oxide; **169**, 1-hydroxypulegone; **170**, 2,3-seco-*p*-methylacetone-3-en-1-ol; **171**, 2-methyl-5-isopropyl-2,5-hexadienoic acid; **172**, 2,5,6-trimethyl-2,4-heptadienoic acid; **173**,

2,5,6-trimethyl-3-heptenoic acid; **174**, 2,5,6-trimethyl-2-heptenoic acid; **175**, 3-hydroxy-2,5,6-trimethyl-3-heptanoic acid; **176**, 3-oxo-2,5,6-trimethyl-3-heptanoic acid; **177**, 3,4-dimethylvaleric acid; **178**, *p*-cymene; **179**, thymol; **180**, 6-hydroxythymol **181**, 6-hydroxymenthenone, 6-hydroxy-4-*p*-menthen-3-one; **182**, 3-hydroxythymo-1,4-quinol; **183**, 2-hydroxythymoquionone; **184**, 2,4-dimethyl-6-oxo-3,7-dimethyl-2,4-octadienoic acid; **185**, 2,4,6-trioxo-3,7-dimethyl octanoic acid; **186**, 2-oxobutanoic acid; **187**, acetic acid; **188**, 8-hydroxy-*p*-cymene; **189**, 9-hydroxy-*p*-cymene; **190**, 2-(4-methylphenyl)-propanoic acid; **191**, carvacrol; **192**, cumin alcohol; **193**, *p*-cumin aldehyde; **194**, cumic acid; **195**, *cis*-2,3-dihydroxy-2, 3-dihydro-*p*-cumic acid; **196**, 3-hydroxycumic acid; **197**, 2,3-dihydroxy-*p*-cumic acid; **198**, 2-hydroxy-6-oxo-7-methyl-2,4-octadien-1,3-dioic acid; **199**, 2-methyl-6-hydroxy-6-carboxy-7-octene; **201**, 4-methyl-2-oxopentanoic acid; **202**, carvacrol methyl ether; **203**, 7-hydroxycarvacrol methyl ether; **204**, 8-hydroxyperillyl alcohol; **205**, 8-hydroxyperillaldehyde; **206**, linalool; **207**, linalyl-6-cation; **208**, linalyl-8-cation; **209**, 6-hydroxymethyl linalool; **210**, linalool-6-al; **211**, 2-methyl-6-hydroxy-6-carboxy-2,7-octadiene; **212**, 2-methyl-6-hydroxy-7-octen-6-oic acid; **213**, phellandric acid-8-cation; **214**, 2,3-epoxylinalool; **215**, 5-methyl-5-vinyltetrahydro-2-furanol; **216**, 5-methyl-5-vinyltetrahydro-2-furanone; **217**, 2,2,6-trimethyl-3-hydroxy-6-vinyltetrahydropyrane; **218**, malonyl ester; **219**, 1-hydroxylinalool (3,7-dimethyl-1,6-octadiene-8-ol); **220**, 2,6-dimethyl-6-hydroxy-*trans*-2,7-octadienoic acid; **221**, 4-methyl-*trans*-3,5-hexadienoic acid; **222**, 4-methyl-trans-3,5-hexadienoic acid; **223**, 4-methyl-*trans*-2-hexenoic acid; **224**, isobutyric acid; **225**, 9-hydroxycamphor; **226**, bornyl acetate; **228**, 6-hydroxycamphor; **229**, 6-oxocamphor; **230**, 4-carboxymethyl-2,3,3-trimethylcyclopentanone; **231**, 4-carboxymethyl-3,5,5-trimethyltetrahydro-2-pyrone; **232**, isohydroxycamphoric acid; **233**, isoketocamphoric acid; **234**, 3,4,4-trimethyl-5-oxo-trans-2-hexenoic acid; **235**, 5-hydroxycamphor; **236**, 5-oxocamphor; **237**, **238**, 6-hydroxy-1, 2-campholide; **239**, 6-oxo-1,2-campholide; **240**, 5-carboxymethyl-3,4,4-trimethyl-2-cyclopentenone; **241**, 6-carboxymethyl-4,5,5-trimethyl-5,6-dihydro-2-pyrone; **242**, 5-hydroxy-3,4,4-trimethyl-2-heptene-1,7-dioic acid; **243**, 3-hydroxycamphor; **244**, camphorquinone; **245**, 2-hydroxyepicamphor; **246**, 2-hydroxy-*p*-menthan-7-oic acid; **247**, 2-oxo-*p*-menthan-7-oic acid; **248**, 3-isopropylheptane-1,7-dioic acid; **249**, 3-isopropylpentane-1,5-dioic acid; **250**, 4-methyl-3-oxopentanoic acid; **251**, methylisopropyl ketone; **252**, *p*-menthane; **253**, 1-hydroxy-*p*-menthane; **254**, *p*-menthane-1,9-diol; **255**, *p*-menthane-1,7-diol; **256**, 1-hydroxy-*p*-menthene-7-al; **257**, 1-hydroxy-*p*-menthene-7-oic acid; **258**, citronellol; **259**, dihydrocitronellol; **260**, 2,8-dihydroxy-2,6-dimethyl octane; **261**, citronellal; **262**, citronellic acid; **263**, pulegol; **264**, 7-hydroxymethyl-6-octene-3-ol; **265**, 3,7-dimethyl-6-octane-1,8-diol; **266**, 3,7-dimethyloctane-1,8-diol; **267**, isopulegol; **268**, 6,7-epoxycitronellol; **269**, citronellyl acetate; **270**, geranyl acetate; **271**, geraniol; **272**, nerol **273**, nerol-6,7-epoxide; **274**, 6,7-epoxygeraniol; **275**, neral; **276**, geranial; **277**, neric acid; **278**, geranic acid; **279**, 2,6-dimethyl-8-hydroxy-7-oxo-2-octene; **280**, 6-methyl-5-heptenoic acid; **281**, 7-methyl-3-carboxymethyl-2,6-octadiene-1-oic acid; **282**, 7-methyl-3-hydroxy-3-carboxymethyl-6-octen-1-oic acid; **283**, 7-methyl-3-oxo-6-octenoic acid; **284**, 6-methyl-5-heptenoic acid; **284**, 4-methyl-3-heptenoic acid; **285**, 4-methyl-3-pentenoic acid; **286**, 3-methyl-2-butenoic acid; **287**, 3-hydroxymethyl-2,6-octadiene-1-ol; **288**, 3-formyl-2,6-octadiene-1-ol; **289**, 3-carboxy-2,6-octadiene-1-ol; **290**, 3-(4-methyl-3-pentenyl)-3-butenolide; **291**, 7-hydroxymethyl-3-methyl-2,6-octadien-1-ol; **292**, 3,7-dimethyl-2,3-dihydroxy-6-octen-1-ol; **293**, 3,7-dimethyl-2-oxo-6-octene-1,3-diol; **294**, 6-methyl-5-hepten-2-one; **295**, 6-methyl-7-hydroxy-5-heptene-2-one; **296**, 2-methyl-γ-butyrolactone; **297**, 6-methyl-5-heptenoic acid; **298**, 6-methyl-5-hepten-2-ol; **299**, 7-methyl-3-oxo-6-octanoic acid; **300**, 3,7-dimethyl-2,6-octadiene-1,8-diol; **301**, isopulegyl acetate; **302**, myrcene; **303**, 2-methyl-6-methylene-7-octene-2,3-diol; **304**, 6-methylene-7-octene-2,3-diol; **305**, myrcenol; **306**, ocimene; **307**, myrcene-3,(10)-epoxide; **308**, myrcene-3,(10)-glycol; **309**, (−)-citronellene; **309′**, (+)-citronellene; **310**, (3*R*)-3,7-dimethyl-6-octene-1,2-diol; **310′**, (3*S*)-3,7-dimethyl-6-octene-1,2-diol; **311**, (*E*)-3,7-dimethyl-5-octene-1,7-diol; **312**, *trans*-rose oxide; **313**, *cis*-rose oxide; **314**, (2*Z*,5*E*)-3,7-dimethyl-2,5-octadiene-1,7-diol; **315**, (Z)-3,7-dimethyl-2,7-octadiene-1,6-diol; **316**, (2*E*,6*Z*)-2,6-dimethyl-2,6-octadiene-1,8-diol; **317**, a cyclization product; **318**, (2*E*,5*E*)-3,7-dimethyl-2,5-octadiene-1,7-diol; **319**, (*E*)-3,7-dimethyl-2,7-octadiene-1,6-diol; **320**, 8-hydroxynerol; **321**,

10-hydroxynerol; **322**, 1-hydroxy-3,7-dimethyl-2*E*,6*E*-octadienal; **323**, 1-hydroxy-3-,7-dimethyl-2*E*,6*E*-octadienoic acid; **324**, 3,9-epoxy-*p*-menth-1-ene; **325**, tetrahydrolinalool; **326**, 3,7-dimethyloctane-3,5-diol; **327**, 3,7-dimethyloctane-3,7-diol; **328**, 3,7-dimethyloctane-3,8-diol; **329**, dihydromyrcenol; **330**, 1,2-epoxydihydromyrcenol; **331**, 3β-hydroxydihydromyrcenol; **332**, dihydromyrcenyl acetate; **333**, 1,2-dihydroxydihydromyrcenyl acetate; **334**, (+)-*p*-menthane-1-β,2α,8-triol; **335**, α-pinene-1,2-epoxide; **336**, 3-carene; **337**, 3-carene-1,2-epoxide; **338**, (1*R*)-*trans*-isolimonene; **338**, (1*R*,4*R*)-*p*-menth-2-ene-8,9-diol; **339**, (1*R*,4*R*)-*p*-menth-2-ene-8,9-diol; **340**, α-terpinene; **341**, α-terpinene-7-oic acid; **342**, (−)-terpinen-4-ol; **343**, *p*-menthane-1,2,4-triol; **344**, γ-terpinene; **345**, γ-terpinene-7-oic acid; **346**, terpinolene; **347**, (1*R*)-8-hydroxy-3-*p*-menthen-2-one; **348**, (1*R*)-1,8-dihydroxy-3-*p*-menthen-2-one; **349**, 6β-hydroxyfenchol; **350**, 5β-hydroxyfenchol (a,5β,b,5α); **351**, terpinolene-1,2-*trans*-diol; **352**, terpinolene-4,8-diol; **353**, terpinolene-9-ol; **354**, terpinolene-10-ol; **355**, α-phellandrene; **356**, α-phellandrene-7-oic acid; **357**, terpinolene-7-oic acid; **358**, thymoquinone; **359**, 1,2-dihydroperillaldehyde; **360**, 1,2-dihydroperillic acid; **361**, 8-hydroxy-1,2-dihydroperillyl alcohol; **362**, tetrahydroperillaldehyde (a *trans*, b *cis*); **363**, tetrahydroperillic acid (a *trans*, b *cis*); **364**, (−)-menthol monoglucoside; **365**, (+)-menthol diglucoside; **366**, (+)-isopulegol; **367**, 7-hydroxy-(+)-isopulegol; **368**, 10-hydroxy-(+)-isopulegol; **369**, 1,2-epoxy-α-terpineol; **370**, bornane-2,8-diol; **371**, *p*-menthane-1α,2β,4β-triol; **372**, 1-*p*-menthene-4β,6-diol; **373**, 1-*p*-menthene-4a,7-diol; **374**, (+)-bottrospicatal; **375**, 1-*p*-menthene-2β,8,9-triol; **376**, (−)-perillyl alcohol monoglucoside; **377**, (−)-perillyl alcohol diglucoside; **378**, 4α-hydroxy-(−)-isodihydrocarvone; **379**, 2-methyl-2-cyclohexenone; **380**, 2-cyclohexenone; **381**, 3-methyl-2-cyclohexenone; **382**, 2-methylcyclohexanone; **383**, 2-methylcyclohexanol (a, *trans*; b, *cis*); **384**, 4-hydroxycarvone; **385**, 2-ethyl-2-cyclohexenone; **386**, 2-ethylcyclohexenone (a1*R*) (b1*S*); **387**, 1-hydroxypulegone; **388**, 5-hydroxypulegone; **389**, 8-hydroxymenthone; **390**, 10-hydroxy-(−)-carvone; **391**, 1,5,5-trimethylcyclopentane-1,4-dicarboxylic acid; **392**, 4β-hydroxy-(−)-menthone; **393**, 1α,4β-dihydroxy-(−)-menthone; **394**, 7-hydroxy-9-carboxymenthone; **395**, 7-hydroxy-1,8-cineole; **396**, methyl ester of 2α-hydroxy-1,8-cineole; **397**, ethyl ester of 2α-hydroxy-1,8-cineole; **398**, *n*-propyl ester of 2α-hydroxy-1,8-cineole; **399**, *n*-butyl ester of 2α-hydroxy-1,8-cineole; **400**, isopropyl ester of 2α-hydroxy-1,8-cineole; **401**, tertiary butyl ester of 2α-hydroxy-1,8-cineole; **402**, methylisopropyl ester of 2α-hydroxy-1,8-cineole; **403**, methyl tertiary butyl ester of 2α-hydroxy-1,8-cineole; **404**, 2α-hydroxy-1,8-cineole monoglucoside (404 and 404'); **405**, 2α-hydroxy-1,8-cineole diglucoside; **406**, *p*-menthane-1,4-diol; **407**, 1-*p*-menthene-4β,6-diol; **408**, (−)-pinane-2α,3α-diol; **409**, (−)-6β-hydroxypinene; **410,** (−)-4α,5-dihydroxypinene; **411**, (−)-4α-hydroxypinene-6-one; **412**, 6β,7-dihydroxyfenchol; **413**, 3-oxo-pinane; **414**, 2α-hydroxy-3-pinanone; **415**, 2α, 5-dihydroxy-3-pinanone; **416**, 2α,10-dihydroxy-3-pinanone; **417**, *trans*-3-pinanol; **418**, pinane-2α,3α-diol; **419**, pinane-2α, 3α, 5-triol; **420**, isopinocampheol (3-pinanol); **421**, pinane-1,3α-diol; **422**, pinane-3α,5-diol; **423**, pinane-3β,9-diol; **424**, pinane-3β,4β,-diol; **425**, **426**, pinane-3α,4β-diol; **427**, pinane-3α,9-diol; **428**, pinane-3α,6-diol; **429**, *p*-menthane-2α,9-diol; **430**, 2-methyl-3α-hydroxy-1-hydroxyisopropyl cyclohexane propane; **431**, 5-hydroxy-3-pinanone; **432**, 2α-methyl,3-(2-methyl-2-hydroxypropyl)-cyclopenta-1β-ol; **433**, 3-acetoxy-2α-pinanol; **434**, 8-hydroxy-α-pinene; **435**, **436**, myrtanic acid; **437**, camphene; **438**, camphene glycol; **439**, (+)-3-carene; **440**, *m*-mentha-4,6-dien-8-ol; **441**, *m*-cymen-8-ol; **442**, 3-carene-9-ol; **443**, 3-carene-9-carboxylic acid; **444**, 3-caren-10-ol-9-carboxylic acid; **445**, 3-carene-9,10-dicarboxylic acid; **446**, (−)-*cis*-carane; **447**, dicarboxylic acid of (−)-*cis*-carane; **448**, (−)-6β-hydroxypinene; **449**, (−)-4α,5-dihydroxypinene; **450**, (−)-4α-hydroxypinene-6-one; **451**, 10-hydroxyverbenol; **452**, (−)-nopol; **453**, 7-hydroxymethyl-1-*p*-menthen-8-ol; **454**, 3-oxonopol; **455**, nopol benzyl ether; **456**, 4-oxonopol-2′,4′-dihydroxybenzyl ether; **457**, 7-hydroxymethyl-1-*p*-menthen-8-ol benzyl ether; **458**, piperitenol; **459**, thymol methyl ether; **460**, 7-hydroxythymol methyl ether; **461**, 9-hydroxythymol methyl ether; **462**, 1,8-cineol-9-oic acid; **463**, 4-hydroxyphellandric acid; **464**, 4-hydroxydihydrophellandric acid; **465**, (+)-8-hydroxyfenchol; **466**, (−)-9-hydroxyfenchol; **467**, (+)-10-hydroxyfenchol, **468**, 4α-hydroxy-6-oxo-α-pinene; **469**, dihydrolinalyl acetate; **470**, 3-hydroxycarvacrol; **471**, 9-hydroxycarvacrol; **472**, carvacrol-9-oic acid; **473**, 8,9 dehydrocarvacrol; **474**, 8-hydroxycarvacrol; **475**, 7-hydroxycarvacrol; **476**, carvacrol-7-oic acid; **477**,

8,9-dihydroxycarvacrol; **478**, 7,9-dihydroxycarvacrol methyl ether; **479**, 7-hydroxythymol; **480**, 9-hydroxythymol; **481**, thymol-7-oic acid; **482**, 7,9-dihydroxythymol; **483**, thymol-9-oic acid; **484,** (1*R*,2*R*,3*S*,4*S*,5*R*)-3,4-pinanediol.

REFERENCES

Abraham, W.-R., H.-A. Arfmann, B. Stumpf, P. Washausen, and K. Kieslich, 1988. Microbial transformations of some terpenoids and natural compounds. In: *Bioflavour'87. Analysis—Biochemistry—Biotechnology*, P. Schreier, ed., pp. 399–414. Berlin, Germany: Walter de Gruyter and Co.

Abraham, W.-R., H.M.R. Hoffmann, K. Kieslich, G. Reng, and B. Stumpf, 1985. Microbial transformation of some monoterpenes and sesquiterpenoids. In: *Enzymes in Organic Synthesis*, R. Porter and S. Clark, eds., *Ciba Foundation Symposium* 111, pp. 146–160. London, U.K.: Pitman Press.

Abraham, W.-R., K. Kieslich, H. Reng, and B. Stumpf, 1984. Formation and production of 1,2-trans-glycols from various monoterpenes with 1-menthene skeleton by microbial transformations with Diplodia gossypina. In: *Third European Congress on Biotechnology*, Vol. 1, pp. 245–248. Verlag Chemie, Weinheim, Germany.

Abraham, W.-R., B. Stumpf, and H.-A. Arfmann, 1990. Chiral intermediates by microbial epoxidations. *J. Essent. Oil Res.*, 2: 251–257.

Abraham, W.-R., B. Stumpf, and K. Kieslich, 1986. Microbial transformation of terpenoids with 1-*p*-menthene skeleton. *Appl. Microbiol. Biotechnol.*, 24: 24–30.

Amsterdam, D. 1997. Susceptibility testing of antimicrobials in liquid media. In: *Antibiotics in Laboratory Medicine*, V. Lorian, ed., 4th ed. Baltimore, MD: Williams & Wilkins, Maple Press.

Asakawa, Y., H. Takahashi, M. Toyota, Y. and Noma, 1991. Biotransformation of monoterpenoids, (−)- and (+)-menthols, terpinolene and carvotanacetone by *Aspergillus* species. *Phytochemistry*, 30: 3981–3987.

Asakawa, Y., M. Toyota, T. Ishida, T. Takemoto, 1983. Metabolites in rabbit urine after terpenoid administration. *Proceedings of the 27th TEAC*, pp. 254–256.

Atta-ur-Rahman, M. Yaqoob, A. Farooq, S. Anjum, F. Asif, and M.I. Choudhary, 1998. Fungal transformation of (1*R*,2*S*,5*R*)-(−)-menthol by *Cephalosporium aphidicola*. *J. Nat. Prod.*, 61: 1340–1342.

Ausgulen, L.T., E. Solheim, and R.R. Scheline, 1987. Metabolism in rats of *p*-cymene derivatives: Carvacrol and thymol. *Pharmacol. Toxicol.*, 61: 98–102.

Babcka, J., J. Volf, J. Czchec, and P. Lebeda, 1956. Patent 56-9686b.

Ballal, N.R., P.K. Bhattacharyya, and P.N. Rangachari, 1967. Microbiological transformation of terpenes. Part XIV. Purification and properties of perillyl alcohol dehydrogenase. *Indian J. Biochem.*, 5: 1–6.

Bauer, K., D. Garbe, and H. Surburg (eds.), 1990. *Common Fragrance and Flavor Materials: Preparation, Properties and Uses*. 2nd revised ed., 218pp. New York: VCH Publishers.

Best, D.J. and K.J. Davis, 1988. *Soap, Perfumery Cosmetics* 4: 47.

Best, D.J., N.C. Floyd, A. Magalhaes, A. Burfield, and P.M. Rhodes, 1987. Initial enzymatic steps in the degradation of alpha-pinene by *Pseudomonas fluorescens* Ncimb 11671. *Biocatal. Biotransform.*, 1: 147–159.

Bhattacharyya, P.K. and K. Ganapathy, 1965. Microbial transformation of terpenes. VI. Studies on the mechanism of some fungal hydroxylation reactions with the aid of model systems. *Indian J. Biochem.*, 2: 137–145.

Bhattacharyya, P.K., B.R. Prema, B.D. Kulkarni, and S.K. Pradhan, 1960. Microbiological transformation of terpenes: Hydroxylation of α-pinene. *Nature*, 187: 689–690.

Bock, G., I. Benda, and P. Schreier, 1986. Biotransformation of linalool by *Botrytis cinerea*. *J. Food Sci.*, 51: 659–662.

Bock, G., I. Benda, and P. Schreier, 1988. Microbial transformation of geraniol and nerol by *Botrytis cinerea*. *Appl. Microbiol. Biotechnol.*, 27: 351–357.

Bouwmester, H.J., J.A.R. Davies, and H. Toxopeus, 1995. Enantiomeric composition of carvone, limonene, and carveols in seeds of dill and annual and biennial caraway varieties. *J. Agric. Food Chem.*, 43: 3057–3064.

Bowen, E.R., 1975. Potential by-products from microbial transformation of *d*-limonene. *Proc. Fla. State Hort. Soc.*, 88: 304–308.

Bradshaw, W.H., H.E. Conrad, E.J. Corey, I.C. Gunsalus, and D. Lednicer, 1959. Microbiological degradation of (+)-camphor. *J. Am. Chem. Soc.*, 81: 5507.

Brit Patent, 1970. No. 1,187,320.

Brunerie, P., I. Benda, G. Bock, and P. Schreier, 1987a. Bioconversion of citronellol by *Botrytis cinerea*. *Appl. Microbiol. Biotechnol.*, 27: 6–10.

Brunerie, P., I. Benda, G. Bock, and P. Schreier, 1987b. Biotransformation of citral by *Botrytis cinerea*. *Z. Naturforsch.*, 42C: 1097–1100.

Brunerie, P., I. Benda, G. Bock, and P. Schreier, 1988. Bioconversion of monoterpene alcohols and citral by Botrytis cinerea. In: *Bioflavour'87. Analysis—Biochemistry—Biotechnology*, P. Schreier, ed., pp. 435–444. Berlin, Germany: Walter de Gruyter and Co.

Busmann, D. and R.G. Berger, 1994. Conversion of myrcene by submerged cultured basidiomycetes. *J. Biotechnol.*, 37: 39–43.

Cadwallander, K.R. and R.J. Braddock, 1992. Enzymatic hydration of (4*R*)-(+)-limonene to (4*R*)-(+)-alpha-terpineol. *Dev. Food Sci.*, 29: 571–584.

Cadwallander, K.R., R.J. Braddock, M.E. Parish, and D.P. Higgins, 1989. Bioconversion of (+)-limonene by *Pseudomonas gladioli*. *J. Food Sci.*, 54: 1241–1245.

Cantwell, S.G., E.P. Lau, D.S. Watt, and R.R. Fall, 1978. Biodegradation of acyclic isoprenoids by *Pseudomonas* species. *J. Bacteriol.*, 135: 324–333.

Chamberlain, E.M. and S. Dagley, 1968. The metabolism of thymol by a *Pseudomonas*. *Biochem. J.*, 110: 755–763.

Chapman, P.J., G. Meerman, and I.C. Gunsalus, 1965. The microbiological transformation of fenchone. *Biochem. Biophys. Res. Commun.*, 20: 104–108.

Chapman, P.J., G. Meerman, I.C. Gunsalus, R. Srinivasan, and K.L. Rinehart Jr., 1966. A new acyclic acid metabolite in camphor oxidation. *J. Am. Chem. Soc.*, 88: 618–619.

Christensen, M. and D.E. Tuthill, 1985. Aspergillus: An overview. In: *Advances in Penicillium and Aspergillus Systematics*, R.A. Samson and J.I. Pitt, eds., pp. 195–209. New York: Plenum Press.

Conrad, H.E., R. DuBus, and I.C. Gunsalus, 1961. An enzyme system for cyclic ketone lactonization. *Biochem. Biophys. Res. Commun.*, 6: 293–297.

Conrad, H.E., R. DuBus, M.J. Mamtredt, and I.C. Gunsalus, 1965a. Mixed function oxidation II. Separation and properties of the enzymes catalyzing camphor ketolactonization. *J. Biol. Chem.*, 240: 495–503.

Conrad, H.E., K. Lieb, and I.C. Gunsalus, 1965b. Mixed function oxidation III. An electron transport complex in camphor ketolactonization. *J. Biol. Chem.*, 240: 4029–4037.

David, L. and H. Veschambre, 1984. Preparation d'oxydes de linalol par bioconversion. *Tetrahadron Lett.*, 25: 543–546.

David, L. and H. Veschambre, 1985. Oxidative cyclization of linalol by various microorganisms. *Agric. Biol. Chem.*, 49: 1487–1489.

DeFrank, J.J. and D.W. Ribbons, 1977a. *p*-Cymene pathway in *Pseudomonas putida*: Initial reactions. *J. Bacteriol.*, 129: 1356–1364.

DeFrank, J.J. and D.W. Ribbons, 1977b. *p*-Cymene pathway in *Pseudomonas putida*: Ring cleavage of 2,3-dihydroxy-*p*-cumate and subsequent reactions. *J. Bacteriol.*, 129: 1365–1374.

Demirci, F., 2000. Microbial transformation of bioactive monoterpenes. PhD thesis, pp. 1–137. Anadolu University, Eskisehir, Turkey.

Demirci, F., H. Berber, and K.H.C. Baser, 2007. Biotransformation of p-cymene to thymoquinone. *Book of Abstracts of the 38th ISEO*, SL-1, p. 6.

Demirci, F., N. Kirimer, B. Demirci, Y. Noma, and K.H.C. Baser, 2001. The biotransformation of thymol methyl ether by different fungi. *Book of Abstracts of the XII Biotechnology Congress*, p. 47.

Demirci, F., Y. Noma, N. Kirimer, and K.H.C. Baser, 2004. Microbial transformation of (–)-carvone. *Z. Naturforsch.*, 59c: 389–392.

Demyttenaere, J.C.R. and H.L. De Pooter, 1996. Biotransformation of geraniol and nerol by spores of *Penicillium italicum*. *Phytochemistry*, 41: 1079–1082.

Demyttenaere, J.C.R. and H.L. De Pooter, 1998. Biotransformation of citral and nerol by spores of *Penicillium digitatum*. *Flav. Fragr. J.*, 13: 173–176.

Demyttenaere, J.C.R., M. del Carmen Herrera, and N. De Kimpe, 2000. Biotransformation of geraniol, nerol and citral by sporulated surface cultures of *Aspergillus niger* and *Penicillium* sp. *Phytochemistry*, 55: 363–373.

Demyttenaere, J.C.R., I.E.I. Koninckx, and A. Meersman, 1996. Microbial production of bioflavours by fungal spores. In: *Flavour Science. Recent Developments*, A.J. Taylor and D.S. Mottram, eds., pp. 105–110. Cambridge, U.K.: The Royal Society of Chemistry.

Demyttenaere, J.C.R. and H.M. Willemen, 1998. Biotransformation of linalool to furanoid and pyranoid linalool oxides by *Aspergillus niger*. *Phytochemistry*, 47: 1029–1036.

Dhavalikar, R.S. and P.K. Bhattacharyya, 1966. Microbial transformation of terpenes. Part VIII. Fermentation of limonene by a soil Pseudomonad. *Indian J. Biochem.*, 3: 144–157.

Dhavalikar, R.S., A. Ehbrecht, and G. Albroscheit, 1974. Microbial transformations of terpenoids: β-pinene. *Dragoco Rep.*, 3: 47–49.

Dhavalikar, R.S., P.N. Rangachari, and P.K. Bhattacharyya, 1966. Microbial transformation of terpenes. Part IX. Pathways of degradation of limonene in a soil Pseudomonad. *Indian J. Biochem.*, 3: 158–164.

Farooq, A., M.I. Choudhary, S. Tahara, T.-U. Rahman, K.H.C. Baser, and F. Demirci, 2002. The microbial oxidation of (−)-p-pinene by *Botrytis cinerea*. *Z. Naturforsch.*, 57c: 686–690.

Fenaroli, G., 1975. Synthetic flavors. In: *Fenaroli's Handbook of Flavor Ingredients*, T.E. Furia and N. Bellanca, eds., Vol. 2, pp. 6–563. Cleveland, OH: CRC Press.

Flynn, T.M. and I.A. Southwell, 1979. 1,3–Dimethyl-2-oxabicyclo [2,2,2]-octane-3-methanol and 1,3-dimethyl-2-oxabicyclo[2,2,2]-octane-3-carboxylic acid, urinary metabolites of 1,8-cineole. *Aust. J. Chem.*, 32: 2093–2095.

Ganapathy, K. and P.K. Bhattacharyya, unpublished data.

Gibbon, G.H., N.F. Millis, and S.J. Pirt, 1972. Degradation of α-pinene by bacteria. *Proc. IV IFS, Ferment. Technol. Today*, pp. 609–612.

Gibbon, G.H. and S.J. Pirt, 1971. The degradation of a-pinene by *Pseudomonas* PX 1. *FEBS Lett.*, 18: 103–105.

Gondai, T., M. Shimoda, and T. Hirata, 1999. Asymmetric reduction of enone compounds by Chlorella miniata. *Proceedings of the 43rd TEAC*, pp. 217–219.

Griffiths, E.T., S.M. Bociek, P.C. Harries, R. Jeffcoat, D.J. Sissons, and P.W. Trudgill, 1987a. Bacterial metabolism of alpha-pinene: Pathway from alpha-pinene oxide to acyclic metabolites in *Nocardia* sp. strain P18.3. *J. Bacteriol.*, 169: 4972–4979.

Griffiths, E.T., P.C. Harries, R. Jeffcoat, and P.W. Trudgill, 1987b. Purification and properties of alpha-pinene oxide lyase from *Nocardia* sp. strain P18.3. *J. Bacteriol.*, 169: 4980–4983.

Gunsalus, I.C., P.J. Chapman, and J.-F. Kuo, 1965. Control of catabolic specificity and metabolism. *Biochem. Biophys. Res. Commun.*, 18: 924–931.

Gyoubu, K. and M. Miyazawa, 2005. Biotransformation of (+)- and (−)-fenchone by liver microsomes. *Proceedings of the 49th TEAC*, pp. 420–422.

Gyoubu, K. and M. Miyazawa, 2006. Biotransformation of (+)- and (−)-camphor by liver microsome. *Proceedings of the 50th TEAC*, pp. 253–255.

Hagiwara, Y. and M. Miyazawa, 2007. Biotransformation of cineole by the larvae of common cutworm (Spodoptera litura) as a biocatalyst. *Proceedings of the 51st TEAC*, pp. 304–305.

Hagiwara, Y., H. Takeuchi, and M. Miyazawa, 2006. Biotransformation of (+)-and (−)-menthone by the larvae of common cutworm (Spodoptera litura) as a biocatalyst. *Proceedings of the 50th TEAC*, pp. 279–280.

Hamada, H. and T. Furuya, 2000. Hydroxylation of monoterpenes by plant suspension cells. *Proceedings of the 44th TEAC*, pp. 167–168.

Hamada, H., T. Furuya, and N. Nakajima, 1996. The hydroxylation and glycosylation by plant catalysts. *Proceedings of the 40th TEAC*, pp. 111–112.

Hamada, H., T. Harada, and T. Furuya, 2001. Hydroxylation of monoterpenes by algae and plant suspension cells. *Proceedings of the 45th TEAC*, pp. 366–368.

Hamada, H., M. Kaji, T. Hirata, T. Furuya, 2003. Enantioselective biotransformation of monoterpenes by Cyanobacterium. *Proceedings of the 47th TEAC*, pp. 162–163.

Hamada, H., Y. Kondo, M. Kaji, and T. Furuta, 2002. Glycosylation of monoterpenes by plant suspension cells. *Proceedings of the 46th TEAC*, pp. 321–322.

Hamada, H., A. Matsumoto, and J. Takimura, 2004. Biotransformation of acyclic monoterpenes by biocatalysts of plant cultured cells and Cyanobacterium. *Proceedings of the 48th TEAC*, pp. 393–395.

Hamada, H. and H. Yasumune, 1995. The hydroxylation of monoterpenoids by plant cell biotransformation. *Proceedings of the 39th TEAC*, pp. 375–377.

Hartline, R.A. and I.C. Gunsalus, 1971. Induction specificity and catabolite repression of the early enzymes in camphor degradation by *Pseudomonas putida*. *J. Bacteriol.*, 106: 468–478.

Hashimoto Y. and M. Miyazawa, 2001. Microbial resolution of esters of racemic 2-endo-hydroxy-1,8-cineole by Glomerella cingulata. *Proceedings of the 45th TEAC*, pp. 363–365.

Hayashi, T., T. Kakimoto, H. Ueda, and C. Tatsumi, 1969. Microbiological conversion of terpenes. Part VI. Conversion of borneol. *J. Agric. Chem. Soc. Jpn.*, 43: 583–587.

Hayashi, T., H. Takashiba, S. Ogura, H. Ueda, and C. Tsutsumi, 1968. *Nippon Nogei-Kagaku Kaishi*, 42: 190–196.

Hayashi, T., H. Takashiba, H. Ueda, and C. Tsutsumi, 1967. *Nippon Nogei Kagaku Kaishi*, 41(254): 79878g.

Hayashi, T., S. Uedono, and C. Tatsumi, 1972. Conversion of a-terpineol to 8,9-epoxy-*p*-menthan-1-ol. *Agric. Biol. Chem.*, 36: 690–691.

Hirata, T., K. Shimoda, and T. Gondai, 2000. Asymmetric hydrogenation of the C–C double bond of enones with the reductases from *Nicotiana tabacum*. *Chem. Lett.*, 29: 850–851.

Hosler, P., 1969. U.S. Patent 3,458,399.

Hudlicky, T., D. Gonzales, and D.T. Gibson, 1999. Enzymatic dihydroxylation of aromatics in enantioselective synthesis: Expanding asymmetric methodology. *Aldrichim. Acta*, 32: 35–61.

Hungund, B.L., P.K. Bhattachayya, and P.N. Rangachari, 1970. Methylisopropyl ketone from a terpene fermentation by the soil Pseudomonad, PL-strain. *Indian J. Biochem.*, 7: 80–81.
Ishida, T., Y. Asakawa, and T. Takemoto, 1981a. Metabolism of myrtenal, perillaldehyde and dehydroabietic acid in rabbits. *Res. Bull. Hiroshima Inst. Technol.*, 15: 79–91.
Ishida, T., Y. Asakawa, T. Takemoto, and T. Aratani, 1979. Terpenoid biotransformation in mammals. II. Biotransformation of *dl*-camphene. *J. Pharm. Sci.*, 68: 928–930.
Ishida, T., Y. Asakawa, T. Takemoto, and T. Aratani, 1981b. Terpenoids biotransformation in mammals. III. Biotransformation of α-pinene, (3–pinene, pinane, 3-carene, carane, myrcene, and *p*-cymene in rabbits. *J. Pharm. Sci.*, 70: 406–415.
Iscan, G., 2005. Unpublished data.
Ismaili-Alaoui, M., B. Benjulali, D. Buisson, and R. Azerad, 1992. Biotransformation of terpenic compounds by fungi I. Metabolism of *R*-(+)-pulegone. *Tetrahedron Lett.*, 33: 2349–2352.
Janssens, L., H.L. De Pooter, N.M. Schamp, and E.J. Vandamme, 1992. Production of flavours by microorganisms. *Process Biochem.*, 27: 195–215.
Joglekar, S.S. and R.S. Dhavlikar, 1969. Microbial transformation of terpenoids. I. Identification of metabolites produced by a Pseudomonad from citronellal and citral. *Appl. Microbiol.*, 18: 1084–1087.
Kaji, M., H. Hamada, and T. Furuya, 2002. Biotransformation of monoterpenes by Cyanobacterium and plant suspension cells. *Proceedings of the 46th TEAC*, pp. 323–325.
Kamino, F. and M. Miyazawa, 2005. Biotransformation of (+)-and (−)-pinane-2,3-diol using plant pathogenic fungus, Glomerella cingulata as a biocatalyst. *Proceedings of the 49th TEAC*, pp. 395–396.
Kamino, F., Y. Noma, Y. Asakawa, and M. Miyazawa, 2004. Biotransformation of (1S,2S,3R,5S)-(+)-pinane-2,3-diol using plant pathogenic fungus, Glomerella cingulata as a biocatalyst. *Proceedings of the 48th TEAC*, pp. 383–384.
Kayahara, H., T. Hayashi, C. and Tatsumi, 1973. Microbiological conversion of (−)-perillaldehyde and *p*-mentha-1,3-dien-7-al. *J. Ferment. Technol.*, 51: 254–259.
Kieslich, K., W.-R. Abraham, and P. Washausen, 1985. Microbial transformations of terpenoids. In: *Topics in Flavor Research*, R.G. Berger, S. Nitz, and P. Schreier, eds., pp. 405–427. Marzling Hangenham, Germany: Eichborn.
Koneman, E.W., S.D. Allen, W.M. Janda, P.C. Schreckenberger, and W.C. Winn, 1997. *Color Atlas and Textbook of Diagnostic Microbiology*, Philadelphia, PA: Lippincott-Raven Publishers.
Kraidman, G., B.B. Mukherjee, and I.D. Hill, 1969. Conversion of limonene into an optically active isomer of α-terpineol by a *Cladosporium species. Bacteriol. Proc.*, 69: 63.
Krasnobajew, V., 1984. Terpenoids. In: *Biotechnology*, K. Kieslich, ed., Vol. 6a, pp. 97–125. Weinheim, Germany: Verlag Chemie.
Kumagae, S. and M. Miyazawa, 1999. Biotransformation of p-menthanes using common cutworm larvae, Spodoptera litura as a biocatalyst. *Proceedings of the 43rd TEAC*, pp. 389–390.
Lassak, E.V., J.T. Pinkey, B.J. Ralph, T. Sheldon, and J.J.H. Simes, 1973. Extractives of fungi. V. Microbial transformation products of piperitone. *Aust. J. Chem.*, 26: 845–854.
Liu, W., A. Goswami, R.P. Steffek, R.L. Chemman, F.S. Sariaslani, J.J. Steffens, and J.P.N. Rosazza, 1988. Stereochemistry of microbiological hydroxylations of 1,4-cineole. *J. Org. Chem.*, 53: 5700–5704.
MacRae, I.C., V. Alberts, R.M. Carman, and I.M. Shaw, 1979. Products of 1,8-cineole oxidation by a Pseudomonad. *Aust. J. Chem.*, 32: 917–922.
Madyastha, K.M. 1984. Microbial transformations of acyclic monoterpenes. *Proc. Indian Acad. Sci. (Chem. Sci.)*, 93: 677–686.
Madyastha, K.M. and P.K. Bhattacharyya, 1968. Microbiological transformation of terpenes. Part XIII. Pathways for degradation of *p*-cymene in a soil pseudomonad (PL-strain). *Indian J. Biochem.*, 5: 161–167.
Madyastha, K.M., P.K. Bhattacharyya, and C.S. Vaidyanathan, 1977. Metabolism of a monoterpene alcohol, linalool, by a soil pseudomonad. *Can. J. Microbiol.*, 23: 230–239.
Madyastha, K.M. and N.S.R. Krishna Murthy, 1988a. Regiospecific hydroxylation of acyclic monoterpene alcohols by *Aspergillus niger. Tetrahedron Lett.*, 29: 579–580.
Madyastha, K.M. and N.S.R. Krishna Murthy, 1988b. Transformations of acetates of citronellol, geraniol, and linalool by *Aspergillus niger*: Regiospecific hydroxylation of citronellol by a cell-free system. *Appl. Microbiol. Biotechnol.*, 28: 324–329.
Madyastha, K.M. and V. Renganathan, 1983. Bio-degradation of acetates of geraniol, nerol and citronellol by *P. incognita*: Isolation and identification of metabolites. *Indian J. Biochem. Biophys.*, 20: 136–140.
Mattison, J.E., L.L. McDowell, and R.H. Baum, 1971. Cometabolism of selected monoterpenoids by fungi associated with monoterpenoid-containing plants. *Bacteriol. Proc.*, 1971: 141.

Miyamoto, Y. and M. Miyazawa, 2001. Biotransformation of (+)- and (−)-borneol by the larvae of common cutworm (Spodoptera litura) as a biocatalyst. *Proceedings of the 45th TEAC*, pp. 377–378.

Miyazato, Y. and M. Miyazawa, 1999. Biotransformation of (+)- and (−)-a-fenchyl acetated using plant parasitic fungus, Glomerella cingulata as a biocatalyst. *Proceedings of the 43rd TEAC*, pp. 213–214.

Miyazawa, M., H. Furuno, and H. Kameoka, 1992a. Biotransformation of thujone by plant pathogenic microorganism, Botrytis allii IFO 9430. *Proceedings of the 36th TEAC*, pp. 197–198.

Miyazawa, M., H. Furuno, K. Nankai, and H. Kameoka, 1991a. Biotransformation of verbenone by plant pathogenic microorganism, Rhizoctonia solani. *Proceedings of the 35th TEAC*, pp. 274–275.

Miyazawa, M., H. Huruno, and H. Kameoka, 1991b. Biotransformation of (+)-pulegone to (−)-1*R*-8-hydroxy-4-*p*-menthen-3-one by *Botrytis allii. Chem. Express*, 6: 479–482.

Miyazawa, M., H. Huruno, and H. Kameoka, 1991c. *Chem. Express*, 6: 873.

Miyazawa, M., H. Kakita, M. Hyakumachi, and H. Kameoka, 1992b. Biotransformation of monoterpenoids having p-menthan-3-one skeleton by Rhizoctonia solani. *Proceedings of the 36th TEAC*, pp. 191–192.

Miyazawa, M., H. Kakita, M. Hyakumachi, K. Umemoto, and H. Kameoka, 1991d. Microbiological conversion of piperitone oxide by plant pathogenic fungi Rhizoctonia solani. *Proceedings of the 35th TEAC*, pp. 276–277.

Miyazawa, M., H. Kakita, M. Hyakumachi, K. Umemoto, and H. Kameoka, 1992c. Microbiological conversion of monoterpenoids containing p-menthan-3-one skeleton by plant pathogenic fungi Rhizoctonia solani. *Proceedings of the 36th TEAC*, pp. 193–194.

Miyazawa, M., S. Kumagae, H. Kameoka, 1997a. Biotransformation of (−)-menthol and (+)-menthol by common cutworm Larvae, Spodoptera litura as a biocatalyst. *Proceedings of the 41st TEAC*, pp. 391–392.

Miyazawa, M., S. Kumagae, H. Kameoka, 1997b. Biotransformation of (+)-trans myrtanol and (−)-trans-myrtanol by common cutworm Larvae, Spodoptera litura as a biocatalyst. *Proceedings of the 41st TEAC*, pp. 389–390.

Miyazawa, M. and Y. Miyamoto, 2004. Biotransformation of (+)-(1*R*, 2*S*)-fenchol by the larvae of common cutworm (*Spodoptera litura*). *Tetrahadron*, 60: 3091–3096.

Miyazawa, M., T. Murata, and H. Kameoka, 1998. Biotransformation of β-myrcene by common cutworm larvae, Spodoptera litura as a biocatalyst. *Proceedings of the 42nd TEAC*, pp. 123–125.

Miyazawa, M., H. Nankai, and H. Kameoka, 1996a. Microbial oxidation of citronellol by *Glomerella cingulata. Nat. Prod. Lett.*, 8: 303–305.

Miyazawa, M., Y. Noma, K. Yamamoto, and H. Kameoka, 1983. Microbiological conversion of d- and l-limonene. *Proceedings of the 27th TEAC*, pp. 147–149.

Miyazawa, M., Y. Noma, K. Yamamoto, and H. Kameoka, 1991e. Biotransformation of 1,4-cineole to 2-*endo*-hydroxy-1,4-cineole by *Aspergillus niger. Chem. Express*, 6: 771–774.

Miyazawa, M., Y. Noma, K. Yamamoto, and H. Kameoka, 1992d. Biohydroxylation of 1,4-cineole to 9-hydroxy-1,4-cineole by *Aspergillus niger. Chem. Express*, 7: 305–308.

Miyazawa, M., Y. Noma, K. Yamamoto, and H. Kameoka, 1992e. Biotransformation of 1,4-cineole to 3-*endo*-hydroxy-1,4-cineole by *Aspergillus niger. Chem. Express*, 7: 125–128.

Miyazawa, M., Y. Suzuki, and H. Kameoka, 1994a. Biotransformation of myrtanol by plant pathogenic microorganism, Glomerella cingulata. *Proceedings of the 38th TEAC*, pp. 96–97.

Miyazawa, M., Y. Suzuki, and H. Kameoka, 1997c. Biotransformation of (−)- and (+)-isopinocampheol by three fungi. *Phytochemistry*, 45: 945–950.

Miyazawa, M., T. Wada, and H. Kameoka, 1995a. Biotransformation of terpinene, limonene and α-phellandrene in common cutworm larvae, Spodoptera litura Fabricius. *Proceedings of the 39th TEAC*, pp. 362–363.

Miyazawa, M., T. Wada, and H. Kameoka, 1996b. Biotransformation of p-menthanes using common cutworm larvae, Spodoptera litura as a biocatalyst. *Proceedings of the 40th TEAC*, pp. 80–81.

Miyazawa, M., K. Yamamoto, Y. Noma, and H. Kameoka, 1990a. Bioconversion of (+)-fenchone to (+)-6-*endo*-hydroxyfenchone by *Aspergillus niger. Chem. Express*, 5: 237–240.

Miyazawa, M., K. Yamamoto, Y. Noma, and H. Kameoka, 1990b. Bioconversion of (+)-fenchone to 5-*endo*-hydroxyfenchone by *Aspergillus niger. Chem. Express*, 5: 407–410.

Miyazawa, M., H. Yanagihara, and H. Kameoka, 1996c. Biotransformation of pinanes by common cutworm larvae, Spodoptera litura as a biocatalyst. *Proceedings of the 40th TEAC*, pp. 84–85.

Miyazawa, M., H. Yanahara, and H. Kameoka, 1995b. Biotransformation of trans-pinocarveol by plant pathogenic microorganism, Glomerella cingulata, and by the larvae of common cutworm, Spodoptera litura Fabricius. *Proceedings of the 39th TEAC*, pp. 360–361.

Miyazawa, M., K. Yokote, and H. Kameoka, 1994b. Biotransformation of linalool oxide by plant pathogenic microorganisms, Glomerella cingulata. *Proceedings of the 38th TEAC*, pp. 101–102.

Miyazawa, M., K. Yokote, and H. Kameoka, 1995c. Biotransformation of 2-endo-hydroxy-1,4-cineole by plant pathogenic microorganism, Glomerella cingulata. *Proceedings of the 39th TEAC*, pp. 352–353.
Mizutani, S., T. Hayashi, H. Ueda, and C. Tstsumom, 1971. Microbiological conversion of terpenes. Part IX. Conversion of linalool. *Nippon Nogei Kagaku Kaishi*, 45: 368–373.
Moroe, T., S. Hattori, A. Komatsu, and Y. Yamaguchi, 1971. Japanese Patent, 2,036,875. No. 98195t.
Mukherjee, B.B., G. Kraidman, and I.D. Hill, 1973. Synthesis of glycols by microbial transformation of some monocyclic terpenes. *Appl. Microbiol.*, 25: 447–453.
Mukherjee, B.B., G. Kraidman, I.D. Hill, 1974. Transformation of 1-menthene by a *Cladosporium*: Accumulation of β-isopropyl glutaric acid in the growth medium. *Appl. Microbiol.*, 27: 1070–1074.
Murakami, T., I. Ichimoto, and C. Tstsumom, 1973. Microbiological conversion of linalool. *Nippon Nogei Kagaku Kaishi*, 47: 699–703.
Murata, T. and M. Miyazawa, 1999. Biotransformation of dihydromyrcenol by common cutworm larvae, Spodoptera litura as a biocatalyst. *Proceedings of the 43rd TEAC*, pp. 393–394.
Nakanishi, K. and M. Miyazawa, 2004. Biotransformation of (−)-menthone by human liver microsomes. *Proceedings of the 48th TEAC*, pp. 401–402.
Nakanishi, K. and M. Miyazawa, 2005. Biotransformation of (+)- and (−)- menthol by liver microsomal humans and rats. *Proceedings of the 49th TEAC*, pp. 423–425.
Nishimura, H., S. Hiramoto, and J. Mizutani, 1983a. Biological activity of bottrospicatol and related compounds produced by microbial transformation of (−)-cis-carveol towards plants. *Proceedings of the 27th TEAC*, pp. 107–109.
Nishimura, H., S. Hiramoto, J. Mizutani, Y. Noma, A. Furusaki, and T. Matsumoto, 1983b. Structure and biological activity of bottrospicatol, a novel monoterpene produced by microbial transformation of (−)-*cis*-carveol. *Agric. Biol. Chem.*, 47: 2697–2699.
Nishimura, H. and Y. Noma, 1996. Microbial transformation of monoterpenes: Flavor and biological activity. In: *Biotechnology for Improved Foods and Flavors*, G.R. Takeoka, R. Teranishi, P.J. Williams, and A. Kobayashi, eds., ACS Symposium Series 637, pp. 173–187. Washington, DC: American Chemical Society.
Nishimura, H., Y. Noma, and J. Mizutani, 1982. *Eucalyptus* as biomass. Novel compounds from microbial conversion of 1,8–cineole. *Agric. Biol. Chem.*, 46: 2601–2604.
Nishimura, H., D.M. Paton, and M. Calvin, 1980. *Eucalyptus radiata* oil as a renewable biomass. *Agric. Biol. Chem.*, 44: 2495–2496.
Noma, Y., 1976. Microbiological conversion of carvone. Biochemical reduction of terpenes, part VI. *Ann. Res. Stud. Osaka Joshigakuen Junior College*, 20: 33–47.
Noma, Y., 1977. Conversion of the analogues of carvone and dihydrocarvone by *Pseudomonas ovalis*, strain 6-1, Biochemical reduction of terpenes, part VII. *Nippon Nogeikagaku Kaishi*, 51: 463–470.
Noma, Y., 1979a. Conversion of (−)-carvone by *Nocardia lurida* A-0141 and *Streptosporangium roseum* IFO3776. Biochemical reduction of terpenes, part VIII. *Nippon Nogeikagaku Kaishi*, 53: 35–39.
Noma, Y., 1979b. On the pattern of reaction mechanism of (+)-carvone conversion by actinomycetes. Biochemical reduction of terpenes, part X, Ann. Res. Stud. *Osaka Joshigakuen Junior College*, 23: 27–31.
Noma, Y., 1980. Conversion of (−)-carvone by strains of *Streptomyces*, A-5–1 and Nocardia, 1-3-11. *Agric. Biol. Chem.*, 44: 807–812.
Noma, Y., 1984. Microbiological conversion of carvone, *Kagaku to Seibutsu*, 22: 742–746.
Noma, Y., 1988. Formation of p-menthane-2,8-diols from (−)-dihydrocarveol and (+)-dihydrocarveol by Aspergillus spp. *The Meeting of Kansai Division of The Agricultural and Chemical Society of Japan*, Kagawa, Japan, p. 28.
Noma, Y., 2000. unpublished data.
Noma, Y., 2007. Microbial production of mosquitocidal (1R,2S,4R)-(+)-menthane- 2,8-diol. In: *Aromatic Plants from Asia their Chemistry and Application in Food and Therapy*, L. Jiarovetz, N.X. Dung, and V.K. Varshney, eds., pp. 169–186. Dehradun, Uttarakhand: Har Krishan Bhalla & Sons.
Noma, Y., E. Akehi, N. Miki, and Y. Asakawa, 1992a. Biotransformation of terpene aldehyde, aromatic aldehydes and related compounds by *Dunaliella tertiolecta. Phytochemistry*, 31: 515–517.
Noma, Y. and Y. Asakawa, 1992. Enantio- and diastereoselectivity in the biotransformation of carveols by *Euglena gracilis* Z. *Phytochem.*, 31: 2009–2011.
Noma, Y. and Y. Asakawa, 1995. Aspergillus spp.: Biotransformation of terpenoids and related compounds. In: *Biotechnology in Agriculture and Forestry, Vol. 33. Medicinal and Aromatic Plants VIII*, Y.P.S. Bajaj, ed., pp. 62–96. Berlin, Germany: Springer.
Noma, Y. and Y. Asakawa, 1998. Euglena gracilis Z: Biotransformation of terpenoids and related compounds. In: *Biotechnology in Agriculture and Forestry, Vol. 41. Medicinal and Aromatic Plants X*, Y.P.S. Bajaj, ed., pp. 194–237. Berlin Heidelberg, Germany: Springer.

Noma, Y. and Y. Asakawa, 2005a. New metabolic pathways of β-pinene and related compounds by Aspergillus niger. *Book of Abstracts of the 36th ISEO*, p. 32.

Noma, Y. and Y. Asakawa, 2005b. Microbial transformation of (−)-myrtenol and (−)-nopol. *Proceedings of the 49th TEAC*, pp. 78–80.

Noma, Y. and Y. Asakawa, 2006a. Biotransformation of (+)-limonene and related compounds by Citrus pathogenic fungi. *Proceedings of the 50th TEAC*, pp. 431–433.

Noma, Y. and Y. Asakawa, 2006b. Biotransformation of β-pinene, myrtenol, nopol and nopol benzyl ether by Aspergillus niger TBUYN-2. *Book of Abstracts of the 37th ISEO*, p. 144.

Noma, Y. and Y. Asakawa, 2006c. Microbial transformation of (−)-nopol benzyl ether. *Proceedings of the 50th TEAC*, pp. 434–436.

Noma, Y. and Y. Asakawa, 2007a. Biotransformation of limonene and related compounds by newly isolated low temperature grown citrus pathogenic fungi and red yeast. *Book of Abstracts of the 38th ISEO*, p. 7.

Noma, Y. and Y. Asakawa, 2007b. Microbial transformation of limonene and related compounds. *Proceedings of the 51st TEAC*, pp. 299–301.

Noma, Y. and Y. Asakawa, 2008. New metabolic pathways of (+)-carvone by Citrus pathogenic Aspergillus niger Tiegh CBAYN and A. niger TBUYN-2. *Proceedings of the 52nd TEAC*, pp. 206–208.

Noma, Y., M. Furusawa, T. Hashimoto, and Y. Asakawa, 2002. Stereoselective formation of (1R, 2S, 4R)-(+)-p-menthane-2,8-diol from α-pinene. *Book of Abstracts of the 33rd ISEO*, p. 142.

Noma, Y., M. Furusawa, T. Hashimoto, and Y. Asakawa, 2004. Biotransformation of (+)- and (−)-3-pinanone by Aspergillus niger. *Proceedings of the 48th TEAC*, pp. 390–392.

Noma, Y., T. Hashimoto, S. Uehara, and Y. Asakawa, 2009. unpublished data.

Noma, Y., T. Higata, T. Hirata, Y. Tanaka, T. Hashimoto, and Y. Asakawa, 1995. Biotransformation of [6-2H]-(−)-carvone by Aspergillus niger, Euglena gracilis Z and Dunaliella tertiolecta. *Proceedings of the 39th TEAC*, pp. 367–368.

Noma, Y., K. Hirata, and Y. Asakawa, 1996. Biotransformation of 1,8-cineole. Why do the biotransformed 2α- and 3α-hydroxy-1,8-cineole by Aspergillus niger have no optical activity? *Proceedings of the 40th TEAC*, pp. 89–91.

Noma, Y. and M. Iwami, 1994. Separation and identification of terpene convertible actinomycetes: *S. bottropensis* SY-2-1, S. ikutamanensis Ya-2–1 and S. humidus Tu-1. *Bull. Tokushima Bunri Univ.*, 47: 99–110.

Noma, Y., F. Kamino, T. Hashimoto, and Y. Asakawa, 2003. Biotransformation of (+)- and (−)-pinane-2,3-diol and related compounds by Aspergillus niger. *Proceedings of the 47th TEAC*, pp. 91–93.

Noma, Y., K. Matsueda, I. Maruyama, and Y. Asakawa, 1997. Biotransformation of terpenoids and related compounds by Chlorella species. *Proceedings of the 41st TEAC*, pp. 227–229.

Noma, Y., N. Miki, E. Akehi, E. Manabe, and Y. Asakawa, 1991a. Biotransformation of monoterpenes by photosynthetic marine algae, Dunaliella tertiolecta. *Proceedings of the 35th TEAC*, pp. 112–114.

Noma, Y., M. Miyazawa, K. Yamamoto, H. Kameoka, T. Inagaki, and H. Sakai, 1984. Microbiological conversion of perillaldehyde. Biotransformation of l- and dl-perillaldehyde by Streptomyces ikutamanensis, Ya-2-1. *Proceedings of the 28th TEAC*, pp. 174–176.

Noma, Y. and H. Nishimura, 1980. Microbiological transformation of 1,8-cineole. Oxidative products from 1,8-cineole by S. bottropensis, SY-2-1. *Book of Abstracts of the Annual Meeting of Agricultural and Biological Chemical Society*, p. 28.

Noma, Y. and H. Nishimura, 1981. Microbiological transformation of 1,8-cineole. Production of 3β-hydroxy-1,8-cineole from 1,8-cineole by S. ikutamanensis, Ya-2-1. *Book of Abstracts of the Annual Meeting of Agricultural and Biological Chemical Society*, p. 196.

Noma, Y. and H. Nishimura, 1982. Biotransformation of carvone. 4. Biotransformation of (+)-carvone by Streptomyces bottropensis, SY-2-1. *Proceedings of the 26th TEAC*, pp. 156–159.

Noma, Y. and H. Nishimura, 1983a. Biotransformation of (−)-carvone and (+)-carvone by S. ikutamanensis Ya-2-1. *Book of Abstracts of the Annual Meeting of Agricultural and Biological Chemical Society*, p. 390.

Noma, Y. and H. Nishimura, 1983b. Biotransformation of carvone. 5. Microbiological transformation of dihydrocarvones and dihydrocarveols. *Proceedings of the 27th TEAC*, pp. 302–305.

Noma, Y. and H. Nishimura, 1984. Microbiological conversion of carveol. Biotransformation of (−)-cis-carveol and (+)-cis-carveol by S. bottropensis, Sy-2-1. *Proceedings of the 28th TEAC*, pp. 171–173.

Noma, Y. and H. Nishimura, 1987. Bottrospicatols, novel monoterpenes produced on conversion of (−)- and (+)-*cis*-carveol by *Streptomyces*. *Agric. Biol. Chem.*, 51: 1845–1849.

Noma, Y., H. Nishimura, S. Hiramoto, M. Iwami, and C. Tstsumi, 1982. A new compound, (4R, 6R)-(+)-6,8-oxidomenth-1-en-9-ol produced by microbial conversion of (−)-cis-carveol. *Agric. Biol. Chem.*, 46: 2871–2872.

Noma, Y., H. Nishimura, and C. Tatsumi, 1980. Biotransformation of carveol by Actinomycetes. 1. Biotransformation of (−)-cis-carveol and (−)-trans-carveol by Streptomyces bottropensis, SY-2-1. *Proceedings of the 24th TEAC*, pp. 67–70.

Noma, Y. and S. Nonomura, 1974. Conversion of (−)-carvone and (+)-carvone by a strain of *Aspergillus niger. Agric. Biol. Chem.*, 38: 741–744.

Noma, Y., S. Nonomura, and H. Sakai, 1974a. Conversion of (−)-carvotanacetone and (+)-carvotanacetone by *Pseudomonas ovalis*, strain 6-1. *Agric. Biol. Chem.*, 38: 1637–1642.

Noma, Y., S. Nonomura, and H. Sakai, 1975. Epimerization of (−)-isodihydrocarvone to (−)-dihydrocarvone by *Pseudomonas fragi* IFO 3458. *Agric. Biol. Chem.*, 39: 437–441.

Noma, Y., S. Nonomura, H. Ueda, H. Sakai, and C. Tstusmi, 1974b. Microbial transformation of carvone. *Proceedings of the 18th TEAC*, pp. 20–23.

Noma, Y., S. Nonomura, H. Ueda, and C. Tatsumi, 1974c. Conversion of (+)-carvone by *Pseudomonas ovalis*, strain 6-1(1). *Agric. Biol. Chem.*, 38: 735–740.

Noma, Y. and H. Sakai, 1984. Investigation of the conversion of (−)-perillyl alcohol, 1,8–cineole, (+)-carvone and (−)-carvone by rare actinomycetes. *Ann. Res. Stud. Osaka Joshigakuen Junior College*, 28: 7–18.

Noma, Y., A. Sogo, S. Miki, N. Fujii, T. Hashimoto, and Y. Asakwawa, 1992b. Biotransformation of terpenoids and related compounds. *Proceedings of the 36th TEAC*, pp. 199–201.

Noma, Y., H. Takahashi, and Y. Asakawa, 1989. Microbiological conversion of menthol. Biotransformation of (+)-menthol by a strain of Aspergillus niger. *Proceedings of the 33rd TEAC*, pp. 124–126.

Noma, Y., H. Takahashi, and Y. Asakawa, 1990. Microbiological conversion of p-menthane 1. Formation of p-menthane-1,9-diol from p-menthane by a strain of Aspergillus niger. *Proceedings of the 34th TEAC*, pp. 253–255.

Noma, Y., H. Takahashi, and Y. Asakawa, 1991b. Biotransformation of terpene aldehyde by *Euglena gracilis. Z. Phytochem.*, 30: 1147–1151.

Noma, Y., H. Takahashi, and Y. Asakawa, 1993. Formation of 8 kinds of p-menthane-2,8-diols from carvone and related compounds by Euglena gracilis Z. Biotransformation of monoterpenes by photosynthetic microorganisms. Part VIII. *Proceedings of the 37th TEAC*, pp. 23–25.

Noma, Y., H. Takahashi, M. Toyota, and Y. Asakawa, 1988a. Microbiological conversion of (−)-carvotanacetone and (+)-carvotanacetone by a strain of Aspergillus niger. *Proceedings of the 32nd TEAC*, pp. 146–148.

Noma, Y., H. Takahashi, T. Hashimoto, and Y. Asakawa, 1992c. Biotransformation of isopiperitenone, 6-gingerol, 6-shogaol and neomenthol by a strain of Aspergillus niger. *Proceedings of the 37th TEAC*, pp. 26–28.

Noma, Y. and C. Tatsumi, 1973. Conversion of (−)-carvone by Pseudomonas ovalis, strain 6-1(1), Microbial conversion of terpenes part XIII. *Nippon Nogeikagaku Kaishi*, 47: 705–711.

Noma, Y., M. Toyota, and Y. Asakawa, 1985a. Biotransformation of (−)-carvone and (+)-carvone by Aspergillus spp. *Annual Meeting of Agricultural and Biological Chemistry*, Sapporo, Japan, p. 68.

Noma, Y., M. Toyota, and Y. Asakawa, 1985b. Biotransformation of carvone. 6. Biotransformation of (−)-carvone and (+)-carvone by a strain of Aspergillus niger. *Proceedings of the 29th TEAC*, pp. 235–237.

Noma, Y., M. Toyota, Y. and Asakawa, 1985c. Microbiological conversion of (−)-carvotanacetone and (+)-carvotanacetone by S. bottropensis SY-2-1. *Proceedings of the 29th TEAC*, pp. 238–240.

Noma, Y., M. Toyota, and Y. Asakawa, 1986. Reduction of terpene aldehydes and epoxidation of terpene alcohols by S. ikutamanensis, Ya-2-1. *Proceedings of the 30th TEAC*, pp. 204–206.

Noma, Y., M. Toyota, and Y. Asakawa, 1988b. Microbial transformation of thymol formation of 2-hydroxy-3-p-menthen-5-one by Streptomyces humidus, Tu-1. *Proceedings of the 28th TEAC*, pp. 177–179.

Noma, Y., J. Watanabe, T. Hashimoto, and Y. Asakawa, 2001. Microbiological transformation of β-pinene. *Proceedings of the 45th TEAC*, pp. 88–90.

Noma, Y., S. Yamasaki, and Asakawa Y. 1992d. Biotransformation of limonene and related compounds by *Aspergillus cellulosae. Phytochemistry*, 31: 2725–2727.

Nonoyama, H., H. Matsui, M. Hyakumachi, and M. Miyazawa, 1999. Biotransformation of (−)-menthone using plant parasitic fungi, Rhizoctonia solani as a biocatalyst. *Proceedings of the 43rd TEAC*, pp. 387–388.

Ohsawa, M. and M. Miyazawa, 2001. Biotransformation of (+)- and (−)-isopulegol by the larvae of common cutworm (Spodoptera litura) as a biocatalyst. *Proceedings of the 45th TEAC*, pp. 375–376.

Omata, T., N. Iwamoto, T. Kimura, A. Tanaka, S. Fukui, 1981. Stereoselective hydrolysis of *dl*-menthyl succinate by gel-entrapped *Rhodotorula minuta* var. *texensis* cells in organic solvent. *Appl. Microbiol. Biotechnol.*, 11: 119–204.

Oosterhaven, K., K.J. Hartmans, and J.J.C. Scheffer, 1995a. Inhibition of potato sprouts growth by carvone enantiomers and their bioconversion in sprouts. *Potato Res.*, 38: 219–230.

Oosterhaven, K., B. Poolman, and E.J. Smid, 1995b. *S*-Carvone as a natural potato sprouts inhibiting, fungistatic and bacteriostatic compound. *Indian Crops Prod.*, 4: 23–31.
Oritani, T. and K. Yamashita, 1973a. Microbial dl-acyclic alcohols. *Agric. Biol. Chem.*, 37: 1923–1928.
Oritani, T. and K. Yamashita, 1973b. Microbial resolution of racemic 2- and 3-alkylcyclohexanols. *Agric. Biol. Chem.*, 37: 1695–1700.
Oritani, T. and K. Yamashita, 1973c. Microbial resolution of *dl*-isopulegol. *Agric. Biol. Chem.*, 37: 1687–1689.
Oritani, T. and K. Yamashita, 1973d. Microbial resolution of racemic carvomenthols. *Agric. Biol. Chem.*, 37: 1691–1694.
Oritani, T. and K. Yamashita, 1974. Microbial resolution of (±)-borneols. *Agric. Biol. Chem.*, 38: 1961–1964.
Oritani, T. and K. Yamashita, 1980. Optical resolution of dl-(β, γ-unsaturated terpene alcohols by biocatalyst of microorganism. *Proceedings of the 24th TEAC*, pp. 166–169.
Pfrunder, B. and Ch. Tamm, 1969a. Mikrobiologische Umwandlung von bicyclischen monoterpenen durch *Absidia orchidis* (Vuill.) Hagem. 2. Teil: Hydroxylierung von Fenchon und Isofenchon. *Helv. Chim. Acta.*, 52: 1643–1654.
Pfrunder, B. and Ch. Tamm, 1969b. Mikrobiologische Umwandlung von bicyclischen monoterpenen durch *Absidia orchidis* (Vuill.) Hagem. 1. Teil: Reduktion von Campherchinon und Isofenchonchinon. *Helv. Chim. Acta.*, 52: 1630–1642.
Prema, B.R. and P.K. Bhattachayya, 1962. Microbiological transformation of terpenes. II. Transformation of a-pinene. *Appl. Microbiol.*, 10: 524–528.
Rama Devi, J., S.G. Bhat, and P.K. Bhattacharyya, 1977. Microbiological transformations of terpenes. Part XXV. Enzymes involved in the degradation of linalool in the Pseudomonas incognita, linalool strain. *Indian J. Biochem. Biophys.*, 15: 323–327.
Rama Devi, J. and P.K. Bhattacharyya, 1977a. Microbiological transformations of terpenes. Part XXIV. Pathways of degradation of linalool, geraniol, nerol and limonene by Pseudomonas incognita, linalool strain. *Indian J. Biochem. Biophys.*, 14: 359–363.
Rama Devi, J. and P.K. Bhattacharyya, 1977b. Microbiological transformation of terpenes. Part XXIII. Fermentation of geraniol, nerol and limonene by soil Pseudomonad, *Pseudomonas incognita* (linalool strain). *Indian J. Biochem. Biophys.*, 14: 288–291.
Rama Devi, J. and P.K. Bhattacharyya, 1978. Molecular rearrangements in the microbiological transformations of terpenes and the chemical logic of microbial processes. *J. Indian Chem. Soc.*, 55: 1131–1137.
Rapp, A. and H. Mandery, 1988. Influence of Botrytis cinerea on the monoterpene fraction wine aroma. In: *Bioflavour'87. Analysis—Biochemistry—Biotechnology*, P. Schreier, ed., pp. 445–452. Berlin, Germany: Walter de Gruyter and Co.
Saeki, M. and N. Hashimoto, 1968. Microbial transformation of terpene hydrocarbons. Part I. Oxidation products of d-limonene and d-pentene. *Proceedings of the 12th TEAC*, pp. 102–104.
Saeki, M. and N. Hashimoto, 1971. Microorganism biotransformation of terpenoids. Part II. Production of cis-terpin hydrate and terpineol from d-limonene. *Proceedings of the 15th TEAC*, pp. 54–56.
Saito, H. and M. Miyazawa, 2006. Biotransformation of 1,8-cineole by Salmonella typhimurium OY1001/3A4. *Proceedings of the 50th TEAC*, pp. 275–276.
Savithiry, N., T.K. Cheong, and P. Oriel, 1997. Production of alpha-terpineol from *Escherichia coli* cells expressing thermostable limonene hydratase. *Appl. Biochem. Biotechnol.*, 63–65: 213–220.
Sawamura, Y., S. Shima, H. Sakai, and C. Tatsumi, 1974. Microbiological conversion of menthone. *Proceedings of the 18th TEAC*, pp. 27–29.
Schwammle, B., E. Winkelhausen, S. Kuzmanova, and W. Steiner, 2001. Isolation of carvacrol assimilating microorganisms. *Food Technol. Biotechnol.*, 39: 341–345.
Seubert, W. and E. Fass, 1964a. Studies on the bacterial degradation of isoprenoids. V. The mechanism of isoprenoid degradation. *Biochem. Z.*, 341: 35–44.
Seubert, W. and E. Fass, 1964b. Studies on the bacterial degradation of isoprenoids. IV. The purification and properties of beta-isohexenylglutaconyl-COA-hydratase and beta-hydroxy-beta-isohexenylglutaryl-COA-lyase. *Biochem. Z.*, 341: 23–34.
Seubert, W., E. Fass, and U. Remberger, 1963. Studies on the bacterial degradation of isoprenoid compounds. III. Purification and properties of geranyl carboxylase. *Biochem. Z.*, 338: 265–275.
Seubert, W. and U. Remberger, 1963. Studies on the bacterial degradation of isoprenoid compounds. II. The role of carbon dioxide. *Biochem. Z.*, 338: 245–246.
Shima, S., Y. Yoshida, Y. Sawamura, and C. Tstsumi, 1972. Microbiological conversion of perillyl alcohol. *Proceedings of the 16th TEAC*, pp. 82–84.
Shimoda, K., T. Hirata, and Y. Noma, 1998. Stereochemistry in the reduction of enones by the reductase from *Euglena gracilis*. *Z. Phytochem.*, 49: 49–53.

Shimoda, K., D.I. Ito, S. Izumi, and T. Hirata, 1996. Novel reductase participation in the syn-addition of hydrogen to the C=C bond of enones in the cultured cells of *Nicotiana tabacum*. *J. Chem. Soc. Perkin Trans.*, 1: 355–358.

Shimoda, K., S. Izumi, and T. Hirata, 2002. A novel reductase participating in the hydrogenation of an exocyclic C–C double bond of enones from *Nicotiana tabacum*. *Bull. Chem. Soc. Jpn.*, 75: 813–816.

Shimoda, K., N. Kubota, H. Hamada, and M. Kaji, 2003. Cyanobacterium catalyzed asymmetric reduction of enones. *Proceedings of the 47th TEAC*, pp. 164–166.

Shindo, M., T. Shimada, and M. Miyazawa, 2000. Metabolism of 1,8-cineole by cytochrome P450 enzymes in human and rat liver microsomes. *Proceedings of the 44th TEAC*, pp. 141–143.

Shukla, O.P., R.C. Bartholomeus, and I.C. Gunsalus, 1987. Microbial transformation of menthol and menthane-3,4-diol. *Can. J. Microbiol.*, 33: 489–497.

Shukla, O.P., and P.K. Bhattacharyya, 1968. Microbiological transformations of terpenes: Part XI—Pathways of degradation of α- & β-pinenes in a soil Pseudomonad (PL-strain). *Indian J. Biochem.*, 5: 92–101.

Shukla, O.P., M.N. Moholay, and P.K. Bhattacharyya, 1968. Microbiological transformation of terpenes: Part X—Fermentation of α- & β-pinenes by a soil Pseudomonad (PL-strain). *Indian J. Biochem.*, 5: 79–91.

Southwell, I.A. and T.M. Flynn, 1980. Metabolism of α- and β-pinene, p-cymene and 1,8–cineole in the brush tail possum. *Xenobiotica*, 10: 17–23.

Suga, T. and T. Hirata, 1990. Biotransformation of exogenous substrates by plant cell cultures. *Phytochemistry*, 29: 2393–2406.

Suga, T., T. Hirata, and H. Hamada, 1986. The stereochemistry of the reduction of carbon–carbon double bond with the cultured cells of *Nicotiana tabacum*. *Bull. Chem. Soc. Jpn.*, 59: 2865–2867.

Sugie, A. and M. Miyazawa, 2003. Biotransformation of (−)-a-pinene by human liver microsomes. *Proceedings of the 47th TEAC*, pp. 159–161.

Swamy, G.K., K.L. Khanchandani, and P.K. Bhattacharyya, 1965. *Symposium on Recent Advances in the Chemistry of Terpenoids*, Natural Institute of Sciences of India, New Delhi, India, p. 10.

Takagi, K., Y. Mikami, Y. Minato, I. Yajima, and K. Hayashi, 1972. Manufacturing method of carvone by microorganisms, Japanese Patent 7,238,998.

Takahashi, H., Y. Noma, M. Toyota, and Y. Asakawa, 1994. The biotransformation of (−)- and (+)-neomenthols and isomenthols by *Aspergillus niger*. *Phytochemistry*, 35: 1465–1467.

Takeuchi, H. and M. Miyazawa, 2004. Biotransformation of nerol by the larvae of common cutworm (Spodoptera litura) as a biocatalyst. *Proceedings of the 48th TEAC*, pp. 399–400.

Takeuchi, H. and M. Miyazawa, 2005. Biotransformation of (−)- and (+)-citronellene by the larvae of common cutworm (Spodoptera litura) as biocatalyst. *Proceedings of the 49th TEAC*, pp. 426–427.

Trudgill, P.W., 1990. Microbial metabolism of terpenes—Recent developments. *Biodegradation* 1: 93–105.

Tsukamoto, Y., S. Nonomura, and H. Sakai, 1975. Formation of *p-cis*-menthan-1-ol from *p*-menthane by *Pseudomonas mendocina* SF. *Agric. Biol. Chem.*, 39: 617–620.

Tsukamoto, Y., S. Nonomura, H. Sakai, and C. Tatsumi, 1974. Microbiological oxidation of p-menthane 1. Formation of formation of p-cis-menthan-1-ol. *Proceedings of the 18th TEAC*, pp. 24–26.

Van der Werf, M.J. and J.A.M. de Bont, 1998a. Screening for microorganisms converting limonene into carvone. In: *New Frontiers in Screening for Microbial Biocatalysts, Proceedings of the International Symposium, Ede, the Netherlands*, K. Kieslich, C.P. Beek, J.A.M. van der Bont, and W.J.J. van den Tweel, eds., Vol. 53, pp. 231–234. Amsterdam, the Netherlands: Studies in Organic Chemistry.

Van der Werf, M.J., J.A.M. de Bont, and D.J. Leak, 1997. Opportunities in microbial biotransformation of monoterpenes. *Adv. Biochem. Eng. Biotechnol.*, 55: 147–177.

Van der Werf, M.J., K.M. Overkamp, and J.A.M. de Bont, 1998b. Limonene-1,2-epoxide hydrolase from *Rhodococcus erythropolis* DCL14 belongs to a novel class of epoxide hydrolases. *J. Bacteriol.*, 180: 5052–5057.

van Dyk, M.S., E. van Rensburg, I.P.B. Rensburg, and N. Moleleki, 1998. Biotransformation of monoterpenoid ketones by yeasts and yeast-like fungi. *J. Mol. Catal. B: Enzym.*, 5: 149–154.

Verstegen-Haaksma, A.A., H.J. Swarts, B.J.M. Jansen, A. de Groot, N. Bottema-MacGillavry, and B. Witholt, 1995. Application of *S*-(+)-carvone in the synthesis of biologically active natural products using chemical transformations and bioconversions. *Indian Crops Prod.*, 4: 15–21.

Watanabe, Y. and T. Inagaki, 1977a. Japanese Patent 77,12,989. No. 187696x.

Watanabe, Y. and T. Inagaki, 1977b. Japanese Patent 77, 122,690. No. 87656g.

Watanabe, T., H. Nomura, T. Iwasaki, A. Matsushima, and T. Hirata, 2007. Cloning of pulegone reductase and reduction of enones with the recombinant reductase. *Proceedings of the 51st TEAC*, pp. 323–325.

Wolf-Rainer, A., 1994. Phylogeny and biotransformation. Part 5. Biotransformation of isopinocampheol. *Z. Naturforsch.*, 49c: 553–560.

Yamada, K., S. Horiguchi, and J. Tatahashi, 1965. Studies on the utilization of hydrocarbons by microorganisms. Part VI. Screening of aromatic hydrocarbon-assimilating microorganisms and cumic acid formation from p-cymene. *Agric. Biol. Chem.*, 29: 943–948.

Yamaguchi, Y., A. Komatsu, and T. Moroe, 1977. Asymmetric hydrolysis of *dl*-menthyl acetate by *Rhodotorula mucilaginosa. J. Agric. Chem. Soc. Jpn.*, 51: 411–416.

Yamamoto, K., M. Miyazawa, H. Kameoka, and Y. Noma, 1984. Biotransformation of d- and l-fenchone by a strain of Aspergillus niger. *Proceedings of the 28th TEAC*, pp. 168–170.

Yamanaka, T. and M. Miyazawa, 1999. Biotransformation of (−)-trans-verbenol by common cutworm larvae, Spodoptera litura as a biocatalyst. *Proceedings of the 43rd TEAC*, pp. 391–392.

Yawata, T., M, Ogura, K. Shimoda, S. Izumi, and T. Hirata, 1998. Epoxidation of monoterpenes by the peroxidase from the cultured cells of Nicotiana tabacum, *Proceedings of the 42nd TEAC*, pp. 142–144.

Yonemoto, N., S. Sakamoto, T. Furuya, and H. Hamada, 2005. Preparation of (−)-perillyl alcohol oligosaccharides. *Proceedings of the 49th TEAC*, pp. 108–110.

23 Biotransformation of Sesquiterpenoids, Ionones, Damascones, Adamantanes, and Aromatic Compounds by Green Algae, Fungi, and Mammals

Yoshinori Asakawa and Yoshiaki Noma

CONTENTS

23.1 INTRODUCTION

Recently, environment-friendly green or clean chemistry is emphasized in the field of organic and natural product chemistry. Noyori's highly efficient production of (−)-menthol using (*S*)-BINAP-Rh catalyst is one of the most important green chemistries (Tani et al., 1982; Otsuka and Tani, 1991), and 1000 ton of (−)-menthol has been produced by this method in 1 year. On the other hand, enzymes of microorganisms and mammals are able to transform a huge variety of organic compounds, such as mono-, sesqui-, and diterpenoids, alkaloids, steroids, antibiotics, and amino acids from crude drugs and spore-forming green plants to produce pharmacologically and medicinally valuable substances.

Since Meyer and Neuberg (1915) studied the microbial transformation of citronellal, there are a great number of reports concerning biotransformation of essential oils, terpenoids, steroids, alkaloids, and acetogenins using bacteria, fungi, and mammals. In 1988, Mikami (1988) reported the review article of biotransformation of terpenoids entitled *Microbial Conversion of Terpenoids*. Lamare and Furstoss (1990) reviewed biotransformation of more than 25 sesquiterpenoids by microorganisms.

In this chapter, the recent advances in the biotransformation of natural and synthetic compounds; sesquiterpenoids, ionones, α-damascone, and adamantanes; and aromatic compounds, using microorganisms including algae and mammals, are described.

23.2 BIOTRANSFORMATION OF SESQUITERPENOIDS BY MICROORGANISMS

23.2.1 Highly Efficient Production of Nootkatone (2) from Valencene (1)

The most important and expensive grapefruit aroma, nootkatone (**2**), decreases the somatic fat ratio (Haze et al., 2002), and therefore, its highly efficient production has been requested by the cosmetic and fiber industrial sectors. Previously, valencene (**1**) from the essential oil of Valencia orange was converted into nootkatone (**2**) by biotransformation using *Enterobacter* species only in 12% yield (Dhavlikar and Albroscheit, 1973), *Rhodococcus* KSM-5706 in 0.5% yield with a complex mixture (Okuda et al., 1994), and cytochrome P450 (CYP450) in 20% yield with other complex products (Sowden et al., 2005). Nootkatone (**2**) was chemically synthesized from valencene (**1**) with $AcOOCMe_3$ in three steps and chromic acid in low yield (Wilson and Saw, 1978) and using surface-functionalized silica supported by metal catalysts such as Co^{2+} and Mn^{2+} with *tert*-butyl hydroperoxide in 75% yield (Salvador and Clark, 2002). However, these synthetic methods are not safe because they involve toxic heavy metals. An environment-friendly method for the synthesis of nootkatone that does not use any heavy metals such as chromium and manganese must be designed. The commercially available and cheap sesquiterpene hydrocarbon (+)-valencene (**1**) ($[\alpha]_D$ + 84.6°,c = 1.0) obtained from Valencia orange oil was very efficiently converted into nootkatone (**2**) by biotransformations using *Chlorella* (Hashimoto et al., 2003b), *Mucor* species (Hashimoto et al., 2003a), *Botryosphaeria dothidea*, and *Botryodiplodia theobromae* (Noma et al., 2001b; Furusawa et al., 2005a,b).

Chlorella fusca var. *vacuolata* IAMC-28 (Figure 23.1) was inoculated and cultivated while stationary under illumination in Noro medium $MgCl_2{\cdot}6H_2O$ (1.5 g), $MgSO_4{\cdot}7H_2O$ (0.5 g), KCl (0.2 g), $CaCl_2{\cdot}2H_2O$ (0.2 g), KNO_3 (1.0 g), $NaHCO_3$ (0.43 g), TRIS (2.45 g), K_2HPO_4 (0.045 g), Fe-EDTA (3.64 mg), EDTA-2Na (1.89 mg), $ZnSO_4{\cdot}7H_2O$ (1.5 g), H_3BO_2 (0.61 mg), $CoCl_2{\cdot}6H_2O$ (0.015 mg), $CuSO_4{\cdot}5H_2O$ (0.06 mg), $MnCl_2{\cdot}4H_2O$ (0.23 mg), and $(NH_4)_6Mo_7O_{24}{\cdot}4H_2O$ (0.38 mg), in distilled

FIGURE 23.1 *Chlorella fusca* var. *vacuolata.*

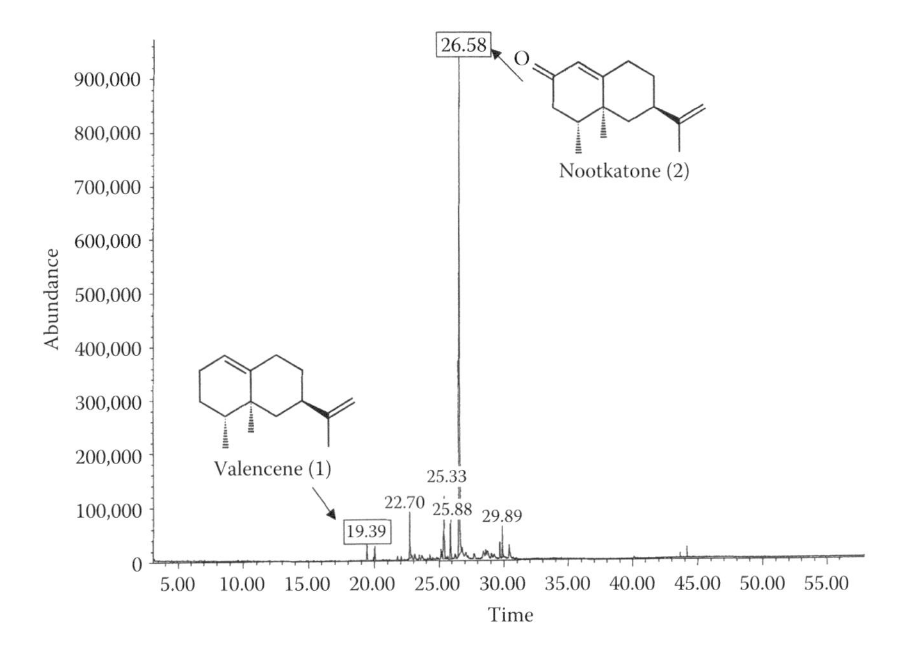

FIGURE 23.2 Total ion chromatogram of metabolites of valencene (**1**) by *Chlorella fusca* var. *vacuolata*.

H_2O 1 L (pH 8.0). Czapek-peptone medium (1.5% sucrose, 1.5% glucose, 0.5% polypeptone, 0.1% K_2HPO_4, 0.05% $MgSO_4 \cdot 7H_2O$, 0.05% KCl, and 0.001% $FeSO_4 \cdot 7H_2O$, in distilled water [pH 7.0]) was used for the biotransformation of substrate by microorganism. *Aspergillus niger* was isolated in our laboratories from soil in Osaka prefecture and was identified according to its physiological and morphological characters.

(+)-Valencene (**1**) (20 mg/50 mL) isolated from the essential oil of Valencia orange was added to the medium and biotransformed by *Chlorella fusca* for a further 18 days to afford nootkatone (**2**) (gas chromatography–mass spectrometry [GC-MS] peak area, 89%; isolated yield, 63%) (Figure 23.2) (Noma et al., 2001b; Furusawa et al., 2005a,b). The reduction of **2** with $NaBH_4$ and $CeCl_3$ gave 2α-hydroxyvalencene (**3**) in 87% yield, followed by Mitsunobu reaction with *p*-nitrobenzoic acid, triphenylphosphine, and diethyl azodicarboxylate to give nootkatol (2β-hydroxyvalencene) (**4**)—possessing calcium antagonistic activity—isolated from *Alpinia oxyphylla* (Shoji et al., 1984) in 42% yield. Compounds **3** and **4** thus obtained were easily biotransformed by *C. fusca* and *Chlorella pyrenoidosa* for only 1 day to give nootkatone (**2**) in good yield (80%–90%), respectively. The biotransformation of compound **1** was further carried out by *C. pyrenoidosa* and *Chlorella vulgaris* (Furusawa et al., 2005a,b) and soil bacteria (Noma et al., 2001a) to give nootkatone in good yield (Table 23.1).

In the time course of the biotransformation of **1** by *C. pyrenoidosa*, the yield of nootkatone (**2**) and nootkatol (**4**) without 2α-hydroxyvalencene (**3**) increased with the decrease in that of **1**,

TABLE 23.1
Conversion of Valencene (1) to Nootkatone (2) by *Chlorella* sp. for 14 Days

Chlorella sp.	Metabolites (% of the Total in GC-MS)				
	Valencene (1)	2α-Nootkatol (3)	2β-Nootkatol (4)	Nootkatone (2)	Conversion Ratio (%)
C. fusca	11	0	0	89	89
C. pyrenoidosa	7	0	0	93	93
C. vulgaris	0	0	0	100	100

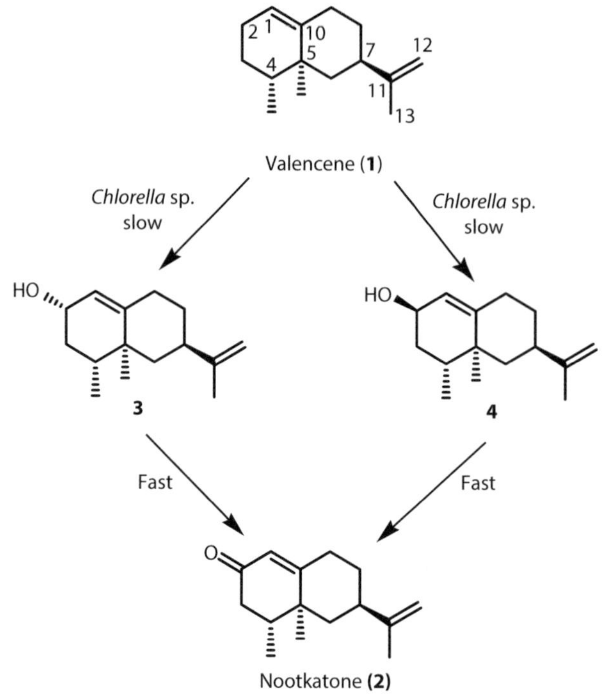

FIGURE 23.3 Biotransformation of valencene (**1**) by *Chlorella* species.

and subsequently, the yield of **2** increased with decrease in that of **3**. In the metabolic pathway of valencene (**1**), **1** was slowly converted into nootkatol (**4**), and subsequently, **4** was rapidly converted into **2**, as shown in Figure 23.3.

A fungus strain from the soil adhering to the thalloid liverwort *Pallavicinia subciliata*, *Mucor* species, was inoculated and cultivated statically in Czapek-peptone medium (pH 7.0) at 30°C for 7 days. Compound **1** (20 mg/50 mL) was added to the medium and incubated for a further 7 days. Nootkatone (**2**) was then obtained in very high yield (82%) (Noma et al., 2001b; Furusawa et al., 2005a).

The biotransformation from **1** to **2** was also examined using the plant pathogenic fungi *B. dothidea* and *B. theobromae* (a total of 31 strains) separated from fungi infecting various types of fruit, and so on. *B. dothidea* and *B. theobromae* were both inoculated and cultivated while stationary in Czapek-peptone medium (pH 7.0) at 30°C for 7 days. The same size of the substrate **1** was added to each medium and incubated for a further 7 days to obtain nootkatone (42%–84%) (Furusawa et al., 2005a).

The expensive grapefruit aromatic nootkatone (**2**) used by cosmetic and fiber manufacturers was obtained in high yield by biotransformation of (+)-valencene (**1**), which can be cheaply obtained from Valencia orange, by *Chlorella* species, fungi such as *Mucor* species, *B. dothidea*, and *B. theobromae*. This is a very inexpensive and clean oxidation reaction, which does not use any heavy metals, and thus, this method is expected to find applications in the industrial production of nootkatone.

23.2.2 Biotransformation of Valencene (1) by *Aspergillus niger* and *Aspergillus wentii*

Valencene (**1**) from Valencia orange oil was cultivated by *A. niger* in Czapek-peptone medium, for 5 days to afford six metabolites: **5** (1.0%), **6** and **7** (13.5%), **8** (1.1%), **9** (1.5%), **10** (2.0%), and **11** (0.7%), respectively. Ratio of compounds **6** (11*S*) and **7** (11*R*) was determined as 1:3 by high-performance liquid chromatography (HPLC) analysis of their thiocarbonates (**12** and **13**) (Noma et al., 2001b) (Figure 23.4).

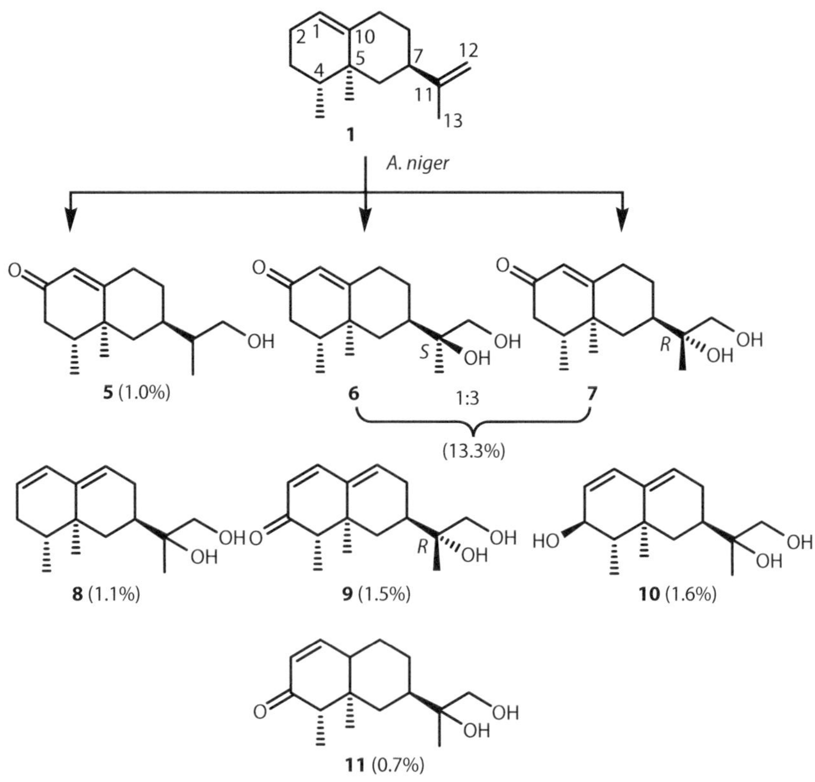

FIGURE 23.4 Biotransformation of valencene (**1**) by *Aspergillus niger*.

Compounds **8–11** could be biosynthesized by elimination of a hydroxy group of 2-hydroxyvalencenes (**3**, **4**). Compound **3** was biotransformed for 5 days by *A. niger* to give three metabolites: **6** and **7** (6.4%), **8** (34.6%), and **9** (5.5%), respectively. Compound **4** was biotransformed for 5 days by *A. niger* to give three metabolites: **6** and **7** (21.8%), **9** (5.5%), and **10** (10.4%), respectively (Figure 23.5).

Both ratios of **6** (11*S*) and **7** (11*R*) obtained from **3** and **4** were 1:3, respectively. From the aforementioned results, plausible metabolic pathways of valencene (**1**) and 2-hydroxyvalencene (**3**, **4**) by *A. niger* are shown in Figure 23.6 (Noma et al., 2001b).

Aspergillus wentii and *Epicoccum purpurascens* converted valencene (**1**) to 11,12-epoxide (**14a**) and the same diol (**6**, **7**) (Takahashi and Miyazawa, 2005) as well as nootkatone (**2**) and 2α-hydroxyvalencene (**3**) (Takahashi and Miyazawa, 2006).

Kaspera et al. (2005) reported that valencene (**1**) was incubated in submerged cultures of the ascomycete *Chaetomium globosum*, to give nootkatone (**2**), 2α-hydroxyvalencene (**3**), and valencene 11,12-epoxide (**14a**), together with a valencene ketodiol, valencene diols, a valencene triol, or valencene epoxydiol that were detected by liquid chromatography–MS (LC-MS) spectra and GC-MS of trimethylsilyl derivatives. These metabolites are accumulated preferably inside the fungal cells (Figure 23.7).

The metabolites of valencene, nootkatone (**2**), (**3**), and (**14a**), indicated grapefruit with sour and citrus with bitter odor, respectively. Nootkatone 11,12-epoxide (**14**) showed no volatile fragrant properties.

23.2.3 Biotransformation of Nootkatone (2) by *Aspergillus niger*

A. niger was inoculated and cultivated rotatory (100 rpm) in Czapek-peptone medium at 30°C for 7 days. (+)-Nootkatone (**2**) ($[\alpha]_D$ + 193.5°,c = 1.0) (80 mg/200 mL), which was isolated from the

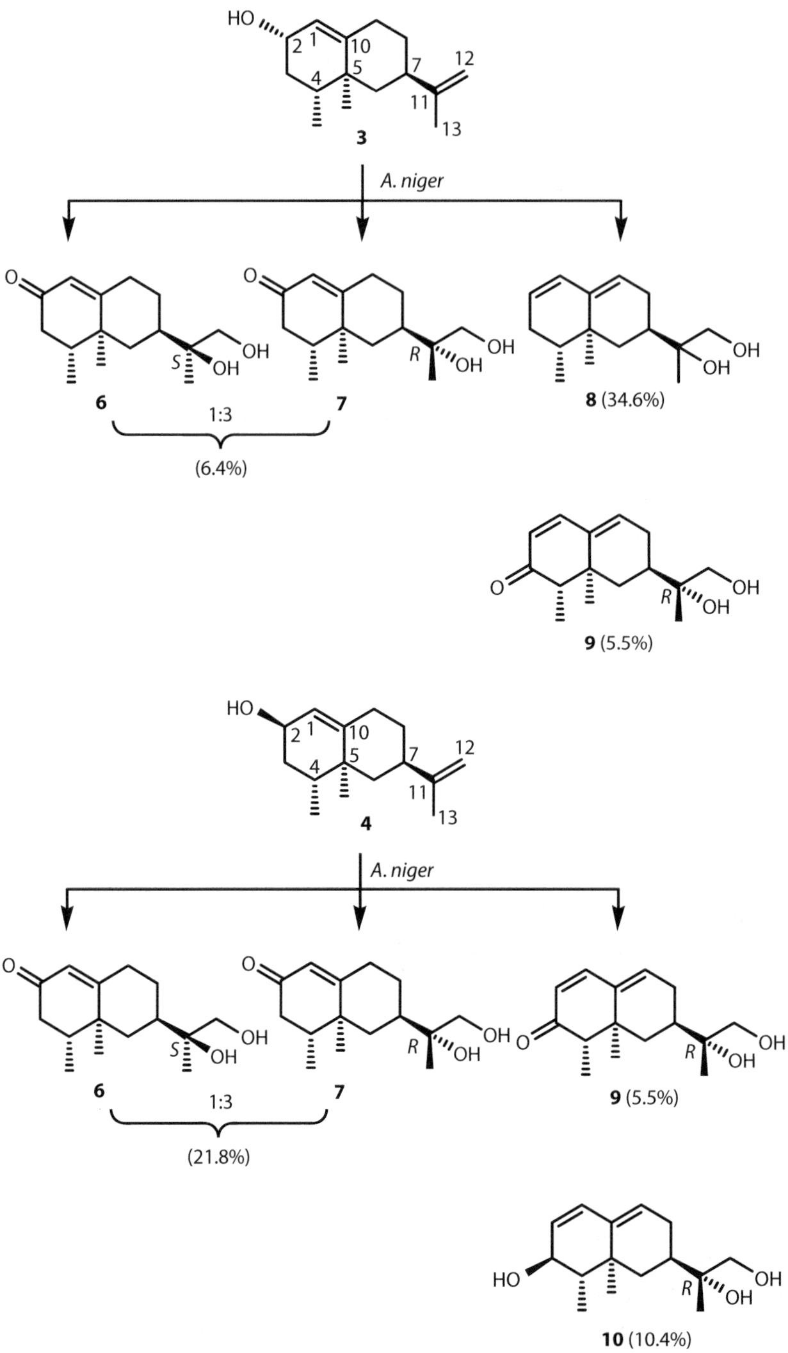

FIGURE 23.5 Biotransformation of 2α-hydroxyvalencene (**3**) and 2β-hydroxyvalencene (**4**) by *Aspergillus niger*.

essential oil of grapefruit, was added to the medium and further cultivated for 7 days to obtain two metabolites, 12-hydroxy-11,12-dihydronootkatone (**5**) (10.6%) and C11 stereo mixtures (51.5%), of nootkatone-11*S*,12-diol (**6**) and its 11*R* isomer (**7**) (11*R*:11*S* = 1:1) (Hashimoto et al., 2000b; Noma et al., 2001b; Furusawa et al., 2003) (Figure 23.8).

11,12-Epoxide (**14**) obtained by epoxidation of nootkatone (**2**) with *meta-chloroperbenzoic acid (mCPBA)* was biotransformed by *A. niger* for 1 day to afford **6** and **7** (11*R*:11*S* = 1:1) in good yield

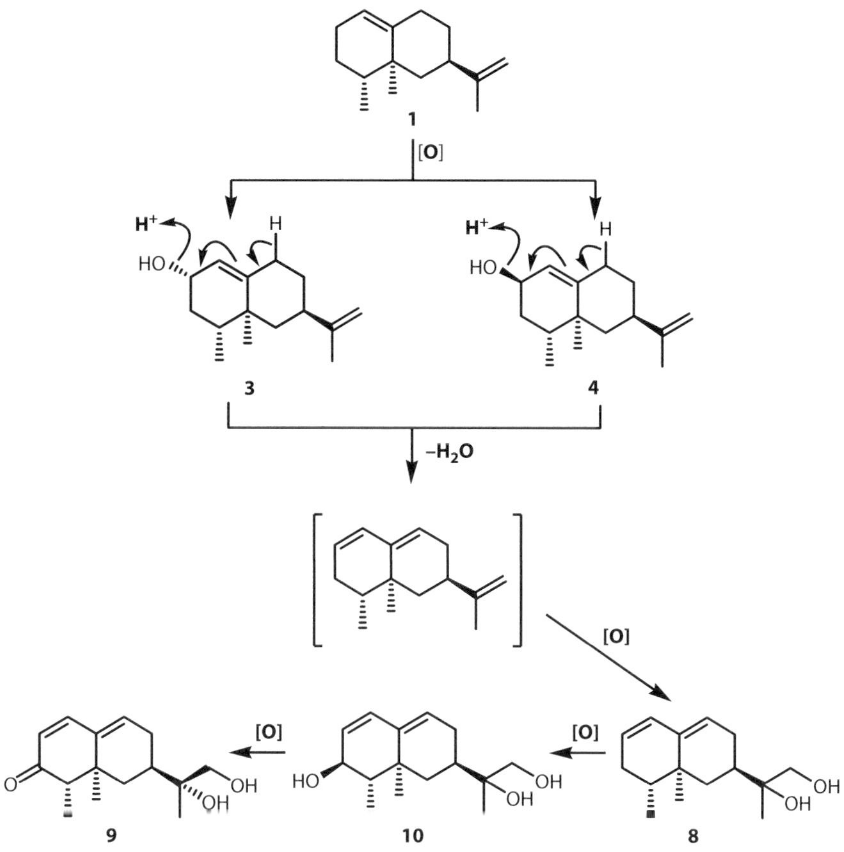

FIGURE 23.6 Possible pathway of biotransformation of valencene (**1**) by *Aspergillus niger*.

(81.4%). 1-Aminobenzotriazole, an inhibitor of CYP450, inhibited the oxidation process of **1** into compounds **5–7** (Noma et al., 2001b). From the aforementioned results, possible metabolic pathways of nootkatone (**2**) by *A. niger* might be considered as shown in Figure 23.9.

The same substrate was incubated with *A. wentii* to produce diol (**6**, **7**) and 11,12-epoxide (**14**) (Takahashi and Miyazawa, 2005).

23.2.4 Biotransformation of Nootkatone (2) by *Fusarium culmorum* and *Botryosphaeria dothidea*

(+)-Nootkatone (**2**) was added to the same medium as mentioned earlier including *Fusarium culmorum* to afford nootkatone-11*R*,12-diol (**7**) (47.2%) and 9β-hydroxynootkatone (**15**) (14.9%) (Noma et al., 2001b).

Compound **7** was stereospecifically obtained at C11 by biotransformation of **1**. Purity of compound **7** was determined as ca. 95% by HPLC analysis of the thiocarbonate (**13**).

The biotransformation of nootkatone (**2**) was examined by the plant pathogenic fungus *B. dothidea* separated from the fungus that infected the peach. (+)-Nootkatone (**2**) was cultivated with *B. dothidea* (Peach PP8402) for 14 days to afford nootkatone diols (**6** and **7**) (54.2%) and 7α-hydroxynootkatone (**16**) (20.9%). Ratio of compounds **6** and **7** was determined as 3:2 by HPLC analysis of the thiocarbonates (**12**, **13**) (Noma et al., 2001b). Nootkatone (**2**) was administered into rabbits to give the same diols (**6**, **7**) (Asakawa et al., 1986; Ishida, 2005).

E. purpurascens also biotransformed nootkatone (**2**) to **5–7**, **14**, and **15a** (Takahashi and Miyazawa, 2006).

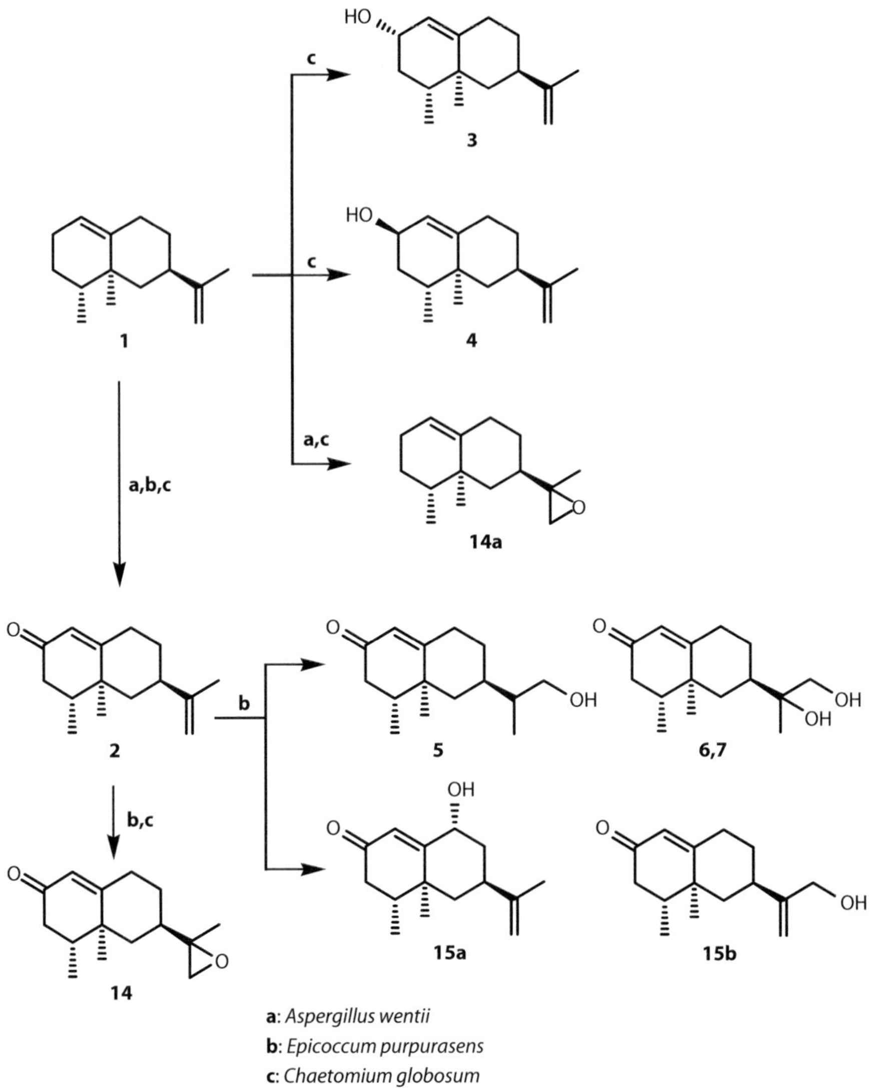

FIGURE 23.7 Biotransformation of valencene (**1**) and nootkatone (**2**) by *Aspergillus wentii*, *Epicoccum purpurascens*, and *Chaetomium globosum*.

The biotransformation of **2** by *A. niger* and *B. dothidea* resembled to that of the oral administration to rabbit since the ratio of the major metabolites 11*S*- (**6**) and 11*R*-nootkatone-11,12-diol (**7**) was similar. It is noteworthy that the biotransformation of **2** by *F. culmorum* affords stereospecifically nootkatone-11*R*,12-diol (**7**) (Noma et al., 2001b) (Figure 23.10).

Metabolites **3–5**, **12**, and **13** from (+)-nootkatone (**2**) and **14–17** from (+)-valencene (**1**) did not show an effective odor.

Dihydronootkatone (**17**), which shows that citrus odor possesses antitermite activity, was also treated in *A. niger* to obtain 11*S*-mono- (**18**) and 11*R*-dihydroxylated products (**19**) (the ratio 11*S* and 11*R* = 3:2). On the other hand, *Aspergillus cellulosae* reduced ketone group at C2 of **17** to give 2α- (**20**) (75.7%) and 2β-hydroxynootkatone (**21**) (0.7%) (Furusawa et al., 2003) (Figure 23.11).

Tetrahydronootkatone (**22**) also shows antitermite and mosquito-repellant activity. It was incubated with *A. niger* to give two similar hydroxylated compounds (**23**, 13.6% and **24**, 9.9%) to those obtained from **17** (Furusawa, 2006) (Figure 23.12).

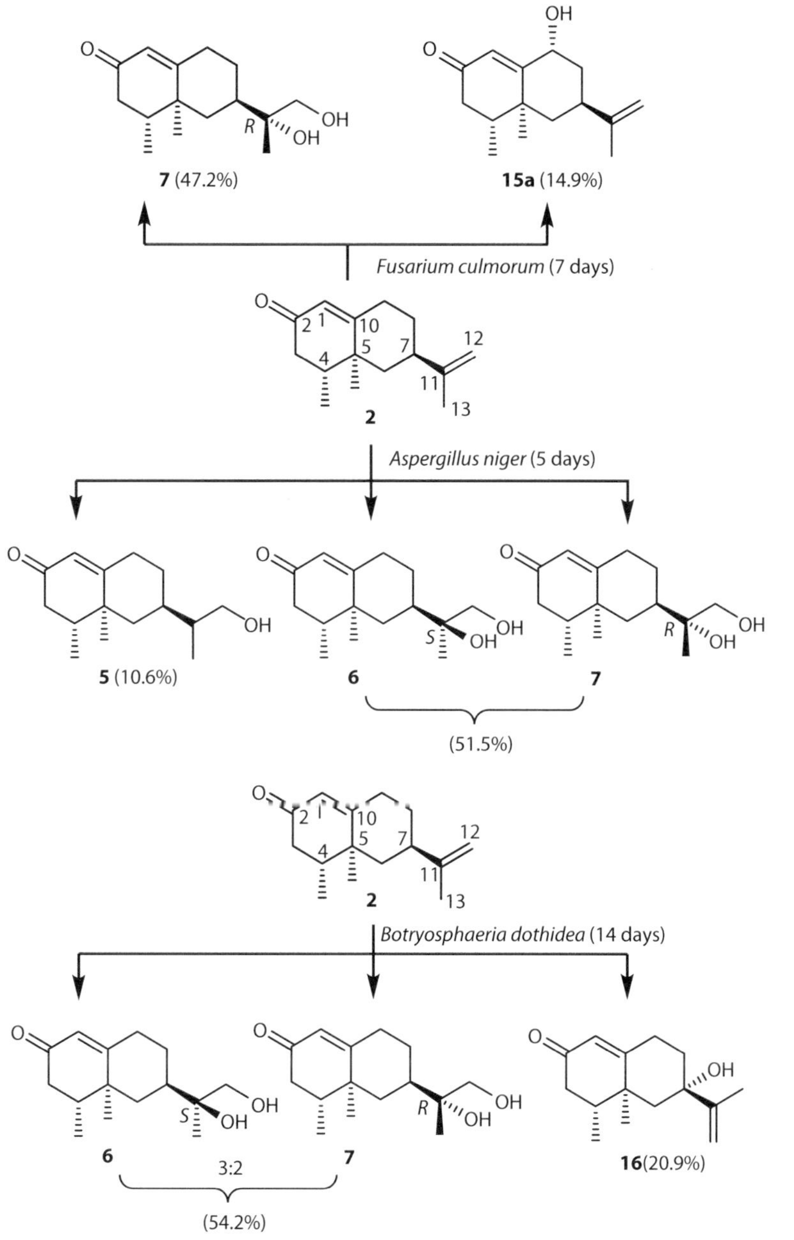

FIGURE 23.8 Biotransformation of nootkatone (**2**) by *Fusarium culmorum*, *Aspergillus niger*, and *Botryosphaeria dothidea*.

8,9-Dehydronootkatone (**25**) was incubated with *A. niger* to give four metabolites, a unique acetonide (**26**, 15.6%), monohydroxylated (**27**, 0.2%), dihydroxylated (**28%**, 69%), and a carboxyl derivative (**29**, 0.8%) (Figure 23.13).

When the same substrate was treated in *Aspergillus sojae* IFO 4389, compound **25** was converted to the different monohydroxylated products (**30**, 15.8%) from that mentioned earlier. *A. cellulosae* is an interesting fungus since it did not give any same products as mentioned earlier; in place, it produced trinorsesquitepene ketone (**31%**, 6%) and nitrogen-containing aromatic compound (**32**) (Furusawa et al., 2003) (Figure 23.14).

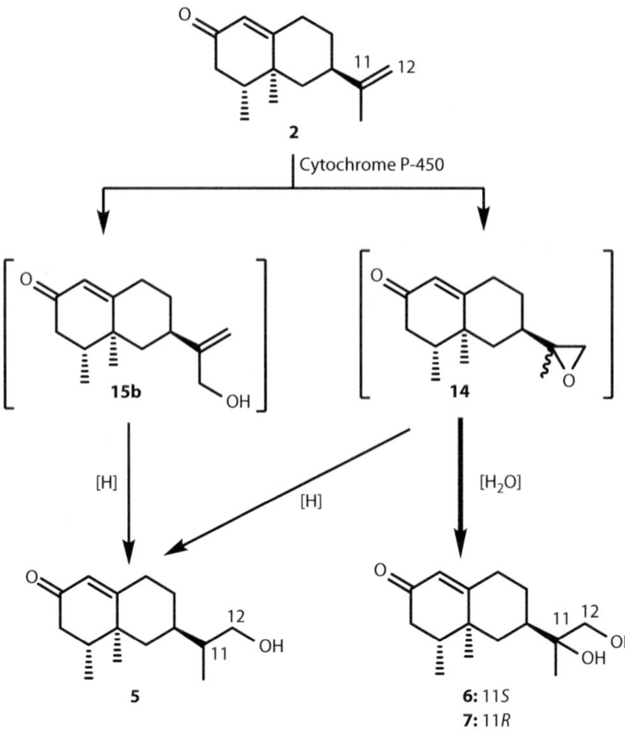

FIGURE 23.9 Possible pathway of biotransformation of valencene (**1**) by cytochrome P-450.

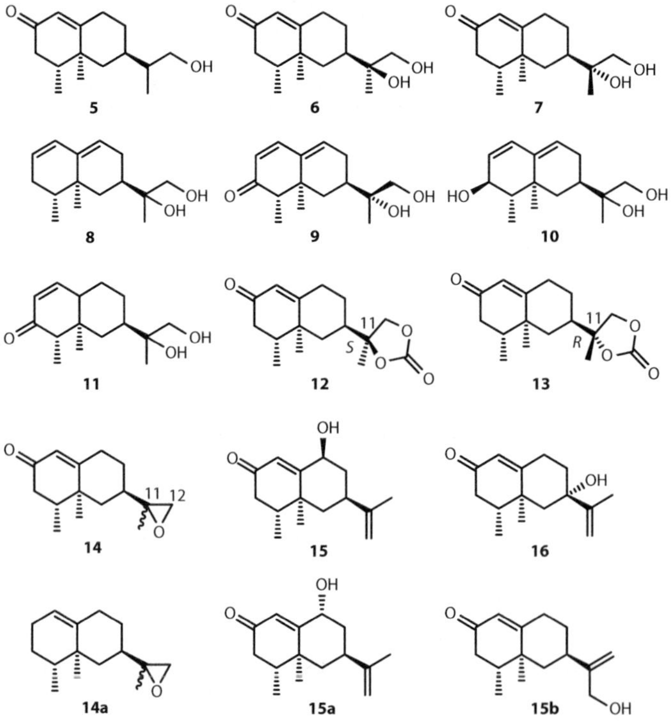

FIGURE 23.10 Metabolites (**5–11**, **14–15b**) from valencene (**1**) and nootkatone (**2**) by various microorganisms.

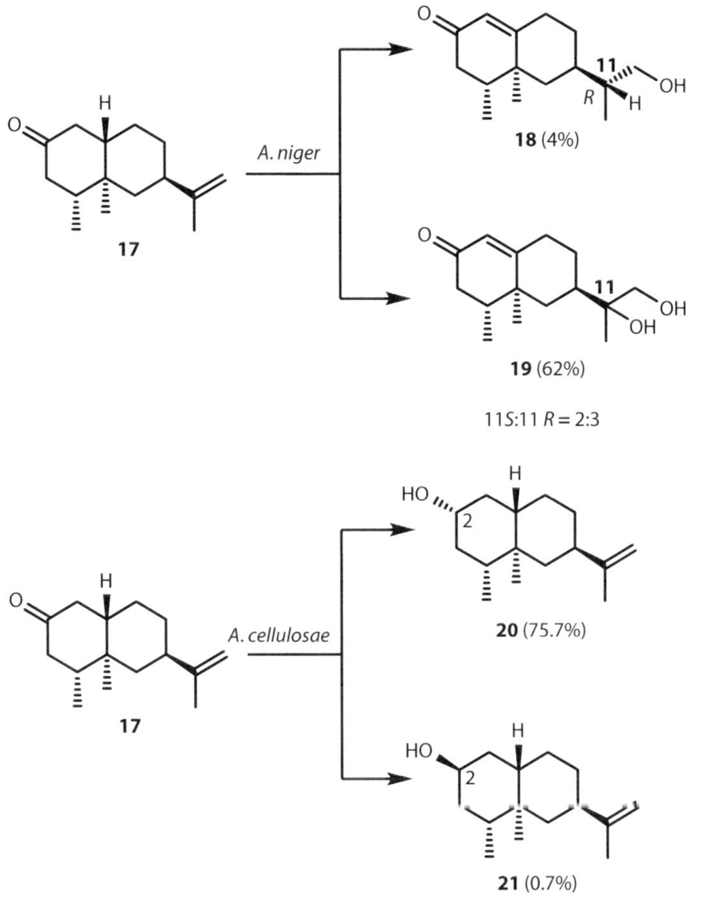

A. niger

17

18 (4%)

19 (62%)

11S:11 R = 2:3

A. cellulosae

17

20 (75.7%)

21 (0.7%)

FIGURE 23.11 Biotransformation of dihydronootkatone (**17**) by *Aspergillus niger* and *Aspergillus cellulosae*.

Mucor species also oxidized compound **25** to give three metabolites, 13-hydroxy-8,9-dehydronootkatone (**33**, 13.2%), an epoxide (**34**, 5.1%), and a diol (**35**, 19.9%) (Furusawa et al., 2003). The same substrate was investigated with cultured suspension cells of the liverwort *Marchantia polymorpha* to afford **33** (Hegazy et al., 2005) (Figure 23.15).

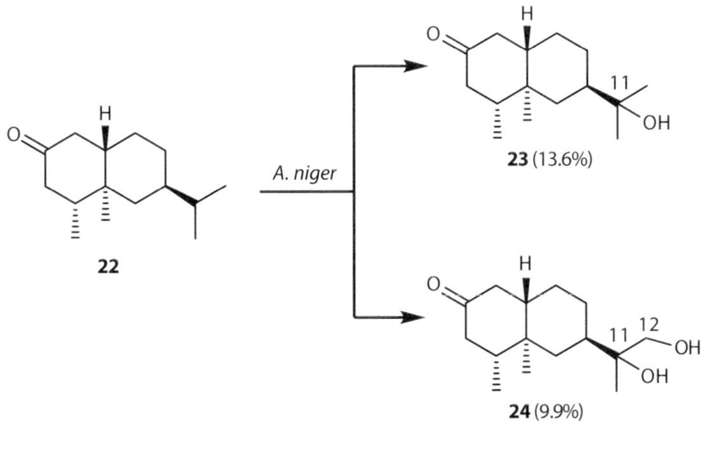

A. niger

22

23 (13.6%)

24 (9.9%)

FIGURE 23.12 Biotransformation of tetrahydronootkatone (**22**) by *Aspergillus niger*.

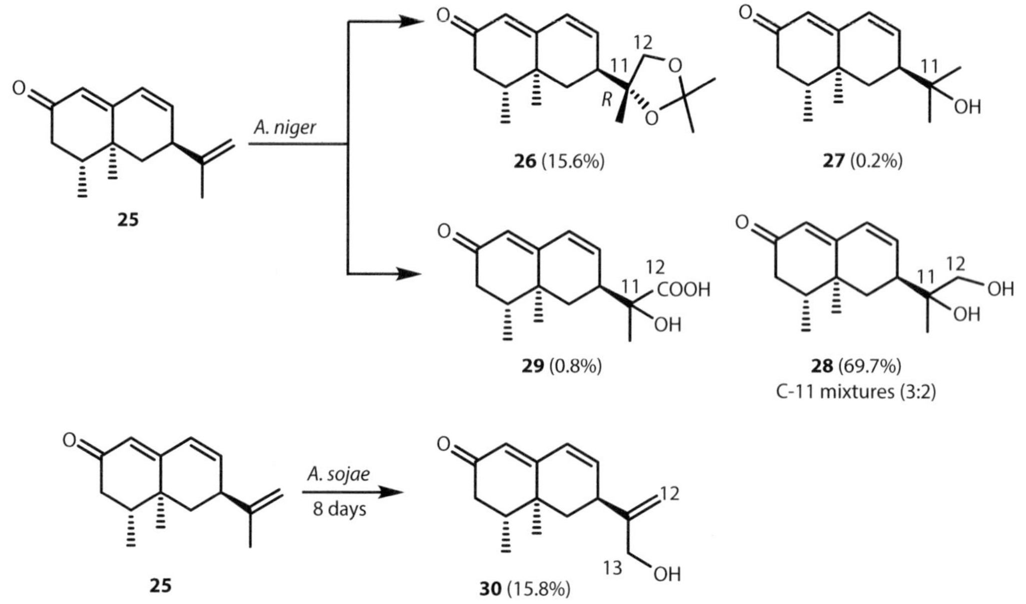

FIGURE 23.13 Biotransformation of 8,9-dehydronootkatone (**25**) by *Aspergillus sojae*.

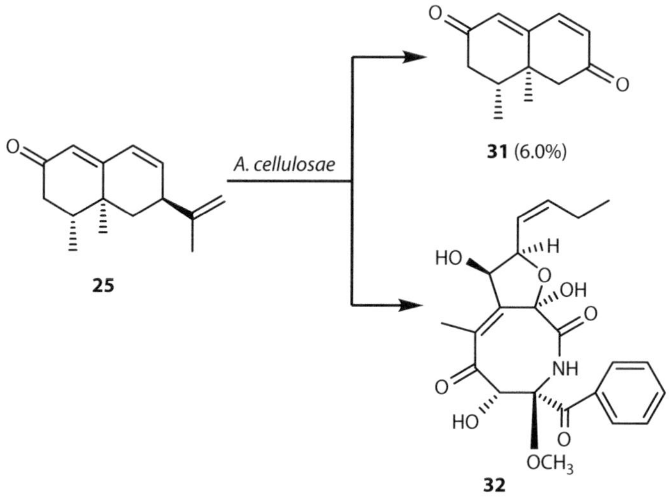

FIGURE 23.14 Biotransformation of 8,9-dehydronootkatone (**25**) by *Aspergillus cellulosae*.

Although *Mucor* species could give nootkatone (**21**) from valencene (**1**), this fungus biotransformed the same substrate (**25**) to the same alcohol (**30**, 13.2%) obtained from the same starting compound (**25**) in *A. sojae*, a new epoxide (**34**, 5.1%), and a diol (**35**, 9.9%).

The metabolites (**3**, **4**, **20**, **21**) inhibited the growth of lettuce stem, and **3** and **4** inhibited germination of the same plant (Hashimoto and Asakawa, 2007).

Valerianol (**35a**), from *Valeriana officinalis* whose dried rhizome is traditionally used for its carminative and sedative properties, was biotransformed by *Mucor plumbeus*, to produce three metabolites, a bridged ether (**35b**), and a triol (**35c**), which might be formed via C1–C10 epoxide, and **35d** arises from double dehydration (Arantes et al., 1999). In this case, allylic oxidative compounds have not been found (Figure 23.16).

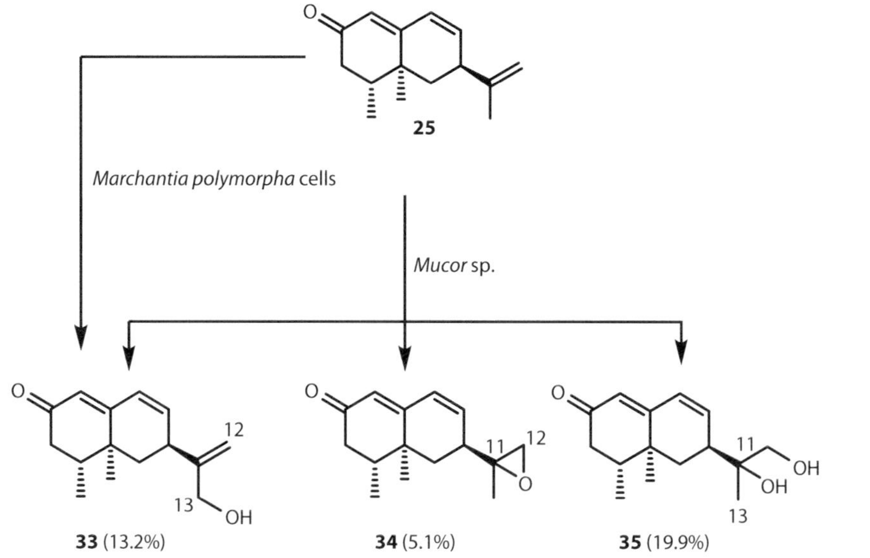

FIGURE 23.15 Biotransformation of 8,9-dehydronootkatone (**25**) by *Marchantia polymorpha* and *Mucor* species.

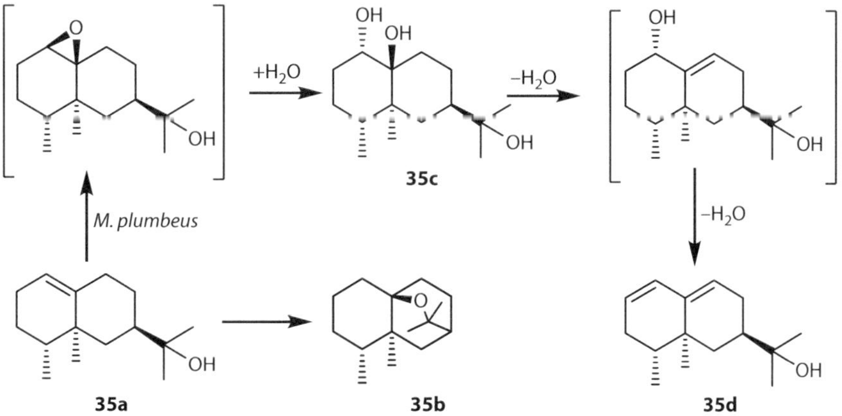

FIGURE 23.16 Biotransformation of valerianol (**35a**) by *Mucor plumbeus.*

23.2.5 Biotransformation of (+)-1(10)-Aristolene (36) from the Crude Drug *Nardostachys chinensis* by *Chlorella fusca*, *Mucor* Species, and *Aspergillus niger*

The structure of sesquiterpenoid (+)-1(10)-aristolene (=calarene) (**36**) from the crude drug *Nardostachys chinensis* was similar to that of nootkatone. 2-Oxo-1(10)-aristolene (**38**) shows antimelanin-inducing activity and excellent citrus fragrance. On the other hand, the enantiomer (**37**) of **36** and (+)-aristolone (**41**) were also found in the liverworts as the natural products. In order to obtain compound **38** and its analogues, compound **36** was incubated with *Chlorella fusca* var. *vacuolata* IAMC-28, *Mucor* species, and *A. niger* (Furusawa et al., 2006a) (Figure 23.17).

C. fusca was inoculated and cultivated stationary in Noro medium (pH 8.0) at 25°C for 7 days, and (+)-1(10)-aristolene (**36**) (20 mg/50 mL) was added to the medium and further incubated for 10–14 days and cultivated stationary under illumination (pH 8.0) at 25°C for 7 days to afford

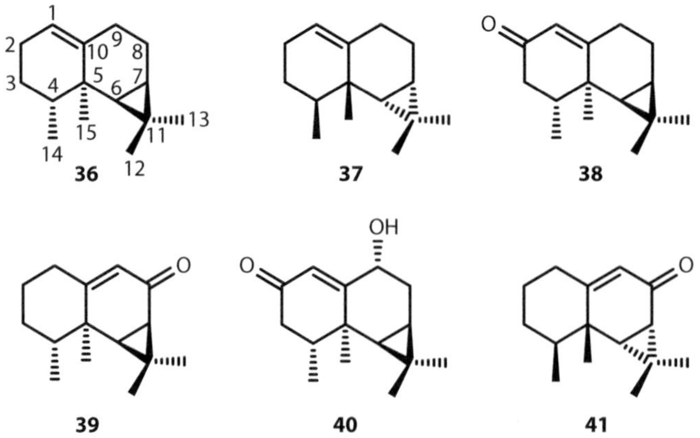

FIGURE 23.17 Naturally occurring aristolane sesquiterpenoids.

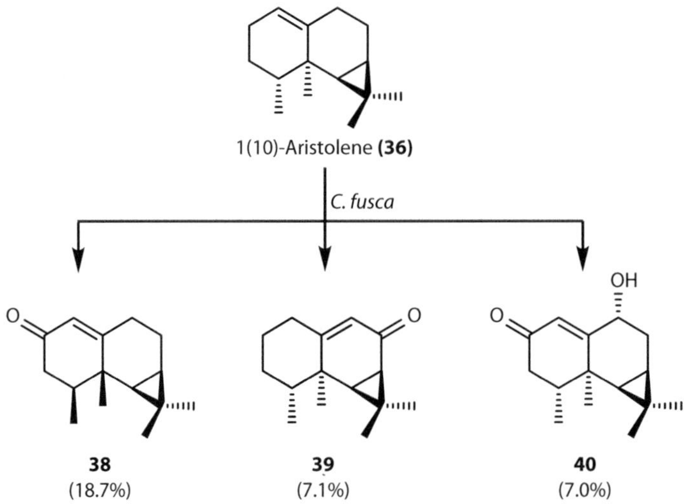

FIGURE 23.18 Biotransformation of 1(10)-aristolene (**36**) by *Chlorella fusca*.

1(10)-aristolen-2-one (**38**, 18.7%), (−)-aristolone (**39**, 7.1%), and 9-hydroxy-1(10)-aristolen-2-one (**40**). Compounds **38** and **40** were found in *Aristolochia* species (Figure 23.18).

Mucor species was inoculated and cultivated rotatory (100 rpm) in Czapek-peptone medium (pH 7.0) at 30°C for 7 days. (+)-1(10)-Aristolene (**36**) (100 mg/200 mL) was added to the medium and further for 7 days. The crude metabolites contained **38** (0.9%) and **39** (0.7%) as very minor products (Figure 23.19).

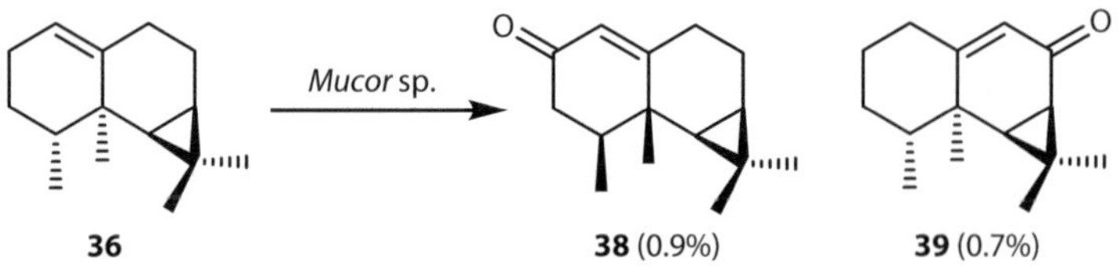

FIGURE 23.19 Biotransformation of 1(10)-aristolene (**36**) by *Mucor* species.

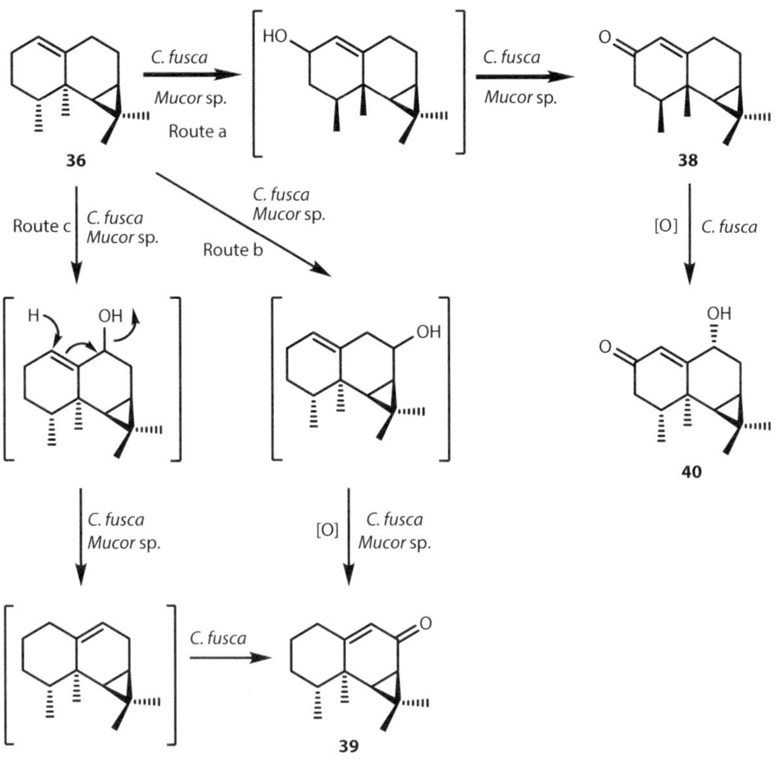

FIGURE 23.20 Possible pathway of biotransformation of 1(10)-aristolene (**36**) by *Chlorella fusca* and *Mucor* species.

Although *Mucor* species produced a large amount of nootkatone (**2**) from valencene (**1**), however, only poor yield of similar products as those from valencene (**1**) was seen in the biotransformation of tricyclic substrate (**36**). Possible biogenetic pathway of (+)-1(10)-aristolene (**36**) is shown in Figure 23.20.

A. niger was inoculated and cultivated rotatory (100 rpm) in Czapek-peptone medium (pH 7.0) at 30°C for 3 days. (+)-1(10)-Aristolene (**36**) (100 mg/200 mL) was added to the medium and further for 7 days. From the crude metabolites, four new metabolic products **42**, 1.3%; **43**, 3.2%; **44**, 0.98%; and **45**, 2.8% were obtained in very poor yields (Figure 23.21). Possible metabolic pathways of **36** by *A. niger* are shown in Figure 23.22.

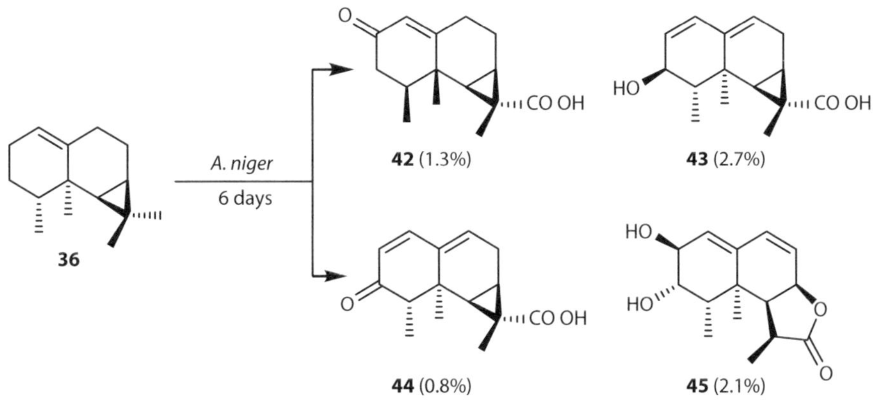

FIGURE 23.21 Biotransformation of 1(10)-aristolene (**36**) by *Aspergillus niger*.

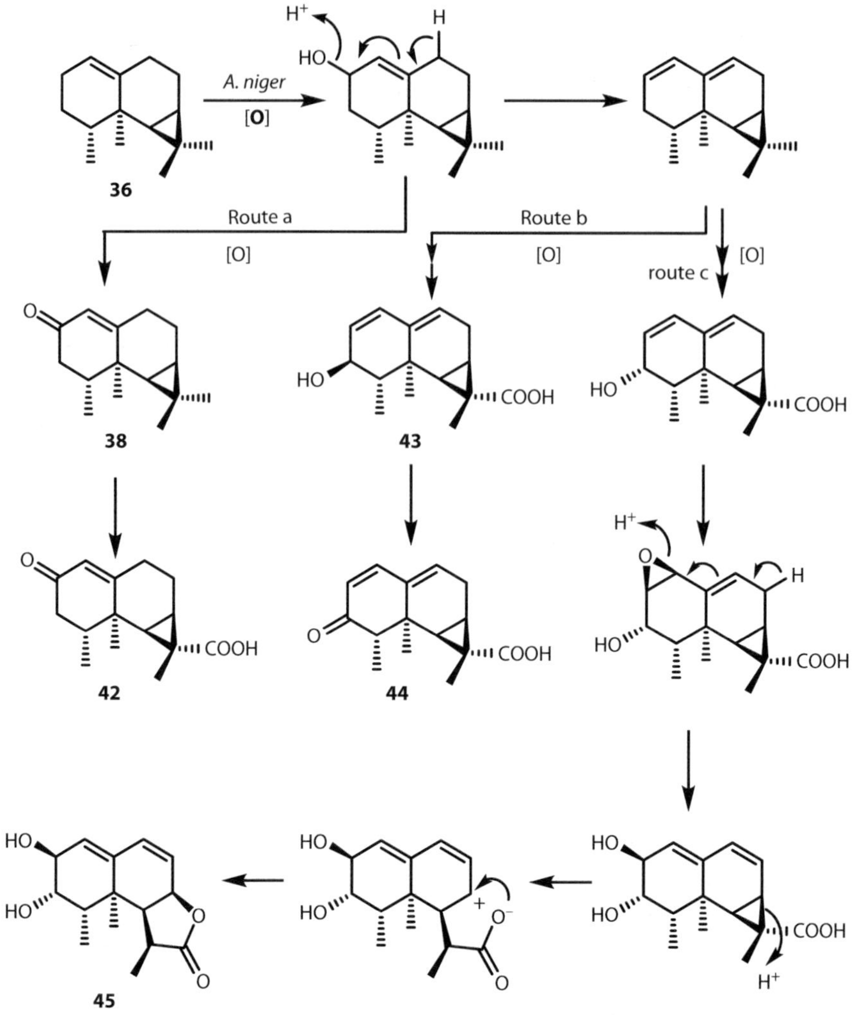

FIGURE 23.22 Possible pathway of biotransformation of 1(10)-aristolene (**36**) by *Aspergillus niger*.

Commercially available (+)-1(10)-aristolene (**36**) was treated with *Diplodia gossypina* and *Bacillus megaterium*. Both microorganisms converted **36**–**4** (**46**–**49**; 0.8%, 1.1%, 0.16%, 0.38%) and six metabolites (**40**, **50**–**55**; 0.75%, 1.0%, 1.0%, 2.0%, 1.1%, 0.5%, 0.87%), together with **40** (0.75%), respectively (Abraham et al., 1992) (Figure 23.23).

It is noteworthy that *Chlorella* and *Mucor* species introduce hydroxyl group at C2 of the substrate (**36**) as seen in the biotransformation of valencene (**1**), while *D. gossypina* and *B. megaterium* oxidize C2, C8, C9, and/or 1,1-dimethyl group on a cyclopropane ring. *A. niger* oxidizes not only C2 but also stereoselectively oxidized one of the gem-dimethyl groups on cyclopropane ring. Stereoselective oxidation of one of gem-dimethyl of cyclopropane and cyclobutane derivatives is observed in biotransformation using mammals (see later).

23.2.6 Biotransformation of Various Sesquiterpenoids by Microorganisms

Aromadendrene-type sesquiterpenoids have been found not only in higher plants but also in liverworts and marine sources. Three aromadendrenes (**56**, **57**, **58**) were biotransformed by *D. gossypina*, *B. megaterium*, and *Mycobacterium smegmatis* (Abraham et al., 1992). Aromadendrene (**56**) (800 mg)

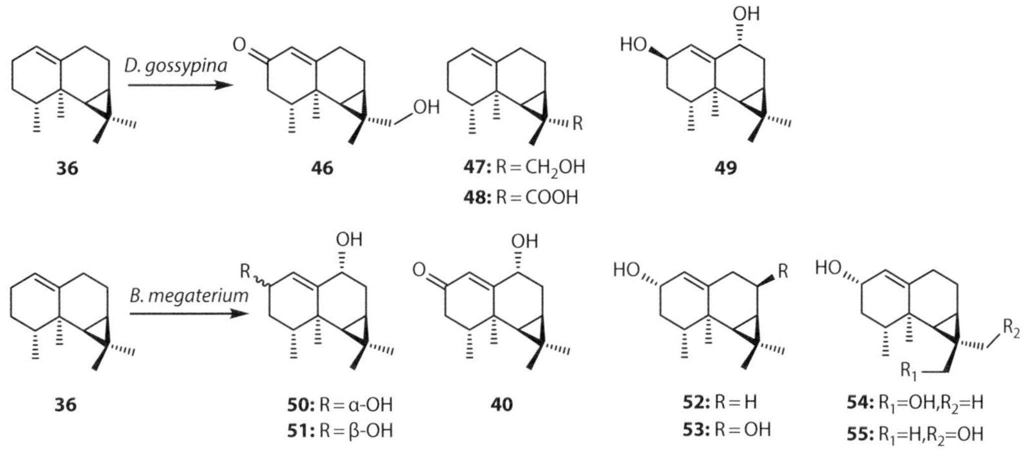

FIGURE 23.23 Biotransformation of 1(10)-aristolene (**36**) by *Diplodia gossypina* and *Bacillus megaterium*.

was converted by *B. megaterium* to afford a diol (**59**) and a triol (**60**) of which **59** (7 mg) was the major product. The triol (**60**) was also obtained from the metabolite of (+)-(1*R*)-aromadendrene (**56**) by the plant pathogen *Glomerella cingulata* (Miyazawa et al., 1995d). *allo*-Aromadendrene (**57**) (1.2 g) was also treated in *M. smegmatis* to afford **61** (10 mg) (Abraham et al., 1992) (Figure 23.24).

The same substrate was also incubated with *G. cingulata* to afford C10 epimeric triol (**62**) (Miyazawa et al., 1995d). Globulol (**58**) (400 mg) was treated in *M. smegmatis* to give only a carboxylic acid (**63**) (210 mg). The same substrate (**58**) (1 g) was treated in *D. gossypina* and *B. megaterium* to give two diols, **64** (182 mg) and **65**, and a triol (**66**) from the former and **67**–**69** from the latter organism among which **64** (60 mg) was predominant (Abraham et al., 1992). *G. cingulata* and

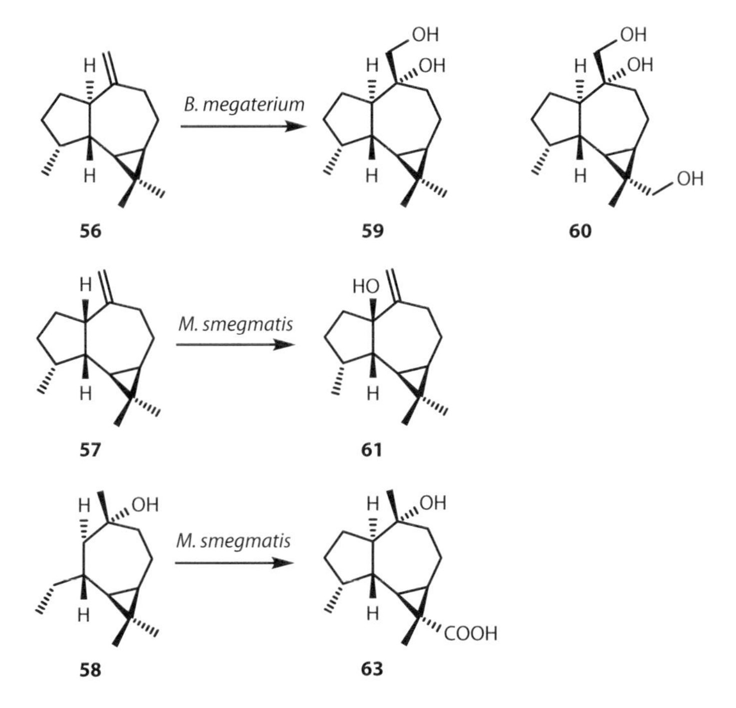

FIGURE 23.24 Biotransformation of aromadendrene (**56**), allo-aromadendrene (**57**), and globulol (**58**) by *Bacillus megaterium* and *Mycobacterium smegmatis*.

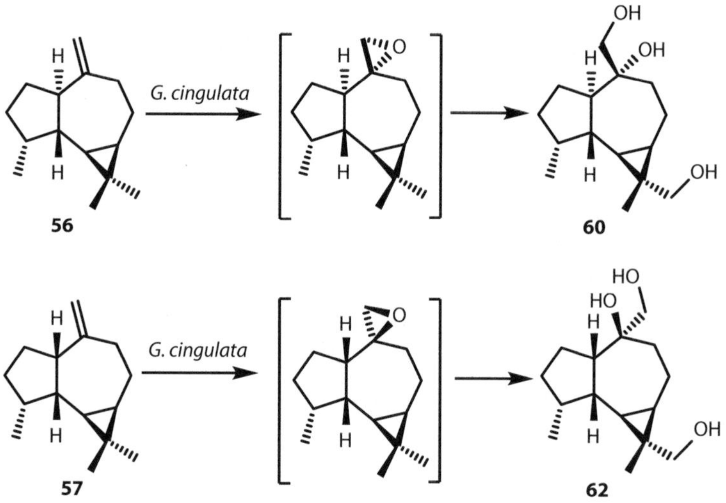

FIGURE 23.25 Biotransformation of aromadendrene (**56**) and allo-aromadendrene (**57**) by *Glomerella cingulata*.

Botrytis cinerea also bioconverted globulol (**58**) to diol (**64**) regio- and stereoselectively (Miyazawa et al., 1994) (Figures 23.25 and 23.26).

Globulol (**58**) (1.5 g) and 10-epiglobulol (**70**) (1.2 mL) were separately incubated with *Cephalosporium aphidicola* in shake culture for 6 days to give the same diol **64** (780 mg) as obtained from the same substrate by *B. megaterium* mentioned earlier and **71** (720 mg) (Hanson et al., 1994). *A. niger* also converted globulol (**58**) and epiglobulol (**70**) to a diol (**64**) and three 13-hydroxylated globulol (**71, 72, 74**) and 4α-hydroxylated product (**73**). The epimerization at C4 is very rare example (Hayashi et al., 1998).

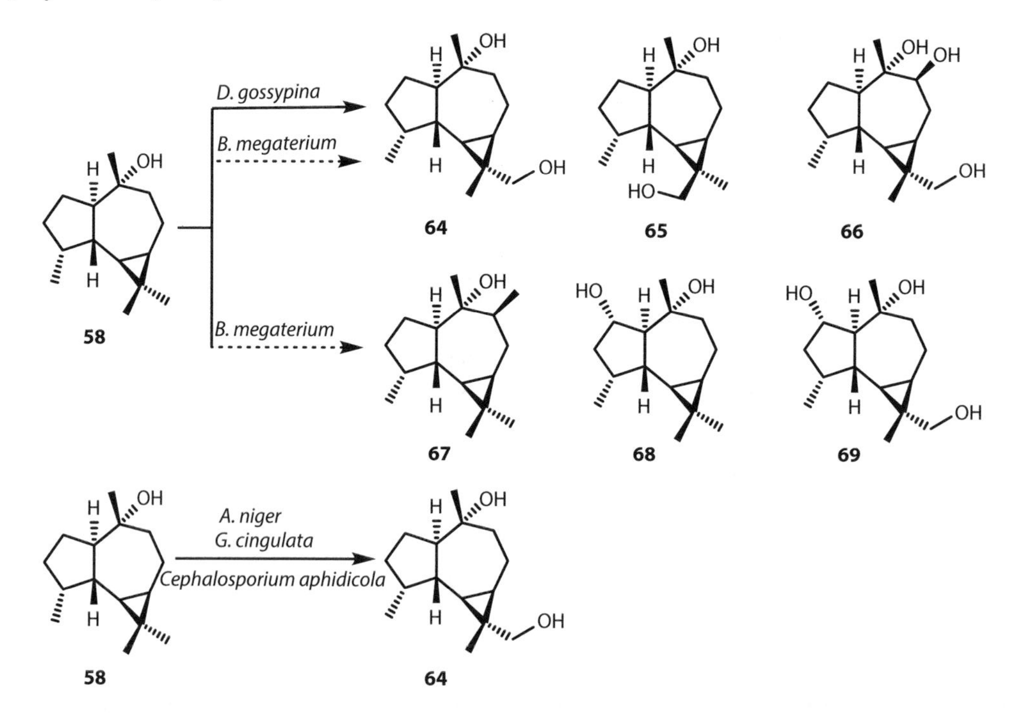

FIGURE 23.26 Biotransformation of globulol (**58**) by various microorganisms.

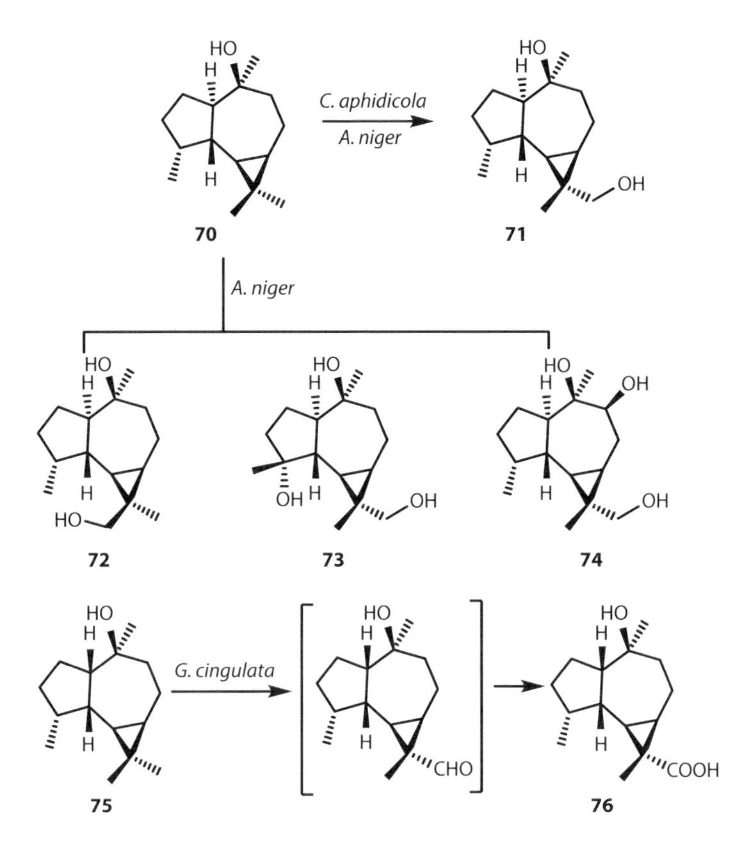

FIGURE 23.27 Biotransformation of 10-epi-globulol (**70**) and ledol (**75**) by *Cephalosporium aphidicola*, *Aspergillus niger*, and *Glomerella cingulata*.

Ledol (**75**), an epimer at C1 of globulol, was incubated with *G. cingulata* to afford C13 carboxylic acid (**76**) (Miyazawa et al., 1994) (Figure 23.27).

Squamulosone (**77**), aromadendr-1(10)-en-9-one isolated from *Hyptis verticillata* (Labiatae), was reduced chemically to give **78**–**82**, which were incubated with the fungus *Curvularia lunata* in two different growth media (Figure 23.28).

From **78**, two metabolites **80** and **83** were obtained. Compound **79** and **80** were metabolized to give ketone **81** as the sole product and **78** and **83**, respectively. From compound **81**, two metabolites, **79** and **84**, were obtained (Figure 23.29). From the metabolite of the substrate (**82**), five products (**84**–**88**) were isolated (Collins et al., 2002a) (Figure 23.30).

Squamulosone (**77**) was treated in the fungus *M. plumbeus* ATCC 4740 to give not only cyclopentanol derivatives (**89**, **90**) but also C12-hydroxylated products (**91**–**93**) (Collins et al., 2002b) (Figure 23.31).

Spathulenol (**94**), which is found in many essential oils, was fed by *A. niger* to give a diol (**95**) (Higuchi et al., 2001). *ent*-10β-Hydroxycyclocolorenone (**96**) and myli-4(15)-en-9-one (**96a**) isolated from the liverwort *Mylia taylorii* were incubated with *A. niger* IFO 4407 to give C10 epimeric product (**97**) (Hayashi et al., 1999) and 12-hydroxylated product (**96b**), respectively (Nozaki et al., 1996) (Figures 23.32 and 23.33).

(+)-*ent*-Cyclocolorenone (**98**) $[\alpha]_D - 405°$ (c = 8.8, EtOH), one of the major compounds isolated from the liverwort *Plagiochila sciophila* (Asakawa, 1982, 1995), was treated by *A. niger* to afford three metabolites, 9-hydroxycyclocolorenone (**99**, 15.9%) 12-hydroxy-(+)-cyclocolorenone (**100**, 8.9%) and a unique cyclopropane-cleaved metabolite, 6β-hydroxy-4,11-guaiadien-3-one (**101**, 35.9%), and 6β,7β-dihydroxy-4,11-guaiadien-3-one (**102**, trace), of which **101** was the major component.

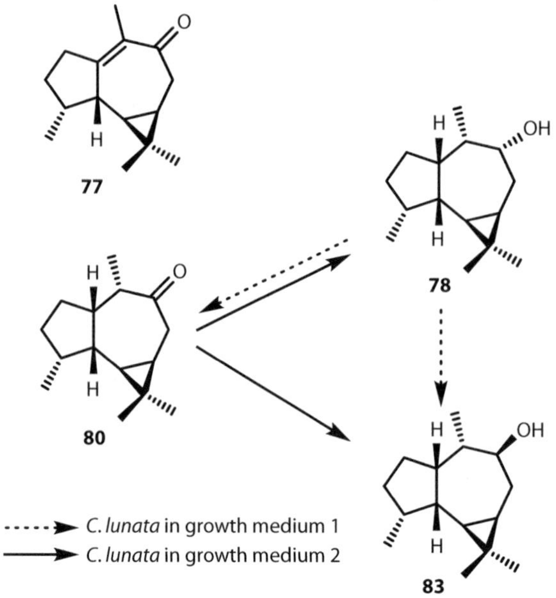

FIGURE 23.28 Biotransformation of aromadendra-9-one (**80**) by *Curvularia lunata*.

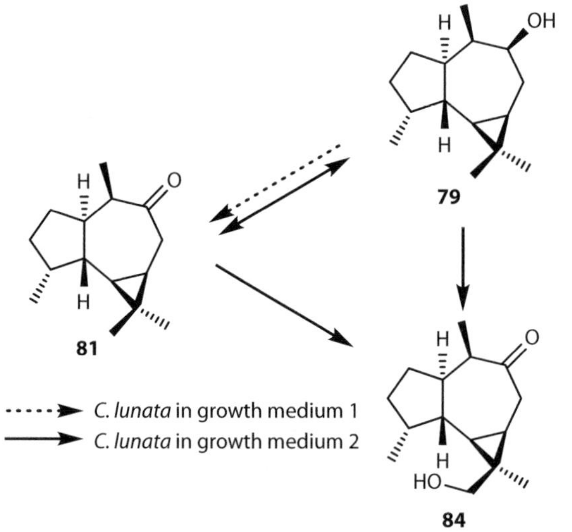

FIGURE 23.29 Biotransformation of 10-epi-aromadendra-9-one (**81**) by *Curvularia lunata*.

The enantiomer (**103**) $[\alpha]_D$ + 402° (c = 8.8, EtOH) of **98** isolated from *Solidago altissima* was biotransformed by the same organism to give 13-hydroxycycolorenone (**103a**, 65.5%), the enantiomer of **100**, 1β,13-dihydroxycyclocolorenone (**103b**, 5.0%), and its C11 epimer (**103c**) (Furusawa et al., 2005c, 2006a). It is noteworthy that no cyclopropane-cleaved compounds from **103** have been detected in the crude metabolites even in GC-MS analysis (Figure 23.34).

Plagiochiline A (**104**) that shows potent insect antifeedant, cytotoxicity, and piscicidal activity is very pungent, 2,3-secoaromadendrane sesquiterpenoids having 1,1-dimethyl cyclopropane ring, isolated from the liverwort *Plagiochila fruticosa*. Plagiochilide (**105**) is the major component of this liverwort. In order to get more pungent component, the lactone (**105**, 101 mg) was incubated with *A. niger* to give two metabolites: **106** (32.5%) and **107** (9.7%). Compound **105** was incubated

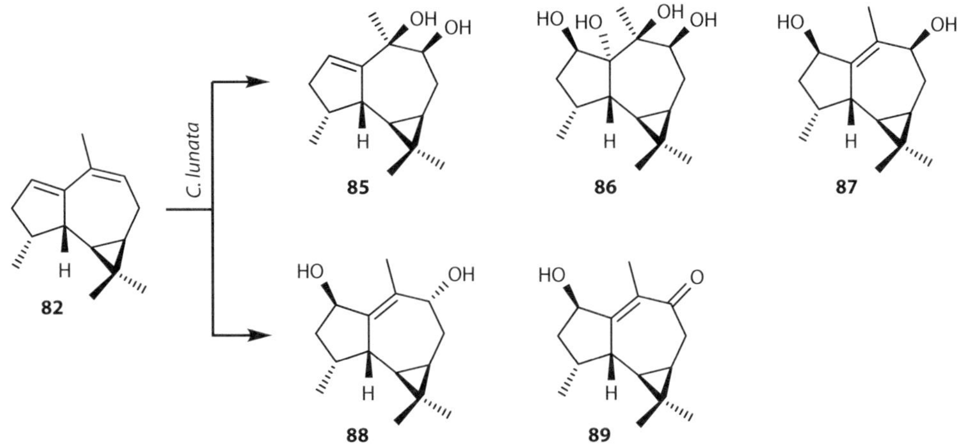

FIGURE 23.30 Biotransformation of aromadendr-1(10),9-diene (**82**) by *Curvularia lunata*.

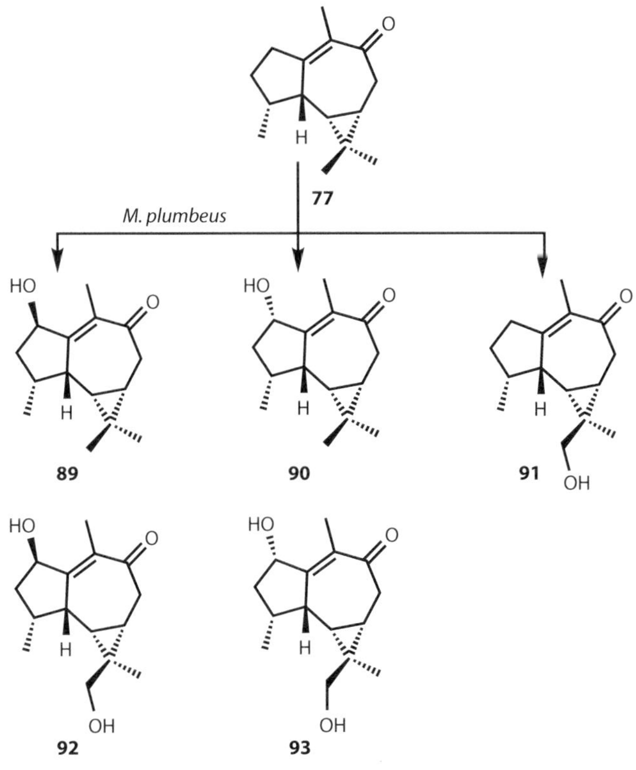

FIGURE 23.31 Biotransformation of squamulosone (**77**) by *Mucor plumbeus*.

in *A. niger* including 1-aminobenzotriazole, the inhibitor of CYP450, to produce only **106**, since this enzyme plays an important role in the formation of carboxylic acid (**107**) from primary alcohol (**106**). Unfortunately, two metabolites show no hot taste (Hashimoto et al., 2003c; Furusawa et al., 2006a,b) (Figure 23.35).

Partheniol, 8α-hydroxybicyclogermacrene (**108**) isolated from *Parthenium argentatum* × *Parthenium tomentosum*, was cultured in the media of *Mucor circinelloides* ATCC 15242 to afford six metabolites, a humulane (**109**), three maaliane (**110**, **112**, **113**), an aromadendrane (**111**), and a tricylohumulane

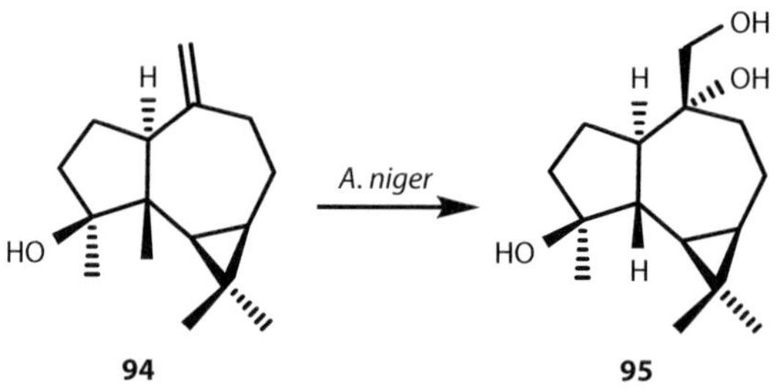

FIGURE 23.32 Biotransformation of spathulenol (**94**) by *Aspergillus niger.*

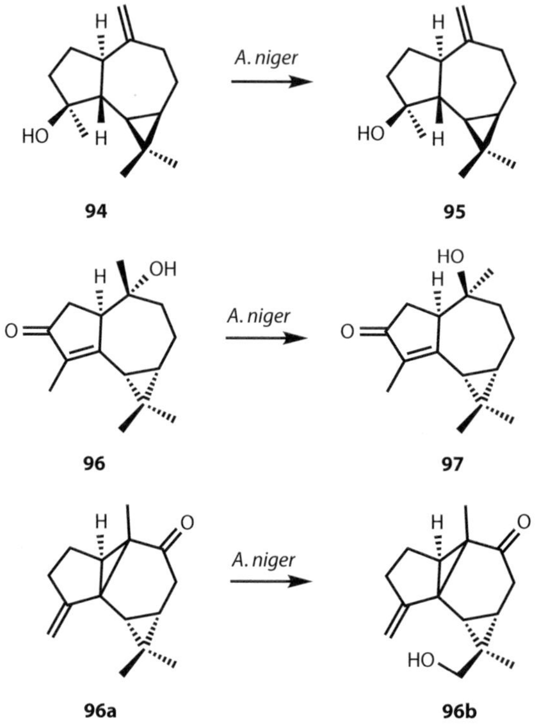

FIGURE 23.33 Biotransformation of spathulenol (**94**), *ent*-10β-hydroxycyclocolorenone (**96**) and myli-4(15)-en-9-one (**96a**) by *Aspergillus niger.*

triol (**114**), the isomer of compound **111**. Compounds **110**, **111**, and **114** were isolated as their acetates (Figure 23.36).

Compounds **110** might originate from the substrate by acidic transannular cyclization since the broth was pH 6.4 just before extraction (Maatooq, 2002a).

The same substrate (**108**) was incubated with the fungus *Calonectria decora* to afford six new metabolites (**108a–108f**). In these reactions, hydroxylation, epoxidation, and transannular cyclization were evidenced (Maatooq, 2002c) (Figure 23.37).

ent-Maaliane-type sesquiterpene alcohol, 1α-hydroxymaaliene (**115**), isolated from the liverwort *M. taylorii*, was treated in *A. niger* to afford two primary alcohols (**116**, **117**) (Morikawa et al., 2000). Such an oxidation pattern of 1,1-dimethyl group on the cyclopropane ring has been found in aromadendrane series as described earlier and mammalian biotransformation of a monoterpene hydrocarbon, D^3-carene (Ishida et al., 1981) (Figure 23.38).

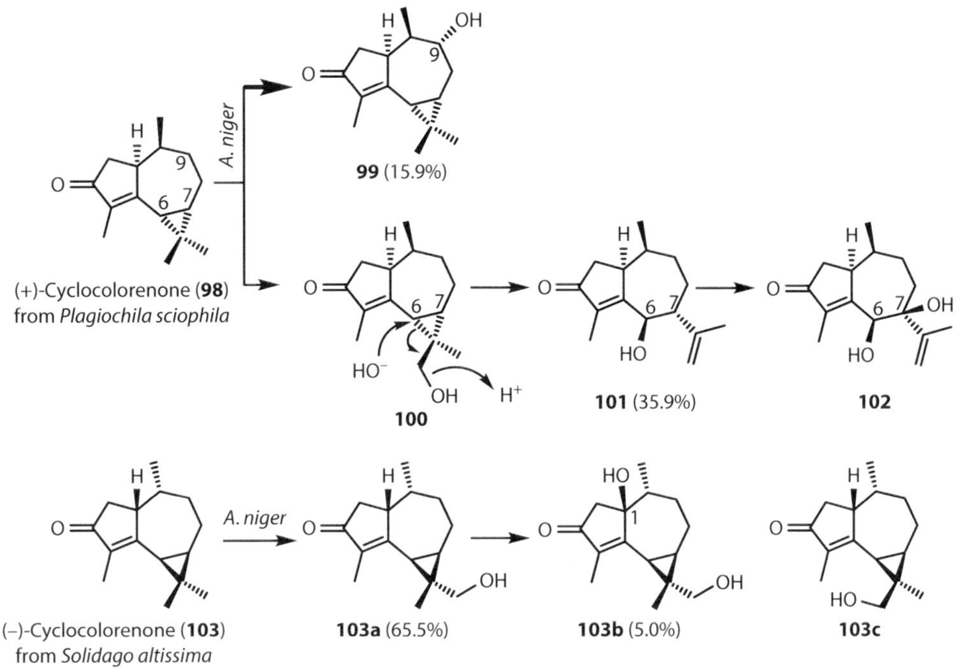

(+)-Cyclocolorenone (**98**) from *Plagiochila sciophila* → *A. niger* → **99** (15.9%); **100** → **101** (35.9%) → **102**

(–)-Cyclocolorenone (**103**) from *Solidago altissima* → *A. niger* → **103a** (65.5%) → **103b** (5.0%); **103c**

FIGURE 23.34 Biotransformation of (+)-cyclocolorenone (**98**) and (–)-cyclocolorenone (**103**) by *Aspergillus niger*.

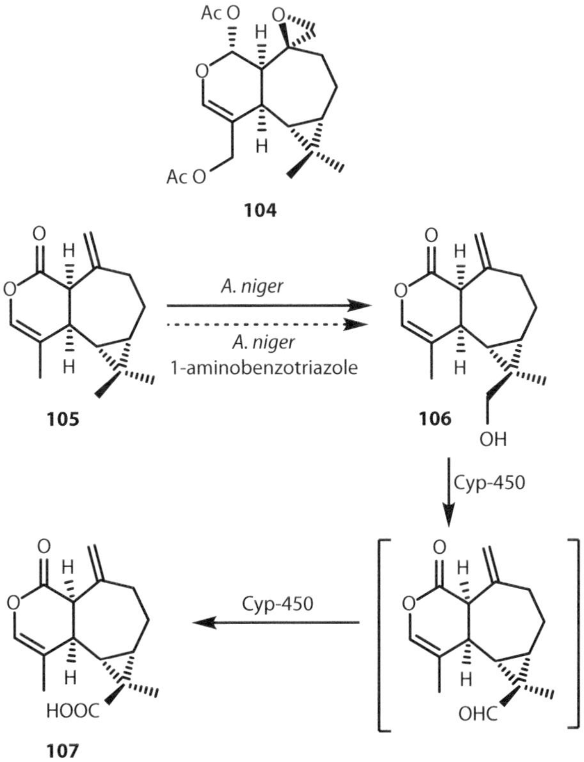

104

105 → *A. niger*; *A. niger* 1-aminobenzotriazole → **106** → Cyp-450 → [aldehyde] → Cyp-450 → **107**

FIGURE 23.35 Biotransformation of plagiochiline C (**104**) by *Aspergillus niger*.

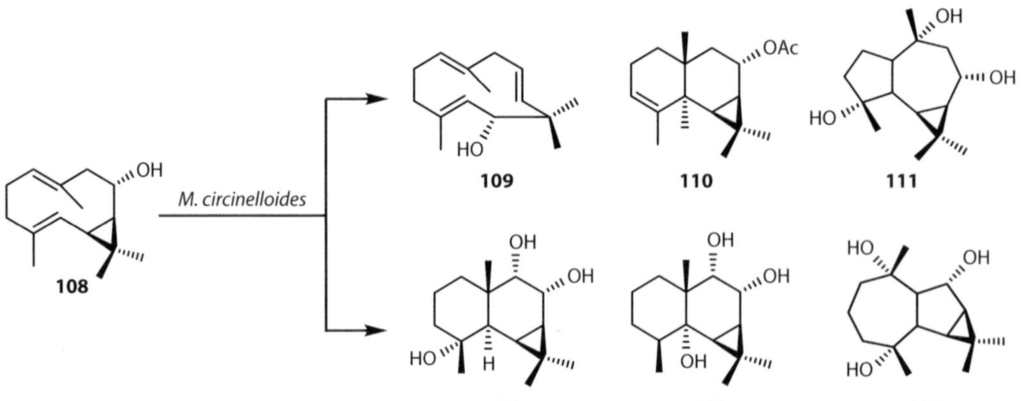

FIGURE 23.36 Biotransformation of 8α-hydroxybicyclogermacrene (**108**) by *Mucor circinelloides.*

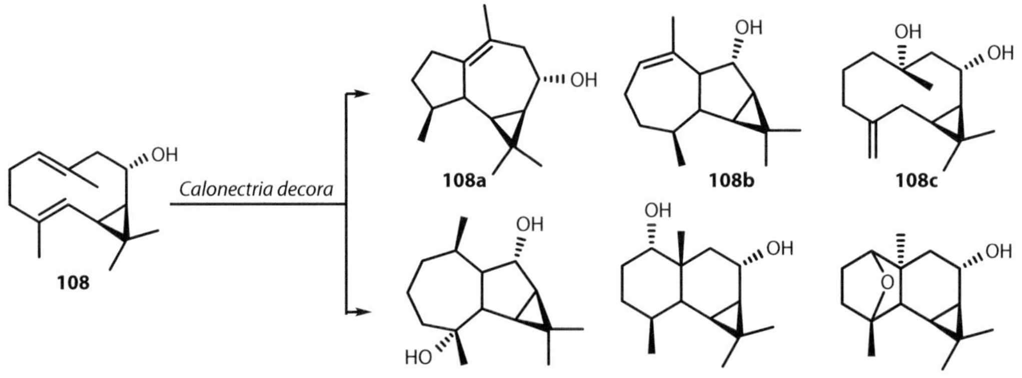

FIGURE 23.37 Biotransformation of 8α-hydroxybicyclogermacrene (**108**) by *Calonectria decora.*

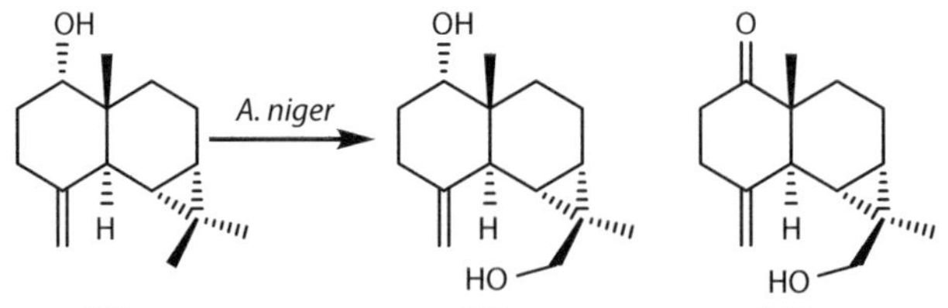

FIGURE 23.38 Biotransformation of 1α-hydroxymaaliene (**115**) *Aspergillus niger.*

9(15)-Africanene (**117a**), a tricyclic sesquiterpene hydrocarbon isolated from marine soft corals of *Simularia* species, was biotransformed by *A. niger* and *Rhizopus oryzae* for 8 days to give 10α-hydroxy (**117b**) and 9α,15-epoxy derivative (**117c**) (Venkateswarlu et al., 1999) (Figure 23.39).

Germacrone (**118**), (+)-germacrone-4,5-epoxide (**119**), and curdione (**120**) isolated from *Curcuma aromatica*, which has been used as crude drug, were incubated with *A. niger.* From compound **119** (700 mg), two naturally occurring metabolites, zedoarondiol (**121**) and isozedoarondiol (**122**), were obtained (Takahashi, 1994). Compound **119** was cultured in callus of *Curcuma zedoaria* and *C. aromatica* to give the same secondary metabolites **121**, **122**, and **124** (Sakui et al., 1988) (Figures 23.40 and 23.41).

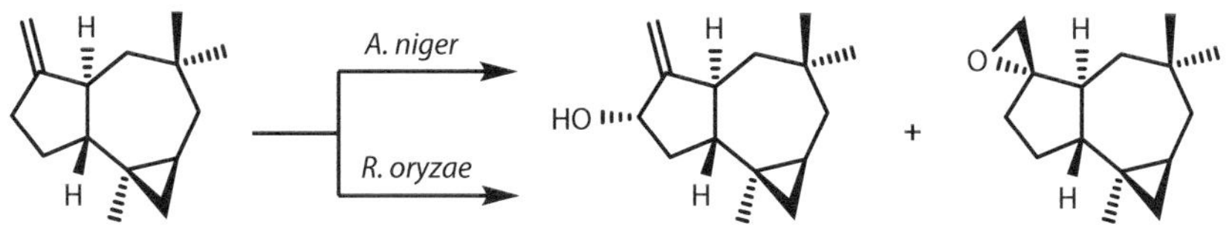

FIGURE 23.39 Biotransformation of 9(15)-africanene (**117a**) by *Aspergillus niger* and *Rhizopus oryzae*.

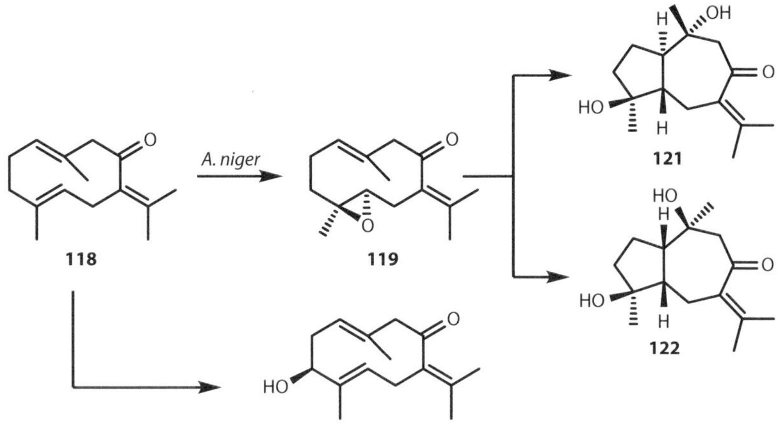

FIGURE 23.40 Biotransformation of germacrone (**118**) by *Aspergillus niger*.

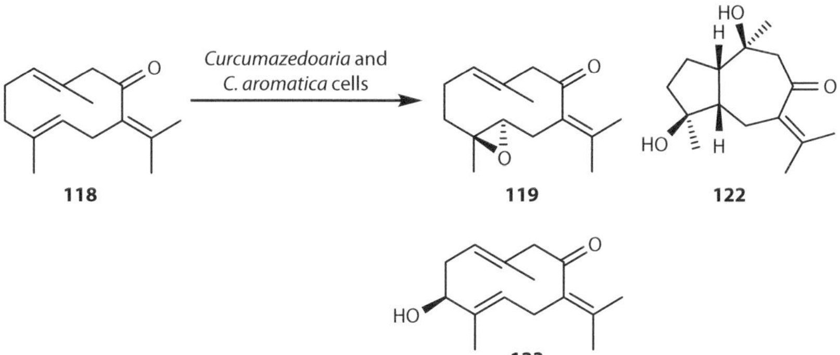

FIGURE 23.41 Biotransformation of germacrone (**118**) by *Curcuma zedoaria* and *Curcuma aromatica* cells.

A. niger biotransformed germacrone (**118**, 3 g) to very unstβ-hydroxygermacrone (**123**) and 4,5-epoxygermacrone (**119**), which was further converted to two guaiane sesquiterpenoids (**121**) and (**122**) through transannular-type reaction (Takahashi, 1994). The same substrate was incubated in the microorganism *Cunninghamella blakesleeana* to afford germacrone-4,5-epoxide (**119**) (Hikino et al., 1971), while the treatment of **118** in the callus of *C. zedoaria* gave four metabolites **121**, **122**, **125**, and **126** (Sakamoto et al., 1994) (Figure 23.42).

The same substrate (**118**) was treated in plant cell cultures of *S. altissima* (Asteraceae) for 10 days to give various hydroxylated products (**121**, **127**, **125**, **128**–**132**) (Sakamoto et al., 1994). Guaiane (**121**) underwent further rearrangement C4–C5, cleavage, and C5–C10 transannular cyclization to the bicyclic hydroxyketone (**128**) and diketone (**129**) (Sakamoto et al., 1994) (Figure 23.43).

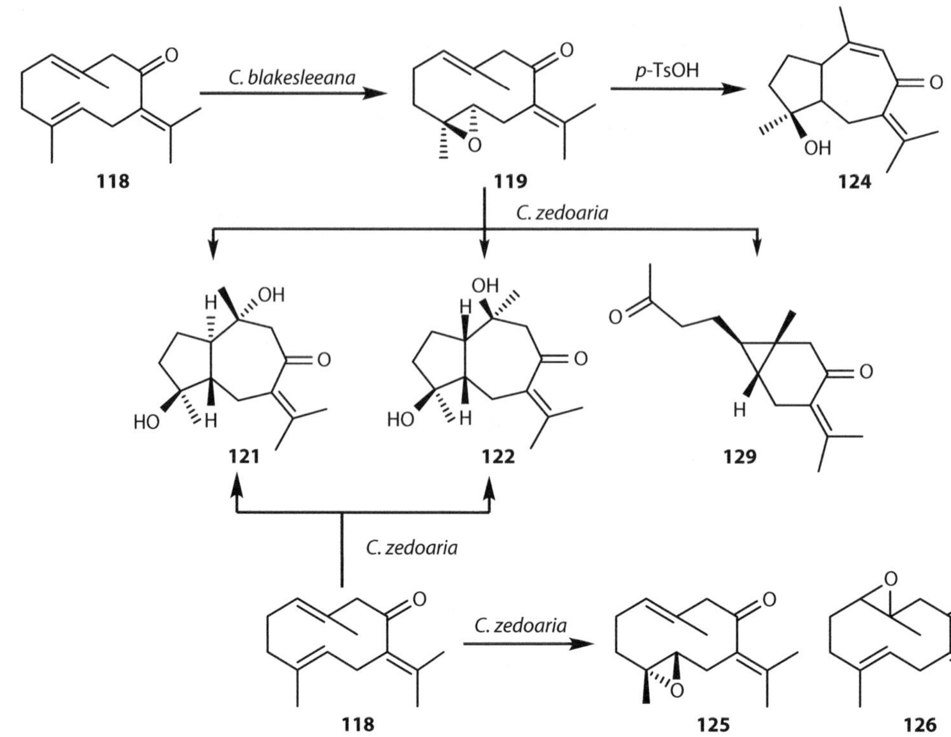

FIGURE 23.42 Biotransformation of germacrone (**118**) by *Cunninghamella blakesleeana* and *Curcuma zedoaria* cells.

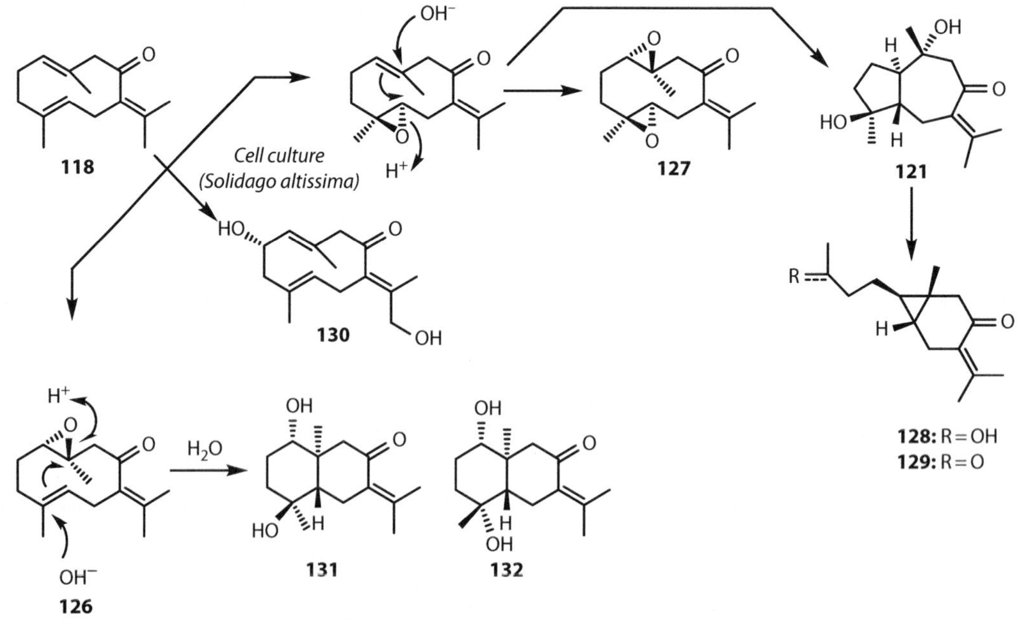

FIGURE 23.43 Biotransformation of germacrone (**118**) by *Solidago altissima* cells.

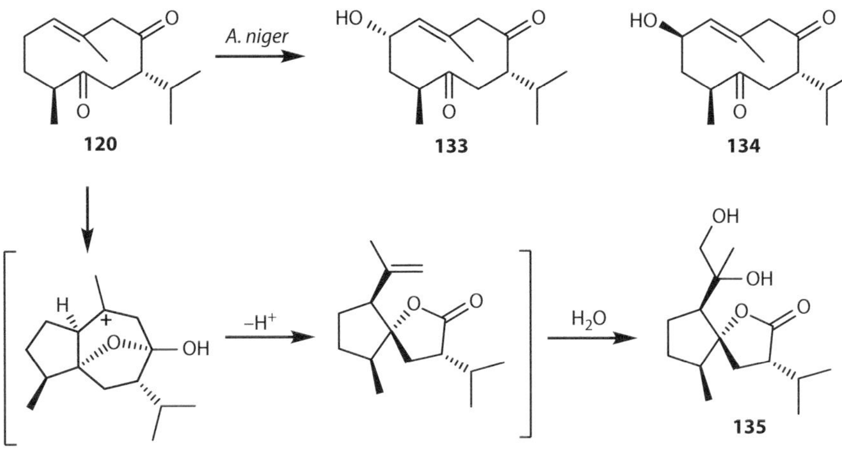

FIGURE 23.44 Biotransformation of curdione (**120**) by *Aspergillus niger.*

Curdione (**120**) was also treated in *A. niger* to afford two allylic alcohols (**133**, **134**) and a spirolactone (**135**). *C. aromatica* and *Curcuma wenyujin* produced spirolactone (**135**), which might be formed from curdione via transannular reaction *in vivo* and was biotransformed to spirolactone diol (**135**) (Asakawa et al., 1991; Sakui et al., 1992) (Figure 23.44).

A. niger also converted shiromodiol diacetate (**136**) isolated from *Neolitsea sericea* to 2β-hydroxy derivative (**137**) (Nozaki et al., 1996) (Figure 23.45).

Twenty strains of filamentous fungi and four species of bacteria were screened initially by thin-layer chromatography for their biotransformation capacity of curdione (**120**). *Mucor spinosus*, *Mucor polymorphosporus*, *Cunninghamella elegans*, and *Penicillium janthinellum* were found to be able to biotransform curdione (**120**) to more polar metabolites. Incubation of curdione with *M. spinosus*, which was most potent strain to produce metabolites, for 4 days using potato medium gave five metabolites (**134**, **134a**–**134d**) among which compounds **134c** and **134d** are new products (Ma et al., 2006) (Figure 23.46).

Many eudesmane-type sesquiterpenoids have been biotransformed by several fungi and various oxygenated metabolites obtained.

β-Selinene (**138**) is ubiquitous sesquiterpene hydrocarbon of seed oil from many species of Apiaceae family, for example, *Cryptotaenia canadensis* var. *japonica*, which is widely used as vegetable for Japanese soup. β-Selinene was biotransformed by plant pathogenic fungus *G. cingulata* to give an epimeric mixtures (1:1) of 1β,11,12-trihydroxy product (**139**) (Miyazawa et al., 1997b). The same substrate was treated in *A. wentii* to give 2α,11,12-trihydroxy derivative (**140**) (Takahashi et al., 2007).

Eudesm-11(13)-en-4,12-diol (**141**) was biotransformed by *A. niger* to give 3β-hydroxy derivative (**142**) (Hayashi et al., 1999).

α-Cyperone (**143**) was fed by *Collectotrichum phomoides* (Lamare and Furstoss, 1990) to afford 11,12-diol (**144**) and 12-manool (**145**) (Higuchi et al., 2001) (Figure 23.47).

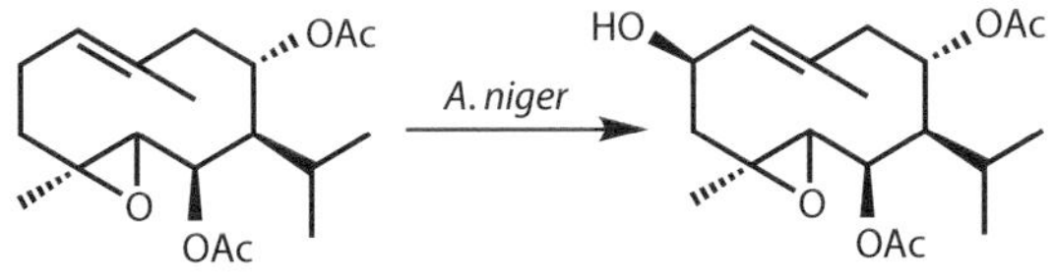

FIGURE 23.45 Biotransformation of shiromodiol diacetate (**136**) by *Aspergillus niger.*

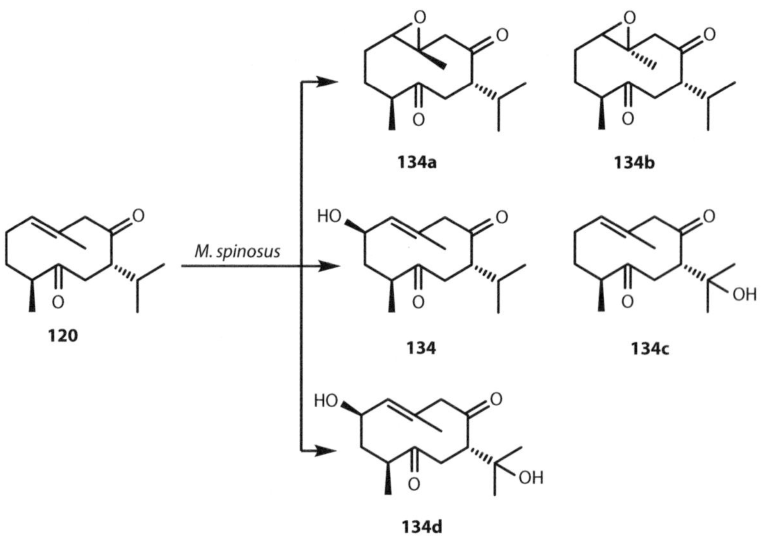

FIGURE 23.46 Biotransformation of curdione (**120**) by *Mucor spinosus.*

The filamentous fungi *Gliocladium roseum* and *Exserohilum halodes* were used as the bioreactors for 4β-hydroxyeudesmane-1,6-dione (**146**) isolated from *Sideritis varoi* subsp. *cuatrecasasii.* The former fungus transformed **146** to 7α-hydroxyl (**147**), 11-hydroxy (**148**), 7α,11-dihydroxy (**149**), 1α,11-dihydroxy (**150**), and 1α,8α-dihydroxy derivatives (**151**), while *E. halodes* gave only 1α-hydoxy product (**152**) (Garcia-Granados et al., 2001) (Figure 23.48).

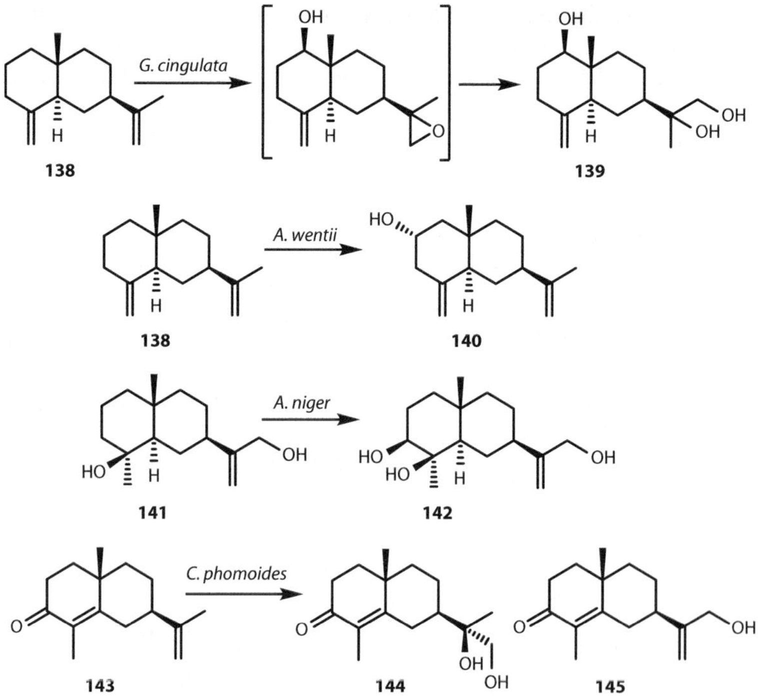

FIGURE 23.47 Biotransformation of eudesmenes (**138**, **141**, **143**) by *Aspergillus wentii*, *Glomerella cingulata*, and *Collectotrium phomoides.*

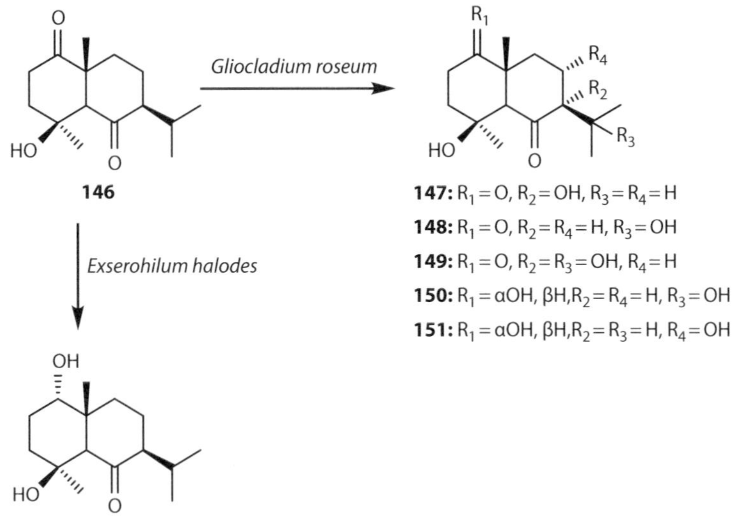

146

147: $R_1=O$, $R_2=OH$, $R_3=R_4=H$

148: $R_1=O$, $R_2=R_4=H$, $R_3=OH$

149: $R_1=O$, $R_2=R_3=OH$, $R_4=H$

150: $R_1=\alpha OH$, βH, $R_2=R_4=H$, $R_3=OH$

151: $R_1=\alpha OH$, βH, $R_2=R_3=H$, $R_4=OH$

152

FIGURE 23.48 Biotransformation of 4β-hydroxy-eudesmane-1,6-dione (**146**) by *Gliocladium roseum* and *Exserohilum halodes*.

Orabi (2000) reported that *Beauveria bassiana* is the most efficient microorganism to metabolize plectanthone (**152a**) among 20 microorganisms, such as *Absidia glauca*, *Aspergillus flavipes*, *B. bassiana*, *Cladosporium resinae*, and *Penicillium frequentans*. The substrate (**152a**) was incubated with *B. bassiana* to give metabolites **152b** (2.1%), **152c** (21.2%), **152d** (2.5%), **152e** (no data), and **152f** (1%) (Figure 23.49).

(–)-α-Eudesmol (**153**) isolated from the liverwort *Porella stephaniana* was treated by *A. cellulosae* and *A. niger* to give 2-hydroxy (**154**) and 2-oxo derivatives (**155**), among which the latter product

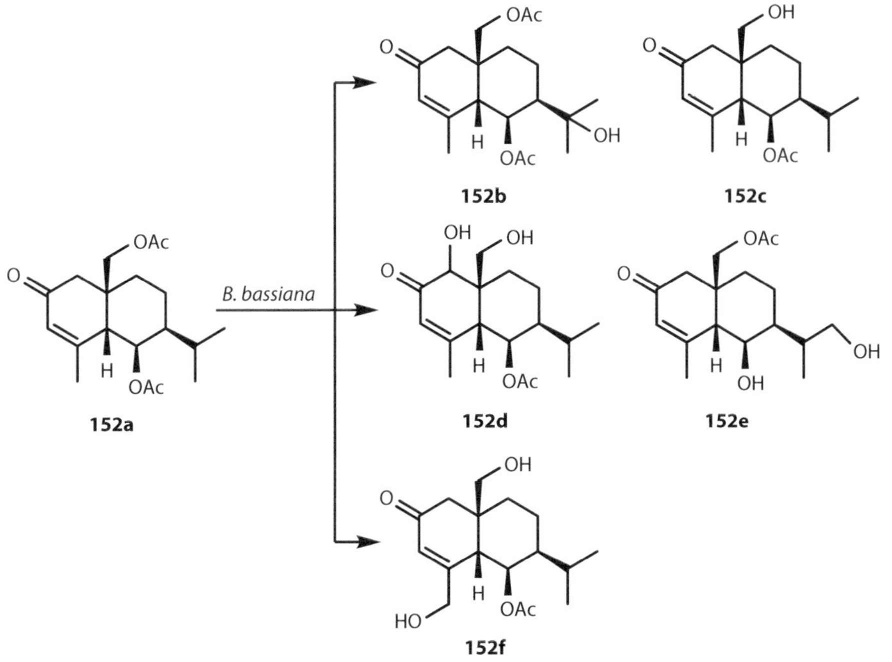

152a 152b 152c 152d 152e 152f

FIGURE 23.49 Biotransformation of eudesmenone (**152a**) by *Beauveria bassiana*.

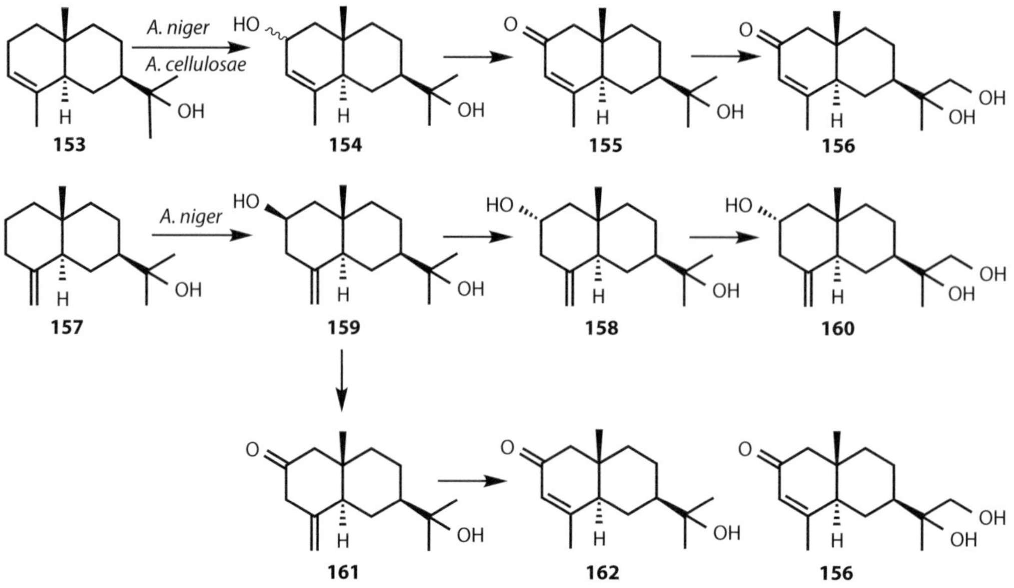

FIGURE 23.50 Biotransformation of α-eudesmol (**153**) and β-eudesmol (**157**) by *Aspergillus niger* and *Aspergillus cellulosae.*

was predominantly obtained. This bioconversion was completely blocked by 1-aminobenzotriazole, CYP450 inhibitor. Compound **155** has been known as natural product, isolated from *Pterocarpus santalinus* (Noma et al., 1996). Biotransformation of α-eudesmol (**153**) isolated from the dried *Atractylodes lancea* was reinvestigated by *A. niger* to give 2-oxo-11,12-dihydro-α-eudesmol (**156**) together with 2-hydroxy- (**154**) and 2-oxo-α-eudesmol (**155**). β-Eudesmol (**157**) was treated in *A. niger*, with the same culture medium, to afford 2α- (**158**) and 2β-hydroxy-α-eudesmol (**159**) and 2α,11,12-trihydroxy-β-eudesmol (**160**) and 2-oxo derivative (**161**), which was further isomerized to compound **162** (Noma et al., 1996, 1997a) (Figure 23.50).

Three new hydroxylated metabolites (**157b–157d**) along with a known **158** and **157e–157 g** were isolated from the biotransformation reaction of a mixture of β- (**157**) and γ-eudesmols (**157a**) by *Gibberella suabinetii.* The metabolites proved a super activity of the hydroxylase, dehydrogenase, and isomerase enzymes. The hydroxylation is a common feature; on the contrary, cyclopropyl ring formation like compound (**158d**) is very rare (Maatooq, 2002b) (Figure 23.51).

A furanosesquiterpene, atractylon (**163**) obtained from *Atractylodes* rhizoma, was treated with the same fungus to yield atractylenolide III (**164**) possessing inhibition of increased vascular permeability in mice induced by acetic acid (Hashimoto et al., 2001a,b).

The biotransformation of sesquiterpene lactones have been carried out by using different microorganisms.

Costunolide (**165**), a very unstable sesquiterpene γ-lactone, from *Saussurea radix*, was treated in *A. niger* to produce three dihydrocostunolides (**166–168**) (Clark and Hufford, 1979). Costunolide is easily converted into eudesmanolides (**169–172**) in diluted acid, thus, **166–168** might be biotransformed after being cyclized in the medium including the microorganisms. If the crude drug including costunolide (**165**) is orally administered, **165** will be easily converted into **169–172** by stomach juice (Figure 23.52).

(+)-Costunolide (**165**), (+)-cnicin (**172a**), and (+)-salonitgenolide (**172b**) were incubated with *Cunninghamella echinulata* and *R. oryzae*.

The former fungus converted compound **165**, to four metabolites, (+)-11β,13-dihydrocostunolide (**165a**), 1β-hydroxyeudesmanolide, (+)-santamarine (**166a**), (+)-reynosin (**166b**), and (+)-1β-hydroxyarbusculin A (**168a**), which might be formed from 1β,10α-epoxide (**166c**). Treatment of **172a** with *C. echinulata* gave (+)-salonitenolide (**172b**) (Barrero et al., 1999) (Figure 23.53).

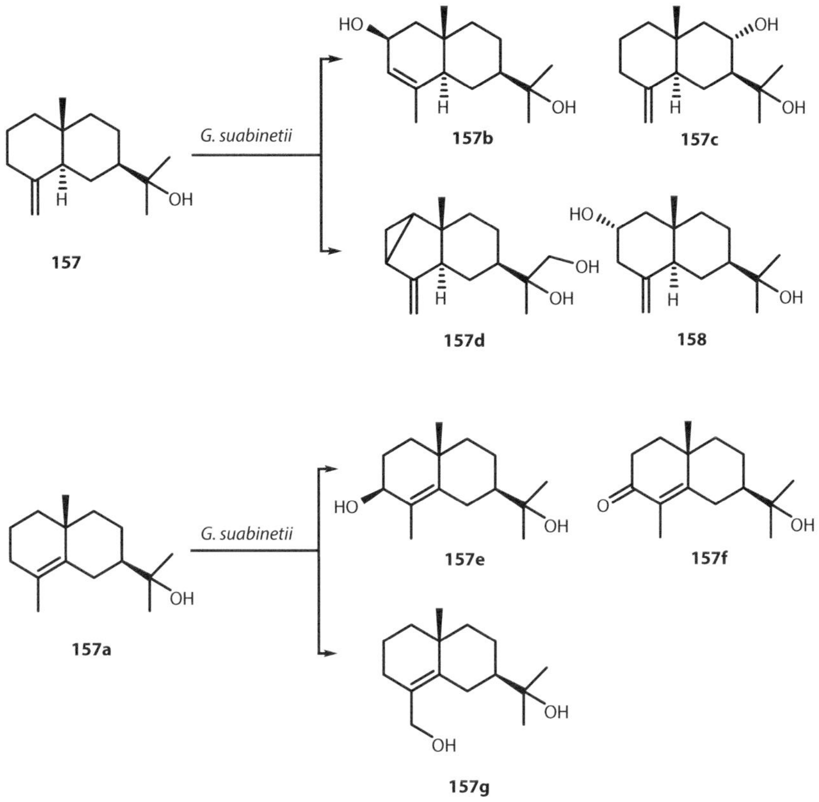
G. suabinetii
157
157b
157c
157d
158
G. suabinetii
157a
157e
157f
157g

FIGURE 23.51 Biotransformation of β-eudesmol (**157**) and γ-eudesmol (**157a**) by *Gibberella suabinetii*.

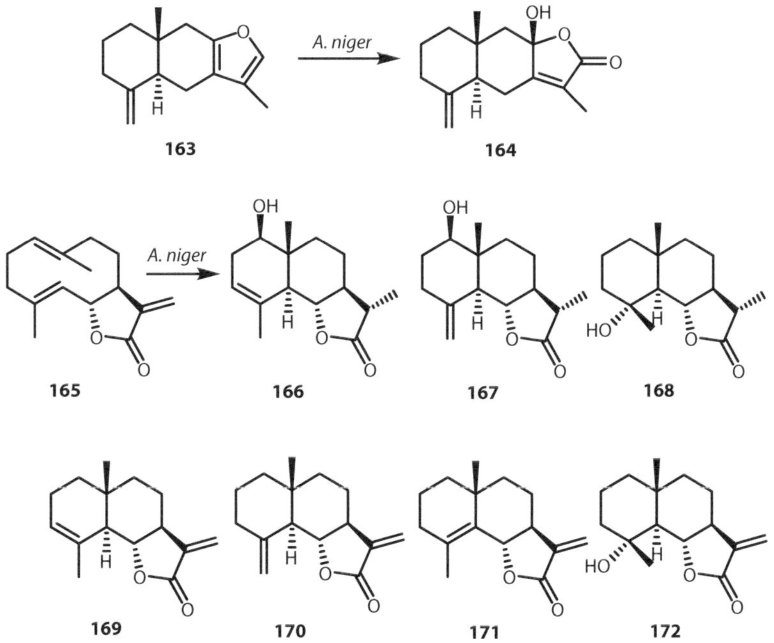
A. niger
163
164
A. niger
165
166
167
168
169
170
171
172

FIGURE 23.52 Biotransformation of atractylon (**163**) and costunolide (**165**) by *Aspergillus niger*.

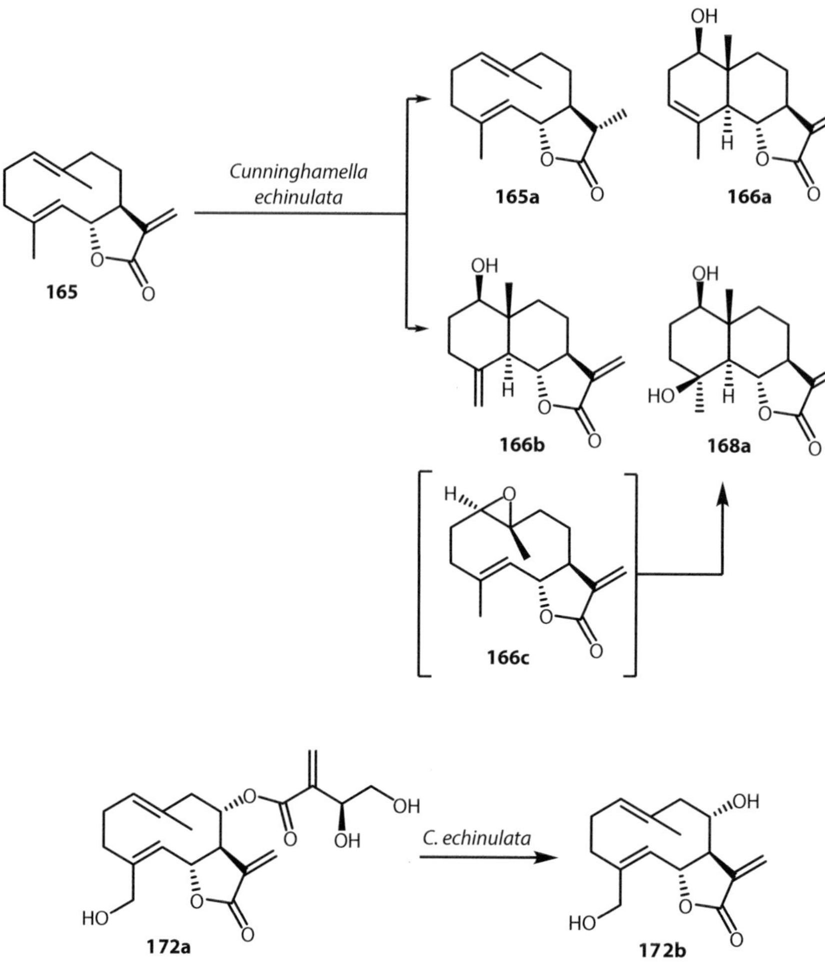

FIGURE 23.53 Biotransformation of costunolide (**165**) and its derivative (**172a**) by *Cunninghamella echinulata* and *Rhizopus oryzae*.

α-Cyclocostunolide (**169**), β-cyclocostunolide (**170**), and γ-cyclocostunolide (**171**) prepared from costunolide were cultivated in *A. niger*. From the metabolite of **169**, four dihydro lactones (**173–176**) were obtained, among which sulfur-containing compound (**176**) was predominant (Figure 23.54).

The same substrate (**169**) was cultivated for 3 days by *A. cellulosae* to afford a sole metabolite, 11β,13-dihydro-α-cyclocostunolide (**177**). Possible metabolic pathways of **169** by both microorganisms were shown in Figure 23.55.

A double bond at C11–C13 of **169** was firstly reduced stereoselectively to afford **177**, followed by oxidation at C2 to give **173**, and then further oxidation occurred to furnish two hydroxyl derivatives (**174**, **175**) in *A. niger*. The sulfide compound (**176**) might be formed from **175** or by Michael condensation of ethyl 2-hydroxy-3-mercaptopropanate, which might originate from Czapek-peptone medium into exomethylene group of α-cyclocostunolide (Hashimoto et al., 1999c, 2001a,b).

A. niger converted β-cyclocostunolide (**170**) to two oxygenated metabolites (**173**, **174**, **178–181**) of which **173** was predominant. It is suggested that compound **173** and **174** might be formed during biotransformation period since metabolite media after 7 days was acidic (pH 2.7). Surprisingly, *A. cellulosae* gave a sole 11β,13-dihydro-β-cyclocostunolide (**182**), which was abnormally folded in the mycelium of *A. cellulosae* as a crystal form after biotransformation of **170**. On the other hand, the

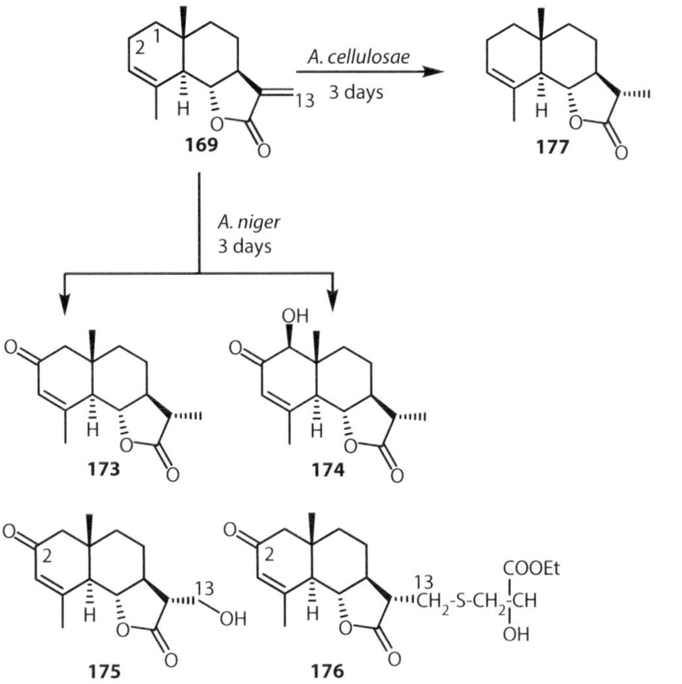

FIGURE 23.54 Biotransformation of α-cyclocostunolide (**169**) by *Aspergillus niger* and *Aspergillus cellulosae.*

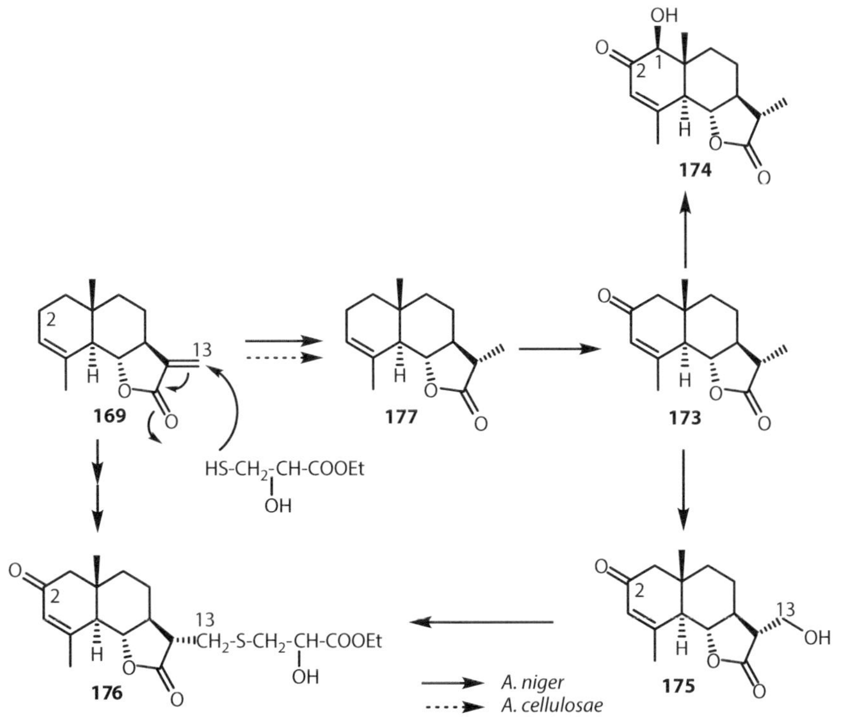

FIGURE 23.55 Possible pathway of biotransformation of α-cyclocostunolide (**169**) by *Aspergillus niger* and *Aspergillus cellulosae.*

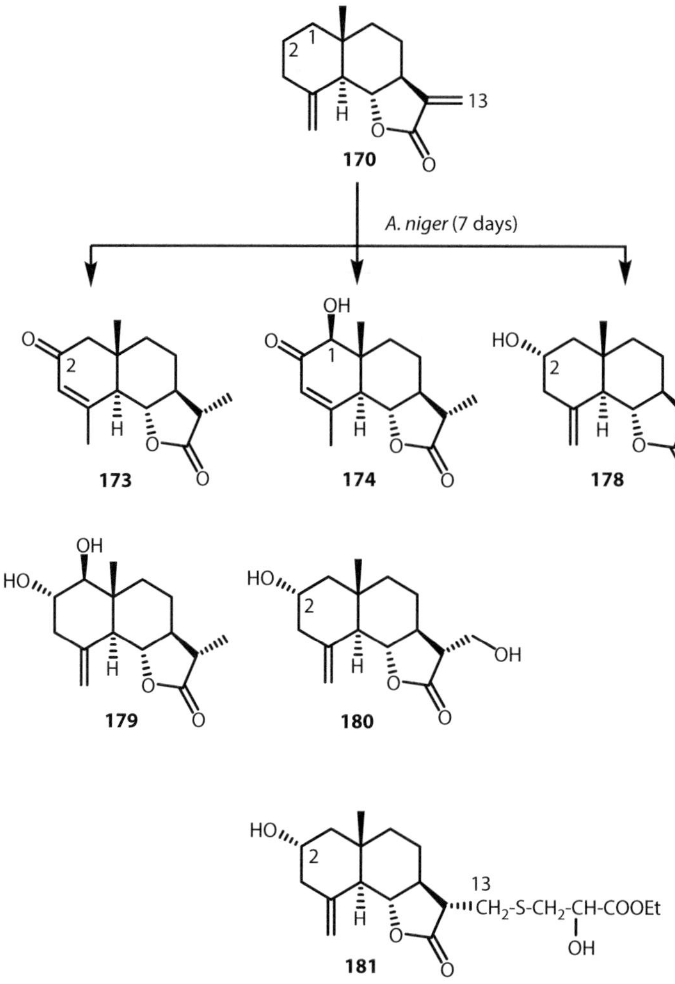

FIGURE 23.56 Biotransformation of β-cyclocostunolide (**170**) by *Aspergillus niger.*

metabolites were normally liberated in medium outside of the mycelium of *A. niger* and *B. dothidea* (Hashimoto et al., 1999c, 2001a,b) (Figure 23.56).

B. dothidea has no stereoselectivity to reduce the C11–C13 double bond of β-cyclocostunolide (**170**) since this organism gave two dihydro derivatives, **182** (16.7%) and **183** (37.8%), respectively, as shown in Figure 23.57.

It is noteworthy that both α- and β-cyclocostunolides were biotransformed by *A. niger* to give the sulfur-containing metabolites (**176, 181**). Possible biogenetic pathway of **170** is shown in Figure 23.58.

When γ-cyclocostunolide (**171**) was cultivated in *A. niger* to give dihydro-α-santonin (**187%**, 25%) and its related C11–C13, dihydro derivatives (**184–186, 188, 189**) were obtained as a small amount. Compound **186** was recultivated for 2 days by the same organism as mentioned earlier to afford **187** (25%) and 5β-hydroxy-α-cyclocostunolide (**189%**, 54%). Recultivation of **185** for 2 days by *A. niger* afforded compound **187** as a sole metabolite. During the biotransformation of **171**, no sulfur-containing product was obtained. Both *A. cellulosae* and *B. dothidea* produced only dihydro-γ-cyclocostunolide (**184**) from the substrate (**171**) (Hashimoto et al., 1999c, 2001a,b) (Figure 23.59).

Santonin (**190**) has been used as vermicide against roundworm. *C. blakesleeana* and *A. niger* converted **190–187** (Atta-ur-Rahman et al., 1998). When **187** was fed by *A. niger* for 1 week to give

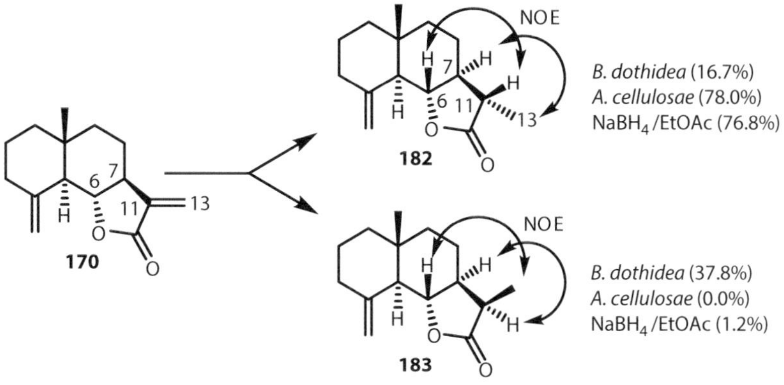

FIGURE 23.57 Biotransformation of β-cyclocostunolide (**170**) by *Aspergillus cellulosae* and *Botryosphaeria dothidea*.

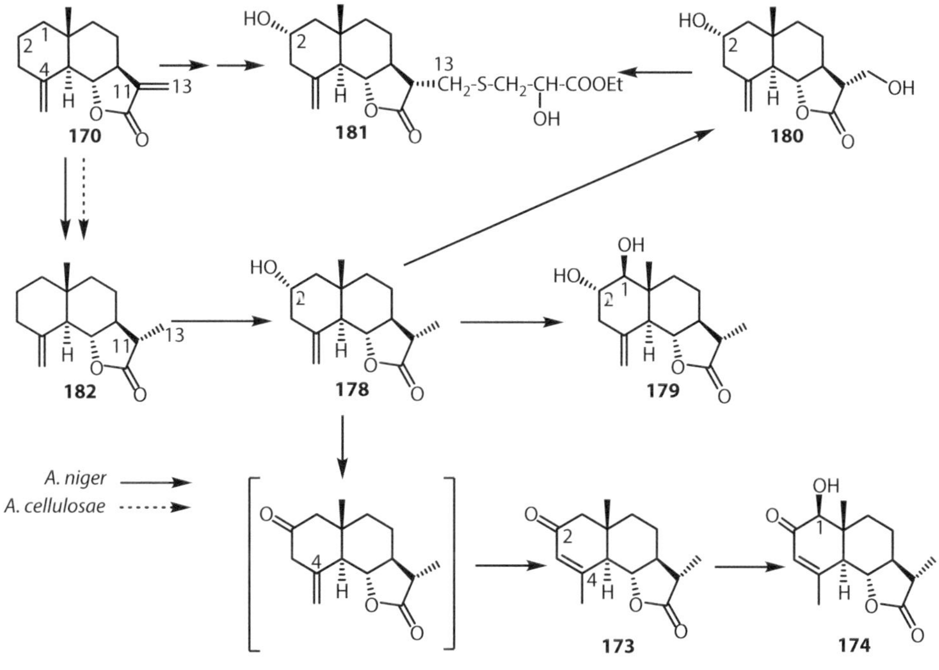

FIGURE 23.58 Possible pathway of biotransformation of β-cyclocostunolide (**170**) by *Aspergillus niger* and *Aspergillus cellulosae*.

2β-hydroxy-1,2-dihydro-α-santonin (**188%**, 39%) as well as 1β-hydroxy-1,2-dihydro-α-santonin (**195**, 6.5%), 9β-hydroxy-1,2-dihydro-α-santonin (**196**, 6.9%), and α-santonin (**190**, 5.4%), which might be obtained from dehydroxylation of **188**, as a minor component (Hashimoto et al., 2001a,b), compound **188** was isolated from the crude metabolite of γ-cyclocostunolide (**171**) by *A. niger* as mentioned earlier (Figure 23.60).

It was treated with *A. niger* for 7 days to give **191** (18.3%), **192** (2.3%), **193** (19.3%), and **194** (3.5%) of which **193** was the major metabolite. Compound **191** was isolated from dog's urine after the oral administration of **190**. The structure of compound **194** was established as lumisantonin obtained by the photoreaction of **190**. α-Santonin **190** was not converted into 1,2-dihydro derivative by *A. niger*, whereas the other strain of *A. niger* gave a single product, 1,2-dihydro-α-santonin (**187**) (Hashimoto et al., 2001a,b) (Figure 23.61).

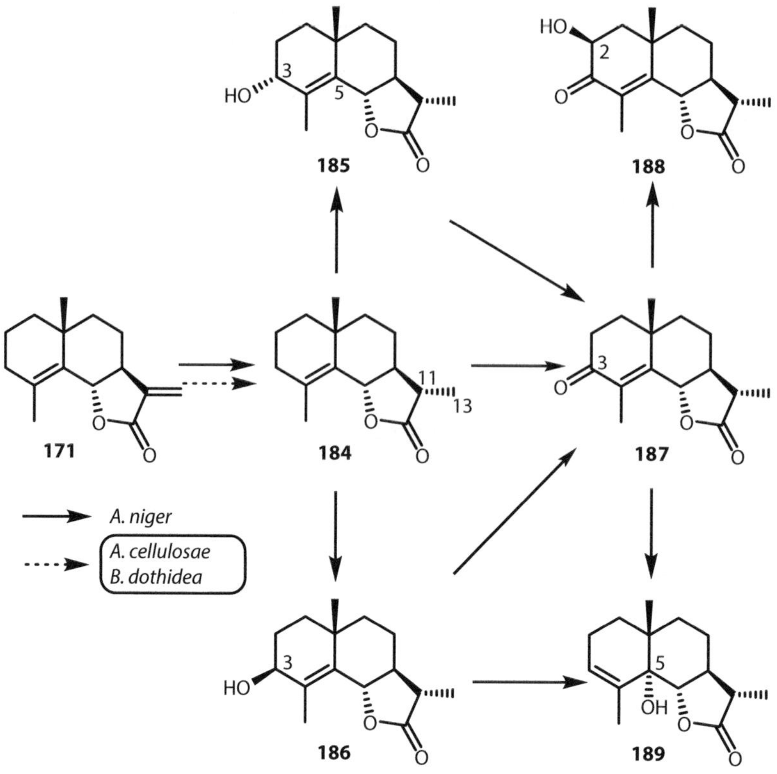

FIGURE 23.59 Biotransformation of γ-cyclocostunolide (**171**) by *Aspergillus niger*, *Aspergillus cellulosae*, and *Botryosphaeria dothidea*.

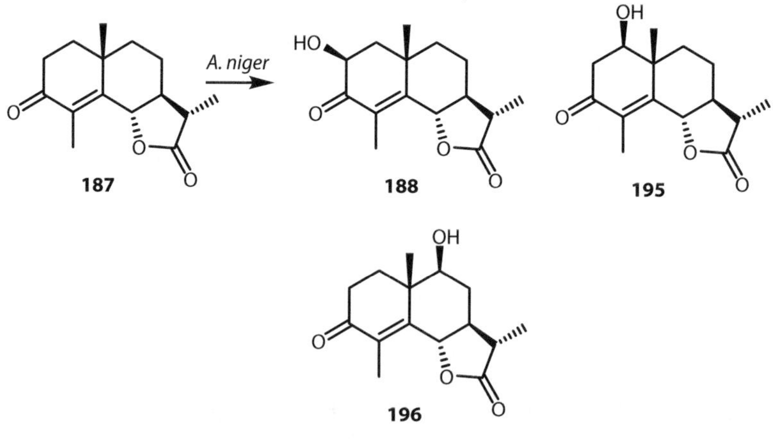

FIGURE 23.60 Biotransformation of dihydro-α-santonin (**187**) by *Aspergillus niger*.

Ata and Nachtigall (2004) reported that α-santonin (**190**) was incubated with *Rhizopus stolonifer* to give (**187a**) and (**183b**), while with *Cunninghamella bainieri*, *C. echinulata*, and *M. plumbeus* to afford the known 1,2-dihydro-α-santonin (**187**) (Figure 23.62).

α-Santonin (**190**) and 6-*epi*-α-santonin (**198**) were cultivated in *Absidia coerulea* for 2 days to give 11β-hydroxy- (**191**, 71.4%) and 8α-hydroxysantonin (**197**, 2.0%), while 6-*epi*-santonin (**198**) afforded four major products (**199–201**, **206**) and four minor analogues (**202**, **203–205**). *Asparagus*

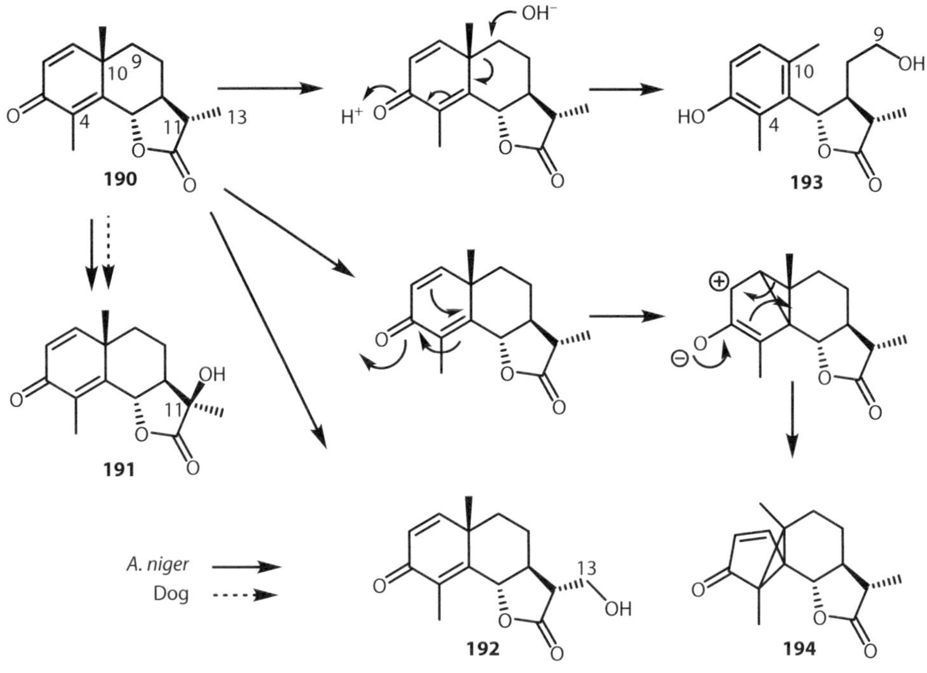

FIGURE 23.61 Biotransformation of α-santonin (**190**) by *Aspergillus niger* and dogs.

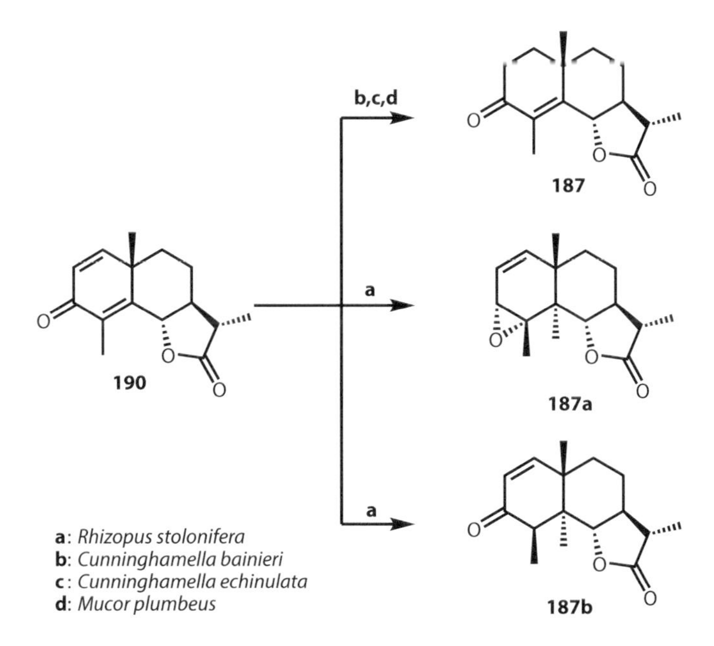

FIGURE 23.62 Biotransformation of α-santonin (**190**) by *Rhizopus stolonifer*, *Cunninghamella bainieri*, *Cunninghamella echinulata*, and *Mucor plumbeus*.

officinalis also biotransformed α-santonin (**190**) into three eudesmanolides (**187**, **207**, **208**) and a guaianolide (**209**) in a small amount. 6-*epi*-Santonin (**198**) was also treated in the same bioreactor as mentioned earlier to give **199** and **206**, the latter of which was obtained as a major metabolite (44.7%) (Yang et al., 2003) (Figure 23.63).

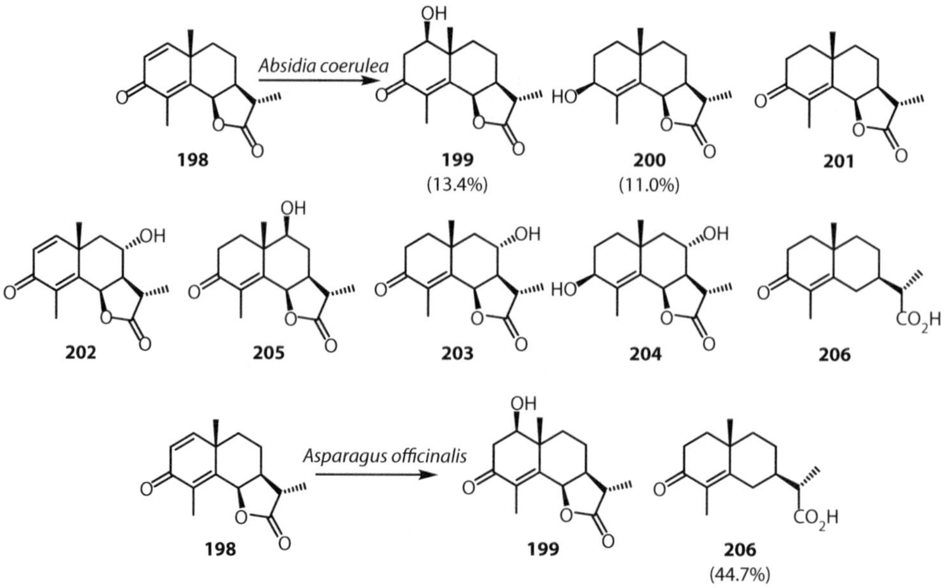

FIGURE 23.63 Biotransformation of α-epi-santonin (**198**) by *Absidia coerulea* and *Asparagus officinalis.*

α-Santonin (**190**) was incubated in the cultured cells of *Nicotiana tabacum* and the liverwort *M. polymorpha*. *N. tabacum* cells gave 1,2-dihydro-α-santonin (**187**) (50%) for 6 days. The latter cells also converted α-santonin to 1,2-dihydro-α-santonin, but conversion ratio was only 28% (Matsushima et al., 2004) (Figure 23.64).

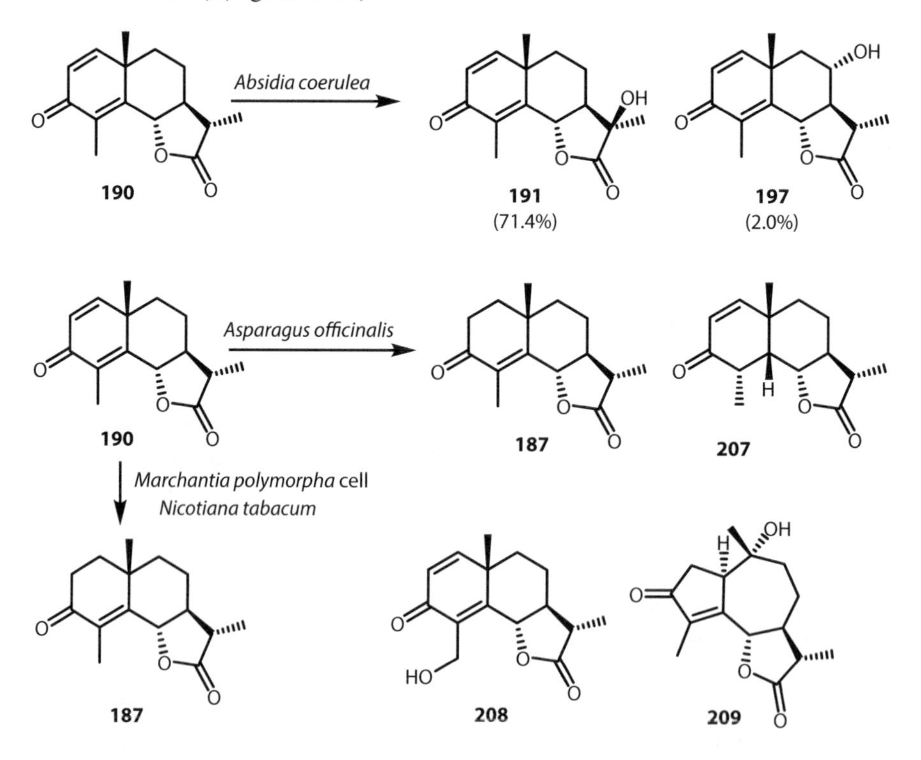

FIGURE 23.64 Biotransformation of 6-epi-α-santonin (**190**) by *Absidia coerulea*, *Asparagus officinalis*, *Marchantia polymorpha*, and *Nicotiana tabacum*.

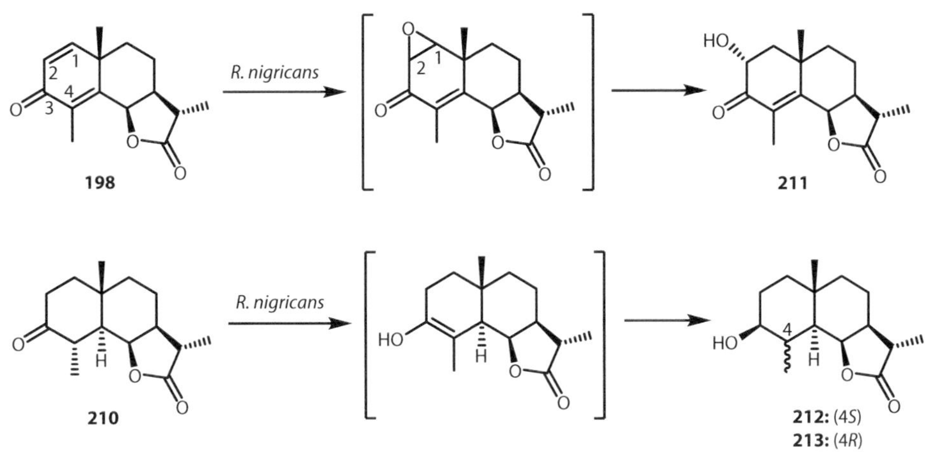

FIGURE 23.65 Biotransformation of α-epi-santonin (**198**) and tetrahydrosantonin (**210**) by *Rhizopus nigricans*.

6-*epi*-α-Santonin (**198**) and its tetrahydro analogue (**210**) were also incubated with fungus *Rhizopus nigricans* to give 2α-hydroxydihydro-α-santonin (**211**) (Amate et al., 1991), the epimer of **188** obtained from the biotransformation of dihydro-α-santonin (**187**) by *A. niger* (Hashimoto et al., 2001a,b). The product **211** might be formed via 1,2-epoxide of **198**. Compound **210** was converted through carbonyl reduction to furnish **212** and **213** under epimerization at C4 (Amate et al., 1991) (Figure 23.65).

1,2,4β,5α-Tetrahydro-α-santonin (**214**) prepared from α-santonin (**190**) was treated with *A. niger* to afford six metabolites (**215**–**220**) of which **219** was the major product (21%). When the substrate (**214**) was treated with CYP450 inhibitor, 1-aminobenzotriazole, only **215** was obtained without its homologues, **216**–**220**, while the C4 epimer (**221**) of **214** was converted by the same microorganism to afford a single metabolite (**222**) (73%). Further oxidation of **222** did not occur. This reason might be considered by the steric hindrance of β-(axial) methyl group at C4 (Hashimoto et al., 2001a,b) (Figure 23.66).

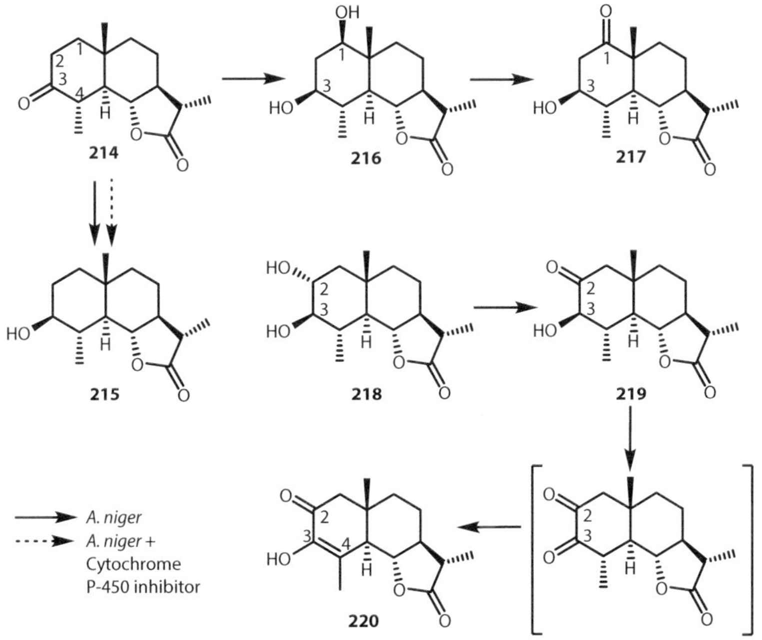

FIGURE 23.66 Biotransformation of 1,2,4β,5α-tetrahydro-α-santonin (**214**) by *Aspergillus niger*.

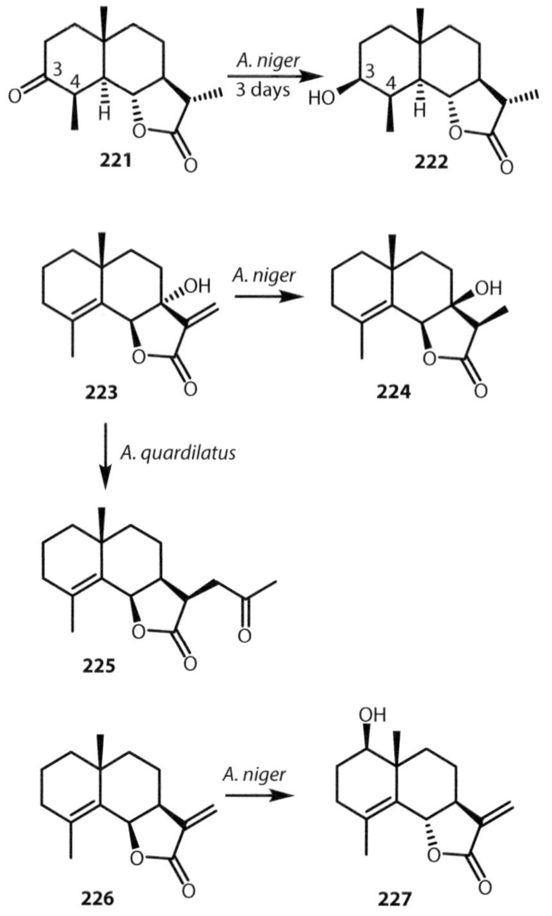

FIGURE 23.67 Biotransformation of C4-epimer (**221**) of **214**, 7α-hydroxyfrullanolide (**223**), and frullanolide (**226**) by *Aspergillus niger* and *Aspergillus quardilatus*.

7α-Hydroxyfrullanolide (**223**) possessing cytotoxicity and antitumor activity, isolated from *Sphaeranthus indicus* (Asteraceae), was bioconverted by *A. niger* to afford 13*R*-dihydro derivative (**224**). The same substrate was also treated in *Aspergillus quardilatus* (wild type) to give 13-acetyl product (**225**) (Atta-ur-Rahman et al., 1994) (Figure 23.67).

Incubation of (−)-frullanolide (**226**), obtained from the European liverwort *Frullania tamarisci* subsp. *tamarisci* that causes a potent allergenic contact dermatitis, was incubated by *A. niger* to give dihydrofrullanolide (**227**), nonallergenic compound in 31.8% yield. In this case, C11–C13 dihydro derivative was not obtained (Hashimoto et al., 2005).

Guaiane-type sesquiterpene hydrocarbon, (+)-γ-gurjunene (**228**), was treated in plant pathogenic fungus *G. cingulata* to give two diols, (1*S*,4*S*,7*R*,10*R*)-5-guaien-11,13-diol (**229**) and (1*S*,4*S*,7*R*,10*S*)-5-guaien-10,11,13-triol (**230**) (Miyazawa et al., 1997a, 1998a) (Figure 23.68).

G. cingulata converted guaiol (**231**) and bulnesol (**232**) to 5,10-dihydroxy (**233**) and 15-hydroxy derivative (**234**), respectively (Miyazawa et al., 1996a) (Figure 23.69).

When *Eurotium rubrum* was used as the bioreactor of guaiene (**235**), rotundone (**236**) was obtained (Sugawara and Miyazawa, 2004). Guaiol (**231**) was also transformed by *A. niger* to give a cyclopentane derivative, pancherione (**237**), and two dihydroxy guaiols (**238**, **239**) (Morikawa et al., 2000), of which **237** was obtained from the same substrate using *E. rubrum* for 10 days (Sugawara and Miyazawa, 2004; Miyazawa and Sugawara, 2006) (Figure 23.70).

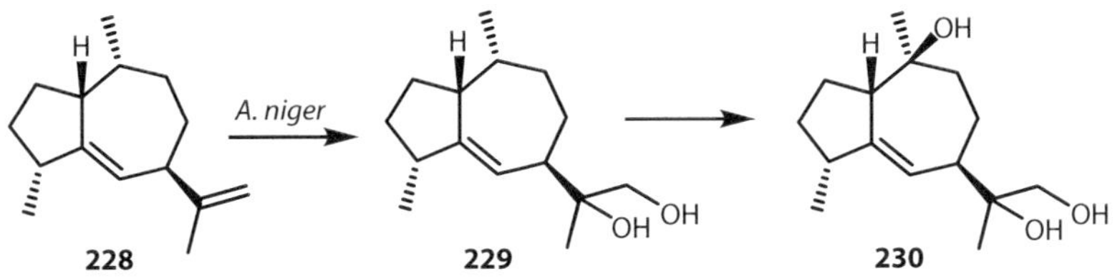

FIGURE 23.68 Biotransformation of (+)-γ-gurjunene (**228**) by *Glomerella cingulata*.

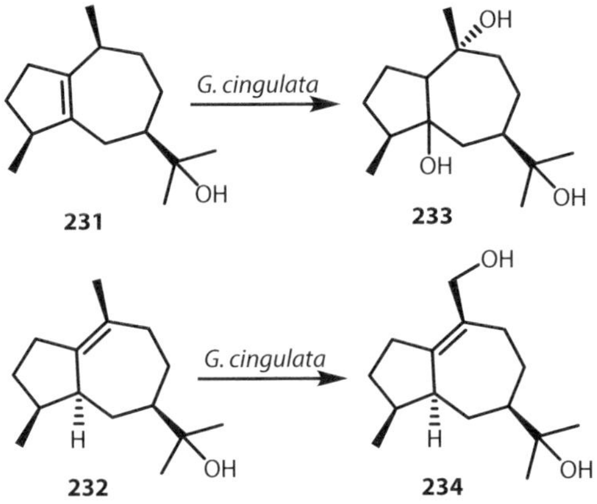

FIGURE 23.69 Biotransformation of guaiol (**221**) and bulnesol (**232**) by *Glomerella cingulata*.

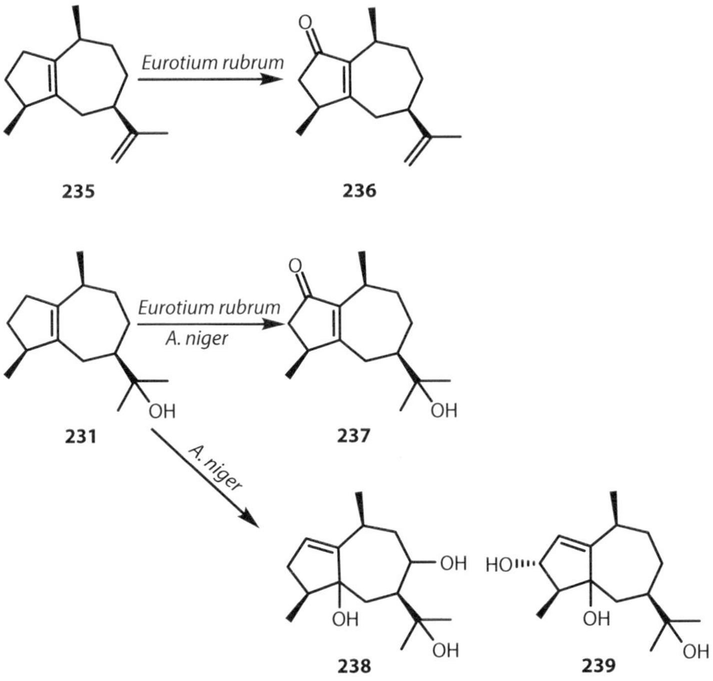

FIGURE 23.70 Biotransformation of guaiene (**235**) by *Eurotium rubrum* and guaiol (**231**) by *Aspergillus niger* and *Eurotium rubrum*.

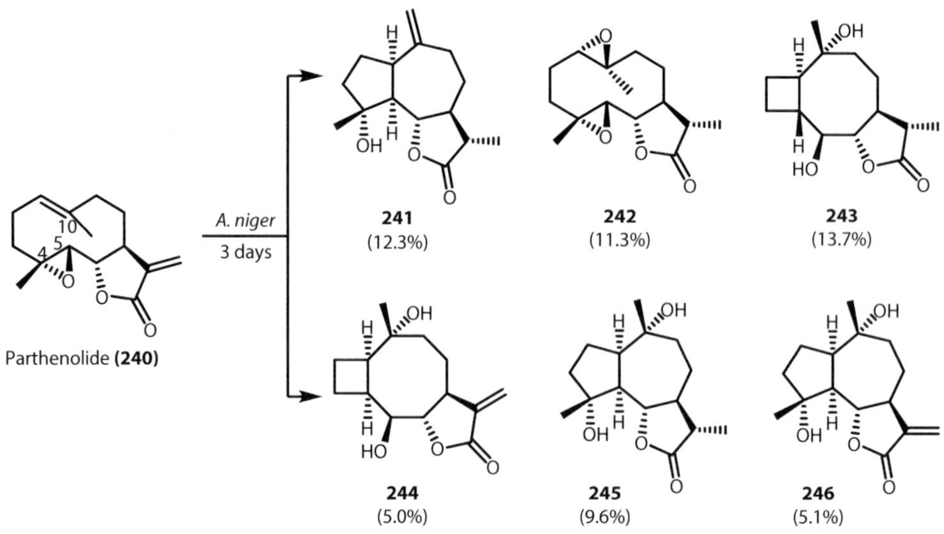

Parthenolide **(240)**
A. niger
3 days
241 (12.3%)
242 (11.3%)
243 (13.7%)
244 (5.0%)
245 (9.6%)
246 (5.1%)

FIGURE 23.71 Biotransformation of parthenolide (**240**) by *Aspergillus niger*.

Parthenolide (**240**), a germacrane-type lactone, isolated from the European feverfew (*Tanacetum parthenium*) as a major constituent, shows cytotoxic, antimicrobial, antifungal, anti-inflammatory, antirheumatic activity, apoptosis inducing, and NF-βB and DNA binding inhibitory activity. This substrate was incubated with *A. niger* in Czapek-peptone medium for 2 days to give six metabolites (**241**, 12.3%; **242**, 11.3%; **243**, 13.7%; **244**, 5.0%; **245**, 9.6%; **246**, 5.1%) (Hashimoto et al., 2005) (Figure 23.71). Compound **244** was a naturally occurring lactone from *Michelia champaca* (Jacobsson et al., 1995). The stereostructure of compound **243** was established by x-ray crystallographic analysis.

When parthenolide (**240**) was treated in *A. cellulosae* for 5 days, two new metabolites, 11β,13-dihydro-(**247**, 43.5%) and 11α,13-dihydroparthenolides (**248**, 1.6%), were obtained together with the same metabolites (**241**, 5.3%; **243**, 11.2%; **245**, 10.4%) as described earlier (Figure 23.72). Possible metabolic root of **240** has been shown in Figure 23.73 (Hashimoto et al., 2005).

Galal et al. (1999) reported that *Streptomyces fulvissimus* or *R. nigricans* converted parthenolide (**240**) into 11α-methylparthenolide (**247**) in 20%–30% yield, while metabolite 11β-hydroxyparthenolide (**248**) was obtained by incubation of **240** with *R. nigricans* and *Rhodotorula rubra*. In addition to the metabolite **247**, *S. fulvissimus* gave a minor polar metabolite, 9β-hydroxy derivative (**248a**) in low yield (3%). The same metabolite (**248a**) was obtained from **247** by fermentation of *S. fulvissimus* as

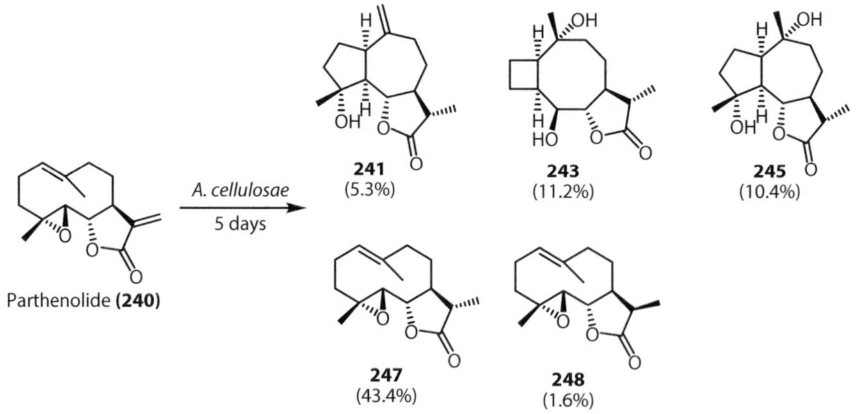

Parthenolide **(240)**
A. cellulosae
5 days
241 (5.3%)
243 (11.2%)
245 (10.4%)
247 (43.4%)
248 (1.6%)

FIGURE 23.72 Biotransformation of parthenolide (**240**) by *Aspergillus cellulosae*.

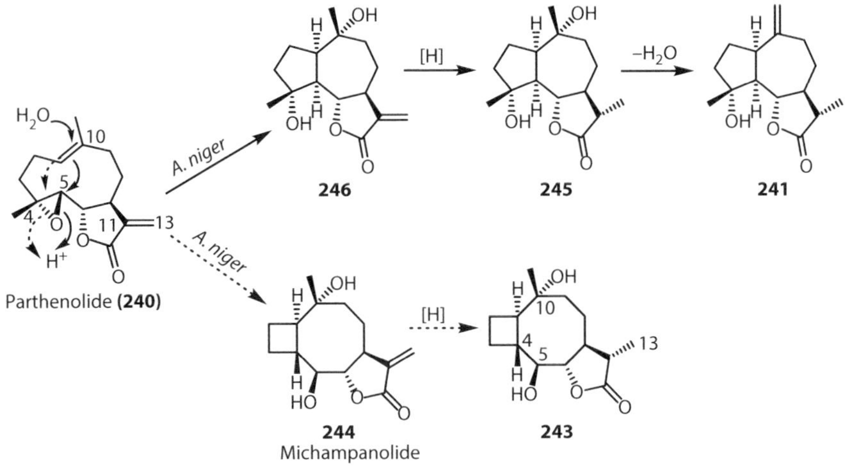

FIGURE 23.73 Possible pathway of biotransformation of parthenolide (**240**).

a minor constituent. Furthermore, 14-hydroxyparthenolide (**248b**) was obtained from **240** and **247** as a minor component (4%) by *R. nigricans* (Figure 23.74).

Pyrethrosin (**248c**), a germacranolide, was treated in the fungus *R. nigricans* to afford five metabolites (**248d–248 h**). Pyrethrosin itself and metabolite **248e** displayed cytotoxic activity against human malignant melanoma with IC_{50} 4.20 and 7.5 mg/mL, respectively. Metabolite **248 h** showed significant *in vitro* cytotoxic activity against human epidermoid carcinoma (KB cells) and against human ovary carcinoma with $IC_{50} < 1.1$ and 8.0 mg/mL, respectively. Compounds **248f** and **248i** were active against *Cryptococcus neoformans* with IC_{50} 35.0 and 25 mg/mL, respectively, while **248a** and **248 g** showed antifungal activity against *Candida albicans* with IC_{50} 30 and 10 mg/mL. Metabolites **248 g** and its acetate (**248i**), derived from **248 g**, showed antiprotozoal activity against *Plasmodium falciparum* with IC_{50} 0.88 and 0.32 mg/mL, respectively, without significant toxicity. Compound **248i** also exhibited pronounced activity against the chloroquine-resistant strain of *P. falciparum* with IC_{50} 0.38 mg/mL (Galal, 2001) (Figure 23.75).

(−)-Dehydrocostus lactone (**249**), inhibitors of nitric oxide synthases and TNF-α, isolated from *Saussurea* radix, was incubated with *C. echinulata* to afford (+)-11α,13-dihydrodehydrocostuslactone (**250a**). The epoxide (**251**) and a C11 reduced compound (**250**) were obtained by the aforementioned microorganisms (Galal, 2001).

C. echinulata and *R. oryzae* bioconverted **249** into C11/C13 dihydrogenated (**250**) and C10/C14 epoxidated product (**251**). Treatment of **252a** in *C. echinulata* and *R. oryzae* gave (–)-16-(1-methyl-1-propenyl)eremantholanolide (**252b**) (Galal, 2001) (Figure 23.76).

The same substrate (**249**) was fed by *A. niger* for 7 days to afford four metabolites, costus lactone (**250**), and their derivatives (**251–253**), of which **251** was the major product (28%); while the same substrate was cultivated with *A. niger* for 10 days, two minor metabolites (**254**, **255**) were newly obtained in addition to **252** and **253** of which the latter lactone was predominant (20.7%) (Hashimoto et al., 2001a,b) (Figure 23.77).

When compound **249** was treated with *A. niger* in the presence of 1-aminobenzotriazole, **249** was completely converted into 11β,13-dihydro derivative (**250**) for 3 days; however, further biodegradation did not occur for 10 days (Hashimoto et al., 1999a, 2001a,b). The same substrate (**249**) was cultivated with *A. cellulosae* IFO to furnish 11,13-dihydro- (**250**) (82%) for only 1 day and then the product (**250**) slowly oxidized into 11,13-dihydro-8β-hydroxycostuslactone (**256**) (1.6%) for 8 days (Hashimoto et al., 1999a, 2001a,b) (Figure 23.78).

The lactone (**249**) was biodegraded by the plant pathogen *B. dothidea* for 4 days to give the metabolites **250** (37.8%) and **257** (8.6%), while *A. niger* IFO-04049 (4 days) and *A. cellulosae* for

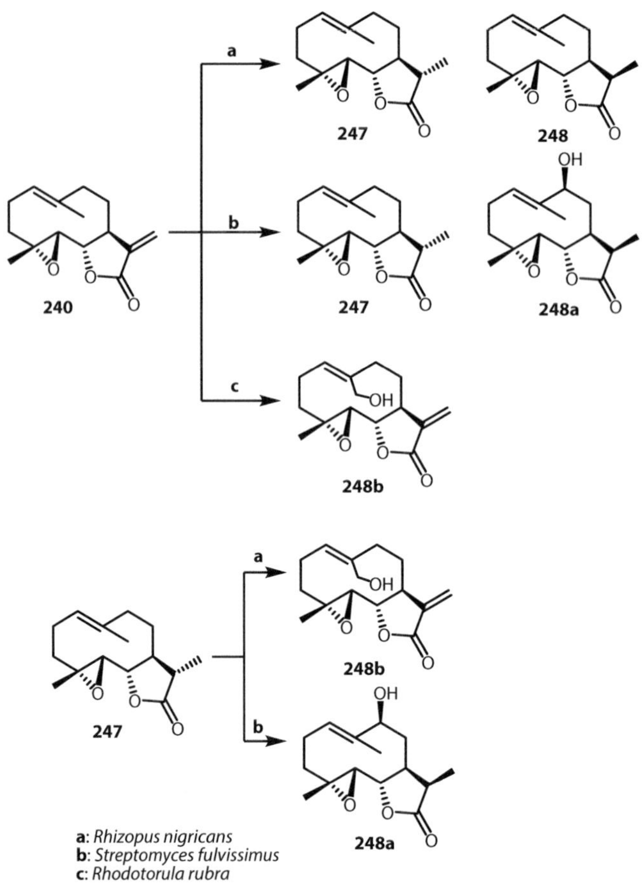

a: *Rhizopus nigricans*
b: *Streptomyces fulvissimus*
c: *Rhodotorula rubra*

FIGURE 23.74 Biotransformation of parthenolide (**240**) and its dihydro derivative (**247**) by *Rhizopus nigricans*, *Streptomyces fulvissimus*, and *Rhodotorula rubra*.

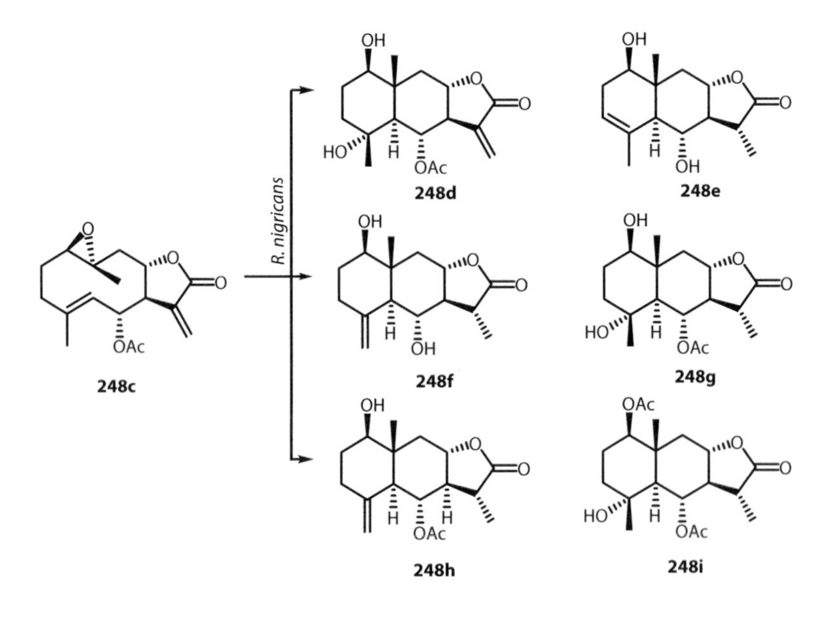

FIGURE 23.75 Biotransformation of pyrethrosin (**248c**) by *Rhizopus nigricans*.

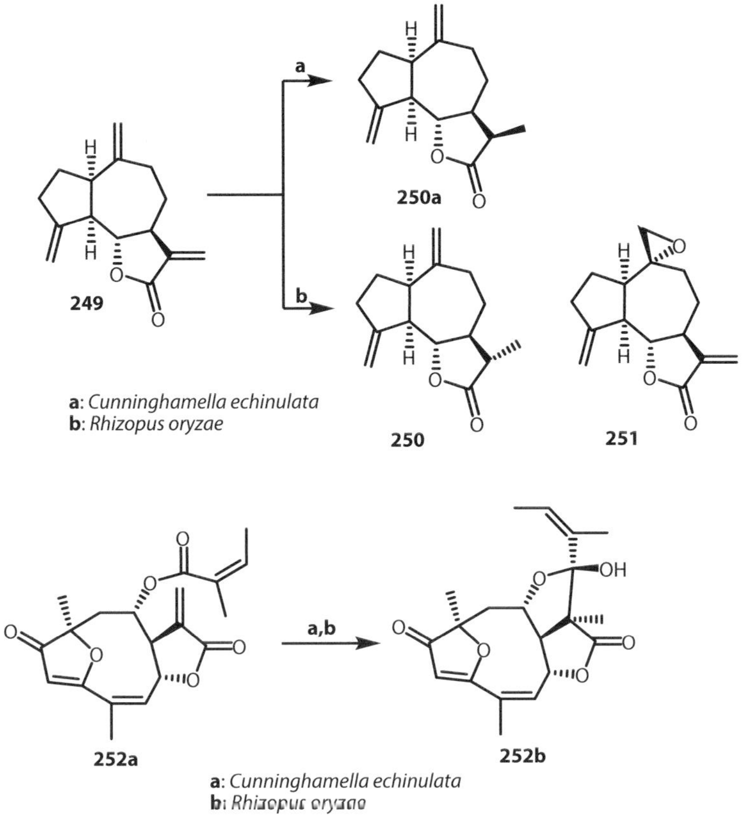

249 **250a** **250** **251**

a: *Cunninghamella echinulata*
b: *Rhizopus oryzae*

252a **252b**

a: *Cunninghamella echinulata*
b: *Rhizopus oryzae*

FIGURE 23.76 Biotransformation of (–)-dehydrocostuslactone (**249**) and rearranged guaianolide (**252a**) by *Cunninghamella echinulata* and *Rhizopus oryzae*.

1 day gave only **250**. Thus, *B. dothidea* demonstrated low stereoselectivity to reduce the C11–C13 double bond (Hashimoto et al., 2001a,b). Furthermore, three *Aspergillus* species, *A. niger* IFO 4034, *Aspergillus awamori* IFO 4033, and *Aspergillus terreus* IFO6123, were used as bioreactors for compounds **249**. *A. niger* IFO 4034 gave three products (**250**–**252**), of which **252** was predominant (56% in GC-MS). *A. awamori* IFO 4033 and *A. terreus* IFO 6123 converted **249** to give **250** (56% from *A. awamori*, 43% from *A. terreus*) and **252** (43% from *A. awamori*, 57% from *A. terreus*), respectively (Hashimoto et al., 2001a,b) (Figure 23.79).

Vernonia arborea (Asteraceae) contains zaluzanin D (**258**) in high content. Ten microorganisms were used for the biotransformation of compound **258**. *B. cinerea* converted **258** into **259** and **260** (85%:15%) and *Fusarium equiseti* gave **259** and **260** (33%:66%). *C. lunata*, *Colletotrichum lindemuthianum*, *Alternaria alternata*, and *Phyllosticta capsici* produced **259** as the sole metabolite in good yield, while *Sclerotinia sclerotiorum* and *Rhizoctonia solani* gave deactyl product (**261**) as a sole product and **260** and **262**–**264**, among which **263** and **264** are the major products, respectively. Reduction of C11–C13 exocylic double bond is the common transformation of α-methylene-γ-butyrolactone (Kumari et al., 2003) (Figure 23.80).

Incubation of parthenin (**264a**) with the fungus *B. bassiana* in modified Richard's medium gave C11–C13 reduced product (**264b**) in 37% yield, while C11 α-hydroxylated product (**264c**) was obtained in 32% yield from the broth of the fungus *Sporotrichum pulverulentum* using the same medium (Bhutani and Thakur, 1991) (Figure 23.81).

Cadina-4,10(15)-dien-3-one (**265**) possessing insecticidal and ascaricidal activity, from the Jamaican medicinal plant *H. verticillata*, was metabolized by *C. lunata* ATCC 12017 in potato dextrose to give its 12-hydoxy- (**266**), 3α-hydroxycadina-4,10(15)-dien (**267**), and

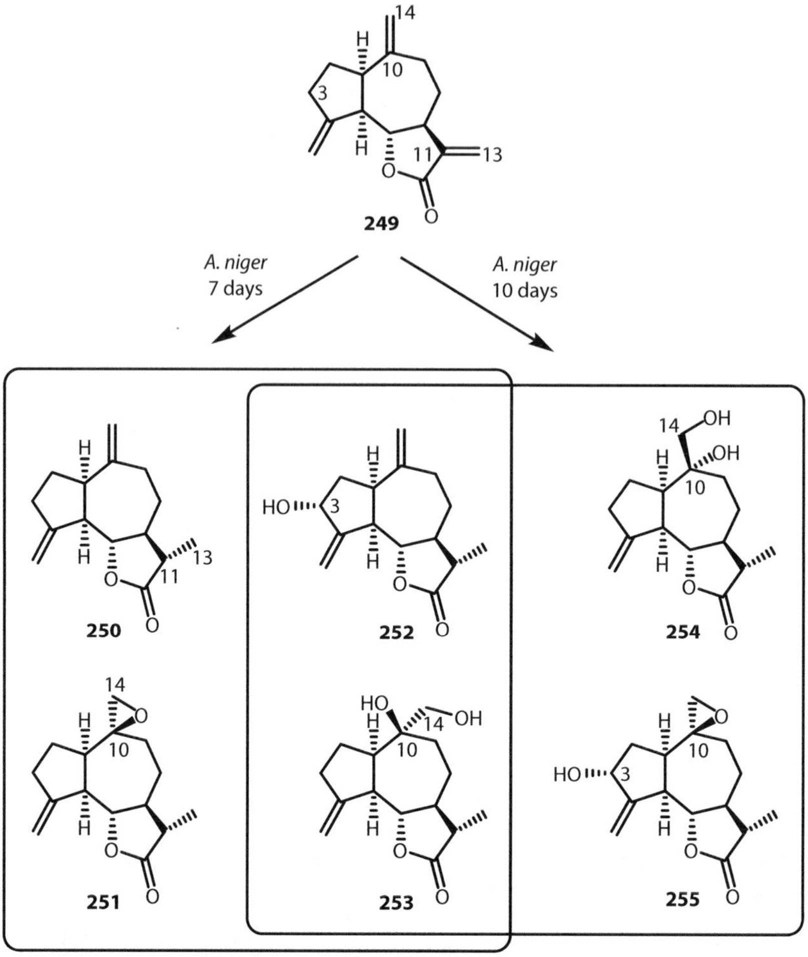

FIGURE 23.77 Biotransformation of (−)-dehydrocostuslactone (**249**) by *Aspergillus niger.*

3α-hydroxy-4,5dihydrocadinenes (**268**), while **265** was incubated by the same fungus in peptone, yeast, and beef extracts and glucose medium, only **267** and **268** were obtained. Compound **267** derived synthetically was treated in the same fungus *C. lunata* to afford three metabolites (**269–271**) (Collins and Reese, 2002) (Figure 23.82).

The incubation of the same substrate (**265**) in *M. plumbeus* ATCC 4740 in high-iron-rich medium gave **270**, which was obtained from *C. lunata* mentioned earlier, **268**, **272**, **273**, **277**, **278**, and **279**. In low-iron medium, this fungus converted the same substrate **265** into three epoxides (**274–276**), a tetraol (**280**) with common metabolites (**268, 273, 277, 278**), and **271**, which was the same metabolite used by *C. lunata* (Collins et al., 2002a). It is interesting to note that only epoxides were obtained from the substrate (**265**) by *Mucor* fungus in low-iron medium (Figure 23.83).

The same substrate (**265**) was incubated with the deuteromycete fungus *B. bassiana*, which is responsible for the muscardine disease in insects, in order to obtain new functionalized analogues with improved biological activity. From compound **265**, nine metabolites were obtained. The insecticidal potential of the metabolites (**267, 268, 268a–268f**) were evaluated against *Cylas formicarius*. The metabolites (**273, 268, 268d**) showed enhanced activity compared with the substrate (**265**). The plant growth regulatory activity of the metabolites against radish seeds was tested. All the compounds showed inhibitory activity; however, their activity was less than colchicine (Buchanan et al., 2000) (Figure 23.84).

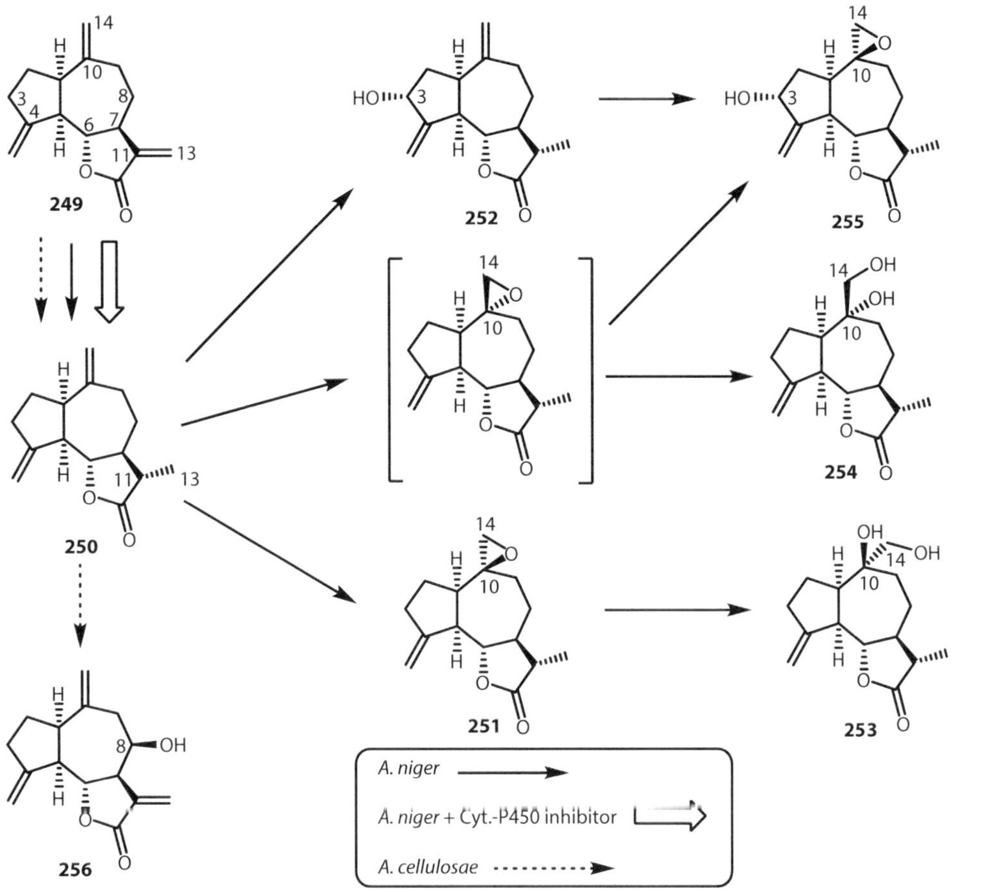

FIGURE 23.78 Possible pathway of biotransformation of (−)-dehydrocostuslactone (**249**) by *Aspergillus niger* and *Aspergillus cellulosae*.

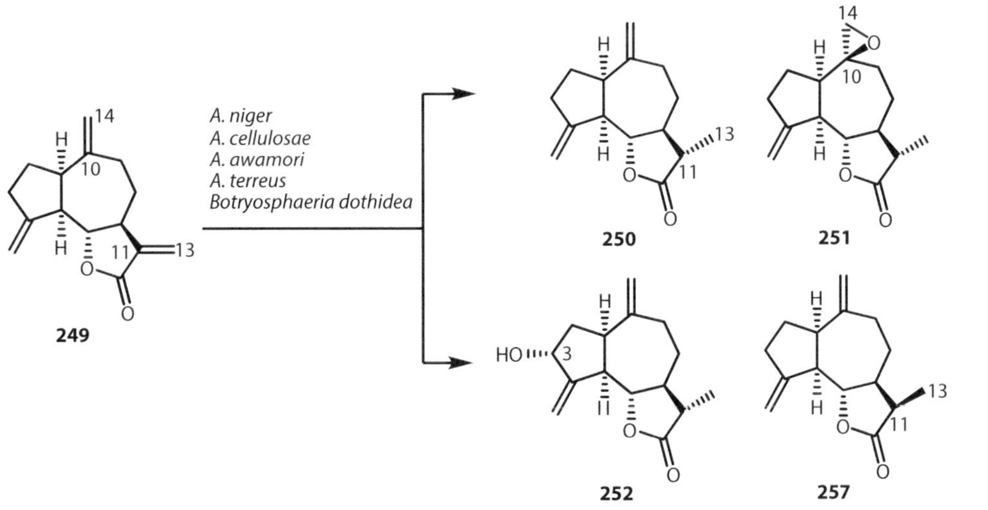

FIGURE 23.79 Biotransformation of (−)-dehydrocostuslactone (**249**) by *Aspergillus* species and *Botryosphaeria dothidea*.

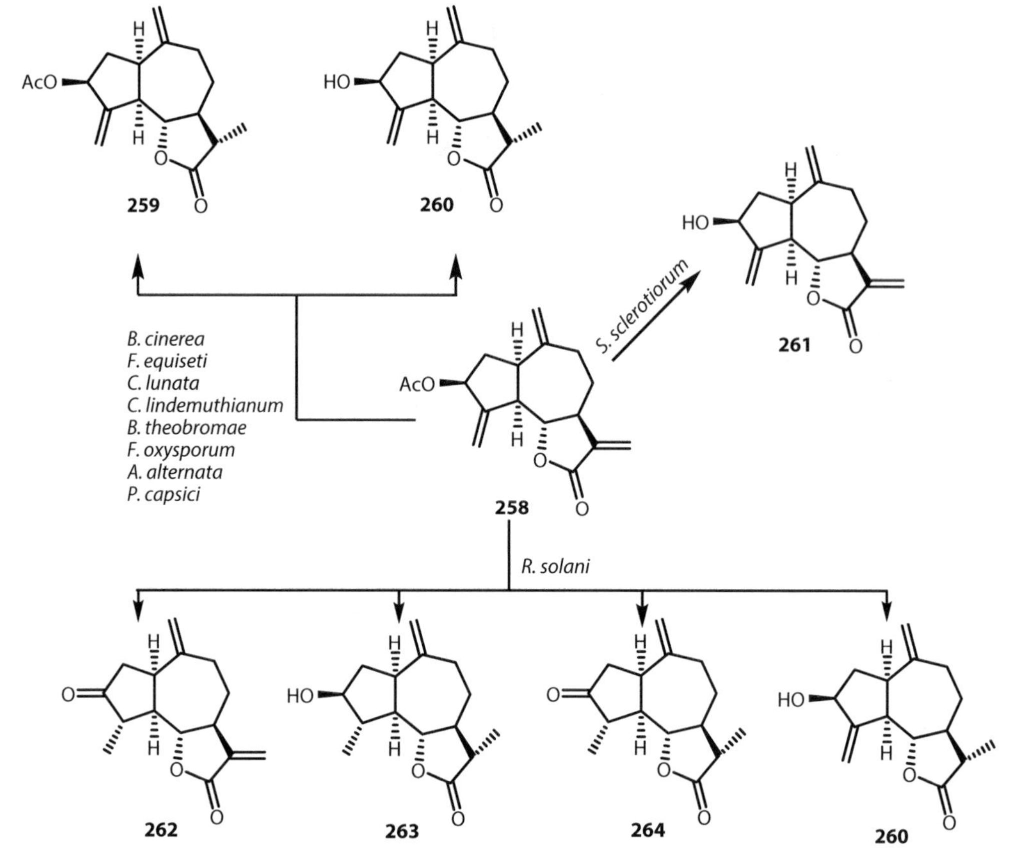

FIGURE 23.80 Biotransformation of zaluzanin D (**258**) by various fungi.

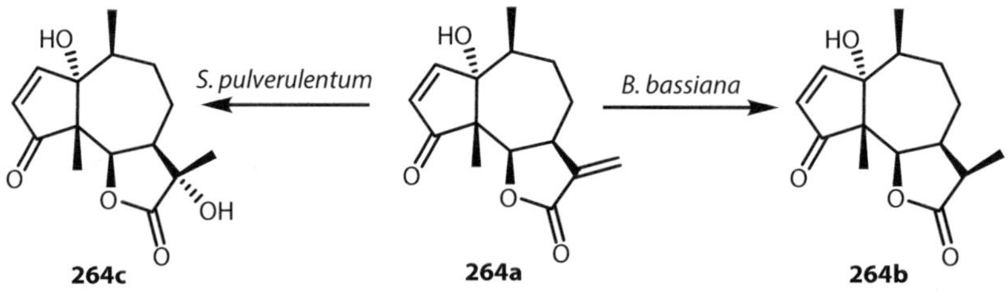

FIGURE 23.81 Biotransformation of parthenin (**264a**) by *Sporotrichum pulverulentum* and *Beauveria bassiana*.

Cadinane-type sesquiterpene alcohol (**281**) isolated from the liverwort *M. taylorii* gave a primary alcohol (**282**) by *A. niger* treatment (Morikawa et al., 2000) (Figure 23.85).

Fermentation of (−)-α-bisabolol (**282a**) possessing an anti-inflammatory activity with the plant pathogenic fungus *G. cingulata* for 7 days yielded oxygenated products (**282b**–**282e**) of which compound **282e** was predominant. 3,4-Dihydroxy products (**282b**, **282d**, **282e**) could be formed by hydrolysis of the 3,4-epoxide from **282a** and **282c** (Miyazawa et al., 1995c) (Figure 23.86).

El Sayed et al. (2002) reported microbial and chemical transformation of (*S*)-(+)-curcuphenol (**282 g**) and curcudiol (**282n**), isolated from the marine sponges *Didiscus axiata*. Incubation of compound **282 g** with *Kluyveromyces marxianus* var. *lactis* resulted in the isolation of six

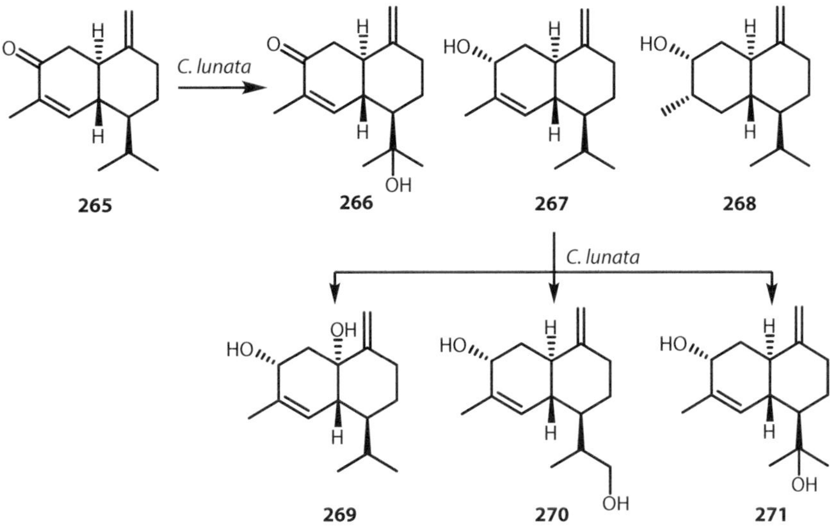

FIGURE 23.82 Biotransformation of cadina-4,10(15)-dien-3-one (**265**) by *Curvularia lunata*.

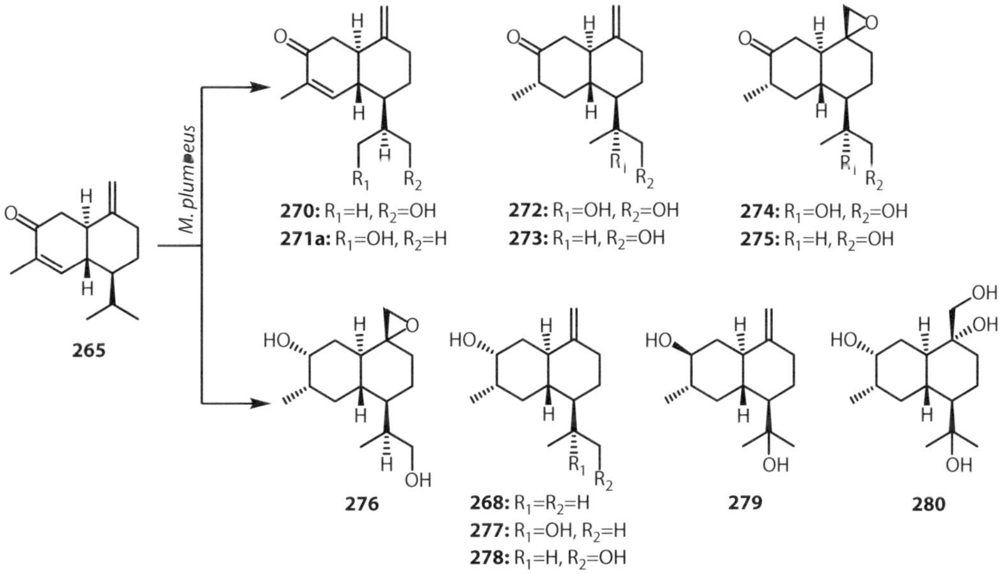

FIGURE 23.83 Biotransformation of cadina-4,10(15)-dien-3-one (**265**) by *Mucor plumbeus*.

metabolites (3–8, **282h–282j**). The same substrate was incubated with *Aspergillus alliaceus* to give the metabolites (**282p**, **282q**, **282 s**) (Figure 23.87).

Compounds **282 g** and **282n** were treated in *Rhizopus arrhizus* and *Rhodotorula glutinus* for 6 and 8 days to afford glucosylated metabolites, 1α-D-glucosides (**282o**) and **282r**, respectively. The substrate itself showed antimicrobial activity against *C. albicans*, *C. neoformans*, and MRSA-resistant *Staphylococcus aureus* and *S. aureus* with MIC and MFC/MBC ranges of 7.5–25 and 12.5–50 mg/mL, respectively. Compounds **282 g** and **282 h** also exhibited *in vitro* antimalarial activity against *P. falciparum* (D6 clone) and *P. falciparum* (W2 clone) of 3600 and 3800 ng/mL (selective index [S.I.] > 1.3), 1800 (S.I. > 2.6), and 2900 (S.I. > 1.6), respectively (El Sayed et al., 2002) (Figure 23.87).

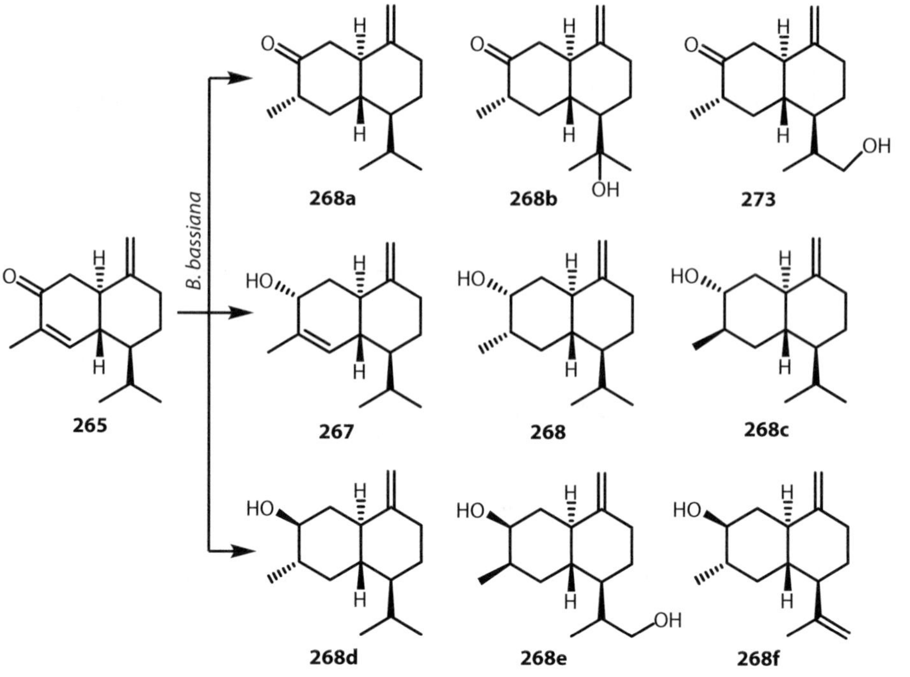

FIGURE 23.84 Biotransformation of cadina-4,10(15)-dien-3-one (**265**) by *Beauveria bassiana*.

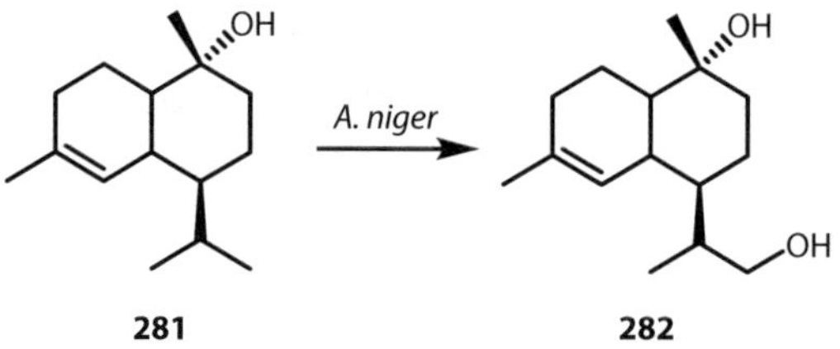

FIGURE 23.85 Biotransformation of cadinol (**281**) by *Aspergillus niger*.

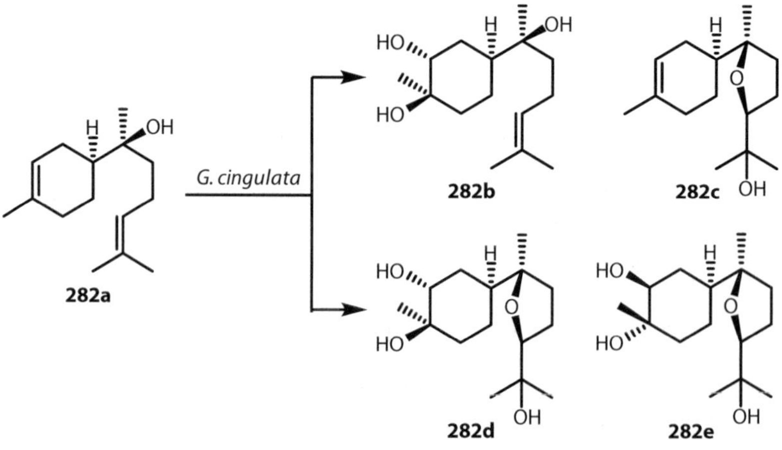

FIGURE 23.86 Biotransformation of β-bisabolol (**282a**) by *Glomerella cingulata*.

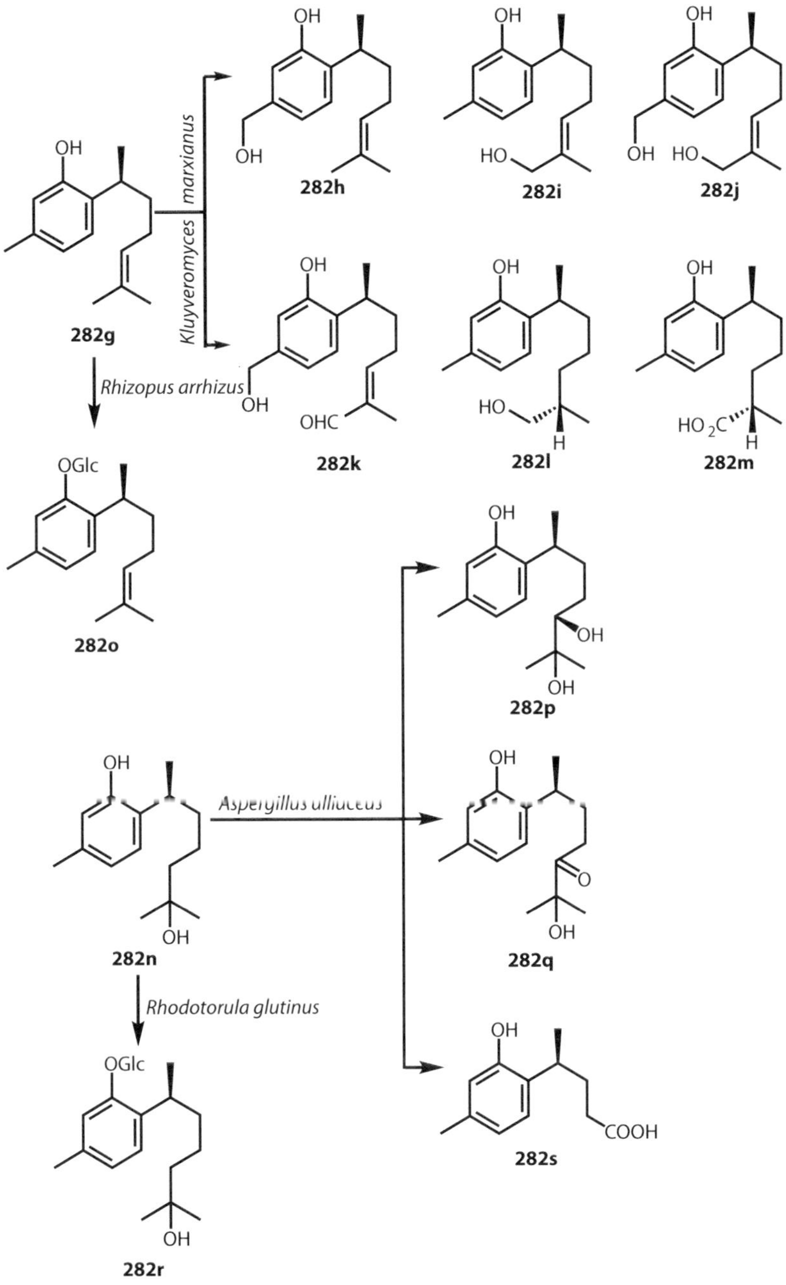

FIGURE 23.87 Biotransformation of (*S*)-(+)-curcuphenol (**282 g**) by *Kluyveromyces marxianus* and *Rhizopus arrhizus* and curcudiol (282n) by *Aspergillus alliaceus* and *Rhodotorula glutinus*.

Artemisia annua is one of the most important Asteraceae species as antimalarial plant. There are many reports of microbial biotransformation of artemisinin (**283**), which is active antimalarial rearranged cadinane sesquiterpene endoperoxide, and its derivatives to give novel antimalarials with increased activities or differing pharmacological characteristics.

Lee et al. (1989) reported that deoxoartemisinin (**284**) and its 3α-hydroxy derivative (**285**) were obtained from the metabolites of artemisinin (**283**) incubated with *Nocardia corallina* and *Penicillium chrysogenum* (Figure 23.88).

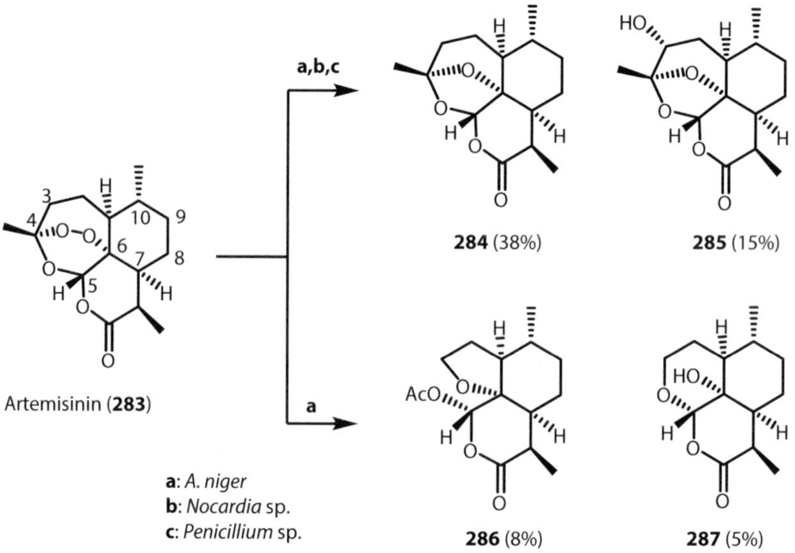

FIGURE 23.88 Biotransformation of artemisinin (**283**) by *Aspergillus niger*, *Nocardia corallina*, and *Penicillium chrysogenum*.

Zhan et al. (2002) reported that incubation of artemisinin (**283**) with *C. echinulata* and *A. niger* for 4 days at 28°C resulted in the isolation of two metabolites, 10β-hydroxyartemisinin (**287a**) and 3α-hydroxydeoxyartemisinin (**285**), respectively.

Compound **283** was also biotransformed by *A. niger* to give four metabolites, deoxyartemisinin (**284%**, 38%), 3α-hydroxydeoxyartemisinin (**285%**, 15%), and two minor products (**286%**, 8% and **287%**, 5%) (Hashimoto et al., 2003d).

Artemisinin (**283**) was also bioconverted by *C. elegans*. During this process, 9β-hydroxyartemisinin (**287b**, 78.6%), 9β-hydroxy-8α-artemisinin (**287c**, 6.0%), 3α-hydroxydeoxoartemisinin (**285**, 5.4%), and 10β-hydroxyartemisinin (**287d**, 6.5%) have been formed. On the basis of quantitative structure–activity relationship and molecular modeling investigations, 9β-hydoxy derivatization of artemisinin skeleton may yield improvement in antimalarial activity and may potentially serve as an efficient means of increasing water solubility (Parshikov et al., 2004) (Figure 23.89).

Albicanal (**288**) and (−)-drimenol (**289**) are simple drimane sesquiterpenoids isolated from the liverwort, *Diplophyllum serrulatum*, and many other liverworts and higher plants. The latter compound was incubated with *M. plumbeus* and *R. arrhizus*. The former microorganism converted **289** to 6,7α-epoxy- (**290**), 3β-hydroxy- (**291**), and 6α-drimenol (**292**) in the yields of 2%, 7%, and 50%, respectively. On the other hand, the latter species produced only 3β-hydroxy derivative (**291**) in 60% yield (Aranda et al., 1992) (Figure 23.90).

(−)-Polygodial (**293**) possessing piscicidal, antimicrobial, and mosquito-repellant activity is the major pungent sesquiterpene dial isolated from *Polygonum hydropiper* and the liverwort *Porella vernicosa* complex. Polygodial was incubated with *A. niger*; however, because of its antimicrobial activity, no metabolite was obtained (Sekita et al., 2005). Polygodiol (**295**) prepared from polygodial (**293**) was also treated in the same manner as described earlier to afford 3β-hyrdoxy (**297**), which was isolated from *Marasmius oreades* as antimicrobial activity (Ayer and Craw, 1989), and 6α-hydroxypolygodiol (**298**) in 66%–70% and 5%–10% yields, respectively (Aranda et al., 1992). The same metabolite (**297**) was also obtained from polygodiol (**295**) as a sole metabolite from the culture broth of *A. niger* in Czapek-peptone medium for 3 days in 70.5% yield (Sekita et al., 2005), while the C9 epimeric product (**296**) from isopolygodial (**294**) was incubated with *M. plumbeus* to afford 3β-hydroxy- (**299**) and 6α-hydroxy derivative (**300**) in low yields, 7% and 13% (Aranda et al.,

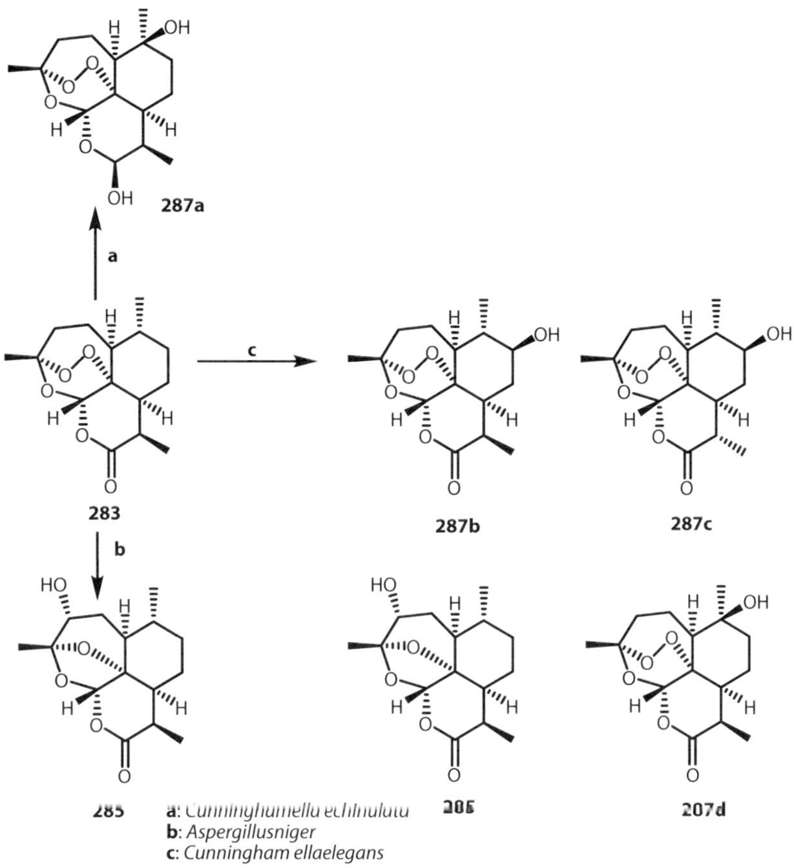

FIGURE 23.89 Biotransformation of artemisinin (**283**) by *Cunninghamella echinulata*, *Cunninghamella elegans*, and *Aspergillus niger*.

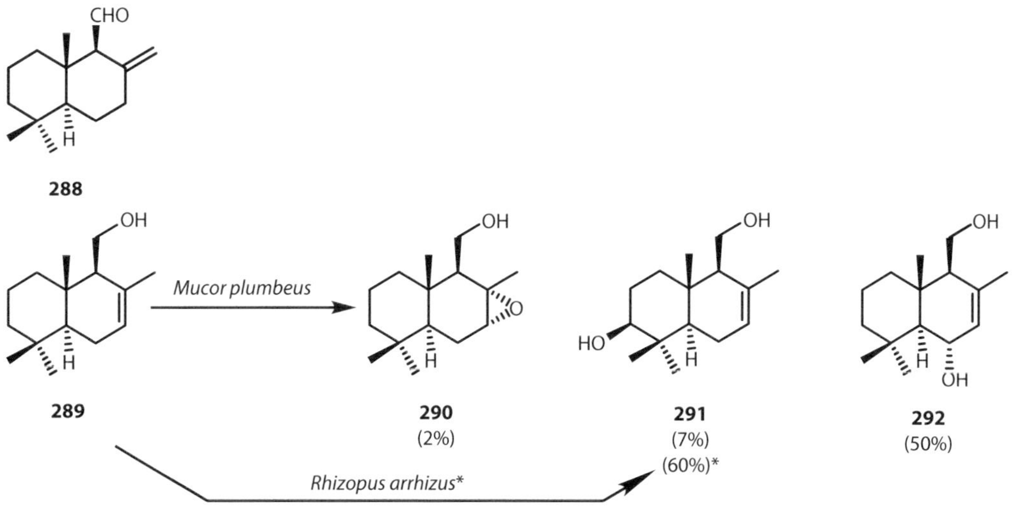

FIGURE 23.90 Biotransformation of drimenol (**289**) by *Mucor plumbeus* and *Rhizopus arrhizus*.

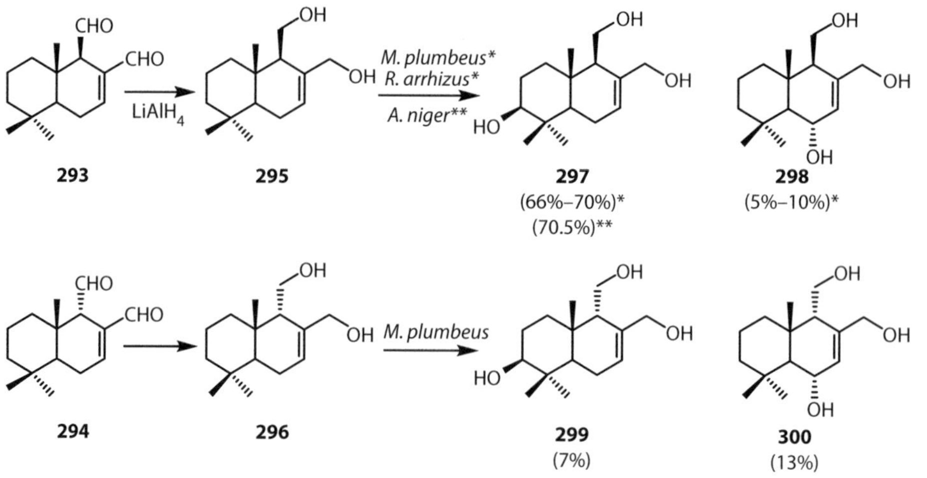

FIGURE 23.91 Biotransformation of polygodiol (**295**) by *Mucor plumbeus*, *Rhizopus arrhizus*, and *Aspergillus niger*.

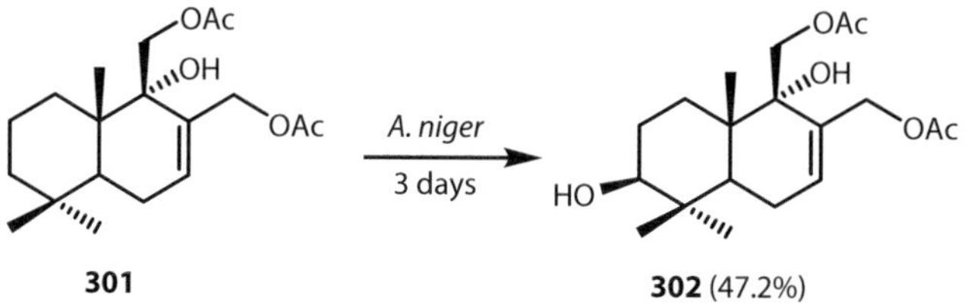

FIGURE 23.92 Biotransformation of drim-9α-hydroxy-11,12-diacetoxy-7-ene (**301**) by *Aspergillus niger*.

1992). Drim-9α-hydroxy-11β,12-diacetoxy-7-ene (**301**) derived from polygodiol (**295**) was treated in the same manner as described earlier to yield its 3β-hydroxy derivative (**302%**, 42%) (Sekita et al., 2005) (Figures 23.91 and 23.92).

Cinnamodial (**303**) from the Malagasy medicinal plant, *Cinnamosma fragrans*, was also treated in the same medium including *A. niger* to furnish three metabolites, respectively, in very law yields (**304**, 2.2%; **305**, 0.05%; and **306**, 0.62%). Compound **305** and **306** are naturally occurring cinnamosmolide, possessing cytotoxicity and antimicrobial activity, and fragrolide. In this case, the introduction of 3β-hydroxy group was not observed (Sekita et al., 2006) (Figure 23.93).

Naturally occurring rare drimane sesquiterpenoids (**307–314**) were biosynthesized by the fungus *Cryptoporus volvatus* with isocitric acids. Among these compounds, in particular, cryptoporic acid E (**312**) possesses antitumor promoter, anticolon cancer, and very strong superoxide anion radical scavenging activities (Asakawa et al., 1992). When the fresh fungus allowed standing in moisture condition, olive fungus *Paecilomyces variotii* grows on the surface of the fruit body of this fungus. *C. volvatus* infected 2 kg of the fresh fungus for 1 month, followed by the extraction of methanol to give the crude extract and then purification using silica gel and Sephadex LH-20 to give five metabolites (**316**, **318–321**), which were not found in the fresh fungus (Takahashi et al., 1993b). Compound **318** was also isolated from the liverworts, *Bazzania* and *Diplophyllum* species (Asakawa, 1982, 1995) (Figure 23.94).

Liverworts produce a large number of enantiomeric mono-, sesqui-, and diterpenoids to those found in higher plants and lipophilic aromatic compounds. It is also noteworthy that some liverworts produce both normal and its enantiomers. The more interesting phenomenon in the chemistry of liverworts is that the different species in the same genus, for example, *Frullania tamarisci* subsp. *tamarisci* and *Frullania dilatata*, produce totally enantiomeric terpenoids. Various sesqui- and

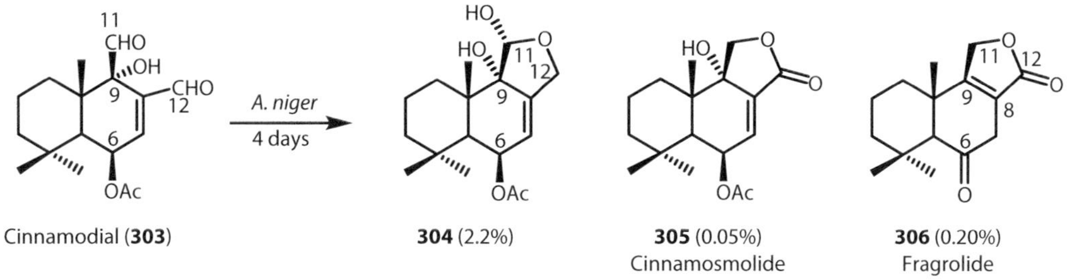

A. niger
4 days

Cinnamodial (**303**) **304** (2.2%) **305** (0.05%) Cinnamosmolide **306** (0.20%) Fragrolide

FIGURE 23.93 Biotransformation of cinnamodial (**303**) by *Aspergillus niger*.

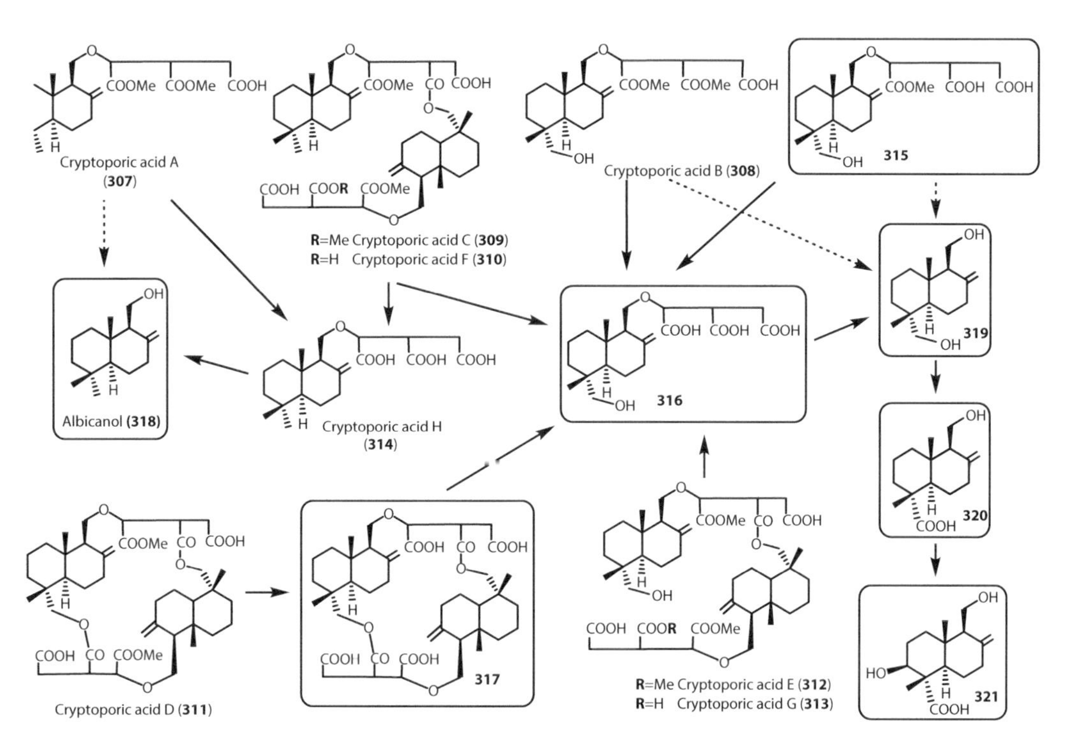

FIGURE 23.94 Biotransformation of cryptoporic acids (**307–317**, **316**) by *Paecilomyces variotii*.

diterpenoids, bibenzyls, and bisbibenzyls isolated from several liverworts show a characteristic fragrant odor, intensely hot and bitter taste, muscle relaxation, allergenic contact dermatitis, and antimicrobial, antifungal, antitumor, insect antifeedant, superoxide anion release inhibitory, piscicidal, and neurotrophic activity (Asakawa, 1982, 1990, 1995, 1999, 2007, 2008; Asakawa and Ludwiczuk, 2008). In order to obtain the different kinds of biologically active products and to compare the metabolites of both normal and enantiomers of terpenoids, several secondary metabolites of specific liverworts were biotransformed by *Penicillium sclerotiorum*, *A. niger*, and *A. cellulosae*.

(−)-Cuparene (**322**) and (−)-2-hydroxycuparene (**323**) have been isolated from the liverworts *Bazzania pompeana* and *M. polymorpha*, while its enantiomer (+)-cuparene (**324**) and (+)-2-hydroxycuparene (**325**) from the higher plant *Biota orientalis* and the liverwort *Jungermannia rosulans*. (*R*)-(−(-α-Cuparenone (**326**) and grimaldone (**327**) demonstrate intense flagrance. In order to obtain such compounds from both cuparene and its hydroxy compounds, both enantiomers mentioned earlier were cultivated with *A. niger* (Hashimoto et al., 2001a) (Figure 23.95).

From (−)-cuparene (**322**), five metabolites (**328–332**) all of which contained cyclopentanediols or hydroxycyclopentanones were obtained. An aryl methyl group was also oxidized to give primary

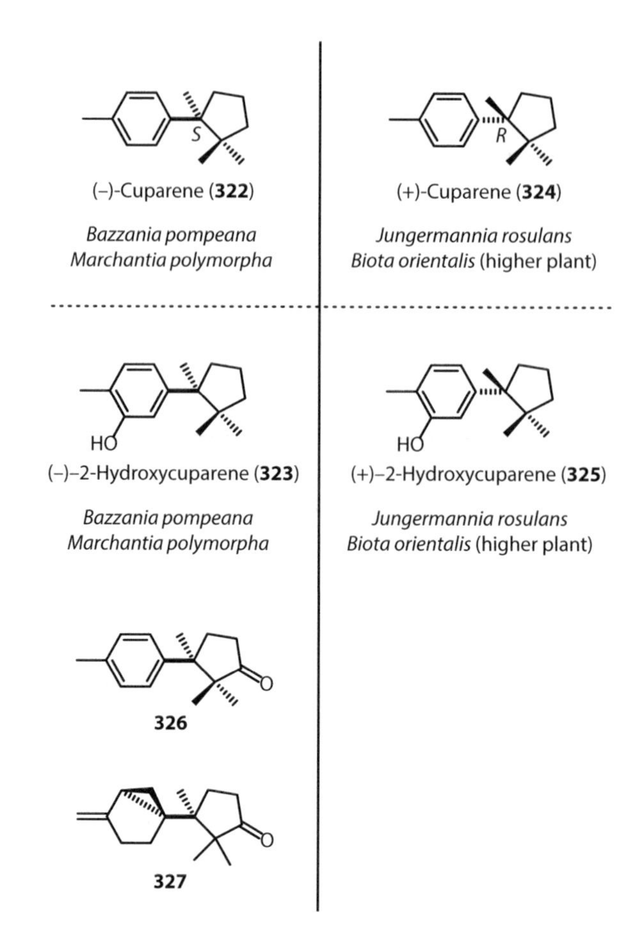

FIGURE 23.95 Naturally occurring cuparene sesquiterpenoids (**322–327**).

alcohol, which was further oxidized to afford carboxylic acids (**329–331**) (Hashimoto et al., 2001a) (Figure 23.96).

From (+)-cuparene, six metabolites (**333–338**) were obtained. These are structurally very similar to those found in the metabolites of (−)-cuparene, except for the presence of an acetonide (**336**), but they are not identical. All metabolites possess benzoic acid moiety.

The possible biogenetic pathways of (+)-cuparene (**324**) have been proposed in Figure 23.97. Unfortunately, none of the metabolites show strong mossy odor (Hashimoto et al., 2006). The presence of an acetonide in the metabolites has also been seen in those of dehydronootkatone (**25**) (Furusawa et al., 2003) (Figure 23.98).

The liverworts *Herbertus aduncus*, *Herbertus sakurai*, and *Mastigophora diclados* produce (−)-herbertene, the C3 methyl isomer of cuparene, with its hydroxy derivatives, for example, herbertanediol (**339**), which shows no production inhibitory activity (Harinantenaina et al., 2007), and herbertenol (**342**). Treatment of compound (**339**) in *P. sclerotiorum* in Czapek-polypeptone medium gave two dimeric products, mastigophorene A (**340**) and mastigophorene B (**341**), which showed neurotrophic activity (Harinantenaina et al., 2005).

When (−)-herbertenol (**342**) was biotransformed for 1 week by the same fungus, no metabolic product was obtained; however, five oxygenated metabolites (**344–348**) wcrc obtained from its methyl ether (**343**). The possible metabolic pathway is shown in Figure 23.99. Except for the presence of the ether (**348**), the metabolites from **342** resemble those found in (−)- and (+)-cuparene (Hashimoto et al., 2006) (Figures 23.100 and 23.101).

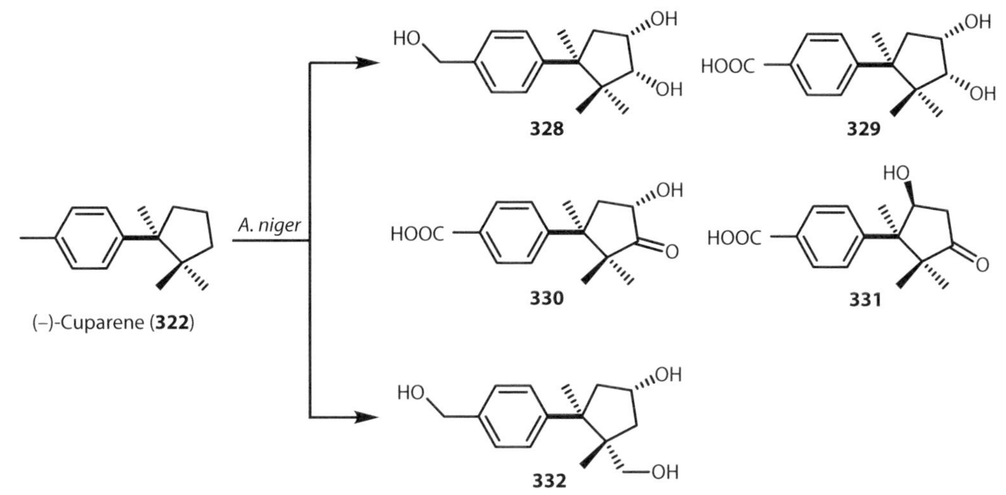

FIGURE 23.96 Biotransformation of (−)-cuparene (**322**) by *Aspergillus niger*.

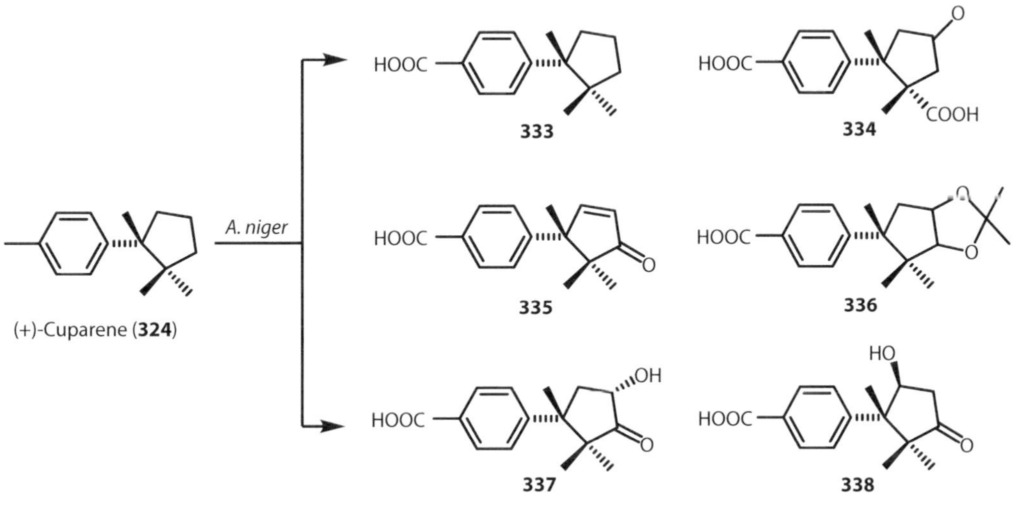

FIGURE 23.97 Biotransformation of (+)-cuparene (**324**) by *Aspergillus niger*.

Maalioxide (**349**), mp 65°–66°, $[\alpha]_D^{21}$ − 34.4°, obtained from the liverwort *P. sciophila* was inoculated and cultivated rotatory (100 rpm) in Czapek-peptone medium (pH 7.0) at 30°C for 2 days. (−)-Maalioxide (**349**) (100 mg/200 mL) was added to the medium and further cultivated for 2 days to afford three metabolites, 1β-hydroxy- (**350**), 1β,9β-dihydroxy- (**351**), and 1β,12-dihydroxymaalioxides (**352**), of which **351** was predominant (53.6%). When the same substrate was cultured with *A. cellulosae* in the same medium for 9 days, 7β-hydroxymaalioxide (**353**) was obtained as a sole product in 30% yield (Hashimoto et al., 2004). The same substrate (**349**) was also incubated with the fungus *M. plumbeus* to obtain a new metabolite, 9β-hydroxymaalioxide (**354**), together with two known hydroxylated products (**350**, **353**) (Wang et al., 2006).

Maalioxide (**349**) was oxidized by *m*-chloroperoxybenzoic acid to give a very small amount of **353** (1.2%), together with 2α-hydroxy- (**355%**, 2%) and 8α-hydroxymaalioxide (**356**, 1.5%), which have not been obtained in the metabolite of **349** in *A. niger* and *A. cellulosae* (Tori et al., 1990) (Figure 23.102).

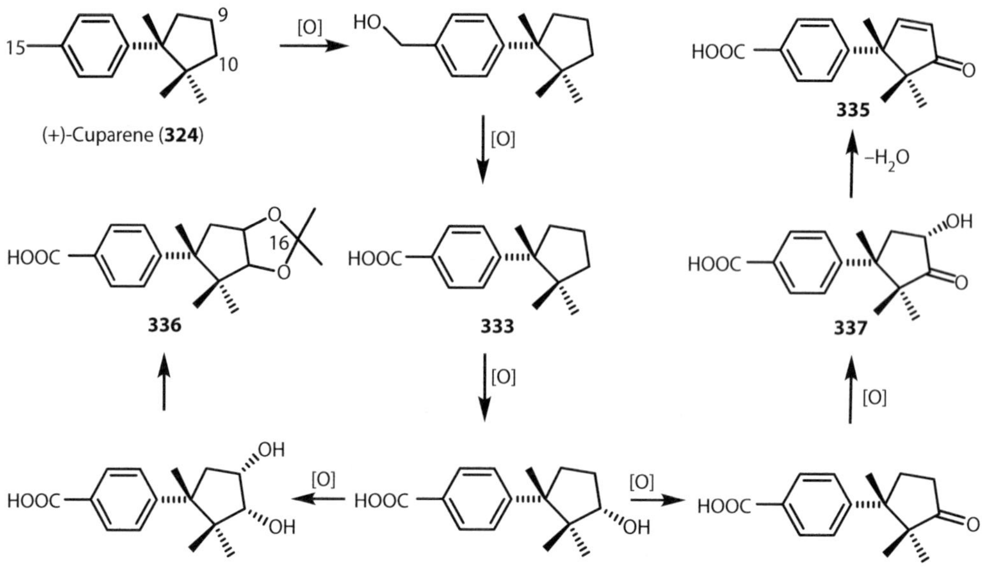

FIGURE 23.98 Possible pathway of biotransformation of (+)-cuparene (**324**) by *Aspergillus niger.*

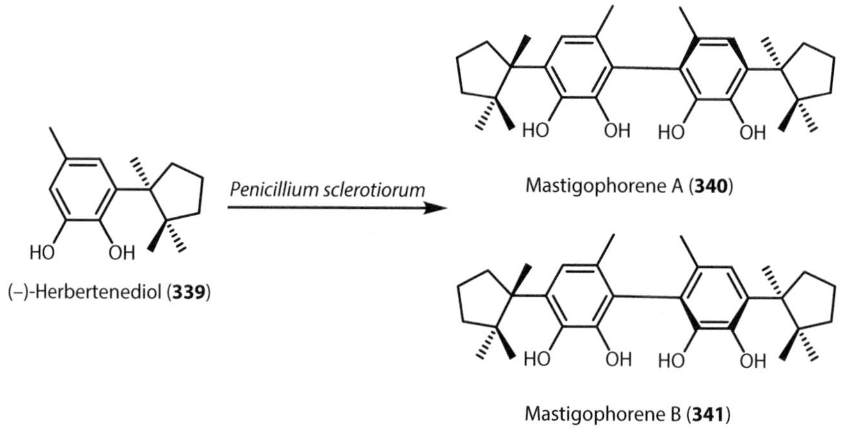

FIGURE 23.99 Biotransformation of (–)-herbertenediol (**339**) by *Penicillium sclerotiorum.*

P. sciophila is one of the most important liverworts, since it produces bicyclohumulenone (**357**), which possesses strong mossy note and is expected to manufacture compounding perfume. In order to obtain much more strong scent, **357** was treated in *A. niger* for 4 days to give 4α,10β-dihydroxybicyclohumulenone (**358**, 27.4%) and bicyclohumurenone-12-oic acid (**359**). An epoxide (**360**) prepared by *m*-chloroperoxybenzoic acid was further treated in the same fungus as described earlier to give 10β-hydoxy derivative (**361**, 23.4%). Unfortunately, these metabolites possess only faint mossy odor (Hashimoto et al., 2003c) (Figure 23.103).

The liverwort *Reboulia hemisphaerica* biosynthesizes cyclomyltaylanoids like **362** and also *ent*-1α-hydroxy-β-chamigrene (**367**). Biotransformation of cyclomyltaylan-5-ol (**362**) in the same medium including *A. niger* gave four metabolites, 9β-hydroxy (**363%**, 27%), 9β,15-dihydroxy (**364**, 1.7%), 10β-hydroxy (**365**, 10.3%), and 9β,15-dihydroxy derivative (**366**, 12.6%). In this case, the stereospecificity of alcohol was observed, but the regiospecificity of alcohol moiety was not seen in this substrate (Furusawa et al., 2005c, 2006b) (Figure 23.104).

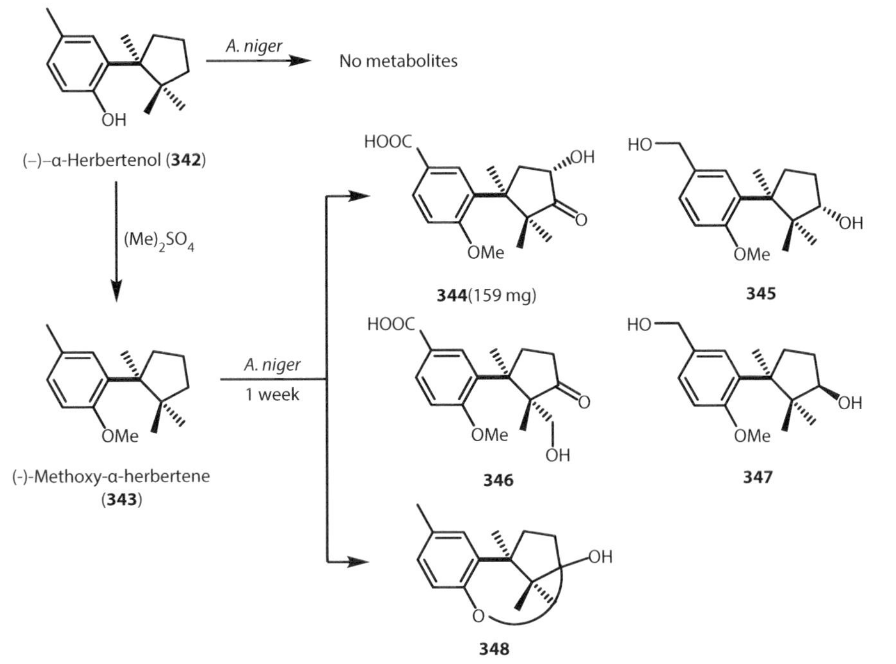

FIGURE 23.100 Biotransformation of (−)-methoxy-α-herbertene (**343**) by *Aspergillus niger.*

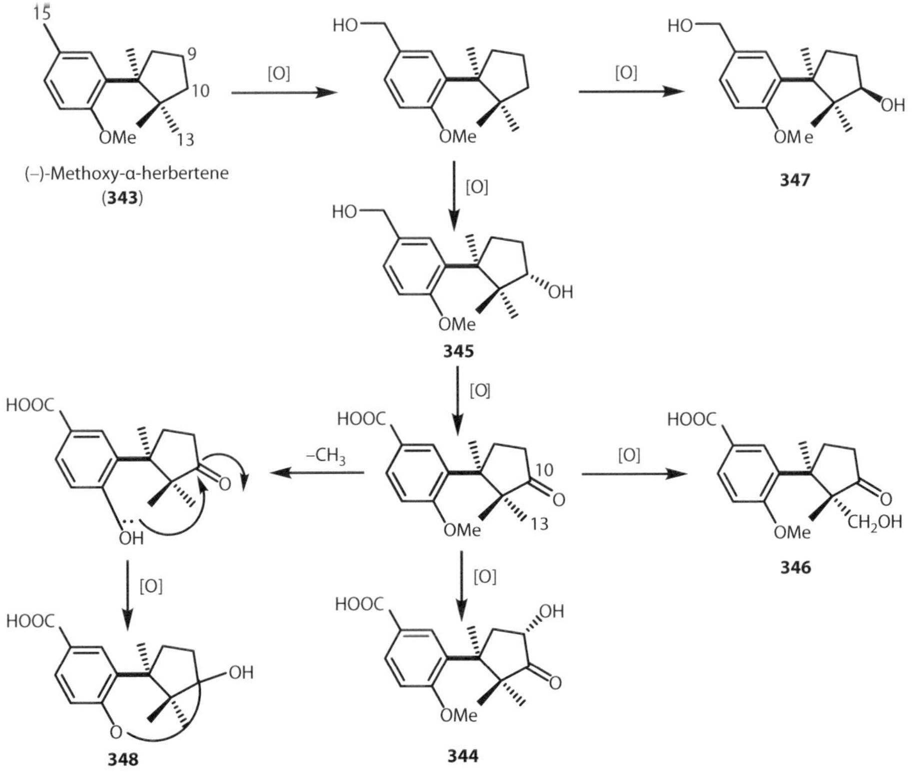

FIGURE 23.101 Possible pathway of biotransformation of (−)-methoxy-α-herbertene (**343**) by *Aspergillus niger.*

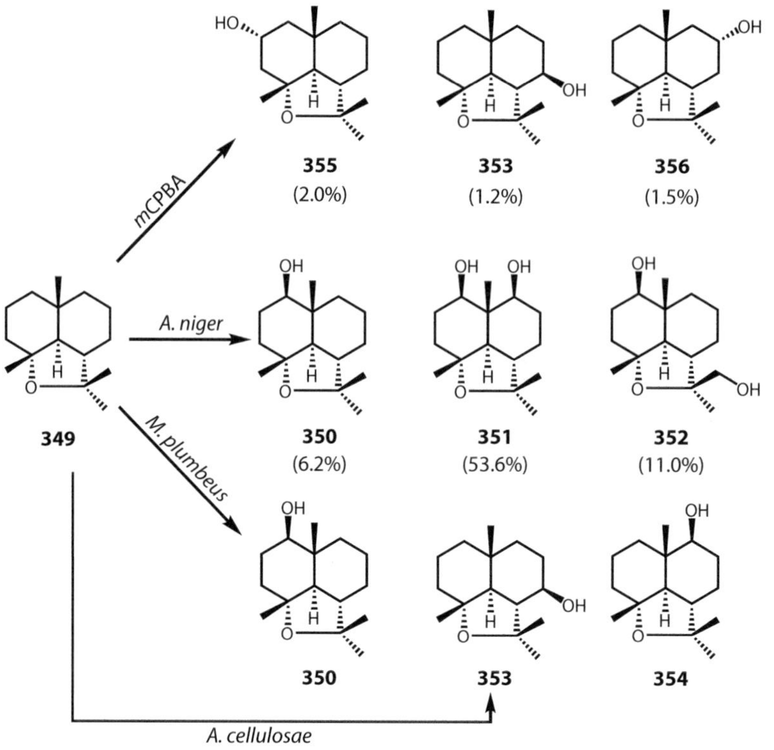

FIGURE 23.102 Biotransformation of maalioxide (**349**) by *Aspergillus niger*, *Aspergillus cellulosae*, and *Mucor plumbeus*.

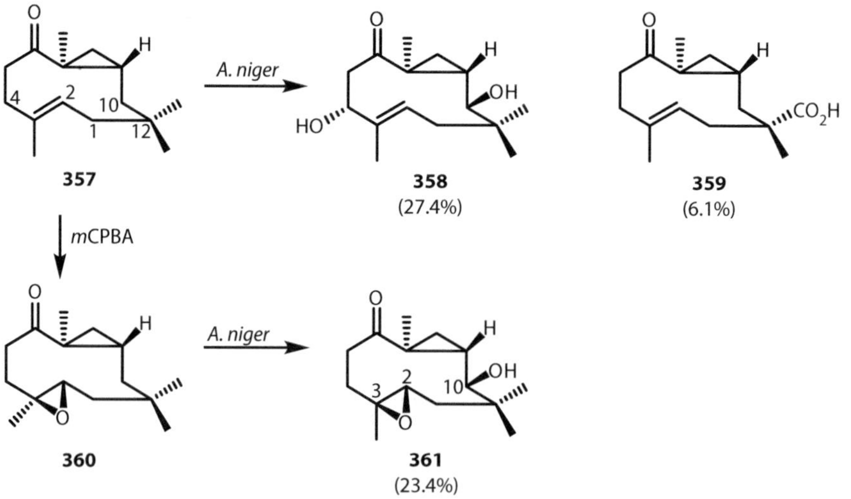

FIGURE 23.103 Biotransformation of bicyclohumulenone (**357**) by *Aspergillus niger*.

The biotransformation of spirostructural terpenoids was not carried out. *ent*-1α-Hydroxy-β-chamigrene (**367**) was inoculated in the same manner as described earlier to give three new metabolites (**368–370**), of which **370** was the major product (46.2% in isolated yield). The hydroxylation of vinyl methyl group has been known to be very common in the case of microbial and mammalian biotransformation (Furusawa et al., 2005a, 2006a,b) (Figure 23.105).

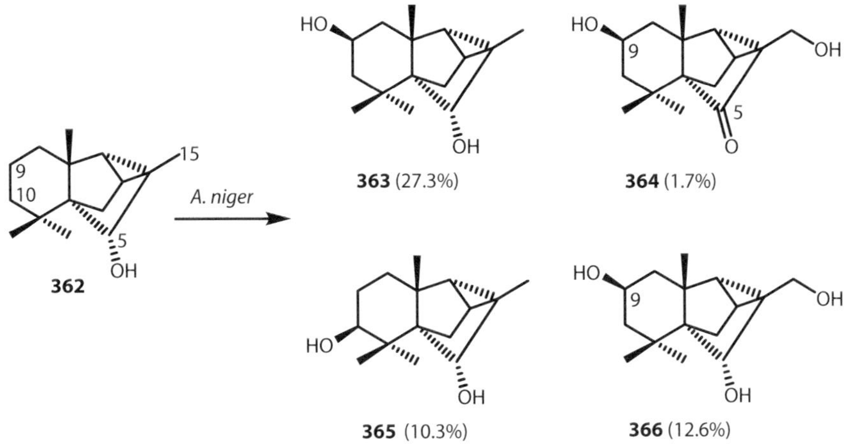

FIGURE 23.104 Biotransformation of cyclomyltaylan-5-ol (**362**) by *Aspergillus niger.*

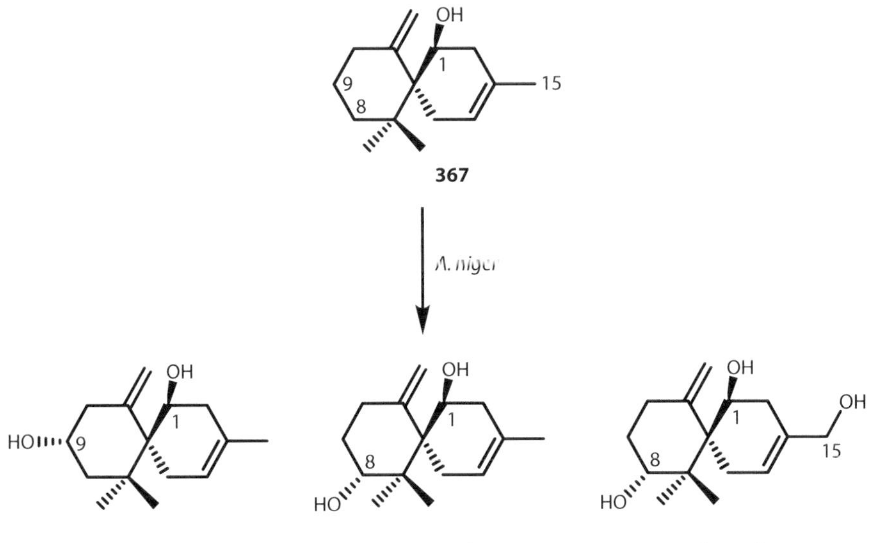

FIGURE 23.105 Biotransformation of *ent*-1α-hydroxy-β-chamigrene (**367**) by *Aspergillus niger.*

β-Barbatene (=gymnomitrene) (**4**), a ubiquitous sesquiterpene hydrocarbon, is from liverwort like *P. sciophila* and many others. Jungermanniales liverworts treated in the same manner using *A. niger* for 1 day gave a triol, 4β,9β,10β-trihydoxy-β-barbatene (**27%**, 8%) (Hashimoto et al., 2003c).

Pinguisane sesquiterpenoids have been isolated from the Jungermanniales, Metzgeriales, and Marchantiales. In particular, the Lejeuneaceae and Porellaceae are rich sources of this unique type of sesquiterpenoids. One of the major furanosesquitepene (**373**) was biodegradated by *A. niger* to afford primary alcohol (**375**), which might be formed from **374** as shown in Figure 23.106 (Lahlou et al., 2000) (Figure 23.107).

In order to obtain more pharmacologically active compounds, the secondary metabolites from crude drugs and animals, for example, nardosinone (**376**) isolated from the crude drug *N. chinensis*, which has been used for headache, stomachache, and diuresis, possesses antimalarial activity. Hinesol (**384**), possessing spasmolytic activity, obtained from *A. lancea* rhizoma, and animal perfume (−)-ambrox (**391**) from ambergris were biotransformed by *A. niger, A. cellulosae, B. dothidea*, and so on.

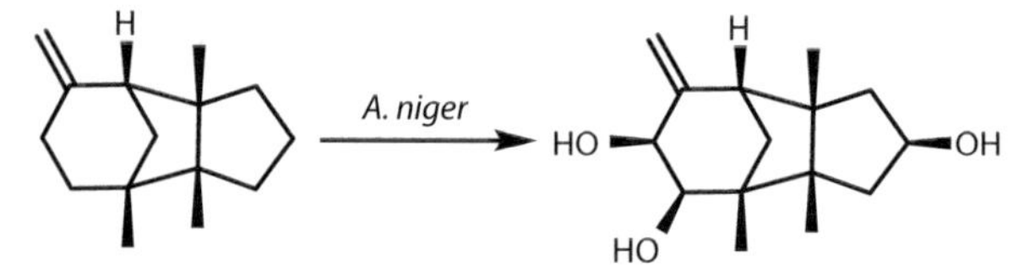

FIGURE 23.106 Biotransformation of β-barbatene (**371**) by *Aspergillus niger*.

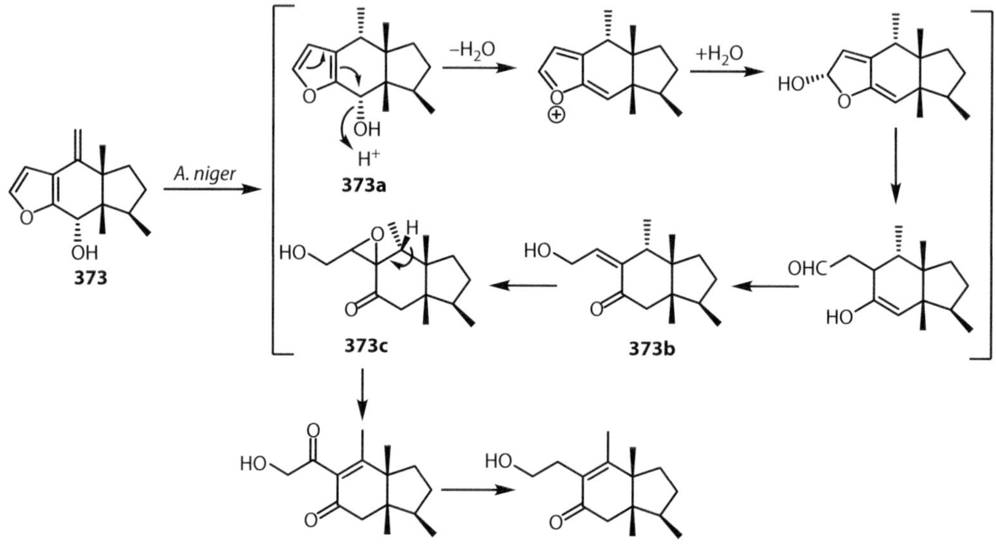

FIGURE 23.107 Biotransformation of pinguisanol (**373**) by *Aspergillus niger*.

Nardosinone (**376**) was incubated in the same medium including *A. niger* as described earlier for 1 day to give six metabolites (**377%**, 45%; **378%**, 3%; **379%**, 2%; **380%**, 5%; **381%**, 6%; and **382%**, 3%). Compounds **380–382** are unique trinorsesquiterpenoids although their yields are very poor. Compound **380** might be formed by the similar manner to that of phenol from cumene (**383**) (Figures 23.108 and 23.109) (Hashimoto et al., 2003d).

From hinesol (**384**), two allylic alcohols (**386**, **387**) and their oxygenated derivative (**385**) and three unique metabolites (**388–390**) having oxirane ring were obtained. The metabolic pathway is very similar to that of oral administration of hinesol since the same metabolites (**395–387**) were obtained from the urine of rabbits (Hashimoto et al., 1998a, 1999b, 2001a,b) (Figure 23.110).

To obtain a large amount of ambrox (**391**), a deterrence, labda-12,14-dien-7α,8-diol obtained from the liverwort *Porella pettottetiana* as a major component, was chemically converted into (−)-ambrox via six steps in relatively high yield (Hashimoto et al., 1998b). Ambrox was added to Czapek-peptone medium including *A. niger*, for 4 days, followed by chromatography of the crude extract to afford four oxygenated products (**392–395**), among which the carboxylic acid (**393**, 52.4%) is the major product (Hashimoto et al., 2001a,b) (Figure 23.111).

When ambrox (**391**) was biotransformed by *A. niger* for 9 days in the presence of 1-aminobenzotriazole, an inhibitor of CYP450, compounds **396** and **397** were obtained instead of the metabolites (**392–395**), which were obtained by incubation of ambrox in the absence of the inhibitor. Ambrox was cultivated by *A. cellulosae* for 4 days in the same medium to afford C1 oxygenated products (**398** and **399**), the former of which was the major product (41.3%) (Hashimoto et al., 2001a,b) (Figure 23.112).

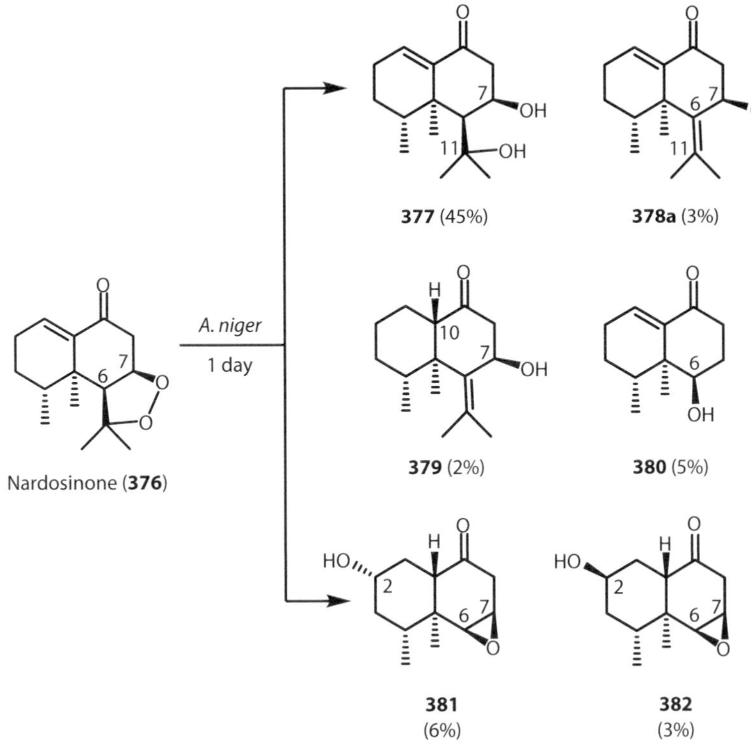

FIGURE 23.108 Biotransformation of nardosinone (**376**) by *Aspergillus niger.*

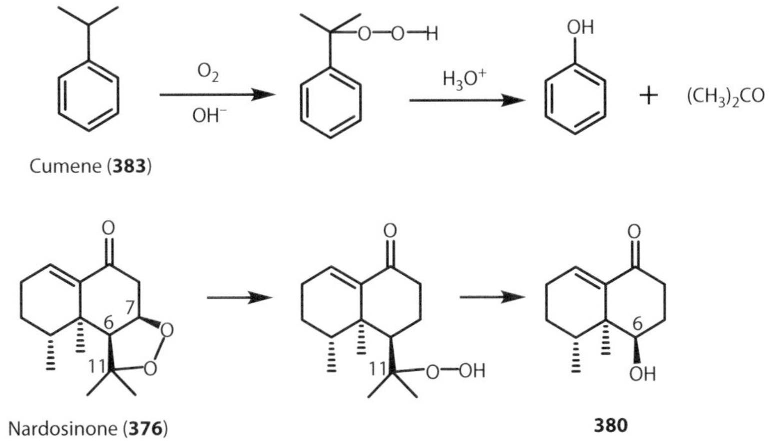

FIGURE 23.109 Possible pathway of biotransformation of nardosinone (**376**) to trinornardosinone (**380**) by *Aspergillus niger.*

The metabolite pathways of ambrox are quite different between *A. niger* and *A. cellulosae.* Oxidation at C1 occurred in *A. cellulosae* to afford **398** and **399**, which was also afforded by John's oxidation of **398**, while oxidation at C3 and C18 and ether cleavage between C8 and C12 occurred in *A. niger* to give **392–395**. Ether cleavage seen in *A. niger* is very rare.

Fragrances of the metabolites (**392–395**) and 7α-hydroxy-(−)-ambrox (**400**) and 7-oxo-(−)-ambrox (**401**) obtained from labdane diterpene diol were estimated. Only **399** demonstrated a similar odor to ambrox (**391**) (Hashimoto et al., 2001a,b) (Figure 23.113).

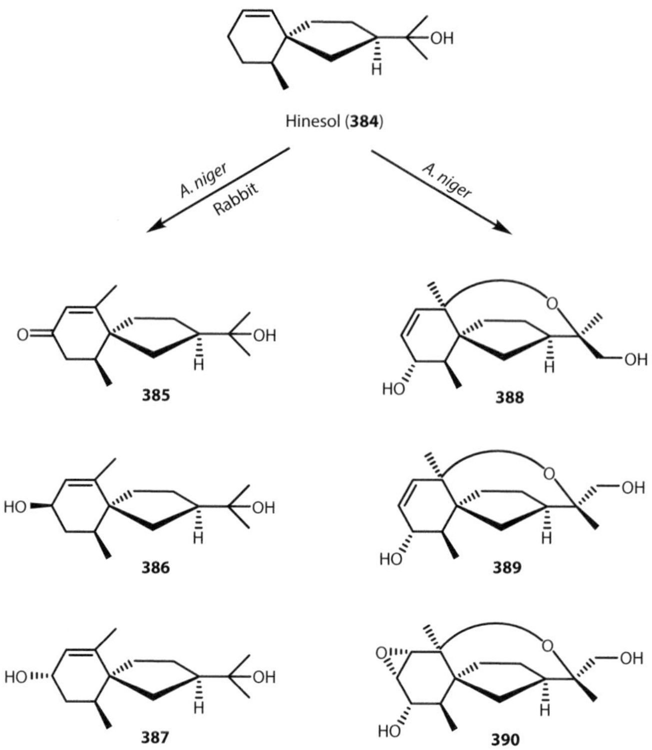

Hinesol (**384**)

A. niger Rabbit → **385**, **386**, **387**

A. niger → **388**, **389**, **390**

FIGURE 23.110 Biotransformation of hinesol (**384**) by *Aspergillus niger.*

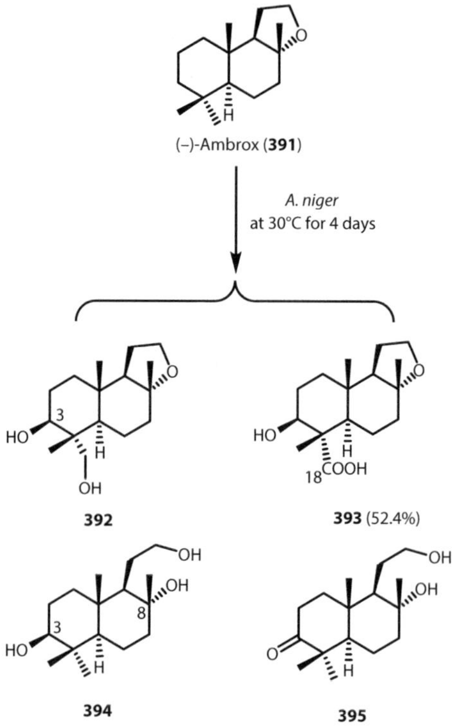

(–)-Ambrox (**391**)

A. niger at 30°C for 4 days

392 **393** (52.4%)

394 **395**

FIGURE 23.111 Biotransformation of (–)-ambrox (**391**) by *Aspergillus niger.*

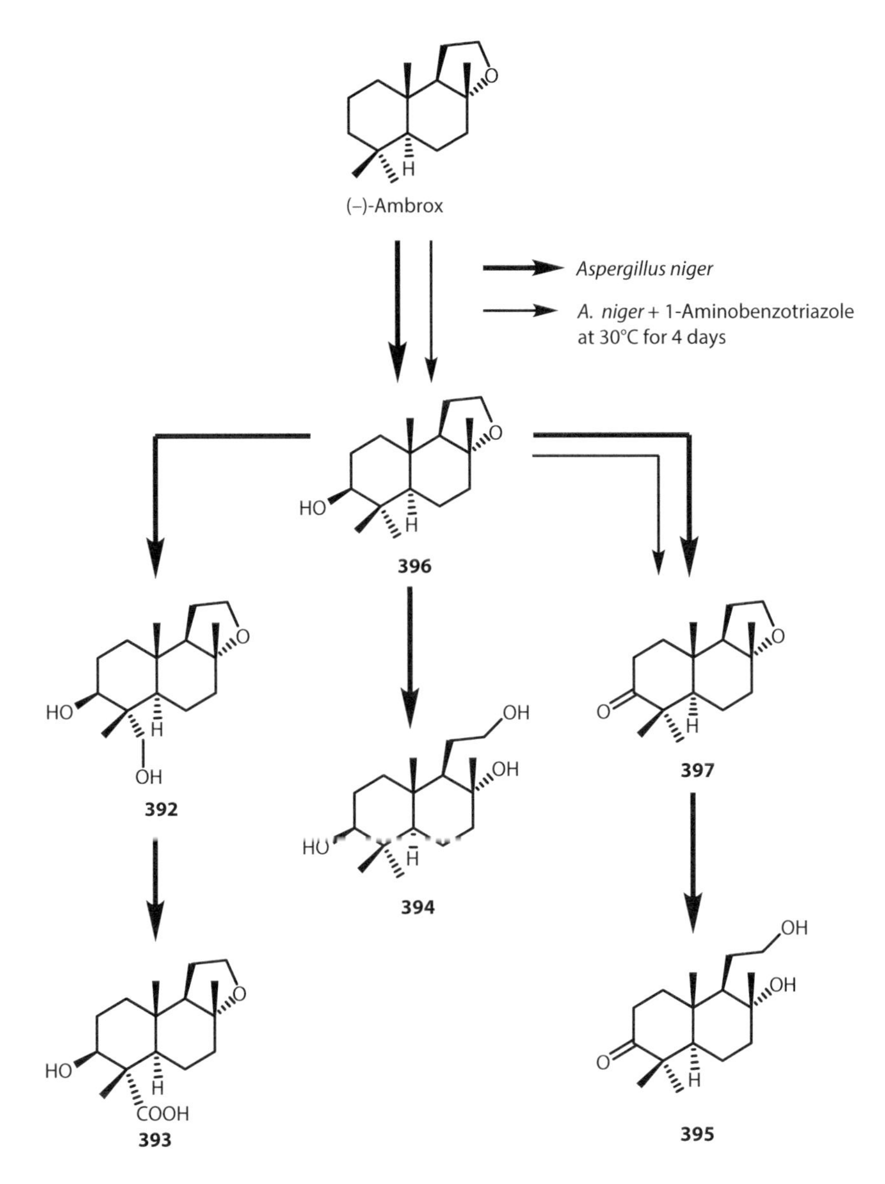

FIGURE 23.112 Possible pathway of biotransformation of (−)-ambrox (**391**) by *Aspergillus niger*.

(−)-Ambrox (**391**) was also microbiotransformed with *Fusarium lini* to give mono-, di-, and trihydroxylated metabolites (**401a**–**401d**), while incubation of the same substrate with *R. stolonifer* afforded two metabolites (**394**, **396**), which were obtained from **391** by *A. niger* as mentioned earlier, together with **397** and **401e** (Choudhary et al., 2004) (Figure 23.114).

The sclareolide (**402**), which is C12 oxo derivative of ambrox, was incubated with *M. plumbeus* to afford three metabolites, 3β-hydroxy- (**403**, 7.9%), 1β-hydroxy- (**404**, 2.5%), and 3-ketosclareolide (**405**, 7.9%) (Aranda et al., 1991) (Figure 23.115).

A. niger in the same medium as mentioned earlier converted sclareolide (**402**) into two new metabolites (**406**, **407**), together with known compounds (**403**, **405**), of which 3β-hydroxy-sclareolide (**403**) is preferentially obtained (Hashimoto et al., 2007) (Figure 23.116).

From the metabolites of sclareolide (**402**) incubated with *C. lunata* and *A. niger*, five oxidized compounds (**403**, **404**, **405**, **405a**, **405b**) were obtained. Fermentation of **402** with *Gibberella fujikuroi* afforded **403**, **404**, **405**, and **405a**. Metabolites **403** and **405a** were formed from the same substrate by

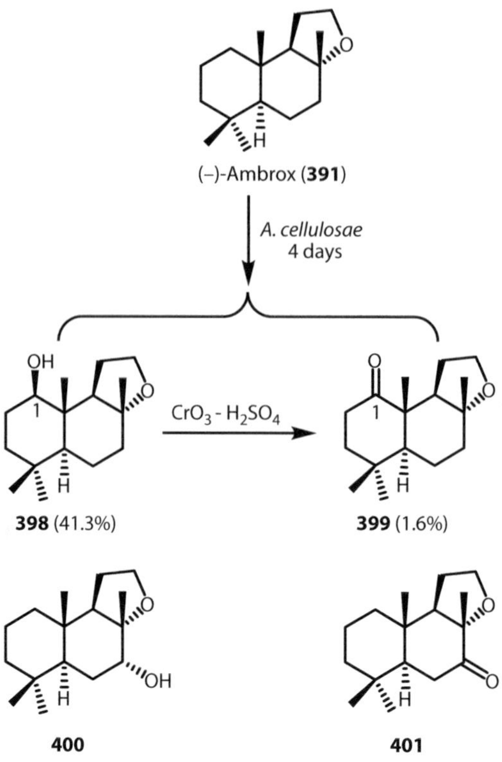

(–)-Ambrox (**391**)

A. cellulosae
4 days

CrO_3 - H_2SO_4

398 (41.3%) **399** (1.6%)

400 **401**

FIGURE 23.113 Biotransformation of (–)-ambrox (**391**) by *Aspergillus cellulosae.*

the incubation of *F. lini*. No microbial transformation of **402** was observed with *Pleurotus ostreatus* (Atta-ur-Rahman et al., 1997) (Figure 23.117).

Compound **391** treated in *C. lunata* gave metabolites **401e** and **396**, while *C. elegans* yielded compounds **401e** and **396** and (+)-sclareolide (**402**) (Figure 23.113). The metabolites (**401a–401e**, **396**) from **391** do not release any effective aroma when compared to **391.** Compound **394** showed a strong sweet odor quite different from the amber-like odor (Choudhary et al., 2004).

Sclareolide (**402**) exhibited phytotoxic and cytotoxic activity against several human cancer cell lines. *C. elegans* gave new oxidized metabolites (**403**, **404**, **405a**, **405c**, **405d**, **405e**), resulting from the enantioselective hydroxylation. Metabolites **403**, **404**, and **405a** have been known as earlier as biotransformed products of **402** by many different fungi and have shown cytotoxicity against various human cancer cell lines. The metabolites (**403**, **404**, and **405a**) indicated significant phytotoxicity at higher dose against *Lemna minor* L. (Choudhary et al., 2004) (Figure 23.117).

Ambrox (**391**) and sclareolide (**402**) were incubated with the fungus *C. aphidicola* for 10 days in shake culture to give 3β-hydroxy- (**396**), 3β,6β-dihydroxy- (**401 g**), 3β,12-dihydroxy- (**401 h**), and sclareolide 3β,6β-diol (**401f**), and 3β-hydroxy- (**403**), 3-keto- (**405**), and sclareolide 3β,6β-diol (**401f**), respectively (Hanson and Truneh, 1996) (Figure 23.118).

Zerumbone (**408**), which is easily isolated from the wild ginger *Zingiber zerumbet*, and its epoxide (**409**) were incubated with *F. culmorum* and *A. niger* in Czapek-peptone medium, respectively. The former fungus gave (1R,2R)-(+)-2,3-dihydrozerumbol (**410**) stereospecifically via either 2,3-dihydrozerumbone (**408a**) or zerumbol (**408b**) or both and accumulated **410** in the mycelium. The facile production of optically active **410** will lead a useful material of woody fragrance, 2,3-dihydrozerumbone. *A. niger* biotransformed **408** via epoxide (**409**) to several metabolites containing zerumbone-6,7-diol as a main product. The same fungus converted the epoxide (**409**) into three major metabolites containing (2*R*,6*S*,7*S*,10*R*,11*S*)-1-oxo7,9-dihydroxyisodaucane (**413**) via

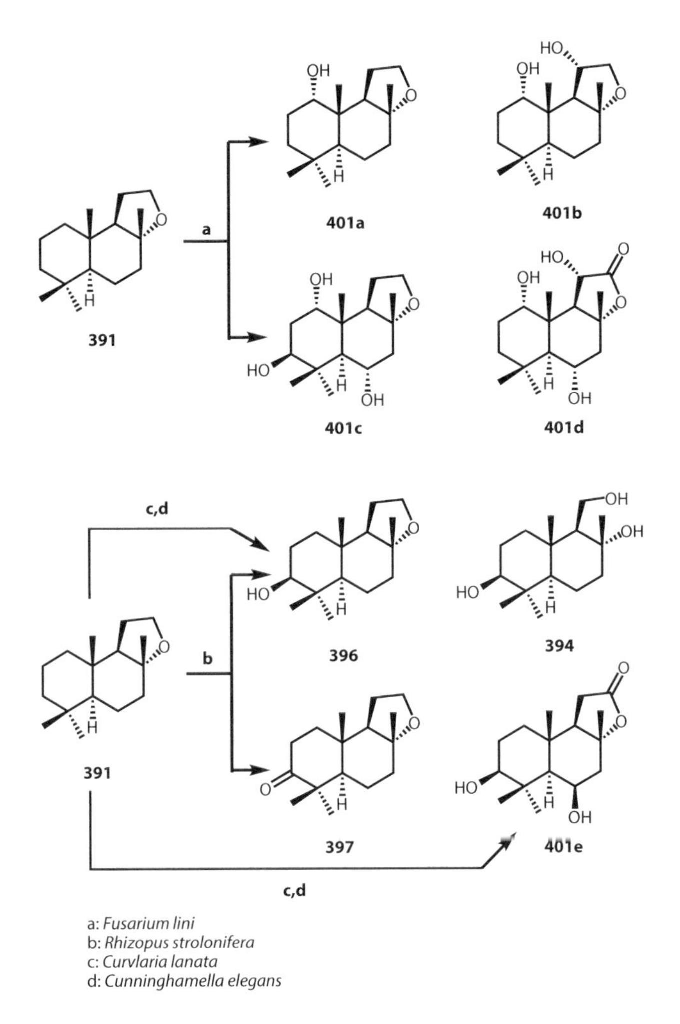

FIGURE 23.114 Biotransformation of (−)-ambrox (**391**) by *Fusarium lini* and *Rhizopus stolonifer*.

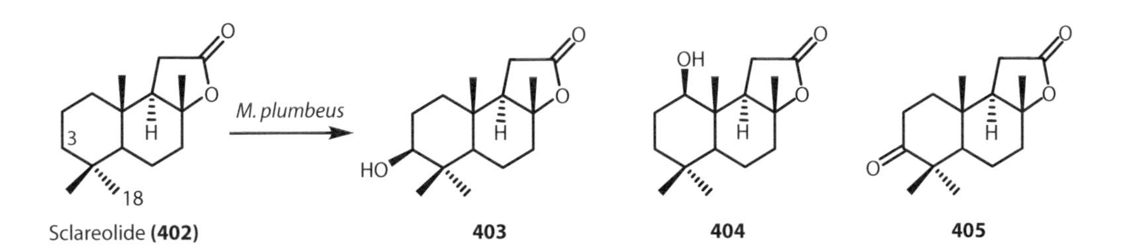

FIGURE 23.115 Biotransformation of (+)-sclareolide (**402**) by *Mucor plumbeus*.

dihydro derivatives (**411**, **412**). However, *A. niger* biotransformed **409** only into **412** in the presence of the CYP450 inhibitor 1-aminobenzotriazole (Noma et al., 2002).

The same substrate was incubated in *A. niger*, *Aspergillus oryzae*, *Candida rugosa*, *Candida tropicalis*, *Mucor mucedo*, *Bacillus subtilis*, and *Schizosaccharomyces pombe*; however, any metabolites have been obtained. All microbes except for the last organism, zerumbone epoxide (**409**), prepared by *m*CPBA, bioconverted into two diastereoisomers, 2*R*,6*S*,7*S*-dihydro- (**411**) and 2*R*,6*R*,7*R*-derivative (**412**), whose ratio was determined by GC, and their enantio-excess was over 99% (Nishida and Kawai, 2007) (Figure 23.119).

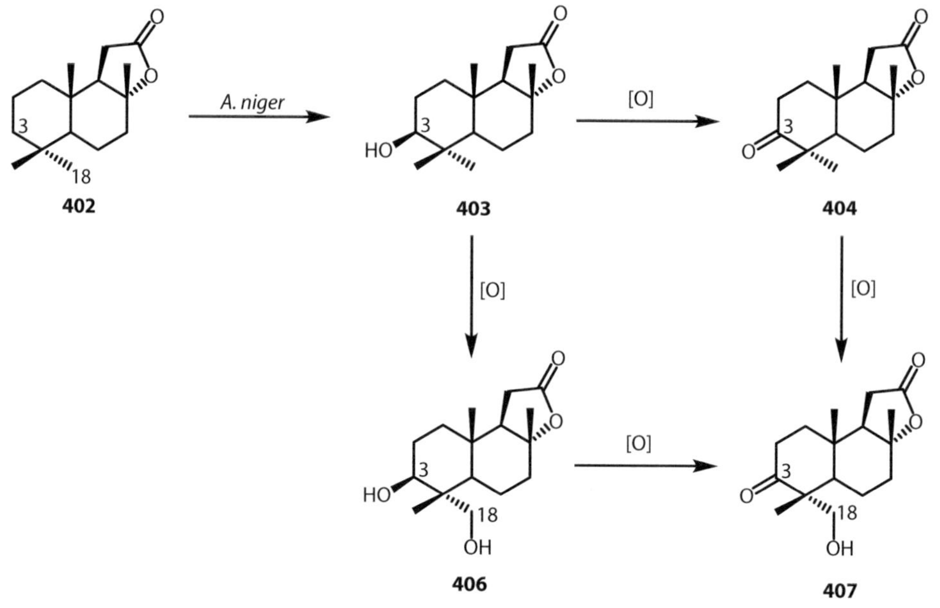

FIGURE 23.116 Biotransformation of (+)-sclareolide (**402**) by *Aspergillus niger*.

Several microorganisms and a few mammals (see later) for the biotransformation of (+)-cedrol (**414**), which is widely distributed in the cedar essential oils, were used. Plant pathogenic fungus *G. cingulata* converted cedrol (**414**) into three diols (**415–417**) and 2α-hydroxycedrene (**418**) (Miyazawa et al., 1995a). The same substrate (**414**) was incubated with *A. niger* to give **416** and **417** together with a cyclopentanone derivative (**419**) (Higuchi et al., 2001). Human skin microbial flora, *Staphylococcus epidermidis* also converted (+)-cedrol into 2α-hydroxycedrol (**415**) (Itsuzaki et al., 2002) (Figure 23.120).

C. aphidicola bioconverted cedrol (**414**) into **417** (Hanson and Nasir, 1993). On the other hand, *Corynespora cassiicola* produced **419** in addition to **417** (Abraham et al., 1987). It is noteworthy that *B. cinerea* that damages many flowers, fruits, and vegetables biotransformed cedrol into different metabolites (**420–422**) from those mentioned earlier (Aleu et al., 1999a).

4α-Hydroxycedrol (**424**) was obtained from the metabolite of cedrol acetate (**423**) by using *G. cingulata* (Matsui et al., 1999) (Figure 23.121).

Patchouli alcohol (**425**) was treated in *B. cinerea* to give three metabolites two tertiary alcohols (**426**, **427**), four secondary alcohols (**428**, **430**, **430a**), and two primary alcohols (**430b**, **430c**) of which compounds **425**, **427**, and **428** are the major metabolites (Aleu et al., 1999a), while plant pathogenic fungus *G. cingulata* converted the same substrate to 5-hydroxy- (**426**) and 5,8-dihydroxy derivative (**429**) (Figure 23.122).

In order to confirm the formation of **429** from **426**, the latter product was reincubated in the same medium including *G. cingulata* to afford **429** (Miyazawa et al., 1997c) (Figure 23.123).

Patchouli acetate (**431**) was also treated in the same medium to give **426** and **429** (Matsui and Miyazawa, 2000). 5-Hydroxy-α-patchoulene (**432**) was incubated with *G. cingulata* to afford 1α-hydroxy derivative (**426**) (Miyazawa et al., 1998b).

(−)-α-Longipinene (**433**) was treated with *A. niger* to afford 12-hydroxylated product (**434**) (Sakata et al., 2007).

Ginsenol (**435**), which was obtained from the essential oil of *Panax ginseng*, was incubated with *B. cinerea* to afford four secondary alcohols (**436–439**) and two cyclohexanone derivatives (**440**) from **437** and **441** from **438** or **439**. Some of the oxygenated products were considered as potential antifungal agents to control *B. cinerea* (Aleu et al., 1999b) (Figures 23.124 and 23.125).

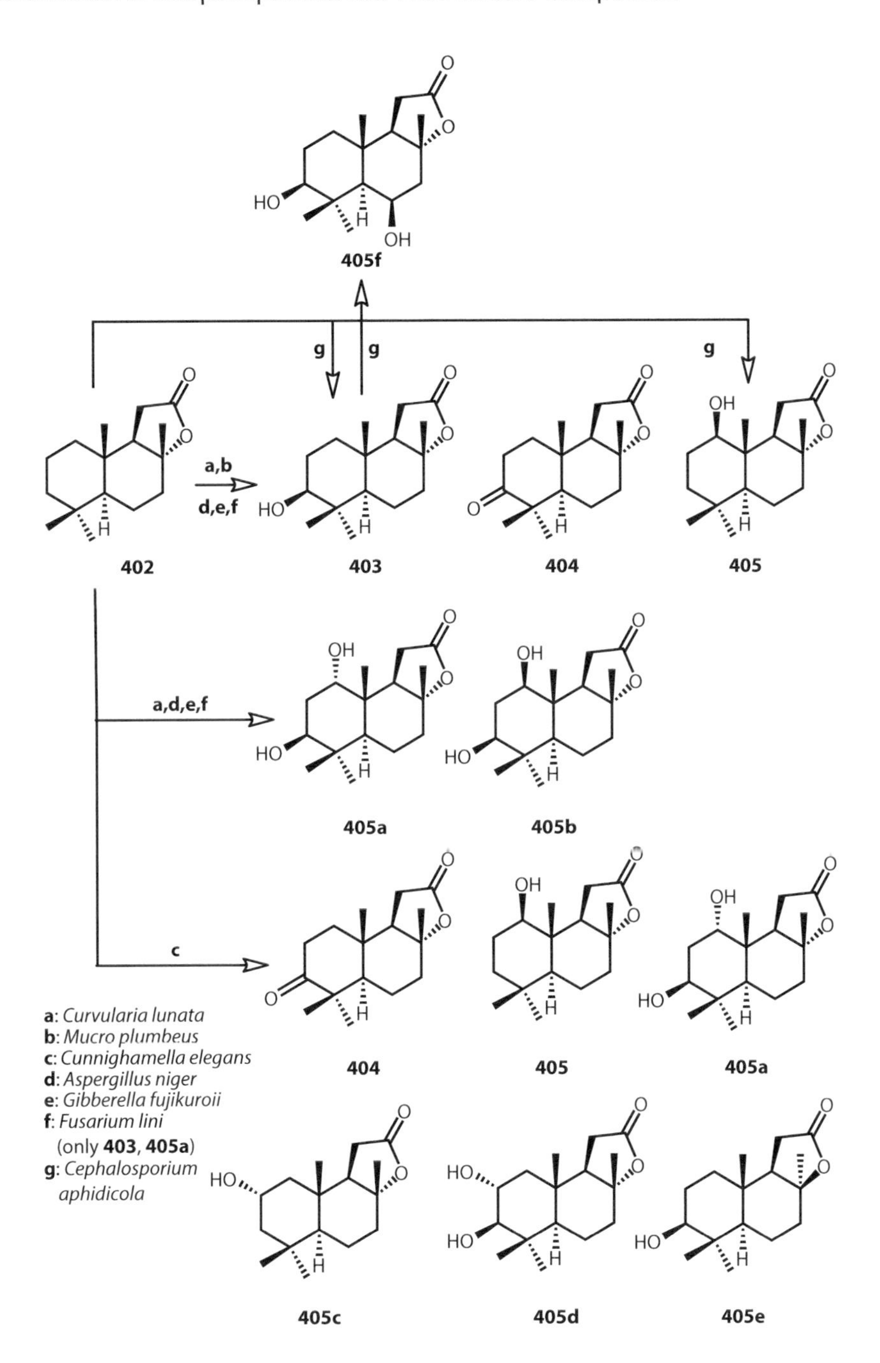

FIGURE 23.117 Biotransformation of (+)-sclareolide (**402**) by various fungi.

(+)-Isolongifolene-9-one (**442**), which was isolated from some cedar trees, was treated in *G. cingulata* for 15 days to afford two primary alcohols (**443**, **444**) and a secondary alcohol (**445**) (Sakata and Miyazawa, 2006) (Figure 23.126).

Choudhary et al. (2005) reported that fermentation of (−)-isolongifolol (**445a**) with *F. lini* resulted in the isolation of three metabolites, 10-oxo- (**445b**), 10α-hydroxy- (**445c**), and 9α-hydroxyisolongifolol (**445d**). Then the same substrate was incubated with *A. niger* to yield the products **445c** and **445d**. Both **445c** and **445d** showed inhibitory activity against butylcholinesterase enzyme in a concentration-dependent manner with IC_{50} 13.6 and 299.5 mM, respectively (Figure 23.127).

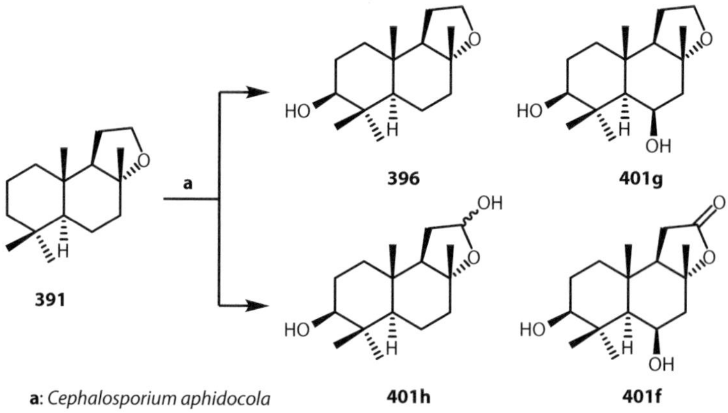

FIGURE 23.118 Biotransformation of (−)-ambrox (**391**) by *Cephalosporium aphidicola.*

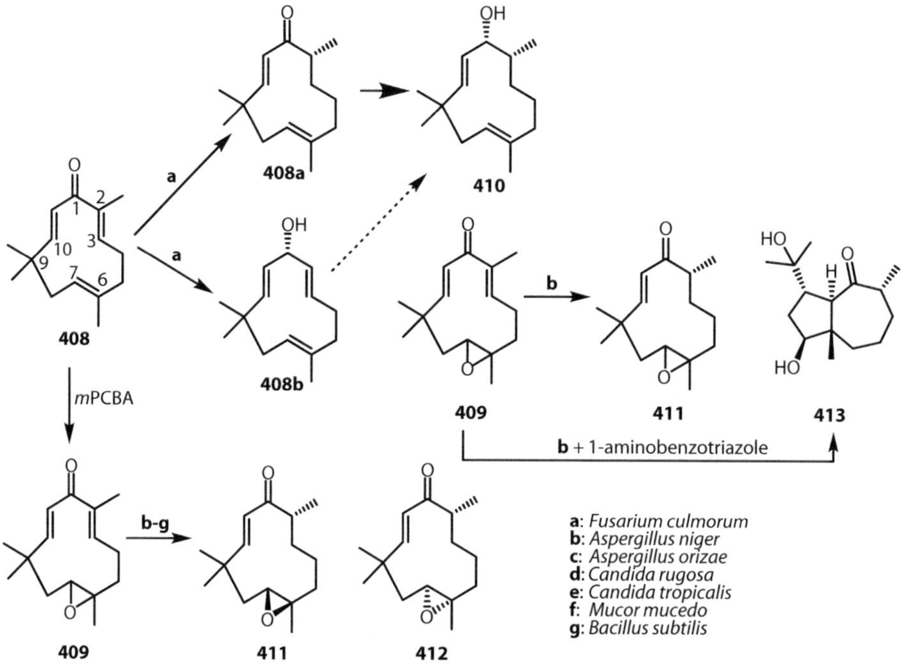

FIGURE 23.119 Biotransformation of zerumbone (**408**) by various fungi.

(+)-Cycloisolongifol-5β-ol (**445e**) was fermented with *C. elegans* to afford three oxygenated metabolites: 11-oxo (**445f**), 3β-hydroxy (**445 g**), and 3β,11α-dihydroxy derivative (**445 h**) (Choudhary et al., 2006a) (Figure 23.128).

A daucane-type sesquiterpene derivative, lancerroldiol *p*-hydroxybenzoate (**446**), was hydrogenated with cultured suspension cells of the liverwort *M. polymorpha* to give 3,4-dihydrolancerodiol (**447**) (Hegazy et al., 2005) (Figure 23.129).

Widdrane sesquiterpene alcohol (**448**) was incubated with *A. niger* to give an oxo and an oxy derivative (**449**, **450**) (Hayashi et al., 1999) (Figure 23.130).

(−)-β-Caryophyllene (**451**), one of the ubiquitous sesquiterpene hydrocarbons found not only in higher plants but also in liverworts, was biotransformed by *Pseudomonas cruciviae*, *D. gossypina*, and *Chaetomium cochliodes* (Lamare and Furstoss, 1990). *P. cruciviae* gave a ketoalcohol (**452**)

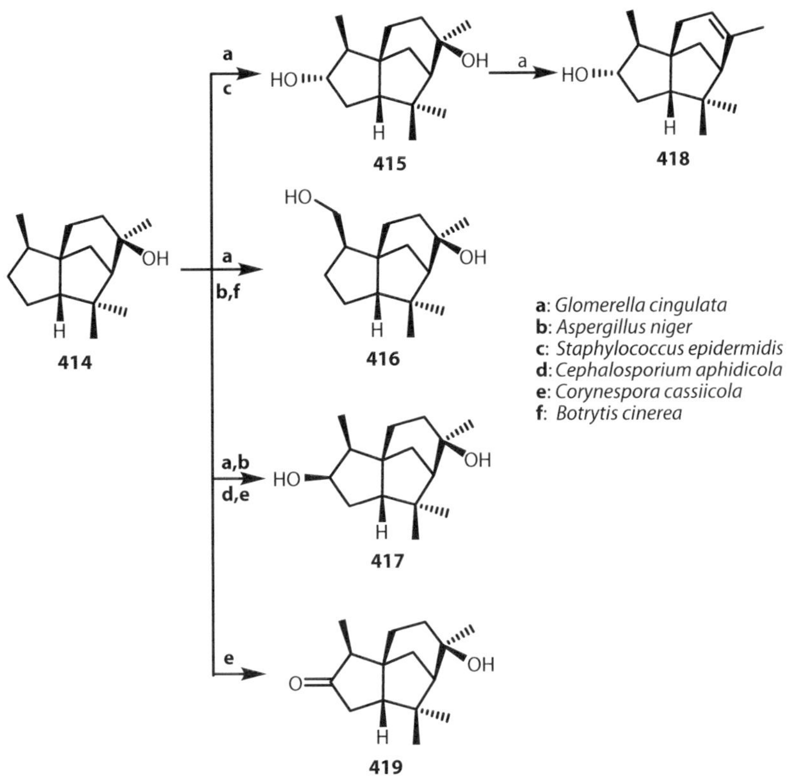

414
415
416
417
418
419
a: Glomerella cingulata
b: Aspergillus niger
c: Staphylococcus epidermidis
d: Cephalosporium aphidicola
e: Corynespora cassiicola
f: Botrytis cinerea

FIGURE 23.120 Biotransformation of cedrol (**414**) by various fungi.

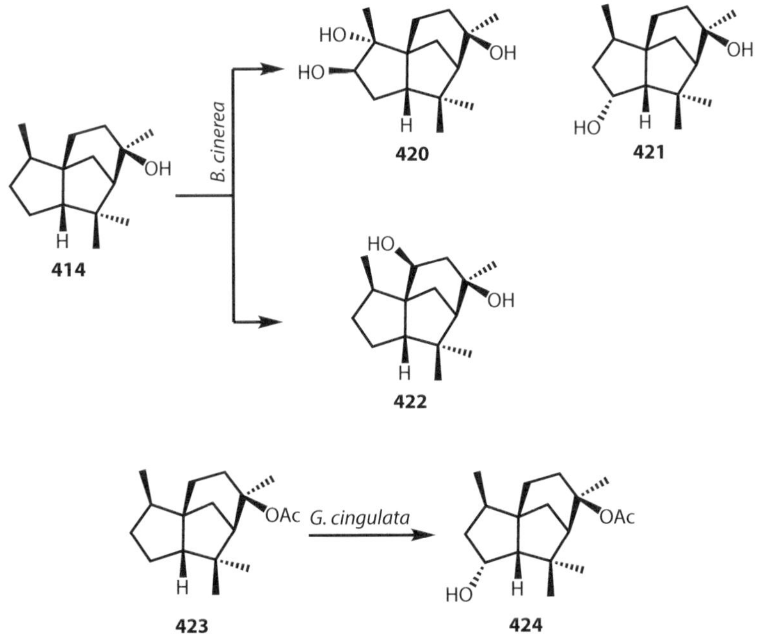

B. cinerea
G. cingulata
414
420
421
422
423
424

FIGURE 23.121 Biotransformation of cedrol (**414**) by *Botrytis cinerea and Glomerella cingulata.*

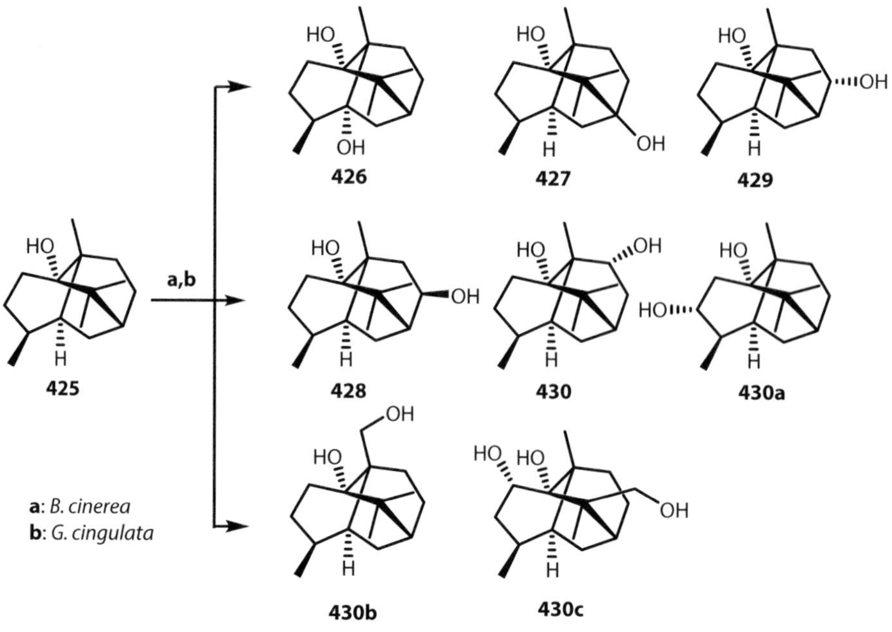

a: *B. cinerea*
b: *G. cingulata*

FIGURE 23.122 Biotransformation of patchoulol (**425**) by *Botrytis cinerea*.

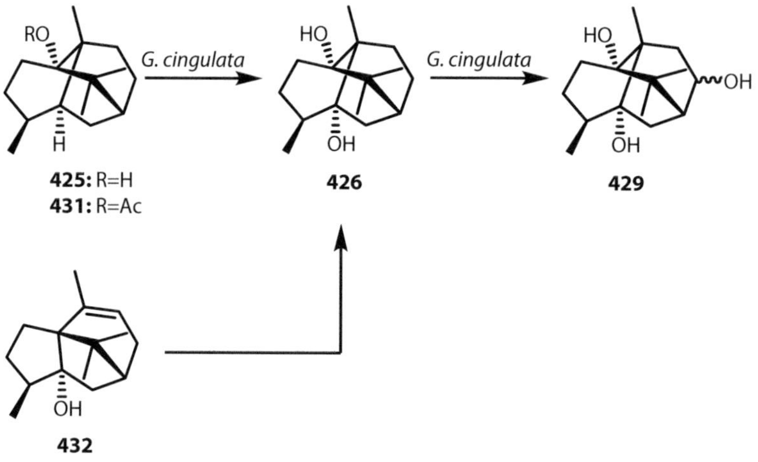

425: R=H
431: R=Ac

FIGURE 23.123 Biotransformation of patchoulol (**425**) by *Glomerella cingulata*.

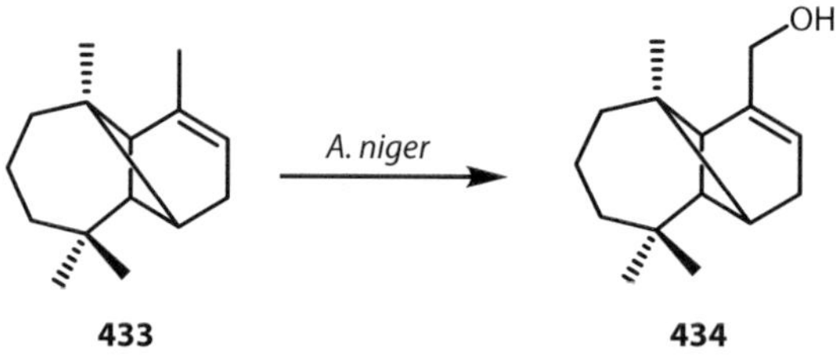

FIGURE 23.124 Biotransformation of α-longipinene (**433**) by *Aspergillus niger*.

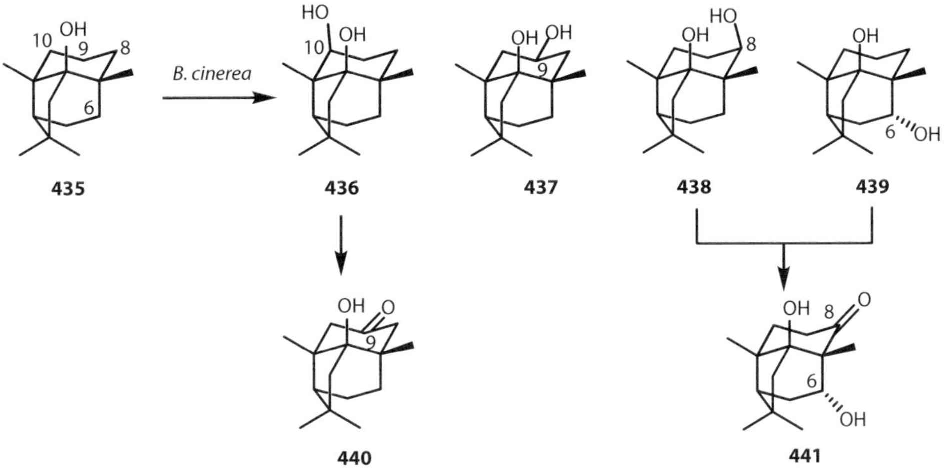

FIGURE 23.125 Biotransformation of ginsenol (**435**) by *Botrytis cinerea.*

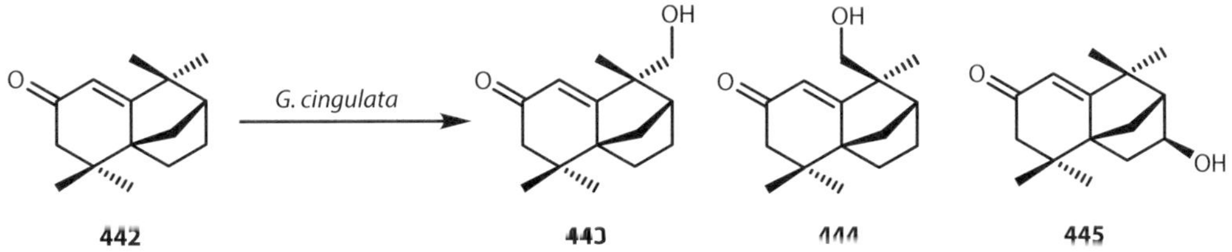

FIGURE 23.126 Biotransformation of (+)-isolongifolene-9-one (**442**) by *Glomerella cingulata.*

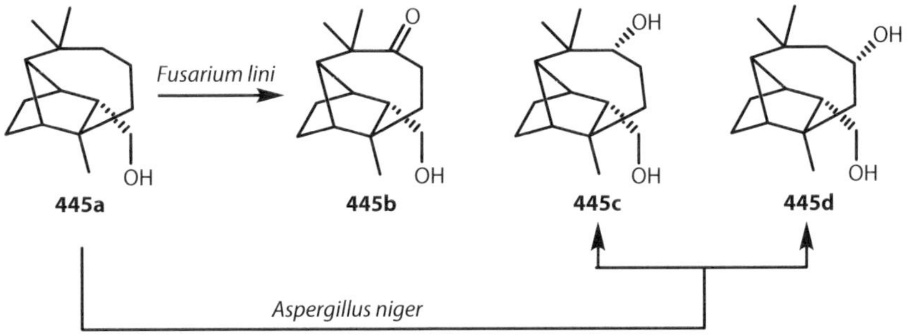

FIGURE 23.127 Biotransformation of (−)-isolongifolol (**445a**) by *Aspergillus niger* and *Fusarium lini.*

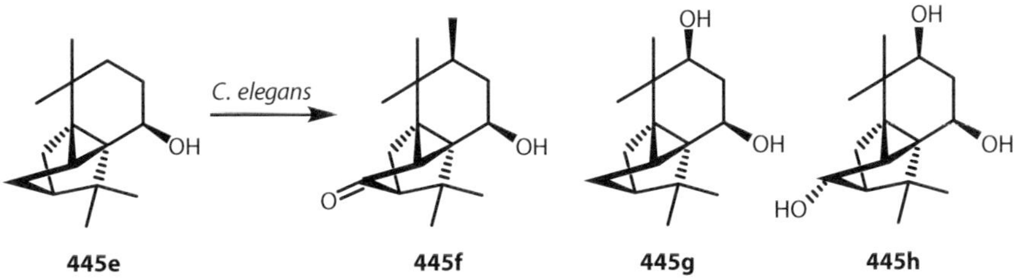

FIGURE 23.128 Biotransformation of (+)-cycloisolongifol-5β-ol (**445e**) by *Cunninghamella elegans.*

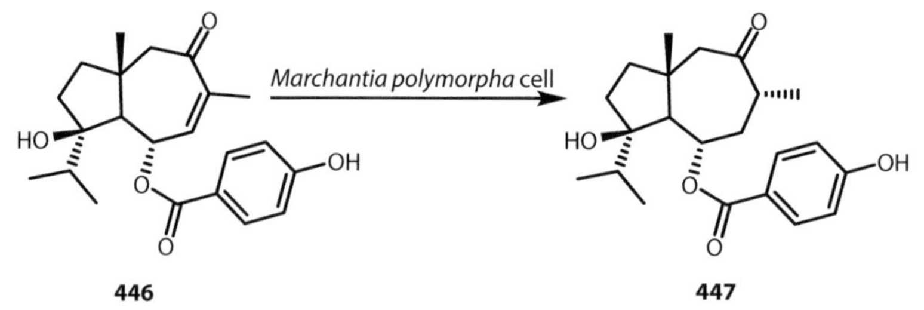

FIGURE 23.129 Biotransformation of lancerodiol *p*-hydroxybenzoate (**446**) by *Marchantia polymorpha* cells.

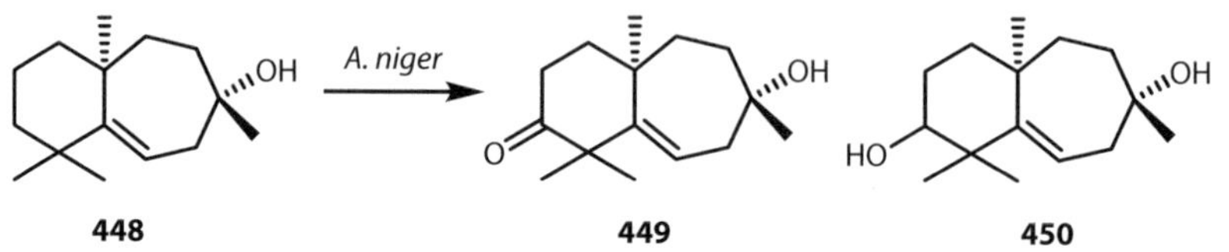

FIGURE 23.130 Biotransformation of widdrol (**448**) by *Aspergillus niger.*

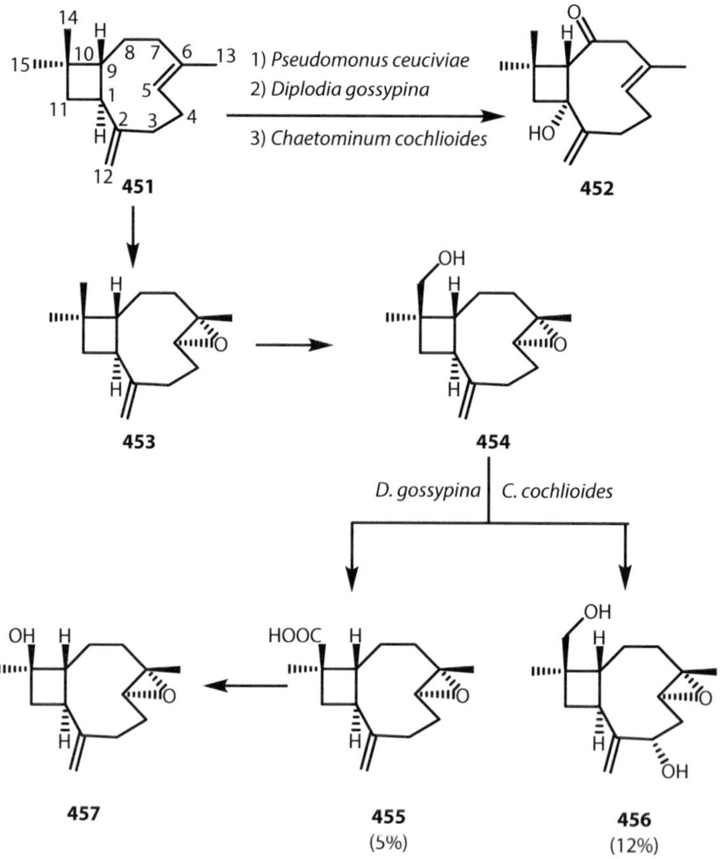

FIGURE 23.131 Biotransformation of (−)-β-caryophyllene (**451**) by *Pseudomonas cruciviae*, *Diplodia gossypina*, and *Chaetomium cochliodes*.

(Devi, 1979), while the latter two species produced the 14-hydroxy-5,6-epoxide (**454**), its carboxylic (**455**), and 3α-hydroxy- (**456**) and norcaryophyllene alcohol (**457**), all of which might be formed from caryophyllene C5,C6-epoxide (**453**). The oxidation pattern of (−)-β-caryophyllene by the fungi is very similar to that by mammals (see later) (Figure 23.131).

Fermentation of (−)-β-caryophyllene (**451**) with *D. gossypina* afforded **14** different metabolites (**453–457j**), among which 14-hydroxy-5,6-epoxide (**454**) and the corresponding acid (**455**) were the major metabolites. Compound **457j** is structurally very rare and found in *Poronia punctata*. The main reaction path is epoxidation at C5, C6 as mentioned earlier and selective hydroxylation at C4 (Abraham et al., 1990) (Figure 23.132).

(−)-β-Caryophyllene epoxide (**453**) was incubated with *C. aphidicola* for 6 days to afford two metabolites (**457 l**, **457 m**), while *Macrophomina phaseolina* biotransformed the same substrate to 14- (**454**) and 15-hydroxy derivatives (**457k**). The same substrate was treated in *A. niger*, *G. fujikuroi*, and *R. stolonifer* for 8 days and *F. lini* for 10 days to afford the metabolites **457n**, **457o**, **457p** and

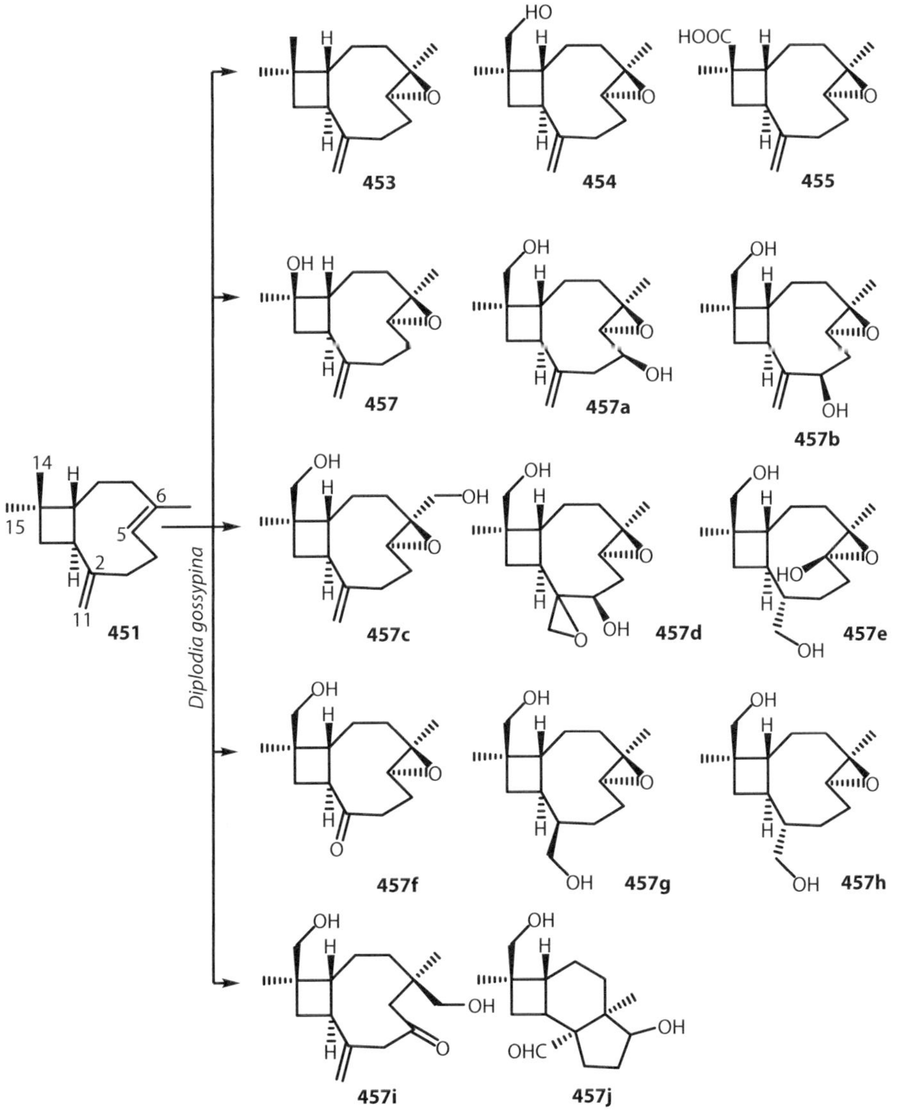

FIGURE 23.132 Biotransformation of (−)-β-caryophyllene (**451**) by *Diplodia gossypina*.

457q, and **457r**, respectively. All metabolites were estimated for butyrylcholine esterase inhibitory activity, and compound **457k** was found to show potency similar activity to galantamine HBr (IC_{50} 10.9 vs. 8.5 mM) (Choudhary et al., 2006b) (Figure 23.133).

The fermentation of (−)-β-caryophyllene oxide (**453**) using *B. cinerea* and the isolation of the metabolites were carried out by Duran et al. (1999). Kobuson (**457w**) was obtained with 14 products (**457s**–**457u**, **457x**). Diepoxides **457t** and **457u** could be the precursors of epimeric alcohols **457q**

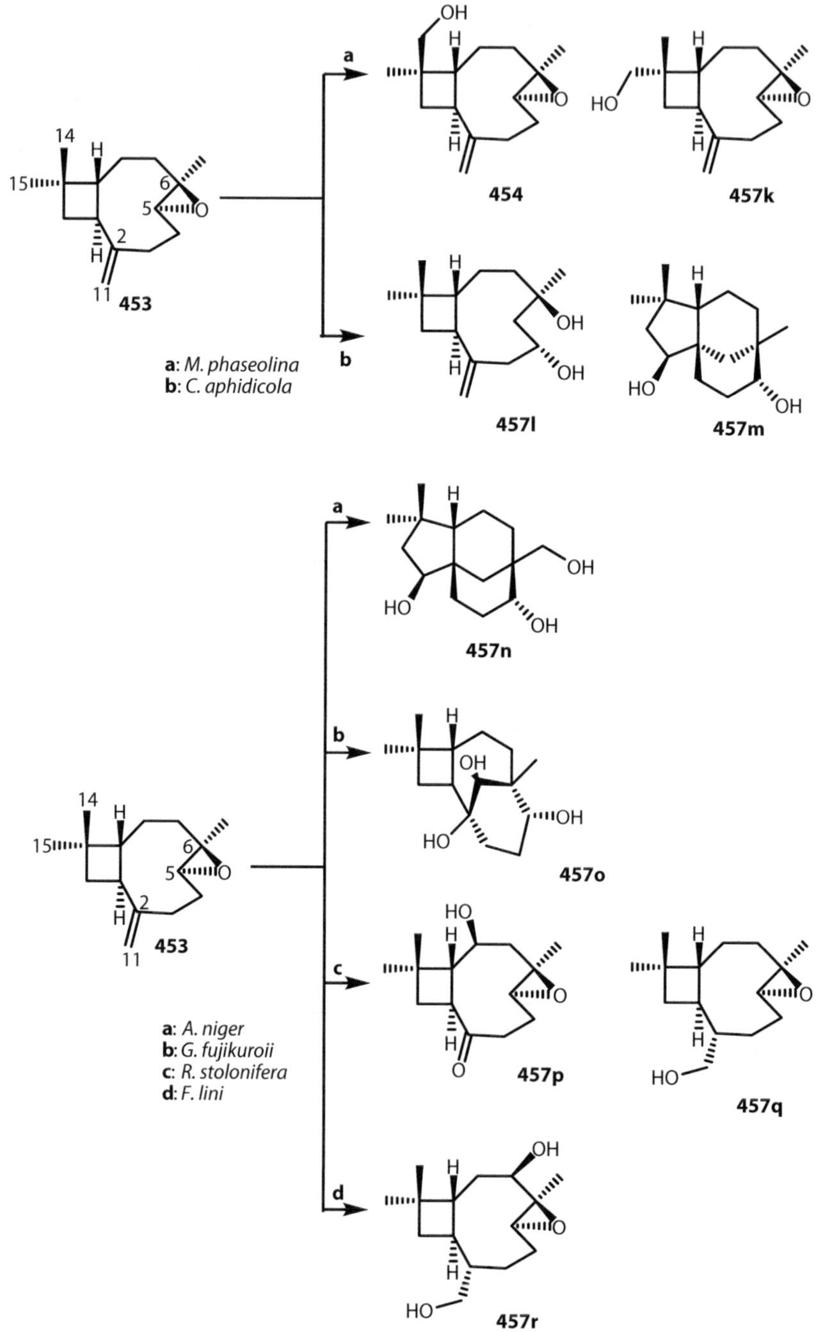

a: *M. phaseolina*
b: *C. aphidicola*

a: *A. niger*
b: *G. fujikuroii*
c: *R. stolonifera*
d: *F. lini*

FIGURE 23.133 Biotransformation of (−)-β-caryophyllene epoxide (**453**) by various fungi.

and **457y** obtained through reductive opening of the C2,C11-epoxide. The major reaction paths are stereoselective epoxidation and introduction of hydroxyl group at C3. Compound **457ae** has a caryolane skeleton (Figure 23.134).

When isoprobotryan-9α-ol (**458**) produced from isocaryophyllene was incubated with *B. cinerea*, it was hydroxylated at tertiary methyl groups to give three primary alcohols (**459–461**) (Aleu et al., 2002) (Figure 23.135).

Acyclic sesquiterpenoids, racemic *cis*-nerolidol (**462**), and nerylacetone (**463**) were treated by the plant pathogenic fungus *G. cingulata* (Miyazawa et al., 1995d). From the former substrate, a triol (**464**) was obtained as the major product. The latter was bioconverted to give the two methyl

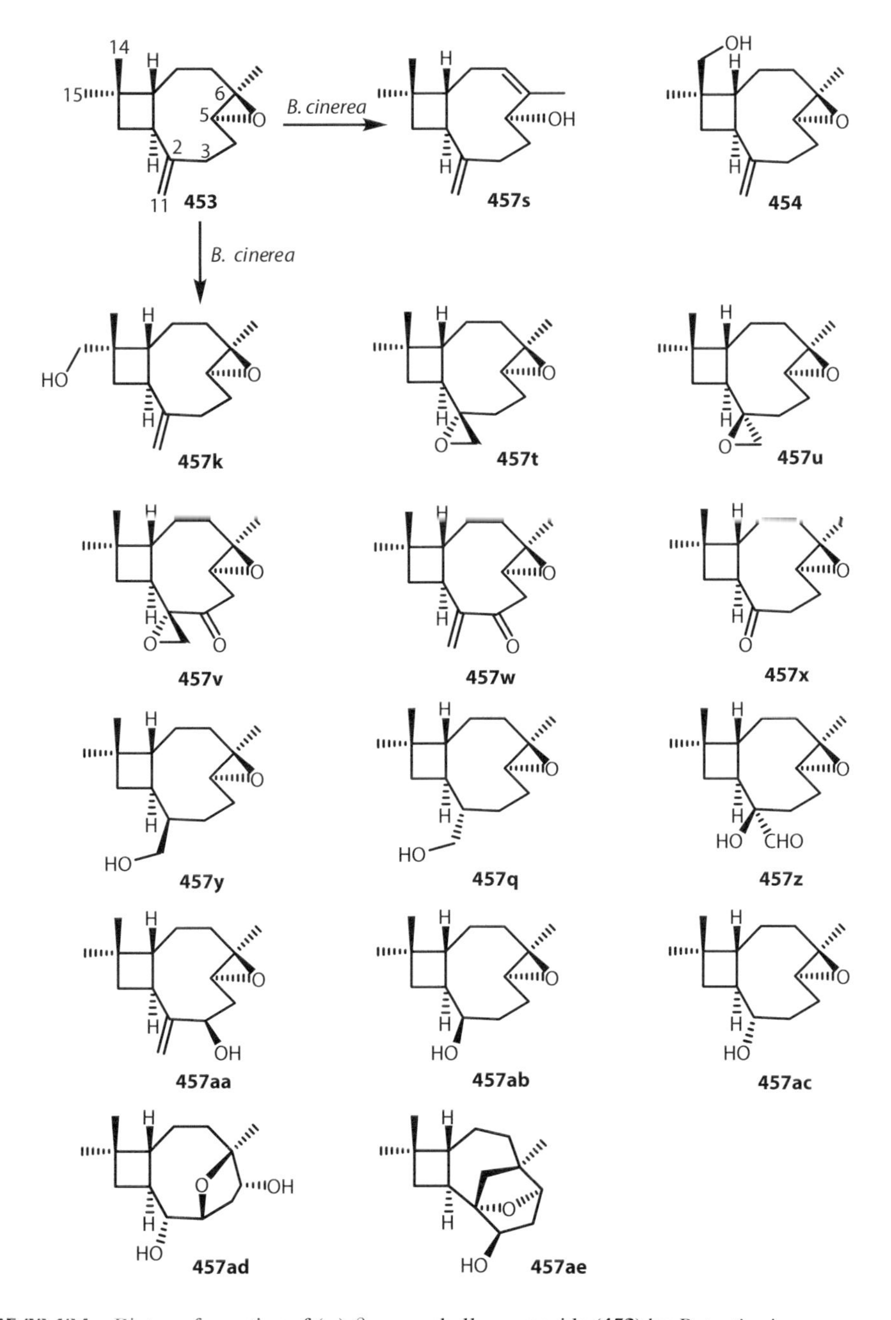

FIGURE 23.134 Biotransformation of (−)-β-caryophyllene epoxide (**453**) by *Botrytis cinerea*.

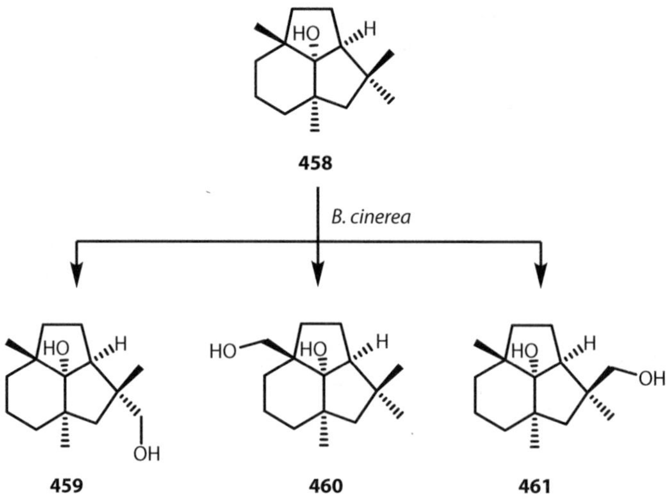

FIGURE 23.135 Biotransformation of isoprobotryan-9α-ol (**458**) by *Botrytis cinerea*.

ketones (**465**, **467**) and a triol (**468**), among which **465** was the predominant. The C10,C11 diols (**464**, **465**) might be formed from both epoxides of the substrates, followed by the hydration although no C10,C11-epoxides were detected (Figure 23.136).

Racemic *trans*-nerolidol (**469**) was also treated in the same fungus to afford w2-hydroxylated product (**471**) and C10,C11-hydroxylated compounds (**472**) as seen in racemic *cis*-nerolidol (**462**) (Miyazawa et al., 1996b) (Figure 23.137).

12-Hydroxy-*trans*-nerolidol (**472a**) is an important precursor in the synthesis of interesting flavor of α-sinensal. Hrdlicka et al. (2004) reported the biotransformation of *trans*- (**469**) and *cis*-nerolidol (**462**) and *cis-/trans*-mixture of nerolidol using repeated batch culture of *A. niger* grown in computer-controlled bioreactors. *Trans*-nerolidol (**469**) gave **472a** and **472** and *cis*-isomer (**462**) afforded **464a** and **464**. From a mixture of *cis*- and *trans*-nerolidol, 12-hydroxy-*trans*-neroridol **472a** (8%) was

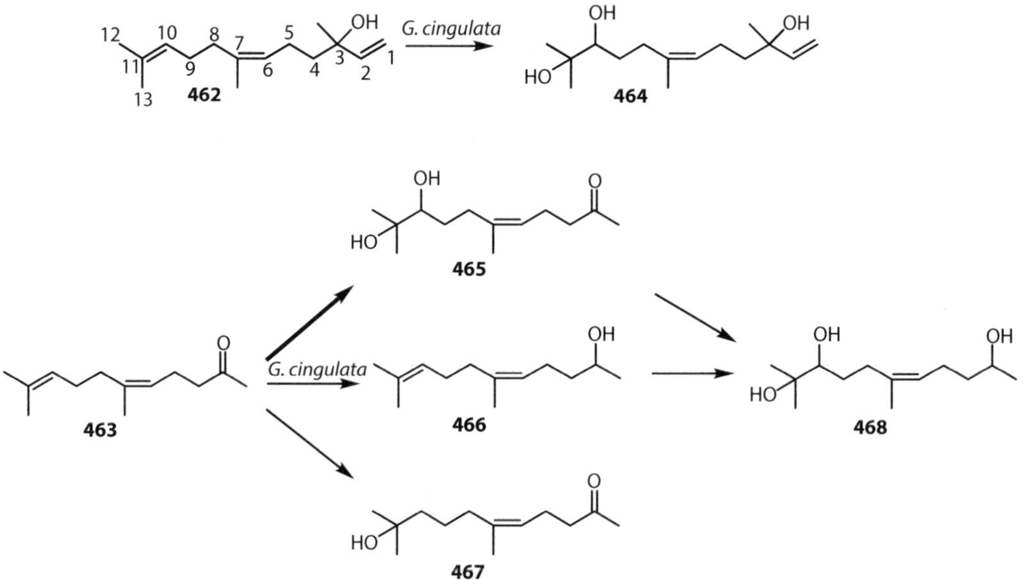

FIGURE 23.136 Biotransformation of *cis*-nerolidol (**462**) and *cis*-geranyl acetone (**463**) by *Glomerella cingulata*.

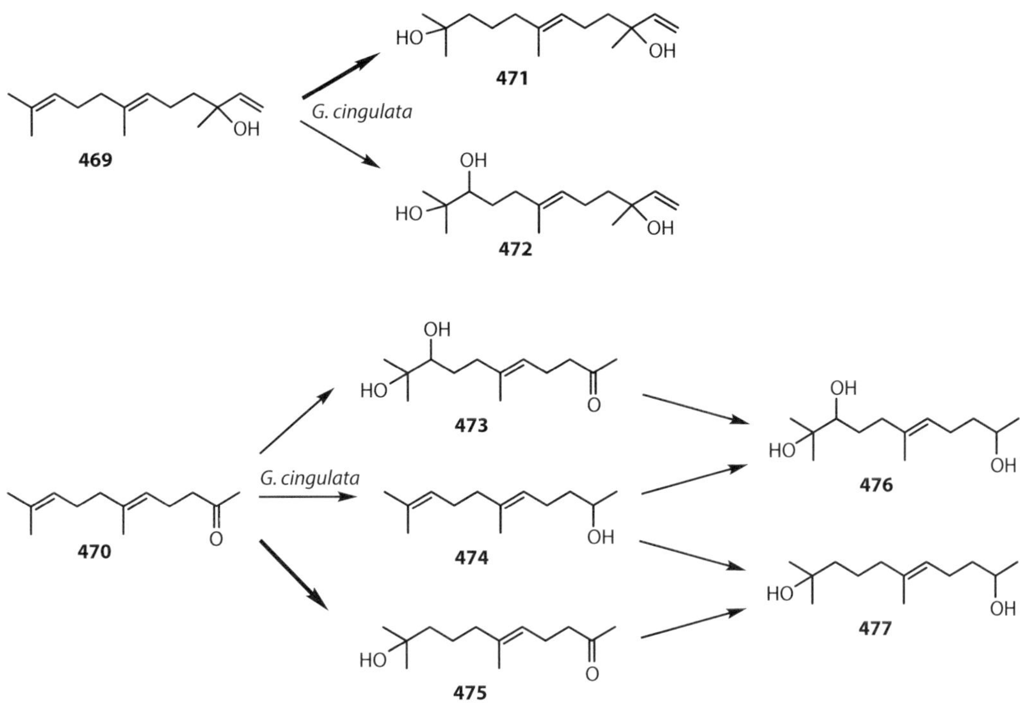

FIGURE 23.137 Biotransformation of *trans*-nerolidol (**469**) and *trans*-geranyl acetone (**470**) by *Glomerella cingulata.*

obtained in postexponential phase at high dissolved oxygen. At low dissolved oxygen condition, the mixture gave **472a** (7%) and **464a** (6%) (Figures 23.138 and 23.139).

From geranyl acetone (**470**) incubated with *G. cingulata*, four products (**473**–**477**) were formed. It is noteworthy that the major compounds from both substrates (**469**, **470**) were w2-hydroxylated

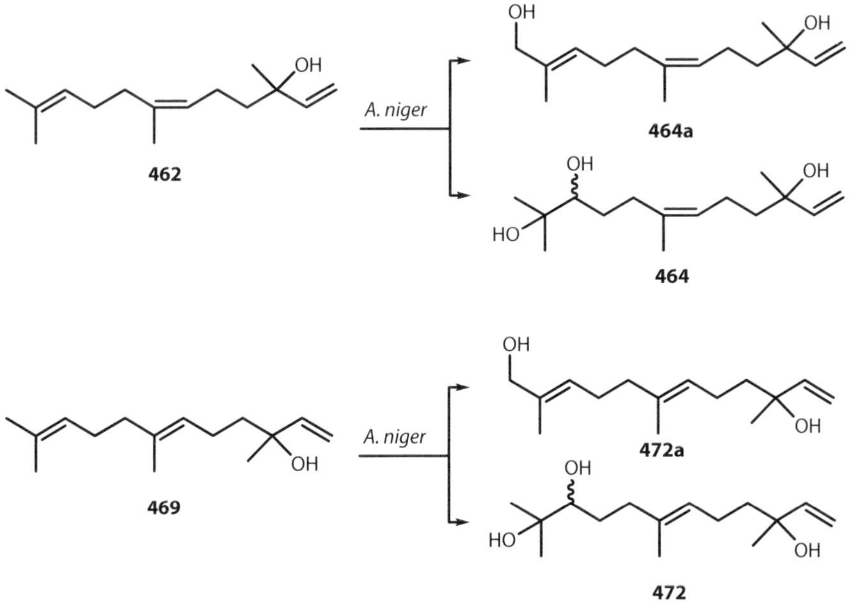

FIGURE 23.138 Biotransformation of *cis*- (**462**) and *trans*-nerolidol (**469**) by *Aspergillus niger.*

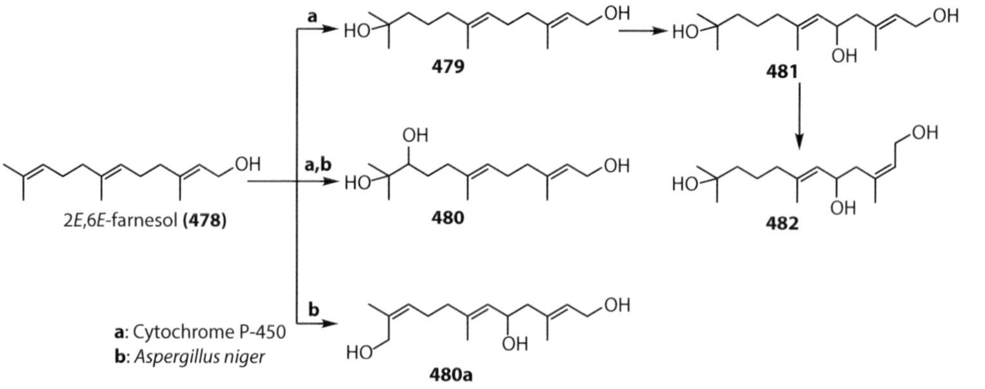

FIGURE 23.139 Biotransformation of *2E,6E*-farnesol (**478**) by cytochrome P-450 and *Aspergillus niger*.

products, but not C10,C11-dihydroxylated products as seen in *cis*-nerolidol (**462**) and nerylacetone (**463**) (Miyazawa et al., 1995b) (Figure 23.136).

The same fungus bioconverted (2*E*,6*E*)-farnesol (**478**) to four products, w2-hydroxylated product (**479**), which was further oxidized to give C10,C11 dihydroxylated compound (**480**) and 5-hydroxy derivative (**481**), followed by isomerization at C2,C3 double bond to afford a triol (**482**) (Miyazawa et al., 1996c) (Figure 23.140).

The same substrate was bioconverted by *A. niger* to afford two metabolites, 10,11-dihydroxy (**480**) and 5,13-hydroxy derivative (**480a**) (Madyastha and Gururaja, 1993).

The same fungus also converted (2*Z*,6*Z*)-farnesol (**483**) to three hydroxylated products: 10,11-dihydroxy-(2*Z*,6*Z*)- (**484**), 10,11-dihydroxy-(2*E*,6*Z*)-farnesol (**485**), and (5*Z*)-9,10-dihydroxy-6,10-dimethyl-5-undecen-2-one (**486**) (Nankai et al., 1996) (Figures 23.140 and 23.141).

A linear sesquiterpene 9-oxonerolidol (**487**) was treated in *A. niger* to give wl-hydroxylated product (**488**) (Higuchi et al., 2001) (Figure 23.142).

Racemic diisophorone (**488a**) dissolved in ethanol was incubated with the Czapek–Dox medium of *A. niger* to afford 8α- (**488b**), 10β- (**488c**), and 17-hydroxydiisophorone (**488d**) (Kiran et al., 2004).

On the other hand, the same substrate was fed with *Neurospora crassa* and *C. aphidicola* to afford only 8β-hydroxydiisophorone (**488e**) in 20% and 10% yield, respectively (Kiran et al., 2005) (Figure 23.143).

From the metabolites of 5β,6β-dihydroxypresilphiperfolane 2β-angelate (**488f**) using the fungus *Mucor ramannianus*, 2,3-epoxyangeloyloxy derivative (**488 g**) was obtained (Orabi, 2001) (Figure 23.144).

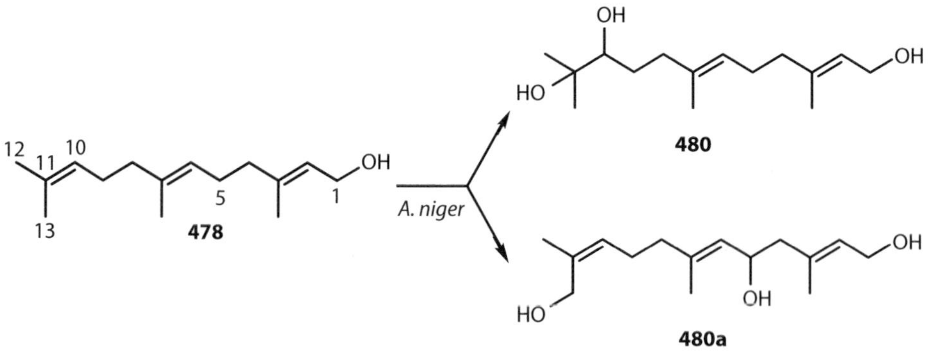

FIGURE 23.140 Biotransformation of *2E,6E*-farnesol (**478**) by *Aspergillus niger*.

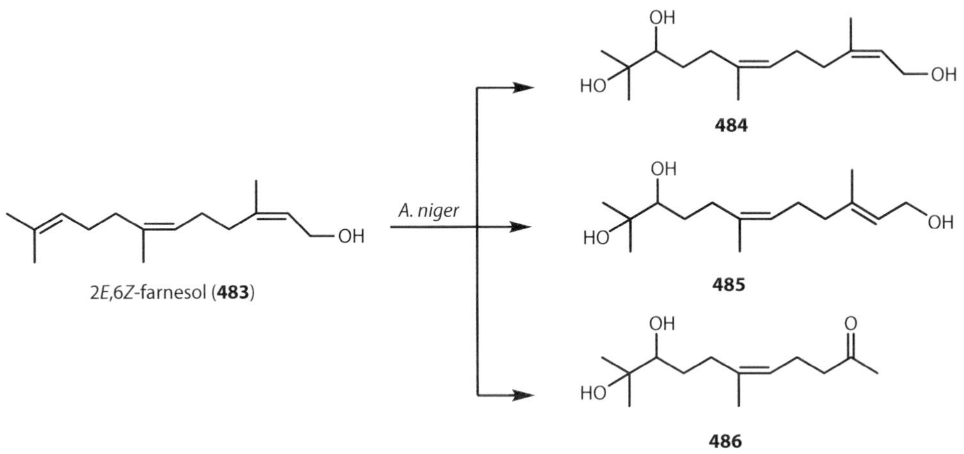

FIGURE 23.141 Biotransformation of *2Z,6Z*-farnesol (**478**) by *Aspergillus niger.*

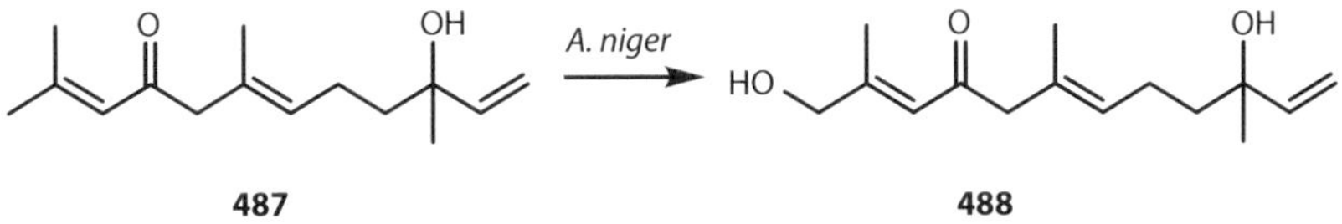

FIGURE 23.142 Biotransformation of 9-oxo-*trans*-nerolidol (**487**) by *Aspergillus niger.*

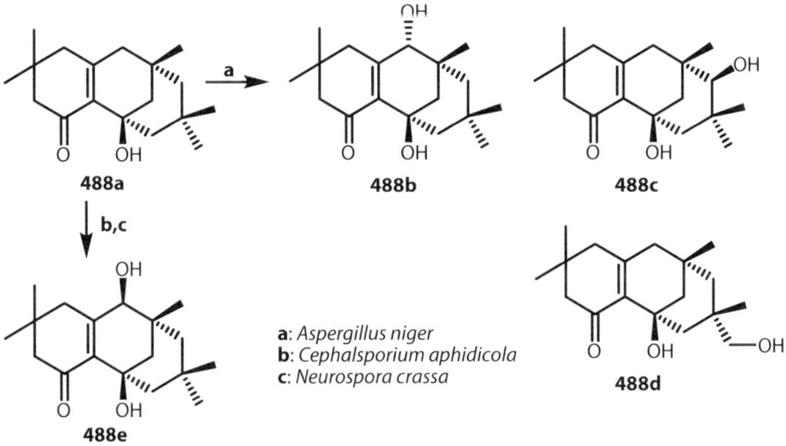

FIGURE 23.143 Biotransformation of diisophorone (**488a**) by *Aspergillus niger, Cephalosporium aphidicola,* and *Neurospora crassa.*

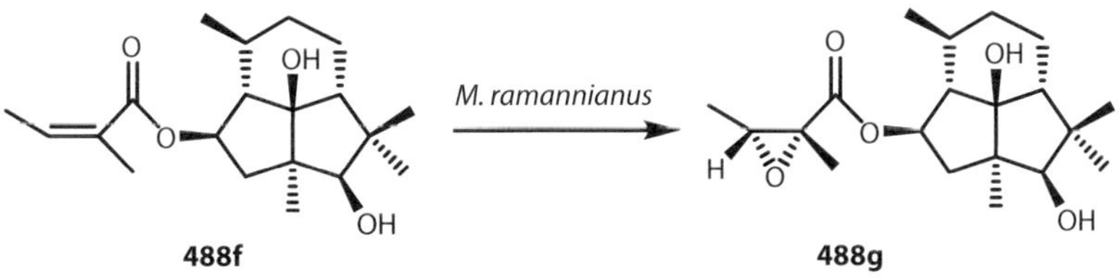

FIGURE 23.144 Biotransformation of 5β,6β-dihydroxypresilphiperfolane 2β-angelate (**488f**) by *Mucor ramannianus.*

23.3 BIOTRANSFORMATION OF SESQUITERPENOIDS BY MAMMALS, INSECTS, AND CYTOCHROME P-450

23.3.1 ANIMALS (RABBITS) AND DOSING

Six male albino rabbits (2–3 kg) were starved for 2 days before experiment. Monoterpenes were suspended in water (100 mL) containing polysorbate 80 (0.1 g) and were homogenized well. This solution (20 mL) was administered to each rabbit through a stomach tube followed by water (20 mL). This dose of sesquiterpenoids corresponds to 400–700 mg/kg. Rabbits were housed in stainless steel metabolism cages and were allowed rabbit food and water ad libitum. The urine was collected daily for 3 days after drug administration and stored at 0°–5°C until the time of analysis. The urine was centrifuged to remove feces and hairs at 0°C and the supernatant was used for the experiments. The urine was adjusted to pH 4.6 with acetate buffer and incubated with β-glucuronidase–arylsulfatase (3/100 mL of fresh urine) at 37°C for 48 h, followed by continuous ether extraction for 48 h. The ether extracts were washed with 5% $NaHCO_{3\%}$ and 5% NaOH to remove the acidic and phenolic components, respectively. The ether extract was dried over $MgSO_4$, followed by evaporation of the solvent to give the neutral crude metabolites (Ishida et al., 1981).

23.3.2 SESQUITERPENOIDS

Wild rabbits (hair) and deer damage the young leaves of *Chamaecyparis obtusa*, one of the most important furniture and house-constructing tree in Japan. The essential oil of the leaves contains a large amount of (−)-longifolene (**489**). Longifolene (36 g) was administered to 18 of rabbits to obtain the metabolites (3.7 g) from which an aldehyde (**490**) (35.5%) was isolated as pure state. In the metabolism of terpenoids having an exomethylene group, glycol formation was often found, but in the case of longifolene such as a diol was not formed. Introduction of an aldehydes group in biotransformation is very remarkable. Stereoselective hydroxylation of the gem-dimethyl group on a seven-membered ring is first time (Ishida et al., 1982) (Figure 23.145).

(−)-β-Caryophyllene (**451**) is one of the ubiquitous sesquiterpene hydrocarbons in the plant kingdom and the main component of beer hops and clove oil and is being used as a culinary ingredient and as a cosmetic in soaps and fragrances. (−)-β-Caryophyllene also has cytotoxic against breast carcinoma cells and its epoxide is toxic to *Planaria* worms. It contains unique 1,1-dimethylcyclobutane skeleton. (−)-β-Caryophyllene (3 g) was treated in the same manner as described earlier to afford the crude metabolite (2.27 g) from which (10*S*)-14-hydroxycaryophyllene-5,6-oxide (**491**) (80%) and a diol (**492**) were obtained (Asakawa et al., 1981). Later, compound **491** was isolated from the Polish mushroom *Lactarius camphoratus* (Basidiomycetes) as a natural product (Daniewski et al., 1981). 14-Hyroxy-β-callyophyllene and 1-hydroxy-8-keto-β-caryophyllene have been found in Asteraceae and *Pseudomonas* species, respectively. In order to confirm that caryophyllene epoxide (**453**) is the intermediate of both metabolites, it was treated in the same manner as described earlier to give the same metabolites, **491** and **492**, of which **491** was predominant (Asakawa et al., 1981, 1986) (Figure 23.146).

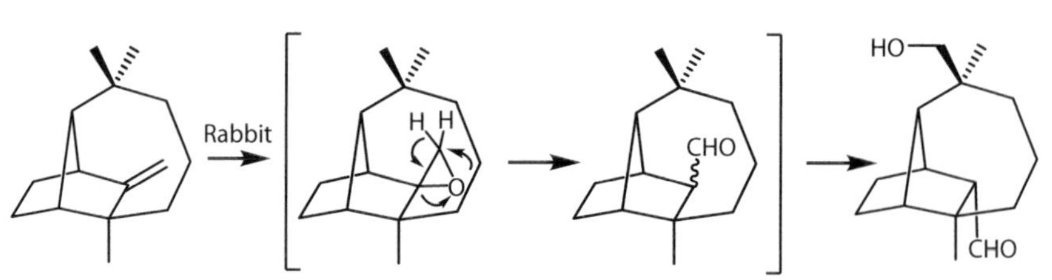

FIGURE 23.145 Biotransformation of longifolene (**489**) by rabbit.

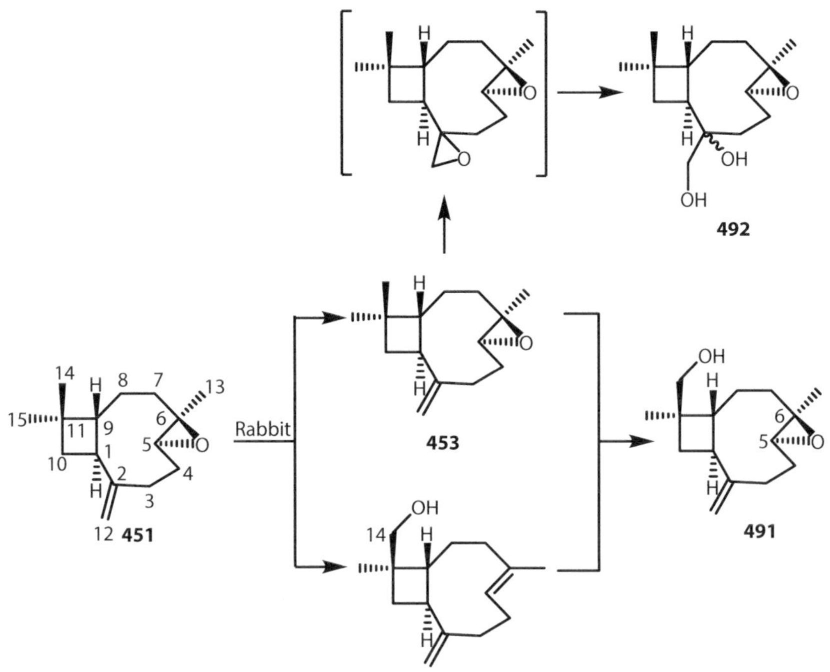

FIGURE 23.146 Biotransformation of (−)-β-caryophyllene (**451**) by rabbit.

The grapefruit aroma, (+)-nootkatone (**2**), was administered into rabbits to give 11,12-diol (**6**, **7**). The same metabolism has been found in that of biotransformation of nootkatone by microorganisms as mentioned in the previous paragraph. Compounds (**6**, **7**) were isolated from the urine of hypertensive subjects and named urodiolenone. The endogenous production of **6** and **7** seem to occur interdentally from the administrative manner of nootkatone or grapefruit. Synthetic racemic nootkatone epoxide (**14**) was incubated with rabbit liver microsomes to give 11,12-diol (**6**, **7**) (Ishida, 2005). Thus, the role of the epoxide was clearly confirmed as an intermediate of nootkatone (**2**).

(+)-*ent*-Cyclocolorenone (**98**) and its enantiomer (**103**) were biotransformed by *Aspergillus* species to give cyclopropane-cleaved metabolites as described in the previous paragraph.

In order to compare the metabolites between mammals and microorganisms, the essential oil (2 g/rabbit) containing (−)-cyclocolorenone (**103**) obtained from *S. altissima* was administered in rabbits to obtain two metabolites: 9β-hydroxycyclocolorenone (**493**) and 10-hydroxycyclocolorenone (**494**) (Asakawa et al., 1986). 10-Hydroxyaromadendrane-type compounds are well known as the natural products. No oxygenated compound of cyclopropane ring was found in the metabolites of cyclocolorenone in rabbit (Figure 23.147).

From the metabolites of elemol (**495**) possessing the same partial structures of monoterpene hydrocarbon, myrcene, and nootkatone, one primary alcohol (**496**) was obtained from rabbit urine after the administration of **495** (Asakawa et al., 1986) (Figure 23.148).

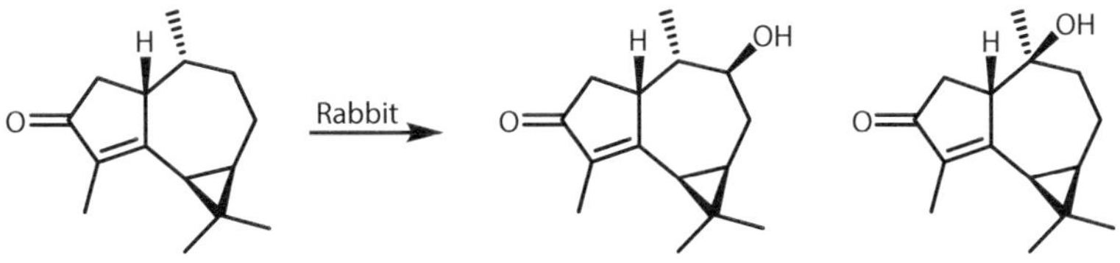

FIGURE 23.147 Biotransformation of (+)-*ent*-cyclocolorenone (**101**) by rabbit.

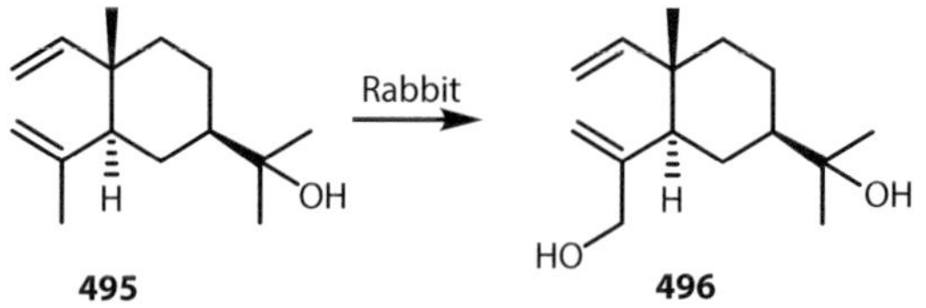

495 496

FIGURE 23.148 Biotransformation of elemol (**495**) by rabbit.

Components of cedar wood such as cedrol (**414**) and cedrene shorten the sleeping time of mice. In order to search for a relationship between scent, olfaction, and detoxifying enzyme induction, (+)-cedrol (**414**) was administered to rabbits and dogs. From the metabolites from rabbits, two C3-hydroxylated products (**418** and **497**) and a diol (**415** or **416**), which might be formed after the hydrogenation of double bond. Dogs converted cedrol (**414**) into different metabolite products, C2- (**498**) and C2/C14-hydroxylated products (**499**), together with the same C3- (**415**) and C15-hydroxylated products (**416**) as those found in the metabolites of microorganisms and rabbits. The aforementioned species-specific metabolism is very remarkable (Bang and Ourisson, 1975).

The microorganisms *C. aphidicola*, *C. cassiicola*, *B. cinerea*, and *G. cingulata* also biotransformed cedrol to various C2-, C3-, C4-, C6-, and C15-hydroxylated products as shown in the previous paragraph. The microbial metabolism of cedrol resembles that of mammals (Figure 23.149).

Patchouli alcohol (**425**) with fungi static properties is one of the important essential oils in perfumery industry. Rabbits and dogs gave two oxidative products (**500**, **501**) and one norpatchoulen-1-one (**502**) possessing a characteristic odor. Plant pathogen *B. cinerea* causes many diseases for vegetables and flowers. This pathogen gave totally different five metabolites (**426**–**430**) from those found in the urine metabolites of mammals as described earlier (Bang et al., 1975) (Figure 23.150).

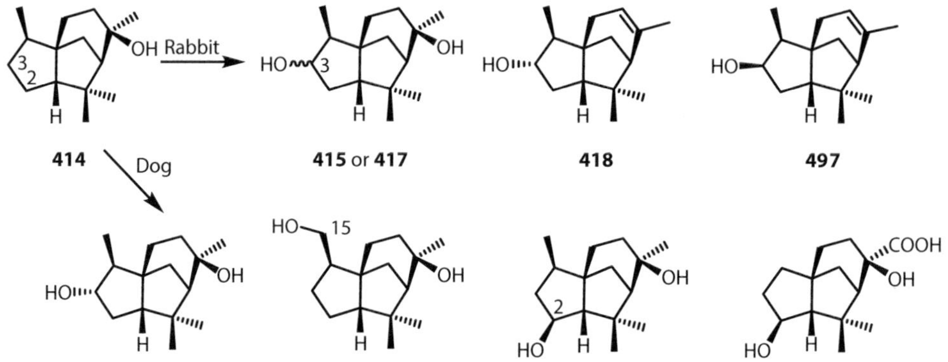

414 415 or 417 418 497 415 416 498 499

FIGURE 23.149 Biotransformation of cedrol (**414**) by rabbits or dogs.

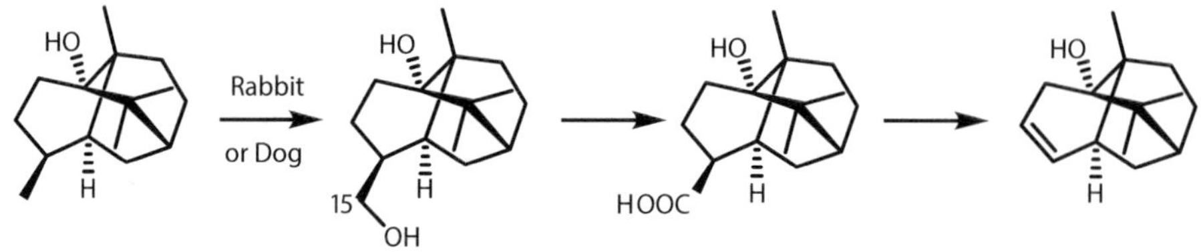

425 500 501 502

FIGURE 23.150 Biotransformation of patchouli alcohol (**425**) by rabbits or dogs.

Sandalwood oil contains mainly α-santalol (**503**) and β-santalol. Rabbits converted α-santalol to three diastereomeric primary alcohols (**504**–**506**) and dogs did carboxylic acid (**507**) (Zundel, 1976) (Figure 23.151).

(2*E*,6*E*)-Farnesol (**478**) was treated in cockroach cytochrome P-450 (CYP4C7) to form regio- and diastereospecifically w-hydroxylated at the C12 methyl group to the corresponding diol (**508**) with 10*E*-configuration (Sutherland et al., 1998) (Figure 23.152).

Juvenile hormone III (**509**) was also treated in cockroach CYP4C7 to the corresponding 12-hydroxylated product (**510**) (Sutherland et al., 1998).

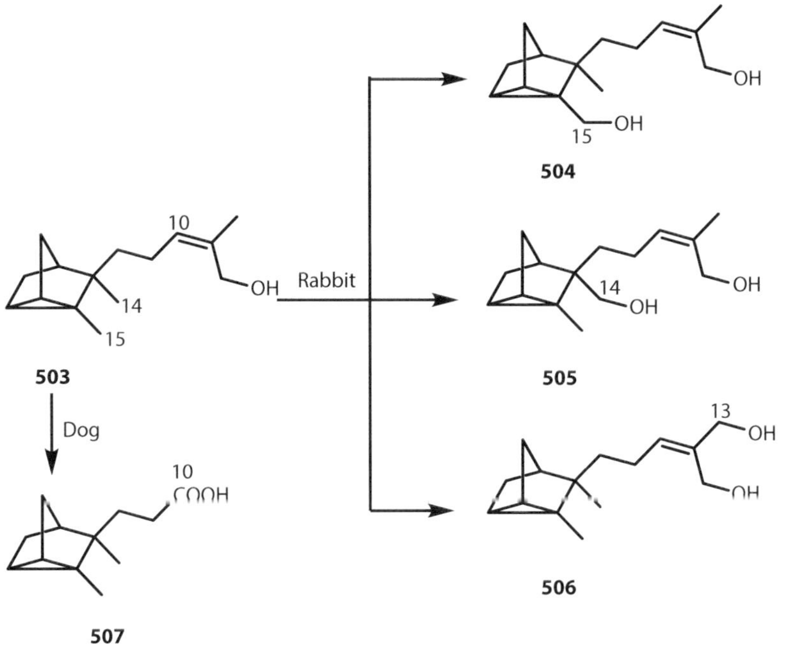

FIGURE 23.151 Biotransformation of santalol (**503**) by rabbits or dogs.

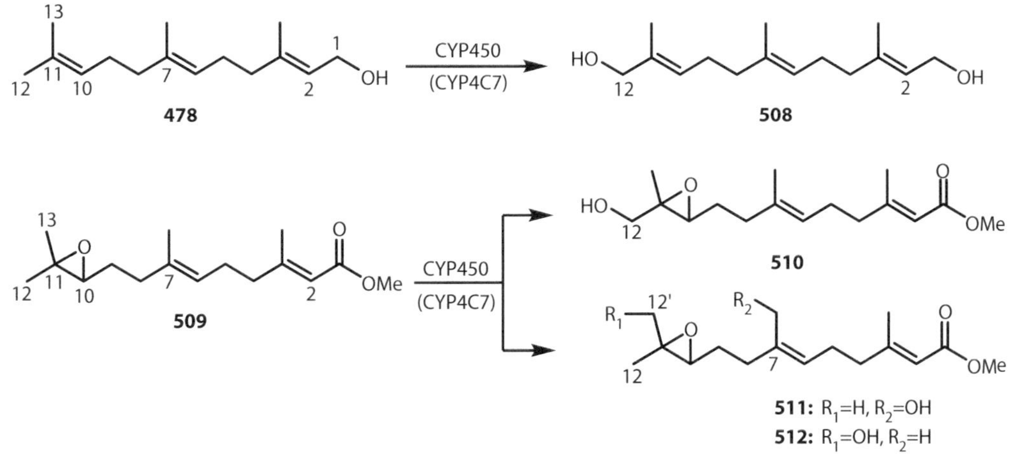

FIGURE 23.152 Biotransformation of *2E*,*6E*-farnesol (**478**) by cockroach cytochrome P-450 and 10,11-epoxyfarnesic acid methyl ester (**509**) by African locust cytochrome P-450.

The African locust converted the same substrate (**509**) into a 7-hydroxy product (**511**) and a 13-hydroxylated product (**512**). It is noteworthy that the African locust and cockroach showed clear species specificity for introduction of oxygen function (Darrouzet et al., 1997).

23.4 BIOTRANSFORMATION OF IONONES, DAMASCONES, AND ADAMANTANES

Racemic α-ionone (**513**) was converted to 4-hydroxy-α-ionone (**514**), which was further dehydrogenated to 4-oxo-α-ionone (**515**) by *Chlorella ellipsoidea* IAMC-27 and *C. vulgaris* IAMC-209. α-Ionone (**513**) was reduced preferentially to α-ionol (**516**) by *Chlorella sorokiniana* and *Chlorella salina* (Noma et al., 1997b).

α-Ionol (**516**) was oxidized by *C. pyrenoidosa* to afford 4-hydroxy-α-ionol (**524**). The same substrate was fed by the same microorganism and *A. niger* to furnish α-ionone (**513**) (Noma and Asakawa, 1998) (Figure 23.153).

4-Oxo-α-ionone (**515**), which is one of the major product of α-ionone (**513**) by *A. niger*, was transformed reductively by *Hansenula anomala*, *Rhodotorula minuta*, *Dunaliella tertiolecta*, *Euglena gracilis*, *C. pyrenoidosa* C28 and other eight kinds of *Chlorella* species, *B. dothidea*, *A. cellulosae* IFO 4040, and *A. sojae* IFO 4389 to give 4-oxo-α-ionol (**517**), 4-oxo-7,8-dihydro-α-ionone (**518**), and 4-oxo-7,8-dihydro-α-ionol (**519**). Compound **515** was also oxidized by *A. niger* and *A. sojae* to give 1-hydroxy-4-oxo-α-ionone (**520**) and 7,11-oxido-4-oxo-7,8-dihydro-α-ionone (**521**). C7–C8 double bond of α-ionone (**513**), 4-oxo-α-ionone (**515**), and 4-oxo-α-ionol (**517**) were easily reduced to their corresponding dihydro products (**522**, **518**, **519**), respectively, by *Euglena*,

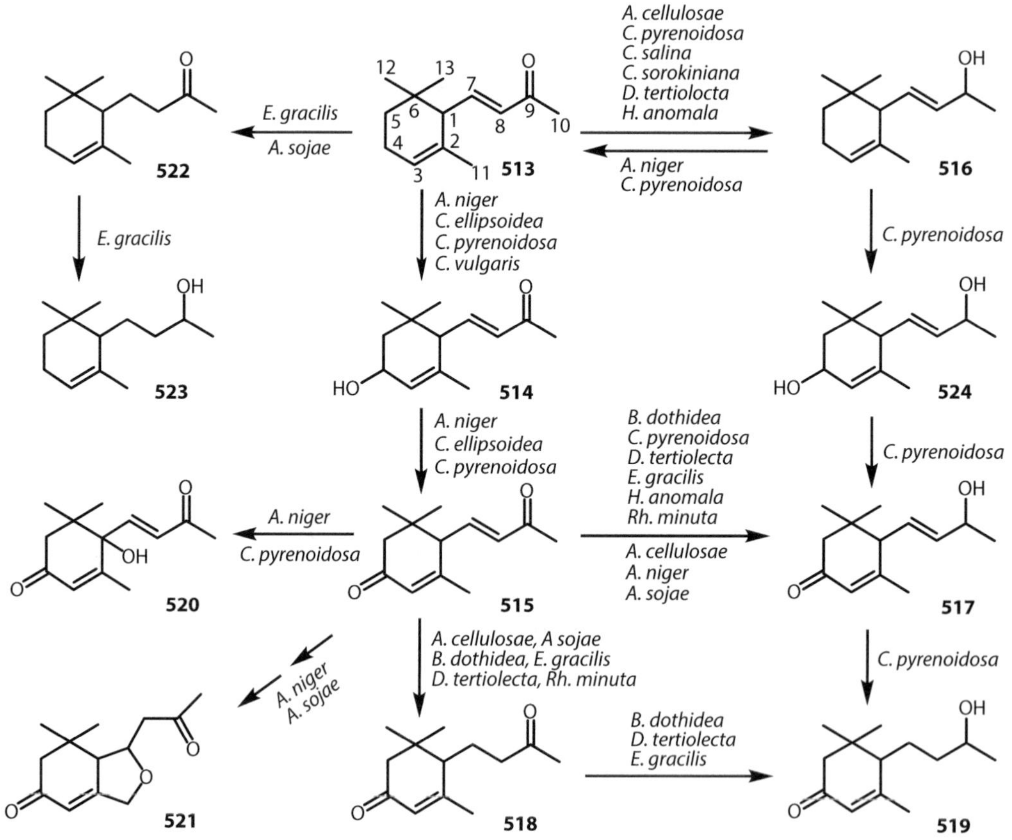

FIGURE 23.153 Biotransformation of α-ionone (**513**) by various microorganisms.

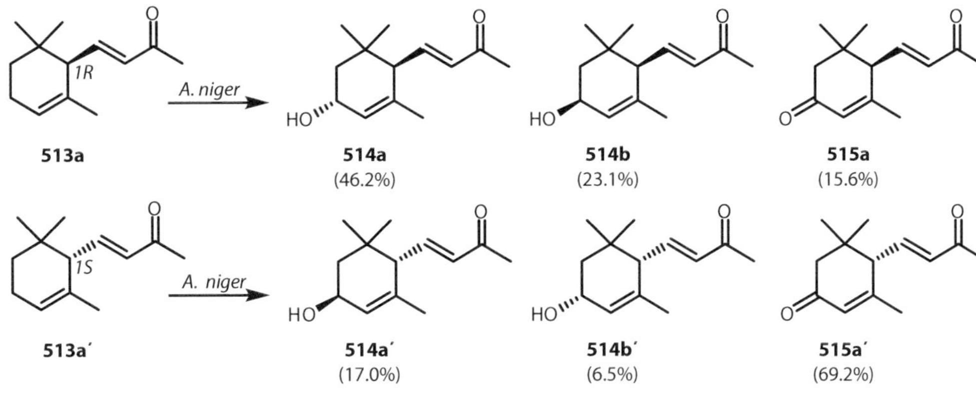

FIGURE 23.154 Biotransformation of (1*R*)-α-ionone (**513a**) and (1*S*)-α-ionone (**513a′**) by *Aspergillus niger.*

Aspergillus, *Botryosphaeria*, and *Chlorella* species. The metabolite (**522**) was further reduced to **523** by *E. gracilis* (Noma and Asakawa, 1998).

Biotransformation of (+)-1*R*-α-ionone (**513a**), $[\alpha]_D$ + 386.5°, 99% ee, and (−)-1*S*-α-ionone (**513a′**), $[\alpha]_D$ 361.6°, 98% ee, which were obtained by optical resolution of racemic α-ionone (**513**), was fed by *A. niger* for 4 days in Czapek-peptone medium. From **513a**, 4α-hydroxy-α-ionone (**514a**), 4β-hydroxy-α-ionone (**514b**), and 4-oxo-α-ionone (**515a**) were obtained, while from compound **513a′**, the enantiomers (**514a′**, **514b′**, **515a′**) of the metabolites from **513a** were obtained; however, the difference of their yields was observed. In case of **513a**, 4α-hydroxy-α-ionone (**514a**) was obtained as the major product, while **515a′** was predominantly obtained from **513a′**. This oxidation was inhibited by 1-aminobenzotriazole, thus, CYP-450 is contributed to this oxidation process (Hashimoto et al., 2000a) (Figure 23.154).

α-Damascone (**525**) was incubated with *A. niger* and *A. terreus*, in Czapek-peptone medium to give *cis*- (**525**) and *trans*-3-hydroxy-α-damascones (**527**) and 3-oxo-α-damscone (**528**), while the latter *Aspergillus* species afforded 3-oxo-8,9-dihydro-α-damascone (**529**). The hydroxylation process of **525**–**527** was inhibited by CYP-450 inhibitor. *H. anomala* reduced α-damascone (**525**) to α-damascol (**530**). *Cis*- (**526**) and *trans*-4-hydroxy-α-damascone (**527**) were fed by *C. pyrenoidosa* in Noro medium to give 4-oxodamascone (**528**) (Noma et al., 2001b) (Figure 23.155).

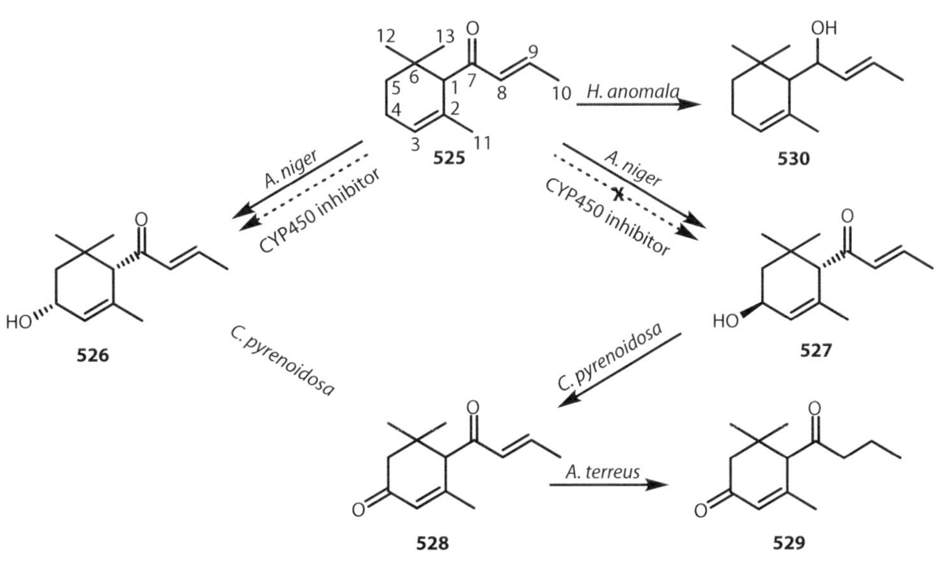

FIGURE 23.155 Biotransformation of α-damascone (**525**) by various microorganisms.

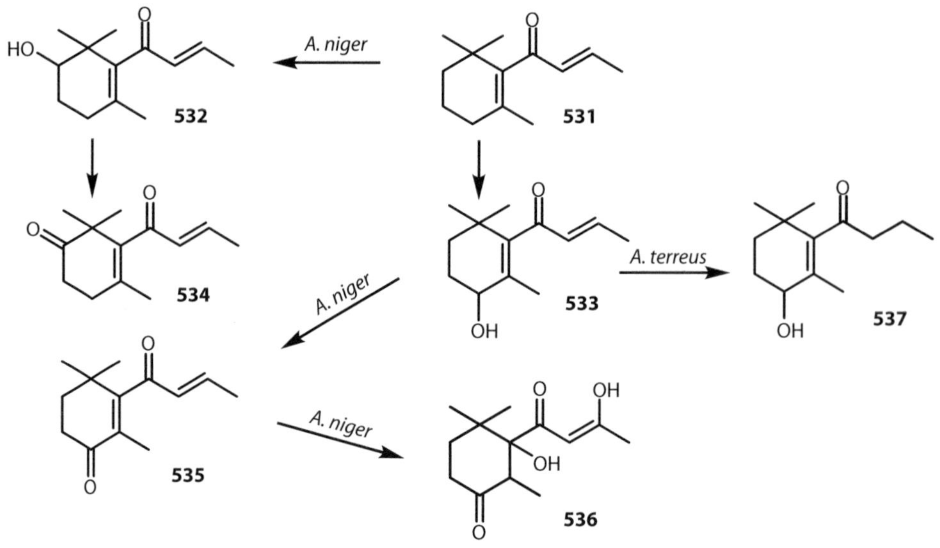

FIGURE 23.156 Biotransformation of β-damascone (**531**) by *Aspergillus niger* and *Aspergillus terreus*.

β-Damascone (**531**) was also treated in *A. niger* to afford 5-hydroxy-β-damascone (**532**), 3-hydroxy-β-damascone (**533**), 5-oxo- (**534**), 3-oxo-β-damscone (**535**), and 3-oxo-1,9-dihydroxy-1,2-dihydro-β-damascone (**536**) as the minor components. In case of *A. terreus*, 3-hydroxy-8,9-dihydro-β-damascone (**537**) was also obtained (Figure 23.156).

Adamantane derivatives have been used as many medicinal drugs. In order to obtain the drugs, adamantanes were incubated by many microorganisms, such as *A. niger*, *A. awamori*, *A. cellulosae*, *Aspergillus fumigatus*, *A. sojae*, *A. terreus*, *B. dothidea*, *C. pyrenoidosa* IMCC-28, *C. sorokiniana*, *F. culmorum*, *E. gracilis*, and *H. anomala* (Figure 23.157).

Adamantane (**538**) was incubated with *A. niger*, *A. cellulosae*, and *B. dothidea* in Czapek-peptone medium. The same substrate was also treated in *C. pyrenoidosa* in Noro medium. Compound **538** was converted into both 1-hydroxy- (**539**) and 9α-hydroxyadamantane (**540**) by all four microorganisms, followed by oxidation oxidized to give 1,9α-dihydroxyadamantanol (**541**) by *A. niger*, which was further oxidized to 1-hydroxyadamantane-9-one (**542**), which was reduced to afford 1,9β-hydroxyadamantane (**544**). *A. niger* gave the metabolite (**541**) as the major product in 80% yield. *A. cellulosae* converted **538**–**539** and **540** in the ratio of 81:19. *C. pyrenoidosa* gave **539**, **540**, and adamantane-9-one (**543**) in the ratio 74:16:10. 4α-Adamantanol (**540**) was directly

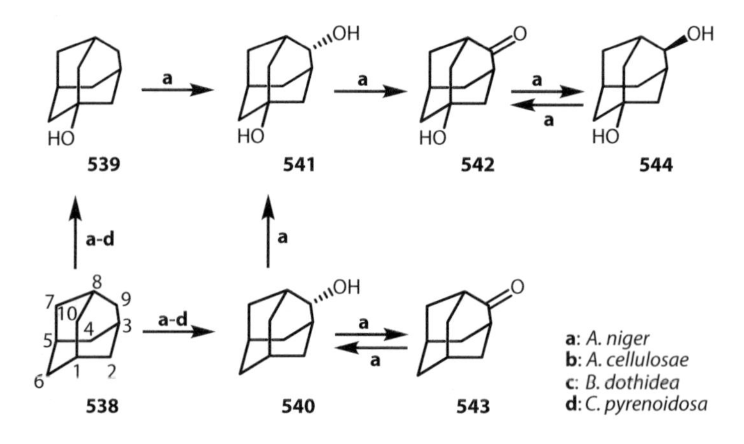

FIGURE 23.157 Biotransformation of adamantane (**538**) by various microorganisms.

converted by *C. pyrenoidosa*, *A. niger*, and *A. cellulosae* to afford **543**, which was also reduced to 9α-adamantanol (**540**) by *A. niger*. The biotransformation of adamantane, however, did not occur by the microorganisms *H. anomala*, *C. sorokiniana*, *D. tertiolecta*, and *E. gracilis* (Noma et al., 1999).

Adamantanes (**538–543**, **542**) were also incubated with various fungi including with *F. culmorum*. 1-Hydroxyadamantane-9-one (**542**) was reduced stereoselectively to **541** by *A. niger*, *A. cellulosae*, *B. dothidea*, and *F. culmorum*. On the contrary, *F. culmorum* reduced **541–542**. *A. cellulosae* and *B. dothidea* bioconverted **542** to 1,9β-hydroxyadamantane (**544**) stereoselectively. Adamantane-9-one (**543**) was treated by *A. niger* to give nonstereoselectively **545–547** that were further converted into diketone (**548**, **549**) and a diol (**550**). It is noteworthy that oxidation and reduction reactions were observed between ketoalcohol (**547**) and diols (**551**, **552**). The same phenomenon was also seen between **546** and **553**. The latter diol was also oxidized by *A. niger* to furnish diketone (**549**) (Noma et al., 2001c, 2003). Direct hydroxylation at C3 of 1-hydroxyadamantane-9-one (**542**) was seen in the incubation of **539** with *A. niger*.

4-Adamantanone (**543**) showed promotion effect of cell division of the fungus, while 1-adamantanol (**539**) and adamantane-9-one (**543**) inhibited germination of lettuce seed. 1-Hydroxyadamantane-9-one (**542**) inhibited the elongation of root of lettuce, while adamantane-1,4-diol (**544**) and adamantane itself (**538**) promoted root elongation (Noma et al., 1999, 2001c) (Figure 23.158).

Stereoselective reduction of racemic bicyclo[3.3.1]nonane-2,6-dione (**555a**, **555a′**) was carried out by *A. awamori*, *A. fumigatus*, *A. cellulosae*, *A. sojae*, *A. terreus*, *A. niger*, *B. dothidea*, and *F. culmorum* in Czapek-peptone, *H. anomala* in yeast, *E. gracilis* in Hunter, and *D. tertiolecta* in Noro

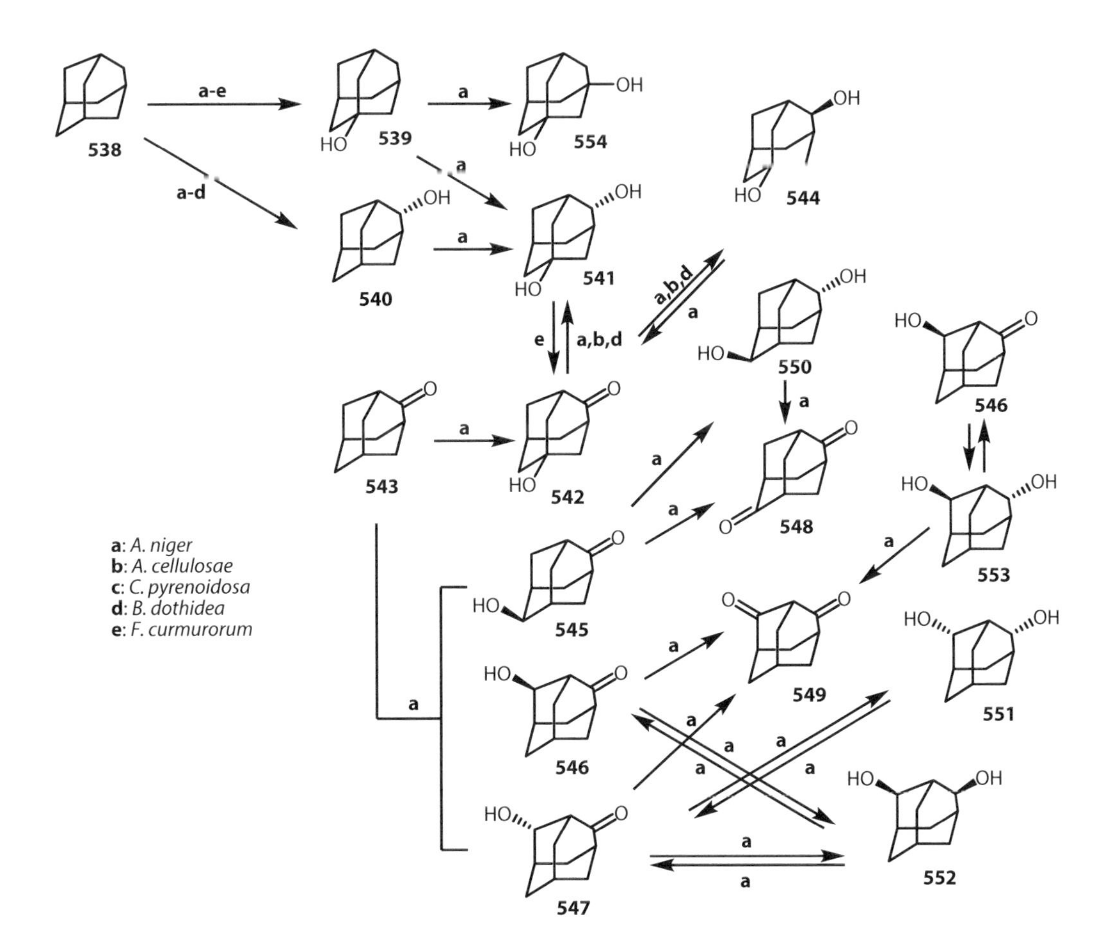

FIGURE 23.158 Biotransformation of adamantane (**538**) and adamantane-9-one by various microorganisms.

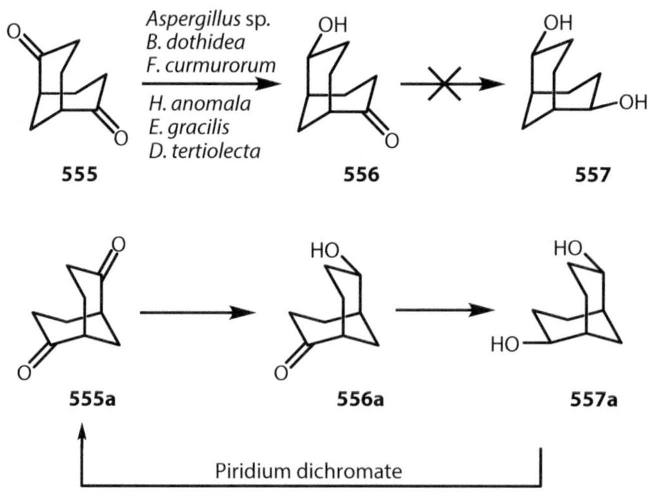

FIGURE 23.159 Biotransformation of bicyclo[3.3.1]nonane-2,6-dione (**555a**, **555a′**) by various microorganisms.

medium. All microorganisms reduced **555** and **555a′** to give corresponding monoalcohol (**556**, **556a**) and optical-active (−)-diol (**557a**) ($[\alpha]_D$ − 71.8° in the case of *A. terreus*), which was formed by enantioselective reduction of racemic monool, **556** and **556a** (Noma et al., 2003) (Figure 23.159).

23.5 BIOTRANSFORMATION OF AROMATIC COMPOUNDS

Essential oils contain aromatic compounds, such as *p*-cymene, carvacrol, thymol, vanillin, cinnamaldehyde, eugenol, chavicol, safrole, and asarone (**558**), among others.

Takahashi (1994) reported that simple aromatic compounds, propylbenzene, hexylbenzene, decylbenzene, *o*- and *p*-hydroxypropiophenones, *p*-methoxypropiophenone, 4-hexylresorcinol, and methyl 4-hexylresorcionol, were incubated with *A. niger*. From hexyl- and decylbenzenes, w1-hydroxylated products were obtained, whereas from propylbenzene, w2-hydroxylated metabolites were obtained (Takahashi, 1994).

Asarone (**558**) and dihydroeugenol (**562**) were not biotransformed by *A. niger*. However, dihydroasarone (**559**) and methyl dihydroeugenol (**563**) were biotransformed by the same fungus to produce a small amount of 2-hydroxy (**560**, **561**) and 2-oxo derivatives (**564**, **565**), respectively. The chirality at C2 was determined to be *R* and *S* mixtures (1:2) by the modified Mosher method (Takahashi, 1994) (Figure 23.160).

Chlorella species are excellent microalgae as oxidation bioreactors as mentioned earlier. Treatment of monoterpene aldehydes and related aldehydes were reduced to the corresponding primary alcohols, indicating that these green algae possess reductase.

A microalgae *E. gracilis* Z. also contains reductase. The following aromatic aldehydes were treated in this organism: benzaldehyde; 2-cyanobenzaldehyde; *o*-, *m*-, and *p*-anisaldehyde; *o*-, *m*-, and *p*-salicylaldehyde; *o*-, *m*-, and *p*-tolualdehyde; *o*-chlorobenzaldehyde; *p*-hydroxybenzaldehyde; *o*-, *m*-, and *p*-nitrobenzaldehyde; 3-cyanobenzaldehyde; vanillin; isovanillin; *o*-vanillin; nicotine aldehyde; 3-phenylpropionaldehyde; and ethyl vanillin. Veratraldehyde, 3-nitrosalicylaldehde, phenylacetaldehyde, and 2-phenylproanaldehyde gave their corresponding primary alcohols. 2-Cyanobenzaldehyde gave its corresponding alcohol with phthalate. *m*- and *p*-Chlorobenzaldehyde gave its corresponding alcohols and *m*- and *p*-chlorobenzoic acids. *o*-Phthalaldehyde and *p*-phthalate and iso- and terephthalaldehydes gave their corresponding monoalcohols and dialcohols. When cinnamaldehyde and α-methyl cinnamaldehyde were incubated in *E. gracilis*, cinnamyl alcohol and 3-phenylpropanol, and 2-methylcinnamyl alcohol, and 2-methyo-3-phenylpropanol were obtained in

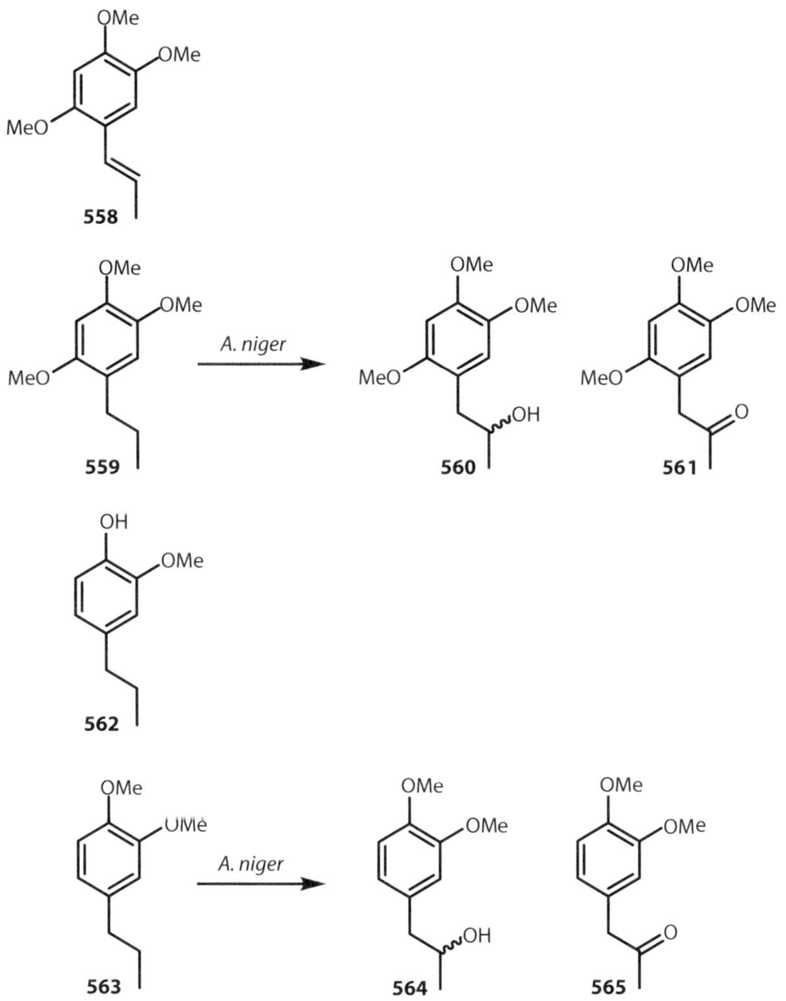

FIGURE 23.160 Biotransformation of dihydroasarone (**559**) and methyl dihydroeugenol (**563**) by *Aspergillus niger*.

good yield. *E. gracilis* could convert acetophenone to 2-phenylethanol; however, its enantio-excess is very poor (10%) (Takahashi, 1994).

Raspberry ketone (**566**) and zingerone (**574**) are the major components of raspberry (*Rubus idaeus*) and ginger (*Zingiber officinale*), and these are used as food additive and spice. Two substrates were incubated with the *Phytolacca americana* cultured cells for 3 days to produce two secondary alcohols (**567**, **568**) as well as five glucosides (**569–572**) from **566** and a secondary alcohol (**576**) and four glycoside products (**575**, **577–579**) from **574**. In the case of raspberry ketone, phenolic hydroxyl group was preferably glycosylated after the reduction of carbonyl group of the substrate occurred. It is interesting to note that one more hydroxyl group was introduced into the benzene ring to give **568**, which were further glycosylated by one of the phenolic hydroxyl groups, and no glycoside of the secondary alcohol at C2 were obtained (Figure 23.161).

On the other hand, zingerone (**574**) was converted into **576**, followed by glycosylation to give both glucosides (**577**, **578**) of phenolic and secondary hydroxyl groups and a diglucoside (**579**) of both phenolic and secondary hydroxyl group in the molecule. It is the first report on the introduction of individual glucose residues onto both phenolic and secondary hydroxyl groups by cultured plant cells (Shimoda et al., 2007a) (Figure 23.162).

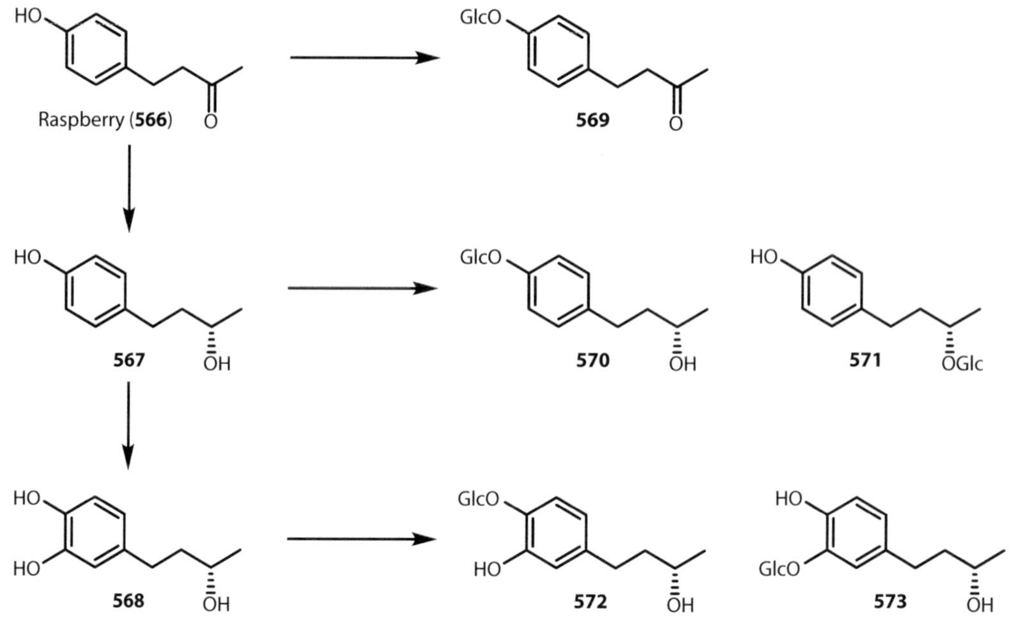

FIGURE 23.161 Biotransformation of raspberry ketone (**566**) by *Phytolacca americana* cells.

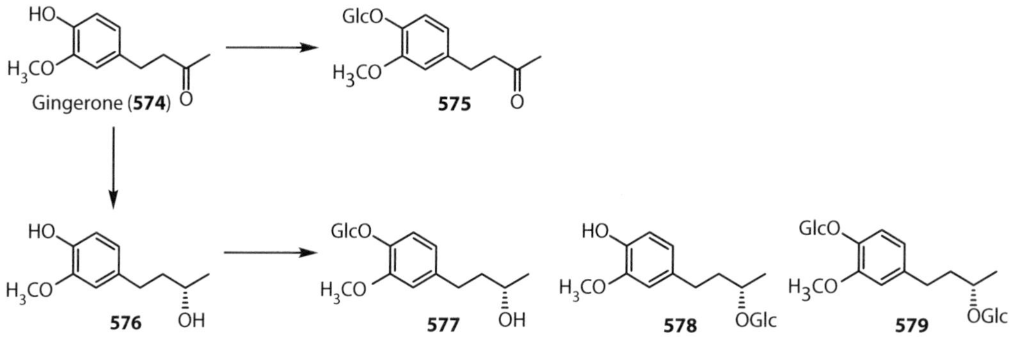

FIGURE 23.162 Biotransformation of zingerone (**574**) by *Phytolacca americana* cells.

Thymol (**580**), carvacrol (**583**), and eugenol (**586**) were glucosylated by glycosyltransferase of cell-cultured *Eucalyptus perriniana* to each glucoside (**581%**, 3%; **584%**, 5%; **587%**, 7%) and gentiobioside (**582%**, 87%; **585%**, 56%; **588%**, 58%). The yield of thymol glycosides was 1.5 times higher than that of carvacrol and 4 times higher than that of eugenol. Such glycosylation is useful to obtain higher water-soluble products from natural and commercially available secondary metabolites for food additives and cosmetic fields (Shimoda et al., 2006) (Figure 23.163).

Hinokitiol (**589**), which is easily obtained from cell suspension cultures of *Thujopsis dolabrata* and possesses potent antimicrobial activity, was incubated with cultured cells of *E. perriniana* for 7 days to give its monoglucosides (**590**, **591%**, 32%) and gentiobiosides (**592**, **593**) (Furuya et al., 1997; Hamada et al., 1998) (Figure 23.164).

(−)-Nopol benzyl ether (**594**) was smoothly biotransformed by *A. niger*, *A. cellulosae*, *A. sojae*, *Aspergillus usami*, and *Penicillium* species in Czapek-peptone medium to give (−)-4-oxonopol-2′,4′-dihydroxybenzyl ether (**595%**, 23% in the case of *A. niger*), which demonstrated antioxidant activity (ID_{50} 30.23 m/M), together with a small amount of nopol (6.3% in *A. niger*). This is very rare direct introduction of oxygen function on the phenyl ring (Noma and Asakawa, 2006) (Figure 23.165).

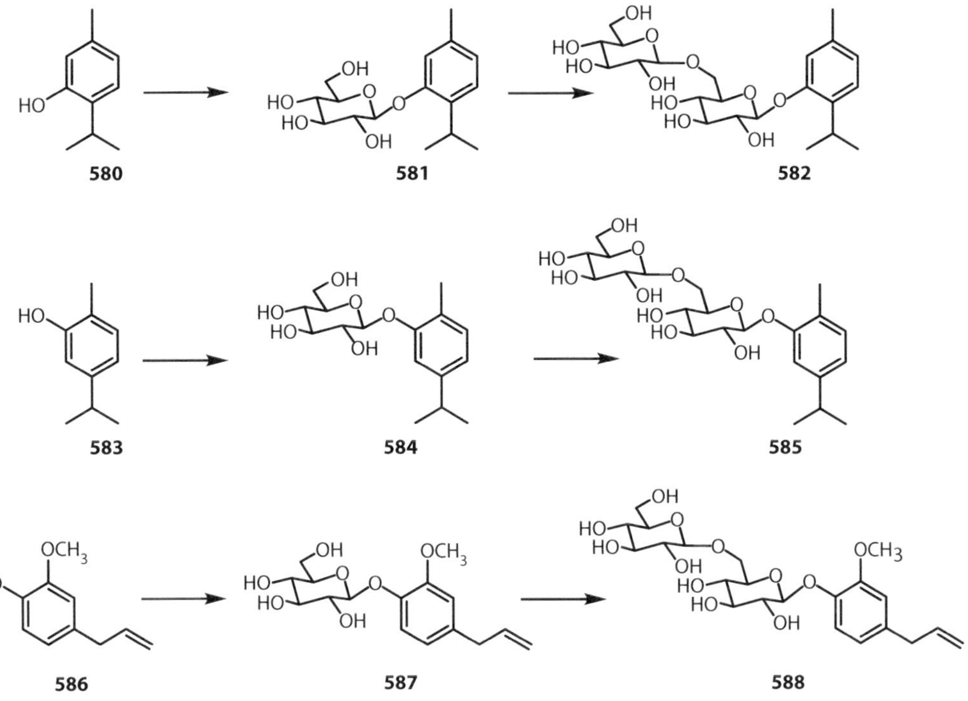

FIGURE 23.163 Biotransformation of thymol (**580**), carvacrol (**583**), and eugenol (**586**) by *Eucalyptus perriniana* cells.

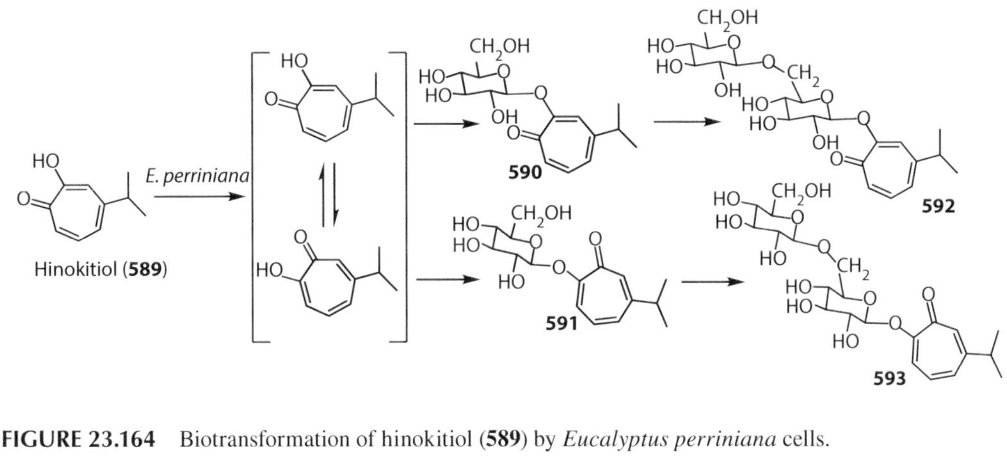

FIGURE 23.164 Biotransformation of hinokitiol (**589**) by *Eucalyptus perriniana* cells.

Capsicum annuum contains capsaicin (**596**), and its homologues having an alkylvanillylamides possess various interesting biological activities such as anti-inflammatory, antioxidant, saliva- and stomach juice–inducing activity, analgesic, antigenotoxic, antimutagenic, anticarcinogenic, antirheumatoid arthritis, and diabetic neuropathy and are used as food additives. On the other hand, because of potent pungency and irritation on skin and mucous membrane, it has not yet been permitted as medicinal drug. In order to reduce this typical pungency and application of nonpungent capsaicin metabolites to the crude drug, capsaicin (**596**) (600 mg) including 30% of dihydrocapsaicin (**600**) was incubated in Czapek-peptone medium including *A. niger* for 7 days to give three metabolites, w1-hydroxylated capsaicin (**597**, 60.9%), 8,9-dihydro-w1-hydroxycapsaicin (**598%**, 16%), and a carboxylic acid (**599**, 13.6%). All of the metabolites do not show pungency (Figure 23.166).

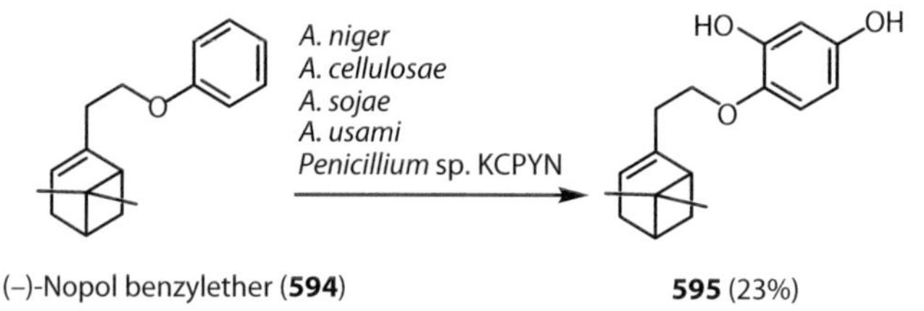

FIGURE 23.165 Biotransformation of nopol benzyl ether (**531**) by *Aspergillus* and *Penicillium* species.

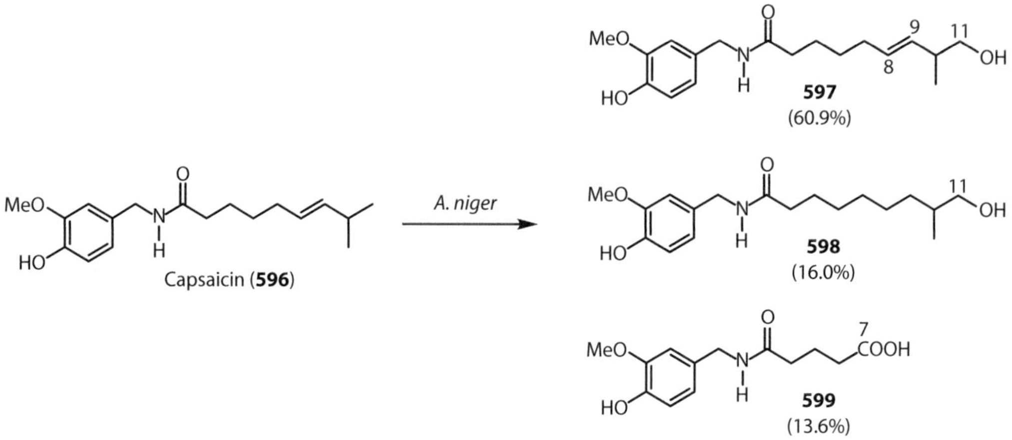

FIGURE 23.166 Biotransformation of capsaicin (**596**) by *Aspergillus niger*.

Dihydrocapsaicin (**600**) was also treated in the same manner as described earlier to afford wl-hydroxydihyrocpsaicin (**598**, 80.9%) in high yield and the carboxylic acid (**599**, 5.0%). Capsaicin itself showed carbachol-induced contraction of 60% in the bronchus at a concentration of 1 mmol/L. 11-Hydroxycapsicin (**85**) retained this activity of 60% at a concentration of 30 mmol/L. Dihydrocapsaicin (**600**) showed the same activity of contraction in the bronchus, at the same concentration as that used in capsaicin. However, the activity of contraction in the bronchus of 11-hydroxy derivative (**598**) showed weaker (50% at 30 mmol/L) than that of the substrate. Since both metabolites (**597** and **598**) are tasteless, these products might be valuable for the crude drug although the contraction in the bronchus is weak. 2,2-Diphenyl-1-picrylhydrazyl radical scavenging activity test of capsaicin and dihydrocapsaicin derivatives was carried out. 11-Hydroxycapsaicin (**597**), 11-dihydrocapsaicin (**598**), and capsaicin (**596**) showed higher activity than (±)-α-tocopherol, and 11-dihydroxycapsaicin (**598**) displayed a strong scavenging activity (IC_{50} 50 mmol/L) (Hashimoto and Asakawa, unpublished results) (Figure 23.167).

Shimoda et al. (2007b) reported the bioconversion of capsaicin (**596**) and 8-nor-dihydrocapsaicin (**601**) by the cultured cell of *Catharanthus roseus* to give more water-soluble capsaicin derivatives. From capsaicin, three glycosides, capsaicin 4-*O*-β-D-glucopyranoside (**602**), which was one of the capsaicinoids in the fruit of *Capsicum* and showed 1/100 weaker pungency than capsaicin, 4-*O*-(6-*O*-β-D-xylopyranosyl)-β-D-glucoside (**603**), and 4-*O*-(6-*O*-α-L-arbinosyl)-β-D-glucopyranoside (**604**), were obtained. 8-Nor-dihydrocapsaicin (**601**) was also incubated with the same cultured cell to afford the similar products (**605–607**) all of which reduced their pungency and enhanced water solubility. Since many synthetic capsaicin glycosides possess remarkable pharmacological activity, such as decrease of liver and serum lipids, the present products will be used for valuable prodrugs (Figure 23.168).

Z. officinale contains various sesquiterpenoids and pungent aromatic compounds such as 6-shogaol (**608**) and 6-gingerol (**613**), and their pungent compounds that possess cardio tonic and sedative

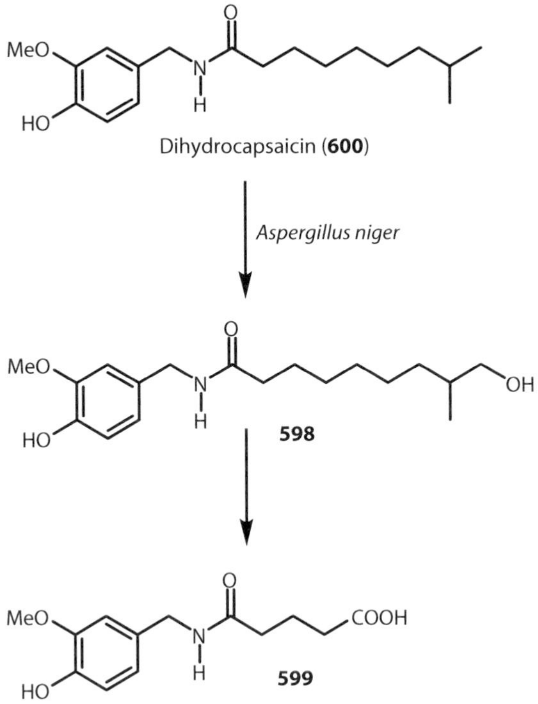

Dihydrocapsaicin (**600**)

Aspergillus niger

598

599

FIGURE 23.167 Biotransformation of dihydrocapsaicin (**600**) by *Aspergillus niger*.

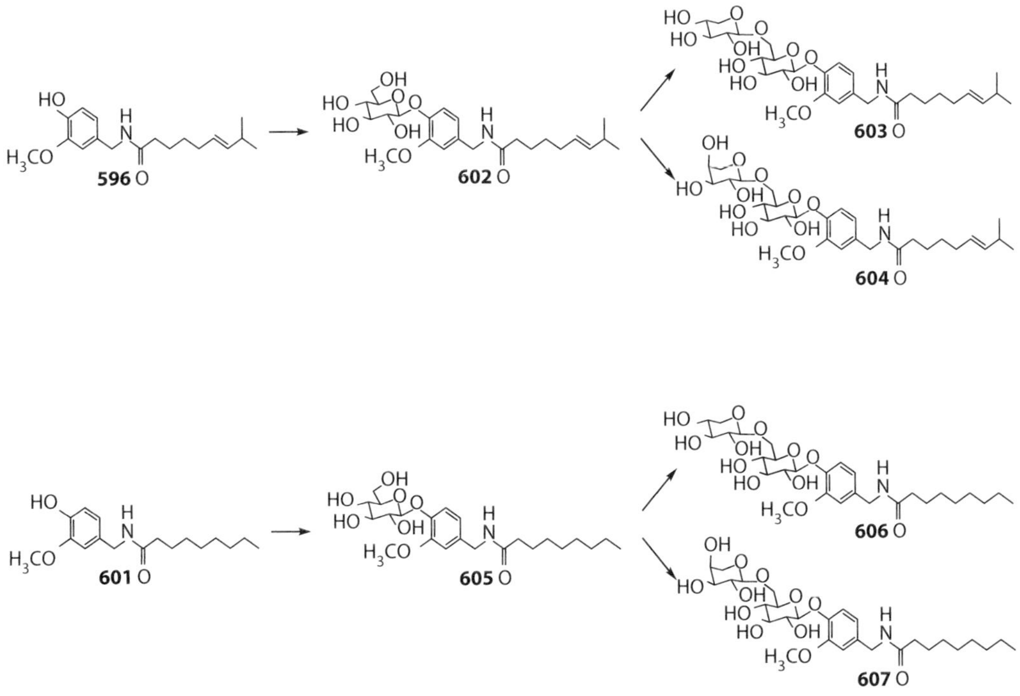

FIGURE 23.168 Biotransformation of capsaicin (**596**) and 8-nor-dihydrocapsaicin (**601**) by *Catharanthus roseus* cells.

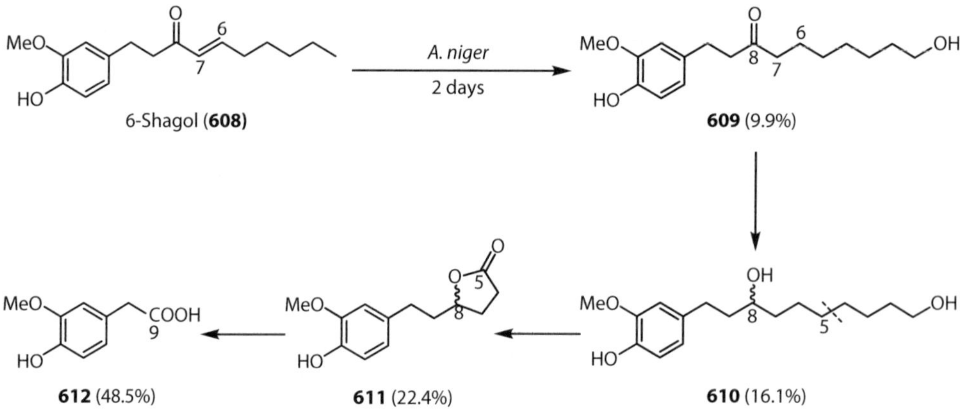

FIGURE 23.169 Biotransformation of 6-shogaol (**608**) by *Aspergillus niger.*

activity. 6-Shogaol (**608**) was incubated with *A. niger* in Czapek-peptone medium for 2 days to afford wl-hydroxy-6-shagaol (**609**, 9.9%), which was further converted to 8-hydroxy derivative (**610**, 16.1%), a γ-lactone (**611**, 22.4%), and 3-methoxy-4-hydroxyphenylacetic acid (**612**, 48.5%) (Figure 23.169).

6-Gingerol (**613**) (1 g) was treated in the same condition as mentioned earlier to yield six metabolites, wl-hydroxy-6-gingerol (**614**, 39.8%), its carboxylic derivative (**616**, 14.5%), a γ-lactone (**618**) (16.9%) that might be formed from (**616**), its 8-hydoxy-γ-lactone (**619**, 12.1%), w2-hydroxy-6-gingerol (**615**, 19.9%), and 6-deoxy-gingerol (**617**, 14.5%) (Takahashi et al., 1993a).

The metabolic pathway of 6-gingerol (**613**) resembles that of 6-shagaol (**608**). That of 6-shogaol and dihydrocapsaicin (**600**) is also similar since both substrates gave carboxylic acids as the final metabolites (Takahashi et al., 1993a) (Figure 23.170).

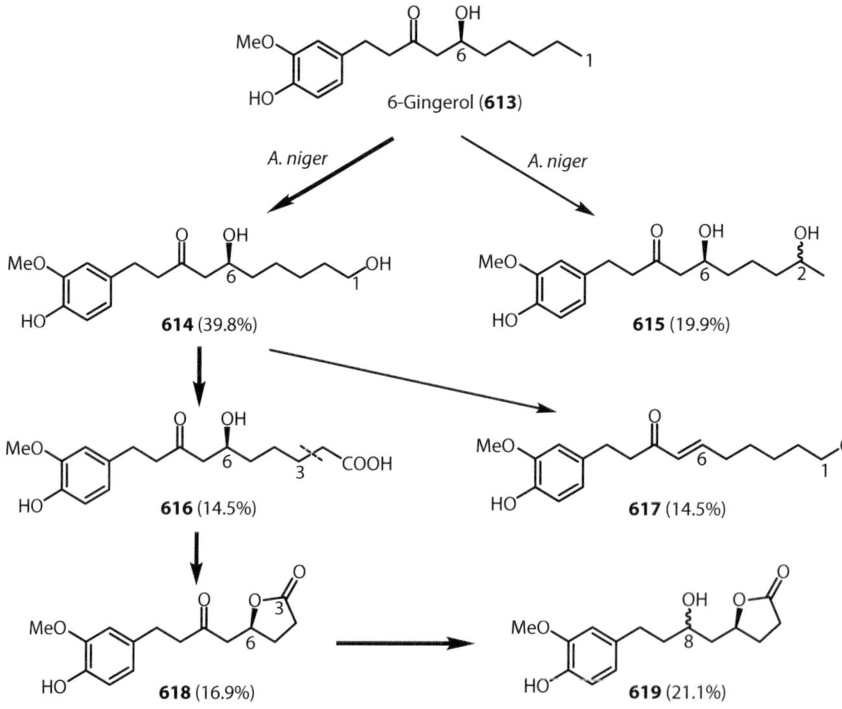

FIGURE 23.170 Biotransformation of 6-gingerol (**613**) by *Aspergillus niger.*

In conclusion, a number of sesquiterpenoids were biotransformed by various fungi and mammals to afford many metabolites, several of which showed antimicrobial and antifungal, antiobesity, cytotoxic, neurotrophic, and enzyme inhibitory activity. Microorganisms introduce oxygen atom at allylic position to give secondary hydroxyl and keto groups. Double bond is also oxidized to give epoxide, followed by hydrolysis to afford a diol. These reactions precede stereo- and regiospecifically. Even at nonactivated carbon atom, oxidation reaction occurs to give primary alcohol. Some fungi like *A. niger* cleave the cyclopropane ring with a 1,1-dimethyl group. It is noteworthy that *A. niger* and *A. cellulosae* produce totally different metabolites from the same substrates. Some fungi occurs reduction of carbonyl group, oxidation of aryl methyl group, phenyl coupling, and cyclization of a 10-membered ring sesquiterpenoids to give C6/C6- and C5/C7-cyclic or spiro compounds. Cytochrome P-450 is responsible for the introduction of oxygen function into the substrates.

The present methods are very useful for the production of medicinal and agricultural drugs as well as fragrant components from commercially available cheap, natural, and unnatural terpenoids or a large amount of terpenoids from higher medicinal plants and spore-forming plants like liverworts and fungi.

The methodology discussed in this chapter is a very simple one-step reaction in water, nonhazard, and very cheap, and it gives many valuable metabolites possessing different properties from those of the substrates.

REFERENCES

Abraham, W.R., L. Ernst, and B. Stumpf, 1990. Biotransformation of caryophyllene by *Diplodia gossypina*. *Phytochemisrty*, 29: 115–120.

Abraham, W.-R., K. Kieslich, B. Stumpf, and L. Ernst, 1992. Microbial oxidation of tricyclic sesquiterpenoids containing a dimethylcyclopropane ring. *Phytochemistry*, 31: 3749–3755.

Abraham, W.-R., P. Washausen, and K. Kieslich, 1987. Microbial hydroxylation of cedrol and cedrene. *Z. Naturforsch.*, 42C: 414–419.

Aleu, J., J.R. Hanson, R. Hernandez Galan, and I.G. Collado, 1999a. Biotransformation of the fungistatic sesquiterpenoid patchoulol by *Botrytis cinerea*. *J. Nat. Prod.*, 62: 437–440.

Aleu, J., R. Hernandez-Galan, and I.G. Collad, 2002. Biotransformation of the fungistatic sesquiterpenoid isobotryan-9α-ol by Botrytis cinerea. *J. Mol. Catal. B*, 16: 249–253.

Aleu, J., R. Hernandez-Galan, J.R. Hanson, P.B. Hitchcock, and I.G. Collado, 1999b. Biotransformation of the fungistatic sesquiterpenoid ginsenol by *Botrytis cinerea*. *J. Chem. Soc. Perkin Trans.*, 1: 727–730.

Amate, A., A. Garcia-Granados, A. Martinez et al. 1991. Biotransformation of 6α-eudesmanolides functionalized at C-3 with Curvularia lunata and Rhizopus nigricans cultures. *Tetrahedron*, 47: 5811–5818.

Aranda G., I. Facon, J.-Y. Lallemand, and M. Leclaire, 1992. Microbiological hydroxylation in the drimane series. *Tetrahedron Lett.*, 33, 7845–7848.

Aranda, G., M.S. Kortbi, J.-Y. Lallemand et al. 1991. Microbial transformation of diterpenes: Hydroxylation of sclareol, manool and derivatives by *Mucor plumbeus*. *Tetrahedron*, 47: 8339–8350.

Arantes, S.F., J.R. Hanson, and P.B. Hitchcok, 1999. The hydroxylation of the sesquiterpenoid valerianol by *Mucor plumbeus*. *Phytochemistry*, 52: 1063–1067.

Asakawa, Y., 1982. Chemical constituents of the Hepaticae. In: *Progress in the Chemistry of Organic Natural Products*, Herz, W., H. Grisebach, and G.W. Kirby (eds.), Vol. 42, pp. 1–285. Vienna, Austria: Springer.

Asakawa, Y., 1990. Terpenoids and aromatic compounds with pharmaceutical activity from bryophytes. In: *Bryophytes: Their Chemistry and Chemical Taxonomy*, Zinsmeister, D.H. and R. Mues (eds.), pp. 369–410. Oxford, U.K.: Clarendon Press.

Asakawa, Y., 1995. Chemical constituents of the bryophytes. In: *Progress in the Chemistry of Organic Natural Products*, Herz, W., G.W. Kirby, R.E. Moore, W. Steglich, and Ch. Tamm (eds.), Vol. 65, pp. 1–618. Vienna, Austria: Springer.

Asakawa, Y., 1999. Phytochemistry of bryophytes. In: *Phytochemicals in Human Health Protection, Nutrition, and Plant Defense*, J. Romeo (ed.), Vol. 33, pp. 319–342. New York: Kluwer Academic, Plenum Publishers.

Asakawa, Y., 2007. Biologically active compounds from bryophytes. *Pure Appl. Chem.*, 75: 557–580.

Asakawa, Y., 2008. Recent advances of biologically active substances from the Marchantiophyta. *Nat. Prod. Commun.*, 3: 77–92.

Asakawa, Y., T. Hashimoto, Y. Mizuno, M. Tori, and Y. Fukuzawa, 1992. Cryptoporic acids A-G, drimane-type sesquiterpenoid ethers of isocitric acid from the fungus *Cryptoporus volvatus*. *Phytochemistry*, 31: 579–592.

Asakawa, Y., T. Ishida, M. Toyota, and T. Takemoto, 1986. Terpenoid biotransformation in mammals IV. Biotransformation of (+)-longifolene, (−)-caryophyllene, (−)-caryophyllene oxide, (−)-cyclocolorenone, (+)-nootkatone, (−)-elemol, (−)-abietic acid and (+)-dehydroabietic acid in rabbits. *Xenobiotica*, 6: 753–767.

Asakawa, Y. and A. Ludwiczuk, 2008. Bryophytes-Chemical diversity, bioactivity and chemosystematics. Part 1. Chemical diversity and bioactivity. *Med. Plants Poland World*, 14: 33–53.

Asakawa, Y., Z. Taira, T. Takemoto, T. Ishida, M. Kido, and Y. Ichikawa, 1981. X-ray crystal structure analysis of 14-hydroxycaryophyllene oxide, a new metabolite of (−)-caryophyllene in rabbits. *J. Pharm. Sci.*, 70: 710–711.

Asakawa, Y., H. Takahashi, and M. Toyota, 1991. Biotransformation of germacrane-type sesquiterpenoids by *Aspergillus niger. Phytochemistry*, 30: 3993–3997.

Ata, A. and J.A. Nachtigall, 2004. Microbial transformation of α-santonin. *Z. Naturforsch.*, 59C: 209–214.

Atta-ur-Rahman, M.I. Choudhary, A. Ata et al., 1994. Microbial transformation of 7α-hydroxyfrullanolide. *J. Nat. Prod.*, 57: 1251–1255.

Atta-ur-Rahman, M.I. Choudhary, F. Shaheen, A. Rauf, and A. Farooq, 1998. Microbial transformation of some bioactive natural products. *Nat. Prod. Lett.*, 12: 215–222.

Atta-ur-Rahman, A. Farooq, and M.I. Choudhary, 1997. Microbial transformation of sclareolide. *J. Nat. Prod.*, 60: 1038–1040.

Ayer, W.A. and P.A. Craw, 1989. Metabolites of fairy ring fungus, *Marasmius oreades*. Part 2. Norsesquiterpenes, further sesquiterpenes, and argocybin. *Can. J. Chem.*, 67: 1371–1380.

Bang, L. and G. Ourisson, 1975. Hydroxylation of cedrol by rabbits. *Tetrahedron Lett.*, 16: 1881–1884.

Bang, L., G. Ourisson, and P. Teisseire, 1975. Hydroxylation of patchoulol by rabbits. Hemisynthesis of nor-patchoulenol, the odour carrier of patchouli oil. *Tetrahedron Lett.*, 16: 2211–2214.

Barrero, A.F., J.E. Oltra, D.S. Raslan, and D.A. Sade, 1999. Microbial transformation of sesquiterpene lactones by the fungi *Cunninghamella echinulata* and *Rhizopus oryzae*. *J. Nat. Prod.*, 62: 726–729.

Bhutani, K.K. and R.N. Thakur, 1991. The microbiological transformation of parthenin by *Beauveria bassiana* and *Sporotrichum pulverulentum*. *Phytochemistry*, 30: 3599–3600.

Buchanan, G.O., L.A.D. Williams, and P.B. Reese, 2000. Biotransformation of cadinane sesquiterpenes by *Beauveria bassiana* ATCC 7159. *Phytochemistry*, 54: 39–45.

Choudhary, M.I., W. Kausar, Z.A. Siddiqui, and Atta-ur-Rahman, 2006a. Microbial metabolism of (+)-cycloisolongifol-5β-ol. *Z. Naturforsch.*, 61B: 1035–1038.

Choudhary, M.I., S.G. Musharraf, S.A. Nawaz et al., 2005. Microbial transformation of (−)-isolongifolol and butyrylcholinesterase inhibitory activity of transformed products. *Bioorg. Med. Chem.*, 13: 1939–1944.

Choudhary, M.I., S.G. Musharraf, A. Sami, and Atta-ur-Rahman, 2004. Microbial transformation of sesquiterpenes, (−)-ambrox® and (+)-sclareolide. *Helv. Chim. Acta*, 87: 2685–2694.

Choudhary, M.I., Z.A. Siddiqui, S.A. Nawaz, and Atta-ur-Rahman, 2006b. Microbial transformation and butyrylcholinesterase inhibitory activity of (−)-caryophyllene oxide and its derivatives. *J. Nat. Prod.*, 69: 1429–1434.

Clark, A.M. and C.D. Hufford, 1979. Microbial transformation of the sesquiterpene lactone costunolide. *J. Chem. Soc. Perkin Trans.*, 1: 3022–3028.

Collins, D.O. and P.B. Reese, 2002. Biotransformation of cadina-4,10(15)-dien-3-one and 3α-hydroxycadina-4,10(15)-diene by Curvularia lunata ATCC 12017. *Phytochemistry*, 59: 489–492.

Collins, D.O., W.F. Reynold, and P.B. Reese, 2002a. Aromadendrane transformations by *Curvularia lunata* ATCC 12017. *Phytochemistry*, 60: 475–481.

Collins, D.O., P.L.D. Ruddock, J. Chiverton, C. de Grasse, W.F. Reynolds, and P.B. Reese, 2002b. Microbial transformation of cadina-4,10(15)-dien-3-one, aromadendr-1(10)-en-9-one and methyl ursolate by *Mucor plumbeus* ATCC 4740. *Phytochemistry*, 59: 479–488.

Daniewski, W.M., P.A. Grieco, J. Huffman, A. Rymkiewicz, and A. Wawrzun, 1981. Isolation of 12-hydroxycaryophyllene-4,5-oxide, a sesquiterpene from *Lactarius camphoratus*. *Phytochemistry*, 20: 2733–2734.

Darrouzet, E., B. Mauchamp, G.D. Prestwich, L. Kerhoas, I. Ujvary, and F. Couillaud 1997. Hydroxy juvenile hormones: New putative juvenile hormones biosynthesized by locust corpora allata *in vitro*. *Biochem. Biophys. Res. Commun.*, 240: 752–758.

Devi, J.R., 1979. Microbiological transformation of terpenes: Part XXVI. Microbial transformation of caryophyllene. *Ind. J. Biochem. Biophys.*, 16: 76–79.

Dhavlikar, R.S. and G. Albroscheit, 1973. Microbiologische Umsetzung von Terpenen: Valencen. *Dragoco Rep.*, 12: 250–258.

Duran, R., E. Corrales, R. Hernandez-Galan, and G. Collado, 1999. Biotransformation of caryophyllene oxide by *Botrytis cinerea*. *J. Nat. Prod.*, 62: 41–44.

El Sayed, K.A., M. Yousaf, M.T. Hamann, M.A. Avery, M. Kelly, and P. Wipf, 2002. Microbial and chemical transformation studies of the bioactive marine sesquiterpenes (*S*)-(+)-curcuphenol and -curcudiol isolated from a deep reef collection of the Jamaican sponge *Didiscus oxeata*. *J. Nat. Prod.*, 65: 1547–1553.

Furusawa, M., 2006. Microbial biotransformation of sesquiterpenoids from crude drugs and liverworts: Production of functional substances. PhD thesis. Tokushima Bunri University, Tokushima, Japan, pp. 1–156.

Furusawa, M., T. Hashimoto, Y. Noma, and Y. Asakawa, 2005a. Biotransformation of Citrus aromatics nootkatone and valencene by microorganisms. *Chem. Pharm. Bull.*, 53: 1423–1429.

Furusawa, M., T. Hashimoto, Y. Noma, and Y. Asakawa 2005b. Highly efficient production of nootkatone, the grapefruit aroma from valencene, by biotransformation. *Chem. Pharm. Bull.*, 53: 1513–1514.

Furusawa, M., T. Hashimoto, Y. Noma, and Y. Asakawa, 2005c. The structure of new sesquiterpenoids from the liverwort *Reboulia hemisphaerica* and their biotransformation. *Proceeding of 49th TEAC*, pp. 235–237.

Furusawa, M., T. Hashimoto, Y. Noma, and Y. Asakawa, 2006a. Biotransformation of aristolene- and 2,3-secoaromadendrane-type sesquiterpenoids having a 1,1-dimethylcyclopropane ring by *Chlorella fusca* var. *vacuolata, Mucor* species, and *Aspergillus niger. Chem. Pharm. Bull.*, 54: 861–868.

Furusawa, M., T. Hashimoto, Y. Noma, and Y. Asakawa, 2006b. Isolation and structures of new cyclomyltaylane and ent-chamigrane-type sesquiterpenoids from the liverwort *Reboulia hemisphaerica. Chem. Pharm. Bull.*, 54: 996–1003.

Furusawa, M., Y. Noma, T. Hashimoto, and Y. Asakawa, 2003. Biotransformation of Citrus oil nootkatone, dihydronootkatone and dehydronootkatone. *Proceedings of 47th TEAC*, pp. 142–144.

Furuya, T., Y. Asada, Y. Matsuura, S. Mizobata, and H. Hamada, 1997. Biotransformation of β-thujaplicin by cultured cells of Eucalyptus perriniana. *Phytochemistry*, 46: 1355–1358.

Galal, A.M., 2001. Microbial transformation of pyrethrosin. *J. Nat. Prod.*, 64: 1098–1099.

Galal, A.M., A.S. Ibrahim, J.S. Mossa, and F.S. El-Feraly, 1999. Microbial transformation of parthenolide. *Phytochemistry*, 51: 761–765.

Garcia-Granados, A., M.C. Gutierrez, F. Rivas, and J.M. Arias, 2001. Biotransformation of 4β-hydroxyeudesmane-1,6-dione by Gliocladium roseum and Exserohilum halodes. *Phytochemistry*, 58: 891–895.

Hamada, H., F. Murakami, and T. Furuya, 1998. The production of hinokitiol glycoside. *Proceedings of 42nd TEAC*, pp. 145–147.

Hanson, J.R. and H. Nasir, 1993. Biotransformation of the sesquiterpenoid, cedrol, by *Cephalosporium aphidicola. Phytochemistry*, 33: 835–837.

Hanson, J.R. and A. Truneh, 1996. The biotransformation of ambrox and sclareolide by *Cephalosporium aphidicola. Phytochemistry*, 42: 1021–1023.

Hanson, R.L., J.M. Wasylyk, V.B. Nanduri, D.L. Cazzulino, R.N. Patel, and L.J. Szarka, 1994. Site-specific enzymatic hydrolysis of tisanes at C-10 and C-13. *J. Biol. Chem.*, 269: 22145–22149.

Harinantenaina, L., Y. Noma, and Y. Asakawa, 2005. Penicillium sclerotiorum catalyzes the conversion of herbertenediol into its dimers: Mastigophorenes A and B. *Chem. Pharm. Bull.*, 53: 256–257.

Harinantenaina, L., D.N. Quang, T. Nishizawa et al., 2007. Bioactive compounds from liverworts: Inhibition of lipopolysaccharide-induced inducible NOS mRNA in RAW 264.7 cells by herbertenoids and cuparenoids. *Phytomedicine*, 14: 486–491.

Hashimoto, T. and Y. Asakawa, 2007. Biological activity of fragrant substances from Citrus and herbs, and production of functional substances using microbial biotransformation. In: *Development of Medicinal Foods*, M. Yoshikawa (ed.), pp. 168–184. Tokyo, Japan: CMC Publisher.

Hashimoto, T., Y. Asakawa, Y. Noma et al., 2003a. Production method of nootkatone. *Jpn. Kokai Tokkyo Koho*, 250591A.

Hashimoto, T., Y. Asakawa, Y. Noma et al., 2003b. Production method of nootkatone. *Jpn. Kokai Tokkyo Koho*, 70492A.

Hashimoto, T., M. Fujiwara, K. Yoshikawa, A. Umeyama, M. Tanaka, and Y. Noma, 2007. Biotransformation of sclareolide and sclareol by microorganisms. *Proceedings of 51st TEAC*, pp. 316–318.

Hashimoto, T., S. Kato, M. Tanaka, S. Takaoka, and Y. Asakawa, 1998a. Biotransformation of sesquiterpenoids by microorganisms (4): Biotransformation of hinesol by *Aspergillus niger. Proceedings of 42nd TEAC*, pp. 127–129.

Hashimoto, T., Y. Noma, Y. Akamatsu, M. Tanaka, and Y. Asakawa, 1999a. Biotransformation of sesquiterpenoids by microorganisms. (5): Biotransformation of dehydrocostuslactone. *Proceedings of 43rd TEAC*, pp. 202–204.

Hashimoto, T., Y. Noma, and Y. Asakawa, 2001a. Biotransformation of terpenoids from the crude drugs and animal origin by microorganisms. *Heterocycles*, 54: 529–559.

Hashimoto, T., Y. Noma, and Y. Asakawa, 2006. Biotransformation of cuparane- and herbertane-type sesquiterpenoids. *Proceedings of 50th TEAC*, pp. 263–265.

Hashimoto, T., Y. Noma, Y. Goto, S. Takaoka, M. Tanaka, and Y. Asakawa, 2003c. Biotransformation of sesquiterpenoids from the liverwort Plagiochila species. *Proceedings of 47th TEAC*, pp. 139–141.

Hashimoto, T., Y. Noma, Y. Goto, M. Tanaka, S. Takaoka, and Y. Asakawa, 2004. Biotransformation of (−)-maalioxide by *Aspergillus niger* and *Aspergillus cellulosae*. *Heterocycles*, 62: 655–666.

Hashimoto, T., Y. Noma, S. Kato, M. Tanaka, S. Takaoka, and Y. Asakawa, 1999b. Biotransformation of hinesol isolated from the crude drug Atractylodes lancea by *Aspergillus niger* and Aspergillus cellulosae. *Chem. Pharm. Bull.*, 47: 716–717.

Hashimoto, T., Y. Noma, H. Matsumoto, Y. Tomita, M. Tanaka, and Y. Asakawa, 2000a. Microbial biotransformation of optically active (+)-α-ionone and (−)-α-ionone. *Proceedings of 44th TEAC*, pp. 154–156.

Hashimoto, T., Y. Noma, Y. Matsumoto, Y. Akamatsu, M. Tanaka, and Y. Asakawa, 1999c. Biotransformation of sesquiterpenoids by microorganisms. (6): Biotransformation of α-, β- and γ-cyclocostunolides. *Proceedings of 43rd TEAC*, pp. 205–207.

Hashimoto, T., Y. Noma, C. Murakami, N. Nishimatsu, M. Tanaka, and Y. Asakawa, 2001b. Biotransformation of valencene and aristolene. *Proceedings of 45th TEAC*, pp. 345–347.

Hashimoto, T., Y. Noma, C. Murakami, M. Tanaka, and Y. Asakawa, 2000b. Microbial transformation of α-santonin derivatives and nootkatone. *Proceedings of 44th TEAC*, pp. 157–159.

Hashimoto, T., Y. Noma, N. Nishimatsu, M. Sekita, M. Tanaka, and Y. Asakawa, 2003d. Biotransformation of antimalarial sesquiterpenoids by microorganisms. *Proceedings of 47th TEAC*, pp. 136–138.

Hashimoto, T., M. Sekita, M. Furusawa, Y. Noma, and Y. Asakawa, 2005. Biotransformation of sesquiterpene lactones, (−)-parthenolide and (−)-frullanolide by microorganisms. *Proceedings of 49th TEAC*, pp. 387–389.

Hashimoto, T., K. Shiki, M. Tanaka, S. Takaoka, and Y. Asakawa, 1998b. Chemical conversion of labdane-type diterpenoid isolated from the liverwort *Porella perrottetiana* into (−)-ambrox. *Heterocycles*, 49: 315–325.

Hayashi, K., H. Morikawa, A. Matsuo, D. Takaoka, and H. Nozaki, 1999. Biotransformation of sesquiterpenoids by *Aspergillus niger* IFO 4407. *Proceedings of 43rd TEAC*, pp. 208–210.

Hayashi, K., H. Morikawa, H. Nozaki, and D. Takaoka, 1998. Biotransformation of globulol and epiglubulol by *Aspergillus niger* IFO 4407. *Proceedings of 42nd TEAC*, pp. 136–138.

Haze, S., K. Sakai, and Y. Gozu, 2002. Effects of fragrance inhalation on sympathetic activity in normal adults. *Jpn. J. Pharmacol.*, 90: 247–253.

Hegazy, M.-E.F., C. Kuwata, Y. Sato et al., 2005. Research and development of asymmetric reaction using biocatalysts-biotransformation of enones by cultured cells of Marchantia polymorpha. *Proceedings of 49th TEAC*, pp. 402–404.

Higuchi, H., R. Tsuji, K. Hayashi, D. Takaoka, A. Matsuo, and H. Nozaki, 2001. Biotransformation of sesquiterpenoids by *Aspergillus niger*. *Proceedings of 45th TEAC*, pp. 354–355.

Hikino, H., T. Konno, T. Nagashima, T. Kohama, and T. Takemoto, 1971. Stereoselective epoxidation of germacrone by *Cunninghamella blakesleeana*. *Tetrahedron Lett.*, 12: 337–340.

Hrdlicka, P.J., A.B. Sorensen, B.R. Poulsen, G.J.G. Ruijter, J. Visser, and J.J.L. Iversen, 2004. Characterization of nerolidol biotransformation based on indirect on-line estimation of biomass concentration and physiological state in bath cultures of *Aspergillus niger*. *Biotechnol. Prog.*, 20: 368–376.

Ishida, T., 2005. Biotransformation of terpenoids by mammals, microorganisms, and plant-cultured cells. *Chem. Biodivers.*, 2: 569–590.

Ishida, T., Y. Asakawa, and T. Takemoto, 1982. Hydroxyisolongifolaldehyde: A new metabolite of (+)-longifolene in rabbits. *J. Pharm. Sci.*, 71: 965–966.

Ishida, T., Y. Asakawa, T. Takemoto, and T. Aratani, 1981. Terpenoid biotransformation in mammals. III. Biotransformation of α-pinene, β-pinene, pinane, 3-carene, carane, myrcene and p-cymene in rabbits. *J. Pharm. Sci.*, 70: 406–415.

Itsuzaki, Y., K. Ishisaka, and M. Miyazawa, 2002. Biotransformation of (+)-cedrol by using human skin microbial flora *Staphylococcus epidermidis*. *Proceedings of 46th TEAC*, pp. 101–102.

Jacobsson, U., V. Kumar, and S. Saminathan, 1995. Sesquiterpene lactones from *Michelia champaca*. *Phytochemistry*, 39: 839–843.

Kaspera, R., U. Krings, T. Nanzad, and R.G. Berger, 2005. Bioconversion of (+)-valencene in submerged cultures of the ascomycete *Chaetomium globosum*. *Appl. Microbiol. Biotechnol.*, 67: 477–583.

Kiran, I., T. Akar, A. Gorgulu, and C. Kazaz, 2005. Biotransformation of racemic diisophorone by *Cephalosporium aphidicola* and *Neurospora crassa*. *Biotechnol. Lett.*, 27: 1007–1010.

Kiran, I., H.N. Yildirim, J.R. Hanson, and P.B. Hitchcock, 2004. The antifungal activity and biotransformation of diisophorone by the fungus *Aspergillus niger. J. Chem. Technol. Biotechnol.*, 79: 1366–1370.

Kumari, G.N.K., S. Masilamani, R. Ganesh, and S. Aravind, 2003. Microbial biotransformation of zaluzanin D. *Phytochemistry*, 62: 1101–1104.

Lahlou, E.L., Y. Noma, T. Hashimoto, and Y. Asakawa, 2000. Microbiotransformation of dehydropinguisenol by *Aspergillus* sp. *Phytochemistry*, 54: 455–460.

Lamare, V. and R. Furstoss, 1990. Bioconversion of sesquiterpenes. *Tetrahedron*, 46: 4109–4132.

Lee, I.-S., H.N. ElSohly, E.M. Coroom, and C.D. Hufford, 1989. Microbial metabolism studies of the antimalarial sesquiterpene artemisinin. *J. Nat. Prod.*, 52: 337–341.

Ma, X.C., M. Ye, L.J. Wu, and D.A. Guo, 2006. Microbial transformation of curdione by *Mucor spinosus. Enzyme Microb. Technol.*, 38: 367–371.

Maatooq, G.A., 2002b. Microbial transformation of a β- and γ-eudesmols mixture. *Z. Naturforsch.*, 57C: 654–659.

Maatooq, G.A., 2002c. Microbial conversion of partheniol by *Calonectria decora. Z. Naturforsch.*, 57C: 680–685.

Maatooq, G.T., 2002a. Microbial metabolism of partheniol by *Mucor circinelloides. Phytochemistry*, 59: 39–44.

Madyastha, K.M. and T.L. Gururaja, 1993. Utility of microbes in organic synthesis: Selective transformations of acyclic isoprenoids by *Aspergillus niger. Ind. J. Chem.*, 32B: 609–614.

Matsui, H., Y. Minamino, and M. Miyazawa, 1999. Biotransformation of (+)-cedryl acetate by *Glomerella cingulata*, parasitic fungus. *Proceedings of 43rd TEAC*, pp. 215–216.

Matsui, H. and M. Miyazawa, 2000. Biotransformation of pathouli acetate using parasitic fungus *Glomerella cingulata* as a biocatalyst. *Proceedings of 44th TEAC*, pp. 149–150.

Matsushima, A., M.-E.F. Hegazy, C. Kuwata, Y. Sato, M. Otsuka, and T. Hirata, 2004. Biotransformation of enones using plant cultured cells—The reduction of α-santonin. *Proceedings of 48th TEAC*, pp. 396–398.

Meyer, P. and C. Neuberg 1915. Phytochemische reduktionen. XII. Die umwandlung von citronellal in citronelol. *Biochem. Z*, 71: 174–179.

Mikami, Y., 1988. *Microbial Conversion of Terpenoids. Biotechnology and Genetic Engineering Reviews*, Vol. 6, pp. 271–320. Wimborne, UK: Intercept Ltd. https://www.nottingham.ac.uk/ncmh/documents/bger/volume-6/bger6-7.pdf Accessed date: 16 March 2020.

Miyazawa, M., S. Akazawa, H. Sakai, and H. Kameoka, 1997a. Biotransformation of (+)-γ-gurjunene using plant pathogenic fungus, *Glomerella cingulata* as a biocatalyst. *Proceedings of 41st TEAC*, pp. 218–219.

Miyazawa, M., Y. Honjo, and H. Kameoka, 1996a. Biotransformation of guaiol and bulnesol using plant pathogenic fungus *Glomerella cingulata* as a biocatalyst. *Proceedings of 40th TEAC*, pp. 82–83.

Miyazawa, M., Y. Honjo, and H. Kameoka, 1997b. Biotransformation of the sesquiterpenoid β-selinene using the plant pathogenic fungus *Glomerella cingulata. Phytochemistry*, 44: 433–436.

Miyazawa, M., Y. Honjo, and H. Kameoka, 1998a. Biotransformation of the sesquiterpenoid (+)-γ-gurjunene using a plant pathogenic fungus *Glomerella cingulata* as a biocatalyst. *Phytochemistry*, 49: 1283–1285.

Miyazawa, M., H. Matsui, and H. Kameoka, 1997c. Biotransformation of patchouli alcohol using plant parasitic fungus *Glomerella cingulata* as a biocatalyst. *Proceedings of 41st TEAC*, pp. 220–221.

Miyazawa, M., H. Matsui, and H. Kameoka, 1998b. Biotransformation of unsaturated sesquiterpene alcohol using plant parasitic fungus, *Glomerella cingulata* as a biocatalyst. *Proceedings of 42nd TEAC*, pp. 121–122.

Miyazawa, M., H. Nakai, and H. Kameoka, 1995a. Biotransformation of (+)-cedrol by plant pathogenic fungus, *Glomerella cingulata. Phytochemistry*, 40: 69–72.

Miyazawa, M., H. Nakai, and H. Kameoka, 1995b. Biotransformation of cyclic terpenoids, (±)-cis-nerolidol and nerylacetone, by plant pathogenic fungus, *Glomerella cingulata. Phytochemistry*, 40: 1133–1137.

Miyazawa, M., H. Nakai, and H. Kameoka, 1996b. Biotransformation of acyclic terpenoid (±)-*trans*-nerolidol and geranylacetone by *Glomerella cingulata. J. Agric. Food Chem.*, 44: 1543–1547.

Miyazawa, M., H. Nakai, and H. Kameoka, 1996c. Biotransformation of acyclic terpenoid (*2E,6E*)-farnesol by plant pathogenic fungus *Glomerella cingulata. Phytochemistry*, 43: 105–109.

Miyazawa, M., H. Nankai, and H. Kameoka, 1995c. Biotransformation of (−)-α-bisabolol by plant pathogenic fungus, *Glomerella cingulata. Phytochemistry*, 39: 1077–1080.

Miyazawa, M. and A. Sugawara, 2006. Biotransformation of (2)-guaiol by *Euritium rubrum. Nat. Prod. Res.*, 20: 731–734.

Miyazawa, M., T. Uemura, and H. Kameoka, 1994. Biotransformation of sesquiterpenoids, (−)-globulol and (+)-ledol by *Glomerella cingulata. Phytochemistry*, 37: 1027–1030.

Miyazawa, M., T. Uemura, and H. Kameoka, 1995d. Biotransformation of sesquiterpenoids, (+)-aromadendrene and (−)-alloaromadendrene by *Glomerella cingulata*. *Phytochemistry*, 40: 793–796.

Morikawa, H., K. Hayashi, K. Wakamatsu et al., 2000. Biotransformation of sesquiterpenoids by *Aspergillus niger* IFO 4407. *Proceedings of 44th TEAC*, pp. 151–153.

Nankai, H., M. Miyazawa, and H. Kameoka, 1996. Biotransformation of (Z,Z)-farnesol using plant pathogenic fungus, *Glomerella cingulata* as a biocatalyst. *Proceedings of 40th TEAC*, pp. 78–79.

Nishida, E. and Y. Kawai, 2007. Bioconversion of zerumbone and its derivatives. *Proceedings of 51st TEAC*, pp. 387–389.

Noma, Y. and Y. Asakawa, 1998. Microbiological transformation of 3-oxo-α-ionone. *Proceedings of 44th TEAC*, pp. 133–135.

Noma, Y. and Y. Asakawa, 2006. Biotransformation of (−)-nopol benzyl ether. *Proceedings of 50th TEAC*, pp. 434–436.

Noma, Y., M. Furusawa, C. Murakami, T. Hashimoto, and Y. Asakawa, 2001a. Formation of nootkatol and nootkatone from valencene by soil microorganisms. *Proceedings of 45th TEAC*, pp. 91–92.

Noma, Y., T. Hashimoto, Y. Akamatsu, S. Takaoka, and Y. Asakawa, 1999. Microbial transformation of adamantane (Part 1). *Proceedings of 43rd TEAC*, pp. 199–201.

Noma, Y., T. Hashimoto, and Y. Asakawa, 2001b. Microbiological transformation of damascone. *Proceedings of 45th TEAC*, pp. 93–95.

Noma, Y., T. Hashimoto, and Y. Asakawa, 2001c. Microbial transformation of adamantane. *Proceedings of 45th TEAC*, pp. 96–98.

Noma, Y., T. Hashimoto, S. Kato, and Y. Asakawa, 1997a. Biotransformation of (+)-β-eudesmol by *Aspergillus niger*. *Proceedings of 41st TEAC*, pp. 224–226.

Noma, Y., T. Hashimoto, A. Kikkawa, and Y. Asakawa, 1996. Biotransformation of (−)-α-eudesmol by *Asp. niger* and *Asp. cellulosae* M-77. *Proceedings of 40th TEAC*, pp. 95–97.

Noma, Y., T. Hashimoto, S. Sawada, T. Kitayama, and Y. Asakawa, 2002. Microbial transformation of zerumbone. *Proceedings of 46th TEAC*, pp. 313–315.

Noma, Y., K. Matsueda, I. Maruyama, and Y. Asakawa, 1997b. Biotransformation of terpenoids and related compounds by *Chlorella* species. *Proceedings of 41st TEAC*, pp. 227–229.

Noma, Y., Y. Takahashi, and Y. Asakawa, 2003. Stereoselective reduction of racemic bicycle[33.1]nonane-2,6-dione and 5-hydroxy-2-adamantanone by microorganisms. *Proceedings of 46th TEAC*, pp. 118–120.

Nozaki, H., K. Asano, K. Hayashi, M. Tanaka, A. Masuo, and D. Takaoka, 1996. Biotransformation of shiromodiol diacetate and myli-4(15)-en-9-one by *Aspergillus niger* IFO 4407. *Proceedings of 40th TEAC*, pp. 108–110.

Okuda, M., K. Sonohara, and H. Takikawa, 1994. Production of natural flavors by laccase catalysis. *Jpn. Kokai Tokkyo Koho*, 303967.

Orabi, K.Y., 2000. Microbial transformation of the eudesmane sesquiterpene plectranthone. *J. Nat. Prod.*, 63: 1709–1711.

Orabi, K.Y., 2001. Microbial epoxidation of the tricyclic sesquiterpene presilphiperfolane angelate ester. *Z. Naturforsch.*, 56C: 223–227.

Otsuka, S. and K. Tani, 1991. Catalytic asymmetric hydrogen migration of ally amines. *Synthesis*, (9) 665–680. https://www.thieme-connect.com/products/ejournals/pdf/10.1055/s-1991-26541.pdf. Accessed date: 16 March 2020.

Parshikov, I.A., K.M. Muraleedharan, and M.A. Avery, 2004. Transformation of artemisinin by Cunninghamella elegans. *Appl. Microbiol. Biotechnol.*, 64: 782–786.

Sakamoto, S., N. Tsuchiya, M. Kuroyanagi, and A. Ueno, 1994. Biotransformation of germacrone by suspension cultured cells. *Phytochemistry*, 35: 1215–1219.

Sakata, K., I. Horibe, and M. Miyazawa, 2007. Biotransformation of (+)-α-longipinene by microorganisms as a biocatalyst. *Proceedings of 51st TEAC*, pp. 321–322.

Sakata, K. and M. Miyazawa, 2006. Biotransformation of (+)-isolongifolen-9-one by *Glomerella cingulata* as a biocatalyst. *Proceedings of 50th TEAC*, pp. 258–260.

Sakui, N., M. Kuroyamagi, Y. Ishitobi, M. Sato, and A. Ueno, 1992. Biotransformation of sesquiterpenes by cultured cells of *Curcuma zedoaria*. *Phytochemistry*, 31: 143–147.

Sakui, N., M. Kuroyanagi, M. Sato, and A. Ueno, 1988. Transformation of ten-membered sesquiterpenes by callus of *Curcuma*. *Proceedings of 32nd TEAC*, pp. 322–324.

Salvador, J.A.R. and J.H. Clark, 2002. The allylic oxidation of unsaturated steroids by *tert*-butyl hydroperoxide using surface functionalized silica supported metal catalysts. *Green Chem.*, 4: 352–356.

Sekita, M., M. Furusawa, T. Hashimoto, Y. Noma, and Y. Asakawa, 2005. Biotransformation of pungent tasting polygodial from *Polygonum hydropiper* and related compounds by microorganisms. *Proceedings of 49th TEAC*, pp. 380–381.

Sekita, M., T. Hashimoto, Y. Noma, and Y. Asakawa, 2006. Biotransformation of biologically active terpenoids, sacculatal and cinnamodial by microorganisms. *Proceedings of 50th TEAC*, pp. 406–408.

Shimoda, K., T. Harada, H. Hamada, N. Nakajima, and H. Hamda, 2007a. Biotransformation of raspberry ketone and zingerone by cultured cells of *Phytolacca americana. Phytochemistry*, 68: 487–492.

Shimoda, K., Y. Kondo, T. Nishida, H. Hamada, N. Nakajima, and H. Hamada, 2006. Biotransformation of thymol, carvacrol, and eugenol by cultured cells of *Eucalyptus perriniana. Phytochemistry*, 67: 2256–2261.

Shimoda, K., S. Kwon, A. Utsuki et al., 2007b. Glycosylation of capsaicin and 8-norhydrocapsaicin by cultured cells of *Catharanthus roseus. Phytochemistry*, 68: 1391–1396.

Shoji, N., A. Umeyama, Y. Asakawa, T. Takeout, K. Nocoton, and Y. Ohizumi, 1984. Structure determination of nootkatol, a new sesquiterpene isolated form *Alpinia oxyphylla* Miquel possessing calcium antagonist activity. *J. Pharm. Sci.*, 73: 843–844.

Sowden, R.J., S. Yasmin, N.H. Rees, S.G. Bell, and L.-L. Wong, 2005. Biotransformation of the sesquiterpene (+)-valencene by cytochrome $P450_{cam}$ and $P450_{BM-3}$. *Org. Biomol. Chem.*, 3: 57–64.

Sugawara, A. and M. Miyazawa, 2004. Biotransformation of guaiene using plant pathogenic fungus, *Eurotium rubrum* as a biocatalyst. *Proceedings of 48th TEAC*, pp. 385–386.

Sutherland, T.D., G.C. Unnithan, J.F. Andersen et al., 1998. A cytochrome P450 terpenoid hydroxylase linked to the suppression of insect juvenile hormone synthesis. *Proc. Natl. Acad. Sci. USA*, 95: 12884–12889.

Takahashi, H., 1994. Biotransformation of terpenoids and aromatic compounds by some microorganisms. PhD thesis. Tokushima Bunri University, Tokushima, Japan, pp. 1–115.

Takahashi, H., T. Hashimoto, Y. Noma, and Y. Asakawa, 1993a. Biotransformation of 6-gingerol, and 6-shogaol by *Aspergillus niger. Phytochemistry*, 34: 1497–1500.

Takahashi, T., I. Horibe, and M. Miyazawa, 2007. Biotransformation of β-selinene by *Aspergillus wentii. Proceedings of 51st TEAC*, pp. 319–320.

Takahashi, T. and M. Miyazawa, 2005. Biotransformation of (+)-nootkatone by *Aspergillus wentii*, as biocatalyst. *Proceedings of 49th TEAC*, pp. 393–394.

Takahashi, T. and M. Miyazawa, 2006. Biotransformation of sesquiterpenes which possess an eudesmane skeleton by microorganisms. *Proceedings of 50th TEAC*, pp. 256–257.

Takahashi, H., M. Toyota, and Y. Asakawa, 1993b. Drimane-type sesquiterpenoids from *Cryptoporus volvatus* infected by *Paecilomyces variotii. Phytochemistry*, 33: 1055–1059.

Tani, K., T. Yamagata, S. Otsuka et al. 1982. Cationic rhodium (I) complex-catalyzed asymmetric isomerization of allylamines to optically active enamines. *J. Chem. Soc. Chem. Commun.*, (11) 600–601.

Tori, M., M. Sono, and Y. Asakawa, 1990. The reaction of three sesquiterpene ethers with *m*-chloroperbenzoic acid. *Bull. Chem. Soc. Jpn.*, 63: 1770–1776.

Venkateswarlu, Y., P. Ramesh, P.S. Reddy, and K. Jamil, 1999. Microbial transformation of $D^{9(15)}$-africane. *Phytochemistry*, 52: 1275–1277.

Wang, Y., T.-K. Tan, G.K. Tan, J.D. Connolly, and L.J. Harrison, 2006. Microbial transformation of the sesquiterpenoid (−)-maalioxide by *Mucor plumbeus. Phytochemistry*, 67: 58–61.

Wilson, C.W. III and P.E. Saw, 1978. Quantitative composition of cold-pressed grapefruit oil. *J. Agric. Food Chem.*, 26: 1430–1432.

Yang, L., K. Fujii, J. Dai, J. Sakai, and M. Ando, 2003. Biotransformation of α-santonin and its C-6 epimer by fungus and plant cell cultures. *Proceedings of 47th TEAC*, pp. 148–150.

Zhan, J., H. Guo, J. Dai, Y. Zhang, and D. Guo, 2002. Microbial transformation of artemisinin by *Cunninghamella echinulata* and *Aspergillus niger. Tetrahedron Lett.*, 43: 4519–4521.

Zundel, J.-L., 1976. PhD thesis. *Universite Louis Pasteur*, Strasbourg, France.

24 Use of Essential Oils in Agriculture

Catherine Regnault-Roger, Susanne Hemetsberger, and Gerhard Buchbauer

CONTENTS

24.1 INTRODUCTION

Essential oils (EOs) play a major role in nature for the plants that produce them in order to protect the plants against bacteria, fungi, and viruses, as well as against herbivores and pests (Bakkali et al., 2008). Since pollution of our environment and intoxication of mammalians with some pesticides and herbicides is an acute problem all over the world, healthier alternatives must be researched. Natural products such as EOs offer some solutions for agricultural pests as far as they are efficient but nontoxic to vertebrates and do not harm our ambience. Even though there are so many advantages of EOs, also some disadvantages must be mentioned, like the slow action and short duration of effectiveness as well as the high quantities needed. The object of this compilation is to review the chances and problems of the use of EOs in agriculture.

Crop protection has been an important topic in agriculture for thousands of years. Many Greek and Roman authors—for example, Theophrastus (371–287 B.C.), Cato the Censor (234–149 B.C.), Varro (116–27 B.C.), Vergil (70–19 B.C.), Columella (4–70 A.D.), and Pliny the elder (23–79 A.D.)—wrote about different substances, such as EOs, to protect plants against pests. Also, Chinese literature, such as a survey of the Shengnong Ben Tsao Jing era (25–220 A.D.) tells about pesticidal activity of plants. Famously, Linnaeus (1752 A.D.) did research about natural pesticides which could be used on caterpillars. During the 19th century, compounds of plant origins were identified and frequently used as repellents or toxic compounds. Among them are alkaloids like nicotine and its isomer anabasine extracted from tobacco, a Solanaceae (*Nicotiana tabacum*, *N. rustica*, and *N. galuca*) or isolated from a plant growing in the Russian steppes and high plateaux of North Africa, *Anabasis aphylla*; nornicotine from *Duboisia hopwoodi* (an Australian plant); veratrine extracted from a Liliaceae of the Balkans, *Veratrum album*; ryanodine identified from *Rynia* spp. (chiefly *R. speciose*) in Amazonia; and also other families of molecules represented by rotenone and rotenoids and by pyrethrines, which were made of *Chrysanthemum* (*Pyrethrum*) *cinerariaefolium* flower heads and *Lonchocarpus nicou* or *Derris elliptica* roots. Most of these compounds are not used today because of toxicity to humans (e.g., neurotoxic activity of nicotine or Parkinson disease linked with rotenone) (Philogène et al., 2005). Consequently and according to the societal claim for a friendly environment fort use of pesticides today, more plant-derived solutions to control agricultural pests have been actively explored since 1960. Among these plant extracts, EOs became important, because they have a wide range of activities against microorganisms, weeds, and insects (insecticides, antifeedants, or repellents), as previously mentioned (Bakkali et al., 2008). Grieneisen and Isman (2018) have counted more than 1300 published articles per year devoted to botanical research efforts since 2012, with 20% focused on EOs, including numerous reviews (as examples, Arnason et al., 2012; Regnault-Roger et al., 2012a,b; Regnault-Roger, 2013; Buchbauer and Hemetberger, 2016; Mossa, 2016; Sparagano et al., 2016; Isman and Tak, 2017a).

EOs uses have a long history. They have been used for many industrial applications for a long time. The first distillation of these products was mentioned during the 13th century in Andalusia (Spain) by Ibn al-Baitar, after which their pharmacological properties induced their inclusion into the very early pharmacopoeias of several European countries (Regnault-Roger, 2013). These oils were first used largely for medicinal or ceremonial purposes, and according to Isman and Tak (2017a), anecdotal records suggest that "*certain aromatic oils were burned during the Black Death (1346–1353) 'to ward off evil spirits' (healingscents.net), but may have inadvertently reduced the spread of bubonic plague by repelling or killing fleas that vectored the pathogen.*" They have been used since ancient times as cosmetics and pharmaceuticals and also since the last century in fine chemistry and aromatics for the food industry. Today, they still play a major role in our life (Ammon et al., 2010). The plant protection in agriculture is now one very important focus of EOs activities.

24.1.1 Essential Oils: Very Complex Natural Mixes

EOs appear to be very complex natural mixes. EOs are biosynthesized by around 17,500 species of aromatic plants belonging to a few families. They are particularly abundant in Conifers, Rutaceae, Umbelliferae, Myrtaceae, Lamiaceae, and Lauraceae. Depending on the species and families, they are localized in specialized histological structures: glandular trichomes (Lamiaceae), secretory canals (Myrtaceae), or resin ducts (Apiaceae). They could be stored in different parts of the plant such as flowers, leaves, wood, roots, or seeds (Bruneton, 2009).

Terpenoids are major constituents of EOs and, to a lesser amount, phenylpropanoids. EO constituents belonging to terpenoids are mainly monoterpenes (ten atoms of carbon) and sesquiterpenes (15 atoms of carbon) of low molecular weight. They generally consist of several tens of constituents, of which the great majority possess an isoprenoid skeleton. Monoterpenes present in EOs may contain terpenes that are hydrocarbons (alpha-pinene), alcohols (menthol, geraniol, linalool, terpinen-4-ol, *p*-menthane-3,8-diol), aldehydes (cinnamaldehyde, cuminaldehyde), ketones (thujone), ethers (1,8-cineole or eucalyptol), and lactones (nepetalactone). As the elongation of the chain to 15 carbons increases the number of possible cyclizations, sesquiterpenes have a wide variety of structures (over 100 skeletons). Aromatic compounds are less common and are derived mainly from the shikimate pathway. Some compounds identified in EOs result from the degradation of fatty acids (jasmonic acid) or are glycosylated volatile compounds (e.g., linalool glucoside) (Regnault-Roger et al., 2012a,b).

The majority of EOs contains a limited number of main compounds, but some of the minor compounds play an important role as vectors of fragrance and make up the richness of an extract. It is thus well established that EO composition is very variable depending of the species and of chemotypes within the species, as well as physiological parameters. Thyme (*Thymus vulgaris* L.) is a species with numerous chemotypes, named according to the major compound; for example, thyme with chemotype thymol or chemotype carvacrol or terpineol or linalool. A typical EO may contain 20–80 phytochemicals. Physiological expression of secondary metabolism of the plant may be different at all stages of its development. The proportions of monoterpenes depend on temperature and circadian rhythm and vary according to plant stage. For example, Gershenzon et al. (2000) showed that limonene and menthone are the major monoterpenes present in the youngest leaves of peppermint, but limonene content declines rapidly with development whereas menthone increases and then declines at later stages; consequently, menthol becomes the dominant constituent. Soil acidity and climate (heat, photoperiod, humidity) directly affect the secondary metabolism of the plant and EO composition (Muller-Riebau et al., 1997). Isman and Machial (2006) reported that rosemary oil extracted from plants harvested in two different areas of Italy contained 1,8-cineole concentrations ranging from 7% to 55% and α-pinene concentrations ranging from 11% to 36%. The variability of the composition of an EO is also impacted by the choice of the method of extraction of EOs. One characteristic of EOs is the volatility of their compounds which allows them to be easily extracted by water vapors, in contrast to fixed lipid oils and essences (concrete, absolute, oleoresins, and resinoids) which are extracted by solvents and alcohol. More recently, extraction and separation methods using supercritical fluids, steam distillation, dry distillation, or mechanical cold pressing of plants are also mentioned (Regnault-Roger et al. 2012a,b). All these factors lead to the complexity of EO phytochemistry and to a certain inconsistency of the chemical composition of an EO.

The biological significance of EOs has long been discussed. First considered as wastes of phytometabolism, it appears today that EOs present multiple actions. They prevent plants from losing water by excessive evaporation and some of their components react as donors of hydrogen in oxydation-reduction reactions. They also seem to be important agents of interspecific communication as they favor pollination by attracting insects and they also play a part in the defense of plants against herbivores, microorganisms, and fungi (Regnault-Roger, 1997). Therefore, all these activities could be taken into account in the development of pest-management strategies.

24.2 ESSENTIAL OILS AS ANTIPESTS

Insects form the largest population in the animal kingdom and many of them are harmful toward human beings since they act as pathogenic vectors and devaster crops. Pest control, both by synthetics and biopesticides, bear a huge market in the regular growth. The massive use of oil-based synthetic pesticides since the middle of the 20th century, with limited knowledge of properties of some compounds, produced unexpected effects with negative sides, such as environmental persistence (organochloride compounds classified as POPs), chronic toxicity toward human or mammalians and increased insect resistance (Regnault-Roger, 1997). Because of that, it is important to find alternatives which do not damage our environment.

24.2.1 Health and Environmental Impact of Botanical Antipests

EOs are natural products, which are excellent alternatives to synthetic products because they reduce negative impacts on human health and the environment (Koul et al., 2008). Even though there are a lot of botanical insecticides available, from 1980 to 2000, only one single product was registered in the United States and Europe, which is called Neem (now considered an endocrine disruptor). It is obtained by the seeds of the Indian tree *Azadirachta indica*, Meliaceae. In the last few years, EOs also obtained influence as botanical insecticides in the United States. But, due to the relatively slow action, variable efficacy, lack of persistence and inconsistent availability of natural products, they still cannot compete against synthetic pesticides. On the other hand, it is not necessary to kill harmful insects since botanical insecticides, like EOs, can also be used as repellents against animals. Moreover, those natural pesticides can also be mixed with synthetic products which could lessen the needed quantities of pesticides and improve the environmental problems which were caused by excessive use of synthetic pesticides. Especially, developing countries could benefit from the increased use of botanical pesticides, such as EOs, since farmers are more familiar with plant extracts and EOs they can do by themselves. They often use pesticides and do not know about the dangers because many of them do not receive any information and they are often unable to read the official languages in which the instructions are written (Isman, 2008). In static water, eugenol appears to be 1500 times less toxic than pyrethrum, which is also a botanical insecticide, and not more than 15,000 times less toxic than the organophosphate insecticide azinphosmethyl. Moreover, eugenol is volatile, which means that its half-life is extremely short. After about two days, there will be no eugenol left, which also avoids rare side effects of EOs (Isman, 2000). Also, the tobacco cutworm, which is a severe problem to vegetable and tobacco crops in Asia, can be killed by compounds of EOs, namely thymol, carvacrol, pulegone, eugenol, and *trans*-anethole, even though it is quite resistant (Hummelbrunner et al., 2001). Many monoterpenes like (+)-limonene, pinene, and Δ3-carene show acaricidal activity. Carvomenthenol and terpinen-4-ol proved to cause highest toxicity against mites (Ibrahim et al., 2008). Eucalyptus EO was tested on several parasites, such as *Varroa destructor* (Varroa mite), *Tetranychus urticae*, *Phytoseiulus perisimilis*, and *Boophilus microplus*. The studies concluded that several *Eucalyptus* sp. (Myrtaceae) EOs can be used as acaricides (Batish et al., 2008). *T. urticae* can be killed by EOs obtained from *Satureja hortensis* L. (Lamiaceae), *Ocimum basilicum* L. (Lamiacae), and *Thymus vulgaris* L. (Lamiaceae) (Aslan et al., 2004). EOs do not only act as deterrents but also as attractants toward insects. Ethanolic extracts of *Rosmarinus officinalis* L. (Lamiaceae) EO attracted the moth *Loberia botrana*, a pest of grape berries (Katerinopoulos et al., 2005).

24.2.2 Pesticidal and Repellent Action of Essential Oils

EOs are neurotoxic to some pests as some of them interfere with the neuromodulator octopamine and others with GABA-gated chloride channels (Isman, 2006). Octopamine "controls and modulates neuronal development, circadian rhythm, locomotion 'fight or flight' responses, as well as learning and memory" (Balfanz et al. 2005). It activates GTP-binding protein G-coupled receptors, which leads to cAMP production or calcium release (Balfanz et al., 2005). Interrupting its function, "results in total break down of nervous system in insects" (Tripathi et al., 2009). $GABA_A$ receptors of insects showed similarities as well as differences to vertebrate ones. Activation of insect GABA channels can increase the chloride ion conductance across cell membranes. Insect GABA receptors are also sensitive to benzodiazepines and barbiturates, but they are insensitive to antagonists like bicuculline and pitrazepin. The differences of insect and vertebrate GABA channels can be used as targets for pesticides (Anthony et al., 1993). At high concentrations of EOs, they can also inhibit the acetylcholinesterase activity, but it cannot explain the low-dose pesticidal effect (Kostyukovsky et al., 2002). Moreover, the acetylcholinesterase-inhibiting effect could not be seen *in vivo* for all monoterpenes (Rajendran et al., 2008). It seems to be a very good idea to use nontoxic substances that only act as deterrents or antifeedants toward insects. This concept obtained influence with the discovery of the antifeedant azadirachtin in the 1970s and 1980s. But even though deterrents and antifeedants were hyped a lot during that time, a few problems remain, such as the variability of response—even closely related insects vary in their behavior toward these substances; some may even get attracted by substances that other insects may find detesting. Furthermore, the deterrent action may change during a period of time, and insects can get used to it (Isman, 2006). The exact mode of action of repellents still cannot be explained completely. It is also unclear if there are common mechanisms in different anthropods. Just as insects do on the antennae, "tick detects repellents on the tarsi of the first pair of legs" (Tripathi et al., 2009). It is known that hairs on the antennae of mosquitos are used to detect temperature and moisture. Repellents target female mosquito olfactory receptors and block them. As for cockroaches and the method of defense, is just known that oleic and linoleic acid cause death recognition and aversion (Tripathi et al., 2009).

Even though single compounds of the EO can be isolated and tested for repellent activity, it has been reported that the synergistic effect of the whole EO is more effective (Nerio et al., 2010) (Table 24.1).

24.2.2.1 Insecticidal Activities of Essential Oils

An abundant literature, more than 2000 scientific papers, is devoted to study the EO–insect relationships over the past 20 years, with some major reviews (Regnault-Roger, 1997; Isman, 2000, 2006; Regnault-Roger and Philogène, 2008; Regnault-Roger et al., 2008; Adorjan and Buchbauer, 2010; Isman et al., 2011; and reviews mentioned above). The majority of this scientific literature focuses on the characterization of bioactivities of oils in the laboratory from previously uninvestigated plant species or untested insect pests. Most papers document the immediate effects (acute toxicity or repellency) of given EOs on a number of arthropod taxa, frequently on the basis of assays lasting less than 48 h, but few papers investigate the modes or mechanisms of action of EOs and the real efficacy in the field and the effective protection of a crop. If the complexity in the way EOs act on insects justifies this rich literature, some more consistent results regarding situations out of laboratories, as well as more papers devoted to a better understanding of mechanisms of action, ought to be produced. In this context, we just intend to give some keys to understand the diversities of EO activities on insects in regard to the routes of exposure and their cellular targets.

Insect biocontrol by EOs results from several kinds of modes of action and depends on the routes of the exposure. An activity observed in earlier studies was the toxicity by ingestion or by contact through cuticle or by inhalation for volatile compounds. Insects are very sensitive to topical applications of EOs; for example, *Citrus* spp. EOs on the maize weevil *Sitophilus zeamaïs* (Motschulsky) or the larger grain borer *Prostephanus Americana* (Horn) (Haubruge et al., 1989) and the bruchid *Acanthoscelides obtectus* (Say) to fumigant toxicity of a large range of Mediterranean

TABLE 24.1
Examples of Harmful Pests

Name Family	Notes	Source
Acanthoscelides obtectus Bean weevil Coleoptera: Bruchidae	Native to Africa, Europe, America; eats fruits of crops	(a)
Acrolepiopsis assectella Leek moth Lepidoptera: Yponomeutidae	Native to Asia, Europe; eats foliage of crops	(a)
Acyrthosiphon pisum Pea aphid Hemiptera: Aphididae	Pest of legumes, e.g., alfalfa, clover and peas	(a)
Aedes aegypti Yellow fever mosquito Diptera: Culicidae	Lives in tropics and subtropics; pathogenous	(a)
Anopheles stephensi Diptera: Anophelinae	Vector of malaria disease	(b)
Bemisia tabaci Sweetpotato white fly Homoptera: Aleyrodidae	Polyphageous, pathogenous vector (i.e. cassava mosaic disease); lives in America, Africa, Asia	(a)
Boophilus microplus Ixodida, Ixodidae		(a)
Callosobruchus chinensis Adzuki bean weevil Coleoptera: Bruchidae	Storage pest; i.e., in Syria	(a)
Callosobruchus maculatus Cowpea weevil Coleoptera: Bruchidae	Damages cowpeas	(a)
Cryptolestes pusillus Flat grain beetle Coleoptera: Laemophloeidae	Stored grain pest; in temperate climate	(a)
Culex pipiens House mosquito Diptera: Culicidae	Pathogenous vector of diseases	(a)
Culex quinquefasciatus Diptera: Culicidae	Tropical regions; pathogenous vector of diseases	(a)
Delia radicum Cabbage fly Diptera: Anthomyiidae	Native to Asia, Europe and North America; damages crucifers	(a)
Dendroctonus rufipennis Spruce beetle Coleoptera: Scolytidae	Damages spruce trees	(c)
Dendroctonus simplex Eastern larch beetle Coleoptera: Scolytidae	Damages spruce trees	(c)
Dermanyssus gallinae Roost mite Gamasida, Dermanyssoidea, Dermanyssidae	Worldwide; poultry pest; can reduce egg production	(a)
Drosophila auraria	Subgroup of *D. melanogaster*	(d)

(Continued)

TABLE 24.1 (*Continued*)
Examples of Harmful Pests

Name Family	Notes	Source
Ephestia kuehniella Mediterranean flour moth Lepidoptera: Pyralidae	Pest of industrial flour mills; lives in temperate climate	(a), (e)
Helicoverpa armigera Cotton bollworm Lepidoptera: Noctuideae	Pest of maize and other crops; native to Asia and Africa	(a)
Hyalomma marginatum Acari: Ixodidae	In Europe and Asia	(a)
Hylobius abietis Large pine weevil Coleoptera: Curculionidae	Pest of conifers; in Europe and Asia	(a)
Lasioderma serricorne Cigarette beetle Coleoptera: Anobiidae	Pest of stored food, hoursehold pest; native to America, Asia, and Africa	(a)
Leptinotarsa decemlineata Colorado potato beetle Coleoptera: Chrysomelidae	Destroys solanaceous crops; in Europe, North America	(a)
Limantria dispar Gypsy moth Lepidoptera: Lymantridae	Pest of cork oak forests	(f)
Listronotus oregonensis Carrot weevil Coleoptera: Curculionidae	Damages mainly carrots	(a)
Lobesia botrana Grape berry moth Lepidoptera: Tortricidae	Pest of grapes; for example, in Jordan and Iran	(a)
Megastigmus pinus Hymenoptera: Torymidae	Destroys white fir trees	(c)
Megastigmus rafini Hymenoptera: Torymidae	Destroys white fir trees	(c)
Musca domestica Housefly Diptera: Muscidae	Feeds on garbage and feces; worldwide	(a)
Oryzaephilus surinamensis Sawtoothed grain beetle Coleoptera: Silvavidae	Stored-product pest	(a)
Papilio demoleus Common lime butterfly Lepidoptera: Papilionidae	Southeast Asia; pest of citrus	(a)
Periplaneta americana American cockroach Blattaria: Blattidae	Household pest; omnivore	(a)
Phytoseiulus persimilis Acarina: Phytoseiidae	Is used to control mite pests in greenhouses; predatory mite	(a)
Pissodes strobi White pine weevil Coleoptera: Curculionidae	Destroys pines, for example, Eastern White pine	(a), (c)

(Continued)

TABLE 24.1 (*Continued*)
Examples of Harmful Pests

Name Family	Notes	Source
Plodia interpunctella Indian meal moth Lepidoptera: Pyralidae	Stored-grain pest; worldwide	(a)
Prays citri (Citrus blossom moth) Lepidoptera: Hyponomentidae	For example, in Egypt	(a)
Prostephanus truncatus Larger grain borer Coleoptera: Bostrichidae	Pest of maize; especially in tropical regions	(a)
Rhyzopertha dominica Lesser grain Borer Coleoptera: Bostrichidae	Native to America, Africa, Asia; pest of grain	(a)
Sitophilus granarius Granary weevil Coleoptera: Curculionidae	Pest of stored grain	(a)
Sitophilus granarius Granary weevil Coleoptera: Curculionidae	Pest of stored grain	(a)
Sitophilus oryzae Rice weevil Coleoptera: Curculionidae	Pest of stored grain	(a)
Sitophilus zeamays Maize weevil Coleoptera: Curculionidae	Pest of stored grain	(a)
Spodoptera frugiperda Fall armyworm Lepidoptera: Noctuideae	Pest of sugarcane fields and other crops; North and South America	(a)
Spodoptera litura Rice cutworm Lepidoptera: Noctuidae	Native to Australia and Asia; damages legumes and solanaceous crops	(a)
Stegobium paniceum Drugstore beetle Coleoptera: Anobiidae	Temperate and subtropical areas; eats pharmaceutical drugs	(a)
Stephanitis pyri Azalea lace bug Heteroptera: Tingidae	Native to Europe; damages trees	(a)
Tenebrio molitor Yellow mealworm Coleoptera: Tenebrionidae	Stored-product pest	(a)
Tetranychus cinnabarinus Carmine mite Acari: Tetranychidae	Damages soybeans, cotton, plums, citrus, vegetables; in Syria, Saudi Arabia, Lebanon, Jordan	(a)
Tetranychus urticae Two-spotted spider mite Acari: Tetranychidae	Worldwide in temperate and subtropical areas; greenhouse pest	(a), (g)

(*Continued*)

TABLE 24.1 (*Continued*)
Examples of Harmful Pests

Name Family	Notes	Source
Thaumetopoea pityocampa Pine processionary Lepidoptera: Thaumetopoeidae	Pest of Pinaceae	(a)
Triatoma infestans Kissing bug Hemiptera: Reduviidae: Triatominae	Pathogeous vectors of Chagas disease in South America	(a)
Tribolium castaneum Red flour beetle Coleoptera: Tenebrionidae	Native to America, Africa and Asia; pest of stored food	(a)
Tribolium confusum Confused flour beetle Coleoptera: Tenebrionidae	Stored food pest, especially flour; difficult to control	(a)
Tyrophagus putrescentiae Mold mite Acaridida: Acaroidea: Acaridae	Eats processed cereals	(a)
Varroa destructor (*Varroa jacobsoni*) Acari: Varroidae	Pest of honey bees	(a)

Source: (a) Capinera (2008), (b) Prajapazi et al. (2005), (c) Ibrahim et al. (2008), (d) Konstantopoulou et al. (1992), (e) Ayvaz et al. (2010), (f) Moretti et al. (2002), (g).

EOs (Regnault-Roger et al., 1993). The first observations were mainly focused on insects of the stored products, but EOs could control a large range of flying insects as well: the Mediterranean fruit fly *Ceratitis capitata*, the greenhouse white fly *Trialeurodes vaporariorum*, and also the green peach aphid *Myzus persicae* (Regnault-Roger, 1997). They also repel or act as deterrents or antifeedants that affect insect fitness (Regnault-Roger and Hamraoui, 1994). All these activities the EOs exert on an insect could occur on several physiological targets at the same or at different stages of the insect development. As examples, EOs of *Artemisia vulgaris* develop a combined activity as repellent and fumigant upon *Tribolium castaneum* (Wang et al., 2006).

Beside these activities on adult, EOs also have an influence on the level of reproduction. EOs disturb oviposition or disrupt the larvae growth or modify the imago's behavior or physiology of the insect. The beetle *Acanthoscelides obtectus* (Say) has been shown to be a convenient model to point out the different ways the EOs impact the fitness of adult insects and their reproduction, in particular, which reproductive stage (oviposition, eggs hatching, larvae) is targeted and the speed of the activity of EOs. The inhibition of reproduction and the toxicity on adults can be quite different. As examples, parsley (*Petroselinum sativum* L., Umbelliferae) did not have significant fumigant toxicity on the beetle adults but inhibited strongly its reproduction, whereas the summer savory (*Satureia hortensis* L., Lamiaceae) presented a high toxicity on adults but inhibited poorly the bruchid reproduction. Mint (*Mentha piperita* L. Lamiaceae) developed fumigant toxicity and very strong inhibition of reproduction but no antifeedant effect; dill (*Anethum graveolens* L., Umbelliferae) has little fumigant toxicity and inhibited poorly the reproduction but had a significant antifeedant effect (Regnault-Roger, 2013). The complexity of the large range of EO activities on insects is reinforced by a pronounced variability of physiological responses of insect species for the same EO. Tansy oil (*Tanacetum vulgare* L.) is attractive and paralyzing for *Rhizoperta dominica*, repulsive for *Tribolium confusum*, and toxic for *Sitophilus americana* (Koul et al., 1990).

From all these observations, it could be deduced that EOs present a widespread range of activities on insects that necessitate to be sharpened on a case-by-case study before application in pest management.

These biological effects on pests involve both physical and chemical mechanisms, as the following considerations underline:

- Creating a physical barrier, a topical application of EOs laid a film on insect cuticle that modifies the physiology of the insect. Because oils are lipophilic, the film changes the conditions of penetration of the air and other substances inside the body of the insect and causes the disruption of lipid bilayers of cell walls.
- Because of the volatility of monoterpenes, several families of proteins including OBPs (odorant binding proteins) and CSPs (chemosensory proteins) are involved in the detection of bouquets of fragrant and chemosensory-active compounds of EOs. OBPs and CSPs are found on the periphery of the sensory receptors and function in the capture and transport of molecular stimuli. In the sensilla of insects like trichoid sensilla of the female silkworm, *Bombyx mori,* specialized OBPs respond to the volatile monoterpene linalool (Picimbon and Regnault-Roger, 2008). In moths, the protein GOBP2, identified in tobacco hornworm, *Manduca sexta*, preferentially interacts with floral aromas and green plant odors such as geraniol, geranyl acetate, and limonene (Feng and Prestwich, 1997). These proteins play an important role in the response of the insect to an EO blend.
- According to the toxicity of monoterpenes, several main compounds identified in EOs (thymol, α-terpineol, linalool, geraniol, eugenol) induce a neurotoxicity through different receptors. Thymol binds to GABA receptors associated with chloride channels located on the membrane of postsynaptic neurons and disrupts the functioning of GABA synapses (Priestley et al., 2003). Eugenol acts through the concentration of cAMP and the octopaminergic system. It mimics the action of octopamine with the consequence of increasing intracellular calcium levels in the cockroach *Periplaneta americana.* Tyramine (a precursor of octopamine) receptors are involved in the recognition of monoterpenes (Enan, 2005). Thymol, carvacrol, and α-terpineol influence the production of cellular cAMP and calcium in *D. melanogaster* (Kostyukovsky et al., 2002).

Another mechanism of neurotoxicity was identified through the transmission of the nervous impulse. Electrophysiological experiments showed that eugenol inhibits deeply neuronal activity, whereas citral and geraniol have a biphasic effect that is dose dependent. At low doses, these compounds induce an increase in spontaneous electrical activity but, at high doses, cause a decrease (Price and Berry, 2006). Using a similar electrophysiological experimentation, Huignard et al. (2008) observed that the *Ocimum basilicum* EO has a complex neurotoxic activity. The EO inhibited fully the neuronal electrical activity by decreasing the magnitude of nerve-action current, then reducing the post-hyperpolarization phase and firing the frequency of nerve-action current. The authors hypothesized that this effect could be the result of the combined action of linalool and estragole, two major components of the *O. basilicum* EO. They demonstrated that the mere application of pure linalool produced a reduction in the amplitude of nerve-action current and decreased the post-hyperpolarization while estragole specifically induced a reduction of post-hyperpolarization.

Another well-known target for neurotoxicity is the enzyme acetylcholinesterase (AChE). Tea tree oil inhibited acetylcholinesterase (Mills et al., 2004). Several mechanisms involving monoterpenes were identified through a reversible competitive inhibition at the enzyme's hydrophobic active site or a mixed inhibition for this enzyme by linking to a different site from the active site where the substrate was bound (Lopez and Pascual-Villalobos, 2010). However, a systematic study of bioactivity of plant EOs and their AChE inhibitory activity *in vitro* led to the conclusion "that AChE is not a principal target site for insecticidal monoterpenoids naturally occurring in plant EOs, nor is inhibition of AChE an important mode of action for these substances in insects" (Isman and Tak, 2017b).

These studies confirm that the insecticidal activity of EOs is the consequence of several mechanisms that affect multiple cellular and physiological targets.

24.2.3 Development and Commercialization of Botanical

The utilization of insecticides, both botanicals and synthetics, are handled differently in various countries (Isman, 2006): In the US, many EOs are registered for agricultural use. Canada has a slightly different law as there are not as many EOs allowed. Mexico has a similar utilization of EOs as the US, but there is no exception of EOs. The European Union allows the utilization of components of EOs. Even though, the individual law differs for countries in the European Union. In India, more biological pesticides are used than in other countries. In South America, pesticide use varies in every region. In Africa there are no real restrictions by law for its local use of various pesticides. In general, countries tend to make laws for use of pesticides stricter than they were, because the majority of the population and media are aware of the environmental and health impacts of pesticides. This is especially the case in wealthier countries; but since developing countries are dependent on selling their products to so-called "first-world countries," they also have to meet up with stricter criteria, even though it might not make sense for them (Isman, 2006). According to Isman, three aspects are important for successful commercialization of pesticides: sustainability of the botanical resource, standardization of the chemically complex extracts, and regulatory approval. Sustainability means that the base of the product must be available all the time and not only seasonally. Standardization means that there must be a method to investigate analytically the product so that the product obtained from plants will not vary. It is a problem if the pesticidal action of a substance differs even though it has been cultivated under the same circumstances. Regulatory control is—among those—the most difficult aspect. The profit made by selling botanicals will not be very high in industrialized countries. That is why "green pesticides" might not even reach the market. In industrialized countries, it is hard to imagine that botanicals such as EOs will play a bigger role in the future since synthetic pesticides are affordable and very potent against various insects (Isman, 2006).

While the global market for synthetic and biopesticides pooled together should reach USD 79.3 billion by 2022 from USD 61.2 billion in 2017 at a compound annual growth rate (CAGR) of 5.3% during 2017–2022 (https://www.bccresearch.com), in 2016, the global biopesticides market was valued at USD 2.83 billion and, in 2017, at USD 3.3 billion with a projection of USD 9.5 billion by 2025, which means a CAGR of 13.9% during 2017–2022. North America, with USD 1300.25 million is the leading region followed by Asia, Europe, Latin America, Middle East, and Africa (TRM, 2018). Botanicals (including EOs) value for 10% of the biopesticides market. This market is dynamic and full of opportunities.

However, to paraphrase Isman (2008) speaking of botanical insecticides, the EO market has been considered for two categories of countries: for richer and for poorer.

For the richer, in developed countries, EOs are used for high-value crops (orchard, greenhouse vegetable production). For example, in California, 37 EOs have been registered as active ingredients. Among them, clove oil against post-harvest potato sprouting, tagetes, and wintergreen oil are newly registered as agricultural insecticides, and many crude EOs as repellent (eucalyptus, cedarwood, citronella, lemon grass, peppermint, rosemary) (Grieneisen and Isman, 2018). Arnason (2011) indicated that about 90 insect-repellent products sold in the US market contain an EO as one of the active ingredients in the formulation. The most commonly used ingredient is citronella oil (45 products), followed by geranium oil (geraniol) (33), lemongrass oil (24), cedar oil (22), peppermint oil (16), rosemary oil (15), soybean oil (15), and eucalyptus oil (14). EOs are now considered to be the most important commercial application of botanical insecticides in North America.

The enhancement of using EOs to control insect pests in orchards and to protect high value crops is probably the result of the regulation rules in the US. Biopesticides are subject to special procedures outlined in Title 40, Code of Federal Regulations, of FIFRA. A number of natural substances, such as EOs of mint, thyme, rosemary, and lemon grass that did not benefit from this simplified procedure, however, were classified as GRAS (generally regarded as safe). They were placed on a list

(FIFRA Section 25[b]) exempting them from the registration process (EPA, 2018). This exemption has become a marketing strategy to promote these products (Isman, 2004; Regnault-Roger et al., 2005). The Company EcoSMART™ was a pioneer, 15 years ago, to develop a line of products on this exemption. These products are named "Ecoexempt® Minimum-Risk & EcoPCO® Products" and are commercialized on a large scale by several suppliers inside the US and abroad. Nowadays, these products are currently approved for use in seven other countries, including Mexico, Peru, Uruguay, Singapore, and the United Arab Emirates (Isman and Tak, 2017a).

In the European Union, the procedure for reevaluation of plant protection products (PPPs) ended in 2008. To be authorized on market, all PPPs derived from biological as well as chemical sources have to be listed. Because they meet the purposes of Directive 2009/128/CE to promote Integrated Pest Management (IPM) obligatory in all European Union member states from the beginning of 2014, several plants oils have been authorized for plant protection on a special procedure. French Ministry of Agriculture listed natural products for biocontrol, among them orange and mint EOs, but also eugenol, geraniol, and thymol (DGAL, 2017). Mint oil is also currently used as a PPP in Belgium, Germany, and Italy, and fennel oil is used in organic farming in Czech Republic and Slovakia, and gillyflower oil in Italy (Szulc and Sobczak, 2017). Nevertheless, Isman and Tak (2017a) underlined that very few PPPs based on EOs are available in the European Union (EU) compared to the US. French Organic Farming Technical Institute (ITAB), looking for new solutions to expand the range of organic pesticides, indicated only two commercial formulations based on orange oil and mint EOs as insecticide and fungicide for legumes, fruits, and vines for the first one and potato sprout inhibition for the second. Some experimental research in the field was conducted between 2014 and 2017 in France to test the efficacy of seven EOs against mildew of potatoes, vines and salads, and apple scab. The EOs and their main compounds were eucalyptus (citronellal), clove (eugenol), mint (D-carvone and L-carvone), oregano (carvacrol), tea tree (terpinen-4-ol), and thyme (thymol). If these EOs showed some efficacy *in vitro* on the two fungi implicated in mildew and apple scab diseases, *Phytophthora infestans* and *Venturia inaequalis,* results were very poor in the field trials; these EOs biocontrol abilities were less effective than synthesized fungicides and also copper hydroxide. Moreover, the beneficial mites *Typhlodromus* spp. were impacted by these EOs (Vidal et al., 2018). This very interesting experiment showed that there is a gap between promising results in the laboratory and the real ones in the field. It also could explain the limited success for registration of new PPP substances based on EOs in the EU. In fact, to be registered as a PPP in the EU, contrary to US registration, the active substance has to be proved to have a real efficacy on the target species.

In other developed or emerging countries like India, Brazil, or Australia, where an abundant plant biodiversity grows and where a "prolific research on botanicals" is developed, Isman and Tak (2017a) also noted the lack of approved active substances based on EOs.

For poorer countries, the situation for using EOs is quite different. The tropical climate develops numerous species with high potentiality for pest management (Arnason et al., 2008). Aromatic plants are traditionally and widely used for stored-product insect biocontrol or to repel harmful insects in fields (Glitho et al., 2008). Currently, there is a move to enhance the use of steam-distilled EOs, but the lack of technologic resources leads to use of crafty solutions like a domestic pressure cooker for extracting EOs by steam distillation. Another point that impedes the proper development of EOs in Africa is that it is unsupported by scientific experimentation (Regnault-Roger, 2013). Isman (2017), analyzing the gap between the abundance of tropical pesticide plants and the lack of practical solutions for farmers in these countries, argued to develop simple methods for local cultivation of a relevant plant species, simple methods for extraction of active substances at minimal cost, and simple methods to do bioassays and then field trials, thus, to demonstrate and to learn strategies to optimize efficacy. He concluded that EOs are "well suited for exploitation by smallholder farmers in sub-Saharan Africa, whether prepared on an individual ('do-it-yourself') basis." According to him, the use of low-cost botanical insecticides with consistent efficacy would enhance local productions. Appropriate regulations and relevant public policies for sustainable agriculture are also needed in these countries.

24.2.3.1 Examples of Essential Oils Used as Antipests

24.2.3.1.1 *Rosmarinus officinalis*

Rosmarinus officinalis L. (Lamiaceae) can be found in the Mediterranean area (Tables 24.2–24.4). The used parts are the fresh flowering tops. The EO can be used as an antipest due to its effect on beetles, caterpillar larvae, and many other insects (Dayan et al., 2009) like *Drosophila auraria* (Konstantopoulou et al., 1992), *Sitophilus oryzae* (Lee et al., 2004), or *Rhyzopertha dominica* (Shaaya et al., 1991). The microencapsulated EO also showed larvicidal effects on *Limantria dispar* (Moretti et al., 2002). In *Acanthoscelides obtectus* males and females, *R. officinalis* volatile oil also caused high mortality rates (Papachristos et al., 2002). Repellent activity was reported against *Listronotus oregonensis* (Niepel, 2000).

TABLE 24.2
Essential Oils that can be Used as Insecticide and Acaricide

Name Family	Constituents
Acorus sp. Acoraceae	β-asarone, acorenone, acoragermacrone
Artemisia sp. Asteraceae	Limonene, myrcene, α-thujone, β-thujone, caryophellene, sabinyl-acetate
Baccharis salicifolia (Ruiz & Parvon) Pers Asteraceae, Garcia et al. (2005)	β-Pinene, pulegone, camphene, limonene, α-pinene, pulegone, pulegol, germacrol, germacrone
Carum sp. Apiaceae	D-Carvone, limonene
Chenopodium ambrosioides L. Amaranthaceae	α-Terpinene, cymol, ***cis***-β-farnesene, acaridole, carvacrol
Cinnamomum sp. Lauraceae	Cinnamaldehyde, linalol, β-caryophyllene, eugenol
Citrus sinensis (L.) Osbeck Rutaceae, Loebe (2001), Njoroge et al. (2005)	Limonene, α-pinene, sabinene, α-terpinene
Coriandrum sativum L. Apiaceae	Linalol, α-terpinyl acetate, 1,8-cineole, linalyacetate, geranyl acetate, camphor
Cuminum cyminum L. Umbelliferae, Li and Jiang (2004)	Cuminal, cuminic alcohol, γ-terpinene, safranal, p-cymene, β-pinene
Cymbopogon sp. Poaceae	Citronellal, geraniol, citronellol, citral, limonene
Elettaria cardamomum (L.) Maton Zingiberaceae	α-Terpinyl acetate, 1,8-cineole, linalyl acetate
Eucalyptus sp. Myrtaceae	1,8-Cineole, limonene, α-pinene
Evodia rutaecarpa Hook f. et Thomas Rutaceae, Liu and Ho (1999)	Limonene, β-elemene, linalool, myrcene, valencene, linalyl acetate, β-caryophyllene
Foeniculum vulgare L. Apiaceae	Anethol, fenchone, estragol

(*Continued*)

TABLE 24.2 (*Continued*)
Essential Oils that can be Used as Insecticide and Acaricide

Name Family	Constituents
Hedomea mandonianum Wedd. Lamiaceae	Eucalyptol, pulegone, menthone
Juniperus sp. Cupressaceae	α-Pinene, myrcene, β-pinene, terpinen-4-ol, germacrene D
Lavandula sp. Lamiaceae	Linalol, linalylacetate, fenchone, camphor
Leptospermum scoparium J.R. et G. Forst. Myrtaceae, Douglas et al. (2004)	*Trans*-calamenene, δ-cadinene, β-caryophyllene, leptospermone
Lippia sp. Verbenaceae	Piperitenone oxide, limonene, thymol, carvacrol, ***p***-cymene, β-caryophyllene and γ-terpinene
Melaleuca sp. Myrtaceae	Terpinene-4-ol, γ-terpinene, α-terpinene
Mentha sp. Lamiaceae	Carvone, pulegone, menthol, menthone, menthylacetate
Micromeria fruticosa L. Lamiaceae	Pulegone, β-caryophyllene, isomenthol, limonene
Minthostachys andina (Britton) Eppling Lamiaceae	Menthone, pulegone, iso-menthone
Myristica fragrans Houtt. Myristicaceae	Sabinene, α-pinene, β-pinene, myristicine
Myrtus communis Myrtaceae	1,8-Cineole, α-pinene, limonene, linalol, myrtenol
Nepeta racemosa L. Lamiaceae	Nepetalactone, myrtenol, terpinen-4-ol
Ocimum sp. Lamiaceae	γ-Terpinene, α-terpinene, thymol, carvacrol, linalool, eugenol
Origanum sp. Lamiaceae	Carvacrol, thymol, γ-terpinene, ***p***-cymene, linalool
Pelargonium graveolens L'Hér Geraniaceae	Geraniol, citronellol
Pimpinella anisum L. Apiaceae	*cis*-Anethole
Piper nigrum L. Piperaceae	β-Caryophyllene, α-pinene, sabinene, limonene
Rosmarinus officinalis L. Lamiaceae	Limonene, cineole, borneol, terpineol
Salvia sp. Lamiaceae	Cineole, thujone, camphor, α-pinene, β-pinene

(*Continued*)

TABLE 24.2 (*Continued*)
Essential Oils that can be Used as Insecticide and Acaricide

Name Family	Constituents		
Satureja sp. Lamiaceae	Carvacrol, thymol, γ-terpinene, *p*-cymene		
Syzygium aromaticum (L.) Merr. Et L.M. Perry Myrtaceae	Eugenol, aceteugenol, β-caryophyllene		
Syzygium aromaticum (L.) Merr. Et L.M. Perry Myrtaceae	Eugenol, aceteugenol, β-caryophyllene		Teuscher et al. (2004)
Tagetes sp. Asteraceae	Limonene, (*Z*)-β-ocimene, (*Z*)- and (*E*)-ocimenone	Toxic to *V. destructor*, no toxicity toward honey bees; causes repellence to *S. zeamais*	Eguaras et al. (2005); Nerio et al. (2009)
Tanacetum vulgare L. Asteraceae	α-Pinene, α-terpinene, γ-terpinene, carvone, borneol, β-thujone, camphor	Antifeedant against *L. decemlineata* and acaricide toward *T. urticae*	Chiasson et al. (2001)
Thymbra sp. Lamiaceae	Carvacrol, thymol, γ-terpinene, *p*-cymene	High mortality rates in *D. gallinae* and *T. cinnabarinus*; Insecticidal activity on *D. auraria*	Konstantopoulou et al. (1992); Ghrabi-Gammar et al. (2009); Sertkaya et al. (2010)
Thymus sp. Lamiaceae	Thymol, carvacrol, *p*-cymene		Teuscher et al. (2004)

24.2.3.1.2 *Thymus* sp.

Thymus vulgaris L. (Lamiaceae) also grows in the Mediterranean area. Parts used are partial dried or fresh aerial parts. The main components are "available for broad-spectrum insect control in organic farming" (Dayan et al., 2009). The EO was tested against *Bemisia tabaci* and caused its death (Aslan et al., 2004). *T. vulgaris* EO also showed toxicity against *Dermanyssus gallinae*, a pathogenic mite on hens (George et al., 2009a, 2010; Ghrabi-Gammar et al., 2009). It also caused high mortality in *Tenebrio molitor* (George et al., 2009b) and *Tyrophagus putrescentiae* (Kim et al., 2003a,b). The EO of *Thymus herba-barona* Loisel was microencapsulated and tested on *Limantria dispar*, on which it caused mortality (Moretti et al., 2002).

24.2.3.1.3 *Syzygium aromaticum*

Syzygium aromaticum (L.) Merr et L. M. Perry (Myrtaceae) (Teuscher et al., 2004) grows in tropical areas. Parts used are flower buds, which have been dried. Its field of application is insect pest management as eugenol is toxic to many common pests (Dayan et al., 2009). In another study conducted by Huang et al. (2002), toxicity of eugenol to *Sitophilus zeamais* and *Tribolium castaneum* could be shown. Acaricidal activity against *Tyrophagus putrescentiae* (Kim et al., 2003a,b) and *Dermanyssus gallinae* (Kim et al., 2007) could also be confirmed for clove EO.

TABLE 24.3
Effect of Limonene Application on Various Pests (Ibrahim et al., 2008)

Insect Species	Host Plant	Limonene Application	Response
		Reduced Activity of Pest Insect	
Acrolepiopsis assectella (Leek moth)	Leek & onion, all Allium crops	Olfactometer with two parallel air currents containing a Y-shaped nylon fibre	Limonene shows repellent activity
Delia radicum (Cabbage fly)	Brassicaceae	Olfactory stimuli for orientation behavior	Limonene from the surface part of plant host acts as a repellent
Dendroctonus rufipennis, D. *simplex* (Spruce beetle, eastern larch beetle)	Spruce spp	Bioassayed for their toxicity	(+)-Limonene at 60 ppm kills 100% of the pests after 24 hours
Hylobius abietis (Large pine weevil)	Pine, Spruce	Exposure to limonene vapors	High limonene concentrations show signs of poisoning within a few hours
Megastigmus pinus, *M. rafini* (Silver fir seed wasp)	White fir	Olfactory responses to pure α-pinene and limonene	Limonene significally acts as repellent
Thaumatopoea pityocampa (Pine processionary)	Pine	Emulsified with water and sprayed on the foliage of pine seedlings	(+)-Limonene reduces the number of egg clusters on plants sprayed with it
		Increased Activity of Pest Insect	
Helicoverpa armigera (Cotton bollworm)	Polyphageous moth (e.g., cotton, tomatoes, conifers, etc)	Electroantennography used to investigate electrophysiological responses	Attractive to 1–2 days old moths
Papilio demoleus (Common lime butterfly)	Fabaceae	Orientation responses to different odors in Olfactory	(−)-Limonene shows Maximum attraction to the larvae
Prays citri (Citrus blossom moth)	Citrus limonum, C. decuminata, C. aurantium	Electroantennogram response to pure limonene	Limonene activates oviposition
		No effect	
Hylobius abietis (Large pine weevil)	Pine	Exposure to limonene vapors	Low limonene levels does not affect feeding activity

TABLE 24.4
Effect of Eucalyptus Essential Oil on Various Pests (Batish et al., 2008)

Eucalyptus sp.	Tested Organism
E. camaldulensis (River red gum E.)	Repels adult females of *Culex pipiens* (common house mosquito); mortality of eggs in *Tribolium confusum* amd *Ephestia kuehniella*
E. citriodora (Lemon scented E.)	Toxicity against *Sitophilus zeamais* (greater rice weevil)
E. globulus (Tasmanian blue gum E.)	Toxic to pupae of *Musca domestica* (housefly) as well as to female *Pediculus humanus* (louse)
E. intertexta *E. Sargentii* *E. camaldulensis*	Causes death of adult *Callosobruchus maculatus*, *Sitophilus oryzae* and *Tribolium castaneum*
E. saligna	Repelled *Sitophilus zeamais* and *Tribolium confusum* (confused flour beetle)
E. tereticornis	Pesticidal activity against *Anopheles stephensi*
Eucalyptus sp.	Toxic to *Thaumetopoea pityocampa* (pine processionary)

24.2.3.1.4 Muña

Muña are Bolivian medical plants and include species of *Satureja*, *Minthostachys*, *Mentha*, and *Hedomea*. They derive their name muña by the Kechuas people, who live in the Andean mountains where these plants grow at an altitude between 2500 and 5000 meters. These indigenous people of the Andes use these plants among others because of their insecticide and repellent activity in order to protect the crops of potatoes against insects and also to prevent Chagas disease of which an insect acts as the pathogenous vector. Plants used in the study against *Triatoma infestans*, which was conducted by Fournet et al. (1996), were *Minthostachys andina* (Britton) Eppling (Lamiaceae) and *Hedomea mandonianum* Wedd. (Lamiaceae). The EO of *M. andina* could significantly decrease the number of insects: after 28 days, 10 out of 20 insects were found dead. *H. mandonianum* had no such effect (Fournet et al., 1996).

24.2.3.1.5 *Eucalyptus* sp.

The EO of *Eucalyptus* is obtained from the Australian tree *Eucalyptus* sp. that belongs to the Myrtaceae family. It has been known for its antibacterial, antifungal, and antiseptic action for hundreds of years (Batish et al., 2008). Toxicity of eucalyptus EO toward *Sitophilus oryzae* was reported by Lee et al. (2001), and it proves to be a promising fumigant to control that pest. Lee et al. (2004) reported in another study that several Eucalyptus species were toxic to S. *oryzae*, namely *E. nicholii*, *E. codonocarpa*, and *E. blakelyi*. The same species cause mortality in *Tribolium castaneum* and *Rhyzopertha dominica*.

24.2.3.1.6 *Satureja* sp.

Aslan et al. tested the EO of *Satureja hortensis* L. (Lamiaceae) against *Bemisia tabaci*. Of the three oils (also, *T. vulgaris* and *O. basilicum*) tested, *S. hortensis* EO was most toxic to it (Aslan et al., 2004). *S. thymbra* L. (Lamiaceae) was tested against *Hyalomma marginatum* and was able to kill almost all of the ticks after 24 hours (Cetin et al., 2010). It also showed toxicity to *Drosophila auraria* (Konstantopoulou et al., 1992). According to Ayvaz et al., savory EO caused mortality against the pests *Ephestia kuehniella* and *Plodia interpunctella* (Ayvaz et al., 2010).

24.2.3.1.7 *Ocimum* sp.

Basil (*Ocimum basilicum* L., Lamiaceae) EO was tested against *B. tabaci* and proved to be effective against it (Aslan et al., 2004). Keita et al. (2001) tested the fumigant and contact toxicity of *O. basilicum* and *O. gratissimum* L. on *Callosobruchus maculatus* beetles. Both EOs showed a high toxicity toward the beetles compared to the control. Also, the egg hatch rate decreased. Another study reports methyleugenol, methylcinnamate, linalool, and estragol as main compounds. Estragole was identified as one of the effective compounds against *Cryptolestes pusillus* and *R. dominica*, while it was variable against *Sitophilus oryzae* (López et al., 2008). Basil EO showed also very promising results in *Oryzaephilus surinamensis* (Shaaya et al., 1991). *O. kilimandscharicum* Wild. EO, with camphor as the major compound, was proved to be toxic to four pests, namely, *Sitophilus granarius*, *Sitophilus zeamais*, *Tribolium castaneum*, and *Prostephanus truncatus*, while it also caused repellency in them. Compared with the other three, *T. castaneum* is the strongest among them in tolerating fumigation, as it has the lowest mortality rate (Obeng-Ofori et al., 1998).

24.2.3.1.8 *Origanum* sp.

Calmasur et al. tested the pesticidal effect of *Origanum vulgare* L. (Lamiaceae) EO on two pests, namely *Tetranychus urticae* and *Bemisia tabaci*. They showed that after 120 hours and at a concentration of 2 μL/L air, the mortality rate on *B. tabaci* was 100% and on *Tetranychus urticae*, it was 95% (Çalmaşur et al., 2006). Oregano EO can also be used as an acaricide against *Tyrophagus putrescentiae* (Kim et al., 2003a,b) and as insecticide against *Oryzaephilus surinamensis* (Shaaya et al., 1991) and *Drosophila auraria*. *O. dictamnus* L. and O. *majorana* L. were also effective against *D. auraria* (Konstantopoulou et al., 1992). *O. acutidens* EO can be used against *Sitophilus granarius* and *Tribolium confusum* (Kordali et al., 2008), and the oil of *O. onites* L. (with carvacrol, linalool, and thymol as main constituents) is successful in controlling *T. cinnabarinus*, as reported by Sertkaya et al. (2010). Oregano EO showed high activity against *Plodia interpunctella* and *Ephestia kuehniella* (Ayvaz et al., 2010).

24.2.3.1.9 *Artemisia* sp.

The main constituents of *Artemisia absinthium* L. (Asteraceae) EO are said to be effective against several fleas, flies, and mosquitoes. In a study conducted by Chiasson et al. (2001), *A. absinthium* EO showed toxicity to *T. urticae* and *T. putrescentiae* (Kim et al., 2003a,b). When used as fumigant, *A. sieberi* Besser EO caused mortality in *Callosobruchus maculatus*, *Sitophilus oryzae*, and *Tribolium castaneum*. Of all three insects, it was most toxic to *C. maculatus*, as it caused 100% mortality at a concentration of 37 μL/L after 12 h (Negahban et al., 2007). In another study, this author group tested the effect of A. *scoparia* Waldst et Kit EO on the same three stored-product pests as mentioned above. At a concentration of 37 μL/L 100% mortality of the pests can be observed after one day. It also causes repellent effects on all three insects (Negahban et al., 2006).

24.2.3.1.10 *Mentha* sp.

Mentha pulegium L. (Lamiaceae) EO was tested on *Dermanyssus gallinae*, it showed high toxicity as fumigant against the mite (George et al., 2009a). The same EO proved to be active as an acaricide against *Tyrophagus putrescentiae*. The same effect can be observed with *M. spicata* L. (Kim et al., 2003a,b). *M. pulegium* L. and *M. spicata* were very effective against *Drosophila auraria* (Konstantopoulou et al., 1992). *Mentha arvensis* var. *piperascens* also showed high mortality rates on *Dermanyssus gallinae* (Kim et al., 2007). *Mentha microphylla* and *M. viridis* proved to be toxic to *Acanthoscelides obtectus*, and they also decreased the number of eggs (Papachristos et al., 2002).

24.2.3.1.11 *Cinnamomum* sp.

Cinnamomum aromaticum Nees, Lauraceae proved to be a contact insecticide to *Tribolium castaneum* and *Sitophilus zeamais* (Huang et al., 1998). *C. cassia* and *C. sieboldii* proved to have pesticidal action against *Sitophilus oryzae* and *Callosobruchus chinensis* (Kim et al., 2003a,b). Cinnamaldehyde causes acaricidal effects on *Tyrophagus putrescentiae*, especially in closed containers (Kim et al., 2004). *C. camphora* (L.) J. Presl causes mortality to *Dermanyssus gallinae* and can be used as an acaricide against them (Kim et al., 2007).

24.2.3.1.12 *Acorus* sp.

Acorus calamus var. *angustatus* L. and *Acorus gramineus* Sol. showed 100% mortality at concentrations of 3.5 mg/cm^3 after three days treatment with *Sitophilus oryzae*. EO obtained from *A. calamus* var. *angustatus* was also very potent against *Callosobruchus chinensis*. A mortality rate of 100% after one day could be realized at a concentration of 3.5 mg/cm^3 (Kim et al., 2003a,b). Another pest that can be controlled by *A. calamus* var. *angustatus* is *Dermanyssus gallinae* (Kim et al., 2007). *A. gramineus* EO shows insecticidal action against *Sitophilus oryzae* and *Callosobruchus chinensis* when directly applied. It can also be used against *Lasioderma serricorne*, but it is more potent on the first two insects. As fumigant, it worked best in closed containers (Park et al., 2003).

24.2.3.1.13 *Foeniculum vulgare*

Fennel EO caused 100% toxicity to *Sitophilus oryzae* after exposition for 3 days at concentrations of 3.5 mg/cm^3. The EO was even more potent against *Callosobruchus chinensis* because 100% mortality can be realized at the same concentration after one day (Kim et al., 2003a,b). Compounds of fennel EO were also toxic, as proved in the previous study, to *S. oryzae* and *C. chinensis* and also to *Lasioderma serricorne*. At concentrations of 0.168 mg/cm^2, estragole was most toxic to *S. oryzae* after one day, followed by (+)-fenchone and (*E*)-anethol. Against *C. chinensis*, (*E*)-anethole was most effective, followed by estragol and (+)-fenchone. After one day, (*E*)-anethole was lethal to *L. serricorne*, whereas estragole and (+)-fenchone showed lower toxicity (Kim et al., 2001). *F. vulgare* EO is also potent in controlling *Dermanyssus gallinae* (Kim et al., 2007). Fennel EO also showed toxic effects to *Tyrophagus putrescentiae* (Lee et al., 2006), which can possibly be attributed to fenchone, as the isolated compound also showed high mortality rates (Sánchez-Ramos and Castañera , 2001).

24.2.3.1.14 *Lavandula* sp.

Lavandula stoechas L. (Lamiaceae) EO is effective in killing *Drosophila auraria* in a study conducted by Konstantopoulou et al. In the same study, the insecticidal effect of *Lavandula angustifolia* Miller (Lamiaceae), with linalool and linalylacetate as main compounds, could be confirmed as well (Konstantopoulou et al., 1992). *Lavandula hybrida* volatile oil caused fumigant toxicity to *Acanthoscelides obtectus* males and females and a decreased number of eggs (Papachristos et al., 2002). Lavender EO also showed high activity against *Rhyzopertha dominica* (Shaaya et al., 1991).

24.2.3.1.15 *Carum* sp.

Carum carvi L. (Apiaceae) EO is able to kill *Sitophilus oryzae*; carvone seems to be the active compound against it. (*E*)-anethole is proved to be effective against *Rhyzopertha dominica*. Limonene and fenchone, for example were active against *Cryptolestes pusillus* (López et al., 2008). *C. copticum* C. B. Clarke volatile oil constituents are thymol, α-terpineol and p-cymene. Especially *S. oryzae* was weak against the fumigant action of the EO, but also mortality on *Tribolium castaneum* can be observed (Sahaf et al., 2007).

24.2.3.1.16 *Chenopodium ambrosioides*

At a concentration of 0.4%, the EO of *Chenopodium ambrosioides* was able to kill *Callosobruchus chinensis*, *Callosobruchus maculatus*, and *Acanthoscelides obtectus* at a high percentage. *Sitophilus granarius* and *Sitophilus zeamais* were totally killed at 6.4%, *Prostephanus truncatus* was the least sensitive, as 56% survived (Tapondjou et al., 2002).

The use of EOs as an antipest is presented in this chapter. It could be shown that several EOs, like *Rosmarinus officinalis*, *Thymus vulgaris*, *Eucalyptus* species, and *Origanum* species show great potential in controlling several pests—for example, *Sitophilus* sp. or *Tetranychus* sp., among many others—both on stored products and in protection of livestock, bees, and crops. Since synthetical pesticides bear a lot of risk toward environmental and mammalian health there is a need for natural and healthier alternatives, just like EOs. Of course, the high volatility is a limiting factor in the use of volatile oils on the field, but in glass houses or on stored food, they bear a great chance to change the current situation.

24.3 ESSENTIAL OILS AS HERBICIDES

Weed control in agriculture is very important as a high percentage of weed causes a high crop reduction, even more than insect pests and pathogens. While in ancient times, people were dependent on removing weeds by hand like several famous painters (Emile Klaus, Vincent Van Gogh, or Jules Breton) illustrated, today's weed control is managed by synthetic and natural substances but also by machine weeding. In order to diversify the approaches to control weeds, natural substances such as EOs have to be considered (Dayan et al., 2009).

24.3.1 Phytotoxicity

Phytotoxicity means the harmful impact of chemical substances on plants which can be seen in many different ways: their parts can appear to be burned; they can suffer from chlorosis, which colors the leaves yellow; or they can also result in a lack of growth or even an excessive growth. When phytotoxicity of various components of EOs was tested, pulegone was least, and D-carvone most, phytotoxic to maize plants. (+)-Limonene appeared most toxic to sugar beet seedlings. Limonene showed toxicity toward strawberry seedlings in preliminary screenings, as well as toward cabbage and carrot seedlings (Ibrahim et al., 2008; de Almeida et al., 2010) found that EOs of balm, caraway, hyssop, thyme, and vervain showed a 100% inhibition of germination of *Lepidium sativum* (Brassicaceae). *Raphanus sativus*' (Brassicaceae) germination was inhibited by 100% by vervain oil at all concentrations. At low concentrations, anise and basil EOs promoted the germination and radicle growth, while caraway, sage, vervain, and marjoram EOs inhibited the growth of radish. *Lactuca sativa's* (Asteraceae) growth was inhibited by thyme EO. Vervain, balm, and caraway also had the ability to suppress the growth of lettuce. Generally said, a high level of oxygenated monoterpenes is most harmful toward weeds. When it comes to inhibit the seed germination and subsequent growth, ketones and alcohol showed the highest activity followed by aldehydes and phenols (de Almeida et al., 2010).

Especially plants from the Lamiaceae family can inhibit the growth of several weeds by releasing phytotoxic monoterpenes, namely α-pinene, β-pinene, camphene, limonene, α-phellandrene, *p*-cymene, 1,8-cineole, borneol, pulegone, and camphor (Angelini et al., 2003). The herbicide effect of 1,4-cineole and 1,8-cineole is also described by Dayan et al. (2012). Plants that are exposed to EOs often metabolize them, and when citral was added, geraniol, nerol, and their acids appeared. When citronellal metabolization was tested, citronellol and citronellic acid were formed, with pulegone (iso)-menthone, isopulegol, and menthofuran found (Dudai et al., (2000).

It was also reported that EOs lead to accumulation of H_2O_2 in other plants. That way they increase oxidative stress, which leads to "disruption of metabolic activities in the cell," (Mutlu et al 2010) Another cause for phytotoxicity is that EO destroys the cell membrane of plants and inhibits their enzymes (Mutlu et al., 2010) (Table 24.5).

24.3.2 Prospects of Organic Weed Control

The problem with organic weed control is the huge quantities that must be applied to harm the weeds. This may also cause a large negative impact on the environment and the microbes in the soil. Also, the volatility is a problem of EOs, on account of which they also have to be applied very often, which could be avoided by using microencapsulation of EOs, because it will decrease their volatility. The third problem is that EOs can also harm the desired crop and lead to part destruction of the culture. Altogether, organic weed control using EOs does not seem too promising as a future replacement of synthetic products (Dayan et al., 2009).

24.3.3 Examples of Essential Oils in Weed Control

24.3.3.1 *Thymus vulgaris*

Thymus vulgaris L. belongs to the Lamiaceae family as well (Angelini et al., 2003; de Almeida et al., 2010) (Table 24.6). The latter author group observed that thyme EO can inhibit the growth of the

TABLE 24.5
Examples of Tested Plants

Name	Common Name	Family	Source
Achyranthes aspera	Prickly chaff flower	Amaranthaceae	(a)
Ageratum conyzoides	Billygoat weed	Asteraceae	(a)
Agrostis stolonifera	Creeping bentgrass	Poaceae	(a)
Alcea pallida	Hollyhock	Malvaceae	(a)
Amaranthus hybridus	Smooth amaranth	Amaranthaceae	(a)
Amaranthus retroflexus	Red-root amaranth	Amaranthaceae	(a)
Amaranthus viridis	Slender amaranth	Amaranthaceae	(a)
Ambrosia artemisiifolia	Common ragweed	Asteraceae	(a), (b)
Arabidopsis thaliana	Thale cress	Brassicaceae	(a)
Bromus danthoniae		Poaceae	(a)
Bromus intermedius		Poaceae	(a)
Cassia occidentalis			(a)
Centaurea solstitialis			(a)
Chenopodium album	Commmon lambsquarters	Amaranthaceae	(a), (b)
Cirsium arvense	Creeping thistle	Asteraceae	(a)
Cynodon dactylon	Bermuda grass	Poaceae	(a)
Cyperus rotundus	Purple nut sedge	Cyperaceae	(a)
Echinochloa crus-galli	Common barnyard grass	Poaceae	(a)
Hordeum spontaneum	Wild barley	Poaceae	(a)
Lactuca sativa	Lettuce	Asteraceae	(a)
Lactuca serriola	Prickly lettuce	Asteraceae	(a)
Lepidium sativum	Garden cress	Brassicaceae	(a)
Lolium multiflorum	Annual ryegrass	Poaceae	(a)
Lolium rigidum	Ryegrass	Poaceae	(a)
Parthenium hysterophorus	Whitetop weed	Asteraceae	(a)
Phalaris canariensis	Canarygrass	Poaceae	(a)
Phalaris minor	Littleseed canarygrass	Poaceae	(a)
Portulaca oleracea	Common purslane	Portulacaceae	(a)
Ranunculus repens	Creeping buttercup	Ranunculaceae	(a)
Raphanus raphanistrum	Wild radish	Brassicaceae	(a)
Raphanus sativus	Radish	Brassicaceae	(a)
Rumex crispus	Curled dock	Polygonaceae	(a)
Rumex nepalensis		Polygonaceae	(a)
Rumex obtusifolius	Broad-leaved dock	Polygonaceae	(a)
Secale cereale	Rye	Poaceae	(a)
Senecio jacobaea	Common ragwort	Asteraceae	(a), (c)
Sinapis arvensis	Wild mustard	Brassicaceae	(a)
Sonchus oleraceus	Common sowthistle	Asteraceae	(a)
Sorghum halepense	Johnsongrass	Poaceae	(a), (b)
Taraxacum sp.	Dandelion	Asteraceae	(a)
Trifolium campestre	Hop trefoil	Fabaceae	(a), Teuscher et al. (2004)
Urtica dioica	Common nettle	Urticaceae	Teuscher et al. (2004)

Source: (a) Plant Encyclopedia (2012), (b) Tworkoski (2002), (c) Clay et al. (2005).

TABLE 24.6
Essential Oils that can be Used in Weed Control

Name Family	Constituents	Notes	Source
Achillea sp. Asteraceae	Camphor, 1,8-cineole, piperitone, borneol, α-terpineol	Inhibitory effect on germination and seedling growth of *A. retroflexus*, *C. arvense*, and *L. serriola*	Kordali et al. (2009)
Ageratum conyzoides L. Asteraceae	Precocene I and II, β-caryophyllene, γ-bisabolene, fenchyl acetate	Causes phytotoxic effects on radish, mungbean, and tomatoes	Kong et al. (1999); Plant Encyclopedia (2012)
Anisomeles indica L. Lamiaceae	Isobornyl-acetate, isothujone, nerolidol, camphene, eugenol	Herbicide against *P. minor,* positive effects on growth of wheat	Batish et al. (2007b); Ushir et al. (2010)
Artemisia scoparia Waldst et Kit. Asteraceae	*p*-Cymene, β-myrcene, (+)-limonene		Kaur et al. (2010)
Callicarpa japonica Thunb. Verbenaceae	Spathulenol, germacrene B, viridiflorol, globulol	Toxic to *A. stolonifera*, but had no such effect on lettuce	Kobaisy et al. (2002)
Carum carvi L. Apiaceae	D-Carvone, limonene	Inhibits germination of *A. retroflexus*, *C. salsotitialis*, *S. arvensis*, *S. oleraceus*, *R. raphanistrum* and *R. nepalensis* and *A. pallida*	Teuscher et al. (2004); Azirek et al. (2008); de Almeida et al. (2010)
Coriandrum sativum L. Apiaceae	Linalol, α-terpinylacetate, 1,8-cineole, linalyacetate	Effective against *C. salsotitialis*, *S. arvensis*, *S. oleraceus*, *R. raphanistrum*, and *R. nepalensis*	
Cymbopogon sp. Poaceae	Citronellal, geraniol, citronellol, citral, limonene		
Eucalyptus sp. Myrtaceae	1,8-Cineole, limonene, α-pinene citronellal, citronellol, linalool, α-terpinene		
Foeniculum vulgare L. Apiaceae	Anethol, fenchone, estragol	Reduces germination rate (under 25%) of *C. salsotitialis*, *S. arvensis* and *R. raphanistrum*	
Hibiscus cannabinus L. Malvaceae	α-Terpineol, myrtenol, limonene, *trans*-carveol and γ-eudesmol	Controls various weeds e.g., *A. retroflexus* and *L. multiflorum*, at higher concentration effective against lettuce, bentgrass, and against one cyanobacterium	
			Teuscher et al. (2004); Azirak et al. (2008)
			Teuscher et al. (2004); Dayan et al. (2009)
			Teuscher et al. (2004); Batish et al. (2006)

(*Continued*)

TABLE 24.6 (*Continued*)
Essential Oils that can be Used in Weed Control

Name Family	Constituents	Notes	Source
			Teuscher et al. (2004); Azirak et al. (2008)
			Kobaisy et al. (2001)
Hyssopus officinalis L. Lamiaceae	β-Pinene, iso-pinocamphone, *trans*-pinocamphone	Inhibitory effect on germination of wheat seeds	Dudai et al. (1999); de Almeida et al. (2010)
Juniperus sp. Cupressaceae	α-Pinene, myrcene, β-pinene, terpinene-4-ol, germacrene D	Effective against *S. arvensis*, *T. campestre*, *L. rigidum* and *P. canariensis* causing electrolyte leakage in them	Teuscher et al. (2004); Ismail et al. (2012)
Lavandula sp. Lamiaceae	Linalol, linalyacetate, fenchone, camphor		Konstantopoulou et al. (1992); de Almeida et al. (2010)
Majorana hortensis L. Lamiaceae	1,8-Cineole, α-pinene, limonene		
Melissa officinalis L. Lamiaceae	Citral, citronellal, carvacrol, iso-menthone	Inhibitory effect on germination of wheat seeds	
Mentha sp. Lamiaceae	Carvone, pulegone, menthol, menthone, menthylacetate		
Micromeria fruticosa Lamiaceae	Pulegone, β-caryophyllene, isomenthol, limonene	Inhibits germination of wheat seedlings	
Nepeta meyeri Benth. Lamiaceae	Nepetalactone, myrtenol, terpinen-4-ol	Inhibits germination of seeds of *B. danthoniae*, *B. intermedius*, *A. retroflexus*, *C. dactylon*, *C. album*, and *L. serriola*	
Ocimum sp. Lamiaceae	γ-Terpinene, α-terpinene, thymol, carvacrol, linalool, eugenol, carvone		
			de Almeida et al. (2010)
			Dudai et al. (1999); Teuscher et al. (2004); de Almeida et al. (2010)
			Konstantopoulou et al. (1992); Teuscher et al. (2004)
			Dudai et al. (1999)
			Mutlu et al. (2010)

(*Continued*)

TABLE 24.6 (*Continued*)
Essential Oils that can be Used in Weed Control

Name Family	Constituents	Notes	Source
		Inhibits germination of wheat seeds	Dudai et al. (1999); Aslan et al. (2004); de Almeida et al. (2010)
Origanum sp. Lamiaceae	Carvacrol, thymol, γ-terpinene, *p*-cymene, linalool		Konstantopoulou et al. (1992); de Almeida et al. (2010)
Peumus boldus Mol. Monimiaceae	Limonene, 1,8-cineole, *p*-cymene, ascaridol	Herbicidal activity against *A. hybridus* and *P. oleracea*	Teuscher et al. (2004); Ibrahim et al. (2008); Verdeguer et al. (2011)
Pimpinella anisum L. Apiaceae	*cis*-anethol	Herbicidal effect against *R. raphanistrum*	Azirak et al. (2008); de Almeida et al. (2010)
Pinus sp. Pinaceae	α-Pinene, β-pinene	Herbicide against various weeds	Teuscher et al. (2004); Dayan et al. (2009)
Pistacia sp. Anacardiaceae	α-Terpinene, limonene	Completely inhibits germination of dicotyles *S. arvensis* and *T. campestre*, partially germination of monocotyles *L. rigidum* and *P. canariensis*	Ismail et al. (2012)
Rosa damascena Mill. Rosaceae	Citronellol, geraniol, nerol, linalool		Aridoğan et al. (2002); Teuscher et al. (2004)
Rosmarinus officinalis L. Lamiaceae	Limonene, 1,8-cineole, borneol, terpineol	Selective toxicity against *C. album*, *P. oleracea*, *E. crus-galli*, *C. annuum*; good inhibition rates against *C. salsotitialis*, *S. arvensis* and *R. raphanistrum* and wheat seeds	Konstantopoulou et al. (1992); Dudai et al. (1999); Angelini et al. (2003); Azirak et al. (2008)
Ruta graveolens L. Rutaceae	α-Pinene, limonene, 1,8-cineole, nonan-2-one, undecan-2-one	Inhibits germination and radicle elongation of radish	de Feo et al. (2002)
Salvia officinalis L. Lamiaceae	1,8-Cineole, thujone, camphor, borneol	Lowers germination rate of *S. arvensis* and *R. raphanistrum*; inhibits germination of wheat seeds	Dudai et al. (1999); Teuscher et al. (2004); Azirak et al. (2008); de Almeida et al. (2010)
Satureja sp. Lamiaceae	Carvacrol, thymol, γ-terpinene, *p*-cymene	Damages several weeds, e.g., dandelion leaves, *C. album*, *A. artemisiifolia*, *S. halepense*	Konstantopoulou et al. (1992); Tworkoski (2002); Angelini et al. (2003)

(*Continued*)

TABLE 24.6 (*Continued*)
Essential Oils that can be Used in Weed Control

Name Family	Constituents	Notes	Source
Syzygium aromaticum (L.) Merr et L.M. Perry Myrtaceae	Eugenol, aceteugenol, β-caryophyllene	Damages *Taraxacum* sp., *S. halepense*, *C. album*, and *A. artemisiifolia*	Tworkoski (2002); Teuscher et al. (2004); Dayan et al. (2009)
Tagetes minuta L. Asteraceae	Limonene, (Z)-β-ocimene, (Z)- and (E)-ocimenone	Inhibits germination of *E. crus-galli* and *C. rotundus*; inhibits root-growth of *Zea mays*	Eguaras et al. (2005); Batish et al. (2007a); Scrivanti et al. (2003)
Tanacetum sp. Asteraceae	α-Pinene, α-terpinene, γ-terpinene, carvone, borneol, β-thujone, camphor		Chiasson et al. (2001)
Thymbra sp. Lamiaceae	Thymol, carvacrol	Lowers germination rate to under 5% of *A. retroflexus*, *S. arvensis*, *S. oleraceus*, and *R. raphanistrum*	Azirak et al. (2008); Ghrabi-Gammar et al. (2009)
Thymus vulgaris L. Lamiaceae	Thymol, *p*-cymene, carvacrol	Inhibits the growth of *P. oleracea*, *E. crus-galli*, and *C. album*, lettuce and pepper crops; damages *Taraxacum* sp., C. *album*, *A. artemisiifolia*, and *S. halepense*	Kim et al. (2001); Tworkoski (2002); Teuscher et al. (2004); de Almeida et al. (2010)
Verbena officinalis L. Verbenaceae	Citral, isobornyl formate, linalol, geranial		de Almeida et al. (2010)
Zataria multiflora Boiss. Lamiaceae	Carvacrole, linalool, α-pinene, *p*-cymene		Saharkhiz et al. (2010)

weeds *Portulaca oleracea, Echinochloa crus-galli*, and *Chenopodium album*, but also of lettuce and pepper crops. *T. vulgaris* EO also causes damage to dandelion (*Taraxacum* sp., Asteraceae) leaves, common lambsquarters (*Chenopodium album* L.), common ragweed (*Ambrosia artemisiifolia*), and Johnsongrass (*Sorghum halepense* L.) (Tworkoski 2002).

24.3.3.2 *Mentha* sp.

2-Phenylethyl propionate can be obtained from peppermint oil (*Mentha x piperita*, L., Lamiaceae); it also contains menthol and menthone (Dayan et al., 2009) *Mentha spicata* L. EO contains (−)-carvone and limonene. *M. spicata* shows good effects against several weeds. It is able to inhibit germination (under 15%) of *Amaranthus retroflexus, Centaurea solstitialis, Sinapis arvensis, Sonchus oleraceus, Raphanus raphanistrum*, and *Rumex nepalensis* Spreng (Azirak et al., 2008). *Mentha longifolia* (L.) Huds and *Mentha officinalis* can be used as herbicides as they are able to inhibit germination of wheat seeds. This must be considered if the crop happens to be wheat (Dudai et al., 1999).

24.3.3.3 *Cymbopogon* sp.

Besides its mosquito-repellent actions, lemongrass oil can also be used as an herbicide and *Cymbopogon citratus* (DC) Stapf (Poaceae) as weed control and an antipest. It is used as contact herbicide, since limonene is able to "remove the waxy cuticular layer from leaves"(Dayan et al. 2009) which leads to death by dehydration. Limonene can be used as a pesticide as well as an herbicide (Dayan et al., 2009). *Cymbopogon winterianus* EO shows toxicity toward several weeds, namely *Senecio jacobaea* L., *Ranunculus repens* L., *Rumex obtusifolius* L., and *Urtica dioica* L. When sprayed on trees, it causes damage to the foliage, but it does not inhibit their growth (Clay et al., 2005). The mode of action of citral is probably that it causes disruption of microtubule polymerization in weeds, as could be proved in *Arabidopsis thaliana* seedlings (Dayan et al., 2012). The effect of *C. citratus* EO was also proved by Dudai et al. The EO proves to be one of the most effective ones, as it is able to inhibit germination sooner (0% germination at 80 nL/mL) than most other EOs tested (Dudai et al., 1999).

24.3.3.4 *Eucalyptus* sp.

Species of *Eucalyptus* (Myrtaceae), such as *Eucalyptus citriodora* and *Eucalyptus tereticornis*, can be used for weed control, because they could reduce growth, chlorophyll and water content when used as fumigants. They also have been shown to reduce the cellular respiration of *Parthenium hysterophorus* (Asteraceae). After about two weeks, injuries like necrosis or wilting could be observed (Batish et al., 2008; Dayan et al., 2009). Moreover, the EO of *E. citriodora* can also inhibit its seed germination. The suspected mode of action is the inhibition of mitosis (Singh et al., 2005). The EO obtained from *E. tereticornis* showed a higher toxicity than that of *E. citriodora*, because of a slightly different composition of the EO. In another study, *E. citriodora* EO was tested on crops—namely *Triticum aestivum*, *Zea mays*, and *Raphanus sativus*—and weeds—namely *Cassia occidentalis*, *Amaranthus viridis*, and *Echinochloa crus-galli*. It could be shown that the EO was more toxic to small-seeded crops such as *A. viridis*. When *Eucalyptus* EO was tested on *P. hysterophorus* and *Phalaris minor*, the herbicidal effect could also be confirmed (Batish et al., 2008). The herbicidal effect of *E. citriodora* on *P. minor* was examined in another study by Batish et al. They found that the concentration of lemon-scented eucalyptus oil is important for inhibiting the growth. The inhibition is higher at higher concentrations. The mode of action is more the inhibition of growth of seedlings than inhibition of germination; *E. citriodora* also lessens the chlorophyll content and the respiratory activity of the weed and causes an ion leakage in membranes (Batish et al., 2007c). The herbicidal effect of *E. citriodora* was also mentioned by Dudai et al. (1999) when they examined the inhibition of germination of wheat seeds. *E. camaldulensis* also showed promising effects in controlling *Amaranthus hybridus* and *Portulaca oleracea*, because it inhibits seedling growth as well as germination in both of them. Its main compound is identified as spathulenol (Verdeguer et al., 2009).

24.3.3.5 *Lavandula* sp.

When tested on various weeds by Azirak et al., *Lavandula stoechas* was not able to significantly inhibit the germination (under 25%) of most of the weed plants. It showed an effect on a few of them, namely *Centaurea solstitialis*, *Sinapis arvensis*, and *Raphanus raphanistrum* (Azirak et al., 2008). *Lavandula x intermedia* cv Grosso was tested on *Lolium rigidum* by Haig et al. It showed a high phytotoxicity when it comes to inhibiting root growth (Haig et al., 2009).

24.3.3.6 *Origanum* sp.

An author group found out that the composition of oregano was different compared to the literature because other authors named *p*-cymene as main compound (de Almeida et al., 2010). *O. onites* L. EO proved to be effective against several weeds, namely *Amaranthus retroflexus*, *Centaurea solstitialis*, *Sinapis arvensis*, *Sonchus oleraceus*, and *Raphanus raphanistrum*, as they lower the germination rate to less than 20% (Azirak et al., 2008). *O. majorana* L. is also able to significantly inhibit germination of wheat seeds (Dudai et al., 1999). *O. acutidens* is able to inhibit germination and seedling growth of *Amaranthus retroflexus*, *Chenopodium album*, and *Rumex crispus* (Kordali et al., 2008).

24.3.3.7 *Artemisia scoparia*

Artemisia scoparia Waldst. and Kit. (Asteraceae) was tested on *Achyranthes aspera* L., *Cassia occidentalis* L., *Echinochloa crus-galli* (L.) P. Beauv., *Parthenium hysterophorus* L. and *Ageratum conyzoides* L. The inhibitory effect on seedling growth was greatest in the latter two weeds. When the EO was used on the weeds, it caused wilting and necrosis of sprayed parts. Also, a decreasing chlorophyll amount and ion leakage can be observed. This effect was greatest in *E. crus-galli* and *P. hysterophorus*. *A. scoparia* EO's main constituents are *p*-cymene, β-myrcene, and (+)-limonene (Kaur et al., 2010).

24.3.3.8 *Zataria multiflora*

Two different ecotypes of *Zataria multiflora* Boiss, Lamiaceae, were tested in the study conducted by Saharkhiz et al. (2010). Ecotype B had similar constituents but in lower concentrations; this also explains why ecotype A was more successful in inhibiting the germination and growth of the tested weeds, which were *Hordeum spontaneum* Koch, *Secale cereale*, *Amaranthus retroflexus*, and *Cynodon dactylon*.

24.3.3.9 *Tanacetum* sp.

Tanacetum aucheranum and *Tanacetum chiliophyllum* var. *chiliophyllum*, Asteraceae, are common in Europe and western Asia. The EOs were tested for their potential to inhibit germination and seedling growth of *Amaranthus retroflexus*, *Chenopodium album*, and *Rumex crispus*, in which they were very successful. Besides their herbicidal effect, they also have antibacterial and antifungal activity (Salamci et al., 2007).

Weeds are a severe problem for agriculture today as they lead to decreased yields in crops. EOs, like those obtained from *Cymbopogon* sp., *Eucalyptus* sp., *Origanum* sp., and *Thymus vulgaris* have proved to be very potent inhibitors of both germination and seedling growth of various weeds. However, two limiting factors must be mentioned: first, the high volatiliy of EOs and, second, the fact that some EOs do not only inhibit the growth of weeds, but also that of crops. More research needs to be done on which weeds and crops are influenced by different volatile oils.

24.4 ESSENTIAL OILS AS INHIBITORS OF VARIOUS PESTS

Volatile oils can also be used in controlling various other pests like bacteria, viruses, fungi, and nematodes. Their application and potential will be discussed in the following section (Table 24.7).

TABLE 24.7
Essential Oils with Effect on Bacteria, Viruses, Fungi, and Nematodes

Name Family	Constituents	Notes	Source
Aloysia triphyllata Verbenaceae	α-Thujone, cis-carveol, carvone, limonene	Nematicide	Duschatzky et al. (2004)
Artemisia sp. Asteraceae	Limonene, myrcene, α-thujone, β-thujone, caryophyllyne, sabinyl-acetate	Fungicide Nematicide	Chiasson et al. (2001)
Calamintha nepeta Lamiaceae	Limonene, menthone, pulegone, menthol	Bactericide	Ibrahim et al. (2008)
Carum carvi L. Apiaceae	D-Carvone, limonene	Nematicide	Teuscher et al. (2004)
Cinnamomum zeylanicum Lauraceae	Cinnamaldehyde, linalool, β-caryophyllene, eugenol	Fungicide	Teuscher et al. (2004)
Coridothymus sp. Lamiaceae	Carvacrol, thymol, γ-terpinene, *p*-cymene	Nematicide	Konstantopoulou et al. (1992)
Cotinus coggygria Scop. Anacardiaceae	Limonene, (*Z*)-β-ocimene, (*E*)-β-ocimene, β- caryophyllene	Bactericide	Demirci et al. (2003)
Cymbopogon sp. Poaceae	Citronellal, geraniol, citronellol	Virucide Bactericide Fungicide	Teuscher et al. (2004)
Ducrosia anethifolia (DC.) Boiss Apiaceae	α-Pinene, myrcene, limonene, terpinolene, (*E*)-β-ocimene	Bactericide	Ibrahim et al. (2008)
Echinophora tenuifolia Apiaceae	α-Phellandrene, eugenol, *p*-cymene	Bactericide	Aridoğan et al. (2002)
Elettaria cardamomum (L.) Maton Zingiberaceae	α-Terpinylacetate, 1,8-cineole, linalylacetate	Fungicide	Teuscher et al. (2004)
Eucalyptus sp. Myrtaceae	1,8-Cineole, limonene, α-pinene	Bactericide Fungicide Nematicide	Teuscher et al. (2004)
Foeniculum vulgare L. Apiaceae	Anethol, fenchone, estragol	Bactericide Fungicide Nematicide	Teuscher et al. (2004)
Hibiscus cannabinus L. Malvaceae	α-Terpineol, myrtenol, limonene, *trans*-carveol and γ-eudesmol	Fungicide	Kobaisy et al. (2001)
Juniperus sp. Cupressaceae	α-Pinene, myrcene, β-pinene, terpinen-4-ol, germacrene D	Bactericide	Teuscher et al. (2004)
Laurus nobilis L. Lauraceae	1,8-Cineole, linalool, terpineol acetate, methyleugenol, linalylacetate, eugenol	Fungicide	de Corato et al. (2010)
Lavandula sp. Lamiaceae	Linalool, linalylacetate, fenchone, camphor	Bactericide	Konstantopoulou et al. (1992)
Lippia sp. Verbenaceae	Piperitenone-oxide, limonene	Nematicide	Duschatzky et al. (2004)
Melaleuca alternifolia (Maiden et Betche) Cheel Myrtaceae	Terpinene-4-ol, γ-terpinene, α-terpinene	Virucide	Teuscher et al. (2004)

(*Continued*)

TABLE 24.7 (*Continued*)
Essential Oils with Effect on Bacteria, Viruses, Fungi, and Nematodes

Name Family	Constituents	Notes	Source
Mentha sp. Lamiaceae	Carvone, pulegone, menthol, menthone, menthylacetate	Nematicide Bactericide	Konstantopoulou et al. (1992); Teuscher et al. (2004)
Micromeria fruticosa Lamiaceae	Pulegone, β-caryophyllene, isomenthol, limonene	Nematicide	Dudai et al. (1999)
Ocimum sp. Lamiaceae	γ-Terpinene, α-terpinene, thymol, carvacrol, linalol, and eugenol	Nematicide	Aslan et al. (2004)
Origanum sp. Lamiaceae	Carvacrol, thymol, γ-terpinene, *p*-cymene, linalool	Fungicide Bactericide Nematicide	Konstantopoulou et al. (1992)
Pelargonium graveolens L'Hér. Geraniaceae	Geraniol, citronellol	Nematicide	Ghrabi-Gammar et al. (2009)
Peumus boldus Mol. Monimiaceae	Limonene, 1,8-cineole, *p*-cymene	Bactericide	Teuscher et al. (2004); Ibrahim et al. (2008)
Pinus sp. Pinaceae	α-Pinene, β-pinene	Fungicide	Teuscher et al. (2004)
Rosa damascena Mill. Rosaceae	Citronellol, geraniol, nerol, linalool	Bactericide	Aridoğan et al. (2002); Teuscher et al. (2004)
Rosmarinus officinalis L. Lamiaceae	Limonene, cineole, borneol, terpineol	Bactericide	Konstantopoulou et al. (1992)
Salvia fruticosa Miller Lamiaceae	Cineole, thujone, camphor, α-pinene, β-pinene	Fungicide	Konstantopoulou et al. (1992)
Satureja thymbra L. Lamiaceae	Carvacrol, thymol, γ-terpinene, *p*-cymene	Fungicide	Konstantopoulou et al. (1992)
Syzygium aromaticum (L.) Merr. et L.M. Perry Myrtaceae	Eugenol, aceteugenol, β-caryophyllene	Fungicide	Teuscher et al. (2004)
Tagetes minuta L. Asteraceae	Limonene, (*Z*)-β-ocimene, (*Z*)- and (*E*)-ocimenone	Bactericide	Eguaras et al. (2005)
Thymbra sp. Lamiaceae	Thymol	Fungicide	Ghrabi-Gammar et al. (2009)
Thymus sp. Lamiaceae	Thymol, carvacrol, *p*-cymene	Bactericide Fungicide	Teuscher et al. (2004)
Xylopia longifolia A. DC. Annonaceae	α-Phellandrene, limonene, *p*-cymene, spathulenol	Bactericide	Fournier et al. (1993)

24.4.1 Effect on Bacteria

EOs also have the potential to kill bacteria. Since they are lipophiles themselves, they can pass through the membrane of cells and cause cytotoxicity (Bakkali et al., 2008). On one hand, they "inhibit the respiration and increase the permeability of bacterial cytoplasmatic membranes" of bacteria, and on the other hand, they cause potassium leakage (Ibrahim et al., 2008). According to Burt (2004), other mechanisms that cause the bactericidal effects are depletion of proton motive force and damage to membrane proteins, as well as to the to the cytoplasmatic membrane and degradation of the cell wall. For example, the EO of *Cymbopogon densiflorus* (Poaceae), whose main compounds are limonene, cymenene, *p*-cymene, carveol, and carvone, were active against Gram-negative, as well as Gram-positive, bacteria (Ibrahim et al., 2008). Another EO with effect against bacteria is that of *Eucalyptus* sp.

(Myrtaceae), which mostly contains 1,8-cineole, linalool, citronellal, and limonene (Batish et al., 2008). The EO of *Calamintha nepeta* (Lamiaceae), which contains limonene, menthone, and pulegone, showed high activity against *Salmonella*. Of all the constituents, pulegone was the most effective. The EO from *Peumus boldus* (Monimiaceae) contains mainly monoterpenes, especially limonene, and was also highly efficient against Gram-negative and Gram-positive bacteria. Also the EOs of *Foeniculum vulgare* (Apiaceae)—mostly sweet fennel, *Rosmarinus officinalis*, and *Thymus vulgaris* (Lamiaceae)—showed high activity against bacteria. Other EOs with activity against bacteria are those obtained from *Tagetes minuta* (Asteraceae), *Xylopia longifolia* (Annonaceae), *Cotinus coggygria* Scop. (Anacardiaceae), and *Ducrosia anethifolia* (Ibrahim et al., 2008). EOs obtained from *Origanum onites* (Lamiaceae), *Rosa damascena* (Rosaceae), *Mentha piperita* (Lamiaceae), *Echinophora tenuifolia* (Apiaceae), *Lavandula hybrida* (Lamiaceae), and *Juniperus exalsa* (Cupressaceae) were tested on various bacteria. The EO of *Origanum onites* was reported to elicit the biggest effect (Aridoğan et al., 2002).

24.4.2 Effect on Fungi

Volatile oils are also capable of killing fungi that harm crops, because they cause the increasing of plasma membrane permeability. Several monoterpenes, such as isopulegol, (*R*)-carvone, and iso-limonene, were very efficient against *Candida albicans*. The EOs obtained from *Foeniculum vulgare* (Apiaceae) proved to be very active against *Aspergillus niger*. Also, EOs of citrus fruits (other than limonene) were tested against several *Penicillum* species, and they were highly efficient. Unlike many other EOs, volatile compounds that can be found in tomato leaves were able to inhibit the growth of hyphes of *Alternaria alternata*. Cardamom oil, linalool, limonene, and cineole showed the highest toxicity against fungi (Ibrahim et al., 2008). Another EO that is able to kill harmful fungi like *Aspergillus* sp., *Penicillum* sp., *Fusarium* sp., and *Mucor* sp. is that of *Eucalyptus* sp., which mostly contains 1,8-cineole, citronellal, limonene, and linalool. It is able to inhibit the growth of mycelium as well as spore production and germination (Batish et al., 2008). *Pinus roxburghii* (Pinaceae) EO was able to kill *Aspergillus* sp. and *Cymbopogon pendulus* (Poaceae) and can inhibit the mycelium growth of *Microsporum gypseum* and *Trichophyton mentagrophytes*. Moreover, EOs of *Eucalyptus amygdalia* (Myrtaceae) and *Eucalyptus panciflora* showed activity against *Erysiphe cichoracearum*. EOs obtained from *Thymbra spicata*, *Satureja thymbra*, *Salvia fruticosa*, *Eucalyptus* sp., and *Origanum minutiflorum* (Lamiaceae) were tested on several fungi. Most toxic to those fungi were the EOs of *T. spicata*, *S. thymbra*, and *O. minutiflorum* (Blaeser et al., 2002). Origanum, cassia, and red thyme EOs are able to inhibit fungi that destroy wood (Chao et al., 2000). The EO of *Origanum acutidens* (Lamiaceae) shows antifungal activity as well, for example, against several fungi species of *Fusarium* and *Botrytis* (Kordali et al., 2008). Against *Botrytis cinereal*, the EO from *Cymbopogon martini* (Poaceae), *Thymus zygis* (Lamiaceae), *Cinnamomum zeylanicum* (Lauraceae), and *Syzygium aromaticum* (Myrtaceae) showed high activity as well. Especially C. *martini* and *T. zygis* were very effective in inhibiting spore germination, followed by *S. aromaticum* EO (Wilson et al., 1997). Laurel oil (*Laurus nobilis* L., Lauraceae) showed antifungal activity against *Rhizoctonia solani* and *Sclerotinia sclerotiorum* (de Corato et al., 2010). EO of *Hibiscus cannabinus* L. (Malvaceae) was effective against fungi of *Colletotrichum* species, even though the antifungal effect was weak (Kobaisy et al., 2001). EOs from *Artemisia dranunculus*, *Artemisia absinthium*, *Artemisia santonicum*, and *Artemisia spicigera*, all of them Asteraceae, were tested against various fungi and shown to be effective against them (Kordali et al., 2005).

24.4.3 Effect on Viruses

EOs were also found to be active against viruses. The tobacco mosaic virus, which is an important pest in agriculture, is weakly resistant to lemongrass EO (Chao et al., 2000). But, also, EO of *Melaleuca alternifolia (Myrtaceae)* resulted in fewer lesions caused by tobacco mosaic virus for at least 10 days (Bishop, 1995).

24.4.4 Effect on Nematodes

The population of nematodes (*Heterodera schachtii*) could be decreased down to less than 3% of the control within 3 months using (+)-limonene. Also, menthol proved to be very efficient against nematodes (Ibrahim et al., 2008). EO of *Ocimum sanctum* L. (Lamiaceae) and eugenol as the main compound were tested on *Caenorhabditis elegans* and showed an anthelmintic effect (Asha et al., 2001). Another EO that is very promising in nematode control is *Eucalyptus* sp. EO. It proved to be toxic to *Meloidogyne incognita* and *Meloidogyne exigua* (Batish et al., 2008). *Ocimum basilicum* L. (Lamiaceae) and *Ocimum sanctum* L. (Lamiaceae), as well as their main compounds linalol and eugenol, are effective in killing larvae of *M. incognita* (Chatterjee et al., 1982). Pandey et al. (2000) reported that several *Eucalyptus* and *Mentha* species, as well as *O. basilicum*, as proved in the previous study, and *Pelargonium graveolens* L'Hér (Geraniaceae), volatile oils act as nematicides on *M. incognita*. Another study conducted on *Meloidogyne* sp. showed that *Aloysia triphyllata* (Verbenaceae), *Lippia juneliana* (Verbenaceae), and *Lippia turbinata* (Verbenaceae) have a nematicidal effect. The main components are α-thujone, *cis*-carveole, carvone, and limonene in *A. triphyllata*; piperitenone oxide, limonene, and camphor in *L. junelia* and limonene and piperitenone oxide in *L. turbinata* (Duschatzky et al., 2004). Oka et al. (2000) examined the effect of EO on *M. javanica*. EOs that significantly inhibited its mobility (more than 80%) were *Artemisia judaica*, *Carum carvi*, *Coridothymus capitatus*, *Coridothymus citratus*, *Foeniculum vulgare*, *Mentha rotundifolia*, *Mentha spicata*, *Micromeria fruticosa*, *Origanum syriacum*, and *Origanum vulgare*. EOs that lowered the egg hatching rate to less than 2% were *Artemisia judaica, Carum carvi, Foeniculum vulgare*, and *Mentha rotundifolia*. *Ocimum gratissimum*, with eugenol and 1,8-cineole as main compounds, was reported to have anthelmintic activity on *Haemonchus contortus*, as it decreases the egg hatch rate (Pessoa et al., 2002) (Table 24.8).

This section gave a short overview about various pests that play a role in agriculture. The bactericidal and fungicidal effect of various EOs, like *Cymbopogon* sp., *Thymus* sp., *Eucalyptus* sp., and *Foeniculum vulgare*, among many others, has been known for quite a long time. This effect has been studied in many different publications, and many facts about the mode of action and the different effects are known already. EOs already show virucidal and nematocidal effects as well. An agrochemical relevant virus is the tobacco mosaic virus, for example, that causes big losses every year. An EO that could be of used in killing the virus is *Cymbopogon* sp. Also, nematodes, like the root-knot nematode, can be controlled by EOs—for example, by several *Mentha* species.

24.5 EFFECT OF ESSENTIAL OILS ON THE CONDITION OF THE SOIL

24.5.1 Effects of Essential Oils on Microorganisms and Soil

Successful biodegradation is influenced by several factors, namely soil water, oxygen, redox potential, pH, nutrient status, and temperature (Holden et al., 1997). EOs are not only capable of destroying microorganisms, but these can also use the EOs as carbon and energy sources if they have been exposed to them recently. Thus, some microorganisms can actually be activated by EOs. This applies mostly to bacteria living in Mediterranean areas where many EO-containing plants also exist (Vokou et al., 1999). Especially, bacteria like *Arthrobacter* sp. and *Nocardia* sp. have been reported to use EOs as a carbon source. Also, several *Pseudomonas* species have been detected to use α-pinene as an energy source (Hassiotis, 2010; Hassiotis and Lazari, 2010b). EOs tested by Vokou et al. (1999) were *Origanum vulgare* subsp. *hirtum* (Lamiaceae), *Rosmarinus officinalis* (Lamiaceae), *Mentha spicata* (Lamiaceae), and *Coridothymus capitatus* (Lamiaceae), as well as *Lavandula angustifolia* (Lamiaceae). In another study, Vokou et al. (2002) observed the effect of EOs obtained from *Lavandula stoechas* (Lamiaceae), *Satureia thymbra* (Lamiaceae), or fenchone. They were all able to increase the CO_2 emissions of the soil and the microbial population, even though some substances had not been in previous contact with the soil sample before (Vokou et al., 1999). According to Hassiotis and Dina (2010a), the deciding factor if EOs act as inhibitors or not is the concentration of the EO applied. At lower concentrations, they can induce bacterial growth; at higher concentrations, most likely they

TABLE 24.8
List of Various Pests in Agriculture, Including Post-Harvest Food Pathogens and Soil Microbes

Name	Notes	Group	Source
Aeromonas hydrophila	Spoils meat	Bacterium	Burt (2004)
Alternaria alternata		Fungus	Burt (2004)
Arthrobacter sp.	Can use EO as carbon source	Bacterium	Hassiotis (2010)
Aspergillus flavus	Storage fungus, can produce toxic aflatoxines	Fungus	Blaeser et al. (2002); Razzaghi-Abyaneh et al. (2008)
Aspergillus niger	Post-harvest pathogen	Fungus	Tzortzakis et al. (2007)
Aspergillus parasiticus	Can produce toxic aflatoxines	Fungus	Razzaghi-Abyaneh et al. (2008)
Bacillus cereus	Food pathogen	Bacterium	Burt (2004)
Botrytis cinerea	Post-harvest pathogen	Fungus	Tzortzakis et al. (2007)
Caenorhabditis elegans		Nematode	Tzortzakis et al. (2007)
Candida albicans		Fungus	Tzortzakis et al. (2007)
Cladosporium herbarum	Post-harvest pathogen	Fungus	Razzaghi-Abyaneh et al. (2008)
Clostridium botulinum	Spoils food, produces neurotoxins	Bacterium	Jobling (2000)
Colletotrichum coccodes	Post-harvest pathogen	Fungus	Tzortzakis et al. (2007)
Erysiphe cichoracearum	Phytopathogenous funghus	Fungus	Blaeser et al. (2002)
Eurotium sp.	Spoils food, i.e., bakery products	Fungus	Guynot et al. (2005)
Fusarium oxyporum f. sp. dianthi	Phytopathogenous fungus	Fungus	Pitarokili et al. (2002)
Haemonchus contortus	Gastrointestinal parasite of ruminants	Nematode	Pessoa et al. (2002)
Heterodera schachtii	Infects various plants, i.e., sugarbeet	Nematode	Ibrahim et al. (2008)
Liseria monocytogenes	Spoils food	Bacterium	Jobling (2000)
Meloidogyne exigua *Root-knot nematode*	Damages plant roots	Nematode	Batish et al. (2008)
Meloidogyne incognita *Root-knot nematode*	Damages plant roots	Nematode	Batish et al., (2008)
Meloidogyne javanica *Root-knot nematode*	Damages plant roots	Nematode	Oka et al. (2000)
Methylosinus trichosporium	Consumes methane in soils	Bacterium	Amaral et al. (1998)
Microsporum gypseum		Fungus	Amaral et al. (1998)
Monilinia laxa	Post-harvest pathogen	Fungus	de Corato et al. (2010)
Mucor sp.		Fungus	de Corato et al. (2010)
Nitrosomonas sp.	Oxydize ammonia	Bacterium	Chalkos et al. (2010)
Nitrosospira sp.	Oxydize ammonia	Bacterium	Chalkos et al. (2010)
Nocardia sp.	Can use EOs as carbon source	Bacterium	Hassiotis (2010)
Penicillium digitatum	Spoils food, i. e. bakery products	Fungus	Guynot et al. (2005)
Polysphondylium pallidum	Cellular slime mold, feeds on bacteria in forest soils	Fungus	Hwang et al. (2004)
Pseudomonas sp.		Bacterium	Hwang et al. (2004)
Rhizoctonia solani	Post-harvest fungus	Fungus	de Corato et al. (2010)
Rhizopus stolonifer	Spoils food	Fungus	Tzortzakis et al. (2007)
Salmonella enteritidis	Spoils food, especially raw food	Bacteria	US-FDA (2000)
Sclerotinia sclerotiorum	Phytopathogenous fungus	Fungus	Pitarokili et al. (2002)
Sclerotium cepivorum	Phytopathogenous fungus	Fungus	Pitarokili et al. (2002)
Trichophyton mentagrophytes		Fungus	Pitarokili et al. (2002)

inhibit bacterial action. The only difference between the EOs was the delay they showed before an effect. While some were immediately effective, others needed more time to be efficient. Gram-positive, as well as Gram-negative, bacteria were able to use monoterpenes or oxygenated products (Vokou et al., 2002). This effect is also enhanced by the ability of EOs to kill harmful fungi and their spores. As one can see, plants that contain EOs may influence the condition of the soil by activating the soil respiration through increasing the microbial growth. The plants have to be native to the soil's place to achieve that effect. If EOs are tested on soil where they do not belong, they appear to be less or non-effective (Vokou et al., 1999). Amaral et al. (1998) found that several monoterpenes, especially (−)-α-pinene, were able to inhibit the atmospheric methane consumption by forest soils. EOs can kill *Methylosinus trichosporium*, a methanotrope bacterium, and they are also said to be able to inhibit the nitrification of forest soils. In the same way as the nitrogen cycle, also the carbon cycle is affected by EOs (Amaral et al., 1998). Generally said, terpenes, alcohols and esters can be degraded by bacteria in a very fast way, while it takes a lot longer for ketones (Hassiotis, 2010). Another interesting question is how EOs affect *Polysphondylium*, a cellular slime mold that feeds on bacteria in forest soils. (*R*)-(−)-limonene, (−)-camphene, and (*S*)-(+)-carvone, as well as (−)-menthone and (+)-β-pinene, inhibited growth of *Polysphondylium pallidum* (Hwang et al., 2004).

24.5.2 Examples of Essential Oils with an Effect on Soil Conditions

24.5.2.1 *Mentha spicata*

The main compounds of *Mentha spicata*(Lamiaceae) are (*R*)-(−)-carvone, limonene, and 1,8-cineole (Vokou et al., 2002) (Table 24.9). Chalkos et al. (2010) evaluated the ability of *M. spicata*-composted plants to increase growth of tomato crops and avoid weed growth. It also has

TABLE 24.9
Essential Oils that can Influence Condition of the Soil

Name Family	Constituents	Notes	Source
Cymbopogon sp. Poaceae	Citronellal, geraniol, citronellol		Teuscher et al. (2004)
Laurus nobilis L. Lauraceae	1,8-Cineole, linalool, terpineol acetate, methyleugenol, linalylacetate, eugenol		de Corato et al. (2010)
Lavandula sp. Lamiaceae	Linalool, linalylacetate, fenchone, camphor		Konstantopoulou et al. (1992)
Mentha spicata L. Lamiaceae	Carvone, pulegone, menthol, menthone, menthylacetate		Konstantopoulou et al. (1992), (Teuscher et al. (2004)
Myrtus communis L. Myrtaceae	1,8-Cineole, α-pinene, limonene, linalool, myrtenol		Teuscher et al. (2004)
Origanum sp. Lamiaceae	Carvacrol, thymol, γ-terpinene, *p*-cymene, linalool	Activates soil respiration	Konstantopoulou et al. (1992); Vokou et al. (1999, 2002)
Rosmarinus officinalis L. Lamiaceae	Limonene, 1,8-cineole, borneol, terpineol, camphor	Activates soil respiration	Konstantopoulou et al. (1992); Vokou et al. (1999, 2002)
Salvia sp. Lamiaceae	Cineole, thujone, camphor, α-pinene, β-pinene		Konstantopoulou et al. (1992)
Satureja thymbra L. Lamiaceae	Carvacrol, thymol, γ-terpinene, *p*-cymene	Activates soil respiration	Konstantopoulou et al. (1992); Vokou et al. (2002)
Thymbra capitata (L.) Cav. Lamiaceae	Carvacrol, thymol, γ-terpinene, *p*-cymene	Activates soil respiration	Vokou et al. (2002); Ghrabi-Gammar et al. (2009)

a positive effect on the bacterial and fungal population in the soil. The bacteria *Nitrosomonas* and *Nitrosospira* are still oxidizing ammonia in the soil despite adding the compost. Among all these effects, the compost of *M. spicata* was able to increase the pH. A long time after adding *M. spicata* to the soil, compounds of the EO can be found, but in another composition than the original EO. Carvone, for example, makes up 50% of the EO in the beginning, but after some time in the soil, the concentration lowers to 1%, while other monoterpenes are gone completely. Sesquiterpenes, which act as allelochemicals, on the other hand, can be found at the same or even higher concentrations than in the original oil.

24.5.2.2 *Lavandula* sp.

Lavender oil (*Lavandula angustifolia* Mill., Lamiaceae) mainly contains linalool and linalyl acetate. Its natural habitat is also the Mediterranean area. Spanish or French lavender oil (*L. stoechas* L., Lamiaceae), as it is also called, is obtained from *L. stoechas* and consists of fenchone (Vokou et al., 2002). Other compounds are camphor, *p*-cymene, and 1,8-cineole. In a study by Hassiotis and Dina (2010a), the bacterial growth-stimulating effect could be shown. In another study, *L. stoechas*' EO degradation was examined. The highest EO degradation was in October, November, and December. During the hot summer months, the EO decrease was not very high. After 17 months, only camphene, 1,8-cineole, and camphor remained (Hassiotis, 2010).

24.5.2.3 *Salvia* sp.

Adding *Salvia fruticosa* Mill. (Lamiaceae) compost to soil decreases the high C:N rates and raises the pH of the soil. Moreover, it increases the number of microbes in the soil, such as bacteria and fungi (Chalkos et al., 2010). *S. sclarea* (Lamiaceae) volatile oil's compounds are linalyl acetate, linalool, geranyl acetate, and α-terpineol. The EO is able to cause fungicidal activity on *Sclerotinia sclerotiorum*, a soil-borne pathogen. It also inhibits growth of *Sclerotium cepivorum* and *Fusarium oxysporum* f. sp. *Dianthi* (Pitarokili et al., 2002).

24.5.2.4 *Myrtus communis*

Hassiotis and Lazari examined the bacterial population growth during degradation of *Myrtus communis* L. (Myrtaceae) EO. It could be shown that the bacterial population was able to grow and use myrtle EO, for which the compounds were decreasing over the time, as an energy source. At the end of the study, only low percentages of 1,8-cineole and camphene could be detected (Hassiotis and Lazari, 2010b).

24.5.2.5 *Laurus nobilis*

Laurus nobilis showed inhibition of bacterial activity and lowered the number of colonies as well as the soil respiration rate. This is because 1,8-cineole and eugenol are known for their bactericidal effects (Hassiotis and Dina, 2010a).

24.5.2.6 *Cymbopogon* sp.

Malkomes (2006) tested the dehydrogenase activity in soil with and without dung and the addition of citronella (*Cymbopogon* sp., Poaceae) oil. A low concentration of citronella oil caused a decreasing stimulation of the activity of dehydrogenases. In the soil in which no dung was added, citronella oil also stimulated the CO_2 production from the very beginning; in the soil in which dung was added, the CO_2 production decreased at first, but after three weeks, it started to increase.

The bactericidal action of EOs has been known for quite a long time, but they can also be used as energy sources for bacteria and activate them. That way, they can increase the quality of the soil because CO_2 production is stimulated and pathogenous microorganisms are killed. Several EOs, namely *Mentha spicata*, *Lavandula* sp., and *Cymbopogon* sp., bear great potential to be used as dung and control of soil microbes in the future, but not expected effects on non-targeted species have to be also carefully explored.

24.6 EOS USED IN POST-HARVEST DISEASE CONTROL

Food safety is a very important topic. Even though there is increased hygiene during food production, still, 30% of people suffer from food poisoning every year, and as a result, two million people worldwide still die. A lot of people nowadays also want their food to have less synthetic food additives in it; therefore, there is a need to find natural substances which can replace them (Burt, 2004).

Insects and mites that damage food and act as pathogens are also a major problem in food storage, both in developed and developing countries. EOs can be used as acaricides, and insecticides are explained in the Section 24.2.

24.6.1 Effects of Essential Oils on Stored-Product Pests

In order to keep the concentration of health-damaging effects low, it is important to find out on the benefits of EOs on stored products. EOs that have proved themselves to be useful against post-harvest pathogens, namely the fungi *Botrytis cinerea*, are red thyme oil, clove oil, and cinnamon oil. EOs of *Monarda citriodora* (Lamiaceae) and *Melaleuca alternifolia* (Myrtaceae) showed an antifungal effect against many different fungi. Tea tree oil also works as an inhibitor on bacteria (Jobling, 2000). The effect of anethole, carvacrol, cinnamaldehyde, eugenol, and safrole were tested against fungi of *Mucor*, *Aspergillus*, and *Rhizopus* species. Against most *Rhizopus* and *Mucor* species, carvacrol proved itself to be most effective. Against *Aspergillus* species, both carvacrol and eugenol showed best results (Thompson, 1989). Carvacrol is also able to inhibit toxins which are produced by *Bacillus cereus* in various dishes (Burt, 2004). Against bacteria that damage stored products, cedar, eucalyptus, thyme, and chamomile oils proved themselves useful, especially when used against pathogenous bacteria *B. cereus*, *Clostridium botulinum*, and *Listeria monocytogenes* (Jobling, 2000). Generally said, EOs with a high percentage of phenolic compounds seem to be more effective against pathogens that damage food. Moreover, EOs have a bigger effect on Gram-positive bacteria than on Gram-negative bacteria (Burt, 2004).

But, it must be mentioned that the effect of EOs was tested in laboratories and not in the commercial field, and that many EOs only show a fungistatic effect, which means that the fungi will start growing again after the EO vapor is gone (Jobling, 2000). Moreover, a higher concentration of EOs—up to 100 times as much in soft cheese—is needed to inhibit bacteria in food than *in vitro*. The chemical environment of the food is also important for the effect; for example, the bactericide effect increases when the pH decreases; also the concentration of the fat in the food plays a major role because EOs lose their ability to harm bacteria with increasing fat concentration (Burt, 2004).

24.6.2 Examples of Essential Oils Used on Stored Products

24.6.2.1 *Thymus zygis*

The main components of thyme (*Thymus zygis* L., Lamiaceae) oil are thymol and *p*-cymene (Table 24.10). Its home is the Mediterranean area (Jobling, 2000). Thyme oil can be used to protect meat products from two bacteria, namely *Listeria monocytogenes* and *Aeromonas hydrophila*. Thymol also showed activity against several *Salmonella* species (Burt, 2004).

24.6.2.2 *Cinnamomum* sp.

Cinnamon oil (*Cinnamomum zeylanicum* J. Presl, Lauraceae) consists mainly of cinnamaldehyde (Jobling, 2000). It can be used to preserve dairy products, but it also proved to be successful in defeating several *Salmonella* bacteria on vegetables (Burt, 2004). Guynot et al. (2005) tested Cinnamon EO on *Eurotium* sp., *Aspergillus* sp., and *Penicillium* sp. to preserve bakery products.

24.6.2.3 *Cymbopogon citratus*

Cymbopogon citratus (Poaceae) was able to kill *Aspergillus flavus*, a fungus that can produce very toxic aflatoxins and harms stored products, to 100%. It has to be applied at about 1000 ppm to

TABLE 24.10
Essential Oils that can be Used to Preserve Stored Products

Name Family	Constituents	Notes	Source
Artemisia sp. Asteraceae	Limonene, myrcene, α-thujone, β-thujone, caryophellene, sabinyl-acetate, camphor, 1,8-cineole, borneol, terpinen-4-ol, bornyl acetate	Very strong antifungal activity	Chiasson et al. (2001); Kordali et al. (2005)
Cinnamomum sp. Lauraceae	Cinnamaldehyde, linalol, β-caryophyllene, eugenol		Teuscher et al. (2004)
Coriandrum sp. Apiaceae	Linalool, α-terpinyl acetate, 1,8-cineole, linalyl acetate	Can be used to preserve meat products from bacteria	Burt (2004); Teuscher et al. (2004)
Cymbopogon sp. Poaceae	Citronellal, geraniol, citronellol		Teuscher et al. (2004)
Eucalyptus sp. Myrtaceae	1,8-Cineole, limonene, α-pinene	Against browning of mushrooms in plastic bags, antibacterial	Teuscher et al. (2004); Jobling (2000)
Laurus nobilis L. Lauraceae	1,8-Cineole, linalool, terpineol acetate, methyleugenol, linalyacetate, eugenol		de Corato et al. (2010)
Melaleuca alternifolia (Maiden et Betche) Cheel Myrtaceae	Terpinene-4-ol, γ-terpinene, α-terpinene	At 500 ppm it is able to kill 100% of the fungi *Botrytis* sp. on grapes	Jobling (2000); Teuscher et al. (2004)
Mentha piperita L. Lamiaceae	Carvone, pulegone, menthol, menthone, menthylacetate	Is able to kill *Salmonella enteritidis* in several foods	Konstantopoulou et al. (1992); Tassou et al. (1995); Teuscher et al. (2004)
Monarda citriodora Lamiaceae	Thymol, thymol methylester, α-terpinene, *p*-cymene	Antifungal activity on most post-harvest fungi, e.g., *Alternaria*, *Fusarium* and *Rhizopus* species	Bishop et al. (1997); Jobling (2000); Rozzi et al. (2002)
Origanum sp. Lamiaceae	Carvacrole, thymol, γ-terpinene, *p*-cymene, linalool	Inhibits bacterial growth on meat, fish and vegetable products	Konstantopoulou et al. (1992), Burt (2004)
Satureja hortensis Lamiaceae	Carvacrol, thymol	Inhibitor of *Aspergillus parasiticus* (can produce aflatoxines)	Razzaghi-Abyaneh et al. (2008)
Syzygium aromaticum (L.) Merr. Et L.M. Perry Myrtaceae	Eugenol, aceteugenol, β-caryophyllene	Eugenol shows high activity against *A. flavus*; is antibacterial; inhibits *L. monocytogenes* and *A. hydrophila*	Jobling (2000); Blaeser et al. (2002); Burt (2004); Teuscher et al. (2004)
Thymus zygis L. Lamiaceae	Thymol, carvacrol, *p*-cymene	Protects meat products against *L. monocytogenes* and *A. hydrophila*; also against *Salmonella* species	Jobling (2000); Burt (2004); Teuscher et al. (2004)
Zizyphus jujuba Rhamnaceae	Eugenol, isoeugenol, caryophyllene, eucalyptol, caryophyllene oxide	Active acainst *Listeria monocytogenes*	Al-Reza et al. (2009)

inhibit *Aspergillus flavus*. Besides *A. flavus*, it has the ability to kill many other fungi, as well, for about 210 days in storage. The EO of *Cymbopogon citratus*, especially if it was produced between May and November, also appeared to be more successful in fighting fungi than several synthetic products (Mishra et al., 1994). Lemongrass EO also showed activity against colony development of *Cladosporium herbarum* and *Rhizopus stolonifer*. A concentration of 500 ppm was lethal for *Colletotrichum coccodes*, *Cladosporium herbarum*, *Rhizopus stolonifera*, and *Aspergillus niger*. The fungal colony growth of *Botrytis cinerea* was only inhibited up to 60%. The fungal spore production could be inhibited in all five fungi (Tzortzakis *et al.*, 2007). Lemongrass EO can also be used against *A. flavus* in parboiled rice. It causes fungistatic and fungicidal effects (Paraganama et al., 2003).

24.6.2.4 *Laurus nobilis*

EO of *Laurus nobilis* L., Lauraceae, showed good effects in the protection of the harvested fruits of kiwi and peach against the post-harvest fungi *Monilinia laxa*, *Botrytis cinerea*, and *Penicillium digitatum*. The fungistatic effect is very high in *M. laxa* and *B. cinerea*, but *P. digitatum* cannot be completely killed (de Corato et al., 2010).

The protection of stored food against pathogens like bacteria, fungi, insects, and mites is a big challenge, especially in developing countries. Especially volatile oils obtained from *Cinnamomum* sp., *Cymbopogon* sp., and *Syzygium aromaticum* (earlier known as *Eugenia caryophyllata*) bear great potential. One thing that should be considered is that they have a strong taste themselves, therefore not every EO can be used in every dish as the flavors might not match very well. But overall, adding EOs to food helps preserve it and seems to be a good idea since many EO plants are already used as spices throughout the world, and most of them do not have any known toxic effects. But of course, studies must be conducted to make sure adding volatile oils does not lead to yet-unknown side effects.

24.7 CONCLUSION

EOs can be used in various fields of application, and one of them is in agriculture. Until now, agriculture has been dominated by synthetic products in order to control various pests, weed, and soil condition, but EOs bear great potential to replace or complement those substances. In insect pest control, EOs proved to be very promising, as they have great potential for killing various harmful insects. Especially in developing countries, they might play a bigger role in the future as they are very cheap and can be combined with synthetic products to minimize their toxicity. When it comes to repellent activity, EOs might not be that useful as only some insects will be deterred while others can be attracted by them. When it comes to weed control, EOs do not seem as promising because of the high quantities needed and the frequent application, which might also lead to negative environmental effects. Moreover, the possibility that the applied EOs might damage the wanted crops cannot be excluded. Volatile oils are also able to kill various pests, namely bacteria, fungi, and viruses, as well as mites and nematodes. Especially in the control of tobacco mosaic virus, EOs might become very important. More research needs to be done in this field of application. Furthermore, EOs are also able to improve the condition of the soil because if exposed to the oils earlier, bacteria can use them as a carbon source. That way, the respiration of the soil can be activated. Finally, a very interesting aspect of EOs is to use them in post-harvest pest management and against fungi and bacteria on stored products. Especially in developing countries, insect-, and other pest-, management on stored products might become very useful. But, also in industrialized countries, there is some use for them; for example, in supermarkets on fruits and vegetables to make them less perishable.

Because they have a natural origin, they are biodegradable and diverse physiological targets to control pests that may delay the evolution of pest resistance, EOs are considered to be friendly environmental products and have been embraced by the public.

However, the development of EOs as PPPs for a sustainable agriculture needs to have more data on environmental features of EOs and their components, and also on plant–pest interrelationships and mechanisms. More fundamental and applied studies are needed to better know the impact of these compounds on non-target organisms (unintended affects) or on the biodiversity if some plant endemic resources are hugely used to provide EOs. It is not because they are natural that the EOs have no adverse effect, even they have no history of adverse effects. So, a benefit-to-risk assessment on a case-by-case basis must to be required to prevent undesirable situations and to have a reasonable management of real risk when there is some.

Most government policies in all the world are now seeking low-risk plant protection products. The US EPA is the first administration that has considered reduced regulatory regime for about 20 years and, as a result, has allowed a significant number of EOs for commercial use. The results of this American position must be checked carefully to determine positive and negative inputs of using EOs in such a way. In EU, the Directive 2009/128/CE makes an obligation to develop Integrated Pest Management (IPM) as the strategy of agricultural pests control. It entered into force in January 2014. Because of their numerous advantages in controlling harmful pests, and because they are especially suited to organic farming as well as to Integrated Pest Management in a complementary approach to synthetic insecticides, EOs certainly have a room in biocontrol approaches to promote sustainable development.

REFERENCES

Adorjan, B., and G. Buchbauer, 2010. Biological properties of EOs: An updated review. *Flavour Fragance J.*, 25: 407–426.

Al-Reza, S.M., V.K. Bajpai, and S.C. Kang, 2009. Antioxidant and antilisterial effect of seed EO and organic extracts from Zizyphus jujuba. *Food Chem. Toxicol.*, 47: 2374–2380.

Amaral, J.A., and R. Knowles, 1998. Inhibition of methane consumption in forest soils by monoterpenes. *J. Chem. Ecol.*, 24: 723–734.

Ammon, H.P.T., C. Hunnius, and A. Bihlmayer, 2010. *Hunnius pharmazeutisches Wörterbuch*. de Gruyter.

Angelini, L.G., G. Carpanese, P.L. Cioni, I. Morelli, M. Macchia, and G. Flamini, 2003. EOs from Mediterranean Lamiaceae as weed germination inhibitors. *J. Agric. Food Chem.*, 51: 6158–6164.

Anthony, N.M., J.B. Harrison, and D.B. Sattelle, 1993. GABA receptor molecules of insects. *EXS*, 63: 172–209.

Aridoğan, B.C., H. Baydar, S. Kaya, M. Demirci, D. Ozbaşar, and E. Mumcu, 2002. Antimicrobial activity and chemical composition of some EOs. *Arch. Pharm. Res.*, 25: 860–864.

Arnason, J.T., 2011. Natural products from plants as insecticides in agriculture and human health (Chapter 13). In *Photochemistry and pharmacognosy in Encyclopedia of Life Support Systems (EOLSS), developed under the Auspices of the UNESCO*, J.M. Pezzuto, and M.J. Kato (eds.). Oxford, UK: Eolss Publishers.

Arnason, J.T., T. Durst, J.R. Philogène, and I.A. Scott, 2008. Prospection d'insecticides phytochimiques de plantes tempérées et tropicales communes ou rares. In *Biopesticides d'origine végétale*, C. Regnault-Roger, B.J.R. Philogène, and C. Vincent (eds.). Paris: Lavoisier.

Arnason, J.T., S.R. Sims, and I. Scott. 2012. Natural products from plants as insecticides. In *Encyclopedia of Life Support Systems (EOLSS) EOLSS-publishers (editors). Developed under the Auspices of the UNESCO*. Oxford, UK: Eolss Publishers, [http://www.eolss.net].

Asha, M.K., D. Prashanth, B. Murali, R. Padmaja, and A. Amit, 2001, Anthelmintic activity of EO of *Ocimum sanctum* and eugenol. *Fitoterapia*, 72: 669–670.

Aslan, İ., H. Özbek, Ö. Çalmaşur, and F. Şahin, 2004. Toxicity of EO vapours to two greenhouse pests, *Tetranychus urticae* Koch and Bemisia tabaci Genn. *Ind. Crops Prod.*, 19: 167–173.

Ayvaz, A., O. Sagdic, S. Karaborklu, and I. Ozturk, 2010. Insecticidal activity of the EOs from different plants against three stored-product insects. *J. Insect Sci.*, 10: 21.

Azirak, S., and S. Karaman, 2008. Allelopathic effect of some EOs and components on germination of weed species. *Acta Agric. Scand. Section B—Soil Plant Sci.*, 58: 88–92.

Bakkali, F., S. Averbeck, D. Averbeck, and M. Idaomar, 2008. Biological effects of EOs—a review. *Food Chem. Toxicol.*, 46: 446–475.

Balfanz, S., T. Strünker, S. Frings, and A. Baumann, 2005. A family of octopamine [corrected] receptors that specifically induce cyclic AMP production or Ca2+ release in *Drosophila melanogaster. J. Neurochem.*, 93: 440–451.

Batish, D.R., K. Arora, H.P. Singh, and R.K. Kohli, 2007a. Potential utilization of dried powder of *Tagetes minuta* as a natural herbicide for managing rice weeds. *Crop Prot.*, 26: 566–571.

Batish, D.R., M. Kaur, H.P. Singh, and R.K. Kohli, 2007b. Phytotoxicity of a medicinal plant, *Anisomeles indica*, against *Phalaris minor* and its potential use as natural herbicide in wheat fields. *Crop Prot.*, 26: 948–952.

Batish, D.R., H.P. Singh, R.K. Kohli, and S. Kaur, 2008. Eucalyptus EO as a natural pesticide. *Forest Ecol. Manage.*, 256: 2166–2174.

Batish, D.R., H.P. Singh, N. Setia, S. Kaur, and R.K. Kohli, 2006. Chemical composition and phytotoxicity of volatile EO from intact and fallen leaves of *Eucalyptus citriodora*. *Z. Naturforsch. C. J. Biosci.*, 61: 465–471.

Batish, D.R., H.P. Singh, N. Setia, R.K. Kohli, S. Kaur, and S.S. Yadav, 2007c. Alternative control of littleseed canary grass using eucalypt oil. *Agron. Sustain. Dev.*, 27: 171–177.

Bishop, C.D., 1995. Antiviral activity of the EO of *Melaleuca alternifolia* (Maiden amp; Betche) Cheel (Tea Tree) against tobacco mosaic virus. *J. Essent. Oil Res.*, 7: 641–644.

Bishop, C.D., and I.B. Thornton, 1997. Evaluation of the antifungal activity of the EOs of *Monarda citriodora* var. citriodora and *Melaleuca alternifolia* on post-harvest pathogens. *J. Essent. Oil Res.*, 9: 77–82.

Blaeser, P., U. Steiner, and H.W. Dehne, 2002. Pflanzeninhaltsstoffe mit fungizider Wirkung. *Schriftenreihe des Lehr- und Forschungsschwerpunktes USL,* Landwirtschaftliche Fakultät der Universität Bonn, 97, 1–143.

Bruneton, J., 2009. *Pharmacognosie*, 4ème éd. Paris: Tec & Doc Lavoisier.

Buchbauer, G., S. Hemetsberger, 2016. Use of EOs in agriculture. In *Handbook of EOs: Science, Technology, and Applications*, 2nd ed, K. Husnu Can Baser and G. Buchbauer (eds.), pp. 669–706. CRC Press Book.

Burt, S., 2004. EOs: Their antibacterial properties and potential applications in foods—a review. *Intern. J. Food Microbiol.*, 94: 223–253.

Çalmaşur, Ö., İ. Aslan, and F. Şahin, 2006. Insecticidal and acaricidal effect of three Lamiaceae plant EOs against *Tetranychus urticae* Koch and Bemisia tabaci Genn. *Indus. Crops Prod.*, 23: 140–146.

Capinera, J.L., 2008. *Encyclopedia of Entomology*. Springer Science & Business Media.

Cavalcanti, S.C.H., E. dos S. Niculau, A.F. Blank, C.A.G. Câmara, I.N. Araújo, and P.B. Alves, 2010. Composition and acaricidal activity of *Lippia sidoides* EO against two-spotted spider mite (*Tetranychus urticae* Koch). *Bioresour. Technol.*, 101: 829–832.

Cetin, H., J.E. Cilek, E. Oz, L. Aydin, O. Deveci, and A. Yanikoglu, 2010. Acaricidal activity of *Satureja thymbra* L. EO and its major components, carvacrol and γ-terpinene against adult *Hyalomma marginatum* (Acari: Ixodidae). *Vet. Parasitol.*, 170: 287–290.

Chalkos, D., K. Kadoglidou, K. Karamanoli, C. Fotiou, A.S. Pavlatou-Ve, I.G. Eleftherohorinos, H.-I.A. Constantinidou, and D. Vokou, 2010. *Mentha spicata* and *Salvia fruticosa* composts as soil amendments in tomato cultivation. *Plant Soil*, 332: 495–509.

Chao, S.C., D.G. Young, and C.J. Oberg, 2000. Screening for inhibitory activity of EOs on selected bacteria, fungi and viruses. *J. Essent. Oil Res.*, 12: 639–649.

Chatterjee, A., N.C. Sukul, S. Laskar, and S. Ghoshmajumdar, 1982. Nematicidal principles from two species of Lamiaceae. *J. Nematol.*, 14: 118–120.

Chiasson, H., A. Bélanger, N. Bostanian, C. Vincent, and A. Poliquin, 2001. Acaricidal properties of *Artemisia absinthium* and *Tanacetum vulgare* (Asteraceae) EOs obtained by three methods of extraction. *J. Econom. Entomol.*, 94: 167–171.

Clay, D.V., F.L. Dixon, and I. Willoughby, 2005. Natural products as herbicides for tree establishment. *Forestry*, 78: 1–9.

Dayan, F.E., C.L. Cantrell, and S.O. Duke, 2009. Natural products in crop protection. *Bioorg. Med. Chem.*, 17: 4022–4034.

Dayan, F.E., D.K. Owens, and S.O. Duke, 2012. Rationale for a natural products approach to herbicide discovery. *Pest Manag. Sci.*, 68: 519–528.

de Almeida, L.F.R., F. Frei, E. Mancini, L. de Martino, and V. de Feo, 2010. Phytotoxic activities of Mediterranean EOs. *Molecules*, 15: 4309–4323.

de Corato, U., O. Maccioni, M. Trupo, and G. Di Sanzo, 2010. Use of EO of Laurus nobilis obtained by means of a supercritical carbon dioxide technique against post harvest spoilage fungi. *Crop Prot.*, 29: 142–147.

de Feo, V., F. de Simone, and F. Senatore, 2002. Potential allelochemicals from the EO of *Ruta graveolens*. *Phytochemistry*, 61: 573–578.

DGAL (Direction générale de l'alimentation du Ministère de l'agriculture et de l'alimentation). 2017. Note de service DGAL/SDQSPV/2017-635 du 19/07/2017.

Demirci, B., F. Demirci, and K.H.C. Başer, 2003. Composition of the EO of *Cotinus coggygria* Scop. from Turkey. *Flavour Fragr. J.*, 18: 43–44.

Douglas, M.H., J.W. van Klink, B.M. Smallfield, N.B. Perry, R.E. Anderson, P. Johnstone, and R.T. Weavers, 2004. EOs from New Zealand manuka: Triketone and other chemotypes of *Leptospermum scoparium*. *Phytochemistry*, 65: 1255–1264.

Dudai, N., O. Larkov, E. Putievsky, H.R. Lerner, U. Ravid, E. Lewinsohn, and A.M. Mayer, 2000. Biotransformation of constituents of EOs by germinating wheat seed. *Phytochemistry*, 55: 375–382.

Dudai, N., A. Poljakoff-Mayber, A.M. Mayer, E. Putievsky, and H.R. Lerner, 1999. EOs as allelochemicals and their potential use as bioherbicides. *J. Chem. Ecol.*, 25: 1079–1089.

Duschatzky, C.B., A.N. Martinez, N.V. Almeida, and S.L. Bonivardo, 2004. Nematicidal activity of the EOs of several Argentina plants against the root-knot nematode. *J. Essent. Oil Res.*, 16: 626–628.

Eguaras, M.J., S. Fuselli, L. Gende, R. Fritz, S.R. Ruffinengo, G. Clemente, A. Gonzalez, P.N. Bailac, and M.I. Ponzi, 2005. An in vitro evaluation of tagetes minuta EO for the control of the honeybee pathogens *Paenibacillus* larvae and ascosphaera apis, and the parasitic mite *Varroa destructor. J. Essent. Oil Res.*, 17: 336–340.

Enan, E.E., 2005. Molecular and pharmacological analysis of an octopamine receptor from American cockroach and fruit fly in response to EOs. *Arch. Insect. Biochem. Physiol.*, 59: 161–171.

Environ Protection Agency (EPA). 2018. Regulating biopesticides. http://www.epa.gov/pesticides/biopesticides (accessed 15 August 2018).

Feng, L., and Prestwich, G.D. 1997. Expression and characterization of a lepidopteran general odorant binding protein. *Insect. Biochem. Mol. Biol.*, 27: 405–412.

Fournet, A., A. Rojas de Arias, B. Charles, and J. Bruneton, 1996. Chemical constituents of EOs of muña, Bolivian plants traditionally used as pesticides, and their insecticidal properties against Chagas' disease vectors. *J. Ethnopharmacol.*, 52: 145–149.

Fournier, G., A. Hadjiakhoondi, M. Leboeuf, A. Cavé, J. Fourniat, and B. Charles, 1993. Chemical and biological studies of *Xylopia longifolia* A. DC. EOs. *J. Essent. Oil Res.*, 5: 403–410.

García, M., O.J. Donadel, C.E. Ardanaz, C.E. Tonn, and M.E. Sosa, 2005. Toxic and repellent effects of *Baccharis salicifolia* EO on *Tribolium castaneum. Pest. Manag. Sci.*, 61: 612–618.

George, D.R., G. Olatunji, J. H. Guy, and O. A. E. Sparagano, 2010. Effect of plant EOs as acaricides against the poultry red mite, *Dermanyssus gallinae*, with special focus on exposure time. *Vet. Parasitol.*, 169: 222–225.

George, D.R., T.J. Smith, R.S. Shiel, O.A.E. Sparagano, and J.H. Guy, 2009a. Mode of action and variability in efficacy of plant EOs showing toxicity against the poultry red mite, *Dermanyssus gallinae. Vet. Parasitol.*, 161: 276–282.

George, D.R., O.A.E. Sparagano, G. Port, E. Okello, R.S. Shiel, and J.H. Guy, 2009b. Repellence of plant EOs to *Dermanyssus gallinae* and toxicity to the non-target invertebrate *Tenebrio molitor. Vet. Parasitol.*, 162: 129–134.

Gershenzon, J., M.E. McConkey, and R.B. Croteau, 2000. Regulation of monoterpene accumulation in leaves of peppermint. *Plant Physiol.*, 122: 205–213.

Ghrabi-Gammar, Z., D.R. George, A. Daoud-Bouattour, I. Ben Haj Jilani, S.B. Saad-Limam, and O.A.E. Sparagano, 2009. Screening of EOs from wild-growing plants in Tunisia for their yield and toxicity to the poultry red mite, *Dermanyssus gallinae. Indus. Crops Prod.*, 30: 441–443.

Glitho, A.I., G.K. Ketoh, P.Y. Nuto, S.K. Amevoin, and J. Huignard, 2008. Approches non Centre et de l'Ouest. In *Biopesticides d'origine végétale*, C. Regnault-Roger, B.J.R. Philogène, and C. Vincent (eds.). Paris: Lavoisier.

Grieneisen, M.L., and M.B. Isman, 2018. Recent developments in the registration and usage of botanical pesticides in California in managing and analyzing pesticide use data for pest management, environmental monitoring, public health, and public policy. *American Chemical Society Symposium Series*. vol. 1283, DOI: 10.1021/bk-2018-1283.ch008

Guynot, M.E., S. MarÍn, L. SetÚ, V. Sanchis, and A.J. Ramos, 2005. Screening for antifungal activity of some EOs against common spoilage fungi of bakery products. *Food Sci. Technol. Intern.*, 11: 25–32.

Haig, T.J., T.J. Haig, A.N. Seal, J.E. Pratley, M. An, and H. Wu, 2009. Lavender as a source of novel plant compounds for the development of a natural herbicide. *J. Chem. Ecol.*, 35: 1129–1136.

Hassiotis, C.N., 2010. Chemical compounds and EO release through decomposition process from *Lavandula stoechas* in Mediterranean region. *Biochem. Systematics Ecol.*, 38: 493–501.

Hassiotis, C.N., and E.I. Dina, 2010a. The influence of aromatic plants on microbial biomass and respiration in a natural ecosystem. *Israel J. Ecol. Evol.*, 56: 181–196.

Hassiotis, C.N., and D.M. Lazari, 2010b. Decomposition process in the Mediterranean region. Chemical compounds and EO degradation from *Myrtus communis. Int. Biodeter. Biodegr.*, 64: 356–362.

Haubruge, E., G. Lognay, M. Marlier, P. Danhier, J.-C. Gilson, and C. Gaspar, 1989. The toxicity of five EOs extracted from *Citrus* species with regards to *Sitophilus zeamais* Motsch (Col., Curculionidae), *Prostephanus truncatus* (Horn) (Col., Bostrychidae) and *Tribolium castaneum* Herbst (Col., Tenebrionidae). *Medelingen van de Fac. Landbouwwet. Rijksuniv. Gent*, 54: 1083–1093.

Holden, P.A., and M.K. Firestone, 1997. Soil microorganisms in soil cleanup: How can we improve our understanding? *J. Environm. Qual.*, 26: 32.

Huang, Y., and S.H. Ho, 1998. Toxicity and antifeedant activities of cinnamaldehyde against the grain storage insects, *Tribolium castaneum* (Herbst) and *Sitophilus zeamais* Motsch. *J. Stored Prod. Res.*, 34: 11–17.

Huang, Y., S.-H. Ho, H.-C. Lee, and Y.-L. Yap, 2002. Insecticidal properties of eugenol, isoeugenol and methyleugenol and their effects on nutrition of *Sitophilus zeamais* Motsch. (Coleoptera: Curculionidae) and *Tribolium castaneum* (Herbst) (Coleoptera: Tenebrionidae). *J. Stored Prod. Res.*, 38: 403–412.

Huang, Y., S.L. Lam, and S.H. Ho, 2000. Bioactivities of EO from *Elletaria cardamomum* (L.) Maton. to *Sitophilus zeamais* Motschulsky and *Tribolium castaneum* (Herbst). *J. Stored Prod. Res.*, 36: 107–117.

Huignard, J., B. Lapied, S. Dugravot, M. Magnin-Robert, G.K. Ketoh, 2008. Modes d'action neurotoxiques des dérivés soufrés et de certaines huiles essentielles et risques liés à leur utilisation. In *Biopesticides d'Origine Végétale*, 2nd ed, C. Regnault-Roger, B.J.R. Philogène, and C. Vincent (eds.). Paris: Lavoisier Tech & Doc.

Hummelbrunner, L.A., and M.B. Isman, 2001. Acute, sublethal, antifeedant, and synergistic effects of monoterpenoid EO compounds on the tobacco cutworm, *Spodoptera litura* (Lep., Noctuidae). *J. Agric. Food Chem.*, 49: 715–720.

Hwang, J.-Y., and J.-H. Kim, 2004. *The effect of monoterpenoids on growth of a cellular slime mold, Polysphondylium pallidum. J. Plant Biol.*, 47: 8–14.

Ibrahim, M.A., P. Kainulainen, and A. Aflatuni, 2008. Insecticidal, repellent, antimicrobial activity and phytotoxicity of EOs: With special reference to limonene and its suitability for control of insect pests. *Agric. Food Sci.*, 10: 243–259.

Ismail, A., H. Lamia, H. Mohsen, and J. Bassem, 2012. Herbicidal potential of EOs from three Mediterranean trees on different weeds. *Curr. Bioactive Compd.*, 8: 3–12.

Isman, M.B., 2000. Plant EOs for pest and disease management. *Crop Prot.*, 19: 603–608.

Isman, M.B., 2004. Plant EOs as green pesticides for pest and diseases management. *ACS Symp. Ser.*, 887: 41–51

Isman, M.B., 2006. Botanical insecticides, deterrents, and repellents in modern agriculture and an increasingly regulated world. *Annu. Rev. Entomol.*, 51: 45–66.

Isman, M.B., 2008. Botanical insecticides: For richer, for poorer. *Pest Manag. Sci.*, 64: 8–11.

Isman, M.B., 2017. Bridging the gap: Moving botanical insecticides from the laboratory to the farm. *Indus. Crops Prod.*, 110: 10–14.

Isman, M.B., and C.M. Machial, 2006. Pesticides based on plant EOs: From traditional practice to commercialization. In *Naturally Occurring Bioactive Compounds*, M. Rai and M.C. Carpinella (eds.). Amsterdam: Elsevier BV.

Isman, M.B., S. Miresmailli, and C. Machial, 2011. Commercial opportunities for pesticides based on plant EOs in agriculture, industry and consumer products. *Phytochem. Rev.*, 10, 197–204.

Isman, M.B., and J.-H. Tak, 2017a. Commercialization of insecticides based on plant EOs: Past, present, and future. In *Green Pesticides Handbook: EOs for pest Control*, L.M.L. Nollet and H.S. Rathore (eds.), pp. 27–39. CRC Press.

Isman, M.B., and J.-H. Tak, 2017b. Inhibition of acetylcholinesterase by EOs and monoterpenoids: A relevant mode of action for insecticidal EOs? *Biopestic. Int.*, 13(2): 71–78.

Jobling, J., 2000. EOs: A new idea for postharvest disease control. *Good Fruit Vegetables Mag.*, 11: 50.

Katerinopoulos, H.E., G. Pagona, A. Afratis, N. Stratigakis, and N. Roditakis, 2005. Composition and insect attracting activity of the EO of *Rosmarinus officinalis. J. Chem. Ecol.*, 31: 111–122.

Kaur, S., H.P. Singh, S. Mittal, D.R. Batish, and R.K. Kohli, 2010. Phytotoxic effects of volatile oil from *Artemisia scoparia* against weeds and its possible use as a bioherbicide. *Indus. Crops Prod.*, 32: 54–61.

Kéita, S.M., C. Vincent, J.-P. Schmit, J.T. Arnason, and A. Bélanger, 2001. Efficacy of EO of *Ocimum basilicum* L. and *O. gratissimum* L. applied as an insecticidal fumigant and powder to control *Callosobruchus maculatus* (Fab.) [Coleoptera: Bruchidae]. *J. Stored Prod. Res.*, 37: 339–349.

Kim, D.-H., and Y.-J. Ahn, 2001. Contact and fumigant activities of constituents of *Foeniculum vulgare* fruit against three coleopteran stored-product insects. *Pest. Manag. Sci.*, 57: 301–306.

Kim, E.-H., H.-K. Kim, and Y.-J. Ahn, 2003a. Acaricidal activity of plant EOs against *Tyrophagus putrescentiae* (Acari: Acaridae). *J. Asia-Pacific Entomol.*, 6: 77–82.

Kim, H.-K., J.-R. Kim, and Y.-J. Ahn, 2004. Acaricidal activity of cinnamaldehyde and its congeners against *Tyrophagus putrescentiae* (Acari: Acaridae). *J. Stored Prod. Res.*, 40: 55–63.

Kim, S.-I., Y.-E. Na, J.-H. Yi, B.-S. Kim, and Y.-J. Ahn, 2007. Contact and fumigant toxicity of oriental medicinal plant extracts against *Dermanyssus gallinae* (Acari: Dermanyssidae). *Vet. Parasitol.*, 145: 377–382.

Kim, S.-I., J.-Y. Roh, D.-H. Kim, H.-S. Lee, and Y.-J. Ahn, 2003b. Insecticidal activities of aromatic plant extracts and EOs against *Sitophilus oryzae* and *Callosobruchus chinensis. J. Stored Prod. Res.*, 39: 293–303.

Kobaisy, M., M.R. Tellez, F.E. Dayan, and S.O. Duke, 2002. Phytotoxicity and volatile constituents from leaves of *Callicarpa japonica* Thunb. *Phytochemistry*, 61: 37–40.
Kobaisy, M., M.R. Tellez, C.L. Webber, F.E. Dayan, K.K. Schrader, and D.E. Wedge, 2001. Phytotoxic and fungitoxic activities of the EO of kenaf (*Hibiscus cannabinus* L.) leaves and its composition. *J. Agric. Food Chem.*, 49: 3768–3771.
Kong, C., F. Hu, T. Xu, and Y. Lu, 1999. Allelopathic potential and chemical constituents of volatile oil from *Ageratum conyzoides*. *J. Chem. Ecol.*, 25: 2347–2356.
Konstantopoulou, I., L. Vassilopoulou, P. Mavragani-Tsipidou, and Z.G. Scouras, 1992. Insecticidal effects of EOs. A study of the effects of EOs extracted from eleven Greek aromatic plants on *Drosophila auraria*. *Experientia*, 48: 616–619.
Kordali, S., A. Cakir, T.A. Akcin, E. Mete, A. Akcin, T. Aydin, and H. Kilic, 2009. Antifungal and herbicidal properties of EOs and n-hexane extracts of *Achillea gypsicola* Hub-Mor. and *Achillea biebersteinii* Afan. (Asteraceae). *Indus. Crops Prod.*, 29: 562–570.
Kordali, S., A. Cakir, H. Ozer, R. Cakmakci, M. Kesdek, and E. Mete, 2008. Antifungal, phytotoxic and insecticidal properties of EO isolated from Turkish *Origanum acutidens* and its three components, carvacrol, thymol and p-cymene. *Bioresour. Technol.*, 99: 8788–8795.
Kordali, S., R. Kotan, A. Mavi, A. Cakir, A. Ala, and A. Yildirim, 2005. Determination of the chemical composition and antioxidant activity of the EO of *Artemisia dracunculus* and of the antifungal and antibacterial activities of Turkish *Artemisia absinthium*, *A. dracunculus*, *Artemisia santonicum*, and *Artemisia spicigera* EOs. *J. Agric. Food Chem.*, 53: 9452–9458.
Kostyukovsky, M., A. Rafaeli, C. Gileadi, N. Demchenko, and E. Shaaya, 2002. Activation of octopaminergic receptors by EO constituents isolated from aromatic plants: Possible mode of action against insect pests. *Pest Manag. Sci.*, 58: 1101–1106.
Koul, O., M.J. Smirle, and M.B. Isman, 1990. Asarones from *Acorus Calamus* oil: Their effect on feeding behavior and dietary utilization in *Peridroma saucia*. *J. Chem. Ecol.*, 16: 1911–1920.
Koul, O., S. Walia, and G. S. Dhaliwal, 2008. EOs as green pesticides: Potential and constraints. *Biopestic. Int.*, 4: 63–84.
Lee, B.-H., P.C. Annis, F. Tumaalii, and W.-S. Choi, 2004. Fumigant toxicity of EOs from the Myrtaceae family and 1,8-cineole against 3 major stored-grain insects. *J. Stored Prod. Res.*, 40: 553–564.
Lee, B.-H., W.-S. Choi, S.-E. Lee, and B.-S. Park, 2001. Fumigant toxicity of EOs and their constituent compounds towards the rice weevil, *Sitophilus oryzae* (L.). *Crop Prot.*, 20: 317–320.
Lee, C.-H., B.-K. Sung, and H.-S. Lee, 2006. Acaricidal activity of fennel seed oils and their main components against *Tyrophagus putrescentiae*, a stored-food mite. *J. Stored Prod. Res.*, 42: 8–14.
Li, R., and Z.-T. Jiang, 2004. Chemical composition of the EO of *Cuminum cyminum* L. from China. *Flavour Fragr. J.*, 19: 311–313.
Liu, Z.L., and S.H. Ho, 1999. Bioactivity of the EO extracted from *Evodia rutaecarpa* Hook f. et Thomas against the grain storage insects, *Sitophilus zeamais* Motsch. and *Tribolium castaneum* (Herbst). *J. Stored Prod. Res.*, 35: 317–328.
Loebe, L., 2001. Olfactory remedies for the evaluation of repellent and attractive properties of EOs against the Pea Aphid Acyrthosiphon pisum. *Master Thesis*, University of Vienna.
López, M.D., M.J. Jordán, and M.J. Pascual-Villalobos, 2008. Toxic compounds in EOs of coriander, caraway and basil active against stored rice pests. *J. Stored Prod. Res.*, 44: 273–278.
Lopez, M.D., and M.J. Pascual-Villalobosa, 2010. Mode of inhibition of acetylcholinesterase by monoterpenoids and implications for pest control. *Indus. Crops Prod.*, 31: 284–288.
Malkomes, H.-P., 2006. Einfluss von herbizidem Citronella-Öl und Neem (Azadirachtin) auf mikrobielle Aktivitäten im Boden. *Gesunde Pflanzen*, 58: 205–212.
Mills, C., B.J. Cleary, J.F. Gilmer, and J.J. Walsh, 2004. Inhibition of acetylcholinesterase by tea tree oil. *J. Pharm. Pharmacol.*, 56: 375–379.
Mishra, A.K., and N.K. Dubey, 1994. Evaluation of some EOs for their toxicity against fungi causing deterioration of stored food commodities. *Appl. Environ. Microbiol.*, 60: 1101–1105.
Moretti, M.D.L., G. Sanna-Passino, S. Demontis, and E. Bazzoni, 2002. EO formulations useful as a new tool for insect pest control. *AAPS Pharm. Sci. Tech.*, 3: E13.
Mossa, A.-T.H., 2016. Green pesticides: EOs as biopesticides in insect-pest management. *J. Environ. Sci. Technol.*, 9: 354–378.
Muller-Riebau, F.J., B.M. Berger, O. Yegen, and C. Cakir, 1997. Seasonal variations in the chemical compositions of EOs of selected aromatic plants growing wild in Turkey. *J. Agric. Food Chem.*, 45: 4821–4825.
Mutlu, S., Ö. Atici, N. Esim, and E. Mete, 2010. EOs of catmint (*Nepeta meyeri* Benth.) induce oxidative stress in early seedlings of various weed species. *Acta Physiol. Plant.*, 33: 943–951.

Negahban, M., S. Moharramipour, and F. Sefidkon, 2006. Chemical composition and insecticidal activity of artemisia scoparia EO against three coleopteran stored-product insects. *J. Asia-Pacific Entomol.*, 9: 381–388.
Negahban, M., S. Moharramipour, and F. Sefidkon, 2007. Fumigant toxicity of EO from *Artemisia sieberi* Besser against three stored-product insects. *J. Stored Prod. Res.*, 43: 123–128.
Nerio, L.S., J. Olivero-Verbel, and E.E. Stashenko, 2009. Repellent activity of EOs from seven aromatic plants grown in Colombia against *Sitophilus zeamais* Motschulsky (Coleoptera). *J. Stored Prod. Res.*, 45: 212–214.
Nerio, L. S., J. Olivero-Verbel, and E. Stashenko, 2010. Repellent activity of EOs: A review. *Bioresour. Technol.*, 101: 372–378.
Niepel, D., 2000. Repellent properties of EOs against the carrot weevil. *Master Thesis*, University of Vienna.
Njoroge, S.M., H. Koaze, P.N. Karanja, and M. Sawamura, 2005. EO constituents of three varieties of Kenyan sweet oranges (*Citrus sinensis*). *Flavour Fragr. J.*, 20: 80–85.
Obeng-Ofori, D., C.H. Reichmuth, A.J. Bekele, and A. Hassanali, 1998. Toxicity and protectant potential of camphor, a major component of EO of *Ocimum kilimandscharicum*, against four stored product beetles. *Intern. J. Pest Manage.*, 44: 203–209.
Oka, Y., S. Nacar, E. Putievsky, U. Ravid, Z. Yaniv, and Y. Spiegel, 2000. Nematicidal activity of EOs and their components against the root-knot nematode. *Phytopathology*, 90: 710–715.
Pandey, R., A. Kalra, S. Tandon, N. Mehrotra, H.N. Singh, and S. Kumar, 2000. EOs as potent source of nematicidal compounds. *J. Phytopathol.*, 148: 501–502.
Papachristos, D.P., and D.C. Stamopoulos, 2002. Repellent, toxic and reproduction inhibitory effects of EO vapours on *Acanthoscelides obtectus* (Say) (Coleoptera: Bruchidae). *J. Stored Prod. Res.*, 38: 117–128.
Paranagama, P.A., K.H.T. Abeysekera, K. Abeywickrama, and L. Nugaliyadde, 2003. Fungicidal and anti-aflatoxigenic effects of the EO of *Cymbopogon citratus* (DC.) Stapf. (lemongrass) against *Aspergillus flavus* Link. isolated from stored rice. *Lett. Appl. Microbiol.*, 37: 86–90.
Park, C., S.-I. Kim, and Y.-J. Ahn, 2003. Insecticidal activity of asarones identified in *Acorus gramineus* rhizome against three coleopteran stored-product insects. *J. Stored Prod. Res.*, 39: 333–342.
Pellati, F., S. Benvenuti, F. Yoshizaki, D. Bertelli, and M.C. Rossi, 2005. Headspace solid-phase microextraction-gas chromatography—mass spectrometry analysis of the volatile compounds of *Evodia* species fruits. *J. Chromatogr. A*, 1087: 265–273.
Pessoa, L.M., S.M. Morais, C.M.L. Bevilaqua, and J.H.S. Luciano, 2002. Anthelmintic activity of EO of *Ocimum gratissimum* Linn. and eugenol against *Haemonchus contortus. Vet. Parasitol.*, 109: 59–63.
Picimbon, J.F., and C. Regnault-Roger, 2008. Composés sémiochimiques volatils, phytoprotection et olfaction: Cibles moléculaires pour la lutte intégrée. In *Biopesticides d'Origine Végétale*, 2nd ed, C. Regnault-Roger, B.J.R. Philogène, and C. Vincent (eds.). Paris: Lavoisier Tech & Doc.
Philogène, B.J.R., C. Regnault-Roger, and C. Vincent, 2005. Botanicals: yesterday and today's promises. In *Biopesticides of Plant Origin*, C. Regnault-Roger, B.J.R. Philogène, and C. Vincent (eds.), pp. 1–16. Andover, UK: Intercept.
Pitarokili, D., M. Couladis, N. Petsikos-Panayotarou, and O. Tzakou, 2002. Composition and antifungal activity on soil-borne pathogens of the EO of *Salvia sclarea* from Greece. *J. Agric. Food Chem.*, 50: 6688–6691.
Plant Encyclopedia, Aden Earth Zone. 2012. https://www.theplantencyclopedia.org
Prajapati, V., A.K. Tripathi, K.K. Aggarwal, and S.P.S. Khanuja, 2005. Insecticidal, repellent and oviposition-deterrent activity of selected EOs against *Anopheles stephensi, Aedes aegypti* and *Culex quinquefasciatus. Bioresour. Technol.*, 96: 1749–1757.
Price, D.N., and M.S. Berry, 2006. Comparison of effects of octopamine and insecticidal EOs on activity in the nerve cord, foregut and dorsal unpaired median neurons of cockroaches. *J. Insect. Physiol.*, 52: 309–319.
Priestley C.M., E.M. Williamson, K.A. Wafford, and D.B. Satelle, 2003. Thymol, a constituent of thyme EOs, is a positive modulator of human GABA and a homo-oligosteric GABA receptor from *Drosophila melanogaster. Br. J. Pharmacol.*, 140: 1363–1372.
Rajendran, S., and V. Sriranjini, 2008. Plant products as fumigants for stored-product insect control. *J. Stored Prod. Res.*, 44: 126–135.
Razzaghi-Abyaneh, M., M. Shams-Ghahfarokhi, T. Yoshinari, M.-B. Rezaee, K. Jaimand, H. Nagasawa, and S. Sakuda, 2008. Inhibitory effects of Satureja hortensis L. EO on growth and aflatoxin production by *Aspergillus parasiticus. Intern. J. Food Microbiol.*, 123: 228–233.
Regnault-Roger, C., 1997. The potential of botanical EOs for insect pest control. *Integr. Pest. Manage. Rev.*, 2: 25–34
Regnault-Roger, C., 2013. EOs in insect control. In *Natural Products*, K.G. Ramawat, and J.M. Merillon (eds.), pp. 4088–4105. Berlin Heidelberg: Springer-Verlag.

Regnault-Roger, C., and A. Hamraoui, 1994. Antifeedant effect of Mediterranean plant EOs upon Acanthoscelides obtectus Say (Coleoptera), bruchid of kidneybeans, Phaseolus vulgaris L. In *Stored Product Protection*, E. Highley, E.J. Wright, H.J. Banks, and B.R. Champs (eds.), vol. 2, pp. 837–840. Wallingford, UK: CAB International.

Regnault-Roger, C., A. Hamraoui, M. Holeman, E. Theron, and R. Pinel, 1993. Insecticidal effect of EOs from mediterranean aromatic plants upon *Acanthoscelides obtectus* Say, Coleopterea, bruchid of kidney bean (*Phaseolus vulgaris* L.). *J. Chem. Ecology*, 19(6): 1233–1244.

Regnault-Roger, C., and B.J.R. Philogène, 2008. Past and current prospects for the use of botanicals and plant allelochemicals in integrated pest management. *Pharm. Biol.*, 46: 1–12.

Regnault-Roger, C., B.J.R. Philogène, and C. Vincent, 2008. *Biopesticides d'Origine Végétale*, 2nd ed. Paris: Lavoisier Tech & Doc.

Regnault-Roger, C., C. Silvy, and C. Alabouvette, 2005. Biopesticides: Réalités et perspectives commerciales. In *Enjeux Phytosanitaires pour l'Agriculture et l'Environnement*, C. Regnault-Roger (ed.). Paris: Lavoisier Tech & Doc.

Regnault-Roger, C., C. Vincent, and J.T. Arnason, 2012a. EOs in insect control: Low-risk products in a high-stakes world. *Annu. Rev. Entomol.*, 57: 405–424.

Regnault-Roger, C., C. Vincent, and J.T. Arnason, 2012b. Low-risk products in a high-stakes world. *Annu. Rev. Entomol.*, 57: 405–424.

Rozzi, N.L., W. Phippen, J.E. Simon, and R.K. Singh, 2002. Supercritical fluid extraction of EO components from lemon-scented botanicals. *LWT – Food Sci. Technol.*, 35: 319–324.

Sahaf, B.Z., S. Moharramipour, and M.H. Meshkatalsadat, 2007. Chemical constituents and fumigant toxicity of EO from *Carum copticum* against two stored product beetles. *Insect Sci.*, 14: 213–218.

Saharkhiz, M.J., S. Smaeili, and M. Merikhi, 2010. EO analysis and phytotoxic activity of two ecotypes of *Zataria multiflora* Boiss. growing in Iran. *Nat. Prod. Res.*, 24: 1598–1609.

Salamci, E., S. Kordali, R. Kotan, A. Cakir, and Y. Kaya, 2007. Chemical compositions, antimicrobial and herbicidal effects of EOs isolated from Turkish *Tanacetum aucheranum* and *Tanacetum chiliophyllum* var.\chiliophyllum. *Biochem. Systematics Ecol.*, 35: 569–581.

Sánchez-Ramos, I., and P. Castañera, 2000. Acaricidal activity of natural monoterpenes on *Tyrophagus putrescentiae* (Schrank), a mite of stored food. *J. Stored Prod. Res.*, 37: 93–101.

Scrivanti, L.R., M.P. Zunino, and J.A. Zygadlo, 2003. *Tagetes minuta* and *Schinus areira* EOs as allelopathic agents. *Biochem. Systematics Ecol.*, 31: 563–572.

Sertkaya, E., K. Kaya, and S. Soylu, 2010. Acaricidal activities of the EOs from several medicinal plants against the carmine spider mite (*Tetranychus cinnabarinus* Boisd.\) (Acarina: Tetranychidae). *Indus. Crops Prod.*, 31: 107–112.

Shaaya, E., U. Ravid, N. Paster, B. Juven, U. Zisman, and V. Pissarev, 1991. Fumigant toxicity of EOs against four major stored-product insects. *J. Chem. Ecol.*, 17: 499–504.

Singh, H.P., D.R. Batish, N. Setia, and R.K. Kohli, 2005. Herbicidal activity of volatile oils from *Eucalyptus citriodora* against *Parthenium hysterophorus*. *Ann. Appl. Biol.*, 146: 89–94.

Sparagano, O.A.E., J. Pritchard, and D. George, 2016. The future of EOs as pest biocontrol method. In *Bio-Based Plant Oil Polymers and Composites*, S.A. Madbouly, C. Zhang, and M.R. Kessler (eds.), pp. 207–211. William Andrew Publishing.

Szulc, M., and J. Sobczak, 2017. Formulations of plant oils used in crop protection in selected member states. *IX International Scientific Symposium Farm Machinery and Processes Management in Sustainable Agriculture*, Lublin, Poland, DOI: 10.24326/fmpmsa.2017.66.

Tapondjou, L.A., C. Adler, H. Bouda, and D.A. Fontem, 2002. Efficacy of powder and EO from *Chenopodium ambrosioides* leaves as post-harvest grain protectants against six-stored product beetles. *J. Stored Prod. Res.*, 38: 395–402.

Tassou, C., K. Koutsoumanis, and G.-J.E. Nychas, 2000. Inhibition of *Salmonella enteritidis* and *Staphylococcus aureus* in nutrient broth by mint EO. *Food Res. Intern.*, 33: 273–280.

Teuscher, E., M.F. Melzig, and U. Lindequist, 2004. Biogene Arzneimittel: Ein Lehrbuch der pharmazeutischen Biologie; 14 Tabellen. *Wiss. Verlag-Ges*: Stuttgart.

Thompson, D.P., 1989. Fungitoxic activity of EO components on food storage fungi. *Mycologia*, 81: 151–153.

Tripathi, A.K., S. Upadhyay, M. Bhuiyan, and P. Bhattacharya, 2009. A review on prospects of EOs as biopesticide in insect-pest management. *J. Pharmacog. Phytother.*, 1: 52–63.

TRM Market Biopesticides, 2018. Global Biopesticides Market: Rising Awareness about Environmentally Viable Agriculture Spikes Demand, TMR finds (Feb 2018), https://www.transparencymarketresearch.com/pressrelease/biopesticides-market.htm (accessed 15 August 2018).

Tunç, İ., B.M. Berger, F. Erler, and F. Dağlı, 2000. Ovicidal activity of EOs from five plants against two stored-product insects. *J. Stored Products Res.*, 36: 161–168.
Tworkoski, T., 2002. Herbicide effects of EOs. *Weed Sci.*, 50: 425–431.
Tzortzakis, N.G., and C.D. Economakis, 2007. Antifungal activity of lemongrass (*Cympopogon citratus* L.\) EO against key postharvest pathogens. *Innov. Food Sci. Emer. Technol.*, 8: 253–258.
United states of Food and Drug Administration (US-FDA). 2012. *Bad Bug Book: Foodborne Pathogenic Microorganisms and Natural Toxins*. Lampel K.A., Al-Khaldi S., Cahill S.M. eds., 2nd edition.
Ushir, Y., A. Tatiya, S. Surana, and U. Patil, 2010. Gas chromatography-mass spectrometry analysis and antibacterial activity of EO from aerial parts and roots of *Anisomeles indica* Linn. *Intern. J. Green Pharmacy*, 4: 98.
Verdeguer, M., M.A. Blázquez, and H. Boira, 2009. Phytotoxic effects of *Lantana camara*, *Eucalyptus camaldulensis* and *Eriocephalus africanus* EOs in weeds of Mediterranean summer crops. *Biochem. Syst. Ecol.*, 37: 362–369.
Verdeguer, M., D. García-Rellán, H. Boira, E. Pérez, S. Gandolfo, and M.A. Blázquez, 2011. Herbicidal activity of *Peumus boldus* and *Drimys winterii* EOs from Chile. *Molecules*, 16: 403–411.
Vidal R., J. Muchembled, C., Deweer, L. Tournant, N. Corroyer, S. Flammier, 2018. Évaluation de l'intérêt de l'utilisation d'huiles essentielles dans des stratégies de protection des cultures. *Innov. Agron.*, 63(2018), 191–210.
Vokou, D., D. Chalkos, G. Karamanlidou, and M. Yiangou, 2002, Activation of soil respiration and shift of the microbial population balance in soil as a response to *Lavandula stoechas* EO. *J. Chem. Ecol.*, 28: 755–768.
Vokou, D., and S. Liotiri, 1999. Stimulation of soil microbial activity by EOs. *Chemo. Ecol.*, 9: 41–45.
Walderhaug, M., 2014. *Bad Bug Book: Foodborne Pathogenic Microorganisms and Natural Toxins Handbook*. Createspace Independent Pub.
Wang, J., F. Zhua, X.M. Zhoua, C.Y. Niua, C.L. Leia, 2006. Repellent and fumigant activity of EO from *Artemisia vulgaris* to *Tribolium castaneum* (Herbst) (Coleoptera: Tenebrionidae). *J. Stored Prod. Res.*, 42(3): 339–347.
Wilson, C.L., J.M. Solar, A. El Ghaouth, and M.E. Wisniewski, 1997. Rapid evaluation of plant extracts and EOs for antifungal activity against botrytis cinerea. *Plant Dis.*, 81: 204–210.

25 Essential Oils Used in Veterinary Medicine

K. Hüsnü Can Başer and Chlodwig Franz

CONTENTS

25.1 INTRODUCTION

Essential oils are volatile constituents of aromatic plants. These liquid oils are generally complex mixtures of terpenoid and/or nonterpenoid compounds. Mono-, sesqui-, and sometimes diterpenoids, phenylpropanoids, fatty acids and their fragments, benzenoids, and so on may occur in various essential oils (Baser and Demirci, 2007).

Except for citrus oils obtained by cold pressing, all other essential oils are obtained by distillation. Products obtained by solvent extraction or supercritical fluid extraction are not technically considered as essential oils (Baser, 1995).

Essential oils are used in perfumery, food flavoring, pharmaceuticals, and sources of aromachemicals.

Essential oils exhibit a wide range of biological activities and 31 essential oils have monographs in the latest edition of the *European Pharmacopoeia* (Table 25.1).

Antimicrobial activities of many essential oils are well documented (Bakkali et al., 2008). Such oils may be used singly or in combination with one or more oils. For the sake of synergism this may be necessary.

Although many are generally regarded as safe, essential oils are generally not recommended for internal use. However, their much diluted forms (e.g., hydrosols) obtained during oil distillation as a by-product may be taken orally.

TABLE 25.1
Essential Oil Monographs in the *European Pharmacopoeia* (6.5 Edition, 2009)

English Name	Latin Name	Plant Name
Anise oil	*Anisi aetheroleum*	*Pimpinella anisum* L. fruits
Bitter-fennel fruit oil	*Foeniculi amari fructus aetheroleum*	*Foeniculum vulgare* Miller subsp. *vulgare* var. *vulgare*
Bitter-fennel herb oil	*Foeniculi amari herba aetheroleum*	*Foeniculum vulgare* Miller subsp. *vulgare* var. *vulgare*
Caraway oil	*Carvi aetheroleum*	*Carum carvi* L.
Cassia oil	*Cinnamomi cassiae aetheroleum*	*Cinnamomum cassia* Blume (*Cinnamomum aromaticum* Nees)
Cinnamon bark oil, Ceylon	*Cinnamomi zeylanici corticis aetheroleum*	*Cinnamomum zeylanicum* Nees
Cinnamon leaf oil, Ceylon	*Cinnamomi zeylanici folium aetheroleum*	*Cinnamomum verum* J.S. Presl.
Citronella oil	*Citronellae aetheroleum*	*Cymbopogon winterianus* Jowitt
Clary sage oil	*Salviae sclareae aetheroleum*	*Salvia sclarea* L.
Clove oil	*Caryophylli aetheroleum*	*Syzygium* aromaticum (L.) Merill et L.M. Perry (*Eugenia caryophyllus* C.S. Spreng. Bull. et Harr.)
Coriander oil	*Coriandri aetheroleum*	*Coriandrum sativum* L.
Dwarf pine oil	*Pini pumilionis aetheroleum*	*Pinus mugo* Turra.
Eucalyptus oil	*Eucalypti aetheroleum*	*Eucalyptus globulus* Labill.
Juniper oil	*Juniperi aetheroleum*	*Juniperus communis* L. meyvesi
Lavender oil	*Lavandulae aetheroleum*	*Lavandula angustifolia* P. Mill. (*Lavandula officinalis* Chaix.)
Lemon oil	*Lemonis aetheroleum*	*Citrus limon* (L.) Burman fil.
Mandarin oil	*Citri reticulatae aetheroleum*	*Citrus reticulata* Blanco
Matricaria oil	*Matricariae aetheroleum*	*Matricaria recutita* L. (*Chamomilla recutita* (L.) Rauschert)
Mint oil, partly dementholized	*Menthae arvensis aetheroleum partim mentholi privum*	*Mentha canadensis* L. (*Mentha arvensis* L. var. *glabrata* (Benth.) Fern, *Mentha arvensis* L. var. *piperascens* Malinv. ex Holmes) *Japanese mint*
Neroli oil (formerly bitter-orange flower oil)	*Neroli aetheroleum* (formerly *Aurantii amari floris aetheroleum*)	*Citrus aurantium* L. subsp. *aurantium* (*Citrus aurantium* L. subsp. *amara* Engl.)
Nutmeg oil	*Myristicae fragrantis aetheroleum*	*Myristica fragrans* Houtt.
Peppermint oil	*Menthae piperitae aetheroleum*	*Mentha* × *piperita* L.
Pine silvestris oil	*Pini silvestris aetheroleum*	*Pinus silvestris* L.
Rosemary oil	*Rosmarini aetheroleum*	*Rosmarinus officinalis* L.
Spanish sage oil	*Salviae lavandulifoliae aetheroleum*	*Salvia lavandulifolia* Vahl.
Spike lavender oil	*Spicae aetheroleum*	*Lavandula latifolia* Medik.
Star anise oil	*Anisi stellati aetheroleum*	*Illicium verum* Hooker fil.
Sweet orange oil	*Aurantii dulcis aetheroleum*	*Citrus sinensis* (L.) Osbeck (*Citrus aurantium* L. var. *dulcis* L.)
Tea tree oil	*Melaleucae aetheroleum*	*Melaleuca alternifolia* (Maiden et Betch) Cheel, *Melaleuca linariifolia* Smith, *Melaleuca dissitiflora* F. Mueller, and other species
Thyme oil	*Thymi aetheroleum*	*Thymus vulgaris* L., *T. zygis* L.
Turpentine oil, *Pinus pinaster* type	*Terebinthini aetheroleum ab pinum pinastrum*	*Pinus pinaster* Aiton. (*Maritime pine*)

Topical applications of some essential oils (e.g., oregano and lavender) in wounds and burns bring about fast recovery without leaving any sign of cicatrix. By inhalation, several essential oils act as a mood changer and have effect especially on respiratory conditions.

Several essential oils (e.g., citronella oil) have been used as pest repellents or as insecticides and such uses are frequently encountered in veterinary applications.

In recent years, especially after the ban on the use of antibiotics in animal feed in the European Union since January 2006, essential oils have emerged as a potential alternative to antibiotics in animal feed.

Essential oils used in veterinary medicine may be classified as follows:

1. Oils attracting animals
2. Oils repelling animals
3. Insecticidal, pest repellent, and antiparasitic oils
4. Oils used in animal feed
5. Oils used in treating diseases in animals.

25.2 OILS ATTRACTING ANIMALS

Valeriana oils (and valerianic and isovalerianic acids) and nepeta oils (and nepetalactones) are well-known feline-attractant oils. Their odor attracts male cats.

Douglas fir oil and its monoterpenes have been claimed to attract deer and wild boar (Buchbauer et al., 1994).

Dogs are normally drawn to floral oils and usually choose to take these by inhalation only. Monoterpene-rich oils are usually too strong for dogs, with the exception of bergamot, *Citrus bergamia*.

Cats also usually select only floral oils for inhalation. Cats do not have metabolic mechanism to break down essential oils due to the lack of the enzyme glucuronidase. Therefore, they should not be taken by mouth and should not be generally applied topically (Ingraham, 2008).

25.3 OILS REPELLING ANIMALS

Peppermint oil (*Mentha piperita*) repels mice. It can be applied under the sink in the kitchen or applied in staples to prevent mice annoying horses and livestock. A few drops of peppermint oil in a bucket of water used to scrub out a stall and sprinkling a few drops around the perimeter and directly on straw or bedding are said to eliminate or severely curtail the habitation of mice (Scents and Sensibility, 2001).

A patent (U.S. Patent 4,961,929) claims that a mixture of methyl salicylate, birch oil, wintergreen oil, eucalyptus oil, pine oil, and pine-needle oil repels dogs.

Another patent (U.S. Patent 4,735,803) claims the same using lemon oil and α-terpinyl methyl ether.

Another similar formulation (U.S. Patent 4,847,292) claims that a mixture of citronellyl nitrile, citronellol, α-terpinyl methyl ether, and lemon oil repels dogs.

A mixture of black pepper and capsicum oils and the oleoresin of rosemary is claimed to repel animals (U.S. Patent 6,159,474).

Citronella oil repels cats and dogs (Moschetti, 2003).

Repellents alleged to repel cats include allyl isothiocyanate (oil of mustard), amyl acetate, anethole, capsaicin, cinnamaldehyde, citral, citronella, citrus oil, eucalyptus oil, geranium oil, lavender oil, lemongrass oil, menthol, methyl nonyl ketone, methyl salicylate, naphthalene, nicotine, paradichlorobenzene, and thymol. Oil of mustard, cinnamaldehyde, and methyl nonyl ketone are said to be the most potent.

Essential oils comprised of 10 g/L solutions of cedarwood, cinnamon, sage, juniper berry, lavender, and rosemary; all of these were potent snake irritants. Brown tree snakes exposed to a 2s burst of aerosol of these oils exhibited prolonged, violent undirected locomotory behavior. In contrast, exposure to a 10 g/L concentration of ginger oil aerosol caused snakes to locomote, but in a deliberate, directed manner. The 10 g/L solutions delivered as aerosols of *m*-anisaldehyde, *trans*-anethole, 1,8-cineole, cinnamaldehyde, citral, ethyl phenylacetate, eugenol, geranyl acetate, or methyl salicylate acted as potent irritants for brown tree snakes (*Boiga irregularis*) (Clark and Shivik, 2002).

25.4 OILS AGAINST PESTS

25.4.1 Insecticidal, Pest Repellent, and Antiparasitic Oils

The essential oil of bergamot (*C. bergamia*), anise (*Pimpinella anisum*), sage (*Salvia officinalis*), tea tree (*Melaleuca alternifolia*), geranium (*Pelargonium* sp.), peppermint (*M. piperita*), thyme (*Thymus vulgaris*), hyssop (*Hyssopus officinalis*), rosemary (*Rosmarinus officinalis*), and white clover (*Trifolium repens*) can be used to control certain pests on plants. They have been shown to reduce the number of eggs laid and the amount of feeding damage by certain insects, particularly lepidopteran caterpillars. Sprays made from tansy (*Tanacetum vulgare*) have demonstrated a repellent effect on imported cabbageworm on cabbage, reducing the number of eggs laid on the plants. Teas made from wormwood (*Artemisia absinthium*) or nasturtiums (*Nasturtium* spp.) are reputed to repel aphids from fruit trees, and sprays made from ground or blended catnip (*Nepeta cataria*), chives (*Allium schoenoprasum*), feverfew (*Tanacetum parthenium*), marigolds (*Calendula*, *Tagetes*, and *Chrysanthemum* spp.), or rue (*Ruta graveolens*) have also been used by gardeners against pests that feed on leaves (Moschetti, 2003).

25.4.2 Fleas and Ticks

Dogs, cats, and horses are plagued by fleas and ticks. One to two drops of citronella or lemongrass oils added to the shampoo will repel these pests. Alternatively, 4–5 drops of cedarwood oil and pine oil is added to a bowl of warm water, and a bristle hairbrush is soaked with this solution to brush the pet down with it. Eggs and parasites gathered in the brush are rinsed out. This is repeated several times. This solution can be used similarly for livestock after adding citronella and lemongrass oils to this mixture.

Flea collar can be prepared by a mixture of cedarwood (*Juniperus virginiana*), lavender (*Lavandula angustifolia*), citronella (*Cymbopogon winterianus* [Java]), thyme oils, and 4–5 garlic (*Allium sativum*) capsules. This mixture is thinned with a teaspoonful of ethanol and soaked with a collar or a cotton scarf. This is good for 30 days (Scents and Sensibility, 2001).

Ticks can be removed by applying 1 drop of cinnamon or peppermint oil on Q-tip by swabbing on it.

Carvacrol-rich oil (64%) of *Origanum onites* and carvacrol was found to be effective against the tick *Rhipicephalus turanicus*. Pure carvacrol killed all the ticks following 6 h of exposure, while 25% and higher concentrations of the oil were effective in killing the ticks by the 24 h posttreatment (Coskun et al., 2008).

25.4.3 Mosquitoes

Catnip oil (*N. cataria*) containing nepetalactones can be used effectively as a mosquito repellent. It is said to be 10 times more effective than DEET (*N,N*-diethyl-meta-toluamide) (Moschetti, 2003). *Juniperus communis* berry oil is a very good mosquito repellent. Ocimum volatile oils including camphor, 1,8-cineole, methyl eugenol, limonene, myrcene, and thymol strongly repelled mosquitoes (Regnault-Roger, 1997).

Citronella oil repels mosquitoes, biting insects, and fleas.

Essential oils of *Zingiber officinale* and *R. officinalis* were found to be ovicidal and repellent, respectively, toward three mosquito species (Prajapati et al., 2005). Root oil of *Angelica sinensis* and ligustilide was found to be mosquito repellent (Wedge et al., 2009).

25.4.4 Moths

Cedarwood oil is used in mothproofing. A large number of patents have been assigned to the preservation of cloths from moths and beetles: application of a solution containing clove (*Syzygium aromaticum*) essential oil on woolen cloth, filter paper containing *Juniperus rigida* oil, and tablets of *p*-dichlorobenzene mixed with essential oils to be placed in wardrobe.

25.4.5 Aphids, Caterpillars, and Whiteflies

25.4.5.1 Garlic Oil

Essential oils are effective in insect pest control (Regnault-Roger, 1997).

25.4.6 Ear Mites

Peppermint oil is applied to a Q-tip and swabbed inside of the ear.

25.4.7 Antiparasitic

There is a patent (U.S. Patent 6,800,294) on an antiparasitic formulation comprising eucalyptus oil (*Eucalyptus globulus*), cajeput oil (*Melaleuca cajuputi*), lemongrass oil, clove bud oil (*S. aromaticum*), peppermint oil (*M. piperita*), piperonyl, and piperonyl butoxide. The formulation can be used for treating an animal body, in the manufacture of a medicament for treating ectoparasitic infestation of an animal, or for repelling parasites.

Two essential oils derived from *L. angustifolia* and *Lavandula* × *intermedia* were investigated for any antiparasitic activity against the human protozoal pathogens *Giardia duodenalis* and *Trichomonas vaginalis* and the fish pathogen *Hexamita inflata*, all of which have significant infection and economic impacts. The study has demonstrated that low ($\leq$1%) concentrations of *L. angustifolia* and *Lavandula* × *intermedia* oil can completely eliminate *T. vaginalis*, *G. duodenalis*, and *H. inflata in vitro*. At 0.1% concentration, *L. angustifolia* oil was found to be slightly more effective than *Lavandula* × *intermedia* oil against *G. duodenalis* and *H. inflata* (Moon et al., 2006).

The antiparasitic properties of essential oils from *A. absinthium*, *Artemisia annua*, and *Artemisia scoparia* were tested on intestinal parasites *Hymenolepis nana*, *Lamblia intestinalis*, *Syphacia obvelata*, and *Trichocephalus muris* (*Trichuris muris*). Infested white mice were injected with 0.01 mL/g of the essential oils (6%) twice a day for 3 days. The effectiveness of the essential oils was observed in 70%–90% of the tested animals (Chobanov et al., 2004).

Parasites, such as head lice and scabies, as well as internal parasites, are repelled by oregano oil (86% carvacrol). The oil can be added to soaps, shampoos, and diluted in olive oil for topical applications. By taking a few drops daily under the tongue, one can gain protection from waterborne parasites, such as *Cryptosporidium* and *Giardia*. Internal dosages also are effective in killing parasites in the body (http://curingherbs.com/wild_oregano_oil.htm) (Foster, 2002).

Essential oils from *Pinus halepensis*, *Pinus brutia*, *Pinus pinaster*, *Pinus pinea*, and *Cedrus atlantica* were tested for molluscicidal activity against *Bulinus truncatus*. The oil from *C. atlantica* was found the most active (LC 50 = 0.47 ppm). Among their main constituents, α-pinene, β-pinene, and myrcene exhibited potent molluscicidal activity (LC 50 = 0.49, 0.54, and 0.56 ppm, respectively). These findings have important application of natural products in combating schistosomiasis (Lahlou, 2003).

Origanum essential oils have exhibited differential degrees of protection against myxosporean infections in gilthead and sharpsnout sea bream tested in land-based experimental facilities (Athanassopoulou et al., 2004a,b).

25.5 ESSENTIAL OILS USED IN ANIMAL FEED

Essential oils can be used in feed as appetite stimulant, stimulant of saliva production, gastric and pancreatic juice production enhancer, and antimicrobial and antioxidant to improve broiler performance. Antimicrobial effects of essential oils are well documented. Essential oils due to their potent nature should be used as low as possible levels in animal nutrition. Otherwise, they can lead to feed intake reduction, gastrointestinal tract (GIT) microflora disturbance, or accumulation in animal tissues and products. Odor and taste of essential oils may contribute to feed refusal; however, encapsulation of essential oils could solve this problem (Gauthier, 2005).

Generally, Gram-positive bacteria are considered more sensitive to essential oils than Gram-negative bacteria because of their less complex membrane structure (Lis-Balchin, 2003).

Carvacrol, the main constituent of oregano oils, is a powerful antimicrobial agent (Baser, 2008). It asserts its effect through the biological membranes of bacteria. It acts through inducing a sharp reduction of the intercellular adenosine triphosphate (ATP) pool through the reduction of ATP synthesis and increased hydrolysis. Reduction of the membrane potential (transmembrane electrical potential), which is the driving force of ATP synthesis, makes the membrane more permeable to protons. A high level of carvacrol (1 mM) decreases the internal pH of bacteria from 7.1 to 5.8 related to ion gradients across the cell membrane. One millimolar of carvacrol reduces the internal potassium (K) level of bacteria from 12 mmol/mg of cell protein to 0.99 mmol/mg in 5 min. K plays a role in the activation of cytoplasmic enzymes and in maintaining osmotic pressure and in the regulation of cytoplasmic pH. K efflux is a solid indication of membrane damage (Ultee et al., 1999).

It has been shown that the mode of action of oregano oils is related to an impairment of a variety of enzyme systems, mainly involved in the production of energy and the synthesis of structural components. Leakage of ions, ATP, and amino acids also explains the mode of action. Potassium and phosphate ion concentrations are affected at levels below the minimum inhibitory concentration (MIC) concentration (Lambert et al., 2001).

25.5.1 RUMINANTS

A recent review compiled information on botanicals including essential oils used in ruminant health and productivity (Rochfort et al., 2008). Unfortunately, there are few reports on the effects of essential oils and natural aromachemicals on ruminants. It was demonstrated that the consumption of terpene volatiles such as camphor and α-pinene in "tarbush" (*Flourensia cernua*) effected feed intake in sheep (Estell et al., 1998). In vitro and *in vivo* antimicrobial activities of essential oils have been demonstrated in ruminants (Elgayyar et al., 2001; Wallace et al., 2002; Cardozo et al., 2005; Moreira et al., 2005). Synergistic antinematodal effects of essential oils and lipids were demonstrated (Ghisalberti, 2002). Other nematocidal volatiles reported are as follows: benzyl isothiocyanate (goat), ascaridole (goat and sheep) (Ghisalberti, 2002; Githiori et al., 2006), geraniol, eugenol (Chitwood, 2002; Githiori et al., 2006), menthol, and 1,8-cineole (Chitwood, 2002).

Methylsalicylate, the main component of the essential oil of *Gaultheria procumbens* (wintergreen), is topically used as emulsion in cattle, horses, sheep, goats, and poultry in the treatment of muscular and articular pain. The recommended dose is 600 mg/kg bw twice a day. The duration of treatment is usually less than 1 week (EMA, 1999). It is included in Annex II of Council Regulation (EEC) No. 2377/90 as a substance that does not need an maximum residue limit (MRL) level. *G. procumbens* should not be used as flavoring in pet food since salicylates are toxic to dogs and cats. As cats metabolize salicylates much more slowly than other species, they are more likely to be overdosed. Use of methylsalicylate in combination with anticoagulants such as warfarin can result in adverse interactions and bleedings (Chow et al., 1989; Yip et al., 1990; Ramanathan, 1995; Tam et al., 1995).

The essential oil of *L. angustifolia* (*Lavandulae aetheroleum*) is used in veterinary medicinal products for topical use together with other plant extracts or essential oils for antiseptic and healing purposes. The product is used in horses, cattle, sheep, goats, rabbits, and poultry. It is included in

Annex II of Council Regulation (EEC) No. 2377/90 as a substance that does not need an MRL level (EMEA, 1999; Franz et al., 2005).

The outcomes of *in vitro* studies investigating the potential of *P. anisum* essential oil as a feed additive to improve nutrient use in ruminants are inconclusive, and more and larger preferably *in vivo* studies are necessary for evaluation of efficacy (Franz et al., 2005).

Carvacrol, carvone, cinnamaldehyde, cinnamon oil, clove bud oil, eugenol, and oregano oil have resulted in a 30%–50% reduction in ammonia N concentration in diluted ruminal fluid with a 50:50 forage concentrate diet during the 24 h incubation (Busquet et al., 2006).

Carvacrol has been suggested as a potential modulator of ruminal fermentation (Garcia et al., 2007).

25.5.2 Poultry

25.5.2.1 Studies with CRINA® Poultry

Dietary addition of essential oils in a commercial blend (CRINA Poultry) showed a decreased *Escherichia coli* population in ileocecal digesta of broiler chickens. Furthermore, in high doses, a significant increase in certain digestive enzyme activities of the pancreas and intestine was observed in broiler chickens (Jang et al., 2007).

In another study, CRINA Poultry was shown to control the colonization of the intestine of broilers with *Clostridium perfringens*, and the stimulation of animal growth was put down to this development (Losa, 2001).

Commercial essential oil blends CRINA Poultry and CRINA Alternate were tested in broilers infected with viable oocysts of mixed *Eimeria* spp. It was concluded that these essential oil blends may serve as an alternative to antibiotics and/or ionophores in mixed *Eimeria* infections in non-cocci-vaccinated broilers, but no benefit of essential oil supplementation was observed for vaccinated broilers against coccidia (Oviedo-Rondon et al., 2006).

25.5.2.1.1 Other Studies

Supplementation of 200 ppm essential oil mixture (EOM) that included oregano, clove, and anise oils (no species name or composition given!) in broiler diets was said to significantly improve the daily live weight gain and feed conversion ratio (FCR) during a growing period of 5 weeks (Ertas et al., 2006). Similar results were obtained with 400 mg/kg anise oil (composition not known!) (Ciftci et al., 2005).

A total of 50 and 100 mg/kg of feed of oregano oil* were tested on broilers. No growth-promoting effect was observed. At 100 mg/kg of feed, antioxidant effect was detected on chicken tissues (Botsoglou et al., 2002a).

Positive results were also reported for oregano oil added in poultry feed (Bassett, 2000).

Antioxidant activities of rosemary and sage oils on lipid oxidation of broiler meat have been shown. Following dietary administration of rosemary and sage oils to the live birds, a significant inhibition of lipid peroxidation was reported in chicken meat stored for 9 days (Lopez-Bote et al., 1998). A dietary supplementation of oregano essential oil (300 mg/kg) showed a positive effect on the performance of broiler chickens experimentally infected with *Eimeria tenella*. Throughout the experimental period of 42 days, oregano essential oil exerted an anticoccidial effect against *E. tenella*, which was, however, lower than that exhibited by lasalocid. Supplementation with dietary oregano oil to *E. tenella*–infected chickens resulted in body weight gains and FCRs not differing from the noninfected group, but higher than those of the infected control group and lower than those of chickens treated with the anticoccidial lasalocid (Giannenas et al., 2003).

Inclusion of oregano oil at 0.005% and 0.01% in chicken diets for 38 days resulted in a significant antioxidant effect in raw and cooked breast and thigh muscle stored up to 9 days in refrigerator (Botsoglou et al., 2002b).

---

* Oregano essential oil was in the form of a powder called Orego-Stim. This product contains 5% oregano essential oil (Ecopharm Hellas S.A., Kilkis, Greece) and 95% natural feed grade inert carrier. The oil of *Origanum vulgare* subsp. *hirtum* used in this product contains 85% carvacrol and thymol.

Oregano oil (55% carvacrol) exhibited a strong bactericidal effect against lactobacilli and following the oral administration of the oil MIC values of amikacin, apramycin, and streptomycin and neomycin against *E. coli* strains increased (Horosova et al., 2006).

An *in vitro* assay measuring the antimicrobial activity of essential oils of *Coridothymus capitatus*, *Satureja montana*, *Thymus mastichina*, *Thymus zygis*, and *Origanum vulgare* was carried out against poultry origin strains of *E. coli*, *Salmonella enteritidis*, and *Salmonella essen* and pig origin strains of enterotoxigenic *E. coli*, *Salmonella choleraesuis*, and *Salmonella typhimurium*. *O. vulgare* (MIC $\leq$ 1% v/v) oil showed the highest antimicrobial activity against the four strains of *Salmonella*. It was followed by *T. zygis* oil (MIC $\leq$ 2% v/v). *T. mastichina* oil inhibited all the microorganisms at the highest concentration, 4% (v/v). Monoterpenic phenols carvacrol and thymol showed higher inhibitory capacity than the monoterpenic alcohol linalool. The results confirmed potential application of such oils in the treatment and prevention of poultry and pig diseases caused by *Salmonella* (Penalver et al., 2005).

In another study, groups of male, 1-day-old Lohmann broilers were given maize–soya bean meal diets, with oils extracted from thyme, mace, and caraway or coriander, garlic, and onion (0, 20, 40, and 80 mg/kg) for 6 weeks. The average daily gain and FCR were not different between the broilers fed with the different oils; meat was not tainted with flavor or smell of the oils (Vogt and Rauch, 1991).

25.5.2.2 Studies with Herbromix®

Essential oils from oregano herb (*O. onites*), laurel leaf (*Laurus nobilis*), sage leaf (*Salvia fruticosa*), fennel fruit (*Foeniculum vulgare*), myrtle leaf (*Myrtus communis*), and citrus peel (rich in limonene) were mixed and formulated as feed additive after encapsulation. It is marketed in Turkey as poultry feed under the name Herbromix.

The following three *in vivo* experiments with this product were recently accomplished.

25.5.2.2.1 In Vivo Experiment 1

In this study, 1250 sexed 1-day-old broiler chicks obtained from a commercial hatchery were randomly divided into five treatment groups of 250 birds each (negative control, antibiotic, and essential oil combination [EOC] at 3 levels). Each treatment group was further subdivided into five replicates of 50 birds (25 males and 25 females) per replicate. Commercial EOC at three different levels (24, 48, and 72 mg) and antibiotic (10 mg avilamycin) per kg were added to the basal diet. There were significant effects of dietary treatments on body weight, feed intake (except at day 42), FCR, and carcass yield at 21 and 42 days. Body weights were significantly different between the treatments. Birds fed on diet containing 48 mg essential oil/kg being the highest, and this treatment was followed by chicks fed on the diet containing 72 mg essential oil/kg, antibiotic, negative control, and 24 mg essential oil/kg at day 42.

Supplementation with 48 mg EOC/kg to the broiler diet significantly improved the body weight gain, FCR, and carcass yield compared to other dietary treatments on 42 days of age. EOC may be considered as a potential growth promoter in the future of the new era, which agrees with producer needs for increased performance and today's consumer demands for environment-friendly broiler production. The EOC can be used cost-effectively when its cost is compared with antibiotics and other commercially available products in the market.

25.5.2.2.2 In Vivo Experiment 2

In this study, 1250 sexed 1-day-old broiler chicks were randomly divided into five treatment groups of 250 birds each (negative control, organic acid, probiotic, and EOC at two levels). Each treatment group was further subdivided into five replicates of 50 birds (25 males and 25 females) per replicate. The oils in the EOC were extracted from different herbs growing in Turkey. The organic acid at 2.5 g/kg diet, the probiotic at 1 g/kg diet, and the EOC at 36 and 48 mg/kg diet were added to the basal diet.

The results obtained from this study indicated that the inclusion of 48 mg EOC/kg broiler diet significantly improved the body weight gain, FCR, and carcass yield of broilers compared to organic acid and probiotic treatments after a growing period of 42 days. The EOC may be considered as a potential growth promoter like organic acids and probiotics for environment-friendly broiler production.

25.5.2.2.3 In Vivo Experiment 3

The aim of this study was to examine the effect of essential oils and breeder age on growth performance and some internal organs' weight of broilers. A total of 1008 unsexed 1-day-old broiler chicks (Ross-308) originating from young (30 weeks) and older (80 weeks) breeder flocks were randomly divided into three treatment groups of 336 birds each, consisting of control and two EOMs at a level of 24 and 48 mg/kg diet. There were no significant effects of dietary treatments on body weight gain of broilers at days 21 and 42.

On the other hand, there were significant differences on the feed intake at days 21 and 42. The addition of 24 or 48 mg/kg EOM to the diet reduced significantly the feed intake compared to the control. The groups fed with the added EOM had significantly better FCR than the control at days 21 and 42. Although there was no significant effect of broiler breeder age on body weight gain at day 21, significant differences were observed on body weight gain at 42 days of age. Broilers originating from young breeder flock had significantly higher body weight gain than those originating from old breeder flock at 42 days of age. No difference was noticed for carcass yield, liver, pancreas, proventriculus, gizzard, and small intestine weight. Supplementation with EOM to the diet in both levels significantly decreased mortality at days 21 and 42.

The results indicated that the Herbromix may be considered as a potential growth promoter. However, more trials are needed to determine the effect of essential oil supplementation to diet on the performance of broilers with regard to variable management conditions including different stress factors, essential oils and their optimal dietary inclusion levels, active substances of oils, dietary ingredients, and nutrient density (Alcicek et al., 2003, 2004; Bozkurt and Baser, 2002a,b; Cabuk et al., 2006a,b).

25.5.3 Pigs

CRINA® Pigs was tested on pigs. The results for the first 21-day period showed that males grew faster, ate less, and exhibited superior FCR compared to females. Although female carcass weight was higher, males had a significantly lower carcass fat than females (Losa, 2001).

The addition of fennel (*F. vulgare*) and caraway (*Carum carvi*) oils was not found beneficial for weaned piglets. In feed choice conditions, fennel oil caused feed aversion (Schoene et al., 2006).

Oregano oil was found to be beneficial for piglets (Molnar and Bilkei, 2005).

In a preliminary investigation, the effects of low-level dietary inclusion of rosemary, garlic, and oregano oils on pig performance and pork quality were carried out. Unfortunately, no information on the species from which the oils were obtained and their composition existed in the paper. The pigs appeared to prefer the garlic-treated diet, and the feed intake and the average daily gain were significantly increased although no difference in the feed efficiency was observed. Carcass and meat quality attributes were unchanged, although a slight reduction of lipid oxidation was noted in oregano-fed pork. Since the composition of the oils is not clear, it is not possible to evaluate the results (Janz et al., 2007).

A study revealed that the inclusion of essential oil of oregano in pigs' diet significantly improved the average daily weight gain and FCR of the pigs. Pigs fed with the essential oils had higher carcass weight, dressing percentage, and carcass length than those fed with the basal and antibiotic-supplemented diet. The pigs that received the essential oil supplementation had a significantly lower fat thickness. Also lean meat and ham portions from these pigs were significantly higher. Therefore, the use of *Origanum* essential oil as feed additive improves the growth of pigs and has greater positive effects on carcass composition than antibiotics (Onibala et al., 2001).

Ropadiar®, an essential oil of the oregano plant, was supplemented in the diet of weaning pigs as alternative for antimicrobial growth promoters (AMGPs), observing its efficacy on the performance of the piglets. Ropadiar liquid contains 10% oregano oil and has been designed to be added to water. Compared to the negative control (without AMGP), Ropadiar improved performance only during the first 14 days after weaning. Based on the results of this trial, it cannot be argued about the usefulness of Ropadiar as an alternative for AMGP in diets of weanling pigs. However, its addition in prestarter diets could improve performance of these animals (Krimpen and Binnendijk, 2001).

The objective of another trial was to ascertain the effect on nutrient digestibilities and N-balance, as well as on parameters of microbial activity in the GIT of weaned pigs after adding oregano oil to the feed. The apparent digestibility of crude nutrients (except fiber) and the N-balance of the weaned piglets in this study were not influenced by feeding piglets restrictively with this feed additive. By direct microbiological methods, no influence of the additive on the gut flora could be found (Moller, 2001).

The inclusion of essential oil of spices in the pigs' diet significantly improved the average daily weight gain and FCR of the pigs in Groups 3, 4, and 5, as compared to Groups 1 and 2 ($P < 0.01$). Furthermore, pigs fed with the essential oils had higher carcass weight ($P < 0.01$), dressing percentage ($P < 0.01$), and carcass length ($P < 0.01$) than those fed with the basal and antibiotic-supplemented diet. In Groups 3, 4, and 5, backfat thickness was significantly lower than those in Groups 1 and 2. Moreover, lean meat and ham portions from pigs in Groups 3, 4, and 5 were significantly higher than those from pigs in Groups 1 and 2. In conclusion, the use of essential oils as feed additives improves the growth of pigs and has greater positive effects on carcass composition than antibiotics (Onibala et al., 2001).

25.6 ESSENTIAL OILS USED IN TREATING DISEASES IN ANIMALS

There is scarce scientific information on the use of essential oils in treating diseases in animals. Generally, the oils used in treating diseases in humans are also recommended for animals.

Internet literature is abound with valid and/or suspicious information in this issue. We have tried to compile relevant information using the reachable resources. The information may not be concise or comprehensive but should be seen as an effort to combine the available information in a short period of time.

The oil of *Ocimum basilicum* has been reported as an expectorant in animals. The combined oils of *Ocimum micranthum* and *Chenopodium ambrosioides* are claimed to treat stomachache and colic in animals (Cornell University Animal Science, 2009).

Bad breath as a result of gum disease and bacterial buildup on the teeth of pets can be treated by brushing their teeth with a mixture of a couple of tablespoons of baking soda, 1 drop of clove oil, and 1 drop of aniseed oil. Lavender, myrrh, and clove oils can also be directly applied to their gums.

For wounds, abscesses, and burns, lavender and tea tree oils are used by topical application. Skin rashes can be treated with tea tree, lavender, and chamomile oils.

Earache of pets can be healed by dripping a mixture of lavender, chamomile, and tea tree oils (1 drop each) dissolved in a teaspoonful of grape-seed or olive oil in the infected ears.

Hoof rot in livestock can be treated with a hot compress made up of 10 drops of chamomile, 15 drops of thyme, and 5 drops of melissa oils diluted in about 100 mL of vegetable oil (e.g., grape-seed oil).

Intestinal worms of horses can be expelled by applying 3–4 drops of thyme oil and tansy leaves to each feed. Melissa oil can be added to feed to increase milk production of both cows and goats (http://scentsnsensibility.com/newsletter/Apr0601.htm).

Aromatic plants such as *Pimpinella isaurica*, *Pimpinella aurea*, and *Pimpinella corymbosa* are used as animal feed to increase milk secretion in Turkey (Tabanca et al., 2003).

To calm horses, chamomile oil is added to their feed. Pneumonia in young elephants caused by *Klebsiella* is claimed to be healed by *Lippia javanica* oil. Rose and yarrow oils bring about emotional release in donkeys by licking them. Wounds in horses are treated with *Achillea millefolium* oil; sweet itch is treated with peppermint oil. *Matricaria recutita* and *A. millefolium* oils are used to heal the skin with inflammatory conditions (Ingraham, 2008).

A study evaluated the effect of dietary oregano etheric oils as nonspecific immunostimulating agents in growth-retarded, low-weight growing–finishing pigs. A group of pigs were fed with commercial fattening diet supplemented with 3000 ppm oregano additive (Oregpig®, Pecs, Hungary), composed of dried leaf and flower of *O. vulgare*, enriched with 500 g/kg cold-pressed essential oils of the leaf and flower of *O. vulgare*, and composed of 60 g carvacrol and 55 g thymol/kg. Dietary oregano improved growth in growth-retarded growing–finishing pigs and had nonspecific immunostimulatory effects on porcine immune cells (Walter and Bilkei, 2004).

Menthol is often used as a repellent against insects and in lotions to cool legs (especially for horses) (Franz et al., 2005).

Milk cows become restless and aggressive each time a group of cows are separated and regrouped. This can last a few days putting cows in more stress resulting in a drop in milk production. Two Auburn University scientists could solve this problem by spraying anise oil (*P. anisum*) on the cows. Treated animals could not distinguish any differences among the cows in new or old groupings. They were mellower and kept their milk production up. Among many other oils tested but only anise seemed to work (HerbalGram, 1990).

Essential oils have been found effective in honeybee diseases (Ozkirim, 2006; Ozkirim et al., 2007).

In this review, we tried to give you an insight into the use of essential oils in animal health and nutrition. Due to the paucity of research in this important area, there is not much to report. Most information on usage exists in the form of not-so-well-qualified reports. We hope that this rather preliminary report can be of use as a starting point for more comprehensive reports.

REFERENCES

Alcicek, A., M. Bozkurt, and M. Cabuk, 2003. The effect of an essential oil combination derived from selected herbs growing wild in Turkey on broiler performance. *S. Afr. J. Anim. Sci.*, 33(2): 89–94.

Alcicek, A., M. Bozkurt, and M. Cabuk, 2004. The effect of a mixture of herbal essential oils, an organic acid or a probiotic on broiler performance. *S. Afr. J. Anim. Sci.*, 34(4): 217–222.

Athanassopoulou, F., E. Karagouni, E. Dotsika, V. Ragias, J. Tavla, and P. Christofilloyani, 2004a. Efficacy and toxicity of orally administrated anticoccidial drugs for innovative treatments of *Polysporoplasma sparis* infection in *Sparus aurata* L. *J. Appl. Ichthyol.*, 20: 345–354.

Athanassopoulou, F., E. Karagouni, E. Dotsika, V. Ragias, J. Tavla, P. Christofilloyanis, and I. Vatsos, 2004b. Efficacy and toxicity of orally administrated anticoccidial drugs for innovative treatments of *Myxobolus* sp. infection in *Puntazzo puntazzo*. *Dis. Aquat. Org.*, 62: 217–226.

Bakkali, F., S. Averbeck, D. Averbeck, and M. Idaomar, 2008. Biological effects of essential oils—A review. *Food Chem. Technol.*, 46: 446–475.

Baser, K.H.C., 1995. Analysis and quality assessment of essential oils. In: *A Manual on the Essential Oil Industry*, K.T. De Silva (ed.), pp. 155–177. Vienna, Austria: UNIDO.

Baser, K.H.C., 2008. Chemistry and biological activities of carvacrol and carvacrol-bearing essential oils. *Curr. Pharm. Des.*, 14: 3106–3120.

Baser, K.H.C. and F. Demirci, 2007. Chemistry of essential oils. In: *Flavours and Fragrances. Chemistry, Bioprocessing and Sustainability*, R.G. Berger (ed.), pp. 43–86. Berlin, Germany: Springer.

Bassett, R., 2000. Oreganos positive impact on poultry production. *World Poultry—Elsevier*, 16: 31–34.

Botsoglou, N.A., E. Christaki, D.J. Fletouris, P. Florou-Paneri, and A.B. Spais, 2002a. The effect of dietary oregano essential oil on lipid oxidation in raw and cooked chicken during refrigerated storage. *Meat Sci.*, 62: 259–265.

Botsoglou, N.A., P. Floron-Paneri, E. Christaki, D.J. Fletouris, and A.B. Spais, 2002b. Effect of dietary oregano essential oil on performance of chickens and on iron-induced lipid peroxidation of breast, thigh and abdominal fat tissues. *Br. Poult. Sci.*, 43(2): 223–230.

Bozkurt, M. and K.H.C. Baser, 2002a. The effect of antibiotic, Mannan oligosaccharide and essential oil mixture on the laying egg performance. In: *First European Symposium on Bioactive Secondary Plant Products in Veterinary Medicine*, Vienna, Austria, October 4–5, 2002.

Bozkurt, M. and K.H.C. Baser, 2002b. The effect of commercial organic acid, probiotic and essential oil mixture at two levels on the performance of broilers. In: *First European Symposium on Bioactive Secondary Plant Products in Veterinary Medicine*, Vienna, Austria, October 4–5, 2002.

Buchbauer, G., L. Jirovetz, M. Wasicky, and A. Nikiforov, 1994. Comparative investigation of Douglas fir headspace samples, essential oils, and extracts (needles and twigs) using GC-FID and GC-FTIR-MS. *J. Agric. Food Chem.*, 42: 2852–2854.

Busquet, M., S. Calsamiglia, A. Ferret, and C. Kamel, 2006. Plant extracts affect *in vitro* rumen microbial fermentation. *J. Dairy Sci.*, 89: 761–771.

Cabuk, M., M. Bozkurt, A. Alcicek, Y. Akbas, and K. Kucukyilmaz, 2006a. Effect of a herbal essential oil mixture on growth and internal organ weight of broilers from young and old breeder flocks. *S. Afr. J. Anim. Sci.*, 36(2): 135–141.

Cabuk, M., M. Bozkurt, A. Alcicek, A.U. Catli, and K.H.C. Baser, 2006b. Effect of dietary essential oil mixture on performance of laying hens in summer season. *S. Afr. J. Anim. Sci.*, 36(4): 215–221.

Cardozo, P.W., S. Calsamiglia, A. Ferret, and C. Kamel, 2005. Screening for the effects of natural plant extracts at different pH on *in vitro* Rumen microbial fermentation of a high-concentrate diet for beef cattle. *J. Anim. Sci.*, 83: 2572–2579.

Chitwood, D.J., 2002. Phytochemical based strategies for nematode control. *Annu. Rev. Phytopathol.*, 40: 221–249.

Chobanov, R.E., A.N. Aleskerova, S.N. Dzhanahmedova, and L.A. Safieva, 2004. Experimental estimation of antiparasitic properties of essential oils of some *Artemisia* (Asteraceae) species of Azerbaijan flora. *Rastitel'nye Resursy*, 40(4): 94–98.

Chow, W.H., K.L. Cheung, H.M. Ling, and T. See, 1989. Potentiation of warfarin anticoagulation by topical methylsalicylate ointment. *J. R. Soc. Med.*, 82(8): 501–502.

Ciftci, M., T. Guler, B. Dalkilic, and O.K. Ertas, 2005. The effect of anise oil (*Pimpinella anisum* L.) *on broiler performance. Int. J. Poultry Sci.*, 4(11): 851–855.

Clark, L. and J. Shivik, 2002. Aerosolized essential oils and individual natural product compounds as brown tree snake repellents. *Pest Manage. Sci.*, 58(8): 775–783.

Cornell University Animal Science, 2009. Treating livestock with medicinal plants: Beneficial or toxic?—*Ocimum basilicum, O. americanum* and *O. micranthum*. http://www.ansci.cornell.edu/plants/medicinal/basil.html. Accessed September 10, 2009.

Coskun, S., O. Grekin, M. Kurkcuoglu, H. Malyer, A.O. Grekin, N. Kirimer, and K.H.C. Baser, 2008. Acaricidal efficacy of *Origanum onites* L. essential oil against *Rhipicephalus turanicus* (Ixodidae). *Parasitol. Res.*, 103: 259–261.

Council Regulation (EEC) No. 2377/90, 1990. Laying down a community procedure for the establishment of maximum residue limits of veterinary medicinal products in foodstuffs of animal origin. http://ec.europa.eu/health/files/eudralex/vol-5/reg_1990_2377/reg_1990_2377_en.pdf. Accessed September 10, 2008.

Elgayyar, M., F.A. Draughon, D.A. Golden, and J.A. Mount, 2001. Antimicrobial activity of essential oils from plants against selected pathogenic and saprophytic microorganisms. *J. Food Prot.*, 64(7): 1019–1024.

EMA (European Medicines Agency), 1999. Salicylic acid, sodium salicylate, aluminium salicylate and methyl salicylate. Committee for Veterinary Medicinal Products (EMEA/MRL/696/99, 1999). www.ema.europa.eu/ema/pages/includes/document/open_document.jsp?webContentId=WC500015823. Accessed May 17, 2008.

Ertas, O.K., T. Guler, M. Ciftci, B. Dalkilic, and U.G. Simsek, 2006. The effect of an essential oil mixture from oregano, clove and anis on broiler performance. *Int. J. Poult. Sci.*, 4(11): 879–884.

Estell, R.E., E.L. Fredrickson, M.R. Tellez, K.M. Havstad, W.L. Shupe, D.M. Anderson, and M.D. Remmenga, 1998. Effects of volatile compounds on consumption of alfalfa pellets by sheep. *J. Anim. Sci.*, 76: 228–233.

Foster, S., 2002. The fighting power of Oregano: This versatile herb packs a powerful punch—Earth medicine. *Better Nutr.*, 1. http://www.oreganocures.com/articles74.html. Accessed May 17, 2015.

Franz, Ch., R. Bauer, R. Carle, D. Tedesco, A. Tubaro, and K. Zitterl-Eglseer, 2005. Study on the assessment of plants/herbs, plant/herb extracts and their naturally or synthetically produced components as "additives" for use in animal production. CFT/EFSA/FEEDAP/2005/01. http://www.agronavigator.cz/UserFiles/File/Agronavigator/Kvasnickova_2/EFSA_feedap_report_plantsherbs.pdf. Accessed September 15, 2008.

Garcia, V., P. Catala-Gregori, J. Madrid, F. Hernandez, M.D. Megias, and H.M. Andrade-Montemayor, 2007. Potential of carvacrol to modify *in vitro* rumen fermentation as compared with monensin. *Animal*, 1: 675–680.

Gauthier, R., 2005. Organic acids and essential oils, a realistic alternative to antibiotic growth promoters in poultry. *I Forum Internacional de Avicultura*. Foz do Iguaçu, PR, Brazil, August 17–19, 2005.

Ghisalberti, E.L., 2002. Secondary metabolites with antinematodal activity. In: *Studies in Natural Products Chemistry*, Atta-ur-Rahman (ed.), Vol. 26, pp. 425–506. Amsterdam, the Netherlands: Elsevier Science BV.

Giannenas, I., P.P. Florou, M. Papazahariadou, E. Christaki, N.A. Botsoglou, and A.B. Spais, 2003. Effect of dietary supplementation with oregano essential oil on performance of broilers after experimental infection with *Eimeria tenella. Arch. Anim. Nutr.*, 57(2): 99–106.

Githiori, J.B., S. Athanasiadou, and S.M. Thamsborg, 2006. Use of plants in novel approaches for control of gastrointestinal helminthes in livestock with emphasis on small ruminants. *Vet. Parasitol.*, 139: 308–320.

HerbalGram, 1990. Bovine aromatherapy: Common herb quells cowcophony. *HerbalGram*, 22: 8.

Horosova, K., D. Bujnakova, and V. Kmet, 2006. Effect of oregano essential oil on chicken lactobacilli and *E. coli. Folia Microbiol.*, 51(4): 278–280.

Ingraham, C, 2008. Zoopharmacognosy—Working with aromatic medicine. http://www.ingraham.co.uk/. Accessed May 17, 2008.

Jang, I.S., Y.H. Ko, S.Y. Kang, and C.Y. Lee, 2007. Effect of commercial essential oil on growth performance digestive enzyme activity and intestinal microflora population in broiler chickens. *Anim. Feed Sci. Technol.*, 134: 304–315.

Janz, J.A.M., P.C.H. Morel, B.H.P. Wilkinson, and R.W. Purchas, 2007. Preliminary investigation of the effects of low-level dietary inclusion of fragrant essential oils and oleoresins on pig performance and pork quality. *Meat Sci.*., 75: 350–355.

Krimpen, M.V. and G.P. Binnendijk, 2001. *Ropadiar® as alternative for anti microbial growth promoter in diets of weanling pigs.* Rapport Praktijkonderzoek Veehouderij, May 15, 2001. ISSN: 0169-3689.

Lahlou, M., 2003. Composition and molluscicidal properties of essential oils of five Moroccan Pinaceae. *Pharm. Biol.*, 41(3): 207–210.

Lambert, R.J.W., P.N. Skandamis, P.J. Coote, and G.-J.E. Nychas, 2001. A study of the minimum inhibitory concentration and mode of action of oregano essential oil, thymol and carvacrol. *J. Appl. Microbiol.*, 91: 453–462.

Lis-Balchin, M., 2003. Feed additives as alternatives to antibiotic growth promoters: Botanicals. In: *Proceedings of the Ninth International Symposium on Digestive Physiology in Pigs*, Vol. 1, pp. 333–352. Banff, Alberta, Canada: University of Alberta.

Lopez-Bote, L.J., J.I. Gray, E.A. Gomaa, and C.I. Flegal, 1998. Effect of dietary administration of oil extracts from rosemary and sage on lipid oxidation in broiler meat. *Br. Poult. Sci.*, 39: 235–240.

Losa, R., 2001. The use of essential oils in animal nutrition. In: *Feed Manufacturing in the Mediterranean Region. Improving Safety: From Feed to Food*, J. Brufau (ed.), pp. 39–44. Zaragoza, Spain: CIHEAM-IAMZ (Cahiers Options Méditerranéennes; v. 54), *Third Conference of Feed Manufacturers of the Mediterranean*, Reus, Spain, March 22–24, 2000.

Moller, T., 2001. Studies on the effect of an oregano-oil-addition to feed towards nutrient digestibilities, N-balance as well as towards the parameters of microbial activity in the alimentary tract of weaned-piglets. Thesis. http://www.agronavigator.cz/UserFiles/File/Agronavigator/Kvasnickova_2/EFSA_feedap_report_plantsherbs.pdf. Accessed July 15, 2009.

Molnar, C. and G. Bilkei, 2005. The influence of an oregano feed additive on production parameters and mortality of weaned piglets, Tieraerzliche Praxis, Ausgabe Grosstiere. *Nutztiere*, 33: 42–47.

Moon, T., J. Wilkinson, and H. Cavanagh, 2006. Antiparasitic activity of two Lavandula essential oils against *Giardia duodenalis*, *Trichomonas vaginalis* and *Hexamita inflata*. *Parasitol. Res.*, 99(6): 722–728.

Moreira, M.R., A.G. Ponze, C.E. del Valle, and S.I. Roura, 2005. Inhibitory parameters of essential oils to reduce a foodborne pathogen. *LWT*, 38: 565–570.

Moschetti, R., 2003. Pesticides made with botanical oils and extracts. http://www.plantoils.in/uses/other/other.html. Accessed September 2009.

Onibala, J.S.I.T., K.D. Gunther, and Ut. Meulen, 2001. Effects of essential oil of spices as feed additives on the growth and carcass characteristics of growing-finishing pigs. In: *Sustainable Development in the Context of Globalization and Locality: Challenges and Options for Networking in Southeast Asia.* EFSA. http://www.efsa.europa.eu/en/scdocs/doc/070828.pdf. Accessed September 10, 2008.

Oviedo-Rondon, E.O., S. Clemente-Hernandez, F. Salvador, R. Williams, and R. Losa, 2006. Essential oils on mixed coccidia vaccination and infection in broilers. *Int. J. Poult. Sci.*, 5(8): 723–730.

Ozkirim, A., 2006. The detection of antibiotic resistance in the American and European Foulbrood diseases of honey bees (*Apis mellifera L.*). PhD thesis, Hacettepe University, Ankara, Turkey.

Ozkirim, A., N. Keskin, M. Kurkcuoglu, and K.H.C. Baser, 2007. Screening alternative antibiotics-essential oils from Seseli spp. against Paenibacillus larvae subsp. larvae strains isolated from different regions of Turkey. In: *40th Apimondia International Apicultural Congress*, Melbourne, Victoria, Australia, September 9–14, 2007.

Penalver, P., B. Huerta, C. Borge, R. Astorga, R. Romero, and A. Perea, 2005. Antimicrobial activity of five essential oils against origin strains of the Enterobacteriaceae family. *APMIS*, 113: 1–6.

Prajapati, V., A.K. Tripathi, K.K. Aggarwal, and S.P.S. Khanuja, 2005. Insecticidal, repellent and oviposition-deterrent activity of selected essential oils against *Anopheles stephensi*, *Aedes aegypti* and *Culex quinquefasciatus*. *Bioresour. Technol.*, 96(16): 1749–1757.

Ramanathan, M., 1995. Warfarin–topical salicylate interactions: Case reports. *Med. J. Malaysia*, 50(3): 278–279.

Regnault-Roger, C., 1997. The potential of botanical essential oils for insect pest control. *Int. Pest Manage. Rev.*, 2: 25–34.

Rochfort, S., A.J. Parker, and F.R. Dunshea, 2008. Plant bioactives for ruminant health and productivity. *Phytochemistry*, 69: 299–322.

Scents & Sensibility, 2001. Pet care & pest control—Using essential oils. *Scents Sensibility Newsl.*, 2(12): 1. http://scentsnsensibility.com/newsletter/Apr0601.htm. Accessed May 17, 2008.

Schoene, F., A. Vetter, H. Hartung, H. Bergmann, A. Biertuempfel, G. Richter, S. Mueller, and G. Breitschuh, 2006. Effects of essential oils from fennel (*Foeniculi aetheroleum*) and caraway (*Carvi aetheroleum*) in pigs. *J. Anim. Physiol. Anim. Nutr.*, 90: 500–510.

Tabanca, N., E. Bedir, N. Kirimer, K.H.C. Baser, S.I. Khan, M.R. Jacob, and I.A. Khan, 2003. Antimicrobial compounds from *Pimpinella* species growing in Turkey. *Planta Med.*, 69: 933.

Tam, L.S., T.Y. Chan, W.K. Leung, and J.A. Critchley, 1995. Warfarin interactions with Chinese traditional medicines: Danshen and methyl salicylate medicated oil. *Aust. N. Z. J. Med.*, 25(3): 258.

Ultee, A., E.P.W. Kets, and E.J. Smid, 1999. Mechanisms of action of carvacrol in the food-borne pathogen *Bacillus cereus*. *Appl. Environ. Microbiol.*, 65: 4606–4610.

U.S. Patent 4,735,803. Naturally-odoriferous animal repellent. http://digitalcommons.unl.edu/cgi/viewcontent.cgi?article=1151&context=icwdm_usdanwrc. Accessed September 10, 2008.

U.S. Patent 4,847,292. Repelling animals with compositions comprising citronellyl nitrile, citronellol, alpha-terpinyl methyl ether and lemon oil. http://www.freepatentsonline.com/4847292.html. Accessed September 10, 2008.

U.S. Patent 4,961,929. Process of repelling dogs and dog repellent material. http://www.freepatentsonline.com/4961929.html. Accessed September 10, 2008.

U.S. Patent 6,159,474. Animal repellant containing oils of black pepper and/or capsicum. http://www.freepatentsonline.com/6159474.html. Accessed September 10, 2008.

U.S. Patent 6,800,294. Antiparasitic formulation. http://www.freepatentsonline.com/6800294.html. Accessed March 16, 2020.

Vogt, H. and H.W. Rauch, 1991. The use of several essential oils in broiler diets. *Landbauforschung Volkenrode*, 41: 94–97.

Wallace, R.J., N.R. McEwan, F.M. McIntosh, B. Teferedegne, and C.J. Newbold, 2002. Natural products as manipulators of rumen fermentation. *Asian-Aust. J. Anim. Sci.*, 15: 1458–1468.

Walter, B.M. and G. Bilkei, 2004. Immunostimulatory effect of dietary oregano etheric oils on lymphocytes from growth-retarded, low-weight growing-finishing pigs and productivity. *Tijdschr. Diergeneeskd.*, 129(6): 178–181.

Wedge, D.E., J.A. Klun, N. Tabanca, B. Demirci, T. Ozek, K.H.C. Baser, Z. Liu, S. Zhang, C.L. Cantrell, and J. Zhang, 2009. Bioactivity-guided fractionation and GC-MS fingerprinting of *Angelica sinensis* and *A. archangelica* root components for antifungal and mosquito deterrent activity. *J. Agric. Food Chem.*, 57: 464–470.

Yip, A.S., W.H. Chow, Y.T. Tai, and K.L. Cheung, 1990. Adverse effect of topical methylsalicylate ointment on warfarin anticoagulation: An unrecognized potential hazard. *Postgrad. Med. J.*, 66(775): 367–369.

26 Encapsulation and Other Programmed/Sustained-Release Techniques for Essential Oils and Volatile Terpenes

*Jan Karlsen**

CONTENTS

26.1 INTRODUCTION

In order to widen the medical and industrial applications of volatiles (essential oils), it is necessary to lower the volatility of the compounds to obtain a longer shelf life of products containing volatiles. However, due to the many contradictory reports on the activity/non-activity of volatile compounds toward a biological system, a prolonged-release formulation of volatiles would also be of great interest to basic research. The short contact time of the volatiles with a biological system may make it difficult to verify a biological effect, and this could be changed by, for instance, encapsulation of the volatile compounds before biological testing. Microencapsulation of essential oils and other volatiles has become an attractive tool and formulation principle for the different applications of volatile compounds. Several new patents have been filed in recent years for new application purposes In short, the encapsulation of volatiles using different materials was found to retain the volatiles in a larger matrix of industrial products. Recently, nano-emulsions have been developed with similar properties, as encapsulation techniques for volatiles.

Volatile compounds like essential oils and terpenes have mainly been used in the perfumery and fragrance industry for the impact on our sense of smelling. Formulations which can lower the volatility or prolong the release of volatiles obviously would be of interest not only to industry developing new products, but also to scientists interested in the biological effects of volatile compounds. Encapsulation of special groups of volatile compounds in a mixture may change the smell of the original mixture of compounds and change the smell of a perfume even though the composition of the essential oils and volatiles are the same.

* Passed away on March 21, 2019, in Norway.

When the first edition of this *Handbook of Essential Oils* was published in 2010, there were several patents describing the encapsulation of essential oils, but the studies into the impact upon the biological effects of encapsulated (or "low volatile ") volatiles were scarce. However, during the last two to three years (2015–2018), many new publications describing the biological effects of volatiles have been published. In these publications, the essential oils were often present in an encapsulated formulation. In the opinion of the author, we can expect many more studies to be carried out where the prolonged release of volatile compounds and the resulting longer interaction with a biological system may give rise to data on biological effects which, until now, have been bypassed.

There are many practical aspects of this use of volatile compounds in which the limited contact time of volatiles can be significantly changed and therefore give rise to a range of new and better applications and industrial products. Encapsulation of essential oils can make a "dry" free-flowing powder easily incorporated into consumer products. Lowering the volatility can also allow incorporating these compounds into textiles, surface films, aerosols for spraying, etc.

To lower the volatility, one needs to encapsulate the volatile into a polymer matrix, utilize a complex formation, use covalent bonding to a matrix, to mention a few techniques. We therefore need to formulate the volatiles and take many of the techniques from areas where controlled release formulations have been in use for many years. Especially, the area of controlled drug delivery has a large number of such formulations. Today, there exists a large number of sustained drug-delivery formulations in both journal publications and in international patents (Deasy, 1984).

However, one of the limitations of encapsulation is still the encapsulated percentage of the volatiles. It is difficult, if not impossible, to increase the degree of encapsulation above 20%. But it must be mentioned that even a 20% encapsulation of volatiles represents a considerable improvement compared to a non-encapsulated product.

So far in the area of volatile terpenes/essential oils, we have seen a large number of investigations that focused on plant selection, volatile isolation techniques, separation of volatiles isolated, identification of isolated compounds, and the biochemical formation of terpenes. The formulation of a volatile into products has been seen as an area of industrial research. This has naturally led to a large number of patents but very few scientific publications on the formulation of essential oils and lower terpenes.

The idea of this chapter has been to give an introduction to the area of making a controlled-release product of volatiles and, in particular, of essential oils and their constituents. It will be impossible to give a total survey of this area in a chapter, but hopefully some ideas are given which inspire the reader to use the internet search in databases for more complete coverage. Several keywords for computer searching are given.

26.2 CONTROLLED RELEASE OF VOLATILES

The main interest of volatile encapsulation is the possibility to extend the biological effects of the compounds. For essential oils, we want to prolong the activity by lowering the evaporation of the volatile compounds (Baranauskiene, 2007). During the last ten years, there are not many publications on this topic in the scientific literature compared to the number of patents that describe various ways of prolonging the effect of volatiles (Sair, 1980; Sair and Sair, 1980; Fulger and Popplewell, 1997, 1998; Zasypkin and Porzio, 2004; McIver, 2005; Porzio, 2008). One reason for this fact is that the prolonging of the evaporation of volatiles is regarded as so close to practical applications that further development of this area will immediately be blocked by a number of patents. However, by 2013, there were signs that this attitude was changing. The change of the impact of a perfume by encapsulation of the ingredients is given in a patent by Budijono (2014). The possibility to change the top note, middle note, and end note in a perfume without changing the composition should be a tempting solution to new perfumes. This area should interest the perfume industry, but the cost of making well-controlled nano-particles with the size less that 100 nm is at the moment too high to be applied to perfume production. In order to look into the lowering of the volatility of essential oils, we have to approach another active area of scientific research—controlled delivery of pharmacologically active ingredients.

The reasons for controlled release of volatiles may be the following:

- Changing and improving the biological impact of volatiles
- Adding fragrance to textiles
- Stabilizing specific compounds
- Tailoring a fragrance to a specific application
- Lowering the volatility to increase the shelf life of a formulated product
- Improving the study of the biological effect of volatiles

The slow or controlled release of volatiles is achieved by:

- Encapsulation
- Solution or dispersion into a polymer matrix
- Complex formation
- Covalent bonding to another molecule or matrix
- Nano-emulsion formulation

For essential oils/terpenes and natural volatiles, the following techniques can be utilized depending upon the volatiles and the intended use of the final product:

- Microcapsule production
- Microparticle formation
- Melt extrusion
- Melt injection
- Complex formation
- Liposomes
- Micelles
- Covalent bonding to a matrix
- Combination of nano-capsules into larger microcapsules
- Nano-emulsion formulation

The making of one of the abovementioned type of products and techniques utilized will influence the activity toward the human biological membranes in one way or another. Therefore, the relevant sizes of biological units are listed in Table 26.1, and the average sizes of units produced in consumer products where volatiles are involved are listed in Table 26.2.

The introduction of encapsulation of volatiles, which ten years ago saw very few publications, has now, by 2018, turned into a very active field of applied research. Encapsulation or prolonged delivery of volatiles give us more predictable and long-lasting effects of the compounds. The areas of application are very varied, and the industries using essential oils and terpenes foresee many prospects for new microencapsulated products.

Application markets for encapsulated essential oils and volatiles are:

- Applied to the surface of various paper products
- Medicine
- Food and household and personal care items
- Biotechnology
- Pharmaceuticals
- Electronics
- Photography
- Chemical industry
- Textile industry
- Cosmetics

TABLE 26.1
Size Diameters of Biological Entities

Human blood cells	7000–8000 nm
Bacteria	800–2000 nm
Human cell nucleus	1000 nm
Nano-particles that can cross biomembranes	60 nm
Virus	17–300 nm
Hemoglobin molecules	3.3 nm
Nano-particles that can cross blood–brain barrier	4 nm
DNA helix	3 nm
Water molecule	0.3 nm

TABLE 26.2
Average Size of Formulation Units, in nm (Sizes below 150 nm may be Invisible to the Naked Eye)

Solutions	0.1 nm
Micellar solutions	0.5 nm
Macromolecular solutions	0.5 nm
Microemulsions	5–20 nm
Liposomes SUV	20–150 nm
Nano-emulsions	20–150 nm
Nano-spheres	100–500 nm
Nano-capsules	100–500 nm
Liposomes LUV	200–500 nm
Liposomes MLV	200–1000 nm
Microcapsules	500–30,000 nm
Simple emulsions	500–5000 nm
Multiple emulsions	10,000–100,000 nm

Abbreviations: SUV, unilamellar vesicles; LUV, large unilamellar vesicles; MLV, multilamellar vesicles.

It is therefore easy to understand that the encapsulation procedures will open up a much larger and different market for essential oil/terpene products. Experience from all the areas mentioned above can be applied to the study of volatile compounds in products.

In the area of essential oils and lower terpenes, simple encapsulation procedures from the area of drug delivery are applied. The essential oils or single active constituents are mixed with a hydrophilic polymer and spray-dried using a commercial spray-drier. Depending on whether we have an emulsion or a solution of the volatile fraction in the polymer, we obtain monolithic particles or normal microcapsules.

The most usual polymers used for encapsulation are:

- Oligosaccharides from α-amylase
- Acacia gum
- Gum arabic
- Alginate
- Chitosan

Many different emulsifiers are used to solubilize the essential oils totally or partly prior to the encapsulation process. This can result in a monolithic particle or a normal capsule, where the essential oil is surrounded by a hydrophilic coating. When the mixture of an essential oil and a hydrophilic polymer is achieved, the application of the spray-drying procedure of the resulting mixture will result in the formation of microcapsules. The technique for achieving an encapsulated product in high efficiency will depend upon many technical parameters and a description of the procedures can be found in the patent literature. Often, a mixture of essential oil/polymer (4:1) can be used, but a successful result will also depend on the type of apparatus being used for the production. The reader is advised to refer to the parameters given for the polymer used in the experiment. To achieve the encapsulated product, a mixture of low pressure and a slightly elevated temperature is used in the spray-drying equipment and a certain loss of volatiles is inevitable. However, a recovery of more than 70% can be achieved by carefully monitoring the production parameters.

26.3 USE OF HYDROPHILIC POLYMERS

In product development, one tends to use cheap derivatives of starches or other low-grade quality polymers. Early studies with protein-based polymers such as gelatine, gelatine-derivatives, soy proteins, silk proteins, and milk-derived proteins gave reasonable technical quality of the products. However, even if these materials show stable emulsification properties with essential oils, they have some unwanted side effects in products. We have seen that a more careful control of the polymer used can result in real high-tech products, where the predictability of the release of the volatiles can be assured like a programmed release of drug molecules in drug-delivery devices. The polymer quality to be used will depend on the intention of the final product. In the cosmetic industry, in which one is looking for an essential oil product, free flowing and dry, to mix with a semisolid or solid matrix, the use of simple starch derivatives will be very good. For other applications, in which the release of the volatile needs to be closely controlled or predicted more accurately, it is recommended that a more thorough selection of a well-characterized polymer is done. One example is the use of chitosan in prolonging the volatile profile of saffron by encapsulation (Chranioti, 2013). Another example of a very good and controllable hydrophilic polymer is alginate. This polymer is available in many qualities and can be tailored to any controlled-release product. The chemistry of alginate is briefly discussed below and will allow the reader to decide whether to opt for an alginate of technical quality, or if a high-tech product is the aim, to choose a better characterized hydrocolloid.

26.4 ALGINATE

Alginates are naturally occurring polycarbohydrates consisting of monomers of α-L-glucuronic acid (G) and β-D-mannuronic acid (M). The relative amounts of these two building blocks will influence the total chemistry of the polymer. The linear polymer is water soluble due to its polarity. Today, alginate can be produced by bacteria that allow us to control the composition of the monomer (G/M) ratio. The chemical composition of the natural occurring polymer is dependent upon the origin of the raw material. Marine species display seasonal differences in the composition and different parts of the plant produce different alginates, and this may make standardization of the isolated polymer difficult. Alginates may undergo epimerization to obtain the preferred chemical composition. This composition (G/M ratio) will determine the diffusion rate through the swollen alginate gel, which surrounds the encapsulated essential oil (Ogston, 1973; Elias, 1997; Amsden, 1998a, 1998b). An important structure parameter is also the distribution of the carboxyl groups along the polymer chain. The molecular weight of the polymer is equally important, and molecular weights between 12,000 and 250,000 are readily available in the market. The alginate polymer can form a swollen gel by hydrophobic interaction or by cross-linking with divalent ions like calcium. The G/M ratio determines the swelling rate and therefore also the release of encapsulated compounds. The diffusion of different substances has been studied and reference can be made to essential oil encapsulation. The size of alginate capsules can vary from 100 μm or more down to the nanometer range depending on the production procedure chosen (Draget et al., 1994, 1997; Tonnesen and Karlsen, 2002; Shilpa, 2003; Donati et al., 2005).

26.5 STABILIZATION OF ESSENTIAL OIL CONSTITUENTS

The encapsulation of essential oils in a hydrophilic polymer may stabilize the constituents of the oil, but a better technique for this purpose will often be to use cyclodextrins in the encapsulation procedure. The use of cyclodextrins will lead to a complexation of the single compounds, which again will stabilize the complexed molecule. Complex formation with cyclodextrins is often used in drug delivery to promote solubility of lipophilic compounds, however, in the case of volatiles containing compounds that may oxidize, the complex formation will definitely prolong the shelf life of the finished product. A good review of the flavor encapsulation advantages is given by Risch and Reineccius (1987). The most important aspect of essential oil encapsulation in a hydrophilic polymer is that the volatility is lowered. Lowering the volatility will result in longer shelf life of products and a better stability of the finished product in this respect. A very interesting product is the development of microcapsules containing essential oils having adherence to keratinous surfaces. This should find applications in cosmetics and the personal-care area (Alden-Danforth, 2013). By modifying the capsule surface, we can change the properties of the essential oil impact—in addition to lowering the volatility, we can also change the adherence of the capsules.

26.6 CONTROLLED RELEASE OF VOLATILES FROM NONVOLATILE PRECURSORS

The time-limited effect of volatiles for olfactive perception has led to the development of encapsulated volatiles and also to the development of covalent-bonded fragrance molecules to matrices. In this way, molecules release their fragrance components by the cleavage of the covalent bond. Mild reaction conditions met in practical life initiated by light, pH, hydrolysis, temperature, oxygen, and enzymes may release the flavors. The production of "profragrances" is a very active field for the industry and has led to numerous patents. The botanical plants producing essential oils have developed means by which the volatiles are produced, stored, and released into the atmosphere related to environmental factors. The making of a "profragrance" involves mimicking these natural procedures into flavor products. However, we are simplifying the process by using only one parameter in this release process, that is, the splitting of a covalent bond. In doing this, we mimic the formation of terpene-glucosides in the plant which are then split, giving off the terpene. In theory, the making of a long-lasting biological impact and the breakdown of a constituent are contradictory reactions. However, for practical purposes, the use of covalent bonding and thereafter the controlled splitting of this bond by the parameters mentioned above (temperature, humidity etc.) can be built into suitable flavor and fragrance products. This technique is only applicable to single essential oil constituents but constitutes a natural follow-up of essential encapsulation (Herrmann, 2004, 2007; Powell, 2006).

26.7 CYCLODEXTRIN COMPLEXATION OF VOLATILES

Cyclodextrin molecules are modified carbohydrates that have been used for many years to modify the solubility properties of drug molecules by complexation. The cyclodextrins can also be applied to volatiles to protect them against environmental hazards and thus prolong the shelf life of these compounds. Cyclodextrin complexation will also modify the volatility of essential oils and prolong the bioactivity (Han, 2011; Na, 2012; Zhu, 2012). Recently, the activity of monoterpenes (linalool, S-carvone, camphor, geraniol, γ-terpinene, and fenchone) as insecticides has been shown (Don-Pedro, 1966., Lopez, 2008). However, the efficient application of these compounds is difficult owing to low-stability and high-evaporation properties. Products made by encapsulation into β-cyclodextrins show a better control of the release rate of the volatiles which therefore facilitates their use in products (Lopez and Pascual-Villalobos, 2010). The cyclodextrins will give a molecular encapsulation by the complexation reaction. There are many cyclodextrins on the market, but β-cyclodextrin is the most popular in use for small molecules. Complexation of volatiles with cyclodextrins may improve the heat stability, improve the stability toward oxygen, and improve the stability against light-induced reactions (Szenta, 1988). Cyclodextrin complexation will also protect enantiomers against racemization. The degree of encapsulation using cyclodextrins will be around 6%. A significant lowering of the volatility has been observed for the complexation with essential oils (Risch and Reineccius, 1988). The complexation with cyclodextrins will result in increased heat stability. This is in contrast with the stability of volatiles that has been adsorbed on a polymer matrix. Cyclodextrin complexation can protect against:

- Loss of volatiles upon storage of finished product
- Light-induced instability
- Heat decomposition
- Oxidation
- Racemization of enantiomers

26.8 ENHANCED BIOLOGICAL EFFECT BY PROLONGED DELIVERY OF VOLATILES AND ESSENTIAL OILS

In cases where a biological effect has been indicated by introductory experiments, a prolonged contact time may give real evidence of the biological effect. This can then be applied to product development. The prolonged delivery has been introduced into a variety of commercial products such as fruit juices (Fujimoto, 1992; Donsi, 2011) and general improvement of microbial effect (Paluch, 2011), to mention a few applications. Several new patents and publications describe the encapsulation or prolonged delivery of essential oils (Trinh, 1994; Principato, 2007; Gaonkar, 2010; Behle, 2011; Bhala, 2012; Ortan, 2013).

26.9 METHODS FOR PRODUCING PROLONGED DELIVERY UNITS OF VOLATILES

The methods for prolonging the volatility of essential oils and other volatiles are varied. Some indications of useful excipient are given in this chapter. However, the efficiency of the encapsulation procedures shows great variation, and a thorough investigation into suitable methods is needed. This is the reason for the many patents being published in this area. The surrounding matrix of the volatiles in a product will greatly influence the chosen method and polymer—whether the volatile will be incorporated into cosmetics, food, and pharmaceuticals or onto surfaces such as paper, wood, lacquer, and textile fibers (Koike, 1992; Habar, 1996; Boh, 2006; Madene, 2006; Shah, 2012; Boardman, 2013; Vella, 2013; Zhang, 2013; Jimenez, 2014).

26.10 PRESENTING VOLATILES IN NANO-EMULSIONS

Recently, the formulation of essential oils and volatiles as nano-emulsions has attracted the interest of scientists and industry. The industrial interest in naturally occurring antimicrobial compounds and preservatives has been the driving force for the studies on nano-emulsions. These types of formulations offer scope for better bioavailability and stability of the volatiles, as well as opening up new, interesting applications. A recent publication by Khan et al. (Khan, 2018) has shown an interesting *in vitro* and *in vivo* potential of a carvacrol nano-emulsion in which the mean droplet size of carvacrol was around 120 nm. The emulsion seems to have potential in lung cancer therapy. Likewise, the incorporation of essential oils in nano-emulsions have shown that the preparations have interesting effects as naturally antimicrobial, as well as acting as preservatives in food products (Donsi, 2016). The production of nano-emulsions seems easier than the encapsulation process and, therefore, at the moment, more attractive to industrial production. It remains to be seen, however, if nano-emulsions will replace encapsulated essential oils in consumer products.

26.11 CONCLUDING REMARKS

The encapsulation/complexation/covalent bonding of essential oils and single volatile compounds will result in significant lowering of the volatility, stabilizing the compounds, improving the shelf life of finished products and prolonging the biological activity. It may also allow a better testing of the biological activity of volatiles and thereby improve basic research on the biological effect of volatile compounds. Until a few years ago, most of the literature on prolonged flavor products was to be found in patent literature. Today, a series of papers describe the use of prolonged effect formulations for volatile constituents, and the research activity in this field of volatile application is very high. The effect of controlled delivery of volatiles opens up different areas of applications which previously were limited due to the volatility of the essential oils and their constituents. The encapsulation or the lowering of the volatility of compounds like the essential oils will allow for more relevant studies into the biological effects of volatile compounds. The readers of this chapter should also be aware that the literature of essential oil encapsulation and other means of lowering the volatility of compounds cannot always be found in the traditional essential oil research journals. It may also be advisable to look into Chinese and Japanese journals and patents as scientists of these countries have a long tradition using hydrocolloids for industrial production, among which many utilize the encapsulated volatiles.

REFERENCES

Alden-Danforth, E., W. Feuer, M. White, and N. Williams. 2013. Aqueous-based personal care product formula that combines friction controlled fragrance technology with a film-forming compound to improve adherence of capsules on keratinous surfaces. Patent WO 2013087549 A1 20130620.

Amsden, B. 1998a. Solute diffusion within hydrogels. *Macromolecules*, 31: 8382–8395.

Amsden, B. 1998b. Solute diffusion in hydrogels. *Polym. Gels. Networks*, 6: 13–43.

Baranauskiene, R., E. Bylaite, J. Zukauskaite, and R. Venskutonis. 2007. Flavour retention of peppermint (*Mentha piperita* L.) essential oil spray-dried in modified starches during encapsulation and storage. *J. Agric. Food Chem.*, 55: 3027–3036.

Bhala, R., V. Dhandania, and A. P. Periyasamy. 2012. Bio-finishing of fabrics. *Asian. Dyer.*, 9: 45–49.

Behle, R. W., L. B. Flor-Weiler, A. Bharadwaj, and K. C. Stafford. 2011. A formulation to encapsulate nootkatone for tick control. *J. Med. Entomol.*, 48: 1120–1127.

Boardman, C., and K. S. Lee. 2013. Fabric treatment using encapsulated phase-change active materials. Patent WO 2013087550 A1 20130620.

Boh, B., and E. Knez. 2006. Microencapsulation of essential oils and phase change materials for application in textile products. *Ind. J. Fibre Text. Res.*, 31: 72–82.

Budijono, S., L. Ouali, V. Normand, J.-Y. Billard De Saint Laumer, and S. Zhang. 2014. WO Patent 2014029695 A1 20140227.

Chranioti, C., S. Popoutsakis, A. Stephanos, and C. Tzia. 2013. Special publication-Royal Society of Chemistry, section nutrition. *Functional and Sensory Properties of Foods*, 344: 111–116.

Deasy, P. B. 1984. *Microencapsulation and Related Drug Processes*. New York: Marcel Dekker.

Don-Pedro, K. N. 1966. Fumigant toxicity is the major route of insecticidal activity of citrus peel essential oils. *Pestic. Sci.*, 46: 71–78.

Donati, I., S. Holtan, Y. A. Morch, M. Borgogna, M. Dentini, and G. Skjåk-Bræk. 2005. New hypothesis on the role of alternating sequences in calcium-alginate gels. *Biomacromolecules*, 6: 1031–1040.

Donsi, F., M. Annunziata, M. Sessa, and G. Ferrari. 2011. Nanoencapsulation of essential oils to enhance their microbial activity in foods. *Food Sci. Techn.*, 44: 1908–1914.

Donsi, F., and G. Ferrari. 2016. Essential oil nanoemulsions as antimicrobial agents in food. *J. Biotechnol.*, 233: 106–120.

Draget, K. I., G. Kjåk-Bræk, and O. Smidsrød. 1994. Alginic acid gels; The effect of alginate chemical composition and molecular weight. *Carbohydr. Polym.*, 25: 31–38.

Draget, K. I., G. Skjåk-Bræk, B. E. Christiansen, O. Gåserød, and O. Smidsrød. 1997. Swelling and partial solubilization of alginic acid beads in acids. *Carbohydr. Polymer*, 29: 209–215.

Elias, H.-G. 1997. *An Introduction to Polymer Science*. Weinheim, Germany: VCH.

Fujimoto, T., and K. Suehiro. 1992. Antimicrobial wax compositions containing essential oils from wood. JP Patent 04328182 A 19921117.

Fulger, C., and M. Popplewell. 1997. Flavour encapsulation. US Patent 5.601.845.
Fulger, C., and M. Popplewell. 1998. Flavour encapsulation. U.S. Patent 5.792.505.
Gaonkar, A. G., A. Akashe, L. Lawrence, A. R. Lopez, R. L. Meibach, D. Sebesta, J. D. White, Y. Wang, and L. G. West. 2010. Delivery of essential oil esters or other functional compounds in an enteric matrix. US Patent 20100310666 A1 20101209.
Habar, G. L., A. Le Pape, and C. Descusse. 1996. Microcapsules containing terpene or abietic acid derivatives as biodegradable solvents for use on chemical copying paper and pressure sensitive papers. Patent EP 714786 A1 19960605.
Han, L., and Y. Zhang. 2011. Study on the preparation of the complex of β-cyclodextrin-clove oil inclusion. *Xibei Yaoxue Zazhi*, 26: 447–449.
Herrmann, A.. 2004. Photochemical fragrance delivery systems based on the Norrish type II reaction-a review. *Spectrum*, 17: 10–13 and 19.
Herrmann, A. 2007. Controlled release of volatiles under mild reaction conditions. From nature to everyday products. *Angew. Chem. Int. Ed.*, 46: 5836–5863.
Jimenez, A., L. Sanchez-Gonzales, S. Desobry, A. Chiralt, and E. A. Tehrany. 2014. Influence of nanoliposomes incorporation on properties of fil forming dispersions and films based on corn starch and sodium caseinate. *Food Colloids*, 35: 159–169.
Khan, I., A. Bahuguna, P. Kumar, V. K. Bajpai, S. C. Kang. 2018. *In vitro* and *in vivo* antitumor potential of carvacrol nanoemulsion against human lung adenocarcinoma A549 cells via mitochondrial mediated apoptosis, *Sci. Rep.* 8:144.
Koike, S., and A. Imai. 1992. Microcapsules containing fragrant coatings. Patent JP 04351678 A 19921207.
López, M.D., M. J. Jordan, and M. J. Pascual-Villalobos. 2008. Toxic compounds in essential oils of coriander, caraway and basil against stored rice pests. *J. Stored Prod. Res.* 44: 273–278.
Lopez, M. D., and M. J. Pascual-Villalobos. 2010. Analysis of monoterpenoids in inclusion complexes with β-cyclodextrin and study on ratio effect in these microcapsules. *10th International Working Conference on Stored Product Protection, Julius-Kahn-Archiv*, 425: 705–709.
Madene, A., M. Jacquot, J. Scher, and S. Desobry. 2006. Flavour encapsulation and controlled release. *Int. J. Food Sci. Technol.*, 41: 1–21.
McIver, B. 2005. Encapsulation of flavour and/or fragrance composition. U.S. Patent 6.932.982.
Na, S., R. Wu, R. Bu, K. Meng, and G. Hexi. 2012. Mongolian medicinal compound preparation for treating bronchitis and its preparation method. Patent CN 102793840 A 20121128.
Ogston, A. G., B. N. Preston, and J. D. Wells. 1973. On the transport of compact particles through solutions of chain polymers. *Proc. R. Soc. (London) A*, 333: 297–309.
Ortan, A., M. Ferdes, S. Rodino, C. Dinu Pirvu, and D. Draganescu. 2013. Topical delivery system of liposomally encapsulated volatile oil of *Anethum graveolens*. *Farmacia (Bucharest, Roumania)*, 61: 361–370.
Paluch, G., R. Bradbury, and S. Bessette. 2011. Development of botanical pesticides for public health. *J. ASTM Int.*, 8: JAI103468/1–JAI103468/7.
Porzio, M. 2008. Melt extrusion and melt injection. *Perfumer Flavorist*, 33: 48–53.
Principato, M. A. 2007. Insecticidal/acaricidal and insectifungal/acarifungal formulation. *Ital. Appl. Pat IT*, 2006BO0699 A1 20070110.
Powell, K., J. Benkhoff, W. Fischer, and K. Fritsche. 2006. Secret sensations: Novel functionalities triggered by light-Part II: Photolatent fragrances. *Eur. Coat. J.*, 9: 40–49.
Risch, S. J., and G. A. Reineccius. 1988. Flavor encapsultation. *ACS Symposium Series 370*, American Chemical Society, Washington.
Risch, S.J., and G. A. Reineccius (Eds.) 1987. Flavor Encapsulation. *ACS Symposium Series 370*. Washington, DC: American Chemical Society.
Sair, L. 1980. Food supplement concentrate in a dense glass house extrudate. US Patent 4.232.047.
Sair, L., and R. Sair. 1980. Encapsulation of active agents as microdispersions in homogenous natural polymers. US Patent 4.230.687.
Shah, B., M. P. Davidson, and Q. Zhong. 2012. Encapsulation of eugenol using Maillard-type conjugates to form transparent and heat-stable nanoscale dispersions. *Food Sci. Technol.*, 49: 139–148.
Shilpa, A., S. S. Agarwal, and A. R. Ray. 2003. Controlled delivery of drugs from alginate matrix. *Macromol. Sci. Polym. Rev.,C.*, 43: 187–221.
Szenta, L., and J. Szejtli. 1988. Stabilization of flavors by cyclodextrins. In *Flavor encapsulation*. S. J. Risch (ed.), pp. 148–157. ACS Symposium 370. Washington DC: American Chemical Society.
Trinh, T., G. F. Brunner, and T. A. Inglin. 1994. Adsorbent articles for odor control with positive scent signal. WO 9422500 A1 19941013.
Tønnesen, H. H., and J. Karlsen. 2002. Alginate in drug delivery systems. *Drug Dev. Ind. Pharm.*, 28: 621–630.

Vella, J., and T. I. Marks. 2013. Micro-encapsulated chemical re-application method for laundering of fabrics. Patent US 20130239429 A1 20130919.

Venskutonis, R. 2007. Encapsulation of essential oils and release of higher volatile compounds at various storage conditions. *Bulletin USAMV-CN*, 63: 2007.

Zasypkin, D. and M. Porzio. 2004. Glass encapsulation of flavours with chemically modified starch blends. *J. Microencapsulation*, 21: 385–397.

Zhang, N., and P. S. Given. 2013. Releasably encapsulated aroma. Patent WO 2013032631 A1 20130307.

Zhu, Y., and D.-J. Yu. 2012. Technology optimization of inclusion compound for volatile oil from *Cnidium monnieri* with hydroxypropyl-β-cyclodextrin. *Zhungguo Shiyan Fangjixue Zashi*, 18: 28–2031.

27 Essential Oils as Carrier Oils

Romana Aichinger and Gerhard Buchbauer

CONTENTS

27.1 INTRODUCTION

Essential oils (EOs) as carrier oils are receiving more and more attention in recent years. As it is already known, EOs can be obtained from aromatic plants and consist of terpenes, phenylpropanoids, and some other components. In comparison to many other drugs, the mono- and sesquiterpenes, the main components of the complex aromatic volatile mixtures, are rather small molecules with low molecular weights, and thus they are able to overcome the skin barrier and penetrate through human skin. As the terpenes reversibly reduce the barrier function of the *stratum corneum* (SC), they enable drugs with higher molar mass to diffuse into lower skin layers. There are different pathways for percutaneous penetration, caused by penetration enhancers, provided that the intercellular lipid domain of the SC is the main pathway, which will be discussed properly afterward. This paper also focuses on the chemical structures and the lipophilicity and hydrophilicity of both penetration enhancers and drugs and how these influence their interaction. Many medical treatments can be improved with transdermal drug delivery (TDD) because of the positive effects TDD provides, such as avoiding the first-pass effect, reducing the gastrointestinal side effects, or decreasing the frequency of administration. For that reason, transdermal drug delivery gains in importance concerning different diseases and therapies (Fung Chye Lim et al., 2009; Herman and Herman 2014; Chen et al., 2016; Jiang et al., 2017).

27.2 ESSENTIAL OILS IN GENERAL

Essential oils are volatile mixtures of many different components with diverse structures and low molecular weights from aromatic plants. They are complex, multi-component systems in which terpenes play the main part beside some other non-terpene components like phenylpropanoids and fatty acids. The natural agents can be used in many different domains, such as anti-inflammatory therapy, wound healing, inhalation, aromatherapy, anticancer therapy, cosmetic usage, and some others, because of their numerous biological activities (antibacterial, antiviral, antifungal, antioxidant, anti-inflammatory, antidiabetic, and so on). Thus, they are applied orally, nasally, or topically as pure essential oils or formulations like emulsions, gels, or solutions to treat but also prevent many human diseases. To extract the essential oils from aromatic plants, the following techniques are possible: steam or water distillation as well as expression under pressure of the peels of citrus fruits (Cal, 2005; Edris, 2007; Pharmacopoeia Europaea, 2008; Herman and Herman et al., 2014).

27.3 ESSENTIAL OILS AS PENETRATION ENHANCERS

Essential oils (EOs) and their active constituents can be used as natural, safe, and clinically acceptable penetration enhancers for hydrophilic and lipophilic drugs because of their physicochemical properties, their structure, their activity on the skin layers, and the advantages they cause. They show a better safety profile in comparison to some other chemical penetration enhancers like azone, sulphoxides, pyrrolidones, alcohols, or surfactants that often cause skin erythema and edema. Terpenes are one of the main constituents but, remaining in the EOs, they penetrate slower through the skin than as pure substances. EOs facilitate drug delivery through the skin and induce, at best, a reversible, temporary reduction in the stratum corneum (SC) barrier function without damaging viable cells. Characteristics that absolutely influence the penetration are concentration of active compounds, polarity, molecular weight (<500 Da), solubility of molecule in water and oil, and composition of formulation. What is also crucial to obtain the optimal penetration enhancement effectiveness is the volatility (0.759–1.67 mg/h/cm^2), which can be determined by weight loss. If the volatility is too high, it is not possible to achieve good interaction with SC lipids because of the rapid evaporation. It can be assumed that enhancers to be classified as promising penetration enhancers with good activity should be easily removable from the skin; should not lead to loss of body fluids or electrolytes; and must not cause irritation, allergy, or toxicity, and should possess good solvent properties, should be combinable with other drugs, and should not bring color, taste, or odor (Cal et al., 2005; Fung Chye Lim et al., 2009; Herman and Herman, 2014; Feng et al., 2015; Jiang et al., 2017).

27.4 TERPENES

27.4.1 Classes and Complexes

Mono- and sesquiterpenes consisting of only carbon, hydrogen, and oxygen atoms are mostly volatile and extracted from medicinal plants. Those terpenes are formed by a number of isoprene units (C_5H_8), which are repeated in an appropriate number. Monoterpenes (C_{10}) are built from two isoprene units, sesquiterpenes (C_{15}) from three isoprene units; those are the two most important penetration enhancer classes. Monoterpenes seem to be a little more effective as they do not possess the bulky structure like sesquiterpenes do. So, they are able to overcome the biological barriers easily as they constitute small, lipophilic, and unionized molecules. It should be noted that many of the terpenes used as penetration enhancers are oxygen-containing terpenes; thus, they can form hydrogen bonds while interacting with the drugs enhanced, and so, they possess higher ability to support drug permeation. The drug-terpene complex formed by hydrocarbon terpenes needs either donor/acceptor interactions, van der Waals forces, or hydrogen-bond-donor-л interactions. These formed complexes increase the SC partition of the drug. Furthermore, terpenes can loosen the hydrogen bonding network of the SC simply by offering a functional group that can donate or accept a hydrogen bond, hence increasing the fluidity of SC lipids (Cal et al., 2005; Fung Chye Lim et al., 2009; Chen et al., 2016; Jiang et al., 2017).

27.4.2 Hydrophilic, Lipophilic, Amphiphilic Terpenes

Natural terpenes in comparison to conventional synthetic penetration enhancers (PEs) possess higher enhancement activity of hydrophilic and lipophilic compounds and show lower skin irritation potential. As the natural terpenes are non-irritating and non-toxic, they can be regarded as promising PEs with a "generally recognized as safe" (GRAS) status. Highly lipophilic drugs require lipophilic terpenes, so-called hydrocarbon terpenes, to increase their penetration through the skin, while terpenes with polar groups show a better enhancement effect for hydrophilic drugs. However, amphiphilic terpenes show the best penetration enhancement effect for most of the drugs, because they are able to disrupt the highly organized lipid packing in the stratum corneum owing to their amphiphilic structure (Herman and Herman, 2014; Chen et al., 2016; Jiang et al., 2017).

27.4.3 Structure of Terpenes

Another factor that influences the penetration enhancement activity is the structure of the terpenes. A long-chain alkyl structure increases the enhancement effect better than a ring structure. Furthermore, farnesol, which represents a chain molecule, in comparison to cyclic terpenes, possesses better enhancement effects for hydrophilic drugs because of its lower vaporization energy as it can be supposed. In contrast, ring-structure-containing terpenes like menthol and camphor show less of an effect. The size and degree of long-chain alkyl functionality correlates to the degree of SC lipid disorder (Chen et al., 2016; Jiang et al., 2017).

27.4.4 Boiling Point of Terpenes

What acts a part as well is the boiling point. It relates to the penetration enhancement effect of the terpene. The lower the boiling point is, the more effective is the terpene. With a low boiling point comes relatively weaker cohesive forces. Thus, it is easier for the functional groups of terpenes to react with the skin ceramides through competitive hydrogen bonding, as the oxygen of the functional groups is mostly free. Therefore, by weak cohesiveness or self-association of the molecules of the terpenes, they are able to interact with the SC more easily and modify the barrier function (Chen et al., 2016; Jiang et al., 2017).

27.4.5 Concentration of Terpenes

Besides, it is important to choose the right terpene concentration, which is arranged between 1% and 5%, whereas 3% is considered to be the most effective concentration for many drugs. If the concentration is chosen to be too low, the optimal effect is not going to be achieved; on the other hand, if the concentration is too high, side effects like skin irritation have to be expected, but this theme will be dealt with in detail in Section 27.6.1.

Therefore, it is necessary to find the optimal balance between potency and safety and to keep it constant (Herman and Herman, 2014; Chen et al., 2016).

27.4.6 Increase of Terpenes

The combination of terpenes can lead to an increase of the permeation enhancement effect, just as physical enhancers do in combination with terpenes. Ultrasound, as well as iontophoresis, shows synergistic effects when combined, or rather pretreated the skin, with (mono-) terpenes (Fung Chye Lim et al., 2009; Jiang et al., 2017).

27.4.7 Vehicles of Terpenes

The vehicle system in which the EO respectively the terpene will be carried also affects the penetration enhancement effect. Based on the physicochemical properties of the vehicles and their interaction with the SC, they cause a difference concerning the penetration. Some organic solvents like ethanol or propylene glycol even show a penetration-enhancing function and, hence, induce synergistic effects, but work as the formulation of the vehicle to dissolve the drug and the EO. Invasomes, which describe novel vesicles for transdermal drug delivery, consist of phospholipids, little amounts of ethanol, and terpene mixtures as enhancers. In comparison to liposomes or ethosomes, they cause a higher penetration rate. These vesicles are proved to possess the positive effects of both liposomes as potential carriers and terpenes as they are able to modify the skin barrier. Vehicles are needed because of the difficulties that occur while dissolving lipophilic EOs in water and the possible skin irritation when directly applying the EOs on the skin. Namely, the accumulation of the terpenes, when applied as pure EOs, is considerably higher than in vehicles, although proportionality to the penetrant concentration is not discovered. When not applied as pure EOs, there are a few possibilities of dermatological formulations like o/w emulsions, suspensions, oily solutions, hydrogels, ointments, or multi-layer transdermal patches. If grape seed oil is used as vehicle for EOs like linalool, linalyl acetate, terpinen-4-ol, citronellol, or α-pinene, the EOs are present completely dissolved compared to the o/w emulsion, where they are either dissolved in the oily internal phase or in micelles of surfactants or emulsified with the surfactants, forming an internal oily phase (Cal et al., 2005; Fung Chye Lim et al., 2009; Herman and Herman, 2014; Chen et al., 2016; Jiang et al., 2017).

Furthermore, in formulations, terpenes can be combined with other active and synergistic compounds and do not only appear as drugs themselves. But not only the formulation ingredients affect the activity of the penetration enhancers, pH values and skin type also are co-dominant (Cal et al., 2005; Chen et al., 2016)

27.5 SKIN

The skin represents the most accessible and biggest organ of the body with a surface area of about 1.72 m^2. There are many functions it has to fulfill, such as to protect the organism from the outer environment, microorganisms, and damage; regulate the body temperature; avoid transepidermal water loss; and act as an organ of perception. However, although the main task of the skin is protecting the inner organism, it is also able to absorb agents that are applied on the skin through hair follicles or reversibly disrupted skin barrier. This causes increasing interest in transdermal drug delivery (Fung Chye Lim et al., 2009; Herman and Herman, 2014)

27.5.1 Layers

There are three main layers in which the skin can be divided, namely, the epidermis, dermis ,and subcutis. The epidermis itself consists of five other layers, which are identified as stratum corneum (SC)—the outermost skin barrier—stratum lucidum, stratum granulosum, stratum spinosum, and stratum basale. For the topical application of drugs, the SC is the most interesting layer which has to be bridged (Fung Chye Lim et al., 2009).

27.5.2 Stratum Corneum (SC)

The *stratum corneum* that belongs to the epidermis and builds the outermost layer of the skin can be considered as a significant transport barrier and rate-limiting layer which hinders the diffusion of compounds into and out of the host. It consists of 15–20 layers of non-viable, flattened cells embedded in a lipid domain, and its arrangement can be described as "brick-and-mortar," where the keratin-filled corneocytes represent the bricks and the intercellular lipid bilayers, which are composed of 50% ceramides, 25% cholesterol, 15% free fatty acids, and some phospholipids, can be seen as the mortar that causes the permeability properties of SC. Due to this structure, the SC is presumed to be a protective and impermeable barrier to drug diffusion, which justifies the usage of penetration enhancers to overcome the barrier (Fung Chye Lim et al., 2009; Chen et al., 2015; Chen et al., 2016).

27.5.3 Skin Models

To test the penetration of active compounds through the skin, some *in vitro* and *in vivo* models are necessary. Although human skin is considered as the most reliable model, animal models are needed because of the low availability, the limited access, and the high costs of human skin. Instead of human skin, animal skin of rat, mouse, guinea pig, snake ,and some others can be used, considering the differences in skin structure (Herman and Herman, 2014).

27.5.4 Franz Cell

For the *in vitro* permeation experiments, Franz-type diffusion cells can be used. Therefore, the prepared skin surface of about 1.7 cm^2 is mounted between the donor and receptor chambers with the SC side facing the donor compartment. The donor compartment is filled with vehicle-containing drugs and penetration enhancers while the receptor phase consists of buffer pH 7.4, maintained at 37°C and stirred at about 500–600 rpm. To find out whether or not the penetration enhancer is helpful for the drug to overcome the SC barrier, the experiment is carried out with and without it, and the drug concentrations of the collected samples from the receptor compartments are compared (Fang et al., 2004; Brito et al., 2009; Feng et al., 2015; Jiang et al., 2017).

27.6 ADVANTAGES

27.6.1 Advantages of Natural Penetration Enhancers

Natural penetration enhancers (PEs) such as EOs, which come into operation in transdermal drug delivery, are an alternative to synthetic PEs like azone, alcohols, pyrrolidones, sulphoxides, fatty acids, solvents, and surfactants. Natural PEs are more and more preferred, on the one hand, because of their relatively low price and, on the other hand, because of their promising enhancement activities and their hardly existent adverse effects when administered at low concentrations (1%–5%). They facilitate the penetration of both hydrophilic and lipophilic drugs, and besides, they can be considered as safe PEs due to their metabolism, which is quite rapid. So the EOs are quickly eliminated and excreted and thus do not accumulate in the organism. Besides, they cause less skin irritation and higher permeation flux determined by the enhancement ratios (Edris et al., 2007; Herman and Herman, 2014; Feng et al., 2015; Chen et al., 2016; Jiang et al., 2017).

27.6.2 Advantages of Transdermal Drug Delivery

Transdermal drug delivery, in general, represents a positive option to the conventional drug administration routes, such as oral and nasal routes, when not administrated invasively. It causes avoidance of the hepatic first-pass effect and differences in gastrointestinal absorption and metabolism, which normally lead to low bioavailability when administrated orally. Thus, an improved bioavailability, as well as a steady-state plasma level with minimal fluctuations, can be attained with transdermal drug application. Beside the reduction of the gastrointestinal side effects, the administration frequency can be decreased, hence the patient's compliance is improved (Fung Chye Lim et al., 2009; Chen et al., 2015; Chen et al., 2016).

27.7 SIDE EFFECTS OF NATURAL PENETRATION ENHANCERS

27.7.1 Skin Irritancy and Toxicity

EOs and their components do not only show the advantages of an increased skin permeability and a higher drug absorption by disturbing the skin barrier, the disruption of the SC can also lead to cytotoxicity, skin irritation, and allergic reaction when applied at too-high concentrations, although natural PEs are less toxic than synthetic PEs. Furthermore, sesquiterpenes seem more potent compared to monoterpenes because of their chemical structure and their potential in interrupting the SC barrier. To reduce or avoid the adverse reactions it is important to apply the lowest possible, yet still efficacious, concentration of diluted EO (Herman and Herman, 2014; Chen et al., 2016).

27.7.2 Transepidermal Water Loss (TEWL)

To check and demonstrate the health of the skin barrier function, TEWL provides an effective index, where the skin irritation can be determined as well. Skin irritancy is a reaction to substances that cannot be tolerated by the skin and provoke inflammation of skin and itchiness. Non-oxidized terpenes and terpenes at an appropriate concentration do not cause any irritation (Chen et al., 2016).

27.8 MECHANISM OF ACTION

As the natural agents enhance the penetration of drugs through the skin, their main task is changing the structure of the SC and interacting with its lipids to reduce the barrier resistance and increase the drug diffusivity. To improve the permeation of drugs that are normally poorly absorbed, incorporation of PEs into drug formulations is conducive. Drug-permeation enhancement can be induced either by increasing the permeability of the drug into the SC or by reducing the tortuous pathway in the SC, as well as using both options (Brito et al., 2009).

PEs use four different mechanisms of action: (a) disrupting the highly ordered intercellular lipid structure between corneocytes in SC via extraction, fluidization, polarity alteration, and phase separation, which leads to higher permeability; (b) interacting with the intercellular domain of keratinized protein to induce their conformational modification; (c) increasing the partitioning—several solvents alter the SC properties and thus force up the partitioning of a drug; and (d) enhancers acting on desmosomal connections between corneocytes or altering metabolic activity within the skin. There are also three possible pathways that can be used, which include intracellular diffusion across the corneocytes of SC, penetration through the SC intercellular lipid spaces, and appendage penetration across hair follicles and sebaceous and sweat glands. The intracellular pathway is normally chosen by hydrophilic compounds while lipophilic permeants prefer the intercellular route. Although most molecules cross the SC via both routes, the intercellular lipid domain of the SC describes the main pathway; thus, the extraction of SC lipids can be seen as one of the key mechanisms (Fung Chye Lim et al., 2009; Herman and Herman, 2014; Chen et al., 2016; Jiang et al., 2017).

To elucidate which mechanisms take place, different analytical techniques wuch as DSC (differential scanning calorimetry) and FTIR (Fourier transform infrared) spectroscopy can be used. These techniques will be examined in detail in Section 27.8.5.

Besides the four main mechanisms for the enhancement of permeability mentioned above, there are some other possible modes of action. Among those ranks the inhibition of detoxifying enzyme CYP 450, which leads to a delayed metabolism and excretion, so the drug remains in the organism and shows its effect longer (Tak and Isman, 2017).

27.8.1 Effect on Stratum Corneum Lipids

As it is known, lipids are composed of a polar head and lipophilic tails. Therefore, interactions between terpenes and SC lipids are possible at two sites, as there are the polar head groups in the polar transcellular pathways and the lipoidal intracellular pathways, as well as the lipophilic tails in the intercellular lipid pathways. Depending on the biophysical alterations of the skin barrier, it is about extraction or fluidization. To elucidate which of the two options occurs, attenuated total reflection-Fourier transform infrared spectroscopy (ATR-FTIR)—explained in detail in 27.8.5—can be applied (Chen et al., 2016).

27.8.2 Effect on Hydrogen Bond Networks

Hydrogen bonding describes how ceramides are held together tightly in the SC. The hydrogen-bond networks are built at the head of ceramides and are the reason for the stability and strength of the lipid bilayer and thus cause the barrier function. Terpenes with functional groups have the ability to loosen the ceramide network as they are able to accept or donate a hydrogen bond. As terpenes build new hydrogen bonds with the ceramides, the existing hydrogen-bond network between ceramide head groups will be disintegrated, and thus, the permeation of the drug will be facilitated. Besides the increase in diffusivity of the SC caused by the hydrogen-bonding potential of terpenes with functional groups, they are also able to affect the conductivity as they build new polar channels near the head groups of SC lipids that can be passed by ions and polar molecules (Fung Chye Lim et al., 2009; Chen et al., 2016).

27.8.3 Effect on SC Partition of Drugs

With transdermal drug delivery comes first the partition of the drug in the SC, and this process is improved when terpenes are existent in the intercellular lipid domain in dissolved form. The terpene uptake correlates with an increased drug partition coefficient and the enhanced partition coefficient is induced by the interaction between terpenes and drugs via hydrogen bonding—mentioned in Section 27.8.2. This procedure builds the basis for penetration enhancement (Chen et al., 2016).

27.8.4 Affecting Factors

There are a lot of factors that affect the penetration enhancement effect of the terpenes, namely the skin type and origin to start with the treated organ. Besides the pretreatment, the vehicle system and the ingredients, their structure, concentration, and polarity, as well as pH values, are crucial. Terpenes themselves can also influence their activity as some of them—for example, menthol—are able to induce physiological reactions in the living skin, such as increase of skin temperature and vasodilatation. They also have to be suitable for the drug they enhance as polar terpenes are beneficial for hydrophilic drugs, and hydrocarbon terpenes go with lipophilic drugs. Not only the terpenes influence their effects; the drugs are co-decisive too. Drug lipophilicity is presumed to be a predominant factor affecting the skin permeability of drugs, not to forget the molecular weight, where the upper limit seems to be about 500 Da and the melting point. To get information whether a molecule is able to penetrate through the lipophilic SC barrier or not and how it can partition between the SC and hydrophilic or lipophilic vehicle, it is important to know the octanol-water partition coefficient, normally presented as logP (logarithm). The optimal logP for percutaneous penetration is in the range of 1–3, obtained by less lipophilic compounds that penetrate into the skin easily, while a logP >4 caused by highly lipophilic compounds that permeate less through the skin is not desired (Cal et al., 2005; Herman and Herman, 2014; Chen et al., 2016).

Physical methods can affect the skin morphology and thus make it easier for the drugs to permeate. There are a few methods, such as low-frequency ultrasound, also known as sonophoresis and electrical current; namely, iontophoresis and electroporation. By increasing the free volume, space in between the lipid lamellae cavitation is induced by ultrasound, while iontophoresis works with current defects, which ionized drug molecules can get through, and electroporation causes aqueous pore formation for enhancing drug permeation (Fung Chye Lim et al., 2009).

27.8.5 Screening-Techniques

To investigate the mechanism of action and the biophysical alterations of the SC barrier, different analytical techniques come into operation.

One of the most frequently used techniques is the ATR-FTIR (attenuated total reflection-Fourier transform infrared) spectroscopy, also known as FTIR spectrometry. It is often used due to its ability of obtaining the information for how SC lipids and keratins are arranged and how their conformation will be changed after application of penetration enhancers. All the alterations can be seen on the infrared spectra bands of the SC, which are referred to the lipid or protein molecular vibrations. After application of terpenes, stretching peaks near 2850 cm^{-1} (C–H symmetric stretching absorbance frequency peak) and 2920 cm^{-1} (C–H asymmetric stretching absorbance frequency peak) can be noticed because of hydrocarbon chains of SC lipids that give rise and near 1540 cm^{-1} (amide 2) and 1640 cm^{-1} (amide 1) because of SC proteins that give rise to CN stretching and NH bending vibrations. The shift to a higher frequency of C–H stretching peaks, which signifies perturbation of SC lipids and further lipid fluidization, can be noticed in the case in which methylene groups of the SC lipid alkyl chains change from a trans to a gauche conformation, that is more energetic. Shifts to a high frequency mean strong perturbation. Also, the two C–H peak areas and heights act proportionally to the SC lipid amount, and thus, the lipid extraction caused by a penetration enhancer results in a decrease of peak area and height. In contrast to the C–H peaks and shifts that stand for the alteration of SC lipids, the amide peaks represent the change in protein conformation. With FTIR findings, it has been proved that partial extraction is a major mechanism for hydrocarbon, alcoholic, and oxide terpenes. Further, it can be seen that increased drug permeation correlates with delipidization (Fung Chye Lim et al., 2009; Chen et al., 2016)

DSC (differential scanning calorimetry) is the other technique that is often applied because of its sensitivity to the thermal effects that come with phase changes like melting of lipids at 65°C, melting of lipid-protein complexes at 75°C, and protein denaturation at 95°C or transitions of the components of the SC layer. Phase separation denotes a weakened SC resistance caused by the formation of interfacial defects in the lamellae. With the phase separation, a decrease in lipid melting transition can be detected (Fung Chye Lim et al., 2009).

27.9 COMBINATION OF PENETRATION ENHANCERS AND DRUGS

27.9.1 Essential Oils and Anti-Inflammatory Drugs

Anti-inflammatory drugs, more precisely non-steroidal anti-inflammatory drugs (NSAID) like ibuprofen, diclofenac, indomethacin, salicylic acid, and so on come into operation when suffering from inflammation and pain based on dysmenorrhea and rheumatic disease, for example, owing to their analgesic, antipyretic and anti-inflammatory activity. Unfortunately, the oral application of NSAIDs can provoke a lot of gastrointestinal adverse effects such as ulceration and perforation of the stomach and intestines, as well as bleeding. Therefore, the topical application of NSAIDs presents a safer potential option to oral therapy with fewer side effects. However, the poor skin permeability of the NSAIDs and the associated low therapeutic blood concentrations build an obstacle to transdermal drug delivery that needs to be conquered with EOs as penetration enhancers. Hydrophilic terpenes are supposed to be the best penetration enhancers for NSAIDs, under which alcohol terpenes are the most effective followed by ketone and oxide terpenes (Akbari et al., 2015; Chen et al., 2015)

27.9.1.1 Chuanxiong Oil and Ibuprofen

Chuanxiong oil is obtained from Rhizoma Chuanxiong, part of *Ligusticum chuanxiong*, which is a member from the Umbelliferae family; it is often used in traditional Chinese medicine to cure gynecological and cardiovascular diseases. The EO is extracted by steam distillation and consists of many phthalide components under which ligustilide with 41.28% forms the major component that is able to attenuate pain. Thus, the EO shows analgesic activity itself and causes synergistic effects beside its penetration enhancement effect. Chuanxiong oil with its phthalides is suggested to be the best appropriate penetration enhancer for the lipophilic ibuprofen—which possesses the poorest bioavailability among NSAIDs—due to the increased steady state flux and the highest obtained permeation rate of 52.05 ± 7.83 $\mu g/cm^2/h$ compared with the control rate of 14.57 ± 3.47 $\mu g/cm^2/h$. To facilitate the penetration of ibuprofen in order to obtain an effective blood level, Chuanxiong oil disrupts and extracts the SC lipid structure and makes it possible for ibuprofen to overcome the barrier through intercellular spaces. So, the EO and the drug make a good combination to treat dysmenorrhea as they reduce cramping pain and writhing times—attributed to decreased Ca^{2+} levels, increased NO levels, and the reduction of the pro-inflammatory prostaglandins—when applied at the abdominal region (Chen et al., 2015).

27.9.1.2 Rosemary Essential Oil and Na-Diclofenac

Rosemary, also known as *Rosmarinus officinalis*, belongs to the Lamiaceae family and is mostly located in Mediterranean and Iran regions. From the evergreen plant, which shows anti-inflammatory, antiseptic, antioxidant, anti-aging, healing, anti-rheumatic, as well as antispasmodic activity, rosemary EO can be obtained. The EO that is rich in monoterpenoids—identified as 1,8-cineol (16.0%), α-pinene (13.4%), camphor (7.9%), verbenone (5.8%), borneol (5.2%), and camphene (5.0%), beside some other co-major constituents like bornyl acetate (6.5%) and (*E*)-caryophyllene (3,8%)—represents a good candidate in enhancing the penetration of Na-diclofenac that counts among NSAIDs. With this formulation, inflammation of skin tissues as well as supporting structures of the body-bones, joints, ligaments, tendons, and muscles can be treated. The best analgesic effect was noticed with the combination of 1% rosemary essential oil and 1% Na-diclofenac; however the most enhancing effect could be seen at the concentration of 0.5% and 1% of the EO. Rosemary EO helps Na-diclofenac permeate through the skin by vasodilatation, increased disorder of the SC lipids, and complex formation between the enhancer and the drug or structures of the SC (Akbari et al., 2015).

27.9.1.3 *Alpinia oxyphylla* Essential Oil and Indomethacin

Alpinia oxyphylla, a plant that belongs to the Zingiberaceae family, was often used in treatment of diarrhea and gastralgia and also in neuroprotection because of its anti-angiogenetic and anti-oxidative effects. With hydro-distillation from fruits and leaves, the EO can be isolated. With this method, both EOs show a better enhancement effect than an extract obtained with organic solvents like acetone because of the synergistic effects between the constituents. Both the fruit oil and the leaf oil can act as penetration enhancers; however, the fruit oil supports the more lipophilic drugs because of its constituents. The fruit oil consists of 14 hydrocarbon terpenes that make 40.4% of the total content and 15 oxygenated terpenes that make 35.6%. Besides, nootkatone was found with 3.9%, a constituent with beneficial properties. In comparison, the leaf oil contains 27.7% hydrocarbon terpenes, 50.8% oxygenated terpenes, and 1.4% nootkatone. Thus, for indomethacin as a lipophilic drug beneath NSAIDs, *A. oxyphylla* fruit oil is the better candidate to enhance the diffusion through the skin because of the higher content of hydrocarbon terpenes. Furthermore, it is proved that with pretreatment with the EO, better results, such as increase in skin flux values of drug and cumulative amount, can be achieved due to directly acting on the skin and avoiding co-solvent effects on the thermodynamic activities of the drug. The concentration of the EO is also crucial, as the enhancement effect and the reduction of lag time depend on it. So, 3% fruit oil with an enhancement ratio (E_r) of 10.16 exhibits the best effect, followed by 5% fruit oil with E_r of 5.25%, and 3% leaf oil with E_r of 4.61, noting that the differences in E_r can be explained by the different boiling points of the terpenes as well as their molecular weights. However, the 5% level shows an increase in lag time while 3% forms the optimum concentration for both EOs to decrease the lag time to reach steady state flux. The enhancement effect of the *A. oxyphylla* EO may be attributed to disruption of the SC barrier (Feng et al., 2015).

27.9.1.4 *Lippia sidoides* Essential Oil and Salicylic Acid

Lippia sidoides Cham belongs to the Verbenaceae family and is known for its antimicrobial and larvicidal activity, as well as for gastrointestinal treatment. The essential oil basically consists of thymol (59.7%), (*E*)-caryophyllene (10.6%), and *p*-cymene (9.1%). The *L. sidoides* EO (LSEO 1%) as enhancer makes a good formulation in combination with propylene glycol (PG:phosphate buffer 1:1) as co-solvent to facilitate the penetration of salicylic acid. Due to propylene glycol enhanced solubility of LSEO and its diffusion into SC is possible, as well as LSEO causes enhanced PG penetration by its incorporation between the intercellular lipids. The enhancer provokes the disruption of the lipid bilayer and the extraction of SC lipids, as it can be demonstrated with FTIR spectroscopy. Thus, the flux and permeability coefficients of salicylic acid are increased when applied in combination with the penetration enhancer (Brito et al., 2009).

27.9.1.5 Sweet Basil Oil and Indomethacin

Sweet basil oil comes from *Ocimum basilicum* that belongs to the Lamiaceae family and is not only a culinary herb but also a medicinal plant that is the host of many EOs. There are two fractions of sweet basil oil, namely the lower-polarity fraction OB-1 and the higher-polarity fraction OB-2. OB-1 consists predominantly of hydrocarbon components, only estragol (as the major constituent with 76.7%) belongs to the oxygenated components. Because of the lower polarity OB-1 at the concentration of 1%, 3%, or 5% is an adequate penetration enhancer for the lipophilic indomethacin—that prefers a non-polar pathway like intercellular lipids of the SC—and further enables the transfer to the circulation. In comparison to that OB-2, it possesses higher polarity due to oxygenated terpenes that outbalance in this fraction, under which phytol presents the major component with 52.3%. OB-2 shows a relatively low permeation enhancement, but it enables the drug to retain within the skin reservoir. What influences the penetration and the skin reservoir of indomethacin is the pretreatment with 3% of OB-1 or OB-2 in 25% EtOH for 1 hour, which results in avoidance of co-solvent effects on the thermodynamic activities of the drug while directly acting on the skin structure. The presence of EtOH beside the enhancer leads to accumulation in the tissue and a higher partitioning of the drug because of the affinity the drug shows to the solvent. The partitioning of the EOs to the SC can lead to decreased polarity of the SC, which further causes an enhanced penetration of indomethacin, a lipophilic drug into the skin. Thus, the partitioning is considered to be the main mechanism beside the disruption of the SC lipids (Fang et al., 2004).

27.9.1.6 *Zanthoxylum bungeanum* Essential Oil and Indomethacin/5-Fluorouracil

See Section 27.9.4.3.

27.9.2 Essential Oils and Antiseptic Drugs

27.9.2.1 1,8-Cineole and Chlorhexidine

1,8-Cineole, a monoterpene cyclic ether is the main constituent of the eucalyptus oil that is known for its penetration enhancement of lipophilic drugs. Because of the ability of the EO to enhance the absorption of chlorhexidine and its antimicrobial activity, but due to the variability of its constitution, a purified solution of 1,8-cineole should represent a potent alternative with a synergistic effect. Chlorhexidine comes to usage with skin antisepsis; however the SC barrier and the poor diffusion into the skin, caused by the large molecular size and its binding to intercellular lipids in the SC, hinder the treatment of endogenous microorganisms in deeper skin layers. Therefore, 1,8-cineole should facilitate the penetration of chlorhexidine to eradicate the pathogens and lower the risk of infection as it interacts with and disorders the lipids of the SC. In comparison to the application of the combination of 2% (w/v) chlorhexidine with 70% (v/v) isopropyl alcohol, but without the penetration enhancer 2% (v/v) 1,8-cineole, it can be seen that with 1,8-cineole, the concentration of chlorhexidine in the skin is on average 33.3% higher, although the size of the effect does not show significant differences demonstrated by the depth of penetration. Thus, according to the higher concentration, it is proved that 1,8-cineole promotes the permeation of chlorhexidine (Casey et al., 2017).

27.9.3 Essential Oils and Vesicular Carriers

Vesicular carriers are promising candidates for transdermal drug delivery as they encapsulate drugs that are not able to achieve the desired effects or cause side effects like gastrointestinal irritation when administered orally. There are different types of vesicular formulations, namely transferosomes, liposomes, ethosomes, and glycerosomes. Liposomes often come with the disadvantage of low encapsulation efficiency, instability, and limited drug delivery efficiency, although they increase the drug accumulation within the tissue very well. Ethosomes correlate to liposomes with addition of high concentration of short-chain alcohols like ethanol. The alcohols lead to improved deformability of vesicles; however, skin irritation is possible. Glycerosomes that are named after the key-component glycerol, a harmless short-chain alcohol, also improve the deformability and enhance the penetration. Combined with EOs, it is easier for the encapsulated drugs to permeate through the skin and release from the vesicles; note that all vesicular formulations are supposed to increase transdermal flux more than tinctures (Rajan and Vasudevan, 2012; Zhang et al., 2017)

27.9.3.1 Terpenes and Ultra-Deformable Liposomes of Sodium Fluorescein

Ultra-deformable liposomes in combination with terpenes attract attention as they are able to encapsulate and deliver drugs—here, sodium fluorescein to detect the effect—through the skin by using follicular pathways to bypass the outermost layer of the skin that forms a barrier to transdermal drug delivery. The fluidity of these flexible vesicles that consist of phospholipids and surfactant is increased as small, lipophilic monoterpenes like 1,8-cineole or limonene are added. The terpenes alter the fluidity at the C16 atom, the lipophilic region of the acyl chain of the phospholipid bilayer. With the increased fluidity and the reduced liposomal size due to the higher amount of terpenes localized in the outer layer, the flux and thus the penetration of drugs is improved as the permeation through hair follicles becomes the preferred pathway of the ultra-deformable liposomes with terpenes (Thirapit and Tanasait, 2015).

27.9.3.2 Terpenes and Liposomes of Antisense Oligonucleotide

As liposomes represent common carriers for transdermal drug delivery, they gain interest respective to a promising method of cancer therapy—namely, gene therapy—as they can deliver antisense oligonucleotide (AsODN) for treating lung cancer. The vesicle enables the uptake via endocytosis and thus avoids the difficult penetration through the SC. The addition of terpenes—in this case, 1,8-cineole—promotes the liposomal gene delivery by fluidizing the bilayer and thus improves the specific activity of AsODN. 1,8-Cineole is able to increase the enhancement effect, which is dependent on concentration and chemical structure of the enhancer, up to 40 times, although the storage of the lipophilic terpene in the lipid bilayers causes reduction in the encapsulation efficiency (Saffari et al., 2016).

27.9.3.3 Eucalyptus Oil and Transferosomes of Ketoconazole

Transferosomes represent synthetic vesicles that surround drugs and imitate cell vesicles so they enable drug delivery by simply crossing barriers that cannot be passed by drugs themselves. To define it, they improve defective transdermal permeation. Transferosomes form a vesicle with an aqueous core covered with complex lipid bilayers, so they are very flexible, ultra-deformable, and stress responsive. If the transferosome incorporates ketoconazole, a broad-spectrum antifungal agent, life-threatening fungal infections can be treated. This turned out to be difficult when applied orally because of its incomplete absorbance. The inclusion of a penetration enhancer into the transferosomal gel formulation, in this case, eucalyptus oil, is advised because it facilitates the release and permeation of ketoconazole as it increases the diffusion rate (Rajan and Vasudevan, 2012).

27.9.3.4 *Speranskia tuberculata* Essential Oil and Glycerosomes of Paeoniflorin

Glycerosomes are carriers with vesicle structure for transdermal drug delivery that possess the ability to encapsulate drugs with a poor diffusion rate and poor bioavailability and transfer them through the skin barrier. They are formed by 5% (w/v) phospholipids (with their concentration increases the encapsulation efficiency), 0.6% (w/v) cholesterol (causing membrane rigidification), and 10% (v/v) glycerol (which influences the glycerosome particle size as it becomes bigger with the amount of glycerol in the water phase). If this vesicular formulation encapsulates paeoniflorin a potent alternative to treat rheumatoid arthritis—a long-lasting autoimmune disease that leads to synovium inflammation and lesions in joints—is built. Paeoniflorin is known for its anti-inflammatory and immune-regulatory effects and derives from *Paeonia lactiflora* Pall (Paeoniaceae). As monoterpene glucoside, it is a hydrophilic drug that shows poor bioavailability with oral application and a low encapsulation rate. To facilitate the transdermal delivery of paeoniflorin in the glycerosomes, it is recommendable to add 2% (v/v) *Speranskia tuberculata* EO (STEO). This EO consists of 14 compounds, predominantly sesquiterpenes that are able to disturb the lipid bilayers. Thus glycerosomes packed with paeoniflorin and STEO—characterized by their spherical shape and uniform size—produce the best results in transdermal performance as they enhance the transdermal flux and increase the accumulation of paeoniflorin in the inflamed synovium and even maintain the drug concentration at a high level after a long time (Zhang et al., 2017).

27.9.4 Essential Oils and Cytostatic Drugs

With minimizing the serious adverse effects, such as gastrointestinal problems, hair loss, leukopenia, cardiotoxicity, nephrotoxicity, loss of body mass, immunosuppression, drug resistance, and many others that come with anticancer therapy, the transdermal application of cytostatic drugs with EOs gains attention. There are some advantages to commend the transdermal usage of cytostatic drugs, namely improved efficacy with decreased dose, upkeep of the same effect with less toxicity, and reduced drug resistance development, as well as potential synergistic effects. Thus, EOs promote the effectiveness of chemotherapeutic drugs (Amaral et al., 2016).

27.9.4.1 *Myrica rubra* Essential Oil and Doxorubicin

Myrica rubra from the Myricaceae family delivers the EO that comes to operation in treating cancer cell lines with the cytostatic doxorubicin. It consists mostly of sesquiterpenes which are known to promote skin permeability, namely β-caryophyllene oxide, α-humulene, trans-nerolidol, and valencene. The constituents not only enhance the penetration of doxorubicin, they possess anticancer and anti-proliferative activity themselves. So, they can act synergistically and improve the effect of doxorubicin as they increase its efficacy and accumulation beside the increase of reactive oxygen species formation. For doxorubicin, it is important to be enhanced by the EO and its components, because many cancer cell lines react with resistance by increasing drug efflux, more precisely, by increasing the expression of ATP-binding cassette transporters that can be inhibited by the sesquiterpenes. However, the effect of the sesquiterpenes from *M. rubra* EO and doxorubicin depends on the cancer cell lines. Sesquiterpenes can only influence doxorubicin efficacy in the sensitive and partly resistant cancer cells, but not in completely resistant cells (Ambrož et al., 2017).

27.9.4.2 *Mentha x villosa* Essential Oil and 5-Fluorouracil

The essential oil from *Mentha x villosa* Hudson, which belongs to the Lamiaceae family, possesses cytotoxic activity beside antimicrobial, antinociceptive, cardiovascular, spasmolytic, and some other effects. Together with 5-fluorouracil, they make a promising combination for treating tumors. In comparison to 5-fluorouracil (5-FU) alone at highest dose (25 mg/kg/d), the combination causes similar effects with less side effects. As it can be seen with the tumor growth inhibition rate, the combination of the EO (50 or 100 mg/kg/d) and 5-FU (10 mg/kg/d) leads to a higher inhibition rate as 5-FU (10 mg/kg/d) alone, while the combination shows the same effect with higher dosed 5-FU (25 mg/kg/d) and less severe leukopenia, which represents a great benefit. Furthermore, the large doses of the chemotherapeutic drug that are required due to the limited availability in cancer tissues and the repeated treatment that leads to resistance can be reduced with the enhancer *M. x villosa*

EO. Also, the severe side effects like the loss of body mass and the alterations of spleen weight, liver aspartate transaminase (AST), and renal function are decreased, despite the higher antitumor activity when treated with the association (Amaral et al., 2016).

27.9.4.3 *Zanthoxylum bungeanum* Essential Oil and 5-Fluorouracil (5-FU)/Indomethacin

The EO from *Zanthoxylum bungeanum* Maxim., which belongs to the Rutaceae family, consists of oxygenated monoterpenes and monoterpene hydrocarbons, exactly 48 compounds among which the following three are the major constituents: terpinen-4-ol, 1,8-cineole, and limonene. These compounds are able to enhance both polar (5-FU) and non-polar drugs (indomethacin) as they work with different mechanisms. For the hydrophilic 5-FU, lower concentrations (1% or 3%) of the EO come to operation and alter the thermodynamic activity of the drug, which leads to higher saturated solubility. In contrast, the lipophilic indomethacin requires a higher EO concentration (10%) to increase the SC/vehicle partition coefficient and the saturated solubility. It should be mentioned that the alterations of the drug properties described above are relatively weak; the main mechanism of improving the drug permeation is suggested to be changing the SC skin barrier by disturbing and extracting SC lipids. Thus, the drug delivery is promoted in a concentration-dependent manner and causes higher flux and shorter T_{lag}, which decreases with increased EO concentration and higher cumulative amount. In summary, *Z. bungeanum* EO exhibits enhancement activity for hydrophilic drugs with a long T_{lag} and lipophilic drugs with a short T_{lag}; however, efficiency for hydrophilic drugs is higher (Lan et al., 2014a,b).

27.9.5 Essential Oils and Cardiovascular Drugs

27.9.5.1 *Zanthoxylum bungeanum* Essential Oil and Osthole/Tetramethylpyrazine/Ferulic Acid/Puerarin/Geniposide

As noted above, the essential oil from *Z. bungeanum* Maxim. enhances polar and non-polar drugs, and to face the enhancement activities from the EO to the major constituents terpinen-4-ol, 1,8-cineole, and limonene, they are tested in combination with drugs for cardiovascular treatment which possess different polarity; namely, osthole, tetramethylpyrazine, ferulic acid, puerarin, and geniposide. The physicochemical properties such as molecular size, solubility, and lipophilicity specify whether a drug permeates the skin well or not, whereat the lipophilicity is the most crucial factor. To facilitate the permeation of the different polar drugs, adequate penetration enhancers have to be chosen as the lipophilicity of the enhancer and the drug should match. *Z. bungeanum* EO—followed by limonene—possesses the best enhancement permeation capacity as it causes the greatest steady state fluxes and cumulative amounts for all five model drugs, although the EO preferably promotes the absorption of hydrophilic drugs, and limonene primarily enhances the moderate lipophilic drugs. In contrast, terpinen-4-ol and 1,8-cineole show enhancement activity for more lipophilic drugs like osthole and tetramethylpyrazine; however, their enhancement activities are relatively low. The drug absorption is mainly facilitated by changing the skin barrier structure as the enhancers disorder and extract the SC lipids, not to forget that with lipophilic drugs, the enhancers also cause alteration of their thermodynamic activities which leads to increased saturation solubilities. Altogether, the EO in total is the best choice for promoting the skin diffusion of the five model drugs (Lan et al., 2014a,b).

27.9.5.2 Eucalyptus Oil and Tetramethylpyrazine

Eucalyptus oil, isolated from *Eucalyptus globulus*, a member of the Myrtaceae family, consists of more than 80% cineole and represents a promising penetration enhancer that acts on the skin barrier as it disrupts the SC bilayers, thus improving the partitioning and permeation of small polar molecules. 2,3,5,6-Tetramethylpyrazine (TMP), a lipophilic calcium-channel antagonist isolated from *Ligusticum wallichii*—a member from the Apiaceae family—and used for treating cardiovascular disorders, exhibits some problems when administered orally. The first-pass metabolism causes low bioavailability besides a short half-life that leads to a high, frequent dosing. To avoid the side effects that come with oral administration, a reservoir-type transdermal delivery system (TDS) that incorporates the drug and the penetration enhancer is produced; namely, a transdermal delivery system patch. For creating such a reservoir-type TDS for TMP, the following compounds are required: Carbopol gel for gelling, an EVA membrane for rate control and silicone adhesive for pressure sensitivity. With these components, an effective TDS can be formed to reach an appropriate clinical concentration. Further, the TMP patch with a clinical surface area of 20 cm^2 consists of 5% eucalyptus oil, which turned out to be the best concentration for promoting drug permeation. The penetration enhancer causes a 17-fold higher permeation rate compared to an application without enhancer, as well as a lower C_{max} needed for a prolonged steady-state concentration, T_{max}, and mean residence time. The reservoir-type transdermal delivery system presents a promising alternative route to oral administration as it improves the permeation and, thus, the compliance (Shen et al., 2013).

27.9.5.3 Basil Oil and Labetolol Hydrochloride

Basil oil, extracted from *Ocimum basilicum*, a plant that belongs to the Lamiaceae family, contains alcoholic terpenes and exhibits a low boiling point, which is good for interacting with the SC lipids and leads to a higher enhancement rate. Thus, it is an adequate penetration enhancer for a hydrophilic drug like labetolol hydrochloride (LHCl), which is a combined alpha- and beta-blocker that comes to operation while treating hypertension. LHCl presents a good candidate for transdermal application due to the high first-pass metabolism and the poor bioavailability that comes with oral administration. Only a few patients fully benefit from the oral medication because of the fluctuated plasma concentration and the uncontrolled drug release, and for that reason, the transdermal route attracts attention for antihypertensive agents. For overcoming the skin barrier, basil oil and 5% (w/v) terpenes, respectively, are crucial, and in combination with the vehicle for LHCl, synergistic effects are provided. Greater steady-state flux and a decreased lag time are achieved due to the disruption of the SC barrier and the increase in partitioning and diffusion coefficient. Thus, a lower activation energy for LHCl is needed for maximum permeation (Jain et al., 2008).

27.9.5.4 Anethol, Menthone, Eugenol, and Valsartan

Anethol, menthone, and eugenol are oxygen-containing terpenes that act as penetration enhancers for valsartan, a lipophilic specific angiotensin 2 receptor blocker used for treating hypertension. Valsartan should be delivered through human skin due to its low oral bioavailability, its low melting point, and its molecular weight. For the terpenes, 1% (w/v) concentration is the optimum to increase the flux by interacting with the SC lipids, whereas anethol, the most lipophilic enhancer, causes the maximum lipid extraction and thereby improves the drug enhancement rate 4.4 times, followed by menthone. It is always important that penetration enhancers and drugs conform with each other in polarity; thus eugenol, the least lipophilic enhancer, leads to the lowest permeation rate of the highly lipophilic valsartan. Anethol and eugenol promote the drug penetration by keratin denaturation and lipid extraction while menthone only works with lipid extraction. Overall, it is supposed that the following terpenes are preferred over the other ones, because of the better enhancement effects they provoke; namely, liquid terpenes and terpenes with higher Log P values due to the better mixture properties from lipophilic terpenes with SC intercellular lipids (Ahad et al., 2016).

27.9.5.5 Basil Oil, Petit Grain Oil, Thyme Oil, and Nitrendipine

The EOs such as basil oil, petit grain oil, and thyme oil are able to enhance the permeation of nitrendipine 10–12 times by altering the solubility properties and improving the partitioning of the drug within the SC. Nitrendipine, also known as lipophilic 1,4-dihydropyridine derivative calcium channel blocker, represents a potent vasodilator able to decrease blood pressure. Because of its high first-pass effect and the following low bioavailability when administered orally, a transdermal patch is a good alternative for nitrendipine medication, noting that transdermal drug delivery is always a good choice to treat chronic disorders that require long-term dosing. Due to its poor skin permeation activity, a penetration enhancer is needed to overcome the SC barrier. Basil oil, followed by petit grain oil and thyme oil, turned out to be the best penetration enhancer among the other essential oils as it exhibits the highest increase in relative activity value due to increased thermodynamic activity and solubility of nitrendipine in the SC. The only enhancer that shows better results in facilitating the permeation of nitrendipine is oleic acid, an unsaturated fatty acid that is more effective than other saturated fatty acids. The similar structure of the oleic acid to the SC enables the fast penetration through the barrier (Mittal et al., 2008).

27.10 ESSENTIAL OILS AND THE INFLUENCE OF TEMPERATURE

27.10.1 Borneol, Osthole, and Increasing Temperature

Borneol is a cyclic terpene alcohol that is isolated from *Cinnamomum camphora*—a member of the Lauraceae family—and represents a penetration enhancer for osthole, a relative lipophilic drug extracted from the fruit of *Cnidium monnieri*, which belongs to the Apiaceae family. Disturbance of the ordered SC lipids is the main mechanism caused by borneol. This permeation mechanism can be influenced by temperature, an external factor. Increase in temperature leads to changes in the lipid bilayers of SC: lipids become shorter and frizzier and therefore they are more flexible and molecular movement is improved. The area per lipid increases while the thickness and order of lipids decrease, which results in synergism, affecting the permeation of osthole. Thus the penetration enhancement of borneol is promoted with increased temperature as the diffusion rate and the speed of permeation raised. However, the temperature has to be increased carefully, because with too high temperature in combination with borneol, the SC structure gets ruined as a water pore is built and the micelle reversed, caused by 5% borneol at 323 K or 10% borneol at 310 K (Yin et al., 2017).

27.11 ESSENTIAL OILS AND THE EFFECT ON CYTOCHROME P450

27.11.1 *Zataria multiflora* Essential Oils with Cancer Chemopreventive Effect

Zataria multiflora Boiss. belongs to the Lamiaceae family and possesses cytotoxicity, antioxidant, and chemopreventive effects, beside some others. 1,2-Dimethylhydrazine is suggested to be a potent colon-specific carcinogen as it is metabolized by cytochrome P450 to active intermediates that cause colon cancer in the broadest sense. *Z. multiflora* EO is able to inhibit the tumor formation as it interacts with and decreases the activity of CYP450 and thus hinders the metabolism induced by the cytochrome. Provided that essential oils can suppress the activity of CYP450—an important enzyme for the drug metabolism—it can be expected that due to this inhibition effect the drug metabolism can be slowed down thus furnishing an increase and a prolongation of the drug plasma level. Thus, EOs not only improve the medication effect by interacting with the SC barrier and enhancing the drug permeation, but also by limiting the degradation of drugs by inhibiting CYP450 activity (Dadkhah et al., 2014).

27.12 ESSENTIAL OILS AND THEIR SYNERGISTIC EFFECTS

27.12.1 1,8-Cineole and Camphor

Rosemary oil, isolated from *Rosmarinus officinalis*—a member of the Lamiaceae family—possesses 1,8-cineole and camphor as major constituents. The two compounds do not only exhibit insecticidal activity but also penetration enhancing effects. In comparison to the application of 1,8-cineole or camphor alone that may cause partial or incomplete activity, the two major compounds administered as a binary mixture show the same or better activity with lesser amounts, as synergy takes place. The synergy mechanism can be declared as a multi-target effect where the two compounds attack different sites, ameliorate solubility or bioavailability that take place caused by pharmacokinetic or physicochemical effects or interactions with resistance mechanisms. Thus, it comes to synergy as 1,8-cineole enables camphor a better permeation by interacting with the lipid layer. The interaction with each other leads to lowered surface tension, increased solubility of camphor, and higher mobility, which finally causes enhanced penetration. As the two compounds possess insecticidal activity, it can be seen that with the described synergy effect, their toxicity over a larvae of the moth *Trichoplusia ni*, the cabbage looper, is improved in contrast to individual application of higher terpene amounts (Tak and Isman, 2015).

27.13 CONCLUSION

EOs and their terpenes cause many effects, among which the function as carrier oils is the most interesting discussed in this overview. Due to their properties, EOs show the ability to cross the SC barrier of the epidermis by disordering and loosening the lipid packing, thus facilitating the delivery of drugs through human skin that cannot sufficiently overcome the barrier themselves, to achieve therapeutic plasma levels. The conformance of polarity between penetration enhancer and drug, as well as the adequate concentration of the enhancer and the right vehicle, are important factors that influence the enhancement effect. Mixing terpenes can lead to synergism, and increased temperature can improve the permeation activity of carriers and drugs. All those involved factors make transdermal drug delivery and the required therapy, whether according inflammation or cancer treatment, possible without inducing adverse effects, in contrast to drug administration via conventional routes. It should be mentioned that the main mechanism of EOs and their constituents is affecting the SC properties, a further possibility to improve and prolong that the drug remaining in the organism is acting on the metabolism via cytochrome P450 inhibition. So, whatever mechanism is induced with the enhancement of EOs, lower drug concentration, as well as less frequent dosing, is required to achieve the same therapeutic effect with reduced side effects. Altogether, the transdermal drug delivery makes medication more acceptable for the patients and therefore improves their compliance.

REFERENCES

Ahad A., Aqil M. and Ali A. 2016. The application of anethol, menthone, and eugenol in transdermal penetration of valsartan: Enhancement and mechanistic investigation. *Pharm Biol.* 54(6), 1042–1051.

Akbari J., Saeedi M., Farzin D., Morteza-Semnani K. and Esmaili Z. 2015. Transdermal absorption enhancing effect of the essential oil of *Rosmarinus officinalis* on percutaneous absorption of Na diclofenac from topical gel. *Pharm Biol.* 53(10), 1442–1447.

Amaral R. G., Andrade L. N., Dória G. A. A., Barbosa-Filho J. M., de Sousa D. P., Carvalho A. A. and Thomazzi S. M. 2016. Antitumor effects of the essential oil from *Mentha x villosa* combined with 5-fluorouracil in mice. *Flavour Fragr J.* 31(3), 250–254.

Ambrož M., Matoušková P., Skarka A., Zajdlová M., Žáková K. and Skálová L. 2017. The Effects of selected sesquiterpenes from *Myrica rubra* essential oil and the efficacy of doxorubicin in sensitive and resistant cancer cell lines. *Molecules.* 22, 1021.

Brito M. B., Barin G. B., Araújo A. A. S., de Sousa D. P., Cavalcanti S. C. H., Lira A. A. M. and Nunes R. S. 2009. The action modes of *Lippia sidoides* (cham) essential oil as penetration enhancers on snake skin. *J Therm Anal Calorim.* 97(1), 323–1327.

Cal K. 2005. Skin penetration of terpenes from essential oils and topical vehicles. *Planta Med.* 2006, 72, 311–316.

Casey A. L., Karpanen T. J., Conway B. R., Worthington. T., Nightingale P., Waters R. and Elliott T. S. J. 2017. Enhanced chlorhexidine skin penetration with 1,8-cineole. *BMC Infect Dis.* 17, 350.

Chen J., Jiang Q.-D., Chai J.-P., Zhang H., Peng P. and Yang X.-X. 2016. Natural terpenes as penetration enhancers for transdermal drug delivery. *Molecules.* 21, 1709.

Chen J., Jiang Q.-D., Wu Y.-M., Liu P., Yao J.-H., Lu Q., Zhang H. and Duan J.-A. 2015. Potential of essential oils as penetration enhancers for transdermal administrationof ibuprofen to treat dysmenorrhea. *Molecules.* 20, 18219–18236.

Dadkhah A., Ffatemi F., Mohammadi Malayeri M. R. and Rasoli A. 2014. Cancer chemopreventive effect of dietary *Zataria multiflora* essential oils. *Turk J Biol.* 38, 930–939.

Edris A. E. 2007. Pharmaceutical and therapeutic potentials of essential oils and their individual volatile constituents: A review. *Phytother Res.* 21, 308–323.

Fang J.-Y., Leu Y.-L., Hwang T.-L. and Cheng H.-C. 2004. Essential oils from sweet basil (*Ocimum basilicum*) as novel enhancers to accelerate transdermal drug delivery. *Biol Pharm Bull.* 27(11), 1819–1825.

Feng H., Luo J., Kong W., Dou X., Wang Y., Zhao X., Zhang W., Li Q. and Yang M. 2015. Enhancement effect of essential oils from the fruits and leaves of *Alpinia oxyphylla* on skin permeation and deposition of indomethacin. *RSC Adv.* 5, 38910–38917.

Fung Chye Lim P., Yang Liu X. and Yung Chan S. 2009. A review on terpenes as skin penetration enhancers in transdermal drug delivery. *J Essent Oil Res.* 21(5), 423–428.

Herman A. and Herman A. P. 2014. Essential oils and their constituents as skin penetration enhancers for transdermal drug delivery: A review. *J Pharm Pharmacol.* 67, 473–485.

Jain R., Aqil M., Ahad A., Ali A. and Khar R. K. 2008. Basil oil is a promising skin penetration enhancer for transdermal delivery of labetolol hydrochloride. *Drug Dev Ind Pharm.* 34(4), 384–389.

Jiang Q., Wu Y., Zhang H., Liu P., Yao J., Yao P., Chen J. and Duan J. 2017. Development of essential oils as skin permeation enhancers: Penetration enhancement effect and mechanism of action. *Pharm Biol.* 55(1), 1592–1600.

Lan Y., Li H., Chen Y.-Y., Zhang Y.-W., Liu N., Zhang Q. and, Wu Q. 2014a. Essential oil from *Zanthoxylum bungeanum* Maxim. and its main components used as transdermal penetration enhancers: A comparative study. *J Zhejiang Univ Sci B.* 15(11), 940–952.

Lan Y., Wu Q., Mao Y.-Q., Wang Q., An J., Chen Y.-Y., Wang W.-P., Zhao B.-C., Liu N. and Zhang Y.-W. 2014b. Cytotoxicity and enhancement activity of essential oil from *Zanthoxylum bungeanum* Maxim. as a natural transdermal penetration enhancer. *J Zhejiang Univ-Sci B (Biomed & Biotechnol)*, 15(2), 153–164.

Mittal A., Singh Sara U., Ali A. and Aqil M. 2008. The effect of penetration enhancers on permeation kinetics of nitrendipine in two different skin models. *Biol Pharm Bull.* 31(9), 1766–1177.

Pharmacopoea Europaea. 2008. Ätherische Öle, Aetherolea, 6.0/2098, Grundwerk 2008, European Directorate for the Quality of medicines and Healthcare (EDQM), Council of Europe, 6th edition, Strasbourg, 2008, p. 957.

Rajan R. and Vasudevan D. T. 2012. Effect of permeation enhancers on the penetration mechanism of transferosomal gel of ketoconazole. *J Adv Pharm Technol Res.* 3(2), 112–116.

Saffari M., Shirazi F. H. and Moghimi H. R. 2016. Terpene-loaded liposomes and isopropyl myristate as chemical permeation enhancers toward liposomal gene delivery in lung cancer cells; A comparative study. *Iran J Pharm Res.* 15(3), 261–267.

Shen T., Xu H., Wenig W. and Zhang J. 2013. Development of a reservoir-type transdermal delivery system containing eucalyptus oil for tetramethylpyrazine. *Drug Deliv.* 20(1), 19–24.

Tak J.-H. and Isman M. B. 2015. Enhanced cuticular penetration as the mechanism for synergy of insecticidal constituents of rosemary essential oil in *Trichoplusia ni. Sci Rep.* 5, 12690.

Tak J.-H. and Isman M. B. 2017. Enhanced cuticular penetration as the mechanism of synergy for the major constituents of thyme essential oil in the cabbage looper, *Trichoplusia ni. Ind Crops Prod.* 101, 29–35.

Thirapit S. and Tanasait N. 2015. Effect of liposomal fluidity on skin permeation of sodium fluorescein entrapped in liposomes. *Int J Nanomedicine.* 10(1), 4581–4592.

Yin Q., Wang R., Yang S., Wu Z., Guo S., Dai X., Qiao Y. and Shi X. 2017. Influence of temperature on transdermal penetration enhancing mechanism of borneol: A multi-scale study. *Int J Mol Sci.* 18, 195.

Zhang K., Zhang Y., Li Z., Li N. and Feng N. 2017. Essential oil-mediated glycerosomes increase transdermal paeoniflorin delivery: Optimization, characterization, and evaluation *in vitro* and *in vivo. Int J Nanomedicine.* 12, 3521–3532.

28 Influence of Light on Essential Oil Constituents

Marie-Christine Cudlik and Gerhard Buchbauer

CONTENTS

28.1 INTRODUCTION

Light is a key foundation for life. It is essentially electromagnetic radiation and, as such, provides energy. Especially the UV range of the electromagnetic spectrum—which is split into UV-A (315–400 nm), UV-B (280–315 nm), and UV-C (100–280 nm)—is rich in energy; however, only UV-A and a fraction of UV-B reach the earth's surface, as UV-C is completely absorbed by the planet's ozone layer and atmosphere (WHO, 2018). Ultraviolet radiation is known to have both positive and negative effects, for it is important for the biosynthesis of vitamin D3 in the human body, but can also be a dangerous mutagenic agent. It has sufficient energy to break chemical bonds, creating cleavage sites that can react with oxygen, causing degradation (Pelzl et al., 2018).

Plants need light. They use the energy coming from the sun for chemical reactions, producing primary metabolites, for instance, amino acids and carbohydrates, and secondary metabolites, such as essential oils (= EOs) (Prins et al., 2010; Heldt and Piechulla, 2011), which serve them as attractants for insects or as defense compounds against herbivores and pathogens (Heldt and Piechulla, 2011). Various genetic and environmental factors affect the composition and yield of the EO; among them, the intensity, duration and wavelength of the irradiation to which the plant is exposed (Maffei et al., 1999; Kumari et al., 2009; Ivanitskikh and Tarakanov, 2014). The amount of volatile compounds produced is lowest during the time of the year with the lowest temperature and least hours of sunlight (Figueiredo et al., 2008).

And not only does irradiation affect the living plant, but also the finished product, the EO, once it is situated outside the plant. When photosensitive EO constituents are submitted to electromagnetic radiation, the molecules may experience photoexcitation, which means they reach an excited state by absorbing the ultraviolet, visible or infrared light (IUPAC, 1997). Usually, they are initially transferred to an excited singlet state with a short lifespan, followed by fast radiationless relaxation, which leads via intersystem crossing to an energetically lower excited triplet state (Van den Bergh, 1986). As a consequence of the molecule's excitation, its oxidative and reductive properties are enhanced, increasing the possibility of electron transfer processes (Ochsner, 1997). Many different photoreactions are possible due to the chemical heterogeneity of EOs. They consist of 20 to 200 or even more single substances (Hänsel and Sticher, 2010), although mostly of two to three main constituents (20%–95%) and others only in trace amounts (Khayyat and Roselin, 2018). The main chemical components of most EOs are terpenoids (like monoterpenes and sesquiterpenes) as well as phenolic compounds, all of them characterized by low molecular weight (Dhifi et al., 2016). In case of oxygen availability, photooxidation is the most important possible photochemical reaction. Furthermore, double-bond isomerization, photopolymerization, Diels-Alder photocycloaddition and photoepoxidation are prevalent transformation pathways for EO constituents under influence of light. Often, neither light nor oxygen alone can trigger chemical reactions in EOs; they work synergistically (Li et al., 2016). Some transformations occur only when energy is provided in the form of radiation, and oxygen can be taken from the air surrounding. This close relationship between these two factors renders it difficult to differentiate between them and to draw a line as to which is more important in triggering reactions.

The question is, which consequences the various photoalterations have on the EOs and their constituents. Due to their wide-ranging effects, they are ubiquitous in their application as odorants, flavoring substances, and antioxidants in the food and liquor industries (Castro et al., 2010) and, furthermore as expectorants, stomachics, aroma correctors, and many more in the pharmaceutical sector. EOs have been found to be antioxidant, anti-inflammatory (Miguel, 2010; Buchbauer and Erkic, 2016), insect repellent (Chellappandian et al., 2018), antimicrobial and antiviral (Bakkali et al., 2008; Lang and Buchbauer, 2012), spasmolytic, carminative, sedative (Lis-Balchin and Hart, 1999; Buchbauer, 2002), hepatoprotective, and anticarcinogenic (Morita et al., 2003; Raut and Karuppayil, 2014); they even exhibit wound-healing (Pérez-Recalde et al., 2018), antidiabetic, and lipid-lowering effects (Habtemariam, 2018). Due to their range of effects and their tolerability, they are widely used, and therefore, an eventual impact on their safety and/or organoleptic properties is of great interest.

Abundant research has been conducted on the manifold useful properties of EOs, but literature on their stability under different storage conditions is scarce, especially on the topic of light-induced transformations.

Therefore, the aim of this treatise was to summarize the physicochemical interactions of light and EO constituents that may occur during production, storage, and use of EOs. Also, a thorough research was conducted in the field of phototoxicity of EO components, because many EOs have dermal applications, and it is therefore important to look into their potential harmfulness when they could undergo photochemical changes when in contact with direct or indirect sunlight.

28.2 MECHANISMS OF PHOTODEGRADATION

According to the IUPAC gold book of definitions (1997), photodegradation is "the photochemical transformation of a molecule into lower molecular weight fragments, usually in an oxidation process." It is one of the operations that can change the chemical composition of EOs under storage, transportation, or use, causing alterations in organoleptic properties, medicative potency like antimicrobial activity, and toxicity patterns (Beltrame et al., 2013). Most chemical reactions occur with the participating molecules residing in the ground state, a stable electronic state. Photochemical reactions, on the other hand, are characterized by a shift to excited singlet states due to absorption of the energy of a light quantum, followed by intersystem crossing to triplet states, which possess modified electron spins, enabling different reaction mechanisms (Kayyat and Roselin, 2018).

Photochemical degradation is, considered quantitatively, one of the most important reactions in nature. When trace substances in the troposphere interact with sunlight, reactive species, mostly radicals, are formed. These can oxidize hydrocarbons and their derivatives, resulting in indirect photochemical degradation, the products of which are ultimately carbon dioxide and water, thus leading to the natural purification of the atmosphere. Direct photochemical degradation, on the other hand, is the absorption of a light quantum followed by immediate oxidation or rearrangement and is relatively rare owing to the high activation energy (Simmler, 2012). Photodegradation can also take effect for EOs and their components. After being extracted from the plant, the substances can undergo the reaction while being processed, transported, stored, or used, thus being afflicted in their organoleptic properties and possibly as well in their safety and effects on the user. Depending on the exact circumstances of product handling, different mechanisms of photodegradation can come into action. For instance, when oxygen is available, photooxidation is most likely to occur, whereas in the case of hydrogen peroxide presence, epoxidation is possible. These mechanisms are being discussed in this chapter.

28.2.1 Double-Bond Isomerization

Back in 1983, Toda et al. investigated the photolysis of both jasmine absolute oil and one of its main constituents, benzyl benzoate, in ethanolic solution. They used a high-pressure mercury lamp (HPML) and a low-pressure mercury lamp (LPML), respectively, for irradiation of the samples and delivered a proposed degradation pathway and detailed data about the changes in composition over time. The compounds containing a chain double-bond experienced a *Z-E* isomerization reaction, resulting in a fairly large amount of *trans*-configured products. This isomeric shift occurred under both HPML and LPML irradiation, but to a greater extent under the latter. The affected compounds were 3-methyl-2-(*cis*-2-pentenyl)-2-cyclopenten-l-one (= *cis*-jasmone, compound 1 in Figure 28.1), *cis*-3-hexenyl benzoate, *cis*-7-decen-5-olide, phytyl acetate, geranyl linalool, and phytol (Toda et al., 1983). Tateba et al. (1993) subjected a solution containing *cis*-jasmone to irradiation with a high-pressure mercury lamp for 20 hours in methanol under nitrogen atmosphere and came to the same result regarding the isomerization reaction, but also detected two di-π-methane rearrangement

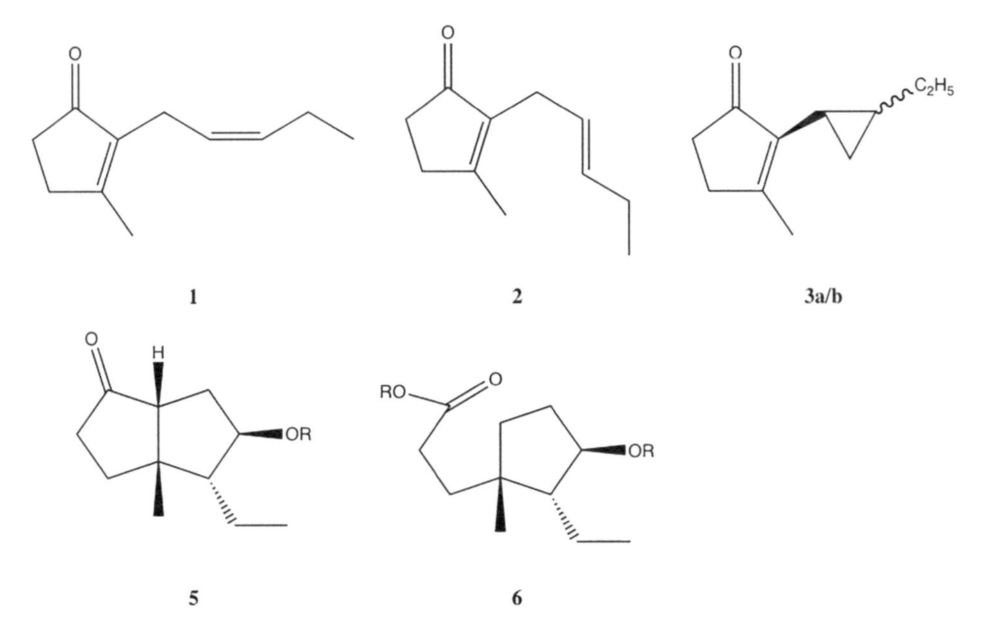

FIGURE 28.1 Photoreaction products of *cis*-jasmone, R=CH_3, C_2H_5 (depending on solvent). (Adapted and newly drawn from Tateba H. et al. 1993. *Biosci. Biotechnol. Biochem.*, 57(2), 220–226.)

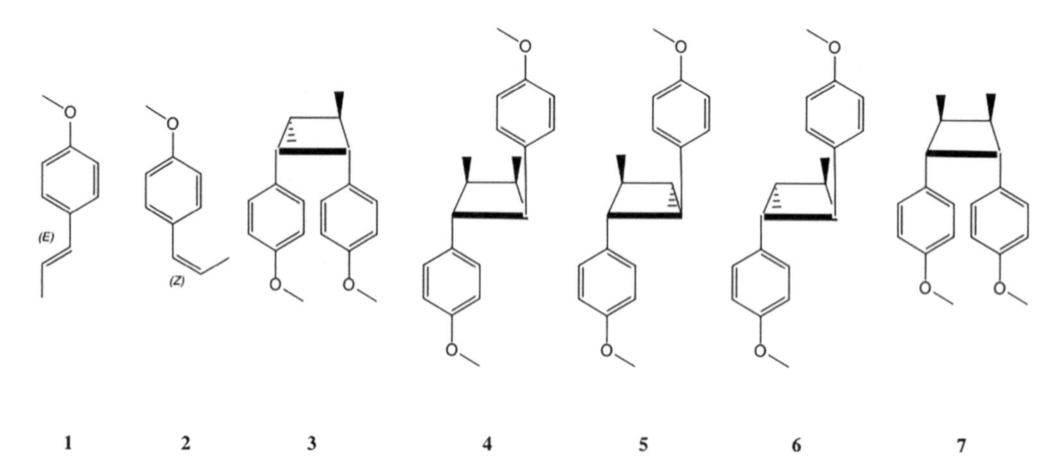

FIGURE 28.2 Photoreaction products of *trans*-anethole. (Adapted and newly drawn from Castro H.T. et al. 2010. *Molecules*, 15, 5012–5030.)

products (compounds 3a/b in Figure 28.1), as well as two intramolecular cyclo-adducts (compounds 4 and 5 in Figure 28.1). When ethyl acetate was used as solvent, only the compounds 2 and 3a and b were generated; however, the yield of products 3a and b (83%) increased considerably compared to ethanol (68%) and methanol (69%).

trans-Anethole (structure 1 in Figure 28.2) is one of the major components of anise, clove, thyme, and cinnamon EOs and often used in food and liquor industries. Its use is controversial for, upon UV irradiation, *cis*-anethole (structure 2 in Figure 28.2) is formed, which is toxic. Castro et al. investigated the transformation of *trans*-anethole in toluene subjected to UV radiation and, with the help of GC-MS, found that (1a,2a,3b,4b)-1,2-bis(4-methoxyphenyl)-3,4-dimethylcyclobutane (structure 7 in Figure 28.2) was the most abundant constituent of the mixture of five methoxyphenyl-disubstituted cyclobutanes found, together with *cis*-anethole. When an excited *trans*-anethole molecule interacts with one in the ground state, a cycloaddition leads to dimers 5 (*anti* head-to-head) and 7 (*syn* head-to-head) in Figure 28.2. For dimers 3, 4, and 6, *cis*-anethole reacted, which was itself formed from *trans*-anethole by photoisomerization. The abundance of dimers 5 and 7 is much higher than of those involving a *cis*-anethole as reaction partner (Castro et al., 2010).

Citral is a mixture of the isomers geranial (= *trans*-citral, substance 1 in Figure 28.3) and neral (= *cis*-citral, substance 2 in Figure 28.3) and one of the most important components of lemon EO, as it conveys the characteristic lemon-like odor. Iwanami et al. (1997) produced a lemon flavor containing lemon oil and irradiated it with UV for 4 days at 30°C under nitrogen atmosphere to block oxidation reactions. The photodegradation led to the photoproducts shown in Figure 28.3: photocitral A (3), epiphotocitral A (4), photocitral B (5), 2-(3-methyl-2-cyclopenten-1-yl)-2-methylpropion aldehyde (6), *trans*-1,3,3-trimethylbicyclo[3.1.0]hexane-1-carboxaldehyde (7), *cis*-1,3,3-trimethylbicyclo[3.1.0]hexane-1-carboxaldehyde (8), (1,2,2-trimethyl-3-cyclopenten-1-yl)-acetaldehyde (9), and α-campholene aldehyde (10). In the case of oxygen availability during irradiation, citral yields other transformation products, as different reactions are feasible in the presence of oxygen (see Chapter 2.2).

The formation of 6 requires a formyl 1,3-migration and is a newly identified photoreaction product. Diethyl acetals 11 and 12 were not obtained in the dark under these conditions, but might be formed by the hydrogen abstraction reaction of excited citral. Limonene, terpinolene, and nonanal decreased in amount, whereas *p*-cymene increased. Other constituents, such as citronellal, linalool, sesquiterpene hydrocarbons, and terpineols, were only insignificantly changed. The fresh, sweet, and characteristically lemon-like odor decreased and a dusty odor became predominant, which is mostly ascribable to compound 6 (Iwanami et al., 1997).

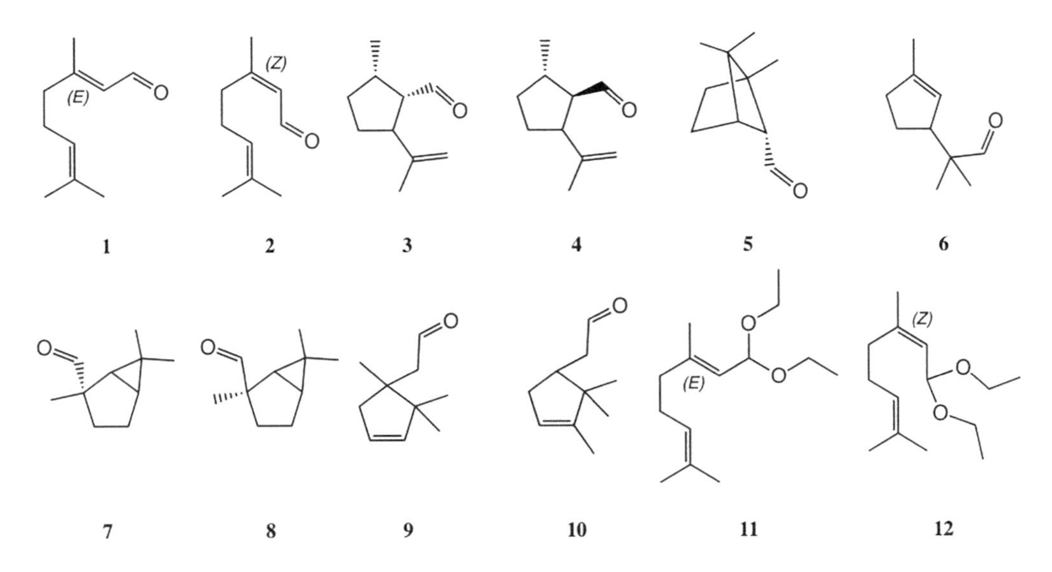

FIGURE 28.3 Photoreaction products of citral. (Adapted and newly drawn from Iwanami Y. et al. 1997. *J. Agric. Food Chem.*, 45, 463–466.)

Krupa et al. (2012) investigated the isomerization reactions of eugenol and isoeugenol, induced by tunable UV laser light. They found that they could prompt photoisomerization in isoeugenol, which contains an asymmetrically substituted exocyclic C=C bond, by irradiation at different wavelengths ranging from 310 to 298 nm. The *E*- and *Z*-isomers and their only practically significant and most stable rotamers are depicted in Figure 28.4.

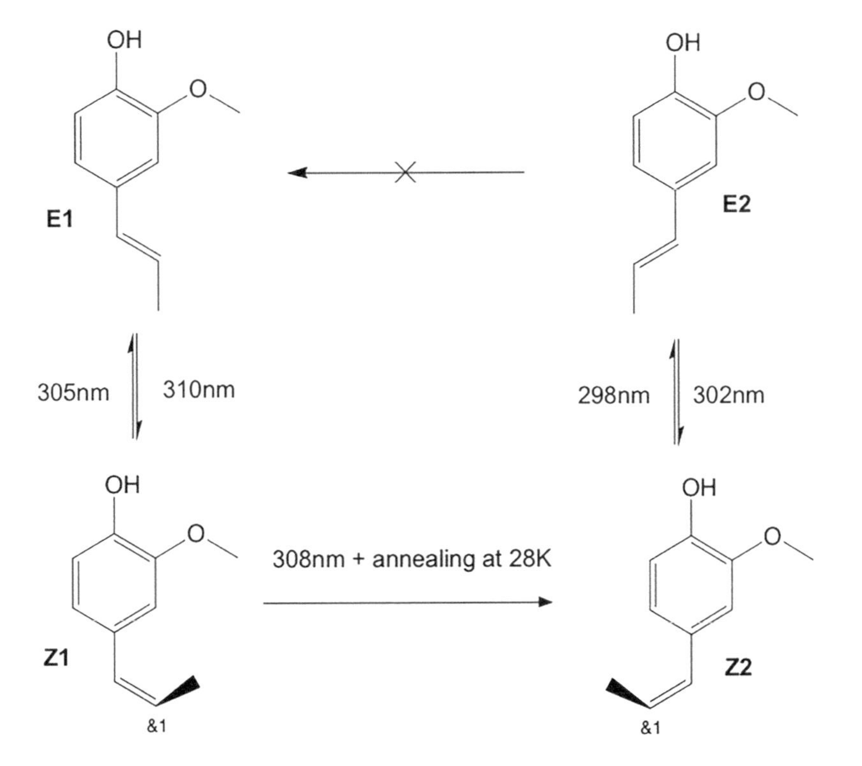

FIGURE 28.4 Interconversion reactions of eugenol. (Adapted and newly drawn from Krupa J. et al. 2012. *J. Phys. Chem. B*, 116, 11148–11158.)

For the conversion of *E1* to *Z1*, a wavelength of 310 nm was applied. This reaction could be reversed by irradiation at $\lambda = 305$ nm. These back and forth reactions between *E1* and *Z1* came to a halt at wavelengths of 306–308 nm, suggesting a photoequilibrium. Irradiation at $\lambda = 302$ nm resulted in a conversion of *E2* to *Z2*. A partial back-transformation could be prompted with $\lambda = 298$ nm, but the spectral changes observed were not as pronounced as for the other interconversions. No changes in the infrared spectrum occurred following the annealing of freshly deposited isoeugenol on Ar matrices up to 30 K, which reveals that the applied temperature and therefore energy does not suffice for *E2* to *E1* transformations. In an additional study combining UV irradiation and matrix isolation technique, isoeugenol on an Ar matrix was first irradiated at $\lambda = 308$ nm to increase the amount of *Z1*, followed by annealing of the matrix in steps of 2 K starting at 15 K. At 28 K (=−245.15°C), the *Z1* bands decreased and *Z2* formed. When *Z1* and *Z2* forms attained equal amounts, the thermally induced partial reaction stagnated, despite further increase of temperature, suggesting they are isoenergetic. The *Z1* and *Z2* forms only occur in populations of about 4%–5% each, since they are the energetically less-convenient forms.

Besides isoeugenol, Krupa et al. (2012) also subjected eugenol to the same irradiation series, but came to the conclusion that no conformational changes were induced. However, aside from interconversions, they also studied photolysis of both eugenol and isoeugenol (see Figure 28.5) using narrow-band UV-irradiation technique. The reaction pathway is very similar for both molecules; therefore, it is illustrated on the example of eugenol. An H-atom shift from the OH-group was found to be the primary photochemical process for either of the compounds.

This hydrogen atom shift had lower threshold energy in isoeugenol than eugenol (308 nm vs. 285 nm). In both cases, it resulted in the generation of two types of long-chain conjugated ketenes, depending on where the hydrogen atom repositioned itself on the ring. In another step of reaction, decarbonylation of the ketenes took place (Krupa et al., 2012).

28.2.2 Photooxidation and Epoxidation Reactions

Mori and Iwahashi (2016) studied the formation of radicals by oxidation of some EOs by measuring the electron spin resonance (ESR) spectra of reaction mixtures of flavin mononucleotide (FMN, an endogenous photosensitizer), EO, acetonitrile, phosphate buffer, α-(4-pyridyl-1-oxide)-*N-tert*-butylnitrone (4-POBN), and $FeSO_4(NH_4)_2SO_4$ irradiated with visible light of 436 nm. Geraniol, being its major constituent, gave similar results to palmarosa EO. In the reaction mixture with geraniol, a new radical, 4-POBN/5-hydroxy-3-methyl-3-pentenyl radical, was identified. Results showed that without light, Fe^{2+}, or FMN, respectively, no reaction occurred. The authors proposed a possible reaction pathway, which is shown in Figure 28.6.

Supposedly, the irradiation with visible light generates the excited singlet state of FMN, ^{1}(FMN)*, which is then transformed by intersystem crossing, resulting in the excited triplet state, ^{3}(FMN)*. Subsequently, ^{3}(FMN)* probably reacts with triplet oxygen $^{3}O_2$, thereby forms $^{1}O_2$, which in turn produces 3,7-dimethyl-6-hydroperoxy-2,7-octadienol (compound 2 in Figure 28.6) following the singlet oxygen *ene*-reaction with geraniol (compound 1 in Figure 28.6). Considering that the reaction does not occur without the ferrous ions, Mori and Iwahashi presume that they catalyze the cleavage of this newly formed compound via β-scission of the alkoxy radical intermediate (compound 3 in Figure 28.6), yielding the newfound radical. Under the same reaction conditions, geranium, clary sage, lavender, petitgrain, and bergamot EO also gave strong ESR signals, a fact which the authors attribute to the autoxidation potential of their constituents, such as geraniol, limonene, and linalool (Mori and Iwahashi, 2016).

Ziegler et al. (1991) studied deterioration products and processes of sweet orange EO (*Citrus sinensis* L. Osbeck) under simulated aging conditions, exposing an aqueous acidic orange oil emulsion to UV light at room temperature. The subsequent GC/MS analysis disclosed an increase in carvone, isopulegol, isomers of carveol, and the limonene and linalool oxides, as well as a significant decrease in neral, geranial, and citronellal. The authors also identified *p*-mentha-1,8-dien-4-ol, α-cyclocitral,

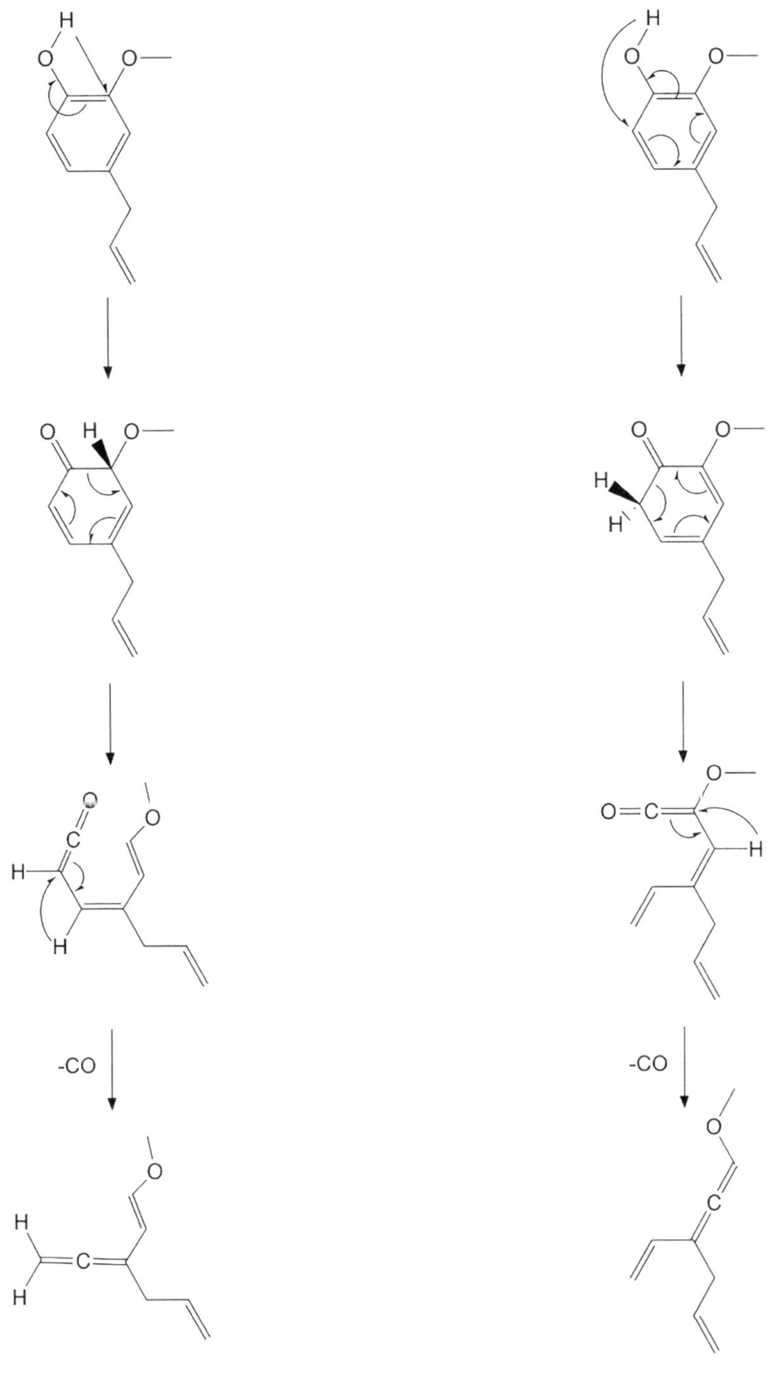

FIGURE 28.5 Photolysis pathways of eugenol. (Adapted and newly drawn from Krupa J. et al. 2012. *J. Phys. Chem. B*, 116, 11148–11158.)

photocitral A, iso(iso)pulegol, carvone, camphor, menthone, isomenthone, isomers of *p*-mentha-1(7),8-dien-2-ol, and isopiperitenol as newly formed secondary constituents (Ziegler et al., 1991).

The effects of different storage conditions on rosemary EO were investigated by Irmak et al. (2010), who stored the EO at 4°C in the dark or at room temperature in indirect daylight for 14 weeks. Considering that they did not only use different light conditions but also varied the temperature, it

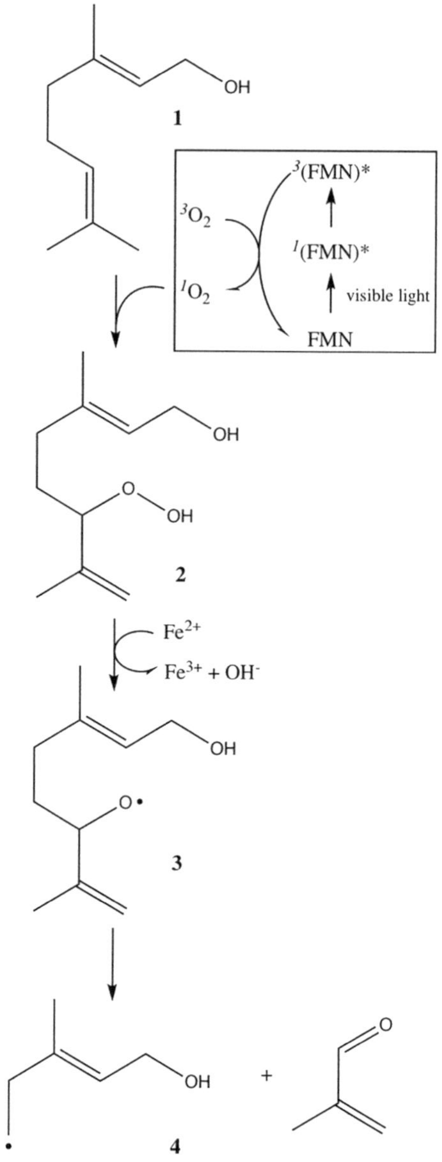

FIGURE 28.6 Geraniol radical formation. (Adapted and newly drawn from Mori H.-M., and H. Iwahashi. 2016. *Free Radic. Res.*, 50(6), 638–644.)

is difficult to say which factor was crucial; but, under the daylight conditions, substantial chemical transformations occurred, while the storage in darkness and low temperature did not have much of an impact. The authors studied the change in total phenolics content and antioxidant properties of the rosemary EO following storage periods of 0, 2, 4, 8, and 14 weeks. The rosemary EO samples showed high antioxidant activity when fresh, largely preventing the bleaching of beta-carotene in the assay. But this capability and the total phenolics content, likewise, diminished after the storage time, especially in the extracts stored in light. The GC-MS peaks of *trans*-caryophyllene and squalene disappeared completely in one of the samples stored under indirect daylight, and in another, there was a reduction of 38% for linalool, 24% for limonene, and 44% for *trans*-caryophyllene (Irmak et al., 2010).

Li et al. (2016) investigated the transformations occurring during UV and air exposure in the EO of white guanxi honey pummelo (*Citrus grandis* (L.) Osbeck, Rutaceae), which is a citrus variety from Southeast Asia. The main constituents of pummelo are (+)-limonene and β-myrcene, germacrene D, geranial, neral, β-pinene, linalool, sabinene, and α-pinene (Sun et al., 2014). The EO was mechanically pressed from the fruit, and one sample was irradiated with UV light for 40 h while being exposed to air at approximately 25°C. In order to identify degradation mechanisms, single standard aldehydes without solvent dilution—that is, octanal (99%), nonanal (96%), citronellal (96%), decanal (95%), citral (97%, mixture of *trans*-citral and *cis*-citral), perilla aldehyde (92%), dodecanal (95%), and dodecenal (93%)—were subjected to air exposure, UV irradiation, or a combination of both (Li et al., 2016).

After UV light and air co-treatment, the concentrations of octanal, nonanal, decanal, dodecanal, dodecenal, perilla aldehyde, *trans*-citral, and *cis*-citral in the pummelo EO decreased by 13.8%, 28.3%, 40.5%, 37.8%, 85.4%, 33.9%, 85.6%, and 82.1% ($p < 0.05$), respectively. The only aldehyde not sticking to this pattern was citronellal, which, in contrast, increased by 84.6%, probably due to precursors existing in the EO. The other aliphatic aldehydes were found to be transformed to their organic acids after the combined exposure to air and UV light, a process in which the UV light grants the energy necessary and the air provides the oxygen for the oxidative reaction (Li et al., 2016). The authors suggested a possible reaction pathway for the aliphatic aldehydes and citral in their article (see Figure 28.7).

The reaction process seems to be the same for all aliphatic aldehydes, starting with the UV-induced loss of a hydrogen radical by the carbon at position 1, generating a free radical (compound 1 in Figure 28.7), followed by an attack by oxygen, resulting in a peroxide radical (compound 2 in Figure 28.7), which then associates with a hydrogen radical and subsequently attacks and oxidizes an original aldehyde molecule, yielding the corresponding aliphatic acid. Neither air nor oxygen alone could trigger the reactions of the aliphatic aldehydes in the study (Li et al., 2016).

trans/cis-Citral was transformed to cyclocitral under exposure to UV light, with or without oxygen availability. Under co-treatment of air and irradiation, citral reacted to form geranic acid and neric acid, in contrast to the study of Schieberle and Grosch (1989), who found citral to be transformed to *p*-methylacetophenone after 4 days in 5% citric acid at 40°C, suggesting that citral undergoes different reaction pathways under different conditions (Li et al., 2016). The formation of cyclocitral is explained by the authors as a succession of steps, starting once more with the loss of a hydrogen radical at the α-position carbon of the carbon at position 3. After a recombination of carbon–carbon double bonds, the position 2 carbon attacks the double bond in position 7, resulting in a cyclic free radical (compound 3 in Figure 28.8), which then associates with a hydrogen radical to form an intermediate (compound 4 in Figure 28.8). Another hydrogen radical loss and double-bond recombination occur, until cyclocitral is formed by hydrogen radical uptake.

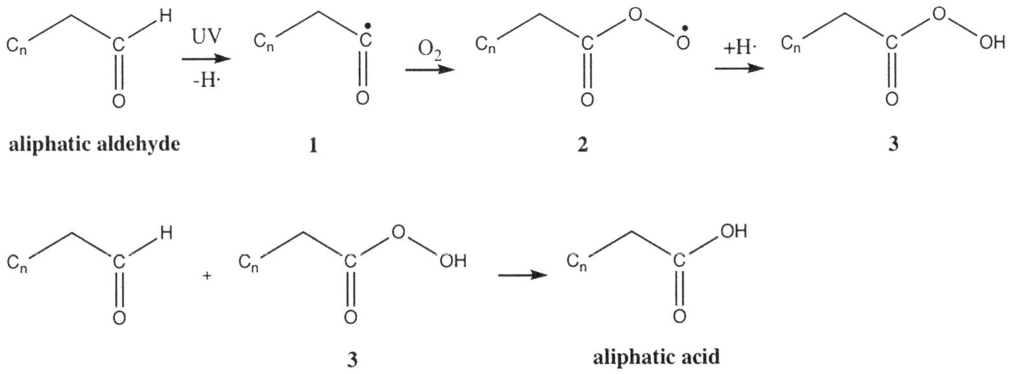

FIGURE 28.7 Aliphatic aldehyde photooxidation, n = 6 (octanal), 7 (nonanal), 8 (decanal), 10 (dodecanal). (Adapted and newly drawn from Li L.J. et al. 2016. *J. Agric. Food Chem.*, 64(24), 5000–5010.)

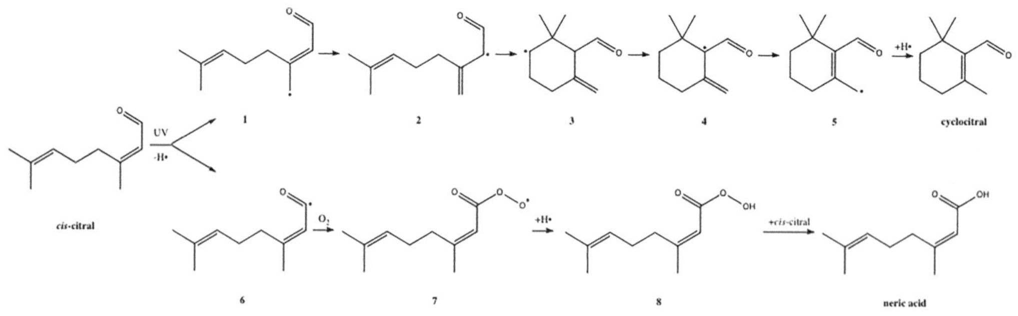

FIGURE 28.8 Citral phototransformations. (Adapted and newly drawn from Li L.J. et al. 2016. *J. Agric. Food Chem.*, 64(24), 5000–5010.)

Another putative reaction pathway leads to the citral acids, also starting with the loss of a hydrogen radical at position 1, generating a free radical (compound 6 in Figure 28.8) ,which is subsequently attacked by oxygen, forming a peroxide radical (compound 7 in Figure 28.8). After forming another intermediate by associating with a hydrogen radical, this compound oxidizes *trans/cis*-citral, accordingly yielding the corresponding citral acid. The transformation pathways are depicted for *cis*-citral in Figure 28.8; for *trans*-citral the mechanism works analogously, ultimately generating also cyclocitral, but geranic acid instead of neric acid.

This change in constituents is followed by a change in odor of pummelo EO. After the treatment with light and oxygen, the minty, herbaceous, and lemon odors decrease, and the oily odor intensifies, due to the concentration reduction of β-myrcene, (+)-limonene, and aldehydes (i.e. octanal, decanal, *cis*-citral, *trans*-citral, and dodecanal).

Interestingly, citral seems to undergo very different reactions when brought into contact with hydrogen peroxide instead of molecular oxygen during irradiation. Elgendy and Khayyat (2008) studied the photooxidations of citral, pulegone, and camphene under irradiation with a sodium lamp under different conditions. In one reaction, 30% hydrogen peroxide was added; in another, 80% *m*-chloroperoxybenzoic acid ($3\text{-}ClC_6H_4CO_3H$) as an oxidant; and in yet another, a singlet oxygen photosensitizer, that is, tetraphenylporphyrin (TPP) or rose bengal or chlorophyll. The mixture of citral and hydrogen peroxide in ethanol was subjected to irradiation for 55 hours, and the reaction yielded (2*E*,*Z*)-5-(3,3-dimethyloxiran-2-yl)-3-methylpent-2-enal (compounds 1a/b in Figure 28.9) and 3-methyl-3-[(3*E*)-4-methylpent-3-en-1-yl]oxirane-2-carbaldehyde (compound 2 in Figure 28.9). Oxidation with the acid in chloroform at room temperature under nitrogen generated the same two isomers 1a and 1b, but no others products were detectable. Interestingly, the photosensitized reaction gave a mixture of (2*E*,5*E*)-7-hydroperoxy-3,7-dimethylocta-2,5-dienal (compound 3 in Figure 28.9) and (2*E*)-6-hydroperoxy-3,7-dimethylocta-2,7-dienal (compound 4 in Figure 28.9). The highest yield was achieved with TPP as a sensitizer. Structure 3 was found to have DNA damaging effects in the study (Elgendy and Khayyat, 2008).

Pulegone is another natural monoterpene found in the EOs of many different *Mentha* species and was isolated for the study of Elgendy and Khayyat (2008) from the leaves of *Mentha pulegium* L. (Lamiaceae). The photoreaction products of pulegone with 30% hydrogen peroxide in ethanol were compounds 1a and 1b, a mixture of isomers with different mutual stereochemical orientations at the oxirane ring and the 7-methyl group, as depicted in Figure 28.10. In relation to the cyclohexane ring, the addition of hydrogen peroxide can occur on either side of the exocyclic double bond; therefore, two isomers are generated after the loss of H_2O by the intermediate oxirane. In the resulting products, the position of the methyl group on C7 and the oxirane oxygen atom are relative to each other in *trans*-position for compound 1a and in *cis*-position for compound 1b. The reaction in the presence of a photosensitizer TPP, rose bengal or chlorophyll, interestingly, produced (besides 1a/b) two additional compounds: 2-(1-hydroperoxy-1-methylethyl)-5-methylcyclohex-2-en-1-one (compound 2 in

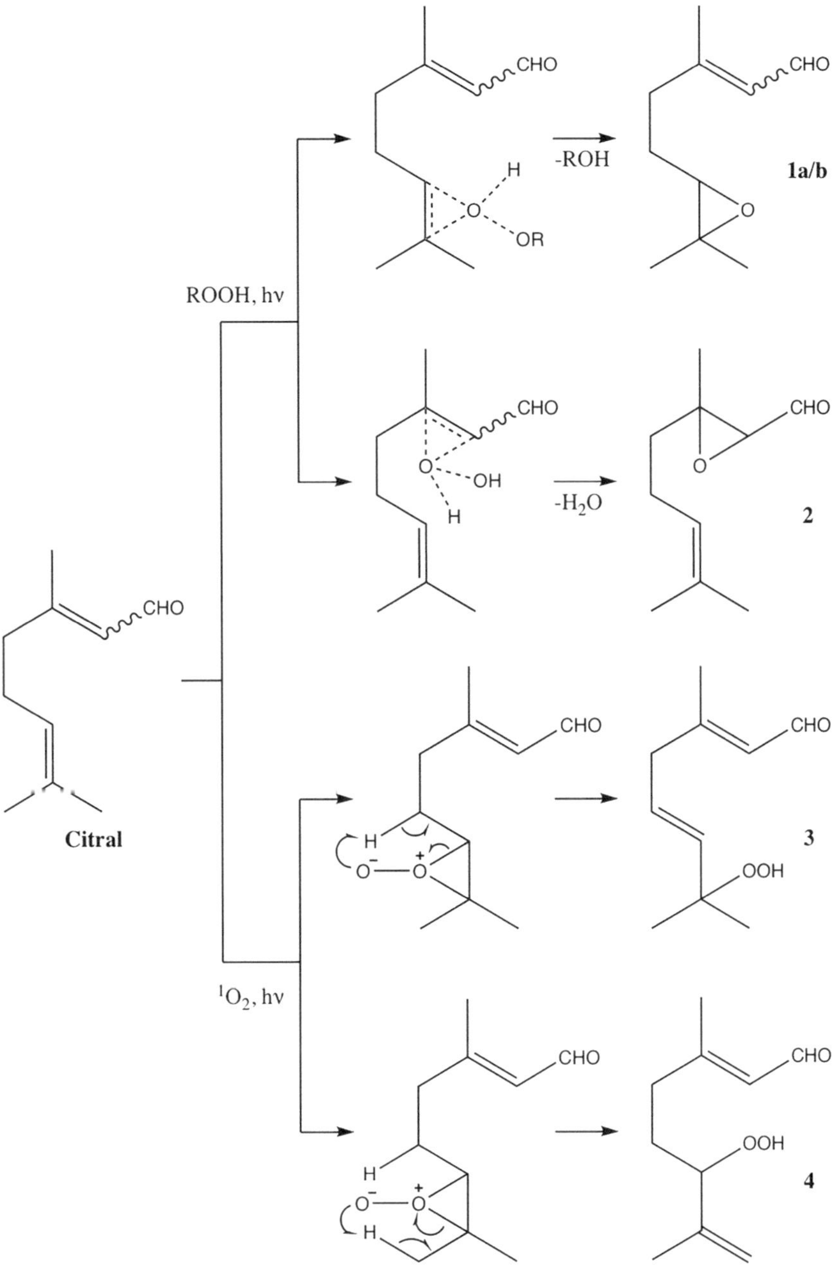

FIGURE 28.9 Photoepoxidation reactions of citral, R=H, 3-$ClC_6H_4CO_3H$. (Adapted and newly drawn from Elgendy E.M., and S.A. Khayyat. 2008. Russ. *J. Org. Chem.*, 44(6), 814–822.)

Figure 28.10) and 2-hydroperoxy-2-isopropenyl-5-methyl-cyclohexan-1-one (compound 3 in Figure 28.10). This oxygenation most likely involves the stabilization of a peroxirane transition state along two different pathways, analogue to the generation of products 3 and 4 in the photosensitized reaction of citral. Compound 2 was found to be genotoxic in a DNA damage assay by the authors (Elgendy and Khayyat, 2008).

The photooxidation of camphene, another monoterpene found, for example, in the EO of camphor, lemongrass, and ginger, was also studied by Elgendy and Khayyat (2008). The reaction with 30%

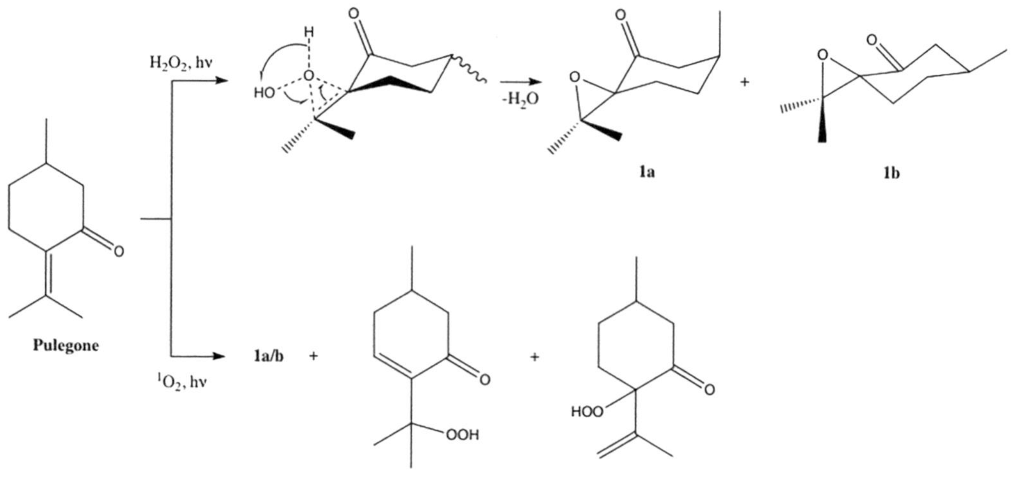

FIGURE 28.10 Photoepoxidation reactions of pulegone. (Adapted and newly drawn from Elgendy E.M., and S.A. Khayyat. 2008. *Russ. J. Org. Chem.*, 44(6), 814–822.)

hydrogen peroxide produced a mixture of *endo*- and *exo*-isomers of 3,3-dimethylspiro[bicyclo[2.2.1]-heptane-2,2′-oxirane] (compounds 1a and b in Figure 28.11) and camphor (compound 2 in Figure 28.11) in approximately 15% yield, while its thermal oxidation with *m*-chloroperoxybenzoic acid gave only the two former products. The generation of camphor in this setting is described as "unusual" by the authors; but nevertheless, they propose a possible formation pathway featuring a photoinitiated rearrangement of camphene followed by the attack of hydrogen peroxide (see Figure 28.11).

Linalyl acetate is an acyclic monoterpene very commonly used in floral scents and poses one of the main constituents of lavender (*Lavandula angustifolia* Mill., Lamiaceae) EO. Khayyat (2018) extracted linalyl acetate from lavender EO and exposed it to 30% hydrogen peroxide and irradiation from a sodium lamp in a nitrogen atmosphere at 0°C in ethanolic medium. The main reaction products were 6,7-epoxy-3,7-dimethyl-1-octene-3-yl acetate (compound 1 in Figure 28.12) and 1,2-epoxy-3,7-dimethyl-6-octene-3-yl acetate (compound 2 in Figure 28.12), which are basically two different epoxides of linalyl acetate, depending on which double bond has been attacked by hydrogen peroxide. The reaction with

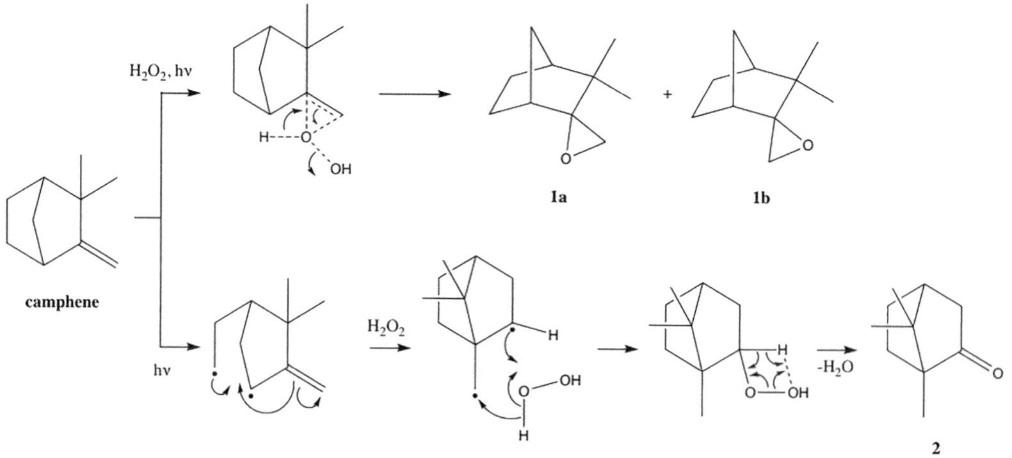

FIGURE 28.11 Photooxidation reactions of camphene. (Adapted and newly drawn from Elgendy E.M., and S.A. Khayyat. 2008. *Russ. J. Org. Chem.*, 44(6), 814–822.)

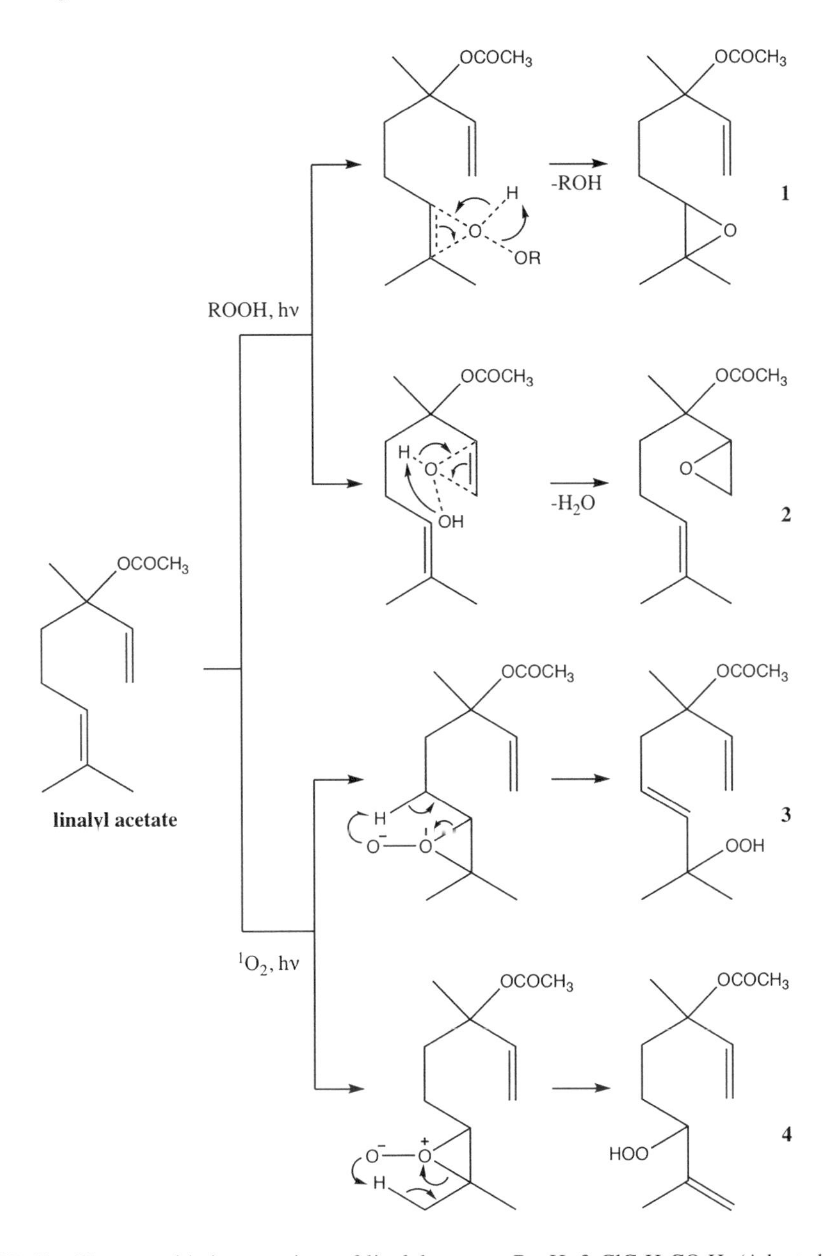

FIGURE 28.12 Photoepoxidation reactions of linalyl acetate, R=H, 3-ClC$_6$H$_4$CO$_3$H. (Adapted and newly drawn from Khayyat, 2018. *Arab. J. Chem.*, Article in Press.)

m-chloroperbenzoic acid instead of hydrogen peroxide yielded only compound 1. When TPP was used as a photosensitizer, a mixture of 7-hydroperoxy-3,7-dimethylocta-1,5-diene-3-yl acetate (compound 3 in Figure 28.12) and 6-hydroperoxy-3,7-dimethylocta-1,7-diene-3-yl acetate (compound 4 in Figure 28.12) was produced, whereas hematoporphyrin (HP) as a sensitizer only gave compound 3 as a product. The reaction pathways proposed by the author are depicted in Figure 28.12 (Khayyat, 2018).

Saffron is, amongst other uses, a spice consisting of the dried stigmas of *Crocus sativus* L. (saffron, Iridaceae) and one of the most expensive spices in the world (Raghavan, 2006). The most abundant constituent of saffron EO, safranal, is mainly responsible for the typical saffron aroma and was found to elicit many effects on the central nervous system, such as antidepressant, anticonvulsive, hypnotic effects, and many more (Rezaee and Hosseinzadeh, 2013). Khayyat and Elgendy studied

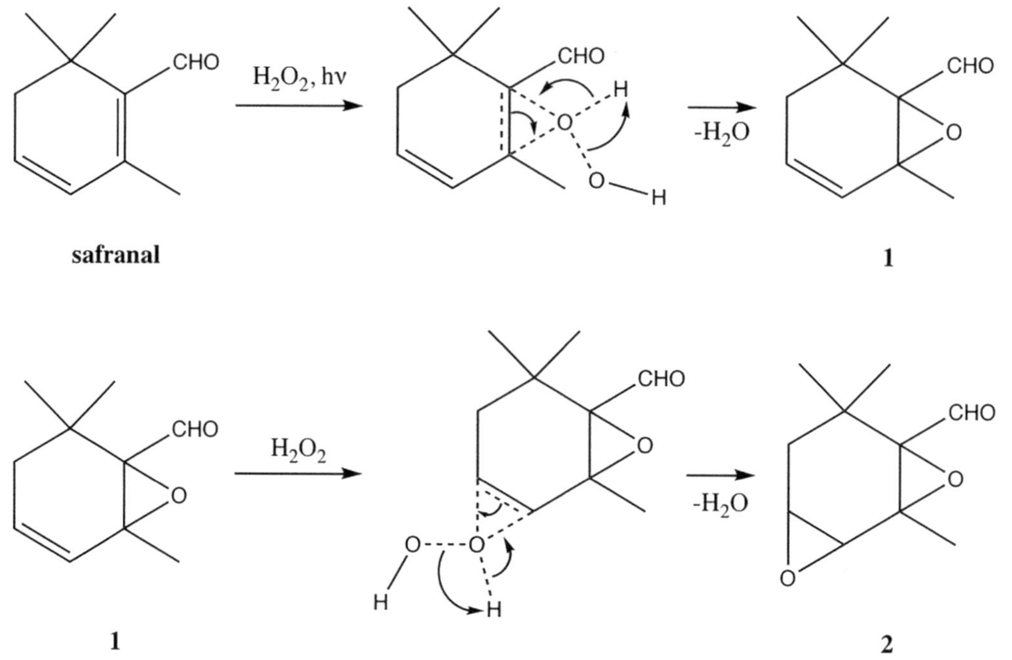

FIGURE 28.13 Photoepoxidation reactions of safranal. (Adapted and newly drawn from Khayyat and Elgendy, 2018. *Saudi. Pharm. J.*, 26(1), 115–119.)

safranal epoxidation, adding 30% hydrogen peroxide and subduing the mixture to 50 hours of irradiation with a sodium lamp under nitrogen atmosphere. The photochemical reaction resulted in 2,2,6-trimethyl-7-oxabicyclo[4.1.0]-hept-4-ene-1-carbaldehyde (compound 1 in Figure 28.13) and diepoxy derivative 2,5,5-Trimethyl-3,8-dioxa-tricyclo[5.1.0.02,4]octane-4-carbaldehyde (compound 2 in Figure 28.13), in yields of 65% and 35%, respectively. The proposed reaction pathways are shown in Figure 28.13. Subsequent analysis of the antibacterial activity of the epoxidation products proved that the monoepoxy and diepoxy derivatives of safranal possess an increased effect against methicillin resistant *Staphylococcus aureus* (MRSA).

Khayyat and Roselin (2018) investigated the photooxidation of geranyl acetate, which is an acyclic monoterpene occurring in the volatile oils of many plant species, such as eucalyptus, cypress, and origanum, for instance. When brought into contact with 30% hydrogen peroxide in ethanolic medium and irradiated with a sodium lamp for 15 hours, geranyl acetate was oxidized at both double bonds and yielded a diepoxy product, 3-(2-(3,3-dimethyloxiran-2-yl)ethyl)-3-methyloxiran-2-yl) methyl acetate. On the other hand, the reaction in the presence of a photosensitizer resulted in a photooxygenation with three different mono- and dihydroperoxide derivatives, acetic acid 2,6-bis-hydroperoxy-7-methyl-3-methylene-oct-7-enyl-ester, acetic acid 7-hydroperoxy-3,7-dimethylocta-2,5-dienyl ester, and acetic acid 3-hydroperoxy-7-methyl-3,7-dimethylocta-1,6-dienyl ester (Khayyat and Sameeh, 2018).

28.2.3 Polymerization Reactions

Some constituents of EOs polymerize when exposed to light and air. Khayyat (2013) studied the photopolymerization of *trans*-cinnamaldehyde, eugenol, and safrole by allowing them to react in chloroform at room temperature with oxygen availability and during irradiation with a sodium lamp. *trans*-Cinnamaldehyde is an aromatic aldehyde, which occurs naturally in the bark of species of the genus *Cinnamomum* (Lauraceae) and poses the main origin of the characteristic cinnamon aroma. Eugenol is the main component of the EO of *Syzygium aromaticum* L. (Myrtaceae), and safrole, on the other hand, is the major constituent of the EO extracted from sassafras (see Section 28.2.4).

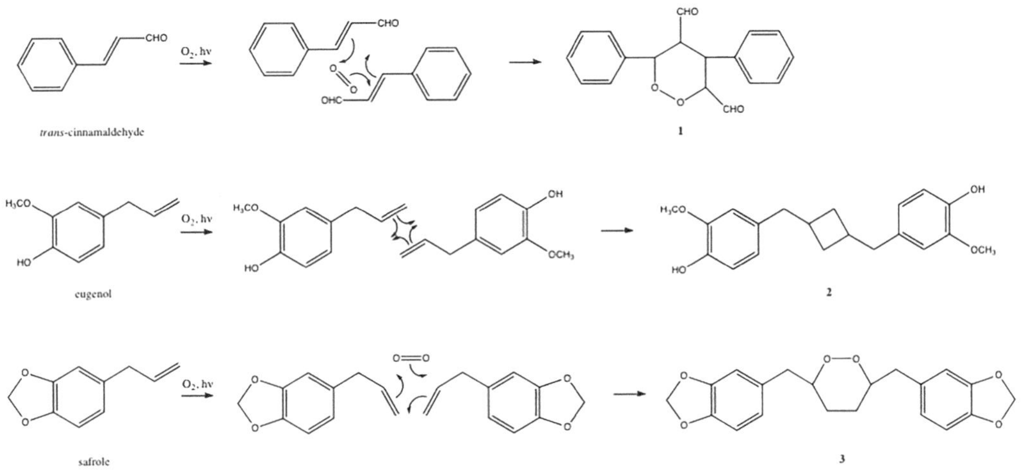

FIGURE 28.14 Photopolymerization of trans-cinnamaldehyde, eugenol, and safrole. (Adapted and newly drawn from Khayyat, 2013. *J. Saudi. Chem. Soc.*, 17(1), 61–65.)

All three produced their corresponding dimers: cinnamaldehyde gave 4,6-diphenyl-1,2-dioxane-3,5-dicarboxaldehyde (compound 1 in Figure 28.14), eugenol gave 4-4′(cyclobutane-1,3-diyl-bis(methylene) bis(2-methoxyphenol)) (compound 2 in Figure 28.14), and safrole gave 3,6-bis(benzo[d][1,3]dioxol-5-ylmethyl)-1,2-dioxane (compound 3 in Figure 28.14). The yields were 73%, 62%, and 55%, respectively.

A probable reaction mechanism for both cinnamaldehyde and safrole is [2+2+2] cycloaddition following singlet oxygen attack on the double bond of the side chain for two molecules each, in both cases resulting in a dimer with a dioxane ring (see Figure 28.14). As for eugenol, this compound yielded a dimer with a cyclobutane ring after a [2+2] cycloaddition reaction. The resulting dimers were tested for their antifungal activity against *Candida albicans*, a very common form of candida yeast which naturally inhabits several moist and warm cavities of the human body but can also cause skin and mucosal infections, in comparison to their respective monomers. The results indicated that the dimers elicit a stronger antifungal effect than their monomer precursors (Khayyat, 2013).

Isoeugenol was also found to undergo dimerization under irradiation with UV light. Chiang and Li (1978) subjected isoeugenol in actone to the light of a high pressure mercury lamp, triggering a photoreaction that results in two different dimers of isoeugenol: diisoeugenol (= 1-ethyl-3-(4′-hydroxy-3′-methoxyphenyl)-6-methoxy-2-methyl-5-indanol) and dehydrodi-isoeugenol (= 4-[2′,3′-dihydro-7′-methoxy-3′-methyl-5′-(1″-propenyl)-2′-benzofuranyl]-2-methoxyphenol) (Chiang and Li, 1978).

Dimerization of *trans*-anethole was described by Lewis and Kojima (1988) as a cycloaddition of a cation radical and a neutral, which yields a mixture of *syn* and *anti* head-to-head dimers. The reaction products are depicted in Figure 28.2 (see above), together with the isomerization products, as irradiation of *trans*-anethole in acetonitrile or toluene gave *cis*-anethole, as well as the various dimers, the concentrations of the latter depending on whether or not a sensitizer was used (Lewis and Kojima, 1988; Castro et al., 2010).

28.2.4 Other Reaction Mechanisms

Moulin et al. (1995) used laser photolysis as an alternative to other separation techniques for purifying complex mixtures such as extracts and EOs by selectively eliminating non-desirable molecules. They monitored the destruction of the molecules by spectral changes and identified the photoproducts by gas chromatography and mass spectrometry. Salvia and bergamot EOs were diluted tenfold and subjected to a laser beam in order to purge the contained amount of toxic thujone and phototoxic bergapten, respectively.

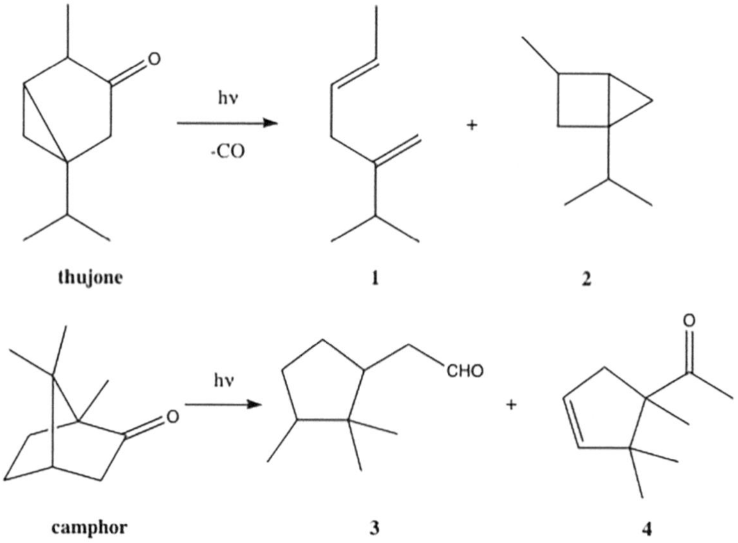

FIGURE 28.15 Thujone and camphor photolysis. (Adapted and newly drawn from Moulin et al., 1995. *J. Photoch. Photobio. A: Chem.*, 85, 165–172.)

Salvia EO was irradiated at 308 nm, and thujone reacted to yield two major reaction products, 1 and 2, which are depicted in Figure 28.15. The supposed driving force of the reaction is ring-strain relief, and the process involves a Norrish type 1 cleavage, a loss of carbon monoxide, and subsequent recombination of the radical fragments. Other molecules in the EO, such as pinene, cineol, and caryophyllene, were not affected by the laser photolysis, except for camphor, which is also a ketone with a very similar structure and absorption spectrum to thujone. The camphor degradation products 3 and 4 are shown in Figure 28.15, although it must be noted that they only posed a small percentage of the original amount of camphor, and the authors suggest that another new reaction product with a higher molecular weight was formed, probably due to rearrangement with the solvent EtOH.

Bergapten (= 5-methoxypsoralen) causes phototoxic reactions when skin treated with bergamot EO is exposed to UV light. The laser photolysis at 10 Hz by Moulin et al. (1995) eliminated 35% of the bergapten in the EO after 1 hour, and 60% after 2 hours of irradiation. Other constituents were not affected, and the process seems to be irreversible due to unchanging absorption spectra several days after the phototreatment (Moulin et al., 1995). In their study, the authors also aimed for the destruction of safrole in the oil of sassafras, which can be obtained from the roots of *Sassafras albidum* (Nutt.) Nees (Lauraceae) or some members of the genus *Ocotea*, which are closely related. Furthermore, it can be sourced from *Piper hispidinervum* L. (Piperaceae), the EO of which also contains high levels of safrole (Rocha and Chau Ming, 1999). Interestingly, the trade of plant extracts rich in safrole is restricted in the European Union, due to the fact that safrole is a precursor in the production of 3,4-methylendioxy-N-methylamphetamin (MDMA) (EMCDDA, 2016). Additionally, safrole is known to have carcinogenic effects. The sassafras EO was subdued to laser photolysis, and with the help of chromatograms before and after, the authors could verify the complete disappearance of the safrole peak. Instead, three new peaks appeared, which can be assumed to be those of short fragments of safrole, because of their short retention times and because laser photolysis generally creates smaller products in comparison to UV photoreactions, a fact attributed to the large difference in energy (Moulin et al., 1995).

Turek and Stintzing (2012) studied the impact of different storage conditions on EOs using the example of rosemary, thyme, pine, and lavender EO. They simulated three different storage circumstances: storage A involved amber glass vials at room temperature in the dark, whereas the alternatives were clear glass vials submitted to cool white light simulating daylight for 24 hours per day at room temperature (23°C ± 3°C, storage B) or elevated temperature (38°C ± 3°C, storage C). The most striking transformation was observed in rosemary (*Rosmarinus officinalis* L., Lamiaceae)

EO, where the amount of the monoterpene α-terpinene did not change under storage condition A, but decreased to under 10% of the original value at elevated temperature under light irradiation after a mere 3 weeks. Both storage experiments under daylight conditions revealed accelerated degradation under irradiation. Phototransformations resulted in a noticeable decrease in α-terpinene and α-phellandrene, together with an increase in *p*-cymene, camphor, 1,8-cineole, and caryophyllene oxide, as well as some yet unidentified oxidation products.

The EO of thyme (*Thymus vulgaris* L., Lamiaceae) showed a distinctive stability under all conditions tested. Minor diminutions of β-myrcene, γ-terpinene, and α-terpinene could be detected in the elevated temperature of storage C, but also, to a lesser extent, in the dark. The high amount of radical-scavenging phenolic structures may be accountable for the resistance of thyme EO to degradation. The EO of *Pinus sylvestris* L. (Pinaceae) showed a peculiar behavior: the highest peroxide values were obtained with storage in the dark, and moreover, the peroxide values were lower at intensified storage conditions C than at room temperature. The authors propose a lability of the hydroperoxide intermediates at higher temperatures to explain these results. The amounts of *p*-methylacetophenone and caryophyllene oxide, as well as a range of newly generated polar compounds, increased significantly, whereas many other unidentified substances decomposed already within 4 to 8 weeks. In summary, pine EO seems more prone to oxidation than the former two EOs, as several constituents such as caryophyllene oxide experienced alterations even in the dark. Lastly, Turek and Stintzing (2012) also investigated the EO of *Lavandula angustifolia* MILL. (Lamiaceae). Already after 4 weeks under daylight conditions (storage B), the peroxide value was higher than after 12 weeks of storage in the dark. However, similar to pine EO, storage under aggravated conditions (38°C, storage C) yielded lower peroxide values than at room temperature, again suggesting thermolabile hydroperoxy intermediates. Unfortunately, most compounds affected by the transformations were not identified by the authors, but it can be stated that imitated daylight conditions forced the total breakdown of an EO constituent and a remarkable decrease of another, while the formation of one structure turned out to be promoted by irradiation (Turek and Stintzing, 2012).

The chemical reactions of juniper EO from *Juniperus communis* L. (Cupressaceae) under different storage conditions was studied by Odak et al. (2018). The EO was stored for 1 year in daylight or dark, at room temperature or in the refrigerator (4°C), with or without oxygen availability. Samples were tested after 1, 2, and 12 months. The most striking transformation was a decline in verbenone content under influence of light, both under nitrogen or oxygen atmosphere. Verbenone is an unsaturated bicyclic monoterpene ketone, which is photochemically reactive. Among the photoreaction products were piperitone and isopiperitone, as well as some small unidentified peaks. Furthermore, β-myrcene decreased under all applied conditions, with the alteration being less conspicuous if the sample was stored in the dark and under nitrogen, and limonene apparently oxidized to form α-terpineol, when oxygen was available, even in the dark. But after all, changes of juniper EO were neither rapid nor quantitatively impressive; therefore, it can be concluded that juniper EO is mostly stable, especially under storage conditions that exclude oxygen (Odak et al., 2018).

Odak et al. (2018) also investigated immortelle EO sourced from *Helychrisum italicum* subsp. *italicum* (Roth) G. Don (Asteraceae) by submitting it to the same storage conditions as they did with juniper EO. When exposed to light, immortelle EO showed photosensitivity, given that the amount of italicene and isoitalicene increased from 4.9% to 6.5% and from 0.4% to 1.0%, respectively, within 12 months, while in the dark, quantities remained unchanged. Transformations were even more distinct under storage in nitrogen atmosphere. Also, a decrease in caryophyllene has been noticed. According to these results, it seems recommendable that immortelle oil be stored in the absence of light (Odak et al., 2018).

28.2.5 Influence of Reaction Conditions

28.2.5.1 Effect of Solvent

Naturally, the solvent medium, in which any photochemical reactions occur, does play a certain role. It could alter the speed or type of reactions due to interactions with the constituents of the EO solved within. Excess vibrational energy is removed by the medium, making it act like a heat sink

(Michl, 1974). Kejlová et al. (2007) assessed the phototoxicity of bergamot oil from four different suppliers, and found that the solvent used had great impact on the results gained by the 3T3 NRU test (see Section 28.3.1.1. For this assay, official guidelines recommend ethanol or DMSO as solvent (OECD, 2004). Utilization of these yielded ambiguous borderline phototoxicity values for two of the four samples, whereas the reaction in aqueous solution (phosphate buffered saline, PBS) gave a clear phototoxic classification. Moreover, in another study, Kejlová et al. (2010) tested the phototoxic potential of EOs of lemon, orange, and *Litsea cubeba* (Lour.) Pers. (Lauraceae) and came to the same conclusion. DMSO poses a typical hydroxyl radical scavenger, and thus it may even attenuate the phototoxic effect of tested substances. Therefore, the phototoxicity assessment without solubilizers might be advisable even for EOs with limited solubility in water to prevent underpredictive classifications due to these radical scavenging effects (Kejlová et al., 2007). In the RBC PT (see Section 28.3.1.2), PBS is being used for sample preparation, and ethanol and DMSO at 10% final test concentration are recommended as vehicles (Pape et al., 2001).

Beltrame et al. (2013) found that the solvent medium of the EO (in their case, marjoram EO) modified the pattern of decomposition by UV radiation. Ethanol and hexane proved to be rather similar in their absorbance levels, indicating the same chemical degradation mechanisms, whereas EO in dichloromethane showed slightly different behavior. In the study of Tateba et al. (1993), the photoconversion of *cis*- and *trans*-jasmone increased in ethyl acetate (aprotic solvent) compared to methanol and ethanol (protic solvents).

28.2.5.2 Effect of Duration and Intensity

The higher the temperature, the higher the yield, naturally (Castro et al., 2010), and according to the Van't Hoff law, a temperature elevation of 10 K doubles the reaction rate, as explained in Glasl (1975). But when it comes to the influence of duration and intensity of light irradiation, opinions differ and no standard procedure seems available.

Beltrame et al. (2013) exposed marjoram EO to UV radiation for only 5 minutes, using a photoreactor with a 125 W mercury lamp, or a 250 W lamp for accelerated photodegradation, both with removed outer bulb to allow UV light emission, and took readings every minute. Castro et al. (2010) left their samples in the photoreactor for 120 minutes, under constant stirring. In the study by Dijoux et al. (2006), the cells underwent UV-A/visible light treatment at 1.6– 1.8 mW/cm^2 for 50 min. Turek and Stintzing (2012) submitted their samples to simulated daylight conditions by means of 24 hours per day irradiation with 5000 lx fluorescent tubes from Osram L36W/840, for up to 24 weeks.

The ICH (2012) stated that UV-A doses ranging from 5 to 20 J/cm^2, that is, UV-A dosage comparable to that procured during extensive outdoor activities on days in summer around noontime in temperate zones at sea level, are effectively used in up-to-date *in vitro* and *in vivo* phototoxicity assays. In the photohemolysis study by Placzek et al. (2007), samples were exposed to 0, 5, 25, 50, or 100 J/cm^2 UV-A (UVASUN 5000) or to 0 (0), 500 (0.2), 1000 (0.4), or 2000 (0.8) mJ/cm^2 UV-B (J/cm^2 UV-A). The RBC photohemolysis test by Pape et al. (2001) demanded an intensity of 15 J/m^2 UV-A and approximately 1 J/m^2 UV-B, since human erythrocytes can be exposed to more-intensive UV-B irradiation than other cell lines due to their specific cellular defense. The protocol of the *in vitro* 3T3 NRU assay stipulates an exposure of 5 J/cm^2 in the UV-A range for 50 minutes for the assessment of phototoxicity (OECD, 2004), suggesting that this duration and intensity should be sufficient for phototoxic effects to be revealed.

28.3 PHOTOTOXICITY

For human skin, solar radiation is not only a source of energy for numerous physiological processes, but also an environmental stressor. Exposure to visible light or UV light can directly cause photodermatoses, and many photoactive chemicals can induce photosensitivity (Maibach and Honari, 2014). UV-A light is able to reach capillary blood, while UV-B only penetrates the epidermis. Hence, UV-A is more relevant for the photochemical activation of systemic drugs, whereas UV-B is of clinical relevance for topical formulations on light-exposed tissues (ICH, 2012).

According to the Organisation for Economic Co-operation and Development (OECD) guidelines for the testing of phototoxic potential of substances, phototoxicity is defined as "a toxic response from a substance applied to the body which is either elicited or increased (apparent at lower close levels) after subsequent exposure to light, or that is induced by skin irradiation after systemic administration of a substance" (OECD, 2004). However, acute skin responses to photosensitizing chemicals can also be photoallergic reactions. Those two processes are distinguished in photochemistry, for phototoxic reactions induce toxic cell damage and are non-immunological, but photoallergic reactions, on the other hand, are T-cell-mediated immunological reactions. Most of the substances that elicit photoallergic responses are also phototoxic (Placzek et al., 2007). It can occur as an adverse reaction to cosmetic products or pharmaceutical drugs, and therefore it is an utmost necessity to evaluate the phototoxic potential of the ingredients of such, so as to not put the patient or user at risk. The use of *trans*-anethole, a major constituent in some EOs, for example, has been the subject of discussion, for when irradiated with UV light, *cis*-anethole is formed, which does not only possess an unpleasant scent and flavor, but is also toxic (Castro et al., 2010).

28.3.1 Methods for the *in vitro* Assessment of Phototoxicity

To determine phototoxicity of chemicals, initially their ability of UV/visible light (290–700 nm) absorption is assessed. For this, the molar extinction coefficient (MEC) is used, a constant for any given molecule under a standard set of conditions like solvent, wavelength, and temperature, which mirrors the photon absorption efficiency of the molecule (ICH, 2012). Subsequently, a chemical assay is utilized to measure reactive oxygen species (ROS), which can help predict phototoxicity, as they are generated by a photoreactive chemical upon exposure to light (Onoue et al., 2008). Apparently, ROS assays have low specificity, giving many false-positive results, but on the other hand, they proving high sensitivity, hence a negative result in this assay would indicate a very low risk of phototoxicity potential (ICH, 2012).

Ultimately, there are several different methods available for the assessment of phototoxicity. All include a certain human or animal cell line, the application of the substance to be tested plus treatment with irradiation, and subsequently a test on cell viability to determine the effect. When the *in vitro* phototoxicity assay shows positive results, further testing is required, utilizing reconstituted 3D human skin models or *in vivo* preclinical trials (Maibach and Honari, 2014).

28.3.1.1 3T3 NRU Assay

The 3T3 cell neutral red uptake (3TS NRU) test is a common colorimetric method of assessing phototoxic potential and currently the only one validated by the European Union Reference Laboratory for alternatives to animal testing (EURL ECVAM, 2018a). It is known to be highly sensitive, specific, and reproducible and identifies substances that either act as photoirritants after dermal application or show phototoxic effects after systemic administration (Dijoux et al., 2006). The 3T3 NRU assay is commonly used in chemical and cosmetics industries as it is regarded as reliable in its capacity to predict acute phototoxicity effects *in vivo* (OECD, 2004). This assay uses monolayer cell cultures of the mouse fibroblast cell line Balb/c 3T3 to assess the cytotoxicity of a compound with or without exposure to a non-cytotoxic dose of simulated sunlight (5 J/cm^2 in the UV-A range for 50 min). The non-diffusion uptake of neutral red (= NR, 3-amino-7-dimethylamino-2-methylphenazine hydrochloride), a weak cationic dye, into the cells and subsequent accumulation in lysosomes is measured by a spectrophotometer. Cells damaged by the phototreatment show a decreased uptake; therefore, viable, damaged, and dead cells can be distinguished. To evaluate the data, the photoirritation factor (PIF, see below) and mean photoeffect (MPE, see below) are calculated and the results of photoexposed samples are compared to the photoprotected ones (Spielmann et al., 1998; OECD, 2004).

Previous studies have found the NRU test and MTT conversion test to be mainly equivalent in results, and rabbit cornea derived SIRC cells or human keratinocytes can be used instead of the murine fibroblastic 3T3 cells (Dijoux et al., 2006); therefore, those variations of the assay could probably be used as well, but the only validated method remains the one utilizing the 3T3 NRU assay.

28.3.1.2 Photohemolysis Test and RBC PT

First ,there was the photohemolysis test, an early protocol, in which the hemoglobin released from cells damaged by phototoxicity was determined by converting it to its stable form by chemical oxidation and then measuring it photometrically. A disadvantage of this test was that it did not include the measurement of met-hemoglobin, the formation of which poses another important endpoint of the red blood cell treatment with phototoxic chemicals under light exposure. Then, Pape et al. (1994) improved the original photohemolysis assay and introduced the red blood cell phototoxicity test (RBC PT). This new approach is designed for the combined testing of the ability of potentially phototoxic substances to hemolyze erythrocyte membranes and/or to oxidize hemoglobin under UV irradiation (Pape et al., 2001).

In the course of this assessment, suspensions of isolated human erythrocytes are subjected to the test compounds and exposed to UV light for 150 minutes. After ensuing 30 minutes of incubation in the dark, changes in optical density are measured at 525 nm and 630 nm to determine the photohemolysis and hemoglobin oxidation, respectively (Pape et al., 2001). Subsequently, the photohemolysis factor (PHF) is calculated as the ratio of the H50 values of the samples incubated in the dark and the ones irradiated. Substances with a PHF $\geq$3 are considered phototoxic. As for hemoglobin oxidation, the met-hemoglobin formation is conveyed as maximal change in the optical density (ΔOD_{MAX}) at 630 nm. If $\Delta OD_{MAX} \geq 0.05$, the substance is assumed to possess phototoxic potential (Pape et al., 2001).

The RBC PT assay passed the pre-validation process during the EU/COLIPA validation program, which was designed to examine the suitability as a regulatory test on phototoxicity (Pape et al., 2001). Erythrocytes have specific cellular defense mechanisms and are therefore not as sensitive to UV-B irradiation as other cell lines (Pape et al., 2001), which is an advantage compared to the 3T3 NRU assay. Also, the RBC PT assay grants additional information about the mechanism of phototoxicity, as compounds will yield a negative result if their cytotoxic mechanism under irradiation does not include the formation of reactive oxygen species. Therefore, this test poses a useful additional *in vitro* test method, especially for mechanistic studies (Liebsch et al., 2005).

28.3.1.3 Reconstituted 3D Human Skin Models

Reconstituted skin models are 3D biostructures consisting of cultured normal human keratinocytes (Maibach and Honari, 2014), and possess a structure very similar to *in vivo* human epidermis. Currently, there are several 3D skin models available, such as EpiDerm™, EpiSkin™, and SkinEthic™. EpiDerm™, for example, has a basal cell layer topped on a support filter, above that spinosum cells, then finally a clear granulosum layer with a stratum corneum at the upper surface (Jones, 2008).

Similar to monolayer assays such as 3T3 NRU, the human skin models work with the premise that phototoxic chemicals will damage cells when exposed to light and can therefore be assessed by measuring the cell viability after irradiation treatment. The cell viability is here scaled by the MTT conversion test (Maibach and Honari, 2014). The 3-(4,5-dimethylthiazol-2-yl)-2,5-diphenyltetrazolium (MTT) assay is based on the uptake of the soluble yellow MTT tetrazolium salt by mitochondrial succinic dehydrogenase and its subsequent reduction to an insoluble blue MTT formazan product. This reaction is dependent on the mitochondrial function and is therefore an indicator of cell viability (Dijoux et al., 2006).

Due to their high cost, they are not part of standard testing operations, but reconstituted human skin models allow for a range of topically applied substances, from single chemicals to final pharmaceutical formulations to be tested (Jones, 2008). In comparison to monolayer cell cultures, there are fewer solubility problems because of the direct application of chemicals to the skin surface; also, higher concentrations can be tested, and the 3D structure of the model allows for higher proportion of UV-B, thus rendering a more accurate solar irradiation model possible (Jones et al., 2001; Jones, 2008). These assays can measure cell viability with or without phototreatment. They might be close to reality, but they can still be less sensitive than human skin *in vivo* (ICH, 2012). EpiDerm™-reconstituted human skin models showed promising results in a pre-validation study (EURL ECVAM, 2018a) and are recommended as further investigations to determine actual hazard when phototoxic effects are solely observed at the highest test concentration in the 3T3 NRU test (OECD, 2004). Also, they are validated as methods for the assessment of skin irritation, that is, a local inflammatory reaction of the skin caused by the non-specific immune system after application of an irritant (EURL ECVAM, 2018b).

28.3.1.4 Photoirritation Factor (PIF) and Mean Photo Effect (MPE)

PIF and MPE are values used to assess the phototoxic potential of chemical substances. The photoirritation factor (PIF) is defined as the difference between IC50 in presence and absence of UV light (Dijoux et al., 2006). The mean photo effect (MPE) is predicated on the comparison of the complete concentration response curves. Concerning the assessment of the results, a PIF<2 or an MPE<0.1 predicts "no phototoxicity." A PIF>2 and < 5 or an MPE > 0.1 and <0.15 suggests "probable phototoxicity" (Spielmann et al., 1998; OECD, 2004), although according to an ICH guideline on the photosafety evaluation of pharmaceuticals, "compounds in this category generally do not warrant further photosafety evaluations" (ICH, 2013). In the case of PIF>5 or an MPE>0.15, "phototoxicity" is forecasted (Spielmann et al., 1998; OECD, 2004).

In a study comparing different systems for the assessment of phototoxic potential, Dijoux et al. (2006) came to the conclusion that the NRU assay and the MTT conversion test on both rabbit-cornea-derived SIRC cells and murine fibroblastic 3T3 cells were all able to differentiate between phototoxic (PIF>5) and non-phototoxic (PIF<5) molecules/oils, but had problems distinguishing mildly or probably phototoxic substances with 2<PIF<5 from non-phototoxic ones with PIF<2 correctly (Dijoux et al., 2006).

28.3.2 Phototoxic Essential Oils and Essential Oil Constituents

Most phototoxic EOs are found in the botanical families of Apiaceae and Rutaceae, probably due to evolutionary divergence. Apart from those, phototoxic EOs may be found prevalently in the Asteraceae and Moraceae families (Tisserand and Young, 2014).

As aforementioned, substances can be phototoxic and/or provoke T-cell-mediated photoallergic reactions, both of which result in skin irritation similar to that of an acute sunburn. Within minutes to hours of sun exposure, agonizing erythema may develop on irradiated skin (Maibach and Honari, 2014). Most phototoxic compounds absorb energy from UV-A radiation leading to the generation of activated derivatives capable of inducing cellular damage (Dijoux et al., 2006). To afflict cutaneous inflammation, chemicals need to be capable of transgression into the epidermis and there must bind to proteins. The allergenic potential of EOs can be accredited to hydroperoxide derivatives of terpenoids, which are generated by (photo-)oxidation (Turek and Stintzing, 2013). The non-oxidized originals, on the other hand, were found to be not or only slightly irritating (Pirilä and Siltanen, 1958; Hausen et al., 1999; Matura et al., 2005; Karlberg et al., 2008; Bråred-Christensson et al., 2009).

A class of substances often brought into connection with phototoxicity of EOs are the furocoumarins. They are synthesized and used by plants as defensive chemicals and are characterized by their coumarin structure conjoined with a furan ring. Depending on their position, they can be differentiated into two subtypes: the linear psoralen type and the angular angelicin type (see Figure 28.16). Furocoumarins like psoralen, bergapten (= 5-MOP), xanthotoxin, and angelicin, which are abundant in the Apiaceae, Rutaceae (e.g., some *Citrus* species), Moraceae. and other families, are known to be phototoxic and also carcinogenic under UV irradiation (Fu et al., 2013). Others, like bergamottin, bergaptol, isobergapten, and isopimpinellin are non-phototoxic. Furocoumarins are larger than most EO constituents, but can pass over during steam-distillation anyway. Still, cold-pressed EOs show much higher content of these

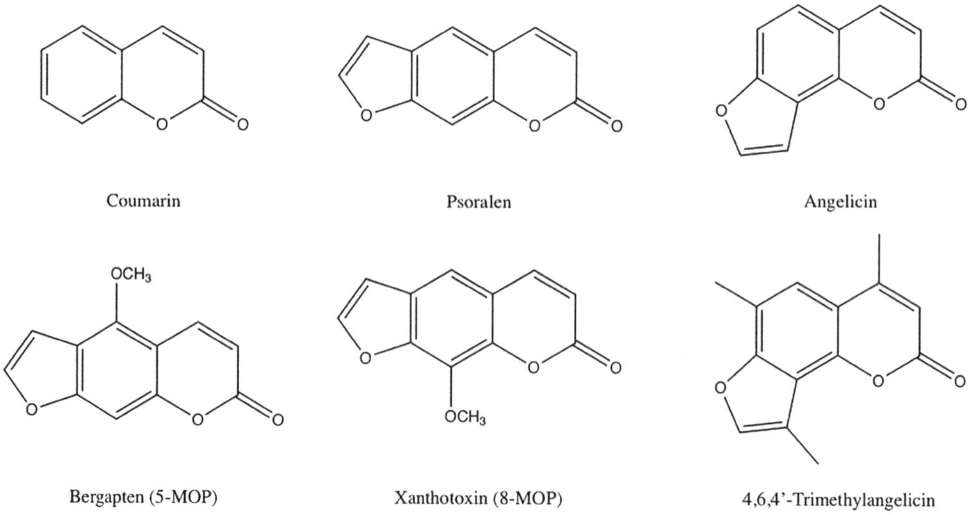

Coumarin Psoralen Angelicin

Bergapten (5-MOP) Xanthotoxin (8-MOP) 4,6,4'-Trimethylangelicin

FIGURE 28.16 Some selected furocoumarins. (Adapted and newly drawn from Fu et al., 2013. *J. Environ. Sci. Heal.*, Part C, 31(3), 213–255; Kinley et al., 1994. *J. Invest. Dermatol.*, 103(1), 97–103.)

compounds than steam-distilled ones. Commonly available EOs that are known to be phototoxic include the EOs of angelica root, bergamot (cold pressed), bitter orange (cold pressed), cumin, fig leaf absolut, grapefruit (cold pressed), lemon (cold pressed), lime (cold pressed), mandarin leaf, opopanax, rue, and tagetes. The following might be phototoxic: clementine (cold pressed), combava fruit, skimmia, angelica root, celery leaf and seed, cumin seed, khella, lovage leaf, and parsnip (Tisserand and Young, 2014).

Three distinct steps of reactions occur between psoralens and DNA: firstly, the non-covalent intercalation of the psoralen between DNA base pairs; secondly, the photochemical reaction of psoralen with a pyrimidine base, resulting in a monoadduct; and lastly, an interstrand cross-link formed by absorption of another photon. The interstrand cross-links in particular are thought to be responsible for photosensitizing effects. For example, 6,4,4′-trimethylangelicin (TMA, see Figure 28.16), a well-known photosensitizing agent, was found to induce interstrand cross-links in mammalian cell DNA (Miolo et al., 1989; Bordin et al., 1994).

The photocycloaddition of a psoralen (8-MOP or 4,5′,8-trimethylpsoralen) to a pyrimidine yields a cyclobutane ring with four asymmetric centers, whereas the stereochemistry is determined by geometrical limitations inflicted by the DNA helix during the formation of the intercalation complex before UV irradiation and cycloaddition. Therefore, all effectively formed adducts have *cis-syn* stereochemistry. But the most abundant products are two diastereomeric thymidine adducts, which are generated by photocycloaddition between the psoralen's 4′,5′ (furan) double-bond and the 5,6 double-bond of the pyrimidine (Kanne et al., 1982). 8-Methoxypsoralen (= 8-MOP, methoxsalen) is infamous for its phototoxic effects, which is put to therapeutic use in photochemotherapy and PUVA (psoralen + UV-A) therapy for patients suffering from psoriasis and other skin diseases. In the process of this therapy, UV-A light energy is absorbed by 8-MOP, and the excited photosensitizer in the triplet state passes energy to molecular oxygen, forming singlet oxygen (type II mechanism), or transfers an electron, which results in the generation of a superoxide anion radical (type I mechanism) (Ochsner, 1997). Either way, ROS are created, increasing the oxidative stress in the cell, which leads to the activation of the complement system, the induction of apoptosis or necrosis, and subsequently to cell death. Ironically, although 8-MOP and psoralen are used to treat skin illnesses, they can also cause such: PUVA therapy has been found to increase the risk of human skin tumors over the years (Stern et al., 1997; Stern, 2001; Katz et al., 2002).

Also, Young et al. (1990) studied the phototumorigenicity of 5-methoxypsoralen (= 5-MOP, bergapten), a constituent of bergamot (*Citrus bergamia* Risso et Poit., Rutaceae) oil, by means of model perfumes containing this oil. They concluded that 5-MOP indeed has phototumorigenic potential already at about 5 ppm. Sunscreens were able to significantly lower the tumorigenicity (Young et al., 1990). The chemical profile and photoinduced cytotoxicity of the EO of *Citrus medica* L. cv. Diamante peel was studied by Menichini et al. (2010). The most abundant compounds were found to be limonene, γ-terpinene, citral, geranial, β-pinene, and α-pinene. The oil also comprised two coumarins, bergapten and citropten. After 100 min of exposure to UV light, the EO showed cytotoxic activity. The phototoxic effect was mainly ascribed to bergapten, as the strong antiproliferative effect of bergapten was not found with citropten (Menichini et al., 2010).

Dijoux et al. (2006) determined in a 3T3 NRU phototoxicity assay that orange EO was "probably" phototoxic, and cytotoxic in the absence of UV light. A study by Binder et al. (2016) supports this result, as they came to the conclusion that orange oil was phototoxic even at low concentrations. Lemongrass EO showed slight phototoxic and cytotoxic properties, whereas sandalwood EO was strongly cytotoxic without UV radiation, but not phototoxic. Carrot and ginger EOs were also not phototoxic, but faintly cytotoxic. To which components the orange and lemongrass EOs owe their phototoxicity remains to be investigated (Dijoux et al., 2006).

The EO of *Anthemis nobilis* L. (or, nowadays, *Chamaemelum nobile L.*, Asteraceae) caused barely perceptible erythema after irradiation in a study of Forbes et al. (1977), where they used hairless mice and miniature swine for testing, and was thus classified as non-phototoxic. 8-Methoxypsoralen, on the other hand, was used as a control and found to be phototoxic (Forbes et al. quoted in Johnson et al., 2017). Gallucci et al. (2010) investigated *Eugenia uniflora* L. (Myrtaceae) leaf EO using the 3T3 NRU test and results showed a PIF <2. They concluded that its use as a fragrance ingredient raises no safety concerns.

A safety assessment study highlighted that undiluted *Mentha piperita* L. (Lamiaceae) oil does evoke moderate to severe reactions in rabbits after repeated intradermal application, but does not appear to be phototoxic. After thoroughly investigating several studies, the CIR expert panel concluded that peppermint oil is safe as used in cosmetic formulations but cautions to keep the concentration of the constituent pulegone under 1% because of its toxicity (CIR expert panel, 2001).

28.4 CONCLUSION

Comprising a range of lipophilic and volatile constituents derived from many different chemical classes, EOs are known to be susceptible to conversion and degradation reactions (Turek and Stintzing, 2013). The stability of EOs depends on several internal factors such as chemical structure and impurities, as well as external factors such as temperature, exposure to humidity, oxygen, or light (Khayyat and Roselin, 2018).

When studying the degradation of EOs, metallic catalysts, molecular oxygen, or photosensitizers are often used to accelerate the occurrence of oxidation reactions; therefore the transformations discovered in such a simulation may not necessarily mirror realistic circumstances. Furthermore, it must be noted that single compounds may not react in the same way as complex mixtures such as EOs, because the different constituents can affect the behavior of the others. On account of this, the findings may not be directly transferable, rendering utilization of single compounds as surrogates or references for EOs questionable (Turek and Stintzing, 2013).

Depending strongly on the exact photoreaction circumstances, photosensitive EOs may be altered in their qualitative and quantitative composition. For instance, if oxygen is available to the reaction mixture, photooxidations are more likely to take place, as with geraniol (Mori and Iwahashi, 2016) or citral (Li et al., 2016). Meanwhile, in the case of hydrogen peroxide presence, photoepoxidation has a higher prevalence, for instance in the photoreactions of pulegone and camphene (Elgendy and Khayyat, 2008). When the irradiated mixture is being deprived of oxygen, there are still other reaction pathways possible, including isomerizations, polymerizations, and cycloadditions. Generally, there is a noticeable influence of the solvent used. DMSO, for example, acts as a hydroxyl radical scavenger, and thus may be able to quench the phototoxic effect of tested substances, as photoreaction progress is inhibited (Kejlová et al., 2007).

Photoinitiation of chemical reactions can be used in positive applications, as for specific elimination by photodegradation of unwanted molecules in complex mixtures such as EOs (Moulin et al., 1995), or as a tool for the drug design of anticancer agents and potent chemoprevention; for example, the photoepoxidation of safranal (Khayyat and Elgendy, 2018). But when transformations are triggered unknowingly, in a product where they should not be, it can be a nuisance, or in the worst case, even potentially dangerous for the consumer. Transformations of EOs and their constituents due to UV light can have different consequences: something as basic as a change in viscosity, odor, and flavor, like in the example of pummelo EO (Li et al., 2016), but also the generation of toxic photoproducts that can harm the consumer, as with the isomerization of *trans*-anethole to toxic *cis*-anethole (Castro et al., 2010) and the phototumorigenic potential of bergapten (= 5-MOP) from bergamot oil (Young et al., 1990). Interestingly, some biological activities of EOs can be enhanced by irradiation with light: the photodimerization products of cinnamaldehyde, safrole, and eugenol elicit stronger inhibiting effects against *Candida albicans* than their corresponding monomers (Khayyat, 2013), and the antibacterial activity of the monoepoxy and diepoxy derivatives of safranal and the hydroperoxides of some monoterpenes (α-pinene, β-pinene, and limonene) are higher than those of their pre-phototransformation parent molecules (Chalchat et al., 2000; Khayyat, 2013; Marqués-Calvo et al., 2017).

In general, there is no risk of phototoxicity to be expected if the EO is used in products which are either not applied to the body or washed off the skin directly after application, such as soaps or shampoos. Moreover, when the EO product is applied to skin that is covered in a way to prevent UV rays from reaching it, there is also no risk of phototoxic reactions. After usage of potentially phototoxic EO products, it is advisable to protect the respective skin area from UV radiation for at least 12–18 hours (Tisserand and Young, 2014). As for the prevention of EO degradation during storage, they should be stored in tightly closed, dark glass vials in a cool place to ensure lasting quality (Buckle, 2003; Clarke, 2008). The headspace should be reduced to a minimum, to avoid contact with oxygen and therefore oxidation reactions, and there should be no water residues (Kaul et a., 1997). Metal contaminants, especially heavy metal copper and ferrous ions, should be avoided, as they are considered to catalyze autoxidation and the formation of singlet oxygen (Choe and Min, 2006). Strategies of EO shielding are being developed, which could improve shelf life of EO products: the nanoemulsion technique, where the EO droplets are encased in propolis (Gismondi et al., 2014) or gum arabic (Bertolini et al., 2001), for instance, or cyclodextrin encapsulation, where inclusion complexes are formed in order to minimize unwanted transformations as well as losses due to evaporation, and increase chemical stability (Marques, 2010).

REFERENCES

Bakkali F., S. Averbeck, D. Averbeck, and M. Idaomar. 2008. Biological effects of essential oils—A review. *Food Chem. Toxicol*, 46(2), 446–475. DOI: 10.1016/j.fct.2007.09.106

Beltrame J.M., R.A. Angnes, and L.U. Rovigatti Chiavelli et al. 2013. Photodegradation of essential oil from marjoram (*Origanum majorana* L.) studied by GC-MS and UV-VIS spectroscopy. *Rev. Latinoamer. Quím.*, 41(2), 81–88.

Bertolini A.C., A. C. Siani, and C. R. Grosso. 2001. Stability of monoterpenes encapsulated in gum arabic by spray-drying. *J. Agric. Food Chem.*, 49(2), 780–785.

Bråred-Christensson J., P. Forsström, and A.-M. Wennberg et al. 2009. Air oxidation increases skin irritation from fragrance terpenes. *Contact Dermatitis*, 60(1), 32–40.

Binder S., A. Hanáková, and K. Tománková et al. 2016. Adverse phototoxic effect of essential plant oils on NIH 3T3 cell line after UV light exposure. *Cent. Eur. J. Public Health*, 24(3), 234–240. DOI: 10.21101/cejph.a4354.

Bordin F., C. Marzano, and C. Gatto et al. 1994. 4,6,4′-Trimethylangelicin induces interstrand cross-links in mammalian cell DNA. *J. Photochem. Photobiol. B. Biology*, 26(2), 197–201. https://doi.org/10.1016/1011-1344(94)07040-7

Buchbauer G. 2002. Lavenderoil and its therapeutic properties. In: M. Lis-Balchin (ed.) *Lavender – The genus Lavender.* London, New York, pp. 124–139.

Buchbauer G., and M. Erkic. 2015. Antioxidative Properties of Essential Oils and Single Fragrance Compounds. In: K.H.C. Başer, and G. Buchbauer (eds.) *Handbook of Essential Oils: Science, Technology and Applications*, 2nd ed., CRC Press, pp. 323–344.

Buckle J. 2003. *Clinical Aromatherapy: Essential Oils in Practice*, 2nd ed., Edinburgh: Churchill Livingstone.

Castro H.T., J.R. Martínez, and E. Stashenko. 2010. Anethole isomerization and dimerization induced by acid sites or UV irradiation. *Molecules*, 15, 5012–5030. DOI: 10.3390/molecules15075012

Chalchat J.C., F. Chiron, R.P. Garry, J. Lacoste, and V. Sautou. 2000. Photochemical hydroperoxidation of terpenes. Antimicrobial activity of α-pinene, β-pinene and limonene hydroperoxides. *J. Essent. Oil Res.*, 12(1), 125–134, DOI: 10.1080/10412905.2000.9712059

Chellappandian M., P. Vasantha-Srinivasan, and S. Senthil-Nathan et al. 2018. Botanical essential oils and uses as mosquitocides and repellents against dengue. *Environ. Int.*, 113, 214–230. DOI: 10.1016/j.envint.2017.12.038

Chiang H., and S. Li. 1978. Studies on the photodimerization of isoeugenol. *Jnl Chinese Chemical Soc.*, 25, 141–147. DOI: 10.1002/jccs.197800024

CIR Cosmetic Ingredient Review Expert Panel. 2001. Final report on the safety assessment of *Mentha piperita* (peppermint) oil, *Mentha piperita* (peppermint) leaf extract, *Mentha piperita* (peppermint) leaf, and *Mentha piperita* (peppermint) leaf water. *Int. J. Toxicol*, 20(Suppl.3), 61–73.

Clarke S. 2008. Chapter 8 – Handling, safety and practical applications for use of essential oils. In S. Clarke (ed.) *Essential Chemistry For Aromatherapy*, 2nd ed., Churchill Livingstone Elsevier, eBook. https://doi.org/10.1016/B978-0-443-10403-9.00008-X

Choe E., and D.B. Min. 2006. Mechanisms and factors for edible oil oxidation. *Compr. Rev. Food. Sci. and Food Safety*, 5, 169–186.

Dhifi W., S. Bellili, and S. Jazi et al. 2016. Essential oils' chemical characterization and investigation of some biological activities: A critical review. *Med. (Basel)*, 3(4), 25.

Dijoux N., Y. Guingand, and C. Bourgeois et al. 2006. Assessment of the phototoxic hazard of some essential oils using modified 3T3 neutral red uptake assay. *Toxicol. In Vitro*, 20, 480–489.

Elgendy E.M., and S.A. Khayyat. 2008. Oxidation studies on some natural monoterpenes: Citral, pulegone, and camphene. *Russ. J. Org. Chem.*, 44(6), 814–822.

European Monitoring Centre for Drugs and Drug Addiction EMCDDA. 2016. EU drug market reports, Chapter 6: Amphetamine, methamphetamine and MDMA—Production and precursors http://www.emcdda.europa.eu/publications/eu-drug-markets/2016/online/amphetamines-ecstasy/production-precursors_en. Accessed December 30, 2018.

European Union Reference Laboratory for alternatives to animal testing EURL ECVAM. 2018a. Alternative methods for toxicity testing – Validated test methods – Phototoxicity https://ec.europa.eu/jrc/en/eurl/ecvam/alternative-methods-toxicity-testing/validated-test-methods/phototoxicity. Accessed November 4, 2018.

European Union Reference Laboratory for alternatives to animal testing EURL ECVAM. 2018b. Alternative methods for toxicity testing – Validated test methods – Skin irritation https://ec.europa.eu/jrc/en/eurl/ecvam/alternative-methods-toxicity-testing/validated-test-methods/skin-irritation. Accessed November 4, 2018.

Figueiredo A.C., J.G. Barroso, L.G. Pedro, and J.J. Scheffer. 2008. Factors affecting secondary metabolite production in plants: Volatile components and essential oils. *Flavour Fragr. J.*, 23, 213–226.

Forbes P.D., F. Urbach, and R.E. Davies. 1977. Phototoxicity testing of fragrance raw materials. *Food and Cosmetics Toxicology*, 15(1), 55–60. https://doi.org/10.1016/S0015-6264(77)80264-2

Fu P.P., Q. Xia, and Y. Zhao et al. 2013. Phototoxicity of herbal plants and herbal products. *J. Environ. Sci. Heal.*, Part C, 31(3), 213–255.

Gallucci S., A. Placeres Neto, and C. Porto et al. 2010. Essential oil of *Eugenia uniflora* L.: An industrial perfumery approach. *J. Essent. Oil Res.*, 22(2), 176–179.

Gismondi A., L. Canuti, M. Grispo, and A. Canini. 2014. Biochemical composition and antioxidant properties of *Lavandula angustifolia* Miller essential oil are shielded by propolis against UV radiations. *Photochem. Photobiol.*, 90, 702–708.

Glasl H. 1975. Über die haltbarkeit von terpenoiden in extrakten und lösungen mit unterschiedlichem alkoholgehalt. *Archiv der Pharmazie*, 308(2), 88–93.

Habtemariam S. 2018. Antidiabetic potential of monoterpenes: A case of small molecules punching above their weight. *Int. J. Mol. Sci.*, 19(1), 4.

Hänsel R., and O. Sticher. 2010. *Pharmakognosie Phytopharmazie*, 9. Aufl, Heidelberg: Springer.

Hausen B.M., J. Reichling, and M. Harkenthal. 1999. Degradation products of monoterpenes are the sensitizing agents in tea tree oil. *Am. J. Cont. Derm.*, 10(2), 68–77.

Heldt H.W., and B. Piechulla. 2011. Chapter 16—Secondary metabolites fulfill specific ecological functions in plants. In: H.W. Heldt, and B. Piechulla (eds.) *Plant Biochemistry*, 4th ed., Academic Press, pp. 399–408.

ICH International Conference on Harmonisation of Technical Requirements for Registration of Pharmaceuticals for Human Use. 2012. Guidance on photosafety evaluation of pharmaceuticals S10 Step 2 version 13. http://www.fda.gov/downloads/Drugs/GuidanceComplianceRegulatoryInformation/Guidances/UCM337572.pdf

ICH International Conference on Harmonisation of Technical Requirements for Registration of Pharmaceuticals for Human Use. 2013. Harmonised tripartite guideline "Photosafety evaluation of pharmaceuticals S10". http://www.ich.org/fileadmin/Public_Web_Site/ICH_Products/Guidelines/Safety/S10/S10_Step_4.pdf. Accessed March 31, 2018.

Irmak S., K. Solakyildirim, A. Hasanoğlu, and O. Erbatur. 2010. Study on the stability of supercritical fluid extracted rosemary (*Rosmarinus Offcinalis* L.) essential oil. *J. Anal. Chem.* , 65: 899–906.

IUPAC International Union of Pure and Applied Chemistry. 1997. *Compendium of Chemical Terminology*, 2nd ed. (the "Gold Book"). Compiled by A. D. McNaught and A. Wilkinson. Blackwell Scientific Publications:Oxford. XML online corrected version: http://goldbook.iupac.org (2006-) created by M. Nic, J. Jirat, B. Kosata; updates compiled by A. Jenkins.

Ivanitskikh A.S., and I.G. Tarakanov. 2014. Effect of light spectral quality on essential oil components in *Ocimum basilicum* and *Salvia officinalis* plants. *Int. J. Second. Metab.*, 1(1), 19.

Iwanami Y., H. Tateba, N. Kodama, and K. Kishino. 1997. Changes of lemon flavor components in an aqueous solution during UV Irradiation. *J. Agric. Food Chem.*, 45, 463–466.

Johnson W., B. Heldreth, and W.F. Bergfeld et al. 2017. Safety assessment of *Anthemis nobilis*-derived ingredients as used in cosmetics. *Int. J. Toxicol.*, 36(Suppl.1), 57–66.

Jones P., W.W. Lovell, A.V. King, and L.K. Earl. 2001. In vitro testing for phototoxic potential using the EpiDerm™ 3-D reconstructed human skin model. *Toxicology Methods*, 11, 1–19.

Jones P. 2008. *In vitro* Phototoxicity Assays. In: R.P. Chilcott, and S. Price (eds.), *Principles and Practice of Skin Toxicology*. DOI: 10.1002/9780470773093.ch10

Kanne D., K. Straub, H. Rapoport, and J.E. Hearst. 1982. The psoralen-DNA photoreaction. Characterization of the monoaddition products from 8–methoxypsoralen and 4,5',8-trimethylpsoralen. *Biochemistry*, 21(5), 861–871.

Karlberg A.-T., M.A. Bergström, and A. Börje et al. 2008. Allergic contact dermatitis – formation, structural requirements, and reactivity of skin sensitizers. *Chem. Res. Toxicol.*, 21(1), 53–69.

Katz K.A., I. Marcil, and R.S. Stern. 2002. Incidence and risk factors associated with a second squamous cell carcinoma or basal cell carcinoma in psoralen + ultraviolet A light-treated psoriasis patients. *J. Invest. Dermatol.*, 118, 1038–1043.

Kaul P.N., B.R. Rajeswara Rao, and A.K. Bhattacharya et al. 1997. Changes in chemical composition of rose-scented geranium (*Pelargonium* sp.) oil during storage. *J. Essent. Oil Res.*, 9(1), 115–117.

Kejlová K., D. Jírová, and H. Bendová et al. 2007. Phototoxicity of bergamot oil assessed by *in vitro* techniques in combination with human patch tests. *Toxicol. In Vitro*, 21(7), 1298–1303.

Kejlová K., D. Jírová, and H. Bendová et al. 2010. Phototoxicity of essential oils intended for cosmetic use. *Toxicol. In Vitro*, 24(8), 2084–2089.

Khayyat S.A. 2013. Photosynthesis of dimeric cinnamaldehyde, eugenol, and safrole as antimicrobial agents. *J. Saudi. Chem. Soc.*, 17(1), 61–65.

Khayyat S.A. 2018. Thermal, photo-oxidation and antimicrobial studies of linalyl acetate as a major ingredient of lavender essential oil. *Arab. J. Chem.*, 13(1), 1575–1581. DOI: 10.1016/j.arabjc.2017.12.008

Khayyat S.A., and E. Elgendy. 2018. Safranal epoxide - A potential source for diverse therapeutic applications. *Saudi. Pharm. J.*, 26(1), 115–119.

Khayyat S.A., and L.S. Roselin. 2018. Recent progress in photochemical reaction on main components of some essential oils. *J. Saudi. Chem. Soc.*, 22(7), 855–875.

Kinley J.S., J. Moan, F. Dall'Aqua, and A. Young. 1994. Quantitative assessment of epidermal melanogenesis in C3H/Tif hr/hr mice treated with topical furocoumarins and UVA radiation. *J. Invest. Dermatol.*, 103(1), 97–103. https://core.ac.uk/download/pdf/81943200.pdf. Accessed January 10, 2019.

Krupa J., A. Olbert-Majkut, and I. Reva et al. 2012. Ultraviolet-tunable laser induced phototransformations of matrix isolated isoeugenol and eugenol. *J. Phys. Chem. B*, 116, 11148–11158.

Kumari R., S.B. Agrawal, S. Singh, and N.K. Dubey. 2009. Supplemental ultraviolet-B induced changes in essential oil composition and total phenolics of *Acorus calamus* L. (sweet flag). *Ecotoxicol. Environ. Saf.*, 72(7), 2013–2019.

Lang G., and G. Buchbauer. 2012. A review on recent research results (2008–2010) on essential oils as antimicrobials and antifungals. *Flavour & Fragrance J.*, 27, 13–39.

Lewis F.D., and M. Kojima. 1988. Electron transfer induced photoisomerization, dimerization, and oxygenation of trans- and cis-anethole - The role of monomer and dimer cation radicals. *J. Am. Chem. Soc.*, 110, 8664–8670.

Li L.J., P. Hong, and F. Chen et al. 2016. Characterization of the aldehydes and their transformations induced by UV irradiation and air exposure of white guanxi honey pummelo (*Citrus Grandis* (L.) Osbeck) essential oil. *J. Agric. Food Chem.*, 64(24), 5000–5010.

Liebsch M., H. Spielmann, and W. Pape et al. 2005. UV-induced effects. *Altern. Lab. Anim.*, 33(Suppl.1), 131–146.

Lis-Balchin M., and S. Hart. 1999. Studies on the mode of action of the essential oil of lavender (*Lavandula angustifolia* P. Miller). *Phytother. Res.*, 13(6), 540–542.

Maffei M., D. Canova, C.M. Bertea, and S. Scannerini. 1999. UVA effects on photomorphogenesis and essential oil composition in *Mentha piperita. J. Photoch. Photobio. B-Biology*, 52, 105–110.

Maibach H., and G. Honari. 2014. Chapter 3 -*Photoirritation (Phototoxicity): Clinical Aspects*. In: H. Maibach, and G. Honari (ed.), *Applied Dermatotoxicology*, pp. 41–56.

Marques H.M. 2010. A review on cyclodextrin encapsulation of essential oils and volatiles. *Flavour Fragr. J.*, 25, 313–326.

Marqués-Calvo M.S., F. Codony, G. Agustí, and C. Lahera. 2017. Visible light enhances the antimicrobial effect of some essential oils. *Photodiagnosis Photody. Ther.*, 17, 180–184.

Matura M., M. Sköld, and A. Börje et al. 2005. Selected oxidized fragrance terpenes are common contact allergens. *Contact Dermatitis*, 52(6), 320–328.

Menichini F., R. Tundis, and M.R. Loizzo et al. 2010. In vitro photo-induced cytotoxic activity of *Citrus bergamia* and *C. medica* L. cv. Diamante peel essential oils and identified active coumarins. *Pharm. Biol.*, 48(9), 1059–1065.

Michl J. 1974. Physical Basis of Qualitative Mo Arguments in Organic Biochemistry. In: A. Davison, M.J.S. Dewar, K. Hafner et al. (eds.), *Topics in current chemistry 46: Photochemistry*, Berlin Heidelberg: Springer.

Miguel M.G. 2010. Antioxidant and anti-inflammatory activities of essential oils: A short review. *Molecules*, 15, 9252–9287.

Miolo G., M. Stefanidis, and R.M. Santella et al. 1989. 6,4,4′-Trimethylangelicin photoadduct formation in DNA: Production and characterization of a specific monoclonal antibody. *J. Photoch. Photobio.y B: Biology*, 3(1), 101–112.

Mori H.-M., and H. Iwahashi. 2016. Characterization of radicals arising from oxidation of commercially-important essential oils. *Free Radic. Res.*, 50(6), 638–644.

Morita T., K. Jinno, and H. Kawagishi et al. 2003. Hepatoprotective effect of myristicin from nutmeg (*Myristica fragrans*) on lipopolysaccharide/d-galactosamine-induced liver injury. *J. Agric. Food Chem.*, 51(6),1560–1565.

Moulin C., A. Petit, and J.C. Baccou. 1995. Selective laser photolysis of organic molecules in complex matrices. *J. Photoch. Photobio. A: Chem.*, 85, 165–172.

Ochsner M. 1997. Photophysical and photobiological processes in the photodynamic therapy of tumours. *J. Photochem. Photobio. B: Biology*, 39(1), 1–18.

Odak I., T. Lukic, and S. Talic. 2018. Impact of storage conditions on alteration of juniper and immortelle essential oils. *J. Essent. Oil Bear. Plants*, 21(3), 614–622.

OECD Organisation for Economic Co-operation and Development. 2004. Guidelines for the testing of chemicals, Section 4. Test no. 432: *in vitro* 3T3 NRU phototoxicity test. DOI: 10.1787/9789264071162-en

Onoue S., K. Kawamura, and N. Igarashi et al. 2008. Reactive oxygen species assay-based risk assessment of drug-induced phototoxicity: Classification criteria and application to drug candidates. *J. Pharm. Biomed. Anal.*, 47(4–5), 967–972.

Pape W.J.W., M. Brandt, and U. Pfannenbecker. 1994. Combined *in vitro* assay for photohaemolysis and haemoglobin oxidation as part of a phototoxicity test system assessed with various phototoxic substances. *Toxicol. In Vitro*, 8, 755–757.

Pape W.J.W., T. Maurer, U. Pfannenbecker, and W. Steiling. 2001. The red blood cell phototoxicity test (photohaemolysis and haemoglobin oxidation). EU/COLIPA Validation programme on phototoxicity (Phase II). *ATLA*, 29, 145–162.

Pelzl B., R. Wolf, and B.L. Kaul. 2018. Plastics, Additives. In: *Ullmann's Encyclopedia of Industrial Chemistry*, Elvers, B (Ed.). DOI: 10.1002/14356007.a20_459.pub2.

Pérez-Recalde M., I.E. Ruiz Ariasa, and É.B. Hermida. 2018. Could essential oils enhance biopolymers performance for wound healing? A systematic review. *Phytomedicine*, 38, 57–65.

Pirilä V., and E. Siltanen. 1958. On the chemical nature of the eczematogenic agent in oil of turpentine. III. *Dermatologica*, 117(1), 1–8.

Placzek M., W. Frömel, and B. Eberlein et al. 2007. Evaluation of phototoxic properties of fragrances. *Acta Derm. Venereol.*, 87, 312–316.

Prins C.L., I.J.C. Vieira, and S.P. Freitas. 2010. Growth regulators and essential oil production. *Braz. J. Plant Physiol.*, 22(2), 91–102.
Raghavan S. 2006. *Handbook Of Spices, Seasonings, and Flavorings, 2nd Edition*, Boca Raton: CRC Press.
Raut J.S., and S.M. Karuppayil. 2014. A status review on the medicinal properties of essential oils. *Ind. Crops Prod.*, 62, 250–264.
Rezaee R., and H. Hosseinzadeh. 2013. Safranal: From an aromatic natural product to a rewarding pharmacological agent. *Iran. J. Basic. Med. Sci.*, 16(1), 12–26.
Rocha S.F.R., and L. Chau Ming. 1999. Piper Hispidinervum: A Sustainable Source of Safrole. In: J. Janick (ed.) *Perspectives on New Crops and New Uses*, Alexandria, VA: ASHS Press, pp. 479–481. https://hort.purdue.edu/newcrop/proceedings1999/v4-479.html. Accessed January 2, 2019.
Schieberle P., and W. Grosch. 1989. Potent odorants resulting from the peroxidation of lemon oil. *Z Lebensm Unters Forsch*, 189(1), 26–31. https://doi.org/10.1007/BF01120443
Simmler W. 2012. *Photochemical Degradation.* In: *Ullmann's Encyclopedia of Industrial Chemistry*, Elvers, B. (Ed.).: 7th edition, Vol. 2, Weinheim: Wiley-VCH Verlag GmbH & Co. KGaA, pp. 127–145. DOI: 10.1002/14356007.o02_o06.
Spielmann H., M. Balls, and J. Dupuis et al. 1998. The international EU/COLIPA *in vitro* phototoxicity validation study: Results of phase II (blind trial). Part 1: The 3T3 NRU phototoxicity test. *Toxicol. In Vitro*, 12(3), 305–327.
Stern R.S., K.T. Nichols, and L.H. Väkevä. 1997. Malignant melanoma in patients treated for psoriasis with methoxsalen (psoralen) and ultraviolet A radiation (PUVA). *New Engl. J. Med.*, 336, 1041–1045.
Stern R.S. 2001. The risk of melanoma in association with long-term exposure to PUVA. *J. Am. Acad. Dermatol.*, 44, 755–761.
Sun H., H. Ni, and Y. Yang et al. 2014. Sensory evaluation and gas chromatography–mass spectrometry (GC-MS) analysis of the volatile extracts of pummelo (*Citrus maxima*) peel. *Flavour Fragrance J.*, 29, 305–312.
Tateba H., K. Morita, W. Kameda, and M. Tada. 1993. Photochemical reaction of (Z)-jasmone under various conditions. *Biosci. Biotechnol. Biochem.*, 57(2), 220–226.
Tisserand R., and R. Young. 2014. Chapter 5 - The skin. In: R. Tisserand, and R. Young (eds.), *Essential Oil Safety*, 2nd ed., Churchill Livingstone, pp. 69–98.
Toda H., S. Mihara, K. Umano, and T. Shibamoto. 1983. Photochemical studies on jasmin oil. *J Agric. Food Chem.*, 31, 554–557.
Turek C., and F.C. Stintzing. 2012. Impact of different storage conditions on the quality of selected essential oils. *Food Res. Int.*, 46(1), 341–353.
Turek C., and F.C. Stintzing. 2013. Stability of essential oils: A review. *Compr. Rev. Food Sci. and Food Safety*, 12, 40–53.
Van den Bergh H. 1986. Light and porphyrins in cancer therapy. *Chem Br*, 22, 430–439.
World Health Organization WHO. 2018. Health topics - Ultraviolet radiation - What is UV radiation? https://www.who.int/uv/uv_and_health/en/. Accessed December 4, 2018.
Young A.R., S.L. Walker, and J.S. Kinley et al. 1990. Phototumorigenesis studies of 5-methoxypsoralen in bergamot oil: Evaluation and modification of risk of human use in an albino mouse skin model. *J. Photoch.Photobio. B: Biology*, 7(2–4): 231–250.
Ziegler M., H. Brandauer, E. Ziegler, and G. Ziegler. 1991. A different aging model for orange oil: Deterioration products. *J. Essent. Oil Res.*, 3(4), 209–220.

29 Influence of Air on Essential Oil Constituents

Darija Gajić and Gerhard Buchbauer

CONTENTS

29.1 INTRODUCTION

Everyday life of a contemporary person shows an ever-increasing trend of using modern, updated synthetic additives and pharmacologically relevant agents, while on the other hand, a worldwide trend toward ever-growing awareness for "natural and natural-like" products speaks in favor of consumers' regained interest for ingredients such essential oils (EOs) that show great popularity when used in food, household, and topical/cosmetic products (Turek and Stintzing,2013). Though overall, the available literature offers many definitions of essential oils, they can be summarized as liquid, odorous mixtures of volatile, organic compounds commonly obtained by steam distillation of aromatic plants and are in general referred to as the "essence" of a plant containing pleasantly scented fragrances (Amorati et al., 2013). The review confirmed a great chemical diversity of EO components, amounting hundreds of compounds (secondary metabolites), which play decisive roles in the oxidative stability of EOs. However, regardless of this variety, the major constituents of EOs are here categorized in two structural families with regard to hydrocarbon skeleton: terpenoids (monoterpenes, sesquiterpenes, diterpenes) and phenyl-propanoids.

This review also stated that an antioxidant quality is one of the major biological activity of EOs that is generally interpreted as the ability of a molecule to react with radicals and thus slow down or retard the oxidation process. In the light of these observations, the study stated that phenolic compounds are ordinarily considered as antioxidants owing to their prominent reactivity with peroxyl radicals (ROO•) and consequently derived phenoxyl radicals (PhO•) that will not further propagate the radical chain owing to its stability. Conversely, a reactive alkyl radical is derived in a rapid reaction between other terpenoid constituents of EOs and peroxyl radicals, so subsequently the alkyl radical containing terpene hydrocarbon skeleton reacts with oxygen, yielding peroxyl radical that further propagates the oxidative chain. It should be noted that phenolic components are to be found in both terpenoid and phenyl-propanoid families. Given that natural EOs contain a mixture of various constituents, it was here also observed that antioxidants and oxidizable terpenoids usually coexist.

The stability aspect is one of the major concerns in terms of handling and storage of EOs. Thus, taking their fragile and volatile nature into account, it has been proved that due to improper management of EOs, they can undergo chemical alterations upon aging which compromise their quality and consequently lead to the loss of therapeutic abilities (Turek et al., 2013). This comprehensive summary stated that there are three crucial factors influencing stability of EOs, thus vigilant storage monitoring of EOs is primarily referred to controlling light, temperature, and oxygen availability.

Terpenes, that is, mono- and sesquiterpenes, are commonly used as fragrance chemicals of natural origin and are widely used in perfumery, pharmaceutical, domestic, and occupational products (Matura et al., 2005). As stated in the study, the main structural feature of this chemical class is the presence of a double bond that is responsible for susceptibility to oxidation upon air exposure, that is, autoxidation which could be shortly referred to as radical chain process that generates numerous primary (hydroperoxides) and secondary oxidation products (epoxides, alcohols, ketones, etc.). Its general mechanism (Turek et al., 2013) could be presented schematically (Figure 29.1). A significant increase of sensitizing potential is one of the essential characteristics of the autoxidation process, whereby the "parent" molecule is transformed to a more or less allergenic oxidation product.

As previously stated, the main focus of this chapter is to investigate the chemical background of air-induced autoxidation processes of the most common fragrance terpenoids regarding their commercial importance, ubiquitous presence, and increasing use of products containing these compounds. Besides from the chemical aspect of terpenoid alteration mechanisms, the nature of adverse effects associated with the oxidation products is discussed as well. Given all the aforementioned, this overview may be considered as an attempt to compile the publications referring autoxidation of individual fragrance chemicals so that an overview of the studies available hitherto could help adequate management of fragrance terpene compounds in the future.

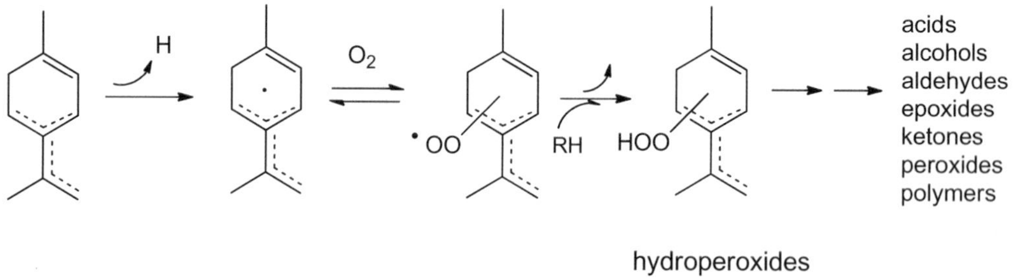

FIGURE 29.1 General autoxidation pattern of single or double unsaturated monocyclic terpenes. (Adapted and newly drawn from Turek C. et al. 2013. *Comprehensive Reviews in Food Science and Food Safety* 12 (1): 40–53.)

29.2 ESSENTIAL OILS AND TERPENOIDS

Growing interest in chemistry of essential oils has been perceived in the recent period so handbooks such as that by Baser and Buchbauer (2016) surely provide comprehensive data in this field. Namely, there are many accorded attempts to establish a definition of essential oils, including the one officially presented by International Organization for Standards (ISO) in 1997 (Geneva), in *Aromatic Natural Raw Materials-Vocabulary*: "Product obtained from vegetable raw material, either by distillation with water or steam, or from the epicarp of Citrus fruits by a mechanical process, or by dry distillation." Many literary sources still use the old term "terpenes" (instead of "terpenoids"), which should generally be restricted to the monoterpenoid hydrocarbons (C10 monoterpenes), as stated in Baser and Buchbauer (2016).

Stability of Essential Oils represents a major concern and is of great significance. Hence, nowadays it is well accepted that EOs can undergo chemical alterations during storage; for example, oxidation in the presence of air. Monitoring of selected quality parameters is a very important aspect of chemical stability of EOs. Turek and Stintzing (2011) established a set of analytical parameters in order to adequately monitor possible changes during the aging process of EOs. This experimental research explored the nature of quality alterations for seven common essential oils under "worst-case conditions" in terms of light, temperature, and in the presence of atmospheric oxygen. After assessing quite a variety of physicochemical parameters, peroxide value (POV), conductivity, and pH were confirmed to be most suitable to evaluate the oxidative deterioration process.

Finally, the study stated that a strict correlation between resulting parameters and detected changes in HPLC fingerprints can only be partially established. Namely, though, alteration changes are similarly reflected by each of the three parameters, direction and extent of these changes may be unequal, which is associated with various oil compositions. This variety in individual composition of EOs was recognized as a decisive factor that determines "the course and the speed" of oxidative alterations which concurrently emphasized the need for further research.

29.3 REACTION MECHANISMS OF TERPENOID OXIDATION PATTERNS

Oxidation patterns of unsaturated hydrocarbons containing double bonds comprise a wide range of chemical reactions. General properties of addition of O_2 to R1-HC = CH-R2, along with establishing a possible link in terms of similarity to terpenoid oxidation, especially in the field of joined experimental and computational studies (Bäcktorp et al., 2006), provides an opportunity for future research and predicting the risk of a compound becoming a contact allergen. The study indicated that simpler molecules—for example, propene—can be used as model compound for terpenoids—for example, linalool—so that a conceptual framework is created. Understanding the reactivity of terpenoids as well as describing the fundamental principles in the oxidation process requires as small a prototype as possible. Modeling can also be perceived as a way to observe the influence of the increasing molecular complexity in the field of the energetic, which also determines the linalool reactivity. Therefore, such a modeling concept can serve two purposes and thus will be more thoroughly analyzed in Section 29.3.1.

A similar form of computational research combined with experimental data was conducted in 2008 (Bäcktorp et al., 2008) where a possible oxidation pattern of another common terpenoid, geraniol, was proposed. In order to fully explore radical chain mechanism, a model concept with smaller and less flexible molecules was used again. The study suggested plausible radical chain reaction with an emphasis upon the primary oxidation products in terms of radical chain process. However, it was also reported that air oxidation process of geraniol was significantly more complex to that of linalool, in terms of forming a mixture of primary oxidation products (such as hydrogen peroxide), aldehydes, (such as geranial and neral), epoxygeraniol, and naturally geraniol hydroperoxide as well as its secondary oxidation product, the allylic alcohol. Additionally, the geranyl formate was observed as a product of post-oxidation bimolecular transformation.

29.3.1 Linalool

In order to further investigate terpenoid oxidation during handling and storage, the first computational study of this kind (Bäcktorp et al., 2006) was conducted in 2006. The main goal was to obtain mechanistic understanding on possible reaction mechanisms of transforming a harmless compound such a linalool into a skin sensitizer. The theoretical framework was established by systematic investigation of smaller unsaturated systems like propene, 2-methyl-2-butene and finally 2-methyl-2-pentene. All the intermediates and products derived from this model concept system were used for comparison with the actual linalool oxidation patterns. The study (Bäcktorp et al., 2006) discussed the addition of O_2 to linalool in terms of an *ene*-type mechanism, the radical mechanism, and the direct reaction pathway. All of the three considered the linalool-O_2 biradical intermediate formation (Figure 29.2) as a branching point for further oxidation. Given that hydroperoxides were confirmed to be the strongest sensitizers of the oxidation products, these compounds were the main focus of the study.

29.3.1.1 The Ene-Type Mechanism

It represents the reaction between the excited singlet state of oxygen molecule $^1O_2$ and an allylic hydrogen atom (Figure 29.3). This results in shifting the double bond to a position adjacent to the original double bond; that is, an allylic hydroperoxide is formed. Notably, the direct excitation (transition of the ground state oxygen $^3O_2$ by light to a singlet form) is highly unlikely. Hence, singlet oxygen (unlike almost all molecules) is uncommon in a standard environment and its excitation usually requires a certain sensitizing molecule.

29.3.1.2 The Free-Radical Chain Reaction

It is described as a slow process including initiation (R1) and a propagation phase (R2; R3) and can be described as initiation phase (R1) preceded with, for instance, hydrogen abstraction by an oxygen (RH+O_2), which results in forming a reactive radical R$^•$ and a hydroperoxyl radical HOO$^•$. It takes a high amount of energy to break a C–H bond which is the reason why this phase is often denoted as the

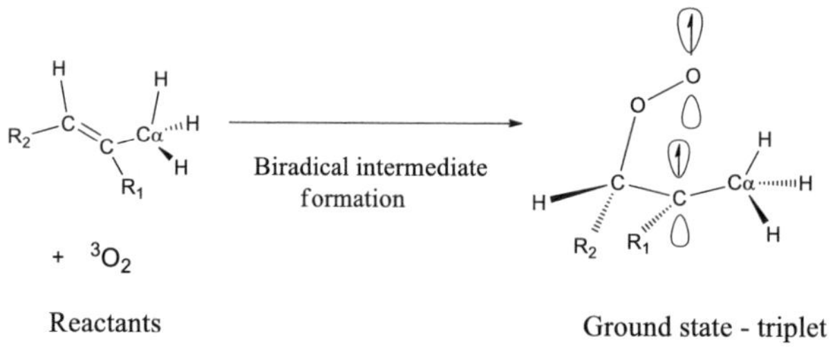

Biradical intermediate formation

Reactants

Ground state - triplet

FIGURE 29.2 Linalool-O_2 biradical intermediate formation. (Adapted and newly drawn from Bäcktorp C. et al. 2006. *The Journal of Physical Chemistry A* 110 (44): 12204–12.)

allylic H atom

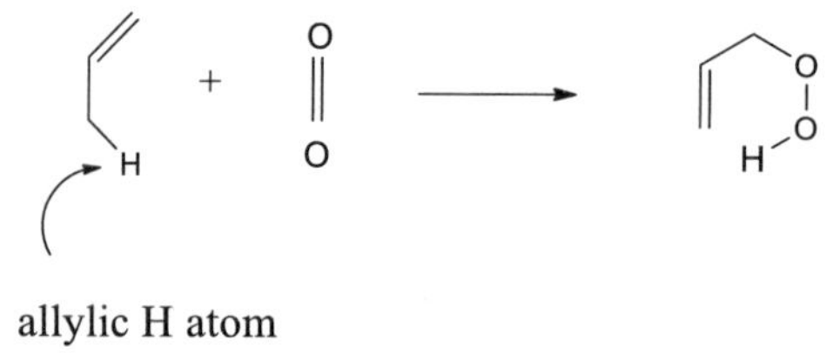

FIGURE 29.3 *Ene*-reaction between singlet oxygen and a double bond leads to allyl hydroperoxide. (Adapted and newly drawn from Bäcktorp, C. et al. 2006. *The Journal of Physical Chemistry A* 110 (44): 12204–12.)

very first bottleneck of the autoxidation. Furthermore, propagation phase (R2) continues with a chain reaction between free radicals and ground state oxygen that leads to the formation of a peroxyl radical ROO•, which in reaction with R′H can further be converted to a hydroperoxide ROOH•. This phase (R3) is often referred to as the slow propagation step wherein a reactive radical R′• is generated. Given is the rate determining step of the propagation phase; it is often denoted as the second bottleneck. Finally, termination phase (R4) results in formation of non-radical products derived from two radicals R′•.

29.3.1.3 The Direct Reaction Pathway

It provides the interpretation of a direct route to hydroperoxy-linalool formation with regard to biradical intermediate. Breaking the C=C π bond is one of the main features of the linalool oxidation, that is, formation of hydroperoxide. Initial formation of an O_2–C=C complex can facilitate interconversion of the internal C=C double bond to a C–C single bond, which allows their π-systems to interact. The ground state oxygen molecule (triplet) is added to one of the carbons in the double bond of linalool ground state (singlet), and a triplet state biradical intermediate is formed. The triplet biradical may undergo a triplet-singlet interconversion (*spinflip,* i.e., change of multiplicity), after which the terminal oxygen atom attacks a hydrogen atom from α-carbon, which leads to the rearrangement of the double bond and structural stabilization. The triplet-singlet transition is considered to be quite slow, whereas the final hydroperoxide production, once a singlet is formed, occurs immediately. Conversely, the triplet-singlet excitation of the *ene*-type reaction takes place prior to the formation of the O_2–C=C complex.

The summary can also be presented in an autoxidation scheme for all three reaction channels (Figure 29.4).

29.3.2 Geraniol

Bäcktorp et al. (2008) explored the theory of geraniol autoxidation in order to complement the theoretical framework of a probable reaction mechanism that follows a radical chain pattern. The study focused on the model concept theory based on the initiation phase; that is, hydrogen abstraction that follows the usual pattern by forming an allylic radical which then interacts with an O_2 molecule, yielding an intermediate peroxyl radical. Propagation of the radical chain reaction by intermolecular hydrogen abstraction was a common feature of the previous studies. It is this subsequent phase that differs from the other patterns studied because an intramolecular rearrangement was shown to be preferred by hydroxy-substituted allylic peroxyl radicals. In turn, aldehydes and hydroperoxyl radicals are yielded to further propagate the reaction (Figure 29.5). This study focused on the theoretical investigation of the primary autoxidation process of geraniol in the presence of triplet oxygen. Enthalpy changes were used to distinguish the radicals that indeed gave no oxidation products. The oxidation products that derived from a radical in a similar pattern to linalool was not discussed due to its analogous nature. Following the geraniol model concept, the study offered the possible radical chain mechanism with the emphasis on the propagation step.

The model molecule—that is, prenol (=3-methyl-2-buten-1-ol, **1**)—used for this research fulfills the requirements necessary to recreate the stability of the main radical—that is, the tri-substituted alkene and the allylic hydroxyl functionality. Notably, these prerequisites exclude the conformationally flexible isoprenoid moiety due to its expected constant nature.

29.3.2.1 Initiation Phase

Hydrogen abstraction phase: Formation of the allylic radicals **2a** and **2b**, depending on the conformation of the model alcohol, prenol (**1**). The *trans* form **2a** is expected to dominate for steric reasons.

29.3.2.2 Propagation Step 1

Radicals interaction with triplet O_2: The study reported no transition, states confirming that the addition is indeed carried out without an energy barrier and it resulted in production of the structure **4**.

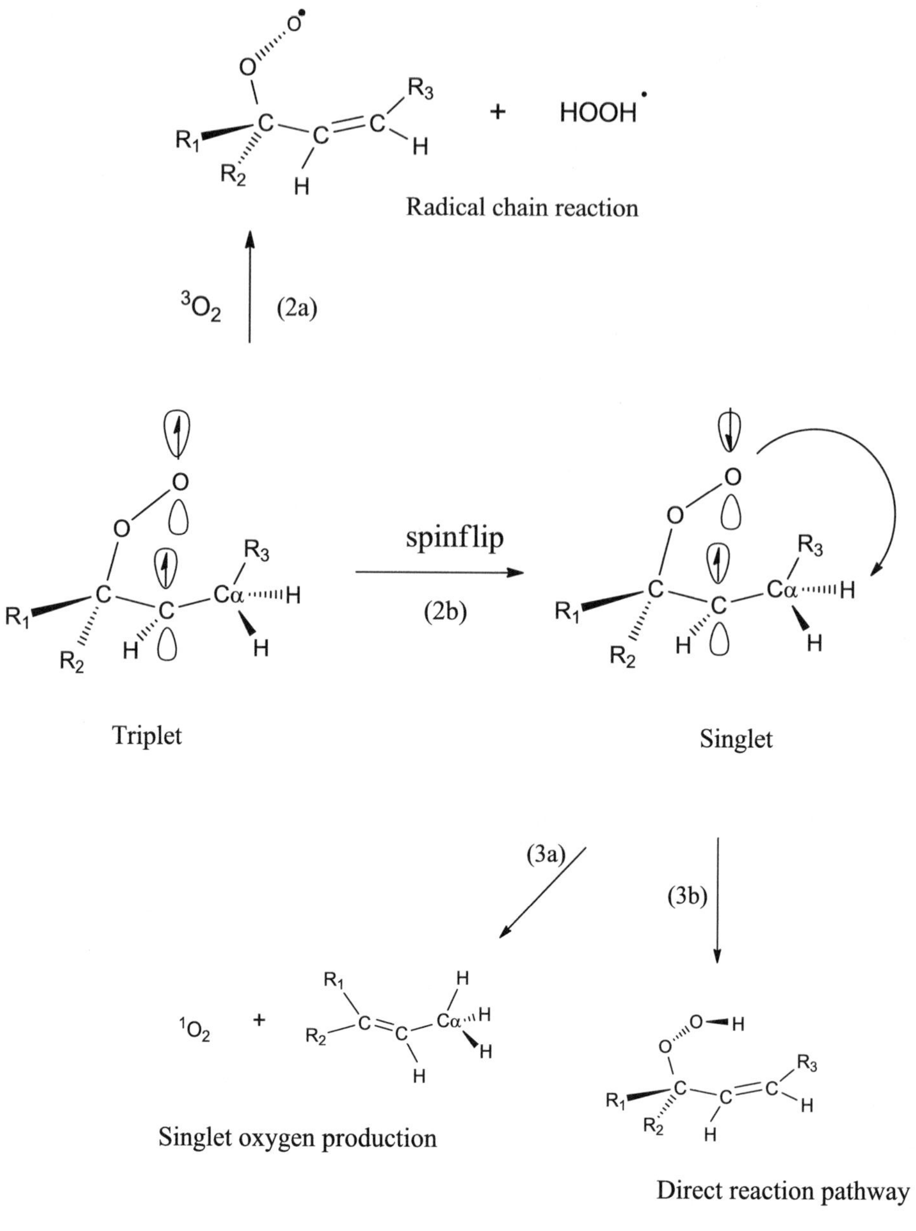

FIGURE 29.4 Triplet biradical, *ene*-type-mechanism, direction pathway. (Adapted and newly drawn from Bäcktorp, C. et al. 2006. *The Journal of Physical Chemistry A* 110 (44): 12204–12.)

29.3.2.3 Propagation Step 2

Another molecule in solution (RH) donates a hydrogen atom to be abstracted by peroxyl radicals **3** and **4**. Hydroperoxides **5** and **6** are formed and are expected to be in equilibrium with aldehyde **7** and H_2O_2, both detected in the autoxidation mixture.

For a further investigation of this phase, a model peroxyl radical $CH_3OO^•$ was used in order to react with prenol as the geraniol model **1** to create **2**. Peroxyl radical *3a* showed no alternative forward reaction (intramolecular hydrogen transfer), and thus it was reported to either revert to **2a** or notably isomerize to **4**. However, **3b** and **4** showed indeed alternative fragmentation paths that could be explained by the proximity of the hydroxy group. Peroxyenoyl radical (Figure 29.6) is

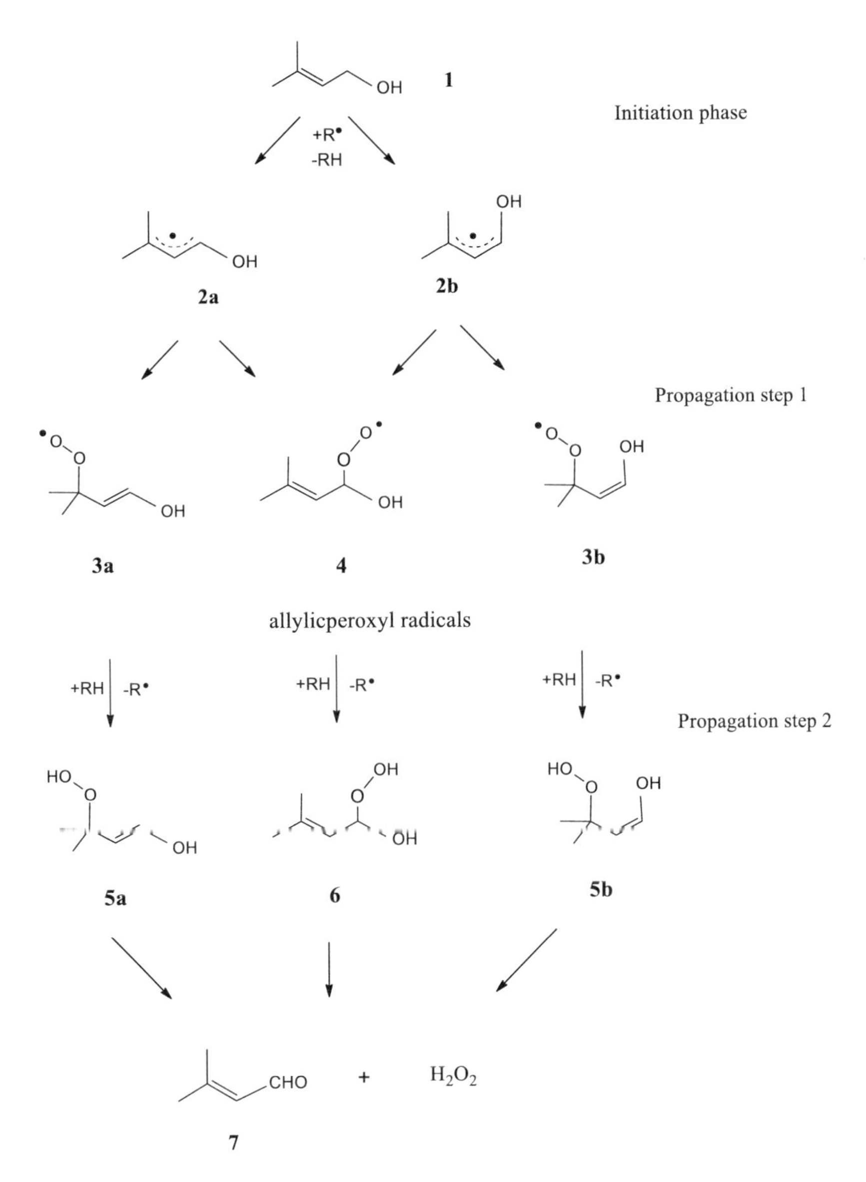

FIGURE 29.5 Prenol as a geraniol model concept: Initiation and propagation phase. (Adapted and newly drawn from Bäcktorp, C. et al. 2008. *Journal of Chemical Theory and Computation* 4 (1): 101–6.)

an intermediate product of the intramolecular hydrogen transfer in **3b** and is characterized with a moderate energy barrier.

It can further eliminate a hydroperoxyl radical forming an aldehyde **7** in an exergonic process. On the other hand, radical **4** produce free aldehyde and hydroperoxyl radical •OOH by intramolecular hydrogen transfer and elimination process. A very low energy barrier is reported on this occasion. Finally, the study stated that fragmentation via **4**, including intramolecular elimination, was considered the preferred path, but once is formed, radical **3b** would rather go via elimination via **8** than revert to allylic radical **2** or isomerize to **4.** The fragmentation product hydroperoxyl radical can further propagate the radical chain reaction by hydrogen abstraction from a prenol molecule or it can participate in the addition reaction to the double bond of the prenol as the model molecule for geraniol (Figure 29.7). Although, the addition process itself is considered endergonic, the final ring closure to epoxy alcohol **9** is extremely exergonic. This secondary autoxidation product is

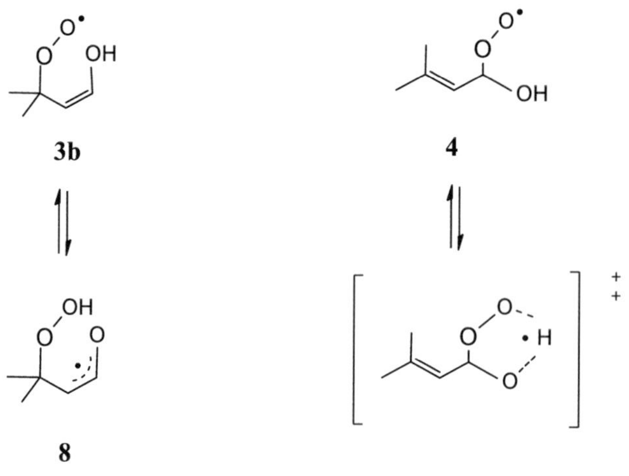

FIGURE 29.6 Peroxy-enolyl radical: Fragmentation path for 3**b** and **4**. (Adapted and newly drawn from Bäcktorp, C. et al. 2008. *Journal of Chemical Theory and Computation* 4 (1): 101–6.)

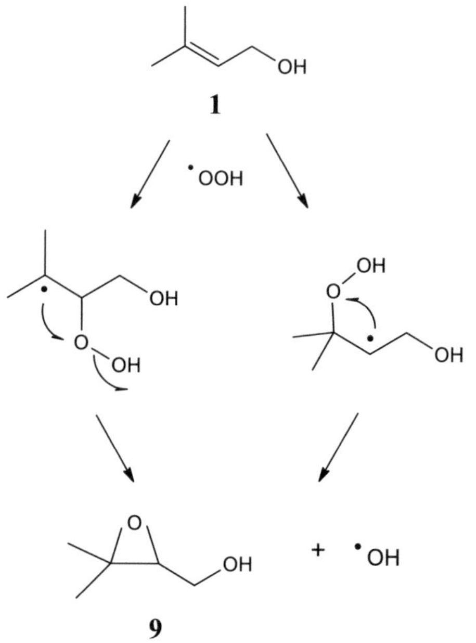

FIGURE 29.7 Epoxidation of prenol. (Adapted and newly drawn from Bäcktorp, C. et al. 2008. *Journal of Chemical Theory and Computation* 4 (1): 101–6.)

considered to be a model compound corresponding to epoxygeraniol, a product found in the oxidation mixture. A highly reactive hydroxyl radical is released and thus responsible for propagation of the radical chain process that occurs as a hydrogen atom abstraction from prenol as a model molecule for geraniol.

The final aspect of this study observed the formation of geranyl formate (**10a**) (Figure 29.8) as a secondary oxidation product. The main characteristic of this product is that it contains one additional carbon atom, which is assumed to have occurred from the fragmentation of another geraniol molecule. Given that no radical process could have been related to forming formates, it

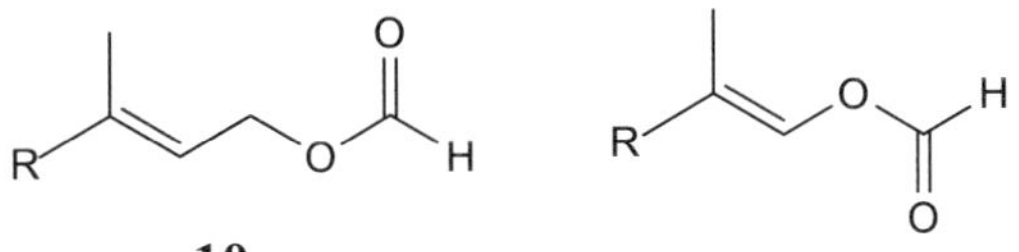

10a **10b**

FIGURE 29.8 Geranyl formate (**10a**) and vinyl formate (**10b**). (Adapted and newly drawn from Bäcktorp, C. et al. 2008. *Journal of Chemical Theory and Computation* 4 (1): 101–6.)

was proposed that formate could be produced from perhydrate **6**, either directly in the radical chain process or by the reversible addition of hydrogen peroxide to geranial or neral. It was here noted that the perhydrate formation is practically a text-book example of an intermediate in the Baeyer–Villiger reaction. Namely, acidic conditions could cause the cleavage of the O–O bond in **6** with simultaneous migration of the vinyl moiety, resulting in a vinyl formate creation (**10b**) (Figure 29.8). Though not detected, a mild acidic environment would expect to cause irreversible trans-esterification of this product with geraniol to finally produce geranyl formate in addition to C_9 aldehyde produced by tautomerization of an enol product. The main deficiency of this theory lies in the fact that no C_9 aldehyde was detected, but geraniol was still considered as the precursor for the formate moiety in **10a**, given it is the only reactant in this process.

In conclusion, the study of prenol as a geraniol model system offered a brief comparison to a linalool oxidation mechanism. The main difference was presented regarding both propagation steps (radical chain transfer and the addition of O_2) that were marked as exergonic. Additionally, besides from the usual chain transfer reaction, peroxyl radicals deriving from geraniol can also undergo intramolecular hydrogen abstraction. This phase is then followed by fragmentation and release of a hydroperoxyl radical. This radical is observed as an alternative chain transfer agent and the process itself was considered to be favored compared to the classical intermolecular hydrogen abstraction. Notably, hydrogen peroxide occurs as a side product of both of these processes. Nevertheless, when a full system of geraniol is used instead of the geraniol model molecule, *cis/trans* isomerization, giving neral and geranial, must be considered too. The expected autoxidation pathway of geraniol will thus be observed further on.

These comprehensive studies showed possible oxidation patterns using some of the most common fragrance terpenoids as an illustration and, therefore, served as a theoretical background for fragrance terpenoid autoxidation research presented in Section 29.4.

29.4 CHEMISTRY OF FRAGRANCE TERPENOID AUTOXIDATION

Stability of EOs is an extremely important aspect of their ever-increasing usage in everyday life. It is a general knowledge that EOs are susceptible to conversion and degradation reactions, so Turek and Stintzing (2013) published a compiled review regarding this matter wherein oxidative degradation is of special interest. Namely, temperature, light, and oxygen availability are outlined as the key factors affecting EO integrity.

It was here emphasized that oxidation, isomerization, cyclization, or dehydrogenation reactions for which the onset is either enzymatically or chemically induced, are the ways of converting essential oil components from one to another owing to structural correlation within one chemical group. With regard to the key features of EOs being volatile, thermolabile, and prone to oxidation as well, it was concluded that chemical composition of EOs is primarily associated with processing and storage routines of the plant material and so, too, to the following handling of the oil itself. Oxidative deterioration, chemical transformation, and polymerization are the most prominent alteration processes of EOs, wherein aging reactions such as these are commonly related with the loss of quality.

The final outcome of these alterations of EOs (including oxidation of terpenoids), characterized by organoleptic and viscosity changes, are generally linked to the increase in skin-sensitizing capacity, clinically manifested as allergic contact dermatitis. The autoxidation reaction was particularly stressed as a spontaneous, air-induced oxidative process. More precisely, aerial oxygen is engaged in a chemical reaction with unsaturated molecules in a free-radical chain mechanism yielding various primary and secondary oxidation products, summarized in the following scheme (Turek and Stintzing, 2013):

Initiation (1) $RH \rightarrow R^{\bullet} + H^{\bullet}$

(2) $ROOH + RH \rightarrow R^{\bullet} + H_2O + RO^{\bullet}$.

(1) Formation of alkyl radicals promoted by heat, catalytic quantities of redox-reactive metals, and exposure to light. (2) An alternative initiation phase with trace levels of hydroperoxides.

Propagation (3) $R^{\bullet} + {}^3O_2 \rightarrow ROO^{\bullet}$.

Formation of peroxyl radical derived from a prompt reaction between alkyl radical and triplet oxygen.

and (4) $ROO^{\bullet} + RH \rightarrow ROOH + R^{\bullet}$.

The creation of hydroperoxides and another alkyl radical, generated by means of a selective reaction of abstraction of a hydrogen atom adjacent to a double bond by a peroxyl radical.

Branching $ROO^{\bullet} + R \rightarrow ROOR^{\bullet}$

(5) $ROOR^{\bullet} \rightarrow R{>}O + RO^{\bullet}$

(6) $RO^{\bullet} + RH \rightarrow ROH + R^{\bullet}$

(7) $RO^{\bullet} + {}^3O_2 \rightarrow R' = O + HO_2^{\bullet}$.

Next to their secondary nature, epoxides (5) (addition of peroxyl radical occurs at a double bond), alcohols (6), and ketones (7) might be produced as primary products in a radical chain mechanism, thus making this reaction competitive with that of formation of hydroperoxides.

(8) $ROOH + ROO^{\bullet} \rightarrow R' = O + HO^{\bullet} + ROOH$

Termination (9) $R^{\bullet} + R^{\bullet} \rightarrow RR$

(10) $ROO^{\bullet} + R^{\bullet} \rightarrow ROOR$

(11) $ROO^{\bullet} + ROO^{\bullet} \rightarrow ROH + R' = O + {}^3O_2$

In a termination phase, radicals combine to alkyl (9) and peroxyl dimers (10) that generally polymerize or decay so that non-radical products are obtained (11).

Molecular structure, oxygen concentration, and energy input, as well as the influence of other participants in further reactions, are the major factors affecting the pathway of the aforementioned reactions. Thus, it is practically a general rule for primary oxidation products like hydroperoxides to deteriorate after some time, and thereby, stable secondary oxidation products such as alcohols, aldehydes, ketones, epoxides, peroxides, and highly viscous polymers are generated in the advanced stages of oxidative degradation.

According to Turek and Stintzing (2013), *oxygen access* was also further confirmed to play a crucial role in maintaining EO integrity and, thus, is considered the most frequent cause of EO deterioration in the sense of changes in composition and physicochemical quality. Besides from ambient temperature, partial pressure of oxygen in the headspace of a container has a major impact on the concentration of dissolved oxygen, which in turn dictates the autoxidation rate in terms of acceleration. Namely, as the study reported, the lower the temperature is, the higher is the oxygen solubility, which is the brief interpretation of Henry's law. Hence, hydroperoxides and peroxyl radicals were presented as the most prominent oxidation products in edible oils at low temperatures. At later oxidation stages, wherein oxygen or oxidizable substrates are consumed, polymers are generated. Presence of alkyl or hydroxyl radicals is mainly associated with elevated temperatures followed by restricted oxygen accessibility. Finally, the individual character of an oil must be taken into account and thus susceptibility to oxidation cannot simply be "copied" from one oil to another. It was here concluded that dissolved atmospheric oxygen will surely initiate the oxidation process, at least to a minor extent. On the other hand, minimizing a container headspace will not necessarily prevent oxidation; so, hence it was here proposed that the remaining air should be replaced with inert gas such argon so that carefully flushed gas could inhibit the formation of peroxides as effectively as possible.

Compound structure and chemical composition are also considered as factors with considerable impact on the oxidation rate of EOs. Several studies were compiled within this review (Turek and Stintzing, 2013) regarding influence of constituents' molecular structures on the autoxidation process. Namely, based on theoretical investigation on linalool oxidation (Bäcktorp et al., 2006), it was here highlighted that the most likely objects of autoxidation are the chemicals rich in allylic hydrogen, given that its abstraction led to formation of resonance-stabilized radicals. Lower activation energy makes this process highly favored. Polyunsaturated terpenic hydrocarbons are also considered extremely liable to autoxidation (Nguyen et al., 2009; Neuenschwander et al., 2010) due to their structural capacity to create a few radicals stabilized by conjugated double bonds or isomerization to tertiary radicals. Hence, chemical alterations in EOs caused by storage take place promptly and are primarily referred to unsaturated mono- and sesquiterpenes (Turek and Stintzing, 2011). Furthermore, a stronger carbon-peroxide bond achieved by means of hyper-conjugative effect is attributed to electron-donating groups and increasing alkyl substitution (Pratt and Porter, 2003). Thus, more stable hydroperoxides are formed, whereas other agents are not properly stabilized and instantly decay. An adequate confirmation for this statement is caryophyllene hydroperoxide, which triggers the immediate formation of corresponding epoxide state (Sköld et al., 2006).

With regard to all of the above presented information, prevention of the autoxidation process has consequently been a subject of research studies as well. A brief insight into possible solutions for preventing oxidative degradation of fragrance terpenes along with an increase of their sensitizing capacity is needed in order to complete this review.

The influence of adding an antioxidant such as butylated hydroxytoluene (BHT) on oxidation of *(R)*-limonene at room temperature was investigated (Karlberg et al., 1994a,b). BHT is commonly regarded as a "true antioxidant" with the ability to hinder oxidation by inhibiting radical chain reactions. Hence, it was here reported that BHT can inhibit the oxidation process, even for a few months. However, after the BHT is fully consumed, autoxidation begins almost instantly and takes place at about the same rate compared to that of preparations with no BHT added. Furthermore, the onset of the autoxidation is determined by the purity of the raw material and storage temperatures. No significant relation was found between the amount of BHT added and the time duration before the oxidation is triggered. Interestingly, in two out of three samples, no decrease in *(R)*-limonene concentration was recorded after storing these in refrigerator (cold and dark surrounding) for 7–12 months before the air-exposure experiment. So, three checkpoints of time could be suggested for controlling the stability of the fragrance terpene known to undergo autoxidation: (a) at production site, (b) in a finalized product, and (c) duration period of use of fragrance product. Handling and storage practice of this product should thus be reviewed, and some new guidelines for instructions of use ought to be implemented for consumers. Initially, it would be advantageous for the manufacturing process itself to involve smaller and air-tight containers and lower storage temperatures during shelf life. Both industrial and consumer products often contain BHT in concentration up to 1% and are generally considered safe in terms of cutaneous allergies.

The following sections contain the presentation of a selection of seven common fragrance terpenoids and one essential oil that were used to illustrate the oxidation pathways, with emphasis on the possible alteration mechanisms and the composition of oxidation mixtures. Simplified oxidation models were used wherein experimental procedures were strictly defined in terms of air exposure, meaning they were strictly defined in terms of adequate preparation of an Erlenmeyer flask, stirring process, exposure to daylight, temperature, and overall duration, as well as storage temperatures, for oxidized samples in order to produce an artificially aged fragrance terpene. Finally, prolonged period of air exposure leads to the increase of the viscosity of the oxidation mixture due to the polymerization process. Furthermore, high molecular weight of polymers excludes the possibility of their skin penetration making them not-so-interesting chemicals in the field of contact allergy.

29.4.1 Linalool

Linalool is a natural ingredient of numerous essential oils extracted from herbs, leaves, flowers, or wood, including oils of rosewood, petit-grain, linaloe seed, bergamot, and jasmine (Sköld et al., 2002). It is also one of the most important constituents of lavender oil, with a maximum concentration range around 40% (Hagvall et al., 2008). The fragrance chemical industry uses linalool as one of the key odorant ingredients in cosmetics and other scented products such as domestic and occupational products, owing to its fresh flowery odor and low cost production (Rastogi et al., 1998a; Rastogi et al., 2001). It is mainly of synthetic origin. Nowadays, linalool is highly valued as a fragrant molecule with significant therapeutic abilities due to its remarkable potential of combining versatility of floral odor and the multifaceted biological profile (Aprotosoaie et al., 2014).

With regard to its chemical structure, linalool (Figure 29.9) represents a doubly unsaturated hydrocarbon—that is, an aliphatic (acyclic) tertiary alcohol. Owing to the chirality of its hydroxylated tertiary carbon (C3), two isomeric forms of linalool can be found naturally, *(S)*-(+) linalool (coriandrol) and *(R)*-(−) linalool (licareol), emphasizing that both stereoisomers are considered equally stable (Bäcktorp et al., 2006). Hence, linalool is generally observed as the racemate of these two forms and thus is denoted as "linalool" in the overall available literature.

Investigation of the compounds formed during air exposure of linalool (Sköld et al., 2002) was the main goal of this research, as well as exploring sensitizing capacity prior to and after oxidation in an animal experiments. Autoxidation of linalool was confirmed to be in agreement with general mechanisms of terpene oxidation pathways. Its purity and storage temperature are among the key factors influencing the autoxidation rate, in general (Aprotosoaie et al., 2014). To the best of the authors' knowledge, this study dated in 2002 identified the major oxidation product linalool hydroperoxide for the first time. Such a curiosity could be explained by inadequate methods (e.g., GC and MS) used for analysis of the oxidation materials in previous studies, due to subjecting these materials to high temperatures and consequently decomposing thermolabile compounds. Although hydroperoxides are commonly marked as unstable chemicals prone to secondary oxidation process, this allylic hydroperoxide was found to be relatively stable, which was then connected to possible stability of the tertiary structure.

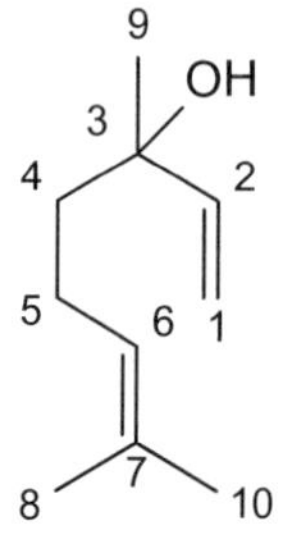

11

FIGURE 29.9 Linalool.

The same group of authors conducted more extensive research in 2004 (Sköld et al., 2004), investigating the autoxidation of linalool "under conditions similar to the room-temperature storage of fragrance products" in order to determine both qualitative and quantitative structure of the hydroperoxides, including the thermolabile ones.

The oxidative degradation showed clearly the decomposition rate of linalool during the period of 80 weeks at room temperature. The experiment proved that the degradation process of linalool starts instantly in the presence of air. Half of the initial amount of linalool was consumed after about 30 weeks, whereas 80 weeks was proved to be sufficient time for 96% of linalool to be decomposed.

To conduct the analysis of the autoxidized linalool, GC with split/splitless injection would be an analytical method of choice because monoterpenes are volatile compounds with low UV absorption. However, this way, only the more stable secondary oxidation products are detected, and due to high temperatures, no hydroperoxides could be identified. Hence, the HPLC was used as a preferred method so that oxidative degradation of linalool could be completely investigated. It was also used to quantify the content of hydroperoxides. Finally, isolation of oxidation products was performed by flash chromatography and preparative HPLC as a preferred method for detection of less stabile primary oxidation products. Hence, together with NMR used for the isolation of impure fractions in low amounts, eight different constituents were confirmed (Figures 29.10 and 29.11).

Quantification results obtained experimentally gave certain insight into the oxidation mixture composition. The etheric component of this mixture contained pyran-derivative **15**, reaching a maximum concentration level of only 4% after 79 weeks, while furan-derivative **14** was produced in a significantly higher amount and thus attained its maximum level of 20% after about 50 weeks.

The main hydroperoxide **12** accumulated in high concentration over time, starting with 4% at the tenth week of air exposure and reaching as much as 15% at the 45 weeks of oxidized sample of linalool. Notably, the hydroperoxide **12** was formed in higher amounts compared to the hydroperoxide **13**; that is, oxidized linalool contained approximately 1% of hydroperoxide **13** after 10 weeks while

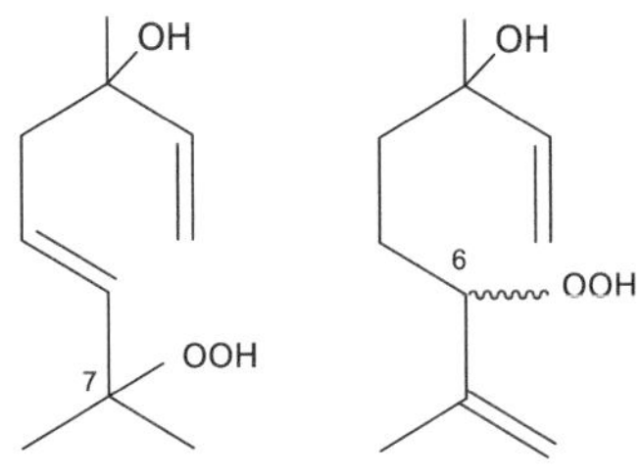

OH OH 6 OOH OOH 7

12 **13**

FIGURE 29.10 Air-exposed linalool: Primary oxidation products **12** and **13**. (Adapted and newly drawn from Sköld, M. et al. 2004. *Chemical Research in Toxicology* 17 (12): 1697–705.)

HO HO O O OH OH OH OH H O OH OH OH

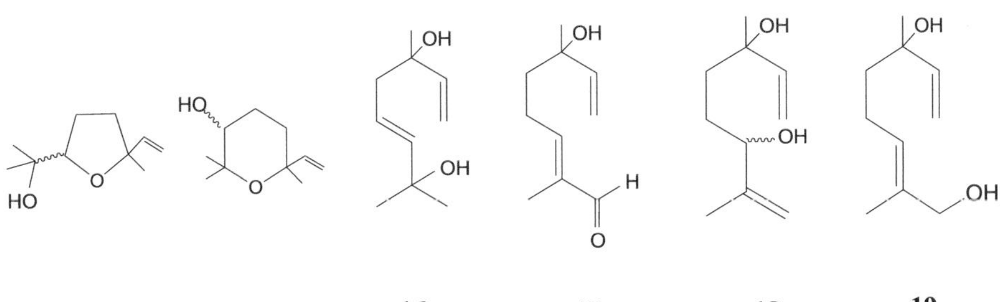

14 **15** **16** **17** **18** **19**

FIGURE 29.11 Air-exposed linalool: Secondary oxidation products **14–19**. (Adapted and newly drawn from Sköld, M. et al. 2004. *Chemical Research in Toxicology* 17 (12): 1697–705.)

4% of hydroperoxide was found in the 45-week air-exposed sample. Hydroperoxide **13** thus belongs to the group of minor oxidation products next to **19.**

Besides the qualitative and quantitative analysis and determination of sensitizing capacity, this experimental study observed a possible autoxidation mechanism of linalool too.

The mechanism can be presented in a few main checkpoints as follows:

- Formation of allylic radicals in the presence of ROO•
- Reversible reaction between radicals and triplet state oxygen, resulting in peroxyl radicals formation
- Hydroperoxides formed in the reversible reaction as products of subsequent hydrogen atom transfer (H•)

Allylic radicals are formed after the abstraction of the allylic hydrogen atoms, which can be achieved in three possible locations of the linalool molecule (Figure 29.12).

- The equilibrium between the alkyl radicals and their matching peroxyl radical can be affected by the difference in alkyl substitution on the peroxyl-bearing carbons. It is implied that the formation of secondary peroxyl radical ***Rb*** takes place in favor of the primary one ***R***a. Tertiary peroxyl radical ***Rd*** would be favored over the secondary one ***R***c for the same reason. Hence, ***R***b and *R*d are considered to be the stable forms of peroxylradicals. This statement is in accordance with the hydroperoxides **12** and **13** being the only ones experimentally detected in the oxidation mixture of linalool. The hydrogen atom abstraction is more easily proceeded with the hydroperoxide **13** compared to the hydroperoxide **12**, which confirms higher stability of the tertiary hydroperoxide **12** and thus explains its larger amount in the oxidation mixture.
- Possible decomposition of hydroperoxides leads to secondary oxidation products

In the group of alcohols (**16, 18**, and **19**), the form **16** is considered to be the most distinguished one, formed as a decomposition product of the main hydroperoxide **12.** Alcohol **18** corresponds to the other hydroperoxide **13**, while identification of **19** indirectly proves the existence of primary hydroperoxide **20** and so does the α,β-unsaturated aldehyde **17**.

This mechanism study suggested that the formation of the pyran and furan derivatives is linked to the ability of tertiary hydroperoxide 12 to form an epoxide. This initial phase is then proceeded with eager attack of a hydroxyl group on one of the two epoxide-carbons, leading to the formation of ethers **14** or **15**. Neither ketones nor hydroperoxide **22** were found in the oxidation mixture. Finally, it is indicated that the dominating reaction pathway leads via the most stable peroxyl radical ***R***d and the hydroperoxide **12**, both formed by the rearrangement process of the peroxyl radical ***R***c or the hydroperoxide **22.**

29.4.2 Linalyl Acetate

Next to linalool, linalyl acetate occurs naturally as the main component in lavender oil, broadly used in massage and aromatherapy treatments (Sköld et al., 2008). The study reported that presently both cosmetic and non-cosmetic products are the sources of dermal exposure to linalyl acetate, so decorative cosmetics, soaps, shampoos, and fine fragrances, as well as household products, are commonly associated with significant frequency of linalyl acetate consumption due to its fresh, fruity scent. This is in accordance with some European studies (Rastogi et al., 1998a; Rastogi et al., 2001) about the utilization rate among fragrance chemicals wherein linalyl acetate was found in nearly 30% of occupational and domestic products and in 70% of deodorants tested. The amount of linalyl acetate used per year reaches up to >1000 metric tons (Letizia et al., 2003).

Regarding its chemical structure, linalyl acetate represents a non-conjugated ester of linalool, that is, the only structural difference compared to linalool is the esterified hydroxyl group (Figure 29.13).

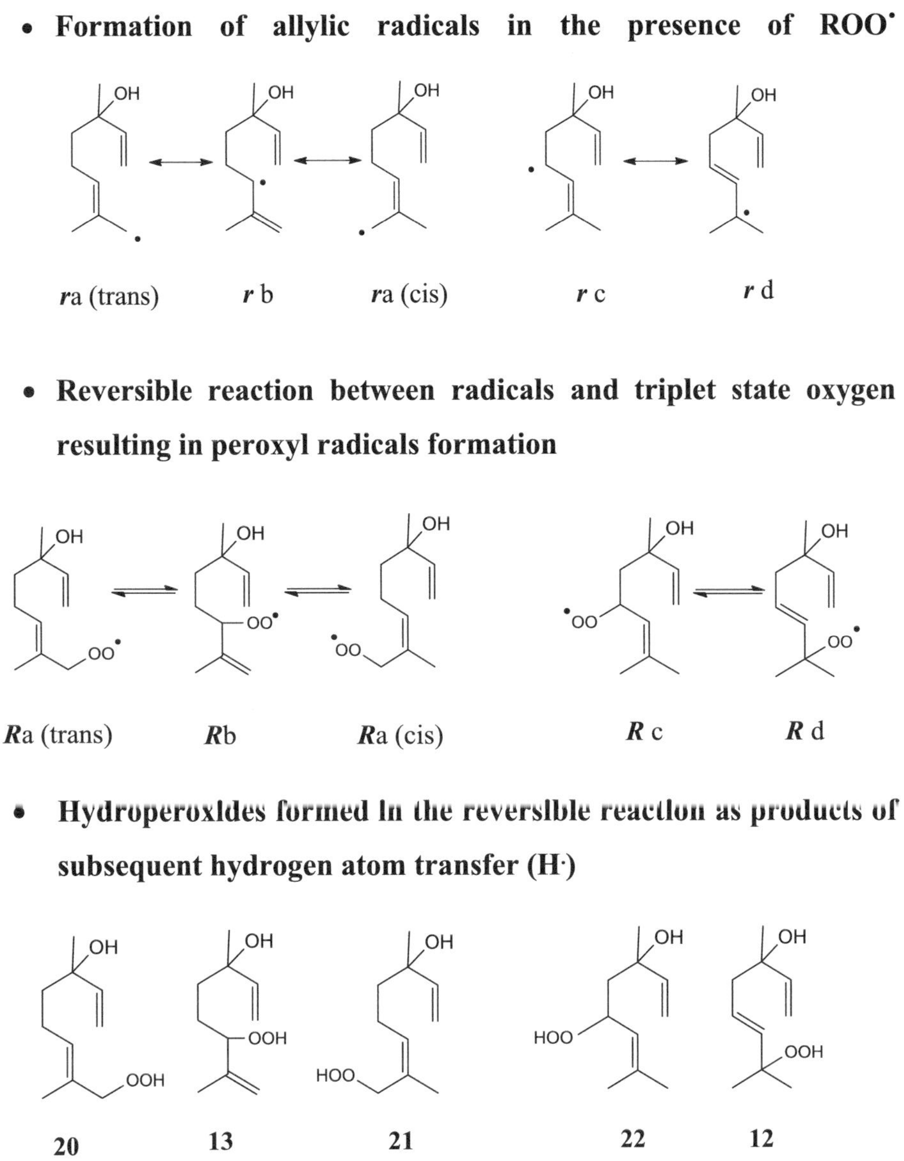

- **Formation of allylic radicals in the presence of ROO$^\bullet$**

*r*a (trans) *r* b *r*a (cis) *r* c *r* d

- **Reversible reaction between radicals and triplet state oxygen resulting in peroxyl radicals formation**

***R*a (trans)** ***R*b** ***R*a (cis)** ***R* c** ***R* d**

- **Hydroperoxides formed in the reversible reaction as products of subsequent hydrogen atom transfer (H·)**

20 **13** **21** **22** **12**

FIGURE 29.12 Allylic radicals of linalool and hydroperoxides 12, 13 and 20–22. (Adapted and newly drawn from Sköld, M. et al. 2004. *Chemical Research in Toxicology* 17 (12): 1697–705.)

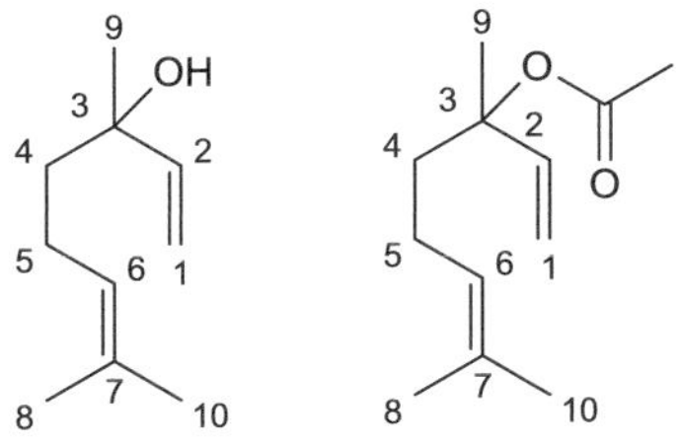

11 **23**

FIGURE 29.13 Linalool (**11**) and linalyl acetate (**23**).

Due to this similarity, it should be expected that linalyl acetate is also prone to oxidation upon air exposure, creating structurally analogous products. Given the aforementioned facts, the studies on linalyl acetate were mostly conducted by comparing this fragrance terpene to linalool.

Sköld et al. (2008) conducted a study in order to explore the nature of the autoxidation process of linalyl acetate and thus to identify the oxidation products. Evaluation of the sensitizing potency of the oxidized linalyl acetate was also in the focus of the research. Therefore, it was here reported that non-conjugated esters are very weak sensitizers, so the allergenic capacity of linalyl acetate might be attributable to the formation of hydroperoxides after linalyl acetate is exposed to air influence. Basic chemical principle of the autoxidation was provided by this study stating that it is a free radical chain reaction. More precisely, peroxy radicals are derived by abstraction of H atoms which is a selective reaction in terms of the position in a molecule. So, the formation of the stable radicals of linalyl acetate is only possible in the allylic position of the double bond, that is, C6–C7. Hence, analogous to linalool, corresponding hydroperoxides of linalyl acetate are expected to be created on air exposure.

The air-exposure procedure was strictly defined regarding plausible contamination, stirring process, and mimicking of daylight. Four different samples after 10, 16, 24, and 28 weeks were used for the experiment and then analyzed with GC to establish decreasing rate of linalyl acetate concentration depending on time exposure. In a 10-week period, 26% of linalyl acetate was consumed, whereas in the end, after 28 weeks, only 24% remained unoxidized. Notably, when compared to linalool (Sköld et al., 2004), a significant similarity can be seen in the oxidation rate of linalyl acetate.

Furthermore, the oxidized samples were then subjected to flash chromatography so that the following single oxidation products could be isolated and identified (Figure 29.14).

From a chemical point of view, it was here observed that autoxidation of linalyl acetate takes place in a way analogous to that of linalool (Sköld et al., 2004), which finally results in creating allergenic hydroperoxides. Thus, the main primary oxidation products of linalyl acetate were identified as **24** and **25**, which is in accordance with the linalool hydroperoxides **12** and **13**. This resemblance could be explained by the same active sites for hydroperoxide production; that is, by the structural similarity of these two chemicals.

One of the main differences is the formation of another secondary oxidation product, the epoxyde **26**. However, it was denoted as a non-sensitizer that is most likely readily converted to the non-sensitizing cyclic agent in the skin, also seen in the oxidation process of linalool (Sköld et al., 2004). Isolation of the linalyl acetate epoxyde is basically in contrast to autoxidation of linalool, wherein two cyclic ethers were identified: 2-(5-methyl-5-vinyltetrahydrofuran-2-yl)-propan-2-ol and 2,2,6-trimethyl-6-vinyltetrahydro-*2H*-pyran-3-ol (Figure 29.15).

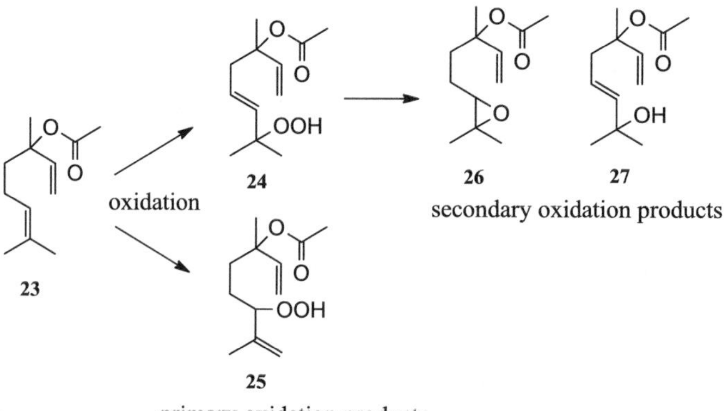

23
oxidation
24
OOH
25
OOH
primary oxidation products
26
27
OH
secondary oxidation products

FIGURE 29.14 Air-exposed linalyl acetate: Oxidation products **24–27**. (Adapted and newly drawn from Sköld, M. et al. 2008. *Contact Dermatitis* 58 (1): 9–14.)

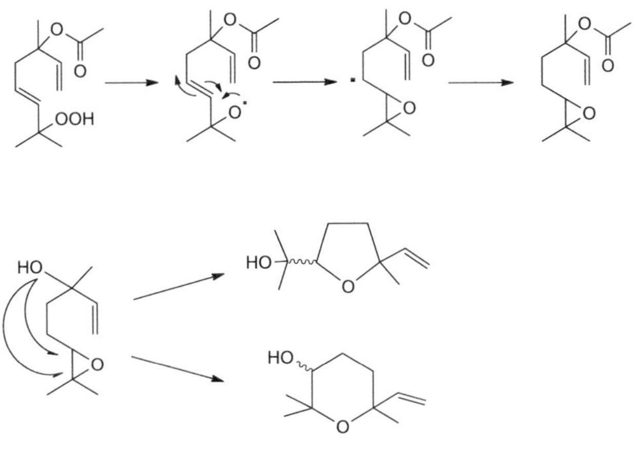

FIGURE 29.15 Linalyl epoxyde and cyclic ethers of linalool. (Adapted and newly drawn from Sköld, M. et al. 2008. *Contact Dermatitis* 58 (1): 9–14.)

The plausible mechanism was proposed as well by this study (Sköld et al., 2008) with the main observation about the stability of the linalyl acetate epoxide which is primarily associated with the acetate group as a key inhibitor of cyclization—that is, it disables the hydroxyl group from any further transformation. Contrary to linalyl acetate, linalool oxides were among secondary oxidation products, but no epoxides were confirmed in the oxidation mixture. Hence, it was here implied that the linalool epoxide actually underwent the attack of the corresponding hydroxyl group and so the cyclic ethers are formed.

Several years later, a similar study on frequency of contact allergy to oxidized linalyl acetate was performed (Hagvall et al., 2015). The aim of the study was to assess whether frequency of allergic reactions are as common as those seen in oxidized linalool. Formation patterns of linalyl acetate hydroperoxides and their sensitizing potential was also examined. The air-exposure procedure was strictly defined in order to determine the concentrations of linalyl acetate and its hydroperoxides during 45 weeks. It was thus noted that the degradation of linalyl acetate took place at an almost constant rate and was comparable to that of linalool, which is in accordance with an earlier study (Sköld et al., 2008). Moreover, the accumulation of linalyl acetate hydroperoxides started after 4 weeks of autoxidation, after which a maximum concentration of 37% was achieved after 42 weeks of time.

A higher concentration of linalyl acetate hydroperoxides was detected compared to that of linalool hydroperoxides which is most likely associated to the stabilized structure of linalyl acetate hydroperoxide by esterification of the alcohol group. Linalool hydroperoxides were shown to be stronger sensitizers than linalyl acetate hydroperoxides compared by their experimental EC3 values.

An interesting observation was provided by this study stating that a similar oxidation process during air exposure is also a common feature of other fragrance terpenes such as geraniol, citronellol, and limonene. Their molecular structure is characterized with "oxidizable positions" that are highly susceptible to oxidation. Those positions are either adjacent to a heteroatom (e.g., oxygen) or contain hydrogen atoms in an allylic position. Secondary oxidation products, the epoxide and the alcohol, were not analyzed given they were denoted as nonsensitizers in a previous study (Sköld et al., 2008).

29.4.3 β-Caryophyllene

β-Caryophyllene belongs to the group of sesquiterpenes and is commonly used as a fragrance chemical (Sköld et al., 2006). The study confirmed that sesquiterpenes are like monoterpenes, volatile compounds of EOs often used as fragrances because of their pleasant scent. As here also reported, the oils of cloves, cinnamon leaves, and copaiba balm are some of the natural sources of β-caryophyllene, and all famous for being natural remedies and fragrances. A study on chemical

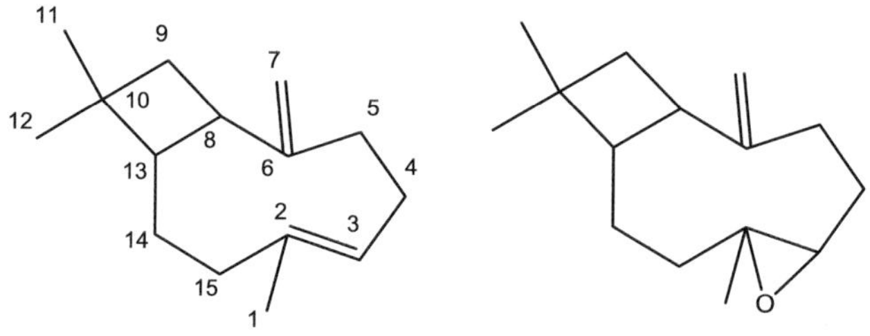

FIGURE 29.16 β-caryophyllene (**28**) and β-caryophyllene oxide (**29**).

analysis of 71 deodorants conducted in 1998 (Rastogi et al., 1998b) discovered that almost half of all the deodorants used on the European market contained caryophyllene which speaks in favor of significant exposure given that these products are frequently applied. For a chemical to act as a hapten, it must be electrophilic or have the ability to form a radical in order to bind covalently with skin proteins (Sköld et al., 2006). The study indicated that β-caryophyllene does not possess any of these characteristics and thus is not likely to act as an allergen. When compared to *(R)*-limonene and linalool, it is discernable that β-caryophyllene, as an unsaturated hydrocarbon, also contains double bonds (Figure 29.16) which makes the former two terpenes prone to autoxidation. So, Sköld et al. (2006) suggested that oxidation caused by influence of air is expected in β-caryophyllene, too. Hence, the main goal of this study was to explore the autoxidation process that took place at room temperature and to identify oxidation products, in parallel.

GC analysis was performed in order to determine autoxidation course after β-caryophyllene was air-exposed for a certain period of time while HPLC was used to analyze air-exposed β-caryophyllene, that is, its oxidation products, by comparing retention times for reference chemicals and isolated compounds. The latter were characterized with NMR and GC-MS. The air oxidation procedure was, as earlier described, strictly defined.

The oxidation process of β-caryophyllene generated only a few oxidation products, despite the fact that autoxidation process was quite fast. The remaining concentrations of air-exposed β-caryophyllene samples were determined after 5 and 48 weeks. The results showed that it took only 5 weeks to consume almost half of the tested concentration of this sesquiterpene—that is, air exposure caused an almost immediate reduction of the content of the original compound. After 48 weeks, there was only 1% of it left.

Oxidative degradation of β-caryophyllene was followed by emergence of a few other peaks in the chromatograms, proving that other compounds were formed beside β-caryophyllene. In the starting phase, the accumulation process of the caryophyllene oxide was found to be just as rapid as the decrease rate of β-caryophyllene. The maximum concentration attained in the oxidation mixture was about 40%. Notably, caryophyllene oxide (**29**) (an epoxide) was the only isolated oxidation product, noting that epoxides are generally secondary oxidation products in a fragrance terpene oxidation mixture. In order to simulate the autoxidation process of β-caryophyllene, photooxidation was conducted and thus investigation of the possible primary oxidation products, that is, hydroperoxides, was enabled. A mixture of hydroperoxides was derived next to high proportions of the epoxide, containing major hydroperoxide **30** and a fraction of hydroperoxides **30** and **31** (Figure 29.17), isolated for the purpose of being the reference compounds.

Given that reactivity of a double bond rises with regard to number of alkyl substituents, the singlet oxygen reacted with the endocyclic double bond so that three different hydroperoxides are formed, in theory. The third hydroperoxide **32** (Figure 29.17) was expected to be present in the oxidation mixture, but it was not experimentally proved. Finally, there was one peak detected in the oxidized sample of β-caryophyllene with the same retention time as hydroperoxide **31**. Unfortunately, isolation of this component was impossible due to the excessively low concentration.

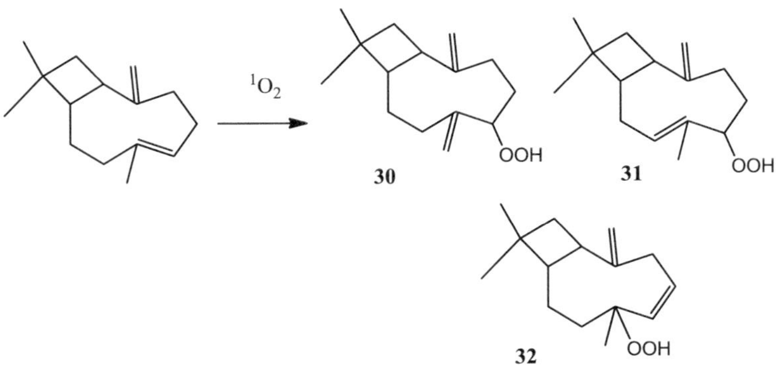

FIGURE 29.17 Photooxidation of β-caryophyllene: Theoretical formation of hydroperoxides **30–32**. (Adapted and newly drawn from Sköld, M. et al. 2006. *Food and Chemical Toxicology* 44 (4): 538–45.)

Furthermore, a stability test was performed on hydroperoxides **30** and **31** using HPLC, after which no decomposition was proved and thus no epoxides could be detected. So, the study indicated that caryophyllene oxide derived from photooxidation was actually yielded by degradation of hydroperoxide **32**. A complete lack of hydroperoxides in the oxidation mixture could thus be explained by the observation here presented that the formation of hydroperoxide **32** is favored and thus caryophyllene oxide is the only oxidation product detected.

In general and as noted in the study (Sköld et al., 2006), autoxidation process of fragrance terpenes results in formation of primary oxidation products (i.e., hydroperoxides) which are then degraded to secondary oxidation products over time; for example, alcohols, epoxides, ketones, aldehydes, and polymeric compounds. However, oxidation of β-caryophyllene after being exposed to air gave no significant amounts of hydroperoxides, whereas caryophyllene oxide was derived to a great extent and was denoted as a stable, crystalline substance. Hence, it was suggested that the plausible alteration process of primary to secondary oxidation products of β-caryophyllene takes place almost immediately.

29.4.4 Lavender Oil

The human population has been familiar with the use of the essential oils distilled from the genus *Lavandula* for centuries so that lavender oil is known to be effective both cosmetically and therapeutically (Cavanagh and Wilkinson, 2002). This review stated that despite its popularity and traditional use that dates back to ancient times, biological activity of lavender oil has become the subject of ongoing research studies just in the recent years. However, most of the claims are linked to its plausible antibacterial, antifungal, carminative, and sedative effects, as well as the positive outcome in treating burns and insect bites. Traditional use of pure lavender oil in folk medicine and cosmetic and hygiene products is a considerable way of exposing population to this herbal remedy due to its pleasant scent and antimicrobial effect. Nowadays, lavender oil is widely used in massage and aromatherapy, as here reported, with constantly growing frequency because many benefits are claimed for these type of treatments.

Today, a pleasant, lavender scent in cosmetics, perfumes, soaps, etc. is mostly mimicked by synthetic substances with the linalool being the most common fragrance chemical in the world used for this purpose (Hagvall et al., 2008). The introduction of the study reported that the contemporary market is supplied with various kinds of lavender oils with different compositions. However, the major, common constituents to all are linalyl acetate and linalool, achieving the concentrations of about 50% and 40%, respectively. The third main terpene component is a sesquiterpene, β-caryophyllene, present in a percentage of approximate 2%–3%. The autoxidation process of linalool and linalyl acetate has already been reviewed in this chapter, so only a few main points will be presented in this section (in accordance with the study): (a) these two compounds show significant structural resemblance, (b) both of them are highly susceptible to oxidation with the hydroperoxides being the major primary oxidation products, and (c) air exposure (i.e., oxidation) greatly influences the sensitizing potential in terms of altering these parent molecules from weak sensitizers to quite potent ones and the hydroperoxides were proved to be the strongest. As for β-caryophyllene, several key features could be summarized: (a) also readily oxidizes on air and (b) sensitizing potential is less affected compared to that of linalool and linalyl acetate because only a moderate allergen, an epoxide, was detected in an oxidation mixture.

Given the above, the aim of this study was to compare the autoxidation outcome of lavender oil to that of previously mentioned individual fragrance chemicals. Likewise, the air-exposure procedure including analytical methods for the analysis was similar to the ones already described. A preparation of a synthetic mix was used for the purpose of this experiment containing linalyl acetate, linalool, and β-caryophyllene in the same proportion as in natural lavender oil, only without any other compounds. Experimental air exposure showed decrease of content of all three main components of lavender oil. The quantitative analysis was conducted by GC on column technique, using two air-exposed samples after 10 weeks and 45 weeks. The autoxidized lavender oil sample, aged for 10 weeks, was used to investigate the content of oxidation products formed after air exposure. The amounts of linalool hydroperoxides (**12** and **13**) and linalyl acetate hydroperoxides (**24** and **25**) were 0.48% and 3.3%, respectively, and were determined by HPLC. Four more compounds were detected in the oxidation mixture but not quantified: **14, 26, 27,** and **29**.

The study confirmed that autoxidation of lavender oil (i.e., its three main components) follows the same pattern in terms of generating the same oxidation products as the one derived from the pure, single synthetic terpene. Corresponding amounts of hydroperoxides were found in oxidized lavender oil and an oxidation mixture of pure terpenes, so it is presumable that they are equally stable. There has been a theory that products extracted from natural essential oils are not as susceptible to autoxidation as synthetical terpenes owing to antioxidant activity of the raw herbal materials. As for the lavender oil, these findings practically denied this speculation. So, the terpenes showed nearly the same rate of autoxidation in lavender oil, in the synthetic mix and as pure, individual chemicals regardless of probable presence of the antioxidant agents. It was finally indicated that the oxidation process caused by air exposure takes place following the identical mechanism in the pure terpenes as in the oil so the presence of antioxidants or scavenging agents was not confirmed. However, it is noteworthy that the oxidation experiment was performed under quite severe conditions in order to promote formation of detectable amounts of hydroperoxides.

29.4.5 Limonene

Scented products, aromatherapy oils, and flavor mixtures contain limonene as a common fragrance ingredient (Kern et al., 2014). Due to its widespread usage, it can be found in all kinds of consumer products from fine fragrances to technical products, often in higher concentrations (0.005%–2%) than other fragrances (Matura et al., 2002). This study reported that industrial degreasing agents and water-free hand cleansers contain even larger amounts of limonene, attaining concentration levels if 20%–100%. Aside from the broad consumer use, limonene is also a common natural ingredient in essential herb oils, such as rosemary, eucalyptus, lavender, peppermint, which are commonly mixed in massage oils and aromatherapy products, leading to the significant increase in exposure to this fragrance chemical (Bråred Christensson et al., 2013). *(R)*-limonene, to be exact, is the main component in peel oil of citrus fruits, that is, a distillate generally contains more than 95% of *(R)*-limonene (Karlberg and Dooms-Goossens, 1997), so just peeling an orange can lead to a significant hand exposure to limonene.

Chemically, limonene represents a monoterpene hydrocarbon *p*-mentha-1,8-diene. Aside from the racemic form of limonene, dipentene, different ratios of two enantiomer forms (*R*)-(+) and (*S*)-(−) can be found in plants (Figure 29.18) which are actually mirror images of one another, whereas the source of essential oils influence exposure of a certain population to one or the other form (Matura et al., 2006).

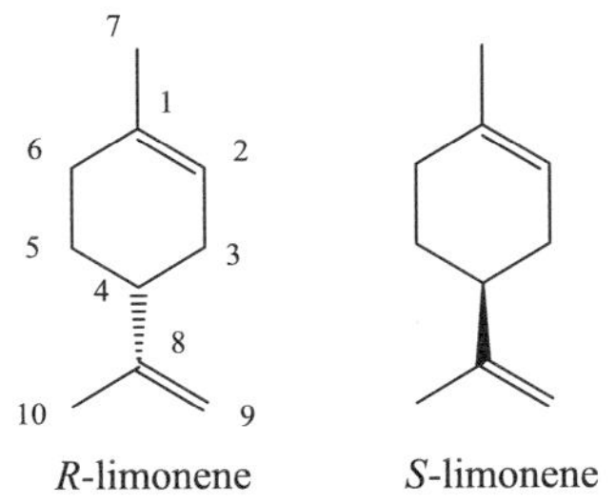

R-limonene *S*-limonene

33

FIGURE 29.18 (*R*)-(+)-limonene and (*S*)-(−)-limonene

Nearly three decades ago, *(R)*-limonene and dipentene joined the industrial market of degreasing and cleaning products as a more environmentally friendly and less toxic agent compared to traditionally used organic solvents (Karlberg and Dooms-Goossens, 1997). Industrial use of *(R)*-limonene is commonly associated with the cutaneous hazards due to its large amounts found in these products, whereas a low concentration of *(R)*-limonene as a flavoring agent, perfume, or an additive in cleaning products rarely cause skin problems (Karlberg et al., 1992). This experimental study confirmed that prolonged air exposure of 8 weeks (at room temperature) induces autoxidation, making *(R)*-limonene quite a potent sensitizer. Conversely, if not air-exposed, *(R)*-limonene caused no significant sensitization reaction on guinea pigs even if the animals are sensitized to oxidized *(R)*-limonene. Five main oxidation products were identified on this occasion: *(R)*-(−)-carvone and a mixture of *cis/trans* isomers of (+)-limonene oxide-(1,2), that is, limonene epoxide, which are described as potent allergens while no significant sensitization response was observed in the animals tested with a mixture of *cis/trans* isomers of (−)-carveol. It should be noted that a GC-MS system was used for the analysis, making primary oxidation products hydroperoxides undetectable due to their instability, which can lead to quick degradation to secondary oxidation products, such as carvone. This problem was solved by a certain modification of GC-MS with chemical ionization in negative ion mode (Nilsson et al., 1996). In addition, single components (i.e., contact allergens) in oxidized *(R)*-limonene were isolated by HPLC with two different stationary phases in normal phase mode. The content of the limonene oxidation mixture was in accordance with the previously mentioned study, including four hydroperoxides as primary oxidation products.

Nowadays, limonene is mostly used as an added fragrance compound, that is, as a by-product from the citrus juice industry. Hence, according to fragrance industry data, it is presumable that the *(R)*-(+)-form is the key source of exposure (Matura et al., 2006). This study actually provided clinical evidence for the significance of limonene oxidation products wherein basic chemical background was briefly investigated as well. Thus, it was confirmed that the oxidation mixture consists of a few oxidation, that is, degradation, products. Carvone, limonene oxide, and limonene hydroperoxide were proved to be major identified compounds with considerable allergenic activity, with the latter being the most important one.

Karlberg and Dooms-Goossens (1997) explored the prevalence of contact allergy to air-exposed limonene, that is, oxidized *(R)*-limonene and *(R)*-limonene hydroperoxide, among dermatitis patients with regard to its high concentrations in industrial, occupational, and domestic products. The results obtained on this occasion confirmed that positive patch test reaction rate was actually comparable to that detected for some of the allergens in the standard series. The study was primarily focused on mimicking the handling conditions of *(R)*-limonene when exposed to daylight and air (open flasks) which resulted in prompt autoxidation yielding various oxygenated terpenes. The experimental part of the study investigated the nature of the degradation process of limonene using samples of *(R)*-limonene that were air-exposed for 10 and 20 weeks. The samples were kept under strict conditions

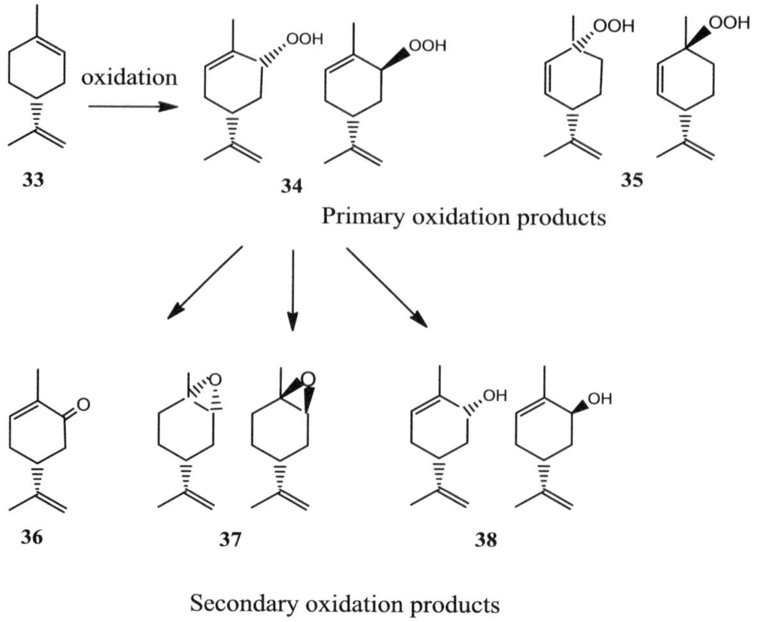

FIGURE 29.19 Autoxidation products of (+)-limonene (**33**). (Adapted and newly drawn from Karlberg, A.-Th., and A. Dooms-Goossens. 1997. Contact Dermatitis 36 (4): 201–6.)

between air-exposure periods and sensitization studies. Notably, frequency of positive reactions between these two forms of oxidized limonene (with the same test concentrations) showed no significant difference. The results of the experiment confirmed decrease in content of *(R)*-limonene. In the first 10 weeks, the amount of *(R)*-limonene was only 40% of its total content. This period, as expected, showed increase in the amount of oxidation products, whereas the maximum of around 15% was attained at 20 weeks. Notably, the amount of oxidation products did not show significant change in the following period. However, the quantity of *(R)*-limonene was continuously decreasing after 10 weeks so that at the end of the experiment (around 50 weeks), less than 10% of *(R)*-limonene was present in the product.

Different kinds of oxidation products (Figure 29.19) were identified in the autoxidation process of limonene (Kern et al. 2014; Karlberg et al., 1994a,b), wherein all oxidation products but carveol were confirmed as potent sensitizers.

A certain discrepancy between the concentrations of 1,2-limonene oxides—that is, limonene epoxide and limonene hydroperoxides (Lim-OOH)—compared to carvone and carveol is discernable. Namely, the former reached the maximum concentrations in the period from 10 to 20 weeks of air exposure, while this could not be depicted for carvone and carveol as stable secondary products. Taken together, none of the single determined allergens exceeded 6% of the oxidized limonene content. Nevertheless, this was still sufficient enough to give an elicitation reaction (Karlberg et al., 1994a,b; Nilsson et al., 1996).

These results are in agreement with another study (Nilsson et al., 1996) indicating that the time of air exposure plays a significant role on accumulating predominant oxidation forms of limonene, which also refers to the level of sensitization induced by industrial limonene products. Further chromatography analysis of limonene hydroperoxide (Karlberg and Dooms-Goossens, 1997) derived from the air-oxidized *(R)*-limonene confirmed the existence of four major hydroperoxide forms. This is in agreement with an earlier investigation (Karlberg et al., 1994a,b) wherein *cis/trans*-limonene-2-hydroperoxide (Lim-2-OOH) was confirmed to be the major autoxidation product formed to the greatest extent. So too, the chromatograms (Nilsson et al., 1996) clearly demonstrate the link between the time of exposure and the difference in composition of oxidation mixture (Table 29.1). Namely, carvone clearly showed high stability given that its concentration reduced only by a few percentage points in 6 months. However, a significant decrease in content was recorded for epoxides and hydroperoxides due to their instability, so only small amounts of these compounds (<4%) could be detected after 9 months. Prolonged air exposure generally leads to the increase of viscosity of the oxidation mixture, which becomes less volatile and more polar and thus is not suitable for GC-MS analysis.

Overall, the study (Karlberg and Dooms-Goossens, 1997) confirmed that autoxidation of *(R)*-limonene (in daylight), like other fragrance terpenes, takes place promptly and results in a variety of oxygenated monocyclic terpenes that are recognized as potent contact allergens using patch testing (in addition to standard series) in consecutive dermatitis patients.

29.4.6 Geraniol

Various plant materials such rose, citronella, and palmarosa contain large amounts of geraniol, which is a common chemical ingredient in Fragrance Mix I (FM I), used in dermatology clinics to screen fragrance allergy among dermatitis patients (Hagvall et al., 2007). Its fresh, flowery odor resembling rose scent made geraniol a widely used fragrance terpene.

Geraniol (**39**) (*trans*-3,7-dimethyl-2,6-octadien-1-ol) belongs to the family of oxygenated monoterpenes and represents an acyclic primary alcohol (Stobiecka, 2015) whose carbon skeleton consists of two isoprene units (see Figure 29.21). This structural feature of geraniol is responsible for the presence of allylic, vinylic, and alkylic sites so, according to the study, different stability and reactivity can be ascribed to these sites. With regard to previously interpreted studies of linalool autoxidation, Stobiecka (2015) also suggested it is exactly the allylic H-atom that determines reactivity toward oxygen as well as the anti-/pro-oxidant qualities of fragrant molecules. As here further accentuated, the flavor and fragrance industry finds geraniol to be one of the most significant compounds owing to its frequent use in the production of fruity pleasant aromas, and furthermore, it is often used as an additive in everyday food products, perfumes, cosmetics, and household products.

Hagvall et al. (2007) published the study on the pattern of geraniol autoxidation in 2007 for which they confirmed it was the first of that kind, to the best of their knowledge. The study was actually based on the potential of geraniol to autoxidize in the presence of air, at normal storage and handling in order to investigate the sensitizing potency of geraniol itself, air-exposed geraniol, and its oxidation products as well. Analogously to previously observed research on linalool, geraniol itself is also a weak allergen and is actually responsible for only 5% of positive patch test reactions toward the single components of FM I (Schnuch et al., 2004). However, geraniol, just like linalool

TABLE 29.1
The Concentration of Oxidation Products of (R)-Limonene, Autoxidized for 3 and 9 Months (Nilsson et al., 1996)

Compound	Concentration after 3 months (μg/mg)	Concentration after 9 months (μg/mg)
cis-limonene-1,2-oxide	18.1	–
trans-limonene-1,2-oxide	32.8	2.8
trans-carveol	26.6	19.0
cis-carveol	11.5	4.7
(R)-(–)-carvone	39.5	35.0
cis-Lim-1-OOH	9.5	3.3
trans-Lim-1-OOH	18.9	3.5
cis-Lim-2-OOH	19.1	2.8
trans-Lim-2-OOH	19.2	2.0

or limonene, has the capacity to spontaneously oxidize during air exposure (Hagvall et al., 2007). The study also stated that for a molecule to act like an antigen and cause contact allergy, it has to act as an electrophilic hapten which reacts with nuclepohilic moieties in amino acid side chains in skin proteins. Geraniol surely does not possess these electrophilic qualities, but when it undergoes autoxidation, sensitizing potency can significantly be altered, that is, increased. Primary oxidation products geranial, neral, and, naturally, hydroperoxide are confirmed to be the most important sensitizing agents and thus make the major contribution to allergenic activity of geraniol. The research (Hagvall et al., 2007) also revealed that the autoxidation pattern of geraniol differs from that of linalool and limonene, stating that it actually follows two paths. Both of these paths are claimed to originate from allylic H abstraction, close to double bonds. Namely, hydrogen peroxide is primarily created from a hydroxyhydroperoxide along with the aldehydes geranial and neral (Figure 29.20). On the other hand, in comparison to the formation of major linalool hydroperoxides, only a small amount of geraniol hydroperoxide is created.

When it comes to the autoxidation process, that is, hydrogen abstraction in radical chain mechanism, the study supported the theory on allylic H atoms being the most probable targets. Moreover, geraniol owns six allylic sites whereas linalool has only three. This structural feature of geraniol explains the fact that geraniol oxidation products derive from allylic H abstraction at the first tri-substituted double bond carrying the allylic alcohol in addition to corresponding hydroperoxides detected with linalool. The air-exposure procedure showed that the rate of autoxidation of geraniol was similar to that of linalool. Namely, after 10 weeks of exposure geraniol was down to 80% of its content while 20% remained at the 45-week period of exposure.

Oxidation products formed after 45 weeks of exposure were not detected and quantified due to polymerization of the oxidation mixture, that is, increased viscosity. Hydrogen peroxide was also proved to be one of the main components in the oxidation mixture of geraniol, next to aldehydes and geraniol hydroperoxide. The first 5 weeks show a similar formation rate of aldehydes (geranial and neral, combined) and hydrogen peroxide. The latter, however, achieves a plateau after this period.

(a)

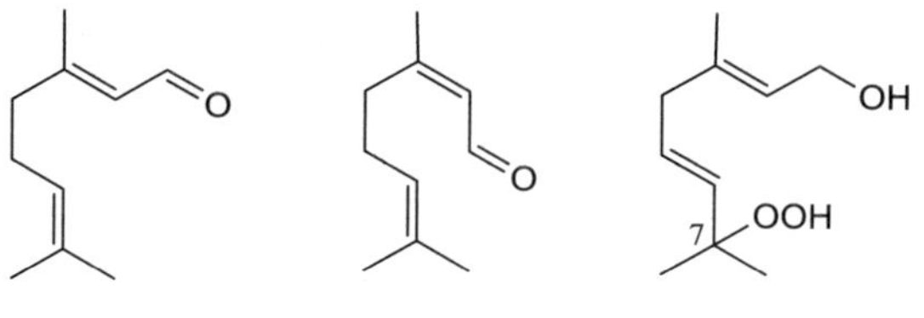

O O OH 7 OOH

40 **41** **42**

Primary oxidation products

(b)

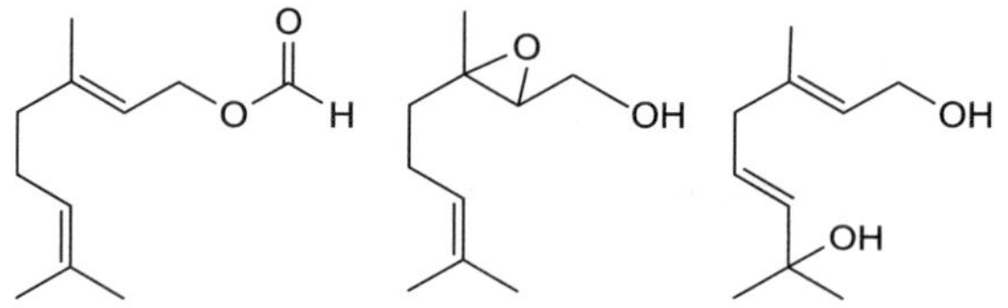

O O H O OH OH OH

43 **44** **45**

Secondary oxidation products

FIGURE 29.20 Oxidation products of geraniol (**39**) geranial (**40**) and neral (**41**), geranylhydroperoxide (**42**), geranylformate (**43**), epoxygeraniol (**44**), and geranyldiol (**45**). (Adapted and newly drawn from Hagvall, L. et al. 2007. *Chemical Research in Toxicology* 20 (5): 807–14.)

Namely, the maximum of hydrogen peroxide formation can be ascribed to week 7 (0.37% w/w). On the other hand, its concentration decreased to 0.31% w/w after 10 weeks.

These experimental data confirmed the assumption that both aldehydes and hydrogen peroxide are indeed derived from hydroxyl-hydroperoxide. According to the graph, the 5-week period is actually a breakpoint for hydrogen peroxide, after which time it is obviously created in slower pace compared to reacting with other components of the mixture. It is a reasonable conclusion that more and more hydrogen peroxide is formed, but it is practically immediately consumed due to its reactivity. The continuous increase in the amount of the aldehydes can confirm this premise. The experimental part of this study was performed in terms of isolation and quantification of primary and secondary oxidation products using NMR, GC-MS, and HPLC. Quantification of hydrogen peroxide in the air-exposed samples was conducted by an FIA method using fluorescence detection. This study finally proposed a mechanism of geraniol alteration caused by air-induced oxidation at room temperature. The focus point of this study begins with the formation of five main radicals (Figure 29.21), and investigation of their relative stability through enthalpy changes relative to the preferential radical A.

Notably, this preferential formation of the radical A was ascribed to the most easily allylic hydrogen abstraction performed on the α-position to the hydroxyl group, which is considered to be the most weakly bonded one in general. Given these basic facts, the possible connection between activation energy and bond strength was suggested. With regard to this suggestion and reported enthalpies, radical A was confirmed to be the main radical product (the most stable one) and thus oxidation products of this radical were considered as the dominant ones in the autoxidation process of geraniol. Lower stabilities of radicals D and E (relative to that of A) were used to explain both small quantities of hydroperoxide **42** (from D) found in this experiment and no hydroperoxide corresponding to E (6-hydroperoxy-3,7-dimethyl-octa-2,7-dien-1-ol) at all. The formation of these two hydroperoxides follow the same mechanisms of the main linalool hydroperoxides given that radicals D and E are analogous to the two main radicals produced in autoxidation of linalool. Hence, they were not further observed. With regard to that previously stated, the study proposed a plausible mechanism of geraniol autoxidation process resulting with the formation of hydrogen peroxide, geranial, and neral starting from radical A (Figure 29.22). The initial addition of a triplet state oxygen molecule takes place in a reversible reaction either at the carbon atom adjacent to the hydroxyl group or at the tri-substituted carbon atom adjacent to the methyl group. Peroxy radicals **46a** and **46b** are produced, respectively.

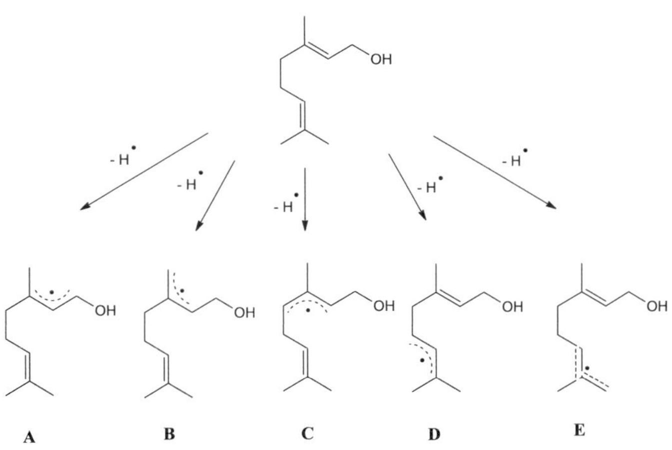

Geraniol

OH

- H•

- H•

- H•

- H•

- H•

OH OH OH OH OH

A B C D E

FIGURE 29.21 Possible radicals from geraniol. (Adapted and newly drawn from Hagvall, L. et al. 2007. *Chemical Research in Toxicology* 20 (5): 807–14.)

In order for the reaction to proceed, the subsequent hydrogen abstraction from a neighboring geraniol takes place, resulting in creation of two hydroperoxide products, **47a** and **47b.**

Notably, **47a** is more stable and thus most likely formed. However, due to their instability in general, these hydroperoxides degrade promptly to the corresponding aldehydes and hydrogen peroxide. Namely, geranial and H_2O_2 are formed via a general acid-catalyzed fragmentation reaction of the hydroperoxides **47a** and **47b**. Additionally, aldehyde neral is formed via **47b** in an analogous way to that of geraniol due to free rotation of the single bond adjacent to the double bond. Hydrogen peroxide is naturally produced in addition to neral. Finally, as far as the product distribution between geranial and neral is concerned, the study stated that geranial is produced to a greater amount due to more favored *E*-configuration. Given all the afore mentioned, it can be stated that this experimental research actually presented the mechanisms for the major reaction pathways of geraniol autoxidation using computational chemistry as a supplemental data source.

29.4.7 Geranial

Geranial is certainly another suitable example of a fragrance terpene compound that was investigated in terms of the nature of the autoxidation process. Namely, from a chemical perspective, it shows a significant level of structural resemblance with other, so far reviewed, monoterpenes such as geraniol, which is actually a corresponding alcohol of geranial.

A few years ago, Hagvall et al. (2011) published research on autoxidation of geranial, the first and only of that kind at the time, to the best of the authors' knowledge. The main focus of the study was to explore the autoxidation mechanism patterns, that is, the nature of the oxidative degradation process including identifying the products derived after geranial was air-exposed as well as determining their sensitizing potential. To fully understand the effects of autoxidation, both theoretical and experimental methods were conducted. Using standard density functional theory (DFT) as a computational method turned out to be quite beneficial in terms of elucidating radical reactions by directly comparing similar (competing) reaction types. On the other hand, experimental

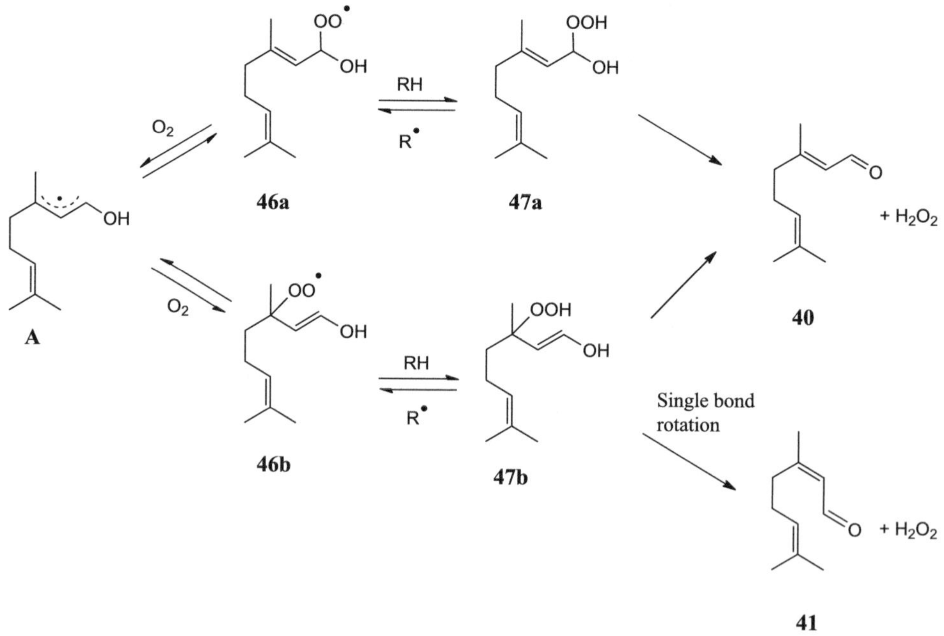

OO• OH RH R• OOH OH O2 O + H2O2 OH 46a 47a A O2 OO• OH OOH OH RH R• 40 Single bond rotation 46b 47b O + H2O2 41

FIGURE 29.22 Mechanism of geraniol oxidation. (Adapted and newly drawn from Hagvall, L. et al. 2007. *Chemical Research in Toxicology* 20 (5): 807–14.)

findings help differentiate these radical reactions as well as closed shell paths. The air-exposure procedure, that is, the way pure geranial was air-exposed to the influence of air in order to simulate artificial oxidation was strictly defined prior to the isolation, identification, and quantification procedures. This is mainly referred to handling and storage of an Erlenmeyer flask, stirring process, and sampling of the geranial preparations. Isolation of geranial oxidation products was conducted using flash chromatography while HPLC-UV was used in order to quantify geranial and its oxidation products after being air-exposed.

The experimental part of the research observed the degradation rate for two samples of air-exposed geranial. The sample 1 showed a decrease in concentration for 40% after 5 weeks, whereas its concentration dropped to 45% after 10 weeks of exposure. The sample 2 was subjected to a longer period of exposure, so the overall concentration of geranial was reduced to 20% after 30 weeks. Further evaluation of the degradation process was disabled by high viscosity, that is, formation of polymers. More precisely, the autoxidation rate began to decrease at 15 weeks, which is easily depicted in the curve slope being less steep. Notably, when compared to the earlier presented oxidation rate of geraniol, the faster degradation rate was seen in geranial.

The oxidation products derived after air exposure of geranial were identified as follows: geranic acid (**48**); 6,7-epoxygeranial (**49**); *(2E)*-5-(2-(*(E)*-2,6-dimethylhepta-1,5-dienyl)-5,5-dimethyl-1,3-dioxolan-4-yl)-3-methylpent-2-enal (**50**); and *(2E)*-5-(2-hydroperoxy-2-(*(E)*-2,6-dimethylhepta-1,5-dienyl)-5,5-dimethyl-1,3-dioxolan-4-yl)-3-methylpent-2-enal (**51**) (Figure 29.23).

The major oxidation product was 6,7-epoxygeranial (**49**), whereas dioxolane hydroperoxide (***51***) was the second most copious, and geranic acid (**48**) was the third. Only after 7 weeks, dioxolane derivative (**50**) was detectable in the oxidation mixture.

Compared to the previously reviewed studies on terpene oxidation, this study reported that geranial actually follows the third pattern of autoxidation. As discussed, stable primary oxidation products, that is, hydroperoxides, are derived from autoxidation of linalool, limonene, and linalyl acetate. However, along with only a small amount of hydroperoxides, autoxidation of geraniol resulted in the formation of aldehydes as the main oxidation products. Hydroperoxides of geranial, structurally analogous to the ones identified in linalool, linalyl acetate, and geraniol, were not detected in the oxidation mixture. The presence of compound **51** was confirmed in concentrations proportional to those of limonene-7- and limonene-6-hydroperoxide. Furthermore, for the first time, dioxolane structures were detected as a product of an autoxidation of fragrance terpenes. DFT calculations also confirmed that formation of the hypothetical geranial hydroperoxides in C6 or C7 is less favored than the formation of dixolane structures **50** and **51.** A thorough study on autoxidation mechanism of geranial was also provided by this research.

The relative stabilities (as calculated energies, ΔG in kJ/ mol) of plausible radicals derived after hydrogen abstraction in geranial were evaluated. Interestingly, a secondary radical at the remote alkene A was proved to have the lowest relative free energy, but no A-oxidation products could be confirmed in this experiment. Nevertheless, oxidation products should emerge from acyl radical C•,

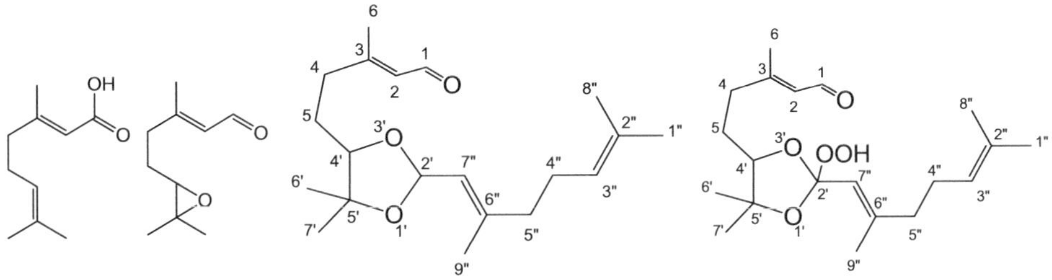

48 **49** **50** **51**

FIGURE 29.23 Oxidation products of geranial (**48–51**). (Adapted and newly drawn from Hagvall, L. et al. 2011. *Chemical Research in Toxicology* 24 (9): 1507–15.)

regardless of its high energy. Though, the second step of the radical chain reaction (production of a peroxyl radical) is usually reversible, results obtained on this occasion showed that the addition of oxygen resulted in formation of a highly stable peroxyl radical C1, with the lowest relative free energy of all peroxyl radicals: A1, A2, B1, B2. Hence, C1 is the only irreversibly formed peroxyl radical. Due to the observed, C1 radical would be yielded to the greatest extent and thus the observed products are expected to originate from C1. These observations were confirmed to be in accordance with the experimental findings as no hydroperoxides nor alcohols and aldehydes originating from allylic radicals $A^{\bullet}$ or $B^{\bullet}$ were detected. According to the proposed mechanism of hydroperoxide formation, radical C1 can abstract a hydrogen atom, which results in formation of peracid.

Still, no peracid was detected in the autoxidation mixture, and two possible scenarios were offered: either peracid is consumed prior to identification of oxidation mixture, or radical C1 reacted directly with some other agent in the solution. Given that peracids are established as epoxidation inducers, it is expected to react with geranial (an olefin) and thus an epoxide **49** and a carboxylic acid **48** are formed (Figure 29.24a). The reaction then proceeds so that the epoxide reacts with acyl radical and the dioxolane hydroperoxide is yielded. It was also suggested that the dioxolane derivative is created in an acid-catalyzed closed shell reaction between 6,7-epoxygeranial and geranial. A more detailed review on dioxolane formation will be proposed later on. Peroxyl radical C1 can, alternatively to hydrogen abstraction, be engaged in a direct reaction with olefins, which can also result in an epoxide and a carboxyl radical creation (Figure 29.24b).

It was here recorded that the results provided by this research are not in agreement with the previously interpreted studies on autoxidation of linalool and geraniol, where stable radicals correlating with A1 were confirmed as well as the formation of its detectable products. The most interesting part of the research was probably the confirmation of the presence of dioxolane structures in the autoxidation mixture which were not previously detected in similar studies on analogous fragrance terpenes. Namely, the first step in dioxolane formation is addition of the carbonyl oxygen to the acid-activated epoxide. As far as this experiment is concerned, epoxide can react with three different carbonyl agents, that is, aldehyde, carboxylic acid, and peracid, with acid catalysis being the main prerequisite. The study confirmed that the formation of dioxolane derivatives with aldehyde is most likely to proceed with remark on large excess of aldehyde. Thus, compound **50** was detected just after the decrease in pH caused by accumulation of carboxylic acid in the oxidation mixture. With regard to the general autoxidation mechanism, it would be expected that **51** is derived from autoxidation of **50**. Surprisingly, the detection of **51** was experimentally confirmed before **50** was identified, which indicated that **51** surely originated from another pathway. Hence, an alternative formation mechanism was suggested by this study, including acyl radical and epoxide (Figure 29.25).

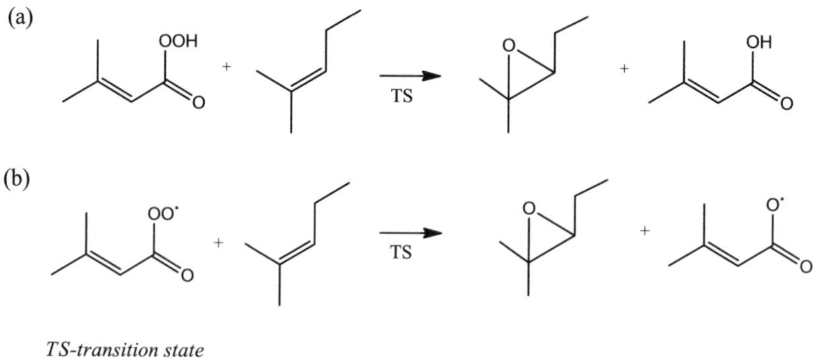

TS-transition state

FIGURE 29.24 Mechanism of formation of secondary products from peracid or peracyl radical. (Adapted and newly drawn from Hagvall, L. et al. 2011. *Chemical Research in Toxicology* 24 (9): 1507–15.)

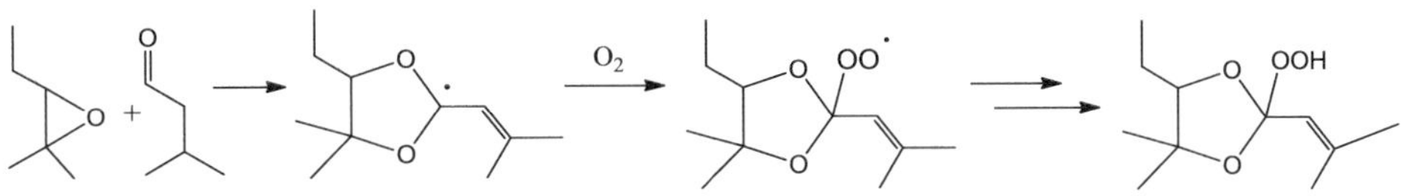

Dioxolan radical intermediate

53

FIGURE 29.25 Dioxolan formation. (Adapted and newly drawn from Hagvall, L. et al. 2011. *Chemical Research in Toxicology* 24 (9): 1507–15.)

Furthermore, it was here emphasized that these two agents form a stable dioxolane radical intermediate in the thermodynamically favored reaction, after which the radical promptly engages into reaction with oxygen, which results in formation of hydroperoxide **53** that corresponds to **51**. In summary, there are two possible mechanisms of dioxolane hydroperoxide: (a) the aforementioned mechanism including the acyl radical that is denoted as the key mechanism and (b) the autoxidation of **50**, after it reaches a certain level of concentration, which was observed as a minor contributor to the overall formation of **51**.

29.4.8 Tea Tree Oil and α-Terpinene

Tea tree oil (TTO) is usually considered a so-called "cure-all" and is very popular as a natural, topical remedy used in the treatment of various cutaneous conditions (infections, psoriasis, etc.), including application to already infected skin (Rubel et al., 1998). This case report stated that the term "tea tree" originates back from the Captain Cook's second trip to Australia when his sailors used tee tree leaves to brew bush tea. It was also reported that the major export industry of TTO is focused on the species *Melaleuca alternifolia* family, which is almost solely spread (in its natural state) around the North Rivers coastal area of New South Wales, Australia. Furthermore, pure oil that is 100% active TTO is becoming more and more available on the market. So too, it was emphasized that out of 100 constituents (mostly monoterpenes, sesquiterpenes, and their corresponding alcohols), terpinen-4-ol and 1,8-cineole (Figure 29.26) are the two most significant compounds of tea tree oil, with the former constituting almost 30% of the TTO volume. Terpinen-4-ol is considered the main active constituent related to the antimicrobial/anti-inflammatory effect of TTO. Aside from these two components, TTO must contain α-terpinene, γ-terpinene, and *p*-cymene (Figure 29.26), according to the ISO standard (ISO 4730:2017).

With regard to the respective structure of certain terpenoids, they may undergo oxidation or hydrolysis as they are volatile and thermolabile species (Yadav et al., 2017). This review stressed that TTO degradation generally takes place in two possible manners: hydrolysis and oxidation, resulting in formation of terpinen-4-ol and *p*-cymene, respectively. Namely, the overall percentage of *p*-cymene serves as an oxidation, that is, degradation, indicator of TTO. The article also presented a study on TTO stability from 2006 wherein the oil sample was subjected to the conditions similar to those

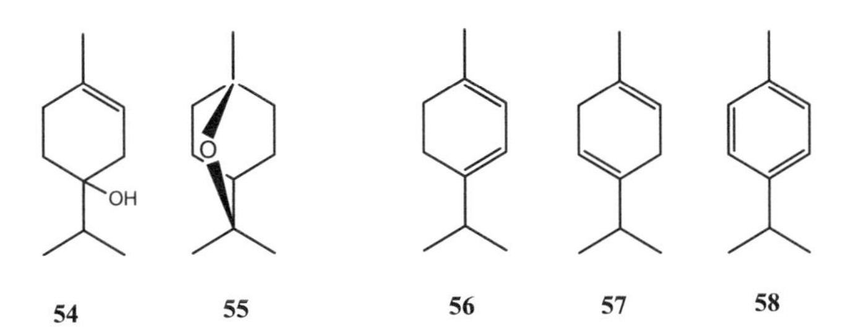

FIGURE 29.26 Major constituents of tea tree oil: terpinene-4-ol (**54**), 1,8-cineol (55), α-terpinene (**56**), γ-terpinene (**57**) and p-cymene (**58**).

during consumers' usage. Levels of *p*-cymene and peroxide value were the main determinants of TTO plausible degradation generated by air, light, or heat. Concurrent increase in peroxide and cymene levels upon oxidation was confirmed. Finally, it was concluded that the official recommendation for formulated TTO in terms of shelf life in Europe amounts to 12 months. Furthermore, maintenance routines were briefly proposed regarding stability of oil in the formulated products, stating that appropriate storage conditions are indispensable. Storing such products in tightly sealed stainless steel containers or amber-colored bottles away from (sun)light, heat, and air was presented as the essential requirement in preserving TTO stability.

α-Terpinene (**56**) (i.e., 1-isopropyl-4-methylcyclohexa-1,3-diene) is often used as a fragrance ingredient that occurs naturally in various essential oils (besides TTO) and is mainly associated with the antioxidant activity of tea tree oil (Rudbäck et al., 2012). As confirmed, structurally it represents a cyclic monoterpene and a conjugated diene for which a certain level of similarity to other monoterpenes (e.g., limonene) can be depicted. With regard to this chemical resemblance, autoxidation at room temperature is expected, which was the main focus of this research performed in order to investigate whether α-terpinene can act as a prehapten. Though chemical analysis proved that α-terpinene is susceptible to a rapid autoxidative degradation, the oxidation pattern is, however, different compared to that of other analogous monoterpenes (e.g., limonene). The study also conducted analysis of four different samples of TTO with regard to its age and concentration. *p*-Cymene, 1,2-epoxide, diol, and *(E)*-3-isopropyl-6-oxohept-2-enal were detected in all of the TTO preparations; thus, plausible allergic reactions to TTO could be attributed to α-terpinene. So, the experimental part of the research confirmed α-terpinene to be a true antioxidant that indeed autoxidizes readily and thus inhibits oxidative degradation of other compounds. However, the autoxidation process of α-terpinene generates allergens so the appropriateness of its topical use in both TTO and dermal preparations becomes questionable. Namely, an "ideal" antioxidant should not just prevent the chain propagation step of oxidative degradation but should also readily decay to a nontoxic agent. The air-exposure procedure used for this experimental research was mainly in accordance with an artificially induced autoxidation process with previously described terpenes. Hence, it was performed in an Erlenmeyer flask, covered with aluminum foil at room temperature, under a daylight lamp. The strict conditions regarding stirring and storage prior to quantification analysis were defined. α-Terpinene was subjected to influence of air for 0 to 17 weeks, after which LC/UV and LC/MS/MS were used to determine its degradation and production of oxidation products. The study performed quantification of α-terpinene and its oxidation products after it was air-exposed at room temperature, and a rapid decrease rate of α-terpinene concentration was easily discernible. It took only 10 days to cut α-terpinene concentration in half and after 66 days of the experiment, no α-terpinene could be detected in the oxidation mixture. Thus, when compared to limonene (structural analogue of α-terpinene), its oxidation rate turned out to be rather quick.

Several compounds were identified in the autoxidation mixture (Figure 29.27), of which allylic epoxides and *p*-cymene were denoted as the major oxidation products. Hydrogen peroxide was also confirmed as one of the main compounds produced after air exposure of α-terpinene. Fraction X

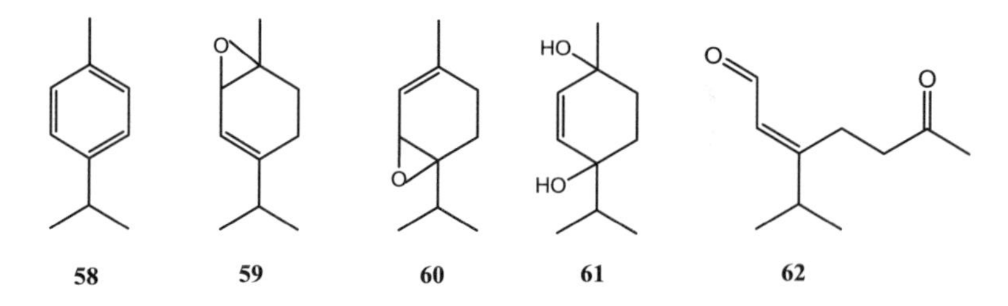

FIGURE 29.27 Autoxidation mixture of α-terpinene (**56**): **58**, 1,2-epoxide (**59**), 3,4-epoxyde (**60**), diol (**61**) and (*E*)-3-sopropyl-6-p-cymene (**62**). (Adapted and newly drawn from Rudbäck, J. et al. 2012. *Chemical Research in Toxicology* 25 (3): 713–21.)

eluted between the diol (**61**) and the 1,2-epoxide (**59**), with two major peaks detected in the LC/MS/MS and LC/UV analyses. The key component and the most stable one was identified as *(E)*-3-isopropyl-6-oxohept-2-enal (**62**).

Notably, no hydroperoxides were detected, so this phenomenon could be considered as the main difference compared to the autoxidation of other common fragrances (e.g., limonene, linalool, and linalyl acetate). As previously shown, their primary oxidation products are mainly allergenic

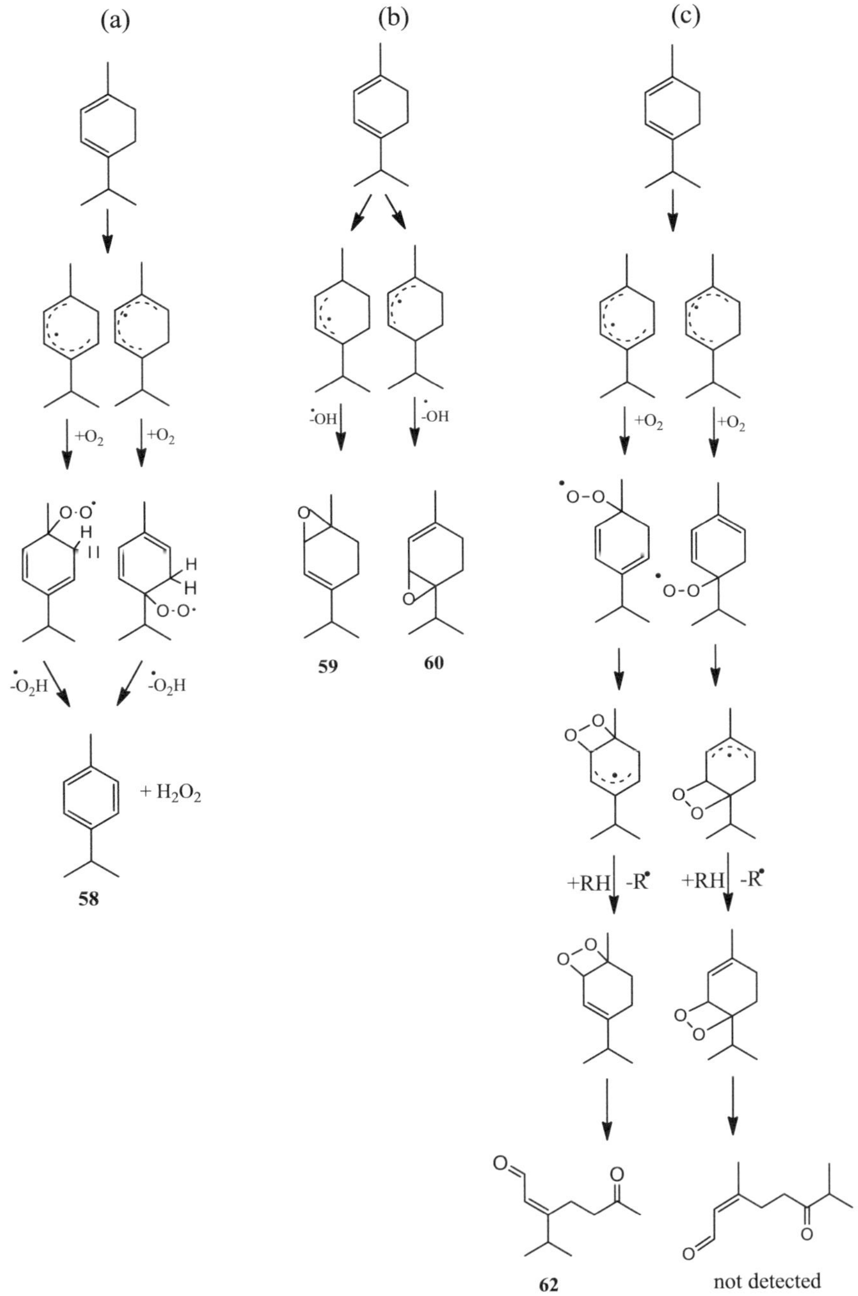

(a) (b) (c)

$+O_2$ $+O_2$ $\dot{}OH$ $\dot{}OH$ $+O_2$ $+O_2$

$\dot{}O_2H$ $\dot{}O_2H$ $+ H_2O_2$

58 59 60

+RH -R• +RH -R•

62 not detected

FIGURE 29.28 α-Terpinene (**56**) products upon air oxidation: (a) **58** + H_2O_2, (b) allylic epoxides, and (c) **62**. (Adapted and newly drawn from Rudbäck, J. et al. 2012. *Chemical Research in Toxicology* 25 (3): 713–21.)

hydroperoxides, which then transform into secondary oxidation products, such as alcohols, epoxides, carbonyl components, with diverse sensitization capacity. However, regardless of their structural and allergenic differences, a common mechanistic starting point can be associated with their oxidation pattern that is the homolytic cleavage of their parent hydroperoxide group. On the other hand, a previous review on geraniol autoxidation pathway in this chapter (Section 29.3.2) confirmed hydrogen peroxide to be the major autoxidation product (along with geranial, neral, and epoxygeraniol). Hydroperoxides were yielded to a significantly lesser extent. Interestingly, the epoxygeraniol could not be related to any of the identified hydroperoxide forms. Therefore, its formation was explained by a computational study wherein addition of a hydroperoxyl radical to the double bond of a new geraniol molecule induced formation of the epoxygeraniol and a hydroxyl radical. Mechanistic considerations were proposed by this study regarding formation of the major oxidation products *p*-cymene and epoxides, and also, α,β-unsaturated aldehyde (**62**) was observed as well (Figure 29.28).

Hence, it was reported that *p*-cymene and hydrogen peroxide are created after α-terpinene was air-exposed (Figure 29.28a). The study then suggested that the chain-carrying reaction in the autoxidation of α-terpinene was probably the same as that investigated for γ-terpinene (Foti and Ingold, 2003), given that no hydroperoxides were detected in the study. Furthermore, it was observed that *p*-cymene and a hydroperoxyl radical are generated from peroxidation of γ-terpinene, wherein the radical can be altered to hydrogen peroxide during the propagation step of the radical chain reaction (Rudbäck et al., 2012). Analogously to the aforementioned and regarding epoxygeraniol, addition of a hydroperoxyl radical to one of the double bonds in α-terpinene might result in ring closure, that is, formation of allylic epoxides (Figure 29.28b). Finally, an oxidative cleavage of the α-terpinene double bond was proposed as a possible explanation for the presence of the α,β-unsaturated aldehyde. With regard to Baldwin's rule, an intramolecular cyclization of the peroxyradical to its double bond was proposed creating 1,2-dioxetane (Figure 29.28c). Due to their extreme chemical instability, they most often cleave to their corresponding carbonyl components.

Extensive data accessible in the scientific literature, such as the ones presented in this treatise, speak with no doubt in favor of the oxidation of fragrance terpenes being the main factor for generating the skin sensitization phenomenon among the consumers that are often exposed to scented products.

29.5 CONCLUSION

EOs are certainly one of the leading examples of the ever-increasing trend in using "natural products" in everyday life instead of those of artificial, that is, industrial, origin. However, this worldwide trend brought a lot of issues in terms of stability during storage and handling with EOs, that is, their constituents, primarily fragrance terpenes. It is of great significance to monitor fragrance terpenes of both natural and synthetic origin not only at the time of production but perhaps even more at the postproduction period after the product is released in the market and used in everyday life. The main focus should be on oxidative degradation process of a parent molecule and composition of an oxidation mixture derived after air exposure, primarily detecting/excluding hydroperoxides. The autoxidation process caused by ambient air, that is, oxygen, may practically be associated with all types of organic chemicals. Fragrance terpenes are especially susceptible to oxidative degradation due to their isoprenoid skeleton that contains numerous allylic positions. Moreover, these allylic sites are easily subjected to radical abstraction of a hydrogen atom after stabilized allylic radicals are yielded. The fragrance terpenes presented in this work are generally denoted as "the (most) common," wherein the review of numerous experimental and clinical studies speaks in favor of autoxidation having the greatest impact on the degradation process and the increase in sensitizing potential of these compounds. Still, no general conclusions can be made on the effect of autoxidation in terms of increasing allergic potential of each fragrance chemical due to (a) different susceptibility to oxidation, (b) various degradation rates of parent molecules and accumulation rates of oxidation products, (c) chemical diversity of yielded primary and secondary oxidation products with (d) varying EC3 values, which classify them from weak to strong sensitizers.

Furthermore, from a public-health perspective, increased sensitizing potential after air exposure of fragrance terpenes gives rise to allergic reactions (ACD), placing them as the second most common allergens after metal salts. Hence, the list of substances for baseline series in clinical patch tests, as well as the chemicals in cosmetic products whose declaration as contact allergens is defined as mandatory by legal authorities in EU, should constantly be updated and revised. Finally, when exploring the nature of air-induced alteration mechanisms, it should be considered that compounded scented products usually contain hundreds of fragrance substances, so more research with consumers products that are already on the market ought to be conducted. This way, a possible correlation between an autoxidation of pure substances and consumer products could/could not be established. Hence, combining the mechanistic chemistry with experimental research and clinical data should be accepted as the basic principle in further investigation of the nature of fragrance terpene autoxidation patterns.

REFERENCES

Amorati, R., M. C. Foti, and L. Valgimigli. 2013. Antioxidant Activity of Essential Oils. *Journal of Agricultural and Food Chemistry* 61 (46): 10835–47.

Aprotosoaie, A. C., M. Hăncianu, I. I. Costache and A. Miron. 2014. Linalool: A Review on a Key Odorant Molecule with Valuable Biological Properties. *Flavour and Fragrance Journal* 29 (4): 193–219.

Bäcktorp, C., L. Hagvall, A. Börje, A.-Th. Karlberg, P.-O. Norrby, and G. Nyman. 2008. Mechanism of Air Oxidation of the Fragrance Terpene Geraniol. *Journal of Chemical Theory and Computation* 4 (1): 101–6.

Bäcktorp, C., J. R. T. Johnson Wass, I. Panas, M. Sköld, A. Börje, and G. Nyman. 2006. Theoretical Investigation of Linalool Oxidation. *The Journal of Physical Chemistry A* 110 (44): 12204–12.

Baser, K. H. Can, and G. Buchbauer. 2016. *Handbook of Essential Oils: Science, Technology, and Applications*. CRC Press, Taylor & Francis, Boca Raton, London, New York.

Bråred Christensson, J., K. E. Andersen, M. Bruze, J. D. Johansen, B. Garcia-Bravo, A. Giménez-Arnau, C.-L. Goh, R. Nixon, and I. R. White. 2013. An International Multicentre Study on the Allergenic Activity of Air-Oxidized R-Limonene. *Contact Dermatitis* 68 (4): 214–23.

Cavanagh, H. M. A., and J. M. Wilkinson. 2002. Biological Activities of Lavender Essential Oil. *Phytotherapy Research* 16 (4): 301–8.

Foti, M. C., and K. U. Ingold. 2003. Mechanism of Inhibition of Lipid Peroxidation by γ-Terpinene, an Unusual and Potentially Useful Hydrocarbon Antioxidant. *Journal of Agricultural and Food Chemistry* 51 (9): 2758–65.

Hagvall, L., C. Bäcktorp, P.-O. Norrby, A.-Th. Karlberg, and A. Börje. 2011. Experimental and Theoretical Investigations of the Autoxidation of Geranial: A Dioxolane Hydroperoxide Identified as a Skin Sensitizer. *Chemical Research in Toxicology* 24 (9): 1507–15.

Hagvall, L., C. Bäcktorp, S. Svensson, G. Nyman, A. Börje, and A.-Th. Karlberg. 2007. Fragrance Compound Geraniol Forms Contact Allergens on Air Exposure. Identification and Quantification of Oxidation Products and Effect on Skin Sensitization. *Chemical Research in Toxicology* 20 (5): 807–14.

Hagvall, L., V. Berglund, and J. Bråred Christensson. 2015. Air-Oxidized Linalyl Acetate – An Emerging Fragrance Allergen? *Contact Dermatitis* 72 (4): 216–23.

Hagvall, L., M. Sköld, J. Bråred-Christensson, A. Börje, and A.-Th. Karlberg. 2008. Lavender Oil Lacks Natural Protection against Autoxidation, Forming Strong Contact Allergens on Air Exposure. *Contact Dermatitis* 59 (3): 143–50.

ISO 4730:2017 – Essential Oil of Melaleuca, Terpinen-4-Ol, Type (Tea Tree Oil). Accessed September 11, 2017, and ISO 9235 Essential Oils, Geneva 1997.

Karlberg, A.-Th., and A. Dooms-Goossens. 1997. Contact Allergy to Oxidized D-Limonene among Dermatitis Patients. *Contact Dermatitis* 36 (4): 201–6.

Karlberg, A.-Th., K. Magnusson, and U. Nilsson. 1992. Air Oxidation of D-Limonene (the Citrus Solvent) Creates Potent Allergens. *Contact Dermatitis* 26 (5): 332–40.

Karlberg, A. T., K. Magnusson, and U. Nilsson. 1994a. Influence of an Anti-Oxidant on the Formation of Allergenic Compounds during Auto-Oxidation of D-Limonene. *The Annals of Occupational Hygiene* 38 (2): 199–207.

Karlberg, A.-T., L. P. Shao, U. Nilsson, E. Gäfvert, and J. L. G. Nilsson. 1994b. Hydroperoxides in Oxidized D-Limonene Identified as Potent Contact Allergens. *Archives of Dermatological Research* 286 (2): 97–103.

Kern, S., T. Granier, H. Dkhil, T. Haupt, G. Ellis, and A. Natsch. 2014. Stability of Limonene and Monitoring of a Hydroperoxide in Fragranced Products: Detection of a Limonene Hydroperoxide. *Flavour and Fragrance Journal* 29 (5): 277–86.

Letizia, C. S., J. Cocchiara, J. Lalko, and A. M. Api. 2003. Fragrance Material Review on Linalyl Acetate. *Food and Chemical Toxicology*, RIFM Toxicologic and Dermatologic Assessments of Linalool and Related Esters when used as Fragrance Ingredients, 41 (7): 965–76.

Matura, M., A. Goossens, O. Bordalo, B. Garcia-Bravo, K. Magnusson, K. Wrangsjö, and A.-Th. Karlberg. 2002. Oxidized Citrus Oil (R-Limonene): A Frequent Skin Sensitizer in Europe. *Journal of the American Academy of Dermatology* 47 (5): 709–14.

Matura, M., M. Sköld, A. Börje, K. E. Andersen, M. Bruze, P. Frosch, A. Goossens et al. 2005. Selected Oxidized Fragrance Terpenes Are Common Contact Allergens. *Contact Dermatitis* 52 (6): 320–28.

Matura, M., M. Sköld, A. Börje, K. E. Andersen, M. Bruze, P. Frosch, A. Goossens et al. 2006. Not Only Oxidized R-(+)- but Also S-(−)-Limonene Is a Common Cause of Contact Allergy in Dermatitis Patients in Europe. *Contact Dermatitis* 55 (5): 274–79.

Neuenschwander, U., F. Guignard, and I. Hermans. 2010. Mechanism of the Aerobic Oxidation of α-Pinene. *ChemSusChem* 3 (1): 75–84.

Nguyen, H., E. M. Campi, W. R. Jackson, and A. F. Patti. 2009. Effect of Oxidative Deterioration on Flavour and Aroma Components of Lemon Oil. *Food Chemistry* 112 (2): 388–93.

Nilsson, U., M. Bergh, L. P. Shao, and A.-Th. Karlberg. 1996. Analysis of Contact Allergenic Compounds in Oxidizedd-Limonene. *Chromatographia* 42 (3–4): 199–205.

Pratt, D. A., and N. A. Porter. 2003. Role of Hyperconjugation in Determining Carbon–Oxygen Bond Dissociation Enthalpies in Alkylperoxyl Radicals. *Organic Letters* 5 (4): 387–90.

Rastogi, S. C., S. Heydorn, J. D. Johansen, and D. A. Basketter. 2001. Fragrance Chemicals in Domestic and Occupational Products. *Contact Dermatitis* 45 (4): 221–25.

Rastogi, S. C., J. D. Johansen, P. Frosch, T. Menne, M. Bruze, J. P. Lepoittevin, B. Dreier, K. E. Andersen, and I. R. White. 1998a. Deodorants on the European Market: Quantitative Chemical Analysis of 21 Fragrances. *Contact Dermatitis* 38 (1): 29–35.

Rastogi, S. C., J.-P. Lepoittevin, J. D. Johansen, P. J. Frosch, T. Menne, M. Bruze, B. Dreier, K. E. Andersen, and I. R. White. 1998b. Fragrances and Other Materials in Deodorants: Search for Potentially Sensitizing Molecules Using Combined GC-MS and Structure Activity Relationship (SAR) Analysis. *Contact Dermatitis* 39 (6): 293–303.

Rubel, D. M., S. Freeman, and I. A. Southwell. 1998. Tea Tree Oil Allergy: What Is the Offending Agent? Report of Three Cases of Tea Tree Oil Allergy and Review of the Literature. *Australasian Journal of Dermatology* 39 (4): 244–47.

Rudbäck, J., M. Andresen Bergström, A. Börje, U. Nilsson, and A.-Th. Karlberg. 2012. α-Terpinene, an Antioxidant in Tea Tree Oil, Autoxidizes Rapidly to Skin Allergens on Air Exposure. *Chemical Research in Toxicology* 25 (3): 713–21.

Schnuch, A., H. Lessmann, J. Geier, P. J. Frosch, and W. Uter. 2004. Contact Allergy to Fragrances: Frequencies of Sensitization from 1996 to 2002. Results of the IVDK*. *Contact Dermatitis* 50 (2): 65–76.

Sköld, M., A. Börje, E. Harambasic, and A.-Th. Karlberg. 2004. Contact Allergens Formed on Air Exposure of Linalool. Identification and Quantification of Primary and Secondary Oxidation Products and the Effect on Skin Sensitization. *Chemical Research in Toxicology* 17 (12): 1697–705.

Sköld, M., A. Börje, M. Matura, and A.-Th. Karlberg. 2002. Studies on the Autoxidation and Sensitizing Capacity of the Fragrance Chemical Linalool, Identifying a Linalool Hydroperoxide. *Contact Dermatitis* 46 (5): 267–72.

Sköld, M., L. Hagvall, and A.-Th. Karlberg. 2008. Autoxidation of Linalyl Acetate, the Main Component of Lavender Oil, Creates Potent Contact Allergens. *Contact Dermatitis* 58 (1): 9–14.

Sköld, M., A.-Th. Karlberg, M. Matura, and A. Börje. 2006. The Fragrance Chemical β-Caryophyllene—air Oxidation and Skin Sensitization. *Food and Chemical Toxicology* 44 (4): 538–45.

Stobiecka, A. 2015. Comparative Study on the Free Radical Scavenging Mechanism Exerted by Geraniol and Geranylacetone Using the Combined Experimental and Theoretical Approach. *Flavour and Fragrance Journal* 30 (5): 399–409.

Turek, C., and F. C. Stintzing. 2011. Evaluation of Selected Quality Parameters to Monitor Essential Oil Alteration during Storage. *Journal of Food Science* 76 (9): C1365–75.

Turek, C., and F. C. Stintzing. 2013. Stability of Essential Oils: A Review. *Comprehensive Reviews in Food Science and Food Safety* 12 (1): 40–53.

Yadav, E., S. Kumar, S. Mahant, S. Khatkar, and R. Rao. 2017. Tea Tree Oil: A Promising Essential Oil. *Journal of Essential Oil Research* 29 (3): 201–13.

30 The Essential Oil Trade

Hugo Bovill

The essential oil industry is an ancient trade complex and fragmented in nature and one that is not easily comprehended by outsiders. Much has been based on perfumers' and flavourists' personal preferences, as odour and flavour are characteristics that are unquantifiable.

Demand is increasing annually for many essential oils, but finding a route to market is not always easy. New producers are sources for many oils, especially those which can be certified organic. This general increase has arisen as a result of two main factors: growth in demand from India and China and the rise of a billion dollar aromatherapy industry.

The usage of essential oils is primarily in flavours and fragrances, but increasingly, aromatherapy. There are in excess of 100 countries who produce essential oils.

Originating countries for specific essential oils can change, as evidenced by these historic examples: peppermint oil Mitcham (*Mentha piperita*) production moved from England to the United States in the 1920s; cornmint (*Mentha arvensis*) (the essential oil source for natural menthol) production originated in Japan, then transferred to China/Brazil/Paraguay and then solely to China but is now produced almost entirely in India. Additionally, there is now adequate production capacity for synthetic menthol to become the dominant quality/source which will once again change the dynamics of cornmint production.

The majority of the essential oil producing countries have been active in the cultivation and processing of these aromatic materials for many decades.

The processes and techniques were, in many cases, introduced during their historical colonization, and even today, some routes to market for producers are through their former colonial powers. For example, clove and ylang oils from Madagascar have traditionally been sold through French intermediaries as a result of historical contact and language. These traditional supply chains are now fragmenting as a result of better communications in producing areas, permitting the next generation of suppliers to look elsewhere for markets.

In most producing countries, there are frequently long supply chains originating from small producers, who may produce just a few kilos, who will then sell their production to a dealer. This collector will visit many different producers in order to purchase various lots that are then bulked together to form a large enough lot of uniform quantity, which would be of interest to an international purchaser. This would then be promoted and, in turn, exported by an enterprise probably located in the commercial capital or main seaport of that country. The exporter will have global commercial relationships and knowledge of international shipping regulations and correct packaging for hazardous goods (UN-certified containers), which is an international legal requirement for many essential oils. The exporter may have a strong financial position or strong banking relationships, enabling them to offer extended terms of payment which are often demanded by many flavour and fragrance houses. The exporter will be able to quote in USD or euros, which is often not possible for small local producers who can only use local currencies.

Producers of essential oils can vary from the very large, such as a sweet orange juice concentrate factory where the oil is a co-product produced in volumes in excess of 100s of tonnes to a small geranium distiller where production can be as low as 100 kilo per annum (Figures 30.1 and 30.2).

The choice of which essential oil to produce is not always so easy.

It is often a misconception that the highest-priced oils such as rose give the best return in terms of profit, but this is not the case. It is important for a potential producer to consider their climate, type of land, availability of plants, labour, the possibility of out-growers and biomass/essential oil

FIGURE 30.1 South American orange juice factory. (Photograph by kind permission of Sucocitrico Cutrale Ltd.)

FIGURE 30.2 Copper still in East Africa.

yield return when compared to competing crops. Then, a review of historical supply and demand with pricing, along with consideration of possible competing countries and their exchange rates, potential usage, and possible legislation such as REACH in EU or Prop 65 in California that might affect long-term consumption of such an oil, and finally the route to buyers.

Prior to planting or committing with out-growers, a producer should seek advice from potential buyers as to their purchase intentions, for the medium and ideally long term. Contracts over 12 months duration are unusual but users or dealers can often provide their best estimates of the volumes they could take. Shortages of many oils are frequent owing to climate changes and increasing demand from wealthier countries. Market prices can move rapidly to reflect increased demand, but new origins can of course bring prices to traditional levels or even lower. There is no internet reference site that is reliable in pricing for essential oils in the somewhat opaque industry because of the long supply chains and varying qualities.

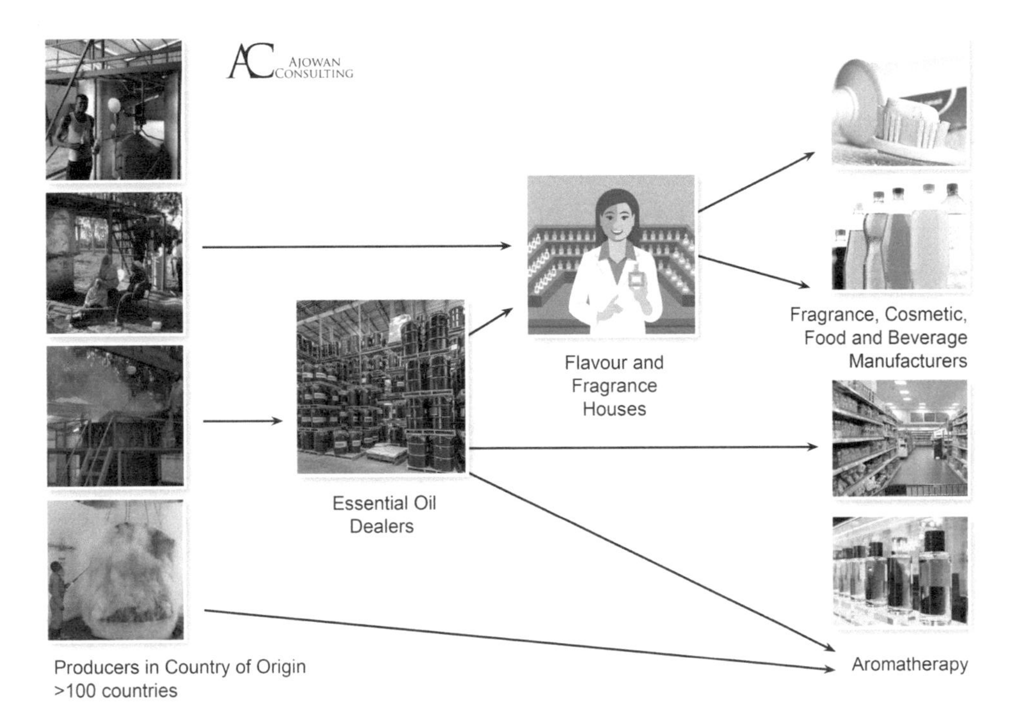

FIGURE 30.3 **(See color insert.)** The essential oil trade flows.

Reliable data and statistics are limited despite the world market being in excess of some billion USD, which is still small in global commodity terms. Much of what is published in costly reports is often misleading as customs tariff numbers are not always specific to a particular botanical.

The primary destinations and markets for essential oils are the United States, Japan, France, Germany, Spain, the United Kingdom, and increasingly India and China.

Within the essential oil market, there are generally four main types of buyers: aromatherapy companies, some of whom may require organically certified oils, the flavour and fragrance companies, and international merchants/dealers. Some major consumer goods houses also purchase some oils but it is rare and extremely difficult to enter (Figure 30.3).

Once an essential oil finally reaches the importer it may be blended and/or resold to other resellers. Essential oils often pass through many hands before reaching end users, which in most cases are the flavour and fragrance houses. Aromatherapy essential oil companies often prefer to purchase direct from source however small their needs. This can lead them to have interruptions in supply when there is a crop failure or in reverse an oversupply scenario of great risk to a producer when the usage of an oil is discontinued and no longer required by the consuming aromatherapy company. Traditional essential oil dealers will have a variety of clients for a product which can ameliorate the risk of a sudden cessation of demand.

All business-to-business essential oil trade is commenced by sending to prospective purchasers "type" samples that are examples of the quality and should be typical of the production that can be supplied. These should not be sent unsolicited as it is an expensive process as often they will need to be sent by international courier, and buyer's laboratories do not analyse samples without good reason.

The samples will need to be accompanied by a Safety Data Sheet and often extensive paperwork as requested by the buyer to qualify a producer as an approved vendor. This paperwork may be concerned with more than just the quality of the essential oil and can cover child labour, site security, Kosher and alal certification, and even the financial strength of the company. The samples will be analysed by gas chromatograph–mass spectrometry and other modern scientific techniques to ensure the oil is of a standard similar to their current purchases and to detect signs of adulteration, deliberate or by contamination with rogue plants. Deliberate adulteration is becoming less common as modern analytical techniques have become cheaper and so are far more practiced.

Samples of oils will additionally be screened by an organoleptic (odour and/or flavour) panel to ensure the oil has the correct odour profile with no "off notes." If present, these could be as a result of many factors, including different botanical plants, terroir, distillation process, storage conditions, and age of the oil.

Oils are frequently tested for pesticides and industrial contaminants such as phthalates at levels as low as PPB (parts per billion). These phthalates can occur in oils; for example, where a harvester or production operative has stored the raw botanicals in contact with plastic for even a few minutes. Pesticide levels should be kept to a minimum and fortunately are not so volatile so they are less likely to appear in volatile oils such as essential oils.

Numbered and traceable lot samples in advance of purchase of a specific quantity will be required by the purchaser to enable them to analyse individual batches/lots of the oil both chemically and organoleptically before a firm purchase order is raised.

It is important to users that the qualities remain fairly constant as differing qualities are not acceptable. They are either the same or similar; better is not considered an acceptable change! The continuity of a standard quality is one of the many keys to building a close and healthy relationship between suppliers and the purchasers.

Should a sample represent an exact quantity for a possible order, the lot should have been bulked in a tank (*made uniform*) prior to sampling. The sample bottle should be made from glass and ideally brown or blue in colour. The oils should never be in contact with plastic during production, storage, or sampling to avoid contamination by phthalates which, if present, can severely reduce the value of an oil to lose much of its value. The sample, as well as drums for storage, should be full or be topped with nitrogen to ensure that there is no oxygen present, in order to prevent oxidation. A few oils such as those containing eugenol and cinnamic aldehyde may be stored in specific types of plastic containers with a special liner. If these are stored in lacquer-lined drums, they can severely discolour.

Some producers try to improve their processes by adapting their equipment and modernizing their processes. In Paraguay, smallholder distillers of petitgrain oil (produced from the leaves of the bitter orange tree) replaced traditional wooden stills with stainless steel stills on the advice of an NGO supported by the British government. This led to a subtle change in quality, but one which was perceived and disliked by perfumers. The quality issues led to dissatisfied users, and in fact, the Paraguayan distillers reverted back to their traditional wooden stills (Figure 30.4).

In China, there are two differing types of oil from cassia leaves (cinnamon leaf oil) producers. The "old" type used primarily for flavouring and a "new" type used for the production of natural

FIGURE 30.4 Petitgrain still.

cinnamic aldehyde. The "new" type was developed in the early 2000s by adapting traditional stills to recover more oil from waste distillation waters. This enabled producers to obtain greater yields of cassia essential oil, rich in cinnamic aldehyde but it was not considered comparable in flavour to the traditional oil made since the early 1900s. This new oil became the dominant quality while the long-term user for flavouring purposes became concerned about the supply of the traditional cassia oil with its unique character. This anxiety was communicated throughout the supply chain which encouraged two types of oil to be produced, old and new.

Conditions of trade are normally FOB or CFR basis, and it is normal practice that the dollar or euro price should be stated before samples are sent. Payment terms are of increasing importance, with the first trade being a matter of trust on both sides. For a new producer to supply a dealer in a weak market, extended terms of payment can be a useful tool to obtain an order rather than a reduction in price. Terms of payment are usually 30 days from receipt of goods, but increasingly, some larger flavour and fragrance houses are demanding and receiving longer terms. Dealers are not so inclined and understand that as they often need new suppliers and so will agree to more favourable payment terms. Letter of credit or payment in advance is rare as the industry was defrauded of many millions of dollars in the 1970s by a criminal Indonesian exporter.

Making a first sale is not an easy task in times of weak markets and oversupply. Sometimes an agent can be used who, in return for a commission of 5%, on an FOB basis, can assist in developing some long-term relationships on an ongoing basis. Two to three years can be the time required to enter a major flavour and fragrance house if fortunate. It is easier for a producer to sell their oil initially if the inventory is lying in the one of the main markets, i.e., Europe or US, rather than in the country of origin.

Once a regular supplier has become established, buyers are loyal as long as qualities remain stable and prices remain in line with pre-agreed terms. Different types of models, such as market pricing (i.e., variable) or medium-term fixed prices can be beneficial for both parties. Fixed pricing requires experience to ensure the model is retained in rising markets when, understandably, producers can be tempted by higher prices and short-term gains. Trust is key and personal knowledge of both buyer and seller by an intermediary can be beneficial in such an arrangement.

There are less than ten major essential oils dealers/merchants remaining in the world, some of whom are also involved in the manufacture of flavours or fragrances. Several of these companies have been established for many years and have good trading histories. Some information about them can be gained from their websites, but without a face-to-face meeting, it is not easy to establish their credentials. Many dealers have product lists which seem almost identical to their competitors so it is important to establish the strengths and weaknesses of their product line which is often best achieved through networking. Some merchants are open to new producers, many are not. Exhibitions and conferences can be a good environment to initiate contact and network, or it may be more cost effective to use an experienced agent or consultant.

Market information, as shared by both producer and buyer is one of the keys to developing long-term mutually beneficial relationships. To better enable the producer to understand market pricing, he should appreciate that when receiving more enquiries than normal for his oil, it is likely that demand is increasing or more likely that supply from elsewhere has been interrupted or reduced. Likewise, if a producer is too enthusiastic in offering their oil to buyers, it can lead clients to consider the market is weak and so they will anticipate a reduction in price. These are examples that both parties may privately consider to establish the direction of the market price, so care should be taken in the frequency of communication.

Producers selling to more than one customer should not "sell out" (*stock out*) unless they have pre-advised their client base. Far too often, new producers have successfully introduced their new oil to buyers who when they return to place a repeat order, learn that the production has been sold elsewhere. Communication from a developing producer is essential even if they have no inventory and do not anticipate a repeat sale until a new crop is available, even if 11 months away, to ensure that possible supply matches potential demand. Once a buyer has undertaken a costly exercise of introducing an oil, he will not wish to learn that his new supplier has sold his production elsewhere without due notice. Therefore, in any business plan, producers should comprehend that the carrying of some inventory for longer than 12 months is important to ensure reliability of supply to existing clients unless it has been made very clear, and even then, it is unwise. Loyalty can be lost rapidly. Most buyers will have two suppliers of an oil but they will expect to be kept informed of their supplier's general inventory position, especially if working with a new vendor.

Essential oils are traditionally used in conjunction with aroma chemicals in fragrances and flavours which can consist of many ingredients. The aromatherapy industry normally uses oils in their pure form or diluted in vegetable oils. A buyer within a flavour or fragrance house can be ultimately responsible for the supply of some thousands of ingredients used as the palate for the flavourists and perfumers, so they would be unlikely to have extensive knowledge of their full supply chain. Some are used in grams per year and others in 100s of tonnes. Prices can range from a few USD per kilo to thousands of dollars. Demand for an oil for aromatherapy use can grow exponentially, whereas in fragrances, this is less likely and flavours less so. Increased demands and requests for oils can be perceived as growth, but often this is due to shortages elsewhere, perhaps caused by a crop failure or perhaps a competitor's batch failing a customer's quality control.

There have been some very unfortunate occurrences of farmers naively following their neighbours into growing essential oil plants such as immortelle (*Helichrysum italicum*). A particular large user may have contracted with one co-operative to ensure their supply for a proprietary product. Other farmers, having seen the change in their local fields, then believe there is potentially a larger global market. They then invest in equipment and plant the same species and then start looking for such a market, which may not in fact exist outside the large user. Once the farmers have the oil whose shelf life is probably less than two years, they then can flood the market with offers. When long-established experienced Essential oil companies read headlines in the mainstream press such as the *NY Times* stating, "Can a Wild Daisy Rejuvenate Croatia's Farming Economy?", they feel the signs are ominous, that is, a market crash is foreseen.

In 1931, V. A. Beckley OBE, MC, Senior Agricultural Chemist, of Kenya, said

> The production of essential oils is perhaps more chancy than most farming propositions; it most certainly requires more attention and supervision than most.

This is still valid today but to a far lesser extent as much of the risk can be minimised with good planning and advice, and certainly the production of many essential oils can be profitable on a medium- to long-term basis.

31 Industrial Uses of Essential Oils

W. S. Brud

CONTENTS

The period when essential oils first were used on an industrial scale is difficult to identify. The 19th century is generally regarded as the commencement of the modern phase of industrial application of essential oils. However, the large-scale usage of essential oils dates back to ancient Egypt. In 1480 B.C., Queen Hatshepsut of Egypt sent an expedition to the country of Punt (now Somalia) to obtain fragrant plants, oils, and resins as ingredients for perfumes, medicaments, flavors, and for the mummification of bodies. Precious fragrances have been found in many Egyptian archeological excavations, as symbols of wealth and social position.

If significant international trade of essential oil-based products is the criterion for industrial use, "Queen of Hungary Water" was the first alcoholic perfume in history. This fragrance, based on rosemary essential oil distillate, was created in the mid-19th century for the Polish-born Queen Elisabeth of Hungary and, following a special presentation to King Charles V The Wise of France in 1350, it became popular in all medieval European courts. The beginning of the 18th century saw the introduction of "Eau de Cologne," based on bergamot and other citrus oils, which remains widely used to this day. This fresh citrus fragrance was the creation of Jean Maria Farina, a descendant of Italian perfumers who came to France with Catherine de Medici and settled in Grasse in XVI century. According to city of Cologne archives, in 1749, Jean Maria Farina and Karl Hieronymus Farina established a factory (Fabriek) of this water, which sounds very "industrial." The "Kölnisch Wasser" became the first unisex fragrance rather than one simply for men, known and used all over Europe, and it has been repeated subsequently in innumerable counter types as fragrance for men.

The history of production of essential oils dates back to approximately 3500 B.C., when the oldest known water-distillation equipment for essential oils was employed and may be seen today in the Texila museum in Pakistan. Ancient India, China, and Egypt were the locations where essential oils were produced and widely used as medicaments, flavors, and fragrances. Perfumes came to Europe most probably from the East at the time of the crusades, and perfumery was accorded a professional status by the approval of a French guild of perfumers in Grasse by King Philippe August in 1190. For centuries, Grasse remained the center of world perfumery and also was the home of the first ever officially registered essential oils producing company—Antoine Chiris—in 1768 (It is worth noting that not much later, in 1798, the first American essential oil company—Dodge and Olcott Inc.—was established in New York.). In November 2018 Perfumery, and its birthplace in Grasse, was awarded World Heritage Status by UNESCO, joining the organization's Intangible Cultural Heritage of Humanity list.

About 150 years earlier, in 1620, an Englishman named Yardley obtained a concession from King Charles I to manufacture soap for the London area. Details of this event are sparse, other than the high fee paid by Yardley for this privilege. Importantly, however, Yardley's soap was perfumed with English lavender, which remains the Yardley trademark today, and it was probably the first case of usage of an essential oil as a fragrance in large-scale soap production.

The use of essential oils as food ingredients has a history dating back to ancient times. There are many examples of the employment of citrus and other squeezed (manually or mechanically expressed) oils for sweets and desserts in ancient Egypt, Greece, and the Roman Empire. Numerous references exist to flavored ice creams in the courts of the Roman Emperor Nero and of China. The reintroduction of recipes in Europe is attributed to Marco Polo on his return from travelling to China. In other stories, Catherine de Medici introduced ice creams in France while Charles I of England served the first dessert in the form of frozen cream. Ice was used for freezing drinks and food in many civilizations and the Eastern practice of employing spices and spice essential oils both as flavoring ingredients and as food conservation agents was adopted centuries ago in Europe.

Whatever may be regarded as the date of their industrial production, essential oils, together with a range of related products—pomades, tinctures, resins, absolutes, extracts, distillates, concretes, etc.—were the only ingredients of flavor and fragrance products until the late 19th century. At this stage, the growth in consumption of essential oils as odoriferous and flavoring materials stimulated the emergence of a great number of manufacturers in France, the United Kingdom, Germany, Switzerland, and the US (see Table 31.1).

The rapid development of the fragrance and flavor industry in the 19th century was generally based on essential oils and related natural products. In 1876, however, Haarman & Reimer started the first production of synthetic aroma chemicals—vanillin, then coumarin, anise aldehyde, heliotropine, and terpineol. Although aroma chemicals made a revolution in fragrances with top discoveries in the 20th century, for many decades, both flavors and fragrances were manufactured with constituents of natural origin, the majority of which were essential oils.

The main reason for the expansion of the essential oils industry and the growing demand for products was the development of the food, soap, and cosmetics industries. Today's multinational

TABLE 31.1
The First Industrial Manufacturers of Essential Oils, Flavors, and Fragrances (Companies Continuing to Operate under Their Original Name Are Printed in Bold)

Company Name	Country	Established
Antoine Chiris	France (Grasse)	1768
Cavallier Freres	France (Grasse)	1784
Dodge & Olcott Inc	US (New York)	1798
Roure Bertrand Fils and Justin Dupont	France (Grasse)	1820
Schimmel & Co	Germany (Leipzig)	1829
J. Mero-Boyveau	France (Grasse)	1832
Stafford Allen and Sons	United Kingdom (London)	1833
Robertet et Cie	**France (Grasse)**	**1850**
W.J.Bush	United Kingdom (London)	1851
Payan-Bertrand et Cie	**France (Grasse)**	**1854**
A. Boake Roberts	United Kingdom (London)	1865
Fritsche-Schimmel Co	US (New York)	1871
V.Mane et Fils	**France (Grasse)**	**1871**
Haarman&Reimer	Germany (Holzminden)	1874
R.C.Treatt Co	**United Kingdom (Bury)**	**1886**
N.V.Polak und Schwartz	Holland (Zaandam)	1889
Ogawa and Co.	**Japan (Osaka)**	**1893**
Firmenich and Cie	**Switzerland (Geneve)**	**1895**
Givaudan S.A.	**Switzerland (Geneve)**	**1895**
Maschmeijer Aromatics	Holland (Amsterdam)	1900

companies, the main users of fragrances and flavors, have evolved directly from the developments during the mid-19th century.

In 1806, William Colgate opened his first store for soaps, candles, and laundry starch on Dutch Street in New York. In 1864, B. J. Johnson in Milwaukee started production of soap, which from 1898, became known as Palmolive. In 1866, Colgate launched its first perfumed soaps and perfumes. In 1873, Colgate launched toothpaste in a glass jug on the market and in the first tube in 1896. In 1926, two soap manufacturers—Palmolive and Peet—merged to create Palmolive-Peet, which two years later merged with Colgate to establish the Colgate-Palmolive-Peet company (renamed as the Colgate-Palmolive Company in 1953).

In October 1837, William Procter and James Gamble signed a formal partnership agreement to develop their production and marketing of soaps (Gamble) and candles (Procter). "Palm oil," "rosin," "toilet," and "shaving" soaps were listed in their advertisements. An "oleine" soap was described as having a violet odor. Only 22 years later, P&G sales reached 1 million dollars. In 1879, a fine but inexpensive "Ivory" white toilet soap was offered to the market with all-purpose applications as a toilet and laundry product. In 1890, Procter & Gamble was selling more than 30 different soaps.

The story of third player started in 1890 when William Hesket Lever created his concept of Sunlight Soap, which revolutionized idea of cleanliness and hygiene in Victorian Britain.

The very beginning of the 20th century marked the next big event when the young French chemist Eugene Schueller prepared his first hair color in 1907 and established what is now L'Oreal. These were the flagships in hundreds of emerging (and disappearing by fusions, takeovers, or bankruptcy) manufacturers of perfumes, cosmetics, toiletries, detergents, household chemicals, and related products, the majority of which were and are perfumed with essential oils.

Over the same time period, another group of users of essential oils entered the markets. In 1790, the term "soda water" for carbon dioxide saturated water as a new drink appeared for the first time in the US and in 1810 the first US patent was issued for manufacture of imitations of natural gaseous mineral waters. Only nine years later, the "soda fountain" was patented by Samuel Fahnestock. In 1833, carbonated lemonade flavored with lemon juice and citric acid was on sale in England. In1835, the first bottled soda water appeared in the US. It is, however, interesting that the first flavored sparkling drink—"Ginger Ale"—was created in Ireland in 1851. The milestones in flavored soft drinks appeared 30 years later: 1881—the first cola-flavored drink in the US; 1885—"Dr Pepper" was invented by Charles Aderton in Waco, Texas; 1886—Coca-Cola by Dr. John S. Pemberton in Atlanta, Georgia; and in 1898—Pepsi-Cola, created by Caleb Bradham, known from 1893 as "Brad's Drink."

"Dr Pepper" was advertised as the "king of beverages, free from caffeine" (which was added to it later on), was flavored with black cherry artificial flavor, and was first sold in the Old Corner Drug Store owned by Wade Morrison. Its market success and position as one of most popular US soft drinks started by a presentation during the St Louis World's Fair where some other important flavor consuming products—ice cream cones, hot dog rolls, and hamburger buns—were shown. All of them remain major users of natural flavors based on essential oils. One-hundred years later, after the merger with another famous lemon-lime drink, "7-up," in 1986, it finally became a part of Cadbury.

Dr. John Pemberton was a pharmacist and mixed up combination of lime, cinnamon, coca leaves, and cola to make the flavor for his famous drink, first as remedy against headache (Pemberton French Wine Coca), reformulated according to prohibition law and used it to add taste to soda water from his "soda fountain." The unique name and logo was created by his bookkeeper Frank Robinson, and Coca-Cola was advertised as "a delicious, exhilarating, refreshing, and invigorating temperance drink." Interestingly, the first year of sales resulted in a $20 loss, as the cost of the flavor syrup used for the drink was higher than total sales of $50. In 1887, another pharmacist—Asa Candler—bought the idea, and with aggressive marketing, in ten years introduced his drink all over the US and Canada by selling syrup to other companies licensed to manufacture and retail the drink. Until 1905, Coca-Cola was known as a tonic drink and contained extracts of cocaine and cola nuts, with flavorings of lime and sugar.

Like Pemberton, Caleb Bradham was a pharmacist, and in his drugstore he offered soda water from his "soda fountain." To promote sales, he flavored that soda with sugar, vanilla, pepsin, cola, and "rare oils"—obviously the essential oils of lemon and lime—and started selling it as cure for dyspepsia, "Brad's Drink" then Pepsi-Cola.

The development of the soft-drink industry is of great importance because it is a major consumer of essential oils, especially those of citrus origin. It is enough to say that nowadays, according to their web pages, only Coca-Cola-produced beverages are consumed worldwide in a quantity exceeding one billion drinks per day. If we consider that the average content of the appropriate essential oil in the final drink is about 0.001%–0.002%, and the standard drink is about 0.3 L (300 g), we approach a daily consumption of essential oils by this company only at the level of 3–6 ton per day, which gives an usage well over 2000 ton annually. All other brands of the food industry, although using substantial quantities of essential oils in ice creams, confectionary, bakery, plus a variety of fast foods (where spice oils are used), together use less oils than the beverage manufacturers.

There is one special range of products that can be situated between the food and cosmetic-toiletries industry sectors and is a big consumer of essential oils, especially of all kinds of mint, eucalyptus, and some other herbal and fruity oils. These are oral-care products, chewing gums, and all kinds of mouth refreshing confectionery. As mentioned above, toothpastes appeared on the market in the late 19th century in the US. Chewing gums or the custom of chewing certain plant secretions was known to the ancient Greeks (e.g., mastic tree resin) and to ancient Mayans (sapodilla tree gum). Chewing gum as we know it now started in America about 1850 when John B. Curtis introduced flavored chewing gum, which was first patented in 1859 by William Semple. In 1892, William Wrigley used chewing gum as a free gift with sales of baking powder in his business in Chicago, and very soon, he realized that chewing gum had real potential. In 1893, "Juicy Fruit" gum came into market and was followed in the same year by "Wrigley's Spearmint"; today, both products are known and consumed worldwide and their names are global trademarks.

This brief and certainly incomplete look into history of industrial usage of essential oils as flavor and fragrance ingredients shows that the real industrial scale of the flavor and fragrance industry developed in the second half of the 19th century, together with the transformation of "manufacture" into "industry."

There are no reliable data on the scale of consumption of essential oils in specific products. On the basis of different sources, it can be estimated that the world market for the flavors and fragrances has a value of €10–12 billion, being equally shared by each group of products. It is very difficult to estimate usage of the essential oils in each of the groups, but it is clear that more oils are used in flavors than in fragrances where, especially in large volume compounds used in detergents and household products, fragrances today are based on aroma chemicals. Table 31.2 presents estimated data on world consumption of major essential oils as available from different sources from 2007 until 2017.

The following oils are used in quantities between 100 and 500 tons p.a.: bergamot, cassia, cinnamon leaf, clary sage, dill, geranium, lemon petitgrain, lemongrass, petitgrain, pine, rosemary, tea tree, and vetivert. It must be emphasized that most of the figures given above on production volume are probably underestimates because no reliable data are available on the domestic consumption of essential oils in major producing areas, such as China, India, Indonesia, and South America. Therefore, the figures given by various sources are sometimes very different. For example, consumption of *Mentha arvensis* is as shown by the above sources given as 5000 and 25,000 tons p.a. The lower value probably relates to direct usage of the oil; the higher includes oil used for production of menthol crystals. Differences in years may be real due to crop availability but also due to lack of reliable data. Considering the above and general figures for flavors and fragrances, it can be estimated that the total production of essential oils in ten years increased over 50% from about 100,000 MT in 2007 to 160,000 MT in 2017. The value of essential oils used worldwide increased from somewhere between €2 and €3 billion in 2007 to almost €6 billion in 2017. Price fluctuations, and many other unpredictable changes, cause any estimation of essential oils consumption value to be very risky and disputable. However general trends in all branches of industry using essential oils and consumers preferences for "natural" and/or "organic" products allow an estimate that by 2020, the production will grow more than two times in tonnage and over €10 billion in value (see comments below). Table 31.2 does not include turpentine, which is sometimes added into essential oils data. It is used mainly as a chemical solvent as it has no

TABLE 31.2
Estimated World Consumption of the Major Essential Oils

Oil Name	Global Production in Metric Tons				Major Applications[b]
	B. Lawrence 2007	V. S. V. Gowda 2011	S. V. Shukla 2014	IFEAT[a] 2017	
Orange	50,000	50,000	30,000	[illegible]5,000	Soft drinks, sweets, fragrances
Cornmint (*M. arvensis*)[c]	25,000	30,000	16,000	[illegible]0,000	Oral care, chewing gum, confectionery, fragrances, menthol crystals
Peppermint	4500	nd	4000	[illegible]300	Oral care, chewing gum, confectionery, liquors, tobacco, fragrances
Eucalyptus (*E. globulus*)	4000	5000	5000	[illegible]1,000	Oral care, chewing gum, confectionery, pharmaceuticals, fragrances
Lemon	3500	nd	5000	[illegible]500	Soft drinks, sweets, diary, fragrances, household chemistry
Citronella	3000	1500	3000	[illegible]100	Perfumery, toiletries, household chemistry
Eucalyptus (*E. citriodora*)	2100	1000	nd	[illegible]00	Confectionery, oral care, chewing gum, pharmaceuticals, fragrances
Clove leaf	2000	4000	nd	>2500	Condiments, sweets, pharmaceuticals, tobacco, toiletries, household chemicals
Spearmint (*M. spicata*)	2000	nd	1500	>3700	Oral care, chewing gum, confectionery
Cedarwood (*Virginia*)	1500	3000	2900	300	Perfumery, toiletries, household chemicals
Lime	1500	nd	1600	1900	Soft drinks, sweets, diary, fragrances
Lavandin	1000	nd	1200	>2200	Perfumery, cosmetics, toiletries
Litsea cubeba	1000	1500	+ lemongrass 2200	1900	Citral for soft drinks, fragrances
Citronella	nd	1500	nd	1100	Perfumery, cosmetics, toiletries
Cedarwood (China)	800	nd	800	1000	Perfumery, toiletries, household chemicals
Camphor	700	nd	800	>1500	Pharmaceuticals,
Coriander	700	nd	700	500	Condiments, pickles, processed food, fragrances
Grapefruit	700	nd	nd	800	Soft drinks, fragrances
Star anise	700	400	400	<800	Liquors, sweets, bakery, household chemicals
Patchouli	600	nd	900	1600	Perfumery, cosmetics, toiletries
Basil	500	500	nd	330	Condiments, processed food, perfumery, toiletries
Mandarin	500	nd	+ tangerine 300	10	Soft drinks, sweets, liquors, perfumery, toiletries
TOTAL with other oils	nd	99,700	nd	>150,000	

[a] Unpublished data.
[b] All of the major oils are used in aromatherapy and pharmacy.
[c] Main source of natural menthol.

practical application as an essential oil, except in some household chemicals. However, both kinds of turpentine—crude sulfate and gum—are the most important raw materials for synthesis of numerous aroma chemicals and many nature identical. Out of approximately 300,000 MT of world production of the turpentine, about 60% is used by the flavor and fragrance industry.

As noted earlier, the largest world consumer of essential oils is the flavor industry, especially for soft drinks. This however is limited to a few essential oils, mainly citrus (orange, lemon, grapefruit, mandarin, lime), ginger, cinnamon, clove, and peppermint. Similar oils are used in confectionery, bakery, desserts, and dairy products, although the range of the oils may be wider and include some fruity products and spices. The spicy oils are widely used in numerous salted chips, which are commonly consumed along with beverages and long drinks. Also, the alcoholic beverage industry is a substantial user of essential oils; for example, anise in numerous specialties of the Mediterranean region, herbal oils in liqueurs, ginger in ginger beer, peppermint in mint liquor, and in many other flavored alcohols.

Next in importance to beverages in the food sector are the sweet, dairy, confectionery, dessert (fresh and powdered), sweet bakery, and cream manufacturing sector, for which the main oils employed are citrus, cinnamon, clove, ginger, and anise. Many other oils are used in an enormous range of very different products in this category.

The fast food and processed food industries are also substantial users of essential oils, although the main demand is for spicy and herbal flavors. Important oils here are coriander (especially popular in the US), pepper, pimento, laurel, cardamom, ginger, basil, oregano, dill, and fennel, which are added to the spices in an aim to strengthen and standardize the flavor.

The major users of essential oils are the big compounders—companies which emerged from historical manufacturers of essential oils and fragrances and flavors and new ones established by various deals between old players in the market or, like IFF (International Flavors & Fragrances, Inc.), were created by talented managers who left their parent companies and started on their own. Today's big ten are listed in Tables 31.3 and 31.4. Although both tables list the same names, there are significant differences in their position in the list. (The first four remained unchanged; however, the merger of IFF and Frutarom in 2018 may change the order.)

Out of the 20 companies listed in Table 31.1, seven were located in France, but by 2018, out of the ten largest, only two are from France. Also, only four of today's big ten are over a century old, with two leaders—Givaudan and Firmenich from Switzerland plus Mane and Robertet from France.

The flavor and fragrance industry is one in which the majority of the oils are introduced into appropriate flavor and fragrance compositions. Created by flavorists and perfumers, an elite of

TABLE 31.3
Leading Producers of Flavors and Fragrances 2007

Pos.	Company Name (Headquarters)	Sales in € million[a]
1	Givaudan S.A. (Vernier, Switzerland)	2550
2	Firmenich S.A. (Geneve, Switzerland)	1620
3	International Flavors and Fragrances (New York, US)	1500
4	Symrise AG (Holzminden, Germany)	1160
5	Takasago International Corporation (Tokyo, Japan)	680
6	Sensient Technologies Flavors&Fragrances (Milwaukee, US)	400
7	T. Hasegawa Co Ltd (Tokyo, Japan)	280
8	Mane S.A. (Le Bar-sur-Loup, France)	260
9	Frutarom Industries Ltd (Haifa, Israel)	220
10	Robertet S.A. (Grasse, France)	210
	Total	8880

[a] Estimated data based on the web pages of the companies, various reports, and journals.

TABLE 31.4
Leading Producers of Flavors and Fragrances 2017

Pos.	Company Name (headquarters)	Sales in € million[a]
1.	Givaudan S.A. (Vernier, Switzerland)	4470
2.	Firmenich S.A. (Geneve, Switzerland)	3120
3.	International Flavors and Fragrances (New York, US)	2890
4.	Symrise AG (Holzminden, Germany)	2890
5.	Mane S.A. (Le Bar-sur-Loup, France)	1110
6.	Frutarom Industries Ltd (Haifa, Israel)	1080
7.	Takasago International Corporation (Tokyo, Japan)	1040
8.	Sensient Technologies Flavors & Fragrances (Milwaukee, US)	530
9.	Robertet S.A. (Grasse, France)	490
10.	T. Hasegawa Co Ltd (Tokyo, Japan)	360
	Total	17,980

[a] Estimated data based on the web pages of the companies, various reports and journals.

professionals in the industry, the compositions, complicated mixtures of natural and nature identical ingredients for flavoring, and natural and synthetic components for fragrances, are offered to end users. The latter are the manufacturers of millions of very different products from luxurious "haute couture" perfumes and top-class flavored liquors and chocolate pralines through cosmetics, household chemicals, sauces, condiments, cleaning products, air fresheners, and aroma marketing.

What is importance in comparison of data in Tables 31.3 and 31.4 is the total value of products. Within 10 years, the consumption of flavors and fragrances doubled, and it is true not only for the big 10 but in the flavors and fragrances industry in general. The global production of the flavors and fragrances industry in 2017 was estimated at €22 billion. This, of course, increased significantly demand for essential oils and their components.

It is important to emphasize that a very wide range of essential oils are used in alternative or "natural" medicine with aromatherapy—treatment of many ailments with use of essential oils as bioactive ingredients—being the leading outlet for the oils and products in which they are applied as major active components. In 2017, the consumption of essential oils in aromatherapy was estimated for over 6500 metric tons. The ideas of aromatherapy from a niche area dominated by lovers of nature and some kind of magic, although based on very old and clinically proved experience, came into mass production appearing as an advertising "hit" in many products, including global ranges. Examples include Colgate-Palmolive liquid soaps, a variety of shampoos, body lotions, creams, etc. by many other producers, and fabric softeners emphasizing the benefits to users' mood and condition from the odors of essential oils (and other fragrant ingredients) remaining on fabrics. Aromatherapy and "natural" products, where essential oils are emphasized as "the natural" ingredients, are a very fast developing segment of the industry and this is a return to what was common practice in ancient and medieval times. According to different sources, "natural" and "organic" cosmetic products production in the last few years increased two times faster than total of the branch.

Until the second half of the 19th century, formulas of perfumes, and flavors (although much less data are available on flavoring products in history) were based on essential oils and some other naturals (musk, civet, amber, resins, pomades, tinctures, extracts, etc.). Now, some 150 years later, old formulations are being taken out of historical books and are advertised as the "back to nature" trend. Perfumery handbooks published until early in the 20th century listed essential oils and none or only one or two aroma chemicals (or isolates from essential oils). A very good illustration of the changes which affected formulation of perfumes in the 20th century is a comparison of rose fragrance as recorded in perfumery handbooks. Dr. Heinrich Hirzel, in his *Die Toiletten Chemie* (1892) gave the following formula for high-quality white-rose perfume:

400 g of rose extract
200 g of violet extract
150 g of acacia extract
100 g of jasmine extract
120 g of iris infusion
25 g of musk tincture
5 g of rose oil
10 drops of patchouli oil

In 1931, Felix Cola's milestone work, *Le Livre de Parfumeur* records a white-rose formula containing only 1% of rose oil, 2% of rose absolute, 7.5% other oils, and aroma chemicals (Cola 1931).

Rose blanche	
Rose oil	10 g
Rose absolute	20 g
Patchouli oil	25 g
Bergamot oil	50 g
Linalool	60 g
Benzyl acetate	75 g
Phenylethyl acetate	75 g
Citronellol	185 g
Geraniol	200 g
Phenylethyl alcohol	300 g

In the mid-20th century, perfumers were educated to consider chemicals as the most convenient, stable, and useful ingredients for fragrance compositions. Several rose fragrance formulas with less than 2% rose oil or absolute can be found in F. V. Wells and M. Billot's *Perfumery Technology*, which was published in 1975, and rose fragrance without any natural rose product is nothing curious in a contemporary perfumers' notebook. However, looking through descriptions of new fragrances launched in the last few years, one can observe a very strong tendency to emphasize the presence of natural ingredients—oils, resinoids, absolutes—in the fragrant mixture. The "back to nature" trend creates another area for essential oils usage in many products.

A very fast growing group of cosmetics and related products today are the so-called "organic products." These are based on plant ingredients obtained from wild harvesting or from "organic cultivation" and which are free of the pesticides, herbicides, synthetic fertilizers, and other chemicals widely used in agriculture. The same "organic raw materials" are becoming more and more popular in the food industry, which in consequence will increase consumption of "organic flavors" based on "organic essential oils." "Organic" certificates, available in many countries (in principle for agricultural products, although there are institutions which certify also cosmetics and related products) are product passports to a higher price level and selective shops or departments in supermarkets. The importance of that segment of essential oils consumption can be illustrated by comparison of the average prices for some standard essential oils and the same oils claimed as "organic". It is worth noting the general price growth within the decade, although in very few cases, the opposite has been observed (see Table 31.5).

Consumption of essential oils in perfumed products varies according to the product (see Table 31.6). From a very high level in perfumes (due to the high concentration of fragrance compounds in perfumes and high content of natural ingredients in perfume fragrances) and in a wide range of "natural" cosmetics and toiletries to relatively low levels in detergents and household chemicals,

TABLE 31.5
Examples of Prices of Some Standard and "Organic" Essential Oils

	Standard Quality €/kg[a]		"Organic" Quality €/kg[a]	
Oil Name	**2007**	**2018**	**2007**	**2018**
Cedarwood (*Virginia*)	15.00	30.00	58.00	74.00
Citronella	11.00	32.00	23.00	44.00
Clove leaf	12.00	35.00	60.00	95.00
Eucalyptus (*Globulus*)	5.50	6.50	26.00	41.00
Lavandin	15.00	51.00	36.00	75.00
Lime	44.00	50.00	92.00	102.00
Litsea cubeba	20.00	47.00	44.00	90.00
Orange	5.50	14.00	35.00	42.00
Patchouli	115.00	72.00	250.00	150.00
Peppermint	27.00	42.00	100.00	107.00

[a] Average prices based on commercial offers in 2007 and 2018.

TABLE 31.6
Average Dosage of Fragrances in Consumer Products and Content of Essential Oils in the Fragrance Compounds

Pos.	Product	Average Dosage of Fragrance Compound in Product (%)	Average Content of Essential Oils in Fragrance (%)
1.	Perfumes	10.0–25.0	5–30[a]
2.	Toilet waters, EdC	3.0–8.0	5–50[a]
3.	Skin care cosmetics	0.1–0.6	0–10
4.	Deodorants (incl. deoparfum)	0.5–5.0	0–10
5.	Shampoos	0.3–2.0	0–5
6.	Body cleansing products (liquid soaps)	0.5–3.0	0–5
7.	Bath preparations	0.5–6.0	0–10
8.	Soaps	0.5–3.0	0–5
8.	Toothpastes	0.5–2.5	10–50[b]
9.	Air fresheners	0.5–30.0	0–20
10.	Washing powders and liquids	0.1–0.5	0–5
11.	Fabric softeners	0.1–0.5	0–10
12.	Home care chemicals	0.5–5.0	0–5
13.	Technical products	0.1–0.5	0–5
14.	Aromatherapy and organic products	0.1–0.5	100

[a] Traditional perfumery products contained more natural oils than modern ones.
[b] Mainly mint oils.

in which fragrances are based on readily available, low-priced aroma chemicals. However, it must be emphasized that although the concentration of essential oils in detergents and related products is low, the large volume sales of these consumer products result in substantial consumption of the oils. According to different sources, production of "natural" and "organic" in the personal care market will reach over $20 billion in 2024. All market research show that a substantial majority of consumers, both women and men, prefer that group of products as more safe, healthy, and beneficial than those with synthetic ingredients.

It should be mentioned here that both terms "natural" and "organic" in reference to cosmetics, their ingredients, and related products created serious difficulties concerning their legal definition, thus proper labelling and description. There is no regulation anywhere concerning natural or organic cosmetic ingredients and cosmetics. As demonstrated by W. S. Brud and I. Konopacka-Brud at the existing level of technology and agriculture, there is no possibility to manufacture on an industrial scale and to sell on the mass market cosmetic products which can be labelled "natural" or "organic" in agreement with the true meaning of this words. Although term "organic" is legally regulated for food products (with numerous differences in different countries) there is lot of freedom of using it for cosmetic ingredients. The same situation is with term "natural" which according to certification system depend mainly on content of "natural" ingredients in final product. Also, "natural" fragrances used in different products are defined as such according to imprecise criteria.

Average values given for fragrance dosage in products and for the content of oils in fragrances are based on literature data and private communications from manufacturers. It should be noted that in many cases, the actual figures for individual products can be significantly different. "Eau Savage" from Dior is a very good example: analytical data indicate a content of essential oils (mainly bergamot) of over 70%. Toothpastes are exceptional in that the content of essential oils in the flavor are in some cases nearly 100% (mainly peppermint, spearmint cooled with natural menthol).

While the average dosage of fragrances in the final product can be very high, flavors in food products are used in very low dosages, well below 1%. The high consumption of essential oils by this sector results from the large volume of sales of flavored foods. Average dosages of flavors and the content of essential oils in the flavors are given in Table 31.7.

As in the case of fragrances, the average figures given in Table 31.7 vary in practice in individual cases, both in the flavor content in the product and much more in the essential oils percentage in the flavor. Again "natural" or "organic" products only contain essential oils, since it is unacceptable to include any synthetic aroma chemicals or so-called "nature identical" food flavors. It should be noted that a substantial number of flavorings are oleoresins: products which are a combination of essential oils and other plant-derived ingredients, which are especially common in hot spices (pepper, chili, pimento, etc.) containing organoleptically important pungent components that do not distill in steam. This group of oleoresin products must be included in the total consumption of essential oils.

For many years after World War II, aroma chemicals were considered the future for the fragrance chemistry and there was strong, if unsuccessful, pressure by manufacturers to get approval for the wide introduction of synthetics (especially those regarded as "nature identical") in food flavors. The very fast development of production and usage of aroma chemicals caused increasing concern over safety issues for human health and for the environment. One by one, certain products were found

TABLE 31.7
Average Content of Flavors in Food Products and of Essential Oils in the Flavors

Pos.	Food Products	Flavor Dosage in Food Product (%)	Essential Oils Content in Flavor (%)
1.	Alcoholic beverages	0.05–0.15	3–100
2.	Soft drinks	0.10–0.15	2–5
3.	Sweets (confectionery, chocolate, etc.)	0.15–0.25	1–100
4.	Bakery (cakes, biscuits, etc.)	0.10–0.25	1–50
5.	Ice creams	0.10–0.30	2–100
6.	Dairy products, desserts	0.05–0.25	1–50
7.	Meat and fish products (also canned)	0.10–0.25	10–20
8.	Sauces, ketchup, condiments	0.10–0.50	2–10
9.	Food concentrates	0.10–0.50	1–25
10.	Snacks	0.10–0.15	2–20

harmful either for human health (e.g., nitro musks) or for nature. This resulted in wide research on the safety of the chemicals and the development of new safe synthetics. Concurrently, the attention of perfumers and producers turned in the direction of essential oils, which as derived from natural sources and known and used for centuries, generally were considered safe. According to recent research, however, this belief is not entirely true, and some, fortunately very few, oils and other fragrance products obtained from plants have been found dangerous, and their use has been banned or restricted. However, these are exceptional cases, and the majority of essential oils are found safe both for use on the human body in cosmetics and related products, as well as for consumption as food ingredients.

It is important to appreciate that the market for "natural," "organic," and "ecological" products both in body care and food industries has changed from a niche area to a boom in recent years, with growth exceeding 30% per annum. The estimated value of sales for "organic" cosmetics and toiletries is 600–800 million Euro in Europe, the US, and Japan and will grow steadily together with organic foods. This creates a very sound future for the essential oils industry, which as such or as isolates derived from the oils, will be widely used for fragrance compounds in cosmetic and related products as well as for flavors.

Furthermore, the modernization of agricultural techniques and the growth of plantation areas result in better economic factors for production of essential oil-bearing plants, creating workplaces in developing countries of Southeast Asia, Africa, and South America, as well as further development of modern farms in the US and Europe (Mediterranean area, Balkans). Despite some regulatory restrictions (EU, REACH, FDA, etc.), essential oils are and will have an important and growing share in the fragrance and flavor industry. The same will be true for usage of essential oils and other products of medicinal plants in pharmaceutical products. It is well known that the big pharmaceutical companies invest substantial resources in studies of folk and traditional medicine, as well as in research on biologically active constituents of plant origin. Both of these areas cover applications of essential oils. The same is observed in cosmetics and toiletries using essential oils as active healing ingredients.

It can be concluded that the industrial use of essential oils is a very promising area and that regular growth shall be observed in the future. Much research work will be undertaken both on the safety of existing products and on development of new oil-bearing plants that are used locally in different regions of the world, both as healing agents and food flavorings. Both directions are equally important. Global exchange of tastes and customs shall not lead to unification by Coca-Cola or McDonalds. With all the positive aspects of these products, there are many local specialties which can became world property, like basil-oregano flavored pizza, curry dishes, spicy kebab, or the universal and always fashionable Eau de Cologne. With growth of the usage of the commonly known essential oils, new ones coming from exotic flowers of the Amazon jungle or from Indian Ayurveda books can add new benefits to the flavor and fragrance industry.

ACKNOWLEDGMENTS

Author is most grateful to Dr. C. Green for his help and assistance in preparation of this chapter.

REFERENCES

Cola, F. 1931. *Le Livre du Parfumeur.* Paris: Casterman.

Hirzel, H. 1892. *Die Toiletten Chemie.* Leipzig: J.J.Weber Verlag.

FURTHER READING

Brud, W.S., and I. Konopacka-Brud. 2018. Natural and organic in cosmetic, related products and ingredients. *SEPAWA Conference*, Berlin.

Dorland, W.F., and J.A. Rogers Jr. 1977. *The Fragrance and Flavor Industry.* New Jersey: V.E.Dorland.

Govda, V.S.V. 2014. *Production of Natural Essential Oils*. Bangalore: CIMAP.
Lawrence, B.M. 2000. *Essential Oils 1995–2000*. Wheaton, IL: Allured Publishing.
Lawrence, B.M. 2004. *Essential Oils 2001–2004*. Wheaton, IL: Allured Publishing.
Lawrence, B.M. 2007. *Essential Oils 2005–2007*. Wheaton, IL: Allured Publishing.
Shukla, S.V. 2014. *Global Scenario and Market of Essential Oils, Fragrance and Flavours*. Kannauj: FFDC.
Wells, W.F. and M. Billot, 1981. *Perfumery Technology*. London: E. Horwood Ltd.

WEB SITES

American Beverage Association: http://www.ameribev.org
The Coca-Cola Company: http://www.thecoca-colacompany.com/heritage/ourheritage.html
Colgate-Palmolive: http://www.colgate.com/app/Colgate/US/Corp/History/1806.cvsp
Procter & Gamble: http://www.pg.com/company/who_we_are/ourhistory.shtml
Unilever: http://www.unilever.com/aboutus

32 Storage, Labeling, and Transport of Essential Oils

Jens Jankowski, Jens-Achim Protzen, and Klaus-Dieter Protzen

CONTENTS

32.1 MARKETING OF ESSENTIAL OILS: THE FRAGRANT GOLD OF NATURE POSTULATES PASSION, EXPERIENCE, AND KNOWLEDGE

Since the publication of the first edition of this book, quite a few changes have taken place regarding the regulations of handling and labeling of essential oils. These are considered and classified by regulatory authorities in most parts of the world not only as *natural* but also as *chemical substances*, abbreviated as NCS (the so-called natural complex substances).

The trade of essential oils is affected more and more by legal regulations related to safety aspects. The knowledge and the compliance with these superseding regulations that affect usual commercial aspects today have become a *conditio sine qua non* (precondition) to ensure trouble-free global business relations when placing essential oils on the market in the European Union (EU) (and other parts of the world) for use as natural flavors and fragrances in food, animal feed, cosmetics, pharmaceuticals, and aromatherapy.

Among others, the following regulations have to be observed (Dueshop, 2015):

- Regulation (EC) No. 1272/2008 on *c*lassification, *l*abeling, and *p*ackaging of substances and mixtures
- Regulation (EC) No. 1907/2006 on *REACH*
- Flavouring Regulation (EC) No. 1334/2008
- Regulation (EU) No. 1169/2011 on the provision of food information to the consumer
- Cosmetic Regulation (EC) No. 1223/2009
- Regulation (EU) No. 528/2012 on biocide products
- Regulation (EC) No. 1831/2003 on additives for use in animal nutrition
- EU Pharmaceutical Legislation—GMP and GDP aspects
- Regulation (EC) No. 648/2004 on detergents

- Regulation (EC) No. 178/2002 on food law
- Novel Food Regulation (EC) No. 258/97
- Regulation (EC) No. 396/2005 on maximum residue level of pesticides

Essential oils are natural substances mainly obtained from vegetable raw materials either by distillation with water or steam or by a mechanical process (expression) from the epicarp of citrus fruits. They are concentrated fragrance and flavor materials of complex composition, in general: volatile alcohols, aldehydes, ethers, esters, hydrocarbons, ketones, and phenols of the group of mono- and sesquiterpenes or phenylpropanes as well as nonvolatile lactones and waxes.

A definition of the term essential oils and related fragrance/aromatic substances is given in the ISO-Norm 9235 Aromatic Natural Raw Materials (International Standard Organization [ISO], Geneva, 1997).

In former times, essential oils were obtained from collected wild-growing plants—in these days, many of them are produced, however, from small-scale plantations and/or manufactured by small individual producers. The reason for this lies probably in the fact that a larger-scale production would require capital investment, which is rarely attracted as investors evidently realize that—if at all—no quick return of money is ensured. The negative factors influencing the market as follows:

- The dependency on weather as climatic conditions may affect the size of a crop over the whole vegetation period.
- Competing crops challenging the acreage.
- A keen global competition striving for market shares.

The general narrow margins do not compensate the involved risks—and last but not least, the adherence to comply with ever-changing administrative regulations is not necessarily the first target of investors of money.

All these aggravating factors also have an impact on the trade of essential oils. This is the reason that particularly the trade of the essential oils is dominated by small-scale and medium-sized family enterprises. Only entrepreneurs with passion, a personal engagement, and a persistent dedication, as well as a long-standing experience, nerve themselves to stay successfully in this business of the liquid gold of nature. A long-term philosophy, a lot of enthusiasm, and hard work together with a broad knowledge of the market situation and also the willingness of spending a lot of time and cost to investigate new ideas of state-of-the-art conditions of processing raw materials affecting the yield and quality and the return of investment.

In the EU, the classification for a chemical substance was laid down in the Council Directive 67/548 and subsequent amendments—Council Directive 79/831/EEC of 18-09-1979 (the famous sixth amendment) is the basis of all existing regulations for dangerous/hazardous chemicals and it earmarked the beginning of a new era that finally resulted in the EU in REACH—a regulation that is superseding and harmonizing the first attempts to regulate the safe handling of dangerous/hazardous materials.

REACH (*Registration, Evaluation, Authorization of Chemicals*) is the consistent continuation of rules to satisfy the EU administration with a perfect system to safeguard absolute security to protect humans and the environment regarding the use of chemicals. The topic of REACH will not be discussed in detail in this chapter because of its complexity and it touches only to a small extent the title of this contribution.

For the trade, that is, the industry and importers and dealers of essential oils, REACH is a heavy and costly burden demanding an unbelievable amount of time to furnish the required product information for an appropriate registration of essential oils as UVCB/NCS substances. In the following, however, a brief introduction to the historic development of the existing regulatory framework is given in order to help to understand the safety aspects, which are the background of the actual regulations and the impediments in connection with REACH.

32.2 IMPACT AND CONSEQUENCES ON THE CLASSIFICATION OF ESSENTIAL OILS AS NATURAL BUT CHEMICAL SUBSTANCES IN REACH

In the EU, the bell for the new era sounded when chemical substances in use within the EC during a reference period of 10 years had to be notified for the European Inventory of Existing Commercial Chemical Substances (EINECS).

At that time, EINECS enabled the EC administration not only to dispose of, for the first time, a survey of all chemical substances that had been in use in the EC between January 1971 and September 1981 but also to distinguish between "known substances" and "new substances."

"Known" substances are all chemicals notified for EINECS, whereas all chemical substances that were not notified (and subsequently had been registered as "known substances" in EINECS) were considered by the EU administration as "new chemicals."

"New" chemical substances could only be marketed in the EU after clearance according to uniform EC standards by competent authorities. Thus, from the beginning, all potential risks of a (new) chemical substance are ascertained for a proper labeling for handling and risks for humans and the environment could be minimized.

In a transitional phase, "known" chemical substances (notified for EINECS) enjoyed a temporary exemption from the obligation to furnish the same safety data required for new chemical substances. The assumption was that, based on the experience gathered during their use for decades and sometimes centuries, the temporary continuation of their use could be tolerated according to the hitherto used older standards of safety (Dueshop, 2007) as a short-term clearance of approximately 100,000 chemical substances registered in EINECS could not be effected overnight.

To make sure that the *known* substances which had been notified for EINECS—and thus became known to the regulative agencies in the EC—also did comply with the new safety standards, they were screened step by step according to the following volume bands either depending on their potential risk or according to the volumes produced or imported:

- >1000 tons
- 100–1000 tons
- 10–100 tons
- 1–10 tons
- 100–1000 kg
- <100 kg

To perform this task, the EU administration—as the United States already did many years ago—made use of the principles of the CAS system and arranged for the majority of essential oils and other "UVCB" (chemical substances of unknown or variable composition, complex reaction products and biological materials) and allocated (new) more precise CAS numbers, which were eventually also published in EINECS.

One should bear in mind, however, that, in principle, the CAS number is an identification number for a defined chemical substance that is allotted by a private enterprise in the United States must not be confused with the EINECS registration number, which is a registration number allocated by the EU administration, that is, ECB/JRC at ISPRA—CAS numbers are assigned by the (private) CAS organization in the United States with the purpose of identification of (defined) chemical substances.

In principle, a CAS number is allocated by the CAS organization to a new (defined) chemical substance only after thorough examination of the product as per the IUPAC rules to make sure that, irrespective of different chemical descriptions and/or coined names that have been given to a product, a substance can be clearly related by the allocated CAS number according to the (CAS) principle "one substance—one number."

Using the CAS number system to also register UVCB substances in EINECS, that is, products that are not defined chemicals, it made it necessary to extend the CAS system also to the so-called UVCBs. Essential oils as NCS eventually are registered by their botanical origin. As, for example,

- *Lavender oil*: lavender—*Lavandula angustifolia*
  - EINECS registration no. 289-995-2—CAS no. (Einecs) 90063-37-9: extractives and their physically modified derivatives such as tinctures, concretes, absolutes, essential oils, terpenes, terpene-free fractions, distillates, and residues from *Lavandula angustifolia*—Labiatae (Lamiaceae)
- *Lavender oil*: lavender—*Lavandula angustifolia*
  - EINECS registration no. 283-994-0—CAS no. (Einecs) 84776-65-8: extractives and their from *Lavandula angustifolia angustifolia*—Labiatae (Lamiaceae)
- *Lavender concrete/absolute*: lavender—*Lavandula angustifolia*
  - EINECS registration no. 289-995-2—CAS no. (Einecs) 90063-37-9 extractives and their physically modified derivatives such as tinctures, *concretes*, *absolutes*, essential oils, terpenes, terpene-free fractions, distillates, and residues from *Lavandula angustifolia*—Labiatae (Lamiaceae)
- *Lavandin oil*: *Lavandula hybrida.*
  - EINECS registration no. 294-470-6—CAS no. (Einecs) 91722-69-9 extractives and their physically modified derivatives such as tinctures, concretes, absolutes, essential oils, terpenes, terpene-free fractions, distillates, and residues from *Lavandula hybrida*—Labiatae (Lamiaceae)
- *Lavandin oil abrialis*: *Lavandula hybrida abrial.*
  - EINECS registration no. 297-384-7—CAS no. (Einecs) 93455-96-0 extractives and their from *Lavandula hybrida abrial*—Labiatae (Lamiaceae)
- *Lavandin oil grosso*: *Lavandula hybrida grosso* ext.
  - EINECS registration no. 297-385-2—CAS no. (Einecs) 93455-97-1 extractives and their from *Lavandula hybrida grosso*—Labiatae (Lamiaceae)

The registration of essential oils under their botanical origin implicated that concretes/absolutes and other natural extractives of the same botanical origin also have the same EINECS and CAS numbers as the essential oil.

In this connection, it should be mentioned that when checking an EINECS number, it is important to investigate the correct number in the official original documentation, as in the secondary literature, there exist many inaccuracies.

Because of the lack of rules for a uniform classification of essential oils as UVCB, it happened that against the principles of the CAS organization in some cases, several CAS numbers had been allocated to essential oils of the same denomination and, in addition,

- An earlier (older) CAS number allocated for the product in the United States
- A new (and more precise) CAS number allocated for registration in the EC for EINECS, respectively (Table 32.1)

32.3 DANGEROUS SUBSTANCES AND DANGEROUS GOODS

There is a significant difference between the similar sounding words and regulations regarding "Dangerous Substances" and "Dangerous (Hazardous) Goods."

Both regulations are targeted to protect humans and the environment, but the term "Dangerous Substance" refers to the risks connected with the properties of the substance, that is, the potential risk of a direct contact with the product during production, packaging, and use.

"Dangerous Goods" refers to dangerous substances that are properly packed and labeled for storage and transport by road, rail, sea, or air (Figure 32.1).

TABLE 32.1
Examples of Different CAS-Numbers Used in USA and EINECS in EU

Article	CAS No. USA	CAS No. EINECS	EC Registration No.
Eucalyptus oil	8000-48-4	84625-32-1	283-406-2
Eucalyptus globulus Lab.—Myrtaceae			
Lavender oil	8000-28-0	90063-37-9	289-995-2
Lavandula angustifolia—Labiatae			
Lavandula angustifolia angustifolia—Labiatae		84776-65-8	283-994-0
Lemon oil	8008-56-8	8028-48-6	284-515-8
Citrus limon L.—Rutaceae		84929-31-7	284-515-8
Orange oil	8008-52-9	8028-48-6	232-433-8
Citrus sinensis—Rutaceae			
Peppermint oil	8006-90-4	98306-02-6	308-770-2
Mentha piperita L.—Lamiaceae			

In 2003, a working group of the United Nations (UN) that was trying to harmonize the hitherto often different national and international existing regulations on the classification and labeling of dangerous/hazardous products published the so-called *purple book*. It contained the results of their first attempts of a global uniform regulation. This publication is updated every 2 years.

In the EU *Regulation (EC) No. 1272 on* ***C****lassification,* ***L****abelling and* ***P****ackaging of substances and mixtures,* in general, often called as CLP-Regulation, was published in 2008. It is based on the Globally Harmonised System (GHS) and became binding in the EU on 2009.01.20 replacing the Regulations (EC) No. 67/548 as well as (EC) No. 1944/45 (Directive 1999/45/EC of the European Parliament and of the Council of 31 May, 1999).

The GHS distinguishes between 16 physical dangers, 10 health risks, and, in addition, the class "aquatic environment."

1. Physical hazards:
   a. Explosives/mixtures and products with explosive properties
   b. Flammable gases
   c. Flammable aerosols
   d. Oxidizing gases
   e. Gases under pressure
   f. Flammable liquids
   g. Flammable solids
   h. Self-reactive substances and mixtures
   i. Pyrophoric liquids

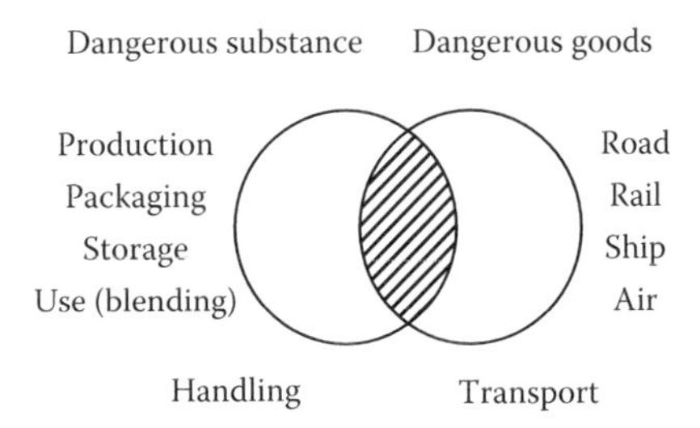

FIGURE 32.1 Interrelationship between dangerous substances and dangerous goods. (Courtesy of Paul Kaders, Hamburg/Germany.)

   j. Pyrophoric solids
   k. Self-heating substances and mixtures
   l. Substances and mixtures which, in contact with water, emit flammable gases
   m. Oxidizing liquids
   n. Oxidizing solids
   o. Organic peroxides
   p. Corrosive to metals
2. Health hazards:
   a. Acute toxicity
   b. Skin corrosion/irritation
   c. Serious eye damage/eye irritation
   d. Respiratory or skin sensitization
   e. Germ cell mutagenicity
   f. Carcinogenicity
   g. Reproductive toxicity
   h. Specific target organ toxicity—single exposure
   i. Specific target organ toxicity—repeated exposure
   j. Aspiration hazard
3. Environmental hazards:
   a. Hazardous to the aquatic environment

In contrast to the CLP regulations—probably for historical reasons—the transport of dangerous goods is divided in nine classes only.

1. Class 1—Explosives
2. Class 2—Gases
3. Class 3—Flammable liquids
4. Class 4—Flammable solids
5. Class 5—Oxidizing substances and organic peroxides
6. Class 6—Toxic and infectious substances
7. Class 7—Radioactive material
8. Class 8—Corrosives
9. Class 9—Miscellaneous dangerous goods

Nowadays, for all ways of transport (road, rail, air, and sea), this uniform classification as per GHS is applicable with the requirement of a correspondent labeling.

32.3.1 Material Safety Data Sheet

As per GHS requirements, for each hazardous substance, a "Material Safety Data Sheet," as per ISO standard (Geneva) and/or REACH regulation, must be furnished.

In Europe, the REACH standard form is binding consisting of the following topics:

1. Identification of the substance/mixture and of the company
2. Hazard identification
3. Composition/information on ingredients
4. First aid measures
5. Fire-fighting measures
6. Accidental release measures
7. Handling and storage

8. Exposure controls/personal protection
9. Physical and chemical properties
10. Stability and reactivity
11. Toxicological information
12. Ecological information
13. Disposal considerations
14. Transport information
15. Regulatory information
16. Other information

The instructions that the supplier has to give in topic 14 of the MSDS permit a correct labeling for the transport of dangerous goods—in case of a doubt, however, the correct classification can also be verified by checking the remarks entered under section 2: "Hazard Identification" of the same MSDS.

In case that no MSDS is available, there exists the possibility to search the required information in the databank of the European Chemicals Agency (ECHA). In the C&L-Inventory, the CAS and/or EINECS numbers, the corresponding classification, required labels and the corresponding "H"-(hazard) and the P (precautionary) statements, and the GHS pictograms of all chemicals registered in the EU can be found. However, this ECHA databank is of help only for orientation, as, for example, there exist for Orange Oil sweet CAS No. 8028-48-6 a total of 56 entries.

Another source for information is the IFRA/IOFI Labelling Manual of the International Fragrance Association. The readings in this Manual are quite simple to understand and the desired information regarding the classification and thus the desired information is easier for use than in the C&L inventory of ECHA. In addition—besides the GHS classification—this manual also still contains a reference to the older Regulation (EC) No. 67/548, which sometimes is a valuable hint.

32.4 PACKAGING OF DANGEROUS GOODS

Those dangerous substances for which the international regulations for transport of dangerous goods apply have to be transported only in UN-approved container. All UN-approved packaging are marked with an immutable code number as, for example (Table 32.2).

According to the risk emanating from a substance, there exist three different packing groups (Table 32.3).

As illustrated, the packing codes "X", "Y," and "Z" refer to the risks emanating from a substance: PC "X" is required for substances with a high risk, and "Z" with a low risk (Table 32.4).

TABLE 32.2
Meaning of the Code Number on Container

	UN1A1/Y/1,4/150/(06)/NL/VL824
1A1	Steel drum—nonremovable head
Y	Allowed for substances with packing group (PG) II and PG III
1.4	Maximum relative density at which the packing has been tested
150	Test pressure
(06)	Year of manufacture
(NL)	State
(VL123)	Code number of manufacturer

TABLE 32.3
Packing Group and Packing Code

Packing Group		Packing Code
PG III	Low risk	Z
PG II	Medium risk	Y
PG I	High risk	X

TABLE 32.4
Correlation between the Packing Code and Packing Group

Packing Code	Packing Group
X	PG I, PG II, PG III
Y	PG II and PG III
Z	PG III

32.5 LABELING

In addition to the correct selection of the packing, the CLP regulations and the guidelines for the transport of dangerous goods require labeling of each piece of packing piece together with "H" (hazard) and "P" (precautionary) statements. These statements must be given in English and national language(s) together with labels of pictograms according to the classification of the potential hazard emanating from the substance. In addition, if applicable, it should also contain the label/pictogram "environmentally hazardous" (Figure 32.2).

The introduction of the label/pictogram "environmentally hazardous" also became part of the present amendments of ADR/RID code/regulations for road and rail traffic in Europe and for the

FIGURE 32.2 "Environmentally Hazardous" label.

IMDG code for sea transport if the hazardous substance falls into the classification (Verordnung zur Änderung der Anlagen A und B zum ADR-Übereinkommen, 2014):

- Aquatic acute 1
- Aquatic chronic or
- Aquatic chronic 2

This label is replacing the former label "Marine Pollutant" for dangerous goods shipped by sea.

For transport by air, as per IATA regulations, this new label is also compulsorily required for goods with the UN numbers UN 3077 and UN 3082—it may, however, also be fixed in those cases where it is required for the transport by other carriers.

In this connection, a few words are due on the so-called UN numbers for dangerous goods. These UN numbers are assigned to dangerous goods according to their hazard classification and composition. These UN numbers should not be confused with the (UN) number for packing. Labels with the UN numbers must be fixed distinctly and visibly on each packing in addition to the already mentioned other labels as per the CLP-Regulation.

UN numbers are listed in all codes for transport of dangerous goods and are identical for all types of transport. Most of the essential oils fall under the numbers:

- UN no. 1169 extracts, aromatic, and liquid
- UN no. 2319 terpene hydrocarbons
- UN no. 3082 environmentally hazardous substances, liquid. n.o.s.
- UN no. 1992 flammable liquid, toxic, n.o.s.
- UN no. 1272 pine oil(s)

N.O.S. (not otherwise specified) requires that the name of the hazardous substance has to be added in braces to the Proper Shipping Name (Figure 32.3).

Symbol	Description
 3	Label 3 Flammable liquids Symbol flame: Black or White Background: Red
 4	Label 4.1 Flammable solid Symbol flame: Black Background: White with seven vertical red stripes
 6	Label 6.1 Toxic Symbol skull and crossbones: Black Background: White
 8	Label 8 Corrosive Symbol liquids spilling from two glass vessels and attacking a hand and a metal: black Background: upper half White, lower half Black with White border
 9	Label 9 Miscellaneous Symbol seven vertical stripes in upper half: Black Background: white

FIGURE 32.3 Important hazard symbols for essential oils.

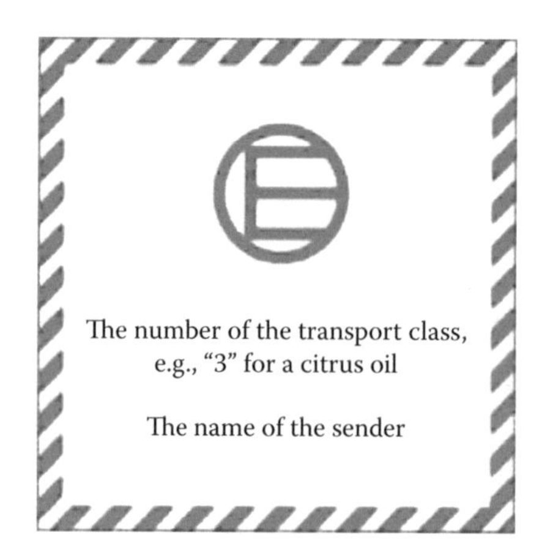

FIGURE 32.4 "Excepted Quantity" label with explanation of the printed.

The aim of dangerous goods regulations is not only to protect persons occupied with the conveyance of dangerous/hazardous substances but also to serve, for example, fire brigades, who in case of an accident or fire are called and have to be aware of special risks.

It should be noted that, at least in Europe, the transport police controls more and more transports of dangerous goods, the accompanying documentation even to the extent of the markings and correct labels (Protzen 1989, 1998).

For the transport of small quantities of hazardous substances, exceptions/exemptions exist that allow simplified procedures in two levels: the "excepted quantity" and the "limited quantity."

For the dispatch of samples of those essential oils classified as dangerous substances, the regulations for documentation are canceled and the requirements for packing have been reduced to a minimum and thus the "excepted quantity" is of invaluable help. However, the dispatch is only allowed in a combined packing and the packaging must be marked with the label "excepted quantity" (Figure 32.4).

In case the quantities to be sent exceed the exemptions for an "excepted quantity," there exists the possibility of a labeling as "limited quantity." As per (different) regulations of the carriers, these packages have to be labeled as shown in Figure 32.5.

For the transport of dangerous goods as "limited quantity," a list is required specifying the net weights of each class of hazardous substances planned for the shipment. This list should be furnished before hand to the forwarder and also has to be attached to the transport documentation. Furthermore, it should be noted that goods have to be packed in a combined packing.

Despite the adherence of aforementioned regulations, it can happen that a carrier might reject the transport of (certain) dangerous substance(s) under the provisions of a "limited quantity." In this case, there only remains the alternative to have the material transported as per the standard regulations of the respective carrier.

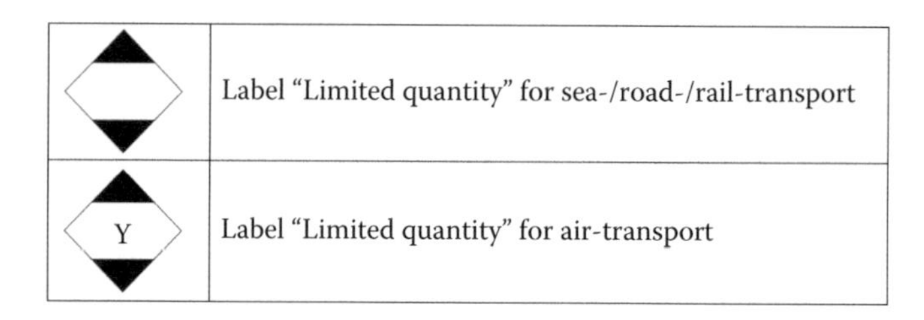

FIGURE 32.5 "Limited Quantity" label.

32.6 LIST OF REGULATIONS FOR THE CONSIDERATION OF DOING BUSINESS IN THE EU

- Regulation (EC) No 1272/2008 of the European Parliament and of the Council of December 16, 2008 on classification, labelling and packaging of substances and mixtures (Text with EEA relevance), OJ. L 353 published December 31, 2008, pp. 1–1355.
- Regulation (EC) No 1907/2006 of the European Parliament and of the Council of December 18, 2006 concerning the Registration, Evaluation, Authorisation and Restriction of Chemicals (REACH), establishing a European Chemicals Agency, OJ. L 396 published December 31, 2006, pp. 1–851.
- Regulation (EC) No. 1334/2008 of the European Parliament and of the Council of December 16, 2008 on flavorings and certain food ingredients with flavouring properties for use in and on foods, OJ L 354 published December 31, 2008, pp. 34–50.
- Regulation (EU) No. 1169/2011 of the European Parliament and of the Council of October 25, 2011 on the provision of food information to consumers, OJ L 304 published November 22, 2011, pp. 18–63.
- Regulation (EC) No. 1223/2009 of the European Parliament and of the Council of November 30, 2009 on cosmetic products, OJ L 342 published December 22, 2009, pp. 59–209.
- Regulation (EU) 528/2012 of the European Parliament and of the Council of May 22, 2012 concerning the making available on the market and use of biocidal products, OJ L 167 published June 27, 2012, pp. 1–123.
- Regulation (EC) No. 1831/2003 of the European Parliament and of the Council on additives for use in animal nutrition, OJ L 268 published October 18, 2003, pp. 29–43.
- Regulation (EC) No. 648/2004 of the European Parliament and of the Council of March 31, 2004 on detergents, OJ L 104 published April 8, 2004, pp. 1–35.
- Regulation (EC) No. 178/2002 of the European Parliament and of the Council of January 28, 2002 laying down the general principles and requirements of food law, establishing the European Food Safety Authority and laying down procedures in matters of food safety, OJ L 31 published February 1, 2002, pp. 1–24.
- Regulation (EC) No. 258/97 of the European Parliament and of the Council of January 27, 1997 concerning novel foods and novel food ingredients, OJ L 43 published February 14, 1997, pp. 1–6.
- Regulation (EC) No. 396/2005 of the European Parliament and of the Council of February 23, 2005 on maximum residue levels of pesticides in or on food and feed of plant and animal origin, OJ L 70 published March 16, 2005, pp. 1–16.
- Directive 2000/13/EC of the European Parliament and of the Council of March 20, 2000 on the approximation of the laws of the Member States relating to the labelling, presentation and advertising of foodstuffs, OJ L 106 published May 6, 2000, pp. 29–42.

ACRONYMS

ADR	Accord europeen relatif au transport international des marchandises dangereuses par route (European agreement for the transport of dangerous goods by road)
CAS	Chemical Abstracts Service
CLP	Classification, Labeling and Packaging of substances and mixtures
ECHA	European Chemicals Agency
EINECS	European Inventory of Existing Commercial Chemical Substances
GDP	Good Distribution Practice
GHS	Globally harmonized system
GMP:	Good Manufacturing Practice
IFRA:	International Fragrance Association

IMDG-Code: International Maritime Dangerous Goods Code
IOFI: International Organization of the Flavor Industry
ISO: International Standard Organization
MSDS: Material Safety Data Sheet
NOS: Not Otherwise Specified
RID: Règlement concernant le transport international ferroviaire des marchandises dangereuses (Order for the international transport of dangerous goods by train)
UVCB: Chemical Substances of Unknown or Variable Composition, Complex Reaction Products and Biological Materials

REFERENCES

C&L Inventory, European chemicals agency, Accessed October 20, 2013 from http://echa.europa.eu/de/regulations/clp/cl-inventory.

Council Directive 67/548/EEC of 27 June 1967 on the approximation of laws, regulations and administrative provisions relating to the classification, packaging and labeling of dangerous substances, OJ. 196 published August 16, 1967, pp. 1–98.

Council Directive 79/831/EEC of 19 September 1979 amending for the sixth time Directive 67/548/EEC on the approximation of the laws, regulations and administrative provisions relating on the classification, packaging and labelling of dangerous substances, OJ L 259, published October 15, 1979, pp. 10–28.

Dangerous Goods Regulations, 56th Edition, International Air Transport Association, Montreal, Quebec, Canada, September 2014.

Directive 1999/45/EC of the European Parliament and of the Council of 31 May 1999 concerning the approximation of the laws, regulations and administrative provisions of the Member States relating to the classification, packaging and labelling of dangerous preparations, OJ. L 200 published July 30, 1999, pp. 1–68.

Dueshop, L., 2007, Personal communications.

Dueshop, L., 2015, Personal communications.

IMDG-Code 2015, *Amdt.*, 37th edn., Storck Verlag, Hamburg, 2014, pp. 37–14.

Protzen, K.-D., October 1989, *Guideline for Classification and Labelling of Essential Oils for Transport and Handling*, distributed during *IFEAT Conference*, London, U.K.

Protzen, K.-D., November 8–12, 1998, Transportation/Safety Regulations Update, *International Conference on Essential Oils and Aromas*, London, U.K., IFRA/IOFI Labeling Manual 2014.

Regulation (EC) No. 178/2002 of the European Parliament and of the Council of 28 January 2002 laying down the general principles and requirements of food law, establishing the European Food Safety Authority and laying down procedures in matters of food safety, OJ L 31 published February 1, 2002, pp. 1–24.

Regulation (EC) No. 258/97 of the European Parliament and of the Council of 27 January 1997 concerning novel foods and novel food ingredients, OJ L 43 published February 14, 1997, pp. 1–6.

Regulation (EC) No. 396/2005 of the European Parliament and of the Council of 23 February 2005 on maximum residue levels of pesticides in or on food and feed of plant and animal origin, OJ L 70 published March 16, 2005, pp. 1–16.

Regulation (EC) No. 648/2004 of the European Parliament and of the Council of 31 March 2004 on detergents, OJ L 104 published April 8, 2004, pp. 1–35.

Regulation (EC) No. 1223/2009 of the European Parliament and of the Council of 30 November 2009 on cosmetic products, OJ L 342 published November 22, 2009, pp. 59–209.

Regulation (EC) No. 1272/2008 of the European Parliament and of the council of 16 December 2008 on classification, labelling and packaging of substances and mixtures (Text with EEA relevance), OJ. L 353 published December 31, 2008, pp. 1–1355.

Regulation (EC) No. 1334/2008 of the European Parliament and of the Council of 16 December 2008 on flavourings and certain food ingredients with flavouring properties for use in and on foods, OJ L 354 published December 31, 2008, pp. 34–50.

Regulation (EC) No. 1831/2003 of the European Parliament and of the Council on additives for use in animal nutrition, OJ L 268 published October 18, 2003, pp. 29–43.

Regulation (EC) No. 1907/2006 of the European Parliament and of the council of 18 December 2006 concerning the Registration, Evaluation, Authorisation and Restriction of Chemicals (REACH), establishing a European Chemicals Agency, OJ. L 396 published December 31, 2006, pp. 1–851.

Regulation (EU) 528/2012 of the European Parliament and of the Council of 22 May 2012 concerning the making available on the market and use of biocidal products, OJ L 167 published June 27, 2012, pp. 1–123.

Regulation (EU) No. 1169/2011 of the European Parliament and of the Council of 25 October 2011 on the provision of food information to consumers, OJ L 304 published November 22, 2011, pp. 18–63.

Verordnung zur Änderung der Anlagen A und B zum ADR-Übereinkommen (24. ADR-Änderungsverordnung—24. ADRÄndV) vom 06. Oktober 2014 (BGBl. II S. 722).

33 Recent EU Legislation on Flavours and Fragrances and Its Impact on Essential Oils

Jan C. R. Demyttenaere

CONTENTS

33.1 INTRODUCTION

In the last two decades, several new European regulations and directives have been adopted and published in relation to flavours and fragrances. As essential oils and extracts are very important ingredients for flavouring and fragrance applications, these new regulations will have a major impact on their trade and use in commerce.

This chapter focuses on a major piece of legislation that is of particular importance for the flavour industry, namely, the flavouring regulation [1] (part of the so-called Food Improvement Agents Package) comprising the flavouring, additives, and enzymes regulation and the Common Authorisation Procedure (CAP). This regulation replaces the former flavouring directive 88/388/EEC [2].

The full title of this regulation is *Regulation (EC) No 1334/2008 of the European Parliament and of the Council on flavourings and certain food ingredients with flavouring properties for use in and on foods and amending Council Regulation (EEC) No 1601/91, Regulations (EC) No 2232/96 and (EC) No 110/2008 and Directive 2000/13/EC.*

In the wake of this new regulation, the former Council Directive 88/388/EEC of June 22, 1988, as well as its amendment directive 91/71/EEC [3] and the commission decision 88/389/EEC [4] have been repealed.

Especially, the impact on labelling, resulting from the difference between the current flavouring regulation and the former "flavour directive" 88/388/EC, is discussed from a B2B (business to business) perspective. Special focus is given to the labelling of "natural flavourings" (including essential oils) and changes in definitions (e.g., what is "natural?") in the current flavouring regulation.

33.2 FORMER FLAVOURING DIRECTIVE AND CURRENT FLAVOURING REGULATION: IMPACT ON ESSENTIAL OILS

In the EU, until 2008, for flavourings, the former flavouring directive 88/388/EC [2] was applied. This was the Council Directive of June 22, 1988 on the approximation of the laws of the Member States (MS) relating to flavourings for use in foodstuffs and to source materials for their production, as published in the Official Journal on July 15, 1988 (OJ L 184, p. 61). It has been amended once by the Commission Directive 91/71/EEC of 16/01/91 (OJ L 42, p. 25, 15/02/91) [3]. As this was a directive, it was up to the EU MS to take the necessary measures to ensure that flavourings may not be marketed or used if they do not comply with the rules stated in Art. 3 of this directive.

However, on December 31, 2008, the current flavouring regulation was published [1], which entered into force on January 20, 2009. This regulation officially came into use on January 20, 2011, and the former flavouring directive was repealed. As many essential oils and extracts either contain flavouring substances or are regarded as "flavouring preparations" or "food ingredients with flavouring properties," this flavouring regulation has an impact on essential oils and their use as flavouring ingredients for food products. Extracts and essential oils contain certain constituents (substances) that according to this regulation "should not be added as such to food" or should be added at a particular level. Especially, the application of maximum levels of these substances will have an impact on how and when extracts, essential oils, and herbs and spices may or can be applied to food. In addition, the definitions for "natural" have drastically changed. The difference between the former directive 88/388/EC and the current flavour regulation in this respect is outlined in the following.

33.2.1 Maximum Levels of "Restricted Substances"

Since 1999, the Scientific Committee on Food (SCF) and the European Food Safety Authority (EFSA)* have expressed their opinions on a number of substances found naturally in source materials for flavourings and food ingredients with flavouring properties, for example, some herbs and spices. According to the Committee of Experts on Flavouring Substances (CEFS) of the Council of Europe (CoE),[†,‡] these substances may raise toxicological concern. Although these substances have, in the past, commonly been referred to, within the food industry, as "biologically active principles" (BAPs), they are referred to as "restricted substances" (RS) in this chapter. Both the former flavouring directive and the current flavouring regulation use the undefined term "certain substances."

33.2.1.1 (Restricted Substances under Former) Flavouring Directive 88/388/EC

In the former flavouring directive 88/388/EC, Annex II set maximum levels (limits) for certain substances obtained from flavourings and other food ingredients with flavouring properties in foodstuffs as consumed in which flavourings have been used. Art. 4 (c) stipulated that

> (c) the use of flavourings and of other food ingredients with flavouring properties does not result in the presence of substances listed in Annex II in quantities greater than those specified therein.

The limits apply to foodstuffs and beverages (mg/kg), with some exceptions, for example, alcoholic beverages and confectionary. In Table 33.1, the maximum levels for these substances for foodstuffs and beverages are given (including the exceptions and/or special restrictions), as per Annex II to the former flavouring directive.

---

* Panel on Food Additives, Flavourings, Processing aids and Flavourings, and Materials in Contact with Food (AFC) and since July 2008 the EFSA Panel on Food Contact Materials, Enzymes, Flavourings and Processing Aids (CEF).

† Within CoE a Committee of Experts on Flavouring Substances has evaluated the safety-in-use of natural flavouring source materials since 1970. The results are published in the "*Blue Book*"—*Flavouring Substances and Natural Sources of Flavourings* [5].

‡ According to CoE "active principles" are chemically defined substances, which occur in certain natural flavouring source materials and preparations and which, on the basis of existing toxicological data, should not be used as flavouring substances in their own right [6].

TABLE 33.1
Maximum Limits for Certain Substances Obtained from Flavourings and Other Food Ingredients with Flavouring Properties Present in Foodstuffs as Consumed in Which Flavourings Have Been Used (Annex II of 88/388/EC)

Substances	Foodstuffs (mg/kg)	Beverages (mg/kg)	Exceptions and/or Special Restrictions
Agaric acid[a]	20	20	100 mg/kg in alcoholic beverages and foodstuffs containing mushrooms
Aloin[a]	0.1	0.1	50 mg/kg in alcoholic beverages
Beta-Asarone[a]	0.1	0.1	1 mg/kg in alcoholic beverages and seasonings used in snack foods
Berberine[a]	0.1	0.1	10 mg/kg in alcoholic beverages
Coumarin[a]	2	2	10 mg/kg in certain types of caramel confectionery 50 mg/kg in chewing gum 10 mg/kg in alcoholic beverages
Hydrocyanic acid[a]	1	1	50 mg/kg in nougat, marzipan, or its substitutes or similar products 1 mg/% volume of alcohol in alcoholic beverages 5 mg/kg in canned stone fruit
Hypericine[a]	0.1	0.1	10 mg/kg in alcoholic beverages 1 mg/kg in confectionery
Pulegone[a]	25	100	250 mg/kg in mint or peppermint-flavoured beverages 350 mg/kg in mint confectionery
Quassine[a]	5	5	10 mg/kg in confectionery in pastille form 50 mg/kg in alcoholic beverages
Safrole and isosafrole[a]	1	1	2 mg/kg in alcoholic beverages with not more than 25% volume of alcohol 5 mg/kg in alcoholic beverages with more than 25% volume of alcohol 15 mg/kg in foodstuffs containing mace and nutmeg
Santonin[a]	0.1	0.1	1 mg/kg in alcoholic beverages with more than 25% volume of alcohol
Thujone (alpha and beta)[a]	0.5	0.5	5 mg/kg in alcoholic beverages with not more than 25% volume of alcohol 10 mg/kg in alcoholic beverages with more than 25% volume of alcohol 25 mg/kg in foodstuffs containing preparations based on sage 35 mg/kg in bitters

[a] May not be added as such to foodstuffs or to flavourings. May be present in a foodstuff either naturally or following the addition of flavourings prepared from natural raw materials.

This means that for essential oils, extracts, complex mixtures containing these "restricted substances" (e.g., nutmeg, cinnamon, peppermint, sage oils, …) and when added to food and flavourings, maximum levels applied. The same applies to herbs and spices containing these "restricted substances" as herbs and spices are also "food ingredients with flavouring properties."

33.2.1.2 (Restricted Substances under) Current Flavouring Regulation 1334/2008/EC

Apart from the fact that the former directive 88/388/EC has now turned into a regulation, there are many changes that have an impact on how essential oils and extracts will be used as source of flavours.

The most important issue is how the restricted substances are addressed.

According to the Regulation, substances for which toxicological concern was confirmed by SCF/EFSA should be regarded as naturally occurring undesirable substances, which should not be added as such to food. Due to their natural occurrence, these substances might be present in flavourings and certain food ingredients with flavouring properties.

This is addressed in Art. 6 of the Regulation: *Presence of certain substances,* which refers to Annex III with the same title. This article clearly states in the first paragraph that *"Substances listed in Part A of Annex III shall not be added as such to food."*

However, when it regards the levels of these substances coming from the use of flavourings and food ingredients with flavouring properties (such as extracts, essential oils, herbs, and spices) the Regulation further specifies (Art. 6.2) as follows:

> 2. Without prejudice to Regulation No 110/2008 maximum levels of certain substances, naturally present in flavourings and/or food ingredients with flavouring properties, in the compound foods listed in Part B of Annex III shall not be exceeded as a result of the use of flavourings and/or food ingredients with flavouring properties in and on those foods.
>
> The maximum levels of the substances set out in Annex III apply to foods as marketed, unless otherwise stated. By way of derogation from this principle, for dried and/or concentrated foods which need to be reconstituted the maximum levels apply to the food as reconstituted according to the instructions on the label, taking into account the minimum dilution factor.

This means that maximum levels of these substances also apply when the substances come from any type of food ingredients with flavouring properties; the only exception is dried and/or concentrated foods, which can have higher levels before they are diluted/reconstituted. Upon dilution/reconstitution, normal maximum levels apply.

The main difference between the former flavouring directive 88/388 and the current flavouring regulation is that in the directive 88/388, there is only one list (Annex II) of substances to which the maximum levels apply—all those substances may not be added *as such* to food. In contrast, in the current flavouring regulation, the Annex III is split in two parts: Part A with "Substances which shall *not* be added *as such* to food" and Part B establishing "Maximum levels of certain substances, naturally present in flavourings and food ingredients with flavouring properties, in certain compound food as consumed to which flavourings and/or food ingredients with flavouring properties have been added."

Part A contains 15 substances, whereas Part B contains 11 substances.

Table 33.2 lists the 15 substances of Part A of Annex III "which shall not be added as such to food," and Table 33.3 lists the 11 substances of Part B with their respective maximum levels in the various compound foods according to the new flavouring regulation.

An important statement in the current flavouring regulation was added only at a final stage of the political discussions of the draft text, as a footnote (asterisk) to the table, applying to three of the substances that are marked with an asterisk: estragol, safrol, and methyleugenol.

This means that the maximum levels do not apply to estragol, safrol, and methyleugenol when only fresh, dried, or frozen herbs and spices are added! However, when "food ingredients with

TABLE 33.2
Annex III, Part A: Substances That May *Not* Be Added *As Such* to Food

Agaric Acid	Aloin	Capsaicin
1,2-Benzopyrone, coumarin	Hypericine	**Beta-Asarone**
1-Allyl-4-methoxybenzene, estragole	**Hydrocyanic acid**	**Menthofuran**
4-Allyl-1,2-dimethoxybenzene, methyleugenol	**Pulegone**	**Quassin**
1-Allyl-3,4-methylene dioxy benzene, safrole	**Teucrin A**	**Thujone (alpha and beta)**

Note: Substances in bold are those that are in both Part A and Part B of Annex III.

TABLE 33.3
Maximum Levels of Certain Substances, Naturally Present in Flavourings and Food Ingredients with Flavouring Properties, in Certain Compound Foods Consumed to Which Flavourings and/or Food Ingredients with Flavouring Properties Have Been Added

Name of the Substance	Compound Food in which the Presence of the Substance is Restricted	Maximum Level (mg/kg)
Beta-Asarone	Alcoholic beverages	1.0
1-Allyl-4-methoxybenzene, estragol[a]	Dairy products	50
	Processed fruits, vegetables (including mushrooms, fungi, roots, tubers, pulses, and legumes), nuts, and seeds	50
	Fish products	50
	Nonalcoholic beverages	10
Hydrocyanic acid	Nougat, marzipan, or its substitutes or similar products	50
	Canned stone fruits	5
	Alcoholic beverages	35
Menthofuran	Mint-/peppermint-containing confectionery, except micro-breath freshening confectionery	500
	Micro-breath freshening confectionery	3000
	Chewing gum	1000
	Mint-/peppermint-containing alcoholic beverages	200
4-Allyl-1,2-dimethoxy-benzene, methyleugenol[a]	Dairy products	20
	Meat preparations and meat products, including poultry and game	15
	Fish preparations and fish products	10
	Soups and sauces	60
	Ready-to-eat savouries	20
	Nonalcoholic beverages	1
Pulegone	Mint-/peppermint-containing confectionery, except micro-breath freshening confectionery	250
	Micro-breath freshening confectionery	2000
	Chewing gum	350
	Mint-/peppermint-containing nonalcoholic beverages	20
	Mint-/peppermint-containing alcoholic beverages	100
Quassin	Nonalcoholic beverages	0.5
	Bakery wares	1
	Alcoholic beverages	1.5
1-Allyl-3,4-methylene dioxy benzene, safrole[a]	Meat preparations and meat products, including poultry and game	15
	Fish preparations and fish products	15
	Soups and sauces	25
	Nonalcoholic beverages	1
Teucrin A	Bitter-tasting spirit drinks or bitter[b]	5
	Liqueurs[c] with a bitter taste	5
	Other alcoholic beverages	2
Thujone (alpha and beta)	Alcoholic beverages, except those produced from *Artemisia* species	10
	Alcoholic beverages produced from *Artemisia* species	35
	Nonalcoholic beverages produced from *Artemisia* species	0.5
Coumarin	Traditional and/or seasonal bakery ware containing cinnamon in the labelling	50
	Breakfast cereals, including muesli	20
	Fine bakery ware with the exception of traditional and/or seasonal bakery ware containing cinnamon in the labelling	15
	Desserts	5

[a] The maximum levels shall not apply where a compound food contains no added flavourings and the only food ingredients with flavouring properties which have been added are fresh, dried or frozen herbs and spices. After consultation with the Member States and the Authority, based on data made available by the Member States and on the newest scientific information, and taking into account the use of herbs and spices and natural flavouring preparations, the Commission, if appropriate, proposes amendments to this derogation.

[b] As defined in Annex II, paragraph 30 of Regulation (EC) No 110/2008.

[c] As defined in Annex II, paragraph 32 of Regulation (EC) No 110/2008.

flavouring properties" such as essential oils are added, or when essential oils and/or other flavourings are added in combination with herbs and spices, the levels *do* apply.

Moreover, as stipulated in the footnote under Annex III (applying to the three substances with an asterisk), amendments to the current derogations for herbs and spices can be expected (see Table 33.3).

Further compared to the former flavouring directive 88/388, some "restricted substances" are new, for example, methyleugenol, estragol, and menthofuran. This is because since the publication (and amendment) of the former directive, some new scientific evidence has become available that suggests that there would be some toxicological concern for these substances. As explained in Recital (8): "Since 1999, the Scientific Committee on Food and subsequently the European Food Safety Authority [EFSA]...have expressed opinions on a number of substances occurring naturally in source materials for flavourings and food ingredients with flavouring properties which, according to the Committee of Experts on Flavouring Substances of the Council of Europe, raise toxicological concern. Substances for which the toxicological concern was confirmed by the Scientific Committee on Food should be regarded as undesirable substances which should not be added as such to food."

It should be noted that for the same reason, as stipulated in Art. 22 of the Flavouring Regulation, the Annex can be amended "to reflect scientific and technical progress...following the opinion of the Authority" (EFSA).

An important example is methyleugenol (4-Allyl-1,2-dimethoxybenzene). In 1999, methyleugenol was evaluated by the CEFS of the council of Europe. The conclusions of this Committee were as follows:

> Available data show that methyleugenol is a naturally-occurring genotoxic carcinogen compound with a DNA-binding potency similar to that of safrole. Human exposure to methyleugenol may occur through the consumption of foodstuffs flavoured with aromatic plants and/or their essential oil fractions which contain methyleugenol. In view of the carcinogenic potential of methyleugenol, it is recommended that absence of methyleugenol in food products be ensured and checked with the most effective available analytical method [7].

Methyleugenol was subsequently evaluated by the SCF, and an opinion on its safety was published in 2001 [8]. The conclusion of the SCF was as follows:

> Methyleugenol has been demonstrated to be genotoxic and carcinogenic. Therefore the existence of a threshold cannot be assumed and the Committee could not establish a safe exposure limit. Consequently, reductions in exposure and restrictions in use levels are indicated.

An equally important but similar example is estragol (1-Allyl-4-methoxybenzene) also known as methylchavicol. In 2000, the CEFS of the council of Europe evaluated estragol, and based on their findings (it was found to be a naturally occurring genotoxic carcinogen in experimental animals), a limit of 0.05 mg/kg (detection limit) was recommended [9].

Estragol was subsequently evaluated by the SCF, and an opinion on its safety was published in 2001 [10]. The conclusion of the SCF was

> Estragole has been demonstrated to be genotoxic and carcinogenic. Therefore the existence of a threshold cannot be assumed and the Committee could not establish a safe exposure limit. Consequently, reductions in exposure and restrictions in use levels are indicated.

As a consequence, both methyleugenol and estragol have been added to Annex III of the current flavouring regulation as "restricted substances."

A particular case is menthofuran, which had already been reviewed first by the SCF together with (*R*)-(+)-pulegone in 2002 [11] and which has more recently been evaluated/considered by EFSA and for which the EFSA CEF Panel raised some concerns [12]. Menthofuran was listed in the register of chemically defined flavouring substances laid down in Commission Decision 1999/217/EC [13], as amended [FL-no: 13.035], but due to these safety concerns, the substance has been introduced in Annex III of the new flavouring regulation and is no longer used/added as such in or on foods as a flavouring substance. It is also not part of the current EU "Union List of Flavourings and Source Materials" Part A: "List of Flavouring Substances," which has been adopted and published recently as part of the "*Implementing Regulation*" [14].

Due to safety/toxicology concerns, some substances should be restricted not only when "added as such" (for which reason they appear in Annex III Part A) but also when they are naturally present in flavourings and food ingredients with flavouring properties (for which reason they appear in Annex III Part B). Others (e.g., capsaicin) are only restricted when added as such (as "chemically defined substance") and appear in Annex III Part A but not when naturally present (hence they do not appear in Annex III Part B). One of the reasons could be that the use of natural sources in which capsaicin is present (e.g., chilli peppers—*Capsicum*) is self-limiting (at least for consumers in the EU), and setting limits for the use of peppers would be extremely difficult from an implementation point of view and for control authorities (to check the maximum levels). An opinion on the safety of capsaicin was published by the SCF in 2002 [15].

It is also important to note that according to Art. 30 of the flavouring regulation (*entry into force*), which was applied 24 months after its entry into force (i.e., 20/01/2011), Art. 22 was applied from the date of its entry into force (i.e., 20/01/2009). Art. 22 concerns the Amendments to Annexes II–V. This means that even before the flavouring regulation was applied, the Annexes could be amended immediately, if necessary. However, between the entry into force and the application date of the regulation, the Annexes are amended, and it is not foreseen that they will be amended in the very near future.

Whereas Art. 6 relates to "certain substances," Art. 7 relates to the "Use of certain source materials," which is even more important in relation to herbs, spices, extracts, and essential oils. This article refers to Annex IV of the regulation, which is a new Annex that was not in the former flavouring directive 88/388/EC entitled: "List of source materials to which restrictions apply for their use in the production of flavourings and food ingredients with flavouring properties." Annex IV consists of two parts:

1. Part A: Source materials that shall not be used for the production of flavourings and food ingredients with flavouring properties.
2. Part B: Conditions of use for flavourings and food ingredients with flavouring properties produced from certain source materials.

The complete Annex IV to the flavouring regulation is given in Appendix 33.A to this chapter. Art. 7 stipulates the following:

1. Source materials listed in Part A of Annex IV shall not be used for the production of flavourings and/or food ingredients with flavouring properties.
2. Flavourings and/or food ingredients with flavouring properties produced from source materials listed in Part B of Annex IV may be used only under the conditions indicated in that Annex.

33.2.2 Definition of "Natural"

33.2.2.1 (Definition of "Natural" under) Former Flavouring Directive 88/388/EC

Also important is how the former flavouring directive addressed "naturalness" of flavours and how "natural" was defined for the purpose of labelling. This was stipulated by Art. 9a.2 (amending the original Art. 9.2 of 88/388/EC by 91/71/EEC) [3]:

> 2. the word "natural", or any other word having substantially the same meaning, may be used only for flavourings in which the flavouring component contains exclusively flavouring substances as defined in Article 1 (2) (b) (i) and/or flavouring preparations as defined in Article 1 (2) (c). If the sales description of the flavourings contains a reference to a foodstuff or a flavouring source, the word "natural" or any other word having substantially the same meaning, may not be used unless the flavouring component has been isolated by appropriate physical processes, enzymatic or microbiological processes or traditional food-preparation processes solely or almost solely from the foodstuff or the flavouring source concerned.

How a "natural flavouring substance" could be obtained was thus defined in Art. 1.2 (b) (i):

> (b) "flavouring substance" means a defined chemical substance with flavouring properties that is obtained:
>
> > (i) by appropriate physical processes (including distillation and solvent extraction) or enzymatic or microbiological processes from material of vegetable or animal origin either in the raw state or after processing for human consumption by traditional food-preparation processes (including drying, torrefaction and fermentation),

How a "flavouring preparation" could be obtained was defined in Art. 1.2 (c):

> (c) "flavouring preparation" means a product, other than the substances defined in (b) (i), whether concentrated or not, with flavouring properties, which is obtained by appropriate physical processes (including distillation and solvent extraction) or by enzymatic or microbiological processes from material of vegetable or animal origin, either in the raw state or after processing for human consumption by traditional food-preparation processes (including drying, torrefaction and fermentation);

This means that a "flavouring preparation" was by default always "natural" and that extracts and essential oils (obtained by appropriate physical processes such as distillation and solvent extraction) from the material of vegetable origin (e.g., plant material) could be considered as "flavouring preparation" and thus "natural."

33.2.2.2 (Definition of "Natural" under) Current Flavouring Regulation 1334/2008/EC

Regarding "naturalness" of flavours and how "natural" is defined for the purpose of labelling the situation has drastically changed since the former flavouring directive 88/388/EC was repealed.

For example, in the past according to 88/388/EC, there were three categories of flavouring substances: natural, nature-identical (NI), and artificial. However, in the new flavouring regulation, there are only two categories: natural and not natural—meaning that the difference between NI and artificial have disappeared, and these two have merged into one category of "non-natural flavouring substances."

In the current Flavouring Regulation, "natural flavouring substance" is defined by Art. 3.2 (c):

> (c) "natural flavouring substance" shall mean a flavouring substance obtained by appropriate physical, enzymatic or microbiological processes from material of vegetable, animal or microbiological origin either in the raw state or after processing for human consumption by one or more of the traditional food preparation processes listed in Annex II. Natural flavouring substances correspond to substances that are naturally present and have been identified in nature;

Important is the last line stating (additional requirement) that a substance has to be identified in nature before it can be regarded as "natural," so it is not only sufficient to produce it "in a natural way" but it should also be identical to something that is present in nature. Of course, "present in nature" includes any natural substance present in any processed foods: for example, cured vanilla beans, roasted coffee, cooked meat, fermented cheese, wine, and beer. This is to avoid that when an enzymatic or microbial process would be developed by which a flavouring substance can be produced "by enzymatic or microbial processes from material of vegetable origin" (i.e., natural source materials) that up to then has never been identified in nature (and is not naturally occurring), such as ethylvanillin, such material would be labelled as a "natural flavouring substance."

Within the global flavour industry, there is a general principle and agreement that in order to determine that a flavouring substance has been "identified in nature," any identification needs to meet the criteria for the validity of identifications in nature as described by IOFI (International Organization of the Flavor Industry) [16].

More important than the definition on "natural" as such (Art. 3.2 (c)) however is how "appropriate physical process" is defined.

Annex II to the flavouring regulation gives a list of "traditional food preparation processes" by which natural flavouring substances and (*natural*) flavouring preparations are obtained. The full list of *traditional food preparation processes* is given in Table 33.4.

TABLE 33.4
List of Traditional Food Preparation Processes (Annex II of Flavouring Regulation (EC) No 1334/2008)

Chopping	Coating
Heating, cooking, baking, frying (up to 240°C at atmospheric pressure) and pressure cooking (up to 120°C)	Cooling
Cutting	Distillation/rectification
Drying	Emulsification
Evaporation	Extraction, including solvent extraction in accordance with Directive 88/344/EEC
Fermentation	Filtration
Grinding	
Infusion	Maceration
Microbiological processes	Mixing
Peeling	Percolation
Pressing	Refrigeration/freezing
Roasting/grilling	Squeezing
Steeping	

The definition of "appropriate physical process" according to the flavour regulation is described in Art. 3.2 (k):

(k) "appropriate physical process" shall mean a physical process which does not intentionally modify the chemical nature of the components of the flavouring, without prejudice to the listing of traditional food preparation processes in Annex II, and does not involve, inter alinea, the use of singlet oxygen, ozone, inorganic catalysts, metal catalysts, organometallic reagents and/or UV radiation.

This definition refers to Annex II, which means that all processes listed in Annex II also fall under the definition of "appropriate physical processes."

"Flavouring preparations" (such as essential oils and extracts) are defined by Art. 3.2(d).

With respect to "natural labelling," according to Art. 16.2, the term "natural" may only be used for the description of a flavouring if the flavouring component comprises only flavouring preparations and/or natural flavouring substances, which means that a "flavouring preparation" is regarded as "natural" by definition. In other words, there is not such a thing as a "synthetic flavouring preparation."

The definition for "flavouring preparation" reads as follows (Art. 3.2 (d)):

(d) "flavouring preparation" shall mean a product, other than a flavouring substance, obtained from:

i. food by appropriate physical, enzymatic or microbiological processes either in the raw state of the material or after processing for human consumption by one or more of the traditional food preparation processes listed in Annex II;

and/or

ii. material of vegetable, animal or microbiological origin, other than food, by appropriate physical, enzymatic or microbiological processes, the material being taken as such or prepared by one or more of the traditional food preparation processes listed in Annex II;

It can be concluded that according to this definition, essential oils and extracts obtained from plant material (*material of vegetable origin*) prepared by distillation (which is a *traditional food preparation process* listed in Annex II), followed by an *appropriate physical process,* can be considered as a "flavouring preparation" and thus natural, as long as the chemical nature of the components is not intentionally modified during the physical process.

However, it can also be argued that if the nature of the components is intentionally modified in the physical process (e.g., extraction, drying, evaporation, and condensation, dilution), which is often the case, then the end product can no longer be regarded as *flavouring preparation* and thus natural, according to the new Flavouring Regulation.

EFFA (the European Flavour Association) has developed a guidance document for the European Flavour Industry on the permissible processes to obtain natural flavouring ingredients (i.e., natural flavouring substances and (natural) flavouring preparations) [17], which is published as part of the General EFFA Guidelines on the flavouring regulation in the EFFA website [18].

33.3 CONCLUSION

As described and outlined earlier, several new European regulations and directives have been adopted and published during the last two decades in relation to flavours and fragrances. NCSs or raw materials (such as essential oils and extracts) are very important ingredients for flavouring and fragrance applications. As a result, these new regulations have a major impact on the trade and use in commerce of these essential oils and extracts, in particular on labelling issues, as has been demonstrated with the new flavouring regulation (labelling of "natural"). The labelling issue is especially important because of its impact on consumer behaviour: with respect to food, consumers prefer natural flavourings to synthetic ones. Good and pragmatic definitions in the flavouring regulation that has recently replaced the former flavouring directive are essential to ensure that all natural raw materials such as essential oils and extracts can continue to be labelled as natural.

33.A APPENDIX

List of source materials to which restrictions apply for their use in the production of flavourings and food ingredients with flavouring properties (Annex IV to Flavouring Regulation (EC) No 1334/2008):

Part A: Source materials which shall not be used for the production of flavourings and food ingredients with flavouring properties

Source Material	
Latin Name	**Common Name**
Tetraploid form of *Acorus calamus*	Tetraploid form of Calamus

Part B: Conditions of use for flavourings and food ingredients with flavouring properties produced from certain source materials

Source Material		
Latin Name	**Common Name**	**Conditions of Use**
Quassia amara L. and *Picrasma excelsa* (Sw)	Quassia	Flavourings and food ingredients with flavouring properties produced from the source material may only be used for the production of beverages and bakery wares
Laricifomes officinales (Vill.: Fr) Kotl. et Pouz or *Fomes officinalis*	White agaric mushroom	Flavourings and food ingredients with flavouring properties produced from the source material may only be used for the production of alcoholic beverages
Hypericum perforatum L.	St John's wort	
Teucrium chamaedrys L.	Wall germander	

REFERENCES

1. Regulation (EC) No 1334/2008 of the European Parliament and of the council on flavourings and certain food ingredients with flavouring properties for use in and on foods and amending Council Regulation (EEC) No 1601/91, Regulations (EC) No 2232/96 and (EC) No 110/2008 and Directive 2000/13/EC (OJ L 354/34, 31.12.2008).
2. Council Directive 88/388/EC of 22 June 1988 on the approximation of the laws of the Member States relating to flavourings for use in foodstuffs and to source materials for their production (OJ L 184, 15.7.1988, p. 61).
3. Commission Directive 91/71/EEC of 16 January, 1991 completing Council Directive 88/388/EEC on the approximation of the laws of the Member States relating to flavourings for use in foodstuffs and to source materials for their production (OJ L42, 15.2.1991, p. 25).
4. 88/389/EEC: Council Decision of 22 June, 1988 on the establishment, by the Commission, of an inventory of the source materials and substances used in the preparation of flavourings (OJ L184, 15.7.1988, p. 67).
5. *Flavouring Substances and Natural Sources of Flavourings ("Blue Book") Council of Europe (CoE)*, 1992.
6. Active principles (constituents of toxicological concern) contained in natural sources of flavourings. Approved by the Committee of Experts on Flavouring Substances, October 2005.
7. Council of Europe—Committee of Experts on Flavouring Substances, 1999. Publication datasheet on Methyleugenol. *Document RD 4.14/2-45 submitted by the delegation of Italy for the 45th meeting in Zurich*, October 1999.
8. Opinion of the Scientific Committee on Food on Methyleugenol (4-Allyl-1,2-dimethoxybenzene). *SCF/CS/FLAV/FLAVOUR/4 ADD1 FINAL*, 2001. European Commission, Health & Consumer Protection Directorate-General. http://ec.europa.eu/food/fs/sc/scf/out102_en.pdf (accessed July 14, 2015).
9. Council of Europe, Committee of Experts on Flavouring Substances, 2000. Final version of the publication datasheet on estragole. *Document RD 4.5/1-47 submitted by Italy for the 47th meeting in Strasbourg*, October 16–20, 2000.
10. Opinion of the Scientific Committee on Food on Estragole (1-Allyl-4-methoxybenzene). *SCF/CS/FLAV/FLAVOUR/6 ADD2 FINAL*, 2001. European Commission, Health & Consumer Protection Directorate-General. http://ec.europa.eu/food/fs/sc/scf/out104_en.pdf (accessed July 14, 2015).
11. Opinion of the Scientific Committee on Food on pulegone and menthofuran. *SCF/CS/FLAV/FLAVOUR/3 ADD2 Final*, 2002. European Commission, Health & Consumer Protection Directorate-General. http://ec.europa.eu/food/fs/sc/scf/out133_en.pdf (accessed July 14, 2015).
12. EFSA Scientific Opinion (FGE.57). Consideration of two structurally related pulegone metabolites and one ester thereof evaluated by JECFA (55th meeting). *The EFSA Journal*, 2009 ON-1079, 1–17.
13. EU Register: EC (European Commission), 1999. Commission Decision 1999/217/EC adopting a register of flavouring substances used in or on foodstuffs drawn up in application of Regulation (EC) No 2232/96 of the European Parliament and of the Council of 28 October 1996 (OJ L84/1-137, 27.3.1999).
14. Commission Implementing Regulation (EU) No 872/2012 of 1 October 2012 adopting the list of flavouring substances provided for by Regulation (EC) No 2232/96 of the European Parliament and of the Council, introducing it in Annex I to Regulation (EC) No 1334/2008 of the European Parliament and of the Council and repealing Commission Regulation (EC) No 1565/2000 and Commission Decision 1999/217/EC Text with EEA relevance (OJ L 267, 2.10.2012, pp. 1–161). http://eur-lex.europa.eu/legal-content/EN/ALL/?uri=CELEX:32012R0872 (accessed July 14, 2015).
15. Opinion of the Scientific Committee on Food on Capsaicin. *SCF/CS/FLAV/FLAVOUR/8 ADD1 Final*, 2002. European Commission, Health & Consumer Protection Directorate-General. http://ec.europa.eu/food/fs/sc/scf/out120_en.pdf (accessed July 14, 2015).
16. IOFI, Statement on the identification in nature of flavouring substances, made by the Working Group on Methods of Analysis of the International Organization of the Flavour Industry (IOFI). *Flavour and Fragrance Journal*, 2006, 21, 185.
17. EFFA Guidance Document for the Production of Natural Flavouring Substances and (Natural) Flavouring Preparations in the EU—2nd Revision (V3.0). 2016. https://www.effa.eu/docs/default-source/guidance-documents/effa-guidance-document-on-the-ec-regulation-on-flavourings---updated-july-2019.pdf?sfvrsn=8 (accessed March 26, 2020).
18. EFFA Guidance Document on the EU Regulation on Flavourings—5th Revision (V6.0). 2018. https://www.effa.eu/docs/default-source/guidance-documents/effa-guidance-document-on-the-ec-regulation-on-flavourings---updated-july-2019.pdf?sfvrsn=8 (accessed March 26, 2020)

Index

A

B

C

D

F

H

I

J

K

L

N

O

Q

R

S

T

U

V